ENDOCRINE SYSTEM
Acts by means of hormones secreted into the blood to regulate processes that require duration rather than speed—e.g., metabolic activities and water and electrolyte balance
See Chapters 4, 18, and 19.

Body systems maintain homeostasis

HOMEOSTASIS
A dynamic steady state of the constituents in the internal fluid environment that surrounds and exchanges materials with the cells
See Chapter 1.
Factors homeostatically maintained:
- Concentration of nutrient molecules
  See Chapters 16, 17, 18, and 19.
- Concentration of O_2 and CO_2
  See Chapter 13.
- Concentration of waste products
  See Chapter 14.
- pH   *See Chapter 15.*
- Concentration of water, salts, and other electrolytes
  See Chapters 14, 15, 18, and 19.
- Temperature   *See Chapter 17.*
- Volume and pressure
  See Chapters 10, 14, and 15.

INTEGUMENTARY SYSTEM
Serves as a protective barrier between the external environment and the remainder of the body; the sweat glands and adjustments in skin blood flow are important in temperature regulation
See Chapters 12 and 17.

Keeps internal fluids in

Keeps foreign material out

IMMUNE SYSTEM
Defends against foreign invaders and cancer cells; paves the way for tissue repair
See Chapter 12.

Protects against foreign invaders

Homeostasis is essential for survival of cells

MUSCULAR AND SKELETAL SYSTEMS
Support and protect body parts and allow body movement; heat-generating muscle contractions are important in temperature regulation; calcium is stored in the bone
See Chapters 8, 17, 18, and 19.

Enables the body to interact with the external environment

Exchanges with all other systems

CELLS
Need homeostasis for their own survival and for performing specialized functions essential for survival of the whole body
See Chapters 1, 2, and 3.
Need a continual supply of nutrients and O_2 and ongoing elimination of acid-forming CO_2 to generate the energy needed to power life-sustaining cellular activities as follows:
Food + $O_2 \rightarrow CO_2 + H_2O$ + energy
See Chapters 13, 15, 16, and 17.

Cells make up body systems

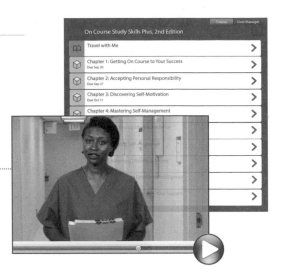
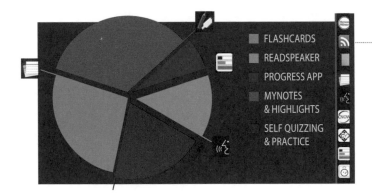

9TH
Edition

Human Physiology
From Cells to Systems

Lauralee Sherwood
Department of Physiology and Pharmacology
School of Medicine
West Virginia University

Australia • Brazil • Mexico • Singapore • United Kingdom • United States

Human Physiology: From Cells to Systems,
  Ninth Edition

Lauralee Sherwood

Product Director: Mary Finch

Product Manager: Yolanda Cossio

Content Developer: Trudy Brown

Associate Content Developers: Casey Lozier,
  Kellie Petruzzelli

Product Assistant: Victor Luu

Media Developer: Lauren Oliveira

Marketing Managers: Tom Ziolkowski,
  Jessica Egbert

Content Project Manager: Tanya Nigh

Art Director: John Walker

Manufacturing Planner: Becky Cross

Compositor and Production Service:
  Graphic World Inc

IP Analyst: Christine Myaskovsky

IP Project Manager: John Sarantakis

Photo and Text Researcher:
  Lumina Datamatics Ltd.

Copy Editor: Graphic World Inc

Text Designer: Lisa Buckley

Cover Designer: John Walker

Cover Image: ©Pete Saloutos/UpperCut
  Images/Getty Images

For product information and technology assistance, contact us at
Cengage Learning Customer & Sales Support, 1-800-354-9706.
For permission to use material from this text or product,

submit all requests online at **www.cengage.com/permissions**.
Further permissions questions can be e-mailed to
permissionrequest@cengage.com.

Library of Congress Control Number: 2014945820

ISBN: 978-1-285-86693-2

Cengage Learning
20 Channel Center Street
Boston, MA 02210
USA

Cengage Learning is a leading provider of customized learning solutions with office locations around the globe, including Singapore, the United Kingdom, Australia, Mexico, Brazil, and Japan. Locate your local office at **www.cengage.com/global.**

Cengage Learning products are represented in Canada by Nelson Education, Ltd.

To learn more about Cengage Learning Solutions,
visit **www.cengage.com.**

Purchase any of our products at your local college store or at our preferred online store **www.cengagebrain.com.**

Printed in the United States
Print Number: 06   Print Year: 2018

With love to my family,
for all that they do for me
and all that they mean to me:

My husband,
Peter Marshall

My daughters and sons-in-law,
Melinda and Mark Marple
Allison Tadros and Bill Krantz

My grandchildren,
Lindsay Marple
Emily Marple
Alexander Tadros
Lauren Krantz

Brief Contents

Contents

Chapter 5 | The Central Nervous System   133

Chapter 6 | The Peripheral Nervous System: Afferent Division; Special Senses   181

Chapter 7 | The Peripheral Nervous System: Efferent Division   233

 Homeostasis Highlights   233

Chapter 8 | Muscle Physiology 251

Chapter 9 | Cardiac Physiology 297

Chapter 10 | The Blood Vessels and Blood Pressure   335

Chapter 13 | The Respiratory System 445

 Homeostasis Highlights 445

Chapter 20 | The Reproductive System   715

 Homeostasis Highlights   715

APPENDIXES

Preface

Goals, Philosophy, and Theme

I am constantly awestruck at the miraculous intricacies and efficiency of body function. No machine can perform even a portion of natural body function as effectively. My goal in writing physiology textbooks is not only to help students learn about how the body works, but also to share my enthusiasm for the subject matter. Most of us, even infants, have a natural curiosity about how our bodies work. When babies first discover they can control their hands, they are fascinated and spend many hours manipulating them in front of their faces. By capitalizing on students' natural curiosity about themselves, I try to make physiology a subject they can enjoy learning.

Even the most tantalizing subject can be difficult to comprehend if not effectively presented, however. Therefore, this book has a logical, understandable format with an emphasis on how each concept is an integral part of the entire subject. Too often, students view the components of a physiology course as isolated entities; by understanding how each component depends on the others, a student can appreciate the integrated functioning of the human body. The text focuses on the mechanisms of body function from cells to systems and is organized around the central theme of *homeostasis*—how the body meets changing demands while maintaining the internal constancy necessary for all cells and organs to function. The text is written in simple, straightforward language, and every effort has been made to ensure smooth reading through good transitions, common-sense reasoning, and integration of ideas throughout.

This text is designed for undergraduate students preparing for health-related careers, but its approach and depth also are appropriate for other undergraduates. Because this book is intended as an introduction and, for most students, may be their only exposure to a formal physiology text, all aspects of physiology receive broad coverage, yet depth, where needed, is not sacrificed. The scope of this text has been limited by judicious selection of pertinent content that a student can reasonably be expected to assimilate in a one-semester physiology course. Materials were selected for inclusion on a "need to know" basis, so the book is not cluttered with unnecessary detail. Instead, content is restricted to relevant information needed to understand basic physiological concepts and to serve as a foundation for future careers in the health professions. Some controversial ideas and hypotheses are presented to illustrate that physiology is a dynamic, changing discipline.

To keep pace with today's rapid advances in the health sciences, students in the health professions must be able to draw on their conceptual understanding of physiology instead of merely recalling isolated facts that soon may be out of date. Therefore, this text is designed to promote understanding of the basic principles and concepts of physiology rather than memorization of details.

In consideration of the clinical orientation of most students, research methodologies and data are not emphasized, although the material is based on up-to-date evidence. New information based on recent discoveries has been included throughout. Students can be assured of the timeliness and accuracy of the material presented. To make room for new, applicable information, I have carefully trimmed content while clarifying, modifying, and simplifying as needed to make this edition fresh, reader-friendly, and current.

Because the function of an organ depends on the organ's construction, enough relevant anatomy is integrated within the text to make the inseparable relation between form and function meaningful.

Hallmark Features and Learning Aids

Implementing the homeostasis theme

Homeostasis is the first word in this text, in the caption for the chapter opener photo for Chapter 1, "Introduction to Physiology and Homeostasis," indicative of the importance placed on homeostasis (see p. 1).

Each chapter begins with **Homeostasis Highlights,** an opening feature that emphasizes the big picture of how the content to come plays a part in homeostasis and functionally fits in with the body as a whole. As an example, see *Homeostasis Highlights* for Chapter 8, "Muscle Physiology," p. 251.

At the close of each chapter, **Homeostasis: Chapter in Perspective** points out specific ways in which the topic covered in the chapter contributes to homeostasis, returning the reader to this central theme, no matter how far the content appears to be removed from playing a role in maintaining internal constancy, as exemplified by *Homeostasis: Chapter in Perspective* for Chapter 3, "The Plasma Membrane and Membrane Potential," p. 84.

A unique, easy-to-follow, pictorial homeostatic model showing the relationship among cells, systems, and homeostasis is developed in the introductory chapter (see pp. 14–15) and presented on the inside front cover as a quick reference.

These opening and closing features and the homeostatic model work together to facilitate students' comprehension of the interactions and interdependency of body systems, even though each system is discussed separately.

Chapter openers

The chapter openers consist of three key components: an eye-catching, informative photo relevant to the chapter; *Chapter at a Glance,* a concise list of contents; and the brief *Homeostasis Highlights* narrative that orients the readers to the homeostatic aspects of the material that follows. Check out the chapter opener for Chapter 13, "The Respiratory System," on p. 445 as an example.

Pedagogical illustrations

Anatomic illustrations, schematic representations, step-by-step descriptions within process-oriented figures, photographs, tables, and graphs complement and reinforce the written material.

Widespread use of integrated descriptions within figures, including numerous **process-oriented figures with incorporated step-by-step descriptions**, allows visually oriented students to review processes through figures. Check out Figure 5-17, p. 161; Figure 8-11, p. 260; and Figure 11-11, p. 396, for examples.

Flow diagrams are used extensively to help students integrate the written information. In the flow diagrams, lighter and darker shades of the same color denote a decrease or increase in a controlled variable, such as blood pressure or the concentration of blood glucose. Physical entities, such as body structures and chemicals, are distinguished visually from actions. Icons of physical entities are incorporated into the flow diagrams. See Figure 15-4, p. 545; Figure 16-12, p. 592; and Figure 20-9, p. 729, for examples.

Most chapters feature one or more **showcase figures**, which are art-enhanced, visually appealing, broad-based foundation figures that draw students' attention to key structural and functional components relevant to the chapter. Examples include the following:

■ Figure 2-1, A diagram of cell structures visible under an electron microscope, p. 23

■ Figure 14-1, The urinary system, p. 493

■ Figure 19-7, Anatomy of and hormonal secretion by the adrenal glands, p. 672

Also, integrated **color-coded figure/table combinations** help students better visualize what part of the body is responsible for what activities. For example, anatomic depiction of the brain is integrated with a table of the functions of the major brain components, with each component shown in the same color in the figure and the table (see Table 5-1, pp. 144–145).

A unique feature of this book is that people depicted in the various illustrations are realistic representatives of a cross-section of humanity. Sensitivity to various races, sexes, and ages should enable all students to identify with the material being presented.

Analogies

Many analogies and frequent references to everyday experiences are included to help students relate to the physiology concepts presented. These useful tools have been drawn in large part from my more than four decades of teaching experience. Knowing which areas are likely to give students the most difficulty, I have tried to develop links that help them relate the new material to something with which they are already familiar. As examples, the lymphatic system as an accessory drainage route for interstitial fluid is compared to a storm sewer that picks up and carries away excess rainwater so that it does not accumulate and flood an area (p. 358); and the effect of sildenafil (Viagra) is likened to pushing a pedal on a piano not causing a note to be played but prolonging a played note (p. 734).

Pathophysiology and clinical coverage

 Another effective way to keep students' interest is to help them realize they are learning worthwhile and applicable material. Because many students using this text will have health-related careers, frequent references to pathophysiology and clinical physiology demonstrate the content's relevance to their professional goals. Clinical Note icons flag clinically relevant material, which is integrated throughout the text.

Boxed features

Two types of boxed features are incorporated within the chapters. Concepts, Challenges, and Controversies boxes expose students to high-interest information on such diverse topics as new technologies involving "seeing" with the tongue or the ear (see p. 210); historical highlights, for example, development of vaccinations (see p. 422); body responses to different environments such as those encountered in mountain climbing and deep-sea diving (see pp. 480–481); and in-depth discussions regarding common diseases such as Alzheimer's disease (see pp. 164–165).

A Closer Look at Exercise Physiology boxes are included for three reasons: increasing national awareness of the importance of physical fitness, increasing recognition of the value of prescribed therapeutic exercise programs for a variety of conditions, and growing career opportunities related to fitness and exercise. As an example, see the exercise physiology box on p. 542 regarding the importance of acclimatization to exercising in the heat.

Major section heads and feedforward statements as subsection titles

Major section heads and subsections logically break up large concepts into smaller, manageable chunks. Instead of traditional short topic titles for each subsection (for example, "Glial cells"), feedforward statements alert students to the main point of the subsection to come (for example, "Glial cells support the interneurons physically, metabolically, and functionally"). As an added bonus, the listing of these headings in the **Contents** at the beginning of the book serves as a set of objectives for each chapter.

Key terms and word derivations

Key terms are defined as they appear in the text. Because physiology is laden with new vocabulary words, many of which are rather intimidating at first glance, word derivations are provided to enhance understanding of new words.

Review and self-evaluation tools in the text

Students are provided opportunities to review and are encouraged to assess their comprehension in a variety of ways.

Check Your Understanding Questions Questions at the end of each major section serve as study breaks for students to test their knowledge before starting the next section. These questions are different than the questions that cover the same content at the end of the chapter. Many of these section questions involve doing something other than copying an answer from a text description, such as drawing and labeling, preparing a chart, predicting based on information provided, and so on. In response to positive feedback regarding the usefulness of this pedagogical tool, which was introduced in the last edition, I have added nearly 100 new Check Your Understanding questions to this edition, bringing them to a total of about 350.

NEW! Figure Focus Questions Designed to check and promote student comprehension, focus questions have been added to specific figures throughout the text. To answer these critical thinking questions correctly, the reader must analyze, interpret, infer, and apply the content of 120 key figures. Check out examples of these new questions in Figures 13-20 and 13-21, p. 467; Figure 14-27, p. 525; and Figure 19-2, p. 667.

NEW! Blooms-Based Organization of Review Exercises The Review Exercises at the end of each chapter are now organized into categories using the educational tool *Bloom's Taxonomy of Learning Domains* as a guide. Questions are grouped in a hierarchy from lower- to higher-order levels as follows:

- **Reviewing Terms and Facts:** The objective-type questions in this exercise are intended for students to self-test their basic knowledge of the chapter by recalling terms and facts.

- **Understanding Concepts:** With this level, students demonstrate their understanding of the concepts presented by describing, explaining, comparing, stating main ideas, and so on in their own words.

- **Solving Quantitative Exercises:** These problem-solving exercises provide students with an opportunity to practice calculations that enhance their understanding of complex relationships.

- **Applying Clinical Reasoning:** This mini case history challenges students to apply acquired knowledge to a patient's specific symptoms, a situation relevant to the health-profession career goals of most students using this textbook.

- **Thinking at a Higher Level:** This section features thought-provoking problems that encourage students to analyze, synthesize, reorganize, or apply in a different way what they have learned in the chapter.

Answers and explanations for these exercises are available in an appendix and online as described in the next section.

Study Cards A tear-out study card is available for each chapter. Each study card presents the major points of the chapter in concise, section-by-section bulleted lists, including cross-references for page numbers, figures, and tables. Students can carry these handy chapter summaries instead of the book to conveniently review key concepts for exams. The tear-out design lets students more efficiently review material even with the book in hand because they can see the written summary and visual information side-by-side without having to flip pages back and forth. This feature enables students to easily review main concepts before moving on.

Appendixes and glossary

Most undergraduate physiology texts have a chapter on chemistry, yet physiology instructors rarely teach basic chemistry concepts. Knowledge of chemistry beyond that introduced in secondary schools is not required for understanding this text. Therefore, I provide instead *Appendix A,* **A Review of Chemical Principles,** as a handy reference for students who need a brief review of basic chemistry concepts that apply to physiology. The following additional review materials are available online at **www.cengagebrain.com** *Storage, Replication, and Expression of Genetic Information* and *Principles of Quantitative Reasoning.*

Appendix B, **Text References to Exercise Physiology,** provides an index of all relevant content on this topic.

Appendix C, **Answers,** provides answers to all objective learning activities, including in-chapter *Check Your Understanding* questions and *Figure Focus* questions and end-of-chapter *Reviewing Terms and Facts,* solutions to the *Solving Quantitative Exercises,* and explanations for *Applying Clinical Reasoning* and *Thinking at a Higher Level* exercises. Answers to *Understanding Concepts* questions can be found at **www.cengagebrain.com.**

The **Glossary,** which offers a way to review the meaning of key terminology, includes phonetic pronunciations of the entries.

Organization

There is no ideal organization of physiologic processes into a logical sequence. In the sequence I chose, most chapters build on material presented in immediately preceding chapters, yet each chapter is designed to stand on its own, allowing the instructor flexibility in curriculum design. This flexibility is facilitated by cross-references to related material in other chapters. The cross-references let students quickly refresh their memory of material already learned or proceed, if desired, to a more in-depth coverage of a particular topic.

The general flow is from introductory background information to cells to excitable tissue (nerve and muscle) to organ systems, with logical transitions from one chapter to the next. For example, Chapter 8, "Muscle Physiology," ends with a discussion of cardiac (heart) muscle, which is carried forward in Chapter 9, "Cardiac Physiology." Even topics that seem unrelated in sequence, such as Chapter 12, "Body Defenses," and Chapter 13, "The Respiratory System," are linked together, in this case by ending Chapter 12 with a discussion of respiratory defense mechanisms.

Several organizational features warrant specific mention. The most difficult decision in organizing this text was placement of the endocrine material. There is merit in placing the chapters on the nervous and the endocrine (hormone-secreting) systems in close proximity because they are the body's two major regulatory systems. However, discussing details of the endocrine system immediately after the nervous system would disrupt the logical flow of material related to excitable tissue. In addition, the endocrine system cannot be covered in the depth its importance warrants if it is discussed before students have the background to understand this system's roles in maintaining homeostasis.

My solution to this dilemma is Chapter 4, "Principles of Neural and Hormonal Communication." This chapter introduces the underlying mechanisms of neural and hormonal action before the nervous system and specific hormones are mentioned in later chapters. It contrasts how nerve cells and endocrine cells communicate with other cells in carrying out their regulatory actions. Building on the different modes of action of nerve and endocrine cells, the last section of this chapter compares, in a general way, how the nervous and endocrine systems differ as regulatory systems. Chapter 5 then begins with the nervous system, providing a good link between Chapters 4 and 5. Chapters 5, 6, and 7 are devoted to the nervous system. Specific hormones are introduced in appropriate chapters, such as hormonal control of the heart and blood vessels in maintaining blood pressure in Chapters 9 and 10 and hormonal control of the kidneys in maintaining fluid balance in Chapters 14 and 15. The body's processing of absorbed energy-rich nutrient molecules is largely under endocrine control, providing a link from digestion (Chapter 16) and energy balance (Chapter 17) to the endocrine system (Chapters 18 and 19). These endocrine chapters pull together the source, functions, and control of specific endocrine secretions and serve as a summarizing and unifying capstone for homeostatic body function. Finally, building on the hormones that control the gonads (testes and ovaries) introduced in the endocrine chapters, the last chapter, Chapter 20, diverges from the theme of homeostasis to focus on reproductive physiology.

Besides the novel handling of hormones and the endocrine system, other organizational features are unique to this book. For example, unlike other physiology texts, the skin is covered in the chapter on defense mechanisms of the body (Chapter 12), in consideration of the skin's recently recognized immune functions. Bone is also covered more extensively in the endocrine chapter than in most undergraduate physiology texts, especially with regard to hormonal control of bone growth and bone's dynamic role in calcium metabolism.

Although there is a rationale for covering the various aspects of physiology in the order given here, it is by no means the only logical way of presenting the topics. Because each chapter is able to stand on its own, especially with the cross-references provided, instructors can vary the sequence of presentation at their discretion. Some chapters may even be omitted, depending on the students' needs and interests and the time constraints of the course. For example, a cursory explanation of the defense role of the leukocytes appears in Chapter 11 on blood, so an instructor can choose to omit the more detailed explanations of immune defense in Chapter 12.

New to the Ninth Edition

This edition has a new look, new pedagogical features, updates, and numerous revisions to make the book as current, relevant, and accessible to students as possible. Every aspect of the text has been upgraded as the following examples illustrate. For a detailed list of all changes, contact your Cengage Learning sales representative.

New look

Not only does this edition have fresh colors but the pages are more visually interesting because of creative wrapping of some of the written material around the art for a contemporary look instead of just being wrapped with a traditional, single 90-degree corner. See pp. 111, 245, and 413 for examples.

New self-check pedagogical tools

Already mentioned, new to this edition are several new or revised self-check features, including the new *Figure Focus* questions, more *Check Your Understanding* questions, and new organization of the end-of-chapter *Review Exercises* into hierarchical learning levels.

New and revised figures

New Art The following exemplify first-time illustrations added in this edition:

- Figure 5-9, Layers of the cerebral cortex, p. 147
- Figure 10-11, Major local chemical and physical means of controlling arteriolar caliber, p. 347
- Figure 16-11, Mechanism of $NaHCO_3$ secretion, p. 591

Revised Art Examples of extensively revised, newly conceptualized, or reorganized figures include the following:

- Figure 8-1, Characteristics of three types of muscle, p. 252
- Figure 10-26, Skeletal muscle pump enhancing venous return and countering effect of gravity on venous pressure, p. 363
- Figure 16-5, Oropharyngeal and esophageal stages of swallowing, p. 577

New Photos More than 50 new photos and replacement photos are incorporated throughout the text, including replacing 45% of the chapter opener photos. For instance, see Chapter 5 opener, a diffusion resonance image (dMRI) scan of the white matter pathways of the brain, p. 133. The following are other examples of content not shown in photos in previous editions:

- Figure 4-7, A micrograph of dendritic spines incorporated into Anatomy of the most common type of neuron, p. 95
- Figure 6-37, A scanning electron micrograph of the tip links between adjacent stereocilia, incorporated into The role of stereocilia in sound transduction, p. 218
- Table 16-3, A scanning electron micrograph of stomach lining showing gastric pits, incorporated into an integrated figure/table featuring The stomach mucosa and the gastric glands, p. 583

New tables

More new tables that group and consolidate information for easier learning have been added to this edition than ever before, as the following samples demonstrate:

- Table 4-2, Major Neurotransmitters, p. 107
- Table 8-3, Motor Control by CNS, p. 280
- Table 20-4, Stages of Follicular Development, p. 743

New boxed features

Several old boxes have been retired and two new boxes regarding timely, relevant content have been added: (1) *The Ups (Causes) and Downs (Treatments) of Hypertension*, in consideration of the fact that one third of all adults in the United States have hypertension (see pp. 372–373); and (2) *Still a Big Question: Why Do We Age?*, which focuses on the current theories of aging, in view of the increased graying of America as baby boomers are reaching old age (see pp. 678–679).

New, updated content

Recent discoveries and hot topics have been incorporated throughout as the following examples illustrate:

- In Chapter 2, inserted a discussion and figure of proteasomes breaking down ubiquinated proteins into recyclable building blocks (p. 27)
- Among the numerous new topics in Chapter 5 is the glymphatic system, a recently identified glia substitute for the lymphatic system in the brain (p. 137)
- Added a comparison of the trichromatic theory and the opponent-process theory of color vision in Chapter 6 (p. 205)
- In Chapter 9, expanded presentation of cardiac autorhythmicity to include both the membrane clock mechanism and the Ca^{2+} mechanism (together, the coupled-clock system) responsible for the pacemaker potential (p. 304)
- Introduced in Chapter 12 newly identified immune cells (innate lymphoid cells [ILCs] and innate response activator [IRA] B cells) that straddle the innate and adaptive immune systems (p. 415)
- Significantly expanded coverage of the microbiota and microbiome in Chapter 16 in light of a torrent of new findings in this hot area of science (pp. 613–614)
- In Chapter 17, updated discussion of brown fat in view of recent studies suggesting that irisin, a newly discovered chemical mediator released from exercising muscles, may promote "browning" of white adipose tissue by stimulating synthesis of uncoupling proteins in mitochondria of white fat cells (p. 632)
- Augmented coverage of the underlying molecular mechanism responsible for the suprachiasmatic nucleus's circadian oscillations in Chapter 18 by adding the interactions of PER and CRY with CLOCK and BMAL-1 (see p. 660)
- Updated discussion of islets of Langerhans in Chapter 19 to include secretion of amylin in addition to insulin by the beta cells and secretion of ghrelin by newly found epsilon cells (pp. 692 and 689)
- Expanded coverage of clinically related issues, such as adding a new discussion of concussions and chronic traumatic encephalopathy (p. 172)

Reorganization

Although the focus of each chapter remains the same as previous editions, I moved some content between and within chapters for better grouping of material, as follows:

- Moved the discussion of eicosanoids from Chapter 20 (in association with male accessory sex gland secretions) to Chapter 4, the chapter devoted primarily to neural and hormonal communication and signal transduction. Eicosanoids and cytokines are now more appropriately grouped together and presented in a new section entitled *Introduction to Paracrine Communication* (pp. 118–120)
- Transferred introduction of the JAK/STAT pathway from Chapter 18, where it was treated more as an aside in the discussion of signal transduction by growth hormone and prolactin, to Chapter 4, where the topic more logically fits in with coverage of other means of signal transduction (pp. 116–117)
- Rearranged and grouped together the material within Chapter 5 related to brain waves and the electroencephalogram for better flow and improved clarity (pp. 168–169)
- Based on reviewer input, relocated presentation of specific somatic reflexes, namely the stretch reflex, withdrawal reflex, and crossed extensor reflex, from Chapter 5, Central Nervous System, where these reflexes were covered in conjunction with the spinal cord, to Chapter 8, Muscle Physiology, where they are now included in the section on Control of Motor Movement (pp. 282, 284–286)

Clearer, more concise coverage

I look at every edition for opportunities to make the writing as clear, concise, well-organized, and relevant for readers as possible. By careful tightening, I was able to shave 22 pages from the text while retaining all essential content and adding more beneficial learning tools and updated content, a win-win for readers.

New and Enhanced Technology for Instructors and Students

NEW! MindTap for Human Physiology

MindTap is a personalized, fully digital learning platform of authoritative content, assignments, and services that engages students with interactivity while also offering instructors their choice in the configuration of coursework and enhancement of the curriculum via Web-based applications known as MindApps. MindApps range from ReadSpeaker (which reads the text out loud to students) to Kaltura (allowing you to insert inline video and audio into your curriculum). MindTap is well beyond an eBook, a homework solution, a digital supplement, a resource center Web site, a course delivery platform, or a Learning Management System. It is the first in a new category—the Personal Learning Experience. MindTap for *Human Physiology* includes an integrated study guide, homework, and clinical case studies, among other valuable learning tools.

NEW! Aplia for Human Physiology

Aplia for *Human Physiology* helps students learn and understand key concepts via focused assignments and active learning opportunities that include randomized, automatically graded questions, exceptional text interaction, and detailed explanations. With Aplia, students move away from just memorizing facts toward a more conceptual understanding of the course materials, In addition, Aplia has a full course management system that can be used independently or in conjunction with other course management systems such as MindTap, D2L, or Blackboard.

NEW! Virtual Physiology Labs 2.0

Virtual Physiology Labs enable students to conduct experiments online without expensive equipment. By acquiring data, performing experiments, and using that data to explain physiology concepts, students become involved in the scientific process—they don't just watch or read about it.

Resources for Instructors

Instructor Companion Site

Everything you need for your course in one place! This collection of book-specific lecture and class tools is available online via www.cengage.com/login. Access and download PowerPoint™ presentations, images, the instructor's manual, videos, and more.

Cognero for Human Physiology, Ninth Edition

Cengage Learning Testing Powered by Cognero is a flexible, online system that allows you to:

- Author, edit, and manage test bank content from multiple Cengage Learning solutions
- Create multiple test versions in an instant
- Deliver tests from your Learning Management System, your classroom, or wherever you want.

Resources for Students

Coloring book for Human Physiology, Ninth Edition

This helpful study tool contains key pieces of art from the book and provides opportunities for students to interact with the material and explain the processes associated with the figures in their own words.

Photo Atlas for Anatomy and Physiology

This full-color atlas (with more than 600 photographs) depicts structures in the same colors as they would appear in real life or in a slide. Labels and color differentiations within each structure are used to facilitate identification of the structure's various components. The atlas includes photographs of tissue and organ slides, the human skeleton, commonly used models, cat dissections, cadavers, some fetal pig dissections, and some physiology materials.

Fundamentals of Physiology Laboratory Manual

This manual, which may be required by the instructor in courses that have a laboratory component, contains a variety of exercises that reinforce concepts covered in *Human Physiology: From Cells to Systems*, Ninth Edition. These laboratory experiences increase students' understanding of the subject matter in a straightforward manner, with thorough directions to guide them through the process and relevant questions for reviewing, explaining, and applying results.

Acknowledgments

I gratefully acknowledge the many people who helped with the first eight editions or this edition of the textbook. Also, I remain indebted to four people who contributed substantially to the original content of the book: Rachel Yeater (West Virginia University), who contributed the original material for the exercise physiology boxes; Spencer Seager (Weber State University), who prepared Appendix A, "A Review of Chemical Principles"; and Kim Cooper (Midwestern University) and John Nagy (Scottsdale Community College), who provided the Solving Quantitative Exercises at the ends of chapters.

In addition to the 184 reviewers who carefully evaluated the forerunner books for accuracy, clarity, and relevance, I express sincere appreciation to the following individuals who served as reviewers for this edition:

Ahmed Al-Assal, West Coast University
Amy Banes-Berceli, Oakland University
Patricia Clark, Indiana University-Purdue University
Elizabeth Co, Boston University
Steve Henderson, Chico State University
James Herman, Texas A&M University
Qingsheng Li, University of Nebraska-Lincoln
Douglas McHugh, Quinnipiac University
Linda Ogren, University of California, Santa Cruz
Roy Silcox, Brigham Young University

Also, I am grateful to the users of the textbook who have taken time to send helpful comments.

I have been fortunate to work with a highly competent, dedicated team from Cengage Learning, along with other capable external suppliers selected by the publishing company. I would like to acknowledge all of their contributions, which collectively made this book possible. It has been a source of comfort and inspiration to know that so many people have been working diligently in so many ways to bring this book to fruition.

From Cengage Learning, Yolanda Cossio, Senior Product Team Manager, deserves warm thanks for her vision, creative ideas, leadership, and ongoing helpfulness. Yolanda's decisions were guided by what is best for the instructors and students who will use the textbook and its learning resources. Thanks also to Product Assistant Victor Luu, who coordinated many tasks for Yolanda during the development process. I appreciate the efforts of Content Developer Alexis Glubka for helping us launch this project and for Managing Developer Trudy Brown for taking over and not missing a beat when Alexis moved to another position. Trudy facilitated most of the development process, which proceeded efficiently and on schedule. Having served as a Production Editor in the past, Trudy was especially helpful in paving the way for a smooth transition from development to production. I am grateful for the biweekly conference calls with Yolanda and Trudy. Their input, expertise, and support were invaluable as collectively we made decisions to make this the best edition yet.

I appreciate the creative insight of Cengage Learning Senior Art Director John Walker, who oversaw the overall artistic design of the text and found the dynamic cover image that displays simultaneous power, agility, and grace in a "body in motion," the theme of our covers. The upward and "over-the-hump" movement of the high jumper symbolizes that this book will move the readers upward and over the hump in their understanding of physiology. I thank John for his patience and perseverance as we sifted through and rejected many photos until we found just what we were looking for.

The technology-enhanced learning tools in the media package were updated under the guidance of Media Developer Lauren Oliveira. These include the online interactive tutorials, media exercises, and other e-Physiology learning opportunities at the CengageBrain Web site. Associate Content Developers Casey Lozier and Kellie Petruzzelli oversaw development of the multiple hard-copy components of the ancillary package, making sure it was a cohesive whole. A hearty note of gratitude is extended to all of them for the high-quality multimedia package that accompanies this edition.

On the production side, I would like to thank Senior Content Project Manager Tanya Nigh, who closely monitored every step of the production process while simultaneously overseeing the complex production process of multiple books. I felt confident knowing that she was making sure that everything was going according to plan. I also thank Photo Researcher Priya Subbrayal and Text Researcher Kavitha Balasundaram for tracking down photos and permissions for the art and other copyrighted materials incorporated in the text. With everything finally coming together, Manufacturing Planner Karen Hunt oversaw the manufacturing process, coordinating the actual printing of the book.

No matter how well a book is conceived, produced, and printed, it would not reach its full potential as an educational tool without being efficiently and effectively marketed. Market Development Manager Julie Schuster played the lead role in marketing this text, for which I am most appreciative.

Cengage Learning also did an outstanding job in selecting highly skilled vendors to carry out particular production tasks. First and foremost, it has been my personal and professional pleasure to work with Cassie Carey, Production Editor at Graphic World, who coordinated the day-to-day management of production. In her competent hands lay the responsibility of seeing that all copyediting, art, typesetting, page layout, and other related details got done right and in a timely fashion. Thanks to her, the production process went smoothly, the best ever. I also want to extend a hearty note of gratitude to compositor Graphic World for their accurate typesetting; execution of the art revisions; and attractive, logical layout. Lisa Buckley deserves thanks for the fresh and attractive, yet space-conscious, appearance of the book's interior and for envisioning the book's visually appealing exterior.

Finally, my love and gratitude go to my family for the sacrifices in family life as this ninth edition was being developed and produced. My husband, Peter Marshall, deserves special appreciation and recognition for assuming extra responsibilities while I was working on the book. I could not have done this, or any of the preceding books, without his help, support, and encouragement.

Thanks to all!

Lauralee Sherwood

Introduction to Physiology and Homeostasis

Oliver Eltinger/Agefotostock

Homeostasis (maintaining internal consistency) in action. Body temperature is maintained as evaporation of sweat cools the body to counterbalance heat gained through exertion on a hot day, and fluid balance is maintained as thirst encourages fluid intake to offset fluid lost in sweat.

Homeostasis Highlights

 Physiology focuses on body functions. This book explores how the various components of the human body function to maintain **homeostasis**, the relatively stable conditions inside the body needed for survival. Each chapter begins with *Homeostasis Highlights* to give you a heads up on how the body part under discussion fits in with the big picture of homeostasis. Each chapter concludes with *Homeostasis: Chapter in Perspective,* which points out specific ways in which the topic covered in the chapter contributes to homeostasis.

1.1 Introduction to Physiology

Look at Figure 1-1. The activities described are a sampling of the body processes that occur all the time just to keep us alive. We usually take these life-sustaining activities for granted and do not really think about "what makes us tick," but that's what physiology is about. **Physiology** is the study of the functions of living things. Specifically, we will focus on how the human body works.

Physiology focuses on mechanisms of action.

Two approaches are used to explain events that occur in the body; one emphasizes the *purpose* of a body process and the other emphasizes the underlying *mechanism* by which this process occurs. In response to the question "Why do I shiver when I am cold?" one answer would be "to help my body warm up, because shivering generates heat." This approach, which explains body functions in terms of meeting a bodily need, emphasizes *why* body processes occur. Physiologists, however, explain *how* processes occur in the body. They view the body as a machine whose mechanisms of action can be explained in terms of cause-and-effect sequences of physical and chemical processes—the same types of processes that occur throughout the universe. A physiologist's explanation of shivering is that when temperature-sensitive nerve cells detect a fall in body temperature, they signal the area in the brain responsible for temperature regulation. In response, this brain area activates nerve pathways that ultimately bring about involuntary, oscillating muscle contractions (that is, shivering).

Structure and function are inseparable.

Physiology is closely related to **anatomy**, the study of the structure of the body. Physiological mechanisms are made possible by the structural design and relationships of the various body parts that carry out each of these functions. Just as the functioning of an automobile depends on the shapes, organization, and interactions of its various parts, the structure and function of the human body are inseparable. Therefore, as we tell the story of how the body works, we provide sufficient anatomic background for you to understand the function of the body part being discussed.

Some structure–function relationships are obvious. For example, the heart is well designed to receive and pump blood, the teeth to tear and grind food, and the hingelike elbow joint to permit bending of the arm. In other situations, the interdependence of form and function is more subtle but equally important. Consider the interface between air and blood in the lungs as an example: The respiratory airways, which carry air from the outside into the lungs, branch extensively when they reach the lungs. Tiny air sacs cluster at the ends of the huge number of airway branches. The branching is so extensive that the lungs contain about 300 million air sacs. Similarly, the vessels carrying blood into the lungs branch extensively and form dense networks of small vessels that encircle each air sac (see ▌Figure 13-2, p. 448). Because of this structural relationship, the total surface area forming an interface between the air in the air sacs and the blood in the small vessels is about the size of one side of a volleyball court. This tremendous interface is crucial for the lungs' ability to efficiently carry out their function: the transfer of needed oxygen (O_2) from the air into the blood and the unloading of the waste product carbon dioxide (CO_2) from the blood into the air. The greater the surface area available for these exchanges, the faster O_2 and CO_2 can move between the air and the blood. This large functional interface packaged within the confines of your lungs is possible only because both the air-containing and blood-containing components of the lungs branch extensively.

Check Your Understanding 1.1

1. Define *physiology*.
2. The nutrient-absorbing intestinal cells have a multitude of fingerlike projections in contact with the digested food (see ▌Figure 16-20, p. 602). Based on your knowledge of structure–function relationships, explain the functional advantage of this structural feature. (Answers are in Appendix C.)

1.2 Levels of Organization in the Body

We now turn to how the body is structurally organized into a total functional unit, from the chemical level to the whole body (▌Figure 1-2). These levels of organization make possible life as we know it.

The chemical level: Various atoms and molecules make up the body.

Like all matter, both living and nonliving, the human body is a combination of specific *atoms,* which are the smallest building blocks of matter. The most common atoms in the body—oxygen, carbon, hydrogen, and nitrogen—make up approximately 96% of the total body chemistry. These common atoms and a few others combine to form the *molecules* of life, such as proteins, carbohydrates, fats, and nucleic acids (genetic material, such as deoxyribonucleic acid, or DNA). These important atoms and molecules are the inanimate raw ingredients from which all living things arise. (See Appendix A for a review of this chemical level.)

The cellular level: Cells are the basic units of life.

The mere presence of a particular collection of atoms and molecules does not confer the unique characteristics of life. Instead, these nonliving chemical components must be arranged and packaged in precise ways to form a living entity. The **cell,** the fundamental unit of both structure and function in a living being, is the smallest unit capable of carrying out the processes associated with life. Cell physiology is the focus of Chapter 2.

During the minute that it will take you to read this page:

Your eyes will convert the image from this page into electrical signals (nerve impulses) that will transmit the information to your brain for processing.

Besides receiving and processing information such as visual input, your brain will provide output to your muscles to help maintain your posture, move your eyes across the page as you read, and turn the page as needed. Chemical messengers will carry signals between your nerves and muscles to trigger appropriate muscle contraction.

Approximately 150 million old red blood cells will die and be replaced by newly produced ones.

Your heart will beat 70 times, pumping 5 liters (about 5 quarts) of blood to your lungs and another 5 liters to the rest of your body.

You will breathe in and out about 12 times, exchanging 6 liters of air between the atmosphere and your lungs.

More than 1 liter of blood will flow through your kidneys, which will act on the blood to conserve the "wanted" materials and eliminate the "unwanted" materials in the urine. Your kidneys will produce 1 mL (about a thimbleful) of urine.

Your cells will consume 250 mL (about a cup) of oxygen and produce 200 mL of carbon dioxide.

You will use about 2 calories of energy derived from food to support your body's "cost of living," and your contracting muscles will burn additional calories.

Your digestive system will be processing your last meal for transfer into your bloodstream for delivery to your cells.

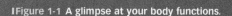

Figure 1-1 A glimpse at your body functions.

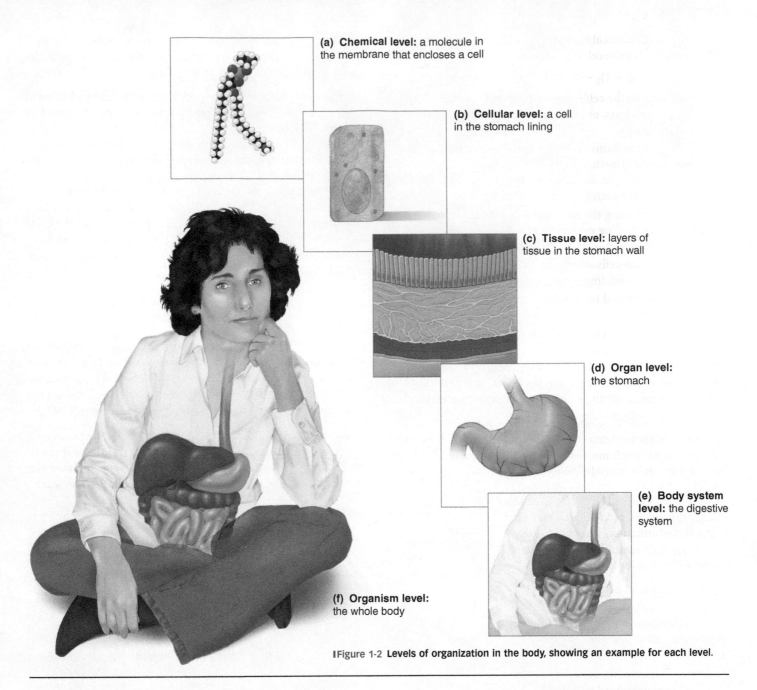

(a) **Chemical level:** a molecule in the membrane that encloses a cell

(b) **Cellular level:** a cell in the stomach lining

(c) **Tissue level:** layers of tissue in the stomach wall

(d) **Organ level:** the stomach

(e) **Body system level:** the digestive system

(f) **Organism level:** the whole body

❙Figure 1-2 **Levels of organization in the body, showing an example for each level.**

An extremely thin, oily, complex barrier, the *plasma membrane*, encloses the contents of each cell and controls movement of materials into and out of the cell. Thus, the cell's interior contains a combination of atoms and molecules that differs from the mixture of chemicals in the environment surrounding the cell. Given the importance of the plasma membrane and its associated functions for carrying out life processes, Chapter 3 is devoted entirely to this structure.

Organisms are independent living entities. The simplest forms of independent life are single-celled organisms such as bacteria and amoebas. Complex multicellular organisms, such as trees and humans, are structural and functional aggregates of trillions of cells (*multi* means "many"). In the simpler multicellular forms of life—for example, a sponge—the cells of the organism are all similar. However, more complex organisms, such as humans, have many kinds of cells, such as muscle cells, nerve cells, and gland cells.

Each human organism begins when an egg and sperm unite to form a single new cell, which multiplies and forms a growing mass through myriad cell divisions. If cell multiplication were the only process involved in development, all body cells would be essentially identical, as in the simplest multicellular life-forms. However, during development of complex multicellular organisms such as humans, each cell also **differentiates**, or becomes specialized to carry out a particular function. As a result of cell differentiation, your body is made up of about 200 specialized types of cells.

Basic Cell Functions All cells, whether they exist as solitary cells or as part of a multicellular organism, perform certain basic functions essential for their own survival, including the following:

1. Obtaining food (nutrients) and O_2 from the environment surrounding the cell.

2. Performing chemical reactions that use nutrients and O_2 to provide energy for the cells, as follows:

$$Food + O_2 \rightarrow CO_2 + H_2O + energy$$

3. Eliminating to the cell's surrounding environment CO_2 and other by-products, or wastes, produced during these chemical reactions.

4. Synthesizing proteins and other components needed for cell structure, for growth, and for carrying out particular cell functions. For example, **enzymes** are specialized proteins that speed up particular chemical reactions in the body.

5. Largely controlling the exchange of materials between the cell and its surrounding environment.

6. Moving materials internally from one part of the cell to another, with some cells also being able to move themselves through their surrounding environment.

7. Being sensitive and responsive to changes in the surrounding environment.

8. In the case of most cells, reproducing. Exceptions are nerve cells and muscle cells, which lose the ability to reproduce soon after they are formed. This is the reason strokes, which result in lost nerve cells in the brain, and heart attacks, which cause death of heart muscle cells, can be so devastating.

Because all cells are remarkably similar in the ways they carry out these basic functions, they share many common characteristics despite their specialization.

Specialized Cell Functions In multicellular organisms, each cell also performs a specialized function, which is usually a modification or elaboration of a basic cell function. Here are a few examples:

■ By taking special advantage of their protein-synthesizing ability, the gland cells of the digestive system secrete digestive enzymes that break down ingested food.

■ Certain kidney cells can selectively retain the substances needed by the body while eliminating unwanted substances in the urine because of their highly specialized ability to control exchange of materials between the cell and its environment.

■ Muscle contraction, which involves selective movement of internal structures to generate tension in the muscle cells, is an elaboration of the inherent ability of these cells to produce intracellular movement (*intra* means "within").

■ Capitalizing on the basic ability of cells to respond to changes in their surrounding environment, nerve cells generate and transmit to other body regions electrical impulses that relay information

about changes to which the nerve cells are responsive. For example, nerve cells in the ear can relay information to the brain about sounds in the body's surroundings.

Each cell performs these specialized activities in addition to carrying on the unceasing, fundamental activities required of all cells. The basic cell functions are essential for survival of individual cells, whereas the specialized contributions and interactions among the cells of a multicellular organism are essential for survival of the whole body.

Just as a machine does not function unless all its parts are properly assembled, the cells of the body must be specifically organized to carry out the life-sustaining processes of the body as a whole, such as digestion, respiration, and circulation. Cells are progressively organized into tissues, organs, body systems, and finally the whole body.

The tissue level: Tissues are groups of cells of similar specialization.

Cells of similar structure and specialized function combine to form **tissues**, of which there are four *primary types:* muscle, nervous, epithelial, and connective (❙ Figure 1-3). Each tissue consists of cells of a single specialized type, along with varying amounts of extracellular material (*extra* means "outside of").

■ **Muscle tissue** consists of cells specialized for contracting, which generates tension and produces movement. The three types of muscle tissue include *skeletal muscle,* which moves the skeleton; *cardiac muscle,* which pumps blood out of the heart;

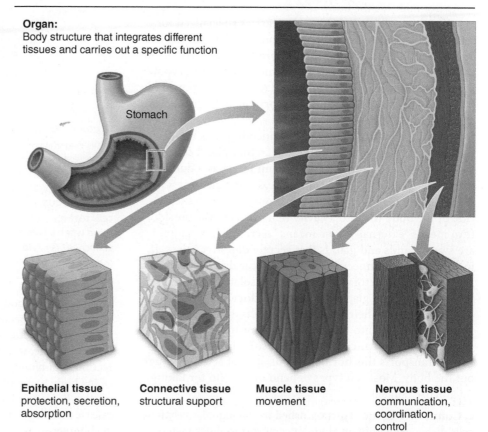

Organ:
Body structure that integrates different tissues and carries out a specific function

Stomach

Epithelial tissue
protection, secretion, absorption

Connective tissue
structural support

Muscle tissue
movement

Nervous tissue
communication, coordination, control

❙ Figure 1-3 **The stomach as an organ made up of all four primary tissue types.**

and *smooth muscle,* which controls movement of contents through hollow tubes and organs, such as movement of food through the digestive tract.

■ **Nervous tissue** consists of cells specialized for initiating and transmitting electrical impulses, sometimes over long distances. These electrical impulses act as signals that relay information from one part of the body to another. Such signals are important in communication, coordination, and control in the body. Nervous tissue is found in the brain, spinal cord, nerves, and special sense organs.

■ **Epithelial tissue** consists of cells specialized for exchanging materials between the cell and its environment. Any substance that enters or leaves the body must cross an epithelial barrier. Epithelial tissue is organized into two general types of structures: *epithelial sheets* and *secretory glands.* Epithelial sheets are layers of tightly joined cells that cover and line various parts of the body. For example, the outer layer of the skin is epithelial tissue, as is the lining of the digestive tract. In general, epithelial sheets serve as boundaries that separate the body from its surroundings and from the contents of cavities that open to the outside, such as the digestive tract lumen. (A **lumen** is the cavity within a hollow organ or tube.) Only selective transfer of materials is possible between regions separated by an epithelial barrier. The type and extent of controlled exchange vary depending on the location and function of the epithelial tissue. For example, the skin can exchange little between the body and outside environment, making it a protective barrier. By contrast the epithelial cells lining the small intestine of the digestive tract are specialized for absorbing nutrients that have come from outside the body.

Glands are epithelial tissue derivatives specialized for secreting. **Secretion** is the release from a cell, in response to appropriate stimulation, of specific products that have been produced by the cell. Glands are formed during embryonic development by pockets of epithelial tissue that dip inward from the surface and develop secretory capabilities. The two categories of glands are *exocrine* and *endocrine* (❙ Figure 1-4). During development, if the connecting cells between the epithelial surface cells and the secretory gland cells within the depths of the pocket remain intact as a duct between the gland and the surface, an exocrine gland is formed. **Exocrine glands** secrete through ducts to the outside of the body (or into a cavity that opens to the outside) (*exo* means "external"; *crine* means "secretion"). Examples are sweat glands and glands that secrete digestive juices. If, in contrast, the connecting cells disappear during development and the secretory gland cells are isolated from the surface, an endocrine gland is formed. **Endocrine glands** lack ducts and release their secretory products, known as *hormones,* internally into the blood (*endo* means "internal"). For example, the pancreas secretes insulin into the blood, which transports this hormone to its sites of action throughout the body. Most cell types depend on insulin for taking up glucose (sugar).

■ **Connective tissue** is distinguished by having relatively few cells dispersed within an abundance of extracellular material. As its name implies, connective tissue connects, supports, and

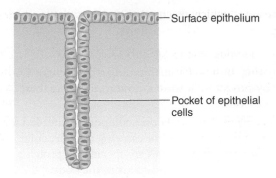

(a) Invagination of surface epithelium during gland formation

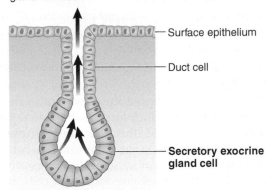

(b) Exocrine gland

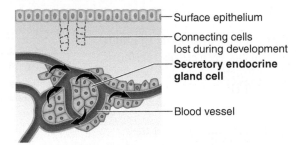

(c) Endocrine gland

❙Figure 1-4 **Exocrine and endocrine glands.** (a) Glands form during development from pocketlike invaginations of surface epithelial cells. (b) Exocrine gland cells release their secretory product through a duct to the outside of the body (or to a cavity in communication with the outside). (c) Endocrine gland cells release their secretory product (a hormone) into the blood.

FIGURE FOCUS: *Milk-secreting glands are surrounded by musclelike cells that squeeze out the milk in response to oxytocin secreted into the blood when a baby breast-feeds. Are milk-secreting glands exocrine or endocrine? Is oxytocin secreted by an exocrine or endocrine gland? (Answers are in Appendix C.)*

anchors various body parts. It includes such diverse structures as the loose connective tissue that attaches epithelial tissue to underlying structures; tendons, which attach skeletal muscles to bones; bone, which gives the body shape, support, and protection; and blood, which transports materials from one part of the body to another. Except for blood, the cells within connective tissue produce specific structural molecules that they release into the extracellular spaces between the cells. One such molecule is the rubber band–like protein fiber *elastin*; its presence facilitates the stretching and recoiling of structures such as the lungs, which alternately inflate and deflate during breathing.

Muscle, nervous, epithelial, and connective tissue are the primary tissues in a classical sense—that is, each is an integrated collection of cells with the same specialized structure and function. The term *tissue* is also often used, as in clinical medicine, to mean the aggregate of various cellular and extracellular components that make up a particular organ (for example, lung tissue or liver tissue).

The organ level: An organ is a unit made up of several tissue types.

Organs consist of two or more types of primary tissue organized to perform particular functions. The stomach, an example of an organ, is made up of all four primary tissue types (see ▮ Figure 1-3). The tissues of the stomach function collectively to store ingested food, move it forward into the rest of the digestive tract, and begin the digestion of protein. The stomach is lined with epithelial tissue that restricts the transfer of harsh digestive chemicals and undigested food from the stomach lumen into the blood. Epithelial gland cells in the stomach include exocrine cells, which secrete protein-digesting juices into the lumen, and endocrine cells, which secrete a hormone that helps regulate the stomach's exocrine secretion and muscle contraction. The stomach wall contains smooth muscle tissue, whose contractions mix ingested food with the digestive juices and push the mixture out of the stomach and into the intestine. The stomach wall also contains nervous tissue, which, along with hormones, controls muscle contraction and gland secretion. Connective tissue binds together all these various tissues.

The body system level: A body system is a collection of related organs.

Groups of organs are further organized into **body systems**. Each system is a collection of organs that perform related functions and interact to accomplish a common activity essential for survival of the whole body. For example, the digestive system consists of the mouth, pharynx (throat), esophagus, stomach, small intestine, large intestine, salivary glands, exocrine pancreas, liver, and gallbladder. These digestive organs cooperate to break food down into small nutrient molecules that can be absorbed into the blood for distribution to all cells.

The human body has 11 systems: circulatory, digestive, respiratory, urinary, skeletal, muscular, integumentary, immune, nervous, endocrine, and reproductive (▮ Figure 1-5). Chapters 4 through 20 cover the details of these systems.

The organism level: The body systems are packaged into a functional whole body.

Each body system depends on the proper functioning of other systems to carry out its specific responsibilities. The whole body of a multicellular organism—a single, independently living individual—consists of the various body systems structurally and functionally linked as an entity that is separate from its sur-

rounding environment. Thus, the body is made up of living cells organized into life-sustaining systems.

The different body systems do not act in isolation from one another. Many complex body processes depend on the interplay among multiple systems. For example, regulation of blood pressure depends on coordinated responses among the circulatory, urinary, nervous, and endocrine systems. Even though physiologists may examine body functions at any level from cells to systems (as indicated in the title of this book), their ultimate goal is to integrate these mechanisms into the big picture of how the entire organism works as a cohesive whole.

Currently, researchers are hotly pursuing several approaches for repairing or replacing tissues or organs that can no longer adequately perform vital functions because of disease, trauma, or age-related changes. (See the boxed feature on pp. 10 and 11, ▮Concepts, Challenges, and Controversies. Each chapter has similar boxed features that explore in greater depth high-interest information on such diverse topics as environmental impact on the body, aging, ethical issues, discoveries regarding common diseases, and historical perspectives.)

We next focus on how the different body systems normally work together to maintain the internal conditions necessary for life.

Check Your Understanding 1.2

1. List and describe the levels of organization in the body.
2. State the basic cell functions.
3. Name the four primary types of tissue and give an example of each.

1.3 Concept of Homeostasis

If each cell has basic survival skills, why can't the body cells live without performing specialized tasks and being organized according to specialization into systems that accomplish functions essential for the whole organism's survival? The cells in a multicellular organism cannot live and function without contributions from the other body cells because most cells are not in direct contact with the external environment. The **external environment** is the surrounding environment in which an organism lives. A single-celled organism such as an amoeba obtains nutrients and O_2 directly from its immediate external surroundings and eliminates wastes back into those surroundings. A muscle cell or any other cell in a multicellular organism has the same need for life-supporting nutrient and O_2 uptake and waste elimination; yet, the muscle cell is isolated from the external environment surrounding the body. How can it make vital exchanges with the external environment with which it has no contact? The key is the presence of a watery internal environment. The **internal environment** is the fluid that surrounds the cells and through which they make life-sustaining exchanges.

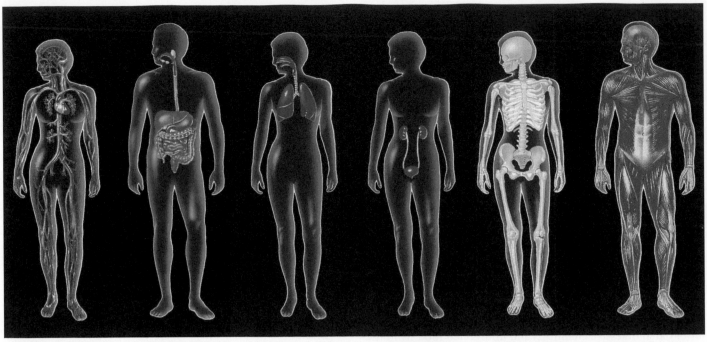

Circulatory system	**Digestive system**	**Respiratory system**	**Urinary system**	**Skeletal system**	**Muscular system**
heart, blood vessels, blood	mouth, pharynx, esophagus, stomach, small intestine, large intestine, salivary glands, exocrine pancreas, liver, gallbladder	nose, pharynx, larynx, trachea, bronchi, lungs	kidneys, ureters, urinary bladder, urethra	bones, cartilage, joints	skeletal muscles

❙Figure 1-5 **Components of the body systems.**

Body cells are in contact with a privately maintained internal environment.

The fluid collectively contained within all body cells is called **intracellular fluid (ICF)**. The fluid outside the cells is called **extracellular fluid (ECF)**. Note that the ECF is outside the cells but inside the body. Thus, the ECF is the internal environment of the body. You live in the external environment; your cells live in the body's internal environment.

ECF is made up of two components: the **plasma**, the fluid portion of the blood, and the **interstitial fluid**, which surrounds and bathes the cells (*inter* means "between"; *stitial* means "that which stands") (❙Figure 1-6).

No matter how remote a cell is from the external environment, it can make life-sustaining exchanges with its surrounding fluid. Particular body systems accomplish the transfer of materials between the external environment and the internal environment so that the composition of the internal environment is appropriately maintained to support the life and functioning of the cells. The digestive system transfers the nutrients required by all body cells from the external environment into the plasma, and the respiratory system transfers O_2 from the external environment into the plasma. The circulatory system distributes these nutrients and O_2 throughout the body. Materials are thoroughly mixed and exchanged between the plasma and the interstitial fluid across the capillaries, the smallest and thinnest of blood vessels. As a result, the nutrients and O_2 originally obtained from the external environment are delivered to the interstitial fluid, from which the body cells pick up these needed supplies. Similarly, wastes produced by the cells are released into the interstitial fluid, picked up by the plasma, and transported to the organs that specialize in eliminating these wastes from the internal environment to the external environment. The lungs remove CO_2 from the plasma and blow out this waste, and the kidneys remove other wastes for elimination in the urine.

Thus, a body cell takes in essential nutrients from its watery surroundings and eliminates wastes into these same surroundings, just as an amoeba does. The main difference is that each body cell must help maintain the composition of the internal environment so that this fluid continuously remains suitable to support the existence of all body cells. In contrast, an amoeba does nothing to regulate its surroundings.

Body systems maintain homeostasis, a dynamic steady state in the internal environment.

Body cells can live and function only when the ECF is compatible with their survival; thus, the chemical composition and physical state of this internal environment must be maintained within narrow limits. As cells take up nutrients and O_2 from the internal environment, these essential materials must constantly be replenished. Likewise, wastes must constantly be removed from the internal environment so that they do not reach toxic

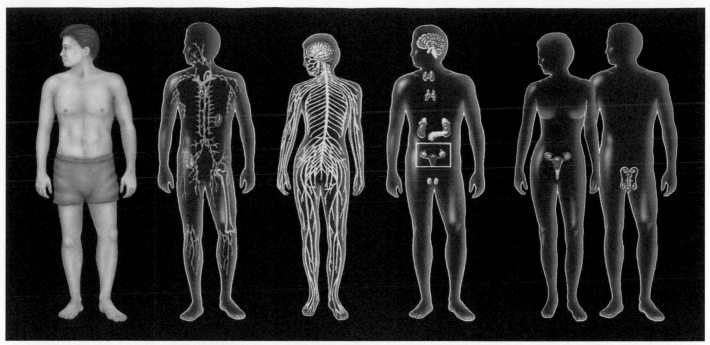

Integumentary system
skin, hair, nails

Immune system
lymph nodes, thymus, bone marrow, tonsils, adenoids, spleen, appendix, and, not shown, white blood cells, gut-associated lymphoid tissue, skin-associated lymphoid tissue

Nervous system
brain, spinal cord, peripheral nerves, and, not shown, special sense organs

Endocrine system
all hormone-secreting tissues, including hypothalamus, pituitary, thyroid, adrenals, endocrine pancreas, gonads, kidneys, pineal, thymus, and, not shown, parathyroids, intestine, heart, skin, adipose tissue

Reproductive system
Male: testes, penis, prostate gland, seminal vesicles, bulbourethral glands, associated ducts

Female: ovaries, oviducts, uterus, vagina, breasts

levels. Other aspects of the internal environment important for maintaining life, such as temperature, also must be kept relatively constant. Maintenance of a relatively stable internal environment is termed **homeostasis** (*homeo* means "similar"; *stasis* means "to stand or stay").

The functions performed by each body system contribute to homeostasis, thereby maintaining within the body the environment required for the survival and function of all cells. Cells, in turn, make up body systems. This is the central theme of physiology and of this book: *Homeostasis is essential for the survival of each cell, and each cell, through its specialized activities as part of a body system, helps maintain the internal environment shared by all cells* (Figure 1-7, p. 12).

The internal environment must be kept relatively stable, but this does not mean that its composition, temperature, and other characteristics are absolutely unchanging. Both external and internal factors continuously threaten to disrupt homeostasis. When any factor starts to move the internal environment away from optimal conditions, the body systems initiate appropriate counter-reactions to minimize the change. For example, when you're exposed to a cold environmental temperature (an external factor), your body temperature tends to fall. In response, the temperature control center in your brain initiates compensatory measures, such as shivering, to raise your body temperature to normal. By contrast, when you exercise, your working muscles produce extra heat (an internal factor) that tends to increase your body temperature. In response, the temperature control

center brings about sweating and other compensatory measures to reduce your body temperature to normal.

Thus, homeostasis is not a rigid, fixed state but a dynamic steady state in which changes that occur are minimized by compensatory physiological responses. The term *dynamic* refers to each homeostatically regulated factor being marked by continuous change, whereas *steady state* implies that these changes do

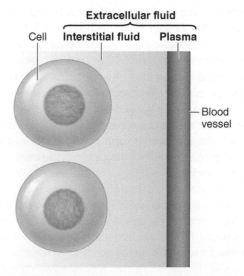

I Figure 1-6 **Components of the extracellular fluid (internal environment).**

Stem Cell Science and Regenerative Medicine: Making Defective Body Parts Like New Again

LIVER FAILURE, PARALYZING SPINAL-CORD INJURY, diabetes mellitus, damaged heart muscle, arthritis, extensive burns, a cancerous breast, an arm mangled in an accident—although our bodies are remarkable and normally serve us well, sometimes a body part is defective, injured beyond repair, or lost in such situations. In addition to the loss of quality of life for affected individuals, the cost of treating patients with lost, permanently damaged, or failing organs accounts for about half of the total health-care expenditures in the United States. Ideally, when the body suffers an irreparable loss, new, permanent replacement parts would be substituted to restore normal function and appearance. Fortunately, this possibility is moving rapidly from the realm of science fiction to the reality of scientific progress.

The Medical Promise of Stem Cells

Stem cells offer exciting medical promise for repairing or replacing organs that are diseased, damaged, or worn out. **Stem cells** are versatile cells that are not specialized for a specific function but can divide to give rise to highly specialized cells while maintaining a supply of new stem cells. Two natural categories of stem cells are under investigation: embryonic stem cells from early embryos and tissue-specific stem cells from adults. **Embryonic stem cells (ESCs)** result from the early divisions of a fertilized egg. These undifferentiated cells ultimately give rise to all mature, specialized cells of the body while simultaneously self-renewing. ESCs are *pluripotent,* meaning they have the potential to generate any of the more than 200 cell types in the body if given the appropriate cues.

During development, the undifferentiated ESCs give rise to many partially differentiated **tissue-specific stem cells**, each of which becomes committed to generating the highly differentiated, specialized cell types that compose a particular tissue. For example, tissue-specific muscle stem cells give rise to specialized muscle cells. Some tissue-specific stem cells remain in adult tissues, where they serve as a continual source of new specialized cells to maintain or repair that particular tissue. Tissue-specific stem cells are even present in adult brain and muscle tissue. Even though mature nerve and muscle cells cannot reproduce themselves, to a limited extent adult brains and muscles can grow new cells throughout life by means of these persisting stem cells. However, this process is too slow to keep pace with major losses, as in a stroke or heart attack.

In 1998, for the first time, researchers succeeded in isolating human ESCs and maintaining them indefinitely in an undifferentiated state in culture. With **cell culture**, cells isolated from a living organism continue to thrive and reproduce in laboratory dishes when supplied with appropriate nutrients and supportive materials.

The medical promise of ESCs lies in their potential to serve as an all-purpose material that can be coaxed into whatever cell types are needed to patch up the body. Studies since their discovery demonstrate that these cells have the ability to differentiate into particular cells when exposed to the appropriate chemical signals. As scientists gradually learn to prepare the right cocktail of chemical signals to direct the undifferentiated cells into the desired cell types, they will have the potential to fill deficits in damaged or dead tissues with healthy cells.

Ethical Concerns and Political Issues

Despite this potential, ESC research is fraught with controversy because of the source of these cells: They are isolated from discarded embryos from abortion clinics and in vitro fertility ("test-tube baby") clinics. Opponents of using ESCs are morally and ethically concerned because embryos are destroyed in the process of harvesting these cells. Proponents argue that these embryos were destined to be destroyed anyway—a decision already made by the parents of the embryos—and that these stem cells have great potential for alleviating human suffering. Thus, ESC science has become inextricably linked with stem cell politics.

Since the mid-1990s, federal funding in the United States for research involving cells derived from embryos has been on-and-off as presidents, Congress, and the courts have grappled with ethical, legal, and public policy challenges. In general, ESC research has been encouraged by Democratic administrations and constrained under Republican leadership. Federally funded research on human ESCs is currently legal as a result of the latest court battle that ended in 2012, but the controversy persists. In the meantime, for more than the past two decades research using ESCs has continued to move forward, funded by private and state resources along with see-saw federal funding.

The Search for Noncontroversial Stem Cells

Because of the controversies and setbacks surrounding use of ESCs, some researchers have searched for alternative ways to obtain stem cells, such as by using tissue-specific stem cells from adult tissues as a substitute for pluripotent ESCs. These adult stem cells were thought to give rise only to the specialized cells of a particular tissue. However, although these partially differentiated adult stem cells do not have the complete developmental potential of ESCs, they have been coaxed into producing a wider variety of cells than originally thought possible. To name a few examples, provided the right supportive environment, stem cells from the brain can give rise to blood cells and fat-tissue stem cells to bone, cartilage, and muscle cells. Although ESCs hold greater potential for developing treatments for a broader range of diseases, adult stem cells are more accessible than ESCs, and their use is not controversial.

Long-running political setbacks have inspired still other scientists to search for new ways to obtain more versatile stem cells without destroying embryos. The newest technique with the greatest potential involves turning back the clock on adult specialized cells, such as easily obtained skin cells, and converting them to their embryonic state.

Like ESCs, these reprogrammed cells, called **induced pluripotent stem cells (iPSCs)**, have the potential to differentiate into any cell type in the body. However, iPSCs are less responsive than ESCs to differentiation signals. iPSCs retain subtle memories of the tissue from which they came, which reduces the efficiency with which they can be converted into new cell types. More recently, investigators have directly converted skin cells into neurons without first reverting them to an embryonic state.

Other researchers have been looking for a nonembryonic source of the more versatile ESCs. One group recently succeeded in creating patient-specific human ESCs through controversial **therapeutic cloning**. With this technique, the nucleus (gene-containing part of a cell) is removed from a patient's skin cell and transferred into a healthy, unfertilized egg cell (donated by another individual) that has been stripped of its own nucleus. When properly cultivated, the egg, which now genetically matches the patient, multiplies into a mass from which ESCs that bear the identical DNA of the patient can be harvested. As a less controversial alternative, another investigator has identified embryoniclike stem cells in breast milk that appear to be pluripotent.

Through rapidly evolving new techniques such as these, soon any mature body cell likely can be converted into any other cell type. Cell reprogramming is one of the hottest areas of investigation in life sciences today.

Whatever the source of the cells, stem cell research promises to revolutionize medicine in the 21st century as profoundly as the impact of vaccines and antibiotics in the 20th century. An estimated 3000 Americans die every day from conditions that may in the future be treatable with stem cell derivatives. As examples of successes, a healthy version of a specific type of ESC-derived eye cell is currently being injected into the eyes of patients with an inherited condition that leads to blindness to save their sight, and stem cells injected into failing hearts have replaced damaged heart muscle thereby improving the hearts' pumping ability. Scientists are even working on growing customized tissues and eventually whole, made-to-order replacement organs. **Regenerative medicine** is the emerging field that involves repairing, replacing, or regenerating cells, tissues, or organs to establish normal function.

The Medical Promise of Regenerative Medicine

The era of regenerative medicine is being ushered in by advances in stem cell science, cell biology, plastic manufacturing, and computer graphics. To make lab-grown replacement parts, using computer-aided designs, *tissue engineers* shape pure, biodegradable plastics into three-dimensional molds or scaffoldings that mimic the structure of a particular tissue or organ. They then "seed" the plastic mold with the desired stem cell types, which they coax, by applying appropriate nourishing and stimulatory chemicals, into multiplying and assembling into the desired body part. After the plastic scaffolding dissolves, only the newly generated tissue remains, ready to be implanted into a patient as a permanent, living replacement part.

Following are some of the tissue engineers' early accomplishments. Engineered skin patches are being used to treat victims of severe burns, and engineered cartilage and bone graft substitutes are already in use. Lab-grown bladders were the first organs successfully implanted in humans, and engineered blood vessels and tracheas (windpipes) have already been built. Tissue-engineered scaffolding to promote nerve regeneration is ready for clinical trials. Progress has been made on growing more complicated organs, including liver, pancreas, heart, kidney, and lung.

Innovative scaffolds have continued to advance transplant technology in recent years. With a newer technique, some investigators are experimenting with three-dimensional *organ printing*. Based on the principle used in desktop printers, organ printing involves computer-aided layer-by-layer deposition of "biological ink." Biological ink consists of cells, scaffold materials, and supportive growth factors that are simultaneously printed in thin layers in a highly organized pattern based on the anatomy of the organ under construction. Fusion of these living layers forms a three-dimensional structure that mimics the body part the printed organ is designed to replace.

The latest scaffolding strategy uses nature's architecture instead of building an intricate scaffold from scratch. Still in the experimental stage, this technique involves chemically stripping all flesh from a donor organ, such as a pig heart, leaving behind a heart-shaped, natural scaffold consisting of collagen, a tough connective tissue fiber. When this naked heart shell is coated with tissue grown from a recipient's stem cells, the cells rebuild a beating heart complete with its blood supply that can be transplanted into the recipient. The goal is to ultimately remove all natural cells from pig hearts (or other organs) and use them as scaffolds for recreating hearts compatible for transplant into humans, thereby overcoming donor shortage and avoiding the risk of transplant rejection.

On the horizon is an emerging capability to produce iPSCs in the body to regenerate damaged parts in place, inside the body itself— the ultimate scaffolding. Another related avenue of exploration is the possibility of promoting regeneration of body parts salamander-style. Some species, including vertebrates such as salamanders and newts, have extraordinary regenerative powers. Whereas humans form scar tissue at the stump end of a severed limb, these species grow entire replacement limbs. Scientists are hopeful that by studying how salamanders regenerate lost tissue, they can duplicate these steps in humans. Most investigators in the field think that humans have the latent capacity to regenerate but have built-in brakes on the process. How to remove the brakes or stimulate the process is the challenge.

Whatever way it happens, replacing or regenerating defective body parts with "the real thing" cannot come soon enough.

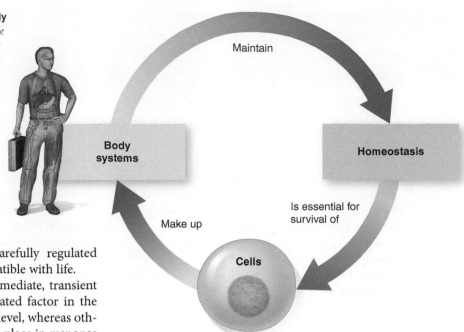

Figure 1-7 Interdependent relationship of cells, body systems, and homeostasis. Homeostasis is essential for the survival of cells, cells make up body systems, and body systems maintain homeostasis. This relationship serves as the foundation for physiology.

Maintain

Body systems

Homeostasis

Is essential for survival of

Make up

Cells

not deviate far from a constant, or steady, level. This situation is comparable to the minor steering adjustments you make as you drive a car along a straight stretch of highway. Small fluctuations around the optimal level for each factor in the internal environment are normally kept, by carefully regulated mechanisms, within the narrow limits compatible with life.

Some compensatory mechanisms are immediate, transient responses to a situation that moves a regulated factor in the internal environment away from the desired level, whereas others are more long-term adaptations that take place in response to prolonged or repeated exposure to a situation that disrupts homeostasis. Long-term adaptations make the body more efficient in responding to an ongoing or repetitive challenge. The body's reaction to exercise includes examples of both short-term compensatory responses and long-term adaptations among the different body systems. (See the accompanying boxed feature, ▮A Closer Look at Exercise Physiology. Most chapters have a boxed feature focusing on exercise physiology. Also, we mention issues related to exercise physiology throughout the text. Appendix B will help you locate all the references to this important topic.)

Homeostatically Regulated Factors Many factors of the internal environment must be homeostatically maintained. They include the following:

1. *Concentration of nutrients.* Cells need a constant supply of nutrient molecules for energy production. Energy, in turn, is needed to support life-sustaining and specialized cell activities.

2. *Concentration of O_2 and CO_2.* Cells need O_2 to carry out energy-yielding chemical reactions. The CO_2 produced during these reactions must be removed so that acid-forming CO_2 does not increase the acidity of the internal environment.

3. *Concentration of waste products.* The end products of some chemical reactions have a toxic effect on body cells if these wastes are allowed to accumulate.

4. *pH.* Changes in the pH (relative amount of acid; see pp. 547–548 and A-8) of the ECF adversely affect nerve cell function and wreak havoc with the enzyme activity of all cells.

5. *Concentrations of water, salt, and other electrolytes.* Because the relative concentrations of salt (NaCl) and water in the ECF influence how much water enters or leaves the cells, these concentrations are carefully regulated to maintain the proper volume of the cells. Cells do not function normally when they are swollen or shrunken. Other electrolytes (chemicals that form ions in solution and conduct electricity; see p. A-3 and A-7) perform a variety of vital functions. For example, the rhythmic

beating of the heart depends on a relatively constant concentration of potassium (K^+) in the ECF.

6. *Volume and pressure.* The circulating component of the internal environment, the plasma, must be maintained at adequate volume and blood pressure to ensure bodywide distribution of this important link between the external environment and the cells.

7. *Temperature.* Body cells function best within a narrow temperature range. If cells are too cold, their functions slow down too much; if they get too hot, their structural and enzymatic proteins are impaired or destroyed.

Body System Contributions to Homeostasis The 11 body systems contribute to homeostasis in the following important ways (▮Figure 1-8):

1. The *circulatory system* (heart, blood vessels, and blood) transports materials such as nutrients, O_2, CO_2, wastes, electrolytes, and hormones from one part of the body to another.

2. The *digestive system* (mouth, esophagus, stomach, intestines, and related organs) breaks down dietary food into small nutrient molecules that can be absorbed into the plasma for distribution to the body cells. It also transfers water and electrolytes from the external environment into the internal environment. It eliminates undigested food residues to the external environment in the feces.

3. The *respiratory system* (lungs and major airways) gets O_2 from and eliminates CO_2 to the external environment. By adjusting the rate of removal of acid-forming CO_2, the respiratory system is also important in maintaining the proper pH of the internal environment.

4. The *urinary system* (kidneys and associated "plumbing") removes excess water, salt, acid, and other electrolytes from the plasma and eliminates them in the urine, along with waste products other than CO_2.

What Is Exercise Physiology?

EXERCISE PHYSIOLOGY is the study of both the functional changes that occur in response to a single session of exercise and the adaptations that occur as a result of regular, repeated exercise sessions. Exercise initially disrupts homeostasis. The changes that occur in response to exercise are the body's attempt to meet the challenge of maintaining homeostasis when increased demands are placed on the body. Exercise often requires prolonged coordination among most body systems, including the muscular, skeletal, nervous, circulatory, respiratory, urinary, integumentary (skin), and endocrine (hormone-producing) systems.

Heart rate is one of the easiest factors to monitor that shows both an immediate response to exercise and long-term adaptation to a regular exercise program. When a person begins to exercise, the ac-tive muscle cells use more O_2 to support their increased energy demands. Heart rate increases to deliver more oxygenated blood to the exercising muscles. The heart adapts to regular exercise of sufficient intensity and duration by increasing its strength and efficiency so that it pumps more blood per beat. Because of increased pumping ability, the heart does not have to beat as rapidly to pump a given quantity of blood as it did before physical training.

Exercise physiologists study the mechanisms responsible for the changes that occur as a result of exercise. Much of the knowledge gained from studying exercise is used to develop appropriate exercise programs that increase the functional capacities of people ranging from athletes to the infirm. The importance of proper and sufficient exercise in disease prevention and rehabilitation is increasingly evident.

---

5. The *skeletal system* (bones and joints) provides support and protection for the soft tissues and organs. It also serves as a storage reservoir for calcium (Ca^{2+}), an electrolyte whose plasma concentration must be maintained within narrow limits. Together with the muscular system, the skeletal system enables the body and its parts to move. Furthermore, the bone marrow—the soft interior portion of some types of bone—is the ultimate source of all blood cells.

6. The *muscular system* (skeletal muscles) moves the bones to which the skeletal muscles are attached. From a purely homeostatic view, this system enables a person to move toward food or away from harm. Furthermore, the heat generated by muscle contraction helps maintain body temperature. In addition, because skeletal muscles are under voluntary control, a person can use them to accomplish myriad other movements by choice. These movements, which range from the fine motor skills required for delicate needlework to the powerful movements involved in weight lifting, are not necessarily directed toward maintaining homeostasis.

7. The *integumentary system* (skin and related structures) serves as an outer protective barrier that prevents internal fluid from being lost from the body and foreign microorganisms from entering. This system is also important in regulating body temperature. The amount of heat lost from the body surface to the external environment can be adjusted by controlling sweat production and by regulating the flow of warm blood through the skin.

8. The *immune system* (white blood cells and lymphoid organs) defends against foreign invaders such as bacteria and viruses and against body cells that have become cancerous. It also paves the way for repairing or replacing injured or worn-out cells.

9. The *nervous system* (brain, spinal cord, nerves, and sense organs) is one of the body's two major regulatory systems. In general, it controls and coordinates body activities that require swift responses. It is especially important in detecting changes in the external environment and initiating reactions to them. Furthermore, it is responsible for higher functions that are not entirely directed toward maintaining homeostasis, such as consciousness, memory, and creativity.

10. The *endocrine system* (all hormone-secreting glands) is the other major regulatory system. In contrast to the nervous system, the endocrine system in general regulates activities that require duration rather than speed, such as growth. It is especially important in controlling the blood concentration of nutrients and, by adjusting kidney function, controlling the volume and electrolyte composition of the ECF.

11. The *reproductive system* (male and female gonads—testes and ovaries, respectively—and related organs) is not essential for homeostasis and therefore is not essential for survival of the individual. It is essential, however, for perpetuating the species.

As we examine each of these systems in greater detail, always keep in mind that the body is a coordinated whole even though each system provides its own special contributions. It is easy to forget that all body parts actually fit together into a functioning, interdependent whole body. Accordingly, each chapter begins with a discussion of how the body system to be described fits into the body as a whole. In addition, each chapter ends with a brief review of the homeostatic contributions of the body system. As a further tool to help you keep track of how all the pieces fit together, Figure 1-8 is duplicated on the inside front cover as a handy reference.

Also be aware that the functioning whole is greater than the sum of its separate parts. Through specialization, cooperation, and interdependence, cells combine to form an integrated, unique, single living organism with more diverse and complex capabilities than are possessed by any of the cells that make it

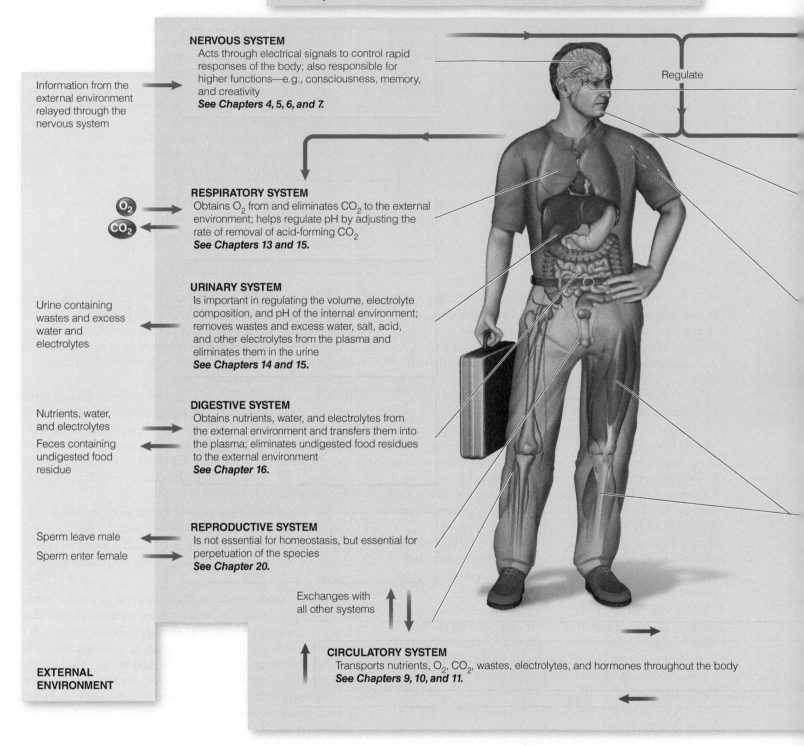

BODY SYSTEMS
Made up of cells organized according to specialization to maintain homeostasis
See Chapter 1.

NERVOUS SYSTEM
Acts through electrical signals to control rapid responses of the body; also responsible for higher functions—e.g., consciousness, memory, and creativity
See Chapters 4, 5, 6, and 7.

Information from the external environment relayed through the nervous system

Regulate

O_2
CO_2

RESPIRATORY SYSTEM
Obtains O_2 from and eliminates CO_2 to the external environment; helps regulate pH by adjusting the rate of removal of acid-forming CO_2
See Chapters 13 and 15.

URINARY SYSTEM
Is important in regulating the volume, electrolyte composition, and pH of the internal environment; removes wastes and excess water, salt, acid, and other electrolytes from the plasma and eliminates them in the urine
See Chapters 14 and 15.

Urine containing wastes and excess water and electrolytes

DIGESTIVE SYSTEM
Obtains nutrients, water, and electrolytes from the external environment and transfers them into the plasma; eliminates undigested food residues to the external environment
See Chapter 16.

Nutrients, water, and electrolytes

Feces containing undigested food residue

REPRODUCTIVE SYSTEM
Is not essential for homeostasis, but essential for perpetuation of the species
See Chapter 20.

Sperm leave male

Sperm enter female

Exchanges with all other systems

EXTERNAL ENVIRONMENT

CIRCULATORY SYSTEM
Transports nutrients, O_2, CO_2, wastes, electrolytes, and hormones throughout the body
See Chapters 9, 10, and 11.

I Figure 1-8 **Role of the body systems in maintaining homeostasis.**
FIGURE FOCUS: *By examining this figure, indicate what four body systems are important in maintaining the proper concentration of the electrolyte calcium in the blood.*

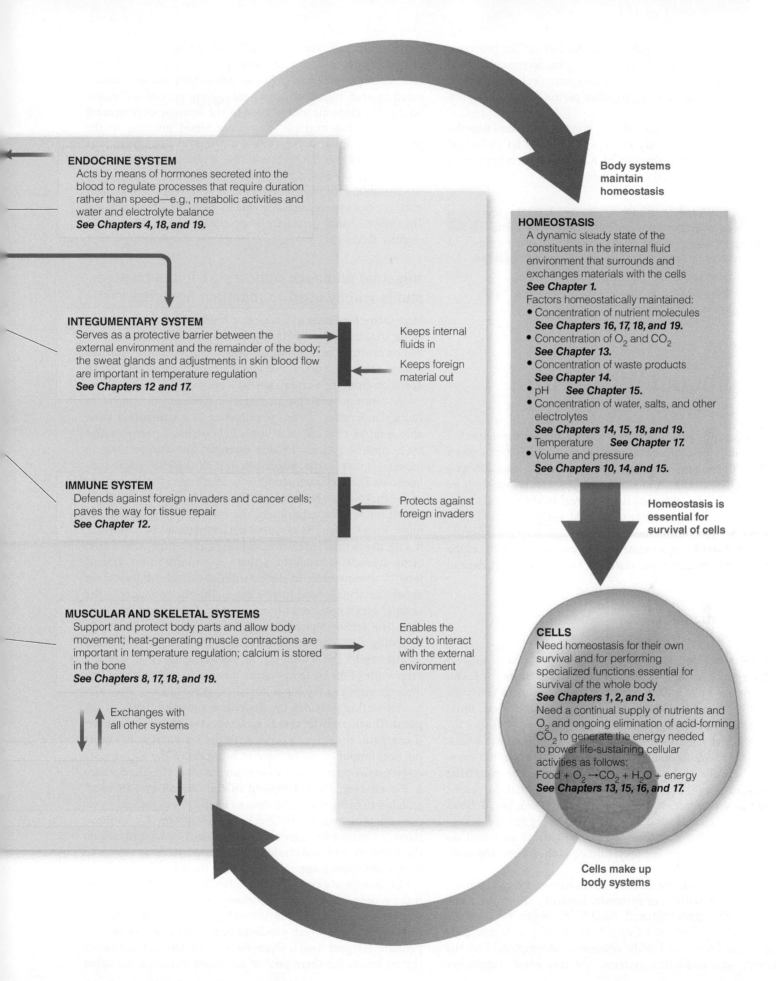

ENDOCRINE SYSTEM
Acts by means of hormones secreted into the blood to regulate processes that require duration rather than speed—e.g., metabolic activities and water and electrolyte balance
See Chapters 4, 18, and 19.

INTEGUMENTARY SYSTEM
Serves as a protective barrier between the external environment and the remainder of the body; the sweat glands and adjustments in skin blood flow are important in temperature regulation
See Chapters 12 and 17.

Keeps internal fluids in

Keeps foreign material out

IMMUNE SYSTEM
Defends against foreign invaders and cancer cells; paves the way for tissue repair
See Chapter 12.

Protects against foreign invaders

MUSCULAR AND SKELETAL SYSTEMS
Support and protect body parts and allow body movement; heat-generating muscle contractions are important in temperature regulation; calcium is stored in the bone
See Chapters 8, 17, 18, and 19.

Enables the body to interact with the external environment

Exchanges with all other systems

Body systems maintain homeostasis

HOMEOSTASIS
A dynamic steady state of the constituents in the internal fluid environment that surrounds and exchanges materials with the cells
See Chapter 1.
Factors homeostatically maintained:
• Concentration of nutrient molecules
  See Chapters 16, 17, 18, and 19.
• Concentration of O_2 and CO_2
  See Chapter 13.
• Concentration of waste products
  See Chapter 14.
• pH   *See Chapter 15.*
• Concentration of water, salts, and other electrolytes
  See Chapters 14, 15, 18, and 19.
• Temperature   *See Chapter 17.*
• Volume and pressure
  See Chapters 10, 14, and 15.

Homeostasis is essential for survival of cells

CELLS
Need homeostasis for their own survival and for performing specialized functions essential for survival of the whole body
See Chapters 1, 2, and 3.
Need a continual supply of nutrients and O_2 and ongoing elimination of acid-forming CO_2 to generate the energy needed to power life-sustaining cellular activities as follows:
Food + $O_2 \rightarrow CO_2 + H_2O$ + energy
See Chapters 13, 15, 16, and 17.

Cells make up body systems

up. For humans, these capabilities go far beyond the processes needed to maintain life. A cell, or even a random combination of cells, cannot create an artistic masterpiece or design a spacecraft, but body cells working together permit those capabilities in an individual.

Now that you have learned what homeostasis is and how the functions of different body systems maintain it, let us look at the regulatory mechanisms by which the body reacts to changes and controls the internal environment.

Check Your Understanding 1.3

1. Distinguish among external environment, internal environment, intracellular fluid, extracellular fluid, plasma, and interstitial fluid.
2. Define *homeostasis*.
3. Draw a figure showing the interdependent relationship of cells, body systems, and homeostasis.

1.4 Homeostatic Control Systems

A **homeostatic control system** is a functionally interconnected network of body components that operates to maintain a given factor in the internal environment at a relatively constant optimal level. To maintain homeostasis, the control system must be able to (1) detect deviations from normal in the internal environmental factor that needs to be held within narrow limits; (2) integrate this information with any other relevant information; and (3) make appropriate adjustments in the activity of the body parts responsible for restoring this factor to its desired value.

Homeostatic control systems may operate locally or bodywide.

Homeostatic control systems can be grouped into two classes—intrinsic and extrinsic controls. **Intrinsic**, or **local, controls** are built into or are inherent in an organ (*intrinsic* means "within"). For example, as an exercising skeletal muscle rapidly uses up O_2 to generate energy to support its contractile activity, the O_2 concentration within the muscle falls. This local chemical change acts directly on the smooth muscle in the walls of the blood vessels supplying the exercising muscle, causing the smooth muscle to relax so that the vessels dilate, or open widely. As a result, increased blood flows through the dilated vessels into the exercising muscle, bringing in more O_2. This local mechanism helps maintain an optimal level of O_2 in the fluid immediately around the exercising muscle's cells.

Most factors in the internal environment are maintained, however, by **extrinsic**, or **systemic, controls**, which are regulatory mechanisms initiated outside an organ to alter the organ's activity (*extrinsic* means "outside of"). Extrinsic control of the organs and body systems is accomplished by the nervous and endocrine systems, the two major regulatory systems. Extrinsic control permits coordinated regulation of several organs toward a common goal; in contrast, intrinsic controls serve only the organ in which they occur. Coordinated, overall regulatory mechanisms are crucial for maintaining the dynamic steady state in the internal environment as a whole. For example, to restore blood pressure to the proper level when it falls too low, the nervous system acts simultaneously on the heart and blood vessels throughout the body to increase blood pressure to normal.

To stabilize the physiological factor being regulated, homeostatic control systems must be able to detect and resist change. They resist change primarily by operating on the principle of negative feedback.

Negative feedback opposes an initial change and is widely used to maintain homeostasis.

In **negative feedback**, a change in a homeostatically controlled factor triggers a response that seeks to restore the factor to normal by moving the factor in the opposite direction of its initial change—that is, a corrective adjustment opposes the original deviation from the normal desired level.

A common example of negative feedback is control of room temperature. Room temperature is a **controlled variable**, a factor that can vary but is held within a narrow range by a control system. In our example, the control system includes a thermostat, a furnace, and all their electrical connections. The room temperature is determined by the activity of the furnace, a heat source that can be turned on or off. To switch on or off appropriately, the control system as a whole must "know" what the *actual* room temperature is, "compare" it with the *desired* room temperature, and "adjust" the output of the furnace to bring the actual temperature to the desired level. A thermometer in the thermostat provides information about the actual room temperature. The thermometer is the **sensor**, which monitors the magnitude of the controlled variable. The sensor typically converts the original information regarding a change into a "language" the control system can "understand." For example, the thermometer converts the magnitude of the air temperature into electrical impulses. This message serves as the input into the control system. The thermostat setting provides the desired temperature level, or **set point**. The thermostat acts as an **integrator**, or **control center**: It compares the sensor's input with the set point and adjusts the heat output of the furnace to bring about the appropriate response to oppose a deviation from the set point. The furnace is the **effector**, the component of the control system commanded to bring about the desired effect. These general components of a negative-feedback control system are summarized in Figure 1-9a. Carefully examine this figure and its key; the symbols and conventions introduced here are used in comparable flow diagrams throughout the text.

Let us look at a typical negative-feedback loop. For example, if the room temperature falls below the set point because it is cold outside, the thermostat, through connecting circuitry, activates the furnace, which produces heat to raise the room temperature (❙Figure 1-9b). Once the room temperature reaches the set point, the thermometer no longer detects a deviation

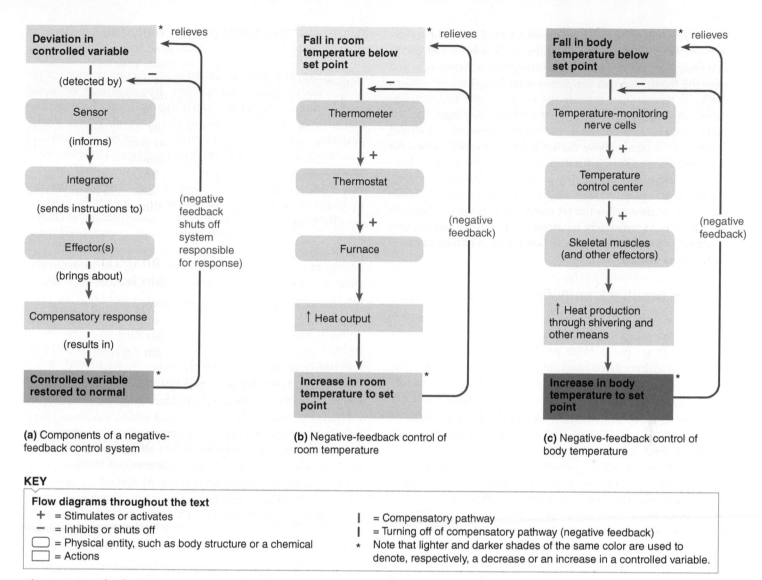

(a) Components of a negative-feedback control system

(b) Negative-feedback control of room temperature

(c) Negative-feedback control of body temperature

KEY

Flow diagrams throughout the text
+ = Stimulates or activates
− = Inhibits or shuts off
⬭ = Physical entity, such as body structure or a chemical
▭ = Actions

| = Compensatory pathway
| = Turning off of compensatory pathway (negative feedback)
* Note that lighter and darker shades of the same color are used to denote, respectively, a decrease or an increase in a controlled variable.

I Figure 1-9 Negative feedback.
FIGURE FOCUS: *The hormone erythropoietin stimulates production of red blood cells. Would the rate of erythropoietin secretion go up, go down, or stay the same after you donate blood?*

from that point. As a result, the activating mechanism in the thermostat and the furnace are switched off. Thus, the heat from the furnace counteracts, or is "negative" to, the original fall in temperature. If the heat-generating pathway was not shut off once the target temperature was reached, heat production would continue and the room would get hotter and hotter. Overshooting the set point does not occur because the heat "feeds back" to shut off the thermostat that triggered its output. Thus, a negative-feedback control system detects a change away from the ideal value in a controlled variable, initiates mechanisms to correct the situation, and then shuts itself off. In this way, the controlled variable does not drift too far above or below the set point.

What if the original deviation is a rise in room temperature above the set point because it is hot outside? A heat-producing furnace is of no use in returning the room temperature to the desired level. An opposing control system involving a cooling air conditioner is needed to reduce the room temperature. In

this case, the thermostat, through connecting circuitry, activates the air conditioner, which cools the room air, the opposite effect from that of the furnace. In negative-feedback fashion, once the set point is reached, the air conditioner is turned off to prevent the room from becoming too cold. Note that if the controlled variable can be deliberately adjusted to oppose a change in one direction only, the variable can move in an uncontrolled fashion in the opposite direction. For example, if the house is equipped only with a furnace that produces heat to oppose a fall in room temperature, no mechanism is available to prevent the house from getting too hot in warm weather. However, the room temperature can be kept relatively constant through two opposing mechanisms, one that heats and one that cools the room, despite wide variations in the temperature of the external environment.

Homeostatic negative-feedback systems in the human body operate in the same way. For example, when temperature-monitoring nerve cells detect a decrease in body temperature below the desired level, these sensors signal the temperature control center in the brain, which begins a sequence of events that ends in responses, such as shivering, that generate heat and raise the temperature to the proper level (I Figure 1-9c). When the body temperature reaches the set point, the temperature-monitoring nerve cells turn off the stimulatory signal to the skeletal muscles. As a result, the body temperature does not continue to increase above the set point. Conversely, when the temperature-monitoring nerve cells detect a rise in body temperature above normal, cooling mechanisms, such as sweating, are called into play to lower the temperature to normal. When the temperature reaches the set point, the cooling mechanisms are shut off. As with body temperature, opposing mechanisms can move most homeostatically controlled variables in either direction as needed.

Positive feedback amplifies an initial change.

In negative feedback, a control system's output is regulated to resist change so that the controlled variable is kept at a relatively steady set point. With **positive feedback**, by contrast, the output enhances or amplifies a change so that the controlled variable continues to move in the direction of the initial change. Such action is comparable to the heat generated by a furnace triggering the thermostat to call for even more heat output from the furnace so that the room temperature continuously rises.

Because the major goal in the body is to maintain stable, homeostatic conditions, positive feedback does not occur nearly as often as negative feedback. Positive feedback does play an important role in certain instances, however, as in the birth of a baby. The hormone oxytocin causes powerful contractions of the uterus (womb). As the contractions push the baby against the cervix (the exit from the uterus), the resultant stretching of the cervix triggers a sequence of events that brings about the release of even more oxytocin, which causes even stronger uterine contractions, triggering the release of more oxytocin, and so on. This positive-feedback cycle does not stop until the cervix is stretched sufficiently for the baby to be pushed through and born. Likewise, all other normal instances of positive-feedback cycles in the body include some mechanism for stopping the cycle.

Feedforward mechanisms initiate responses in anticipation of a change.

In addition to feedback mechanisms, which bring about a reaction to a change in a regulated variable, the body less frequently uses feedforward mechanisms, which respond in anticipation of a change in a regulated variable. For example, when a meal is still in the digestive tract, a feedforward mechanism increases secretion of a hormone (insulin) that promotes the cellular uptake and storage of ingested nutrients after they have been absorbed from the digestive tract. This anticipatory response helps limit the rise in blood nutrient concentration after nutrients have been absorbed.

Disruptions in homeostasis can lead to illness and death.

Clinical Note Despite control mechanisms, when one or more of the body's systems malfunction, homeostasis is disrupted and all cells suffer because they no longer have an optimal environment in which to live and function. Various pathophysiological states develop, depending on the type and extent of the disruption. The term **pathophysiology** refers to the abnormal functioning of the body (altered physiology) associated with disease. When a homeostatic disruption becomes so severe that it is no longer compatible with survival, death is the result.

Physiology serves as an important underpinning of clinical medicine. All health professionals must comprehend basic physiological principles to be able to understand what's happening in the body when things go wrong and to determine what needs to be done to correct the situation by whatever means, if possible. Many diagnostic tests rely heavily on principles learned by physiologists; examples include the electrocardiogram and lung function tests. Treatments for a number of pathophysiological conditions, such as high blood pressure, diabetes mellitus, and erectile dysfunction, are likewise based on knowledge acquired through physiological research. Thus physiology is at the heart of clinical practice.

Check Your Understanding 1.4

1. Distinguish between intrinsic controls and extrinsic controls.
2. Compare negative feedback and positive feedback.
3. Draw a flow diagram showing the relationships among the components of a negative-feedback control system.

Homeostasis: Chapter in Perspective

In this chapter you learned what homeostasis is: a dynamic steady state of the constituents in the internal fluid environment (the extracellular fluid) that surrounds and exchanges materials with the cells. Maintenance of homeostasis is essential for survival and normal functioning of cells. Each cell, through its specialized activities, contributes as part of a body system to the maintenance of homeostasis.

This relationship is the foundation of physiology and the central theme of this book. We have described how cells are organized according to specialization into body systems. How homeostasis is essential for cell survival and how body systems maintain this internal constancy are the topics covered in the rest of this book. Each chapter concludes with this capstone feature to facilitate your understanding of how the system under discussion contributes to homeostasis and of the interactions and interdependency of the body systems.

Review Exercises Answers begin on p. A-19

Reviewing Terms and Facts

1. Which of the following activities is *not* carried out by every cell in the body?
   a. obtaining O_2 and nutrients
   b. performing chemical reactions to acquire energy for the cell's use
   c. eliminating wastes
   d. largely controlling exchange of materials between the cell and its external environment
   e. reproducing

2. Which of the following is the proper progression of the levels of organization in the body?
   a. chemicals, cells, organs, tissues, body systems, whole body
   b. chemicals, cells, tissues, organs, body systems, whole body
   c. cells, chemicals, tissues, organs, whole body, body systems
   d. cells, chemicals, organs, tissues, whole body, body systems
   e. chemicals, cells, tissues, body systems, organs, whole body

3. Which of the following is *not* a type of connective tissue?
   a. bone
   b. blood
   c. the spinal cord
   d. tendons
   e. the tissue that attaches epithelial tissue to underlying structures

4. The term *tissue* can apply either to one of the four primary tissue types or to a particular organ's aggregate of cellular and extracellular components. (*True or false?*)

5. Cells in a multicellular organism have specialized to such an extent that they have little in common with single-celled organisms. (*True or false?*)

6. Cell specializations are usually a modification or elaboration of one of the basic cell functions. (*True or false?*)

7. The four primary types of tissue are _____, _____, _____, and _____.

8. The term _____ refers to the release from a cell, in response to appropriate stimulation, of specific products that have been synthesized largely by the cell.

9. _____ glands secrete through ducts to the outside of the body, whereas _____ glands release their secretory products, known as _____, internally into the blood.

10. _____ controls are inherent to an organ, whereas _____ controls are regulatory mechanisms initiated outside an organ that alter the activity of the organ.

11. Match the following:

1. circulatory system	(a) obtains O_2 and eliminates CO_2
2. digestive system	(b) supports, protects, and moves body parts
3. respiratory system	(c) controls, via hormones it secretes, processes that require duration
4. urinary system	
5. muscular and skeletal systems working together	(d) acts as the transport system
6. integumentary system	(e) removes wastes and excess water, salt, and other electrolytes
7. immune system	(f) perpetuates the species
8. nervous system	(g) obtains nutrients, water, and electrolytes
9. endocrine system	(h) defends against foreign invaders and cancer
10. reproductive system	(i) acts through electrical signals to control the body's rapid responses
	(j) serves as an outer protective barrier

Understanding Concepts
(Answers at www.cengagebrain.com)

1. Compare physiology and anatomy.

2. Explain the difference between basic cell functions and specialized cell functions and indicate in what way each of these categories of functions is essential for life in a multicellular organism.

3. Distinguish between the external environment and the internal environment. What constitutes the internal environment? Distinguish between the intracellular fluid (ICF) and the extracellular fluid (ECF). Discuss the relationship between the internal environment and the ECF. What fluid compartments make up the ECF?

4. State the central theme of physiology and of this book.

5. What factors must be homeostatically maintained, and which body systems contribute to maintaining each of these factors?

6. Define and describe the components of a homeostatic control system.

7. Why is negative feedback important physiologically?

Applying Clinical Reasoning

Jennifer R. has the "stomach flu" that is going around campus and has been vomiting profusely for the past 24 hours. Not only has she been unable to keep down fluids or food, but she also has lost the acidic digestive juices secreted by the stomach that are normally reabsorbed back into the blood farther down the digestive tract. In what ways might this condition threaten to disrupt homeostasis in Jennifer's internal environment? That is, what homeostatically maintained factors are moved away from normal by her profuse vomiting? What body systems respond to resist these changes?

Thinking at a Higher Level

1. Considering the nature of negative-feedback control and the function of the respiratory system, what effect do you predict that a decrease in CO_2 in the internal environment would have on how rapidly and deeply a person breathes?

2. Would the O_2 levels in the blood be (a) normal, (b) below normal, or (c) elevated in a patient with severe pneumonia resulting in impaired exchange of O_2 and CO_2 between the air and blood in the lungs? Would the CO_2 levels in the same patient's blood be (a) normal, (b) below normal, or (c) elevated? Because CO_2 reacts with H_2O to form carbonic acid (H_2CO_3), would the patient's blood (a) have a normal pH, (b) be too acidic, or (c) not be acidic enough (that is, be too alkaline), assuming that other compensatory measures have not yet had time to act?

3. The hormone insulin enhances the transport of glucose (sugar) from the blood into most body cells. Its secretion is controlled by a negative-feedback system between the concentration of glucose in the blood and the insulin-secreting cells. Therefore, which of the following statements is correct?

a. A decrease in blood glucose concentration stimulates insulin secretion, which in turn further lowers blood glucose concentration.

b. An increase in blood glucose concentration stimulates insulin secretion, which in turn lowers blood glucose concentration.

c. A decrease in blood glucose concentration stimulates insulin secretion, which in turn increases blood glucose concentration.

d. An increase in blood glucose concentration stimulates insulin secretion, which in turn further increases blood glucose concentration.

e. None of the preceding is correct.

4. Given that most patients with AIDS die from overwhelming infections or rare types of cancer, what body system do you think the human immunodeficiency virus (or HIV, the AIDS virus) impairs?

5. Body temperature is homeostatically regulated around a set point. Given your knowledge of negative feedback and homeostatic control systems, predict whether narrowing or widening of the blood vessels of the skin will occur when a person exercises strenuously. (*Hints:* Muscle contraction generates heat. Narrowing of the vessels supplying an organ decreases blood flow through the organ, whereas vessel widening increases blood flow through the organ. The more warm blood flowing through the skin, the greater is the loss of heat from the skin to the surrounding environment.)

To access the course materials and companion resources for this text, please visit www.cengagebrain.com

Cell Physiology

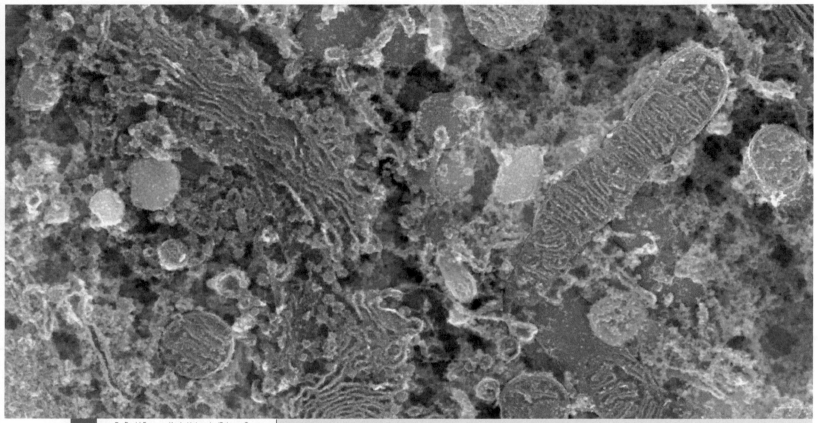

Dr. David Furness, Keele University/Science Source

A scanning electron micrograph of organelles within a cell. Mitochondria (*red*) generate the cell's energy. The Golgi complex (*blue*) processes proteins and lipids produced by the endoplasmic reticulum (small part in *yellow*, bottom left) for secretion and membrane construction. Cytosol (*green*) surrounds the organelles.

CHAPTER AT A GLANCE

Homeostasis Highlights

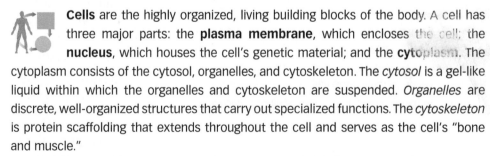

Cells are the highly organized, living building blocks of the body. A cell has three major parts: the **plasma membrane**, which encloses the cell; the **nucleus**, which houses the cell's genetic material; and the **cytoplasm**. The cytoplasm consists of the cytosol, organelles, and cytoskeleton. The *cytosol* is a gel-like liquid within which the organelles and cytoskeleton are suspended. *Organelles* are discrete, well-organized structures that carry out specialized functions. The *cytoskeleton* is protein scaffolding that extends throughout the cell and serves as the cell's "bone and muscle."

Through the coordinated action of these components, every cell performs certain basic functions essential to its survival and a specialized task that helps maintain homeostasis. Cells are organized according to their specialization into body systems that maintain the stable internal environment essential for the whole body's survival. All body functions ultimately depend on the activities of the individual cells that make up the body.

2.1 Cell Theory and Discovery

Although the same chemicals that make up living cells are found in nonliving matter, researchers have not been able to organize these chemicals into living cells in a laboratory. Life stems from the unique and complex organization and interactions of these inanimate chemicals within the cell. Cells, the smallest living entities, are the living building blocks for the immensely complicated whole body. Thus, cells are the bridge between chemicals and humans (and all other living organisms). All body functions of a multicellular organism ultimately depend on the collective structural and functional capabilities of its individual cells. Furthermore, all new cells and all new life arise from the division of preexisting cells, not from nonliving sources. Because of this continuity of life, the cells of all organisms are fundamentally similar in structure and function. ▌Table 2-1 summarizes these principles, which are known collectively as the **cell theory**. By probing deeper into the molecular structure and organization of the cells that make up the body, modern physiologists are unraveling many of the broader mysteries of how the body works.

Body cells are so small they cannot be seen by the unaided eye. The smallest visible particle is 5 to 10 times larger than a typical human cell, which averages about 10 to 20 micrometers (μm) in diameter (1 μm = 1/1,000,000 of a meter). About 100 average-sized cells lined up side by side would stretch a distance of only 1 mm (1 mm = 1/1000 of a meter; 1 m = 39.37 in.).

Until the microscope was invented in the middle of the 17th century, scientists did not know that cells existed. With the development of better light microscopes in the early 19th century, they learned that all plant and animal tissues consist of individual cells. The cells of a hummingbird, a human, and a whale are all about the same size. Larger species have more cells, not larger cells. These early investigators also discovered that cells are filled with a fluid that, given the microscopic capabilities of the time, appeared to be a rather uniform, soupy mixture believed to be the elusive "stuff of life." In the 1940s, when researchers first used electron microscopes to observe living

matter, they began to realize the great diversity and complexity of the internal structure of cells. (Electron microscopes are about 100 times more powerful than light microscopes.) Now that scientists have even more sophisticated microscopes, biochemical techniques, cell culture technology, and genetic engineering, the concept of the cell as a microscopic bag of formless fluid has given way to our current understanding of the cell as a complex, highly organized, compartmentalized structure.

2.2 An Overview of Cell Structure

The trillions of cells in a human body are classified into about 200 types based on specific variations in structure and function. Despite their diverse structural and functional specializations, however, different cells share many features. Most cells have three major subdivisions: the *plasma membrane,* which encloses the cells; the *nucleus,* which contains the cell's genetic material; and the *cytoplasm,* the portion of the cell's interior not occupied by the nucleus (▌Figure 2-1).

The plasma membrane bounds the cell.

The **plasma membrane** is a thin membranous structure that encloses each cell and is composed mostly of lipid (fat) molecules and studded with proteins. This barrier separates the cell's contents from its surroundings; it keeps the *intracellular fluid (ICF)* within the cells from mingling with the *extracellular fluid (ECF)* outside the cells. The plasma membrane is not simply a mechanical barrier to hold in the cell contents; its proteins selectively control movement of molecules between the ICF and ECF. Through this structure, the cell controls entry of nutrients and other needed supplies and export of products manufactured within, while at the same time guarding against unwanted traffic into or out of the cell. The plasma membrane is discussed thoroughly in Chapter 3.

The nucleus contains the DNA.

The two major parts of the cell's interior are the nucleus and the cytoplasm. The **nucleus**, which is typically the largest single organized cell component, can be seen as a distinct spherical or oval structure, usually located near the center of the cell. It is surrounded by a double-layered membrane, the **nuclear envelope**, which separates the nucleus from the rest of the cell. The nuclear envelope is pierced by many **nuclear pores** that allow necessary traffic to move between the nucleus and the cytoplasm.

The nucleus houses the cell's genetic material, **deoxyribonucleic acid (DNA)**, which, along with associated nuclear proteins, is organized into **chromosomes**. Each chromosome con-

▌TABLE 2-1 Principles of the Cell Theory

- The cell is the smallest structural and functional unit capable of carrying out life processes.
- The functional activities of each cell depend on the specific structural properties of the cell.
- Cells are the living building blocks of all multicellular organisms.
- An organism's structure and function ultimately depend on the collective structural characteristics and functional capabilities of its cells.
- All new cells and new life arise only from preexisting cells.
- Because of this continuity of life, the cells of all organisms are fundamentally similar in structure and function.

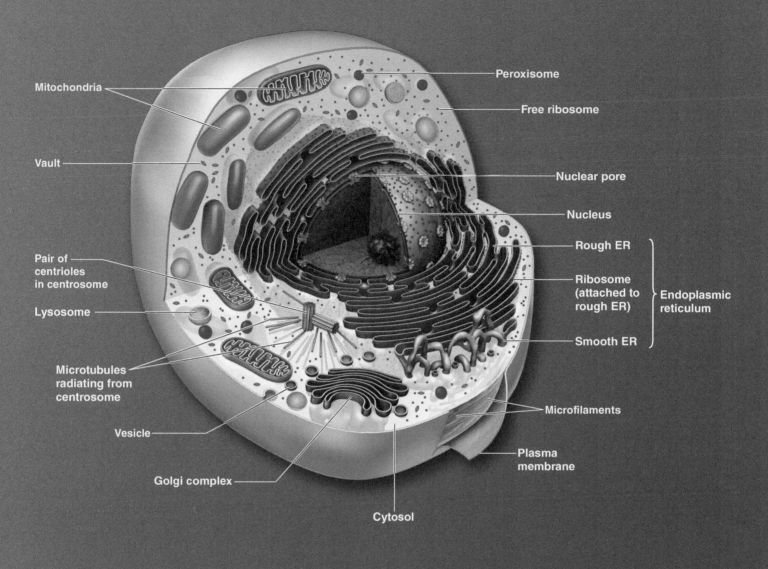

Mitochondria

Vault

Pair of
centrioles
in centrosome

Lysosome

Microtubules
radiating from
centrosome

Vesicle

Golgi complex

Cytosol

Peroxisome

Free ribosome

Nuclear pore

Nucleus

Rough ER

Ribosome
(attached to
rough ER)

Endoplasmic
reticulum

Smooth ER

Microfilaments

Plasma
membrane

IFigure 2-1 Diagram of cell structures visible under an electron microscope.

sists of a different DNA molecule that contains a unique set of genes. Body cells contain 46 chromosomes that can be sorted into 23 pairs on the basis of their distinguishing features. DNA has two important functions:

1. *Serving as a genetic blueprint during cell replication.* Through this role, DNA ensures that the cell produces additional cells just like itself, thus continuing the identical type of cell line within the body. Furthermore, in the reproductive cells (eggs and sperm), the DNA blueprint passes on genetic characteristics to future generations.

2. *Directing protein synthesis.* DNA provides codes, or "instructions," for directing synthesis of specific structural and enzymatic proteins within the cell. Proteins are the main structural component of cells, and enzymes govern the rate of all chemical reactions in the body. By specifying the kinds and amounts of proteins that are produced, the nucleus indirectly governs most cell activities and serves as the cell's control center.

We next examine interactions of RNA with DNA in protein synthesis.

Roles of RNA Three types of **ribonucleic acid (RNA)** play roles in protein synthesis (see p. A-14). First, DNA's genetic code for a particular protein is transcribed into a **messenger RNA (mRNA)** molecule, which exits the nucleus through the nuclear pores. Within the cytoplasm, mRNA delivers the coded message to *ribosomes,* which "read" the code and translate it into the appropriate amino acid sequence for the designated protein being synthesized. **Ribosomal RNA (rRNA)** is an essential component of ribosomes. (We will discuss the structure and function of ribosomes in more detail later.) **Transfer RNA (tRNA)** delivers the appropriate amino acids within the cytoplasm to their designated site in the protein under construction at the ribosome. **Gene expression** refers to the multistepped process by which information encoded in a gene is used to direct the synthesis of a protein molecule. (For further information regarding this topic, see the supplemental resource *Storage, Replication, and Expression of Genetic Information* on the text's Web site at www.cengagebrain.com.)

In addition to these three well-established forms of RNA, *microRNA (miRNA)* and *small interfering RNA (siRNA)* are recently discovered regulatory RNAs that can bind to mRNA and block the production of this mRNA's protein product, a process known as **RNA interference (RNAi)**. By silencing gene expression, these minuscule RNA snippets influence a variety of developmental, differentiation, and cellular processes throughout the body. Much of the DNA in humans is transcribed into regulatory RNA. Only 1.5% of DNA codes for protein synthesis. Scientists long thought that most of the DNA that did not code for proteins was "junk" or "nonsense" DNA because they did not understand its purpose. However, the revolutionary discovery of RNAi has changed that misperception and led to an explosion of fundamental and clinical research in this field. More than 1000 distinct miRNAs have been discovered, and each type of body cell contains a different collection of regulatory RNAs. Defects in RNAi have been connected to a variety of diseases, including cancer. For example, *chronic lymphocytic leukemia* is linked to mutations in two miRNA genes. Investigators are searching for ways to exploit RNAi for clinical application, such as by interfering with expression of genes that are misregulated in cancer.

Human Genome and Proteome The human **genome** is all of the genetic information coded in a complete single set of DNA in a typical body cell. The *Human Genome Project* identified and sequenced the entire genetic code through an international collaborative effort among public and private researchers that began in 1990 and was completed in 2003. The human genome mapped the composition and sequence of the 3.2 billion chemical units (nucleotides; see p. A-14) organized into about 20,000 protein-coding genes (representing only 1.5% of the genome), along with extensive intervening stretches of DNA that are involved in various ways with gene regulation. Noncoding regions also affect how DNA is folded and packaged into chromosomes and carry out yet-to-be determined actions. With this complete genetic map in hand, scientists are now scrambling to identify the functions and regulation of the genes and other parts of the genome.

The term **proteome** refers to the complete set of proteins that can be expressed by the protein-coding genes in the genome. Even though all body cells have an identical DNA blueprint, they do not all produce the same proteins. Because of cell-specific regulatory factors, different types of cells express different sets of genes and thus synthesize different sets of structural and enzymatic proteins. For example, only red blood cells can synthesize hemoglobin, even though all body cells carry the DNA instructions for hemoglobin synthesis. A *cellular proteome* is the collection of proteins found in a particular cell type under a particular set of environmental conditions. The cellular proteome, dictated by regulation of gene expression, determines the structure and function of a given cell at a particular time.

By studying the genome and related proteome of healthy and sick individuals, researchers are beginning to yield genome-based diagnostics and therapies. Although clinical applications have been slow to arrive, more than 1800 disease-related genes have been identified and several thousand genetic tests are now available. The genomic revolution will continue to grow as scientists gradually decipher all of the data becoming available to them.

Epigenetics DNA and genetics are not the whole story controlling protein synthesis. In addition to the gene-determined instructions that govern synthesis of structural and enzymatic proteins, the emerging science of **epigenetics** studies environmentally induced modifications of a gene's activity that do not involve a change in the gene's DNA code (*epi* means "on top of"). Environmental factors such as smoking, high-fat diets, and stress can alter the way genes are expressed so that even identical twins are different. Epigenetic modifications arise from chemical tags added to DNA or its associated proteins that influence gene activity without altering the information the gene itself contains. For example, epigenetic tags, such as attachment of methyl groups, may limit the ability of mRNA to access DNA, thereby ultimately reducing synthesis of proteins coded by this stretch of DNA. Epigenetic modifications are dynamic and can change how genes behave over a lifetime. Furthermore, when epigenetic changes occur in the DNA of sperm or egg cells, they can get passed along to future generations.

Lipidome So as not to leave the mistaken impression that proteins are the only important structural and functional molecules in the cell, we digress briefly to mention the other major type of organic molecule in the body, lipids, which exert myriad roles in physiological and disease processes. To mention a few, lipids are the main structural component of the plasma membrane and store excess ingested food energy as body fat. Impaired lipid handling plays a central role in disorders such as heart disease and Alzheimer's disease. Tens of thousands of different types of lipids are included in the **lipidome**, the full roster of lipids in the body cells. Unlike proteins, lipids are not genetically encoded. However, lipid synthesis is governed by enzymes, which are produced under gene control. The formation of some body lipids can be altered by environmental factors such as diet and exercise. For example, differences in these lifestyle patterns affect the extent of cholesterol deposition in blood vessel linings.

The cytoplasm consists of various organelles, the cytoskeleton, and the cytosol.

The **cytoplasm** is that portion of the cell interior not occupied by the nucleus. It contains a number of discrete, specialized *organelles* (the cell's "little organs") and the *cytoskeleton* (a scaffolding of proteins) dispersed within the *cytosol* (a complex, gel-like liquid).

Organelles are distinct, highly organized structures that perform specialized functions within the cell. On average, nearly half of the total cell volume is occupied by two categories of organelles—*membranous organelles* and *nonmembranous organelles*. Each **membranous organelle** is a separate compartment within the cell that is enclosed by a membrane similar to the plasma membrane. Thus, the contents of a membranous organelle are separated from the surrounding cytosol and from the contents of other organelles. Nearly all human cells contain five main types of membranous organelles—the *endoplasmic reticulum, Golgi complex, lysosomes, peroxisomes,* and *mitochondria.* Membranous organelles are like intracellular "specialty shops." Each is a separate internal compartment that contains a specific set of chemicals for carrying out a particular cellular function. This compartmentalization permits chemical activities that would not be compatible with one another to occur simultaneously within the cell. For example, enzymes that destroy unwanted proteins operate within the protective confines of the lysosomes without the risk of destroying essential cell proteins. The **nonmembranous organelles** are not surrounded by membrane and thus are in direct contact with the cytosol. They include *ribosomes, proteasomes, vaults,* and *centrioles.* Like membranous organelles, nonmembranous organelles are organized structures that carry out specific functions within the cell. Organelles are similar in all cells, although some variations occur depending on the specialized capabilities of each cell type. Just as each organ plays a role essential for survival of the whole body, each organelle performs a specialized activity necessary for survival of the whole cell.

The **cytoskeleton** is an interconnected system of protein fibers and tubes that extends throughout the cytosol. This elaborate protein network gives the cell its shape, provides for its internal organization, and regulates its various movements, thus serving as the cell's "bone and muscle."

The remainder of the cytoplasm not occupied by organelles and cytoskeleton consists of the **cytosol** ("cell liquid"). The cytosol is a semiliquid, gel-like mass. Many of the chemical reactions that are compatible with one another are conducted in the cytosol. (For clarification, the ICF encompasses all fluid inside the cell, including that within the cytosol, the organelles, and the nucleus.)

The rest of this chapter examines each of the cytoplasmic components in more detail, concentrating first on organelles. We begin with the endoplasmic reticulum, a membranous organelle.

Check Your Understanding 2.2

1. State the functions of DNA and the different types of RNA.
2. Define *genome, proteome,* and *epigenetics.*
3. Distinguish among *cytoplasm, organelles, cytosol,* and *cytoskeleton.*

2.3 Endoplasmic Reticulum and Segregated Synthesis

The **endoplasmic reticulum (ER)** is an elaborate fluid-filled membranous system distributed extensively throughout the cytosol. It is primarily a protein- and lipid-producing factory. Two distinct types of ER—rough and smooth—can be distinguished. The **rough ER** consists of stacks of relatively flattened interconnected sacs, whereas the **smooth ER** is a meshwork of tiny interconnected tubules (❙Figure 2-2a and b). Even though these two regions differ considerably in appearance and function, they are connected to each other, making the ER one continuous organelle. The relative amount of rough and smooth ER varies among cells, depending on the activity of the cell.

The rough ER synthesizes proteins for secretion and membrane construction.

The outer surface of the rough ER membrane is studded with small particles that give it a "rough" or granular appearance under a light microscope. These particles are *ribosomes,* the "workbenches" where protein synthesis takes place. Not all ribosomes in the cell are attached to the rough ER. Unattached or "free" ribosomes are dispersed throughout the cytosol.

Ribosomes, nonmembranous organelles, carry out protein synthesis by translating mRNA into chains of amino acids in the ordered sequence dictated by the original DNA code. Ribosomes bring together all components that participate in protein synthesis—mRNA, tRNA, and amino acids—and provide the enzymes and energy required for linking the amino acids together. The nature of the protein synthesized by a given ribosome is determined by the mRNA being translated. Each mRNA serves as a code for only one protein.

A finished ribosome is about 20 nm in diameter and is made up of two parts of unequal size, a *large* and a *small ribosomal subunit* (❙Figure 2-2c) (1 nanometer [nm] = 1/1,000,000,000 of a meter). Each subunit is composed of rRNA and ribosomal proteins. These subunits are brought together when a protein is being synthesized. When the two subunits unite, a groove is formed. In translation, mRNA moves through this groove.

The rough ER, in association with its ribosomes, synthesizes and releases various new proteins into the ER lumen, the fluid-filled space enclosed by the ER membrane. These proteins serve one of two purposes: (1) Some proteins are destined for export to the cell's exterior as secretory products, such as protein hormones or enzymes, and (2) other proteins are used in constructing new cellular membrane (either plasma membrane or organelle membrane) or other cell structures, such as lysosomes. The plasma membrane consists mostly of proteins and lipids (fats). The membranous wall of the ER also contains enzymes essential for synthesis of the lipids needed to produce new membranes. These newly synthesized lipids enter the ER lumen along with the proteins. Predictably, the rough ER is most abundant in cells specialized for protein secretion (for example, cells that secrete digestive enzymes) or in cells that require extensive membrane synthesis (for example, rapidly growing cells such as immature egg cells).

After being released into the ER lumen, a newly synthesized protein is folded into its final conformation (see p. A-14); it may also be modified in other ways, such as being pruned or having carbohydrates attached to it. After this processing, a new protein cannot pass out through the ER membrane on its own and therefore becomes permanently separated from the cytosol as soon as it has been synthesized. In contrast to the rough ER

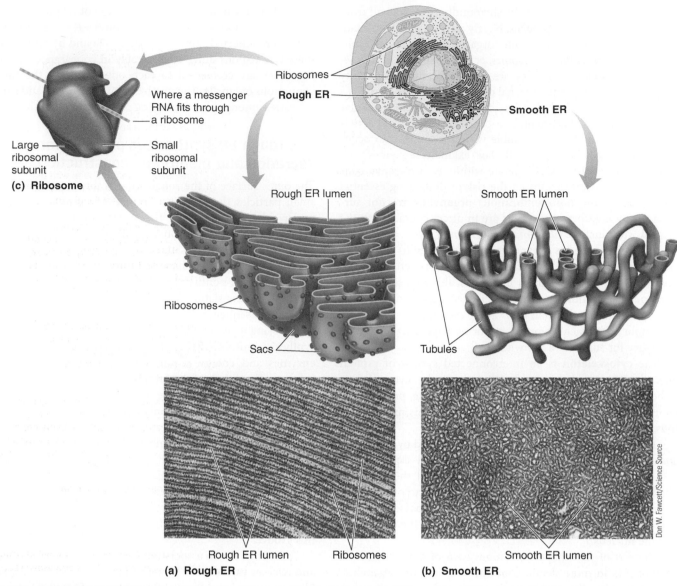

(c) Ribosome
Ribosomes
Where a messenger RNA fits through a ribosome
Large ribosomal subunit
Small ribosomal subunit
Rough ER
Smooth ER
Rough ER lumen
Smooth ER lumen
Ribosomes
Sacs
Tubules
Rough ER lumen
Ribosomes
Smooth ER lumen

Don W. Fawcett/Science Source

(a) Rough ER
(b) Smooth ER

I Figure 2-2 Endoplasmic reticulum (ER). (a) Diagram and electron micrograph of the rough ER, which consists of stacks of relatively flattened interconnected sacs studded with ribosomes. (b) Diagram and electron micrograph of the smooth ER, which is a meshwork of tiny interconnected tubules. The rough ER and smooth ER are connected, making one continuous organelle. (c) Diagram of an assembled ribosome.

ribosomes, free ribosomes synthesize proteins for use within the cytosol. In this way, newly produced molecules destined for export out of the cell or for synthesis of new membrane or other cell components (those synthesized by the ER) are physically separated from those that belong in the cytosol (those produced by the free ribosomes). About one third of the proteome is typically synthesized in the ER.

How do newly synthesized molecules within the ER lumen get to their destinations if they cannot pass out through the ER membrane? They do so by action of the smooth ER.

The smooth ER packages new proteins in transport vesicles.

The smooth ER does not contain ribosomes, so it is "smooth." Lacking ribosomes, it is not involved in protein synthesis. Instead, it serves other purposes that vary in different cell types.

In most cells, the smooth ER is rather sparse and serves primarily as a central packaging and discharge site for molecules to be transported from the ER. Newly synthesized proteins and lipids move within the continuous lumen from the rough ER to gather at specialized exit sites in the smooth ER. These exit sites then "bud off" (that is, balloon outward on the surface and then are pinched off), forming **transport vesicles** that enclose the new molecules (I Figure 2-3). (A **vesicle** is a fluid-filled, membrane-enclosed intracellular cargo container.)

How does the ER exit site bud off? *Coat proteins* of the type known as *coat protein II (COPII)* from the cytosol bind with specific proteins facing the outer surface of the smooth ER membrane at the exit site. The linking of these coat proteins into a cagelike assembly or "coat" causes the surface membrane at the site to curve outward to form a dome-shaped bud around the newly synthesized products to be exported out of the smooth ER. Eventually the surface membrane closes and

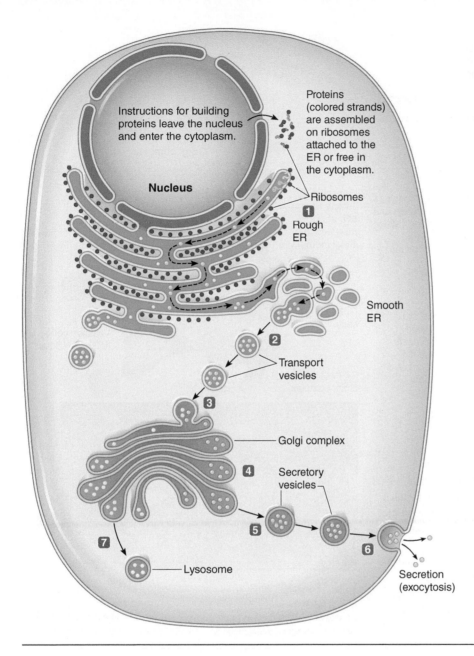

Proteins (colored strands) are assembled on ribosomes attached to the ER or free in the cytoplasm.

Nucleus

Ribosomes

1

Rough ER

Smooth ER

2

Transport vesicles

3

Golgi complex

4

Secretory vesicles

5

6

7

Lysosome

Secretion (exocytosis)

1 The rough ER synthesizes proteins to be secreted to the exterior or to be incorporated into plasma membrane or other cell components.

2 The smooth ER packages the secretory product into transport vesicles, which bud off and move to the Golgi complex.

3 The transport vesicles fuse with the Golgi complex, open up, and empty their contents into the closest Golgi sac.

4 The newly synthesized proteins from the ER travel by vesicular transport through the layers of the Golgi complex, which modifies the raw proteins into final form and sorts and directs the finished products to their final destination by varying their wrappers.

5 Secretory vesicles containing the finished protein products bud off the Golgi complex and remain in the cytosol, storing the products until signaled to empty.

6 On appropriate stimulation, the secretory vesicles fuse with the plasma membrane, open, and empty their contents to the cell's exterior. Secretion has occurred by exocytosis, with the secretory products never having come into contact with the cytosol.

7 Lysosomes also bud from the Golgi complex.

❙Figure 2-3 **Overview of the secretion process for proteins synthesized by the endoplasmic reticulum.** **FIGURE FOCUS**: *Compare the contents of a transport vesicle and a secretory vesicle.*

pinches off a transport vesicle. Transport vesicles move to the Golgi complex, described in the next section, for further processing of their cargo.

In contrast to the sparseness of the smooth ER in most cells, some specialized cell types have an extensive smooth ER, which has additional functions as follows:

- The smooth ER is abundant in cells that specialize in lipid metabolism—for example, cells that secrete lipid-derived steroid hormones. The membranous wall of the smooth ER, like that of the rough ER, contains enzymes for synthesis of lipids. These cells have an expanded smooth-ER compartment that houses the additional enzymes necessary to keep pace with demands for hormone secretion.

- In liver cells, the smooth ER contains enzymes specialized for detoxifying harmful substances produced within the body by metabolism or substances that enter the body from the outside in the form of drugs or other foreign compounds.

- Muscle cells have an elaborate, modified smooth ER known as the *sarcoplasmic reticulum,* which stores calcium used in the process of muscle contraction (see p. 258).

Misfolded proteins are destroyed by the ubiquitin–proteasome pathway.

Like any good factory, the rough ER has a quality control system in place to remove defective products. Misfolded proteins are tagged with **ubiquitin**, a small protein "doom tag" that labels those flawed proteins for destruction. Ubiquitin directs the tagged protein out of the ER to one of many proteasomes located throughout the cell. A **proteasome**, a nonmembranous organelle, is a protein degradation machine: It is a cylinder-shaped complex about the size of a ribosomal subunit that contains multiple protein-digesting enzymes that break down ubiquinated proteins into recyclable building blocks (❙Figure 2-4). A proteasome consists of a hollow *core particle* capped at

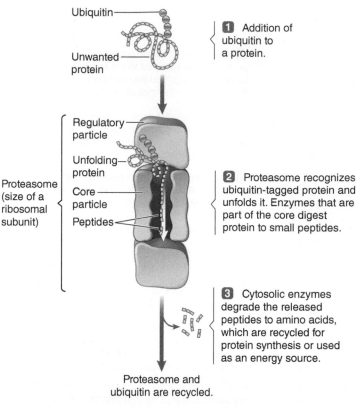

1. Addition of ubiquitin to a protein.

Ubiquitin

Unwanted protein

Regulatory particle

Unfolding protein

Proteasome (size of a ribosomal subunit)

Core particle

Peptides

2. Proteasome recognizes ubiquitin-tagged protein and unfolds it. Enzymes that are part of the core digest protein to small peptides.

3. Cytosolic enzymes degrade the released peptides to amino acids, which are recycled for protein synthesis or used as an energy source.

Proteasome and ubiquitin are recycled.

❚Figure 2-4 **Proteasome.**

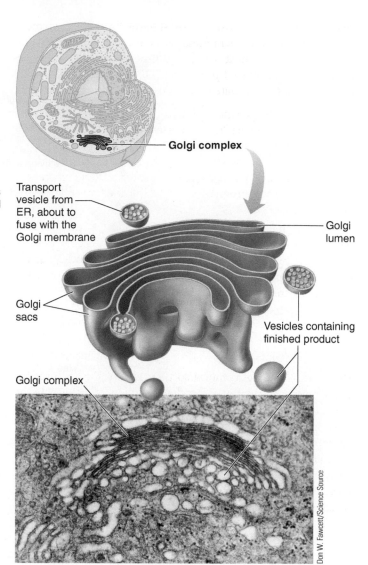

Golgi complex

Transport vesicle from ER, about to fuse with the Golgi membrane

Golgi lumen

Golgi sacs

Vesicles containing finished product

Golgi complex

❚Figure 2-5 **Golgi complex.** Diagram and electron micrograph of a Golgi complex, which consists of a stack of slightly curved, membrane-enclosed sacs. The vesicles at the dilated edges of the sacs contain finished protein products packaged for distribution to their final destination.

Don W. Fawcett/Science Source

each end with a *regulatory particle*. The regulatory particle recognizes the ubiquitin-tagged protein, unfolds it, and feeds it into the core. The core cavity is lined with a variety of enzymatic sites that break down the protein into peptides (small chains of amino acids). After the peptides are released from the proteasome, cytosolic enzymes finish digesting them into their component amino acids. Ubiquitin is released and recycled.

In addition to tagging misfolded proteins in the ER, ubiquitin also labels other damaged or unneeded intracellular proteins for degradation in proteasomes.

2.4 Golgi Complex and Exocytosis

The **Golgi complex**, a membranous organelle, is closely associated with the ER. Each Golgi complex consists of a stack of flattened, slightly curved, membrane-enclosed sacs (❚Figure 2-5 and chapter opener photo). The sacs within each Golgi stack do not come into physical contact with one another. Note that the

flattened sacs are thin in the middle but have dilated, or bulging, edges. The number of Golgi complexes varies, depending on the cell type. Some cells have only one Golgi stack, whereas cells specialized for protein secretion may have hundreds of stacks.

Transport vesicles carry their cargo to the Golgi complex for further processing.

Most newly synthesized molecules that have just budded off from the smooth ER enter a Golgi stack. When a transport vesicle reaches a Golgi stack, the vesicle membrane fuses with the membrane of the sac closest to the center of the cell. The vesicle membrane opens up and becomes integrated into the Golgi membrane, and the contents of the vesicle are released to the interior of the sac (see ❚Figure 2-3).

These newly synthesized raw materials from the ER travel by means of vesicle formation through the layers of the Golgi stack, from the innermost sac closest to the ER to the outermost sac near the plasma membrane. Vesicular transport from one Golgi sac to the next is accomplished through action of membrane-curving *coat protein I (COPI)*. During this transit, two important, interrelated functions take place:

1. *Processing the raw materials into finished products.* Within the Golgi complex, the "raw" proteins from the ER are modified into their final form (for example, by having a carbohydrate attached). The biochemical pathways that the proteins follow during their passage through the Golgi complex are elaborate, precisely programmed, and specific for each final product.

2. *Sorting and directing the finished products to their final destinations.* The Golgi complex is responsible for sorting and segregating products according to their function and destination, such as products to be secreted to the cell's exterior or to be used for constructing new plasma membrane.

The Golgi complex packages secretory vesicles for release by exocytosis.

How does the Golgi complex sort and direct finished proteins to the proper destinations? Finished products collect within the dilated edges of the Golgi complex's sacs. The edge of the outermost sac then pinches off to form a membrane-enclosed vesicle containing a selected product. For each type of product to reach its appropriate site of function, each distinct type of vesicle takes up a specific product before budding off (like a particular piece of mail being placed in an envelope). Vesicles with their selected cargo destined for different sites are wrapped in membranes containing distinct surface proteins. Each distinct surface protein serves as a specific **docking marker** (like an address on an envelope). A vesicle can "dock" lock-and-key fashion and "unload" its selected cargo only at the appropriate **docking-marker acceptor**, a protein located only at the proper destination within the cell (like a house address). Thus, Golgi products reach their appropriate site of function because they are sorted and delivered like addressed envelopes containing particular pieces of mail being delivered only to the appropriate house addresses.

As an example, let us look at secretory cells. **Secretion** is the release to the cell's exterior of a specific product made by the cell for a particular function (see p. 6). Specialized secretory cells include endocrine cells, which secrete protein hormones, and digestive gland cells, which secrete digestive enzymes. In secretory cells, numerous large **secretory vesicles**, which contain proteins to be secreted, bud off from the Golgi stacks. Secretory vesicles, which are about 200 times larger than transport vesicles, store the secretory proteins until the cell is stimulated by a specific signal that indicates a need for release of that particular secretory product. On the appropriate signal, a vesicle moves to the cell's periphery, fuses with the plasma membrane, opens, and empties its contents to the outside (see ❙Figure 2-3 and Figure 2-6a). This mechanism—release to the exterior of substances originating within the cell—is referred to as **exocytosis** (*exo* means "out of"; *cyto* means "cell"). Exocytosis is the primary mechanism

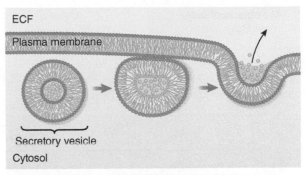

ECF
Plasma membrane
Secretory vesicle
Cytosol

(a) Exocytosis: A secretory vesicle fuses with the plasma membrane, releasing the vesicle contents to the cell exterior. The vesicle membrane becomes part of the plasma membrane.

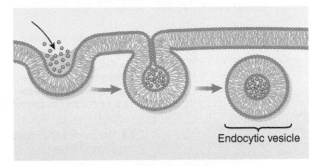

Endocytic vesicle

(b) Endocytosis: Materials from the cell exterior are enclosed in a segment of the plasma membrane that pockets inward and pinches off as an endocytic vesicle.

❙Figure 2-6 **Exocytosis and endocytosis.**
FIGURE FOCUS: *What happens to the surface area of the plasma membrane as a result of exocytosis and of endocytosis?*

for accomplishing secretion. Secretory vesicles fuse only with the plasma membrane and not with any organelle membranes, thereby preventing wasteful or even dangerous discharge of secretory products into organelles.

Let us now examine how secretory vesicles take up specific products in the Golgi stacks for release into the ECF and why they are able to dock only at the plasma membrane (❙Figure 2-7):

■ The newly finished proteins destined for secretion contain a unique sequence of amino acids at one end known as a *sorting signal*, and the interior surface of the Golgi membrane contains *recognition markers,* proteins that recognize and attract specific sorting signals. Recognition of the right protein's sorting signal by the complementary membrane marker ensures that the proper cargo is captured and packaged into the secretory vesicle.

■ *Coat proteins* called *coatomer* from the cytosol bind with another specific protein on the outer surface of the Golgi membrane and cause a bud to form around the captured cargo.

■ After budding off, the vesicle sheds its coat proteins and exposes the docking markers, known as *v-SNAREs,* which face the outer surface of the vesicle membrane.

■ A v-SNARE can link lock-and-key fashion only with its docking-marker acceptor, called a *t-SNARE,* on the targeted membrane. In the case of secretory vesicles, the targeted membrane is the plasma membrane, the designated site for

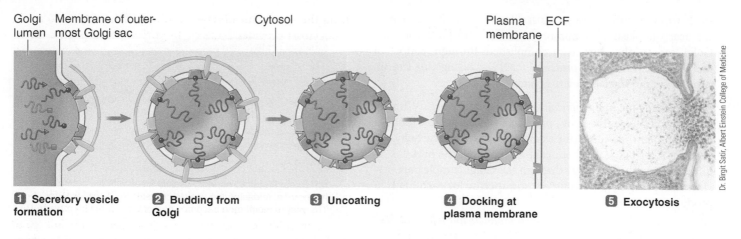

Golgi lumen | Membrane of outer-most Golgi sac | Cytosol | Plasma membrane | ECF

1 Secretory vesicle formation

2 Budding from Golgi

3 Uncoating

4 Docking at plasma membrane

5 Exocytosis

Dr. Birgit Satir, Albert Einstein College of Medicine

KEY

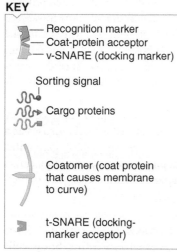

— Recognition marker
— Coat-protein acceptor
— v-SNARE (docking marker)

Sorting signal

Cargo proteins

Coatomer (coat protein that causes membrane to curve)

t-SNARE (docking-marker acceptor)

1 Recognition markers in the membrane of the outermost Golgi sac capture the appropriate cargo from the Golgi lumen by binding only with the sorting signals of the protein molecules to be secreted. The membrane that will wrap the vesicle is coated with coatomer, which causes the membrane to curve, forming a bud.

2 The membrane closes beneath the bud, pinching off the secretory vesicle.

3 The vesicle loses its coating, exposing v-SNARE docking markers on the vesicle surface.

4 The v-SNAREs bind only with the t-SNARE docking-marker acceptors of the targeted plasma membrane, ensuring that secretory vesicles empty their contents to the cell's exterior.

❙Figure 2-7 **Packaging, docking, and release of secretory vesicles.** The diagram series illustrates secretory vesicle formation and budding with the aid of a coat protein and docking with the plasma membrane by means of v-SNAREs and t-SNAREs. The transmission electron micrograph shows secretion by exocytosis.

secretion to take place. Thus, the v-SNAREs of secretory vesicles fuse only with the t-SNAREs of the plasma membrane. Once a vesicle has docked at the appropriate membrane by means of matching SNAREs, the two membranes completely fuse; then the vesicle opens up and empties its contents at the targeted site.

Note that the contents of secretory vesicles never come into contact with the cytosol. From the time these products are first synthesized in the ER until they are released from the cell by exocytosis, they are always wrapped in membrane and thus isolated from the remainder of the cell. By manufacturing its particular secretory protein ahead of time and storing this product in secretory vesicles, a secretory cell has a readily available reserve from which to secrete large amounts of this product on demand. If a secretory cell had to synthesize all its product on the spot as needed for export, the cell would be more limited in its ability to meet varying levels of demand.

Secretory vesicles are formed only by secretory cells. However, the Golgi complexes of these and other cell types also sort and package newly synthesized products for different destinations within the cell in a similar manner: A particular vesicle captures a specific kind of cargo from among the many proteins in the Golgi lumen, then addresses each shipping container for a distinct destination.

2.5 Lysosomes and Endocytosis

Lysosomes are small, membrane-enclosed, degradative organelles that break down organic molecules (*lys* means "break down"; *some* means "body"). Instead of having a uniform structure, as is characteristic of all other organelles, lysosomes vary in size and shape, depending on the contents they are digesting. Most commonly, lysosomes are small (0.2 to 0.5 μm in diameter) oval or spherical bodies (❙Figure 2-8). On average, a cell contains about 300 lysosomes.

Lysosomes digest extracellular material brought into the cell by phagocytosis.

Lysosomes are formed by budding from the Golgi complex. A lysosome contains about 40 different powerful **hydrolytic enzymes** that are synthesized in the ER and then transported to

Unless otherwise noted, all content on this page is © Cengage Learning.

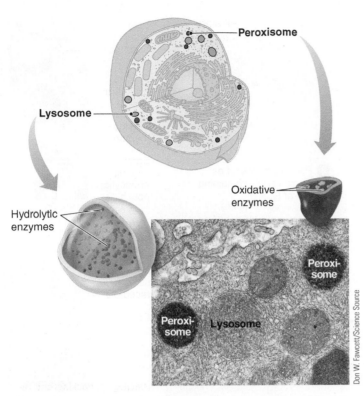

Peroxisome

Lysosome

Oxidative enzymes

Hydrolytic enzymes

Peroxisome

Peroxisome   Lysosome

Don W. Fawcett/Science Source

Figure 2-8 **Lysosomes and peroxisomes.** Diagram and electron micrograph of lysosomes, which contain hydrolytic enzymes, and peroxisomes, which contain oxidative enzymes.

the Golgi complex for packaging in the budding lysosome (see Figure 2-3). These enzymes catalyze **hydrolysis,** reactions that break down organic molecules by the addition of water at a bond site (*hydrolysis* means "splitting with water"; see p. A-14). In lysosomes, the organic molecules are cell debris and foreign material, such as bacteria, that have been brought into the cell. Lysosomal enzymes are similar to the hydrolytic enzymes that the digestive system secretes to digest food. Thus, lysosomes serve as the intracellular "digestive system." Note that lysosomes mostly degrade extracellular proteins brought into the cell, whereas most unwanted intracellular proteins are degraded by the ubiquitin–proteasome pathway.

Extracellular material to be attacked by lysosomal enzymes is brought into the cell through the process of phagocytosis, a type of endocytosis. **Endocytosis,** the reverse of exocytosis, refers to the internalization of extracellular material within a cell (*endo* means "within") (see Figure 2-6b). Endocytosis can be accomplished in three ways—*pinocytosis, receptor-mediated endocytosis,* and *phagocytosis*—depending on the contents of the internalized material.

Pinocytosis In **pinocytosis** ("cell drinking"), a droplet of ECF is taken up nonselectively. First, the plasma membrane dips inward, forming a pouch that contains a small bit of ECF (Figure 2-9a). The plasma membrane then seals at the surface of the pouch, trapping the contents in a small, intracellular **endocytic vesicle,** or **endosome.** *Dynamin,* the protein responsible for pinching off an endocytic vesicle, forms rings that wrap around and "wring the neck" of the pouch, severing the vesicle

from the surface membrane. Besides bringing ECF into a cell, pinocytosis provides a means to retrieve extra plasma membrane that has been added to the cell surface during exocytosis.

Receptor-Mediated Endocytosis Unlike pinocytosis, which involves the nonselective uptake of the surrounding fluid, **receptor-mediated endocytosis** is a highly selective process that enables cells to import specific large molecules that it needs from its environment. Receptor-mediated endocytosis is triggered by the binding of a specific target molecule such as a protein to a surface membrane receptor specific for that molecule (Figure 2-9b). This binding causes the plasma membrane at that site to pocket inward and then seal at the surface, trapping the bound molecule inside the cell. The pouch is formed by the linkage of *clathrin* molecules, which are membrane-deforming coat proteins on the inner surface of the plasma membrane that bow inward, in contrast to the outward-curving coat proteins that form buds. The resulting pouch is known as a *coated pit* because it is coated with clathrin. Cholesterol complexes, vitamin B_{12}, the hormone insulin, and iron are examples of substances selectively taken into cells by receptor-mediated endocytosis.

Clinical Note Unfortunately, some viruses can sneak into cells by exploiting this mechanism. For instance, flu viruses and HIV, the virus that causes AIDS (see p. 426), gain entry to cells via receptor-mediated endocytosis. They do so by binding with membrane receptors normally designed to trigger the internalization of a needed molecule.

Phagocytosis During **phagocytosis** ("cell eating"), large multimolecular particles are internalized. Most body cells perform pinocytosis, many carry out receptor-mediated endocytosis, but only a few specialized cells are capable of phagocytosis, the most notable being certain types of white blood cells that play an important role in the body's defense mechanisms. When a white blood cell encounters a large particle, such as a bacterium or tissue debris, it extends surface projections known as **pseudopods** ("false feet") that surround or engulf the particle and trap it within an internalized vesicle known as a **phagosome** (Figure 2-9c). A lysosome fuses with the membrane of the phagosome and releases its hydrolytic enzymes into the vesicle, where they safely attack the bacterium or other trapped material without damaging the remainder of the cell. The enzymes largely break down the engulfed material into raw ingredients, such as amino acids, glucose, and fatty acids, which the cell can use.

Lysosomes remove worn-out organelles.

Cells typically live longer than many of their internal components. Lysosomes can fuse with aged or damaged organelles to remove these useless parts of the cell. Lysosomal enzymes degrade the dysfunctional organelle, making its building blocks available for reuse by the cell. This selective self-digestion, known as **autophagy** (*auto* means "self"; *phag* means "eating") makes way for new replacement parts. In most cells, all organelles are renewable.

When cells are starving, they often induce autophagy of healthy cellular components that can be spared. This self-

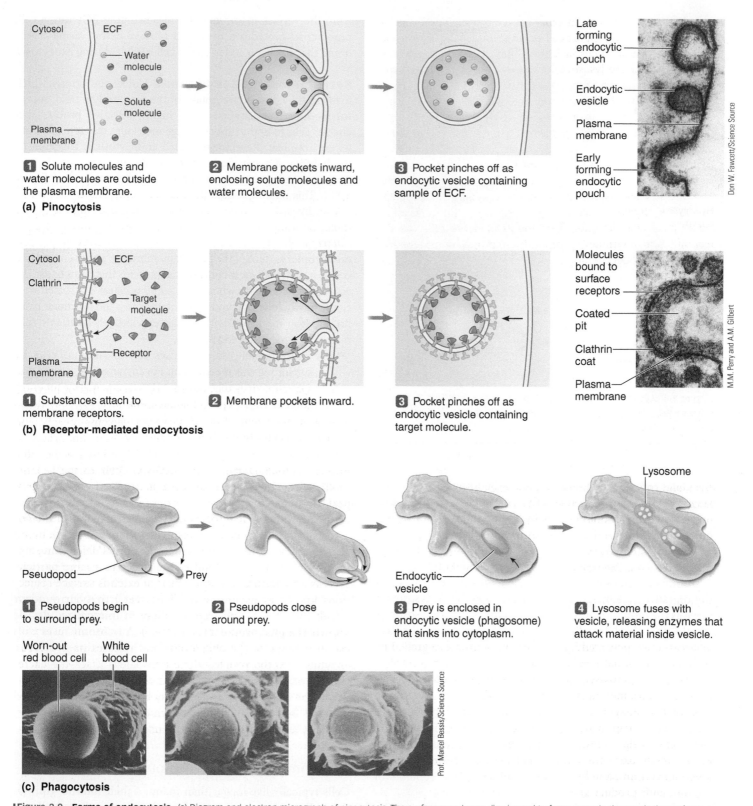

1 Solute molecules and water molecules are outside the plasma membrane.

2 Membrane pockets inward, enclosing solute molecules and water molecules.

3 Pocket pinches off as endocytic vesicle containing sample of ECF.

Cytosol ECF
Water molecule
Solute molecule
Plasma membrane

Late forming endocytic pouch
Endocytic vesicle
Plasma membrane
Early forming endocytic pouch

Don W. Fawcett/Science Source

(a) Pinocytosis

1 Substances attach to membrane receptors.

2 Membrane pockets inward.

3 Pocket pinches off as endocytic vesicle containing target molecule.

Cytosol ECF
Clathrin
Target molecule
Receptor
Plasma membrane

Molecules bound to surface receptors
Coated pit
Clathrin coat
Plasma membrane

M.M. Perry and A.M. Gilbert

(b) Receptor-mediated endocytosis

1 Pseudopods begin to surround prey.

2 Pseudopods close around prey.

3 Prey is enclosed in endocytic vesicle (phagosome) that sinks into cytoplasm.

4 Lysosome fuses with vesicle, releasing enzymes that attack material inside vesicle.

Pseudopod — Prey
Endocytic vesicle
Lysosome

Worn-out red blood cell   White blood cell

Prof. Marcel Bessis/Science Source

(c) Phagocytosis

▮Figure 2-9 **Forms of endocytosis.** (a) Diagram and electron micrograph of pinocytosis. The surface membrane dips inward to form a pouch, then seals the surface, forming an intracellular endocytic vesicle that nonselectively internalizes a bit of ECF. (b) Diagram and electron micrograph of receptor-mediated endocytosis. When a large molecule such as a protein attaches to a specific surface receptor, the membrane pockets inward with the aid of a coat protein, forming a coated pit, then pinches off to selectively internalize the molecule in an endocytic vesicle. (c) Diagram and scanning electron micrograph series of phagocytosis. White blood cells internalize multimolecular particles such as bacteria or old red blood cells by extending pseudopods that wrap around and seal in the targeted material. A lysosome fuses with and degrades the vesicle contents.

cannibalism provides an internal source of raw ingredients to provide energy and support cell survival for as long as possible. *Clinical Note* Some individuals lack the ability to synthesize one or more of the lysosomal enzymes. The result is massive accumulation within the lysosomes of the compound normally digested by the missing enzyme. Clinical manifestations often accompany such disorders because the engorged lysosomes interfere with normal cell activity. More than 50 of these so-called *lysosomal storage diseases* have been identified, and they all differ. The nature and severity of the symptoms depend on the type of substance accumulating, which in turn depends on what lysosomal enzyme is missing. An example is **Tay-Sachs disease**, which is characterized by abnormal accumulation of complex molecules found in nerve cells. As the accumulation continues, profound symptoms of progressive nervous system degeneration result.

Check Your Understanding 2.5

1. State the function of hydrolytic enzymes.
2. Illustrate the three types of endocytosis.
3. Define *autophagy*.

2.6 Peroxisomes and Detoxification

Peroxisomes are membranous organelles that produce and decompose hydrogen peroxide (H_2O_2) in the process of degrading potentially toxic molecules (*peroxi* refers to "hydrogen peroxide"). Typically, several hundred small peroxisomes about one third to one half the average size of lysosomes are present in a cell (see Figure 2-8). They too arise from the ER and Golgi complex.

Peroxisomes house oxidative enzymes that detoxify various wastes.

Like lysosomes, peroxisomes are membrane-enclosed sacs containing enzymes, but unlike lysosomes, which contain hydrolytic enzymes, peroxisomes house several powerful *oxidative enzymes* and contain most of the cell's *catalase*.

Oxidative enzymes, as the name implies, use oxygen (O_2), in this case to strip hydrogen from certain organic molecules. This reaction helps detoxify various wastes produced within the cell or foreign toxic compounds that have entered the cell, such as alcohol consumed in beverages.

The major product generated in the peroxisome, H_2O_2, is formed by molecular oxygen and the hydrogen atoms stripped from the toxic molecule. H_2O_2 is potentially destructive if allowed to accumulate or escape from the confines of the peroxisome. However, peroxisomes also contain an abundance of **catalase**, an enzyme that decomposes potent H_2O_2 into harmless H_2O and O_2. This latter reaction is an important safety mechanism that destroys the potentially deadly H_2O_2 at its site of production, thereby preventing its possible devastating escape into the cytosol.

Check Your Understanding 2.6

1. Discuss the function of the oxidative enzymes in peroxisomes.
2. Name the major product generated in peroxisomes.

2.7 Mitochondria and ATP Production

Mitochondria are the energy organelles, or "power plants," of the cell; they extract energy from the nutrients in food and transform it into a usable form for cell activities. Mitochondria generate about 90% of the energy that cells—and, accordingly, the whole body—need to survive and function. A single cell may contain as few as a hundred or as many as several thousand mitochondria, depending on the energy needs of each particular cell type.

Mitochondria are enclosed by two membranes.

Mitochondria are rod-shaped or oval structures about the size of bacteria. In fact, mitochondria are descendants of bacteria that invaded or were engulfed by primitive cells early in evolutionary history and that subsequently became permanent organelles. As part of their separate heritage, mitochondria possess their own DNA, distinct from the DNA housed in the cell's nucleus. **Mitochondrial DNA (mtDNA)** contains the genetic codes for producing many of the molecules the mitochondria need to generate energy.

Clinical Note *Mitochondrial diseases* affect an estimated 1 in 4000 people. Mitochondrial diseases primarily affect children, often because of inherited defects in mtDNA. However, adult onset of mitochondrial disease is becoming increasingly common. Flaws gradually accumulate in mtDNA over a person's lifetime; these flaws have been implicated in aging and in an array of disorders. The clinical manifestations of mitochondrial diseases are highly diverse, depending on the location and extent of mtDNA mutations. Typically several organ systems are involved. The more common symptoms of mitochondrial diseases include a mixture from among the following: chronic fatigue, incoordination, mental decline, muscle weakness and pain, digestive disturbances, heart disease, liver disease, respiratory complications, blindness, and hearing loss. Mitochondrial diseases get progressively worse and may be fatal, with the only treatments available being aimed at managing symptoms. An example of a mitochondrial disease is *Leigh syndrome*, which begins in infants and leads to early death following a rapid loss of mental and movement abilities and severe breathing problems.

Each mitochondrion, a membranous organelle, is enclosed by a double membrane—a smooth outer membrane that surrounds the mitochondrion and an inner membrane that forms a series of infoldings or shelves called **cristae**, which project into an inner cavity filled with a gel-like solution known as the **matrix** (Figure 2-10a and chapter opener photo). The two membranes are separated by a narrow **intermembrane space**. The cristae contain proteins that ultimately use O_2 to convert

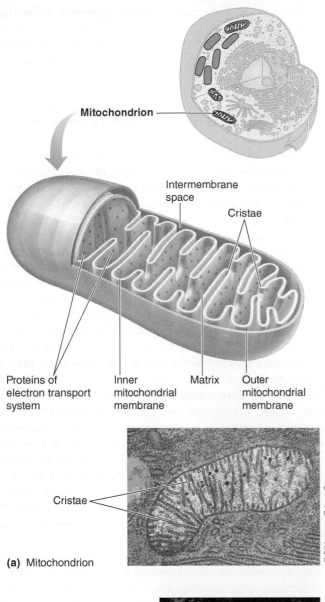

Mitochondrion

Intermembrane space

Cristae

Proteins of electron transport system

Inner mitochondrial membrane

Matrix

Outer mitochondrial membrane

Cristae

(a) Mitochondrion

© Bill Longcore/Science Source

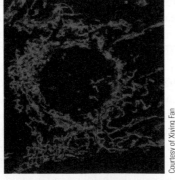

(b) Mitochondrial reticulum

Courtesy of Xiying Fan

I Figure 2-10 Mitochondria. (a) Diagram and electron micrograph of a mitochondrion. Note that the outer membrane is smooth, whereas the inner membrane forms folds known as cristae that extend into the matrix. An intermembrane space separates the outer and inner membranes. The electron transport proteins embedded in the cristae are ultimately responsible for converting much of the energy of food into a usable form. (b) Fluorescent micrograph of the mitochondrial reticulum in a muscle cell, with only the mitochondria being stained (*in red*). The unstained circle in the center of the cell is the nucleus.

much of the energy in food into a usable form. The generous folds of the inner membrane greatly increase the surface area available for housing these important proteins. The matrix consists of a concentrated mixture of hundreds of different dissolved enzymes that prepare nutrient molecules for final extraction of usable energy by the cristae proteins.

Mitochondria form a mitochondrial reticulum in some cell types.

In skeletal muscle and many other cell types, mitochondria rarely exist separately but instead are interconnected in a network, the **mitochondrial reticulum** (I Figure 2-10b). This organized system efficiently distributes materials essential for generating energy—for example O_2 and food derivatives such as fatty acids—from the cell surface to deep within the cell. O_2 and fatty acids, which are poorly soluble in water, are more soluble in and thus can move more quickly through the oily membranes surrounding the interconnected mitochondria than through the watery cytosol.

The mitochondrial reticulum is dynamic; it continuously changes through ongoing rounds of fusion and fission as individual mitochondria join to or split from the network, depending on the cell's energy needs. For example, the mitochondrial network expands in response to contractile activity (exercise) in skeletal muscle.

Mitochondria play a major role in generating ATP.

The source of energy for the body is the chemical energy stored in the carbon bonds of ingested food. Body cells are not equipped to use this energy directly. Instead, the cells must extract energy from food nutrients and convert it into a form they can use—namely, the high-energy phosphate bonds of **adenosine triphosphate (ATP)**, which consists of adenosine with three phosphate groups attached (*tri* means "three") (see p. A-15). When the high-energy bond that binds the terminal phosphate to adenosine is split, a substantial amount of energy is released.

ATP is the universal energy carrier—the common energy "currency" of the body. Cells can "cash in" ATP to pay the energy "price" for running the cell machinery. To obtain immediate usable energy, cells split the terminal phosphate bond of ATP, which yields **adenosine diphosphate (ADP)**—adenosine with two phosphate groups attached (*di* means "two")—plus inorganic phosphate (P_i) plus energy:

$$ATP \xrightarrow{\text{splitting}} ADP + P_i + \text{energy for use by the cell}$$

In this energy scheme, food can be thought of as the "crude fuel" and ATP as the "refined fuel" for operating the body's machinery. Food is digested by the digestive system into small absorbable units that can be transferred from the digestive tract lumen into the blood. For example, dietary carbohydrates are broken down primarily into glucose, which is absorbed into the blood. No usable energy is released during the digestion of food. When delivered to the cells by the blood, the nutrient molecules are transported across the plasma membrane into the

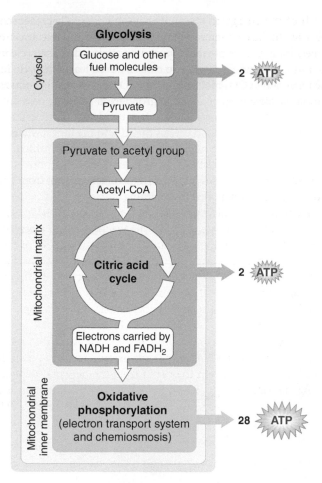

Figure 2-11 Stages of cellular respiration. The three stages of cellular respiration are (1) glycolysis, (2) citric acid cycle, and (3) oxidative phosphorylation

six-carbon sugar molecule, into two **pyruvate** molecules, each with three carbons (*glyc-* means "sweet"; *lysis* means "breakdown") (I Figure 2-12). During this process, two hydrogens are released and transferred to two NADH molecules for later use. (You will learn more about NADH shortly.) Some energy from the broken chemical bonds of glucose is used directly to convert ADP into ATP. However, glycolysis is not efficient in terms of energy extraction: The net yield is only two molecules of ATP per glucose molecule processed. Much of the energy originally contained in the glucose molecule is still locked in the chemical bonds of the pyruvate molecules. The low-energy yield of glycolysis is not enough to support the body's demand for ATP. This is where the mitochondria come into play.

Citric Acid Cycle The pyruvate produced by glycolysis in the cytosol is selectively transported into the mitochondrial matrix. Here, one of its carbons is enzymatically removed in the form of CO_2 (I Figure 2-13). Also, another hydrogen is released and transferred to another NADH. The two-carbon molecule remaining after the breakdown process, an **acetyl group**, combines with coenzyme A (CoA), a derivative of pantothenic acid (a B vitamin), to produce the compound **acetyl coenzyme A (acetyl-CoA)**.

Acetyl-CoA then enters the **citric acid cycle**, a cyclical series of eight biochemical reactions that are catalyzed by the enzymes of the mitochondrial matrix. This cycle of reactions can be compared to one revolution around a Ferris wheel, except that the molecules themselves are not physically moved around in a cycle. On the top of the Ferris wheel, acetyl-CoA, a two-carbon molecule, enters a seat already occupied by oxaloacetate, which has four carbons. These two molecules link to form a six-carbon citrate molecule (at intracellular pH, citric acid exists in an ionized form, citrate), and the trip around the citric acid cycle begins. (This cycle is alternatively known as the **Krebs cycle**, in honor of its principal discoverer, or the **tricarboxylic acid cycle**, because citrate contains three carboxylic acid groups.) At each step in the cycle, matrix enzymes modify the passenger molecule to form a slightly different molecule

cytosol. (Details of how materials cross the membrane are covered in Chapter 3.)

We now turn attention to the steps involved in ATP production within the cell and the role of the mitochondria in these steps. **Cellular respiration** refers collectively to the intracellular reactions in which energy-rich molecules are broken down to form ATP, using O_2 and producing carbon dioxide (CO_2) in the process. In most cells, ATP is generated from the sequential dismantling of absorbed nutrient molecules in three stages: *glycolysis* in the cytosol, the *citric acid cycle* in the mitochondrial matrix, and *oxidative phosphorylation* at the mitochondrial inner membrane (I Figure 2-11). (Muscle cells use an additional cytosolic pathway for immediately generating energy at the onset of exercise; see p. 269.) We use glucose as an example to describe these stages.

Glycolysis Among the thousands of enzymes in the cytosol are the 10 responsible for **glycolysis**, a chemical process involving 10 sequential reactions that break down glucose, a

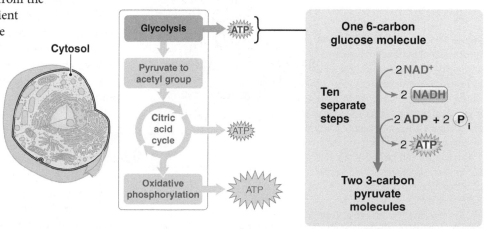

Figure 2-12 Glycolysis in the cytosol. Glycolysis splits glucose (six carbons) into two pyruvate molecules (three carbons each), with a net yield of 2 ATP plus 2 NADH (available for further energy extraction by the electron transport system).

(shown in Figure 2-13). These molecular alterations have the following important consequences:

1. Two carbons are "kicked off the ride"—released one at a time from six-carbon citrate, converting it back into four-carbon oxaloacetate, which is now available at the top of the cycle to pick up another acetyl-CoA for another revolution through the cycle. CoA is recycled too; it is released at the beginning of the cycle, making it available to bind with a new acetyl group to form another acetyl-CoA.

2. The released carbon atoms, which were originally present in the acetyl-CoA that entered the cycle, are converted into two molecules of CO_2. Note that two carbon atoms enter the cycle in the form of acetyl-CoA and two carbon atoms leave the cycle in the form of two CO_2 molecules. This CO_2 and the CO_2 produced during the formation of an acetyl group from pyruvate pass out of the mitochondrial matrix and subsequently out of the cell to enter the blood. The blood carries the CO_2 to the lungs, where it is eliminated as a waste to the atmosphere through breathing. The oxygen used to make CO_2 from these released carbons is derived from the molecules involved in the reactions, not from free molecular oxygen supplied by breathing.

3. Hydrogens are also "bumped off" during the cycle at four of the chemical conversion steps. The key purpose of the citric

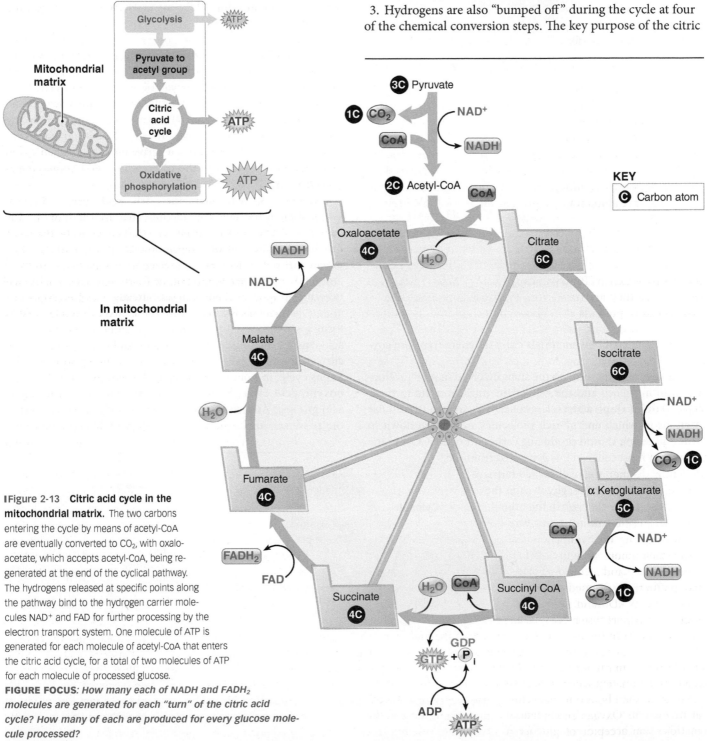

I Figure 2-13 **Citric acid cycle in the mitochondrial matrix.** The two carbons entering the cycle by means of acetyl-CoA are eventually converted to CO_2, with oxaloacetate, which accepts acetyl-CoA, being regenerated at the end of the cyclical pathway. The hydrogens released at specific points along the pathway bind to the hydrogen carrier molecules NAD+ and FAD for further processing by the electron transport system. One molecule of ATP is generated for each molecule of acetyl-CoA that enters the citric acid cycle, for a total of two molecules of ATP for each molecule of processed glucose.

FIGURE FOCUS: *How many each of NADH and FADH₂ molecules are generated for each "turn" of the citric acid cycle? How many of each are produced for every glucose molecule processed?*

acid cycle is to produce these hydrogens for entry into the electron transport system in the inner mitochondrial membrane. The hydrogens are transferred to two different hydrogen carrier molecules—**nicotinamide adenine dinucleotide (NAD$^+$)**, a derivative of the B vitamin niacin, and **flavine adenine dinucleotide (FAD)**, a derivative of the B vitamin riboflavin. The transfer of hydrogen converts these compounds to NADH and FADH$_2$, respectively. Three NADH and one FADH$_2$ are produced for each turn of the citric acid cycle.

4. One more molecule of ATP is produced for each molecule of acetyl-CoA processed. ATP is not directly produced by the citric acid cycle. The released energy directly links inorganic phosphate to **guanosine diphosphate (GDP)**, forming **guanosine triphosphate (GTP)**, a high-energy molecule similar to ATP. The energy from GTP is then transferred to ATP as follows:

$$ADP + GTP \rightleftharpoons ATP + GDP$$

Because each glucose molecule is converted into two acetyl-CoA molecules, fueling two turns of the citric acid cycle, two more ATP molecules are produced from each glucose molecule.

So far, the cell still does not have much of an energy profit. However, the citric acid cycle is important in preparing the hydrogen carrier molecules for their entry into the final stage, oxidative phosphorylation, which produces far more energy than the sparse amount of ATP produced by the cycle itself.

Oxidative Phosphorylation Considerable untapped energy is still stored in the released hydrogens, which contain electrons at high energy levels. **Oxidative phosphorylation** refers to the process by which ATP is synthesized using energy released by electrons as they are transferred to O$_2$. This process involves two groups of proteins at the inner mitochondrial membrane: the *electron transport system* and *ATP synthase.*

The "big payoff" in energy capture begins when NADH and FADH$_2$ carry the hydrogens to the electron transport system. The **electron transport system** consists of electron carriers found in four large stationary protein complexes, numbered *I, II, III,* and *IV,* along with two small highly mobile electron carriers, *cytochrome c* and *ubiquinone* (also known as *coenzyme Q* or *CoQ*), which shuttle electrons between the major complexes (❙ Figure 2-14).

The high-energy electrons are extracted from the hydrogens held in NADH and FADH$_2$ and enter the electron transport chain for transfer through a series of steps from one electron-carrier molecule to another in an assembly line (step **1**). As a result of giving up hydrogen ions (H$^+$) and electrons at the electron transport system, NADH and FADH$_2$ are converted back to NAD$^+$ and FAD (step **2**), freeing them to pick up more hydrogen atoms released during glycolysis and the citric acid cycle. Thus, NAD$^+$ and FAD link the citric acid cycle and the electron transport system. The electron carriers are arranged in a specific order in the inner membrane so that the high-energy electrons fall to successively lower energy levels as they are transferred from carrier to carrier through a chain of reactions (step **3**). Ultimately, when they are in their lowest energy state, the electrons are bound to molecular oxygen derived from the air we breathe. Oxygen enters the mitochondria to serve as the final electron acceptor of the electron transport system. This

negatively charged oxygen (O$_2^-$)—negative because it has acquired additional electrons—then combines with the positively charged hydrogen ions (H$^+$)—positive because they have donated electrons at the beginning of the electron transport system—to form water, H$_2$O (step **4**).

As electrons move through this chain of reactions, they release free energy. Part of the released energy is lost as heat, but some is harnessed by the mitochondrion to synthesize ATP. At three sites in the electron transport system (Complexes I, III, and IV), the energy released during the transfer of electrons is used to transport hydrogen ions across the inner mitochondrial membrane from the matrix to the intermembrane space between the inner and the outer mitochondrial membranes (step **5**). As a result, hydrogen ions are more heavily concentrated in the intermembrane space than in the matrix. This H$^+$ gradient generated by the electron transport system (step **6**) supplies the energy that drives ATP synthesis by the membrane-bound mitochondrial enzyme ATP synthase.

ATP synthase consists of a *basal unit* embedded in the inner membrane, connected by a stalk to a *headpiece* located in the matrix, with the *stator* bridging the basal unit and headpiece. Because H$^+$ ions are more heavily concentrated in the intermembrane space than in the matrix, they have a strong tendency to flow back into the matrix through the inner membrane via channels formed between the basal units and stators of the ATP synthase complexes (step **7**). This flow of H$^+$ ions activates ATP synthase and powers ATP synthesis by the headpiece, a process called **chemiosmosis**. Passage of H$^+$ ions through the channel makes the headpiece and stalk spin like a top (step **8**), similar to the flow of water making a waterwheel turn. As a result of the changes in its shape and position as it turns, the headpiece is able to sequentially pick up ADP and P$_i$, combine them, and release the ATP product (step **9**).

Oxidative phosphorylation encompasses the entire process by which ATP synthase synthesizes ATP by phosphorylating (adding a phosphate to) ADP using the energy released by electrons as they are transferred to O$_2$ by the electron transport system. The harnessing of energy into a useful form as the electrons tumble from a high-energy state to a low-energy state can be likened to a power plant converting the energy of water tumbling down a waterfall into electricity.

When activated, ATP synthase provides a rich yield of 28 more ATP molecules for each glucose molecule processed (from the 10 NADH and 2 FADH$_2$ molecules formed during glucose processing) (❙ Figure 2-15). Approximately 2.5 ATP are synthesized as a pair of electrons released by NADH travels through the entire electron transport system to oxygen, for a total of 25 ATPs from NADH. The shorter pathway followed by an electron pair released from FADH$_2$ (see ❙ Figure 2-14) synthesizes about 1.5 ATP, for a total of 3 ATPs from FADH$_2$. This means a total of 32 molecules of ATP are produced when a glucose molecule is completely dismantled in cellular respiration: 2 during glycolysis, 2 during the citric acid cycle, and 28 during oxidative phosphorylation. The ATP is transported out of the mitochondrion into the cytosol for use as the cell's energy source.

The steps leading to oxidative phosphorylation might at first seem an unnecessary complication. Why not just directly oxidize, or "burn," food molecules to release their energy?

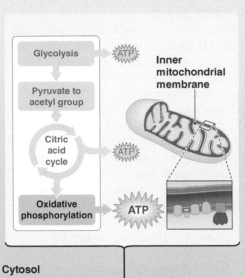

1. The high-energy electrons extracted from the hydrogens in NADH and FADH$_2$ are transferred from one electron-carrier molecule to another.

2. The NADH and FADH$_2$ are converted to NAD$^+$ and FAD, which frees them to pick up more hydrogen atoms released during glycolysis and the citric acid cycle.

3. The high-energy electrons fall to successively lower energy levels as they are transferred from carrier to carrier through the electron transport system.

4. The electrons are passed to O$_2$, the final electron acceptor of the electron transport system. This oxygen, now negatively charged because it has acquired additional electrons, combines with H$^+$ ions, which are positively charged because they donated electrons at the beginning of the electron transport system, to form H$_2$O.

5. As electrons move through the electron transport system, they release free energy. Part of the released energy is lost as heat, but some is harnessed to transport H$^+$ across the inner mitochondrial membrane from the matrix to the intermembrane space at Complexes I, III, and IV.

6. As a result, H$^+$ ions are more heavily concentrated in the intermembrane space than in the matrix. This H$^+$ gradient supplies the energy that drives ATP synthesis by ATP synthase.

7. Because of this gradient, H$^+$ ions have a strong tendency to flow into the matrix across the inner membrane via channels between the basal units and stators of the ATP synthase complexes.

8. This flow of H$^+$ ions activates ATP synthase and powers ATP synthesis by the headpiece, a process called **chemiosmosis**. Passage of H$^+$ ions through the channel makes the headpiece and stalk spin like a top.

9. As a result of changes in its shape and position as it turns, the headpiece picks up ADP and P$_i$, combines them, and releases the ATP product.

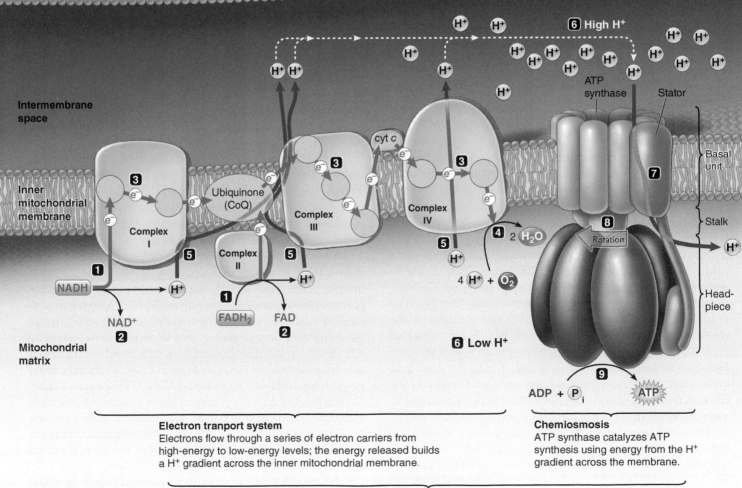

Electron transport system
Electrons flow through a series of electron carriers from high-energy to low-energy levels; the energy released builds a H$^+$ gradient across the inner mitochondrial membrane.

Chemiosmosis
ATP synthase catalyzes ATP synthesis using energy from the H$^+$ gradient across the membrane.

Oxidative phosphorylation

Figure 2-14 Oxidative phosphorylation at the mitochondrial inner membrane. Oxidative phosphorylation involves the electron transport system (steps 1–6) and chemiosmosis by ATP synthase (steps 7–9). The pink circles in the electron transport system represent specific electron carriers.

When this process occurs outside the body, all the energy stored in a food molecule is released explosively in the form of heat (Figure 2-16). Think about what happens when a marshmallow you are roasting accidentally catches on fire. The burning marshmallow gets hot quickly as a result of the rapid oxidation of sugar. In the body, food molecules are oxidized within the mitochondria in many small, controlled steps so that their chemical energy is gradually released in small quantities that can be more efficiently captured in ATP bonds and stored in a form that is useful to the cell. In this way, much less of the energy is converted to heat. The heat that is produced is not completely wasted energy; it is used to help maintain body temperature, with any excess heat being eliminated to the environment.

The cell generates more energy in aerobic than in anaerobic conditions.

The cell is a more efficient energy converter when O_2 is available. In an **anaerobic** ("lack of air," specifically "lack of O_2") condition, the degradation of glucose cannot proceed beyond glycolysis, which takes place in the cytosol and yields only two molecules of ATP per molecule of glucose. The untapped energy of the glucose molecule remains locked in the bonds of the pyruvate molecules, which are eventually converted to **lactate** if they do not enter the pathway that ultimately leads to oxidative phosphorylation.

When sufficient O_2 is present—an **aerobic** ("with air" or "with O_2") condition—mitochondrial processing (that is, the citric acid cycle in the matrix and the electron transport system and ATP synthase at the inner membrane) harnesses enough energy to generate 30 more molecules of ATP, for a total net yield of 32 ATPs per molecule of glucose processed. (For a description of aerobic exercise, see the boxed feature on p. 41, A Closer Look at Exercise Physiology.) The overall reaction for the oxidation of food molecules to yield energy during cellular respiration is as follows:

$$\text{Food} + \underset{\substack{\text{(necessary}\\\text{for oxidative}\\\text{phosphorylation)}}}{O_2} \rightleftharpoons \underset{\substack{\text{(produced}\\\text{primarily}\\\text{by the citric}\\\text{acid cycle)}}}{CO_2} + \underset{\substack{\text{(produced by}\\\text{the electron}\\\text{transport system)}}}{H_2O} + \underset{\substack{\text{(produced}\\\text{primarily by}\\\text{ATP synthase)}}}{\text{ATP}}$$

Note that the oxidative reactions within the mitochondria generate energy, unlike the oxidative reactions controlled by the peroxisome enzymes. Both organelles use O_2, but they do so for different purposes.

◄ FIGURE FOCUS: Where does the O_2 that serves as the final electron acceptor at Step 4 come from?

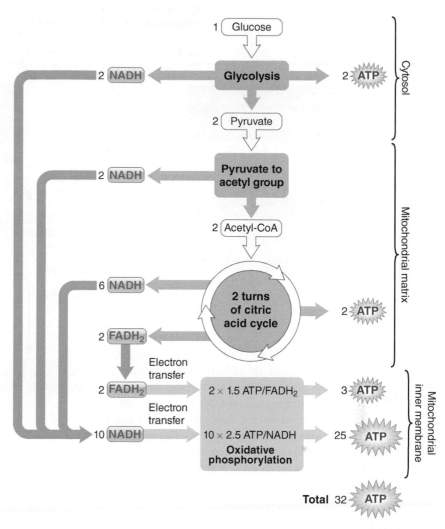

Figure 2-15 **Summary of ATP production from the complete oxidation of one molecule of glucose.** The total of 32 ATP assumes that electrons carried by each NADH yield 2.5 ATP and those carried by each $FADH_2$ yield 1.5 ATP during oxidative phosphorylation.

FIGURE FOCUS: *If no O_2 is available, pyruvate is converted to lactate instead of an acetyl group. Compare the energy yield of glucose degradation with O_2 and without O_2.*

Glucose, the principal nutrient derived from dietary carbohydrates, is the fuel preference of most cells. However, nutrient molecules derived from fats (fatty acids) and, if necessary, from proteins (amino acids) can also participate at specific points in this overall chemical reaction to eventually produce energy. Amino acids are usually used for protein synthesis rather than energy production, but they can be used as fuel if insufficient glucose and fat are available (see Chapter 17). Fatty acids are sequentially broken down in the mitochondrial matrix through the process of **beta (β) oxidation**, which cleaves off blocks of two-carbon units one at a time. Each two-carbon unit is used to form an acetyl-CoA molecule that enters the citric acid cycle. Different fatty acids contain a different number of carbons. Most fatty acids in the body are between 14 and 22 carbons long. Because every two carbons in a long fatty acid chain forms an acetyl-CoA, a single fatty acid can yield 7 to 11 acetyl-CoA molecules, depending on the fatty acid length, compared to the 2 acetyl-CoA molecules resulting from the breakdown of a glucose molecule. For this

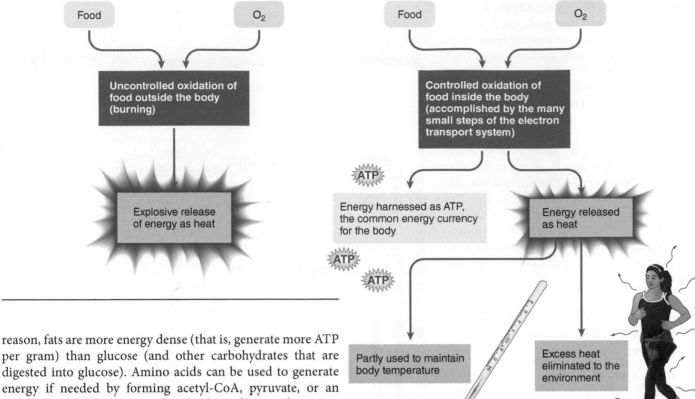

I Figure 2-16 **Uncontrolled versus controlled oxidation of food.** Part of the energy released as heat when food undergoes uncontrolled oxidation (burning) outside the body is instead harnessed and stored in useful form when controlled oxidation of food occurs inside the body.

reason, fats are more energy dense (that is, generate more ATP per gram) than glucose (and other carbohydrates that are digested into glucose). Amino acids can be used to generate energy if needed by forming acetyl-CoA, pyruvate, or an intermediate in the citric acid cycle, depending on the amino acid. Because fatty acids and amino acids generate energy only by the citric acid cycle and oxidative phosphorylation, fats and proteins can be used only aerobically, whereas glucose can be used anaerobically (via glycolysis alone) and aerobically when the product of glycolysis (pyruvate) is further broken down via the citric acid cycle and oxidative phosphorylation.

The energy stored within ATP is used for synthesis, transport, and mechanical work.

Once formed, ATP is transported out of the mitochondria and is then available as an energy source in the cell. Cell activities that require energy expenditure fall into three main categories:

1. *Synthesis of new chemical compounds,* such as protein synthesis by the ER. Some cells, especially cells with a high rate of secretion, use up to 75% of the ATP they generate just to synthesize new chemical compounds.

2. *Membrane transport,* such as the selective transport of molecules across the kidney tubules during the process of urine formation. Kidney cells can expend as much as 80% of their ATP currency to operate their selective membrane-transport mechanisms.

3. *Mechanical work,* such as contraction of the heart muscle to pump blood or contraction of skeletal muscles to lift an object. These activities require tremendous quantities of ATP.

As a result of cell energy expenditure to support these various activities, large quantities of ADP are produced. These energy-depleted molecules enter the mitochondria for "recharging" and then cycle back into the cytosol as energy-rich ATP molecules after participating in oxidative phosphorylation. In this recharging–expenditure cycle, a single ADP/ATP molecule

may shuttle between mitochondria and cytosol thousands of times per day. On average a person recycles the equivalent of his or her body weight of ATP every day.

The high demands for ATP make glycolysis alone an insufficient, and inefficient, supplier of power for most cells. Were it not for the mitochondria, which house the metabolic machinery for oxidative phosphorylation, the body's energy capability would be limited. However, glycolysis does provide cells with a sustenance mechanism that can produce ATP under anaerobic conditions. Skeletal muscle cells in particular take advantage of this ability during short bursts of strenuous exercise, when energy demands for contractile activity outstrip the body's ability to bring adequate O_2 to the exercising muscles to support oxidative phosphorylation.

Mitochondria play a key role in programmed cell death.

In addition to their central role in generating most of the ATP for cell use, mitochondria play a key, unrelated role in deliberate cell suicide, a process called **apoptosis**. Every cell has a built-in biochemical pathway that, if triggered, causes the cell to execute itself as a result of mitochondrial leakage of cytochrome *c* (one of the components of the electron transport system), which out-

AEROBIC ("WITH O₂") EXERCISE involves large muscle groups and is performed at a low-enough intensity and for a long-enough period that fuel sources can be converted to adenosine triphosphate (ATP) by using the citric acid cycle and oxidative phosphorylation as the predominant metabolic pathway. Aerobic exercise can be sustained from 15 to 20 minutes to several hours at a time. Short-duration, high-intensity activities, such as weight training and the 100-meter dash, which last for a matter of seconds and rely solely on energy stored in the muscles and on glycolysis, are forms of **anaerobic** ("without O₂") **exercise**.

Inactivity is associated with increased risk of developing both hypertension (high blood pressure) and coronary artery disease (blockage of the arteries that supply the heart). The American College of Sports Medicine recommends that an individual participate in aerobic exercise a minimum of three times per week for 20 to 60 minutes to reduce the risk of hypertension and coronary artery disease and to improve physi-

cal work capacity. The same health benefits are derived whether the exercise is accomplished in one long stretch or is broken into multiple shorter stints. This is good news because many individuals find it easier to stick with brief bouts of exercise sprinkled throughout the day.

The intensity of the exercise should be based on a percentage of the individual's maximal capacity to work. The easiest way to establish the proper intensity of exercise and to monitor intensity levels is by checking the heart rate. The estimated maximal heart rate is determined by subtracting the person's age from 220. Significant benefits can be derived from aerobic exercise performed between 70% and 80% of maximal heart rate. For example, the estimated maximal heart rate for a 20 year old is 200 beats per minute. If this person exercised three times per week for 20 to 60 minutes at an intensity that increased the heart rate to 140 to 160 beats per minute, the participant should significantly improve his or her aerobic work capacity and reduce the risk of cardiovascular disease.

---

side the confines of the mitochondria exerts the entirely different function of activating intracellular protein-snipping enzymes that slice the cell into small, disposable pieces. Apoptosis is a natural part of the life of an organism: It eliminates cells that are no longer needed or are disordered. (The term *apoptosis* means "dropping off," in reference to the dropping off of cells that are no longer useful, much as autumn leaves drop off trees.) (See the boxed feature on pp. 42 and 43, ▮Concepts, Challenges, and Controversies, for further discussion of apoptosis.)

Check Your Understanding 2.7

1. Draw and label a mitochondrion.
2. List the stages of cellular respiration, and state where each is accomplished.
3. Compare the amount of ATP produced from one glucose molecule in anaerobic and aerobic conditions.
4. Contrast apoptosis and necrosis.

2.8 Vaults as Cellular Trucks

Vaults, which are nonmembranous organelles, are shaped like octagonal barrels (▮Figure 2-17, p. 44). Their name comes from their multiple arches, which reminded their discoverers of vaulted or cathedral ceilings. Just like barrels, vaults have a hollow interior. When open, they appear like pairs of unfolded flowers with each half of the vault bearing eight "petals" attached to a central ring. A cell may contain thousands of vaults, which are three times as large as ribosomes.

Vaults may serve as cellular transport vehicles.

Currently, the function of vaults is uncertain, but their octagonal shape and their hollow interior provides clues. Nuclear pores are also octagonal and the same size as vaults, leading to speculation that vaults may be cellular "trucks." According to this proposal, vaults would dock at or enter nuclear pores, pick up molecules synthesized in the nucleus, and deliver their cargo elsewhere in the cell. Ongoing research supports the role of vaults in nucleus-to-cytoplasm transport, but their cargo has not been determined. One possibility is that vaults carry mRNA from the nucleus to the ribosomal sites of protein synthesis within the cytoplasm. Another possibility is that vaults transport the two subunits that make up ribosomes from the nucleus, where they are produced, to their sites of action—either attached to the rough ER or in the cytosol. The interior of a vault is the right size to accommodate these ribosomal subunits.

Clinical Note Vaults may play an undesirable role in bringing about the multidrug resistance sometimes displayed by cancer cells. Chemotherapy drugs designed to kill cancer cells tend to accumulate in the nuclei of these cells, but some cancer cells develop resistance to a wide variety of these drugs. This broad resistance is a major cause of cancer treatment failure. Researchers have shown that some cancer cells resistant to chemotherapy produce up to 16 times more than normal quantities of the major vault protein. If further investigation confirms that vaults play a role in drug resistance—perhaps by transporting the drugs from the nucleus to sites for exocytosis from the cancer cells—the exciting possibility exists that interference with this vault activity could improve the sensitivity of cancer cells to chemotherapeutic drugs.

Apoptosis: Programmed Cell Suicide

APOPTOSIS is intentional suicide of a cell that is no longer useful. Apoptosis is a normal part of life—individual cells that have become superfluous or disordered are triggered to self-destruct for the greater good of maintaining the whole body's health.

Roles of Apoptosis

Here are examples of the vital roles played by this intrinsic sacrificial program:

■ *Predictable self-elimination of selected cells is a normal part of development.* Certain unwanted cells produced during development are programmed to kill themselves as the body is sculpted into its final form. During the development of a female, for example, apoptosis deliberately removes the embryonic ducts capable of forming a male reproductive tract. Likewise, apoptosis carves fingers from a mitten-shaped developing hand by eliminating the weblike membranes between them.

■ *Apoptosis is important in tissue turnover in the adult body.* Optimal functioning of most tissues depends on a balance between controlled production of new cells and regulated cell self-destruction. This balance maintains the proper number of cells in a given tissue while ensuring a controlled supply of fresh cells that are at their peak of performance.

■ *Programmed cell death plays an important role in the immune system.* Apoptosis provides a means to remove cells infected with harmful viruses. Furthermore, infection-fighting white blood cells that have finished their prescribed function and are no longer needed execute themselves.

■ *Undesirable cells that threaten homeostasis are typically culled from the body by apoptosis.* Included in this hit list are aged cells, cells that have suffered irreparable damage by exposure to radiation or other poisons, and cells that have somehow gone awry. Many mutated cells are eliminated by this means before they become fully cancerous.

Comparison of Apoptosis and Necrosis

Apoptosis is not the only means by which a cell can die, but it is the neatest way. Apoptosis is a controlled, intentional, tidy way of removing individual cells that are no longer needed or that pose a threat to the body. The other form of cell death, **necrosis** (meaning "making dead"), is uncontrolled, accidental, messy murder of useful cells that have been severely injured by an agent external to the cell, as by a physical blow, O_2 deprivation, or disease. For example, heart muscle cells deprived of their O_2 supply by complete blockage of the blood vessels supplying them during a heart attack die as a result of necrosis (see p. 314). Apoptosis is an energy-dependent process, whereas necrosis does not require energy.

Even though necrosis and apoptosis both result in cell death, the steps involved are different. In necrosis the dying cells are passive victims, whereas in apoptosis the cells actively participate in their deaths. In necrosis, the injured cell loses the integrity of its plasma membrane and cannot pump out Na^+ as usual. As a result, water streams in by osmosis (see p. 66), causing the cell to swell and rupture. Typically, in necrosis the insult that prompted cell death injures many cells in the vicinity, so many neighboring cells swell and rupture together. Release of intracellular contents into the surrounding tissues initiates an inflammatory response at the damaged site (see p. 408). Unfortunately, this inflammatory response may harm healthy neighboring cells.

By contrast, apoptosis targets individual cells for destruction, leaving the surrounding cells intact. A condemned cell signaled to commit suicide detaches itself from its neighbors and then shrinks instead of swelling and bursting. The cell's mitochondria become leaky, permitting cytochrome *c* to leak out into the cytosol. **Cytochrome *c***, a component of the electron transport system, usually participates in oxidative phosphorylation to produce ATP. Outside its typical mitochondrial environment, however, cytochrome *c* activates normally inactive intracellular protein-cutting enzymes, the **caspases**, which kill the cell from within. The unleashed caspases act like molecular scissors to systematically dismantle the cell. Snipping protein after protein, they chop up the nucleus, disassembling its life-essential DNA, then break down the internal shape-holding cytoskeleton, and finally fragment the cell into disposable membrane-enclosed packets (see accompanying photo). Of note, the contents of the dying cell remain wrapped by plasma membrane throughout the entire self-execution process, thus avoiding the spewing of poten-

Check Your Understanding 2.8

1. Describe the structure of a vault.
2. Discuss the speculated functions of vaults.

2.9 Cytosol: Cell Gel

Occupying about 55% of the total cell volume, the cytosol is the semiliquid portion of the cytoplasm that surrounds the organelles. Its nondescript appearance under an electron microscope gives the false impression that the cytosol is a liquid mixture of uniform consistency, but it is actually a highly organized, gel-like mass with differences in composition and consistency from one part of the cell to another.

The cytosol is important in intermediary metabolism, ribosomal protein synthesis, and nutrient storage.

Three general categories of activities are associated with the cytosol: (1) enzymatic regulation of intermediary metabolism; (2) ribosomal protein synthesis; and (3) storage of fat, carbohydrate, and secretory vesicles.

tially harmful intracellular contents characteristic of necrosis. No inflammatory response is triggered, so no neighboring cells are harmed. Instead, cells in the vicinity swiftly engulf and destroy the apoptotic cell fragments by phagocytosis. The breakdown products are recycled for other purposes as needed. The tissue as a whole has continued to function normally, while the targeted cell has unobtrusively killed itself.

By comparison to apoptosis and necrosis, both means of cell death, the self-eating process of autophagy (see p. 31) actually promotes cell survival by removing outdated or damaged cell components, thus permitting the cell to refresh itself with healthy new replacement parts.

Control of Apoptosis

If every cell contains caspases, what normally keeps these powerful self-destructive enzymes under control (that is, in inactive form) in cells that are useful to the body and deserve to live? Likewise, what activates the death-wielding caspases in unwanted cells destined to eliminate themselves? Given the importance of these life-or-death decisions, it is not surprising that multiple internal control pathways tightly regulate whether a cell is "to be or not to be." A cell normally receives a constant stream of "survival signals," which reassure the cell that it is useful to the body, that all is right in the internal environment surrounding the cell, and that everything is in good working order within the cell. These signals include tissue-specific growth factors, certain hormones, and appropriate contact with neighboring cells and surrounding connective tissue. These extracellular survival signals trigger intracellular pathways that block activation of the caspases, thus restraining the cell's death ma-

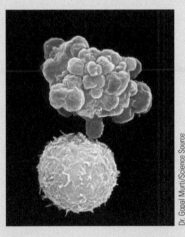

A normal cell (bottom) and a cell undergoing apoptosis (top).

Dr. Gopal Murti/Science Source

chinery. Most cells are programmed to commit suicide if they do not receive their normal reassuring survival signals. With the usual safeguards removed, the lethal protein-snipping enzymes are unleashed. For example, withdrawal of growth factors or detachment from the surrounding connective tissue causes a cell to promptly execute itself.

Furthermore, cells display "death receptors" in their outer plasma membrane that receive specific extracellular "death signals," such as a particular hormone or a specific chemical messenger from white blood cells that arrive at the cell via the blood. Activation of death pathways by these signals can override the life-saving pathways triggered by the survival signals. The death-signal pathway swiftly ignites the internal apoptotic machinery, driving the cell to its demise. Likewise, the self-execution machinery is set in motion when a cell suffers irreparable intracellular damage. Thus, some signals block apoptosis and others promote it. Whether a cell lives or dies depends on which of these competing signals dominates at any given time. Although all cells have the same death machinery, they vary in the specific signals that induce them to commit suicide.

Considering that every cell's life hangs in delicate balance at all times, it is not surprising that faulty control of apoptosis—resulting in either too much or too little cell suicide—appears to participate in many major diseases. Excessive apoptotic activity is believed to contribute to the brain cell death associated with Alzheimer's disease, Parkinson's disease, and stroke and to the premature demise of important infection-fighting cells in AIDS. Conversely, too little apoptosis most likely plays a role in cancer. Evidence suggests that cancer cells fail to respond to the normal extracellular signals that promote cell death. Because these cells neglect to die on cue, they grow in unchecked fashion, forming a chaotic, out-of-control mass.

Apoptosis is currently one of the hottest topics of investigation in the field. Researchers are scrambling to sort out the multiple factors involved in the mechanisms controlling this process. Their hope is to find ways to tinker with the apoptotic machinery to find badly needed new therapies for treating a variety of big killers.

Enzymatic Regulation of Intermediary Metabolism The term **intermediary metabolism** refers collectively to the large set of chemical reactions inside the cell that involve the degradation, synthesis, and transformation of small organic molecules such as simple sugars, amino acids, and fatty acids. These reactions are critical for ultimately capturing the energy used for cell activities and for providing the raw materials needed to maintain the cell's structure, function, and growth. Intermediary metabolism occurs in the cytoplasm, with most of it being accomplished in the cytosol. The cytosol contains thousands of enzymes involved in intermediary biochemical reactions.

Ribosomal Protein Synthesis Also dispersed throughout the cytosol are the free ribosomes, which synthesize proteins for use in the cytosol itself. In contrast, recall that the rough-ER ribosomes synthesize proteins for secretion and for construction of new cell components.

Storage of Fat, Glycogen, and Secretory Vesicles Excess nutrients not immediately used for ATP production are converted in the cytosol into storage forms that are readily visible under a light microscope. Such nonpermanent masses of stored material are known as **inclusions**. Inclusions are not surrounded by membrane, and they may or may not be present,

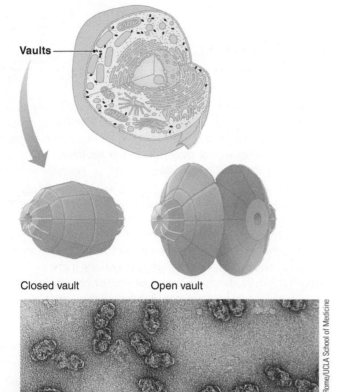

Vaults

Closed vault   Open vault

Figure 2-17 Vaults. Diagram of closed and open vaults and electron micrograph of vaults, which are octagonal, barrel-shaped, nonmembranous organelles believed to transport either messenger RNA or the ribosomal subunits from the nucleus to cytoplasmic ribosomes.

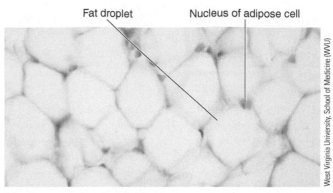

Fat droplet   Nucleus of adipose cell

(a) Fat storage in adipose cells

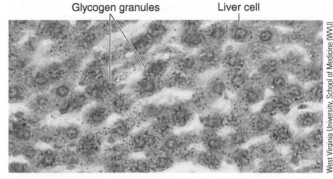

Glycogen granules   Liver cell

(b) Glycogen storage in liver cells

Figure 2-18 Inclusions. (a) Light micrograph showing fat storage in adipose cells. A fat droplet occupies almost the entire cytosol of each cell. (b) Light micrograph showing glycogen storage in liver cells. The red-stained granules throughout the cytosol of each liver cell are glycogen deposits.

depending on the type of cell and the circumstances. The largest and most important storage product is fat. Small fat droplets are present within the cytosol in various cells. In **adipose tissue**, the tissue specialized for fat storage, stored fat molecules can occupy almost the entire cytosol, where they merge to form one large fat droplet (Figure 2-18a). The other storage product is **glycogen**, the storage form of glucose, which appears as clusters or granules dispersed throughout the cell (Figure 2-18b). Cells vary in their ability to store glycogen, with liver and muscle cells having the greatest stores. When food is not available to provide fuel for the citric acid cycle and electron transport system, stored glycogen and fat are broken down to release glucose and fatty acids, respectively, which can feed the mitochondrial energy-producing machinery. An average adult human stores enough glycogen to provide energy for about a day of normal activities and typically has enough fat stored to provide energy for two months.

Secretory vesicles that have been processed and packaged by the ER and Golgi complex also remain in the cytosol, where they are stored until signaled to empty their contents to the outside. In addition, transport and endocytic vesicles move through the cytosol.

2.10 Cytoskeleton: Cell "Bone and Muscle"

Different cell types in the body have distinct shapes, structural complexities, and functional specializations. These unique characteristics are maintained by the **cytoskeleton**, an elaborate protein scaffolding dispersed throughout the cytosol that acts as the "bone and muscle" of the cell by supporting and organizing the cell components and controlling their movements.

The cytoskeleton has three distinct elements: (1) *microtubules,* (2) *microfilaments,* and (3) *intermediate filaments.* These elements are structurally linked and functionally coordinated to provide certain integrated functions for the cell. These functions, along with the functions of all other components of the cytoplasm, are summarized in Table 2-2.

Cytoplasm Component	Structure	Function
Membranous organelles		
Endoplasmic reticulum	Extensive, continuous membranous network of fluid-filled tubules and flattened sacs, partially studded with ribosomes	Forms new cell membrane and other cell components and manufactures products for secretion
Golgi complex	Sets of stacked, flattened, membranous sacs	Modifies, packages, and distributes newly synthesized proteins
Lysosomes	Membranous sacs containing hydrolytic enzymes	Serve as cell's digestive system, destroying foreign substances and cellular debris
Peroxisomes	Membranous sacs containing oxidative enzymes	Perform detoxification activities
Mitochondria	Rod- or oval-shaped bodies enclosed by two membranes, with the inner membrane folded into cristae that project into an interior matrix	Act as energy organelles; major site of ATP production; contain enzymes for citric acid cycle, proteins of electron transport system, and ATP synthase
Nonmembranous organelles		
Ribosomes	Granules of RNA and proteins—some attached to rough ER, some free in cytosol	Serve as workbenches for protein synthesis
Proteasomes	Cylindrical protein complexes consisting of a hollow core particle capped on both ends by a regulatory particle	Degrade unwanted intracellular proteins that have been tagged for destruction by ubiquitin
Vaults	Shaped like hollow octagonal barrels	Serve as cellular trucks for transport from nucleus to cytoplasm
Centrioles	A pair of cylindrical structures at right angles to each other	Form and organize microtubules during assembly of the mitotic spindle during cell division and form cilia and flagella
Cytosol		
Intermediary metabolism enzymes	Dispersed within the cytosol	Facilitate intracellular reactions involving degradation, synthesis, and transformation of small organic molecules
Transport, secretory, and endocytic vesicles	Transiently formed, membrane-enclosed products synthesized within or engulfed by the cell	Transport or store products being moved within, out of, or into the cell, respectively
Inclusions	Glycogen granules, fat droplets	Store excess nutrients
Cytoskeleton		As an integrated whole, serves as the cell's "bone and muscle"
Microtubules	Long, slender, hollow tubes composed of tubulin molecules	Maintain asymmetric cell shapes and coordinate complex cell movements, specifically serving as highways for transport of secretory vesicles within cell, serving as main structural and functional component of cilia and flagella, and assembling into mitotic spindle
Microfilaments	Intertwined helical chains of actin molecules; microfilaments composed of myosin molecules also present in muscle cells	Play a vital role in various cellular contractile systems, including muscle contraction and amoeboid movement; serve as a mechanical stiffener for microvilli
Intermediate filaments	Irregular, threadlike proteins	Help resist mechanical stress

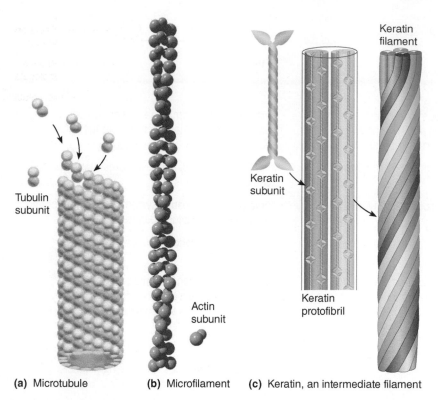

Figure 2-19 **Components of the cytoskeleton.** (a) Microtubules, the largest of the cytoskeletal elements, are long, hollow tubes formed by two slightly different variants of globular-shaped tubulin molecules. (b) Most microfilaments, the smallest of the cytoskeletal elements, consist of two chains of actin molecules wrapped around each other. (c) The intermediate filament keratin, found in skin, is made of four keratin protofibrils twisted together. A protofibril consists of two strands, each made up of two staggered rows of keratin subunits. The composition of intermediate filaments, which are intermediate in size between the microtubules and microfilaments, varies among different cell types.

(a) Microtubule   (b) Microfilament   (c) Keratin, an intermediate filament

Microtubules help maintain asymmetric cell shapes and play a role in complex cell movements.

Microtubules are the largest of the cytoskeletal elements. They are slender (22 nm in diameter), long, hollow, unbranched tubes composed primarily of **tubulin**, a small, globular, protein molecule (Figure 2-19a).

Microtubules arise from the centrosome and its associated centrioles. The **centrosome**, or **cell center**, located near the nucleus, consists of the centrioles surrounded by an amorphous mass of proteins. The **centrioles**, which are nonmembranous organelles, are a pair of short cylindrical structures that lie at right angles to each other at the centrosome's center (Figure 2-20). The centrosome is the cell's main microtubule organizing center. When a cell is not dividing, microtubules are formed from the amorphous mass and radiate outward in all directions from the centrosome (see Figure 2-1, p. 23). Centrioles form microtubules under special circumstances, as you will see as we turn attention to microtubule functions.

Microtubules position many of the cytoplasmic organelles, such as the ER, Golgi complex, lysosomes, and mitochondria. They are also essential for maintaining the shape of asymmetric cells, such as nerve cells, whose elongated axons may extend up to a meter in length from where the cell body originates in the spinal cord to where the axon ends at a muscle (Figure 2-21a).

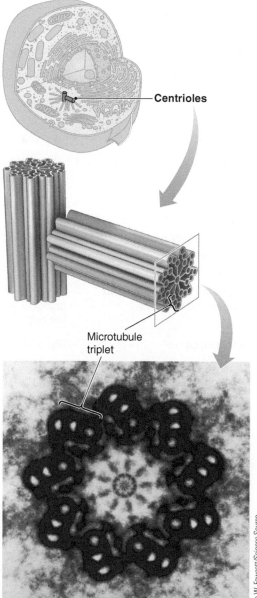

Centrioles

Microtubule triplet

Don W. Fawcett/Science Source

Figure 2-20 **Centrioles.** The two cylindrical centrioles lie at right angles to each other as shown in the diagram. The electron micrograph shows a centriole in cross section. Note that a centriole is made up of nine microtubule triplets that form a ring.

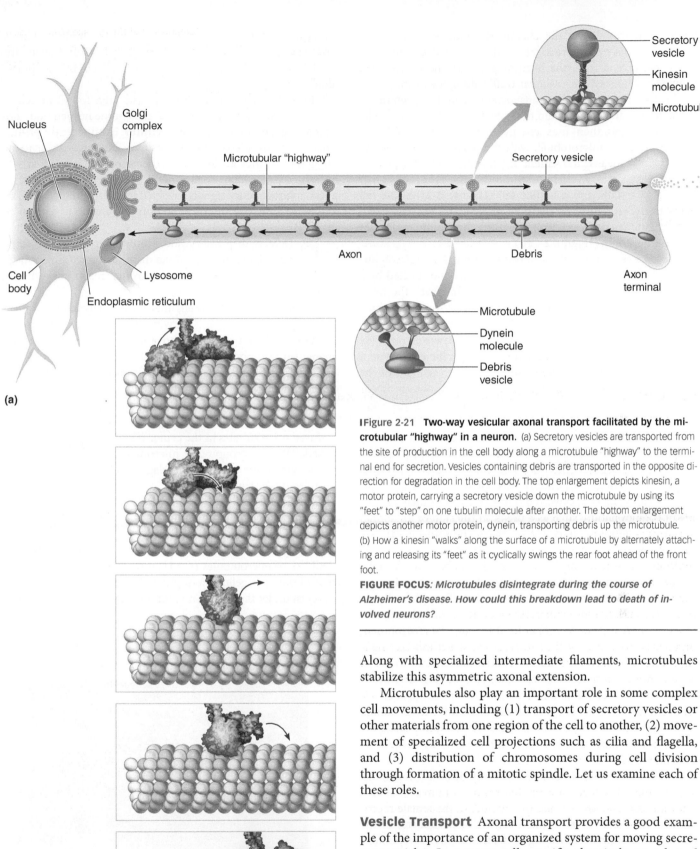

(a)

(b)

Secretory vesicle

Kinesin molecule

Microtubule

Nucleus

Golgi complex

Microtubular "highway"

Secretory vesicle

Axon

Debris

Axon terminal

Cell body

Lysosome

Endoplasmic reticulum

Microtubule

Dynein molecule

Debris vesicle

❚Figure 2-21 Two-way vesicular axonal transport facilitated by the microtubular "highway" in a neuron. (a) Secretory vesicles are transported from the site of production in the cell body along a microtubule "highway" to the terminal end for secretion. Vesicles containing debris are transported in the opposite direction for degradation in the cell body. The top enlargement depicts kinesin, a motor protein, carrying a secretory vesicle down the microtubule by using its "feet" to "step" on one tubulin molecule after another. The bottom enlargement depicts another motor protein, dynein, transporting debris up the microtubule. (b) How a kinesin "walks" along the surface of a microtubule by alternately attaching and releasing its "feet" as it cyclically swings the rear foot ahead of the front foot.

FIGURE FOCUS: *Microtubules disintegrate during the course of Alzheimer's disease. How could this breakdown lead to death of involved neurons?*

Along with specialized intermediate filaments, microtubules stabilize this asymmetric axonal extension.

Microtubules also play an important role in some complex cell movements, including (1) transport of secretory vesicles or other materials from one region of the cell to another, (2) movement of specialized cell projections such as cilia and flagella, and (3) distribution of chromosomes during cell division through formation of a mitotic spindle. Let us examine each of these roles.

Vesicle Transport Axonal transport provides a good example of the importance of an organized system for moving secretory vesicles. In a nerve cell, specific chemicals are released from the terminal end of the elongated axon to influence a muscle or another structure that the nerve cell controls. These chemicals are largely produced within the cell body where the nuclear DNA blueprint, endoplasmic reticular factory, and Golgi packaging and distribution outlet are located. If they had to diffuse on their own from the cell body to a distant axon terminal, it would take the chemicals about 50 years to get there—obviously an impractical solution. Instead, the microtubules that extend from the beginning to the end of the axon provide a "highway" for vesicular traffic along the axon.

Motor proteins are the transporters. A **motor protein**, or **molecular motor**, is a protein that attaches to the particle to be transported and then uses energy harnessed from ATP to "walk" along the microtubule, with the particle riding "piggyback" (*motor* means "movement"). **Kinesin**, the motor protein that carries secretory vesicles to the end of the axon, consists of two "feet," a stalk, and a fanlike tail (▌Figure 2-21a). The tail binds to the secretory vesicle to be moved, and the feet swing forward one at a time, as if walking, using the tubulin molecules as stepping stones (▌Figure 2-21b).

Reverse vesicular traffic also occurs along these microtubular highways. Vesicles that contain debris are transported by a different ATP-driven motor protein, **dynein**, from the axon terminal to the cell body for degradation by lysosomes, which are confined within the cell body. The two ends of a microtubule are different, and each motor protein can travel in only one direction along the microtubule toward a specific end. Dynein always moves toward the centrosome (or "minus") end of the microtubule and kinesin always walks toward the outermost (or "plus") end, ensuring that their cargo is moved in the right direction.

Clinical Note Reverse axonal transport may also serve as a pathway for the movement of some infectious agents, such as herpes viruses (the ones that cause cold sores, genital herpes, and shingles), poliomyelitis virus, and rabies virus. These viruses travel backward along nerves from their surface site of contamination, such as a break in the skin or an animal bite, to the central nervous system (brain and spinal cord).

Movement of Cilia and Flagella Microtubules are also the dominant structural and functional components of cilia and flagella. These specialized motile protrusions from the cell surface allow a cell to move materials across its surface (in the case of a stationary cell) or to propel itself through its environment (in the case of a motile cell). **Cilia** (meaning "eyelashes"; singular, *cilium)* are short, tiny, hairlike protrusions usually found in large numbers on the surface of a ciliated cell. **Flagella** (meaning "whips"; singular, *flagellum)* are long, whiplike appendages; typically, a cell has one or a few flagella at most. Even though they project from the surface of the cell, cilia and flagella are both intracellular structures—they are covered by the plasma membrane.

Cilia beat or stroke in unison in a given direction, much like the coordinated efforts of a rowing team. In humans, ciliated cells line the respiratory tract, the oviduct of the female reproductive tract, and the fluid-filled ventricles (chambers) of the brain. The coordinated stroking of the thousands of respiratory cilia help keep foreign particles out of the lungs by sweeping outward dust and other inspired (breathed-in) particles (▌Figure 2-22). In the female reproductive tract, the sweeping action of the cilia lining the oviduct draws the egg (ovum) released from the ovary during ovulation into the oviduct and then guides it toward the uterus (womb). In the brain, the ciliated cells lining the ventricles produce cerebrospinal fluid, which

flows through the ventricles and around the brain and spinal cord, cushioning and bathing these fragile neural structures. Beating of the cilia helps promote circulation of this supportive fluid.

In addition to the multiple motile cilia found in cells in these specific locations, almost all cells in the human body possess a single nonmotile **primary cilium**. Until recently, primary cilia were considered useless vestiges, but growing evidence suggests that they act as microscopic sensory organs that sample the extracellular environment. They may be critical for receiving regulatory signals involved in controlling growth, cell differentiation, and cell proliferation (expansion of a given cell type).

Clinical Note Defects in primary and motile cilia have been implicated in a range of human disorders, including polycystic kidney disease and chronic respiratory disease, respectively.

In humans, the only cells that have a flagellum are sperm (see ▌Figure 20-9, p. 728). The whiplike motion of the flagellum or "tail" enables a sperm to move through its environment to maneuver into position to fertilize the female ovum.

Cilia and flagella have the same basic internal structure, their only difference being their length: Cilia are short, and flagella are long. Both consist of nine fused pairs of microtubules (doublets) arranged in an outer ring around two single unfused microtubules in the center (▌Figure 2-23). This characteristic "9 + 2" grouping of microtubules extends throughout the length of the motile appendage. Spokelike accessory proteins hold the structure together.

Cilia and flagella arise from centrioles. Each cylinder of the centriole pair contains a bundle of microtubules similar to the 9 + 2 complex, except that the central pair of single microtubules is missing and the outer ring has nine fused triplets rather than doublets of microtubules (see ▌Figure 2-20). During formation of a cilium or flagellum, a duplicated centriole moves to a position just under the plasma membrane, where microtubules grow

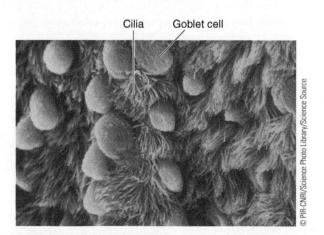

▌Figure 2-22 **Cilia in the respiratory tract**. Scanning electron micrograph of cilia on cells lining the human respiratory tract. The respiratory airways are lined by goblet cells, which secrete sticky mucus that traps inspired particles, and epithelial cells that bear numerous hairlike cilia. The cilia all beat in the same direction to sweep inspired particles up and out of the airways.

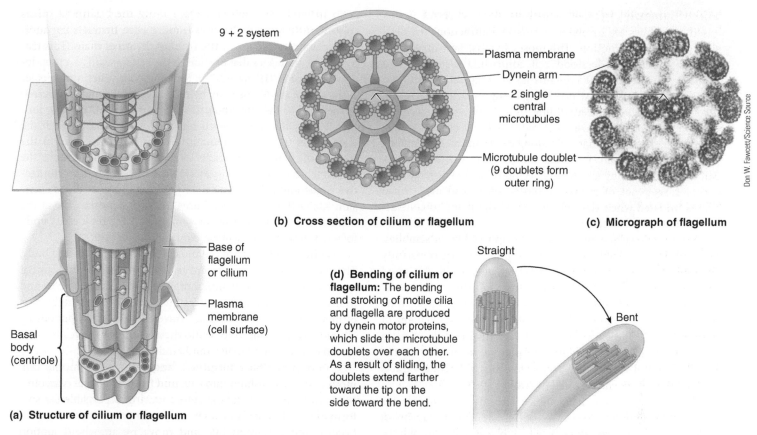

9 + 2 system

(b) Cross section of cilium or flagellum

Plasma membrane
Dynein arm
2 single central microtubules
Microtubule doublet (9 doublets form outer ring)

(c) Micrograph of flagellum

Don W. Fawcett/Science Source

Base of flagellum or cilium

Plasma membrane (cell surface)

Basal body (centriole)

(a) Structure of cilium or flagellum

(d) Bending of cilium or flagellum: The bending and stroking of motile cilia and flagella are produced by dynein motor proteins, which slide the microtubule doublets over each other. As a result of sliding, the doublets extend farther toward the tip on the side toward the bend.

Straight

Bent

Figure 2-23 **Internal structure of a cilium or flagellum.** (a) The relationship between the microtubules and the centriole-turned-basal body of a cilium or flagellum. (b) Diagram of a cilium or flagellum in cross section showing the characteristic "9 + 2" arrangement of microtubules along with the dynein arms and other accessory proteins that hold the system together. (c) Electron micrograph of a flagellum in cross section; individual tubulin molecules are visible in the microtubule walls. (d) Depiction of bending of a cilium or flagellum caused by microtubule sliding brought about by dynein "walking."

outward from the centriole in an orderly pattern to form the motile appendage. The centriole remains at the base of the developed cilium or flagellum as the **basal body** of the structure.

In addition to the accessory proteins that maintain the microtubule's organization, another accessory protein, the motor protein dynein, plays an essential part in the microtubular movement that causes the entire structure to bend. Dynein forms armlike projections from each doublet of microtubules (see Figure 2-23b and c). These dynein arms walk along the adjacent microtubule doublets, causing the doublets to slide past each other, bringing about the bending and stroking (Figure 2-23d). Groups of cilia working together are oriented to beat in the same direction and contract in a synchronized manner through poorly understood control mechanisms involving the single microtubules at the cilium's center.

Formation of the Mitotic Spindle Cell division involves two discrete but related activities: *mitosis* (nuclear division), which depends on microtubules, and *cytokinesis* (cytoplasmic division), which depends on microfilaments and is described in the next section. During **mitosis**, the DNA-containing chromosomes of the nucleus are replicated, resulting in two identical sets. These duplicate sets of chromosomes are separated and drawn apart to opposite sides of the cell so that the genetic material is evenly distributed in the two halves of the cell.

The replicated chromosomes are pulled apart by motor proteins that move them along a cellular apparatus called the **mitotic spindle**, which is transiently assembled from microtubules only during cell division. The microtubules of the mitotic spindle are formed by the centrioles. As part of cell division, the centrioles first duplicate themselves; then, the new centriole pairs move to opposite ends of the cell and form the spindle apparatus between them through a precisely organized assemblage of microtubules.

Microfilaments are important to cellular contractile systems and as mechanical stiffeners.

Microfilaments are the smallest (6 nm in diameter) elements of the cytoskeleton. The most obvious microfilaments in most cells are those composed of **actin**, a protein molecule that has a globular shape similar to that of tubulin. Unlike tubulin, which forms a hollow tube, actin assembles into two strands, which twist around each other to form a microfilament (see Figure 2-19b). In muscle cells, the protein **myosin** forms a different kind of microfilament (see Figure 8-4, p. 255). In most cells, myosin is not as abundant and does not form such distinct filaments.

Microfilaments serve two functions: (1) They play a vital role in various cell contractile systems, and (2) they act as mechanical stiffeners for several specific cell projections.

Microfilaments in Cell Contractile Systems Actin-based assemblies are involved in muscle contraction, cell division, and cell locomotion. The most obvious, best-organized, and most clearly understood cell contractile system is that found in muscle. Muscle contains an abundance of actin and myosin microfilaments, which bring about muscle contraction through the ATP-powered sliding of actin microfilaments in relation to stationary myosin microfilaments. Myosin is a motor protein that has heads that walk along the actin microfilaments, pulling them inward between the myosin microfilaments. Microfilament sliding and force development are triggered by a complex sequence of electrical, biochemical, and mechanical events initiated when the muscle cell is stimulated to contract (see Chapter 8 for details).

Nonmuscle cells may also contain "musclelike" assemblies. Some of these microfilament contractile systems are transiently assembled to perform a specific function when needed. A good example is the contractile ring that forms during **cytokinesis**, the process by which the two halves of a dividing cell separate into two new daughter cells, each with a full complement of chromosomes. The ring consists of a beltlike bundle of actin filaments located just beneath the plasma membrane in the middle of the dividing cell. When this ring of fibers contracts and tightens, it pinches the cell in two (IFigure 2-24).

Complex actin-based assemblies are also responsible for most cell locomotion. Four types of human cells are capable of moving on their own—sperm, white blood cells, fibroblasts, and skin cells. Flagella propel sperm. The other motile cells move via **amoeboid movement**, a cell-crawling process that depends on the activity of their actin filaments, in a mechanism similar to that used by amoebas to maneuver through their environment. When crawling, the motile cell forms footlike protrusions, or *pseudopods*, at the "front" or leading edge of the cell in the direction of the target. For example, the target that triggers amoeboid movement might be the proximity of food in the case of an amoeba or a bacterium in the case of a white blood cell (see IFigure 2-9c, p. 32). Pseudopods are formed as a result of the organized assembly and disassembly of branching actin networks. During amoeboid movement, actin filaments continuously grow at the cell's leading edge through the addition of actin molecules at the front of the actin chain. This filament growth pushes that portion of the cell forward as a pseudopod protrusion (IFigure 2-25). Simultaneously, actin molecules at the rear of the filament are being disassembled and transferred to the front of the line. Thus, the filament does not get any longer; it stays the same length but moves forward through the continuous transfer of actin molecules from the rear to the front of the filament in what is termed *treadmilling* fashion. The cell attaches the advancing pseudopod to surrounding connective tissue and at the same time detaches from its older adhesion site at the rear. The cell uses the new adhesion site at the leading edge as a point of traction to pull the bulk of its body forward through cytoskeletal contraction.

White blood cells are the most active crawlers in the body. These cells leave the blood and travel by amoeboid movement to areas of infection or inflammation, where they engulf and destroy microorganisms and cellular debris. Amazingly, it is estimated that the total distance traveled collectively per day by all your white blood cells while they roam the tissues in their search-and-destroy tactic would circle Earth twice.

Fibroblasts ("fiber formers"), another type of motile cell, move amoeboid fashion into a wound from adjacent connective tissue to help repair the damage; they are responsible for scar formation. Skin cells, which are ordinarily stationary, can become moderately mobile and move by amoeboid motion toward a cut to restore the skin surface.

Microfilaments as Mechanical Stiffeners Besides their role in cellular contractile systems, actin filaments serve as mechanical supports or stiffeners for several cellular extensions, of which the most common are microvilli. **Microvilli** are microscopic, nonmotile, hairlike projections from the surface of epithelial cells lining the small intestine and kidney tubules (IFigure 2-26). Their presence greatly increases the surface area available for transferring material across the plasma membrane. In the small intestine, the microvilli increase the area available for absorbing digested nutrients. In the kidney tubules, microvilli enlarge the absorptive surface that salvages useful sub-

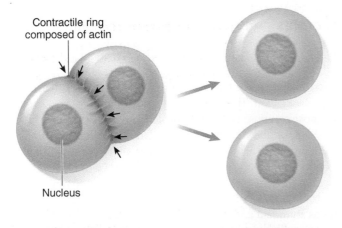

IFigure 2-24 **Cytokinesis.** Diagram of a cell undergoing cytokinesis, in which a contractile ring composed of actin filaments tightens, squeezing apart the two duplicate cell halves formed by mitosis.

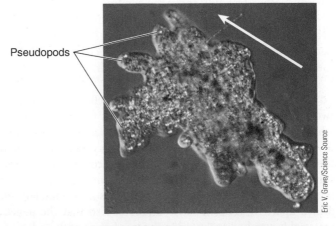

IFigure 2-25 **An amoeba undergoing amoeboid movement.**

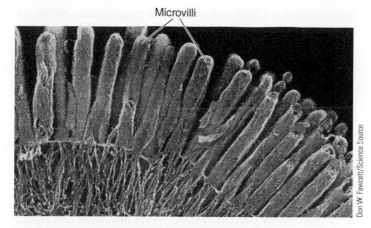

Microvilli

IFigure 2-26 **Microvilli in the small intestine.** Scanning electron micrograph showing microvilli on the surface of a small-intestine epithelial cell.

Don W. Fawcett/Science Source

stances passing through the kidney so that these materials are saved for the body instead of being eliminated in the urine. Within each microvillus, a core of parallel linked actin filaments forms a rigid mechanical stiffener that keeps these valuable surface projections intact.

Intermediate filaments are important in cell regions subject to mechanical stress.

Intermediate filaments are intermediate in size between microtubules and microfilaments (7 to 11 nm in diameter)—hence their name. The proteins that compose the intermediate filaments vary among cell types, but in general they appear as irregular, threadlike molecules. These proteins form tough, durable fibers that play a central role in maintaining the structural integrity of a cell and in resisting mechanical stresses externally applied to a cell.

Intermediate filaments are tailored to suit their structural or tension-bearing role in specific cell types. In general, only one class of intermediate filament is found in a particular cell type. Two important examples follow:

- *Neurofilaments* are intermediate filaments found in nerve cell axons. Together with microtubules, neurofilaments strengthen and stabilize these elongated cellular extensions.
- Skin cells contain irregular networks of intermediate filaments made of the protein **keratin** (see IFigure 2-19c). These intracellular filaments connect with the extracellular filaments that tie adjacent cells together, creating a continuous filamentous network that extends throughout the skin and gives it strength. When surface skin cells die, their tough keratin skeletons persist, forming a protective, waterproof outer layer. Hair and nails are also keratin structures.

Intermediate filaments account for up to 85% of the total protein in nerve cells and keratin-producing skin cells, whereas these filaments constitute only about 1% of other cells' total protein on average.

 Clinical Note Neurofilament abnormalities contribute to some neurological disorders. An important example is **amyotrophic lateral sclerosis (ALS)**, more familiarly known as **Lou Gehrig's disease**. ALS is characterized by progressive degeneration and death of motor neurons, the type of nerve cells that control skeletal muscles. This adult-onset condition leads to gradual loss of control of skeletal muscles, including the muscles of breathing, and ultimately to death, as it did for baseball legend Lou Gehrig. One underlying problem may be an abnormal accumulation and disorganization of neurofilaments. Motor neurons, which have the most neurofilaments, are the most affected. The disorganized neurofilaments are believed to block the axonal transport of crucial materials along the microtubular highways, thus choking off vital supplies from the cell body to the axon terminal.

The cytoskeleton functions as an integrated whole and links other parts of the cell.

Collectively, the cytoskeletal elements and their interconnections support the plasma membrane and are responsible for the particular shape, rigidity, and spatial geometry of each different cell type. Furthermore, growing evidence suggests that the cytoskeleton serves as a lattice to organize groups of enzymes for many cellular activities. This internal framework thus acts as the cell's "skeleton."

New studies hint that the cytoskeleton as a whole is not merely a supporting structure that maintains the tensional integrity of the cell, but also may serve as a mechanical communications system. Various components of the cytoskeleton behave as if they are structurally connected or "hardwired" to each other and to the plasma membrane and the nucleus. This force-carrying network may serve as the mechanism by which mechanical forces acting on the cell surface reach all the way from the plasma membrane through the cytoskeleton to ultimately influence gene regulation in the nucleus.

Furthermore, as you have learned, the coordinated action of the cytoskeletal elements is responsible for directing intracellular transport and for regulating numerous cellular movements and thereby also serves as the cell's "muscle."

Check Your Understanding 2.10

1. List the three types of cytoskeletal elements and state one function of each.

2. Explain how motor proteins transport proteins along a cytoskeletal "highway."

3. Discuss the role of centrioles in the formation of cilia and flagella.

4. Describe how treadmilling forms pseudopods during amoeboid movement.

Homeostasis: Chapter in Perspective

 The ability of cells to perform functions essential for their survival, in addition to specialized tasks that help maintain homeostasis within the body, ultimately depends on the successful, cooperative operation of the intracellular components. For example, to support life-sustaining activities, all cells must generate energy, in a usable form, from nutrient molecules. Energy is generated intracellularly by chemical reactions in the cytosol and mitochondria.

In addition to being essential for basic cell survival, the organelles and cytoskeleton participate in many cells' specialized tasks that contribute to homeostasis. Here are several examples:

■ Both nerve and endocrine cells release protein chemical messengers (neurotransmitters in nerve cells and hormones in endocrine cells) that are important in regulatory activities aimed at maintaining homeostasis. For example, neurotransmitters stimulate the respiratory muscles, which accomplish life-sustaining exchanges of oxygen and carbon dioxide between the body and the atmosphere through breathing. These protein chemical messengers are all produced by the endoplasmic reticulum and Golgi complex and are released by exocytosis from the cell when needed.

■ The ability of muscle cells to contract depends on their highly developed cytoskeletal microfilaments sliding past one another. Muscle contraction is responsible for many homeostatic activities, including (1) contracting the heart muscle, which pumps life-supporting blood throughout the body; (2) contracting the muscles attached to bones, which enables the body to procure food; and (3) contracting the muscle in the walls of the stomach and intestine, which moves the food along the digestive tract so that ingested nutrients can be progressively broken down into a form that can be absorbed into the blood for delivery to the cells.

■ White blood cells help the body resist infection by making extensive use of lysosomal destruction of engulfed particles as they police the body for microbial invaders. These white blood cells are able to roam the body by means of amoeboid movement, a cell-crawling process accomplished by coordinated assembly and disassembly of actin, one of their cytoskeletal components.

As we begin to examine the various organs and systems, keep in mind that proper cell functioning is the foundation of all organ activities.

Review Exercises Answers begin on p. A-21

Reviewing Terms and Facts

1. The barrier that separates and controls movement between the cell contents and the extracellular fluid is the _____.

2. The chemical that directs protein synthesis and serves as a genetic blueprint is _____, which is found in the _____ of the cell.

3. The three major subdivisions of a cell are _____, _____, and _____.

4. The cytoplasm consists of _____, which are discrete, specialized, intracellular compartments; a gel-like mass known as _____; and elaborate protein scaffolding called the _____.

5. Transport vesicles from the _____ fuse with and enter the _____ for modification and sorting.

6. The (what kind of) _____ enzymes within the peroxisomes primarily detoxify various wastes produced within the cell or foreign compounds that have entered the cell.

7. The universal energy carrier of the body is _____.

8. The largest cells in the human body can be seen by the unaided eye. (True or false?)

9. Amoeboid movement is accomplished by the coordinated assembly and disassembly of microtubules. (True or false?)

10. Using the answer code on the right, indicate which form of energy production is being described:

1. takes place in the mitochondrial matrix	(a) glycolysis
2. produces H_2O as a by-product	(b) citric acid cycle
3. results in a rich yield of ATP	(c) oxidative phosphorylation
4. takes place in the cytosol	
5. processes acetyl-CoA	
6. takes place in the mitochondrial inner-membrane cristae	
7. converts glucose into two pyruvate molecules	
8. uses molecular oxygen	
9. accomplished by the electron transport system and ATP synthase	

11. Using the answer code on the right, indicate which type of ribosome is being described:

1. synthesizes proteins used to construct new cell membrane
2. synthesizes proteins used intracellularly within the cytosol
3. synthesizes secretory proteins such as enzymes or hormones
4. synthesizes the hydrolytic enzymes incorporated in lysosomes

(a) free ribosome
(b) rough ER-bound ribosome

Understanding Concepts

(Answers at www.cengagebrain.com)

1. State an advantage of organelle compartmentalization.

2. Distinguish between membranous organelles and non-membranous organelles. List the five types of membranous organelles and the four types of nonmembranous organelles.

3. Describe the structure of the endoplasmic reticulum, distinguishing between rough and smooth. What is the function of each?

4. Compare exocytosis and endocytosis. Define *secretion, pinocytosis, receptor-mediated endocytosis,* and *phagocytosis.*

5. Which organelles serve as the intracellular "digestive system"? What type of enzymes do they contain? What functions do these organelles serve?

6. Compare lysosomes with peroxisomes.

7. Distinguish among *cellular respiration, oxidative phosphorylation,* and *chemiosmosis.*

8. Describe the structure of mitochondria, and explain their role in cellular respiration.

9. Distinguish between the oxidative enzymes found in peroxisomes and those found in mitochondria.

10. Cells expend energy on what three categories of activities?

11. List and describe the functions of each component of the cytoskeleton.

Solving Quantitative Exercises

1. Each "turn" of the citric acid cycle
   a. generates 3 NAD^+, 1 $FADH_2$, and 2 CO_2
   b. generates 1 GTP, 2 CO_2, and 1 $FADH_2$
   c. consumes 1 pyruvate and 1 oxaloacetate
   d. consumes an amino acid

2. Let us consider how much ATP you synthesize in a day. Assume that you consume 1 mole of O_2 per hour or 24 moles per day. (A mole is the number of grams of a chemical equal to its molecular weight; see p. A-6.) About 6 moles of ATP are produced per mole of O_2 consumed. The molecular weight of ATP is 507. How many grams of ATP do you produce per day at this rate? Given that 1000 g equal 2.2 pounds, how many pounds of ATP do you produce per day at this rate? (This is under relatively inactive conditions!)

3. Under resting circumstances a person produces about 144 moles of ATP per day (73,000 g ATP/day). The amount of *free energy* represented by this amount of ATP can be calculated as follows: Cleavage of the terminal phosphate bond from ATP results in a decrease in free energy of approximately 7300 cal/mole. This is a crude measure of the energy available to do work that is contained in the terminal phosphate bond of the ATP molecule. How many calories, in the form of ATP, are produced per day by a resting individual, crudely speaking?

4. Calculate the number of cells in the body of an average 68-kg (150-lb) adult. (This will only be accurate to about 1 part in 10 but should give you an idea how scientists estimate this commonly quoted number.) Assume all cells are spheres 20 μm in diameter. The volume of a sphere can be determined by the equation $v = 4/3 \ \pi^3$. (*Hint:* We know that about two thirds of the water in the body is intracellular and the density of cells is nearly 1 g/mL. The proportion of the mass made up of water is about 60%.)

5. If sucrose is injected into the bloodstream, it tends to stay out of the cells (cells do not use sucrose directly). If it does not go into cells, where does it go? In other words, how much "space" is in the body that is not inside some cell? Sucrose can be used to determine this space. Suppose 150 mg of sucrose is injected into a 55-kg (121-lb) woman. If the concentration of sucrose in her blood is 0.015 mg/mL, what is the volume of her extracellular space, assuming that no metabolism is occurring and that the blood sucrose concentration is equal to the sucrose concentration throughout the extracellular space?

Applying Clinical Reasoning

Kevin S. and his wife have been trying to have a baby for the past 3 years. On seeking the help of a fertility specialist, Kevin learned that he has a hereditary form of male sterility involving nonmotile sperm. His condition can be traced to defects in the cytoskeletal components of the sperm's flagella. As a result of this finding, the physician suspected that Kevin also has a long history of recurrent respiratory tract disease. Kevin confirmed that indeed he has had colds, bronchitis, and influenza more frequently than his friends. Why would the physician suspect that Kevin probably had a history of frequent respiratory disease based on his diagnosis of sterility from nonmotile sperm?

Thinking at a Higher Level

1. The stomach has two types of exocrine secretory cells: *chief cells,* which secrete an inactive form of the protein-digesting enzyme *pepsinogen,* and *parietal cells,* which secrete *hydrochloric acid (HCl)* that activates pepsinogen. Both cell types have an abundance of mitochondria for ATP production—the chief cells need energy to synthesize pepsinogen, and the parietal cells need energy to transport hydrogen ions (H^+) and chloride ions (Cl^-) from the blood into the stomach lumen. Only one of these cell types also has an extensive rough ER and abundant Golgi stacks. Would this type be the chief cells or the parietal cells? Why?

2. The poison *cyanide* acts by binding irreversibly to one component of the electron transport system, blocking its action. As a result, the entire electron-transport process comes to a screeching halt, and the cells lose more than 94% of their ATP-producing capacity. Considering the types of cell activities that depend on energy expenditure, what would be the consequences of cyanide poisoning?

3. Hydrogen peroxide, which belongs to a class of very unstable compounds known as *free radicals,* can bring about drastic, detrimental changes in a cell's structure and function by reacting with almost any molecule with which it comes in contact, including DNA. The resultant cellular changes can lead to genetic mutations, cancer, or other serious consequences. Furthermore, some researchers speculate that cumulative effects of more subtle cellular damage resulting from free radical reactions over a period of time might contribute to the gradual deterioration associated with aging. Related to this speculation, studies have shown that longevity decreases in fruit flies in direct proportion to a decrease in a specific chemical found in one of the cellular organelles. Based on your knowledge of how the body rids itself of dangerous hydrogen peroxide, what do you think this chemical in the organelle is?

4. Why do you think a person is able to perform anaerobic exercise (such as lifting and holding a heavy weight) only briefly but can sustain aerobic exercise (such as walking or swimming) for long periods? (*Hint:* Muscles have limited energy stores.)

5. One type of the affliction *epidermolysis bullosa* is caused by a genetic defect that results in production of abnormally weak keratin. Based on your knowledge of the role of keratin, what part of the body do you think would be affected by this condition?

To access the course materials and companion resources for this text, please visit
www.cengagebrain.com

The Plasma Membrane and Membrane Potential

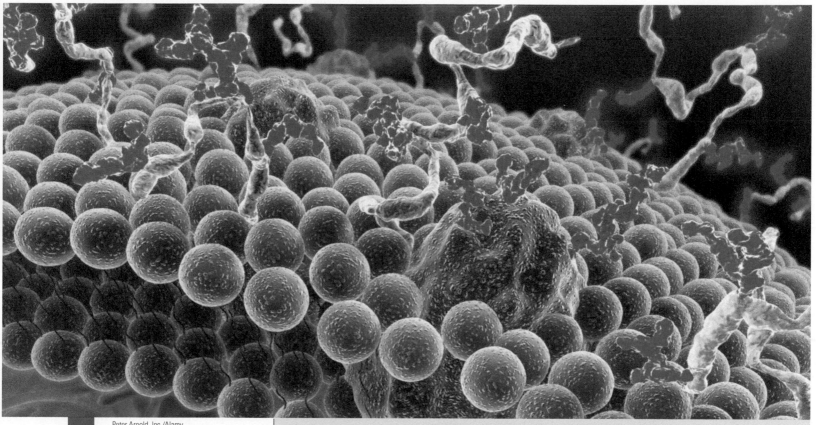

Peter Arnold, Inc./Alamy

A computer model of the plasma membrane. All cells are enclosed by a plasma membrane consisting of a lipid bilayer (*green*) with proteins (*blue*) embedded or attached and carbohydrates (*red*) protruding from the outer surface.

Homeostasis Highlights

All cells are enveloped by a **plasma membrane**, a thin, flexible, lipid barrier that separates the contents of the cell from its surroundings. To carry on life-sustaining and specialized activities, each cell must exchange materials across this membrane with the homeostatically maintained internal fluid environment that surrounds the cell. This discriminating barrier contains specific proteins, some of which enable selective passage of materials. Other membrane proteins are receptors for interaction with specific chemical messengers in the cell's environment. These messengers control many cell activities crucial to homeostasis.

Cells have a **membrane potential**, a slight excess of negative charges lined up along the inside of the membrane and a slight excess of positive charges on the outside. The specialization of nerve and muscle cells depends on the ability of these cells to alter their potential on appropriate stimulation. Much of nerve and muscle function is geared toward maintaining homeostasis.

3.1 Membrane Structure and Functions

To survive, every cell must maintain a specific composition of its contents unique for that cell type despite the remarkably different composition of the extracellular fluid (ECF) surrounding it. This difference in fluid composition inside and outside a cell is maintained by the **plasma membrane**, the extremely thin layer that forms the outer boundary of every cell and encloses the intracellular contents. Besides acting as a mechanical barrier that traps needed molecules within the cell, the plasma membrane helps determine the cell's composition by selectively permitting specific substances to pass between the cell and its environment. The plasma membrane controls the entry of nutrient molecules and the exit of secretory and waste products. In addition, it maintains differences in ion concentrations inside and outside the cell, which are important in the membrane's electrical activity. The plasma membrane also participates in the joining of cells to form tissues and organs. Finally, it plays a key role in enabling a cell to respond to signals from chemical messengers in the cell's environment; this ability is important in communication among cells. No matter what the cell type, these common membrane functions are crucial to the cell's survival, to its ability to perform specialized homeostatic activities, and to its ability to coordinate its functions with those of other cells. Many of the functional differences among cell types are a result of subtle variations in the composition of their plasma membranes, which in turn enable different cells to interact in different ways with essentially the same ECF environment.

The plasma membrane is a fluid lipid bilayer embedded with proteins.

The plasma membrane consists mostly of lipids and proteins plus small amounts of carbohydrate. It is too thin to be seen under an ordinary light microscope, but with an electron microscope it appears as a trilaminar structure consisting of two dark layers separated by a light middle layer (*tri* means "three"; *lamina* means "layer") (❙Figure 3-1). The specific arrangement of the molecules that make up the plasma membrane is responsible for this "sandwich" appearance.

The most abundant membrane lipids are phospholipids, with lesser amounts of cholesterol. An estimated 1 billion phospholipid molecules are present in the plasma membrane of a typical human cell. **Phospholipids** have a polar (electrically charged; see p. A-4) head containing a negatively charged phosphate group and two nonpolar (electrically neutral) fatty acid chain tails (❙Figure 3-2a). The polar end is hydrophilic (meaning "water loving") because it can interact with water molecules, which are also polar; the nonpolar end is hydrophobic (meaning "water fearing") and will not mix with water. In water, phospholipids self-assemble into a **lipid bilayer**, a double layer of lipid molecules (❙Figure 3-2b) (*bi* means "two"). The hydrophobic tails bury themselves in the center of the bilayer away from the water, and the hydrophilic heads line up on both sides in contact with the water. The outer surface of the bilayer is exposed to ECF, whereas the inner surface is in contact with the intracellular fluid (ICF) (❙Figure 3-2c).

The lipid bilayer is fluid, not rigid, with a consistency more like cooking oil than stick butter. The phospholipids, which are not held together by strong chemical bonds, are constantly moving. They can twirl, vibrate, and move around within their half of the bilayer, exchanging places millions of times a second. This phospholipid movement largely accounts for membrane fluidity.

Cholesterol contributes to both the fluidity and the stability of the membrane. Cholesterol molecules are tucked between the phospholipid molecules, where they prevent the fatty acid chains from packing together and crystallizing, a process that would drastically reduce membrane fluidity. Through their spatial relationship with phospholipid molecules, cholesterol molecules also help stabilize the phospholipids' position. Because of its fluidity, the plasma membrane has structural integrity but at the same time is flexible, enabling the cell to change shape. For example, muscle cells change shape as they contract.

Membrane proteins are inserted within or attached to the lipid bilayer (❙Figure 3-3 and chapter opener photo). **Integral proteins** are embedded in the lipid bilayer, with most extending through the entire thickness of the membrane, in which case they are alternatively called **transmembrane proteins** (*trans* means "across"). Like phospholipids, integral proteins have hydrophilic and hydrophobic regions. **Peripheral proteins** are polar molecules that do not penetrate the membrane. They only stud the membrane surface, anchored by weak chemical bonds with the polar parts of integral membrane proteins or membrane lipids. Peripheral proteins are found more commonly on the inner than on the outer surface. The plasma membrane has about 50 times more lipid molecules than protein molecules. However, proteins account for nearly half of the membrane's mass because they are much larger than lipids. The fluidity of the lipid bilayer enables many membrane proteins to float freely like "icebergs" in a moving "sea" of lipids. This view of membrane structure is known as the **fluid mosaic model**, in reference to the membrane fluidity and the ever-changing mosaic pattern of the proteins embedded in the lipid bilayer. (A mosaic is a surface decoration made by inlaying small pieces of variously colored tiles to form patterns or pictures.)

Despite the plasma membrane's generally fluid nature and randomly arranged proteins, researchers recently identified

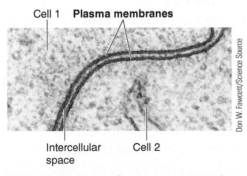

Cell 1   **Plasma membranes**

Intercellular space   Cell 2

Don W. Fawcett/Science Source

❙Figure 3-1 **Trilaminar appearance of a plasma membrane in an electron micrograph.** Depicted are the plasma membranes of two adjacent cells. Note that each membrane appears as two dark, thin layers separated by a light middle layer.

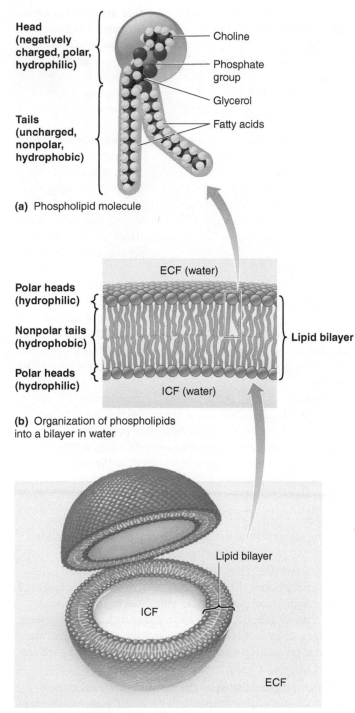

Head (negatively charged, polar, hydrophilic)

Choline

Phosphate group

Glycerol

Tails (uncharged, nonpolar, hydrophobic)

Fatty acids

(a) Phospholipid molecule

ECF (water)

Polar heads (hydrophilic)

Nonpolar tails (hydrophobic)

Polar heads (hydrophilic)

Lipid bilayer

ICF (water)

(b) Organization of phospholipids into a bilayer in water

Lipid bilayer

ICF

ECF

(c) Separation of ICF and ECF by the lipid bilayer

❙Figure 3-2 Structure and organization of phospholipid molecules in a lipid bilayer. (a) Phospholipid molecule. (b) In water, phospholipid molecules organize themselves into a lipid bilayer with the polar heads interacting with the polar water molecules at each surface and the nonpolar tails all facing the interior of the bilayer. (c) An exaggerated view of the plasma membrane enclosing a cell, separating the ICF from the ECF.

specialized membrane patches known as **lipid rafts** made up mostly of sphingolipids (instead of phospholipids), extra cholesterol, and an abundance of particular proteins. Lipid rafts are more highly organized, more tightly packed, and a little thicker than the remainder of the plasma membrane. The rafts are

thicker because the fatty acid tails of sphingolipids are longer than those of phospholipids. The proteins are anchored in place by the sphingolipids or cytoskeletal elements in the cytoplasm.

The proteins gathered in the lipid rafts are *receptors* specialized to interact with specific *extracellular chemical messengers,* or *signal molecules,* in the cell's environment that dictate specific intracellular responses. An example of an extracellular signal is a hormone directing secretion of digestive enzymes by exocrine gland cells of the pancreas in response to food in the small intestine. Lipid rafts exist either as flat platforms on the smooth parts of the plasma membrane or in tiny (50 to 100 nm in diameter) cavelike indentations in the membrane surface aptly called **caveolae** ("tiny caves").

A small amount of **membrane carbohydrate** is located on the outer surface of cells, "sugar coating" them. Short carbohydrate chains protrude like tiny antennas from the outer surface, bound primarily to membrane proteins and, to a lesser extent, to lipids. These sugary combinations are known as *glycoproteins* and *glycolipids,* respectively (❙Figure 3-3 and chapter opener photo), and the coating they form is called the *glycocalyx* (*glyco* means "sweet"; *calyx* means "husk").

This proposed structure accounts for the trilaminar appearance of the plasma membrane. When stains are used to help visualize the plasma membrane under an electron microscope (as in ❙Figure 3-1), the two dark lines represent the hydrophilic polar regions of the lipid and protein molecules that have taken up the stain. The light space between corresponds to the poorly stained hydrophobic core formed by the nonpolar regions of these molecules.

The different components of the plasma membrane carry out a variety of functions. The lipid bilayer forms the primary barrier to diffusion, the proteins perform most of the specific membrane functions, and the carbohydrates play an important role in "self-recognition" processes and cell-to-cell interactions. We now examine these functions in more detail.

The lipid bilayer forms the basic structural barrier that encloses the cell.

The lipid bilayer serves the following functions related to its role as a barrier between a cell's contents and its surroundings:

1. It forms the basic structure of the membrane. The phospholipids can be visualized as the "pickets" that form the "fence" around the cell.

2. Its hydrophobic interior is a barrier to passage of water-soluble substances between the ICF and ECF. Water-soluble substances cannot dissolve in and pass through the lipid bilayer. By means of this barrier, the cell can maintain different mixtures and concentrations of solutes (dissolved substances) inside and outside the cell.

3. It is responsible for the fluidity of the membrane.

In addition to these barrier roles, the lipid bilayer is also a source of lipid signal molecules (although most signal molecules in the body are secreted proteins). In response to specific controlling mechanisms, a portion of specific fatty acid tails of the phospholipid molecules can be cleaved off and used for either intracellular or extracellular communication. An exam-

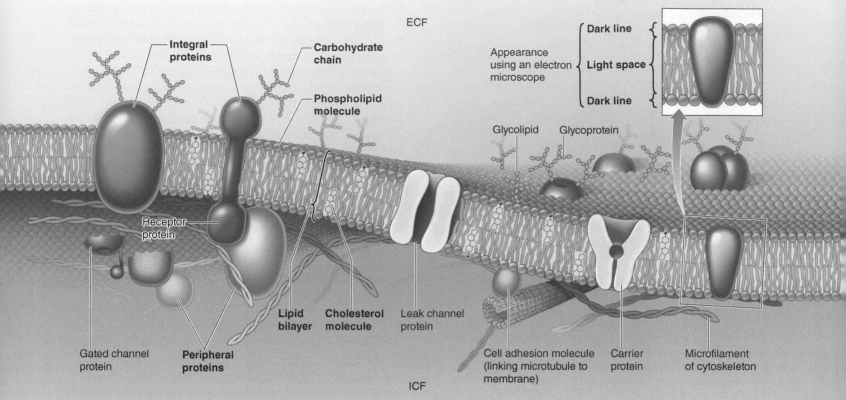

Figure 3-3 Fluid mosaic model of plasma membrane structure. The plasma membrane is composed of a lipid bilayer embedded with proteins. Short carbohydrate chains attach to proteins or lipids on the outer surface only.

ple is the prostaglandin released in response to an infection that turns up the temperature-controlling thermostat in the brain to bring about a fever.

The membrane proteins perform various specific membrane functions.

Different types of membrane proteins serve the following specialized functions:

1. Some transmembrane proteins form water-filled pathways, or **channels**, through the lipid bilayer (▌Figure 3-3). Water-soluble substances small enough to enter a channel can pass through the membrane by this means without coming into direct contact with the hydrophobic lipid interior. Channels are highly selective. The small diameter of channels prevents particles larger than 0.8 nm (40 billionths of an inch) in diameter from entering. Only small ions can fit through channels. Furthermore, a given channel selectively admits particular ions. For example, only sodium ions (Na^+) can pass through Na^+ channels, and only potassium ions (K^+) can pass through K^+ channels. This channel selectivity is a result of specific arrangements of chemical groups on the interior surfaces of the channels. Some channels are **leak channels** that always permit passage of their selected ion. Others are **gated channels** that may be open or closed to their specific ion as a result of changes in channel shape in response to controlling mecha-

nisms, described later. This is a good example of function depending on structural details. Cells vary in the number, kind, and activity of channels they possess.

Clinical Note Some drugs target channels—for example, *Ca^{2+} channel blockers* that are widely used in the management of high blood pressure and abnormal heart rhythms. More than 60 genetic mutations in channels have been linked to human diseases. To learn how a specific channel defect can lead to a devastating disease, see the accompanying boxed feature, ▌Concepts, Challenges, and Controversies.

2. Other proteins that span the membrane are **carrier**, or **transport, molecules**; they transfer across the membrane specific substances that are unable to cross on their own. The means by which carriers accomplish this transport is described later. Each carrier can transport only a particular molecule (or ion) or group of closely related molecules. Cells of different types have different kinds of carriers. As a result, they vary as to which substances they can selectively transport across their membranes. For example, thyroid gland cells are the only cells to use iodine. Appropriately, only the plasma membranes of thyroid gland cells have carriers for iodine, so only these cells can transport iodine from the blood into the cell interior.

3. Other proteins, located on the inner membrane surface, serve as **docking-marker acceptors**; they bind lock-and-key fashion with the docking markers of secretory vesicles (see p. 29). Secretion is initiated as stimulatory signals trigger fusion

Cystic Fibrosis: A Fatal Defect in Membrane Transport

CYSTIC FIBROSIS (CF), THE MOST common fatal genetic disease in the United States, strikes 1 in every 2000 Caucasian children. It is characterized by the production of abnormally thick, sticky mucus. Most dramatically affected are the respiratory airways and the pancreas.

Respiratory Problems

The presence of thick, sticky mucus in the respiratory airways makes it difficult to get adequate air in and out of the lungs. Also, because bacteria thrive in the accumulated mucus, patients with CF experience repeated respiratory infections. They are especially susceptible to *Pseudomonas aeruginosa,* an "opportunistic" bacterium that is often present in the environment but usually causes infection only when some underlying problem handicaps the body's defenses. Gradually, the involved lung tissue becomes scarred (fibrotic), making the lungs harder to inflate. This complication increases the work of breathing beyond the extra effort required to move air through the clogged airways.

Pancreatic Problems

In patients with CF, the pancreatic duct, which carries secretions from the pancreas to the small intestine, becomes plugged with thick mucus. Because the pancreas produces enzymes important in the digestion of food, malnourishment eventually results. In addition, as the pancreatic digestive secretions accumulate behind the blocked duct, fluid-filled cysts form in the pancreas, with the affected pancreatic tissue gradually degenerating and becoming fibrotic.

The name *cystic fibrosis* aptly describes long-term changes that occur in the pancreas and lungs as the result of a single genetic flaw in chloride (Cl^-) channels.

Underlying Cause

CF is caused by one of a number of genetic defects that lead to production of a flawed version of a protein known as *cystic fibrosis transmembrane conductance regulator (CFTR)*. CFTR normally forms the Cl^- channels in the plasma membrane. With CF, the defective CFTR gets "stuck" in the endoplasmic reticulum–Golgi system, which normally manufactures and processes this product and then ships it to the plasma membrane (see pp. 25–30). Thus, in patients with CF, the mutated version of CFTR is only partially processed and never makes it to the cell surface. The resultant absence of CFTR protein in the plasma membrane makes the membrane impermeable to Cl^-.

What are the consequences of this impermeability? Looking at the development of the respiratory problems as an example, because Cl^- transport across the membrane is closely linked to Na^+ transport, cells lining the respiratory airways cannot absorb NaCl (salt) properly. As a result, salt accumulates in the fluid lining the airways.

How does this Cl^- channel defect and resultant salt accumulation lead to the excess mucus problem? The following discoveries provide a plausible answer, although research into other possible mechanisms continues to be pursued. The airway cells produce a natural antibiotic, *defensin,* which normally kills most inhaled airborne bacteria. Defensin cannot function properly in a salty environment. Bathed in the excess salt associated with CF, the disabled antibiotic cannot rid the lungs of inhaled bacteria. In addition, the opportunistic *P. aeruginosa* bacteria disable the white blood cells in the lung that normally fight infectious agents. Besides causing infection, *P. aeruginosa* triggers the airway cells to produce unusually large amounts of abnormal, thick, sticky mucus, which serves as a breeding ground for even more bacterial growth. Furthermore, the mucus is thick and sticky in part because it is underhydrated (has too little water), a problem linked to the defective salt transport. To make matters worse, because the excess mucus is thick and sticky, the normal ciliary defense mechanisms of the lungs have a difficult time sweeping up the bacteria-laden mucus (see pp. 48 and 441). As a result, repeated respiratory infections occur. The vicious cycle continues as the lung-clogging mucus accumulates and lung infections become more frequent.

Treatment and New Research Directions

Treatment consists of physical therapy and mucus-thinning aerosols to help clear the airways of excess mucus and antibiotic therapy to combat respiratory infections, plus special diets and administration of pancreatic digestive enzymes to maintain adequate nutrition. Despite this supportive treatment, most people with CF do not survive beyond their late 30s, with most dying from lung complications.

Researchers are investigating gene therapy approaches to cure CF. Another treatment underway is development of drugs that induce the mutated CFTR to be "finished off" and inserted in the plasma membrane.

of the secretory vesicle membrane with the inner surface of the plasma membrane through interactions between these matching labels. The secretory vesicle subsequently opens up and empties its contents to the outside by exocytosis (see p. 30).

4. Some proteins located on either the inner or the outer cell surface function as **membrane-bound enzymes** that control specific chemical reactions. Cells are specialized in the types of membrane-bound enzymes they have. For example, a special-ized area of the outer plasma membrane surface of skeletal muscle cells contains an enzyme that destroys the chemical messenger responsible for triggering muscle contraction, thus allowing the muscle to relax.

5. Many proteins on the outer surface are **receptors**, sites that "recognize" and bind with specific extracellular chemical messengers (signal molecules) in the cell's environment. This binding initiates a series of membrane and intracellular events (to

be described later) that alter the activity of the particular cell. In this way, chemical messengers in the blood, such as water-soluble hormones, influence only the specific cells that have receptors for a given messenger. Even though every cell is exposed to the same messenger via the circulating blood, a given messenger has no effect on cells lacking receptors for this specific messenger. To illustrate, the anterior pituitary gland secretes into the blood thyroid-stimulating hormone (TSH), which attaches only to the surface of thyroid gland cells to stimulate secretion of thyroid hormone. No other cells have receptors for TSH, so TSH influences only thyroid cells despite its widespread distribution.

6. Still other proteins are **cell adhesion molecules (CAMs)**. Many CAMs protrude from the outer membrane surface and form loops or hooks by which cells grip one another or grasp the connective tissue fibers between cells. For example, *cadherins,* a type of CAM found on the surface of adjacent cells, interlock in zipper fashion to help hold the cells within tissues and organs together. Other CAMs, such as *integrins*, span the plasma membrane, where they serve as a structural link between the outer membrane surface and its extracellular surroundings and connect the inner membrane surface to the intracellular cytoskeletal scaffolding. Besides mechanically linking the cell's external environment and intracellular components, integrins also relay regulatory signals through the plasma membrane in either direction. Some CAMs participate in signaling cells to grow or in signaling immune system cells to interact with the right kind of other cells in inflammatory responses and wound healing, among other things.

7. Still other proteins on the outer membrane surface, especially in conjunction with carbohydrates (as glycoproteins), are important in the cells' ability to recognize "self" (that is, cells of the same type).

The membrane carbohydrates serve as self-identity markers.

The short carbohydrate chains on the outer membrane surface serve as self-identity markers that enable cells to identify and interact with one another in the following ways:

1. Different cell types have different markers. The unique combination of sugar chains projecting from the membrane surface serves as the "trademark" of a particular cell type, enabling a cell to recognize others of its kind. To exemplify, differences in specific membrane carbohydrates are responsible for the variations in human blood group types (A, B, AB, and O). Membrane carbohydrate chains play an important role in recognition of "self" and in cell-to-cell interactions. Cells can recognize other cells of the same type and join to form tissues. This is especially important during embryonic development.

2. Carbohydrate-containing surface markers are also involved in tissue growth, which is normally held within certain limits of cell density. Cells do not "trespass" across the boundaries of neighboring tissues—that is, they do not overgrow their own territory. The exception is the uncontrolled spread of cancer cells, which have been shown to bear abnormal surface carbohydrate markers.

3.2 Cell-to-Cell Adhesions

In multicellular organisms such as humans, the plasma membrane not only is the outer boundary of all cells, but it also participates in cell-to-cell adhesions. These adhesions bind groups of cells together into tissues and package them further into organs. The life-sustaining activities of the body systems depend not only on the functions of the individual cells of which they are made but also on how these cells live and work together in tissue and organ communities.

Cells organized into appropriate groupings are held together by three different means: (1) CAMs, (2) the extracellular matrix, and (3) specialized cell junctions. You are already familiar with CAMs. We now examine the extracellular matrix and then specialized junctions.

The extracellular matrix serves as biological "glue."

Tissues are not made up solely of cells, and many cells within a tissue are not in direct contact with neighboring cells. Instead, they are held together by a biological "glue" called the **extracellular matrix (ECM)**. The ECM is an intricate meshwork of fibrous proteins embedded in a watery, gel-like substance composed of complex carbohydrates. The watery gel, usually called the interstitial fluid (see p. 8), provides a pathway for diffusion of nutrients, wastes, and other water-soluble traffic between the blood and tissue cells. The three major types of protein fibers woven through the gel are collagen, elastin, and fibronectin.

1. **Collagen** forms flexible but nonelastic fibers or sheets that provide tensile strength (resistance to being stretched lengthwise). Collagen is the most abundant protein in the body, making up nearly half of total body protein by weight.

Clinical Note In **scurvy**, a condition caused by vitamin C deficiency, collagen fibers are not properly formed. As a result, the tissues, especially those of the skin and blood vessels, become fragile. This leads to bleeding in the skin and mucous membranes, which is especially noticeable in the gums.

2. **Elastin** is a rubbery protein fiber most plentiful in tissues that must easily stretch and then recoil after the stretching force is removed. It is found, for example, in the lungs, which stretch and recoil as air moves in and out of them.

3. **Fibronectin** promotes cell adhesion and holds cells in position. Reduced amounts of this protein have been found within certain types of cancerous tissue, possibly accounting for cancer cells' inability to adhere well to one another; in-

stead, they tend to break loose and metastasize (spread elsewhere in the body).

The ECM is secreted by local cells present in the matrix. The relative amount of ECM compared to cells varies greatly among tissues. For example, the ECM is scant in epithelial tissue but is the predominant component of connective tissue. Most of this abundant matrix in connective tissue is secreted by **fibroblasts** ("fiber formers"). The exact composition of the ECM also varies for different tissues, thus providing distinct local environments for the various cell types in the body. In some tissues, the matrix becomes highly specialized to form such structures as cartilage or tendons or, on appropriate calcification, the hardened structures of bones and teeth.

Contrary to long-held belief, the ECM is not just passive scaffolding for cellular attachment; it also helps regulate the behavior and functions of the cells with which it interacts. Cells are able to function normally and indeed even to survive only when associated with their normal matrix components. The matrix is especially influential in cell growth and differentiation. In the body, only circulating blood cells are designed to survive and function without attaching to the ECM.

Some cells are directly linked by specialized cell junctions.

In tissues where the cells lie close to one another, CAMs provide some tissue cohesion as they "Velcro" adjacent cells together. In addition, some cells within given types of tissues are directly linked by one of three types of specialized cell junctions: (1) *desmosomes* (adhering junctions), (2) *tight junctions* (impermeable junctions), or (3) *gap junctions* (communicating junctions).

Desmosomes **Desmosomes** act like "spot rivets" that anchor together two adjacent but nontouching cells. A desmosome consists of two components: (1) a pair of dense, buttonlike cytoplasmic thickenings known as *plaques* located on the inner surface of each of the two adjacent cells, and (2) strong filaments containing cadherins (a type of CAM) that extend across the space between the two cells and attach to the plaque on both sides (❙Figure 3-4). These intercellular filaments bind adjacent plasma membranes together so that they resist being pulled apart. Thus, desmosomes are adhering junctions. They are the strongest cell-to-cell connections.

Desmosomes are most abundant in tissues that are subject to considerable stretching, such as those found in the skin, the heart, and the uterus. In these tissues, functional groups of cells are riveted together by desmosomes. Furthermore, intermediate cytoskeletal filaments, such as tough keratin filaments in the skin (see p. 51), stretch across the interior of these cells and attach to the desmosome plaques located on opposite sides of the cells' interior. This arrangement forms a continuous network of strong fibers throughout the tissue, both through and between cells, much like a continuous line of people firmly holding hands. This interlinking fibrous network provides tensile strength, reducing the chances of the tissue being torn when stretched.

Tight Junctions At **tight junctions**, adjacent cells bind firmly with each other at points of direct contact to seal off the pas-

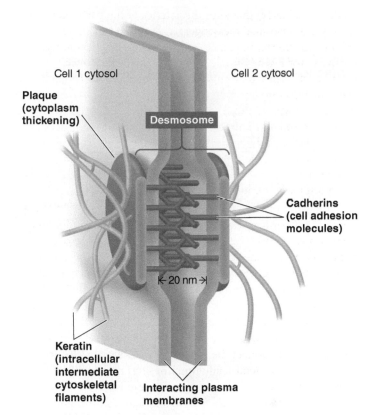

❙Figure 3-4 **Desmosome.** Desmosomes are adhering junctions that spot-rivet cells, anchoring them together in tissues subject to considerable stretching.

sageway between the two cells. Tight junctions are found primarily in sheets of epithelial tissue, which cover the surface of the body and line its internal cavities. All epithelial sheets are highly selective barriers between two compartments with considerably different chemical compositions. For example, the epithelial sheet lining the digestive tract separates the food and potent digestive juices within the inner cavity (lumen) from the blood vessels on the other side. Only completely digested food particles and not undigested food particles or digestive juices must be permitted to move across the epithelial sheet from the lumen to the blood. Accordingly, the lateral (side) edges of adjacent cells in the epithelial sheet are joined in a tight seal near their luminal border by "kiss" sites, at which strands of proteins known as *claudins* on the outer surfaces of the two interacting plasma membranes fuse directly (❙Figure 3-5) (*claudin* means "to close," indicative of the barrier role of these proteins). These tight junctions are impermeable and thus prevent materials from passing between the cells. Passage across the epithelial barrier, therefore, must take place *through* the cells, not *between* them. This **transcellular transport** across the cell (*trans* means "across") is regulated by channel and carrier proteins. If the cells were not joined by tight junctions, uncontrolled exchange of molecules could take place between the compartments by means of unpoliced traffic through the spaces between adjacent cells. Tight junctions thus prevent undesirable leaks within epithelial sheets.

Despite their generally tight nature, some tight junctions are a bit "leaky," enabling water molecules and some small ions to

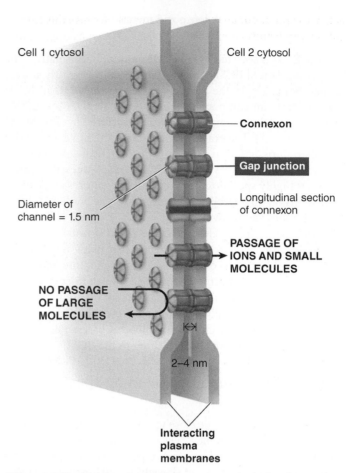

IFigure 3-6 Gap junction. Gap junctions are communicating junctions made up of connexons, which form tunnels that permit movement of charge-carrying ions and small molecules between two adjacent cells.

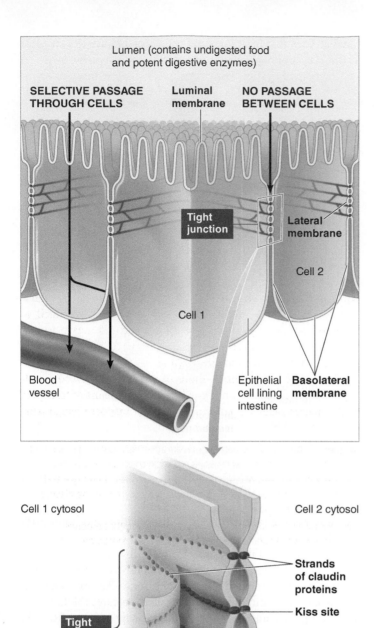

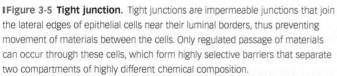

IFigure 3-5 Tight junction. Tight junctions are impermeable junctions that join the lateral edges of epithelial cells near their luminal borders, thus preventing movement of materials between the cells. Only regulated passage of materials can occur through these cells, which form highly selective barriers that separate two compartments of highly different chemical composition.

FIGURE FOCUS: *Different types of specialized membrane proteins such as channels, carriers, or enzymes are localized at the luminal membrane or at the basolateral membrane. What keeps these proteins from migrating to the wrong part of the membrane?*

pass between the cells—for example, in the small intestine during absorption of a meal. This between-cell transport is called **paracellular transport** (*para* means "beside" in reference to this transport occurring beside the adjacent cells).

Gap Junctions At a **gap junction**, as the name implies, a gap exists between adjacent cells, which are linked by small, connecting tunnels formed by connexons. A **connexon** is made up of six protein subunits (called *connexins*) arranged in a hollow, tubelike structure that extends through the thickness of the plasma membrane. Two connexons, one from each of the plasma membranes of two adjacent cells, extend outward and join end-to-end to form a connecting tunnel between the two cells (IFigure 3-6). Gap junctions are communicating junctions. The small diameter of the tunnels permits small, water-soluble particles to pass between the connected cells but precludes passage of large molecules, such as vital intracellular proteins. Ions and small molecules can be directly exchanged between interacting cells through gap junctions without ever entering the ECF.

Gap junctions are especially abundant in cardiac muscle and smooth muscle. In these tissues, movement of ions (charge-carrying particles) through gap junctions transmits electrical activity throughout an entire muscle mass. Because this electrical activity brings about contraction, the presence of gap junctions enables synchronized contraction of a whole muscle mass, such as the pumping chamber of the heart.

Gap junctions are also found in some nonmuscle tissues, where they permit unrestricted passage of small nutrient molecules between cells. For example, glucose, amino acids, and other nutrients pass through gap junctions to a developing egg cell from surrounding cells within the ovary, thus helping the egg stockpile these essential nutrients.

Gap junctions also are avenues for the direct transfer of small signal molecules from one cell to the next. Such transfer permits cells connected by gap junctions to communicate with each other directly.

We now turn to the topic of membrane transport, focusing on how the plasma membrane selectively controls what enters and exits the cell.

Check Your Understanding 3.2

1. Describe the extracellular matrix.
2. List the three types of specialized cell junctions and indicate their primary role.
3. Draw a desmosome.

3.3 Overview of Membrane Transport

Anything that passes between a cell and the surrounding ECF must be able to penetrate the plasma membrane. If a substance can cross the membrane, the membrane is **permeable** to that substance; if a substance cannot pass, the membrane is **impermeable** to it. The plasma membrane is **selectively permeable**: It permits some particles to pass through while excluding others.

Lipid-soluble substances and small water-soluble substances can permeate the plasma membrane unassisted.

Two properties of particles influence whether they can permeate the plasma membrane without assistance: (1) the relative solubility of the particle in lipid and (2) the size of the particle. Highly lipid-soluble particles of any size can dissolve in the lipid bilayer and pass through the membrane. Uncharged or nonpolar molecules, such as oxygen (O_2), carbon dioxide (CO_2), and fatty acids, are highly lipid soluble and readily permeate the membrane. Charged particles (ions such as Na^+ and K^+) and polar molecules (such as glucose and proteins) have low lipid solubility but are very soluble in water. The lipid bilayer is an impermeable barrier to particles poorly soluble in lipid. For water-soluble (and thus lipid-insoluble) ions less than 0.8 nm in diameter, the protein channels are an alternative route for passage across the membrane. Only ions for which specific channels are available and open can permeate the membrane.

Particles that have low lipid solubility and are too large for channels cannot permeate the membrane on their own. Yet some of these particles—for example, glucose—must cross the membrane for the cell to survive and function. (Most cells use glucose as their fuel of choice to produce adenosine triphosphate, or ATP.) Cells have several means of assisted transport to move particles that must cross the membrane but cannot do so unaided, as you will learn shortly.

Active forces use energy to move particles across the membrane, but passive forces do not.

Even if a particle can permeate the membrane because of its lipid solubility or its ability to fit through a channel, some force is needed to move it across the membrane. Two general types of forces accomplish transport of substances across the membrane: (1) **passive forces**, which do not require the cell to expend energy to produce movement, and (2) **active forces**, which do require the cell to expend energy (ATP) in transporting a substance across the membrane.

We now examine the methods of membrane transport, noting whether each is an unassisted or assisted means of transport and whether each is a passive- or active-transport mechanism.

Check Your Understanding 3.3

1. Explain how both highly lipid-soluble substances of any size and small water-soluble substances are able to permeate the plasma membrane without assistance.
2. Distinguish between passive and active forces that produce movement of substances across the plasma membrane.

3.4 Unassisted Membrane Transport

Particles that can penetrate the plasma membrane on their own are passively driven across the membrane by one or both of two forces: diffusion down a concentration gradient or movement along an electrical gradient. We first examine diffusion down a concentration gradient.

Particles that can permeate the membrane diffuse passively down their concentration gradient.

All molecules and ions are in continuous random motion at temperatures above absolute zero as a result of thermal (heat) energy. This motion is most evident in liquids and gases, where the individual molecules (or ions) have more room to move before colliding with another molecule. Each molecule moves separately and randomly in any direction. As a consequence of this haphazard movement, the molecules often collide, bouncing off one another in different directions like billiard balls striking.

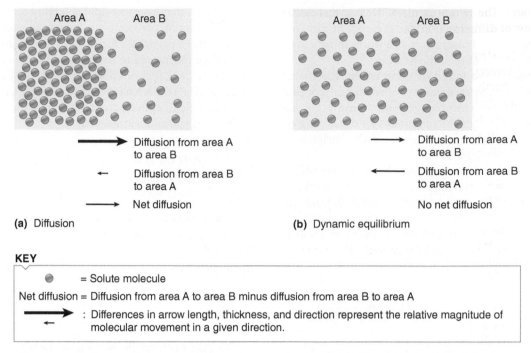

(a) Diffusion

Diffusion from area A to area B

Diffusion from area B to area A

Net diffusion

(b) Dynamic equilibrium

Diffusion from area A to area B

Diffusion from area B to area A

No net diffusion

KEY

● = Solute molecule

Net diffusion = Diffusion from area A to area B minus diffusion from area B to area A

⟶ / ← : Differences in arrow length, thickness, and direction represent the relative magnitude of molecular movement in a given direction.

❙Figure 3-7 **Diffusion.** (a) Diffusion down a concentration gradient. (b) Dynamic equilibrium, with no net diffusion occurring.

Simple Diffusion **Solutions** are homogeneous mixtures containing a relatively large amount of one substance called the **solvent** (the dissolving medium, which in the body is water) and smaller amounts of one or more dissolved substances called **solutes**. The **concentration** of a solution refers to the amount of solute dissolved in a specific amount of solution. The greater the concentration of solute molecules (or ions), the greater the likelihood is of collisions. Consequently, molecules within a particular space tend to become evenly distributed over time. Such uniform spreading out of molecules as a result of their random intermingling is known as **simple diffusion**, or **diffusion** for short (*diffusere* means "to spread out").

To illustrate simple diffusion, in Figure 3-7a, the concentration of the solute in a solution differs between area A and area B. Such a difference in concentration between two adjacent areas is called a **concentration gradient** (or **chemical gradient**). Random molecular collisions occur more frequently in area A because of its greater concentration of solute molecules. For this reason, more molecules bounce from area A into area B than in the opposite direction. In both areas, individual molecules move randomly and in all directions, but the net movement of molecules by diffusion is from the area of higher concentration to the area of lower concentration.

Net Diffusion The term **net diffusion** refers to the difference between two opposing movements. If 10 molecules move from area A to area B while 2 molecules simultaneously move from B to A, the net diffusion is 8 molecules moving from A to B. Molecules spread in this way until the substance is uniformly distributed between the two areas and a concentration gradient no longer exists (❙Figure 3-7b). At this point, even though movement is still taking place, no net diffusion is occurring because the opposing movements exactly counterbalance each other—

that is, they are in **dynamic equilibrium** (*dynamic* in reference to the continuous movement, *equilibrium* in reference to the exact balance between opposing forces). Movement of molecules from area A to area B is exactly matched by movement of molecules from B to A.

What happens if a plasma membrane separates different concentrations of a substance? If the substance can permeate the membrane, net diffusion of the substance occurs through the membrane down its concentration gradient from the area of high concentration to the area of low concentration until the concentration gradient is abolished, unless there's some opposing force (❙Figure 3-8a). No energy is required for this movement, so it is a passive means of membrane transport. The process of diffusion is crucial to the survival of every cell and plays an important role in many specialized homeostatic activities. As an example, O_2 is transferred across the lung membrane by diffusion. The blood carried to the lungs is low in O_2, having given up O_2 to the body tissues for cell metabolism. The air in the lungs, in contrast, is high in O_2 because it is continuously exchanged with fresh air during breathing. Because of this concentration gradient, net diffusion of O_2 occurs from the lungs into the blood as blood flows through the lungs. Thus, as blood leaves the lungs for delivery to the tissues, it is high in O_2.

If the membrane is impermeable to the substance, no diffusion can take place across the membrane, even though a concentration gradient may exist (❙Figure 3-8b). For example, because the plasma membrane is impermeable to the vital intracellular proteins, they are unable to escape from the cell, even though they are in greater concentration in the ICF than in the ECF.

Fick's Law of Diffusion Several factors, in addition to the concentration gradient, influence the rate of net diffusion across a membrane. The effects of these factors collectively make up **Fick's law of diffusion** (Table 3-1):

1. *The magnitude (or steepness) of the concentration gradient.* If a substance can permeate the membrane, its rate of simple diffusion is always directly proportional to its concentration gradient—that is, the greater the difference in concentration, the faster the rate of net diffusion (see Figure 3-15, p. 71). For example, during exercise the working muscles produce CO_2 more rapidly than usual because they are burning additional fuel to produce the extra ATP they need to power the stepped-up, energy-demanding contractile activity. The increase in CO_2 level in the muscles creates a greater-than-normal difference in CO_2 between the muscles and the blood supplying the muscles. Because of this steeper gradient, more CO_2 than usual enters the blood. When this blood with its increased CO_2 load reaches the lungs, a greater-than-normal CO_2 gradient exists between the blood and the air sacs in the lungs. Accordingly, more CO_2 than normal diffuses from the blood into the air sacs. This extra CO_2 is subsequently breathed out to the environment. Thus, any additional CO_2 produced by exercising muscles is eliminated from the body through the lungs as a result of the increase in CO_2 concentration gradient.

2. *The surface area of the membrane across which diffusion is taking place.* The larger the surface area available, the greater the rate of diffusion it can accommodate. Various strategies are used throughout the body for increasing the membrane surface area across which diffusion and other types of transport take place. For example, absorption of nutrients in the small intestine is enhanced by the presence of microvilli, which greatly increase the available absorptive surface in contact with the nutrient-rich contents of the small-intestine lumen (see p. 50). Conversely, abnormal loss of membrane surface area decreases the rate of net diffusion. For example, in *emphysema*, O_2 and CO_2 exchange between air and blood in the lungs is reduced because the walls of the air sacs break down,

TABLE 3-1 Factors Influencing the Rate of Net Diffusion of a Substance across a Membrane (Fick's Law of Diffusion)

Factor	Effect on Rate of Net Diffusion
↑ Concentration gradient of substance (ΔC)	↑
↑ Surface area of membrane (A)	↑
↑ Lipid solubility (β)	↑
↑ Molecular weight of substance (MW)	↓
↑ Distance (thickness) (ΔX)	↓

Modified Fick's equation:

$$\text{Net rate of diffusion } (Q) = \frac{\Delta C \cdot A \cdot \beta}{\sqrt{MW} \cdot \Delta X}$$

$$\left[\text{diffusion constant } (D) \propto \frac{\beta}{\sqrt{MW}} \right]$$

$$\left[\text{permeability } (P) = \frac{D}{\Delta X} \right]$$

$$\text{Restated } Q \propto \Delta C \cdot A \cdot P$$

resulting in less surface area available for diffusion of these gases.

3. *The lipid solubility of the substance.* The greater the lipid solubility of a substance, the more rapidly the substance can diffuse through the membrane's lipid bilayer down its concentration gradient.

4. *The molecular weight of the substance.* Heavier molecules do not bounce as far on collision as lighter molecules such as O_2 and CO_2 do. Consequently, O_2 and CO_2 diffuse rapidly, permitting rapid exchanges of these gases across the lung membranes. As molecular weight increases, the rate of diffusion decreases.

5. *The distance through which diffusion must take place.* The greater the distance, the slower the rate of diffusion. Accordingly, membranes across which diffusing particles must travel are normally relatively thin, such as the membranes separating air and blood in the lungs. Thickening of this air–blood interface (as in *pneumonia*, for example) slows exchange of O_2 and CO_2. Furthermore, diffusion is efficient only for short distances between cells and their surroundings. It becomes an inappropriately slow process for distances of more than a few millimeters. To illustrate, it would take months or even years for O_2 to diffuse from the surface of the body to the cells in the interior. Instead, the circulatory system provides a network of tiny vessels that deliver and pick up materi-

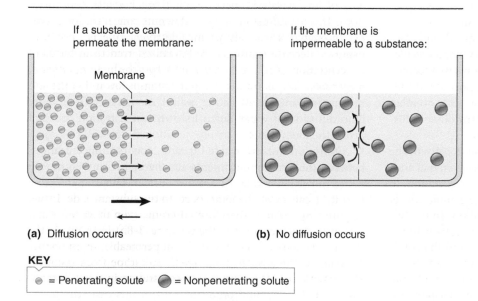

If a substance can permeate the membrane:	If the membrane is impermeable to a substance:

Membrane

(a) Diffusion occurs **(b)** No diffusion occurs

KEY

= Penetrating solute   = Nonpenetrating solute

Figure 3-8 Diffusion through a membrane. (a) Net diffusion of a penetrating solute across the membrane down a concentration gradient. (b) No diffusion of a nonpenetrating solute through the membrane despite the presence of a concentration gradient.

als at every "block" of a few cells, with diffusion accomplishing short local exchanges between blood and surrounding cells.

Ions that can permeate the membrane also move passively along their electrical gradient.

In addition to their concentration gradient, movement of ions is affected by their electrical charge. Ions with like charges (those with the same kind of charge) repel each other, and ions with opposite charges attract each other. If a relative difference in charge exists between two adjacent areas, positively charged ions (*cations*) tend to move toward the more negatively charged area and negatively charged ions (*anions*) tend to move toward the more positively charged area. A difference in charge between two adjacent areas thus produces an **electrical gradient** that promotes movement of ions toward the area of opposite charge. Because a cell does not have to expend energy for ions to move into or out of it along an electrical gradient, this method of membrane transport is passive. When an electrical gradient exists between the ICF and the ECF, only ions that can permeate the plasma membrane can move along this gradient.

Both an electrical and a concentration (chemical) gradient may be acting on a particular ion at the same time. The net effect of simultaneous electrical and concentration gradients on this ion is called an **electrochemical gradient**. Later in this chapter you will learn how electrochemical gradients contribute to the electrical properties of the plasma membrane.

Osmosis is the net diffusion of water down its own concentration gradient.

Water molecules can readily permeate the plasma membrane. Even though water molecules are strongly polar, they are small enough to slip through momentary spaces created between the phospholipid molecules' tails as they sway and move within the lipid bilayer. However, this type of water movement across the membrane is relatively slow. In many cell types, membrane proteins form **aquaporins**, which are channels specific for the passage of water (*aqua* means "water"). This avenue greatly increases membrane permeability to water. About a billion water molecules can pass in single file through an aquaporin channel in a second. Different cell types vary in their density of aquaporins and thus in their water permeability. The driving force for net movement of water across the membrane is the same as for any other diffusing molecule—namely, its concentration gradient.

The term *concentration* usually refers to the density of the solute in a given volume of water. Recognize, however, that adding a solute to pure water decreases the water concentration. In general, one molecule of a solute displaces one molecule of water. Compare the water and solute concentrations in the two containers in Figure 3-9. The container in Figure 3-9a is full of pure water, so the water concentration is 100% and the solute concentration is 0%. In Figure 3-9b, solute has replaced 10% of the water molecules. The water concentration is now 90%, and the solute concentration is 10%—a lower water concentration and a higher solute concentration than in Figure 3-9a. As the solute concentration increases, the water concentration decreases correspondingly.

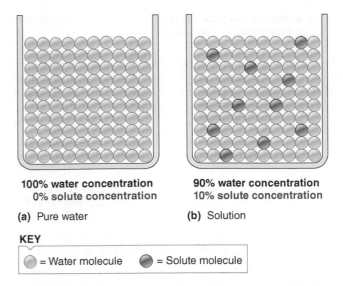

100% water concentration
0% solute concentration

(a) Pure water

90% water concentration
10% solute concentration

(b) Solution

KEY

 = Water molecule   = Solute molecule

❚Figure 3-9 **Relationship between solute and water concentration in a solution.**

We now examine what water movement takes place when a selectively permeable membrane separates two fluid compartments under different circumstances, beginning with pure water being separated from a solution by a membrane permeable to water but not to the solute.

Movement of Water When a Selectively Permeable Membrane Separates Pure Water from a Solution of a Nonpenetrating Solute If as in Figure 3-10 pure water (side 1) and a solution containing a nonpenetrating solute (side 2) are separated by a selectively permeable membrane that permits passage of water but not of solute, water will move passively down its own concentration gradient from the area of higher water concentration (lower solute concentration) to the area of lower water concentration (higher solute concentration). This net diffusion of water down its concentration gradient through a selectively permeable membrane is known as **osmosis**. Because solutions are always referred to in terms of concentration of solute, *water moves by osmosis to the area of higher solute concentration*. Despite the impression that the solutes are "pulling," or attracting, water, osmosis is nothing more than diffusion of water down its own concentration gradient across the membrane.

Osmosis occurs from side 1 to side 2, but the concentrations between the two compartments can never become equal. No matter how dilute side 2 becomes because of water diffusing into it, it can never become pure water, nor can side 1 ever acquire any solute. Therefore, does net diffusion of water (osmosis) continue until all the water has left side 1? No. As the volume expands in side 2, a difference in hydrostatic pressure between the two sides is created, and it opposes osmosis. **Hydrostatic (fluid) pressure** is the pressure exerted by a standing, or stationary, fluid on an object—in this case, the membrane (*hydro* means "fluid"; *static* means "standing"). The hydrostatic pressure exerted by the larger volume of fluid on side 2 is greater than the hydrostatic pressure exerted on side 1.

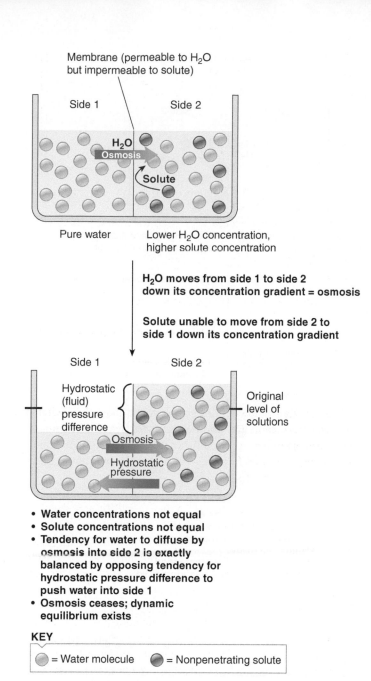

Figure 3-10 **Osmosis when pure water is separated from a solution containing a nonpenetrating solute.**

This difference in hydrostatic pressure tends to push fluid from side 2 to side 1.

The **osmotic pressure** of a solution (a "pulling" pressure) is a measure of the tendency for osmotic flow of water into that solution because of its relative concentration of nonpenetrating solutes and water. Net movement of water by osmosis continues until the opposing hydrostatic pressure (a "pushing" pressure) exactly counterbalances the osmotic pressure. The magnitude of the osmotic pressure is equal to the magnitude of the opposing hydrostatic pressure necessary to completely stop osmosis. The greater the concentration of nonpenetrating solute → the lower the concentration of water → the greater the drive for water to move by osmosis from pure water into the solution → the greater the opposing pressure required to stop the osmotic

flow → the greater the osmotic pressure of the solution. Therefore, a solution with a high concentration of nonpenetrating solute exerts greater osmotic pressure than a solution with a lower concentration of nonpenetrating solute does.

Osmotic pressure is an indirect measure of solute concentration, expressed in units of pressure. A more direct means of expressing solute concentration is the **osmolarity** of a solution, which is a measure of its total solute concentration given in terms of the *number* of particles (molecules or ions). Osmolarity is expressed in *osmoles per liter* (or *Osm/L*), the number of moles of solute particles in 1 liter of solution (see p. A-7). Because glucose remains as an intact molecule when in solution, 1 mole of glucose equals 1 osmole—that is, 1 mole of solute particles. By contrast, because a molecule of NaCl dissociates (separates) into 2 ions—Na^+ and Cl^-—when in solution, 1 mole of NaCl equals 2 osmoles—1 mole of Na^+ and 1 mole of Cl^-, or 2 moles of solute particles. The osmolarity of body fluids is typically expressed in *milliosmoles per liter* (*mOsm/L*) (1/1000 of an osmole) because the solutes in body fluids are too dilute to conveniently use the osmole unit. Because osmolarity depends on the number, not the nature, of particles, any mixture of particles can contribute to the osmolarity of a solution. The normal osmolarity of body fluids is 300 mOsm/L.

Thus far in our discussion of osmosis, we have considered movement of water when pure water is separated from a solution by a membrane permeable to water but not to nonpenetrating solutes. However, in the body, the plasma membrane separates the ICF and the ECF, and both of these contain solutes, some that can and others that cannot penetrate the membrane. Let us compare the results of water movement when solutions of differing osmolarities are separated by a selectively permeable membrane that permits movement of water and only some solutes.

Movement of Water and Solute When a Membrane Separates Unequal Solutions of a Penetrating Solute

Assume that solutions of *unequal* concentration of *penetrating* solute (differing osmolarities) are separated by a membrane that is permeable to both water and solute (Figure 3-11). In this situation, the solute moves down its own concentration gradient in the opposite direction of the net water movement. The movement continues until both solute and water are evenly distributed across the membrane. With all concentration gradients gone, net movement ceases. The final volume of each side when dynamic equilibrium is achieved and no further net movement occurs is the same as at the onset. Water and solute molecules merely exchange places between the two sides until their distributions are equalized—that is, an equal number of water molecules move from side 1 to side 2 as solute molecules move from side 2 to side 1. Therefore, solutes that can penetrate the plasma membrane do not contribute to osmotic differences between the ICF and the ECF and do not affect cell volume.

Movement of Water When a Membrane Separates Equal or Unequal Solutions of a Nonpenetrating Solute

If solutions of *equal* concentration of *nonpenetrating* solute (the same osmolarities) are separated by a membrane that is permeable to water but impermeable to the solute, no concentration

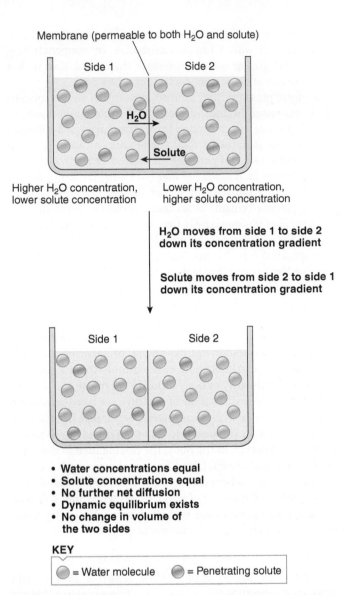

Membrane (permeable to both H₂O and solute)

Side 1   **Side 2**

H_2O →

← **Solute**

Higher H₂O concentration, lower solute concentration

Lower H₂O concentration, higher solute concentration

H₂O moves from side 1 to side 2 down its concentration gradient

Solute moves from side 2 to side 1 down its concentration gradient

Side 1   **Side 2**

- **Water concentrations equal**
- **Solute concentrations equal**
- **No further net diffusion**
- **Dynamic equilibrium exists**
- **No change in volume of the two sides**

KEY

◯ = Water molecule   ◯ = Penetrating solute

▎Figure 3-11 **Movement of water and a penetrating solute unequally distributed across a membrane.**

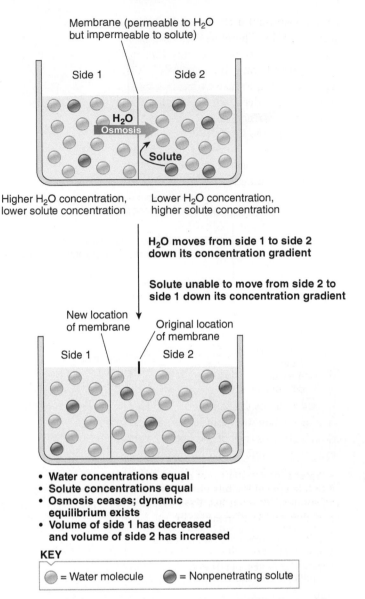

Membrane (permeable to H₂O but impermeable to solute)

Side 1   **Side 2**

H_2O
Osmosis →

Solute

Higher H₂O concentration, lower solute concentration

Lower H₂O concentration, higher solute concentration

H₂O moves from side 1 to side 2 down its concentration gradient

Solute unable to move from side 2 to side 1 down its concentration gradient

New location of membrane   Original location of membrane

Side 1   **Side 2**

- **Water concentrations equal**
- **Solute concentrations equal**
- **Osmosis ceases; dynamic equilibrium exists**
- **Volume of side 1 has decreased and volume of side 2 has increased**

KEY

◯ = Water molecule   ● = Nonpenetrating solute

▎Figure 3-12 **Osmosis in the presence of an unequally distributed nonpenetrating solute.**

differences exist and thus no net movement of water occurs across the membrane. Of course, the solute does not move because the membrane is impermeable to it and no concentration gradient exists for it. This is the usual situation in body fluids. Body cells normally do not experience any net gain (swelling) or loss (shrinking) of volume because the concentration of nonpenetrating solutes in the ECF is normally carefully regulated (primarily by the kidneys) so that the ECF osmolarity is the same as the osmolarity within the cells. Intracellular osmolarity is normally 300 mOsm/L, and all intracellular solutes are assumed to be nonpenetrating.

Now assume that solutions of *unequal* concentration of *nonpenetrating* solute (differing osmolarities) are separated by a membrane that is permeable to water but impermeable to the solute (▎Figure 3-12). Osmotic movement of water across the membrane is driven by the difference in osmotic pressure of the two solutions. At first, the concentration gradients are identical to those in Figure 3-11. Net diffusion of water takes place

from side 1 to side 2, but the solute cannot cross the membrane down its concentration gradient. As a result of water movement alone, the volume of side 2 increases while the volume of side 1 correspondingly decreases. Loss of water from side 1 increases the solute concentration on side 1, whereas addition of water to side 2 reduces the solute concentration on that side. If the membrane is free to move so that side 2 can expand without an opposing hydrostatic pressure developing, eventually the concentrations of water and solute on the two sides of the membrane become equal and net diffusion of water ceases. This situation is similar to what happens across plasma membranes in the body. Within the slight range of changes in ECF osmolarity that occur physiologically, if water moves by osmosis into the cells, their plasma membranes normally accommodate the increase in cell volume with no significant change in hydrostatic pressure inside the cells. Likewise, in the reverse situation, if water moves by osmosis out of the cells, the ECF compartment expands without a change in its hydrostatic pressure. Therefore, osmosis is the major force responsible for the net movement of water into or out of cells, without having to take hydrostatic pressure into consideration. At the endpoint, when osmosis ceases, the volume has increased on the side that originally had the higher solute concentration and the volume has decreased on the side with the lower solute concentration. Therefore, osmotic movement of water across the plasma membrane always results in a change in cell volume, and cells, especially brain cells, do not function properly when they swell or shrink.

Tonicity refers to the effect the concentration of nonpenetrating solutes in a solution has on cell volume.

The **tonicity** of a solution is the effect the solution has on cell volume—whether the cell remains the same size, swells, or shrinks—when the solution surrounds the cell. The tonicity of a solution has no units and is a reflection of its concentration of nonpenetrating solutes relative to the cell's concentration of nonpenetrating solutes. (By contrast, the osmolarity of a solution is a measure of its total concentration of both penetrating and nonpenetrating solutes expressed in units of osmoles per liter.) The easiest way to demonstrate this phenomenon is to place red blood cells in solutions with varying concentrations of a nonpenetrating solute (❙ Figure 3-13).

Normally, the plasma in which red blood cells are suspended has the same osmotic activity as the fluid inside these cells, so the cells maintain a constant volume. An **isotonic solution** (*iso* means "equal") has the same concentration of nonpenetrating solutes as normal body cells do. When a cell is bathed in an isotonic solution, no water enters or leaves the cell by osmosis, so cell volume remains constant. For this reason, the ECF is normally maintained isotonic so that no net diffusion of water occurs into or out of body cells.

If red blood cells are placed in a dilute or **hypotonic solution** (*hypo* means "below"), a solution with a below-normal concentration of nonpenetrating solutes (and therefore a higher concentration of water), water enters the cells by osmosis. Net gain of water by the cells causes them to swell, perhaps to the point of rupturing, or *lysing*. If, in contrast, red blood cells are placed in a concentrated or **hypertonic solution** (*hyper* means "above"), a solution with an above-normal concentration of nonpenetrating solutes (and therefore a lower concentration of water), the cells shrink as they lose water by

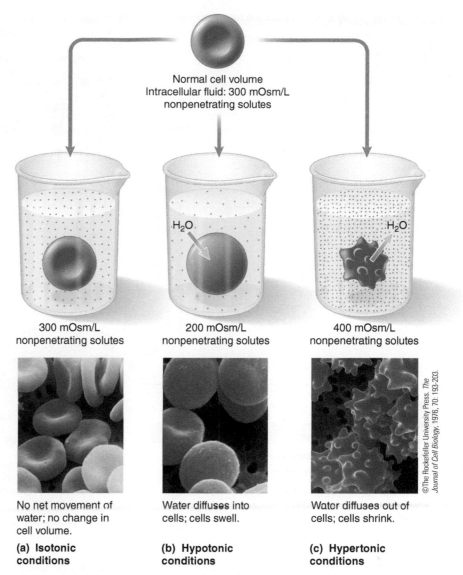

Normal cell volume
Intracellular fluid: 300 mOsm/L
nonpenetrating solutes

300 mOsm/L
nonpenetrating solutes

200 mOsm/L
nonpenetrating solutes

400 mOsm/L
nonpenetrating solutes

No net movement of water; no change in cell volume.

Water diffuses into cells; cells swell.

Water diffuses out of cells; cells shrink.

(a) Isotonic conditions

(b) Hypotonic conditions

(c) Hypertonic conditions

©The Rockefeller University Press. *The Journal of Cell Biology,* 1976, 70: 193-203.

❙Figure 3-13 **Tonicity and osmotic water movement.**
FIGURE FOCUS: *If red blood cells are placed in a 300 mOsm/L mixture of both nonpenetrating and penetrating solutes, what happens to cell volume? What is the tonicity of this solution?*

osmosis. When a red blood cell decreases in volume, its surface area does not decrease correspondingly, so the cell assumes a *crenated,* or spiky, shape (❙ Figure 3-13c). Because cells change volume when surrounded by fluid that is not isotonic, it is crucial that the concentration of nonpenetrating solutes in the ECF quickly be restored to normal should the ECF become hypotonic (as with ingesting too much water) or hypertonic (as with losing too much water through severe diarrhea). (See pp. 540–547 for further details about the important homeostatic mechanisms that maintain the normal concentration of nonpenetrating solutes in the ECF.) For the same reason, fluids injected intravenously should be isotonic to prevent unwanted movement of water into or out of the cells. For example, isotonic saline (0.9% NaCl solution) is used as a vehicle for delivering drugs intravenously or for expanding plasma volume without affecting the cells. (Sometimes hypotonic or hypertonic fluids are injected therapeutically to correct osmotic imbalances.)

▮Figure 3-14 **Facilitated diffusion, a passive form of carrier-mediated transport.**

1 Carrier protein takes conformation in which solute binding site is exposed to region of higher concentration.

ECF — Solute molecule to be transported

— Carrier protein

— Binding site

Concentration gradient

(High)

(Low)

Plasma membrane

ICF

Direction of transport

4 Transported solute is released and carrier protein returns to conformation in step 1.

2 Solute molecule binds to carrier protein.

3 Carrier protein changes conformation so that binding site is exposed to region of lower concentration.

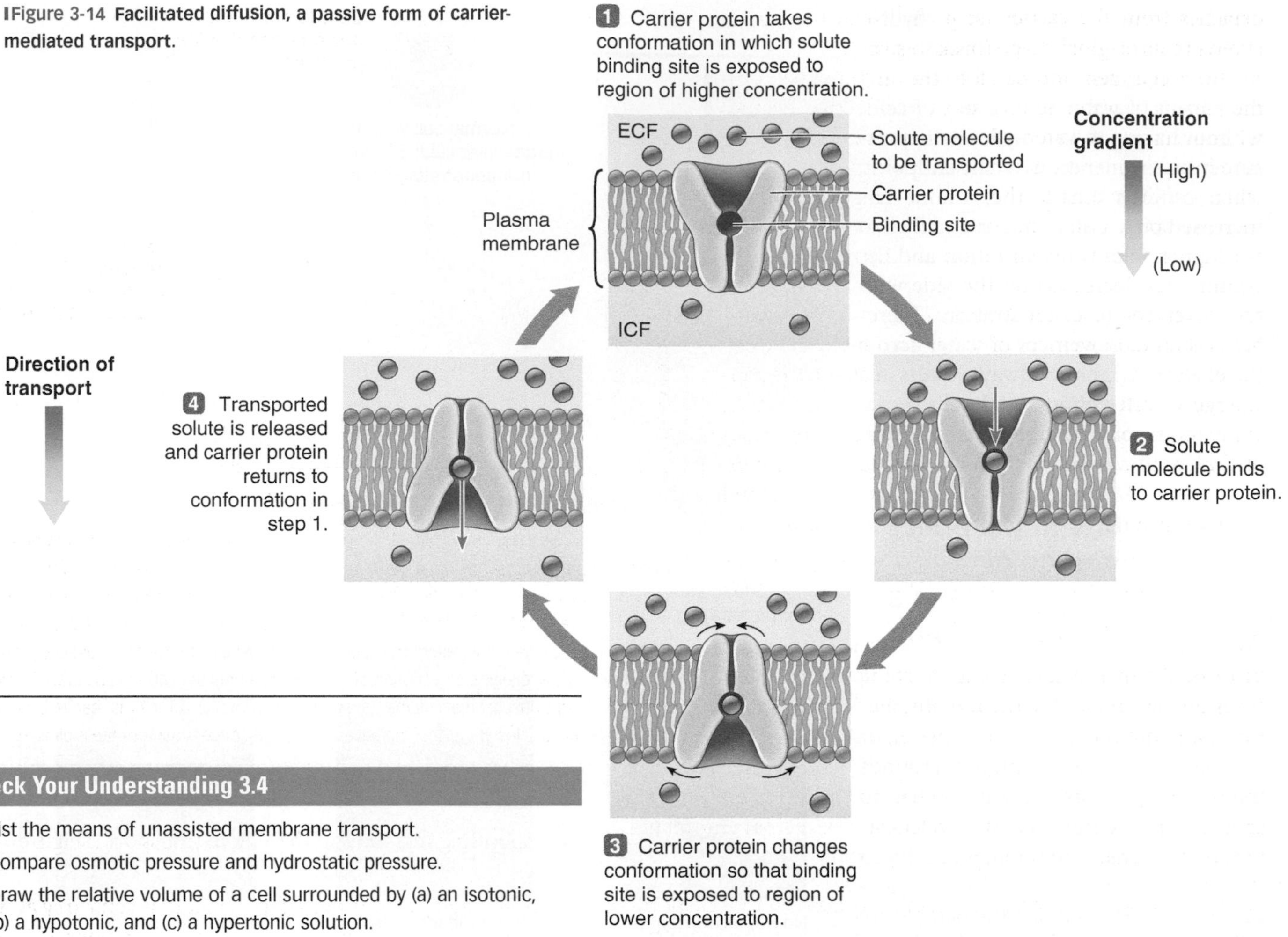

Check Your Understanding 3.4

1. List the means of unassisted membrane transport.
2. Compare osmotic pressure and hydrostatic pressure.
3. Draw the relative volume of a cell surrounded by (a) an isotonic, (b) a hypotonic, and (c) a hypertonic solution.

3.5 Assisted Membrane Transport

All the kinds of transport we have discussed thus far—diffusion down concentration gradients, movement along electrical gradients, and osmosis—produce net movement of particles capable of permeating the plasma membrane because of their lipid solubility (nonpolar molecules of any size) or their ability to fit through channels (selected ions and water). Poorly lipid-soluble polar molecules that are too big for channels, such as proteins, glucose, and amino acids, cannot cross the plasma membrane on their own no matter what forces are acting on them. This impermeability ensures that large, polar intracellular proteins stay in the cell where they belong and can carry out their life-sustaining functions—for example, serving as metabolic enzymes.

However, because poorly lipid-soluble molecules cannot cross the plasma membrane on their own, the cell must provide mechanisms for deliberately transporting these types of molecules into or out of the cell as needed. For example, the cell must usher in essential nutrients, such as glucose for energy and amino acids for synthesis of proteins, and transport out metabolic wastes and secretory products, such as water-soluble pro-

tein hormones. Furthermore, passive diffusion alone cannot always account for the movement of ions. Some ions move through the membrane passively in one direction and actively in the other direction. Cells use two different mechanisms to accomplish these selective transport processes: *carrier-mediated transport* for transfer of small to moderate-sized water-soluble substances across the membrane and *vesicular transport* for movement of large water-soluble molecules and multimolecular particles between the ECF and the ICF. We examine each of these methods of assisted membrane transport in turn.

Carrier-mediated transport is accomplished by a membrane carrier changing its shape.

A carrier protein spans the thickness of the plasma membrane and can change its conformation (shape) so that specific binding sites within the carrier are alternately exposed to the ECF and the ICF. Figure 3-14 shows how this **carrier-mediated transport** works. Step **1** shows the carrier open to the ECF. The molecule to be transported attaches to a carrier's binding site on one side of the membrane—in this case, on the ECF side (step **2**). Then the carrier changes shape, exposing the same site to the other side of the membrane (step **3**). Having been moved in this way from one side of the membrane to the other, the bound molecule

70 CHAPTER 3

Unless otherwise noted, all content on this page is © Cengage Learning.

detaches from the carrier (step **4**). Next, the carrier reverts to its original shape (back to step **1**).

Both channels and carriers are proteins that span the plasma membrane and serve as selective avenues for movement of water-soluble substances across the membrane, but there are notable differences between them: (1) Only ions fit through the narrow channels, whereas small polar molecules such as glucose and amino acids are transported across the membrane by carriers. (2) Channels can be open or closed, but carriers are always "open for business" (although the number and kinds of carriers in the plasma membrane can be regulated). (3) Movement through channels is considerably faster than carrier-mediated transport is. When open for traffic, channels are open at both sides of the membrane at the same time, permitting continuous, rapid movement of ions between the ECF and the ICF through these nonstop passageways. By contrast, carriers are never open to both the ECF and the ICF simultaneously. They must change shape to alternately pick up passenger molecules on one side and drop them off on the other side, a time-consuming process. Whereas a carrier may move up to 5000 particles per second across the membrane, 5 million ions may pass through an open channel in 1 second.

Carrier-mediated transport systems display three important characteristics that determine the kind and amount of material that can be transferred across the membrane: *specificity, saturation,* and *competition*.

1. **Specificity.** Each carrier protein is specialized to transport a specific substance or, at most, a few closely related chemical compounds. For example, amino acids cannot bind to glucose carriers, although several similar amino acids may be able to use the same carrier. Cells vary in the types of carriers they have, thus permitting transport selectivity among cells.

A number of inherited diseases involve defects in transport

Clinical Note systems for a particular substance. *Cysteinuria* (cysteine in the urine) is such a disease involving defective cysteine carriers in the kidney membranes. This transport system normally removes cysteine from the fluid destined to become urine and returns this essential amino acid to the blood. When this carrier malfunctions, large quantities of cysteine remain in the urine, where it is relatively insoluble and tends to precipitate. This is one cause of urinary stones.

2. **Saturation.** A limited number of carrier binding sites are available within a particular plasma membrane for a specific substance. Therefore, the amount of a substance carriers can transport across the membrane in a given time is limited. This limit is known as the **transport maximum (T_m)**. Until the T_m is reached, the number of carrier binding sites occupied by a substance and, accordingly, the substance's rate of transport across the membrane is directly related to its concentration. The more of a substance available for transport, the more transported. When the T_m is reached, the carriers are saturated (all binding sites are occupied) and the rate of the substance's transport across the membrane is maximal. Further increases in the substance's concentration are no longer accompanied by

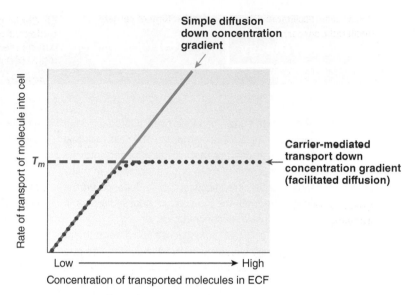

Simple diffusion down concentration gradient

Carrier-mediated transport down concentration gradient (facilitated diffusion)

Low → High
Concentration of transported molecules in ECF

❙Figure 3-15 Comparison of carrier-mediated transport and simple diffusion down a concentration gradient. With simple diffusion of a molecule down its concentration gradient, the rate of transport of the molecule into the cell is directly proportional to the extracellular concentration of the molecule. With carrier-mediated transport of a molecule down its concentration gradient, the rate of transport of the molecule into the cell is directly proportional to the extracellular concentration of the molecule until the carrier is saturated, at which time the rate of transport reaches the transport maximum (T_m). After T_m is reached, the rate of transport levels off despite further increases in the ECF concentration of the molecule.

corresponding increases in the rate of transport (❙Figure 3-15).

As an analogy, consider a ferry boat that can carry at most 100 people across a river during one trip in an hour. If 25 people are on hand to board the ferry, 25 will be transported that hour. Doubling the number of people on hand to 50 will double the rate of transport to 50 people that hour. Such a direct relationship will exist between the number of people waiting to board (the concentration) and the rate of transport until the ferry is fully occupied (its T_m is reached). Even if 150 people are waiting to board, only 100 can be transported per hour.

Saturation of carriers is a critical rate-limiting factor in the transport of selected substances across the kidney membranes during urine formation and across the intestinal membranes during absorption of digested foods. Furthermore, it is sometimes possible to regulate the rate of carrier-mediated transport by varying the affinity (attraction) of the binding site for its passenger or by varying the number of binding sites. For example, the hormone insulin greatly increases the carrier-mediated transport of glucose into most cells of the body by promoting an increase in the number of glucose carriers in the cell's plasma membrane. Deficient insulin action (*diabetes mellitus*) drastically impairs the body's ability to take up and use glucose as the primary energy source.

3. **Competition.** Several closely related compounds may compete for a ride across the membrane on the same carrier. If a given binding site can be occupied by more than one type of molecule, the rate of transport of each substance is less when both molecules are present than when either is present by itself. To illustrate, assume the ferry has 100 seats (binding sites)

CELLS TAKE UP GLUCOSE FROM the blood by facilitated diffusion via glucose carriers in the plasma membrane. The cells maintain an intracellular pool of these carriers that can be inserted into the plasma membrane as the need for glucose uptake increases. In many cells, including resting muscle cells, glucose uptake depends on the hormone insulin, which promotes the insertion of glucose carriers in the plasma membranes of insulin-dependent cells.

During exercise, muscle cells use more glucose and other nutrient fuels than usual to power their increased contractile activity. The rate of glucose transport into exercising muscle may increase more than 10-fold during moderate or intense physical activity. Insulin is not responsible for the increased transport of glucose into exercising muscles, however, because blood insulin levels fall during exercise. Instead muscle cells insert more glucose carriers in their plasma membranes in direct response to exercise.

Exercise influences glucose transport into cells in yet another way. Regular aerobic exercise (see p. 39) has been shown to increase both the affinity (degree of attraction) and number of plasma membrane receptors that bind specifically with insulin. This adaptation results in an increase in insulin sensitivity—that is, the cells are more responsive than normal to a given level of circulating insulin.

Because insulin enhances the facilitated diffusion of glucose into most cells, an exercise-induced increase in insulin sensitivity is one of the factors that makes exercise a beneficial therapy for controlling diabetes mellitus. In this disorder, glucose entry into most cells is impaired as a result of inadequate insulin action (see Chapter 19). Plasma levels of glucose become elevated because glucose remains in the plasma instead of being transported into the cells. In Type 1 diabetes, too little insulin is produced to meet the body's need for glucose uptake. Regular aerobic exercise reduces the amount of insulin that must be injected to promote glucose uptake and lower the blood glucose level toward normal. In Type 2 diabetes, insulin is produced, but insulin's target cells have decreased sensitivity to its presence. By increasing the cells' responsiveness to the insulin available, regular aerobic exercise helps drive glucose into the cells, where it can be used for energy production, instead of remaining in the plasma, where it leads to detrimental consequences for the body.

---

that can be occupied by either men or women. If only men are waiting to board, up to 100 men can be transported during each trip; the same holds true if only women are waiting to board. If both men and women are waiting to board, however, they will compete for the available seats. Fifty of each might make the trip, although the total number of people transported will still be the same, 100 people. In other words, when a carrier can transport two closely related substances, such as the amino acids glycine and alanine, the presence of both diminishes the rate of transfer for either.

Facilitated diffusion is passive carrier-mediated transport.

Carrier-mediated transport takes two forms, depending on whether energy must be supplied to complete the process: facilitated diffusion (not requiring energy) and active transport (requiring energy). **Facilitated diffusion** uses a carrier to facilitate (assist) the transfer of a particular substance across the membrane "downhill" from high to low concentration. This process is passive and does not require energy because movement occurs naturally down a concentration gradient. **Active transport**, however, requires the carrier to expend energy to transfer its passenger "uphill" against a concentration gradient, from an area of lower concentration to an area of higher concentration. An analogous situation is a car on a hill. To move the car downhill requires no energy; it will coast from the top down. Driving the car uphill, however, requires the use of energy (generated by the burning of gasoline).

The most notable example of facilitated diffusion is the transport of glucose into cells. Glucose is in higher concentration in the blood than in the tissues. Fresh supplies of this nutrient are regularly added to the blood by eating and by using reserve energy stores in the body. Simultaneously, the cells metabolize glucose almost as rapidly as it enters from the blood. As a result, a continuous gradient exists for net diffusion of glucose into the cells. However, glucose cannot cross plasma membranes on its own because it is not lipid soluble and is too large to fit through a channel. Without glucose carrier molecules (called **glucose transporters**, or **GLUTs**; see p. 691) to facilitate membrane transport of glucose, cells would be deprived of their preferred source of fuel. (The accompanying boxed feature, ❙ A Closer Look at Exercise Physiology, describes the effect of exercise on glucose carriers in skeletal muscle cells.)

The binding sites on facilitated diffusion carriers can bind with their passenger molecules when exposed on either side of the membrane. As a result of thermal energy, these carriers undergo spontaneous changes in shape, alternately exposing their binding sites to the ECF or the ICF. After picking up the passenger on one side, when the carrier changes its conformation, it drops off the passenger on the opposite side of the membrane. Because passengers are more likely to bind with the carrier on the high-concentration side than on the low-concentration side, the net movement always proceeds down the concentration gradient from higher to lower concentration (see ❙ Figure 3-14). As is characteristic of all types of mediated transport, the rate of facilitated diffusion is limited by saturation of the carrier binding sites—unlike the rate of simple diffusion, which is always directly proportional to the concentration gradient (see Figure 3-15).

Active transport is carrier mediated and uses energy.

Active transport also uses a carrier protein to transfer a specific substance across the membrane, but in this case the carrier transports the substance uphill *against* its concentration gradient. Active transport comes in two forms. In **primary active transport**, energy is *directly* required to move a substance against its concentration gradient; the carrier splits ATP to power the transport process. In **secondary active transport**, energy is required in the entire process, but it is *not directly* used to produce uphill movement. That is, the carrier does not split ATP; instead, it moves a molecule uphill by using "secondhand" energy stored in the form of an **ion concentration gradient** (most commonly a Na^+ gradient). This ion gradient is built up by primary active transport of the ion by a different carrier.

Primary Active Transport In primary active transport, energy in the form of ATP is required for the carrier to change shape and alternately expose the carrier's binding sites for passengers (always ions) to opposite sides of the membrane, with the affinity of the binding sites for their passenger ions being different when the sites are open to the ICF side than when open to the ECF side. The binding sites have a greater affinity for the passenger ion on the low-concentration side where the ion is picked up and a lower affinity on the high-concentration side where the ion is dropped off. In this way, transported ions are moved uphill from an area of low concentration to an area of higher concentration. These active-transport mechanisms are often called "**pumps**," analogous to water pumps that require energy to lift water against the downward pull of gravity. In contrast, in facilitated diffusion, the affinity of the binding site is the same when exposed to either the outside or the inside of the cell and the transported ions are moved downhill from high to low concentration.

All primary-active transport carriers act as enzymes that have ATPase activity, which means they split the terminal phosphate from an ATP molecule to yield adenosine diphosphate (ADP) and inorganic phosphate (P_i) plus free energy (see p. 34). (Do not confuse *ATPase*, which splits ATP, with *ATP synthase*, which synthesizes ATP.) Let us see how ATP is used by primary-active transport carriers to move transported ions against their concentration gradients by considering one of the most important pumps, the **Na^+–K^+ ATPase pump** (**Na^+–K^+ pump** for short) found in the plasma membrane of all cells. This carrier transports Na^+ out of the cell, concentrating it in the ECF, and picks up K^+ from the outside, concentrating it in the ICF. The Na^+–K^+ pump has three binding sites for Na^+ and two binding sites for K^+. When exposed to the cell interior, the pump has high affinity for Na^+ and low affinity for K^+ (Figure 3-16, step **1**). Attachment of three Na^+ to the carrier's high-affinity Na^+ binding sites on the ICF side (where Na^+ is in low concentration) activates the pump's ATPase activity, triggering the splitting of ATP and subsequent *phosphorylation* of the carrier (that is, binding of the inorganic phosphate resulting from ATP splitting) on the intracellular side

(step **2**). This phosphorylation causes the carrier to change shape, which exposes the bound Na^+ to the exterior. The change in carrier shape sharply reduces the affinity of the binding sites for Na^+, so this ion is released to its high concentration side in the ECF (step **3**). Simultaneously, the change in shape increases the carrier's affinity for K^+ on the ECF side (where this ion is in low concentration) (step **4**). Binding of two K^+ leads to *dephosphorylation* of the carrier (that is, the inorganic phosphate detaches from the carrier), inducing a second change in carrier shape, reverting the carrier back to its original conformation, and exposing the bound K^+ to the cell's interior (step **5**). On this side, the affinity of the K^+ binding sites markedly decreases, so K^+ is released to its high concentration side in the ICF (step **6**). Simultaneously, the affinity of the Na^+ binding sites greatly increases as they are exposed to the cell interior once again, so the pump is ready to repeat the cycle (back to step **1**). Thus, the Na^+–K^+ pump moves three Na^+ out of the cell for every two K^+ it pumps in, with both ions moving against their concentration gradients at the expense of energy (ATP splitting). (To appreciate the magnitude of active Na^+–K^+ pumping, consider that a single nerve cell membrane contains roughly 1 million Na^+–K^+ pumps capable of transporting about 200 million ions per second.)

The Na^+–K^+ pump plays three important roles:

1. It establishes Na^+ and K^+ concentration gradients across the plasma membrane of all cells; these gradients are critically important in the ability of nerve and muscle cells to generate electrical signals essential to their functioning (a topic discussed more thoroughly later).

2. It helps regulate cell volume by controlling the concentrations of solutes inside the cell and thus minimizing osmotic effects that would induce swelling or shrinking of the cell.

3. The energy used to run the Na^+–K^+ pump also indirectly serves as the energy source for secondary active transport.

The Na^+–K^+ pump is not the only primary active-transport carrier. Primary active-transport pumps all move positively charged ions—namely, Na^+, K^+, hydrogen ion (H^+), or calcium ion (Ca^{2+})—across the membrane. The simplest primary active-transport systems pump a single type of passenger. For example, the **Ca^{2+} pump** in the plasma membrane transports Ca^{2+} out of the cell, keeping the Ca^{2+} concentration in the cytosol low. These Ca^{2+} transporters are particularly abundant in the plasma membrane of neuron (nerve cell) terminals that store chemical messengers (neurotransmitters) in secretory vesicles (see p. 103). An electrical signal in a neuron terminal triggers the opening of Ca^{2+} channels in the terminal's plasma membrane. Entry of Ca^{2+} down its concentration gradient through these open channels promotes the secretion of neurotransmitter by exocytosis of the secretory vesicles. By keeping the intracellular Ca^{2+} concentration low, the active Ca^{2+} pump helps maintain a large concentration gradient for the entry of secretion-inducing Ca^{2+} from the ECF into the neuron terminal.

More complicated primary active-transport mechanisms involve the transfer of two different passengers in opposite directions, the most important example being the Na^+–K^+ pump. Let us now turn attention to how this primary pump indirectly serves as the energy source for secondary active transport.

Secondary Active Transport With secondary active transport, the carrier does not directly split ATP to move a substance against its concentration gradient. Instead, the movement of Na^+ into the cell down its concentration gradient (established by the ATP-splitting Na^+–K^+ pump) drives the uphill transport of another solute by a secondary active-transport carrier. This is efficient because Na^+ must be pumped out anyway to maintain the electrical and osmotic integrity of the cell.

In secondary active transport, the transfer of the solute across the membrane is always coupled (occurs together) with the transfer of the ion that supplies the driving force. We use Na^+ as the main example. Secondary active transport carriers have two binding sites: one for the solute being moved and one for Na^+. Secondary active transport occurs by two mechanisms—symport and antiport—depending on the direction the transported solute moves in relation to Na^+ movement. In **symport** (also called **cotransport**), the solute and Na^+ move through the membrane in the same direction—that is, into the cell (*sym* means "together"; *co* means "with"). Glucose and amino acids are examples of molecules transported by symport in intestinal and kidney cells. We discuss the importance of these carriers in more detail shortly. In **antiport** (also known as **countertransport** or **exchange**), the solute and Na^+ move through the membrane in opposite directions—that is, Na^+ into and the solute out of the cell (*anti* means "opposite"; *counter* means "against") (❚ Figure 3-17). For example, cells exchange Na^+ and H^+ by means of anti-

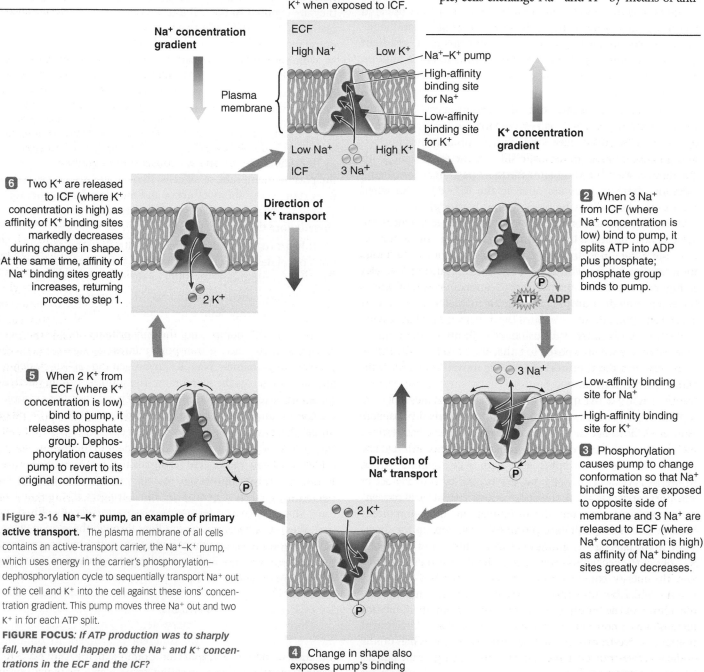

① Pump has 3 high-affinity sites for Na^+ and 2 low-affinity sites for K^+ when exposed to ICF.

Na^+ concentration gradient

ECF
High Na^+ Low K^+ Na^+–K^+ pump
                                        High-affinity binding site for Na^+
Plasma membrane   Low-affinity binding site for K^+

Low Na^+ High K^+
ICF   3 Na^+ **K^+ concentration gradient**

⑥ Two K^+ are released to ICF (where K^+ concentration is high) as affinity of K^+ binding sites markedly decreases during change in shape. At the same time, affinity of Na^+ binding sites greatly increases, returning process to step 1.

Direction of K^+ transport

2 K^+

② When 3 Na^+ from ICF (where Na^+ concentration is low) bind to pump, it splits ATP into ADP plus phosphate; phosphate group binds to pump.

ATP ADP

⑤ When 2 K^+ from ECF (where K^+ concentration is low) bind to pump, it releases phosphate group. Dephosphorylation causes pump to revert to its original conformation.

Direction of Na^+ transport

3 Na^+
Low-affinity binding site for Na^+
High-affinity binding site for K^+

③ Phosphorylation causes pump to change conformation so that Na^+ binding sites are exposed to opposite side of membrane and 3 Na^+ are released to ECF (where Na^+ concentration is high) as affinity of Na^+ binding sites greatly decreases.

❚ **Figure 3-16 Na^+–K^+ pump, an example of primary active transport.** The plasma membrane of all cells contains an active-transport carrier, the Na^+–K^+ pump, which uses energy in the carrier's phosphorylation–dephosphorylation cycle to sequentially transport Na^+ out of the cell and K^+ into the cell against these ions' concentration gradient. This pump moves three Na^+ out and two K^+ in for each ATP split.

FIGURE FOCUS: *If ATP production was to sharply fall, what would happen to the Na^+ and K^+ concentrations in the ECF and the ICF?*

2 K^+

④ Change in shape also exposes pump's binding sites for K^+ to ECF and greatly increases affinity of K^+ sites.

port. This carrier plays an important role in maintaining the appropriate pH inside the cells (a fluid becomes more acidic as its H^+ concentration rises).

Let us examine Na^+ and glucose symport in more detail as an example of secondary active transport. Unlike most body cells, the intestinal and kidney cells actively transport glucose by moving it uphill from low to high concentration. The intestinal cells transport this nutrient from the intestinal lumen into the blood, concentrating it there, until none is left in the lumen to be lost in the feces. The kidney cells save this nutrient for the body by transporting it out of the fluid that is to become urine, moving it against a concentration gradient into the blood. The symport carriers that transport glucose against its concentration gradient from the lumen in the intestine and kidneys are distinct from the glucose facilitated-diffusion carriers that transport glucose down its concentration gradient into most cells.

Here, we focus specifically on the symport carrier that cotransports Na^+ and glucose in intestinal epithelial cells. This carrier, known as the **sodium and glucose cotransporter** or **SGLT**, is located in the luminal membrane (the membrane facing the intestinal lumen) (❙Figure 3-18). The Na^+–K^+ pump in these cells is located in the basolateral membrane (the membrane on the side of the cell opposite the lumen and along the lateral edge of the cell below the tight junction; see Figure 3-5, p. 62). More Na^+ is present in the lumen than inside the cells because the energy-requiring Na^+–K^+ pump transports Na^+ out of the cell at the basolateral membrane, keeping the intracellular Na^+ concentration low (❙Figure 3-18, step **1**). Because of this Na^+ concentration difference, more Na^+ binds to the SGLT when it is exposed to the lumen than when it is exposed to the ICF. Binding of Na^+ to this carrier increases the carrier's affinity for glucose, so glucose binds to the SGLT when it is open to the lumen side where glucose concentration is low (step **2a**). When both Na^+ and glucose are bound to it, the SGLT changes shape and opens to the inside of the cell (step **2b**). Both Na^+ and glucose are released to the interior—Na^+ because of the lower intracellular Na^+ concentration and glucose because of the reduced affinity of the binding site on release of Na^+ (step **2c**). The movement of Na^+ into the cell by this cotransport carrier is downhill because the intracellular Na^+ concentration is low, but the movement of glucose is uphill because glucose becomes concentrated in the cell.

The released Na^+ is quickly pumped out by the active Na^+–K^+ transport mechanism, keeping the level of intracellular Na^+ low. The energy expended in this process is not used directly to run the SGLT because phosphorylation is not required to alter the affinity of the binding site to glucose. Instead, the establishment of a Na^+ concentration gradient by the Na^+–K^+ pump (a primary active-transport mechanism) drives the SGLT (a secondary active-transport mechanism) to move glucose against its concentration gradient.

The glucose carried across the luminal membrane into the cell by secondary active transport then moves passively out of

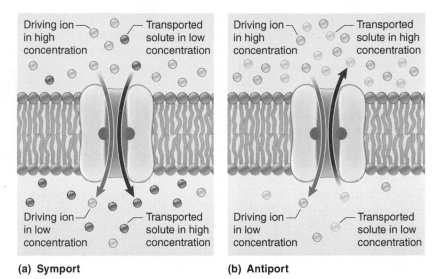

(a) **Symport**   (b) **Antiport**

❙Figure 3-17 **Secondary active transport.** With secondary active transport, an ion concentration gradient (established by primary active transport) is used as the energy source to transport a solute against its concentration gradient. (Usually the driving ion is Na^+, whose concentration gradient is established by the Na^+–K^+ pump.) Note that for convenience in using arrows to depict the direction in which the carrier moves the transported solute and driving ion, the carrier is shown as being open to both sides of the membrane at the same time, which is never the case in reality. (a) In symport, the transported solute moves in the same direction as the gradient of the driving ion. (b) In antiport, the transported solute moves in the direction opposite from the gradient of the driving ion.

the cell by facilitated diffusion across the basolateral membrane and into the blood (step **3**). This facilitated diffusion, which moves glucose down its concentration gradient, is mediated by a passive GLUT identical to the one that transports glucose into other cells, but in intestinal and kidney cells it transports glucose out of the cell. The difference depends on the direction of the glucose concentration gradient. In the case of intestinal and kidney cells, the glucose concentration is higher inside the cells. Note that in this sequence of events, secondary active transport refers only to the cotransport of glucose uphill across the luminal membrane driven by the Na^+ concentration gradient—that is, the transport accomplished by SGLT.

Before leaving the topic of carrier-mediated transport, think about all the activities that rely on carrier assistance. All cells depend on carriers for the uptake of glucose and amino acids, which serve as the major energy source and the structural building blocks, respectively. Na^+–K^+ pumps are essential for generating cellular electrical activity and for ensuring that cells have an appropriate intracellular concentration of osmotically active solutes. Both primary active transport and secondary active transport are used extensively to accomplish the specialized functions of the nervous and digestive systems and those of the kidneys and all types of muscle.

With vesicular transport, material is moved into or out of the cell wrapped in membrane.

The special carrier-mediated transport systems embedded in the plasma membrane selectively transport ions and small polar molecules. But how do large polar molecules, such as the protein hormones secreted by endocrine cells, or even multimolecular

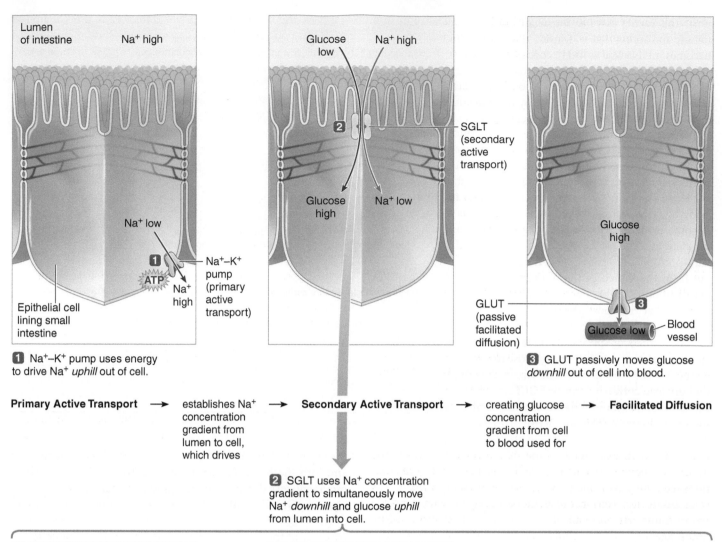

1 Na⁺–K⁺ pump uses energy to drive Na⁺ *uphill* out of cell.

3 GLUT passively moves glucose *downhill* out of cell into blood.

Primary Active Transport → establishes Na⁺ concentration gradient from lumen to cell, which drives → **Secondary Active Transport** → creating glucose concentration gradient from cell to blood used for → **Facilitated Diffusion**

2 SGLT uses Na⁺ concentration gradient to simultaneously move Na⁺ *downhill* and glucose *uphill* from lumen into cell.

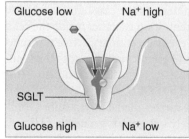

2a Binding of Na⁺ on luminal side, where Na⁺ concentration is higher, increases affinity of SGLT for glucose. Therefore, glucose also binds to SGLT on luminal side, where glucose concentration is lower.

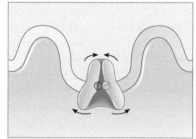

2b When both Na⁺ and glucose are bound, SGLT changes shape, opening to cell interior.

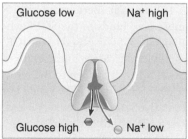

2c SGLT releases Na⁺ to cell interior, where Na⁺ concentration is lower. Because affinity of SGLT for glucose decreases on release of Na⁺, SGLT also releases glucose to cell interior, where glucose concentration is higher.

I Figure 3-18 Symport of glucose. Glucose is transported across intestinal and kidney cells against its concentration gradient by means of secondary active transport mediated by the sodium and glucose cotransporter (SGLT) at the cells' luminal membrane.

FIGURE FOCUS: *By what chain of events does SGLT lead to water absorption from the digestive tract lumen into the blood to promote rehydration when a child with dehydrating diarrhea sips a salt and glucose solution such as Pedialyte or a homemade version?*

materials, such as the bacteria ingested by white blood cells, leave or enter the cell? These materials are unable to cross the plasma membrane, even with assistance: They are much too big for channels, and no carriers exist for them (they would not even fit into a carrier molecule). These large particles are transferred between the ICF and the ECF not by crossing the membrane but by being wrapped in a membrane-enclosed vesicle, a process known as **vesicular transport**. Vesicular transport requires energy expenditure by the cell, so this is an active method of membrane transport. Energy is needed to accomplish vesicle formation and vesicle movement within the cell. Transport into the cell in this manner is termed *endocytosis,* whereas transport out of the cell is called *exocytosis* (see ❙Figure 2-6, p. 29).

Endocytosis To review, in **endocytosis** the plasma membrane surrounds the substance to be ingested and then fuses over the surface, pinching off a membrane-enclosed vesicle so that the engulfed material is trapped within the cell. Recall that there are three forms of endocytosis, depending on the material internalized: *pinocytosis* (nonselective uptake of a sample of ECF), *receptor-mediated endocytosis* (selective uptake of a large molecule), and *phagocytosis* (selective uptake of a multimolecular particle) (see ❙Figure 2-9, p. 32). Once inside the cell, an engulfed vesicle has two possible destinies:

1. In most instances, lysosomes fuse with the vesicle, degrading and releasing its contents into the intracellular fluid.
2. In some cells, the endocytic vesicles bypass the lysosomes where they are normally degraded and instead travel to the opposite side of the cell, where they release their contents by exocytosis. This provides a pathway, termed **transcytosis,** to shuttle large intact molecules through the cell. Such vesicular traffic is one means by which materials are transferred through the thin cells lining the capillaries, the smallest of blood vessels, across which exchanges are made between the blood and the surrounding tissues.

Exocytosis In exocytosis, almost the reverse of endocytosis occurs. A membrane-enclosed vesicle formed within the cell fuses with the plasma membrane, then opens up and releases its contents to the exterior. Materials packaged for export by the endoplasmic reticulum and Golgi complex are externalized by exocytosis.

Exocytosis serves two different purposes:

1. It provides a mechanism for secreting large polar molecules, such as protein hormones and enzymes that are unable to cross the plasma membrane. In this case, the vesicular contents are highly specific and are released only on receipt of appropriate signals.
2. It enables the cell to add specific components to the membrane, such as selected carriers, channels, or receptors, depending on the cell's needs. In such cases, the composition of the membrane surrounding the vesicle is important and the contents may be merely a sampling of ICF.

Balance Between Endocytosis and Exocytosis The rates of endocytosis and exocytosis must be kept in balance to maintain a constant membrane surface area. In a cell actively involved in endocytosis, more than 100% of the plasma membrane may be used in an hour to wrap internalized vesicles, necessitating rapid replacement of surface membrane by exocytosis. In contrast, when a secretory cell is stimulated to secrete, it may temporarily insert up to 30 times its surface membrane through exocytosis. This added membrane must be specifically retrieved by an equivalent level of endocytic activity. Thus, through exocytosis and endocytosis, portions of the membrane are constantly being restored, retrieved, and generally recycled.

Our discussion of membrane transport is now complete; ❙Table 3-2 summarizes the pathways by which materials can pass between the ECF and the ICF. Cells are differentially selective in what enters or leaves because they have varying numbers and kinds of channels, carriers, and mechanisms for vesicular transport. Large polar molecules (too large for channels and not lipid soluble) that have no special transport mechanisms are unable to permeate.

The selective transport of K^+ and Na^+ is responsible for the electrical properties of cells. We turn to this topic next.

Check Your Understanding 3.5

1. Draw a graph comparing simple diffusion down a concentration gradient and carrier-mediated transport.
2. Describe what causes the carrier to change shape to expose binding sites for passengers to opposite sides of the membrane in facilitated diffusion, primary active transport, and secondary active transport.
3. Distinguish between symport and antiport.

3.6 Membrane Potential

The plasma membranes of all living cells have a membrane potential, or are polarized electrically.

Membrane potential is a separation of opposite charges across the plasma membrane.

The term **membrane potential** refers to a separation of opposite charges across the membrane or to a difference in the relative number of cations and anions in the ICF and ECF. Recall that opposite charges tend to attract each other and like charges tend to repel each other. Work must be performed (energy expended) to separate opposite charges after they have come together. Conversely, when oppositely charged particles have been separated, the electrical force of attraction between them can be harnessed to perform work when the charges are permitted to come together again. This is the basic principle underlying electrically powered devices. A separation of charges across the membrane is called a *membrane potential* because separated charges have the potential to do work. Potential is measured in volts (the same unit used for the voltage in electrical devices), but because the membrane potential is relatively low, the unit used is the **millivolt** (**mV**; 1/1000 of a volt).

Because the concept of potential is fundamental to understanding much of physiology, especially nerve and muscle physiology, it is important to understand clearly what this term means. The membrane in ❙Figure 3-19a is electrically neutral; with an equal number of positive (+) and negative (−) charges on each side of the membrane, no membrane potential exists. In ❙Figure 3-19b, some of the positive charges from the right side have been moved to the left. Now the left side has an excess of positive charges, leaving an excess of negative charges on the right. In other words, opposite charges are separated across the

Method of Transport	Substances Involved	Energy Requirements and Force Producing Movement	Limit to Transport
Simple Diffusion			
Diffusion through lipid bilayer	Nonpolar molecules of any size (e.g., O_2, CO_2, fatty acids)	Passive; molecules move down concentration gradient (from high to low concentration)	Continues until gradient is abolished (dynamic equilibrium with no net diffusion)
Diffusion through protein channel	Specific small ions (e.g., Na^+, K^+, Ca^{2+}, Cl^-)	Passive; ions move down electrochemical gradient through open channels (from high to low concentration and by attraction of ion to area of opposite charge)	Continues until there is no net movement and dynamic equilibrium is established
Osmosis	Water only	Passive; water moves down its own concentration gradient (to area of lower water concentration—that is, higher solute concentration)	Continues until concentration difference is abolished or until stopped by opposing hydrostatic pressure or until cell is destroyed
Carrier-Mediated Transport			
Facilitated diffusion	Specific polar molecules for which carrier is available (e.g., glucose)	Passive; molecules move down concentration gradient (from high to low concentration)	Displays a transport maximum (T_m); carrier can become saturated
Primary active transport	Specific cations for which carriers are available (e.g., Na^+, K^+, H^+, Ca^{2+})	Active; ions move against concentration gradient (from low to high concentration); requires ATP	Displays a transport maximum; carrier can become saturated
Secondary active transport (symport or antiport)	Specific polar molecules and ions for which coupled transport carriers are available (e.g., glucose, amino acids for symport; some ions for antiport)	Active; substance moves against concentration gradient (from low to high concentration); driven directly by ion gradient (usually Na^+) established by ATP-requiring primary pump. In symport, cotransported molecule and driving ion move in same direction; in antiport, transported solute and driving ion move in opposite directions	Displays a transport maximum; coupled transport carrier can become saturated
Vesicular Transport			
Endocytosis			
Pinocytosis	Small volume of ECF fluid; also important in membrane recycling	Active; plasma membrane dips inward and pinches off at surface, forming internalized vesicle	Control poorly understood
Receptor-mediated endocytosis	Specific large polar molecule (e.g., protein)	Active; plasma membrane dips inward and pinches off at surface, forming internalized vesicle	Necessitates binding to specific receptor on membrane surface
Phagocytosis	Multimolecular particles (e.g., bacteria and cellular debris)	Active; cell extends pseudopods that surround particle, forming internalized vesicle	Necessitates binding to specific receptor on membrane surface
Exocytosis	Secretory products (e.g., hormones and enzymes) as well as large molecules that pass through cell intact; also important in membrane recycling	Active; increase in cytosolic Ca^{2+} induces fusion of secretory vesicle with plasma membrane; vesicle opens up and releases contents to outside	Secretion triggered by specific neural or hormonal stimuli; other controls involved in transcellular traffic and membrane recycling not known

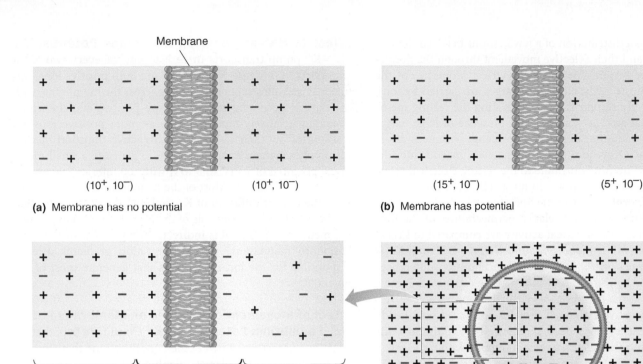

Membrane

$(10^+, 10^-)$ $(10^+, 10^-)$

(a) Membrane has no potential

$(15^+, 10^-)$ $(5^+, 10^-)$

(b) Membrane has potential

Remainder of fluid electrically neutral   **Separated charges**   Remainder of fluid electrically neutral

(c) Separated charges responsible for potential

Plasma membrane

(d) Separated charges forming a layer along plasma membrane

⎮Figure 3-19 Determination of membrane potential by unequal distribution of positive and negative charges across the membrane.
(a) When positive and negative charges are equally balanced on each side of the membrane, no membrane potential exists. (b) When opposite charges are separated across the membrane, membrane potential exists. (c) The unbalanced charges responsible for the potential accumulate in a thin layer along opposite surfaces of the membrane. (d) Most of the fluid in the ECF and ICF is electrically neutral. (e) The greater the separation of charges across the membrane, the larger the potential.

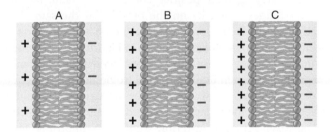

A   B   C

(e) Magnitude of potential: membrane B has more potential than membrane A and less potential than membrane C

membrane, or the relative number of positive and negative charges differs between the two sides. A membrane potential now exists. The attractive force between the separated charges causes them to accumulate in a thin layer along the outer and inner surfaces of the plasma membrane (⎮Figure 3-19c). These separated charges represent only a fraction of the total number of charged particles (ions) present in the ICF and ECF, however, and most fluid inside and outside the cells is electrically neutral (⎮Figure 3-19d). The electrically balanced ions can be ignored because they do not contribute to membrane potential. Thus, an almost insignificant fraction of the total number of charged particles present in the body fluids is responsible for the membrane potential.

Note that the membrane itself is not charged. The term *membrane potential* refers to the difference in charge between the wafer-thin regions of ICF and ECF lying next to the inside and outside of the membrane, respectively. The magnitude of the potential depends on the number of opposite charges separated: The greater the number of charges separated, the larger

the potential. Therefore, in ⎮Figure 3-19e membrane B has more potential than A and less potential than C.

Membrane potential results from differences in the concentration and permeability of key ions.

All cells have membrane potential. The cells of **excitable tissues**—namely, nerve cells and muscle cells—have the ability to produce rapid, transient changes in their membrane potential when excited. These brief fluctuations in potential serve as electrical signals. The constant membrane potential present in the cells of nonexcitable tissues and those of excitable tissues when they are at rest—that is, when they are not producing electrical signals—is known as the **resting membrane potential**. Here, we concentrate on the generation and maintenance of the resting membrane potential; in later chapters, we examine the changes that take place in excitable tissues during electrical signaling.

The unequal distribution of a few key ions between the ICF and the ECF and their selective movement through the plasma membrane are responsible for the electrical properties of the membrane. In the body, electrical charges are carried by ions. The ions directly responsible for generating the resting membrane potential are Na^+ and K^+. The presence of large, negatively charged (anionic) intracellular proteins, written as A^-, is also important. Other ions (calcium, magnesium, and chloride, to name a few) do not contribute to the resting electrical properties of the plasma membrane in most cells, even though they play other important roles in the body.

The concentrations and relative permeabilities of the ions critical to membrane electrical activity are compared in Table 3-3. Note that *Na$^+$ is more concentrated in the ECF and K$^+$ is more concentrated in the ICF.* These concentration differences are maintained by the Na^+–K^+ pump at the expense of energy. Because the plasma membrane is virtually impermeable to A^-, these large, negatively charged proteins are found only inside the cell. After they have been synthesized from amino acids transported into the cell, they remain trapped within the cell.

In addition to the active carrier mechanism, Na^+ and K^+ can passively cross the membrane through protein channels specific for them. It is usually much easier for K^+ than for Na^+ to get through the membrane because the membrane typically has many more leak channels always open for passive K^+ traffic than channels open for passive Na^+ traffic. At resting potential in a nerve cell, the membrane is typically about 25 to 30 times more permeable to K^+ than to Na^+.

Armed with knowledge of the relative concentrations and permeabilities of these ions, we can analyze the forces acting across the plasma membrane. We consider (1) the direct contributions of the Na^+–K^+ pump to membrane potential, (2) the effect that the movement of K^+ alone would have on membrane potential, (3) the effect of Na^+ alone, and (4) the situation that exists in the cells when both K^+ and Na^+ effects are taking place concurrently. Remember throughout this discussion that *the concentration gradient for K$^+$ will always be outward* and *the concentration gradient for Na$^+$ will always be inward* because the Na^+–K^+ pump maintains a higher concentration of K^+ inside the cell and a higher concentration of Na^+ outside the cell. Also, note that because K^+ and Na^+ are both cations (positively charged), *the electrical gradient for both of these ions will always be toward the negatively charged side of the membrane.*

Effect of Na$^+$–K$^+$ Pump on Membrane Potential The Na^+–K^+ pump transports three Na^+ out for every two K^+ it transports in. Because Na^+ and K^+ are both positive ions, this unequal transport separates charges across the membrane, with the outside becoming relatively more positive and the inside becoming relatively more negative as more positive ions are transported out than in. However, this active-transport mechanism only separates enough charges to generate a small membrane potential of 1 mV to 3 mV, with the interior negative to the exterior of the cell. Most of the membrane potential results from the passive diffusion of K^+ and Na^+ down concentration gradients. Thus, the main role of the Na^+–K^+ pump in producing membrane potential is indirect, through its critical contribution to maintaining the concentration gradients directly responsible for the ion movements that generate most of the potential.

Effect of Movement of K$^+$ Alone on Membrane Potential: Equilibrium Potential for K$^+$ Consider a hypothetical situation characterized by (1) the concentrations that exist for K^+ and A^- across the plasma membrane, (2) free permeability of the membrane to K^+ but not to A^-, and (3) no potential as yet present. The concentration gradient for K^+ would tend to move these ions out of the cell (Figure 3-20). Because the membrane is permeable to K^+, these ions would readily pass through, carrying their positive charge with them, so more positive charges would be on the outside. At the same time, negative charges in the form of A^- would be left behind on the inside, similar to the situation shown in Figure 3-19b. (Remember that A^- cannot diffuse out, despite a tremendous concentration gradient.) A membrane potential would now exist. Because an electrical gradient would also be present, K^+ would be attracted toward the negatively charged interior and repelled by the positively charged exterior. Thus, two opposing forces would now be acting on K^+: the concentration gradient tending to move K^+ out of the cell and the electrical gradient tending to move these same ions into the cell.

Initially, the concentration gradient would be stronger than the electrical gradient, so net movement of K^+ out of the cell would continue and the membrane potential would increase. As more and more K^+ moved out of the cell, however, the opposing electrical gradient would become stronger as the outside became increasingly positive and the inside became increasingly negative. One might think that the outward concentration gradient for K^+ would gradually decrease as K^+ leaves the cell down this gradient. It is surprising, however, that the K^+ concentration gradient would remain essentially constant despite the outward movement of K^+. The reason is that even infinitesimal movement of K^+ out of the cell would bring about rather large changes in membrane potential. Accordingly, such an extremely small number of K^+ ions would have to leave the cell to establish an opposing electrical gradient that the K^+ concentration inside and outside the cell would remain essentially unaltered. As K^+ continued to move out down its unchanging concentration gradient, the inward electrical gradient would continue to increase in strength. Net outward movement would gradually be reduced as the strength of the electrical gradient approached that of the concentration gradient.

TABLE 3-3 Concentration and Permeability of Ions Responsible for Membrane Potential in a Resting Nerve Cell

Ion	Extracellular Concentration*	Intracellular Concentration*	Relative Permeability
Na^+	150	15	1
K^+	5	150	25–30
A^-	0	65	0

*Concentration expressed in millimoles per liter, mM

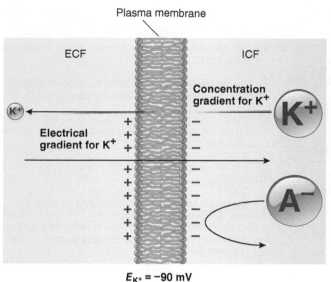

Plasma membrane

ECF   ICF

Concentration gradient for K+

Electrical gradient for K+

1 The concentration gradient for K+ tends to move this ion out of the cell.

2 The outside of the cell becomes more positive as K+ moves to the outside down its concentration gradient.

3 The membrane is impermeable to the large intracellular protein anion (A−). The inside of the cell becomes more negative as K+ moves out, leaving behind A−.

4 The resulting electrical gradient tends to move K+ into the cell.

5 No further net movement of K+ occurs when the inward electrical gradient exactly counterbalances the outward concentration gradient. The membrane potential at this equilibrium point is the equilibrium potential for K+ (E_{K^+}) at −90 mV.

$$E_{K^+} = -90 \text{ mV}$$

❙Figure 3-20 **Equilibrium potential for K+.**

FIGURE FOCUS: *If the ECF concentration of K+ decreases, does E_{K^+} become more negative, less negative, or stay the same?*

Finally, when these two forces exactly balanced each other (that is, when they were in dynamic equilibrium), no further net movement of K+ would occur. The potential that would exist at this equilibrium is known as the **equilibrium potential for K+** (E_{K^+}). At this point, a large concentration gradient for K+ would still exist, but no more net movement of K+ out of the cell would occur down this concentration gradient because of the exactly equal opposing electrical gradient (❙Figure 3-20).

The membrane potential at E_{K^+} is −90 mV. By convention, *the sign always designates the polarity of the excess charge on the inside of the membrane.* A membrane potential of −90 mV means that the potential is of a magnitude of 90 mV, with the inside being negative relative to the outside. A potential of +90 mV would have the same strength, but the inside would be more positive than the outside. For convenience in depicting the charge separation responsible for potential, each pair of separated charges in a figure represents 10 mV of potential. (This is not technically correct because in reality many separated charges must be present to account for a potential of 10 mV.) Thus E_{K^+} in Figure 3-20 is represented by nine separated pairs of charges, with the negative charges lined up along the intracellular side of the membrane.

The equilibrium potential for a given ion with differing concentrations across a membrane can be calculated by means of the **Nernst equation** as follows:

$$E_{\text{ion}} = \frac{61}{z} \log \frac{C_o}{C_i}$$

where

E_{ion} = equilibrium potential for ion in mV

61 = a constant that incorporates the universal gas constant (R), absolute temperature (T), and an electrical constant known as Faraday (F), along with the conversion of the natural logarithm (*ln*) to the logarithm to base 10 (*log*); 61 = RT/F.

z = the ion's valence; $z = 1$ for K+ and Na+, the ions that contribute to membrane potential

C_o = concentration of the ion outside the cell in millimoles/liter (millimolars; mM)

C_i = concentration of the ion inside the cell in mM

Given that the ECF concentration of K+ is 5 mM and the ICF concentration is 150 mM (see ❙Table 3-3),

$$E_{K^+} = 61 \log \frac{5 \text{ mM}}{150 \text{ mM}}$$

$$= 61 \log \frac{1}{30}$$

Because the log of 1/30 = −1.477,

$$E_{K^+} = 61(-1.477) = -90 \text{ mV}$$

Because 61 is a constant, the equilibrium potential is essentially a measure of the membrane potential (that is, the magnitude of the electrical gradient) that exactly counterbalances the concentration gradient for the ion (that is, the ratio between the ion's concentration outside and inside the cell). The larger the concentration gradient is for an ion, the greater the ion's equilibrium potential. A comparably greater opposing electrical gradient would be required to counterbalance the larger concentration gradient.

Effect of Movement of Na+ Alone on Membrane Potential: Equilibrium Potential for Na+ A similar hypothetical situation could be developed for Na+ alone (❙Figure 3-21). The concentration gradient for Na+ would move these ions into the cell, producing a buildup of positive charges on the interior of the membrane and leaving negative charges unbalanced outside (primarily in the form of chloride, Cl−; Na+ and Cl−—that is, salt—are the predominant ECF ions). Net inward movement would continue until equilib-

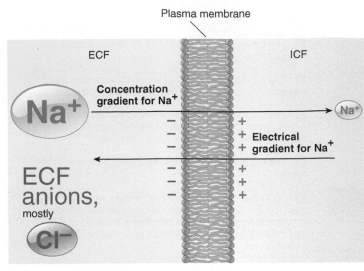

Plasma membrane

ECF | ICF

Concentration gradient for Na⁺

Na⁺ → Na⁺

Electrical gradient for Na⁺

ECF anions, mostly

Cl⁻

E_{Na^+} = +60 mV

1 The concentration gradient for Na⁺ tends to move this ion into the cell.

2 The inside of the cell becomes more positive as Na⁺ moves to the inside down its concentration gradient.

3 The outside becomes more negative as Na⁺ moves in, leaving behind in the ECF unbalanced negatively charged ions, mostly Cl⁻.

4 The resulting electrical gradient tends to move Na⁺ out of the cell.

5 No further net movement of Na⁺ occurs when the outward electrical gradient exactly counterbalances the inward concentration gradient. The membrane potential at this equilibrium point is the equilibrium potential for Na⁺ (E_{Na^+}) at +60 mV.

I Figure 3-21 Equilibrium potential for Na⁺.

rium was established by the development of an opposing electrical gradient that exactly counterbalanced the concentration gradient. At this point, given the concentrations for Na⁺, the **equilibrium potential for Na⁺ (E_{Na^+})** as calculated by the Nernst equation would be +61 mV. Given that the ECF concentration of Na⁺ is 150 mM and the ICF concentration is 15 mM,

$$E_{Na^+} = 61 \log \frac{150 \text{ mM}}{15 \text{ mM}}$$

$$= 61 \log 10$$

Because the log of 10 = 1,

$$E_{Na^+} = 61(1) = 61 \text{ mV}$$

In this case, the inside of the cell would be positive, in contrast to the equilibrium potential for K⁺. The magnitude of E_{Na^+} is somewhat less than that for E_{K^+} (61 mV compared to 90 mV) because the concentration gradient for Na⁺ is not as large (see I Table 3-3); thus, the opposing electrical gradient (membrane potential) is not as great at equilibrium. (For convenience in representing the magnitude of separated charges in figures, we round E_{Na^+} to +60 mV.)

Simultaneous K⁺ and Na⁺ Effects on Membrane Potential Neither K⁺ nor Na⁺ exists alone in the body fluids, so equilibrium potentials are not present in body cells. They exist only in hypothetical or experimental conditions. In a living cell, the effects of both K⁺ and Na⁺ must be taken into account. *The greater the permeability of the plasma membrane for a given ion, the greater the tendency for that ion to drive the membrane potential toward the ion's equilibrium potential.* Because the membrane at rest is 25 to 30 times more permeable to K⁺ than to Na⁺, K⁺ passes through more readily than Na⁺; thus, K⁺ influences the resting membrane potential to a much greater extent than Na⁺ does. Recall that K⁺ acting

alone would establish an equilibrium potential of −90 mV. The membrane is somewhat permeable to Na⁺, however, so some Na⁺ enters the cell in a limited attempt to reach its equilibrium potential. This Na⁺ entry neutralizes, or cancels, some of the potential that would have been produced by K⁺ alone if Na⁺ were not present.

To better understand this concept, examine Figure 3-22, where nine separated pluses and minuses, with the minuses on the inside, represent the E_{K^+} of −90 mV. Superimposing the slight influence of Na⁺ on this K⁺-dominated membrane, assume that two Na⁺ enter the cell down the Na⁺ concentration and electrical gradients. (Note that the electrical gradient for Na⁺ is now inward in contrast to the outward electrical gradient for Na⁺ at E_{Na^+}. At E_{Na^+}, the inside of the cell is positive as a result of the inward movement of Na⁺ down its concentration gradient. In a resting nerve cell, however, the inside is negative because of the dominant influence of K⁺ on membrane potential. Thus, both the concentration and the electrical gradients now favor the inward movement of Na⁺.) The inward movement of these two positively charged Na⁺ neutralizes some of the potential established by K⁺, so now only seven pairs of charges are separated and the potential is −70 mV. This is the resting membrane potential of a typical nerve cell. The resting potential is much closer to E_{K^+} than to E_{Na^+} because of the greater permeability of the membrane to K⁺, but it is slightly less than E_{K^+} (−70 mV is a lower potential than −90 mV) because of the weak influence of Na⁺.

Membrane potential can be measured directly in experimental conditions by recording the voltage difference between the inside and outside of the cell, or it can be calculated using the **Goldman-Hodgkin-Katz equation (GHK equation)**, which takes into account the relative permeabilities and concentration gradients of all permeable ions. The stable, resting membrane is permeable to K⁺, Na⁺, and Cl⁻, but for reasons to be described later Cl⁻ does not directly contribute to potential

1 The Na+–K+ pump actively transports Na+ out of and K+ into the cell, keeping the concentration of Na+ high in the ECF and the concentration of K+ high in the ICF.

2 Given the concentration gradients that exist across the plasma membrane, K+ tends to drive membrane potential to the equilibrium potential for K+ (−90 mV), whereas Na+ tends to drive membrane potential to the equilibrium potential for Na+ (+60 mV).

3 However, K+ exerts the dominant effect on resting membrane potential because the membrane is more permeable to K+. As a result, resting potential (−70 mV) is much closer to E_{K+} than to E_{Na+}.

4 During the establishment of resting potential, the relatively large net diffusion of K+ outward does not produce a potential of −90 mV because the resting membrane is slightly permeable to Na+ and the relatively small net diffusion of Na+ inward neutralizes (in *gray* shading) some of the potential that would be created by K+ alone, bringing resting potential to −70 mV, slightly less than E_{K+}.

5 The negatively charged intracellular proteins (A−) that cannot cross the membrane remain unbalanced inside the cell during the net outward movement of the positively charged ions, so the inside of the cell is more negative than the outside.

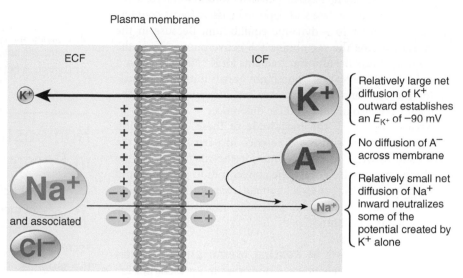

Plasma membrane

ECF · ICF

Relatively large net diffusion of K+ outward establishes an E_{K+} of −90 mV

No diffusion of A− across membrane

Relatively small net diffusion of Na+ inward neutralizes some of the potential created by K+ alone

Na+ and associated Cl−

Resting membrane potential = −70 mV

IFigure 3-22 Effect of concurrent K+ and Na+ movement on establishing the resting membrane potential.

in most cells. Therefore, we can ignore it when calculating membrane potential, making the simplified GHK equation:

$$V_m = 61 \log \frac{P_{K+}[K^+]_o + P_{Na+}[Na^+]_o}{P_{K+}[K^+]_i + P_{Na+}[Na^+]_i}$$

where

$V_m =$ membrane potential in mV

$61 =$ a constant representing RT/zF, when z = 1, as it does for K+ and Na+

$P_{K+}, P_{Na+} =$ permeabilities for K+ and Na+, respectively

$[K^+]_o, [Na^+]_o =$ concentration of K+ and Na+ outside the cell in mM, respectively

$[K^+]_i, [Na^+]_i =$ concentration of K+ and Na+ inside the cell in mM, respectively.

The GHK equation is basically an expanded version of the Nernst equation. The Nernst equation can only be used to calculate the potential generated by a specific ion, but the GHK equation takes into account the combined contributions to potential of all ions moving across the membrane. Assuming the resting membrane is 25 times more permeable to K+ than to Na+, then the relative permeabilities are P_{K+} = 1.0 and P_{Na+} = 0.04 (1/25 of 1.0). Given these permeabilities and the concentrations for K+ and Na+ in the ECF and ICF listed in ITable 3-3,

$$V_m = 61 \log \frac{(1)(5) + (0.04)(150)}{(1)(150) + (0.04)(15)}$$

$$= 61 \log \frac{5 + 6}{150 + 0.6}$$

$$= 61 \log 0.073$$

Because the log of 0.073 is − 1.137,

$$V_m = 61 (−1.137) = −69 \text{ mV}$$

Adding −1 mV of potential generated directly by the Na+–K+ pump to this value totals −70 mV for the resting membrane potential.

Balance of Passive Leaks and Active Pumping at Resting Membrane Potential At resting potential, neither K+ nor Na+ is at equilibrium. A potential of −70 mV does not exactly counterbalance the concentration gradient for K+; it takes a potential of −90 mV to do that. Thus, K+ slowly continues to passively exit through its leak channels down this small concentration gradient. In the case of Na+, the concentration and electrical gradients do not even oppose each other; they both favor the inward movement of Na+. Therefore, Na+ continually leaks inward down its electrochemical gradient, but only slowly, because of its low permeability—that is, because of the scarcity of Na+ leak channels.

Such leaking goes on all the time, so why doesn't the intracellular concentration of K+ continue to fall and the concentration of Na+ inside the cell progressively increase? The reason is that the Na+–K+ pump counterbalances the rate of passive leakage. At resting potential, this pump transports back into the cell essentially the same number of potassium ions that have leaked out and simultaneously transports to the outside the sodium ions that have leaked in. At this point, a **steady state** exists: No net movement of any ions takes place because all passive leaks are exactly balanced by active pumping. No net change takes place in either a steady state or a dynamic equilibrium, but in a *steady state* energy must be used to maintain the constancy, whereas in a *dynamic equilibrium* no energy is needed to maintain the constancy. That is, opposing passive and active forces counterbalance each other in a steady state and opposing passive forces counterbalance each other in a dynamic equilibrium. Because in the steady state across the membrane the active pump offsets the passive leaks, the concentration gradients for K^+ and Na^+ remain constant. Thus, the Na^+–K^+ pump not only is initially responsible for the Na^+ and K^+ concentration differences across the membrane but also maintains these differences.

As just discussed, the magnitude of these concentration gradients, together with the difference in permeability of the membrane to these ions, accounts for the magnitude of the membrane potential. Because the concentration gradients and permeabilities for Na^+ and K^+ remain constant in the resting state, the resting membrane potential established by these forces remains constant.

Chloride Movement at Resting Membrane Potential

Thus far, we have largely ignored one other ion present in high concentration in the ECF: Cl^-. Chloride is the principal ECF anion. Its equilibrium potential is -70 mV, exactly the same as the resting membrane potential. Movement alone of negatively charged Cl^- into the cell down its concentration gradient would produce an opposing electrical gradient, with the inside negative compared to the outside. When physiologists were first examining the ionic effects that could account for the membrane potential, they were tempted to think that Cl^- movements and establishment of the Cl^- equilibrium potential could be solely responsible for producing the identical resting membrane potential. Actually, the reverse is the case. The membrane potential is responsible for driving the distribution of Cl^- across the membrane.

Most cells are highly permeable to Cl^- but have no active-transport mechanisms for this ion. With no active forces acting on it, Cl^- passively distributes itself to achieve an individual state of equilibrium. In this case, Cl^- is driven out of the cell, establishing an inward concentration gradient that exactly counterbalances the outward electrical gradient (that is, the resting membrane potential) produced by K^+ and Na^+ movement. Thus, the concentration difference for Cl^- between the ECF and ICF is brought about passively by the presence of the membrane potential rather than maintained by an active pump, as is the case for K^+ and Na^+. Therefore, in most cells Cl^- does not influence resting membrane potential; instead, membrane potential passively influences the Cl^- distribution.

Specialized Use of Membrane Potential in Nerve and Muscle Cells

Nerve and muscle cells have developed a specialized use for membrane potential. They can rapidly and transiently alter their membrane permeabilities to the involved ions in response to appropriate stimulation, thereby bringing about fluctuations in membrane potential. The rapid fluctuations in potential are responsible for producing nerve impulses in nerve cells and for triggering contraction in muscle cells. These activities are the focus of the next five chapters. Even though all cells display a membrane potential, its significance in most other cells is uncertain; however, changes in membrane potential of some secretory cells—for example, insulin-secreting cells—have been linked to their level of secretory activity.

Homeostasis: Chapter in Perspective

 All body cells must obtain vital materials, such as nutrients and O_2, from the surrounding ECF; they must also eliminate wastes to the ECF and release secretory products, such as chemical messengers and digestive enzymes. Thus, transport of materials across the plasma membrane between the ECF and the ICF is essential for cell survival, and the constituents of the ECF must be homeostatically maintained to support these life-sustaining exchanges.

Many cell types use membrane transport to carry out their specialized activities geared toward maintaining homeostasis. Here are several examples:

1. Absorption of nutrients from the digestive tract lumen involves the transport of these energy-giving molecules across the membranes of the cells lining the tract.

2. Exchange of O_2 and CO_2 between air and blood in the lungs involves the transport of these gases across the membranes of the cells lining the air sacs and blood vessels of the lungs.

3. Urine is formed by the selective transfer of materials between the blood and the fluid within the kidney tubules across the membranes of the cells lining the tubules.

4. The beating of the heart is triggered by cyclic changes in the transport of Na^+, K^+, and Ca^{2+} across the heart cells' membranes.

5. Secretion of chemical messengers such as neurotransmitters from nerve cells and hormones from endocrine cells involves the transport of these regulatory products to the ECF on appropriate stimulation.

In addition to providing selective transport of materials between the ECF and the ICF, the plasma membrane contains receptors for binding with specific chemical messengers that regulate various cell activities, many of which are specialized activities aimed toward maintaining homeostasis. For example, the hormone vasopressin, which is secreted in response to a water deficit in the body, binds with receptors in the plasma membrane of a specific type of kidney cell. This binding triggers the cells to conserve water during urine formation by promoting the insertion of additional aquaporins (water channels) in the plasma membrane of these cells, thus helping alleviate the water deficit that initiated the response.

All living cells have a membrane potential, with the cell's interior being slightly more negative than the fluid surrounding the cell when the cell is electrically at rest. The specialized activities of nerve and muscle cells depend on these cells' ability to change their membrane potential rapidly on appropriate stimulation. The transient, rapid changes in potential in nerve cells serve as electrical signals or nerve impulses, which provide a means to transmit information along nerve pathways. This information is used to accomplish homeostatic adjustments, such as restoring blood pressure to normal when signaled that it has fallen too low.

Rapid changes in membrane potential in muscle cells trigger muscle contraction, the specialized activity of muscle. Muscle contraction contributes to homeostasis in many ways, including the pumping of blood by the heart and the movement of food through the digestive tract.

Review Exercises Answers begin on p. A-22

Reviewing Terms and Facts

1. The nonpolar tails of the phospholipid molecules bury themselves in the interior of the plasma membrane. *(True or false?)*

2. Cells shrink when in contact with a hypertonic solution. *(True or false?)*

3. Channels are open to both sides of the membrane at the same time, but carriers are open to only one side of the membrane at a time. *(True or false?)*

4. At resting membrane potential, there is a slight excess of _____ (positive/negative) charges on the inside of the membrane, with a corresponding slight excess of _____ (positive/negative) charges on the outside.

5. Using the answer code on the right, indicate which membrane component is responsible for the function in question:

   1. channel formation   (a) lipid bilayer
   2. barrier to passage of water-   (b) proteins
      soluble substances   (c) carbohydrates
   3. receptor sites
   4. membrane fluidity
   5. recognition of "self"
   6. membrane-bound enzymes
   7. structural boundary
   8. carriers

6. Using the answer code on the right, indicate the direction of net movement in each case:

   1. simple diffusion   (a) movement from high
   2. facilitated diffusion   to low concentration
   3. primary active transport   (b) movement from low
   4. Na^+ during symport or   to high concentration
      antiport
   5. transported solute during
      symport or antiport
   6. water with regard to the
      water concentration gradient
      during osmosis
   7. water with regard to the
      solute concentration
      gradient during osmosis

7. Using the answer code on the right, indicate the type of cell junction described:

   1. adhering junction   (a) gap junction
   2. impermeable junction   (b) tight junction
   3. communicating junction (c) desmosome
   4. made up of connexons, which
      permit passage of ions and small
      molecules between cells
   5. consisting of interconnecting fibers,
      which spot-rivet adjacent cells
   6. formed by an actual fusion of
      proteins on the outer surfaces of
      two interacting cells
   7. important in tissues subject to
      mechanical stretching
   8. important in synchronizing
      contractions within heart and
      smooth muscle by allowing spread
      of electrical activity between the
      cells composing the muscle mass
   9. important in preventing passage
      between cells in epithelial sheets
      that separate compartments of two
      different chemical compositions

Understanding Concepts
(Answers at www.cengagebrain.com)

1. Describe the fluid mosaic model of membrane structure.

2. Discuss the functions of the three major types of protein fibers in the extracellular matrix.

3. What two properties of a particle influence whether it can permeate the plasma membrane?

4. List and describe the methods of membrane transport. Indicate what types of substances are transported by each method, and state whether each means of transport is passive or active and unassisted or assisted.

5. According to Fick's law of diffusion, what factors influence the rate of net diffusion across a membrane?

6. State three important roles of the Na^+–K^+ pump.

7. Describe the contribution of each of the following to establishing and maintaining membrane potential: (a) the Na^+–K^+ pump, (b) passive movement of K^+ across the membrane, (c) passive movement of Na^+ across the membrane, and (d) the large intracellular anions.

Solving Quantitative Exercises

1. Using the Nernst equation, calculate the equilibrium potential for Ca^{2+} and for Cl^- from the following sets of data:

a. Given $[Ca^{2+}]_o = 1$ mM, $[Ca^{2+}]_i = 100$ nM, find $E_{Ca^{2+}}$

b. Given $[Cl^-]_o = 110$ mM, $[Cl^-]_i = 10$ mM, find E_{Cl^-}

2. One of the important uses of the Nernst equation is in describing the flow of ions across plasma membranes. Ions move under the influence of two forces: the concentration gradient (given in electrical units by the Nernst equation) and the electrical gradient (given by the membrane voltage). This is summarized by *Ohm's law*:

$$I_x = G_x (V_m - E_x)$$

which describes the movement of ion x across the membrane. I is the current in amperes (A); G is the conductance, a measure of the permeability of x, in Siemens (S), which is $\Delta I/\Delta V$; V_m is the membrane voltage; and E_x is the equilibrium potential of ion x. Not only does this equation tell how large the current is, but it also tells what direction the current is flowing. By convention, a negative value of the current represents either a positive ion entering the cell or a negative ion leaving the cell. The opposite is true of a positive value of the current.

a. Using the following information, calculate the magnitude of I_{Na^+}.

$[Na^+]_o = 145$ mM, $[Na^+]_i = 15$ mM, $G_{Na^+} = 1$ nS, $V_m = -70$ mV

b. Is Na^+ entering or leaving the cell?

c. Is Na^+ moving with or against the concentration gradient? Is it moving with or against the electrical gradient?

3. Using the Goldman-Hodgkin-Katz equation, determine what happens to the resting membrane potential if the ECF K^+ concentration doubles to 10 mM.

Applying Clinical Reasoning

When William H. was helping victims after a devastating earthquake in a region not prepared to swiftly set up adequate temporary shelter, he developed severe diarrhea. He was diagnosed as having *cholera*, a disease transmitted through unsanitary water supplies contaminated by fecal material from infected individuals. The toxin produced by cholera bacteria causes Cl^- channels in the luminal membranes of the intestinal cells to stay open, thereby increasing the secretion of Cl^- from the cells into the intestinal tract lumen. By what mechanisms would Na^+ and water be secreted into the lumen in conjunction with Cl^- secretion? How does this secretory re-

sponse account for the severe diarrhea that is characteristic of cholera?

Thinking at a Higher Level

1. Which of the following methods of transport is being used to transfer the substance into the cell in the accompanying graph?

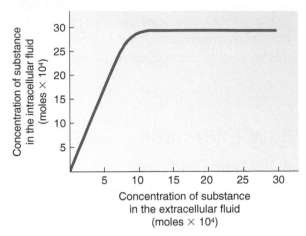

a. diffusion down a concentration gradient

b. osmosis

c. facilitated diffusion

d. active transport

e. vesicular transport

f. It is impossible to tell with the information provided.

2. Assume that a membrane permeable to Na^+ but not to Cl^- separates two solutions. The concentration of sodium chloride on side 1 is higher than on side 2. Which of the following ionic movements would occur?

a. Na^+ would move until its concentration gradient is dissipated (until the concentration of Na^+ on side 2 is the same as the concentration of Na^+ on side 1).

b. Cl^- would move down its concentration gradient from side 1 to side 2.

c. A membrane potential, negative on side 1, would develop.

d. A membrane potential, positive on side 1, would develop.

e. None of the preceding is correct.

3. A solution may have the same osmolarity as normal body fluids yet it may not be isotonic. Explain why.

4. Compared to resting potential, would the membrane potential become more negative or more positive if the membrane were more permeable to Na^+ than to K^+?

5. Colostrum, the first milk that a mother produces, contains an abundance of antibodies, large protein molecules. These maternal antibodies help protect breast-fed infants from infections until the babies are capable of producing their own antibodies. By what means would you suspect these maternal antibodies are transported across the cells lining a newborn's digestive tract into the bloodstream?

Principles of Neural and Hormonal Communication

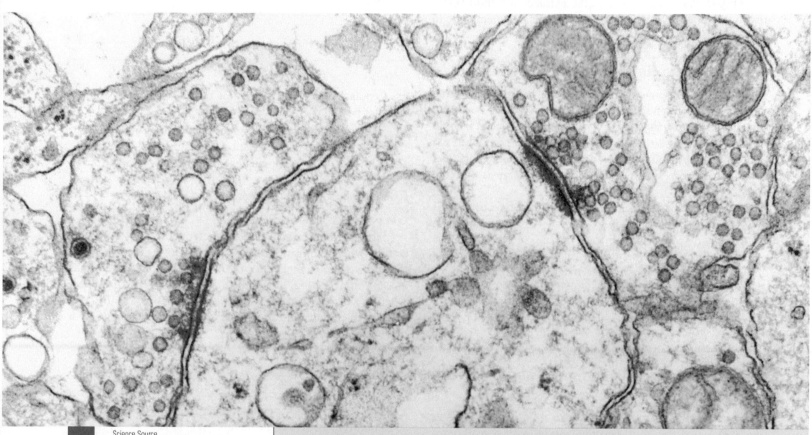

Science Source

A transmission electron micrograph of two adjacent synapses. A synapse is a junction between two neurons separated by a narrow space (*highlighted in purple*). The presynaptic neuron (*light blue, top left or right*) transmits information across the space to the postsynaptic neuron (*darker blue, middle*) by means of a chemical messenger (a neurotransmitter) released from synaptic vesicles (*small orange spheres*).

Homeostasis Highlights

To maintain homeostasis, cells must work in a coordinated fashion toward common goals. The two major regulatory systems of the body that help ensure life-sustaining coordinated responses are the nervous and endocrine systems. **Neural communication** is accomplished by means of nerve cells, or neurons, which are specialized for rapid electrical signaling and for secreting neurotransmitters, short-distance chemical messengers that act on nearby target organs. The nervous system exerts rapid control over most of the body's muscles and exocrine secretions. **Hormonal communication** is accomplished by hormones, which are long-distance chemical messengers secreted by the endocrine glands into the blood. The blood carries the hormones to distant target sites, where they regulate processes that require duration rather than speed, such as metabolic activities, water and electrolyte balance, and growth.

4.1 Introduction to Neural Communication

All body cells display a membrane potential, which is a separation of positive and negative charges across the membrane, as discussed in the preceding chapter (see pp. 77–84). This potential is related to the uneven distribution of sodium (Na^+), potassium (K^+), and large intracellular protein anions between the intracellular fluid (ICF) and the extracellular fluid (ECF) and to the differential permeability of the plasma membrane to these ions.

Nerve and muscle are excitable tissues.

The constant membrane potential present when a cell is electrically at rest—that is, not producing electrical signals—is referred to as the *resting membrane potential*. Two types of cells, *neurons (nerve cells)* and *muscle cells*, have developed a specialized use for membrane potential. They can undergo transient, rapid fluctuations in their membrane potentials, which serve as electrical signals.

Nerve and muscle are considered **excitable tissues** because they produce electrical signals when excited. Neurons use these electrical signals to receive, process, initiate, and transmit messages. In muscle cells, these electrical signals initiate contraction. Thus, electrical signals are critical to the function of the nervous system and all muscles. In this chapter, we examine how neurons undergo changes in potential to accomplish their function. Muscle cells are discussed in later chapters.

Membrane potential becomes less negative during depolarization and more negative during hyperpolarization.

Before you can understand what electrical signals are and how they are created, you must become familiar with several terms used to describe changes in potential, which are graphically represented in ❙Figure 4-1:

1. **Polarization.** Charges are separated across the plasma membrane, so the membrane has potential. Any time membrane potential is other than 0 millivolts (mV), in either the positive or the negative direction, the membrane is in a state of polarization. Recall that the magnitude of the potential is directly proportional to the number of positive and negative charges separated by the membrane and that the sign of the potential (+ or −) always designates whether excess positive or excess negative charges are present, respectively, on the inside of the membrane. At resting potential, the membrane is polarized at −70 mV in a typical neuron (see p. 82).

2. **Depolarization.** The membrane becomes less polarized; the inside becomes less negative than at resting potential, with the potential moving closer to 0 mV (for example, a change from −70 to −60 mV); fewer charges are separated than at resting potential. This term also refers to the inside even becoming positive as it does during an action potential (a major type of electrical signal) when the membrane potential reverses itself (for example, becoming +30 mV).

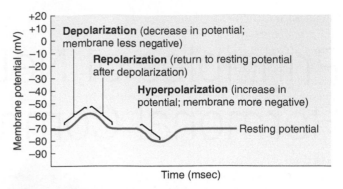

❙Figure 4-1 **Types of changes in membrane potential.**

3. **Repolarization.** The membrane returns to resting potential after having been depolarized.

4. **Hyperpolarization.** The membrane becomes more polarized; the inside becomes more negative than at resting potential, with the potential moving even farther from 0 mV (for instance, a change from −70 to −80 mV); more charges are separated than at resting potential.

One possibly confusing point should be clarified. On the device used for recording rapid changes in potential, during a depolarization, when the inside becomes less negative than at resting, this *decrease* in the magnitude of the potential is represented as an *upward* deflection. By contrast, during a hyperpolarization, when the inside becomes more negative than at resting, this *increase* in the magnitude of the potential is represented by a *downward* deflection.

Electrical signals are produced by changes in ion movement across the plasma membrane.

Changes in membrane potential are brought about by changes in ion movement across the membrane. For example, if the net inward flow of positively charged ions increases compared to the resting state (such as more Na^+ moves in), the membrane depolarizes (becomes less negative inside). By contrast, if the net outward flow of positively charged ions increases compared to the resting state (such as more K^+ moves out), the membrane hyperpolarizes (becomes more negative inside).

Changes in ion movement are brought about by changes in membrane permeability in response to triggering events. A **triggering event** triggers a change in membrane potential by altering membrane permeability and consequently altering ion flow across the membrane. These ion movements redistribute charge, thus changing the potential.

Because the water-soluble ions responsible for carrying charges cannot penetrate the plasma membrane's lipid bilayer, these charges can cross the membrane only through channels specific for them or by carrier-mediated transport. Membrane channels may be either *leak channels* or *gated channels*. As described in Chapter 3, leak channels, which are open all the time, permit unregulated leakage of their specific ion across the membrane through the channels. **Gated channels**, in contrast, have gates that can be open or closed, permitting ion passage through the channels when open and preventing ion passage through the channels when closed. Gate opening and closing occurs in response to a triggering event that causes a change in the conformation (shape) of the protein that forms the gated channel. There are four kinds of gated channels, depending on the factor that causes the channel to change shape: (1) **Voltage-gated channels** open or close in response to changes in membrane potential, (2) **chemically gated channels** change shape in response to binding of a specific extracellular chemical messenger to a surface membrane receptor, (3) **mechanically gated channels** respond to stretching or other mechanical deformation, and (4) **thermally gated channels** respond to local changes in temperature (heat or cold).

There are two basic forms of electrical signals: (1) *graded potentials,* which serve as short-distance signals, and (2) *action potentials,* which signal over long distances. We next examine these types of signals in more detail.

Check Your Understanding 4.1

1. Name the two types of excitable tissue.
2. Draw a graph depicting the changes in potential during depolarization, repolarization, and hyperpolarization as compared to resting membrane potential.
3. State the factor responsible for triggering gate opening and closing in each of the four types of gated channels.

4.2 Graded Potentials

Graded potentials are local changes in membrane potential that occur in varying grades or degrees of magnitude or strength. For example, membrane potential could change from -70 to -60 mV (a 10-mV graded potential) or from -70 to -50 mV (a 20-mV graded potential).

The stronger a triggering event, the larger the resultant graded potential.

Graded potentials are usually produced by a specific triggering event that causes gated ion channels to open in a specialized region of the excitable cell membrane. The resultant ion movement produces the graded potential, which most commonly is a depolarization resulting from net Na^+ entry. The graded potential is confined to this small, specialized region of the total plasma membrane.

The magnitude of the initial graded potential (that is, the difference between the new potential and the resting potential) is related to the magnitude of the triggering event. *The stronger the triggering event, the larger the resultant graded potential.* Here's why, using gated channels that permit net Na^+ entry as a common example: The stronger the triggering event is, the more gated Na^+ channels open. As more gated Na^+ channels open, more positive charges in the form of Na^+ enter the cell. The more positive charges that enter the cell, the less negative (more depolarized) the inside becomes at this specialized region. This depolarization is the graded potential.

Also, the duration of the graded potential varies, depending on how long the triggering event keeps the gated channels open. *The longer the duration of the triggering event, the longer the duration of the graded potential.*

Graded potentials spread by passive current flow.

When a graded potential occurs locally in a nerve or muscle cell membrane, the rest of the membrane remains at resting potential. The temporarily depolarized region is called an *active area*. Note from ▮Figure 4-2b that, inside the cell, the active area is relatively more positive than the neighboring *inactive areas* that are still at resting potential. Outside the cell, the active area is relatively less positive than adjacent inactive areas. Because of this difference in potential, electrical charges, which are carried by ions, passively flow between the active and the adjacent resting regions on both the inside and the outside of the membrane. Any flow of electrical charges is called a **current**. By convention, the direction of current flow is always expressed as the direction in which the positive charges are moving (▮Figure 4-2c). Inside the cell, positive charges flow through the ICF away from the relatively more positive depolarized active region toward the more negative adjacent resting regions. Outside the cell, positive charges flow through the ECF from the more positive adjacent inactive regions toward the relatively more negative active region. Ion movement (that is, current) is occurring *along* the membrane between regions next to each other on the same side of the membrane. This flow is in contrast to ion movement *across* the membrane through ion channels or by means of carriers.

As a result of local current flow between an active area and an adjacent inactive area, the potential changes in the previously inactive area. Because positive charges have flowed simultaneously into the adjacent inactive area on the inside and out of this area on the outside, the adjacent area is now more positive (or less negative) on the inside than before and less positive (or more negative) on the outside (▮Figure 4-2c). Stated differently, the previously inactive adjacent area has been depolarized, so the graded potential has spread. This area's potential now differs from that of the inactive region immediately next to it on the other side, inducing further current flow at this new site, and so on. In this manner, current spreads in both directions away from the initial site of the change in potential.

The amount of current that flows between two areas depends on the difference in potential between the areas and on the resistance of the material through which the charges are moving. **Resistance** is the hindrance to electrical charge movement. The *greater* the difference in potential, the *greater* the current flow; by contrast, the *lower* the resistance, the *greater* the current flow. *Conductors* have low resistance, providing little hindrance to current flow. Electrical wires and the ICF and ECF are all good conductors, so current readily flows through them. *Insulators* have high resistance and greatly hinder movement of charge. The plastic surrounding electrical wires has high resistance, as do body lipids. Thus, current does not flow directly through the plasma membrane's lipid bilayer. Current, carried by ions, can move across the membrane only through ion channels.

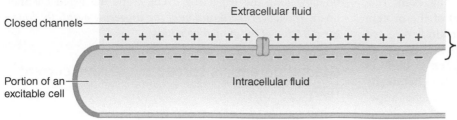

Closed channels

Extracellular fluid

Unbalanced charges distributed across the plasma membrane that are responsible for membrane potential

Portion of an excitable cell

Intracellular fluid

(a) Entire membrane at resting potential

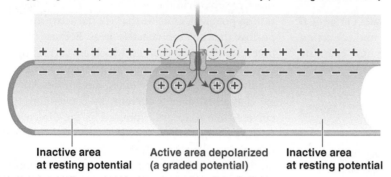

Triggering event opens ion channels, most commonly permitting net Na⁺ entry

Inactive area at resting potential

Active area depolarized (a graded potential)

Inactive area at resting potential

(b) Inward movement of Na⁺ depolarizes membrane, producing a graded potential

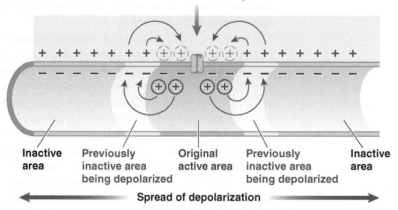

Current flows between the active and adjacent inactive areas

Inactive area

Previously inactive area being depolarized

Original active area

Previously inactive area being depolarized

Inactive area

← Spread of depolarization →

(c) Depolarization spreads by local current flow to adjacent inactive areas, away from point of origin

❚Figure 4-2 **Current flow during a graded potential.** (a) The membrane of an excitable cell at resting potential. (b) A triggering event opens ion channels, usually leading to net Na⁺ entry that depolarizes the membrane at this site. The adjacent inactive areas are still at resting potential. (c) Local current flows between the active and adjacent inactive areas, resulting in depolarization of the previously inactive areas. In this way, the depolarization spreads away from its point of origin.

Graded potentials die out over short distances.

The passive current flow between active and adjacent inactive areas is similar to the means by which current is carried through electrical wires. We know from experience that current leaks out of an electrical wire with dangerous results unless the wire is covered with an insulating material such as plastic. (People can get an electric shock if they touch a bare wire.) Likewise, current is lost across the plasma membrane as

charge-carrying ions in the form of K⁺ leak out through the "uninsulated" parts of the membrane—that is, by diffusing outward down their electrochemical gradient through open K⁺ leak channels. Because of this current loss, the magnitude of the local current—and thus the magnitude of the graded potential—progressively diminishes the farther it moves from the initial active area (❚Figure 4-3a). Another way of saying this is that the spread of a graded potential is *decremental* (gradually decreases) (❚Figure 4-3b). Note that in this example the magnitude of the initial change in potential is 15 mV (a change from the resting state of −70 to −55 mV); the change in potential decreases as it moves along the membrane to a potential of 10 mV (from −70 to −60 mV) and continues to diminish the farther it moves away from the initial active area until there is no longer a change in potential. In this way, local currents die out within micrometers (less than 1 mm) as they move away from the initial site of change in potential and consequently can function as signals for only very short distances.

Although graded potentials have limited signaling distance, they are critically important to the body's function, as explained in later chapters. The following are all graded potentials: *postsynaptic potentials, receptor potentials, end-plate potentials, pacemaker potentials,* and *slow-wave potentials.* These terms are unfamiliar to you now, but you will become well acquainted with them as we continue discussing nerve and muscle physiology. We are including this list here because it is the only place all these types of graded potentials are listed together. For now it's enough to say that most excitable cells produce one of these types of graded potentials in response to a triggering event. In turn, graded potentials can initiate action potentials, the long-distance signals, in an excitable cell.

1. Discuss how the magnitude and duration of a graded potential vary with the magnitude and duration of a triggering event.

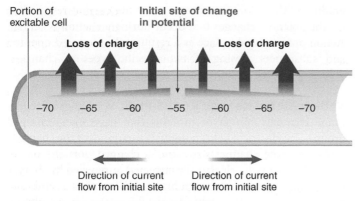

Portion of
excitable cell

Initial site of change
in potential

Loss of charge   **Loss of charge**

−70   −65   −60   −55   −60   −65   −70

Direction of current   Direction of current
flow from initial site   flow from initial site

* Numbers refer to the local potential in mV
at various points along the membrane.

(a) Current loss across the membrane

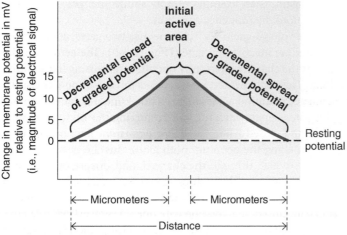

(b) Decremental spread of graded potentials

❙Figure 4-3 **Current loss across the plasma membrane leading to decremental spread of a graded potential.** (a) Leakage of charge-carrying ions across the plasma membrane results in progressive loss of current with increasing distance from the initial site of the change in potential. (b) Because of leaks in current, the magnitude of a graded potential continues to decrease as it passively spreads from the initial active area. The potential dies out altogether within micrometers (less than 1 mm) of its site of initiation.

FIGURE FOCUS: *How is charge being lost across the membrane?*

2. Compare how an increase in the difference in potential and an increase in resistance would affect current flow.

3. Explain why the spread of graded potentials is decremental.

4.3 Action Potentials

Action potentials are brief, rapid, large (100-mV) changes in membrane potential during which the potential actually reverses so that the inside of the excitable cell transiently becomes more positive than the outside. As with a graded potential, a single action potential involves only a small portion of the total excitable cell membrane. Unlike graded potentials, however, action potentials are conducted, or propagated, throughout the entire membrane *nondecrementally*—that is, they do not diminish in

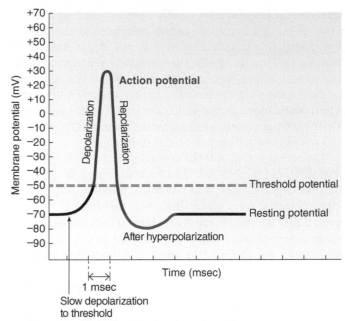

❙Figure 4-4 **Changes in membrane potential during an action potential.**
FIGURE FOCUS: *The magnitude of potential at the peak of an action potential is greater than the magnitude at resting potential.* (True or false?)

strength as they travel from their site of initiation throughout the remainder of the cell membrane. Thus, action potentials can serve as faithful long-distance signals.

Think about the neuron that causes the muscle cells in your big toe to contract (see ❙Figure 4-7, p. 95). If you want to wiggle your big toe, commands are sent from your brain down your spinal cord to initiate an action potential at the beginning of this neuron, which is located in the spinal cord. The action potential travels all the way down the neuron's long axon, which runs through your leg to terminate on your big-toe muscle cells. The signal does not weaken or die off, being instead preserved at full strength from beginning to end.

Let us now consider the changes in potential during an action potential before we see how action potentials spread throughout the cell membrane without diminishing.

During an action potential, the membrane potential rapidly, transiently reverses.

If a graded potential is large enough, it can initiate an action potential before the graded change dies off. (Later you will discover how this initiation is accomplished for the various types of graded potentials.) Typically, the region of the excitable membrane where graded potentials are produced in response to a triggering event does not undergo action potentials. Instead, passive current flow from the region where a graded potential is taking place depolarizes adjacent portions of the membrane where action potentials can occur.

Depolarization from the resting potential of −70 mV proceeds slowly until it reaches a critical level known as **threshold potential**, typically between −50 and −55 mV (❙Figure 4-4). At threshold potential, an explosive depolarization takes place. A recording of the potential at this time shows a sharp upward deflection as the potential rapidly reverses itself so that the inside of the cell becomes positive compared to the outside. Peak potential is usually +30 to +40 mV, depending on the excitable cell. Just as rapidly, the membrane repolarizes, dropping back to resting potential. Often, the forces that repolarize the membrane push the potential too far, causing a brief **after hyperpolarization**, during which the inside of the membrane briefly becomes even more negative than normal (for example, −80 mV) before resting potential is restored.

The action potential is the entire rapid change in potential from threshold to peak and then back to resting. If the initial triggered depolarization does not reach threshold potential, no action potential takes place. Therefore, threshold is a critical point: Either the membrane is depolarized to threshold and an action potential takes place, or threshold is not reached in response to the depolarizing event and no action potential occurs.

Unlike the variable duration of a graded potential, the duration of an action potential is always the same in a given excitable cell. In a neuron, an action potential lasts for only 1 msec (0.001 second). It lasts longer in muscle, with the duration depending on the muscle type.

Often an action potential is referred to as a **spike** because of its spikelike recorded appearance. Alternatively, when an excitable membrane is triggered to undergo an action potential, it is said to **fire**. Thus, the terms *action potential*, *spike*, and *firing* all refer to the same phenomenon of rapid reversal of membrane potential.

Marked changes in membrane permeability and ion movement lead to an action potential.

How is the membrane potential, which is usually maintained at a constant resting level, altered to such an extent as to produce an action potential? Recall that K^+ makes the greatest contribution to the establishment of the resting potential because the membrane at rest is considerably more permeable to K^+ than to Na^+ (see pp. 80 and 82). During an action potential, marked changes in membrane permeability to Na^+ and K^+ take place, permitting rapid fluxes of these ions down their electrochemical gradients. These ion movements carry the current responsible for the potential changes that occur during an action potential. Action potentials take place as a result of the triggered opening and subsequent closing of two specific types of channels: voltage-gated Na^+ channels and voltage-gated K^+ channels.

Voltage-Gated Na^+ and K^+ Channels Voltage-gated membrane channels consist of proteins that have many charged groups. The electrical field (potential) surrounding the channels can distort the channel structure as charged portions of the channel proteins are electrically attracted or repelled by charges in the fluids around the membrane. Unlike most membrane proteins, which remain stable despite fluctuations in membrane potential, voltage-gated channel proteins are especially sensitive to voltage changes. Small distortions in shape induced by changes in potential can cause channel gates to open or close. Here, again, is an example of how subtle changes in structure can profoundly influence function.

The voltage-gated Na^+ channel has two gates: an activation gate and an inactivation gate (Figure 4-5). The *activation gate* guards the channel interior by opening and closing like a sliding door. The *inactivation gate* consists of a ball-and-chain-like sequence of amino acids at the channel opening facing the ICF. This gate is open when the ball is hanging free on the end of its chain and closed when the ball binds to the channel opening, thus blocking the opening. Both gates must be open to permit passage of Na^+ through the channel, and closure of either gate prevents passage. This voltage-gated Na^+ channel can exist in three conformations: (1) *closed but capable of opening* (activation gate closed, inactivation gate open; Figure 4-5a); (2) *open, or activated* (both gates open, Figure 4-5b); and (3) *closed and not capable of opening*, or *inactivated* (activation gate open, inactivation gate closed, Figure 4-5c). The channel moves through these various conformations as a result of voltage changes that take place during an action potential, as described shortly. When the action potential is over and the membrane has returned to resting potential, the channel reverts back to its "closed but capable of opening" conformation.

The voltage-gated K^+ channel is simpler. It has only an activation gate, which can be either closed (Figure 4-5d) or open

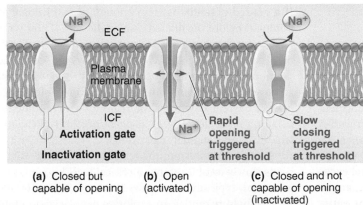

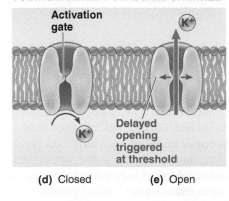

VOLTAGE-GATED SODIUM CHANNEL

ECF
Plasma membrane
ICF
Activation gate
Inactivation gate

Na^+
Na^+
Rapid opening triggered at threshold
Slow closing triggered at threshold
Na^+

(a) Closed but capable of opening

(b) Open (activated)

(c) Closed and not capable of opening (inactivated)

VOLTAGE-GATED POTASSIUM CHANNEL

Activation gate
K^+
K^+
Delayed opening triggered at threshold

(d) Closed

(e) Open

Figure 4-5 **Conformations of voltage-gated sodium and potassium channels.**

(❙Figure 4-5e). These voltage-gated Na^+ and K^+ channels exist in addition to the Na^+–K^+ pump and the leak channels for these ions (described in Chapter 3).

Changes in Permeability and Ion Movement During an Action Potential

At resting potential (-70 mV), all voltage-gated channels for both Na^+ and K^+ are closed, with the activation gates of the Na^+ channels being closed and their inactivation gates being open—that is, the voltage-gated Na^+ channels are in their "closed but capable of opening" conformation. Therefore, Na^+ and K^+ cannot pass through these voltage-gated channels at resting potential. However, because many K^+ leak channels and few Na^+ leak channels are present, the resting membrane is 25 to 30 times more permeable to K^+ than to Na^+.

When current spreads passively from an adjacent site already depolarized (such as from a site undergoing a graded potential) into a new region still at resting potential, the new region of membrane starts to depolarize toward threshold. This depolarization causes the activation gates of some voltage-gated Na^+ channels in the new region to open so that both gates of these activated channels are now open. Because both the concentration and the electrical gradients for Na^+ favor its movement into the cell, Na^+ starts to move in. The inward movement of positively charged Na^+ depolarizes the membrane further, opening even more voltage-gated Na^+ channels and allowing more Na^+ to enter, and so on, in a positive-feedback cycle.

At threshold potential, Na^+ permeability, which is symbolized as P_{Na^+}, increases explosively as the membrane swiftly becomes about 600 times more permeable to Na^+ than to K^+. Each channel is either closed or open and cannot be partially open. However, the delicately poised gating mechanisms of the various voltage-gated Na^+ channels are jolted open by slightly different voltage changes. During the early depolarizing phase, more and more Na^+ channels open as the potential progressively decreases. At threshold, enough Na^+ gates have opened to set off the positive-feedback cycle that rapidly causes the remaining Na^+ gates to open. Now Na^+ permeability dominates the membrane, in contrast to the K^+ domination at resting potential. Thus, at threshold Na^+ rushes into the cell, rapidly eliminating the internal negativity and even making the inside of the cell more positive than the outside in an attempt to drive the membrane potential to the Na^+ equilibrium potential (which is $+61$ mV; see p. 82). The potential reaches $+30$ mV, close to the Na^+ equilibrium potential. The potential does not become any more positive because, at the peak of the action potential, the Na^+ channels start to close to the inactivated state and P_{Na^+} starts to fall to its low resting value (❙Figure 4-6).

What causes the Na^+ channels to close? When the membrane potential reaches threshold, two closely related events take place in each Na^+ channel's gates. First, the activation gate is triggered to *open rapidly* in response to the depolarization, converting the channel to its open (activated) conformation. Surprisingly, the conformational change that opens the channel also allows the inactivation gate's ball to bind to the channel opening, thereby physically blocking the mouth of the channel. However, this closure process takes time, so the inactivation gate *closes*

slowly compared to the rapidity of channel opening (see ❙Figure 4-5c). Meanwhile, during the 0.5-msec delay after the activation gate opens and before the inactivation gate closes, both gates are open and Na^+ rushes into the cell through these open channels, bringing the action potential to its peak. Then the inactivation gate closes, membrane permeability to Na^+ plummets to its low resting value, and further Na^+ entry is prevented. The channel remains in this inactivated conformation until the membrane potential has been restored to resting.

Simultaneous with inactivation of Na^+ channels, the voltage-gated K^+ channels start to slowly open at the peak of the action potential. Opening of the K^+ channel gate is a delayed voltage-gated response triggered by the initial depolarization to threshold (see ❙Figures 4-5e and 4-6). Thus, three action potential–related events occur at threshold: (1) rapid opening of the Na^+ activation gates, which permits Na^+ to enter, moving the potential from threshold to its positive peak; (2) slow closing of the Na^+ inactivation gates, which halts further Na^+ entry after a brief time delay, thus keeping the potential from rising any further; and (3) slow opening of the K^+ gates, which is in large part responsible for the potential plummeting from its peak back to resting.

The membrane potential would gradually return to resting after closure of the Na^+ channels as K^+ continued to leak out but no further Na^+ entered. However, the return to resting is hastened by the opening of K^+ gates at the peak of the action potential. Opening of the voltage-gated K^+ channels greatly increases K^+ permeability (designated P_{K^+}) to about 300 times the resting P_{Na^+}. This marked increase in P_{K^+} causes K^+ to rush out of the cell down its electrochemical gradient, carrying positive charges back to the outside. Note that at the peak of the action potential, the positive potential inside the cell tends to repel the positive K^+ ions, so the electrical gradient for K^+ is outward, unlike at resting potential. Of course, the concentration gradient for K^+ is always outward. The outward movement of K^+ rapidly restores the negative resting potential.

To review (see ❙Figure 4-6), *the rising phase of the action potential* (from threshold to $+30$ mV) *is due to Na^+ influx* (Na^+ entering the cell) induced by an explosive increase in P_{Na^+} at threshold. The *falling phase* (from $+30$ mV to resting potential) *is brought about largely by K^+ efflux* (K^+ leaving the cell) caused by the marked increase in P_{K^+} occurring at the peak of the action potential.

As the potential returns to resting, the changing voltage shifts the Na^+ channels to their "closed but capable of opening" conformation, with the activation gate closed and the inactivation gate open. Now the channel is reset, ready to respond to another triggering event. The newly opened voltage-gated K^+ channels also close, so the membrane returns to the resting number of open K^+ leak channels. Typically, the voltage-gated K^+ channels are slow to close. As a result of this persistent increased permeability to K^+, more K^+ may leave than is necessary to bring the potential to resting. This slightly excessive K^+ efflux makes the interior of the cell transiently even more negative than resting potential, causing the after hyperpolarization. When the voltage-gated K^+ channels all close, the membrane returns to resting potential, where it remains until another triggering event alters the gated Na^+ and K^+ channels.

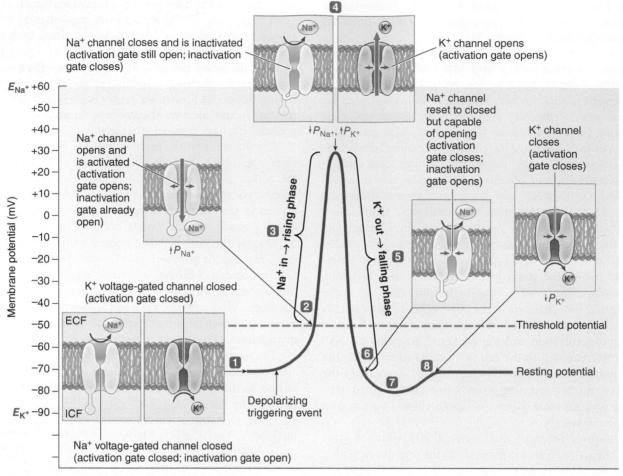

E_Na+ +60
+50
+40
+30
+20
+10
0
−10
−20
−30
−40
−50
−60
−70
−80
E_K+ −90

Membrane potential (mV)

Na$^+$ channel closes and is inactivated (activation gate still open; inactivation gate closes)

K$^+$ channel opens (activation gate opens)

Na$^+$ channel reset to closed but capable of opening (activation gate closes; inactivation gate opens)

K$^+$ channel closes (activation gate closes)

Na$^+$ channel opens and is activated (activation gate opens; inactivation gate already open)

↓P_Na+, ↑P_K+

↑P_Na+

Na$^+$ in → rising phase

K$^+$ out → falling phase

↓P_K+

K$^+$ voltage-gated channel closed (activation gate closed)

ECF

ICF

Na$^+$ voltage-gated channel closed (activation gate closed; inactivation gate open)

Threshold potential

Depolarizing triggering event

Resting potential

Time (msec)

1 Resting potential: all voltage-gated channels closed.

2 At threshold, Na$^+$ activation gate opens and P_{Na^+} rises.

3 Na$^+$ enters cell, causing explosive depolarization to +30 mV, which generates rising phase of action potential.

4 At peak of action potential, Na$^+$ inactivation gate closes and P_{Na^+} falls, ending net movement of Na$^+$ into cell. At the same time, K$^+$ activation gate opens and P_{K^+} rises.

5 K$^+$ leaves cell, causing its repolarization to resting potential, which generates falling phase of action potential.

6 On return to resting potential, Na$^+$ activation gate closes and inactivation gate opens, resetting channel to respond to another depolarizing triggering event.

7 Further outward movement of K$^+$ through still-open K$^+$ channel briefly hyperpolarizes membrane, which generates after hyperpolarization.

8 K$^+$ activation gate closes, and membrane returns to resting potential.

I Figure 4-6 **Permeability changes and ion fluxes during an action potential.**
FIGURE FOCUS: *Compare the status of the voltage-gated Na$^+$ and K$^+$ channels during the rising and falling phases of an action potential.*

The Na$^+$–K$^+$ pump gradually restores the concentration gradients disrupted by action potentials.

At the completion of an action potential, the membrane potential has been restored to resting, but the ion distribution has been altered slightly. Na$^+$ entered the cell during the rising phase, and K$^+$ left during the falling phase. The Na$^+$–K$^+$ pump restores these ions to their original locations in the long run, but not after each action potential.

The active pumping process takes much longer to restore Na$^+$ and K$^+$ to their original locations than it takes for the passive fluxes of these ions during an action potential. However, the membrane does not need to wait until the concentration gradients are slowly restored before it can undergo another action potential. Actually, the movement of relatively few Na$^+$

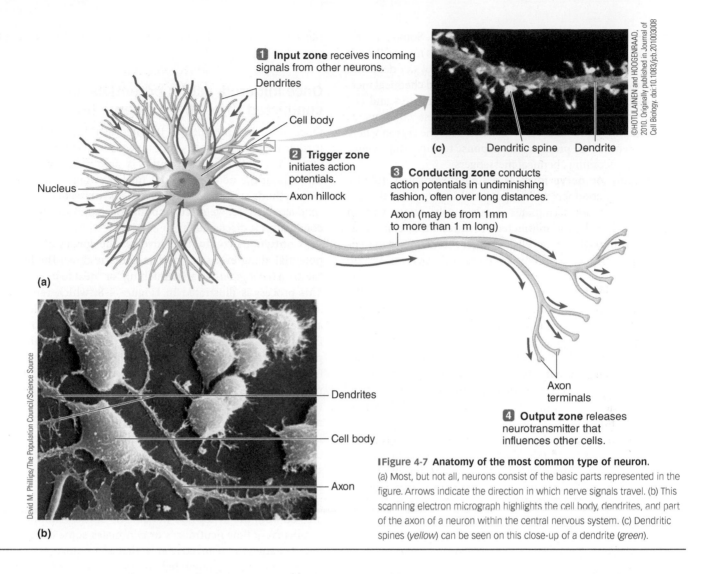

1 **Input zone** receives incoming signals from other neurons.

Dendrites

Cell body

2 **Trigger zone** initiates action potentials.

Nucleus

Axon hillock

©HOTULAINEN and HOOGENRAAD, 2010. Originally published in Journal of Cell Biology. doi:10.1063/jcb.201003008

(c)   Dendritic spine   Dendrite

3 **Conducting zone** conducts action potentials in undiminishing fashion, often over long distances.

Axon (may be from 1mm to more than 1 m long)

(a)

David M. Phillips/The Population Council/Science Source

Dendrites

Cell body

Axon

(b)

Axon terminals

4 **Output zone** releases neurotransmitter that influences other cells.

❙Figure 4-7 Anatomy of the most common type of neuron. (a) Most, but not all, neurons consist of the basic parts represented in the figure. Arrows indicate the direction in which nerve signals travel. (b) This scanning electron micrograph highlights the cell body, dendrites, and part of the axon of a neuron within the central nervous system. (c) Dendritic spines (*yellow*) can be seen on this close-up of a dendrite (*green*).

---

and K^+ ions causes the large swings in membrane potential that occur during an action potential. Only about 1 out of 100,000 K^+ ions present in the cell leaves during an action potential, while a comparable number of Na^+ ions enter from the ECF. The movement of this extremely small proportion of the total Na^+ and K^+ during a single action potential produces dramatic 100-mV changes in potential (between -70 and $+30$ mV) but only infinitesimal changes in the ICF and ECF concentrations of these ions. Much more K^+ is still inside the cell than outside, and Na^+ is still predominantly extracellular. Consequently, the Na^+ and K^+ concentration gradients still exist, so repeated action potentials can occur without the pump having to keep pace restoring the gradients.

Were it not for the pump, even tiny fluxes accompanying repeated action potentials would eventually "run down" the concentration gradients so that further action potentials would be impossible. If the concentrations of Na^+ and K^+ were equal between the ECF and the ICF, changes in permeability to these ions would not bring about ion fluxes, so no change in potential would occur. Thus, the Na^+–K^+ pump is critical to maintaining the concentration gradients in the long run. However, it does not have to perform its role between action potentials, nor is it directly involved in the ion fluxes or potential changes that occur during an action potential.

Action potentials are propagated from the axon hillock to the axon terminals.

A single action potential involves only a small patch of the total surface membrane of an excitable cell. But if action potentials are to serve as long-distance signals, they cannot be merely isolated events occurring in a limited area of a nerve or muscle cell membrane. Mechanisms must exist to conduct or spread the action potential throughout the entire cell membrane. Furthermore, the signal must be transmitted from one cell to the next (for example, along specific nerve pathways). To explain these mechanisms, we begin with a brief look at neuronal structure. Then we examine how an action potential (nerve impulse) is conducted throughout a neuron before we turn to how the signal is passed to another cell.

A single **neuron** typically consists of three basic parts—the *cell body,* the *dendrites,* and the *axon*—although the structure varies depending on the location and function of the neuron (❙Figure 4-7). The nucleus and organelles are housed in the **cell body,** from which numerous extensions known as **dendrites** typically project like antennae to increase the surface area available for receiving signals from other neurons. Often tiny spike-like or knoblike projections, known as *dendritic spines* arise from the dendrites, increasing even further the surface area available for reception of incoming signals (Figure 4-7c). Some neurons have up to 400,000 dendrites, which carry signals *toward* the cell body. In most neurons, the plasma membrane of the dendrites and cell body contains protein receptors that bind chemical messengers from other neurons. Therefore, the dendrites and cell body are the neuron's *input zone* because these components receive and integrate incoming signals. This is the region where graded potentials are produced in response to triggering events, in this case, incoming chemical messengers.

The **axon**, or **nerve fiber**, is a single, elongated, tubular extension that conducts action potentials *away from* the cell body and eventually terminates at other cells. Axons vary in length from less than a millimeter in neurons that communicate only with neighboring cells to longer than a meter in neurons that communicate with distant parts of the nervous system or with peripheral organs. For example, the axon of the neuron innervating your big toe must traverse the distance from the origin of its cell body within the spinal cord of your lower back all the way down your leg to your toe.

The first portion of the axon plus the region of the cell body from which the axon leaves are known collectively as the **axon hillock** or **initial segment**. The axon hillock is the neuron's *trigger zone* because it is the site where action potentials are triggered, or initiated, by a graded potential of sufficient magnitude. The action potentials are then conducted along the axon from the axon hillock to what is typically the highly branched ending at the **axon terminals**. These terminals release chemical messengers that simultaneously influence numerous other cells with which they come into close association. Functionally, therefore, the axon is the *conducting zone* of the neuron, and the axon terminals constitute its *output zone*. (The major exceptions to this typical neuronal structure and functional organization are neurons specialized to carry sensory information, a topic described in a later chapter.)

Action potentials can be initiated only in portions of the membrane with abundant voltage-gated Na^+ channels that can be triggered to open by a depolarizing event. Typically, regions of excitable cells where graded potentials take place do not undergo action potentials because voltage-gated Na^+ channels are sparse there. Therefore, sites specialized for graded potentials do not undergo action potentials, even though they might be considerably depolarized. However, before dying out, graded potentials can trigger action potentials in adjacent portions of the membrane by bringing these more sensitive regions to threshold through local current flow spreading from the site of the graded potential. In a typical neuron, for example, graded potentials are generated in the dendrites and cell body in response to incoming chemical signals. If these graded potentials have sufficient magnitude by the time they have spread to the axon hillock, they initiate an action potential at this trigger zone. The axon hillock has the lowest threshold in the neuron because this region has a much higher density of voltage-gated Na^+ channels than anywhere else in the neuron. For this reason, the axon hillock is considerably more responsive than the dendrites or remainder of the cell body to changes in potential and is the first to reach threshold (the dendrites and cell body at the same potential are still considerably below their much higher thresholds). Therefore, an action potential originates in the axon hillock and is propagated from there to the end of the axon.

Once initiated, action potentials are conducted throughout a nerve fiber.

Once an action potential is initiated at the axon hillock, no further triggering event is necessary to activate the remainder of the nerve fiber. The impulse is automatically conducted throughout the neuron without further stimulation by one of two methods of propagation: *contiguous conduction* or *saltatory conduction*. Here, we discuss contiguous conduction. Saltatory conduction is discussed later.

Contiguous conduction involves the spread of the action potential along every patch of membrane down the length of the axon (*contiguous* means "touching" or "next to in sequence"). This process is illustrated in Figure 4-8, which represents a longitudinal section of the axon hillock and the portion of the axon immediately beyond it. The membrane at the axon hillock is at the peak of an action potential. The inside of the cell is positive in this active area because Na^+ has already rushed in here. The remainder of the axon, still at resting potential and negative inside, is considered inactive. For the action potential to spread from the active to the inactive areas, the inactive areas must somehow be depolarized to threshold. This depolarization is accomplished by local current flow between the area already undergoing an action potential and the adjacent inactive area, similar to the current flow responsible for the spread of graded potentials. Because opposite charges attract, current can flow locally between the active area and the neighboring inactive area on both the inside and the outside of the membrane. This local current flow neutralizes or eliminates some of the unbalanced charges in the inactive area; that is, it reduces the number of opposite charges separated across the membrane, reducing the potential in this area. This depolarizing effect quickly brings the involved inactive area to threshold, at which time the voltage-gated Na^+ channels in this region of the membrane are all thrown open, leading to an action potential in this previously inactive area. Meanwhile, the original active area returns to resting potential as a result of K^+ efflux.

Beyond the new active area is another inactive area, so the same thing happens again. This cycle repeats itself in a chain reaction until the action potential has spread to the end of the axon. *Once an action potential is initiated in one part of a neuron's cell membrane, a self-perpetuating cycle is initiated so that the action potential is propagated along the rest of the fiber automatically.* In this way, the axon is like a firecracker fuse that needs to be lit at only one end. Once ignited, the fire spreads down the fuse; it is not necessary to hold a match to every separate section of the fuse. Note that the original action potential does not travel along the membrane. Instead, it triggers an identical new action potential in the bordering area of the membrane, with this process being repeated along the axon's length. An analogy is the "wave" at a stadium. Each section of spectators stands up (the rising phase of an action potential) and then sits down (the falling phase) in sequence one after another as the wave moves around the stadium. The wave, not individual spectators, travels around the stadium. Similarly,

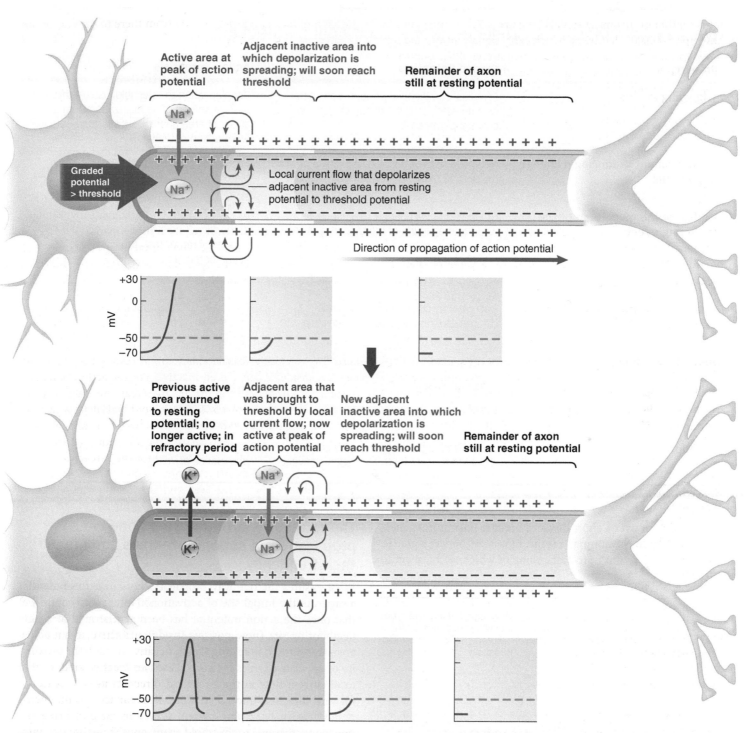

IFigure 4-8 Contiguous conduction. Local current flow between the active area at the peak of an action potential and the adjacent inactive area still at resting potential reduces the potential in this contiguous inactive area to threshold, which triggers an action potential in the previously inactive area. The original active area returns to resting potential, and the new active area induces an action potential in the next adjacent inactive area by local current flow as the cycle repeats itself down the length of the axon.

FIGURE FOCUS: *What initiates an action potential at the axon hillock? What initiates the action potential at each subsequent adjacent inactive area as the action potential propagates along the axon?*

new action potentials arise sequentially down the axon. Each new action potential is a fresh local event that depends on induced permeability changes and electrochemical gradients that are virtually identical down the length of the axon. Therefore, the last action potential at the end of the axon is identical to the original one, no matter how long the axon is. In this way, action potentials can serve as long-distance signals without weakening or distortion.

This nondecremental propagation of an action potential contrasts with the decremental spread of a graded potential,

Property	Graded Potentials	Action Potentials
Triggering events	Stimulus, combination of neurotransmitter with receptor, or self-induced changes in channel permeability	Depolarization to threshold, usually through passive spread of depolarization from an adjacent area undergoing a graded potential or an action potential
Ion movement producing a change in potential	Net movement of Na^+, K^+, Cl^-, or Ca^{2+} across the plasma membrane by various means	Sequential movement of Na^+ into and K^+ out of the cell through voltage-gated channels
Coding of the magnitude of the triggering event	Graded potential change; magnitude varies with the magnitude of the triggering event	All-or-none membrane response; magnitude of the triggering event is coded in the frequency rather than the amplitude of action potentials
Duration	Varies with the duration of the triggering event	Constant
Magnitude of the potential change with distance from the initial site	Decremental conduction; magnitude diminishes with distance from the initial site	Propagated throughout the membrane in an undiminishing fashion; self-regenerated in neighboring inactive areas of the membrane
Refractory period	None	Relative, absolute
Summation	Temporal, spatial	None
Direction of potential change	Depolarization or hyperpolarization	Always depolarization and reversal of charges
Location	Specialized regions of the membrane designed to respond to the triggering event	Regions of the membrane with an abundance of voltage-gated channels

which dies out over a short distance because it cannot regenerate itself. ▌Table 4-1 summarizes the differences between graded potentials and action potentials, some of which we have yet to discuss.

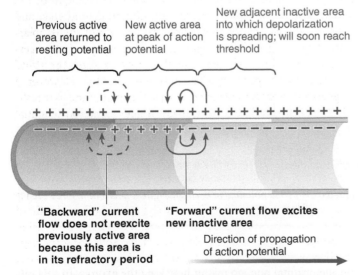

Previous active area returned to resting potential   New active area at peak of action potential   New adjacent inactive area into which depolarization is spreading; will soon reach threshold

"Backward" current flow does not reexcite previously active area because this area is in its refractory period

"Forward" current flow excites new inactive area

Direction of propagation of action potential

▌Figure 4-9 **Value of the refractory period.** The refractory period prevents "backward" current flow. During an action potential and slightly afterward, an area cannot be restimulated by normal events to undergo another action potential. Thus, the refractory period ensures that an action potential can be propagated only in the forward direction along the axon.

The refractory period ensures one-way propagation of action potentials and limits their frequency.

What ensures the one-way propagation of an action potential away from the initial site of activation? Note from ▌Figure 4-9 that once the action potential has been regenerated at a new neighboring site (now positive inside) and the original active area has returned to resting (again negative inside), the proximity of opposite charges between these two areas is conducive to local current flow in the backward direction as well as in the forward direction into as-yet-unexcited portions of the membrane. If such backward current flow were able to bring the previous active area to threshold again, another action potential would be initiated here, which would spread both forward and backward, initiating still other action potentials, and so on. But if action potentials were to move in both directions, the situation would be chaotic, with numerous action potentials bouncing back and forth along the axon until the neuron eventually fatigued. Fortunately, neurons are saved from this fate of oscillating action potentials by the **refractory period**, during which a new action potential cannot be initiated in a region that has just undergone an action potential.

Because of the changing status of the voltage-gated Na^+ and K^+ channels during and after an action potential, the refractory period has two components: the *absolute refractory period* and the *relative refractory period* (▌Figure 4-10). When a particular

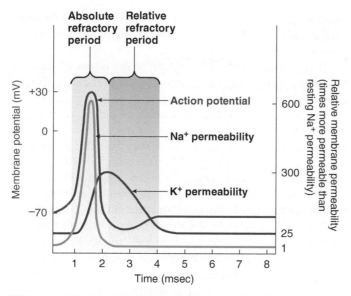

Absolute
refractory
period

Relative
refractory
period

Action potential

Na⁺ permeability

K⁺ permeability

Membrane potential (mV)

+30

0

−70

Relative membrane permeability
(times more permeable than
resting Na⁺ permeability)

600

300

25
1

1 2 3 4 5 6 7 8
Time (msec)

ǀFigure 4-10 **Absolute and relative refractory periods.** During the absolute refractory period, the portion of the membrane that has just undergone an action potential cannot be restimulated. This period corresponds to the time during which the Na⁺ gates are not in their resting conformation. During the relative refractory period, the membrane can be restimulated only by a stronger stimulus than is usually necessary. This period corresponds to the time during which the K⁺ gates opened during the action potential have not yet closed, coupled with lingering inactivation of the voltage-gated Na⁺ channels.

patch of axonal membrane is undergoing an action potential, it cannot initiate another action potential, no matter how strong the depolarizing triggering event is. This period when a recently activated patch of membrane is completely refractory (meaning "stubborn" or "unresponsive") to further stimulation is known as the **absolute refractory period**. Once the voltage-gated Na⁺ channels are triggered to open at threshold, they cannot open again in response to another depolarizing triggering event, no matter how strong, until they pass through their "closed and not capable of opening" conformation and then are reset to their "closed and capable of opening" conformation when resting potential is restored. Accordingly, the absolute refractory period lasts the entire time from threshold, through the action potential, and until return to resting potential. Only then can the voltage-gated Na⁺ channels respond to another depolarization with an explosive increase in P_{Na^+} to initiate another action potential. Because of the absolute refractory period, one action potential must be over before another can be initiated at the same site. Action potentials cannot overlap or be added one on top of another "piggyback" fashion.

Following the absolute refractory period is a **relative refractory period**, during which a second action potential can be produced only by a triggering event considerably stronger than usual. The relative refractory period occurs after the action potential is completed because of a twofold effect. First, the voltage-gated Na⁺ channels that opened during the action potential do not all reset at once when resting potential is reached. Some take a little longer to be restored to their capable of opening conformation. As a result, fewer voltage-gated Na⁺ channels are in a position to be jolted open in

response to another depolarizing triggering event. Second, the voltage-gated K⁺ channels that opened at the peak of the action potential are slow to close. During this time, the resultant less-than-normal Na⁺ entry in response to another triggering event is opposed by K⁺ still leaving through its slow-to-close channels during the after hyperpolarization. Thus, a greater depolarizing triggering event than normal is needed to offset the persistent hyperpolarizing outward movement of K⁺ and bring the membrane to threshold during the relative refractory period.

By the time the original site has recovered from its refractory period and is capable of being restimulated by normal current flow, the action potential has been propagated in the forward direction only and is so far away that it can no longer influence the original site. Thus, *the refractory period ensures the one-way propagation of the action potential down the axon away from the initial site of activation.*

The refractory period is also responsible for setting an upper limit on the frequency of action potentials—that is, it determines the maximum number of new action potentials that can be initiated and propagated along a fiber in a given period. The original site must recover from its refractory period before a new action potential can be triggered to follow the preceding action potential. The length of the refractory period varies for different types of neurons. The longer the refractory period, the greater the delay before a new action potential can be initiated and the lower the frequency with which a neuron can respond to repeated or ongoing stimulation.

Action potentials occur in all-or-none fashion.

If any portion of the neuronal membrane is depolarized to threshold, an action potential is initiated and relayed along the membrane in an undiminished fashion. Furthermore, once threshold has been reached, the resultant action potential always goes to maximal height. The reason for this effect is that the changes in voltage during an action potential result from ion movements down concentration and electrical gradients, and these gradients are not affected by the strength of the depolarizing triggering event. A triggering event stronger than necessary to bring the membrane to threshold (a **suprathreshold** event) does not produce a larger action potential. However, a triggering event that fails to depolarize the membrane to threshold (a **subthreshold** event) does not trigger an action potential at all. Thus, *an excitable membrane either responds to a triggering event with a maximal action potential that spreads nondecrementally throughout the membrane or does not respond with an action potential at all.* This property is called the **all-or-none law**.

The all-or-none concept is analogous to firing a gun. Either the trigger is not pulled sufficiently to fire the bullet (threshold is not reached), or it is pulled hard enough to elicit the full firing response of the gun (threshold is reached). Squeezing the trigger harder does not produce a greater explosion. Just as it is not possible to fire a gun halfway, it is not possible to cause a halfway action potential.

The threshold phenomenon allows some discrimination between important and unimportant stimuli or other triggering events. Stimuli too weak to bring the membrane to threshold do not initiate action potentials and therefore do not clutter up the nervous system by transmitting insignificant signals.

The strength of a stimulus is coded by the frequency of action potentials.

How is it possible to differentiate between two stimuli of varying strengths when both stimuli bring the membrane to threshold and generate action potentials of the same magnitude? For example, how can we distinguish between touching a warm object and touching a hot object if both trigger identical action potentials in a nerve fiber relaying information about skin temperature to the central nervous system (CNS) (the brain and spinal cord). The answer partly lies in the *frequency* with which the action potentials are generated. A stronger stimulus does not produce a *larger* action potential, but it does trigger a greater *number* of action potentials per second. For an illustration, see Figure 10-31, p. 367, in which changes in blood pressure are coded by corresponding changes in the frequency of action potentials generated in the neurons monitoring blood pressure.

In addition, a stronger stimulus in a region causes more neurons to reach threshold, increasing the total information sent to the CNS. For example, lightly touch this page with your finger and note the area of skin in contact with the page. Now, press down more firmly and note that a larger surface area of skin is in contact with the page. Therefore, more neurons are brought to threshold with this stronger touch stimulus.

Once initiated, the velocity, or speed, with which an action potential travels down the axon depends on two factors: (1) whether the fiber is myelinated and (2) the diameter of the fiber. Contiguous conduction occurs in unmyelinated fibers. In this case, as you just learned, each action potential initiates an identical new action potential in the next contiguous segment of the axon membrane so that every portion of the membrane undergoes an action potential as this electrical signal is conducted from the beginning to the end of the axon. A faster method of propagation, saltatory conduction, takes place in myelinated fibers. We show next how a myelinated fiber compares with an unmyelinated fiber and then how saltatory conduction compares with contiguous conduction.

Myelination increases the speed of conduction of action potentials.

Myelinated fibers are axons covered with **myelin**, a thick layer composed primarily of lipids, at regular intervals along their length (Figure 4-11a). Because the water-soluble ions responsible for carrying current across the membrane cannot permeate this myelin coating, it acts as an insulator, just like plastic around an electrical wire, to prevent leakage of current across the myelinated portion of the membrane. Myelin is not actually a part of the neuron but consists of separate myelin-forming cells that wrap themselves around the axon in jelly-roll fashion. These myelin-forming cells are **Schwann cells** in the peripheral nervous system (PNS) (Figure 4-11b) (the nerves running between the CNS and the various regions of the body), and **oligodendrocytes** in the CNS (Figure 4-11c). Each patch of lipid-rich myelin consists of multiple layers of the myelin-forming cell's plasma membrane (predominantly the lipid bilayer) as the cell repeatedly wraps itself around the axon. A patch of myelin might be made up of as many as 300 wrapped lipid bilayers.

Between the myelinated regions, at the **nodes of Ranvier**, the axonal membrane is bare and exposed to the ECF. Current can flow across the membrane only at these bare spaces to produce action potentials. Voltage-gated Na^+ and K^+ channels are concentrated at the nodes, whereas the myelin-covered regions are almost devoid of these special passageways (Figure 4-11d). By contrast, an unmyelinated fiber has a high density of these voltage-gated channels along its entire length. As you now know, action potentials can be generated only at portions of the membrane furnished with an abundance of these channels.

The distance between the nodes is short enough that local current can flow between an active node and an adjacent inactive node before dying off. When an action potential occurs at one node, local current flow between this node and the oppositely charged adjacent resting node reduces the adjacent node's potential to threshold so that it undergoes an action potential, and so on. Consequently, in a myelinated fiber, the impulse "jumps" from node to node, skipping over the myelinated sections of the axon (Figure 4-12); this process is called **saltatory conduction** (*saltare* means "to jump"). Saltatory conduction propagates action potentials more rapidly than contiguous conduction does, because the action potential does not have to be regenerated at myelinated sections but must be regenerated within every section of an unmyelinated axon from beginning to end. Myelinated fibers conduct impulses about 50 times faster than unmyelinated fibers of comparable size. You can think of myelinated fibers as the "superhighways" and unmyelinated fibers as the "back roads" of the nervous system when it comes to the speed with which information can be transmitted.

Besides permitting action potentials to travel faster, myelination also conserves energy. Because the ion fluxes associated with action potentials are confined to the nodal regions, the energy-consuming Na^+-K^+ pump must restore fewer ions to their respective sides of the membrane following propagation of an action potential.

The boxed feature on p. 103, Concepts, Challenges, and Controversies, examines the myelin-destroying disease *multiple sclerosis*.

Fiber diameter also influences the velocity of action potential propagation.

Besides the effect of myelination, fiber diameter influences the speed with which an axon can conduct action potentials. The magnitude of current flow (that is, the amount of charge that moves) depends not only on the difference in potential between two adjacent electrically charged regions, but also on the resistance to electrical charge movement between the two regions. When fiber diameter increases, the resistance to local current decreases. Thus, the larger the fiber diameter, the faster action potentials can be propagated.

Large myelinated fibers, such as those supplying skeletal muscles, can conduct action potentials at a velocity of up to 120 meters/second (268 miles/hour), compared with a conduction velocity of 0.7 meters/second (2 miles/hour) in small unmy-

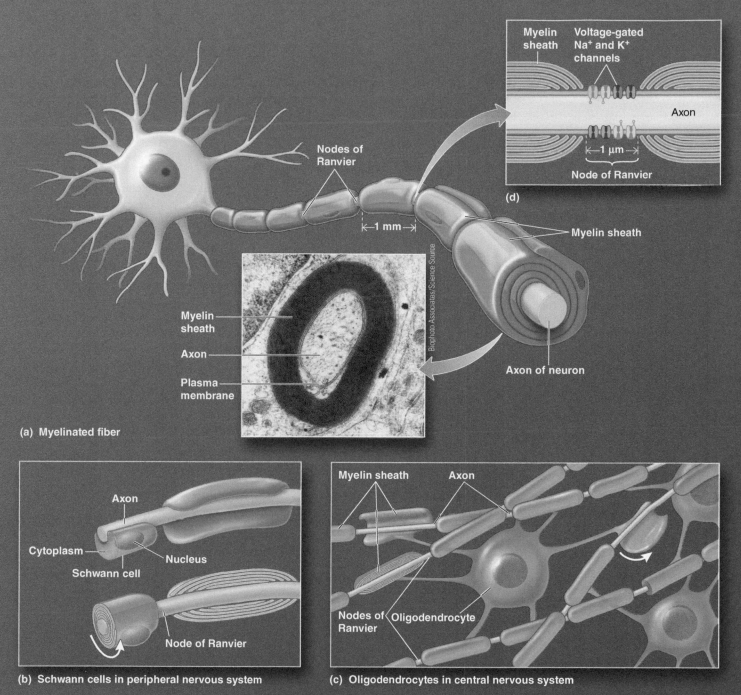

I Figure 4-11 Myelinated fibers. (a) A myelinated fiber is surrounded by myelin at regular intervals. The intervening bare, unmyelinated regions are known as nodes of Ranvier. The electron micrograph shows a myelinated fiber in cross section at a myelinated region. (b) In the PNS each patch of myelin is formed by a separate Schwann cell that wraps itself jelly-roll fashion around the nerve fiber. (c) In the CNS each of several processes ("arms") of a myelin-forming oligodendrocyte forms a patch of myelin around a separate nerve fiber.

elinated fibers such as those supplying the digestive tract. This difference in speed of propagation is related to the urgency of the information being conveyed. A signal to skeletal muscles to execute a particular movement (for example, to prevent you from falling as you trip on something) must be transmitted more rapidly than a signal to modify a slow-acting digestive process. Without myelination, axon diameters within urgent nerve pathways would have to be very large and cumbersome to achieve

the necessary conduction velocities. Indeed, many invertebrates have large axons. In the course of vertebrate evolution, an efficient alternative to very large nerve fibers was development of the myelin sheath, which allows economic, rapid, long-distance signaling. For example, in humans the optic nerve leading from the eye to the brain is only 3 mm in diameter but is packed with more than a million myelinated axons. If those axons were unmyelinated, each would have to be about 100 times thicker to

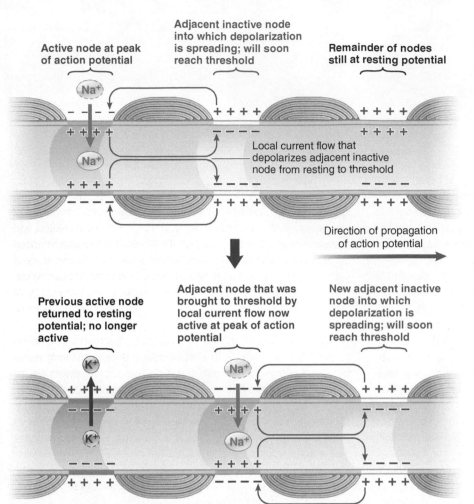

Active node at peak
of action potential

Adjacent inactive node
into which depolarization
is spreading; will soon
reach threshold

Remainder of nodes
still at resting potential

Na⁺

Na⁺

Local current flow that
depolarizes adjacent inactive
node from resting to threshold

Direction of propagation
of action potential

Previous active node
returned to resting
potential; no longer
active

Adjacent node that was
brought to threshold by
local current flow now
active at peak of action
potential

New adjacent inactive
node into which
depolarization is
spreading; will soon
reach threshold

K⁺

K⁺

Na⁺

Na⁺

Figure 4-12 Saltatory conduction. The impulse "jumps" from node to node in a myelinated fiber.

4.4 Synapses and Neuronal Integration

When an action potential reaches the axon terminals, they release a chemical messenger that alters the activity of the cells on which the neuron terminates. A neuron may terminate on one of three structures: a muscle, a gland, or another neuron. Therefore, depending on where a neuron terminates, it can cause a muscle cell to contract, a gland cell to secrete, another neuron to convey an electrical message along a nerve pathway, or some other function. When a neuron terminates on a muscle or a gland, the neuron is said to **innervate**, or supply, the structure. The junctions between the nerves and the muscles and glands that they innervate are described later. For now, we concentrate on the junction between two neurons—a **synapse** (*synapsis* means "juncture"). (Sometimes the term *synapse* is used to describe a junction between any two excitable cells, but we reserve this term for the junction between two neurons.)

conduct impulses at the same velocity, resulting in an optic nerve about 300 mm (12 inches) in diameter.

The presence of myelinating cells can be either a tremendous benefit or a tremendous detriment when an axon is cut, depending on whether the damage occurs in a peripheral nerve or in the CNS. See the boxed feature on p. 104, ▌Concepts, Challenges, and Controversies, to learn more about the regeneration of damaged nerve fibers, a matter of crucial importance in spinal-cord injuries.

You have now seen how an action potential is propagated along the axon and learned about the factors influencing the speed of this propagation. But what happens when an action potential reaches the end of the axon?

Check Your Understanding 4.3

1. Draw and label an action potential, indicating the ion movements responsible for the rising phase and the falling phase.

2. Describe the three conformations of a voltage-gated Na⁺ channel and indicate the membrane potential at which each of these conformations exists.

3. Draw and label the most common type of neuron and identify its four functional zones.

4. Explain why saltatory conduction propagates action potentials more rapidly than contiguous conduction does.

Synapses are typically junctions between presynaptic and postsynaptic neurons.

There are two types of synapses: *electrical synapses* and *chemical synapses*, depending on how information is transferred between the two neurons.

Electrical Synapses In an **electrical synapse**, two neurons are connected by gap junctions (see p. 62), which allow charge-carrying ions to flow directly between the two cells in either direction. Although electrical synapses lead to unbroken transmission of electrical signals and are extremely rapid, this type of connection is essentially "on" or "off" and is unregulated. At an electrical synapse, an action potential in one neuron always leads to an action potential in the connected neuron.

Electrical synapses are not as common as chemical synapses in the human nervous system. Gap junctions are more numerous in smooth muscle and cardiac muscle, where their function is better understood. Until recently electrical synapses in the nervous system were thought to be confined to specialized locations such as the retina of the eye and the pulp of a tooth. Neuroscientists now know that these synapses are present in widespread locations within the CNS. Electrical synapses are typically found among populations of neurons

Multiple Sclerosis: Myelin—Going, Going, Gone

MULTIPLE SCLEROSIS (MS) IS A pathophysiologic condition in which nerve fibers in various locations throughout the nervous system lose their myelin. MS is an autoimmune disease (*auto* means "self"; *immune* means "defense against") in which the body's defense system erroneously attacks the myelin sheath surrounding myelinated nerve fibers. The condition afflicts about 1 in 1000 people in the United States. MS typically begins between the ages of 20 and 40.

Many investigators believe that MS arises from a combination of genetic and environmental factors. Relatives of those with MS have a 6 to 10 times greater chance of developing the disease themselves than the general population does. Because of their genetic predisposition, these relatives have increased susceptibility to environmental factors that may trigger the disease. Various environmental triggers have been proposed, including viral infections, environmental toxins, and vitamin D deficiency, but no evidence has been conclusive.

Loss of myelin as a result of a misguided immune attack slows transmission of impulses in the affected neurons. A hardened scar known as a *sclerosis* (meaning "hard") forms at the multiple sites of myelin damage. These scars interfere with and can eventually block the propagation of action potentials in the underlying axons. Furthermore, the inflammatory phase characterized by myelin destruction sets off a subsequent degenerative phase characterized by deterioration of the affected axons.

The symptoms of MS vary considerably, depending on the extent and location of myelin damage and axon degeneration. The most common symptoms include fatigue, visual problems, tingling and numbness, muscle weakness, impaired balance and coordination, and gradual paralysis. The early stage of the disease is often characterized by cycles of relapse and recovery, whereas the later chronic stage is marked by slow, progressive worsening of symptoms. MS can be debilitating, but is not generally fatal, although the life expectancy of those with the condition averages 5 to 10 years less than the unaffected population.

Currently there is no cure for MS, but researchers have been scrambling to find means to treat attacks, reduce debilitating symptoms, and improve the course of the disease. The symptoms are so mild in some people that no treatment is necessary. For those with more pronounced symptoms, current treatments include drugs that suppress the immune attack on myelin in various ways, along with physical therapy and muscle relaxants. Some patients respond better than others to current drug therapies. Among recent efforts to thwart MS are development of an experimental vaccine that calms the myelin-attacking immune cells, strategies to promote remyelination, and use of neuroprotective drugs.

where synchronization of activity is paramount. For example, electrical synapses interconnect a cluster of neurosecretory neurons in the brain that all secrete the same neurohormone, *GnRH,* which is at the head of an endocrine chain of command that governs reproductive function. Because of these electrical synapses, these neurons all fire and secrete GnRH in synchrony, resulting in coordinated bursts of secretion once every 2 to 3 hours, with no secretion occurring between these bursts. The target cells of GnRH in the chain of command are programmed to respond only to the normal pulsatile pattern of GnRH, so control of the reproductive system ultimately depends on electrical synapses enabling this synchronized secretion.

Chemical Synapses Most synapses in the human nervous system are **chemical synapses** at which a chemical messenger transmits information one way across a space separating the two neurons. A chemical synapse typically involves a junction between an axon terminal of one neuron, known as the *presynaptic neuron,* and the dendrites or cell body of a second neuron, known as the *postsynaptic neuron.* (*Pre* means "before," and *post* means "after"; the presynaptic neuron lies before the synapse, and the postsynaptic neuron lies after the synapse.) The dendrites and, to a lesser extent, the cell body of most neurons receive thousands of synaptic inputs, which are axon terminals from many other neurons. Some neurons in the CNS receive as many as 100,000 synaptic inputs (■ Figure 4-13).

The anatomy of a chemical synapse is shown in ■ Figure 4-14 and in the chapter opener photo. The axon terminal of the **presynaptic neuron**, which conducts its action potentials *toward* the synapse, ends in a slight swelling, the **synaptic knob**. The synaptic knob contains **synaptic vesicles**, which store a specific chemical messenger, a **neurotransmitter** that has been synthesized and packaged by the presynaptic neuron. The synaptic knob comes close to, but does not touch, the **postsynaptic neuron**, whose action potentials are propagated *away from* the synapse. The space between the presynaptic and postsynaptic neurons is called the **synaptic cleft**. Fingerlike CAMs (see p. 60) extend partway across the synaptic cleft from the surfaces of both the presynaptic and postsynaptic neurons. These projections are "Velcroed" together where they meet and overlap in the middle of the cleft, much as if you interlock the extended fingers from both hands together. This physical tethering stabilizes the close proximity of the presynaptic and postsynaptic neurons at the synapse.

Regeneration: PNS Axons Can Do It, But CNS Axons Cannot

NERVE FIBERS MAY BE DAMAGED by being severed or crushed (as during a traumatic event, such as a vehicle wreck, a gunshot wound, or a diving accident) or by being deprived of their blood supply (as during a stroke). When damaged, the affected axons can no longer conduct action potentials to convey messages. Using a severed axon as an example, the portion of the axon farthest from the cell body degenerates. Whether the lost portion of the axon regenerates depends on its location. Cut axons in the peripheral nervous system (PNS) can regenerate, whereas those in the central nervous system (CNS) cannot.

Regeneration of Peripheral Axons

In the case of a cut axon in a peripheral nerve, when the detached part of the axon degenerates, the surrounding Schwann cells phagocytize the debris. The Schwann cells themselves remain, then form a **regeneration tube** that guides the regenerating nerve fiber to its proper destination. The remaining portion of the axon connected to the cell body starts to grow and move forward within the Schwann cell column by amoeboid movement (see p. 50). The growing axon tip "sniffs" its way forward in the proper direction, guided by a chemical secreted into the regeneration tube by the Schwann cells. Successful fiber regeneration is responsible for the eventual return of sensation and movement after traumatic peripheral nerve injuries, although regeneration is not always successful.

Inhibited Regeneration of Central Axons

Fibers in the CNS, which are myelinated by oligodendrocytes, do not have this regenerative ability. Actually, the axons themselves have the ability to regenerate, but the oligodendrocytes surrounding them synthesize proteins that inhibit axonal growth, in sharp contrast to the nerve growth–promoting action of the Schwann cells that myelinate peripheral axons. Nerve growth in the brain and spinal cord is controlled by a delicate balance between *nerve growth–enhancing* and *nerve growth–inhibiting proteins.* During fetal development, nerve growth in the CNS is possible as the brain and spinal cord are being formed. Researchers speculate that nerve growth inhibitors, which are produced late in fetal development in the myelin sheaths surrounding central nerve fibers, may normally serve as "guardrails" that keep new nerve endings from straying outside their proper paths. The growth-inhibiting action of oligodendrocytes may thus serve to stabilize the enormously complex structure of the CNS.

The synaptic cleft is too wide for the direct spread of current from one cell to the other and therefore prevents action potentials from electrically passing between the neurons. Instead, an action potential in the presynaptic neuron alters the postsynaptic neuron's potential by chemical means. Synapses operate in one direction only—that is, the presynaptic neuron brings about changes in the membrane potential of the postsynaptic neuron, but the postsynaptic neuron does not directly influence the potential of the presynaptic neuron. The reason for this becomes readily apparent when you examine the events that occur at a synapse.

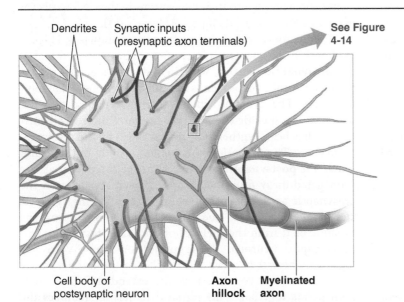

Dendrites — Synaptic inputs (presynaptic axon terminals) — See Figure 4-14

Cell body of postsynaptic neuron — Axon hillock — Myelinated axon

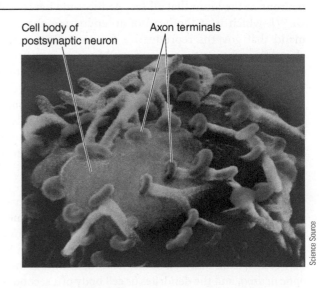

Cell body of postsynaptic neuron — Axon terminals

Science Source

❙ Figure 4-13 Synaptic inputs (presynaptic axon terminals) to the cell body and dendrites of a single postsynaptic neuron. The drying process used to prepare the neuron for the electron micrograph has toppled the presynaptic axon terminals and pulled them away from the postsynaptic cell body.

Unless otherwise noted, all content on this page is © Cengage Learning.

Growth inhibition is a disadvantage, however, when CNS axons need to be mended, as when the spinal cord has been severed accidentally. Damaged central fibers show immediate signs of repairing themselves after an injury, but within several weeks they start to degenerate, and scar tissue forms at the site of injury, halting any recovery. Therefore, damaged neuronal fibers in the brain and spinal cord never regenerate.

Research on Regeneration of Central Axons

In the future, however, it may be possible to promote significant regeneration of damaged fibers in the CNS. Investigators are exploring several promising ways of spurring repair of central axonal pathways, with the goal of enabling victims of spinal-cord injuries to walk again and to regain control of bladder emptying. Here are some current lines of research:

■ Scientists have been able to induce significant nerve regeneration in rats with severed spinal cords by *chemically blocking the nerve growth inhibitor* dubbed *Nogo,* thereby allowing nerve growth enhancers to promote abundant sprouting of new nerve fibers at the site of injury.

■ Other researchers are experimentally *using peripheral nerve grafts* to bridge the defect at an injured spinal-cord site. These grafts contain the nurturing Schwann cells, which release nerve growth–enhancing proteins.

■ Another avenue of hope is the discovery of *neural stem cells* (see pp. 10 and 138). These cells might someday be implanted into a damaged spinal cord and coaxed into multiplying and differentiating into mature, functional neurons that will replace those lost.

■ Yet another new strategy under investigation is *enzymatically breaking down inhibitory components in the scar tissue* that naturally forms at the injured site and prevents sprouting nerve fibers from crossing this barrier.

■ Other researchers are taking a *bionic approach,* trying to create electronic devices that plug into the nervous system to bypass a broken connection in the spinal cord. The idea is to ultimately implant brain chips that could pick up the electrical message intended to command muscle movement, then to relay this message to a second device implanted in the spinal cord below the level of damage. This second device would stimulate the motor neurons to produce the commanded movement.

1 Action potential reaches axon terminal of presynaptic neuron.

2 Ca^{2+} enters synaptic knob (presynaptic axon terminal).

3 Neurotransmitter is released by exocytosis into synaptic cleft.

4 Neurotransmitter binds to receptors that are an integral part of chemically gated channels on subsynaptic membrane of postsynaptic neuron.

5 Binding of neurotransmitter to receptor-channel opens that specific channel.

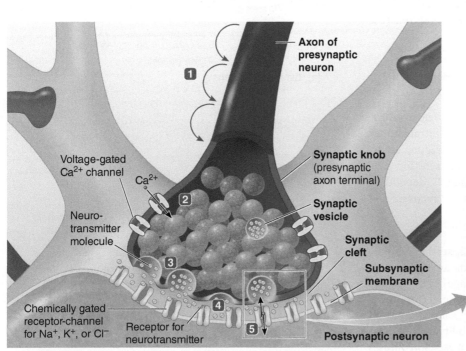

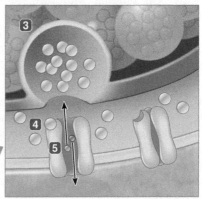

IFigure 4-14 **Structure and function of a single synapse.** The numbered steps designate the sequence of events that take place at a synapse. The blowup depicts the release by exocytosis of neurotransmitter from the presynaptic axon terminal and its subsequent binding with receptor-channels specific for it on the subsynaptic membrane of the postsynaptic neuron.

A neurotransmitter carries the signal across a synapse.

When an action potential in a presynaptic neuron has been propagated to the axon terminal (Figure 4-14, step **1**), this local change in potential triggers the opening of voltage-gated calcium (Ca^{2+}) channels in the synaptic knob. Because Ca^{2+} is more highly concentrated in the ECF (see p. 73), this ion flows into the synaptic knob through the open channels (step **2**). Ca^{2+} promotes the release of a neurotransmitter from some synaptic vesicles into the synaptic cleft (step **3**). The release is accomplished by exocytosis (see p. 29). The released neurotransmitter diffuses across the cleft and binds with specific protein receptors on the **subsynaptic membrane**, the portion of the postsynaptic membrane immediately underlying the synaptic knob (*sub* means "under") (step **4**). These receptors are an integral part of specific ion channels. These combined receptor and channel units are appropriately known as **receptor-channels**. Binding of neurotransmitter to the receptor-channels causes the channels to open, changing the ion permeability and thus the potential of the postsynaptic neuron (step **5**). These are chemically gated channels, in contrast to the voltage-gated channels responsible for the action potential and for Ca^{2+} influx into the synaptic knob. Because the presynaptic terminal releases the neurotransmitter and the subsynaptic membrane of the postsynaptic neuron has receptor-channels for the neurotransmitter, the synapse can operate only in the direction from presynaptic to postsynaptic neuron.

Conversion of the electrical signal in the presynaptic neuron (an action potential) to an electrical signal in the postsynaptic neuron by chemical means (via the neurotransmitter–receptor combination) takes time. This **synaptic delay** is usually about 0.5 to 1 msec. In a neural pathway, chains of neurons often must be traversed. The more complex the pathway, the more synaptic delays and the longer the *total reaction time* (the time required to respond to a particular event).

Some synapses excite, whereas others inhibit, the postsynaptic neuron.

Each presynaptic neuron typically releases only one neurotransmitter; however, different neurons vary in the neurotransmitter they release. On binding with their subsynaptic receptor-channels, different neurotransmitters cause different ion permeability changes. There are two types of synapses, depending on the resultant permeability changes: *excitatory synapses* and *inhibitory synapses*.

Excitatory Synapses At an **excitatory synapse**, the receptor-channels to which the neurotransmitter binds are nonspecific cation channels that permit simultaneous passage of Na^+ and K^+ through them. (These are a different type of channel from those you have encountered before.) When these channels open in response to neurotransmitter binding, permeability to both these ions is increased at the same time. How much of each ion diffuses through an open nonspecific cation channel depends on the ions' electrochemical gradients. At resting potential, both the concentration and the electrical gradients for Na^+ favor its movement

into the postsynaptic neuron, whereas only the concentration gradient for K^+ favors its movement outward. Therefore, the permeability change induced at an excitatory synapse results in the movement of a few K^+ ions out of the postsynaptic neuron, whereas a larger number of Na^+ ions simultaneously enter this neuron. The result is net movement of positive ions into the cell. This makes the inside of the membrane slightly less negative than at resting potential, thus producing a *small depolarization* of the postsynaptic neuron.

Activation of one excitatory synapse cannot depolarize the postsynaptic neuron enough to bring it to threshold. Too few channels are involved at a single synaptic site to permit adequate ion flow to reduce the potential to threshold. This small depolarization, however, does bring the postsynaptic neuron's membrane closer to threshold, increasing the likelihood that threshold will be reached (in response to further excitatory input) and that an action potential will occur—that is, the membrane is now more excitable (easier to bring to threshold) than when at rest. Accordingly, the change in postsynaptic potential occurring at an excitatory synapse is called an **excitatory postsynaptic potential**, or EPSP (Figure 4-15a).

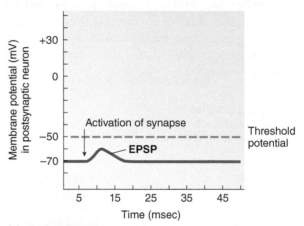

(a) Excitatory synapse

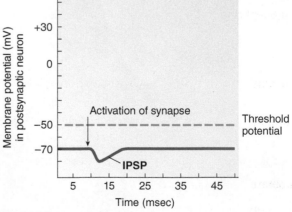

(b) Inhibitory synapse

Figure 4-15 Postsynaptic potentials. (a) An excitatory postsynaptic potential (EPSP) brought about by activation of an excitatory presynaptic input brings the postsynaptic neuron closer to threshold potential. (b) An inhibitory postsynaptic potential (IPSP) brought about by activation of an inhibitory presynaptic input moves the postsynaptic neuron farther from threshold potential.

Inhibitory Synapses At an **inhibitory synapse**, binding of a neurotransmitter with its receptor-channels increases the permeability of the subsynaptic membrane to either K^+ or chloride (Cl^-), depending on the synapse. The resulting ion movements bring about a *small hyperpolarization* of the postsynaptic neuron—that is, greater internal negativity. In the case of increased P_{K^+}, more positive charges leave the cell via K^+ efflux, leaving more negative charges behind on the inside. In the case of increased P_{Cl^-}, because the concentration of Cl^- is higher in the ECF, negative charges enter the cell in the form of Cl^-. In either case, this small hyperpolarization moves the membrane potential even farther from threshold (Figure 4-15b), reducing the likelihood that the postsynaptic neuron will reach threshold (in response to excitatory input) and undergo an action potential—that is, the membrane is now less excitable (harder to bring to threshold by excitatory input) than when it is at resting potential. The membrane is said to be inhibited under these circumstances, and the small hyperpolarization of the postsynaptic cell is called an **inhibitory postsynaptic potential**, or **IPSP**.

In cells where the equilibrium potential for Cl^- exactly equals the resting potential (see p. 84), an increased P_{Cl^-} does not result in a hyperpolarization because there is no driving force to produce Cl^- movement. Opening of Cl^- channels in these cells tends to hold the membrane at resting potential, reducing the likelihood that threshold will be reached.

Note that EPSPs and IPSPs are produced by opening of chemically gated channels, unlike action potentials, which are produced by opening of voltage-gated channels.

Each neurotransmitter–receptor combination always produces the same response.

Many different molecules serve as neurotransmitters. The chemical classes and functions of the major neurotransmitters are found in Table 4-2. You will learn about the details of these neurotransmitter actions as we discuss their roles later on.

Even though neurotransmitters vary from synapse to synapse, the same neurotransmitter is always released at a particular synapse. Furthermore, at a given synapse, binding of a neurotransmitter with its appropriate subsynaptic receptor-channels always leads to the same change in permeability and resultant change in potential of the postsynaptic membrane. Thus, the response to a given neurotransmitter–receptor combination is always the same; the combination does not generate an EPSP under one circumstance and an IPSP under another. Some neurotransmitters (for example, *glutamate*, the most common excitatory neurotransmitter in the brain) typically bring about EPSPs, whereas others (for

TABLE 4-2 Major Neurotransmitters

Classes and Examples of Neurotransmitters	Chemical Structure	Functions
Choline derivative		
Acetylcholine	Synthesized from choline and acetyl CoA	Major neurotransmitter in PNS*: released from motor nerves that supply skeletal muscle and from parasympathetic nerves that supply smooth muscle, cardiac muscle, and exocrine glands; also acts in CNS**
Biogenic amines (Monoamines)	Amines each derived from a single amino acid	
Norepinephrine	Made from tyrosine; is a catecholamine	Important neurotransmitter in PNS: released from sympathetic nerves that supply smooth muscle, cardiac muscle, and exocrine glands; also acts in CNS in pathways involved with memory, mood, emotions, behavior, sensory perception, sleep, and muscle movements
Dopamine	Made from tyrosine; is a catecholamine	Acts in CNS in many pathways similar to norepinephrine; especially important in "pleasure" pathways and muscle movements
Serotonin	Made from tryptophan; is an indoleamine	Acts in CNS in pathways involving mood, emotions, behavior, appetite, states of consciousness, and muscle movements
Amino acids	Are single amino acids	Most abundant neurotransmitters
Glutamate		Primary excitatory neurotransmitter in CNS; important in pathways involved with memory and learning
Gamma-aminobutyric acid (GABA)		Primary inhibitory neurotransmitter in brain; often acts in same circuits as glutamate
Glycine		Primary inhibitory neurotransmitter in spinal cord and brain stem

*PNS refers to the peripheral nervous system.
**CNS refers to the central nervous system.

example, *gamma-aminobutyric acid,* or GABA, the brain's main inhibitory neurotransmitter) always produce IPSPs. Still other neurotransmitters (for example, *norepinephrine*) can produce EPSPs at one synapse and IPSPs at a different synapse, because different permeability changes occur in response to the binding of this same neurotransmitter to different postsynaptic neurons. Yet the response at a given norepinephrine-influenced synapse is always either excitatory or inhibitory.

Most of the time, each axon terminal releases only one neurotransmitter. However, in some cases two different neurotransmitters can be released simultaneously from a single axon terminal. For example, *glycine* and GABA, both of which produce inhibitory responses, can be packaged and released from the same synaptic vesicles. Scientists speculate that the fast-acting glycine and more slowly acting GABA may complement each other in the control of activities that depend on precise timing—for example, coordination of complex movements.

Neurotransmitters are quickly removed from the synaptic cleft.

As long as the neurotransmitter remains bound to the receptor-channels, the alteration in membrane permeability responsible for the EPSP or IPSP continues. For the postsynaptic neuron to be ready to receive additional messages from the same or other presynaptic inputs, the neurotransmitter must be inactivated or removed from the postsynaptic cleft after it has produced the appropriate response in the postsynaptic neuron—that is, the postsynaptic "slate" must be "wiped clean." Thus, after combining with the postsynaptic receptor-channel, chemical transmitters are removed and the response is terminated.

Several mechanisms can remove the neurotransmitter: It may diffuse away from the synaptic cleft, be inactivated by specific enzymes within the subsynaptic membrane, or be actively taken back up into the axon terminal by transport mechanisms in the presynaptic membrane. Once the neurotransmitter is taken back up, it can be stored and released another time (recycled) in response to a subsequent action potential or destroyed by enzymes within the synaptic knob. The method used depends on the particular synapse.

Clinical Note Some drugs work by interfering with removal of specific neurotransmitters from synapses. For example, **selective serotonin reuptake inhibitors (SSRIs)**, as their name implies, selectively block the reuptake of *serotonin* into presynaptic axon terminals, thereby prolonging the action of this neurotransmitter at synapses that use this messenger. SSRIs, such as *Prozac,* are prescribed to treat depression, which is characterized by a deficiency of serotonin, among other things. Serotonin is involved in neural pathways that regulate mood and behavior.

The grand postsynaptic potential depends on the sum of the activities of all presynaptic inputs.

EPSPs and IPSPs are graded potentials. Unlike action potentials, which behave according to the all-or-none law, graded potentials can be of varying magnitude, have no refractory period, and can be summed (added on top of one another). What are the mechanisms and significance of summation?

The events that occur at a single synapse result in either an EPSP or an IPSP at the postsynaptic neuron. But if a single EPSP is inadequate to bring the postsynaptic neuron to threshold and an IPSP moves it even farther from threshold, how can an action potential be initiated in the postsynaptic neuron? The answer lies in the thousands of presynaptic inputs that a typical neuronal cell body receives from many other neurons. Some of these presynaptic inputs may be carrying sensory information from the environment; some may be signaling internal changes in homeostatic balance; others may be transmitting signals from control centers in the brain; and still others may arrive carrying other bits of information. At any given time, any number of these presynaptic neurons (probably hundreds) may be firing and thus influencing the postsynaptic neuron's level of activity. The total potential in the postsynaptic neuron, the **grand postsynaptic potential (GPSP)**, is a composite of all EPSPs and IPSPs occurring around the same time.

The postsynaptic neuron can be brought to threshold by either *temporal summation* or *spatial summation.* To illustrate these methods of summation, we examine the possible interactions of three presynaptic inputs—two excitatory inputs (Ex1 and Ex2) and one inhibitory input (In1)—on a hypothetical postsynaptic neuron (❚Figure 4-16). The recording shown in the figure represents the potential in the postsynaptic cell. Bear in mind during our discussion of this simplified version that many thousands of synapses are actually interacting in the same way on a single cell body and its dendrites.

Temporal Summation Suppose that Ex1 has an action potential that causes an EPSP in the postsynaptic neuron. After this EPSP has died off, if another action potential occurs in Ex1, an EPSP of the same magnitude takes place before dying off (❚Figure 4-16a). Next, suppose that Ex1 has two action potentials in close succession (❚Figure 4-16b). The first action potential in Ex1 produces an EPSP in the postsynaptic membrane. While the postsynaptic membrane is still partially depolarized from this first EPSP, the second action potential in Ex1 produces a second EPSP. Because graded potentials do not have a refractory period, the second EPSP can add to the first, bringing the membrane to threshold and initiating an action potential in the postsynaptic neuron. EPSPs can add together or sum because an EPSP lasts longer than the action potential that caused it. The presynaptic neuron (Ex1) can recover from its refractory period following the first action potential and have a second action potential, causing a second EPSP in the postsynaptic neuron, before the first EPSP is finished.

The summing of several EPSPs occurring very close together in time because of successive firing of a single presynaptic neuron is known as **temporal summation** (*tempus* means "time"). In reality, up to 50 EPSPs might be needed to bring the postsynaptic membrane to threshold. Each action potential in a single presynaptic neuron triggers the emptying of a certain number of synaptic vesicles. The amount of neurotransmitter released and the resultant magnitude of the change in postsynaptic potential are thus directly related to the frequency of presynaptic action potentials. One way in which the postsynaptic mem-

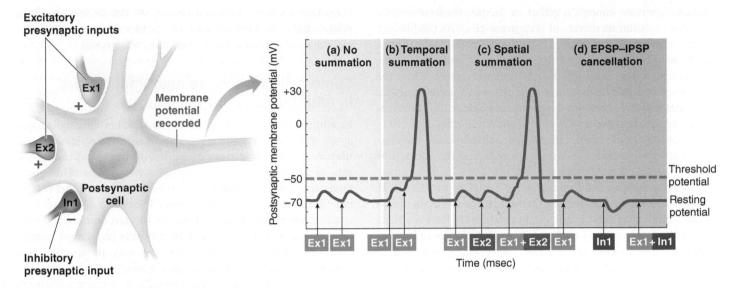

Excitatory presynaptic inputs

Ex1

Ex2

Postsynaptic cell

Membrane potential recorded

In1

Inhibitory presynaptic input

(a) No summation

(b) Temporal summation

(c) Spatial summation

(d) EPSP–IPSP cancellation

Postsynaptic membrane potential (mV)

+30

0

−50 — — — Threshold potential

−70 — Resting potential

Ex1 | Ex1 | Ex1 | Ex1 | Ex1 | Ex2 | Ex1 + Ex2 | Ex1 | In1 | Ex1 + In1

Time (msec)

(a) If an excitatory presynaptic input (Ex1) is stimulated a second time after the first EPSP in the postsynaptic cell has died off, a second EPSP of the same magnitude will occur.

(b) If, however, Ex1 is stimulated a second time before the first EPSP has died off, the second EPSP will add onto, or sum with, the first EPSP, resulting in *temporal summation*, which may bring the postsynaptic cell to threshold.

(c) The postsynaptic cell may also be brought to threshold by *spatial summation* of EPSPs that are initiated by simultaneous activation of two (Ex1 and Ex2) or more excitatory presynaptic inputs.

(d) Simultaneous activation of an excitatory (Ex1) and inhibitory (In1) presynaptic input does not change the postsynaptic potential, because the resultant EPSP and IPSP cancel each other out.

I Figure 4-16 Determination of the grand postsynaptic potential by the sum of activity in the presynaptic inputs. Two excitatory (Ex1 and Ex2) and one inhibitory (In1) presynaptic inputs terminate on this hypothetical postsynaptic neuron. The potential of the postsynaptic neuron is being recorded. For simplicity in the figure, summation of two EPSPs brings the postsynaptic neuron to threshold, but in reality many EPSPs must sum to reach threshold.

brane can be brought to threshold, then, is through rapid, repetitive excitation from a single persistent input.

Spatial Summation Let us now see what happens in the postsynaptic neuron if both excitatory inputs are stimulated simultaneously (I Figure 4-16c). An action potential in either Ex1 or Ex2 produces an EPSP in the postsynaptic neuron; however, neither of these alone brings the membrane to threshold to elicit a postsynaptic action potential. But simultaneous action potentials in Ex1 and Ex2 produce EPSPs that add to each other, bringing the postsynaptic membrane to threshold, so an action potential does occur. The summation of EPSPs originating simultaneously from several presynaptic inputs (that is, from different points in "space") is known as **spatial summation**. A second way to elicit an action potential in a postsynaptic cell, therefore, is through concurrent activation of several excitatory inputs. Again, in reality, up to 50 simultaneous EPSPs are required to bring the postsynaptic membrane to threshold.

Similarly, IPSPs can undergo both temporal and spatial summation. As IPSPs add together, however, they progressively move the potential farther from threshold.

Cancellation of Concurrent EPSPs and IPSPs If an excitatory and an inhibitory input are simultaneously activated, the concurrent EPSP and IPSP more or less cancel each other

out. The extent of cancellation depends on their respective magnitudes. In most cases, the postsynaptic membrane potential remains close to resting potential (I Figure 4-16d).

Importance of Postsynaptic Integration The magnitude of the GPSP depends on the sum of activity in all presynaptic inputs and, in turn, determines whether or not the postsynaptic neuron will undergo an action potential to pass information on to the cells on which the neuron terminates. The following oversimplified real-life example demonstrates the benefits of this neuronal integration. The explanation is not completely accurate technically, but the principles of summation are accurate.

Assume for simplicity's sake that urination is controlled by a postsynaptic neuron that innervates the urinary bladder. When this neuron fires, the bladder contracts. (Actually, voluntary control of urination is accomplished by postsynaptic integration at the neuron controlling the external urethral sphincter rather than the bladder itself.) As the bladder fills with urine and becomes stretched, a reflex is initiated that ultimately produces EPSPs in the postsynaptic neuron responsible for causing bladder contraction. Partial filling of the bladder does not cause enough excitation to bring the neuron to threshold, so urination does not take place—that is, action potentials do not occur frequently enough in presynaptic neuron Ex1, which fires reflexly in response to the degree of bladder stretching, to generate EPSPs close enough together in the postsynaptic neuron to bring the latter to threshold (▪Figure 4-16a). As the bladder progressively fills, the frequency of action potentials progressively increases in presynaptic neuron Ex1, leading to more rapid formation of EPSPs in the postsynaptic neuron. Thus, the frequency of EPSP formation arising from Ex1 activity signals the postsynaptic neuron of the extent of bladder filling. When the bladder becomes sufficiently stretched that the Ex1-generated EPSPs are temporally summed to threshold, the postsynaptic neuron undergoes an action potential that stimulates bladder contraction (▪Figure 4-16b).

What if the time is inopportune for urination to take place? Presynaptic inputs originating in higher levels of the brain responsible for voluntary control can produce IPSPs at the bladder postsynaptic neuron (In1 in ▪Figure 4-16d). These "voluntary" IPSPs in effect cancel out the "reflex" EPSPs triggered by stretching of the bladder. Thus, the postsynaptic neuron remains at resting potential and does not have an action potential, so the bladder is prevented from contracting and emptying even though it is full.

What if a person's bladder is only partially filled, so that the presynaptic input from this source (Ex1) is insufficient to bring the postsynaptic neuron to threshold to cause bladder contraction, yet the individual needs to supply a urine specimen for laboratory analysis? The person can voluntarily activate another excitatory presynaptic neuron originating in higher brain levels (Ex2 in ▪Figure 4-16c). The "voluntary" EPSPs arising from Ex2 activity and the "reflex" EPSPs arising from Ex1 activity are spatially summed to bring the postsynaptic neuron to threshold. This achieves the action potential necessary to stimulate bladder contraction, even though the bladder is not full.

This example illustrates the importance of postsynaptic neuronal integration. Each postsynaptic neuron in a sense "computes" all the input it receives and "decides" whether to pass the information on (that is, whether threshold is reached and an action potential is transmitted down the axon). In this way, neurons serve as complex computational devices, or integrators. The dendrites function as the primary processors of incoming information. They receive and tally the signals from all presynaptic neurons. Each neuron's output in the form of frequency of action potentials to other cells (muscle cells, gland cells, or other neurons) reflects the balance of activity in the inputs it receives via EPSPs or IPSPs from the thousands of other neurons that terminate on it. Each postsynaptic neuron filters out information that is not significant enough to bring it to threshold and does not pass it on. If every action potential in every presynaptic neuron that impinges on a particular postsynaptic neuron were to cause an action potential in the postsynaptic neuron, the neuronal pathways would be overwhelmed with trivia. Only if an excitatory presynaptic signal is reinforced by other supporting signals through summation will the information be passed on. Furthermore, interaction of EPSPs and IPSPs provides a way for one set of signals to offset another, allowing a fine degree of discrimination and control in determining what information will be passed on. Thus, unlike an electrical synapse, a chemical synapse is more than a simple on–off switch because many factors can influence the generation of a new action potential in the postsynaptic cell. Whether the postsynaptic neuron has an action potential depends on the relative balance of information coming in via presynaptic neurons at all of its excitatory and inhibitory synapses.

Some neurons secrete neuromodulators in addition to neurotransmitters.

In addition to the classical neurotransmitters released at synapses, some neurons also release *neuromodulators*. **Neuromodulators** are chemical messengers that do not cause the formation of EPSPs or IPSPs but instead act slowly to bring about long-term changes that subtly *modulate* (that is, depress or enhance) the action of the synapse. The neuronal receptors to which neuromodulators bind are not located on the subsynaptic membrane, and they do not directly alter membrane permeability and potential. Neuromodulators may act at either presynaptic or postsynaptic sites. For example, a neuromodulator may influence the level of an enzyme involved in the synthesis of a neurotransmitter by a presynaptic neuron, or it may alter the sensitivity of the postsynaptic neuron to a particular neurotransmitter by causing long-term changes in the number of subsynaptic receptors for the neurotransmitter. Thus, neuromodulators delicately fine-tune the synaptic response. The effect may last for days, months, or even years. Whereas neurotransmitters are involved in rapid communication between neurons, neuromodulators are involved with more long-lasting events, such as learning and motivation.

A variety of chemicals serve as neuromodulators. (1) *Neuropeptides* are the largest class of neuromodulators. Other neuromodulators include some novel chemical messengers such as (2) *adenosine triphosphate (ATP)*, which normally serves as the primary energy carrier but can be released into the synaptic cleft; (3) the short-lived gas *nitric oxide (NO)*, which serves multiple other roles in the body (see p. 346); and (4) *endocannabinoids*, a group of lipid messengers that act in a way similar to the active component of *cannabis*, or *marijuana*. Let us examine the most abundant of the neuromodulators, neuropeptides, in more detail.

Neuropeptides differ from classical neurotransmitters in several important ways. Classical neurotransmitters are small, rapid-acting molecules that typically trigger the opening of specific ion channels to bring about a change in potential in the postsynaptic neuron (an EPSP or an IPSP) within a few milliseconds or less. Most classical neurotransmitters are synthesized and packaged locally in synaptic vesicles in the cytosol of the axon terminal. These chemical messengers are primarily amino acids or closely related compounds. By contrast, neuropeptides are larger molecules made up of anywhere from 2 to about 40 amino acids. They are synthesized in the endoplasmic reticulum and Golgi complex (see ▪Figure 2-3, p. 27) of the neuronal cell body and are moved by axonal transport along the microtubular highways to the axon terminal (see p. 47). Neuropeptides are not stored in small synaptic vesicles with the classical neurotransmitters but instead are packaged in large **dense-core vesicles**, which are also present in the axon terminal [see the chapter opener photo on p. 87. A dense-core vesicle is larger deep blue vesicle among the smaller orange synaptic vesicles in the synaptic knob on the top left. (The large green spheres are mitochondria.)]. Dense-core vesicles undergo Ca^{2+}-induced exocytosis and release neuropeptides at the same time that neurotransmitter is released from synaptic vesicles. An axon terminal typically releases only a single classical neurotransmitter, but the same terminal may also contain one or more neuropeptides that are co-secreted simultaneously with the neurotransmitter.

Most but not all neuropeptides function as neuromodulators. (An example of a neuropeptide that has no effect on neuronal activity and thus does not function as a neuromodulator is a neurohormone, which is secreted by specialized neurons into the blood instead of being released into a synaptic cleft.) *Endogenous opioids* are examples of neuropeptides that do function as neuromodulators. Endogenous opioids are internally produced morphinelike substances that dampen the sensation of pain by exerting effects similar to the opiate drugs morphine and codeine (see p. 192). Neuropeptides that serve as neuromodulators also include many substances that in addition function as hormones released into the blood from endocrine tissues. *Cholecystokinin (CCK)* is an example. As a hormone, CCK is released from the small intestine following a meal and causes the gallbladder to contract and release bile into the intestine, among other digestive actions (see Chapter 16). CCK also functions in the brain as a neuromodulator, causing the sensation of no longer being hungry. A number of chemical messengers are versatile and can assume different roles, depending on their source, distribution, and interaction with different cell types.

Presynaptic inhibition or facilitation can selectively alter the effectiveness of a presynaptic input.

Besides neuromodulation, another means of depressing or enhancing synaptic effectiveness is presynaptic inhibition or facilitation. Sometimes, a third neuron influences activity between a presynaptic ending and a postsynaptic neuron. The presynaptic axon terminal (labeled A in ▌Figure 4-17) may itself be innervated by another axon terminal (labeled B). Note that this axon-to-axon synapse is different than the usual axon-to-dendrite (or cell body) synapse. The neurotransmitter released from modulatory terminal B binds with receptors on terminal A. This binding alters the amount of neurotransmitter released from terminal A in response to action potentials. If the amount of neurotransmitter released from A is reduced, the phenomenon is known as **presynaptic inhibition**. If the release of neurotransmitter is enhanced, the effect is called **presynaptic facilitation**.

Let us look more closely at how this process works. You know that Ca^{2+} entry into an axon terminal causes the release of neurotransmitter by exocytosis of synaptic vesicles. The amount of neurotransmitter released from terminal A depends on how much Ca^{2+} enters this terminal in response to an action potential. Ca^{2+} entry into terminal A, in turn, can be influenced by activity in modulatory terminal B. We will use presynaptic inhibition to illustrate (▌Figure 4-17). The amount of neurotransmitter released from presynaptic terminal A, an excitatory input in our example, influences the potential in the postsynaptic neuron at which it terminates (labeled C in the figure). Firing of A by itself generates an EPSP in postsynaptic neuron C. Now consider that B is stimulated simultaneously with A. Neurotransmitter from terminal B binds on terminal A, reducing Ca^{2+} entry into terminal A. Less Ca^{2+} entry means less neurotransmitter release from A. Note that modulatory neuron B can suppress neurotransmitter release from A only when A is firing. If this presynaptic inhibition by B prevents A from releasing its neurotransmitter, the formation of EPSPs on postsynaptic membrane C from input A is specifically prevented. As a result, no change in the potential of the postsynaptic neuron occurs despite action potentials in A.

Would the simultaneous production of an IPSP through activation of an inhibitory input to negate an EPSP produced by excitatory input A achieve the same result? Not quite. Activation of an inhibitory input to cell C would produce an IPSP in cell C, but this IPSP could cancel out not only an EPSP from excitatory input A but also any EPSPs produced by other excitatory terminals, such as terminal D in the figure. The entire postsynaptic membrane is hyperpolarized by IPSPs, thereby negating (canceling) excitatory information fed into any part of the cell from any presynaptic input. By contrast, presynaptic inhibition (or presynaptic facilitation) works in a much more specific way. Presynaptic inhibition provides a means by which certain inputs to the postsynaptic neuron can be *selectively*

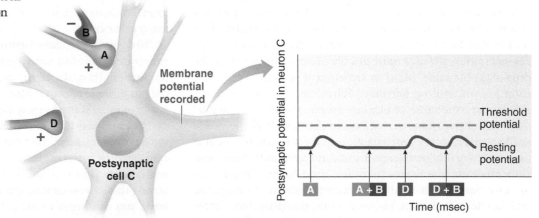

▌Figure 4-17 **Presynaptic inhibition.** A, an excitatory terminal ending on postsynaptic cell C, is itself innervated by inhibitory terminal B. Stimulation of terminal A alone produces an EPSP in cell C, but simultaneous stimulation of terminal B prevents the release of excitatory neurotransmitter from terminal A. Consequently, no EPSP is produced in cell C despite the fact that terminal A has been stimulated. Such presynaptic inhibition selectively depresses activity from terminal A without suppressing any other excitatory input to cell C. Stimulation of excitatory terminal D produces an EPSP in cell C even though inhibitory terminal B is simultaneously stimulated because terminal B only inhibits terminal A.

FIGURE FOCUS: *Assume excitatory terminal D is itself innervated by an excitatory terminal E. What would happen to the potential in postsynaptic cell C if terminal D and terminal E were simultaneously stimulated?*

inhibited without affecting the contributions of any other inputs. For example, firing of B specifically prevents the formation of an EPSP in the postsynaptic neuron from excitatory presynaptic neuron A but does not influence other excitatory presynaptic inputs. Excitatory input D can still produce an EPSP in the postsynaptic neuron even when B is firing. This type of neuronal integration is another means by which electrical signaling between neurons can be carefully fine-tuned.

Drugs and diseases can modify synaptic transmission.

Clinical Note Most drugs that influence the nervous system function by altering synaptic mechanisms. Synaptic drugs may block an undesirable effect or enhance a desirable effect. Possible drug actions include (1) altering the synthesis, storage, or release of a neurotransmitter; (2) modifying neurotransmitter interaction with the postsynaptic receptor; (3) influencing neurotransmitter reuptake or destruction; and (4) replacing a deficient neurotransmitter with a substitute transmitter.

You already learned about SSRIs. As another example, the socially abused drug **cocaine** blocks the reuptake of the neurotransmitter *dopamine* at presynaptic terminals. It does so by binding competitively with the dopamine reuptake transporter, which is a protein molecule that picks up released dopamine from the synaptic cleft and shuttles it back to the axon terminal. With cocaine occupying the dopamine transporter, dopamine remains in the synaptic cleft longer than usual and continues to interact with its postsynaptic receptors. The result is prolonged activation of neural pathways that use this chemical as a neurotransmitter, especially pathways that play a role in feelings of pleasure. In essence, when cocaine is present, the neural switches in the pleasure pathway are locked in the "on" position.

Cocaine is addictive because the involved neurons become *desensitized* to the drug. After the postsynaptic cells are incessantly stimulated for an extended time, they can no longer transmit normally across synapses without increasingly larger doses of the drug. Specifically, with prolonged use of cocaine, the number of dopamine receptors in the brain is reduced in response to the glut of the abused substance. As a result of this desensitization, the user must steadily increase the dosage of the drug to get the same "high," or sensation of pleasure, a phenomenon known as *drug tolerance*. When the cocaine molecules diffuse away, the sense of pleasure evaporates because the normal level of dopamine activity no longer "satisfies" the overly needy demands of the postsynaptic cells for stimulation. Cocaine users reaching this low become frantic and profoundly depressed. Only more cocaine makes them feel good again. But repeated use of cocaine often modifies responsiveness to the drug; the user no longer derives pleasure from the drug but suffers unpleasant *withdrawal symptoms* once its effect has worn off. The user typically becomes **addicted** to the drug, compulsively seeking it out, first to experience the pleasurable sensations and later to avoid the negative withdrawal symptoms. Cocaine is abused by millions who have become addicted to its mind-altering properties, with devastating social and economic effects.

Synaptic transmission is also vulnerable to neural toxins, which may cause nervous system disorders by acting at either presynaptic or postsynaptic sites. For example, two different neural poisons, tetanus toxin and strychnine, act at different synaptic sites to block inhibitory impulses while leaving excitatory inputs unchecked. Tetanus toxin prevents the presynaptic release of a specific inhibitory neurotransmitter, whereas strychnine blocks specific postsynaptic inhibitory receptors.

Tetanus toxin prevents release of GABA from inhibitory presynaptic inputs terminating at neurons that supply skeletal muscles. Unchecked excitatory inputs to these neurons result in uncontrolled muscle spasms. These spasms occur especially in the jaw muscles early in the disease, giving rise to the common name of *lockjaw* for this condition. Later, they progress to the muscles responsible for breathing, at which point death occurs.

Strychnine competes with another inhibitory neurotransmitter, glycine, at the postsynaptic receptor. This poison combines with the receptor but does not directly alter the potential of the postsynaptic cell in any way. Instead, it blocks the receptor so that it is not available for interaction with glycine when the latter is released from the inhibitory presynaptic ending. Thus, strychnine abolishes postsynaptic inhibition (formation of IPSPs) in nerve pathways that use glycine as an inhibitory neurotransmitter. Unchecked excitatory pathways lead to convulsions, muscle spasticity, and death.

Many other drugs and diseases influence synaptic transmission, but as these examples illustrate, any site along the synaptic pathway is vulnerable to interference.

Neurons are linked through complex converging and diverging pathways.

Two important relationships exist between neurons: convergence and divergence. A given neuron may have many other neurons synapsing on it. Such a relationship is known as **convergence** (Figure 4-18). Through converging input, a single cell is influenced by thousands of other cells. This single cell, in turn, influences the level of activity in many other cells by divergence of output. The term **divergence** refers to the branching of axon terminals so that a single cell synapses with and influences many other cells.

Note that a particular neuron is postsynaptic to the neurons converging on it but presynaptic to the other cells at which it terminates. Thus, the terms *presynaptic* and *postsynaptic* refer only to a single synapse. Most neurons are presynaptic to one group of neurons and postsynaptic to another group.

An estimated 100 billion neurons and 10^{14} (100 quadrillion) synapses are found in the brain alone! A single neuron may be connected to between 5000 and 10,000 other neurons. When you consider the vast and intricate interconnections possible among these neurons through converging and diverging pathways, you can begin to imagine the complexity of the wiring mechanism of our nervous system. Even the most sophisticated computers are far less complex than the human brain. The "language" of the nervous system—that is, all communication among neurons—is in the form of graded potentials, action potentials, neurotransmitter signaling across synapses, and other nonsynaptic forms of chemical chatter. All activities for which the nervous system is responsible—every sensation, every command to move a muscle, every thought, every emo-

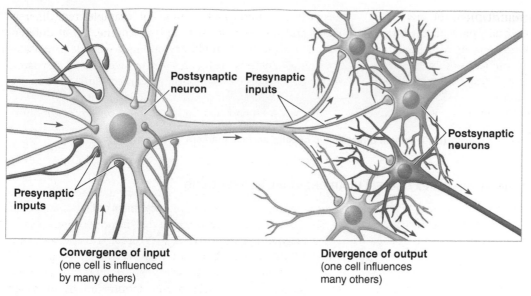

Postsynaptic neuron

Presynaptic inputs

Postsynaptic neurons

Presynaptic inputs

Convergence of input
(one cell is influenced by many others)

Divergence of output
(one cell influences many others)

❙Figure 4-18 **Convergence and divergence.** Arrows indicate the direction in which information is being conveyed.

tion, every memory, every spark of creativity—depend on the patterns of electrical and chemical signaling among neurons along these complexly wired neural pathways.

A neuron communicates with the cells it influences by releasing a neurotransmitter, but this is only one means of intercellular ("between cell") communication. We now consider all the ways by which cells can "talk" with one another.

Check Your Understanding 4.4

1. Explain why synapses operate only in the direction from presynaptic to postsynaptic neurons.
2. Draw a graph of an EPSP and an IPSP, showing the relative distance between each of these and threshold potential.
3. Distinguish between temporal summation and spatial summation.
4. Compare neurotransmitters and neuromodulators.

4.5 Intercellular Communication and Signal Transduction

Coordination of the diverse activities of cells throughout the body to accomplish life-sustaining and other desired responses depends on the ability of cells to communicate with one another.

Communication among cells is largely orchestrated by extracellular chemical messengers.

Intercellular communication can take place either directly or indirectly (❙Figure 4-19). *Direct* intercellular communication involves physical contact between the interacting cells:

1. *Through gap junctions and possibly through tunneling nanotubes.* The most intimate means of intercellular communication is through gap junctions, the minute tunnels that bridge the cytoplasm of neighboring cells in some types of tissues. Through gap junctions, ions and small molecules are directly exchanged between closely associated interacting cells without ever entering the ECF.

Scientists recently discovered a possible new route for direct exchange of materials between cells—long, thin, hollow filaments called **tunneling nanotubes (TNTs)**— that transiently form between laboratory-grown cells of a variety of types (❙Figure 4-20) and have now been confirmed to exist in living tissue. Studies suggest that these intercellular bridges serve as a route for selective, relatively long transfer from one cell to another of rather large cargo, including proteins or even organelles such as mitochondria. Whereas cells connected by gap junctions are in close proximity (being 2 to 4 nm apart), TNTs may extend up to 150 μm (150,000 nm) between cells. Furthermore, the opening in a gap junction is 1.5 nm in diameter compared to the much larger 50 to 200 nm diameter opening in a TNT. Because of these major differences between gap junctions and TNTs, TNTs can transfer larger cargo considerably longer distances than gap junctions can. Researchers have identified motor proteins (see p. 48) in TNTs that are believed to help move substances through these long connecting tunnels. Evidence further suggests that viruses, including HIV, the AIDS virus, can hijack TNTs to move directly between cells without entering the ECF.

2. *Through transient direct linkup of surface markers.* Some cells, such as those of the immune system, have specialized markers on the surface membrane that allow them to directly link with certain other cells that have compatible markers for transient interactions. This is how cell-destroying immune cells specifically recognize and selectively destroy only undesirable cells, such as cancer cells, while leaving the body's healthy cells alone (see p. 423).

The most common means by which cells communicate with one another is *indirectly* through **extracellular chemical messengers**, or **signal molecules**, of which there are four types: *paracrines/autocrines, neurotransmitters, hormones,* and *neurohormones.* In each case, a specific chemical messenger, the signal molecule, is synthesized by specialized controlling cells to serve a designated purpose. On being released into the ECF by appropriate stimulation, these extracellular chemical messengers act on other particular cells, the messenger's **target cells**, in a prescribed manner. To exert its effect, an extracellular chemi-

DIRECT INTERCELLULAR COMMUNICATION

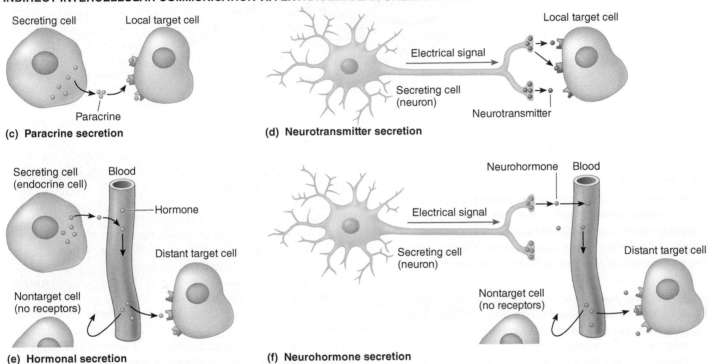

(a) Gap junctions

(b) Transient direct linkup of cells' surface markers

INDIRECT INTERCELLULAR COMMUNICATION VIA EXTRACELLULAR CHEMICAL MESSENGERS

(c) Paracrine secretion

(d) Neurotransmitter secretion

(e) Hormonal secretion

(f) Neurohormone secretion

❙Figure 4-19 **Types of intercellular communication.** Gap junctions and transient direct linkup of cells by means of complementary surface markers are both means of direct communication between cells. Paracrines, neurotransmitters, hormones, and neurohormones are all extracellular chemical messengers that accomplish indirect communication between cells. These chemical messengers differ in their source and the distance they travel to reach their target cells.

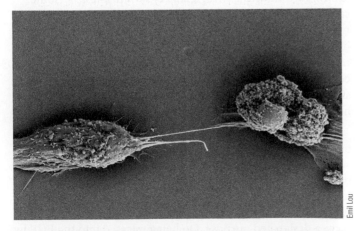

❙Figure 4-20 **Tunneling nanotubes.** These recently discovered long, hollow filaments, shown here in cells grown in a laboratory, are believed to be a new route for direct transfer of relatively large cargo between cells connected by these bridges in the body.

cal messenger must bind with target cell receptors specific for it. A given cell may have thousands to as many as a few million receptors, of which hundreds to as many as 100,000 may be for the same chemical messenger. Different cell types have distinct combinations of receptors, allowing them to react individually to various regulatory extracellular chemical messengers. Nearly 5% of all genes in humans code for synthesis of these membrane receptors, indicative of the importance of this means of intercellular communication.

The four types of extracellular chemical messengers differ in their source and the distance to and means by which they get to their site of action.

1. **Paracrines** are local chemical messengers whose effect is exerted only on neighboring cells in the immediate environment of their site of secretion. An **autocrine** is even more localized—after being secreted, it acts only on the cell that secreted it. For convenience, we implicitly include autocrines in

future discussions regarding paracrines. Because paracrines are distributed by simple diffusion within the interstitial fluid, their action is restricted to short distances. They do not gain entry to the blood in any significant quantity because they are rapidly inactivated by locally existing enzymes. One example of a paracrine is *histamine*, which is released from a specific type of connective tissue cell during an inflammatory response within an invaded or injured tissue (see p. 409). Among other things, histamine dilates (opens more widely) the blood vessels in the vicinity to increase blood flow to the tissue. This action brings additional blood-borne combat supplies into the affected area.

2. As you just learned, neurons communicate directly with the cells they innervate (their target cells) by releasing **neurotransmitters**, which are short-range chemical messengers, in response to electrical signals (action potentials). Like paracrines, neurotransmitters diffuse from their site of release across a narrow extracellular space to act locally on an adjoining target cell, which may be another neuron, a muscle, or a gland. Neurons themselves may carry electrical signals long distances (the length of the axon), but the chemical messenger released at the axon terminal acts at short range—just across the synaptic cleft.

3. **Hormones** are long-range chemical messengers specifically secreted into the blood by endocrine glands in response to an appropriate signal. The blood carries the messengers to other sites in the body, where they exert their effects on their target cells some distance from their site of release. Only the target cells of a particular hormone have membrane receptors for binding with this hormone. Nontarget cells are not influenced by any blood-borne hormones that reach them.

4. **Neurohormones** are hormones released into the blood by *neurosecretory neurons*. Like ordinary neurons, neurosecretory neurons can respond to and conduct electrical signals. Instead of directly innervating target cells and releasing a neurotransmitter into the synaptic cleft, however, a neurosecretory neuron releases its chemical messenger, a neurohormone, into the blood when an action potential reaches the axon terminals. The neurohormone is then distributed through the blood to distant target cells. An example is *vasopressin*, a neurohormone produced by nerve cells in the brain that promotes water conservation by the kidneys during urine formation. In the future, the general term *hormone* will tacitly include both blood-borne hormonal and neurohormonal messengers.

We now turn to how these chemical messengers cause the right cellular response.

Extracellular chemical messengers bring about cell responses by signal transduction.

The term **signal transduction** refers to the process by which incoming signals (instructions from extracellular chemical messengers) are conveyed into the target cell, where they are transformed into the dictated cellular response. Binding of the extracellular chemical messenger amounts to a signal for the cell to get a certain job done. During signal transduction, the extracellular signal is transduced, or changed into a form

necessary to modify intracellular activities to accomplish the desired outcome. (A *transducer* is a device that receives energy from one system and transmits it in a different form to another system. For example, your radio receives radio waves sent out from the broadcast station and transmits these signals in the form of sound waves that can be detected by your ears.)

Signal transduction occurs by different mechanisms, depending on the messenger and the receptor type (▮Table 4-3):

1. *Lipid-soluble extracellular chemical messengers,* such as cholesterol-derived steroid hormones, gain entry into the cell by dissolving in and passing through the lipid bilayer of the target cell's plasma membrane. Thus, these extracellular chemical messengers bind to receptors inside the target cell to initiate the desired intracellular response themselves, usually by *changing gene activity,* either turning on or suppressing transcription of specific genes. In this way the messenger controls the level of activity of the transcribed protein, such as an enzyme.

2. *Water-soluble extracellular chemical messengers,* by contrast, cannot gain entry to the target cell because they are poorly soluble in lipid and cannot dissolve in the plasma membrane. The major water-soluble extracellular messengers are peptide (protein) hormones delivered by the blood, neurotransmitters released from nerve endings, and paracrines released locally. These water-soluble messengers signal the target cell to perform a certain response by first binding with receptors specific for that given messenger on the outer surface of the plasma membrane. This binding triggers a sequence of intracellular events that controls a particular cellular activity,

---

▮**TABLE 4-3** Signal Transduction Pathways Used by Extracellular Chemical Messengers

I. Pathway used by lipid-soluble extracellular messengers that bind to intracellular receptors

A. Function in nucleus to change specific gene activity (example: steroid hormones)

II. Pathways used by water-soluble extracellular messengers that bind to surface membrane receptors

A. Bind to and open or close chemically gated receptor-channels (example: neurotransmitters)

B. Bind to and activate receptor–enzyme complexes

1. Use tyrosine kinase pathway, where the receptor itself functions as an enzyme (examples: insulin, growth factors)

2. Use JAK/STAT pathway, where the receptor and attached enzymes function as a unit (examples: prolactin, immune cytokines)

C. Bind to G-protein-coupled receptors (GPCRs) and activate second-messenger pathways (examples: eicosanoids and most peptide hormones)

---

such as membrane transport, secretion, metabolism, or contraction.

Despite the wide range of possible responses, binding of a water-soluble extracellular messenger (also known as the **first messenger**) to its matching surface membrane receptor brings about the desired intracellular response by one of three general means, depending on receptor type.

1. Messenger binding to a *chemically gated receptor-channel* opens or closes the channel, with the resultant ion movement leading to the cell's response.

2. Messenger binding to a *receptor–enzyme complex* activates tyrosine kinase, which phosphorylates designated proteins that lead to the cell's response.

3. Messenger binding to a *G-protein-coupled receptor* activates a second-messenger pathway that carries out the cell's response.

Because of the universal nature of these events, let us examine each more closely.

Some water-soluble extracellular messengers open chemically gated receptor-channels.

Some extracellular messengers carry out the assigned task by opening or closing specific chemically gated receptor-channels to regulate movement of particular ions across the membrane. In this case, *the receptor itself serves as an ion channel*. When the appropriate extracellular messenger binds to the **receptor-channel**, the channel opens or closes, depending on the signal. (In the future, for convenience when discussing receptor-channels in general we refer only to the more common opening of channels.) An example is the opening of chemically gated receptor-channels in the subsynaptic membrane in response to neurotransmitter binding (see ▮Figure 4-14). The resultant small, short-lived movement of given charge-carrying ions across the membrane through these open channels generates electrical signals—in this example, EPSPs and IPSPs.

On completion of the response, the extracellular messenger is removed from the receptor site and the chemically gated channels close once again. The ions that moved across the membrane through opened channels to trigger the response are returned to their original location by special membrane carriers.

Some water-soluble extracellular messengers activate receptor-enzymes.

Most water-soluble extracellular messengers that cannot enter their target cells issue their orders by triggering a "Psst, pass it on" process. In most instances, on binding to a surface membrane receptor, the extracellular messenger relays its message inside the cell by activating intracellular **protein kinases**, the name for any enzyme that transfers a phosphate group from ATP to a particular intracellular protein. As a result of phosphorylation, these proteins alter their shape and function, that is, are activated, to ultimately accomplish the cellular response dictated by the first messenger. Transduction may occur in a single step, although typically phosphorylation of a single kind of protein does not get the job done. Usually, protein kinases act in a chain of reactions, called a **cascade**, to pass along the signal to the final designated proteins capable of carrying out the desired effect. A body cell contains an estimated 1 billion protein molecules. The ultimate cellular response to receptor binding by an extracellular messenger hinges on which of these proteins are activated by phosphorylation, which depends on the messenger.

Protein kinases are activated on binding of the signal molecule to the surface receptor in one of two ways: by activating tyrosine kinase on binding to a receptor–enzyme complex or by activating a second-messenger pathway on binding to a G-protein-coupled receptor. We next describe these mechanisms.

The protein kinase **tyrosine kinase** is a key participant in two different signaling pathways—the tyrosine kinase pathway and the JAK/STAT pathway (▮Table 4-3). In both of these pathways, protein activation on binding of the extracellular messenger to a receptor–enzyme complex is accomplished by specifically phosphorylating tyrosine, a type of amino acid within the protein, hence the name *tyrosine kinase.*

Tyrosine Kinase Pathway In the **tyrosine kinase pathway**, the simplest of the pathways, the *receptor itself functions as an enzyme*, a so-called **receptor-enzyme**, which has a receptor portion facing the ECF and protein kinase (tyrosine kinase) site on its portion that faces the cytosol (▮Figure 4-21). To activate tyrosine kinase, appropriate extracellular messengers must bind with two of these receptor-enzymes, which assemble into a pair. On activation, the tyrosine kinase site adds phosphate groups to the tyrosines on the cytosolic side of the receptor-enzyme. Designated proteins inside the cell recognize and bind to the phosphorylated receptor-enzyme. Then the receptor-enzyme's tyrosine kinase adds phosphate groups to the tyrosines in the bound proteins. As a result of phosphorylation, the designated proteins change shape and function (are activated), enabling them to bring about the desired cellular response. The hormone insulin, which plays a major role in maintaining glucose homeostasis, exerts its effects via the tyrosine kinase pathway. Also, many growth factors that help regulate cell growth and division, such as *nerve growth factor* and *epidermal growth factor,* act via this pathway.

JAK/STAT Pathway In the **JAK/STAT pathway**, instead of the receptor itself having tyrosine kinase activity, the tyrosine kinase activity resides in a family of separate cytosolic enzymes called *Janus family tyrosine kinases,* better known as *JAKs,* two of which are attached, one on each side, to the receptor. *The receptor and attached enzymes function as a unit.* Binding of an extracellular messenger to the receptor on the ECF side causes a conformational change in the receptor that activates the JAKs bound to the cytosolic side of the receptor. Activated JAKs phosphorylate *signal transducers and activators of transcription (STAT)* within the cytosol. Phosphorylated STAT moves to the nucleus and turns on transcription of selected genes, resulting in synthesis of new proteins that carry out the cellular response. Some hormones, for example prolactin, the hormone that

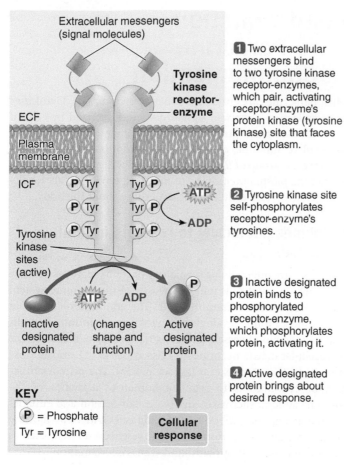

1 Two extracellular messengers bind to two tyrosine kinase receptor-enzymes, which pair, activating receptor-enzyme's protein kinase (tyrosine kinase) site that faces the cytoplasm.

2 Tyrosine kinase site self-phosphorylates receptor-enzyme's tyrosines.

3 Inactive designated protein binds to phosphorylated receptor-enzyme, which phosphorylates protein, activating it.

4 Active designated protein brings about desired response.

I Figure 4-21 **Tyrosine kinase pathway.**

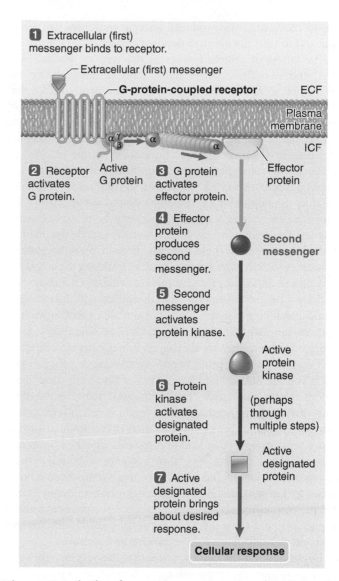

I Figure 4-22 **Activation of a second-messenger pathway via binding of a first messenger to a G-protein-coupled receptor.** Binding of an extracellular (first) messenger to the extracellular side of a G-protein-coupled receptor activates a membrane-bound effector protein by means of a G-protein intermediary. The effector protein produces an intracellular second messenger, which ultimately leads to the cellular response.

stimulates milk secretion in lactating mothers, as well as cytokines, chemical mediators of the immune system, use this JAK/ STAT signal transduction pathway.

Most water-soluble extracellular chemical messengers activate second-messenger pathways via G-protein-coupled receptors.

The **second-messenger pathway** is initiated by binding of the first messenger (alias the extracellular chemical messenger, alias the signal molecule) to a surface membrane receptor specific for it. In this pathway, *the receptor is coupled with a G protein,* appropriately called a **G-protein-coupled receptor (GPCR)**, which snakes through the membrane (I Figure 4-22). Binding of the first messenger to the receptor activates the **G protein,** which is a membrane-bound intermediary. On activation, a portion of the G protein shuttles along the membrane to alter the activity of a nearby membrane protein called the **effector protein**. Once altered, the effector protein leads to an increased concentration of an intracellular messenger, known as the **second messenger**. The second messenger relays the orders through a cascade of chemical reactions inside the cell that cause a change in the shape and function of designated proteins. Once activated, these designated proteins accomplish the cellular response dictated by the first messenger. Most commonly,

the second messenger activates an intracellular protein kinase, which leads to phosphorylation and thereby altered function of designated proteins. The intracellular pathways activated by a second messenger are remarkably similar among different cells despite the diversity of ultimate responses. The variability in response depends on the specialization of the cell, not on the mechanism used.

Clinical Note About half of all drugs prescribed today act on G-protein-coupled receptors. These receptors participate in some way in most body functions, so they are important targets for a variety of drugs used to treat diverse disorders. For example, they include drugs used to reduce high blood pressure, to treat congestive heart failure, to suppress stomach acid, to open airways in asthmatics, to ease symptoms of enlarged prostate, to block histamine-induced allergic responses, to relieve pain, and to treat hormone-dependent cancers.

The effects of protein kinases in the tyrosine kinase, JAK/STAT, and second-messenger signal transduction pathways are reversed by another group of enzymes called **protein phosphatases**, which remove phosphate groups from the designated proteins. Unlike protein kinases, which are active only when an extracellular messenger binds to a surface membrane receptor, most protein phosphatases are continuously active in cells. By continually removing phosphate groups from designated proteins, protein phosphatases quickly shut off a signal transduction pathway if its signal molecule is no longer bound at the cell surface. Thus, kinases activate a signaling pathway by phosphorylating designated proteins, whereas phosphatases inactivate the pathway by dephosphorylating these proteins. Protein phosphorylation/dephosphorylation plays a central role in regulating the activity of proteins and thus their extensive roles in cellular physiology.

Some neurotransmitters function through intracellular second-messenger systems. Most, but not all, neurotransmitters function by changing the conformation of chemically gated receptor-channels, thereby altering membrane permeability and ion fluxes across the postsynaptic membrane, a process with which you are already familiar. Synapses involving these rapid responses are considered **"fast" synapses**. However, another mode of synaptic transmission used by some neurotransmitters, such as *serotonin,* involves the activation of intracellular second messengers. Synapses that lead to responses mediated by second messengers are known as **"slow" synapses** because these responses take longer and often last longer than those accomplished by fast synapses. For example, neurotransmitter-activated second messengers may trigger long-term postsynaptic cellular changes believed to be linked to neuronal growth and development and possibly play a role in learning and memory.

Depending on the cell type, the first messenger either can be released from the target cell and ultimately degraded by the liver and excreted in the urine, or alternatively the first messenger and receptor complex can be removed from action by receptor-mediated endocytosis, in which both the receptor and extracellular chemical messenger are internalized by the target cell (see p. 31).

Second-messenger pathways are widely used throughout the body, including being the key means by which paracrines and most water-soluble hormones ultimately bring about their effects. Let us now turn attention to paracrine communication before focusing on hormonal communication, where we will examine specific second-messenger systems in more detail.

4.6 Introduction to Paracrine Communication

Most paracrines are either cytokines or eicosanoids. **Cytokines** are a collection of *protein* signal molecules secreted by cells of the immune system and other cell types that largely act locally to regulate immune responses. **Eicosanoids** (eye-KOH-sah-noydz) are a group of *lipid* signal molecules derived from a fatty acid in the plasma membrane of most cell types that act locally to regulate diverse cellular processes throughout the body.

Cytokines act locally to regulate immune responses.

Cytokines are intercellular regulatory proteins secreted primarily by white blood cells and other cells of the immune system but also by some nonimmune cells. The nonimmune cells that are the most prolific producers of cytokines are endothelial cells that line the blood vessels (see p. 346), fibroblasts that form the extracellular matrix in connective tissue (see p. 61), and adipose tissue cells that store excess fat (see p. 623). The primary function of cytokines is to regulate numerous activities of the immune system, such as mediating inflammation and enhancing the activity of antibody-producing cells and virus-fighting cells. Some cytokines exert nonimmune effects as well, such as by influencing cell growth and cell differentiation during embryonic development. Cytokines important in development are typically referred to specifically as various **growth factors**, but growth factors have other functions not related to development as well. For example, growth factors as well as other cytokines both help regulate wound healing. More than 100 cytokines have been identified. You will learn about the specific functions of major cytokines in later chapters.

Cytokines typically act locally as paracrines but some travel in the blood, similar to hormones, to distant target cells to produce systemic (bodywide) responses. For example, specific cytokines are responsible for the generally miserable way you feel when you have the "flu." The boundary between whether a given signal molecule should be classified as a cytokine, a growth factor, or a hormone sometimes gets blurry.

Cytokines function primarily by binding to receptor–enzyme complexes, with immune cytokines largely employing the JAK/STAT pathway and growth factors mainly using the tyrosine kinase pathway.

Eicosanoids are locally acting chemical messengers derived from plasma membrane.

Eicosanoids are a group of lipid signal molecules that act locally as paracrines to regulate a plethora of physiological processes. Eicosanoids are modified 20-carbon fatty acids derived from **arachidonic acid**, a 20-carbon polyunsaturated fatty acid constituent of the phospholipids within the plasma membrane (see p. 56) (*eikos* means "20" in Greek). On appropriate stimulation, arachidonic acid is split from the plasma membrane by a membrane-bound enzyme, **phospholipase A_2**, and then is converted

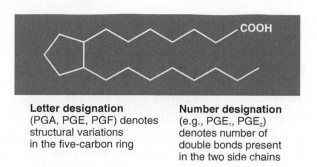

Letter designation
(PGA, PGE, PGF) denotes
structural variations
in the five-carbon ring

Number designation
(e.g., PGE$_1$, PGE$_2$)
denotes number of
double bonds present
in the two side chains

Figure 4-23 **Structure and nomenclature of prostaglandins.**

into one of three main classes of eicosanoids—**prostaglandins**, **thromboxanes**, and **leukotrienes**—depending on how it is further processed. The enzyme **cyclooxygenase (COX)** initiates a pathway leading to formation of prostaglandins and thromboxanes, whereas the enzyme **lipooxygenase (LOX)** sets off another pathway that results in generation of leukotrienes.

Each eicosanoid class has multiple members, depending on slight structural variations driven by further enzymatic processing. Using prostaglandins as an example, prostaglandins are designated as belonging to one of three groups—PGA, PGE, or PGF—according to structural variations in the five-carbon ring that they contain at one end (Figure 4-23). Within each group, prostaglandins are further identified by the number of double bonds present in the two side chains that project from the ring structure (for example, PGE$_1$ has one double bond and PGE$_2$ has two double bonds). Which eicosanoids are produced by a particular cell depends on what specific enzymes the cell has at its disposal for processing arachidonic acid.

Once synthesized, eicosanoids diffuse out of the cell to serve as local extracellular messengers. They exert their effects by binding with surface membrane receptors of their neighboring target cells and initiating second messenger pathways. After eicosanoids act, they are rapidly inactivated by local enzymes before they gain access to the blood; if they do reach the circulatory system, they are swiftly degraded on their first pass through the lungs so that they are not dispersed through the systemic arterial system.

Eicosanoids are the most ubiquitous extracellular chemical messengers and are among the most biologically active compounds known. They exert a tremendous number of very specific actions in almost every tissue of the body, as made evident by the following discussion of the functions of each of the main classes of eicosanoids:

1. *Prostaglandins* were first identified in the semen and were believed to be of prostate gland origin (hence their name, even though they are actually secreted into the semen by another male accessory sex gland; see p. 731). Thus they were first known for enhancing sperm transport in the male and female reproductive systems. However, their production and actions are not limited to the reproductive system. Prostaglandins also are produced by and exert a bewildering variety of effects in the respiratory, urinary, digestive, nervous, endocrine, circulatory, and immune systems, in addition to affecting fat metabolism, as can be seen in Table 4-4. Slight variations in prostaglandin structure are accompanied by profound differences in biological action.

TABLE 4-4 Actions of Prostaglandins

Body System Affected	Actions of Prostaglandins
Reproductive system	Promote sperm transport by action on smooth muscle in the male and female reproductive tracts
	Play a role in ovulation
	Play important role in menstruation
	Contribute to preparation of the maternal portion of the placenta
	Promote contractions of the uterus
Respiratory system	Some promote dilation, others constriction, of respiratory airways
Urinary system	Increase renal blood flow
	Increase excretion of water and salt
Digestive system	Inhibit HCl secretion by the stomach
	Stimulate intestinal motility
Nervous system	Influence neurotransmitter release and action
	Act at the hypothalamic "thermostat" to increase body temperature
	Intensify the sensation of pain
Endocrine system	Enhance cortisol secretion
	Influence tissue responsiveness to hormones in many instances
Circulatory system	Influence platelet aggregation
	Some decrease, others increase, blood pressure
Immune system	Promote many aspects of inflammation, including development of fever
Fat metabolism	Inhibit fat breakdown

Clinical Note Prostaglandins' various actions can be manipulated therapeutically. A classic example is the use of *aspirin* and other *nonsteroidal anti-inflammatory drugs (NSAIDs),* such as *ibuprofen,* which all inhibit COX, thus blocking the conversion of arachidonic acid into prostaglandins, for pain relief and fever reduction. (Note in Table 4-4 that prostaglandins intensify pain and play a key role in development of a fever.)

2. *Thromboxanes* were originally discovered as secretory products of thrombocytes (alias blood platelets), accounting for their name. Thromboxanes promote platelet aggregation (to help stop bleeding) and constrict blood vessels.

3. *Leukotrienes* were initially encountered as secretory products from leukocytes (white blood cells), thus their name. These local messengers are involved in inflammatory responses and cause the profound airway constriction characteristic of asthma.

Check Your Understanding 4.6

1. Distinguish between cytokines and eicosanoids.
2. Discuss the roles of phospholipase A₂, cyclooxygenase, and lipooxygenase.
3. Explain how NSAIDs provide pain relief.

4.7 Introduction to Hormonal Communication

Endocrinology is the study of homeostatic chemical adjustments and other activities accomplished by hormones, which are secreted into the blood by endocrine glands. Earlier we described the underlying molecular and cellular mechanisms of the nervous system—electrical signaling within neurons and chemical transmission of signals between neurons. We now focus on the molecular and cellular features of hormonal action and compare the similarities and differences in how neurons and endocrine cells communicate with other cells in carrying out their regulatory actions.

Hormones are classified chemically as hydrophilic or lipophilic.

Hormones fall into two chemical groups based on their solubility: *hydrophilic* and *lipophilic* hormones. Hormones can also be classified according to their chemical structure (namely, *peptides, amines,* and *steroids*) as follows (Table 4-5):

1. **Hydrophilic** ("water-loving") **hormones** are highly water soluble and have low lipid solubility. Most hydrophilic hormones are peptide or protein hormones consisting of specific amino acids arranged in a chain of varying length. The shorter chains are peptides, and the longer ones are proteins. For convenience, we refer to this entire category as **peptides**. Insulin from the pancreas is a peptide hormone. The **amines** are so called because they are amino acid derivatives. The amine hor-

▮TABLE 4-5 Chemical Classification of Hormones

Properties	Peptides	AMINES		Steroids
		Catecholamines and Indoleamines	Thyroid Hormone	
Solubility	Hydrophilic	Hydrophilic	Lipophilic	Lipophilic
Structure	Chains of specific amino acids	Tyrosine derivative (catecholamines) or tryptophan derivative (indoleamines)	Iodinated tyrosine derivative	Cholesterol derivative
Synthesis	In rough endoplasmic reticulum; packaged in Golgi complex	In cytosol	In colloid within thyroid gland (see p. 666)	Stepwise modification of cholesterol molecule in various intracellular compartments
Storage	Large amounts in secretory vesicles	In secretory vesicles	In colloid	Not stored; cholesterol precursor stored in lipid droplets
Secretion	Exocytosis of secretory vesicles	Exocytosis of secretory vesicles	Endocytosis of colloid	Simple diffusion
Transport in blood	As free hormone	Half bound to plasma proteins	Mostly bound to plasma proteins	Mostly bound to plasma proteins
Receptor site	Surface of target cell	Surface of target cell	Inside target cell	Inside target cell
Mechanism of action	Activation of second-messenger pathway to alter activity of preexisting proteins that produce the effect	Activation of second-messenger pathway to alter activity of preexisting proteins that produce the effect	Activation of specific genes to make new proteins that produce the effect	Activation of specific genes to make new proteins that produce the effect
Hormones of this type	Most hormones	Catecholamines: hormones from the adrenal medulla, dopamine from hypothalamus. Indoleamines: melatonin from pineal gland	Only hormones from the follicular cells of the thyroid gland	Hormones from the adrenal cortex and gonads and some placental hormones; vitamin D (a hormone) is steroidlike

mones include two types of hydrophilic hormones (catecholamines and indoleamines) and one type of lipophilic hormone (thyroid hormone). *Catecholamines* are derived from the amino acid tyrosine and are largely secreted by the adrenal medulla. The adrenal gland consists of an inner adrenal medulla surrounded by an outer adrenal cortex. (You will learn more about the location and structure of the endocrine glands and the functions of specific hormones in later chapters.) Epinephrine is the major catecholamine hormone. *Indoleamines* are derived from the amino acid tryptophan and are secreted by the pineal gland. Melatonin is the only indoleamine hormone. Some neurotransmitters are also amines, such as dopamine (a catecholamine) and serotonin (an indoleamine). Dopamine also acts as a neurohormone, and serotonin is the precursor for melatonin, serving as examples of the overlapping activities of the nervous and endocrine systems.

2. **Lipophilic** ("lipid-loving") **hormones** have high lipid solubility and are poorly soluble in water. Lipophilic hormones include thyroid hormone and the steroid hormones. *Thyroid hormone*, as its name implies, is secreted exclusively by the thyroid gland; it is an iodinated tyrosine derivative. Even though catecholamines and thyroid hormone are both derived from tyrosine, they behave differently because of their solubility properties. **Steroids** are neutral lipids derived from cholesterol. They include hormones secreted by the adrenal cortex, such as cortisol, and the sex hormones (testosterone in males and estrogen in females) secreted by the reproductive organs.

Minor differences in chemical structure among hormones within each category often result in profound differences in biological response. For example, in ❚Figure 4-24, note the subtle difference between the steroid hormone testosterone, the male sex hormone responsible for inducing the development of masculine characteristics, and the steroid hormone estradiol, a form of estrogen, which is the feminizing female sex hormone.

The mechanisms of synthesis, storage, and secretion of hormones vary according to their chemical differences.

The solubility properties of a hormone determine (1) how the hormone is synthesized, stored, and secreted by the endocrine cell; (2) how the hormone is transported in the blood; and (3) how the hormone exerts its effects at the target cell. We first consider the different ways in which hydrophilic and lipophilic hormones are processed at their site of origin, the endocrine cell, before comparing their means of transport and their mechanisms of action.

Processing of Hydrophilic Peptide Hormones Peptide hormones are synthesized and secreted by an endocrine cell via the same steps used for manufacturing any protein that is exported from a cell (see ❚Figure 2-3, p. 27). From the time peptide hormones are synthesized until they are secreted, they are always segregated from intracellular proteins within membrane-enclosed compartments. Here is a brief outline of these steps:

1. Large precursor proteins, or **preprohormones**, are synthesized by ribosomes on the rough endoplasmic reticulum (ER).

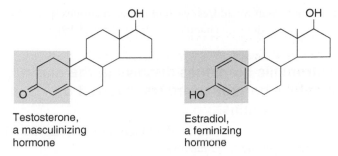

❚Figure 4-24 **Comparison of two steroid hormones: testosterone and estradiol.**

They then migrate to the Golgi complex in membrane-enclosed transport vesicles that pinch off from the smooth ER.

2. During their journey through the ER and Golgi complex, the preprohormones are pruned to active hormones.

3. The Golgi complex packages the finished hormones into secretory vesicles that are pinched off and stored in the cytoplasm until an appropriate signal triggers their secretion.

4. On appropriate stimulation, the secretory vesicles fuse with the plasma membrane and release their contents to the outside by exocytosis. Such secretion usually does not go on continuously; it is triggered only by specific stimuli. The blood then picks up the secreted hormones for distribution.

Processing of Lipophilic Steroid Hormones All steroidogenic (steroid-producing) endocrine cells perform the following steps to produce and release their hormonal product:

1. Cholesterol is the precursor for all steroid hormones.

2. Synthesis of the various steroid hormones requires a series of enzymatic reactions that modify the basic cholesterol molecule—for example, by varying the type and position of side groups attached to the cholesterol framework. Each conversion from cholesterol to a specific steroid hormone requires the help of particular enzymes limited to certain steroidogenic organs. Thus, each steroidogenic organ can produce only the steroid hormone or hormones for which it has a complete set of appropriate enzymes. For example, a key enzyme necessary for producing cortisol is found only in the adrenal cortex, so no other steroidogenic organ can produce this hormone.

3. Unlike peptide hormones, steroid hormones are not stored. Once formed, the lipid-soluble steroid hormones immediately diffuse through the steroidogenic cell's lipid plasma membrane to enter the blood. Only the hormone precursor cholesterol is stored within steroidogenic cells. Accordingly, the rate of steroid hormone secretion is controlled entirely by the rate of hormone synthesis. In contrast, peptide hormone secretion is controlled primarily by regulating the release of presynthesized stored hormone.

4. Following their secretion into the blood, some steroid hormones (and thyroid hormone) undergo further interconversions within the blood or other organs, where they are changed into more potent or different hormones.

The adrenomedullary catecholamines and thyroid hormone have unique synthetic and secretory pathways that are described when we address each of these hormones specifically in the endocrine chapters (Chapters 18 and 19).

Hydrophilic hormones dissolve in the plasma; lipophilic hormones are transported by plasma proteins.

All hormones are carried by the blood, but they are not all transported in the same manner:

1. The hydrophilic peptide hormones simply dissolve in the blood.

2. Lipophilic steroids and thyroid hormone, which are poorly soluble in water, cannot dissolve to any extent in the watery blood. Instead, most lipophilic hormones circulate to their target cells reversibly bound to plasma proteins in the blood. Some plasma proteins carry only one type of hormone, whereas others, such as albumin, indiscriminately pick up any "hitchhiking" hormone.

Only the small, unbound, freely dissolved fraction of a lipophilic hormone is biologically active (that is, free to cross capillary walls and bind with target cell receptors to exert an effect). The bound form of steroid and thyroid hormones provides a large reserve of these lipophilic hormones that can be used to replenish the active free pool. To maintain normal endocrine function, the magnitude of the small, free, effective pool, rather than the total blood concentration of a particular lipophilic hormone, is monitored and adjusted.

3. Catecholamines are unusual in that only about 50% of these hydrophilic hormones circulate as free hormone; the other 50% are loosely bound to albumin. Because catecholamines are water soluble, the importance of this binding to plasma proteins is unclear.

Clinical Note The chemical properties of a hormone dictate not only how it is transported in the blood, but also the way in which it can be artificially introduced into the blood for therapeutic purposes. Because the digestive system does not secrete enzymes that can digest steroid and thyroid hormones, hormones such as the sex steroids contained in birth control pills can, when swallowed, be absorbed intact from the digestive tract into the blood. No other type of hormones can be taken orally (by mouth) because protein-digesting enzymes would attack and convert them into inactive fragments. Therefore, these hormones must be administered by non-oral routes; for example, insulin deficiency is treated with daily injections of insulin.

Next, we examine how the hydrophilic and lipophilic hormones vary in their mechanisms of action at their target cells.

Hormones generally produce their effect by altering intracellular proteins.

To induce their effect, hormones must bind with target cell receptors specific for them. Each interaction between a particular hormone and a target cell receptor produces a highly characteristic response that differs among hormones and among different target cells influenced by the same hormone. Both the location of the receptors within the target cell and the mechanism by which binding of the hormone with the receptors brings about a response vary, depending on the hormone's solubility characteristics.

Location of Receptors for Hydrophilic and Lipophilic Hormones Hormones can be grouped into two categories based on the location of their receptors:

1. The hydrophilic peptides and catecholamines, which are poorly soluble in lipid, cannot pass through the lipid membrane barriers of their target cells. Instead, they bind with specific receptors on the outer plasma membrane surface of the target cell.

2. The lipophilic steroids and thyroid hormone easily pass through the surface membrane to bind with specific receptors located inside the target cell.

General Means of Hydrophilic and Lipophilic Hormone Action Even though hormones cause a wide variety of responses, they ultimately influence their target cells by altering the cell's proteins in one of two major ways:

1. Surface-binding hydrophilic hormones function largely by activating second-messenger pathways within the target cell. This activation directly *alters the activity of preexisting intracellular proteins,* usually enzymes, to produce the desired effect.

2. Lipophilic hormones function mainly by activating specific genes in the target cell to cause *formation of new intracellular proteins,* which in turn produce the desired effect. The new proteins may be enzymatic or structural.

Let us examine the two major mechanisms of hormonal action (activation of second-messenger pathways and activation of genes) in more detail.

Hydrophilic hormones alter preexisting proteins via second-messenger systems.

Most hydrophilic hormones (peptides and catecholamines) bind to G-protein-coupled surface membrane receptors and produce their effects in their target cells by acting through a second-messenger pathway to alter the activity of preexisting proteins. There are two major second-messenger pathways: One uses **cyclic adenosine monophosphate (cyclic AMP, or cAMP)** as a second messenger, and the other uses Ca^{2+} in this role.

Both pathways use a G protein, which is found on the inner surface of the plasma membrane, as an intermediary between the receptor and the effector protein (see Figure 4-22). G proteins are so named because they are bound to guanine nucleotides—*guanosine triphosphate (GTP)* when active or *guanosine diphosphate (GDP)* when inactive. An inactive G protein consists of a complex of alpha (α), beta (β), and gamma (γ) subunits, with a GDP molecule bound to the α subunit. A number of different G proteins with varying α subunits have been identified. The different G proteins are activated in response to binding of various first messengers to surface receptors. When an appropriate extracellular messenger (a first messenger) binds with its receptor, the receptor attaches to the associated G pro-

FIGURE FOCUS: *Does active protein kinase A phosphorylate the same inactive designated proteins in all cells that use the cAMP second-messenger pathway?*

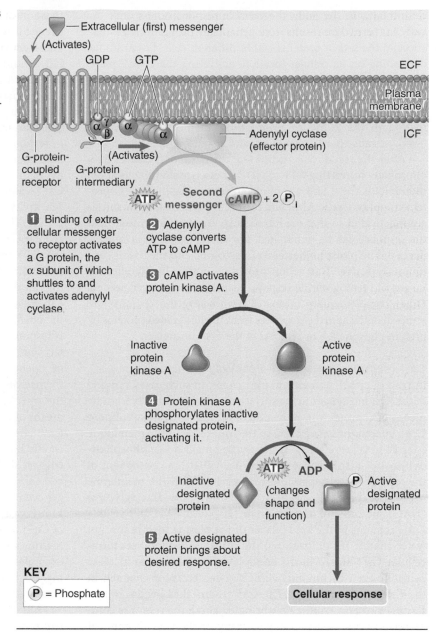

tein, resulting in release of GDP from the G-protein complex. GTP then attaches to the α subunit, an action that activates the G protein. Once activated, the α subunit breaks away from the G-protein complex and moves along the inner surface of the plasma membrane until it reaches an effector protein, which is typically either an enzyme or an ion channel within the membrane. The α subunit links up with the effector protein and alters its activity. Researchers have identified more than 300 different receptors that convey instructions of extracellular messengers through the membrane to various effector proteins by means of G proteins. We next examine the cAMP pathway in more detail as an example of what happens after an effector protein is activated.

Cyclic AMP Second-Messenger Pathway

Cyclic AMP is the most widely used second messenger. In the following description of the cAMP pathway, the numbered steps correlate to the numbered steps in Figure 4-25. When the appropriate extracellular messenger binds to its surface membrane receptor and activates the associated G protein, the G protein in turn activates the effector protein—in this case, the enzyme **adenylyl cyclase** (step **1**), which is located on the cytoplasmic side of the plasma membrane. Adenylyl cyclase converts intracellular ATP to cAMP by cleaving off two of the phosphates (step **2**). (This is the same ATP used as energy currency in the body.) Acting as the intracellular second messenger, cAMP triggers a preprogrammed series of chemical steps within the cell to bring about the response dictated by the first messenger. To begin, cAMP activates the intracellular enzyme, **protein kinase A (PKA)** (step **3**). Protein kinase A, in turn, phosphorylates a designated preexisting intracellular protein, such as an enzyme important in a particular metabolic pathway. Phosphorylation causes the protein to change its shape and function, thereby activating it (step **4**). This activated protein brings about the target cell's ultimate response to the first messenger (step **5**). For example, the activity of a particular enzymatic protein that regulates a specific metabolic event may be increased or decreased.

Note that in this signal transduction pathway the steps involving the extracellular first messenger, the receptor, the G-protein complex, and the effector protein occur *in the plasma membrane* and lead to activation of the second messenger. The extracellular messenger cannot enter the cell to "personally"

deliver its message to the proteins that carry out the desired response. Instead, it initiates membrane events that activate an intracellular second messenger, cAMP. The second messenger then triggers a chain reaction of biochemical events *inside the cell* that leads to the cellular response.

Different types of cells have different designated proteins available for phosphorylation and modification by PKA. Therefore, a *common second messenger such as cAMP can cause widely differing responses in different cells*, depending on what proteins are modified. Cyclic AMP can be thought of as an intracellular molecular "switch" that can "turn on" (or "turn off") different cell events, depending on the kinds of protein activity ultimately modified in the various target cells. The type of proteins altered by a second messenger depends on the unique specialization of a particular cell type. This can be likened to being able to either illuminate or cool a room depending on whether the wall switch you flip on is wired to a device specialized to shed light (a chandelier) or one specialized to create air movement (a ceiling fan). In the body, the variable responsiveness once the switch is turned on results from genetically programmed differences in the sets of proteins within different cells. For example, depending on its cellular location, activating the cAMP pathway can modify heart rate in the heart, stimulate the formation of female sex hormones in the ovaries, break down stored glucose in the liver, control water conservation during urine formation in the kidneys, create simple memory traces in the brain, or cause perception of a sweet taste by a taste bud.

After the response is completed, the α subunit cleaves off a phosphate, converting GTP to GDP, in essence shutting itself off, then rejoins the β and γ subunits to restore the inactive G-protein complex. Cyclic AMP and the other participating chemicals are inactivated so that the intracellular message is "erased" and the response can be terminated. For example, cAMP is quickly degraded by **phosphodiesterase**, a cytosolic enzyme that is continuously active. This action provides another highly effective means of turning off the response when it is no longer needed. Other complementary means of terminating the response are removal of the added phosphates from the designated proteins by protein phosphatase or removal of the first messenger.

Ca²⁺ Second-Messenger Pathway

Some cells use Ca^{2+} instead of cAMP as a second messenger. In such cases, binding of the first messenger to the surface receptor eventually leads by means of G proteins to activation of the enzyme **phospholipase C**, an effector protein bound to the inner side of the membrane (step **1** in Figure 4-26). This enzyme breaks down **phosphatidylinositol bisphosphate** (abbreviated **PIP₂**), a component of the tails of the phospholipid molecules within the membrane itself. The products of PIP_2 breakdown are **diacylglycerol (DAG)** and **inositol trisphosphate (IP₃)** (step **2**). Lipid-soluble DAG remains in the lipid bilayer of the plasma membrane, but water-soluble IP_3 diffuses into the cytosol. IP_3 mobilizes intracellular Ca^{2+} stored in the endoplasmic reticulum to increase cytosolic Ca^{2+} by binding with IP_3-gated receptor-channels in the ER membrane (step **3a**). Ca^{2+} then takes on the role of second messenger, ultimately bringing about the response commanded by the first messenger. Many of the Ca^{2+}-dependent cellular events are triggered by activation of **calmodulin**, an intracellular Ca^{2+}-binding protein (step **4a**). The Ca^{2+}–calmodulin complex activates **Ca²⁺–calmodulin dependent protein kinase (CaM kinase)** (or activates another kinase) (step **5a**). Activation of CaM kinase by the Ca^{2+}–calmodulin complex is similar to activation of PKA by cAMP. From here the patterns of the two pathways are similar. The activated CaM kinase phosphorylates the designated proteins (perhaps through multiple steps), thereby causing these proteins to change their shape and function (activating them) (step **6a**). The active designated proteins bring about the ultimate desired cellular response (step **7a**). For example, the Ca^{2+}–calmodulin pathway is the means by which chemical messengers can activate smooth muscle contraction.

Simultaneous to the IP_3 pathway, the other PIP_2 breakdown product, DAG, sets off another second-messenger pathway. (IP_3 and DAG themselves are sometimes considered to be second messengers.) DAG activates **protein kinase C (PKC)** (step **3b**), which phosphorylates designated proteins, different from

those phosphorylated by calmodulin (step **4b**). The resultant change in shape and function of these proteins activates them. These active proteins produce another cellular response (step **5b**). Although currently the subject of considerable investigation, the DAG pathway is not yet understood as well as the other signaling pathways. IP_3 and DAG typically trigger complementary actions inside a target cell to accomplish a common goal because both of these products are formed at the same time in response to the same first messenger. For example, extracellular chemical messengers promote increased contractile activity of blood-vessel smooth muscle via the IP_3–intracellular Ca^{2+}–calmodulin pathway, and the DAG pathway enhances the sensitivity of the contractile apparatus to Ca^{2+}.

The IP_3 pathway is not the only means of increasing intracellular Ca^{2+}. Intracellular Ca^{2+} can be increased by entry from the ECF or by release from Ca^{2+} stores in the endoplasmic reticulum via means other than the IP_3 pathway. Ca^{2+} channels in both the surface membrane and in the ER may be opened by either electrical or chemical means. For example, Ca^{2+} entry on the opening of voltage-gated surface-membrane Ca^{2+} channels is responsible for exocytosis of neurotransmitter from the axon terminal. Alternatively, surface-membrane Ca^{2+} channels may be opened via activation of receptors that serve as channels themselves or via activation of GPCRs. In yet another pathway, the opening of surface-membrane Na^+ and K^+ channels by means of receptor-channels leads to electrical signals that open Ca^{2+} channels in the endoplasmic reticulum. The latter pathway is how neurotransmitter released from neuron terminals triggers skeletal muscle contraction. The resultant rise in intracellular Ca^{2+} turns on the contractile apparatus. The pathways get even more complex than this. Ca^{2+} entering from the ECF can serve as a second messenger to trigger an even larger release of Ca^{2+} from intracellular stores, as it does to bring about contraction in cardiac muscle. This all sounds confusing, but these examples are meant to illustrate the complexity of Ca^{2+} signaling, not to overwhelm. You will learn more about the details of these pathways when appropriate in later chapters.

Although the cAMP and Ca^{2+} pathways are the most prevalent second-messenger systems, they are not the only ones. For example, a few cells use **cyclic guanosine monophosphate (cyclic GMP)** as a second messenger in a system analogous to the cAMP system. In other cells, the second messenger is still unknown. Remember that activation of second messengers is a universal mechanism used by a variety of extracellular messengers in addition to hydrophilic hormones.

Amplification by a Second-Messenger Pathway

Several remaining points about receptor activation and the ensuing events merit attention. First, considering the number of steps in a second-messenger relay chain, you might wonder why so many cell types use the same complex system to accomplish such a wide range of functions. The multiple steps of a second-messenger pathway are actually advantageous because the cascading (multiplying) effect of these pathways greatly amplifies the initial signal (Figure 4-27). *Amplification* means that the output of a system is much greater than the input. Using the cAMP pathway as an example, binding of one extracellular messenger molecule to a receptor activates a number of

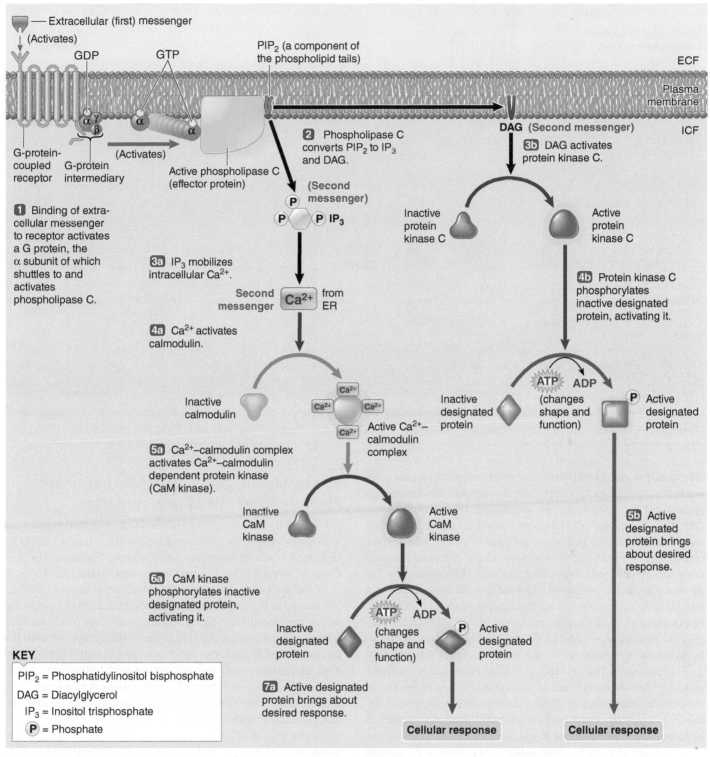

Figure 4-26 Mechanism of action of hydrophilic hormones via concurrent activation of the IP₃–Ca²⁺ second-messenger pathway and the DAG pathway.

The following text appears within the figure:

— Extracellular (first) messenger
(Activates)

GDP   GTP   PIP₂ (a component of the phospholipid tails)

ECF

Plasma membrane

α γ β   α   α   DAG (Second messenger)   ICF

G-protein-coupled receptor   G-protein intermediary   (Activates)   Active phospholipase C (effector protein)

2 Phospholipase C converts PIP₂ to IP₃ and DAG.

3b DAG activates protein kinase C.

(Second messenger)

1 Binding of extracellular messenger to receptor activates a G protein, the α subunit of which shuttles to and activates phospholipase C.

P   P   P IP₃

Inactive protein kinase C   Active protein kinase C

3a IP₃ mobilizes intracellular Ca²⁺.

Second messenger Ca²⁺ from ER

4a Ca²⁺ activates calmodulin.

4b Protein kinase C phosphorylates inactive designated protein, activating it.

Inactive calmodulin   Ca²⁺ Ca²⁺ Ca²⁺ Ca²⁺   Active Ca²⁺–calmodulin complex

ATP ADP (changes shape and function)

Inactive designated protein   P Active designated protein

5a Ca²⁺–calmodulin complex activates Ca²⁺–calmodulin dependent protein kinase (CaM kinase).

Inactive CaM kinase   Active CaM kinase

5b Active designated protein brings about desired response.

6a CaM kinase phosphorylates inactive designated protein, activating it.

ATP ADP (changes shape and function)

Inactive designated protein   P Active designated protein

KEY

PIP₂ = Phosphatidylinositol bisphosphate

DAG = Diacylglycerol

IP₃ = Inositol trisphosphate

P = Phosphate

7a Active designated protein brings about desired response.

Cellular response   **Cellular response**

---

adenylyl cyclase molecules (let's arbitrarily say 10), each of which activates many (in our hypothetical example, let's say 100) cAMP molecules. Each cAMP molecule then acts on a single protein kinase A, which phosphorylates and thereby influences many (again, let's say 100) specific proteins, such as enzymes. Each enzyme, in turn, is responsible for producing many (perhaps 100) molecules of a particular product, such as

a secretory product. The result of this cascade, with one event triggering the next in sequence, is a tremendous amplification of the initial signal. In our hypothetical example, one extracellular messenger molecule has been responsible for inducing a yield of 10 million molecules of a secretory product. In this way, very low concentrations of hormones and other chemical messengers can trigger pronounced cell responses.

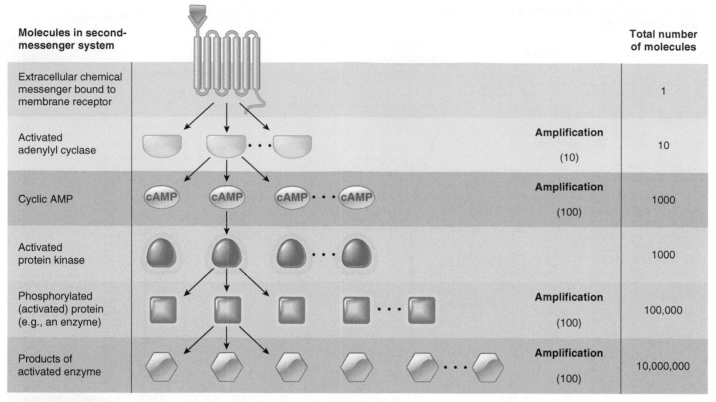

Molecules in second-messenger system		Total number of molecules
Extracellular chemical messenger bound to membrane receptor		1
Activated adenylyl cyclase	Amplification (10)	10
Cyclic AMP	Amplification (100)	1000
Activated protein kinase		1000
Phosphorylated (activated) protein (e.g., an enzyme)	Amplification (100)	100,000
Products of activated enzyme	Amplification (100)	10,000,000

❙Figure 4-27 **Amplification of the initial signal by a second-messenger pathway.** Through amplification, very low concentrations of extracellular chemical messengers, such as hormones, can trigger pronounced cellular responses.

Regulation of Receptors Although membrane receptors are links between extracellular first messengers and intracellular second messengers in the regulation of specific cellular activities, the receptors themselves are also often subject to regulation. In many instances, receptor number and affinity (attraction of a receptor for its extracellular chemical messenger) can be altered, depending on the circumstances. For example, a chronic elevation in blood insulin levels leads to a reduction in the number of insulin receptors, thus reducing the responsiveness of this hormone's target cells to its high levels.

Clinical Note Many diseases can be linked to malfunctioning receptors or to defects in the ensuing signal transduction pathways. For example, defective receptors are responsible for **Laron dwarfism**. In this condition, the person is abnormally short, despite having normal levels of growth hormone, because the tissues cannot respond normally to growth hormone. This is in contrast to the more usual type of dwarfism in which the person is abnormally short because of growth hormone deficiency. As another example, the toxins released by some infecting bacteria, such as those that cause cholera and whooping cough, keep second-messenger pathways "turned on" at a high level. **Cholera toxin** prevents the involved G protein from converting GTP to GDP, thus keeping the G protein in its active state. **Pertussis (whooping cough) toxin** blocks the inhibition of adenylyl cyclase, thereby keeping the ensuing second-messenger pathway continuously active.

Having examined the means by which hydrophilic hormones alter their target cells, we now focus on the mechanism of lipophilic hormone action.

By stimulating genes, lipophilic hormones promote synthesis of new proteins.

All lipophilic hormones (steroids and thyroid hormone) bind with intracellular receptors and primarily produce effects in their target cells by activating specific genes that cause the synthesis of new proteins, as summarized in ❙Figure 4-28.

Free lipophilic hormone (hormone not bound with its plasma–protein carrier) diffuses through the plasma membrane of the target cell (step ❶ in ❙Figure 4-28) and binds with its specific receptor inside the cell, either in the cytoplasm or in the nucleus (step ❷). Each receptor has a specific region for binding with its hormone and another region for binding with DNA. The receptor cannot bind with DNA unless it first binds with the hormone. Once the hormone is bound to the receptor, the hormone receptor complex binds with DNA at a specific attachment site on the DNA known as the **hormone response element (HRE)** (step ❸). Different steroid hormones and thyroid hormone, once bound with their respective receptors, attach at different HREs on DNA. For example, the estrogen receptor complex binds at DNA's *estrogen response element*.

Binding of the hormone receptor complex with DNA "turns on" or activates a specific gene within the target cell (step ❹). This gene contains a code for synthesizing a given protein. The code of the activated gene is transcribed into complementary messenger RNA (mRNA) (step ❺). The new mRNA leaves the nucleus and enters the cytoplasm (step ❻), where it binds to a ribosome, the "workbench" that mediates the assembly of new proteins (see pp. 23 and 25). Here, mRNA directs the synthesis

FIGURE FOCUS: *Which steps of this pathway by which lipophilic hormones bring about their effects occur in the cytoplasm and which take place in the nucleus?*

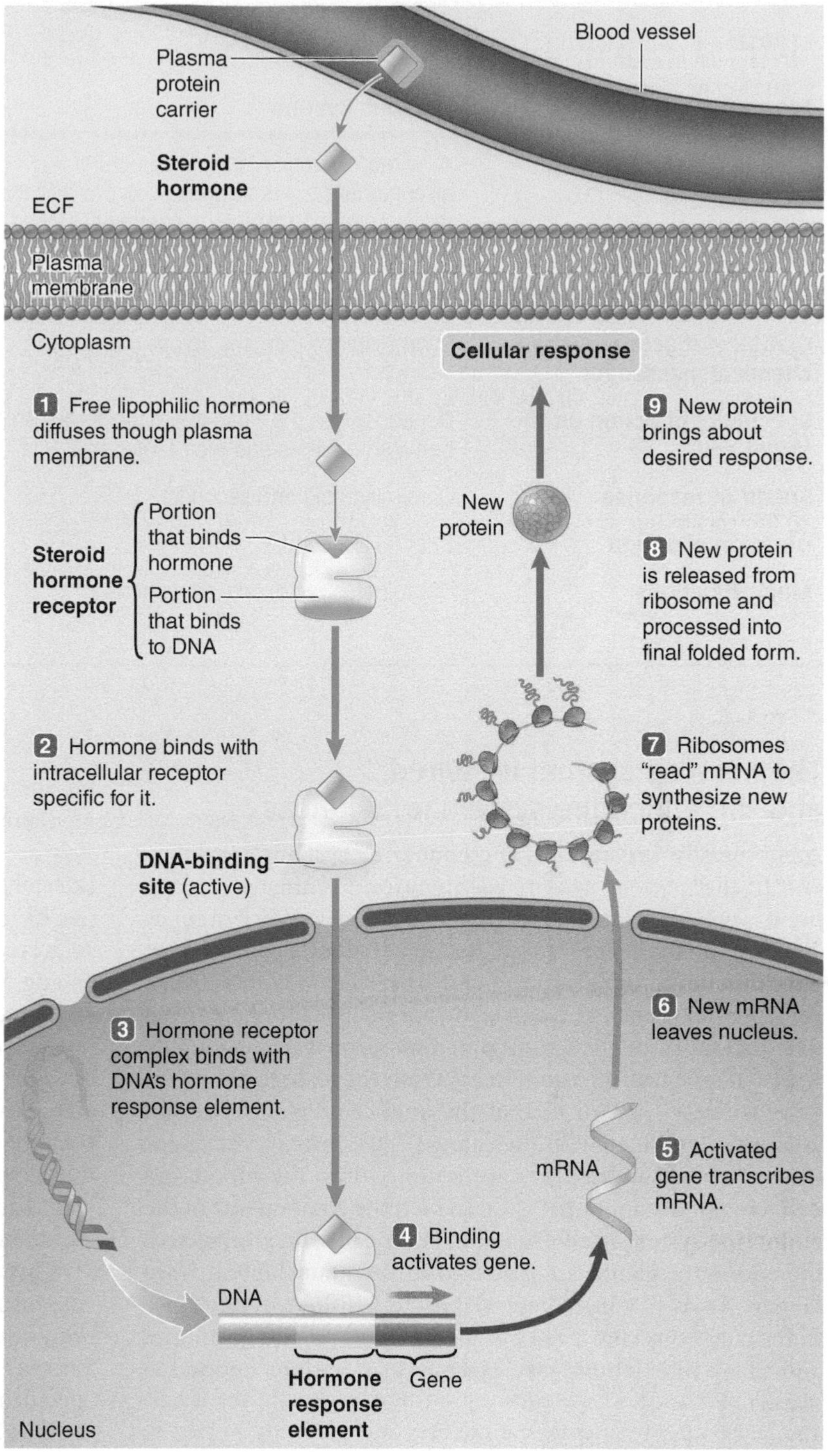

Blood vessel

Plasma protein carrier

Steroid hormone

ECF

Plasma membrane

Cytoplasm

Cellular response

1 Free lipophilic hormone diffuses though plasma membrane.

Steroid hormone receptor { Portion that binds hormone / Portion that binds to DNA

New protein

9 New protein brings about desired response.

8 New protein is released from ribosome and processed into final folded form.

2 Hormone binds with intracellular receptor specific for it.

DNA-binding site (active)

7 Ribosomes "read" mRNA to synthesize new proteins.

6 New mRNA leaves nucleus.

3 Hormone receptor complex binds with DNA's hormone response element.

5 Activated gene transcribes mRNA.

mRNA

4 Binding activates gene.

DNA

Hormone response element   Gene

Nucleus

of the designated new proteins according to the DNA code in the activated genes (step **7**). The newly synthesized protein, either enzymatic or structural, is released from the ribosome (step **8**) and produces the target cell's ultimate response to the hormone (step **9**). Different genes are activated by different lipophilic hormones, resulting in different biological effects.

Even though most steroid actions are accomplished by hormonal binding with intracellular receptors that leads to gene activation, recent studies have unveiled another mechanism by which steroid hormones induce effects that occur too rapidly to be mediated by gene transcription. Some steroid hormones, most notably some of the sex hormones, bind with unique steroid receptors in the plasma membrane, in addition to binding with the traditional steroid receptors in the nucleus. This membrane binding leads to **nongenomic steroid receptor actions**—that is, actions accomplished by something other than altering gene activity, such as by inducing changes in ionic flux across the membrane or by altering activity of cellular enzymes.

Next, we compare the similarities and differences between neural and hormonal responses at the system level.

Check Your Understanding 4.7

1. Prepare a chart comparing the synthesis, storage, secretion, and transport in the blood of peptide hormones and steroid hormones.

2. Explain how a common second messenger such as cAMP can induce widely differing responses in different cells.

3. Describe the role of a hormone response element.

4.8 Comparison of the Nervous and Endocrine Systems

The nervous and endocrine systems are the two main regulatory systems of the body. The nervous system swiftly transmits electrical impulses to the skeletal muscles and the exocrine glands that it innervates. The endocrine system secretes hor-

mones into the blood for delivery to distant sites of action. Although these two systems differ in many respects, they have much in common (I Table 4-6). They both alter their target cells (their sites of action) by releasing chemical messengers (neurotransmitters in the case of neurons, hormones in the case of endocrine cells) that bind with specific receptors of the target cells. This binding triggers the cellular response dictated by the regulatory system.

Now let us examine the anatomic distinctions between these two systems and the different ways in which they accomplish specificity of action.

ITABLE 4-6 Comparison of the Nervous System and the Endocrine System

Property	Nervous System	Endocrine System
Anatomic arrangement	A "wired" system: A specific structural arrangement exists between neurons and their target cells, with structural continuity in the system	A "wireless" system: Endocrine glands are widely dispersed and not structurally related to one another or to their target cells
Type of chemical messenger	Neurotransmitters released into the synaptic cleft	Hormones released into the blood
Distance of action of the chemical messenger	Short distance (diffuses across the synaptic cleft)	Long distance (carried by the blood)
Specificity of action on the target cell	Dependent on the close anatomic relationship between neurons and their target cells	Dependent on the specificity of target cell binding and responsiveness to a particular hormone
Speed of response	Generally rapid (milliseconds)	Generally slow (minutes to hours)
Duration of action	Brief (milliseconds)	Long (minutes to days or longer)
Major functions	Coordinates rapid, precise responses	Controls activities that require long duration rather than speed

The nervous system is "wired," and the endocrine system is "wireless."

Anatomically, the nervous and endocrine systems are different. In the nervous system, each neuron terminates directly on its specific target cells—that is, the nervous system is "wired" into highly organized, distinct anatomic pathways for transmission of signals from one part of the body to another. Information is carried along chains of neurons to the desired destination through action potential propagation coupled with synaptic transmission. In contrast, the endocrine system is a "wireless" system in that the endocrine glands are not anatomically linked with their target cells. Instead, the endocrine chemical messengers are secreted into the blood and delivered to distant target sites. In fact, the components of the endocrine system itself are not anatomically interconnected; the endocrine glands are scattered throughout the body (see IFigure 18-1, p. 639). These glands constitute a system in a functional sense, however, because they all secrete hormones and many interactions take place among various endocrine glands.

Neural specificity is a result of anatomic proximity, and endocrine specificity is a result of receptor specialization.

Because of their anatomic differences, the nervous and endocrine systems accomplish specificity of action by distinctly different means. Specificity of neural communication depends on neurons having a close anatomic relationship with their target cells, so each neuron has a narrow range of influence. A neurotransmitter is released only to specific adjacent target cells and then is swiftly inactivated or removed before it can enter the blood. The target cells for a particular neuron have receptors for the neurotransmitter, but so do many other cells in other locations, and they could respond to this same mediator if it were delivered to them. For example, the entire system of neurons (called motor neurons) supplying your skeletal muscles uses the same neurotransmitter, *acetylcholine (ACh)*, and all your skeletal muscles bear complementary ACh receptors. Yet you can wiggle your big toe without influencing any of your other muscles because ACh can be discretely released from the motor neurons specifically wired to the muscles controlling your toe. If ACh were indiscriminately released into the blood, as are hormones, all the skeletal muscles would simultaneously respond by contracting because they all have identical receptors for ACh. This does not happen because of the precise wiring patterns that provide direct lines of communication between motor neurons and their target cells.

This specificity sharply contrasts to the way specificity of communication is built into the endocrine system. Because hormones travel in the blood, they reach virtually all tissues. Yet only specific target cells can respond to each hormone. Specificity of hormonal action depends on specialization of target cell receptors. For a hormone to exert its effect, the hormone must first bind with receptors specific for it that are located only on or in the hormone's target cells. Target cell receptors are highly selective in their binding function. A receptor recognizes a specific hormone because a portion of its conformation matches a unique portion of its binding hormone in "lock-and-key" fashion. Binding of a hormone with target cell receptors initiates a reaction that culminates in the hormone's final effect. The hormone cannot influence any other cells because nontarget cells lack the right binding receptors. Likewise, a given target cell has receptors that are "tuned" to recognize only one or a few of the many hormones that circulate in its vicinity. Other signals pass by without effect because the cell has no receptors for them.

The nervous and endocrine systems have their own realms of authority but interact functionally.

The nervous and endocrine systems are specialized for controlling different types of activities. In general, the nervous system governs the coordination of rapid, precise responses. It is especially important in the body's interactions with the external environment. Neural signals in the form of action potentials are rapidly propagated along neuronal fibers, resulting in the release at the axon terminal of a neurotransmitter that must diffuse only a microscopic distance to its target cell before a response is affected. A neurally mediated response is rapid but brief; the action is quickly halted as the neurotransmitter is swiftly removed from the target site. This permits ending the response, almost immediately repeating the response, or rapidly initiating an alternate response as circumstances demand (for example, the swift changes in commands to muscle groups needed to coordinate walking). This mode of action makes neural communication extremely rapid and precise. The target tissues of the nervous system are the muscles and glands, especially exocrine glands, of the body.

The endocrine system, in contrast, is specialized to control activities that require duration rather than speed, such as regulating organic metabolism; maintaining water and electrolyte balance; promoting smooth, sequential growth and development; and controlling reproduction. The endocrine system responds more slowly to its triggering stimuli than the nervous system does for several reasons. First, the endocrine system must depend on blood flow to convey its hormonal messengers over long distances. Second, hormones typically have a more complex mechanism of action at their target cells than neurotransmitters do; thus, they require more time before a response occurs. The ultimate effect of some hormones cannot be detected until a few hours after they bind with target cell receptors. Also, because of the receptors' high affinity for their respective hormone, the hormones often remain bound to receptors for some time, thus prolonging their biological effectiveness. Furthermore, unlike the brief, neurally induced responses that stop almost immediately after the neurotransmitter is removed, endocrine effects usually last for some time after the hormone's withdrawal. Neural responses to a single burst of neurotransmitter release usually last only milliseconds to seconds, whereas the alterations that hormones induce in target cells range from minutes to days or, in the case of growth-promoting effects, even a lifetime. Thus, hormonal action is relatively slow and prolonged, making endocrine control particularly suitable for regulating metabolic activities that require long-term stability.

Although the endocrine and nervous systems have their areas of specialization, they are intimately interconnected functionally. Some neurons do not release neurotransmitters at synapses but instead end at blood vessels and release their chemical messengers (neurohormones) into the blood, where these chemicals act as hormones. A given messenger may even be a neurotransmitter when released from a nerve ending and a hormone when secreted by an endocrine cell. An example is *norepinephrine* (see p. 640). The nervous system directly or indirectly controls the secretion of many hormones (see Chapter 18). At the same time, many hormones act as neuromodulators, altering synaptic effectiveness and thereby influencing the excitability of the nervous system. The presence of certain key hormones is even essential for the proper development and maturation of the brain during fetal life. Furthermore, in many instances the nervous and endocrine systems both influence the same target cells in supplementary fashion. For example, these two major regulatory systems both help regulate the circulatory and digestive systems. Thus, many important regulatory interfaces exist between the nervous and the endocrine systems. The study of these relationships is known as **neuroendocrinology**.

In the next three chapters, we concentrate on the nervous system. We examine the endocrine system in more detail in later chapters. Throughout the text, we continue to point out the numerous ways in which these two regulatory systems interact so that the body is a coordinated whole, even though each system has its own realm of authority.

Check Your Understanding 4.8

1. Compare how neural and endocrine specificity of action is accomplished.
2. What regulatory system enables you to turn the pages of this book and what regulatory system is maintaining your blood glucose (sugar) level?

Homeostasis: Chapter in Perspective

To maintain homeostasis, cells must communicate so that they work together to accomplish life-sustaining activities. To bring about desired responses, the two major regulatory systems of the body, the nervous system and the endocrine system, must communicate with the target cells they are controlling. Neural and hormonal communication is therefore critical in maintaining a stable internal environment and in coordinating nonhomeostatic activities.

Neurons are specialized to receive, process, encode, and rapidly transmit information from one part of the body to another. The information is transmitted over intricate neuronal pathways by propagation of action potentials along the neuron's length and by chemical transmission of the signal from neuron to neuron at synapses and from neuron to muscles and glands through other neurotransmitter–receptor interactions at these junctions.

Neurons are the key functional components of the nervous system. Many activities controlled by the nervous system are geared toward maintaining homeostasis. Some neuronal electrical signals convey information about changes to which the body must rapidly respond to maintain homeostasis—for example, information about a fall in blood pressure. Other neuronal electrical signals swiftly convey messages to muscles and glands stimulating appropriate responses to counteract these changes—for example, adjustments in heart and blood vessel activity that restore blood pressure to normal when it starts to fall. Furthermore, the nervous system directs many activities not geared toward maintaining homeostasis, many of which are subject to voluntary control, such as playing basketball or browsing the Internet.

The endocrine system secretes hormones into the blood, which carries these chemical messengers to distant target cells where they bring about their effect by changing the activity of enzymatic or structural proteins within these cells. Through its relatively slow-acting hormonal messengers, the endocrine system generally regulates activities that require

duration rather than speed. Most of these activities are directed toward maintaining homeostasis. For example, hormones help maintain the proper concentration of nutrients in the internal environment by directing chemical reactions involved in the cellular uptake, storage, release, and use of these molecules. Also, hormones help maintain the proper water and electrolyte balance in the internal environment. Unrelated to homeostasis, hormones direct growth and control most aspects of the reproductive system.

Together, the nervous and the endocrine systems orchestrate a range of adjustments that help the body maintain homeostasis in response to stress. Likewise, these systems work in concert to control the circulatory and digestive systems, which carry out many homeostatic activities.

Review Exercises Answers begin on p. A-24

Reviewing Terms and Facts

1. Conformational changes in channel proteins brought about by voltage changes are responsible for opening and closing Na^+ and K^+ gates during the generation of an action potential. (*True or false?*)

2. The $Na^+–K^+$ pump restores the membrane to resting potential after it reaches the peak of an action potential. (*True or false?*)

3. After an action potential, the K^+ concentration is greater outside the cell than inside the cell because of the efflux of K^+ during the falling phase. (*True or false?*)

4. Postsynaptic neurons can either excite or inhibit presynaptic neurons. (*True or false?*)

5. Second-messenger systems ultimately bring about the desired cell response by inducing a change in the shape and function of particular designated intracellular proteins. (*True or false?*)

6. Each steroidogenic organ has all the enzymes necessary to produce any steroid hormone. (*True or false?*)

7. The two types of excitable tissue are _____ and _____.

8. The one-way propagation of action potentials away from the original site of activation is ensured by the _____.

9. The _____ is the site of action potential initiation in most neurons because it has the lowest threshold.

10. A junction in which electrical activity in one neuron influences the electrical activity in another neuron by means of a neurotransmitter is called a _____.

11. Summing of EPSPs occurring very close together in time as a result of repetitive firing of a single presynaptic input is known as _____.

12. Summing of EPSPs occurring simultaneously from several different presynaptic inputs is known as _____.

13. The neuronal relationship in which synapses from many presynaptic inputs act on a single postsynaptic cell is called _____, whereas the relationship in which a single presynaptic neuron synapses with and thereby influences the activity of many postsynaptic cells is known as _____.

14. A common membrane-bound intermediary between the receptor and the effector protein within the plasma membrane is the _____.

15. The three types of receptors with regard to mode of action in signal transduction pathways are _____, _____, and _____.

16. The three classes of eicosanoids are _____, _____, and _____.

17. Using the answer code on the right, indicate which potential is being described:

1. behaves in all-or-none fashion	(a) graded potential
2. has a magnitude of potential change that varies with the magnitude of the triggering event	(b) action potential
3. spreads decrementally away from the original site	
4. spreads nondecrementally throughout the membrane	
5. serves as a long-distance signal	
6. serves as a short-distance signal	

18. Using the answer code on the right, indicate which characteristics apply to peptide and steroid hormones:

1. are hydrophilic
2. are lipophilic
3. are synthesized by the ER
4. are synthesized by modifying cholesterol
5. include epinephrine from the adrenal medulla
6. include cortisol from the adrenal cortex
7. bind to plasma proteins
8. bind to intracellular receptors
9. bind to surface membrane receptors
10. activate genes to promote synthesis of new proteins
11. act via second messenger to alter preexisting proteins
12. are secreted into blood by endocrine glands and carried to distant target sites

(a) peptide hormones
(b) steroid hormones
(c) both peptide and steroid hormones
(d) neither peptide nor steroid hormones

Understanding Concepts

(Answers at www.cengagebrain.com)

1. Define the following terms: *polarization, depolarization, hyperpolarization, repolarization, resting membrane potential, threshold potential, action potential, refractory period,* and *all-or-none law.*

2. Describe the permeability changes and ion fluxes that occur during an action potential.

3. Compare contiguous conduction and saltatory conduction.

4. Compare the events at excitatory and inhibitory synapses.

5. Compare the four kinds of gated channels in terms of the factor that opens or closes them.

6. Distinguish among a neurotransmitter, a neuromodulator, and a neurohormone.

7. Discuss the possible outcomes of the GPSP brought about by interactions between EPSPs and IPSPs.

8. Distinguish between presynaptic inhibition and an inhibitory postsynaptic potential.

9. List and describe the types of intercellular communication.

10. Define *signal transduction.*

11. Compare the tyrosine kinase and JAK/STAT pathways.

12. Distinguish between first and second messengers.

13. Compare cytokines and hormones.

14. Describe how arachidonic acid is converted into prostaglandins, thromboxanes, and leukotrienes.

15. Describe the sequence of events in the cAMP second-messenger pathway.

16. Describe the sequence of events in the Ca^{2+} second-messenger pathway.

17. Explain how the cascading effect of hormonal pathways amplifies the response.

18. Compare the nervous and endocrine systems.

Solving Quantitative Exercises

1. Answer the following questions regarding conduction of action potentials using the velocities given on p. 100:

a. How long would it take for an action potential to travel 0.6 m along the axon of an unmyelinated neuron of the digestive tract?

b. How long would it take for an action potential to travel the same distance along the axon of a large myelinated neuron innervating a skeletal muscle?

c. Suppose there were two synapses in a 0.6 m nerve tract and the delay at each synapse is 1 msec. How long would it take an action potential and chemical signal to travel the 0.6 m now, for both the myelinated and unmyelinated neurons?

d. What if there were five synapses?

2. Suppose point A is 1 m from point B. Compare the following situations:

i. A single axon spans the distance from A to B, and its conduction velocity is 60 m/sec.

ii. Three neurons span the distance from A to B, all three neurons have the same conduction velocity, and the synaptic delay at both synapses (draw a picture) is 1 msec. What are the conduction velocities of the three neurons in this second situation if the total conduction time in both cases is the same?

3. One can predict what the Na^+ current produced by the Na^+–K^+ pump is with the following equation:[1]

$$p = \frac{kT}{q}\left(\frac{G_{Na^+}G_{K^+}}{G_{Na^+}+G_{K^+}}\right)\log\frac{G_{K^+}[Na^+]_o}{G_{Na^+}[K^+]_i}$$

where p is the Na^+ pump current; G is membrane conductance to the indicated ion expressed in $\mu S/cm^2$ (S = Siemens); $[x]_o$ and $[x]_i$ are the concentrations of ion x outside and inside the cell, respectively; k is Boltzmann's constant; T is the temperature in kelvins; and q is the elementary charge constant. Suppose $kT/q = 25$ mV, $G_{Na^+} = 3.3$ $\mu S/cm^2$, $G_{K^+} = 240$ $\mu S/cm^2$, $[Na^+]_o = 145$ mM, and $[K^+]_i = 4$ mM. What is the pump current for Na^+, in $\mu A/cm^2$ (A = amperes, an expression of current)?

Applying Clinical Reasoning

Becky N. was apprehensive as she sat in the dentist's chair awaiting the placement of her first silver amalgam (the "filling" in a cavity in a tooth). Before preparing the tooth for the amalgam by drilling away the decayed portion of the tooth, the dentist injected a local anesthetic in the nerve pathway

---

[1]F. C. Hoppensteadt and C. S. Peskin, *Mathematics in Medicine and the Life Sciences* (New York: Springer, 1992), equation 7.4.35, p. 178.

supplying the region. As a result, Becky, much to her relief, did not feel any pain during the drilling and filling procedure. Local anesthetics block voltage-gated Na^+ channels. Explain how this action prevents the transmission of pain impulses to the brain.

Thinking at a Higher Level

1. The rate at which the Na^+–K^+ pump operates is not constant but is controlled by a combined effect of changes in ICF Na^+ concentration and ECF K^+ concentration. Do you think the changes in both ICF Na^+ and ECF K^+ concentration following a series of action potentials in a neuron would accelerate, slow down, or have no effect on the Na^+–K^+ pumps in this cell?

2. Which of the following would occur if a neuron were experimentally stimulated simultaneously at both ends?

   a. The action potentials would pass in the middle and travel to the opposite ends.

   b. The action potentials would meet in the middle and then be propagated back to their starting positions.

   c. The action potentials would stop as they met in the middle.

   d. The stronger action potential would override the weaker action potential.

   e. Summation would occur when the action potentials met in the middle, resulting in a larger action potential.

3. Assume you touched a hot stove with your finger. Contraction of the biceps muscle causes flexion (bending) of the elbow, whereas contraction of the triceps muscle causes extension (straightening) of the elbow. What pattern of postsynaptic potentials would you expect to be initiated as a reflex in the cell bodies of the neurons controlling these muscles to pull your hand away from the painful stimulus: excitatory postsynaptic potentials (EPSPs) or inhibitory postsynaptic potentials (IPSPs)?

   Now assume your finger is being pricked to obtain a blood sample. The same withdrawal reflex would be initiated. What pattern of postsynaptic potentials would you voluntarily produce in the neurons controlling the biceps and triceps to keep your arm extended despite the painful stimulus?

4. Assume presynaptic excitatory neuron A terminates on a postsynaptic cell near the axon hillock and presynaptic excitatory neuron B terminates on the same postsynaptic cell on a dendrite located on the side of the cell body opposite the axon hillock. Explain why rapid firing of presynaptic neuron A could bring the postsynaptic neuron to threshold through temporal summation, thus initiating an action potential, whereas firing of presynaptic neuron B at the same frequency and the same magnitude of EPSPs may not bring the postsynaptic neuron to threshold.

5. Sometimes patients are treated for a number of years following surgical removal of a breast because of estrogen-dependent breast cancer with *selective estrogen receptor modulators (SERMs)*. Speculate how this drug might be beneficial. Indicate by what route this drug would be administered and explain why.

To access the course materials and companion resources for this text, please visit www.cengagebrain.com

The Central Nervous System

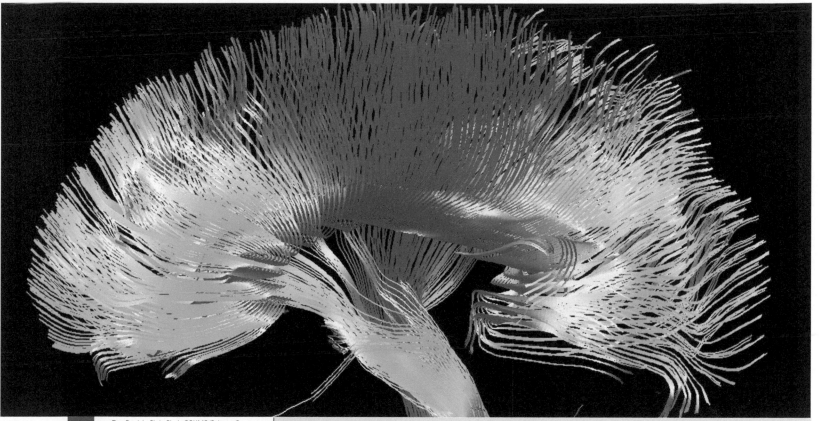

Tom Barrick, Chris Clark, SGHMS/Science Source

A diffusion magnetic resonance image (dMRI) scan of the white matter pathways of the brain. White matter is composed of myelinated nerve fibers that carry information between neurons in different parts of the brain. Blue represents neural pathways from top to bottom, green delineates pathways between front (left) and back, and red shows pathways between the left and right halves of the brain.

Homeostasis Highlights

The nervous system is one of the two major regulatory systems of the body; the other is the endocrine system. The three basic functional types of neurons—afferent neurons, efferent neurons, and interneurons—form a complex interactive network of excitable cells. Ninety percent of the cells of the nervous system are nonexcitable glial cells, which interact extensively both structurally and functionally with neurons. The **central nervous system (CNS)**, which consists of the brain and spinal cord, receives input about the external and internal environment from the afferent neurons. The CNS sorts and processes this input via interneurons and then initiates appropriate directions in the efferent neurons, which carry the instructions to glands or muscles to bring about the desired response—some type of secretion or movement. Many of these neurally controlled activities are directed toward maintaining homeostasis. In general, the nervous system acts by means of its electrical signals (action potentials) and neurotransmitter release to control the rapid responses of the body.

5.1 Organization and Cells of the Nervous System

The way humans act and react depends on complex, organized, discrete neuronal processing. Many basic life-supporting neuronal patterns, such as those controlling respiration and circulation, are similar in all individuals. However, there must be subtle differences in neuronal integration between someone who is a talented composer and someone who cannot carry a tune or between someone who is a math wizard and someone who struggles with long division. Some differences in the nervous systems of individuals are genetically endowed. The rest, however, are a result of environmental encounters and experiences. When the immature nervous system develops according to its genetic plan, an overabundance of neurons and synapses is formed. Depending on external stimuli and the extent to which these pathways are used, some are retained, firmly established, and even enhanced, whereas others are eliminated.

Clinical Note A case in point is **amblyopia** (lazy eye), in which the weaker of the two eyes is not used for vision. A lazy eye that does not get appropriate visual stimulation during a critical developmental period will almost completely and permanently lose the power of vision. The functionally blind eye itself is normal; the defect lies in the lost neuronal connections in the brain's visual pathways. However, if the weak eye is forced to work by covering the stronger eye with a patch during the sensitive developmental period, the weaker eye will retain full vision.

Maturation of the nervous system involves many instances of "use it or lose it." Once the nervous system has matured, modifications still occur as we continue to learn from our unique set of experiences. For example, the act of reading this page is somehow altering the neuronal activity of your brain as you (it is hoped) tuck the information away in your memory.

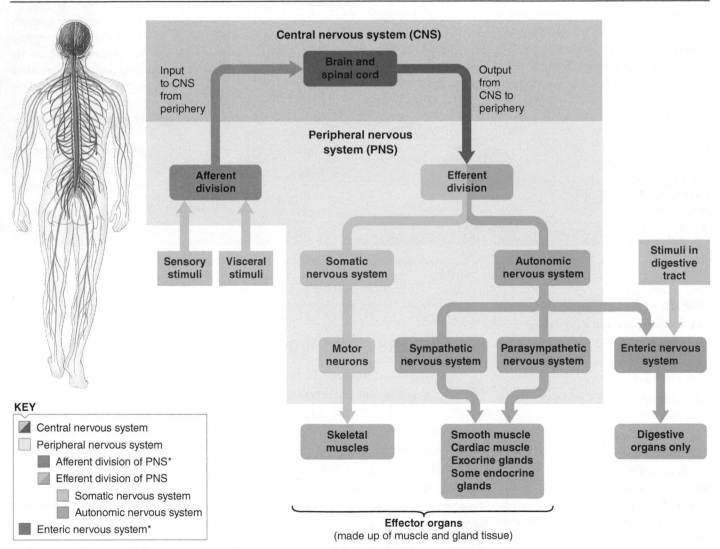

KEY

- Central nervous system
- Peripheral nervous system
  - Afferent division of PNS*
  - Efferent division of PNS
    - Somatic nervous system
    - Autonomic nervous system
- Enteric nervous system*

I Figure 5-1 Organization of the nervous system. *The afferent division of the peripheral nervous system (PNS) and enteric nervous system are not shown in the human figure. Afferent fibers travel within the same nerves as efferent fibers but in the opposite direction. The enteric nervous system lies entirely within the wall of the digestive tract.

FIGURE FOCUS: *What parts of the nervous system come into play when you are taking a walk? How about when you are digesting a meal?*

The nervous system is organized into the central nervous system and the peripheral nervous system.

The nervous system is organized into the **central nervous system (CNS)**, consisting of the brain and spinal cord, and the **peripheral nervous system (PNS)**, consisting of nerve fibers that carry information between the CNS and the other parts of the body (the periphery) (❙ Figure 5-1). The PNS is further subdivided into afferent and efferent divisions. The **afferent division** carries information *to* the CNS, apprising it of the external environment and providing status reports on internal activities being regulated by the nervous system (*a* is from *ad*, meaning "toward," as in *advance; ferent* means "carrying"; thus, *afferent* means "carrying toward"). Instructions *from* the CNS are transmitted via the **efferent division** to **effector organs**—the muscles or glands that carry out the orders to bring about the desired effect (*e* is from *ex*, meaning "from," as in *exit*; thus, *efferent* means "carrying from"). The efferent nervous system is divided into the **somatic nervous system**, which consists of the fibers of the motor neurons that supply the skeletal muscles; and the **autonomic nervous system**, which consists of fibers that innervate smooth muscle, cardiac muscle, and glands. The latter system is further subdivided into the **sympathetic nervous system** and the **parasympathetic nervous system**, both of which innervate most of the organs supplied by the autonomic system. In addition to the CNS and PNS, the **enteric nervous system** is an extensive nerve network in the wall of the digestive tract. Digestive activities are controlled by the autonomic nervous system, the enteric nervous system, and by hormones. The enteric nervous system can act independently of the rest of the nervous system but is also influenced by autonomic fibers that terminate on the enteric neurons. Sometimes the enteric nervous system is considered a third component of the autonomic nervous system, one that supplies the digestive organs only.

All these "nervous systems" are really subdivisions of a single, integrated nervous system. These subdivisions are based on differences in the structure, location, and functions of the various diverse parts of the whole nervous system.

The three functional classes of neurons are afferent neurons, efferent neurons, and interneurons.

Three functional classes of neurons make up the nervous system: *afferent neurons, efferent neurons,* and *interneurons.* The afferent division of the PNS consists of **afferent neurons**, which are shaped differently from efferent neurons and interneurons (❙ Figure 5-2). At its peripheral ending, a typical afferent neuron has a **sensory receptor** that generates action potentials in response to a particular type of stimulus (a change detectable by the neuron). (This stimulus-sensitive afferent neuronal receptor should not be confused with the special

protein receptors that bind chemical messengers and are found in the plasma membrane of all cells.) The afferent neuron cell body, which is devoid of dendrites and presynaptic inputs, is adjacent to the spinal cord. A long *peripheral axon,* commonly called the *afferent fiber,* extends from the receptor to the cell body, and a short *central axon* passes from the cell body into the spinal cord. Action potentials are initiated at the receptor end of the peripheral axon in response to a stimulus and are propagated along the peripheral axon and the central axon toward the spinal cord. The terminals of the central axon diverge and synapse with other neurons within the spinal cord, thus disseminating information about the stimulus. Afferent neurons lie primarily within the PNS. Only a small portion of their central axon endings projects into the spinal cord to relay signals from the periphery to the CNS.

Efferent neurons also lie primarily in the PNS. Efferent neuron cell bodies originate in the CNS, where many centrally located presynaptic inputs converge on them to influence their outputs to the effector organs. Efferent axons (*efferent fibers*) leave the CNS to course their way to the muscles or glands they innervate, conveying their integrated output for the effector organs to put into effect. (An autonomic nerve pathway consists of a two-neuron chain between the CNS and the effector organ.)

About 99% of all neurons are **interneurons**, which lie entirely within the CNS. As their name implies, interneurons lie between the afferent and the efferent neurons and are important in integrating peripheral information to peripheral responses (*inter* means "between"). For example, on receiving information through afferent neurons that you are touching a hot object, appropriate interneurons signal efferent neurons that transmit

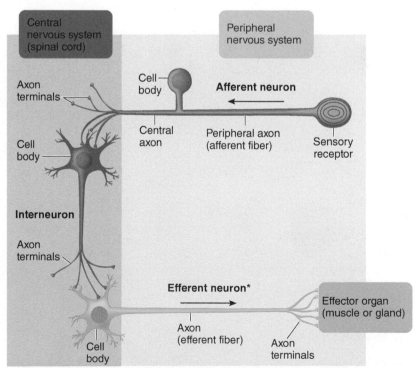

* Efferent autonomic nerve pathways consist of a two-neuron chain between the CNS and the effector organ.

❙Figure 5-2 **Structure and location of the three functional classes of neurons.**

to your hand and arm muscles the message, "Pull the hand away from the hot object!" The more complex the required action, the greater the number of interneurons interposed between the afferent message and the efferent response. In addition, interconnections between interneurons themselves are responsible for the abstract phenomena associated with the "mind," such as thoughts, emotions, memory, creativity, intellect, and motivation. These activities are the least understood functions of the nervous system.

Glial cells support the interneurons physically, metabolically, and functionally.

About 90% of the cells within the CNS are not neurons but **glial cells** or **neuroglia**. Despite their large numbers, glial cells occupy only about half the volume of the brain because they do not branch as extensively as neurons do.

Unlike neurons, glial cells do not initiate or conduct nerve impulses. However, they do communicate with neurons and among themselves by means of chemical signals. For much of the time since glial cells were discovered in the 19th century, they were considered passive "mortar" that physically supported the functionally important neurons. In the past three decades, however, the varied and important roles of these dynamic cells have become apparent. Glial cells help support the neurons both physically and metabolically. They also maintain the composition of the specialized extracellular environment surrounding the neurons within the narrow limits optimal for normal neuronal function. Furthermore, they actively modulate (depress or enhance) synaptic function and are considered nearly as important as neurons to learning and memory. There are four major types of glial cells in the CNS—*astrocytes, oligodendrocytes, microglia,* and *ependymal cells*—each with specific roles (❙Figure 5-3).

Astrocytes Named for their starlike shape (*astro* means "star"; *cyte* means "cell") (❙Figure 5-4), **astrocytes** are the most abundant glial cells. They fill several critical functions:

1. As the main "glue" (*glia* means "glue") of the CNS, astrocytes hold the neurons together in proper spatial relationships.

2. Astrocytes serve as a scaffold that guides neurons to their proper final destination during fetal brain development.

3. These glial cells induce the small blood vessels (capillaries) of the brain to undergo the anatomic and functional changes that establish the blood–brain barrier, a highly selective, protective barricade between the blood and brain.

4. They help transfer nutrients from the blood to the neurons.

5. Astrocytes form neural scars to help repair brain injuries.

6. They take up and degrade some locally released neurotransmitters, thus bringing the actions of these chemical messengers to a halt.

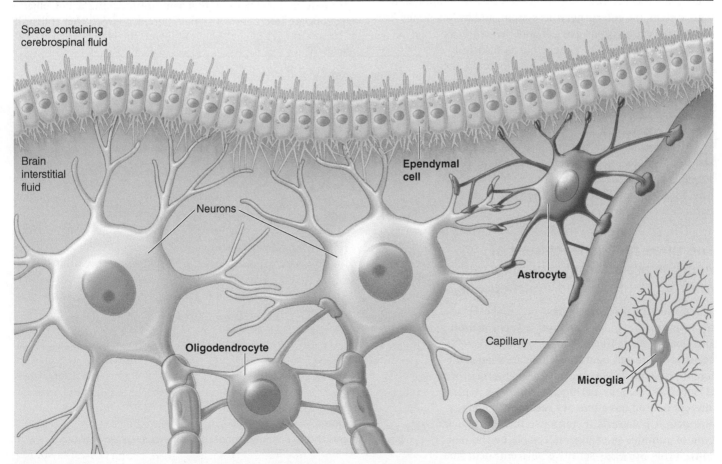

❙Figure 5-3 **Glial cells of the central nervous system.** The glial cells include the astrocytes, oligodendrocytes, microglia, and ependymal cells.

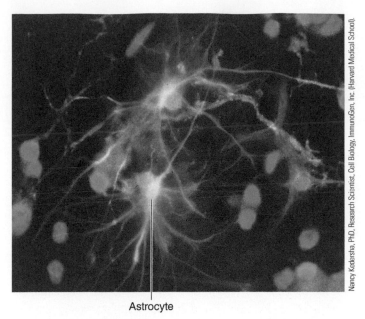

Astrocyte

Figure 5-4 Astrocytes. Note the starlike shape of these astrocytes, which have been grown in tissue culture.

7. Astrocytes take up excess K^+ from the brain extracellular fluid (ECF) when high action potential activity outpaces the ability of the $Na^+–K^+$ pump to return to the neurons the K^+ that leaves during the falling phase of an action potential. By taking up excess K^+, astrocytes help maintain the optimal ion conditions around neurons to sustain normal neural excitability.

8. Astrocytes along with other glial cells enhance synapse formation and modify synaptic transmission.

9. Astrocytes communicate with neurons and with one another by means of chemical signals passing locally in both directions between these cells in two ways. First, chemical signals pass directly among astrocytes and between astrocytes and neurons through gap junctions (see p. 62) without entering the ECF. Second, chemical signals pass extracellularly between these cells. Astrocytes have receptors for the common neurotransmitter glutamate. Furthermore, firing of neurons in the brain in some instances triggers the release of adenosine triphosphate (ATP) along with the classical neurotransmitter from the axon terminal. Binding of glutamate to an astrocyte's receptors or detection of extracellular ATP by the astrocyte leads to Ca^{2+} influx into this glial cell. The resultant rise in intracellular Ca^{2+} prompts the astrocyte itself to release ATP, thereby activating adjacent glial cells. In this way, astrocytes can share information about action potential activity in a nearby neuron. In addition, astrocytes and other glial cells can also release glutamate and other chemical signals. These extracellular chemical signals from glial cells, collectively called **gliotransmitters**, can affect neuronal excitability and strengthen synaptic activity, such as by increasing neuronal release of neurotransmitter or promoting the formation of new synapses. Glial modulation of synaptic activity is likely important in memory and learning. Also, astrocytes are thought to coordinate and integrate synaptic activity among networks of neurons working together.

Scientists are trying to sort out the two-directional chatter that takes place between and among these glial cells and neurons because this dialogue plays an important role in information processing in the brain. In fact, some neuroscientists suggest that synapses should be considered "three-party" junctures involving the glial cells and the presynaptic and postsynaptic neurons. This point of view is indicative of the increasingly important role being placed on astrocytes in synapse function. Thus, astrocytes have come a long way from their earlier reputation as "support staff" for neurons; these glial cells might turn out to be the "board members" commanding the neurons.

10. The most recently identified role of astrocytes is their role in clearing toxic metabolic byproducts from the brain by means of the **glymphatic system**, a glia substitute for the lymphatic system (hence, this system is dubbed "glymphatic.") The brain is not supplied by the lymphatic system, a system of lymph-carrying vessels that transports excess interstitial fluid from the tissues into the blood, proteins that have leaked into the interstitial fluid, and toxic metabolic wastes that have accumulated in the interstitial fluid (see p. 358). The glymphatic system serves as a functional waste clearance pathway in the brain, similar to the lymphatic system in peripheral tissues. Astrocytes facilitate a cleansing fluid exchange within spaces that lie between microscopic brain blood vessels and the long astrocyte projections. This "brainwashing" increases during sleep.

Oligodendrocytes **Oligodendrocytes** form the insulative myelin sheaths around axons in the CNS. An oligodendrocyte has several elongated projections, each of which wraps jelly-roll fashion around a section of an interneuronal axon to form a patch of myelin (see **Figure 4-11c, p. 101; and Figure 5-3).

Microglia **Microglia** are the immune defense cells of the CNS. These scavengers are similar to monocytes, a type of white blood cell that leaves the blood and sets up residence as a front-line defense agent in various tissues throughout the body. Microglia are derived from the same bone-marrow tissue that gives rise to monocytes. During embryonic development, microglia migrate to the CNS, where they remain stationary until activated by an infection or injury.

In the resting state, microglia are wispy cells with many long branches that radiate outward. Resting microglia are not just waiting watchfully, however. In addition to providing surveillance, they release low levels of growth factors, such as *nerve growth factor,* which help neurons and other glial cells survive and thrive. Also, emerging evidence suggests that microglia may play an important role in synaptic pruning (eliminating unneeded synapses) during development and memory processing. When trouble occurs in the CNS, microglia retract their branches, round up, and become highly mobile and move toward the affected area to remove any foreign invaders or tissue debris by phagocytosis (see p. 31). Activated microglia also release destructive chemicals for assault against their target.

 Microglia are the only CNS cell type that can be infected by HIV, the virus that causes AIDS. Microglia dysfunction ultimately leads to AIDS-related

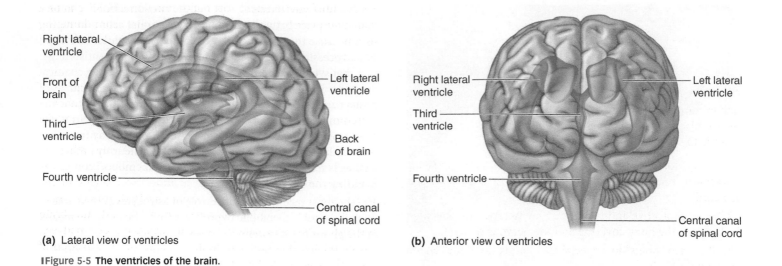

Right lateral ventricle

Front of brain

Left lateral ventricle

Third ventricle

Back of brain

Fourth ventricle

Central canal of spinal cord

(a) Lateral view of ventricles

Right lateral ventricle

Left lateral ventricle

Third ventricle

Fourth ventricle

Central canal of spinal cord

(b) Anterior view of ventricles

❙Figure 5-5 **The ventricles of the brain.**

dementia (mental failing). Furthermore, researchers suspect that excessive release of destructive chemicals from overzealous microglia may damage the neurons they are meant to protect, thus contributing to the insidious neuronal damage seen in stroke, Alzheimer's disease, multiple sclerosis, and other **neurodegenerative diseases** in which nerve cells are destroyed.

Ependymal Cells **Ependymal cells** line the internal, fluid-filled cavities of the CNS. As the nervous system develops embryonically from a hollow neural tube, the original central cavity of this tube is maintained and modified to form the ventricles and central canal. The four **ventricles** are interconnected chambers within the brain that are continuous with the narrow, hollow **central canal** through the middle of the spinal cord (❙Figure 5-5). The ependymal cells lining the ventricles help form cerebrospinal fluid, a topic to be discussed shortly. Ependymal cells are one of the few cell types with cilia (see p. 48). Beating of ependymal cilia contributes to the flow of cerebrospinal fluid through the ventricles.

Ependymal cells also have a different role: They serve as neural stem cells with the potential of forming not only other glial cells, but also new neurons, especially after injury (see p. 10). The long held traditional view was that new neurons are not produced in the mature brain. Then, in the late 1990s, scientists discovered that new neurons are produced in the **hippocampus**, a brain structure important for learning and memory (see ❙Figure 5-16, p. 155). Astrocyte-like neural stem cells continuously renew a subpopulation of hippocampal neurons, whereas the rest of the neurons in this structure do not turn over. The nonrenewable neurons are produced during embryonic development, and most of them survive the life of the individual, although some succumb to head trauma, stroke, or neurodegenerative disease. In contrast, the new neurons generated in adults "grow up," function as mature cells for a period of time, then die and are exchanged for new neurons, thus maintaining a mix of young and old cells in the renewable hippocampal subpopulation. Young and mature neurons process

information in different ways, both of which are critical for hippocampal function. Ongoing **neurogenesis** (production of new neurons) is crucial for maintaining the forever-young pool of neurons. The rate of hippocampal neurogenesis declines with age and chronic stress and is increased in response to physical exercise.

Neurons in the rest of the brain are considered irreplaceable. But the discovery of ependymal cells as a reservoir of precursors for new neurons suggests that the adult brain has more potential for repairing damaged regions than previously assumed. Currently, no evidence shows that the brain spontaneously repairs itself following neuron-losing insults. Apparently, most brain regions cannot activate this mechanism for replenishing neurons, probably because the appropriate "cocktail" of supportive chemicals is not present. Researchers hope that probing into why these ependymal cells are dormant and how they might be activated will lead to the possibility of unlocking the brain's undeveloped capacity for self-repair.

Clinical Note Unlike neurons, glial cells do not lose the ability to undergo cell division, so most brain tumors of neural origin consist of glial cells (**gliomas**). Neurons themselves do not form tumors because they are unable to divide and multiply. Brain tumors of non-neural origin are of two types: (1) those that metastasize (spread) to the brain from other sites and (2) **meningiomas**, which originate from the meninges, the protective membranes covering the CNS. We next examine the meninges and other means by which the CNS is protected.

Check Your Understanding 5.1

1. Draw a flow diagram showing the organization of the subdivisions of the human nervous system.

2. Compare the structure, location, and function of the functional classes of neurons.

3. List the four types of glial cells.

5.2 Protection and Nourishment of the Brain

Central nervous tissue is delicate. Because of this characteristic, and because damaged nerve cells cannot be replaced, this fragile, irreplaceable tissue must be well protected. Four major features help protect the CNS from injury:

1. It is enclosed by hard, bony structures. The *cranium (skull)* encases the brain, and the *vertebral column* surrounds the spinal cord.

2. Three protective and nourishing membranes, the *meninges,* lie between the bony covering and the nervous tissue.

3. The brain "floats" in a special cushioning fluid, the *cerebrospinal fluid (CSF).*

4. A highly selective *blood–brain barrier* limits access of blood-borne materials into the vulnerable brain tissue.

The role of the first of these protective devices, the bony covering, is self-evident. The latter three protective mechanisms warrant further discussion.

Three meningeal membranes wrap, protect, and nourish the central nervous system.

Three membranes, the **meninges**, wrap the CNS. From the outermost to the innermost layer, they are the *dura mater,* the *arachnoid mater,* and the *pia mater* (Figure 5-6). (*Mater* means "mother," indicative of these membranes' protective and supportive role.)

The **dura mater** is a tough, inelastic covering that consists of two layers (*dura* means "tough"). Usually, these layers adhere closely, but in some regions they are separated to form blood-filled cavities, **dural sinuses**, or in the case of the larger cavities, **venous sinuses**. Venous blood draining from the brain empties into these sinuses to be returned to the heart. CSF also reenters the blood at these sinus sites.

The **arachnoid mater** is a delicate, richly vascularized layer with a "cobwebby" appearance (*arachnoid* means "spiderlike"). The space between the arachnoid layer and the underlying pia mater, the **subarachnoid space**, is filled with CSF. Protrusions of arachnoid tissue, the **arachnoid villi**, penetrate through gaps in the overlying dura and project into the dural sinuses (Figure 5-6b). CSF is reabsorbed across the surfaces of these villi into the blood circulating within the sinuses.

The innermost meningeal layer, the **pia mater**, is the most fragile (*pia* means "gentle"). It is highly vascular and closely adheres to the surfaces of the brain and spinal cord, following every ridge and valley. In certain areas, the pia mater and ependymal cells form a special relationship important in the formation of CSF, a topic to which we now turn attention.

The brain floats in its own special cerebrospinal fluid.

Cerebrospinal fluid (CSF) surrounds and cushions the brain and spinal cord. The CSF has about the same density as the brain itself, so the brain essentially floats or is suspended in this special fluid environment. The major function of CSF is to be a shock-absorbing fluid to prevent the brain from bumping against the interior of the hard skull when the head is subjected to sudden, jarring movements.

In addition to protecting the delicate brain from mechanical trauma, the CSF plays an important role in the exchange of materials between the neural cells and the interstitial fluid surrounding the brain. Only the brain interstitial fluid—not the blood or CSF—comes into direct contact with the neurons and glial cells. Because the brain interstitial fluid directly bathes the neural cells, its composition is critical. The composition of the brain interstitial fluid is influenced more by changes in the composition of the CSF than by alterations in the blood. Materials are exchanged fairly freely between the CSF and brain interstitial fluid, whereas only limited exchange occurs between the blood and brain interstitial fluid. Thus, the composition of the CSF must be carefully regulated.

CSF is formed primarily by the **choroid plexuses**, which are specialized structures in the walls of the ventricles that protrude into the ventricular cavity in particular regions (Figure 5-6a). Choroid plexuses consist of richly vascularized, cauliflowerlike masses of pia mater tissue that dip into pockets formed by ependymal cells (Figure 5-6c). CSF forms as the ependymal cells of the choroid plexuses selectively transport materials from the pia capillary blood into the ventricular cavity. The composition of CSF differs from that of blood. For example, CSF is lower in K^+ and slightly higher in Na^+, making the brain interstitial fluid an ideal environment for movement of these ions down concentration gradients, a process essential for conduction of nerve impulses (see p. 96). The biggest difference is the presence of plasma proteins in the blood but almost no proteins normally present in the CSF. Plasma proteins cannot exit the brain capillaries to leave the blood during formation of CSF.

Once CSF is formed, it flows through the four interconnected ventricles of the brain and through the spinal cord's narrow central canal, which is continuous with the last ventricle. CSF also escapes through small openings from the fourth ventricle at the base of the brain to enter the subarachnoid space and subsequently flows between the meningeal layers over the entire surface of the brain and spinal cord (Figure 5-6). When the CSF reaches the upper regions of the brain, it is reabsorbed from the subarachnoid space into the venous blood through the arachnoid villi. Flow of CSF through this system is facilitated by ciliary beating along with circulatory and postural factors that result in a CSF pressure of about 10 mm Hg. Reduction of this pressure by removal of even a few milliliters (mL) of CSF during a spinal tap for laboratory analysis may produce severe headaches.

Clinical Note Through the ongoing processes of formation, circulation, and reabsorption, the entire CSF volume of about 125 to 150 mL is replaced more than three times a day. If any one of these processes is defective so that excess CSF accumulates (for example, as with obstruction of the CSF pathways by a malformation or a tumor), **hydrocephalus** ("water on the brain") occurs. The resulting increase in CSF pressure can lead to brain damage if untreated.

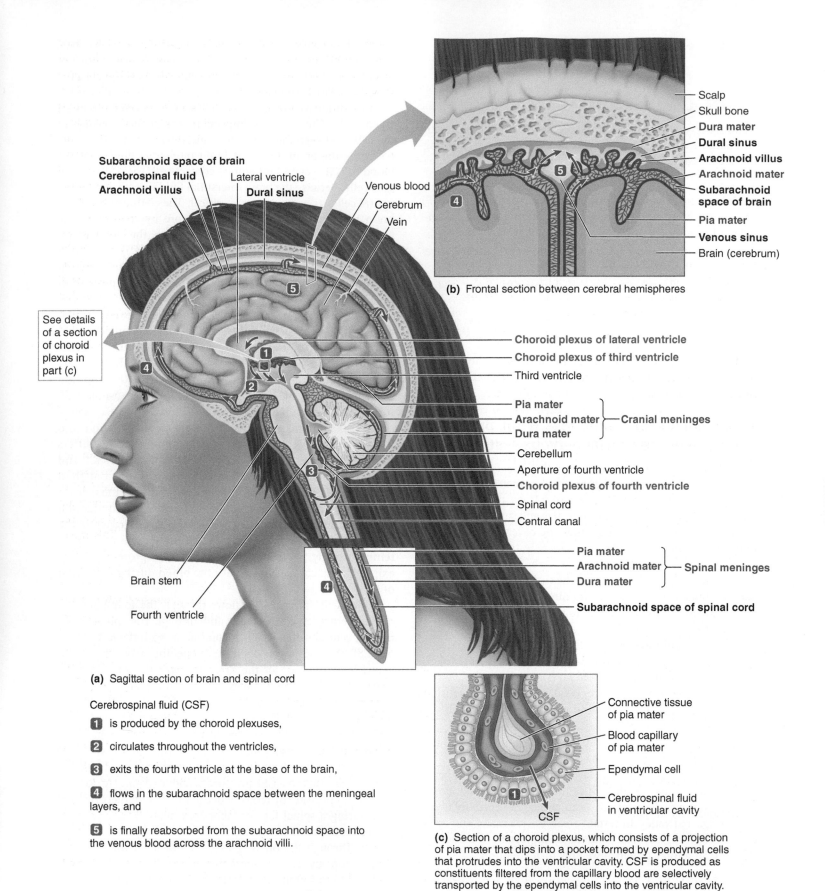

Subarachnoid space of brain
Cerebrospinal fluid
Arachnoid villus

Lateral ventricle
Dural sinus

Venous blood
Cerebrum
Vein

Scalp
Skull bone
Dura mater
Dural sinus
Arachnoid villus
Arachnoid mater
Subarachnoid space of brain
Pia mater
Venous sinus
Brain (cerebrum)

(b) Frontal section between cerebral hemispheres

See details of a section of choroid plexus in part (c)

Choroid plexus of lateral ventricle
Choroid plexus of third ventricle
Third ventricle

Pia mater
Arachnoid mater } **Cranial meninges**
Dura mater

Cerebellum
Aperture of fourth ventricle
Choroid plexus of fourth ventricle
Spinal cord
Central canal

Pia mater
Arachnoid mater } **Spinal meninges**
Dura mater

Subarachnoid space of spinal cord

Brain stem

Fourth ventricle

(a) Sagittal section of brain and spinal cord

Cerebrospinal fluid (CSF)

1 is produced by the choroid plexuses,

2 circulates throughout the ventricles,

3 exits the fourth ventricle at the base of the brain,

4 flows in the subarachnoid space between the meningeal layers, and

5 is finally reabsorbed from the subarachnoid space into the venous blood across the arachnoid villi.

Connective tissue of pia mater
Blood capillary of pia mater
Ependymal cell
Cerebrospinal fluid in ventricular cavity

CSF

(c) Section of a choroid plexus, which consists of a projection of pia mater that dips into a pocket formed by ependymal cells that protrudes into the ventricular cavity. CSF is produced as constituents filtered from the capillary blood are selectively transported by the ependymal cells into the ventricular cavity.

∎Figure 5-6 **Relationship of the meninges and cerebrospinal fluid (CSF) to the brain and spinal cord.** (a) Brain, spinal cord, and meninges in sagittal section. The *blue* arrows and numbered steps indicate the direction of flow of CSF (in *yellow*). (b) Frontal section in the region between the two cerebral hemispheres of the brain, depicting the meninges and arachnoid villi in greater detail. CSF is reabsorbed into the blood across the arachnoid villi. (c) Close-up of a portion of a choroid plexus showing the relationship between the pia mater capillaries and the ependymal cells in the formation of CSF.

A highly selective blood–brain barrier regulates exchanges between the blood and brain.

The brain is carefully shielded from harmful changes in the blood by a highly selective **blood–brain barrier (BBB)** that limits access of blood-borne materials into the vulnerable brain tissue. Throughout the body, materials can be exchanged between the blood and the interstitial fluid only across the walls of capillaries. Capillary walls are formed by a single layer of *endothelial cells*. The holes or pores usually present between the endothelial cells permit rather free exchange across capillaries elsewhere. However, the cells that form the walls of a brain capillary are joined by tight junctions (see p. 61). These impermeable junctions seal the capillary wall so that nothing can be exchanged across the wall by passing between the cells. The only permissible exchanges occur through the endothelial cells themselves. Lipid-soluble substances such as O_2, CO_2, alcohol, and steroid hormones penetrate these cells easily by dissolving in their lipid plasma membrane. Small water molecules also readily diffuse through by passing between the phospholipid molecules of the plasma membrane or through aquaporins (water channels) (see p. 66). All other substances exchanged between the blood and brain interstitial fluid, including such essential materials as glucose, amino acids, and ions, are transported by highly selective membrane-bound carriers. Thus, transport across brain capillary walls *between* the wall-forming cells is *anatomically prevented* and transport *through* the cells is *physiologically restricted*. Together, these mechanisms constitute the BBB. By strictly limiting exchange between the blood and brain, the BBB protects the delicate brain from chemical fluctuations in the blood. For example, even if the K^+ level in the blood is doubled, little change occurs in the K^+ concentration of the fluid bathing the central neurons. This is beneficial because alterations in interstitial fluid K^+ would be detrimental to neuronal function. Also, the BBB minimizes the possibility that potentially harmful blood-borne substances might reach the central neural tissue. It further prevents certain circulating hormones that could act as neurotransmitters from reaching the brain, where they could produce uncontrolled nervous activity.

Clinical Note On the negative side, the BBB limits the use of drugs for the treatment of brain and spinal cord disorders because many drugs cannot penetrate this barrier. Researchers are seeking ways to safely and temporarily breech the barrier to permit drugs to get through, such as by sneaking them through normal gateways for entry of essentials such as iron; by temporarily shrinking the endothelial cells via injection of hyperosmotic solutions that draw water out of the cells, thus creating minuscule gaps in the BBB by pulling apart the tight junctions between the cells; or by injecting microscopic gas bubbles and using ultrasound to vibrate them against a precise location of the BBB, thereby briefly forcing open the tight junctions in the bombarded section.

Pericytes and astrocytes both contribute to formation and maintenance of the BBB. **Pericytes** are contractile cells that wrap around capillary endothelial cells throughout the body. The precise roles of brain pericytes in BBB function are still being investigated. Highly complex signaling cascades take place among pericytes, brain capillary endothelial cells, and astrocytes. Brain capillaries are surrounded by astrocyte processes, which lie outside of the pericytes. Astrocytes play three roles in the BBB: (1) They signal the cells forming the brain capillaries to "get tight." Endothelial cells do not have an inherent ability to form tight junctions; they do so only at the command of a signal within their neural environment. (2) Astrocytes promote the production of specific carrier proteins and ion channels that regulate the transport of selected substances through the endothelial cells. (3) These glial cells participate in the cross-cellular transport of some substances, such as K^+.

Certain areas of the brain, most notably a portion of the hypothalamus, are not subject to the BBB. Functioning of the hypothalamus depends on its "sampling" the blood and adjusting its controlling output accordingly to maintain homeostasis. Part of this output is in the form of water-soluble hormones that must enter hypothalamic capillaries to be transported to their sites of action. Appropriately, these hypothalamic capillaries are not sealed by tight junctions, the presence of which would prevent entry of these hormones into the blood.

The brain depends on constant delivery of oxygen and glucose by the blood.

Even though many substances in the blood never come in contact with the brain tissue, the brain depends more than any other tissue on a constant blood supply. Unlike most tissues, which can resort to anaerobic metabolism to produce ATP in the absence of O_2 for at least short periods (see p. 39), the brain cannot produce ATP without O_2. Scientists recently discovered an O_2-binding protein, **neuroglobin**, in the brain. This molecule, which is similar to hemoglobin, the O_2-carrying protein in red blood cells (see p. 383), is thought to play a key role in O_2 handling in the brain, although its exact function remains to be determined. Also in contrast to most tissues, which can use other sources of fuel for energy production in lieu of glucose, the brain normally uses only glucose but does not store any of this nutrient. Because of its high rate of demand for ATP, under resting conditions the brain uses 20% of the O_2 and 50% of the glucose consumed in the body. Therefore, the brain depends on a continuous, adequate blood supply of O_2 and glucose. Although it constitutes only 2% of body weight, the brain receives 15% of the blood pumped out by the heart. (Instead of using glucose during starvation, the brain can resort to using ketone bodies produced by the liver, but this alternate nutrient source also must be delivered by the blood to the brain.)

Clinical Note Brain damage results if this organ is deprived of its critical O_2 supply for more than 4 to 5 minutes or if its glucose supply is cut off for more than 10 to 15 minutes. The most common cause of inadequate blood supply to the brain is a stroke. (See the accompanying boxed feature, Concepts, Challenges, and Controversies, for details.)

Check Your Understanding 5.2

1. Name the meninges, from outermost to innermost.
2. Discuss the function of cerebrospinal fluid.

Strokes: A Deadly Domino Effect

THE MOST COMMON CAUSE OF brain damage is a **cerebrovascular accident (CVA** or **stroke)**. When a cerebral (brain) blood vessel is blocked by a clot (which accounts for more than 80% of strokes) or ruptures, the brain tissue supplied by that vessel loses its vital O_2 and glucose supply. The result is damage and usually death of the deprived tissue. Furthermore, neural damage (and the subsequent loss of neural function) extends well beyond the blood-deprived area as a result of a neurotoxic effect that leads to the death of additional nearby cells. The initial blood-deprived cells die by necrosis (unintentional cell death), but the doomed neighbors undergo apoptosis (deliberate cell suicide; see p. 40). In a process known as **excitotoxicity**, the initial O_2-starved cells release excessive amounts of glutamate, a common excitatory neurotransmitter. The excitatory overdose of glutamate from the damaged brain cells binds with and overexcites surrounding neurons. Specifically, glutamate binds with excitatory receptors known as NMDA receptors, which function as calcium (Ca^{2+}) channels. As a result of toxic activation of these receptor-channels, they remain open for too long, permitting too much Ca^{2+} to rush into the affected neighboring neurons. This elevated intracellular Ca^{2+} triggers these cells to self-destruct. Cell-damaging **free radicals** are produced during this process. These highly reactive, electron-deficient particles cause further cell damage by snatching electrons from other molecules. Adding to the injury, researchers speculate that the Ca^{2+} apoptotic signal may spread from these dying cells to abutting healthy cells through gap junctions, cell-to-cell conduits that allow Ca^{2+} and other small ions to diffuse freely between cells (see p. 62). This action kills even more neuronal victims. Thus, most neurons that die following a stroke are originally unharmed cells that commit suicide in response to the chain of reactions unleashed by the toxic release of glutamate from the initial site of O_2 deprivation.

Until late last century, physicians could do nothing to halt the inevitable neuronal loss following a stroke, leaving patients with an unpredictable mix of neural deficits. Treatment was limited to rehabilitative therapy after the damage was already complete. In recent years, armed with new knowledge about the underlying factors in stroke-related neuronal death, the medical community has been seeking ways to halt the cell-killing domino effect. The goal is to limit the extent of neuronal damage and thus minimize or even prevent clinical symptoms such as paralysis. If a stroke is caught in time, doctors now administer clot-dissolving drugs within the first 3 hours after onset to restore blood flow through blocked cerebral vessels. Clot busters were the first drugs used to treat strokes, but they are only the beginning of new stroke therapies. Other methods are under investigation to prevent adjacent neurons from succumbing to the neurotoxic release of glutamate. These include blocking the NMDA receptor-channels that initiate the death-wielding chain of events in response to glutamate, halting the apoptosis pathway that results in self-execution, and blocking the gap junctions that permit the Ca^{2+} death messenger to spread to adjacent cells. Evidence in a recent rat study even suggests that sensory stimulation, such as rubbing a stroke victim's face or fingers, may represent a cheap, simple way to help minimize disability while the person is on the way to medical attention. These tactics hold much promise for treating strokes, which are the most common cause of adult disability and the third leading cause of death in the United States. However, to date, no new neuroprotective drugs have been found that do not cause serious side effects.

3. Explain by what means the blood–brain barrier anatomically prevents transport between the cells that form the walls of brain capillaries and physiologically restricts transport through these cells.

5.3 Overview of the Central Nervous System

The CNS consists of the brain and spinal cord. The human brain is the consistency of tofu and weighs just three pounds. The estimated 85 billion neurons in your brain are joined together by an estimated quadrillion synaptic connections and are assembled into complex networks that enable you to (1) subconsciously regulate your internal environment by neural means, (2) experience emotions, (3) voluntarily control your movements, (4) perceive (be consciously aware of) your body and your surroundings, and (5) engage in other higher cognitive processes such as thought and memory. The term **cognition** refers to the act or process of "knowing," including both awareness and judgment.

No part of the brain acts in isolation from other brain regions because networks of neurons are anatomically linked by synapses and neurons throughout the brain communicate extensively with one another by electrical and chemical means. However, neurons that work together to ultimately accomplish a given function tend to be organized within a discrete location. Therefore, even though the brain operates as a whole, it is organized into regions. The parts of the brain can be grouped in various ways based on anatomic distinctions, functional specialization, and evolutionary development. We use the following grouping:

1. Brain stem
2. Cerebellum

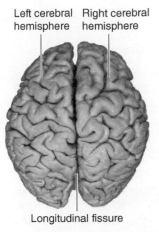

Left cerebral hemisphere   Right cerebral hemisphere

Longitudinal fissure

(a) Brain, superior (top) view

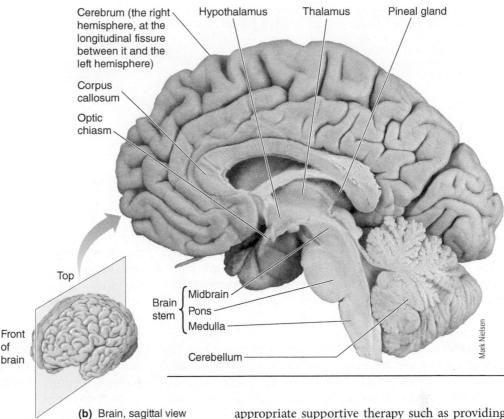

Cerebrum (the right hemisphere, at the longitudinal fissure between it and the left hemisphere)

Corpus callosum

Optic chiasm

Hypothalamus   Thalamus   Pineal gland

Brain stem { Midbrain   Pons   Medulla

Cerebellum

Mark Nielsen

Top

Front of brain

(b) Brain, sagittal view

❘Figure 5-7 **Brain of a human cadaver.** (a) Superior view looking down on the top of the brain. Note that the deep longitudinal fissure divides the cerebrum into the right and left cerebral hemispheres. (b) Sagittal view of the right half of the brain. All major brain regions are visible from this midline interior view. The corpus callosum serves as a neural bridge between the two cerebral hemispheres.

3. Forebrain
   a. Diencephalon
      (1) Hypothalamus
      (2) Thalamus
   b. Cerebrum
      (1) Basal nuclei
      (2) Cerebral cortex

The order in which these components are listed generally represents both their anatomic location (from bottom to top) and their complexity and sophistication of function (from the least specialized, oldest evolutionary level to the newest, most specialized level).

A primitive nervous system consists of comparatively few interneurons interspersed between afferent and efferent neurons. During evolutionary development, the interneuronal component progressively expanded, formed more complex interconnections, and became localized at the head end of the nervous system, forming the brain. Newer, more sophisticated layers of the brain were added on to the older, more primitive layers. The human brain represents the present peak of development.

The *brain stem,* the oldest region of the brain, is continuous with the spinal cord (❘Table 5-1 and ❘Figure 5-7b). It consists of the midbrain, pons, and medulla. The brain stem controls many life-sustaining processes, such as respiration, circulation, and digestion, common to all vertebrates. These processes are often referred to as *vegetative functions,* meaning functions performed unconsciously or involuntarily. With the loss of higher brain functions, these lower brain levels, in conjunction with

appropriate supportive therapy such as providing adequate nourishment, can still sustain the functions essential for survival, but the person has no awareness or control of that life.

Attached at the top rear portion of the brain stem is the *cerebellum,* which is concerned with maintaining proper position of the body in space and subconscious coordination of motor activity (movement). The cerebellum also plays a key role in learning skilled motor tasks, such as a dance routine.

On top of the brain stem, tucked within the interior of the cerebrum, is the *diencephalon.* It houses two brain components: the *hypothalamus,* which controls many homeostatic functions important in maintaining stability of the internal environment; and the *thalamus,* which begins sensory processing.

Using an ice cream cone as an analogy, on top of this "cone" of lower brain regions is the *cerebrum,* whose "scoop" gets progressively larger and more highly convoluted (that is, has tortuous ridges delineated by deep grooves or folds) the more evolutionarily advanced the vertebrate species is. The cerebrum is most highly developed in humans, where it constitutes about 80% of the total brain weight. The outer layer of the cerebrum is the highly convoluted *cerebral cortex,* which caps an inner core that houses the *basal nuclei.* The myriad convolutions of the human cerebral cortex give it the appearance of a much-folded walnut (❘Figure 5-7a). In more ancestral mammal groups, the cortex is smooth. Without these surface wrinkles, the human cortex would take up to three times the area it does and, thus, would not fit like a cover over the underlying structures. The increased neural circuitry housed in the extra cerebral cortical area not found in less highly developed species is responsible for many of our unique human abilities. The cerebral cortex plays a key role in the most sophisticated neural functions, such as voluntary initiation of movement, final sen-

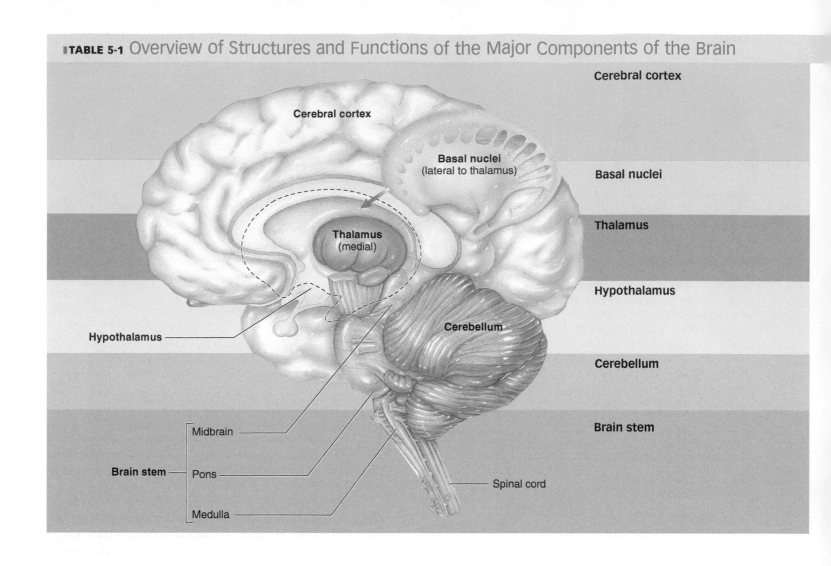

sory perception (the brain's interpretation of the body and its surroundings based on sensory input), conscious thought, language, personality traits, and other factors we associate with the mind or intellect. It is the highest, most complex, integrating area of the brain.

Each of these regions of the CNS is now discussed in turn, starting with the highest level, the cerebral cortex, and moving down to the lowest level, the spinal cord.

Check Your Understanding 5.3

1. Define *cognition*.
2. Outline the components of the brain from the least specialized, oldest evolutionary level to the newest, most specialized level.
3. Discuss the significance of the human cerebral cortex being highly convoluted.

5.4 Cerebral Cortex

The **cerebrum** is divided into two halves, the right and left **cerebral hemispheres** (Figure 5-7a). They are connected to each other by the **corpus callosum**, a thick band consisting of an

estimated 300 million neuronal axons that connect the two hemispheres (Figure 5-7b; also see Figure 5-14, p. 153). The corpus callosum is the body's "information superhighway." The two hemispheres communicate and cooperate with each other by means of constant information exchange through this neural connection.

The cerebral cortex is an outer shell of gray matter covering an inner core of white matter.

Each hemisphere is composed of a thin outer shell of *gray matter*, the **cerebral cortex**, covering a thick central core of *white matter* (see Figure 5-14). Several other masses of gray matter that collectively constitute the basal nuclei are located deep within the white matter. Throughout the entire CNS, **gray matter** consists mostly of densely packaged neuronal cell bodies and their dendrites, in addition to most glial cells. Bundles or tracts of myelinated nerve fibers (axons) constitute **white matter**; its white appearance is a result of the lipid composition of the myelin. Gray matter and white matter each make up about half of the brain.

Gray matter can be viewed as the "computers" of the CNS and white matter as the "wires" that connect the computers to

1. Sensory perception
2. Voluntary control of movement
3. Language
4. Personality traits
5. Sophisticated mental events, such as thinking, memory, decision making, creativity, and self-consciousness

1. Inhibition of muscle tone
2. Coordination of slow, sustained movements
3. Suppression of useless patterns of movement

1. Relay station for all synaptic input
2. Crude awareness of sensation
3. Some degree of consciousness
4. Role in motor control

1. Regulation of many homeostatic functions, such as temperature control, thirst, urine output, and food intake
2. Important link between nervous and endocrine systems
3. Extensive involvement with emotion and basic behavioral patterns
4. Role in sleep–wake cycle

1. Maintenance of balance
2. Enhancement of muscle tone
3. Coordination and planning of skilled voluntary muscle activity

1. Origin of majority of peripheral cranial nerves
2. Cardiovascular, respiratory, and digestive control centers
3. Regulation of muscle reflexes involved with equilibrium and posture
4. Reception and integration of all synaptic input from spinal cord; arousal and activation of cerebral cortex
5. Role in sleep–wake cycle

one another. White matter contains collectively an estimated quarter million miles of fibers, enough that if lined up end to end, they would stretch from Earth to the moon. Integration of neural input and initiation of neural output take place at synapses within gray matter. The fiber tracts in white matter transmit signals from one part of the cerebral cortex to another or between the cortex and other regions of the CNS. Such communication between different areas of the cortex and elsewhere facilitates integration of their activity. This integration is essential for even a relatively simple task such as picking a flower. Vision of the flower is received by one area of the cortex, reception of its fragrance takes place in another area, and movement is initiated by still another area. More subtle neuronal responses, such as appreciation of the flower's beauty and the urge to pick it, are poorly understood but undoubtedly involve extensive interconnection of fibers among different cortical regions.

Our knowledge about the complex connections that underlie brain function is expanding at a rapid rate owing to the **Human Connectome Project** launched by the National Institutes of Health in 2009. Through this ambitious project, scientists across the country are collecting and sharing data using cutting-edge brain imaging techniques on healthy adults to map the entire white matter fiber circuitry in the human brain. One imaging tool being used is **diffusion magnetic resonance imaging (dMRI)**, which maps the orientation in space of the brain's white matter tracts by tracking how the organized bundles of fibers affect the diffusion of water. Because diffusion is faster in the direction parallel to the fibers than in the perpendicular direction, the technique can be used to reveal the patterns of white matter connectivity (see chapter opener photo).

Our knowledge about gray matter function has been leaping forward in recent decades too through use of modern technologies. In the early 1980s, new minuscule glass electrodes made it possible to directly record electrical activity of individual neurons in experimental animals engaged in particular motor tasks or encountering various sensations. The first pictures of the human brain at work were also snapped in the 1980s through use of **positron emission tomography (PET) scans** that depend on injection of small amounts of radioactive material that accumulates in the part of the body under study and leads to release of a tiny burst of gamma-ray energy that can be detected by the PET equipment. Increased gamma activity can be correlated with increased neural activity in the brain region under study (❙ Figure 5-8). In the 1990s, **functional magnetic resonance imaging (fMRI)** was used for the first time to detect functionally induced changes in regional cerebral blood flow and O_2 use by taking advantage of the fact that the magnetic properties of hemoglobin, the O_2-carrying molecule in the blood, are affected by the amount of O_2 it is carrying. By using a *blood oxygen-level dependent (BOLD)* signal, the fMRI method highlights neural areas that are more active. The latest technology for studying neurons, **optogenetics**, debuted in the naughts (*opto* refers to "light," *genetics* refers to the genetic engineering involved in the technique). By inserting genes that code for light-responsive molecules into experimental animals' neurons, scientists can use flashes of light to turn the neurons on or off at will (by depolarizing or hyperpolarizing these cells via light-induced changes in ion channels). By being able to manipulate neurons on command in living animals going about their routine activities, scientists hope to shed further light on complex information processing in the brain.

Neurons in different regions of the cerebral cortex may fire in rhythmic synchrony.

Neither whole-brain imaging techniques or single-neuron recordings can identify concurrent changes in electrical activity in a group of neurons working together to accomplish a particular activity. As an analogy, imagine trying to record a concert by using a single microphone that could pick up only the sounds produced by one musician. You would get a very limited impression of the performance by hearing only the changes in notes and tempo as played by this one individual. You would miss the richness of the melody and rhythm being performed in synchrony by the entire orchestra. Similarly, in recording from single neurons, scientists have overlooked a parallel informa-

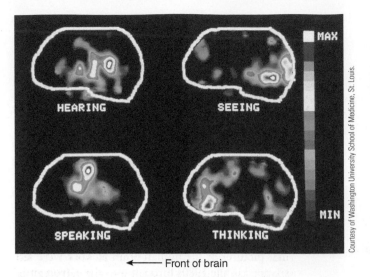

<— Front of brain

Figure 5-8 PET scans of cerebral cortex during different tasks. Different areas of the brain "light up" on positron emission tomography (PET) scans as a person performs different tasks. PET scans detect the magnitude of blood flow in various regions of the brain. Because more blood flows into a particular region of the brain when it is more active, neuroscientists can use PET scans to "take pictures" of the brain at work on various tasks.

tion mechanism involving changes in the relative timing of action potential discharges among a functional group of neurons, called a **neural network** or **assembly**. Studies involving simultaneous recordings from multiple neurons show that interacting neurons may transiently fire together for fractions of a second. Many neuroscientists believe that the brain encodes information not just by changing the firing rates of individual neurons but also by changing the patterns of these brief neural synchronizations. That is, groups of neurons communicate, or send messages about what is happening, by changing their pattern of synchronous firing. Recall that astrocytes also help coordinate synaptic activity among neural networks.

For example, when you view a bouncing ball, different visual units initially process different aspects of this object—its shape, its color, its movement, and so on. Somehow all these separate processing pathways must be integrated, or "bound together," for you to "see" the bouncing ball as a whole unit without stopping to contemplate its many separate features. The solution to the longtime mystery of how the brain accomplishes this integration might lie in the synchronous firing of neurons in separate regions of the brain that are functionally linked by virtue of being responsive to different aspects of the same objects, such as the bouncing ball.

Two new ambitious projects, both launched in 2013, will greatly expand our knowledge of brain function by studying via different approaches how neurons are functionally linked. The **BRAIN Initiative (Brain Research through Advancing Innovative Neurotechnologies)**, commonly referred to as the **Brain Activity Map Project**, is a United States–led collaborative 10-year research project with the goal of developing new technologies to advance our understanding of brain activity by real-time mapping of thousands or millions of neurons working together in coordinated networks. The European Commission launched the **Human Brain Project**, a 10-year initiative involv-

ing about 130 universities around the world to create a supercomputer simulation of the human brain. To mimic the complexity of the human brain, this digital brain will eventually require computers thousands of times more powerful than those available today.

The cerebral cortex is organized into layers and functional columns.

The cerebral cortex is organized into six well-defined layers based on varying distributions of several distinctive cell types (∎Figure 5-9). These layers are organized into functional vertical columns that extend perpendicularly about 2 mm from the cortical surface down through the thickness of the cortex to the underlying white matter. The neurons within a given column function as a "team," with each cell being involved in different aspects of the same specific activity—for example, perceptual processing of the same stimulus from the same location.

The functional differences between various areas of the cortex result from different layering patterns within the columns and from different input–output connections, not from the presence of unique cell types or different neuronal mechanisms. For example, those regions of the cortex responsible for sensory perception have an expanded layer IV, a layer rich in **stellate cells**, which are neurons responsible for initial processing of sensory input to the cortex. In contrast, cortical areas that control output to skeletal muscles have a thickened layer V, which contains an abundance of large neurons known as **pyramidal cells**. These nerve cells send fibers down the spinal cord from the cortex to terminate on efferent motor neurons that innervate skeletal muscles (∎Figure 5-9).

The four pairs of lobes in the cerebral cortex are specialized for different activities.

We now consider the locations of the major functional areas of the cerebral cortex. Throughout this discussion, keep in mind that even though a discrete activity is ultimately attributed to a particular region of the brain, each part depends on complex interplay among numerous other regions for both incoming and outgoing messages.

The anatomic landmarks used in cortical mapping are specific deep folds that divide each half of the cortex into four major lobes: the *occipital, temporal, parietal,* and *frontal lobes* (∎Figure 5-10). The **occipital lobes**, located posteriorly (at the back of the head), carry out initial processing of visual input. Auditory (sound) sensation is initially received by the **temporal lobes**, located laterally (on the sides of the head) (see ∎Figures 5-8 and 5-11). You will learn more about the functions of these regions in Chapter 6 when we discuss vision and hearing. The parietal lobes and frontal lobes, located on the top of the head, are separated by a deep infolding, the **central sulcus**, which runs roughly down the middle of the lateral surface of each hemisphere. The **parietal lobes** lie to the rear of the central sulcus on each side, and the **frontal lobes** lie in front of it. The parietal lobes are primarily responsible for receiving and processing sensory input. The frontal lobes are responsible for three main functions: (1) voluntary motor activity, (2) speaking

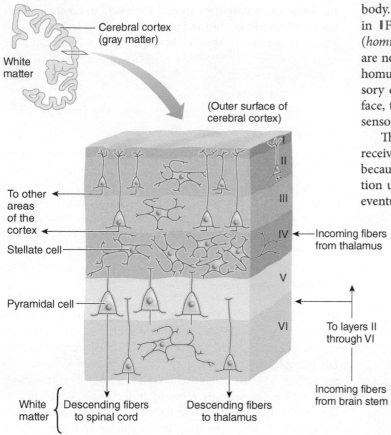

Figure 5-9 Layers of the cerebral cortex. Layer I is mostly glial cells and axons that run laterally. Layers II through VI contain different proportions of two main classes of cortical neurons: *pyramidal cells,* which are shaped like upside down pyramids and are the major output neurons, and *stellate cells,* which are shaped like stars. Stellate cells primarily receive input to the cortex and process local information.

FIGURE FOCUS: *(1) Which type of cortical neuron is most abundant in regions of the cortex that control output to skeletal muscles? (2) Which type is most abundant in cortical regions responsible for sensory perception?*

body. This distribution of cortical sensory processing is depicted in ❚Figure 5-12b. Note that on this **sensory homunculus** (*homunculus* means "little man") the different parts of the body are not equally represented. The size of each body part in this homunculus indicates the relative proportion of the somatosensory cortex devoted to that area. The exaggerated size of the face, tongue, hands, and genitalia indicates the high degree of sensory perception associated with these body parts.

The somatosensory cortex on each side of the brain mostly receives sensory input from the opposite side of the body because most ascending pathways that carry sensory information up the spinal cord cross over to the opposite side before eventually terminating in the cortex (see ❚Figure 5-26a, p. 175).

Thus, damage to the somatosensory cortex in the left hemisphere produces sensory deficits on the right side of the body, whereas sensory losses on the left side are associated with damage to the right half of the cortex.

Simple awareness of touch, pressure, temperature, or pain is detected by the thalamus, a lower level of the brain, but the somatosensory cortex goes beyond mere recognition of sensations to fuller sensory perception. The thalamus makes you aware that something hot versus something cold is touching your body, but it does not tell you where or of what intensity. The somatosensory cortex localizes the source of sensory input and perceives the level of intensity of the stimulus. It also is capable of spatial discrimination, so it can discern shapes of objects being held and can distinguish subtle differences in similar objects that come into contact with the skin.

The somatosensory cortex, in turn, projects this sensory input via white matter fibers to adjacent higher sensory areas for even further elaboration, analysis, and integration of sensory information. These higher areas are important in perceiving com-

ability, and (3) elaboration of thought. We next examine the role of the parietal lobes in sensory perception and then turn to the functions of the frontal lobes in more detail.

The parietal lobes accomplish somatosensory processing.

Sensations from the surface of the body, such as touch, pressure, heat, cold, and pain, are collectively known as **somesthetic sensations** (*somesthetic* means "body feelings"). Somesthetic information is detected by sensory receptors in the skin and relayed along afferent fibers to the CNS. Within the CNS, this information is **projected** (transmitted along specific neural pathways to higher brain levels) to the **somatosensory cortex**. The somatosensory cortex is located in the front portion of each parietal lobe immediately behind the central sulcus (❚Figures 5-11 and 5-12a). It is the site for initial cortical processing and perception of both somesthetic and proprioceptive input. **Proprioception** is the awareness of body position.

Each region within the somatosensory cortex receives somesthetic and proprioceptive input from a specific area of the

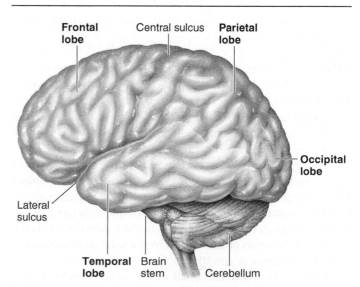

Figure 5-10 Cortical lobes. Each half of the cerebral cortex is divided into the occipital, temporal, parietal, and frontal lobes, as depicted in this lateral view of the left side of the brain.

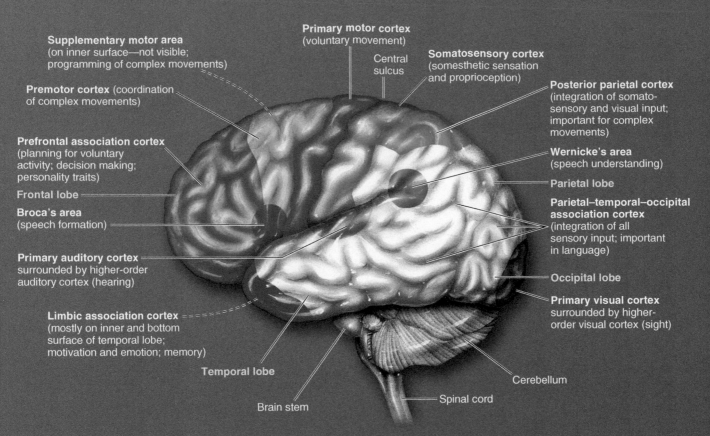

┃Figure 5-11 Functional areas of the cerebral cortex. Various regions of the cerebral cortex are primarily responsible for various aspects of neural processing, as indicated in this lateral view of the left side of the brain.

FIGURE FOCUS: *Assume you are petting a cat. Indicate which cortical lobe processes each of the following parts of this action: (1) seeing the cat, (2) commanding hand and arm movements, (3) feeling the soft fur, and (4) hearing the cat purr.*

plex patterns of somatosensory stimulation—for example, simultaneous appreciation of the texture, firmness, temperature, shape, position, and location of an object you are holding.

The primary motor cortex located in the frontal lobes controls the skeletal muscles.

The area in the rear portion of the frontal lobe immediately in front of the central sulcus and next to the somatosensory cortex is the **primary motor cortex** (see ┃Figures 5-11 and 5-12a). It confers voluntary control over movement produced by skeletal muscles. As in sensory processing, the motor cortex on each side of the brain primarily controls muscles on the opposite side of the body. Neuronal tracts originating in the motor cortex of the left hemisphere cross over before passing down the spinal cord to terminate on efferent motor neurons that trigger skeletal muscle contraction on the right side of the body (see ┃Figure 5-26b, p. 175). Accordingly, damage to the motor cortex on the left side of the brain produces paralysis on the right side of the body; the converse is also true.

Stimulation of different areas of the primary motor cortex brings about movement in different regions of the body. Like the sensory homunculus for the somatosensory cortex, the **motor homunculus**, which depicts the location and relative

amount of motor cortex devoted to output to the muscles of each body part, is distorted (┃Figure 5-12c). The fingers, thumbs, and muscles important in speech, especially those of the lips and tongue, are grossly exaggerated, indicating the fine degree of motor control these body parts have. Compare this to how little brain tissue is devoted to the trunk, arms, and lower extremities, which are not capable of such complex movements. Thus, the extent of representation in the motor cortex is proportional to the precision and complexity of motor skills required of the respective part.

Higher motor areas are also important in motor control.

Even though signals from the primary motor cortex terminate on the efferent neurons that trigger voluntary skeletal muscle contraction, the motor cortex is not the only region of the brain involved with motor control. First, lower brain regions and the spinal cord control involuntary skeletal muscle activity, such as in maintaining posture. Some of these same regions play an important role in monitoring and coordinating voluntary motor activity that the primary motor cortex has set in motion. Second, although fibers originating from the motor cortex can activate motor neurons to bring about muscle contraction, the motor cortex itself does not *initiate* voluntary movement. The motor cortex is activated by a widespread pattern of neuronal discharge, the **readiness potential**, which occurs about 750 msec before specific electrical activity is detectable in the motor cortex. Three higher motor areas of the cortex are involved in this voluntary decision-making period. These higher areas, which all command the primary motor cortex, include the *supplementary motor area,* the *premotor cortex,* and the *posterior parietal cortex* (see ❙Figure 5-11). Furthermore, a subcortical region of the brain, the *cerebellum,* plays an important role in planning, initiating, and timing certain kinds of movement by sending input to the motor areas of the cortex.

The three higher motor areas of the cortex and the cerebellum carry out different, related functions that are all important in programming and coordinating complex movements that involve simultaneous contraction of many muscles. Even though electrical stimulation of the primary motor cortex brings about contraction of particular muscles, no purposeful coordinated movement can be elicited, just as pulling on isolated strings of a puppet does not produce any meaningful movement. A puppet displays purposeful movements only when a skilled puppeteer manipulates the strings in a coordinated manner. In the same way, these four regions (and perhaps other areas as yet undetermined) develop a **motor program** for the specific voluntary task and then "pull" the appropriate pattern of "strings" in the primary motor cortex to produce the sequenced contraction of appropriate muscles that accomplishes the desired complex movement.

The **supplementary motor area** lies on the medial (inner) surface of each hemisphere anterior to (in front of) the primary motor cortex. It plays a preparatory role in programming complex sequences of movement. Stimulation of various regions of this motor area brings about complex patterns of movement, such as opening or closing the hand. Lesions here do not result in paralysis, but they do interfere with performance of more complex, useful integrated movements.

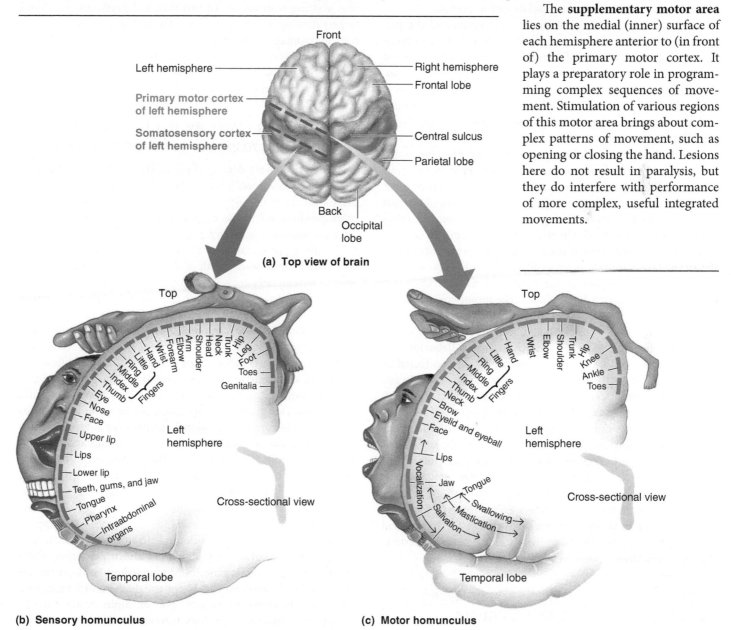

(a) Top view of brain

(b) Sensory homunculus

(c) Motor homunculus

❙Figure 5-12 **Somatotopic maps of the somatosensory cortex and primary motor cortex.** (a) Top view of cerebral hemispheres showing somatosensory cortex and primary motor cortex. (b) Sensory homunculus showing the distribution of sensory input to the somatosensory cortex from different parts of the body. The distorted graphic representation of the body parts indicates the relative proportion of the somatosensory cortex devoted to reception of sensory input from each area. (c) Motor homunculus showing the distribution of motor output from the primary motor cortex to different parts of the body. The distorted graphic representation of the body parts indicates the relative proportion of the primary motor cortex devoted to controlling skeletal muscles in each area.

The **premotor cortex**, located on the lateral surface of each hemisphere in front of the primary motor cortex, is important in orienting the body and arms toward a specific target. To command the primary motor cortex to produce the appropriate skeletal muscle contraction for accomplishing the desired movement, the premotor cortex must be informed of the body's momentary position in relation to the target. The premotor cortex is guided by sensory input processed by the **posterior parietal cortex**, a region that lies posterior to (in back of) the primary somatosensory cortex. These two higher motor areas have many anatomic interconnections and are closely related functionally. When either of these areas is damaged, the person cannot process complex sensory information to accomplish purposeful movement in a spatial context; for example, the person cannot successfully manipulate eating utensils.

Even though these higher motor areas command the primary motor cortex and are important in preparing for execution of deliberate, meaningful movement, researchers cannot say that voluntary movement is actually initiated by these areas. This pushes the question of how and where voluntary activity is initiated one step further. Probably no single area is responsible; undoubtedly, numerous pathways can ultimately bring about deliberate movement.

Think about the neural systems called into play, for example, during the simple act of picking up an apple to eat. Your memory tells you the fruit is in a bowl on the kitchen counter. Sensory systems, coupled with your knowledge based on past experience, enable you to distinguish the apple from the other kinds of fruit in the bowl. On receiving this integrated sensory information, motor systems issue commands to the exact muscles of the body in the proper sequence to enable you to move to the fruit bowl and pick up the targeted apple. During execution of this act, minor adjustments in the motor command are made as needed, based on continual updating provided by sensory input about the position of your body relative to the goal. Then there is the issue of motivation and behavior. Are you reaching for the apple because you are hungry (detected by a neural system in the hypothalamus) or because of a more complex behavioral scenario (for example, you started to think about food because you just saw someone eating on television)? Why did you choose an apple rather than a banana when both are in the fruit bowl and you like the taste of both, and so on? Thus, initiating and executing purposeful voluntary movement actually include a complex neuronal interplay that involves output from the motor regions guided by integrated sensory information and ultimately depends on motivational systems and elaboration of thought. All this plays against a background of memory stores from which you can make meaningful decisions about desirable movements.

Somatotopic maps vary slightly between individuals and are dynamic, not static.

Although the general organizational pattern of sensory and motor somatotopic ("body representation") maps of the cortex is similar in all people, the precise distribution is unique for each individual. Just as each of us has two eyes, a nose, and a mouth and yet no two faces have these features arranged in exactly the same way, so it is with brains. Furthermore, an individual's somatotopic mapping is not "carved in stone" but is subject to constant subtle modifications based on use. The general pattern is governed by genetic and developmental processes, but the individual cortical architecture can be influenced by **use-dependent competition** for cortical space. For example, when monkeys were encouraged to use their middle fingers instead of their other fingers to press a bar for food, after only several thousand bar presses the "middle finger area" of the motor cortex was greatly expanded and encroached on territory previously devoted to the other fingers. Similarly, modern neuroimaging techniques reveal that the left hand of a right-handed string musician is represented by a larger area of the somatosensory cortex than is the left hand of a person who does not play a string instrument. In this way, the musician's left-hand fingers develop a greater "feel" for the instrument as they skillfully manipulate the strings.

Other regions of the brain besides the somatosensory cortex and motor cortex can also be modified by experience. We now turn our attention to this plasticity of the brain.

Because of its plasticity, the brain can be remodeled in response to varying demands.

The brain displays a degree of **plasticity**—that is, an ability to change or be functionally remodeled in response to the demands placed on it. The ability of the brain to modify as needed is more pronounced in the early developmental years, but even adults retain some plasticity. When one area of the brain associated with a particular activity is destroyed, other areas may gradually assume some or all of the functions of the damaged region. Researchers are only beginning to unravel the molecular mechanisms responsible for the brain's plasticity. Current evidence suggests that the formation of new neural pathways (not new neurons, but new connections between existing neurons) in response to changes in experience are mediated in part by alterations in dendritic shape resulting from modifications in certain cytoskeletal elements (see p. 44). As its dendrites become more branched and elongated and more dendritic spines form (see ❚ Figure 4-7, p. 95), a neuron becomes able to receive and integrate more signals from other neurons. Thus, the precise synaptic connections between neurons are not fixed but can be modified by experience. The gradual modification of each person's brain by a unique set of experiences provides a biological basis for individuality. Even though the particular architecture of your own rather plastic brain has been and continues to be influenced by your unique experiences, it is important to realize that what you do and do not do cannot totally shape the organization of your cortex and other parts of the brain. Some limits are genetically established, and others are developmental limits on the extent to which modeling can be influenced by patterns of use. For example, some cortical regions maintain their plasticity throughout life, especially the ability to learn and to add new memories, but other cortical regions can be modified by use for only a specified time after birth before becoming permanently fixed. The length of this critical developmental period varies for different cortical regions.

Different regions of the cortex control different aspects of language.

Language ability is an excellent example of early cortical plasticity coupled with later permanence. Unlike the sensory and motor regions of the cortex, which are present in both hemispheres, in most people the areas of the brain responsible for language ability are found in only one hemisphere—the left hemisphere.

Clinical Note However, if a child younger than the age of two accidentally suffers damage to the left hemisphere, language functions are transferred to the right hemisphere with no delay in language development but at the expense of less obvious nonverbal abilities for which the right hemisphere is normally responsible. Up to about the age of 10, after damage to the left hemisphere, language ability can usually be reestablished in the right hemisphere following a temporary period of loss. If damage occurs beyond the early teens, however, language ability is permanently impaired, even though some limited restoration may be possible. The regions of the brain involved in comprehending and expressing language apparently are permanently assigned before adolescence.

Even in normal individuals, there is evidence for early plasticity and later permanence in language development. Infants can distinguish between and articulate the entire range of speech sounds, but each language uses only a portion of these sounds. As children mature, they often lose the ability to distinguish between or express speech sounds that are not important in their native language. For example, Japanese children can distinguish between the sounds of *r* and *l*, but many Japanese adults cannot perceive the difference between them.

Roles of Broca's Area and Wernicke's Area **Language** is a complex form of communication in which written or spoken words (or hand gestures in the case of sign language) symbolize objects and convey ideas. It involves the integration of two distinct capabilities—namely, *expression* (speaking ability) and *comprehension*—each of which is related to a specific area of the cortex. The primary areas of cortical specialization for language are Broca's area and Wernicke's area. **Broca's area**, which governs speaking ability, is located in the left frontal lobe in close association with the motor areas of the cortex that control the muscles necessary for speaking (see ❙ Figures 5-8, 5-11, and 5-13). **Wernicke's area**, located in the left cortex at the juncture of the parietal, temporal, and occipital lobes, is concerned with language comprehension. It plays a critical role in understanding both spoken and written messages. Furthermore, it is responsible for formulating coherent patterns of speech that are transferred via a bundle of fibers to Broca's area, which in turn controls the act of speaking. Wernicke's area receives input from the visual cortex in the occipital lobe, a pathway important in reading comprehension and in describing objects seen, and from the auditory cortex in the temporal lobe, a pathway essential for understanding spoken words. Wernicke's area also receives input from the somatosensory cortex, a pathway important in the ability to read Braille. Precise interconnecting pathways between these localized cortical areas are involved in the various aspects of speech (❙ Figure 5-13).

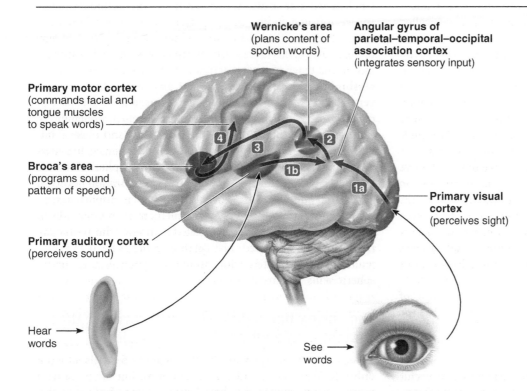

Wernicke's area (plans content of spoken words)

Angular gyrus of parietal–temporal–occipital association cortex (integrates sensory input)

Primary motor cortex (commands facial and tongue muscles to speak words)

Broca's area (programs sound pattern of speech)

Primary auditory cortex (perceives sound)

Primary visual cortex (perceives sight)

Hear words

See words

1a To speak about something seen, the brain transfers the visual information from the primary visual cortex to the angular gyrus of the parietal–temporal–occipital association cortex, which integrates inputs such as sight, sound, and touch.

1b To speak about something heard, the brain transfers the auditory information from the primary auditory cortex to the angular gyrus.

2 The information is transferred to Wernicke's area, where the choice and sequence of words to be spoken are formulated.

3 This language command is then transmitted to Broca's area, which translates the message into a programmed sound pattern.

4 This sound program is conveyed to the precise areas of the primary motor cortex that activate the appropriate facial and tongue muscles for causing the desired words to be spoken.

❙ Figure 5-13 **Cortical pathway for speaking a word seen or heard.** The arrows and numbered steps describe the pathway used to speak about something seen or heard. Similarly, appropriate muscles of the hand can be commanded to write the desired words.
FIGURE FOCUS: *(1) Will a person who has damage to Broca's area be able to speak clearly? (2) Will the person still be able to comprehend spoken messages and write coherent responses?*

Clinical Note · Language Disorders Because various aspects of language are localized in different regions of the cortex, damage to specific regions of the brain can result in selective disturbances of language. Damage to Broca's area results in a failure of word formation, although the patient can still understand the spoken and written word. Such people know what they want to say but cannot express themselves. Even though they can move their lips and tongue, they cannot establish the proper motor command to say the desired words. In contrast, patients with a lesion in Wernicke's area cannot understand words they see or hear. They can speak fluently, but their perfectly spoken words make no sense. They cannot attach meaning to words or choose appropriate words to convey their thoughts. Such language disorders caused by damage to specific cortical areas are known as **aphasias**, most of which result from strokes. Aphasias should not be confused with **speech impediments**, which are caused by a defect in the mechanical aspect of speech, such as weakness or incoordination of the muscles controlling the vocal apparatus.

Dyslexia, another language disorder, is a difficulty in learning to read despite normal intelligence. Emerging evidence suggests that dyslexia stems from a deficit in phonological processing, meaning an impaired ability to break down written words into their underlying phonetic components. People with dyslexia have difficulty decoding and thus identifying and assigning meaning to words.

The association areas of the cortex are involved in many higher functions.

The motor, sensory, and language areas account for only about half of the total cerebral cortex. The remaining areas, called **association areas**, are involved in higher functions. There are three association areas: (1) the *prefrontal association cortex,* (2) the *parietal–temporal–occipital association cortex,* and (3) the *limbic association cortex* (see ❘ Figure 5-11). At one time the association areas were called "silent" areas, because stimulation does not produce any observable motor response or sensory perception. (During brain surgery, typically the patient remains awake and only local anesthetic is used along the cut scalp. This is possible because the brain itself is insensitive to pain. Before cutting into this nonregenerative tissue, the neurosurgeon explores the exposed region with a tiny stimulating electrode. The patient is asked to describe what happens with each stimulation—the flick of a finger, a prickly feeling on the bottom of the foot, nothing? In this way, the surgeon can ascertain the appropriate landmarks on the neural map before making an incision.)

The **prefrontal association cortex** is the front portion of the frontal lobe just anterior to the premotor cortex. This is the part of the brain that "brainstorms" or thinks (see ❘ Figure 5-8). Specifically, the roles attributed to this region are (1) planning for voluntary activity, (2) decision making (that is, weighing consequences of future actions and choosing among options for various social or physical situations), (3) creativity, and (4) personality traits. To carry out these highest of neural functions, the prefrontal cortex is the site of operation of *working memory,* where the brain temporarily stores and actively manipulates information used in reasoning and planning. You will learn more about working memory later.

The **parietal–temporal–occipital association cortex** lies at the interface of the three lobes for which it is named. In this strategic location, it pools and integrates somatic, auditory, and visual sensations projected from these three lobes for complex perceptual processing. It enables you to "get the complete picture" of the relationship of various parts of your body with the external world. For example, it integrates visual information with proprioceptive input to let you place what you are seeing in proper perspective, such as realizing that a bottle is in an upright position despite the angle from which you view it (that is, whether you are standing up, lying down, or hanging upside down from a tree branch). This region is also involved in the language pathway connecting Wernicke's area to the visual and auditory cortices.

The **limbic association cortex** is located mostly on the bottom and adjoining inner portion of each temporal lobe. This area is concerned primarily with motivation and emotion and is extensively involved in memory.

The association areas are all interconnected by bundles of fibers within the cerebral white matter. Collectively, they integrate diverse information for purposeful action.

The cerebral hemispheres have some degree of specialization.

The cortical areas described thus far appear to be equally distributed in both the right and the left hemispheres, except for the language areas, which are found only on one side, usually the left. The left side is also most commonly the dominant hemisphere for fine motor control. Thus, most people are right handed, because the left side of the brain controls the right side of the body. Furthermore, each hemisphere is somewhat specialized in the types of mental activities it carries out best. The **left cerebral hemisphere** excels in logical, analytical, sequential, and verbal tasks, such as math, language forms, and philosophy. In contrast, the **right cerebral hemisphere** excels in nonlanguage skills, especially spatial perception and artistic and musical talents. The left hemisphere tends to process information in a fine-detail, fragmentary way, whereas the right hemisphere views the world in a big-picture, holistic way. This specialization is known as **cerebral lateralization**. Normally, the two hemispheres share so much information that they complement each other, but in many individuals the skills associated with one hemisphere are more strongly developed. Left cerebral hemisphere dominance tends to be associated with "thinkers," whereas right hemispheric skills dominate in "creators."

The cortex has a default mode network that is most active when the mind wanders.

New imaging techniques such as PET scans and fMRIs have not only helped researchers identify the brain regions involved with performing specific tasks. These cutting-edge technologies also led to discovery of a previously unrecognized brain system, the **default mode network (DMN)**, which is more active during resting states such as when you're daydreaming than during focused tasks such as when you're reading this page. Surprisingly, an estimated 60% to 80% of the brain's energy expenditure is used by the DMN circuits unrelated to any externally cued tasks. Most people spend an estimated 30% of their waking hours lost in spontaneous thoughts or "spacing out." The major DMN brain hubs lay midline in the medial prefrontal cortex and the medial parietal cortex. Evidence suggests that these areas communicate with one another and have functional connectivity while the brain is focused on internal signals rather than on external stimuli. Neuroscientists are unsure what purpose this DMN activity serves. Alternative proposed roles include enabling creativity, preparing the brain for conscious activity, constituting internally generated thoughts, retrieving and manipulating memories, or establishing a sense of self.

We now shift attention to the **subcortical regions** of the brain, which interact extensively with the cortex in the performance of their functions (*subcortical* means "under the cortex"). These regions include the *basal nuclei*, located in the cerebrum, and the *thalamus* and *hypothalamus*, located in the diencephalon.

5.5 Basal Nuclei, Thalamus, and Hypothalamus

The **basal nuclei** (also known as **basal ganglia**) consist of several masses of gray matter located deep within the cerebral white matter (see ▌Table 5-1 and ▌Figure 5-14). In the CNS, a **nucleus** (plural, **nuclei**) is a functioning group of neuron cell bodies.

The basal nuclei play an important inhibitory role in motor control.

The basal nuclei play a complex role in controlling movement. In particular, they are important in (1) inhibiting muscle tone throughout the body (proper muscle tone is normally maintained by a balance of excitatory and inhibitory inputs to the neurons that innervate skeletal muscles); (2) selecting and maintaining purposeful motor activity while suppressing useless or unwanted patterns of movement; and (3) helping monitor and coordinate slow, sustained contractions, especially those related to posture and support. The basal nuclei do not directly influence the efferent motor neurons that bring about muscle contraction but act instead by modifying ongoing activity in motor pathways.

To accomplish these complex integrative roles, the basal nuclei receive and send out a large volume of information, as is indicated by the tremendous number of fibers linking them to other regions of the brain. One important pathway consists of strategic interconnections that form a complex feedback loop linking the motor regions of the cerebral cortex, the basal nuclei, and the thalamus. The thalamus positively reinforces voluntary motor behavior initiated by the cortex, whereas the basal nuclei modulate this activity

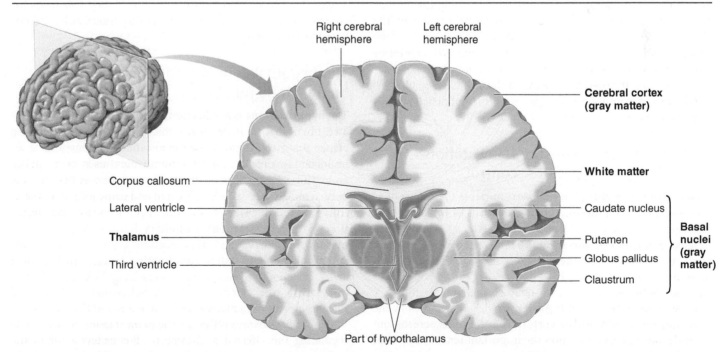

▌Figure 5-14 **Frontal section of the brain.** The cerebral cortex, an outer shell of gray matter, surrounds an inner core of white matter. Deep within the cerebral white matter are several masses of gray matter, the basal nuclei. The ventricles are cavities in the brain through which the cerebrospinal fluid flows. The thalamus forms the walls of the third ventricle. For comparison, the colors used for the thalamus and basal nuclei are the same as those used in the lateral view depicted in ▌Table 5-1, p. 144.

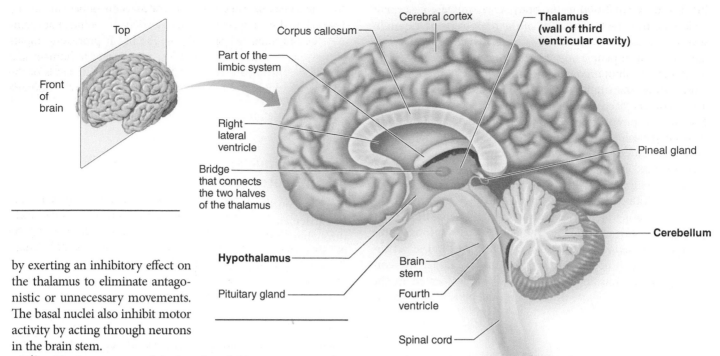

by exerting an inhibitory effect on the thalamus to eliminate antagonistic or unnecessary movements. The basal nuclei also inhibit motor activity by acting through neurons in the brain stem.

Clinical Note The importance of the basal nuclei in motor control is evident in **Parkinson's disease (PD)**. This condition is associated with a gradual destruction of neurons that release the neurotransmitter dopamine in the basal nuclei. Because the basal nuclei lack enough dopamine to exert their normal roles, three types of motor disturbances characterize PD: (1) increased muscle tone, or rigidity; (2) involuntary, useless, or unwanted movements, such as *resting tremors* (for example, hands rhythmically shaking, making it difficult or impossible to hold a cup of coffee); and (3) slowness in initiating and carrying out different motor behaviors. People with PD find it difficult to stop ongoing activities. If sitting down, they tend to remain seated, and if they get up, they do so slowly. The standard treatment for PD is the administration of *levodopa (L-dopa)*, a precursor of dopamine. Dopamine itself cannot be given because it is unable to cross the BBB, but L-dopa can enter the brain from the blood. Once inside the brain, L-dopa is converted into dopamine, thus substituting for the deficient neurotransmitter.

The thalamus is a sensory relay station and is important in motor control.

Deep within the brain near the basal nuclei is the **diencephalon**, a midline structure that forms the walls of the third ventricular cavity, one of the spaces through which CSF flows (see ❙Figure 5-5, p. 138). The diencephalon consists of two main parts, the *thalamus* and the *hypothalamus* (see ❙Table 5-1 and ❙Figures 5-7b, 5-14, and 5-15).

The **thalamus** serves as a "relay station" for preliminary processing of sensory input. All sensory input synapses in the thalamus on its way to the cortex. The thalamus screens out insignificant signals and routes the important sensory impulses to appropriate areas of the somatosensory cortex, and to other regions of the brain. Along with the brain stem and cortical association areas, the thalamus helps direct attention to stimuli

❙Figure 5-15 **Location of the thalamus, hypothalamus, and cerebellum in sagittal section**.

of interest. For example, parents can sleep soundly through the noise of outdoor traffic but become instantly aware of their baby's slightest whimper. The thalamus is also capable of crude awareness of various sensations but cannot distinguish their location or intensity. Some degree of consciousness resides here as well. Finally, the thalamus plays an important role in motor control by positively reinforcing voluntary motor behavior initiated by the cortex.

The hypothalamus regulates many homeostatic functions.

The **hypothalamus** is a collection of specific nuclei and associated fibers that lie beneath the thalamus. It is an integrating center for many important homeostatic functions and is an important link between the autonomic nervous system and the endocrine system. Specifically, the hypothalamus (1) controls body temperature; (2) controls thirst and urine output; (3) controls food intake; (4) controls anterior pituitary hormone secretion; (5) produces posterior pituitary hormones; (6) controls uterine contractions and milk ejection; (7) serves as a major autonomic nervous system coordinating center, which in turn affects all smooth muscle, cardiac muscle, and exocrine glands; (8) plays a role in emotional and behavioral patterns; and (9) participates in the sleep–wake cycle.

The hypothalamus is the brain area most involved in directly regulating the internal environment. For example, when the body is cold, the hypothalamus initiates internal responses to increase heat production (such as shivering) and to decrease heat loss (such as constricting the skin blood vessels to reduce the flow of warm blood to the body surface, where heat could be lost to the external environment). Other areas of the brain, such as the cerebral cortex, act more indirectly to regulate the internal environment. For example, a person who feels cold is motivated to voluntarily put on warmer clothing, close the window, turn up the thermostat, and so on. Even these voluntary behavioral activities are strongly influenced by the hypothalamus, which, as a part of the limbic system, functions with the cortex in controlling emotions and motivated behavior. We now turn to the limbic system and its functional relations with the higher cortex.

Check Your Understanding 5.5

1. State the functions of the basal nuclei.
2. Describe how the thalamus serves as a sensory relay station.
3. Name the brain area most involved directly in regulating homeostatic functions.

5.6 Emotion, Behavior, and Motivation

The **limbic system** is not a separate structure but a functional system consisting of a ring of forebrain structures that surround the brain stem and are interconnected by intricate neuron pathways (Figure 5-16). It includes portions of each of the following: the lobes of the cerebral cortex (especially the limbic association cortex), the basal nuclei, the thalamus, and the hypothalamus. This complex interacting network is associated with emotions, basic behavioral patterns, motivation, learning, and memory. Let us examine each of these brain functions further.

The limbic system plays a key role in emotion.

The concept of **emotion** encompasses subjective emotional feelings and moods (such as anger, fear, sadness, and joy) plus the overt physical responses associated with these feelings. These responses include specific behavioral patterns (for example, preparing for attack or defense when angered by an adversary) and observable emotional expressions (for example, laughing, crying, or blushing). Emotions are highly subjective and can vary among individuals in response to an identical circumstance. The limbic system plays a central role in all aspects of emotion. Stimulating specific regions of the limbic system during brain surgery produces vague subjective sensations

that the patient may describe as joy, satisfaction, or pleasure in one region and discouragement, fear, or anxiety in another. For example, the **amygdala** is especially important in processing inputs that give rise to the sensation of fear and anxiety. In humans and to an undetermined extent in other species, higher levels of the cortex are also crucial for conscious awareness of emotional feelings.

The limbic system and higher cortex participate in controlling basic behavioral patterns.

Basic behavioral patterns controlled at least in part by the limbic system include those aimed at individual survival (attack, searching for food) and those directed toward perpetuating the species (sociosexual behaviors conducive to mating). These behaviors are inborn and shared among members of a species. In experimental animals, stimulating the limbic system brings about complex and even bizarre behaviors. For example, stimulation in one area can elicit responses of anger and rage in a normally docile animal, whereas stimulation in another area results in placidity and tameness, even in an otherwise vicious animal. Stimulation in yet another limbic area can induce sexual behaviors such as copulatory movements.

The extensive involvement of the hypothalamus in the limbic system governs the involuntary internal responses of various body systems in preparation for appropriate action to accompany a particular emotional state. For example, the hypothalamus controls the increase of heart rate and respiratory rate, elevation of blood pressure, and diversion of blood to skeletal muscles that occur in anticipation of attack or when angered. These preparatory changes in the internal state require no conscious control.

In executing complex behavioral activities such as attacking, fleeing, or mating, the individual (humans and other animals) must interact with the external environment. Higher cortical mechanisms are called into play to connect the limbic system and hypothalamus with the outer world so that appropriate overt behaviors are manifested. At the simplest level, the cortex provides the neural mechanisms necessary for implementing the appropriate skeletal muscle activity required to approach or avoid an adversary, participate in sexual activity, or display emotional expression. For example, the stereotypical sequence of movement for the universal human emotional expression of smiling is preprogrammed in the cortex and can be called forth by the limbic system. One can also voluntarily call

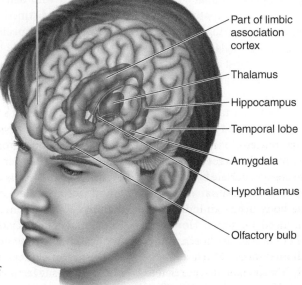

Frontal lobe

Part of limbic association cortex

Thalamus

Hippocampus

Temporal lobe

Amygdala

Hypothalamus

Olfactory bulb

Figure 5-16 Limbic system. The structures located in the interior of the brain that constitute the limbic system (in *pink*) are revealed in this partially transparent view of the brain.

forth the smile program, as when posing for a picture. Even individuals blind from birth have normal facial expressions—that is, they do not learn to smile by observation. Smiling means the same thing in every culture, despite widely differing environmental experiences.

Higher cortical levels also can reinforce, modify, or suppress basic behavioral responses so that actions can be guided by planning, strategy, and judgment based on an understanding of the situation. Even if you are angry at someone and your body is internally preparing for attack, you can usually judge whether an attack is appropriate and can consciously suppress the external manifestation of this basic emotional behavior. Thus, the higher levels of the cortex, particularly the prefrontal and limbic association areas, are important in conscious, learned control of innate behavioral patterns. Using fear as an example, exposure to an aversive experience calls two parallel tracks into play for processing this emotional stimulus: a fast track in which the lower-level amygdala plays a key role and a slower track mediated primarily by the higher-level prefrontal cortex. The fast track permits a rapid, rather crude, instinctive response ("gut reaction") and is essential for the "feeling" of being afraid. The slower track involving the prefrontal cortex permits a more refined response to the aversive stimulus based on a rational analysis of the current situation compared to stored past experiences. The prefrontal cortex formulates plans and guides behavior, suppressing impulsive amygdala-induced responses that may be inappropriate for the situation at hand.

Motivated behaviors are goal directed.

An individual tends to reinforce behaviors that have proved gratifying and to suppress behaviors that have been associated with unpleasant experiences. Certain regions of the limbic system have been designated as **"reward"** and **"punishment" centers** because stimulation in these respective areas gives rise to pleasant or unpleasant sensations. When a self-stimulating device is implanted in a reward center, an experimental animal will self-deliver up to 5000 stimulations per hour and continue self-stimulation in preference to food, even when starving. When the device is implanted in a punishment center, animals avoid stimulation at all costs. Reward centers are found most abundantly in regions involved in mediating the highly motivated behavioral activities of eating, drinking, and sexual activity.

Motivation is the ability to direct behavior toward specific goals. Some goal-directed behaviors are aimed at satisfying specific identifiable physical needs related to homeostasis. **Homeostatic drives** represent the subjective urges associated with specific bodily needs that motivate appropriate behavior to satisfy those needs. As an example, the sensation of thirst accompanying a water deficit in the body drives an individual to drink to satisfy the homeostatic need for water. However, whether water, a soft drink, or another beverage is chosen as the thirst quencher is unrelated to homeostasis. Much human behavior does not depend on purely homeostatic drives related to simple tissue deficits such as thirst. Human behavior is influenced by experience, learning, and habit, shaped in a complex framework of unique personal gratifications blended with cultural expectations. To what extent, if any, motivational drives unrelated to homeostasis, such as the drive to pursue a particular career or win a certain race, are involved with the reinforcing effects of the reward and punishment centers is unknown. Indeed, some individuals motivated toward a particular goal may even deliberately "punish" themselves in the short term to achieve their long-range gratification (for example, the temporary pain of training in preparation for winning a competitive athletic event).

Norepinephrine, dopamine, and serotonin are neurotransmitters in pathways for emotions and behavior.

The underlying neurophysiological mechanisms responsible for the psychological observations of emotions and motivated behavior largely remain a mystery, although the neurotransmitters *norepinephrine, dopamine,* and *serotonin* all have been implicated. Norepinephrine and dopamine, both chemically classified as *catecholamines* (see p. 121), are known transmitters in the regions that elicit the highest rates of self-stimulation in animals equipped with do-it-yourself devices. Numerous **psychoactive drugs** affect moods in humans, and some of these drugs have been shown to influence self-stimulation in experimental animals. For example, increased self-stimulation is observed after the administration of drugs that increase catecholamine synaptic activity, such as *amphetamine,* an "upper" drug. Amphetamine stimulates the release of dopamine from dopamine-secreting neurons.

Although most psychoactive drugs are used therapeutically to treat various mental disorders, others, unfortunately, are abused. Many abused drugs act by enhancing the effectiveness of dopamine in the "pleasure" pathways, thus initially giving rise to an intense sensation of pleasure. As you have already learned, an example is cocaine, which blocks the reuptake of dopamine at synapses (see p. 112).

Clinical Note **Depression** is among the psychiatric disorders associated with defects in limbic system neurotransmitters. (As a distinction, *psychiatric disorders* involve abnormal activity in specific neurotransmitter pathways in the absence of detectable brain lesions, whereas *neurological disorders* are associated with specific lesions of the brain and may or may not involve abnormalities in neurotransmission. Examples of neurological disorders include Parkinson's disease and Alzheimer's disease.) Depression is a psychiatric mood disorder. In psychiatry, **mood** is an abstract term referring to a person's prolonged subjective emotional state that influences his or her behavior and perception of external events. Depression is characterized by a pervasive negative mood accompanied by a generalized loss of interests, an inability to experience pleasure, and suicidal tendencies. A functional deficiency of serotonin, norepinephrine, or both is implicated in depression. These neurotransmitters are synaptic messengers in regions of the brain involved in pleasure and motivation, suggesting that the pervasive sadness and lack of interest (no motivation) in patients who are depressed are related at least partly to deficiencies of these neurotransmitters. Areas of the brain that play a significant role in depression are the limbic system, especially the amygdala and hippocampus, both of which are located in the medial temporal lobe (see ▌Figure 5-16), and the medial prefrontal cortex. All of these brain regions are central to emotion and behavior. For example, research shows that the hippocampus is 9% to 13% smaller in depressed women compared with women who are not depressed. Events perceived as being stressful can trigger depression, but the underlying link has not been determined. Recall that stress suppresses production of new neurons in the hippocampus.

All effective antidepressant drugs increase the available concentration of these two neurotransmitters in the CNS. There are four main classes of antidepressants:

■ **Selective serotonin reuptake inhibitors (SSRIs).** As implied by the name, these drugs selectively block the reuptake of released serotonin into the presynaptic terminal, thus prolonging serotonin activity at synapses (see p. 108). *Prozac (fluoxetine)*, the most widely prescribed drug in American psychiatry, is illustrative.

■ **Serotonin norepinephrine reuptake inhibitors (SNRIs).** By blocking the reuptake of both released serotonin and norepinephrine, SNRIs cause these neurotransmitters to linger longer at synapses. An example of an SNRI is *Cymbalta (duloxetine)*. SSRIs and SNRIs are newer classes of antidepressants.

■ **Tricyclic antidepressants (TCAs). TCAs** are an older class of drug that also blocks reuptake of norepinephrine and serotonin, especially norepinephrine, thus prolonging action of these neurotransmitters at synapses, but unlike SNRIs, TCAs also affect activity of the autonomic nervous system and thus have more undesirable side effects. *Pamelor (nortriptyline)* belongs to this class of antidepressants.

■ **Monoamine oxidase inhibitors (MAOIs).** The oldest class of antidepressants, MAOIs elevate levels of norepinephrine, serotonin, and dopamine by inhibiting monoamine oxidase, an enzyme that breaks down these three neurotransmitters. MAOIs tend to exert more side effects than more selective antidepressants, especially when combined with certain foods and other medicines. *Emsam (selegiline)* is an MAOI administered as a patch worn on the skin.

Antidepressants immediately boost the concentration of affected neurotransmitters at synapses, yet a reduction in symptoms of depression typically does not take place until several weeks after medication begins. Some experts believe that people do not start to feel better as soon as the neurotransmitter levels increase because improvement of mood further depends on growth of new neurons and formation of new connections in the hippocampus, a process that takes several weeks and is stimulated by antidepressants. Furthermore, antidepressants trigger formation of new astrocytes, which are also reduced in numbers in areas of the brain affected by depression.

Researchers are optimistic that as understanding of the molecular mechanisms of mental disorders is expanded in the future, many psychiatric problems can be corrected or better managed through drug or other therapeutic intervention, a hope of great medical significance.

Check Your Understanding 5.6

1. State the brain functions associated with the limbic system.
2. Name the brain area most important in processing the sensation of fear.
3. List the three neurotransmitters involved in pathways that process emotions and motivated behavior.

5.7 Learning and Memory

In addition to their involvement in emotion and basic behavioral patterns, the limbic system and higher cortex are involved in learning and memory. The cerebellum also plays a key role in some types of learning and memory.

Learning is the acquisition of knowledge as a result of experiences.

Learning is the acquisition of knowledge or skills as a consequence of experience, instruction, or both. Rewards and punishments are integral parts of many types of learning. If an animal is rewarded on responding in a particular way to a stimulus, the likelihood increases that the animal will respond in the same way again to the same stimulus as a consequence of this experience. Conversely, if a particular response is accompanied by punishment, the animal is less likely to repeat the same response to the same stimulus. When behavioral responses that give rise to pleasure are reinforced or those accompanied by punishment are avoided, learning has taken place. Housebreaking a puppy is an example. If the puppy is praised when it urinates outdoors but scolded when it wets the carpet, it will soon learn the acceptable place to empty its bladder. Thus, learning is a change in behavior that occurs as a result of experiences. It highly depends on the organism's interaction with its environment. The only limits to the effects that environmental influences can have on learning are the biological constraints imposed by species-specific and individual genetic endowments.

Memory is laid down in stages.

Memory is the storage of acquired knowledge for later recall. Learning and memory form the basis by which individuals adapt their behavior to their particular external circumstances. Without these mechanisms, it would be impossible for individuals to plan for successful interactions and to intentionally avoid predictably disagreeable circumstances.

The neural change responsible for retention or storage of knowledge is known as the **memory trace**, or **engram**. Generally, concepts, not verbatim information, are stored. As you read this page, you are storing the concept discussed, not the specific words. Later, when you retrieve the concept from memory, you will convert it into your own words. It is possible, however, to memorize bits of information word by word.

Storage of acquired information is accomplished in at least two stages: short-term memory and long-term memory (▌Table 5-2). **Short-term memory** lasts for seconds to hours, whereas **long-term memory** is retained for days to years. The process of

| TABLE 5-2 Comparison of Short-Term and Long-Term Memory

Characteristic	Short-Term Memory	Long-Term Memory
Time of storage after acquisition of new information	Immediate	Later; must be transferred from short-term to long-term memory through consolidation; enhanced by practice or recycling of information through short-term mode
Duration	Lasts for seconds to hours	Retained for days to years
Capacity of storage	Limited	Very large
Retrieval time (remembering)	Rapid retrieval	Slower retrieval, except for thoroughly ingrained memories, which are rapidly retrieved
Inability to retrieve (forgetting)	Permanently forgotten; memory fades quickly unless consolidated into long-term memory	Usually only transiently unable to access; relatively stable memory trace
Mechanism of storage	Involves transient modifications in functions of preexisting synapses, such as altering amount of neurotransmitter released	Involves relatively permanent functional or structural changes between existing neurons, such as formation of new synapses; synthesis of new proteins plays a key role

transferring and fixing short-term memory traces into long-term memory stores is known as **consolidation**.

Working memory, or what has been called "the erasable blackboard of the mind," is a complex type of short-term memory you use on an ongoing basis to carry out daily activities. Working memory temporarily holds and interrelates various pieces of information relevant to a current mental task. Through your working memory, you briefly hold and process data for immediate use—both newly acquired information and related, previously stored knowledge that is transiently brought forth into working memory—so that you can evaluate the incoming data in context. This integrative function is crucial to your ability to reason, plan, and make judgments. By comparing and manipulating new and old information within your working memory, you can comprehend what you are reading, carry on a conversation, calculate a restaurant tip in your head, find your way home, and know that you should put on warm clothing if you see snow outside. In short, working memory enables people to string thoughts together in a logical sequence and plan for future action.

Recent findings suggest that once an established memory is actively recalled, it becomes labile (unstable or subject to change) and must be **reconsolidated** into a restabilized, inactive state. New information may be incorporated into the old memory trace during reconsolidation. Thus, an old memory may actually be changed each time it is recalled.

Comparison of Short-Term and Long-Term Memory

Newly acquired information is initially deposited in short-term memory, which has a limited capacity for storage. Information in short-term memory has one of two eventual fates. Either it is soon forgotten (for example, forgetting a telephone number after you have looked it up and finished dialing), or it is trans-ferred into the more permanent long-term memory mode through active practice or rehearsal. The recycling of newly acquired information through short-term memory increases the likelihood that the information is consolidated into long-term memory. (Therefore, when you cram for an exam, your long-term retention of the information is poor!) The original short-term memory rapidly fades unless it is consolidated to provide a more enduring long-term memory. Sometimes only parts of memories are fixed, while other parts fade away. Information of interest or importance to the individual is more likely to be recycled and fixed in long-term stores, whereas less important information is quickly erased.

The storage capacity of the long-term memory bank is much larger than the capacity of short-term memory. Different informational aspects of long-term memory traces seem to be processed, codified, and then stored with other memories of the same type; for example, visual memories are stored separately from auditory memories. This organization facilitates future searching of memory stores to retrieve desired information. For example, in remembering a woman you once met, you may use various recall cues from different storage pools, such as her name, her appearance, the fragrance she wore, an incisive comment she made, or the song playing in the background.

Stored knowledge is of no use unless it can be retrieved and used to influence current or future behavior. Because long-term memory stores are larger, it often takes longer to retrieve information from long-term memory than from short-term memory. *Remembering* is the process of retrieving specific information from memory stores; *forgetting* is the inability to retrieve stored information. Information lost from short-term memory is permanently forgotten, but information in long-term storage is often forgotten only temporarily. For example, you may be transiently unable to remember an acquaintance's name and then have it suddenly "come to you" later. Some forms of long-term memory involving information or skills used daily are essentially never forgotten and are rapidly accessible, such as knowing your name or being able to write.

Amnesia Occasionally, individuals suffer from a lack of memory that involves whole portions of time rather than isolated bits of information. This condition, known as **amnesia**, occurs in two forms. The most common form, *retrograde* (meaning "going backward") *amnesia,* is the inability to recall recent past events. It usually follows a traumatic event that interferes with electrical activity of the brain, such as a concussion or stroke. If a person is knocked unconscious, the content of short-term memory is essentially erased, resulting in loss of memory about activities that occurred within about the last half hour before the event. Severe trauma may interfere with access to recently acquired information in long-term stores also.

Anterograde (meaning "going forward") *amnesia,* conversely, is the inability to consolidate memory in long-term storage for later retrieval. It is usually associated with lesions of the medial portions of the temporal lobes, which are generally considered critical regions for memory consolidation. People suffering from this condition may be able to recall things they learned before the onset of their problem, but they cannot establish new permanent memories. New information is lost as quickly as it fades from short-term memory. In one case study, the person could not remember where the bathroom was in his new home but still had total recall of his old home.

Interestingly, the processes of human memory are adapting to new communication technology. Through widespread use of computers and smartphones, the Internet has become a ubiquitous presence in most of our lives and serves as a readily accessible external memory source. Having immediate access to vast stores of information online thanks to databases and search engines has reduced the need to commit as much information to memory. Studies show that people who expect to be able to easily access needed information are less apt to consolidate short-term memory into the long-term mode. On a simple level, have you memorized the phone numbers of your friends and family, a common practice in the past, or do you rely on pushing their numbers on speed dial?

Short-term memory and long-term memory involve different molecular mechanisms.

Obviously, some change must take place within the neural circuitry of the brain to account for the altered behavior that follows learning. Different mechanisms are responsible for short-term and long-term memory. Short-term memory involves transient modifications in the function of preexisting synapses, such as a temporary change in the amount of neurotransmitter released in response to stimulation or temporary increased responsiveness of the postsynaptic cell to the neurotransmitter within affected nerve pathways. Long-term memory, in contrast, involves relatively permanent functional or structural changes between existing neurons in the brain. Let us look at each of these types of memory in more detail.

Short-term memory involves transient changes in synaptic activity.

Ingenious experiments in the sea snail *Aplysia* have shown that two forms of short-term memory—habituation and sensitization—result from modification of different channel proteins in presynaptic terminals of specific afferent neurons involved in the pathway that mediates the behavior being modified. This modification, in turn, brings about changes in neurotransmitter release. **Habituation** is a decreased responsiveness to repetitive presentations of an indifferent stimulus—that is, one that is neither rewarding nor punishing. **Sensitization** is increased responsiveness to mild stimuli following a strong or noxious stimulus. *Aplysia* reflexly withdraws its gill when its siphon, a breathing organ at the top of its gill, is touched. Afferent (presynaptic) neurons responding to touch of the siphon directly synapse on efferent (postsynaptic) motor neurons controlling gill withdrawal. The snail becomes habituated when its siphon is repeatedly touched—that is, it learns to ignore the stimulus and no longer withdraws its gill in response. Sensitization, a more complex form of learning, takes place in *Aplysia* when it is given a hard bang on the siphon. Subsequently, the snail withdraws its gill more vigorously in response to even mild touch. These different forms of learning affect the same site—the synapse between a siphon afferent and a gill efferent—in opposite ways. Habituation depresses this synaptic activity, whereas sensitization enhances it. These transient modifications persist for as long as the memory.

Mechanism of Habituation As a result of habituation, the voltage-gated Ca^{2+} channels in a siphon afferent axon terminal do not open as readily when an action potential arrives, reducing entry of exocytosis-inducing Ca^{2+}, which leads to a decrease in neurotransmitter release. As a consequence of less neurotransmitter binding at the membrane of the motor neuron to the gill, the postsynaptic potential is reduced compared to normal, resulting in a diminished behavioral response controlled by the gill efferent (decreased gill withdrawal). Thus, the memory for habituation in *Aplysia* is stored in the form of modification of specific Ca^{2+} channels. With no further training, this reduced responsiveness lasts for several hours. A similar process is responsible for short-term habituation in other species studied, although in higher species the involvement of intervening interneurons makes the process somewhat more complicated. Habituation is probably the most common form of learning and is believed to be the first learning process to take place in human infants. By learning to ignore indifferent stimuli, the animal or person is free to attend to other more important stimuli.

Mechanism of Sensitization Sensitization in *Aplysia,* which results from increased Ca^{2+} influx into the siphon afferent terminal, does not have a direct effect on presynaptic Ca^{2+} channels. Instead, it indirectly enhances Ca^{2+} entry via presynaptic facilitation (see p. 111). The neurotransmitter serotonin is released from a facilitating interneuron that synapses on the presynaptic terminal to bring about increased release of presynaptic neurotransmitter in response to an action potential. Serotonin does so by triggering activation of a cAMP second-messenger pathway (see p. 123) that causes blockage of K^+ channels, thereby prolonging the action potential in the presynaptic terminal. (Remember that K^+ efflux is responsible for returning the potential to resting during an action potential.) Because the Ca^{2+} channels are kept open longer as a result of the prolonged action potential, more Ca^{2+} enters the terminal. The subsequent increase in neurotransmitter release produces a larger postsynaptic potential, resulting in a more vigorous gill-withdrawal response.

Thus, existing synaptic pathways may be functionally interrupted (habituated) or enhanced (sensitized) during simple learning. Scientists speculate that much of short-term memory is similarly a temporary modification of already existing processes. Several lines of research suggest that the cAMP cascade, especially activation of protein kinase, plays an important role, at least in elementary forms of learning and memory.

Memories more complex than habituation and sensitization that involve conscious awareness are initially stored by means of long-term potentiation, a mechanism that involves more persistent changes in activity of existing synapses.

Mechanism of Long-Term Potentiation With **long-term potentiation (LTP)**, modifications take place as a result of increased use at a given preexisting synapse that enhance the future ability of the presynaptic neuron to excite the postsynaptic neuron—that is, this connection gets stronger the more often it is used. Such strengthening of synaptic activity results in the formation of more EPSPs in the postsynaptic neuron in response to chemical signals from this particular excitatory presynaptic input. The increased excitatory responsiveness is ultimately translated into more action potentials being sent along this postsynaptic cell to other neurons. LTP lasts for days or even weeks—long enough for this short-term memory to be consolidated into more permanent long-term memory. LTP is especially prevalent in the hippocampus, a site critical for converting short-term memories into long-term memories. Less commonly, poorly understood **long-term depression (LTD)**, or weakening of synaptic transmission, has been demonstrated.

Enhanced synaptic transmission with LTP could theoretically result from either changes in the postsynaptic neuron (such as increased responsiveness to the neurotransmitter via insertion of more receptors for this messenger in the postsynaptic membrane) or in the presynaptic neuron (such as increased release of neurotransmitter). The underlying mechanisms for LTP are still the subject of much research and debate. Most likely, multiple mechanisms are involved in this complex phenomenon. It appears that there are several forms of LTP, some arising from changes only in the postsynaptic neuron and others also having a presynaptic component. Based on current scientific evidence, the following is a plausible mechanism for LTP involving both a postsynaptic change and a presynaptic modification (Figure 5-17).

LTP begins when a presynaptic neuron releases the common excitatory neurotransmitter glutamate in response to an action potential. Glutamate binds to two types of receptors on the postsynaptic neuron: *AMPA receptors* and *NMDA receptors*. An **AMPA receptor** is a chemically mediated receptor-channel

that opens on binding of glutamate and permits net entry of Na^+ ions, leading to formation of an EPSP at the postsynaptic neuron (see pp. 116 and 106). This is the ordinary receptor at excitatory synapses about which you already learned. An **NMDA receptor** is a receptor-channel that permits Ca^{2+} entry when it is open. This receptor-channel is unusual because it is both chemically gated and voltage dependent. It is closed by both a gate and by a magnesium ion (Mg^{2+}) that physically blocks the channel opening at resting potential. Two events must happen almost simultaneously to open an NMDA receptor-channel: presynaptic glutamate release and postsynaptic depolarization by other inputs. The gate opens on binding of glutamate, but this action alone does not permit Ca^{2+} entry. Additional depolarization of the postsynaptic neuron beyond that produced by the EPSP resulting from glutamate binding to the AMPA receptor is needed to depolarize the postsynaptic neuron enough to force Mg^{2+} out of the channel. Thus, even though glutamate binds with the NMDA receptor, the channel does not open unless the postsynaptic cell is sufficiently depolarized as a result of other excitatory activity. The postsynaptic cell can be sufficiently depolarized to expel Mg^{2+} in two ways: by repeated input from this single excitatory presynaptic neuron, resulting in temporal summation of EPSPs from this source, or by additional excitatory input from another presynaptic neuron at about the same time, resulting in spatial summation of EPSPs (see p. 108). When the NMDA receptor-channel opens as a result of simultaneous gate opening and Mg^{2+} expulsion, Ca^{2+} enters the postsynaptic cell. The entering Ca^{2+} activates a Ca^{2+} second-messenger pathway in this neuron. This second-messenger pathway leads to the physical insertion of additional AMPA receptors in the postsynaptic membrane. Because of the increased availability of AMPA receptors, the postsynaptic cell exhibits a greater EPSP response to subsequent release of glutamate from the presynaptic cell. This heightened sensitivity of the postsynaptic neuron to glutamate from the presynaptic cell helps maintain LTP.

Furthermore, at some synapses, activation of the Ca^{2+} second-messenger pathway in the postsynaptic neuron causes this cell to release a retrograde ("going backward") paracrine that diffuses to the presynaptic neuron (see p. 114). Here, the retrograde paracrine activates a second-messenger pathway in the presynaptic neuron, ultimately enhancing the release of glutamate from the presynaptic neuron. This positive feedback strengthens the signaling process at this synapse, also helping sustain LTP. Note that in this mechanism, a chemical factor from the postsynaptic neuron influences the presynaptic neuron, just the opposite direction of neurotransmitter activity at a synapse. Most investigators believe that the retrograde messenger is **nitric oxide**, a chemical that performs a bewildering array of other functions in the body. These other functions range from dilation of blood vessels in the penis during erection to destruction of foreign invaders by the immune system (see pp. 346, 411 and 733).

The modifications that take place during the development of LTP are sustained long after the activity that led to these changes has ceased. Therefore, information can be transmitted along this same synaptic pathway more efficiently when activated in the future—that is, the synapse "remembers." LTP is specific for the activated pathway. Pathways between other inactive presynaptic inputs and the same postsynaptic cell are not affected. Note that LTP develops in response to frequent activity across a synapse as a result of repetitive, intense firing of a given input (as with repeatedly going over a particular fact during studying) or to the linking of one input with another input firing at the same time. For example, when you smell a pie baking in the oven, your mouth waters in anticipation of the imminent arrival of a tasty treat you have come to associate with this aroma. The taste and feel of food in the mouth is the built-in trigger for salivation. However, through experience, neurons in the pathway that control salivation link input arising from the smell of pie with input from its delicious taste. After the smell-input pathway is strengthened through development of LTP and ultimate consolidation into long-term storage, the smell of pie alone can cause salivation.

Ethanol in alcoholic beverages blocks NMDA receptors while facilitating GABA function. Ethanol's blockage of NMDA receptors is likely the reason people have difficulty remembering what happened during a time of heavy drinking. Furthermore, by enhancing the actions of GABA, the brain's major inhibitory neurotransmitter, ethanol depresses overall CNS activity.

Studies suggest a regulatory role for the cAMP second-messenger pathway in the development and maintenance of LTP in addition to the Ca^{2+} second-messenger pathway. Participation of cAMP may hold a key to linking short-term memory to long-term memory consolidation.

Long-term memory involves formation of new, permanent synaptic connections.

Whereas short-term memory involves transient strengthening of preexisting synapses, long-term memory storage requires activation of specific genes that control synthesis of proteins, needed for lasting structural or functional changes at specific synapses. Examples of such changes include formation of new synaptic connections or permanent changes in presynaptic or postsyn-

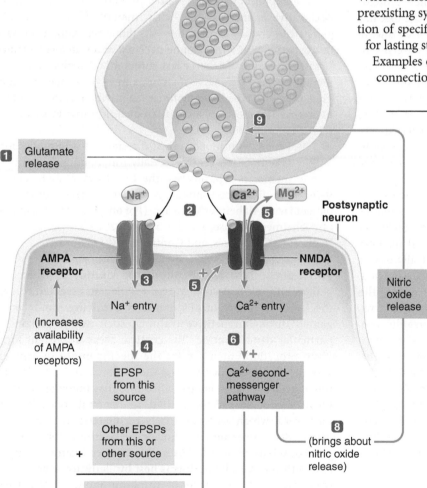

1. Glutamate is released from activated presynaptic neuron.

2. Glutamate binds with both AMPA and NMDA receptors.

3. Binding opens AMPA receptor-channel.

4. Na^+ entry through open AMPA channel depolarizes postsynaptic neuron, producing EPSP.

5. Binding opens gate of NMDA receptor-channel but Mg^{2+} still blocks channel. Sufficient depolarization from this AMPA opening plus other EPSPs drives Mg^{2+} out.

6. Ca^{2+} entry through open NMDA channel activates Ca^{2+} second-messenger pathway.

7. Second-messenger pathway promotes insertion of additional AMPA receptors in postsynaptic membrane, increasing its sensitivity to glutamate.

8. Second-messenger pathway also triggers release of retrograde paracrine (likely nitric oxide).

9. Nitric oxide stimulates long-lasting increase in glutamate release by presynaptic neuron.

⏐Figure 5-17 **Possible pathways for long-term potentiation**.
FIGURE FOCUS: *Would simultaneous binding of glutamate from a presynaptic neuron to both AMPA and NMDA receptors at a postsynaptic neuron automatically trigger a chain of events leading to long-term potentiation? Explain why or why not.*

aptic membranes. Thus, long-term memory storage involves permanent physical changes in the brain. Long-term memory holds the sum total of all you can remember about what you have done, who you have known, where you have been, when something took place, and what you have learned.

Studies comparing the brains of experimental animals reared in a sensory-deprived environment with those raised in a sensory-rich environment demonstrate readily observable microscopic differences. The animals afforded more environmental interactions—and therefore, more opportunity to learn—displayed greater branching, elongation of dendrites, and more dendritic spines in nerve cells in regions of the brain suspected to be involved with memory storage. Greater dendritic surface area presumably provides more sites for synapses. Thus, long-term memory may be stored at least in part by a particular pattern of dendritic branching and synaptic contacts.

No one knows for sure how the transient short-term memory is converted to the permanent long-term mode, but many researchers believe that cAMP and *immediate early genes* play critical roles in memory consolidation. cAMP can switch on **cAMP responsive element binding protein (CREB)**, which acts on DNA to ultimately influence synthesis of new proteins important for maintaining long-term memory. (Recall that this second messenger also plays a regulatory role in LTP and in sensitization.) The **immediate early genes (IEGs)** govern synthesis of proteins that encode long-term memory. The exact role that these critical newly synthesized long-term memory proteins might play remains speculative. They may be needed for structural changes in dendrites or used for synthesis of more neurotransmitters or additional receptor sites. Alternatively, they may accomplish long-term modification of neurotransmitter release by sustaining biochemical events initially activated by short-term memory processes.

Most investigation of learning and memory has focused on changes in synaptic connections within the brain's gray matter. To complicate the issue further, scientists now have evidence that white matter also changes during learning and memory formation as more myelin surrounds axons, especially during adolescence, speeding up transmission between connected neurons. Neurons produce chemical signals such as **neuregulin** that regulate the extent to which myelin-forming cells wrap themselves around the axon. The amount of neuregulin produced is correlated with the extent of action potential propagation within the axon. Accordingly, researchers propose that conduction velocity can be increased by means of further myelination in more active pathways, and that these changes support learning and memory. In addition to a probable role of white matter, numerous hormones and neuropeptides have also been shown to affect learning and memory processes.

Memory traces are present in multiple regions of the brain.

Another question besides the "how" of memory is the "where" of memory. What parts of the brain are responsible for memory? There is no single "memory center" in the brain. Instead, the neurons involved in memory traces are widely distributed throughout the subcortical and cortical regions of the brain.

The regions of the brain most extensively implicated in memory include the hippocampus, the cerebellum, the prefrontal cortex, and other areas of the cerebral cortex. Despite widespread circuitry for all memories, recent evidence indicates that a particular memory concept may be stored in a sparse number of neurons in a specific location. For example, a small set of neurons might store the concept of "grandmother" and fire in response to all inputs relevant to her, such as seeing her or her picture from various angles and distances, reading her written name, hearing her voice, and so on. According to this controversial proposal, other small sets of neurons would each store other specific concepts.

The Hippocampus and Declarative Memories The hippocampus is a prominent site where LTP takes place and is also crucial for consolidation into long-term memory. The hippocampus is believed to store new long-term memories only temporarily and then transfer them to other cortical sites for more permanent storage. The sites for long-term storage of various types of memories are only beginning to be identified by neuroscientists.

The hippocampus plays an especially important role in **declarative memories**—the "what" memories of specific people, places, objects, facts, and events that often result after only one experience and that can be declared in a statement such as "I saw the Statue of Liberty last summer" or conjured up in a mental image. Declarative memories involve conscious recall. This memory type is sometimes subdivided into **semantic memories** (memories of facts) and **episodic memories** (memories of events in our lives).

Clinical Note People with hippocampal damage are profoundly forgetful of facts critical to daily functioning. Declarative memories typically are the first to be lost. Extensive damage in the hippocampus is evident during autopsy in patients with **Alzheimer's disease**. (For an expanded discussion of Alzheimer's disease, see the boxed feature on pp. 164–165, Concepts, Challenges, and Controversies.)

The Cerebellum and Procedural Memories The cerebellum and relevant cortical regions play an essential role in the "how to" **procedural memories** involving motor skills gained through repetitive training, such as memorizing a particular dance routine. The cortical areas important for a given procedural memory are the specific motor or sensory systems engaged in performing the routine. For example, different groups of muscles are called into play to tap dance than those needed to execute a dive. In contrast to declarative memories, which are consciously recollected from previous experiences, procedural memories can be brought forth without conscious effort. To exemplify, an ice skater during a competition typically performs best by "letting the body take over" the routine instead of thinking about exactly what needs to be done next.

Clinical Note The distinct localization in different parts of the brain of declarative and procedural memory is apparent in people who have hippocampal lesions. They can perform a skill, such as playing a piano, but the next day they have no recollection of having done so.

The Prefrontal Cortex and Working Memory The major orchestrator of the complex reasoning skills associated with *working memory* is the prefrontal association cortex. The prefrontal cortex serves as a temporary storage site for holding relevant data "online" and is largely responsible for the so-called **executive functions** involving manipulation and integration of this information for planning, juggling competing priorities, problem solving, making choices, organizing activities, and inhibiting impulses. Executive functions allow a person to decide what to do instead of just reacting to the situation at hand. The prefrontal cortex carries out these complex reasoning functions in cooperation with all the brain's sensory regions, which are linked to the prefrontal cortex through neural connections.

Researchers have identified different storage bins in the prefrontal cortex, depending on the nature of the current relevant data. For example, working memory involving spatial cues is in a prefrontal location distinct from working memory involving verbal cues or cues about an object's appearance. One recent fascinating proposal suggests that a person's intelligence may be determined by the capacity of working memory to temporarily hold and relate a variety of relevant data.

Check Your Understanding 5.7

1. Define *consolidation*.
2. Compare the molecular mechanisms for short-term and long-term memories.
3. Define long-term potentiation.
4. Distinguish among *declarative memories, procedural memories,* and *working memory* and indicate the brain area primarily associated with each.

5.8 Cerebellum

The **cerebellum** is a highly folded, baseball-sized part of the brain that lies underneath the occipital lobe of the cortex and is attached to the back of the upper portion of the brain stem (see ▮Table 5-1, p. 144 and ▮Figures 5-7b, p. 143, and 5-15, p. 154).

The cerebellum is important in balance and in planning and executing voluntary movement.

About four times as many individual neurons are found in the cerebellum than in the entire rest of the brain, indicative of the importance of this structure. The cerebellum consists of three functionally distinct parts with different roles concerned primarily with subconscious control of motor activity (▮Figure 5-18, p. 166). Specifically, the different parts of the cerebellum perform the following functions:

1. The **vestibulocerebellum** is important for maintaining balance and controls eye movements.
2. The **spinocerebellum** enhances muscle tone and coordinates skilled, voluntary movements. This brain region is especially important in ensuring the accurate timing of various muscle contractions to coordinate movements involving mul-

tiple joints. For example, the movements of your shoulder, elbow, and wrist joints must be synchronized even during the simple act of reaching for a pencil. When cortical motor areas send messages to muscles for executing a particular movement, the spinocerebellum is informed of the intended motor command. This region also receives input from peripheral receptors that inform it about the body movements and positions that are actually taking place. The spinocerebellum essentially acts as "middle management," comparing the "intentions" or "orders" of the higher centers with the "performance" of the muscles and then correcting any "errors" or deviations from the intended movement. The spinocerebellum even seems able to predict the position of a body part in the next fraction of a second during a complex movement and to make adjustments accordingly. If you are reaching for a pencil, for example, this region "puts on the brakes" soon enough to stop the forward movement of your hand at the intended location rather than letting you overshoot your target. These ongoing adjustments, which ensure smooth, precise, directed movement, are especially important for rapidly changing (phasic) activities such as typing, playing the piano, or running.

3. The **cerebrocerebellum** plays a role in planning and initiating voluntary activity by providing input to the cortical motor areas. This is also the region that stores procedural memories.

Recent discoveries suggest that in addition to these well-established functions, the cerebellum has even broader responsibilities, such as perhaps coordinating the brain's acquisition of sensory input. Researchers are currently trying to make sense of new and surprising findings that do not fit with the cerebellum's traditional roles in motor control.

Clinical Note All the following symptoms of cerebellar disease result from a loss of the cerebellum's established motor functions: poor balance; "drunken sailor" gait with wide stance and unsteady walking; nystagmus (rhythmic, oscillating eye movements); reduced muscle tone but no paralysis; inability to perform rapid alternating movements smoothly, such as being unable to swiftly slap the open palm of one hand alternately with the palm or back of the other hand; and inability to stop and start skeletal muscle action quickly. The latter gives rise to an *intention tremor* characterized by oscillating to-and-fro movements of a limb as it approaches its intended destination. A person with cerebellar damage who tries to pick up a pencil may overshoot the pencil and then rebound excessively, repeating this to-and-fro process until success is finally achieved. No tremor is observed except in performing intentional activity, in contrast to the resting tremor associated with disease of the basal nuclei, most notably PD.

The cerebellum and basal nuclei both monitor and adjust motor activity commanded from the motor cortex, and like the basal nuclei, the cerebellum does not directly influence the efferent motor neurons. Although they perform different roles (for example, the cerebellum enhances muscle tone, whereas the basal nuclei inhibit it), both function indirectly by modifying the output of major motor systems in the brain. The motor command for a particular voluntary activity arises from the motor cortex, but the actual execution of that activity is coordinated subconsciously by these subcortical regions. To illustrate, you voluntarily

Alzheimer's Disease: A Tale of Beta Amyloid Plaques, Tau Tangles, and Dementia

"**I CAN'T REMEMBER WHERE I PUT** my keys. I must be getting Alzheimer's." The incidence and awareness of **Alzheimer's disease (AD)**, which is characterized in its early stages by loss of recent memories, have become so commonplace that people sometimes jest about having it when they can't remember something. AD is no joking matter, however.

Incidence

AD is the most common and most costly neurological disorder of the CNS. About 5.4 million Americans currently have AD, but because it is an age-related condition and the population is aging, the incidence is expected to climb. The number of affected individuals is expected to swell to 7 million as "baby boomers" age. About 0.1% of those between 60 and 65 years of age are afflicted with the disease, but the incidence rises to 47% among those older than age 85. A small percentage of people with AD are in their 40s and 50s.

Symptoms

AD accounts for about two thirds of the cases of *senile dementia,* which is a generalized age-related diminution of mental abilities. In the earliest stages of AD, only short-term memory is impaired, but as the disease progresses, even firmly entrenched long-term memories, such as recognition of family members, are lost. Confusion, disorientation, and personality changes characterized by irritability and emotional outbursts are common. Higher mental abilities gradually deteriorate as the patient inexorably loses the ability to read, write, and calculate. Language ability and speech are often impaired. In later stages, patients with AD become childlike and are unable to feed, dress, and groom themselves. Patients usually die in a severely debilitated state 4 to 12 years after onset of the disease.

Characteristic Brain Lesions

The characteristic brain lesions of the condition are extracellular *neuritic (senile) plaques* and intracellular *neurofibrillary tangles,* which are dispersed throughout the cerebral cortex and are especially abundant in the hippocampus. A **neuritic plaque** consists of a central core of extracellular, waxy, fibrous protein known as **beta amyloid (Aβ)** surrounded by degenerating dendritic and axonal nerve endings. **Neurofibrillary tangles** are dense bundles of abnormal, paired helical filaments that accumulate in the cell bodies of affected neurons. AD is characterized by degeneration of the cell bodies of certain neurons in the basal forebrain. The acetylcholine-releasing axons of these neurons normally terminate in the cerebral cortex and hippocampus, so the loss of these neurons results in a deficiency of acetylcholine in these areas. Neuron death and loss of synaptic communication are responsible for the ensuing dementia.

Underlying Pathology

Much progress has been made in understanding the pathology underlying the condition in recent years. **Amyloid precursor protein (APP),** a structural component of all neuronal plasma membranes, is especially abundant in presynaptic terminal endings. APP can be cleaved at different locations to produce different products. Cleavage of APP at one site yields a product believed to play a role in learning and memory. Cleavage of APP at an alternative site yields Aβ. Depending on the exact site of cleavage, two different variants of Aβ are produced and released from the neuron. Normally about 90% of the Aβ is a soluble and harmless form of this product. The other 10% is the dangerous, plaque-forming version, which forms thin, insoluble filaments that readily aggregate into Aβ plaques and also appears to be neurotoxic. Furthermore, some researchers propose a controversial new theory that free-floating, short-chain, potentially toxic soluble molecules (known as *oligomers*) derived from Aβ that diffuse through the brain instead of aggregating into plaques are the real culprits responsible for the symptoms of AD. The balance between these APP products can be shifted by mutations in APP, other genetic defects, age-related or pathological changes in the brain, or perhaps environmental factors. The end result is increased production of toxic Aβ. Some evidence suggests that increased levels of Aβ may arise not from overproduction of this product but from failure to adequately clear from the brain the amount of Aβ normally produced.

Aβ formation is seen early in the course of the disease, with neurofibrillary tangles developing somewhat later. AD does not "just happen" in old age. Instead, it results from a host of gradual, insidious processes that occur over the course of years or decades. Although some pieces of the puzzle have not been identified, the following is a possible scenario based on the findings to date. The deposited Aβ is directly toxic to neurons. Furthermore, the gradual buildup of Aβ plaques attracts microglia to the plaque sites. These immune cells of the brain launch an inflammatory attack on the plaque, releasing toxic chemicals that can damage surrounding "innocent bystander" neurons in the process.

These inflammatory assaults, along with the direct toxicity of the deposited Aβ, also cause changes in the neuronal cytoskeleton that lead to formation of nerve-cell clogging neurofibrillary tangles. The protein **tau** normally associates with tubulin molecules in the formation of microtubules, which serve as axonal "highways" for transport of materials back and forth between the cell body and the axon terminal (see p. 46). Tau molecules act like "railroad ties" anchoring the "rails" of tubulin molecules within the microtubule. If tau molecules become hyperphosphorylated (have too many phosphate groups attached), they cannot interact with tubulin. Research suggests that Aβ binds to receptors on the surface of nerve cells, triggering a chain of intracellular events that leads to tau hyperphosphorylation. When not bound to tubulin, the incapacitated tau molecules intertwine, forming paired helical filaments that aggregate to form neurofibrillary tangles. More important, just as train tracks start to fall apart if too many ties are missing, the microtubules start to break down as increasing numbers of tau molecules can no longer do their job. The resultant loss of the neuron's transport system can lead to death of the cell.

Other factors also play a role in the complex story of AD, but exactly where they fit in is unclear. According to a leading proposal, Aβ causes excessive influx of Ca^{2+}, which triggers a chain of biochemical events that kills the cells. Brain cells that have an abundance of glutamate NMDA receptors, most notably the hippocampal cells involved in long-term potentiation (see p. 160), are especially vulnerable to glutamate toxicity. Loss of hippocampal memory-forming capacity is a hallmark feature of AD. Other studies suggest that cell-damaging free radicals (see p. 142) are produced during the course of the disease. One of the most startling recent discoveries is that Aβ proteins might act much like *prions,* the infectious proteins that lead to brain damage in mad cow disease. Prions are misfolded proteins that wreak havoc by causing other normal similar proteins to also misfold, leading to further misfolding in a toxic chain reaction. These misshapen proteins clump together and ultimately kill nerve cells. The Aβ proteins clumped in plaques are misfolded proteins. This prionlike behavior of aberrant brain proteins may play a role in the neurodegeneration seen not only in AD but also in Parkinson's disease (see p. 154) and amyotrophic lateral sclerosis (see p. 51). Unlike with infectious prions associated with mad cow disease, the normal brain proteins turned prionlike are not transmitted from person to person.

Possible Causes

The underlying trigger for abnormal Aβ formation in AD is unknown in most cases. Many investigators believe the condition has many underlying causes. Both genetic and environmental factors have been implicated in an increased risk of acquiring AD. About 15% of cases are linked to specific, known genetic defects that run in families and cause *early onset,* or *familial Alzheimer's disease.* Individuals with this form of the condition typically develop clinical symptoms in their 40s or 50s.

The other 85% of patients with AD do not begin to manifest symptoms until later in life, somewhere between 65 and 85 years of age. Specific gene traits have also been identified that increase an individual's vulnerability of acquiring *late-onset Alzheimer's disease.* However, not everyone with genetic tendencies for AD develops the disease. Furthermore, many develop the illness with no apparent genetic predisposition. Obviously, other factors must also be at play in producing the condition. Hormonal imbalances may play a role. In particular, research findings suggest that *cortisol,* the stress hormone, increases the propensity to develop the condition. In addition, investigators have been searching for possible environmental triggers, but none have been found to date.

Diagnosis

AD can only be confirmed at autopsy on finding the characteristic brain lesions associated with the disease—Aβ plaques and neurofibrillary tangles. Currently, AD is clinically diagnosed before death by a process of elimination—that is, all other disorders that could produce dementia, such as a stroke or brain tumor, must be ruled out. In 2011, diagnostic criteria for AD were changed for the first time in 25 years, although these new guidelines are being used now primarily in research settings. Researchers hope to confirm that earlier diagnosis of AD via the revised guidelines will lead to interventions early in the disease before dementia occurs. Neurological damage can begin up to 20 years before symptoms appear.

The new diagnostic criteria depend on the presence of *biomarkers* (short for biological marker, anything that can be objectively measured as an indicator of a particular physiological or disease state). The two newly accepted biomarkers for AD are (1) imaging of Aβ plaques in the living brain using a PET scan (see p. 145) following injection of a radioactive compound that binds to Aβ (a recently discovered technique) and (2) measuring Aβ and tau in a cerebrospinal fluid (CSF) sample obtained via a spinal tap.

Treatment

The *National Alzheimer's Project Act* was established in the United States in 2011 with the goal of providing a comprehensive approach for ensuring quality care of patients with AD, family and caregiver support, much needed new treatments, and most ambitiously prevention of the condition by 2025.

Currently available drugs can transiently reduce symptoms in some patients but do nothing to slow down or halt progression of the disease. Two classes of drugs are specifically approved for treatment of AD. One class raises the levels of acetylcholine (the deficient neurotransmitter) in the brain. For example, *Aricept (donepezil),* the most commonly prescribed drug for AD, inhibits the enzyme that normally clears released acetylcholine from the synapse. The second, newer class of approved drugs, an example being *Nemenda (memantine),* interferes with the NMDA receptors, blocking the toxic effects of excess glutamate release. Several over-the-counter agents are also used to treat AD. Antioxidants hold some promise of thwarting free-radical damage. Aspirin and other anti-inflammatory drugs may slow the course of AD by blocking the inflammatory components of the condition.

As researchers continue to unravel the underlying factors, the likelihood of finding various means to block the gradual, relentless progression of AD increases. For example, the search is on for new drugs that might block the cleavage of plaque-forming Aβ from APP or might inhibit the aggregation of Aβ into dangerous plaques, thus halting AD in its tracks in the earliest stages. Another thrust is to clear from the brain toxic Aβ fragments. Right now more than 800 therapeutic approaches targeted at different steps of the AD pathway are under development.

The financial payoff for the drug companies that come up with successful products will be huge, and the broader effect will be even more significant. Prevention or treatment of AD cannot come too soon in view of the tragic toll the condition takes on its victims, their families, and society. The cost of custodial care for patients with AD is currently estimated at $183 billion annually and will continue to rise as a greater percentage of our population ages and becomes afflicted with the condition.

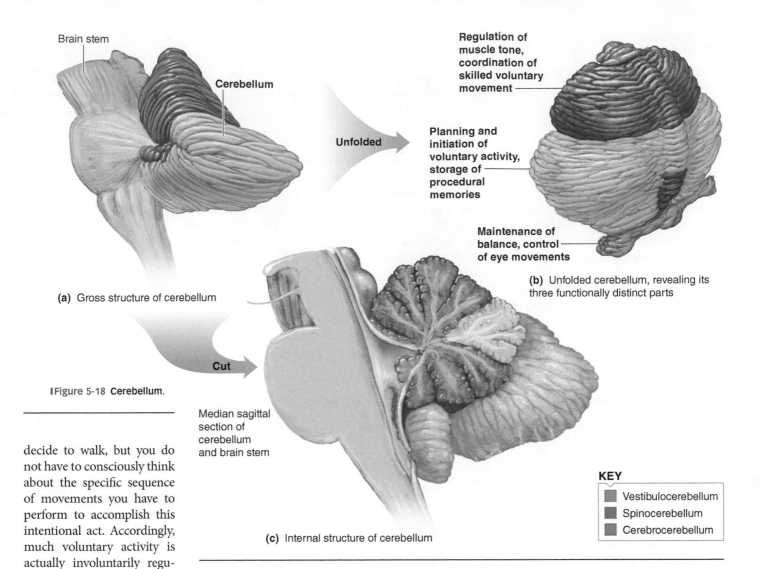

Brain stem

Cerebellum

Unfolded

Regulation of
muscle tone,
coordination of
skilled voluntary
movement

Planning and
initiation of
voluntary activity,
storage of
procedural
memories

Maintenance of
balance, control
of eye movements

(b) Unfolded cerebellum, revealing its
three functionally distinct parts

(a) Gross structure of cerebellum

Cut

Median sagittal
section of
cerebellum
and brain stem

(c) Internal structure of cerebellum

❙Figure 5-18 **Cerebellum**.

KEY

▨	Vestibulocerebellum
▨	Spinocerebellum
▨	Cerebrocerebellum

decide to walk, but you do not have to consciously think about the specific sequence of movements you have to perform to accomplish this intentional act. Accordingly, much voluntary activity is actually involuntarily regulated.

You will learn more about motor control when we discuss skeletal muscle physiology in Chapter 8. For now, we move on to the remaining part of the brain, the brain stem.

Check Your Understanding 5.8

1. State the functions of the three parts of the cerebellum.
2. Compare resting tremors and intention tremors.

5.9 Brain Stem

The **brain stem** consists of the **medulla**, **pons**, and **midbrain** (see ❙Table 5-1 and Figure 5-7b).

The brain stem is a vital link between the spinal cord and higher brain regions.

All incoming and outgoing fibers traversing between the periphery and the higher brain centers must pass through the brain stem, with incoming fibers relaying sensory information to the brain and outgoing fibers carrying command signals from the brain for efferent output. Most of these fibers synapse within the brain stem for important processing. Thus, the brain stem is a critical connecting link between the rest of the brain and the spinal cord.

The functions of the brain stem include the following:

1. Most of the 12 pairs of **cranial nerves** arise from the brain stem. With one major exception, these nerves supply structures in the head and neck with both sensory and motor fibers (❙Table 5-3). They are important in sight, hearing, equilibrium, taste, smell, sensation of the face and scalp, eye movement, chewing, swallowing, facial expressions, and salivation. The major exception is cranial nerve X, the **vagus nerve**. Instead of innervating regions in the head, most branches of the vagus nerve supply organs in the thoracic (chest) and abdominal (belly) cavities. The vagus is the major nerve of the parasympathetic nervous system.

2. Collected within the brain stem are neuronal clusters or **centers** that control heart and blood vessel function, respiration, and many digestive activities. A functional collection of neuronal cell bodies within the CNS is alternately known as a *center,* such as the respiratory control center in the brain stem, or as a *nucleus* (plural *nuclei*), such as the basal nuclei. You will learn about these centers when we discuss the body systems controlled by their activity.

TABLE 5-3 Functions of Cranial Nerves

Number	Name	Fiber Types	Functions
I	Olfactory	Sensory	■ Smell
II	Optic	Sensory	■ Vision
III	Oculomotor	Mixed (mainly motor)	● Eyeball and eyelid movement, pupil constriction, change of lens shape for near vision.
			■ Proprioception (awareness of position of body parts)
IV	Trochlear	Mixed (mainly motor)	● Eyeball movement
			■ Proprioception
V	Trigeminal	Mixed	● Chewing
			■ Somatic sensations (touch, pressure, pain, and temperature) of face and mouth
VI	Abducens	Mixed (mainly motor)	● Eyeball movement
			■ Proprioception
VII	Facial	Mixed	● Facial expression, secretion of saliva and tears
			■ Taste from front of tongue
VIII	Vestibulocochlear	Sensory	■ Hearing, sense of equilibrium
IX	Glossopharyngeal	Mixed	● Swallowing, secretion of saliva
			■ Taste from back of tongue, somatic sensation of oral cavity, blood-pressure monitoring
X	Vagus	Mixed	● Efferent output for skeletal muscles of pharynx (throat) and larynx (voice box) and for smooth muscle and glands of thoracic and abdominal organs and for cardiac muscle of heart
			■ Afferent input from thoracic and abdominal organs, blood-pressure monitoring
XI	Accessory	Motor	● Efferent output for skeletal muscles of pharynx, larynx, neck, and shoulder
XII	Hypoglossal	Motor	● Tongue movement

■ Carried by afferent fibers ● Carried by efferent fibers

3. The brain stem helps regulate muscle reflexes involved in equilibrium and posture.

4. A widespread network of interconnected neurons called the **reticular formation** runs throughout the entire brain stem and into the thalamus. This network receives and integrates all incoming sensory synaptic input. Ascending fibers originating in the reticular formation carry signals upward to arouse and activate the cerebral cortex (▮Figure 5-19). These fibers compose the **reticular activating system (RAS)**, which controls

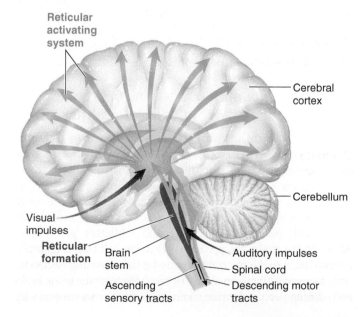

▮Figure 5-19 **The reticular activating system.** The reticular formation, a widespread network of neurons within the brain stem (in *red*), receives and integrates all synaptic input. The reticular activating system, which promotes cortical alertness and helps direct attention toward specific events, consists of ascending fibers (in *blue*) that originate in the reticular formation and carry signals upward to arouse and activate the cerebral cortex.

FIGURE FOCUS: *Describe the pathway by which your alarm going off wakes you up.*

the overall degree of cortical alertness and is important in the ability to direct attention. In turn, fibers descending from the cortex, especially its motor areas, can activate the RAS.

5. The centers that govern sleep are housed within the brain stem and the hypothalamus.

We now examine sleep and the other states of consciousness.

Consciousness refers to awareness of one's existence, thoughts, and surroundings.

The term **consciousness** refers to subjective awareness of the external world and self, including awareness of the private inner world of one's mind—that is, awareness of thoughts, perceptions, dreams, and so on. Even though the final level of awareness resides in the cerebral cortex and a crude sense of awareness is detected by the thalamus, conscious experience depends on the integrated functioning of many parts of the nervous system.

The cellular and molecular basis underlying consciousness is one of the greatest unanswered questions in neuroscience. One proposal that is gaining increasing support is the **global workspace theory**, which suggests that conscious experience depends on the brain functioning as a "brainweb" in which some of the separate bits of subconscious information that are being processed locally at the same time are momentarily broadcast throughout the brain (that is, to a global workspace). This highly coordinated, widespread information exchange among much of the cortex gives rise to subjective experience of the information. That is, we become conscious of what we are experiencing only when information received through specialized channels (such as sensory information) is distributed to much of the cortex, creating a unity of mind.

Normal states of consciousness are wakefulness and sleep. The **sleep–wake cycle** is a normal cyclic variation in awareness of surroundings. In the **waking state** people are alert and aware of their surroundings and consciously engage in coherent thoughts and actions. Wakefulness depends on attention-getting sensory input that "energizes" the RAS and subsequently the activity level of the CNS as a whole. Wakefulness is not a constant level of arousal but varies from maximum alertness to drowsiness, depending on the extent of interaction between peripheral stimuli and the brain. Different arousal and activity states are characterized by different brain wave activity as recorded on an electroencephalogram.

An electroencephalogram is a record of postsynaptic activity in cortical neurons.

Extracellular current flow arising from electrical activity within the cerebral cortex can be detected by placing recording electrodes on the scalp to produce a graphic record known as an **electroencephalogram**, or **EEG.** These "brain waves" for the most part are not the result of action potentials but instead represent the momentary collective postsynaptic potential activity (that is, excitatory postsynaptic potentials, or EPSPs, and inhibitory postsynaptic potentials, or IPSPs; see pp. 106–107) in the cell bodies and dendrites located in the cortical layers under the recording electrode.

Electrical activity can always be recorded from the living brain, even during sleep and unconscious states, but the waveforms vary, depending on the degree of activity in the cerebral cortex. Often the waveforms appear irregular, but sometimes distinct patterns in wave frequency (number of waves per second) and amplitude (height of each wave) can be observed. There are five categories of brain waves, depending on the mental state. The higher the frequency of the brain waves, the faster the brain activity. Ranging from the most to the least activity, these waveforms are:

- **gamma waves**, which are the fastest brain waves with the smallest amplitude and are the most recently identified brain waves. (Gamma waves in the brain are distinctly different from gamma rays released from radioactive material, as used in PET scans; see p. 145.) Gamma waves are associated with peak concentration; the highest levels of cognition; and simultaneous processing of information from different brain areas, as when the brain is actively tying together the sights and sounds of a current experience.

- **beta waves**, which have a high frequency and low amplitude and are prominent when you are fully awake, focused, and alert. This is the dominant waveform during much of the day, such as when you are focused on your surroundings, actively thinking, or engaged in conversation.

- **alpha waves**, which have a lower frequency and greater amplitude than beta waves and are present when you are awake,

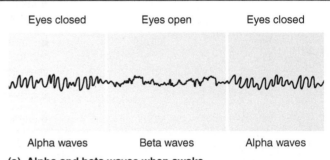

Eyes closed	Eyes open	Eyes closed
Alpha waves	Beta waves	Alpha waves

(a) Alpha and beta waves when awake

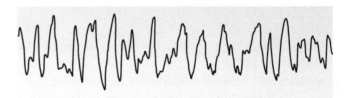

(b) Delta waves when in slow-wave sleep, stage 4

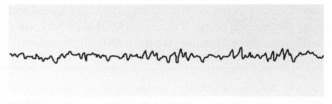

(c) Waves during paradoxical sleep

❙Figure 5-20 **Electroencephalogram (EEG) patterns under different circumstances.** (a) An alpha rhythm when the eyes are closed is replaced by a beta rhythm when the eyes are opened during the waking state. (b) A delta rhythm is associated with deep stage-4 slow-wave sleep. (c) Note that the EEG pattern during paradoxical sleep is similar to the beta rhythm of an alert, awake person.

but relaxed, calm, and not processing much information. ▮Figure 5-20 illustrates how the EEG waveform recorded over the occipital (visual) cortex dramatically shifts between alpha and beta waves in response to simply closing and opening the eyes.

■ **theta waves**, which have an even slower frequency and greater amplitude than alpha waves and dominate when you are extremely relaxed, drowsy, or are in light sleep.

■ **delta waves**, which have the greatest amplitude and slowest frequency and occur when you are in deep, dreamless sleep (▮Figure 5-20).

Clinical Note The EEG has three major uses:

1. The EEG is often used as a *clinical tool in the diagnosis of cerebral dysfunction*. Diseased or damaged cortical tissue often gives rise to altered EEG patterns. One of the most common neurological diseases accompanied by a distinctively abnormal EEG is **epilepsy**. Epileptic seizures occur when a large collection of neurons undergo abnormal, synchronous action potentials that produce stereotypical, involuntary spasms and alterations in behavior. Different underlying problems, including genetic defects and traumatic brain injuries, can lead to the neuronal hyperexcitability that characterizes epilepsy. Typically there is too little inhibitory compared to excitatory activity, as with compromised functioning of the inhibitory neurotransmitter GABA or prolonged action of the excitatory neurotransmitter glutamate. The seizures may be partial or generalized, depending on the location and extent of the abnormal neuronal discharge. Each type of seizure displays different EEG features.

2. The EEG is also used in the *legal determination of brain death*. Even though a person may have stopped breathing and the heart may have stopped pumping blood, it is often possible to restore and maintain respiratory and circulatory activity if resuscitative measures are instituted soon enough. Yet because the brain is susceptible to O_2 deprivation, irreversible brain damage may occur before lung and heart function can be reestablished, resulting in the paradoxical situation of a dead brain in a living body. The determination of whether a comatose patient being maintained by artificial respiration and other supportive measures is alive or dead has important medical, legal, and social implications. The need for viable organs for modern transplant surgery has made the timeliness of such life-or-death determinations of utmost importance. Physicians, lawyers, and Americans in general have accepted the notion of brain death—that is, a brain that is not functioning, with no possibility of recovery—as the determinant of death under such circumstances. The most widely accepted indication of brain death is electrocerebral silence—a flat EEG.

3. The EEG is also used to *distinguish various stages of sleep*.

Sleep is an active process consisting of alternating periods of slow-wave and paradoxical sleep.

In contrast to being awake, sleeping people are not consciously aware of the external world, but they do have inward conscious experiences such as dreams. Furthermore, they can be aroused by external stimuli, such as an alarm going off. **Sleep** is an active process, not just the absence of wakefulness. The brain's overall level of activity is not reduced during sleep. During certain stages of sleep, O_2 uptake by the brain is even increased above normal waking levels. There are two types of sleep, characterized by different EEG patterns and different behaviors: *slow-wave sleep* and *paradoxical*, or REM, *sleep* (▮Table 5-4).

EEG Patterns During Sleep Slow-wave sleep occurs in four stages, each displaying progressively slower EEG waves of higher amplitude (hence, "slow-wave" sleep) (▮Figure 5-20). At the onset of sleep, you move from the light sleep of stage 1 to the deep sleep of stage 4 of slow-wave sleep during a period of 30 to 45 minutes; then you reverse through the same stages in the same amount of time. A 10- to 15-minute episode of **paradoxical sleep** punctuates the end of each slow-wave sleep cycle.

▮TABLE 5-4 Comparison of Slow-Wave and Paradoxical Sleep

Characteristic	TYPE OF SLEEP	
	Slow-Wave Sleep	Paradoxical Sleep
EEG	Displays slow waves	Similar to EEG of alert, awake person
Motor activity	Considerable muscle tone; frequent shifting	Abrupt inhibition of muscle tone; no movement
Heart rate, respiratory rate, blood pressure	Minor reductions	Irregular
Dreaming	Rare (mental activity is extension of waking-time thoughts)	Common
Arousal	Sleeper easily awakened	Sleeper hard to arouse but apt to wake up spontaneously
Percentage of sleeping time	80%	20%
Other important characteristics	Has four stages; sleeper must pass through this type of sleep first	Rapid eye movements

Paradoxically, your EEG pattern during this time abruptly becomes similar to that of a wide-awake, alert individual, even though you are still asleep (hence, "paradoxical" sleep) (▮Figure 5-20). After the paradoxical episode, the stages of slow-wave sleep repeat. You cyclically alternate between the two types of sleep throughout the night. Brief periods of wakefulness occasionally occur. Most stage 4 deep sleep occurs during the first several hours of sleep, with paradoxical sleep occupying an increasingly greater share of sleep time as morning approaches (▮Figure 5-21). Because of the resemblance of this graphic representation of the cyclical sleep pattern to a city skyline, the pattern of sleep is sometimes referred to as *sleep architecture*.

In a normal sleep cycle, you always pass through slow-wave sleep before entering paradoxical sleep. On average, paradoxical sleep occupies 20% of total sleeping time throughout adolescence and most of adulthood. Infants spend considerably more time in paradoxical sleep. In contrast, paradoxical and deep stage 4 slow-wave sleep declines in the elderly.

Behavioral Patterns During Sleep In addition to distinctive EEG patterns, the two types of sleep are distinguished by behavioral differences. It is difficult to pinpoint exactly when an individual drifts from drowsiness into slow-wave sleep. In this type of sleep, the person still has considerable muscle tone and often shifts body position. Respiratory rate, heart rate, and blood pressure remain regular. During this time, the sleeper can be easily awakened and rarely dreams. The mental activity associated with slow-wave sleep is less visual than dreaming. It is more conceptual and plausible—like an extension of waking-time thoughts concerned with everyday events—and it is less likely to be recalled. The major exception is nightmares, which occur during stages 3 and 4.

The behavioral pattern accompanying paradoxical sleep is marked by abrupt inhibition of muscle tone throughout the body. The muscles are completely relaxed, with no movement taking place except in the eye muscles. Paradoxical sleep is characterized by *rapid eye movements,* hence the alternative name, **REM sleep**. Heart rate and respiratory rate become irregular, and blood pressure may fluctuate. Another characteristic of REM sleep is dreaming. Recent evidence indicates that at least part of the rapid eye movements are related to "watching" the dream imagery, although traditionally scientists thought that the eye movements were caused by an automatic, rhythmic pattern of discharge not influenced by dream content.

Brain imaging of volunteers during REM sleep shows heightened activity in the higher-level visual processing areas and limbic system (seat of emotions), coupled with reduced activity in the prefrontal cortex (seat of reasoning). This activity pattern lays the groundwork for the characteristics of dreaming: internally generated visual imagery reflecting activation of the person's "emotional memory bank" with little guidance or interpretation from the complex thinking areas. As a result, dreams are often charged with intense emotions, a distorted sense of time, and bizarre content that is uncritically accepted as real, with little reflection about all the strange happenings.

The sleep–wake cycle is controlled by interactions among three neural systems.

The sleep–wake cycle, and the various stages of sleep, result from the cyclic interplay of three neural systems: (1) an **arousal system** involving the RAS in the brain stem, which is commanded by a specialized group of neurons in the hypothalamus; (2) a **slow-wave sleep center** in the hypothalamus that contains *sleep-on neurons,* which bring on slow-wave sleep; and (3) a **paradoxical sleep center** in the brain stem that houses REM *sleep-on neurons,* which switches to paradoxical sleep. The patterns of interaction among these three neural regions, which bring about the fairly predictable cyclical sequence between being awake and passing alternately between the two types of sleep, are the subject of intense investigation. A growing body of evidence suggests the following relationships:

1. A group of neurons in the hypothalamus is at the top of the chain of command for regulating the arousal system. These neurons secrete the excitatory neurotransmitter **hypocretin** (also known as **orexin**). Surprisingly, hypocretin is better known as an appetite-enhancing signal, but it is now known to play an important role in arousal too. These hypocretin-secreting neurons fire autonomously (on their own) and continuously and keep you awake and alert by stimulating the RAS. They must be inhibited to induce sleep, as perhaps by IPSPs generated by input from the sleep-on neurons.

2. The **sleep-on neurons** in the slow-wave sleep center appear to be responsible for bringing on sleep, likely by inhibiting the arousal-promoting neurons by releasing the inhibitory neurotransmitter GABA. This mechanism would explain why we enter slow-wave sleep first when we fall asleep. The sleep-on neurons are inactive when a person is awake and are maximally active only during slow-wave sleep. Scientists do not know much about the factors that activate the sleep-on neurons to induce sleep.

3. The **REM sleep-on neurons** in the paradoxical sleep center become very active during REM sleep. It appears that they

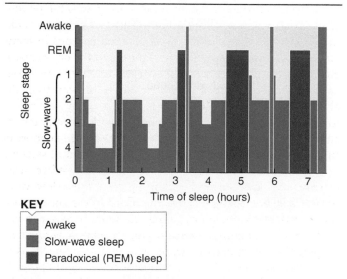

▮Figure 5-21 **Typical cyclical sleep pattern in a young adult.**
FIGURE FOCUS: *Describe how the cyclical pattern of slow-wave sleep, REM sleep, and awake periods changes as the person moves from beginning to end of the sleep period depicted.*

can turn off the sleep-on neurons and switch the sleep pattern from slow-wave sleep to REM sleep. The underlying molecular mechanisms responsible for the cyclical interplay between the two types of sleep remain poorly understood.

The normal cycle can easily be interrupted, with the arousal system more readily overriding the sleep systems than vice versa—that is, it is easier to stay awake when you are sleepy than to fall asleep when you are wide awake. The arousal system can be activated by afferent sensory input (for example, a person has difficulty falling asleep when it is noisy) or by input descending to the brain stem from higher brain regions. Intense concentration or strong emotional states, such as anxiety or excitement, can keep a person from falling asleep, just as motor activity, such as getting up and walking around, can arouse a drowsy person. However, you can override the urge to sleep for just so long before the pressure to sleep becomes irresistible. Sleep on a regular basis is an absolute necessity of life, even though scientists are not sure what purpose sleep serves.

The function of sleep is unclear.

Even though humans spend about a third of their lives sleeping, why sleep is needed largely remains a mystery. Sleep is by the brain and for the brain, not for other parts of the body. It is not accompanied by a reduction in neural activity (that is, the brain cells are not "resting"), as once was suspected, but rather by a profound *change* in activity.

One widely accepted proposal holds that sleep provides "catch-up" time for the brain to restore biochemical or physiological processes that have progressively degraded during wakefulness.

The most direct evidence supporting this proposal is the potential role of **adenosine** as a neural sleep factor. Adenosine, the backbone of adenosine triphosphate (ATP), the body's energy currency, is generated during the awake state by metabolically active neurons and glial cells. Thus, the brain's extracellular concentration of adenosine continues to rise the longer a person has been awake. Adenosine, which acts as a neuromodulator, has been shown experimentally to inhibit the arousal center. This action can bring on slow-wave sleep, during which restoration and recovery activities are believed to take place. Injections of adenosine induce apparently normal sleep, whereas **caffeine**, which blocks adenosine receptors in the brain, revives drowsy people by removing adenosine's inhibitory influence on the arousal center. Adenosine levels diminish during sleep, presumably because the brain uses this adenosine as a raw ingredient for replenishing its limited energy stores. Thus, the body's need for sleep may stem from the brain's periodic need to replenish diminishing energy stores. Because adenosine reflects the level of brain cell activity, the concentration of this chemical in the brain may serve as a gauge of how much energy has been depleted.

Another "restoration and recovery" proposal suggests that slow-wave sleep provides time for the brain to repair damage caused by toxic free radicals (see p. 142) produced as by-products of the stepped-up metabolism during the waking state. Other organs can sacrifice and replace cells damaged by free radicals, but this is not an option for the nonregenerative brain.

A related, new "restoration and recovery" role for sleep involves the recently identified cleansing of the brain's intersti-tial fluid by the glymphatic system whose activity increases during sleep (see p. 137).

One more possible "restoration and recovery" function of REM sleep is to let some of the neural pathways regain full sensitivity. When a person is awake, brain neurons that release the neurotransmitters norepinephrine and serotonin are maximally and continuously active. Release of these neurotransmitters ceases during REM sleep. Studies suggest that constant release of norepinephrine and serotonin can desensitize their receptors. Perhaps REM sleep is needed to restore receptor sensitivity for optimal functioning during the next period of wakefulness.

A different leading theory has nothing to do with restoration and recovery. Instead, other researchers believe that sleep is necessary to allow the brain to "shift gears" to accomplish the long-term structural and chemical adjustments necessary for learning and memory. This theory might explain why infants need so much sleep. Their highly plastic brains are rapidly undergoing profound synaptic modifications in response to environmental stimulation. In contrast, mature individuals, in whom neural changes are less dramatic, sleep less. Some evidence suggests that the different types of sleep might be involved in consolidation of different kinds of memories, with declarative memories being consolidated during slow-wave sleep and procedural memories during REM sleep.

A recent memory-related theory is that sleep, especially slow-wave sleep, is a time for replaying the events of the day, not only to help consolidate memories but perhaps to make recent experiences more meaningful by catching information missed on first pass and by "connecting the dots" between new pieces of information. This information-processing proposal could explain why people with an important decision to make sometimes say they will "sleep on it" before arriving at a conclusion.

The latest, highly debated proposal for sleep's role in learning and memory is the **synaptic homeostasis hypothesis**, which is based on preventing brain overload. According to this proposal, sleep provides off-line time needed for synaptic downscaling to offset the increases in synaptic activity that accompany all kinds of stimulation during wakeful periods (that is, to maintain synaptic homeostasis). During wakefulness, LTP strengthens connections between neurons in many circuits throughout the cortex in response to the day's experiences. The resultant increase in synaptic activity is metabolically costly and eventually day after day would constrain the ability to learn further (for example, only so many new receptors can be inserted into the postsynaptic membrane to strengthen synaptic activity). Evidence gathered in support of this hypothesis suggests that the greater the extent of synaptic potentiation during wakefulness, the more slow-wave sleep activity takes place during the following sleep. These slow waves depress synaptic strength across the board in the cortex. As a result of this synaptic downscaling, the more weakly potentiated synaptic connections made during the day are eliminated, keeping the cortex from filling up with useless connections of inadequate strength. Only the strongest of the newly potentiated memory bonds are preserved (in a sort of "survival of the fittest"). This downsizing restores total synaptic activity to a sustainable baseline level that preserves resources needed for the next day's round of synaptic strengthening and learning.

Another highly controversial new proposal is that local neural networks in the brain may go to sleep at different times, depending on how much they have been used recently. When a particular neuronal circuit is exhausted from prolonged or intense use, it can "fall asleep" even though the rest of the brain remains awake. The characteristic behavioral sleep state kicks in only when most of the brain's neurons are in the sleep mode. Furthermore, portions of the brain that have been relatively inactive during the day may remain "awake" even after the person falls asleep. This scattered pattern of sleep is known as **local use-dependent sleep**.

The various sleep theories are not mutually exclusive. Sleep might serve multiple purposes.

Little is known about the brain's need for cycling between the two types of sleep, although a specified amount of paradoxical sleep appears to be required. Individuals experimentally deprived of paradoxical sleep for a night or two by being aroused every time the paradoxical EEG pattern appeared suffered hallucinations and spent proportionally more time in paradoxical sleep during subsequent undisturbed nights, as if to make up for lost time.

Clinical Note An unusual sleep disturbance is **narcolepsy**. It is characterized by brief (5- to 30-minute), irresistible sleep attacks during the day. A person suffering from this condition suddenly falls asleep during any ongoing activity, often without warning. Patients with narcolepsy typically enter into paradoxical sleep directly without the normal prerequisite passage through slow-wave sleep. Investigators recently learned that narcolepsy is linked to a deficiency of hypocretin as a result of selective autoimmune destruction of the hypocretin-secreting neurons in the hypothalamus.

Impaired states of consciousness are associated with minimal or no awareness.

Impaired states of consciousness include the minimally conscious state, the vegetative state, and coma. A person in a **minimally conscious state** inconsistently shows signs of minimal conscious awareness of self or environment and can move purposefully, such as deliberately making a hand response to a simple command. The **vegetative state** is characterized by periodic sleep–wake cycles but without detectable awareness. The person may appear wide awake but is unaware of surroundings or self. **Coma** is the total unresponsiveness of a living person to external stimuli. These altered states of consciousness are caused either by brain stem damage that interferes with the RAS or by widespread depression of the cerebral cortex, such as accompanies O_2 deprivation or traumatic brain injury. When under **anesthesia**, a person's state of consciousness is more like being in a drug-induced coma than in a deep sleep.

A **concussion** is a transient change in mental status caused by a traumatic brain injury. It is typified by the following range of symptoms, some but not necessarily all of which may be present: temporary loss of consciousness, headache, confusion, dizziness, and amnesia. Recent studies have raised concerns that repeated blows to the head, especially before the brain has healed from a previous concussion, as can occur in boxing or football, can lead to increased risk in later life for dementia, *chronic traumatic encephalopathy* (progressive degeneration of the brain similar to amyotrophic lateral sclerosis, or Lou Gehrig's disease), Parkinson's disease, or depression. One worrisome finding was that players retired from the National Football League had a 19 times higher incidence of severe memory problems than other men of comparable age.

We have finished discussing the brain and now shift attention to the other CNS component, the spinal cord.

Check Your Understanding 5.9

1. List the functions of the brain stem.
2. Define *consciousness*.
3. Discuss the location and functions of the three neural systems that play a role in the sleep–wake cycle.

5.10 Spinal Cord

The **spinal cord** is a long, slender cylinder of nerve tissue that extends from the brain stem. It is about 45 cm (18 in.) long and 1 to 1.5 cm wide (about the width of your finger).

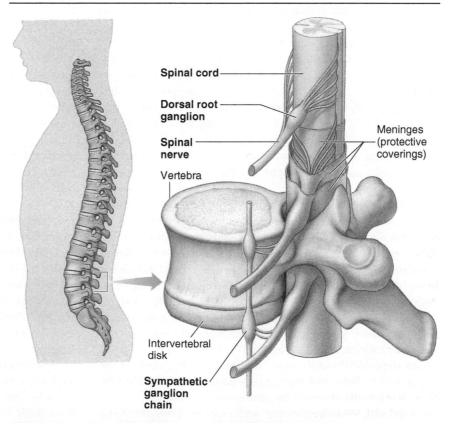

Spinal cord
Dorsal root ganglion
Spinal nerve
Vertebra
Meninges (protective coverings)
Intervertebral disk
Sympathetic ganglion chain

❙Figure 5-22 **Location of the spinal cord relative to the vertebral column.**

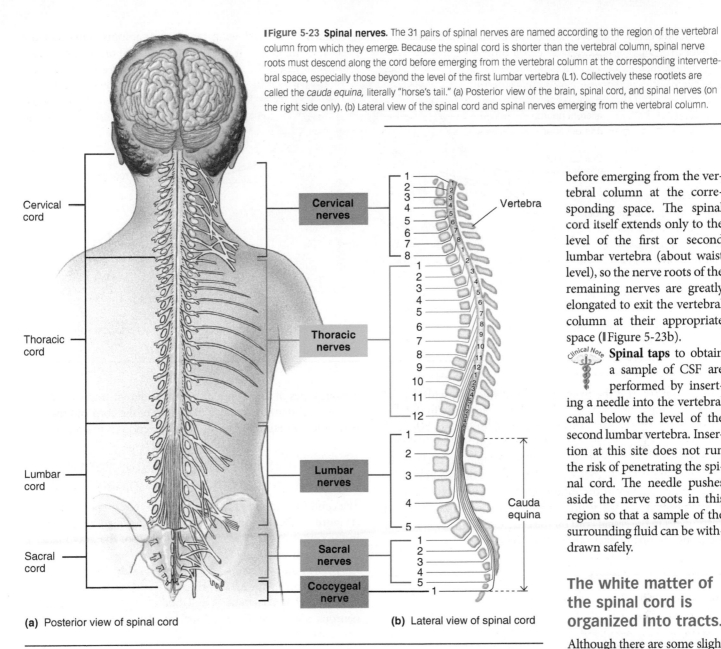

Figure 5-23 Spinal nerves. The 31 pairs of spinal nerves are named according to the region of the vertebral column from which they emerge. Because the spinal cord is shorter than the vertebral column, spinal nerve roots must descend along the cord before emerging from the vertebral column at the corresponding intervertebral space, especially those beyond the level of the first lumbar vertebra (L1). Collectively these rootlets are called the *cauda equina,* literally "horse's tail." (a) Posterior view of the brain, spinal cord, and spinal nerves (on the right side only). (b) Lateral view of the spinal cord and spinal nerves emerging from the vertebral column.

Cervical cord

Thoracic cord

Lumbar cord

Sacral cord

Cervical nerves

Thoracic nerves

Lumbar nerves

Sacral nerves

Coccygeal nerve

Vertebra

Cauda equina

(a) Posterior view of spinal cord

(b) Lateral view of spinal cord

before emerging from the vertebral column at the corresponding space. The spinal cord itself extends only to the level of the first or second lumbar vertebra (about waist level), so the nerve roots of the remaining nerves are greatly elongated to exit the vertebral column at their appropriate space (Figure 5-23b).

Clinical Note **Spinal taps** to obtain a sample of CSF are performed by inserting a needle into the vertebral canal below the level of the second lumbar vertebra. Insertion at this site does not run the risk of penetrating the spinal cord. The needle pushes aside the nerve roots in this region so that a sample of the surrounding fluid can be withdrawn safely.

The white matter of the spinal cord is organized into tracts.

Although there are some slight regional variations, the cross-sectional anatomy of the spinal cord is generally the same throughout its length (Figure 5-24). In contrast to the brain, where the gray matter forms an outer shell capping an inner white core, the gray matter in the spinal cord forms an inner butterfly-shaped region surrounded by the outer white matter. As in the brain, the cord gray matter consists primarily of neuronal cell bodies and their dendrites, and glial cells. The white matter is organized into **tracts**, which are bundles of nerve fibers (axons of long interneurons) with a similar function. The bundles are grouped into columns that extend the length of the cord. Each of these tracts begins or ends within a particular area of the brain, and each transmits a specific type of information. Some are **ascending** (cord to brain) **tracts** that transmit to the brain signals derived from afferent input. Others are **descending** (brain to cord) **tracts** that relay messages from the brain to efferent neurons (Figure 5-25).

The tracts are generally named for their origin and termination. For example, the **ventral spinocerebellar tract** is an ascending pathway that originates in the spinal cord and runs up

The spinal cord extends through the vertebral canal and is connected to the spinal nerves.

Exiting through a large hole in the base of the skull, the spinal cord is enclosed by the protective vertebral column as it descends through the vertebral canal (Figure 5-22). Paired **spinal nerves** emerge from the spinal cord through spaces formed between the bony, winglike arches of adjacent vertebrae. The spinal nerves are named according to the region of the vertebral column from which they emerge (Figure 5-23): There are 8 pairs of *cervical (neck) nerves* (namely, C1 to C8), 12 *thoracic (chest) nerves,* 5 *lumbar (abdominal) nerves,* 5 *sacral (pelvic) nerves,* and 1 *coccygeal (tailbone) nerve.*

During development, the vertebral column grows about 25 cm longer than the spinal cord. Because of this differential growth, segments of the spinal cord that give rise to various spinal nerves are not aligned with the corresponding intervertebral spaces. Most spinal nerve roots must descend along the cord

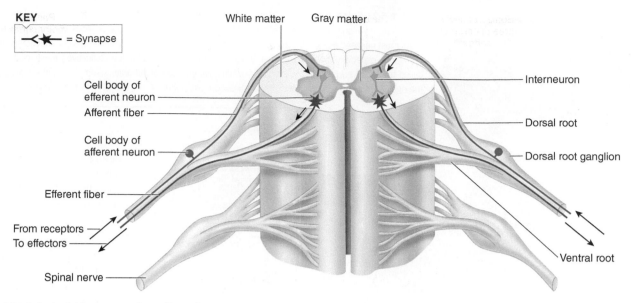

KEY

─<─*── = Synapse

White matter

Gray matter

Cell body of efferent neuron

Afferent fiber

Cell body of afferent neuron

Efferent fiber

From receptors

To effectors

Spinal nerve

Interneuron

Dorsal root

Dorsal root ganglion

Ventral root

▮Figure 5-24 **Spinal cord in cross section.** Afferent fibers enter through the dorsal root, and efferent fibers exit through the ventral root. Afferent and efferent fibers are enclosed together within a spinal nerve.

the ventral (toward the front) margin of the cord with several synapses along the way until it eventually terminates in the cerebellum (▮Figure 5-26a). This tract carries information derived from muscle stretch receptors that has been delivered to the spinal cord by afferent fibers for use by the spinocerebellum. In contrast, the **ventral corticospinal tract** is a descending pathway that originates in the motor region of the cerebral cortex, then travels down the ventral portion of the spinal cord, and terminates in the spinal cord on the cell bodies of efferent motor neurons supplying skeletal muscles (▮Figure 5-26b). Because

various types of signals are carried in different tracts within the spinal cord, damage to particular areas of the cord can interfere with some functions, whereas other functions remain intact.

Each horn of the spinal cord gray matter houses a different type of neuronal cell body.

The centrally located gray matter is also functionally organized (▮Figure 5-27). The central canal, which is filled with CSF, lies in the center of the gray matter. Each half of the gray matter is arbitrarily divided into a *dorsal (posterior) horn*, a *ventral (anterior) horn*, and a *lateral horn*. The **dorsal horn** contains cell bodies of interneurons on which afferent neurons terminate. The **ventral horn** contains cell bodies of the efferent motor neurons supplying skeletal muscles. Autonomic nerve fibers

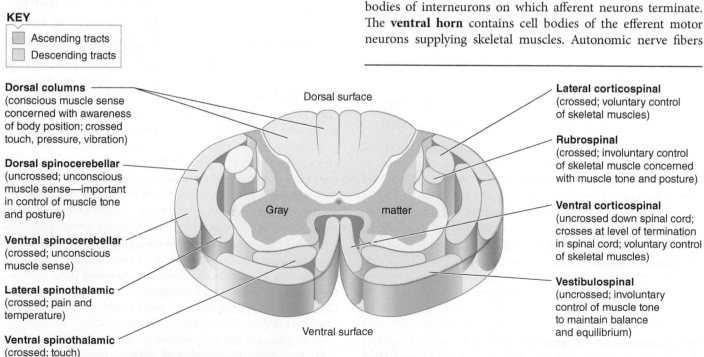

KEY

▢ Ascending tracts
▢ Descending tracts

Dorsal columns (conscious muscle sense concerned with awareness of body position; crossed touch, pressure, vibration)

Dorsal spinocerebellar (uncrossed; unconscious muscle sense—important in control of muscle tone and posture)

Ventral spinocerebellar (crossed; unconscious muscle sense)

Lateral spinothalamic (crossed; pain and temperature)

Ventral spinothalamic (crossed; touch)

Dorsal surface

Gray   matter

Ventral surface

Lateral corticospinal (crossed; voluntary control of skeletal muscles)

Rubrospinal (crossed; involuntary control of skeletal muscle concerned with muscle tone and posture)

Ventral corticospinal (uncrossed down spinal cord; crosses at level of termination in spinal cord; voluntary control of skeletal muscles)

Vestibulospinal (uncrossed; involuntary control of muscle tone to maintain balance and equilibrium)

▮Figure 5-25 **Ascending and descending tracts in the white matter of the spinal cord in cross section.**

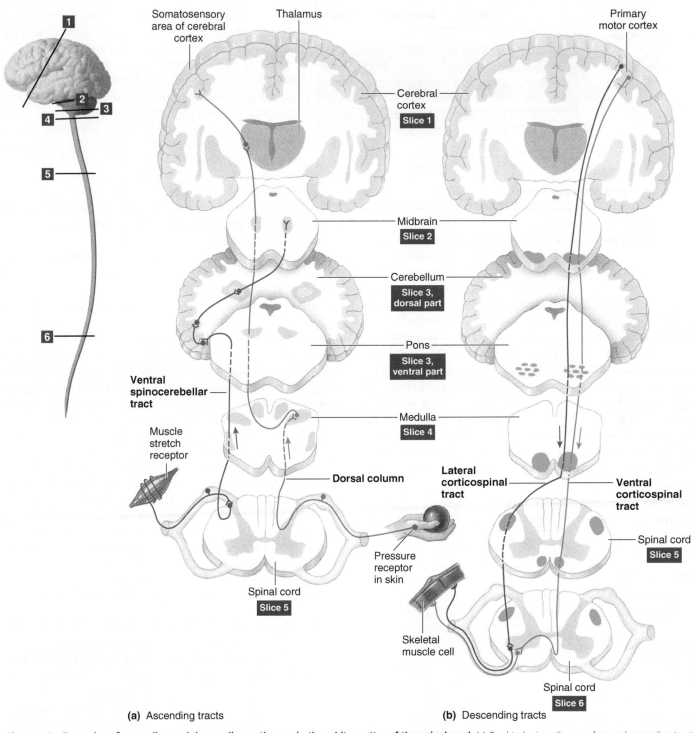

(a) Ascending tracts   **(b)** Descending tracts

IFigure 5-26 Examples of ascending and descending pathways in the white matter of the spinal cord. (a) Cord-to-brain pathways of several ascending tracts (a dorsal column tract and ventral spinocerebellar tract). (b) Brain-to-cord pathways of several descending tracts (lateral corticospinal and ventral corticospinal tracts).

supplying cardiac and smooth muscle and exocrine glands originate at cell bodies found in the **lateral horn**.

Spinal nerves carry both afferent and efferent fibers.

Spinal nerves connect with each side of the spinal cord by a *dorsal root* and a *ventral root* (see IFigure 5-24). Afferent fibers carrying incoming signals from peripheral receptors

enter the spinal cord through the **dorsal root**. The cell bodies for the afferent neurons at each level are clustered together in a **dorsal root ganglion**. (A collection of neuronal cell bodies located outside the CNS is called a *ganglion*, whereas a functional collection of cell bodies within the CNS is referred to as a *nucleus* or a *center*.) The cell bodies for the efferent neurons originate in the gray matter, and the efferent fibers carrying outgoing signals to muscles and glands exit through the **ventral root**.

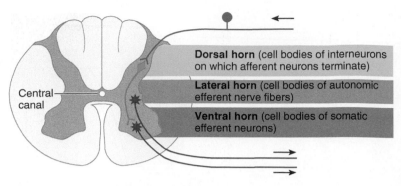

Figure 5-27 Regions of the gray matter.

Labels in figure:
- Central canal
- **Dorsal horn** (cell bodies of interneurons on which afferent neurons terminate)
- **Lateral horn** (cell bodies of autonomic efferent nerve fibers)
- **Ventral horn** (cell bodies of somatic efferent neurons)

The dorsal and ventral roots at each level join to form a spinal nerve that emerges from the vertebral column (see ❚Figure 5-24). A spinal nerve contains both afferent and efferent fibers traversing between the spinal cord and a particular body region. Note the relationship between a *nerve* and a *neuron*. A **nerve** is a bundle of peripheral neuronal axons, some afferent and some efferent, enclosed by a connective tissue covering and following the same pathway (❚Figure 5-28). A nerve does not contain a complete nerve cell, only the axonal portions of many neurons. (By this definition, there are no nerves in the CNS! Bundles of axons in the CNS are called tracts.) The individual fibers within a nerve generally do not have any direct influence on one another. They travel together for convenience, just as many individual land telephone lines are carried within a telephone cable yet any particular landline phone connection can be private without interference from other lines in the cable.

The 31 pairs of spinal nerves and the 12 pairs of cranial nerves that arise from the brain stem constitute the peripheral nervous system. After they emerge, the spinal nerves progressively branch to form a vast network of peripheral nerves that supply the tissues. Because each segment of the spinal cord gives rise to a pair of spinal nerves that ultimately supplies a particular region of the body with both afferent and efferent fibers, the location and extent of sensory and motor deficits associated with spinal-cord injuries can be clinically important in determining the level and extent of the cord injury.

Each spinal nerve carries afferent sensory fibers from a particular region on the body surface called a **dermatome**. The body surface can be mapped with multiple dermatomes, each one associated with a different spinal nerve (❚Figure 5-29a).

Clinical Note **Shingles**, an infection of a sensory nerve fiber and the area of skin supplied by this fiber, is caused by *varicella-zoster virus*, the same virus that causes chicken pox. After a bout of chicken pox, some of the virus may survive and travel in sensory axons to dorsal root ganglia, where it can remain dormant and produce no symptoms for years. When the immune system weakens, as it can with aging, debilitating diseases, or stress, the virus may reactivate and travel back through the sensory axon to the skin. Here the virus causes pain and a blistery rash along a band of skin supplied by the affected nerve; that is, along a dermatome (❚Figure 5-29b). Typically an episode of shingles is limited to a single nerve fiber and dermatome. Shingles occurs in about 30% of those who have had chicken pox and is most common in people older than 50. With development of the chicken pox vaccine in 1995 and development of the shingles vaccine in 2006, the incidence of shingles will continue to decline.

Spinal nerves also carry fibers that branch off to supply internal organs, and sometimes pain originating from one of these organs is "referred" to the corresponding dermatome (surface region) supplied by the same spinal nerve. **Referred pain** originating in the heart, for example, may appear to come from the left shoulder and arm. The mechanism responsible for referred pain is not completely understood. Inputs arising from the heart presumably share a pathway to the brain in common with inputs from the left upper extremity. The higher perception levels, being more accustomed to receiving sensory input from the left arm than from the heart, may interpret the input from the heart as having arisen from the left arm.

The spinal cord is responsible for the integration of many innate reflexes.

The spinal cord is strategically located between the brain and the afferent and efferent fibers of the PNS; this location enables the spinal cord to fulfill

Figure 5-28 Structure of a nerve. Neuronal axons (both afferent and efferent fibers) are bundled together into connective tissue–wrapped fascicles. A nerve consists of a group of fascicles enclosed by a connective tissue covering and following the same pathway. The photograph is a light micrograph of a nerve in cross section.

Labels in figure 5-28:
- Axon
- Myelin sheath
- Connective tissue around the axon
- Connective tissue around a fascicle
- **Nerve fascicle** (many axons bundled in connective tissue)
- Blood vessels
- Connective tissue around the nerve
- **Nerve**

Manfred Kage/Science Source

its two primary functions: (1) serving as a link for transmission of information between the brain and rest of the body and (2) integrating reflex activity between afferent input and efferent output without involving the brain.

Reflex Arc A **reflex** is any response that occurs automatically without conscious effort. The neural pathway involved in accomplishing reflex activity is known as a **reflex arc**, which typically includes five basic components:

1. Sensory receptor
2. Afferent pathway
3. Integrating center
4. Efferent pathway
5. Effector organ

The *sensory receptor* (*receptor* for short) responds to a stimulus, which is a detectable change in the environment of the receptor. In response to the stimulus, the receptor produces an action potential that is relayed by the *afferent pathway* to the *integrating center* (usually the CNS) for processing. The integrating center processes all information available to it from this receptor, and from all other inputs, and then "makes a decision" about the appropriate response. The instructions from the integrating center are transmitted via the *efferent pathway* to the *effector organ*—a muscle or gland—that carries out the desired response. Unlike conscious behavior, in which any one of numerous responses is possible, a reflex response is predictable, because the pathway is always the same.

Not all reflex activity involves a clear-cut reflex arc, although the basic principles of a reflex (that is, an automatic response to a detectable change) are present.

(a) Distribution of dermatomes

Pathways for unconscious responsiveness digress from the typical reflex arc in two general ways:

1. *Responses at least partly mediated by hormones.* A particular reflex may be mediated solely by either neurons or hormones or may involve a pathway using both.

2. *Local responses that do not involve either nerves or hormones.* For example, the blood vessels in an exercising muscle dilate because of local metabolic changes, thereby increasing blood flow to match the active muscle's metabolic needs.

Reflex Categories Reflexes can be categorized in the following variable ways:

1. As *spinal* or *cranial reflexes,* depending on the CNS level at which the reflex is integrated. **Spinal reflexes** are integrated by the spinal cord, an example of which is the withdrawal reflex, such as automatically withdrawing your hand from a hot object (see p. 284). To a certain extent, the brain may consciously

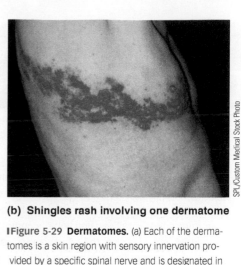

SPL/Custom Medical Stock Photo

(b) Shingles rash involving one dermatome

▮Figure 5-29 **Dermatomes.** (a) Each of the dermatomes is a skin region with sensory innervation provided by a specific spinal nerve and is designated in the figure by the name of the nerve supplying this area. (b) A photograph of shingles, a painful blistery rash caused by infection with varicella-zoster virus of a single spinal nerve and its associated dermatome.

override a spinal reflex, such as when you voluntarily prevent urination if the time is not convenient when the micturition (bladder-emptying) reflex is called forth when your bladder begins to be stretched as it fills with urine (see p. 531). This modulatory influence of the brain is above and beyond the level of the spinal reflex. In the case of a **cranial reflex**, the reflex itself is subconsciously integrated by the brain at levels lower than the cortex, such as by the brain stem or hypothalamus. An example of a cranial reflex is constriction of the pupils of your eyes in response to bright light (see p. 194).

2. As *innate* or *conditioned reflexes,* depending on whether the reflex is inborn or learned. **Innate** (or **simple** or **basic**) **reflexes** are built-in, unlearned responses. The withdrawal reflex, the micturition reflex, and the pupillary constriction reflex are all examples of innate reflexes. **Conditioned** (or **acquired**) **reflexes** are a result of learning, such as increased secretion of saliva on smelling a favorite food being prepared. Increased salivation occurs via the innate salivary reflex (a cranial reflex) on tasting a favorite food, but your mouth waters (increased salivation) via a conditioned salivary reflex once you have learned to associate the smell with the anticipation of getting to eat the tasty food (see p. 575). The spinal cord and brain stem integrate innate reflexes, whereas higher brain levels usually process acquired reflexes. (For a discussion of the role of conditioned reflexes in many sports skills, see the accompanying boxed feature, ▮A Closer Look at Exercise Physiology.)

3. As *somatic* or *autonomic,* depending on which efferent division of the peripheral nervous system and which effector organs are involved. Output in a **somatic reflex** is transmitted by motor neurons to skeletal muscles, an example being the withdrawal reflex. Output in an **autonomic** (or **visceral**) **reflex** is

Swan Dive or Belly Flop:
It's a Matter of CNS Control

SPORT SKILLS MUST BE LEARNED. Much of the time, strong innate reflexes must be overridden to perform the skill. Learning to dive into water, for example, is very difficult initially. Strong head-righting reflexes controlled by sensory organs in the neck and ears initiate a straightening of the neck and head before the beginning diver enters the water, causing what is commonly known as a "belly flop." In a backward dive, the head-righting reflex causes the beginner to land on his or her back or even in a sitting position. To perform any motor skill that involves body inversions, somersaults, back flips, or other abnormal postural movements, the person must learn to consciously inhibit basic postural reflexes. This is accomplished by having the person concentrate on specific body positions during the movement. For example, to perform a somersault, the person must concentrate on keeping the chin tucked and grabbing the knees. After the skill is performed repeatedly, new synaptic patterns are formed in the CNS, and the new or conditioned response substitutes for the natural innate reflex responses. Sport skills must be practiced until the movement becomes automatic; then the athlete is free during competition to think about strategy or the next move to be performed in a routine.

carried via the autonomic nervous system to smooth muscle, cardiac muscle, or glands. The micturition, pupillary constriction, and salivary reflexes are all autonomic reflexes. Another example of an autonomic reflex is the baroreceptor (blood-pressure regulating) reflex (see p. 367).

4. As *monosynaptic* or *polysynaptic*, depending on how many synapses are in the reflex arc.

The simplest reflex is the stretch reflex, in which an afferent neuron originating at a stretch-detecting receptor in a skeletal muscle terminates directly on the efferent neuron supplying the same skeletal muscle to cause it to contract and counteract the stretch (see p. 282). In this reflex, the integrating center is the single synapse within the spinal cord between the afferent and the efferent pathways. The output of this system (whether or not the muscle contracts in response to passive stretch) depends on the extent of summation of EPSPs at the cell body of the efferent neuron arising from the frequency of afferent input (determined by the extent of stretch detected by the receptor). Integration in this case simply involves summation of EPSPs from a single source. The stretch reflex is a **monosynaptic** ("one synapse") **reflex** because the only synapse in the reflex arc is the one between the afferent neuron and the efferent neuron. All other reflexes are **polysynaptic** ("many synapses") because interneurons are interposed in the reflex pathway and, therefore, a number of synapses are involved.

You will learn more about specific reflexes in later chapters devoted to the involved effector organs.

Check Your Understanding 5.10

1. Draw a cross-section of a spinal cord and a pair of spinal nerves, showing the location of an afferent neuron, efferent neuron, and interneuron. Label the gray matter, white matter, dorsal root, ventral root, and spinal nerve.

2. Distinguish among a *tract, ganglion, nucleus, center,* and *nerve.*

3. List the components of a reflex arc.

4. Describe the ways in which reflexes can be categorized.

Homeostasis: Chapter in Perspective

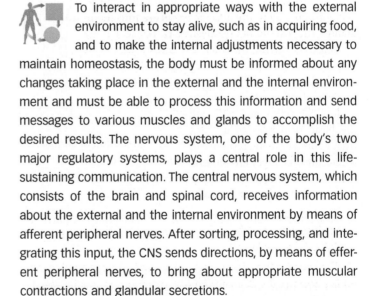 To interact in appropriate ways with the external environment to stay alive, such as in acquiring food, and to make the internal adjustments necessary to maintain homeostasis, the body must be informed about any changes taking place in the external and the internal environment and must be able to process this information and send messages to various muscles and glands to accomplish the desired results. The nervous system, one of the body's two major regulatory systems, plays a central role in this life-sustaining communication. The central nervous system, which consists of the brain and spinal cord, receives information about the external and the internal environment by means of afferent peripheral nerves. After sorting, processing, and integrating this input, the CNS sends directions, by means of efferent peripheral nerves, to bring about appropriate muscular contractions and glandular secretions.

With its swift electrical signaling system, the nervous system is especially important in controlling the rapid responses of the body. Many neurally controlled muscular and glandular activities are aimed toward maintaining homeostasis. The CNS is the main site of integration between afferent input and efferent output. It links the appropriate response to a particular input so that conditions compatible with life are maintained in the body. For example, when informed by the afferent nervous system that blood pressure has fallen, the CNS sends appropriate commands via the efferent nervous system to the heart and blood vessels to increase blood pressure to normal. Likewise, when informed that the body is overheated, the CNS promotes secretion of sweat, among other cooling responses. Evaporation of sweat helps cool the body to normal temperature. Were it not for this processing and integrating ability of the CNS, maintaining homeostasis in an organism as complex as a human would be impossible.

At the simplest level, the spinal cord integrates many basic protective and evacuative reflexes that do not require conscious participation, such as withdrawing from a painful stimulus and emptying of the urinary bladder. In addition to serving as a more complex integrating link between afferent input and efferent output, the brain is responsible for the initiation of all voluntary movement; complex perceptual awareness of the external environment; self-awareness; language; and abstract neural phenomena such as thinking, learning, remembering, consciousness, emotions, creativity, and personality traits. All neural activity—from the most private thoughts to commands for motor activity, from enjoying a concert to retrieving memories from the distant past—is ultimately attributable to propagation of action potentials along individual nerve cells and chemical transmission between cells.

During evolutionary development, the nervous system has become progressively more complex. Newer, more complicated, and more sophisticated layers of the brain have been piled on top of older, more primitive regions. Mechanisms for governing many basic activities necessary for survival are built into the older parts of the brain. The newer, higher levels progressively modify, enhance, or nullify actions coordinated by lower levels in a hierarchy of command; they also add new capabilities. Many of these higher neural activities are not aimed at maintaining life, but they add immeasurably to the quality of being alive.

Review Exercises Answers begin on p. A-26

Reviewing Terms and Facts

1. The major function of the CSF is to nourish the brain. (*True or false?*)

2. In emergencies when O_2 supplies are low, the brain can perform anaerobic metabolism. (*True or false?*)

3. Stellate cells initially process sensory input to the cortex, whereas pyramidal cells send fibers from the cortex to terminate on efferent motor neurons. (*True or false?*)

4. Damage to the left cerebral hemisphere brings about paralysis and loss of sensation on the left side of the body. (*True or false?*)

5. The hands and structures associated with the mouth have a disproportionately large share of representation in both the sensory and the motor cortexes. (*True or false?*)

6. The left cerebral hemisphere specializes in artistic and musical ability, whereas the right side excels in verbal and analytical skills. (*True or false?*)

7. The specific function a particular cortical region will carry out is permanently determined during embryonic development. (*True or false?*)

8. The amygdala is the brain area where long-term potentiation and memory consolidation take place. (*True or false?*)

9. _____ is a decreased responsiveness to an indifferent stimulus that is repeatedly presented.

10. The process of transferring and fixing short-term memory traces into long-term memory stores is known as _____.

11. Afferent fibers enter through the _____ root of the spinal cord, and efferent fibers leave through the _____ root.

12. Using the answer code on the right, indicate which neurons are being described (a characteristic may apply to more than one class of neurons):

1. have receptor at peripheral endings
2. lie entirely within the CNS
3. lie primarily within the peripheral nervous system
4. innervate muscles and glands
5. cell body is devoid of presynaptic inputs
6. predominant type of neuron
7. responsible for thoughts, emotions, memory, etc.

(a) afferent neurons
(b) efferent neurons
(c) interneurons

13. Match the following:

1. consists of nerves carrying information between the periphery and the CNS
2. consists of the brain and spinal cord
3. division of the peripheral nervous system that transmits signals to the CNS
4. division of the peripheral nervous system that transmits signals from the CNS
5. supplies skeletal muscles
6. supplies smooth muscle, cardiac muscle, and glands

(a) somatic nervous system
(b) autonomic nervous system
(c) central nervous system
(d) peripheral nervous system
(e) efferent division
(f) afferent division

Understanding Concepts

(Answers at www.cengagebrain.com)

1. Discuss the function of each of the following: astrocytes, oligodendrocytes, ependymal cells, microglia, cranium, vertebral column, meninges, cerebrospinal fluid, and blood–brain barrier.

2. Compare the composition of white and gray matter.

3. Draw and label the major functional areas of the cerebral cortex, indicating the functions attributable to each area.

4. Define *cognition*.

5. Describe the circumstances in which the default mode network is active.

6. Discuss the function of each of the following parts of the brain: thalamus, hypothalamus, basal nuclei, limbic system, cerebellum, and brain stem.

7. Define *somesthetic sensations* and *proprioception*.

8. Discuss the roles of Broca's area and Wernicke's area in language.

9. Compare short-term and long-term memory.

10. Discuss the difference between AMPA and NMDA glutamate receptors and their roles in long-term potentiation.

11. What is the reticular activating system?

12. Describe the appearance and consciousness level associated with each waveform on an electroencephalogram.

13. Compare slow-wave and paradoxical (REM) sleep.

14. Discuss what types of neuronal cell bodies are located in the dorsal, ventral, and lateral horns of the spinal cord.

15. Distinguish between a monosynaptic and a polysynaptic reflex.

Applying Clinical Reasoning

Julio D., who had recently retired, was enjoying an afternoon of playing golf when suddenly he experienced a severe headache and dizziness. These symptoms were quickly followed by numbness and partial paralysis on the upper right side of his body, accompanied by an inability to speak. After being rushed to the emergency room, Julio was diagnosed as having suffered a stroke. Given the observed neurological impairment, what areas of his brain were affected?

Thinking at a Higher Level

1. Special studies designed to assess the specialized capacities of each cerebral hemisphere have been performed on "split-brain" patients. In these people, the corpus callosum—the bundle of fibers that links the two halves of the brain—has been surgically cut to prevent the spread of epileptic seizures from one hemisphere to the other. Even though no overt changes in behavior, intellect, or personality occur in these patients, because both hemispheres individually receive the same information, deficits are observable with tests designed to restrict information to one brain hemisphere at a time. One such test involves limiting a visual stimulus to only half of the brain. Because of a crossover in the nerve pathways from the eyes to the occipital cortex, the visual information to the right of a midline point is transmitted to only the left half of the brain, whereas visual information to the left of this point is received by only the right half of the brain. A split-brain patient presented with a visual stimulus that reaches only the left hemisphere accurately describes the object seen, but when a visual stimulus is presented to only the right hemisphere, the patient denies having seen anything. The right hemisphere does receive the visual input, however, as demonstrated by nonverbal tests. Even though a split-brain patient denies having seen anything after an object is presented to the right hemisphere, the patient can correctly match the object by picking it out from among several objects, usually to the patient's surprise. What is your explanation of this finding?

2. Which of the following symptoms are most likely to occur as the result of a severe blow to the back of the head?
   a. paralysis
   b. hearing impairment
   c. visual disturbances
   d. burning sensations
   e. personality disorders

3. The hormone insulin enhances the carrier-mediated transport of glucose into most of the body's cells but not into brain cells. The uptake of glucose from the blood by neurons does not depend on insulin. Knowing the brain's need for a continuous supply of blood-borne glucose, predict the effect that insulin excess would have on the brain.

4. Give examples of conditioned reflexes you have acquired.

5. Under what circumstances might it be inadvisable to administer a clot-dissolving drug to a stroke victim?

To access the course materials and companion resources for this text, please visit www.cengagebrain.com

The Peripheral Nervous System: Afferent Division; Special Senses

6

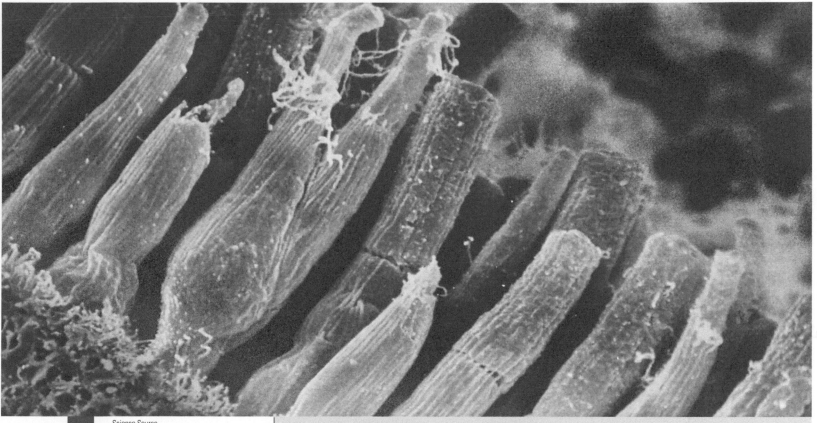

A scanning electron micrograph of rods and cones. Rods and cones are the photoreceptors (light detectors) in the eye. Their outer segments, which are rod shaped in rods (*blue*) and cone shaped in cones (*green*), contain photopigments that absorb light in the initial step of vision.

CHAPTER AT A GLANCE

Homeostasis Highlights

The nervous system, one of the two major regulatory systems of the body, consists of the central nervous system (CNS), composed of the brain and spinal cord, and the **peripheral nervous system (PNS)**, composed of the afferent and efferent fibers that relay signals between the CNS and the periphery (other parts of the body).

The **afferent division** of the PNS detects, encodes, and transmits peripheral signals to the CNS, thus informing the CNS about the internal and the external environment. This afferent input to the controlling centers of the CNS is essential in maintaining homeostasis. To make appropriate adjustments in effector organs via efferent output, the CNS has to "know" what is going on. Afferent input is also used to plan for voluntary actions unrelated to homeostasis.

6.1 Receptor Physiology

The peripheral nervous system consists of nerve fibers that carry information between the CNS and other parts of the body. The afferent division of the PNS sends information about the external and the internal environment to the CNS.

A **stimulus** is a change detectable by the body. Stimuli exist in various energy forms, or **modalities**, such as heat, light, sound, pressure, and chemical changes. Afferent neurons have **sensory receptors** (*receptors* for short) at their peripheral endings that respond to stimuli in both the external world and the internal environment. (Although both are called *receptors*, stimulus sensitive sensory receptors are distinctly different from the plasma-membrane protein receptors that bind with extracellular chemical messengers; see p. 59.) Because the only way afferent neurons can transmit information to the CNS about stimuli is via action potential propagation, receptors must convert these other forms of energy into electrical signals. Stimuli bring about graded potentials known as **receptor potentials** in the receptor. The conversion of stimulus energy into a receptor potential is known as **sensory transduction**. Receptor potentials in turn trigger action potentials in the afferent fiber.

Receptors have differential sensitivities to various stimuli.

Each type of receptor is specialized to respond to one type of stimulus, its **adequate stimulus**. For example, receptors in the eye are sensitive to light, receptors in the ear to sound waves, and heat receptors in the skin to heat energy. Because of this differential sensitivity of receptors, we cannot see with our ears or hear with our eyes. Some receptors can respond weakly to stimuli other than their adequate stimulus, but even when activated by a different stimulus, a receptor still gives rise to the sensation usually detected by that receptor type. As an example, the adequate stimulus for eye receptors (photoreceptors) is light, to which they are exquisitely sensitive, but these receptors can also be activated to a lesser degree by mechanical stimulation. When hit in the eye, a person often "sees stars" because the mechanical pressure stimulates the photoreceptors.

Types of Receptors According to Their Adequate Stimulus Depending on the type of energy to which they ordinarily respond, receptors are categorized as follows:

- **Photoreceptors** are responsive to visible light.
- **Mechanoreceptors** are sensitive to mechanical energy. Examples include skeletal muscle receptors sensitive to stretch, the receptors in the ear containing fine hairs that are bent as a result of sound waves, and blood pressure–monitoring baroreceptors.
- **Thermoreceptors** are sensitive to heat and cold.
- **Osmoreceptors** detect changes in the concentration of solutes in the extracellular fluid (ECF) and the resultant changes in osmotic activity (see p. 67).
- **Chemoreceptors** are sensitive to specific chemicals. Chemoreceptors include the receptors for taste and smell and those located deeper within the body that detect O_2 and CO_2 concentrations in the blood or the chemical content of the digestive tract.
- **Nociceptors**, or **pain receptors**, are sensitive to tissue damage such as cutting or burning. Intense stimulation of any receptor is also perceived as painful.

Although not yet fully accepted as a receptor category, **itch-specific receptors** were recently discovered in the skin. This finding is contrary to the long-held belief that itch is a mild manifestation of pain elicited by stimulation of nociceptors on exposure to itch-inducing substances, for example histamine released in response to a mosquito bite. The urge to scratch can also be a symptom of some systemic conditions as a result of central processing unrelated to input from skin itch receptors, as may occur in kidney or liver failure for example.

Some sensations are compound sensations in that their perception arises from central integration of several simultaneously activated primary sensory inputs. For instance, the perception of wetness comes from touch, pressure, and thermal receptor input; there is no such thing as a "wetness receptor."

Uses for Information Detected by Receptors The information detected by receptors is conveyed via afferent neurons to the CNS, where it is used for various purposes:

- Afferent input is essential for control of efferent output, both for regulating motor behavior in accordance with external circumstances and for coordinating internal activities directed at maintaining homeostasis. At the most basic level, afferent input provides information (of which the person may or may not be consciously aware) for the CNS to use in directing activities necessary for survival. On a broader level, we could not interact successfully with our environment or with one another without sensory input.
- Processing of sensory input by the reticular activating system in the brain stem is critical for cortical arousal and consciousness (see p. 167).
- Central processing of sensory information gives rise to our perceptions of the world around us.
- Selected information delivered to the CNS may be stored for future reference.
- Sensory stimuli can have a profound effect on our emotions. The smell of just-baked apple pie, the sensuous feel of silk, the sight of a loved one, the sound of someone sharing bad news—sensory input can gladden, sadden, arouse, calm, anger, frighten, or evoke a range of other emotions.

We next examine how adequate stimuli initiate action potentials that ultimately are used for these purposes.

A stimulus alters the receptor's permeability, leading to a graded receptor potential.

A receptor may be either (1) a specialized ending of the afferent neuron or (2) a separate receptor cell closely associated with the peripheral ending of the neuron. Stimulation of a receptor alters its membrane permeability, usually by opening channels that permit an inward flux of Na^+, which depolarizes the receptor membrane (see p. 88). (There are exceptions; for example, photoreceptors are hyperpolarized on stimulation.) This local depolarization, the receptor potential, is a graded potential. As is true of all graded potentials, the stronger the stimulus, the greater the permeability change and the larger the receptor potential (see p. 89). Also, receptor potentials have no refractory period, so summation in response to rapidly successive stimuli is possible. Because the receptor region has few to no voltage-gated Na^+ channels and thus has a high threshold, action potentials do not take place at the receptor itself. (The channels in the receptor region that open in response to a stimulus are not voltage-gated Na^+ channels and vary depending on the receptor type.) For long-distance transmission, the receptor potential must be converted into action potentials that can be propagated along the afferent fiber.

Receptor potentials may initiate action potentials in the afferent neuron.

If a receptor potential is large enough, it may trigger an action potential in the afferent neuron membrane next to the receptor by promoting the opening of voltage-gated Na^+ channels in this

adjacent region. In myelinated afferent fibers, this trigger zone is the node of Ranvier closest to the receptor. The means by which the Na^+ channels are opened differ depending on whether the receptor is a specialized afferent ending or a separate cell.

- In the case of a specialized afferent ending, local current flow between the activated receptor ending undergoing a receptor potential and the cell membrane next to the receptor depolarizes this adjacent region (▌Figure 6-1a). If the region is depolarized to threshold, voltage-gated Na^+ channels open here, triggering an action potential that is conducted along the afferent fiber to the CNS.

- In the case of a separate receptor cell, the receptor cell synapses with the ending of the afferent neuron (▌Figure 6-1b). A receptor potential promotes the opening of voltage-gated Ca^{2+} channels in the receptor cell. The resultant Ca^{2+} entry causes the release by exocytosis of a neurotransmitter that diffuses across the synaptic cleft and binds with specific protein receptors on the afferent neuron membrane. This binding opens chemically gated Na^+ receptor-channels (see p. 116). If the re-

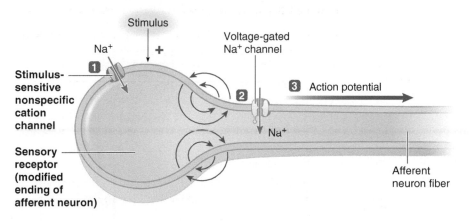

1 In sensory receptors that are specialized afferent neuron endings, stimulus opens stimulus-sensitive channels, permitting net Na^+ entry that produces receptor potential.

2 Local current flow between depolarized receptor ending and adjacent region opens voltage-gated Na^+ channels.

3 Na^+ entry initiates action potential in afferent fiber that self-propagates to CNS.

(a) Receptor potential in specialized afferent ending

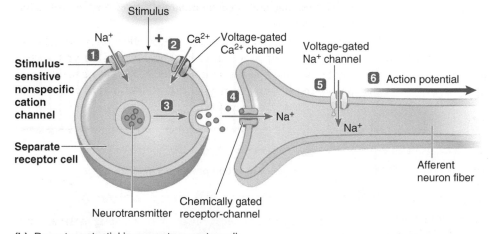

1 In sensory receptors that are separate cells, stimulus opens stimulus-sensitive channels, permitting net Na^+ entry that produces receptor potential.

2 This local depolarization opens voltage-gated Ca^{2+} channels.

3 Ca^{2+} entry triggers exocytosis of neurotransmitter.

4 Neurotransmitter binding opens chemically gated receptor-channels at afferent ending, permitting net Na^+ entry.

5 Resultant depolarization opens voltage-gated Na^+ channels in adjacent region.

6 Na^+ entry initiates action potential in afferent fiber that self-propagates to CNS.

(b) Receptor potential in separate receptor cell

▌Figure 6-1 **Conversion of receptor potential into action potentials.** (a) Specialized afferent ending as sensory receptor. Local current flow between a depolarized receptor ending undergoing a receptor potential and the adjacent region initiates an action potential in the afferent fiber by opening voltage-gated Na^+ channels. (b) Separate receptor cell as sensory receptor. The depolarized receptor cell undergoing a receptor potential releases a neurotransmitter that binds with chemically gated channels in the afferent fiber ending. This binding leads to a depolarization that opens voltage-gated Na^+ channels, initiating an action potential in the afferent fiber.

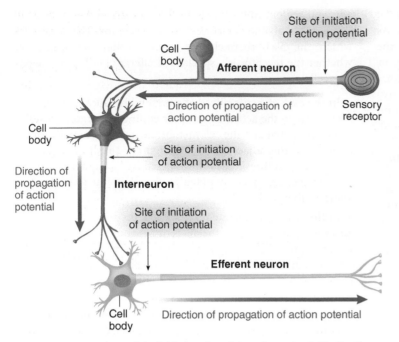

Cell body

Afferent neuron

Site of initiation of action potential

Cell body

Sensory receptor

Direction of propagation of action potential

Cell body

Site of initiation of action potential

Direction of propagation of action potential

Interneuron

Site of initiation of action potential

Efferent neuron

Cell body

Direction of propagation of action potential

ǀFigure 6-2 **Comparison of the initiation site of an action potential in the three types of neurons.**

many pressure receptors in the skin as does a more forceful touch applied to the same area. Stimulus intensity is therefore distinguished both by the frequency of action potentials generated in the afferent neuron and by the number of receptors and thus afferent fibers activated within the area.

Receptors may adapt slowly or rapidly to sustained stimulation.

Stimuli of the same intensity do not always result in receptor potentials of the same magnitude in the same receptor. Some receptors diminish the extent of their depolarization despite sustained stimulus strength, a phenomenon called **adaptation**. Subsequently, the frequency of action potentials generated in the afferent neuron decreases—that is, the receptor "adapts" to the stimulus by no longer responding to it to the same degree.

Types of Receptors According to Their Speed of Adaptation Depending on their speed of adaptation,

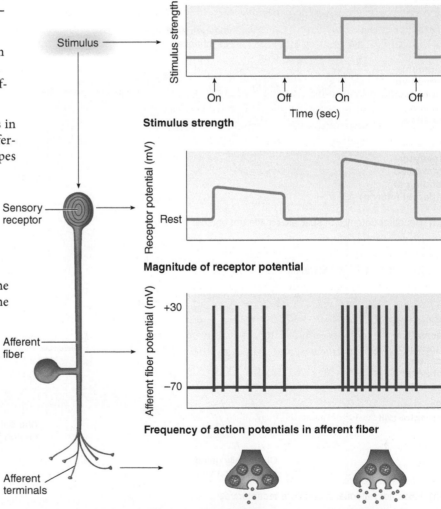

Stimulus

Stimulus strength

On   Off   On   Off

Time (sec)

Stimulus strength

Sensory receptor

Receptor potential (mV)

Rest

Magnitude of receptor potential

Afferent fiber

Afferent fiber potential (mV)

+30

−70

Frequency of action potentials in afferent fiber

Afferent terminals

Rate of neurotransmitter release at afferent terminals

ǀFigure 6-3 **Coding of stimulus strength by an afferent neuron.**
FIGURE FOCUS: *How would an acid-monitoring neuron's activity change compared to normal if body fluids became too acidic? If they became too alkaline?*

sultant Na$^+$ entry depolarizes the afferent neuron ending to threshold, voltage-gated Na$^+$ channels open here, triggering an action potential that self-propagates to the CNS.

Note that the initiation site of action potentials in an afferent neuron differs from the site in an efferent neuron or interneuron. In the latter two types of neurons, action potentials are initiated at the axon hillock located at the start of the axon next to the cell body (see p. 96). By contrast, action potentials are initiated at the peripheral end of an afferent nerve fiber next to the receptor, a long distance from the cell body (ǀFigure 6-2).

The intensity of the stimulus is reflected by the magnitude of the receptor potential. The larger the receptor potential, the greater the frequency of action potentials generated in the afferent neuron (ǀFigure 6-3). A larger receptor potential cannot bring about a larger action potential (because of the all-or-none law), but it can induce more rapid firing of action potentials (see p. 99). The more rapidly an afferent fiber fires, the more neurotransmitter it releases. This neurotransmitter influences the next cell in the neural pathway, passing on information about stimulus strength. Stimulus strength is also reflected by the size of the area stimulated. Stronger stimuli usually affect larger areas, so correspondingly more receptors respond. For example, a light touch does not activate as

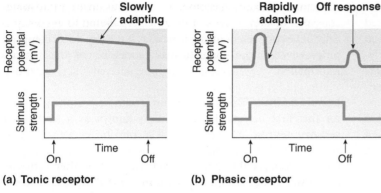

(a) Tonic receptor

(b) Phasic receptor

I Figure 6-4 Tonic and phasic receptors. (a) A tonic receptor does not adapt at all or adapts slowly to a sustained stimulus and thus provides continuous information about the stimulus. (b) A phasic receptor adapts rapidly to a sustained stimulus and frequently exhibits an off response when the stimulus is removed. Thus, the receptor signals changes in stimulus intensity rather than relaying status quo information.

receptors are classified as *tonic receptors* or *phasic receptors*. **Tonic receptors** do not adapt or adapt slowly (I Figure 6-4a). These receptors are useful when it is valuable to maintain information about a stimulus. Examples of tonic receptors are muscle stretch receptors, which monitor muscle length, and joint proprioceptors, which measure the degree of joint flexion. To maintain posture and balance, the CNS must continually get information about the degree of muscle length and joint position. It is important, therefore, that these receptors do *not* adapt to a stimulus but continue to generate action potentials to relay this information to the CNS.

Phasic receptors are rapidly adapting receptors. The receptor quickly adapts by no longer responding to a maintained stimulus. Some phasic receptors, most notably the *Pacinian corpuscle,* respond with a slight depolarization called the **off response** when the stimulus is removed (I Figure 6-4b). Phasic receptors are useful when it is important to signal a change in stimulus intensity rather than to relay status quo information. Many *tactile (touch) receptors* that signal changes in pressure on the skin surface are phasic receptors. Because these receptors adapt rapidly, you are not continually conscious of wearing your watch, rings, and clothing. When you put something on, you soon become accustomed to it because of these receptors' rapid adaptation. When you take the item off, you are aware of its removal because of the off response.

Tactile Receptors Tactile (touch) receptors in the skin are mechanoreceptors. The mechanical forces of a stimulus distort non-specific cation channel proteins in the plasma membrane of these receptors, leading to net Na^+ entry, which causes a receptor potential that triggers an action potential in the afferent fiber. Sensory input from these receptors

informs the CNS of the body's contact with objects in the external environment. Tactile receptors include the following (I Figure 6-5):

- A **hair receptor** is rapidly adapting and senses hair movement and very gentle touch, such as stroking the hair on your arm with a wisp of cotton.
- A **Merkel's disc** is slowly adapting and detects light, sustained touch and texture, such as reading Braille.
- A **Pacinian corpuscle** is rapidly adapting and responds to vibrations and deep pressure.
- **Ruffini endings** are slowly adapting and respond to deep, sustained pressure and stretch of the skin, such as during a massage.
- A **Meissner's corpuscle** is rapidly adapting and sensitive to light, fluttering touch, such as tickling with a feather.

Mechanism of Adaptation in the Pacinian Corpuscle The mechanism by which adaptation is accomplished varies for different receptors and is not fully understood for all receptor types. Many receptors adapt as a result of inactivation of channels that opened in response to the stimulus. Adaptation in the well-studied Pacinian corpuscle depends on the physical properties of this receptor. A Pacinian corpuscle is

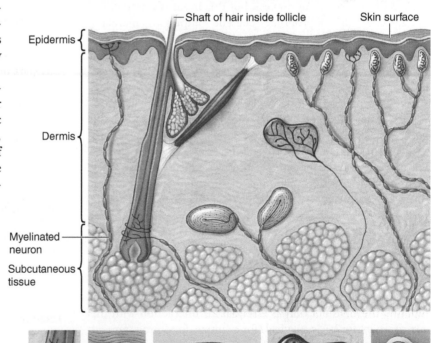

Hair receptor: hair movement and very gentle touch

Merkel's disc: light, sustained touch

Pacinian corpuscle: vibrations and deep pressure

Ruffini endings: deep pressure

Meissner's corpuscle: light, fluttering touch

I Figure 6-5 Tactile receptors in the skin.

a specialized receptor ending that consists of concentric layers of connective tissue resembling layers of an onion wrapped around the peripheral terminal of an afferent neuron. When pressure is first applied to the Pacinian corpuscle, the underlying terminal responds with a receptor potential of a magnitude that reflects the intensity of the stimulus. As the stimulus continues, the pressure energy is dissipated because it causes the receptor layers to slip (just as steady pressure on a peeled onion causes its layers to slip). Because this physical effect filters out the steady component of the applied pressure, the underlying neuronal ending no longer responds with a receptor potential—that is, adaptation has occurred.

Adaptation should not be confused with habituation (see p. 159). Although both these phenomena involve decreased neural responsiveness to repetitive stimuli, they operate at different points in the neural pathway. Adaptation is a receptor adjustment in the PNS, whereas habituation involves a modification in synaptic effectiveness in the CNS.

Visceral afferents carry subconscious input; sensory afferents carry conscious input.

Action potentials generated by receptors in afferent fibers in response to stimuli are propagated to the CNS. Afferent information about the internal environment, such as blood pressure and the concentration of CO_2 in the body fluids, never reaches the level of conscious awareness, but this input is essential for determining the appropriate efferent output to maintain homeostasis. The incoming pathway for information derived from the internal viscera (organs in the body cavities, such as the abdominal cavity) is called a **visceral afferent**. Even though mostly subconscious information is transmitted via visceral afferents, people do become aware of pain signals arising from viscera. Afferent input derived from receptors located at the body surface or in the muscles or joints typically reaches the level of conscious awareness. This input is known as *sensory information,* and the incoming pathway is considered a **sensory afferent**. Sensory information is categorized as (1) **somatic** (body sense) **sensation** arising from the body surface, including *somesthetic sensation* from the skin and *proprioception* from the muscles, joints, skin, and inner ear (see p. 147); or (2) **special senses**, including *vision, hearing, equilibrium, taste,* and *smell*. (See the accompanying boxed feature, A Closer Look at Exercise Physiology, for a description of the usefulness of proprioception in athletic performance.) Final processing of sensory input by the CNS not only is essential for interaction with the environment for basic survival (for example, food procurement and defense from danger), but also adds immeasurably to the richness of life.

Each somatosensory pathway is "labeled" according to modality and location.

On reaching the spinal cord, afferent information has two possible destinies: (1) it may become part of a reflex arc, bringing about an appropriate effector response, or (2) it may be relayed upward to the brain via ascending pathways for further processing and possible conscious awareness. Pathways conveying conscious somatic sensation, the **somatosensory pathways**, consist of discrete chains of neurons, or **labeled lines**, synaptically interconnected in a particular sequence to accomplish progressively more sophisticated processing of the sensory information.

Labeled Lines The afferent neuron with its peripheral receptor that first detects the stimulus is known as a **first-order sensory neuron**. It synapses on a **second-order sensory neuron**, either in the spinal cord or the medulla, depending on which sensory pathway is involved. This neuron then synapses on a **third-order sensory neuron** in the thalamus, and so on. With each step, the input is processed further. A particular sensory modality detected by a specialized receptor type is sent over a specific afferent and ascending pathway (a neural pathway committed to that modality) to excite a defined area in the somatosensory cortex—that is, a particular sensory input is **projected** to a specific region of the cortex (see Figure 5-26a, p. 175, for an example). Thus, different types of incoming information are kept separated within specific labeled lines between the periphery and the cortex. In this way, even though all information is propagated to the CNS via the same type of signal (action potentials), the brain can decode the type and location of the stimulus. Table 6-1 summarizes how the CNS is informed of the type (what), location (where), and intensity (how much) of a stimulus.

Clinical Note **Phantom Pain** Activation of a sensory pathway at any point gives rise to the same sensation that would be produced by stimulation of the receptors in the body part itself. This phenomenon is the traditional explanation for **phantom pain**—for example, pain perceived as originating in the foot by a person whose leg has been amputated at the knee. Irritation of the severed endings of the afferent

TABLE 6-1 Coding of Sensory Information

Stimulus Property	Mechanism of Coding
Type of stimulus (stimulus modality)	Distinguished by the type of receptor activated and the specific pathway over which this information is transmitted to a particular area of the cerebral cortex
Location of stimulus	Distinguished by the location of the activated receptive field and the pathway that is subsequently activated to transmit this information to the area of the somatosensory cortex representing that particular location
Intensity of stimulus (stimulus strength)	Distinguished by the frequency of action potentials initiated in an activated afferent neuron and the number of receptors (and afferent neurons) activated

Back Swings and Prejump Crouches: What Do They Share in Common?

PROPRIOCEPTION, THE SENSE OF THE body's position in space, is critical to any movement and is especially important in athletic performance, whether it be a figure skater performing triple jumps on ice, a gymnast performing a difficult floor routine, or a football quarterback throwing perfectly to a spot 60 yards downfield. To control skeletal muscle contraction to achieve the desired movement, the CNS must be continuously apprised of the results of its action through sensory feedback.

A number of receptors provide proprioceptive input. Muscle proprioceptors provide feedback information on muscle tension and length. Joint proprioceptors provide feedback on joint acceleration, angle, and direction of movement. Skin proprioceptors inform the CNS of weight-bearing pressure on the skin.

Proprioceptors in the inner ear, along with those in neck muscles, provide information about head and neck position so that the CNS can orient the head correctly. For example, neck reflexes facilitate essential trunk and limb movements during somersaults, and divers and tumblers use strong movements of the head to maintain spins.

The most complex and probably one of the most important proprioceptors is the muscle spindle (see p. 281). Muscle spindles are found throughout a muscle but tend to be concentrated in its center. Each spindle lies parallel to the muscle fibers within the muscle. The spindle is sensitive to both the muscle's rate of change in length and the final length achieved. If a muscle is stretched, each muscle spindle

within the muscle is also stretched, and the afferent neuron whose peripheral axon terminates on the muscle spindle is stimulated. The afferent fiber passes into the spinal cord and synapses directly on the motor neurons that supply the same muscle. Stimulation of the stretched muscle as a result of this stretch reflex causes the muscle to contract sufficiently to relieve the stretch.

Older persons or those with weak quadriceps (thigh) muscles unknowingly take advantage of the muscle spindle by pushing on the center of the thighs when they get up from a sitting position. Contraction of the quadriceps muscle extends the knee joint, thus straightening the leg. The act of pushing on the center of the thighs when getting up slightly stretches the quadriceps muscle in both limbs, stimulating the muscle spindles. The resultant stretch reflex aids in contraction of the quadriceps muscles and helps the person assume a standing position.

In sports, people use the muscle spindle to advantage all the time. To jump high, as in basketball jump balls, an athlete starts by crouching down. This action stretches the quadriceps muscles and increases the firing rate of their spindles, thus triggering the stretch reflex that reinforces the quadriceps muscles' contractile response so that these extensor muscles of the legs gain additional power. The same is true for crouch starts in running events. The backswing in tennis, golf, and baseball similarly provides increased muscular excitation through reflex activity initiated by stretched muscle spindles.

pathways in the stump can trigger action potentials that, on reaching the foot region of the somatosensory cortex, are interpreted as pain in the missing foot. New evidence suggests that in addition, the sensation of phantom pain may arise from extensive remodeling of the brain region that originally handled sensations from the severed limb. This "remapping" of the "vacated" area of the brain is speculated to somehow lead to signals from elsewhere being misinterpreted as pain arising from the missing extremity.

Acuity is influenced by receptive field size and lateral inhibition.

Each somesthetic sensory neuron responds to stimulus information only within a circumscribed region of the skin surface surrounding it; this region is called its **receptive field**. The size of a receptive field varies inversely with the density of receptors in the region; the more closely receptors of a particular type are spaced, the smaller the area of skin each monitors. The smaller the receptive field is in a region, the greater its **acuity** or **discriminative ability**. Compare the tactile discrimination in your fingertips with that in your calf by "feeling" the same object with both. You can sense more precise information about the

object with your richly innervated fingertips because the receptive fields there are small; as a result, each neuron signals information about small, discrete portions of the object's surface. An estimated 17,000 tactile mechanoreceptors are present in the fingertips and palm of each hand. In contrast, the skin over the calf is served by relatively few sensory endings with larger receptive fields. Subtle differences within each large receptive field cannot be detected (IFigure 6-6). The distorted cortical representation of various body parts in the sensory homunculus (see p. 149) corresponds precisely with the innervation density; more cortical space is allotted for sensory reception from areas with smaller receptive fields and, accordingly, greater tactile discriminative ability.

Besides receptor density, a second factor influencing acuity is *lateral inhibition*. You can appreciate the importance of this phenomenon by slightly indenting the surface of your skin with the point of a pencil (IFigure 6-7a). The receptive field is excited immediately under the center of the pencil point where the stimulus is most intense, but the surrounding receptive fields are also stimulated, only to a lesser extent because they are less distorted. If information from these marginally excited afferent fibers in the fringe of the stimulus area were to reach the cortex, localization of the pencil point would

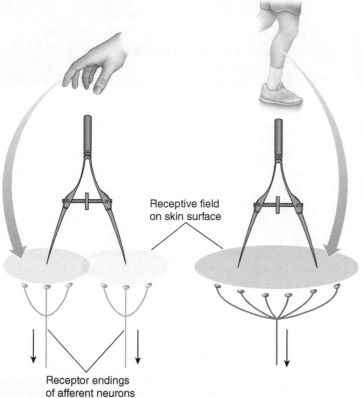

Receptive field
on skin surface

Receptor endings
of afferent neurons

Two receptive fields stimulated by
the two points of stimulation:
Two points felt

Only one receptive field stimulated
by the two points of stimulation
the same distance apart as in (a):
One point felt

(a) Region with small receptive fields

(b) Region with large receptive field

❚Figure 6-6 **Comparison of discriminative ability of regions with small versus large receptive fields.** The relative tactile acuity of a given region can be determined by the *two-point threshold-of-discrimination test.* If the two points of a pair of calipers applied to the surface of the skin stimulate two different receptive fields, two separate points are felt. If the two points touch the same receptive field, they are perceived as only one point. By adjusting the distance between the caliper points, one can determine the minimal distance at which the two points can be recognized as two rather than one, which reflects the size of the receptive fields in the region. With this technique, it is possible to plot the discriminative ability of the body surface. The two-point threshold ranges from 2 mm in the fingertip (enabling a person to read Braille, where the raised dots are spaced 2.5 mm apart) to 48 mm in the poorly discriminative skin of the calf of the leg.

be blurred. To facilitate localization and sharpen contrast, **lateral inhibition** occurs within the CNS (❚Figure 6-7b, p. 189). With lateral inhibition, each activated signal pathway inhibits the pathways next to it by stimulating inhibitory interneurons that pass laterally between ascending fibers serving neighboring receptive fields. The most strongly activated pathway originating from the center of the stimulus area inhibits the less excited pathways from the fringe areas to a greater extent than the weakly activated pathways in the fringe areas inhibit the more excited central pathway. Blockage of further transmission in the weaker inputs increases the contrast between wanted and unwanted information so that the pencil point can be precisely localized. The extent of lateral

inhibitory connections within sensory pathways varies for different modalities. Those with the most lateral inhibition— touch and vision—bring about the most accurate localization.

Perception is the conscious awareness of surroundings derived from interpretation of sensory input.

Perception is our conscious interpretation of the external world as created by the brain from a pattern of nerve impulses delivered to it from receptors. Is the world, as we perceive it, reality? The answer is a resounding no. Our perception is different from what is really "out there" for several reasons. First, humans have receptors that detect only a limited number of existing energy forms. We perceive sounds, colors, shapes, textures, smells, tastes, and temperature but are not informed of magnetic forces, polarized light waves, radio waves, or X-rays because we do not have receptors to respond to the latter energy forms. What is not detected by receptors, the brain will never know. Our response range is limited even for the energy forms for which we do have receptors. For example, dogs can hear a whistle whose pitch is above our level of detection. Second, the information channels to our brains are not high-fidelity recorders. During precortical processing of sensory input, some features of stimuli are accentuated and others are suppressed or ignored, as through lateral inhibition. Third, the cerebral cortex further manipulates the data, comparing the sensory input with other incoming information and with memories of past experiences to extract the significant features—for example, sifting out a friend's words from the hubbub of sound in a school cafeteria. In the process, the cortex often fills in or distorts the information to abstract a logical perception—that is, it "completes the picture." Much of our daily perceptual experience is made by taking what we know and using that information to fill in the blanks to imagine what we do not know. As a simple example, you "see" a white square in ❚Figure 6-8 even though there is not a white square but right-angle wedges taken out of four red circles. Optical illusions illustrate how the brain interprets reality according to its own rules. Thus, our perceptions do not replicate reality. Other species, equipped with different types of receptors, sensitivities, and neural processing, perceive a markedly different world from what we perceive.

Having completed our general discussion of receptor physiology, we now examine one important somatic sensation in greater detail—pain.

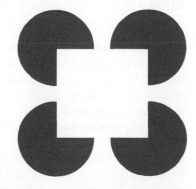

❚Figure 6-8 **Do you "see" a white square that is not really there?**

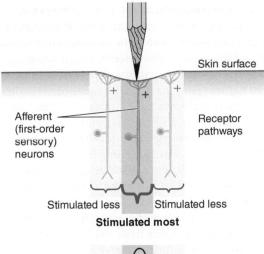

Skin surface

Afferent (first-order sensory) neurons

Receptor pathways

Stimulated less — Stimulated less
Stimulated most

Frequency of action potentials

Baseline level of activity

Location on skin

(a) Activity in afferent neurons

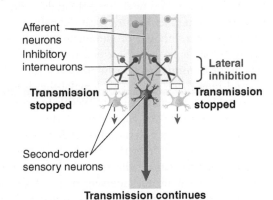

Afferent neurons

Inhibitory interneurons

} Lateral inhibition

Transmission stopped

Transmission stopped

Second-order sensory neurons

Transmission continues

Frequency of action potentials

Baseline level of activity

Area of sensation on skin

(b) Lateral inhibition

I Figure 6-7 Lateral inhibition. (a) The receptor at the site of most intense stimulation is activated to the greatest extent. Surrounding receptors are also stimulated but to a lesser degree. (b) The most intensely activated receptor pathway halts transmission of impulses in the less intensely stimulated pathways through lateral inhibition. This process facilitates localization of the site of stimulation.

Check Your Understanding 6.1

1. Define *stimulus, receptor potential, labeled line,* and *perception.*
2. Draw the response of a tonic receptor and of a phasic receptor to a stimulus of sustained strength.
3. Compare the receptive field size for a sensory neuron on your tongue and a sensory neuron on your back.

6.2 Pain

Pain is primarily a protective mechanism triggered on stimulation of danger-sensing nociceptors (pain receptors) that brings to conscious awareness tissue damage that is occurring or about to occur. Because of their value to survival, nociceptors do not adapt to sustained or repetitive stimulation. Storage of painful experiences in memory helps us avoid potentially harmful events in the future.

Stimulation of nociceptors elicits the perception of pain plus motivational and emotional responses.

Pain is more than a direct response to a stimulus. Unlike other somatosensory modalities, the sensation of pain is accompanied by motivated behavioral responses (such as withdrawal or defense) and emotional reactions (such as crying or fear). Also, unlike other sensations, the subjective perception of pain can be influenced by other past or present experiences (for example, heightened pain perception accompanying fear of the dentist or lowered pain perception in an injured athlete during a competitive event). Therefore pain is a personal, multidimensional experience.

Categories of Pain Receptors There are three categories of nociceptors: **Mechanical nociceptors** respond to mechanical damage such as cutting, crushing, or pinching; **thermal nociceptors** respond to temperature extremes, especially heat; and **polymodal nociceptors** respond equally to all kinds of damaging stimuli, including irritating chemicals released from injured tissues.

Clinical Note All nociceptors can be sensitized by *prostaglandins,* which greatly enhance the receptor response to noxious stimuli (that is, it hurts more when prostaglandins are present). Prostaglandins are a type of eicosanoid, a lipid signal molecule derived from a fatty acid in the plasma membrane that acts locally where released (see p. 119). Tissue injury, among other things, can lead to local release of prostaglandins. These paracrines act on nearby nociceptors' peripheral endings to lower their threshold for activation. Aspirin-like drugs inhibit the synthesis of prostaglandins, accounting at least in part for the pain-relieving properties of these drugs.

Fast and Slow Afferent Pain Fibers Pain impulses originating at nociceptors are transmitted to the CNS via one of two types of afferent fibers (I Table 6-2). Signals arising from noci-

| TABLE 6-2 Characteristics of Pain

Fast Pain	Slow Pain
Occurs on stimulation of mechanical and thermal nociceptors	Occurs on stimulation of polymodal nociceptors
Carried by small, myelinated A-delta fibers	Carried by small, unmyelinated C fibers
Produces sharp, prickling sensation	Produces dull, aching, burning sensation
Easily localized	Poorly localized
Occurs first	Occurs second; persists for longer time; more unpleasant

ceptors that respond to mechanical damage such as cutting or to thermal damage such as burning are transmitted over small, myelinated **A-delta fibers** at rates of up to 30 m/sec (the **fast pain pathway**). Impulses from polymodal nociceptors that respond to chemicals released into the ECF from damaged tissue are carried by small, unmyelinated **C fibers** at a slower rate of 12 m/sec or less (the **slow pain pathway**). Think about the last time you cut or burned your finger. You undoubtedly felt a sharp twinge of pain at first, with a more diffuse, disagreeable pain commencing shortly thereafter. Pain typically is perceived initially as a brief, sharp, prickling sensation that is easily localized; this is fast pain originating from specific mechanical or heat nociceptors. This feeling is followed by a dull, aching, poorly localized sensation that persists for a longer time and is more unpleasant; this is slow pain triggered by chemicals, especially **bradykinin**, a normally inactive substance that is activated by enzymes released into the ECF from damaged tissue. Bradykinin and related compounds not only provoke pain by stimulating the polymodal nociceptors, but they also contribute to the inflammatory response to tissue injury (see Chapter 12). This slow, aching pain is activated for a prolonged time because of the persistence of released chemicals at the site long after removal of the mechanical or thermal stimulus that caused the tissue damage.

Clinical Note Interestingly, the peripheral receptors of afferent C fibers are activated by **capsaicin**, the ingredient in hot chili peppers that gives them their fiery zing. (In addition to binding with pain receptors, capsaicin binds with heat receptors—hence the burning sensation when eating hot peppers.) Ironically, local application of capsaicin can reduce clinical pain, most likely by overstimulating and damaging the nociceptors with which it binds.

Higher-Level Processing of Pain Input Multiple structures are involved in pain processing—primary afferent pain fibers, ascending pain pathways in the spinal cord, and brain regions involved with pain perception. The primary afferent pain fibers synapse with specific second-order excitatory interneurons in the dorsal horn of the spinal cord. In response to stimulus-induced action potentials, afferent pain fibers release neurotransmitters that influence these next neurons in line. The two best known pain neurotransmitters are *substance P* and *glutamate*. **Substance P**, which is unique to pain fibers, activates ascending pathways that transmit nociceptive signals to higher levels for further processing (I Figure 6-9a). Ascending pain pathways have different destinations in the *cortex*, the *thalamus*, and the *reticular formation*. Cortical somatosensory processing areas localize the pain, whereas other cortical areas participate in other conscious components of the pain experience, such as deliberation about the incident. Pain can still be perceived at the level of the thalamus in the absence of the cortex. The reticular formation increases the level of alertness associated with the noxious encounter. Interconnections from the thalamus and reticular formation to the *hypothalamus* and *limbic system* elicit the behavioral and emotional responses accompanying the painful experience. The limbic system is especially important in perceiving the unpleasant aspects of pain.

Glutamate, the other neurotransmitter released from primary afferent pain terminals, is a major excitatory neurotransmitter (see p. 107). Glutamate acts on two different plasma membrane receptors on the dorsal horn excitatory interneurons, with two different outcomes (see p. 160). First, binding of glutamate with its *AMPA receptors* leads to permeability changes that ultimately result in generation of action potentials in the dorsal horn cells. These action potentials transmit the pain message to higher centers. Second, binding of glutamate with its *NMDA receptors* leads to Ca^{2+} entry into these neurons. This pathway is not involved in the transmission of pain messages. Instead, Ca^{2+} initiates second-messenger systems that make the dorsal horn cells more excitable than usual (see p. 117). This hyperexcitability contributes in part to the exaggerated sensitivity of an injured area to subsequent exposure to painful or even normally nonpainful stimuli, such as a light touch. Think about how exquisitely sensitive your sunburned skin is, even to clothing. Other mechanisms also contribute to supersensitivity of an injured area. For example, responsiveness of the pain-sensing peripheral receptors can be boosted so that they react more vigorously to subsequent stimuli. This exaggerated sensitivity presumably serves a useful purpose by discouraging activities that could cause further damage or interfere with healing of the injured area. Usually this hypersensitivity resolves as the injury heals.

Clinical Note Persistent, **chronic pain**, sometimes excruciating, can occur in the absence of tissue injury. In contrast to the acute pain accompanying peripheral injury, which serves as a normal protective mechanism to warn of impending or actual damage to the body, abnormal chronic pain results from prolonged hypersensitivity within the pain transmission pathways in the peripheral nerves or in the CNS—that is, pain is perceived because of abnormal signaling within the pain pathways in the absence of typical painful stimuli. Recent evidence suggests that the persistent, abnormal excitability among neurons in the pain pathways that leads to chronic pain is the result of a complex interplay among the involved neurons, glial cells (especially microglia and astrocytes; see p. 136), and immune cells. These cells release many types of intercellular chemical messengers, such as inflammatory cytokines (see

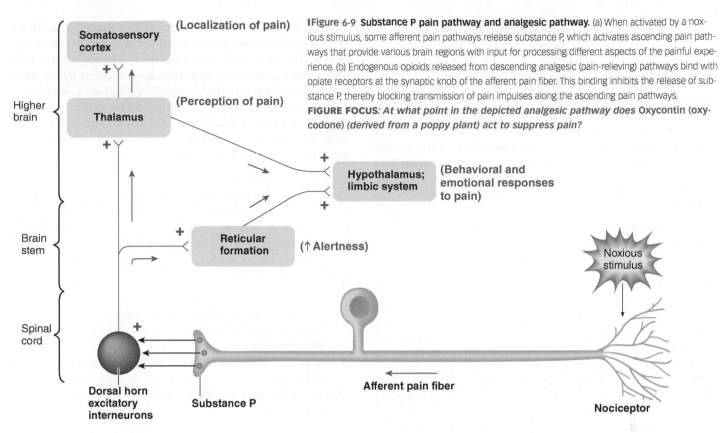

Figure 6-9 Substance P pain pathway and analgesic pathway. (a) When activated by a noxious stimulus, some afferent pain pathways release substance P, which activates ascending pain pathways that provide various brain regions with input for processing different aspects of the painful experience. (b) Endogenous opioids released from descending analgesic (pain-relieving) pathways bind with opiate receptors at the synaptic knob of the afferent pain fiber. This binding inhibits the release of substance P, thereby blocking transmission of pain impulses along the ascending pain pathways.
FIGURE FOCUS: *At what point in the depicted analgesic pathway does* Oxycontin (oxycodone) *(derived from a poppy plant) act to suppress pain?*

(a) Substance P pain pathway

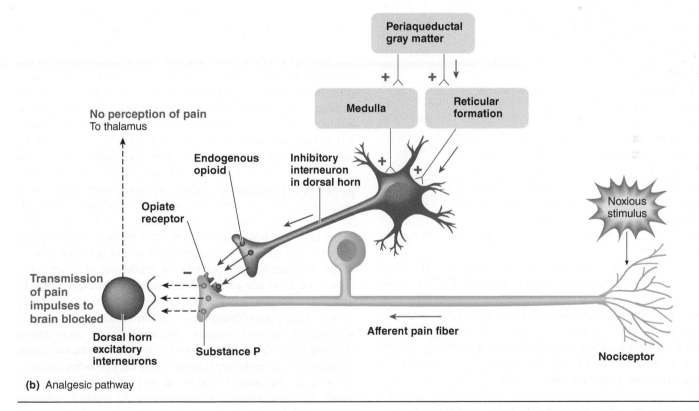

(b) Analgesic pathway

p. 118) that are meant to be helpful, such as by promoting healing in response to the original tissue insult. However, many of these molecules increase the excitability of involved neurons via long-term potentiation (LTP) (see p. 160), a state that can last long after the initial damage is healed. By unleashing exaggerated reactions to stimuli that are ordinarily too mild to trigger a response, the overly sensitive neurons continue to fire and transmit pain signals in the absence of obvious tissue damage. Chronic pain is sometimes categorized as **neuropathic pain**. Worldwide, 15% to 20% of adults suffer from this affliction.

The brain has a built-in analgesic system.

In addition to the chain of neurons connecting peripheral nociceptors with higher CNS structures for pain perception, the CNS contains a built-in pain-suppressing or **analgesic system** that suppresses transmission in the pain pathways as they enter the spinal cord. Three brain-stem regions are part of this descending analgesic pathway: the *periaqueductal gray matter* (gray matter surrounding the cerebral aqueduct, a narrow canal that connects the third and fourth ventricular cavities) and specific nuclei in the *medulla* and *reticular formation*. Electrical stimulation of any of these parts of the brain produces profound analgesia.

The periaquaductal gray matter stimulates particular neurons whose cell bodies lie in the medulla and reticular formation and that terminate on inhibitory interneurons in the dorsal horn of the spinal cord (❙Figure 6-9b). These inhibitory interneurons release *enkephalin,* an endogenous opioid that binds with **μ opiate receptors** at the afferent pain-fiber terminal. People have long known that **morphine**, a component of the opium poppy, is a powerful analgesic. Researchers considered it unlikely that the body has been endowed with opiate receptors only to interact with chemicals derived from a flower. They therefore began to search for the substances that normally bind with these opiate receptors. The result was the discovery of **endogenous opioids** (morphinelike substances)—the **endorphins**, **enkephalins**, and **dynorphin**—which are important in the body's natural analgesic system. These endogenous opioids serve as analgesic neurotransmitters. Binding of enkephalin from the dorsal-horn inhibitory interneuron with the afferent pain-fiber terminal suppresses the release of substance P via presynaptic inhibition, thereby blocking further transmission of the pain signal (see p. 111). Morphine binds to these same opiate receptors, which largely accounts for its analgesic properties. Furthermore, injection of morphine into the periaqueductal gray matter and medulla causes profound analgesia, suggesting that endogenous opioids also are released centrally to block pain.

It is not clear how this natural pain-suppressing mechanism is normally activated. Factors known to modulate pain include exercise, stress, and acupuncture.

■ Endorphins are released during prolonged exercise and are natural painkillers and mood enhancers. Endorphins are believed to be a major player in "runner's high," a feeling of euphoria some avid runners (or other vigorous exercisers) get during high-intensity exertion that gets them hooked on exercise. Recent evidence suggests that endocannibinoids, bodymade marijuanalike chemical messengers, also may contribute to runner's high (see p. 110).

■ Some types of stress also induce analgesia. It is sometimes disadvantageous for a stressed organism to display the normal reaction to pain. For example, when two male lions are fighting for dominance of the group, withdrawing, escaping, or resting when injured would mean certain defeat.

■ **Acupuncture analgesia (AA)** is the technique of relieving pain by inserting and manipulating fine needles at key points. The overwhelming body of evidence supports the *acupuncture*

endorphin hypothesis as the primary mechanism of AA's action. According to this hypothesis, the needle twirling activates specific afferent nerve fibers, which send impulses to the CNS. Here, the incoming impulses cause analgesia by blocking pain transmission at both the spinal-cord and the brain level through use of endorphins and other endogenous opioids. Several other neurotransmitters, such as serotonin and norepinephrine, as well as cortisol, the major hormone released during stress, are implicated as well.

This completes our discussion of somatic sensation. Whereas somatic sensation is detected by widely distributed receptors that provide information about the body's interactions with the environment in general, each of the special senses has highly localized, extensively specialized receptors that respond to unique environmental stimuli. We now turn attention to the special senses, starting with vision.

Check Your Understanding 6.2

1. Explain why pain is considered a multidimensional experience.
2. Compare the type of pain signals transmitted via A-delta fibers and C fibers.
3. Describe the role of endogenous opioids in the body's natural analgesic system.

6.3 Eye: Vision

For vision, the eyes capture the patterns of illumination in the environment as an "optical picture" on a layer of light-sensitive cells, the *retina*. The coded image on the retina is transmitted through the steps of visual processing until it is finally consciously perceived as a visual likeness of the original image. Before considering the steps involved in the process of vision, we first examine how the eyes are protected from injury.

Protective mechanisms help prevent eye injuries.

Several mechanisms help protect the eyes from injury. Except for its anterior (front) portion, the eyeball is sheltered by the bony socket in which it is positioned. The **eyelids** act like shutters to protect the exposed part of the eye from environmental insults. They close reflexly to cover the eye under threatening circumstances, such as rapidly approaching objects, dazzling light, and when the eye or eyelashes are touched. Frequent spontaneous blinking of the eyelids helps disperse the lubricating, cleansing, bactericidal ("germ-killing") **tears**. Tears are produced continuously by the **lacrimal gland** in the upper lateral corner under the eyelid. This eye-washing fluid flows across the anterior surface of the eye and drains into tiny canals in the medial corner of each eye (❙Figure 6-10a), eventually emptying into the back of the nasal passageway. This drainage system cannot handle the profuse tear production during crying, so the tears overflow from the eyes. (The tears associated with crying have a different composition than the ongoing tear production.) The eyes are also

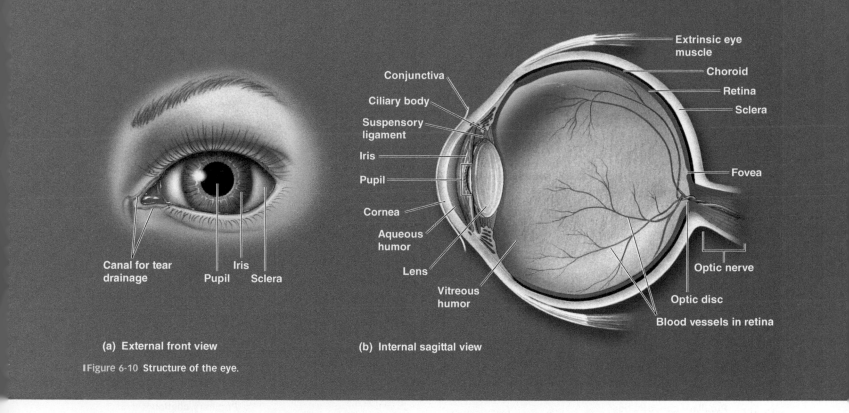

Conjunctiva
Ciliary body
Suspensory ligament
Iris
Pupil
Cornea
Aqueous humor
Lens
Vitreous humor

Extrinsic eye muscle
Choroid
Retina
Sclera
Fovea
Optic nerve
Optic disc
Blood vessels in retina

Canal for tear drainage
Pupil
Iris
Sclera

(a) External front view   (b) Internal sagittal view

❙Figure 6-10 Structure of the eye.

equipped with protective **eyelashes**, which trap fine, airborne debris such as dust before it can fall into the eye.

The eye is a fluid-filled sphere enclosed by three specialized tissue layers.

Each eye is a spherical, fluid-filled structure enclosed by three layers. From outermost to innermost, these are (1) the *sclera/cornea;* (2) the *choroid/ciliary body/iris;* and (3) the *retina* (❙Figure 6-10b). Most of the eyeball is covered by a tough outer layer of connective tissue, the **sclera**, which forms the visible white part of the eye (❙Figure 6-10a). Anteriorly, the outer layer consists of the transparent **cornea**, through which light rays pass into the interior of the eye. The middle layer underneath the sclera is the highly pigmented **choroid**, which contains many blood vessels that nourish the retina. The choroid layer becomes specialized anteriorly to form the ciliary body and iris, which we describe shortly. The innermost coat under the choroid is the **retina**, which consists of an outer pigmented layer and an inner nervous-tissue layer. The latter contains the *rods* and *cones,* the photoreceptors that convert light energy into nerve impulses. Like the black walls of a photographic studio, the pigment in the choroid and retina absorbs light after it strikes the retina to prevent reflection or scattering of light within the eye.

The interior of the eye consists of two fluid-filled cavities, separated by an elliptical **lens**, all of which are transparent to permit light to pass through the eye from the cornea to the retina. In an adult, the lens is about 10 mm in diameter, the size of a shirt button. The larger posterior cavity between the lens and the retina contains a clear, jellylike substance, the **vitreous humor**. The vitreous humor helps maintain the spherical shape of the eyeball. The anterior cavity between the cornea and the lens contains a clear, watery fluid, the **aqueous humor**. The aqueous humor carries nutrients for the cornea and lens, both of which lack a blood supply. Blood vessels in these structures would impede the passage of light to the photoreceptors.

The aqueous humor is produced at a rate of about 5 mL/day by a capillary network within the **ciliary body**, a specialized anterior derivative of the choroid layer. This fluid drains into a canal at the edge of the cornea and eventually enters the blood. *Clinical Note* If the aqueous humor is not drained as rapidly as it forms (for example, because of a blocked drainage canal), the excess accumulates in the anterior cavity, causing the pressure to rise within the eye. This condition is known as **glaucoma**. The excess aqueous humor pushes the lens backward into the vitreous humor, which in turn pushes against the inner neural layer of the retina. This compression causes retinal and optic nerve damage that can lead to blindness if the condition is not treated.

The amount of light entering the eye is controlled by the iris.

Not all light passing through the cornea reaches the light-sensitive photoreceptors because of the presence of the **iris**, a thin, pigmented smooth muscle that forms a visible ringlike structure within the aqueous humor (❙Figure 6-10a and b). The pigment in the iris is responsible for eye color. The varied flecks, lines, and other nuances of the iris are unique for each individual, making the iris the basis of the latest identification technology. Recognition of iris patterns by a video camera that captures iris images and translates the landmarks into a computerized code is more foolproof than fingerprinting or even DNA testing.

The round opening in the center of the iris through which light enters the interior portions of the eye is the **pupil**. The size

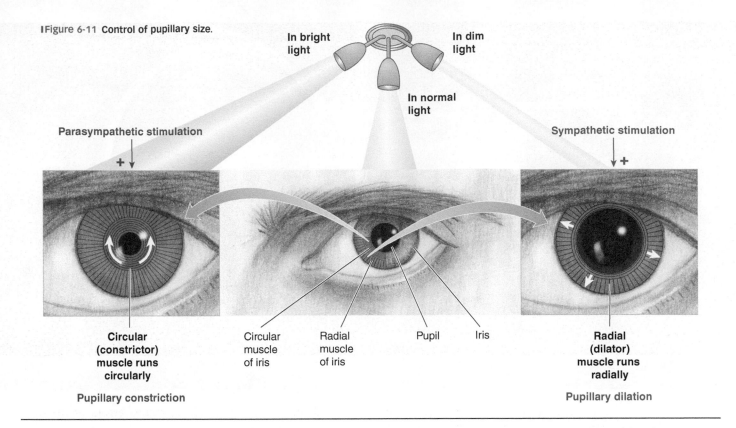

Figure 6-11 Control of pupillary size.

In bright light

In dim light

In normal light

Parasympathetic stimulation

Sympathetic stimulation

+

+

Circular (constrictor) muscle runs circularly

Circular muscle of iris

Radial muscle of iris

Pupil

Iris

Radial (dilator) muscle runs radially

Pupillary constriction

Pupillary dilation

of this opening can be adjusted by variable contraction of the iris smooth muscles to admit more or less light as needed. The iris contains two sets of smooth muscle networks, one *circular* (the muscle fibers run in a ringlike fashion within the iris) and the other *radial* (the fibers project outward from the pupillary margin like bicycle spokes) (Figure 6-11). Because muscle fibers shorten when they contract, the pupil gets smaller when the **circular** (or **constrictor**) **muscle** contracts and forms a smaller ring. This reflex pupillary constriction occurs in bright light to decrease the amount of light entering the eye. When the **radial** (or **dilator**) **muscle** shortens, the size of the pupil increases. Such pupillary dilation occurs in dim light to allow the entrance of more light. Iris muscles are controlled by the autonomic nervous system. Parasympathetic nerve fibers innervate the circular muscle (causing pupillary constriction), and sympathetic fibers supply the radial muscle (causing pupillary dilation).

The eye refracts entering light to focus the image on the retina.

Light is a form of electromagnetic radiation composed of particlelike individual packets of energy called **photons** that travel in wavelike fashion. The distance between two wave peaks is known as the *wavelength* (Figure 6-12). The wavelengths in the electromagnetic spectrum range from 10^{-14} m (quadrillionths of a meter, as in the extremely short cosmic rays) to 10^4 m (10 km, as in long radio waves) (Figure 6-13). The photoreceptors in the eye are sensitive only to wavelengths between 400 and 700 nanometers (nm; billionths of a meter). Thus, **visible light** is only a small portion of the total electromagnetic spectrum. Light of different wavelengths in this visible band is perceived as different color sensations. The shorter visible wavelengths are sensed as violet and blue; the longer wavelengths are interpreted as orange and red.

In addition to having variable wavelengths, light energy varies in intensity—that is, the amplitude, or height, of the wave (see Figure 6-12). Dimming a bright red light does not change its color; it just becomes less intense or less bright.

Light waves *diverge* (radiate outward) in all directions from every point of a light source. The forward movement of a light wave in a particular direction is known as a **light ray**. Divergent light rays reaching the eye must be bent inward to be focused back into a point (the **focal point**) on the light-sensitive retina and provide an accurate image of the light source (Figure 6-14).

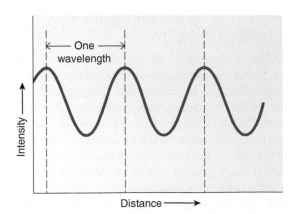

One wavelength

Intensity

Distance

Figure 6-12 Properties of an electromagnetic wave. A wavelength is the distance between two wave peaks. The intensity is the amplitude of the wave.

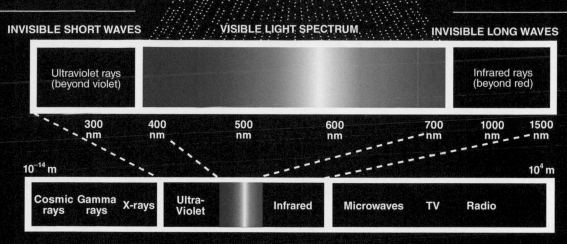

INVISIBLE SHORT WAVES   VISIBLE LIGHT SPECTRUM   INVISIBLE LONG WAVES

Ultraviolet rays (beyond violet)		Infrared rays (beyond red)

300 nm   400 nm   500 nm   600 nm   700 nm   1000 nm   1500 nm

10^{-14} m 10^{4} m

Cosmic rays	Gamma rays	X-rays	Ultra-Violet		Infrared	Microwaves	TV	Radio

IFigure 6-13 Electromagnetic spectrum. The wavelengths in the electromagnetic spectrum range from less than 10^{-14} m to 10^{4} m. The visible spectrum includes wavelengths ranging from 400 to 700 nanometers (nm).

Process of Refraction Light travels faster through air than through other transparent media such as water and glass. When a light ray enters a medium of greater density, it is slowed down (the converse is also true). The course of direction of the ray changes if it strikes the surface of the new medium at any angle other than perpendicular (I Figure 6-15). The bending of a light ray is known as **refraction**. With a curved surface such as a lens, the greater the curvature, the greater is the degree of bending and the stronger the lens. When a light ray strikes the curved surface of any object of greater density, the direction of refraction depends on the angle of the curvature (I Figure 6-16). A **convex surface** curves outward (like the outer surface of a ball), whereas a **concave surface** curves inward (like a cave). Convex surfaces converge light rays, bringing them closer together. Because convergence is essential for bringing an image to a focal point, refractive surfaces of the eye are convex. Concave surfaces diverge light rays (spread them farther apart). A concave lens is useful for correcting certain refractive errors of the eye, such as nearsightedness.

The Eye's Refractive Structures The two structures most important in the eye's refractive ability are the *cornea* and the *lens*. The curved corneal surface, the first structure light passes through as it enters the eye, contributes most extensively to the eye's total refractive ability because the difference in density at the air–cornea interface is greater than the differences in density between the lens and the fluids surrounding it. In **astigmatism**, the curvature of the cornea is uneven, so light rays are unequally refracted. The refractive ability of a person's cornea remains constant because the curvature of the cornea never changes. In contrast, the refractive ability of the lens can be adjusted by changing its curvature as needed for near or far vision.

Rays from light sources more than 20 feet away are considered parallel by the time they reach the eye. Light rays originating from near objects are still diverging when they reach the eye. For a given refractive ability of the eye, the diverging rays of a near source come to a focal point a greater distance behind the lens than the parallel rays of a far source come to a focal point (I Figure 6-17a and b). However, in a particular eye, the distance between the lens and the retina always remains the same. Therefore, a greater distance beyond the lens is not available for bringing near objects into focus. Yet for clear vision, the refractive structures of the eye must bring both near and far light sources into focus on the retina. If an image is focused before it reaches the retina or is not yet focused

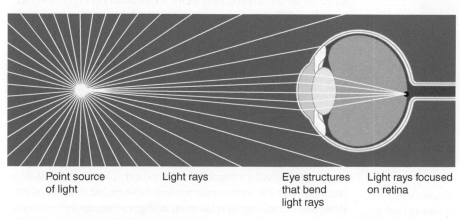

Point source of light   Light rays   Eye structures that bend light rays   Light rays focused on retina

IFigure 6-14 Focusing of diverging light rays. Diverging light rays must be bent inward to be focused.

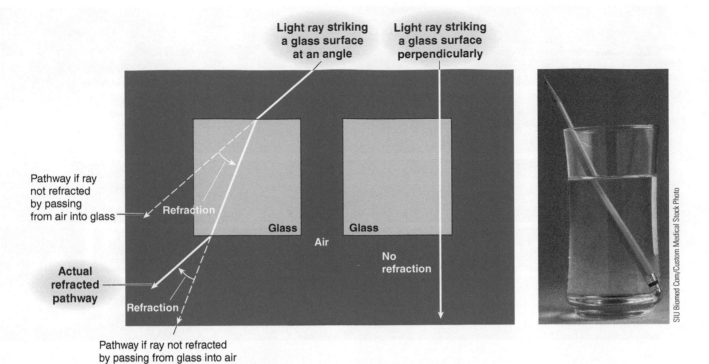

Light ray striking a glass surface at an angle

Light ray striking a glass surface perpendicularly

Pathway if ray not refracted by passing from air into glass

Refraction

Glass

Glass

Air

No refraction

Actual refracted pathway

Refraction

Pathway if ray not refracted by passing from glass into air

▮Figure 6-15 **Refraction.** A light ray is bent (refracted) when it strikes the surface of a medium of different density from the one in which it had been traveling (for example, moving from air into glass) at any angle other than perpendicular to the new medium's surface. Thus, the pencil in the glass of water appears to bend. What is happening, though, is that the light rays coming to the camera (or your eyes) are bent as they pass through the water, then the glass, and then the air. Consequently, the pencil appears distorted.

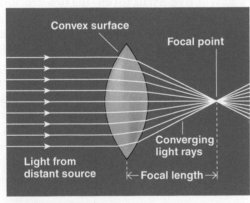

(a) Convex lens

Convex surface

Focal point

Converging light rays

Light from distant source

Focal length

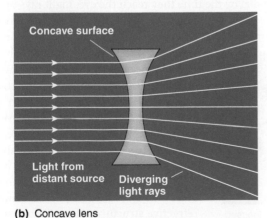

Concave surface

Light from distant source

Diverging light rays

(b) Concave lens

▮Figure 6-16 **Refraction by convex and concave lenses.** (a) A lens with a convex surface converges the rays (brings them closer together). (b) A lens with a concave surface diverges the rays (spreads them farther apart).

when it reaches the retina, it will be blurred. To bring both near and far light sources into focus on the retina (that is, in the same distance), a stronger lens must be used for the near source (▮Figure 6-17c). Let us see how the strength of the lens can be adjusted as needed.

Accommodation increases the strength of the lens for near vision.

The ability to adjust the strength of the lens is known as **accommodation**. The strength of the lens depends on its shape, which in turn is regulated by the ciliary muscle. The **ciliary muscle** is part of the ciliary body, an anterior specialization of the choroid layer. The ciliary body has two major components: the ciliary muscle and the capillary network that produces the aqueous humor. The ciliary muscle is a circular ring of smooth muscle attached to the lens by **suspensory ligaments** (▮Figure 6-18a).

When the ciliary muscle is relaxed, the suspensory ligaments are taut, and they pull the lens into a flattened, weakly refractive shape (▮Figure 6-18b). As the muscle contracts, its circumference decreases, slackening the tension in the suspensory ligaments (▮Figure 6-18c). When the suspensory ligaments are not pulling on the lens, it becomes more spherical because of its inherent elasticity. The greater curvature of the more rounded lens increases its strength, further bending light rays. In the normal eye, the ciliary muscle is relaxed and the lens is flat for far vision but the muscle contracts to let the lens become more convex and stronger for near vision. The ciliary muscle is controlled by the autonomic nervous system, with sympathetic stimulation causing its relaxation and parasympathetic stimulation causing its contraction.

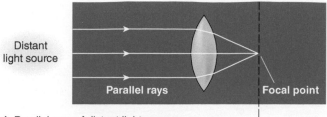

Distant light source

Parallel rays   **Focal point**

(a) Parallel rays of distant light source

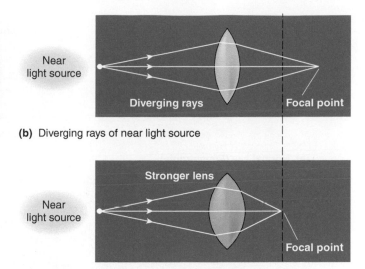

Near light source

Diverging rays   **Focal point**

(b) Diverging rays of near light source

Stronger lens

Near light source

Focal point

(c) Stronger lens needed to focus near light source

❙Figure 6-17 **Focusing of distant and near sources of light.** (a) The rays from a distant (far) light source (more than 20 feet from the eye) are parallel by the time the rays reach the eye. (b) The rays from a near light source (less than 20 feet from the eye) are still diverging when they reach the eye. A longer distance is required for a lens of a given strength to bend the diverging rays from a near light source into focus than to bend the parallel rays from a distant light source into focus. (c) To focus both a distant and a near light source in the same distance (the distance between the lens and retina), a stronger lens must be used for the near source.

ᵍˡⁱⁿⁱᶜᵃˡ ᴺᵒᵗᵉ The lens is made up of about 1000 layers of cells that destroy their nucleus and organelles during development so that the cells are perfectly transparent. Lacking DNA and protein-synthesizing machinery, mature lens cells cannot regenerate or repair themselves. Cells in the center of the lens are in double jeopardy. Not only are they oldest, but they also are farthest from the aqueous humor, the lens's nutrient source. With advancing age, these non-renewable central cells die and become stiff. With loss of elasticity, the lens can no longer assume the spherical shape required to accommodate for near vision. This age-related reduction in accommodative ability, **presbyopia**, affects most people by middle age (45 to 50 years), requiring them to resort to corrective lenses for near vision (reading).

❙Figure 6-18 **Mechanism of accommodation.** (a) Suspensory ligaments extend from the ciliary muscle to the outer edge of the lens. (b) When the ciliary muscle is relaxed, the suspensory ligaments are taut, putting tension on the lens so that it is flat and weak. (c) When the ciliary muscle is contracted, the suspensory ligaments become slack, reducing the tension on the lens, allowing it to assume a stronger, rounder shape because of its elasticity.

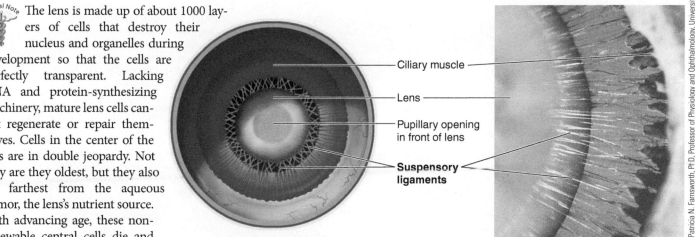

Ciliary muscle

Lens

Pupillary opening in front of lens

Suspensory ligaments

Patricia N. Farnsworth, Ph.D. Professor of Physiology and Ophthalmology, University of Medicine and Dentistry of New Jersey, New Jersey Medical School

(a) Anterior view of suspensory ligaments extending from ciliary muscles to lens

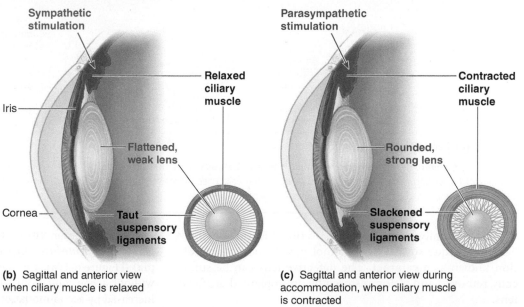

Sympathetic stimulation

Parasympathetic stimulation

Relaxed ciliary muscle

Contracted ciliary muscle

Iris

Flattened, weak lens

Rounded, strong lens

Cornea

Taut suspensory ligaments

Slackened suspensory ligaments

(b) Sagittal and anterior view when ciliary muscle is relaxed

(c) Sagittal and anterior view during accommodation, when ciliary muscle is contracted

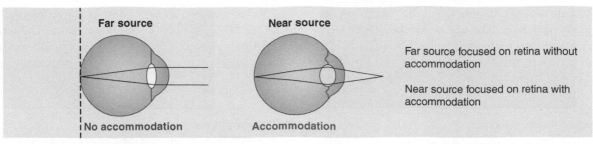

(a) Normal eye (Emmetropia)

Far source
Near source
Far source focused on retina without accommodation
Near source focused on retina with accommodation
No accommodation
Accommodation

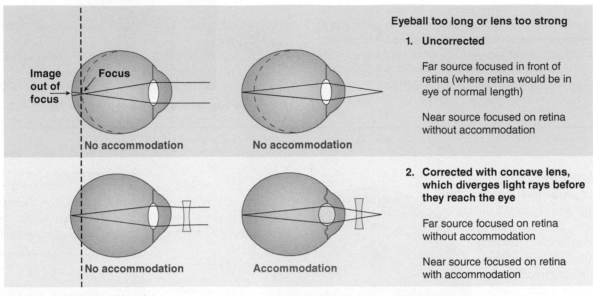

(b) Nearsightedness (Myopia)

Eyeball too long or lens too strong

1. **Uncorrected**

   Far source focused in front of retina (where retina would be in eye of normal length)

   Near source focused on retina without accommodation

2. **Corrected with concave lens, which diverges light rays before they reach the eye**

   Far source focused on retina without accommodation

   Near source focused on retina with accommodation

Image out of focus · Focus

No accommodation · No accommodation · No accommodation · Accommodation

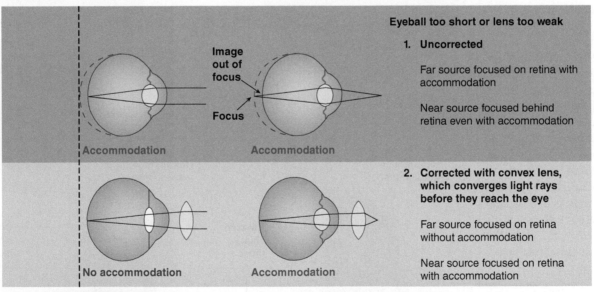

(c) Farsightedness (Hyperopia)

Eyeball too short or lens too weak

1. **Uncorrected**

   Far source focused on retina with accommodation

   Near source focused behind retina even with accommodation

2. **Corrected with convex lens, which converges light rays before they reach the eye**

   Far source focused on retina without accommodation

   Near source focused on retina with accommodation

Image out of focus · Focus

Accommodation · Accommodation · No accommodation · Accommodation

⏐Figure 6-19 Emmetropia, myopia, and hyperopia. This figure compares far vision and near vision (a) in the normal eye with (b) nearsightedness and (c) farsightedness in both their (1) uncorrected and (2) corrected states. The vertical dashed line represents the normal distance of the retina from the cornea—that is, the site at which an image is brought into focus by the refractive structures in a normal eye.

The normally transparent elastic fibers in the lens may become opaque so that light cannot pass through, a condition known as a **cataract**. The defective lens can be surgically removed and vision restored by an implanted artificial lens.

Other common vision disorders are *nearsightedness (myopia)* and *farsightedness (hyperopia)*. In a normal eye (**emmetropia**) (⏐Figure 6-19a), a far light source is focused on the retina without accommodation, whereas the strength of the lens is increased by accommodation to bring a near source into focus.

In **myopia** (❙Figure 6-19b1) because the eyeball is too long or the lens is too strong, a near light source is brought into focus on the retina without accommodation (even though accommodation is normally used for near vision), whereas a far light source is focused in front of the retina and is blurry. Thus, a myopic individual has better near vision than far vision, a condition that can be corrected by a concave lens (❙Figure 6-19b2). With **hyperopia** (❙Figure 6-19c1), either the eyeball is too short or the lens is too weak. Far objects are focused on the retina only with accommodation, whereas near objects are focused behind the retina even with accommodation and, accordingly, are blurry. Thus, a hyperopic individual has better far vision than near vision, a condition that can be corrected by a convex lens (❙Figure 6-19c2). Instead of using corrective eyeglasses or contact lenses, many people are now opting to compensate for refractive errors with laser eye surgery (such as LASIK) to permanently change the shape of the cornea.

Light must pass through several retinal layers before reaching the photoreceptors.

The major function of the eye is to focus light rays from the environment on the rods and cones, the photoreceptor cells of the retina. The photoreceptors then transform the light energy into electrical signals for transmission to the CNS.

The receptor-containing portion of the retina is actually an anatomic extension of the CNS, not a separate peripheral organ. During embryonic development, the retinal cells "back out" of the nervous system, so the retinal layers, surprisingly, are facing backward. The neural portion of the retina consists of three layers of excitable cells (❙Figure 6-20): (1) the outermost layer (closest to the choroid) containing the **rods** and **cones**, whose light-sensitive ends face the choroid (away from the incoming light); (2) a middle layer of **bipolar cells** and associated interneurons; and (3) an inner layer of **ganglion cells**. Axons of the ganglion cells join to form the **optic nerve**, which leaves the retina slightly off center. The point on the retina at which the optic nerve leaves and through which blood vessels pass is the **optic disc** (see ❙Figure 6-10b, p. 193). This region is often called the **blind spot**; no image can be detected in this area because it has no rods and cones (❙Figure 6-21). We are normally not aware of the blind spot because central processing somehow "fills in" the missing spot. You can discover the existence of your own blind spot by a simple demonstration (❙Figure 6-22).

Light must pass through the ganglion and bipolar layers before reaching the photoreceptors in all areas of the retina except the fovea. In the **fovea**, which is a pinhead-sized depression located in the exact center of the retina (see ❙Figure 6-10b), the bipolar and ganglion cell layers are pulled aside so that light strikes the photoreceptors directly. Because of this feature, and because only cones

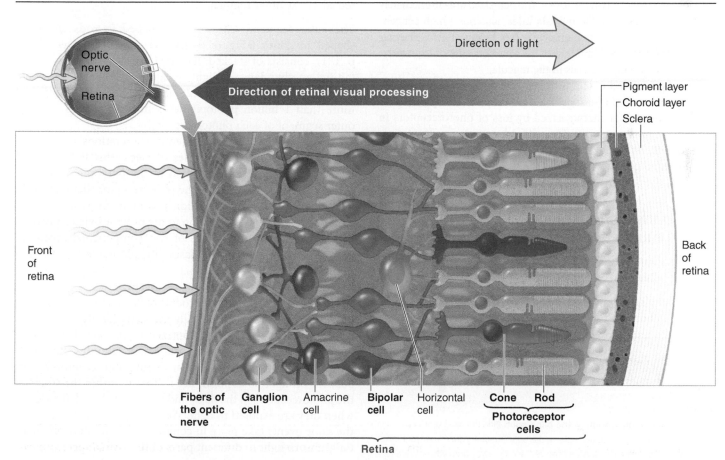

❙Figure 6-20 **Retinal layers.** The retinal visual pathway extends from the photoreceptor cells (rods and cones, whose light-sensitive ends face the choroid away from the incoming light) to the bipolar cells to the ganglion cells. The horizontal and amacrine cells are interneurons that act locally for retinal processing of visual input.
FIGURE FOCUS: *Does convergence exist in the pathway from rods to bipolar cells to ganglion cells? How about in the pathway from cones to bipolar cells to ganglion cells? Remember these wiring patterns. They will be important in a later discussion.*

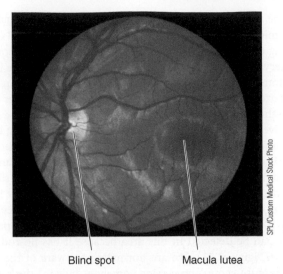

Blind spot   Macula lutea

I Figure 6-21 View of the retina seen through an ophthalmoscope. With an ophthalmoscope, a lighted viewing instrument, it is possible to view the optic disc (blind spot) and macula lutea within the retina at the rear of the eye.

I Figure 6-23 View with macular degeneration.

(which have greater acuity or discriminative ability than the rods) are found here, the fovea is the point of most distinct vision. In fact, the fovea has the greatest concentration of cones in the retina. Thus, we turn our eyes so that the image of the object at which we are looking is focused on the fovea. The pea-sized area immediately surrounding the fovea, the **macula lutea**, also has a high concentration of cones and fairly high acuity (see I Figure 6-21). Macular acuity, however, is less than that of the fovea because of the overlying ganglion and bipolar cells in the macula.

Clinical Note **Age-related macular degeneration (AMD)** is the leading cause of blindness in the Western Hemisphere. This condition is characterized by loss of photoreceptors in the macula lutea in association with advancing age. Its victims have "doughnut" vision. They suffer a loss in the middle of their visual field, which normally has the highest acuity, and are left with only the less distinct peripheral vision (I Figure 6-23).

Phototransduction by retinal cells converts light stimuli into neural signals.

Photoreceptors (rod and cone cells) consist of three parts (I Figure 6-24a):

1. An *outer segment,* which lies closest to the eye's exterior, facing the choroid. It detects the light stimulus.

I Figure 6-22 Demonstration of the blind spot. Find the blind spot in your left eye by closing your right eye and holding the book about 4 inches from your face. While focusing on the cross, gradually move the book away from you until the circle vanishes from view. At this time, the image of the circle is striking the blind spot of your left eye. You can similarly locate the blind spot in your right eye by closing your left eye and focusing on the circle. The cross disappears when its image strikes the blind spot of your right eye.

2. An *inner segment,* which lies in the middle of the photoreceptor's length. It contains the metabolic machinery of the cell.

3. A *synaptic terminal,* which lies closest to the eye's interior, facing the bipolar cells. It varies its rate of neurotransmitter release, depending on the extent of dark or light exposure detected by the outer segment.

The outer segment, which is rod shaped in rods and cone shaped in cones (I Figure 6-24a and chapter opener photo, p. 181), consists of stacked, flattened, membranous discs containing an abundance of light-sensitive **photopigments**. Each retina contains more than 125 million photoreceptors, and more than 1 billion photopigments may be packed into the outer segment of each photoreceptor.

Photopigments undergo chemical alterations when activated by light. Through a series of steps, this light-induced change and subsequent activation of the photopigment bring about a receptor potential in the photoreceptor that ultimately leads to the generation of action potentials in ganglion cells, which transmit this information to the brain for visual processing. A photopigment consists of two components: **opsin**, an integral protein in the disc plasma membrane; and **retinal**, a derivative of vitamin A. Retinal is the light-absorbing part of the photopigment.

Phototransduction, the process of converting light stimuli into electrical signals, is basically the same for all photoreceptors, but the mechanism is contrary to the usual means by which receptors respond to their adequate stimulus. Receptors typically *depolarize* when stimulated, but photoreceptors *hyperpolarize* on light absorption. Let us first examine the status of the photoreceptors in the dark, and then consider what happens when they are exposed to light. We use rods as an example, but the same events take place in cones, except that they preferentially absorb light in different parts of the visible spectrum.

Photoreceptor Activity in the Dark The photopigment in rods is **rhodopsin**. Retinal exists in different conformations in the dark and light. In the dark, it exists as 11-*cis* retinal,

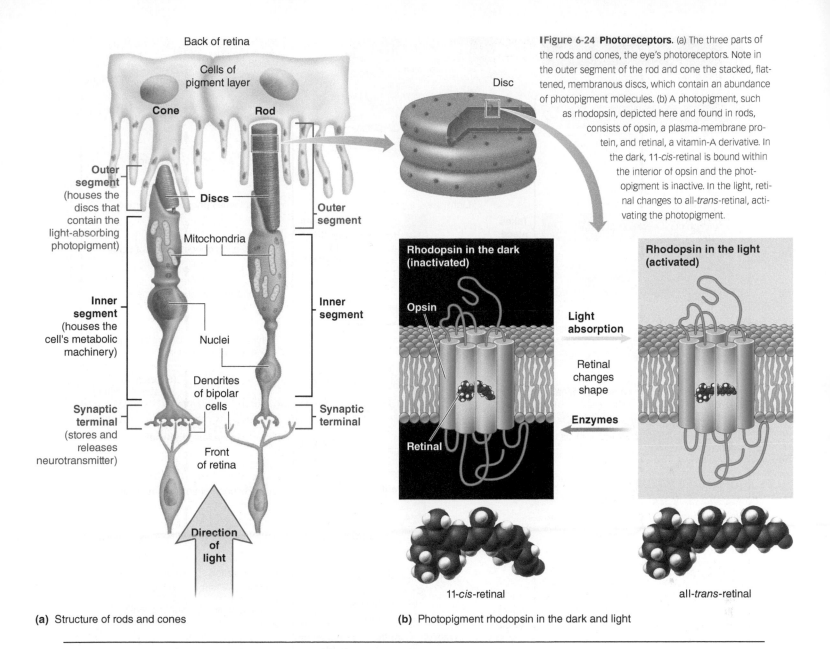

I Figure 6-24 Photoreceptors. (a) The three parts of the rods and cones, the eye's photoreceptors. Note in the outer segment of the rod and cone the stacked, flattened, membranous discs, which contain an abundance of photopigment molecules. (b) A photopigment, such as rhodopsin, depicted here and found in rods, consists of opsin, a plasma-membrane protein, and retinal, a vitamin-A derivative. In the dark, 11-*cis*-retinal is bound within the interior of opsin and the photopigment is inactive. In the light, retinal changes to all-*trans*-retinal, activating the photopigment.

Back of retina

Cells of pigment layer

Cone   Rod

Outer segment (houses the discs that contain the light-absorbing photopigment)

Discs

Outer segment

Mitochondria

Inner segment (houses the cell's metabolic machinery)

Nuclei

Inner segment

Dendrites of bipolar cells

Synaptic terminal (stores and releases neurotransmitter)

Synaptic terminal

Front of retina

Direction of light

Disc

Rhodopsin in the dark (inactivated)

Opsin

Retinal

Light absorption

Retinal changes shape

Enzymes

Rhodopsin in the light (activated)

11-*cis*-retinal

all-*trans*-retinal

(a) Structure of rods and cones

(b) Photopigment rhodopsin in the dark and light

which fits into a binding site within the interior of the opsin portion of rhodopsin (I Figure 6-24b). The plasma membrane of a photoreceptor's outer segment contains chemically gated Na^+ channels. Unlike other chemically gated channels that respond to extracellular chemical messengers, these channels respond to an internal second messenger, **cyclic GMP**, or **cGMP** (cyclic guanosine monophosphate). Binding of cGMP to these Na^+ channels keeps them open. In the absence of light, the concentration of cGMP is high (I Figure 6-25a). (Light absorption leads to the breakdown of cGMP.) Therefore, the Na^+ channels of a photoreceptor, unlike most receptors, are open in the absence of stimulation, that is, in the dark. The resultant passive inward Na^+ leak, the so-called *dark current*, depolarizes the photoreceptor. The passive spread of this depolarization from the outer segment (where the Na^+ channels are located) to the synaptic terminal (where the photoreceptor's neurotransmitter is stored) keeps the synaptic terminal's voltage-gated Ca^{2+} channels open. Ca^{2+} entry triggers the release by exocytosis of the neurotransmitter glutamate from the synaptic terminal while in the dark.

Photoreceptor Activity in the Light On exposure to light, the concentration of cGMP is decreased through a series of biochemical steps triggered by photopigment activation (I Figure 6-25b). When 11-*cis* retinal absorbs light, it changes to the all-*trans* retinal conformation (see I Figure 6-24b). This is the only light-dependent step in the entire process of phototransduction. As a result of this change in shape, retinal no longer fits snugly in its binding site in opsin, causing opsin to also change its conformation, which activates the photopigment. Membrane-bound opsin is similar in shape and behavior to G-protein-coupled receptors (GPCRs; see p. 117), except that instead of being activated by binding with an extracellular chemical messenger, photopigments are activated in response to light absorption by retinal. Rod and cone cells contain a G protein called **transducin**. The activated photopigment

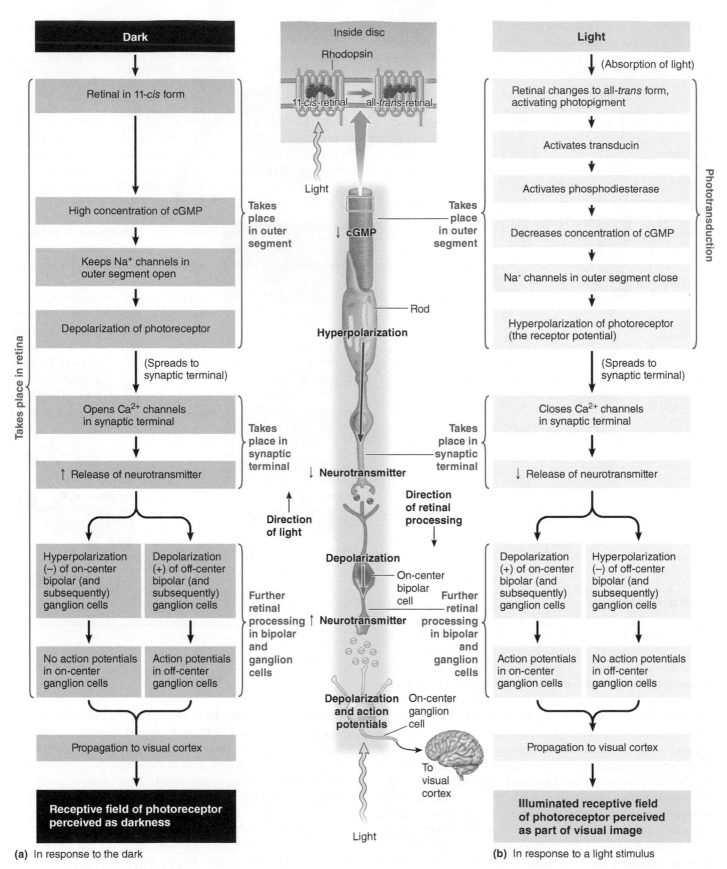

(a) In response to the dark

(b) In response to a light stimulus

I Figure 6-25 Phototransduction, further retinal processing, and initiation of action potentials in the visual pathway. (a) Events occurring in the retina and visual pathway in response to the dark. (b) Events occurring in the retina and visual pathway in response to a light stimulus.

FIGURE FOCUS: *How are photoreceptors depolarized in the absence of a stimulus and hyperpolarized in response to their adequate stimulus? How does a hyperpolarizing receptor potential lead to propagation of action potentials in on-center ganglion cells?*

activates transducin, which in turn activates the intracellular enzyme *phosphodiesterase*. This enzyme degrades cGMP, thus decreasing the concentration of this second messenger in the photoreceptor. During the light excitation process, the reduction in cGMP permits the chemically gated Na^+ channels to close. This channel closure stops the depolarizing Na^+ leak, thereby causing hyperpolarization. The hyperpolarization, which is the receptor potential, passively spreads from the outer segment to the synaptic terminal of the photoreceptor. Here the potential change leads to closure of the voltage-gated Ca^{2+} channels and a subsequent reduction in glutamate release from the synaptic terminal. Thus, photoreceptors are *inhibited by their adequate stimulus* (hyperpolarized by light) and *excited in the absence of stimulation* (depolarized by darkness). The hyperpolarizing potential and subsequent decrease in neurotransmitter release are graded according to the light intensity. The brighter the light is, the greater the hyperpolarizing response and the greater the reduction in glutamate release.

The short-lived active form of the photopigment quickly dissociates into opsin and retinal. The retinal is converted back into its 11-*cis* form. In the dark, enzyme-mediated mechanisms rejoin opsin and this recycled retinal to restore the photopigment to its original inactive conformation (see ❙Figure 6-24b).

Further Retinal Processing of Light Input How does the retina signal the brain about light stimulation through such an inhibitory response? Further retinal processing involves different influences of glutamate on two parallel pathways. Each photoreceptor synapses with two side-by-side bipolar cells, one an *on-center bipolar cell* and the other an *off-center bipolar cell*. These cells, in turn, terminate respectively on *on-center ganglion cells* and *off-center ganglion* cells, whose axons collectively form the optic nerve for transmission of signals to the brain.

The receptive field of a bipolar or ganglion cell is determined by the field of light detection by the photoreceptor with which it is linked. (Of course light is not directly detected by the bipolar or ganglion cells; light stimulates the photoreceptors, which signal the bipolar cells that in turn send the message to the ganglion cells.) **On-center** and **off-center cells** respond in opposite ways, depending on the relative comparison of illumination between the center and periphery (surround) of their receptive fields. Think of the receptive field as a doughnut. An on-center cell increases its rate of firing when light is most intense at the center of its receptive field (that is, when the doughnut hole is lit up) and stops firing when its surround is most intensely illuminated. In contrast, an off-center cell increases its firing rate when light is the brightest in the periphery of its receptive field (that is, when the doughnut itself is lit up) and stops firing when light is most intense in its center (❙Figure 6-26a). Thus, on-center cells are "turned on" and off-center cells are "turned off" when light shines most intensely on their centers. Both cells respond only weakly when light shines evenly on both their centers and surrounds. This response pattern is useful for enhancing the difference in light level between one small area at the center of a receptive field and the illumination immediately around it. By emphasizing differences in relative brightness, this mechanism helps define contours of images,

but in so doing, information about absolute brightness is sacrificed (❙Figure 6-26b).

Glutamate released from the photoreceptor terminal in the dark has opposite effects on the two types of bipolar cells because they have different types of receptors that lead to different channel responses on binding with this neurotransmitter. Glutamate hyperpolarizes (inhibits) on-center bipolar cells and depolarizes (excites) off-center bipolar cells. When glutamate secretion decreases on light exposure, this reduction depolarizes (stimulates) the hyperpolarized on-center bipolar cells and hyperpolarizes (inhibits) the depolarized off-center bipolar cells. The bipolar cells pass on the information about patterns of illumination to the next neurons in the processing chain, the ganglion cells, by changing their rate of neurotransmitter release in accordance with their state of polarization—increased neurotransmitter release on depolarization and decreased neurotransmitter release on hyperpolarization.

Bipolar cells, similar to the photoreceptors, display graded potentials. Action potentials do not originate until the ganglion cells (the first neurons in the chain that must propagate the visual message over long distances to the brain) are stimulated. As the firing rates of the on-center and off-center ganglion cells change in response to the changing pattern of illumination, the brain is informed about the rapidity and extent of change in contrast within the visual image.

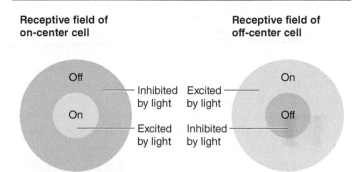

| Receptive field of on-center cell | | Receptive field of off-center cell |

Off / On — Inhibited by light / Excited by light — Excited by light / Inhibited by light — On / Off

Both types of cells are weakly stimulated by uniform light on both center and surround.

(a) Receptive fields of on-center and off-center cells

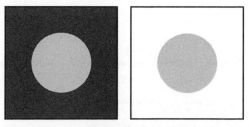

(b) Outcome of retinal processing by on-center and off-center cells

❙Figure 6-26 **On-center and off-center cells in retina.** (a) On-center cells are excited and off-center cells are inhibited by bright light in the centers of their receptive fields. (b) Retinal processing by on-center and off-center ganglion cells is largely responsible for enhancing differences in relative (rather than absolute) brightness, which helps define contours. Note that the gray circle surrounded by black appears brighter than the one surrounded by white, even though the two circles are identical (same shade and size).

Rods provide indistinct gray vision at night; cones provide sharp color vision during the day.

The retina contains 20 times more rods than cones (120 million rods compared to 6 million cones per eye). Cones are most abundant in the macula lutea in the center of the retina. From this point outward, the concentration of cones decreases and the concentration of rods increases. Rods are most abundant in the periphery. We have examined the similar way in which phototransduction takes place in rods and cones. Now we focus on the differences between these photoreceptors (Table 6-3).

Rods Have High Sensitivity; Cones Have Lower Sensitivity The outer segments are longer in rods than in cones, so they contain more photopigments and thus can absorb light more readily. Also, as you will see shortly, the way in which rods connect with other neurons in their processing pathway further increases the sensitivity of rod vision. Rods have high sensitivity, so they can respond to the dim light of night. Cones, by contrast, have lower sensitivity to light, being activated only by bright daylight. Thus, rods are specialized for night vision and cones for day vision.

Cone Vision Has High Acuity; Rod Vision Has Low Acuity The pathways by which cones are "wired" to the other retinal neuronal layers confer high acuity (sharpness, or the ability to distinguish between two nearby points). Thus, cones provide sharp vision with high resolution for fine detail during the day. By contrast, the wiring pathways of rods provide low acuity, so you can see at night but at the expense of distinctness. Let us see how wiring patterns influence sensitivity and acuity.

Little convergence of neurons takes place in the retinal pathways for cone output (see p. 112). Each cone generally has a private line connecting it to a particular ganglion cell. In contrast, much convergence occurs in rod pathways. Output from more than 100 rods may converge via bipolar cells on a single ganglion cell. Before a ganglion cell can have an action potential, the cell must be brought to threshold through influence of the graded potentials in the photoreceptors to which it is wired.

Because a single-cone ganglion cell is influenced by only one cone, only bright daylight is intense enough to induce a sufficient receptor potential in the cone to ultimately bring the ganglion cell to threshold. The abundant convergence in the rod visual pathways, in contrast, offers good opportunities for summation of subthreshold events in a rod ganglion cell (see p. 108). Whereas a small receptor potential induced by dim light in a single cone would not be sufficient to bring its ganglion cell to threshold, similar small receptor potentials induced by the same dim light in multiple rods converging on a single ganglion cell would have an additive effect to bring the rod ganglion cell to threshold. Because rods can bring about action potentials in response to small amounts of light, they are much more sensitive than cones. (Rods are also more sensitive than cones because they have more photopigment.) However, because cones have dedicated lines into the optic nerve, each cone transmits information about an extremely small receptive field on the retinal surface. Cones are thus able to provide highly detailed vision at the expense of sensitivity. With rod vision, acuity is sacrificed for sensitivity. Because many rods share a single ganglion cell, once an action potential is initiated, it is impossible to discern which of the multiple rod inputs were activated to bring the ganglion cell to threshold. Objects appear fuzzy when rod vision is used because of this poor ability to distinguish between two nearby points.

Cones Provide Color Vision; Rods Provide Vision in Shades of Gray There are four different photopigments, one in the rods and one in each of three types of cones—**red**, **green**, and **blue cones**. Each photopigment has the same retinal but a different opsin. Because each opsin binds retinal in a unique way, each of the four photopigments absorbs different wavelengths of light in the visible spectrum to varying degrees. Each photopigment maximally absorbs a particular wavelength but also absorbs a range of wavelengths shorter and longer than this peak absorption. The farther a wavelength is from the peak wavelength absorbed, the less strongly the photopigment responds. Rods absorb the greatest range of wavelengths. The absorption curves for the three cone types overlap so that two or three cones may respond to a given wavelength but to a different extent (Figure 6-27). Because the photopigments in the three types of cones each respond selectively to a different part of the visible light spectrum, the brain can compare the responses of the three cone types, making color vision in daylight possible. In contrast, the brain cannot discriminate among various wavelengths when using visual input from the rods. The rhodopsin in every rod responds in the same way to a given wavelength, so no comparison among rod inputs is possible. Therefore, rods provide vision at night only in shades of gray by detecting different intensities, not different colors. We now examine color vision in further detail.

Color vision depends on the ratios of stimulation of the three cone types.

Vision depends on stimulation of photoreceptors by light. Certain objects in the environment, such as the sun, fire, and light bulbs, emit light. But how do you see objects such as chairs,

TABLE 6-3 Properties of Rod Vision and Cone Vision

Rods	Cones
120 million per retina	6 million per retina
More numerous in periphery	Concentrated in fovea
High sensitivity	Low sensitivity
Night vision	Day vision
Low acuity	High acuity
Much convergence in retinal pathways	Little convergence in retinal pathways
Vision in shades of gray	Color vision

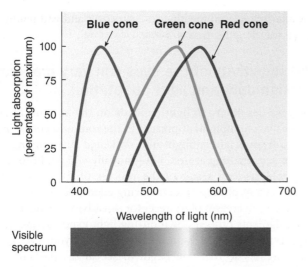

Figure 6-27 Sensitivity of the three types of cones to different wavelengths.

trees, and people, which do not emit light? The pigments in various objects selectively absorb particular wavelengths of light transmitted to them from light-emitting sources, and the unabsorbed wavelengths are reflected from the objects' surfaces. These reflected light rays enable you to see the objects. An object perceived as blue absorbs the longer red and green wavelengths of light and reflects the shorter blue wavelengths, which can be absorbed by the photopigment in your blue cones, thereby activating them.

Each cone type is most effectively activated by a particular wavelength of light in the range of color indicated by its name. The *S-type photopigment* in blue cones absorbs light maximally in the *s*hort-wavelength (blue) part of the visible spectrum, whereas the *M-type photopigment* in green cones is most sensitive to the *m*edium wavelengths (green) of visible light, and the *L-type photopigment* in red cones responds best to the *l*ong (red) wavelengths. However, cones also respond in varying degrees to other wavelengths (Figure 6-27). According to the **trichromatic theory** of color vision, the perception of the many colors of the world depends on the three cone types' various *ratios of stimulation* in response to different wavelengths (*tri* means "three;" *chroma* refers to "color"). A wavelength perceived as blue does not stimulate red or green cones but excites blue cones maximally. (The percentage of maximal stimulation for red, green, and blue cones, respectively, is 0:0:100.) The sensation of yellow, in comparison, arises from a stimulation ratio of 83:83:0—that is, red and green cones are each stimulated 83% of maximum, while blue cones are not excited. The ratio for green is 31:67:36, and so on, with various combinations giving rise to the sensation of all the different colors. White is a mixture of all wavelengths of light, whereas black is the absence of light.

The extent to which each of the cone types is excited is coded and transmitted in separate parallel pathways to the brain. Distinct color processing pathways in the primary visual cortex in the occipital lobe of the brain (see Figure 5-11, p. 148) combine and process these inputs to generate the perception of color. The concept of color is in the mind of the beholder, but most of us agree on what color we see because we have the same types of cones and use similar neural pathways for comparing their output. Occasionally, however, individuals lack a particular cone type, so their color vision is a product of the differential sensitivity of only two types of cones, a condition known as **color blindness**. Not only do color-defective individuals perceive certain colors differently, but they are also unable to distinguish as many varieties of colors (Figure 6-28). For example, people with certain color defects cannot distinguish between red and green. At a traffic light, they can tell which light is "on" by its intensity, but they must rely on the position of the bright light to know whether to stop or go. Because the defective gene associated with red–green color blindness is on the X sex chromosome, the incidence of this condition is greater in males than in females (affecting 8% of men and less than 1% of women). Females have XX and males have XY sex chromosomes; see p. 718. A female who has a defective copy of the gene on one X chromosome usually has a good copy of the gene on her other X chromosome and so has normal color vision, but a male who has a defective copy of the gene on his X chromosome has no comparable gene on his Y chromosome as a backup and so is color blind.

In addition to the trichromatic theory, which applies to the way the retina detects colors with the three cone types, the complementary opponent-process theory applies to further visual processing of information from the cones at a point where inputs from these photoreceptors are interconnected neurally. The **opponent-process theory** of color vision states that subsequent neural handling of trichromatic signals from the cones involves antagonist responses to opponent color pairs within three separate processing channels, namely red versus green, blue versus yellow, and black versus white. Specialized ganglion cells known as **opponent color cells**, which display baseline action potential activity, have opposite responses to the members of a color pair. For example, the firing rate of a red–green opponent cell, which receives both red-cone and green-cone input from the same vicinity of the retina, increases above baseline when it processes a signal from a red

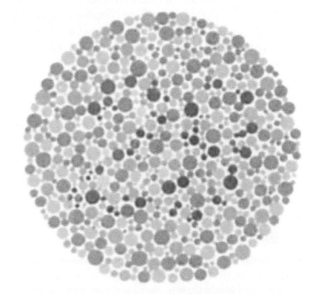

Figure 6-28 Color blindness chart. People with red–green color blindness cannot detect the number 29 in this chart.

light and decreases below baseline when it processes a signal from a green light. Opponent colors in a pair are in competition; that is, stimulation of this opponent color cell by red input inhibits green input, and vice versa, because the firing rate of this cell cannot be increased and decreased at the same time. Thus red and green are mutually antagonistic. This explains why we never see both members of an opponent color pair at the same place and at the same time. That is, we don't see reddish green or bluish yellow. However, we do see combinations of colors that are in different pairs, for example reddish blue (purple) or yellowish red (orange). There is no yellow cone, but blue–yellow opponent cells respond to input from blue cones compared to combined information from red and green cones. (Remember that the sensation of yellow comes from near maximum stimulation of both red and green cones, with no excitation of blue cones.) The black–white opponent processing channel distinguishes darkness and lightness, not specific colors. According to the opponent-process theory, color vision depends on the differential response of these various opponent color cells, which enhances our perception of color contrast, similar to the way in which the differential response of on-center and off-center ganglion cells enhances differences in relative brightness. Further processing of opponent colors takes place in the thalamus and visual cortex.

The opponent-process theory can account for the visual phenomena of opponent color afterimages. When you intently stare at one color of an opponent pair for 30 seconds, then look away at a white space, you briefly perceive a ghostly image of the opposite color (❙Figure 6-29). This optical illusion is an opponent color **afterimage**. When you stare at the blue square, the blue cones fatigue from overstimulation and lose their sensitivity. When you shift your gaze to the white space, the cones that detect the opponent color (yellow, detected by combined activation of red and green cones) are not fatigued and have just been removed from prolonged inhibition by blue cone input. Thus these "fresh" cones can respond to the medium and long wavelengths of light in the white light, whereas the fatigued blue cones briefly are unable to respond to the short wavelengths in the white light. (Recall that white light contains all wavelengths of color.) Because the light reflected off the white paper is able to excite only the yellow-detecting cones, you briefly see a yellow afterimage until the blue cones recover and you once again see all wavelengths, that is, white.

The sensitivity of the eyes can vary markedly through dark and light adaptation.

The eyes' sensitivity to light depends on the amount of light-responsive photopigment present in the rods and cones. When you go from bright sunlight into darkened surroundings, you cannot see anything at first, but gradually you begin to distinguish objects as a result of the process of **dark adaptation**. Breakdown of photopigments during exposure to sunlight tremendously decreases photoreceptor sensitivity. In the dark, the photopigments broken down during light exposure are gradually regenerated. As a result, the sensitivity of your eyes gradually increases so that you begin to see in the darkened surroundings. However, only the highly sensitive, rejuvenated rods are "turned on" by the dim light.

Conversely, when you move from the dark to the light (for example, leaving a movie theater and entering bright sunlight), at first your eyes are very sensitive to the dazzling light. With little contrast between lighter and darker parts, the entire image appears bleached. As some of the photopigments are rapidly broken down by the intense light, the sensitivity of the eyes decreases and normal contrasts can again be detected, a process known as **light adaptation**. The rods are so sensitive to light that enough rhodopsin is broken down in bright light to essentially "burn out" the rods—that is, after the rod photopigments have already been broken down by the bright light, they no longer respond to the light. Therefore, only the less sensitive cones are used for day vision.

Our eyes' sensitivity can change as much as 1 million times as they adjust to various levels of illumination through dark and light adaptation. These adaptive measures are also enhanced by pupillary reflexes that adjust the amount of available light permitted to enter the eye.

Clinical Note Because retinal is a derivative of vitamin A, adequate amounts of this nutrient must be available for synthesis of photopigments. **Night blindness** occurs as a result of dietary deficiencies of vitamin A. Although photopigment concentrations in both rods and cones are reduced in this condition, there is still enough cone photopigment to respond to the intense stimulation of bright light, except in the most severe cases. However, even modest reductions in rhodopsin content can decrease the sensitivity of rods so much that they cannot respond to dim light. The person can see in the day using cones but cannot see at night because the rods are no longer functional. Thus, carrots are "good for your eyes" because they are rich in vitamin A.

Visual information is modified and separated before reaching the visual cortex.

The field of view that can be seen without moving the head is known as the **visual field**. The information that reaches the **primary visual cortex** in the occipital lobe is not a replica of the visual field for several reasons:

1. The image detected on the retina at the onset of visual processing is upside down and backward because of bending of

❙Figure 6-29 **Demonstration of opponent color afterimage.** Focus on the small black circle in the middle of the blue square while you slowly count to 30. Then quickly shift your gaze to the small black triangle in the middle of the white space to the right of the blue square and watch for an afterimage in the white space. What color is the afterimage?

the light rays. Once it is projected to the brain, the inverted image is interpreted as being in its correct orientation.

2. The information transmitted from the retina to the brain is not merely a point-to-point record of photoreceptor activation. Before the information reaches the brain, the retinal neuronal layers beyond the rods and cones reinforce selected information and suppress other information to enhance contrast. The differential activity of on-center and off-center cells along with the contributions of specialized retinal interneurons, the *horizontal cells* and *amacrine cells* (see ❙ Figure 6-20), are responsible for much of this retinal processing. For example, horizontal cells participate in lateral inhibition, by which strongly excited cone pathways suppress activity in surrounding pathways of weakly stimulated cones. This increases the dark–bright contrast to enhance the sharpness of boundaries.

3. Various aspects of visual information, such as shape, color, and motion, are separated and projected in parallel pathways to different regions of the cortex. Only when these separate bits of processed information are integrated by higher visual regions is a reassembled picture of the visual scene perceived.

Clinical Note Patients with lesions in specific visual-processing regions of the brain may be unable to completely combine components of a visual impression. For example, a person may be unable to discern movement of an object but have reasonably good vision for shape, pattern, and color. Sometimes the defect can be remarkably specific, like being unable to recognize familiar faces while retaining the ability to recognize inanimate objects.

4. Because of the pattern of wiring between the eyes and the visual cortex, the left half of the cortex receives information only from the right half of the visual field as detected by both eyes, and the right half receives input only from the left half of the visual field of both eyes.

As light enters the eyes, light rays from the left half of the visual field fall on the right half of the retina of both eyes (the medial or inner half of the left retina and the lateral or outer half of the right retina) (❙ Figure 6-30a). Similarly, rays from the right half of the visual field reach the left half of each retina (the lateral half of the left retina and the medial half of the right retina). Each optic nerve exiting the retina carries information from both halves of the retina it serves. This information is separated as the optic nerves meet at the **optic chiasm** located underneath the hypothalamus (*chiasm* means "cross") (see ❙ Figure 5-7b, p. 143). Within the optic chiasm, the fibers from the medial half of each retina cross to the opposite side, but those from the lateral half remain on the original side. The reorganized bundles of fibers leaving the optic chiasm are known as **optic tracts**. Each optic tract carries information from the lateral half of one retina and the medial half of the other retina. Therefore, this partial crossover brings together, from the two eyes, fibers that carry information from the same half of the visual field. Each optic tract, in turn, delivers to the half of the brain on its same side information about the opposite half of the visual field.

Clinical Note Knowledge of these pathways can facilitate diagnosis of visual defects arising from interruption of the visual pathway at various points (❙ Figure 6-30b).

Before we move on to how the brain processes visual information, take a look at ❙ Table 6-4, which summarizes the functions of the various components of the eyes.

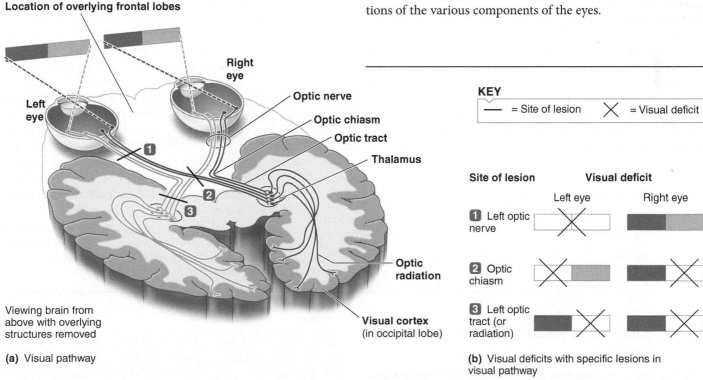

Location of overlying frontal lobes

Right eye

Left eye

Optic nerve

Optic chiasm

Optic tract

Thalamus

① ② ③

Optic radiation

Viewing brain from above with overlying structures removed

Visual cortex (in occipital lobe)

(a) Visual pathway

KEY

— = Site of lesion   ✕ = Visual deficit

Site of lesion	Visual deficit	
	Left eye	Right eye
① Left optic nerve		
② Optic chiasm		
③ Left optic tract (or radiation)		

(b) Visual deficits with specific lesions in visual pathway

❙ **Figure 6-30 The visual pathway and visual deficits associated with lesions in the pathway.** (a) Note that the left half of the visual cortex in the occipital lobe receives information from the right half of the visual field of both eyes (in *green*), and the right half of the cortex receives information from the left half of the visual field of both eyes (in *red*). (b) Each visual deficit illustrated is associated with a lesion at the corresponding numbered point of the visual pathway in part (a).

TABLE 6-4 Functions of the Major Components of the Eye

Structures	Location	Function
Aqueous humor	Anterior cavity between the cornea and lens	Clear, watery fluid that is continually formed and carries nutrients to the cornea and lens
Bipolar cells	Middle layer of nerve cells in the retina	Important in retinal processing of light stimulus
Blind spot	Point slightly off center on the retina where the optic nerve exits; is devoid of photoreceptors (also known as *optic disc*)	Route for passage of the optic nerve and blood vessels
Choroid	Middle layer of the eye	Pigmented to prevent scattering of light rays in the eye; contains blood vessels that nourish the retina; anteriorly specialized to form the ciliary body and iris
Ciliary body	Specialized anterior derivative of the choroid layer; forms a ring around the outer edge of the lens	Produces aqueous humor and contains the ciliary muscle
Ciliary muscle	Circular muscular component of the ciliary body; attached to the lens by means of suspensory ligaments	Important in accommodation
Cones	Photoreceptors in the outermost layer of the retina	Responsible for high acuity, color, and day vision
Cornea	Anterior, clear, outermost layer of the eye	Contributes most extensively to the eye's refractive ability
Fovea	Exact center of the retina	Region with greatest acuity
Ganglion cells	Inner layer of nerve cells in the retina	Important in retinal processing of light stimulus; form the optic nerve
Iris	Visible pigmented ring of muscle within the aqueous humor	Varies size of the pupil by variable contraction; responsible for eye color
Lens	Between the aqueous humor and vitreous humor; attaches to the ciliary muscle by suspensory ligaments	Provides variable refractive ability during accommodation
Macula lutea	Area immediately surrounding the fovea	Has high acuity because of abundance of cones
Optic disc	*(See entry for blind spot)*	
Optic nerve	Leaves each eye at the optic disc (blind spot)	First part of the visual pathway to the brain
Pupil	Anterior round opening in the middle of the iris	Permits variable amounts of light to enter the eye
Retina	Innermost layer of the eye	Contains the photoreceptors (rods and cones)
Rods	Photoreceptors in the outermost layer of the retina	Responsible for high-sensitivity, black-and-white, and night vision
Sclera	Tough outer layer of the eye	Protective connective tissue coat; forms the visible white part of the eye; anteriorly specialized to form the cornea
Suspensory ligaments	Suspended between ciliary muscle and lens	Important in accommodation
Vitreous humor	Between the lens and retina	Semiliquid, jellylike substance that helps maintain spherical shape of the eye

The thalamus and visual cortex elaborate the visual message.

The first stop in the brain for information in the visual pathway is the *lateral geniculate nucleus* of the thalamus (▮Figure 6-30a). It separates information received from the eyes and relays it via fiber bundles known as **optic radiations** to different zones in the primary visual cortex located in the occipital lobes. Each zone processes different aspects of the visual stimulus (for example, form, movement, color, and depth). This sorting process is no small task because each optic nerve contains more than a million fibers carrying information from the photoreceptors in one retina. This is more than all the afferent fibers carrying somatosensory input from all other regions of the body! Researchers estimate that hundreds of millions of neurons occupying about 40% of the cortex participate in visual processing, compared to 8% devoted to touch perception and 3% to hearing. Yet the connections in the visual pathways are precise. The lateral geniculate nucleus and each of the zones in the cortex that processes visual information have a topographical map representing the retina point for point. As with the somatosensory cortex, the neural maps of the retina are distorted. The fovea, the retinal region capable of greatest acuity, has much greater representation in the neural map than the more peripheral regions of the retina do.

Depth Perception Although each half of the visual cortex receives information simultaneously from the same part of the visual field as received by both eyes, the messages from the two eyes are not identical. Each eye views an object from a slightly different vantage point, even though the overlap is tremendous. The overlapping area seen by both eyes at the same time is known as the **binocular** ("two-eyed") field of vision, which is important for **depth perception**. Like other areas of the cortex, the primary visual cortex is organized into functional columns, each processing information from a small region of the retina. Independent alternating columns are devoted to information about the same point in the visual field from the right and left eyes. The brain uses the slight disparity in the information received from the two eyes to estimate distance, allowing you to perceive three-dimensional objects in spatial depth. Some depth perception is possible using only one eye, based on experience and comparison with other cues. For example, if your one-eyed view includes a car and a building and the car is larger, you correctly interpret that the car must be closer to you than the building is.

Sometimes the two views are not successfully merged. This condition may occur for two reasons: (1) The eyes are not both focused on the same object simultaneously because of defects of the external eye muscles that make fusion of the two eyes' visual fields impossible (for example, being cross-eyed); or (2) the binocular information is improperly integrated during visual processing. The result is double vision, or **diplopia**, a condition in which the disparate views from both eyes are seen simultaneously.

Hierarchy of Visual Cortical Processing Within the cortex, visual information is first processed in the primary visual cortex and then is sent to surrounding higher-level visual areas for even more complex processing and abstraction. The visual cortex is precisely organized both vertically and horizontally. Vertical columns extend through the thickness of the cortex from its outer surface to the white matter. Each column is made up of cells that process the same bit of visual input. There are three types of columns, based on the type of visual input they process: (1) as discussed in the preceding section, an alternating system of *ocular dominance columns* devoted to input from the left or the right eye is important for binocular interaction and depth perception; (2) *orientation columns* related to axis of orientation of visual stimuli play a key role in perceiving form and movement; and (3) shortened columns known as *blobs* process color.

The orientation columns contain a hierarchy of visual cells that respond to increasingly complex stimuli. Three types of visual cortical neurons have been identified based on the complexity of stimulus requirements needed for the cell to respond; these are called **simple**, **complex**, and **hypercomplex cells**. All of the cells within a given orientation column process input arising from visual stimuli in the same axis of orientation, such as a slit of light oriented vertically, horizontally, or at some oblique angle. The primary visual cortex has orientation columns for every possible axis of orientation. Dissection of visual input into these various orientations is important for discerning form and movement. The visual cortex is also organized into six layers, with each layer consisting of specific cell types. For example, simple cells are found in layer IV. Simple and complex cells are stacked on top of one another in a specific way within each orientation column. Hypercomplex cells are found only in higher visual-processing areas. Horizontal connections within the layers link vertical columns that carry out similar functions. Each layer has different inputs and outputs and is specialized to perform a particular task.

Unlike a retinal cell that responds to the amount of light, a cortical cell fires only when it receives a particular pattern of illumination for which it is programmed. These patterns are built up by converging connections that originate from closely aligned photoreceptor cells in the retina. For example, in the orientation columns some simple cells fire only when a bar is viewed vertically in a specific location, others when a bar is horizontal, and others at various oblique orientations. Movement of a critical axis of orientation becomes important for response by some of the complex cells. Hypercomplex cells add a new dimension to visual processing by responding only to particular edges, corners, and curves. Each level of cortical visual neurons has increasingly greater capacity for abstraction of information built up from the increasing convergence of input from lower-level neurons. In this way, the cortex transforms the dotlike pattern of photoreceptors stimulated to varying degrees by varying light intensities in the retinal image into information about position, orientation, movement, contour, and length. Other aspects of visual information, such as depth perception and color perception, are processed simultaneously by the other vertical and horizontal organizational systems. How and where the entire image is finally put together is still unresolved. This is similar to the blobs of paint on an artist's palette versus the finished portrait; the separate pigments do not represent a portrait of a face until they are appropriately integrated on a canvas.

Visual input goes to other areas of the brain not involved in vision perception.

Not all fibers in the visual pathway terminate in the visual cortex. Some are projected to other regions of the brain for purposes other than direct vision perception. Examples of nonsight activities dependent on input from the rods and cones include (1) contribution to cortical alertness and attention (for example, you get drowsy in a dimly lit room), (2) control of pupil size (for example, your pupils constrict in bright light), and (3) control of eye movements (for example, input from your

"Seeing" with the Tongue or the Ear

ALTHOUGH EACH TYPE OF SENSORY input is received primarily by a distinct brain region responsible for perception of that modality, the regions of the brain involved with perceptual processing receive sensory signals from a variety of sources. Thus, the visual cortex receives sensory input not only from the eyes, but also from the body surface and ears. One group of scientists exploited, in an unusual but exciting way, this sharing of sensory input by multiple regions of the brain. They developed *BrainPort,* a noninvasive device that enables the blind to crudely perceive shapes and motion in space by means of a *tongue display unit (TDU),* a flat, 9-cm², lollipop-like unit consisting of a grid of electrodes that is positioned against the tongue (see the accompanying figure). A miniature camera mounted on glasses sends visual data to a handheld base unit that converts light input into a pattern of electrical signals that are sent to the TDU, where they activate touch receptors on the tongue. The pattern of "tingling" on the tongue (similar to the feeling of effervescent champagne) as a result of the light-induced electrical signals corresponds with the image recorded by the camera. With practice, the visual cortex interprets this alternate sensory input as a visual image. As one of the investigators who developed this technique claims, a person

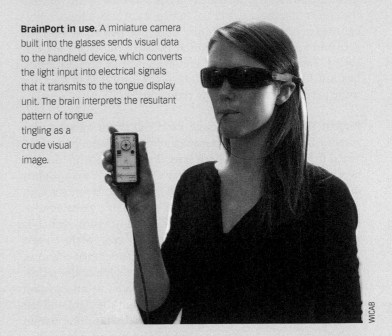

BrainPort in use. A miniature camera built into the glasses sends visual data to the handheld device, which converts the light input into electrical signals that it transmits to the tongue display unit. The brain interprets the resultant pattern of tongue tingling as a crude visual image.

WICAB

photoreceptors is used to guide contraction of your external eye muscles to enable you to read this page). Each eye is equipped with a set of six **external eye muscles** that position and move the eye so that it can better locate, see, and track objects. Eye movements are among the fastest, most discretely controlled movements of the body.

About 3% of the eyes' ganglion cells are not wired to rods and cones and are not involved in visual processing. Instead, they make **melanopsin,** a light-sensitive pigment that plays a key role in setting the body's "biological clock" to march in step with the light–dark cycles (see p. 661).

Some sensory input may be detected by multiple sensory-processing areas in the brain.

Before shifting gears to another sense—hearing—we should mention a new theory regarding the senses that challenges the prevailing view that the separate senses feed into distinct brain regions that handle only one sense. A growing body of evidence suggests that the brain regions devoted almost exclusively to a certain sense, such as the visual cortex for visual input and the somatosensory cortex for touch input, actually receive a variety of sensory signals. No sense works alone. For example, tactile and auditory signals also arrive in the visual cortex. One study using new brain-imaging techniques showed that people blind from birth use the visual cortex

when they read Braille, even though they are not "seeing" anything. The tactile input from their fingers reaches the visual area of the brain as well as the somatosensory cortex. This input helps them "visualize" the patterns of the Braille bumps. More and more evidence supports this multisensory revolution, the notion that our senses "eavesdrop" on one another and that the brain makes sense of the world by deriving information from as many avenues as possible and blending the diverse forms of sensory perception.

Also reinforcing the notion that central processing of different types of sensory input overlaps to some extent, scientists recently discovered *multisensory neurons*—brain cells that react to multiple sensory inputs instead of just to one. No one knows whether these cells are rare or commonplace in the brain. (See the accompanying boxed feature, ■Concepts, Challenges, and Controversies, for one way in which researchers exploited this sharing of sensory input by multiple regions of the brain.)

Such cross-sensory experiences may also be the basis of **synaesthesia** (meaning "joining of senses"), a poorly understood condition in which two or more senses are connected. For example, those with synaesthesia may "see" letters, words, and numbers as colors, "taste" shapes, or "hear" colors. Synaesthetic abilities are very specific for the individual. One person with synaesthesia may always see "M" as indigo blue; another may always see this letter as red. The blending of senses occurs in one direction only; even though "M" is always red, red can be seen without conjuring up the letter "M."

sees with the visual cortex, not with the eyes. Any means of sending signals to the visual cortex can be perceived as a visual image. For example, with this device, the blind can see the shapes of furniture, track the movement of people, identify doorways and elevator buttons, read letters and numbers, pick up a cup without groping, or enjoy the flickering of a candle flame. Although using the tongue as a surrogate eye can never provide anywhere near the same vision as a normal eye, even this limited visual input can enable a sightless person to get around more easily and enjoy a better quality of life.

The tongue is a better choice than the skin for receiving this light-turned-tactile input because the saliva is an electrically conducive fluid that readily conducts the current generated in the device by the visual input. Furthermore, the tongue is densely populated with tactile receptors, opening up the possibility that the tongue can provide higher acuity of visual input than the skin could. In fact, researchers plan to improve the resolution of the device by increasing the number of in-the-mouth electrodes.

In a related approach, other researchers have created an iPhone app, *EyeMusic*, that can scan and convert images into sequences of sound that allow the blind to "see" with sound. With this technology,

sound frequencies correspond to particular aspects of the scene. For example, higher frequencies of sound denote higher locations in the image. With training, the user can learn to "hear" everyday visual scenes, such as identifying individuals by their general facial characteristics (for instance by hair color, whether or not they have a beard or are wearing glasses, and so on).

Still other research groups have alternately developed microelectronic chips for implant in the eye to bypass defective photoreceptors or, much farther out, in the visual cortex (to circumvent deficits in the visual pathway) to enable the blind to "see the light" at least to some extent. Several different versions of retinal implants in early stages of development have enabled blind people to make out the basic outlines of people and objects, detect movement in front of them, and even to string together words by reading high-contrast letters. Several other promising avenues under investigation for halting or even reversing the loss of sight in degenerative eye diseases are injecting functional genes into eyes blinded by genetic mutations in retinal cells (already here), regenerating the retina through fetal retinal transplants, growing replacement retinas in the laboratory, and stem cell therapy (see p. 10).

For the remainder of the chapter, we will concentrate on the mainstream function of the other special senses. We next shift attention from the eyes to the ears.

Check Your Understanding 6.3

1. Draw two sagittal sections of an emmetropic eye, one for far vision and one accommodated for near vision.
2. Explain how light absorption by a photopigment leads to a hyperpolarizing receptor potential.
3. Compare rod and cone vision.
4. Distinguish among optic nerve, optic chiasm, optic tract, and optic radiation.

6.4 Ear: Hearing and Equilibrium

Each **ear** consists of three parts: the *external*, the *middle*, and the *inner ear* (❙ Figure 6-31). The external and middle portions of the ear transmit airborne sound waves to the fluid-filled inner ear, amplifying sound energy in the process. The inner ear houses two sensory systems: the *cochlea*, which contains the receptors for conversion of sound waves into nerve impulses,

making hearing possible, and the *vestibular apparatus*, which is necessary for the sense of equilibrium.

Sound waves consist of alternate regions of compression and rarefaction of air molecules.

Hearing is the neural perception of sound energy. It helps a person interact with both the external environment and with human society. Hearing involves two aspects: the identification of the sounds (what) and their localization (where). We first examine the characteristics of sound waves and then explore how the ears and brain process sound input to accomplish hearing.

Sound waves are traveling vibrations of air. They consist of regions of high pressure, caused by compression of air molecules, alternating with regions of low pressure, caused by rarefaction of the molecules (❙ Figure 6-32a). Any device capable of producing such a disturbance pattern in air molecules is a source of sound. A simple example is a tuning fork. When a tuning fork is struck, its prongs vibrate. As a prong of the fork moves in one direction (❙ Figure 6-32b), air molecules ahead of it are pushed closer together, or compressed, increasing the pressure in this area. Simultaneously, as the prong moves forward, the air molecules behind the prong spread out, or are rarefied, lowering the pressure in that region. As the prong moves in the opposite direction, an opposite wave of compression and rarefaction is created. Even though individual molecules are moved only short distances as the tuning fork vibrates, alternating waves of com-

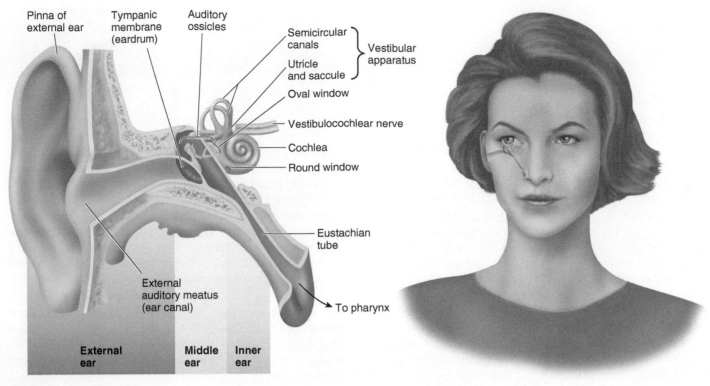

Figure 6-31 Anatomy of the ear.

pression and rarefaction spread out considerable distances in a rippling fashion. Disturbed air molecules disturb other molecules in adjacent regions, setting up new regions of compression and rarefaction, and so on (∎Figure 6-32c). Sound energy gradually dissipates as sound waves travel farther from the original sound source; it finally dies out when the last sound wave is too weak to disturb the air molecules around it.

Sound waves can also travel through media other than air, such as water. They do so less efficiently, however; greater pressures are required to cause movements of fluid than movements of air because of the fluid's greater inertia (resistance to change).

Sound is characterized by its pitch (tone), intensity (loudness), and timbre (quality) (∎Figure 6-33):

- The **pitch**, or **tone**, of a sound (for example, whether it is a C or a G note) is determined by the *frequency* of vibrations. The greater the frequency of vibration, the higher the pitch. Human ears can detect sound waves with frequencies from 20 to 20,000 cycles per second, or **hertz (Hz)**, but are most sensitive to frequencies between 1000 and 4000 Hz.

- The **intensity**, or **loudness**, of a sound depends on the *amplitude* of the sound waves, or the pressure differences between a high-pressure region of compression and a low-pressure region of rarefaction. Within the hearing range, the greater the amplitude, the louder the sound. Human ears can detect a wide range of sound intensities, from the slightest whisper to the painfully loud takeoff of a jet. Loudness is measured in **decibels (dB)**, which are a logarithmic measure of intensity compared with the faintest sound that can be heard—the **hearing threshold**. Because of the logarithmic relationship, every 10 dB indicates a 10-fold increase in loudness. A few ex-

amples of common sounds illustrate the magnitude of these increases (∎Table 6-5). Note that the rustle of leaves at 10 dB is 10 times louder than hearing threshold but the sound of a jet taking off at 150 dB is 1 quadrillion (1 million billion) times, not 150 times, louder than the faintest audible sound. Sounds greater than 100 dB can permanently damage the sensitive sensory apparatus in the cochlea.

- The **timbre**, or **quality**, of a sound depends on its *overtones*, which are additional frequencies superimposed on the fundamental pitch or tone. A tuning fork has a pure tone, but most sounds lack purity. For example, complex mixtures of overtones impart different sounds to different instruments playing the same note (a C note on a trumpet sounds different from C on a piano). Overtones are likewise responsible for characteristic differences in voices. Timbre enables the listener to distinguish the source of sound waves because each source produces a different pattern of overtones. Thanks to timbre, you can tell whether it is your mother or girlfriend calling on the phone before you say the wrong thing.

Next we describe how sound waves are detected by our ears, and then converted into neural signals that are sent to the brain for interpretation.

The external ear plays a role in sound localization.

The specialized receptor cells for sound are located in the fluid-filled inner ear. Airborne sound waves must therefore be channeled toward and transferred into the inner ear, compensating for the loss in sound energy that naturally occurs as sound

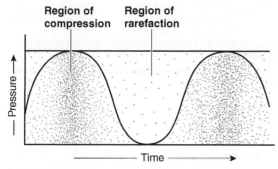

Region of compression   Region of rarefaction

(a) Sound waves

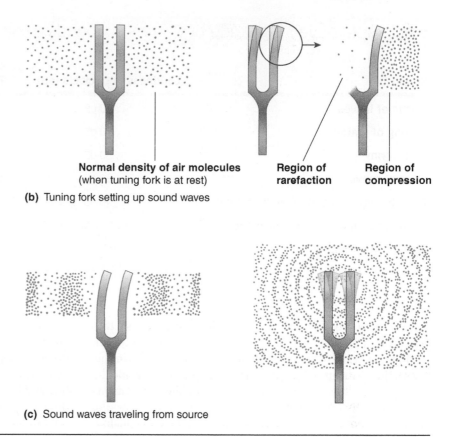

Normal density of air molecules (when tuning fork is at rest)   Region of rarefaction   Region of compression

(b) Tuning fork setting up sound waves

(c) Sound waves traveling from source

▮Figure 6-32 **Formation of sound waves.** (a) Sound waves are alternating regions of compression and rarefaction of air molecules. (b) A vibrating tuning fork sets up sound waves as the air molecules ahead of the advancing arm of the tuning fork are compressed while the molecules behind the arm are rarefied. (c) Disturbed air molecules bump into molecules beyond them, setting up new regions of air disturbance more distant from the original source of sound. In this way, sound waves travel progressively farther from the source, even though each individual air molecule travels only a short distance when it is disturbed. The sound wave dies out when the last region of air disturbance is too weak to disturb the region beyond it.

waves pass from air into water. This function is performed by the external ear and the middle ear.

The **external ear** (see ▮Figure 6-31) consists of the *pinna* (ear), *external auditory meatus* (ear canal), and *tympanic membrane* (eardrum). The **pinna**, a prominent skin-covered flap of cartilage, collects sound waves and channels them down the ear canal. The **ear canal** tunnels through the temporal bone from the exterior to the **tympanic membrane**, a thin membrane that separates the external ear and the middle ear.

Many species (dogs, for example) can cock their ears in the direction of sound to collect more sound waves, but human ears are relatively immobile. Because of its shape, the pinna partially shields sound waves that approach the ear from the rear and thus helps a person distinguish whether a sound is coming from directly in front or behind. Sound localization for sounds approaching from the right or left depends on the sound wave reaching the ear closer to the sound source slightly before it arrives at the farther ear. The auditory cortex integrates all these cues to determine the location of the sound source. It is difficult to localize sound with only one functional ear.

The tympanic membrane vibrates in unison with sound waves in the external ear.

The tympanic membrane, which is stretched across the entrance to the middle ear, vibrates when struck by sound waves. The alternating higher- and lower-pressure regions of a sound wave cause the exquisitely sensitive eardrum to bow inward and outward in unison with the wave's frequency.

For the membrane to be free to move as sound waves strike it, the resting air pressure on both sides of the tympanic membrane must be equal. The outside of the eardrum is exposed to atmospheric pressure that reaches it through the ear canal. The inside of the eardrum facing the middle ear cavity is also exposed to atmospheric pressure via the **eustachian (auditory) tube**, which connects the middle ear to the **pharynx** (back of the throat) (see ▮Figure 6-31). The eustachian tube is normally closed, but it can be pulled open by yawning, chewing, and swallowing. Such opening permits air pressure within the

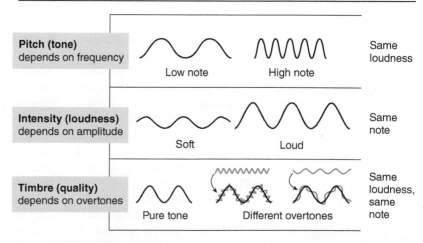

Pitch (tone) depends on frequency		Same loudness
Low note	High note	
Intensity (loudness) depends on amplitude		Same note
Soft	Loud	
Timbre (quality) depends on overtones		Same loudness, same note
Pure tone	Different overtones	

▮Figure 6-33 **Properties of sound waves.**

TABLE 6-5 Relative Magnitude of Common Sounds

Sound	Loudness in Decibels (dB)	Comparison to Faintest Audible Sound (Hearing Threshold)
Rustle of leaves	10 dB	10 times louder
Ticking of watch	20 dB	100 times louder
Whispering	30 dB	1 thousand times louder
Normal conversation	60 dB	1 million times louder
Food blender, lawn mower, hair dryer	90 dB	1 billion times louder
Loud rock concert, ambulance siren	120 dB	1 trillion times louder
Takeoff of jet plane	150 dB	1 quadrillion times louder

middle ear to equilibrate with atmospheric pressure so that pressures on both sides of the tympanic membrane are equal. During rapid external pressure changes (for example, during air flight), the eardrum bulges painfully as the pressure outside the ear changes while the pressure in the middle ear remains unchanged. Opening the eustachian tube by yawning allows the pressure on both sides of the tympanic membrane to equalize, relieving the pressure distortion as the eardrum "pops" back into place.

Clinical Note Infections originating in the throat sometimes spread through the eustachian tube to the middle ear. The resulting fluid accumulation in the middle ear not only is painful but also interferes with sound conduction across the middle ear.

The middle ear bones convert tympanic membrane vibrations into fluid movements in the inner ear.

The **middle ear** transfers the vibrating movements of the tympanic membrane to the fluid of the inner ear. This transfer is facilitated by a movable chain of three small bones, or **ossicles** (the **malleus, incus,** and **stapes**), that extend across the middle ear (Figure 6-34a). The first bone, the malleus, is attached to the tympanic membrane, and the last bone, the stapes, is attached to the **oval window**, the entrance into the fluid-filled cochlea. As the tympanic membrane vibrates in response to sound waves, the chain of bones is set into motion at the same frequency, transmitting this frequency of movement from the tympanic membrane to the oval window. The resulting pressure on the oval window with each vibration produces wavelike movements in the inner ear fluid at the same frequency as the original sound waves. Recall that it takes greater pressure to set fluid in motion than needed to move air, but the ossicular system amplifies the pressure of the airborne sound waves sufficiently by two mechanisms to produce fluid movements in the cochlea. First, because the surface area of the tympanic membrane is much larger than that of the oval window, pressure is increased as force exerted on the tympanic membrane is con-

veyed by the ossicles to the oval window (pressure = force/unit area). Second, the lever action of the ossicles provides an additional mechanical advantage. Together, these mechanisms increase the force exerted on the oval window by 20 times what it would be if the airborne sound wave struck the oval window directly. This additional pressure is sufficient to set the cochlear fluid in motion.

Several tiny muscles in the middle ear contract reflexly in response to loud sounds (greater than 70 dB), causing the tympanic membrane to tighten and limiting movement of the ossicular chain. This reduced movement of middle ear structures diminishes the transmission of loud sound waves to the inner ear to protect the delicate sensory apparatus from damage. This reflex response is relatively slow, however, happening at least 40 msec after exposure to a loud sound. It thus provides protection only from prolonged loud sounds, not from sudden sounds like an explosion. Taking advantage of this reflex, World War II antiaircraft guns were designed to make a loud prefiring sound to protect the gunner's ears from the much louder boom of the actual firing.

The cochlea contains the organ of Corti, the sense organ for hearing.

The pea-sized, snail-shaped **cochlea**, the "hearing" portion of the inner ear, is a coiled tubular system lying deep within the temporal bone (see Figure 6-31) (*cochlea* means "snail"). It is easier to understand the functional components of the cochlea by "uncoiling" it, as shown in Figure 6-34a. The cochlea is divided throughout most of its length into three fluid-filled longitudinal compartments. A blind-ended **cochlear duct**, which is also known as the **scala media**, constitutes the middle compartment. It tunnels lengthwise through the center of the cochlea, not quite reaching its end. The upper compartment, the **scala vestibuli**, follows the inner contours of the spiral, and the **scala tympani**, the lower compartment, follows the outer contours (Figure 6-34a and b). The fluid within the scala vestibuli and scala tympani is called **perilymph**. The cochlear duct contains a slightly different fluid, the **endolymph** (Figure

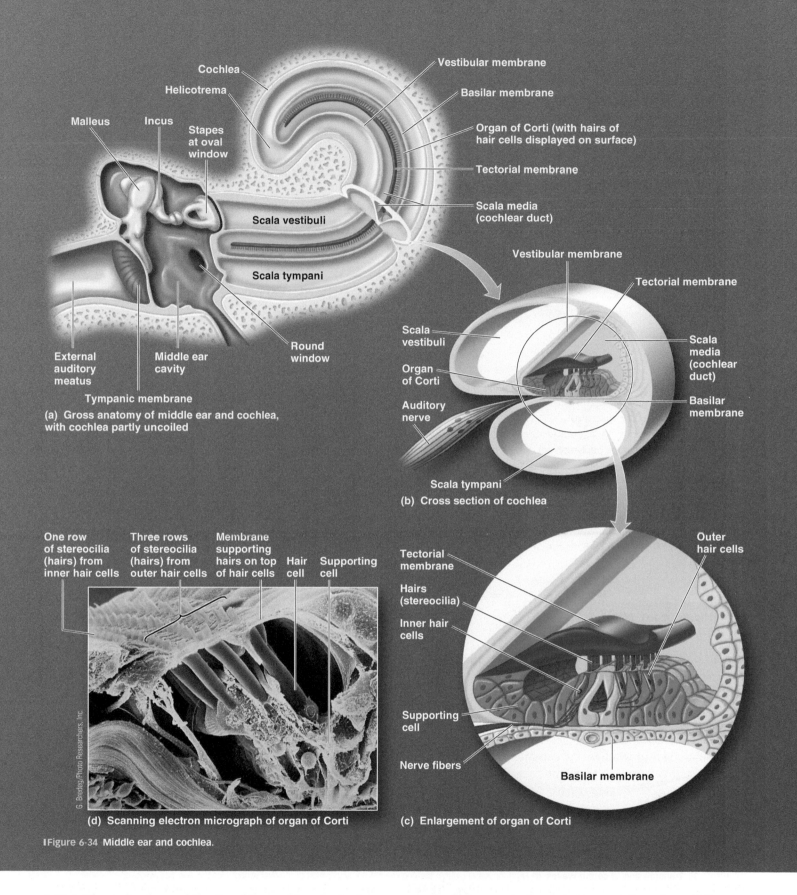

Cochlea

Helicotrema

Vestibular membrane

Basilar membrane

Malleus   Incus

Stapes at oval window

Organ of Corti (with hairs of hair cells displayed on surface)

Tectorial membrane

Scala vestibuli

Scala media (cochlear duct)

Scala tympani

External auditory meatus

Middle ear cavity

Round window

Tympanic membrane

(a) Gross anatomy of middle ear and cochlea, with cochlea partly uncoiled

Vestibular membrane

Tectorial membrane

Scala vestibuli

Scala media (cochlear duct)

Organ of Corti

Auditory nerve

Basilar membrane

Scala tympani

(b) Cross section of cochlea

One row of stereocilia (hairs) from inner hair cells

Three rows of stereocilia (hairs) from outer hair cells

Membrane supporting hairs on top of hair cells

Hair cell

Supporting cell

G. Bredeg/Photo Researchers, Inc.

(d) Scanning electron micrograph of organ of Corti

Tectorial membrane

Hairs (stereocilia)

Inner hair cells

Outer hair cells

Supporting cell

Nerve fibers

Basilar membrane

(c) Enlargement of organ of Corti

❙Figure 6-34 Middle ear and cochlea.

6-35a). The region beyond the tip of the cochlear duct where the fluid in the upper and lower compartments is continuous is called the **helicotrema**. The scala vestibuli is sealed from the middle ear cavity by the oval window, to which the stapes is attached. Another small membrane-covered opening, the

round window, seals the scala tympani from the middle ear. The thin **vestibular membrane** forms the ceiling of the cochlear duct and separates it from the scala vestibuli. The **basilar membrane** forms the floor of the cochlear duct, separating it from the scala tympani. The basilar membrane is especially

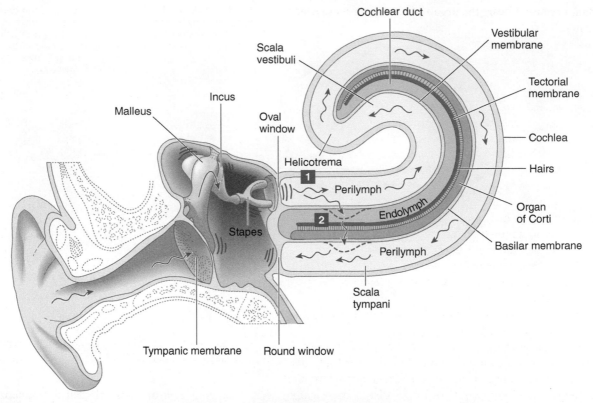

Fluid movement within the perilymph set up by vibration of the oval window follows two pathways:

Pathway 1: Through the scala vestibuli, around the heliocotrema, and through the scala tympani, causing the round window to vibrate. This pathway just dissipates sound energy.

Pathway 2: A "shortcut" from the scala vestibuli through the basilar membrane to the scala tympani. This pathway triggers activation of the receptors for sound by bending the hairs of hair cells as the organ of Corti on top of the vibrating basilar membrane is displaced in relation to the overlying tectorial membrane.

(a) Fluid movement in cochlea

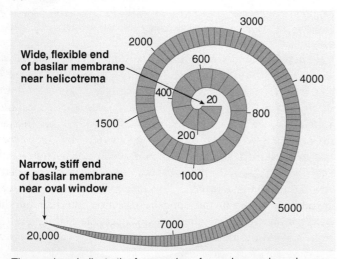

The numbers indicate the frequencies of sound waves in cycles per second (hertz) with which different regions of the basilar membrane maximally vibrate.

(b) Basilar membrane, partly uncoiled

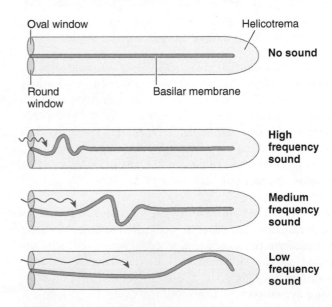

(c) Basilar membrane, completely uncoiled

‖Figure 6-35 Transmission of sound waves. (a) Fluid movement within the cochlea set up by vibration of the oval window follows two pathways, one dissipating sound energy and the other initiating the receptor potential. (b) Different regions of the basilar membrane vibrate maximally at different frequencies. (c) The narrow, stiff end of the basilar membrane nearest the oval window vibrates best with high-frequency pitches. The wide, flexible end of the basilar membrane near the helicotrema vibrates best with low-frequency pitches.

FIGURE FOCUS: *The spiral shape of the cochlea steers low-frequency sound waves (bass sounds) toward the tightest turn at its center. What is the functional significance of this phenomenon?*

important because it bears the **organ of Corti**, the sense organ for hearing.

Hair cells in the organ of Corti transduce fluid movements into neural signals.

The organ of Corti, which rests on top of the basilar membrane throughout its full length, contains **auditory hair cells** that are the receptors for sound. The 15,000 hair cells within each cochlea are arranged in four parallel rows along the length of the basilar membrane: one row of **inner hair cells** and three rows of **outer hair cells** (see ❙Figure 6-34c and d). Protruding from the surface of each hair cell are about 100 hairs known as **stereocilia**, which are actin-stiffened microvilli, not true cilia (see p. 50). Hair cells are mechanoreceptors; they generate neural signals when their surface hairs are mechanically deformed by fluid movements in the inner ear. These stereocilia contact the **tectorial membrane**, an awninglike projection overhanging the organ of Corti throughout its length (see ❙Figure 6-34b and c).

The pistonlike action of the stapes against the oval window sets up pressure waves in the upper compartment. Because fluid is incompressible, pressure is dissipated in two ways as the stapes causes the oval window to bulge inward: (1) displacement of the round window and (2) deflection of the basilar membrane (❙Figure 6-35a). In the first of these pathways, the pressure wave pushes the perilymph forward in the upper compartment, around the helicotrema, and into the lower compartment, where it causes the round window to bulge outward into the middle ear cavity to compensate for the pressure increase. As the stapes rocks backward and pulls the oval window outward toward the middle ear, the perilymph shifts in the opposite direction, displacing the round window inward. This pathway does not result in sound reception; it just dissipates pressure.

Pressure waves of frequencies associated with sound reception take a "shortcut" (❙Figure 6-35a). Pressure waves in the upper compartment are transferred through the thin vestibular membrane, into the cochlear duct and then through the basilar membrane into the lower compartment. Transmission of pressure waves through the basilar membrane causes this membrane to move up and down, or vibrate, in synchrony with the pressure wave. Because the organ of Corti rides on the basilar membrane, the hair cells also move up and down.

Role of the Inner Hair Cells The inner and outer hair cells differ in function. The inner hair cells are the ones that "hear": They transform the mechanical forces of sound (cochlear fluid vibration) into the electrical impulses of hearing (action potentials propagating auditory messages to the cerebral cortex). Because the stereocilia of these receptor cells are connected to the stiff, stationary tectorial membrane, they are bent back and forth when the oscillating basilar membrane shifts their position in relationship to the tectorial membrane (❙Figure 6-36). This back-and-forth mechanical deformation of the hairs alternately opens and closes mechanically gated cation channels (see p. 89) in the hair cell, resulting in alternating depolarizing and hyperpolarizing potential changes—the receptor potential—at the same frequency as the original sound stimulus.

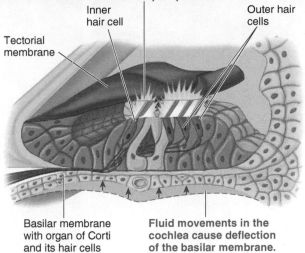

The stereocilia (hairs) from the hair cells of the basilar membrane contact the overlying tectorial membrane. These hairs are bent when the basilar membrane is deflected in relation to the stationary tectorial membrane. This bending of the inner hair cells' hairs opens mechanically gated channels, leading to ion movements that result in a receptor potential.

❙Figure 6-36 **Bending of hairs on deflection of the basilar membrane.**

The stereocilia of each hair cell are organized into three rows of increasing heights in a precise staircaselike pattern resembling organ pipes (❙Figure 6-37a). **Tip links**, which are CAMs (cell adhesion molecules; see p. 60), link the tip of one stereocilium to the side of the next taller stereocilium in the next row up (❙Figure 6-37b). The tallest stereocilia in the top row is directly linked with the overlying tectorial membrane. When the basilar membrane moves upward, the bundle of stereocilia bends toward its tallest membrane that is connected to the stationary tectorial membrane, stretching the tip links. Stretched tip links tug open the mechanically gated cation channels to which they are attached (❙Figure 6-37c). The resultant ion movement is unusual because of the unique composition of the endolymph that bathes the stereocilia. In sharp contrast to ECF elsewhere, endolymph has a higher concentration of K^+ than found inside the hair cell. Some cation channels are open in a resting hair cell, allowing low-level K^+ entry down its concentration gradient. When more cation channels are pulled open, more K^+ enters the hair cell. This additional entering K^+ depolarizes (excites) the hair cell. When the basilar membrane moves in the opposite direction, the hair bundle bends away from the tallest stereocilium, slackening the tip links and closing all the channels. As a result, K^+ entry ceases, hyperpolarizing the hair cell.

Like photoreceptors, hair cells do not undergo action potentials. The inner hair cells communicate via a chemical synapse with the terminals of afferent nerve fibers making up the **auditory (cochlear) nerve**. Because of low-level K^+ entry, the inner hair cells spontaneously release some neurotransmitter (glutamate) via Ca^{2+}-induced exocytosis in the absence of stimulation. Depolarization of these hair cells opens more voltage-gated Ca^{2+} channels. The resultant additional Ca^{2+} entry increases their rate of neurotransmitter secretion, which steps up the rate of firing in the afferent fibers with which the inner

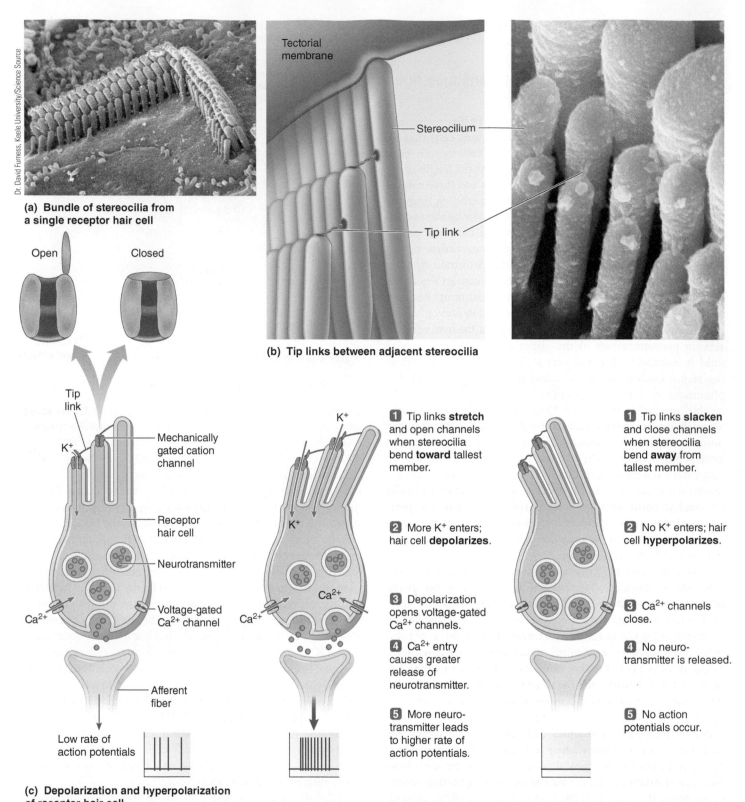

(a) Bundle of stereocilia from a single receptor hair cell

Open   Closed

Tectorial membrane

Stereocilium

Tip link

(b) Tip links between adjacent stereocilia

Tip link

K⁺

Mechanically gated cation channel

Receptor hair cell

Neurotransmitter

Voltage-gated Ca²⁺ channel

Ca²⁺

Afferent fiber

Low rate of action potentials

K⁺

K⁺

Ca²⁺

Ca²⁺

1 Tip links **stretch** and open channels when stereocilia bend **toward** tallest member.

2 More K⁺ enters; hair cell **depolarizes**.

3 Depolarization opens voltage-gated Ca²⁺ channels.

4 Ca²⁺ entry causes greater release of neurotransmitter.

5 More neuro-transmitter leads to higher rate of action potentials.

1 Tip links **slacken** and close channels when stereocilia bend **away** from tallest member.

2 No K⁺ enters; hair cell **hyperpolarizes**.

3 Ca²⁺ channels close.

4 No neuro-transmitter is released.

5 No action potentials occur.

(c) Depolarization and hyperpolarization of receptor hair cell

❙Figure 6-37 **The role of stereocilia in sound transduction.**

FIGURE FOCUS: *K⁺ passage through open channels in stereocilia depolarizes the receptor hair cell, yet K⁺ passage through open channels at an inhibitory synapse hyperpolarizes the postsynaptic cell. What is responsible for the different outcomes in potential in these two situations?*

hair cells synapse. Conversely, the firing rate decreases below resting level as these hair cells release less neurotransmitter when they are hyperpolarized on displacement in the opposite direction.

To summarize, the ear converts sound waves in the air into oscillating movements of the basilar membrane that bends the hairs of the receptor cells back and forth. This shifting mechanical deformation of the hairs alternately opens and closes the receptor cells' channels, bringing about graded potential changes in the receptor that lead to changes in the rate of action potentials propagated to the brain. These neural signals are perceived by the brain as sound sensations (Figure 6-38).

Role of the Outer Hair Cells Whereas the inner hair cells send auditory signals to the brain over afferent fibers, the outer hair cells do not signal the brain about incoming sounds. Instead, the outer hair cells actively and rapidly change length in response to changes in membrane potential, a behavior known as **electromotility**. The outer hair cells shorten on depolarization and lengthen on hyperpolarization. These changes in length mechanically amplify or accentuate the motion of the basilar membrane. An analogy would be a person deliberately pushing the pendulum of a grandfather clock in time with its swing to accentuate its motion. Such modification of basilar membrane movement improves and tunes stimulation of the inner hair cells. Thus, the outer hair cells enhance the response of the inner hair cells, the real auditory sensory receptors, making them exquisitely sensitive to sound intensity and highly discriminatory between various pitches of sound.

Pitch discrimination depends on the region of the basilar membrane that vibrates.

Pitch discrimination (that is, the ability to distinguish among various frequencies of incoming sound waves) depends on the shape and properties of the basilar membrane, which is narrow and stiff at its oval window end and wide and flexible at its helicotrema end (see Figure 6-35b). Different regions of the basilar membrane naturally vibrate maximally at different frequencies—that is, each frequency displays peak vibration at a different position along the membrane. The narrow end nearest the oval window vibrates best with high-frequency pitches, whereas the wide end nearest the helicotrema vibrates maximally with low-frequency tones (see Figure 6-35c). The pitches in between are sorted out precisely along the length of the membrane from higher to lower frequency. As a sound wave of a particular frequency is set up in the cochlea by oscillation of the stapes, the wave travels to the region of the basilar membrane that naturally responds maximally to that frequency. The energy of the pressure wave is dissipated with this vigorous membrane oscillation, so the wave dies out at the region of maximal displacement.

The hair cells in the region of peak vibration of the basilar membrane undergo the most mechanical deformation and accordingly are the most excited. You can think of the organ of Corti as a piano with 15,000 keys (represented by the 15,000 hair cells) rather than the usual 88 keys. Each hair cell is "tuned" to an optimal sound frequency, determined by its location on the organ of Corti. Different sound waves promote maximal movement of different regions of the basilar membrane and thus activate differently tuned hair cells (that is, different sound waves "strike" different "piano keys"). This information is propagated to the CNS, which interprets the pattern of hair cell stimulation as a sound of a particular frequency. The basilar membrane is so fine-tuned that the peak membrane response to a single pitch probably extends no more than the width of a few hair cells.

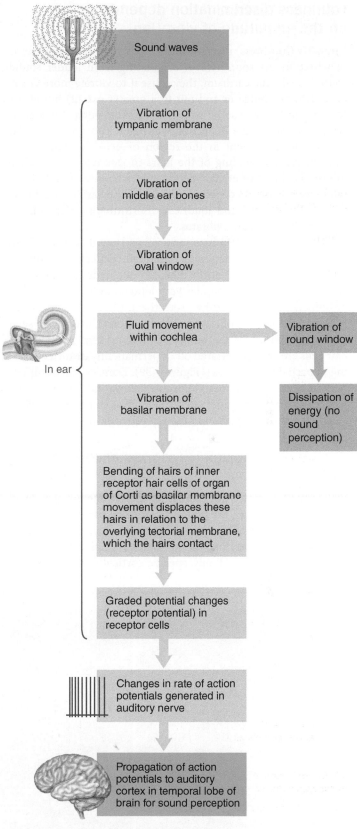

Figure 6-38 **Pathway for sound transduction.**

Overtones of varying frequencies cause many points along the basilar membrane to vibrate simultaneously but less intensely than the fundamental tone, enabling the CNS to distinguish the timbre of the sound (**timbre discrimination**).

Loudness discrimination depends on the amplitude of vibration.

Intensity (loudness) discrimination depends on the amplitude of vibration. As sound waves originating from louder sound sources strike the eardrum, they cause it to vibrate more vigorously (that is, bulge in and out to a greater extent) but at the same frequency as a softer sound of the same pitch. The greater tympanic membrane deflection translates into greater basilar membrane movement in the region of peak responsiveness, causing greater bending of the hairs in this region. The CNS interprets this greater hair bending as a louder sound. Thus, pitch discrimination depends on "where" the basilar membrane maximally vibrates and loudness discrimination depends on "how much" this place vibrates.

The auditory system is so sensitive and can detect sounds so faint that the distance of basilar membrane deflection is comparable to only a fraction of the diameter of a hydrogen atom, the smallest of atoms. No wonder very loud sounds, which cannot be sufficiently attenuated by protective middle ear reflexes (for example, the sounds of a typical rock concert), can set up such violent vibrations of the basilar membrane that irreplaceable hair cells are actually sheared off or permanently distorted, leading to partial hearing loss (Figure 6-39). Damage can occur not only from brief exposure to high-intensity sounds but also from frequent exposure to moderately loud noises (those greater than 75 dB), something common in today's environment.

The auditory cortex is mapped according to tone.

Just as various regions of the basilar membrane are associated with particular tones, the **primary auditory cortex** in the temporal lobe is also *tonotopically* organized. Each region of the basilar membrane is linked to a specific region of the primary auditory cortex. Accordingly, specific cortical neurons are activated only by particular tones—that is, each region of the auditory cortex becomes excited only in response to a specific tone detected by a selected portion of the basilar membrane.

The afferent neurons that pick up the auditory signals from the inner hair cells exit the cochlea via the auditory nerve. The neural pathway between the organ of Corti and the auditory cortex involves several synapses en route, the most notable of which are in the brain stem and *medial geniculate nucleus* of the thalamus. The brain stem uses the auditory input for alertness and arousal. The thalamus sorts and relays the signals upward. Unlike signals in the visual pathways, auditory signals from each ear are transmitted to both temporal lobes because the fibers partially cross over in the brain stem. For this reason, a disruption of the auditory pathways on one side beyond the brain stem does not affect hearing in either ear to any extent.

The primary auditory cortex perceives discrete sounds, whereas the surrounding higher-order auditory cortex integrates the separate sounds into a coherent, meaningful pattern. Think about the complexity of the task accomplished by your auditory system. When you are at a concert, your organ of Corti responds to the simultaneous mixture of the instruments, the applause and hushed talking of the audience, and the background noises in the theater. You can distinguish these separate parts of the many sound waves reaching your ears and can pay attention to those of importance to you.

Deafness is caused by defects in either conduction or neural processing of sound waves.

Clinical Note Loss of hearing, or **deafness**, may be temporary or permanent, partial or complete. Deafness is classified into two types—*conductive deafness* and *sensorineural deafness*—depending on the part of the hearing mechanism that fails to function adequately. **Conductive deafness** occurs when sound waves are not adequately conducted through the external and middle portions of the ear to set the fluids in the inner ear in motion. Possible causes include physical blockage of the ear canal with earwax, rupture of the eardrum, middle ear infections with accompanying fluid accumulation, or restriction of ossicular movement because of bony adhesions. In **sensorineural deafness**, sound waves are transmitted to the inner ear, but they are not translated into nerve signals that are interpreted by the brain as sound sensations. The defect can lie in the organ of Corti, in the auditory nerves or, rarely, in the ascending auditory pathways or auditory cortex.

One of the most common causes of partial hearing loss, **neural presbycusis**, is a degenerative, age-related process that occurs as hair cells "wear out" with use. Over time, exposure to even ordinary modern-day sounds eventually damages hair cells so that, on average, adults have lost more than 40% of their cochlear hair cells by age 65. Hearing loss is the second most common physical disability in the United

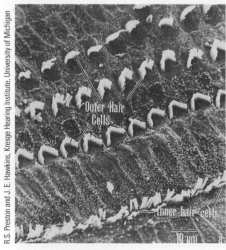

(a) Normal hair cells   **(b)** Damaged hair cells

Figure 6-39 **Loss of hair cells caused by loud noises.** The scanning electron micrographs show portions of the organ of Corti, with its three rows of outer hair cells and one row of inner hair cells, from the inner ear of (a) a normal guinea pig and (b) a guinea pig after a 24-hour exposure to noise at 120 decibels SPL (sound pressure level), a level approached by loud rock music.

States. Currently, more than 48 million Americans have some degree of hearing loss, and this number is expected to climb to 78 million by 2030. Unfortunately, partial hearing loss caused by excessive exposure to loud noises is affecting people at younger ages than in the past because we live in an increasingly noisy environment. An estimated 6.5 million children between 6 and 19 years of age in the United States already have some hearing damage resulting from amplified music and other noise pollution. Hair cells that process high-frequency sounds are the most vulnerable to destruction.

Hearing aids are helpful in conductive deafness but are less beneficial for sensorineural deafness. These devices increase the intensity of airborne sounds and may modify the sound spectrum and tailor it to the person's particular pattern of hearing loss at higher or lower frequencies. For the sound to be perceived, however, the receptor cell–neural pathway system must still be intact.

The first **cochlear implant** went on the market in 1972. These electronic devices, which are surgically implanted, transduce sound signals into electrical signals that can directly stimulate the auditory nerve, thus bypassing a defective cochlear system. Cochlear implants cannot restore normal hearing, but they do permit recipients to recognize sounds. Success ranges from an ability to "hear" a phone ringing to being able to carry on a conversation over the phone.

Recent findings suggest that in the future it may be possible to restore hearing by stimulating an injured inner ear to repair itself. Scientists have long considered the hair cells of the inner ear irreplaceable. Thus, hearing loss resulting from hair cell damage caused by the aging process or exposure to loud noises is considered permanent. Encouraging new studies suggest, to the contrary, that hair cells in the inner ear have the latent ability to regenerate in response to an appropriate chemical signal. Researchers are currently trying to develop a drug that spurs regrowth of hair cells, thus repairing inner ear damage and hopefully restoring hearing. Other investigators are employing a gene therapy approach to prompt growth of replacement hair cells. Still others are using neural growth factors to coax auditory nerve cell endings to resprout in the hopes of reestablishing lost neural pathways.

The vestibular apparatus is important for equilibrium by detecting head position and motion.

In addition to its cochlear-dependent role in hearing, the inner ear has another specialized component, the **vestibular apparatus**, which provides information essential for the sense of equilibrium and for coordinating head movements with eye and postural movements (❚ Figure 6-40). **Equilibrium** is the sense of body orientation and motion. The vestibular apparatus consists of two sets of structures lying within a tunneled-out region of the temporal bone near the cochlea—the *semicircular canals* and the *otolith organs*.

The vestibular apparatus detects changes in position and motion of the head. As in the cochlea, all components of the vestibular apparatus contain endolymph and are surrounded by perilymph. Also, similar to the organ of Corti, the vestibular components each contain hair cells that respond to mechanical deformation triggered by specific movements of the endolymph. And like the auditory hair cells, the vestibular receptors may be either depolarized or hyperpolarized, depending on the direction of the fluid movement. Unlike information from the auditory system, much of the information provided by the vestibular apparatus does not reach the level of conscious awareness.

Role of the Semicircular Canals The **semicircular canals** detect rotational or angular acceleration or deceleration of the head, such as when turning the head, starting or stopping spinning, or somersaulting. Each ear contains three Hula-Hoop-shaped semicircular canals arranged three-dimensionally in planes that lie at right angles to each other. The receptor hair cells of each semicircular canal are situated on top of a saddle-shaped ridge located in the **ampulla**, a swelling at the base of the canal (❚ Figure 6-40a and b). The hairs are embedded in an overlying, caplike, gelatinous layer, the **cupula**, which protrudes into the endolymph and stretches to the roof of the ampulla. The force of moving endolymph pushes against the cupula, causing it to bow so that the embedded hairs are bent.

Acceleration or deceleration during rotation of the head in any direction causes endolymph movement in at least one of the semicircular canals because of their three-dimensional arrangement. As you start to move your head, the bony canal and the ridge of hair cells embedded in the cupula move with your head. Initially, however, the fluid within the canal, not being attached to your skull, does not move in the direction of the rotation but lags behind because of its inertia. (Because of inertia, a resting object remains at rest and a moving object continues to move in the same direction unless the object is acted on by some external force that induces change.) When the endolymph is left behind as you start to rotate your head, the fluid that is in the same plane as the head movement is in effect shifted in the opposite direction from the movement (similar to your body tilting to the right as the car in which you are riding suddenly turns to the left) (❚ Figure 6-40c). This fluid movement causes the cupula to lean in the opposite direction from the head movement, bending the sensory hairs embedded in it. If your head movement continues at the same rate in the same direction, the endolymph catches up and moves in unison with your head so that the hairs return to their unbent position. When your head slows down and stops, the reverse situation occurs. The endolymph briefly continues to move in the direction of the rotation while your head decelerates to a stop. As a result, the cupula and its hairs are transiently bent in the direction of the preceding spin, which is opposite to the way they were bent during acceleration.

The hairs of a **vestibular hair cell** consist of one cilium, the **kinocilium**, along with a tuft of 20 to 50 microvilli—the **stereocilia**—arranged in rows of decreasing height from the taller kinocilium (❚ Figure 6-40d) (see p. 48). As in the auditory hair cell, the stereocilia are linked by tip links. When the stereocilia are deflected by endolymph movement, the resultant tension on the tip links pulls on mechanically gated ion channels in the hair cell. Depending on whether the ion channels are mechanically opened or closed by hair bundle displacement, the hair cell either depolarizes or hyperpolarizes. Each hair cell is oriented so

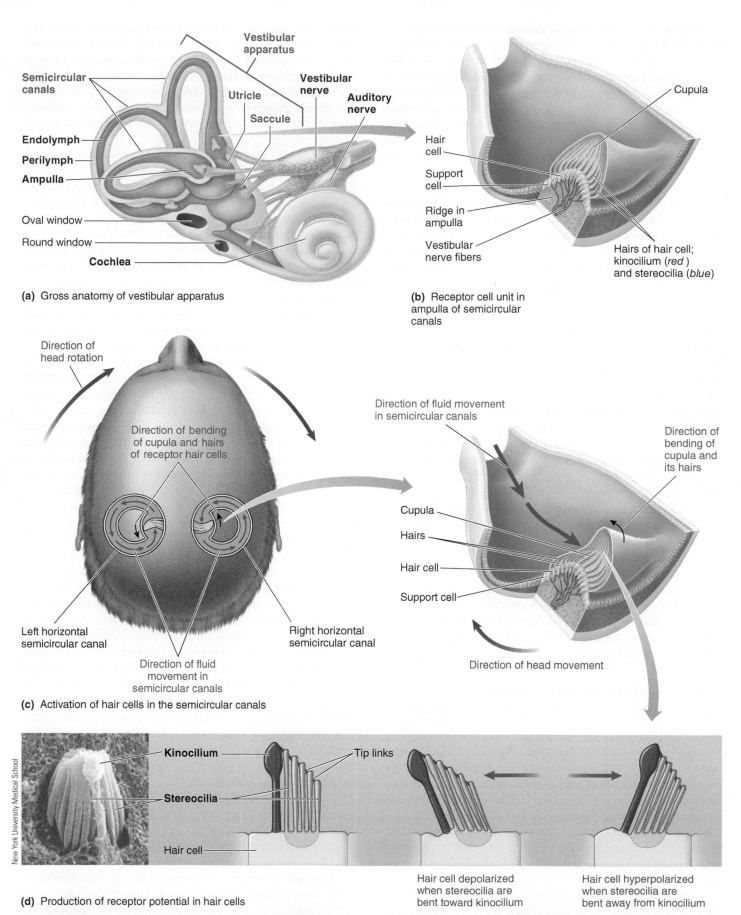

(a) Gross anatomy of vestibular apparatus

(b) Receptor cell unit in ampulla of semicircular canals

(c) Activation of hair cells in the semicircular canals

(d) Production of receptor potential in hair cells

New York University Medical School

I**Figure 6-40 Structure and activation of vestibular apparatus.** The scanning electron micrograph shows the kinocilium and stereocilia on the hair cells within the vestibular apparatus.

FIGURE FOCUS: *What direction will the endolymph in your semicircular canals move as you do a forward somersault?*

that it depolarizes when its stereocilia are bent toward the kinocilium and hyperpolarizes when the stereocilia are bent away from the kinocilium. The hair cells form a chemically mediated synapse with terminal endings of afferent neurons whose axons join with those of the other vestibular structures to form the **vestibular nerve**. This nerve unites with the auditory nerve from the cochlea to form the **vestibulocochlear nerve**. Depolarization increases the release of neurotransmitter from the hair cells, thereby bringing about an increased rate of firing in the afferent fibers; conversely, hyperpolarization reduces neurotransmitter release from the hair cells, in turn decreasing the frequency of action potentials in the afferent fibers. When the fluid gradually comes to a halt, the hairs straighten again. Thus, the semicircular canals detect changes in the rate of rotational movement (rotational acceleration or deceleration) of your head. They do not respond when your head is motionless or when it is moving in a circle at a constant speed.

Role of the Otolith Organs The **otolith organs** provide information about the position of the head relative to gravity (that is, static head tilt) and detect changes in the rate of lin-

ear motion (moving in a straight line regardless of direction). The otolith organs, the **utricle** and the **saccule**, are saclike structures housed within a bony chamber situated between the semicircular canals and the cochlea (▮Figure 6-40a). The hairs (kinocilium and stereocilia) of the receptor hair cells in these sense organs also protrude into an overlying gelatinous sheet, whose movement displaces the hairs and results in changes in hair cell potential. Many tiny crystals of calcium carbonate—the **otoliths** ("ear stones")—are suspended within the gelatinous layer, making it heavier and giving it more inertia than the surrounding fluid (▮Figure 6-41a). When a person is in an upright position, the hairs within the utricle are oriented vertically and the saccule hairs are lined up horizontally.

Let us look at the *utricle* as an example. Its otolith-embedded, gelatinous mass shifts positions and bends the hairs in two ways:

1. When you tilt your head in any direction so that it is no longer vertical (that is, when your head is not straight up and down), the hairs are bent in the direction of the tilt because of the gravitational force exerted on the top-heavy gelatinous

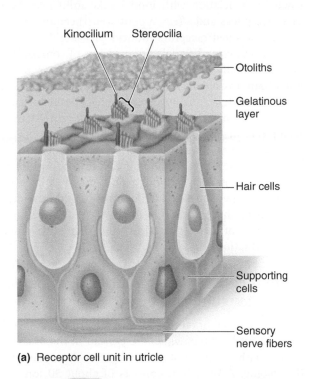

(a) Receptor cell unit in utricle

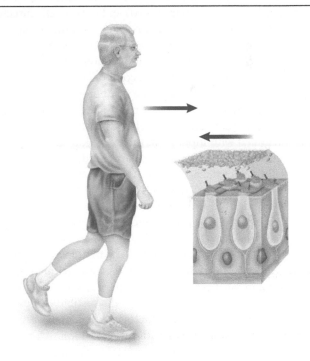

(c) Activation of utricle by horizontal linear acceleration

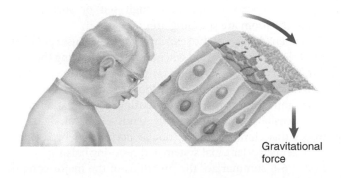

(b) Activation of utricle by change in head position

▮Figure 6-41 **Structure and activation of a receptor cell unit in the utricle.**

layer (Figure 6-41b). This bending produces depolarizing or hyperpolarizing receptor potentials depending on the tilt of your head. The CNS thus receives different patterns of neural activity depending on head position with respect to gravity.

2. The utricle hairs are also displaced by any change in horizontal linear motion (such as moving straight forward, backward, or to the side). As you start to walk forward (Figure 6-41c), the top-heavy otolith membrane at first lags behind the endolymph and hair cells because of its greater inertia. The hairs are thus bent to the rear, in the opposite direction of the forward movement of your head. If you maintain your walking pace, the gelatinous layer soon catches up and moves at the same rate as your head so that the hairs are no longer bent. When you stop walking, the otolith sheet continues to move forward briefly as your head slows and stops, bending the hairs toward the front. Thus, the hair cells of the utricle detect horizontally directed linear acceleration and deceleration, but they do not provide information about movement in a straight line at constant speed.

The saccule functions similarly to the utricle, except that it responds selectively to tilting of the head away from a horizontal position (such as getting up from bed) and to vertically directed linear acceleration and deceleration (such as jumping up and down or riding in an elevator). Together the otolith organs let you know which way is up and what direction you are heading.

Signals arising from the various components of the vestibular apparatus are carried through the vestibulocochlear nerve to the **vestibular nuclei**, a cluster of neuronal cell bodies in the brain stem, and to the cerebellum. Here, the vestibular information is integrated with input from the eyes, skin surface, joints, and muscles for (1) maintaining balance and desired posture; (2) controlling the external eye muscles so that the eyes remain fixed on the same point, despite movement of the head; and (3) perceiving motion and orientation.

Clinical Note Some people, for poorly understood reasons, are especially sensitive to particular motions that activate the vestibular apparatus and cause symptoms of dizziness and nausea; this sensitivity is called **motion sickness**. Occasionally, fluid imbalances within the inner ear lead to **Ménière's disease**. Not surprisingly because both the vestibular apparatus and cochlea contain the same inner ear fluids, both vestibular and auditory symptoms occur with this condition. An afflicted individual suffers transient attacks of severe vertigo (dizziness) accompanied by pronounced ringing in the ears and some loss of hearing. During these episodes, the person cannot stand upright and reports feeling as though self or surrounding objects in the room are spinning around.

Permanent damage to the semicircular canals causes poor balance and shaky, blurred vision when the head is moving (because the person cannot keep the eyes on target during the motion). Now researchers are working on a bionic ear implant that incorporates a miniature gyroscope for sensing head rotation in all three dimensions, which ultimately sends electrical signals to electrodes that stimulate the vestibular nerve, thus bypassing a defective semicircular canal system and restoring balance.

Table 6-6 summarizes the functions of the major components of the ear.

6.5 Chemical Senses: Taste and Smell

Unlike the eyes' photoreceptors and the ears' mechanoreceptors, the receptors for taste and smell are chemoreceptors, which generate neural signals on binding with particular chemicals in their environment. The sensations of taste and smell in association with food intake influence the flow of digestive juices and affect appetite. Furthermore, stimulation of taste or smell receptors induces pleasurable or objectionable sensations and signals the presence of something to seek (a nutritionally useful, good-tasting food) or to avoid (a potentially toxic, bad-tasting substance). Thus, the chemical senses provide a "quality-control" checkpoint for substances available for ingestion. In lower animals, smell also plays a major role in finding direction, in seeking prey or avoiding predators, and in sexual attraction to a mate. The sense of smell is less sensitive in humans and less important in influencing our behavior (although millions of dollars are spent annually on perfumes and deodorants to make us smell better and appear more socially attractive). We first examine the mechanism of taste (**gustation**) and then turn attention to smell (**olfaction**).

Taste receptor cells are located primarily within tongue taste buds.

The chemoreceptors for **taste** sensation are packaged in taste buds, about 10,000 of which are present in the oral cavity and throat, with the greatest percentage on the upper surface of the tongue. A **taste bud** consists of about 50 long, spindle-shaped *taste receptor cells* packaged with *supporting cells* in an arrangement like slices of an orange (Figure 6-42). Each taste bud has a small opening, the **taste pore**, through which fluids in the mouth come into contact with the surface of its receptor cells. **Taste receptor cells** are modified epithelial cells with many surface folds, or microvilli, that protrude slightly through the taste pore, greatly increasing the surface area exposed to the oral contents. The plasma membrane of the microvilli contains receptor sites that bind selectively with chemical molecules in the environment. Only chemicals in solution—either ingested liquids or solids that have been dissolved in saliva—can attach to receptor cells and evoke the

Structure	Location	Function
External ear		Collects and transfers sound waves to the middle ear
Pinna (ear)	Skin-covered flap of cartilage located on each side of the head	Collects sound waves and channels them down the ear canal; contributes to sound localization
External auditory meatus (ear canal)	Tunnel from the exterior through the temporal bone to the tympanic membrane	Directs sound waves to the tympanic membrane
Tympanic membrane (eardrum)	Thin membrane that separates the external ear and the middle ear	Vibrates in synchrony with sound waves that strike it, setting middle ear bones in motion
Middle ear		Transfers vibrations of the tympanic membrane to the fluid in the cochlea
Malleus, incus, stapes	Movable chain of bones that extends across the middle ear cavity; malleus attaches to the tympanic membrane, and stapes attaches to the oval window	Oscillate in synchrony with tympanic membrane vibrations and set up wavelike movements in the cochlear perilymph at the same frequency
Inner ear: cochlea		Houses sensory system for hearing
Oval window	Thin membrane at the entrance to the cochlea; separates the middle ear from the scala vestibuli	Vibrates in unison with movement of the stapes, to which it is attached; oval window movement sets cochlear perilymph in motion
Scala vestibuli	Upper compartment of the cochlea, a snail-shaped tubular system that lies deep within the temporal bone	Contains perilymph that is set in motion by oval window movement driven by oscillation of middle ear bones
Scala tympani	Lower compartment of the cochlea	Contains perilymph that is continuous with the scala vestibuli
Cochlear duct (scala media)	Middle compartment of the cochlea; a blind-ended tubular compartment that tunnels through the center of the cochlea	Contains endolymph; houses the basilar membrane
Basilar membrane	Forms the floor of the cochlear duct	Vibrates in unison with perilymph movements; bears the organ of Corti, the sense organ for hearing
Organ of Corti	Rests on top of the basilar membrane throughout its length	Contains hair cells, the receptors for sound; inner hair cells undergo receptor potentials when their hairs are bent as a result of fluid movement in the cochlea
Tectorial membrane	Stationary membrane that overhangs the organ of Corti and contacts the surface hairs of the receptor hair cells	Serves as the stationary site against which the hairs of the receptor cells are bent and undergo receptor potentials as the vibrating basilar membrane moves in relation to this overhanging membrane
Round window	Thin membrane that separates the scala tympani from the middle ear	Vibrates in unison with fluid movements in perilymph to dissipate pressure in the cochlea; does not contribute to sound reception
Inner ear: vestibular apparatus		Houses sensory systems for equilibrium and provides input essential for maintaining posture and balance
Semicircular canals	Three semicircular canals arranged three-dimensionally in planes at right angles to each other near the cochlea	Detect rotational or angular acceleration or deceleration
Utricle	Saclike structure in a bony chamber between the cochlea and semicircular canals	Detects changes in head position away from vertical and horizontally directed linear acceleration and deceleration
Saccule	Lies next to the utricle	Detects changes in head position away from horizontal and vertically directed linear acceleration and deceleration

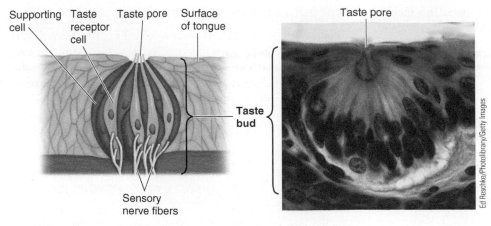

Figure 6-42 **Structure of the taste buds.** The receptor cells and supporting cells of a taste bud are arranged like slices of an orange.

Supporting cell · Taste receptor cell · Taste pore · Surface of tongue

Taste bud

Sensory nerve fibers

Taste pore

Taste discrimination is coded by patterns of activity in various taste bud receptors.

We can discriminate among thousands of taste sensations, yet all tastes are varying combinations of five **primary tastes**: *salty, sour, sweet, bitter,* and *umami*. Umami, a meaty or savory taste, has recently been added to the list of primary tastes.

The five established primary taste sensations are elicited by the following stimuli:

- **Salty taste** is stimulated by chemical salts, especially NaCl (table salt). Direct entry of positively charged Na^+ ions through specialized Na^+ channels in the receptor cell membrane, a movement that reduces the cell's internal negativity, is responsible for receptor depolarization in response to salt. Saltiness signals the presence of electrolytes, necessary components of a healthy diet.

- **Sour taste** is caused by acids, which contain a free hydrogen ion, H^+. The citric acid content of lemons, for example, accounts for their distinctly sour taste. Depolarization of the receptor cell by sour tastants occurs because H^+ blocks K^+ channels in the receptor cell membrane. The resultant decrease in the passive movement of positively charged K^+ ions out of the cell reduces the internal negativity, producing a depolarizing receptor potential. For animals and early humans using taste as a guide to evaluate the nutritional value of a potential food source, a sour taste could indicate spoiled food.

- **Sweet taste** is a pleasurable sensation evoked by the particular configuration of glucose. From an evolutionary perspective, we crave sweet foods because they supply necessary calories in a readily usable form. However, other organic molecules with similar structures but no calories, such as saccharin, aspartame, sucralose, and other artificial sweeteners, can also interact with "sweet" receptor binding sites. Binding of glucose or another chemical with a sweet taste receptor activates a G protein, which acts through a cAMP second-messenger pathway (see p. 123) in the taste cell to ultimately cause phosphorylation and blockage of K^+ channels in the receptor cell membrane, leading to a depolarizing receptor potential.

- **Bitter taste** is elicited by a more chemically diverse group of tastants than the other taste sensations. For example, alkaloids (such as caffeine, nicotine, strychnine, morphine, and other toxic plant derivatives) and poisonous substances, all taste bitter, presumably as a protective mechanism to discourage ingestion of these potentially dangerous compounds (the tendency is to spit out something bitter). Taste cells that detect bitter flavors possess about 30 bitter receptor types, all of which are G-protein-coupled receptors, each of which responds to a different bitter flavor. (By comparison, there appears to be only one receptor type each for the other primary tastes.) Because each bitter taste receptor cell has a diverse family of bitter receptors, a wide variety of unrelated chemi-

sensation of taste. Binding of a taste-provoking chemical, a **tastant**, with a receptor cell alters the cell's ionic channels to produce a depolarizing receptor potential. Like the other special sense receptors, a depolarizing receptor potential opens voltage-gated Ca^{2+} channels, leading to the entry of Ca^{2+}, which promotes release of neurotransmitter. This neurotransmitter (serotonin or ATP [adenosine triphosphate], depending on the taste sensation), in turn, initiates action potentials within terminal endings of afferent nerve fibers with which the receptor cell synapses. Each taste receptor cell responds to only one tastant.

Most receptors are carefully sheltered from direct exposure to the environment, but the taste receptor cells, by virtue of their task, frequently come into contact with potent chemicals. Unlike the eye or ear receptors, which are irreplaceable, taste receptors have a life span of about 10 days. Epithelial cells surrounding the taste bud differentiate first into supporting cells and then into receptor cells to constantly renew the taste bud components.

Terminal afferent endings of several cranial nerves terminate on taste buds in various regions of the mouth. Signals in these sensory inputs are conveyed via synaptic stops in the brain stem and thalamus to the **primary gustatory** (or **taste**) **cortex**, where the taste is perceived. The primary gustatory cortex is found in the *insula*, a cortical region that lays hidden from the surface in an especially deep fold, the lateral sulcus, that separates the temporal lobes from the overlying parietal and frontal lobes (see Figure 5-10, p. 147). Contrary to the long-held belief that each primary taste is encoded by a unique ensemble of neurons dispersed throughout the gustatory cortex without any spatial clustering, recent evidence suggests that each taste modality is represented in its own separate territory in the gustatory cortex. This taste map is similar to the way in which brain regions that process tactile, visual, and auditory input are organized into spatial maps. Unlike most sensory input, the gustatory pathways are primarily uncrossed. Taste signals are also sent to the hypothalamus and limbic system to add affective dimensions, such as whether the taste is pleasant or unpleasant, and to process behavioral aspects associated with taste and smell.

cals all taste bitter despite their diverse structures. This mechanism expands the ability of a bitter taste receptor cell to detect a wide range of potentially harmful chemicals. The first G protein in taste—**gustducin**—was identified in one of the bitter signaling pathways. This G protein, which sets off a second-messenger pathway in the taste cell, is very similar to the visual G protein, transducin. (Gustducin is also the G protein in the sweet and umami signaling pathways.)

- **Umami taste**, a pleasant savory taste that was first identified and named by a Japanese researcher, is triggered by amino acids, especially glutamate (*umami* means "pleasant savory taste"). The presence of amino acids, as found in meat, for example, is a marker for a desirable, nutritionally protein-rich food. Glutamate binds to a GPCR and acts via a second-messenger pathway. In addition to giving us our sense of meaty flavors, this pathway is responsible for the distinctive taste of the flavor additive monosodium glutamate (MSG), which is especially popular in Asian dishes.

Note that transduction of salt and sour taste is mediated by ion channels, whereas transduction of the other three primary tastes depends on GPCRs. The different GPCRs for detection of sweet, umami, and bitter have been identified. Bitter is detected by T2R receptors (of which there are about 30 different variants), whereas sweet and umami are detected by specific combinations of two of the three different T1R receptors. Sweet taste results from binding of a tastant to the receptor combination of T1R2 + T1R3; umami taste is triggered by tastant binding to the receptor combination of T1R1 + T1R3.

Another new taste sensation has also been proposed—**fat taste**. Scientists have identified a sensor in the mouth for long-chain fatty acids, which could explain our fondness for fat-rich foods (think full-fat ice cream compared to the less satisfactory fat-free version). Early evidence suggests that people who have a higher sensitivity to the taste of fat (that is, they can detect fat at lower concentrations) tend to consume less fat and are not as likely to be overweight as people whose sensitivity to fat taste is lower (that is, they detect fat only at higher concentrations). This finding could be relevant in the fight against obesity.

Each receptor cell is preferentially responsive to one of the taste modalities. The richness of fine taste discrimination beyond the primary tastes depends on subtle differences in the stimulation patterns of all taste buds in response to various substances, similar to the variable stimulation of the three cone types that gives rise to the range of color sensations. For example, eating seasoned lemon chicken simultaneously stimulates salt, sour, and umami taste receptors.

Taste perception is also influenced by information derived from other receptors, especially odor. Complex flavors beyond the five primary tastes depend on smell. When you temporarily lose your sense of smell during a cold because of swollen nasal passageways, your sense of taste is also markedly reduced, even though your taste receptors are unaffected by the cold. Other factors affecting taste include temperature and texture of the food and psychological factors associated with past experiences with the food. How the gustatory cortex accomplishes the complex perceptual processing of taste sensation is not yet known.

The gut and airways "taste" too.

Interestingly, scientists have discovered cells in the stomach and intestine that have the same GPCRs and gustducin-activated pathways for "tasting" sweet, umami, and bitter as those in the tongue taste buds. The gut taste cells sense the chemical composition of the contents in the lumen of the digestive tract and are believed to trigger physiological responses important in dealing with the food. For example, when the gut taste cells detect something sweet (indicative of a nutritive substance), they initiate a cascade of events leading to production of molecules that stimulate gut motility, enhance absorption of glucose by the digestive tract, stimulate secretion of insulin (a hormone that promotes cell uptake and storage of glucose) in anticipation of the blood-borne arrival of absorbed sweet food, and contribute to the sensation of being full. By contrast, detection by the gut taste cells of something bitter (suggestive of being potentially toxic) slows absorption or spurs vomiting.

More recently the same GPCRs used by the taste buds for detecting tastants, especially the T2R bitter-detecting receptors, have also been identified in epithelial cells lining the respiratory airways. Here the T2R receptors respond to inhaled toxic dusts and aerosols and produce the sensation of irritation, which may trigger protective reflexes such as sneezing or coughing to help expel the potentially dangerous irritant.

The olfactory receptors in the nose are specialized endings of renewable afferent neurons.

The **olfactory** ("smell") **mucosa**, a 3-cm^2 patch of mucosa in the ceiling of the nasal cavity, contains three cell types: *olfactory receptor cells, supporting cells,* and *basal cells* (❙ Figure 6-43). The supporting cells secrete mucus, which coats the nasal passages. The basal cells are precursors for new olfactory receptor cells, which are replaced about every 2 months. The sense of **smell** depends on the **olfactory receptor cells** detecting odors, or scents. An olfactory receptor cell is an afferent neuron whose receptor portion lies in the olfactory mucosa in the nose and whose afferent axon traverses into the brain. The axons of the olfactory receptor cells collectively form the **olfactory nerve**.

The receptor portion of an olfactory receptor cell consists of an enlarged knob bearing several long cilia that extend like a tassel to the surface of the mucosa (❙ Figure 6-43). These cilia contain the receptors for binding of **odorants**, molecules that can be smelled. During quiet breathing, odorants typically reach the sensitive receptors only by diffusion because the olfactory mucosa is above the normal path of airflow. The act of sniffing enhances this process by drawing the air currents upward within the nasal cavity so that a greater percentage of the odoriferous molecules in the air come into contact with the olfactory mucosa. Odorants also reach the olfactory mucosa during eating by wafting up to the nose from the mouth through the pharynx (back of the throat).

To be smelled, a substance must be (1) sufficiently volatile (easily vaporized) so that some of its molecules can enter the nose in the inspired air and (2) sufficiently water soluble so that it can dissolve in the mucus coating the olfactory mucosa. As

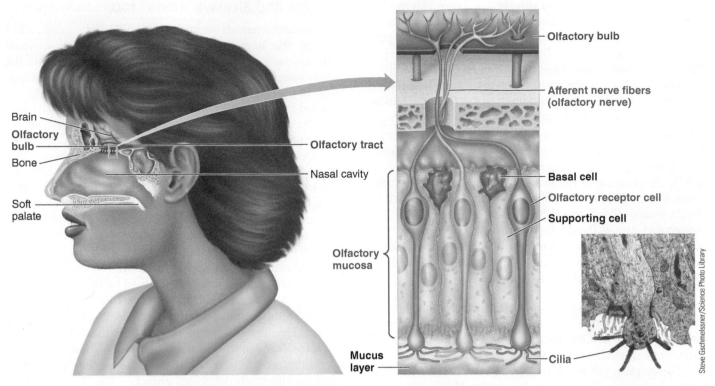

Brain
Olfactory bulb
Bone
Soft palate

Olfactory tract
Nasal cavity

Olfactory mucosa

Mucus layer

Olfactory bulb

Afferent nerve fibers (olfactory nerve)

Basal cell
Olfactory receptor cell
Supporting cell

Cilia

Steve Gschmeissner/Science Photo Library

┃Figure 6-43 **Location and structure of the olfactory receptor cells.** The photo is an electron micrograph of the tassel of cilia at the sensory ending of an olfactory receptor.

with taste receptors, molecules must be dissolved to be detected by olfactory receptors.

Various parts of an odor are detected by different olfactory receptors and sorted into "smell files."

The human nose contains 5 million olfactory receptors, of which there are 1000 types. During smell detection, an odor is "dissected" into various components. Each receptor responds to only one discrete component of an odor rather than to the whole odorant molecule. Accordingly, each of the various parts of an odor is detected by one of the thousand different receptors, and a given receptor can respond to a particular odor component that may be shared by different scents. Compare this to the three cone types for coding color vision and the taste buds that respond differentially to only five (maybe six) primary tastes to accomplish coding for taste discrimination.

Binding of an appropriate scent signal to an olfactory receptor activates a specific G protein, G_{olf}, triggering a cascade of cAMP-dependent intracellular reactions that leads to opening of an olfactory-specific cAMP-gated channel. This channel opening leads to net Na^+ and Ca^{2+} entry, which causes a depolarizing receptor potential that generates action potentials in the afferent fiber. The frequency of the action potentials depends on the concentration of the stimulating chemical molecules.

The afferent fibers arising from the receptor endings in the nose pass through tiny holes in the flat bone plate separating the olfactory mucosa from the overlying brain tissue (┃Figure

6-43). They immediately synapse in the **olfactory bulb,** a complex neural structure containing several layers of cells that are functionally similar to the retinal layers of the eye. The twin olfactory bulbs, one on each side, are about the size of small grapes (see ┃Figure 5-16, p. 155). Each olfactory bulb is lined by small, ball-like neural junctions known as **glomeruli** (meaning "little balls") (┃Figure 6-44). Within each glomerulus, the terminals of receptor cells carrying information about a particular scent component synapse with the next cells in the olfactory pathway, the **mitral cells.** Because each glomerulus receives signals only from receptors that detect a particular odor component, the glomeruli serve as "smell files." The separate components of an odor are sorted into different glomeruli, one component per file. Thus, the glomeruli, which are the first relay station in the brain for processing olfactory information, play a key role in organizing scent perception.

The mitral cells on which the olfactory receptors terminate in the glomeruli refine the smell signals and relay them to the brain for further processing. Fibers leaving the olfactory bulb travel in two routes:

1. A route going primarily to the lower medial sides of the temporal lobes, especially to regions of the limbic system. A group of structures in this area are collectively considered the **primary olfactory cortex,** the largest component of which is the *piriform* (or *pyriform*) *cortex.* This route, which includes hypothalamic involvement, permits close coordination between smell and behavioral reactions associated with feeding, mating, and direction orienting.

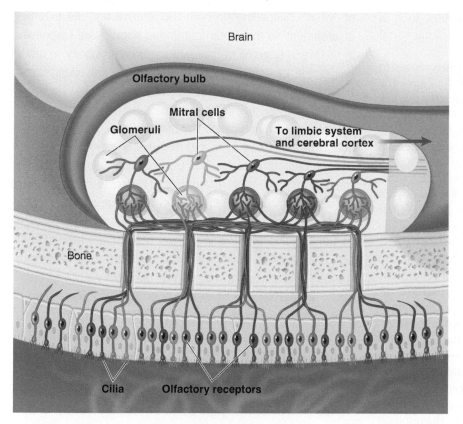

Brain

Olfactory bulb

Mitral cells

Glomeruli

To limbic system
and cerebral cortex

Bone

Cilia   Olfactory receptors

IFigure 6-44 Processing of scents in the olfactory bulb. Each of the glomeruli lining the olfactory bulb receives synaptic input from only one type of olfactory receptor, which, in turn, responds to only one discrete component of an odorant. Thus, the glomeruli sort and file the various components of an odoriferous molecule before relaying the smell signal to the mitral cells and higher brain levels for further processing. **FIGURE FOCUS:** *When you smell your favorite cologne, how are the different components of that odor processed by your olfactory receptors and olfactory bulb?*

2. A route through the thalamus to higher centers, especially the *orbitofrontal cortex*, located on the medial ventral surface of the frontal lobe above the boney orbits that house the eyes. As with other senses, this cortical route is important for conscious perception and fine discrimination of smell.

Odor discrimination is coded by patterns of activity in the olfactory bulb glomeruli.

Because each odorant activates multiple receptors and glomeruli in response to its various odor components, odor discrimination is based on different patterns of glomeruli activated by various scents. In this way, the cortex can distinguish more than 10,000 scents. Studies indicate that a particular scent may be variably perceived by different people. What is a pleasant odor for one person may be unpleasant for another.

This mechanism for sorting out and distinguishing odors is very effective. A noteworthy example is our ability to detect methyl mercaptan (garlic odor) at a concentration of 1 molecule per 50 billion molecules in the air. This substance is added to odorless natural gas to enable us to detect potentially lethal gas leaks. Despite this impressive sensitivity, humans have a poor sense of smell compared to other species. By comparison, dogs' sense of smell is hundreds of times more sensitive than

that of humans. Bloodhounds, for example, have about 4 billion olfactory receptor cells compared to our 5 million such cells, accounting for bloodhounds' superior scent-sniffing ability.

The olfactory system adapts quickly, and odorants are rapidly cleared.

Although the olfactory system is sensitive and highly discriminating, it is also quickly adaptive. Sensitivity to a new odor diminishes rapidly after a short period of exposure to it, even though the odor source continues to be present. This reduced sensitivity does not involve receptor adaptation, as researchers thought for years; actually, the olfactory receptors themselves adapt slowly. It apparently involves some sort of adaptation process in the CNS. Adaptation is specific for a particular odor, and responsiveness to other odors remains unchanged.

What clears the odorants away from their binding sites on the olfactory receptors so that the sensation of smell doesn't "linger" after the source of the odor is removed? Several "odor-eating" enzymes in the olfactory mucosa serve as molecular janitors, clearing away the odoriferous molecules so that they do not continue to stimulate the olfactory receptors. Interestingly, these odorant-clearing enzymes are very similar chemically to detoxification enzymes found in the liver. (These liver enzymes inactivate potential toxins absorbed from the digestive tract; see p. 27.) This resemblance may not be coincidental. Researchers speculate that the nasal enzymes may serve the dual purpose of clearing the olfactory mucosa of old odorants and transforming potentially harmful chemicals into harmless molecules. Such detoxification would serve a very useful purpose, considering the open passageway between the olfactory mucosa and the brain.

The vomeronasal organ detects pheromones.

In addition to the olfactory mucosa, the nose contains another sense organ, the **vomeronasal organ (VNO)**, which is common in mammals but until recently was thought nonexistent in humans. The VNO is located about half an inch inside the human nose next to the vomer bone, hence its name. It detects **pheromones**, nonvolatile chemical signals passed subconsciously between individuals of the same species. In animals, binding of a pheromone to its receptor on the surface of a neuron in the VNO triggers an action potential that travels through nonolfactory pathways to the limbic system, the brain region that governs emotional responses and sociosexual behaviors. These signals never reach the higher levels of conscious awareness. In animals, the VNO is known as the "sexual nose" for its role in governing reproductive and social behaviors, such as identifying and attracting a mate and communicating social status.

Some scientists now claim the existence of pheromones in humans, although many skeptics doubt these findings. The major goal of investigators in the field is to identify the key chemicals that serve as pheromones in humans (one candidate, for instance, is the nonsmelly steroid compound *androstadienone* found in male sweat) and to learn how the body processes these surreptitious signals. Although the role of the VNO in human behavior has not been validated, some researchers suspect that it is responsible for spontaneous "feelings" between people, either "good chemistry," such as "love at first sight," or "bad chemistry," such as "getting bad vibes" from someone you just met. They speculate that pheromones in humans subtly influence sexual activity, compatibility with others, or group behavior, similar to the role they play in other mammals, although this messenger system is nowhere as powerful or important in humans as in animals. For example, a recent study suggests that a chemical signal (that is, a pheromone) in women's emotional tears reduces sexual arousal in males. Not only did men who sniffed these tear drops suddenly feel less sexually interested than those who sniffed saline drops did, but the tear sniffers experienced a small but measurable decrease in testosterone (male sex hormone) levels. Because messages conveyed by the VNO seem to bypass cortical consciousness, the response to the largely odorless pheromones is not a distinct, discrete perception, such as smelling a favorite fragrance, but more like an inexplicable impression.

Check Your Understanding 6.5

1. List the five established primary tastes and the stimuli that evoke each of these taste sensations.
2. Discuss the role of gustducin in taste signaling.
3. Describe how odor discrimination is accomplished.

Homeostasis: Chapter in Perspective

To maintain a life-sustaining stable internal environment, the body must constantly make adjustments to compensate for myriad internal and external factors that continuously threaten to disrupt homeostasis, such as internal acid production or external exposure to cold. Many of these adjustments are directed by the nervous system, one of the body's two major regulatory systems. The central nervous system (CNS), the integrating and decision-making component of the nervous system, must continuously be informed of "what's happening" in both the internal and the external environment so that it can command appropriate responses in the organ systems to maintain the body's viability. In other words, the CNS must know what changes are taking place before it can respond to these changes.

The afferent division of the peripheral nervous system (PNS) is the communication link by which the CNS is informed about the internal and the external environment. The afferent division detects, encodes, and transmits peripheral signals to the CNS for processing. Afferent input is necessary for arousal, perception, and determination of efferent output.

Afferent information about the internal environment, such as the CO_2 level in the blood, never reaches the level of conscious awareness, but this input to the controlling centers of the CNS is essential for maintaining homeostasis. Afferent input that reaches the level of conscious awareness, called sensory information, includes somesthetic and proprioceptive sensation (body sense) and special senses (vision, hearing, equilibrium, taste, and smell).

The body sense receptors are distributed over the entire body surface and throughout the joints and muscles. Afferent signals from these receptors provide information about what's happening directly to each specific body part in relation to the external environment (that is, the "what," "where," and "how much" of stimulatory inputs to the body's surface and the momentary position of the body in space). In contrast, each special sense organ is restricted to a single site in the body. Rather than provide information about a specific body part, a special sense organ provides a specific type of information about the external environment that is useful to the body as a whole. For example, through their ability to detect, extensively analyze, and integrate patterns of illumination in the external environment, the eyes and visual processing system enable you to see your surroundings. The same integrative effect could not be achieved if photoreceptors were scattered over your entire body surface, as are touch receptors.

Sensory input (both body sense and special senses) enables a complex multicellular organism such as a human to interact in meaningful ways with the external environment in procuring food, defending against danger, and engaging in other behavioral actions geared toward maintaining homeostasis. In addition to providing information essential for interactions with the external environment for basic survival, the perceptual processing of sensory input adds immeasurably to the richness of life, such as enjoyment of a good book, concert, or meal.

Review Exercises Answers begin on p. A-28

Reviewing Terms and Facts

1. Conversion of the energy forms of stimuli into electrical energy by the receptors is known as _____.

2. The type of stimulus to which a particular receptor is most responsive is called its _____.

3. All afferent information is sensory information. *(True or false?)*

4. Off-center ganglion cells increase their rate of firing when a beam of light strikes the periphery of their receptive field. *(True or false?)*

5. During dark adaptation, rhodopsin is gradually regenerated to increase the sensitivity of the eyes. *(True or false?)*

6. An optic nerve carries information from the lateral and medial halves of the same eye, whereas an optic tract carries information from the lateral half of one eye and the medial half of the other. *(True or false?)*

7. Displacement of the round window generates neural impulses perceived as sound sensations. *(True or false?)*

8. Stereocilia of the inner hair cells hyperpolarize when they bend toward and depolarize when they bend away from their tallest member. *(True or false?)*

9. Hair cells in different regions of the organ of Corti and neurons in different regions of the auditory cortex are activated by different tones. *(True or false?)*

10. The receptor potential in a sour taste receptor cell occurs when H^+ in an acid blocks K^+ channels in the receptor cell membrane. *(True or false?)*

11. Rapid adaptation to odors results from adaptation of the olfactory receptors. *(True or false?)*

12. Match the following:

1. layer that contains photoreceptors	(a) choroid
2. point from which optic nerve leaves retina	(b) aqueous humor
3. forms white part of eye	(c) fovea
4. thalamic structure that processes visual input	(d) lateral geniculate nucleus
5. colored diaphragm of muscle that controls amount of light entering eye	(e) cornea
	(f) retina
	(g) lens
6. contributes most to refractive ability	(h) optic disc; blind spot
7. supplies nutrients to lens and cornea	(i) iris
8. produces aqueous humor	(j) ciliary body
9. contains vascular supply for retina and a pigment that minimizes scattering of light within eye	(k) optic chiasm
10. has adjustable refractive ability	(l) sclera

11. portion of retina with greatest acuity

12. point at which fibers from medial half of each retina cross to opposite side

13. Using the answer code on the right, indicate which properties apply to taste and/or smell:

1. Receptors are separate cells that synapse with terminal endings of afferent neurons.	(a) applies to taste
2. Receptors are specialized endings of afferent neurons.	(b) applies to smell
	(c) applies to both taste and smell

3. Receptors are regularly replaced.

4. Specific chemicals in the environment attach to special binding sites on the receptor surface, leading to a depolarizing receptor potential.

5. There are two processing pathways: a limbic system route and a thalamic–cortical route.

6. Discriminative ability is based on patterns of receptor stimulation by five (maybe six) different modalities.

7. A thousand different receptor types are used.

8. Information from receptor cells is filed and sorted by neural junctions called glomeruli.

Understanding Concepts
(Answers at www.cengagebrain.com)

1. List and describe the receptor types according to their adequate stimulus.

2. Compare tonic and phasic receptors.

3. Explain how acuity is influenced by receptive field size and by lateral inhibition.

4. Compare the fast and slow pain pathways.

5. Describe the built-in analgesic system of the brain.

6. Describe the process of phototransduction by photoreceptors and further retinal processing by bipolar and ganglion cells.

7. Compare the functional characteristics of rods and cones.

8. Discuss the trichromatic theory and the opponent-process theory of color vision.

9. What are sound waves? What is responsible for the pitch, intensity, and timbre of a sound?

10. Describe the function of each of the following parts of the ear: pinna, ear canal, tympanic membrane, ossicles, oval window, and various parts of the cochlea. Include a discussion of how sound waves are transduced into action potentials.

11. Discuss the functions of the semicircular canals, the utricle, and the saccule.

12. Describe the location, structure, and general means of activation of the receptors for taste and smell.

13. Compare the processes of color vision, hearing, taste, and smell discrimination.

Solving Quantitative Exercises

1. Calculate the difference in the time it takes for an action potential to travel 1.3 m between the slow (12 m/sec) and fast (30 m/sec) pain pathways.

2. Have you ever noticed that humans have circular pupils, whereas cats' pupils are more elongated from top to bottom? For simplicity in calculation, assume the cat's pupil is rectangular. The following calculations will help you understand the implication of this difference. For simplicity, assume a constant intensity of light.

   a. If the diameter of a human's circular pupil were decreased by half on contraction of the constrictor muscle of the iris, by what percentage would the amount of light allowed into the eye be decreased?

   b. If a cat's rectangular pupil were decreased by half along one axis only, by what percentage would the amount of light allowed into the eye be decreased?

   c. Comparing these calculations, do humans or cats have more precise control over the amount of light falling on the retina?

3. A decibel is the unit of sound level, β, defined as follows:

$$\beta = (10 \text{ dB}) \log_{10}(I/I_0)$$

where I is *sound intensity*, or the rate at which sound waves transmit energy per unit area. The units of I are watts per square meter (W/m^2). I_0 is a constant intensity close to the human hearing threshold, namely, 10^{-12} W/m^2.

   a. For the following sound levels, calculate the corresponding sound intensities:

     (1) 20 dB (a ticking watch)

     (2) 70 dB (a car horn)

     (3) 120 dB (a loud rock concert)

     (4) 170 dB (a space shuttle launch)

   b. Explain why the sound levels of these sounds increase by the same increment (that is, each sound is 50 dB higher than the one preceding it), yet the incremental increases in sound intensities you calculated are so different. What implications does this have for performance of the human ear?

Applying Clinical Reasoning

Suzanne J. complained to her physician of bouts of dizziness. The physician asked her whether by "dizziness" she meant a feeling of lightheadedness, as if she were going to faint (a condition known as **syncope**), or a feeling that she or surrounding objects in the room were spinning around (a condition known as **vertigo**). Why is this distinction important in the differential diagnosis of her condition? What are some possible causes of each of these symptoms?

Thinking at a Higher Level

1. Patients with certain nerve disorders are unable to feel pain. Why is this disadvantageous?

2. Ophthalmologists often instill eye drops in their patients' eyes to bring about pupillary dilation, which makes it easier for the physician to view the eye's interior. In what way would the drug in the eye drops affect autonomic nervous system activity in the eye to cause the pupils to dilate?

3. A patient complains of not being able to see the right half of the visual field with either eye. At what point in the patient's visual pathway does the defect lie?

4. *Retinitis pigmentosa* is a hereditary eye disease characterized by gradual accumulation of excess pigment in the retina, which leads to slow degeneration of photoreceptors, especially rods. What symptoms would occur as a result of rod deterioration?

5. Explain how middle ear infections interfere with hearing. Of what value are the "tubes" that are sometimes surgically placed in the eardrums of patients with a history of repeated middle ear infections accompanied by chronic fluid accumulation?

To access the course materials and companion resources for this text, please visit www.cengagebrain.com

The Peripheral Nervous System: Efferent Division

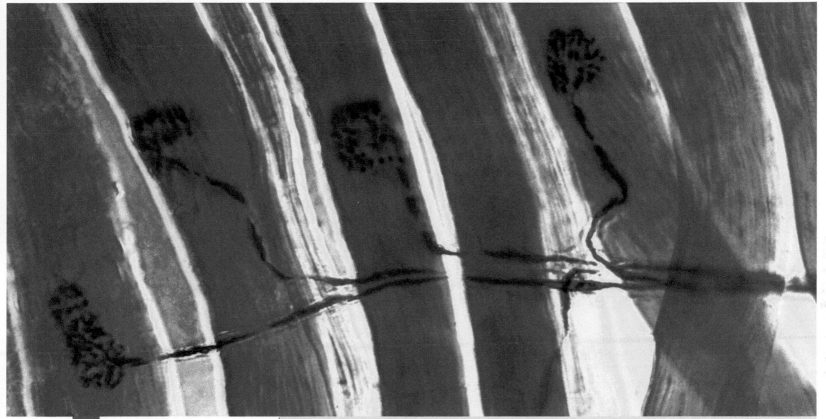

Wood/Custom Medical Stock Photo/Getty Images

A light micrograph of a motor neuron innervating skeletal muscle cells. When a motor neuron (*black*) reaches a skeletal muscle, it divides into many terminal branches, each of which forms a neuromuscular junction (*enlarged ending*) with a single long, cylindrical muscle cell (*red*). Release of neurotransmitter from knoblike terminal buttons (*small black dots*) in the neuromuscular junction excites the muscle cell to trigger contraction.

CHAPTER AT A GLANCE

Homeostasis Highlights

The nervous system, one of the two major regulatory systems of the body, consists of the central nervous system (CNS), composed of the brain and spinal cord, and the **peripheral nervous system (PNS)**, composed of the afferent and efferent fibers that relay signals between the CNS and the periphery (other parts of the body).

Once informed by the afferent division of the PNS that a change in the internal or the external environment is threatening homeostasis, the CNS makes appropriate adjustments to maintain homeostasis. The CNS makes these adjustments by controlling the activities of effector organs (muscles and glands), transmitting signals from the CNS to these organs through the **efferent division** of the PNS.

7.1 Autonomic Nervous System

The efferent division of the PNS is the communication link by which the CNS controls muscles and glands, the effector organs that carry out the intended effects or actions (typically contraction or secretion, respectively). The CNS regulates these effectors by initiating action potentials in the cell bodies of efferent neurons whose axons terminate on these organs. Cardiac muscle, smooth muscle, most exocrine glands, some endocrine glands, and adipose tissue (fat) are innervated by the **autonomic nervous system**, the involuntary branch of the peripheral efferent division. Skeletal muscle is innervated by the **somatic nervous system**, the branch of the efferent division subject to voluntary control. The following are examples of the effects of neural control on various effectors composed of different types of muscle and gland tissue:

- *Heart (cardiac muscle)*: increased pumping of blood by the heart when blood pressure falls too low
- *Stomach (smooth muscle)*: delayed emptying of the stomach until the intestine is ready to process the food
- *Respiratory muscles (skeletal muscle)*: augmented breathing in response to exercise
- *Sweat glands (exocrine glands)*: initiation of sweating on exposure to a hot environment
- *Endocrine pancreas (endocrine gland)*: increased secretion of insulin, a hormone that puts excess nutrients in storage following a meal

As these examples illustrate, much of efferent output is directed toward maintaining homeostasis. The efferent output to skeletal muscles is also directed toward voluntarily controlled nonhomeostatic activities, such as riding a bicycle or texting a message. (Many effector organs are also subject to hormonal control or to intrinsic control; see p. 16.)

Almost all neurally controlled effector organ responses are directly mediated by one of two neurotransmitters: acetylcholine or norepinephrine. Acting independently, these neurotransmitters bring about such diverse effects as salivary secretion, bladder contraction, and voluntary motor movements. These effects are a prime example of how the same chemical messenger may cause different responses in various tissues, depending on specialization of the effector organs.

An autonomic nerve pathway consists of a two-neuron chain.

Each autonomic nerve pathway extending from the CNS to an innervated organ is a two-neuron chain (❙Figure 7-1). The cell body of the first neuron in the series is located in the CNS. Its axon, the **preganglionic fiber**, synapses with the cell body of the second neuron, which lies within a ganglion. (Recall that a ganglion is a cluster of neuronal cell bodies outside the CNS.) The axon of the second neuron, the **postganglionic fiber**, innervates the effector organ.

The autonomic nervous system has two subdivisions—the **sympathetic** and the **parasympathetic nervous systems**[1] (❙Figure 7-2). Sympathetic nerve fibers originate in the lateral horn of the thoracic (chest) and lumbar (abdominal) regions of the spinal cord (see pp. 173 and 175). Most sympathetic preganglionic fibers are very short, synapsing with cell bodies of postganglionic neurons within ganglia that lie in a **sympathetic ganglion chain** (also called the **sympathetic trunk**) located along either side of the spinal cord (see ❙Figure 5-22, p. 172). Long postganglionic fibers originate in the ganglion chain and end on the effector organs. Some preganglionic fibers pass through the ganglion chain without synapsing. Instead, they end later in sympathetic **collateral ganglia** about halfway between the CNS and the innervated organs, with postganglionic fibers traveling the rest of the distance.

Parasympathetic preganglionic fibers arise from the cranial (brain) and sacral (lower spinal cord) areas of the CNS. These fibers are longer than sympathetic preganglionic fibers because they do not end until they reach **terminal ganglia** that lie in or near the effector organs. Very short postganglionic fibers end on the cells of an organ itself.

---

[1]Some physiologists include the enteric nervous system in the autonomic nervous system, but we consider it as a separate entity (see pp. 135 and 572).

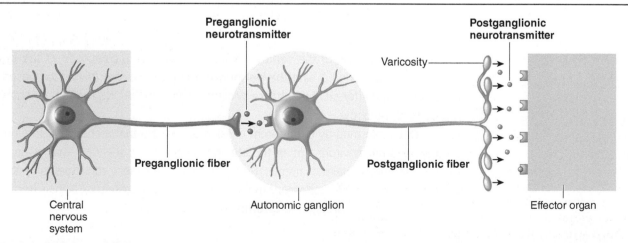

Preganglionic neurotransmitter

Postganglionic neurotransmitter

Varicosity

Preganglionic fiber

Postganglionic fiber

Central nervous system

Autonomic ganglion

Effector organ

❙Figure 7-1 **Autonomic nerve pathway.**

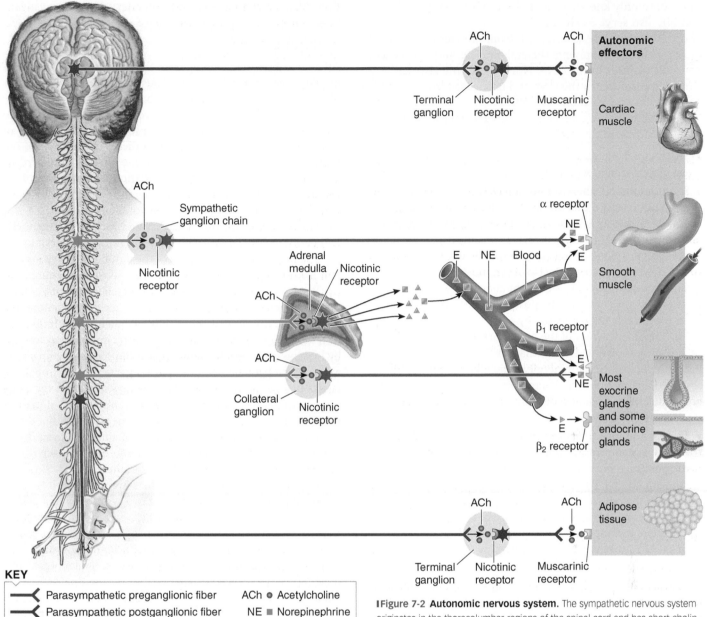

Parasympathetic preganglionic fiber	ACh ●	Acetylcholine
Parasympathetic postganglionic fiber	NE ■	Norepinephrine
Sympathetic preganglionic fiber	E ▲	Epinephrine
Sympathetic postganglionic fiber		

❙Figure 7-2 Autonomic nervous system. The sympathetic nervous system originates in the thoracolumbar regions of the spinal cord and has short cholinergic preganglionic fibers and long adrenergic postganglionic fibers. The parasympathetic nervous system originates in the brain and sacral region of the spinal cord and has long cholinergic preganglionic fibers and short cholinergic postganglionic fibers. Sympathetic and parasympathetic postganglionic fibers typically both innervate the same effector organs. The adrenal medulla is a modified sympathetic ganglion, which releases E and NE into the blood. Nicotinic cholinergic receptors respond to ACh released by all autonomic preganglionic fibers. Muscarinic cholinergic receptors respond to ACh released by parasympathetic postganglionic fibers. α_1, α_2, β_1, and β_2 adrenergic receptors are variably located at the autonomic effectors and differentially respond to NE released by sympathetic postganglionic fibers and to E released by the adrenal medulla.

FIGURE FOCUS: *(1) Which of the types of autonomic fibers release acetylcholine (ACh)? (2) Would a drug that interferes with ACh action at nicotinic receptors block the influence of the parasympathetic nervous system, the sympathetic nervous system, or both, at effector organs? (3) How about for a drug that interferes with ACh action at muscarinic receptors?*

Parasympathetic postganglionic fibers release acetylcholine; sympathetic ones release norepinephrine.

Sympathetic and parasympathetic preganglionic fibers release the same neurotransmitter, **acetylcholine (ACh)**, but the postganglionic endings of these two systems release different neurotransmitters (the neurotransmitters that influence the autonomic effectors). Parasympathetic postganglionic fibers release ACh. Accordingly, they, along with all autonomic preganglionic fibers, are called **cholinergic fibers**. Most sympathetic postganglionic fibers, in contrast, are called **adrenergic fibers** because they release **noradrena-**

line, commonly known as **norepinephrine (NE)**.[2] Both ACh and NE also serve as chemical messengers elsewhere in the body. Acetylcholine is released from the terminals of all motor neurons supplying skeletal muscle and serves as a neurotransmitter in the CNS. Norepinephrine also serves as a neurotransmitter in the CNS and is released from the adrenal medulla as a hormone as well.

The following are exceptions to this general pattern of autonomic neurotransmitter release:

- The sympathetic postganglionic fibers that supply most sweat glands secrete ACh rather than NE.

- Some autonomic fibers are **nonadrenergic, noncholinergic**: they do not release either NE or ACh. Instead they use other chemical mediators as neurotransmitters. For example, *adenosine triphosphate (ATP)* is secreted by some sympathetic fibers supplying blood-vessel smooth muscle to cause vasoconstriction (vessel narrowing). (However, most sympathetically induced vasoconstriction is brought about by NE, which is released from most sympathetic postganglionic fibers.) By contrast, *nitric oxide (NO)* released by parasympathetic fibers supplying the blood vessels of the penis contributes to the vasodilation (vessel widening) that leads to erection (hardening of the penis). Nonadrenergic, noncholinergic fibers are scattered among traditional autonomic fibers that supply not only blood vessels but also digestive, respiratory, urinary, and reproductive organs.

- Many autonomic fibers release **cotransmitters** along with the classical neurotransmitters. For example, certain sympathetic postganglionic fibers cosecrete *neuropeptide Y (NPY)* along with NE. NPY functions as a neuromodulator; that is, it modulates the release and actions of NE instead of exerting a direct action on the effector organ (see p. 110). Autonomic cotransmitters include ATP, dopamine, and various peptides such as NPY, vasoactive intestinal peptide (VIP), endogenous opioids (see p. 192), and others.

Postganglionic autonomic fibers do not end in a single terminal swelling like a synaptic knob. Instead, the terminal branches of autonomic fibers have numerous swellings, or **varicosities**, that simultaneously release neurotransmitter over a large area of the innervated organ rather than on single cells (see Figures 7-1 and 8-33, p. 292). Because of this diffuse release of neurotransmitter, and because any resulting change in electrical activity is spread throughout a smooth or cardiac muscle mass via gap junctions (see p. 62), autonomic activity typically influences whole organs instead of discrete cells.

---

[2]*Noradrenaline (norepinephrine)* is chemically similar to *adrenaline (epinephrine)*, the primary hormone product secreted by the adrenal medulla (an endocrine gland). Because a United States pharmaceutical company marketed this product for use as a drug under the trade name Adrenalin, the scientific community in this country prefers the alternative name "epinephrine" as a generic term for this chemical messenger, and accordingly, "noradrenaline" is known as "norepinephrine." In most other English-speaking countries, however, "adrenaline" and "noradrenaline" are the terms of choice.

The sympathetic and parasympathetic nervous systems dually innervate most visceral organs.

Afferent information coming from the viscera (internal organs) usually does not reach the conscious level (see p. 186). Examples of visceral afferent information include input from the baroreceptors that monitor blood pressure and input from the chemoreceptors that monitor the protein or fat content of ingested food. This input is used to direct the activity of the autonomic efferent neurons. Autonomic efferent output regulates visceral activities such as circulation and digestion. Like visceral afferent input, autonomic efferent output operates outside the realm of consciousness and voluntary control.

Most visceral organs are innervated by both sympathetic and parasympathetic nerve fibers (Figure 7-3). Innervation of a single organ by both branches of the autonomic nervous system is known as **dual innervation** (*dual* means "pertaining to two"). Table 7-1 summarizes the major effects of these autonomic branches. Although the details of this array of autonomic responses are described more fully later when the individual effector organs are discussed, you can infer one general concept by looking over the table now. As you can see, the sympathetic and parasympathetic nervous systems generally exert opposite effects in a particular organ. Sympathetic stimulation increases the heart rate, whereas parasympathetic stimulation decreases it; sympathetic stimulation slows movement within the digestive tract, whereas parasympathetic stimulation enhances digestive motility. Note that both systems increase the activity of some organs and reduce the activity of others.

Rather than memorize a list such as in Table 7-1, it is better to logically deduce the actions of the two systems by first understanding the circumstances under which each system dominates. Usually, both systems are partially active—that is, normally some level of action potential activity exists in both the sympathetic and the parasympathetic fibers supplying a particular organ. This ongoing activity is called **sympathetic** or **parasympathetic tone**. Under given circumstances, activity of one division can dominate the other. *Sympathetic dominance* to a particular organ exists when the sympathetic fibers' rate of firing to that organ increases above tone level, coupled with a simultaneous decrease below tone level in the parasympathetic fibers' frequency of action potentials to the same organ. The reverse situation is true for *parasympathetic dominance*. The balance between sympathetic and parasympathetic activity can be shifted separately for individual organs to meet specific demands (for example, sympathetically induced dilation of the pupil in dim light; see p. 194), or a more generalized, widespread discharge of one autonomic system in favor of the other can be elicited to control bodywide functions. Massive widespread discharges take place more often in the sympathetic system. The value of massive sympathetic discharge is clear, considering the circumstances during which this system usually dominates.

Times of Sympathetic Dominance The sympathetic system promotes responses that prepare the body for strenuous

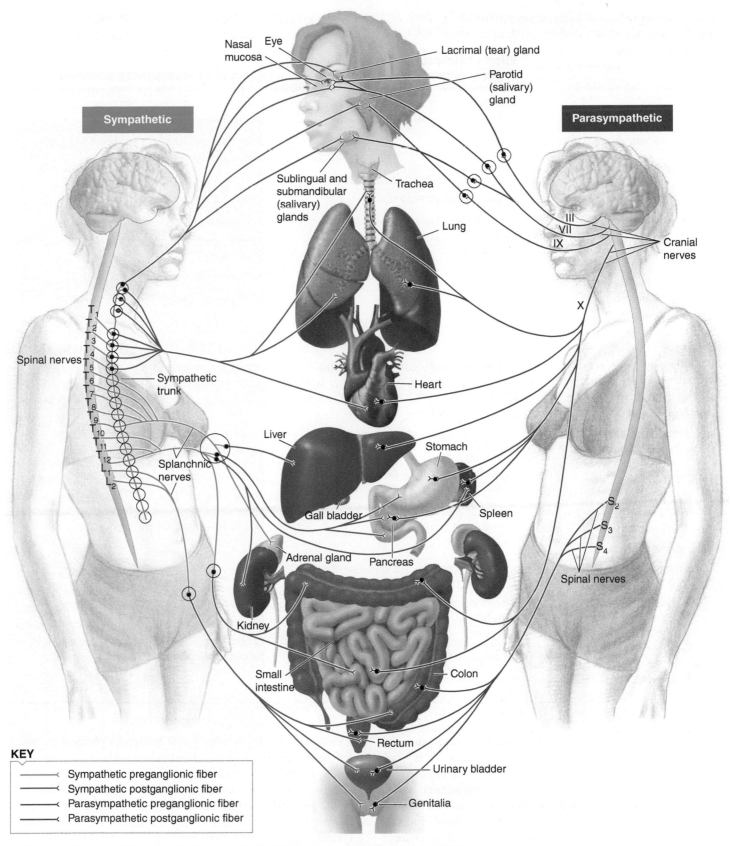

Sympathetic

Parasympathetic

Nasal mucosa
Eye
Lacrimal (tear) gland
Parotid (salivary) gland

Sublingual and submandibular (salivary) glands
Trachea

Lung

III
VII
IX
Cranial nerves

X

Spinal nerves

T₁
T₂
T₃
T₄
T₅
T₆
T₇
T₈
T₉
T₁₀
T₁₁
T₁₂
L₁
L₂

Sympathetic trunk

Heart

Liver
Stomach

Splanchnic nerves

Gall bladder
Spleen

Adrenal gland
Pancreas

S₂
S₃
S₄

Spinal nerves

Kidney

Small intestine
Colon

Rectum

Urinary bladder

Genitalia

KEY

⊸	Sympathetic preganglionic fiber
⊸	Sympathetic postganglionic fiber
⊸	Parasympathetic preganglionic fiber
⊸	Parasympathetic postganglionic fiber

I Figure 7-3 **Structures innervated by the sympathetic and the parasympathetic nervous systems.**

TABLE 7-1 Effects of Autonomic Nervous System on Various Organs

Organ	Effect of Sympathetic Stimulation (and Types of Adrenergic Receptors)	Effect of Parasympathetic Stimulation
Heart	Increases heart rate and increases force of contraction of the whole heart) (β_1)	Decreases heart rate and decreases force of contraction of the atria only
Most innervated blood vessels	Constricts (α_1)	Dilates vessels supplying the penis and clitoris only
Lungs	Dilates the bronchioles (airways) (β_2) Inhibits mucus secretion (α)	Constricts the bronchioles Stimulates mucus secretion
Digestive tract	Decreases motility (movement) (α_2, β_2) Contracts sphincters (to prevent forward movement of contents) (α_1) Inhibits digestive secretions (α_2)	Increases motility Relaxes sphincters (to permit forward movement of contents) Stimulates digestive secretions
Urinary bladder	Relaxes (β_2)	Contracts (emptying)
Eye	Dilates the pupil (contracts radial muscle) (α_1) Adjusts the eye for far vision (β_2)	Constricts the pupil (contracts circular muscle) Adjusts the eye for near vision
Liver (glycogen stores)	Glycogenolysis (glucose is released) (β_2)	None
Adipose cells (fat stores)	Lipolysis (fatty acids are released) (β_2)	None
Exocrine glands		
Exocrine pancreas	Inhibits pancreatic exocrine secretion (α_2)	Stimulates pancreatic exocrine secretion (important for digestion)
Sweat glands	Stimulates secretion by sweat glands; important in cooling the body (α_1; most are cholinergic)	None
Salivary glands	Stimulates a small volume of thick saliva rich in mucus (α_1)	Stimulates a large volume of watery saliva rich in enzymes
Endocrine glands		
Adrenal medulla	Stimulates epinephrine and norepinephrine secretion (cholinergic)	None
Endocrine pancreas	Inhibits insulin secretion; stimulates glucagon secretion (α_2)	Stimulates insulin and glucagon secretion
Genitals	Controls ejaculation (males) and orgasmic contractions (both sexes) (α_1)	Controls erection (penis in males and clitoris in females)
Brain activity	Increases alertness (receptors unknown)	None

physical activity in emergency or stressful situations, such as a physical threat from the outside. This response is typically referred to as a **"fight-or-flight"** response (some physiologists even throw in **fright**, too), because the sympathetic system readies the body to fight against or flee from (and be frightened by) the threat. Think about the body resources needed in such circumstances. The heart beats more rapidly and more forcefully, blood pressure is elevated by generalized constriction of the blood vessels, respiratory airways dilate to permit maximal airflow, glycogen (stored sugar) and fat stores are broken down to release extra fuel into the blood, and blood vessels supplying skeletal muscles dilate. All these responses are aimed at providing increased flow of oxygenated, nutrient-rich blood to the

skeletal muscles in anticipation of strenuous physical activity. Furthermore, the pupils dilate and the eyes adjust for far vision, letting the person visually assess the entire threatening scene. Sweating is promoted in anticipation of excess heat production by the physical exertion. Because digestive and urinary activities are not essential in meeting the threat, the sympathetic system inhibits these activities.

Times of Parasympathetic Dominance The parasympathetic system dominates in quiet, relaxed situations. Under such nonthreatening circumstances, the body can be concerned with its "general housekeeping" activities, such as digestion. The parasympathetic system promotes such **"rest-and-digest"**

bodily functions while slowing down those activities that are enhanced by the sympathetic system. For example, the heart does not need to beat rapidly and forcefully when the person is in a tranquil setting.

Advantage of Dual Autonomic Innervation Dual inner-vation of organs with nerve fibers whose actions oppose each other enables precise control over an organ's activity, like having both an accelerator and a brake to control the speed of a car. If an animal suddenly darts across the road as you are driving, you could eventually stop if you just took your foot off the accelerator, but you might stop too slowly to avoid hitting the animal. If you simultaneously apply the brake as you lift up on the accelerator, however, you can come to a more rapid, controlled stop. In a similar manner, a sympathetically accelerated heart rate could gradually be reduced to normal following a stressful situation by decreasing the firing rate in the cardiac sympathetic nerve (let-ting up on the accelerator). However, the heart rate can be reduced more rapidly by simultaneously increasing activity in the parasympathetic supply to the heart (applying the brake). Indeed, the two divisions of the autonomic nervous system are usually reciprocally controlled; increased activity in one division is accompanied by a corresponding decrease in the other.

There are several exceptions to the general rule of dual reciprocal innervation by the two branches of the autonomic nervous system; the most notable are the following:

■ *Innervated blood vessels* receive only sympathetic nerve fibers. (Of the blood vessel types, most arterioles and veins are innervated; arteries and capillaries are not [see pp. 343 and 361].) Regulation of vessel caliber in innervated blood vessels (that is, promoting vasoconstriction or vasodilation) is accom-plished by increasing or decreasing the firing rate above or be-low tone level, respectively, in these sympathetic fibers. The only blood vessels to receive both sympathetic and parasym-pathetic fibers are those supplying the penis and clitoris. The precise vascular control this dual innervation affords these or-gans is important in accomplishing erection.

■ *Sweat glands* are innervated only by sympathetic nerves. Re-member that the postganglionic fibers of most of these nerves are unusual because they secrete ACh rather than NE.

■ *Salivary glands* are innervated by both autonomic divisions, but unlike elsewhere, sympathetic and parasympathetic activi-ties are not antagonistic. Both stimulate salivary secretion, but the saliva's volume and composition differ, depending on which autonomic branch is dominant.

You will learn more about these exceptions in later chapters. We now turn to the adrenal medulla, a unique endocrine compo-nent of the sympathetic nervous system.

The adrenal medulla is a modified part of the sympathetic nervous system.

The two *adrenal glands* lie above the kidneys, one on each side (*ad* means "next to"; *renal* means "kidney"). The adrenal glands are endocrine glands, each with an outer portion, the *adrenal cortex,* and an inner portion, the *adrenal medulla* (see pp. 672 and 681–682). The **adrenal medulla** is a modified sympathetic

ganglion that does not give rise to postganglionic fibers. Instead, on stimulation by the preganglionic fiber that originates in the CNS, it secretes catecholamine hormones (see p. 121) into the blood (see ❚Figure 7-2). Not surprisingly, the hormones are identical or similar to postganglionic sympathetic neurotrans-mitters. About 20% of the adrenal medullary hormone output is norepinephrine, and the remaining 80% is the closely related **epinephrine (E) (adrenaline)** (see footnote 2, p. 236). These hormones, in general, reinforce activity of the sympathetic ner-vous system.

Several receptor types are available for each autonomic neurotransmitter.

Because each autonomic neurotransmitter and medullary hor-mone stimulates activity in some tissues but inhibits activity in others, the particular responses must depend on specialization of the tissue cells rather than on properties of the chemicals themselves. Responsive tissue cells have one or more of several types of plasma membrane receptor proteins for these chemical messengers. Binding of a neurotransmitter to a receptor induces the tissue-specific response.

Cholinergic Receptors Researchers have identified two types of ACh (cholinergic) receptors—*nicotinic* and *musca-rinic*—on the basis of their response to particular drugs. **Nico-tinic receptors** are activated by the tobacco plant derivative nicotine, whereas **muscarinic receptors** are activated by the mushroom poison muscarine (❚Table 7-2).

Nicotinic receptors are found on the postganglionic cell bodies in all autonomic ganglia. These receptors respond to ACh released from both sympathetic and parasympathetic preganglionic fibers. These receptors are nonspecific cation receptor-channels that permit passage of both Na^+ and K^+ when ACh binds to them (see p. 116). Because the permeability of the postganglionic membrane to Na^+ and K^+ on opening of these channels is essentially equal, the relative movement of these ions through the channels depends on their electrochem-ical driving forces. Recall that at resting potential the net driv-ing force for Na^+ is much greater than that for K^+ because the resting potential is much closer to the K^+ equilibrium potential than to the Na^+ equilibrium potential. Both the concentration and the electrical gradients for Na^+ are inward, whereas the outward concentration gradient for K^+ is almost, but not quite, balanced by the opposing inward electrical gradient (see p. 82). As a result, when ACh triggers the opening of these receptor-channels, considerably more Na^+ moves inward than K^+ moves outward, bringing about a depolarization that leads to initiation of an action potential in the postganglionic cell.

Muscarinic receptors are found on effector cell membranes (cardiac muscle, smooth muscle, and glands). They bind with ACh released from parasympathetic postganglionic fibers. The five subtypes of muscarinic receptors are all linked to G pro-teins that activate second-messenger pathways leading to the target cell response (see p. 117).

Adrenergic Receptors The two major classes of adrenergic receptors for norepinephrine and epinephrine are **alpha (α)**

Receptor Type	Neurotransmitter Affinity	Effector(s) with Receptor Type	Mechanism of Action at Effector	Effect on Effector
Nicotinic	ACh from autonomic preganglionic fibers	All autonomic postganglionic cell bodies; adrenal medulla	Opens nonspecific cation receptor-channels	Excitatory
	ACh from motor neurons	Motor end plates of skeletal muscle fibers	Opens nonspecific cation receptor-channels	Excitatory
Muscarinic	ACh from parasympathetic postganglionic fibers	Cardiac muscle, smooth muscle, glands	Activates various G-protein-coupled receptor pathways, depending on effector	Excitatory or inhibitory, depending on effector
α_1	Greater affinity for NE (from sympathetic postganglionic fibers) than for E (from the adrenal medulla)	Most sympathetic target tissues	Activates IP_3–Ca^{2+} second-messenger pathway	Excitatory
α_2	Greater affinity for NE than for E	Digestive organs	Inhibits cAMP	Inhibitory
β_1	Equal affinity for NE and for E	Heart	Activates cAMP	Excitatory
β_2	Affinity for E only	Smooth muscles of arterioles and bronchioles	Activates cAMP	Inhibitory

and **beta (β) receptors**, which are further subclassified into α_1 and α_2 and into β_1 and β_2 **receptors,** respectively (ITable 7-2). These various receptor types are distinctly distributed among sympathetically controlled effector organs as follows:

- α_1 receptors are present on most sympathetic target tissues.
- α_2 receptors are located mainly on digestive organs.
- β_1 receptors are restricted to the heart.
- β_2 receptors are found on smooth muscles of arterioles and bronchioles (small blood vessels and airways).

Different receptor types also have different affinities (attraction) for norepinephrine and epinephrine:

- α receptors of both subtypes have a greater affinity for NE than for E.
- β_1 receptors have about equal affinities for NE and E.
- β_2 receptors bind only with E.

All adrenergic receptors are coupled to G proteins, but the ensuing pathway activated on binding of a catecholamine differs for the various receptor types:

- Activation of both β_1 and β_2 receptors brings about the target cell response by activating the cyclic adenosine monophosphate (cAMP) second-messenger pathway (see p. 123).
- Stimulation of α_1 receptors elicits the desired response via the IP_3–Ca^{2+} second-messenger pathway (see p. 124).
- By contrast, binding of a neurotransmitter to an α_2 receptor inhibits cAMP production in the target cell.

The effector organ response also varies depending on the adrenergic receptor type:

- Activation of α_1 receptors usually brings about an excitatory response in the effector organ—for example, arteriolar constriction caused by increased contraction of smooth muscle in the walls of these blood vessels.
- Activation of α_2 receptors, in contrast, brings about an inhibitory response in the effector, such as decreased smooth muscle contraction in the digestive tract.
- Stimulation of β_1 receptors, which are found only in the heart, causes an excitatory response—namely, increased rate and force of cardiac contraction.
- The response to β_2 receptor activation is generally inhibitory, such as arteriolar or bronchiolar dilation caused by relaxation of the smooth muscle in the walls of these tubular structures.

As a quick rule, activation of the subscript "1" versions of adrenergic receptors leads to excitatory responses, and activation of the subscript "2" versions leads to inhibitory responses.

Autonomic Agonists and Antagonists Drugs are available that selectively alter autonomic responses at each of the receptor types. An **agonist** binds to the neurotransmitter's receptor and causes the same response as the neurotransmitter would. An **antagonist**, by contrast, binds with the receptor, preventing the neurotransmitter from binding and causing a response, yet the antagonist itself produces no response. Thus, an agonist mimics the neurotransmitter's response, and an antagonist blocks the neurotransmitter's response. Some of these drugs are only of experimental interest, but others are important therapeutically. For example, *atropine* blocks the effect of ACh at muscarinic receptors but does not affect nicotinic receptors. Because ACh released at both parasympathetic and sympathetic preganglionic fibers combines with nicotinic receptors, blockage at nicotinic synapses would knock out both these autonomic branches. By acting selectively to interfere with ACh action only at muscarinic junctions, which are the sites of parasympathetic postganglionic action, atropine blocks parasympathetic effects but does not influence sympathetic activity. Doctors use this principle to suppress salivary and bronchial secretions before surgery and thus reduce the risk of a patient inhaling these secretions into the lungs.

Likewise, drugs that act selectively at α- and β-adrenergic receptor sites to either activate or block specific sympathetic effects are widely used. Following are several examples. *Salbutamol* selectively activates β_2-adrenergic receptors at low doses, making it possible to dilate the bronchioles in the treatment of asthma without undesirably stimulating the heart (which has mostly β_1 receptors). By contrast, *metoprolol* selectively blocks β_1-adrenergic receptors and is prescribed to treat high blood pressure because it decreases the amount of blood the heart pumps into the blood vessels. Metoprolol does not affect β_2 receptors and so has no effect on the bronchioles.

Many regions of the CNS are involved in the control of autonomic activities.

Messages from the CNS are delivered to cardiac muscle, smooth muscle, and glands via autonomic nerves, but what CNS regions regulate autonomic output? Autonomic control of these effectors is mediated by reflexes and through centrally located control centers. Going back one step further, ultimately information carried to the CNS via visceral afferents is used to determine the appropriate output via autonomic efferents to the effectors to maintain homeostasis. (Some physiologists regard visceral affer-

ents as part of the autonomic nervous system, whereas others consider the sympathetic and parasympathetic efferents as being the only components of the autonomic nervous system.)

■ Some autonomic reflexes, such as urination, defecation, and erection, are integrated at the spinal-cord level, but all these spinal reflexes are subject to control by higher levels of consciousness.

■ The medulla within the brain stem is the region most directly responsible for autonomic output. Centers for controlling cardiovascular, respiratory, and digestive activity via the autonomic system are located there.

■ The hypothalamus plays an important role in integrating the autonomic, somatic, and endocrine responses that automatically accompany various emotional and behavioral states. For example, the increased heart rate, blood pressure, and respiratory activity associated with anger or fear are brought about by the hypothalamus acting through the medulla.

■ Autonomic activity can also be influenced by the prefrontal association cortex through its involvement with emotional expression. An example is blushing when embarrassed, which is caused by dilation of blood vessels supplying the skin of the cheeks. Such responses are mediated through hypothalamic–medullary pathways.

▌Table 7-3 summarizes the main distinguishing features of the sympathetic and parasympathetic nervous systems.

Check Your Understanding 7.1

1. Illustrate the origin, termination, fiber length, and neurotransmitter released for parasympathetic and sympathetic preganglionic fibers and postganglionic fibers.

2. Compare the times of sympathetic and of parasympathetic dominance.

3. Discuss the relationship of the adrenal medulla to the autonomic nervous system.

▌TABLE 7-3 Comparison of the Sympathetic and the Parasympathetic Nervous System

Feature	Sympathetic System	Parasympathetic System
Origin of preganglionic fiber	Thoracic and lumbar regions of the spinal cord	Brain and sacral region of the spinal cord
Origin of postganglionic fiber	Sympathetic ganglion chain (near the spinal cord) or collateral ganglia (about halfway between spinal cord and effector organs)	Terminal ganglia (in or near effector organs)
Fiber length	Short preganglionic fibers, long postganglionic fibers	Long preganglionic fibers, short postganglionic fibers
Neurotransmitter released	Preganglionic: ACh Postganglionic: NE	Preganglionic: ACh Postganglionic: ACh
Types of receptors for neurotransmitters	For preganglionic neurotransmitter: nicotinic For postganglionic neurotransmitter: α_1, α_2, β_1, β_2	For preganglionic neurotransmitter: nicotinic For postganglionic neurotransmitter: muscarinic
Dominance	Dominates in "fight-or-flight" situations	Dominates in "rest-and-digest" situations

7.2 Somatic Nervous System

Motor neurons supply skeletal muscle.

Motor neurons, whose axons constitute the somatic nervous system, supply skeletal muscles and bring about movement (*motor* means "movement"). (Sometimes all efferent neurons are referred to as *motor neurons*, but we reserve this term for the efferent somatic fibers that supply skeletal muscles.) The cell bodies of almost all motor neurons are within the ventral horn of the spinal cord (see p. 175). The only exception is that the cell bodies of motor neurons supplying muscles in the head are in the brain stem. Unlike the two-neuron chain of autonomic nerve fibers, the axon of a motor neuron is continuous from its origin in the CNS to its ending on skeletal muscle. Motor-neuron axon terminals release ACh, which brings about excitation and contraction of the innervated muscle cells. Motor neurons can only stimulate skeletal muscles, in contrast to autonomic fibers, which can either stimulate or inhibit their effector organs. Inhibition of skeletal muscle activity can be accomplished only within the CNS through inhibitory synaptic input to the dendrites and cell bodies of the motor neurons supplying that particular muscle.

Motor neurons are the final common pathway.

Motor-neuron dendrites and cell bodies are influenced by many converging presynaptic inputs, both excitatory and inhibitory. Some of these inputs are part of spinal reflex pathways originating with peripheral sensory receptors. Others are part of descending pathways originating within the brain. Areas of the brain that exert control over skeletal muscle movements include the motor regions of the cortex, the basal nuclei, the cerebellum, and the brain stem (see pp. 148–150, 153, 163, and 166–167; also see Table 8-3, p. 280, for a summary of motor control and Figure 5-26b, p. 175, for specific examples of these descending motor pathways).

Motor neurons are considered the **final common pathway** because the only way any other parts of the nervous system can influence skeletal muscle activity is by acting on these motor neurons. The level of activity in a motor neuron and its subsequent output to the skeletal muscle fibers it innervates depend on the relative balance of excitatory postsynaptic potentials (EPSPs) and inhibitory postsynaptic potentials (IPSPs) (see pp. 106–107) brought about by its presynaptic inputs originating from these diverse sites in the brain.

The somatic system is under voluntary control, but much of skeletal muscle activity involving posture, balance, and stereotypical movements is subconsciously controlled. You may decide you want to start walking, but you do not have to consciously bring about the alternate contraction and relaxation of the involved muscles because these movements are involuntarily coordinated by lower brain centers.

Clinical Note The cell bodies of motor neurons may be selectively destroyed by **poliovirus**. The result is paralysis of the muscles innervated by the affected neurons. **Amyotrophic lateral sclerosis (ALS)**, also known as **Lou Gehrig's disease**, is the most common motor-neuron disease. This

---

ITABLE 7-4 Comparison of the Autonomic and the Somatic Nervous System

Feature	Autonomic Nervous System	Somatic Nervous System
Site of origin	Sympathetic: lateral horn of thoracic and lumbar spinal cord Parasympathetic: brain and sacral spinal cord	Ventral horn of spinal cord for most; those supplying muscles in head originate in brain
Number of neurons from CNS to effector organ	Two-neuron chain (preganglionic and postganglionic)	Single neuron (motor neuron)
Organs innervated	Cardiac muscle, smooth muscle, most exocrine and some endocrine glands	Skeletal muscle
Type of innervation	Most effector organs dually innervated by the two antagonistic branches of this system (sympathetic and parasympathetic)	Effector organs innervated only by motor neurons
Neurotransmitter at effector organs	May be ACh (parasympathetic terminals) or NE (sympathetic terminals)	Only ACh
Effects on effector organs	Either stimulation or inhibition (antagonistic actions of two branches)	Stimulation only (inhibition possible only centrally through IPSPs on dendrites and cell body of motor neuron)
Type of control	Under involuntary control	Subject to voluntary control; much activity subconsciously coordinated
Higher centers involved in control	Spinal cord, medulla, hypothalamus, prefrontal association cortex	Spinal cord, motor cortex, basal nuclei, cerebellum, brain stem

incurable condition is characterized by degeneration and eventual death of motor neurons. The result is gradual loss of motor control, progressive paralysis, and finally death within 3 to 5 years of onset. The exact cause is uncertain, although researchers are investigating a variety of potential underlying problems. Among these are pathological changes in neurofilaments that block axonal transport of crucial materials (see p. 51), extracellular accumulation of toxic levels of the excitatory neurotransmitter glutamate, aggregation of misfolded intracellular proteins, mitochondrial dysfunction leading to reduced energy production, and activation of protein-cutting enzymes (caspases, the ones involved in apoptosis; see p. 43) that selectively cut up the neuronal cell body and nucleus.

Before turning to the junction between a motor neuron and the muscle cells it innervates, we pull together in table form two groups of information we have been examining in this and preceding nervous system chapters. Table 7-4 summarizes the features of the two branches of the efferent division of the PNS: the autonomic nervous system and the somatic nervous system. Table 7-5 compares the three functional types of neurons: afferent neurons, efferent neurons, and interneurons.

TABLE 7-5 Comparison of Types of Neurons

Feature	Afferent Neuron	Efferent Neuron in Autonomic Nervous System	Efferent Neuron in Somatic Nervous System	Interneuron
Origin, structure, location	Receptor at peripheral ending; elongated peripheral axon travels in the peripheral nerve; cell body located in the dorsal root ganglion; short central axon enters the spinal cord	Two-neuron chain; first neuron (preganglionic fiber) originates in the CNS and terminates on a ganglion; second neuron (postganglionic fiber) originates in the ganglion and terminates on the effector organ	Cell body of motor neuron in the spinal cord; long axon travels in the peripheral nerve and terminates on the effector organ	Lies entirely within the CNS; some cell bodies originate in the brain, with long axons traveling down the spinal cord in descending pathways; some originate in the spinal cord, with long axons traveling up the cord to the brain in ascending pathways; others form short local connections
Termination	Interneurons*	Effector organs (cardiac muscle, smooth muscle, glands)	Effector organs (skeletal muscle)	Other interneurons and efferent neurons
Function	Carries information about the external and the internal environment to the CNS	Carries instructions from the CNS to the effector organs	Carries instructions from the CNS to the effector organs	Processes and integrates afferent input; initiates and coordinates efferent output; is responsible for thought and other higher mental functions
Convergence of input on cell body	No (only input is through the receptor)	Yes	Yes	Yes
Effect of input to neuron	Can only be excited (through a receptor potential induced by a stimulus; must reach threshold for an action potential)	Can be excited or inhibited (through EPSPs and IPSPs at the first neuron; must reach threshold for an action potential)	Can be excited or inhibited (through EPSPs and IPSPs; must reach threshold for an action potential)	Can be excited or inhibited (through EPSPs and IPSPs; must reach threshold for an action potential)
Site of action potential initiation	First excitable portion of the membrane adjacent to the receptor	Axon hillock	Axon hillock	Axon hillock
Divergence of output	Yes	Yes	Yes	Yes
Effect of output on effector organ	Only excites	Postganglionic fiber either excites or inhibits	Only excites	Either excites or inhibits

*Except in stretch reflex where afferent neuron terminates directly on efferent neuron; see p. 178.

7.3 Neuromuscular Junction

Motor neurons and skeletal muscle fibers are chemically linked at neuromuscular junctions.

An action potential in a motor neuron is rapidly propagated from the cell body within the CNS to the skeletal muscle along the large myelinated axon (efferent fiber) of the neuron. As the axon approaches a muscle, it divides and loses its myelin sheath. Each of these axon terminals forms a special junction, a **neuromuscular junction**,[3] with one of the many muscle cells that compose the whole muscle (❙ Figure 7-4 and chapter opener photo, p. 233). Each branch innervates only one muscle cell; therefore, each muscle cell has only one neuromuscular junc-

---

[3]Many scientists refer to a **synapse** as any junction between two cells that handle information electrically. According to this broad point of view, *chemical synapses* include junctions between two neurons and those between a neuron and an effector cell (such as muscle cells of any type or gland cells), and *electrical synapses* include gap junctions between smooth muscle cells, between cardiac muscle cells, or between some neurons. We narrowly reserve the term *synapse* specifically for neuron-to-neuron junctions and use different terms for other types of junctions, such as the term *neuromuscular junction* for a junction between a motor neuron and a skeletal muscle cell.

tion. Both the neural and muscular components make up the neuromuscular junction, just as a synapse includes both presynaptic and postsynaptic components. A single muscle cell, called a **muscle fiber**, is long and cylindrical. Within a neuromuscular junction, the axon terminal splits into multiple fine branches, each of which ends in an enlarged knoblike structure called the **terminal button**, or **bouton**. The entire axon terminal ending (all the fine branches with terminal buttons) fits into a shallow depression, or groove, in the underlying muscle fiber. This specialized underlying portion of the muscle cell membrane is called the **motor end plate** (❙ Figure 7-5).

ACh is the neuromuscular junction neurotransmitter.

Nerve and muscle cells do not come into direct contact at a neuromuscular junction. The space, or cleft, between these two structures is too large for electrical transmission of an impulse between them (that is, an action potential cannot "jump" that far). Just as at a neuronal chemical synapse (see p. 103), a chemical messenger carries the signal between a terminal button and the muscle fiber. This neurotransmitter is ACh.

Release of ACh at the Neuromuscular Junction Each terminal button contains thousands of vesicles that store ACh. Propagation of an action potential to the axon terminal (❙ Figure 7-5, step **1**) triggers the opening of voltage-gated calcium (Ca^{2+}) channels in all of its terminal buttons (see p. 89). We focus on one terminal button, but the same events take place concurrently at all terminal buttons of a given neuromuscular junction. When Ca^{2+} channels open, Ca^{2+} diffuses into the terminal button from its higher extracellular concentration

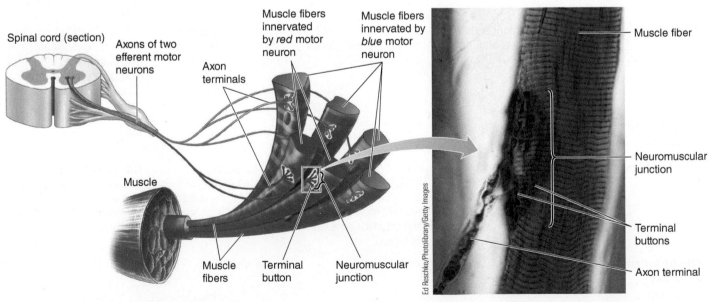

❙ Figure 7-4 **Motor neuron innervating skeletal muscle cells.** The cell body of a motor neuron originates in the ventral horn of the spinal cord. The axon (somatic efferent fiber) exits through the ventral root and travels through a spinal nerve to the skeletal muscle it innervates. When the axon reaches a skeletal muscle, it divides into many axon terminals, each of which forms a neuromuscular junction with a single muscle cell (muscle fiber). The axon terminal within a neuromuscular junction further divides into fine branches, each of which ends in an enlarged terminal button. Note that the muscle fibers innervated by a single axon terminal are dispersed throughout the muscle, but for simplicity they are grouped together in this figure.

FIGURE FOCUS: *What is the relationship among* axon terminal, neuromuscular junction, *and* terminal button?

(step **2**), which in turn causes release of ACh by exocytosis from several hundred vesicles into the cleft (step **3**).

Formation of an End-Plate Potential The released ACh diffuses across the cleft and binds with chemically gated receptor-channels of the cholinergic nicotinic type on the motor end-plate portion of the muscle fiber membrane (step **4**). Binding with ACh causes these receptor-channels to open. They are nonspecific cation channels that permit both Na^+ and K^+ traffic through them (step **5**). Because of the greater electrochemical gradient for Na^+ than for K^+, considerably more Na^+ moves inward than K^+ moves outward, depolarizing the motor end plate. The collective potential change resulting from these ion movements across all of the terminal buttons within a neuromuscular junction is called the **end-plate potential (EPP)**. It is a graded potential similar to an EPSP, except that an EPP is much larger for the following reasons: (1) A neuromuscular junction consists of multiple terminal buttons, each of which simultaneously releases ACh on activation of the axon terminal; (2) more neurotransmitter is released from a terminal button than from a presynaptic knob in response to an action potential; (3) the motor end plate has a larger surface area and a higher density of neurotransmitter receptor-channels and thus has more sites for binding with neurotransmitter than a subsynaptic membrane has; and (4) accordingly, many more receptor-channels are opened in response to neurotransmitter release at a neuromuscular junction than at a synapse. This permits a greater net influx of positive ions and a larger depolarization

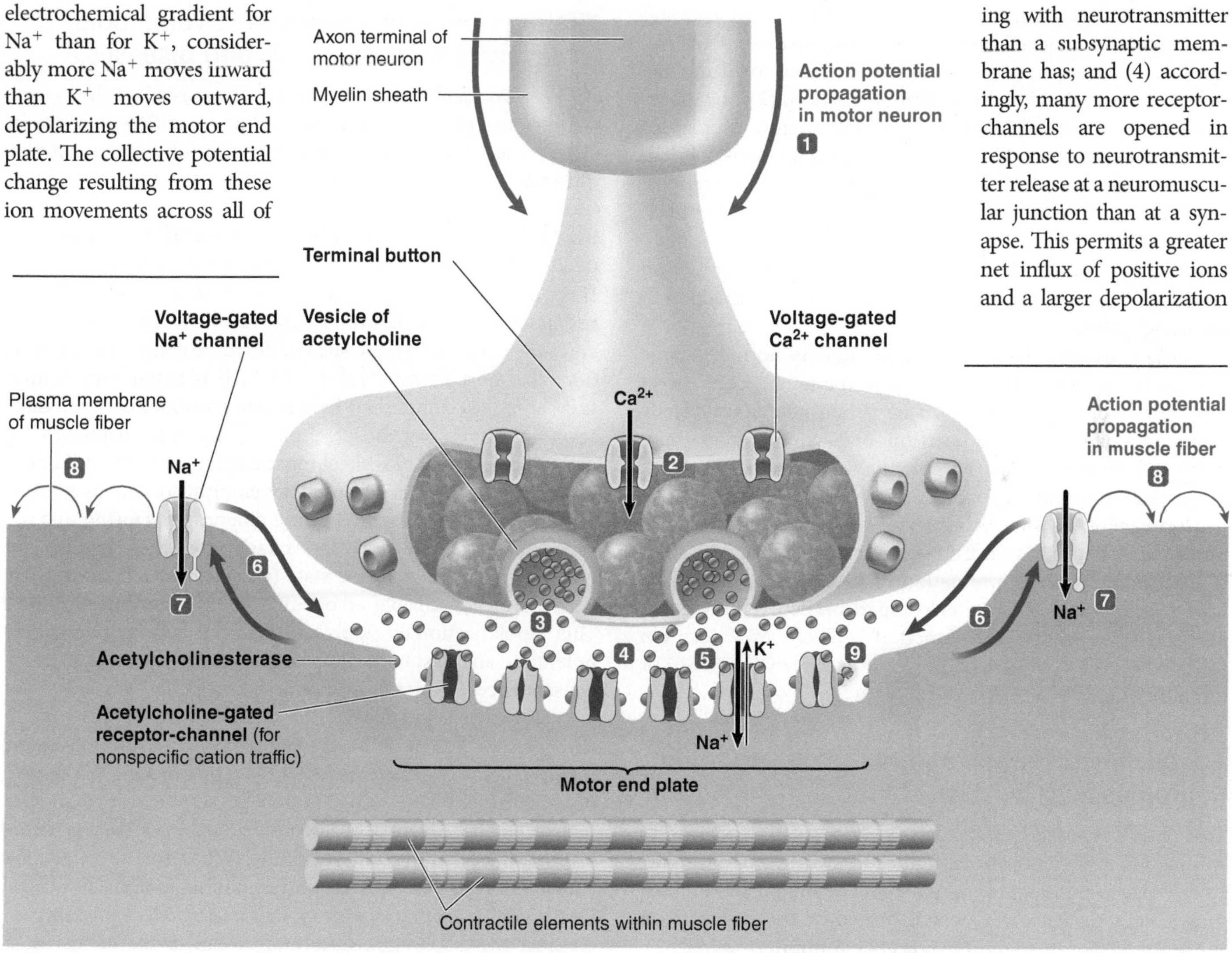

Figure labels:
Axon terminal of motor neuron
Myelin sheath
Action potential propagation in motor neuron **1**
Terminal button
Voltage-gated Na^+ channel
Vesicle of acetylcholine
Voltage-gated Ca^{2+} channel
Plasma membrane of muscle fiber
Ca^{2+}
2
Action potential propagation in muscle fiber **8**
Na^+ **8**
6
7
Acetylcholinesterase
Acetylcholine-gated receptor-channel (for nonspecific cation traffic)
3
4 **5**
K^+
Na^+
6 Na^+ **7**
9
Motor end plate
Contractile elements within muscle fiber

1 An action potential in a motor neuron is propagated to the terminal button.

2 This local action potential triggers the opening of voltage-gated Ca^{2+} channels and the subsequent entry of Ca^{2+} into the terminal button.

3 Ca^{2+} triggers the release of acetylcholine (ACh) by exocytosis from a portion of the vesicles.

4 ACh diffuses across the space separating the nerve and muscle cells and binds with receptor-channels specific for it on the motor end plate of the muscle cell membrane.

5 This binding brings about the opening of these nonspecific cation channels, leading to a relatively large movement of Na^+ into the muscle cell compared to a smaller movement of K^+ outward.

6 The result is an end-plate potential. Local current flow occurs between the depolarized end plate and the adjacent membrane.

7 This local current flow opens voltage-gated Na^+ channels in the adjacent membrane.

8 The resultant Na^+ entry reduces the potential to threshold, initiating an action potential, which is propagated throughout the muscle fiber.

9 ACh is subsequently destroyed by acetylcholinesterase, an enzyme located on the motor end-plate membrane, terminating the muscle cell's response.

Figure 7-5 Events at a neuromuscular junction.
FIGURE FOCUS: *By studying this figure, what are three ways that detrimental chemical agents could interfere with the normal action of ACh at the neuromuscular junction?*

(an EPP) at a neuromuscular junction. As with an EPSP, an EPP is a graded potential, whose magnitude depends on the amount and duration of ACh at the end plate.

Initiation of an Action Potential The motor end-plate region itself does not have a threshold potential, so an action potential cannot be initiated at this site. However, an EPP brings about an action potential in the rest of the muscle fiber as follows: The neuromuscular junction is usually in the middle of the long, cylindrical muscle fiber. When an EPP takes place, local current flow occurs between the depolarized end plate and the adjacent, resting cell membrane in both directions step ❻), opening voltage-gated Na^+ channels and thus reducing the potential to threshold in the adjacent areas (step ❼). The subsequent action potential initiated at these sites propagates throughout the muscle fiber membrane by contiguous conduction (step ❽) (see p. 96). The spread runs in both directions, away from the motor end plate toward both ends of the fiber. This electrical activity triggers contraction of the muscle fiber. Thus, by means of ACh, an action potential in a motor neuron brings about an action potential and subsequent contraction in the muscle fiber.

Unlike synaptic transmission, an EPP is normally large enough to cause an action potential in the muscle cell. Therefore, one-to-one transmission of an action potential typically occurs at a neuromuscular junction; one action potential in a nerve cell triggers one action potential in a muscle cell that it innervates. At a synapse, one action potential in a presynaptic neuron cannot by itself bring about an action potential in a postsynaptic neuron. An action potential in a postsynaptic neuron occurs only when summation of EPSPs brings the membrane to threshold. As another comparison between these two junctions, a neuromuscular junction is always excitatory (an EPP), whereas a synapse may be either excitatory (an EPSP) or inhibitory (an IPSP).

Acetylcholinesterase ends ACh activity at the neuromuscular junction.

To ensure purposeful movement, a muscle cell's response to stimulation by its motor neuron must be switched off promptly when there is no longer a signal from the motor neuron. The muscle cell's electrical response is turned off by an enzyme in the motor end-plate membrane, **acetylcholinesterase (AChE)**, which inactivates ACh.

As a result of diffusion, many of the released ACh molecules come into contact with and bind to receptor-channels on the surface of the motor end-plate membrane. However, some of the ACh molecules bind with AChE, which is also at the end-plate surface. Being quickly inactivated, this ACh never contributes to the EPP. The ACh that does bind with receptor-channels does so briefly (for about 1 millionth of a second) and then detaches. Some of the detached ACh molecules quickly rebind with receptor-channels, keeping these end-plate channels open, but some randomly contact AChE instead and are inactivated (step ❾). As this process repeats, more ACh is inactivated until all of it has been removed from the cleft within a few milliseconds after its release. ACh removal ends the EPP, so

the remainder of the muscle cell membrane returns to resting potential. Now the muscle cell can relax. Or, if sustained contraction is essential for the desired movement, another motor-neuron action potential leads to the release of more ACh, which keeps the contractile process going. By removing contraction-inducing ACh from the motor end plate, AChE permits the choice of allowing relaxation to take place (no more ACh released) or keeping the contraction going (more ACh released), depending on the body's momentary needs.

The neuromuscular junction is vulnerable to several chemical agents and diseases.

 Several chemical agents and diseases affect the neuromuscular junction by acting at different sites in the transmission process, as the following examples illustrate.

Black Widow Spider Venom Causes Explosive Release of ACh The venom of black widow spiders exerts its deadly effect by triggering explosive release of ACh from the storage vesicles, not only at neuromuscular junctions but at all cholinergic sites. All cholinergic sites undergo prolonged depolarization, the most harmful result of which is respiratory failure. Breathing is accomplished by alternate contraction and relaxation of respiratory muscles, particularly the diaphragm, a dome-shaped sheet of skeletal muscle that forms the floor of the thoracic (chest) cavity. Respiratory paralysis occurs as a result of prolonged depolarization of the diaphragm. During this so-called *depolarization block,* the voltage-gated Na^+ channels are trapped in their inactivated state (that is, they remain in their closed and not capable of opening conformation; see p. 92). This depolarization block prohibits the initiation of new action potentials and resultant contraction of the diaphragm. As a consequence, the victim cannot breathe.

Botulinum Toxin Blocks Release of ACh Botulinum toxin, in contrast, exerts its lethal blow by blocking the release of ACh from the terminal button in response to a motor-neuron action potential. *Clostridium botulinum* toxin causes **botulism**, a form of food poisoning. When this toxin is consumed, it prevents muscles from responding to nerve impulses. Death results from respiratory failure caused by inability to contract the diaphragm. Botulinum toxin is one of the most lethal poisons known; ingesting less than 0.0001 mg can kill an adult human. (See the accompanying boxed feature, Concepts, Challenges, and Controversies, to learn about a wrinkle in the botulinum toxin story.)

Curare Blocks Action of ACh at Receptor-Channels Other chemicals interfere with neuromuscular junction activity by blocking the effect of released ACh. The best-known example is the antagonist **curare**, which reversibly binds to the ACh receptor-channels on the motor end plate. Unlike ACh, however, curare does not alter membrane permeability, nor is it inactivated by AChE. When curare occupies ACh receptor-channels, ACh cannot combine with and open these channels to permit the ionic movement responsible for an EPP. Conse-

Botulinum Toxin's Reputation Gets a Facelift

THE POWERFUL TOXIN PRODUCED BY *Clostridium botulinum* causes the deadly food poisoning botulism. Yet this dreaded, highly lethal poison has been put to use as a treatment for alleviating specific movement disorders and, more recently, has been added to the list of tools that cosmetic surgeons use to fight wrinkles.

During the past several decades, botulinum toxin, marketed in therapeutic doses as *Botox,* has offered welcome relief to people with painful, disruptive neuromuscular diseases known categorically as **dystonias**. These conditions are characterized by spasms (excessive, sustained, involuntarily produced muscle contractions) that result in involuntary twisting or abnormal postures, depending on the body part affected. For example, painful neck spasms that twist the head to one side result from *spasmodic torticollis* (*tortus* means "twisted"; *collum* means "neck"), the most common dystonia. The problem is believed to arise from too little inhibitory input compared to excitatory input to the motor neurons that supply the affected muscle. The reasons for this imbalance in motor-neuron input are unknown. The end result of excessive motor-neuron activation is sustained, disabling contraction of the muscle supplied by the overactive motor neurons. Fortunately, injecting minuscule amounts of botulinum toxin into the affected muscle causes a reversible, partial paralysis of the muscle. Botulinum toxin interferes with the release of muscle-contraction-causing ACh from the overactive motor neurons at the neuromuscular junctions in the treated muscle. The goal is to inject just enough botulinum toxin to alleviate the troublesome spasmodic contractions but not enough to eliminate the normal contractions needed for ordinary movements. The therapeutic dose is considerably less than the amount of toxin needed to induce even mild symptoms of botulinum poisoning. Botulinum toxin is eventually cleared away, so its muscle-relaxing effects wear off after 3 to 6 months, at which time the treatment must be repeated.

The first dystonia for which Botox was approved as a treatment by the U.S. Food and Drug Administration (FDA) was *blepharospasm* (*blepharo* means "eyelid"). In this condition, sustained and involuntary contractions of the muscles around the eye nearly permanently close the eyelids.

Botulinum toxin's potential as a treatment option for cosmetic surgeons was accidentally discovered when physicians noted that injections used to counter abnormal eye muscle contractions also smoothed the appearance of wrinkles in the treated areas. It turns out that frown lines, crow's feet, and furrowed brows are caused by facial muscles that have become overactivated, or permanently contracted, as a result of years of performing certain repetitive facial expressions. By relaxing these muscles, botulinum toxin temporarily smoothes out these age-related wrinkles. Botox now has FDA approval as an antiwrinkle treatment. The agent is considered an excellent alternative to facelift surgery for combating lines and creases. This treatment is among the most rapidly growing cosmetic procedures in the United States, especially in the entertainment industry and in high-fashion circles. However, as with its therapeutic use to treat dystonias, the costly injections of botulinum toxin must be repeated every 3 to 6 months to maintain the desired effect in appearance. Furthermore, Botox does not work against the fine, crinkly wrinkles associated with years of excessive sun exposure because these wrinkles are caused by skin damage, not by contracted muscles.

---

quently, because muscle action potentials cannot occur in response to nerve impulses to these muscles, paralysis ensues. When enough curare is present to block a significant number of ACh receptor-channels, the person dies from respiratory paralysis caused by inability to contract the diaphragm. In the past, some peoples used curare as a deadly arrowhead poison.

Organophosphates Prevent Inactivation of ACh **Organophosphates** are a group of chemicals that modify neuromuscular junction activity in yet another way—namely, by irreversibly inhibiting AChE. Inhibition of AChE prevents the inactivation of released ACh. Death from organophosphates also results from respiratory failure because the diaphragm remains in a depolarization block and cannot repolarize and return to resting conditions then be stimulated to contract again to bring in a fresh breath of air. In addition to organophosphates' lethal action at nicotinic receptors of respiratory muscles, other symptoms occur related to the effect of ACh buildup at muscarinic

receptors, such as drooling, pupillary constriction, vomiting, and diarrhea, as well as from ACh overstimulation in the brain, which can lead to seizures. These toxic agents are used in some pesticides (for example, *malathion* used as a mosquito control product) and as military nerve gases (such as the chemical weapon *sarin gas*).

Myasthenia Gravis Inactivates ACh Receptor-Channels **Myasthenia gravis**, a disease involving the neuromuscular junction, is characterized by extreme muscular weakness (*myasthenia* means "muscular weakness"; *gravis* means "severe"). It is an autoimmune (meaning "immunity against self") condition in which the body erroneously produces antibodies against its motor end-plate ACh receptor-channels. Thus, not all released ACh molecules can find a functioning receptor-channel with which to bind. As a result, AChE destroys much of the ACh before it ever has a chance to interact with a receptor-channel and contribute to the EPP. Treatment consists of administering a drug such as *neostigmine* that inhibits AChE temporarily (in contrast to the toxic organophosphates, which irreversibly block this enzyme). This drug prolongs the action of ACh at the neuromuscular junction by permitting it to build up for the short term. The resultant EPP is of sufficient magnitude to initiate an action potential and subsequent contraction in the muscle fiber, as it normally would.

Check Your Understanding 7.3

1. Discuss the role of ACh and of AChE at a neuromuscular junction.
2. Compare the magnitude of an EPP and an EPSP and explain the functional significance of this difference.
3. Describe the cause, symptoms, and treatment of myasthenia gravis.

Homeostasis: Chapter in Perspective

The nervous system, along with the other major regulatory system, the endocrine system, controls most muscle contractions and gland secretions. Whereas the afferent division of the PNS detects and carries information to the CNS for processing and decision making, the efferent division of the PNS carries directives from the CNS to the effector organs (muscles and glands), which carry out the intended response. Much of this efferent output is directed toward maintaining homeostasis.

The autonomic nervous system, which is the efferent branch that innervates smooth muscle, cardiac muscle, and glands, plays a major role in the following homeostatic activities, among others:

- Regulating blood pressure
- Controlling digestive juice secretion and digestive tract contractions that mix ingested food with the digestive juices
- Controlling sweating to help maintain body temperature

The somatic nervous system, the efferent branch that innervates skeletal muscle, contributes to homeostasis by stimulating the following activities:

- Skeletal muscle contractions that enable the body to move in relation to the external environment, contributing to homeostasis by moving the body toward food or away from harm
- Contractions that accomplish breathing to maintain appropriate levels of O_2 and CO_2 in the body
- Shivering, which is important in maintaining body temperature

In addition, efferent output to skeletal muscles accomplishes many movements that are not aimed at maintaining a stable internal environment but nevertheless enrich our lives and enable us to engage in activities that contribute to society, such as dancing, building bridges, or performing surgery.

Review Exercises Answers begin on p. A-29

Reviewing Terms and Facts

1. Sympathetic preganglionic fibers begin in the thoracic and lumbar segments of the spinal cord. (*True or false?*)

2. Action potentials are transmitted on a one-to-one basis at both a neuromuscular junction and a synapse. (*True or false?*)

3. The sympathetic nervous system
   a. is always excitatory.
   b. innervates only tissues concerned with protecting the body against challenges from the outside environment.
   c. has short preganglionic and long postganglionic fibers.
   d. is part of the afferent division of the PNS.
   e. is part of the somatic nervous system.

4. Acetylcholinesterase
   a. is stored in vesicles in the terminal button.
   b. combines with receptor-channels on the motor end plate to bring about an end-plate potential.
   c. is inhibited by organophosphates.
   d. is the chemical transmitter at the neuromuscular junction.
   e. paralyzes skeletal muscle by strongly binding with acetylcholine (ACh) receptor-channels.

5. The two divisions of the autonomic nervous system are the _____ nervous system, which dominates in "fight-or-flight" situations, and the _____ nervous system, which dominates in "rest-and-digest" situations.

6. The _____ is a modified sympathetic ganglion that does not give rise to postganglionic fibers but instead secretes hormones similar or identical to sympathetic postganglionic neurotransmitters into the blood.

7. The _____ is the specialized portion of muscle cell membrane that underlies the terminal button at a neuromuscular junction.

8. Using the answer code on the right, identify the autonomic neurotransmitter being described:

1. is secreted by all preganglionic fibers	(a) acetylcholine
2. is secreted by sympathetic postganglionic fibers	(b) norepinephrine
3. is secreted by parasympathetic postganglionic fibers	
4. is secreted by the adrenal medulla	
5. is secreted by motor neurons	
6. binds to muscarinic or nicotinic receptors	
7. binds to α or β receptors	

9. Using the answer code on the right, indicate which type of efferent output is being described:

1. is composed of two-neuron chains	(a) characteristic of the somatic nervous system
2. innervates cardiac muscle, smooth muscle, and glands	(b) characteristic of the autonomic nervous system
3. innervates skeletal muscle	
4. consists of the axons of motor neurons	
5. exerts either an excitatory or an inhibitory effect on its effector organs	
6. dually innervates its effector organs	
7. exerts only an excitatory effect on its effector organs	

10. Using the answer code on the right, indicate what types of receptors are present for each of the organs listed (more than one answer may apply).

1. heart	(a) α_1
2. arteriolar smooth muscle	(b) α_2
3. bronchiolar smooth muscle	(c) β_1
4. skeletal muscle fibers	(d) β_2
5. adrenal medulla	(e) nicotinic
6. digestive glands	(f) muscarinic

Understanding Concepts

(Answers at www.cengagebrain.com)

1. Distinguish between preganglionic and postganglionic fibers.

2. What is the advantage of dual innervation of many organs by both branches of the autonomic nervous system?

3. Distinguish among the following types of receptors in terms of mechanism of action and effect at the effector organ: nicotinic receptors, muscarinic receptors, α_1 receptors, α_2 receptors, β_1 receptors, and β_2 receptors.

4. Compare *agonists* and *antagonists*.

5. What regions of the CNS regulate autonomic output?

6. Describe the sequence of events that occurs at a neuromuscular junction.

7. Explain why an end-plate potential (EPP) has a larger magnitude than an excitatory postsynaptic potential (EPSP).

8. Discuss the effect each of the following has at the neuromuscular junction: black widow spider venom, botulinum toxin, curare, myasthenia gravis, and organophosphates.

Solving Quantitative Exercises

1. When a muscle fiber is activated at the neuromuscular junction, tension does not begin to rise until about 1 msec after initiation of the action potential in the muscle fiber. Many things occur during this delay, one time-consuming event being diffusion of ACh across the neuromuscular junction. The following equation can be used to calculate how long this diffusion takes:

$$t = x^2/2D$$

In this equation, x is the distance covered, D is the diffusion coefficient, and t is the time it takes for the substance to diffuse across the distance x. In this example, x is the width of the cleft between the neuronal terminal button and the muscle fiber at the neuromuscular junction (assume 200 nm), and D is the diffusion coefficient of ACh (assume 1×10^{-5} cm^2/sec). How long does it take ACh to diffuse across the neuromuscular junction?

Applying Clinical Reasoning

Christopher K. experienced chest pains when he climbed the stairs to his fourth-floor office or played tennis, but he had no symptoms when not physically exerting himself. His condition was diagnosed as *angina pectoris* (*angina* means "pain"; *pectoris* means "chest"), heart pain that occurs whenever the blood supply to the heart muscle cannot meet the muscle's need for oxygen delivery. This condition usually is caused by narrowing of the blood vessels supplying the heart by cholesterol-containing deposits. Most people with this condition do not have any pain at rest but experience bouts of pain whenever the heart's need for oxygen increases, such as during exercise or emotionally stressful situations that increase sympathetic nervous activity. Christopher obtains immediate relief of angina attacks by promptly taking a vasodilator drug such as *nitroglycerin,* which relaxes the smooth muscle in the walls of his narrowed heart vessels. Consequently, the vessels open more widely and more blood can flow through them. For prolonged treatment, his doctor has indicated that Christopher will experience fewer and less severe angina attacks if he takes a β_1-blocker drug, such as *metoprolol,* regularly. Explain why.

Thinking at a Higher Level

1. Explain why epinephrine, which causes arteriolar constriction (narrowing) in most tissues, is often administered in conjunction with local anesthetics.

2. Would skeletal muscle activity be affected by atropine (see p. 241)? Why or why not?

3. Considering that you can voluntarily control the emptying of your urinary bladder by contracting (preventing emptying) or relaxing (permitting emptying) your external urethral sphincter, a ring of muscle that guards the exit from the bladder, of what type of muscle is this sphincter composed and what branch of the nervous system supplies it?

4. The venom of certain poisonous snakes contains α-bungarotoxin, which binds tenaciously to ACh receptor sites on the motor end-plate membrane. What would the resultant symptoms be?

5. Explain how destruction of motor neurons by poliovirus or amyotrophic lateral sclerosis can be fatal.

To access the course materials and companion resources for this text, please visit www.cengagebrain.com

Muscle Physiology

CNRI/Science Source

A scanning electron micrograph of a skeletal muscle fiber. Skeletal muscle cells, or muscle fibers, are long, cylindrical, and striated (striped). A muscle fiber is packed with myofibrils, which are cylinder-shaped contractile structures that run the length of the fiber and have alternating dark and light bands responsible for the muscle fiber's striations.

Homeostasis Highlights

Muscles are contraction specialists. The three types of muscle are skeletal, smooth, and cardiac. **Skeletal muscle** attaches to the skeleton. Contraction of skeletal muscles moves bones to which they are attached, allowing the body to perform a variety of motor activities. Skeletal muscles that support homeostasis include those important in acquiring, chewing, and swallowing food and those essential for breathing. Also, heat-generating muscle contractions help regulate body temperature. Skeletal muscles are further used to move the body away from harm. Skeletal muscle contractions are also important for nonhomeostatic activities, such as dancing or operating a computer. **Smooth muscle** is found in the walls of hollow organs and tubes. Controlled contraction of smooth muscle regulates movement of blood through blood vessels, food through the digestive tract, air through respiratory airways, and urine to the exterior. **Cardiac muscle** is found only in the walls of the heart, whose contraction pumps life-sustaining blood throughout the body.

8.1 Structure of Skeletal Muscle

By moving specialized intracellular components, muscle cells can develop tension and shorten (contract). Through their highly developed ability to contract, groups of muscle cells working together within a muscle can produce movement and do work. Controlled contraction of muscles allows (1) purposeful movement of the whole body or parts of the body (such as walking or waving your hand), (2) manipulation of external objects (such as driving a car or moving a piece of furniture), (3) propulsion of contents through hollow internal organs (such as circulation of blood or movement of a meal through the digestive tract), and (4) emptying the contents of certain organs to the external environment (such as urination or giving birth).

Muscle comprises the largest group of tissues in the body, accounting for approximately half of body weight. Skeletal muscle alone makes up about 40% of body weight in men and 32% in women, with smooth and cardiac muscle making up another 10% of total weight. Although these three muscle types are structurally and functionally distinct, they can be classified in two ways according to common characteristics (Figure 8-1). First, muscles are categorized as *striated* (skeletal and cardiac muscle) or *unstriated* (smooth muscle), depending on whether alternating dark and light bands, or striations (stripes), can be seen when the muscle is viewed under a light microscope. Second, muscles are categorized as *voluntary* (skeletal muscle) or *involuntary* (cardiac and smooth muscle), depending on whether they are innervated by the somatic nervous system and subject to voluntary control or are innervated by the autonomic nervous system and not subject to voluntary control, respectively (see p. 234). Although skeletal muscle is categorized as voluntary, because it can be consciously controlled, much skeletal muscle activity is also subject to subconscious, involuntary regulation, such as that related to posture, balance, and rhythmic movements like walking.

Most of this chapter is a detailed examination of the most abundant and best understood muscle: skeletal muscle. Skeletal muscles make up the muscular system. The chapter concludes with a discussion of the unique properties of smooth and cardiac muscle in comparison to skeletal muscle.

Skeletal muscle fibers are striated by a highly organized internal arrangement.

A single skeletal muscle cell, known as a **muscle fiber**, is relatively large, elongated, and cylinder shaped, measuring from 10 to 100 micrometers (μm) in diameter and up to 750,000 μm, or 2.5 feet, in length (1 μm = 1 millionth of a meter). A skeletal muscle consists of a number of muscle fibers lying parallel to one another and bundled together by connective tissue (Figure 8-2a). The fibers usually extend the entire length of the muscle. During embryonic development, the huge skeletal muscle fibers are formed by the fusion of many smaller cells called **myoblasts** (*myo* means "muscle"; *blast* refers to a primitive cell that forms more specialized cells); thus, one striking feature is the presence of multiple nuclei dispersed just beneath the plasma membrane in a single muscle cell (see Figure 8-1a). Another feature is the abundance of mitochondria, the energy-generating organelles, as would be expected with the high energy demands of a tissue as active as skeletal muscle.

A skeletal muscle fiber contains numerous **myofibrils**, which are cylindrical intracellular structures 1 μm in diameter that extend the entire length of the muscle fiber (Figure 8-2b

(a) Skeletal muscle

Multiple nuclei in single cell

Skeletal muscle cell (muscle fiber)

Classification: Striated muscle, voluntary muscle

Description: Bundles of long, thick, cylindrical, striated, contractile, multinucleate cells that extend the length of the muscle

Typical location: Attached to bones of skeleton

Function: Movement of body in relation to external environment

(b) Cardiac muscle

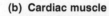
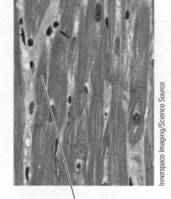
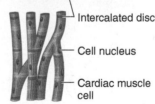

Intercalated disc

Cell nucleus

Cardiac muscle cell

Classification: Striated muscle, involuntary muscle

Description: Interlinked network of short, slender, cylindrical, striated, branched, contractile cells connected cell to cell by intercalated discs

Location: Wall of heart

Function: Pumping of blood out of heart

(c) Smooth muscle

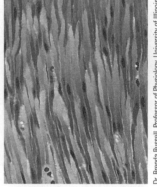
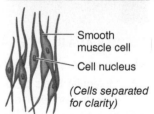

Smooth muscle cell

Cell nucleus

(Cells separated for clarity)

Classification: Unstriated muscle, involuntary muscle

Description: Loose network of short, slender, spindle-shaped, unstriated, contractile cells that are arranged in sheets

Typical location: Walls of hollow organs and tubes, such as stomach and blood vessels

Function: Movement of contents within hollow organs

Innerspace Imaging/Science Source

Dr. Brenda Russell, Professor of Physiology, University of Illinois

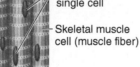

Figure 8-1 Characteristics of three types of muscle. The photos in (a), (b), and (c) are light micrographs of longitudinal sections of skeletal, cardiac, and smooth muscle, respectively.

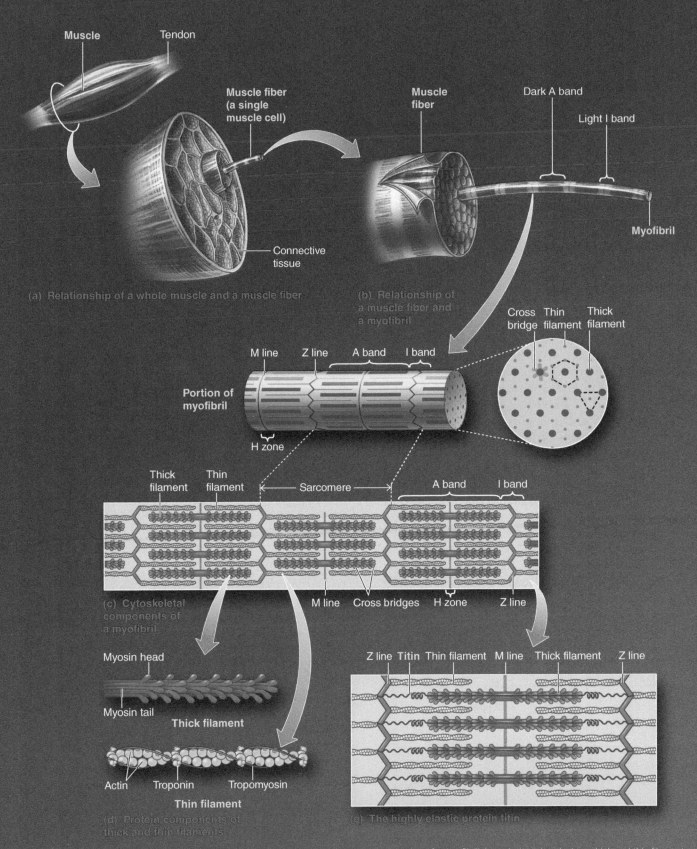

(a) Relationship of a whole muscle and a muscle fiber

(b) Relationship of a muscle fiber and a myofibril

(c) Cytoskeletal components of a myofibril

(d) Protein components of thick and thin filaments

(e) The highly elastic protein titin

Muscle   **Tendon**

Muscle fiber (a single muscle cell)

Connective tissue

Muscle fiber

Dark A band

Light I band

Myofibril

Cross bridge   Thin filament   Thick filament

M line   Z line   A band   I band

Portion of myofibril

H zone

Thick filament   Thin filament   Sarcomere   A band   I band

M line   Cross bridges   H zone   Z line

Myosin head

Myosin tail

Thick filament

Actin   Troponin   Tropomyosin

Thin filament

Z line   **Titin**   Thin filament   M line   Thick filament   Z line

❙Figure 8-2 Levels of organization in a skeletal muscle. Note in part (c) in the cross section of a myofibril through the A band where thick and thin filaments overlap that each thick filament is surrounded by six thin filaments and each thin filament is surrounded by three thick filaments.

FIGURE FOCUS: *How would a cross section through the H zone (outside of the M line) and one through the I band (outside of the Z line) differ from the cross section through the A band (outside of the H zone) shown in part (c)?*

and chapter opener photo). Myofibrils are specialized contractile elements that constitute 80% of the volume of the muscle fiber. Each myofibril consists of a regular arrangement of highly organized cytoskeletal microfilaments (see p. 44)—the thick and the thin filaments (∎Figure 8-2c). The **thick filaments**, which are 12 to 18 nm in diameter and 1.6 μm in length, are special assemblies of the protein *myosin*, whereas the thin filaments, which are 5 to 8 nm in diameter and 1.0 μm long, are made up primarily of the protein *actin* (∎Figure 8-2d). The levels of organization in a skeletal muscle can be summarized as follows:

Whole muscle	→	muscle fiber	→	myofibril	→	thick and thin filament	→	myosin and actin
(an organ)		(a cell)		(a specialized intracellular structure)		(cytoskeletal elements)		(protein molecules)

A and I Bands Viewed with an electron microscope, a myofibril displays alternating dark bands (the A bands) and light bands (the I bands) (∎Figure 8-3). The bands of all the myofibrils lined up parallel to one another collectively produce the striated appearance of a skeletal muscle fiber visible under a light microscope. Alternate stacked sets of thick and thin filaments that slightly overlap one another are responsible for the A and I bands (see ∎Figure 8-2c). Several cytoskeletal proteins maintain this precise filament geometry.

An **A band** is made up of a stacked set of thick filaments along with the portions of the thin filaments that overlap on both ends of the thick filaments. The thick filaments lie only within the A band and extend its entire width. The lighter area within the middle of the A band, where the thin filaments do not reach, is the **H zone**. Only the central portions of the thick filaments are found in this region. Supporting proteins that hold the thick filaments together vertically within each stack can be seen as the **M line**, which extends vertically down the middle of the A band within the center of the H zone.

An **I band** consists of the remaining portion of the thin filaments that do not project into the A band. Visible in the middle of each I band is a dense, vertical **Z line**. The area between two Z lines is called a **sarcomere**, which is the functional unit of skeletal muscle. A **functional unit** of any organ is the smallest component that can perform all functions of that organ. Accordingly, a sarcomere is the smallest component of a muscle fiber that can contract. The Z line is a flat, cytoskeletal disc that connects the thin filaments of two adjoining sarcomeres. Each relaxed sarcomere is about 2 μm in width and consists of one whole A band and half of each of the two I bands located on either side. An I band contains only thin filaments from two adjacent sarcomeres but not the entire length of these filaments. During growth, a muscle increases in length by adding new sarcomeres on the ends of the myofibrils, not by increasing the size of each sarcomere.

Single strands of a giant, highly elastic protein known as **titin** extend in both directions from the M line along the length of the thick filament to the Z lines at opposite ends of the sarcomere (see ∎Figure 8-2e). Titin is the largest protein in the body, being made up of nearly 30,000 amino acids. It serves three important roles:

1. *Serving as scaffolding.* Along with the M-line proteins, titin helps stabilize the position of the thick filaments in relation to the thin filaments, thus contributing to sarcomere stability.

2. *Acting as an elastic spring.* By acting like a spring, titin greatly augments a muscle's elasticity. That is, titin helps a muscle stretched by an external force passively recoil to its resting length when the stretching force is removed, much like a stretched spring. Because it behaves like an elastic spring and lies parallel to the thick and thin filaments, titin (along with the elastic connective tissue surrounding the muscle fibers) constitutes the **parallel-elastic component** of muscle.

3. *Participating in signal transduction.* Titin is also involved in diverse signaling pathways, such as the complex pathway involved in muscle enlargement in response to weight lifting.

Cross Bridges With an electron microscope, fine **cross bridges** can be seen extending from each thick filament toward the surrounding thin filaments in the areas where the thick and thin filaments overlap (see the longitudinal view in ∎Figure 8-2c). Three-dimensionally, the thin filaments are arranged hexagonally around the thick filaments. Cross bridges project from each thick filament in all six directions toward the six surrounding thin filaments. Each thin filament, in turn, is surrounded by three thick filaments (see the cross-section view through an A band in ∎Figure 8-2c). A single muscle fiber may contain an estimated 16 billion thick and 32 billion thin filaments, all arranged in this precise pattern within the myofibrils.

Myosin forms the thick filaments.

Each thick filament has several hundred myosin molecules packed together in a specific arrangement. A **myosin** molecule is a protein consisting of two identical subunits, each shaped somewhat like a golf club (∎Figure 8-4a). The tail ends of the two subunits are intertwined around each other like golf-club shafts twisted together, with the two globular heads projecting

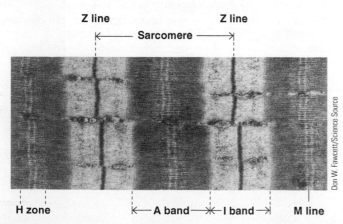

∎Figure 8-3 **Electron micrograph of a relaxed myofibril.** Note the A and I bands.
Reprinted with permission from the Sydney Schochet Jr., M. D. Collection, Diagnostic Pathology of Skeletal Muscle and Nerve, Fig. 1-13 (Stamford, CT: Appleton & Lan, 1986).

Z line   Z line
← Sarcomere →

Don W. Fawcett/Science Source

H zone   ←A band→   ←I band→   M line

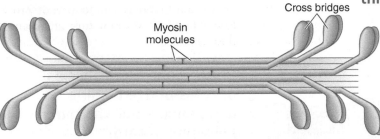

Actin-binding site • **Myosin ATPase site**

Hinges

Heads

Tail

← 100 nm →

(a) Myosin molecule

Cross bridges

Myosin molecules

(b) Thick filament

❙Figure 8-4 **Structure of myosin molecules and their organization within a thick filament.** (a) Each myosin molecule consists of two identical, golf-club–shaped subunits with
their tails intertwined and their globular heads, each of which contains an actin-binding site
and a myosin ATPase site, projecting out at one end. Each myosin subunit has two hinge
points: one along the tail and the other at the junction of the tail with the head. (b) A thick filament is made up of myosin molecules lying lengthwise parallel to one another. Half are oriented in one direction and half in the opposite direction. The globular heads, which protrude
at regular intervals along the thick filament, form the cross bridges.

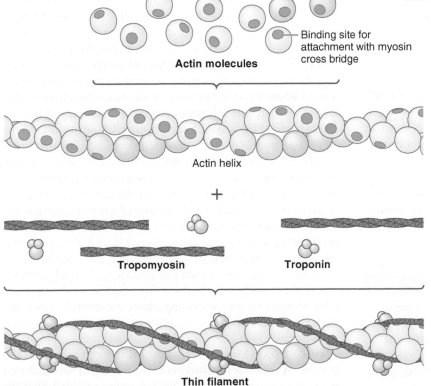

**Binding site for
attachment with myosin
cross bridge**

Actin molecules

Actin helix

+

Tropomyosin   **Troponin**

Thin filament

out at one end. Myosin can bend at hinge points in two
locations: one along the tail and the other at the "neck" or
junction of the tail with each head. The two halves of each
thick filament are mirror images made up of myosin molecules
lying lengthwise in a regular, staggered array, with their tails
oriented toward the center of the filament and their globular
heads protruding outward at regular intervals (❙Figure 8-4b).
The heads form the cross bridges between the thick and thin
filaments. Each cross bridge has two sites crucial to the contractile process: (1) an actin-binding site and (2) a myosin ATPase
(ATP-splitting) site.

Actin is the main structural component of the thin filaments.

Thin filaments consist of three proteins: *actin, tropomyosin,* and *troponin* (❙Figure 8-5). **Actin** molecules,
the primary structural proteins of the thin filament, are
spherical. The thin filament's backbone is formed by
actin molecules joined into two strands and twisted
together, like two intertwined strings of pearls. Each
actin molecule has a binding site for attaching with a
myosin cross bridge. Binding of myosin and actin at
the cross bridges leads to contraction of the muscle
fiber, by a mechanism to be described shortly. Myosin
and actin are not unique to muscle cells (see p. 49), but
these proteins are more abundant and more highly
organized in muscle cells.

In a relaxed muscle fiber, contraction does not take
place; actin cannot bind with cross bridges because of
the way the two other types of protein—tropomyosin
and troponin—are positioned within the thin filament.
Tropomyosin molecules are threadlike proteins that lie
end to end alongside the groove of the actin spiral. In
this position, tropomyosin covers the actin sites
that bind with the cross bridges, blocking the
interaction that leads to muscle contraction. The
other thin filament component, **troponin**, is a
protein complex made of three polypeptide units:
one binds to tropomyosin, one binds to actin,
and a third can bind with Ca^{2+}.

When troponin is not bound to Ca^{2+}, this
protein stabilizes tropomyosin in its blocking

❙Figure 8-5 **Composition of a thin filament.** The main
structural component of a thin filament is two chains of spherical actin molecules that are twisted together. Troponin molecules (which consist of three small, spherical subunits) and
threadlike tropomyosin molecules are arranged to form a rib-
bon that lies alongside the groove of the actin helix and physically covers the binding sites on actin molecules for attachment
with myosin cross bridges. (The thin filaments shown here are
not drawn in proportion to the thick filaments in Figure 8-4.
Thick filaments are several times larger in diameter than thin
filaments.)

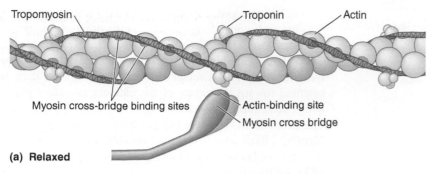

Tropomyosin Troponin Actin

Myosin cross-bridge binding sites — Actin-binding site
Myosin cross bridge

(a) Relaxed

1 No excitation.

2 No cross-bridge binding because cross-bridge binding site on actin is physically covered by troponin–tropomyosin complex.

3 Muscle fiber is relaxed.

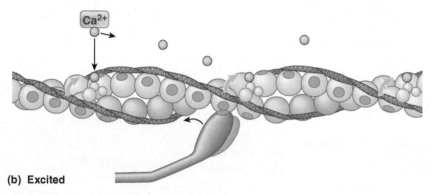

Ca²⁺

(b) Excited

1 Muscle fiber is excited and Ca²⁺ is released.

2 Released Ca²⁺ binds with troponin, pulling troponin–tropomyosin complex aside to expose cross-bridge binding site.

3 Cross-bridge binding occurs.

4 Binding of actin and myosin cross bridge triggers power stroke that pulls thin filament inward during contraction.

❙Figure 8-6 **Role of calcium in turning on cross bridges.**

position over actin's cross-bridge binding sites (❙Figure 8-6a). When Ca²⁺ binds to troponin, the shape of this protein is changed in such a way that tropomyosin slips away from its blocking position (❙Figure 8-6b). With tropomyosin out of the way, actin and myosin can bind and interact at the cross bridges, resulting in muscle contraction. Tropomyosin and troponin are often called **regulatory proteins** because of their role in covering (preventing contraction) or exposing (permitting contraction) the binding sites for cross-bridge interaction between actin and myosin.

Check Your Understanding 8.1

1. Compare the relationship among myofibrils, muscle fibers, and a whole muscle.

2. Illustrate the relative positions of the cytoskeletal structures that make up a sarcomere.

3. Describe how actin, tropomyosin, and troponin are organized in a relaxed muscle fiber.

8.2 Molecular Basis of Skeletal Muscle Contraction

Several important links in the contractile process remain to be discussed. How does cross-bridge interaction between actin and myosin bring about muscle contraction? How does a muscle action potential trigger this contractile process? What is the source of the Ca²⁺ that physically repositions troponin and tropomyosin to permit cross-bridge binding? We now turn attention to these topics.

During contraction, cycles of cross-bridge binding and bending pull the thin filaments inward.

Cross-bridge interaction between actin and myosin brings about muscle contraction by means of the sliding filament mechanism.

Sliding Filament Mechanism The thin filaments on each side of a sarcomere slide inward over the stationary thick filaments toward the A band's center during contraction (❙Figure 8-7). As they slide inward, the thin filaments pull the Z lines to which they are attached closer together, so the sarcomere shortens. As all sarcomeres throughout the muscle fiber's length shorten simultaneously, the entire fiber shortens. This is the **sliding filament mechanism** of muscle contraction. The H zone, in the center of the A band where the thin filaments do not reach, becomes smaller as the thin filaments approach each other when they slide more deeply inward. The I band, which consists of the portions of the thin filaments that do not overlap with the thick filaments, narrows as the thin filaments further overlap the thick filaments during their inward slide. The thin filaments themselves do not change length during muscle fiber shortening. The width of the A band remains unchanged during contraction because its width is determined by the length of the thick filaments, and the thick filaments do not change length during the shortening process. Note that neither the thick nor the thin filaments decrease in length to shorten the sarcomere. Instead, contraction is accomplished by the thin filaments from the opposite sides of each sarcomere sliding closer together between the thick filaments.

Power Stroke During contraction, with the tropomyosin and troponin "chaperones" pulled out of the way by Ca²⁺, the myosin heads or cross bridges from a thick filament can bind with the

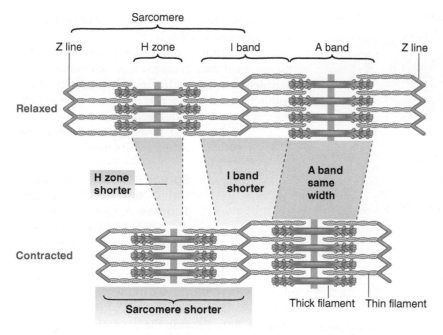

Sarcomere

Z line   H zone   I band   A band   Z line

Relaxed

H zone shorter   I band shorter   A band same width

Contracted

Sarcomere shorter

Thick filament   Thin filament

▮Figure 8-7 Changes in banding pattern during shortening. During muscle contraction, each sarcomere shortens as the thin filaments slide closer together between the thick filaments so that the Z lines are pulled closer together. The width of the A bands does not change as a muscle fiber shortens, but the I bands and H zones become shorter.

FIGURE FOCUS: *Predict what happens to the H zone when the thin filaments from opposite sides of an A band touch each other as they slide inward toward the center of the sarcomere.*

actin molecules in the surrounding thin filaments. Myosin is a motor protein, similar to kinesin and dynein. Recall that kinesin and dynein have little feet that "walk" along microtubules to transport specific products (as within a neuronal axon; see p. 47) or to move microtubules in relation to one another (as to accomplish beating of cilia or flagella; see p. 49). In the same way, myosin cross bridges "walk" along an actin filament to pull it inward relative to the stationary thick filament. Let us concentrate on a single cross-bridge interaction (▮Figure 8-8a). The two myosin heads of each myosin molecule act independently, with only one head attaching to actin at a given time. When the binding site on an actin molecule is exposed, the myosin molecule tilts at the hinge point on the tail, elevating the myosin head to facilitate the binding of this cross bridge to the nearest actin molecule. On binding, the myosin head tilts 45 degrees inward. Bending at this neck hinge point creates a "stroking" motion that pulls the thin filament toward the center of the sarcomere, like the stroking of a boat oar. This action is known as the **power stroke** of a cross bridge. A single power stroke pulls the thin filament inward only a small percentage of the total shortening distance. Repeated cycles of cross-bridge binding and bending complete the shortening.

At the end of one cross-bridge cycle, the link between the myosin cross bridge and actin molecule breaks. The cross bridge returns to its original angle and binds to an actin molecule behind its previous actin partner. The cross bridge tilts inward again to pull the thin filament in farther, then detaches and repeats the cycle. Repeated cycles of cross-bridge power

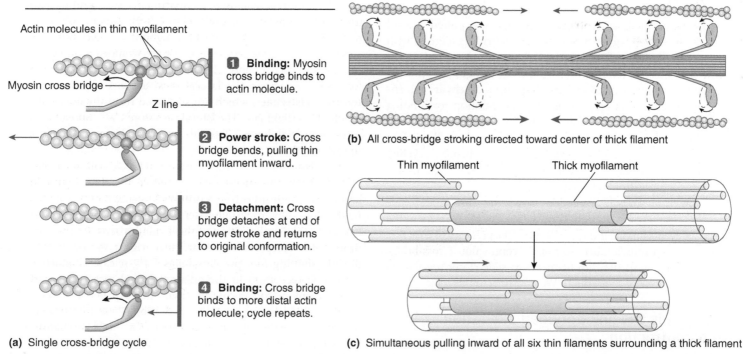

Actin molecules in thin myofilament

Myosin cross bridge

Z line

1 Binding: Myosin cross bridge binds to actin molecule.

2 Power stroke: Cross bridge bends, pulling thin myofilament inward.

3 Detachment: Cross bridge detaches at end of power stroke and returns to original conformation.

4 Binding: Cross bridge binds to more distal actin molecule; cycle repeats.

(a) Single cross-bridge cycle

(b) All cross-bridge stroking directed toward center of thick filament

Thin myofilament   Thick myofilament

(c) Simultaneous pulling inward of all six thin filaments surrounding a thick filament

▮Figure 8-8 Cross-bridge activity. (a) During each cross-bridge cycle, the cross bridge binds with an actin molecule, bends to pull the thin filament inward during the power stroke, then detaches and returns to its resting conformation, ready to repeat the cycle. (b) The power strokes of all cross bridges extending from a thick filament are directed toward the center of the thick filament. (c) All six thin filaments surrounding each end of a thick filament are pulled inward simultaneously through cross-bridge cycling during muscle contraction.

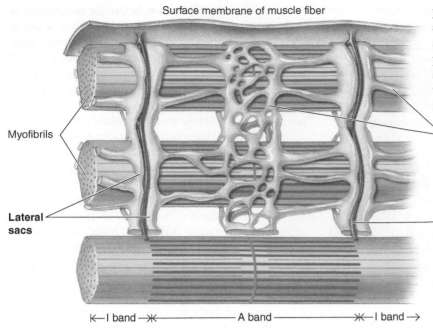

Surface membrane of muscle fiber

Myofibrils

Segments of
sarcoplasmic
reticulum

Lateral
sacs

Transverse (T)
tubule

|← I band →|← A band →|← I band →|

Figure 8-9 The T tubules and sarcoplasmic reticulum in relationship to the myofibrils. The transverse (T) tubules are membranous, perpendicular extensions of the surface membrane that dip deep into the muscle fiber at the junctions between the A and I bands of the myofibrils. The sarcoplasmic reticulum (SR) is a fine, membranous network that runs longitudinally and surrounds each myofibril, with separate segments encircling each A band and I band. The ends of each segment are expanded to form lateral sacs that lie next to the adjacent T tubules.

strokes successively pull in the thin filaments, much like pulling in a rope hand over hand.

Because of the way myosin molecules are oriented within a thick filament (**Figure 8-8b**), all cross bridges stroke toward the center of the sarcomere so that all six of the surrounding thin filaments on each end of the sarcomere are pulled inward simultaneously (**Figure 8-8c**). The cross bridges aligned with given thin filaments do not all stroke in unison, however. At any time during contraction, part of the cross bridges are attached to the thin filaments and are stroking, while others are returning to their original conformation in preparation for binding with another actin molecule. Thus, some cross bridges are "holding on" to the thin filaments, whereas others "let go" to bind with new actin. Were it not for this asynchronous cycling of the cross bridges, the thin filaments would slip back toward their resting position between strokes.

How does muscle excitation switch on this cross-bridge cycling? The term **excitation–contraction coupling** refers to the series of events linking muscle excitation (the presence of an action potential in a muscle fiber) to muscle contraction (cross-bridge activity that causes the thin filaments to slide closer together to produce sarcomere shortening). We now turn to this topic.

Calcium is the link between excitation and contraction.

Skeletal muscles are stimulated to contract by release of acetylcholine (ACh) at neuromuscular junctions between motor-neuron terminal buttons and muscle fibers. Recall that binding of ACh with the motor end plate of a muscle fiber brings about

permeability changes in the muscle fiber, resulting in an action potential that is conducted over the entire surface of the muscle cell membrane (see p. 245). The plasma membrane in muscle is sometimes called the **sarcolemma**. Two other membranous structures within the muscle fiber play important roles in linking this excitation to contraction—*transverse tubules* and *sarcoplasmic reticulum*. We examine them next.

Spread of the Action Potential Down the Transverse Tubules

At each junction of an A band and I band, the surface membrane dips into the muscle fiber to form a **transverse tubule** (**T tubule**), which runs perpendicularly from the surface of the muscle cell membrane into the central portions of the muscle fiber (**Figure 8-9**). Because the T tubule membrane is continuous with the sarcolemma, an action potential on the surface membrane spreads down into the T tubule, rapidly transmitting the surface electrical activity into the interior of the fiber. The presence of a local action potential in the T tubules leads to permeability changes in a separate membranous network within the muscle fiber, the sarcoplasmic reticulum.

Release of Calcium from the Sarcoplasmic Reticulum

The **sarcoplasmic reticulum (SR)** is a modified endoplasmic reticulum (see p. 27) that consists of a fine network of interconnected membrane-enclosed compartments surrounding each myofibril like a mesh sleeve (**Figure 8-9**). This membranous network encircles the myofibril throughout its length but is not continuous. Separate segments of SR are wrapped around each A band and each I band. The ends of each segment expand to form saclike regions, the **lateral sacs** (alternatively known as **terminal cisternae**), which are separated from the adjacent T tubules by a slight gap. The lateral sacs store Ca^{2+}. Spread of an action potential down a T tubule triggers release of Ca^{2+} from the SR into the cytosol.

How is a change in T tubule potential linked with the release of Ca^{2+} from the lateral sacs? T tubule membrane proteins known as **dihydropyridine receptors** (because they are blocked by the drug dihydropyridine) serve as voltage sensors (**Figure 8-10a**). Local depolarization of the T tubules activates the dihydropyridine receptors, which in turn trigger the opening of directly abutting foot proteins (alias Ca^{2+}-release channels or ryanodine receptors) in the adjacent lateral sacs. An orderly arrangement of **foot proteins** spans the gap between the T tubule and the lateral sac (**Figure 8-10b**). These foot proteins not only bridge the gap, but also serve as Ca^{2+}-**release channels** and are also known as **ryanodine receptors** because they are locked in the open position by the plant chemical ryanodine.

When these Ca^{2+}-release channels are opened in the presence of a local action potential in the adjacent T tubule, Ca^{2+} is released into the cytosol from the lateral sacs (**Figure 8-10c**).

By slightly repositioning the troponin and tropomyosin molecules, this released Ca^{2+} exposes the binding sites on the actin molecules so that they can link with the myosin cross bridges at their complementary binding sites. Excitation–contraction coupling is summarized in ❙Figure 8-11.

ATP-Powered Cross-Bridge Cycling

A myosin cross bridge has two special sites: an actin-binding site and an ATPase site (see ❙Figure 8-4a). The latter is an enzymatic site that can bind the energy carrier *adenosine triphosphate (ATP)* and split it into *adenosine diphosphate (ADP)* and *inorganic phosphate (P_i)*, yielding energy in the process. The breakdown of ATP occurs on the myosin cross bridge before the bridge ever links with an actin molecule (❙Figure 8-12 step ❶). The ADP and P_i remain tightly bound to the myosin, and the generated energy is stored within the cross bridge to produce a high-energy form of myosin. To use an analogy, the cross bridge is "cocked" like a gun, ready to be fired when the trigger is pulled. When the muscle fiber is excited, Ca^{2+} pulls the troponin–tropomyosin complex out of its blocking position so that the energized (cocked) myosin cross bridge can bind with an actin molecule (step ❷ᵃ). This contact between myosin and actin "pulls the trigger," causing the cross-bridge bending that produces the power stroke (step ❸). P_i is released from the cross bridge during the power stroke. After the power stroke is complete, ADP is released.

When the muscle is not excited and Ca^{2+} is not released, troponin and tropomyosin remain in their blocking position so that actin and the myosin cross bridges do not bind and no power stroking takes place (step ❷ᵇ).

When P_i and ADP are released from myosin following contact with actin and the subsequent power stroke, the myosin ATPase site is free for attachment of another ATP molecule. The actin and myosin remain linked at the cross bridge until a fresh molecule of ATP attaches to myosin at the end of the power stroke. Attachment of the new ATP molecule reduces the binding affinity between the myosin head and actin, thus allowing the cross bridge to detach (step ❹ᵃ) and return to its unbent form. The newly attached ATP is then split by myosin ATPase, recocking and energizing the myosin cross bridge once again so that it is ready to start another cycle (step ❶). On binding with another actin molecule, the energized cross bridge again bends, and so on, successively pulling the thin filament inward to accomplish contraction.

Clinical Note **Rigor Mortis** Note that fresh ATP must attach to myosin to permit the cross-bridge link between myosin and actin to break at the end of a cycle, even though the ATP is not split during this dissociation process. The need for ATP in separating myosin and actin is amply shown in **rigor mortis**. This "stiffness of death" is a generalized locking in place of the skeletal muscles that begins 3 to 4 hours after death and completes in about 12 hours. Following death, the cytosolic concentration of Ca^{2+} begins to rise, most likely because the inactive muscle cell membrane cannot keep out extracellular Ca^{2+} and perhaps because Ca^{2+} leaks out of the lateral sacs. This Ca^{2+} moves troponin and tropomyosin aside, letting actin bind with the myosin cross bridges, which were already charged with ATP before death. Dead cells cannot produce any more ATP, so actin and myosin, once bound, cannot detach because they lack fresh ATP. The thick and thin filaments

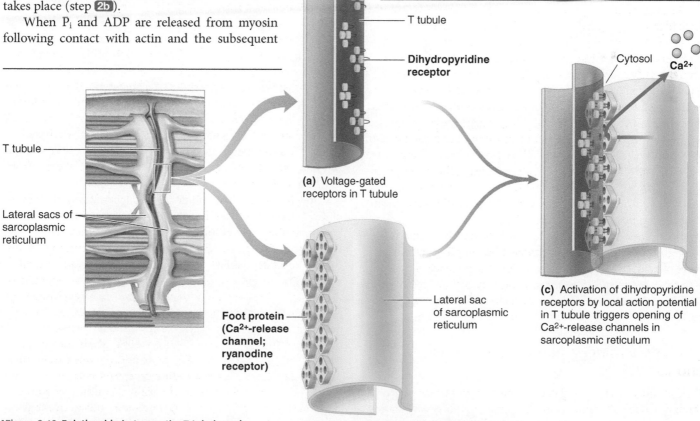

(a) Voltage-gated receptors in T tubule

(b) Foot proteins that serve as Ca^{2+}-release channels in sarcoplasmic reticulum

(c) Activation of dihydropyridine receptors by local action potential in T tubule triggers opening of Ca^{2+}-release channels in sarcoplasmic reticulum

❙Figure 8-10 **Relationship between the T tubule and adjacent lateral sacs of the sarcoplasmic reticulum.**

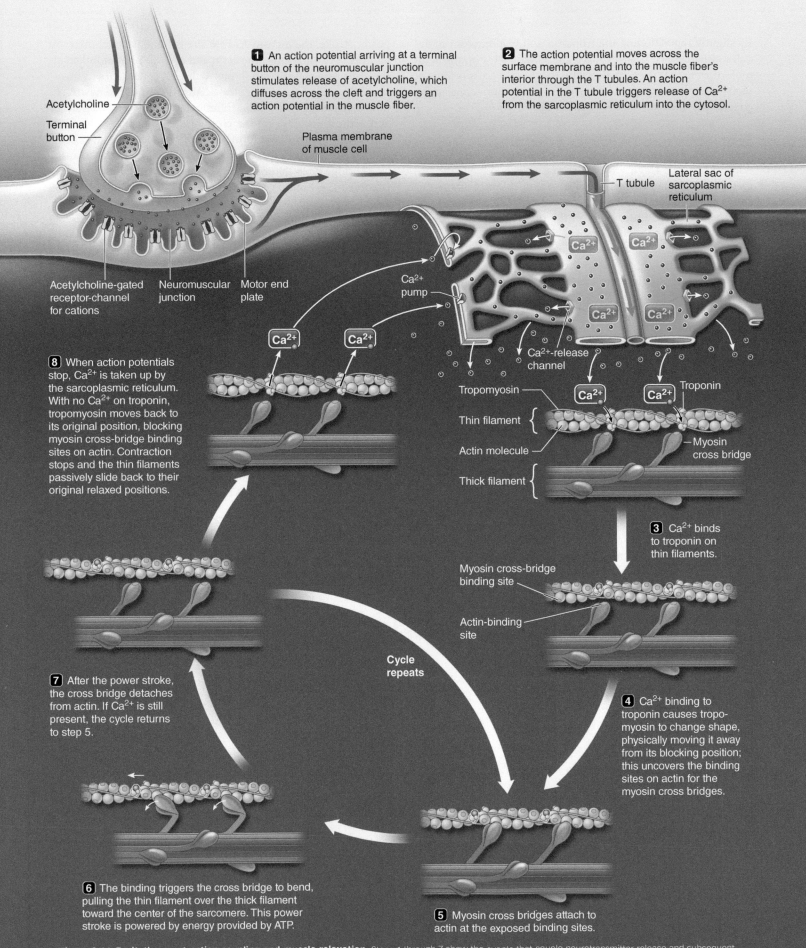

1 An action potential arriving at a terminal button of the neuromuscular junction stimulates release of acetylcholine, which diffuses across the cleft and triggers an action potential in the muscle fiber.

2 The action potential moves across the surface membrane and into the muscle fiber's interior through the T tubules. An action potential in the T tubule triggers release of Ca²⁺ from the sarcoplasmic reticulum into the cytosol.

Acetylcholine

Terminal button

Plasma membrane of muscle cell

T tubule

Lateral sac of sarcoplasmic reticulum

Acetylcholine-gated receptor-channel for cations

Neuromuscular junction

Motor end plate

Ca²⁺ pump

Ca²⁺-release channel

8 When action potentials stop, Ca²⁺ is taken up by the sarcoplasmic reticulum. With no Ca²⁺ on troponin, tropomyosin moves back to its original position, blocking myosin cross-bridge binding sites on actin. Contraction stops and the thin filaments passively slide back to their original relaxed positions.

Tropomyosin

Thin filament

Actin molecule

Thick filament

Troponin

Myosin cross bridge

3 Ca²⁺ binds to troponin on thin filaments.

Myosin cross-bridge binding site

Actin-binding site

Cycle repeats

7 After the power stroke, the cross bridge detaches from actin. If Ca²⁺ is still present, the cycle returns to step 5.

4 Ca²⁺ binding to troponin causes tropomyosin to change shape, physically moving it away from its blocking position; this uncovers the binding sites on actin for the myosin cross bridges.

6 The binding triggers the cross bridge to bend, pulling the thin filament over the thick filament toward the center of the sarcomere. This power stroke is powered by energy provided by ATP.

5 Myosin cross bridges attach to actin at the exposed binding sites.

▮Figure 8-11 **Excitation–contraction coupling and muscle relaxation.** Steps 1 through 7 show the events that couple neurotransmitter release and subsequent electrical excitation of the muscle cell with muscle contraction. At step 7, if Ca²⁺ is still present, the cross-bridge cycle returns to step 5 for another power stroke. If Ca²⁺ is no longer present as a consequence of step 8, relaxation occurs.

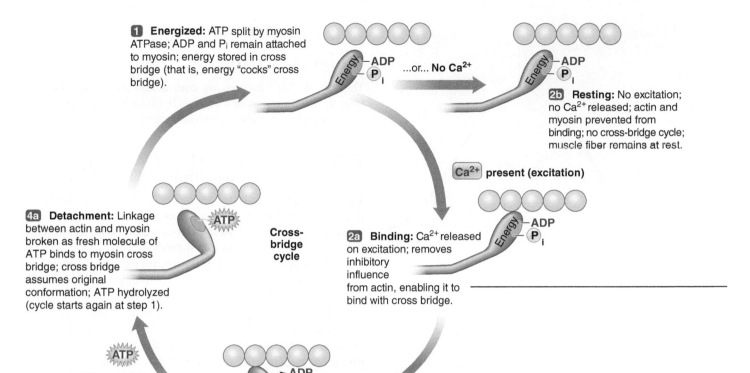

1 Energized: ATP split by myosin ATPase; ADP and P_i remain attached to myosin; energy stored in cross bridge (that is, energy "cocks" cross bridge).

...or... **No Ca²⁺**

2b Resting: No excitation; no Ca²⁺ released; actin and myosin prevented from binding; no cross-bridge cycle; muscle fiber remains at rest.

Ca²⁺ present (excitation)

4a Detachment: Linkage between actin and myosin broken as fresh molecule of ATP binds to myosin cross bridge; cross bridge assumes original conformation; ATP hydrolyzed (cycle starts again at step 1).

Cross-bridge cycle

2a Binding: Ca²⁺ released on excitation; removes inhibitory influence from actin, enabling it to bind with cross bridge.

Fresh ATP available

Energy

...or...

No ATP (after death)

3 Bending: Power stroke of cross bridge triggered on contact between myosin and actin; P_i released during and ADP released after power stroke.

4b Rigor complex: If no fresh ATP available (after death), actin and myosin remain bound in rigor complex.

Figure 8-12 **Cross-bridge cycle.**
FIGURE FOCUS: *How many ATP molecules are used for each cross-bridge cycle, keeping in mind that splitting of ATP energizes the cross bridge and binding of ATP to myosin allows detachment of the cross bridge from actin.*

thus stay linked by the immobilized cross bridges, leaving dead muscles stiff (step **4b**). During the next several days, rigor mortis gradually subsides as the proteins involved in the rigor complex begin to degrade.

Relaxation How is relaxation normally accomplished in a living muscle? Just as an action potential in a muscle fiber turns on the contractile process by triggering release of Ca²⁺ from the lateral sacs into the cytosol, the contractile process is turned off and **relaxation** occurs when Ca²⁺ is returned to the lateral sacs when local electrical activity stops. The SR has an energy-consuming carrier, the **sarcoplasmic/endoplasmic reticulum Ca²⁺–ATPase (SERCA) pump,** which actively

transports Ca²⁺ from the cytosol and concentrates it in the lateral sacs (see Figure 8-11). Recall that the end-plate potential and resultant muscle-fiber action potential stop when the membrane-bound enzyme acetylcholinesterase removes ACh from the neuromuscular junction (see p. 246). When a local action potential is no longer in the T tubules to trigger release of Ca²⁺, the ongoing activity of the SERCA pump returns released Ca²⁺ back into the lateral sacs. Removing cytosolic Ca²⁺ lets the troponin–tropomyosin complex slip back into its blocking position, so actin and myosin can no longer bind at the cross bridges. The thin filaments, freed from cycles of cross-bridge attachment and pulling, return passively to their resting position. The muscle fiber has relaxed. Next we see how long the contractile activity initiated by a single action potential lasts before relaxation occurs.

Contractile Activity Far Outlasts the Electrical Activity that Initiated It. A single action potential in a skeletal muscle fiber lasts only 1 to 2 msec. The onset of the resulting contractile response lags behind the action potential because the entire excitation–contraction coupling must occur before cross-bridge activity begins. In fact, the action potential is over before the contractile apparatus even becomes operational. This time delay of a few milliseconds between stimulation and onset of contraction is called the **latent period** (Figure 8-13). Time is also needed for generating tension within the muscle fiber by means of the sliding interactions between the thin and the thick filaments through cross-bridge activity. The time from contraction onset until peak tension develops—**contraction time**—varies from 15 to 50 msec, depending on muscle fiber type. The contractile response does not end until the lateral sacs have taken up all Ca²⁺ released in response to the action potential. This reuptake of Ca²⁺ is also time-consuming. As Ca²⁺ is pumped back into the lateral sacs, cytosolic Ca²⁺ is reduced

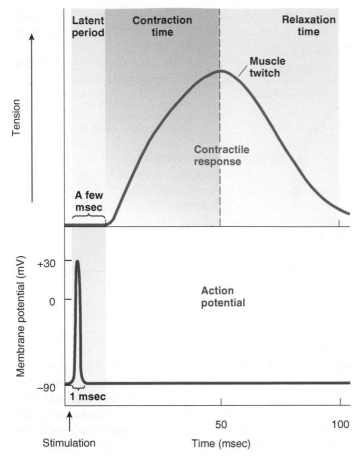

| Latent period | Contraction time | Relaxation time |

Muscle twitch

Contractile response

A few msec

Tension

+30
0
−90
Membrane potential (mV)

1 msec

Action potential

Stimulation

50   100

Time (msec)

Figure 8-13 Relationship of an action potential to the resultant muscle twitch in a slow contracting fiber. The duration of the action potential is not drawn to scale but is exaggerated. Note that the resting potential of a skeletal muscle fiber is −90 mV, compared to a resting potential of −70 mV in a neuron.

and, as a result, cross-bridge activity declines and overall contractile force decreases. The time from peak tension until relaxation is complete—**relaxation time**—varies from 15 to 50 msec, again depending on muscle fiber type. Consequently, the entire contractile response to a single action potential may last from 30 msec in fast contracting fibers to 100 msec or more in slow contracting fibers. This is much longer than the duration of the action potential that initiates contraction (30 to 100 msec as compared to 1 to 2 msec). This fact is important in the body's ability to produce muscle contractions of variable strength, as you will discover in the next section.

Check Your Understanding 8.2

1. Illustrate the relationship between the thick and the thin filaments in a relaxed sarcomere and in a contracted sarcomere.

2. Explain the role of the dihydropyridine and the ryanodine receptors in the process of excitation–contraction coupling.

3. Describe the cross-bridge cycle and indicate whether ATP, ADP, or ADP and P$_i$ are bound to the myosin head during the various stages of the cycle.

4. Explain why muscles are stiff during rigor mortis.

8.3 Skeletal Muscle Mechanics

Thus far we have focused on contraction in a single muscle fiber. In the body, groups of muscle fibers are organized into whole muscles. We now turn to contraction of whole muscles.

Whole muscles are groups of muscle fibers bundled together and attached to bones.

The about 600 skeletal muscles in the body range in size from delicate external eye muscles that control eye movements and contain only a few hundred fibers to large, powerful leg muscles that contain several hundred thousand fibers.

Each muscle is sheathed by connective tissue that penetrates from the surface into the muscle to envelop each individual fiber and divide the muscle into bundles. The connective tissue extends beyond the ends of the muscle to form tough, collagenous **tendons** that attach the muscle to bones. A tendon may be quite long, attaching to a bone some distance from the fleshy part of the muscle. For example, many muscles involved in finger movement are in the forearm, with long tendons extending down to attach to the bones of the fingers. (You can readily see these tendons move on the top of your hand when you wiggle your fingers.) This arrangement permits greater dexterity; the fingers would be thicker and more awkward if all the muscles used in finger movement were actually in the fingers.

Muscle tension is transmitted to bone as the contractile component tightens the series-elastic component.

Tension is produced internally within the sarcomeres, considered the **contractile component** of the muscle, as a result of cross-bridge activity and the resulting sliding of filaments. However, the sarcomeres are not attached directly to the bones. Instead, the tension generated by these contractile elements must be transmitted to the bone via a tendon before the bone can be moved. Tendons have a certain degree of passive elasticity. This noncontractile elastic tissue is in series with the contractile component (being *in series* means that one component is positioned after another in a row) and thus is called the **series-elastic component** of the muscle (unlike titin, which is a major part of the parallel-elastic component and contributes to the muscle's passive elastic recoil). The series-elastic component behaves like a stiff spring placed between the internal tension-generating elements and the bone that is to be moved against an external **load**, or opposing force (**Figure 8-14**). Shortening of the sarcomeres stretches the series-elastic component (tendon). Muscle tension is transmitted to the bone by this tightening of the series-elastic component. This force applied to the bone moves the bone against a load.

A muscle is typically attached to at least two bones across a joint by means of tendons that extend from each end of the muscle. When the muscle shortens during contraction, the position of the joint changes as one bone is moved in relation to the other—for example, *flexion* (bending) of the elbow joint by

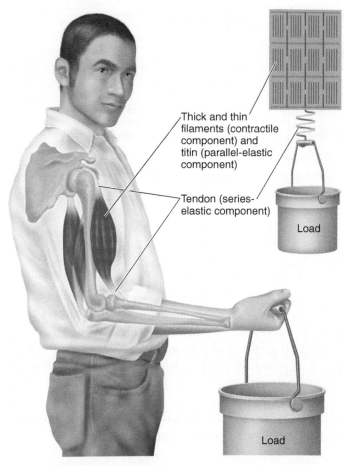

Thick and thin filaments (contractile component) and titin (parallel-elastic component)

Tendon (series-elastic component)

Load

Load

|Figure 8-14 Relationship between the contractile component and the series-elastic component in transmitting muscle tension to bone. Muscle tension is transmitted to the bone by means of the stretching and tightening of the muscle's elastic tendon as a result of sarcomere shortening brought about by cross-bridge cycling.

contraction of the biceps and *extension* (straightening) of the elbow by contraction of the triceps (**I**Figure 8-15). The end of the muscle attached to the more stationary part of the skeleton is called the **origin**, and the end attached to the skeletal part that moves is the **insertion**. Note that because muscle contraction can only pull and not push bone, two different antagonistic muscles or muscle groups are situated to pull on opposite sides of the joint. For example, the biceps can pull the joint in one direction (flexion) and the triceps can pull the joint in the other direction (extension).

The three primary types of contraction are isotonic, isokinetic, and isometric.

Not all muscle contractions shorten muscles and move bones. For a muscle to shorten during contraction, the tension developed in the muscle must exceed the forces that oppose movement of the bone to which the muscle's insertion is attached. In the case of elbow flexion, the opposing force or load is the weight of an object being lifted. When you

flex your elbow without lifting any external object, there is still a load, albeit a minimal one—the weight of your forearm being moved against the force of gravity.

There are three primary types of contraction. In an **isotonic contraction**, the load remains constant as the muscle changes length. In an **isokinetic contraction**, the velocity of shortening remains constant as the muscle changes length. In an **isometric contraction**, the muscle is prevented from shortening, so tension develops at constant muscle length. The same internal events occur in isotonic, isokinetic, and isometric contractions: Muscle excitation turns on the tension-generating contractile process; the cross bridges start cycling; and filament sliding shortens the sarcomeres, which stretches the series-elastic components to exert forces on the bones at the sites of the muscle's origin and insertion.

Considering your biceps as an example, assume you are going to lift an object. When the tension developing in your biceps becomes great enough to overcome the weight of the object in your hand, you can lift the object, with the whole muscle shortening in the process. Because the weight of the object does not change as it is lifted, this type of contraction is an *isotonic* (literally, "constant tension") *contraction*. Owing to the mechanical arrangement of the joint, as the angle of the joint changes while the object is lifted, muscle tension must also change to counterbalance the load. Thus, muscle tension does not remain constant throughout the period of shortening during an isotonic contraction (despite its name) even though the load remains constant.

Isokinetic (literally, "constant motion") *contractions* occur when muscle fibers shorten at a constant velocity, or speed. Isokinetic contractions do not take place normally but can be achieved using special exercise machines set up to require muscle contraction at a constant velocity throughout the entire range in motion. One of the proposed advantages of isokinetic exercise is more rapid development of muscle strength.

What happens if you try to lift an object too heavy for you (that is, if the tension you can develop in your arm muscles is less than required to lift the load)? In this case, the muscle cannot shorten and lift the object but remains at constant length despite the development of tension, so an *isometric* ("constant length") *contraction* occurs. In addition to occurring when the load is too great, isometric contractions take place when the tension developed in the muscle is deliberately less than needed to move the load, with the goal of keeping the muscle at a fixed

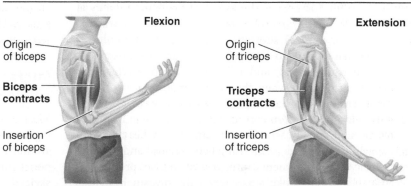

Flexion

Extension

Origin of biceps

Biceps contracts

Insertion of biceps

Origin of triceps

Triceps contracts

Insertion of triceps

IFigure 8-15 **Flexion and extension of the elbow joint.**

length. These submaximal isometric contractions are important for maintaining posture (such as keeping the legs stiff while standing) and for supporting objects in a fixed position (such as holding a beverage between sips).

Muscle contractions often are not of one pure primary type. Muscle tension, length, and velocity of shortening may vary throughout a range of motion. Think about pulling back a bow and arrow. The tension of your biceps muscle continuously increases to overcome the progressively increasing resistance as you stretch the bow further. At the same time, the joint angle changes and the muscle progressively shortens as your elbow bends to draw the bow farther back. Such a contraction does not occur at constant tension, length, or velocity.

Concentric and Eccentric Contractions There are also two other descriptors of muscle contraction—*concentric* and *eccentric*. In **concentric contractions** the muscle shortens, whereas with **eccentric contractions** the muscle lengthens. An example of an eccentric contraction is lowering a book to place it on a desk. During this action, the muscle fibers in the biceps are lengthening but are still actively contracting in opposition to being passively stretched by the load. The contraction itself does not lengthen the muscle; the contraction is resisting the stretch of the muscle imposed externally by the weight of the book.

Other Contractions Some skeletal muscles do not attach to bones at both ends but still produce movement. For example, tongue muscles are not attached at the free end. Contractions of tongue muscles maneuver the free, unattached portion of the tongue to facilitate speech and eating. External eye muscles attach to the skull at their origin and to the eye, not another bone, at their insertion. Contractions of these muscles produce eye movements that enable us to track moving objects or read. A few skeletal muscles known as **sphincters** are not attached to bone at all and actually prevent movement. Sphincters are voluntarily controlled rings of skeletal muscles that, when contracted, close an opening, thereby guarding movement of material through the opening, such as the exit of urine and feces from the body.

The velocity of shortening is related to the load.

The load is also an important determinant of the velocity with which a muscle changes length (❙Figure 8-16). During a concentric contraction, the greater the load, the lower the velocity at which a single muscle fiber (or a constant number of contracting fibers within a muscle) shortens. The speed of shortening is maximal when there is no external load, progressively decreases with an increasing load, and falls to zero (no shortening—isometric contraction) when the load cannot be overcome by maximal tension. You have frequently experienced this **load–velocity relationship**. You can lift light objects requiring little muscle tension quickly, whereas you can lift very heavy objects only slowly, if at all. This relationship between load and shortening velocity is a fundamental property of muscle, presumably because the power stroke slows when the myosin head tilts against a greater load.

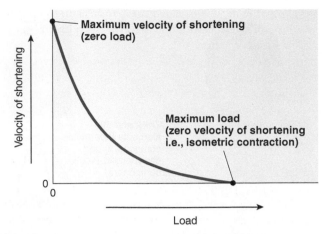

❙Figure 8-16 **Load–velocity relationship in concentric contractions.** The velocity of shortening decreases as the load increases.

Whereas load and velocity for shortening are *inversely* related for *concentric* contractions, load and velocity for lengthening are *directly* related for *eccentric* contractions. An external force (load) greater than a muscle's maximal contraction force causes the muscle to lengthen, with the velocity of lengthening being directly dependent on the load.

Although muscles can accomplish work, much of the energy is converted to heat.

Muscle accomplishes work in a physical sense only when an object is moved. **Work** is defined as force multiplied by distance. **Force** can be equated to the muscle tension required to overcome the load (the weight of the object). The amount of work accomplished by a contracting muscle therefore depends on how much an object weighs and how far it is moved. In an isometric contraction when no object is moved, the muscle contraction's efficiency as a producer of external work is zero. All energy consumed by the muscle during the contraction is converted to heat. In an isotonic or isokinetic contraction, the muscle's efficiency is about 25%. Of the energy consumed by the muscle during the contraction, 25% is realized as external work, whereas the remaining 75% is converted to heat.

Much of this heat is not wasted energy because it is used in maintaining body temperature. In fact, shivering—a form of involuntarily induced skeletal muscle contraction—is a well-known means of increasing heat production on a cold day. Heavy exercise on a hot day, in contrast, may overheat the body because the normal heat-loss mechanisms may not be able to compensate for this increase in heat production.

Interactive units of skeletal muscles, bones, and joints form lever systems.

Most skeletal muscles are attached to bones across joints, forming lever systems. A **lever** is a rigid structure capable of moving around a pivot point known as a **fulcrum**. In the body, the bones function as levers, the joints serve as fulcrums, and the skeletal muscles provide the force to move the bones. The portion of a lever between the fulcrum and the point where a force is applied by the muscle is called the **power arm**; the portion between the fulcrum and the force exerted by a load is known as the **load arm** (Figure 8-17a).

The most common type of lever system in the body is exemplified by flexion of the elbow joint on contraction of the biceps. Skeletal muscles, such as the biceps, consist of many parallel (side-by-side) tension-generating fibers that can exert a large force at their insertion but shorten only a small distance and at relatively slow velocity. The lever system of the elbow joint amplifies the slow, short movements of the biceps to produce more rapid movements of the hand that cover a greater distance. Consider how an object weighing 5 kg is lifted by the hand (Figure 8-17b). When the biceps contracts, it exerts an upward force at the point where it inserts on the forearm bone about 5 cm away from the elbow joint, the fulcrum. Thus, the power arm of this lever system is 5 cm long. The length of the load arm, the distance from the elbow joint to the hand, averages 35 cm. In this case, the load arm is seven times longer than the power arm, which enables the load to be moved a distance seven times greater than the shortening distance of the muscle (while the biceps shortens a distance of 1 cm, the hand moves the load a distance of 7 cm) and at a velocity seven times greater (the hand moves 7 cm during the time that the biceps shortens 1 cm).

The disadvantage of this lever system is that at the point of insertion the biceps muscle must exert a force seven times greater than the load. To keep from dropping the 5-kg load, the product of the length of the power arm times the force in the biceps must equal the product of the length of the load arm times the force exerted by the load. These products are referred to as **moments** (force times power arm or force times load arm). The moment for the load is 5 kg (force) times 35 cm (load arm). This must be matched by the moment for the muscle;

35 kg (force) times 5 cm (power arm). To lift the 5-kg load, the biceps muscle must generate a force greater than 35 kg. As shown by this example, skeletal muscles typically work at a mechanical disadvantage in that they must exert a considerably greater force than the actual load to be moved. Nevertheless, the amplification of velocity and distance afforded by the lever arrangement enables muscles to move loads faster over greater distances than would otherwise be possible. This amplification provides valuable maneuverability and speed.

Next we examine the means by which muscle tension can be graded or varied.

Contractions of a whole muscle can be of varying strength.

A single action potential in a muscle fiber produces a brief, weak contraction called a **twitch**, which is too short and not strong enough to be useful and thus rarely occurs. Muscle fibers are arranged into whole muscles, where they function cooperatively to produce contractions of variable grades of strength stronger than a twitch. You can vary the force you exert by the same muscle, depending on whether you are picking up a piece of paper, a book, or a 50-pound weight. Two primary factors can be adjusted to accomplish gradation of whole-muscle tension: (1) *the number of muscle fibers contracting within a muscle* and (2) *the tension developed by each contracting fiber*. We discuss each of these factors in turn.

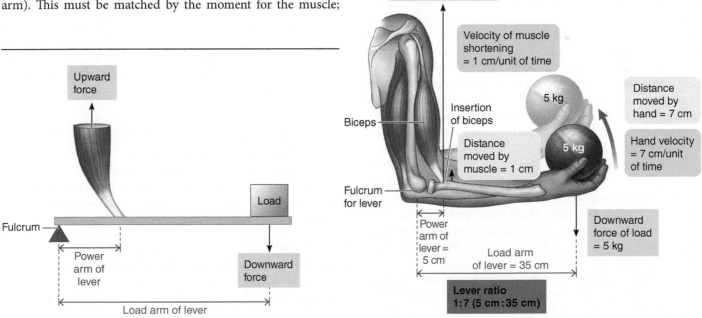

(a) Most common type of lever system in body

(b) Flexion of elbow joint as example of body lever action

Figure 8-17 Lever systems of muscles, bones, and joints. Note that the lever ratio (length of the power arm to length of the load arm) is 1:7 (5 cm:35 cm), which amplifies the distance and velocity of movement seven times (distance moved by the muscle [extent of shortening] =1 cm, distance moved by the hand = 7 cm, velocity of muscle shortening = 1 cm/unit of time, hand velocity = 7 cm/unit of time), but at the expense of the muscle having to exert seven times the force of the load (muscle force = 35 kg, load = 5 kg).

FIGURE FOCUS: *If the biceps of a child inserts 4 cm from the elbow and the length of the arm from the elbow to the hand is 28 cm, how much force must the biceps generate for the child to lift a 6-kg backpack with one hand?*

The number of fibers contracting within a muscle depends on the extent of motor unit recruitment.

The greater the number of fibers contracting, the greater the total muscle tension. Therefore, larger muscles consisting of more muscle fibers can generate more tension than smaller muscles with fewer fibers can.

Each whole muscle is innervated by a number of different motor neurons. When a motor neuron enters a muscle, it branches, with each axon terminal supplying a single muscle fiber (Figure 8-18). One motor neuron innervates a number of muscle fibers, but each muscle fiber is supplied by only one motor neuron. When a motor neuron is activated, all the muscle fibers it supplies are stimulated to contract simultaneously. This team of concurrently activated components—one motor neuron plus all the muscle fibers it innervates—is called a **motor unit**. The muscle fibers that compose a motor unit are dispersed throughout the whole muscle; thus, their simultaneous contraction results in an evenly distributed, although weak, contraction of the whole muscle. Each muscle consists of numerous intermingled motor units. For a weak contraction of the whole muscle, only one or a few of its motor units are activated. For stronger and stronger contractions, more and more motor units are recruited, or stimulated to contract simultaneously, a phenomenon known as **motor unit recruitment**.

How much stronger the contraction is with the recruitment of each additional motor unit depends on motor unit size (that is, the number of muscle fibers controlled by a single motor neuron). The number of muscle fibers per motor unit and the number of motor units per muscle vary widely, depending on the specific function of the muscle. For muscles that produce precise, delicate movements, such as external eye muscles and hand muscles, a single motor unit may contain as few as a dozen muscle fibers. Because so few muscle fibers are involved with each motor unit, recruitment of each additional motor unit adds only a small increment to the whole muscle's strength of contraction. These small motor units allow fine control over muscle tension. In contrast, in muscles designed for powerful, coarsely controlled movement, such as those of the legs, a single motor unit may contain 1500 to 2000 muscle fibers. Recruitment of motor units in these muscles results in large incremental increases in whole-muscle tension. More powerful contractions occur at the expense of less precisely controlled gradations. Thus, the number of muscle fibers participating in the whole muscle's total contractile effort depends on the number of motor units recruited and the number of muscle fibers per motor unit in that muscle.

To delay or prevent **fatigue** (inability to maintain muscle tension at a given level) during a sustained contraction involving only a portion of a muscle's motor units, as is necessary in muscles supporting the weight of the body against the force of gravity, **asynchronous recruitment** of motor units takes place. The body alternates motor unit activity, like shifts at a factory, to give motor units that have been active an opportunity to rest while others take over. Changing of the shifts is carefully coordinated, so the sustained contraction is smooth rather than jerky. Asynchronous recruitment is possible only for submaxi-

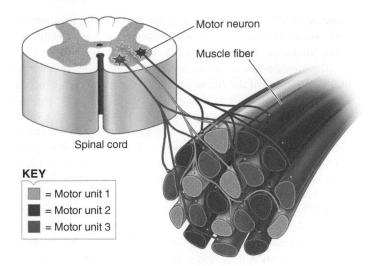

Motor neuron
Muscle fiber
Spinal cord

KEY

▨	= Motor unit 1
■	= Motor unit 2
▦	= Motor unit 3

Figure 8-18 **Motor units in a skeletal muscle.**

mal contractions, during which only some of the motor units must maintain the desired level of tension. During maximal contractions, when all muscle fibers must participate, it is impossible to alternate motor unit activity to prevent fatigue. This is one reason you cannot support a heavy object as long as you can support a light one.

Furthermore, the type of muscle fiber activated varies with the extent of gradation. Most muscles consist of a mixture of fiber types that differ metabolically, some being more resistant to fatigue than others. During weak or moderate endurance-type activities (aerobic exercise), the motor units most resistant to fatigue are recruited first. The last fibers to be called into play in the face of demands for further increases in tension are those that fatigue rapidly. An individual can therefore engage in endurance activities for prolonged periods but can only briefly maintain bursts of all-out, powerful effort. Of course, even the muscle fibers most resistant to fatigue eventually tire if required to maintain a certain level of sustained tension.

The frequency of stimulation can influence the tension developed by each muscle fiber.

Whole-muscle tension depends not only on the number of muscle fibers contracting, but also on the tension developed by each contracting fiber. Various factors influence the extent to which tension can be developed, including the following:

1. Frequency of stimulation
2. Length of the fiber at the onset of contraction
3. Extent of fatigue
4. Thickness of the fiber

We now examine the effect of frequency of stimulation; we discuss the other factors in later sections.

Twitch Summation and Tetanus Even though a single action potential in a muscle fiber produces only a twitch, contractions with longer duration and greater tension can be achieved by repeated stimulation of the fiber. Let us see what happens when a second action potential occurs in a muscle fiber. If the muscle fiber has completely relaxed before the next action potential takes place, a second twitch of the same magnitude as the first occurs (⏐Figure 8-19a). The same excitation–contraction events take place each time, resulting in identical twitch responses. If, however, the muscle fiber is stimulated a second time before it has completely relaxed from the first twitch, a second action potential causes a second contractile response, which is added "piggyback" on top of the first twitch (⏐Figure 8-19b). The two twitches from the two action potentials add together, or sum, to produce greater tension in the fiber than that produced by a single action potential, a process known as **twitch summation**.

Twitch summation is possible only because the duration of the action potential (1 to 2 msec) is much shorter than the duration of the resulting twitch (30 msec to 100 msec). Once an action potential has been initiated, a brief refractory period occurs during which another action potential cannot be initiated (see p. 98). It is therefore impossible to achieve summation of action potentials. The membrane must return to resting potential and recover from its refractory period before another action potential can occur. However, because the action potential and refractory period are over long before the resulting muscle twitch is completed, the muscle fiber may be restimulated while some contractile activity still exists to produce summation of the mechanical response.

If the muscle fiber is stimulated so rapidly that it does not have a chance to relax at all between stimuli, a smooth, sustained contraction of maximal strength known as **tetanus**

occurs (⏐Figure 8-19c). A tetanic contraction is usually three to four times stronger than a single twitch. (Don't confuse this normal muscle tetanus with the disease tetanus; see p. 112.)

Twitch summation results primarily from a sustained elevation in cytosolic Ca²⁺.

What is the mechanism of twitch summation and tetanus at the cell level? The tension produced by a contracting muscle fiber increases as a result of greater cross-bridge cycling. The series-elastic component (tendon) must be stretched to transmit the tension generated in the muscle fibers to the bone, and it takes time to stretch this elastic element. Accordingly, two factors contribute to twitch summation: (1) sustained elevation in cytosolic Ca²⁺ permitting greater cross-bridge cycling, and (2) more time to stretch the series-elastic component.

The most important factor in the development of twitch summation is sustained elevation in cytosolic Ca²⁺ as the frequency of action potentials increases. Enough Ca²⁺ is released in response to a single action potential to interact with all the troponin within the cell. As a result, all cross bridges are free to participate in the contractile response. How, then, can repetitive action potentials bring about a greater contractile response? The difference depends on how long enough Ca²⁺ is available. The cross bridges remain active and continue to cycle as long as enough Ca²⁺ is present to keep the troponin–tropomyosin complexes away from the cross-bridge binding sites on actin. Each troponin–tropomyosin complex spans a distance of seven actin molecules. Thus, binding of Ca²⁺ to one troponin mole-

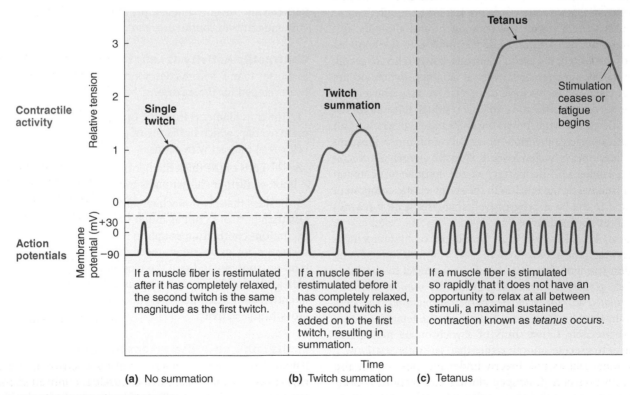

⏐Figure 8-19 **Twitch summation and tetanus.**

FIGURE FOCUS: *Compare the relative tension developed in a single twitch and during tetanus. What accounts for this difference in tension?*

cule leads to the uncovering of only seven cross-bridge binding sites on the thin filament.

As soon as Ca^{2+} is released in response to an action potential, the SERCA pump starts pumping Ca^{2+} back into the lateral sacs. As the cytosolic Ca^{2+} concentration subsequently declines, less Ca^{2+} is present to bind with troponin, so some of the troponin–tropomyosin complexes slip back into their blocking positions. Consequently, not all cross-bridge binding sites remain available to participate in the cycling process during a single twitch induced by a single action potential. Because not all cross bridges find a binding site, the resulting contraction during a single twitch is not of maximal strength.

If action potentials and twitches occur far enough apart in time for all released Ca^{2+} from the first contractile response to be pumped back into the lateral sacs between the action potentials, an identical twitch response occurs as a result of the second action potential. If, however, a second action potential occurs and more Ca^{2+} is released while the Ca^{2+} that was released in response to the first action potential is being taken back up, the cytosolic Ca^{2+} concentration remains high and might even be elevated further. This prolonged availability of Ca^{2+} in the cytosol permits more of the cross bridges to continue participating in the cycling process for a longer time. As a result, tension development increases correspondingly. As the frequency of action potentials increases, the duration of elevated cytosolic Ca^{2+} concentration increases, and contractile activity likewise increases until a maximum tetanic contraction is reached. With tetanus, the maximum number of cross-bridge binding sites remains uncovered so that cross-bridge cycling, and consequently tension development, is at its peak.

The second factor contributing to twitch summation is related to the elastic structures of the muscle fiber. During a single twitch, the contraction does not last long enough to completely stretch the series-elastic component and allow the full sarcomere-generated tension to be transmitted to the bone. At the end of the twitch, the elastic elements slowly relax, or recoil, to their initial nonstretched state. If another twitch occurs before the elastic elements have completely relaxed, the tension from the second twitch adds to the residual tension in the series-elastic component remaining from the first twitch. With greater frequencies of action potentials and more frequent twitches, less time is available for the elastic elements to recoil between twitches. Consequently, as the frequency of action potentials increases, the tension in the series-elastic component transmitted to the bone progressively increases until it reaches its maximum during tetanus.

Because skeletal muscle must be stimulated by motor neurons to contract, the nervous system plays a key role in regulating contraction strength. The two main factors subject to control to accomplish gradation of contraction are the *number of motor units stimulated* and the *frequency of their stimulation*. The areas of the brain that direct motor activity combine tetanic contractions and precisely timed shifts of asynchronous motor unit recruitment to execute smooth rather than jerky contractions.

Additional factors not directly under nervous control also influence the tension developed during contraction. Among these is the length of the fiber at the onset of contraction, to which we now turn attention.

At the optimal muscle length, maximal tension can be developed.

A relationship exists between the length of the muscle before the onset of contraction and the tetanic tension that each contracting fiber can subsequently develop at that length. Every muscle has an **optimal length** (l_o) at which maximal force can be achieved during a tetanic contraction beginning at that length—that is, more tension can be achieved during tetanus when beginning at l_o than can be achieved when the contraction begins with the muscle longer or shorter than l_o. This **length–tension relationship** can be explained by the sliding filament mechanism of muscle contraction.

Contractile Activity at l_o At l_o, when maximum tension can be developed (point A in I Figure 8-20), the thin filaments optimally overlap the regions of the thick filaments where the cross bridges are located. At this length, a maximal number of cross bridges and actin molecules are accessible to each other for cycles of binding and bending. The central region of thick filaments, where the thin filaments do not overlap at l_o, lacks cross bridges; only myosin tails are found here.

Contractile Activity at Lengths Greater Than l_o At greater lengths, as when a muscle is passively stretched (point B), the thin filaments are pulled out from between the thick filaments, decreasing the number of actin sites available for cross-bridge binding—that is, some of the actin sites and cross bridges no longer "match up," so they "go unused." When less cross-bridge activity can occur, less tension can develop. In fact, when the muscle is stretched to about 70% longer than its l_o (point C) the thin filaments are completely pulled out from between the thick filaments, preventing cross-bridge activity; consequently, no contraction can occur.

Contractile Activity at Lengths Less Than l_o If a muscle is shorter than l_o before contraction (point D), less tension can be developed for three reasons:

1. The thin filaments from the opposite sides of the sarcomere overlap, which limits the opportunity for the cross bridges to interact with actin.

2. The ends of the thick filaments become forced against the Z lines, so further shortening is impeded.

3. Besides these two mechanical factors, at muscle lengths less than 80% of l_o, not as much Ca^{2+} is released during excitation–contraction coupling for reasons unknown. Furthermore, by an unknown mechanism, the ability of Ca^{2+} to bind to troponin and pull the troponin–tropomyosin complex aside is reduced at shorter muscle lengths. Consequently, fewer actin sites are uncovered for participation in cross-bridge activity.

Limitations on Muscle Length The extremes in muscle length that prevent development of tension occur only under experimental conditions, when a muscle is removed and stimulated at various lengths. Attachment of muscles to the skeleton imposes limits on muscle shortening and lengthening. Muscles

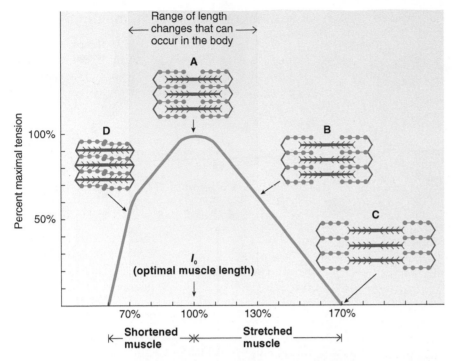

Range of length changes that can occur in the body

Shortened muscle ← → **Stretched muscle**

Muscle fiber length compared with optimal length

❙Figure 8-20 Length–tension relationship. Maximal contraction strength can be achieved when a muscle fiber is at its optimal length (l_o) before the onset of contraction because this is the point of optimal overlap of thick-filament cross bridges and thin-filament cross-bridge binding sites (point A). The percentage of maximal contraction strength that can be achieved decreases when the muscle fiber is longer or shorter than l_o before contraction. When it is longer, fewer thin-filament binding sites are accessible for binding with thick-filament cross bridges because the thin filaments are pulled out from between the thick filaments (points B and C). When the fiber is shorter, fewer thin-filament binding sites are exposed to thick-filament cross bridges because the thin filaments overlap (point D). Also, further shortening and tension development are impeded as the thick filaments become forced against the Z lines (point D). In the body, the resting muscle length (that is, when the muscle is not actively contracting or passively positioned) is near l_o. Furthermore, because of restrictions imposed by skeletal attachments, muscles cannot vary beyond 30% of their l_o in either direction (the range screened in light green). At the outer limits of this range, muscles still can achieve about 50% of their maximal contraction strength.

FIGURE FOCUS: *What percentage of maximal tension can a muscle generate if it is positioned at 80% of its optimal length at the onset of contraction?*

Check Your Understanding 8.3

1. Explain how muscle tension is transmitted to bone.
2. Compare how the contraction force needed to support a 5-kg load and the velocity of moving the hand would be affected if the tendon of insertion of the biceps muscle were 10 cm instead of 5 cm from the elbow (fulcrum) in ❙Figure 8-17.
3. Describe the means by which the strength of contraction of a skeletal muscle can be changed to generate greater force.
4. Describe the role of Ca^{2+} in twitch summation.

8.4 Skeletal Muscle Metabolism and Fiber Types

Four steps in the excitation, contraction, and relaxation processes require ATP:

1. Splitting of ATP by myosin ATPase provides the energy for the power stroke of the cross bridge.
2. Binding (but not splitting) of a fresh molecule of ATP to myosin lets the cross bridge detach from the actin filament at the end of a power stroke so that the cycle can be repeated. This ATP is later split to provide energy for the next stroke of the cross bridge.
3. Active transport of Ca^{2+} back into the lateral sacs of the SR during relaxation depends on energy derived from the breakdown of ATP.
4. The ATP-dependent Na^+–K^+ pump actively returns the ions (Na^+ back out of the cell and K^+ back into the cell) that moved during the generation of a contraction-inducing action potential in the muscle cell.

are positioned so that their relaxed length (the length when the muscle is not actively contracting or passively positioned) is approximately at l_o; thus, they can achieve near maximal tetanic contraction most of the time. (The sarcomeres are between 2.0 and 2.2 μm wide at l_o, and the relaxed width of a sarcomere averages about 2.0 μm.) Furthermore, because of skeletal constraints, muscles cannot be stretched or shortened more than 30% of their optimal length. Even at the outer limits (130% and 70% of l_o), the muscles still can generate half their maximum tension.

The factors covered thus far that influence how much tension a contracting muscle fiber can develop—the frequency of stimulation and the muscle length at onset of contraction—can vary from contraction to contraction. Other determinants of muscle fiber tension—how resistant the muscle fiber is to fatigue and how thick the fiber is—do not vary from contraction to contraction but depend on the fiber type and can be modified over time. We consider these other factors next.

Muscle fibers have alternate pathways for forming ATP.

Because ATP is the only energy source that can be directly used for these activities, for contractile activity to continue, ATP must constantly be supplied. Only limited stores of ATP are immediately available in muscle tissue, enough to power the first few seconds of exercise. However, three pathways supply additional ATP as needed during muscle contraction: (1) transfer of a high-energy phosphate from creatine phosphate to ADP, (2) oxidative phosphorylation (the electron transport system and chemiosmosis), and (3) glycolysis.

Creatine Phosphate **Creatine phosphate** is the first energy storehouse tapped at the onset of contractile activity (❙Figure 8-21, step **3a**). Like ATP, creatine phosphate contains a high-

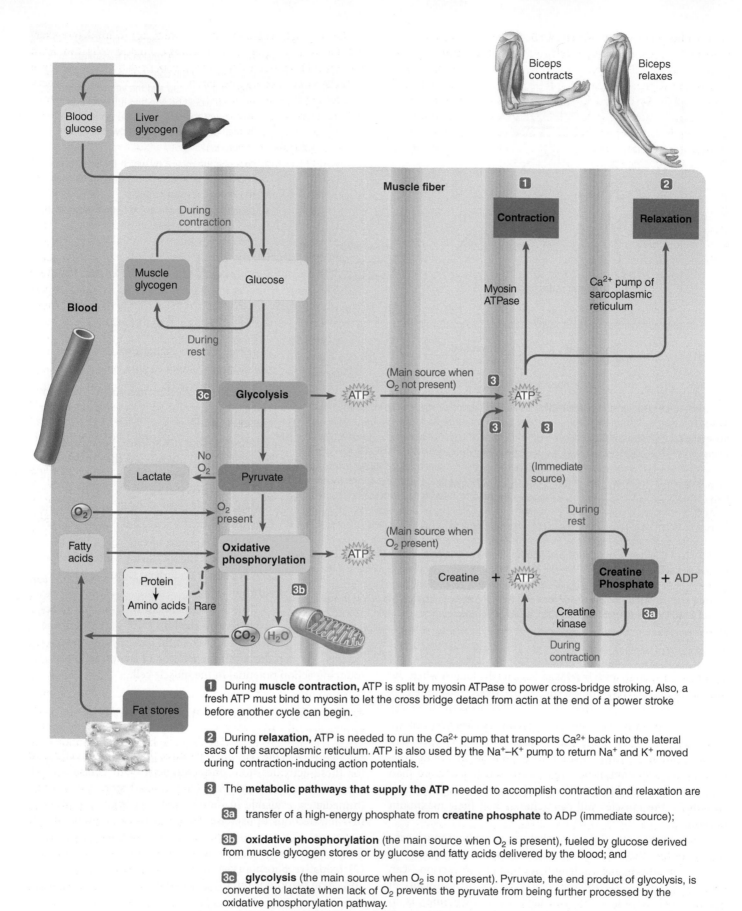

Biceps contracts

Biceps relaxes

Muscle fiber

1 **Contraction**

2 **Relaxation**

Blood glucose

Liver glycogen

During contraction

Muscle glycogen

Glucose

Blood

During rest

Myosin ATPase

Ca²⁺ pump of sarcoplasmic reticulum

(Main source when O₂ not present) **3**

3c **Glycolysis** → ATP → ATP

No O₂

Lactate ← Pyruvate

O₂ present

O₂

(Immediate source)

During rest

Fatty acids

(Main source when O₂ present)

Oxidative phosphorylation → ATP

Creatine + ATP

Creatine Phosphate + ADP

Protein ↓ Amino acids Rare

3b

CO_2 H_2O

Creatine kinase

3a

During contraction

Fat stores

1 During **muscle contraction,** ATP is split by myosin ATPase to power cross-bridge stroking. Also, a fresh ATP must bind to myosin to let the cross bridge detach from actin at the end of a power stroke before another cycle can begin.

2 During **relaxation,** ATP is needed to run the Ca²⁺ pump that transports Ca²⁺ back into the lateral sacs of the sarcoplasmic reticulum. ATP is also used by the Na⁺–K⁺ pump to return Na⁺ and K⁺ moved during contraction-inducing action potentials.

3 The **metabolic pathways that supply the ATP** needed to accomplish contraction and relaxation are

3a transfer of a high-energy phosphate from **creatine phosphate** to ADP (immediate source);

3b **oxidative phosphorylation** (the main source when O₂ is present), fueled by glucose derived from muscle glycogen stores or by glucose and fatty acids delivered by the blood; and

3c **glycolysis** (the main source when O₂ is not present). Pyruvate, the end product of glycolysis, is converted to lactate when lack of O₂ prevents the pyruvate from being further processed by the oxidative phosphorylation pathway.

❙Figure 8-21 **Metabolic pathways producing ATP used during muscle contraction and relaxation.**

energy phosphate group, which can be donated directly to ADP to form ATP. Just as energy is released when the terminal phosphate bond in ATP is split, energy is released when the bond between phosphate and creatine is broken. The energy released from the hydrolysis of creatine phosphate, along with the phosphate, can be donated directly to ADP to form ATP. This reaction, which is catalyzed by the muscle cell enzyme **creatine kinase**, is reversible; energy and phosphate from ATP can be transferred to creatine to form creatine phosphate:

$$\text{Creatine phosphate} + \text{ADP} \overset{\text{creatine kinase}}{\rightleftharpoons} \text{creatine} + \text{ATP}$$

As energy reserves are built up in a resting muscle, the increased concentration of ATP favors transfer of the high-energy phosphate group from ATP to form creatine phosphate. By contrast, at the onset of contraction when myosin ATPase splits the meager reserves of ATP, the resultant fall in ATP favors transfer of the high-energy phosphate group from stored creatine phosphate to form more ATP. A rested muscle contains about five times as much creatine phosphate as ATP. Thus, most energy is stored in muscle in creatine phosphate pools. Because only one enzymatic reaction is involved in this energy transfer, ATP can be formed rapidly (within a fraction of a second) by using creatine phosphate.

Thus, creatine phosphate is the first source for supplying additional ATP when exercise begins. Muscle ATP levels actually remain fairly constant early in contraction, but creatine phosphate stores become depleted. In fact, short bursts of high-intensity contractile effort, such as high jumps, sprints, or weight lifting, are supported primarily by ATP derived at the expense of creatine phosphate. Creatine phosphate stores typically power 5 to 10 seconds of exercise before the stores run out.

Some athletes hoping to gain a competitive edge take oral creatine supplements to boost their performance in short-term, high-intensity activities. (We naturally get creatine in our diets, especially in meat.) Loading the muscles with extra creatine means larger creatine phosphate stores—that is, larger energy stores that can translate into a small edge in performance of activities requiring short, explosive bursts of energy. Yet creatine supplements should be used with caution because the long-term health effects are unknown. Also, extra creatine stores are of no use in activities of longer duration that rely on more long-term energy-supplying mechanisms.

Oxidative Phosphorylation The multistep oxidative phosphorylation pathway produces ATP at a relatively slow rate when compared to the transfer of a high-energy phosphate from creatine phosphate to ADP or the process of glycolysis. Oxidative phosphorylation takes place within the muscle mitochondria if sufficient O_2 is present (see p. 37). Oxygen is required to support the mitochondrial electron transport system, which, together with chemiosmosis by ATP synthase, efficiently harnesses energy captured from the breakdown of nutrient molecules and uses it to generate ATP. This pathway is fueled by glucose or fatty acids, depending on the intensity and duration of the activity (❙ Figure 8-21, step **3b**). Although it provides a rich yield of 32 ATP molecules for each glucose molecule processed, oxidative phosphorylation is relatively slow because of the number of enzymatic steps involved.

During light exercise (such as walking) to moderate exercise (such as jogging or swimming), muscle cells can form enough ATP through oxidative phosphorylation to keep pace with the modest energy demands of the contractile machinery for prolonged periods. To sustain ongoing oxidative phosphorylation, exercising muscles depend on delivery of adequate O_2 and nutrients to maintain their activity. Activity supported in this way is **aerobic** ("with O_2") or **endurance-type exercise**.

Most O_2 required for oxidative phosphorylation is delivered by the blood. Increased O_2 is made available to muscles during exercise by several means: Deeper, more rapid breathing brings more O_2 into the blood; the heart contracts more rapidly and forcefully to pump more oxygenated blood to the tissues; more blood is diverted to exercising muscles by dilation of the blood vessels supplying them; and hemoglobin molecules that carry O_2 in the blood release more O_2 in exercising muscles. (These mechanisms are discussed further in later chapters.) Furthermore, some types of muscle fibers have an abundance of **myoglobin**, which is similar to hemoglobin. Myoglobin can store small amounts of O_2, but more important, it increases the rate of O_2 transfer from the blood into muscle fibers.

Glucose and fatty acids, ultimately derived from food eaten, are also delivered to muscle cells by the blood. In addition, muscle cells are able to store limited quantities of glucose in the form of glycogen (chains of glucose). Furthermore, up to a point the liver can store excess ingested carbohydrates as glycogen, which can be broken down to release glucose into the blood for use between meals. *Carbohydrate loading*—increasing carbohydrate intake before a competition—is a tactic used by some athletes in hopes of boosting performance in endurance events such as marathons. However, once muscle and liver glycogen stores are filled, excess ingested carbohydrates (or any other energy-rich nutrient) are converted to body fat.

Glycolysis There are limits as to how much O_2 the lungs can pick up and the circulatory system can deliver to exercising muscles. Furthermore, in near-maximal contractions, the powerful contraction almost squeezes closed the blood vessels that course through the muscle, severely limiting O_2 availability to the muscle fibers. Even when O_2 is available, the relatively slow oxidative phosphorylation system may not be able to produce ATP rapidly enough to meet the muscle's needs during intense activity. A skeletal muscle's energy consumption may increase up to 100-fold when going from rest to high-intensity exercise. When O_2 delivery or oxidative phosphorylation cannot keep pace with the demand for ATP formation as the intensity of exercise increases, the muscle fibers rely increasingly on glycolysis to generate ATP (❙ Figure 8-21, step **3c**) (see p. 35). The chemical reactions of glycolysis yield products for ultimate entry into the oxidative phosphorylation pathway, but glycolysis can also proceed alone in the absence of further processing of its products by oxidative phosphorylation. During glycolysis, a glucose molecule is broken down into two pyruvate molecules, yielding two ATP molecules in the process. Pyruvate can be further degraded by oxidative phosphorylation to extract more energy. However, glycolysis alone has two advantages over the oxidative phosphorylation pathway: (1) glycolysis can form ATP in the absence of O_2 (operating *anaerobically*—that is, "without O_2"), and (2) it can proceed more rapidly than oxidative phosphorylation. Although glycolysis extracts considerably fewer ATP molecules from each nutrient molecule processed, because of its speed its rate of ATP production can exceed the rate of generation of ATP by oxidative phosphorylation as long as glucose is present. Activity that can be supported in this way is anaerobic or **high-intensity exercise**.

Lactate Production Even though anaerobic glycolysis provides a means of performing intense exercise when O_2 delivery or the oxidative phosphorylation capacity is exceeded, using this pathway has two consequences. First, large amounts of nutrient fuel must be processed because glycolysis is less efficient than oxidative phosphorylation in converting nutrient energy into the energy of ATP. (Glycolysis yields a net of 2 ATPs for each glucose molecule degraded, whereas the oxidative phosphorylation pathway can extract 32 ATPs from each glucose molecule.) Muscle cells can store limited quantities of glucose as glycogen, but anaerobic glycolysis rapidly depletes these glycogen supplies. Therefore, anaerobic high-intensity exercise can be sustained for only a short duration, in contrast to the body's prolonged ability to sustain aerobic, endurance-type activities. Anaerobic glycolysis can support muscle contractile activity for less than 2 minutes.

Second, when the end product of anaerobic glycolysis, pyruvate, cannot be further processed by oxidative phosphorylation, it is converted to **lactate**. Lactate accumulation has been implicated in the acute muscle soreness (burning sensation) that occurs when intense exercise is actually taking place. (The delayed-onset muscle pain and stiffness that begin the day after unaccustomed muscular exertion, however, are caused by reversible structural damage.) Furthermore, lactate (lactic acid) picked up by the blood produces the metabolic acidosis accompanying intense exercise.

Fatigue may be of muscle or central origin.

Contractile activity in a particular skeletal muscle cannot be maintained at a high level indefinitely. Eventually, the tension in the muscle declines as fatigue sets in. There are two types of fatigue: muscle fatigue and central fatigue.

Muscle fatigue occurs when an exercising muscle can no longer respond to stimulation with the same degree of contractile activity. Muscle fatigue is a defense mechanism that protects a muscle from reaching a point at which it can no longer produce ATP. An inability to produce ATP would result in rigor mortis (obviously not an acceptable outcome of exercise). The underlying causes of muscle fatigue are unclear. The primary implicated factors include the following:

- The *local increase in inorganic phosphate* from ATP breakdown is considered the primary cause of muscle fatigue. Increased levels of P_i reduce the strength of contraction by interfering with the power stroke of the myosin heads. In addition, increased P_i appears to decrease the sensitivity of the regulatory proteins to Ca^{2+} and to decrease the amount of Ca^{2+} released from the lateral sacs.

- *Inappropriate leakage of Ca^{2+}* through the SR's Ca^{2+}-release channels is the latest factor implicated in muscle fatigue after

long and intense exercise. Some of the leaked Ca^{2+} exits the cell and cannot be returned to the SR by the SERCA pump. This Ca^{2+} loss from the cell depletes the SR Ca^{2+} supply needed to sustain contractile activity, leading to weaker contractions. Furthermore, exposure to leaked Ca^{2+} during fatiguing exercise activates proteases, protein-snipping enzymes that cause transient muscle damage, which likely contributes to weakening of contractions.

- *Depletion of glycogen energy reserves* may also lead to muscle fatigue in exhausting exercise.

The time of onset of muscle fatigue varies with the type of muscle fiber (some fibers being more resistant to fatigue than others) and with the intensity of the exercise (more rapid onset of fatigue being associated with high-intensity activities).

Central fatigue occurs when the central nervous system (CNS) no longer adequately activates the motor neurons supplying the working muscles. The person slows down or stops exercising even though the muscles are still able to perform. Central fatigue often is psychologically based. During strenuous exercise, central fatigue may stem from discomfort associated with the activity; it takes strong motivation (a will to win) to deliberately persevere when in pain. In less strenuous activities, central fatigue may reduce physical performance in association with boredom and monotony (such as assembly-line work) or tiredness (lack of sleep). The mechanisms involved in central fatigue are poorly understood. In some cases, central fatigue may stem from increased levels of serotonin (a neurotransmitter) and tryptophan (an amino acid from which serotonin is made) within the brain.

Increased O_2 consumption is necessary to recover from exercise.

A person continues to breathe deeply and rapidly for some time after exercising. The need for elevated O_2 uptake during recovery from exercise (**excess postexercise oxygen consumption,** or **EPOC**) results from various factors. The best known is repayment of an **oxygen deficit** incurred during exercise, when contractile activity was being supported by ATP derived from nonoxidative sources such as creatine phosphate and anaerobic glycolysis. During exercise, the creatine phosphate stores of active muscles are reduced, lactate may accumulate, and glycogen stores may be tapped; the extent of these effects depends on the intensity and duration of the activity. Oxygen is needed for recovery of the energy systems. During the recovery period, fresh supplies of ATP are formed by oxidative phosphorylation using the newly acquired O_2, which is provided by the sustained increase in breathing after exercise has stopped. Most of this ATP is used to resynthesize creatine phosphate to restore its reserves, which can be accomplished in a few minutes. Any accumulated lactate is converted back into pyruvate, part of which is used by the oxidative phosphorylation system for ATP production. The remainder of the pyruvate is converted back into glucose by the liver. Most of this glucose is used to replenish the glycogen stores drained from the muscles and liver during exercise. These biochemical reactions involving pyruvate require O_2 and take several hours for completion. Thus, EPOC provides the O_2 needed to restore the creatine phosphate system, remove lactate, and at least partially replenish glycogen stores.

Unrelated to increased O_2 uptake is the need to restore nutrients after grueling exercise, such as marathon races, in which glycogen stores are severely depleted. In such cases, long-term recovery can take a day or more, because the exhausted energy stores require nutrient intake for full replenishment. Therefore, depending on the type and duration of activity, recovery can be complete within a few minutes or can require more than a day.

Part of EPOC is not directly related to repayment of energy stores but instead results from a general metabolic disturbance following exercise. For example, the local increase in muscle temperature arising from heat-generating contractile activity speeds up the rate of all chemical reactions in the muscle tissue, including those dependent on O_2. Likewise, body temperature rises several degrees Fahrenheit during exercise. A rise in temperature speeds up O_2-consuming chemical reactions. Until body temperature returns to preexercise levels, the increased speed of these chemical reactions partly accounts for EPOC. Furthermore, the secretion of epinephrine, a hormone that increases O_2 consumption by the body, is elevated during exercise. Until the circulating level of epinephrine returns to its preexercise state, O_2 uptake is increased above normal.

We have been looking at the contractile and metabolic activities of skeletal muscle fibers in general. However, not all skeletal muscle fibers use these mechanisms to the same extent. We next examine the types of muscle fibers based on their speed of contraction and how they are metabolically equipped to generate ATP.

The three types of skeletal muscle fibers differ in ATP hydrolysis and synthesis.

Classified by their biochemical capacities, there are three major types of muscle fibers (Table 8-1):

1. Slow-oxidative (type I) fibers
2. Fast-oxidative (type IIa) fibers
3. Fast-glycolytic (type IIx) fibers

As their names imply, the two main differences among these fiber types are their speed of contraction (slow or fast) and the type of enzymatic machinery they primarily use for ATP formation (oxidative or glycolytic).

Fast Versus Slow Fibers Fast fibers have higher myosin ATPase (ATP-splitting) activity than slow fibers do. The higher the ATPase activity, the more rapidly ATP is split and the faster the rate at which energy is made available for cross-bridge cycling. The result is a fast twitch,

compared to the slower twitches of those fibers that split ATP more slowly. The time to peak twitch tension for fast fibers is 15 to 40 msec compared to 50 to 100 msec for slow fibers (Figure 8-22a). Thus, two factors determine the speed with which a muscle contracts: the load (load–velocity relationship) and the myosin ATPase activity of the contracting fibers (fast or slow twitch).

Oxidative Versus Glycolytic Fibers Fiber types also differ in ATP-synthesizing ability. Those with a greater capacity to form ATP are more resistant to fatigue. Some fibers are better equipped for oxidative phosphorylation, whereas others rely primarily on anaerobic glycolysis for synthesizing ATP. Because oxidative phosphorylation yields considerably more ATP from each nutrient molecule processed, it does not readily deplete energy stores. Furthermore, it does not result in lactate accumulation. Oxidative types of muscle fibers are therefore more resistant to fatigue than glycolytic fibers are.

Other related characteristics distinguishing these three fiber types are summarized in Table 8-1. As you would expect, the oxidative fibers, both slow and fast, contain an abundance of mitochondria, the organelles that house the enzymes involved in oxidative phosphorylation. Because adequate oxygenation is essential to support this pathway, these fibers are richly supplied with capillaries. Oxidative fibers also have high myoglobin content. Myoglobin not only helps support oxidative fibers' O_2 dependency, but also gives them a red color, just as oxygenated hemoglobin produces the red color of arterial blood. Accordingly, these muscle fibers are called **red fibers**.

In contrast, the fast fibers specialized for glycolysis contain few mitochondria but have a high content of glycolytic enzymes instead. Also, to supply the large amounts of glucose needed for glycolysis, they contain a lot of stored glycogen. Because the glycolytic fibers need relatively less O_2 to function, they have

TABLE 8-1 Characteristics of Skeletal Muscle Fibers

Characteristic	Slow-Oxidative (Type I) Fiber	Fast-Oxidative (Type IIa) Fiber	Fast-Glycolytic (Type IIx) Fiber
Myosin–ATPase activity	Low	High	High
Speed of contraction	Slow	Fast	Fast
Resistance to fatigue	High	Intermediate	Low
Oxidative phosphorylation capacity	High	High	Low
Enzymes for anaerobic glycolysis	Low	Intermediate	High
Mitochondria	Many	Many	Few
Capillaries	Many	Many	Few
Myoglobin content	High	High	Low
Color of fiber	Red	Red	White
Glycogen content	Low	Intermediate	High

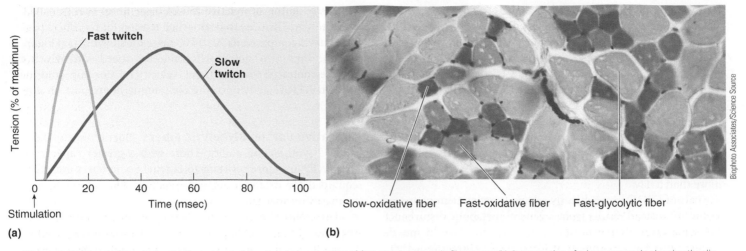

IFigure 8-22 Muscle fiber types. (a) Comparison of the speed of contraction of fast and slow muscle fiber types. (b) Cross section of a human muscle showing the distribution of slow-oxidative, fast-oxidative, and fast-glycolytic muscle fiber types.

only a meager capillary supply compared with the oxidative fibers. The glycolytic fibers contain little myoglobin and therefore are pale in color, so they are sometimes called **white fibers**. (The most readily observable comparison between red and white fibers is the dark and white meat in poultry; muscles of the legs consist primarily of red fibers and the breast muscles consist primarily of white fibers.)

Genetic Endowment of Muscle Fiber Types In humans, most muscles contain a mixture of all three fiber types (IFigure 8-22b); the percentage of each type is largely determined by the type of activity for which the muscle is specialized. Accordingly, a high proportion of slow-oxidative fibers are found in muscles specialized for maintaining low-intensity contractions for long periods without fatigue, such as the muscles of the back and legs that support the body's weight against the force of gravity. A preponderance of fast-glycolytic fibers are found in the arm muscles, which are adapted for performing rapid, forceful movements such as lifting heavy objects.

The percentage of these various fibers not only differs among muscles within an individual but also varies considerably among individuals. Athletes genetically endowed with a higher percentage of the fast-glycolytic fibers are good candidates for power and sprint events, whereas those with a greater proportion of slow-oxidative fibers are more likely to succeed in endurance activities such as marathon races.

Of course, success in any event depends on many factors other than genetic endowment, such as the extent and type of training and the level of dedication. Indeed, the mechanical and metabolic capabilities of muscle fibers can change a lot in response to the patterns of demands placed on them. Let us see how.

Muscle fibers adapt considerably in response to the demands placed on them.

Different types of exercise produce different patterns of neuronal discharge to the muscle involved. Depending on the pattern of neural activity, long-term adaptive changes occur in the muscle fibers, enabling them to respond most efficiently to the types of demands placed on the muscle. Two types of changes can be induced in muscle fibers: changes in their oxidative capacity and changes in their diameter.

Improvement in Oxidative Capacity Regular aerobic endurance exercise, such as long-distance jogging or swimming, promotes metabolic changes within the oxidative fibers, which are the ones primarily recruited during aerobic exercise. For example, the number of mitochondria and the number of capillaries supplying blood to these fibers both increase. Muscles so adapted can use O_2 more efficiently and therefore can better endure prolonged activity without fatiguing. However, they do not change in size.

Muscle Hypertrophy The actual size of the muscles can be increased by regular bouts of anaerobic, short-duration, high-intensity resistance training, such as weight lifting. The resulting muscle enlargement comes primarily from an increase in diameter (**hypertrophy**) of the fast-glycolytic fibers called into play during such powerful contractions. Most fiber thickening results from increased synthesis of myosin and actin filaments, which permits a greater opportunity for cross-bridge interaction and consequently increases the muscle's contractile strength. The mechanical stress that resistance training exerts on a muscle fiber triggers signaling proteins, which turn on genes that direct the synthesis of more myosin and actin. Vigorous weight training can double or triple a muscle's size. The resultant bulging muscles are better adapted to activities that require intense strength for brief periods, but endurance has not been improved.

Influence of Testosterone Men's muscle fibers are thicker, and accordingly, their muscles are larger and stronger than those of women, even without weight training, because of the actions of testosterone, a steroid hormone secreted primarily in males. Testosterone promotes the synthesis and assembly of myosin and actin. This fact has led some athletes, both males and females, to the dangerous practice of taking this or closely related steroids to increase their athletic performance. (To explore this topic further, see the boxed feature (on pp. 276–277, A Closer Look at Exercise Physiology.)

Interconversion Between Fast Muscle Types All the muscle fibers within a single motor unit are of the same fiber type. This pattern usually is established early in life, but the two types of fast-twitch fibers are interconvertible, depending on training efforts—that is, fast-glycolytic fibers can be converted to fast-oxidative fibers, and vice versa, depending on the types of demands repetitively placed on them. Adaptive changes in skeletal muscle gradually reverse to their original state over a period of months if the regular exercise program that induced these changes is discontinued.

Slow and fast fibers are not interconvertible, however. Although training can induce changes in muscle fibers' metabolic support systems, whether a fiber is fast or slow twitch depends on the fiber's nerve supply. Slow-twitch fibers are supplied by motor neurons that exhibit a low-frequency pattern of electrical activity, whereas fast-twitch fibers are innervated by motor neurons that display intermittent rapid bursts of electrical activity. Experimental switching of motor neurons supplying slow muscle fibers with those supplying fast fibers gradually reverses the speed at which these fibers contract.

 Muscle Atrophy At the other extreme, if a muscle is not used, its actin and myosin content decreases, its fibers become smaller, and the muscle accordingly **atrophies** (decreases in mass) and becomes weaker. Muscle atrophy can take place in three ways. (1) **Disuse atrophy** occurs when a muscle is not used for a long period even though the nerve supply is intact, as when a cast or brace must be worn or during prolonged bed confinement. (2) **Denervation atrophy** occurs after the nerve supply to a muscle is lost. If the muscle is stimulated electrically until innervation can be reestablished, such as during regeneration of a severed peripheral nerve, atrophy can be diminished but not entirely prevented. Contractile activity itself obviously plays an important role in preventing atrophy; however, poorly understood factors released from active nerve endings, perhaps packaged with the ACh vesicles, apparently contribute to the integrity and growth of muscle tissue. (3) **Age-related atrophy**, or **sarcopenia**, occurs naturally with aging. Beginning at approximately 40 years of age, people progressively lose motor neurons, particularly those that innervate the fast-glycolytic fiber types. As a result, a gradual loss of muscle mass, strength, and speed of muscle contraction occurs in aging individuals. Reduced rates of protein synthesis and lowered hormone levels (growth hormone, testosterone, and insulin-like growth factor-I) contribute to this loss of muscle mass. On average, most people lose about a quarter of a pound of muscle per year starting in their 40s. Although age-related muscle atrophy is inevitable, resistance training exercise and proper diet can slow the rate of development of sarcopenia.

 Limited Repair of Muscle When a muscle is damaged, limited repair is possible, even though muscle cells cannot divide mitotically to replace lost cells. A small population of inactive muscle-specific stem cells called **satellite cells** are located close to the muscle surface (see p. 10). When a muscle fiber is damaged, locally released factors activate the satellite cells, which divide to give rise to myoblasts, the same undifferentiated cells that formed the muscle during embryonic development. A group of myoblasts fuse to form a large, multinucleated cell, which immediately begins to synthesize and assemble the intracellular machinery characteristic of the muscle, ultimately differentiating completely into a mature muscle fiber. With extensive injury, this limited mechanism is not adequate to completely replace all the lost fibers. In that case, the remaining fibers often hypertrophy to compensate.

Transplantation of satellite cells or myoblasts provides one of several glimmers of hope for victims of **muscular dystrophy**, a hereditary pathological condition characterized by progressive degeneration of contractile elements, which are ultimately replaced by fibrous tissue. (See the boxed feature on pp. 278–279, Concepts, Challenges, and Controversies, for further information on this devastating condition.)

We have now completed our discussion of all the determinants of whole-muscle tension in a skeletal muscle, which are summarized in Table 8-2. Next, we examine the central and local mechanisms involved in regulating the motor activity performed by these muscles.

Check Your Understanding 8.4

1. State four functions related to muscle excitation, contraction, and relaxation that require ATP.

2. Describe the differences between muscle fibers in a turkey drumstick (slow-oxidative) and muscles fibers in turkey breast meat (fast-glycolytic).

3. Discuss the relative contributions of creatine phosphate, glycolysis, and oxidative phosphorylation to the production of ATP during running of a marathon.

TABLE 8-2 Determinants of Whole-Muscle Tension in Skeletal Muscle

Number of Fibers Contracting

Number of motor units recruited*

Number of muscle fibers per motor unit

Number of muscle fibers available to contract (size of muscle)

Tension Developed by Each Contracting Fiber

Frequency of stimulation (twitch summation and tetanus)*

Length of fiber at onset of contraction (length–tension relationship)

Extent of fatigue

Type of fiber (fatigue-resistant oxidative or fatigue-prone glycolytic)

Thickness of fiber (strength training and testosterone)

*Factors controlled to accomplish gradation of contraction.

A Closer Look at Exercise Physiology

Are Athletes Who Use Steroids to Gain Competitive Advantage Really Winners or Losers?

ATHLETES INVOLVED IN ELITE SPORTS or competitions such as the Olympics are tested for use of performance-enhancing drugs and those found to be using substances outlawed by sports federations are banned from participating or lose awards earned in the event. One such group of drugs is **anabolic androgenic steroids** (*anabolic* means "buildup of tissues," *androgenic* means "male producing," and *steroids* are a class of hormone). These agents are closely related to testosterone, the natural male sex hormone, which is responsible for promoting the increased muscle mass characteristic of males.

Although their use is outlawed (possessing anabolic steroids without a prescription became a federal offense in 1991), these agents are taken by many athletes who specialize in power events such as weight lifting and sprinting in the hopes of increasing muscle mass and, accordingly, muscle strength. Both male and female athletes have resorted to using these substances in an attempt to gain a competitive edge. Bodybuilders also take anabolic steroids. Furthermore, these performance enhancers are widely used in professional sports. There are an estimated 1 million anabolic steroid abusers in the United States. Compounding the problem, underground chemists have created new synthetic performance-enhancing steroids undetectable by standard drug tests.

Studies have confirmed that steroids can increase muscle mass when used in large amounts and coupled with heavy exercise. The adverse effects of these drugs, however, outweigh any benefits derived.

Adverse Effects on the Reproductive System

In males, testosterone secretion and sperm production by the testes are normally controlled by hormones from the anterior pituitary gland. In negative-feedback fashion, testosterone inhibits secretion of these controlling hormones so that a constant level of testosterone is maintained. The anterior pituitary is similarly inhibited by androgenic steroids taken as a drug. Because the testes do not receive their normal stimulatory input from the anterior pituitary, testosterone secretion and sperm production decrease and the testes shrink. This hormone abuse also may set the stage for testicular and prostate cancer.

In females, who normally lack potent androgenic hormones, anabolic steroid drugs not only promote "male-type" muscle mass and strength but also "masculinize" the users in other ways, such as by inducing growth of facial hair and by deepening the voice. Furthermore, inhibition of the anterior pituitary by androgenic drugs suppresses the hormonal output that controls ovarian function. The result is failure to ovulate, menstrual irregularities, and decreased secretion of "feminizing" female sex hormones. Their decline diminishes breast size and other female characteristics.

Adverse Effects on the Cardiovascular System

Use of anabolic steroids induces cardiovascular changes that increase the risk of developing atherosclerosis, which in turn is associated with an increased incidence of heart attacks and strokes (see p. 327).

8.5 Control of Motor Movement

The nervous system controls motor movement by activating motor neurons, the final common pathway for motor processing (see p. 242). Each activated motor neuron triggers contraction of all of the skeletal muscle fibers within its motor unit.

Motor activity can be classified as reflex, voluntary, or rhythmic.

Particular patterns of motor unit output govern motor activity, which can be divided into three broad, overlapping classes: somatic reflex responses, voluntary movements, and rhythmic activities. These movements differ in their complexity and in the nervous system level at which they are integrated.

■ **Somatic reflex responses** are automatic responses brought about by skeletal muscle contraction that take place without conscious effort (see p. 177). Somatic reflexes are the least complex type of purposeful motor movement. They include protective reflexes and postural reflexes. Examples of *protective reflexes* are withdrawing from a painful stimulus and coughing. *Postural reflexes* maintain the desired position of the head, trunk, and limbs and stabilize our balance against gravity and other external forces as we stand and move about. Examples of postural reflexes include the stretch reflex, which is the simplest reflex and involves involuntary contraction to counteract passive stretching of a muscle; and the more complicated vestibular reflexes, which bring the body and head in proper alignment when a person gets off balance. Postural reflexes are elicited in response to input from the following sources: (1) proprioceptors in the muscles and joints that provide information regarding the relative position and movement of the body and its different parts (see p. 187), (2) the vestibular apparatus in the inner ear, which detects changes in position and motion of the head (see p. 221), (3) touch receptors that monitor pressure on the skin from interaction with the environment (such as standing on

Adverse Effects on the Liver

Liver dysfunction is common with high steroid intake because the liver, which normally inactivates steroid hormones and prepares them for urinary excretion, is overloaded by the excess steroid intake. The incidence of liver cancer is also increased.

Adverse Effects on Behavior

Anabolic steroid use promotes aggressive, even hostile behavior—the so-called 'roid rages.

Addictive Effects

Another concern is addiction to anabolic steroids of some who abuse these drugs. This apparent tendency to become chemically dependent on steroids is alarming because the potential for adverse effects on health increases with long-term, heavy use—the kind of use that would be expected from someone hooked on the drug.

Thus, for health reasons, without even taking into account the legal and ethical issues, people should not use anabolic steroids. However, the problem appears to be worsening. Currently, the international black market for anabolic steroids is estimated at $1 billion per year.

Other Cheating Ways to Build Muscle Mass

Athletes seeking an artificial competitive edge have resorted to other illicit measures besides taking anabolic steroids, such as using the hormone *erythropoietin,* see p. 385) to promote production of extra O_2-carrying red blood cells or using *human growth hormone* to spur muscle buildup. More worrisome, scientists predict the next illicit frontier will be gene doping. **Gene doping** refers to gene therapy aimed at improving athletic performance, such as by promoting production of naturally occurring muscle-building chemicals (for example, *insulin-like growth factor-I*); by blocking production of *myostatin,* a natural body chemical that puts the brakes on muscle growth; or by boosting endurance by tinkering with the nuclear receptor *PPAR-δ,* which regulates genes involved in energy use, insulin action, and muscle metabolism. Because these chemicals occur naturally in the body, detection of gene doping will be a challenge.

More than 100 drugs are currently banned by the World Anti-Doping Agency (WADA). Furthermore, to keep pace in the race between regulators and athlete dopers, WADA introduced guidelines in 2009 for the **athlete biological passport (ABP)**, which is based on blood tests administered multiple times per year that look for physiological consequences of doping rather than for abused substances themselves. Any suspicious change from the athlete's baseline pattern raises the alarm for further testing. The ABP is used in addition to traditional testing for specific drugs. WADA continues to expand the ABP by striving to identify new biological markers to indirectly detect more types of doping.

---

the ground), and (4) the eyes, which provide visual input about body position with relation to surroundings. Somatic reflexes may be integrated in the spinal cord (spinal reflexes, for example the withdrawal reflex) or at the brain stem level (cranial reflexes, for example vestibular reflexes). Most somatic reflexes can be modulated by conscious input from the cerebral cortex. To exemplify, a diver can learn to override neck-righting reflexes to execute a complicated dive involving abnormal postural movements (see p. 277).

■ **Voluntary movements** are the most complex type of motor activity. They are goal-directed movements initiated and terminated at will and are integrated by the cerebral cortex. Examples range from simple acts such as picking up a cup of coffee to highly skilled movements such as gymnastics or playing a musical instrument. Unconscious postural adjustments are integrated as necessary with all voluntary motor tasks. With repetitive practice, learned voluntary movements like gymnastic or dance routines become almost reflexlike: The learned actions get stored as procedural memory in the cerebellum and their execution can be brought forth by unconscious brain levels without deliberate thought of each maneuver (see p. 162). "Muscle memory" is the term applied to having the body unconsciously reproduce a memorized voluntary motor routine.

■ **Rhythmic activities** are stereotypical movements repeated in a general pattern, like walking or chewing. The cerebral cortex consciously starts and stops rhythmic activities but the details of their execution are accomplished in reflexlike fashion by lower CNS levels without conscious effort. For example, networks of specific interconnected excitatory and inhibitory interneurons and their associated motor neurons in the spinal cord function as **central pattern generators**. Once activated, central pattern generators autonomously bring about rhythmic patterned outputs such as walking by commanding precisely timed alternate contraction and relaxation of multiple muscles of the legs in a cyclical, coordinated fashion to accomplish repetitive stepping movements. Output from these intrinsic oscillatory networks can be modulated by higher motor areas in the cortex, such as if you decide to pick up your walking speed or turn a corner.

Muscular Dystrophy: When One Small Step Is a Big Deal

HOPE OF TREATMENT IS ON the horizon for **muscular dystrophy (MD)**, a fatal muscle-wasting disease that primarily strikes boys and relentlessly leads to their death before age 30.

Symptoms

Muscular dystrophy encompasses more than 30 distinct hereditary pathological conditions, which have in common a progressive degeneration of contractile elements and their replacement by fibrous tissue. The gradual muscle wasting is characterized by progressive weakness over a period of years. Typically, a patient with MD begins to show symptoms of muscle weakness at about 2 to 3 years of age, becomes wheelchair bound when he is 10 to 12 years old, and dies within the next 15 or so years, either from respiratory failure when his respiratory muscles become too weak or from heart failure when his heart becomes too weak.

Cause

The disease is caused by a recessive genetic defect on the X sex chromosome, of which males have only one copy. (Males have XY sex chromosomes; females have XX sex chromosomes.) If a male inherits from his mother an X chromosome bearing the defective dystrophic gene, he is destined to develop the disease, which affects 1 out of every 3500 boys worldwide. To acquire the condition, females must inherit a dystrophic-carrying X gene from both parents, a much rarer occurrence.

The defective gene responsible for *Duchenne muscular dystrophy (DMD)*, the most common and most devastating form of the disease, was pinpointed in 1986. The gene normally produces **dystrophin**, a large protein that provides structural stability to the muscle cell's plasma membrane. Dystrophin is part of a complex of membrane-associated proteins that form a mechanical link between actin, a major component of the muscle cell's internal cytoskeleton, and the extracellular matrix, an external support network. This mechanical reinforcement of the plasma membrane enables the muscle cell to withstand the stresses and strains encountered during repeated cycles of contraction and stretching.

Dystrophic muscles are characterized by a lack of dystrophin. Although this protein represents only 0.002% of the total amount of skeletal muscle protein, its presence is crucial in maintaining the integrity of the muscle cell membrane. The absence of dystrophin permits a constant leakage of Ca^{2+} into the muscle cells. This Ca^{2+} activates proteases that harm the muscle fibers. The resultant ongoing damage leads to the muscle wasting and ultimate fibrosis that characterize the disorder.

With the discovery of the dystrophin gene and its deficiency in DMD came the hope that scientists could somehow replenish this missing protein in the muscles of the disease's young victims. Although the disease is still considered untreatable and fatal, several lines of research are being pursued vigorously to intervene in the relentless muscle loss.

Gene-Therapy Approach

One approach is a possible "gene fix." With gene therapy, healthy genes are usually delivered to the defective cells by means of viruses. Viruses operate by invading a body cell and micromanaging the cell's genetic machinery. In this way, the virus directs the host cell to synthesize the proteins needed for viral replication. With gene therapy, the desired gene is inserted into an incapacitated virus that cannot cause disease but can still enter the target cell and take over genetic commands.

One of the big challenges for gene therapy for DMD is the enormous size of the dystrophin gene. This gene, being more than 3 million base pairs long, is the largest gene ever found. It does not fit inside the viruses usually used to deliver genes to cells—they only have enough space for a gene one thousandth the size of the dystrophin gene. Therefore, researchers have created a minigene that is one thousandth of the size of the dystrophin gene but still contains the essential components for directing the synthesis of dystrophin. This

Multiple neural inputs influence motor unit output.

Control of any motor movement, regardless of its level of complexity, depends on converging input to the motor neurons of specific motor units. Three levels of input to the motor neurons control their output to the muscle fibers they innervate:

1. *Input from afferent neurons.* This input, usually through intervening interneurons, occurs at the level of the spinal cord. Afferent neuronal input to motor neurons, which is typically proprioceptive or painful in nature, elicits somatic spinal reflexes. Spinal reflexes initiated by input from afferent neurons are important in maintaining posture and in executing basic protective movements, such as the withdrawal reflex. Afferent information may also be transmitted via ascending pathways to higher cortical levels for perception, planning voluntary movements, and so on, but this pathway is above and beyond input to motor neurons in the spinal cord for reflex responses.

2. *Input from the primary motor cortex.* Fibers originating from neuronal cell bodies known as **pyramidal cells** within the primary motor cortex (see p. 148) descend directly without synaptic interruption to terminate on motor neurons (or

stripped-down minigene can fit inside the viral carrier. Injection of these agents has stopped and even reversed the progression of MD in experimental animals. Gene therapy clinical trials in humans have not been completed.

Cell-Transplant Approach

Another approach involves injecting cells that can functionally rescue the dystrophic muscle tissue. *Myoblasts* are undifferentiated cells that fuse to form the large, multinucleated skeletal muscle cells during embryonic development. After development, a small group of stem cells known as *satellite cells* remain close to the muscle surface. Satellite cells can be activated to form myoblasts, which can fuse together to form a new skeletal muscle cell to replace a damaged cell. When the loss of muscle cells is extensive, however, as in MD, this limited mechanism is not adequate to replace all the lost fibers.

One therapeutic approach for MD under study involves the transplantation of dystrophin-producing myoblasts harvested from muscle biopsies of healthy donors into the patient's dwindling muscles. Other researchers are pinning their hopes on delivery of satellite cells or partially differentiated adult stem cells that can be converted into healthy muscle cells (see p. 10).

Utrophin Approach

An alternative strategy that holds considerable promise for treating MD is upregulation of **utrophin**, a naturally occurring protein in muscle that is closely related to dystrophin. Eighty percent of the amino acid sequence for dystrophin and utrophin is identical, but these two proteins normally have different functions. Whereas dystrophin is dispersed throughout the muscle cell's surface membrane, where it contributes to the membrane's structural stability, utrophin is concentrated at the motor end plate. Here, utrophin plays a role in anchoring the acetylcholine receptors.

When researchers genetically engineered dystrophin-deficient mice that produced extra amounts of utrophin, this utrophin upregulation compensated in large part for the absent dystrophin—that is, the additional utrophin dispersed throughout the muscle cell membrane, where it assumed dystrophin's responsibilities. The result was improved intracellular Ca^{2+} homeostasis, enhanced muscle strength, and a marked reduction in the microscopic signs of muscle degeneration. Researchers are now scrambling to find a drug that will entice muscle cells to overproduce utrophin in humans, in the hopes of preventing or even repairing the muscle wasting that characterizes this devastating condition.

RNA "Bandage" Approach

The most recent promising approach is circumventing the portions of messenger RNA that carry the faulty message from mutated DNA that leads to failure to synthesize dystrophin. When specially prepared nucleotide snippets that bind to the parts of messenger RNA transcribed from the defective parts of DNA are injected intramuscularly, the snippets cover up the faulty parts of the message. These "bandaged" parts of the modified messenger RNA are skipped over during translation when the protein is being synthesized. The result is a modified version of the dystrophin protein that is shortened but often still functional. The investigators are encouraged by animal studies and early studies in humans that suggest dystrophin is being produced in muscles following injection with these RNA bandages.

Anti-Myostatin Approach

Still other groups are exploring different tactics, such as interventions with newly designed drugs that increase the size of dwindling muscle fibers to counter the functional decline of dystrophic muscles. As an example, scientists have learned that **myostatin**, a protein produced in muscle cells, normally inhibits skeletal muscle growth in a check-and-balance fashion. They are working on ways to inhibit this inhibitor in patients with MD, thereby stimulating muscle growth.

These steps toward an eventual treatment mean that hopefully one day the boys affected by MD will be able to take steps on their own instead of being destined to wheelchairs and early death.

on local interneurons that terminate on motor neurons) in the spinal cord. These fibers make up the **corticospinal (or pyramidal) motor system**.

3. *Input from the brain stem as part of the multineuronal motor system.* The pathways composing the **multineuronal (or extrapyramidal or brain stem) motor system** include synapses that involve many regions of the brain (*extra* means "outside of"; *pyramidal* refers to the pyramidal system). The final link in multineuronal pathways is the brain stem (see p. 166), which in turn is influenced by motor regions of the cortex, the cerebellum, and the basal nuclei. In addition, the primary motor cortex itself is interconnected with the thalamus, as well as with premotor and supplementary motor areas; these are all part of the multineuronal system.

The corticospinal system plays a key role in mediating performance of fine, discrete, voluntary movements of the hands and fingers, such as those required to send a text message. Premotor and supplementary motor areas, with input from the cerebrocerebellum, plan the voluntary motor command that is issued to the appropriate motor neurons by the primary motor cortex through this descending system. The multineuronal system, in contrast, primarily regulates overall body posture involving involuntary movements of large muscle

CNS Region	Involvement in Motor Control
Motor neurons* in the spinal cord	Receive direct input from afferent neurons, brain stem, and primary motor cortex
	Send output to skeletal muscle fibers
	Serve as final common pathway; are efferent component for somatic spinal reflexes, voluntary motor activities, and rhythmic activities
Brain stem	Receives input from motor cortex, cerebellum, basal nuclei, and afferent neurons
	Serves as final link in multineuronal motor system
	Directly influences motor neurons in spinal cord
	Origin of cranial nerves that supply motor fibers to head; mediates cranial somatic reflexes and rhythmic activities
	Shares information with cerebellum; routes sensory information to thalamus
	Regulates postural reflexes and coordinates eye and head movements
Primary motor cortex	Receives input from higher motor areas (premotor cortex and supplementary motor area) and from thalamus
	Fibers from pyramidal cells make up corticospinal motor system
	Directly influences motor neurons
	Triggers voluntary movement
	Provides output to brain stem, cerebellum, basal nuclei, and thalamus for motor control
Higher motor areas (premotor cortex and supplementary motor area)	Receive input from cerebellum and sensory areas of cortex
	Are part of multineuronal motor system; do not directly influence motor neurons; act through primary motor cortex
	Important for planning and coordination of complex movement sequences
	Send voluntary motor command to primary motor cortex for execution
Cerebellum	Shares information with brain stem; monitors motor commands from motor cortex and sensory feedback from muscle
	Does not directly influence motor neurons; modulates corticospinal and multineuronal motor systems by acting on higher cortical motor areas and brain stem
	Improves accuracy of rapidly changing movements by comparing cortical command with muscle execution and makes adjustments as needed; important in balance; enhances muscle tone
	Plans skilled muscle activity
	Serves as site for procedural ("muscle sense") memory
Basal nuclei	Receive input from cortical motor and sensory areas
	Do not act directly on motor neurons; modulate multineuronal and corticospinal motor systems by acting through the cortex, brain stem, and thalamus
	Inhibit muscle tone and useless patterns of movement; coordinate slow, sustained movement; contribute to motor planning
Thalamus	Receives motor-related input from primary motor cortex and basal nuclei; receives sensory input from brain stem
	Part of multineuronal motor system; does not act directly on motor neurons; is part of a complex feedback loop linking motor cortex, basal nuclei, and thalamus
	Positively reinforces voluntary motor activity initiated by cortex; basal nuclei modulate this activity by inhibiting thalamus to eliminate unnecessary movements

*Includes interneurons associated with the motor neurons

groups of the trunk and limbs. The corticospinal and multineuronal systems show considerable complex interaction and overlapping of function. To voluntarily manipulate your thumbs to text message, for example, you subconsciously assume a particular posture of your arms that lets you hold your phone in the proper position.

The only brain regions that directly influence motor neurons are the primary motor cortex and brain stem; the other involved brain regions indirectly regulate motor activity by adjusting motor output from the motor cortex and brain stem. Numerous complex interactions take place among these various brain regions; the most important are presented in ▮Table 8-3, which summarizes the major involvement of different regions of the CNS in motor control. (See Chapter 5 for further discussion of the specific roles of these regions.)

Muscle tone refers to an ongoing, involuntary, low-level state of tension in a muscle even at rest. Skeletal muscle tone is important in maintaining postural stability. Without it, you would be floppy, like a rag doll. The skeleton itself cannot hold your body in position. Tension in muscles that are attached to the bones holds the bones in place. Skeletal muscle tone is due to both elastic properties of the muscle that resist passive stretching and continuous, minimal stimulation by motor neurons that produces a constant state of partial muscle contraction. Muscle tone is regulated by postural reflexes and output from the multineuronal motor system, namely the spinocerebellum, basal nuclei, and brain stem. Of course more forceful purposeful contractions above tonic level can be commanded at any time.

Clinical Note Some inputs converging on motor neurons are excitatory, whereas others are inhibitory. Coordinated movement depends on an appropriate balance of activity in these inputs. The following types of motor abnormalities result from defective motor control:

■ Loss of descending inhibitory inputs on motor neurons may result in **spastic paralysis**, a condition characterized by increased muscle tone (rigidity) and augmented limb reflexes.

■ In contrast, loss of excitatory input from higher centers brings about **flaccid paralysis**. In this condition, the muscles are limp with reduced or no muscle tone and the person cannot voluntarily contract them, although spinal reflex activity is still present. Damage to the primary motor cortex on one side of the brain, as with a stroke, leads to flaccid paralysis on the opposite half of the body (**hemiplegia**, or paralysis of one side of the body). Disruption of all descending pathways, as in traumatic severance of the spinal cord, produces flaccid paralysis below the level of the damaged region—**quadriplegia** (paralysis of all four limbs) in upper spinal cord damage and **paraplegia** (paralysis of the legs) in lower spinal cord injury.

■ Destruction of motor neurons—either their cell bodies or efferent fibers (as with amyotrophic lateral sclerosis [Lou Gehrig's disease]; see p. 242)—causes flaccid paralysis and lack of reflex responsiveness in the affected muscles.

■ Disorders of the cerebellum (such as cerebellar ataxia; *ataxia* means "without coordination") or basal nuclei (for example, Parkinson's disease; see p. 154) results not in paralysis but in uncoordinated, clumsy activity and inappropriate patterns of movement. These regions normally smooth out activity initiated voluntarily. Furthermore, cerebellar damage results in decreased muscle tone, whereas impairment of the basal nuclei leads to increased muscle tone.

■ Damage to higher cortical regions (premotor cortex or supplementary motor area) involved in planning motor activity (for instance brought about by traumatic brain injury or by surgical removal of a tumor in the vicinity) results in the inability to establish appropriate motor commands to accomplish desired goals.

Muscle receptors provide afferent information needed to control skeletal muscle activity.

To plan coordinated, purposeful skeletal muscle activity, the brain regions that direct motor output depend on afferent input from various sources. For example, if you are going to catch a ball, the motor systems must program sequential motor commands that move and position your body correctly for the catch, using predictions of the ball's direction and rate of movement provided by visual input. Many muscles acting simultaneously or alternately at different joints are called into play to shift your body's location and position rapidly while maintaining your balance. To appropriately program muscle activity, your CNS must know the starting position of your body. Furthermore, it must be constantly informed by proprioceptive input about the progression of movement it has initiated so that it can make adjustments as needed.

Proprioceptors are found in the joints and the muscles themselves. They provide information about the location of body parts relative to one another and about body movement. You can demonstrate your joint and muscle proprioceptors in action by closing your eyes and bringing the tips of your right and left index fingers together at any point in space. You can do so without seeing where your hands are because your brain is informed of the position of your hands and other body parts at all times by afferent input from the joint and muscle receptors.

For effective control of motor output, the CNS needs continual information regarding ongoing changes in muscle length and tension. Two types of muscle proprioceptors—*muscle spindles* and *Golgi tendon organs*—provide this input. Muscle length is monitored by muscle spindles; changes in muscle tension are detected by Golgi tendon organs. Both these receptor types are activated by muscle stretch, but they convey different types of information. Let us see how.

Muscle Spindle Structure **Muscle spindles**, which are distributed throughout the fleshy part of a skeletal muscle, consist of collections of specialized muscle fibers known as **intrafusal fibers**, which lie within spindle-shaped connective tissue capsules parallel to the "ordinary" **extrafusal fibers** (*fusus* means "spindle") (▮Figure 8-23a). Unlike an ordinary extrafusal skeletal muscle fiber, which contains contractile elements (myofibrils) throughout its entire length, an intrafusal fiber has a noncontractile central portion, with the contractile elements being limited to both ends.

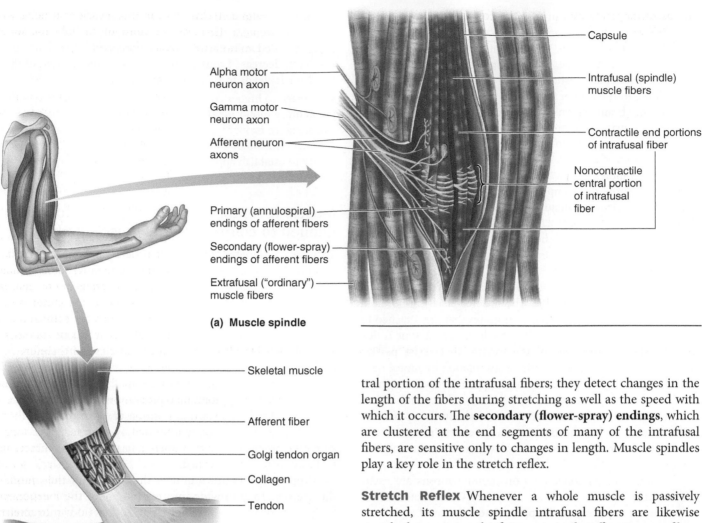

Alpha motor neuron axon

Gamma motor neuron axon

Afferent neuron axons

Primary (annulospiral) endings of afferent fibers

Secondary (flower-spray) endings of afferent fibers

Extrafusal ("ordinary") muscle fibers

Capsule

Intrafusal (spindle) muscle fibers

Contractile end portions of intrafusal fiber

Noncontractile central portion of intrafusal fiber

(a) Muscle spindle

Skeletal muscle

Afferent fiber

Golgi tendon organ

Collagen

Tendon

Bone

(b) Golgi tendon organ

▮Figure 8-23 **Muscle receptors.** (a) A muscle spindle consists of a collection of specialized intrafusal fibers that lie within a connective tissue capsule parallel to the ordinary extrafusal skeletal muscle fibers. The muscle spindle is innervated by its own gamma motor neuron and is supplied by two types of afferent sensory terminals, the primary (annulospiral) endings and the secondary (flower-spray) endings, both of which are activated by stretch. (b) The Golgi tendon organ is entwined with the collagen fibers in a tendon and monitors changes in muscle tension transmitted to the tendon.

Each muscle spindle has its own private efferent and afferent nerve supply. The efferent neuron that innervates a muscle spindle's intrafusal fibers is known as a **gamma motor neuron**, whereas the motor neurons that supply the extrafusal fibers are called **alpha motor neurons**. Two types of afferent sensory endings terminate on the intrafusal fibers and serve as muscle spindle receptors, both of which are activated by stretch. The **primary (annulospiral) endings** are wrapped around the cen-

tral portion of the intrafusal fibers; they detect changes in the length of the fibers during stretching as well as the speed with which it occurs. The **secondary (flower-spray) endings**, which are clustered at the end segments of many of the intrafusal fibers, are sensitive only to changes in length. Muscle spindles play a key role in the stretch reflex.

Stretch Reflex Whenever a whole muscle is passively stretched, its muscle spindle intrafusal fibers are likewise stretched, increasing the firing rate in the afferent nerve fibers whose sensory endings terminate on the stretched spindle fibers. The afferent neuron directly synapses on the alpha motor neuron that innervates the extrafusal fibers of the same muscle, resulting in contraction of that muscle (▮Figure 8-24a, ❶ and ❷). This monosynaptic (one-synapse) **stretch reflex** (see p. 178) serves as a local negative-feedback mechanism to sense and resist changes in muscle length when an additional load is applied.

The classic example of the stretch reflex is the **patellar tendon**, or **knee-jerk**, **reflex** (▮Figure 8-25). The extensor muscle of the knee is the *quadriceps femoris*, which forms the anterior (front) portion of the thigh and is attached just below the knee to the tibia (shinbone) by the *patellar tendon*. Tapping this tendon with a rubber mallet passively stretches the quadriceps muscle, activating its spindle receptors. The resulting stretch reflex brings about contraction of this extensor muscle, causing the knee to extend and raise the foreleg in the well-known knee-jerk fashion.

Clinical Note This test is routinely done as a preliminary assessment of nervous system function. A normal knee jerk indicates that a number of neural and muscular components—muscle spindle, afferent input, motor neurons, efferent output, neuromuscular junctions, and the muscles themselves—are functioning normally. It also indicates an appropriate balance of excitatory and inhibitory input to the motor neurons

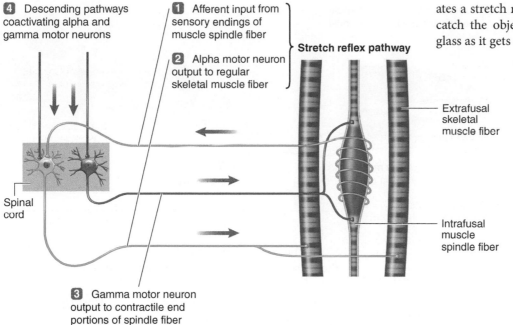

4 Descending pathways coactivating alpha and gamma motor neurons

1 Afferent input from sensory endings of muscle spindle fiber

Stretch reflex pathway

2 Alpha motor neuron output to regular skeletal muscle fiber

Extrafusal skeletal muscle fiber

Intrafusal muscle spindle fiber

Spinal cord

3 Gamma motor neuron output to contractile end portions of spindle fiber

(a) Pathways involved in monosynaptic stretch reflex and coactivation of alpha and gamma motor neurons

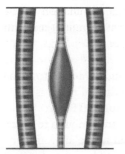

Relaxed muscle; spindle fiber sensitive to stretch of muscle

(b) Relaxed muscle

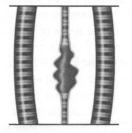

Contracted muscle in hypothetical situation of no spindle coactivation; slackened spindle fiber not sensitive to stretch of muscle

(c) Contracted muscle with no spindle coactivation

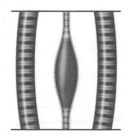

Contracted muscle in normal situation of spindle coactivation; contracted spindle fiber sensitive to stretch of muscle

(d) Contracted muscle with spindle coactivation

❙Figure 8-24 **Muscle spindle function**.

ates a stretch reflex in this muscle that helps you catch the object or continue to hold the water glass as it gets heavier during filling.

Coactivation of Gamma and Alpha Motor Neurons

Gamma motor neurons initiate contraction of the muscular end regions of intrafusal fibers (see ❙Figure 8-24a, 3). This contractile response is too weak to have any influence on whole-muscle tension, but it does have an important localized effect on the muscle spindle itself. If there were no compensating mechanisms, shortening of the whole muscle by alpha motor-neuron stimulation of extrafusal fibers would slacken the spindle fibers so that they would be less sensitive to stretch and therefore not as effective as muscle length detectors (❙Figure 8-24b and c). **Alpha-gamma coactivation** (simultaneous stimulation of the gamma motor-neuron system and the alpha motor-neuron system) during reflex and voluntary contractions (❙Figure 8-24a, 4) takes the slack out of the spindle fibers as the whole muscle shortens, letting these receptor structures maintain their high sensitivity to stretch over a wide range of muscle lengths. Gamma motor-neuron stimulation triggers simultaneous contraction of both contractile ends of the intrafusal fibers, causing tightening of their central (noncontractile) portions to remove slack in the muscle spindle (❙Figure 8-24d). Whereas the extent of alpha motor-neuron activation depends on the intended strength of the motor response, the extent of simultaneous gamma motor-neuron activity to the same muscle depends on the anticipated distance of shortening. When shortening of the overall muscle is less than expected (for example, the load is greater than anticipated), the muscle spindle receptors signal the alpha motor neurons to increase their rate of firing and, thereby, compensate for the additional load.

Golgi Tendon Organs In contrast to muscle spindles, which lie within the belly of the muscle, Golgi tendon organs are in the tendons of the muscle, where they can respond to changes in the muscle's tension rather than to changes in its length. Because a number of factors determine the tension developed in the whole muscle during contraction (for example, frequency of stimulation or length of the muscle at the onset of contraction), it is essential that motor control systems be apprised of

from higher brain levels. Muscle jerks may be absent or depressed with loss of higher-level excitatory inputs, or may be greatly exaggerated with loss of inhibitory input to the motor neurons from higher brain levels.

The stretch reflex is a postural reflex. The primary purpose of the patellar tendon reflex is to react to loads that tend to stretch the extensor muscles of the legs. Whenever your knee joint tends to buckle while standing, walking, running or jumping, your quadriceps muscle is stretched. The resulting enhanced contraction of this extensor muscle brought about by the stretch reflex quickly straightens out the knee, keeping your leg extended so that you remain upright. A similar stretch reflex involving the biceps muscle comes into play when you catch an object with your hand or fill a water glass that you are holding. The resultant stretch of the muscle spindle in your biceps initi-

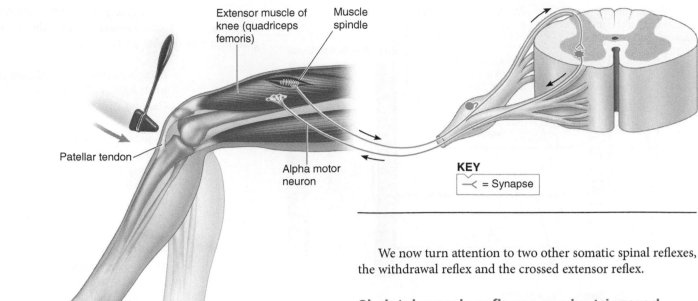

Extensor muscle of
knee (quadriceps
femoris)

Muscle
spindle

Patellar tendon

Alpha motor
neuron

KEY

≺ = Synapse

Figure 8-25 Patellar tendon reflex (a stretch reflex). Tapping the patellar tendon with a rubber mallet stretches the muscle spindles in the quadriceps femoris muscle. The resultant monosynaptic stretch reflex results in contraction of this extensor muscle, causing the characteristic knee-jerk response.

the tension actually achieved so that adjustments can be made if necessary.

A Golgi tendon organ consists of endings of an afferent fiber entwined within bundles of connective tissue (collagen) fibers that make up the tendon (see Figure 8-23b). When the extrafusal muscle fibers contract, the resulting pull on the tendon tightens the collagen bundles, which in turn increase the tension exerted on the bone to which the tendon is attached. In the process, the entwined Golgi organ afferent receptor endings are stretched, causing the afferent fibers to fire; the frequency of firing is directly related to the tension developed. This afferent information is sent to the brain for processing. Much of this information is used subconsciously for smoothly executing motor activity, but unlike afferent information from the muscle spindles, afferent information from the Golgi tendon organ reaches the level of conscious awareness. You are aware of the tension within a muscle but not of its length.

Scientists once thought the Golgi tendon organ triggered a protective spinal reflex that halted further contraction and brought about sudden reflex relaxation when the muscle tension became great enough, thus helping prevent damage to the muscle or tendon from excessive, tension-developing muscle contractions. Scientists now believe, however, that this receptor is a pure sensor and does not initiate any reflexes. Other unknown mechanisms are apparently involved in inhibiting further contraction to prevent tension-induced damage.

We now turn attention to two other somatic spinal reflexes, the withdrawal reflex and the crossed extensor reflex.

Skeletal muscle reflexes can be triggered by painful stimulation of the skin.

All other reflexes besides the monosynaptic stretch reflex are polysynaptic. Two common polysynaptic spinal reflexes can be elicited in response to painful cutaneous stimulation of an extremity: the withdrawal reflex (a protective reflex) and the crossed extensor reflex (a postural reflex).

Withdrawal Reflex When a person touches a hot stove (or receives another painful stimulus), a withdrawal reflex is initiated to withdraw from the painful stimulus (Figure 8-26). When a pain receptor in the skin is stimulated enough to reach threshold, an action potential is generated in the afferent neuron. Once the afferent neuron enters the spinal cord, it diverges to synapse with the following interneurons (the letters correspond to those in step ❸ of Figure 8-26).

1. An excited afferent neuron stimulates excitatory interneurons that in turn stimulate the efferent motor neurons supplying the biceps (**3a**). The resultant contraction of the biceps causes flexion of the elbow joint, which leads to withdrawal of the hand from the injurious stimulus.

2. The afferent neuron also stimulates inhibitory interneurons that inhibit the efferent neurons supplying the triceps (**3b**) to prevent it from contracting. When the biceps contracts to flex the elbow, it would be counterproductive for the triceps, which extends the elbow joint, to contract. Therefore, built into the withdrawal reflex is inhibition of the muscle that antagonizes (opposes) the desired response. This type of connection involving stimulation of the nerve supply to one muscle and simultaneous inhibition of the nerves to its antagonistic muscle is known as **reciprocal innervation**.

3. The afferent neuron stimulates still other interneurons that carry the signal up the spinal cord to the brain via an ascending pathway (**3c**). Only when the impulse reaches the sensory area of the cortex is the person aware of the pain, its location, and the type of stimulus. Also, when the impulse reaches the brain, the information can be stored as memory, and the person can start thinking about the situation—how it happened,

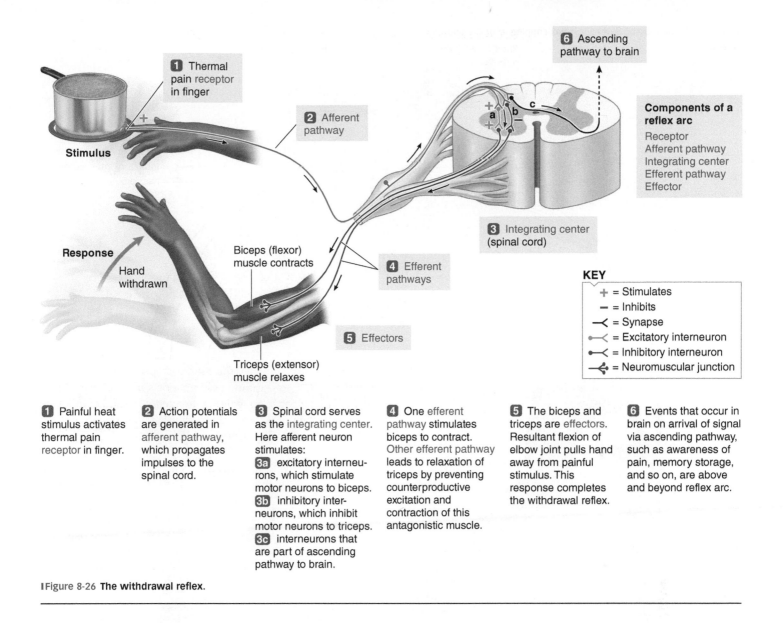

1 Thermal pain receptor in finger

Stimulus

2 Afferent pathway

6 Ascending pathway to brain

Components of a reflex arc
Receptor
Afferent pathway
Integrating center
Efferent pathway
Effector

3 Integrating center (spinal cord)

4 Efferent pathways

Response

Hand withdrawn

Biceps (flexor) muscle contracts

5 Effectors

Triceps (extensor) muscle relaxes

KEY
+ = Stimulates
− = Inhibits
< = Synapse
•—< = Excitatory interneuron
•—< = Inhibitory interneuron
•—<• = Neuromuscular junction

1 Painful heat stimulus activates thermal pain receptor in finger.

2 Action potentials are generated in afferent pathway, which propagates impulses to the spinal cord.

3 Spinal cord serves as the integrating center. Here afferent neuron stimulates:
3a excitatory interneurons, which stimulate motor neurons to biceps.
3b inhibitory interneurons, which inhibit motor neurons to triceps.
3c interneurons that are part of ascending pathway to brain.

4 One efferent pathway stimulates biceps to contract. Other efferent pathway leads to relaxation of triceps by preventing counterproductive excitation and contraction of this antagonistic muscle.

5 The biceps and triceps are effectors. Resultant flexion of elbow joint pulls hand away from painful stimulus. This response completes the withdrawal reflex.

6 Events that occur in brain on arrival of signal via ascending pathway, such as awareness of pain, memory storage, and so on, are above and beyond reflex arc.

Figure 8-26 **The withdrawal reflex.**

what to do about it, and so on. All this activity at the conscious level is beyond the basic reflex.

As with all spinal reflexes, the brain can modify the withdrawal reflex. Impulses may be sent down descending pathways to the efferent motor neurons supplying the involved muscles to override the input from the receptors, actually preventing the biceps from contracting despite the painful stimulus. When your finger is being pricked to obtain a blood sample, pain receptors are stimulated, initiating the withdrawal reflex. Knowing that you must be brave and not pull your hand away, you can consciously override the reflex by sending IPSPs via descending pathways to the motor neurons supplying the biceps and EPSPs to those supplying the triceps. The activity in these efferent neurons depends on the sum of activity of all their synaptic inputs. Because the neurons supplying your biceps are now receiving more IPSPs from your brain (voluntary) than EPSPs from the afferent pain pathway (reflex), these neurons are inhibited and do not reach threshold. Therefore, the biceps is not stimulated to contract and withdraw your hand. Simultaneously, the neu-

rons to your triceps are receiving more EPSPs from your brain than IPSPs via the reflex arc, so they reach threshold, fire, and consequently stimulate the triceps to contract. In this way, you voluntarily override the withdrawal reflex and keep your arm extended despite the painful stimulus.

Crossed Extensor Reflex Spinal reflex action is not necessarily limited to motor responses on the side of the body to which the stimulus is applied. Assume that a person steps on a tack instead of burning a finger. A reflex arc is initiated to withdraw the injured foot from the painful stimulus, while the opposite leg simultaneously prepares to suddenly bear all the weight so that the person does not lose balance or fall (Figure 8-27). Unimpeded bending of the injured extremity's knee is accomplished by concurrent reflex stimulation of the muscles that flex the knee and inhibition of the muscles that extend the knee. This response is a typical withdrawal reflex. At the same time, unimpeded extension of the opposite limb's knee is accomplished by activation of pathways that cross over to the

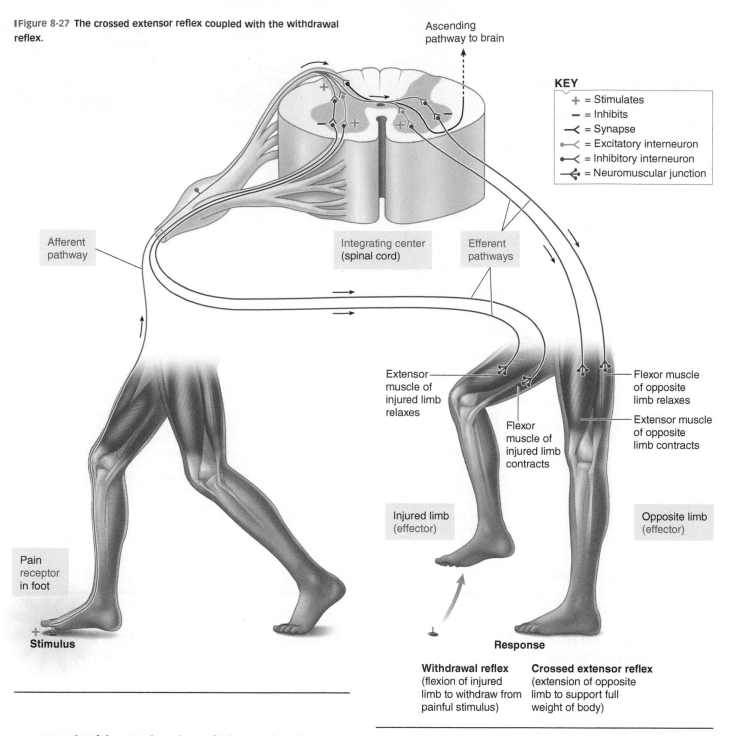

⎮Figure 8-27 The crossed extensor reflex coupled with the withdrawal reflex.

Ascending pathway to brain

KEY

+	= Stimulates
−	= Inhibits
⌐<	= Synapse
●—<	= Excitatory interneuron
●—<	= Inhibitory interneuron
⟜<	= Neuromuscular junction

Afferent pathway

Integrating center (spinal cord)

Efferent pathways

Extensor muscle of injured limb relaxes

Flexor muscle of injured limb contracts

Flexor muscle of opposite limb relaxes

Extensor muscle of opposite limb contracts

Injured limb (effector)

Opposite limb (effector)

Pain receptor in foot

Stimulus

Response

Withdrawal reflex (flexion of injured limb to withdraw from painful stimulus)

Crossed extensor reflex (extension of opposite limb to support full weight of body)

opposite side of the spinal cord to reflexly stimulate this knee's extensors and inhibit its flexors. This **crossed extensor reflex** is a postural reflex that ensures that the opposite limb is in a position to bear the weight of the body as the injured limb is withdrawn from the stimulus.

Having completed our discussion of skeletal muscle, we now examine smooth and cardiac muscle.

Check Your Understanding 8.5

1. Distinguish among the three classes of motor activity.
2. Describe the three levels of direct input to motor neurons.
3. Diagram the patellar tendon reflex.

8.6 Smooth and Cardiac Muscle

The two other types of muscle—smooth muscle and cardiac muscle—share some basic properties with skeletal muscle, but each also displays unique contractile characteristics (⎮Table 8-4). The three muscle types all have a specialized contractile apparatus made up of thin actin filaments that slide relative to stationary thick myosin filaments in response to a rise in cytosolic Ca^{2+} to accomplish contraction. Also, they all directly use ATP as the energy source for cross-bridge cycling. However, the

Characteristic	Skeletal Muscle	Multiunit Smooth Muscle	Single-Unit Smooth Muscle	Cardiac Muscle
Mechanism of contraction	Sliding filament mechanism	Sliding filament mechanism	Sliding filament mechanism	Sliding filament mechanism
Innervation	Somatic nervous system	Autonomic nervous system	Autonomic nervous system	Autonomic nervous system
Level of control	Under voluntary control; also subject to subconscious regulation	Under involuntary control	Under involuntary control	Under involuntary control
Initiation of contraction	Neurogenic	Neurogenic	Myogenic (pacemaker potentials and slow-wave potentials)	Myogenic (pacemaker potentials)
Role of nervous stimulation	Initiates contraction; accomplishes gradation	Initiates contraction; contributes to gradation	Modifies contraction; can excite or inhibit; contributes to gradation	Modifies contraction; can excite or inhibit; contributes to gradation
Modification by hormones	No	Yes	Yes	Yes
Presence of myosin and actin filaments	Yes	Yes	Yes	Yes
Presence of troponin and tropomyosin	Yes	Tropomyosin only	Tropomyosin only	Yes
Presence of T tubules	Yes	No	No	Yes
Development of sarcoplasmic reticulum	Well developed	Poorly developed	Poorly developed	Moderately developed
Source of increased cytosolic Ca^{2+}	Sarcoplasmic reticulum	ECF and sarcoplasmic reticulum	ECF and sarcoplasmic reticulum	ECF and sarcoplasmic reticulum
Mechanism of Ca^{2+} action to permit cross-bridge binding	Physically repositions troponin–tropomyosin complex in thin filaments to uncover actin cross-bridge binding sites	Chemically brings about phosphorylation of myosin cross bridges in thick filaments so that they can bind with actin	Chemically brings about phosphorylation of myosin cross bridges in thick filaments so that they can bind with actin	Physically repositions troponin–tropomyosin complex in thin filaments to uncover actin cross-bridge binding sites
Presence of gap junctions	No	Yes (very few)	Yes	Yes
Speed of contraction	Fast or slow, depending on type of fiber	Very slow	Very slow	Slow
Means by which gradation is accomplished	Varying number of motor units contracting (motor unit recruitment) and frequency at which they are stimulated (twitch summation)	Varying number of muscle fibers contracting and varying cytosolic Ca^{2+} concentration in each fiber by autonomic and hormonal influences	Varying cytosolic Ca^{2+} concentration through myogenic activity and influences of the autonomic nervous system, mechanical stretch, hormones, and local metabolites	Varying length of fibers (extent of filling of heart chambers) and varying cytosolic Ca^{2+} concentration through autonomic, hormonal, and local metabolite influences
Clear-cut length–tension relationship	Yes	No	No	Yes

structure and organization of fibers within these muscle types vary, as do their mechanisms of excitation and the means by which excitation and contraction are coupled. Furthermore, important distinctions occur in the contractile response itself. We spend the rest of this chapter highlighting unique features of smooth and cardiac muscle as compared with skeletal muscle, saving more detailed discussion of their function for chapters on organs containing these muscle types.

Smooth muscle cells are small and unstriated.

Most smooth muscle cells are found in the walls of hollow organs and tubes. Their contraction exerts pressure on and regulates forward movement of the contents of these structures.

Both smooth and skeletal muscle cells are elongated, but in contrast to their large, cylindrical skeletal muscle counterparts, smooth muscle cells are spindle shaped (tapered at both ends), have a single nucleus, and are considerably smaller (2 to 10 μm in diameter and 50 to 400 μm long). Also unlike skeletal muscle cells, a single smooth muscle cell does not extend the full length of a muscle. Instead, groups of smooth muscle cells are typically arranged in sheets (see Figure 8-1c).

A smooth muscle cell has three types of filaments: (1) thick myosin filaments, which are longer than those in skeletal muscle; (2) thin actin filaments, which contain tropomyosin but lack troponin; and (3) filaments of intermediate size, which do not directly participate in contraction but are part of the cytoskeletal framework that supports the cell shape. Smooth muscle filaments do not form myofibrils and are not arranged in the sarcomere pattern found in skeletal muscle. Thus, smooth muscle cells do not show the banding or striation of skeletal muscle, hence the term *smooth* for this muscle type.

Lacking sarcomeres, smooth muscle does not have Z lines, but it does have **dense bodies** containing the same protein constituent found in Z lines (Figure 8-28). Dense bodies are positioned throughout the smooth muscle cell, as well as attached to the internal surface of the plasma membrane. Dense bodies are held in place by a scaffold of intermediate filaments. The actin filaments are anchored to the dense bodies. Considerably more actin is present in smooth muscle cells than in skeletal muscle cells, with 10 to 15 thin filaments for each thick myosin filament in smooth muscle compared to 2 thin filaments for each thick filament in skeletal muscle.

The thick- and thin-filament contractile units are oriented slightly diagonally from side to side within the smooth muscle cell in an elongated, diamond-shaped lattice, rather than running parallel with the long axis as myofibrils do in skeletal muscle (Figure 8-29a). Relative sliding of the thin filaments past the thick filaments during contraction causes the filament lattice to shorten and expand from side to side. As a result, the whole cell shortens and bulges out between the points where the thin filaments are attached to the inner surface of the plasma membrane (Figure 8-29b).

Unlike in skeletal muscle, myosin molecules are arranged in a smooth-muscle thick filament so that cross bridges are present along the entire filament length (that is, there is no bare portion in the center of a smooth-muscle thick filament). As a result, the surrounding thin filaments can be pulled along the

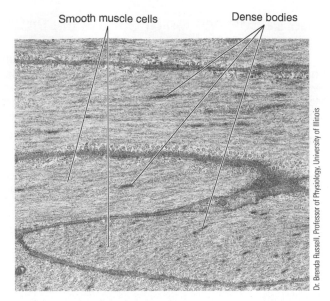

Smooth muscle cells   Dense bodies

Figure 8-28 **Electron micrograph of smooth muscle cells.** Note the presence of dense bodies and lack of banding.

Dr. Brenda Russell, Professor of Physiology, University of Illinois

thick filaments for longer distances than in skeletal muscle. Also dissimilar to skeletal muscle (in which all thin filaments surrounding a thick filament are pulled toward the center of the stationary thick filament), the myosin proteins in smooth-muscle thick filaments are organized so that half of the surrounding thin filaments are pulled toward one end of the stationary thick filament and the other half are pulled toward the opposite end (Figure 8-29b).

Smooth muscle cells are turned on by Ca^{2+}-dependent phosphorylation of myosin.

The thin filaments of smooth muscle cells do not contain troponin, and tropomyosin does not block actin's cross-bridge binding sites. What, then, prevents actin and myosin from binding at the cross bridges in the resting state, and how is cross-bridge activity switched on in the excited state? Lightweight chains of proteins are attached like "necklaces" to the heads of myosin molecules, near the "neck" region. These so-called **light chains** are only of secondary importance in skeletal muscle, but they have a crucial regulatory function in smooth muscle. Smooth muscle myosin can interact with actin only when the light chain is *phosphorylated* (that is, has an inorganic phosphate from ATP attached to it). During excitation, the increased cytosolic Ca^{2+} acts as an intracellular messenger, initiating a chain of biochemical events that results in phosphorylation of the myosin light chain (Figure 8-30). Smooth muscle Ca^{2+} binds with **calmodulin**, an intracellular protein found in most cells that is structurally similar to troponin (see p. 124). This Ca^{2+}–calmodulin complex binds to and activates another protein, **myosin light chain kinase (MLC kinase)**, which in turn phosphorylates the myosin light chain. This phosphate on the myosin light chain is in addition to the phosphate accompanying ADP on the myosin cross-bridge ATPase site during the energy-supplying cycle that powers cross-bridge bending. The P_i on the light chain per-

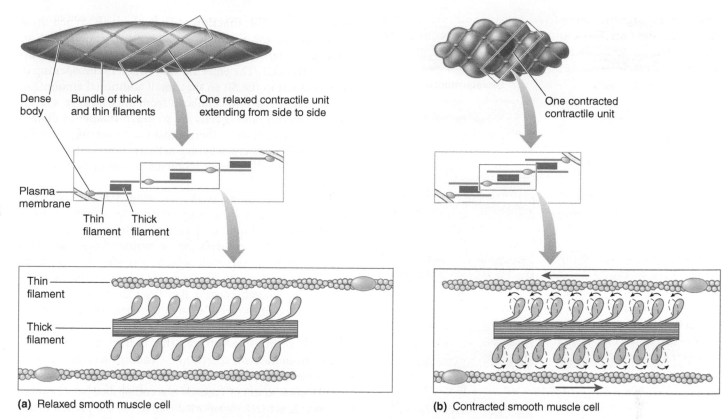

(a) Relaxed smooth muscle cell

(b) Contracted smooth muscle cell

❙Figure 8-29 **Arrangement of thick and thin filaments in a smooth muscle cell in relaxed and contracted states.**

mits the myosin cross bridge to bind with actin so that cross-bridge cycling can begin. Therefore, smooth muscle is triggered to contract by a rise in cytosolic Ca^{2+}, similar to what happens in skeletal muscle, but in smooth muscle, Ca^{2+} ultimately turns on the cross bridges by inducing a *chemical* change in myosin in the *thick* filaments (phosphorylation), whereas in skeletal muscle it exerts its effects by causing a *physical* change at the *thin* filaments (moving troponin and tropomyosin from their blocking positions) (❙Figure 8-31).

Phasic smooth muscle contracts in bursts; tonic smooth muscle maintains tone.

Smooth muscle can be grouped into two categories depending on its pattern of contractile activity and how its cytosolic Ca^{2+} concentration increases: *phasic smooth muscle* and *tonic smooth muscle*. **Phasic smooth muscle** contracts in bursts, triggered by action potentials that lead to increased cytosolic Ca^{2+}. These bursts in contraction are characterized by pronounced increases in contractile activity. Phasic smooth muscle is most abundant in the walls of hollow organs that push contents through them, such as digestive organs. Phasic digestive contractions mix food with digestive juices and propel the mass forward for further processing. **Tonic smooth muscle** is usually partially contracted at all times; that is, it exhibits smooth muscle tone. Tone exists because this type of smooth muscle has a relatively low resting potential of -55 to -40 mV. Some surface-membrane voltage-gated Ca^{2+} channels are open at these potentials. The resultant Ca^{2+} entry maintains a state of

❙Figure 8-30 **Calcium activation of myosin cross bridge in smooth muscle.**

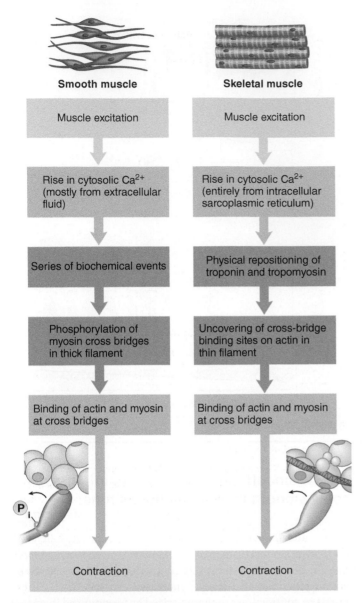

Smooth muscle

Muscle excitation

↓

Rise in cytosolic Ca²⁺ (mostly from extracellular fluid)

↓

Series of biochemical events

↓

Phosphorylation of myosin cross bridges in thick filament

↓

Binding of actin and myosin at cross bridges

↓

Contraction

Skeletal muscle

Muscle excitation

↓

Rise in cytosolic Ca²⁺ (entirely from intracellular sarcoplasmic reticulum)

↓

Physical repositioning of troponin and tropomyosin

↓

Uncovering of cross-bridge binding sites on actin in thin filament

↓

Binding of actin and myosin at cross bridges

↓

Contraction

❙Figure 8-31 **Comparison of the role of Ca2⁺ in bringing about contraction in smooth muscle and skeletal muscle.**

partial contraction, or tone, in the absence of action potentials. Tonic smooth muscle does not display bursts of contractile activity but instead incrementally varies its extent of contraction above or below this tonic level in response to regulatory factors, which alter the cytosolic Ca^{2+} concentration. The smooth muscle in walls of arterioles is an example of tonic smooth muscle. The ongoing tonic contraction in these small blood vessels squeezes down on the blood flowing through them and is one of the major contributing factors to maintenance of blood pressure.

A smooth muscle cell has no T tubules and a poorly developed SR. In phasic smooth muscle, the increased cytosolic Ca^{2+} that triggers contraction comes from two sources: Most Ca^{2+} enters from the extracellular fluid (ECF), but some is released intracellularly from the sparse SR stores. Unlike their role in skeletal muscle cells, voltage-sensitive dihydropyridine receptors in the plasma membrane of

smooth muscle cells function as Ca^{2+} channels. When these surface-membrane channels are opened in response to an action potential, Ca^{2+} enters down its concentration gradient from the ECF. The entering Ca^{2+} triggers the opening of Ca^{2+} channels in the SR so that small additional amounts of Ca^{2+} are released intracellularly from this meager source. Because smooth muscle cells are so much smaller in diameter than skeletal muscle fibers, most Ca^{2+} entering from the ECF can influence cross-bridge activity, even in the central portions of the cell, without requiring an elaborate T tubule–SR mechanism.

One of the major means of increasing cytosolic Ca^{2+} concentration and thus increasing contractile activity in tonic smooth muscle is binding of an extracellular chemical messenger, such as norepinephrine or various hormones, to a G-protein-coupled receptor, which activates the IP_3–Ca^{2+} second-messenger pathway (see p. 124). The membrane of the SR in tonic smooth muscle has IP_3 receptors, which like ryanodine receptors, are Ca^{2+}-release channels. IP_3 binding leads to release of contractile-inducing Ca^{2+} from this intracellular store into the cytosol. This is how norepinephrine released from the sympathetic nerve endings acts on arterioles to increase blood pressure.

Relaxation in smooth muscle is accomplished by removal of Ca^{2+} as it is actively transported out across the plasma membrane or back into the SR, depending on its source. When Ca^{2+} is removed, myosin is dephosphorylated (the phosphate is removed) and can no longer interact with actin, so the muscle relaxes.

We still have not addressed the question of how action potentials are initiated in smooth muscle. Smooth muscle is grouped in another way into two categories—*multiunit* and *single-unit smooth muscle*—based on differences in how the muscle fibers become excited. Let us compare them.

Multiunit smooth muscle is neurogenic.

Multiunit smooth muscle exhibits properties partway between those of skeletal muscle and those of single-unit smooth muscle. As the name implies, a multiunit smooth muscle consists of multiple discrete units that function independently of one another and must be separately stimulated by nerves to undergo action potentials and contract, similar to skeletal muscle motor units. Thus, contractile activity in both skeletal muscle and multiunit smooth muscle is **neurogenic** ("nerve produced"). That is, contraction in these muscle types is initiated only in response to stimulation by the nerves supplying the muscle. All multiunit smooth muscle is phasic, contracting only when neurally stimulated. Whereas skeletal muscle is innervated by the voluntary somatic nervous system (motor neurons), multiunit (as well as single-unit) smooth muscle is supplied by the involuntary autonomic nervous system.

Multiunit smooth muscle is found (1) in the walls of large blood vessels; (2) in small airways to the lungs; (3) in the muscle of the eye that adjusts the lens for near or far vision; (4) in the iris of the eye, which alters the pupil size to adjust the amount of light entering the eye; and (5) at the base of hair follicles, contraction of which causes "goose bumps."

Single-unit smooth muscle cells form functional syncytia.

Most smooth muscle is **single-unit smooth muscle**, alternatively called **visceral smooth muscle**, because it is found in the walls of the hollow organs or viscera (for example, the digestive, reproductive, and urinary tracts and small blood vessels). The term *single-unit smooth muscle* derives from the muscle fibers that make up this type of muscle becoming excited and contracting as a single unit. The muscle fibers in single-unit smooth muscle are electrically linked by gap junctions (see p. 62). When an action potential occurs anywhere within a sheet of single-unit smooth muscle, it is quickly propagated via these special points of electrical contact throughout the entire group of interconnected cells, which then contract as a single, coordinated unit. Such a group of interconnected muscle cells that function electrically and mechanically as a unit is known as a **functional syncytium** (plural, *syncytia; syn* means "together"; *cyt* means "cell").

Thinking about the role of the uterus during labor can help you appreciate the significance of this arrangement. Muscle cells composing the uterine wall act as a functional syncytium. They repetitively become excited and contract as a unit during labor, exerting a series of coordinated "pushes" that eventually deliver the baby. Independent, uncoordinated contractions of individual muscle cells in the uterine wall could not exert the uniformly applied pressure needed to expel the baby.

Single-unit smooth muscle is myogenic.

Single-unit smooth muscle is **self-excitable**, so it does not require nervous stimulation for contraction. Single-unit smooth muscle may be of the phasic or tonic type. In phasic single-unit smooth muscle, clusters of specialized cells within a functional syncytium display spontaneous electrical activity; that is, they can undergo action potentials without any external stimulation. In contrast to the other excitable cells we have been discussing (such as neurons, skeletal muscle fibers, and multiunit smooth muscle), the self-excitable cells of phasic single-unit smooth muscle do not maintain a constant resting potential. Instead, their membrane potential inherently fluctuates without any influence by factors external to the cell. Two major types of spontaneous depolarizations displayed by self-excitable cells are *pacemaker potentials* and *slow-wave potentials*.

Pacemaker Potentials With **pacemaker potentials**, the membrane potential gradually depolarizes on its own because of shifts in passive ionic fluxes accompanying automatic changes in ion channel permeability (Figure 8-32a). When the membrane has depolarized to threshold, an action potential is initiated. After repolarizing, the membrane potential again depolarizes to threshold, cyclically continuing in this manner to repetitively self-generate action potentials.

Self-excitable smooth muscle pacemaker cells are specialized to initiate action potentials, but they are not equipped to contract. Only a few of all the cells in a functional syncytium are noncontractile, pacemaker cells. Most smooth muscle cells are specialized to contract but cannot self-initiate action potentials. However, once an action potential is initiated by a self-

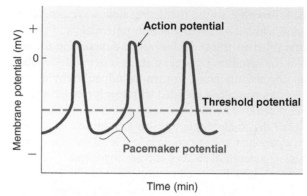

(a) Pacemaker potential

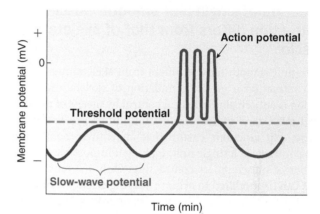

(b) Slow-wave potential

Figure 8-32 **Self-generated electrical activity in smooth muscle.** (a) With pacemaker potentials, the membrane gradually depolarizes to threshold on a regular periodic basis without any nervous stimulation. These regular depolarizations cyclically trigger self-induced action potentials. (b) In slow-wave potentials, the membrane gradually undergoes self-induced hyperpolarizing and depolarizing swings in potential. A burst of action potentials occurs if a depolarizing swing brings the membrane to threshold.

FIGURE FOCUS: *(1) A depolarizing pacemaker potential always initiates an action potential. (True or false?) (2) A depolarizing slow-wave potential always initiates an action potential. (True or false?)*

excitable pacemaker cell, it is conducted to the remaining contractile, nonpacemaker cells of the functional syncytium via gap junctions, so the entire group of connected cells contracts as a unit without any nervous input. Such nerve-independent contractile activity initiated by the muscle itself is called **myogenic activity** ("muscle-produced" activity), in contrast to the neurogenic activity of skeletal muscle and multiunit smooth muscle.

Slow-Wave Potentials **Slow-wave potentials** are spontaneous, gradually alternating depolarizing and hyperpolarizing swings in potential (Figure 8-32b) brought about by unknown means. They occur only in smooth muscle of the digestive tract. Slow-wave potentials are initiated by specialized clusters of nonmuscle pacemaker cells within the digestive tract wall and spread to the adjacent smooth muscle cells via gap junctions. If threshold is reached at the peak of a depolarizing swing, a burst of action potentials occurs. These action potentials bring about myogenically induced contraction. Threshold is not always reached, however, so the oscillating slow-wave potentials can continue without generating action potentials and contractile activity. Whether threshold is reached depends on the starting point of the membrane potential at the onset of its depolarizing swing. The starting point, in turn, is influenced by neural and local factors typically associated with meals (see Chapter 16 for further detail).

Recall that tonic single-unit smooth muscle cells have sufficient cytosolic Ca^{2+} to maintain low level tension even without action potentials, so they too are myogenic. (Thus, multiunit smooth muscles are all neurogenic and phasic; single-unit smooth muscles are all myogenic and may be phasic or tonic.)

Gradation of single-unit smooth muscle contraction differs from that of skeletal muscle.

Single-unit smooth muscle differs from skeletal muscle in the way contraction is graded. Gradation of skeletal muscle contraction is entirely under neural control by means of motor unit recruitment and twitch summation. In single-unit smooth muscle, gap junctions ensure that an entire smooth muscle sheet contracts as a single unit, making it impossible to vary the number of muscle fibers contracting. Only the tension of the fibers can be modified to achieve varying strengths of contraction of the whole organ. The portion of cross bridges activated and the tension subsequently developed in single-unit smooth muscle can be graded by varying the cytosolic Ca^{2+} concentration. A single excitation in smooth muscle does not cause all cross bridges to switch on, in contrast to skeletal muscles, where a single action potential triggers release of enough Ca^{2+} to permit all cross bridges to cycle. As Ca^{2+} concentration increases in smooth muscle, more cross bridges are brought into play and greater tension develops.

Modification of Smooth Muscle Activity by the Autonomic Nervous System Smooth muscle is typically innervated by both branches of the autonomic nervous system. In single-unit smooth muscle (both phasic and tonic), this nerve supply does not *initiate* contraction, but it can *modify* the rate and strength of contraction, either enhancing or retarding the inherent contractile activity of a given organ. Recall that the isolated motor end-plate region of a skeletal muscle fiber interacts with ACh released from a single axon terminal of a motor neuron. In contrast, the receptors that bind with autonomic neurotransmitters are dispersed throughout the entire surface membrane of a smooth muscle cell. Smooth muscle cells are sensitive to varying degrees and in varying ways to autonomic neurotransmitters, depending on the cells' distribution of cholinergic and adrenergic receptors (see pp. 239–240).

Each terminal branch of a postganglionic autonomic fiber travels across the surface of one or more smooth muscle cells, releasing neurotransmitter from the vesicles within its multiple varicosities (bulges) as an action potential passes along the terminal (Figure 8-33). The neurotransmitter diffuses to the many receptors specific for it on the cells underlying the terminal. Thus, in contrast to the discrete one-to-one relationship at motor end plates, a given smooth muscle cell can be influenced by more than one type of neurotransmitter, and each autonomic terminal can influence more than one smooth muscle cell.

Other Factors Influencing Smooth Muscle Activity

Other factors (besides autonomic neurotransmitters) can influence the rate and strength of both multiunit and single-unit smooth muscle contraction, including mechanical stretch, certain hormones, local metabolites, and specific drugs. The smooth muscle of digestive organs is also influenced by the enteric nervous system, which is a specialized network of nerve fibers built into the wall of the digestive tract (see pp. 135 and 572). Some smooth muscle is poorly innervated, an example being the uterus, in which case the rate and strength of contraction is regulated entirely by circulating and locally released chemical messengers, which vary with the stage of the menstrual cycle and with the stage of pregnancy. All these factors ultimately act by modifying the permeability of Ca^{2+} channels in the plasma membrane, the SR, or both, through a variety of mechanisms. Thus, smooth muscle is subject to more external influences than skeletal muscle is, even though smooth muscle can contract on its own and skeletal muscle cannot.

Next, as we look at the length–tension relationship in smooth muscle, we consider the effect of mechanical stretch (as occurs during filling of a hollow organ) on smooth muscle contractility. We examine the extracellular chemical influences (certain hormones and local metabolites) on smooth muscle

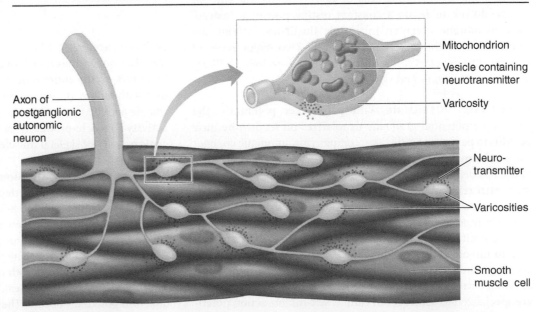

Axon of postganglionic autonomic neuron

Mitochondrion

Vesicle containing neurotransmitter

Varicosity

Neurotransmitter

Varicosities

Smooth muscle cell

Figure 8-33 **Innervation of smooth muscle by autonomic postganglionic nerve terminals.**

contractility in later chapters when we discuss regulation of the various organs that contain smooth muscle.

Smooth muscle can still develop tension yet inherently relaxes when stretched.

The relationship between muscle fiber length before contraction and the tension that can be developed on a subsequent contraction is less closely linked in smooth muscle than in skeletal muscle. The range of lengths over which a smooth muscle fiber can develop near-maximal tension is greater than the range for skeletal muscle. Smooth muscle can still develop considerable tension even when stretched up to 2.5 times its resting length, for two reasons. First, in contrast to skeletal muscle, in which the resting length is near its l_o, in smooth muscle the resting (nonstretched) length is much shorter than its l_o. Therefore, smooth muscle can be stretched considerably before reaching its optimal length. Second, the thin filaments still overlap the longer thick filaments even in the stretched-out position, so cross-bridge interaction and tension development can still take place. In contrast, when skeletal muscle is stretched only three-fourths longer than its resting length, the thick and thin filaments are completely pulled apart and can no longer interact (see Figure 8-20, p. 269).

The ability of a considerably stretched smooth muscle fiber to still develop tension is important, because the smooth muscle fibers within the wall of a hollow organ are progressively stretched as the volume of the organ's contents expands. Consider the urinary bladder as an example. Even though the muscle fibers in the urinary bladder are stretched as the bladder gradually fills with urine, they still maintain their tone and can even develop further tension in response to inputs that regulate bladder emptying. If considerable stretching prevented tension development, as in skeletal muscle, a filled bladder would not be capable of emptying.

Stress Relaxation Response When a smooth muscle is suddenly stretched, it initially increases its tension, much like the tension created in a stretched rubber band. The muscle quickly adjusts to this new length, however, and inherently relaxes to the tension level before the stretch, probably as a consequence of rearrangement of cross-bridge attachments. Smooth muscle cross bridges detach comparatively slowly. On sudden stretching, it is speculated that any attached cross bridges would strain against the stretch, contributing to a passive (not actively generated) increase in tension. As these cross bridges detach, the filaments would be permitted to slide into an unstrained stretched position, restoring the tension to its original level. This inherent property of smooth muscle is called the **stress relaxation response**.

Advantages of the Smooth Muscle Length–Tension Relationship These two responses of smooth muscle to being stretched—being able to develop tension even when considerably stretched and inherently relaxing when stretched—are highly advantageous. They enable smooth muscle to exist at a variety of lengths with little change in tension. As a result, a hollow organ enclosed by smooth muscle can accommodate variable volumes of contents with little change in the pressure exerted on the contents except when the contents are to be pushed out of the organ. At that time, the tension is deliberately increased by fiber shortening. Smooth muscle fibers can contract to half their normal length, enabling hollow organs to dramatically empty their contents on increased contractile activity; thus, smooth-muscled viscera can easily accommodate large volumes but can empty to practically zero volume. This length range in which smooth muscle normally functions (anywhere from 0.5 to 2.5 times the normal length) is greater than the limited length range within which skeletal muscle remains functional.

Smooth muscle contains a lot of connective tissue, which helps prevent a hollow organ from being overstretched. Unlike skeletal muscle, in which the skeletal attachments restrict how far the muscle can be stretched, this connective tissue puts an upper limit on how much a smooth-muscled hollow organ can hold.

Smooth muscle is slow and economical.

A smooth muscle contractile response proceeds more slowly than a skeletal muscle twitch. ATP splitting by myosin ATPase is much slower in smooth muscle, so cross-bridge activity and filament sliding occur about 10 times more slowly in smooth muscle than in skeletal muscle. A single smooth muscle contraction may last as long as 3 seconds (3000 msec), compared to the maximum of 100 msec for a single contractile response in skeletal muscle. Smooth muscle also relaxes more slowly because of slower Ca^{2+} removal. Slowness should not be equated with weakness, however. Smooth muscle can generate the same contractile tension per unit of cross-sectional area as skeletal muscle, but it does so more slowly and at considerably less energy expense. Because of slow cross-bridge cycling during smooth muscle contraction, cross bridges stay attached for more time during each cycle, compared with skeletal muscle; that is, the cross bridges "latch onto" the thin filaments for a longer time each cycle. This **latch phenomenon** enables smooth muscle to maintain tension with comparatively less ATP consumption because each cross-bridge cycle uses up one molecule of ATP. The duration of force maintained by a single cross-bridge interaction lasts about eight times longer in smooth muscle than in skeletal muscle. Smooth muscle is therefore an economical contractile tissue, making it well suited for long-term sustained contractions with little energy consumption and without fatigue. In contrast to the rapidly changing demands placed on your skeletal muscles as you maneuver through and manipulate your external environment, your smooth muscle activities are geared for long-term duration and slower adjustments to change. Because of its slowness and the less ordered arrangement of its filaments, smooth muscle has often been mistakenly viewed as a poorly developed version of skeletal muscle. Actually, smooth muscle is just as highly specialized for the demands placed on it. It is an extremely adaptive, efficient tissue.

Nutrient and O_2 delivery are generally adequate to support the smooth muscle contractile process. Smooth muscle can use a wide variety of nutrient molecules for ATP production. There are no energy storage pools comparable to creatine phosphate in smooth muscle; they are not necessary. Oxygen delivery is usually adequate to keep pace with the low rate of oxidative phosphorylation needed to provide ATP for the energy-efficient smooth muscle. If necessary, anaerobic glycolysis can sustain adequate ATP production if O_2 supplies are diminished.

Cardiac muscle blends features of both skeletal and smooth muscle.

Cardiac muscle, found only in the heart, shares structural and functional features with both skeletal and single-unit smooth muscle. Like skeletal muscle, cardiac muscle is striated, with its thick and thin filaments highly organized into a banding pattern. Cardiac thin filaments contain troponin and tropomyosin, which constitute the site of Ca^{2+} action in switching on cross-bridge activity, as in skeletal muscle. Also like skeletal muscle, cardiac muscle has a clear length–tension relationship. Like the oxidative skeletal muscle fibers, cardiac muscle cells have lots of mitochondria and myoglobin. They also have T tubules and a moderately well-developed SR.

Like smooth muscle, cardiac muscle fibers are slender and short (10 to 20 μm in diameter and 50 to 100 μm long). Like single-unit smooth muscle, the heart displays pacemaker (but not slow-wave) activity, initiating its action potentials without any external influence. Cardiac cells are interconnected by gap junctions found in *intercalated discs* that join cells together (see ▮Figure 8-1b). Gap junctions enhance the spread of action potentials throughout the heart, just as in single-unit smooth muscle. As in smooth muscle, Ca^{2+} enters the cytosol from both the ECF and the SR during cardiac excitation. Ca^{2+} entry from the ECF occurs through voltage-gated dihydropyridine receptors, which also act as Ca^{2+} channels in the T tubule membrane. This Ca^{2+} entry from the ECF triggers release of Ca^{2+} intracellularly from the SR. Also similarly, the heart is innervated by the autonomic nervous system, which, along with certain hormones and local factors, can modify the rate and strength of contraction.

Unique to cardiac muscle, cardiac fibers are joined in a branching network, and cardiac muscle action potentials last much longer before repolarizing. Further details of cardiac muscle's features are addressed in the next chapter.

Check Your Understanding 8.6

1. Compare the thick and thin filaments of skeletal muscle and smooth muscle.
2. Describe the differences between multiunit and single-unit smooth muscle.
3. Contrast the sources and roles of Ca^{2+} in skeletal, smooth, and cardiac muscle.

Homeostasis: Chapter in Perspective

 Skeletal muscles comprise the muscular system itself. Cardiac muscle and smooth muscle are part of organs that make up other body systems. Cardiac muscle is found only in the heart, which is part of the circulatory system. Smooth muscle is found in the walls of hollow organs and tubes, including blood vessels in the circulatory system, airways in the respiratory system, bladder in the urinary system, stomach and intestines in the digestive system, and tubular components of the reproductive system (an example being the uterus in females).

Contraction of skeletal muscles accomplishes movement of body parts in relation to one another and movement of the whole body in relation to the external environment. Thus, these muscles permit you to move through and manipulate your external environment. At a general level, some of these movements are aimed at maintaining homeostasis, such as moving the body toward food or away from harm. Examples of more specific homeostatic functions accomplished by skeletal muscles include chewing and swallowing food for further breakdown in the digestive system into usable energy-producing nutrient molecules (the mouth and throat muscles are all skeletal muscles), and breathing to obtain O_2 and get rid of CO_2 (the respiratory muscles are all skeletal muscles). Contracting skeletal muscles also are the major source of heat production in maintaining body temperature. Skeletal muscles further accomplish many nonhomeostatic activities that enable us to work and play—for example, operating a piece of equipment or riding a bicycle—so that we can contribute to society and enjoy ourselves.

All other systems of the body, except the immune (defense) system, depend on their nonskeletal muscle components to enable them to accomplish their homeostatic functions. For example, contraction of cardiac muscle in the heart pushes life-sustaining blood forward into the blood vessels, and contraction of smooth muscle in the stomach and intestines pushes ingested food through the digestive tract at a rate appropriate for the digestive juices secreted along the route to break down the food into usable units.

Review Exercises Answers begin on p. A-31

Reviewing Terms and Facts

1. When an action potential in a muscle fiber is completed, the contractile activity initiated by the action potential stops. (*True or false?*)

2. The velocity at which a muscle shortens depends entirely on the ATPase activity of its fibers. (*True or false?*)

3. When a skeletal muscle is maximally stretched, it can develop maximal tension on contraction because the actin filaments can slide in a maximal distance. (*True or false?*)

4. Smooth muscle can develop tension even when considerably stretched because the thin filaments still overlap with the long, thick filaments. (*True or false?*)

5. The muscle shortens in a(n) _____ contraction, whereas the muscle lengthens in a(n) _____ contraction.

6. _____ motor neurons supply extrafusal muscle fibers, whereas intrafusal fibers are innervated by _____ motor neurons.

7. The three types of atrophy are _____, _____, and _____.

8. An ongoing, involuntary, low-level state of muscle tension is known as _____.

9. A group of interconnected muscle cells that function electrically and mechanically as a unit because of the presence of gap junctions is called a _____.

10. Prolonged attachment of the cross bridge to actin during each cross-bridge cycle in smooth muscle is known as the _____.

11. Which of the following is *not* involved in bringing about muscle relaxation?

    a. reuptake of Ca^{2+} by the sarcoplasmic reticulum
    b. no more ATP
    c. no more action potential
    d. removal of ACh at the end plate by acetylcholinesterase
    e. filaments sliding back to their resting position

12. Match the following (with reference to skeletal muscle):

    1. Ca^{2+}
    2. T tubule
    3. ATP
    4. lateral sac of the sarcoplasmic reticulum
    5. myosin
    6. troponin–tropomyosin complex
    7. actin

    (a) cyclically binds with the myosin cross bridges during contraction
    (b) has ATPase activity
    (c) supplies energy for the power stroke of a cross bridge
    (d) rapidly transmits the action potential to the central portion of the muscle fiber
    (e) stores Ca^2
    (f) pulls the troponin–tropomyosin complex out of its blocking position
    (g) prevents actin from interacting with myosin when the muscle fiber is not excited

13. Indicate which of the following shorten during muscle contraction. (*Indicate all correct answers.*)

    a. thick filament
    b. thin filament
    c. A band
    d. I band
    e. H zone
    f. sarcomere

Understanding Concepts
(Answers at www.cengagebrain.com)

1. Describe the levels of organization in a skeletal muscle.

2. What produces the striated appearance of skeletal muscles? Describe or draw the arrangement of thick and thin filaments that gives rise to the banding pattern.

3. Explain what a functional unit is and name the functional unit of skeletal muscle.

4. Describe the composition of thick and thin filaments.

5. Describe the sliding filament mechanism of muscle contraction. How do cross-bridge power strokes bring about shortening of the muscle fiber?

6. Compare the excitation–contraction coupling process in skeletal muscle with that in smooth muscle.

7. How can gradation of skeletal muscle contraction be accomplished?

8. What is a motor unit? Compare the size of motor units in finely controlled muscles with those specialized for coarse, powerful contractions. Describe motor unit recruitment.

9. Explain twitch summation and tetanus.

10. How does a skeletal muscle fiber's length at the onset of contraction affect the strength of the subsequent contraction?

11. Compare isotonic, isokinetic, and isometric contractions.

12. Describe the role of each of the following in powering skeletal muscle contraction: ATP, creatine phosphate, oxidative phosphorylation, and glycolysis. Distinguish between aerobically and anaerobically supported exercises.

13. Compare the three types of skeletal muscle fibers.

14. What are the roles of the corticospinal system and multineuronal system in controlling motor movement?

15. Describe the structure and function of muscle spindles and Golgi tendon organs.

16. Contrast the arrangement and stroking pattern of cross bridges in smooth muscle with that in skeletal muscle.

17. What activates myosin light chain kinase in smooth muscle? What does activated myosin light chain kinase do?

18. Distinguish between phasic and tonic smooth muscle.

19. Distinguish between multiunit and single-unit smooth muscle.

20. Differentiate between neurogenic and myogenic muscle activity.

21. How can smooth muscle contraction be graded?

22. Compare the contractile speed and relative energy expenditure of skeletal muscle with that of smooth muscle.

23. In what ways is cardiac muscle functionally similar to skeletal muscle and to single-unit smooth muscle?

Solving Quantitative Exercises

1. Consider two individuals each throwing a baseball, one a weekend athlete and the other a professional pitcher.

   a. Given the following information, calculate the velocity of the ball as it leaves the amateur's hand:
   - The distance from his shoulder socket (humeral head) to the ball is 70 cm.
   - The distance from his humeral head to the points of insertion of the muscles moving his arm forward (we must simplify here because the shoulder is such a complex joint) is 9 cm.
   - The velocity of muscle shortening is 2.6 msec.

   b. The professional pitcher throws the ball 85 miles per hour. If his points of insertion are also 9 cm from the humeral head and the distance from his humeral head to the ball is 90 cm, how much faster did the professional pitcher's muscles shorten compared to the amateur's?

2. The velocity at which a muscle shortens is related to the force that it can generate in the following way:[1]

$$v = b(F_0 - F) / (F + a)$$

where v is the velocity of shortening, and F_0 can be thought of as an "upper load limit," or the maximum force a muscle can generate against a resistance. The parameter a is inversely proportional to the cross-bridge cycling rate, and b is proportional to the number of sarcomeres in line in a muscle. Draw the resistance (load)–velocity curve predicted by this equation by plotting the points $F = 0$ and $F = F_0$. Values of v are on the vertical axis; values of F are on the horizontal axis; a, b, and F_0 are constants.

   a. Notice that the curve generated from this equation is the same as that in Figure 8-16, p. 264. Why does the curve have this shape? That is, what does the shape of the curve tell you about muscle performance in general?

   b. What happens to the resistance (load)–velocity curve when F_0 is increased? When the cross-bridge cycling rate is increased? When the size of the muscle is increased? How will each of these changes affect the performance of the muscle?

[1] F. C. Hoppensteadt and C. S. Peskin, *Mathematics in Medicine and the Life Sciences* (New York: Springer, 1992), equation 9.1.1, p. 199.

Applying Clinical Reasoning

Jason W. is waiting impatiently for the doctor to finish removing the cast from his leg, which Jason broke the last day of school 6 weeks ago. Summer vacation is half over, and he hasn't been able to swim, play baseball, or ride his bike. When the cast is finally off, Jason's excitement gives way to concern when he sees that the injured limb is noticeably smaller in diameter than his normal leg. What explains this reduction in size? How can the leg be restored to its normal size and functional ability?

Thinking at a Higher Level

1. Why does regular aerobic exercise provide more cardiovascular benefit than weight training does? (*Hint:* The heart responds to the demands placed on it in a way similar to that of skeletal muscle.)

2. Put yourself in the position of the scientists who discovered the sliding filament mechanism of muscle contraction by considering what molecular changes must be involved to account for the observed alterations in the banding pattern during contraction. If you were comparing a relaxed and contracted muscle fiber under an electron microscope (see Figure 8-3, p. 254), how could you determine that the thin filaments do not change in length during muscle contraction? You cannot see or measure a single thin filament at this magnification. (*Hint:* What landmark in the banding pattern represents each end of the thin filament? If these landmarks are the same distance apart in a relaxed and contracted fiber, then the thin filaments must not change in length.) How could you determine that thick filaments remain the same length?

3. What type of off-the-snow training would you recommend for a competitive downhill skier versus a competitive cross-country skier? What adaptive skeletal muscle changes would you hope to accomplish in each case?

4. Explain how the rate of firing of the muscle spindle receptors (primary and secondary endings) would change if (a) the gamma motor neurons are activated, but the alpha motor neurons are not activated and (b) the gamma motor neurons are not activated, but the alpha motor neurons are activated.

5. When the bladder is filled and the micturition (urination) reflex is initiated, the nervous supply to the bladder promotes contraction of the bladder and relaxation of the external urethral sphincter, a ring of muscle that guards the exit from the bladder. If the time is inopportune for bladder emptying when the micturition reflex is initiated, the external urethral sphincter can be voluntarily tightened to prevent urination even though the bladder is contracting. Using your knowledge of the muscle types and their innervation, of what types of muscle are the bladder and the external urethral sphincter composed, and what branch of the efferent division of the peripheral nervous system supplies each of these muscles?

To access the course materials and companion resources for this text, please visit
www.cengagebrain.com

Cardiac Physiology

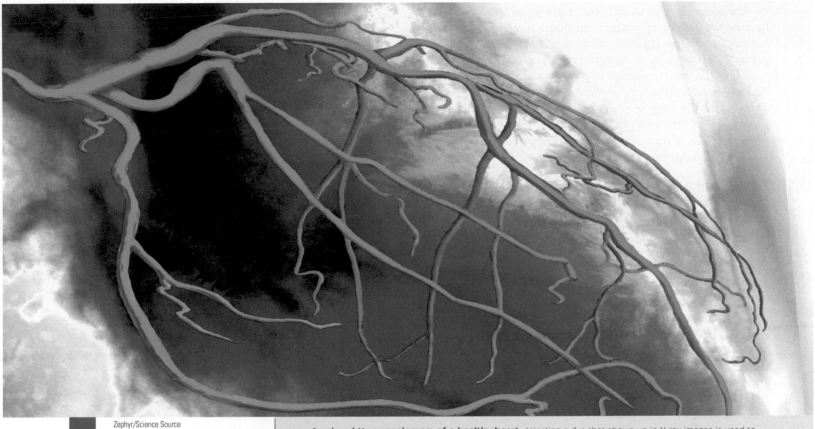

Zephyr/Science Source

A colored X-ray angiogram of a healthy heart. Injecting a dye that shows up in X-ray images is used to examine the blood vessels supplying the heart muscle. Shown here is a normal left coronary artery (*blue*) that supplies much of the heart.

CHAPTER AT A GLANCE

Homeostasis Highlights

To maintain homeostasis, essential materials such as O_2 and nutrients must continually be picked up from the external environment and delivered to the cells, and waste products must continually be removed. Furthermore, excess heat generated by muscles must be transported to the skin where it can be lost from the body surface to help maintain body temperature. Homeostasis also depends on the transfer of hormones, which are important regulatory chemical messengers, from their site of production to their site of action. The **circulatory system**, which contributes to homeostasis by serving as the body's transport system, consists of the heart, blood vessels, and blood.

All body tissues constantly depend on the life-supporting blood flow the **heart** provides them by contracting, or beating. The heart drives blood through the blood vessels for delivery to the tissues in sufficient amounts, whether the body is at rest or engaging in vigorous exercise.

9.1 Anatomy of the Heart

From just days following conception until death, the beat goes on. Throughout an average human life span, the heart contracts about 3 billion times, never stopping except for a fraction of a second to fill between beats. Within about 3 weeks after conception, the heart of the developing embryo starts to function. It is the first organ to become functional. At this time, the human embryo is only a few millimeters long, about the size of a capital letter on this page.

The heart develops so early and is so crucial throughout life because the circulatory system is the body's transport system. A human embryo, having little yolk available as food, depends on promptly establishing a circulatory system that can interact with the mother's circulation to pick up and distribute to the developing tissues the supplies needed for survival and growth. Thus begins the story of the circulatory system, which serves throughout life as a vital pipeline for transporting materials on which the cells of the body absolutely depend.

The **circulatory system** has three components:

1. The **heart** is the pump that imparts pressure to the blood to establish the pressure gradient needed for blood to flow to the tissues. Like all liquids, blood flows down a pressure gradient from an area of higher pressure to an area of lower pressure. This chapter focuses on cardiac physiology (*cardia* means "heart").

2. The **blood vessel**s are the passageways through which blood is directed and distributed from the heart to all parts of the body and subsequently returned to the heart. The smallest of the blood vessels are designed for rapid exchange of materials between the surrounding tissues and the blood within the vessels (see Chapter 10).

3. **Blood** is the transport medium within which materials being transported long distances in the body, such as O_2, CO_2, nutrients, wastes, electrolytes, and hormones, are dissolved or suspended (see Chapter 11).

Blood travels continuously through the circulatory system to and from the heart through two separate vascular (blood vessel) loops, both originating and terminating at the heart (Figure 9-1). The **pulmonary circulation** consists of a closed loop of vessels carrying blood between the heart and the lungs (*pulmo* means "lung"). The **systemic circulation** is a circuit of vessels carrying blood between the heart and all body systems (except for the air sacs of the lungs, which are supplied by the pulmonary circulation). Each of these vascular loops forms a figure "8." The pulmonary circulation simultaneously loops through the right lung and the left lung; the systemic circulation simultaneously loops through the upper half and the lower half of the body.

The heart is positioned in the middle of the thoracic cavity.

The heart is a hollow, muscular organ about the size of a clenched fist. It lies in the **thoracic** (chest) **cavity** about midline between the **sternum** (breastbone) anteriorly and the **vertebrae**

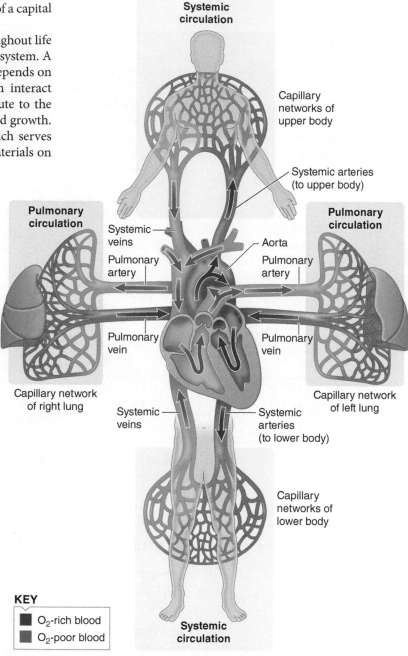

KEY

■ O_2-rich blood
■ O_2-poor blood

Figure 9-1 **Pulmonary and systemic circulations in relation to the heart.** The circulatory system consists of two separate vascular loops: the pulmonary circulation, which carries blood between the heart and lungs, and the systemic circulation, which carries blood between the heart and organ systems. Each of these loops forms a figure "8," with the pulmonary circulation simultaneously supplying the right and left lungs and the systemic circulation simultaneously supplying the upper body and lower body.

FIGURE FOCUS: *Examine this figure to correctly complete the following description of blood flow through the heart. The (right/left) side of the heart receives O_2-poor blood from the (systemic/pulmonary) circulation and pumps it into the (systemic/pulmonary) circulation. The (right/left) side of the heart receives O_2-rich blood from the (systemic/pulmonary) circulation and pumps it into the (systemic/pulmonary) circulation.*

(backbone) posteriorly. Place your hand over your heart. People usually put their hand on the left side of the chest, even though the heart is actually in the middle. The heart has a broad **base** at the top and tapers to a pointed tip, the **apex**, at the bottom. It is situated at an angle under the sternum so that its base lies predominantly to the right and the apex lies to the left of the sternum. When the heart beats forcefully, the apex thumps against the inside of the chest wall on the left. Because we are aware of the beating heart through the apex beat on the left, we tend to think—erroneously—that the entire heart is on the left. The heart's position between bony structures anteriorly and posteriorly makes it possible to manually drive blood out of the heart when it is not pumping effectively. Rhythmically depressing the sternum compresses the heart between the sternum and the vertebrae so that blood is squeezed out into the blood vessels, maintaining blood flow to the tissues. This *external cardiac compression,* which is part of **cardiopulmonary resuscitation (CPR),** may be lifesaving until appropriate therapy can restore the heart to normal function.

The heart is a dual pump.

Although anatomically the heart is a single organ, the right and left sides of the heart function as two separate pumps. The heart is divided into right and left halves and has four chambers, an upper and a lower chamber within each half (Figure 9-2a). The upper chambers, the **atria** (singular, *atrium*), receive blood returning to the heart and transfer it to the lower chambers, the **ven-**

tricles, which pump blood from the heart. The vessels that return blood from the tissues to the atria are **veins,** and those that carry blood away from the ventricles to the tissues are **arteries.** The two halves of the heart are separated by the **septum,** a continuous muscular partition that prevents blood mixing from the two sides of the heart. This separation is extremely important because the right side of the heart receives and pumps O_2-poor blood, whereas the left side of the heart receives and pumps O_2-rich blood.

The Complete Circuit of Blood Flow Let us look at how the heart functions as a dual pump by tracing a drop of blood through one complete circuit (Figure 9-2a). Blood returning from the systemic circulation enters the right atrium via two large veins, the **venae cavae,** one returning blood from above and the other returning blood from below heart level. The drop of blood entering the right atrium has returned from the body tissues, where O_2 has been taken from it and CO_2 has been added to it. This partially deoxygenated blood flows from the right atrium into the right ventricle, which pumps it out through the **pulmonary artery.** This artery immediately forms two branches, one going to each of the two lungs. Thus, *the right side of the heart receives deoxygenated blood from the systemic circulation and pumps it into the pulmonary circulation.*

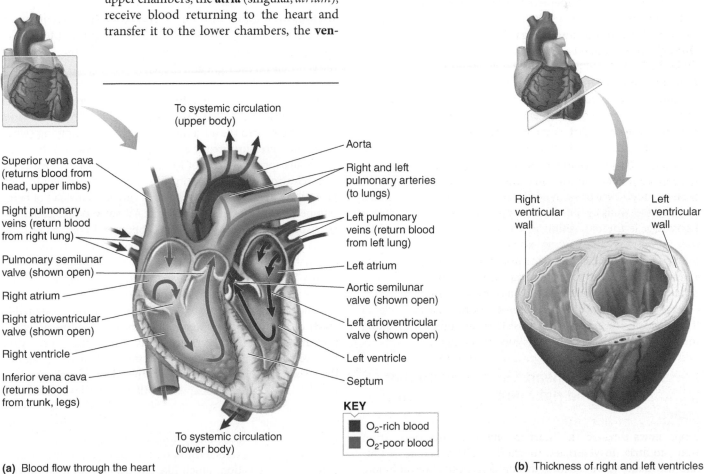

(a) Blood flow through the heart

To systemic circulation (upper body)

Superior vena cava (returns blood from head, upper limbs)

Right pulmonary veins (return blood from right lung)

Pulmonary semilunar valve (shown open)

Right atrium

Right atrioventricular valve (shown open)

Right ventricle

Inferior vena cava (returns blood from trunk, legs)

To systemic circulation (lower body)

Aorta

Right and left pulmonary arteries (to lungs)

Left pulmonary veins (return blood from left lung)

Left atrium

Aortic semilunar valve (shown open)

Left atrioventricular valve (shown open)

Left ventricle

Septum

KEY
- O_2-rich blood
- O_2-poor blood

(b) Thickness of right and left ventricles

Right ventricular wall

Left ventricular wall

Figure 9-2 **Blood flow through and pump action of the heart.** (a) The arrows indicate the direction of blood flow. To illustrate the direction of blood flow through the heart, all of the heart valves are shown open, which is never the case. (b) Note that the left ventricular wall is much thicker than the right wall.

Within the lungs, the drop of blood loses its extra CO_2 and picks up a fresh supply of O_2 by means of gas exchange with the air sacs before being returned to the left atrium via the **pulmonary veins** coming from both lungs. This O_2-rich blood returning to the left atrium subsequently flows into the left ventricle, the pumping chamber that propels the blood to the body systems—that is, *the left side of the heart receives oxygenated blood from the pulmonary circulation and pumps it into the systemic circulation.* The single large artery carrying blood away from the left ventricle is the **aorta**. Major arteries branch from the aorta to bring blood to the various organs.

In contrast to the pulmonary circulation, in which all the blood flows through the lungs, the systemic circulation may be viewed as a series of parallel pathways. Part of the blood pumped out by the left ventricle goes to the digestive system, part to the kidneys, part to the brain, part to the muscles, and so on (see ❙ Figure 10-1, p. 336). Even the heart muscle itself and the lung tissue other than the air sacs, namely the airways and lung connective tissue, receive blood from the left ventricle. Thus, the output of the left ventricle is distributed so that each part of the body receives a fresh blood supply. Accordingly, the drop of blood we are tracing goes to only one of the systemic organs. Tissue cells within the organ take O_2 from the blood and use it to oxidize nutrients for energy production; in the process, the tissue cells form CO_2 as a waste product that is added to the blood (see pp. 5 and 39). The drop of blood, now partially depleted of O_2 content and increased in CO_2 content, returns to the right side of the heart, which again pumps it to the lungs for gas exchange with the air sacs. One circuit is complete.

Comparison of the Right and Left Pumps Both sides of the heart simultaneously pump equal amounts of blood. The volume of O_2-poor blood being pumped to the lungs by the right side of the heart soon becomes the same volume of O_2-rich blood being delivered to the tissues by the left side of the heart. The pulmonary circulation is a low-pressure, low-resistance system, whereas the systemic circulation is a high-pressure, high-resistance system. Pressure is the force exerted on the vessel walls by the blood pumped into them by the heart. Resistance is the opposition to blood flow, largely caused by friction between the flowing blood and the vessel wall. Even though the right and left sides of the heart pump the same amount of blood, the left side works harder because it pumps an equal volume of blood at a higher pressure into a higher-resistance and longer system. Accordingly, the heart muscle on the left side is thicker than the muscle on the right side, making the left side a stronger pump (❙ Figure 9-2a and b).

Pressure-operated heart valves ensure that blood flows in the right direction through the heart.

Blood flows through the heart in one fixed direction—from veins, to atria, to ventricles, to arteries. The presence of four one-way **heart valves** ensures this unidirectional flow of blood. The valves are positioned so that they open and close passively because of pressure differences, similar to a one-way door

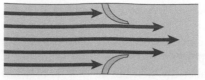

When pressure is greater behind the valve, it opens.

Valve opened

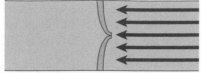

When pressure is greater in front of the valve, it closes. Note that when pressure is greater in front of the valve, it does not open in the opposite direction; that is, it is a one-way valve.

Valve closed; does not open in opposite direction

❙Figure 9-3 **Mechanism of valve action.**

(❙ Figure 9-3). A forward pressure gradient (that is, a greater pressure behind the valve) forces the valve open, much as you open a door by pushing on one side of it, whereas a backward pressure gradient (that is, a greater pressure in front of the valve) forces the valve closed, just as you apply pressure to the opposite side of the door to close it. Note that a backward gradient can force the valve closed but cannot force it to swing open in the opposite direction—that is, heart valves are not like swinging, saloon-type doors.

Atrioventricular Valves Between the Atria and Ventricles Two of the heart valves, the **right** and **left atrioventricular (AV) valves**, are positioned between the atrium and the ventricle on the right and the left sides, respectively (❙ Figure 9-4a). These valves let blood flow from the atria into the ventricles during ventricular filling (when atrial pressure exceeds ventricular pressure) but prevent the backflow of blood from the ventricles into the atria during ventricular emptying (when ventricular pressure greatly exceeds atrial pressure). If the rising ventricular pressure did not force the AV valves to close as the ventricles contracted to empty, much of the blood would inefficiently be forced back into the atria and veins instead of being pumped into the arteries. The right AV valve is also called the **tricuspid valve** (*tri* means "three") because it consists of three cusps or leaflets (❙ Figure 9-4b). Likewise, the left AV valve, which has two cusps, is often called the **bicuspid valve** (*bi* means "two") or the **mitral valve** (because of its physical resemblance to a mitre, or a bishop's traditional hat).

The edges of the AV valve leaflets are fastened by tough, thin cords of tendinous-type tissue, the **chordae tendineae**, which prevent the valve from *everting* (that is, from being forced by the high ventricular pressure to open in the opposite direction into the atria). These cords extend from the edges of each cusp and attach to small, nipple-shaped **papillary muscles**, which protrude from the inner surface of the ventricular walls (*papilla* means "nipple"). When the ventricles contract, the papillary muscles also contract, pulling downward on the chordae tendineae. This pulling exerts tension on the closed AV valve cusps to hold them in position, much like tethering ropes hold down a hot-air balloon. This action helps keep the valve tightly sealed in the face of a strong backward pressure gradient (❙ Figure 9-4c).

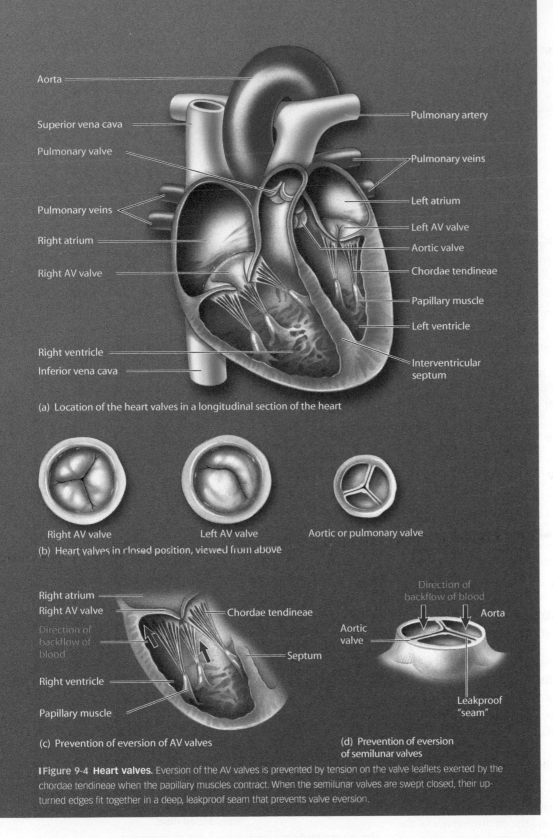

Aorta

Superior vena cava

Pulmonary valve

Pulmonary veins

Right atrium

Right AV valve

Right ventricle

Inferior vena cava

Pulmonary artery

Pulmonary veins

Left atrium

Left AV valve

Aortic valve

Chordae tendineae

Papillary muscle

Left ventricle

Interventricular septum

(a) Location of the heart valves in a longitudinal section of the heart

Right AV valve

Left AV valve

Aortic or pulmonary valve

(b) Heart valves in closed position, viewed from above

Right atrium
Right AV valve

Direction of backflow of blood

Right ventricle

Papillary muscle

Chordae tendineae

Septum

(c) Prevention of eversion of AV valves

Direction of backflow of blood

Aortic valve

Aorta

Leakproof "seam"

(d) Prevention of eversion of semilunar valves

Figure 9-4 Heart valves. Eversion of the AV valves is prevented by tension on the valve leaflets exerted by the chordae tendineae when the papillary muscles contract. When the semilunar valves are swept closed, their upturned edges fit together in a deep, leakproof seam that prevents valve eversion.

Semilunar Valves Between the Ventricles and Major Arteries The two remaining heart valves, the **aortic** and **pulmonary valves**, lie at the juncture where the major arteries leave the ventricles (**Figure 9-4a**). They are known as **semilunar valves** because they have three cusps, each resembling a shallow half-moon-shaped pocket (*semi* means "half"; *lunar* means "moon") (**Figure 9-4b**). These valves are forced open when the left and right ventricular pressures exceed the pres-

sure in the aorta and pulmonary artery, respectively, during ventricular contraction and emptying. Closure results when the ventricles relax and ventricular pressures fall below the aortic and pulmonary artery pressures. The closed valves prevent blood from flowing from the arteries back into the ventricles from which it has just been pumped.

The semilunar valves are prevented from everting by the anatomic structure and positioning of the cusps. When on ventricular relaxation a backward pressure gradient is created, the back surge of blood fills the pocketlike cusps and sweeps them into a closed position, with their unattached upturned edges fitting together in a deep, leakproof seam (**Figure 9-4d**).

No Valves Between the Atria and Veins Even though there are no valves between the atria and veins, backflow of blood from the atria into the veins usually is not a significant problem for two reasons: (1) Atrial pressures usually are not much higher than venous pressures, and (2) the sites where the venae cavae enter the atria are partially compressed during atrial contraction.

Fibrous Skeleton Surrounding the Valves Four interconnecting rings of dense connective tissue known as the **fibrous skeleton** of the heart surround and support the four heart valves, similar to the way the interconnected plastic rings hold together a six pack of beverage cans (**Figure 9-5**). The fibrous skeleton also separates the atria from the ventricles and provides a fairly rigid structure for attachment of the cardiac muscle. The atrial muscle mass is anchored above the rings, and the ventricular muscle mass is attached to the bottom of the rings.

It might seem surprising that the inlet valves to the ventricles (the AV valves) and the outlet valves from the ventricles (the semilunar valves) all lie on the same plane through the heart, as delineated by the fibrous skeleton. This relationship comes about because the heart forms from a single tube that

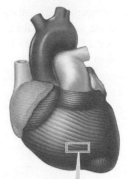

Figure 9-5 Fibrous skeleton of the heart. A view of the heart from above, with the atria and major vessels removed to show the heart valves and fibrous rings. Note that the inlet and outlet valves all lie on the same plane through the heart.

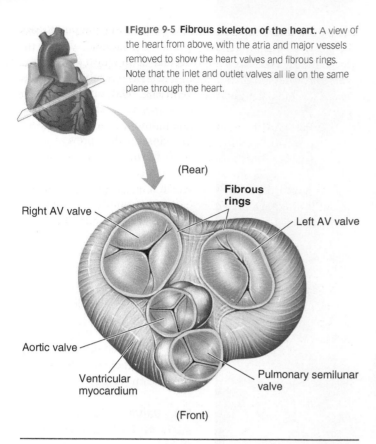

(Rear)

Fibrous rings

Right AV valve

Left AV valve

Aortic valve

Ventricular myocardium

Pulmonary semilunar valve

(Front)

bends on itself and twists on its axis during embryonic development. Another outcome of developmental twisting is that the cardiac muscle fibers run spirally around the blood-filled chambers, an arrangement that helps the heart pump blood more efficiently. Let us see how.

The heart walls are composed primarily of spirally arranged cardiac muscle fibers.

The heart wall has three distinct layers:

■ A thin, inner layer, the **endothelium**, a unique type of epithelial tissue that lines the entire circulatory system

■ A middle layer, the **myocardium**, which is composed of cardiac muscle and constitutes the bulk of the heart wall (*myo* means "muscle")

■ A thin, external layer, the **epicardium**, that covers the heart (*epi* means "on")

The myocardium consists of interlacing bundles of cardiac muscle fibers arranged spirally around the circumference of the heart (▌Figure 9-6a). As a result of this arrangement, when the ventricular muscle contracts and shortens, the diameter of the ventricular chambers is reduced while the apex is simultaneously pulled upward toward the top of the heart in a rotating manner. This exerts a "wringing" effect, efficiently exerting pressure on the blood within the enclosed chambers and directing it upward toward the openings of the major arteries that exit at the base of the ventricles.

(a) Bundles of cardiac muscle are arranged spirally around the ventricle. When they contract, they "wring" blood from the apex to the base where the major arteries exit.

Intercalated discs

Manfred Kage/Science Source

(b) Cardiac muscle fibers branch and are interconnected by intercalated discs.

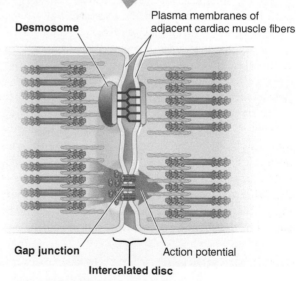

Desmosome

Plasma membranes of adjacent cardiac muscle fibers

Gap junction

Action potential

Intercalated disc

(c) Intercalated discs contain two types of membrane junctions: mechanically important desmosomes that hold the cardiac cells together and electrically important gap junctions that link the cells of each chamber into a functional syncytium.

▌Figure 9-6 **Organization of cardiac muscle fibers.**

To support their ongoing, rhythmic, contractile activity, cardiac muscle cells have an abundance of energy-generating mitochondria, and they receive a rich blood supply, which is delivered by about one capillary for each myocardial fiber.

Cardiac muscle fibers are interconnected by intercalated discs and form functional syncytia.

The individual cardiac muscle cells are interconnected to form branching fibers, with adjacent cells joined end to end at specialized structures called **intercalated discs**[1] (❚ Figure 9-6b). Two types of membrane junctions are present within an intercalated disc: desmosomes and gap junctions (❚ Figure 9-6c). A *desmosome,* a type of adhering junction that mechanically holds cells together, is particularly abundant in tissues such as the heart that are subject to considerable mechanical stress (see p. 61). At intervals along the intercalated disc, the opposing membranes approach each other closely to form *gap junctions,* which are areas of low electrical resistance that allow action potentials to spread from one cardiac cell to adjacent cells (see p. 62). Some specialized cardiac muscle cells can generate action potentials without any nervous stimulation. When one of the cardiac cells spontaneously undergoes an action potential, the electrical impulse spreads to all the other cells that are joined by gap junctions in the surrounding muscle mass so that they become excited and contract as a single, *functional syncytium* (see p. 291). The atria and the ventricles each form a functional syncytium and contract as separate units. The synchronous contraction of the muscle cells that make up the walls of each of these chambers produces the force needed to eject the enclosed blood.

No gap junctions join the atrial and ventricular contractile cells; furthermore, the atria and the ventricles are separated by the electrically nonconductive fibrous skeleton that surrounds and supports the valves. However, an important, specialized conduction system facilitates and coordinates transmission of electrical excitation from the atria to the ventricles to ensure synchronization between atrial and ventricular pumping.

Heart muscle performs an endocrine function in addition to pumping blood. The atria and ventricles each secrete a hormone involved in regulation of blood pressure. These related hormones both act on the kidneys to promote elimination of water-retaining salt into the urine. The resultant loss of water in the urine reduces blood volume and blood pressure accordingly. This hormonal action of cardiac muscle cells will be discussed further in a later chapter.

The heart is enclosed by the pericardial sac.

The heart is enclosed in the double-walled, membranous **pericardial sac** (*peri* means "around"). The sac consists of two layers—a tough, fibrous covering and a secretory lining. The outer fibrous covering of the sac attaches to the connective tis-

sue partition that separates the lungs. This attachment anchors the heart so that it remains properly positioned within the chest. The sac's secretory lining secretes a thin **pericardial fluid**, which provides lubrication to prevent friction between the pericardial layers as they glide over each other with every beat of the heart.

 Pericarditis, an inflammation of the pericardial sac that results in painful friction between the two pericardial layers, occurs occasionally because of viral or bacterial infection.

We next explain how action potentials are initiated and spread throughout the heart, followed by a discussion of how this electrical activity brings about coordinated pumping.

Check Your Understanding 9.1

1. Schematically draw the relationship of the pulmonary circulation and the systemic circulation to the chambers of the heart.

2. Name and discuss the functions of the four heart valves.

3. Describe an intercalated disc.

9.2 Electrical Activity of the Heart

Contraction of cardiac muscle cells to eject blood is triggered by action potentials sweeping across the muscle cell membranes. The heart contracts, or beats, rhythmically as a result of action potentials that it generates by itself, a property called **autorhythmicity**, or **automaticity** (*auto* means "self"). There are two specialized types of cardiac muscle cells:

1. **Contractile cells**, which are 99% of the cardiac muscle cells, do the mechanical work of pumping. These working cells normally do not initiate their action potentials.

2. **Autorhythmic cells**, the small but extremely important remainder of the cardiac cells, do not contract but instead are specialized for initiating and conducting the action potentials responsible for contraction of the working cells.

Cardiac autorhythmic cells display pacemaker activity.

In contrast to nerve and skeletal muscle cells, in which the membrane remains at constant resting potential unless the cell is stimulated, the cardiac autorhythmic cells do not have a resting potential. Instead, they display *pacemaker activity*—that is, their membrane potential slowly depolarizes, or drifts, between action potentials until threshold is reached, at which time the membrane fires or has an action potential. An autorhythmic cell membrane's slow drift to threshold is called the **pacemaker potential** (❚ Figure 9-7; see also p. 291). Through repeated cycles of drift and fire, these autorhythmic cells cyclically initiate action potentials, which then spread throughout the heart to trigger rhythmic beating without any nervous stimulation.

---

[1]Many sources use the spelling *disk* in *intercalated disc*. However, the Federative International Committee on Anatomical Terminology indicates that *disc* is the preferred spelling for anatomical references.

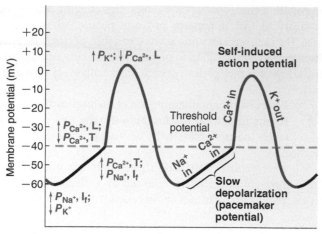

KEY

I_f	= Funny channels
T	= Transient-type Ca^{2+} channels
L	= Long-lasting Ca^{2+} channels

Figure 9-7 Pacemaker activity of cardiac autorhythmic cells. The first half of the pacemaker potential is the result of simultaneous opening of unique funny channels, which permits inward Na^+ current, and closure of K^+ channels, which reduces outward K^+ current. The second half of the pacemaker potential is the result of opening of T-type Ca^{2+} channels. Once threshold is reached, the rising phase of the action potential is the result of opening of L-type Ca^{2+} channels, whereas the falling phase is the result of opening of K^+ channels.

FIGURE FOCUS: *How would a low ECF K^+ concentration affect the rate of slow depolarization to threshold in an autorhythmic cell?*

Pacemaker Potential in Autorhythmic Cells Complex interactions of several different ionic mechanisms are responsible for the pacemaker potential. The most important changes in ion movement that give rise to the pacemaker potential are (1) an increased inward Na^+ current, (2) a decreased outward K^+ current, and (3) an increased inward Ca^{2+} current.

The initial phase of the slow depolarization to threshold is caused by net Na^+ entry through a type of voltage-gated channel found only in cardiac pacemaker cells. Typically, voltage-gated channels open when the membrane becomes less negative (depolarizes), but these unique channels open when the potential becomes more negative (hyperpolarizes) at the end of repolarization from the previous action potential. Because of their unusual behavior, they are called **funny, or I_f, channels.** When one action potential ends and the I_f channels open, the resultant depolarizing net inward Na^+ current through these open channels starts immediately moving the pacemaker cell's membrane potential toward threshold once again.

The second mechanism contributing to this pacemaker potential is a progressive reduction in the passive outward flux of K^+. In cardiac autorhythmic cells, permeability to K^+ does not remain constant between action potentials as it does in nerve and skeletal muscle cells. The K^+ channels that opened during the falling phase of the preceding action potential slowly close at negative potentials. This slow closure gradually diminishes the outflow of K^+ down its concentration gradient. The resultant slow decline in the rate of K^+ efflux occurring simul-

taneous with the slow inward leak of Na^+ through the open I_f channels further contributes to the early drift toward threshold.

The third ionic contribution to pacemaker potential is increased Ca^{2+} entry. In the second half of the pacemaker potential, the I_f channels close and transient Ca^{2+} channels (**T-type Ca^{2+} channels**), one of two types of voltage-gated Ca^{2+} channels, open before the membrane reaches threshold. ("T" stands for *transient*.) The resultant brief influx of Ca^{2+} further depolarizes the membrane, bringing it to threshold, at which time the transient Ca^{2+} channels close.

These permeability changes in surface membrane ion channels that cyclically bring the membrane of autorhythmic cells to threshold are collectively termed the **membrane clock mechanism**, which was long thought to be solely responsible for the pacemaker potential. Recent evidence, however, suggests that another clock, the **Ca^{2+} clock mechanism**, acts concurrently and independently of the membrane clock as a redundant means of self-depolarizing the membrane of autorhythmic cells to threshold. The Ca^{2+} clock depends on local Ca^{2+} recycling within a pacemaker cell. Spontaneous, rhythmic local releases of Ca^{2+} from the sarcoplasmic reticulum (SR) (see p. 258) increase the cytosolic Ca^{2+} concentration. Each time the cytosolic Ca^{2+} concentration rises, the *Na^+–Ca^{2+} exchanger (NCX)*, which is a plasma membrane antiport carrier (see p. 74), repetitively transports one intracellular Ca^{2+} ion out for every three extracellular Na^+ ions it moves in. This exchange leads to net inward movement of positive ions, a process that gradually depolarizes the autorhythmic cell to threshold. These two clocks act cooperatively to ensure periodic generation of the pacemaker potential that drives cardiac rhythmicity. This cooperative relationship between the membrane clock and Ca^{2+} clock is known as the **coupled-clock system**. The same regulatory mechanisms influence the "ticking speed" of both clocks so that they are synchronized to bring the membrane of pacemaker cells to threshold at a frequency befitting the body's momentary needs, such as speeding up the heart rate during exercise.

Next we see what happens when the pacemaker potential brings the membrane to threshold.

Action Potential in Autorhythmic Cells Once threshold is reached, the rising phase of the action potential occurs in response to activation of a long-lasting, voltage-gated Ca^{2+} channel (**L-type Ca^{2+} channel**; "L" standing for *long-lasting*) and a subsequently large influx of Ca^{2+}. The Ca^{2+}-induced rising phase of a cardiac pacemaker cell differs from that in nerve and skeletal muscle cells, where Na^+ influx rather than Ca^{2+} influx swings the potential in the positive direction.

The falling phase is the result, as usual, of the K^+ efflux that occurs when K^+ permeability increases on activation of voltage-gated K^+ channels, coupled with closure of the L-type Ca^{2+} channels. After the action potential is over, slow closure of these K^+ channels contributes to the next slow depolarization to threshold.

Ca^{2+} influx during the rising phase, combined with active reuptake of Ca^{2+} by the sarcoplasmic/endoplasmic reticulum Ca^{2+} ATPase pump (SERCA pump; see p. 261), refills the intracellular Ca^{2+} stores. The Na^+–K^+ pump returns to the extracellular fluid (ECF) the Na^+ that moved into the pacemaker cell during the early phase of slow depolarization as well as the Na^+ that enters via NCX. Simultaneously the pump moves back into the cell the K^+ that exited during the falling phase. These actions reset the Ca^{2+} clock and membrane clock for another cycle of slow depolarization to threshold and initiation of another self-induced action potential.

The sinoatrial node is the normal pacemaker of the heart.

The specialized noncontractile cardiac cells capable of autorhythmicity lie in the following specific sites (Figure 9-8):

1. The **sinoatrial node (SA node)**, a small, specialized region in the right atrial wall near the opening of the superior (upper) vena cava.

2. The **atrioventricular node (AV node)**, a small bundle of specialized cardiac muscle cells located at the base of the right atrium near the septum, just above the junction of the atria and ventricles.

3. The **bundle of His** (pronounced *Hiss*) or **atrioventricular bundle**, a tract of specialized cells that originates at the AV node and enters the septum between the ventricles. Here, it divides to form the right and left bundle branches that travel down the septum, curve around the tip of the ventricular chambers, and travel back toward the atria along the outer walls.

4. **Purkinje fibers**, small terminal fibers that extend from the bundle of His and spread throughout the ventricular myocardium, much like small twigs of a tree branch.

Normal Pacemaker Activity Because these various autorhythmic cells have different rates of slow depolarization to threshold, the rates at which they can generate action potentials also differ. The number of action potentials per minute each type of autorhythmic cell can generate under resting conditions is as follows:

- The rate for the SA node is 70–80
- The rate for the AV node is 40–60
- The rate for the Bundle of His and Purkinje fibers is 20–40

The heart cells with the fastest rate of action potential initiation are localized in the SA node. Once an action potential occurs in any cardiac muscle cell, it is propagated throughout the rest of the myocardium via gap junctions and the specialized conduction system. Therefore, the SA node, which normally has the fastest rate of autorhythmicity, at 70 to 80 action potentials per minute, drives the rest of the heart at this rate and thus is known as the **pacemaker** of the heart—that is, the entire heart becomes excited, triggering the contractile cells to contract and the heart to beat at the pace set by SA node autorhythmicity, normally at 70 to 80 beats per minute. The other autorhythmic tissues cannot assume their own naturally slower rates because they are activated by action potentials originating in the SA node before they can reach threshold at their slower rhythm.

The following analogy shows how the SA node drives the rest of the heart at its own pace. Suppose a train has 100 cars, 3 of which are engines capable of moving on their own; the other 97 cars must be pulled (Figure 9-9a). One engine (the SA node) can travel at 70 miles per hour (mph) on its own, another engine (the AV node) at 50 mph, and the last engine (the Purkinje fibers) at 30 mph. If all these cars are joined, the engine that travels at 70 mph pulls the rest of the cars at that speed. The engines that can travel at lower speeds on their own are pulled at a faster speed by the fastest engine and therefore cannot assume their slower rate as long as a faster engine drives them.

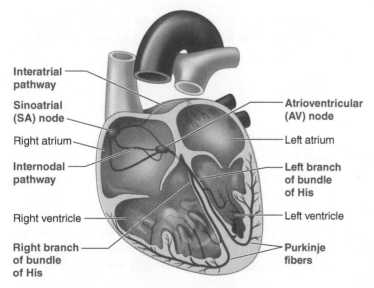

(a) Specialized conduction system of the heart

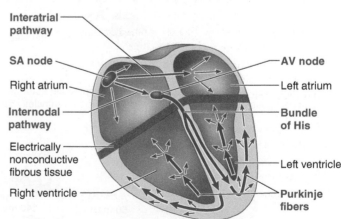

(b) Spread of cardiac excitation

Figure 9-8 **Specialized conduction system of the heart and spread of cardiac excitation.** An action potential initiated at the SA node first spreads throughout both atria. Two specialized atrial conduction pathways facilitate its spread: the interatrial and internodal pathways. The AV node is the only point where an action potential can spread from the atria to the ventricles. From the AV node, the action potential spreads rapidly throughout the ventricles, hastened by a specialized ventricular conduction system consisting of the bundle of His and Purkinje fibers.

The other 97 cars (nonautorhythmic, contractile cells), being unable to move on their own, likewise travel at the speed the fastest engine pulls them.

Abnormal Pacemaker Activity If for some reason the fastest engine breaks down (SA node damage), the next-fastest engine (AV node) takes over and the entire train travels at 50 mph—that is, if the SA node becomes nonfunctional such as from a heart attack, the AV node assumes pacemaker activity (Figure 9-9b). If impulse conduction becomes blocked between the atria and the ventricles, the atria continue at the typical rate of 70 beats per minute, and the ventricular tissue, not being driven by the faster SA nodal rate, assumes its own, slower autorhythmic rate of about 30 beats per minute, initiated by the Purkinje fibers. This situation is like a breakdown of the second engine (AV node) so that the lead engine (SA node) becomes disconnected from the slow third engine (Purkinje fibers) and the rest of the cars (Figure 9-9c). The lead engine (and cars connected directly to it—that is, the atrial cells) continues at 70 mph while the rest of the train proceeds at 30 mph. This **complete heart block** occurs when the conducting tissue between the atria and the ventricles is damaged and becomes nonfunctional. A ventricular rate of 30 beats per minute supports only a very sedentary existence; in fact, the patient usually becomes comatose.

When a person has an abnormally low heart rate, as in SA node failure or heart block, an **artificial pacemaker** can be used. Such an implanted device rhythmically generates impulses that spread throughout the heart to drive both the atria and the ventricles at the typical rate of 70 beats per minute.

Occasionally, an area of the heart, such as a Purkinje fiber, becomes overly excitable and depolarizes more rapidly than the

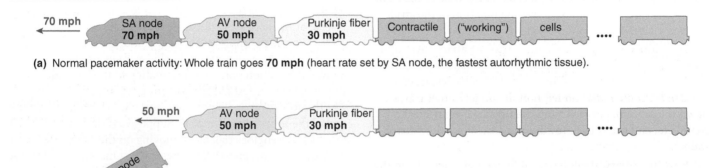

(a) Normal pacemaker activity: Whole train goes **70 mph** (heart rate set by SA node, the fastest autorhythmic tissue).

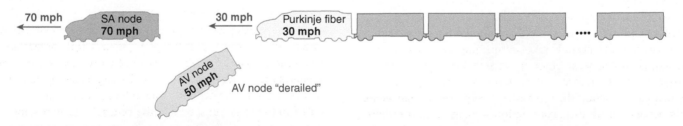

(b) Takeover of pacemaker activity by AV node when the SA node is nonfunctional: Train goes **50 mph** (the next fastest autorhythmic tissue, the AV node, sets the heart rate).

(c) Takeover of ventricular rate by the slower ventricular autorhythmic tissue in complete heart block: First part of train goes **70 mph**; last part goes **30 mph** (atria are driven by SA node; ventricles assume own, much slower rhythm).

(d) Takeover of pacemaker activity by an ectopic focus: Train is driven by ectopic focus, which is now going faster than the SA node (the whole heart is driven more rapidly by an abnormal pacemaker).

Figure 9-9 Analogy of pacemaker activity. In complete heart block (c), when ventricular rate is taken over by the slower ventricular autorhythmic tissue the atrial rate (not shown) is still driven by the SA node.

SA node. (The slow engine suddenly goes faster than the lead engine; Figure 9-9d). This abnormally excitable area, an **ectopic focus**, initiates a premature action potential that spreads throughout the rest of the heart before the SA node can initiate a normal action potential (*ectopic* means "out of place"). The resultant contraction is called a **premature ventricular contraction (PVC)**. If the ectopic focus continues to discharge at its more rapid rate, pacemaker activity shifts from the SA node to the ectopic focus. The heart rate abruptly becomes greatly accelerated and continues this rapid rate for a variable period until the ectopic focus returns to normal. Such overly irritable areas may be associated with heart disease, but more frequently they occur in response to anxiety; lack of sleep; or excess caffeine, nicotine, or alcohol consumption.

We now turn to how an action potential, once initiated, is conducted throughout the heart.

The spread of cardiac excitation is coordinated to ensure efficient pumping.

Once initiated in the SA node, an action potential spreads throughout the rest of the heart. For efficient cardiac function, the spread of excitation should satisfy three criteria:

1. *Atrial excitation and contraction should be complete before the onset of ventricular contraction.* Complete ventricular filling requires that atrial contraction precede ventricular contraction. During cardiac relaxation, the AV valves are open, so venous blood entering the atria continues to flow directly into the ventricles. Almost 80% of ventricular filling occurs by this means before atrial contraction. When the atria do contract, more blood is squeezed into the ventricles to complete ventricular filling. Ventricular contraction then occurs to eject blood from the heart into the arteries.

If the atria and ventricles were to contract simultaneously, the AV valves would close immediately because ventricular pressures would greatly exceed atrial pressures. The ventricles have thicker walls and, accordingly, can generate more pressure. Atrial contraction would be unproductive because the atria could not squeeze blood into the ventricles through closed valves. Therefore, to ensure complete filling of the ventricles—to obtain the remaining 20% of ventricular filling that occurs during atrial contraction—the atria must become excited and contract before ventricular excitation and contraction. During a normal heartbeat, atrial contraction occurs about 160 msec before ventricular contraction.

2. *Excitation of cardiac muscle fibers should be coordinated to ensure that each heart chamber contracts as a unit to pump efficiently.* If the muscle fibers in a heart chamber became excited and contracted randomly rather than contracting simultaneously in a coordinated fashion, they would be unable to eject blood. A smooth, uniform ventricular contraction is essential to squeeze out the blood. As an analogy, assume you have a basting syringe full of turkey pan juice. If you merely poke a finger here or there into the rubber bulb of the syringe, you will not eject much juice. However, if you compress the bulb in a smooth, coordinated fashion, you can squeeze out the juice to baste the turkey.

Clinical Note In a similar manner, contraction of isolated cardiac muscle fibers cannot successfully pump blood. Such random, uncoordinated excitation and contraction of cardiac cells is known as **fibrillation**. Ventricular fibrillation is more serious than atrial fibrillation. Ventricular fibrillation rapidly causes death because of the heart's inability to pump blood. This condition can often be corrected by *electrical defibrillation,* in which a strong electrical current is applied on the chest wall. When this current reaches the heart, it depolarizes all parts of the heart simultaneously and serves as a "reset button." Usually the first part of the heart to recover is the SA node, which takes over pacemaker activity, again initiating impulses that trigger the synchronized contraction of the rest of the heart.

3. *The pair of atria and pair of ventricles should be functionally coordinated so that both members of the pair contract simultaneously.* This coordination permits synchronized pumping of blood into the pulmonary and systemic circulations.

The normal spread of cardiac excitation is carefully orchestrated to ensure that these criteria are met and the heart functions efficiently, as follows (see Figure 9-8b).

Atrial Excitation An action potential originating in the SA node first spreads throughout both atria, primarily from cell to cell via gap junctions. In addition, several specialized conduction pathways speed up conduction of the impulse through the atria.

- The *interatrial pathway* extends from the SA node within the right atrium to the left atrium. Because this pathway rapidly transmits the action potential from the SA node to the pathway's termination in the left atrium, a wave of excitation can spread across the gap junctions throughout the left atrium at the same time as excitation is similarly spreading throughout the right atrium. This ensures that both atria become depolarized to contract simultaneously.

- The *internodal pathway* extends from the SA node to the AV node. The AV node is the only point of electrical contact between the atria and the ventricles; in other words, because the atria and the ventricles are structurally connected by electrically nonconductive fibrous tissue, the only way an action potential in the atria can spread to the ventricles is by passing through the AV node. The internodal conduction pathway directs the spread of an action potential originating at the SA node to the AV node to ensure sequential contraction of the ventricles following atrial contraction. Hastened by this pathway, the action potential arrives at the AV node within 30 msec of SA node firing.

Conduction Between the Atria and the Ventricles The action potential is conducted relatively slowly through the AV node. This slowness is advantageous because it allows time for complete ventricular filling. The impulse is delayed about 100 msec (the **AV nodal delay**), which enables the atria to become completely depolarized and to contract, emptying their contents into the ventricles, before ventricular depolarization and contraction occur.

Ventricular Excitation After the AV nodal delay, the impulse travels rapidly down the septum via the right and left branches of the bundle of His and throughout the ventricular myocardium via the Purkinje fibers. The network of fibers in this ventricular conduction system is specialized for rapid propagation of action potentials. Its presence hastens and coordinates the spread of ventricular excitation to ensure that the ventricles contract as a unit. The action potential is transmitted through the entire Purkinje fiber system within 30 msec.

Although this system carries the action potential rapidly to a large number of cardiac muscle cells, it does not terminate on every cell. The impulse quickly spreads from the excited cells to the rest of the ventricular muscle cells by means of gap junctions.

The ventricular conduction system is more highly organized and more important than the atrial conduction pathways. Because the ventricular mass is so much larger than the atrial mass, the ventricular conduction system is crucial for hastening the spread of excitation in the ventricles. Purkinje fibers can transmit an action potential six times faster than the ventricular syncytium of contractile cells could. If the entire ventricular depolarization process depended on cell-to-cell spread of the impulse via gap junctions, the ventricular tissue immediately next to the AV node would become excited and contract before the impulse had even passed to the heart apex. This, of course, would not allow efficient pumping. Rapid conduction of the action potential down the bundle of His and its swift, diffuse distribution throughout the Purkinje network lead to almost simultaneous activation of the ventricular myocardial cells in both ventricular chambers, which ensures a single, smooth, coordinated contraction that can efficiently eject blood into the systemic and pulmonary circulations at the same time.

The action potential of cardiac contractile cells shows a characteristic plateau.

The action potential in cardiac contractile cells, although initiated by the nodal pacemaker cells, varies considerably in ionic mechanisms and shape from the SA node potential (compare Figures 9-7 and 9-10). Unlike the membrane of autorhythmic cells, the membrane of contractile cells remains essentially at rest, about −90 millivolts (mV), until excited by electrical activity propagated from the pacemaker. Myocardial contractile cells have several subclasses of K^+ channels. At resting potential, the type of K^+ channel that is open is especially leaky.[2] The resultant outward movement of K^+ through these channels keeps the resting potential close to the K^+ equilibrium potential at −90 mV. Once the membrane of a ventricular myocardial contractile cell is depolarized to threshold via current flow through gap junctions, an action potential is generated by a complicated interplay of changes in membrane permeability and potential as follows (Figure 9-10):

1. During the rising phase of the action potential, the membrane potential rapidly reverses to a positive value of about

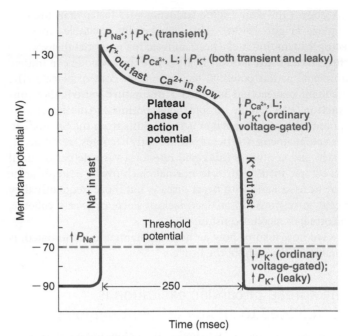

Figure 9-10 Action potential in cardiac contractile cells. The action potential in cardiac contractile cells differs considerably from the action potential in cardiac autorhythmic cells (compare with Figure 9-7, p. 304). The rapid rising phase of the action potential in contractile cells is the result of Na^+ entry on opening of fast Na^+ channels at threshold. The early, brief repolarization after the potential reaches its peak is because of limited K^+ efflux on opening of transient K^+ channels, coupled with inactivation of the Na^+ channels. The prolonged plateau phase is the result of slow Ca^{2+} entry on opening of L-type Ca^{2+} channels, coupled with reduced K^+ efflux on closure of several types of K^+ channels. The rapid falling phase is the result of K^+ efflux on opening of ordinary voltage-gated K^+ channels, as in other excitable cells. Resting potential is maintained by opening of leaky K^+ channels.

FIGURE FOCUS: *Compare the role of L-type Ca^{2+} channels in development of an action potential in cardiac contractile cells and in autorhythmic cells.*

+20 to +30 mV (depending on the myocardial cell) as a result of activation of voltage-gated Na^+ channels and Na^+ subsequently rapidly entering the cell, as it does in other excitable cells undergoing an action potential (see p. 93). These are the same type of double-gated Na^+ channels found in nerve and skeletal muscle cells. At peak potential, the Na^+ permeability then rapidly plummets to its low resting value.

2. At peak potential, another subclass of K^+ channels transiently opens. The resultant fast, limited efflux of K^+ through these transient channels brings about a brief, small repolarization as the membrane becomes slightly less positive.

3. Unique to the cardiac contractile cells, however, the membrane potential is maintained close to this peak positive level for several hundred milliseconds, producing a **plateau phase** of the action potential. In contrast, the short action potential of neurons and skeletal muscle cells lasts 1 to 2 msec. This plateau is maintained by two voltage-dependent permeability changes: activation of "slow" L-type Ca^{2+} channels and a marked decrease in K^+ permeability in the cardiac contractile cell membrane. These permeability changes occur in response to the sudden change in voltage during the rising phase of the action

---

[2]For reasons beyond the scope of this book, this channel type is officially known as an *inward rectifier K^+ channel*, and the "ordinary" K^+ channel discussed shortly is called a *delayed rectifier K^+ channel*.

potential. Opening of the L-type Ca^{2+} channels results in a slow, inward diffusion of Ca^{2+} because Ca^{2+} is in greater concentration in the ECF. This continued influx of positively charged Ca^{2+} prolongs the positivity inside the cell and is primarily responsible for the plateau part of the action potential. This effect is enhanced by the concomitant decrease in K^+ permeability on closure of both the briefly opened transient K^+ channels and the leaky K^+ channels open at resting potential. The resultant reduction in outward movement of positively charged K^+ prevents rapid repolarization of the membrane and thus contributes to prolongation of the plateau phase.

4. The rapid falling phase of the action potential results from inactivation of the Ca^{2+} channels and delayed activation of "ordinary" voltage-gated K^+ channels, yet another subclass of K^+ channels identical to the ones responsible for repolarization in neurons and skeletal muscle cells. The decrease in Ca^{2+} permeability diminishes the slow, inward movement of positive Ca^{2+}, whereas the sudden increase in K^+ permeability simultaneously promotes rapid outward diffusion of positive K^+. Thus, as in other excitable cells, the cell returns to resting potential as K^+ leaves the cell. At resting potential, the ordinary voltage-gated K^+ channels close and the leaky K^+ channels open once again.

Next we see how this action potential initiates contraction.

Calcium Entry From the ECF Induces a Much Larger Ca^{2+} Release From the Sarcoplasmic Reticulum. In cardiac contractile cells, the L-type Ca^{2+} channels lie mostly in the transverse (T) tubules. (In fact, these channels are modified dihydropyridine receptors found in skeletal muscle T tubules; see p. 258.) As you just learned, these voltage-gated channels open during a local action potential. Thus, unlike in skeletal muscle, Ca^{2+} diffuses into the cytosol from the ECF across the T tubule membrane during a cardiac action potential. This entering Ca^{2+} triggers the opening of nearby ryanodine Ca^{2+}-release channels in the adjacent lateral sacs of the SR (see p. 258). By means of this action, termed **Ca^{2+}-induced Ca^{2+} release**, Ca^{2+} entering the cytosol from the ECF induces a much larger release of Ca^{2+} into the cytosol from the intracellular stores (❙Figure 9-11). The resultant local bursts of Ca^{2+} release, known as **Ca^{2+} sparks**, from the SR collectively increase the cytosolic Ca^{2+} pool sufficiently to turn on the contractile machinery. Ninety percent of the Ca^{2+} needed for muscle contraction comes from the SR. This extra supply of Ca^{2+}, coupled with the slow Ca^{2+} removal processes, is responsible for the long period of cardiac contraction, which lasts about three times longer than the contraction of a single skeletal muscle fiber (300 msec compared to 100 msec). This increased contractile time ensures adequate time to eject the blood.

As in skeletal muscle, the role of cytosolic Ca^{2+} is to bind with the troponin–tropomyosin complex and physically pull it aside to allow cross-bridge cycling and contraction (❙Figure 9-11 and see p. 256). However, unlike in skeletal muscle, in which sufficient Ca^{2+} is always released to turn on all the cross bridges, in cardiac muscle the extent of cross-bridge activity varies with the amount of cytosolic Ca^{2+}. As we will show, various regulatory factors can alter the amount of cytosolic Ca^{2+}.

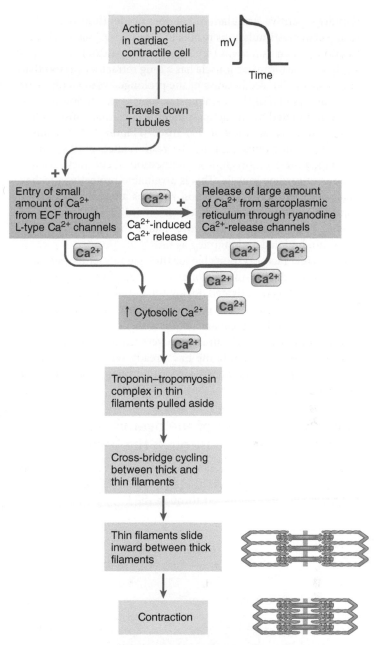

❙Figure 9-11 **Excitation–contraction coupling in cardiac contractile cells.**

Removal of Ca^{2+} from the cytosol by energy-dependent mechanisms in both the plasma membrane (primarily by means of the Na^+–Ca^{2+} exchanger, or NCX) and the SR (via the SERCA pump) restores the blocking action of troponin and tropomyosin, so contraction ceases and the heart muscle relaxes.

A long refractory period prevents tetanus of cardiac muscle.

Like other excitable tissues, cardiac muscle has a refractory period. During the refractory period, a second action potential cannot be triggered until the membrane has recovered from the preceding action potential. In skeletal muscle, the refractory period is very short compared with the duration of the resulting contraction, so the fiber can be restimulated before the first contraction is complete to produce summation of contractions.

Rapidly repetitive stimulation that does not let the muscle fiber relax between stimulations results in a sustained, maximal contraction known as *tetanus* (see ❚Figure 8-19, p. 267).

In contrast, cardiac muscle has a long refractory period that lasts about 250 msec because of the prolonged plateau phase of the action potential. This is almost as long as the period of contraction initiated by the action potential; a cardiac muscle fiber contraction averages about 300 msec (❚Figure 9-12). Consequently, cardiac muscle cannot be restimulated until contraction is almost over, precluding summation of contractions and tetanus of cardiac muscle. This is a valuable protective mechanism because pumping of blood requires alternate periods of contraction (emptying) and relaxation (filling). A prolonged tetanic contraction would prove fatal: The heart chambers could not be filled and emptied again.

The chief factor responsible for the long refractory period is inactivation, during the prolonged plateau phase, of the Na^+ channels that were activated during the initial Na^+ influx of the rising phase—that is, the double-gated Na^+ channels are in their closed and not capable of opening conformation (see p. 92). Not until the membrane recovers from this inactivation process (when the membrane has already repolarized to resting), can the Na^+ channels be activated once again to begin another action potential.

The ECG is a record of the overall spread of electrical activity through the heart.

The electrical currents generated by cardiac muscle during depolarization and repolarization spread into the tissues around the heart and are conducted through the body fluids. A small part of this electrical activity reaches the body surface, where it can be detected using recording electrodes. The record produced is an **electrocardiogram,** or **ECG**. (Alternatively, the abbreviation EKG is often used, from the ancient Greek word *kardia*, instead of the Latin *cardia*, for "heart.")

When considering what an ECG represents, remember these three important points:

1. An ECG is a recording of that part of the electrical activity present in body fluids from the cardiac impulse that reaches the body surface, not a direct recording of the actual electrical activity of the heart.

2. The ECG is a complex recording representing the *overall* spread of activity throughout the heart during depolarization and repolarization. It is not a recording of a *single* action potential in a single cell at a single point in time. The record at any given time represents the sum of electrical activity in all the cardiac muscle cells, some of which may be undergoing action potentials while others may not yet be activated. For example, immediately after the SA node fires, the atrial cells are undergoing action potentials while the ventricular cells are still at resting potential. At a later point, the electrical activity spreads to the ventricular cells while the atrial cells are repolarizing. Therefore, the overall pattern of cardiac electrical activity varies with time as the impulse passes throughout the heart.

3. The recording represents comparisons in voltage detected by electrodes at two points on the body surface, not the actual potential. For example, the ECG does not record a potential when the ventricular muscle is either completely depolarized or completely repolarized; both electrodes are "viewing" the same potential, so no difference in potential between the two electrodes is recorded.

The exact pattern of electrical activity recorded from the body surface depends on the orientation of the recording electrodes. Electrodes may be loosely thought of as "eyes" that "see" electrical activity and translate it into a visible recording, the ECG record. Whether an upward or downward deflection is recorded is determined by the way the electrodes are oriented with respect to the current flow in the heart. For example, the spread of excitation across the heart is "seen" differently from the right arm, from the left leg, or from a recording directly over the heart. Even though the same electrical events are occurring in the heart, different waveforms representing the same electrical activity result when electrodes at different points on the body record this activity.

To provide a common basis for comparison and for recognizing deviations from normal, the same 12 conventional electrode arrangements, or leads, are routinely used for all ECG recordings (❚Figure 9-13). When an electrocardiograph machine is connected between recording electrodes at two points on the body, the specific arrangement of each pair of connections is called a **lead**. The 12 leads each record electrical activity in the heart from different locations—six different electrode arrangements from the limbs and six chest leads at various sites around the heart.

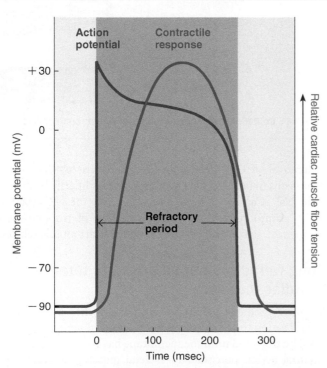

❚Figure 9-12 **Relationship of an action potential and the refractory period to the duration of the contractile response in cardiac muscle.**

Different parts of the ECG record can be correlated to specific cardiac events.

Interpretation of the wave configurations recorded from each lead depends on a thorough knowledge of the sequence of cardiac excitation spread and the position of the heart relative to electrode placement. A normal ECG has three distinct waveforms: the P wave, the QRS complex, and the T wave (❚Figure 9-14). (The letters only indicate the orderly sequence of the waves. The inventor of the technique started in mid-alphabet when naming the waves.)

- The **P wave** represents atrial depolarization.
- The **QRS complex** represents ventricular depolarization.
- The **T wave** represents ventricular repolarization.

These waves of depolarization and repolarization bring about alternating contraction and relaxation of the heart, respectively.

Note the following points about the ECG record:

1. Firing of the SA node does not generate enough electrical activity to reach the body surface, so no wave is recorded for SA nodal depolarization. Therefore, the first recorded wave, the P wave, occurs when the wave of depolarization spreads across the atria.

2. In a normal ECG, no separate wave for atrial repolarization is visible. The electrical activity associated with atrial repolarization normally takes place simultaneously with ventricular depolarization and is masked by the QRS complex.

3. The P wave is smaller than the QRS complex because the atria have a smaller muscle mass than the ventricles and consequently generate less electrical activity.

4. At the following three points in time, no net current flow is occurring in the heart muscle, so the ECG remains at baseline:

   a. *During the AV nodal delay.* This delay is represented by the interval of time between the end of P and the onset of QRS; this segment of the ECG is known as the **PR segment**. (It is called the "PR segment" rather than the "PQ segment" because the Q deflection is small and sometimes absent, whereas the R deflection is the dominant wave of the complex.) Current is flowing through the AV node, but the magnitude is too small for the ECG electrodes to detect.

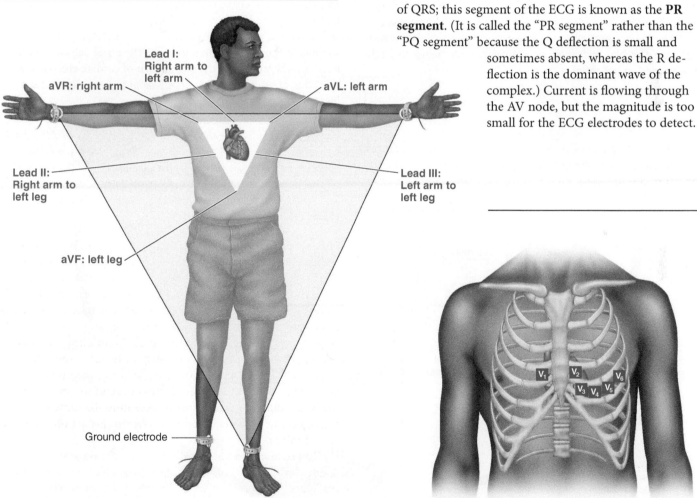

(a) Limb leads

(b) Chest leads

❚Figure 9-13 **Electrocardiogram leads.** (a) The six limb leads include leads I, II, III, aVR, aVL, and aVF. Leads I, II, and III are bipolar leads because two recording electrodes are used. The tracing records the *difference* in potential between the two electrodes. For example, lead I records the difference in potential detected at the right arm and left arm. The electrode placed on the right leg serves as a ground and does not record. The aVR, aVL, and aVF leads are unipolar leads. Even though two electrodes are used, only the actual potential under one "exploring" electrode is recorded. The other electrode is set at zero potential and serves as a neutral reference point. For example, the aVR lead records the potential reaching the right arm in comparison to the rest of the body. (b) The six chest leads, V_1 through V_6, are also unipolar leads. The exploring electrode records the potential of the cardiac muscle immediately beneath the electrode in six different locations surrounding the heart.

b. *When the ventricles are completely depolarized and the cardiac contractile cells are undergoing the plateau phase of their action potential before they repolarize*, represented by the **ST segment**. This segment lies between QRS and T; it coincides with the time during which ventricular activation is complete and the ventricles are contracting and emptying. Note that the ST segment is not a record of cardiac contractile activity. The ECG is a measure of the electrical activity that triggers the subsequent mechanical activity.

c. *When the heart muscle is completely repolarized*, after the T wave and before the next P wave. This period is called the **TP segment**, which coincides with the time when the ventricles are at rest and ventricular filling is taking place.

The ECG can detect abnormal heart rates and rhythms and heart muscle damage.

Clinical Note Because electrical activity triggers mechanical activity, abnormal electrical patterns are usually accompanied by abnormal contractile activity of the heart. Thus, evaluation of ECG patterns can provide useful information about the heart's status. The main deviations from normal that can be found through an ECG are (1) abnormalities in rate, (2) abnormalities in rhythm, and (3) cardiac myopathies (▌Figure 9-15). (For use of the ECG in stress tests, see the boxed feature on p. 314, ▌A Closer Look at Exercise Physiology.)

Abnormalities in Rate Heart rate can be determined from the distance between two consecutive

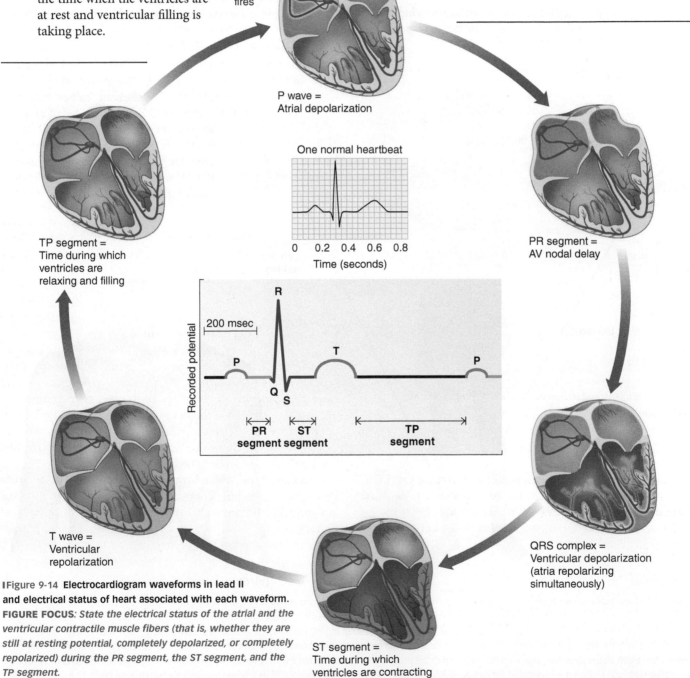

SA node fires

P wave = Atrial depolarization

One normal heartbeat

0   0.2   0.4   0.6   0.8
Time (seconds)

PR segment = AV nodal delay

TP segment = Time during which ventricles are relaxing and filling

Recorded potential

200 msec

R

P   T   P

Q   S

PR ST   TP
segment segment   segment

QRS complex = Ventricular depolarization (atria repolarizing simultaneously)

T wave = Ventricular repolarization

▌**Figure 9-14 Electrocardiogram waveforms in lead II and electrical status of heart associated with each waveform.**
FIGURE FOCUS: *State the electrical status of the atrial and the ventricular contractile muscle fibers (that is, whether they are still at resting potential, completely depolarized, or completely repolarized) during the PR segment, the ST segment, and the TP segment.*

ST segment = Time during which ventricles are contracting and emptying

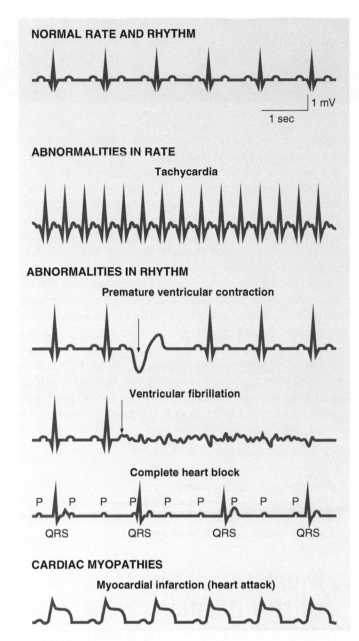

NORMAL RATE AND RHYTHM

1 mV

1 sec

ABNORMALITIES IN RATE

Tachycardia

ABNORMALITIES IN RHYTHM

Premature ventricular contraction

Ventricular fibrillation

Complete heart block

P P P P P P P P P

QRS QRS QRS QRS

CARDIAC MYOPATHIES

Myocardial infarction (heart attack)

Figure 9-15 **Representative heart conditions detectable through electrocardiography.**

QRS complexes on the calibrated paper used to record an ECG. A rapid heart rate of more than 100 beats per minute is called **tachycardia** (*tachy* means "fast"), whereas a slow heart rate of fewer than 60 beats per minute is called **bradycardia** (*brady* means "slow").

Abnormalities in Rhythm *Rhythm* refers to the regularity or spacing of the ECG waves. Any variation from the normal rhythm and sequence of excitation of the heart is termed an **arrhythmia**. It may result from ectopic foci, alterations in SA node pacemaker activity, or interference with conduction. Heart rate is also often altered. *Premature ventricular contractions* originating from an ectopic focus are common deviations from normal rhythm. Other abnormalities in rhythm easily

detected on an ECG include atrial flutter, atrial fibrillation, ventricular fibrillation, and heart block.

Atrial flutter is characterized by a rapid but regular sequence of atrial depolarizations at rates between 200 and 380 beats per minute. The ventricles rarely keep pace with the racing atria. Because the conducting tissue's refractory period is longer than that of the atrial muscle, the AV node is unable to respond to every impulse that converges on it from the atria. Maybe only one out of every two or three atrial impulses successfully passes through the AV node to the ventricles. Such a situation is referred to as a *2:1* or *3:1 rhythm*. The fact that not every atrial impulse reaches the ventricle in atrial flutter is important because it precludes a rapid ventricular rate of more than 200 beats per minute. Such a high rate would not allow adequate time for ventricular filling between beats. In such a case, the output of the heart would be reduced to the extent that loss of consciousness or even death could result because of decreased blood flow to the brain.

Atrial fibrillation is characterized by rapid, irregular, uncoordinated depolarizations of the atria with no definite P waves. Accordingly, atrial contractions are chaotic and asynchronized. Because impulses reach the AV node erratically, the ventricular rhythm is also very irregular. The QRS complexes are normal in shape but occur sporadically. Variable lengths of time between ventricular beats are available for ventricular filling. Some ventricular beats come so close together that little filling can occur between beats. When less filling occurs, the subsequent contraction is weaker. In fact, some of the ventricular contractions may be too weak to eject enough blood to produce a palpable wrist pulse. In this situation, if the heart rate is determined directly, either by the apex beat or via the ECG, and the pulse rate is taken concurrently at the wrist, the heart rate exceeds the pulse rate. Such a difference in heart rate and pulse rate is known as a **pulse deficit**. Normally, the heart rate coincides with the pulse rate because each cardiac contraction initiates a pulse wave as it ejects blood into the arteries.

Ventricular fibrillation is a very serious rhythmic abnormality in which multiple impulses travel erratically in all directions around the ventricles. The ECG tracing is very irregular with no detectable pattern or rhythm. The resultant chaotic contractions are so disorganized that the ventricles are ineffectual as pumps. If circulation is not restored in less than four minutes through external cardiac compression or electrical defibrillation, irreversible brain damage occurs, and death is imminent.

Another type of arrhythmia, **heart block**, arises from defects in the cardiac conducting system. The atria still beat regularly, but the ventricles occasionally fail to be stimulated and thus do not contract following atrial contraction. Impulses between the atria and ventricles can be blocked to varying degrees. In some forms of heart block, only every second or third atrial impulse is passed to the ventricles. This is known as *2:1* or *3:1 block,* which can be distinguished from the 2:1 or 3:1 rhythm associated with atrial flutter by the rates involved. In heart block, the atrial rate is normal but the ventricular rate is considerably below normal, whereas in atrial flutter the atrial rate is very high, in accompaniment with a normal or above-normal ventricular rate. *Complete heart block* is characterized by complete dissociation between atrial and ventricular activity,

The What, Who, and When of Stress Testing

STRESS TESTS, OR GRADED EXERCISE TESTS, are conducted primarily to aid in diagnosing or quantifying heart or lung disease and to evaluate the functional capacity of asymptomatic individuals. The tests are usually given on motorized treadmills or bicycle ergometers (stationary, variable-resistance bicycles). Workload intensity (how hard the subject is working) is adjusted by progressively increasing the speed and incline of the treadmill or by progressively increasing the pedaling frequency and resistance on the bicycle. The test starts at a low intensity and continues until a prespecified workload is achieved, physiological symptoms occur, or the subject is too fatigued to continue.

During diagnostic testing, the patient is monitored with an ECG and blood pressure is taken each minute. A test is considered positive if ECG abnormalities occur (such as ST segment depression, inverted T waves, or dangerous arrhythmias) or if physical symptoms such as chest pain develop. A test that is interpreted as positive in a person who does not have heart disease is called a *false-positive test.* In men, false positives occur only about 10% to 20% of the time, so the diagnostic stress test for men has a *specificity* of 80% to 90%. Women

have a greater frequency of false-positive test results, with a corresponding lower specificity of about 70%.

The *sensitivity* of a test means that people with disease are correctly identified and there are few false negatives. The sensitivity of the stress test is 60% to 80%—that is, if 100 people with heart disease were tested, 60 to 80 would be correctly identified, but 20 to 40 would have a *false-negative test*. Although stress testing is now an important diagnostic tool, it is just one of several tests used to determine the presence of coronary artery disease.

Stress tests are also conducted on people not suspected of having heart or lung disease to determine their present functional capacity. These functional tests are administered in the same way as diagnostic tests, but exercise physiologists conduct them and a physician need not be present. These tests are used to establish safe exercise prescriptions, to aid athletes in establishing optimal training programs, and to serve as research tools to evaluate the effectiveness of a particular training regimen. Functional stress testing is becoming more prevalent as more people are joining hospital- or community-based wellness programs for disease prevention.

---

with impulses from the atria not being conducted to the ventricles at all. The SA node continues to govern atrial depolarization, but the ventricles generate their own impulses at a rate slower than that of the atria. On the ECG, the P waves exhibit a normal rhythm. The QRS and T waves also occur regularly but more slowly than the P waves and independently of P wave rhythm. Because atrial activity and ventricular activity are not synchronized, waves for atrial repolarization may appear, no longer masked by the QRS complex.

Cardiac Myopathies Abnormal ECG waves are also important in recognizing **cardiac myopathies** (damage of the heart muscle). **Myocardial ischemia** is inadequate delivery of oxygenated blood to the heart tissue. Actual death, or **necrosis**, of heart muscle cells occurs when a blood vessel supplying that area of the heart becomes blocked or ruptured. This condition is **acute myocardial infarction**, commonly called a **heart attack**. Abnormal QRS waveforms appear when part of the heart muscle becomes necrotic. Furthermore, damaged heart muscle cells release characteristic enzymes into the blood that can be measured to provide a further index of the extent of myocardial damage.

1. Draw two graphs comparing the electrical activity in a cardiac autorhythmic cell and in a cardiac contractile cell. Label the ion movement responsible for each change in potential.

2. List the autorhythmic tissues of the heart and indicate the normal rate of action potential discharge of each.

3. Define *Ca²⁺-induced Ca²⁺ release*.

4. Draw and label a normal ECG and state the electrical event associated with each waveform. Explain why no separate wave for atrial repolarization is visible on a normal ECG.

9.3 Mechanical Events of the Cardiac Cycle

The mechanical events of the cardiac cycle—contraction, relaxation, and the resultant changes in blood flow through the heart—are brought about by the rhythmic changes in cardiac electrical activity.

The heart alternately contracts to empty and relaxes to fill.

The **cardiac cycle** consists of alternate periods of **systole** (contraction and emptying) and **diastole** (relaxation and filling). Contraction results from the spread of excitation across the heart, whereas relaxation follows the subsequent repolarization of the cardiac muscle. The atria and ventricles go through separate cycles of systole and diastole. Unless qualified, the terms *systole* and *diastole* refer to what is happening with the ventricles.

The following discussion and corresponding ❙Figure 9-16 correlate various events that occur concurrently during the cardiac cycle, including ECG features, pressure changes, volume changes, valve activity, and heart sounds. This integrated diagram is known as **Wigger's diagram**. Only the events on the left side of the heart are described, but keep in mind that identical events are occurring on the right side of the heart, except that the pressures are lower. To complete one full cardiac cycle, our discussion begins and ends with ventricular diastole.

Mid-Ventricular Diastole During most of ventricular diastole, the atrium is still also in diastole. This stage corresponds to the TP segment on the ECG—the interval after ventricular repolarization and before another atrial depolarization. Because of the continuous inflow of blood from the venous system into the atrium, atrial pressure slightly exceeds ventricular pressure even though both chambers are relaxed (❙ Figure 9-16, point ❶). Because of this pressure differential, the AV valve is open and blood flows directly from the atrium into the ventricle throughout the filling phase of ventricular diastole (❙ Figure 9-16, heart a). As a result of this passive filling, the ventricular volume slowly continues to rise even before atrial contraction takes place (point ❷).

Late Ventricular Diastole Late in ventricular diastole, the SA node reaches threshold and fires. The impulse spreads throughout the atria, which appears on the ECG as the P wave (point ❸). Atrial depolarization brings about atrial contraction, raising the atrial pressure curve (point ❹) and squeezing more blood into the ventricle. The excitation–contraction coupling process takes place during the short delay between the P wave and the rise in atrial pressure. The corresponding rise in ventricular pressure (point ❺) that occurs simultaneously with the rise in atrial pressure results from the additional volume of blood added to the ventricle by atrial contraction (point ❻ and heart b). Throughout atrial contraction, atrial pressure still slightly exceeds ventricular pressure, so the AV valve remains open.

End of Ventricular Diastole Ventricular diastole ends at the onset of ventricular contraction. By this time, atrial contraction and ventricular filling are completed. The volume of blood in the ventricle at the end of diastole (point ❼) is known as the **end-diastolic volume (EDV)**, which averages about 135 mL. No more blood is added to the ventricle during this cycle. Therefore, the end-diastolic volume is the maximum amount of blood that the ventricle contains during this cycle.

Onset of Ventricular Systole After atrial excitation, the impulse travels through the AV node and specialized conduction system to excite the ventricle. Simultaneously, the atria are contracting. By the time ventricular activation is complete, atrial contraction is already over. The QRS complex represents this ventricular excitation (point ❽), which induces ventricular contraction. The ventricular pressure curve sharply increases shortly after the QRS complex (that is, after excitation–contraction coupling has occurred), signaling the onset of ventricular systole (point ❾). As ventricular contraction begins, ventricular pressure immediately exceeds atrial pressure. This backward pressure differential forces the AV valve closed (point ❾).

Isovolumetric Ventricular Contraction After ventricular pressure exceeds atrial pressure and the AV valve has closed, to open the aortic valve, the ventricular pressure must continue to increase until it exceeds aortic pressure. Therefore, after the AV valve closes and before the aortic valve opens, the ventricle briefly remains a closed chamber (point ❿). Because all valves are closed, no blood can enter or leave the ventricle during this time, which is termed the period of **isovolumetric ventricular contraction** (*isovolumetric* means "constant volume and length") (heart c). Because no blood enters or leaves the ventricle, the ventricular chamber stays at constant volume, and the muscle fibers stay at constant length (point ⓫) while ventricular pressure continues to rise.

Ventricular Ejection When ventricular pressure exceeds aortic pressure (point ⓬), the aortic valve is forced open and ejection of blood begins (heart d). The amount of blood pumped out of each ventricle with each contraction is called the **stroke volume (SV)**. The aortic pressure curve rises as blood is forced into the aorta from the ventricle faster than blood is draining off into the smaller vessels at the other end (point ⓭). The ventricular volume decreases substantially as blood is rapidly pumped out (point ⓮). Ventricular systole includes both isovolumetric ventricular contraction and ventricular ejection.

End of Ventricular Systole The ventricle does not empty completely during ejection. Normally, only about half the blood within the ventricle at the end of diastole is pumped out during the subsequent systole. The amount of blood left in the ventricle at the end of systole when ejection is complete is the **end-systolic volume (ESV)** (point ⓯), which averages about 65 mL. This is the least amount of blood that the ventricle contains during this cycle.

The difference between the volume of blood in the ventricle before contraction and the volume after contraction is the amount of blood ejected during the contraction, that is, EDV − ESV = SV. In our example, EDV is 135 mL, ESV is 65 mL, and SV is 70 mL.

Onset of Ventricular Diastole The T wave signifies ventricular repolarization at the end of ventricular systole (point ⓰). When the ventricle repolarizes and starts to relax, ventricular pressure falls below aortic pressure and the aortic valve closes (point ⓱). Closure of the aortic valve produces a disturbance or notch on the aortic pressure curve, the **dicrotic notch** (point ⓲). No more blood leaves the ventricle during this cycle because the aortic valve has closed.

Isovolumetric Ventricular Relaxation When the aortic valve closes, the AV valve is not yet open because ventricular pressure still exceeds atrial pressure, so no blood can enter the ventricle from the atrium. Therefore, all valves are again closed for a brief period known as **isovolumetric ventricular relaxation** (point ⓳ and heart e). The muscle fiber length and chamber volume (point ⓴) remain constant. No blood leaves or enters as the ventricle continues to relax and the pressure steadily falls.

Ventricular Filling When ventricular pressure falls below atrial pressure, the AV valve opens (point ㉑), and ventricular filling occurs again. Ventricular diastole includes both isovolumetric ventricular relaxation and ventricular filling.

Figure 9-16 Cardiac cycle. This diagram depicts various events that occur concurrently during the cardiac cycle. Follow each horizontal strip across to see the changes that take place in the electrocardiogram; aortic, ventricular, and atrial pressures; ventricular volume; and heart sounds throughout the cycle. The last half of diastole, one full systole and diastole (one full cardiac cycle), and another systole are shown for the left side of the heart. Follow each vertical strip downward to see what happens simultaneously with each of these factors during each phase of the cardiac cycle. See the text (pp. 315 and 317) for a detailed explanation of the numbered points. The sketches of the heart illustrate the flow of O₂-poor (*dark blue*) and O₂-rich (*dark pink*) blood in and out of the ventricles during the cardiac cycle.

FIGURE FOCUS: *If the length of the diastolic filling phase is reduced by one half because the heart rate increases, would only half as much blood enter the ventricles? Use this figure to defend your answer.*

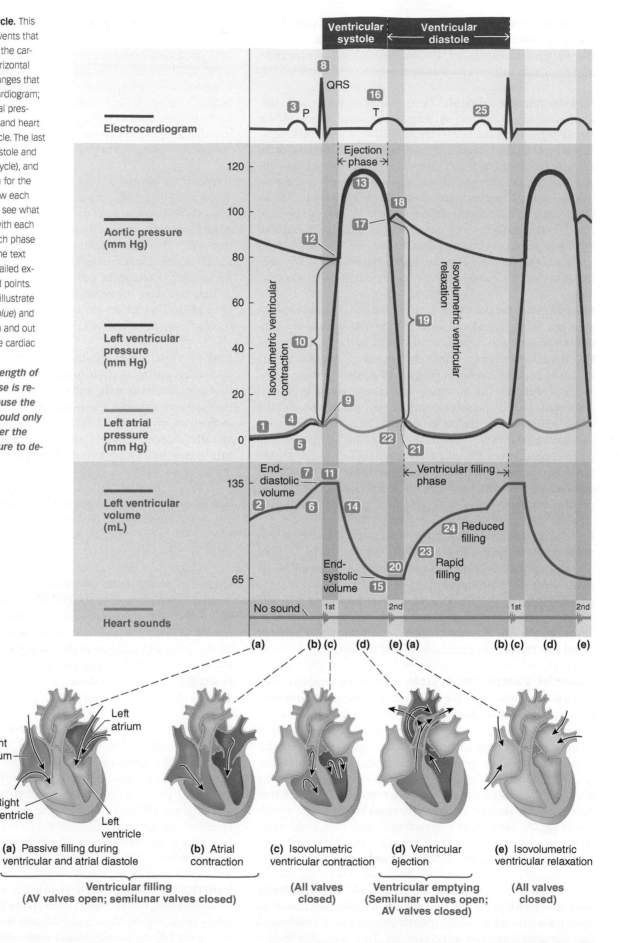

(a) Passive filling during ventricular and atrial diastole

(b) Atrial contraction

(c) Isovolumetric ventricular contraction

(d) Ventricular ejection

(e) Isovolumetric ventricular relaxation

Ventricular filling (AV valves open; semilunar valves closed)

(All valves closed)

Ventricular emptying (Semilunar valves open; AV valves closed)

(All valves closed)

Atrial repolarization and ventricular depolarization occur simultaneously, so the atria are in diastole throughout ventricular systole. Blood continues to flow from the pulmonary veins into the left atrium. As this incoming blood pools in the atrium, atrial pressure rises continuously (point 22). When the AV valve opens at the end of ventricular systole, blood that accumulated in the atrium during ventricular systole pours rapidly into the ventricle (heart a again). Ventricular filling thus occurs rapidly at first (point 23) because of the increased atrial pressure resulting from the accumulation of blood in the atria. Then ventricular filling slows down (point 24) as the accumulated blood has already been delivered to the ventricle. During this period of reduced filling, blood continues to flow from the pulmonary veins into the left atrium and through the open AV valve into the left ventricle. During late ventricular diastole, when the ventricle is filling slowly, the SA node fires again, and the cardiac cycle starts over (point 25).

Another way besides Wigger's diagram of viewing the relationship between pressure and volume throughout the cardiac cycle is a **pressure–volume loop**, which does not consider the element of time but is an excellent tool for visualizing pressure and volume changes as valves open and close and blood flows in and out of the ventricle (∎Figure 9-17).

When the body is at rest, one complete cardiac cycle lasts 800 msec, with 300 msec devoted to ventricular systole and 500 msec taken up by ventricular diastole. Significantly, much of ventricular filling occurs early in diastole during the rapid-filling phase. During times of rapid heart rate, diastole length is shortened more than systole length is. For example, if the heart rate increases from 75 to 180 beats per minute, the duration of diastole decreases about 75%, from 500 msec to 125 msec. This greatly reduces the time available for ventricular relaxation and filling. However, because much ventricular filling is accomplished during early diastole, filling is not seriously impaired during periods of increased heart rate, such as during exercise. There is a limit, however, to how rapidly the heart can beat without decreasing the period of diastole to the point that ventricular filling is severely impaired. At heart rates greater than 200 beats per minute, diastolic time is too short to allow adequate ventricular filling. With inadequate filling, the resultant cardiac output is deficient. Normally, ventricular rates do not exceed 200 beats per minute because the relatively long refractory period of the AV node will not allow impulses to be conducted to the ventricles more frequently than this.

Two normal heart sounds are associated with valve closures.

Two major heart sounds normally can be heard with a stethoscope during the cardiac cycle. The **first heart sound** is low-pitched, soft, and relatively long; it sounds like "lub." The **second heart sound** has a higher pitch and is shorter and sharper; it sounds like "dup." Thus, one normally hears "lub-dup-lub-dup-lub-dup . . . ." The first heart sound is associated with closure of the AV valves, whereas the second sound is associated with closure of the semilunar valves (see the "Heart sounds" line at the bottom of the graphs in ∎Figure 9-16). Opening of valves does not produce any sound.

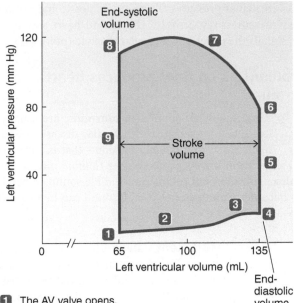

1 The AV valve opens.

2 Passive ventricular filling occurs. Volume increases considerably and pressure increases slightly as blood enters.

3 Atrial contraction completes ventricular filling. End-diastolic volume is reached at the end of this phase.

4 The AV valve closes.

5 Isovolumetric ventricular contraction occurs. Volume remains constant; pressure increases markedly.

6 The aortic valve opens.

7 A stroke volume of blood is ejected. As blood leaves, volume decreases considerably as pressure peaks, then falls more slowly until end-systolic volume is reached at the end of this phase.

8 The aortic valve closes.

9 Isovolumetric ventricular relaxation occurs. Volume remains constant; pressure falls sharply.

Back to step 1.

∎Figure 9-17 **Left-ventricular pressure–volume loop for a single cardiac cycle.**
FIGURE FOCUS: *How would the difference in pressure between point 4 (AV valve closure) and point 6 (semilunar valve opening) in a right-ventricular pressure–volume loop compare with the difference in pressure between these same two points in this left-ventricular pressure–volume loop (taking into account that the pulmonary valve instead of the aortic valve is opening at 6 and closing at 8 in the right-sided loop)? How would the change in volume between points 6 and 8 compare in these two pressure–volume loops?*

The sounds are caused by vibrations set up within the walls of the ventricles and major arteries during valve closure, not by the valves snapping shut. Because the AV valves close at the onset of ventricular contraction, when ventricular pressure first exceeds atrial pressure, the first heart sound signals the onset of ventricular systole (∎Figure 9-16, point 9). The semilunar valves close at the onset of ventricular relaxation, as the left and right ventricular pressures fall below the aortic and pulmonary artery pressures, respectively. The second heart sound, therefore, signals the onset of ventricular diastole (point **17**).

Turbulent blood flow produces heart murmurs.

Clinical Note Abnormal heart sounds, or **murmurs**, are usually (but not always) associated with cardiac disease. Blood normally flows in a *laminar* fashion—that is, layers of the fluid slide smoothly over one another (*lamina* means "layer"). Laminar flow does not produce an audible sound. When blood flow becomes turbulent, however, a sound can be heard (**I**Figure 9-18).

Stenotic and Insufficient Valves The most common cause of turbulence is valve malfunction, either a stenotic or an insufficient valve. A **stenotic valve** is a stiff, narrowed valve that does not open completely. Blood must be forced through the constricted opening at tremendous velocity, resulting in turbulence that produces an abnormal whistling sound similar to the sound produced when you force air rapidly through narrowed lips to whistle. An **insufficient**, or **incompetent**, **valve** is one that cannot close completely because the valve edges do not fit together properly. Turbulence is produced when blood flows backward through the insufficient valve and collides with blood moving in the opposite direction, creating a swishing or gurgling murmur. Such backflow of blood is known as **regurgitation**. An insufficient heart valve is often called a **leaky valve** because it lets blood leak back through when the valve should be closed.

Most often, both valvular stenosis and insufficiency are caused by **rheumatic fever**, an autoimmune ("immunity against self") disease triggered by a streptococcus bacterial infection. Antibodies formed against toxins produced by these bacteria interact with many of the body's tissues, resulting in immunological damage. The heart valves are among the most susceptible tissues in this regard. Large, hemorrhagic, fibrous lesions form along the inflamed edges of an affected heart valve, causing the valve to become thickened, stiff, and scarred. Sometimes the leaflet edges permanently adhere to each other. Depending on the extent and specific nature of the lesions, the valve may become either stenotic or insufficient or some degree of both. Rheumatic fever is much less common since antibiotics that can treat "strep throat" became available. On occasion children are born with malfunctioning valves.

Timing of Murmurs The valve involved and the type of defect can usually be detected by the timing and location of the murmur. The *timing* of the murmur refers to the part of the cardiac cycle during which the murmur is heard. Recall that the first heart sound signals the onset of ventricular systole and the second heart sound signals the onset of ventricular diastole. Thus, a murmur between the first and the second heart sounds ("lub-murmur-dup—lub-murmur-dup") is a **systolic murmur**. A **diastolic murmur**, in contrast, occurs between the second and the first heart sounds ("lub-dup-murmur—lub-dup-murmur"). The sound of the murmur characterizes it as either a stenotic (whistling) murmur or an insufficient (swishy) mur-

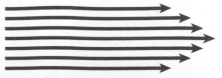

(a) Laminar flow (does not create any sound)

(b) Turbulent flow (can be heard)

IFigure 9-18 **Comparison of laminar and turbulent flow.**

mur. Armed with these facts, one can determine the cause of a valvular murmur. For example, a whistling murmur (denoting a stenotic valve) occurring between the first and the second heart sounds (denoting a systolic murmur)—that is, *lub-whistle-dup*—indicates stenosis in a valve that should be open during systole. It could be either the aortic or the pulmonary semilunar valve through which blood is being ejected. Identifying which of these valves is stenotic is accomplished by finding where the murmur is best heard. Each heart valve can be heard best at a specific location on the chest. Noting where a murmur is loudest helps the diagnostician tell which valve is involved. Let's analyze another example for practice. A swishy diastolic murmur—*lub-dup-swish*—signifies that a valve that should be closed during diastole (a semilunar valve) does not close completely, that is, is insufficient. Using the same line of reasoning, you can figure out that a *lub-dup-whistle* murmur denotes a stenotic AV valve, whereas a *lub-swish-dup* murmur is a sign of an insufficient AV valve.

The main concern with a heart murmur is not the murmur itself but the harmful circulatory results of the defect. An *echocardiogram* can be used to further evaluate a valve defect. With this noninvasive technique, ultrasound waves transmitted to the heart bounce off the valves and heart chambers. Returning echoes of the waves are processed by a computer and assembled into moving pictures of the beating heart that can be viewed on a monitor. The resultant images are used to identify various abnormalities in valves and heart muscle.

Check Your Understanding 9.3

1. Define *systole* and *diastole*.

2. State the pressure relationships among the aortic, atrial, and ventricular pressures (for example, ventricular pressure > aortic pressure > atrial pressure) during each of these phases of the cardiac cycle: (1) ventricular filling, (2) isovolumetric ventricular contraction, (3) ventricular ejection, and (4) isovolumetric ventricular relaxation (think what pressure relationships must exist for the valves to be opened or closed as appropriate in each phase).

3. Distinguish among end-diastolic volume, end-systolic volume, and stroke volume.

9.4 Cardiac Output and Its Control

Cardiac output (CO) is the volume of blood pumped by *each ventricle* per minute (not the total amount of blood pumped by the heart). During any period, the volume of blood flowing through the pulmonary circulation is the same as the volume flowing through the systemic circulation. Therefore, the cardiac output from each ventricle normally is the same, although minor variations may occur on a beat-to-beat basis.

Cardiac output depends on heart rate and stroke volume.

The two determinants of cardiac output are **heart rate (HR)** (beats per minute) and **stroke volume (SV)** (volume of blood pumped per beat or stroke). The average resting HR is 70 beats per minute, established by SA node rhythmicity; the average resting SV is 70 mL per beat, producing an average CO of 4900 mL per minute, or close to 5 L per minute:

$$CO = HR \times SV$$
$$= 70 \text{ beats/min} \times 70 \text{ mL/beat}$$
$$= 4900 \text{ mL/min} \approx 5 \text{ L/min}$$

Because the body's total blood volume averages 5 to 5.5 liters, each half of the heart pumps the equivalent of the entire blood volume each minute. In other words, each minute the right ventricle normally pumps 5 liters of blood through the lungs and the left ventricle pumps 5 liters through the systemic circulation. At this rate, each half of the heart would pump about 2.5 million liters of blood in just 1 year. Yet this is only the resting CO; during exercise, CO can increase to 20 to 25 liters per minute (and even more in trained athletes during heavy endurance-type exercise). The difference between the cardiac output at rest and the maximum volume of blood the heart can pump per minute is called the **cardiac reserve**.

How can CO vary so tremendously, depending on the demands of the body? You can readily answer this question by thinking about how your heart pounds rapidly (increased heart rate) and forcefully (increased stroke volume) when you engage in strenuous physical activities (when you need increased cardiac output). Thus, regulation of CO depends on control of both HR and SV.

Heart rate is determined primarily by autonomic influences on the SA node.

The heart is innervated by both divisions of the autonomic nervous system, which can modify the rate (and the strength) of contraction, even though nervous stimulation is not required to initiate contraction. The parasympathetic nerve to the heart, the *vagus nerve*, primarily supplies the atrium, especially the SA and AV nodes. Parasympathetic innervation of the ventricles is sparse. The cardiac sympathetic nerves also supply the atria, including the SA and AV nodes, and richly innervate the ventricles as well.

Both the parasympathetic and sympathetic nervous system bring about their effects on the heart primarily by altering the activity of the cyclic adenosine monophosphate (cAMP) second-messenger pathway in the innervated cardiac cells. Acetylcholine (ACh) released from the vagus nerve binds to a muscarinic cholinergic receptor and is coupled to an inhibitory G protein that reduces activity of the cAMP pathway (see pp. 117, 123, and 239). By contrast, the sympathetic neurotransmitter norepinephrine binds with a β_1-adrenergic receptor and is coupled to a stimulatory G protein that accelerates the cAMP pathway in the target cells (see p. 240). The cAMP pathway leads to phosphorylation and altered activity of various proteins within cardiac muscle, for example, keeping channels open longer. Let us examine the specific effects that parasympathetic and sympathetic stimulation have on the heart (Table 9-1).

TABLE 9-1 Effects of the Autonomic Nervous System on Heart Activity

Area Affected	Effect of Parasympathetic Stimulation	Effect of Sympathetic Stimulation
SA node	Decreases the rate of depolarization to threshold; decreases the heart rate	Increases the rate of depolarization to threshold; increases the heart rate
AV node	Decreases excitability; increases the AV nodal delay	Increases excitability; decreases the AV nodal delay
Ventricular conduction pathway	No effect	Increases excitability; hastens conduction through the bundle of His and Purkinje cells
Atrial muscle	Decreases contractility; weakens contraction	Increases contractility; strengthens contraction
Ventricular muscle	No effect	Increases contractility; strengthens contraction
Adrenal medulla (an endocrine gland)	No effect	Promotes secretion of epinephrine, a hormone that augments sympathetic nervous system actions
Veins	No effect	Increases venous return, which increases the strength of cardiac contraction via intrinsic control

- - - = Inherent SA node pacemaker activity
——— = SA node pacemaker activity on parasympathetic stimulation
——— = SA node pacemaker activity on sympathetic stimulation

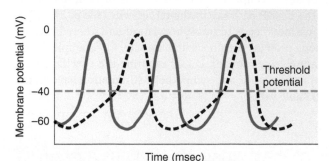

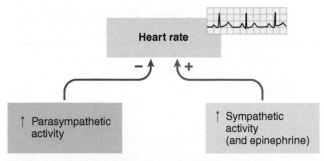

(a) Autonomic influence on SA node potential

Heart rate

– ↗ ↖ +

↑ Parasympathetic activity

↑ Sympathetic activity (and epinephrine)

(b) Control of heart rate by autonomic nervous system

❙Figure 9-19 **Autonomic control of SA node activity and heart rate.**
(a) Parasympathetic stimulation decreases the rate of SA nodal depolarization so that the membrane reaches threshold more slowly and has fewer action potentials, whereas sympathetic stimulation increases the rate of depolarization of the SA node so that the membrane reaches threshold more rapidly and has more frequent action potentials. (b) Because each SA node action potential ultimately leads to a heartbeat, increased parasympathetic activity decreases the heart rate, whereas increased sympathetic activity increases the heart rate.

Effect of Parasympathetic Stimulation on the Heart

The parasympathetic nervous system's influence on the SA node is to decrease HR (❙Figure 9-19). In a different mechanism than its usual reduction in cAMP activity, ACh slows heart rate primarily by increasing K^+ permeability of the pacemaker cells in the SA node by binding with muscarinic cholinergic receptors that are coupled directly to ACh-regulated K^+ channels by a G protein. This action augments opening of these K^+ channels. As a result, the rate at which spontaneous action potentials are initiated is reduced through a twofold effect:

1. Enhanced K^+ permeability hyperpolarizes the SA node membrane because more positive K^+ ions leave than normal, making the inside even more negative. Because the "resting" potential starts even farther away from threshold, it takes longer to reach threshold.

2. The enhanced K^+ permeability induced by vagal stimulation also opposes the automatic reduction in K^+ permeability that contributes to development of the pacemaker potential. This countering effect decreases the SA node's rate of spontaneous depolarization, prolonging the time required to drift to threshold. ACh, by inhibiting the cAMP pathway, also depresses both the inward movement of Na^+ and Ca^{2+} through the I_f and T-type channels, respectively, further slowing the depolarization to threshold. Therefore, the SA node reaches threshold and fires less frequently, decreasing the heart rate.

■ Parasympathetic stimulation decreases the AV node's excitability, prolonging transmission of impulses to the ventricles even longer than the usual AV nodal delay. This effect is brought about by increasing K^+ permeability, which hyperpolarizes the membrane, thereby retarding the initiation of excitation in the AV node.

■ Parasympathetic stimulation of the atrial contractile cells shortens the plateau phase of the action potential by reducing the slow inward current carried by Ca^{2+}. As a result, atrial contraction is weakened.

■ The parasympathetic system has little effect on ventricular contraction because of the sparseness of parasympathetic innervation to the ventricles.

Thus, the heart is more "leisurely" under parasympathetic influence—it beats less rapidly, the time between atrial and ventricular contraction is stretched out, and atrial contraction is weaker. These actions are appropriate, considering that the parasympathetic system controls heart action in quiet, relaxed situations when the body is not demanding enhanced cardiac output.

Effect of Sympathetic Stimulation on the Heart

In contrast, the sympathetic nervous system, which controls heart action in emergency or exercise situations that require greater blood flow, "revs up" the heart.

■ The main effect of sympathetic stimulation on the SA node is to speed up depolarization so that threshold is reached more rapidly. In pacemaker cells, the rate of depolarization increases as a result of greater inward movement of Na^+ and Ca^{2+} through augmented I_f and T-type Ca^{2+} channels. This swifter drift to threshold under sympathetic influence permits more frequent action potentials and a correspondingly faster heart rate (❙Figure 9-19 and ❙Table 9-1).

■ Sympathetic stimulation of the AV node reduces the AV nodal delay by increasing conduction velocity, as a result enhancing the slow, inward Ca^{2+} current.

■ Similarly, sympathetic stimulation speeds up spread of the action potential throughout the specialized conduction pathway.

■ In the atrial and ventricular contractile cells, both of which have many sympathetic nerve endings, sympathetic stimulation increases contractile strength so that the heart beats more forcefully and squeezes out more blood. This effect is produced by increasing Ca^{2+} permeability through prolonged opening of L-type Ca^{2+} channels. The resultant enhanced Ca^{2+} influx strengthens contraction by intensifying Ca^{2+} participation in excitation–contraction coupling.

- Sympathetic stimulation not only increases the speed of contraction by allowing greater influx of Ca^{2+} into the cell through L-type Ca^{2+} channels, but it also speeds up relaxation by enhancing the SERCA pump that removes Ca^{2+} from the cytosol (see p. 261).

The overall effect of sympathetic stimulation on the heart, therefore, is to improve its effectiveness as a pump by increasing HR, decreasing the delay between atrial and ventricular contraction, decreasing conduction time throughout the heart, increasing the force of contraction, and speeding up the relaxation process so that more time is available for filling.

Control of Heart Rate Thus, as is typical of the autonomic nervous system, parasympathetic and sympathetic effects on heart rate are antagonistic (oppose each other). At any given moment, HR is determined largely by the balance between inhibition of the SA node by the vagus nerve and stimulation by the cardiac sympathetic nerves. Under resting conditions, parasympathetic discharge dominates because ACh (the parasympathetic neurotransmitter) suppresses sympathetic activity by inhibiting the release of norepinephrine (the sympathetic neurotransmitter) from neighboring sympathetic nerve endings. If all autonomic nerves to the heart were blocked, the resting HR would increase from its average value of 70 beats per minute to about 100 beats per minute, which is the inherent rate of the SA node's spontaneous discharge when not subjected to any nervous influence. (We use 70 beats per minute as the normal rate of SA node discharge because this is the average rate under normal resting conditions when parasympathetic activity dominates.) HR can be altered beyond this resting level in either direction by shifting the balance of autonomic nervous stimulation. HR is speeded up by simultaneously increasing sympathetic and decreasing parasympathetic activity; HR is slowed by a concurrent rise in parasympathetic activity and decline in sympathetic activity. The relative level of activity in these two autonomic branches to the heart in turn is primarily coordinated by the *cardiovascular control center* in the brain stem.

Although autonomic innervation is the primary means by which HR is regulated, other factors affect it as well. The most important is epinephrine, a hormone secreted into the blood from the adrenal medulla on sympathetic stimulation. Epinephrine acts in a manner similar to norepinephrine to increase HR, thus reinforcing the direct effect that the sympathetic nervous system has on the heart.

Stroke volume is determined by the extent of venous return and by sympathetic activity.

The other component besides heart rate that determines cardiac output is stroke volume, the amount of blood pumped out by each ventricle during each beat. Two types of control influence

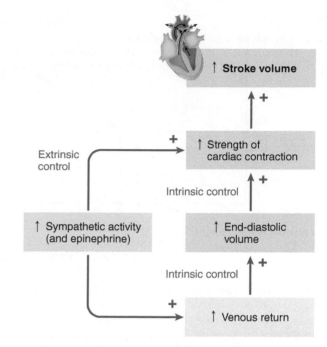

❙Figure 9-20 Intrinsic and extrinsic control of stroke volume.

stroke volume: (1) *intrinsic control* related to the extent of venous return and (2) *extrinsic control* related to the extent of sympathetic stimulation of the heart. Both factors increase SV by increasing the strength of heart contraction (❙Figure 9-20). Let us examine each of these mechanisms in detail.

Increased end-diastolic volume results in increased stroke volume.

Intrinsic control of stroke volume, which refers to the heart's inherent ability to vary SV, depends on the direct correlation between end-diastolic volume (EDV) and SV. As more blood returns to the heart, the heart pumps out more blood, but the relationship is not as simple as it might seem because the heart does not eject all the blood it contains. This intrinsic control depends on the length–tension relationship of cardiac muscle, which is similar to that of skeletal muscle. For skeletal muscle, the resting muscle length is approximately the optimal length (l_o) at which maximal tension can be developed during a subsequent contraction. When the skeletal muscle is longer or shorter than l_o, the subsequent contraction is weaker (see ❙Figure 8-20, p. 269). For cardiac muscle, the resting cardiac muscle fiber length is less than l_o. Therefore, the length of cardiac muscle fibers normally varies along the ascending limb of the length–tension curve. An increase in cardiac muscle fiber length, by moving closer to l_o, increases the contractile tension of the heart on the following systole (❙Figure 9-21).

Unlike in skeletal muscle, the length–tension curve of cardiac muscle normally does not operate at lengths that fall within the region of the descending limb. That is, within physiologic limits, cardiac muscle does not get stretched beyond its l_o to the point that contractile strength diminishes with further stretching.

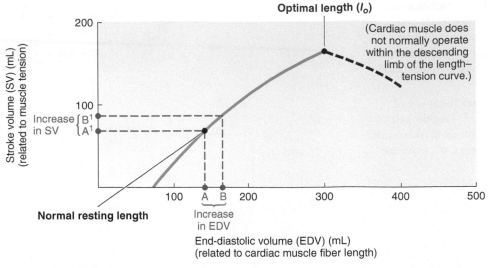

Figure 9-21 Intrinsic control of stroke volume (Frank–Starling curve). The cardiac muscle fiber's length, which is determined by the extent of venous filling, is normally less than the optimal length (l_o) for developing maximal tension. Therefore, an increase in end-diastolic volume (EDV) (that is, an increase in venous return), by moving the cardiac muscle fiber length closer to l_o, increases the contractile tension of the fibers on the next systole. A stronger contraction squeezes out more blood. Thus, as more blood is returned to the heart and EDV increases, the heart automatically pumps out a correspondingly larger stroke volume (SV).

Frank–Starling Law of the Heart What causes cardiac muscle fibers to vary in length before contraction? Skeletal muscle length can vary before contraction because of the positioning of the skeletal parts to which the muscle is attached, but cardiac muscle is not attached to any bones. The main determinant of cardiac muscle fiber length is the degree of diastolic filling. An analogy is a balloon filled with water—the more water you put in, the larger the balloon becomes, and the more it is stretched. Likewise, the greater the diastolic filling, the larger the EDV, and the more the heart is stretched. The more the heart is stretched, the longer the cardiac fibers before contraction. The increased length results in a greater force on the subsequent cardiac contraction and thus in a greater SV. This intrinsic relationship between EDV and SV is known as the **Frank–Starling law of the heart**. Stated simply, the law says that the heart normally pumps out during systole the volume of blood returned to it during diastole; increased venous return results in increased SV. In ❙Figure 9-21, assume that EDV increases from point A to point B. You can see that this increase in EDV is accompanied by a corresponding increase in SV from point A^1 to point B^1. The extent of filling is referred to as the **preload** because it is the workload imposed on the heart before contraction begins.

Advantages of the Cardiac Length–Tension Relationship The built-in relationship matching SV with venous return has two important advantages. First, this intrinsic mechanism equalizes output between the right and the left sides of the heart so that blood pumped out by the heart is equally distributed between the pulmonary and systemic circulations. If, for example, the right side of the heart ejects a larger SV, more blood enters the pulmonary circulation, so venous return to the left side of the heart increases accordingly. The increased EDV of the left side of the heart causes it to contract more forcefully, so it too pumps out a larger SV. In this way, output of the two ventricular chambers is kept equal. If such equalization did not happen, too much blood would be dammed up in the venous system before the ventricle with the lower output.

Second, when a larger CO is needed, as during exercise, venous return is increased by the sympathetic nervous system constricting the veins to drive blood forward and by the contracting muscles compressing the veins, which squeezes more blood toward the heart. The resulting increase in EDV automatically increases SV correspondingly. Because exercise also increases HR, these two factors act together to increase CO so that more blood can be delivered to the exercising muscles.

Mechanism of the Cardiac Length–Tension Relationship

Although the length–tension relationship in cardiac muscle fibers depends to a degree on the extent of overlap of thick and thin filaments, similar to the length–tension relationship in skeletal muscle, the key factor relating cardiac muscle fiber length to tension development is the dependence of myofilament Ca^{2+} sensitivity on the fiber's length. Specifically, as a cardiac muscle fiber is stretched as a result of greater ventricular filling, its myofilaments are pulled closer together side by side. As a result of this reduction in distance between the thick and thin filaments, more cross-bridge interactions between myosin and actin can take place when Ca^{2+} pulls the troponin–tropomyosin complex away from actin's cross-bridge binding sites—that is, myofilament Ca^{2+} sensitivity increases. Thus, the length–tension relationship in cardiac muscle depends not on muscle fiber length per se but on the resultant variations in the lateral spacing between the myosin and actin filaments.

We now shift from intrinsic to extrinsic control of SV.

Sympathetic stimulation increases the contractility of the heart.

In addition to intrinsic control, SV is subject to **extrinsic control** by factors originating outside the heart, the most important of which are actions of the cardiac sympathetic nerves and epinephrine (see ❙Table 9-1). Sympathetic stimulation and epinephrine enhance the heart's **contractility**, which is the strength of contraction at any given EDV. In other words, on sympathetic stimulation the heart contracts more forcefully and squeezes out a greater percentage of the blood it contains, leading to more complete ejection. This increased contractility results from the increased Ca^{2+} influx triggered by norepinephrine and epinephrine. The extra cytosolic Ca^{2+} lets the myocardial fibers generate more force through greater cross-bridge cycling than they would without sympathetic influence. Normally, the EDV is 135 mL and the end-systolic volume (ESV) is 65 mL for a SV of 70 mL (Figure 9-22a). Under sympathetic influence, for the same EDV of 135 mL, the ESV might be 35 mL and the SV 100 mL (Figure 9-22b). In effect, sympathetic stimulation shifts the Frank–Starling curve to the left (Figure 9-23). Depending on the extent of sympathetic stimulation, the curve can be shifted to varying degrees, up to a maximal increase in contractile strength of about 100% greater than normal.

Clinical Note The **ejection fraction** is the ratio of stroke volume to end-diastolic volume (ejection fraction = SV/EDV); that is, it is the proportion of the blood in the ventricle that is pumped out. The ejection fraction is often used clinically as an indication of contractility. A healthy heart normally has an ejection fraction of 50% to 75% under resting conditions and may go as high as 90% during strenuous exercise, but a failing heart may pump out 30% or less.

Sympathetic stimulation increases SV not only by strengthening cardiac contractility but also by enhancing venous return (see Figure 9-22c). Sympathetic stimulation constricts the veins, which squeezes more blood forward from the veins to the heart, increasing the EDV and subsequently increasing SV even further.

Summary of Factors Affecting Stroke Volume and Cardiac Output

The strength of cardiac muscle contraction and, accordingly, stroke volume can thus be graded by (1) varying the initial length of the muscle fibers, which in turn depends on the degree of ventricular filling before contraction (intrinsic control), and (2) varying the extent of sympathetic stimulation (extrinsic control) (see Figure 9-20). This is in contrast to gradation of skeletal muscle, in which twitch summation and recruitment of motor units play key roles in producing variable strength of muscle contraction. These mechanisms do not apply to cardiac muscle. In cardiac muscle, twitch summation is impossible because of the long refractory period. Recruitment of motor units is not possible because heart muscle cells are arranged into functional syncytia where all contractile cells become excited and contract with every beat, instead of into distinct motor units that can be discretely activated. Therefore, unlike skeletal muscle, where graded contractions can be produced by varying the number of muscle cells contracting within the muscle, either all cardiac muscle fibers contract or none do. A "halfhearted" contraction is not possible. Cardiac contraction is thus graded by varying the strength of contraction of all the cardiac muscle cells by intrinsic and extrinsic control mechanisms.

All the factors that determine CO by influencing HR or SV are summarized in Figure 9-24. Note that sympathetic stimulation increases CO by increasing both HR and SV. Sympathetic activity to the heart increases, for example, during exercise when the

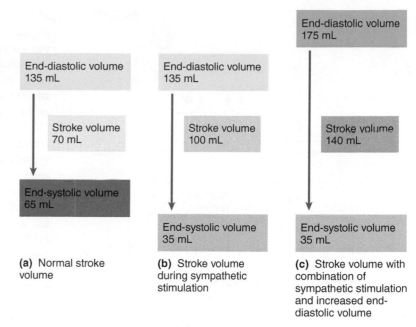

(a) Normal stroke volume

(b) Stroke volume during sympathetic stimulation

(c) Stroke volume with combination of sympathetic stimulation and increased end-diastolic volume

Figure 9-22 **Effect of sympathetic stimulation on stroke volume.**

working skeletal muscles need increased delivery of O_2-laden blood to support their high rate of ATP consumption.

We next examine how the afterload influences the ability of the heart to pump out blood.

High blood pressure increases the workload of the heart.

Clinical Note When the ventricles contract, to force open the semilunar valves they must generate sufficient pressure to exceed the blood pressure in the major arteries. The arterial blood pressure is called the **afterload** because it is the workload imposed

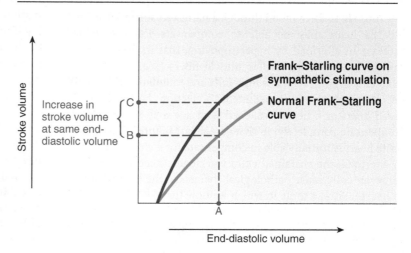

Figure 9-23 **Shift of the Frank–Starling curve to the left by sympathetic stimulation.** For the same end-diastolic volume (point A), a larger stroke volume (from point B to point C) is ejected on sympathetic stimulation as a result of increased contractility of the heart. The Frank–Starling curve is shifted to the left by variable degrees, depending on the extent of sympathetic stimulation.

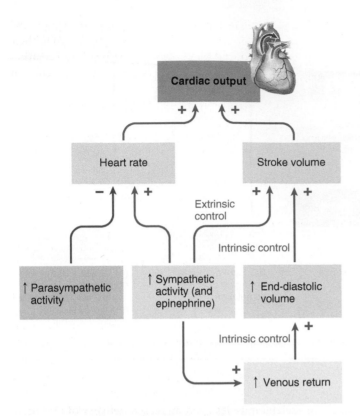

Figure 9-24 Control of cardiac output. Because cardiac output equals heart rate times stroke volume, this figure is a composite of Figure 9-19b (control of heart rate) and Figure 9-20 (control of stroke volume).

FIGURE FOCUS: *Through what regulatory mechanisms can a transplanted heart, which does not have any innervation, adjust cardiac output to meet the body's changing needs?*

on the heart after contraction has begun. If arterial blood pressure is chronically elevated (high blood pressure) or if the exit valve is stenotic, the ventricle must generate more pressure to eject blood. For example, instead of generating the normal pressure of 120 mm Hg, the ventricular pressure may need to rise as high as 400 mm Hg to force blood through a narrowed aortic valve.

The heart may be able to compensate for a sustained increase in afterload by hypertrophying, that is, by increasing the thickness of the cardiac muscle fibers (see p. 274). This enables it to contract more forcefully and maintain a normal SV despite an abnormal impediment to ejection. However, a diseased heart or a heart weakened with age may not be able to compensate completely; in that case, heart failure ensues. Even if the heart is initially able to compensate for a chronic increase in afterload, the sustained extra workload placed on the heart can eventually cause pathological changes in the heart that lead to heart failure, a topic to which we now turn.

A failing heart cannot pump out enough blood.

Heart failure (HF) is the inability of CO to keep pace with the body's demands for supplies and removal of wastes. Heart failure can be of two types: *systolic HF,* in

which the heart has difficulty pumping blood out, or *diastolic HF,* in which the heart has trouble filling. Traditionally all HF was considered systolic, but improvements in cardiac imaging technology, such as use of echocardiograms, have revealed that nearly half of all cases are diastolic HF. Heart failure presently affects 5 million Americans, with this number expected to rise as the population ages.

Defect in Systolic Heart Failure The prime defect in **systolic HF** is decreased cardiac contractility—that is, weakened cardiac muscle cells contract less effectively, resulting in a greatly reduced ejection fraction. Either one or both ventricles may progressively weaken and fail. Systolic HF may occur for a variety of reasons, but the two most common are (1) damage to the heart muscle as a result of a heart attack or impaired circulation to the cardiac muscle and (2) prolonged pumping against a chronically increased afterload, as with a sustained elevated blood pressure or a stenotic semilunar valve.

With systolic HF, the intrinsic ability of the heart to develop pressure and eject a SV is reduced so that the heart operates on a lower length–tension curve (Figure 9-25a). The Frank–Starling curve shifts downward and to the right such that, for a given EDV, a failing heart pumps out a smaller SV than a normal healthy heart does.

Compensatory Measures for Systolic Heart Failure In the early stages of systolic HF, two major compensatory measures help restore SV to normal. First, sympathetic activity to the heart is reflexly increased, which increases heart contractility toward normal (Figure 9-25b). Sympathetic stimulation can help compensate only for a limited time, however, because the heart becomes less responsive to norepinephrine after prolonged exposure, and furthermore, norepinephrine stores in the heart's sympathetic nerve terminals become depleted. Second, when CO is reduced, the kidneys, in a compensatory attempt to improve their reduced blood flow, retain extra salt and water in the body during urine formation to expand the blood volume. The increase in circulating blood volume increases the EDV. The resultant stretching of the cardiac muscle fibers enables the weakened heart to pump out a normal SV (Figure 9-25b). The heart is now pumping out the blood returned to it but is operating at a greater cardiac muscle fiber length.

Decompensated Systolic Heart Failure As the disease progresses and heart contractility deteriorates further, the heart reaches a point at which it can no longer pump out a normal SV despite compensatory measures. At this point, the heart slips from compensated HF into a state of decompensated HF. Now the cardiac muscle fibers are stretched to the point that they are operating in the descending limb of the length–tension curve. *Forward failure* occurs as the heart fails to pump an adequate amount of blood forward to the tissues because the SV becomes progressively smaller. *Backward failure* occurs simultaneously as the failing heart cannot pump out all of the blood returned to it (SV cannot keep pace with

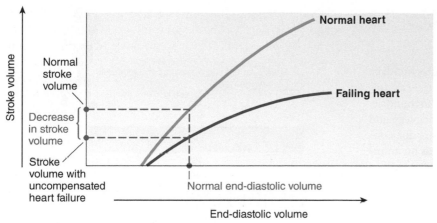

(a) Reduced contractility in a failing heart

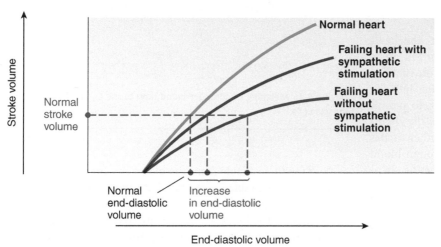

(b) Compensation for heart failure

▮Figure 9-25 Compensated heart failure. (a) The Frank–Starling curve shifts downward and to the right in a failing heart. Because its contractility is decreased, the failing heart pumps out a smaller stroke volume at the same end-diastolic volume than a normal heart does. (b) During compensation for heart failure, reflex sympathetic stimulation shifts the Frank–Starling curve of a failing heart to the left, increasing the contractility of the heart toward normal. A compensatory increase in end-diastolic volume as a result of blood volume expansion further increases the strength of contraction of the failing heart. Operating at a longer cardiac muscle fiber length, a compensated failing heart is able to eject a normal stroke volume.

venous return) so that the "backlogged" returning blood continues to dam up in the venous system. Congestion of blood in the venous system behind a failing ventricle is the reason this condition is sometimes termed **congestive heart failure**.

Left-sided failure has more serious consequences than right-sided failure. Backward failure of the left side leads to pulmonary edema (excess tissue fluid in the lungs) because blood dams up in the lungs. This fluid accumulation in the lungs reduces exchange of O_2 and CO_2 between air and blood in the lungs, reducing arterial oxygenation and elevating levels of acid-forming CO_2 in the blood. In addition, one of the more serious consequences of left-sided forward failure is inadequate blood flow to the kidneys, which causes a twofold problem. First, vital kidney function is depressed; second, the kidneys retain even more salt and water in the body during

urine formation as they try to expand the blood volume even further to improve their reduced blood flow. Excessive fluid retention worsens the already existing problems of venous congestion.

Treatment of congestive heart failure therefore includes measures that reduce salt and water retention and increase urinary output and drugs that enhance the contractile ability of the weakened heart—digitalis, for example. *Digitalis* increases cardiac contractility by causing accumulation of cytosolic Ca^{2+}.

Diastolic Heart Failure With the more recently recognized **diastolic HF**, the ventricles do not fill normally either because the heart muscle does not adequately relax between beats or because it stiffens and cannot expand as much as usual. The heart can pump properly and the ejection fraction is normal. For this reason, diastolic HF is alternatively called **heart failure with preserved ejection fraction**. However, even though the ejection fraction is normal (that is, the ventricles pump out a normal percentage of the blood present in their chambers), less blood than normal is pumped out by a diastolic failing heart because the ventricles are inadequately filled with blood. Both forward and backward failure, including congestive heart failure, can result, so symptoms are similar to systolic HF.

The abnormal ventricular relaxation and stiffness associated with diastolic HF are attributed to (1) excess collagen deposition in the ventricular muscle's extracellular matrix (see p. 60); (2) increased passive tension of titin, the giant elastic protein found in striated muscle (see p. 254); and (3) elevated resting cardiac muscle tension caused by slow or incomplete removal of cytosolic Ca^{2+} following an action potential.

No drugs are available yet that reliably help the heart relax, so treatment of diastolic HF is aimed at relieving symptoms, halting underlying causes, or lessening aggravating factors, such as by controlling high blood pressure.

Check Your Understanding 9.4

1. Indicate the effect (increase or decrease) of parasympathetic stimulation and sympathetic stimulation on heart rate and stroke volume.

2. Draw a graph showing the relationship between end-diastolic volume and stroke volume, according to the Frank–Starling law of the heart.

3. Define *preload* and *afterload*.

9.5 Nourishing the Heart Muscle

Cardiac muscle cells contain an abundance of mitochondria, the O_2-dependent energy organelles. Up to 40% of the cell volume of cardiac muscle cells is occupied by mitochondria, indicative of how much the heart depends on O_2 delivery and aerobic metabolism to generate the energy necessary for contraction (see pp. 34 and 39). Cardiac muscle also has an abundance of myoglobin, which stores limited amounts of O_2 within the heart for immediate use (see p. 271).

The heart receives most of its blood supply through the coronary circulation during diastole.

Although all the blood passes through the heart, the heart muscle cannot extract O_2 or nutrients from the blood within its chambers for two reasons. First, the watertight endothelial lining does not permit blood to pass from the chamber into the myocardium. Second, the heart walls are too thick to permit diffusion of O_2 and other supplies from the blood in the chamber to the individual cardiac cells. Therefore, like other tissues of the body, heart muscle must receive blood through blood vessels, specifically via the **coronary circulation** (see the chapter opener photo, p. 297 and Figure 9-29, p. 331). The coronary arteries branch from the aorta just beyond the aortic valve, and the coronary veins empty into the right atrium.

The heart muscle receives most of its blood supply during diastole. Blood flow to the heart muscle cells is substantially reduced during systole for two reasons: (1) The contracting myocardium compresses the major branches of the coronary arteries and (2) the open aortic valve partially blocks the entrance to the coronary vessels. Thus, about 70% of coronary arterial flow occurs during diastole, driven by the aortic blood pressure, with only 30% occurring during systole, driven by ventricular contraction.

This limited time for coronary blood flow becomes especially important during rapid heart rates, when diastolic time is much reduced. Just when increased demands are placed on the heart to pump more rapidly, it has less time to provide O_2 and nourishment to its own musculature to accomplish the increased workload.

Matching of Coronary Blood Flow to Heart Muscle's O_2 Needs Nevertheless, under normal circumstances, the heart muscle receives adequate blood flow to support its activities—even during exercise, when the rate of coronary blood flow increases up to five times its resting rate. Extra blood is delivered to the cardiac cells primarily by vasodilation, or enlargement, of the coronary vessels, which lets more blood flow through them, especially during diastole. The increased coronary blood flow is necessary to meet the heart's increased O_2 requirements because the heart, unlike most other tissues, is unable to remove much additional O_2 from the blood passing through its vessels to support increased metabolic activities. Most other tissues under resting conditions extract only about

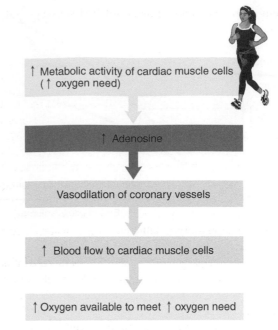

Figure 9-26 Matching of coronary blood flow to the O_2 need of cardiac muscle cells.

25% of the O_2 available from the blood flowing through them, leaving a considerable O_2 reserve that can be drawn on when a tissue has increased O_2 needs—that is, the tissue can immediately increase the O_2 available to it by removing a greater percentage of O_2 from the blood passing through it. In contrast, the heart, even under resting conditions, removes up to 65% of the O_2 available in the coronary vessels, far more than is withdrawn by other tissues. This leaves little O_2 in reserve in the coronary blood should cardiac O_2 demands increase. Therefore, the primary means by which more O_2 can be made available to the heart muscle is by increasing coronary blood flow.

Coronary blood flow is adjusted primarily in response to changes in the heart's O_2 requirements. Among the proposed links between blood flow and O_2 needs is *adenosine,* which is formed from adenosine triphosphate (ATP) during cardiac metabolic activity. Cardiac cells form and release more adenosine when cardiac activity increases, and the heart accordingly needs more O_2 and uses more ATP as an energy source. The released adenosine, acting as a paracrine (see p. 114), induces dilation of the coronary blood vessels, allowing more O_2-rich blood to flow to the more active cardiac cells to meet their increased O_2 demand (Figure 9-26). This matching of O_2 delivery to O_2 needs is crucial because heart muscle depends on oxidative processes to generate energy. The heart cannot get enough ATP through anaerobic metabolism (see p. 39).

Nutrient Supply to the Heart Although the heart must rely heavily on its O_2 supply to generate ATP, it can tolerate wide variations in its nutrient supply. As fuel sources, the heart primarily uses fatty acids and, to a lesser extent, glucose and lactate, depending on their availability. Because cardiac muscle is adaptable and can shift metabolic pathways to use whatever nutrient is available, the primary danger of insufficient coronary blood flow is not fuel shortage but O_2 deficiency.

Atherosclerotic coronary artery disease can deprive the heart of essential O_2.

Clinical Note Adequacy of coronary blood flow is relative to the heart's O_2 demands at any moment. In the normal heart, coronary blood flow increases correspondingly as O_2 demands rise. With coronary artery disease, coronary blood flow may not be able to keep pace with rising O_2 needs. The term **coronary artery disease (CAD)** refers to pathological changes within the coronary artery walls that diminish blood flow through these vessels. A given rate of coronary blood flow may be adequate at rest but insufficient in physical exertion or other stressful situations.

CAD is the underlying cause of about 50% of all deaths in the United States and worldwide. More people die of complications of CAD than from all cancers combined. CAD can cause myocardial ischemia and possibly lead to a heart attack by three mechanisms: (1) profound vascular spasm of the coronary arteries, (2) formation of atherosclerotic plaques, and (3) thromboembolism. We discuss each in turn.

Vascular Spasm **Vascular spasm** is an abnormal spastic constriction that transiently narrows the coronary vessels. Vascular spasms are associated with the early stages of CAD and are most often triggered by exposure to cold, physical exertion, or anxiety. The condition is reversible and usually does not last long enough to damage the cardiac muscle.

When too little O_2 is available in the coronary vessels, the endothelium (blood vessel lining) releases *platelet-activating factor (PAF)*. PAF, which exerts a variety of actions, was named for its first discovered effect, activating platelets. Among its other effects, PAF, once released from the endothelium, diffuses to the underlying vascular smooth muscle and causes it to contract, bringing about vascular spasm.

Development of Atherosclerosis **Atherosclerosis** is a progressive, degenerative arterial disease that leads to occlusion (gradual blockage) of affected vessels, reducing blood flow through them. Atherosclerosis is characterized by plaques forming beneath the vessel lining within arterial walls. An **atherosclerotic plaque** consists of a lipid-rich core covered by an abnormal overgrowth of smooth muscle cells, topped off by a collagen-rich connective tissue cap. As the plaque forms, it bulges into the vessel lumen (I Figure 9-27).

Although all the contributing factors have not yet been identified, in recent years investigators have sorted out the following complex sequence of events in the gradual development of atherosclerosis:

1. Atherosclerosis starts with injury to the blood vessel wall, which triggers an *inflammatory response* that sets the stage for plaque buildup. Normally, inflammation is a protective response that fights infection and promotes repair of damaged tissue (see p. 408). However, when the cause of the injury persists within the vessel wall, the sustained, low-grade inflammatory response over a course of decades can insidiously lead to

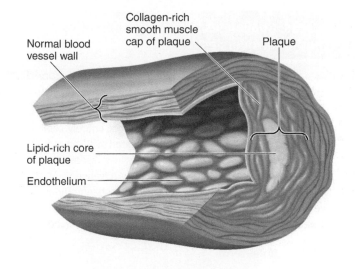

Collagen-rich smooth muscle cap of plaque
Normal blood vessel wall
Plaque
Lipid-rich core of plaque
Endothelium

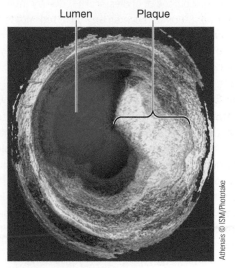

Lumen   Plaque

Athenais © ISM/Phototake

I Figure 9-27 **Atherosclerotic plaque in a coronary vessel.**

arterial plaque formation and heart disease. Plaque formation likely has many causes. Suspected artery-abusing agents that may set off the vascular inflammatory response include oxidized cholesterol, free radicals, high blood pressure, homocysteine, chemicals released from fat cells, or even bacteria and viruses that damage blood vessel walls. The most common triggering agent appears to be oxidized cholesterol. (For further discussion of the role of cholesterol and other factors in the development of atherosclerosis, see the boxed feature on pp. 328–329, I Concepts, Challenges, and Controversies.)

2. Typically, the initial stage of atherosclerosis is characterized by accumulation beneath the endothelium of excessive amounts of *low-density lipoprotein (LDL)*, the so-called bad cholesterol in combination with a protein carrier. As LDL accumulates within the vessel wall, this cholesterol product becomes oxidized, primarily by oxidative wastes produced by the blood vessel cells. These wastes are *free radicals*, very unstable electron-deficient particles that are highly reactive and cause cell damage by snatching electrons from other molecules. Antioxidant vitamins that prevent LDL oxidation, such as *vitamin E, vitamin C,* and *beta-carotene,* slow plaque deposition.

Atherosclerosis: Cholesterol and Beyond

THE CAUSE OF ATHEROSCLEROSIS IS still not entirely clear. Certain high-risk factors have been associated with an increased incidence of atherosclerosis and coronary artery disease. Included among them are genetic predisposition, obesity, advanced age, smoking, hypertension, lack of exercise, high blood concentrations of C-reactive protein, elevated levels of homocysteine, infectious agents, and (most notoriously) elevated cholesterol levels in the blood.

Sources of Cholesterol

The body has two sources of cholesterol: (1) dietary intake of cholesterol, with animal products such as egg yolk, red meats, and butter being especially rich in this lipid (animal fats contain cholesterol, whereas plant fats typically do not), and (2) manufacture of cholesterol by cells, especially liver cells.

"Good" versus "Bad" Cholesterol

Actually, the amount of cholesterol bound to various plasma protein carriers, not the total blood cholesterol, is most important to the risk of developing atherosclerotic heart disease. Because cholesterol is a lipid, it is not very soluble in blood. Most cholesterol in the blood is attached to specific plasma protein carriers in the form of lipoprotein complexes, which are soluble in blood. The three major lipoproteins are named for their density of protein as compared to lipid: (1) **high-density lipoproteins (HDL)**, which contain the most protein and least cholesterol; (2) **low-density lipoproteins (LDL)**, which have less protein and more cholesterol; and (3) **very-low-density lipoproteins (VLDL)**, which have the least protein and most lipid, but the lipid they carry is neutral fat, not cholesterol.

Cholesterol carried in LDL complexes has been termed "bad" cholesterol because cholesterol is transported *to* the cells, including those lining the blood vessel walls, by LDL. The propensity toward developing atherosclerosis substantially increases with elevated levels of LDL. The presence of oxidized LDL within an arterial wall is a major trigger for the inflammatory process that leads to development of atherosclerotic plaques (see p. 327).

In contrast, cholesterol carried in HDL complexes has been dubbed "good" cholesterol because HDL removes cholesterol *from* the cells and transports it to the liver for partial elimination from the body. HDL also helps protect against formation of atherosclerotic plaques by inhibiting oxidation of LDL. Furthermore, HDL has anti-inflammatory action, helps stabilize atherosclerotic plaques so that they are less prone to rupture, and reduces clot formation, all actions that counter progressive development of atherosclerosis. The risk of atherosclerosis is inversely related to HDL concentration in the blood—that is, higher HDL levels are associated with a lower incidence of atherosclerotic heart disease.

Some other factors known to influence atherosclerotic risk can be related to HDL levels; for example, cigarette smoking lowers HDL, whereas regular exercise raises HDL.

Cholesterol Uptake by Cells

Unlike most lipids, cholesterol is not used as metabolic fuel by cells. Instead, it is an essential component of plasma membranes. In addition, a few special cell types use cholesterol as a precursor for synthesis of secretory products, such as steroid hormones and bile salts. Although most cells can synthesize some of the cholesterol needed for their plasma membranes, they cannot manufacture sufficient amounts and therefore must rely on supplemental cholesterol being delivered by the blood. Cells accomplish cholesterol uptake from the blood by synthesizing receptor proteins specifically capable of binding LDL and inserting these receptors into the plasma membrane. When an LDL particle binds to one of the membrane receptors, the cell engulfs the particle by receptor-mediated endocytosis, receptor and all (see p. 31). Within the cell, lysosomal enzymes break down the LDL to free the cholesterol, making it available to the cell for synthesis of new cellular membrane. The LDL receptor, which is also freed within the cell, is recycled back to the surface membrane.

If too much free cholesterol accumulates in the cell, the cell shuts down synthesis of LDL receptor proteins so that it takes up less cholesterol from the blood. Faced with a cholesterol shortage, in contrast, the cell makes more LDL receptors so that it can engulf more cholesterol from the blood.

Maintenance of Blood Cholesterol Level and Cholesterol Metabolism

Maintaining a constant blood-borne cholesterol supply to the cells involves an interaction between dietary cholesterol and synthesis of

3. In response to the presence of oxidized LDL or other irritants, the endothelial cells produce chemicals that attract *monocytes*, a type of white blood cell, to the site. These immune cells trigger a local inflammatory response.

4. Once they leave the blood and enter the vessel wall, monocytes settle down permanently, enlarge, and become large phagocytic cells called *macrophages*. Macrophages voraciously phagocytize (see p. 31) the oxidized LDL until these cells become so packed with fatty droplets that they appear foamy under a microscope. Now called *foam cells*, these engorged macrophages accumulate beneath the vessel lining and form a visible *fatty streak*, the beginning of an atherosclerotic plaque.

5. Thus, the earliest stage of a plaque is accumulation beneath the endothelium of a cholesterol-rich deposit. The disease progresses as smooth muscle cells within the blood vessel wall migrate from the muscular layer of the blood vessel to a position on top of the lipid accumulation, just beneath the endothelium. This migration is triggered by chemicals released at

cholesterol by the liver. When the amount of dietary cholesterol is increased, hepatic (liver) synthesis of cholesterol is turned off because cholesterol in the blood directly inhibits a hepatic enzyme essential for cholesterol synthesis. Thus, as more cholesterol is ingested, the liver produces less. Conversely, when cholesterol intake from food is reduced, the liver synthesizes more of this lipid because the inhibitory effect of cholesterol on the crucial hepatic enzyme is removed. In this way, the blood concentration of cholesterol is maintained at a fairly constant level despite changes in cholesterol intake; thus, it is difficult to significantly reduce cholesterol levels in the blood by decreasing cholesterol intake.

HDL transports cholesterol to the liver. The liver secretes cholesterol, and cholesterol-derived bile salts, into the bile. Bile enters the intestinal tract, where bile salts participate in the digestive process. Most of the secreted cholesterol and bile salts are subsequently reabsorbed from the intestine into the blood to be recycled to the liver. However, the cholesterol and bile salts not reclaimed by absorption are eliminated in the feces and lost from the body. Thus, the liver has a primary role in determining total blood cholesterol levels, and the interplay between LDL and HDL determines the traffic flow of cholesterol between the liver and other cells.

Varying the intake of dietary fatty acids may alter total blood cholesterol levels by influencing one or more of the mechanisms involving cholesterol balance. The blood cholesterol level tends to be raised by ingesting saturated fatty acids found predominantly in animal fats because these fatty acids stimulate cholesterol synthesis and inhibit its conversion to bile salts. In contrast, ingesting polyunsaturated fatty acids, the predominant fatty acids of most plants, tends to reduce blood cholesterol levels by enhancing elimination of both cholesterol and cholesterol-derived bile salts in the feces.

Risk Factors Besides Cholesterol

Despite the strong links between cholesterol and heart disease, more than half of all patients with heart attacks have a normal cholesterol profile and no other well-established risk factors. Clearly, other factors are involved in the development of coronary artery disease in these people. These same factors may also contribute to development of atherosclerosis in people with unfavorable cholesterol levels. The following are among the leading other possible risk factors:

■ Elevated blood levels of **homocysteine** have been implicated as a strong predictor for heart disease, independent of the person's cholesterol or lipid profile. Homocysteine is formed as an intermediate product during metabolism of the essential dietary amino acid *methionine*. Investigators believe homocysteine contributes to atherosclerosis by promoting proliferation of vascular smooth muscle cells, an early step in development of this condition. Furthermore, homocysteine appears to damage endothelial cells and cause oxidation of LDL, both of which contribute to plaque formation. Three B vitamins—*folic acid, vitamin B_{12}* and *vitamin B_6*—all are important in pathways that clear homocysteine from the blood.

■ People with elevated levels of **C-reactive protein (CRP)**, a blood-borne marker of inflammation, have a higher risk for developing coronary artery disease. About half of all people who have a heart attack have high CRP, whereas high CRP is much less common in those without heart disease. Because inflammation plays a crucial role in the development of atherosclerosis, anti-inflammatory drugs such as aspirin help prevent heart attacks. Furthermore, aspirin protects against heart attacks through its role in inhibiting clot formation. Also, *statin* drugs not only lower LDL, but also have anti-inflammatory effects.

■ Accumulating data implicate an infectious agent as the underlying culprit in some cases of atherosclerotic disease. Among the leading suspects are respiratory infection–causing *Chlamydia pneumoniae,* cold sore–causing herpes virus, and gum disease–causing bacteria. If a link between infections and coronary artery disease can be confirmed, antibiotics may be added as a heart-disease prevention strategy.

As you can see, the relationships among atherosclerosis, cholesterol, and other factors are far from clear. Much research on this complex disease is currently in progress because the incidence of atherosclerosis is so high and its consequences are potentially fatal.

the inflammatory site. At their new location, the smooth muscle cells continue to divide and enlarge. Together the lipid-rich core and overlying smooth muscle form a maturing plaque.

6. As it continues to develop, the plaque progressively bulges into the lumen of the vessel, narrowing the opening through which blood can flow.

7. Further contributing to vessel narrowing, oxidized LDL inhibits release of *nitric oxide* from the endothelial cells. Nitric oxide is a local chemical messenger that relaxes the underlying layer of normal smooth muscle cells within the vessel wall. Relaxation of these smooth muscle cells dilates the vessel. Because of reduced nitric oxide release, vessels damaged by developing plaques cannot dilate as readily as normal.

8. A thickening plaque also interferes with nutrient exchange for cells located within the involved arterial wall, leading to degeneration of the wall in the vicinity of the plaque. The damaged area is invaded by *fibroblasts* (scar-forming cells), which form a collagen-rich connective tissue cap over the plaque.

9. Late in the disease, Ca^{2+} often precipitates in the plaque. A vessel so afflicted becomes hard and cannot distend easily.

Thromboembolism and Other Complications of Atherosclerosis Atherosclerosis attacks arteries throughout the body, but the most serious consequences involve damage to the vessels of the brain and heart. Atherosclerosis in regions other than the brain and heart is known as **peripheral artery disease (PAD)**, which most commonly reduces blood flow to the legs. The resultant intermittent leg pain when circulation in the legs becomes inadequate during muscular exertion may be a warning that atherosclerosis is present in the brain and heart, too. In the brain, atherosclerosis is the prime cause of strokes, whereas in the heart it brings about myocardial ischemia and its complications. The following are potential complications of coronary atherosclerosis:

■ *Angina pectoris.* Gradual enlargement of a protruding plaque continues to narrow the vessel lumen and progressively diminishes coronary blood flow, triggering increasingly frequent bouts of transient myocardial ischemia as the ability to match blood flow with cardiac O_2 needs becomes more limited. Although the heart cannot normally be "felt," pain is associated with myocardial ischemia. Such cardiac pain, known as **angina pectoris** (meaning "pain of the chest"), can be felt beneath the sternum and is often referred to (or appears to come from) the left shoulder and down the left arm (see p. 176). The symptoms of angina pectoris recur whenever cardiac O_2 demands become too great in relation to the coronary blood flow—for example, during exertion or emotional stress. The pain is thought to result from stimulation of cardiac nerve endings by the accumulation of lactate when the heart shifts to its limited ability to perform anaerobic metabolism. The ischemia associated with the characteristically brief angina attacks is usually temporary and reversible and can be relieved by rest, taking vasodilator drugs such as *nitroglycerin,* or both. Nitroglycerin brings about coronary vasodilation by being metabolically converted to nitric oxide, which in turn relaxes the vascular smooth muscle.

■ *Thromboembolism.* The enlarging atherosclerotic plaque can rupture, or break through the weakened endothelial lining that covers it, a process that can trigger clot formation. Foam cells release chemicals that weaken the fibrous cap of a plaque by breaking down the connective tissue fibers. Plaques with thick fibrous caps are considered stable because they are not likely to rupture. However, plaques with thinner fibrous caps are unstable because they are likely to rupture and trigger clot formation.

When a plaque ruptures through the endothelium, blood is exposed to collagen in the plaque's collagen-rich connective tissue cap. Blood platelets (formed elements of the blood involved in plugging vessel defects and in clot formation) normally do not adhere to smooth, healthy vessel linings. However, when platelets contact collagen at the site of vessel damage, they stick to the site and help promote the formation of a blood clot. Furthermore, foam cells produce a potent clot promoter. Such an abnormal clot attached to a vessel wall is called a **thrombus**. The thrombus may enlarge gradually until it completely blocks the vessel at that site, or the continued flow of blood past the thrombus may break it loose. As it heads downstream, such a freely floating clot, or **embolus**, may completely plug a smaller vessel (Figure 9-28). Thus, through **thromboembolism**, atherosclerosis can result in a gradual or sudden occlusion of a coronary vessel (or any other vessel).

■ *Heart attack.* When a coronary vessel is completely plugged, the cardiac tissue served by the vessel soon dies from O_2 deprivation and a heart attack occurs, unless the area can be supplied with blood from nearby vessels.

Sometimes a deprived area of cardiac tissue is lucky enough to receive blood from more than one pathway. **Collateral circulation** exists when small terminal branches from adjacent blood vessels nourish the same area. These accessory vessels cannot develop suddenly after an abrupt blockage but may be lifesaving if already developed. Such alternate vascular pathways often develop over a period of time when an atherosclerotic constriction progresses slowly, or they may be induced by sustained demands on the heart through regular aerobic exercise.

In the absence of collateral circulation, the extent of the damaged area during a heart attack depends on the size of the blocked vessel: The larger the vessel occluded, the greater the area deprived of blood supply. As Figure 9-29 illustrates, a blockage at point A in the coronary circulation would cause more extensive damage than a blockage at point B would. Because there are only two major coronary arteries, complete blockage of either one of these main branches results in extensive myocardial damage. Left coronary artery blockage is most devastating because this vessel supplies blood to 85% of the cardiac tissue.

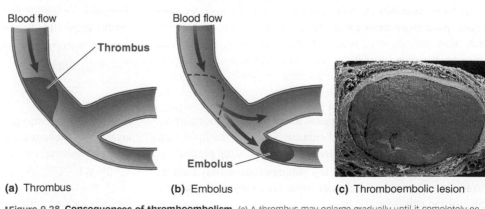

(a) Thrombus   **(b)** Embolus   **(c)** Thromboembolic lesion

Figure 9-28 **Consequences of thromboembolism.** (a) A thrombus may enlarge gradually until it completely occludes the vessel at that site. (b) A thrombus may break loose from its attachment, forming an embolus that may completely occlude a smaller vessel downstream. (c) Scanning electron micrograph of a vessel completely occluded by a thromboembolic lesion.

Moredun Animal Health Ltd./Science Source

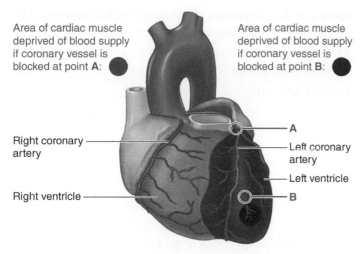

Area of cardiac muscle deprived of blood supply if coronary vessel is blocked at point **A**:

Area of cardiac muscle deprived of blood supply if coronary vessel is blocked at point **B**:

Right coronary artery

Right ventricle

A

Left coronary artery

Left ventricle

B

I Figure 9-29 **Extent of myocardial damage as a function of the size of the occluded vessel.**

The four possible outcomes of a heart attack are:

1. *Immediate death* resulting from (a) acute systolic heart failure occurring because the heart is too weakened to pump effectively to support the body tissues, or (b) fatal ventricular fibrillation brought about by damage to the specialized conducting tissue or induced by O_2 deprivation

2. *Delayed death from complications* because of (a) fatal rupture of the dead, degenerating area of the heart wall affected by the acute infarction, or (b) slowly progressing congestive heart failure because the pumping ability of the weakened heart is unable to keep up with venous return

3. *Full functional recovery* owing to replacement of the damaged area with a strong scar, accompanied by hypertrophy (enlargement) of the remaining normal contractile tissue to compensate for the lost cardiac musculature

4. *Recovery with impaired function* on account of persistence of permanent functional defects, such as bradycardia or conduction blocks, caused by destruction of irreplaceable autorhythmic or conductive tissues

The discovery in 2006 of cardiac stem cells, coupled with earlier studies demonstrating that injecting damaged hearts with stem cells derived from other sources improves cardiac function, has generated hopes of future regenerative therapies. Even though the heart has the potential to produce new myocardial cells on its own through activity of its stem cells, this regenerative ability is minimally functional. Only 1% of heart muscle cells are replaced each year in young adults, and the yearly turnover rate drops to less than half of that in the elderly. For this reason, damaged heart muscle cells are replaced with scar tissue, not new muscle cells. Scientists hope to find a way to spur the latent stem cells into action so that myocardial cells can be replaced if they are lost. Another approach under investigation is to genetically reprogram scar tissue cells (fibroblasts) to transform them into heart muscle cells after a heart attack.

1. Explain why heart muscle receives most of its blood supply during diastole.
2. Discuss how coronary blood flow varies to match the heart muscle's O_2 needs.
3. Describe an atherosclerotic plaque.

Homeostasis: Chapter in Perspective

Survival depends on continual delivery of needed supplies to all body cells and on ongoing removal of wastes generated by the cells. Furthermore, regulatory chemical messengers, such as hormones, must be transported from their production site to their action site, where they control a variety of activities, most of which are directed toward maintaining a stable internal environment. Finally, to maintain normal body temperature, excess heat produced during muscle contraction must be carried to the skin, where the heat can be lost from the body surface.

The circulatory system contributes to homeostasis by serving as the body's transport system. It provides a way to rapidly move materials from one part of the body to another. Without the circulatory system, materials would not get quickly enough to where they need to be to support life-sustaining activities. For example, O_2 would take months to years to diffuse from the body surface to internal organs, yet through the heart's swift pumping action the blood can pick up and deliver O_2 and other substances to all the cells in a few seconds.

The heart is a dual pump that continuously circulates blood between the lungs, where O_2 is picked up, and the other body tissues, which use O_2 to support their energy-generating chemical reactions. As blood is pumped through the various tissues, other substances besides O_2 are exchanged between the blood and the tissues. For example, blood picks up nutrients as it flows through the digestive organs, and other tissues remove nutrients from blood as it flows through them. Even excess heat is transported by blood from exercising muscles to the skin surface, where it is lost to the external environment.

Although all body tissues constantly depend on the life-supporting blood flow provided to them by the heart, the heart itself is quite an independent organ. It can take care of many of its needs without any outside influence. Contraction of this magnificent muscle is self-generated through a carefully orchestrated interplay of changing ionic permeabilities. Local mechanisms within the heart ensure that blood flow to the cardiac muscle normally meets the heart's need for O_2. In addition, the heart has built-in capabilities to vary its strength of contraction, depending on the amount of blood returned to it. The heart does not act entirely autonomously, however. It is innervated by the autonomic nervous system and is influenced by the hormone epinephrine, both of which can vary heart rate and contractility, depending on the body's needs for blood delivery. Furthermore, as with all tissues, the cells that make up the heart depend on the other body systems to maintain a stable internal environment in which they can survive and function.

Review Exercises <small>Answers begin on p. A-32</small>

Reviewing Terms and Facts

1. The heart lies in the left half of the thoracic cavity. (*True or false?*)

2. The left ventricle is a stronger pump than the right ventricle because more blood is needed to supply the body tissues than to supply the lungs. (*True or false?*)

3. The only point of electrical contact between the atria and the ventricles is the fibrous skeletal rings that surround and support the heart valves. (*True or false?*)

4. Entrance of Ca^{2+} through funny channels is responsible for the unique plateau phase of the action potential in cardiac contractile cells. (*True or false?*)

5. The atria and the ventricles each act as a functional syncytium. (*True or false?*)

6. Contraction of the papillary muscles tugs on the chordae tendinae to pull open the AV valves during diastole. (*True or false?*)

7. The three components of the circulatory system are _____, _____, and _____.

8. The three layers of the heart wall are _____, _____, and _____.

9. Adjacent cardiac muscle cells are joined end to end at specialized structures known as _____, which contain two types of membrane junctions: _____ and _____.

10. _____% of ventricular filling is normally accomplished before atrial contraction begins.

11. The link that coordinates coronary blood flow with myocardial oxygen needs is _____.

12. Which of the following is the proper sequence of cardiac excitation?

   a. SA node → AV node → atrial myocardium → bundle of His → Purkinje fibers → ventricular myocardium

   b. SA node → atrial myocardium → AV node → bundle of His → ventricular myocardium → Purkinje fibers

   c. SA node → atrial myocardium → ventricular myocardium → AV node → bundle of His → Purkinje fibers

   d. SA node → atrial myocardium → AV node → bundle of His → Purkinje fibers → ventricular myocardium

13. Sympathetic stimulation of the heart

   a. increases the heart rate

   b. increases the contractility of the heart muscle

   c. shifts the Frank–Starling curve to the left

   d. both (a) and (b)

   e. all of the above

14. Match the following:

   1. receives O_2-poor blood from the venae cavae

   2. prevents backflow of blood from the ventricles to the atria

   3. pumps O_2-rich blood into the aorta

   4. prevents backflow of blood from the arteries into the ventricles

   5. pumps O_2-poor blood into the pulmonary artery

   6. receives O_2-rich blood from the pulmonary veins

   (a) AV valves
   (b) semilunar valves
   (c) left atrium
   (d) left ventricle
   (e) right atrium
   (f) right ventricle

15. Circle the correct choice in each instance to complete the statement: The first heart sound is associated with closing of the (*AV/semilunar*) valves and signals the onset of (*systole/diastole*), whereas the second heart sound is associated with closing of the (*AV/semilunar*) valves and signals the onset of (*systole/diastole*).

16. Use the following answer code to compare the relative magnitudes of the pair of items in question:

(a) = Item A is greater than item B

(b) = Item B is greater than item A

(c) = Item A and item B are approximately equal

   1. A. Resistance and pressure in pulmonary circulation

      B. Resistance and pressure in systemic circulation

   2. A. Volume of blood pumped out by left side of heart

      B. Volume of blood pumped out by right side of heart

3. A. Spontaneous rate of depolarization to threshold in SA node

   B. Spontaneous rate of depolarization to threshold in ventricular Purkinje fibers

4. A. Velocity of impulse conduction through AV node

   B. Velocity of impulse conduction through bundle of His and Purkinje fibers

5. A. Rate of ventricular filling in early diastole

   B. Rate of ventricular filling in late diastole

6. A. Stroke volume when EDV equals 130 mL

   B. Stroke volume when EDV equals 160 mL

7. A. Normal stroke volume

   B. Stroke volume on sympathetic stimulation

8. A. Normal stroke volume

   B. Stroke volume on parasympathetic stimulation

9. A. Volume of blood in ventricles at onset of isovolumetric ventricular contraction

   B. Volume of blood in ventricles at end of isovolumetric ventricular contraction

10. A. Volume of blood in left ventricle at the time aortic valve opens

    B. Volume of blood in left ventricle at the time aortic valve closes

11. A. Volume of blood in left ventricle at the time left AV valve opens

    B. Volume of blood in left ventricle at the time left AV valve closes

12. A. Duration of refractory period in cardiac muscle

    B. Duration of contraction in cardiac muscle

Understanding Concepts

(Answers at www.cengagebrain.com)

1. Trace a drop of blood through one complete circuit of the circulatory system.

2. Describe the location and function of each of the four heart valves. What keeps each of these valves from everting?

3. Describe the structure and arrangement of cardiac muscle cells. What are the two specialized types of cardiac muscle cells?

4. Discuss the ionic movements involved in the membrane clock mechanism and Ca^{2+} clock mechanism that are collectively responsible for the pacemaker potential.

5. Why is the SA node the pacemaker of the heart?

6. Describe the normal spread of cardiac excitation. What is the significance of the AV nodal delay? Why is the ventricular conduction system important?

7. Compare the changes in permeability, ionic movements, and membrane potential associated with an action potential in a nodal pacemaker cell with those in a myocardial contractile cell.

8. Why is tetanus of cardiac muscle impossible? Why is this inability advantageous?

9. What electrical event does each component of the ECG represent?

10. Describe the mechanical events (that is, pressure changes, volume changes, valve activity, and heart sounds) of the cardiac cycle. Correlate the mechanical events of the cardiac cycle with the changes in electrical activity.

11. Compare the defect, murmur associated with, and circulatory consequences of a stenotic and an insufficient valve.

12. Distinguish among cardiac output, ejection fraction, and cardiac reserve.

13. Discuss autonomic nervous system control of heart rate.

14. Describe intrinsic and extrinsic control of stroke volume.

15. What are the pathological changes and consequences of coronary artery disease?

16. Discuss the sources, transport, and elimination of cholesterol in the body. Distinguish between "good" cholesterol and "bad" cholesterol.

Solving Quantitative Exercises

1. During heavy exercise, the CO of a trained athlete may increase to 40 liters per minute. If SV could not increase above the normal value of 70 mL, what HR would be necessary to achieve this CO? Is such a HR physiologically possible?

2. How much blood remains in the heart after systole if the SV is 85 mL and the EDV is 125 mL?

3. Calculate the ejection fraction in each of the three circumstances illustrated in Figure 9-22, p. 323.

Applying Clinical Reasoning

In a physical exam Rachel B.'s heart rate was rapid and very irregular. Furthermore, her heart rate, determined directly by listening to her heart with a stethoscope, exceeded the pulse rate taken concurrently at her wrist. No definite P waves could be detected on Rachel's ECG. The QRS complexes were normal in shape but occurred sporadically. Given these findings, what is the most likely diagnosis of Rachel's condition? Explain why the condition is characterized by a rapid, irregular heartbeat. Would CO be seriously impaired by this condition? Why or why not? What accounts for the pulse deficit?

Thinking at a Higher Level

1. The stroke volume ejected on the next heartbeat after a premature ventricular contraction (PVC) is usually larger than normal. Can you explain why? (*Hint:* At a given heart rate, the interval between a PVC and the next normal beat is longer than the interval between two normal beats.)

2. Trained athletes usually have lower resting heart rates than normal (for example, 50 beats per minute in an athlete compared to 70 beats per minute in a sedentary individual). Considering that the resting CO is 5000 mL per minute in both trained athletes and sedentary people, what is responsible for the bradycardia of trained athletes?

3. During fetal life, because of the tremendous resistance offered by the collapsed, nonfunctioning lungs, the pressures in the right half of the heart and pulmonary circulation are higher than those in the left half of the heart and systemic circulation, a situation that reverses after birth. Also in the fetus, a vessel called the **ductus arteriosus** connects the pulmonary artery and aorta as these major vessels both leave the heart. The blood pumped out by the heart into the pulmonary circulation is shunted from the pulmonary artery into the aorta through the ductus arteriosus, bypassing the nonfunctional lungs. What force is driving blood to flow in this direction through the ductus arteriosus?

At birth, the ductus arteriosus normally collapses and eventually degenerates into a thin, ligamentous strand. On occasion, this fetal bypass fails to close properly at birth, leading to a patent (open) ductus arteriosus. In what direction would blood flow through a patent ductus arteriosus? What possible outcomes would you predict might occur as a result of this blood flow?

4. There are two branches of the bundle of His, the right and left bundle branches, each of which travels down its respective side of the ventricular septum (see ❙ Figure 9-8, p. 305). Occasionally, conduction through one of these branches becomes blocked (so-called *bundle-branch block*). In this case, the wave of excitation spreads out from the terminals of the intact branch and eventually depolarizes the whole ventricle, but the normally stimulated ventricle completely depolarizes a considerable time before the ventricle on the side of the defective bundle branch. For example, if the left bundle branch is blocked, the right ventricle will be completely depolarized two to three times more rapidly than the left ventricle. How would this defect affect the heart sounds?

5. Occasionally a child is born with a defective aortic valve that is both stenotic and insufficient. The abnormally shaped valve leaflets neither open nor close properly. List the sequence of sounds that would be heard when listening to the heart with a stethoscope, taking into account the timing and type of murmur(s).

To access the course materials and companion resources for this text, please visit www.cengagebrain.com

The Blood Vessels and Blood Pressure

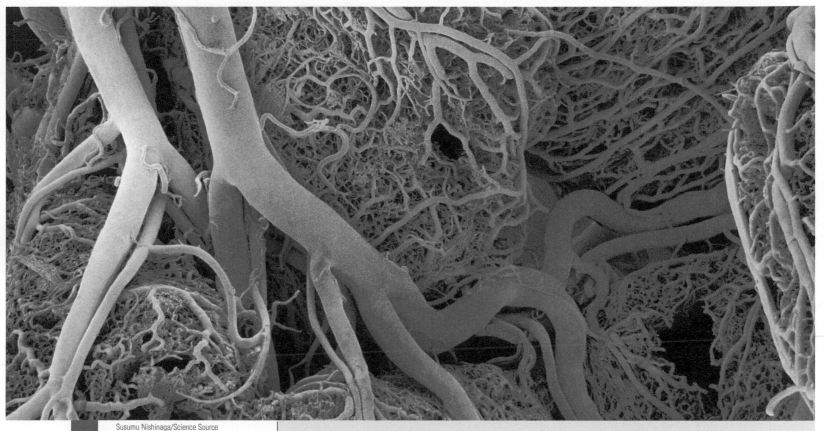

Susumu Nishinaga/Science Source

A scanning electron micrograph of a resin cast of blood vessels supplying the small intestine. After liquid resin injected into the vessels hardens, the tissues are chemically digested leaving a cast of the highly branched vessels. The myriad smallest blood vessels are capillaries across which materials are exchanged between the blood and surrounding cells.

Homeostasis Highlights

The circulatory system contributes to homeostasis by serving as the body's transport system. The **blood vessels** transport and distribute blood pumped through them by the heart to meet the body's needs for O_2 and nutrient delivery, waste removal, and hormonal signaling. The highly elastic **arteries** transport blood from the heart to the organs and serve as a pressure reservoir to continue driving blood forward when the heart is relaxing and filling. The **mean arterial blood pressure** is closely regulated to ensure adequate blood delivery to the organs. The amount of blood that flows through a given organ depends on the caliber (internal diameter) of the highly muscular **arterioles** that supply the organ. Arteriolar caliber is subject to control so that flow to particular organs can be variably adjusted to best serve the body's needs at the moment. The thin-walled **capillaries** are the actual site of exchange between blood and surrounding tissue cells. The highly distensible **veins** return blood from the organs to the heart and serve as a blood reservoir.

10.1 Patterns and Physics of Blood Flow

Most body cells are not in direct contact with the external environment, yet these cells must make exchanges with this environment, such as picking up O_2 and nutrients and eliminating wastes. Furthermore, chemical messengers must be transported between cells to accomplish integrated activity. To achieve these long-distance exchanges, cells are linked with one another and with the external environment by vascular (blood vessel) highways. Blood is transported to all parts of the body through a system of vessels that brings fresh supplies to the vicinity of all cells while removing their wastes.

To review, all blood pumped by the right side of the heart passes through the pulmonary circulation to the lungs for O_2 pickup and CO_2 removal. The blood pumped by the left side of the heart into the systemic circulation is distributed in various proportions to the systemic organs through a parallel arrangement of vessels that branch from the aorta (❙Figure 10-1). This arrangement ensures that all organs receive blood of the same composition—that is, one organ does not receive "leftover" blood that has passed through another organ. Because of this parallel arrangement, blood flow through each systemic organ can be independently adjusted as needed.

We first examine some general principles regarding blood flow patterns and the physics of blood flow. Then we turn attention to the various types of blood vessels through which blood flows. We end by discussing how blood pressure is regulated to ensure adequate delivery of blood to the tissues.

To maintain homeostasis, reconditioning organs receive blood flow in excess of their needs.

Blood is constantly "reconditioned" so that its composition remains relatively constant despite an ongoing drain of supplies to support metabolic activities and despite continual addition of wastes from the tissues. Organs that **recondition** the blood normally receive much more blood flow than is necessary to meet their basic metabolic needs, so they can adjust the extra blood to achieve homeostasis. For example, large percentages of the cardiac output (CO) are distributed to the digestive tract (to pick up nutrients), to the kidneys (to eliminate metabolic wastes and adjust water and electrolyte composition), and to the skin (to eliminate heat). Blood flow to the other organs—heart, skeletal muscles, and so on—is solely for filling these organs' metabolic needs and can be adjusted according to their level of activity. For example, during exercise, additional blood is delivered to the active muscles to meet their increased metabolic needs.

Because reconditioning organs—digestive organs, kidneys, and skin—receive blood flow in excess of their needs, they can withstand temporary reductions in blood flow much better than other organs can that do not have this extra margin of blood supply. The brain in particular suffers irreparable damage when transiently deprived of blood supply. After only 4 minutes without O_2, permanent brain damage occurs. Therefore, con-

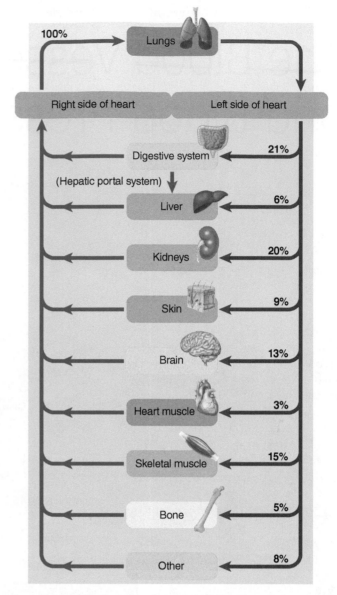

❙Figure 10-1 **Distribution of cardiac output (CO) at rest.** The lungs receive all the blood pumped out by the right side of the heart, whereas the systemic organs each receive some of the blood pumped out by the left side of the heart. The percentage of pumped blood received by the various organs under resting conditions is indicated. This distribution of CO can be adjusted as needed.

stant delivery of adequate blood to the brain, which can least tolerate disrupted blood supply, is a high priority in the overall operation of the circulatory system.

In contrast, the reconditioning organs can tolerate significant reductions in blood flow for quite a long time, and often do. For example, during exercise some of the blood that normally flows through the digestive organs and kidneys is diverted to the skeletal muscles. Likewise, to conserve body heat, blood flow through the skin is markedly restricted during exposure to cold.

Later in the chapter, you will see how distribution of CO is adjusted according to the body's current needs. For now, we concentrate on the factors that influence blood flow through a given blood vessel.

Blood flow through a vessel depends on the pressure gradient and vascular resistance.

The **flow rate** of blood through a vessel (that is, the volume of blood passing through per unit of time) is directly proportional to the pressure gradient (as the pressure gradient increases, flow rate increases) and inversely proportional to vascular resistance (as resistance increases, flow rate decreases):

$$F = \Delta P/R$$

where

F = flow rate of blood through a vessel
ΔP = pressure gradient
R = resistance of blood vessel

Pressure Gradient The **pressure gradient** is the difference in pressure between the beginning and the end of a vessel. Blood flows from an area of higher pressure to an area of lower pressure down a pressure gradient. Contraction of the heart imparts pressure to the blood, which is the main driving force for flow through a vessel. Because of frictional losses (resistance), the pressure drops as blood flows throughout the vessel's length. Accordingly, pressure is higher at the beginning than at the end of the vessel, establishing a pressure gradient for forward flow of blood through the vessel. The greater the pressure gradient forcing blood through a vessel, the greater the flow rate through that vessel (**I** Figure 10-2a). Think of a garden hose attached to a faucet. If you turn on the faucet slightly, a small stream of water flows out of the end of the hose because the pressure is slightly greater at the beginning than at the end of the hose. If you open the faucet all the way, the pressure gradient increases tremendously so that water flows through the hose faster and spurts from the end of the hose. Note that the *difference* in pressure between the two ends of a vessel, not the absolute pressures within the vessel, determines flow rate (**I** Figure 10-2b).

Resistance The other factor influencing flow rate through a vessel is **resistance**, which is a measure of the hindrance or opposition to blood flow through the vessel, caused by friction between the moving fluid and the stationary vascular walls. As resistance to flow increases, it is more difficult for blood to pass through the vessel, so flow rate decreases (as long as the pressure gradient remains unchanged). When resistance

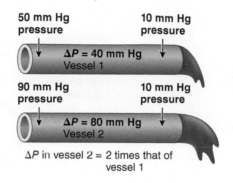

50 mm Hg pressure   **10 mm Hg pressure**

ΔP = 40 mm Hg
Vessel 1

90 mm Hg pressure   **10 mm Hg pressure**

ΔP = 80 mm Hg
Vessel 2

ΔP in vessel 2 = 2 times that of vessel 1

Flow in vessel 2 = 2 times that of vessel 1

Flow ∝ ΔP

(a) Comparison of flow rate in vessels with a different ΔP

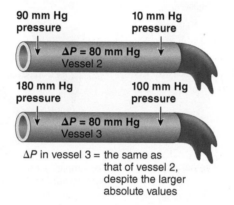

90 mm Hg pressure   **10 mm Hg pressure**

ΔP = 80 mm Hg
Vessel 2

180 mm Hg pressure   **100 mm Hg pressure**

ΔP = 80 mm Hg
Vessel 3

ΔP in vessel 3 = the same as that of vessel 2, despite the larger absolute values

Flow in vessel 3 = the same as that of vessel 2

Flow ∝ ΔP

(b) Comparison of flow rate in vessels with the same ΔP

I Figure 10-2 **Relationship of flow to the pressure gradient in a vessel.** (a) As the difference in pressure (ΔP) between the two ends of a vessel increases, the flow rate increases proportionately. (b) Flow rate is determined by the *difference* in pressure between the two ends of a vessel, not the magnitude of the pressures at each end.

increases, the pressure gradient must increase correspondingly to maintain the same flow rate. Accordingly, when the vessels offer more resistance to flow, the heart must work harder to maintain adequate circulation.

Resistance to blood flow is (1) directly proportional to viscosity of the blood, (2) directly proportional to vessel length, and (3) inversely proportional to vessel radius, which is by far the most important:

$$R \propto \eta\, L/r^4$$

where

η = viscosity
L = vessel length
r = vessel radius

Viscosity refers to the friction developed between the molecules of a fluid as they slide over each other during flow of the fluid. The thicker a liquid is, the greater its viscosity, the greater the resistance to flow. For example, molasses flows more slowly than water because molasses has greater viscosity. Blood viscosity is determined primarily by the number of circulating red blood cells. Normally, this factor is relatively constant and not important in controlling resistance. Occasionally, however, blood viscosity and resistance to flow are increased because an excessive number of red blood cells are present, in which case blood flow is more sluggish than normal.

Because blood "rubs" against the lining of the vessels as it flows past, the greater the vessel surface area in contact with the blood, the greater the resistance to flow. Surface area is determined by both the length and the radius of the vessel. At a constant radius, the longer the vessel is, the greater the surface area and the greater the resistance to flow. Because vessel length remains constant in the body, it is not a variable factor in the control of vascular resistance.

Therefore, the major determinant of resistance to flow is the vessel's radius. Fluid passes more readily through a large vessel than through a smaller vessel. The reason is that a given volume of blood comes into contact with more of the surface area of a small-radius vessel than of a larger-radius vessel, resulting in greater resistance (**I** Figure 10-3a).

Furthermore, a slight change in the radius of a vessel brings about a substantial change in flow because, as can be noted in the preceding equation for R, resistance is inversely proportional to the fourth power of the radius (multiplying the radius by itself four times; $R \propto 1/r^4$). Thus, doubling the radius reduces

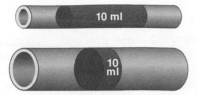

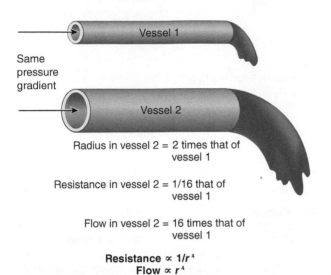

(a) Comparison of contact of a given volume of blood with the surface area of a small-radius vessel and a large-radius vessel

Same pressure gradient → Vessel 1

→ Vessel 2

Radius in vessel 2 = 2 times that of vessel 1

Resistance in vessel 2 = 1/16 that of vessel 1

Flow in vessel 2 = 16 times that of vessel 1

Resistance ∝ 1/r^4
Flow ∝ r^4

(b) Influence of vessel radius on resistance and flow

❙Figure 10-3 **Relationship of resistance and flow to the vessel radius.** (a) The smaller-radius vessel offers more resistance to blood flow because the blood "rubs" against a larger surface area. (b) Doubling the radius decreases the resistance to 1/16 and increases the flow 16 times because the resistance is inversely proportional to the fourth power of the radius.

FIGURE FOCUS: *How would the flow rate in vessel 3, which has the same radius as vessel 2 but double the pressure gradient, compare with the flow rate in vessel 1 and in vessel 2? How would the flow rate in vessel 4, which has double the radius of vessel 2 but the same pressure gradient as vessels 1 and 2, compare with the flow rate in vessel 1 and in vessel 2?*

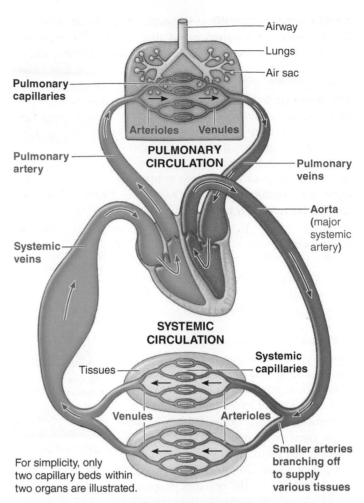

For simplicity, only two capillary beds within two organs are illustrated.

❙Figure 10-4 **Basic organization of the cardiovascular system.** Arteries progressively branch as they carry blood from the heart to the organs. A separate small arterial branch delivers blood to each of the various organs. As a small artery enters the organ it is supplying, it branches into arterioles, which further branch into an extensive network of capillaries. The capillaries rejoin to form venules, which further unite to form small veins that leave the organ. The small veins progressively merge as they carry blood back to the heart.

the resistance to 1/16th its original value ($r^4 = 2 \times 2 \times 2 \times 2 = 16$; $R = 1/16$) and therefore increases flow through the vessel 16-fold (at the same pressure gradient) (❙Figure 10-3b). The converse is also true: Only 1/16th as much blood flows through a vessel at the same driving pressure when its radius is halved. Importantly, the radius of arterioles can be regulated and is the key factor in controlling resistance to blood flow throughout the vascular circuit.

Poiseuille's Law The factors that affect flow rate through a vessel are integrated in **Poiseuille's law** as follows:

$$\text{Flow rate} = \frac{\pi \Delta P r^4}{8 \eta L}$$

The significance of the relationships among flow, pressure, and resistance will become even more apparent as we embark on a voyage through the vessels.

The vascular tree consists of arteries, arterioles, capillaries, venules, and veins.

The systemic and pulmonary circulations each consist of a closed system of vessels (❙Figure 10-4). These vascular loops each are made up of a continuum of different blood vessel types that begins and ends with the heart.

Looking specifically at the systemic circulation, **arteries**, which carry blood from the heart to the organs, branch into a "tree" of progressively smaller vessels, with the various branches delivering blood to different regions of the body. When a small artery reaches the organ it is supplying, it branches into numerous **arterioles**. The volume of blood flowing through an organ can be adjusted by regulating the caliber (internal diameter) of the organ's arterioles. Arterioles branch further within the organs into **capillaries**, the smallest of vessels, across which all exchanges are made with surrounding cells. Capillary exchange is the entire purpose of the circulatory system; all other activities of the system are directed toward ensuring an adequate distribution of replenished blood to capillaries for exchange with all cells. Capillaries rejoin to form small **venules**, which further merge to form small **veins** that leave the organs. The small veins progressively unite to form larger veins that eventually empty into the heart. The arterioles, capillaries, and venules are collectively referred to as the **microcirculation** because they are only visible through a microscope. The microcirculatory vessels are all located within the organs.

The pulmonary circulation consists of the same vessel types, but all the blood in this loop goes between the heart and the lungs. If all of the vessels in the body were strung end to end, they could circle the circumference of Earth twice. In discussing the vessel types in this chapter, we refer to their roles in the systemic circulation, starting with systemic arteries.

Check Your Understanding 10.1

1. Name the reconditioning organs.
2. Give the equation showing the relationship among flow rate of blood through a vessel, the pressure gradient, and the resistance to flow and give the equation for the relationship between resistance and vessel radius.
3. List the blood vessel types, state their functions, and indicate which comprise the microcirculation.

10.2 Arteries

The consecutive segments of the vascular tree are specialized to perform specific tasks (▮Table 10-1).

▮TABLE 10-1 Features of Blood Vessels

Feature	Arteries	Arterioles	Capillaries	Veins
Number	Several hundred*	Half a million	10 billion	Several hundred*
Special features	Thick, highly elastic, walls; large radii*	Highly muscular, well-innervated walls; small radii	Very thin walled; large total cross-sectional area	Thin walled compared to arteries; highly distensible; large radii*
Functions	Passageway from the heart to organs; pressure reservoir	Primary resistance vessels; determine distribution of cardiac output	Site of exchange; determine distribution of extracellular fluid between plasma and interstitial fluid	Passageway to the heart from organs; blood reservoir
Structure				

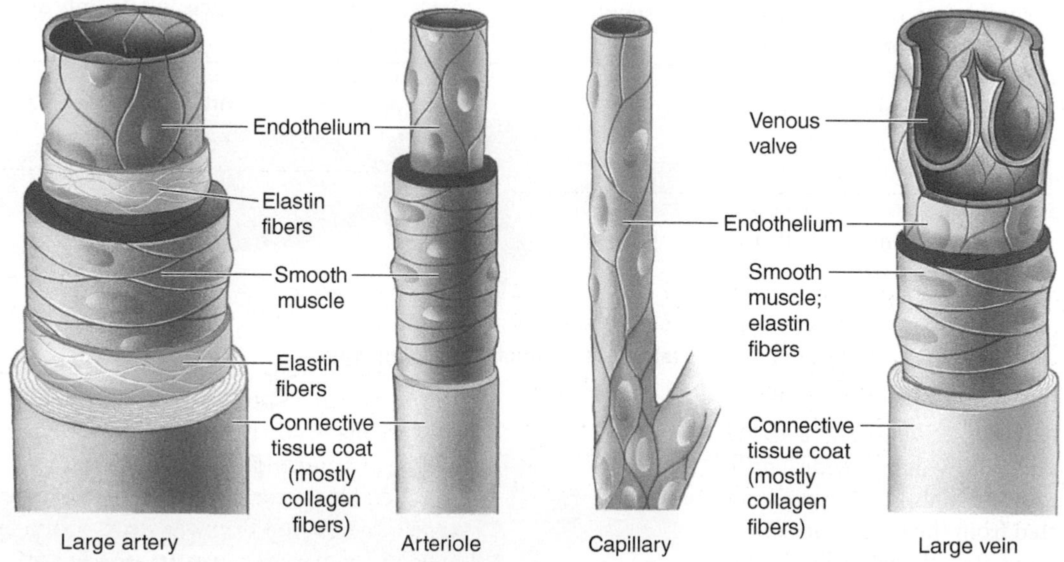
Endothelium
Elastin fibers
Smooth muscle
Elastin fibers
Connective tissue coat (mostly collagen fibers)
Large artery
Arteriole
Capillary
Venous valve
Endothelium
Smooth muscle; elastin fibers
Connective tissue coat (mostly collagen fibers)
Large vein

Relative thickness of layers in wall

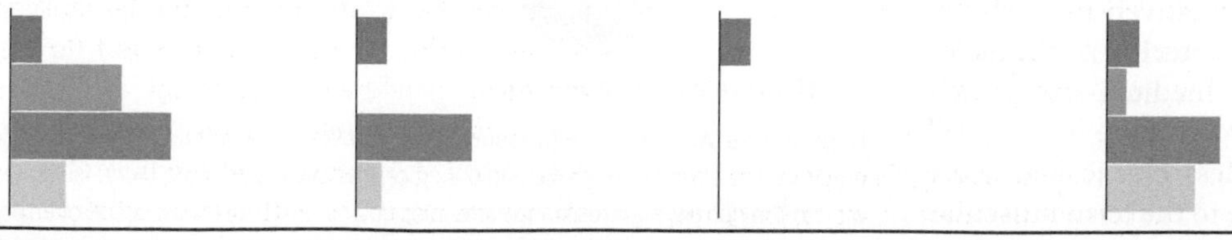
Endothelium
Elastin fibers
Smooth muscle
Collagen fibers

*These numbers and special features refer to the large arteries and veins, not to the smaller arterial branches or venules.

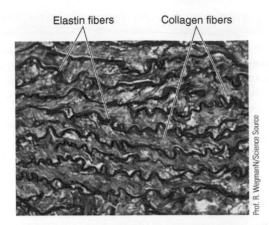

Elastin fibers   Collagen fibers

Prof. R. Wegmann/Science Source

❙Figure 10-5 **Elastin and collagen fibers in an artery.** Light micrograph of a portion of the aorta wall in cross section, showing numerous wavy elastin fibers (*dark red*) among the collagen fibers (*dark yellow*), common to all arteries.

Arteries serve as rapid-transit passageways to the organs and as a pressure reservoir.

Arteries are specialized (1) to serve as rapid-transit passageways for blood from the heart to the organs (because of their large radius, arteries offer little resistance to blood flow) and (2) to act as a **pressure reservoir** to provide the driving force for blood when the heart is relaxing.

How do the arteries act as a pressure reservoir? The heart alternately contracts to pump blood into the arteries and then relaxes to refill with blood from the veins. When the heart is relaxing and refilling, no blood is pumped out. However, capillary flow does not fluctuate between cardiac systole and diastole—that is, blood flow is continuous through the capillaries supplying the organs. The driving force for the continued flow of blood to the organs during cardiac relaxation is provided by elastic recoil of the walls of large arteries. Let us see how.

All vessels are lined with an *endothelium*, a single layer of smooth, flat endothelial cells, which is continuous with the endothelial lining of the heart (see p. 302). A thick wall made up of smooth muscle and connective tissue surrounds the arteries' endothelial lining (❙Table 10-1). Arterial connective tissue contains an abundance of two types of connective tissue fibers: *collagen fibers,* which provide tensile strength against the high, driving pressure of blood ejected from the heart, and *elastin fibers,* which give the arterial walls elasticity (❙Figure 10-5). The large arteries nearest the heart contain relatively more elastin and are called **elastic arteries**. As the major vessels branch into medium-size vessels approaching the organs, the vessel wall becomes relatively less elastic and more muscular, giving rise to the term **muscular arteries** for these vessels.

As the heart pumps blood into the arteries during ventricular systole, a greater volume of blood enters the arteries from the heart than leaves them to flow into the smaller arterioles downstream because these smaller vessels have a greater resistance to flow than the arteries do. However, because of the abundance of elastin in the arterial walls, they behave much like a balloon does when you blow it up. The highly elastic large arteries expand to temporarily hold the excess volume of ejected blood, storing some of the pressure energy imparted by cardiac contraction in their stretched walls—just as a balloon expands to accommodate the extra volume of air you blow into it (❙Figure 10-6a). When the heart relaxes and temporarily stops pumping blood into the arteries, the stretched arterial walls passively recoil, like an inflated balloon that is released. This **elastic recoil** exerts pressure on the blood in the large arteries during diastole. The pressure pushes the excess blood contained in the arteries into the vessels downstream, ensuring continued blood flow to the organs when the heart is relaxing and not pumping blood into the system (❙Figure 10-6b).

Arterial pressure fluctuates in relation to ventricular systole and diastole.

Blood pressure, the force exerted by the blood against a vessel wall, depends on the volume of blood contained within the vessel and the **compliance**, or **distensibility**, of the vessel walls (how easily they can be stretched). If the volume of blood entering the arteries were equal to the volume of blood leaving the arteries during the same period, arterial blood pressure would remain constant. This is not the case, however. During ventricular systole, a stroke volume of blood enters the arteries from the ventricle, while only about one third as much

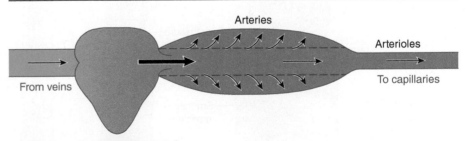

(a) Heart contracting and emptying

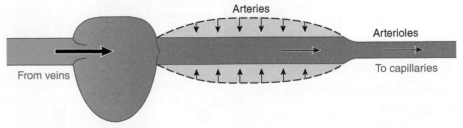

(b) Heart relaxing and filling

❙Figure 10-6 **Arteries as a pressure reservoir.** Because of their elasticity, arteries act as a pressure reservoir. (a) The elastic arteries distend during cardiac systole as more blood is ejected into them than drains off into the narrow, high-resistance arterioles downstream. (b) Elastic recoil of arteries during cardiac diastole continues driving blood forward when the heart is not pumping.

blood leaves the arteries to enter the arterioles. During diastole, no blood enters the arteries, while blood continues to leave, driven by elastic recoil. The maximum pressure exerted in the arteries when blood is ejected into them during systole, the **systolic pressure**, averages 120 mm Hg. The minimum pressure within the arteries when blood is draining off into the rest of the vessels during diastole, the **diastolic pressure**, averages 80 mm Hg. Although ventricular pressure falls to 0 mm Hg during diastole, arterial pressure does not fall to 0 mm Hg because the next cardiac contraction refills the arteries before all the blood drains off (Figure 10-7a; also see Figure 9-16, p. 316).

Clinically, arterial blood pressure is expressed as systolic pressure over diastolic pressure, with desirable blood pressure being 120/80 (read "120 over 80") mm Hg or slightly less.

When you palpate (feel with your fingers) an artery lying close to the surface of the skin (such as at your wrist or neck), you can feel the artery expand as the pressure rises during systole when blood is ejected into the arterial system by the left ventricle. What you feel when you "take a pulse" is the difference between systolic and diastolic pressures; you don't feel anything during diastole, but you feel the surge in pressure during systole. This pressure difference is known as the **pulse pressure**. When blood pressure is 120/80, pulse pressure is 40 mm Hg (120 minus 80 mm Hg). Because the pulse can be felt each time the ventricles pump blood into the arteries, the pulse rate is a measure of the heart rate.

Blood pressure can be measured indirectly by using a sphygmomanometer.

The changes in arterial pressure throughout the cardiac cycle can be measured directly by connecting a pressure-measuring device to a needle inserted in an artery. However, it is more convenient and reasonably accurate to measure the pressure indirectly with a **sphygmomanometer**, an externally applied inflatable cuff attached to a pressure gauge. When the cuff is wrapped around the upper arm and then inflated with air, the pressure of the cuff is transmitted through the tissues to the underlying brachial artery, the main vessel carrying blood to the forearm (Figure 10-7b). The technique involves balancing the pressure in the cuff against the pressure in the artery. When cuff pressure is greater than the pressure in the vessel, the vessel is pinched closed so that no blood flows through it. When blood pressure is greater than cuff pressure, the vessel is open and blood flows through.

During the determination of blood pressure, a stethoscope is placed over the brachial artery at the inside bend of the elbow just below the cuff. No sound can be detected either when blood is not flowing through the vessel or when blood is flowing in the normal, smooth laminar flow (see p. 318). Turbulent blood flow, in contrast, creates vibrations that can be heard. The sounds heard when determining blood pressure, known as **Korotkoff sounds**, are distinct from the heart sounds associated with valve closure heard when listening to the heart with a stethoscope.

At the onset of a blood pressure determination, the cuff is inflated to a pressure greater than systolic blood pressure so

that the brachial artery collapses. Because the externally applied pressure is greater than the peak internal pressure, the artery remains completely pinched closed throughout the entire cardiac cycle; no sound can be heard because no blood is passing through (point **1** in Figure 10-7c). As air in the cuff is slowly released, the pressure in the cuff is gradually reduced. When the cuff pressure falls to just below the peak systolic pressure, the artery transiently opens a bit when the blood pressure reaches this peak. Blood escapes through the partially occluded artery for a brief interval before the arterial pressure falls below the cuff pressure and the artery collapses again. This spurt of blood is turbulent, so it can be heard. Thus, the highest cuff pressure at which the *first sound* can be heard indicates the *systolic pressure* (point **2**). As the cuff pressure continues to fall, blood intermittently spurts through the artery and produces a sound with each subsequent cardiac cycle whenever the arterial pressure exceeds the cuff pressure (point **3**).

When the cuff pressure finally falls below diastolic pressure, the brachial artery is no longer pinched closed during any part of the cardiac cycle, and blood can flow uninterrupted through the vessel (point **5**). With the return of nonturbulent blood flow, no further sounds can be heard. Therefore, the lowest cuff pressure at which the *last sound* can be detected indicates the *diastolic pressure* (point **4**).

Mean arterial pressure is the main driving force for blood flow.

The **mean arterial pressure (MAP)** is the *average pressure* driving blood forward into the tissues throughout the cardiac cycle. MAP, not the systolic or diastolic pressure, is the pressure that is monitored and regulated. Contrary to what you might expect, MAP is not the halfway value between systolic and diastolic pressure (for example, with a blood pressure of 120/80, MAP is not 100 mm Hg). The reason is that arterial pressure remains closer to diastolic than to systolic pressure for a longer portion of each cardiac cycle. At resting heart rate, about two thirds of the cardiac cycle is spent in diastole and only one third in systole. As an analogy, if a race car traveled 80 miles per hour (mph) for 40 minutes and 120 mph for 20 minutes, its average speed would be 93 mph, not the halfway value of 100 mph. Similarly, a good approximation of MAP can be determined using the following equation:

$$\text{MAP} = \text{diastolic pressure} + 1/3 \text{ pulse pressure}$$

At 120/80,

$$\text{MAP} = 80 + (1/3)\ 40 = 93 \text{ mm Hg}$$

Because arteries offer little resistance to flow, only a negligible amount of pressure energy is lost in them because of friction. Therefore, arterial pressure—systolic, diastolic, pulse, or mean—is essentially the same in all arteries (Figure 10-8).

Blood pressure exists throughout the entire vascular tree, but when discussing a person's "blood pressure" without qualifying which blood vessel type is being referred to, the term is tacitly understood to mean the pressure in the arteries.

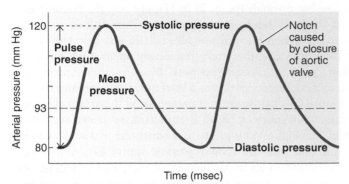

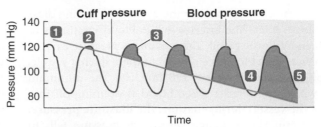

(a) Arterial blood pressure

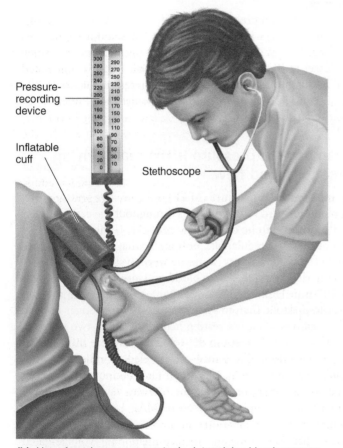

(b) Use of a sphygmomanometer in determining blood pressure

When blood pressure is 120/80:

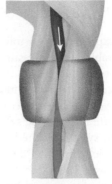

When cuff pressure is greater than 120 mm Hg and exceeds blood pressure throughout the cardiac cycle:

No blood flows through the vessel.

1 No sound is heard because no blood is flowing.

When cuff pressure is between 120 and 80 mm Hg:

Blood flow through the vessel is turbulent whenever blood pressure exceeds cuff pressure.

2 The first sound is heard at peak systolic pressure.

3 Intermittent sounds are produced by turbulent spurts of flow as blood pressure cyclically exceeds cuff pressure.

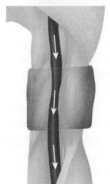

When cuff pressure is less than 80 mm Hg and is below blood pressure throughout the cardiac cycle:

Blood flows through the vessel in smooth, laminar fashion.

4 The last sound is heard at minimum diastolic pressure.

5 No sound is heard thereafter because of uninterrupted, smooth, laminar flow.

(c) Blood flow through the brachial artery in relation to cuff pressure and sounds

▐Figure 10-7 Arterial blood pressure and its measurement. (a) The systolic pressure is the peak pressure exerted in the arteries when blood is pumped into them during ventricular systole. The diastolic pressure is the lowest pressure exerted in the arteries when blood is draining off into the vessels downstream during ventricular diastole. The pulse pressure is the difference between systolic and diastolic pressure. The mean pressure is the average pressure throughout the cardiac cycle. (b) During measurement of blood pressure, the pressure in the sphygmomanometer (inflatable cuff) can be varied to prevent or permit blood flow in the underlying brachial artery. Turbulent blood flow can be detected with a stethoscope, whereas smooth laminar flow and no flow are inaudible. (c) The *red shaded areas* in the graph are the times during which blood is flowing in the brachial artery.

FIGURE FOCUS: *Assume a person has a blood pressure recording of 125/77. (1) What is the systolic pressure? (2) What is the diastolic pressure? (3) What is the pulse pressure? (4) What is the mean arterial pressure? (5) Would any sound be heard when the pressure in an external cuff around the arm was 130 mm Hg? (Yes or no?) (6) Would any sound be heard when cuff pressure was 118 mm Hg? (7) Would any sound be heard when cuff pressure was 75 mm Hg?*

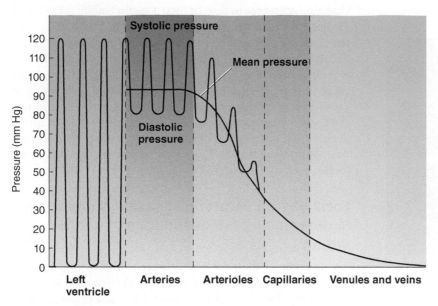

Figure 10-8 Pressures throughout the systemic circulation. Left ventricular pressure swings between a low pressure of 0 mm Hg during diastole to a high pressure of 120 mm Hg during systole. Arterial blood pressure, which fluctuates between a peak systolic pressure of 120 mm Hg and a low diastolic pressure of 80 mm Hg each cardiac cycle, is of the same magnitude throughout the arteries. Because of the arterioles' high resistance, the pressure drops precipitously and the systolic-to-diastolic swings in pressure are converted to a nonpulsatile pressure when blood flows through the arterioles. The pressure continues to decline but at a slower rate as blood flows through the capillaries and venous system.

Check Your Understanding 10.2

1. Indicate what structural feature enables arteries to serve as a pressure reservoir.

2. Draw a graph of average arterial blood pressure throughout the cardiac cycle, labeling the systolic pressure, diastolic pressure, and pulse pressure.

3. Calculate MAP if blood pressure is 135/90.

10.3 Arterioles

When an artery reaches the organ it is supplying, it branches into numerous arterioles within the organ.

Arterioles are the major resistance vessels.

Arterioles are the main resistance vessels in the vascular tree because their radius is small enough to offer considerable resistance to flow. (Even though capillaries have a smaller radius than arterioles, you will see later how collectively the capillaries do not offer as much resistance to flow as the arterioles do.) In contrast to the low resistance of the arteries, the high degree of arteriolar resistance causes a marked drop in mean pressure as blood flows through these small vessels. On average, the pressure falls from the MAP of 93 mm Hg (the pressure of the blood entering the arterioles from the arteries), to 37 mm Hg, the pressure of the blood leaving the arterioles and entering the capillaries (Figure 10-8). This decline in pressure helps estab-

lish the pressure differential that encourages the flow of blood from the heart to the various organs downstream. Arteriolar resistance also converts the pulsatile systolic-to-diastolic pressure swings in the arteries into the nonfluctuating pressure present in the capillaries.

The radius (and, accordingly, the resistances) of arterioles supplying individual organs can be adjusted independently to accomplish two functions: (1) to variably distribute the cardiac output among the systemic organs, depending on the body's momentary needs, and (2) to help regulate arterial blood pressure. Before considering how such adjustments are important in accomplishing these two functions, we discuss the mechanisms involved in adjusting arteriolar resistance.

Vasoconstriction and Vasodilation Unlike arteries, arteriolar walls contain little elastic connective tissue. However, they do have a thick layer of smooth muscle that is richly innervated by sympathetic nerve fibers (see Table 10-1). The smooth muscle is also sensitive to many local chemical changes, to a few circulating hormones, and to mechanical factors such as stretch. The smooth muscle layer runs circularly around the arteriole (Figure 10-9a); so when the smooth muscle layer contracts, the vessel's circumference (and its radius) becomes smaller, increasing resistance and decreasing flow through that vessel.

Vasoconstriction is the term applied to such narrowing of a vessel (Figure 10-9c). In contrast, the term **vasodilation** refers to enlargement in the circumference and radius of a vessel as a result of its smooth muscle layer relaxing (Figure 10-9d). Vasodilation leads to decreased resistance and increased flow through that vessel.

Vascular Tone The extent of contraction of arteriolar smooth muscle depends on the cytosolic Ca^{2+} concentration. Arteriolar smooth muscle normally displays a state of partial constriction known as **vascular tone**, which establishes a baseline of arteriolar resistance (Figure 10-9b). Two factors are responsible for vascular tone. First, arteriolar smooth muscle is tonic smooth muscle that has sufficient surface-membrane voltage-gated Ca^{2+} channels open even at resting potential to trigger partial contraction (see p. 289). This myogenic activity is independent of any neural or hormonal influences, leading to self-induced contractile activity (see p. 291). Second, the sympathetic fibers supplying most arterioles continually release norepinephrine, which further enhances vascular tone.

This ongoing tone makes it possible to either increase or decrease contractile activity to accomplish vasoconstriction or vasodilation, respectively. Were it not for tone, it would be impossible to reduce tension in an arteriolar wall to accomplish vasodilation; only varying degrees of vasoconstriction would be possible.

A variety of factors can influence the level of contractile activity in arteriolar smooth muscle, thereby substantially

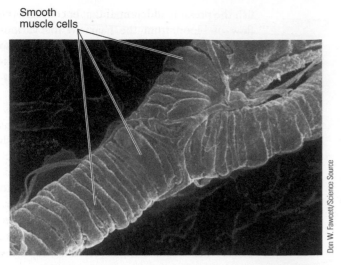

Smooth muscle cells

(a) Scanning electron micrograph of an arteriole showing how the smooth muscle cells run circularly around the vessel wall

Cross section of arteriole

(b) Normal arteriolar tone

Caused by:
↑ Myogenic activity
↑ Oxygen (O$_2$)
↓ Carbon dioxide (CO$_2$) and other metabolites
↑ Endothelin
↑ Sympathetic stimulation
  Vasopressin; angiotensin II
  Cold

(c) Vasoconstriction (increased contraction of circular smooth muscle in the arteriolar wall, which leads to increased resistance and decreased flow through the vessel)

Caused by:
↓ Myogenic activity
↓ O$_2$
↑ CO$_2$ and other metabolites
↑ Nitric oxide
↓ Sympathetic stimulation
  Histamine release
  Heat

(d) Vasodilation (decreased contraction of circular smooth muscle in the arteriolar wall, which leads to decreased resistance and increased flow through the vessel)

❙Figure 10-9 **Arteriolar vasoconstriction and vasodilation.**

changing resistance to flow in these vessels. Unlike skeletal and cardiac muscle in which action potentials trigger muscle contraction, vascular smooth muscle can undergo graded changes in force in response to chemical, physical, and neural factors without undergoing action potentials. These agents largely act via second-messenger pathways (see p. 117). The factors that cause arteriolar vasoconstriction or vasodilation fall into two categories: *local (intrinsic) controls,* which are important in determining distribution of cardiac output, and *extrinsic controls,* which are important in blood pressure regulation. We look at each of these controls in turn.

Local control of arteriolar radius is important in determining the distribution of cardiac output.

The fraction of the total CO delivered to each organ is not always constant; it varies, depending on the demands for blood at the time. The share of CO received by each organ is determined by the number and caliber of the arterioles supplying that area. Recall that $F = \Delta P/R$. Because blood is delivered to all organs at the same mean arterial pressure, the driving pressure gradient for flow is identical for each organ. Therefore, differences in flow to various organs are determined by differences in the extent of vascularization and by differences in the resistance offered by the arterioles supplying each organ. From moment to moment, the distribution of CO can be varied by differentially adjusting arteriolar resistance in the various vascular beds.

As an analogy, consider a pipe carrying water, with several adjustable valves located throughout its length (❙Figure 10-10). Assuming that water pressure in the pipe is constant, differences in the amount of water flowing into a beaker under each valve depend entirely on which valves are open and to what extent. No water enters beakers under closed valves (high resistance), and more water flows into beakers under valves that are opened completely (low resistance) than into beakers under valves that are only partially opened (moderate resistance).

Similarly, more blood flows to areas whose arterioles offer the least resistance to its passage. During exercise, for example, not only is CO increased, but also because of vasodilation in skeletal muscle and in the heart, a greater percentage of the pumped blood is diverted to these organs to support their increased metabolic activity. Simultaneously, blood flow to the digestive tract and kidneys is reduced as a result of arteriolar vasoconstriction in these organs, leaving more blood available for diversion to the active muscles (see pp. 370–371 for details). Only the blood supply to the brain remains remarkably constant no matter what the person is doing, be it vigorous physical activity, intense mental concentration, or sleep. Although total blood flow to the brain remains constant, new imaging techniques demonstrate that regional blood flow varies within the brain in close correlation with local neural activity patterns (see ❙Figure 5-8, p. 146).

Local (intrinsic) controls are changes within an organ that adjust blood flow through the organ by affecting the smooth muscle of the organ's arterioles to alter their caliber and resistance. Local influences may be either chemical or physical.

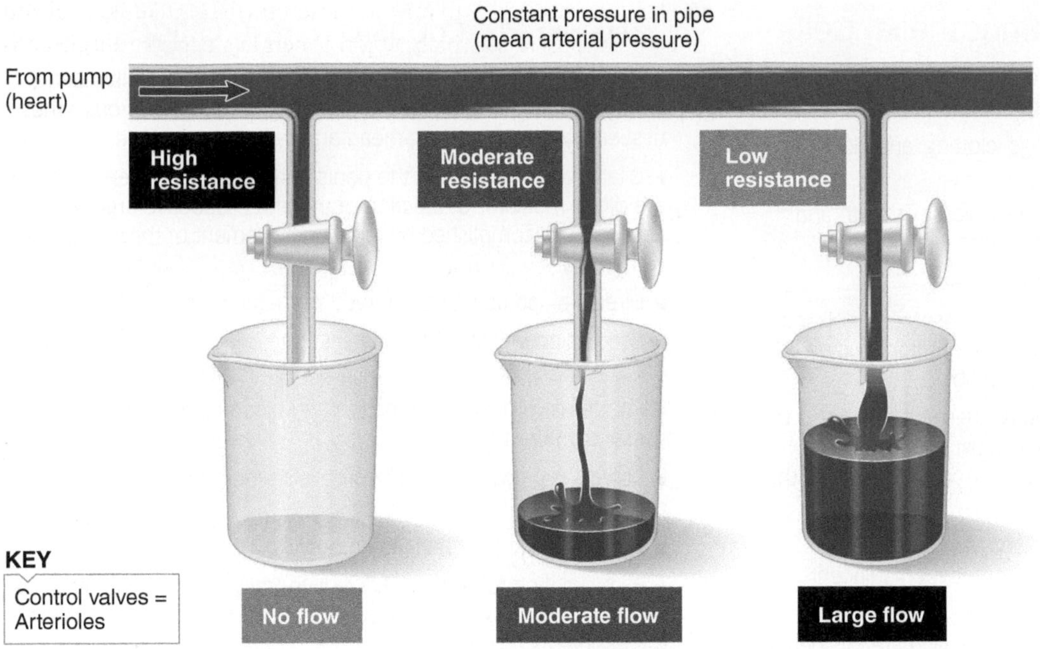

Constant pressure in pipe
(mean arterial pressure)

From pump
(heart)

High
resistance

Moderate
resistance

Low
resistance

KEY

Control valves =
Arterioles

No flow

Moderate flow

Large flow

Figure 10-10 **Flow rate as a function of resistance.**

Local chemical influences on arteriolar radius include (1) local metabolic changes and (2) histamine release. Local physical influences include (1) how much the vessel is stretched, (2) the extent of shear stress, and (3) local application of heat or cold. Let us examine the role and mechanism of each of these local influences.

Local metabolic influences on arteriolar radius help match blood flow with the organs' needs.

The most important local chemical influences on arteriolar smooth muscle are related to metabolic changes within a given organ. The influence of these local changes on arteriolar radius is important in matching blood flow through an organ with the organ's metabolic needs. Local metabolic controls are especially important in skeletal muscle and in the heart, the organs whose metabolic activity and need for blood supply normally vary most extensively, and in the brain, whose overall metabolic activity and need for blood supply remain constant. Local controls help maintain the constancy of blood flow to the brain.

Active Hyperemia Arterioles lie within the organ they are supplying and can be acted on by local factors within the organ. During increased metabolic activity, such as when a skeletal muscle is contracting during exercise, local concentrations of several of the organ's chemicals change. For example, the local O_2 concentration decreases as the actively metabolizing cells use up more O_2 to support oxidative phosphorylation for ATP production (see p. 37). This and other local chemical changes produce local arteriolar dilation by triggering relaxation of the arteriolar smooth muscle in the vicinity. Local arteriolar vaso-

dilation then increases blood flow to that particular area. This increased blood flow in response to enhanced tissue activity is called **active hyperemia** (*hyper* means "above normal"; *emia* means "blood"). When cells are more active metabolically, they need more blood to bring in O_2 and nutrients and to remove metabolic wastes.

Conversely, when an organ, such as a relaxed muscle, is less active metabolically and thus has reduced needs for blood delivery, the resultant local chemical changes (for example, increased local O_2 concentration) bring about local arteriolar vasoconstriction and a subsequent reduction in blood flow to the area. Local metabolic changes can thus adjust blood flow as needed without involving nerves or hormones.

Local Metabolic Changes that Influence Arteriolar Radius A variety of local chemical changes act together in a cooperative, redundant manner to bring about these "selfish" local adjustments in arteriolar caliber that match a tissue's blood flow with its needs. Specifically, the following local chemical factors produce relaxation of arteriolar smooth muscle:

- *Decreased O_2.*

- *Increased CO_2.* More CO_2 is generated as a by-product during the stepped-up pace of oxidative phosphorylation that accompanies increased activity.

- *Increased acid.* More carbonic acid is generated from the increased CO_2 produced as the metabolic activity of a cell increases. Also, lactate (lactic acid) accumulates if the glycolytic pathway is used for ATP production (see pp. 39 and 272).

- *Increased K^+.* Repeated action potentials that outpace the ability of the Na^+–K^+ pump to restore the resting concentration gradients (see p. 94) result in an increase in K^+ in the interstitial fluid of a more active tissue.

- *Increased osmolarity.* Osmolarity (the concentration of osmotically active solutes) increases during elevated cellular metabolism.

- *Adenosine release.* Especially in cardiac muscle, adenosine is released in response to increased metabolic activity or O_2 deprivation (see p. 326).

Endothelial Derived Vasoactive Paracrines These local chemical changes do not act directly on vascular smooth muscle to change its contractile state. Instead, they act on arteriolar endothelial cells. **Endothelial cells** are the single layer of spe-

TABLE 10-2 Functions of Endothelial Cells

- Line the blood vessels and heart chambers; serve as a physical barrier between the blood and the remainder of the vessel wall.

- Influence formation of platelet plugs, clotting, and clot dissolution.

- Secrete substances that stimulate new vessel growth and proliferation of smooth muscle cells in vessel walls.

- Secrete cytokines during immune responses.

- Secrete paracrines that promote blood vessel remodeling in response to long-term changes in shear stress (longitudinal force on vessel lining caused by forward blood flow).

- In arterioles, secrete vasoactive paracrines in response to local chemical and physical changes; these substances cause relaxation (vasodilation) or contraction (vasoconstriction) of the underlying smooth muscle.

- In capillaries, help determine capillary permeability by contracting to vary the size of the pores between adjacent endothelial cells.

- Also in capillaries, participate in exchange of materials between the blood and surrounding tissue cells through vesicular transport.

TABLE 10-3 Functions of Nitric Oxide (NO)

- Causes relaxation of arteriolar smooth muscle. By this means, NO plays an important role in controlling blood flow through the tissues and in maintaining mean arterial blood pressure.

- Dilates the arterioles of the penis and clitoris, thus serving as the direct mediator of erection of these reproductive organs. Erection is accomplished by rapid engorgement of these organs with blood.

- Directs blood flow to O_2-starved tissues.

- Used as chemical warfare against bacteria and cancer cells by macrophages, large phagocytic cells of the immune system.

- Interferes with platelet function and blood clotting at sites of vessel damage.

- Serves as a novel type of neurotransmitter in the brain and elsewhere.

- Plays a role in the changes underlying memory.

- By promoting relaxation of digestive-tract smooth muscle, helps regulate peristalsis, a type of contraction that pushes digestive tract contents forward.

- Relaxes the smooth muscle cells in the airways of the lungs, helping keep these passages open to facilitate movement of air in and out of the lungs.

- Modulates the filtering process involved in urine formation.

- May play a role in relaxation of skeletal muscle.

cialized epithelial cells which form the endothelium that lines the lumen of all vessels. Arteriolar endothelial cells release paracrines (locally acting chemical messengers; see p. 114) in response to chemical changes in the cells' environment (such as a reduction in O_2) or physical changes (such as an increase in the frictional force of blood as it flows over the surface of the vessel lining). These vasoactive ("acting on vessels") paracrines act on the underlying smooth muscle to alter its state of contraction, thus locally regulating arteriolar caliber.

Scientists used to regard endothelial cells as little more than a passive barrier between the blood and the rest of the vessel wall. However they have discovered that endothelial cells are active participants in a variety of vessel-related activities in addition to secreting vasoactive paracrines (Table 10-2).

Among the best studied of vasoactive paracrines is **nitric oxide (NO)**, which brings about local arteriolar vasodilation by causing relaxation of arteriolar smooth muscle in the vicinity. It does so by increasing the concentration of the intracellular second messenger cyclic GMP, which leads to activation of an enzyme that reduces phosphorylation of myosin. Remember that smooth-muscle myosin can bind with actin and promote filament sliding through cycles of power strokes only when myosin is phosphorylated (see p. 288). NO plays a role in regulating mean arterial pressure by exerting an ongoing vasodilatory effect. Release of additional NO in response to local metabolic changes promotes further vasodilation in the area. NO is a small, highly reactive, short-lived gas molecule that once was known primarily as a toxic air pollutant. Yet studies have revealed an astonishing number of biological roles for NO, which is produced in many tissues besides endothelial cells. NO is one of the body's most important messenger molecules, as shown by its range of functions listed in Table 10-3. As you can

see, most areas of the body are influenced by this versatile intercellular messenger molecule.

Endothelial cells release other important paracrines besides NO. As an example, **endothelin** causes arteriolar smooth muscle contraction and is one of the most potent vasoconstrictors yet identified. Still other chemicals, released from the endothelium in response to chronic changes in blood flow to an organ, trigger long-term vascular changes that permanently influence blood flow to a region. For instance, **vascular endothelial growth factor (VEGF)** stimulates new vessel growth, a process known as **angiogenesis**.

Reactive Hyperemia Active hyperemia takes place in response to changes in local chemical composition resulting from changes in local metabolic activity. When the blood supply to a region is completely occluded (for example, by means of a tourniquet applied to the upper arm when a blood sample is being drawn), many of the same chemical changes occur in the blood-deprived tissue that occur during metabolically induced active hyperemia. In the case of reactive hyperemia, the imbalance between blood supply and metabolic activity occurs because the blood supply is cut off while metabolic activity remains constant. When a tissue's blood supply is blocked, O_2 levels decrease in the deprived tissue; the tissue continues to consume O_2, but no fresh supplies are being delivered. Meanwhile, the concentrations of CO_2, acid, and other metabolites rise. Even though their production does not increase as it does when a tissue is more active metabolically,

346 | CHAPTER 10

these substances accumulate in the tissue when the normal amounts produced are not "washed away" by blood. As a result of these local chemical changes, the arterioles in the blood-deprived area dilate. Despite this arteriolar vasodilation, the occlusion prevents blood flow through these dilated vessels.

However, after the occlusion is removed, blood flow to the previously deprived tissue is transiently much higher than normal because the arterioles are widely dilated. This post-occlusion increase in blood flow is called **reactive hyperemia**. Such a response is beneficial for rapidly restoring the local chemical composition to normal. Of course, prolonged blockage of blood flow (such as by a blood clot that completely blocks a coronary artery during a heart attack; see p. 314) leads to irreversible damage in the deprived tissue.

The same mechanisms are responsible for both active hyperemia and reactive hyperemia—release from the endothelial cells of vasodilating paracrines in response to decreased O_2 and other associated local chemicals that relax the underlying smooth muscle (Figure 10-11a and b). The difference lies in the cause responsible for these local chemical changes: increased local metabolic activity in the case of active hyperemia and local blockage of the blood supply to the area in the case of reactive hyperemia.

Local histamine release pathologically dilates arterioles.

Histamine is another local chemical mediator that influences arteriolar smooth muscle, but it is not released in response to local metabolic changes and is not derived from endothelial

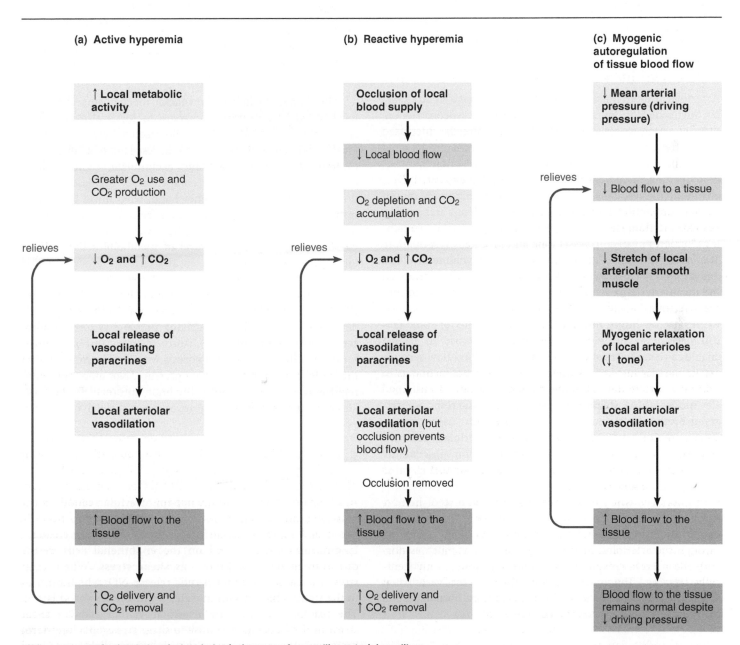

I Figure 10-11 **Major local chemical and physical means of controlling arteriolar caliber.**
FIGURE FOCUS: *Using this figure, identify the triggering cause, the local mediators that produce arteriolar vasodilation, and the compensatory result for (a) active hyperemia, (b) reactive hyperemia, and (c) autoregulation of tissue blood flow.*

cells. Although histamine normally does not participate in controlling blood flow, it is important in certain pathological conditions.

Clinical Note Histamine is synthesized and stored within special connective tissue cells in many organs and in certain types of circulating white blood cells. When organs are injured or during allergic reactions, histamine is released and acts as a paracrine in the damaged region. By promoting relaxation of arteriolar smooth muscle, histamine is the major cause of vasodilation in an injured area. The resultant increase in blood flow into the area produces the redness and contributes to the swelling seen with inflammatory responses (see Chapter 12 for further details).

We now shift from local chemical influences to local physical influences on arteriolar radius, the most important of which is the myogenic response to stretch.

The myogenic response of arterioles to stretch helps tissues autoregulate their blood flow.

Arteriolar smooth muscle responds to being passively stretched by myogenically increasing its tone via vasoconstriction, thereby acting to resist the initial passive stretch.[1] Increased vessel stretching brings about opening of mechanically gated cation channels, which leads to a small depolarization that triggers opening of more surface-membrane voltage-gated Ca^{2+} channels. The resultant Ca^{2+} entry promotes increased smooth muscle contraction, boosting myogenic vessel tone and causing vasoconstriction. Conversely, a reduction in arteriolar stretching decreases myogenic vessel tone and promotes vasodilation. The extent of passive stretch varies with the volume of blood delivered to the arterioles from the arteries, which depends on the mean arterial pressure (the pressure that drives blood into the arterioles). Mean arterial pressure is normally maintained within narrow limits, but if this driving pressure for some reason becomes abnormal, the myogenic response to stretch enables a tissue to resist changes in its own blood flow secondary to changes in MAP by making appropriate adjustments in arteriolar radius. For example, in the presence of sustained elevations in MAP (hypertension), the myogenic response triggered by the initial increased flow of blood to tissues brings about vasoconstriction, which increases arteriolar tone and resistance. This greater degree of vasoconstriction subsequently reduces tissue blood flow toward normal despite this elevated blood pressure (Figure 10-12).

Conversely, when MAP falls (such as because of hemorrhage or a weakened heart), the driving force is reduced, so blood flow to organs decreases. Because less blood is flowing through the arterioles, they are not stretched as much as normal. The arterioles respond to this reduced stretch by myogenically relaxing. The increased flow through the vasodilated arterioles helps restore tissue blood flow toward normal despite the reduced driving pressure (see Figure 10-11c).

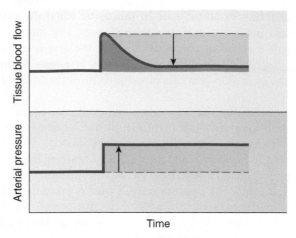

Figure 10-12 **Autoregulation of tissue blood flow.** Even though blood flow through a tissue immediately increases in response to a rise in arterial pressure, the tissue blood flow is reduced gradually as a result of autoregulation within the tissue, despite a sustained increase in arterial pressure.

Simultaneous with these myogenic mechanisms, changes in local blood flow driven by changes in MAP also alter local chemical factors, kind of like a mild reactive hyperemia. When local blood flow initially decreases in response to a fall in MAP, O_2 levels decline and metabolites accumulate, leading to local arteriolar vasodilation and a compensatory increase in blood flow to the tissue. On the other hand, when local blood flow initially increases in response to a rise in MAP, O_2 levels increase and metabolites are washed away more quickly as the blood supply outpaces the level of metabolic activity. These local chemical changes cause local arteriolar vasoconstriction and a compensatory decrease in blood flow.

These local arteriolar myogenic and chemical mechanisms that keep tissue blood flow fairly constant despite rather wide deviations in mean arterial driving pressure is termed **autoregulation** ("self-regulation"). Autoregulatory responses bring local tissue flow back toward normal within a few minutes after the initial change in the driving pressure. Not all organs autoregulate equally. As examples, the brain autoregulates best, the kidneys are good at autoregulation, and skeletal muscle has poor autoregulatory abilities.

Arterioles release vasodilating NO in response to an increase in shear stress.

Another local physical influence on arteriolar caliber is the vessel's response to changes in shear stress. Due to friction, blood flowing over the surface of the vessel lining creates a longitudinal force applied on the endothelial cells in the direction of the flow known as **shear stress**. When shear stress increases, endothelial cells release NO, which diffuses to the underlying smooth muscle and promotes vasodilation. The resultant increase in arteriolar caliber reduces shear stress in the vessel. In response to shear stress on a long-term basis, endothelial cells orient themselves parallel to the direction of blood flow (that is, they line up their long axis to "go with the flow.")

---

[1]Because of the continuum of vessels, the small muscular arteries behave similarly to arterioles by responding myogenically to varying degrees of stretch.

Local heat application dilates arterioles and cold application constricts them.

Clinical Note The effect of temperature changes, another local physical influence, on arterioles can be exploited clinically. Heat application, by causing localized arteriolar vasodilation, is a useful therapeutic agent for promoting increased blood flow to an area. Conversely, applying an ice pack to an inflamed area produces vasoconstriction, which reduces swelling by counteracting histamine-induced vasodilation.

This completes our discussion of local control of arteriolar radius. We now shift to extrinsic control of arteriolar radius.

Extrinsic control of arteriolar radius is important in regulating blood pressure.

Extrinsic control of arteriolar radius includes both neural and hormonal influences, the effects of the sympathetic nervous system being the most important. Sympathetic nerve fibers supply arteriolar smooth muscle everywhere in the systemic circulation except in the brain. Recall that a certain level of ongoing sympathetic activity contributes to vascular tone. Increased sympathetic activity produces generalized arteriolar vasoconstriction, whereas decreased sympathetic activity leads to generalized arteriolar vasodilation. These widespread changes in arteriolar resistance bring about changes in mean arterial pressure because of their influence on total peripheral resistance.

Influence of Total Peripheral Resistance on Mean Arterial Pressure To find the effect of changes in arteriolar resistance on MAP, the equation $F = \Delta P/R$ applies to the entire circulation as well as to a single vessel:

- *F*. Looking at the circulatory system as a whole, flow (*F*) through all the vessels in either the systemic or the pulmonary circulation is equal to the cardiac output (CO).
- ΔP. The pressure gradient (ΔP) for the entire systemic circulation is the mean arterial pressure. (ΔP equals the difference in pressure between the beginning and the end of the systemic circulatory system. The beginning pressure is the MAP as the blood leaves the left ventricle at an average of 93 mm Hg. The end pressure in the right atrium is 0 mm Hg. Therefore, $\Delta P = 93$ minus $0 = 93$ mm Hg, which is equivalent to the MAP.)
- *R*. The total resistance (*R*) offered by all the systemic peripheral vessels together is the **total peripheral resistance (TPR)**. By far, the greatest percentage of the TPR is caused by arteriolar resistance because arterioles are the primary resistance vessels.

Therefore, for the entire systemic circulation, rearranging

$$F = \Delta P/R$$

to

$$\Delta P = F \times R$$

gives us the equation

$$MAP = CO \times TPR$$

(Do not confuse this equation, which indicates the factors that *determine* MAP, with the equation used to *calculate* mean arterial pressure, namely, MAP = diastolic pressure + 1/3 pulse pressure.)

Thus, the extent of TPR offered collectively by all the systemic arterioles influences MAP immensely. A dam provides an analogy to this relationship. At the same time a dam restricts the flow of water downstream, it increases the pressure upstream by elevating the water level in the reservoir behind the dam. Similarly, generalized, sympathetically induced vasoconstriction reflexly reduces blood flow downstream to the organs while elevating the upstream mean arterial pressure, thereby increasing the main driving force for blood flow to all the organs.

These effects seem counterproductive. Why increase the driving force for flow to the organs by increasing MAP while reducing flow to the organs by narrowing the vessels supplying them? In effect, the sympathetically induced arteriolar responses help maintain the appropriate driving pressure head (that is, the MAP) to all organs. The extent to which each organ actually receives blood flow is determined by local arteriolar adjustments that override the sympathetic constrictor effect. If all arterioles were dilated, blood pressure would fall substantially, so there would not be an adequate driving force for blood flow. An analogy is the pressure head for water in the pipes in your home. If the water pressure is adequate, you can selectively obtain satisfactory water flow at any of the faucets by turning the appropriate handle to the open position. If the water pressure in the pipes is too low, however, you cannot obtain satisfactory flow at any faucet, even if you turn the handle to the maximally open position. Tonic sympathetic activity thus constricts most vessels (with the exception of those in the brain) to help maintain a pressure head on which organs can draw as needed through local mechanisms that control arteriolar radius.

Norepinephrine's Influence on Arteriolar Smooth Muscle The norepinephrine released from sympathetic nerve endings combines with α_1-adrenergic receptors on arteriolar smooth muscle to bring about vasoconstriction (see p. 240). Cerebral (brain) arterioles are the only ones that do not have α_1 receptors, so no vasoconstriction occurs in the brain. It is important that cerebral arterioles are not reflexly constricted by neural influences because brain blood flow must remain constant to meet the brain's continuous need for O_2, no matter what is going on elsewhere in the body. Cerebral vessels are almost entirely controlled by local mechanisms that maintain a constant blood flow to support a constant level of brain metabolic activity. In fact, reflex vasoconstrictor activity in the remainder of the cardiovascular system is aimed at maintaining an adequate pressure head for blood flow to the vital brain.

Thus, sympathetic activity contributes in an important way to maintaining MAP, ensuring an adequate driving force for blood flow to the brain at the expense of organs that can better withstand reduced blood flow. Other organs that really need additional blood, such as active muscles (including active heart muscle), obtain it through local controls that override the sympathetic effect.

Local Controls Overriding Sympathetic Vasoconstriction Skeletal and cardiac muscles have the most powerful local control mechanisms with which to override generalized sympathetic vasoconstriction. For example, if you are pedaling a bicycle, the increased activity in the skeletal muscles of your legs brings about an overriding local, metabolically induced vasodilation in those particular muscles, despite the generalized sympathetic vasoconstriction that accompanies exercise. As a result, more blood flows through your leg muscles but not through your inactive arm muscles.

No Parasympathetic Innervation to Arterioles Arterioles have no significant parasympathetic innervation, with the exception of the abundant parasympathetic vasodilator supply to the arterioles of the penis and clitoris. The rapid, profuse vasodilation induced by parasympathetic stimulation in these organs (by means of promoting release of NO) is largely responsible for accomplishing erection. Vasodilation elsewhere is produced primarily by decreasing sympathetic vasoconstrictor activity below its normal tone level. When MAP rises above normal, reflex reduction in sympathetic vasoconstrictor activity accomplishes generalized arteriolar vasodilation to help bring the driving pressure down toward normal. (Also, the hormone epinephrine causes vasodilation in arteriolar smooth muscle specifically in the skeletal muscles and heart by a mechanism described shortly.)

The cardiovascular control center and several hormones regulate blood pressure.

The main region of the brain that adjusts sympathetic output to the arterioles is the *cardiovascular control center* in the medulla of the brain stem. This is the integrating center for blood pressure regulation. Several other brain regions also influence blood distribution, the most notable being the hypothalamus, which, as part of its temperature-regulating function, controls blood flow to the skin to adjust heat loss to the environment.

In addition to neural reflex activity, several hormones extrinsically influence arteriolar radius. These hormones include the adrenal medullary hormones *epinephrine* and *norepinephrine*, which generally reinforce the sympathetic nervous system in most organs, and *vasopressin* and *angiotensin II*, which are important in controlling fluid balance and blood volume.

Influence of Epinephrine and Norepinephrine Sympathetic stimulation of the adrenal medulla causes this endocrine gland to release epinephrine and norepinephrine. Adrenal medullary norepinephrine combines with the same α_1 receptors as sympathetically released norepinephrine to produce generalized vasoconstriction. However, epinephrine, the more abundant of the adrenal medullary hormones, combines with both β_2 and α_1 receptors but has a much greater affinity for the β_2 receptors. Activation of β_2 receptors produces vasodilation, but not all tissues have β_2 receptors; they are most abundant in the arterioles of the skeletal muscles and heart. During sympathetic discharge, the released epinephrine combines with the β_2 receptors in the skeletal muscles and heart to reinforce local vasodilatory mechanisms in these tissues. Arterioles in digestive organs and kidneys, in contrast, are equipped only with α_1 receptors. Therefore, the

arterioles of these organs undergo more profound vasoconstriction during generalized sympathetic discharge than those in the skeletal muscles and heart do. Lacking β_2 receptors, the digestive organs and kidneys do not experience an overriding vasodilatory response on top of the α_1 receptor-induced vasoconstriction.

Influence of Vasopressin and Angiotensin II The two other hormones that extrinsically influence arteriolar tone are vasopressin and angiotensin II. Vasopressin is primarily involved in maintaining water balance by regulating the amount of water the kidneys retain for the body during urine formation (see pp. 524 and 544). Angiotensin II is part of a hormonal pathway, the *renin–angiotensin–aldosterone system*, which is important in regulating the body's salt balance. This pathway promotes salt conservation during urine formation and leads to water retention because salt exerts a water-holding osmotic effect in the ECF (see p. 508). Thus, both these hormones play important roles in maintaining the body's fluid balance, which in turn is an important determinant of blood volume and blood pressure.

In addition, both vasopressin and angiotensin II are potent vasoconstrictors. Their role in this regard is especially crucial during hemorrhage. A sudden loss of blood reduces the blood volume, which triggers increased secretion of both these hormones to help restore blood volume. Their vasoconstrictor effect also helps maintain blood pressure despite abrupt loss of blood volume. (The functions and control of these hormones are discussed more thoroughly in later chapters.)

This completes our discussion of the factors that affect TPR, the most important of which are controlled adjustments in arteriolar radius. These factors are summarized in ❙ Figure 10-13. We now turn to the next vessels in the vascular tree, the capillaries.

Check Your Understanding 10.3

1. Draw cross sections of an arteriole (a) with normal arteriolar tone, (b) during vasoconstriction, and (c) during vasodilation.

2. Discuss the mechanism and purpose of active hyperemia.

3. Define *autoregulation*.

4. Write the equation showing the determinants of MAP and the equation used to calculate MAP.

10.4 Capillaries

Capillaries, the sites for exchange of materials between blood and tissue cells,[2] branch extensively to bring blood within the reach of essentially every cell (see chapter opener photo, p. 335).

---

[2]Actually, some exchange takes place across the other microcirculatory vessels, especially the postcapillary venules. The entire vasculature is a continuum and does not abruptly change from one vascular type to another. When the term *capillary exchange* is used, it tacitly refers to all exchange at the microcirculatory level, the majority of which occurs across the capillaries.

Capillaries are ideally suited to serve as sites of exchange.

There are no carrier-mediated transport systems across capillaries, with the exception of those in the brain that play a role in the blood–brain barrier (see p. 141). Materials are exchanged across capillary walls mainly by diffusion.

Factors that Enhance Diffusion Across Capillaries
Capillaries are ideally suited to enhance diffusion, in accordance with Fick's law of diffusion (see p. 65). They minimize diffusion distances while maximizing surface area and time available for exchange, as follows:

1. Diffusing molecules have only a short distance to travel between blood and surrounding cells because of the thin capillary wall and small capillary diameter, coupled with the proximity of every cell to a capillary. This short dis-

tance is important because the rate of diffusion slows down as the diffusion distance increases.

a. Capillary walls are very thin (1 μm in thickness; in contrast, the diameter of a human hair is 100 μm). Capillaries consist of only a single layer of flat endothelial cells—essentially the lining of the other vessel types. No smooth muscle or connective tissue is present (Figure 10-14a; also see Table 10-1, p. 339). The endothelial

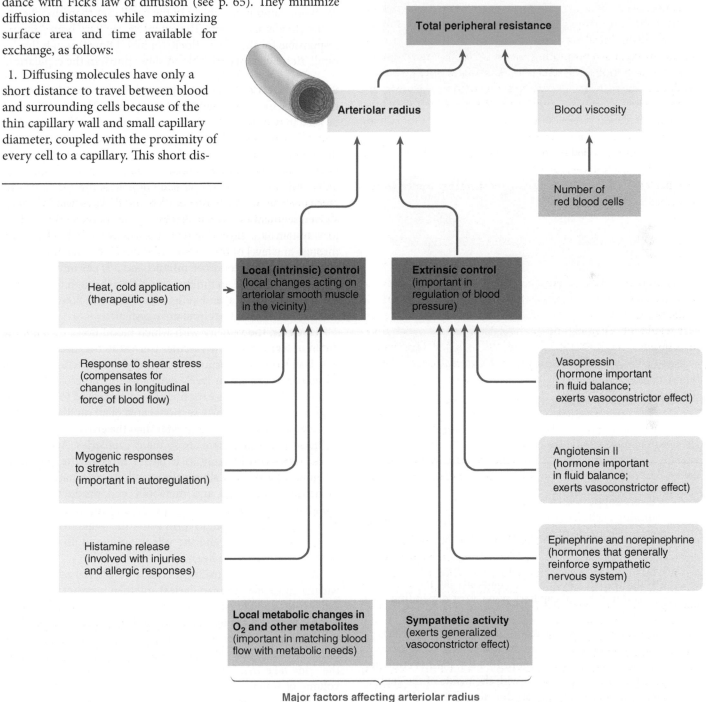

Major factors affecting arteriolar radius

Figure 10-13 Factors affecting total peripheral resistance (TPR). The primary determinant of TPR is the adjustable arteriolar radius. Two major categories of factors influence arteriolar radius: (1) local (intrinsic) control, which is primarily important in matching blood flow through a tissue with the tissue's metabolic needs and is mediated by local factors acting on the arteriolar smooth muscle; and (2) extrinsic control, which is important in regulating blood pressure and is mediated primarily by sympathetic influence on arteriolar smooth muscle.

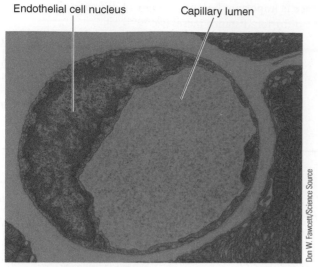

Endothelial cell nucleus   Capillary lumen

Don W. Fawcett/Science Source

(a) Cross section of a capillary

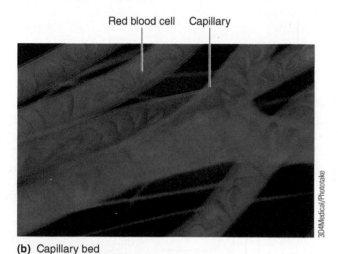

Red blood cell   Capillary

3D4Medical/Phototake

(b) Capillary bed

❙Figure 10-14 **Capillary anatomy.** (a) Electron micrograph showing that the capillary wall consists of a single layer of endothelial cells. The nucleus of one of these cells is shown. (b) The capillaries are so narrow that red blood cells must pass through the capillary bed in single file.

cells are supported by a thin *basement membrane,* a surrounding acellular (lacking cells) layer of extracellular matrix consisting of glycoproteins and collagen. Materials entering or leaving the capillaries pass freely through the basement membrane. Capillaries also have pores through which water-soluble materials can pass. The size and number of capillary pores vary, depending on the tissue.

b. Each capillary is so narrow (7 μm average diameter) that red blood cells (8 μm diameter) have to squeeze through in single file (❙Figure 10-14b). Plasma contents either are in direct contact with the inside of the capillary wall or are only a short diffusing distance from it.

c. Because of extensive capillary branching, scarcely any cell is farther than 0.1 mm (4/1000 in.) from a capillary.

2. Because capillaries are distributed in such incredible numbers (estimates range from 10 to 40 billion capillaries), a tre-

mendous total surface area is available for exchange (an estimated 600 m²). Despite this large number of capillaries, at any point in time these microscopic vessels contain only 5% of the total blood volume (250 mL out of a total of 5000 mL). As a result, a small volume of blood is exposed to an extensive surface area. If all capillary surfaces were stretched out in a flat sheet and the volume of blood contained within the capillaries was spread over the top, this would be roughly equivalent to spreading a half pint of paint over the floor of a high school gymnasium. Imagine how thin the paint layer would be!

3. Blood flows more slowly in the capillaries than elsewhere in the circulatory system. The extensive capillary branching is responsible for this slow velocity of blood flow through the capillaries. Let us see why blood slows down in the capillaries.

Slow Velocity of Flow Through Capillaries First, we need to clarify a potentially confusing point. The term *flow* can be used in two contexts—flow rate and velocity of flow. The *flow rate* refers to the *volume* of blood per unit of time flowing through a given segment of the circulatory system (this is the flow we have been talking about in relation to the pressure gradient and resistance). The *velocity of flow* is the *speed,* or distance per unit of time, with which blood flows forward through a given segment of the circulatory system. Because the circulatory system is a closed system, the volume of blood flowing through any level of the system must equal the CO. If the heart pumps out 5 L of blood per minute, and 5 L per minute return to the heart, then 5 L per minute must flow through the arteries, arterioles, capillaries, and veins. Therefore, the flow rate is the same at all levels of the circulatory system.

However, the velocity with which blood flows through the different segments of the vascular tree varies because velocity of flow is inversely proportional to the total cross-sectional area of all vessels at any given level. Even though the cross-sectional area of each capillary is extremely small compared to that of the large aorta, the total cross-sectional area of all capillaries added together is about 750 times greater than the cross-sectional area of the aorta because there are so many capillaries. Accordingly, blood slows considerably as it passes through the capillaries (❙Figure 10-15). This slow velocity allows adequate time for exchange of nutrients and metabolic end products between blood and tissue cells—the sole purpose of the circulatory system. As capillaries rejoin to form veins, the total cross-sectional area is again reduced, and the velocity of blood flow increases as blood returns to the heart.

As an analogy, consider a river (the arterial system) that widens into a lake (the capillaries) and then narrows into a river again (the venous system) (❙Figure 10-16). The flow rate is the same throughout the length of this body of water—that is, identical volumes of water are flowing past all points along the bank of the river and lake. However, the velocity of flow is slower in the wide lake than in the narrow river because the identical volume of water, now spread out over a larger cross-sectional area, moves forward a shorter distance in the wide lake than in the narrow river during a given period. You could readily observe the forward movement of water in the swift-flowing river, but the forward motion of water in the lake would be unnoticeable.

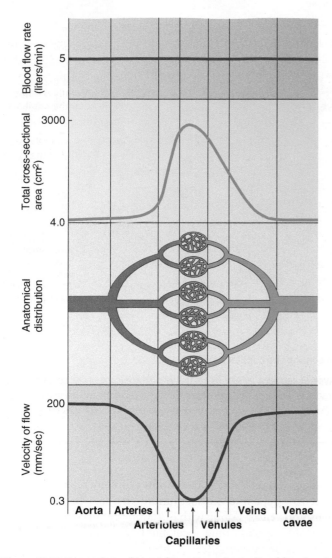

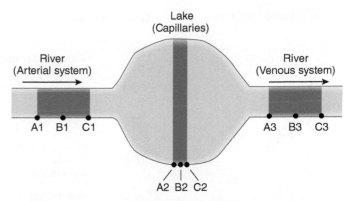

Figure 10-16 **Relationship between total cross-sectional area and velocity of flow.** The three dark blue areas represent equal volumes of water. During 1 minute, this volume of water moves forward from points A to points C. Therefore, an identical volume of water flows past points B1, B2, and B3 during this minute—that is, the flow rate is the same at all points along the length of this body of water. However, during that minute the identical volume of water moves forward a much shorter distance in the wide lake (A2 to C2) than in the much narrower river (A1 to C1 and A3 to C3). Thus, velocity of flow is much slower in the lake than in the river. Similarly, velocity of flow is much slower in the capillaries than in the arterial and venous systems.

Figure 10-15 **Comparison of blood flow rate and velocity of flow in relation to total cross-sectional area.** The blood flow rate (*red curve*) is identical through all levels of the circulatory system and is equal to the cardiac output (5 L/ min at rest). The velocity of flow (*purple curve*) varies throughout the vascular tree and is inversely proportional to the total cross-sectional area (*green curve*) of all the vessels at a given level. Note that the velocity of flow is slowest in the capillaries, which have the largest total cross-sectional area.

Also, because of the capillaries' tremendous total cross-sectional area, the resistance offered by all capillaries is lower than that offered by all arterioles, even though each capillary has a smaller radius than each arteriole. For this reason, the arterioles contribute more to TPR. Furthermore, arteriolar caliber (and, accordingly resistance) is subject to control, whereas capillary caliber cannot be adjusted.

Water-filled capillary pores permit passage of small, water-soluble substances.

Diffusion across capillary walls also depends on the walls' permeability to the materials being exchanged. Endothelial cells forming the capillary walls fit together like a jigsaw puzzle, but the closeness of the fit varies considerably among organs. In most capillaries, endothelial cells are *continuous,* or closely joined, with only narrow, water-filled clefts, or **pores**, at the junctions between the cells (Figure 10-17). The size of capillary pores varies from organ to organ. Following is an examination of capillaries of different porosity, from the tightest to the leakiest:

- At one extreme, the endothelial cells in brain capillaries are joined by tight junctions so that pores are nonexistent (see p. 51). These junctions prevent transcapillary ("across capillary") passage of materials between these cells and thus constitute part of the protective blood–brain barrier.

- In most capillaries (for example, in skeletal muscle and in lung tissue), small, water-soluble substances such as ions, glucose, and amino acids can readily pass through the water-filled pores, which are about 4 nm wide, but large, water-soluble materials such as plasma proteins are kept from passing through. Lipid-soluble substances, such as O_2 and CO_2, can readily pass through the endothelial cells themselves by dissolving in the lipid bilayer barrier of the plasma membrane surrounding the cells.

- In addition to having the narrow clefts between endothelial cells, the leakier capillaries of the kidneys and intestines have larger 20- to 100-nm holes known as **fenestrations** (*fenestra* means "window") that extend through the thickness of the endothelial cells. These through-the-cell passageways are important in the rapid movement of fluid across the capillaries in these organs during the formation of urine and during the absorption of a digested meal, respectively.

- At the other extreme, the endothelial cells of liver cells are *discontinuous*—that is, they are not in such close contact as in continuous capillaries. The gaps between adjacent cells in discontinuous capillaries range from 10 nm to 1000 nm, creating very large pores compared to the 4 nm clefts found in continuous capillaries. Discontinuous capillaries form large channels

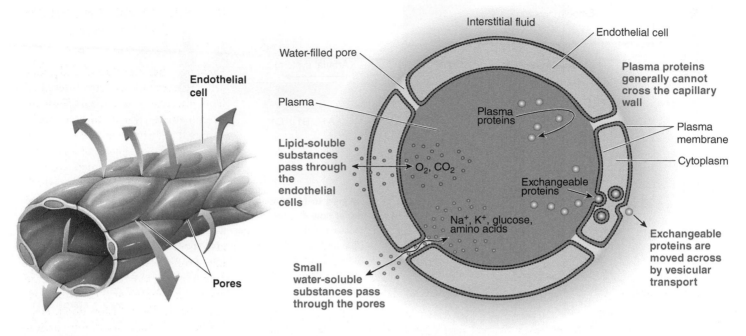

Endothelial cell

Water-filled pore

Plasma

Lipid-soluble substances pass through the endothelial cells

Pores

Small water-soluble substances pass through the pores

Interstitial fluid

Endothelial cell

Plasma proteins generally cannot cross the capillary wall

Plasma proteins

O_2, CO_2

Na$^+$, K$^+$, glucose, amino acids

Exchangeable proteins

Plasma membrane

Cytoplasm

Exchangeable proteins are moved across by vesicular transport

(a) Continuous capillary

(b) Transport across a continuous capillary wall

IFigure 10-17 Exchanges across a continuous capillary wall, the most common type of capillary. (a) Slitlike gaps between adjacent endothelial cells form pores within the capillary wall. (b) As depicted in this cross section of a capillary wall, small water-soluble substances are exchanged between the plasma and the interstitial fluid by passing through the water-filled pores, whereas lipid-soluble substances are exchanged across the capillary wall by passing through the endothelial cells. Proteins to be moved across are exchanged by vesicular transport. Plasma proteins generally cannot escape from the plasma across the capillary wall.

known as *sinusoids* that are five times wider than traditional capillaries. Liver sinusoids have fenestrations and such large intercellular pores that even proteins pass through readily. This is adaptive because the liver's functions include synthesis of plasma proteins and the metabolism of protein-bound substances such as cholesterol. These proteins must all pass through the liver's capillary (sinusoid) walls.

The leakiness of various capillary beds is therefore a function of how tightly the endothelial cells are joined (how wide the intercellular spaces are) and whether fenestrations are present, which varies according to the different organs' needs. For convenience, in the future we lump the between-cell pores and the through-the-cell fenestrations into the single category of *capillary pores.*

Scientists traditionally considered the capillary wall a passive sieve, like a brick wall with permanent gaps in the mortar acting as pores. However, they now know that endothelial cells can actively change to regulate capillary permeability— that is, in response to appropriate signals, the "bricks" can readjust themselves to vary the size of the holes between them. Thus, the degree of leakiness does not necessarily remain constant for a given capillary bed. For example, histamine increases capillary permeability by triggering contractile responses in endothelial cells to widen the intercellular gaps. This is not a muscular contraction because no smooth muscle cells are present in capillaries; it is the result of an actin–myosin contractile apparatus in the nonmuscular capillary endothelial cells. Because of these enlarged pores, the affected

capillary wall is leakier. As a result, normally retained plasma proteins escape into the surrounding tissue, where they exert an osmotic effect. Along with histamine-induced vasodilation, the resulting additional local fluid retention contributes to inflammatory swelling (see p. 409).

Vesicular transport also plays a limited role in passage of materials across the capillary wall. Large molecules that are not lipid-soluble, such as protein hormones that must be exchanged between blood and surrounding tissues, are transported from one side of the capillary wall to the other in endocytic–exocytic vesicles, a process called *transcytosis* (see p. 77).

Many capillaries are not open under resting conditions.

The branching and reconverging arrangement within capillary beds varies somewhat, depending on the tissue. Capillaries typically branch either directly from an arteriole or from a thoroughfare channel known as a **metarteriole**, which runs between an arteriole and a venule. Likewise, capillaries may rejoin at either a venule or a metarteriole (IFigure 10-18a).

Wisps of spiraling smooth muscle cells form **precapillary sphincters**, each of which consists of a ring of smooth muscle around the entrance to a capillary as it arises from a metarteriole or an arteriole.

Role of Precapillary Sphincters Precapillary sphincters are not innervated, but they have a high degree of myogenic tone and are sensitive to local metabolic changes. They act as

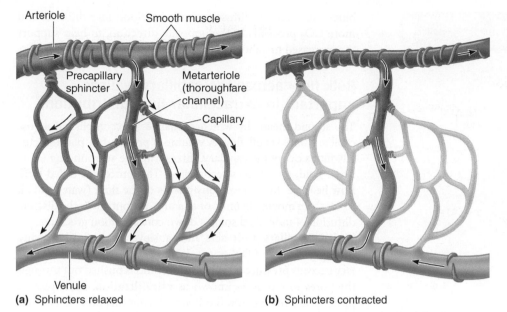

Arteriole

Smooth muscle

Precapillary
sphincter

Metarteriole
(thoroughfare
channel)

Capillary

Venule

(a) Sphincters relaxed

(b) Sphincters contracted

▌Figure 10-18 **Capillary bed.** Capillaries branch either directly from an arteriole or from a metarteriole, a thoroughfare channel between an arteriole and venule. Capillaries rejoin at either a venule or a metarteriole. Smooth muscle cells form precapillary sphincters that encircle capillaries as they arise from a metarteriole or an arteriole. (a) When the precapillary sphincters are relaxed, blood flows through the entire capillary bed. (b) When the precapillary sphincters are contracted, blood flows only through the metarteriole, bypassing the capillary bed.

Interstitial fluid is a passive intermediary between blood and cells.

Exchanges between blood and tissue cells are not made directly. Interstitial fluid, the true internal environment in immediate contact with the cells, acts as the go-between. Only 20% of the extracellular fluid (ECF) circulates as plasma. The remaining 80% consists of interstitial fluid, which bathes all cells in the body. Cells exchange materials directly with interstitial fluid, with the type and extent of exchange being governed by the properties of cellular plasma membranes. Movement across the plasma membrane may be either passive (that is, by diffusion down electrochemical gradients or by carrier-mediated facilitated diffusion) or active (that is, by carrier-mediated active transport or by vesicular transport) (see ▌Table 3-2, p. 78).

In contrast, exchanges across the capillary wall between the plasma and the interstitial fluid are largely passive. The only transport across this barrier that requires energy is the limited vesicular transport. Because capillary walls are highly permeable, exchange is so thorough that the interstitial fluid takes on essentially the same composition as incoming arterial blood, with the exception of the large plasma proteins that usually do not escape from the blood. Therefore, when we speak of exchanges between blood and tissue cells, we tacitly include interstitial fluid as a passive intermediary.

Exchanges between blood and surrounding tissues across capillary walls are accomplished in two ways: (1) passive diffusion down concentration gradients, the primary mechanism for exchanging individual solutes, and (2) bulk flow, a process that fills the different function of determining the distribution of the ECF volume between the vascular and the interstitial fluid compartments. We now examine each of these mechanisms in more detail.

stopcocks to control blood flow through the particular capillary that each one guards. Capillaries themselves have no smooth muscle, so they cannot actively participate in regulating their own blood flow. When tissue metabolic activity increases, the local metabolic changes bring about relaxation of precapillary sphincters in the vicinity, thereby increasing the number of open capillaries (▌Figure 10-18a). When tissue activity decreases, the local precapillary sphincters contract. As a result, blood bypasses the capillary bed and flows only through the metarteriole (▌Figure 10-18b).

Generally, tissues that are more metabolically active have a greater density of capillaries. Muscles, for example, have relatively more capillaries than their tendinous attachments. Only about 10% or less of the precapillary sphincters in a resting muscle are open at any moment, however, so blood is flowing through only about 10% of the muscle's capillaries or flowing directly through the metarteriole without entering the remaining capillary bed. When the muscle becomes more active, a greater percentage of the precapillary sphincters relax in response to the local chemical changes, simultaneously opening up more capillary beds. Concurrently, arteriolar vasodilation increases total flow to the organ. As a result of more blood flowing through more open capillaries, the total volume and surface area available for exchange increase, and the diffusion distance between the cells and an open capillary decreases. Thus, blood flow through a particular tissue (assuming a constant blood pressure) is regulated by (1) the degree of resistance offered by the arterioles in the organ, controlled by sympathetic activity and local metabolic factors; and (2) the number of open capillaries, controlled by action of the same local factors on precapillary sphincters.

Diffusion across capillary walls is important in solute exchange.

Because most capillary walls have no carrier-mediated transport systems, solutes cross primarily by diffusion down concentration gradients. The chemical composition of arterial blood is carefully regulated to maintain the concentrations of individual solutes at levels that promote each solute's movement in the appropriate direction across the capillary walls. The reconditioning organs continuously add nutrients and O_2 and remove CO_2 and other wastes as blood passes through them. Meanwhile, cells constantly use up supplies and generate metabolic wastes. As cells use up O_2 and glucose, the blood constantly brings in fresh supplies of these

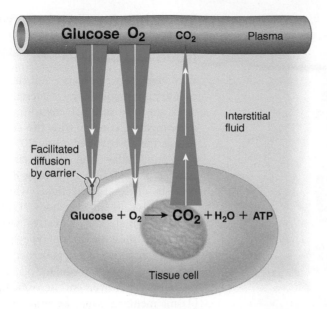

Facilitated
diffusion
by carrier

Glucose O_2 CO_2 Plasma

Interstitial
fluid

Glucose + O_2 ⟶ CO_2 + H_2O + ATP

Tissue cell

❙Figure 10-19 **Independent exchange of individual solutes down their own concentration gradients across the capillary wall.**

vital materials, maintaining concentration gradients that favor the net diffusion of these substances from blood to cells. Simultaneously, ongoing net diffusion of CO_2 and other metabolic wastes from cells to blood is maintained by the continual production of these wastes at the cell level and by their constant removal by the circulating blood (❙Figure 10-19).

Because the capillary wall does not limit passage of any constituent except plasma proteins, the extent of exchanges for each solute is independently determined by the magnitude of its concentration gradient between blood and surrounding cells. As cells increase their level of activity, they use up more O_2 and produce more CO_2, among other things. This creates larger concentration gradients for O_2 and CO_2 between these cells and

blood, so more O_2 diffuses out of the blood into the cells and more CO_2 proceeds in the opposite direction, to help support the increased metabolic activity.

Bulk flow across the capillary walls is important in extracellular fluid distribution.

The second means by which exchange is accomplished across capillary walls is bulk flow. A volume of protein-free plasma actually filters out of the capillary, mixes with the surrounding interstitial fluid, and then is reabsorbed. This process is called **bulk flow** because the various constituents of the fluid (water and all solutes) are moving in bulk, or as a unit, in contrast to the discrete diffusion of individual solutes down concentration gradients.

The capillary wall acts like a sieve, with fluid moving through its water-filled pores. When pressure inside the capillary exceeds pressure on the outside, fluid is pushed out through the pores in a process known as **ultrafiltration**. Most plasma proteins are retained on the inside during this process because of the pores' filtering effect, although a few do escape. Because all other constituents in plasma are dragged along as a unit with the volume of fluid leaving the capillary, the filtrate is essentially protein-free plasma. When inward-driving pressures exceed outward pressures across the capillary wall, net inward movement of fluid from the interstitial fluid into the capillaries takes place through the pores, a process known as **reabsorption**.

Forces Influencing Bulk Flow Bulk flow occurs because of differences in hydrostatic and colloid osmotic pressures between plasma and interstitial fluid. Even though pressure differences exist between plasma and surrounding fluid elsewhere in the circulatory system, only capillaries have pores that let fluids pass through. Four forces influence fluid movement across the capillary wall (❙Figure 10-20):

1. **Capillary blood pressure** (P_C) is the fluid or hydrostatic pressure exerted on the inside of the capillary walls by blood.

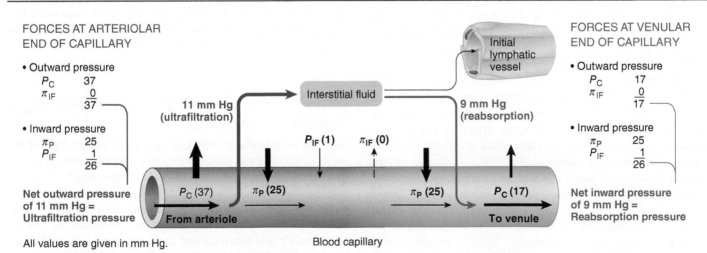

FORCES AT ARTERIOLAR
END OF CAPILLARY

• Outward pressure

P_C	37
π_{IF}	0
	37

• Inward pressure

π_P	25
P_{IF}	1
	26

Net outward pressure
of 11 mm Hg =
Ultrafiltration pressure

11 mm Hg
(ultrafiltration)

Interstitial fluid

Initial
lymphatic
vessel

9 mm Hg
(reabsorption)

P_{IF} (1) π_{IF} (0)

P_C (37) π_P (25) π_P (25) P_C (17)

From arteriole   To venule

FORCES AT VENULAR
END OF CAPILLARY

• Outward pressure

P_C	17
π_{IF}	0
	17

• Inward pressure

π_P	25
P_{IF}	1
	26

Net inward pressure
of 9 mm Hg =
Reabsorption pressure

All values are given in mm Hg.   Blood capillary

❙Figure 10-20 **Bulk flow across the capillary wall.** Ultrafiltration occurs at the arteriolar end and reabsorption occurs at the venule end of the capillary as a result of imbalances in the physical forces acting across the capillary wall.

FIGURE FOCUS: *If π_P decreases to 23 mm Hg as a result of a reduced plasma protein concentration, calculate the rates of ultrafiltration and reabsorption across the capillaries. What would happen to the volume of interstitial fluid compared to normal?*

This pressure tends to force fluid *out of* the capillaries into the interstitial fluid. By the level of the capillaries, blood pressure has dropped substantially because of frictional losses in pressure in the high-resistance arterioles upstream. On average, the hydrostatic pressure is 37 mm Hg at the arteriolar end of a tissue capillary (compared to a mean arterial pressure of 93 mm Hg). It declines even further, to 17 mm Hg, at the capillary's venular end because of further frictional loss coupled with the exit of fluid through ultrafiltration along the capillary's length (see ▌Figure 10-8, p. 343).

2. **Plasma-colloid osmotic pressure (π_P)**, also known as *oncotic pressure,* is a force caused by colloidal dispersion of plasma proteins (see p. A-7); it encourages fluid movement *into* the capillaries. Because plasma proteins remain in the plasma rather than entering the interstitial fluid, a protein concentration difference exists between plasma and interstitial fluid. Accordingly, a water concentration difference also exists between these two regions. Plasma has a higher protein concentration and a lower water concentration than interstitial fluid does. This difference exerts an osmotic effect that tends to move water from the area of higher water concentration in interstitial fluid to the area of lower water concentration in plasma (see p. 66). The other plasma constituents do not exert an osmotic effect because they readily pass through the capillary wall, so their concentrations are equal in plasma and interstitial fluid. π_P averages 25 mm Hg.

3. **Interstitial fluid hydrostatic pressure (P_{IF})** is the fluid pressure exerted on the outside of the capillary wall by interstitial fluid. This pressure tends to force fluid *into* the capillaries. Because of difficulties measuring P_{IF}, the actual value is controversial. It is either at, slightly above, or slightly below atmospheric pressure. For purposes of illustration, we will say it is 1 mm Hg above atmospheric pressure.

4. **Interstitial fluid–colloid osmotic pressure (π_{IF})** is another force that does not normally contribute significantly to bulk flow. The small fraction of plasma proteins that leak across the capillary walls into the interstitial spaces are normally returned to the blood by the lymphatic system. Therefore, the protein concentration in the interstitial fluid is extremely low, and the interstitial fluid–colloid osmotic pressure is essentially zero. If plasma proteins pathologically leak into the interstitial fluid, however, as they do when histamine widens the capillary pores during tissue injury, the leaked proteins exert an osmotic effect that tends to promote movement of fluid *out of* the capillaries into the interstitial fluid.

Therefore, the two pressures that tend to force fluid out of the capillary are P_C and π_{IF}. The two opposing pressures that tend to force fluid into the capillary are π_P and P_{IF}. Now let us analyze the fluid movement that occurs across a capillary wall because of imbalances in these opposing physical forces (▌Figure 10-20).

Net Exchange of Fluid Across the Capillary Wall Net exchange at a given point across the capillary wall can be calculated using the following equation:

$$\textbf{Net exchange pressure} = \underbrace{(P_C + \pi_{IF})}_{\text{(outward pressure)}} - \underbrace{(\pi_P + P_{IF})}_{\text{(inward pressure)}}$$

A positive net exchange pressure (when outward pressure exceeds inward pressure) represents an ultrafiltration pressure. A negative net exchange pressure (when inward pressure exceeds outward pressure) represents a reabsorption pressure.

At the arteriolar end of the capillary, outward pressure totals 37 mm Hg, whereas inward pressure totals 26 mm Hg, for a net outward pressure of 11 mm Hg. Ultrafiltration takes place at the beginning of the capillary as this outward pressure gradient forces a protein-free filtrate through the capillary pores.

By the time the venular end of the capillary is reached, capillary blood pressure has dropped but the other pressures have remained essentially constant. At this point, outward pressure has fallen to a total of 17 mm Hg, whereas the total inward pressure is still 26 mm Hg, for a net inward pressure of 9 mm Hg. Reabsorption of fluid takes place as this inward pressure gradient forces fluid back into the capillary at its venular end.

Ultrafiltration and reabsorption, collectively known as *bulk flow,* are thus the result of a shift in the balance between the passive physical forces acting across the capillary wall. No active forces or local energy expenditures are involved in bulk exchange of fluid between plasma and surrounding interstitial fluid. With only minor contributions from interstitial fluid forces, ultrafiltration occurs at the beginning of the capillary because capillary blood pressure exceeds plasma-colloid osmotic pressure, whereas by the end of the capillary, reabsorption takes place because blood pressure has fallen below osmotic pressure.

Note that we have taken "snapshots" at two points—at the beginning and at the end—in a hypothetical capillary. Actually, blood pressure gradually diminishes along the length of the capillary so that progressively diminishing quantities of fluid are filtered out in the first half of the vessel and progressively increasing quantities of fluid are reabsorbed in the last half. Even this situation is idealized. The pressures used in this figure are average values. Some capillaries have such high hydrostatic pressure that filtration actually occurs throughout their entire length, whereas others have such low hydrostatic pressure that reabsorption takes place throughout their length.

Role of Bulk Flow Bulk flow does not play an important role in exchange of individual solutes between blood and tissues because the quantity of solutes moved across the capillary wall by bulk flow is extremely small compared to the larger transfer of solutes by diffusion. The composition of the fluid filtered out of the capillary is essentially the same as the composition of the fluid that is reabsorbed. Thus, ultrafiltration and reabsorption are not important in exchange of nutrients and wastes. Bulk flow is extremely important, however, in regulating the distribution of ECF between plasma and interstitial fluid. Maintenance of proper arterial blood pressure depends in part on an appropriate volume of circulating blood. If plasma volume is reduced (for example, by hemorrhage), blood pressure falls. The resultant lowering of capillary blood pressure alters the balance of forces across the capillary walls. Because net outward pressure is decreased while net inward pressure remains unchanged, extra fluid is shifted from the interstitial compartment into the plasma as a result of reduced filtration and increased reabsorption. The extra fluid soaked up from the interstitial fluid provides additional fluid for the plasma, temporarily compensating for the blood loss. Meanwhile, reflex mechanisms acting on the heart and blood vessels (described later) also come into play to help maintain blood pressure until long-term mechanisms, such as thirst (and its satisfaction) and reduced urinary output, can restore the fluid volume to completely compensate for the loss.

Conversely, if the plasma volume becomes overexpanded, as with excessive fluid intake, the resulting rise in capillary blood pressure forces extra fluid from the capillaries into the interstitial fluid, temporarily relieving the expanded plasma volume until the excess fluid can be eliminated from the body by long-term measures, such as increased urinary output.

These internal fluid shifts between the two ECF compartments occur automatically and immediately whenever the balance of forces acting across the capillary walls is changed; they provide a temporary mechanism to help keep plasma volume fairly constant. In the process of restoring plasma volume to an appropriate level, interstitial fluid volume fluctuates, but it is more important that plasma volume be kept constant to ensure that the circulatory system functions effectively.

The lymphatic system is an accessory route for return of interstitial fluid to the blood.

Even under normal circumstances, slightly more fluid is filtered out of the capillaries into the interstitial fluid than is reabsorbed from the interstitial fluid back into the plasma. On average, the net ultrafiltration pressure starts at 11 mm Hg at the beginning of the capillary, whereas the net reabsorption pressure only reaches 9 mm Hg by the vessel's end (see ▌Figure 10-20). Because of this pressure differential, on average more fluid is filtered out of the first half of the capillary than is reabsorbed in its last half. The extra fluid filtered out as a result of this filtration–reabsorption imbalance is picked up by the **lymphatic system**. This extensive network of one-way vessels provides an accessory route by which fluid can be returned from the interstitial fluid to the blood. The lymphatic system functions much like a storm sewer that picks up and carries away excess rainwater so that it does not accumulate and flood an area.

Pickup and Flow of Lymph Small, blind-ended terminal lymph vessels known as **initial lymphatics** permeate almost every tissue of the body (▌Figure 10-21a). The endothelial cells forming the walls of initial lymphatics slightly overlap like shingles on a roof, with their overlapping edges being free instead of attached to the surrounding cells. This arrangement creates one-way, valvelike openings in the vessel wall. Fluid pressure on the outside of the vessel pushes the innermost edge of a pair of overlapping edges inward, creating a gap between the edges (that is, opening the valve). This opening permits interstitial fluid to enter (▌Figure 10-21b). Once interstitial fluid enters a lymphatic vessel, it is called **lymph**. Fluid pressure on the inside forces the overlapping edges together, closing the valves so that lymph does not escape. These lymphatic valvelike openings are larger than the pores in blood capillaries. Consequently, large particles in the interstitial fluid, such as escaped plasma proteins and bacteria, can gain access to initial lymphatics but are excluded from blood capillaries.

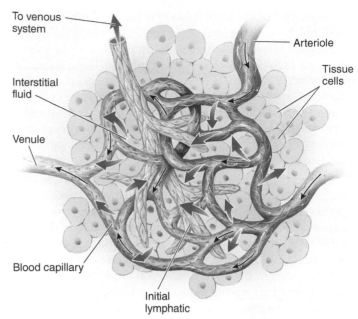

(a) Relationship between initial lymphatics and blood capillaries

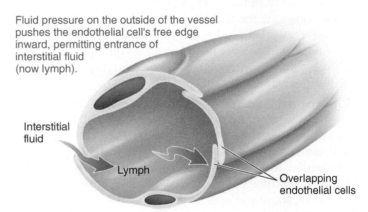

Fluid pressure on the outside of the vessel pushes the endothelial cell's free edge inward, permitting entrance of interstitial fluid (now lymph).

Fluid pressure on the inside of the vessel forces the overlapping edges together so that lymph cannot escape.

(b) Arrangement of endothelial cells in an initial lymphatic

▌Figure 10-21 **Initial lymphatics.** (a) Blind-ended initial lymphatics pick up excess fluid filtered by blood capillaries and return it to the venous system in the chest. (b) Note that the overlapping edges of the endothelial cells create valvelike openings in the vessel wall.

Initial lymphatics converge to form larger and larger **lymph vessels**, which eventually empty into the venous system near where blood enters the right atrium (▌Figure 10-22a). There is no "lymphatic heart" to provide driving pressure, but lymph is directed from the tissues toward the venous system in the thoracic cavity by two mechanisms. First, lymph vessels beyond the initial lymphatics are surrounded by smooth muscle, which contracts rhythmically as a result of myogenic activity. When this muscle is stretched because the vessel is distended with lymph, the muscle inherently contracts more forcefully, pushing the lymph through the vessel. This intrinsic "lymph pump" is the major force for propelling lymph. Stimulation of lymphatic smooth muscle by the sympathetic nervous system further increases the pumping activity of the lymph vessels. Sec-

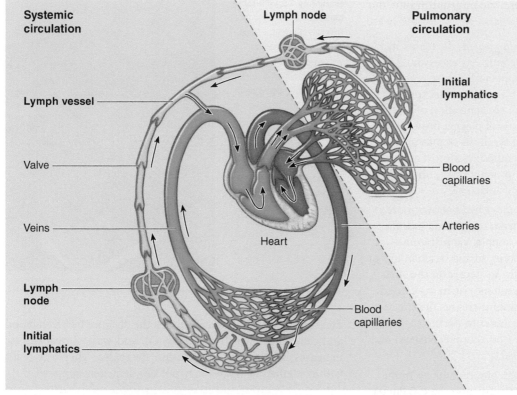

(a) Relationship of lymphatic system to circulatory system

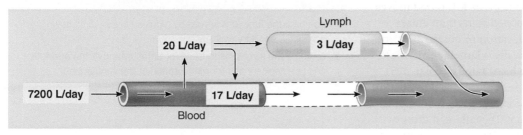

(b) Comparison of blood flow and lymph flow per day

I Figure 10-22 **Lymphatic system.** (a) Lymph empties into the venous system near its entrance to the right atrium. (b) Lymph flow averages 3 liters per day, whereas blood flow averages 7200 liters per day.

beat results in the equivalent of more than the entire plasma volume being left behind in the interstitial fluid each day. Obviously, this fluid must be returned to the circulating plasma, and the lymph vessels accomplish this task. The average rate of flow through the lymph vessels is 3 liters per day, compared with 7200 liters per day through the circulatory system.

■ *Defense against disease.* Lymph percolates through **lymph nodes** located en route within the lymphatic system. Passage of this fluid through the lymph nodes is important in the body's defense against disease. For example, bacteria picked up from the interstitial fluid are destroyed by special phagocytes within the lymph nodes (see Chapter 12).

■ *Transport of absorbed fat.* The lymphatic system transports fat absorbed from the digestive tract. The end products of dietary fat digestion are packaged by cells lining the digestive tract into fatty particles that are too large to gain access to the blood capillaries but can easily enter the initial lymphatics (see Chapter 16).

■ *Return of filtered protein.* Most capillaries permit leakage of a small amount of plasma proteins during filtration. These proteins cannot readily be reabsorbed back into the blood capillaries but can easily gain access to the initial lymphatics. If the proteins were allowed to accumulate in the interstitial fluid rather than being returned to the circulation via the lymphatics, π_{IF} (an outward pressure) would progressively increase while π_P (an inward pressure) would progressively fall. As a result, ultrafiltration forces would gradually increase and reabsorption forces would gradually decrease, resulting in progressive accumulation of fluid in the interstitial spaces at the expense of loss of plasma volume.

ond, because lymph vessels lie between skeletal muscles, contraction of these muscles squeezes the lymph out of the vessels. One-way valves spaced at intervals within the lymph vessels direct the flow of lymph toward its venous outlet in the chest.

Functions of the Lymphatic System Here are the most important functions of the lymphatic system:

■ *Return of excess filtered fluid.* Normally, capillary filtration exceeds reabsorption by about 3 liters per day (20 liters filtered, 17 liters reabsorbed) (I Figure 10-22b). Yet the entire blood volume is only 5 liters, and only 2.75 liters of that is plasma. (Blood cells make up the rest of the blood volume.) With an average CO of 5 L/min, 7200 liters of blood pass through the capillaries daily under resting conditions (more when CO increases). Even though only a small fraction of the filtered fluid is not reabsorbed by the blood capillaries, the cumulative effect of this process being repeated with every heart-

Edema occurs when too much interstitial fluid accumulates.

Clinical Note Occasionally, excessive interstitial fluid does accumulate when one of the physical forces acting across the capillary walls becomes abnormal for some reason. Swelling of the tissues because of excess interstitial fluid is known as **edema**. The causes of edema can be grouped into four categories:

1. *A reduced concentration of plasma proteins* decreases π_P. Such a drop in the major inward pressure lets excess fluid filter out, whereas less-than-normal amounts of fluid are reabsorbed; hence, extra fluid remains in the interstitial spaces. Edema can be caused by a decreased concentration of plasma proteins in several ways: excessive loss of plasma proteins in urine, from kidney disease; reduced synthesis of plasma proteins, from liver disease (the liver synthesizes almost all plasma proteins); a diet deficient in protein; or significant loss of plasma proteins from large burned surfaces.

2. *Increased permeability of the capillary walls* allows more plasma proteins than usual to pass from the plasma into the surrounding interstitial fluid—for example, via histamine-induced widening of the capillary pores during tissue injury or allergic reactions. The resultant fall in π_P decreases the effective inward pressure, whereas the resultant rise in π_{IF} caused by excess protein in the interstitial fluid increases the effective outward force. This imbalance contributes in part to the localized edema associated with injuries (for example, blisters) and allergic responses (for example, hives).

3. *Increased venous pressure*, as when blood dams up in the veins, is accompanied by increased P_C. Because the capillaries drain into the veins, damming of blood in the veins leads to a "backlog" of blood in the capillaries because less blood moves out of the capillaries into the overloaded veins than enters from the arterioles. The resultant elevation in outward hydrostatic pressure across the capillary walls is largely responsible for the edema seen with congestive heart failure (see p. 325). Regional edema can also occur because of localized restriction of venous return. An example is the swelling often occurring in the legs and feet during pregnancy. The enlarged uterus compresses the major veins that drain the lower extremities as these vessels enter the abdominal cavity. The resultant damming of blood in these veins raises blood pressure in the capillaries of the legs and feet, which promotes regional edema of the lower extremities.

4. *Blockage of lymph vessels* produces edema because the excess filtered fluid is retained in the interstitial fluid rather than returned to the blood through the lymphatics. Protein accumulation in the interstitial fluid compounds the problem through its osmotic effect. Local lymph blockage can occur, for example, in the arms of women whose major lymphatic drainage channels from the arm have been blocked as a result of lymph node removal during surgery for breast cancer. More widespread lymph blockage occurs with *filariasis*, a mosquito-borne parasitic disease found predominantly in tropical coastal regions. In this condition, small, threadlike filaria worms infect the lymph vessels, where their presence prevents proper lymph drainage. The affected body parts, particularly the scrotum and extremities, become grossly edematous. The condition is often called *elephantiasis* because of the elephant-like appearance of the swollen extremities (❙Figure 10-23).

Whatever the cause of edema, an important consequence is reduced exchange of materials between blood and cells. As

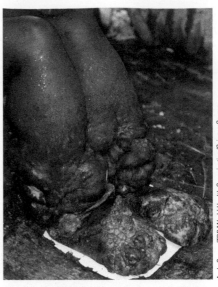

❙Figure 10-23 **Elephantiasis.** Note the pronounced edema in the legs, ankles, and feet of this person with elephantiasis. This tropical condition is caused by a mosquito-borne parasitic worm that invades the lymph vessels. As a result of the interference with lymph drainage, the affected body parts, usually the extremities, become grossly edematous, appearing elephantlike.

Andy Crump/TDR/World Health Organization/Science Source

excess interstitial fluid accumulates, the distance between blood and cells across which nutrients, O_2, and wastes must diffuse increases, so the rate of diffusion decreases. Therefore, cells within edematous tissues may not be adequately supplied.

Check Your Understanding 10.4

1. Compare the functions served by diffusion and by bulk flow across the capillary walls.
2. Describe the forces responsible for ultrafiltration at the arteriolar end of a capillary and for reabsorption at the venular end.
3. Define *lymph*.

10.5 Veins

The venous system completes the circulatory circuit. Blood leaving the capillary beds enters the venous system for transport back to the heart.

Venules communicate chemically with nearby arterioles.

At the microcirculatory level, capillaries drain into **venules**, which progressively converge to form small veins that exit the organ. In contrast to arterioles, venules have little tone and resistance. Extensive communication takes place via chemical signals between venules and nearby arterioles. This venuloarteriolar signaling is vital to matching capillary inflow and outflow within an organ.

Veins serve as a blood reservoir and as passageways back to the heart.

Veins have a large radius, so they offer little resistance to flow. Furthermore, because the total cross-sectional area of the venous system gradually decreases as smaller veins converge into progressively fewer but larger vessels, blood flow speeds up as blood approaches the heart.

In addition to serving as low-resistance passageways to return blood from the tissues to the heart, systemic veins also serve as a *blood reservoir*. Because of their storage capacity, veins are often called **capacitance vessels**. Veins have thinner walls with less smooth muscle than arteries do. Also, in contrast to arteries, veins have little elasticity because venous connective tissue contains considerably more collagen fibers than elastin fibers (see ▌Table 10-1, p. 339). Unlike arteriolar smooth muscle, venous smooth muscle has little inherent myogenic tone. Because of these features, veins are highly distensible, or stretchable, and have little elastic recoil. They easily distend to accommodate additional volumes of blood with only a small increase in venous pressure. Arteries stretched by an excess volume of blood recoil because of the elastin fibers in their walls, driving the blood forward. Veins containing an extra volume of blood simply stretch to accommodate the additional blood without tending to recoil. In this way veins serve as a **blood reservoir**—that is, when demands for blood are low, the veins can store extra blood in reserve because of their passive distensibility. Under resting conditions, the veins contain more than 60% of the total blood volume (▌Figure 10-24).

Contrary to a common misconception, blood stored in the veins is not being held in a stagnant holding tank. Normally all the blood is circulating all the time. When the body is at rest and many of the capillary beds are closed, the capacity of the venous reservoir is increased as extra blood bypasses the closed capillaries and enters the veins. When this extra volume of blood stretches the veins, the blood moves forward through the veins more slowly because the total cross-sectional area of the veins has been increased as a result of the stretching. Therefore, the blood spends more time in the veins. As a result of this slower transit time through the veins, the veins are essentially storing the extra volume of blood because it is not moving forward as quickly to the heart to be pumped out again.

When the stored blood is needed, such as during exercise, extrinsic factors (soon to be described) reduce the capacity of the venous reservoir and drive the extra blood from the veins to the heart so that it can be pumped to the tissues. Increased venous return leads to an increased stroke volume, in accordance with the Frank–Starling law of the heart (see p. 322). In contrast, if too much blood pools in the veins instead of being returned to the heart, CO is abnormally diminished. Thus, a delicate balance exists among the capacity of the veins, the extent of venous return, and CO. We now turn attention to the factors that affect venous capacity and contribute to venous return.

Venous return is enhanced by several extrinsic factors.

Venous capacity (the volume of blood that the veins can accommodate) depends on the distensibility of the vein walls (how much they can stretch to hold blood) and the influence of any externally applied pressure squeezing inwardly on the veins. At a constant blood volume, as venous capacity increases, more blood remains in the veins instead of being returned to

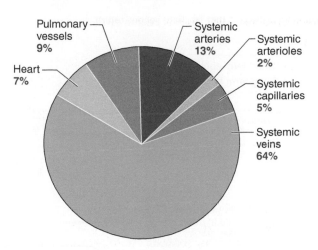

▌Figure 10-24 **Percentage of total blood volume in different parts of the circulatory system**.

the heart. Such venous storage decreases the **effective circulating blood volume**, the volume of blood being returned to and pumped out of the heart. Conversely, when venous capacity decreases, more blood is returned to the heart and is subsequently pumped out. Thus, changes in venous capacity directly influence the magnitude of venous return, which in turn is an important (although not the only) determinant of effective circulating blood volume. The effective circulating blood volume is also influenced short term by passive shifts in bulk flow between plasma and interstitial fluid and long term by factors that control total ECF volume, such as salt and water balance.

The term **venous return** refers to the volume of blood per minute entering each atrium from the veins. Recall that the magnitude of flow through a vessel is directly proportional to the pressure gradient. Much driving pressure imparted to the blood by cardiac contraction has been lost by the time the blood reaches the venous system because of frictional losses along the way, especially during passage through the high-resistance arterioles. By the time the blood enters the venous system, blood pressure averages only 17 mm Hg (see ▌Figure 10-8, p. 343). However, because atrial pressure is near 0 mm Hg, a small but adequate driving pressure still exists to promote flow of blood through the large-radius, low-resistance veins.

In addition to the driving pressure imparted by cardiac contraction, five other factors enhance venous return: sympathetically induced venous vasoconstriction, skeletal muscle pump, venous valves, respiratory pump, and cardiac suction (▌Figure 10-25). Most of these secondary factors affect venous return by increasing the pressure gradient between the veins and the heart. We examine each in turn.

Effect of Sympathetic Activity on Venous Return
Veins are not very muscular and have little inherent tone, but venous smooth muscle is abundantly supplied with sympathetic nerve fibers. Sympathetic stimulation produces venous vasoconstriction, which modestly elevates venous pressure; this, in turn, increases the pressure gradient to drive more of the stored blood from the veins into the right atrium, thus enhancing venous

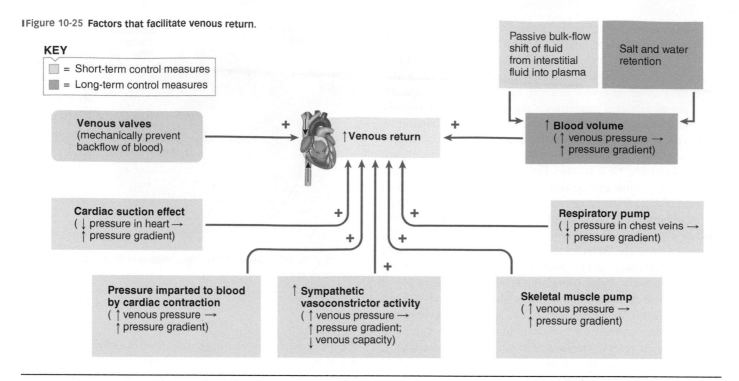

❙ Figure 10-25 Factors that facilitate venous return.

KEY
☐ = Short-term control measures
▓ = Long-term control measures

Passive bulk-flow shift of fluid from interstitial fluid into plasma

Salt and water retention

Venous valves (mechanically prevent backflow of blood)

+ **↑ Venous return**

+ **↑ Blood volume** (↑ venous pressure → ↑ pressure gradient)

Cardiac suction effect (↓ pressure in heart → ↑ pressure gradient)

Respiratory pump (↓ pressure in chest veins → ↑ pressure gradient)

Pressure imparted to blood by cardiac contraction (↑ venous pressure → ↑ pressure gradient)

↑ Sympathetic vasoconstrictor activity (↑ venous pressure → ↑ pressure gradient; ↓ venous capacity)

Skeletal muscle pump (↑ venous pressure → ↑ pressure gradient)

return. Veins normally have such a large radius that the moderate vasoconstriction from sympathetic stimulation has little effect on resistance to flow. Even when constricted, veins still have a relatively large radius and are still low-resistance vessels.

Note the different outcomes of vasoconstriction in arterioles and veins. Arteriolar vasoconstriction immediately *reduces* flow through these vessels because of their increased resistance (less blood can enter and flow through a narrowed arteriole), whereas venous vasoconstriction immediately *increases* flow through these vessels because of their decreased capacity (narrowing of veins squeezes out more of the blood already in the veins, increasing blood flow through these vessels). In addition to mobilizing the stored blood, venous vasoconstriction sustains increased venous return. With the filling capacity of the veins reduced, less blood draining from the capillaries remains in the veins but continues to flow instead toward the heart.

The increased venous return initiated by sympathetic stimulation leads to increased CO because of the increase in end-diastolic volume. Sympathetic stimulation of the heart also increases CO by increasing the heart rate and increasing the heart's contractility (see pp. 320 and 322). Thus, as long as sympathetic activity remains elevated, as during exercise, the heart pumps out more blood than usual for use by the exercising muscles.

Effect of Skeletal Muscle Activity on Venous Return
Many large veins in the extremities lie between skeletal muscles, so muscle contraction compresses the veins. This external venous compression decreases venous capacity and increases venous pressure, in effect squeezing blood in the veins forward toward the heart (❙ Figure 10-26a). This pumping action, known as the **skeletal muscle pump**, is another way extra blood stored

in the veins is returned to the heart during exercise. The skeletal muscle pump also counters the effect of gravity on the venous system. Let us see how.

Countering the Effects of Gravity on the Venous System The average pressures mentioned thus far for various regions of the vascular tree are for a person in the horizontal position. When a person is lying down, the force of gravity is uniformly applied, so it need not be considered. When a person stands up, however, gravitational effects are not uniform. In addition to the usual pressure from cardiac contraction, vessels below heart level are subject to pressure from the weight of the column of blood extending from the heart to the level of the vessel (❙ Figure 10-26b).

This increased pressure has two major consequences. First, the distensible veins yield under the increased hydrostatic pressure, further expanding so that their capacity is increased. Even though the arteries are subject to the same gravitational effects, they are not nearly as distensible and do not expand like the veins. Much of the blood entering from the capillaries tends to pool in the expanded lower-leg veins instead of returning to the heart. Because venous return is reduced, CO decreases and the effective circulating volume shrinks. Second, the marked increase in P_C resulting from the effect of gravity causes excessive fluid to filter out of capillary beds in the lower extremities, producing localized edema (that is, swollen feet and ankles) (❙ Figure 10-26c).

Two compensatory measures normally counteract these gravitational effects. First, the resultant fall in MAP that occurs when a person moves from a lying-down to an upright position triggers sympathetically induced venous vasoconstriction, which drives some of the pooled blood forward. Second, when

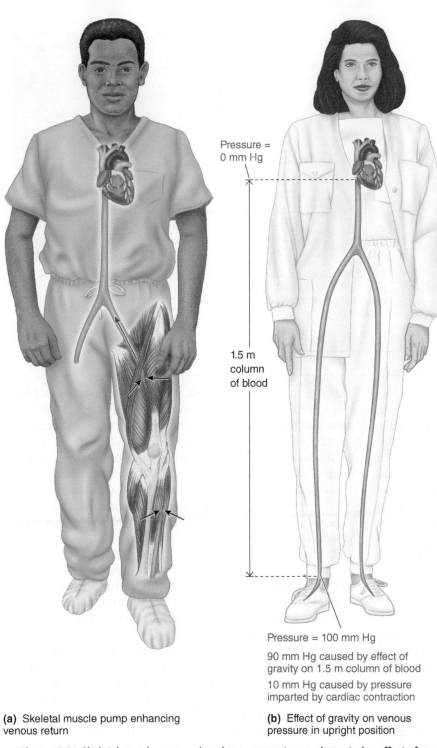

Pressure =
0 mm Hg

1.5 m
column
of blood

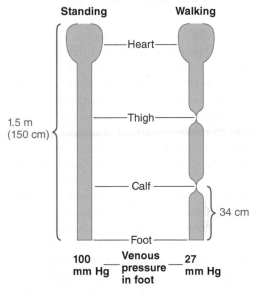

↓ **Venous return**

**Pooling of
blood in
distended
veins**

Venous
pressure =
100 mm Hg

Capillary blood
pressure averages
137 mm Hg

↑ **Ultrafiltration →
swelling of ankles and feet**

(c) Results of increased pressure caused by
gravity in upright position

Standing		Walking
	Heart	
1.5 m (150 cm)	Thigh	
	Calf	
		34 cm
	Foot	
100 mm Hg	Venous pressure in foot	27 mm Hg

Foot vein supporting
column of blood 1.5 m
(150 cm) in height

Foot vein supporting
column of blood
34 cm in height

Pressure = 100 mm Hg

90 mm Hg caused by effect of
gravity on 1.5 m column of blood

10 mm Hg caused by pressure
imparted by cardiac contraction

(a) Skeletal muscle pump enhancing
venous return

(b) Effect of gravity on venous
pressure in upright position

(d) Effect of contraction of skeletal muscles of legs
in counteracting the effects of gravity

I**Figure 10-26 Skeletal muscle pump enhancing venous return and countering effect of gravity on venous pressure.** (a) Compression of veins by contraction
of nearby skeletal muscles squeezes extra blood out of the veins, increasing venous return. (b) In an upright adult, the blood in the vessels extending between the heart
and the foot is equivalent to a 1.5-m column of blood. The pressure exerted by this column of blood as a result of the effect of gravity is 90 mm Hg. The pressure imparted
to the blood by the heart has declined to about 10 mm Hg in the lower-leg veins because of frictional losses in preceding vessels. Together these pressures produce a venous pressure of 100 mm Hg in the ankle and foot veins. The capillaries in the region are subjected to these same gravitational effects. (c) Because of the increased pressure caused by gravity, blood pools in the distended veins, resulting in decreased venous return. Ultrafiltration also increases across capillary walls, resulting in swollen ankles and feet, unless compensatory measures can counteract the effect of gravity. (d) Contraction of skeletal muscles (as in walking) completely empties given vein
segments, interrupting the column of blood that the lower veins must support.

(Source: Part (d) adapted from *Physiology of the Heart and Circulation*, 4th ed., by R. C. Little and W. C. Little. Copyright © 1989 Year Book Medical Publishers, Inc., with permission from Elsevier.)

FIGURE FOCUS: *Put in order from highest to lowest the venous pressure in your ankle and foot veins when you are (a) sitting in class (b) walking to
class, (c) standing at a concert, and (d) taking a nap.*

a person moves around, the skeletal muscle pump "interrupts" the column of blood by completely emptying given vein segments intermittently so that a particular portion of a vein is not subjected to the weight of the entire venous column from the heart to that portion's level (Figure 10-26d). Reflex venous vasoconstriction cannot completely compensate for gravitational effects without skeletal muscle activity. When a person stands still for a long time, therefore, blood flow to the brain is reduced because of the decline in effective circulating volume, despite reflexes aimed at maintaining MAP. Reduced flow of blood to the brain, in turn, leads to fainting, which returns the person to a horizontal position, eliminating the gravitational effects on

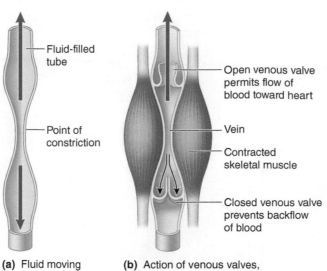

(a) Fluid moving in both directions on squeezing a fluid-filled tube

- Fluid-filled tube
- Point of constriction

(b) Action of venous valves, permitting flow of blood toward heart and preventing backflow of blood

- Open venous valve permits flow of blood toward heart
- Vein
- Contracted skeletal muscle
- Closed venous valve prevents backflow of blood

(c) Photograph of a closed venous valve

Photodisc/GettyImages

Figure 10-27 **Function of venous valves.**

the vascular system and restoring effective circulation. For this reason, it is counterproductive to try to hold upright someone who has fainted. Fainting is a remedy to the problem, not the problem itself.

Because the skeletal muscle pump facilitates venous return and helps counteract the detrimental effects of gravity on the circulatory system, when you are working at a desk it's a good idea to get up periodically and, when you are on your feet, you should move around. The mild muscular activity "gets the blood moving." It is further recommended that people who must be on their feet for long periods wear elastic stockings that apply a continuous gentle external compression, similar to the effect of skeletal muscle contraction, to further counter the effect of gravitational pooling of blood in the leg veins.

Effect of Venous Valves on Venous Return Venous vasoconstriction and external venous compression both drive blood toward the heart. Yet if you squeeze a fluid-filled tube in the middle, fluid is pushed in both directions from the point of constriction (Figure 10-27a). Then why isn't blood driven backward, as well as forward, by venous vasoconstriction and the skeletal muscle pump? Blood can only be driven forward because the large veins are equipped with one-way valves spaced at 2- to 4-cm intervals; these valves let blood move forward toward the heart but keep it from moving back toward the tissues (Figure 10-27b and c). These venous valves also play a role in counteracting the gravitational effects of upright posture by helping minimize the backflow of blood that tends to occur when a person stands up and by temporarily supporting portions of the column of blood when the skeletal muscles are relaxed.

Clinical Note **Varicose veins** occur when the venous valves become incompetent and can no longer support the column of blood above them. People predisposed to this condition usually have inherited an overdistensibility and weakness of their vein walls. Aggravated by frequent, prolonged standing,

the veins become so distended as blood pools in them that the edges of the valves can no longer meet to form a seal. Varicosed superficial leg veins become visibly overdistended and tortuous. Contrary to what might be expected, chronic pooling of blood in the pathologically distended veins does not reduce CO because of a compensatory increase in total circulating blood volume. Instead, the most serious consequence of varicose veins is the possibility of abnormal clot formation in the sluggish, pooled blood. Particularly dangerous is the risk that these clots may break loose and block small vessels elsewhere, especially the pulmonary capillaries.

Effect of Respiratory Activity on Venous Return As a result of respiratory activity, the pressure within the chest cavity averages 5 mm Hg less than atmospheric pressure. As the venous system returns blood to the heart from the lower regions of the body, it travels through the chest cavity, where it is exposed to this subatmospheric pressure. Because the venous system in the limbs and abdomen is subject to normal atmospheric pressure, an externally applied pressure gradient exists between the lower veins (at atmospheric pressure) and the chest veins (at less than atmospheric pressure). This pressure difference pushes blood from the lower veins to the chest veins, promoting increased venous return (Figure 10-28). This mechanism of facilitating venous return is called the **respiratory pump** because it results from respiratory activity. Increased respiratory activity, the skeletal muscle pump, and venous vasoconstriction, all enhance venous return during exercise.

Effect of Cardiac Suction on Venous Return The extent of cardiac filling does not depend entirely on factors affecting the veins. The heart plays a role in its filling. During ventricular contraction, the AV valves are drawn downward, enlarging the atrial cavities. As a result, atrial pressure transiently drops below 0 mm Hg, thus increasing the vein-to-atria pressure gradient so

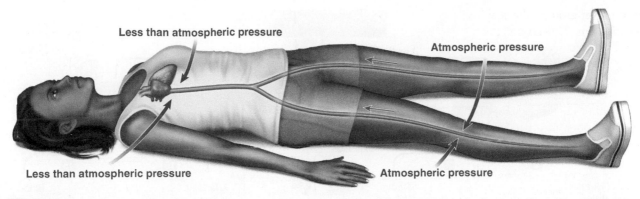

Less than atmospheric pressure

Atmospheric pressure

Less than atmospheric pressure

Atmospheric pressure

Figure 10-28 **Respiratory pump enhancing venous return.** As a result of respiratory activity, the pressure surrounding the chest veins is lower than the pressure surrounding the veins in the extremities and abdomen. This establishes an externally applied pressure gradient on the veins, which drives blood toward the heart.

that venous return is enhanced. In addition, rapid expansion of the ventricular chambers during ventricular relaxation creates a transient negative pressure in the ventricles so that blood is "sucked in" from the atria and veins—that is, the negative ventricular pressure increases the vein-to-atria-to-ventricle pressure gradient, further enhancing venous return. Thus, the heart functions as a "suction pump" to facilitate cardiac filling.

Check Your Understanding 10.5

1. Explain how veins have the capacity to store an extra volume of blood with little change in venous pressure.
2. Define *effective circulating blood volume*.
3. List the factors that enhance venous return.

10.6 Blood Pressure

Mean arterial pressure is the blood pressure that is monitored and regulated in the body. Routine blood pressure measurements record the arterial systolic and diastolic pressures, which can be used as a yardstick for assessing MAP.

Blood pressure is regulated by controlling cardiac output, total peripheral resistance, and blood volume.

Mean arterial pressure is the main driving force for propelling blood to the tissues. This pressure must be closely regulated for two reasons. First, it must be high enough to ensure sufficient driving pressure; without this pressure, the brain and other organs do not receive adequate flow, no matter what local adjustments are made in the resistance of the arterioles supplying them. Second, the pressure must not be so high that it creates extra work for the heart and increases the risk of vascular damage and possible rupture of small blood vessels.

Determinants of Mean Arterial Pressure Elaborate mechanisms involving the integrated action of the various components of the circulatory system and other body systems are vital in regu-

lating this all-important mean arterial pressure (Figure 10-29). Remember that the two determinants of MAP are CO and TPR:

$$MAP = CO \times TPR$$

Recall that a number of factors, in turn, determine CO and TPR (see Figure 9-24, p. 324; Figure 10-13, p. 351; and Figure 10-25, p. 362). Thus, you can quickly appreciate the complexity of blood pressure regulation. Let us work through Figure 10-29, reviewing all the factors that affect MAP. Even though we have covered all these factors before, it is useful to pull them together. The numbers in the text correspond to the numbers in the figure.

- MAP depends on CO and TPR (**1** on Figure 10-29).
- CO depends on heart rate and stroke volume **2**.
- Heart rate depends on the relative balance of parasympathetic activity **3**, which decreases heart rate, and sympathetic activity **4** (including epinephrine throughout this discussion), which increases heart rate.
- Stroke volume increases in response to sympathetic activity **5** (extrinsic control of stroke volume).
- Stroke volume also increases as venous return increases **6** (intrinsic control of stroke volume according to the Frank–Starling law of the heart).
- Venous return is enhanced by sympathetically induced venous vasoconstriction **7**, the skeletal muscle pump **8**, the respiratory pump **9**, and cardiac suction **10**.
- The effective circulating blood volume also influences how much blood is returned to the heart **11** and therefore ultimately on how much blood is pumped out by the heart. Blood volume depends in the short term on the size of passive bulk-flow fluid shifts between the plasma and the interstitial fluid across the capillary walls **12**. In the long term, blood volume depends on salt and water balance **13**, which are hormonally controlled by the renin–angiotensin–aldosterone system and vasopressin, respectively **14**.
- The other major determinant of mean arterial pressure, TPR, depends on the radius of all arterioles and on blood viscosity **15**. The main factor determining blood viscosity is the number of red blood cells **16**. However, arteriolar radius is the more important factor determining TPR.

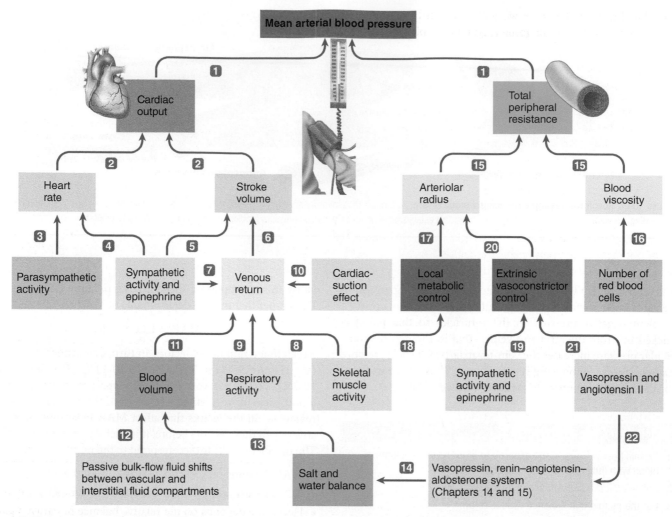

IFigure 10-29 Determinants of mean arterial blood pressure. Note that this figure is basically a composite of Figure 9-24, p. 324, "Control of cardiac output"; Figure 10-13, p. 351, "Factors affecting total peripheral resistance"; and Figure 10-25, p. 362, "Factors that facilitate venous return." See the text for a discussion of the numbers.

FIGURE FOCUS: *State the pathway on this flow diagram by which drugs that interfere with the salt-conserving renin–angiotensin–aldosterone system act to bring down blood pressure in patients with high blood pressure.*

■ Arteriolar radius is influenced by local (intrinsic) metabolic controls that match blood flow with metabolic needs **17**. For example, local changes that take place in active skeletal muscles cause local arteriolar vasodilation and increased blood flow to these muscles **18**.

■ Arteriolar radius is also influenced by sympathetic activity **19**, an extrinsic control mechanism that causes arteriolar vasoconstriction **20** to increase TPR and subsequently MAP.

■ Arteriolar radius is also extrinsically controlled by the hormones vasopressin and angiotensin II, which are potent vasoconstrictors **21** as well as being important in salt and water balance **22**.

Altering any of the pertinent factors that influence MAP changes this pressure, unless a compensatory change in another variable keeps the blood pressure constant. Blood flow to any given organ depends on the driving force of the MAP and on the degree of vasoconstriction of the organ's arterioles. Because MAP depends on CO and the degree of arteriolar vasoconstriction, if the arterioles in one organ

dilate, the arterioles in other organs must constrict to maintain an adequate MAP. An adequate pressure is needed to provide a driving force to push blood not only to the vasodilated organ but also to the brain, which depends on a constant blood supply. Thus, the cardiovascular variables must be continuously juggled to maintain a constant blood pressure despite organs' varying needs for blood.

Short-Term and Long-Term Control Measures Mean arterial pressure is constantly monitored by **baroreceptors** (pressure sensors) within the circulatory system. When deviations from normal are detected, multiple reflex responses are initiated to return MAP to its normal value. *Short-term* (within seconds) adjustments are made by alterations in CO and TPR, mediated by means of autonomic nervous system influences on the heart, veins, and arterioles. *Long-term* (requiring minutes to days) control involves adjusting total blood volume by restoring normal salt and water balance through mechanisms that regulate urine output and thirst (see Chapters 14 and 15). The size of the total blood volume, in turn, has a profound effect on CO and, thereby, MAP. For now we will focus on the short-term mechanisms involved in ongoing regulation of blood pressure.

The baroreceptor reflex is a short-term mechanism for regulating blood pressure.

Any change in MAP triggers an autonomically mediated **baroreceptor reflex** that influences the heart and blood vessels to adjust CO and TPR in an attempt to restore blood pressure toward normal. Like any reflex, the baroreceptor reflex includes a receptor, an afferent pathway, an integrating center, an efferent pathway, and effector organs.

The most important receptors involved in the moment-to-moment regulation of blood pressure, the **carotid sinus** and **aortic arch baroreceptors**, are mechanoreceptors sensitive to changes in MAP. These baroreceptors are strategically located (IFigure 10-30) to provide critical information about arterial blood pressure in the vessels leading to the brain (the carotid sinus baroreceptor) and in the major arterial trunk before it gives off branches that supply the rest of the body (the aortic arch baroreceptor).

The baroreceptors constantly provide information about MAP; in other words, they continuously generate action potentials in response to the ongoing pressure within the arteries. When MAP increases, the receptor potential of these baroreceptors increases, thus increasing the rate of firing in the corresponding afferent neurons. Conversely, a decrease in MAP slows the rate of firing generated in the afferent neurons by the baroreceptors (IFigure 10-31).

The integrating center that receives the afferent impulses about the state of the mean arterial pressure is the **cardiovascular control center**, located in the medulla within the brain stem. The efferent pathway is the autonomic nervous system. The cardiovascular control center alters the ratio between sympathetic and parasympathetic activity to the effector organs (the heart and blood vessels). To review how autonomic changes alter arterial blood pressure, study IFigure 10-32, which summarizes the major effects of parasympathetic and sympathetic stimulation on the heart and blood vessels.

Let us fit all the pieces of the baroreceptor reflex together by tracing the reflex activity that compensates for an elevation or fall in arterial blood pressure. If for any reason MAP transiently rises above normal (for example, caused by anxiety or taking vasoconstricting decongestants), the carotid sinus and aortic arch baroreceptors increase the rate of firing in their respective afferent neurons (IFigure 10-33a). On being informed by increased afferent firing that blood pressure has become too high, the cardiovascular control center responds by decreasing sympathetic and increasing parasympathetic activity to the cardiovascular system. These efferent signals decrease heart rate, decrease stroke volume, and produce arteriolar and venous vasodilation, which in turn lead to a decrease in CO and a decrease in TPR, with a subsequent fall in blood pressure back toward normal.

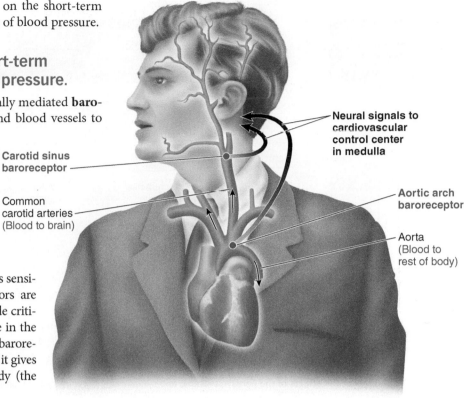

Carotid sinus baroreceptor

Common carotid arteries (Blood to brain)

Neural signals to cardiovascular control center in medulla

Aortic arch baroreceptor

Aorta (Blood to rest of body)

IFigure 10-30 **Location of the arterial baroreceptors.** The arterial baroreceptors are strategically located to monitor the mean arterial blood pressure in the arteries that supply blood to the brain (carotid sinus baroreceptor) and to the rest of the body (aortic arch baroreceptor).

Conversely, when blood pressure falls below normal (for example caused by blood pooling in the leg veins when a person gets up from bed or by blood loss from a traumatic injury), baroreceptor activity decreases, inducing the cardiovascular center to increase sympathetic cardiac and vasoconstrictor nerve activity while decreasing its parasympathetic output (IFigure 10-33b). This efferent pattern of activity leads to an increase in heart rate and stroke volume, coupled with arteriolar and venous vasoconstriction. These changes increase both CO and TPR, raising blood pressure back toward normal.

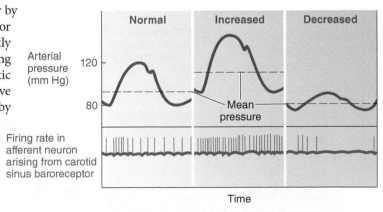

IFigure 10-31 **Firing rate in the afferent neuron from the carotid sinus baroreceptor in relation to the magnitude of mean arterial pressure.**

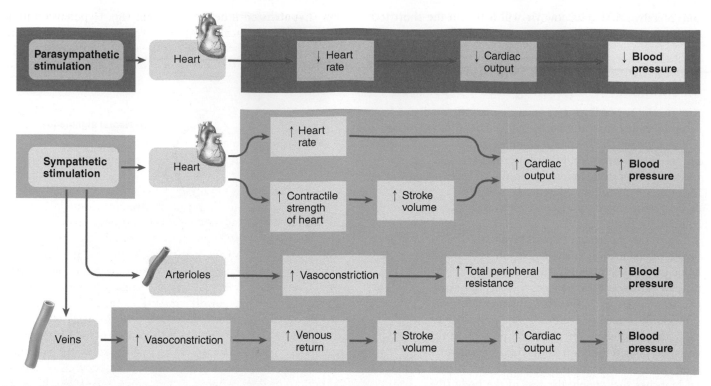

I Figure 10-32 **Summary of the effects of the parasympathetic and sympathetic nervous systems on factors that influence mean arterial blood pressure.**

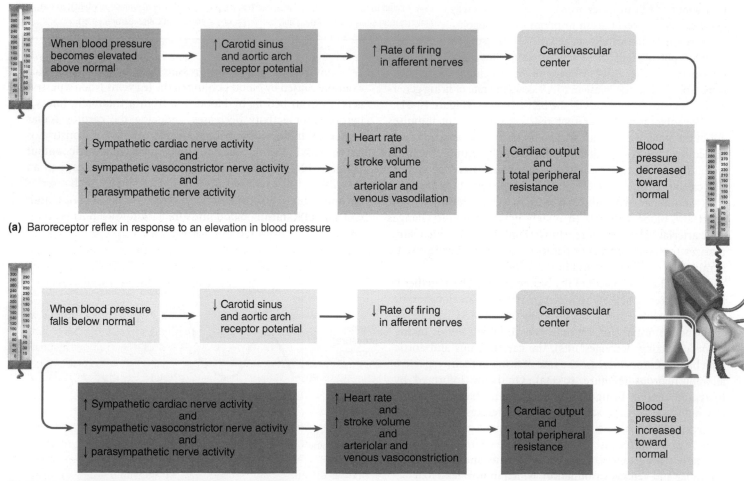

(a) Baroreceptor reflex in response to an elevation in blood pressure

(b) Baroreceptor reflex in response to a fall in blood pressure

I Figure 10-33 **Baroreceptor reflexes to restore blood pressure to normal.**

Other reflexes and responses influence blood pressure.

Besides the baroreceptor reflex, whose sole function is blood pressure regulation, several other reflexes and responses influence the cardiovascular system and blood pressure even though they primarily regulate other body functions. These factors include the following:

1. Left atrial volume receptors and hypothalamic osmoreceptors are primarily important in water and salt balance in the body; they thus affect long-term regulation of blood pressure by controlling blood volume.

2. Chemoreceptors located in the carotid and aortic arteries, in close association with but distinct from the baroreceptors, are sensitive to low O_2 or high acid levels in the blood. These chemoreceptors' main function is to reflexly increase respiratory activity to bring in more O_2 or to blow off more acid-forming CO_2, but they also reflexly increase blood pressure by sending excitatory impulses to the cardiovascular center.

3. Cardiovascular responses associated with certain behaviors and emotions are mediated through the cerebral cortex–hypothalamic pathway and appear preprogrammed. These responses include the widespread changes in cardiovascular activity accompanying the generalized sympathetic fight-or-flight response, the characteristic marked increase in heart rate and blood pressure associated with sexual orgasm, and the localized cutaneous (skin) vasodilation characteristic of blushing.

4. Hypothalamic control over cutaneous arterioles for the purpose of temperature regulation takes precedence over control that the cardiovascular center has over these same vessels for the purpose of blood pressure regulation. As a result, blood pressure can fall when the skin vessels are widely dilated to eliminate excess heat from the body, even though the baroreceptor responses are calling for cutaneous vasoconstriction to help maintain adequate TPR.

5. Marked cardiovascular changes occur in response to exercise, mediated by exercise centers in the brain at the onset of exercise or even in anticipation of exercise. In moderate exercise, CO more than doubles (and increases up to fivefold in heavy exercise). CO not only is larger but also is redistributed compared to resting; whereas skeletal muscles receive 15% of the CO at rest, during moderate exercise their share of the CO goes up to 64%. As a result, exercising skeletal muscles receive more than a 1000% increase in blood flow. (Details of these and other cardiovascular changes during exercise can be found in the boxed feature on pp. 370–371, A Closer Look at Exercise Physiology.)

We next examine blood pressure abnormalities.

Hypertension is a national public-health problem, but its causes are largely unknown.

Clinical Note Sometimes blood pressure control mechanisms do not function properly or are unable to completely compensate for changes that have taken place. Blood pressure may be too high (**hypertension** if above 140/90 mm Hg) or too low (**hypotension** if below 90/60 mm Hg). Hypotension in its extreme form is *circulatory shock*. We first examine hypertension, which is by far the most common of blood pressure abnormalities, and then conclude this chapter with a discussion of hypotension and shock.

Types of Hypertension There are two broad types of hypertension, secondary hypertension and primary hypertension, depending on the cause. A definite cause for hypertension can be established in only 10% of cases. Hypertension that occurs secondary to another known primary problem is called **secondary hypertension**. For example, hypertension can occur secondary to kidney disease. If the kidneys are unable to eliminate the normal salt load, retention of salt and accompanying water expands the blood volume, thus chronically increasing blood pressure. The underlying cause is unknown in the remaining 90% of hypertension cases. Such hypertension is known as **primary**, or **essential**, **hypertension**. Primary hypertension is a catchall category for blood pressure elevated by a variety of unknown causes rather than by a single disease entity. People show a strong genetic tendency to develop primary hypertension, which can be hastened or worsened by contributing factors such as obesity, chronic stress, smoking, excessive alcohol consumption, or dietary habits. For example, excessive salt intake can contribute to development of primary hypertension, especially in salt-sensitive individuals. (For further discussion of the range of known causes for secondary hypertension and potential causes for primary hypertension, see the boxed feature on pp. 372–373, Concepts, Challenges, and Controversies.)

Whatever the underlying defect, once initiated, hypertension appears to be self-perpetuating. Constant exposure to elevated blood pressure damages vessel walls and predisposes them to development of atherosclerosis (see p. 327). The resultant narrowing of vessel lumens by atherosclerotic plaques increases TPR, which further elevates blood pressure. Thus a detrimental positive-feedback cycle ensues where hypertension and atherosclerosis each promote development of the other.

Baroreceptor Adaptation During Hypertension The baroreceptors do not respond to bring blood pressure back to normal during hypertension because they adapt, or are "reset," to operate at a higher level. In the presence of chronically elevated blood pressure, the baroreceptors still function to regulate blood pressure, but they maintain it at a higher mean pressure.

Complications of Hypertension Hypertension imposes stresses on both the heart and the blood vessels. The heart has an increased workload because it is pumping blood out against an increased TPR, and the high internal pressure may damage blood vessels, particularly when the vessel wall is weakened by the degenerative process of atherosclerosis. Complications of hypertension include (1) *left ventricular hypertrophy* in the early stages as the heart muscle thickens to pump a normal stroke volume against an elevated blood pressure, followed in later stages by *systolic heart failure* as the heart weakens and becomes unable to pump continuously against a sustained elevation in arterial pressure (see p. 324); (2) *strokes* caused by rupture of brain vessels (see p. 142); and (3) *heart attacks* caused by

The Body Gets a Jump on Jogging: Cardiovascular Changes during Exercise

PRONOUNCED CARDIOVASCULAR CHANGES ACCOMPANY EXERCISE. Not only does cardiac output (CO) increase significantly during exercise but the distribution of CO is adjusted to support the heightened physical activity. For example, CO jumps from its resting value of 5 L/min to 12.5 L/min during moderate exercise and may increase up to 25 L/min (or more in trained athletes) during heavy exercise. The exercising skeletal muscles receive a substantially greater percentage of the larger-than-normal CO, thereby obtaining the extra O_2 and nutrients needed to support these muscles' stepped-up rate of ATP consumption (see the accompanying figure for the magnitude and distribution of CO at rest and during moderate exercise). Heart muscle likewise receives a greater proportion of the CO than usual to support its increased contractile activity during exercise. The percentage going to the skin also increases as a way to eliminate from the body surface the extra heat generated by the exercising muscles. The share of the CO going to most other organs shrinks. Only the magnitude of blood flow to the brain remains unchanged as the distribution of CO is readjusted during exercise. Other noteworthy cardiovascular changes during exercise are a fall in total peripheral resistance (TPR) (because of widespread vasodilation in skeletal muscles despite generalized arteriolar vasoconstriction in most organs), and a modest increase in mean arterial pressure (see the accompanying table for a summary of these cardiovascular changes).

Discrete exercise centers yet to be identified in the brain induce the appropriate cardiac and vascular changes at the onset of exercise or even in feedforward fashion in anticipation of exercise. These effects are then reinforced by afferent inputs to the medullary cardiovascular center from chemoreceptors in exercising muscles and by local mechanisms important in maintaining vasodilation in active muscles. The baroreceptor reflex further modulates these cardiovascular responses.

Cardiovascular Changes during Exercise

Cardiovascular Variable	Change	Comment
Heart rate	Increases	Occurs as a result of increased sympathetic and decreased parasympathetic activity to the SA node
Venous return	Increases	Occurs as a result of sympathetically induced venous vasoconstriction and increased activity of the skeletal muscle pump and respiratory pump
Stroke volume	Increases	Occurs both as a result of increased venous return by means of the Frank-Starling mechanism and as a result of a sympathetically induced increase in myocardial contractility
Cardiac output (CO)	Increases	Occurs as a result of increases in both heart rate and stroke volume
Blood flow to active skeletal muscles and heart muscle	Increases	Occurs as a result of locally controlled arteriolar vasodilation, which is reinforced by the vasodilatory effects of epinephrine and overpowers the weaker sympathetic vasoconstrictor effect
Blood flow to the brain	Unchanged	Occurs because sympathetic stimulation has no effect on brain arterioles; local control mechanisms maintain constant cerebral blood flow
Blood flow to the skin	Increases	Occurs because the hypothalamic temperature control center induces vasodilation of skin arterioles; increased skin blood flow brings heat produced by exercising muscles to the body surface where the heat can be lost to the external environment
Blood flow to the digestive system, kidneys, and other organs	Decreases	Occurs via generalized sympathetically induced arteriolar vasoconstriction
Total peripheral resistance (TPR)	Decreases	Occurs because resistance in the skeletal muscles, heart, and skin decreases to a greater extent than resistance in the other organs increases
Mean arterial blood pressure	Increases (modest)	Occurs because CO increases more than TPR decreases

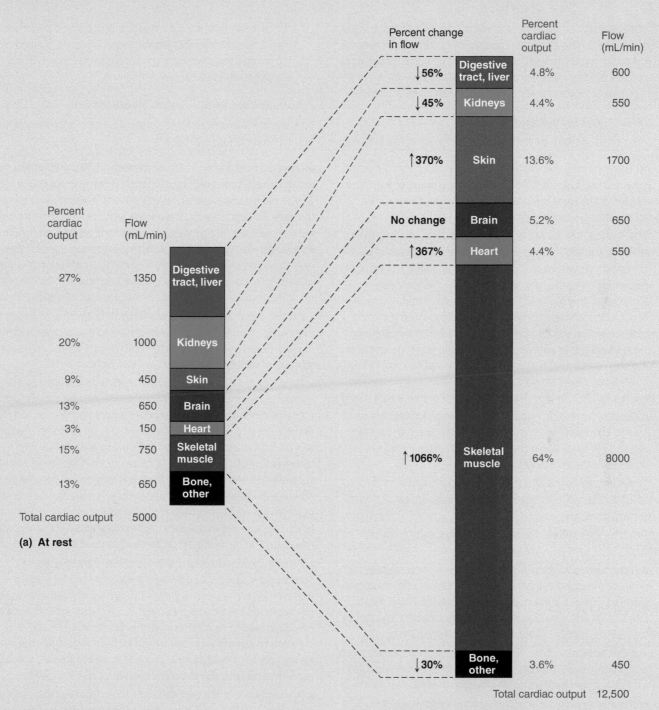

Percent change in flow

		Percent cardiac output	Flow (mL/min)
↓56%	Digestive tract, liver	4.8%	600
↓45%	Kidneys	4.4%	550
↑370%	Skin	13.6%	1700
No change	Brain	5.2%	650
↑367%	Heart	4.4%	550
↑1066%	Skeletal muscle	64%	8000
↓30%	Bone, other	3.6%	450

Total cardiac output 12,500

(b) During moderate exercise

Percent cardiac output	Flow (mL/min)	
27%	1350	Digestive tract, liver
20%	1000	Kidneys
9%	450	Skin
13%	650	Brain
3%	150	Heart
15%	750	Skeletal muscle
13%	650	Bone, other

Total cardiac output 5000

(a) At rest

Magnitude and distribution of cardiac output at rest and during moderate exercise.

The Ups (Causes) and Downs (Treatments) of Hypertension

HYPERTENSION, OR HIGH BLOOD PRESSURE, falls into two types, depending on the cause. Ten percent of hypertension cases are classified as *secondary hypertension* because they occur secondary to an identifiable primary cause. For the other 90% of cases, known as *primary hypertension,* a specific underlying cause cannot be established.

Causes of Secondary Hypertension

The pathological causes of secondary hypertension fall into four categories:

1. *Renal (kidney) hypertension* may occur as a result of either partial occlusion of the renal arteries that supply the kidneys or diseases of the kidney tissue itself.

   ■ Atherosclerotic lesions protruding into the lumen of a renal artery (see p. 327) or external compression of the vessel by a tumor may reduce blood flow through the kidney. The kidney responds by initiating the renin–angiotensin–aldosterone system (RAAS). This pathway promotes salt and water retention during urine formation, thus increasing blood volume to compensate for the reduced renal blood flow. Recall that angiotensin II, a part of this pathway, is also a powerful vasoconstrictor. Although these effects are compensatory mechanisms to improve blood flow through the narrowed renal artery, they also are responsible for elevating arterial blood pressure as a whole.

   ■ Renal hypertension also occurs if diseased kidneys are unable to eliminate salt normally in the urine. The resultant salt retention induces water retention, which expands blood volume and leads to hypertension.

2. *Cardiovascular hypertension* is usually associated with chronically elevated total peripheral resistance (TPR) caused by advanced atherosclerosis.

3. *Endocrine hypertension* arises from the following endocrine disorders:

   ■ A *pheochromocytoma* is a tumor of the adrenal medulla that secretes excessive epinephrine and norepinephrine. Abnormally elevated levels of these hormones bring about a high cardiac output (CO) and generalized peripheral vasoconstriction that increases TPR, both of which contribute to hypertension.

   ■ With *Conn's syndrome,* the adrenal cortex produces too much aldosterone. Being part of the salt- and water-conserving RAAS, excess aldosterone leads to an elevation of blood pressure.

4. *Neurogenic hypertension* occurs secondary to neural lesions.

   ■ The problem may be erroneous blood-pressure control caused by a defect in the cardiovascular control center or in the baroreceptors.

   ■ Neurogenic hypertension may also occur as a compensatory response to a reduction in cerebral blood flow—for example, because a major cerebral vessel is compressed by a tumor. Reflexes are initiated that elevate blood pressure in an attempt to provide sufficient driving pressure to adequately supply the O_2-dependent brain tissue with blood.

Potential Causes of Primary Hypertension

Consider the following range of potential causes for primary hypertension currently being investigated:

■ *Defects in salt management by the kidneys.* Many identified gene variations associated with hypertension in humans are linked in some way to the blood-pressure raising RAAS. For example, a variation in the gene that encodes for angiotensinogen, the precursor for angiotensin II, was the first gene–hypertension link discovered in humans. Individuals with hypertension-producing defects in this pathway appear to be salt sensitive—that is, they do not eliminate salt in the urine as they should, leading to gradual accumulation of salt and water in the body, resulting in progressive elevation of arterial pressure.

■ *Excessive salt intake.* Because salt osmotically retains water, thus expanding blood volume and contributing to the long-term control of blood pressure, excessive ingestion of salt can contribute to hypertension, especially in salt-sensitive individuals.

■ *Diets low in fruits, vegetables, and dairy products (that is, low in K^+ and Ca^{2+}).* Dietary factors other than salt have been shown to markedly affect blood pressure. The DASH (Dietary Approaches to Stop Hypertension) studies found that a low-fat diet rich in fruits, vegetables, and dairy products lowers blood pressure in people with mild hypertension as much as any single drug treatment. The high K^+ intake associated with eating abundant fruits and vegetables may lower blood pressure by relaxing arteries. Furthermore, inadequate Ca^{2+} intake from dairy products has been identified as the most prevalent dietary pattern among individuals with untreated hypertension, although the role of Ca^{2+} in regulating blood pressure is unclear.

■ *Plasma membrane abnormalities such as defective Na^+–K^+ pumps.* Such defects, by altering the electrochemical gradient across plasma membranes, could change the excitability and contractility of the heart and the smooth muscle in blood vessel walls in such a way as to lead to high blood pressure. In addition, the Na^+–K^+ pump is crucial to salt management by the kidneys. A genetic defect in the Na^+–K^+ pump of hypertensive-prone laboratory rats was the first gene–hypertension link to be discovered.

■ *Abnormalities in NO, endothelin, or other locally acting vasoactive paracrines.* For example, a shortage of NO has been discovered in the blood vessel walls of some hypertensive patients, leading to an impaired ability to accomplish blood-pressure-lowering vasodilation. Furthermore, an underlying abnormality in the gene that codes for endothelin, a locally acting vasoconstrictor, has been strongly implicated as a possible cause of hypertension.

- *Excess vasopressin.* Evidence suggests that hypertension may result from a malfunction of the vasopressin-secreting cells of the hypothalamus. Vasopressin is a potent vasoconstrictor and also promotes water retention.

Lifestyle Changes for Treatment of Hypertension

The following lifestyle modifications are recommended for reducing blood pressure in all patients with hypertension and prehypertension:

- *Lose weight if overweight.* Because most obese people are hypertensive, losing weight helps bring blood pressure down in these individuals. Even losing 10 pounds helps improve blood pressure, but the more excess weight lost, the better the improvement. Weight loss is the most important behavioral step in reducing hypertension.

- *Restrict salt intake.* Because cutting down on salt intake can reduce blood pressure, and because a high salt intake, independent of its effect on blood pressure, is linked with an increased lifetime risk of heart attack, strokes, and kidney disease, experts now recommend no more than 1.5 g of sodium (Na^+) (equivalent to 3.8 g of salt) per day for those who already have high blood pressure or are at high risk (African American or older than 40 years old) and an average of 2.3 g of Na^+ (5.8 g of salt) per day for everyone else. Yet Americans currently consume 3.4 to 4 g of Na^+ (or 8.5 to 10 g of salt) per day. One teaspoon of salt has 2.3 g of Na^+. Most salt consumption is not from our saltshakers but hidden in processed foods (such as lunch meats, canned soups, frozen dinners, and potato chips); fast foods; and many restaurant meals.

- *Follow the DASH eating plan.*

- *Exercise regularly.* Moderate aerobic exercise performed for 30 minutes per day on at least four days per week can help reduce blood pressure. If more convenient, the total exercise time on a given day may even be split up into smaller sessions and still provide the same benefits.

- *Limit alcohol consumption.* Limiting alcohol intake to one drink per day for women and men older than 65, or two per day for men younger than 65, is recommended for optimal blood pressure control.

- *Avoid tobacco products.* Smokers are advised to quit smoking and everyone is urged to avoid secondhand smoke. Nicotine in smoke raises blood pressure by activating the sympathetic nervous system and promoting epinephrine release.

- *Reduce stress.* Preventing, reducing, and/or successfully coping with stressful situations (such as through relaxation techniques) can move blood pressure toward normal by cutting down on stress-induced sympathetic discharge.

Antihypertensive Drugs for Treatment of Hypertension

Blood pressure depends on CO and TPR. When needed, a variety of drugs that manipulate salt and water management (to reduce blood volume and venous return, thus reducing CO) or that modify autonomic activity on the heart (to reduce heart rate and cardiac contractility, thus reducing CO) and on the blood vessels (to reduce TPR) can be used to treat hypertension, as follows:

- *Diuretic drugs that increase urinary output* (for example, hydrochlorothiazide) lower blood pressure by increasing urinary output, thereby reducing blood volume. Salt and water that normally would have been retained in the plasma are excreted in the urine.

- *Drugs that block Ca^{2+} channels* (for example, verapamil) reduce the entry of contraction-inducing Ca^{2+} into vascular smooth muscle cells from the ECF in response to excitatory input. Even though cardiac muscle also depends on Ca^{2+} entry to trigger contraction, vascular smooth muscle is more sensitive than cardiac muscle to Ca^{2+} channel blockers. Because the level of contractile activity in vascular smooth muscle cells depends on their cytosolic Ca^{2+} concentration, drugs that block Ca^{2+} channels reduce the contractile activity of these cells, thereby decreasing TPR.

- *Drugs called ACE inhibitors that interfere with production of angiotensin II* (for example, benazepril) block action of RAAS, resulting in more salt and water being lost in the urine and less fluid being retained in the plasma. The resultant reduction in blood volume lowers blood pressure. Also, angiotensin II, a vasoconstrictor, is not formed, thereby decreasing TPR and further reducing blood pressure.

- *Drugs that block angiotensin receptors* (for example, losartan) prevent angiotensin II from causing arteriolar vasoconstriction, thereby decreasing TPR, and also reduce the action of RAAS, thereby decreasing blood volume.

- *Drugs that block β_1-adrenergic receptors* (for example, metoprolol) act by decreasing CO. Because activation of β_1-adrenergic receptors, which are found primarily in the heart (see p. 240), increases the rate and strength of cardiac contraction, drugs that block these receptors decrease heart rate and stroke volume.

- *Drugs that block α_1-adrenergic receptors* (for example, phentolamine) reduce blood pressure by decreasing TPR. Because activation of α_1-adrenergic receptors in vascular smooth muscle brings about vasoconstriction, blockage of α_1-adrenergic receptors reduces arteriolar vasoconstrictor activity, thereby lowering TPR.

- *Drugs that directly relax arteriolar smooth muscle* (for example, hydralazine) promote arteriolar vasodilation, thus reducing TPR.

- *Drugs that block release of norepinephrine from sympathetic endings* (for example, guanethidine) prevent sympathetically induced generalized arteriolar vasoconstriction, thereby lowering TPR.

- *Drugs that act on the brain to reduce sympathetic output* (for example, clonidine) prevent sympathetically induced arteriolar vasoconstriction, thereby precluding the resultant increase in TPR.

rupture of coronary vessels, especially those damaged by atherosclerosis (see p. 314). (Recall that heart attacks can also occur as a result of blocked coronary vessels through thromboembolism; see p. 330.) Hypertension underlies an estimated 47% of all coronary artery disease and 54% of all strokes worldwide. Another serious complication of hypertension is (4) *kidney failure* caused by progressive impairment of blood flow through damaged renal blood vessels. Furthermore, retinal damage from changes in the blood vessels supplying the eyes may result in (5) *progressive loss of vision.*

Until complications occur, hypertension is symptomless because the tissues are adequately supplied with blood. Therefore, unless blood pressure measurements are made routinely, the condition can go undetected until a precipitous complicating event occurs. When you become aware of these potential complications of hypertension and consider that one third of all adults in America are estimated to have chronic elevated blood pressure, you can appreciate the magnitude of this national health problem.

Treatment of Hypertension Once hypertension (or prehypertension) is detected, therapeutic intervention can reduce the course and severity of the problem. By making lifestyle modifications that eliminate or minimize contributing factors, people with hypertension may be able to avoid, delay, or reduce the need for antihypertensive drugs. The most common nondrug therapies to reduce blood pressure are weight reduction, salt restriction, and exercise. Antihypertensive drugs may be employed as needed. No matter what the original cause, agents that reduce blood volume or TPR (or both) decrease blood pressure toward normal. Sometimes a combination of several drugs that act by different mechanisms may be required to achieve normal blood pressure. (To learn more about treatment of hypertension, see the boxed feature on pp. 372–373, ▮Concepts, Challenges, and Controversies.)

Prehypertension In its recent guidelines, the National Institutes of Health identified **prehypertension** as a new category for blood pressures in the range between normal (120/80) and hypertension (140/90). Blood pressures in the prehypertension range can usually be reduced by appropriate dietary and exercise measures, whereas those in the hypertension range typically must be treated with blood pressure medication in addition to changing health habits. The goal in managing blood pressures in the prehypertension range is to take action before the pressure climbs into the hypertension range, where serious complications may develop.

We now examine the other extreme, hypotension, looking first at transient orthostatic hypotension, then at (the more serious) circulatory shock.

Orthostatic hypotension results from transient inadequate sympathetic activity.

Hypotension, or low blood pressure, occurs either when there is a disproportion between vascular capacity and blood volume (in essence, too little blood to fill the vessels) or when the heart is too weak to drive the blood.

The most common situation in which hypotension occurs transiently is orthostatic hypotension. **Orthostatic (postural) hypotension** results from insufficient compensatory responses to the gravitational shifts in blood when a person moves from a horizontal to a vertical position. When a person moves from lying down to standing up, pooling of blood in the leg veins from gravity reduces venous return, decreasing stroke volume and thus lowering CO and blood pressure. This fall in blood pressure is normally detected by the baroreceptors, which initiate immediate compensatory responses to restore blood pressure to its proper level. In some people this reflex adaptation to standing is impaired, as in those taking certain antihypertension drugs that interfere with the reflex or in long-bedridden patients in whom the reflex is temporarily reduced because of disuse. When someone with impaired reflex adaptation first stands up, sympathetic control of the leg veins is inadequate. Consequently, blood pools in the lower extremities without sufficient compensatory responses coming into play to counter the gravity-induced fall in blood pressure. The resultant orthostatic hypotension and decrease in blood flow to the brain cause dizziness or actual fainting.

Circulatory shock can become irreversible.

When blood pressure falls so low that adequate blood flow to the tissues can no longer be maintained, the condition known as **circulatory shock** occurs. Circulatory shock may result from (1) extensive loss of blood volume as through hemorrhage *(hypovolemic shock)*; (2) failure of a weakened heart to pump blood adequately *(cardiogenic shock)*; (3) widespread arteriolar vasodilation *(vasogenic shock)* triggered by vasodilator substances (such as extensive histamine release in severe allergic reactions); or (4) neurally defective vasoconstrictor tone *(neurogenic shock)* (▮Figure 10-34).

We now examine the consequences of and compensations for shock, using hemorrhage as an example (▮Figure 10-35). This figure may look intimidating, but we will work through it step by step. It is an important example that pulls together many of the principles discussed in this and the preceding chapter. As before, the numbers in the text and the figure correspond.

Consequences and Compensations of Shock Following severe loss of blood, the resultant reduction in circulating blood volume leads to a decrease in venous return ❶ and stroke volume and a subsequent fall in CO and mean arterial blood pressure. (Note the *blue* boxes, which indicate consequences of hemorrhage.)

Compensatory measures immediately attempt to maintain adequate blood flow to the brain by increasing blood pressure toward normal, followed by longer-range measures aimed at restoring plasma volume and replacing lost red blood cells, as follows: (Note the *pink* boxes, which indicate compensations for hemorrhage.)

▪ In the short term, the baroreceptor reflex response to the fall in blood pressure brings about decreased parasympathetic activity to the heart and increased sympathetic activity to the heart and the innervated vessels (arterioles and veins) ❷. The resultant increase in heart rate ❸ offsets the reduced stroke

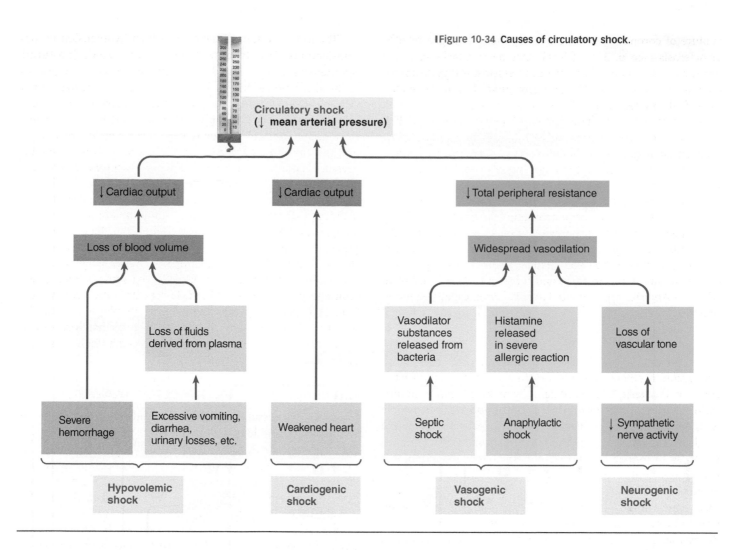

|Figure 10-34 **Causes of circulatory shock.**

volume **4** brought about by the loss of blood volume. With severe fluid loss, the pulse is weak because of reduced stroke volume but rapid because of increased heart rate.

■ Increased sympathetic activity to the veins produces generalized venous vasoconstriction **5**, increasing venous return and thereby stroke volume by means of the Frank–Starling mechanism **6**.

■ Simultaneously, sympathetic stimulation of the heart increases the heart's contractility **7** so that it beats more forcefully and ejects a greater volume of blood, likewise increasing the stroke volume.

■ The increase in heart rate and in stroke volume collectively increase CO **8**.

■ Sympathetically induced generalized arteriolar vasoconstriction **9** leads to an increase in TPR **10**.

■ Together, the increase in CO and TPR lead to a compensatory increase in arterial pressure **11**.

■ The original fall in arterial pressure is accompanied by a fall in capillary blood pressure **12**, which results in immediate fluid shifts from the interstitial fluid into the capillaries to expand the plasma volume **13**. This response is sometimes termed *autotransfusion* because it restores the plasma volume as a transfusion does.

■ This ECF shift is enhanced by plasma protein synthesis by the liver during the next few days following hemorrhage **14**. The plasma proteins exert a colloid osmotic pressure that helps retain extra fluid in the plasma.

■ Urinary output is reduced, thereby conserving water that normally would have been lost from the body **15**. This additional fluid retention helps expand the reduced plasma volume **16**. Expansion of plasma volume further augments the increase in CO brought about by the baroreceptor reflex **17**. Reduction in urinary output results from decreased renal blood flow caused by compensatory renal arteriolar vasoconstriction **18**. The initial reduction in plasma/blood volume resulting from hemorrhage also triggers increased secretion of the hormone vasopressin and activation of the salt- and water-conserving renin–angiotensin–aldosterone hormonal pathway, which further reduces urinary output **19**.

■ Increased thirst is also stimulated by the initial fall in plasma/blood volume that occurs with hemorrhage **20**. The resultant increased fluid intake helps restore plasma volume.

■ Over a longer course (a week or more), lost red blood cells are replaced through increased red blood cell production triggered by a reduction in O_2 delivery to the kidneys **21** (see p. 385 for further details).

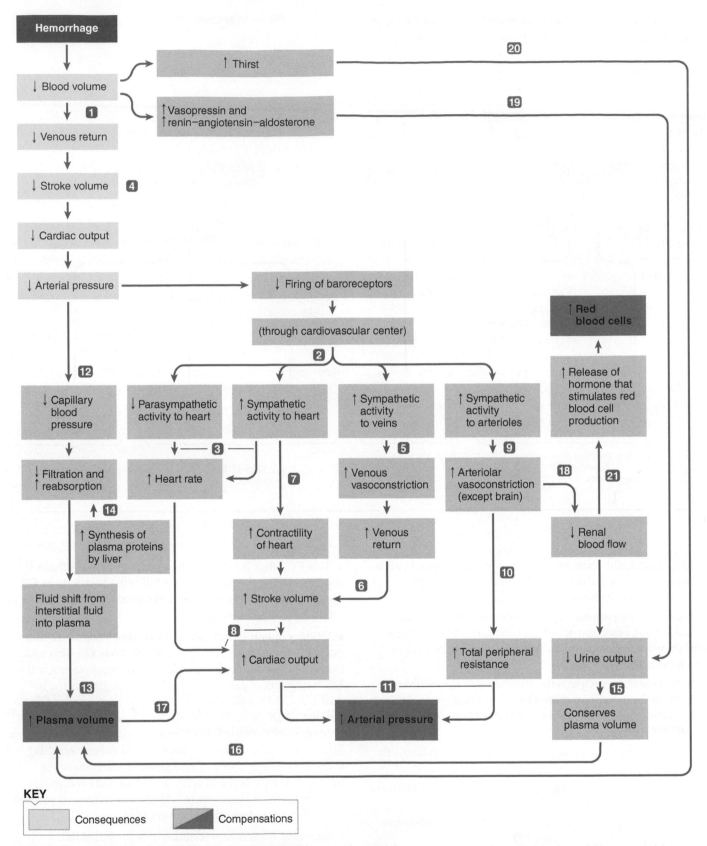

IFigure 10-35 Consequences and compensations of hemorrhage. The reduction in blood volume resulting from hemorrhage leads to a fall in mean arterial pressure. (Note the *blue boxes,* representing consequences of hemorrhage.) A series of compensations ensue (*light pink boxes*) that ultimately restore plasma volume, arterial pressure, and the number of red blood cells toward normal (*dark pink boxes*). See the text for an explanation of the numbers and a detailed discussion of the compensations.

FIGURE FOCUS: *Describe all pathways that lead to a compensatory increase in cardiac output in response to hemorrhage.*

Irreversible Shock These compensatory mechanisms are often not enough to counteract substantial blood loss. Even if they can maintain an adequate blood pressure level, the short-term measures cannot continue indefinitely. Ultimately, fluid volume must be replaced from the outside through drinking, transfusion, or a combination of both. Blood supply to the kidneys, digestive tract, skin, and other organs can be compromised to maintain blood flow to the brain only so long before organ damage begins to occur. A point may be reached at which blood pressure continues to drop rapidly because of tissue damage, despite vigorous therapy. This condition is often termed *irreversible shock*, in contrast to *reversible shock*, which can be corrected by compensatory mechanisms and effective therapy.

Although the exact mechanism underlying irreversibility is not currently known, many logical possibilities could contribute to the unrelenting, progressive circulatory deterioration that characterizes irreversible shock. Metabolic acidosis arises when lactate (lactic acid) production increases as blood-deprived tissues resort to anaerobic metabolism. Acidosis deranges the enzymatic systems responsible for energy production, limiting the capability of the heart and other tissues to produce ATP. Prolonged depression of kidney function results in electrolyte imbalances that may lead to cardiac arrhythmias. The blood-deprived pancreas releases a chemical that is toxic to the heart (*myocardial toxic factor*), further weakening the heart. Vasodilator substances build up within ischemic organs, inducing local vasodilation that overrides the generalized reflex vasoconstriction. As CO progressively declines because of the heart's diminishing effectiveness as a pump and TPR continues to fall, hypotension becomes increasingly severe. This causes further cardiovascular failure, which leads to a further decline in blood pressure. Thus, when shock progresses to the point that the cardiovascular system itself starts to fail, a vicious positive-feedback cycle ensues that ultimately results in death.

Check Your Understanding 10.6

1. Prepare a flow diagram showing the baroreceptor reflex response to a fall in MAP.

2. Define *secondary hypertension, primary hypertension, orthostatic hypertension,* and *circulatory shock.*

3. Explain why the baroreceptors do not reduce blood pressure to normal during hypertension.

Homeostasis: Chapter in Perspective

 Homeostatically, the blood vessels are passageways to transport blood to and from the cells for O_2 and nutrient delivery, waste removal, distribution of fluid and electrolytes, elimination of excess heat, and hormonal signaling, among other things. Cells soon die if deprived of their blood supply; brain cells succumb within 4 minutes. Blood is constantly recycled and reconditioned as it travels through the various organs via the vascular highways. Hence, the body needs only a small volume of blood to maintain the appropriate chemical composition of the entire internal fluid environment on which the cells depend for their survival. For example, O_2 is continually picked up by blood in the lungs and constantly delivered to all body cells.

The smallest blood vessels, the capillaries, are the actual site of exchange between blood and surrounding cells. Capillaries bring homeostatically maintained blood within 0.1 mm of every cell in the body; this proximity is critical because beyond a few millimeters materials cannot diffuse rapidly enough to support life-sustaining activities. O_2 that would take months to years to diffuse from the lungs to all the cells of the body is continuously delivered to the "doorstep" of every cell, where diffusion can efficiently accomplish short local exchanges between capillaries and surrounding cells. Likewise, hormones must be rapidly transported through the circulatory system from their sites of production in endocrine glands to their sites of action in other parts of the body. These chemical messengers could not diffuse nearly rapidly enough to their target organs to effectively exert their controlling effects, many of which are aimed toward maintaining homeostasis.

The rest of the circulatory system is designed to transport blood to and from the capillaries. The arteries and arterioles distribute blood pumped by the heart to the capillaries for life-sustaining exchanges to take place, and the venules and veins collect blood from the capillaries and return it to the heart, where the process is repeated.

Review Exercises Answers begin on p. A-34

Reviewing Terms and Facts

1. In general, the parallel arrangement of the vascular system enables each organ to receive its separate arterial blood supply. (*True or false?*)

2. More blood flows through the capillaries during cardiac systole than during diastole. (*True or false?*)

3. The capillaries contain only 5% of the total blood volume at any point in time. (*True or false?*)

4. The same volume of blood passes through the capillaries in a minute as passes through the aorta, even though blood flow is much slower in the capillaries. (*True or false?*)

5. Because capillary walls have no carrier transport systems, all capillaries are equally permeable. *(True or false?)*

6. Because of gravitational effects, venous pressure in the lower extremities is greater when a person is standing up than when the person is lying down. *(True or false?)*

7. _____ is a measure of the hindrance to blood flow through a vessel caused by friction between the moving fluid and the stationary vascular walls.

8. Local mechanisms that keep tissue blood flow fairly constant despite wide deviations in mean arterial driving pressure is termed _____.

9. _____ refers to the volume of blood per unit of time flowing through a given segment of the circulatory system, whereas _____ is the speed, or distance per unit of time, with which the blood flows forward through a given segment of the circulatory system.

10. _____ is the term applied to vessel narrowing that increases resistance to flow, whereas _____ is the term applied to vessel widening that decreases resistance to flow.

11. Which of the following functions is or are attributable to arterioles? *(Indicate all correct answers.)*

   a. produce a significant decline in mean pressure, which helps establish the driving pressure gradient between the heart and the organs

   b. serve as the site of exchange of materials between blood and surrounding tissue cells

   c. act as the main determinant of total peripheral resistance

   d. determine the pattern of distribution of cardiac output

   e. help regulate mean arterial blood pressure

   f. convert the pulsatile nature of arterial blood pressure into a smooth, nonfluctuating pressure in the vessels farther downstream

   g. act as a pressure reservoir

12. Using the answer code on the right, indicate whether the following factors increase or decrease venous return:

   1. sympathetically induced venous vasoconstriction   (a) increases venous return
   2. skeletal muscle activity   (b) decreases venous return
   3. gravitational effects on the venous system   (c) has no effect on venous return
   4. respiratory activity
   5. increased atrial pressure associated with a leaky AV valve
   6. ventricular pressure change associated with diastolic recoil

Understanding Concepts

(Answers at www.cengagebrain.com)

1. Compare blood flow through reconditioning organs and through organs that do not recondition the blood.

2. Discuss the relationships among flow rate, pressure gradient, and vascular resistance. What is the major determinant of resistance to flow?

3. Describe the structure and major functions of each segment of the vascular tree.

4. How do the arteries serve as a pressure reservoir?

5. Describe the indirect technique of measuring arterial blood pressure by means of a sphygmomanometer.

6. Compare the consequences of vasoconstriction and vasodilation each on distribution of cardiac output and on control of mean arterial blood pressure.

7. Discuss the local and extrinsic controls that regulate arteriolar resistance.

8. What is the primary means by which individual solutes are exchanged across capillary walls? What forces produce bulk flow across capillary walls? Of what importance is bulk flow?

9. How is lymph formed? What are the functions of the lymphatic system?

10. Define *edema*, and discuss its possible causes.

11. How do veins serve as a blood reservoir?

12. Compare the effect of vasoconstriction on the rate of blood flow in arterioles and in veins.

13. Discuss the factors that determine mean arterial pressure.

14. Review the effects on the cardiovascular system of parasympathetic and sympathetic stimulation.

15. Differentiate between secondary hypertension and primary hypertension. What are the potential consequences of hypertension?

16. Define circulatory shock. What are its consequences and compensations? What is irreversible shock?

Solving Quantitative Exercises

1. Recall that the flow rate of blood equals the pressure gradient divided by the total peripheral resistance (TPR) of the vascular system. The conventional unit of resistance in physiological systems is expressed in PRU (peripheral resistance unit), which is defined as (1 L/min)/(1 mm Hg). At rest, Tom's TPR is about 20 PRU. Last week while playing racquetball, his cardiac output increased to 30 L/min and his mean arterial pressure (MAP) increased to 120 mm Hg. What was his TPR during the game?

2. Systolic pressure rises as a person ages. By age 85, an average male (untreated for hypertension) has a systolic pressure of 180 mm Hg and a diastolic pressure of 90 mm Hg.

   a. What is the MAP of this average 85-year-old male?

   b. Considering capillary dynamics, predict the result at the capillary level of this age-related change in MAP if no homeostatic mechanisms were operating. (Recall that MAP is about 93 mm Hg at age 20.)

3. Compare the flow rates in the systemic and the pulmonary circulations of an individual with the following measurements:

    a. systemic mean arterial pressure = 95 mm Hg

    b. systemic resistance = 19 PRU

    c. pulmonary mean arterial pressure = 20 mm Hg

    d. pulmonary resistance = 4 PRU

4. Which of the following changes would increase the resistance in an arteriole? Explain.

    a. a longer length

    b. a smaller caliber

    c. increased sympathetic stimulation

    d. increased blood viscosity

    e. all of the above

Applying Clinical Reasoning

Li-Ying C. has just been diagnosed as having hypertension secondary to a *pheochromocytoma*, a tumor of the adrenal medulla that secretes excessive epinephrine. Explain how this condition leads to secondary hypertension by describing the effect that excessive epinephrine would have on various factors that determine arterial blood pressure.

Thinking at a Higher Level

1. During coronary bypass surgery, a piece of vein is often removed from the patient's leg and surgically attached within the coronary circulatory system so that blood detours, through the vein, around an occluded coronary artery segment. Why must the patient wear, for an extended period after surgery, an elastic support stocking on the limb from which the vein was removed?

2. A classmate who has been standing still for several hours working on a laboratory experiment suddenly faints. What is the probable explanation? What would you do if the person next to him tried to get him up?

3. A drug applied to a piece of excised arteriole causes the vessel to relax, but an isolated piece of arteriolar muscle stripped from the other layers of the vessel fails to respond to the same drug. What is the probable explanation?

4. Children who suffer from protein malnutrition because their diets are high in starch and low in protein (as in poor countries with limited food supply) often develop *kwashiorkor*. This condition is characterized among other things by a pronounced, protruding belly caused by a fluid-filled abdominal cavity (called *ascites*), although the rest of the body is "skin and bones." What causes the markedly distended abdomen?

5. Why is the risk for developing *deep venous thrombosis (DVT)* (abnormal formation of blood clots, especially in the deep veins of the legs) increased during a long airplane flight? What can you do while onboard to decrease your risk?

To access the course materials and companion resources for this text, please visit
www.cengagebrain.com

The Blood

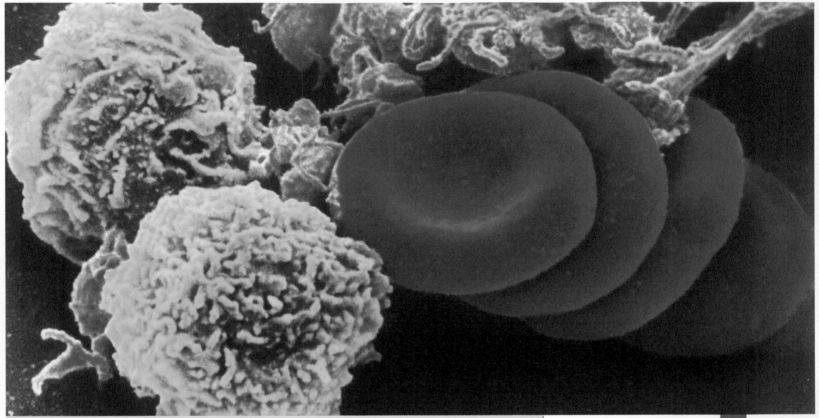

A scanning electron micrograph of cellular elements in the blood. Erythrocytes (*red*) carry oxygen. Leukocytes (*blue*) are vital to immune defense. Platelets (*pink*) help stop bleeding. The light brown threads are fibrin, which is not normally present in the blood but, once produced during vessel injury, forms the meshwork of a clot.

Dr. Yorgos Nikas/Science Source

CHAPTER AT A GLANCE

Homeostasis Highlights

Blood is the vehicle for long-distance, mass transport of materials between the cells and external environment or between the cells themselves. Such transport is essential for maintaining homeostasis. Blood consists of a complex liquid **plasma** in which the cellular elements—*erythrocytes, leukocytes,* and *platelets*—are suspended. **Erythrocytes (red blood cells,** or **RBCs)** are essentially plasma membrane–enclosed bags of hemoglobin that transport O_2 in the blood. **Leukocytes (white blood cells,** or **WBCs),** the immune system's mobile defense units, are transported in the blood to sites of injury or of invasion by disease-causing microorganisms. **Platelets (thrombocytes)** are important in hemostasis, the stopping of bleeding from an injured vessel.

11.1 Plasma

The hematocrit is the packed cell volume of blood; the rest of the volume is plasma.

Blood represents about 8% of total body weight and has an average volume of 5 liters in women and 5.5 liters in men. It consists of three types of specialized cellular elements, *erythrocytes (red blood cells), leukocytes (white blood cells),* and *platelets (thrombocytes),* suspended in the complex liquid *plasma* (▌Figure 11-1, chapter opener photo, and ▌Table 11-1). Erythrocytes and leukocytes are both whole cells, whereas platelets are cell fragments.

The constant movement of blood as it flows through the blood vessels keeps the cellular elements rather evenly dispersed within the plasma. However, if you put a sample of whole blood in a test tube and treat it to prevent clotting, the heavier cells slowly settle to the bottom and the lighter plasma rises to the top. This process can be hastened by centrifuging, which quickly packs the cells in the bottom of the tube (▌Figure 11-1). Because more than 99% of the cells are erythrocytes, the **hematocrit**, or **packed cell volume**, essentially represents the percentage of erythrocytes in the total blood volume. The hematocrit averages about 42% for women and slightly higher, 45%, for men. Plasma accounts for the remaining volume. Accordingly, the average volume of plasma in the blood is about 58% for women and 55% for men. White blood cells and platelets, which are colorless and less dense than red cells, are packed in a thin, cream-colored layer, the *buffy coat,* on top of the packed red cell column. They are less than 1% of the total blood volume.

Let us first consider the largest portion of the blood, the plasma, before turning to the cellular elements.

Plasma water is a transport medium for many inorganic and organic substances.

Plasma, being a liquid, consists of 90% water. Plasma water is a medium for materials being carried in the blood. Also, plasma absorbs and distributes much of the heat generated metabolically within tissues, with only minimal changes in the temperature of the blood itself. As blood travels close to the surface of the skin, heat energy not needed to maintain body temperature is eliminated to the environment.

A large number of inorganic and organic substances are dissolved in the plasma. Inorganic constituents account for about 1% of plasma weight. The most abundant electrolytes (ions) in the plasma are Na^+ and Cl^-, the components of common salt. Smaller amounts of HCO_3^-, K^+, Ca^{2+}, and other ions are present. The most notable functions of these ions are their roles in membrane excitability, osmotic distribution of fluid between the extracellular fluid (ECF) and the cells, and buffering of pH changes; these functions are discussed elsewhere.

The most plentiful organic constituents by weight are the plasma proteins, which make up 6% to 8% of plasma weight. We examine them more thoroughly in the next section. The remaining small percentage of plasma consists of other organic substances, including nutrients (such as glucose, amino acids, lipids, and vitamins), waste products (creatinine, bilirubin, and nitrogenous substances such as urea), dissolved gases (O_2 and CO_2), and hormones. Most of these substances are merely being transported in the plasma. For example, endocrine glands secrete hormones into the plasma, which transports these chemical messengers to their sites of action.

Many of the functions of plasma are carried out by plasma proteins.

Plasma proteins are the one group of plasma constituents that are not along just for the ride. These important components normally stay in the plasma, where they perform many valuable functions. Here are the most important of these functions, which are elaborated on elsewhere in the text:

1. Unlike other plasma constituents that are dissolved in the plasma water, plasma proteins are dispersed as a colloid (see p. A-7). Furthermore, because they are the largest of the plasma constituents, plasma proteins usually do not exit through the narrow pores in the capillary walls to enter the interstitial fluid. By their presence as a colloidal dispersion in the plasma and their absence in the interstitial fluid, plasma proteins establish

Plasma
(55% of whole blood)

Buffy coat: platelets and leukocytes (<1% of whole blood)

Platelets

Leukocytes (white blood cells)

Erythrocytes (45% of whole blood)

Erythrocytes (red blood cells)

Packed cell volume, or hematocrit

National Cancer Institute/Science Source

▌Figure 11-1 **Hematocrit and types of blood cells.** The values given are for men. The average hematocrit for women is 42%, with plasma occupying 58% of the blood volume. Note the biconcave shape of the erythrocytes.

FIGURE FOCUS: *If the hematocrit is 37, what percentage of the whole blood is occupied by plasma? Would O_2-carrying capacity be at, above, or below normal?*

Constituent	Functions
Plasma	
Water	Acts as a transport medium; carries heat
Electrolytes	Are important in membrane excitability; distribute fluid by osmosis between ECF and ICF; buffer pH changes
Nutrients, wastes, gases, hormones	Are transported in blood; blood CO_2 plays a role in acid–base balance
Plasma proteins	In general, exert an osmotic effect important in the distribution of the ECF between the vascular and interstitial compartments; buffer pH changes
Albumins	Transport many substances; contribute most to colloid osmotic pressure
Globulins	
Alpha and beta	Transport many water-insoluble substances; include clotting factors and inactive precursor molecules
Gamma	Are antibodies
Fibrinogen	Is an inactive precursor for a clot's fibrin meshwork
Cellular elements	
Erythrocytes	Transport O_2 and CO_2 (mainly O_2)
Leukocytes	
Neutrophils	Engulf bacteria and debris
Eosinophils	Attack parasitic worms; play a role in allergic reactions
Basophils	Release histamine, which is important in allergic reactions, and heparin, which helps clear fat from the blood
Monocytes	Are in transit to become tissue macrophages
Lymphocytes	
B lymphocytes	Produce antibodies
T lymphocytes	Produce cell-mediated immune responses
Platelets	Contribute to hemostasis

an osmotic gradient between the blood and the interstitial fluid. This colloid osmotic pressure is the primary force preventing excessive loss of plasma from the capillaries into the interstitial fluid and thus helps maintain plasma volume (see p. 357).

2. Plasma proteins are partially responsible for plasma's capacity to buffer changes in pH (see p. 552).

3. The three groups of plasma proteins—*albumins, globulins,* and *fibrinogen*—are classified according to their various physical and chemical properties. In addition to the general functions just listed, each type of plasma protein performs specific tasks as follows:

a. **Albumins**, the most abundant plasma proteins, contribute most extensively to the colloid osmotic pressure by virtue of their numbers. They also nonspecifically bind substances that are poorly soluble in plasma (such as bilirubin, bile salts, and penicillin) for transport in the plasma.

b. There are three subclasses of **globulins: alpha (α), beta (β),** and **gamma (γ)**.

(1) Like albumins, some α and β globulins bind poorly water-soluble substances for transport in the plasma, but these globulins are highly specific as to which passenger they bind and carry. Examples of substances carried by specific globulins include thyroid hormone (see p. 668), cholesterol (see p. 328), and iron (see p. 608).

(2) Many of the factors involved in the blood-clotting process are α or β globulins.

(3) Some plasma proteins are inactive, circulating precursor molecules, which are activated as needed by specific regulatory inputs. For example, the α glob-

ulin *angiotensinogen* is activated to *angiotensin*, which plays an important role in regulating salt balance in the body (see p. 508).

(4) The γ globulins are *antibodies*, or *immunoglobulins*, which are crucial to the body's defense mechanism (see p. 417).

c. **Fibrinogen** is a key factor in blood clotting.

Plasma proteins are synthesized by the liver, with the exception of antibodies, which are produced by lymphocytes, one of the types of white blood cells.

Check Your Understanding 11.1

1. Draw a centrifuged tube of whole blood, showing the packed cell volume and indicating the percentage in a normal man occupied by plasma, by erythrocytes, and by the buffy coat.
2. Define *hematocrit*.
3. List the functions of plasma proteins.

11.2 Erythrocytes

Each milliliter of blood on average contains about 5 billion **erythrocytes (red blood cells,** or **RBCs)**, commonly reported clinically in a **red blood cell count** as 5 million cells per cubic millimeter (mm³).

Erythrocytes are well designed for their main function of O₂ transport in the blood.

The shape and content of erythrocytes are ideally suited to carry out their primary function, transporting O_2 in the blood.

Erythrocyte Structure Three anatomic features of erythrocytes contribute to the efficiency with which they transport O_2. First, erythrocytes are flat, disc-shaped cells indented in the middle on both sides, like a doughnut with a flattened center instead of a hole (that is, they are biconcave discs 8 μm in diameter, 2 μm thick at the outer edges, and 1 μm thick in the center) (see ❙ Figure 11-1 and chapter opener photo). The biconcave shape provides a larger surface area for diffusion of O_2 from the plasma across the membrane into the erythrocyte than a spherical cell of the same volume would. Also, the thinness of the cell enables O_2 to diffuse rapidly between the exterior and innermost regions of the cell.

A second structural feature that facilitates RBCs' transport function is their flexible membrane. Red blood cells, whose diameter is normally 8 μm, can deform amazingly as they squeeze single file through capillaries as narrow as 3 μm in diameter. Because they are extremely pliant, RBCs can travel through the narrow, tortuous capillaries to deliver their O_2 cargo at the tissue level without rupturing in the process.

The third and most important anatomic feature that enables RBCs to transport O_2 is the hemoglobin they contain.

Role of Hemoglobin Hemoglobin is found only in red blood cells. A **hemoglobin** molecule has two parts: (1) the **globin** portion, a protein made up of four highly folded polypeptide chains (two α subunits and two β subunits), and (2) four iron-containing, nonprotein groups known as **heme groups**, each of which is bound to one of the polypeptides (❙ Figure 11-2). Each of the four iron atoms can combine reversibly with one molecule of O_2; thus, each hemoglobin molecule can pick up four O_2 passengers in the lungs. Because O_2 is poorly soluble in the plasma, 98.5% of the O_2 carried in the blood is bound to hemoglobin (see p. 472).

Hemoglobin is a pigment (that is, it is naturally colored). Because of its iron content, it appears reddish when combined with O_2 and bluish when deoxygenated. Thus, fully oxygenated

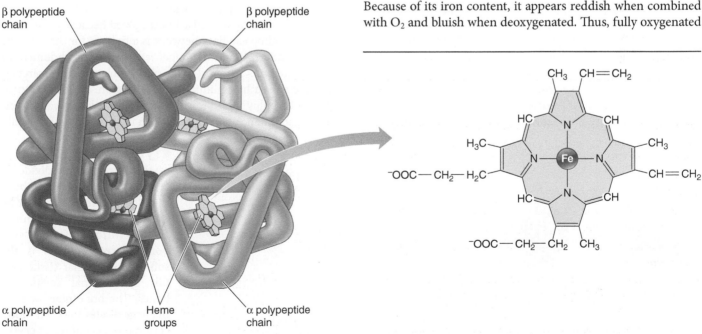

β polypeptide chain

β polypeptide chain

α polypeptide chain

Heme groups

α polypeptide chain

(a) Hemoglobin molecule

(b) Iron-containing heme group

❙Figure 11-2 **Hemoglobin molecule.** A hemoglobin molecule consists of four highly folded polypeptide chains (the globin portion) and four iron-containing heme groups.

arterial blood is red, and venous blood, which has lost some of its O_2 load at the tissue level, has a bluish cast.

In addition to carrying O_2, hemoglobin can combine with the following:

1. *Carbon dioxide (CO_2).* Hemoglobin helps transport this gas from the tissue cells back to the lungs (see p. 476).

2. *The acidic hydrogen-ion portion (H^+) of ionized carbonic acid,* which is generated at the tissue level from CO_2. Hemoglobin buffers this acid so that it minimally alters the pH of the blood (see p. 477).

3. *Carbon monoxide (CO).* This gas is not normally in the blood, but if inhaled, it preferentially occupies the O_2-binding sites on hemoglobin, causing CO poisoning (see p. 475).

4. *Nitric oxide (NO).* In the lungs, the vasodilator nitric oxide binds to hemoglobin. This NO is released at the tissues, where it relaxes and dilates the local arterioles (see p. 346). Vasodilation helps ensure that the O_2-rich blood can make its vital rounds and helps stabilize blood pressure.

Therefore, hemoglobin plays the key role in O_2 transport while contributing significantly to CO_2 transport and the pH-buffering capacity of blood. Furthermore, by toting along its own vasodilator, hemoglobin helps deliver the O_2 it is carrying.

Lack of Nucleus and Organelles To maximize its hemoglobin content, a single erythrocyte is stuffed with more than 250 million hemoglobin molecules, excluding almost everything else. (That means each RBC can carry more than 1 billion O_2 molecules.) Red blood cells contain no nucleus or organelles. During the cell's development, these structures are extruded to make room for more hemoglobin (❙Figure 11-3). Thus, an RBC is mainly a plasma membrane–enclosed sac full of hemoglobin.

Key Erythrocyte Enzymes Only a few crucial, nonrenewable enzymes remain within a mature erythrocyte: *glycolytic enzymes* and *carbonic anhydrase*. The **glycolytic enzymes** are necessary for generating the energy needed to fuel the active-transport mechanisms involved in maintaining proper ionic concentrations within the cell. Ironically, even though erythrocytes are the vehicles for transporting O_2 to all other tissues of the body, for energy production erythrocytes themselves can-

not use the O_2 they are carrying. Lacking the mitochondria that house the enzymes for oxidative phosphorylation, erythrocytes must rely entirely on glycolysis for ATP formation (see p. 35).

The other important enzyme within RBCs, **carbonic anhydrase**, is critical in CO_2 transport. This enzyme catalyzes a key reaction that leads to the conversion of metabolically produced CO_2 into **bicarbonate ion (HCO_3^-)**, which is the primary form in which CO_2 is transported in the blood. Thus, erythrocytes contribute to CO_2 transport in two ways—by means of its carriage on hemoglobin and its carbonic anhydrase–induced conversion to HCO_3^-.

The bone marrow continuously replaces worn-out erythrocytes.

Each of us has a total of 25 trillion to 30 trillion RBCs streaming through our blood vessels at any given time (100,000 times more in number than the entire U.S. population). Yet these vital gas-transport vehicles are short-lived and must be replaced at the average rate of 2 million to 3 million cells per second.

Erythrocytes' Short Life Span The price erythrocytes pay for their generous content of hemoglobin to the exclusion of the usual specialized intracellular machinery is a short life span. Without DNA, RNA, and ribosomes, red blood cells cannot synthesize proteins for cell repair, growth, and division or for renewing enzyme supplies. Equipped only with initial supplies synthesized before they extrude their nucleus and organelles, RBCs survive an average of only 120 days, in contrast to nerve and muscle cells, which last a person's entire life. During its short life span of 4 months, each erythrocyte travels about 700 miles as it circulates through the vasculature.

As a red blood cell ages, its plasma membrane, which cannot be repaired, becomes fragile and prone to rupture as the cell squeezes through tight spots in the vascular system. Most old RBCs meet their final demise in the **spleen** because this organ's narrow, winding capillary network is a tight fit for these fragile cells. The spleen lies in the upper left part of the abdomen. In addition to removing most of the old erythrocytes from circulation, the spleen has a limited ability to store healthy erythrocytes in its pulpy interior, is a reservoir for platelets, and contains an abundance of lymphocytes, a type of white blood cell.

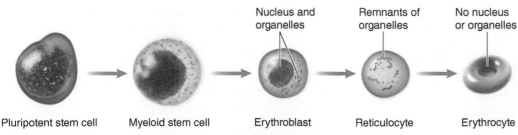

Nucleus and organelles — Remnants of organelles — No nucleus or organelles

Pluripotent stem cell   Myeloid stem cell   Erythroblast   Reticulocyte   Erythrocyte

❙Figure 11-3 **Major steps in erythrocyte production (erythropoiesis).** Erythrocytes are derived in the red bone marrow from pluripotent stem cells that give rise to all the types of blood cells. Myeloid stem cells are partially differentiated cells that give rise to erythrocytes and several other types of blood cells. Nucleated erythroblasts are committed to becoming mature erythrocytes. These cells extrude their nucleus and organelles, making more room for hemoglobin. Reticulocytes are immature red blood cells that contain organelle (mostly ribosome) remnants. Mature erythrocytes are released into the abundant capillaries in the bone marrow.

Erythropoiesis Because erythrocytes cannot divide to replenish their own numbers, the old ruptured cells must be replaced by new cells produced in an erythrocyte factory—the **bone marrow**—which is the soft, highly cellular tissue that fills the internal cavities of bones. The bone marrow normally generates new red blood cells, a process known as **erythropoiesis**, at a rate to keep pace with the demolition of old cells.

During intrauterine development, RBCs are produced first by the yolk sac and then by the developing liver and spleen, until the bone marrow is formed and takes over erythrocyte production exclusively. In children, most bones are filled with **red bone marrow** capable of blood cell production. As a person matures, fatty **yellow bone marrow** incapable of erythropoiesis gradually replaces red marrow, which remains only in a few places, such as the sternum (breastbone), ribs, pelvis, and upper ends of the limb bones. These are sites where bone marrow is extracted for examination or for bone marrow transplants.

Red marrow not only produces RBCs but also is the ultimate source for leukocytes and platelets. Undifferentiated **pluripotent stem cells**, the source of all blood cells, reside in the red marrow, where they continuously divide and differentiate to give rise to each of the types of blood cells (▮Figure 11-3; also see ▮Figure 11-9, p. 394). The different types of immature blood cells, along with the stem cells, are intermingled in the red marrow at various stages of development. Once mature, the blood cells are released into the rich supply of capillaries that permeate the red marrow. Bone marrow capillaries are of the less common discontinuous type that has large gaps between the endothelial cells (see p. 353). Mature red blood cells can pass through these very large pores to enter the blood, but, once circulating, these cells cannot exit the blood through the much narrower pores in ordinary capillaries. Regulatory factors act on the *hemopoietic* ("blood-producing") red marrow to govern the type and number of cells generated and discharged into the blood. The mechanism for regulating RBC production is the best understood. We consider it next.

Erythropoiesis is controlled by erythropoietin from the kidneys.

Because O_2 transport in the blood is the erythrocytes' main function, you might logically suspect that the primary stimulus for increased erythrocyte production would be reduced O_2 delivery to the tissues. You would be correct, but low O_2 levels do not stimulate erythropoiesis by acting directly on the red marrow. Instead, reduced O_2 delivery to the kidneys stimulates them to secrete the hormone **erythropoietin (EPO)** into the blood, and this hormone in turn stimulates erythropoiesis by the red marrow (▮Figure 11-4). The magnitude of EPO release reflects the degree of O_2 insufficiency detected by the kidneys.

Undifferentiated stem cells continuously generate derivatives that are already committed to becoming RBCs (see ▮Figure 11-3). Erythropoietin acts on these erythrocyte forerunners, stimulating their proliferation and maturation into mature erythrocytes that are released into the blood. This increased erythropoietic activity elevates the number of circulating RBCs, thereby increasing O_2-carrying capacity of the blood and restoring O_2 delivery to the tissues to normal. Once normal O_2 delivery to the kidneys is achieved, EPO secretion is turned down until needed again. In this way, erythrocyte production is normally balanced against destruction or loss of these cells so that O_2-carrying capacity in the blood stays fairly constant. In severe loss of RBCs, as in hemorrhage or abnormal destruction of young circulating erythrocytes, the rate of erythropoiesis can be increased to more than six times the normal level. (For a discussion of erythropoietin abuse by some athletes, see the boxed feature on p. 386, ▮A Closer Look at Exercise Physiology.)

The preparation of an erythrocyte for its departure from the marrow involves several steps, such as synthesis of hemoglobin and extrusion of the nucleus and organelles. Cells closest to maturity need a few days to be "finished off" and released into the blood in response to erythropoietin, and less developed and newly proliferated cells may take up to several weeks to reach maturity. Therefore, the time required for complete replacement of lost RBCs depends on how many are needed

1 Kidneys detect reduced O_2-carrying capacity of blood.

2 When less O_2 is delivered to the kidneys, they secrete erythropoietin into blood.

3 Erythropoietin stimulates erythropoiesis by red bone marrow.

4 Additional circulating erythrocytes increase O_2-carrying capacity of blood.

5 Increased O_2-carrying capacity relieves initial stimulus that triggered erythropoietin secretion.

▮Figure 11-4 **Control of erythropoiesis.**
FIGURE FOCUS: *What effect would kidney failure have on the rate of erythropoiesis?*

Blood Doping: Is More of a Good Thing Better?

EXERCISING MUSCLES NEED A CONTINUAL supply of O_2 for generating energy to sustain endurance activities (see p. 271). **Blood doping** is a technique used to temporarily increase the O_2-carrying capacity of blood in an attempt to gain a competitive advantage. Blood doping involves removing blood from an athlete, then promptly reinfusing the plasma but freezing the RBCs for reinfusion one to seven days before a competitive event. One to four units of blood (one unit equals 450 mL) are usually withdrawn at three- to eight-week intervals before the competition. In the periods between blood withdrawals, increased erythropoietin activity restores the RBC count to normal. Reinfusion of the stored RBCs artificially increases the RBC count and hemoglobin level above normal before competing. Theoretically, blood doping would benefit endurance athletes by improving the blood's O_2-carrying capacity. If too many red cells were infused, however, performance could suffer because the increased blood viscosity would decrease blood flow.

Research indicates that athletes who have used blood doping may realize a 5% to 13% increase in aerobic capacity; a reduction in heart rate during exercise compared to the rate during the same exercise in the absence of blood doping; improved performance; and reduced lactate levels in the blood. (Lactate is produced when muscles resort to less efficient anaerobic glycolysis for energy production; see page 272.)

Blood doping, although effective, is illegal in collegiate athletics and Olympic competition for ethical and medical reasons. Of concern, as with use of any banned performance-enhancing product, is loss of fair competition. Furthermore, the practice can lead to high blood pressure and has been implicated in the deaths of some athletes. The development of synthetic erythropoietin (EPO) exacerbates the problem of blood doping. Injection of this product stimulates RBC production and thus temporarily increases the O_2-carrying capacity of the blood. Injected EPO may improve an endurance athlete's performance by 7% to 10%. Although formally banned, a black market for EPO developed among sports cheats when this product became available as a drug to treat anemia. Erythropoietin is now widely used among competitors in cycling, cross-country skiing, and long-distance running and swimming. This practice is ill advised, however, not only because of legal and ethical implications, but also because of the dangers of increasing blood viscosity. Synthetic EPO is believed responsible for the deaths of a number of cyclists. Unfortunately, too many athletes are willing to take the risks.

It is hoped that recent development of tests to detect blood doping and illegal erythropoietin use, coupled with recent public scorn and medal stripping related to their abuse, will curb their use in the future.

---

to return the number to normal. (When you donate blood, your circulating erythrocyte supply is replenished in less than a week.)

Reticulocytes When demands for RBC production are high (for example, following hemorrhage), the bone marrow may release large numbers of immature erythrocytes, known as **reticulocytes**, into the blood to quickly meet the need (see ▮Figure 11-3). These immature cells are recognized by staining techniques that make visible the residual organelle remnants (mostly ribosomes) that have not yet been extruded. Their presence above the normal level of 0.5% to 1.5% of the total number of circulating erythrocytes indicates a high rate of erythropoietic activity. At very rapid rates, more than 30% of circulating RBCs may be at the immature reticulocyte stage.

Clinical Note **Laboratory-produced Erythropoietin** Synthetic erythropoietin (*Epogen, Procrit*) has currently become biotechnology's single-biggest moneymaker, with sales exceeding $1 billion annually. This hormone is often used to boost RBC production in patients with suppressed erythropoietic activity, such as those with kidney failure or those undergoing chemotherapy for cancer. (Chemotherapy drugs interfere with the rapid cell division characteristic of both cancer cells and developing RBCs.)

Anemia can be caused by a variety of disorders.

Clinical Note Despite control measures, O_2-carrying capacity cannot always be maintained to meet tissue needs. A below-normal O_2-carrying capacity of the blood is known as **anemia**, which is characterized by a low hematocrit (▮Figure 11-5a and b). Anemia can be brought about by a decreased rate of erythropoiesis, excessive losses of erythrocytes, or a deficiency in the hemoglobin content of erythrocytes. The various causes of anemia can be grouped into six categories:

1. **Nutritional anemia** is caused by a dietary deficiency of a factor needed for erythropoiesis, which depends on an adequate supply of essential raw ingredients, some of which are not synthesized in the body, so must be provided by food intake. For example, *iron deficiency anemia* occurs when not enough iron is available for synthesizing hemoglobin.

2. **Pernicious anemia** is caused by an inability to absorb enough ingested vitamin B_{12} from the digestive tract. Vitamin B_{12} is essential for normal RBC production and maturation. It is abundant in various foods and thus is rarely deficient in the diet. The problem is a deficiency of *intrinsic factor*, a special substance secreted by the lining of the stomach (see p. 585). Vitamin B_{12} can be absorbed from the intestinal tract only when this nutrient is bound to intrinsic factor.

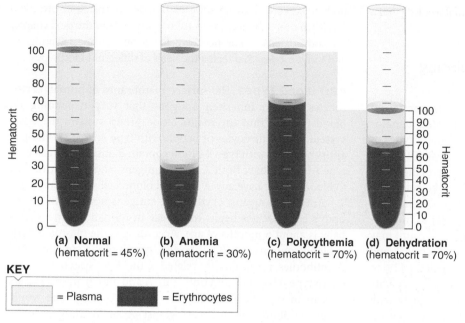

KEY

☐ = Plasma   ■ = Erythrocytes

I Figure 11-5 **Hematocrit under various circumstances.** (a) Normal hematocrit. (b) The hematocrit is lower than normal in anemia because of too few circulating erythrocytes. (c) The hematocrit is above normal in polycythemia because of excess circulating erythrocytes. (d) The hematocrit can also be elevated in dehydration when the normal number of circulating erythrocytes is concentrated in a reduced plasma volume.

3. **Aplastic anemia** is caused by the bone marrow failing to produce enough RBCs, even though all ingredients needed for erythropoiesis are available. Reduced erythropoietic capability can be caused by destruction of red marrow by toxic chemicals (such as benzene), heavy exposure to ionizing radiation, invasion of the marrow by cancer cells, or chemotherapy for cancer. The destructive process may selectively reduce the marrow's output of erythrocytes, or it may also reduce the productive capability for leukocytes and platelets. The anemia's severity depends on the extent to which erythropoietic tissue is destroyed; severe losses are fatal.

4. **Renal anemia** may result from kidney disease. Because erythropoietin from the kidneys is the primary stimulus for promoting erythropoiesis, inadequate erythropoietin secretion by diseased kidneys leads to insufficient RBC production.

5. **Hemorrhagic anemia** is caused by losing a lot of blood. The loss can be either acute, such as a bleeding wound, or chronic, such as excessive menstrual flow.

6. **Hemolytic anemia** is caused by rupture of too many circulating erythrocytes. **Hemolysis**, the rupture of RBCs, occurs either because an external factor brings about rupture of otherwise normal cells, as in invasion of RBCs by malaria parasites, or because the cells are defective, as in sickle cell disease.

Malaria is caused by protozoan parasites introduced into a victim's blood by the bite of a carrier mosquito (hence the use of mosquito nets over beds at night in tropical regions to reduce the incidence of malaria). These parasites selectively invade RBCs, where they multiply to the point that the mass of malarial organisms ruptures the cells, releasing hundreds of new active parasites that quickly invade other RBCs. As this cycle continues and more erythrocytes are destroyed, the anemic

condition progressively worsens and leads to death if not treated.

Sickle cell disease is the best-known example among various hereditary abnormalities of erythrocytes that make these cells fragile. It affects about 1 in 650 African Americans. Sickle cell disease is caused by a genetic mutation that changes a single amino acid in the 146-long amino acid chain that makes up the β chain of hemoglobin (valine replaces glutamate at position 6 in this amino acid chain). This defective type of hemoglobin joins together to form rigid chains that make the RBC stiff and unnaturally shaped, like a crescent or sickle (I Figure 11-6). Unlike normal erythrocytes, these deformed RBCs tend to clump together. The resultant "logjam" blocks blood flow through small vessels, leading to pain and tissue damage at the affected site. Furthermore, the defective erythrocytes are fragile and prone to rupture, even as young cells, as they travel through the narrow splenic capillaries. Despite an accelerated rate of erythropoiesis triggered by the constant excessive loss of RBCs, production may not be able to keep pace with the rate of destruction, and anemia may result. Interestingly, patients who have sickle cell anemia are more likely to survive malaria because defective cells invaded by malaria parasites are more apt to be destroyed as they travel through the spleen, eliminating the infected cells before the parasites have a chance to multiply and spread. This advantage may explain the increased incidence of sickle cell disease among descendants of people who lived in tropical regions where malaria is prevalent. In an evolutionary

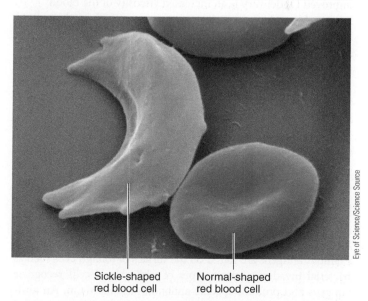

Sickle-shaped   Normal-shaped
red blood cell   red blood cell

I Figure 11-6 **Sickle-shaped red blood cell.** A scanning electron micrograph comparing a sickle cell and normal red blood cell.

"trade-off," the benefit conferred by surviving malaria may have offset the disadvantages of sickle cell disease.

Polycythemia is an excess of circulating erythrocytes.

Clinical Note **Polycythemia**, in contrast to anemia, is characterized by too many circulating RBCs and an elevated hematocrit (see Figure 11-5c). There are two general types of polycythemia, depending on what causes the excess RBC production: primary polycythemia and secondary polycythemia.

Primary polycythemia is caused by a tumorlike condition of the bone marrow in which erythropoiesis proceeds at an excessive, uncontrolled rate instead of being subject to the normal erythropoietin regulatory mechanism. The RBC count may reach 11 million cells/mm^3 (normal is 5 million cells/mm^3), and the hematocrit may be as high as 70% to 80% (normal is 42% to 45%). No benefit is derived from the extra O_2-carrying capacity of the blood, because O_2 delivery is more than adequate with normal RBC numbers. Inappropriate polycythemia has harmful effects, however. The excessive number of red cells increases blood's viscosity up to five to seven times normal (that is, makes the blood "thicker"), causing the blood to flow sluggishly, which may actually reduce O_2 delivery to the tissues (see p. 337). The increased viscosity also increases total peripheral resistance, which may elevate blood pressure, thus increasing the workload of the heart, unless blood-pressure control mechanisms can compensate (see Figure 10-13, p. 351).

Secondary polycythemia, in contrast, is an appropriate erythropoietin-induced adaptive mechanism to improve the blood's O_2-carrying capacity in response to a prolonged reduction in O_2 delivery to the tissues. It occurs normally in people living at high altitudes, where less O_2 is available in the air, or in people for whom O_2 delivery to the tissues is impaired by chronic lung disease or cardiac failure. The red cell count in secondary polycythemia is usually lower than that in primary polycythemia, typically averaging 6 million to 8 million cells/mm^3. The price paid for improved O_2 delivery is an increased viscosity of the blood.

An elevated hematocrit can occur when the body loses fluid but not erythrocytes, as in dehydration accompanying heavy sweating or profuse diarrhea (see Figure 11-5d). This is not a true polycythemia, however, because the number of circulating RBCs is not increased. A normal number of erythrocytes are simply concentrated in a smaller plasma volume. This condition is sometimes termed **relative polycythemia**.

We next shift attention to blood types, which depend on erythrocyte antigens.

Blood types depend on surface antigens on erythrocytes.

An **antigen** is a large, complex molecule that triggers a specific immune response against itself when it gains entry to the body. For example, antigens are found on the surface of foreign cells such as bacterial invaders. Certain types of white blood cells recognize antigens and produce specific antibodies against them. An **antibody** binds with the specific antigen against which it is produced and leads to the antigen's destruction by various means. Thus, the

body rejects cells bearing antigens that do not match its own. You will learn more about these immune responses in the next chapter on body defenses. For now, we focus on the special antigen–antibody reaction that forms the basis of different blood types.

ABO Blood Types The surface membranes of human erythrocytes contain inherited antigens that vary depending on blood type. Within the major blood group system, the **ABO system**, the erythrocytes of people with type A blood contain A antigens, those with type B blood contain B antigens, those with type AB blood have both A and B antigens, and those with type O blood do not have any A or B red blood cell surface antigens.

Antibodies against erythrocyte antigens not present on the body's erythrocytes begin to appear in human plasma after a baby is about 6 months of age. Accordingly, the plasma of type A blood contains anti-B antibodies, type B blood contains anti-A antibodies, no antibodies related to the ABO system are present in type AB blood, and both anti-A and anti-B antibodies are present in type O blood (Table 11-2). Typically, one would expect antibody production against A or B antigen to be induced only if blood containing the alien antigen were injected into the body. However, high levels of these antibodies are found in the plasma of people who have never been exposed to a different type of blood. Consequently, these were considered naturally occurring antibodies—that is, produced without any known exposure to the antigen. Scientists now know that people are routinely exposed at an early age to small amounts of A- and B-like antigens associated with common, harmless intestinal bacteria. Antibodies produced against these foreign antigens coincidentally also interact with a nearly identical antigen for a foreign blood group, even on first exposure to it.

Clinical Note **Transfusion Reaction** If a person is given blood of an incompatible type, two antigen–antibody interactions take place. By far, the more serious consequences arise from the effect of the antibodies in the recipient's plasma on the incoming donor erythrocytes. The effect of the donor's antibodies on the recipient's erythrocyte-bound antigens is less important unless a large amount of blood is transfused, because the donor's antibodies are so diluted by the recipient's plasma that little RBC damage takes place in the recipient.

Antibody interaction with an erythrocyte-bound antigen may result in agglutination (clumping) or hemolysis (rupture) of the attacked RBCs. Agglutination and hemolysis of donor RBCs

TABLE 11-2 ABO Blood Types

Blood Type	Antigens on Erythrocytes	Antibodies in Plasma
A	A	Anti-B
B	B	Anti-A
AB	A and B	No antibodies
O	No antigens	Both anti-A and anti-B

by antibodies in the recipient's plasma can lead to a sometimes fatal **transfusion reaction** (❙Figure 11-7). Agglutinated clumps of incoming donor cells can plug small blood vessels. In addition, one of the most lethal consequences of mismatched transfusions is acute kidney failure caused by the release of large amounts of hemoglobin from ruptured donor erythrocytes. If the free hemoglobin in the plasma rises above a critical level, it precipitates in the kidneys and blocks the urine-forming structures, leading to acute kidney shutdown.

Universal Blood Donors and Recipients

Because type O individuals have no A or B antigens, their erythrocytes are not attacked by either anti-A or anti-B antibodies, so they are considered **universal donors**. Their blood can be transfused into people of any blood type. However, type O individuals can receive only type O blood, because the anti-A and anti-B antibodies in their plasma attack either A or B antigens in incoming blood. In contrast, type AB individuals are called **universal recipients**. Lacking both anti-A and anti-B antibodies, they can accept donor blood of any type, although they can donate blood only to other AB people. Because their erythrocytes have both A and B antigens, their cells would be attacked if transfused into individuals with antibodies against either of these antigens.

The terms *universal donor* and *universal recipient* are misleading, however. In addition to the ABO system, many other erythrocyte antigens and plasma antibodies can cause transfusion reactions, the most important of which is the Rh factor.

Donor with type B blood

Recipient with type A blood

Antigen B

Antibody to type A blood

Antibody to type B blood

Antigen A

Red blood cells from donor agglutinate

Red blood cells usually rupture

Clumping blocks blood flow in capillaries

Hemoglobin precipitates in kidney, interfering with kidney function

Oxygen and nutrient flow to cells and tissues is reduced

❙Figure 11-7 **Transfusion reaction.** A transfusion reaction resulting from type B blood being transfused into a recipient with type A blood.
FIGURE FOCUS: *(1) Describe what would happen to donor red blood cells if type B blood were transfused into a recipient with type O blood? (2) How about if type O donor blood were transfused into a recipient with type B blood?*

Rh Blood-Group System

People who have the **Rh factor** (an erythrocyte antigen first observed in rhesus monkeys, hence the designation Rh) are said to have *Rh-positive* blood, whereas those lacking the Rh factor are considered *Rh-negative*. In contrast to the ABO system, no naturally occurring antibodies develop against the Rh factor.

According to the American Association of Blood Banking, following are the percentages of blood types in the ABO and Rh blood systems in the United States population: A+, 34%; A−, 6%; B+, 9%; B−, 2%; AB+, 3%; AB−, 1%; O+, 38%; and O−, 7%.

Clinical Note Anti-Rh antibodies are produced only by Rh-negative individuals when (and if) such people are first exposed to the foreign Rh antigen present in Rh-positive blood. A subsequent transfusion of Rh-positive blood could produce a transfusion reaction in such a sensitized Rh-negative person. Rh-positive individuals, in contrast, never produce antibodies against the Rh factor that they themselves possess. Therefore, Rh-negative people should be given only Rh-negative blood, whereas Rh-positive people can safely receive either Rh- negative or Rh-positive blood. The Rh factor is of particular medical importance when an Rh-negative mother develops antibodies against the erythrocytes of an Rh-positive fetus she is carrying, a condition known as **erythroblastosis fetalis**, or **hemolytic disease of the newborn**. Because the maternal antibodies destroy many fetal RBCs, the fetal bone marrow cannot keep pace with the rate of destruction and releases immature precursors of erythrocytes, such as reticulocytes and eventually erythroblasts (see ❙Figure 11-3), hence the name of the condition. (See Applying Clinical Reasoning on p. 402 for further discussion of this disorder.)

Except in extreme emergencies, it is safest to individually cross-match blood before a transfusion is undertaken even though the ABO and Rh typing is already known, because there are approximately 23 other minor human erythrocyte antigen systems, with hundreds of subtypes. Compatibility is determined by mixing the RBCs from the potential donor with plasma from the recipient. If no clumping occurs, the blood is considered an adequate match for transfusion. (See the boxed feature on pp. 390–391, ❙Concepts, Challenges, and Controversies, for an update on alternatives to whole-blood transfusions under investigation.)

In Search of a Blood Substitute

ONE OF THE HOTTEST MEDICAL contests of the past three decades has been the race to develop a universal substitute for human blood that is safe, inexpensive, and disease free and that has a long shelf life.

Need for a Blood Transfusion

A blood transfusion is administered, on average, every two seconds in the United States alone. As a general rule, a transfusion is warranted when the hemoglobin concentration falls below 7 g/100 mL whole blood (normal range is 13 to 18 g/100 mL for men and 12 to 16 g/100 mL for women) or when another specific component of the blood, such as platelets or clotting factors, is in short supply. The need can arise with victims of traumatic emergencies, such as automobile accidents or gunshot wounds; with surgery patients; with cancer patients on chemotherapy that suppresses blood cell production; with people who have blood disorders like leukemia or sickle cell anemia; with patients undergoing bone marrow transplants; and with other illnesses. Some patients need ongoing transfusions to survive, such as those with hemophilia. Sometimes whole blood is separated so that different components may be used for multiple recipients. Collection specialists have even developed a technique known as *apheresis* in which blood is collected into an instrument that separates the blood into separate portions, then the rest of the blood is returned to the donor after a selected portion, such as platelets, is recovered.

Need for a Blood Substitute

With only about 5% of the population now donating blood, regional shortages of particular blood types necessitate the shipping and sharing of blood among areas. (Almost everyone older than 17 years of age who weighs more than 110 pounds, is healthy, and has not donated blood in the last 8 weeks is eligible to be a blood donor, but most people do not take this opportunity to give the gift of life.) Medical personnel anticipate widespread shortages of blood in the near future because the number of blood donors continues to decline at the same time that the number of senior citizens, the group of people who most often need transfusions, continues to grow. The benefits for society of a safe blood substitute that could be administered without regard for the recipient's blood type are great, as will be the profits for the manufacturer of the first successful product. Experts estimate that the world market for a good blood substitute, more accurately called an *oxygen therapeutic,* may be $10 billion per year.

Researchers began working on blood substitutes in the 1960s, but the search for an alternative to whole-blood transfusions was given new impetus in the 1980s by the rising incidence of AIDS and the concomitant concern over the safety of the nation's blood supply. Infectious diseases such as AIDS, viral hepatitis, and West Nile virus infection can be transmitted from infected blood donors to recipients of blood transfusions. Although careful screening of our blood supply minimizes the possibility that infectious diseases will be transmitted through transfusion, the public remains wary and would welcome a safe substitute.

Eliminating the risk of disease transmission is only one advantage of an alternative to whole-blood transfusion. Whole blood must be kept refrigerated, and even then its shelf life is only 42 days. Also, transfusion of whole blood requires blood typing and cross matching, which cannot be done at the scene of an accident or on a battlefield.

Major Approaches

The goal is not to find a replacement for whole blood but to duplicate its O_2-carrying capacity. The biggest need for blood transfusions is to replace acute blood loss in surgical patients, accident victims, and wounded soldiers. These individuals require short-term replenishment of blood's O_2-carrying capacity until their bodies synthesize replacement erythrocytes. The many other important elements in blood are not as immediately critical in sustaining life as the hemoglobin within the RBCs is. Unfortunately, RBCs are the whole-blood component that requires refrigeration, has a short shelf life, and bears the markers for the various blood types.

Therefore, the search for a blood substitute has focused on two major possibilities: (1) hemoglobin products that exist outside an RBC and can be stored at room temperature for up to a year, and (2) chemically synthesized products that serve as artificial hemoglobin by dissolving large amounts of O_2 when O_2 levels are high (as in the lungs) and releasing it when O_2 levels are low (as in the tissues). A variety of potential blood substitutes are in various stages of development and clinical trials. No products have yet reached the market, although they are getting close. Let us examine each of the major approaches.

Hemoglobin Products

By far the greatest number of research efforts have focused on manipulating the structure of hemoglobin so that it can be effectively and safely administered as a substitute for whole-blood transfusions. If appropriately stabilized and suspended in saline solution, hemoglobin could be injected to bolster O_2-carrying capacity of the recipients' blood no matter what their blood type. The following strategies are among those being pursued to develop a hemoglobin product:

■ One problem is that hemoglobin behaves differently when it is outside RBCs. "Naked" hemoglobin splits into halves that do not release O_2 for tissue use as readily as normal hemoglobin does. Also, these hemoglobin fragments can cause kidney damage. A cross-binding reagent has been developed that keeps hemoglobin molecules intact when they are outside RBCs, thus surmounting one major obstacle to administering free hemoglobin.

■ Some products under investigation are derived from outdated, donated human blood. Instead of the blood being discarded, its hemoglobin is extracted, purified, sterilized, and chemically stabilized. However, this strategy relies on the continued practice of collecting human blood donations.

■ Several products use cows' blood as a starting point. Bovine hemoglobin is readily available from slaughterhouses, is cheap, and can be treated for administration to humans. A big concern with these products is the potential of introducing into humans unknown disease-causing microbes that might be lurking in the bovine products.

■ A potential candidate as a blood substitute is genetically engineered hemoglobin, which bypasses the ongoing need for human donors or the risk of spreading disease from cows to humans. Genetic engineers can insert the gene for human hemoglobin into bacteria, which act as a "factory" to produce the desired hemoglobin product. A drawback for genetically engineered hemoglobin is the high cost involved in operating the facilities.

■ One promising strategy encapsulates hemoglobin within liposomes—membrane-wrapped containers—similar to real hemoglobin-stuffed, plasma membrane–enclosed RBCs. These so-called neo red cells (*neo* means "new") await further investigation.

Synthetic O_2 Carriers

Other researchers are pursuing chemical-based strategies that rely on *perfluorocarbons (PFCs)*, which are artificial O_2-carrying compounds composed of carbon and fluorine. PFCs are completely inert, chemically synthesized molecules that can dissolve large quantities of O_2 in direct proportion to the amount of O_2 breathed in. Because they are derived from a nonbiological source, PFCs cannot transmit disease. This, coupled with their low cost, makes them attractive as a blood substitute. Yet use of PFCs is not without risk. Their administration can cause flulike symptoms, and because of poor excretion, they may be retained and accumulate in the body. Ironically, PFC administration poses the danger of causing O_2 toxicity by delivering too much O_2 to the tissues in uncontrolled fashion.

A more recent strategy under development is a water-soluble, synthetic plastic version of hemoglobin. The molecule has the same size and shape of hemoglobin, reversibly binds with O_2, and is made from a substance known to be safe in the body, but it has not yet undergone biological testing.

Red Blood Cells Produced from Stem Cells

The newest technique being explored is using hemopoietic stem cells to "grow" a supply of transfusable red blood cells. One rich source of hemopoietic stem cells under investigation is umbilical cord blood after birth. The umbilical cord is the connection between mother and fetus during gestation but is cut when the baby is born. Hemopoietic stem cells from cord blood that normally would have been discarded can be used to produce mass quantities of new red blood cells of the O negative type (universal donor type; see p. 389) in a machine that mimics the bone marrow environment. This process is called *blood pharming*. These laboratory-produced RBCs are functionally indistinguishable from body-produced erythrocytes. However, as of now the process is cost prohibitive for general use and is being explored for more limited battlefield application.

Tactics to Reduce Need for Donated Blood

Other tactics aimed toward reducing the need for donated blood include the following:

■ By changing surgical practices, the medical community has reduced the need for transfusions. These blood-saving methods include recycling a patient's blood during surgery (collecting lost blood, then reinfusing it); using less invasive and therefore less bloody surgical techniques; and treating the patient with blood-enhancing erythropoietin before surgery.

■ The necessity of matching blood types for transfusions is a major reason for waste at blood banks. Transfusion of mismatched blood causes a serious, even fatal reaction (see p. 389). Therefore, a blood bank may be discarding stocks of one blood type that has gone unused, while facing a serious shortage of another type. Researchers have zeroed in on a way to use enzymes to cleave A and B antigens from RBCs, thus converting all donated blood into "universal donor" type O blood. If clinical trials using this enzyme-converted blood are successful, such a product would reduce the current waste and would be highly beneficial to battlefield surgeons who often do not have time to type blood.

■ Other investigators are seeking ways to prolong the life of RBCs, either in a blood bank or in patients, thus reducing the need for fresh blood that can be transfused.

As this list of strategies attests, considerable progress has been made toward developing a safe, effective alternative to whole-blood transfusions. After three decades of intense effort, however, considerable challenges remain, and no ideal solution has been found.

1. Describe the anatomic features of erythrocytes that contribute to the efficiency with which they transport O_2.

2. List the chemicals with which hemoglobin can combine.

3. Discuss the source, control, and function of erythropoietin.

11.3 Leukocytes

Leukocytes (white blood cells, or WBCs) are the mobile units of the body's immune defense system. **Immunity** is the body's ability to resist or eliminate potentially harmful foreign materials or abnormal cells. Leukocytes and their derivatives, along with a variety of plasma proteins, make up the **immune system**, an internal defense system that recognizes and either destroys or neutralizes materials that are foreign to the "normal self." Specifically, the immune system (1) defends against invading disease-producing microorganisms (such as bacteria and viruses); (2) functions as a "cleanup crew" that removes worn-out cells (such as aged red blood cells) and tissue debris (for example, tissue damaged by trauma or disease), paving the way for wound healing and tissue repair; and (3) identifies and destroys cancer cells that arise in the body.

Leukocytes primarily function as defense agents outside the blood.

To carry out their functions, WBCs largely use a "seek out and attack" strategy—that is, they go to sites of invasion or tissue damage. The main reason WBCs are in the blood is for rapid transport from their site of production or storage to wherever they are needed. Unlike erythrocytes, leukocytes are able to exit the blood by assuming amoebalike behavior to wriggle through narrow capillary pores and crawl to assaulted areas (see ▌ Figure 12-2, p. 408). As a result, the immune system's effector cells are widely dispersed throughout the body and can defend in any location. For this reason, we introduce the specific circulating WBCs here to round out our discussion of blood but leave the details of their phagocytic and immunologic functions, which take place primarily in the tissues, for the next chapter.

There are five types of leukocytes.

Leukocytes lack hemoglobin, so they are colorless (that is, "white") unless specifically stained for microscopic visibility. Unlike red blood cells, which are of uniform structure, identical function, and constant number, white blood cells vary in structure, function, and number. There are five types of circulating WBCs—neutrophils, eosinophils, basophils, monocytes, and lymphocytes—each with a characteristic structure and function. They are all somewhat larger than RBCs.

Granulocytes and Agranulocytes The five types of leukocytes fall into two main categories, depending on the appearance of their nuclei and the presence or absence of granules in their cytoplasm when viewed microscopically (▌ Figure 11-8). Neutrophils, eosinophils, and basophils are categorized as **poly-** morphonuclear (meaning "many-shaped nucleus") **granulocytes** (meaning "granule-containing cells"). Their nuclei are segmented into several lobes of varying shapes, and their cytoplasm contains an abundance of membrane-enclosed granules. The granules contain preformed, stored chemicals that are released by exocytosis on appropriate stimulation for carrying out the granulocyte's functions. The three types of granulocytes are distinguished on the basis of the varying affinity of their granules for dyes: *Eosinophils* have an affinity for the red dye eosin, *basophils* preferentially take up a basic blue dye, and *neutrophils* are neutral, showing no dye preference. Monocytes and lymphocytes are known as **mononuclear** (meaning "single nucleus") **agranulocytes** (meaning "cells lacking granules"). Both have a single, large, nonsegmented nucleus and few granules. *Monocytes* are the larger of the two and have an oval or kidney-shaped nucleus. *Lymphocytes,* the smallest of the leukocytes, characteristically have a large spherical nucleus that occupies most of the cell.

Functions and Life Spans of Leukocytes The following are the functions and life spans of the granulocytes:

■ **Neutrophils** are phagocytic specialists; they engulf and destroy bacteria intracellularly (see ▌ Figure 2-9c, p. 32). Neutrophil granules, which contain an arsenal of antimicrobial proteins, fuse with invading bacteria ingested by phagocytosis and kill them inside the cell. Neutrophils can also release the bacteria-killing chemicals into the ECF by exocytosis of granule contents to the cell exterior, a process called *degranulation*. As a further assault, neutrophils can act like "suicide bombers" by undergoing an unusual type of programmed cell death called *NETosis,* during which they kill nearby bacteria. During NETosis, neutrophils use vital cellular materials to prepare a web of fibers dubbed **neutrophil extracellular traps (NETs)**, which they release into the ECF on their death. These fibers consist of chromatin from the neutrophil's nucleus studded with antimicrobial proteins from its cytoplasmic granules. NETs bind with bacteria, trapping and then destroying these foreign invaders extracellularly. Neutrophils invariably are the first defenders on the scene of bacterial invasion. Furthermore, they scavenge to clean up debris.

Clinical Note As might be expected in view of these functions, an increase in circulating neutrophils (**neutrophilia**) typically accompanies acute bacterial infections. In fact, a differential WBC count (a determination of the proportion of each type of leukocyte present) can be used to make an immediate, reasonably accurate prediction of whether an infection, such as pneumonia or meningitis, is of bacterial or viral origin. Obtaining a definitive answer as to the causative agent by culturing a sample of the infected tissue's fluid takes several days. Because an elevated neutrophil count is highly indicative of bacterial infection, antibiotic therapy can be initiated long before the causative agent is actually known. (Bacteria generally succumb to antibiotics, whereas viruses do not.)

■ **Eosinophils** are specialists of another type. An increase in circulating eosinophils (**eosinophilia**) is associated with allergic conditions (such as asthma and hay fever) and with internal parasite infestations (for example, worms). Eosinophils ob-

	Leukocytes					Erythrocyte	Platelets
	Polymorphonuclear granulocytes			Mononuclear agranulocytes			
	Neutrophil	Eosinophil	Basophil	Monocyte	Lymphocyte		
	60%–70%	1%–4%	0.25%–0.5%	2%–6%	25%–33%	Erythrocyte concentration = 5 billion/ mL blood	Platelet concentration = 250 million/ mL blood
	Differential WBC count (percentage distribution of types of leukocytes)						
	Leukocyte concentration = 7 million/mL blood					RBC count = 5,000,000/mm^3	Platelet count = 250,000/mm^3
	WBC count = 7000/mm^3						

▌Figure 11-8 **Normal blood cellular elements and typical human blood cell count.**
Cell types: Courtesy and copyright of the Clinical Chemistry and Hematology Laboratory, Wadsworth Center, NY State Department of Health (http://www.wadsworth.org); platelets: © Peter Arnold, Inc./Alamy

viously cannot engulf a larger parasitic worm, but they do attach to the worm and secrete substances that kill it.

■ **Basophils** are the least numerous of the leukocytes. They are quite similar structurally and functionally to *mast cells,* which never circulate in the blood but instead are dispersed in connective tissue throughout the body. Both basophils and mast cells synthesize and store *histamine* and *heparin,* powerful chemical substances that can be released on appropriate stimulation. Histamine release is important in allergic reactions, whereas heparin speeds up removal of fat particles from the blood after a fatty meal and plays a role in certain immune responses. Heparin can also prevent clotting (coagulation) of blood samples drawn for clinical analysis and is used extensively as an anticoagulant drug, but it does not appear to play a physiologic role as an anticoagulant.

Once released into the blood from the bone marrow, a granulocyte usually stays in transit in the blood for less than a day before leaving to enter the tissues, where it survives another 3 to 4 days unless it dies sooner in the line of duty.

By comparison, the functions and life spans of the agranulocytes are as follows:

■ **Monocytes,** like neutrophils, become professional phagocytes. They emerge from the bone marrow while still immature and circulate for only a day or two before settling down in various tissues throughout the body. At their new residences, monocytes continue to mature and greatly enlarge, becoming large tissue phagocytes known as **macrophages** (*macro* means "large"; *phage* means "eater"). A macrophage's life span ranges from months to years unless it dies sooner while performing phagocytosis. A phagocytic cell can ingest only a limited amount of foreign material before it succumbs.

■ **Lymphocytes** provide immune defense against targets for which they are specifically programmed. There are two types of lymphocytes, B and T lymphocytes (B and T cells), which look alike. **B lymphocytes** produce antibodies, which circulate in the blood and are responsible for *antibody-mediated,* or *hu-*

moral, immunity. An antibody binds with and marks for destruction (by phagocytosis or other means) the specific kinds of antigen-containing foreign matter, such as bacteria, that induced production of the antibody. **T lymphocytes** do not produce antibodies; instead, they directly destroy their specific target cells by releasing chemicals that punch holes in the victim cell, a process called *cell-mediated immunity.* The target cells of T cells include body cells invaded by viruses and cancer cells. Lymphocytes live for about 100 to 300 days. Only a small part of the total lymphocytes are in transit in the blood at any given moment. Most continually recycle among the blood, lymph and lymphoid tissues, spending only a few hours at a time in the blood. *Lymphoid tissues* are lymphocyte-containing tissues such as the lymph nodes and tonsils.

Leukocytes are produced at varying rates depending on the body's changing needs.

All leukocytes ultimately originate from common precursor undifferentiated pluripotent stem cells in bone marrow that also give rise to erythrocytes and platelets (▌Figure 11-9). The cells destined to become WBCs eventually differentiate into various committed cell lines and proliferate under the influence of appropriate stimulating factors. Granulocytes and monocytes are produced only in bone marrow, which releases these mature leukocytes into the blood. Lymphocytes are originally derived from precursor cells in bone marrow, but most new lymphocytes actually come from lymphocyte colonies already in lymphoid tissues originally populated by cells derived from bone marrow.

The total number of white blood cells normally ranges from 5 million to 10 million cells per milliliter of blood, with an average of 7 million/mL, expressed as an average **white blood cell count** of 7000/mm^3. Leukocytes are the least numerous of the blood cells (about 1 white blood cell for every 700 red blood cells), not because fewer are produced but because they are merely in transit while in the blood. Normally, about two thirds of the circulating leukocytes are granulocytes, mostly neutrophils, whereas one third are agranulocytes, predominantly lymphocytes (see ▮Figure 11-8). However, the total number of white cells and the percentage of each type may vary considerably to meet changing defense needs. Depending on the type and extent of assault the body is combating, different types of leukocytes are selectively produced at varying rates, as in neutrophilia for example. Chemical messengers arising from invaded or damaged tissues or from activated leukocytes themselves govern the rates of production of the various WBCs. Specific messengers analogous to erythropoietin direct the differentiation and proliferation of each cell type. Some of these messengers have been identified and can be produced in the laboratory; an example is **granulocyte colony–stimulating factor**, which stimulates increased replication and release of granulocytes, especially neutrophils, from bone marrow. Marketed under the name *Neulasta,* this synthetic agent is a powerful new therapeutic tool that can be used to bolster defense and thus decrease the incidence of infection in cancer patients being treated with chemotherapy drugs. These drugs suppress all rapidly dividing cells, including hemopoietic cells in bone marrow, and the targeted cancer cells.

Clinical Note **Abnormalities in Leukocyte Production** Even though levels of circulating WBCs may vary, changes in these levels are normally controlled and adjusted according to the body's needs. However, abnormalities in leukocyte production can occur that are not subject to control—that is, either too few or too many WBCs can be produced. Bone marrow can greatly slow down or even stop production of white blood cells when it is exposed to certain toxic chemical agents (such as benzene and anticancer drugs) or to excessive radiation. The most serious consequence is the reduction in professional phagocytes (neutrophils and macrophages), which greatly diminishes the body's defense capabilities against invading microorganisms. The only defense still available when bone marrow fails is the immune capabilities of the lymphocytes produced by lymphoid tissues.

Surprisingly, one of the major consequences of **leukemia**, a cancerous condition involving uncontrolled proliferation of WBCs, is inadequate defense capabilities against foreign invasion. In leukemia, the WBC count may reach as high as 500,000/mm³, compared with the normal 7000/mm³, but because most of these cells are abnormal or immature, they cannot perform their normal defense functions. Another devastating consequence of leukemia is displacement of the other blood cell lines in bone marrow. This results in anemia because of reduced erythropoiesis and in internal bleeding because of a platelet deficit. Platelets play a critical role in preventing bleeding from myriad tiny breaks that normally occur in small blood vessel walls. Consequently, overwhelming infections or hemorrhage are the most common causes of death in leukemic patients. The next section examines platelets' role in minimizing the threat of bleeding.

Check Your Understanding 11.3

1. Define *immunity* and list the functions of the immune system.
2. Name the polymorphonuclear granulocytes and the mononuclear agranulocytes.
3. State the function of each type of leukocyte.

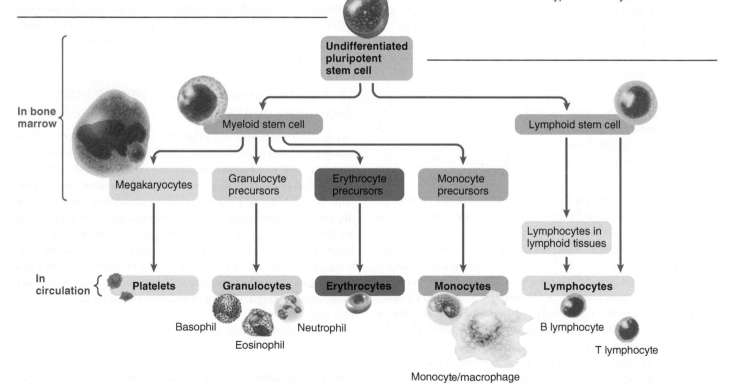

▮Figure 11-9 **Blood cell production (hemopoiesis).** All the blood cell types ultimately originate from the same undifferentiated pluripotent stem cells in the red bone marrow.

FIGURE FOCUS: *Name all of the blood cellular elements derived from myeloid stem cells.*

11.4 Platelets and Hemostasis

An average of 250 million platelets are normally present in each milliliter of blood (range of 150,000 to 350,000/mm³).

Platelets are cell fragments shed from megakaryocytes.

Platelets, or **thrombocytes**, are not whole cells but small cell fragments (about 2 to 4 μm in diameter) shed from the outer edges of extraordinarily large (up to 60 μm in diameter) bone marrow–bound cells known as **megakaryocytes** (▮Figure 11-10). A single megakaryocyte typically produces about 1000 platelets. Megakaryocytes are derived from the same undifferentiated stem cells that give rise to the erythrocytic and leukocytic cell lines (see ▮Figure 11-9). Platelets are essentially detached vesicles containing pieces of megakaryocyte cytoplasm wrapped in plasma membrane.

Platelets remain functional for an average of 10 days, at which time they are removed from circulation by tissue macrophages, especially those in the spleen and liver, and are replaced by newly released platelets from bone marrow. The hormone **thrombopoietin**, produced by the liver, increases the number of megakaryocytes in bone marrow and stimulates each megakaryocyte to produce more platelets as needed. The factors that control thrombopoietin secretion and regulate the platelet level are currently under investigation.

Because platelets are cell fragments, they lack nuclei. However, they have organelles and cytosolic enzyme systems for generating energy and synthesizing secretory products, which they store in numerous granules dispersed throughout the cytosol. Furthermore, platelets contain high concentrations of actin and myosin, which enable them to contract. Their secretory and contractile abilities are important in hemostasis, a topic to which we now turn.

Hemostasis prevents blood loss from damaged small vessels.

Hemostasis is the arrest of bleeding from a broken blood vessel—that is, the stopping of hemorrhage (*hemo* means "blood"; *stasis* means "standing"). For bleeding to take place from a vessel, a break must be present in the vessel wall and the pressure inside must be greater than the pressure outside the vessel to force blood out through the defect.

The small capillaries, arterioles, and venules are often ruptured by minor traumas of everyday life; such traumas are the most common source of bleeding, although we often are not even aware that any damage has taken place. The body's inherent hemostatic mechanisms normally are adequate to seal defects and stop blood loss through these small microcirculatory vessels. *Clinical Note* The rarer occurrence of bleeding from medium to large vessels usually cannot be stopped by hemostatic mechanisms alone. Bleeding from a severed artery is more profuse and therefore more dangerous than venous bleeding, because the outward driving pressure is greater in arteries (that is, arterial blood pressure is higher than venous pressure). First-

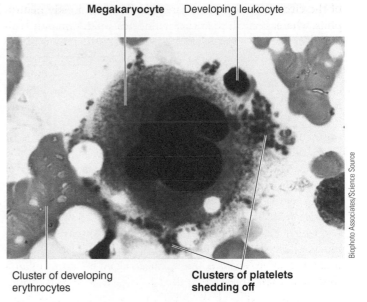

Megakaryocyte Developing leukocyte

Cluster of developing erythrocytes

Clusters of platelets shedding off

Biophoto Associates/Science Source

▮Figure 11-10 **A megakaryocyte forming platelets.**

aid measures for a severed artery include applying external pressure to the wound that is greater than the arterial pressure to temporarily halt bleeding until the torn vessel can be surgically closed. Hemorrhage from a severed vein can often be stopped simply by elevating the bleeding body part to reduce gravity's effects on pressure in the vein (see p. 362). If the accompanying drop in venous pressure is not enough to stop bleeding, mild external compression is usually adequate.

Hemostasis involves three major steps: (1) *vascular spasm,* (2) *formation of a platelet plug,* and (3) *blood coagulation (clotting).* Platelets play a pivotal role in hemostasis. They obviously play a major part in forming a platelet plug, but they also contribute significantly to the other two steps.

Vascular spasm reduces blood flow through an injured vessel.

A cut or torn blood vessel immediately constricts. The underlying mechanism is thought to be an intrinsic response triggered by a paracrine released locally from the injured vessel's endothelial lining (see p. 345). This constriction, or **vascular spasm**, slows blood flow through the defect and thus minimizes blood loss. Also, as the opposing endothelial surfaces of the vessel are pressed together by this initial vascular spasm, they become sticky and adhere to each other, further sealing off the damaged vessel. These physical measures alone cannot completely prevent further blood loss, but they minimize blood flow through the break in the vessel until other hemostatic measures can actually plug the hole.

Platelets aggregate to form a plug at a vessel injury.

Platelets normally do not adhere to the smooth endothelial lining of blood vessels, but they do stick to damaged vessels. When the endothelial lining is disrupted because of vessel injury, *von Willebrand factor (vWF)*, a plasma protein secreted by mega-

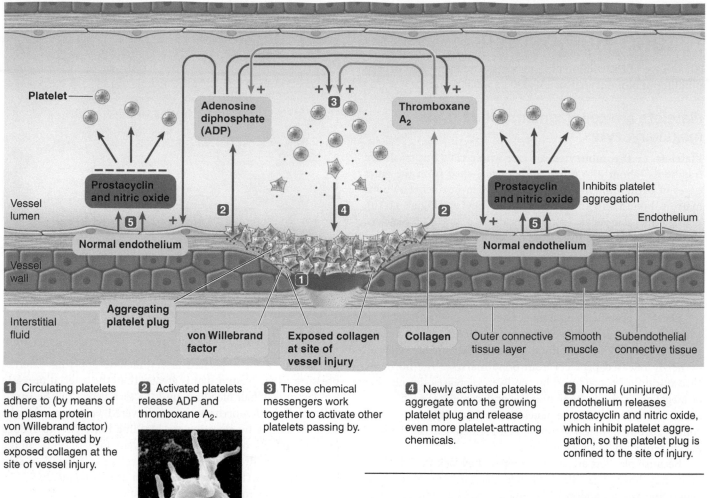

Platelet	Adenosine diphosphate (ADP)		Thromboxane A₂	

Prostacyclin and nitric oxide

Prostacyclin and nitric oxide — Inhibits platelet aggregation

Vessel lumen

Normal endothelium

Normal endothelium

Endothelium

Vessel wall

Interstitial fluid

Aggregating platelet plug

von Willebrand factor

Exposed collagen at site of vessel injury

Collagen

Outer connective tissue layer

Smooth muscle

Subendothelial connective tissue

1 Circulating platelets adhere to (by means of the plasma protein von Willebrand factor) and are activated by exposed collagen at the site of vessel injury.

2 Activated platelets release ADP and thromboxane A₂.

3 These chemical messengers work together to activate other platelets passing by.

4 Newly activated platelets aggregate onto the growing platelet plug and release even more platelet-attracting chemicals.

5 Normal (uninjured) endothelium releases prostacyclin and nitric oxide, which inhibit platelet aggregation, so the platelet plug is confined to the site of injury.

Activated platelet

Circulating (inactive) platelet

SPL/Science Source

❙Figure 11-11 **Platelet activation, formation of a platelet plug, and pre-vention of platelet aggregation at adjacent normal vessel lining.**

karyocytes, platelets, and endothelial cells and always present in the plasma, adheres to the exposed collagen. *Collagen* is a fibrous protein in the connective tissue underlying the endothelial lining (see p. 60 and p. 340). vWF has binding sites to which the fast-moving platelets attach by means of their cell-surface receptors specific for this plasma protein. Thus vWF serves as a bridge between platelets and the injured vessel wall. This adhesion prevents these platelets from being swept forward in the circulation. This layer of stuck platelets forms the foundation of a hemostatic **platelet plug** at the site of the defect. Collagen activates the bound platelets. Normally platelets are disc-shaped and have a smooth surface, but activated platelets quickly reorganize their actin cytoskeletal elements to develop spiky processes, which help them adhere to the collagen and to other platelets (❙Figure 11-11). Activated platelets also release several important chemicals from their storage granules. Among these

chemicals is *adenosine diphosphate (ADP)*, which causes the surfaces of nearby circulating platelets to become sticky so that they adhere to the first layer of aggregated platelets and are activated. These newly aggregated platelets release more ADP, which causes more platelets to pile on, and so on; thus, a plug of platelets is rapidly built up at the defect site in a positive-feedback fashion (❙Figure 11-11). This aggregating process is reinforced by the ADP-stimulated formation of an eicosanoid paracrine similar to prostaglandins, *thromboxane A₂*, from a component of the platelet plasma membrane (see p. 119). Thromboxane A₂ directly promotes platelet aggregation and further enhances it indirectly by triggering the release of even more ADP from the platelet granules. Thus, formation of a platelet plug involves the three successive, closely integrated events of adhesion, activation, and aggregation.

Given the self-perpetuating nature of platelet aggregation, why does the platelet plug not continue to expand over the surface of the adjacent normal vessel lining? The reason is that ADP discharged by activated platelets stimulates release of *prostacyclin* and *nitric oxide* from the adjacent normal endothelium. Both these chemicals profoundly inhibit platelet aggregation. Thus, the platelet plug is limited to the defect and does not spread to the nearby undamaged vascular tissue (❙Figure 11-11).

The aggregated platelet plug not only physically seals the break in the vessel but also performs three other important roles. (1) The actin–myosin complex within the aggregated platelets contracts to compact and strengthen what was originally a fairly loose plug. (2) The platelet plug releases several powerful vasoconstrictors that induce profound constriction of the affected vessel to reinforce the initial vascular spasm. (3) The platelet plug releases other chemicals that enhance blood coagulation, the next step of hemostasis. Although the platelet-plugging mechanism alone is often enough to seal the myriad minute tears in capillaries and other small vessels that occur many times daily, larger holes in these vessels require the formation of a blood clot to completely stop the bleeding.

Scientists continue to discover other platelet functions besides thwarting blood loss, such as releasing growth factors to help damaged tissue rebuild, inducing inflammation, serving as detectors of disease-causing microorganisms, and promoting NET release by neutrophils. Thus, platelets are not mere "band-aids" in the blood.

Clot formation results from a triggered chain reaction involving plasma clotting factors.

Blood coagulation, or **clotting**, is the transformation of blood from a liquid into a solid gel. Formation of a clot on top of the platelet plug reinforces the seal over a break in a vessel. Furthermore, as blood in the vicinity of the vessel defect solidifies, it can no longer flow. Clotting is the body's most powerful hemostatic mechanism. It is required to stop bleeding from all but the minutest defects.

Clot Formation The ultimate step in clot formation is conversion of **fibrinogen**, a large, soluble plasma protein produced by the liver and normally always present in the plasma, into **fibrin**, an insoluble, threadlike molecule. This conversion into fibrin is catalyzed by the enzyme **thrombin** at the site of the injury. Fibrin molecules adhere to the damaged vessel surface, forming a loose, netlike meshwork that traps blood cells, including aggregating platelets. The resulting mass, or **clot**, typically appears red because of the abundance of trapped RBCs, but the foundation of the clot is formed of fibrin derived from the plasma (❙Figure 11-12). Except for platelets, which help bring about the conversion of fibrinogen to fibrin, clotting can take place in the absence of all other blood cells.

Fibrin is the stretchiest natural protein that scientists have ever studied. On average fibrin fibers can be passively stretched nearly 3 times their original length and still snap back to their starting length and can be stretched more than 4 times their original length before they break. This highly elastic property accounts for the extraordinary stretchiness of blood clots.

The original fibrin web is rather weak, because the fibrin strands are only loosely interlaced. However, chemical linkages rapidly form between adjacent strands to strengthen and stabilize the clot meshwork. This cross-linkage is catalyzed by a clotting factor known as *factor XIII (fibrin-stabilizing factor)*, which normally is present in the plasma in inactive form.

Roles of Thrombin Multitasking thrombin, in addition to (1) converting fibrinogen into fibrin, also (2) activates factor XIII to stabilize the resultant fibrin mesh, (3) acts in a positive-feedback fashion to facilitate its own formation, and (4) enhances platelet aggregation, which in turn is essential to the clotting process.

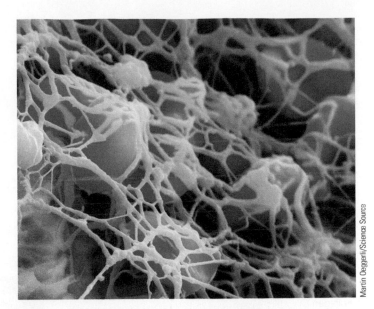

❙Figure 11-12 **Erythrocytes trapped in the fibrin meshwork of a clot.**

Martin Oeggerli/Science Source

Because thrombin converts the ever-present fibrinogen molecules in the plasma into a blood-stanching clot, thrombin must normally be absent from the plasma except in the vicinity of vessel damage. Otherwise, blood would always be coagulated—a situation incompatible with life. How can thrombin normally be absent from the plasma, yet be readily available to trigger fibrin formation when a vessel is injured? The answer is that thrombin exists in the plasma in the form of an inactive precursor called **prothrombin**, which is converted into thrombin by the clotting cascade when blood clotting is desirable.

The Clotting Cascade Yet another activated plasma clotting factor, **factor X**, converts prothrombin to thrombin; factor X itself is normally present in the blood in inactive form and must be converted into its active form by still another activated factor, and so on. Altogether, 12 plasma clotting factors participate in essential steps that lead to the final conversion of fibrinogen into a stabilized fibrin mesh (❙Figure 11-13). These factors are designated by roman numerals in the order in which the factors were discovered, not the order in which they participate in the clotting process.[1] Most of these clotting factors are plasma proteins synthesized by the liver. Normally, they are always present in the plasma in an inactive form, such as fibrinogen and prothrombin. In contrast to fibrinogen, which is converted into insoluble fibrin strands, prothrombin and the other precursors, when converted to their active form, act as proteolytic (protein-splitting) enzymes. These enzymes activate another specific factor in the clotting sequence. Once the first factor in the sequence is activated, it in turn activates the next factor, and so on, in a series of sequential reactions known as the **clotting cascade**, until thrombin catalyzes the final conversion of fibrinogen into fibrin. Several of these steps require the presence of plasma Ca^{2+} and *platelet factor 3 (PF3)*, a chemical secreted by the aggregated platelet plug. Thus, platelets also contribute to clot formation.

---

[1] The term *factor VI* is no longer used. What once was considered a separate factor VI has now been determined to be an activated form of factor V.

Intrinsic and Extrinsic Pathways The clotting cascade may be triggered by the *intrinsic pathway* or the *extrinsic pathway*:

■ The **intrinsic pathway** precipitates clotting within damaged vessels, and clotting of blood samples in test tubes. All elements necessary to bring about clotting by means of the intrinsic pathway are present in the blood. This pathway, which involves seven separate steps (shown in *blue* in ❙Figure 11-13), is set off when **factor XII (Hageman factor)** is activated by coming into contact with either exposed collagen in an injured vessel or a foreign surface such as a glass test tube. Remember

that exposed collagen also initiates platelet aggregation. Thus, formation of a platelet plug and the chain reaction leading to clot formation are simultaneously set in motion when a vessel is damaged. Furthermore, these complementary hemostatic mechanisms reinforce each other. The aggregated platelets secrete PF3, which is essential for the clotting cascade that in turn enhances further platelet aggregation.

■ The **extrinsic pathway** takes a shortcut and requires only four steps (shown in *gray* in ❙Figure 11-13). This pathway, which requires contact with tissue factors external to the blood, initiates clotting of blood that has escaped into the tissues. When a tissue is traumatized, it releases a protein complex known as **tissue thromboplastin**. Tissue thromboplastin directly activates factor X, thereby bypassing all preceding steps of the intrinsic pathway. From this point on, the two pathways are identical.

The intrinsic and extrinsic mechanisms usually operate simultaneously. When a blood vessel ruptures during tissue injury, the intrinsic mechanism stops blood in the injured vessel, and the extrinsic mechanism clots blood that escaped into the tissue before the vessel was sealed off. Typically, clots are fully formed in 3 to 6 minutes.

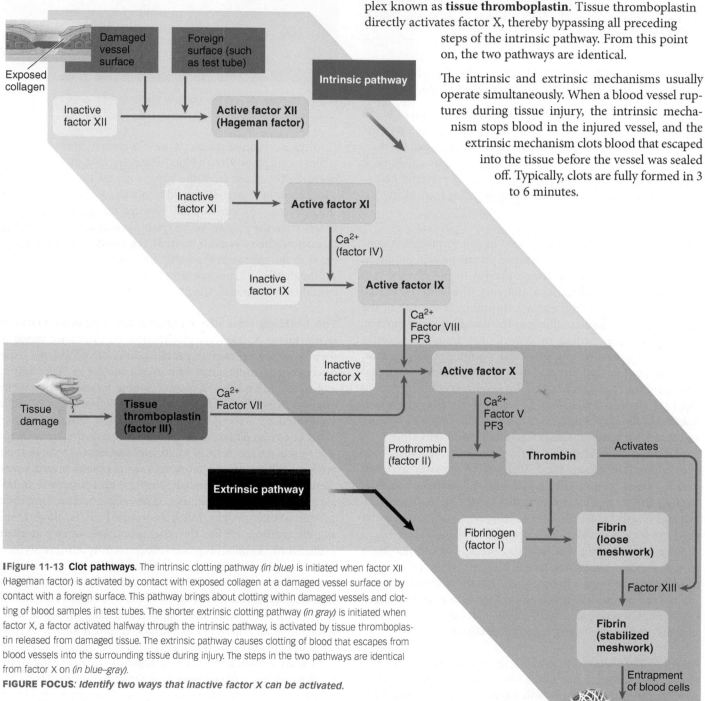

❙Figure 11-13 **Clot pathways.** The intrinsic clotting pathway *(in blue)* is initiated when factor XII (Hageman factor) is activated by contact with exposed collagen at a damaged vessel surface or by contact with a foreign surface. This pathway brings about clotting within damaged vessels and clotting of blood samples in test tubes. The shorter extrinsic clotting pathway *(in gray)* is initiated when factor X, a factor activated halfway through the intrinsic pathway, is activated by tissue thromboplastin released from damaged tissue. The extrinsic pathway causes clotting of blood that escapes from blood vessels into the surrounding tissue during injury. The steps in the two pathways are identical from factor X on *(in blue–gray)*.

FIGURE FOCUS: *Identify two ways that inactive factor X can be activated.*

Clot Retraction Once a clot is formed, **clot retraction** occurs as platelets trapped within the clot contract and shrink the fibrin mesh, pulling the edges of the damaged vessel closer together. During clot retraction, fluid is squeezed from the clot. This fluid, which is essentially plasma minus fibrinogen and other clotting precursors that have been removed during the clotting process, is called **serum**.

Fibrinolytic plasmin dissolves clots.

A clot is not meant to be a permanent solution to vessel injury. It is a transient device to stop bleeding until the vessel can be repaired.

Vessel Repair The aggregated platelets secrete a chemical that helps promote the invasion of fibroblasts ("fiber formers") from the surrounding connective tissue into the wounded area of the vessel. Fibroblasts form a scar at the vessel defect.

Clot Dissolution Simultaneous with the healing process, the clot, which is no longer needed to prevent hemorrhage, is slowly dissolved by a fibrinolytic (fibrin-splitting) enzyme called **plasmin**. If clots were not removed after they performed their hemostatic function, their accumulation would eventually obstruct the vessels, especially the small ones that regularly endure tiny ruptures. Plasmin, like the clotting factors, is a plasma protein produced by the liver and present in the blood in an inactive precursor form, **plasminogen**. Plasmin is activated in a fast cascade of reactions involving many factors, among them factor XII (Hageman factor), which also triggers the rapid chain reaction leading to clot formation (Figure 11-14). Fast-activated plasmin becomes trapped in the clot and later dissolves it by slowly breaking down the fibrin meshwork.

Phagocytic white blood cells gradually remove the products of clot dissolution. You have observed the slow removal of blood that has clotted after escaping into the tissue layers of your skin following an injury. The black-and-blue marks of bruised skin result from deoxygenated clotted blood within the skin; this blood is eventually cleared by plasmin action, followed by the phagocytic "cleanup crew."

Preventing Inappropriate Clot Formation In addition to removing clots that are no longer needed, plasmin functions continually to prevent clots from forming inappropriately. Throughout the vasculature, small amounts of fibrinogen are constantly being converted into fibrin, triggered by unknown mechanisms. Clots do not develop, however, because the fibrin is quickly disposed of by plasmin activated by **tissue plasminogen activator (tPA)** from the tissues, especially the lungs. Normally, the low level of fibrin formation is counterbalanced by a low level of fibrinolytic activity, so inappropriate clotting does not occur. Only when a vessel is damaged do additional factors precipitate the explosive chain reaction that leads to more extensive fibrin formation and results in local clotting at the site of injury.

Clinical Note Genetically engineered tPA and other similar chemicals that trigger clot dissolution are frequently used to limit damage to cardiac muscle during heart attacks. Administering a clot-busting drug within the first hours after a clot has blocked a coronary (heart) vessel often dissolves the clot in time to restore blood flow to the cardiac muscle supplied by the blocked

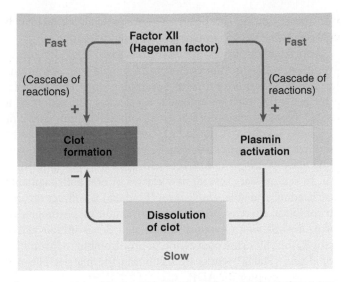

Figure 11-14 Role of factor XII in clot formation and dissolution. Activation of factor XII (Hageman factor) simultaneously initiates a fast cascade of reactions that result in clot formation and a fast cascade of reactions that result in plasmin activation. Plasmin, which is trapped in the clot, subsequently slowly dissolves the clot. This action removes the clot when it is no longer needed after the vessel has been repaired.

vessel before the muscle dies of O_2 deprivation. tPA and related drugs can also be administered to promptly dissolve a stroke-causing clot within a cerebral (brain) blood vessel within 3 hours of onset of symptoms to minimize loss of irreplaceable brain tissue.

Inappropriate clotting produces thromboembolism.

Clinical Note Despite protective measures, clots occasionally form in intact vessels. Abnormal or excessive clot formation within blood vessels—that is, hemostasis in the wrong place—can compromise blood flow to vital organs. The body's clotting and anticlotting systems normally function in a check-and-balance manner. Acting in concert, they permit prompt formation of "good" blood clots, thus minimizing blood loss from damaged vessels, while preventing "bad" clots from forming and blocking blood flow in intact vessels. To review, an abnormal intravascular clot attached to a vessel wall is known as a **thrombus**, and freely floating clots are called **emboli** (singular, *embolus*). An enlarging thrombus narrows and can eventually completely occlude the vessel in which it forms. By entering and completely plugging a smaller vessel, a circulating embolus can suddenly block blood flow (see Figure 9-28, p. 330).

Several factors, acting independently or simultaneously, can cause *thromboembolism*: (1) Roughened vessel surfaces associated with atherosclerosis can lead to thrombus formation (see p. 327). (2) Imbalances in the clotting–anticlotting systems can trigger clot formation. (3) Slow-moving blood is more apt to clot, probably because small quantities of fibrin accumulate in the stagnant blood, for example, in blood pooled in varicosed leg veins (see p. 364) or pooled in ineffectively pumping atria during atrial fibrillation (see p. 313). (4) Widespread clotting is occasionally triggered by release of tissue thromboplastin into the blood from large amounts of traumatized tissue. Similar widespread clotting can occur in **septicemic shock**, in which bacteria or their toxins initiate the clotting cascade.

Anticoagulant drugs ("blood thinners") are sometimes given to prevent thromboembolism in people with conditions that make them more prone to developing clots. For example, *heparin*, which must be injected, accelerates the action of a normal blood-borne inhibitor of thrombin, halting this clot promoter. *Warfarin* (Coumadin), which can be taken orally, interferes with vitamin K's action. Vitamin K, commonly known as the *blood-clotting vitamin*, is essential for normal clot formation. In recent years several new classes of oral anticoagulants have become available. *Dabigatran* (Pradaxa) is a direct thrombin inhibitor, thereby preventing conversion of fibrinogen to fibrin. *Rivaroxaban* (Xarelto) and related drugs inhibit active factor X, thus blocking conversion of prothrombin to thrombin. *Clopidogrel* (Plavix) works by inhibiting the platelet plasma-membrane receptor for ADP.

Hemophilia is the primary condition that produces excessive bleeding.

Clinical Note In contrast to inappropriate clot formation in intact vessels, the opposite hemostatic disorder is failure of clots to form promptly in injured vessels, resulting in life-threatening hemorrhage from even relatively mild traumas. The most common cause of excessive bleeding is **hemophilia** resulting from a deficiency of one of the factors in the clotting cascade (low or no factor VIII in 80% of cases).

People with a platelet deficiency, in contrast to the more profuse bleeding that accompanies defects in the clotting mechanism, continuously develop hundreds of small, confined hemorrhagic areas throughout the body as blood leaks from tiny breaks in the small blood vessels before coagulation takes place. Platelets normally are the primary sealers of these ever-occurring minute ruptures. In the skin of a platelet-deficient person, the diffuse capillary hemorrhages are visible as small, purplish blotches, giving rise to the term **thrombocytopenia purpura** ("the purple of thrombocyte deficiency") for this condition. (Recall that *thrombocyte* is another name for *platelet*.)

Vitamin K deficiency can also cause a bleeding tendency because of incomplete activation of vitamin K–dependent clotting factors.

Check Your Understanding 11.4

1. Describe how a platelet plug forms at the site of a vessel defect.
2. Draw a flow diagram showing the intrinsic and extrinsic pathways of the clotting cascade.
3. Explain how a clot is dissolved after it is no longer needed.

Homeostasis: Chapter in Perspective

 Blood contributes to homeostasis in a variety of ways. First, the composition of interstitial fluid, the true internal environment that surrounds and directly exchanges materials with the cells, depends on the composition of blood plasma. Because of the thorough exchange that occurs between the interstitial and vascular compartments, interstitial fluid has the same composition as plasma with the exception of plasma proteins, which cannot escape through the capillary walls. Thus, blood serves as the vehicle for rapid, long-distance, mass transport of materials to and from the cells, and interstitial fluid serves as the go-between.

Homeostasis depends on the blood carrying essential supplies such as O_2 and nutrients to the cells as rapidly as the cells consume them and carrying materials such as metabolic wastes away from the cells as rapidly as the cells produce these products. It also depends on the blood carrying hormonal messengers from their site of production to their distant site of action. Once a substance enters the blood, it can be transported throughout the body within seconds, whereas diffusion of the substance over long distances in a large multicellular organism such as a human would take months to years—a situation incompatible with life. Diffusion can, however, effectively accomplish short local exchanges of materials between blood and surrounding cells through the intervening interstitial fluid.

Blood has special transport capabilities that enable it to move its cargo efficiently throughout the body. For example, life-sustaining O_2 is poorly soluble in water, but blood is equipped with O_2-carrying specialists, the erythrocytes (red blood cells), which are stuffed full of hemoglobin, a complex molecule that transports O_2. Likewise, homeostatically important water-insoluble hormonal messengers are shuttled in the blood by plasma protein carriers.

Specific components of the blood perform the following additional homeostatic activities unrelated to blood's transport function:

- Blood helps maintain the proper pH in the internal environment by buffering changes in the body's acid–base load.

- Blood helps maintain body temperature by absorbing heat produced by heat-generating tissues such as contracting skeletal muscles and distributing it throughout the body. Excess heat is carried by blood to the body surface for elimination to the external environment.

- Electrolytes (mostly Na^+ and K^+) carried by plasma are important in membrane excitability, which is critical for nerve and muscle function.

- Plasma electrolytes (mostly Na^+ and Cl^-) are important in osmotic distribution of water between the ECF and ICF. Plasma proteins play a critical role in distributing the ECF between the plasma and interstitial fluid.

- Through their hemostatic functions the platelets and clotting factors minimize the loss of life-sustaining blood after vessel injury.

- The leukocytes (white blood cells), their secretory products, and certain types of plasma proteins, such as antibodies, constitute the immune defense system. This system defends the body against invading disease-causing agents, paves the way for wound healing and tissue repair by clearing away debris from dead or injured cells, and destroys cancer cells. These actions indirectly contribute to homeostasis by helping organs that directly maintain homeostasis stay healthy. We could not survive beyond early infancy were it not for the body's defense mechanisms.

Review Exercises Answers begin on p. A-35

Reviewing Terms and Facts

1. Blood can absorb metabolic heat while undergoing only small changes in temperature. (True or false?)

2. Hemoglobin can carry only O_2. (True or false?)

3. Erythrocytes, leukocytes, and platelets all originate from the same undifferentiated stem cells. (True or false?)

4. Erythrocytes are unable to use the O_2 they contain for their own ATP formation. (True or false?)

5. White blood cells spend most of their time in the blood. (True or false?)

6. The type of leukocyte produced primarily in lymphoid tissue is _____.

7. Most clotting factors are synthesized by the _____.

8. Which of the following is *not* a function of plasma proteins?
   a. facilitating retention of fluid in the blood vessels
   b. playing an important role in blood clotting
   c. transporting water-insoluble substances in the blood
   d. transporting O_2 in the blood
   e. serving as antibodies
   f. contributing to the buffering capacity of the blood

9. Which of the following is *not* directly triggered by exposed collagen in an injured vessel?
   a. initial vascular spasm
   b. platelet aggregation
   c. activation of the clotting cascade
   d. activation of plasminogen

10. Match the following (an answer may be used more than once):
   1. causes platelets to aggregate in positive-feedback fashion
   2. activates prothrombin
   3. fibrinolytic enzyme
   4. inhibits platelet aggregation
   5. first factor activated in intrinsic clotting pathway
   6. forms meshwork of the clot
   7. stabilizes the clot
   8. activates fibrinogen
   9. activated by tissue thromboplastin

   (a) prostacyclin
   (b) plasmin
   (c) ADP
   (d) fibrin
   (e) thrombin
   (f) factor X
   (g) factor XII
   (h) factor XIII

11. Match the following blood abnormalities with their causes:
   1. deficiency of intrinsic factor
   2. insufficient amount of iron to synthesize adequate hemoglobin
   3. destruction of bone marrow
   4. abnormal loss of blood
   5. tumorlike condition of bone marrow
   6. inadequate erythropoietin secretion
   7. excessive rupture of circulating erythrocytes
   8. associated with living at high altitudes

   (a) hemolytic anemia
   (b) aplastic anemia
   (c) nutritional anemia
   (d) hemorrhagic anemia
   (e) pernicious anemia
   (f) renal anemia
   (g) primary polycythemia
   (h) secondary polycythemia

Understanding Concepts
(Answers at www.cengagebrain.com)

1. What is the average blood volume in women and in men? What is the normal percentage of blood occupied by erythrocytes and by plasma in women and in men? What is the hematocrit? What is the buffy coat?

2. What is the composition of plasma?

3. List the three major groups of plasma proteins, and state their functions.

4. Describe the structure and functions of erythrocytes.

5. Why can erythrocytes survive for only about 120 days?

6. Describe the process and control of erythropoiesis.

7. List the antigens and antibodies present in the blood of the four major ABO blood types. Describe what happens in a transfusion reaction. Does the recipient have to have had prior exposure to incompatible blood of the ABO type for a transfusion reaction to occur? Does the recipient have to have had prior exposure to incompatible blood of the Rh type for a transfusion reaction to occur? Explain.

8. Compare the structure, functions, and life spans of the five types of leukocytes.

9. Discuss the derivation of platelets.

10. Describe the three steps of hemostasis, including a comparison of the intrinsic and extrinsic pathways by which the clotting cascade is triggered.

11. Compare plasma and serum.

12. What normally prevents inappropriate clotting in vessels?

Solving Quantitative Exercises

1. The normal concentration of hemoglobin in blood (as measured clinically) is 15 g/100 mL of blood.

   a. Given that 1 mole of hemoglobin weighs 66,000 grams, what is the concentration of hemoglobin in millimoles (mM)?

   b. Each hemoglobin molecule can bind four molecules of O_2. What is the concentration of O_2 bound to hemoglobin at maximal saturation (in mM)?

   c. Given that 1 mole of an ideal gas occupies 22.4 liters, what is the maximal carrying capacity of normal blood for O_2 (usually expressed in mL of O_2/liter of blood)?

2. Assume that the blood sample in ❙Figure 11-5b, p. 387, is from a patient with hemorrhagic anemia. Given a normal blood volume of 5 liters, a normal red blood cell concentration of 5 billion/mL, and an RBC production rate of 3 million cells/second, how long will it take the body to return the hematocrit to normal?

3. Note that in the blood sample in ❙Figure 11-5c from a patient with polycythemia, the hematocrit has increased to 70%. An increased hematocrit increases blood viscosity, which in turn increases total peripheral resistance and increases the workload on the heart. The effect of hematocrit (h) on relative blood viscosity (v, viscosity relative to that of water) is given approximately by the following equation:

$$v = 1.5 \times \exp(2h)$$

Note that in this equation, h is the hematocrit as a fraction, not a percentage. Given a normal hematocrit of 0.40, what percent increase in viscosity would result from the polycythemia in ❙Figure 11-5c? What percent change would this cause in total peripheral resistance?

Applying Clinical Reasoning

Heather L., who has Rh-negative blood, has just given birth to her first child, who has Rh-positive blood. Both mother and baby are fine, but the doctor administers an Rh immunoglobulin preparation (short-lived antibodies against the Rh factor antigen) so that any future Rh-positive babies Heather has will not suffer from erythroblastosis fetalis (hemolytic disease of the newborn) (see p. 389). During gestation (pregnancy), fetal and maternal blood do not mix. Instead, materials are exchanged between these two circulatory systems across the placenta, a special organ that develops during gestation from both maternal and fetal structures (see p. 757). Red blood cells are unable to cross the placenta, but antibodies can cross. During the birthing process, a small amount of the infant's blood may enter the maternal circulation.

1. Why did Heather's first-born child not have erythroblastosis fetalis—that is, why didn't maternal antibodies against the Rh factor attack the fetal Rh-positive RBCs during gestation?

2. Why would any subsequent Rh-positive babies Heather might carry be likely to develop erythroblastosis fetalis if she were not treated with Rh immunoglobulin?

3. How would administering Rh immunoglobulin immediately following Heather's first pregnancy with an Rh-positive child prevent erythroblastosis fetalis in a later pregnancy with another Rh-positive child? Similarly, why must Rh immunoglobulin be given to Heather after the birth of each Rh-positive child she bears?

4. Suppose Heather were not treated with Rh immunoglobulin after the birth of her first Rh-positive child, and a second Rh-positive child developed erythroblastosis fetalis. Would administering Rh immunoglobulin to Heather immediately after the second birth prevent this condition in a third Rh-positive child? Why or why not?

Thinking at a Higher Level

1. A person has a hematocrit of 62. Can you conclude from this finding that the person has polycythemia? Explain.

2. There are different forms of hemoglobin. *Hemoglobin A* is normal adult hemoglobin, which has two α chains and two β chains. The abnormal form *hemoglobin S*, which has a single "typographical error" in the genetic code for the β chains, causes RBCs to warp into fragile, sickle-shaped cells (see p. 387). Fetal RBCs contain *hemoglobin F*, the production of which stops soon after birth. Hemoglobin F contains two γ chains instead of two β chains, along with the two α chains. This substitution increases hemoglobin's affinity for O_2. Now researchers are trying to goad the genes that direct hemoglobin F synthesis back into action as a means of treating sickle cell anemia. Explain how turning on these fetal genes could be a useful remedy. (Indeed, the first effective drug therapy for treating sickle cell anemia, *hydroxyurea*, acts on the bone marrow to boost production of fetal hemoglobin.)

3. What types of blood in the ABO and Rh blood systems could safely be transfused into a person with Type A+ blood?

4. Low on the list of popular animals are vampire bats, leeches, and ticks, yet these animals may someday indirectly save your life. Scientists are currently examining the "saliva" of these blood-sucking creatures in search of new chemicals that might limit cardiac muscle damage in heart attack victims. What do you suspect the nature of these sought-after chemicals is?

5. *Porphyria* is a genetic disorder that shows up in about one in every 25,000 individuals. Affected individuals lack certain enzymes that are part of a metabolic pathway leading to formation of heme, which is the iron-containing group of hemoglobin. An accumulation of porphyrins, which are intermediates of the pathway, causes a variety of symptoms, especially after exposure to sunlight. Lesions and scars form on the skin. Hair grows thickly on the face and hands. As gums retreat from teeth, the elongated canine teeth take on a fanglike appearance. Symptoms worsen on exposure to a variety of substances, including garlic and alcohol. Affected individuals avoid sunlight and aggravating substances and get injections of heme from normal red blood cells. If you are familiar with vampire stories, which date from the Middle Ages or earlier, speculate on how they may have evolved among superstitious folk who did not have medical knowledge of porphyria.

To access the course materials and companion resources for this text, please visit
www.cengagebrain.com

12 Body Defenses

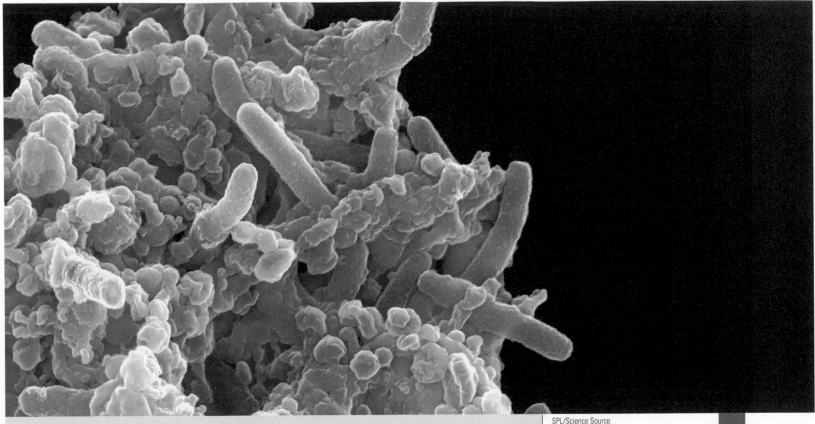

A scanning electron micrograph of a macrophage at work. A macrophage (*green*), a large tissue phagocyte, extends projections to surround and engulf tuberculosis bacteria (*purple*). After internalizing the bacteria, the macrophage will destroy them by lysosomal enzymes. The macrophage will also present bacterial antigens (protein fragments) on its surface that will activate specific helper lymphocytes, setting in motion events leading to production of antibodies against the tuberculosis bacteria.

Homeostasis Highlights

Humans constantly come into contact with external agents that could be harmful if they entered the body. The most serious are disease-causing microorganisms. If bacteria or viruses do gain entry, the body is equipped with a complex, multifaceted internal defense system—the **immune system**—that provides continual protection against invasion by foreign agents. Furthermore, body surfaces exposed to the external environment, such as the **integumentary system (skin)**, serve as a first line of defense to resist penetration by foreign microorganisms. The immune system also protects against cancer and paves the way for repair of damaged tissues.

The immune system indirectly contributes to homeostasis by helping maintain the health of organs that directly contribute to homeostasis.

12.1 Immune System: Targets, Effectors, Components

Immunity is the body's ability to protect itself by resisting or eliminating potentially harmful foreign invaders (such as bacteria and viruses) or abnormal cells (such as cancer cells). The first line of defense against foreign invaders is the epithelial barriers that surround the outer surface of the body (the skin) and line the body cavities (such as the digestive tract and lungs) that are in contact with the external environment. These epithelial barriers are not part of the immune system. We will discuss their roles in the body's overall defense mechanisms after examining the immune system in detail. As an overview, the following activities are attributable to the immune system, an internal defense system that plays a key role in recognizing and either destroying or neutralizing things that are not "normal self":

1. Defending against invading **pathogens** (disease-producing microorganisms)

2. Removing worn-out cells and tissue damaged by trauma or disease, paving the way for wound healing and tissue repair

3. Identifying and destroying abnormal cancer cells that have originated in the body, a function termed *immune surveillance*

Pathogenic bacteria and viruses are the major targets of the immune system.

The primary foreign enemies against which the immune system defends are bacteria and viruses. Comparing their relative sizes, if an average bacterium were the size of a pitcher's mound, a virus would be the size of a baseball. **Bacteria** are nonnucleated, single-celled microorganisms self-equipped with all the machinery essential for their survival and reproduction. Pathogenic bacteria that invade the body cause tissue damage and produce disease largely by releasing enzymes or toxins that physically injure or functionally disrupt affected cells and organs. The disease-producing power of a pathogen is known as its **virulence**.

In contrast to bacteria, **viruses** are not self-sustaining cellular entities. They consist only of nucleic acids (genetic material—DNA or RNA) enclosed by a protein coat. Because they lack cellular machinery for energy production and protein synthesis, viruses cannot carry out metabolism and reproduce unless they invade a **host cell** (a body cell of the infected individual) and take over the cell's biochemical facilities for their own uses. Not only do viruses sap the host cell's energy resources, but the viral nucleic acids also direct the host cell to synthesize proteins needed for viral replication.

When a virus becomes incorporated into a host cell, the body's defense mechanisms may destroy the cell because they no longer recognize it as a normal-self cell. Other ways in which viruses can lead to cell damage or death are by depleting essential cell components, dictating that the cell produce substances toxic to the cell, or transforming the cell into a cancer cell.

Leukocytes are the effector cells of the immune system.

Leukocytes (white blood cells, or WBCs) and their derivatives, along with a variety of plasma proteins, are responsible for immune defense.

Leukocyte Functions As a brief review, the functions of the five types of leukocytes are as follows (see pp. 392–393 and ▮Figure 11-8):

1. **Neutrophils** are highly mobile phagocytic specialists that engulf and destroy unwanted materials.

2. **Eosinophils** secrete chemicals that destroy parasitic worms and are involved in allergic reactions.

3. **Basophils** release histamine and heparin and also are involved in allergic reactions.

4. **Monocytes** are transformed into **macrophages**, which are large, tissue-bound phagocytic specialists.

5. **Lymphocytes** are of two types:

   a. **B lymphocytes (B cells)** produce antibodies that indirectly lead to the destruction of foreign material (antibody-mediated immunity).

   b. **T lymphocytes (T cells)** directly destroy virus-invaded cells and mutant cells by releasing chemicals that punch lethal holes in the victim cells (cell-mediated immunity).

A given leukocyte is present in the blood only transiently. Most leukocytes are out in the tissues on defense missions.

Lymphoid Tissues Almost all leukocytes originate from common precursor stem cells in the bone marrow and are subsequently released into the blood. The only exception is lymphocytes, which arise in part from lymphocyte colonies in various lymphoid tissues originally populated by cells derived from bone marrow (see ▮Figure 11-9, p. 394).

Lymphoid tissues, collectively, are the tissues that produce, store, or process lymphocytes. These include the bone marrow, lymph nodes, spleen, thymus, tonsils, adenoids, appendix, and aggregates of lymphoid tissue in the lining of the digestive tract called **Peyer's patches** or **gut-associated lymphoid tissue (GALT)** (▮Figure 12-1 and ▮Table 12-1). Lymphoid tissues are strategically located to intercept invading microorganisms before they have a chance to spread very far. For example, lymphocytes populating the *tonsils* and *adenoids* are situated advantageously to respond to inhaled microbes, whereas microorganisms invading through the digestive system immediately encounter lymphocytes in the *appendix* and GALT. Potential pathogens that gain access to lymph are filtered through *lymph nodes,* where they are exposed to lymphocytes as well as to macrophages that line lymphatic passageways. The *spleen*, the largest lymphoid tissue, performs immune functions on blood similar to those that lymph nodes perform on lymph. Through actions of its lymphocyte and macrophage population, the spleen clears blood that passes through it of microorganisms and other foreign matter and also removes worn-out red blood cells (see p. 384). The *thymus* and *bone marrow* play important roles in processing T and B lymphocytes, respectively, to prepare them to carry out their specific immune strategies.

We now turn attention to the two major components of the immune system's response to foreign invaders and other targets—innate and adaptive immune responses. In the process, we will further examine the roles of each type of leukocyte.

Immune responses may be either innate and nonspecific or adaptive and specific.

Protective immunity is conferred by the complementary actions of two separate but interdependent components of the immune system: the innate immune system and the adaptive, or acquired, immune system. The responses of these two systems differ in timing and in the selectivity of the defense mechanisms.

The **innate immune system** encompasses the body's *nonspecific* immune responses that come into play immediately on exposure to a threatening agent. These nonspecific responses are inherent (innate or built-in) defense mechanisms that nonselectively defend against foreign or abnormal material of any type, even on initial exposure to it. Such responses provide a first line of internal

defense against a range of threats, including infectious agents, chemical irritants, and tissue injury. Everyone is born with essentially the same innate immune-response mechanisms. The **adaptive**, or **acquired**, **immune system**, in contrast, relies on *specific* immune responses selectively targeted against a particular foreign material to which the body has already been exposed and has had an opportunity to prepare for a discriminating attack on the enemy. The adaptive immune system thus takes considerably more time to mount and takes on specific foes. The innate and adaptive immune systems work in harmony to contain, and then eliminate, harmful agents.

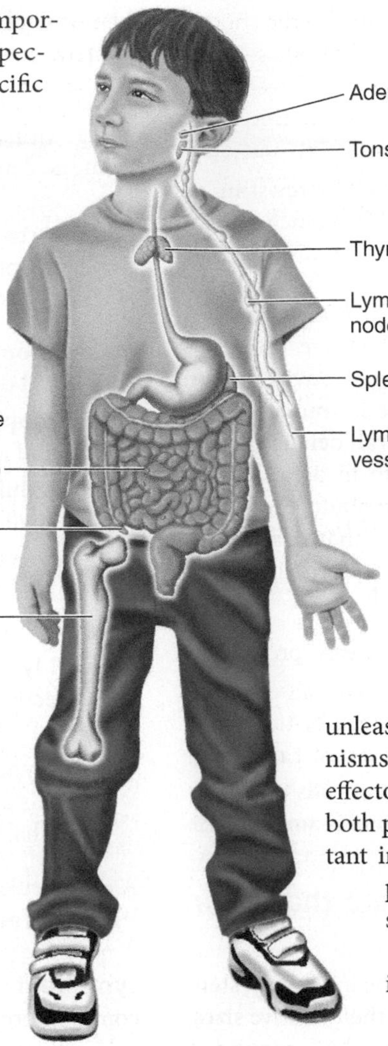

Innate Immune System The innate system is always on guard, ready to unleash a limited repertoire of defense mechanisms at any and every invader. Of the immune effector cells, the neutrophils and macrophages—both phagocytic specialists—are especially important in innate defense. Several groups of plasma proteins also play key roles, as you will see shortly.

Nonspecific immune responses are set in motion on exposure to generic molecular patterns associated with threatening agents. Two categories of patterns call forth the innate

Figure 12-1 **Lymphoid tissues.**

Labels: Adenoid · Tonsil · Thymus · Lymph node · Spleen · Lymphatic vessel · Peyer's patches in small intestine (gut-associated lymphoid tissue) · Appendix · Bone marrow

▮TABLE 12-1 Functions of Lymphoid Tissues

Lymphoid Tissue	Functions
Bone marrow	Origin of all blood cells
	Site of maturational processing for B lymphocytes
Lymph nodes, tonsils, adenoids, appendix, gut-associated lymphoid tissue	Exchange lymphocytes with the lymph (remove, store, produce, and add them)
	Resident lymphocytes produce antibodies and activated T cells, which are released into the lymph
	Resident macrophages remove microbes and other particulate debris from the lymph
Spleen	Exchanges lymphocytes with the blood (removes, stores, produces, and adds them)
	Resident lymphocytes produce antibodies and activated T cells, which are released into the blood
	Resident macrophages remove microbes and other particulate debris, most notably worn-out red blood cells, from the blood
	Stores a small percentage of red blood cells, which can be added to the blood by splenic contraction as needed
Thymus	Site of maturational processing for T lymphocytes
	Secretes the hormone thymosin

response: exogenous (originating from outside the body) **pathogen-associated molecular patterns (PAMPs)**, such as the carbohydrates typically found in bacterial cell walls but not found in human cells, and endogenous (originating from within the body) **damage-associated molecular patterns (DAMPs;** also known as **alarmins)**, such as extracellular adenosine triphosphate (ATP) released from trauma-damaged cells. Both patterns trigger identical innate pathways leading to inflammation, a multistep process involving phagocytic removal of offending agents and tissue debris and promoting tissue repair. In addition, the response to PAMPs includes augmenting adaptive immunity. We will focus on the response called forth by PAMPs. The responding phagocytic cells possess *pattern recognition receptors* on their surface membrane or in their cytosol for detecting the patterns associated with threatening agents. For example, phagocytes are studded with plasma membrane proteins known as **toll-like receptors (TLRs)**, which recognize PAMPs. About a dozen distinct TLRs each recognize a different specific set of molecular patterns. For example, some recognize gram-positive bacteria (bacteria that can be stained with Gram stain), others recognize gram-negative bacteria (bacteria whose cell wall does not pick up Gram stain), still others recognize viral DNA or RNA, and so on. TLRs have been dubbed the "eyes of the innate immune system" because these immune sensors recognize and bind with the unique, telltale pathogen markers, allowing the effector cells of the innate system to "see" pathogens as distinct from normal-self cells.

A TLR's recognition of a pathogen triggers the phagocyte to engulf and destroy the infectious microorganism. Moreover, activation of the TLR induces the phagocytic cell to secrete chemicals, some of which contribute to inflammation. TLRs link the innate and adaptive branches of the immune system because still other chemicals secreted by the phagocytes recruit cells of the adaptive immune system. Because of their pivotal role in the immune system, TLRs are targets for many new drugs and vaccines under development.

As another link between the innate and adaptive branches of the immune system, foreign particles are deliberately marked for phagocytic ingestion by being coated with antibodies produced by the B cells of the adaptive immune system. These are but a few examples of how various components of the immune system are highly interactive and interdependent. The most significant cooperative relationships among the immune effectors are pointed out throughout this chapter.

TLRs function at the cell surface to recognize pathogens in the extracellular fluid (ECF), but most viruses are hidden inside host cells instead of being free in the ECF. Scientists recently discovered intracellular pattern recognition receptors, including several **RLRs,**[1] which recognize viral DNA or RNA within the cytosol, and multiple **NLRs,**[2] which distinguish intracellular PAMPs such as bits of bacterial cell wall engulfed by a phago-

cyte or parasites that have invaded a nonimmune cell. Appropriately, activation of RLRs by viral genetic material triggers synthesis of interferon, an innate mechanism to be described later that defends against viral invasion. Activated NLRs trigger formation of cytosolic, multiprotein complexes termed **inflammasomes**, of which the NLRs themselves are a part. Inflammasomes bring about a potent inflammatory response, complementing the actions triggered by activated TLRs. Thus, the trinity of pathogen receptors—TLRs, RLRs, and NLRs—cooperate to ensure efficient innate immune responses against pathogenic interlopers.

The innate mechanisms give us all a rapid but limited and nonselective response to unfriendly challenges of all kinds, much like medieval guardsmen lashing out with brute-force weapons at any enemy approaching the walls of the castle they are defending. Innate immunity largely contains and limits the spread of infection. These nonspecific responses are important for keeping the foe at bay until the adaptive immune system, with its highly selective weapons, can be prepared to take over and mount strategies to eliminate the villain.

Adaptive Immune System Unlike the innate immune system, the adaptive immune system customizes its defenses for specific pathogens. The responses of the adaptive immune system are mediated by the B and T lymphocytes. Each B and T cell can recognize and defend against only one particular type of foreign material, such as one kind of bacterium. Among the millions of B and T cells in the body, only the ones specifically equipped to recognize the unique molecular features of a particular infectious agent are called into action to discriminatingly defend against this agent. This specialization is similar to modern, specially trained military personnel called into active duty to accomplish a specific task. The chosen lymphocytes multiply, expanding the pool of specialists that can launch a highly targeted attack against the invader.

The adaptive immune system is the ultimate weapon against most pathogens. The repertoire of activated and expanded B and T cells constantly changes, or adapts, to wage battle against the specific pathogens encountered in each person's environment. Thus, the targets of the adaptive immune system vary among people, depending on the types of immune assaults each individual meets. Furthermore, this system acquires an ability to more efficiently eradicate a particular foe when rechallenged by the same pathogen in the future. It does so by establishing a pool of memory cells as a result of an encounter with a given pathogen so that, when later exposed to the same agent, it can more swiftly defend against the invader.

Next we examine in more detail the innate immune responses before looking more closely at adaptive immunity.

Check Your Understanding 12.1

1. Define *immunity*.

2. Explain why innate immune responses are considered nonspecific and adaptive immune responses are said to be specific.

3. Distinguish among PAMPs, DAMPs, TLRs, RLRs, and NLRs.

---

[1] RLRs stands for *retinoic acid inducible gene I (RIG-I)-like receptors*
[2] NLRs stands for *nucleotide-binding oligomerization domain (NOD)-like receptors*

12.2 Innate Immunity

Innate defenses include the following:

1. *Inflammation,* a nonspecific response to tissue injury in which the phagocytic specialists—neutrophils and macrophages—play a major role

2. *Interferon,* a family of proteins that nonspecifically defend against viral infection

3. *Natural killer cells,* a special class of lymphocytelike cells that spontaneously and nonspecifically lyse (rupture) and thereby destroy virus-infected host cells and cancer cells

4. The *complement system,* a group of inactive plasma proteins that, when sequentially activated, bring about destruction of foreign cells by attacking their plasma membranes

We now elaborate on each of these mechanisms in order.

Inflammation is a nonspecific response to foreign invasion or tissue damage.

Inflammation is an innate, nonspecific series of interrelated events set into motion in response to foreign invasion, tissue damage, or both. The goal of inflammation is to bring to the invaded or injured area phagocytes and plasma proteins that can (1) isolate, destroy, or inactivate invaders; (2) remove debris; and (3) prepare for subsequent healing and repair. The overall inflammatory response is remarkably

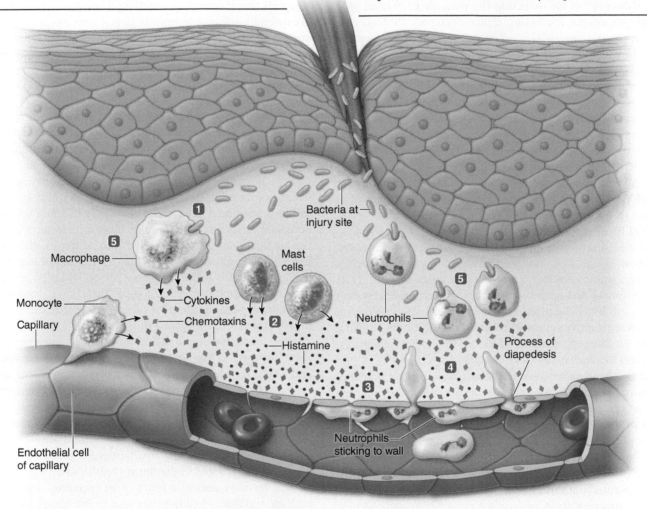

1 A break in the skin introduces bacteria, which reproduce at the wound site. Activated resident macrophages engulf the pathogens and secrete chemotaxins and other cytokines.

2 Activated mast cells release histamine.

3 Histamine dilates local blood vessels and widens capillary pores. Some cytokines cause neutrophils and monocytes to stick to the blood vessel wall.

4 Chemotaxins attract neutrophils and monocytes, which squeeze out between cells of the blood vessel wall, a process called diapedesis, and migrate to the infection site.

5 Monocytes enlarge into macrophages. Newly arriving macrophages and neutrophils engulf the pathogens and destroy them.

❙ Figure 12-2 Steps producing inflammation. Chemotaxins released at the site of damage attract phagocytes to the scene. Note the leukocytes emigrating from the blood into the tissues by assuming amoebalike behavior and squeezing through the capillary pores. Mast cells secrete vessel-dilating, pore-widening histamine. Macrophages secrete cytokines that exert multiple local and systemic effects.

FIGURE FOCUS: *Use the figure (1) to identify the first immune cells to defend at a site of injury and/or infection, and (2) to describe how neutrophils and monocytes exit the blood to enter the site.*

similar no matter what the triggering event (bacterial invasion, chemical injury, or mechanical trauma), although some subtle differences may be evident, depending on the injurious agent or the site of damage. The following sequence of events typically occurs during inflammation. As an example, we use bacterial entry into a break in the skin (Figure 12-2).

Defense by Resident Tissue Macrophages When bacteria invade through a break in the external barrier of skin (or enter through another avenue), macrophages already in the area immediately begin phagocytizing the foreign microbes, defending against infection during the first hour or so, before other mechanisms can be mobilized (see chapter opener photo). Resident macrophages also secrete chemical mediators, or *cytokines,* that exert various immune responses, as will be described shortly (Figure 12-2, step **1**).

Localized Vasodilation Almost immediately on microbial invasion, arterioles within the area dilate, increasing blood flow to the site of injury. This localized vasodilation is mainly induced by **histamine** released from mast cells in the area of tissue damage (the connective tissue–bound "cousins" of circulating basophils; see p. 393) (steps **2** and **3**). Increased local delivery of blood brings to the site more phagocytic leukocytes and plasma proteins, both crucial to the defense response.

Increased Capillary Permeability Released histamine also increases the capillaries' permeability by enlarging the capillary pores (the spaces between the endothelial cells) so that plasma proteins normally prevented from leaving the blood can escape into the inflamed tissue (see p. 354).

Localized Edema Accumulation of leaked plasma proteins in the interstitial fluid raises the local interstitial fluid–colloid osmotic pressure. Furthermore, the increased local blood flow elevates capillary blood pressure. Because both these pressures tend to move fluid out of the capillaries, these changes favor enhanced ultrafiltration and reduced reabsorption of fluid across the involved capillaries. The end result of this shift in fluid balance is localized edema (see p. 359). Thus, the familiar

swelling that accompanies inflammation is the result of histamine-induced vascular changes. Likewise, the other well-known gross manifestations of inflammation, such as redness and heat, are largely caused by the enhanced flow of warm arterial blood to the damaged tissue (*inflammare* means "to set on fire"). Pain is caused both by local distension within the swollen tissue and by the direct effect of locally produced substances on the receptor endings of afferent neurons that supply the area. These observable characteristics of the inflammatory process (swelling, redness, heat, and pain) are coincidental to the primary purpose of the vascular changes in the injured area—to increase the number of leukocytic phagocytes and crucial plasma proteins in the area (Figure 12-3).

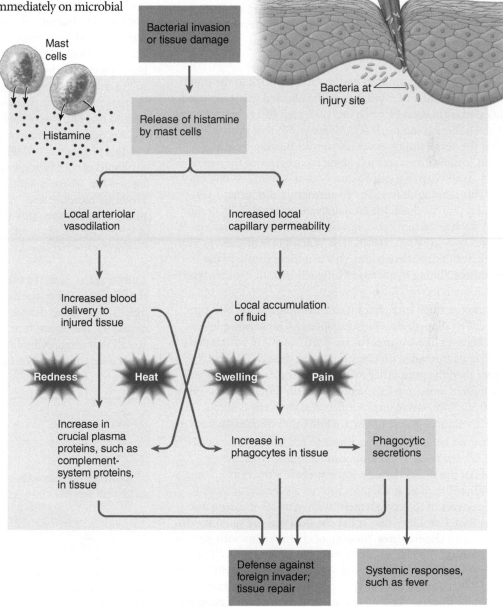

Figure 12-3 Gross manifestations and outcomes of inflammation.
FIGURE FOCUS: *Trace the pathways by which histamine leads to the undesirable gross manifestations of inflammation. Which box in the diagram designates the desirable outcome of inflammation?*

Walling Off the Inflamed Area The leaked plasma proteins most critical to the immune response are those in the complement system and clotting and anticlotting factors. On exposure to tissue thromboplastin in the injured tissue and to specific chemicals secreted by phagocytes on the scene, fibrinogen—the final factor in the clotting system—is converted into fibrin (see p. 397). Fibrin forms interstitial fluid clots in the spaces around the bacterial invaders and damaged cells. This walling off the injured region from the surrounding tissues prevents, or at least delays, the spread of bacterial invaders and their toxic products. Later, the more slowly acting anticlotting factors gradually dissolve the clots after they are no longer needed (see p. 399).

Emigration of Leukocytes Within an hour after injury, the area is teeming with leukocytes that have left the vessels. Neutrophils arrive first, followed during the next 8 to 12 hours by the slower-moving monocytes. The latter swell and mature into macrophages during another 8- to 12-hour period. Once neutrophils or monocytes leave the bloodstream, they never recycle back to the blood. Leukocytes can emigrate from the blood into the tissues via the following steps:

- Blood-borne neutrophils and monocytes stick to the inner endothelial lining of capillaries in the affected tissue, an event called **margination** (see ∎Figure 12-2, step ❸). *Selectins*, a type of cell adhesion molecule (CAM; see p. 60) that protrudes from the vessel lining, cause leukocytes flowing by in the blood to slow down and roll along the interior of the vessel, much as the nap of a carpet slows down a child's rolling toy car. This slowing-down allows neutrophils and monocytes enough time to check for local activating factors—"SOS signals", such as certain cytokines released by resident macrophages in nearby infected tissues. When present, these activating factors cause these leukocytes to adhere firmly to the endothelial lining by means of interaction with another type of CAM, the *integrins*.

- Soon the stuck leukocytes start leaving the vessel by a process called **diapedesis**. During diapedesis, an adhered leukocyte behaves like an amoeba (see p. 50), forming pseudopods to wriggle through a capillary pore even though it is much larger than the pore. After exiting, a leukocyte crawls toward the injured area (step ❹). Neutrophils, the "foot soldiers" on the front lines, arrive on the inflammatory scene earliest, within minutes, because they are more mobile than monocytes.

- How do leukocytes know where to go? These phagocytic cells are attracted to **chemotaxins**, chemical mediators released at the site of damage (*taxis* means "attraction"). Some cytokines function as chemotaxins, as do certain activated components of the complement system, another innate defense tool. Cytokines that act as chemotaxins are often specifically called **chemokines**. Binding of chemotaxins with protein receptors on the plasma membrane of a phagocytic cell increases Ca^{2+} entry into the cell. Calcium, in turn, switches on the cellular contractile apparatus that leads to amoebalike crawling. Because the concentration of chemotaxins progressively increases toward the site of injury, phagocytic cells move unerringly toward this site along a chemotaxin concentration gradient.

Leukocyte Proliferation Resident tissue macrophages and leukocytes that exited from the blood and migrated to the site are soon joined by new phagocytic recruits from the bone marrow. Within a few hours after onset of the inflammatory response, the number of neutrophils in the blood may increase up to five times that of normal. Mobilization of stored neutrophils and proliferation of new neutrophils and monocytes–macrophages are stimulated by various chemical mediators (*colony–stimulating factors;* see p. 394) released from the inflamed region.

Marking of Bacteria for Destruction by Opsonins Phagocytes must be able to distinguish between normal-self cells and foreign or abnormal cells before accomplishing their destructive mission. Otherwise, they could not selectively engulf and destroy only unwanted materials. First, phagocytes, by means of their TLRs, recognize and subsequently engulf infiltrators that have standard bacterial cell wall components not found in human cells. Second, foreign particles are deliberately marked for phagocytic ingestion by being coated with chemical mediators generated by the immune system. Such body-produced chemicals that make bacteria more susceptible to phagocytosis are known as **opsonins** (*opsonin* means "to prepare for eating"). The most important opsonins are antibodies and one of the activated proteins of the complement system.

An opsonin enhances phagocytosis by linking the foreign cell to a phagocytic cell (∎Figure 12-4). One portion of an opsonin molecule binds nonspecifically to the surface of an invading bacterium, whereas another portion of the opsonin molecule binds to receptor sites specific for it on the phagocytic cell's plasma membrane. This link ensures that the bacterial victim does not have a chance to "get away" before the phagocyte can perform its lethal attack.

Leukocytic Destruction of Bacteria Neutrophils and macrophages clear the inflamed area of infectious and toxic agents, and tissue debris, largely by phagocytosis; this clearing action is the main function of the inflammatory response. Recall that phagocytosis involves engulfment of a bacterium or other foreign particle or tissue debris, which becomes enclosed

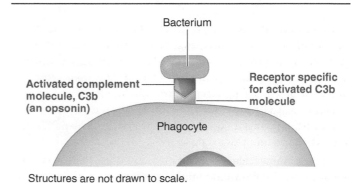

Bacterium

Activated complement molecule, C3b (an opsonin)

Receptor specific for activated C3b molecule

Phagocyte

Structures are not drawn to scale.

∎Figure 12-4 **Mechanism of opsonin action.** One of the activated complement molecules, C3b, links a foreign cell, such as a bacterium, and a phagocytic cell by nonspecifically binding with the foreign cell and specifically binding with a receptor on the phagocyte. This link ensures that the foreign victim does not escape before it can be engulfed by the phagocyte.

in a membrane-bound intracellular vesicle called a *phagosome*. The phagosome fuses with a lysosome and the entrapped material degrades intracellularly by hydrolytic enzymes within the confines of this organelle (see p. 31). Macrophages can engulf a bacterium in less than 0.01 second. Phagocytes eventually die from accumulation of toxic by-products from foreign particle degradation or from inadvertent release of destructive lysosomal chemicals into the cytosol. Neutrophils usually succumb after phagocytizing from 5 to 25 bacteria, whereas macrophages survive much longer and can engulf up to 100 or more bacteria. Indeed, the longer-lived macrophages even clear the area of dead neutrophils in addition to other tissue debris. The **pus** that forms in an infected wound is a collection of these phagocytic cells, both living and dead; necrotic (dead) tissue liquefied by lysosomal enzymes released from the phagocytes; and bacteria.

Mediation of Inflammation by Cytokines Microbe-stimulated phagocytes release many chemicals that function as mediators of inflammation. All chemicals other than antibodies that leukocytes secrete are categorized as **cytokines**, which are protein signal molecules whose function is to regulate immune responses (see p. 118). Macrophages, monocytes, neutrophils, a type of T cell called a helper T cell, and some nonimmune cells such as endothelial cells and fibroblasts (fiber-formers in the connective tissue) all secrete cytokines. More than 100 cytokines have been identified, and the list continues to grow as researchers unravel the complicated chemical means by which immune effector cells communicate with one another to coordinate their activities. Unlike antibodies, cytokines do not interact directly with the antigen (foreign material) that induces their production. Instead, cytokines largely spur other immune cells into action to help ward off the invader. Cytokines typically act locally as paracrines on cells in the vicinity, but some circulate in the blood to exert endocrine effects at distant sites. Some cytokines have names related to their first identified or most important function, examples being specific *colony–stimulating factors*. Other cytokines are designated as specific numbered *interleukins*, such as interleukin 1 (IL-1), interleukin 2 (IL-2), up to IL-38, numbered in order of their discovery (*interleukin* means "between leukocytes"). Cytokines secreted by phagocytes orchestrate a wide variety of independent and overlapping immune activities, the following being among the most important:

1. Some phagocytic secretions are very destructive and directly kill microbes by nonphagocytic means. For example, macrophages secrete *nitric oxide (NO)*, a multipurpose chemical that is toxic to nearby microbes (see p. 346). As a more subtle means of destruction, neutrophils secrete **lactoferrin**, a protein that tightly binds with iron, making it unavailable for use by invading bacteria. Bacterial multiplication depends on high concentrations of available iron.

2. Several chemicals released by macrophages, namely **interleukin 1 (IL-1)**, **interleukin 6 (IL-6)**, and **tumor necrosis factor (TNF)** collectively act to bring about a diverse array of effects locally and throughout the body, all of which are geared toward defending the body against infection or tissue injury. They promote inflammation and are largely responsible for systemic (body-wide) manifestations, such as malaise (not feeling well), accompanying an infection. (*Tumor necrosis factor* was named for its first identified role in killing cancer cells.)

3. The same cytokine trio function together as **endogenous pyrogen (EP)**, which induces development of fever (*endogenous* means "from within the body"; *pyro* means "fire" or "heat"; *gen* means "production"). This response occurs especially when invading organisms have spread into the blood. Endogenous pyrogen causes release within the hypothalamus of *prostaglandins*, locally acting lipid signal molecules (see p. 119) that "turn up" the hypothalamic "thermostat" that regulates body temperature. The function of the resulting elevation in body temperature in fighting infection remains unclear. Fever is a common systemic manifestation of inflammation, suggesting the raised temperature plays an important beneficial role, as supported by recent evidence. Higher temperatures appear to augment phagocytosis, increase the rate of the many enzyme-dependent inflammatory activities, and interfere with bacterial multiplication by increasing bacterial requirements for iron. Resolving the controversial issue of whether a fever can be beneficial is extremely important, given the widespread use of drugs that suppress fever.

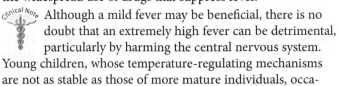

Although a mild fever may be beneficial, there is no doubt that an extremely high fever can be detrimental, particularly by harming the central nervous system. Young children, whose temperature-regulating mechanisms are not as stable as those of more mature individuals, occasionally have convulsions in association with high fevers.

4. IL-1, IL-6, and TNF also decrease the plasma concentration of iron by altering iron metabolism within the liver, spleen, and other tissues. This action reduces the amount of iron available to support bacterial multiplication.

5. Furthermore, the same three cytokines stimulate release of **acute phase proteins** from the liver. This collection of proteins, which have not yet been sorted out by scientists, exerts a multitude of wide-ranging effects associated with the inflammatory process, tissue repair, and immune cell activities. One of the better-known acute phase proteins is **C-reactive protein**, considered clinically as a blood-borne marker of inflammation (see p. 329). C-reactive protein serves as a nonspecific opsonin that binds to the surface of many kinds of bacteria.

6. TNF stimulates release of histamine from mast cells in the vicinity. Histamine, in turn, promotes the local vasodilation and increased capillary permeability of inflammation.

7. IL-1 enhances the proliferation and differentiation of both B and T lymphocytes, which, in turn, are responsible for antibody production and cell-mediated immunity, respectively.

8. *Granulocyte-macrophage colony–stimulating factor* from macrophages and other immune and nonimmune cells on the scene stimulates the bone marrow to synthesize and release neutrophils and monocytes–macrophages, an effect especially prominent in response to bacterial infections.

9. Still other phagocytic chemical mediators trigger both the clotting and anticlotting systems to first enhance the walling-off process and then facilitate gradual dissolution of the fibrin clot after it is no longer needed.

10. A chemical secreted by neutrophils, **kallikrein**, converts specific plasma protein precursors produced by the liver into ac-

tivated **kinins**. Activated kinins augment a variety of inflammatory events. For example, the end product of the kinin cascade, **bradykinin**, activates nearby pain receptors and thus partially produces the soreness associated with inflammation. Bradykinin also dilates blood vessels in the area, reinforcing the effects of histamine. In positive-feedback fashion, kinins also act as powerful chemotaxins to entice more neutrophils to join the battle.

This list of events augmented by phagocyte-secreted chemicals is not complete, but it illustrates the diversity and complexity of responses these mediators elicit. Furthermore, other important interactions between macrophages and lymphocytes that do not depend on release of chemicals from phagocytic cells are described later. Thus, the effect that phagocytes, especially macrophages, ultimately have on microbial invaders far exceeds their "engulf and destroy" tactics.

Tissue Repair The ultimate purpose of the inflammatory process is to isolate and destroy injurious agents and to clear the area for tissue repair. In some tissues (for example, skin, bone, and liver), the healthy organ-specific cells surrounding the injured area undergo cell division to replace the lost cells, often repairing the wound perfectly. In typically nonregenerative tissues such as nerve and muscle, however, lost cells are replaced by **scar tissue**. Connective-tissue fibroblasts start to divide rapidly in the vicinity and secrete large quantities of the protein collagen, which fills in the region vacated by the lost cells and results in the formation of scar tissue (see p. 60). Even in a tissue as readily replaceable as skin, scars sometimes form when complex underlying structures, such as hair follicles and sweat glands, are permanently destroyed by deep wounds.

Inflammation is an underlying culprit in many common, chronic illnesses.

Clinical Note Acute (short-term) inflammatory responses serve a useful purpose for eliminating pathogens from the body, but scientists are increasingly becoming aware that chronic (long-term), low-grade inflammation is a unifying theme for many chronic diseases. Chronic inflammation occurs when the triggering agent persists long term, either because it is not entirely eliminated or because it is constantly present or continually renewed. Chronic inflammation has an important role in Alzheimer's disease (p. 164), atherosclerosis and coronary artery disease (see p. 327), asthma (see p. 457), rheumatoid arthritis (see p. 420), obesity (see p. 623), diabetes (see p. 696), and possibly cancer, among other health problems. Collectively, these conditions are responsible for most morbidity (illness) and mortality (death). Reining in this underlying inflammation could have a huge effect on the quality and quantity of life for much of the world's population.

Nonsteroidal anti-inflammatory drugs and glucocorticoids suppress inflammation.

Clinical Note Many drugs can suppress inflammation; the most commonly used are the *nonsteroidal anti-inflammatory drugs,* or *NSAIDs* (aspirin, ibuprofen, and related compounds) and *glucocorticoids* (drugs similar to the steroid hormone cortisol, which is secreted by the adrenal cortex and exerts anti-inflammatory actions; see p. 674). For example, aspirin interferes with the inflammatory response by decreasing histamine release, thus reducing pain, swelling, and redness. Furthermore, aspirin reduces fever by inhibiting production of prostaglandins, the local mediators of endogenous pyrogen–induced fever.

Glucocorticoids, which are potent anti-inflammatory drugs, suppress almost every aspect of the inflammatory response. In addition, they destroy lymphocytes within lymphoid tissue and reduce antibody production. These therapeutic agents are useful for treating undesirable immune responses, such as allergic reactions (for example, poison ivy rash and asthma) and the inflammation associated with arthritis. However, by suppressing inflammatory and other immune responses that localize and eliminate bacteria, such therapy also reduces the body's ability to resist infection. For this reason, glucocorticoids should be used discriminatingly.

Other newer, more specialized classes of anti-inflammatory agents have recently appeared on the market, such as drugs that specifically inhibit TNF, such as adalimumab *(Humira)*. These new drugs are used, for example, in the treatment of rheumatoid arthritis.

Now let us shift from inflammation to interferon, another component of innate immunity.

Interferon transiently inhibits multiplication of viruses in most cells.

Interferon, a group of three related cytokines, is released from virus-infected cells and briefly provides nonspecific resistance to viral infections by transiently interfering with replication of the same or unrelated viruses in other host cells. Interferon was named for its ability to "interfere" with viral replication.

Antiviral Effect of Interferon When a virus invades a cell, the cell synthesizes and secretes interferon in response to being exposed to viral nucleic acid. Once released into the ECF from a virus-infected cell, interferon binds with receptors on the plasma membranes of healthy neighboring cells or even distant cells that it reaches through the blood, signaling these cells to prepare for possible viral attack. Interferon thus acts as a "whistle-blower," forewarning healthy cells of potential viral attack and helping them prepare to resist. Interferon does not have a direct antiviral effect; instead, it triggers production of virus-blocking enzymes by potential host cells. When interferon binds with these other cells, they synthesize enzymes that can break down viral messenger RNA (see p. 23) and inhibit protein synthesis. Both these processes are essential for viral replication. Although viruses are still able to invade these forewarned cells, the pathogens are unable to govern cellular protein synthesis for their own replication (▌Figure 12-5).

The newly synthesized inhibitory enzymes remain inactive within the tipped-off potential host cell unless it is actually invaded by a virus, at which time the enzymes are activated by the presence of viral nucleic acid. This activation requirement protects the cell's own messenger RNA and protein-synthesizing machinery from unnecessary inhibition by these enzymes should viral invasion not occur. Because activation can take place only during a limited time span, this is a short-term defense mechanism.

Interferon is released nonspecifically from any cell infected by any virus and, in turn, can induce temporary self-protective activity against many different viruses in any other cells that it reaches. Thus, it provides a general, rapidly responding defense strategy against viral invasion until more specific but slower-responding immune mechanisms come into play.

In addition to facilitating inhibition of viral replication, interferon reinforces other immune activities. For example, it enhances macrophage phagocytic activity, stimulates production of antibodies, and boosts the power of killer cells.

Anticancer Effects of Interferon Interferon exerts anticancer as well as antiviral effects. It markedly enhances the actions of cell-killing cells—the *natural killer cells,* the component of innate immunity we describe next—and a special type of T lymphocyte, *cytotoxic T cells*—that attack and destroy both virus-infected cells and cancer cells. Furthermore, interferon itself slows cell division and suppresses tumor growth.

Natural killer cells destroy virus-infected cells and cancer cells on first exposure to them.

Natural killer (NK) cells are naturally occurring, lymphocyte-like cells that nonspecifically destroy virus-infected cells and cancer cells by releasing chemicals that directly lyse the membranes of such cells on first exposure to them. NK cells recognize general features of virus-infected cells and cancer cells. Their mode of action and major targets are similar to those of cytotoxic T cells, but the latter can fatally attack only the specific types of virus-infected cells and cancer cells to which they have been previously exposed. Furthermore, after exposure cytotoxic T cells require a maturation period before they can launch their lethal assault. NK cells provide an immediate, non-

IFigure 12-5 Mechanism of action of interferon in preventing viral replication. Interferon, which is released from virus-infected cells, binds with other uninvaded host cells and induces these cells to produce inactive enzymes capable of blocking viral replication. The inactive enzymes are activated only if a virus subsequently invades one of these prepared cells.
FIGURE FOCUS: *Explain how interferon secreted by virus-infected cells transiently protects all cells against all kinds of viruses.*

specific defense against virus-invaded cells and cancer cells before the more specific and abundant cytotoxic T cells become functional. Antibodies produced as part of the adaptive immune response enhance the killing action of NK cells.

The complement system punches holes in microorganisms.

The **complement system** is another defense mechanism brought into play nonspecifically in response to invading organisms. This system can be activated in two major ways (IFigure 12-6a):

1. By exposure to particular carbohydrate chains present on the surfaces of microorganisms but not found on human cells, a nonspecific innate immune response known as the **alternate complement pathway**

2. By exposure to antibodies produced against a specific foreign invader, an adaptive immune response known as the **classical complement pathway**

The system derives its name from its ability to "complement" the action of antibodies; it is the primary mechanism activated by antibodies to kill foreign cells. The complement system

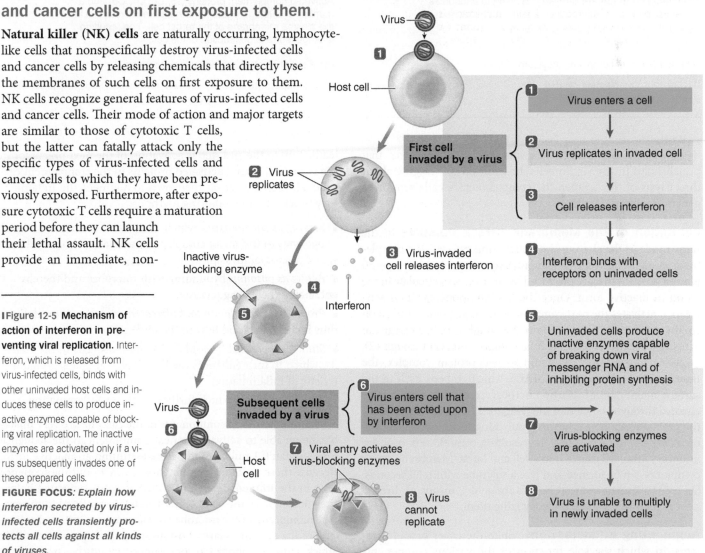

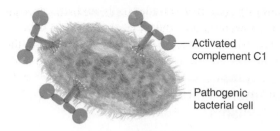

Activated
complement C1

Pathogenic
bacterial cell

Alternate complement pathway: Binding directly
to a foreign invader nonspecifically activates the
complement cascade (an innate immune response).

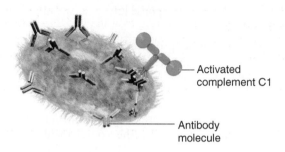

Activated
complement C1

Antibody
molecule

Classical complement pathway: Binding to antibodies
(Y-shaped molecules) produced against and attached
to a particular foreign invader specifically activates the
complement cascade (an adaptive immune response).

(a) Activation of complement system

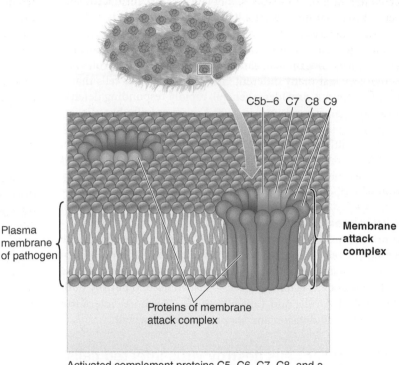

C5b–6 C7 C8 C9

Plasma
membrane
of pathogen

Membrane
attack
complex

Proteins of membrane
attack complex

Activated complement proteins C5, C6, C7, C8, and a
number of C9s aggregate to form a porelike channel in
the plasma membrane of the target cell. The resulting
leakage leads to destruction of the cell.

(b) Formation of membrane attack complex

❙Figure 12-6 **Complement system.** (a) Activation of complement system via the alternate pathway and the classical pathway. (b) Formation of membrane attack complex of the complement system.

---

destroys cells by forming *membrane attack complexes* that punch holes in the victim cells. In addition to bringing about direct lysis of the invader, the complement cascade reinforces general inflammatory tactics.

Formation of the Membrane Attack Complex In the same mode as the clotting and anticlotting systems, the complement system consists of an interactive group of more than 30 plasma proteins that are produced by the liver and circulate in the blood in inactive form. Once the first component, C1, is activated, it activates the next component in the sequence, and so on (in the order of C4, C2, C3, then C5 through C9), in a cascade of activation reactions. The five final components, C5 through C9, assemble into a large, doughnut-shaped protein complex, the **membrane attack complex (MAC)**, which embeds itself in the surface membrane of nearby microorganisms, creating a large channel through the membrane (❙Figure 12-6b). In other words, the parts make a hole. This hole-punching technique makes the membrane extremely leaky; the resulting osmotic flux of water into the victim cell causes it to swell and burst. Membrane attack complexes can be formed and lyse a microbe within 30 seconds. Such complement-induced lysis is the major means of directly killing microbes without phagocytizing them.

Augmenting Inflammation Unlike the other cascade systems, in which the sole function of the various components

leading up to the end step is activation of the next precursor in the sequence, several activated proteins in the complement cascade additionally act on their own to augment the inflammatory process by the following methods:

- *Serving as chemotaxins,* which attract and guide professional phagocytes to the site of complement activation (that is, the site of microbial invasion)

- *Acting as opsonins* by binding with microbes and thereby enhancing their phagocytosis

- *Promoting vasodilation and increased vascular permeability,* thus increasing blood flow to the invaded area

- *Stimulating release of histamine* from mast cells in the vicinity, which in turn enhances the local vascular changes characteristic of inflammation

- *Activating kinins,* which further reinforce inflammation

Several activated components in the cascade are very unstable, being able to activate the next component in the sequence for less than 0.1 millisecond before assuming an inactive form. Because these unstable components can carry out the sequence only in the immediate area in which they are activated before they decompose, the complement attack is confined to the surface membrane of the microbe whose presence initiated activation of the system. Nearby host cells are thus spared from lytic attack. However, abnormal complement activity has been implicated in a number of diseases, including osteoarthritis and age-related macular degeneration (see p. 200).

This completes our discussion of innate immunity. Before turning to adaptive immunity, we draw attention to recently found immune effector cells that blur the boundary between innate and adaptive immunity.

Newly discovered immune cells straddle innate and adaptive defenses.

In recent years scientists have identified new immune cell types that have characteristics belonging to both branches of the immune system. For example, **innate lymphoid cells (ILCs)** are innate mirror images of *helper T cells,* a specialized class of adaptive-system T lymphocytes that spurs other immune cells into action. The always-ready ILCs swiftly but less powerfully perform many of the same duties as the slower but stronger tailor-made helper T cells. In the opposite direction, **innate response activator (IRA) B cells** from the adaptive system recognize bacteria by means of TLRs and act quickly to carry out functions that overlap with the innate system, such as producing cytokines that activate other innate immune cells. By contrast, typical B cells do not respond to TLRs and act slowly to produce antibodies.

Check Your Understanding 12.2

1. List and briefly describe the four types of innate defense.
2. Define *chemotaxin* and *opsonin.*
3. Describe the formation and function of a membrane attack complex.

12.3 Adaptive Immunity: General Concepts

A specific adaptive immune response is a selective attack aimed at limiting or destroying a particular offending target for which the body has been specially prepared after exposure to it.

Adaptive immune responses include antibody-mediated immunity and cell-mediated immunity.

There are two classes of adaptive immune responses: **antibody-mediated**, or **humoral**, **immunity**, involving production of antibodies by B lymphocyte derivatives known as *plasma cells,* and **cell-mediated immunity**, involving production of *activated T lymphocytes,* which directly attack unwanted cells. (The term *humoral immunity* is a reference to the ancient Greek classification of body fluids into four *humors,* blood being one of them. Antibodies responsible for humoral immunity are carried by the blood.)

Lymphocytes can specifically recognize and selectively respond to an almost limitless variety of foreign agents, as well as cancer cells. The recognition and response processes are different in B and in T cells. B cells recognize free-existing foreign invaders such as bacteria and their toxins and a few viruses, which they combat by secreting antibodies specific for the invaders. T cells specialize in recognizing and destroying body cells gone awry, including virus-infected cells and cancer cells. Before we examine each of these processes in detail, we explore the different life histories of B and T cells.

Origins of B and T Cells Both types of lymphocytes, like all blood cells, are derived from common stem cells in the bone marrow (see ❙Figure 11-9, p. 394). Whether a lymphocyte and all its progeny are destined to be B or T cells depends on the site of final differentiation and maturation of the original cell in the lineage (❙Figure 12-7). B cells differentiate and mature in the bone marrow. As for T cells, during fetal life and early childhood, some immature lymphocytes from the bone marrow migrate through the blood to the thymus, where they undergo further processing to become T lymphocytes (named for their site of maturation). The **thymus** is a lymphoid tissue located

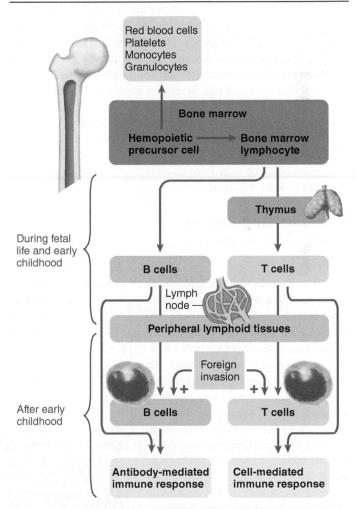

❙Figure 12-7 **Origins of B and T cells.** B cells are derived from lymphocytes that matured and differentiated in the bone marrow, whereas T cells are derived from lymphocytes that originated in the bone marrow but matured and differentiated in the thymus. After early childhood, new B and T cells are produced primarily by colonies of B and T cells established in peripheral lymphoid tissues during fetal life and early childhood.

midline within the chest cavity above the heart in the space between the lungs (see Figure 12-1, p. 406).

On being released into the blood from either the bone marrow or the thymus, mature B and T cells take up residence and establish lymphocyte colonies in the peripheral lymphoid tissues. Here, on appropriate stimulation, they undergo cell division to produce new generations of either B or T cells, depending on their ancestry. After early childhood, most new lymphocytes are derived from these peripheral lymphocyte colonies rather than from the bone marrow.

Each of us has an estimated 2 trillion lymphocytes, which, if aggregated in a mass, would be about the size of the brain. At any one time, most of these lymphocytes are concentrated in the various strategically located lymphoid tissues, but both B and T cells continually circulate among the lymph, blood, and body tissues, where they remain on constant surveillance.

Role of Thymosin Because most of the migration and differentiation of T cells occurs early in development, the thymus gradually atrophies and becomes less important as a person matures. It does, however, continue to produce **thymosin**, a hormone important in maintaining the T-cell lineage. Thymosin enhances proliferation of new T cells within the peripheral lymphoid tissues and augments the immune capabilities of existing T cells. Thymosin secretion starts to decline after about 30 to 40 years of age. Scientists speculate that diminishing T-cell capacity with advancing age may be linked to increased susceptibility to viral infections and cancer because T cells play an especially important role in defense against viruses and cancer.

Let us now see how lymphocytes detect their selected target.

An antigen induces an immune response against itself.

Both B and T cells must be able to specifically recognize unwanted cells and other material to be destroyed as being distinct from the body's normal cells. The presence of antigens enables lymphocytes to make this distinction. An **antigen** is a large, unique molecule that triggers a specific immune response against itself, such as generation of antibodies that lead to its destruction, when it gains entry into the body (*antigen* means *anti*body *gen*erator, although some antigens trigger cell-mediated immune responses instead of antibody production). In general, the more complex a molecule is, the greater its antigenicity. Foreign proteins are the most common antigens because of their size and structural complexity, although other macromolecules, such as large

polysaccharides (carbohydrates) and lipids (fats), can also act as antigens. Antigens may exist as isolated molecules, such as bacterial toxins, or they may be an integral part of a multimolecular structure, as when they are on the surface of an invading foreign microbe.

We'll first see how B cells respond to their targeted antigen, after which we'll look at T cells' response to their antigen.

Check Your Understanding 12.3

1. State the two classes of adaptive immunity and indicate which type of lymphocyte accomplishes each.

2. Define *antigen*.

12.4 B Lymphocytes: Antibody-Mediated Immunity

Each B and T cell has receptors—**B-cell receptors (BCRs)** and **T-cell receptors (TCRs)**—on its surface for binding with one particular type of the multitude of possible antigens (Figure 12-8). These receptors are the "eyes of the adaptive immune system," although a given lymphocyte can "see" only one unique antigen. This is in contrast to the TLRs of the innate effector cells, which recognize generic "trademarks" found on the surface of all microbial invaders.

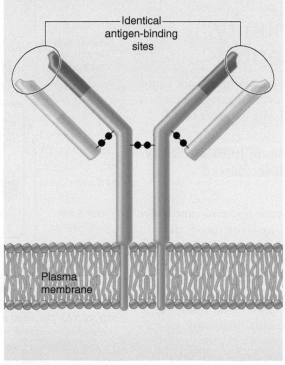
(a) B-cell receptor (BCR)

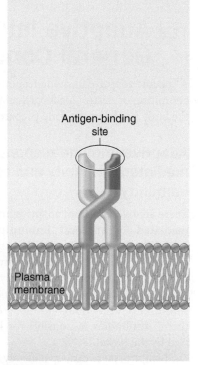
(b) T-cell receptor (TCR)

Figure 12-8 **B-cell and T-cell receptors.**

The antigens to which B cells respond can be T-independent or T-dependent.

T-independent antigens stimulate production of antibody without any assistance from T cells. For example, B cells can bind with and be directly activated by T-independent polysaccharide antigens without T-cell help. By contrast, **T-dependent antigens**, which are typically protein antigens, do not directly stimulate production of antibody without the help of a *helper T cell*. Most antigens to which B cells respond are T-dependent antigens. For now we will consider activation of B cells by binding with antigen without regard to whether helper T cells must be present. We will discuss how helper T cells are involved in the response of B cells to T-dependent antigens when we consider the different types of T cells later.

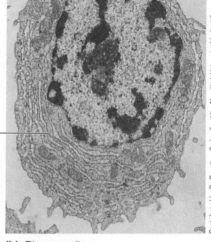

(a) Unactivated B cell

(b) Plasma cell

Endoplasmic reticulum

Contributed by Dr. Dorothea Zucker-Franklin, New York University Medical Center

❙Figure 12-9 **Comparison of an unactivated B cell and a plasma cell.** Electron micrographs at the same magnification of (a) an unactivated B cell, or small lymphocyte, and (b) a plasma cell. A plasma cell is an activated B cell. It is filled with an abundance of rough endoplasmic reticulum distended with antibody molecules.

Antigens stimulate B cells to convert into plasma cells that produce antibodies.

When BCRs (❙Figure 12-8a) bind with an antigen, most B cells differentiate into active *plasma cells,* whereas others become dormant *memory cells.* We first examine the role of plasma cells and their antibodies and then later in the chapter turn attention to memory cells.

Plasma Cells A **plasma cell** produces **antibodies** that can combine with the specific type of antigen that stimulated activation of the plasma cell. During differentiation into a plasma cell, a B cell swells as the rough endoplasmic reticulum (the site for synthesis of proteins to be exported; see p. 25) greatly expands (❙Figure 12-9). Because antibodies are proteins, plasma cells essentially become prolific protein factories, producing up to 2000 antibody molecules per second. So great is the commitment of a plasma cell's protein-synthesizing machinery to antibody production that it cannot maintain protein synthesis for its own viability and growth. Consequently, it dies after a brief (5- to 7-day), highly productive life span.

Antibody Subclasses Antibodies (also known as γ *globulins* or *immunoglobulins*; see p. 383) are grouped into five subclasses based on differences in their biological activity:

- **IgM** serves as the BCR for antigen attachment and is produced in the early stages of plasma cell response.
- **IgG**, the most abundant immunoglobulin in the blood, is produced and secreted copiously when the body is subsequently re-exposed to the same antigen. IgG produces most specific immune responses against bacterial invaders.
- **IgE** helps protect against parasitic worms and is the antibody mediator for common allergic responses, such as hay fever, asthma, and hives.
- **IgA** is found in secretions of the digestive, respiratory, and urogenital (urinary and reproductive) systems and in milk and tears.
- **IgD** is present on the surface of many B cells, but its function is unclear.

Recognize that within each of these five functional subclasses are millions of different antibodies, each able to bind with only one specific antigen.

Antibodies are Y shaped and classified according to properties of their tail portion.

Antibodies of all five subclasses are composed of four interlinked polypeptide chains—two long, heavy chains and two short, light chains—arranged in the shape of a Y (❙Figure 12-10). Characteristics of the arm regions of the Y determine the *specificity* of the antibody (that is, with what antigen the antibody can bind). Properties of the tail portion of the antibody determine the *functional properties* of the antibody (what the antibody does once it binds with an antigen).

An antibody has two identical antigen-binding sites, one at the tip of each arm. These **antigen-binding fragments (Fab)** are unique for each antibody, so each antibody can interact only with an antigen that specifically matches it, much like a lock and key. The tremendous variation in the antigen-binding fragments of different antibodies leads to the extremely large number of unique antibodies that can bind specifically with only one of millions of different antigens.

In contrast to these variable Fab regions at the arm tips, the tail portion of every antibody within each immunoglobulin subclass is identical. The tail, the antibody's **constant (Fc) region**, contains binding sites for particular mediators of antibody-induced activities, which vary among the subclasses.

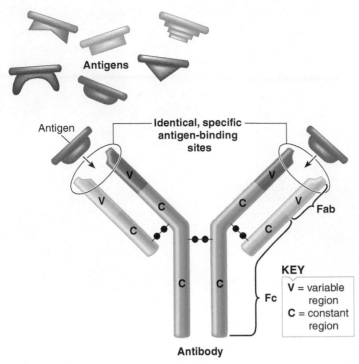

Antigens

Identical, specific antigen-binding sites

Antigen

Fab

KEY

V = variable region
C = constant region

Fc

Antibody

ǀFigure 12-10 **Antibody structure.** An antibody is Y-shaped. It is able to bind only with the specific antigen that "fits" its antigen-binding sites (Fab) on the arm tips. The tail region (Fc) binds with particular mediators of antibody-induced activities.

In fact, differences in the constant region are the basis for distinguishing among the immunoglobulin subclasses. For example, the constant tail region of IgG antibodies, when activated by antigen binding in the Fab region, binds with phagocytic cells and serves as an opsonin to enhance phagocytosis. In comparison, the constant tail region of IgE antibodies attaches to mast cells and basophils, even in the absence of antigens. When the appropriate antigen gains entry to the body and binds with the attached antibodies, this triggers the release of histamine from the affected mast cells and basophils. Histamine, in turn, induces the allergic manifestations that follow.

Antibodies largely amplify innate immune responses to promote antigen destruction.

Antibodies cannot directly destroy foreign organisms or other unwanted materials on binding with antigens on their surfaces. Instead, they exert their protective influence by physically hindering antigens or, more commonly, by amplifying innate immune responses.

Physical Hindrance of an Antigen Through neutralization and agglutination, antibodies can physically hinder some antigens from exerting their detrimental effects.

■ In **neutralization**, antibodies combine with bacterial toxins, preventing these harmful chemicals from interacting with susceptible cells. Similarly, antibodies can neutralize some types of viruses by binding with their surface antigens and preventing these viruses from entering cells, where they could exert their damaging effects.

■ In **agglutination**, multiple antibody molecules cross-link numerous antigen molecules into chains or lattices of antigen–antibody complexes (ǀFigure 12-11a). Through this means, foreign cells, such as bacteria or mismatched transfused red blood cells, bind together in a clump (see p. 389). When soluble antigens such as tetanus toxin are involved instead of foreign cells that can clump together, the dissolved antigen–antibody lattices become so large that they *precipitate* (separate from solution).

Within the body, these physical hindrance mechanisms play only a minor protective role against invading agents. *Clinical Note* However, the tendency for certain antigens to agglutinate or precipitate on forming large complexes with antibodies specific for them is useful for detecting the presence of particular antigens or antibodies. Pregnancy diagnosis tests, for example, use this principle to detect, in urine, the presence of a hormone secreted soon after conception.

Amplification of Innate Immune Responses The most important function of antibodies is to profoundly augment the innate immune responses already initiated by the invaders. Antibodies mark foreign material as targets for actual destruction by the complement system, phagocytes, or NK cells as follows:

1. *Activating the complement system.* When an appropriate antigen binds with an antibody, receptors on the tail portion of the antibody bind with and activate C1, the first component of the complement system. This sets off the cascade of events leading to formation of the membrane attack complex, which is specifically directed at the membrane of the invading cell that bears the antigen that initiated the activation process (ǀFigure 12-11b). Antibodies are the most powerful activators of the complement system (the classical complement pathway). The biochemical attack subsequently unleashed against the invader's membrane is the most important means by which antibodies exert their protective influence. Furthermore, various activated complement components enhance every aspect of inflammation. The same complement system is activated by an antigen–antibody complex regardless of the type of antigen. Although binding of antigen to antibody is highly specific, the outcome, which is determined by the antibody's constant tail region, is identical for all activated antibodies within a given subclass; for example, all IgG antibodies activate the same complement system.

2. *Enhancing phagocytosis.* Antibodies, especially IgG, act as opsonins. The tail portion of an antigen-bound IgG antibody binds with a receptor on the surface of a phagocyte and subsequently promotes phagocytosis of the antigen-containing victim attached to the antibody (ǀFigure 12-11c).

3. *Stimulating NK cells.* Binding of antibody to antigen also induces attack of the antigen-bearing target cell by NK cells. NK cells have receptors for the constant tail portion of antibodies. In this case, when the target cell is coated with antibodies, the tail portions of the antibodies link the target cell to NK cells, which destroy the target cell by releasing chemicals that lyse its plasma membrane (ǀFigure 12-11d). This process is known as **antibody-dependent cellular cytotoxicity (ADCC)**.

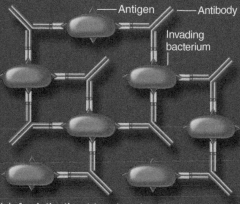

(a) **Agglutination** (clumping of antigenic cells)

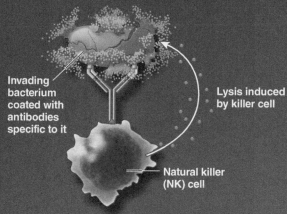

(c) **Enhancement of phagocytosis (opsonization)**

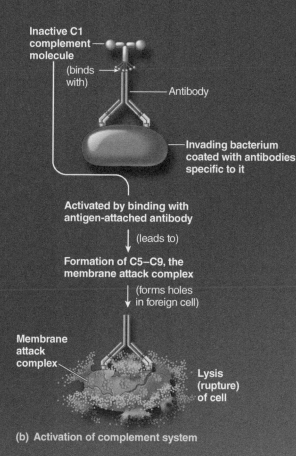

(b) **Activation of complement system**

(d) **Stimulation of natural killer (NK) cells:** antibody-dependent cellular cytotoxicity

Structures are not drawn to scale.

❙Figure 12-11 How antibodies help eliminate invading microbes. Antibodies may physically hinder antigens, such as through neutralization or (a) agglutination. More commonly, antibodies amplify innate immune responses by (b) activating the complement system, (c) enhancing phagocytosis by acting as opsonins, and (d) stimulating natural killer cells.

FIGURE FOCUS: *A friend has bacterial pneumonia. Describe three means by which antibodies against the causative microorganism indirectly lead to destruction of the bacteria.*

By amplifying other nonspecific lethal defenses in these ways, antibodies, although unable to directly destroy invading bacteria or other undesirable material, bring about destruction of the antigens to which they are specifically attached.

Immune Complex Disease Occasionally, an over-zealous antigen–antibody response inadvertently causes damage to normal cells, as well as to invading foreign cells. Typically, antigen–antibody complexes, formed in response to foreign invasion, are removed by phagocytic cells after having revved up nonspecific defense strategies. If large numbers of these complexes are continuously produced, how-ever, the phagocytes cannot clear away all the immune complexes formed. Antigen–antibody complexes that are not removed continue to activate the complement system, among other things. Excessive amounts of activated complement and other inflammatory agents may "spill over," damaging the surrounding normal cells, as well as the unwanted cells. Furthermore, destruction is not necessarily restricted to the initial site of inflammation. Antigen–antibody complexes may circulate freely and become trapped in the kidneys, joints, brain, small vessels of the skin, and elsewhere, causing widespread inflammation and tissue damage. Such damage produced by immune complexes is referred to as an **immune complex disease**, which can be a complicating outcome of bacterial, viral, or parasitic infection.

More insidiously, immune complex disease can stem from overzealous inflammatory activity prompted by immune complexes formed by "self-antigens" (proteins synthesized by the person's body) and antibodies erroneously produced against them. *Rheumatoid arthritis* develops in this way.

Clonal selection accounts for the specificity of antibody production.

Consider the diversity of foreign molecules a person can potentially encounter during a lifetime. Despite this, each B cell is preprogrammed to respond to only 1 of probably more than 100 million different antigens. Other antigens cannot combine with the same B cell and induce it to secrete different antibodies. The astonishing implication is that each of us is equipped with about 100 million kinds of preformed B lymphocytes, at least one "matching" B lymphocyte for every possible antigen that we might ever encounter.

Early researchers in immunologic theory believed antibodies were "made to order" whenever a foreign antigen gained entry to the body. In contrast, the currently accepted **clonal selection theory** proposes that diverse B lymphocytes are produced during fetal development, each capable of synthesizing an antibody against a particular antigen before ever being exposed to it. All offspring of a particular ancestral B lymphocyte form a family of identical cells, or a **clone**, committed to producing the same specific antibody. B cells remain dormant, not actually secreting their particular antibody product or undergoing rapid division until (or unless) they come into contact with the appropriate antigen. Lymphocytes that have not yet been exposed to their specific antigen are known as **naïve lymphocytes**. When an antigen gains entry to the body, the particular clone of B cells that bear receptors (BCRs) on their surface uniquely specific for that antigen is activated or "selected" by the antigen binding with the BCRs, hence the term *clonal selection theory* (❙Figure 12-12).

The first antibodies produced by a newly formed B cell are IgM immunoglobulins, which are inserted into the cell's plasma membrane rather than secreted. Here they serve as BCRs for binding with a specific kind of antigen, almost like "advertisements" for the kind of antibody that the cell can produce. Binding of the appropriate antigen to a B cell amounts to "placing an order" for the manufacture and secretion of large quantities of that particular antibody.

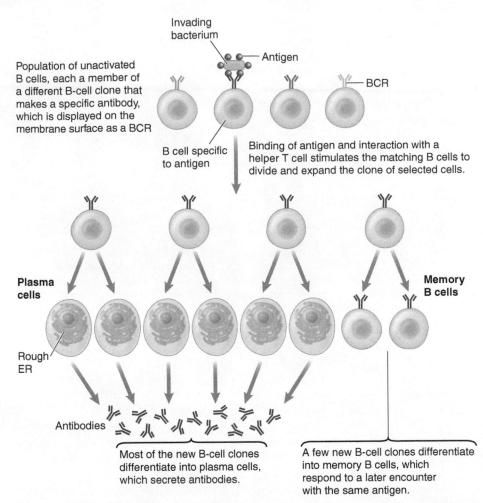

❙Figure 12-12 **Clonal selection theory.** The B-cell clone specific to the antigen proliferates and differentiates into plasma cells and memory cells. Plasma cells secrete antibodies that bind with free antigen not attached to B cells. Memory cells are primed and ready for subsequent exposure to the same antigen.

Selected clones differentiate into active plasma cells and dormant memory cells.

Antigen binding causes the activated B-cell clone to multiply (the clone expands at the rate of two to three cell divisions per day) and differentiate into two cell types—the previously described plasma cells and memory cells, which we focus on next. Most progeny are transformed into active plasma cells, which are prolific producers of customized antibodies that contain the same antigen-binding sites as the surface receptors. However, plasma cells switch to producing IgG antibodies, which are secreted rather than remaining membrane bound. In the blood, the secreted antibodies combine with the invading, free extracellular antigen (such as on the surface of a bacterium), marking it for destruction by the complement system, phagocytic ingestion, or NK cells.

Memory Cells Not all new B lymphocytes produced by the specifically activated clone differentiate into short-lived antibody-secreting plasma cells. A small proportion become long-lived **memory cells**, which do not participate in the current immune attack against the antigen but instead remain dormant and expand this specific clone. If the person is ever re-exposed to the same antigen, these memory cells are primed and ready for even more immediate action than the original lymphocytes in the clone were.

Even though each of us has essentially the same original pool of different naive B-cell clones, the pool gradually becomes appropriately biased to respond most efficiently to each person's particular antigenic environment. Those clones specific for antigens to which a person is never exposed remain dormant for life, whereas those specific for antigens in the individual's environment typically become expanded and enhanced by forming highly responsive memory cells. The different naive clones provide protection against new pathogens, and the evolving populations of memory cells protect against the recurrence of infections encountered in the past.

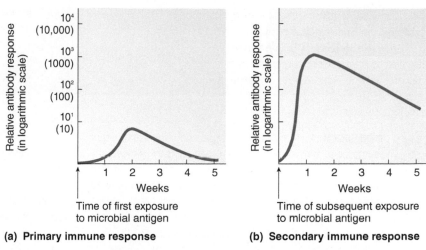

(a) **Primary immune response**   (b) **Secondary immune response**

Figure 12-13 Primary and secondary immune responses. (a) Primary response on first exposure to a microbial antigen. (b) Secondary response on subsequent exposure to the same microbial antigen. The primary response does not peak for a couple of weeks, whereas the secondary response peaks in a week. The magnitude of the secondary response is 100 times that of the primary response.

Primary and Secondary Responses During initial contact with a microbial antigen, the antibody response is delayed for several days until plasma cells are formed and does not reach its peak for a couple of weeks. This response is known as the **primary response**, which is mediated by IgM antibodies (Figure 12-13a). Meanwhile, symptoms characteristic of the particular microbial invasion persist until either the invader succumbs to the mounting specific immune attack against it or the infected person dies. After reaching the peak, the antibody levels gradually decline over time. If the same antigen ever reappears, the long-lived memory cells launch a more rapid, more potent, and longer-lasting **secondary response** than occurred during the primary response (Figure 12-13b). This swifter, more powerful immune attack, which is mediated by IgM antibodies, is frequently adequate to prevent overt infection on subsequent exposures to the same microbe, forming the basis of long-term immunity against a specific disease.

Clinical Note The original antigenic exposure that induces the formation of memory cells can occur through the person either having the disease or being vaccinated. **Vaccination (immunization)** deliberately exposes the person to a pathogen that has been stripped of its disease-inducing capability (that is, the pathogen is *attenuated*) but that can still induce antibody formation against itself. (For the early history of vaccination development, see the boxed feature on p. 422, Concepts, Challenges, and Controversies.)

Active immunity is self-generated; passive immunity is "borrowed."

The production of antibodies as a result of exposure to an antigen is referred to as **active immunity** against that antigen. A second way in which an individual can acquire antibodies is by direct transfer of antibodies actively formed by another person (or animal). The immediate "borrowed" immunity conferred on receipt of preformed antibodies is known as **passive immunity**. Such transfer of antibodies of the IgG class normally

occurs from the mother to the fetus across the placenta during intrauterine development. In addition, a mother's colostrum (first milk) contains IgA antibodies that provide further protection for breast-fed babies. Passively transferred antibodies are usually broken down in less than a month, but meanwhile the newborn is provided important immune protection (essentially the same as the mother's) until he or she can begin actively mounting immune responses. Antibody-synthesizing ability does not develop for about a month after birth.

Clinical Note Passive immunity is sometimes used clinically to provide immediate protection or to bolster resistance against an extremely virulent infectious agent or potentially lethal toxin to which a person has been exposed (for example, rabies virus, tetanus toxin in nonimmunized individuals, and poisonous snake venom). The administered preformed antibodies are harvested from another source (often nonhuman) that has been exposed to an attenuated form of the antigen. Frequently, horses or sheep are used in the deliberate production of antibodies to be collected for passive immunizations. Although injection of serum containing these antibodies (**antiserum** or **antitoxin**) is beneficial in providing immediate protection against the specific disease or toxin, the recipient may develop an immune response against the injected antibodies themselves because they are foreign proteins. The result may be a severe allergic reaction to the treatment, a condition known as **serum sickness**.

The huge repertoire of B cells is built by reshuffling a small set of gene fragments.

How is it possible for each person to have such a tremendous diversity of B lymphocytes that all individuals (with healthy immune systems) have the ability to produce specific antibodies against every one of the 100 million different antigens that we potentially might encounter? Antibodies are proteins synthesized in accordance with a nuclear DNA blueprint. Because all cells of the body, including the antibody-producing cells,

Vaccination: A Victory Over Many Dreaded Diseases

MODERN SOCIETY HAS COME TO hope and even expect that vaccines can be developed to protect us from almost any dreaded infectious disease. This expectation has been brought into sharp focus by our current frustration over the inability to date to develop a successful vaccine against HIV, the virus that causes AIDS.

Nearly 2500 years ago, our ancestors were aware of the existence of immune protection. Writing about a plague that struck Athens in 430 BC, Thucydides observed that the same person was never attacked twice by this disease. However, the ancients did not understand the basis of this protection, so they could not manipulate it to their advantage.

Early attempts to deliberately acquire lifelong protection against smallpox, a dreaded disease that was highly infectious and frequently fatal (up to 40% of the sick died), consisted of intentionally exposing oneself by coming into direct contact with a person suffering from a milder form of the disease. The hope was to protect against a future fatal bout of smallpox by deliberately inducing a mild case of the disease. By the beginning of the 17th century, this technique had evolved into using a needle to extract small amounts of pus from active smallpox pustules (the fluid-filled bumps on the skin, which leave a characteristic depressed scar or "pock" mark after healing) and introducing this infectious material into healthy individuals. This inoculation process was done by applying the pus directly to slight cuts in the skin or by inhaling dried pus.

Edward Jenner, an English physician, was the first to demonstrate that immunity against cowpox, a disease similar to but less serious than smallpox, could also protect humans against smallpox. Having observed that milkmaids who got cowpox seemed to be protected from smallpox, Jenner in 1796 inoculated a healthy boy with pus he had extracted from cowpox boils (*vacca*, as in *vaccination*, means

"cow"). After the boy recovered, Jenner (not being restricted by modern ethical standards of research on human subjects) deliberately inoculated him with what was considered a normally fatal dose of smallpox infectious material. The boy survived.

Jenner's results were not taken seriously, however, until a century later when, in the 1880s, Louis Pasteur, the first great experimental immunologist, extended Jenner's technique. Pasteur demonstrated that the disease-inducing capability of organisms could be greatly reduced (attenuated) so that they could no longer produce disease but would still induce antibody formation when introduced into the body—the basic principle of modern vaccines. His first vaccine was against anthrax, a deadly disease of sheep and cows. Pasteur isolated and heated anthrax bacteria and then injected these attenuated organisms into a group of healthy sheep. A few weeks later, at a gathering of fellow scientists, Pasteur injected these vaccinated sheep and a group of unvaccinated sheep with fully potent anthrax bacteria. The result was dramatic—all the vaccinated sheep survived, but all the unvaccinated sheep died. Pasteur's notorious public demonstrations such as this, coupled with his charismatic personality, caught the attention of physicians and scientists of the time, sparking the development of modern immunology.

Today preventive vaccines save an estimated 3 million lives annually. Now scientists are striving to develop therapeutic vaccines aimed not at preventing disease but at squelching established diseases and addictions. New kinds of vaccines at various stages of development include those for the following conditions: cancer, coronary artery disease (by targeting LDL, the "bad" cholesterol); allergies; obesity (by targeting ghrelin, the "hunger hormone"); and addictions, such as for nicotine and cocaine.

---

contain the same nuclear DNA, it is hard to imagine how enough DNA could be packaged within the nuclei of every cell to code for the 100 million different antibodies (a different portion of the genetic code being used by each B-cell clone), along with all the other genetic instructions used by other cells. Actually, only a relatively small number of gene fragments code for antibody synthesis, but during B-cell development these fragments are cut, reshuffled, and spliced in a vast number of different combinations. Each different combination gives rise to a unique B-cell clone. Antibody genes of already-formed B cells are later even further diversified by mutations, to which the region that codes for the antibody's variable antigen-binding sites are highly prone. Each different mutant cell in turn gives rise to a new clone. In this way, a huge antibody repertoire is possible using only a modest share of the genetic blueprint.

We now turn our attention to T cells.

Check Your Understanding 12.4

1. Describe the structure and function of the Fab and Fc regions of an antibody.

2. List the means by which antibodies help eliminate invading microbes.

3. According to the clonal selection theory, draw a flow diagram showing the formation of plasma cells and memory cells on binding of an antigen to its specific B-cell receptor.

12.5 T Lymphocytes: Cell-Mediated Immunity

As important as B lymphocytes and their antibody products are in specific defense against invading bacteria and other foreign material, they represent only half of the body's specific immune defenses. The T lymphocytes are equally important in defense against most viral infections and also play an important regulatory role in immune mechanisms.

T cells bind directly with their targets.

Whereas B cells defend against conspicuous invaders in the ECF, T cells defend against covert invaders that hide inside cells where antibodies and the complement system cannot reach them. Unlike B cells, which secrete antibodies that can attack antigens at long distances, T cells must directly contact their targets, a process known as *cell-mediated immunity*. T cells of the killer type release chemicals that destroy the targeted cells they contact, such as virus-infected cells and cancer cells.

Like B cells, T cells are clonal and exquisitely antigen specific. On its plasma membrane, each T cell bears unique receptor proteins called *T-cell receptors (TCRs)*, similar although not identical to the surface receptors on B cells (see Figure 12-8b, p. 416). Immature lymphocytes acquire their TCRs in the thymus during their differentiation into T cells. Unlike B cells, T cells are activated by a foreign antigen only when it is on the surface of a cell that also carries a marker of the individual's identity; that is, both foreign antigens and **self-antigens** known as **major histocompatibility complex (MHC) molecules** must be on a cell's surface before a T cell can bind with it. During thymic education, T cells learn to recognize foreign antigens only in combination with the person's own tissue antigens—a lesson passed on to all T cells' future progeny. The importance of this dual antigen requirement and the nature of the MHC self-antigens are described shortly.

A delay of a few days generally follows exposure to the appropriate antigen before **activated T cells** are prepared to launch a cell-mediated immune attack. When exposed to a specific antigen combination, cells of the complementary T-cell clone proliferate and differentiate for several days, yielding large numbers of activated effector T cells that carry out various cell-mediated responses.

Like B cells, T cells form a memory pool and display both primary and secondary responses. Primary responses tend to be initiated in the lymphoid tissues. During a few-week period after the infection is cleared, more than 90% of the huge number of effector T cells generated during the primary response die by means of *apoptosis* (cell suicide; see p. 42). To stay alive, activated T lymphocytes require the continued presence of their specific antigen and appropriate stimulatory signals. Once the foe succumbs, most of the now superfluous T lymphocytes commit suicide because their supportive antigen and stimulatory signals are withdrawn. Elimination of most of the effector T cells following a primary response is essential to prevent congestion in the lymphoid tissues. (Such paring down is not needed for B cells—those that become plasma cells and not memory B cells on antigen stimulation rapidly work themselves to death producing antibodies.) The remaining surviving effector T cells become long-lived memory T cells that migrate to all areas of the body, where they are poised for a swift secondary response to the same pathogen in the future.

The three types of T cells are cytotoxic, helper, and regulatory T cells.

The three subpopulations of T cells can be designated in two ways: by their roles when activated by an antigen or by specific proteins associated with their outer membrane. By role, the three types of T cells are *cytotoxic T cells*, *helper T cells*, and *regulatory T cells*. By specific membrane protein type, these same cells are *CD8+ T cells*, *CD4+ T cells*, and *CD4+CD25+ T cells*, respectively. The various types of immune cells each have an assortment of specific immune-related surface membrane proteins given official *cluster designation (CD) numbers* that help characterize them.

- **Cytotoxic**, or **killer**, **T cells** destroy host cells harboring anything foreign and thus bearing a foreign antigen, such as body cells invaded by viruses, cancer cells that have mutated proteins resulting from malignant transformations, and transplanted cells. The T-cell receptors for cytotoxic T cells are associated with **coreceptors** designated CD8, which are inserted into the plasma membrane as these cells pass through the thymus. Therefore, these cells are also known as **CD8+ T cells**.

- **Helper T cells** do not directly participate in immune destruction of invading pathogens. Instead, they modulate activities of other immune cells. Because of the important role they play in "turning on" the full power of all the other activated lymphocytes and macrophages, helper T cells constitute the immune system's "master switch." Helper T cells are by far the most numerous T cells, making up 60% to 80% of circulating T cells. The T-cell receptors for helper T cells are associated with coreceptors designated CD4. Accordingly, helper T cells are also called **CD4+ cells**.

- **Regulatory T cells** (T_{regs}) are a recently identified small subset of CD4+ cells. They have the same CD4 coreceptors as helper T cells, but in addition they also have CD25, a component of a receptor for IL-2, which promotes T_{reg} activities. Thus, these cells are also called **CD4+CD25+ T cells**.

We now further examine these T cell types.

Cytotoxic T cells secrete chemicals that destroy target cells.

Cytotoxic T cells are microscopic "hit men." The targets of these destructive cells most frequently are host cells infected with viruses. When a virus invades a body cell, as it must to survive (Figure 12-14, step **1**), the cell breaks down the envelope of proteins surrounding the virus and loads a fragment of this viral antigen piggyback onto a newly synthesized MHC self-antigen. This self-antigen–viral antigen complex is inserted into the host cell's surface membrane, where it serves as a red flag indicating the cell is harboring the invader (step **2**). To attack the intracellular virus, cytotoxic T cells must destroy the infected host cell in the process. Cytotoxic T cells of the clone specific for this particular virus recognize and bind to the viral antigen and self-antigen on the surface of an infected cell (step **3**). Thus activated by the viral antigen, a cytotoxic T cell can kill the infected cell by either direct or indirect means, depending

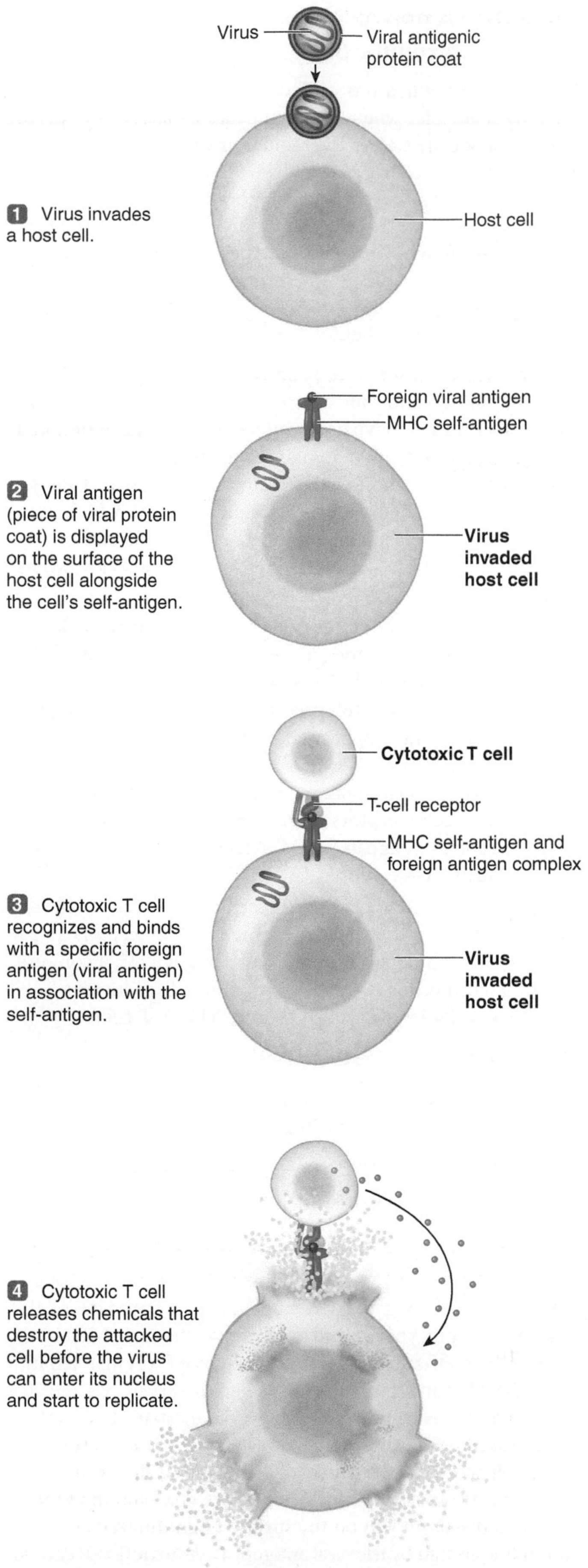

1 Virus invades a host cell.

Virus — Viral antigenic protein coat

Host cell

2 Viral antigen (piece of viral protein coat) is displayed on the surface of the host cell alongside the cell's self-antigen.

Foreign viral antigen
MHC self-antigen

Virus invaded host cell

3 Cytotoxic T cell recognizes and binds with a specific foreign antigen (viral antigen) in association with the self-antigen.

Cytotoxic T cell

T-cell receptor

MHC self-antigen and foreign antigen complex

Virus invaded host cell

4 Cytotoxic T cell releases chemicals that destroy the attacked cell before the virus can enter its nucleus and start to replicate.

Figure 12-14 A cytotoxic T cell lysing a virus-invaded cell.

on the type of lethal chemicals the activated T cell releases. Let us elaborate.

■ An activated cytotoxic T cell may directly kill the victim cell by releasing chemicals that lyse the attacked cell before viral replication can begin (step **4**). Specifically, cytotoxic T cells, as well as NK cells, destroy a targeted cell by releasing **perforin** molecules, which penetrate the target cell's surface membrane and join to form porelike channels (■ Figure 12-15). This technique of killing a cell by punching holes in its membrane is similar to the method employed by the membrane attack complex of the complement cascade. This contact-dependent mechanism of killing has been nicknamed the "kiss of death."

■ A cytotoxic T cell can also indirectly bring about death of an infected host cell by releasing **granzymes**, which are enzymes similar to digestive enzymes. Granzymes enter the target cell through the perforin channels. Once inside, these chemicals trigger the virus-infected cell to self-destruct through apoptosis.

The virus released on destruction of the host cell by either of these methods is directly destroyed in the ECF by phagocytic cells, neutralizing antibodies, and the complement system. Meanwhile, the cytotoxic T cell, which has not been harmed in the process, can move on to kill other infected host cells. Recall that other defense mechanisms including interferon and NK cells also come into play to combat viral infections. As usual, an intricate interplay exists among the immune defenses that are launched against viruses, as summarized in ■ Table 12-2.

When virus-infected cells are destroyed, surrounding healthy cells replace the lost cells by means of cell division. Usually, to halt a viral infection, only some of the host cells must be destroyed. If the virus has had a chance to multiply, however, with replicated virus leaving the original cell and spreading to other host cells, the cytotoxic T-cell defense mechanism may sacrifice so many of the host cells that serious malfunction may ensue.

Hybrid Natural Killer T Cells A recently discovered subclass of cytotoxic T cells, **natural killer T (NKT) cells**, have properties of both NK cells and cytotoxic T cells. Constituting only 0.2% of all circulating T cells, the hybrid NKT cells have both NK-cell markers and T-cell receptors and thus are another cell type that sits at the crossroads of innate and adaptive immunity. Scientists are scrambling to identify the physiological roles of these multifunctional cells. They seem to be important in suppression of autoimmunity (erroneous production of antibodies against self-antigens) and in tumor rejection.

Defense in the Nervous System The usual method of destroying virus-infected host cells is not appropriate for the nervous system. If cytotoxic T cells destroyed virus-infected neurons, the lost cells could not be replaced because neurons cannot reproduce. Fortunately, virus-infected neurons are spared from extermination by the immune system, but how, then, are neurons protected from viruses? Immunologists long thought that the only antiviral defenses for neurons were those aimed at free viruses in the ECF. Surprising new research has revealed, however, that antibodies not only target viruses for destruction in the ECF but can also eliminate viruses inside neurons. It is unclear whether antibodies actually enter the

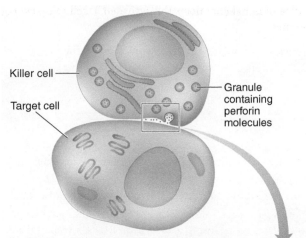

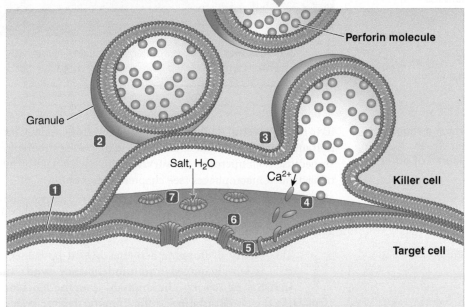

1 Killer cell binds to its target.

2 As a result of binding, killer cell's perforin-containing granules fuse with plasma membrane.

3 Granules disgorge their perforin by exocytosis into a small pocket of intercellular space between killer cell and its target.

4 On exposure to Ca^{2+} in this space, individual perforin molecules change from spherical to cylindrical shape.

5 Remodeled perforin molecules bind to target cell membrane and insert into it.

6 Individual perforin molecules group together like staves of a barrel to form pores.

7 Pores admit salt and H_2O, causing target cell to swell and burst.

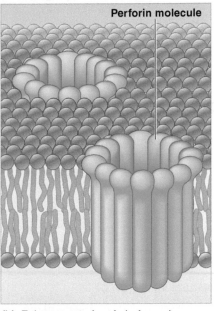

(a) Details of the killing process for cytotoxic T cells and NK cells

(b) Enlargement of perforin-formed pores in a target cell

I Figure 12-15 **Mechanism of killing by killer cells.**

Source: Adapted from the illustration by Dana Burns-Pizer in "How Killer Cells Kill," by John Ding-E Young and Zanvil A. Cohn in *Scientific American*, 1988.

FIGURE FOCUS: *Describe the similarities and differences between the membrane-attack complex formed by the complement system (see I Figure 12-6b, p. 414) and the lethal pores in the victim cell created by killer cells.*

neurons and interfere directly with viral replication (neurons have been shown to take up antibodies near their synaptic endings) or bind with the surface of nerve cells and trigger intracellular changes that stop viral replication.

Clinical Note The fact that some viruses, such as the herpes virus, persist for years in nerve cells, occasionally "flaring up" to produce symptoms, demonstrates that the antibodies' intraneuronal mechanism does not provide a foolproof antiviral defense for neurons.

Helper T cells secrete chemicals that amplify the activity of other immune cells.

In contrast to cytotoxic T cells, helper T cells are not killer cells. Instead, helper T cells secrete cytokines that "help," or augment, nearly all aspects of the immune response. Most cytokines are

produced by helper T cells. The following are among the best known of helper T-cell cytokines:

1. Helper T cells secrete several interleukins (*IL-4, IL-5,* and *IL-6*) that serve collectively as a **B-cell growth factor**, which contributes to B-cell function in concert with the IL-1 secreted by macrophages. Antibody secretion is greatly reduced or absent without the assistance of helper T cells, especially in defense against T-dependent antigen.

2. Helper T cells similarly secrete **T-cell growth factor** (*IL-2*), which augments the activity of cytotoxic T cells and even of other helper T cells responsive to the invading antigen. In typical interplay fashion, IL-1 secreted by macrophages not only enhances the activity of both the appropriate B- and T-cell clones but also stimulates secretion of IL-2 by activated helper T cells.

| TABLE 12-2 Defenses Against Viral Invasion

When the virus is free in the ECF,

Macrophages

Destroy the free virus by phagocytosis.

Process and present the viral antigen to helper T cells.

Secrete IL-1, which activates B- and T-cell clones specific to the viral antigen.

Antibodies

Are secreted by plasma cells derived from B cells specific to the viral antigen.

Neutralize the virus to prevent its entry into a host cell.

Act as opsonins to enhance phagocytosis of the virus.

Activate the lethal complement cascade as part of adaptive immunity.

The Complement System

Directly destroys the free virus by forming a hole-punching membrane attack complex.

Acts as an opsonin to enhance phagocytosis of the virus.

When the virus has entered a host cell (which it must do to survive and multiply, with the replicated viruses leaving the original host cell to enter the ECF in search of other host cells),

Interferon

Is secreted by virus-infected cells.

Binds with and prevents viral replication in other host cells.

Enhances the killing power of macrophages, natural killer cells, and cytotoxic T cells.

Natural Killer (NK) Cells

Nonspecifically lyse virus-infected host cells.

Cytotoxic T Cells

Are specifically activated by the viral antigen and lyse the infected host cells before the virus has a chance to replicate.

Helper T Cells

Secrete cytokines, which enhance cytotoxic T-cell activity and B-cell antibody production.

When a virus-infected cell is destroyed, the free virus is released into the ECF, where it is attacked directly by macrophages, antibodies, and the activated complement components.

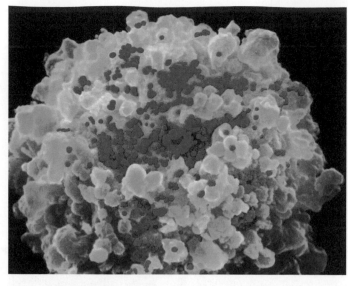

Figure 12-16 AIDS virus. Human immunodeficiency virus (HIV) (*red*), the AIDS-causing virus, on a helper (CD4+) T lymphocyte, HIV's primary target.

NIBSC/Science Source

3. Some helper T-cell cytokines act as *chemotaxins* to lure more neutrophils and macrophages-to-be to the invaded area.

4. Once macrophages are attracted to the area, **macrophage-migration inhibition factor**, another cytokine released from helper T cells, keeps these large phagocytic cells in the region by inhibiting their outward migration. As a result, many chemotactically attracted macrophages accumulate in the infected area. This factor also confers greater phagocytic power on the gathered macrophages. These so-called **angry macrophages**

have more powerful destructive ability. They are especially important in defending against the bacteria that cause tuberculosis because such microbes can survive simple phagocytosis by nonactivated macrophages (see chapter opener photo).

5. One cytokine secreted by helper T cells (*IL-5*) activates eosinophils, and another (*IL-4*) promotes the development of IgE antibodies for defense against parasitic worms.

Clinical Note This variety of immune activities helped by helper T cells is why **acquired immunodeficiency syndrome (AIDS)**, caused by the **human immunodeficiency virus (HIV)**, is so devastating to the immune defense system. The AIDS virus selectively invades helper (CD4+) T cells, destroying or incapacitating the cells that normally orchestrate much of the immune response (|Figure 12-16). The virus also invades macrophages, further crippling the immune system, and sometimes enters brain cells, leading to the dementia (severe impairment of intellectual capacity) noted in some AIDS victims.

HIV infections are still incurable, and efforts to develop a vaccine against HIV have been only slightly successful to date because the virus rapidly mutates. However, a variety of drugs are now available to control HIV. For example, some classes of drugs disable HIV from making copies of itself, whereas others block HIV entry into CD4+ cells. The most successful therapy is a three-drug "cocktail" that typically changes the course of the disease from a short-term death sentence to a long-term, often manageable condition. The triple combination of drugs from different classes slows progression of the condition and allows the body to replenish its CD4+ population while reducing the likelihood that the virus develops drug resistance. A recent novel treatment with promising preliminary results involves disabling by means of new gene-editing technology the receptor on CD4+ cells that HIV uses to gain entry into these cells. Another new tactic under investigation to thwart the spread of HIV entails treatment as prevention (that is, giving drugs to individuals not infected) for those at high risk of acquiring the virus.

T Helper 1 and T Helper 2 Cells Not all helper cells secrete the same cytokines. Two subsets of helper T cells—**T helper 1 (TH1)** and **T helper 2 (TH2)** cells—augment different patterns of immune responses by secreting different types of cytokines. TH1 cells rally a cell-mediated (cytotoxic T-cell) response, which is appropriate for infections with intracellular microbes, such as viruses. By contrast, TH2 cells promote antibody-mediated immunity, especially IgE production, by B cells and rev up eosinophil activity for defense against parasitic worms.

Helper T cells produced in the thymus are in a naive state until they encounter the antigen they are primed to recognize. Whether a naive helper T cell becomes a TH1 or TH2 cell depends on which cytokines are secreted by macrophages and dendritic cells (macrophagelike cells) of the innate immune system that "present" the antigen to the uncommitted T cell. You will learn about antigen presentation in an upcoming section. **IL-12** drives a naive T cell specific for the antigen to become a TH1 cell, whereas **IL-4** favors the development of a naive cell into a TH2 cell. Thus, the antigen-presenting cells of the nonspecific immune system can influence the whole tenor of the specific immune response by determining whether the TH1 or TH2 cellular subset dominates. In the usual case, the secreted cytokines promote the appropriate specific immune response against the particular threat at hand.

Scientists have also recently discovered much smaller subsets of helper T cells: TH17 cells and T follicular helper (TFH) cells. **TH17 cells** produce *IL-17* (accounting for the name of this subset). They promote inflammation and are effector cells in the development of inflammatory autoimmune diseases (immune attack against a body tissue), such as the attack against myelin in multiple sclerosis (see p. 103). The newest discovered **TFH cells** specifically interact with B cells in lymph node follicles (hence the name of this subset) to help them secrete antibodies in response to T-dependent antigens and also promote differentiation of memory B cells.

Regulatory T cells suppress immune responses.

Representing 5% to 10% of CD4+ T cells, regulatory T cells suppress immune responses, thus keeping the rest of the immune system under tight control. They are specialized to inhibit both innate and adaptive immune responses in a check-and-balance fashion to minimize harmful immune pathology. T_{regs} contain large amounts of the intracellular protein **Foxp3**, which is essential for turning developing T cells into T_{regs} and gives these suppressor cells the ability to quiet other immune cells. Researchers hope that the ability of T_{regs} to put the brakes on helper T cells, B cells, NK cells, and dendritic cells can somehow be used therapeutically to curb autoimmune diseases and prevent rejection of transplanted organs.

A newly discovered subset of B cells, **regulatory B cells (B_{regs})**, performs a similar role in suppressing harmful immune responses instead of turning into plasma cells that churn out antibodies as most B cells do. Constituting only about 1% to 2% of all B cells, B_{regs} help squelch autoimmunity and dampen innate-immunity directed inflammatory responses.

We next examine how T cells are activated by antigen-presenting cells and the roles of MHC molecules.

T cells respond only to antigens presented to them by antigen-presenting cells.

T cells cannot perform their tasks without assistance from antigen-presenting cells. That is, relevant T cells cannot recognize "raw" foreign antigens entering the body; before reacting to it, a T-cell clone must be formally "introduced" to the antigen. **Antigen-presenting cells (APCs)** handle the formal introduction; they engulf, then process and present antigens, complexed with MHC self-antigen molecules, on their surface to the T cells. APCs include macrophages and closely related dendritic cells, both of which not only are phagocytic effector cells of the innate defense system but also play a key role in activating cell-mediated adaptive immunity. You are already familiar with macrophages. **Dendritic cells** are specialized APCs that act as sentinels in almost every tissue. They are so named because they have many surface projections, or branches, that resemble the dendrites of neurons (*dendros* means "tree") (❚Figure 12-17). Dendritic cells are especially abundant in the skin and mucosal linings of the lungs and digestive tract—strategic locations where microbes are likely to enter the body. After exposure to the appropriate antigen, dendritic cells leave their tissue home and migrate through the lymphatic system to nearby lymph nodes, where they cluster and activate T cells.

We will use a dendritic cell engulfing a bacterium and presenting it to a helper T cell as an example of an APC. Invading bacteria are phagocytized by dendritic cells (or macrophages) (❚Figure 12-18, step ❶). Within the dendritic cell, the phagosome containing the engulfed bacterium fuses with a lysosome, which enzymatically breaks down the bacterium's proteins into antigenic peptides (small protein fragments) (step ❷). Each antigenic peptide then binds to an MHC molecule, which has been newly synthesized within the endoplasmic reticulum—

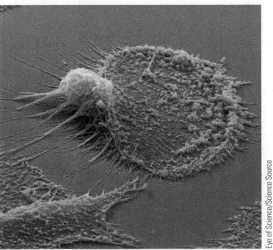

❚Figure 12-17 **Dendritic cell.**

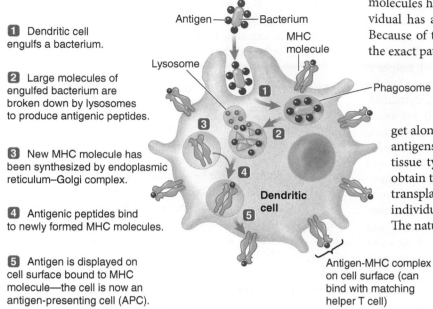

1 Dendritic cell engulfs a bacterium.

2 Large molecules of engulfed bacterium are broken down by lysosomes to produce antigenic peptides.

3 New MHC molecule has been synthesized by endoplasmic reticulum–Golgi complex.

4 Antigenic peptides bind to newly formed MHC molecules.

5 Antigen is displayed on cell surface bound to MHC molecule—the cell is now an antigen-presenting cell (APC).

Antigen—Bacterium

MHC molecule

Lysosome

Phagosome

Dendritic cell

Antigen-MHC complex on cell surface (can bind with matching helper T cell)

⏐Figure 12-18 **Generation of an antigen-presenting cell when a dendritic cell engulfs a bacterium.**

Golgi complex (steps 3 and 4). An MHC molecule has a deep groove into which a variety of antigenic peptides can bind, depending on what the macrophage has engulfed. The MHC molecule then transports the bound antigen to the cell surface, where it is presented to passing T lymphocytes (step 5). The dendritic cell is now an antigen-presenting cell. Once displayed at the cell surface, the combined presence of these self- and nonself-antigens alerts the immune system to the presence of an undesirable agent within the cell. Unlike B cells, T cells cannot bind with foreign antigen that is not in association with self-antigen. It would be futile for T cells to bind with free, extracellular antigen—they cannot defend against foreign material unless it is intracellular. Only a specific helper T cell with a T-cell receptor that fits the displayed antigen-MHC complex in complementary fashion can bind with the APC.

When the helper T cell binds to the APC, the APC secretes cytokines that activate the associated T cell, among other effects. The activated T cell then secretes various other cytokines, which act in an autocrine manner to stimulate clonal expansion of this particular helper T cell and act in a paracrine manner on adjacent B cells, cytotoxic T cells, and NK cells to enhance their responses to the same foreign antigen.

What is the nature of the MHC self-antigens that the immune system learns to recognize as markers of a person's cells? That is the topic of the next section.

The major histocompatibility complex is the code for self-antigens.

Self-antigens are plasma membrane–bound glycoproteins (proteins with sugar attached) known as **MHC molecules** because their synthesis is directed by a group of genes called the **major histocompatibility complex**, or **MHC.** The MHC genes are the most variable ones in humans. More than 100 different MHC molecules have been identified in human tissue, but each individual has a code for only 3 to 6 of these possible antigens. Because of the tremendous number of combinations possible, the exact pattern of MHC molecules varies from one individual to another, much like a "biochemical fingerprint" or "molecular identification card."

The major histocompatibility complex (*histo* means "tissue"; *compatibility* means "ability to get along") was so named because these genes and the self-antigens they encode were first discerned in relation to tissue typing (similar to blood typing), which is done to obtain the most compatible matches for tissue grafting and transplantation. However, the transfer of tissue from one individual to another does not normally occur in nature. The natural function of MHC antigens lies in their ability to direct the responses of T cells, not in their artificial role in rejecting transplanted tissue.

Class I and Class II MHC Glycoproteins

T cells become active only when they match a given MHC–foreign antigen combination. In addition to having to fit a specific foreign antigen, the T-cell receptor must match the appropriate MHC molecule. Each person has two main classes of MHC-encoded molecules that are differentially recognized by cytotoxic and helper T cells—*class I* and *class II MHC glycoproteins*, respectively (⏐Figure 12-19). The class I and II markers serve as signposts to guide cytotoxic and helper T cells to the precise cellular locations where their immune capabilities can be most effective. The coreceptors (either CD8 or CD4) on the T cells bind with the MHC molecules on the target molecule, linking the two cells together.

Cytotoxic (CD8+) T cells can respond to foreign antigens only in association with **class I MHC glycoproteins**, which are found on the surface of all nucleated body cells. This binding specificity occurs because the cytotoxic T cell's CD8 coreceptor can interact only with class I MHC molecules. To carry out their role of dealing with pathogens that have invaded host cells, it is appropriate that cytotoxic T cells bind only with body cells that viruses have infected—that is, with foreign antigens in association with self-antigens. Furthermore, these deadly T cells can link up with any cancerous body cell because class I MHC molecules also display mutated (and thus "foreign" appearing) cellular proteins characteristic of these abnormal cells. Because any nucleated body cell can be invaded by viruses or become cancerous, essentially all cells display class I MHC glycoproteins, enabling cytotoxic T cells to attack any virus-invaded host cell or any cancer cell. In the case of cytotoxic T cells, the outcome of this binding is destruction of the infected body cell. Because cytotoxic T cells do not bind to MHC self-antigens in the absence of foreign antigens, normal body cells are protected from lethal immune attack.

In contrast, **class II MHC glycoproteins**, which are recognized by helper (CD4+) T cells, are restricted to the surface of a few special types of immune cells. The CD4 coreceptor associated with these helper T cells can interact only with class II MHC molecules. A helper T cell can bind with a foreign antigen only when it is found on the surfaces of immune cells with

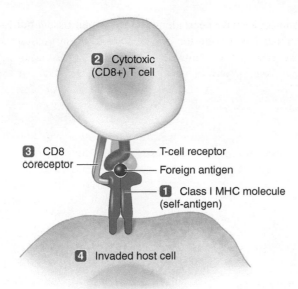

2 Cytotoxic (CD8+) T cell

3 CD8 coreceptor

T-cell receptor

Foreign antigen

1 Class I MHC molecule (self-antigen)

4 Invaded host cell

1 Class I MHC molecules are found on the surface of all cells.

2 They are recognized only by cytotoxic (CD8+) T cells.

3 CD8 coreceptor links the two cells together.

4 Linked in this way, cytotoxic T cells can destroy body cells if invaded by foreign (viral) antigen.

(a) Class I MHC self-antigens

2 Helper (CD4+) T cell

3 CD4 coreceptor

T-cell receptor

Foreign antigen

1 Class II MHC molecule (self-antigen)

4 Dendritic cell (APC)

1 Class II MHC molecules are found on the surface of immune cells with which helper T cells interact: dendritic cells, macrophages, and B cells.

2 They are recognized only by helper (CD4+) T cells.

3 CD4 coreceptor links the two cells together.

4 To be activated, helper T cells must bind with a class II MHC-bearing APC (dendritic cell or macrophage). To activate B cells, helper T cell must bind with a class II MHC-bearing B cell with displayed foreign antigen.

(b) Class II MHC self-antigens

▮Figure 12-19 **Distinctions between class I and class II major histocompatibility complex (MHC) glycoproteins.** Specific binding requirements for the two types of T cells ensure that these cells bind only with the target cells with which they can interact. Cytotoxic (CD8+) T cells can recognize and bind with foreign antigen only when the antigen is in association with class I MHC glycoproteins, which are found on the surface of all body cells. This requirement is met when a virus invades a body cell, whereupon the cell is destroyed by the cytotoxic T cells. Helper (CD4+) T cells, which are activated by or enhance the activities of dendritic cells, macrophages, and B cells, can recognize and bind with foreign antigen only when it is in association with class II MHC glycoproteins, which are found only on the surface of these other immune cells. The CD8+ or CD4+ T-cell's coreceptor CD8 or CD4 links these cells to the target cell's class I or class II MHC molecules, respectively.

which the helper T cell directly interacts—*macrophages, dendritic cells,* and *B cells.* Class II MHC molecules are found on macrophages and dendritic cells, which present antigens to helper T cells and on B cells, whose activities are enhanced by cytokines secreted by helper T cells. B cells do not phagocytize antigenic particles like macrophages and dendritic cells do, but they can internalize T-dependent antigen bound to their surface receptor by receptor-mediated endocytosis (see p. 31), then display the antigens complexed with class II MHC on their surface. Binding of the antigen-bearing B cell with the matching helper T cell causes the T cell to secrete cytokines that activate this specific B cell, leading to clonal expansion and conversion of this B-cell clone into antibody-producing plasma cells and memory cells (▮Figure 12-20). The capabilities of helper T cells would be squandered if these cells were able to bind with body cells other than these special APC and B immune cells. This is the primary pathway by which the adaptive immune system fights bacteria. ▮Table 12-3 summarizes

the innate and adaptive immune strategies that defend against bacterial invasion.

Clinical Note **Transplant Rejection** T cells bind with MHC antigens present on the surface of transplanted cells in the absence of a foreign viral antigen. The ensuing destruction of the transplanted cells triggers the rejection of transplanted or grafted tissues. Presumably, some of the recipient's T cells "mistake" the MHC antigens of the donor cells for a closely resembling combination of a conventional viral foreign antigen complexed with the recipient's MHC self-antigens.

To minimize the rejection phenomenon, technicians match the tissues of donor and recipient according to MHC antigens as closely as possible. Therapeutic procedures to suppress the immune system then follow. In recent years, new therapeutic agents have become extremely useful in selectively depressing T-cell-mediated immune activity while leaving B-cell antibody-mediated immunity essentially intact. For example, *cyclosporin*

Activation of helper T cells by antigen presentation

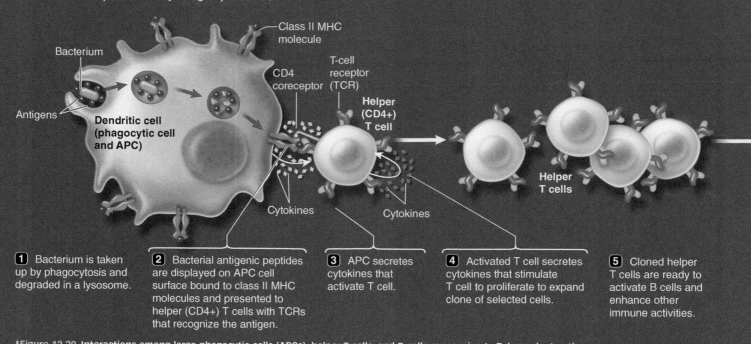

1 Bacterium is taken up by phagocytosis and degraded in a lysosome.

2 Bacterial antigenic peptides are displayed on APC cell surface bound to class II MHC molecules and presented to helper (CD4+) T cells with TCRs that recognize the antigen.

3 APC secretes cytokines that activate T cell.

4 Activated T cell secretes cytokines that stimulate T cell to proliferate to expand clone of selected cells.

5 Cloned helper T cells are ready to activate B cells and enhance other immune activities.

Figure 12-20 **Interactions among large phagocytic cells (APCs), helper T cells, and B cells responsive to T-dependent antigen.**
FIGURE FOCUS: *In this sequence of events, with what two immune cell types do helper T cells bind and what is the outcome of each binding?*

TABLE 12-3 Innate and Adaptive Immune Responses to Bacterial Invasion

Innate Immune Mechanisms	Adaptive Immune Mechanisms
Roles of Inflammation	**Roles of B Cells and Helper T Cells**
Resident tissue macrophages engulf invading bacteria.	B cells specific to T-independent antigen are activated on binding with the antigen.
Histamine-induced vascular responses increase blood flow to the area, bringing in additional immune effector cells and plasma proteins.	B cells specific to T-dependent antigen present antigen to helper T cells. On binding with the B cells, helper T cells activate the B cells.
A fibrin clot walls off the invaded area.	The activated B-cell clone proliferates and differentiates into plasma cells and memory cells.
Neutrophils and monocytes–macrophages migrate from the blood to the area to engulf and destroy foreign invaders and to remove cell debris.	Plasma cells secrete customized antibodies, which specifically bind to the invading bacteria. Plasma cell activity is enhanced by
Phagocytic cells secrete cytokines, which enhance both innate and adaptive immune responses and induce local and systemic symptoms associated with an infection.	IL-1 secreted by macrophages.
	Helper T cells, which have been activated by the same bacterial antigen processed and presented to them by macrophages or dendritic cells.
Roles of the Complement System	
The complement system is nonspecifically activated by exposure to PAMPs (alternate complement pathway).	Antibodies bind to invading bacteria and enhance innate mechanisms that lead to the bacteria's destruction. Specifically, antibodies
Complement components form a hole-punching membrane attack complex that lyses bacterial cells.	Act as opsonins to enhance phagocytic activity.
Complement components enhance many steps of inflammation.	Activate the lethal complement system (classical complement pathway).
	Stimulate natural killer (NK) cells, which directly lyse bacteria.
	Memory cells persist that are capable of responding more rapidly and more forcefully should the same bacteria be encountered again.

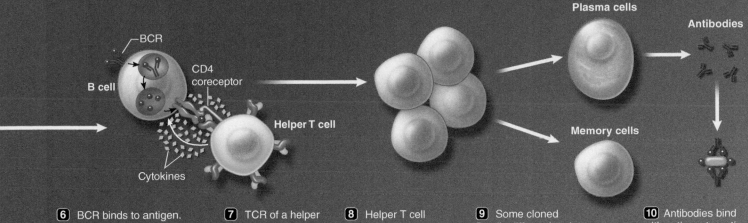

Activation of B cells responsive to T-dependent antigen

BCR
CD4 coreceptor
B cell
Helper T cell
Cytokines
Plasma cells
Antibodies
Memory cells

6 BCR binds to antigen. Antigen is internalized by receptor-mediated endocytosis and its macromolecules degraded. Antigenic peptides produced are displayed on cell surface bound to class II MHC molecules.

7 TCR of a helper T cell recognizes specific antigen on B cell, and CD4 coreceptor links the two cells together.

8 Helper T cell secretes cytokines that stimulate B cell proliferation to produce clone of selected cells.

9 Some cloned B cells differentiate into plasma cells, which secrete antibodies specific for the antigen, while a few differentiate into memory B cells.

10 Antibodies bind with antigen, targeting antigenic invader for destruction by the innate immune system.

---

blocks IL-2, the cytokine secreted by helper T cells that is required for expansion of the selected cytotoxic T-cell clone.

What factors normally prevent the adaptive immune system from unleashing its powerful defense capabilities against the body's self-antigens? We examine this issue next.

The immune system is normally tolerant of self-antigens.

Self-tolerance refers to preventing the immune system from attacking the person's own tissues. During the genetic "cut, shuffle, and paste process" that goes on during lymphocyte development, some B and T cells are by chance formed that could react against the body's own tissue antigens. If these lymphocyte clones were allowed to function, they would destroy the individual's body. Fortunately, the immune system normally does not produce antibodies or activated T cells against the body's self-antigens. A variety of mechanisms are involved in self-tolerance, including the following:

1. *Clonal deletion.* In response to continuous exposure to body antigens early in development, B and T lymphocyte clones specifically capable of attacking these self-antigens in most cases are permanently destroyed within the thymus. This **clonal deletion** is accomplished by triggering apoptosis of immature cells that would react with the body's proteins. This physical elimination is the major mechanism by which self-tolerance is developed.

2. *Clonal anergy.* Self-reactive lymphocyte clones that escape destruction in the thymus are rendered nonfunctional in the periphery by various means, the most important of which is clonal anergy. The premise of clonal anergy is that

a T lymphocyte must receive two specific simultaneous signals to be activated ("turned on"), one from its compatible antigen and a stimulatory cosignal molecule found only on the surface of an APC. Both signals are present for foreign antigens, which are introduced to T cells by APCs. In contrast, these dual signals—antigen plus cosignal—never are present for self-antigens because these antigens are not handled by cosignal-bearing APCs. The first exposure to a single signal from a self-antigen "turns off" the compatible T cell, rendering the cell unresponsive to further exposure to the antigen. This reaction is referred to as **clonal anergy** (*anergy* means "lack of energy") because T cells are inactivated (that is, "become lazy") rather than activated by their antigens. Anergized T lymphocyte clones survive, but they cannot function.

3. *Active suppression by regulatory cells.* T_{regs} (and maybe B_{regs}) play a role in self-tolerance by inhibiting throughout life some lymphocyte clones specific for the body's tissues.

4. *Receptor editing.* With **receptor editing**, once a B cell that bears a receptor for one of the body's antigens encounters the self-antigen, the B cell escapes death by swiftly changing its antigen receptor to a nonself version. In this way, an originally self-reactive B cell survives but is "rehabilitated" so that it never targets the body's tissues again.

5. *Immunological ignorance,* alternatively known as *antigen sequestering.* Some self-molecules are normally hidden from the immune system because they never come into direct contact with the ECF in which the immune cells and their products circulate. An example is thyroglobulin, a complex protein sequestered within the hormone-secreting structures of the thyroid gland (see p. 666).

Autoimmune diseases arise from loss of tolerance to specific self-antigens.

Clinical Note Occasionally, the immune system fails to distinguish between self-antigens and foreign antigens and unleashes its deadly powers against one or more of the body's tissues. A condition in which the immune system fails to recognize and tolerate self-antigens associated with particular tissues is known as an **autoimmune disease** (*auto* means "self"). Autoimmunity underlies more than 80 diseases, many of which are well known. Examples include *multiple sclerosis, rheumatoid arthritis, Type 1 diabetes mellitus,* and *psoriasis* (a skin condition). About 50 million Americans suffer from some type of autoimmune disease, with the incidence being about three times higher in females than in males.

Autoimmune diseases may arise from various causes:

1. Normal-self antigens may be modified by factors such as drugs, environmental chemicals, viruses, or genetic mutations so that they are no longer recognized and tolerated by the immune system.

2. Exposure of normally inaccessible self-antigens sometimes induces an immune attack against these antigens. Because the immune system is usually never exposed to hidden self-antigens, it does not "learn" to tolerate them. Inadvertent exposure of these normally inaccessible antigens to the immune system because of tissue disruption caused by injury or disease can lead to a rapid immune attack against the affected tissue, just as if these self-proteins were foreign invaders. *Hashimoto's disease,* which involves production of antibodies against thyroglobulin and destruction of the thyroid gland's hormone-secreting capacity, is one such example.

3. Exposure of the immune system to a foreign antigen structurally almost identical to a self-antigen may induce the production of antibodies or activated T lymphocytes that not only interact with the foreign antigen but also cross-react with the closely similar body antigen. An example of this molecular mimicry is the streptococcal bacteria responsible for "strep throat." The bacteria possess antigens structurally very similar to self-antigens in the tissue covering the heart valves of some individuals, in which case the antibodies produced against the streptococcal organisms may also bind with this heart tissue. The resultant inflammatory response is responsible for the heart-valve lesions associated with *rheumatic fever.*

4. New studies hint at a possible trigger of autoimmune diseases that could explain why a whole host of these disorders are more common in women than in men. Scientists have long speculated that the sex bias of autoimmune diseases was somehow related to hormonal differences. Recent findings suggest, however, that the higher incidence of these self-destructive conditions in females may be a legacy of pregnancy. Researchers have learned that fetal cells, which often gain access to the mother's bloodstream during the trauma of labor and delivery, sometimes linger in the mother for decades after the pregnancy. The immune system typically clears these cells from the mother's body following childbirth, but studies have demonstrated that women with autoimmune conditions are more likely than healthy women to have persistent fetal cells in their blood. The persistence of similar but not identical fetal antigens that were not wiped out early on as being foreigners may somehow trigger a gradual, more subtle immune attack that eventually turns against the mother's closely related antigens. For example, this phenomenon might be the underlying trigger for *systemic lupus erythematosus,* an autoimmune attack against DNA, a condition that can affect many organs, most commonly skin, joints, and kidneys.

We now turn to T cells' role in defending against cancer.

An interplay among immune cells and interferon defends against cancer.

Besides destroying virus-infected host cells, an important function of the T-cell system is recognizing and destroying newly arisen, potentially cancerous tumor cells before they have a chance to multiply and spread, a process known as **immune surveillance**. At least once a day, on average, your immune system destroys a mutated cell that could potentially become cancerous. Any normal cell may be transformed into a cancer cell if mutations occur within its genes that govern cell division and growth. Such mutations may occur by chance or, more frequently, by exposure to **carcinogenic** (cancer-causing) **factors** such as ionizing radiation, certain environmental chemicals, or physical irritants. Alternatively, a few cancers are caused by tumor viruses, which turn the cells they invade into cancer cells, an example being the *human papillomavirus* that causes cervical cancer. The immune system recognizes cancer cells because they bear new and different surface antigens alongside the cell's normal-self antigens because of either genetic mutation or invasion by a tumor virus.

Clinical Note **Benign and Malignant Tumors** Cell multiplication and growth are normally under strict control by largely unknown regulatory mechanisms. Cell multiplication in an adult is generally restricted to replacing lost cells. Furthermore, cells normally respect their place and space in the body's society of cells. If a cell that has transformed into a tumor cell manages to escape immune destruction, however, it defies the normal controls on its proliferation and position. Unrestricted multiplication of a single tumor cell results in a **tumor** that consists of a clone of cells identical to the original mutated cell. If the mass is slow growing, stays put in its original location, and does not infiltrate the surrounding tissue, it is considered a **benign tumor**. In contrast, the transformed cell may multiply rapidly and form an invasive mass that lacks the "altruistic" behavior characteristic of normal cells. Such invasive tumors are **malignant tumors**, or **cancer**. Malignant tumor cells usually do not adhere well to the neighboring normal cells, so often some of the cancer cells break away from the parent tumor. These "emigrant" cancer cells are transported through the blood to new territories, where they continue to proliferate, forming multiple malignant tumors. The term **metastasis** is applied to this spreading of cancer to other parts of the body.

If a malignant tumor is detected early, before it has metastasized, it can be removed surgically. Once cancer cells have dispersed and seeded multiple cancerous sites, surgical elimination of the malignancy is impossible. In this case, agents that interfere with rapidly dividing and growing cells, such as certain *chemotherapy drugs*, are used in an attempt to destroy the malignant cells. Unfortunately, these agents also harm normal body cells, especially rapidly proliferating cells such as blood cells and the cells lining the digestive tract.

Untreated cancer is eventually fatal in most cases, for several interrelated reasons. The uncontrollably growing malignant mass crowds out normal cells by vigorously competing with them for space and nutrients, yet the cancer cells cannot take over the functions of the cells they are destroying. Cancer cells typically remain immature and do not become specialized, often resembling embryonic cells instead (Figure 12-21). Such poorly differentiated malignant cells lack the ability to perform the specialized functions of the normal cell type from which they mutated. Affected organs gradually become disrupted to the point that they can no longer perform their life-sustaining functions, and the person dies.

Genetic Mutations that Do Not Lead to Cancer
Even though many body cells undergo mutations throughout a person's lifetime, most of these mutations do not result in malignancy for three reasons:

1. Only a fraction of the mutations involve loss of control over the cell's growth and multiplication. More frequently, other facets of cellular function are altered.

2. A cell usually becomes cancerous only after an accumulation of multiple independent mutations. This requirement contributes at least in part to the much higher incidence of cancer in older individuals, in whom mutations have had more time to accumulate in a single cell lineage.

3. Potentially cancerous cells that do arise are usually destroyed by the immune system early in their development.

Effectors of Immune Surveillance
Immune surveillance against cancer depends on interplay among three types of immune cells—*cytotoxic T cells, NK cells,* and *macrophages*—and *interferon.* These three immune cell types not only can attack and destroy cancer cells directly but also secrete interferon. Interferon, in turn, inhibits multiplication of cancer cells and increases the killing ability of the immune cells (Figure 12-22). Because NK cells do not require prior exposure and activation in response to a cancer cell before being able to launch a lethal attack, they are the first line of defense against cancer. Cytotoxic T cells take aim at cancer cells after being activated by mutated surface proteins alongside normal class I MHC molecules. On contacting a cancer cell, both these killer cells release perforin and other toxic chemicals that destroy the targeted mutant cell. Macrophages, in addition to clearing away the remains of the dead victim cell, can engulf and destroy cancer cells intracellularly.

Still, cancer does sometimes occur because cancer cells occasionally escape these immune

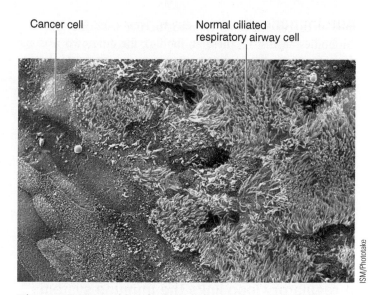

Cancer cell

Normal ciliated respiratory airway cell

ISM/Phototake

Figure 12-21 **Comparison of normal and cancerous cells in the large respiratory airways.** The normal cells display specialized cilia (*blue*), which constantly contract in whiplike motion to sweep debris and microorganisms from the respiratory airways so that they do not gain entrance to the deeper portions of the lungs. The cancerous cells are not ciliated, so they are unable to perform this specialized defense task.

mechanisms. Some cancer cells are believed to survive by evading immune detection—for example, by failing to display identifying antigens on their surface or by being surrounded by counterproductive **blocking antibodies** that interfere with T-cell function. Although B cells and antibodies are not believed to play a direct role in cancer defense, B cells, on viewing a

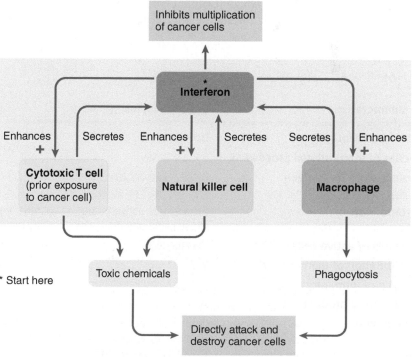

Figure 12-22 **Immune surveillance against cancer.** Anticancer interactions of cytotoxic T cells, natural killer cells, macrophages, and interferon.

mutant cancer cell as an alien to "normal self," may produce antibodies against it. These antibodies, for unknown reasons, do not activate the complement system, which could destroy the cancer cells. Instead, the antibodies bind with the antigenic sites on the cancer cell, "hiding" these sites from recognition by cytotoxic T cells. The coating of a tumor cell, by blocking antibodies, thus protects the harmful cell from attack by deadly T cells. Still other successful cancer cells thwart immune attack by turning on their pursuers. They induce the cytotoxic T cells that bind with them to commit suicide. In addition, some cancer cells secrete large amounts of a specific chemical messenger that recruits T_{regs} and programs them to suppress cytotoxic T cells.

We have finished discussing B cells and T cells, the effector cells of adaptive immunity. ▌Table 12-4 compares these two types of lymphocytes.

A regulatory loop links the immune system with the nervous and endocrine systems.

Even though complex controlling factors operate within the immune system itself, scientists traditionally thought that the immune system functioned independently of other control systems in the body. However, they now know that the immune system both influences and is influenced by the two major regulatory systems, the nervous and endocrine systems. Neurons have receptors for specific cytokines. For example, IL-1 can turn on the stress response by activating a sequence of nervous and endocrine events that result in secretion of cortisol, one of the major hormones released during stress. This linkage between a mediator of the immune response and a mediator of the stress response is appropriate. Cortisol mobilizes the body's nutrient stores so that metabolic fuel is readily available to keep pace with the body's energy demands at a time when the person is sick and may not be eating enough (or, in the case of an animal, may not be able to search for food). Furthermore, cortisol

mobilizes amino acids, which serve as building blocks to repair any tissue damage sustained during the encounter that triggered the immune response.

In the reverse direction, lymphocytes and macrophages are responsive to blood-borne signals from the nervous system and from certain endocrine glands. These immune cells possess receptors for a wide variety of neurotransmitters, hormones, and other chemical mediators. For example, cortisol and other chemical mediators of the stress response have a profound immunosuppressive effect, inhibiting many functions of lymphocytes and macrophages and decreasing the production of cytokines. Thus, a negative-feedback loop exists among the immune system and the nervous and endocrine systems. Cytokines released by immune cells enhance the neurally and hormonally controlled stress response, whereas cortisol and related chemical mediators released during the stress response suppress the immune system. For example, the anti-inflammatory effect of cortisol modulates stress-activated immune responses, preventing them from overshooting, and thus protecting us against damage by potentially overreactive defense mechanisms.

However, in large part because stress suppresses the immune system, stressful physical, psychological, and social life events are sometimes linked with increased susceptibility to infections and cancer. Thus, the body's resistance to disease can be influenced by the person's mental state—a case of "mind over matter."

Other important links exist between the immune and nervous systems besides the cortisol connection. For example, many immune system organs, such as the thymus, spleen, and lymph nodes, are innervated by the sympathetic nervous system, the branch of the nervous system called into play during stress-related "fight-or-flight" situations (see p. 238). Also, a parasympathetic anti-inflammatory pathway helps rein in the inflammatory response. The vagus nerve, the main nerve of the parasympathetic nervous system, supplies the spleen and other

▌**TABLE 12-4** B versus T Lymphocytes

Characteristic	B Lymphocytes	T Lymphocytes
Ancestral origin	Bone marrow	Bone marrow
Site of maturational processing	Bone marrow	Thymus
Receptors for antigen	B-cell receptors are antibodies inserted in the plasma membrane; highly specific	T-cell receptors present in the plasma membrane are not the same as antibodies; highly specific
Bind with	Extracellular antigens such as bacteria, free viruses, and other circulating foreign material	Foreign antigen in association with self-antigen, such as virus-infected cells
Types of active cells	Plasma cells	Cytotoxic T cells, helper T cells, regulatory T cells
Formation of memory cells	Yes	Yes
Type of immunity	Antibody-mediated immunity	Cell-mediated immunity
Secretory product	Antibodies	Cytokines
Functions	Help eliminate free foreign invaders (mostly bacteria) by enhancing innate immune responses against them	Lyse virus-infected cells and cancer cells; aid B cells in antibody production; modulate immune responses
Life span	Short	Long

Exercise: A Help or Hindrance to Immune Defense?

FOR YEARS PEOPLE WHO ENGAGE in moderate exercise regimens have claimed they have fewer colds when they are in good aerobic condition. In contrast, elite athletes and their coaches have often complained about the number of upper respiratory infections that the athletes seem to contract at the height of their competitive seasons. The results of recent scientific studies lend support to both these claims. The effect of exercise on immune defense depends on the intensity of the exercise.

Animal studies have shown that high-intensity exercise after experimentally induced infection results in more severe infection. Moderate exercise performed before infection or to tumor implantation, in contrast, results in less severe infection and slower tumor growth in experimental animals.

Studies on humans lend further support to the hypothesis that exhaustive exercise suppresses immune defense, whereas moderate exercise stimulates the immune system. A survey of 2300 runners competing in a major marathon indicated that those who ran more than 60 miles a week had twice the number of respiratory infections as those who ran less than 20 miles a week in the two months preceding the race. In another study, 10 elite athletes were asked to run on a treadmill for 3 hours at the same pace they would run in competition. Blood tests after the run indicated that natural killer (NK) cell activity had decreased by 25% to 50%, and this decrease lasted for 6 hours. The runners also showed a 60% increase in the stress hormone cortisol, which is known to suppress immunity. Other studies have shown that athletes have lower resting salivary IgA levels compared with control subjects and that their respiratory mucosal immunoglobulins are decreased after prolonged exhaustive exercise. These results suggest a lower resistance to respiratory infection following high-intensity exercise. Because of these results, researchers in the field recommend that athletes keep exposure to respiratory viruses to a minimum by avoiding crowded places or anyone with a cold or flu for the first 6 hours after strenuous competition.

However, a study evaluating the effects of a moderate exercise program in which a group of women walked 45 minutes a day, 5 days a week, for 15 weeks found that the walkers' antibody levels and NK cell activity increased throughout the exercise program. Other studies using moderate exercise on stationary bicycles in subjects older than age 65 showed increases in NK cell activity as large as those found in young people.

---

sites where macrophages are abundant. Cytokine-secreting macrophages, the orchestrators of many aspects of inflammation, have receptors for acetylcholine (ACh). Even though ACh is typically the parasympathetic postganglionic neurotransmitter, vagal nerve endings in the spleen are adrenergic, not cholinergic. What is the source then of the ACh to which splenic macrophages respond? In an interesting twist, vagal stimulation of a population of memory T cells in the spleen causes these immune cells to produce the ACh that inhibits cytokine secretion by the splenic macrophages, in this way suppressing inflammation. In the reverse direction (immune system influencing nervous system), immune system secretions act on the brain to produce fever and other general symptoms that accompany infections. Furthermore, immune cells secrete some traditional hormones once thought to be produced only by the endocrine system. For example, many of the hormones secreted by the pituitary gland are produced by lymphocytes as well. Scientists are only beginning to sort out the mechanisms and implications of the many complex neuro–endocrine–immune interactions. (For a discussion of the possible effects of exercise on immune defense, see the accompanying boxed feature, ▮A Closer Look at Exercise Physiology.)

Check Your Understanding 12.5

1. List the three major types of T cells and state their functions.
2. Describe the role of antigen-presenting cells.
3. By what means is self-tolerance achieved?
4. Define *immune surveillance* and discuss how it is accomplished.

12.6 Immune Diseases

Abnormal functioning of the immune system can lead to immune diseases in two general ways: *immunodeficiency diseases* (too little immune response) and *inappropriate immune attacks* (too much or mistargeted immune response).

Immunodeficiency diseases result from insufficient immune responses.

Clinical Note **Immunodeficiency diseases** occur when the immune system fails to respond adequately to foreign invasion. The condition may be congenital (present at birth) or acquired (nonhereditary), and it may involve impairment of antibody-mediated or cell-mediated immunity, or both.

In a rare hereditary condition known as **severe combined immunodeficiency**, both B and T cells are lacking. Such people have extremely limited defenses against pathogenic organisms and die in infancy unless maintained in a germ-free environment (that is, live in a "bubble"). However, that verdict has changed with recent successes using gene therapy to cure the disease in some patients.

Acquired (nonhereditary) immunodeficiency states can arise from inadvertent destruction of lymphoid tissue during prolonged therapy with anti-inflammatory agents, such as cortisol derivatives, or from cancer therapy aimed at destroying rapidly dividing cells (which unfortunately include lymphocytes as well as cancer cells). The most recent and tragically the most common acquired immunodeficiency disease is AIDS, which, as described earlier, is caused by HIV, a virus that invades and incapacitates the critical helper T cells.

Let us now look at inappropriate immune attacks.

Allergies are inappropriate immune attacks against harmless environmental substances.

Inappropriate immune attacks cause reactions harmful to the body. These include (1) *autoimmune diseases,* in which the immune system turns against one of the body's tissues; (2) *immune complex diseases,* which involve overexuberant antibody responses that "spill over" and damage normal tissue; and (3) allergies. The first two conditions have been described earlier in this chapter, so we now concentrate on allergies.

An **allergy** is the acquisition of an inappropriate specific immune reactivity, or **hypersensitivity**, to a normally harmless environmental substance, such as dust or pollen. The offending agent is known as an **allergen**. Subsequent re-exposure of a sensitized individual to the same allergen elicits an immune attack, which may vary from a mild, annoying reaction to a severe, body-damaging reaction that may even be fatal.

Allergic responses are classified into two categories: immediate hypersensitivity and delayed hypersensitivity. In **immediate hypersensitivity**, the allergic response appears within about 20 minutes after a sensitized person is exposed to an allergen. In **delayed hypersensitivity**, the reaction does not generally show up until a day or so following exposure. The difference in timing is the result of the different mediators involved. A particular allergen may activate either a B- or a T-cell response. Immediate allergic reactions involve B cells and are elicited by antibody interactions with an allergen; delayed reactions involve T cells and the more slowly responding process of cell-mediated immunity against the allergen. Let us examine the causes and consequences of each of these reactions in more detail.

Triggers for Immediate Hypersensitivity

In immediate hypersensitivity, the antibodies involved and the events that ensue on exposure to an allergen differ from the typical antibody-mediated response to bacteria. The most common allergens that provoke immediate hypersensitivities are pollen grains, bee stings, penicillin, certain foods, molds, dust, feathers, and animal fur. (Actually, people allergic to cats are not allergic to the fur itself. The true allergen is in the cat's saliva, which is deposited on the fur during licking. Likewise, people are not allergic to dust or feathers per se but to tiny mites that inhabit the dust or feathers.) For unclear reasons, these allergens bind to and elicit the synthesis of IgE antibodies rather than the IgG antibodies associated with bacterial antigens. IgE antibodies are the least plentiful immunoglobulin, but their

presence spells trouble. Without IgE antibodies, there would be no immediate hypersensitivity. When a person with an allergic tendency is first exposed to a particular allergen, compatible helper T cells secrete IL-4, a cytokine that prods compatible B cells to synthesize IgE antibodies specific for the allergen. During this initial **sensitization period** no symptoms are evoked, but memory cells form that are primed for a more powerful response on subsequent re-exposure to the same allergen.

In contrast to the antibody-mediated response elicited by bacterial antigens, IgE antibodies do not freely circulate. Instead, their tail portions attach to mast cells and basophils, both of which produce and store an arsenal of potent inflammatory chemicals, such as histamine, in preformed granules. Mast cells are most plentiful in regions that come into contact with the external environment, such as the skin, the outer surface of the eyes, and the linings of the respiratory system and digestive tract. Binding of an appropriate allergen with the outreached arm regions of the IgE antibodies that are lodged tail-first in a mast cell or basophil triggers the rupture of the cell's granules. As a result, histamine and other chemical mediators spew forth into the surrounding tissue.

A single mast cell (or basophil) may be coated with several different IgE antibodies, each able to bind with a different allergen. Thus, the mast cell can be triggered to release its chemical products by any one of several allergens (∎Figure 12-23).

Chemical Mediators of Immediate Hypersensitivity

Following are among the most important reaction-causing chemicals released during immediate allergic reactions:

1. *Histamine,* which brings about vasodilation and increased capillary permeability and increased mucus production.

2. **Slow-reactive substance of anaphylaxis (SRS-A)**, which induces prolonged and profound contraction of smooth muscle, especially of the small respiratory airways. SRS-A is a collection of three related *leukotrienes,* locally acting eicosanoids similar to prostaglandins (see p. 119).

3. **Eosinophil chemotactic factor**, which specifically attracts eosinophils to the area. Interestingly, eosinophils release enzymes that inactivate SRS-A and may inhibit histamine, perhaps serving as an "off switch" to limit the allergic response.

Symptoms of Immediate Hypersensitivity

Symptoms of immediate hypersensitivity vary depending on the site, allergen, and mediators involved. Most frequently, the reaction is localized to the body site in which the IgE-bearing cells first come into contact with the allergen. If the reaction is limited to the upper respiratory passages after a person inhales an allergen such as ragweed pollen, the released chemicals bring about the symptoms characteristic of **hay fever**—nasal congestion caused by histamine-induced localized edema and sneezing and runny nose caused by increased mucus secretion. If the reaction is concentrated primarily within the bronchioles (the small respiratory airways that lead to the tiny air sacs within the lungs), **asthma** results. Contraction of the smooth muscle in the walls of the bronchioles in response to SRS-A narrows or constricts these passageways, making breathing difficult. Localized swelling in

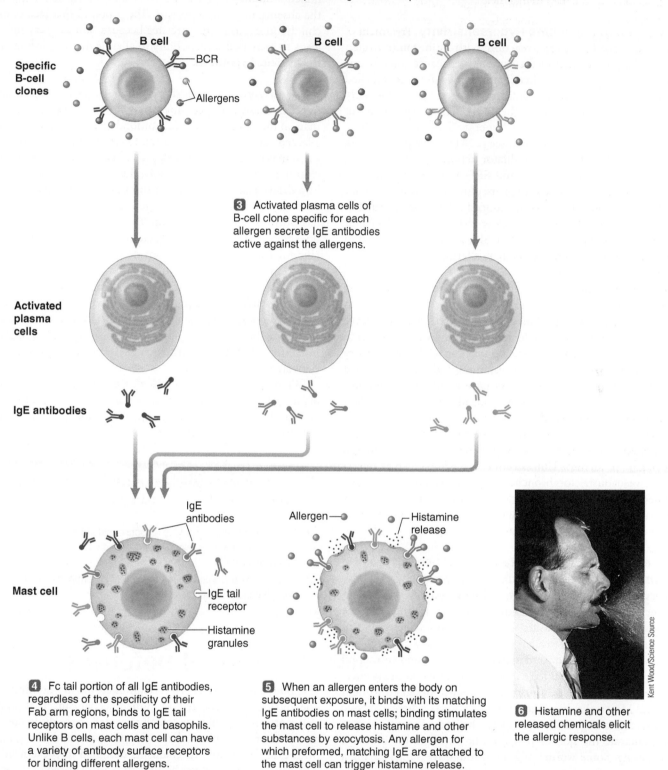

1 Allergens (antigens) enter the body for the first time.

2 Allergens bind to matching BCRs; B cells now process the allergens and, with stimulation by helper T cells (not shown), proceed through the steps leading to clonal expansion of activated plasma cells.

Specific B-cell clones

B cell

BCR

Allergens

B cell

B cell

3 Activated plasma cells of B-cell clone specific for each allergen secrete IgE antibodies active against the allergens.

Activated plasma cells

IgE antibodies

Mast cell

IgE antibodies

IgE tail receptor

Histamine granules

Allergen

Histamine release

Kent Wood/Science Source

4 Fc tail portion of all IgE antibodies, regardless of the specificity of their Fab arm regions, binds to IgE tail receptors on mast cells and basophils. Unlike B cells, each mast cell can have a variety of antibody surface receptors for binding different allergens.

5 When an allergen enters the body on subsequent exposure, it binds with its matching IgE antibodies on mast cells; binding stimulates the mast cell to release histamine and other substances by exocytosis. Any allergen for which preformed, matching IgE are attached to the mast cell can trigger histamine release.

6 Histamine and other released chemicals elicit the allergic response.

Figure 12-23 Role of IgE antibodies and mast cells in immediate hypersensitivity. B-cell clones are converted into plasma cells, which secrete IgE antibodies on contact with the allergen for which they are specific. All IgE antibodies, regardless of their antigen specificity, bind to mast cells or basophils. When an allergen combines with the IgE receptor specific for it on the surface of a mast cell, the mast cell releases histamine and other chemicals from preformed granules by exocytosis. These chemicals elicit the allergic response.

the skin because of allergy-induced histamine release causes **hives**. An allergic reaction in the digestive tract in response to an ingested allergen can lead to diarrhea.

Treatment of Immediate Hypersensitivity Treatment of localized immediate allergic reactions with antihistamines often offers only partial relief of the symptoms because some manifestations are invoked by other chemical mediators not blocked by these drugs. For example, antihistamines are not particularly effective in treating asthma, the most serious symptoms of which are invoked by SRS-A. Adrenergic drugs (which mimic the sympathetic nervous system; see p. 241) are helpful through their vasoconstrictor–bronchodilator actions in counteracting the effects of both histamine and SRS-A. Anti-inflammatory drugs such as cortisol derivatives are often used as the primary treatment for ongoing allergen-induced inflammation, such as that associated with asthma. Newer drugs such as *Singulair* that inhibit leukotrienes, including SRS-A, have been added to the arsenal for combating immediate allergies.

Anaphylactic Shock A life-threatening systemic (involving the entire body) reaction known as **anaphylactic shock** can occur if the allergen becomes blood borne or if very large amounts of chemicals are released from the localized site into the circulation. Severe hypotension that can lead to circulatory shock (see p. 374) results from histamine-induced widespread vasodilation and a massive shift of plasma fluid into the interstitial spaces as a result of a generalized increase in capillary permeability. Concurrently, pronounced bronchiolar constriction occurs via SRS-A action and can lead to respiratory failure. The person may suffocate from an inability to move air through the narrowed airways. Unless countermeasures, such as injecting a vasoconstrictor–bronchodilator drug, are undertaken immediately, anaphylactic shock is often fatal. This reaction is the reason even a single bee sting or eating a peanut can be so dangerous in people sensitized to these allergens.

Immediate Hypersensitivity and Absence of Parasitic Worms Although the immediate hypersensitivity response differs considerably from the typical IgG antibody response to bacterial infections, it is strikingly similar to the immune response elicited by parasitic worms. Shared characteristics of the immune reactions to allergens and parasitic worms include IgE antibody production and increased basophil and eosinophil activity. This finding has led to the proposal that harmless allergens somehow trigger an immune response designed to fight worms. Mast cells are concentrated in areas where parasitic worms (and allergens) could contact the body. Parasitic worms can penetrate the skin or penetrate or attach to the digestive tract lining. Some worms migrate through the lungs during a part of their life cycle. Scientists suspect the IgE response helps ward off these invaders as follows. The inflammatory response in the skin could wall off parasitic worms attempting to burrow in. Coughing and sneezing could expel worms that migrated to the lungs. Diarrhea could help flush out worms before they could penetrate or attach to the digestive tract lining. Epidemiological studies suggest that the incidence of allergies in a country rises as the presence of parasites decreases. Thus,

superfluous immediate hypersensitivity responses to normally harmless allergens may represent a pointless marshaling of a honed immune-response system "with nothing better to do" in the absence of parasitic worms. The proposal that lack of early childhood exposure to parasites because of a sanitary lifestyle leads to increased susceptibility to allergic diseases is known as the **hygiene hypothesis.**

Delayed Hypersensitivity Some allergens invoke delayed hypersensitivity, a T-cell-mediated immune response, rather than an immediate, B cell–IgE antibody response. Among these allergens are poison ivy toxin and certain chemicals to which the skin is frequently exposed, such as cosmetics and household cleaning agents. Most commonly, the response is characterized by a delayed skin eruption that reaches its peak intensity 1 to 3 days after contact with an allergen to which the T-cell system has previously been sensitized. To illustrate, poison ivy toxin does not harm the skin on contact, but it activates T cells specific for the toxin, including formation of a memory component. On subsequent exposure to the toxin, activated T cells diffuse into the skin within a day or two, combining with the poison ivy toxin that is present. The resulting interaction gives rise to the tissue damage and discomfort typical of the condition. The best relief is obtained from application of anti-inflammatory preparations, such as those containing cortisol derivatives.

▐ Table 12-5 summarizes the distinctions between immediate and delayed hypersensitivities. This completes our discussion of the immune defense system. We now turn to external defenses that thwart entry of foreign invaders.

Check Your Understanding 12.6

1. Name the three categories of inappropriate adaptive immune attacks.

2. Compare the type of immune response and immune effectors involved in immediate hypersensitivity and delayed hypersensitivity.

3. Discuss how the mechanism of action of IgE antibodies differs from that of IgG antibodies.

12.7 External Defenses

The body's defenses against foreign microbes are not limited to the intricate, interrelated immune mechanisms that destroy microorganisms that have actually invaded the body. In addition to the internal immune defense system, the body is equipped with external defense mechanisms designed to prevent microbial penetration wherever body tissues are exposed to the external environment. All of our epithelial surfaces, namely the skin and the linings of the digestive tract, the urogenital (urinary and reproductive) tracts, and respiratory airways and lungs, are protected by antimicrobial peptides called **defensins**. Epithelial cells of these surfaces secrete defensins on attack by microbial pathogens, thereby killing the would-be invaders by disrupting their membranes.

TABLE 12-5 Immediate versus Delayed Hypersensitivity Reactions

Characteristic	Immediate Hypersensitivity Reaction	Delayed Hypersensitivity Reaction
Time of onset of symptoms after exposure to the allergen	Within 20 minutes	Within 1 to 3 days
Type of immune response	Antibody-mediated immunity against the allergen	Cell-mediated immunity against the allergen
Immune effectors involved	B cells, IgE antibodies, mast cells, basophils, histamine, slow-reactive substance of anaphylaxis, and eosinophil chemotactic factor	T cells
Allergies commonly involved	Hay fever, asthma, hives, and (in extreme cases) anaphylactic shock	Contact allergies, such as allergies to poison ivy, cosmetics, and household cleaning agents

The most obvious external defense is the **skin**, or **integument**, which covers the outside of the body (*integere* means "to cover").

The skin consists of an outer protective epidermis and an inner, connective tissue dermis.

The skin, which is the largest organ of the body, not only is a mechanical barrier between the external environment and the underlying tissues but is dynamically involved in defense mechanisms and other important functions as well. The skin in an average adult weighs 9 pounds and covers a surface area of about 20 square feet (1.86 m²). Its deeper layer contains an abundance of blood vessels, which if laid end to end would extend more than 11 miles. The skin consists of two layers, an outer *epidermis* and an inner *dermis* (Figure 12-24).

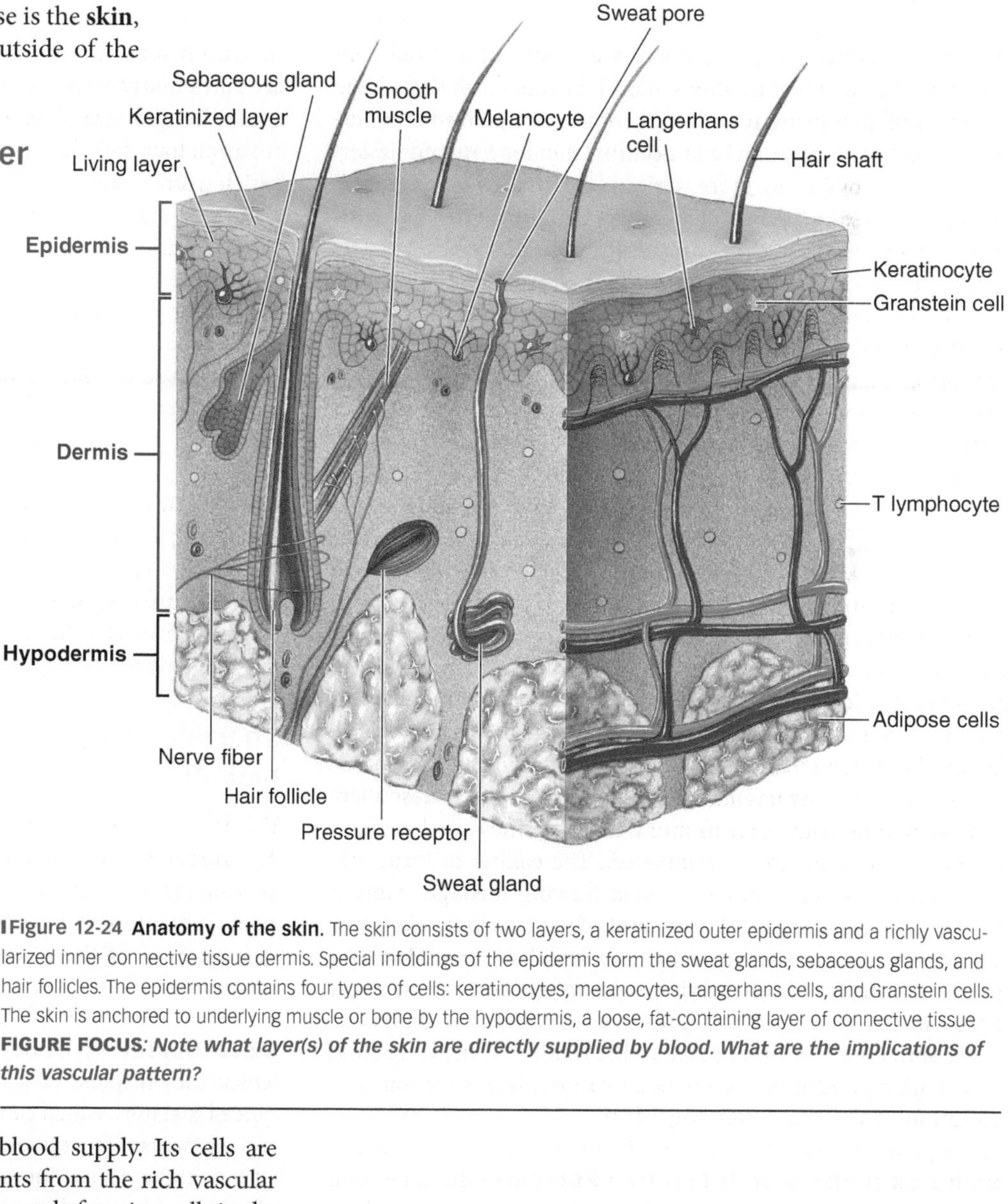

Epidermis The **epidermis** consists of numerous layers of epithelial cells. On average, the epidermis replaces itself about every 2.5 months. The inner epidermal layers are composed mostly of cube-shaped cells that are living and rapidly dividing, whereas the cells in the outer layers are dead and flattened. The epidermis has no direct blood supply. Its cells are nourished only by diffusion of nutrients from the rich vascular network in the underlying dermis. The newly forming cells in the

Figure 12-24 Anatomy of the skin. The skin consists of two layers, a keratinized outer epidermis and a richly vascularized inner connective tissue dermis. Special infoldings of the epidermis form the sweat glands, sebaceous glands, and hair follicles. The epidermis contains four types of cells: keratinocytes, melanocytes, Langerhans cells, and Granstein cells. The skin is anchored to underlying muscle or bone by the hypodermis, a loose, fat-containing layer of connective tissue **FIGURE FOCUS**: *Note what layer(s) of the skin are directly supplied by blood. What are the implications of this vascular pattern?*

inner layers constantly push the older cells closer to the surface and ever farther from their nutrient supply. This, coupled with the continuous subjection of the outer layers to pressure and "wear and tear," causes these older cells to die and become flattened. Epidermal cells are riveted tightly together by desmosomes (see p. 61), which interconnect with intracellular keratin filaments (see p. 51) to form a strong, cohesive covering. During maturation of a keratin-producing cell, keratin filaments progressively accumulate and cross-link with one another within the cytosol. As the outer cells die, this fibrous keratin core remains, forming flattened, hardened scales that provide a tough, protective **keratinized layer**. As the scales of the outermost keratinized layer slough or flake off through abrasion, they are continuously replaced by means of cell division in the deeper epidermal layers. The rate of cell division, and consequently the thickness of this keratinized layer, varies in different regions of the body. It is thickest in the areas where the skin is subjected to the most pressure, such as the bottom of the feet. The keratinized layer is airtight, fairly waterproof, and impervious to most substances. It resists anything passing in either direction between the body and the external environment.

Clinical Note The skin minimizes loss of water and other vital constituents from the body. This protective layer's value in holding in body fluids becomes obvious after severe burns. Bacterial infections can occur in the unprotected underlying tissue, but even more serious are the systemic consequences of losing body water and plasma proteins from the exposed, burned surface. The resulting circulatory disturbances can be life threatening.

Likewise, the skin barrier impedes passage into the body of most materials that come into contact with the body surface, including bacteria and toxic chemicals. In many instances, the skin modifies compounds that come into contact with it. For example, epidermal enzymes can convert many potential carcinogens into harmless compounds. Some materials, however, especially lipid-soluble substances, can penetrate intact skin through the lipid bilayers of the plasma membranes of the epidermal cells. Drugs that can be absorbed by the skin, such as nicotine or estrogen, are sometimes used in the form of a cutaneous "patch" impregnated with the drug.

Dermis Under the epidermis is the **dermis**, a connective tissue layer that contains many elastin fibers (for stretch) and collagen fibers (for strength) and an abundance of blood vessels and specialized nerve endings. The dermal blood vessels not only supply both the dermis and the epidermis but also play a major role in temperature regulation. The caliber of these vessels, and hence the volume of blood flowing through them, is subject to control to vary the amount of heat exchange between these skin surface vessels and the external environment (see p. 632). Receptors at the peripheral endings of afferent nerve fibers in the dermis detect pressure, temperature, pain, and other somatosensory input (see p. 182). Efferent nerve endings in the dermis control blood vessel caliber, hair erection, and secretion by the skin's exocrine glands.

Skin's Exocrine Glands and Hair Follicles Special infoldings of the epidermis into the underlying dermis form the skin's exocrine glands—the sweat glands and the sebaceous glands—as well as the hair follicles. **Sweat glands** of the most common type, which are distributed over most of the body, release a dilute salt solution through small openings, the sweat pores, onto the skin surface (❙ Figure 12-24). Evaporation of this sweat cools the skin and is important in regulating temperature.

The amount of sweat produced is subject to regulation and depends on the environmental temperature, the amount of heat-generating skeletal muscle activity, and various emotional factors (for example, a person often sweats when nervous). A special type of sweat gland located in the axilla (armpit) and pubic region produces a protein-rich sweat that supports the growth of surface bacteria, which give rise to a characteristic odor. In contrast, most sweat, as well as the secretions from the sebaceous glands, contains chemicals that are generally highly toxic to bacteria.

The cells of the **sebaceous glands** produce **sebum**, an oily secretion released into adjacent hair follicles. From there sebum flows to the skin surface, oiling both the hairs and the outer keratinized layers of the skin, helping to waterproof them and prevent them from drying and cracking. Chapped hands or lips indicate insufficient protection by sebum. The sebaceous glands are particularly active during adolescence, causing the oily skin common among teenagers.

Each **hair follicle** is lined by special keratin-producing cells, which secrete keratin and other proteins that form the hair shaft. Hairs increase the sensitivity of the skin's surface to tactile (touch) stimuli. In some other species, this function is more finely tuned. For example, the whiskers on a cat are exquisitely sensitive in this regard. An even more important role of hair in hairier species is heat conservation, but this function is not significant in us relatively hairless humans. Like hair, the **nails** are another special keratinized product derived from living epidermal structures, the nail beds.

Hypodermis The skin is anchored to the underlying tissue (muscle or bone) by the **hypodermis** (*hypo* means "below"), also known as **subcutaneous tissue** (*sub* means "under"; *cutaneous* means "skin"), a loose layer of connective tissue. Most fat cells are housed within the hypodermis. These fat deposits throughout the body are collectively referred to as **adipose tissue**.

Specialized cells in the epidermis produce melanin, keratin, and vitamin D and participate in immune defense.

The epidermis contains four resident cell types— *melanocytes, keratinocytes, Langerhans cells,* and *Granstein cells*—plus transient T lymphocytes scattered throughout the epidermis and dermis. Each of these resident cell types performs specialized functions.

Melanocytes **Melanocytes** produce the pigment **melanin**, which they disperse to surrounding skin cells. The amount and type of melanin, which can vary among black, brown, yellow, and red pigments, are responsible for the shades of skin color of the various races. Fair-skinned people have about the same number of melanocytes as dark-skinned people; the difference in skin color depends on the amount of melanin produced by each melanocyte. Melanin is produced through complex biochemical pathways in which the melanocyte enzyme *tyrosinase* plays a key role. Most people, regardless of skin color, have enough tyrosinase that, if fully functional, could result in enough melanin to make their skin very black. In those with lighter skin, however, two genetic factors prevent this melanocyte enzyme from functioning at full capacity: (1) much of the tyrosinase produced is in an inactive form, and (2) various inhibitors that block tyrosinase action are produced. As a result, less melanin is produced.

In addition to hereditary determination of melanin content, the amount of this pigment can be increased transiently in response to exposure to ultraviolet (UV) light rays from the sun. On exposure to UV light, keratinocytes secrete **α-melanocyte stimulating hormone (α-MSH)**, which acts as a paracrine on neighboring melanocytes to darken the skin. This additional melanin, the outward appearance of which constitutes a "tan," performs the protective function of absorbing harmful UV rays.

Keratinocytes By far the most abundant epidermal cells are the **keratinocytes**, which, as the name implies, are specialists in keratin production. As they die, they form the outer protective keratinized layer. They also generate keratin-rich hair and nails.

Keratinocytes also synthesize vitamin D in the presence of sunlight. Vitamin D, which is derived from a precursor molecule closely related to cholesterol, promotes the absorption of Ca^{2+} from the digestive tract into the blood (Chapter 16). Dietary supplements of vitamin D are usually required because typically the skin is not exposed to sufficient sunlight to produce adequate amounts of this essential chemical.

Furthermore, keratinocytes are important immunologically. They secrete IL-1 (a product also secreted by macrophages), which influences the maturation of T cells that tend to localize in the skin. Interestingly, the epithelial cells of the thymus bear anatomic, molecular, and functional similarities to those of the skin. Some post-thymic steps in T-cell maturation take place in the skin under keratinocyte guidance.

Other Immune Cells of the Skin The two other epidermal cell types also play a role in immunity. **Langerhans cells**, which migrate to the skin from bone marrow, are dendritic cells that serve as antigen-presenting cells. Thus, the skin not only is a mechanical barrier but actually alerts lymphocytes if the barrier is breached by invading microorganisms. Langerhans cells present antigen to helper T cells, facilitating their responsiveness to skin-associated antigens. In contrast, **Granstein cells** act as a "brake" on skin-activated immune responses. They are antigen-presenting cells to skin-associated T$_{regs}$, thus exerting an immune suppressor effect. Significantly, Langerhans cells are more susceptible to damage by UV radiation than Granstein cells are. Losing Langerhans cells as a result of exposure to UV radiation can detrimentally lead to a predominant suppressor signal rather than the normally dominant helper signal, leaving the skin more vulnerable to microbial invasion and cancer cells.

The various epidermal components of the immune system are collectively termed **skin-associated lymphoid tissue**, or **SALT**. SALT plays an important protective role, because the skin is a major interface with the external environment.

Protective measures within body cavities discourage pathogen invasion into the body.

The human body's defense system must guard against entry of potential pathogens not only through the outer surface of the body but also through the internal cavities that communicate directly with the external environment—namely, the digestive system, the urogenital system, and the respiratory system. These systems use various tactics to destroy microorganisms entering through these routes.

Defenses of the Digestive System Saliva secreted into the mouth at the entrance of the digestive system contains an enzyme that lyses certain ingested bacteria. "Friendly" bacteria that live on the back of the tongue convert food-derived nitrate into nitrite, which is swallowed. Acidification of nitrite on reaching the highly acidic stomach generates nitric oxide, which is toxic to a variety of microorganisms. Furthermore, many of the surviving bacteria that are swallowed are killed directly by the strongly acidic gastric juice in the stomach. Farther down the tract, the intestinal lining is endowed with gut-associated lymphoid tissue. These defensive mechanisms are not 100% effective, however. Some bacteria do manage to survive and reach the large intestine (the last portion of the digestive tract), where they continue to flourish. Surprisingly, this normal microbial population provides a natural barrier against infection within the lower intestine. These harmless resident microbes competitively suppress the growth of potential pathogens that have managed to escape the antimicrobial measures of earlier parts of the digestive tract.

Clinical Note Occasionally, orally administered antibiotic therapy against an infection elsewhere within the body may induce an intestinal infection. By knocking out some normal intestinal bacteria, an antibiotic may permit an antibiotic-resistant pathogenic species to overgrow in the intestine.

Defenses of the Urogenital System Within the urogenital system, would-be invaders encounter hostile conditions in the acidic urine and acidic vaginal secretions. The urogenital organs also produce a sticky mucus, which, like flypaper, entraps small invading particles. Subsequently, the particles are either engulfed by phagocytes or are swept out as the organ empties (for example, they are flushed out with urine flow).

Defenses of the Respiratory System The respiratory system is likewise equipped with several important defense mechanisms against inhaled particulate matter. The respiratory system is the largest surface of the body that comes into direct contact with the increasingly polluted external environment. The surface area of the respiratory system exposed to the air is about 40 times that of the skin. Larger airborne particles are filtered out of the inhaled air by hairs at the entrance of the nasal passages. Lymphoid tissues, the *tonsils* and *adenoids,* provide immunological protection against inhaled pathogens near the beginning of the respiratory system. Farther down in the respiratory airways, millions of tiny hairlike projections known as *cilia* constantly beat in an outward direction (see p. 48). The respiratory airways are coated with a layer of thick, sticky mucus secreted by epithelial cells within the airway lining. This mucus sheet, laden with any inspired particulate debris (such as dust) that adheres to it, is constantly moved upward to the throat by ciliary action. This moving "staircase" of mucus is known as the **mucus escalator**. The dirty mucus is either expectorated (spit out) or, in most cases, is swallowed without the person even being aware of it; any indigestible foreign particulate matter is later eliminated in the feces. Besides keeping the lungs clean, this mechanism is an important defense against bacterial infection because many bacteria enter the body on dust particles. Also contributing to defense against respiratory infections are IgA antibodies secreted in the mucus. In addition, an abundance of phagocytic specialists called **alveolar macrophages** scavenge within the air sacs (alveoli) of the lungs. Further respiratory defenses include coughs and sneezes. These commonly experienced reflex mechanisms involve forceful outward expulsion of material in an attempt to remove irritants from the trachea (*coughs*) or nose (*sneezes*). A sneeze can propel air and particles from the respiratory airways up to 30 feet away at rates more than 90 miles per hour.

Clinical Note Cigarette smoking suppresses these normal respiratory defenses. The smoke from a single cigarette can paralyze the cilia for several hours, with repeated exposure eventually leading to ciliary destruction. Failure of ciliary activity to sweep out a constant stream of particulate-laden mucus enables inhaled carcinogens to remain in contact with the respiratory airways for prolonged periods. Furthermore, cigarette smoke incapacitates alveolar macrophages. In addition, noxious agents in tobacco smoke irritate the mucous linings of the respiratory tract, resulting in excess mucus production, which may partially obstruct the airways. "Smoker's cough" is an attempt to dislodge this excess stationary mucus. These and other direct toxic effects on lung tissue lead to the increased incidence of lung cancer and chronic respiratory diseases associated with cigarette smoking. Air pollutants include some of the same substances found in cigarette smoke and can similarly affect the respiratory system.

We examine the respiratory system in greater detail in the next chapter.

Check Your Understanding 12.7

1. Describe the epidermis, dermis, and hypodermis.
2. List and state the functions of the four resident cell types in the skin.
3. Tell how the mucus escalator works.

Homeostasis: Chapter in Perspective

 We could not survive beyond early infancy were it not for the body's defense mechanisms. These mechanisms resist and eliminate potentially harmful foreign agents with which we continuously come into contact in our hostile external environment and also destroy abnormal cells that often arise within the body. Homeostasis can be optimally maintained, and thus life sustained, only if the body cells are not physically injured or functionally disrupted by pathogenic microorganisms or are not replaced by abnormally functioning cells, such as traumatized cells or cancer cells. The immune defense system—a complex, multifaceted, interactive network of leukocytes, their secretory products, and plasma proteins—contributes indirectly to homeostasis by keeping other cells alive so that they can perform their specialized activities to maintain a stable internal environment. The immune system protects the other healthy cells from foreign agents that have gained entrance to the body, clears away dead and injured cells to pave the way for replacement with healthy new cells, and eliminates newly arisen cancer cells.

The skin contributes indirectly to homeostasis by serving as a protective barrier between the external environment and the rest of the body cells. It helps prevent harmful foreign agents such as pathogens and toxic chemicals from entering the body and helps prevent the loss of precious internal fluids from the body. The skin also contributes directly to homeostasis by helping maintain body temperature by means of the sweat glands and adjustments in skin blood flow. The amount of heat carried to the body surface for dissipation to the external environment is determined by the volume of warmed blood flowing through the skin.

Other systems that have internal cavities in contact with the external environment, such as the digestive, urogenital, and respiratory systems, also have defense capabilities to prevent harmful external agents from entering the body through these avenues.

Review Exercises   Answers begin on p. A-38

Reviewing Terms and Facts

1. The complement system can be activated only by antibodies. (*True or false?*)

2. Specific adaptive immune responses are accomplished by neutrophils. (*True or false?*)

3. Damaged tissue is always replaced by scar tissue. (*True or false?*)

4. Active immunity against a particular disease can be acquired only by actually having the disease. (*True or false?*)

5. A secondary response has a more rapid onset, is more potent, and has a longer duration than a primary response. (*True or false?*)

6. _____ are receptors on the plasma membrane of phagocytes that recognize and bind with telltale molecular patterns present on the surface of microorganisms but absent from human cells.

7. The complement system's _____ forms a doughnut-shaped complex that embeds in a microbial surface membrane, causing osmotic lysis of the victim cell.

8. _____ is a collection of phagocytic cells, necrotic tissue, and bacteria.

9. _____ is the localized response to microbial invasion or tissue injury that is accompanied by swelling, heat, redness, and pain.

10. A chemical that enhances phagocytosis by serving as a link between a microbe and the phagocytic cell is known as a(n) _____.

11. All the chemical messengers other than antibodies secreted by lymphocytes that regulate immune process are categorized as _____.

12. _____ molecules released by killer cells form porelike channels that punch holes in a victim cell, and _____ released by killer cells trigger apoptosis of victim cells.

13. Which of the following statements concerning leukocytes is/are *incorrect*?
   a. Monocytes are transformed into macrophages.
   b. T lymphocytes are transformed into plasma cells that secrete antibodies.
   c. Neutrophils are highly mobile phagocytic specialists.
   d. Basophils release histamine.
   e. Lymphocytes arise in large part from lymphoid tissues.

14. Match the following:
   1. a family of proteins that nonspecifically defend against viral infection
   2. a response to tissue injury in which neutrophils and macrophages play a major role
   3. a group of plasma proteins that, when activated, bring about destruction of foreign cells by attacking their plasma membranes
   4. lymphocytelike entities that spontaneously lyse tumor cells and virus-infected host cells

   (a) complement system
   (b) natural killer (NK) cells
   (c) interferon
   (d) inflammation

15. Using the answer code on the right, indicate whether the numbered characteristics of the adaptive immune system apply to antibody-mediated immunity or cell-mediated immunity (or both):
   1. involves secretion of antibodies
   2. mediated by B cells
   3. mediated by T cells
   4. accomplished by thymus-educated lymphocytes
   5. triggered by the binding of specific antigens to complementary lymphocyte receptors
   6. involves formation of memory cells in response to initial exposure to an antigen
   7. primarily aimed against virus-infected host cells
   8. protects primarily against bacterial invaders
   9. directly destroys targeted cells
   10. involved in rejection of transplanted tissue
   11. requires binding of a lymphocyte to a free extracellular antigen
   12. requires dual binding of a lymphocyte with both foreign antigen and self-antigens present on the surface of a host cell

   (a) antibody-mediated immunity
   (b) cell-mediated immunity
   (c) both antibody-mediated and cell-mediated immunity

16. Using the answer code on the right, indicate whether the numbered characteristics apply to the epidermis or dermis:
   1. is the inner layer of skin
   2. has layers of epithelial cells that are dead and flattened
   3. has no direct blood supply
   4. contains sensory nerve endings
   5. contains keratinocytes
   6. contains melanocytes
   7. contains rapidly dividing cells
   8. is mostly connective tissue

   (a) epidermis
   (b) dermis

Understanding Concepts
(Answers at www.cengagebrain.com)

1. Distinguish between bacteria and viruses.
2. Summarize the functions of each of the lymphoid tissues.

3. Distinguish between innate and adaptive immune responses.

4. Compare the life histories of B cells and T cells.

5. Describe the structure of an antibody. List and describe the five subclasses of immunoglobulins.

6. In what ways do antibodies exert their effect?

7. Describe the clonal selection theory.

8. Compare the functions of B cells and T cells. What are the roles of the three types of T cells?

9. Summarize the functions of macrophages in immune defense.

10. What mechanisms are involved in self-tolerance?

11. What is the importance of class I and class II MHC glycoproteins?

12. Describe the factors that contribute to immune surveillance against cancer cells.

13. Distinguish among immunodeficiency disease, autoimmune disease, immune complex disease, immediate hypersensitivity, and delayed hypersensitivity.

14. What are the immune functions of the skin?

Solving Quantitative Exercises

1. As a result of the innate immune response to an infection—for example, from a cut on the skin—capillary walls near the site of infection become very permeable to plasma proteins that normally remain in the blood. These proteins diffuse into the interstitial fluid, raising the interstitial fluid–colloid osmotic pressure (π_{IF}). Increased π_{IF} causes fluid to leave the circulation and accumulate in the tissue, forming a welt, or wheal. This process is referred to as the *wheal response*. The wheal response is mediated in part by histamine secreted from mast cells in the area of infection. The histamine binds to receptors, called *H-1 receptors*, on capillary endothelial cells. The histamine signal is transduced via the Ca^{2+} second-messenger pathway involving phospholipase C (see p. 124). In response to this signal, the capillary endothelial cells contract (via internal actin–myosin interaction), which causes a widening of the intercellular gaps (pores) between the capillary endothelial cells (see p. 354). In addition, substance P (see p. 190) also contributes to pore widening. Plasma proteins can pass through these widened pores and leave the capillaries. Looking at ❙Figure 10-20,

p. 356, compare the magnitude of the wheal response (that is, the extent of localized edema) if π_{IF} were raised (a) from 0 mm Hg to 5 mm Hg and (b) from 0 mm Hg to 10 mm Hg. In both cases, compare the net exchange pressure (NEP) at the arteriolar end of the capillary, the venular end of the capillary, and the average NEP. (Assume the other forces acting across the capillary wall remain unchanged.)

Applying Clinical Reasoning

Linda P. is allergic to molds and dust mites. She takes allergy medication as needed to keep her symptoms under control, but in the hopes of ridding herself of the troublesome allergies, she has been taking allergy (desensitization) shots. Allergy shots consist of weekly injections of minute, gradually increasing doses of the offending allergens. How can deliberately injecting the offending agent lead to a reduction in allergic response to the allergen? The leading theory regarding the mechanism of action of allergy shots is that the immune response to the allergen is gradually shifted from production of IgE antibodies to production of IgG antibodies against this antigen. How would this switch lead to a reduction in allergic symptoms on environmental exposure to higher doses of the antigen?

Thinking at a Higher Level

1. Compare the defense mechanisms that come into play in response to bacterial and viral pneumonia.

2. Three decades have passed since the first cases of AIDS were reported in the United States and millions of research dollars have been spent studying this disease. Much has been learned, and drugs have been developed that delay or manage the condition, but no AIDS vaccine has been approved despite many unsuccessful attempts. Why does the frequent mutation of HIV (the AIDS virus) make it difficult to develop a vaccine against this virus?

3. What effect would failure of the thymus to develop embryonically have on the immune system after birth?

4. Medical researchers are currently working on ways to "teach" the immune system to view foreign tissue as "self." What useful clinical application will the technique have?

5. When someone looks at you, are the cells of your body that person is viewing dead or alive?

To access the course materials and companion resources for this text, please visit www.cengagebrain.com

The Respiratory System

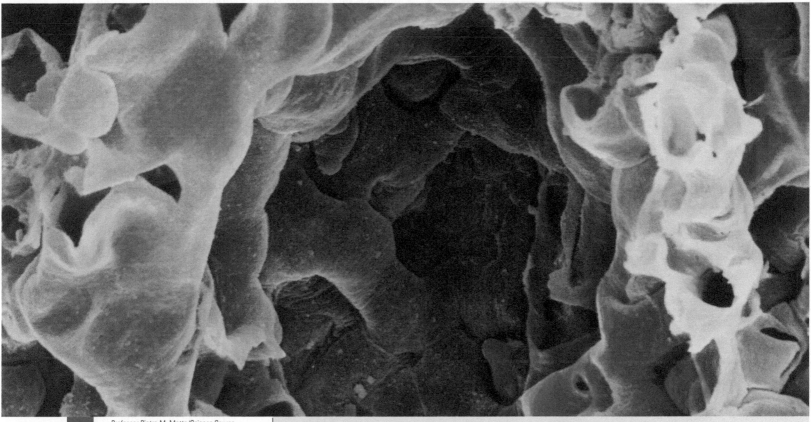

Professor Pietro M. Motta/Science Source

A scanning electron micrograph of an alveolus in the lung. An alveolus (air sac; *blue cavity*) enclosed by an extremely thin membrane (*white*) is surrounded by pulmonary capillaries (*red*). O_2 and CO_2 are exchanged across this thin barrier between air and blood. The large blue ridges at the rear of the alveolus denote the position of underlying capillaries.

Homeostasis Highlights

Energy is essential for sustaining life-supporting cellular activities, such as protein synthesis and active transport across plasma membranes. Body cells need a continuous supply of O_2 to support their energy-generating chemical reactions. The CO_2 produced during these reactions must be eliminated from the body at the same rate as it is produced to prevent dangerous fluctuations in pH (that is, to maintain acid–base balance) because CO_2 generates carbonic acid.

Respiration involves the sum of the processes that accomplish ongoing passive movement of O_2 from the atmosphere to the tissues to support cell metabolism and the continual passive movement of metabolically produced CO_2 from the tissues to the atmosphere. The **respiratory system** contributes to homeostasis by exchanging O_2 and CO_2 between the atmosphere and blood. The blood transports O_2 and CO_2 between the respiratory system and the tissues.

13.1 Respiratory Anatomy

The primary function of respiration is to obtain O_2 for use by the body cells and to eliminate the CO_2 the cells produce.

The respiratory system does not participate in all steps of respiration.

Most people think of respiration as the process of breathing in and breathing out. In physiology, however, respiration has a broader meaning. Respiration encompasses two separate but related processes: cellular respiration and external respiration.

Cellular Respiration The term **cellular respiration** refers to the intracellular metabolic processes carried out within the mitochondria, which use O_2 and produce CO_2 while deriving energy from nutrient molecules (see p. 35). The **respiratory quotient (RQ)**, the ratio of CO_2 produced to O_2 consumed, varies depending on the foodstuff consumed. When carbohydrate is being used, the RQ is 1—that is, for every molecule of O_2 consumed, one molecule of CO_2 is produced. For fat utilization, the RQ is 0.7; for protein, it is 0.8. On a typical American diet consisting of a mixture of these three nutrients, resting O_2 consumption averages about 250 mL/min, and CO_2 production averages about 200 mL/min, for an average RQ of 0.8:

$$RQ = \frac{CO_2 \text{ produced}}{O_2 \text{ consumed}} = \frac{(200 \text{ mL/min})}{(250 \text{ mL/min})} = 0.8$$

External Respiration The term **external respiration** refers to the entire sequence of events in exchange of O_2 and CO_2 between the external environment and tissue cells. External respiration, the topic of this chapter, encompasses four steps (Figure 13-1):

Step **1** Air is alternately moved into and out of the lungs so that air can be exchanged between the atmosphere (external environment) and air sacs (*alveoli*) of the lungs. This exchange is accomplished by the mechanical act of **breathing**, or **ventilation**. The rate of ventilation is regulated to adjust the flow of air between the atmosphere and alveoli according to the body's metabolic needs for O_2 uptake and CO_2 removal.

Step **2** O_2 and CO_2 are exchanged between air in the alveoli and blood within the pulmonary (*pulmonary* means "lung") capillaries by the process of diffusion.

Step **3** The blood transports O_2 and CO_2 between the lungs and the tissues.

Step **4** O_2 and CO_2 are exchanged between the tissue cells and blood by the process of diffusion across the systemic (tissue) capillaries.

The respiratory system does not accomplish all steps of respiration; it is involved only

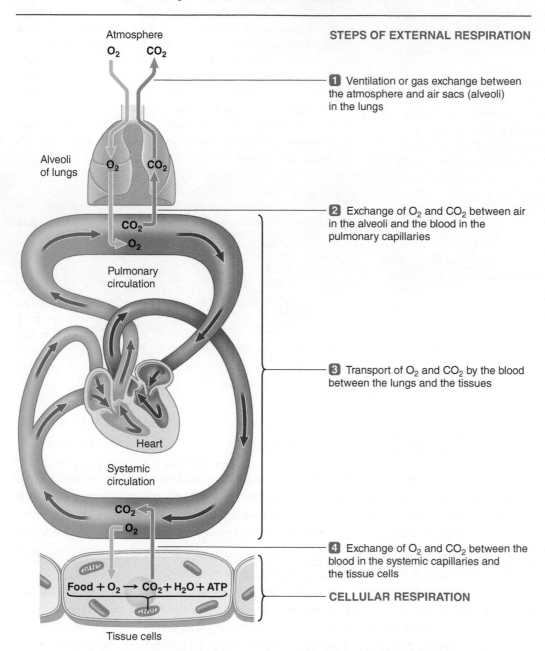

STEPS OF EXTERNAL RESPIRATION

Atmosphere
O_2 CO_2

1 Ventilation or gas exchange between the atmosphere and air sacs (alveoli) in the lungs

Alveoli of lungs
O_2 CO_2

CO_2
O_2

Pulmonary circulation

2 Exchange of O_2 and CO_2 between air in the alveoli and the blood in the pulmonary capillaries

Heart

Systemic circulation

3 Transport of O_2 and CO_2 by the blood between the lungs and the tissues

CO_2
O_2

4 Exchange of O_2 and CO_2 between the blood in the systemic capillaries and the tissue cells

Food + O_2 → CO_2 + H_2O + ATP

CELLULAR RESPIRATION

Tissue cells

Figure 13-1 External and cellular respiration. External respiration encompasses the steps involved in exchange of O_2 and CO_2 between the external environment and tissue cells (steps 1 through 4). Cellular respiration encompasses the intracellular metabolic reactions involving the use of O_2 to derive energy (ATP) from food, producing CO_2 as a by-product.

with ventilation and exchange of O_2 and CO_2 between the lungs and blood (steps ❶ and ❷). The circulatory system is involved in step ❷ and carries out steps ❸ and ❹.

Nonrespiratory Functions of the Respiratory System

The respiratory system also fills these nonrespiratory functions:

■ It is a route for water loss and heat elimination. Inspired (inhaled) atmospheric air is humidified and warmed by the respiratory airways before it is expired. Moistening of inspired air is essential to prevent the alveolar linings from drying out. O_2 and CO_2 cannot diffuse through dry membranes.

■ It enhances venous return (see "respiratory pump," p. 364).

■ It helps maintain normal acid–base balance by altering the amount of H^+-generating CO_2 exhaled (see p. 553).

■ It enables speech, singing, and other vocalization.

■ It defends against inhaled foreign matter (see p. 441).

■ It removes, modifies, activates, or inactivates various materials passing through the pulmonary circulation. All blood returning to the heart from the tissues must pass through the lungs before being returned to the systemic circulation. Thus the lungs are uniquely situated to act on specific materials that have been added to the blood at the tissue level before these substances have a chance to reach other parts of the body by means of the arterial system. For example, prostaglandins, a collection of chemical messengers released in many tissues to mediate particular local responses (see p. 119), may spill into the blood, but they are *inactivated* during passage through the lungs so that they cannot exert systemic effects. By contrast, the lungs *activate* angiotensin II, which is part of the renin–angiotensin–aldosterone hormonal pathway that plays an important role in regulating Na^+ concentration in the ECF (see p. 508).

■ The nose, a part of the respiratory system, is the organ of smell (see p. 227).

The respiratory airways conduct air between the atmosphere and alveoli.

The **respiratory system** includes the respiratory airways leading into the lungs, the lungs themselves, and the respiratory muscles of the thorax (chest) and abdomen involved in producing movement of air through the airways into and out of the lungs. The **respiratory airways** are tubes that carry air between the atmosphere and the air sacs, the latter being the only site where gases can be exchanged between air and blood. If all the airways were lined up end to end, they would extend 1500 miles. The airways begin with the **nasal passages (nose)** (❙ Figure 13-2a). The nasal passages open into the **pharynx (throat)**, which serves as a common passageway for both the respiratory and digestive systems. Two tubes lead from the pharynx—the **trachea (windpipe)**, through which air is conducted to the lungs, and the **esophagus**, the tube through which food passes to the stomach. Air normally enters the pharynx through the nose, but it can enter by the mouth as well when the nasal passages are congested—that is, you can breathe through your mouth when you have a cold. Because the pharynx serves as a common passageway for food and air, reflex mechanisms close

off the trachea during swallowing so that food does not enter the airways. The esophagus stays closed except during swallowing to keep air from entering the stomach during breathing.

The **larynx**, or **voice box**, is located at the entrance of the trachea. The anterior protrusion of the larynx forms the "Adam's apple." The **vocal folds**, two bands of elastic tissue that lie across the opening of the larynx, can be stretched and positioned in different shapes by laryngeal muscles (❙ Figure 13-3a). Air passes into the larynx through the space between the vocal folds. This laryngeal opening is known as the **glottis**. As air moves through the open glottis past the variably positioned, taut vocal folds, they vibrate to produce the many sounds of speech. The lips, tongue, and soft palate modify the sounds into recognizable sound patterns. During swallowing, the vocal folds assume a function not related to speech: They close the glottis. That is, laryngeal muscles bring the vocal folds into tight apposition to each other to close off the entrance to the trachea so that food does not get into the airways (❙ Figure 13-3b).

Beyond the larynx, the trachea divides into two main branches, the right and left **bronchi**, which enter the right and left lungs, respectively. Within each lung the bronchus continues to branch into progressively narrower, shorter, and more numerous airways, like the branching of a tree. The smaller branches are known as **bronchioles**. Clustered at the ends of the terminal bronchioles are the **alveoli**, the tiny air sacs where gases are exchanged between air and blood (see ❙ Figure 13-2b).

To permit airflow into and out of the gas-exchanging portions of the lungs, the continuum of conducting airways from the entrance through the terminal bronchioles to the alveoli must remain open. The trachea and larger bronchi are fairly rigid, nonmuscular tubes encircled by a series of cartilaginous rings that prevent these tubes from compressing. The smaller bronchioles have no cartilage to hold them open. Their walls contain smooth muscle that is innervated by the autonomic nervous system and is sensitive to certain hormones and local chemicals. These factors, by varying the degree of contraction of bronchiolar smooth muscle and hence the caliber of these small terminal airways, regulate the amount of air passing between the atmosphere and each cluster of alveoli.

The gas-exchanging alveoli are thin-walled air sacs encircled by pulmonary capillaries.

The lungs are ideally structured for gas exchange. According to Fick's law of diffusion, the shorter the distance and the greater the surface area through which diffusion takes place, the greater the rate of diffusion (see p. 65).

The alveoli are clusters of thin-walled, inflatable, grapelike sacs at the terminal branches of the conducting airways. The alveolar walls consist of a single layer of flattened, **Type I alveolar cells** (❙ Figure 13-4a and b). Each alveolus is surrounded by a network of pulmonary capillaries, the walls of which are also only one cell thick (❙ Figure 13-4a, b, and c and chapter opener photo). The interstitial space between an alveolus and the surrounding capillary network is extremely thin. Thus, only a 0.5 μm barrier known as the **alveolar–capillary membrane** separates air in the alveoli from blood in the pulmonary capillaries. (A sheet of tracing paper is about 50 times thicker than

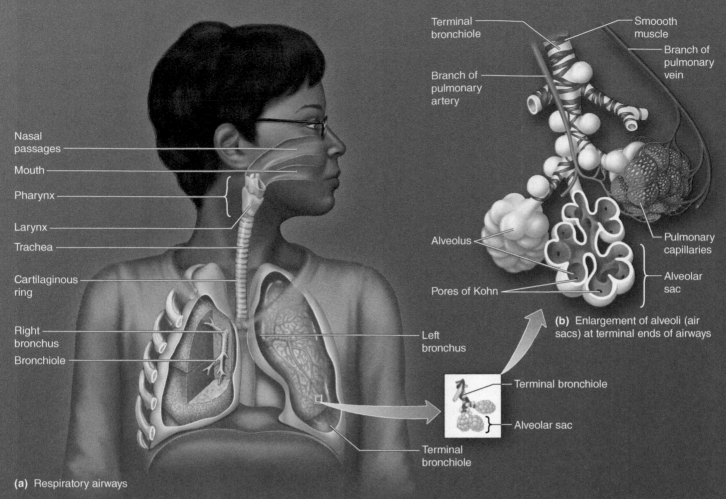

Nasal passages

Mouth

Pharynx

Larynx

Trachea

Cartilaginous ring

Right bronchus

Bronchiole

Left bronchus

Terminal bronchiole

(a) Respiratory airways

Terminal bronchiole

Branch of pulmonary artery

Alveolus

Pores of Kohn

Smoooth muscle

Branch of pulmonary vein

Pulmonary capillaries

Alveolar sac

(b) Enlargement of alveoli (air sacs) at terminal ends of airways

Terminal bronchiole

Alveolar sac

❙Figure 13-2 **Anatomy of the respiratory system.** (a) The respiratory airways include the nasal passages, pharynx, larynx, trachea, bronchi, and bronchioles. (b) Most alveoli (air sacs) are clustered in grapelike arrangements at the end of the terminal bronchioles.

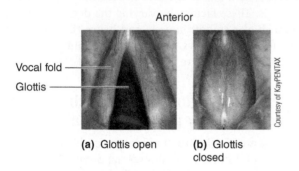

Anterior

Vocal fold

Glottis

(a) Glottis open

(b) Glottis closed

❙Figure 13-3 **Vocal folds.** Photographs showing the vocal folds (a) positioned apart when the glottis is open and (b) in tight apposition when the glottis is closed.

this air-to-blood barrier.) The thinness of this barrier facilitates gas exchange.

Furthermore, the alveolar air–pulmonary blood interface presents a tremendous surface area for exchange. The lungs contain about 500 million alveoli, each about 200 μm to 300 μm in diameter. So dense are the pulmonary capillary networks that an almost continuous sheet of blood encircles each alveolus. The pulmonary capillaries if laid end to end would

extend 620 miles. The total surface area thus exposed between alveolar air and pulmonary capillary blood is about 75 m² (about the size of one side of a volleyball court). In contrast, if the lungs consisted of a single hollow chamber of the same dimensions instead of being divided into a multitude of alveolar units, the total surface area would be only about 0.01 m².

In addition to the thin, wall-forming Type I cells, 5% of the alveolar surface epithelium is covered by **Type II alveolar cells** (❙Figure 13-4a and b). These cells secrete *pulmonary surfactant,* a phospholipoprotein complex that facilitates lung expansion (described later). Furthermore, defensive alveolar macrophages stand guard within the lumen of the air sacs (see p. 442).

Minute **pores of Kohn** exist in the walls between adjacent alveoli (see ❙Figure 13-2b). Their presence permits airflow between adjoining alveoli. These passageways are especially important in allowing fresh air to enter an alveolus whose terminal conducting airway is blocked because of disease.

The lungs occupy much of the thoracic cavity.

The two **lungs** are each divided into several lobes and each is supplied by one of the bronchi. The lung tissue itself consists of the series of highly branched airways, the alveoli, the pulmonary blood vessels, and large quantities of elastic connective tissue. The only muscle within the lungs is the smooth muscle in the walls of the arterioles and the walls of the bronchioles, both of which are subject to control. No muscle is present within the alveolar walls to cause them to inflate and deflate during the breathing process. Instead, changes in lung volume (and accompanying changes in alveolar volume) are brought about through changes in the dimensions of the thoracic (chest) cavity. You will learn about this mechanism after we finish discussing respiratory anatomy.

The lungs occupy most of the volume of the **thoracic (chest) cavity**, the only other structures in the chest being the heart and associated vessels, the esophagus, the thymus, and some nerves. The outer chest wall **(thorax)** is formed by 12 pairs of curved **ribs**, which join the **sternum** (breastbone) anteriorly and the **thoracic vertebrae** (backbone) posteriorly. The rib cage provides bony protection for the lungs and heart. Skeletal muscles connect these bony structures and enclose the thoracic cavity. The **diaphragm**, which forms the floor of the thoracic cavity, is a large, dome-shaped sheet of skeletal muscle that separates the thoracic cavity from the abdominal cavity. It is penetrated only by the esophagus and blood vessels traversing the thoracic and abdominal cavities. At the neck, muscles and connective tissue enclose the thoracic cavity. The only communication between the atmosphere and the thoracic cavity is through the respiratory airways into the alveoli.

A pleural sac separates each lung from the thoracic wall.

A double-walled, closed sac called the **pleural sac** separates each lung from the thoracic wall and other surrounding structures (Ⅰ Figure 13-5). The interior of the pleural sac is known as the **pleural cavity**. In the illustration, the dimensions of the pleural cavity are greatly exaggerated to aid visualization; in reality, the layers of the pleural sac are in close contact with one another. The surfaces of the pleura secrete a thin **intrapleural fluid** (*intra* means "within"), which lubricates the pleural surfaces as they slide past each other during respiratory movements.

 Pleurisy, an inflammation of the pleural sac, is accompanied by painful breathing because each inflation and each deflation of the lungs cause a "friction rub."

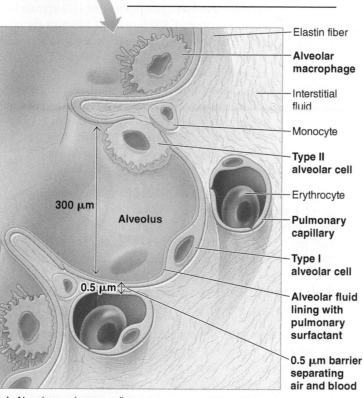

300 μm

Alveolus

0.5 μm

- Elastin fiber
- **Alveolar macrophage**
- Interstitial fluid
- Monocyte
- **Type II alveolar cell**
- Erythrocyte
- **Pulmonary capillary**
- **Type I alveolar cell**
- **Alveolar fluid lining with pulmonary surfactant**
- **0.5 μm barrier separating air and blood**

(a) Alveolus and surrounding pulmonary capillaries

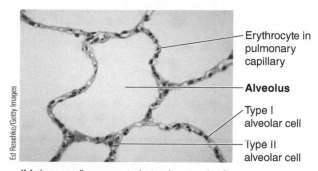

- Erythrocyte in pulmonary capillary
- **Alveolus**
- Type I alveolar cell
- Type II alveolar cell

(b) Immunofluorescent photomicrograph of several alveoli

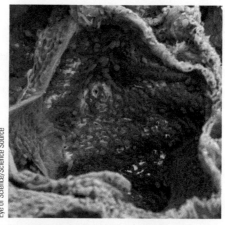

(c) Scanning electron micrograph of a network of pulmonary capillaries surrounding an alveolus cut open for visibility

Ⅰ **Figure 13-4 Alveolus and associated pulmonary capillaries.** (a) A single layer of flattened Type I alveolar cells forms the alveolar walls. Type II alveolar cells embedded within the alveolar wall secrete pulmonary surfactant. Wandering alveolar macrophages are found within the alveolar lumen. The size of the cells and respiratory membrane is exaggerated compared to the size of the alveolar and pulmonary capillary lumens. The diameter of an alveolus is actually about 600 times larger (300 μm) than the intervening space between air and blood (0.5 μm). (b) Note the thin layer separating air in the alveolus and blood in the surrounding pulmonary capillaries. (c) Each alveolus is encircled with a dense network of pulmonary capillaries.

FIGURE FOCUS: *List the layers through which O$_2$ diffuses to move from alveolar air into pulmonary capillary blood.*

(a) Analogy of relationship between lung and pleural sac

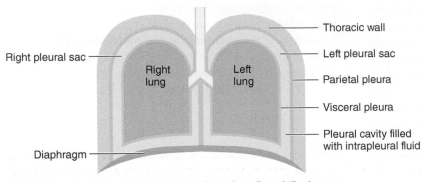

(b) Relationship of lungs to pleural sacs, thoracic wall, and diaphragm

▮ **Figure 13-5 Pleural sac.** (a) Pushing a lollipop into a small water-filled balloon produces a relationship analogous to that between each double-walled, closed pleural sac and the lung that it surrounds and separates from the thoracic wall. (b) One layer of the pleural sac, the *visceral pleura,* closely adheres to the surface of the lung (*viscus* means "organ") and then reflects back on itself to form another layer, the *parietal pleura,* which lines the interior surface of the thoracic wall (*paries* means "wall"). The relative size of the pleural cavity between these two layers is grossly exaggerated for the purpose of visualization.

Check Your Understanding 13.1

1. List the steps of external respiration accomplished by the respiratory system and those carried out by the circulatory system.
2. Discuss how the alveolar air–pulmonary blood interface is ideally structured for gas exchange.
3. State the functions of Type I alveolar cells, Type II alveolar cells, and alveolar macrophages.

13.2 Respiratory Mechanics

Air tends to move from a region of higher pressure to a region of lower pressure—that is, down a **pressure gradient**.

Interrelationships among pressures inside and outside the lungs are important in ventilation.

Air flows into and out of the lungs during the act of breathing by moving down alternately reversing pressure gradients established between the alveoli and the atmosphere by cyclic respira-

tory muscle activity. Three pressure considerations are important in ventilation (▮Figure 13-6):

1. **Atmospheric (barometric) pressure** is the pressure exerted by the weight of the air in the atmosphere on objects on Earth's surface. At sea level it equals 760 mm Hg. Atmospheric pressure diminishes with increasing altitude above sea level as the layer of air above Earth's surface correspondingly decreases in thickness. Minor fluctuations in atmospheric pressure occur at any height because of changing weather conditions (that is, when barometric pressure is rising or falling).

2. **Intra-alveolar pressure** is the pressure within the alveoli. Because the alveoli communicate with the atmosphere through the conducting airways, air quickly flows down its pressure gradient any time intra-alveolar pressure differs from atmospheric pressure; air flow continues until the two pressures equilibrate (become equal).

3. **Intrapleural pressure** is the pressure within the pleural sac. It is the pressure exerted outside the lungs within the thoracic cavity. The intrapleural pressure is usually less than atmospheric pressure, averaging 756 mm Hg at rest. Just as blood pressure is recorded using atmospheric pressure as a reference point (that is, a systolic blood pressure of 120 mm Hg is 120 mm Hg greater than the atmospheric pressure of 760 mm Hg or, in reality, 880 mm Hg), 756 mm Hg is sometimes referred to as a pressure of -4 mm Hg. However, there is really no such thing as an absolute negative pressure. A pressure of -4 mm Hg is just negative when compared with the normal atmospheric pressure of 760 mm Hg. To avoid confusion, we use absolute positive values throughout our discussion of respiration.

Intrapleural pressure does not equilibrate with atmospheric or intra-alveolar pressure because the pleural sac is a closed sac with no openings, so air cannot enter or leave despite any pressure gradients that might exist between the pleural cavity and the atmosphere or lungs.

The transmural pressure gradient stretches the lungs to fill the larger thoracic cavity.

The thoracic cavity is larger than the unstretched lungs because the thoracic wall grows more rapidly than the lungs during development. However, a *transmural pressure gradient* across the lung wall holds the lungs and thoracic wall in close apposition, stretching the lungs to fill the larger thoracic cavity.

Transmural Pressure Gradient The intra-alveolar pressure, equilibrated with atmospheric pressure at 760 mm Hg, is greater than the intrapleural pressure of 756 mm Hg, so a greater pressure is pushing outward than is pushing inward across the lung wall. This net outward pressure differential, the **transmural pressure gradient**, pushes out on the lungs, stretching, or distending, them (*trans* means "across"; *mural* means "wall") (▮Fig-

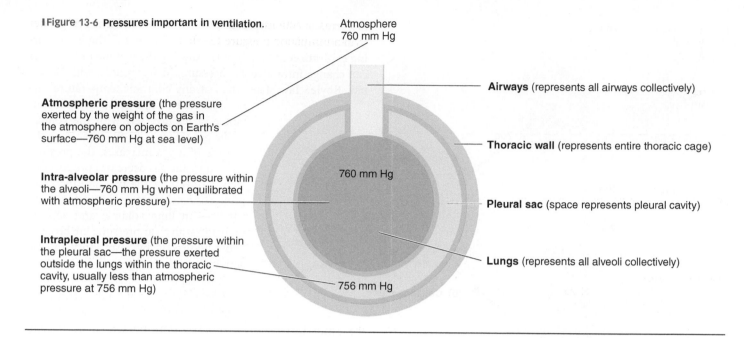

Figure 13-6 Pressures important in ventilation.

Atmosphere
760 mm Hg

Atmospheric pressure (the pressure exerted by the weight of the gas in the atmosphere on objects on Earth's surface—760 mm Hg at sea level)

Intra-alveolar pressure (the pressure within the alveoli—760 mm Hg when equilibrated with atmospheric pressure)

Intrapleural pressure (the pressure within the pleural sac—the pressure exerted outside the lungs within the thoracic cavity, usually less than atmospheric pressure at 756 mm Hg)

Airways (represents all airways collectively)

Thoracic wall (represents entire thoracic cage)

Pleural sac (space represents pleural cavity)

Lungs (represents all alveoli collectively)

760 mm Hg

756 mm Hg

ure 13-7). Because of this pressure gradient, the lungs are always forced to expand to fill the thoracic cavity, no matter its size. As the thoracic cavity enlarges, the lungs enlarge—that is, the lungs follow the movements of the chest wall.

Why the Intrapleural Pressure Is Subatmospheric

Because of the lungs' elasticity, they try to pull inward away from the thoracic wall as they are stretched to fill the larger thoracic cavity. The transmural pressure gradient, however, prevents the lungs from pulling away except to the slightest degree. The resultant ever-so-slight expansion of the pleural cavity is sufficient to drop the pressure in this cavity by 4 mm Hg, bringing the intrapleural pressure to the subatmospheric level of 756 mm Hg. This pressure drop occurs because the pleural cavity is filled with fluid, which cannot expand to fill the slightly larger volume. Therefore, a vacuum exists in the infinitesimal space in the slightly expanded pleural cavity not occupied by intrapleural fluid, producing a small drop in intrapleural pressure below atmospheric pressure.

Note the interrelationship between the transmural pressure gradient and the subatmospheric intrapleural pressure. The lungs are stretched by the transmural pressure gradient that exists across their walls because the intrapleural pressure is less than atmospheric pressure. The intrapleural pressure, in turn, is subatmospheric because the stretched lungs tend to pull away from the larger thoracic wall, slightly expanding the pleural cavity and dropping the intrapleural pressure below atmospheric pressure.

Pneumothorax Normally, air does not enter the pleural cavity because there is no communication between the cavity and either the atmosphere or the alveoli. However, if the chest wall is punctured (for example, by a stab wound or a broken rib), air flows down its pressure gradient from the higher atmospheric pressure and rushes into the cavity (Figure

13-8a). The abnormal condition of air in the pleural cavity is known as **pneumothorax** ("air in the chest"). Intrapleural and intra-alveolar pressure are now both equilibrated with atmospheric pressure, so a transmural pressure gradient no longer exists across the lung wall. With no force present to stretch the lung, it collapses to its unstretched size (Figure 13-8b). Similarly, pneumothorax and lung collapse can occur if air enters the pleural cavity through a hole in the lung produced, for

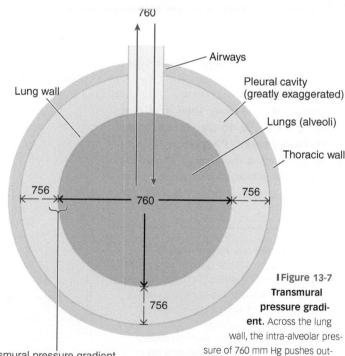

Transmural pressure gradient across lung wall = intra-alveolar pressure minus intrapleural pressure

Numbers are mm Hg pressure.

Figure 13-7 Transmural pressure gradient. Across the lung wall, the intra-alveolar pressure of 760 mm Hg pushes outward, while the intrapleural pressure of 756 mm Hg pushes inward. This 4 mm Hg difference in pressure constitutes a transmural pressure gradient that pushes out on the lungs, stretching them to fill the larger thoracic cavity.

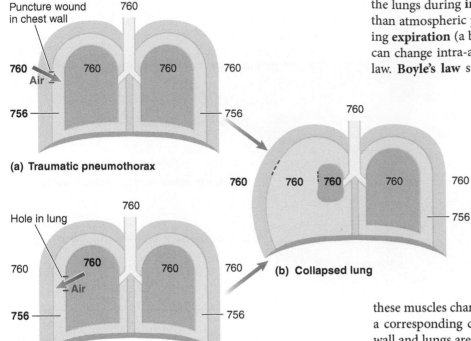

Puncture wound in chest wall
760

760
Air
760   760
756
756

(a) Traumatic pneumothorax

760

760
760   760   760   760
760

756

(b) Collapsed lung

760

Hole in lung
760
760   760
760   Air
756
756

(c) Spontaneous pneumothorax

Numbers are mm Hg pressure.

I Figure 13-8 Pneumothorax. (a) In *traumatic pneumothorax,* a puncture in the chest wall permits air from the atmosphere to flow down its pressure gradient and enter the pleural cavity, abolishing the transmural pressure gradient. (b) When the transmural pressure gradient is abolished, the lung collapses to its unstretched size, and the chest wall springs outward. (c) In *spontaneous pneumothorax,* a hole in the lung wall permits air to move down its pressure gradient and enter the pleural cavity from the lungs, abolishing the transmural pressure gradient. As with traumatic pneumothorax, the lung collapses to its unstretched size.

the lungs during **inspiration** (a breath in) and must be greater than atmospheric pressure for air to flow out of the lungs during **expiration** (a breath out). Altering the volume of the lungs can change intra-alveolar pressure, in accordance with Boyle's law. **Boyle's law** states that, at any constant temperature, the pressure exerted by a gas in a closed container varies inversely with the volume of the gas (I Figure 13-9)—that is, as the volume of a gas increases, the pressure exerted by the gas decreases proportionately. Conversely, the pressure increases proportionately as the volume decreases. Changes in lung volume, and accordingly intra-alveolar pressure, are brought about indirectly by respiratory muscle activity.

The respiratory muscles that accomplish breathing do not act directly on the lungs to change their volume. Instead, these muscles change the volume of the thoracic cavity, causing a corresponding change in lung volume because the thoracic wall and lungs are linked by the transmural pressure gradient.

Let us follow the changes that occur during one respiratory cycle—that is, one inspiration and expiration.

Onset of Inspiration: Contraction of Inspiratory Muscles The major **inspiratory muscles**—the muscles that contract to accomplish an inspiration during quiet breathing—include the *diaphragm* and *external intercostal muscles* (I Figure 13-10). Before the beginning of inspiration, all respiratory muscles are

example, by a disease process (I Figure 13-8c).

Air Alternately Flows Into and Out of the Lungs Due to Cyclic Changes in Intra-Alveolar Pressure. Because air flows down a pressure gradient, the intra-alveolar pressure must be less than atmospheric pressure for air to flow into

I Figure 13-9 Boyle's law. Each container has the same number of gas molecules. Given the random motion of gas molecules, the likelihood of a gas molecule striking the interior wall of the container and exerting pressure varies inversely with the volume of the container at any constant temperature. The gas in container B exerts more pressure than the same gas in larger container C but less pressure than the same gas in smaller container A. This relationship is stated as Boyle's law: $P_1V_1 = P_2V_2$. As the volume of a gas increases, the pressure of the gas decreases proportionately; conversely, the pressure increases proportionately as the volume decreases.

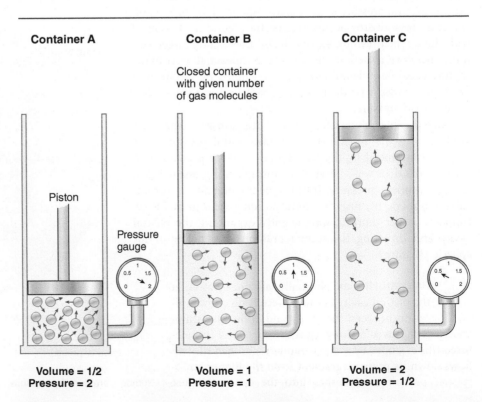

Container A

Piston

Pressure gauge

Volume = 1/2
Pressure = 2

Container B

Closed container with given number of gas molecules

Volume = 1
Pressure = 1

Container C

Volume = 2
Pressure = 1/2

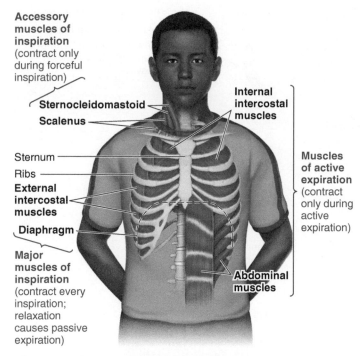

Accessory muscles of inspiration (contract only during forceful inspiration)

Sternocleidomastoid

Scalenus

Sternum

Ribs

External intercostal muscles

Diaphragm

Major muscles of inspiration (contract every inspiration; relaxation causes passive expiration)

Internal intercostal muscles

Muscles of active expiration (contract only during active expiration)

Abdominal muscles

❚ Figure 13-10 **Anatomy of the respiratory muscles.**

relaxed (❚ Figure 13-11a). At the onset of inspiration, contraction of the inspiratory muscles enlarges the thoracic cavity. The major inspiratory muscle is the **diaphragm**, which is innervated by the **phrenic nerve**. The relaxed diaphragm has a dome shape that protrudes upward into the thoracic cavity. The diaphragm descends downward when it contracts on stimulation by the phrenic nerve, enlarging the volume of the thoracic cavity by increasing its vertical (top-to-bottom) dimension (❚ Figure 13-11b). During quiet breathing, the diaphragm descends about 1 cm during inspiration, but during heavy breathing, it may descend as much as 10 cm. The abdominal wall, if relaxed, bulges outward during inspiration as the descending diaphragm pushes the abdominal contents downward and forward. Diaphragm contraction accomplishes 75% of thoracic cavity enlargement during quiet inspiration.

Two sets of **intercostal muscles** lie between the ribs (*inter* means "between"; *costa* means "rib"). The *external intercostal muscles* lie on top of the *internal intercostal muscles*. Contraction of the **external intercostal muscles**, whose fibers run downward and forward between adjacent ribs, enlarges the thoracic cavity in both the lateral (side-to-side) and the anteroposterior (front-to-back) dimensions. When the external intercostals contract, they elevate the ribs and subsequently the sternum upward and outward (❚ Figure 13-11b). **Intercostal nerves** activate these intercostal muscles during inspiration.

Before inspiration, at the end of the preceding expiration, intra-alveolar pressure is equal to atmospheric pressure, so no air is flowing into or out of the lungs (❚ Figure 13-12a). As the thoracic cavity enlarges during inspiration on contraction of the diaphragm and external intercostals, the lungs are also forced to expand to fill the larger thoracic cavity. As the lungs enlarge, the intra-alveolar pressure drops because the same number of air molecules now occupies a larger lung volume. In

a typical inspiratory excursion, the intra-alveolar pressure drops 1 mm Hg to 759 mm Hg (❚ Figure 13-12b). Because the intra-alveolar pressure is now less than atmospheric pressure, air flows into the lungs down this pressure gradient. Air continues to enter the lungs until no further gradient exists—that is, until intra-alveolar pressure equals atmospheric pressure. Thus, movement of air into the lungs does not cause lung expansion; instead, air flows into the lungs because of the fall in intra-alveolar pressure brought about by lung expansion.

During inspiration, the intrapleural pressure falls to 754 mm Hg because the more highly stretched lungs tend to pull away a bit more from the thoracic wall.

Role of Accessory Inspiratory Muscles Deeper inspirations (more air breathed in) can be accomplished by contracting the diaphragm and external intercostal muscles more forcefully and by bringing the **accessory inspiratory muscles** into play to further enlarge the thoracic cavity. Contracting these accessory muscles, which are in the neck (see ❚ Figure 13-10), raises the sternum and elevates the first two ribs, enlarging the upper portion of the thoracic cavity. As the thoracic cavity increases even further in volume than under resting conditions, the lungs likewise expand more, dropping the intra-alveolar pressure further. Consequently, a larger inward flow of air occurs before equilibration with atmospheric pressure is achieved—that is, a deeper breath occurs.

Onset of Expiration: Relaxation of Inspiratory Muscles At the end of inspiration, the inspiratory muscles relax. The diaphragm assumes its original dome-shaped position when it relaxes. The elevated rib cage falls because of gravity when the external intercostals relax (see ❚ Figure 13-11c). With no forces expanding the chest wall (and accordingly, expanding the lungs), the chest wall and stretched lungs recoil to their preinspiratory size because of their elastic properties, much as a stretched balloon would on release. As the lungs recoil and become smaller in volume, the intra-alveolar pressure rises because the greater number of air molecules contained within the larger lung volume at the end of inspiration are now compressed into a smaller volume. In a resting expiration, the intra-alveolar pressure increases about 1 mm Hg above atmospheric pressure to 761 mm Hg (❚ Figure 13-12c), and air leaves the lungs down this pressure gradient. Outward flow of air ceases when intra-alveolar pressure becomes equal to atmospheric pressure and a pressure gradient no longer exists. ❚ Figure 13-13 summarizes the intra-alveolar and intrapleural pressure changes that take place during one respiratory cycle.

Clinical Note Because the diaphragm is the major inspiratory muscle and its relaxation also causes expiration, paralysis of the intercostal muscles alone does not seriously influence quiet breathing. Disruption of diaphragm activity caused by nerve or muscle disorders, however, leads to respiratory paralysis. Fortunately, the phrenic nerve arises from the spinal cord in the neck region (cervical segments 3, 4, and 5) and then descends to the diaphragm at the floor of the thorax, instead of arising from the thoracic region of the cord as might be expected. For this reason, individuals completely paralyzed below the neck by traumatic severance of the spinal cord are

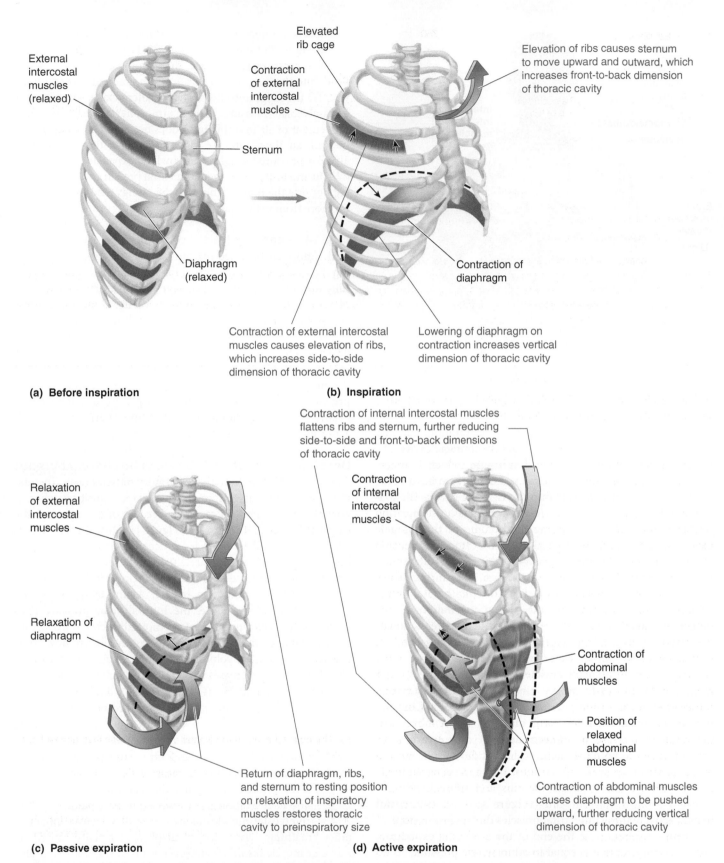

External intercostal muscles (relaxed)

Sternum

Diaphragm (relaxed)

(a) Before inspiration

Elevated rib cage

Contraction of external intercostal muscles

Elevation of ribs causes sternum to move upward and outward, which increases front-to-back dimension of thoracic cavity

Contraction of diaphragm

Contraction of external intercostal muscles causes elevation of ribs, which increases side-to-side dimension of thoracic cavity

Lowering of diaphragm on contraction increases vertical dimension of thoracic cavity

(b) Inspiration

Relaxation of external intercostal muscles

Relaxation of diaphragm

Return of diaphragm, ribs, and sternum to resting position on relaxation of inspiratory muscles restores thoracic cavity to preinspiratory size

(c) Passive expiration

Contraction of internal intercostal muscles flattens ribs and sternum, further reducing side-to-side and front-to-back dimensions of thoracic cavity

Contraction of internal intercostal muscles

Contraction of abdominal muscles

Position of relaxed abdominal muscles

Contraction of abdominal muscles causes diaphragm to be pushed upward, further reducing vertical dimension of thoracic cavity

(d) Active expiration

I **Figure 13-11 Respiratory muscle activity during inspiration and expiration.** (a) Before inspiration, all respiratory muscles are relaxed. (b) During *inspiration,* the diaphragm descends on contraction, increasing the vertical dimension of the thoracic cavity. Contraction of the external intercostal muscles elevates the ribs and subsequently the sternum to enlarge the thoracic cavity from front to back and from side to side. (c) During *quiet passive expiration,* the diaphragm relaxes, reducing the volume of the thoracic cavity from its peak inspiratory size. As the external intercostal muscles relax, the elevated rib cage falls because of the force of gravity. This also reduces the volume of the thoracic cavity. (d) During *active expiration,* contraction of the abdominal muscles increases the intra-abdominal pressure, exerting an upward force on the diaphragm. This reduces the vertical dimension of the thoracic cavity further than it is reduced during quiet passive expiration. Contraction of the internal intercostal muscles decreases the front-to-back and side-to-side dimensions by flattening the ribs and sternum.

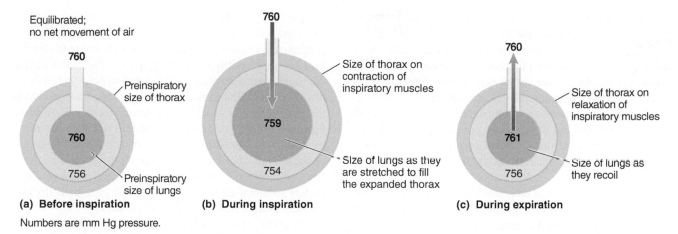

Equilibrated; no net movement of air

760

- Preinspiratory size of thorax

760

756 — Preinspiratory size of lungs

(a) Before inspiration

760

- Size of thorax on contraction of inspiratory muscles

759

754 — Size of lungs as they are stretched to fill the expanded thorax

(b) During inspiration

760

- Size of thorax on relaxation of inspiratory muscles

761

756 — Size of lungs as they recoil

(c) During expiration

Numbers are mm Hg pressure.

❙Figure 13-12 **Changes in lung volume and intra-alveolar pressure during inspiration and expiration.** (a) Before inspiration, at the end of the preceding expiration, intra-alveolar pressure is equilibrated with atmospheric pressure, and no air is flowing. (b) As the lungs increase in volume during inspiration, the intra-alveolar pressure decreases, establishing a pressure gradient that favors the flow of air into the alveoli from the atmosphere—that is, an inspiration occurs. (c) As the lungs recoil to their preinspiratory size on relaxation of the inspiratory muscles, the intra-alveolar pressure increases, establishing a pressure gradient that favors the flow of air out of the alveoli into the atmosphere—that is, an expiration occurs.

FIGURE FOCUS: *Compare the airflow into the lungs if intra-alveolar pressure is reduced to 758 mm Hg instead of the typical 759 mm Hg as a result of more forceful contraction of inspiratory muscles.*

still able to breathe, even though they have lost use of all other skeletal muscles in their trunk and limbs.

Forced Expiration: Contraction of Expiratory Muscles

During quiet breathing, expiration is normally a passive process because it is accomplished by elastic recoil of the lungs on relaxation of the inspiratory muscles, with no muscular exertion or energy expenditure required. In contrast, inspiration is *always* active because it is brought about only by contraction of inspiratory muscles at the expense of energy use. Expiration can become active to empty the lungs more completely and more rapidly than is accomplished during quiet breathing, as during the deeper breaths accompanying exercise. To force more air out, the intra-alveolar pressure must be increased even further above atmospheric pressure than can be accomplished by simple relaxation of the inspiratory muscles and elastic recoil of the lungs. To produce such a **forced**, or **active**, **expiration**, expiratory muscles must contract to further reduce the volume of the thoracic cavity and lungs (see ❙Figures 13-10 and 13-11d). The most important **expiratory muscles** are (unbelievable as it may seem at first) the *muscles of the abdominal wall.* As the abdominal muscles contract, the resultant increase in intra-abdominal pressure exerts an upward force on the diaphragm, pushing it farther up into the thoracic cavity than its relaxed position, thus decreasing the vertical dimension of the thoracic cavity even more. The other expiratory muscles are the **internal intercostal muscles**, whose contraction pulls the ribs downward and inward, flattening the chest wall and further decreasing the size of the thoracic cavity; this action is just the opposite of that of the external intercostal muscles.

As active contraction of the expiratory muscles further reduces the volume of the thoracic cavity, the lungs also become further reduced in volume because they do not have to be stretched as much to fill the smaller thoracic cavity—that is, they are permitted to recoil to an even smaller volume. The intra-alveolar pressure increases further as the air in the lungs

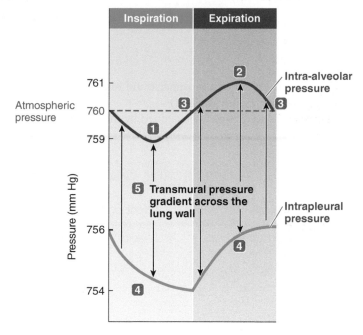

1 During inspiration, intra-alveolar pressure is less than atmospheric pressure.

2 During expiration, intra-alveolar pressure is greater than atmospheric pressure.

3 At the end of both inspiration and expiration, intra-alveolar pressure is equal to atmospheric pressure because the alveoli are in direct communication with the atmosphere, and air continues to flow down its pressure gradient until the two pressures equilibrate.

4 Throughout the respiratory cycle, intrapleural pressure is less than intra-alveolar pressure.

5 Thus, a transmural pressure gradient always exists, and the lung is always stretched to some degree, even during expiration.

❙Figure 13-13 **Intra-alveolar and intrapleural pressure changes throughout the respiratory cycle.**

is confined within this smaller volume. The differential between intra-alveolar and atmospheric pressure is even greater now than during passive expiration, so more air leaves down the pressure gradient before equilibration is achieved. In this way, the lungs are emptied more completely during forceful, active expiration than during quiet, passive expiration.

During forceful expiration, the intrapleural pressure exceeds atmospheric pressure, but the lungs do not collapse. Because the intra-alveolar pressure is also increased correspondingly, a transmural pressure gradient still exists across the walls of the lungs, keeping them stretched to fill the thoracic cavity. For example, if the pressure within the thorax increases 10 mm Hg, the intrapleural pressure becomes 766 mm Hg and the intra-alveolar pressure becomes 770 mm Hg—still a 4–mm Hg pressure difference.

Airway resistance influences airflow rates.

Thus far, we have discussed airflow in and out of the lungs as a function of the magnitude of the pressure gradient between the alveoli and the atmosphere. However, just as flow of blood through the blood vessels depends not only on the pressure gradient but also on the resistance to the flow offered by the vessels, so it is with airflow:

$$F = \frac{\Delta P}{R}$$

where

F = airflow rate
ΔP = difference between atmospheric and intra-alveolar pressure (pressure gradient)
R = resistance of airways, determined by their radius

The primary determinant of resistance to airflow is the radius of the conducting airways. We ignored airway resistance in our preceding discussion of pressure gradient–induced airflow rates because, in a healthy respiratory system, the radius of the conducting system is large enough that resistance remains extremely low. Therefore, the pressure gradient between the alveoli and the atmosphere is usually the primary factor determining airflow rate. Indeed, the airways normally offer such low resistance that only small pressure gradients of 1 to 2 mm Hg are needed to achieve adequate rates of airflow into and out of the lungs. (By comparison, it would take a pressure gradient 250 times greater to move air through a smoker's pipe than through the respiratory airways at the same flow rate.)

Normally, modest adjustments in airway size can be accomplished by autonomic nervous system regulation to suit the body's needs. Parasympathetic stimulation, which occurs in quiet, relaxed situations when the demand for airflow is low, promotes bronchiolar smooth muscle contraction, which increases airway resistance by producing **bronchoconstriction** (a decrease in the radius of bronchioles). In contrast, sympathetic stimulation and to a greater extent its associated hormone, epinephrine, bring about **bronchodilation** (an increase in bronchiolar radius) and decreased airway resistance by promoting bronchiolar smooth muscle relaxation (Table 13-1). Thus, during periods of sympathetic domination, when increased demands for O_2 uptake are actually or potentially placed on the body, bronchodilation ensures that the pressure gradients established by respiratory muscle activity can achieve maximum airflow rates with minimum resistance. Because of this bronchodilator action, epinephrine or related drugs (such as *albuterol,* which is a more selective β_2-adrenergic agonist that relaxes bronchiolar smooth

ITABLE 13-1 Factors Affecting Airway Resistance

Status of Airways	Effect on Resistance	Factors Producing the Effect
Bronchoconstriction	↓ Radius, ↑ resistance to airflow	*Physiological control factors:*
		Neural control: parasympathetic stimulation
		Local chemical control: ↓ CO_2 concentration
		Pathological factors:
		Allergy-induced spasm of the airways caused by
		Slow-reactive substance of anaphylaxis (leukotrienes)
		Histamine
		Physical blockage of the airways caused by
		Excess mucus
		Edema of the walls
		Airway collapse
Bronchodilation	↑ Radius, ↓ resistance to airflow	*Physiological control factors:*
		Neural control: sympathetic stimulation (minimal effect)
		Hormonal control: epinephrine
		Local chemical control: ↑ CO_2 concentration
		Pathological factors: none

muscle without affecting β_1-adrenergic targets like the heart; see p. 240) are useful therapeutic tools to counteract airway constriction in patients with bronchial spasms.

Resistance becomes an extremely important impediment to airflow when airway lumens become abnormally narrowed by disease. We have all transiently experienced the effect that increased airway resistance has on breathing when we have a cold. We know how difficult it is to produce an adequate airflow rate through a "stuffy nose" when the nasal passages are narrowed by swelling and mucus accumulation. More serious is chronic obstructive pulmonary disease, to which we now turn attention.

Airway resistance is abnormally increased with chronic obstructive pulmonary disease.

Clinical Note **Chronic obstructive pulmonary disease (COPD)** is a group of lung diseases characterized by increased airway resistance resulting from narrowing of the lumen of the lower airways. When airway resistance increases, a larger pressure gradient must be established to maintain even a normal airflow rate. For example, if resistance is doubled by narrowing of airway lumens, ΔP must be doubled through increased respiratory muscle exertion to induce the same flow rate of air in and out of the lungs as a healthy person accomplishes during quiet breathing. Accordingly, patients with COPD must work harder to breathe.

COPD, which is the fourth leading cause of death worldwide, encompasses three chronic (long-term) diseases: *chronic bronchitis, asthma,* and *emphysema.*

Chronic Bronchitis Chronic bronchitis is a long-term inflammatory condition of the lower respiratory airways, generally triggered by frequent exposure to irritating cigarette smoke, polluted air, or allergens. In response to the chronic irritation, the airways become narrowed by prolonged edematous thickening of the airway linings, coupled with overproduction of thick mucus. Despite frequent coughing associated with the chronic irritation, the plugged mucus often cannot be satisfactorily removed, especially because the irritants immobilize the ciliary mucus escalator (see p. 442). Pulmonary bacterial infections frequently occur because the accumulated mucus is an excellent medium for bacterial growth.

Asthma In **asthma**, airway obstruction is the result of (1) thickening of airway walls, brought about by inflammation and histamine-induced edema (see p. 409); (2) plugging of the airways by excessive secretion of thick mucus; and (3) airway hyperresponsiveness, characterized by profound constriction of the smaller airways caused by trigger-induced spasm of the smooth muscle in the walls of these airways (see p. 436). In severe asthmatic attacks, pronounced clogging and narrowing of the airways can cut off all airflow, leading to death. Asthma has a variety of causes. Triggers that lead to these inflammatory changes, the exaggerated bronchoconstrictor response, or both, include repeated exposure to allergens (such as dust mites or pollen), irritants (as in cigarette smoke or polluted air), respiratory infections, and vigorous exercise. Long-term infections with *Chlamydia pneumoniae,* a common cause of lung infections, may underlie up to half of the adult cases of asthma. An estimated 15 million people in the United States have asthma, with the number steadily climbing. Asthma is the most common chronic childhood disease; 9.5% of children in the United States as a whole have asthma, with that number rising to 22% in heavily polluted cities.

Emphysema Emphysema is characterized by (1) collapse of the smaller airways and (2) breakdown of alveolar walls. This irreversible condition can arise in two ways. Most commonly, emphysema results from excessive release of protein-digesting enzymes such as *trypsin* from alveolar macrophages as a defense mechanism in response to chronic exposure to inhaled cigarette smoke or other irritants. The lungs are normally protected from damage by these enzymes by α_1-*antitrypsin,* a protein that inhibits trypsin. Excessive secretion of these destructive enzymes in response to chronic irritation, however, can overwhelm the protective capability of α_1-antitrypsin so that these enzymes destroy not only foreign materials but lung tissue as well. Loss of lung tissue leads to the breakdown of alveolar walls and collapse of small airways, which characterize emphysema.

Less frequently, emphysema arises from a genetic inability to produce α_1-antitrypsin so that the lung tissue has no protection from trypsin. The unprotected lung tissue gradually disintegrates under the influence of even small amounts of macrophage-released enzymes, in the absence of chronic exposure to inhaled irritants.

Difficulty in Expiring When COPD of any type increases airway resistance, expiration is more difficult than inspiration. The smaller airways, lacking the cartilaginous rings that hold the larger airways open, are kept open by the same transmural pressure gradient that distends the alveoli. Expansion of the thoracic cavity during inspiration indirectly dilates the airways even further than their expiratory dimensions, like alveolar expansion, so airway resistance is lower during inspiration than during expiration. In a healthy individual, the airway resistance is always so low that the slight variation between inspiration and expiration is not noticeable. When airway resistance has substantially increased, however, as during an asthmatic attack, the difference is quite noticeable. Thus, a person with asthma has more difficulty expiring than inspiring, giving rise to the characteristic "wheeze" as air is forced out through the narrowed airways.

Normally the smaller airways stay open during quiet breathing and even during active expiration when intrapleural pressure is elevated, as during exercise. During normal quiet breathing, because airway resistance is low, little frictional loss of pressure occurs within the airways. Airway pressure is the same as intra-alveolar pressure as air exits the alveoli and drops only slightly as air flows outward through airways. Thus, airway pressure remains higher than intrapleural pressure throughout the length of the airways, so the airways remain open. Even though intrapleural pressure is elevated during the active expiration accompanying routine vigorous activity, intra-alveolar pressure is also elevated and airway resistance is still low, so pressure in the small airways does not fall below the elevated intrapleural pressure. Therefore, collapse of the smaller airways does not occur under typical circumstances.

The smaller airways do collapse and further outflow of air is halted only during maximal forced expiration in people without COPD. Because of this airway collapse, the lungs can never be emptied completely. What causes airway collapse? During maximal forced expiration, both intra-alveolar and intrapleural pressures are markedly increased. The slight frictional losses of pressure as air exits through the airways cause the airway pressure to fall below the surrounding elevated intrapleural pressure before the exiting air reaches the level at which the airways are held open by cartilaginous rings. As a result, the small nonrigid airways are compressed closed by the surrounding higher intrapleural pressure, blocking further expiration of air. In healthy individuals, this dynamic compression of airways occurs only at very low lung volumes during maximal forced expiration. By contrast, in people who have COPD, especially emphysema, the smaller airways often collapse at high lung volumes during routine expiration, preventing further outflow of air through these passageways. This detrimental premature airway collapse occurs for two reasons: (1) the pressure drop along the airways is magnified as a result of increased airway resistance, and (2) the intrapleural pressure is higher than normal because of the loss, as in emphysema, of lung tissue that is responsible for the stretched lung's tendency to recoil and thereby lower the intrapleural pressure. Excessive air trapped in the alveoli behind the compressed bronchiolar segments in COPD reduces the amount of gas exchanged between the alveoli and the atmosphere. Therefore, less alveolar air is "freshened" with each breath when airways collapse at higher lung volumes. Furthermore, because of the extra air trapped behind these collapsed airways, people with emphysema have enlarged lungs and tend to appear barrel-chested. Compared to normal, those with pronounced emphysema suffer an ironic twist of having more air in the lungs but less gas exchange (because of the loss of surface area resulting from the breakdown of alveolar walls) and decreased ability to bring in fresh air (because the chest wall cannot expand as much from its larger-than-normal resting position to reduce the intra-alveolar pressure enough to accomplish adequate inspiration).

The lungs' elastic behavior results from elastin fibers and alveolar surface tension.

During the respiratory cycle, the lungs alternately expand during inspiration and recoil during expiration. What properties of the lungs enable them to behave like balloons, being stretchable and then snapping back to their resting position when the stretching forces are removed? Two interrelated concepts are involved in pulmonary elasticity: compliance and elastic recoil.

Compliance and Elastic Recoil The term **compliance** refers to how much effort is required to stretch or distend the lungs and is analogous to how easy or hard it is to blow up a balloon. (By comparison, 100 times more distending pressure is required to inflate a toy balloon than to inflate the lungs.) Specifically, compliance is a measure of how much change in lung volume results from a given change in the transmural pressure gradient, the force that stretches the lungs. A highly compliant lung stretches further for a given increase in the pressure difference than a less compliant lung does. Stated another way, the lower the compli-

ance of the lungs, the larger the transmural pressure gradient that must be created during inspiration to produce normal lung expansion. In turn, a greater-than-normal transmural pressure gradient during inspiration can be achieved only by making the intrapleural pressure more subatmospheric than usual. This is accomplished by greater expansion of the thoracic cavity through more vigorous contraction of the inspiratory muscles. Therefore, the less compliant the lungs are, the more work is required to produce a given degree of inflation. A poorly compliant lung is referred to as a "stiff" lung because it lacks normal stretchability.

Clinical Note Respiratory compliance can be decreased by several factors, as in _pulmonary fibrosis_, where normal lung tissue is replaced with scar-forming fibrous connective tissue as a result of chronically breathing in asbestos fibers or similar irritants.

The term **elastic recoil** refers to how readily the lungs rebound after having been stretched. It is responsible for the lungs returning to their preinspiratory volume when the inspiratory muscles relax at the end of inspiration.

Pulmonary elastic behavior depends mainly on two factors: large quantities of stretchy _elastin fibers_ in the lung connective tissue (see Figure 13-4a, p. 449; also see p. 60) and, more important, _alveolar surface tension_.

Alveolar Surface Tension The thin liquid film that lines each alveolus displays **alveolar surface tension**. At an air–water interface, the water molecules at the surface are more strongly attracted to other surrounding water molecules (by means of hydrogen bonds; see p. A-5) than to the air above the surface. This unequal attraction produces a force known as _surface tension_ at the surface of the liquid. Surface tension has a twofold effect. First, the liquid layer resists any force that increases its surface area—that is, it opposes expansion of the alveolus because the surface water molecules oppose being pulled apart. Accordingly, the greater the alveolar surface tension is, the less compliant the lungs are. Second, the liquid surface area tends to shrink as small as possible because the surface water molecules, being preferentially attracted to one another, try to get as close together as possible. Thus, the surface tension of the liquid lining an alveolus tends to reduce alveolus size, squeezing in on the air inside. This property, along with the rebound of the stretched elastin fibers, produces the lungs' elastic recoil back to their preinspiratory size when inspiration is over.

Clinical Note With emphysema, elastic recoil is decreased by loss of elastin fibers and the reduction in alveolar surface tension resulting from the breakdown of alveolar walls. The decrease in elastic recoil contributes, along with the increased airway resistance, to the patient's difficulty expiring.

Pulmonary surfactant decreases surface tension and contributes to lung stability.

The cohesive forces between water molecules are so strong that if the alveoli were lined with water alone, alveolar surface tension would be so great that the lungs would collapse. The recoil force attributable to the elastin fibers and high surface tension would exceed the opposing stretching force of the transmural pressure gradient. Furthermore, the lungs would be poorly compliant, so exhausting muscular efforts would be required to accomplish stretching and inflation of the alveoli. The tremendous surface tension of pure water is normally counteracted by pulmonary surfactant.

Pulmonary Surfactant **Pulmonary surfactant** is a complex mixture of lipids and proteins secreted by Type II alveolar cells (see ▮Figure 13-4a and b, p. 449). It intersperses between the water molecules in the fluid lining the alveoli and lowers alveolar surface tension (the term *surfactant* is derived from "*surface active agent*"). Because the cohesive force between a water molecule and an adjacent pulmonary surfactant molecule is very low, pulmonary surfactant decreases the extent of hydrogen bonding between molecules at the alveolar air–water interface. By lowering alveolar surface tension, pulmonary surfactant provides two important benefits: (1) It increases pulmonary compliance, reducing the work of inflating the lungs, and (2) it reduces the lungs' tendency to recoil so that they do not collapse as readily.

Pulmonary surfactant's role in reducing the tendency of alveoli to recoil, thereby discouraging alveolar collapse, is important in helping maintain lung stability. The division of the lung into myriad tiny air sacs provides the advantage of a tremendously increased surface area for exchange of O_2 and CO_2, but it also presents the problem of maintaining the stability of all these alveoli. Recall that the pressure generated by alveolar surface tension is directed inward, squeezing in on the air in the alveoli. If you visualize the alveoli as spherical bubbles, according to the **law of LaPlace**, the magnitude of the inward-directed collapsing pressure is directly proportional to the surface tension and inversely proportional to the radius of the bubble:

$$P = \frac{2T}{r}$$

where

P = inward-directed collapsing pressure
T = surface tension
r = radius of bubble (alveolus)

Law of LaPlace:
Magnitude of inward-directed pressure (P) in a bubble (alveolus) = $\dfrac{2 \times \text{Surface tension } (T)}{\text{Radius } (r) \text{ of bubble (alveolus)}}$

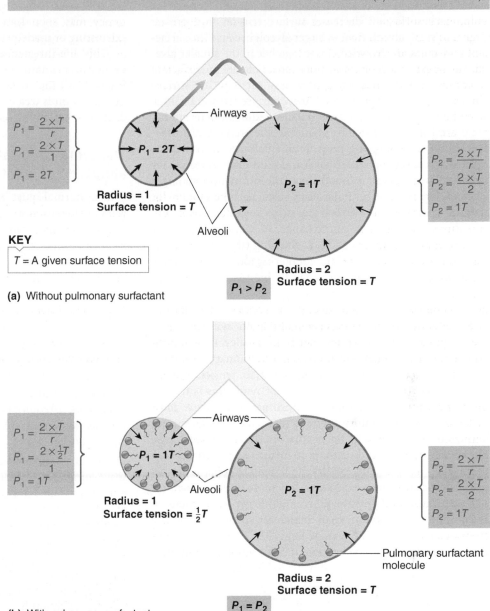

$P_1 = \dfrac{2 \times T}{r}$
$P_1 = \dfrac{2 \times T}{1}$
$P_1 = 2T$

$P_1 = 2T$

Radius = 1
Surface tension = T

KEY

T = A given surface tension

(a) Without pulmonary surfactant

Airways

Alveoli

$P_2 = 1T$

Radius = 2
Surface tension = T

$P_1 > P_2$

$P_2 = \dfrac{2 \times T}{r}$
$P_2 = \dfrac{2 \times T}{2}$
$P_2 = 1T$

$P_1 = \dfrac{2 \times T}{r}$
$P_1 = \dfrac{2 \times \frac{1}{2}T}{1}$
$P_1 = 1T$

$P_1 = 1T$

Radius = 1
Surface tension = $\frac{1}{2}T$

Airways

Alveoli

$P_2 = 1T$

Pulmonary surfactant molecule

Radius = 2
Surface tension = T

$P_1 = P_2$

$P_2 = \dfrac{2 \times T}{r}$
$P_2 = \dfrac{2 \times T}{2}$
$P_2 = 1T$

(b) With pulmonary surfactant

▮Figure 13-14 **Role of pulmonary surfactant in counteracting the tendency for small alveoli to collapse into larger alveoli.** (a) According to the law of LaPlace, if two alveoli of unequal size but the same surface tension are connected by the same terminal airway, the smaller alveolus—because it generates a larger inward-directed collapsing pressure—has a tendency (without pulmonary surfactant) to collapse and empty its air into the larger alveolus. (b) Pulmonary surfactant reduces the surface tension of a smaller alveolus more than that of a larger alveolus. This reduction in surface tension offsets the effect of the smaller radius in determining the inward-directed pressure. Consequently, the collapsing pressures of the small and large alveoli are comparable. Therefore, in the presence of pulmonary surfactant a small alveolus does not collapse and empty its air into the larger alveolus.

Because the collapsing pressure is inversely proportional to the radius, the smaller the alveolus, the smaller its radius and the greater its tendency to collapse at a given surface tension. Accordingly, if two alveoli of unequal size but the same surface tension are connected by the same terminal airway, the smaller alveolus—because it generates a larger collapsing pressure—has a tendency to collapse and empty its air into the larger alveolus (▮Figure 13-14a).

Small alveoli normally do not collapse and blow up larger alveoli, however, because pulmonary surfactant reduces the surface tension of small alveoli more than that of larger alveoli. Pulmonary surfactant decreases surface tension to a greater degree in small alveoli than in larger alveoli because the surfactant molecules are crowded closer together in the smaller alveoli. The larger an alveolus, the more spread out are its surfactant molecules and the less effect they have on reducing surface tension. The surfactant-induced lower surface tension of small alveoli offsets the effect of their smaller radius in determining the inward-directed pressure. Therefore, the presence of surfactant causes the collapsing pressure of small alveoli to become comparable to that of larger alveoli and minimizes the tendency for small alveoli to collapse and empty their contents into larger alveoli (Figure 13-14b). Pulmonary surfactant therefore helps stabilize the sizes of the alveoli and helps keep them open and available to participate in gas exchange.

The opposing forces acting on the lung (that is, the forces keeping the alveoli open and the countering forces that promote alveolar collapse) are summarized in Table 13-2.

Newborn Respiratory Distress Syndrome Developing fetal lungs normally cannot synthesize pulmonary surfactant until late in pregnancy. Especially in an infant born prematurely, not enough pulmonary surfactant may be produced to reduce the alveolar surface tension to manageable levels. The resulting collection of symptoms is termed **newborn respiratory distress syndrome**. The infant must make strenuous inspiratory efforts to overcome the high surface tension in an attempt to inflate the poorly compliant lungs. Moreover, the work of breathing is further increased because the alveoli, in the absence of surfactant, tend to collapse almost completely during each expiration. It is more difficult (requires a greater transmural pressure differential) to expand a collapsed alveolus by a given volume than to increase an already partially expanded alveolus by the same volume. The situation is analogous to blowing up a new balloon. It takes more effort to blow in that first breath of air when starting to blow up a new balloon than to blow additional breaths into the already partially expanded balloon. With newborn respiratory distress syndrome, it is as though with every breath the infant must start blowing up a new balloon. Lung expansion may require transmural pressure gradients of 20 to 30 mm Hg (compared to the normal 4 to 6 mm Hg) to overcome the tendency of surfactant-deprived alveoli to collapse. Worse yet, the newborn's muscles are still weak. The respiratory distress from surfactant deficiency may soon lead to death if breathing efforts become exhausting or inadequate to support sufficient gas exchange.

This life-threatening condition affects 30,000 to 50,000 newborns, primarily premature infants, each year in the United States. Until the surfactant-secreting cells mature sufficiently, the condition is treated by surfactant replacement. In addition, drugs can hasten the maturation process.

The work of breathing normally requires only about 3% of total energy expenditure.

During normal quiet breathing, the respiratory muscles must work during inspiration to expand the lungs against their elastic forces and to overcome airway resistance, whereas expiration is passive. Normally, the lungs are highly compliant and airway resistance is low, so only about 3% of the total energy expended by the body is used for quiet breathing.

 The work of breathing may be increased in four different situations:

1. *When pulmonary compliance is decreased,* such as with pulmonary fibrosis, more work is required to expand the lungs.

2. *When airway resistance is increased,* such as with COPD, more work is required to achieve the greater pressure gradients necessary to overcome the resistance so that adequate airflow can occur.

3. *When elastic recoil is decreased,* as with emphysema, passive expiration may be inadequate to expel the volume of air normally exhaled during quiet breathing. Thus, the abdominal muscles must work to aid in emptying the lungs, even when the person is at rest.

4. *When there is a need for increased ventilation,* such as during exercise, more work is required to accomplish both a greater depth of breathing (a larger volume of air moving in and out with each breath) and a faster rate of breathing (more breaths per minute).

During strenuous exercise, the amount of energy required to power pulmonary ventilation may increase up to 25-fold. However, because total energy expenditure by the body increases up to 15- to 20-fold during heavy exercise, the energy used for increased ventilation still represents only about 5% of total energy expended. In contrast, in patients with poorly compliant lungs or obstructive lung disease, the energy required for breathing even at rest may be as much as 30% of total energy expenditure. In such cases, the individual's exercise ability is severely limited, as breathing itself becomes exhausting.

The lungs normally operate about "half full."

On average, in healthy young adults, the maximum air that the lungs can hold is about 5.7 liters in males (4.2 liters in females). Anatomic build, age, the distensibility of the lungs, and the presence or absence of respiratory disease affect this total lung

TABLE 13-2 Opposing Forces Acting on the Lung	
Forces Keeping the Alveoli Open	**Forces Promoting Alveolar Collapse**
Transmural pressure gradient	Elasticity of stretched elastin fibers in lung connective tissue
Pulmonary surfactant (which opposes alveolar surface tension)	Alveolar surface tension

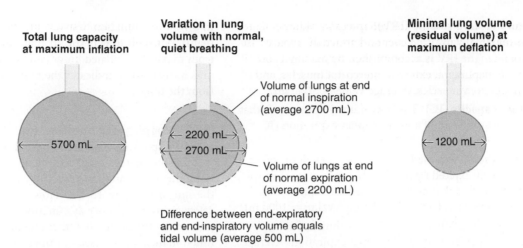

Figure 13-15 Normal lung volumes and capacities of a healthy young adult male. A lung capacity is the sum of two or more lung volumes. Values for females are somewhat lower. (Note that residual volume cannot be measured with a spirometer but must be determined by another means.)

Total lung capacity at maximum inflation

5700 mL

Variation in lung volume with normal, quiet breathing

Volume of lungs at end of normal inspiration (average 2700 mL)

2200 mL
2700 mL

Volume of lungs at end of normal expiration (average 2200 mL)

Difference between end-expiratory and end-inspiratory volume equals tidal volume (average 500 mL)

Minimal lung volume (residual volume) at maximum deflation

1200 mL

(a) Normal range and extremes of lung volume in a healthy young adult male

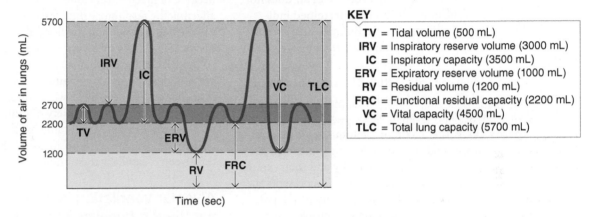

KEY

TV = Tidal volume (500 mL)
IRV = Inspiratory reserve volume (3000 mL)
IC = Inspiratory capacity (3500 mL)
ERV = Expiratory reserve volume (1000 mL)
RV = Residual volume (1200 mL)
FRC = Functional residual capacity (2200 mL)
VC = Vital capacity (4500 mL)
TLC = Total lung capacity (5700 mL)

(b) Normal variations in lung volume in a spirogram in a healthy young adult male

capacity. Normally, during quiet breathing, the lungs are nowhere near maximally inflated, nor are they deflated to their minimum volume. Thus, the lungs normally remain moderately inflated throughout the respiratory cycle. At the end of a normal quiet expiration, the lungs still contain about 2200 mL of air. During each typical breath under resting conditions, about 500 mL of air are inspired and the same quantity is expired, so during quiet breathing the lung volume varies from 2200 mL at the end of expiration to 2700 mL at the end of inspiration (Figure 13-15a). During maximal expiration, lung volume can decrease to 1200 mL in males (1000 mL in females), but the lungs can never be completely deflated because the small airways collapse during forced expirations at low lung volumes, blocking further outflow.

A beneficial outcome of not being able to empty the lungs completely is that even during maximal expiratory efforts, gas exchange can continue between the blood flowing through the lungs and the remaining alveolar air. As a result, the gas content of the blood leaving the lungs for delivery to the tissues normally remains remarkably constant throughout the respiratory cycle. By contrast, if the lungs completely filled and emptied with each breath, the amount of O_2 taken up and CO_2 dumped off by the blood would fluctuate widely. Another advantage of the lungs not completely emptying with each breath is the

reduced work of breathing. Recall that it takes less effort to inflate a partially inflated alveolus than a totally collapsed one.

The changes in lung volume that occur with different respiratory efforts can be determined by a spirometer. A traditional wet **spirometer** consists of an air-filled drum floating in a water-filled chamber. As the person breathes air in and out of the drum through a tube connecting the mouth to the air chamber, the drum rises and falls in the water chamber. This rise and fall can be recorded as a **spirogram**, which is calibrated to changes in lung volume. Inspiration is recorded as an upward deflection and expiration as a downward deflection. Today, less cumbersome computerized spirometers have replaced the wet spirometer for clinical use, but the principles of the lung volumes and capacities determined by the older instrument are the same.

Lung Volumes and Capacities Figure 13-15b is a hypothetical example of a spirogram in a healthy young adult male. Generally, the values are lower for females. The following lung volumes and capacities (a lung capacity is the sum of two or more lung volumes) can be determined:

■ **Tidal volume (TV):** The volume of air entering or leaving the lungs during a single breath. Average value under resting conditions = 500 mL.

- **Inspiratory reserve volume (IRV):** The extra volume of air that can be maximally inspired over and above the typical resting tidal volume. The IRV is accomplished by maximal contraction of the diaphragm, external intercostal muscles, and accessory inspiratory muscles. Average value = 3000 mL.

- **Inspiratory capacity (IC):** The maximum volume of air that can be inspired at the end of a normal quiet expiration (IC = IRV + TV). Average value = 3500 mL.

- **Expiratory reserve volume (ERV):** The extra volume of air that can be actively expired by maximally contracting the expiratory muscles beyond that normally passively expired at the end of a typical resting tidal volume. Average value = 1000 mL.

- **Residual volume (RV):** The minimum volume of air remaining in the lungs even after a maximal expiration. Average value = 1200 mL. The residual volume cannot be measured directly with a spirometer because this volume of air does not move into and out of the lungs. It can be determined indirectly, however, through gas-dilution techniques involving inspiration of a known quantity of a harmless tracer gas such as helium.

- **Functional residual capacity (FRC):** The volume of air in the lungs at the end of a normal passive expiration (FRC = ERV + RV). Average value = 2200 mL.

- **Vital capacity (VC):** The maximum volume of air that can be moved out during a single breath following a maximal inspiration. The subject first inspires maximally and then expires maximally (VC = IRV + TV + ERV). The VC represents the maximum volume change possible within the lungs (IFigure 13-16). It is rarely used because the maximal muscle contractions involved become exhausting, but it is valuable in determining the functional capacity of the lungs. Average value = 4500 mL.

- **Total lung capacity (TLC):** The maximum volume of air that the lungs can hold (TLC = VC + RV). Average value = 5700 mL.

- **Forced expiratory volume in 1 second (FEV$_1$):** The volume of air that can be expired during the first second of expiration

in a VC determination. Usually, FEV$_1$ is about 80% of VC—that is, normally 80% of the air that can be forcibly expired from maximally inflated lungs can be expired within 1 second. This measurement indicates the maximal airflow rate possible from the lungs.

Respiratory Dysfunction Two general categories of respiratory dysfunction yield abnormal results during spirometry—*obstructive lung disease* (difficulty in emptying the lungs) and *restrictive lung disease* (difficulty in filling the lungs) (IFigure 13-17). However, these are not the only categories of respiratory dysfunction, nor is spirometry the only pulmonary function test. Other conditions affecting respiratory function include (1) diseases impairing diffusion of O_2 and CO_2 across the pulmonary membranes; (2) reduced ventilation because of mechanical failure, as with neuromuscular disorders affecting the respiratory muscles; (3) inadequate perfusion (failure of adequate pulmonary blood flow); or (4) ventilation–perfusion imbalances involving a poor matching of air and blood so that efficient gas exchange cannot occur. Some lung diseases are actually a complex mixture of different types of functional disturbances. To determine what abnormalities are present, the diagnostician relies on a variety of pulmonary function tests in addition to spirometry, including X-ray examination, blood-gas determinations, and tests to measure the diffusion capacity of the alveolar–capillary membrane.

Alveolar ventilation is less than pulmonary ventilation because of dead space.

Various changes in lung volume represent only one factor in determining **pulmonary**, or **minute**, **ventilation**, which is the volume of air breathed in and out in 1 minute. The other important factor is **respiratory rate**, which averages 12 breaths per minute:

$$\text{Pulmonary ventilation} = \text{tidal volume} \times \text{respiratory rate}$$
$$\text{(mL/min)} \qquad \text{(mL/breath)} \qquad \text{(breaths/min)}$$

At an average tidal volume of 500 mL/breath and a respiratory rate of 12 breaths/min, pulmonary ventilation is 6000 mL, or 6 liters, of air breathed in and out in 1 minute under resting conditions. For a brief period, a healthy young adult male can voluntarily increase his total pulmonary ventilation 25-fold, to 150 liters/min. To increase pulmonary ventilation, both tidal volume and respiratory rate increase, but depth of breathing increases more than frequency of breathing. It is usually more advantageous to have a greater increase in tidal volume than in respiratory rate because of anatomic dead space, discussed next.

Anatomic Dead Space Not all the inspired air gets down to the site of gas exchange in the alveoli. Part remains in the conducting airways, where it is not available for gas exchange. The volume of the conducting passages in an adult averages about 150 mL. This volume is considered **anatomic dead space** because air within these conducting airways is useless for exchange. Anatomic dead space greatly affects the efficiency of pulmonary ventilation. In effect, even though 500 mL of air are moved in and out with each breath, only 350 mL are actually

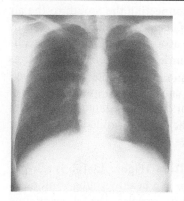 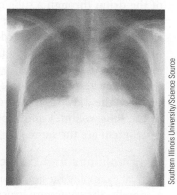

(a) Maximum volume of lungs at maximum inspiration

(b) Minimum volume of lungs at maximum expiration

IFigure 13-16 **X-rays of lungs showing maximum volume change.** The difference between (a) the maximum lung volume at maximum inspiration and (b) the minimum lung volume at maximum expiration is the vital capacity, the maximum volume change possible during a single breath.

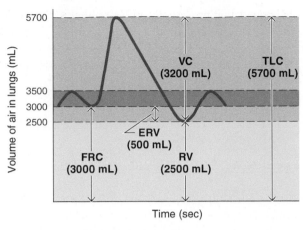

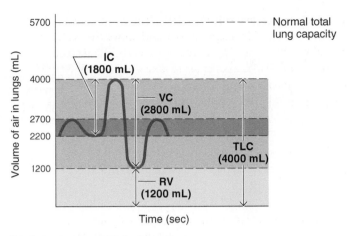

(a) Spirogram in obstructive lung disease

(b) Spirogram in restrictive lung disease

▮Figure 13-17 **Abnormal lung volumes and capacities with obstructive and restrictive lung diseases.** (a) Because a patient with obstructive lung disease experiences more difficulty emptying the lungs than filling them, the total lung capacity (TLC) is essentially normal, but the functional residual capacity (FRC) and the residual volume (RV) are elevated as a result of the additional air trapped in the lungs following expiration. Because the RV is increased, the vital capacity (VC) is reduced. With more air remaining in the lungs, less of the TLC is available to be used in exchanging air with the atmosphere. Another common finding is a markedly reduced forced expiratory volume in one second (FEV₁) because the airflow rate is reduced by the airway obstruction. Even though both the VC and the FEV₁ are reduced, the FEV₁ is reduced more markedly than the VC. As a result, the FEV₁/VC% is much lower than the normal 80%—that is, much less than 80% of the reduced VC can be blown out during the first second. (b) In restrictive lung disease, the lungs are less compliant than normal. Total lung capacity, inspiratory capacity, and VC are reduced because the lungs cannot be expanded as normal. The percentage of the VC that can be exhaled within one second is the normal 80% or an even higher percentage because air can flow freely in the airways. Therefore, the FEV₁/VC% is particularly useful in distinguishing between obstructive and restrictive lung disease. Also, in contrast to obstructive lung disease, the RV is usually normal in restrictive lung disease.

FIGURE FOCUS: *Use the graphs to explain why vital capacity is reduced in obstructive lung disease and in restrictive lung disease compared to normal in Figure 13-15.*

---

exchanged between the atmosphere and the alveoli because of the 150 mL occupying the anatomic dead space.

Looking at Figure 13-18, note that at the end of inspiration the respiratory airways are filled with 150 mL of fresh atmospheric air from the inspiration. During the subsequent expiration, 500 mL of air are expired to the atmosphere. The first 150 mL expired are the fresh air that was retained in the airways and never used. The remaining 350 mL expired are "old" alveolar air that has participated in gas exchange with the blood. During the same expiration, 500 mL of gas also leave the alveoli. The first 350 mL are expired to the atmosphere; the other 150 mL of old alveolar air never reach the outside but remain in the conducting airways.

On the next inspiration, 500 mL of gas enter the alveoli. The first 150 mL to enter the alveoli are the old alveolar air that remained in the dead space during the preceding expiration. The other 350 mL entering the alveoli are fresh air inspired from the atmosphere. Simultaneously, 500 mL of air enter from the atmosphere. The first 350 mL of atmospheric air reach the alveoli; the other 150 mL remain in the conducting airways to be expired without benefit of being exchanged with the blood, as the cycle repeats itself.

Alveolar Ventilation Because the amount of atmospheric air that reaches the alveoli and is actually available for exchange with blood is more important than the total amount breathed in and out, **alveolar ventilation**—the volume of air exchanged between the atmosphere and the alveoli per minute—is more important than pulmonary ventilation. In determining alveolar

ventilation, the amount of wasted air moved in and out through the anatomic dead space must be taken into account, as follows:

Alveolar ventilation = (tidal volume − dead space volume)
× respiratory rate

With average resting values,

Alveolar ventilation = (500 mL/breath − 150 mL
dead space volume) × 12 breaths/min = 4200 mL/min

Thus, with quiet breathing, alveolar ventilation is 4200 mL/min, whereas pulmonary ventilation is 6000 mL/min.

Effect of Breathing Patterns on Alveolar Ventilation
To understand how important dead space volume is in determining the magnitude of alveolar ventilation, examine the effect of various breathing patterns on alveolar ventilation in ▮Table 13-3. If a person deliberately breathes deeply (for example, a tidal volume of 1200 mL) and slowly (for example, a respiratory rate of 5 breaths/min), pulmonary ventilation is 6000 mL/min, the same as during quiet breathing at rest, but alveolar ventilation increases to 5250 mL/min compared to the resting rate of 4200 mL/min. In contrast, if a person deliberately breathes shallowly (for example, a tidal volume of 150 mL) and rapidly (a frequency of 40 breaths/min), pulmonary ventilation would still be 6000 mL/min; however, alveolar ventilation would be 0 mL/min. In effect, the person would only be drawing air in and out of the anatomic dead space without any atmospheric air being exchanged with the alveoli, where it could be useful. The individual could voluntarily maintain such a breath-

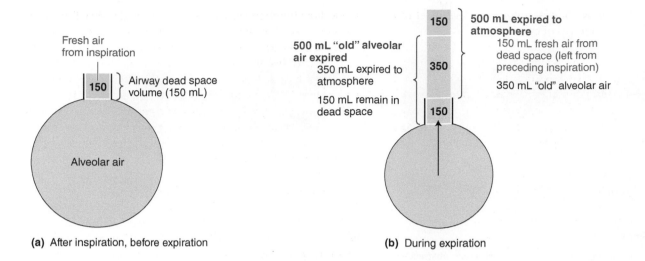

(a) After inspiration, before expiration

(b) During expiration

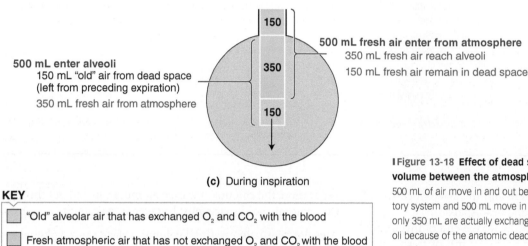

500 mL fresh air enter from atmosphere
350 mL fresh air reach alveoli
150 mL fresh air remain in dead space

500 mL enter alveoli
150 mL "old" air from dead space (left from preceding expiration)
350 mL fresh air from atmosphere

(c) During inspiration

KEY

| | "Old" alveolar air that has exchanged O_2 and CO_2 with the blood |
| | Fresh atmospheric air that has not exchanged O_2 and CO_2 with the blood |

❙Figure 13-18 **Effect of dead space volume on exchange of tidal volume between the atmosphere and the alveoli.** Even though 500 mL of air move in and out between the atmosphere and the respiratory system and 500 mL move in and out of the alveoli with each breath, only 350 mL are actually exchanged between the atmosphere and the alveoli because of the anatomic dead space (the volume of air in the respiratory airways).

❙**TABLE 13-3** Effect of Different Breathing Patterns on Alveolar Ventilation

Breathing Pattern	Tidal Volume (mL/breath)	Respiratory Rate (breaths/min)	Dead Space Volume (mL)	Pulmonary Ventilation (mL/min)*	Alveolar Ventilation (mL/min)†
Quiet breathing at rest	500	12	150	6000	4200
Deep, slow breathing	1200	5	150	6000	5250
Shallow, rapid breathing	150	40	150	6000	0

*Equals tidal volume × respiratory rate.
†Equals (tidal volume − dead space volume) × respiratory rate.

ing pattern for only a few minutes before losing consciousness, at which time normal breathing would resume.

The value of reflexly bringing about a larger increase in depth of breathing than in rate of breathing when pulmonary ventilation increases during exercise should now be apparent. It is the most efficient means of elevating alveolar ventilation. When tidal volume increases, the entire increase in pulmonary ventilation goes toward elevating alveolar ventilation. In contrast, when pulmonary ventilation increases as a result of an increase in respiratory rate, the frequency with which air is wasted in the dead space also increases because a portion of *each breath* must move in and out of the dead space. Thus not all of the increase in pulmonary ventilation goes toward elevating alveolar ventilation. As needs vary, ventilation is normally adjusted to a tidal volume and respiratory rate that meet those needs most efficiently in terms of energy cost.

Alveolar Dead Space We have assumed that all the atmospheric air entering the alveoli participates in exchanges of O_2 and CO_2 with pulmonary blood. However, the match between air and blood is not always perfect because not all alveoli are equally ventilated with air and perfused with blood. Any ventilated alveoli that do not participate in gas exchange with blood because they are inadequately perfused are considered **alveolar dead space**. In healthy people, alveolar dead space is quite small and of little importance, but it can increase to even lethal levels in several types of pulmonary disease.

Next you will learn why alveolar dead space is minimal in healthy individuals.

Local controls act on bronchiolar and arteriolar smooth muscle to match airflow to blood flow.

When discussing the role of airway resistance in determining airflow rate into and out of the lungs, we were referring to the overall resistance of all the airways collectively. However, the resistance of individual airways supplying specific alveoli can be adjusted independently in response to changes in the airway's local environment. This situation is comparable to the control of systemic arterioles. Recall that overall systemic arteriolar resistance (that is, total peripheral resistance) is an important determinant of the blood pressure gradient that drives blood flow throughout the systemic circulatory system (see p. 349). Yet the radius of individual arterioles supplying various tissues can be adjusted locally to match the tissues' differing metabolic needs (see p. 345).

Effect of CO_2 on Bronchiolar Smooth Muscle Similar to arteriolar smooth muscle, bronchiolar smooth muscle is sensitive to local changes within its immediate environment, particularly to local CO_2 levels. If an alveolus is receiving too little airflow in comparison to its blood flow, CO_2 levels will increase in the alveolus and surrounding tissue as the blood drops off more CO_2 than is exhaled into the atmosphere. This local increase in CO_2 directly promotes relaxation of the bronchiolar smooth muscle, bringing about dilation of the airway supplying the underaerated alveolus. The resultant decrease in airway resistance leads to an increased airflow (for the same ΔP) to the involved alveolus, so its airflow now matches its blood supply (❙ Figure 13-19a). The converse is also true. A localized decrease in CO_2 associated with an alveolus that is receiving too much air for its blood supply directly increases contractile activity of the airway smooth muscle involved, constricting the airway supplying this overaerated alveolus. The result is reduced airflow to the overaerated alveolus (❙ Figure 13-19b).

Effect of O_2 on Pulmonary Arteriolar Smooth Muscle Simultaneously, a similar locally induced effect on pulmonary vascular smooth muscle takes place to maximally match blood flow to airflow. The two mechanisms for matching airflow and blood flow function concurrently, so normally very little air or blood is wasted in the lung. In the pulmonary circulation, just as in the systemic circulation, distribution of the cardiac output to different alveolar capillary networks can be controlled by adjusting the resistance to blood flow through specific pulmonary arterioles. If blood flow is greater than airflow to a given alveolus, the O_2 level in the alveolus and surrounding tissues falls below normal as the overabundance of blood extracts more O_2 than usual from the alveolus. The local decrease in O_2 concentration causes vasoconstriction of the pulmonary arteriole supplying this particular capillary bed, thus reducing blood flow to match the smaller airflow (❙ Figure 13-19a). Conversely, an increase in alveolar O_2 concentration caused by a mismatched large airflow and small blood flow brings about pulmonary vasodilation, which increases blood flow to match the larger airflow (❙ Figure 13-19b).

This local effect of O_2 on pulmonary arteriolar smooth muscle is, appropriately, just the opposite of its effect on systemic arteriolar smooth muscle (❙ Table 13-4). In the systemic circulation, a decrease in O_2 in a tissue causes localized vasodilation to increase blood flow to the deprived area, and vice versa, which is important in matching blood supply to local metabolic needs.

Because of gravitational effects, some regional differences in ventilation and perfusion exist from the top to the bottom of the lung. When a person is standing upright, ventilation and perfusion are both less at the top of the lung and greater at the bottom of the lung, but gravity exerts a more marked effect on blood flow than on airflow. Therefore, the ventilation–perfusion ratio (the rate of airflow compared to the rate of blood flow) decreases from the top to the bottom of the lung (❙ Figure 13-20). In other words, the top of the lung receives less air and blood than the bottom of the lung, but it receives relatively more air than blood; the bottom of the lung receives more air and blood than the top of the lung, but it receives relatively less air than blood. In healthy lungs, the effect of this mismatch between air and blood has a negligible effect on overall O_2 uptake and CO_2 elimination. Airflow and blood flow at a particular alveolar interface are usually matched as much as possible by local controls to accomplish efficient exchange of O_2 and CO_2. However, in pathologic conditions, ventilation–perfusion mismatches can exceed the capability of local controls to compensate, so the effect on O_2 uptake and CO_2 elimination can become significant, as with widespread plugging of airways with inflammatory mucus secretion or widespread damage to pulmonary vessels.

We have now finished discussing respiratory mechanics—all the factors involved in ventilation. We next examine gas exchange between alveolar air and blood and then between blood and systemic tissues.

Check Your Understanding 13.2

1. Compare the muscles involved, the intra-alveolar pressure changes, and the air movement that takes place during normal quiet breathing and breathing during strenuous exercise.

2. Define *compliance* and *elastic recoil*.

The Respiratory System | **465**

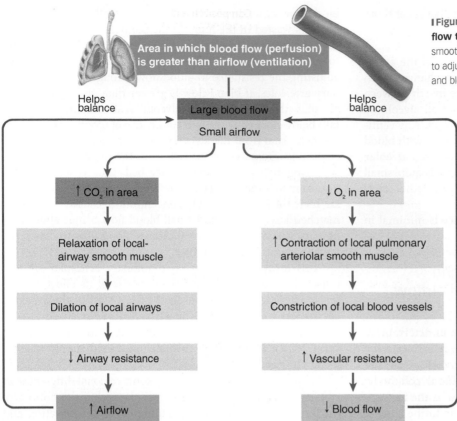

Area in which blood flow (perfusion) is greater than airflow (ventilation)

Helps balance

Large blood flow
Small airflow

Helps balance

↑ CO₂ in area

Relaxation of local-airway smooth muscle

↓ Dilation of local airways

↓ Airway resistance

↑ Airflow

↓ O₂ in area

↑ Contraction of local pulmonary arteriolar smooth muscle

Constriction of local blood vessels

↑ Vascular resistance

↓ Blood flow

(a) Local controls to adjust ventilation and perfusion to lung area with large blood flow and small airflow

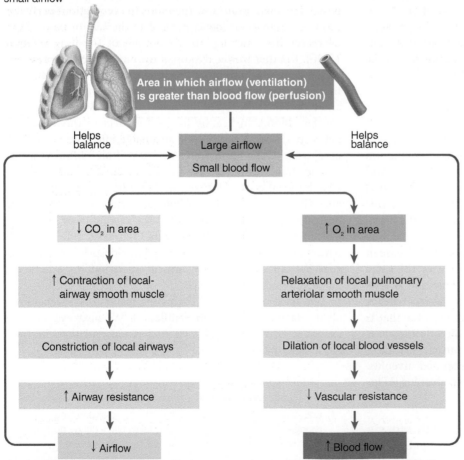

Area in which airflow (ventilation) is greater than blood flow (perfusion)

Helps balance

Large airflow
Small blood flow

Helps balance

↓ CO₂ in area

↑ Contraction of local-airway smooth muscle

Constriction of local airways

↑ Airway resistance

↓ Airflow

↑ O₂ in area

Relaxation of local pulmonary arteriolar smooth muscle

Dilation of local blood vessels

↓ Vascular resistance

↑ Blood flow

(b) Local controls to adjust ventilation and perfusion to a lung area with large airflow and small blood flow

Figure 13-19 Local controls to match airflow and blood flow to an area of the lung. CO_2 acts locally on bronchiolar smooth muscle and O_2 acts locally on arteriolar smooth muscle to adjust ventilation and perfusion, respectively, to match airflow and blood flow to an area of the lung.

---

3. State the forces that keep the alveoli open and those that promote alveolar collapse.
4. Draw and label the lung volumes and capacities in a typical spirogram in a healthy young adult male.

13.3 Gas Exchange

The purpose of breathing is to provide a continual supply of fresh O_2 for pickup by blood and to constantly remove CO_2 unloaded from blood. Blood acts as a transport system for O_2 and CO_2 between the lungs and the tissues, with the tissue cells extracting O_2 from blood and eliminating CO_2 into it.

Gases move down partial pressure gradients.

Gas exchange at both the pulmonary capillary and the tissue capillary levels involves simple passive diffusion of O_2 and CO_2 down *partial pressure gradients*. No active transport mechanisms exist for these gases. Let us see what partial pressure gradients are and how they are established.

Partial Pressures Atmospheric air is a mixture of gases; typical dry air contains about 79% nitrogen (N_2) and 21% O_2, with almost negligible percentages of CO_2, H_2O vapor, other gases, and pollutants. Altogether, these gases exert a total atmospheric pressure of 760 mm Hg at sea level. This total pressure is equal to the sum of the pressures that each gas in the mixture partially contributes. The pressure exerted by a particular gas is directly proportional to the percentage of that gas in the total air mixture. Every gas molecule, no matter what its size, exerts the same amount of pressure; for example, a N_2 molecule exerts the same

TABLE 13-4 Effects of Local Changes in O_2 on Pulmonary and Systemic Arterioles

| Vessels | EFFECT OF A LOCAL CHANGE IN O_2 | |
	Decreased O_2	Increased O_2
Pulmonary arterioles	Vasoconstriction	Vasodilation
Systemic arterioles	Vasodilation	Vasoconstriction

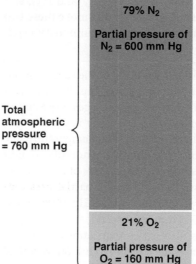

Composition and partial pressures in atmospheric air

79% N₂

Partial pressure of N₂ = 600 mm Hg

Total atmospheric pressure = 760 mm Hg

21% O₂

Partial pressure of O₂ = 160 mm Hg

Partial pressure of N₂ in atmospheric air:
P_{N_2} = 760 mm Hg × 0.79 = 600 mm Hg

Partial pressure of O₂ in atmospheric air:
P_{O_2} = 760 mm Hg × 0.21 = 160 mm Hg

pressure as an O_2 molecule. Because 79% of the air consists of N_2 molecules, 79% of the 760 mm Hg atmospheric pressure, or 600 mm Hg, is exerted by the N_2 molecules. Similarly, because O_2 represents 21% of the atmosphere, 21% of the 760 mm Hg atmospheric pressure, or 160 mm Hg, is exerted by O_2 (Figure 13-21). The individual pressure exerted independently by a particular gas within a mixture of gases is known as its **partial pressure**, designated by P_{gas}. Thus, the partial pressure of O_2 in atmospheric air, P_{O_2}, is normally 160 mm Hg. The atmospheric partial pressure of CO_2, P_{CO_2}, is negligible at 0.23 mm Hg.

Gases dissolved in a liquid such as blood or another body fluid also exert a partial pressure. The greater the partial pressure of a gas in a liquid is, the more of that gas is dissolved.

Partial Pressure Gradients A difference in partial pressure between the capillary blood and the surrounding structures is known as a **partial pressure gradient**. Partial pressure gradients exist between the alveolar air and the pulmonary capillary

Figure 13-21 Concept of partial pressures. The partial pressure exerted by each gas in a mixture equals the total pressure times the fractional composition of the gas in the mixture.

FIGURE FOCUS: *If a person lives 1 mile above sea level in Denver, Colorado (the "mile-high city"), where the atmospheric pressure is 630 mm Hg, what would the P_{O_2} of inspired air be?*

blood. Similarly, partial pressure gradients exist between the systemic capillary blood and the surrounding tissues. A gas always diffuses down its partial pressure gradient from the area of higher partial pressure to the area of lower partial pressure, similar to diffusion down a concentration gradient.

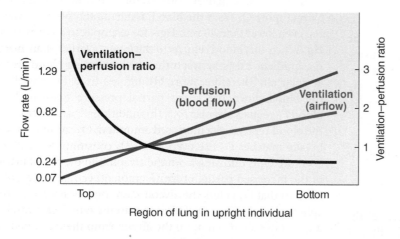

(a) Regional ventilation and perfusion rates and ventilation–perfusion ratios in the lungs

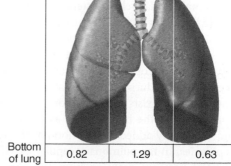

	Ventilation (airflow) (L/min)	Perfusion (blood flow) (L/min)	Ventilation–perfusion ratio
Top of lung	0.24	0.07	3.40
Bottom of lung	0.82	1.29	0.63

(b) Ventilation and perfusion rates and ventilation–perfusion ratios at top and bottom of lungs

Figure 13-20 Differences in ventilation, perfusion, and ventilation–perfusion ratios at the top and bottom of the lungs as a result of gravitational effects. Note that the top of the lungs receives less air and blood than the bottom of the lungs, but the top of the lungs receives relatively more air than blood and the bottom of the lungs receives relatively less air than blood.

FIGURE FOCUS: *(1) Is the top or bottom of the lung better ventilated? (2) Which region is better perfused? (3) Which has a higher ventilation–perfusion ratio?*

O_2 enters and CO_2 leaves the blood in the lungs down partial pressure gradients.

We first consider the magnitude of alveolar P_{O_2} and P_{CO_2} and then look at the partial pressure gradients that move these two gases between the alveoli and the incoming pulmonary capillary blood.

Alveolar P_{O_2} and P_{CO_2} Alveolar air is not of the same composition as inspired atmospheric air, for two reasons. First, as soon as atmospheric air enters the respiratory passages, exposure to the moist airways saturates it with H_2O. Like any other gas, water vapor exerts a partial pressure. At body temperature, the partial pressure of H_2O vapor is 47 mm Hg. Humidification of inspired air in effect "dilutes" the partial pressure of the inspired gases by 47 mm Hg because the sum of the partial pressures must total the atmospheric pressure of 760 mm Hg. In moist air, $P_{H_2O} = 47$ mm Hg, $P_{N_2} = 563$ mm Hg, and $P_{O_2} = 150$ mm Hg.

Second, alveolar P_{O_2} is also lower than atmospheric P_{O_2} because fresh inspired air (average equals 350 mL out of a tidal volume of 500 mL) is mixed with the large volume of old air that remained in the lungs at the end of the preceding expiration (average functional residual capacity equals 2200 mL). At the end of inspiration, only about 13% of the air in the alveoli is fresh air. As a result of humidification and the small turnover of alveolar air, the average alveolar P_{O_2} is 100 mm Hg, compared to the atmospheric P_{O_2} of 160 mm Hg.

You might logically think that alveolar P_{O_2} would increase during inspiration with the arrival of fresh air and would decrease during expiration. Only small fluctuations in alveolar P_{O_2} occur, however, for two reasons. First, only a small proportion of the total alveolar air is exchanged with each breath. The relatively small volume of inspired, high-P_{O_2} air is quickly mixed with the larger volume of retained alveolar air, which has a lower P_{O_2}. Thus, the O_2 in the inspired air can only slightly elevate the level of the total alveolar P_{O_2}. Even this potentially small elevation of P_{O_2} is diminished for another reason. O_2 continually moves by passive diffusion down its partial pressure gradient from the alveoli into the blood. The O_2 arriving in the alveoli in the newly inspired air simply replaces the O_2 diffusing out of the alveoli into the pulmonary capillaries. Therefore, alveolar P_{O_2} remains relatively constant around 100 mm Hg throughout the respiratory cycle. Because pulmonary blood P_{O_2} equilibrates with alveolar P_{O_2}, the P_{O_2} of the blood leaving the lungs likewise remains fairly constant at this same value. Accordingly, the amount of O_2 in the blood available to the tissues varies only slightly during the respiratory cycle.

A similar situation exists in reverse for CO_2, which is continuously produced by the body tissues as a metabolic waste product and constantly added to the blood at the level of the systemic capillaries. In the pulmonary capillaries, CO_2 diffuses down its partial pressure gradient from the blood into the alveoli and is removed from the body during expiration. As with O_2, alveolar P_{CO_2} remains fairly constant throughout the respiratory cycle but at a lower value of 40 mm Hg.

P_{O_2} and P_{CO_2} Gradients Across the Pulmonary Capillaries As blood passes through the lungs, it picks up O_2 and gives up CO_2 by diffusion down partial pressure gradients between blood and alveoli. Ventilation constantly replenishes alveolar O_2 and removes CO_2, thus maintaining the appropriate gradients to ensure this diffusion. Blood entering the pulmonary capillaries is systemic venous blood pumped to the lungs through the pulmonary arteries. This blood, having just returned from the body tissues, is relatively low in O_2, with a P_{O_2} of 40 mm Hg, and is relatively high in CO_2, with a P_{CO_2} of 46 mm Hg. As this blood flows through the pulmonary capillaries, it is exposed to alveolar air (▌Figure 13-22). Because the alveolar P_{O_2} at 100 mm Hg is higher than the P_{O_2} of 40 mm Hg in the blood entering the lungs, O_2 diffuses down its partial pressure gradient from the alveoli into the blood until no further gradient exists. As blood leaves the pulmonary capillaries, it has a P_{O_2} equal to alveolar P_{O_2} at 100 mm Hg.

The partial pressure gradient for CO_2 is in the opposite direction. Blood entering the pulmonary capillaries has a P_{CO_2} of 46 mm Hg, whereas alveolar P_{CO_2} is only 40 mm Hg. CO_2 diffuses from the blood into the alveoli until blood P_{CO_2} equilibrates with alveolar P_{CO_2}. Thus, blood leaving the pulmonary capillaries has a P_{CO_2} of 40 mm Hg. After leaving the lungs, the blood, which now has a P_{O_2} of 100 mm Hg and a P_{CO_2} of 40 mm Hg, is returned to the heart and then pumped out to the body tissues as systemic arterial blood.

Note that blood returning to the lungs from the tissues still contains O_2 (P_{O_2} of systemic venous blood = 40 mm Hg) and that blood leaving the lungs still contains CO_2 (P_{CO_2} of systemic arterial blood = 40 mm Hg). The extra O_2 carried in the blood beyond that normally given up to the tissues represents an immediately available O_2 reserve that can be tapped by tissue cells whenever their O_2 demands increase. The CO_2 remaining in the blood even after passage through the lungs plays an important role in the acid–base balance of the body because CO_2 generates carbonic acid. Furthermore, arterial P_{CO_2} is important in driving respiration, as described later.

The amount of O_2 picked up in the lungs matches the amount extracted and used by the tissues. When the tissues metabolize more actively (for example, during exercise), they extract more O_2 from the blood, reducing the systemic venous P_{O_2} even lower than 40 mm Hg—for example, to a P_{O_2} of 30 mm Hg. When this blood returns to the lungs, a larger-than-normal P_{O_2} gradient exists between the newly entering blood and the alveolar air. Therefore, more O_2 diffuses from the alveoli into the blood down the larger partial pressure gradient before blood P_{O_2} equals alveolar P_{O_2}. This additional transfer of O_2 into the blood replaces the increased amount of O_2 consumed, so O_2 uptake matches O_2 use even when O_2 consumption increases. As more O_2 is diffusing from the alveoli into the blood because of the increased partial pressure gradient, ventilation is stimulated so that O_2 enters the alveoli more rapidly from the atmosphere to replace the O_2 diffusing into the blood. Similarly, the amount of CO_2 given up to the alveoli from the blood matches the amount of CO_2 picked up at the tissues.

Factors other than the partial pressure gradient influence the rate of gas transfer.

We have been discussing diffusion of O_2 and CO_2 between alveoli and blood as if these gases' partial pressure gradients were the sole determinants of their rates of diffusion. According

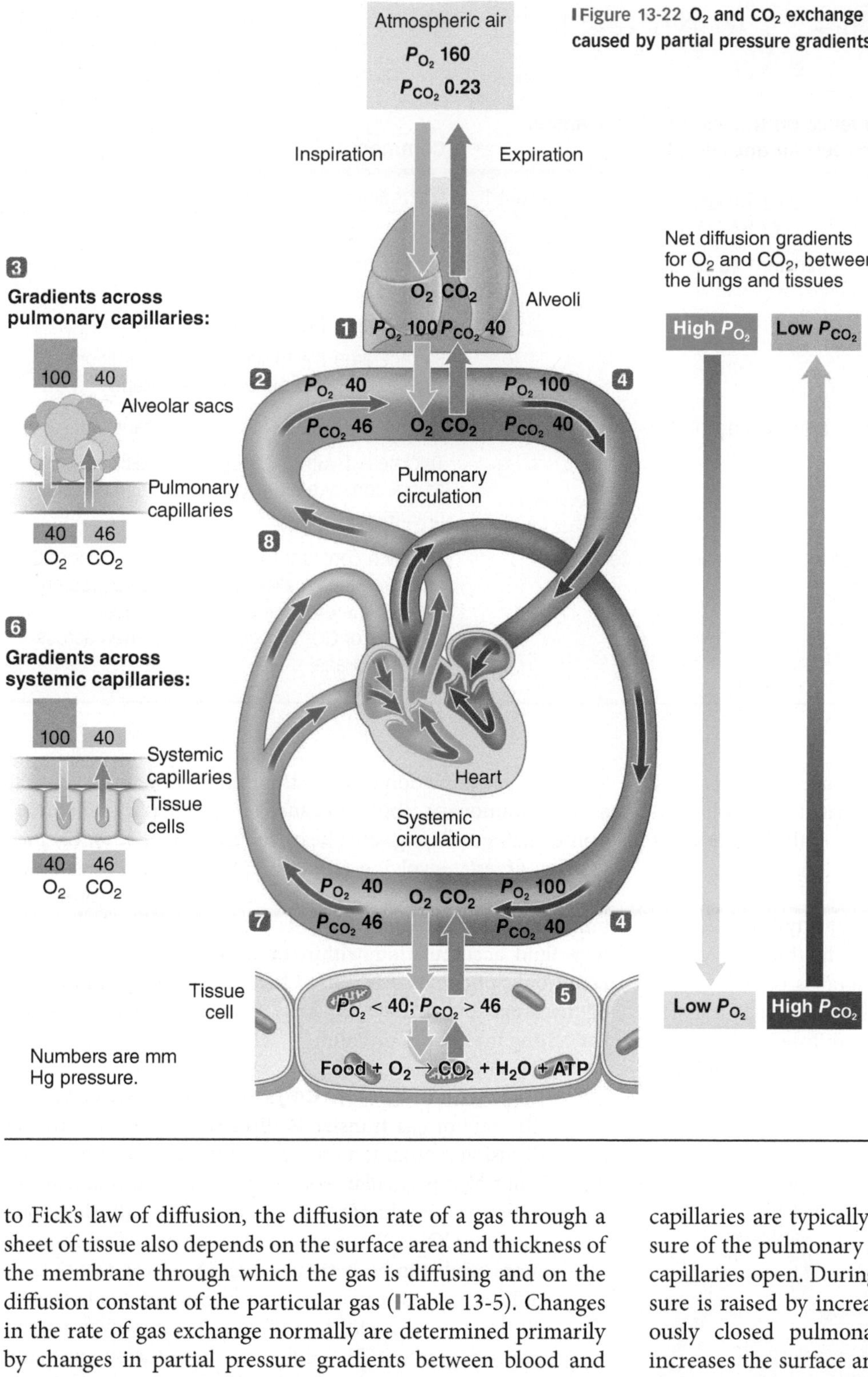

I Figure 13-22 O_2 and CO_2 exchange across pulmonary and systemic capillaries caused by partial pressure gradients.

Atmospheric air
P_{O_2} 160
P_{CO_2} 0.23

Inspiration   Expiration

O_2 CO_2 Alveoli
P_{O_2} 100 P_{CO_2} 40

3
Gradients across pulmonary capillaries:

100 40   Alveolar sacs

P_{O_2} 40 P_{O_2} 100
P_{CO_2} 46 O_2 CO_2 P_{CO_2} 40

Pulmonary circulation

Pulmonary capillaries

40 46
O_2 CO_2

6
Gradients across systemic capillaries:

100 40   Systemic capillaries
               Tissue cells

40 46
O_2 CO_2

Heart

Systemic circulation

P_{O_2} 40 P_{O_2} 100
P_{CO_2} 46 O_2 CO_2 P_{CO_2} 40

Tissue cell

$P_{O_2} < 40; P_{CO_2} > 46$

Numbers are mm Hg pressure.

Food + O_2 → CO_2 + H_2O + ATP

Net diffusion gradients for O_2 and CO_2, between the lungs and tissues

High P_{O_2} Low P_{CO_2}

Low P_{O_2} High P_{CO_2}

1 Alveolar P_{O_2} remains relatively high and alveolar P_{CO_2} remains relatively low because a portion of the alveolar air is exchanged for fresh atmospheric air with each breath.

2 In contrast, the systemic venous blood entering the lungs is relatively low in O_2 and high in CO_2, having given up O_2 and picked up CO_2 at the systemic capillary level.

3 The partial pressure gradients established between the alveolar air and pulmonary capillary blood induce passive diffusion of O_2 into the blood and CO_2 out of the blood until the blood and alveolar partial pressures become equal.

4 The blood leaving the lungs is thus relatively high in O_2 and low in CO_2. It arrives at the tissues with the same blood-gas content as when it left the lungs.

5 The partial pressure of O_2 is relatively low and that of CO_2 is relatively high in the O_2-consuming, CO_2-producing tissue cells.

6 Consequently, partial pressure gradients for gas exchange at the tissue level favor the passive movement of O_2 out of the blood into cells to support their metabolic requirements and also favor the simultaneous transfer of CO_2 into the blood.

7 Having equilibrated with the tissue cells, the blood leaving the tissues is relatively low in O_2 and high in CO_2.

8 The blood then returns to the lungs to once again fill up on O_2 and dump off CO_2.

to Fick's law of diffusion, the diffusion rate of a gas through a sheet of tissue also depends on the surface area and thickness of the membrane through which the gas is diffusing and on the diffusion constant of the particular gas (I Table 13-5). Changes in the rate of gas exchange normally are determined primarily by changes in partial pressure gradients between blood and alveoli because these other factors are relatively constant under resting conditions. However, under circumstances when these other factors do change, these changes alter the rate of gas transfer in the lungs.

Effect of Surface Area on Gas Exchange The rate of gas exchange is directly proportional to the surface area across which exchange takes place. During exercise, the surface area available for exchange can be increased to enhance the rate of gas transfer. During resting conditions, some of the pulmonary capillaries are typically closed because the normally low pressure of the pulmonary circulation is inadequate to keep all the capillaries open. During exercise, when pulmonary blood pressure is raised by increased cardiac output, many of the previously closed pulmonary capillaries are forced open. This increases the surface area of blood available for exchange. Furthermore, the alveolar walls are stretched further than normal during exercise because of the larger tidal volumes (deeper breathing). Such stretching increases the alveolar surface area and decreases the thickness of the alveolar walls. Collectively, these changes expedite gas exchange across the alveolar–capillary membrane during exercise.

Clinical Note By contrast, several pathological conditions can markedly reduce pulmonary surface area and, in turn, decrease the rate of gas exchange. Most notably, in *emphysema* surface area is reduced because many alveolar walls are lost,

Factor	Influence on the Rate of Gas Transfer Between Air and Blood	Comments
Partial pressure gradients of O_2 and CO_2	Rate of transfer ↑ as partial pressure gradient ↑	Major determinant of the rate of transfer
Surface area of the alveolar–capillary membrane	Rate of transfer ↑ as surface area ↑	Surface area remains constant under resting conditions Surface area ↑ during exercise Surface area ↓ with pathological conditions such as emphysema and lung collapse
Thickness of the alveolar–capillary membrane	Rate of transfer ↓ as thickness ↑	Thickness normally remains constant Thickness ↑ with pathological conditions such as pulmonary edema, pulmonary fibrosis, and pneumonia
Diffusion constant	Rate of transfer ↑ as diffusion constant ↑	Diffusion constant for CO_2 is 20 times that of O_2, offsetting the smaller partial pressure gradient for CO_2; therefore, approximately equal amounts of CO_2 and O_2 are transferred across the membrane

resulting in larger but fewer chambers (Figure 13-23). Loss of surface area for exchange is likewise associated with collapsed regions of the lung and also results when part of the lung tissue is surgically removed—for example, in treating lung cancer.

Clinical Note Effect of Thickness on Gas Exchange Inadequate gas exchange can also occur when the thickness of the barrier separating the air and blood is pathologically increased. As the thickness increases, the rate of gas transfer decreases because a gas takes longer to diffuse through the greater thickness. Thickness increases in (1) *pulmonary edema,*

an excess accumulation of interstitial fluid between the alveoli and the pulmonary capillaries caused by pulmonary inflammation or left-sided congestive heart failure (see p. 325); (2) *pulmonary fibrosis,* involving replacement of delicate lung tissue with thick, fibrous tissue in response to certain chronic irritants; and (3) *pneumonia,* which is characterized by inflammatory fluid accumulation within or around the alveoli. Most commonly, pneumonia is caused by bacterial or viral infection of the lungs, but it may also arise from accidental *aspiration* (breathing in) of food or vomit.

Effect of Diffusion Constant on Gas Exchange The rate of gas transfer is directly proportional to the diffusion constant, a constant value related to the solubility of a particular gas in the lung tissues and to its molecular weight ($D \propto sol/\sqrt{MW}$). The diffusion constant for CO_2 is 20 times that of O_2 because CO_2 is much more soluble in body tissues than O_2 is. The rate of CO_2 diffusion across the respiratory membranes is therefore 20 times more rapid than that of O_2 for a given partial pressure gradient. This difference in diffusion constants is normally offset by the difference in partial pressure gradients that exist for O_2 and CO_2 across the alveolar–capillary membrane. The CO_2 partial pressure gradient is 6 mm Hg (P_{CO_2} of 46 mm Hg in the blood; P_{CO_2} of 40 mm Hg in the alveoli), compared to the O_2 gradient of 60 mm Hg (P_{O_2} of 100 mm Hg in the alveoli; P_{O_2} of 40 mm Hg in the blood).

Normally, approximately equal amounts of O_2 and CO_2 are exchanged—a respiratory quotient's worth. Even though a given volume of blood spends three fourths of

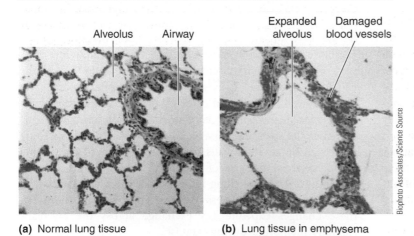

Alveolus   Airway

Expanded alveolus   Damaged blood vessels

(a) Normal lung tissue

(b) Lung tissue in emphysema

Biophoto Associates/Science Source

Figure 13-23 Comparison of normal and emphysematous lung tissue. (a) Each of the smallest clear spaces is an alveolar lumen in normal lung tissue. (b) Note the loss of alveolar walls in the emphysematous lung tissue, resulting in larger but fewer alveolar chambers.

a second passing through the pulmonary capillary bed, P_{O_2} and P_{CO_2} are usually both equilibrated with alveolar partial pressures by the time the blood has traversed only one third the length of the pulmonary capillaries. This means that the lung normally has enormous diffusion reserves, a fact that becomes extremely important during heavy exercise. The time the blood spends in transit in the pulmonary capillaries is decreased as pulmonary blood flow increases with the greater cardiac output that accompanies exercise. Even when less time is available for exchange, blood P_{O_2} and P_{CO_2} are normally able to equilibrate with alveolar levels because of the lungs' diffusion reserves.

Clinical Note In a diseased lung in which diffusion is impeded because the surface area is decreased or the blood–air barrier is thickened, O_2 transfer is usually more seriously impaired than CO_2 transfer because of the larger CO_2 diffusion constant. By the time the blood reaches the end of the pulmonary capillary network, it is more likely to have equilibrated with alveolar P_{CO_2} than with alveolar P_{O_2} because CO_2 can diffuse more rapidly through the respiratory barrier. In milder conditions, diffusion of both O_2 and CO_2 might remain adequate at rest, but during exercise, when pulmonary transit time is decreased, the blood gases, especially O_2, may not have completely equilibrated with the alveolar gases before the blood leaves the lungs.

Gas exchange across the systemic capillaries also occurs down partial pressure gradients.

Just as they do at the pulmonary capillaries, O_2 and CO_2 move between the systemic capillary blood and the tissue cells by simple passive diffusion down partial pressure gradients. Refer again to Figure 13-22. The arterial blood that reaches the systemic capillaries is essentially the same blood that left the lungs by means of the pulmonary veins because the only two places in the entire circulatory system at which gas exchange can take place are the pulmonary and the systemic capillaries. The arterial P_{O_2} is 100 mm Hg, and the arterial P_{CO_2} is 40 mm Hg, the same as alveolar P_{O_2} and P_{CO_2}.

P_{O_2} and P_{CO_2} Gradients Across the Systemic Capillaries
Cells constantly consume O_2 and produce CO_2 through oxidative metabolism. Cellular P_{O_2} averages about 40 mm Hg and P_{CO_2} about 46 mm Hg, although these values are highly variable, depending on the level of cellular metabolic activity. O_2 moves by diffusion down its partial pressure gradient from the entering systemic capillary blood (P_{O_2} = 100 mm Hg) into the adjacent cells (P_{O_2} = 40 mm Hg) until equilibrium is reached. Therefore, the P_{O_2} of venous blood leaving the systemic capillaries is equal to the tissue P_{O_2} at an average of 40 mm Hg.

The reverse situation exists for CO_2, which rapidly diffuses out of the cells (P_{CO_2} = 46 mm Hg) into the entering capillary blood (P_{CO_2} = 40 mm Hg) down the partial pressure gradient created by the ongoing production of CO_2. Transfer of CO_2 continues until blood P_{CO_2} equilibrates with tissue P_{CO_2}.[1] Accordingly, blood leaving the systemic capillaries has an average P_{CO_2} of 46 mm Hg. This systemic venous blood, which is relatively low in O_2 (P_{CO_2} = 40 mm Hg) and relatively high in

CO_2 (P_{CO_2} = 46 mm Hg), returns to the heart and is subsequently pumped to the lungs as the cycle repeats itself.

The more actively a tissue is metabolizing, the lower the cellular P_{O_2} falls and the higher the cellular P_{CO_2} rises. As a consequence of the larger blood-to-cell partial pressure gradients, more O_2 diffuses from blood into cells, and more CO_2 moves from cells into blood before blood P_{O_2} and P_{CO_2} achieve equilibrium with the surrounding cells. Thus, the amount of O_2 transferred to the cells and the amount of CO_2 carried away from the cells both depend on the rate of cellular metabolism.

Net Diffusion of O_2 and CO_2 Between the Alveoli and Tissues Net diffusion of O_2 occurs first between alveoli and blood and then between blood and tissues because of the O_2 partial pressure gradients created by continuous replenishment of fresh alveolar O_2 provided by alveolar ventilation and continuous use of O_2 in the cells. Net diffusion of CO_2 occurs in the reverse direction, first between tissues and blood and then between blood and alveoli because of the CO_2 partial pressure gradients created by continuous production of CO_2 in the cells and continuous removal of alveolar CO_2 through the process of alveolar ventilation (see Figure 13-22).

Now let us see how O_2 and CO_2 are transported in the blood.

Check Your Understanding 13.3

1. Define *partial pressure*.
2. Make a sketch showing the P_{O_2} and P_{CO_2} gradients and the direction of O_2 and CO_2 movement between the alveoli and pulmonary capillaries and between the tissue cells and systemic capillaries.
3. Discuss the factors that influence the rate of gas transfer across the alveolar–capillary membrane.

13.4 Gas Transport

Oxygen picked up by the blood at the lungs must be transported to the tissues for cell use. Conversely, CO_2 produced at the cell level must be transported to the lungs for elimination.

Most O_2 in the blood is transported bound to hemoglobin.

Oxygen is present in the blood in two forms: physically dissolved and chemically bound to hemoglobin (Table 13-6).

Physically Dissolved O_2 Little O_2 physically dissolves in plasma water because O_2 is poorly soluble in body fluids. The amount dissolved is directly proportional to the P_{O_2} of the

---

[1]Actually, the partial pressures of the systemic blood gases never completely equilibrate with tissue P_{O_2} and P_{CO_2}. Because cells constantly consume O_2 and produce CO_2, the tissue P_{O_2} is always slightly less than the P_{O_2} of the blood leaving the systemic capillaries, and the tissue P_{CO_2} always slightly exceeds the systemic venous P_{CO_2}.

TABLE 13-6 Methods of Gas Transport in the Blood

Gas	Method of Transport in Blood	Percentage Carried in This Form
O_2	Physically dissolved	1.5
	Bound to hemoglobin	98.5
CO_2	Physically dissolved	10
	Bound to hemoglobin	30
	As bicarbonate (HCO_3^-)	60

blood: The higher the P_{O_2}, the more O_2 dissolved. At a normal arterial P_{O_2} of 100 mm Hg, only 3 mL of O_2 can dissolve in 1 liter of blood. Thus, only 15 mL of O_2 can dissolve per minute in the normal pulmonary blood flow of 5 liters/min (the resting cardiac output). Even under resting conditions, the cells consume 250 mL of O_2 per minute, and consumption may increase up to 25-fold during strenuous exercise. To deliver the O_2 needed by the tissues even at rest, the cardiac output would have to be 83.3 liters/min if O_2 could only be transported in dissolved form. Obviously, there must be an additional mechanism for transporting O_2 to the tissues. This mechanism is *hemoglobin (Hb)*. Only 1.5% of the O_2 in the blood is dissolved; the remaining 98.5% is transported in combination with Hb. *The O_2 bound to Hb does not contribute to the P_{O_2} of the blood*; thus, blood P_{O_2} is not a measure of the total O_2 content of the blood but only of the dissolved portion of O_2.

Oxygen Bound to Hemoglobin Hemoglobin, an iron-bearing protein molecule contained within the red blood cells, can form a loose, easily reversible combination with O_2 (see p. 383). When not combined with O_2, Hb is referred to as **reduced hemoglobin**, or **deoxyhemoglobin**; when combined with O_2, it is called **oxyhemoglobin (HbO_2)**:

$$\underset{\text{reduced hemoglobin}}{Hb} + O_2 \rightleftharpoons \underset{\text{oxyhemoglobin}}{HbO_2}$$

We need to answer several questions about the role of Hb in O_2 transport. What determines whether O_2 and Hb are combined or dissociated (separated)? Why does Hb combine with O_2 in the lungs and release O_2 at the tissues? How can a variable amount of O_2 be released at the tissues, depending on the level of tissue activity? How can we talk about O_2 transfer between blood and surrounding tissues in terms of O_2 partial pressure gradients when 98.5% of the O_2 is bound to Hb and thus does not contribute to the P_{O_2} of the blood?

The P_{O_2} is the primary factor determining the percent hemoglobin saturation.

Each of the four atoms of iron within the heme portions of a Hb molecule can combine with an O_2 molecule, so each Hb molecule can carry up to four molecules of O_2. Hemoglobin is considered *fully saturated* when all the Hb present is carrying its maximum O_2 load. The **percent hemoglobin (% Hb) saturation**, a measure of the extent to which the Hb present is combined with O_2, can vary from 0% to 100%.

The most important factor determining the % Hb saturation is the P_{O_2} of the blood, which in turn is related to the concentration of O_2 physically dissolved in the blood. According to the **law of mass action**, if the concentration of one substance involved in a reversible reaction is increased, the reaction is driven toward the opposite side. Conversely, if the concentration of one substance is decreased, the reaction is driven toward that side. Applying this law to the reversible reaction involving Hb and O_2 ($Hb + O_2 \rightleftharpoons HbO_2$), when blood P_{O_2} increases, as in the pulmonary capillaries, the reaction is driven toward the right side of the equation, increasing formation of HbO_2 (increased % Hb saturation). When blood P_{O_2} decreases, as in the systemic capillaries, the reaction is driven toward the left side of the equation, and oxygen is released from Hb as HbO_2 dissociates (decreased % Hb saturation). Thus, because of the difference in P_{O_2} at the lungs and other tissues, Hb automatically "loads up" on O_2 in the lungs, where ventilation is continually providing fresh supplies of O_2, and "unloads" it in the tissues, which are constantly using up O_2.

O_2–Hb Dissociation Curve The relationship between blood P_{O_2} and % Hb saturation is not linear, however, a point that is important physiologically. Doubling the partial pressure does not double the % Hb saturation. Rather, the relationship between these variables follows an S-shaped curve, the **O_2–Hb dissociation (or saturation) curve** (Figure 13-24). At the upper end, between a blood P_{O_2} of 60 mm Hg and one of 100 mm Hg, the curve flattens off, or plateaus. Within this pressure range, a rise in P_{O_2} produces only a small increase in the extent to which Hb is bound with O_2. In the P_{O_2} range of 0 to 60 mm Hg, in contrast, a small change in P_{O_2} results in a large change in the extent to which Hb is combined with O_2, as shown by the steep lower part of the curve. Both the upper plateau and the lower steep portion of the curve have physiological significance.

Significance of the Plateau Portion of the O_2–Hb Curve The plateau portion of the curve is in the blood P_{O_2} range at the pulmonary capillaries where O_2 is being loaded onto Hb. The systemic arterial blood leaving the lungs, having equilibrated with alveolar P_{O_2}, normally has a P_{O_2} of 100 mm Hg. Looking at the O_2–Hb curve, note that at a blood P_{O_2} of 100 mm Hg, Hb is 97.5% saturated. Therefore, Hb in the systemic arterial blood normally is almost fully saturated.

If the alveolar P_{O_2} and, consequently, the arterial P_{O_2} fall below normal, there is little reduction in the total amount of O_2 transported by the blood until the P_{O_2} falls below 60 mm Hg. This is because of the plateau region of the curve. If the arterial P_{O_2} falls 40%, from 100 to 60 mm Hg, the concentration of dissolved O_2 as reflected by the P_{O_2} is likewise reduced 40%. At a blood P_{O_2} of 60 mm Hg, however, the % Hb saturation is still remarkably high, at 90%. Accordingly, the total O_2 content of the blood is only slightly decreased despite the 40% reduction in P_{O_2} because Hb is still carrying an almost full load of O_2, and, as mentioned before, most O_2 is transported by Hb rather than dissolved. However, even if the blood P_{O_2} is greatly increased—

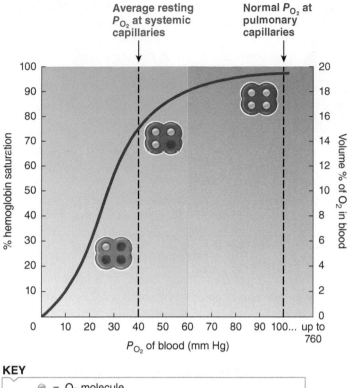

KEY

● = O_2 molecule

= Partially saturated hemoglobin molecule

= Fully saturated hemoglobin molecule

▌Figure 13-24 **Oxygen–hemoglobin (O_2–Hb) dissociation (saturation) curve.** The % hemoglobin saturation (the scale on the left side of the graph) depends on the P_{O_2} of the blood. The relationship between these two variables is depicted by an S-shaped curve with a plateau region between a blood P_{O_2} of 60 and 100 mm Hg and a steep portion between 0 and 60 mm Hg. Another way of expressing the effect of blood P_{O_2} on the amount of O_2 bound with hemoglobin is the volume % of O_2 in the blood (mL of O_2 bound with hemoglobin in each 100 mL of blood). The scale on the right side of the graph represents that relationship.
FIGURE FOCUS: *Looking at the graph, compare the changes in % Hb saturation if P_{O_2} decreases by 25 mm Hg (1) from its normal value of 100 mm Hg in the pulmonary capillaries and (2) from its normal value of 40 mm Hg in the systemic capillaries.*

say, to 600 mm Hg—by breathing pure O_2, little additional O_2 is added to the blood. A small extra amount of O_2 dissolves, but the % Hb saturation can be maximally increased by only another 2.5%, to 100% saturation. Therefore, in the P_{O_2} range between 60 and 600 mm Hg or even higher, there is only a 10% difference in the amount of O_2 carried by Hb. Thus, the plateau portion of the O_2–Hb curve provides a good margin of safety in O_2-carrying capacity of the blood.

Clinical Note Arterial P_{O_2} may be reduced by pulmonary diseases accompanied by inadequate ventilation or defective gas exchange or by circulatory disorders that result in inadequate blood flow to the lungs. It may also fall in healthy people under two circumstances: (1) at high altitudes, where total atmospheric pressure and hence the P_{O_2} of the inspired air are reduced, or (2) in O_2-deprived environments at sea level, such

as if someone were accidentally locked in a vault. Unless the arterial P_{O_2} becomes markedly reduced (falls below 60 mm Hg) in either pathological conditions or abnormal circumstances, near-normal amounts of O_2 can still be carried to the tissues.

Significance of the Steep Portion of the O_2–Hb Curve
The steep portion of the curve between 0 and 60 mm Hg is in the blood P_{O_2} range at the systemic capillaries, where O_2 is unloaded from Hb. In the systemic capillaries, the blood equilibrates with the surrounding tissue cells at an average P_{O_2} of 40 mm Hg. Note in ▌Figure 13-24 that at a P_{O_2} of 40 mm Hg the % Hb saturation is 75%. The blood arrives in the tissue capillaries at a P_{O_2} of 100 mm Hg with 97.5% Hb saturation. Because Hb can only be 75% saturated at the P_{O_2} of 40 mm Hg in the systemic capillaries, nearly 25% of the HbO_2 must dissociate, yielding reduced Hb and O_2. This released O_2 is free to diffuse down its partial pressure gradient from the red blood cells through the plasma and the interstitial fluid into the tissue cells.

The Hb in the venous blood returning to the lungs is still normally 75% saturated. If the tissue cells are metabolizing more actively, the P_{O_2} of the systemic capillary blood falls (for example, from 40 to 20 mm Hg) because the cells are consuming O_2 more rapidly. Note on the curve that this drop of 20 mm Hg in P_{O_2} decreases the % Hb saturation from 75% to 30%—that is, about 45% more of the total HbO_2 than normal gives up its O_2 for tissue use. The normal 60 mm Hg drop in P_{O_2} from 100 to 40 mm Hg in the systemic capillaries causes about 25% of the total HbO_2 to unload its O_2. In comparison, a further drop in P_{O_2} of only 20 mm Hg results in an additional 45% of the total HbO_2 unloading its O_2 because the O_2 partial pressures in this range are operating in the steep portion of the curve. In this range, only a small drop in systemic capillary P_{O_2} can automatically make large amounts of O_2 immediately available to meet the O_2 needs of more actively metabolizing tissues, such as exercising muscles. As much as 85% of the Hb may give up its O_2 to active muscle cells during strenuous exercise. In addition to this more thorough withdrawal of O_2 from the blood, even more O_2 is made available to actively metabolizing cells by circulatory and respiratory adjustments that increase the flow rate of oxygenated blood through the active tissues.

Hemoglobin promotes the net transfer of O_2 at both the alveolar and the tissue levels.

We still have not really clarified the role of Hb in gas exchange. Because blood P_{O_2} depends entirely on the concentration of *dissolved* O_2, we could ignore the O_2 bound to Hb in our earlier discussion of O_2 being driven from the alveoli to the blood by a P_{O_2} gradient. However, Hb does play a crucial role in permitting the transfer of large quantities of O_2 before blood P_{O_2} equilibrates with the surrounding tissues (▌Figure 13-25).

Role of Hb at the Alveolar Level Hemoglobin acts as a "storage depot" for O_2, removing O_2 from solution as soon as it enters the blood from the alveoli. Because only dissolved O_2 contributes to P_{O_2}, the O_2 stored in Hb cannot contribute to blood P_{O_2}. When systemic venous blood enters the pulmonary capillaries, its P_{O_2} is considerably lower than alveolar P_{O_2}, so O_2

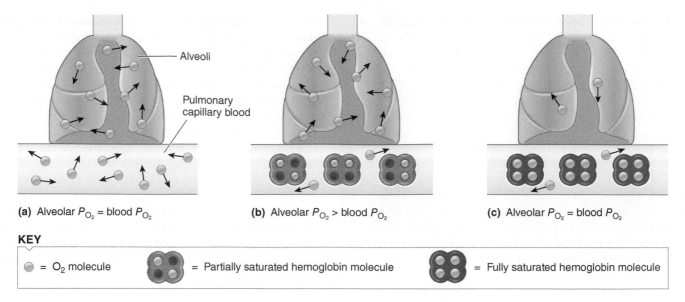

KEY

⊙ = O_2 molecule · ⊛ = Partially saturated hemoglobin molecule · ⊛ = Fully saturated hemoglobin molecule

Figure 13-25 Hemoglobin facilitating a large net transfer of O_2 by acting as a storage depot to keep P_{O_2} low. (a) In the hypothetical situation in which no Hb is present in the blood, the alveolar P_{O_2} and the pulmonary capillary blood P_{O_2} are at equilibrium. (b) Hemoglobin has been added to the blood. As the Hb starts to bind with O_2, it removes O_2 from solution. Because only dissolved O_2 contributes to blood P_{O_2}, the blood P_{O_2} falls below that of the alveoli, even though the same number of O_2 molecules are present in the blood as in part (a). By "soaking up" some of the dissolved O_2, Hb favors the net diffusion of more O_2 down its partial pressure gradient from the alveoli to the blood. (c) Hemoglobin is fully saturated with O_2 and the alveolar and blood P_{O_2} are at equilibrium again. The blood P_{O_2} resulting from dissolved O_2 is equal to the alveolar P_{O_2}, despite the fact that the total O_2 content in the blood is much greater than in part (a) when blood P_{O_2} was equal to alveolar P_{O_2} in the absence of Hb.

immediately diffuses into the blood, raising blood P_{O_2}. As soon as the blood P_{O_2} increases, the percentage of Hb that can bind with O_2 likewise increases, as indicated by the O_2–Hb curve. Consequently, most of the O_2 that has diffused into the blood combines with Hb and no longer contributes to blood P_{O_2}. As O_2 is removed from solution by combining with Hb, blood P_{O_2} falls to about the same level it was when the blood entered the lungs, even though the total quantity of O_2 in the blood actually has increased. Because blood P_{O_2} is again considerably below alveolar P_{O_2}, more O_2 diffuses from the alveoli into the blood, only to be soaked up by Hb again.

Even though we have considered this process stepwise for clarity, net diffusion of O_2 from alveoli to blood occurs continuously until Hb becomes as saturated with O_2 as it can be at that particular P_{O_2}. At a normal P_{O_2} of 100 mm Hg, Hb is 97.5% saturated. Thus, by soaking up O_2, Hb keeps blood P_{O_2} low and prolongs the existence of a partial pressure gradient so that a large net transfer of O_2 into the blood can take place. Not until Hb can store no more O_2 (that is, Hb is maximally saturated for that P_{O_2}) does all the O_2 transferred into the blood remain dissolved and directly contribute to the P_{O_2}. Only now does blood P_{O_2} rapidly equilibrate with alveolar P_{O_2} and bring further O_2 transfer to a halt, but this point is not reached until Hb is already loaded to the maximum extent possible. Once blood P_{O_2} equilibrates with alveolar P_{O_2}, no further O_2 transfer can take place, no matter how little or how much total O_2 has already been transferred.

Role of Hb at the Tissue Level The reverse situation occurs at the tissue level. Because the P_{O_2} of blood entering the systemic capillaries is considerably higher than the P_{O_2} of the surrounding tissue, O_2 immediately diffuses from the blood into the tissues, lowering blood P_{O_2}. When blood P_{O_2} falls, Hb must unload some stored O_2 because the % Hb saturation is reduced. As the O_2 released from Hb dissolves in the blood, blood P_{O_2} increases and again exceeds the P_{O_2} of the surrounding tissues. This favors further movement of O_2 out of the blood, although the total quantity of O_2 in the blood has already fallen. Only when Hb can no longer release any more O_2 into solution (when Hb is unloaded to the greatest extent possible for the P_{O_2} existing at the systemic capillaries) can blood P_{O_2} fall as low as in surrounding tissue. At this time, further transfer of O_2 stops. Hemoglobin, because it stores a large quantity of O_2 that can be freed by a slight reduction in P_{O_2} at the systemic capillary level, permits the transfer of tremendously more O_2 from the blood into the cells than would be possible in its absence.

Thus, Hb plays an important role in the *total quantity* of O_2 that the blood can pick up in the lungs and drop off in the tissues. If Hb levels fall to one half of normal, as in a severely anemic patient (see p. 386), the O_2-carrying capacity of the blood falls by 50% even though the arterial P_{O_2} is the normal 100 mm Hg with 97.5% Hb saturation. Only half as much Hb is available to be saturated, emphasizing again how critical Hb is in determining how much O_2 can be picked up at the lungs and made available to tissues.

Factors at the tissue level promote unloading of O_2 from hemoglobin.

Even though the main factor determining the % Hb saturation is the P_{O_2} of the blood, other factors can affect the affinity, or bond strength, between Hb and O_2 and, accordingly, can shift the O_2–Hb curve (that is, change the % Hb saturation at a given P_{O_2}). These other factors are CO_2, acidity, temperature, and

2,3-bisphosphoglycerate, which we examine separately. The O_2–Hb dissociation curve with which you are already familiar (see ▮Figure 13-24) is a typical curve at normal arterial CO_2 and acidity levels, normal body temperature, and normal 2,3-bisphosphoglycerate concentration.

Effect of CO_2 on % Hb Saturation An increase in P_{CO_2} shifts the O_2–Hb curve to the right (▮Figure 13-26). The % Hb saturation still depends on the P_{O_2}, but for any given P_{O_2} less O_2 and Hb can be combined. This effect is important because the P_{CO_2} of the blood increases in the systemic capillaries as CO_2 diffuses down its gradient from cells into blood. The presence of this additional CO_2 in the blood in effect decreases the affinity of Hb for O_2, so Hb unloads even more O_2 at the tissue level than it would if the reduction in P_{O_2} in the systemic capillaries was the only factor affecting % Hb saturation.

Effect of Acid on % Hb Saturation An increase in acidity also shifts the curve to the right. Because CO_2 generates carbonic acid, the blood becomes more acidic at the systemic capillary level as it picks up CO_2 from the tissues. The resulting reduction in Hb affinity for O_2 in the presence of increased acidity aids in releasing even more O_2 at the tissue level for a given P_{O_2}. In actively metabolizing cells, such as exercising muscles, not only is more carbonic acid–generating CO_2 produced, but lactate (lactic acid) also may be produced if the cells resort to anaerobic metabolism (see pp. 39 and 272). The resultant local elevation of acid in the working muscles facilitates further unloading of O_2 in the very tissues that need the most O_2.

Bohr Effect The influence of CO_2 and acid on the release of O_2 is known as the **Bohr effect**. Both CO_2 and the hydrogen ion (H^+) component of acids can combine reversibly with Hb at sites other than the O_2-binding sites. The result is a change in the molecular structure of Hb that reduces its affinity for O_2. (Note that the % Hb saturation refers only to the extent to which Hb is combined with O_2, not the extent to which it is bound with CO_2, H^+, or other molecules. Indeed, the % Hb saturation decreases when CO_2 and H^+ bind with Hb, because their presence on Hb facilitates increased release of O_2 from Hb.)

Effect of Temperature on % Hb Saturation In a similar manner, a rise in temperature shifts the O_2–Hb curve to the right, resulting in more unloading of O_2 at a given P_{O_2}. An exercising muscle or other actively metabolizing cell produces heat. The resulting local rise in temperature enhances O_2 release from Hb for use by more active tissues.

Comparison of These Factors at the Tissue and Pulmonary Levels As you just learned, increases in CO_2, acidity, and temperature at the tissue level, all of which are associated with increased cellular metabolism and increased O_2 consumption, enhance the effect of a drop in P_{O_2} in facilitating the release of O_2 from Hb. These effects are largely reversed at the pulmonary level, where the extra acid-forming CO_2 is blown off and the local aerated environment is cooler. Appropriately, therefore, Hb has a higher affinity for O_2 in the pulmonary capillary environment, enhancing the effect of raised P_{O_2} in loading O_2 onto Hb.

Effect of 2,3-Bisphosphoglycerate on % Hb Saturation The preceding changes take place in the *environment* of the red blood cells, but a factor *inside* the red blood cells can also affect the degree of O_2–Hb binding: **2,3-bisphosphoglycerate (BPG)**. This erythrocyte constituent, which is produced during red blood cell metabolism, can bind reversibly with Hb and reduce its affinity for O_2, just as CO_2 and H^+ do. Thus, an increased level of BPG, like the other factors, shifts the O_2–Hb curve to the right, enhancing O_2 unloading as the blood flows through the tissues.

BPG production by red blood cells gradually increases whenever Hb in the arterial blood is chronically undersaturated—that is, when arterial HbO_2 is below normal. This condition may occur in people living at high altitudes or in those suffering from certain types of circulatory or respiratory diseases or anemia. By helping unload O_2 from Hb at the tissue level, increased BPG helps maintain O_2 availability for tissue use even though arterial O_2 supply is chronically reduced.

Hemoglobin has a much higher affinity for carbon monoxide than for O_2.

Clinical Note **Carbon monoxide (CO)** and O_2 compete for the same binding sites on Hb, but Hb's affinity for CO is 240 times that of its affinity for O_2. The combination of CO and Hb is known as **carboxyhemoglobin (HbCO)**. Because Hb preferentially latches onto CO, even small amounts of CO can tie up a disproportionately large share of Hb, making Hb unavailable for

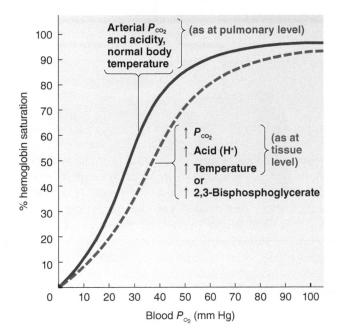

▮Figure 13-26 **Effect of increased P_{CO_2}, H^+, temperature, and 2,3-bisphosphoglycerate on the O_2–Hb curve.** Increased P_{O_2}, acid, and temperature, as found at the tissue level, shift the O_2–Hb curve to the right. As a result, less O_2 and Hb can be combined at a given P_{O_2} so that more O_2 is unloaded from Hb for use by the tissues. Similarly, 2,3-bisphosphoglycerate, whose production is increased in red blood cells when arterial HbO_2 is chronically below normal, shifts the O_2–Hb curve to the right, making more of the limited O_2 available at the tissue level.

O_2 transport. Even though the Hb concentration and P_{O_2} are normal, the O_2 content of the blood is seriously reduced.

Fortunately, CO is not a normal constituent of inspired air. It is a poisonous gas produced during the incomplete combustion (burning) of carbon products such as automobile gasoline, coal, wood, and tobacco. Carbon monoxide is especially dangerous because it is so insidious. If CO is being produced in a closed environment so that its concentration continues to increase (for example, in a parked car with the motor running and windows closed), it can reach lethal levels without the victim ever being aware of the danger. Because it is odorless, colorless, tasteless, and nonirritating, CO is not detectable. Furthermore, for reasons described later, the victim has no sensation of breathlessness and makes no attempt to increase ventilation, even though the cells are O_2 starved.

Most CO_2 is transported in the blood as bicarbonate.

When arterial blood flows through the tissue capillaries, CO_2 diffuses down its partial pressure gradient from the tissue cells into the blood. CO_2 is transported in the blood in three ways (as described in the following list and shown with corresponding numbers in I Figure 13-27; see also I Table 13-6, p. 472):

1. *Physically dissolved.* As with dissolved O_2, the amount of CO_2 physically dissolved in the blood depends on the P_{CO_2}. Because CO_2 is more soluble than O_2 in plasma water, a greater proportion of the total CO_2 than of O_2 in the blood is physically dissolved. Even so, only 10% of the blood's total CO_2 content is carried this way at the normal systemic venous P_{CO_2} level.

2. *Bound to hemoglobin.* Another 30% of the CO_2 combines with Hb to form **carbamino hemoglobin (HbCO$_2$)**. CO_2 binds with the globin portion of Hb, in contrast to O_2, which combines with the heme portions. Reduced Hb has a greater affinity for CO_2 than HbO_2 does. The unloading of O_2 from Hb in the tissue capillaries therefore facilitates the picking up of CO_2 by Hb.

3. *As bicarbonate.* By far the most important means of CO_2 transport is as **bicarbonate (HCO$_3^-$)**, with 60% of the CO_2 being converted into HCO_3^- by the following chemical reaction:

$$CO_2 + H_2O \rightleftharpoons H_2CO_3 \rightleftharpoons H + HCO_3^-$$

In the first step, CO_2 combines with H_2O to form carbonic acid (H_2CO_3). As is characteristic of acids, some of the H_2CO_3 mol-

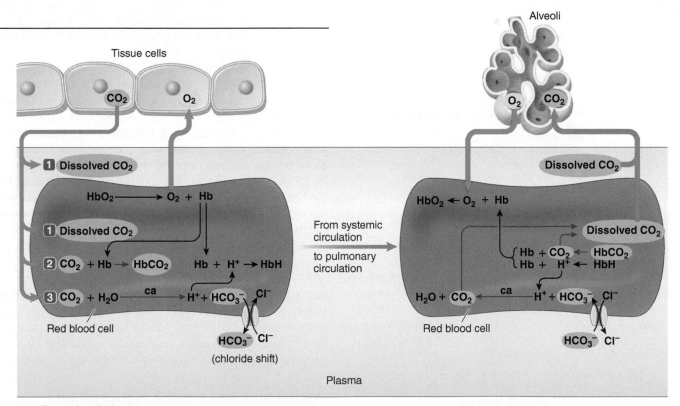

ca = Carbonic anhydrase

I Figure 13-27 **Carbon dioxide transport in the blood.** Carbon dioxide picked up at the tissue level is transported in the blood to the lungs in three ways: (1) physically dissolved, (2) bound to hemoglobin (Hb), and (3) as bicarbonate ion (HCO$_3^-$). Hemoglobin is present only in the red blood cells, as is carbonic anhydrase, the enzyme that catalyzes the production of HCO$_3^-$. The H$^+$ generated during the production of HCO$_3^-$ also binds to Hb. HCO$_3^-$ moves by facilitated diffusion down its concentration gradient out of the red blood cell into the plasma, and chloride (Cl$^-$) moves by means of the same passive carrier into the red blood cell down the electrical gradient created by the outward diffusion of HCO$_3^-$. The reactions that occur at the tissue level are reversed at the pulmonary level, where CO$_2$ diffuses out of the blood to enter the alveoli.

FIGURE FOCUS: *Study the figure to identify three chemicals with which hemoglobin binds.*

ecules spontaneously dissociate into hydrogen ions (H^+) and bicarbonate ions (HCO_3^-). The one carbon and two oxygen atoms of the original CO_2 molecule are thus present in the blood as an integral part of HCO_3^-. This is beneficial because HCO_3^- is more soluble in the blood than CO_2 is.

This reaction takes place slowly in the plasma, but it proceeds swiftly within the red blood cells because of the presence of the erythrocyte enzyme **carbonic anhydrase**, which catalyzes (speeds up) the reaction. In fact, under the influence of carbonic anhydrase, the reaction proceeds directly from $CO_2 + H_2O$ to $H^+ + HCO_3^-$ without the intervening H_2CO_3 step:

$$CO_2 + H_2O \xrightarrow{\text{carbonic anhydrase}} H^+ \; HCO_3^-$$

Chloride Shift As this reaction proceeds, HCO_3^- and H^+ start to accumulate within the red blood cells in the systemic capillaries. The red blood cell membrane has a $HCO_3^- – Cl^-$ carrier that passively facilitates the diffusion of these ions in opposite directions across the membrane. The membrane is relatively impermeable to H^+. Consequently, HCO_3^-, but not H^+, diffuses down its concentration gradient out of the erythrocytes into the plasma. Because HCO_3^- is a negatively charged ion, the efflux of HCO_3^- unaccompanied by a comparable outward diffusion of positively charged ions creates an electrical gradient (see p. 66). Chloride ions (Cl^-), the dominant plasma anions, diffuse into the red blood cells down this electrical gradient to restore electric neutrality. This inward shift of Cl^- in exchange for the efflux of CO_2-generated HCO_3^- is known as the **chloride (Cl^-) shift**.

Haldane Effect Hemoglobin binds with most of the H^+ formed within the erythrocytes. As with CO_2, reduced Hb has a greater affinity for H^+ than HbO_2 does. Therefore, unloading of O_2 facilitates Hb pickup of CO_2 and CO_2-generated H^+, an influence known as the **Haldane effect**. Because only free, dissolved H^+ contributes to the acidity of a solution, venous blood would be considerably more acidic than arterial blood if Hb did not mop up most of the H^+ generated at the tissue level.

Note how Hb's unloading of O_2 and its uptake of CO_2 and CO_2-generated H^+ at the tissue level work in synchrony. Increased CO_2 and H^+ cause increased O_2 release from Hb by the Bohr effect; increased O_2 release from Hb in turn causes increased CO_2 and H^+ uptake by Hb through the Haldane effect. The entire process is efficient. Reduced Hb must be carried back to the lungs to refill on O_2 anyway. After O_2 is released, Hb picks up new passengers—CO_2 and H^+—that are going in the same direction to the lungs.

The reactions at the tissue level as CO_2 enters the blood from the tissues are reversed once the blood reaches the lungs and CO_2 leaves the blood to enter the alveoli (❙Figure 13-27).

Various respiratory states are characterized by abnormal blood-gas levels.

❙Table 13-7 is a glossary of terms used to describe various states associated with respiratory abnormalities, most of which are discussed in more detail here or later in the chapter.

❙**TABLE 13-7** Mini Glossary of Clinically Important Respiratory States

Apnea   Transient cessation of breathing

Asphyxia O_2 starvation of tissues, caused by a lack of O_2 in the air, respiratory impairment, or inability of the tissues to use O_2

Cyanosis   Blueness of the skin resulting from insufficiently oxygenated blood in the arteries

Dyspnea   Difficult or labored breathing

Eupnea   Normal breathing

Hypercapnia   Excess CO_2 in the arterial blood

Hyperpnea   Increased pulmonary ventilation that matches increased metabolic demands, as in exercise

Hyperventilation   Increased pulmonary ventilation in excess of metabolic requirements, resulting in decreased P_{CO_2} and respiratory alkalosis

Hypocapnia   Below-normal CO_2 in the arterial blood

Hypoventilation   Underventilation in relation to metabolic requirements, resulting in increased P_{CO_2} and respiratory acidosis

Hypoxia   Insufficient O_2 at the cellular level

Anemic hypoxia   Reduced O_2-carrying capacity of the blood

Circulatory hypoxia   Too little oxygenated blood delivered to the tissues; also known as stagnant hypoxia

Histotoxic hypoxia   Inability of the cells to use available O_2

Hypoxic hypoxia   Low arterial blood P_{O_2} accompanied by inadequate Hb saturation

Respiratory arrest   Permanent cessation of breathing (unless clinically corrected)

Suffocation O_2 deprivation as a result of an inability to breathe oxygenated air

 Abnormalities in Arterial P_{O_2}   The term **hypoxia** refers to the condition of having insufficient O_2 at the cell level. Following are the four general categories of hypoxia:

1. *Hypoxic hypoxia* is characterized by a low arterial blood P_{O_2} accompanied by inadequate Hb saturation. It is caused by (a) a respiratory malfunction involving inadequate gas exchange, typified by a normal alveolar P_{O_2} but a reduced arterial P_{O_2}, or (b) exposure to high altitude or to a suffocating environment where atmospheric P_{O_2} is reduced so that alveolar and arterial P_{O_2} are likewise reduced.

2. *Anemic hypoxia* is a reduced O_2-carrying capacity of the blood. It can result from (a) a decrease in circulating red blood cells, (b) an inadequate amount of Hb within the red blood cells, or (c) CO poisoning. In all cases of anemic hypoxia, arterial P_{O_2} is normal but the O_2 content of the arterial blood is lower than normal because of inadequate available Hb.

3. *Circulatory hypoxia* arises when too little oxygenated blood is delivered to the tissues. Circulatory hypoxia can be restricted to a limited area by a local vascular blockage. Or the body may experience circulatory hypoxia in general from con-

gestive heart failure or circulatory shock. Arterial P_{O_2} and O_2 content are typically normal, but too little oxygenated blood reaches the cells.

4. In *histotoxic hypoxia*, O_2 delivery to the tissues is normal, but the cells cannot use the O_2 available to them. The classic example is *cyanide poisoning*. Cyanide blocks enzymes essential for cellular respiration (enzymes in the electron transport system; see p. 37).

Hyperoxia, an above-normal arterial P_{O_2}, cannot occur when a person is breathing atmospheric air at sea level. However, breathing supplemental O_2 can increase alveolar and consequently, arterial P_{O_2}. Because more of the inspired air is O_2, more of the total pressure of the inspired air is attributable to the O_2 partial pressure, so more O_2 dissolves in the blood before arterial P_{O_2} equilibrates with alveolar P_{O_2}. Even though arterial P_{O_2} increases, the total blood O_2 content does not significantly increase because Hb is nearly fully saturated at the normal arterial P_{O_2}. In certain pulmonary diseases associated with a reduced arterial P_{O_2}, however, breathing supplemental O_2 can help establish a larger alveoli-to-blood driving gradient, improving arterial P_{O_2}. Far from being advantageous, a markedly elevated arterial P_{O_2} can be dangerous. If arterial P_{O_2} is too high, **oxygen toxicity** can occur. Even though the total O_2 content of the blood is only slightly increased, exposure to a high P_{O_2} can cause brain damage and blindness-causing damage to the retina. Therefore, O_2 therapy must be administered cautiously.

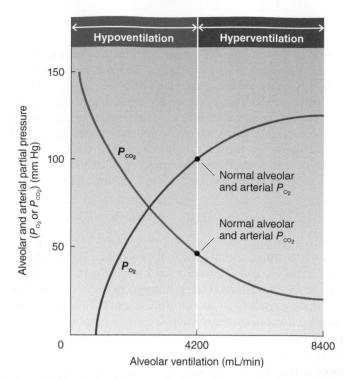

I Figure 13-28 **Effects of hyperventilation and hypoventilation on arterial P_{O_2} and P_{CO_2}.**

Clinical Note **Abnormalities in Arterial P_{CO_2}** The term **hypercapnia** refers to the condition of having excess CO_2 in arterial blood; it is caused by **hypoventilation** (ventilation inadequate to meet metabolic needs for O_2 delivery and CO_2 removal). With most lung diseases, CO_2 accumulates in arterial blood concurrently with an O_2 deficit because both O_2 and CO_2 exchange between lungs and atmosphere are affected (I Figure 13-28).

Hypocapnia, below-normal arterial P_{CO_2} levels, is brought about by hyperventilation. **Hyperventilation** occurs when a person "overbreathes"—that is, when the rate of ventilation exceeds the body's metabolic needs for CO_2 removal. As a result, CO_2 is blown off to the atmosphere more rapidly than it is produced in the tissues, and arterial P_{CO_2} falls. Hyperventilation can be triggered by anxiety states, fever, and aspirin poisoning. Alveolar P_{O_2} increases during hyperventilation as more fresh O_2 is delivered to the alveoli from the atmosphere than the blood extracts from the alveoli for tissue consumption, and arterial P_{O_2} increases correspondingly (I Figure 13-28). However, because Hb is almost fully saturated at the normal arterial P_{O_2}, very little additional O_2 is added to the blood. Except for the small extra amount of dissolved O_2, blood O_2 content remains essentially unchanged during hyperventilation.

Increased ventilation is not synonymous with hyperventilation. Increased ventilation that matches an increased metabolic demand, such as the increased need for O_2 delivery and CO_2 elimination during exercise, is termed **hyperpnea**. During exercise, alveolar and arterial P_{O_2} and P_{CO_2} remain constant, with the increased atmospheric exchange just keeping pace with the increased O_2 consumption and CO_2 production.

Clinical Note **Consequences of Abnormalities in Arterial Blood Gases** The consequences of reduced O_2 availability to the tissues during hypoxia are apparent. The cells need adequate O_2 to sustain energy-generating metabolic activities. The consequences of abnormal blood CO_2 levels are less obvious. Changes in blood CO_2 concentration primarily affect acid–base balance. Hypercapnia elevates production of CO_2-generated H^+. The subsequent generation of excess H^+ produces an acidic condition termed *respiratory acidosis*. Conversely, less-than-normal amounts of H^+ are generated from CO_2 in conjunction with hypocapnia. The resultant alkalotic (less acidic than normal) condition is called *respiratory alkalosis* (see Chapter 15). (To learn about the effects of mountain climbing and deep sea diving on blood gases, see the boxed feature on pp. 480–481, I Concepts, Challenges, and Controversies.)

Check Your Understanding 13.4

1. State the percentage of O_2 and CO_2 carried in the blood by each of the methods of transport.

2. Draw a graph showing the O_2–Hb dissociation curve and label the parts of the curve that operate in the blood P_{O_2} range at the pulmonary capillaries and at the systemic capillaries.

3. Discuss the effect of hypoventilation and of hyperventilation on acid–base balance.

13.5 Control of Respiration

Like the heartbeat, breathing must occur in a continuous, cyclic pattern to sustain life processes. Cardiac muscle must rhythmically contract and relax to alternately pump blood from the heart and fill it again. Similarly, inspiratory muscles must rhythmically contract and relax to alternately fill the lungs with air and empty them. Both these activities are accomplished automatically, without conscious effort. However, the underlying mechanisms and control of these two systems are remarkably different.

Respiratory centers in the brain stem establish a rhythmic breathing pattern.

Whereas the heart can generate its own rhythm by means of its intrinsic pacemaker activity, the respiratory muscles, being skeletal muscles, contract only when stimulated by their nerve supply. The rhythmic pattern of breathing is established by cyclic neural activity to the respiratory muscles. In other words, the pacemaker activity that establishes breathing rhythm resides in the respiratory control centers in the brain, not in the lungs or respiratory muscles themselves. The nerve supply to the heart, not being needed to initiate the heartbeat, only modifies the rate and strength of cardiac contraction. In contrast, the nerve supply to the respiratory system is essential in maintaining breathing and in reflexly adjusting the level of ventilation to match changing needs for O_2 uptake and CO_2 removal. Furthermore, unlike cardiac activity, which is not subject to voluntary control, respiratory activity can be voluntarily modified, as when you are talking.

Components of Neural Control of Respiration Neural control of respiration involves three distinct components: (1) factors that generate the alternating inspiration–expiration rhythm, (2) factors that regulate the magnitude of ventilation (that is, the rate and depth of breathing) to match body needs, and (3) factors that modify respiratory activity to serve other purposes. The latter modifications may be either involuntary, as in the respiratory maneuvers involved in a cough, or voluntary, as in the breath control required for speech.

Respiratory control centers housed in the brain stem generate the rhythmic pattern of breathing. The primary respiratory control center, the *medullary respiratory center*, consists of several aggregations of neuronal cell bodies within the medulla that provide output to the respiratory muscles. In addition, two other respiratory centers lie higher in the brain stem in the pons—the *pneumotaxic center* and the *apneustic center*. These pontine centers influence output from the medullary respiratory center (▮Figure 13-29). Here is how these various regions interact to establish respiratory rhythmicity.

Inspiratory and Expiratory Neurons in the Medullary Center We rhythmically breathe in and out during quiet breathing because of alternate contraction and relaxation of the major inspiratory muscles (the diaphragm and external intercostal muscles). Contraction and relaxation of these muscles in turn is commanded by the **medullary respiratory center**,

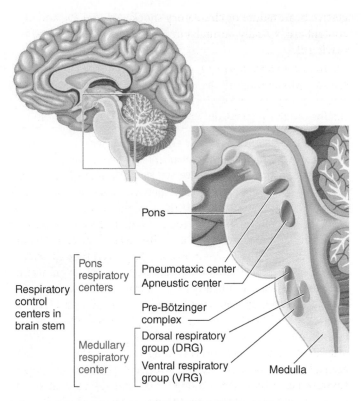

Pons

Respiratory control centers in brain stem
- Pons respiratory centers
  - Pneumotaxic center
  - Apneustic center
- Medullary respiratory center
  - Pre-Bötzinger complex
  - Dorsal respiratory group (DRG)
  - Ventral respiratory group (VRG)

Medulla

▮Figure 13-29 **Respiratory control centers in the brain stem.**

which sends impulses to the cell bodies (located in the spinal cord) of the motor neurons supplying these muscles.

The medullary respiratory center consists of two neuronal clusters known as the *dorsal respiratory group* and the *ventral respiratory group* (▮Figure 13-29).

- The **dorsal respiratory group (DRG)** consists mostly of *inspiratory neurons* whose descending fibers terminate on the motor neurons that supply the inspiratory muscles. When the DRG inspiratory neurons fire, they stimulate the inspiratory muscles and inspiration takes place; when they cease firing, the inspiratory muscles relax and passive expiration occurs. Expiration is brought to an end as the inspiratory neurons fire again. The DRG has important interconnections with the ventral respiratory group.

- The **ventral respiratory group (VRG)** is composed of *inspiratory neurons* and *expiratory neurons,* both of which remain inactive during normal quiet breathing. This region is called into play by the DRG as an "overdrive" mechanism during periods when demands for ventilation are increased. It is especially important in active expiration. No impulses are generated in the descending pathways from the expiratory neurons during quiet breathing. Only during active expiration do the expiratory neurons stimulate the motor neurons supplying the expiratory muscles (the abdominal and internal intercostal muscles). Furthermore, the VRG inspiratory neurons, when stimulated by the DRG, rev up inspiratory activity when demands for ventilation are high.

Generation of Respiratory Rhythm Contrary to a long-held belief, the DRG does not generate the basic rhythm of

Effects of Heights and Depths on the Body

OUR BODIES ARE OPTIMALLY EQUIPPED for existence at normal atmospheric pressure. Ascent into mountains high above sea level or descent into the depths of the ocean can have adverse effects on the body.

Effects of High Altitude on the Body

Atmospheric pressure progressively declines as altitude increases. At 18,000 feet above sea level, atmospheric pressure is only 380 mm Hg—half of its normal sea-level value. Because the proportion of O_2 and N_2 in the air remains the same, the P_{O_2} of inspired air at this altitude is 21% of 380 mm Hg, or 80 mm Hg, with alveolar P_{O_2} being even lower at 45 mm Hg. At any altitude above 10,000 feet, the arterial P_{O_2} falls into the steep portion of the O_2–Hb curve, below the safety range of the plateau region. As a result, the % Hb saturation in arterial blood declines precipitously with further increases in altitude.

People who rapidly ascend to altitudes of 10,000 feet or more experience symptoms of **acute mountain sickness** attributable to hypoxic hypoxia and the resultant hypocapnia-induced alkalosis. The increased ventilatory drive to obtain more O_2 causes respiratory alkalosis because acid-forming CO_2 is blown off more rapidly than it is produced. Symptoms of mountain sickness include fatigue, nausea, loss of appetite, labored breathing, rapid heart rate (triggered by hypoxia as a compensatory measure to increase circulatory delivery of available O_2 to the tissues), and nerve dysfunction characterized by poor judgment, dizziness, and incoordination.

Despite these acute responses to high altitude, millions of people live at elevations above 10,000 feet, with some villagers even residing in the Andes at altitudes higher than 16,000 feet. How do they live and function normally? They do so through the process of **acclimatization**. When a person remains at high altitude, the acute compensatory responses of increased ventilation and increased cardiac output are gradually replaced over a period of days by more slowly developing compensatory measures that permit adequate oxygenation of the tissues and restoration of normal acid–base balance. Red blood cell (RBC) production increases, stimulated by erythropoietin in response to reduced O_2 delivery to the kidneys (see p. 385). The rise in the number of RBCs increases the O_2-carrying capacity of the blood. Hypoxia also promotes synthesis of BPG within the RBCs so that O_2 is unloaded from Hb more easily at the tissues (see p. 475). The number of capillaries within the tissues increases, reducing the distance that O_2 must diffuse from the blood to reach the cells. Furthermore, acclimatized cells are able to use O_2 more efficiently by increasing the number of mitochondria, the energy organelles. In addition, the kidneys restore arterial pH to nearly normal by conserving acid that normally would have been lost in the urine (see p. 554).

These compensatory measures come with undesirable trade-offs. For example, the greater number of RBCs increases blood viscosity (makes the blood "thicker"), thereby increasing resistance to blood

---

ventilation. Generation of respiratory rhythm is now known to lie in the **pre-Bötzinger complex**, a region located in the upper (head) end of the VRG (❚ Figure 13-29). A network of neurons in this region display pacemaker activity, undergoing self-induced action potentials similar to those of the SA node of the heart. The rate at which the DRG inspiratory neurons rhythmically fire is driven by synaptic input from this complex.

Influences From the Pneumotaxic and Apneustic Centers The respiratory centers in the pons exert "fine-tuning" influences over the medullary center to help produce normal, smooth inspirations and expirations. The **pneumotaxic center** sends impulses to the DRG that help "switch off" the inspiratory neurons, limiting the duration of inspiration. In contrast, the **apneustic center** prevents the inspiratory neurons from being switched off, thus providing an extra boost to the inspiratory drive. In this check-and-balance system, the pneumotaxic center dominates over the apneustic center, helping halt inspiration and letting expiration occur normally. Without the pneumotaxic brakes, the breathing pattern consists of prolonged inspiratory gasps abruptly interrupted by brief expirations. This abnormal breathing pattern is known as **apneusis**; hence, the center that promotes this type of breathing is the apneustic center. Apneusis occurs in certain types of severe brain damage.

Hering–Breuer Reflex When the tidal volume is large (greater than 1 liter), as during exercise, the **Hering–Breuer reflex** is triggered to prevent overinflation of the lungs. **Pulmo-**

flow (see p. 337). As a result, the heart has to work harder to pump blood through the vessels.

Recent studies suggest that genetic selection has led to healthier normal red blood cell and hemoglobin levels among populations that have lived at high altitudes for thousands of years, thus bypassing the risks of thick blood. Instead, these peoples have evolved to rely on other means to live with permanently low arterial P_{O_2}. For example, the high-altitude inhabitants' endothelial cells release up to 10 times more nitric oxide (NO) than is released in those dwelling near sea level (see p. 346). This extra NO dilates the arterioles, more than doubling blood flow to the tissues, helping these high dwellers move sufficient O_2 to their tissues despite sustained low arterial O_2 levels.

Effects of Deep-Sea Diving on the Body

When a deep-sea diver, with the help of a *self-contained underwater breathing apparatus (scuba)*, descends underwater, the body is exposed to greater than atmospheric pressure. Pressure rapidly increases with sea depth as a result of the weight of the water. Pressure is already doubled about 30 feet below sea level. The air provided by scuba equipment is delivered to the lungs at these high pressures. Recall that (1) the amount of a gas in solution is directly proportional to the partial pressure of the gas and (2) air is composed of 79% N_2. Nitrogen is poorly soluble in body tissues, but the high P_{N_2} that occurs during deep-sea diving causes more of this gas than normal to dissolve in the body tissues. The small amount of N_2 dissolved in the tissues at sea level has no known effect, but as more N_2 dissolves at greater depths, **nitrogen narcosis,** or **"rapture of the deep,"** develops. Nitrogen narcosis is believed to result from reduced excitability of neurons when the lipid-soluble N_2 dissolves in their lipid membranes. At 150 feet underwater, divers experience a feeling of euphoria and become drowsy, similar to the effect of having a few cocktails. At lower depths, divers become weak and clumsy, and at 350 to 400 feet, they lose consciousness. Oxygen toxicity resulting from the high P_{O_2} is another possible detrimental effect of descending deep underwater.

Another problem associated with deep-sea diving occurs during ascent. If a diver who has been submerged long enough for a significant amount of N_2 to dissolve in the tissues suddenly ascends to the surface, the rapid reduction in P_{N_2} causes N_2 to quickly come out of solution and form bubbles of gaseous N_2 in the body, much as bubbles of gaseous CO_2 form in a bottle of champagne when the cork is popped. The consequences depend on the amount and location of N_2 bubble formation in the body. This condition is called **decompression sickness** or **"the bends"** because the victim often bends over in pain. Decompression sickness can be prevented by ascending slowly to the surface or by decompressing gradually in a decompression tank so that the excess N_2 can slowly escape through the lungs without bubble formation.

nary stretch receptors within the smooth muscle layer of the small airways are activated by stretching of the lungs at large tidal volumes. Action potentials from these stretch receptors travel through afferent nerve fibers to the medullary center and inhibit the inspiratory neurons. This negative feedback from the highly stretched lungs helps cut inspiration short before the lungs become overinflated.

Ventilation magnitude is adjusted in response to three chemical factors: P_{O_2}, P_{CO_2}, and H^+.

No matter how much O_2 is extracted from the blood or how much CO_2 is added to it at the tissue level, the P_{O_2} and P_{CO_2} of the systemic arterial blood leaving the lungs are normally held remarkably constant, indicating that arterial blood-gas content is precisely regulated. Arterial blood gases are maintained within the normal range by varying the magnitude of ventilation (rate and depth of breathing) to match the body's needs for O_2 uptake and CO_2 removal. If the blood extracts more O_2 from the alveoli and drops off more CO_2 because the tissues are metabolizing more actively, ventilation increases correspondingly to bring in more fresh O_2 and blow off more CO_2.

The medullary respiratory center receives inputs that provide information about the body's needs for gas exchange. It responds by sending appropriate signals to the motor neurons supplying the respiratory muscles, to adjust the rate and depth of ventilation to meet those needs. The two most obvious signals to increase ventilation are a decreased arterial P_{O_2} or an

| TABLE 13-8 Influence of Chemical Factors on Respiration

Chemical Factor	Effect on the Peripheral Chemoreceptors	Effect on the Central Chemoreceptors
$\downarrow P_{O_2}$ in the arterial blood	Stimulates only when the arterial P_{O_2} has fallen to the point of being life threatening (<60 mm Hg); an emergency mechanism	Directly depresses the central chemoreceptors and the respiratory center itself when <60 mm Hg
$\uparrow P_{CO_2}$ in the arterial blood ($\uparrow H^+$ in the brain ECF)	Weakly stimulates	Strongly stimulates; is the dominant control of ventilation (Levels >70–80 mm Hg directly depress the respiratory center and central chemoreceptors)
$\uparrow H^+$ in the arterial blood	Stimulates; important in acid–base balance	Does not affect; cannot penetrate the blood–brain barrier

increased arterial P_{CO_2}. These two factors do indeed influence the magnitude of ventilation, but not to the same degree nor through the same pathway. Also, a third chemical factor, H^+, notably influences the level of respiratory activity. We examine the role of each of these important chemical factors in the control of ventilation (Table 13-8).

Decreased arterial P_{O_2} increases ventilation only as an emergency mechanism.

Arterial P_{O_2} is monitored by **peripheral chemoreceptors** known as the **carotid bodies** and **aortic bodies**, which lie at the fork of the common carotid arteries (that supply the brain) on both the right and the left sides and in the arch of the aorta, respectively (Figure 13-30). These chemoreceptors respond to specific changes in the chemical content of the arterial blood that bathes them. They are distinctly different from the carotid sinus and aortic arch baroreceptors located in the same vicinity. The latter monitor pressure changes rather than chemical changes and are important in regulating systemic arterial blood pressure (see p. 367).

Effect of a Large Decrease in P_{O_2} on the Peripheral Chemoreceptors The peripheral chemoreceptors are not sensitive to modest reductions in arterial P_{O_2}. Arterial P_{O_2} must fall below 60 mm Hg (>40% reduction) before the peripheral chemoreceptors respond by sending afferent impulses to the medullary inspiratory neurons, thereby reflexly increasing ventilation. Because arterial P_{O_2} falls below 60 mm Hg only in the unusual circumstances of severe pulmonary disease or reduced atmospheric P_{O_2}, it does not play a role in the normal ongoing regulation of respiration. This fact may seem surprising at first because a primary function of ventilation is to provide enough O_2 for uptake by the blood. However, there is no need to increase ventilation until arterial P_{O_2} falls below 60 mm Hg because of the safety margin in % Hb saturation afforded by the plateau portion of the O_2–Hb curve. Hemoglobin is still 90% saturated at an arterial P_{O_2} of 60 mm Hg, but the % Hb saturation drops precipitously when P_{O_2} falls below this level. Therefore, reflex stimulation of respiration by the peripheral chemoreceptors is an important emergency mechanism in dangerously

low arterial P_{O_2} states. Indeed, this reflex mechanism is a lifesaver because a low arterial P_{O_2} directly depresses the respiratory center, as it does all the rest of the brain.

Clinical Note Because the peripheral chemoreceptors respond to the P_{O_2} of the blood, *not* the total O_2 content of the blood, O_2 content in the arterial blood can fall to dangerously low or even fatal levels without the peripheral chemoreceptors ever responding to reflexly stimulate respiration. Remember that only physically dissolved O_2 contributes to blood P_{O_2}. The total O_2 content in the arterial blood can be reduced in anemic

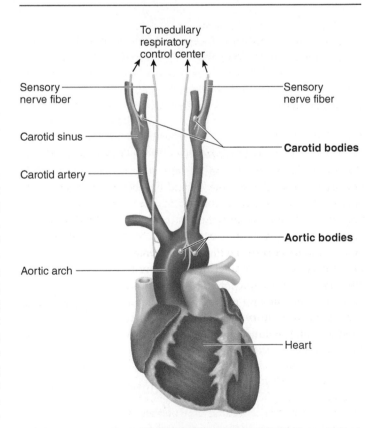

To medullary respiratory control center

Sensory nerve fiber

Sensory nerve fiber

Carotid sinus

Carotid bodies

Carotid artery

Aortic bodies

Aortic arch

Heart

| Figure 13-30 **Location of the peripheral chemoreceptors.** The carotid bodies are located in the carotid sinus, and the aortic bodies are located in the aortic arch.

states, in which O_2-carrying Hb is reduced, or in CO poisoning, when Hb preferentially binds to this molecule rather than to O_2. In both cases, arterial P_{O_2} is normal, so respiration is not stimulated, even though O_2 delivery to the tissues may be so reduced that the person dies from cellular O_2 deprivation.

Direct Effect of a Large Decrease in P_{O_2} on the Respiratory Center Except for the peripheral chemoreceptors, the activity level in all nervous tissue falls in O_2 deprivation. Were it not for stimulatory intervention of the peripheral chemoreceptors when arterial P_{O_2} falls threateningly low, a vicious cycle ending in cessation of breathing would ensue. Direct depression of the respiratory center by the markedly low arterial P_{O_2} would further reduce ventilation, leading to an even greater fall in arterial P_{O_2}, which would even further depress the respiratory center until ventilation ceased and death occurred.

CO_2-generated H^+ in the brain is normally the main regulator of ventilation.

In contrast to arterial P_{O_2}, which does not contribute to the minute-to-minute regulation of respiration, arterial P_{CO_2} is the most important input regulating the magnitude of ventilation under resting conditions. This role is appropriate because changes in alveolar ventilation have an immediate and pronounced effect on arterial P_{CO_2}. By contrast, changes in ventilation have little effect on % Hb saturation and O_2 availability to the tissues until arterial P_{O_2} falls by more than 40%. Even slight alterations from normal in arterial P_{CO_2} bring about a significant reflex effect on ventilation. An increase in arterial P_{CO_2} reflexly stimulates the respiratory center, with the resultant increase in ventilation promoting elimination of the excess CO_2 to the atmosphere. Conversely, a fall in arterial P_{CO_2} reflexly reduces the respiratory drive. The subsequent decrease in ventilation lets metabolically produced CO_2 accumulate so that P_{CO_2} can return normal.

Effect of Increased P_{CO_2} on the Central Chemoreceptors It is surprising that, given the key role of arterial P_{CO_2} in regulating respiration, no important receptors monitor arterial P_{CO_2} per se. The carotid and aortic bodies are only weakly responsive to changes in arterial P_{CO_2}, so they play only a minor role in reflexly stimulating ventilation in response to an elevation in arterial P_{CO_2}. More important in linking changes in arterial P_{CO_2} to compensatory adjustments in ventilation are the **central chemoreceptors**, located in the medulla near the respiratory center. These central chemoreceptors do not monitor CO_2 itself; however, they are sensitive to changes in CO_2-induced H^+ concentration in the brain extracellular fluid (ECF) that bathes them.

Movement of materials across the brain capillaries is restricted by the blood–brain barrier (see p. 141). Because this barrier is readily permeable to CO_2, any increase in arterial P_{CO_2} causes a similar rise in brain-ECF P_{CO_2} as CO_2 diffuses down its pressure

gradient from the cerebral blood vessels into the brain ECF. Under the influence of carbonic anhydrase, the increased P_{CO_2} within the brain ECF correspondingly raises the concentration of H^+ according to the law of mass action as it applies to this reaction: $CO_2 + H_2O \rightleftharpoons H^+ \, HCO_3^-$. An elevation in H^+ concentration in the brain ECF directly stimulates the central chemoreceptors, which in turn increase ventilation by stimulating the respiratory center through synaptic connections (❙ Figure 13-31). As the excess CO_2 is subsequently blown off, arterial P_{CO_2} and brain-ECF P_{CO_2} and H^+ concentration return to normal. Conversely, a decline in arterial P_{CO_2} below normal is paralleled by a fall in P_{CO_2} and H^+ in the brain ECF, the result of which is a central chemoreceptor–mediated decrease in ventilation. As CO_2 produced by cell metabolism is consequently allowed to accumulate, arterial P_{CO_2} and brain-ECF P_{CO_2} and H^+ are restored toward normal.

Unlike CO_2, H^+ cannot readily permeate the blood–brain barrier, so H^+ in the plasma cannot gain access to the central chemoreceptors. Accordingly, the central chemoreceptors respond only to H^+ generated within the brain ECF itself as a result of CO_2 entry. Thus, the major mechanism controlling ventilation under resting conditions is specifically aimed at regulating the brain-ECF H^+ concentration, which in turn directly reflects the arterial P_{CO_2}. Unless there are extenuating circumstances such as reduced availability of O_2 in the inspired

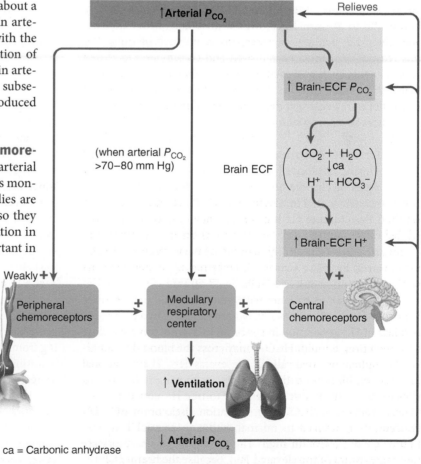

❙ Figure 13-31 **Effect of increased arterial P_{CO_2} on ventilation.**

air, arterial P_{O_2} is coincidentally also maintained at its normal value by the brain-ECF H^+ ventilatory driving mechanism.

The powerful influence of the central chemoreceptors on the respiratory center is widely believed to be responsible for your inability to deliberately hold your breath for more than about a minute. While you hold your breath, metabolically produced CO_2 continues to accumulate in your blood and then to build up the H^+ concentration in your brain ECF. Finally, the increased P_{CO_2}–H^+ stimulant to respiration becomes so powerful that central chemoreceptor excitatory input overrides voluntary inhibitory input to respiration, so breathing resumes despite deliberate attempts to prevent it. Breathing resumes long before arterial P_{O_2} falls to the threateningly low levels that trigger the peripheral chemoreceptors. Therefore, you cannot deliberately hold your breath long enough to create a dangerously high level of CO_2 or low level of O_2 in the arterial blood. (As an alternate theory, recent evidence suggests that the break point at which a breath-holding person is driven to gasp for air might depend on signals resulting from prolonged contraction of the diaphragm, not on elevated P_{CO_2}–H^+.)

Direct Effect of a Large Increase in P_{CO_2} on the Respiratory Center In contrast to the normal reflex stimulatory effect of the increased P_{CO_2}–H^+ mechanism on respiratory activity, very high levels of CO_2 directly depress the entire brain, including the respiratory center, just as very low levels of O_2 do. Up to a P_{CO_2} of 70 to 80 mm Hg, progressively higher P_{CO_2} levels promote correspondingly more vigorous respiratory efforts in an attempt to blow off the excess CO_2. A further increase in P_{CO_2} beyond 70 to 80 mm Hg, however, does not further increase ventilation but actually depresses the respiratory neurons. For this reason, CO_2 must be removed and O_2 supplied in closed environments such as closed-system anesthesia machines, submarines, or space capsules. Otherwise, CO_2 could reach lethal levels, not only because it depresses respiration but also because it produces severe respiratory acidosis.

Clinical Note **Loss of Sensitivity to P_{CO_2} with Lung Disease** During prolonged hypoventilation caused by certain types of chronic lung disease, an elevated P_{CO_2} occurs simultaneously with a markedly reduced P_{O_2}. In most cases, the elevated P_{CO_2} (acting via the central chemoreceptors) and the reduced P_{O_2} (acting via the peripheral chemoreceptors) are *synergistic*—that is, the combined stimulatory effect on respiration exerted by these two inputs together is greater than the sum of their independent effects.

However, some patients with severe chronic lung disease lose their sensitivity to an elevated arterial P_{CO_2}. In a prolonged increase in H^+ generation in the brain ECF, from long-standing CO_2 retention, enough HCO_3^- may cross the blood–brain barrier to buffer, or "neutralize," the excess H^+. The additional HCO_3^- combines with the excess H^+, removing it from solution so that it no longer contributes to free H^+ concentration. When brain-ECF HCO_3^- concentration rises, brain-ECF H^+ concentration returns to normal, although arterial P_{CO_2} and brain-ECF P_{CO_2} remain high. The central chemoreceptors are no longer aware of the elevated P_{CO_2} because the brain-ECF H^+

is normal. Because the central chemoreceptors no longer reflexly stimulate the respiratory center in response to the elevated P_{CO_2}, the drive to eliminate CO_2 is blunted in such patients—that is, their level of ventilation is abnormally low considering their high arterial P_{CO_2}. In these patients, the hypoxic drive to ventilation becomes their primary respiratory stimulus, in contrast to normal individuals, in whom the arterial P_{CO_2} level is the dominant factor governing the magnitude of ventilation. Ironically, administering O_2 to such patients to relieve the hypoxic condition can markedly depress their drive to breathe by elevating the arterial P_{O_2} and removing the primary driving stimulus for respiration. Thus, O_2 therapy must be administered cautiously in patients with long-term pulmonary diseases.

Adjustments in ventilation in response to changes in arterial H^+ are important in acid–base balance.

Changes in arterial H^+ concentration cannot influence the central chemoreceptors because H^+ does not readily cross the blood–brain barrier. However, the aortic and carotid body peripheral chemoreceptors are highly responsive to fluctuations in arterial H^+ concentration, in contrast to their weak sensitivity to deviations in arterial P_{CO_2} and their unresponsiveness to arterial P_{O_2} until it falls 40% below normal.

Any change in arterial P_{CO_2} brings about a corresponding change in the H^+ concentration of the blood as well as of the brain ECF. These CO_2-induced H^+ changes in the arterial blood are detected by the peripheral chemoreceptors; the result is reflexly stimulated ventilation in response to increased arterial H^+ concentration and depressed ventilation in association with decreased arterial H^+ concentration. However, these changes in ventilation mediated by the peripheral chemoreceptors are far less important than the powerful central-chemoreceptor mechanism in adjusting ventilation in response to changes in CO_2-generated H^+ concentration.

The peripheral chemoreceptors do play a major role in adjusting ventilation in response to alterations in arterial H^+ concentration unrelated to fluctuations in P_{CO_2}. In many situations, even though P_{CO_2} is normal, arterial H^+ concentration is changed by the addition or loss of non-CO_2-generated acid from the body. For example, arterial H^+ concentration increases during untreated diabetes mellitus because excess H^+-generating keto acids are abnormally produced and added to the blood. A rise in arterial H^+ concentration reflexly stimulates ventilation by means of the peripheral chemoreceptors. Conversely, the peripheral chemoreceptors reflexly suppress respiratory activity in response to a fall in arterial H^+ concentration resulting from nonrespiratory causes, such as occurs during excessive vomiting. During vomiting, the acid-rich digestive juice that is secreted into the stomach and subsequently reabsorbed back into the blood when digestion is complete is instead lost from the body. Changes in ventilation by this mechanism are extremely important in regulating the body's acid–base balance. Changing the magnitude of ventilation can vary the amount of H^+-generating CO_2 eliminated. The resulting adjustment in the amount of H^+ added to the blood from CO_2 can compensate for the nonrespiratory-induced abnormality in arterial H^+ concentration that first elicited the respiratory response. (See Chapter 15 for further details.)

Exercise profoundly increases ventilation by unclear mechanisms.

Alveolar ventilation may increase up to 20-fold during heavy exercise to keep pace with the increased demand for O_2 uptake and CO_2 output. (Table 13-9 highlights changes in O_2- and CO_2-related variables during exercise.) The cause of increased ventilation during exercise is still largely speculative. It would seem logical that changes in the "big three" chemical factors—decreased P_{O_2}, increased P_{CO_2}, and increased H^+—could account for the increase in ventilation. This does not appear to be the case, however.

- Despite the marked increase in O_2 use during exercise, arterial P_{O_2} does not decrease but remains normal or may actually increase slightly because the increase in alveolar ventilation keeps pace with or even slightly exceeds the stepped-up rate of O_2 consumption.

- Likewise, despite the marked increase in CO_2 production during exercise, arterial P_{CO_2} does not increase but remains normal or decreases slightly because the extra CO_2 is removed as rapidly or even more rapidly than it is produced by the increase in ventilation.

- During mild or moderate exercise, H^+ concentration does not increase because H^+-generating CO_2 is held constant. During heavy exercise, H^+ concentration does increase somewhat from release of H^+-generating lactate (lactic acid) into the blood by anaerobic metabolism in the exercising muscles. Even so, the elevation in H^+ concentration resulting from lactic acid formation is not enough to account for the large increase in ventilation accompanying exercise.

Some investigators argue that the constancy of the three chemical regulatory factors during exercise shows that ventilatory responses to exercise are actually being controlled by these factors—particularly by P_{CO_2} because it is normally the dominant control during resting conditions. According to this reasoning, how else could alveolar ventilation be increased in exact proportion to CO_2 production, thereby keeping the P_{CO_2} constant? This proposal, however, cannot account for the observation that during heavy exercise, alveolar ventilation may increase relatively more than CO_2 production increases, thereby actually causing a slight decline in P_{CO_2}. Also, ventilation increases abruptly at the onset of exercise (within seconds), long before changes in arterial blood gases could become important influences on the respiratory center (which requires a matter of minutes).

TABLE 13-9	Changes in Variables Related to O_2 and CO_2 during Exercise	
Variable	**Change**	**Comment**
O_2 use	Marked ↑	Active muscles oxidize nutrient molecules more rapidly to meet their increased energy needs.
CO_2 production	Marked ↑	More actively metabolizing muscles produce more CO_2.
Alveolar ventilation	Marked ↑	By mechanisms not completely understood, alveolar ventilation keeps pace with or even slightly exceeds the increased metabolic demands during exercise.
Arterial P_{O_2}	Normal or slight ↑	Despite a marked increase in O_2 use and CO_2 production during exercise, alveolar ventilation keeps pace with or even slightly exceeds the stepped-up rate of O_2 consumption and CO_2 production.
Arterial P_{CO_2}	Normal or slight ↓	
O_2 delivery to muscles	Marked ↑	Although arterial P_{O_2} remains normal, O_2 delivery to muscles greatly increases as a result of the increased blood flow to exercising muscles accomplished by increased cardiac output coupled with local vasodilation of active muscles.
O_2 extraction by muscles	Marked ↑	Increased use of O_2 lowers the P_{O_2} at the tissue level, which results in more O_2 unloading from hemoglobin; this is enhanced by ↑ P_{CO_2}, ↑ H^+, and ↑ temperature.
CO_2 removal from muscles	Marked ↑	The increased blood flow to exercising muscles removes the excess CO_2 produced by these more actively metabolizing tissues.
Arterial H^+ Concentration		
Mild to moderate exercise	Normal	Because H^+-generating CO_2 is held constant in arterial blood, arterial H^+ concentration does not change.
Heavy exercise	Modest ↑	In heavy exercise, when muscles resort to anaerobic metabolism, lactic acid is added to the blood.

Researchers have suggested that a number of other factors, including the following, play a role in the ventilatory response to exercise:

1. *Reflexes originating from body movements.* Joint and muscle receptors excited during muscle contraction reflexly stimulate the respiratory center, abruptly increasing ventilation. Even passive movement of the limbs (for example, someone else alternately flexing and extending a person's knee) may increase ventilation several-fold through activation of these receptors, although no actual exercise is occurring. Thus, the mechanical events of exercise are believed to play an important role in coordinating respiratory activity with the increased metabolic requirements of the active muscles.

2. *Increase in body temperature.* Much of the energy generated during muscle contraction is converted to heat rather than to actual mechanical work. Heat-loss mechanisms such as sweating frequently cannot keep pace with the increased heat production that accompanies increased physical activity, so body temperature often rises slightly during exercise (see p. 632). Because raised body temperature stimulates ventilation, this exercise-related heat production undoubtedly contributes to the respiratory response to exercise. For the same reason, increased ventilation often accompanies a fever.

3. *Epinephrine release.* The adrenal medullary hormone epinephrine also stimulates ventilation. The level of circulating epinephrine rises during exercise in response to the sympathetic nervous system discharge that accompanies increased physical activity.

4. *Impulses from the cerebral cortex.* Especially at the onset of exercise, the motor areas of the cerebral cortex are believed to simultaneously stimulate the medullary respiratory neurons and activate the motor neurons of the exercising muscles. This is similar to the cardiovascular adjustments initiated by the motor cortex at the onset of exercise. In this way, the motor region of the brain calls forth increased ventilatory and circulatory responses to support the increased physical activity it is about to orchestrate. These anticipatory adjustments are feedforward regulatory mechanisms—that is, they occur before any homeostatic factors actually change (see p. 18).

None of these factors or combinations of factors are fully satisfactory in explaining the abrupt and profound effect exercise has on ventilation, nor can they completely account for the high degree of correlation between respiratory activity and the body's needs for gas exchange during exercise. (For a discussion of how O_2 consumption during exercise can be measured to determine a person's maximum work capacity, see the accompanying boxed feature, ▮A Closer Look at Exercise Physiology.)

Ventilation can be influenced by factors unrelated to the need for gas exchange.

Ventilation can be modified for reasons other than the need to supply O_2 or remove CO_2. Here are some examples of involuntary influences in this category:

■ Protective reflexes such as sneezing and coughing temporarily govern respiratory activity in an effort to expel irritant materials from respiratory passages.

■ The respiratory center is reflexly inhibited during swallowing, when the airways are closed to prevent food from entering the lungs (see p. 576).

■ Pain originating anywhere in the body reflexly stimulates the respiratory center (for example, one "gasps" with pain).

■ Involuntary modification of breathing also occurs during the expression of various emotional states, such as laughing, crying, sighing, and groaning. The emotionally induced modifications are mediated through connections between the limbic system in the brain (which is responsible for emotions) and the respiratory center.

■ Hiccups occur when involuntary, spasmodic contractions of the diaphragm take place, each causing rapid intake of air, which is suddenly halted by abrupt closure of the glottis, resulting in the "hic" sound. The underlying trigger for hiccups is not known.

Humans also have considerable voluntary control over ventilation. Voluntary control of breathing is accomplished by the cerebral cortex, which does not act on the respiratory center in the brain stem but instead sends impulses directly to the motor neurons in the spinal cord that supply the respiratory muscles. We deliberately control our breathing to perform such voluntary acts as speaking, singing, whistling, playing wind instruments, or swimming. Also, we can voluntarily hyperventilate ("overbreathe") or, at the other extreme, hold our breath, but only for a brief period before reflex control mechanisms take over.

During apnea, a person "forgets to breathe"; during dyspnea, a person feels "short of breath."

Apnea is the transient interruption of ventilation, with breathing resuming spontaneously. If breathing does not resume, the condition is called **respiratory arrest**. Because ventilation is normally decreased and the central chemoreceptors are less sensitive to the arterial P_{CO_2} drive during sleep, especially paradoxical sleep (see p. 169), apnea is most likely to occur during this time. Victims of sleep apnea may stop breathing for a few seconds or up to 1 or 2 minutes as many as 500 times a night. Mild sleep apnea is not dangerous unless the sufferer has pulmonary or circulatory disease, which can be worsened by recurrent bouts of apnea.

Clinical Note **Sudden Infant Death Syndrome** In exaggerated cases of sleep apnea, the victim may be unable to recover from an apneic period, and death results. This is the case in **sudden infant death syndrome (SIDS)**, or "crib death," the leading cause of death in the first year of life. With this tragic form of sleep apnea, a previously healthy 2- to 4-month-old infant is found dead in his or her crib for no apparent reason. The underlying cause of SIDS is the subject of intense investigation. Most evidence suggests that the baby "forgets to breathe" because the respiratory control mechanisms are immature, either in the brain stem or in the chemoreceptors that monitor the body's respiratory status. For example, on autopsy more than half the victims have poorly developed

How to Find Out How Much Work You're Capable of Doing

THE BEST SINGLE PREDICTOR OF a person's work capacity is the determination of **maximal O_2 consumption**, or **max VO_2**, which is the maximum volume of O_2 the person is capable of using per minute to oxidize nutrient molecules for energy production. Max VO_2 is measured by having the person engage in exercise, usually on a treadmill or bicycle ergometer (a stationary bicycle with variable resistance). The workload is incrementally increased until the person becomes exhausted. Expired air samples collected during the last minutes of exercise, when O_2 consumption is at a maximum because the person is working as hard as possible, are analyzed for the percentage of O_2 and CO_2 they contain. Furthermore, the volume of air expired is measured. Equations are then used to determine the amount of O_2 consumed, taking into account the percentages of O_2 and CO_2 in the inspired air, the total volume of air expired, and the percentages of O_2 and CO_2 in the exhaled air.

Maximal O_2 consumption depends on three systems. The respiratory system is essential for ventilation and exchange of O_2 and CO_2 between air and blood in the lungs. The circulatory system is required to deliver O_2 to the working muscles. Finally, the muscles must have the oxidative enzymes available to use the O_2 once it has been delivered.

Regular aerobic exercise can improve max VO_2 by making the heart and respiratory system more efficient, thereby delivering more O_2 to the working muscles. Exercised muscles themselves become better equipped to use O_2 once it is delivered. The number of functional capillaries increases, as do the number and size of mitochondria, which contain the oxidative enzymes.

Maximal O_2 consumption is measured in liters per minute and then converted into milliliters per kilogram of body weight per minute so that large and small people can be compared. As would be expected, athletes have the highest values for maximal O_2 consumption. The max VO_2 for male cross-country skiers has been recorded to be as high as 94 mL O_2/kg/min. Distance runners maximally consume between 65 and 85 mL O_2/kg/min, and football players have max VO_2 values between 45 and 65 mL O_2/kg/min, depending on the position they play. Sedentary young men maximally consume between 25 and 45 mL O_2/kg/min. Female values for max VO_2 are 20% to 25% lower than for males when expressed as mL/kg/min of total body weight. The difference in max VO_2 between females and males is only 8% to 10% when expressed as mL/kg/min of lean body weight, however, because females generally have a higher percentage of body fat (the female sex hormone estrogen promotes fat deposition).

Available norms are used to classify people as being low, fair, average, good, or excellent in aerobic capacity for their age group. Exercise physiologists use max VO_2 measurements to prescribe or adjust training regimens to help people achieve their optimal level of aerobic conditioning.

---

carotid bodies, the more important of the peripheral chemoreceptors. Abnormal lung development has been suggested as being responsible for at least some cases. Some infants who died of SIDS were found to have reduced levels of a growth factor critical for the lungs. Alternatively, some researchers believe the condition may be triggered by an initial cardiovascular failure rather than by an initial cessation of breathing. Still other investigators propose that some cases may result from aspiration of stomach juice containing the bacterium *Helicobacter pylori*. In one study, this microorganism was present in 88% of infants who died of SIDS. Scientists speculate that *H. pylori* may lead to the production of ammonia, which can be lethal if it gains access to the blood from the lungs. Perhaps a combination of factors might be involved, or maybe SIDS is a collection of early infancy deaths from a variety of causes.

Whatever the underlying cause, certain risk factors make babies more vulnerable to SIDS. Among them are sleeping position (an almost 40% higher incidence of SIDS is associated with sleeping on the abdomen rather than on the back or side) and exposure to nicotine during fetal life or after birth. Infants whose mothers smoked during pregnancy or who breathe cigarette smoke in the home are three times more likely to die of SIDS than those not exposed to smoke.

Clinical Note **Dyspnea** People who have dyspnea have the subjective sensation that they are not getting enough air—that is, they feel "short of breath." **Dyspnea** is the mental anguish associated with the unsatiated desire for more adequate ventilation. It often accompanies the labored breathing characteristic of obstructive lung disease or the pulmonary edema associated with congestive heart failure. In contrast, during exercise a person can breathe very hard without experiencing dyspnea because such exertion is not accompanied by a sense of anxiety over the adequacy of ventilation. Surprisingly, dyspnea is not directly related to chronic elevation of arterial P_{CO_2} or reduction of P_{O_2}. The subjective feeling of air hunger may occur even when alveolar ventilation and the blood gases are normal. Some people experience dyspnea when they *perceive* that they are short of air even though this is not actually the case, such as in a crowded elevator.

Check Your Understanding 13.5

1. Briefly describe how the following brain regions contribute to control of respiration: the medullary respiratory center (including the roles of the DRG and VRG), the pneumotaxic center, the apneustic center, and the pre-Bötzinger complex.

2. Discuss the role of the peripheral chemoreceptors.

3. Tell how the magnitude of ventilation is regulated under resting conditions.

Homeostasis: Chapter in Perspective

The respiratory system contributes to homeostasis by obtaining O_2 from and eliminating CO_2 to the external environment. All body cells ultimately need an adequate supply of O_2 to use in oxidizing nutrient molecules to generate ATP. Brain cells, which especially depend on a continual supply of O_2, die if deprived of O_2 for more than 4 minutes. Even cells that can resort to anaerobic ("without O_2") metabolism for energy production, such as strenuously exercising muscles, can do so only transiently by incurring an O_2 deficit that must be made up during the period of excess postexercise O_2 consumption (see p. 272).

As a result of these energy-yielding metabolic reactions, the body produces large quantities of CO_2 that must be eliminated. Because CO_2 and H_2O form carbonic acid, adjustments in the rate of CO_2 elimination by the respiratory system are important in regulating acid–base balance in the internal environment. Cells can survive only within a narrow pH range.

Review Exercises Answers begin on p. A-39

Reviewing Terms and Facts

1. Breathing is accomplished by alternate contraction and relaxation of muscles within the lung tissue. *(True or false?)*

2. Normally, the alveoli empty completely during maximal expiratory efforts. *(True or false?)*

3. Alveolar ventilation does not always increase when pulmonary ventilation increases. *(True or false?)*

4. O_2 and CO_2 have equal diffusion constants. *(True or false?)*

5. Hemoglobin has a higher affinity for O_2 than for any other substance. *(True or false?)*

6. Rhythmicity of breathing is brought about by pacemaker activity displayed by the respiratory muscles. *(True or false?)*

7. The expiratory neurons send impulses to the motor neurons controlling the expiratory muscles during normal quiet breathing. *(True or false?)*

8. The two forces that tend to keep the alveoli open are _____ and _____.

9. The two forces that promote alveolar collapse are _____ and _____.

10. _____ is a measure of the magnitude of change in lung volume accomplished by a given change in the transmural pressure gradient.

11. _____ is the phenomenon of the lungs snapping back to their resting size after having been stretched.

12. _____ is the erythrocytic enzyme that catalyzes the conversion of CO_2 into HCO_3^-.

13. Which of the following reactions take(s) place at the pulmonary capillaries?

  a. $Hb + O_2 \rightarrow HbO_2$

  b. $CO_2 + H_2O \rightarrow H^+ + HCO_3^-$

  c. $Hb + CO_2 \rightarrow HbCO_2$

  d. $Hb + H^+ \rightarrow HbH$

14. Indicate the O_2 and CO_2 partial pressure relationships important in gas exchange by circling > (greater than), < (less than), or = (equal to) as appropriate in each of the following statements:

  a. P_{O_2} in blood entering the pulmonary capillaries is (>, <, or =) P_{O_2} in the alveoli.

  b. P_{CO_2} in blood entering the pulmonary capillaries is (>, <, or =) P_{CO_2} in the alveoli.

  c. P_{O_2} in the alveoli is (>, <, or =) P_{O_2} in blood leaving the pulmonary capillaries.

  d. P_{CO_2} in the alveoli is (>, <, or =) P_{CO_2} in blood leaving the pulmonary capillaries.

  e. P_{O_2} in blood leaving the pulmonary capillaries is (>, <, or =) P_{O_2} in blood entering the systemic capillaries.

  f. P_{CO_2} in blood leaving the pulmonary capillaries is (>, <, or =) P_{CO_2} in blood entering the systemic capillaries.

  g. P_{O_2} in blood entering the systemic capillaries is (>, <, or =) P_{O_2} in the tissue cells.

  h. P_{CO_2} in blood entering the systemic capillaries is (>, <, or =) P_{CO_2} in the tissue cells.

  i. P_{O_2} in the tissue cells is (>, <, or approximately =) P_{O_2} in blood leaving the systemic capillaries.

  j. P_{CO_2} in the tissue cells is (>, <, or approximately =) P_{CO_2} in blood leaving the systemic capillaries.

  k. P_{O_2} in blood leaving the systemic capillaries is (>, <, or =) P_{O_2} in blood entering the pulmonary capillaries.

  l. P_{CO_2} in blood leaving the systemic capillaries is (>, <, or =) P_{CO_2} in blood entering the pulmonary capillaries.

15. Using the answer code on the right, indicate which chemo-receptors are being described:

1. stimulated by an arterial P_{O_2} of 80 mm Hg
2. stimulated by an arterial P_{O_2} of 55 mm Hg
3. directly depressed by an arterial P_{O_2} of 55 mm Hg
4. weakly stimulated by an elevated arterial P_{O_2}
5. strongly stimulated by an elevated brain-ECF H^+ concentration induced by an elevated arterial P_{CO_2}
6. stimulated by an elevated arterial H^+ concentration

(a) peripheral chemoreceptors
(b) central chemoreceptors
(c) both peripheral and central chemoreceptors
(d) neither peripheral nor central chemoreceptors

Understanding Concepts

(Answers at www.cengagebrain.com)

1. Distinguish between cellular and external respiration. List the steps in external respiration.

2. Describe the components of the respiratory system. What is the site of gas exchange?

3. Compare atmospheric, intra-alveolar, and intrapleural pressures.

4. Why are the lungs normally stretched even during expiration?

5. Explain why air enters the lungs during inspiration and leaves during expiration.

6. Why is inspiration normally active and expiration normally passive?

7. Why does airway resistance become an important determinant of airflow rates in chronic obstructive pulmonary disease?

8. Explain pulmonary elasticity in terms of compliance and elastic recoil.

9. State the source and function of pulmonary surfactant.

10. Define the various lung volumes and capacities.

11. Compare pulmonary and alveolar ventilation. What is the consequence of anatomic and alveolar dead space?

12. Compare ventilation, perfusion, and the ventilation–perfusion ratio at the top and the bottom of the lung. Explain what accounts for these differences.

13. What determines the partial pressures of a gas in air and in blood?

14. List the methods of O_2 transport and CO_2 transport in the blood.

15. What is the primary factor that determines the percent hemoglobin saturation? What are the significances of the plateau and the steep portions of the O_2–Hb dissociation curve?

16. How does hemoglobin promote the net transfer of O_2 from the alveoli to the blood?

17. Explain the Bohr and Haldane effects.

18. Define the following: *hypoxic hypoxia, anemic hypoxia, circulatory hypoxia, histotoxic hypoxia, hypercapnia, hypocapnia, hyperventilation, hypoventilation, hyperpnea, apnea,* and *dyspnea.*

19. What are the locations and functions of the three respiratory control centers? Distinguish between the dorsal respiratory group (DRG) and the ventral respiratory group (VRG).

20. What brain region establishes the rhythmicity of breathing?

Solving Quantitative Exercises

1. The two curves in ▌Figure 13-28 (p. 478) show partial pressures for O_2 and CO_2 at various alveolar ventilation rates. These curves can be calculated from the following two equations:

$$P_{AO_2} = P_{IO_2} - (V_{O_2}/V_A) \, 863 \text{ mm Hg}$$

$$P_{ACO_2} = (V_{CO_2}/V_A) \, 863 \text{ mm Hg}$$

In these equations, P_{AO_2} = the partial pressure of O_2 in the alveoli, P_{ACO_2} = the partial pressure of CO_2 in the alveoli, P_{IO_2} = the partial pressure of O_2 in the inspired air, V_{O_2} = the rate of O_2 consumption by the body, V_{CO_2} = the rate of CO_2 production by the body, V_A = the rate of alveolar ventilation, and 863 mm Hg is a constant that accounts for atmospheric pressure and temperature.

John is in training for a marathon tomorrow and just ate a meal of pasta (assume this is pure carbohydrate, which is metabolized with an RQ of 1). His alveolar ventilation rate is 3.0 L/min, and he is consuming O_2 at a rate of 300 mL/min. What is the value of John's P_{ACO_2}?

2. Assume you are flying in an airplane that is cruising at 18,000 feet, where the pressure outside the plane is 380 mm Hg.

a. Calculate the partial pressure of O_2 in the air outside the plane, ignoring water vapor pressure.

b. If the plane depressurized, what would be the value of your P_{AO_2}? Assume that the ratio of your O_2 consumption to ventilation was not changed (that is, equaled 0.06), and note that under these conditions the constant in the equation that accounts for atmospheric pressure and temperature decreases from 863 mm Hg to 431.5 mm Hg.

c. Calculate your P_{ACO_2}, assuming that your CO_2 production and ventilation rates remained unchanged at 200 mL/min and 4.2 L/min, respectively.

3. A student has a tidal volume of 350 mL. While breathing at a rate of 12 breaths/min, her alveolar ventilation is 80% of her pulmonary ventilation. What is her anatomic dead space volume?

Applying Clinical Reasoning

Keith M., a former heavy cigarette smoker, has severe emphysema. How does this condition affect his airway resistance? How does this change in airway resistance influence Keith's inspiratory and expiratory efforts? Describe how his respiratory muscle activity and intra-alveolar pressure changes compare to normal to accomplish a normal tidal volume. How would his spirogram compare to normal? What influence would Keith's condition have on gas exchange in his lungs? What blood-gas abnormalities are likely to be present? Would it be appropriate to administer O_2 to Keith to relieve his hypoxic condition?

Thinking at a Higher Level

1. Why is it important that airplane interiors are pressurized (that is, the pressure is maintained at sea-level atmospheric pressure even though the atmospheric pressure surrounding the plane is substantially lower)? Explain the physiological value of using O_2 masks if the pressure in the airplane interior cannot be maintained.

2. If a severely anemic person has normal lungs, indicate whether each of the following factors will be normal, below normal, or above normal: (a) alveolar P_{O_2}, (b) arterial P_{O_2}, (c) percent hemoglobin saturation, (d) total O_2 content of arterial blood, (e) tissue P_{O_2}, (f) total O_2 transferred from blood to tissues

3. Would hypercapnia accompany the hypoxia produced in each of the following situations? Why or why not?

   a. cyanide poisoning

   b. pulmonary edema

   c. restrictive lung disease

   d. high altitude

   e. severe anemia

   f. congestive heart failure

   g. obstructive lung disease

4. Based on what you know about the control of respiration, explain why it is dangerous to voluntarily hyperventilate to lower the arterial P_{CO_2} before going underwater. The purpose of the hyperventilation is to stay under longer before P_{CO_2} rises above normal and drives the swimmer to surface for a breath of air.

5. If a person whose alveolar–capillary membranes are thickened by disease has an alveolar P_{O_2} of 100 mm Hg and an alveolar P_{CO_2} of 40 mm Hg, which of the following values of systemic arterial blood gases are most likely to exist?

   a. $P_{O_2} = 105$ mm Hg, $P_{CO_2} = 35$ mm Hg

   b. $P_{O_2} = 100$ mm Hg, $P_{CO_2} = 40$ mm Hg

   c. $P_{O_2} = 80$ mm Hg, $P_{CO_2} = 45$ mm Hg

If the person is administered 100% O_2, will the arterial P_{O_2} increase, decrease, or remain the same? Will the arterial P_{CO_2} increase, decrease, or remain the same?

To access the course materials and companion resources for this text, please visit www.cengagebrain.com

The Urinary System

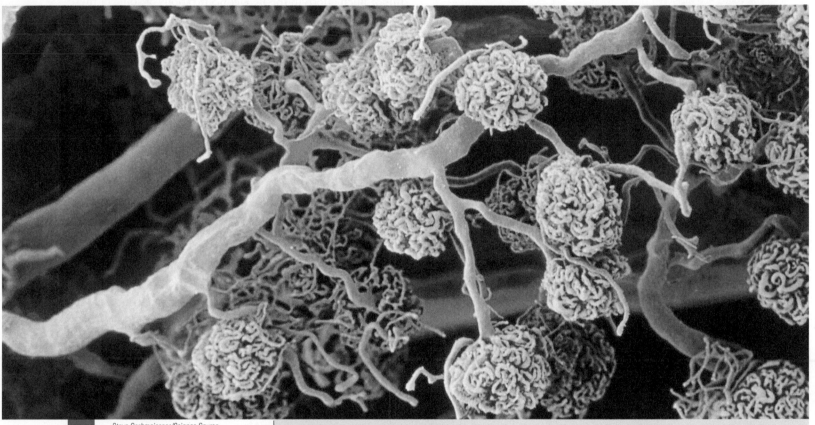

Steve Gschmeissner/Science Source

A scanning electron micrograph of glomeruli and blood vessels in the kidney. The glomeruli (*yellow*) are balls of highly coiled capillaries through which protein-free plasma is filtered as the first step in urine formation. The kidney tubules (stripped away to reveal the glomeruli) collect the filtered fluid and convert it into urine by making selected exchanges with peritubular capillaries that wrap around the tubules.

Homeostasis Highlights

The survival and proper functioning of cells depend on maintaining stable concentrations of salt, acids, and other electrolytes in the internal fluid environment. Cell survival also depends on continuous removal of toxic metabolic wastes that cells produce as they perform life-sustaining chemical reactions. The **kidneys** play a major role in maintaining homeostasis by regulating the concentration of many plasma constituents, especially electrolytes and water, and by eliminating all metabolic wastes (except CO_2, which is removed by the lungs). As plasma repeatedly filters through the kidneys, they retain constituents of value for the body and eliminate undesirable or excess materials in the urine. Of special importance is the kidneys' ability to regulate the volume and osmolarity (solute concentration) of the internal fluid environment by controlling salt and water balance. Also crucial is their ability to help regulate pH by controlling elimination of acid and base in the urine.

14.1 Kidneys: Functions, Anatomy, and Basic Processes

The composition of the fluid bathing all the cells could be notably altered by exchanges between the cells and this internal fluid environment if mechanisms did not exist to keep the extracellular fluid (ECF) stable.

The kidneys perform a variety of functions aimed at maintaining homeostasis.

The **kidneys**, in concert with hormonal and neural inputs that control their function, are primarily responsible for maintaining stable volume, electrolyte composition, and osmolarity (solute concentration) of the ECF. By adjusting the quantity of water and various plasma constituents that are either conserved for the body or eliminated in the urine, the kidneys can maintain water and electrolyte balance within the narrow range compatible with life despite a wide range of intake and losses of these constituents through other avenues. The kidneys not only adjust for variations in ingestion of water, salt, and other electrolytes, but also adjust urinary output of these ECF constituents to compensate for abnormal losses through heavy sweating, vomiting, diarrhea, or hemorrhage. Thus, as the kidneys do what they can to maintain homeostasis, urine composition varies greatly.

When the ECF has a surplus of water or a particular electrolyte such as salt, the kidneys can eliminate the excess in the urine. If a deficit exists, the kidneys cannot provide additional quantities of the depleted constituent, but can limit urinary losses of the material in short supply and thus conserve it until the person can take in more of the depleted substance. Accordingly, the kidneys can compensate more efficiently for excesses than for deficits. In fact, in some instances the kidneys cannot halt the loss of a valuable substance in the urine, even though the substance may be in short supply. A prime example is a water deficit. Even if a person is not consuming any water, the kidneys must put out about half a liter of water in the urine each day to fill another major role as the body's "cleaners."

In addition to the kidneys' important regulatory role in maintaining fluid and electrolyte balance, they are the main route for eliminating potentially toxic metabolic wastes and foreign compounds from the body. These wastes cannot be eliminated as solids; they must be excreted in solution, thus obligating the kidneys to produce a minimum volume of around 500 mL of waste-filled urine per day. Because water eliminated in the urine is derived from the blood plasma, a person stranded without water eventually urinates to death: The plasma volume falls to a fatal level as water is unavoidably removed to accompany the wastes.

Overview of Kidney Functions The kidneys perform the following specific functions, most of which help preserve constancy of the internal fluid environment and most of which will be discussed in this and the next chapter:

1. *Maintaining water (H$_2$O) balance in the body.*

2. *Maintaining the proper osmolarity of body fluids, primarily through regulating H$_2$O balance.* This function prevents osmotic fluxes into or out of the cells, which could lead to detrimental swelling or shrinking of the cells, respectively. Brain cells are particularly sensitive to volume changes.

3. *Regulating the quantity and concentration of most ECF ions,* including sodium (Na$^+$), chloride (Cl$^-$), potassium (K$^+$), calcium (Ca^{2+}), hydrogen ion (H$^+$), bicarbonate (HCO$_3^-$), phosphate (PO$_4^{3-}$), sulfate (SO$_4^{2-}$), and magnesium (Mg^{2+}). Even minor fluctuations in the ECF concentrations of some of these electrolytes can have profound influences. For example, changes in the ECF concentration of K$^+$ can potentially lead to fatal cardiac dysfunction.

4. *Maintaining proper plasma volume,* which is important in the long-term regulation of arterial blood pressure. This function is accomplished through the kidneys' regulatory role in salt (NaCl) and H$_2$O balance.

5. *Helping maintain the proper acid–base balance* of the body by adjusting urinary output of H$^+$ and HCO$_3^-$.

6. *Excreting (eliminating) the end products (wastes) of bodily metabolism,* such as urea (from proteins), uric acid (from nucleic acids), creatinine (from muscle creatine), bilirubin (from hemoglobin), and hormone metabolites. If allowed to accumulate, many of these wastes are toxic, especially to the brain.

7. *Excreting many foreign compounds,* such as drugs, food additives, pesticides, and other exogenous nonnutritive materials that have entered the body.

8. *Producing renin,* an enzymatic hormone that triggers a chain reaction important in salt conservation by the kidneys.

9. *Producing erythropoietin,* a hormone that stimulates red blood cell production (see Chapter 11).

10. *Converting vitamin D into its active form* (see Chapter 19).

The kidneys form urine; the rest of the urinary system carries it to the outside.

The **urinary system** consists of the urine-forming organs—the kidneys—and the structures that carry the urine from the kidneys to the outside for elimination from the body (▮Figure 14-1a). The kidneys are a pair of bean-shaped organs about 4 to 5 inches long that lie behind the abdominal cavity (between the abdominal cavity and the back muscles), one on each side of the vertebral column, slightly above the waistline. Each kidney is supplied by a **renal artery** and a **renal vein**, which, respectively, enters and leaves the kidney at the medial indentation that gives this organ its beanlike form. The kidney acts on the plasma flowing through it to produce urine, conserving materials to be retained in the body and eliminating unwanted materials into the urine.

After urine is formed, it drains into a central collecting cavity, the **renal pelvis,** located at the medial inner core of each kidney (▮Figure 14-1b). From there urine is channeled into the **ureter,** a duct that exits at the medial border close to the renal artery and vein. There are two ureters, one carrying urine from each kidney to the single urinary bladder.

The **urinary bladder,** which temporarily stores urine, is a hollow, distensible, smooth muscle–walled sac. Periodically,

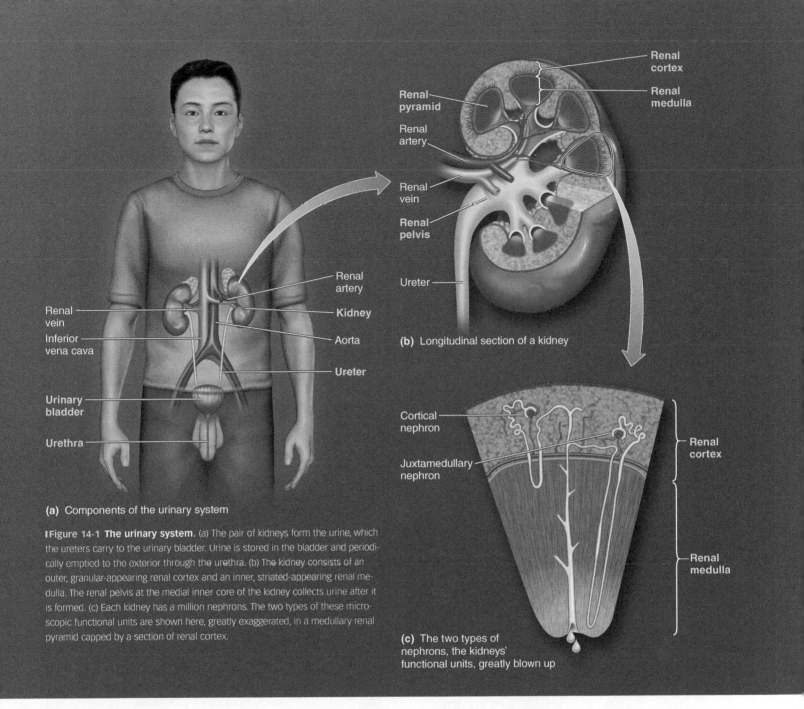

(a) Components of the urinary system

Renal artery
Kidney
Aorta
Ureter
Renal vein
Inferior vena cava
Urinary bladder
Urethra

▮Figure 14-1 The urinary system. (a) The pair of kidneys form the urine, which the ureters carry to the urinary bladder. Urine is stored in the bladder and periodically emptied to the exterior through the urethra. (b) The kidney consists of an outer, granular-appearing renal cortex and an inner, striated-appearing renal medulla. The renal pelvis at the medial inner core of the kidney collects urine after it is formed. (c) Each kidney has a million nephrons. The two types of these microscopic functional units are shown here, greatly exaggerated, in a medullary renal pyramid capped by a section of renal cortex.

Renal cortex
Renal medulla
Renal pyramid
Renal artery
Renal vein
Renal pelvis
Ureter

(b) Longitudinal section of a kidney

Cortical nephron
Juxtamedullary nephron
Renal cortex
Renal medulla

(c) The two types of nephrons, the kidneys' functional units, greatly blown up

urine is emptied from the bladder to the outside through another tube, the **urethra**, as a result of bladder contraction. The urethra in females is straight and short, passing directly from the neck of the bladder to the outside (▮Figure 14-2a; see also ▮Figure 20-2, p. 719). In males, the urethra is longer and follows a curving course from the bladder to the outside, passing through both the prostate gland and the penis (see ▮Figures 14-1a and 14-2b; see also ▮Figure 20-1, p. 717). The male urethra serves the dual function of providing both a route for eliminating urine from the bladder and a passageway for semen from the reproductive organs. The prostate gland lies below the neck of the bladder and completely encircles the urethra.

 Clinical Note Prostatic enlargement, which often occurs during middle to older age, can partially or completely occlude the urethra, impeding the flow of urine.

The parts of the urinary system beyond the kidneys merely serve as "ductwork" to transport urine to the outside. Once formed by the kidneys, urine is not altered in composition or volume as it moves downstream through the rest of the tract.

The nephron is the functional unit of the kidney.

Each kidney consists of about 1 million microscopic functional units known as **nephrons**, which are bound together by connective tissue (see ▮Figure 14-1c). Recall that a functional unit is the smallest unit within an organ capable of performing all of that organ's functions. Because the main function of the kidneys is to produce urine and, in so doing, maintain constancy in the

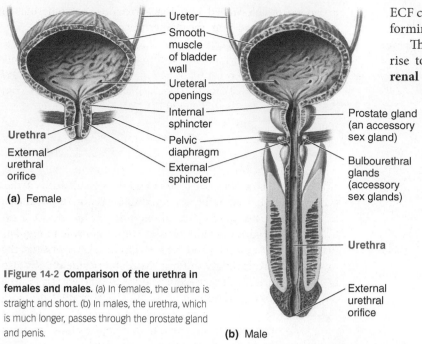

Ureter
Smooth muscle of bladder wall
Ureteral openings
Internal sphincter
Urethra
Pelvic diaphragm
External sphincter
External urethral orifice

(a) Female

Prostate gland (an accessory sex gland)
Bulbourethral glands (accessory sex glands)
Urethra
External urethral orifice

(b) Male

❙Figure 14-2 Comparison of the urethra in females and males. (a) In females, the urethra is straight and short. (b) In males, the urethra, which is much longer, passes through the prostate gland and penis.

ECF composition, a nephron is the smallest unit capable of forming urine.

The arrangement of nephrons within the kidneys gives rise to two distinct regions—an outer region called the **renal cortex**, which looks granular, and an inner region, the **renal medulla**, which is made up of striated triangles, the **renal pyramids** (see ❙Figure 14-1b and c).

Knowledge of the structural arrangement of an individual nephron is essential for understanding the distinction between the cortical and the medullary regions of the kidney and, more important, for understanding renal function. Each nephron consists of a *vascular component* and a *tubular component,* which are intimately related structurally and functionally (❙Figure 14-3).

Vascular Component of the Nephron The dominant part of the nephron's vascular component is the **glomerulus**, a ball-like tuft of capillaries through which part of the water and solutes is filtered from blood passing through (❙Figure 14-4 and chapter opener photo). This filtered

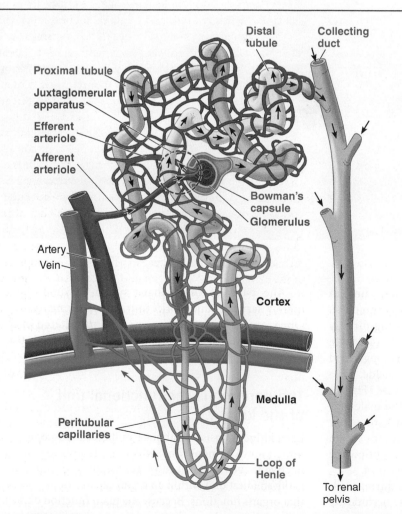

Distal tubule
Collecting duct
Proximal tubule
Juxtaglomerular apparatus
Efferent arteriole
Afferent arteriole
Artery
Vein
Bowman's capsule
Glomerulus
Cortex
Medulla
Peritubular capillaries
Loop of Henle
To renal pelvis

Overview of Functions of Parts of a Nephron

Vascular component
- Afferent arteriole—carries blood to the glomerulus
- Glomerulus—a tuft of capillaries that filters a protein-free plasma into the tubular component
- Efferent arteriole—carries blood from the glomerulus
- Peritubular capillaries—supply the renal tissue; involved in exchanges with the fluid in the tubular lumen

Tubular component
- Bowman's capsule—collects the glomerular filtrate
- Proximal tubule—uncontrolled reabsorption and secretion of selected substances occur here
- Loop of Henle (of juxtamedullary nephrons only; not shown)—establishes an osmotic gradient in the renal medulla that is important in the kidney's ability to produce urine of varying concentration
- Distal tubule and collecting duct— variable, controlled reabsorption of Na^+ and H_2O and secretion of K^+ and H^+ occur here; fluid leaving the collecting duct is urine, which enters the renal pelvis

Combined vascular/tubular component
- Juxtaglomerular apparatus—produces substances involved in the control of kidney function

❙Figure 14-3 A nephron. Components of a cortical nephron, the most abundant type of nephron in humans.

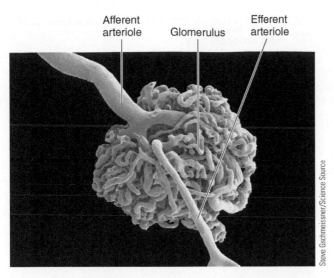

Afferent arteriole   Glomerulus   Efferent arteriole

Steve Gschmeissner/Science Source

▌Figure 14-4 **Scanning electron micrograph of a glomerulus and associated arterioles.**

FIGURE FOCUS: *Compare the diameter of the afferent arteriole and the efferent arteriole. Remember this size discrepancy; it plays a key role in kidney function, as you will learn later in this chapter.*

fluid, which is almost identical in composition to plasma, then passes through the nephron's tubular component, where various transport processes convert it into urine.

On entering the kidney, the renal artery subdivides to ultimately form many small vessels known as **afferent arterioles**, one of which supplies each nephron. The afferent arteriole delivers blood to the glomerulus. The glomerular capillaries rejoin to form another arteriole, the **efferent arteriole**, through which blood that was not filtered into the tubular component leaves the glomerulus (see ▌Figures 14-3 and 14-4). The efferent arterioles are the only arterioles in the body that drain from capillaries. Typically, arterioles break up into capillaries that rejoin to form venules. At the glomerular capillaries, no O_2 or nutrients are extracted from the blood for use by the kidney tissues, nor are waste products picked up from the surrounding tissue. Thus, arterial blood enters the glomerular capillaries through the afferent arteriole, and arterial blood leaves the glomerulus through the efferent arteriole.

The efferent arteriole subdivides into a second set of capillaries, the **peritubular capillaries**, which supply the renal tissue with blood and are important in exchanges between the tubular system and blood during conversion of the filtered fluid into urine. These peritubular capillaries, as their name implies, are intertwined around the tubular system (*peri* means "around"). The peritubular capillaries rejoin to form venules that ultimately drain into the renal vein, by which blood leaves the kidney.

Tubular Component of the Nephron The nephron's tubular component is a hollow, fluid-filled tube formed by a single layer of epithelial cells. Even though the tubule is continuous from its beginning near the glomerulus to its ending at the renal pelvis, it is arbitrarily divided into various segments based on differences in structure and function along its length (see ▌Figure 14-3). The tubular component begins with **Bowman's capsule**, an expanded, double-walled "cup" that surrounds the glomerulus to collect fluid filtered from the glomerular capillaries.

From Bowman's capsule, the filtered fluid passes into the **proximal tubule**, which lies within the cortex and is highly coiled or convoluted throughout much of its course. The next segment, the **loop of Henle**, forms a sharp U-shaped or hairpin loop that dips into the renal medulla. The *descending limb* of the loop of Henle plunges from the cortex into the medulla; the *ascending limb* traverses back up into the cortex. The ascending limb returns to the glomerular region of the same nephron, where it passes through the fork formed by the afferent and efferent arterioles. Both the tubular and the vascular cells at this point are specialized to form the **juxtaglomerular apparatus**, a structure that lies next to the glomerulus (*juxta* means "next to"). This specialized region plays an important role in regulating kidney function. Beyond the juxtaglomerular apparatus, the tubule again coils tightly to form the **distal tubule**, which also lies entirely within the cortex. The distal tubule empties into a **collecting duct** or **tubule**, with each collecting duct draining fluid from up to eight separate nephrons. Each collecting duct plunges down through the medulla to empty its fluid contents (now converted into urine) into the renal pelvis.

Cortical and Juxtamedullary Nephrons Two types of nephrons—*cortical nephrons* and *juxtamedullary nephrons*—are distinguished by the location and length of some of their structures (▌Figures 14-1 and 14-5). All nephrons originate in the cortex, but the glomeruli of **cortical nephrons** lie in the outer layer of the cortex, whereas the glomeruli of **juxtamedullary nephrons** lie in the inner layer of the cortex, next to the medulla. The presence of all glomeruli and associated Bowman's capsules in the cortex is responsible for this region's granular appearance. These two nephron types differ most markedly in their loops of Henle. The hairpin loop of cortical nephrons dips only slightly into the medulla. In contrast, the loop of juxtamedullary nephrons plunges through the entire depth of the medulla. Furthermore, the peritubular capillaries of juxtamedullary nephrons form hairpin vascular loops known as **vasa recta** ("straight vessels"), which run in close association with the long loops of Henle. In cortical nephrons, the peritubular capillaries do not form vasa recta but instead entwine around these nephrons' short loops of Henle in the same manner as the peritubular capillaries wrap around the proximal and distal tubules in both types of nephrons. As they course through the medulla, the collecting ducts of both cortical and juxtamedullary nephrons run parallel to the ascending and descending limbs of the juxtamedullary nephrons' long loops of Henle and vasa recta. The parallel arrangement of tubules and vessels in the medulla creates this region's striated appearance. More important, as you will see, this arrangement—coupled with the permeability and transport characteristics of the long loops of Henle and vasa recta—plays a key role in the kidneys' ability to produce urine of varying concentrations, depending on the needs of the body. About 80% of the nephrons in humans are of the cortical type. Species with greater urine-concentrating abilities than humans, such as the desert rat, have a greater proportion of juxtamedullary nephrons.

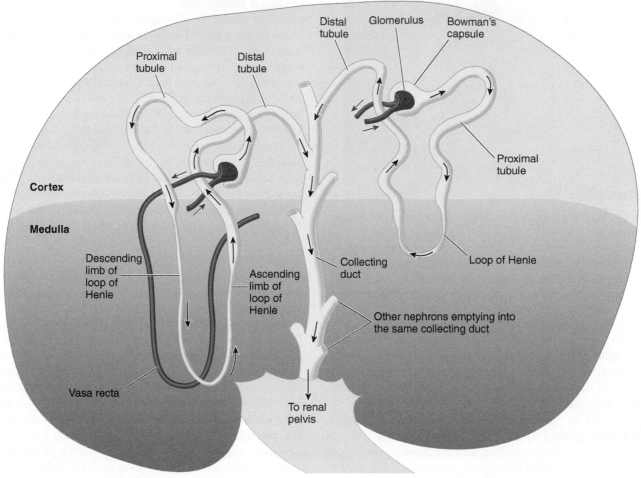

Juxtamedullary nephron:
long-looped nephron important in
establishing the medullary vertical
osmotic gradient (20% this type)

Cortical nephron:
most abundant type
of nephron (80% this type)

Distal tubule

Glomerulus

Bowman's capsule

Proximal tubule

Distal tubule

Proximal tubule

Cortex

Medulla

Descending limb of loop of Henle

Ascending limb of loop of Henle

Collecting duct

Loop of Henle

Other nephrons emptying into the same collecting duct

Vasa recta

To renal pelvis

❙Figure 14-5 Comparison of juxtamedullary and cortical nephrons. The glomeruli of cortical nephrons lie in the outer cortex, whereas the glomeruli of juxtamedullary nephrons lie in the inner part of the cortex next to the medulla. The loops of Henle of cortical nephrons dip only slightly into the medulla, but the juxtamedullary nephrons have long loops of Henle that plunge deep into the medulla. The juxtamedullary nephrons' peritubular capillaries form hairpin loops known as *vasa recta*. (For better visualization, the kidney is rotated 90° from its normal position in an upright person, the nephrons are grossly exaggerated in size, and the peritubular capillaries have been omitted, except for the vasa recta.)

The three basic renal processes are glomerular filtration, tubular reabsorption, and tubular secretion.

Three basic processes are involved in forming urine: *glomerular filtration, tubular reabsorption,* and *tubular secretion.* To aid in visualizing the relationships among these renal processes, it is useful to unwind the nephron schematically, as in ❙Figure 14-6.

Glomerular Filtration As blood flows through the glomerulus, protein-free plasma filters through the glomerular capillaries into Bowman's capsule. Normally, about 20% of the plasma that enters the glomerulus is filtered. This process, known as **glomerular filtration**, is the first step in urine formation. On average, 125 mL of glomerular filtrate (filtered fluid) are formed collectively through all the glomeruli each minute. This amounts to 180 liters (about 47.5 gallons) each day. Considering that the

average plasma volume in an adult is 2.75 liters, this means that the kidneys filter the entire plasma volume about 65 times per day. If everything filtered passed out in the urine, the total plasma volume would be urinated in less than half an hour! This does not happen, however, because the kidney tubules and peritubular capillaries are intimately related throughout their lengths so that the tubular cells can transfer materials as needed between the fluid inside the tubules and the blood within the peritubular capillaries.

Tubular Reabsorption As the filtrate flows through the tubules, substances of value to the body are returned to the peritubular capillary plasma. This selective movement of substances from inside the tubule (the tubular lumen) into the blood is called **tubular reabsorption**. Reabsorbed substances are not lost from the body in the urine but instead are carried by the peritubular capillaries to the venous system and then to

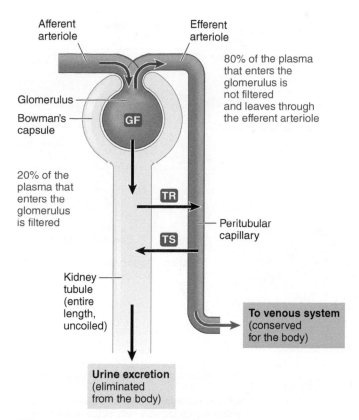

Afferent arteriole

Efferent arteriole

Glomerulus

Bowman's capsule

80% of the plasma that enters the glomerulus is not filtered and leaves through the efferent arteriole

GF

20% of the plasma that enters the glomerulus is filtered

TR

TS

Peritubular capillary

Kidney tubule (entire length, uncoiled)

To venous system (conserved for the body)

Urine excretion (eliminated from the body)

GF = **Glomerular filtration**—nondiscriminant filtration of a protein-free plasma from the glomerulus into Bowman's capsule

TR = **Tubular reabsorption**—selective movement of filtered substances from the tubular lumen into the peritubular capillaries

TS = **Tubular secretion**—selective movement of nonfiltered substances from the peritubular capillaries into the tubular lumen

❙Figure 14-6 **Basic renal processes.** Anything filtered or secreted but not reabsorbed is excreted in the urine and lost from the body. Anything filtered and subsequently reabsorbed, or not filtered at all, enters the venous blood and is saved for the body.

FIGURE FOCUS: *Name two ways that substances can enter and two ways that substances can leave the tubular fluid.*

the heart to be recirculated. Of the 180 liters of plasma filtered per day, 178.5 liters, on average, are reabsorbed. The remaining 1.5 liters of filtered fluid left in the tubules pass into the renal pelvis to be eliminated as urine. In general, substances the body needs to conserve are selectively reabsorbed, whereas unwanted substances that must be eliminated stay in the tubular fluid, which becomes urine after tubular modification is complete.

Tubular Secretion The third renal process, **tubular secretion**, is the selective transfer of substances from the peritubular capillary blood into the tubular lumen. It provides a second route for substances to enter the renal tubules from the blood, the first being by glomerular filtration. Only about 20% of the plasma flowing through the glomerular capillaries is filtered into Bowman's capsule; the remaining 80% flows on through the efferent arteriole into the peritubular capillaries. Tubular secretion provides a mechanism for more rapidly eliminating selected substances from the plasma by extracting an additional

quantity of a particular substance from the 80% of unfiltered plasma in the peritubular capillaries and adding it to the quantity of the substance already present in the tubule as a result of filtration.

Urine Excretion **Urine excretion** is the elimination of substances from the body in the urine. It is not a separate process but the result of the first three processes. All plasma constituents filtered or secreted but not reabsorbed remain in the tubules and pass into the renal pelvis to be excreted as urine and eliminated from the body (❙Figure 14-6). (Do not confuse excretion with secretion.) Note that anything filtered and subsequently reabsorbed, or not filtered at all, enters the venous blood from the peritubular capillaries and thus is conserved for the body instead of being excreted in the urine, despite passing through the kidneys.

The Big Picture of the Basic Renal Processes Glomerular filtration is largely an indiscriminate process. With the exception of blood cells and plasma proteins, all constituents within the blood—H_2O, nutrients, electrolytes, wastes, and so on—nonselectively enter the tubular lumen as a bulk unit during filtration—that is, of the 20% of the plasma filtered at the glomerulus, everything in that part of the plasma enters Bowman's capsule except for the plasma proteins. The highly discriminating tubular processes then work on the filtrate to return to the blood a fluid of the composition and volume necessary to maintain constancy of the internal fluid environment. The unwanted filtered material is left behind in the tubular fluid to be excreted as urine. Glomerular filtration can be thought of as pushing a part of the plasma, with all its essential components and those that need to be eliminated from the body, onto a tubular "conveyor belt" that terminates at the renal pelvis, which is the collecting point for urine within the kidney. All plasma constituents that enter this conveyor belt and are not subsequently returned to the plasma by the end of the line are spilled out of the kidney as urine. It is up to the tubular system to salvage by reabsorption the filtered materials that need to be preserved for the body while leaving behind substances that must be excreted. In addition, some substances not only are filtered, but also are secreted onto the tubular conveyor belt, so the amounts of these substances excreted in the urine are greater than the amounts that were filtered. For many substances, these renal processes are subject to physiologic control. Thus, the kidneys handle each constituent in the plasma by a particular combination of filtration, reabsorption, and secretion.

The kidneys act only on the plasma, yet the ECF consists of both plasma and interstitial fluid. The interstitial fluid is the true internal fluid environment of the body because it is the only component of the ECF that comes into direct contact with the cells. However, because of the free exchange between plasma and interstitial fluid across the capillary walls (with the exception of plasma proteins), interstitial fluid composition reflects the composition of plasma. Thus, by performing their regulatory and excretory roles on plasma, the kidneys maintain the proper interstitial fluid environment for optimal cell function. Most of the rest of this chapter is devoted to considering how the basic renal processes are accomplished and the mecha-

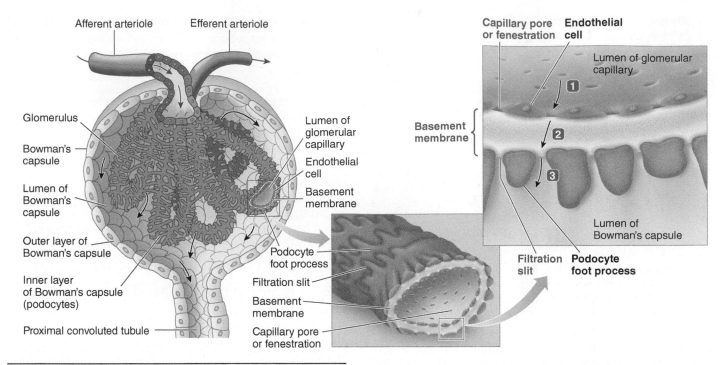

Afferent arteriole   *Efferent arteriole*

Glomerulus

Bowman's
capsule

Lumen of
Bowman's
capsule

Outer layer of
Bowman's capsule

Inner layer
of Bowman's capsule
(podocytes)

Proximal convoluted tubule

Lumen of
glomerular
capillary

Endothelial
cell

Basement
membrane

Podocyte
foot process

Filtration slit

Basement
membrane

Capillary pore
or fenestration

Capillary pore   **Endothelial
or fenestration   cell**

Lumen of glomerular
capillary

**Basement
membrane**

Lumen of
Bowman's capsule

**Filtration   Podocyte
slit   foot process**

To be filtered, a substance must pass through

1 the pores between and the fenestrations within the endothelial cells of the glomerular capillary

2 an acellular basement membrane

3 the filtration slits between the foot processes of the podocytes in the inner layer of Bowman's capsule

❙Figure 14-7 Layers of the glomerular membrane.

nisms by which they are carefully regulated to help maintain homeostasis.

Check Your Understanding 14.1

1. Name and describe the functional unit of the kidneys.
2. Schematically draw an "unwound" nephron and use arrows to show the direction of movement between its vascular and tubular components during the three basic renal processes.
3. Distinguish between cortical and juxtamedullary nephrons.

14.2 Glomerular Filtration

Fluid filtered from the glomerulus into Bowman's capsule must pass through the three layers that make up the **glomerular membrane** (❙Figure 14-7): (1) the glomerular capillary wall, (2) the basement membrane, and (3) the inner layer of Bowman's capsule. Collectively, these layers function as a fine molecular sieve that retains the blood cells and plasma proteins but permits H_2O and solutes of small molecular dimension to filter through. Let us consider each layer in more detail.

The glomerular membrane is considerably more permeable than capillaries elsewhere.

The *glomerular capillary wall* consists of a single layer of flattened endothelial cells. It is perforated by many large pores that make it more than 100 times more permeable to H_2O and solutes than capillaries elsewhere in the body. The glomerular capillaries not only have the traditional pores found between the endothelial cells that form the capillary walls, but the endothelial cells themselves also are perforated by large holes or fenestrations (see p. 353).

The *basement membrane* is an acellular (lacking cells) gelatinous layer composed of collagen and glycoproteins that is sandwiched between the glomerulus and Bowman's capsule. The collagen provides structural strength, and the glycoproteins discourage the filtration of small plasma proteins. The larger plasma proteins cannot be filtered because they cannot fit through the capillary pores, but the pores are just barely large enough to permit passage of albumin, the smallest of plasma proteins. However, because the glycoproteins are negatively charged, they repel albumin and other plasma proteins, which are also negatively charged. Therefore, plasma proteins are almost completely excluded from the filtrate, with less than 1% of the albumin molecules escaping into Bowman's capsule. The small proteins that do slip into the filtrate are picked up by the proximal tubule by receptor-mediated endocytosis (see p. 31), then degraded into constituent amino acids that are returned to the blood. Thus, urine is normally protein free.

Clinical Note Some renal diseases characterized by excessive albumin in the urine (*albuminuria*) are the result of disruption of the negative charges within the basement membrane, which makes the glomerular membrane more permeable to albumin even though the size of the capillary pores remains constant. (Urinary loss of proteins can also fol-

Unless otherwise noted, all content on this page is © Cengage Learning.

When Protein in the Urine Does Not Mean Kidney Disease

URINARY LOSS OF PROTEINS (MOSTLY albumin) usually signifies kidney disease *(nephritis)*. However, a urinary protein loss similar to that of nephritis often occurs following exercise, but the condition is harmless, transient, and reversible. The term *athletic pseudonephritis* (*pseudo* means "false") is used to describe this postexercise (after exercise) proteinuria (protein in the urine). Studies indicate that 70% to 80% of athletes have proteinuria after very strenuous exercise. This condition occurs in participants in both noncontact and contact sports, so it does not arise from physical trauma to the kidneys.

Usually, only a very small fraction of the plasma proteins that enter the glomerulus is filtered; these filtered plasma proteins are subsequently reabsorbed in the tubules, so normally no plasma proteins appear in the urine. Two basic mechanisms can cause proteinuria: (1) increased glomerular permeability with no change in tubular reabsorption or (2) impairment of tubular reabsorption. Research has shown that the proteinuria occurring during mild to moderate exercise results from changes in glomerular permeability, whereas the proteinuria occurring during short-term exhaustive exercise is caused by both increased glomerular permeability and tubular dysfunction.

This reversible kidney dysfunction is believed to result from circulatory and hormonal changes that occur with exercise. Renal blood flow is reduced during exercise as the renal vessels are constricted and blood is diverted to the exercising muscles. This reduction is positively correlated with exercise intensity. With intense exercise, renal blood flow may be reduced to 20% of normal. As a result, glomerular blood flow is also reduced, but not to the same extent as renal blood flow, because of autoregulation (see p. 501). Investigators propose that decreased glomerular blood flow enhances diffusion of proteins into the tubular lumen because as the more slowly flowing blood spends more time in the glomerulus, a greater proportion of the plasma proteins have time to escape through the glomerular membrane. Hormonal changes that occur with exercise may also affect glomerular permeability. For example, injection of the kidney hormone renin is a well-recognized way to experimentally induce proteinuria. Plasma renin activity increases during strenuous exercise and may contribute to postexercise proteinuria. Researchers also hypothesize that maximal tubular reabsorption is reached during severe exercise, which could impair protein reabsorption.

low exercise, but it is transient and harmless. For further discussion, see the accompanying boxed feature, ▮A Closer Look at Exercise Physiology.)

The final layer of the glomerular membrane is the *inner layer of Bowman's capsule*. It consists of **podocytes**, octopuslike epithelial cells that encircle the glomerular tuft. A podocyte bears multiple elongated primary foot processes (*podo* means "foot"; a *process* is a projection or appendage), each of which has many side branches, or secondary foot processes, protruding from it to the right and to the left, similar to the fronds of a fern plant. The secondary foot processes of one podocyte interdigitate with the secondary foot processes of adjacent podocytes as they cup around a glomerular capillary, much as you interlace your fingers when you cup your hands around a ball (▮Figure 14-8). The narrow slits between the interdigitating secondary foot processes of adjacent podocytes are known as **filtration slits**, which provide a pathway through which fluid leaving the glomerular capillaries can enter the lumen of Bowman's capsule.

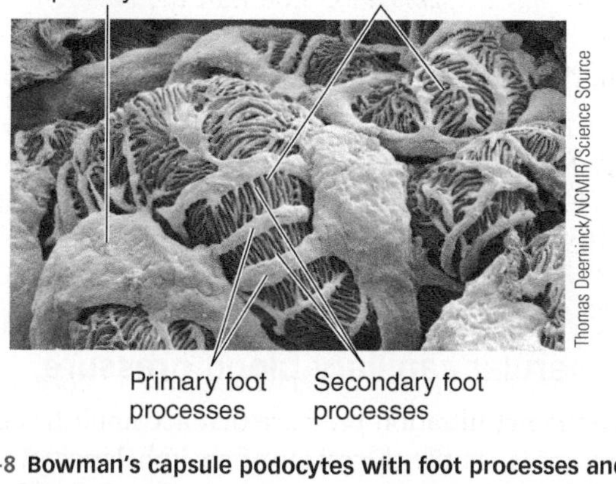

Thus, the route that filtered substances take across the glomerular membrane is completely extracellular—first through capillary pores, then through the acellular basement membrane, and finally through capsular filtration slits (see ▮Figure 14-7).

Glomerular capillary blood pressure is the major force that causes glomerular filtration.

To accomplish glomerular filtration, a force must drive a part of the plasma in the glomerulus through the openings in the glomerular membrane. No local energy is used to move fluid from the plasma across the glomerular membrane into Bowman's capsule. Passive physical forces similar to those acting across capillaries elsewhere accomplish glomerular filtration.

Because the glomerulus is a tuft of capillaries, the same principles of fluid dynamics apply here that cause ultrafiltration across other capillaries (see p. 356), except for two important differences: (1) The glomerular capillaries are more permeable than capillaries elsewhere, so more fluid is filtered for a given

Cell body of podocyte

Filtration slits

Primary foot processes

Secondary foot processes

Thomas Deerinck/NCMIR/Science Source

▮Figure 14-8 **Bowman's capsule podocytes with foot processes and filtration slits.** Note the filtration slits between the fine secondary foot processes of adjacent podocytes on this scanning electron micrograph. The podocytes and their foot processes encircle the glomerular capillaries.

filtration pressure, and (2) the balance of forces across the glomerular membrane is such that filtration occurs the entire length of the capillaries. In contrast, the balance of forces in other capillaries shifts so that filtration occurs in the beginning part of the vessel but reabsorption occurs toward the vessel's end (see Figure 10-20, p. 356).

Forces Involved in Glomerular Filtration Three physical forces are involved in glomerular filtration: glomerular capillary blood pressure, plasma-colloid osmotic pressure, and Bowman's capsule hydrostatic pressure. Glomerular capillary blood pressure *favors* filtration, whereas the two other forces acting across the glomerular membrane *oppose* filtration, as follows (Table 14-1):

1. *Glomerular capillary blood pressure* is the fluid (hydrostatic) pressure exerted by the blood within the glomerular capillaries. It ultimately depends on contraction of the heart (the source of energy that produces glomerular filtration) and the resistance to blood flow offered by the afferent and efferent arterioles. Glomerular capillary blood pressure, at an estimated average value of 55 mm Hg, is higher than capillary blood pressure elsewhere. The reason for the higher pressure is the larger diameter of the afferent arteriole compared to that of the efferent arteriole (see Figure 14-4, p. 495). Because blood can flow more rapidly into the glomerulus through the wide afferent arteriole than it can leave through the narrower efferent arteriole, glomerular capillary blood pressure is maintained high as a result of blood damming up in the glomerular capillaries. Also, because of the high resistance offered by the efferent arterioles, blood pressure does not fall along the length of the glomerular capillaries as it does along other capillaries. This elevated, nondecremental glomerular blood pressure tends to push fluid out of the glomerulus into Bowman's capsule along the glomerular capillaries' entire length, and it is the major force producing glomerular filtration.

2. *Plasma-colloid osmotic pressure* (π_P) is caused by the unequal distribution of plasma proteins across the glomerular membrane. Because plasma proteins cannot be filtered, they are in the glomerular capillaries but not in Bowman's capsule. Accordingly, the concentration of H_2O is higher in Bowman's capsule than in the glomerular capillaries. The resulting tendency for H_2O to move by osmosis down its concentration gradient from Bowman's capsule into the glomerulus opposes glomerular filtration. This opposing osmotic force averages 30 mm Hg, which is slightly higher than across other capillaries. It is higher because more H_2O is filtered out of the glomerular blood, so the concentration of plasma proteins is higher than elsewhere.

3. *Bowman's capsule hydrostatic pressure,* the pressure exerted by the fluid in this initial part of the tubule, is estimated to be about 15 mm Hg. This pressure, which tends to push fluid out of Bowman's capsule, opposes the filtration of fluid from the glomerulus into Bowman's capsule.

Glomerular Filtration Rate As can be seen in Table 14-1, the forces acting across the glomerular membrane are not in balance. The total force favoring filtration is the glomerular capillary blood pressure at 55 mm Hg. The total of the two forces opposing filtration is 45 mm Hg. The net difference favoring filtration (10 mm Hg of pressure) is called the **net filtration pressure**. This modest pressure forces large volumes of fluid from the blood through the highly permeable glomerular membrane. The actual rate of filtration, the **glomerular filtration rate (GFR)**, depends not only on the net filtration pressure, but also on how much glomerular surface area is available for penetration and how permeable the glomerular membrane is (that is, how "holey" it is). These properties of the glomerular membrane are collectively referred to as the **filtration coefficient (K_f)**. Accordingly,

$$\text{GFR} = K_f \times \text{net filtration pressure}$$

Normally, about 20% of the plasma that enters the glomerulus is filtered at the net filtration pressure of 10 mm Hg, producing collectively through all glomeruli 180 liters of glomerular filtrate each day for an average GFR of 125 mL/min in males (160 L/day, 115 mL/min in females).

Changes in GFR result mainly from changes in glomerular capillary blood pressure.

Because the net filtration pressure that accomplishes glomerular filtration is simply the result of an imbalance of opposing physical forces between the glomerular capillary plasma and Bowman's capsule fluid, alterations in any of these physical forces can affect the GFR, as discussed next.

TABLE 14-1 Forces Involved in Glomerular Filtration

Force	Effect	Magnitude (mm Hg)
Glomerular capillary blood pressure	Favors filtration	55
Plasma-colloid osmotic pressure	Opposes filtration	30
Bowman's capsule hydrostatic pressure	Opposes filtration	15
Net filtration pressure (difference between force favoring filtration and forces opposing filtration)	Favors filtration	10

Unregulated Influences on the GFR Plasma-colloid osmotic pressure and Bowman's capsule hydrostatic pressure normally do not vary much and cannot be regulated.

Clinical Note However, these forces can change pathologically and thus inadvertently affect the GFR. Because π_P opposes filtration, a decrease in plasma protein concentration, by reducing this pressure, leads to an increased GFR. Plasma protein concentration might uncontrollably drop, for example, in severely burned patients who lose a large quantity of protein-rich, plasma-derived fluid through the exposed burned surface of their skin. Conversely, when π_P rises, such as in cases of dehydrating diarrhea, the GFR falls.

Bowman's capsule hydrostatic pressure can become uncontrollably elevated, and filtration subsequently can decrease, given a urinary tract obstruction, such as a kidney stone or enlarged prostate. The damming up of fluid behind the obstruction elevates capsular hydrostatic pressure.

Controlled Adjustments in the GFR Unlike plasma-colloid osmotic pressure and Bowman's capsule hydrostatic pressure—which may be uncontrollably altered in various disease states and thereby may inappropriately alter the GFR—glomerular capillary blood pressure can be controlled to adjust the GFR to suit the body's needs. Assuming that all other factors stay constant, as glomerular capillary blood pressure rises, net filtration pressure goes up and the GFR increases correspondingly. The magnitude of the glomerular capillary blood pressure depends on the rate of blood flow within each of the glomeruli. The amount of blood flowing into a glomerulus per minute is determined largely by the magnitude of the mean systemic arterial blood pressure and the resistance offered by the afferent arteriole. If afferent arteriolar resistance increases, less blood flows into the glomerulus, decreasing the GFR. Conversely, if afferent arteriolar resistance decreases, more blood flows into the glomerulus and the GFR increases. Two major control mechanisms regulate the GFR, both directed at adjusting glomerular blood flow by regulating the radius and thus the resistance of the afferent arteriole. These mechanisms are (1) autoregulation, which is aimed at preventing spontaneous changes in GFR; and (2) extrinsic sympathetic control, which is aimed at long-term regulation of arterial blood pressure.

Mechanisms Responsible for Autoregulation of the GFR Because arterial blood pressure is the force that drives blood into the glomerulus, the glomerular capillary blood pressure and, accordingly, the GFR would increase in direct proportion to an increase in arterial pressure if everything else remained constant (Figure 14-9). Similarly, a fall in arterial blood pressure would cause a decline in GFR. Such spontaneous, inadvertent changes in GFR are largely prevented by intrinsic regulatory mechanisms initiated by the kidneys themselves, a process known as **autoregulation** (*auto* means "self"). The kidneys can, within limits, maintain a constant blood flow into the glomerular capillaries (and thus a constant glomerular capillary blood pressure and a stable GFR) despite changes in the driving arterial pressure. They do so primarily by altering

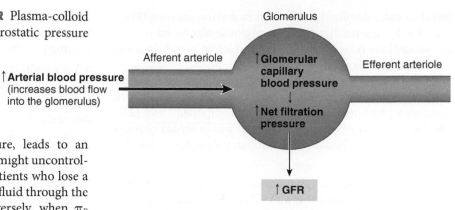

Figure 14-9 Direct effect of arterial blood pressure on the glomerular filtration rate (GFR).

afferent arteriolar caliber, thereby adjusting resistance to flow through these vessels. For example, if the GFR increases as a direct result of a rise in arterial pressure, the net filtration pressure and GFR can be reduced to normal by constriction of the afferent arteriole, which decreases the flow of blood into the glomerulus (Figure 14-10a). This local adjustment lowers glomerular blood pressure and GFR to normal.

Conversely, when GFR falls in the presence of a decline in arterial pressure, glomerular pressure can be increased to normal by vasodilation of the afferent arteriole, which allows more

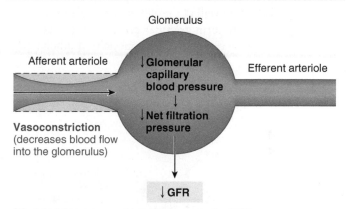

(a) Arteriolar vasoconstriction decreases the GFR

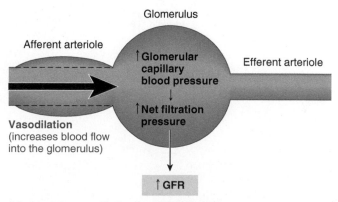

(b) Arteriolar vasodilation increases the GFR

Figure 14-10 Adjustments of afferent arteriole caliber to alter the GFR.

blood to enter despite the reduction in driving pressure (Figure 14-10b). The resultant buildup of glomerular blood volume increases glomerular blood pressure, which in turn brings the GFR back up to normal.

Two mechanisms contribute to autoregulation of the GFR: (1) a *myogenic* mechanism, which responds to changes in pressure within the nephron's vascular component; and (2) a *tubuloglomerular feedback* mechanism, which senses changes in salt level in the fluid flowing through the nephron's tubular component.

- The **myogenic** mechanism is a common property of vascular smooth muscle (*myogenic* means "muscle produced"). Arteriolar vascular smooth muscle contracts inherently in response to the stretch accompanying increased pressure within the vessel (see p. 348). Accordingly, the afferent arteriole automatically constricts on its own when it is stretched because of an increased arterial driving pressure. This response helps limit blood flow into the glomerulus to normal despite the elevated arterial pressure. Conversely, inherent relaxation of an unstretched afferent arteriole when pressure within the vessel is reduced increases blood flow into the glomerulus despite the fall in arterial pressure.

- The **tubuloglomerular feedback (TGF)** mechanism involves the *juxtaglomerular apparatus,* which is the specialized combination of tubular and vascular cells where the tubule, after having bent back on itself, passes through the angle formed by the afferent and efferent arterioles as they join the glomerulus (Figure 14-11; see also Figure 14-3, p. 494). The smooth muscle cells within the wall of the afferent arteriole in this region are specialized to form **granular cells**, so called because they contain many secretory granules. Specialized tubular cells

in this region are collectively known as the **macula densa**. The macula densa cells detect changes in the salt level of the fluid flowing past them through the tubule.

If the GFR is increased secondary to an elevation in arterial pressure, more fluid than normal is filtered and flows through the distal tubule. In response to the resultant rise in salt delivery to the distal tubule, the macula densa cells release *ATP* and *adenosine,* both of which act locally as a paracrine on the adjacent afferent arteriole, causing it to constrict, thus reducing glomerular blood flow and returning GFR to normal. In the opposite situation, when less salt is delivered to the distal tubule because of a spontaneous decline in GFR accompanying a fall in arterial pressure, less ATP and adenosine are released by the macula densa cells. The resultant afferent arteriolar vasodilation increases the glomerular flow rate, restoring the GFR to normal. To exert even more exquisite control over tubuloglomerular feedback, the macula densa cells also secrete the vasodilator *nitric oxide,* which puts the brakes on the action of ATP and adenosine at the afferent arteriole. By means of the TGF mechanism, the tubule of a nephron is able to monitor the salt level in the fluid flowing through it and adjust the rate of filtration through the glomerulus of the same nephron accordingly to keep the early distal tubular fluid and salt delivery constant.

Importance of Autoregulation of the GFR The myogenic and tubuloglomerular feedback mechanisms work in unison to autoregulate the GFR within the mean arterial blood

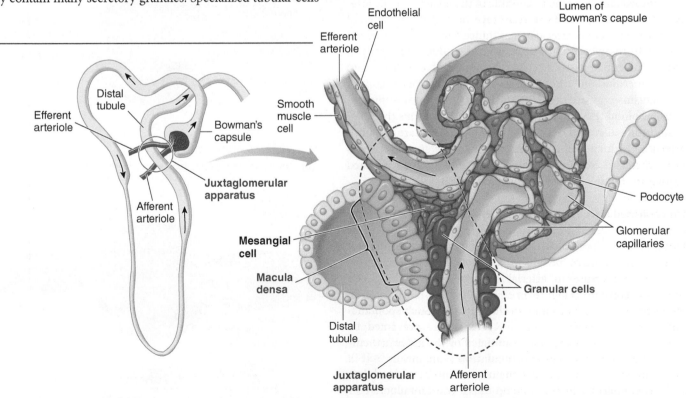

Figure 14-11 **The juxtaglomerular apparatus.** The juxtaglomerular apparatus consists of specialized vascular cells (the granular cells) and specialized tubular cells (the macula densa) at a point where the distal tubule passes through the fork formed by the afferent and efferent arterioles of the same nephron.

pressure range of 80 to 180 mm Hg. Within this wide range, intrinsic autoregulatory adjustments of afferent arteriolar resistance can compensate for changes in arterial pressure, thus preventing inappropriate fluctuations in GFR, even though glomerular pressure tends to change in the same direction as arterial pressure. Normal mean arterial pressure is 93 mm Hg, so this range encompasses the transient changes in blood pressure that accompany daily activities unrelated to the need for the kidneys to regulate H_2O and salt excretion, such as the normal elevation in blood pressure accompanying exercise. Autoregulation is important because unintentional shifts in GFR could lead to dangerous imbalances of fluid, electrolytes, and wastes. Because at least a certain portion of the filtered fluid is always excreted, the amount of fluid excreted in the urine is automatically increased as the GFR increases. If autoregulation did not occur, the GFR would increase and H_2O and solutes would be lost needlessly as a result of the rise in arterial pressure accompanying heavy exercise. If, by contrast, the GFR were too low, the kidneys could not eliminate enough wastes, excess electrolytes, and other materials that should be excreted. Autoregulation thus greatly blunts the direct effect that changes in arterial pressure would otherwise have on GFR and subsequently on H_2O, solute, and waste excretion.

When changes in mean arterial pressure fall outside the autoregulatory range, these mechanisms cannot compensate. Therefore, dramatic changes in mean arterial pressure (<80 mm Hg or >180 mm Hg) directly cause the glomerular capillary pressure and, accordingly, the GFR to decrease or increase in proportion to the change in arterial pressure.

Importance of Extrinsic Sympathetic Control of the GFR In addition to the intrinsic autoregulatory mechanisms designed to keep the GFR constant in the face of fluctuations in arterial blood pressure, the GFR can be *changed on purpose*—even when the mean arterial blood pressure is within the autoregulatory range—by extrinsic control mechanisms that override the autoregulatory responses. Extrinsic control of GFR, which is mediated by sympathetic nervous system input to the afferent arterioles, is aimed at long-term regulation of arterial blood pressure. The parasympathetic nervous system does not exert any influence on the kidneys.

If plasma volume is decreased—for example, by hemorrhage—the resulting fall in arterial blood pressure is detected by the arterial carotid sinus and aortic arch baroreceptors, which initiate neural reflexes to raise blood pressure toward normal (see p. 367). These reflex responses are coordinated by the cardiovascular control center in the brain stem and are mediated primarily through increased sympathetic activity to the heart and blood vessels. Although the resulting increase in both cardiac output and total peripheral resistance helps raise blood pressure toward normal in the short term, plasma volume is still

reduced. In the long term, plasma volume must be restored to normal. One compensation for a depleted plasma volume is reduced urine output so that more fluid than normal is conserved for the body. Urine output is reduced in part by reducing the GFR; if less fluid is filtered, less is available to excrete.

Role of the Baroreceptor Reflex in Extrinsic Control of the GFR No new mechanism is needed to decrease the GFR. It is reduced by the baroreceptor reflex response to a fall in blood pressure (▮Figure 14-12). During this reflex, sympathetically induced vasoconstriction occurs in most arterioles throughout the body (including the afferent arterioles) as a compensatory mechanism to increase total peripheral resistance. The afferent arterioles have α_1-adrenergic receptors (see p. 240) and are innervated with sympathetic vasoconstrictor

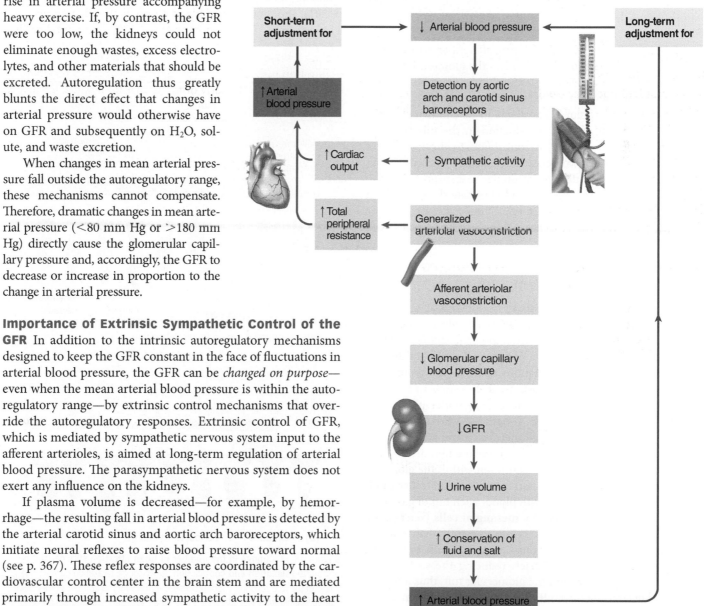

▮Figure 14-12 **Baroreceptor reflex influence on GFR in long-term regulation of blood pressure.**

fibers to a far greater extent than the efferent arterioles are. When the afferent arterioles carrying blood to the glomeruli constrict from increased sympathetic activity, less blood flows into the glomeruli than normal, lowering glomerular capillary blood pressure (see Figure 14-10a). The resulting decrease in GFR, in turn, reduces urine volume. In this way, some of the H_2O and salt that would otherwise have been lost in the urine are saved for the body, helping restore plasma volume to normal in the long term so that short-term cardiovascular adjustments that have been made are no longer necessary. Other mechanisms, such as increased tubular reabsorption of H_2O and salt, and increased thirst (described more thoroughly elsewhere), also contribute to long-term maintenance of blood pressure, despite a loss of plasma volume, by helping restore plasma volume.

Conversely, if blood pressure is elevated (for example, because of an expansion of plasma volume following ingestion of excessive fluid), the opposite responses occur. When the baroreceptors detect a rise in blood pressure, sympathetic vasoconstrictor activity to the arterioles, including the renal afferent arterioles, is reflexly reduced, allowing afferent arteriolar vasodilation to occur. As more blood enters the glomeruli through the dilated afferent arterioles, glomerular capillary blood pressure rises, increasing the GFR (see Figure 14-10b). As more fluid is filtered, more fluid is available to be eliminated in the urine. A hormonally adjusted reduction in the tubular reabsorption of H_2O and salt also contributes to the increase in urine volume. These two renal mechanisms—increased glomerular filtration and decreased tubular reabsorption of H_2O and salt—increase urine volume and eliminate the excess fluid from the body. Reduced thirst and fluid intake also help restore an elevated blood pressure to normal in the long term.

The GFR can be influenced by changes in the filtration coefficient.

Thus far we have discussed changes in the GFR as a result of changes in net filtration pressure. The rate of glomerular filtration, however, depends on the filtration coefficient (K_f) as well as on the net filtration pressure. For years K_f was considered a constant, except in disease situations in which the glomerular membrane becomes leakier than usual. Research to the contrary indicates that K_f is subject to change under physiologic control. Both factors on which K_f depends—the surface area and the permeability of the glomerular membrane—can be modified by contractile activity within the membrane.

The surface area available for filtration within the glomerulus is represented by the inner surface of the glomerular capillaries that comes into contact with blood. Each tuft of glomerular capillaries is held together by **mesangial cells** (see Figure 14-11). These cells contain contractile elements (that is, actinlike filaments). Contraction of these mesangial cells closes off a portion of the filtering capillaries, reducing the surface area available for filtration within the glomerular tuft, thus lowering the K_f and decreasing the GFR. Sympathetic stimulation causes the mesangial cells to contract, thereby providing a second mechanism (besides promoting afferent arteriolar vasoconstriction) by which sympathetic activity can decrease the GFR.

Podocytes also possess actinlike filaments, whose contraction or relaxation can, respectively, decrease or increase the number of filtration slits open in the inner membrane of Bowman's capsule by changing the shapes and proximities of the secondary foot processes (Figure 14-13). The number of slits is a determinant of permeability; the more open slits, the greater the permeability. Contractile activity of the podocytes, which in turn affects permeability and the K_f, is under physiologic control by poorly understood mechanisms.

The kidneys normally receive 20% to 25% of the cardiac output.

At the average net filtration pressure and K_f, 20% of the plasma that enters the kidneys is converted into glomerular filtrate. That means at an average GFR of 125 mL/min, the total renal plasma flow must average about 625 mL/min. Because 55% of whole blood consists of plasma (that is, hematocrit = 45; see p. 381), the total flow of blood through the kidneys averages 1140 mL/min. This quantity is about 22% of the total cardiac

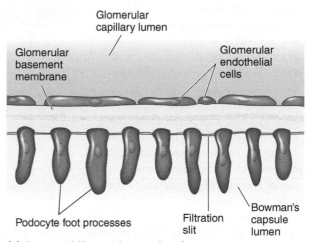

(a) Increased K_f on podocyte relaxation

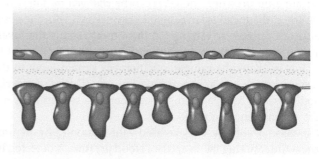

(b) Decreased K_f on podocyte contraction

Figure 14-13 **Change in the number of open filtration slits caused by podocyte relaxation and contraction.** (a) Podocyte relaxation narrows the bases of the fine secondary foot processes, increasing the number of fully open intervening filtration slits spanning a given area. (b) Podocyte contraction flattens the foot process branches and thus decreases the number of intervening filtration slits.

(Source: Adapted from *Federation Proceedings*, vol. 42, pp. 3046–3052, 1983. Reprinted by permission.)

output of 5 liters (5000 mL)/min, although the kidneys compose less than 1% of total body weight.

The kidneys receive such a seemingly disproportionate share of the cardiac output because they must continuously perform their regulatory and excretory functions on the huge volumes of plasma delivered to them to maintain stability in the internal fluid environment. Most of the blood goes to the kidneys not to supply the renal tissue but to be adjusted and purified by the kidneys. On average, 20% to 25% of the blood pumped out by the heart each minute "goes to the cleaners" instead of serving its normal purpose of exchanging materials with the tissues. Only by continuously processing such a large proportion of the blood can the kidneys precisely regulate the volume and electrolyte composition of the internal environment and adequately eliminate the large quantities of metabolic waste products that are constantly produced.

We next consider how the tubules act on this large volume of filtered plasma, first considering the process of tubular reabsorption.

Check Your Understanding 14.2

1. Prepare a table showing the effect and magnitude of the physical forces involved in glomerular filtration.
2. Discuss the mechanisms and importance of autoregulation of the GFR.
3. Discuss the mechanism and importance of extrinsic control of the GFR.

14.3 Tubular Reabsorption

All plasma constituents except the plasma proteins are indiscriminately filtered together through the glomerular capillaries. In addition to waste products and excess materials that the body must eliminate, the filtered fluid contains nutrients, electrolytes, and other substances that the body cannot afford to lose in the urine. Indeed, through ongoing glomerular filtration, greater quantities of these materials are filtered per day than are even present in the entire body. The essential materials that are filtered are returned to the blood by *tubular reabsorption,* the discrete transfer of substances from the tubular lumen into the peritubular capillaries.

Tubular reabsorption is tremendous, highly selective, and variable.

Tubular reabsorption is a highly selective process. All constituents except plasma proteins are at the same concentration in the glomerular filtrate as in plasma. In most cases, the quantity reabsorbed of each substance is the amount required to maintain the proper composition and volume of the internal fluid environment. In general, the tubules have a high reabsorptive capacity for substances needed by the body and little or no reabsorptive capacity for substances of no value. Accordingly, only a small percentage, if any, of filtered plasma constituents that are useful to the body are present in the urine, most having

been reabsorbed and returned to the blood. Of the 125 mL of fluid filtered per minute, typically 124 mL/min are reabsorbed. Considering the magnitude of glomerular filtration, the extent of tubular reabsorption is tremendous: The tubules typically reabsorb 99% of the filtered water (47 gallons per day), 100% of the filtered sugar (0.4 pound per day), and 99.5% of the filtered salt (3.5 pounds per day). Only excess amounts of essential materials such as electrolytes are excreted in the urine. For the essential plasma constituents regulated by the kidneys, absorptive capacity may vary depending on the body's needs. In contrast, a large percentage of filtered waste products are present in the urine. These wastes, which are useless or even potentially harmful to the body if allowed to accumulate, generally are not reabsorbed. Instead, they stay in the tubules to be eliminated in the urine. For example, creatinine, a waste produced during muscle metabolism, is not reabsorbed at all, so 100% of filtered creatinine is excreted in the urine. As H_2O and other valuable constituents are reabsorbed, the waste products remaining in the tubular fluid become highly concentrated.

Tubular reabsorption involves transepithelial transport.

Throughout its length, the tubule wall is one cell thick and is close to a surrounding peritubular capillary (❙Figure 14-14). Adjacent tubular cells do not come into contact with each other except where they are joined by tight junctions (see p. 61) at their lateral edges near their *luminal membranes,* which face the tubular lumen. Interstitial fluid lies in the gaps between adjacent cells—the **lateral spaces**—as well as between the tubules and the capillaries. The *basolateral membrane* faces the interstitial fluid at the base and lateral edges of the cell. The tight junctions largely prevent substances from moving *between* the cells, so materials must pass *through* the cells to leave the tubular lumen and gain entry to the blood.

Transepithelial Transport To be reabsorbed, a substance must go across the following five distinct barriers (the numbers correspond to the numbered barriers in ❙Figure 14-14):

1 Leave the tubular fluid by crossing the luminal membrane of the tubular cell.

2 Pass through the cytosol from one side of the tubular cell to the other.

3 Cross the basolateral membrane of the tubular cell to enter the interstitial fluid.

4 Diffuse through the interstitial fluid.

5 Penetrate the capillary wall to enter the plasma.

This entire sequence of steps is known as **transepithelial transport** (*transepithelial* means "across the epithelium").

Passive Versus Active Reabsorption The two types of tubular reabsorption—passive and active—depend on whether local energy expenditure is needed for reabsorbing a particular substance. In **passive reabsorption**, all steps in the transepithelial transport of a substance from the tubular lumen to the plasma are passive—that is, no energy is spent for the substance's net movement, which occurs down electrochemical or

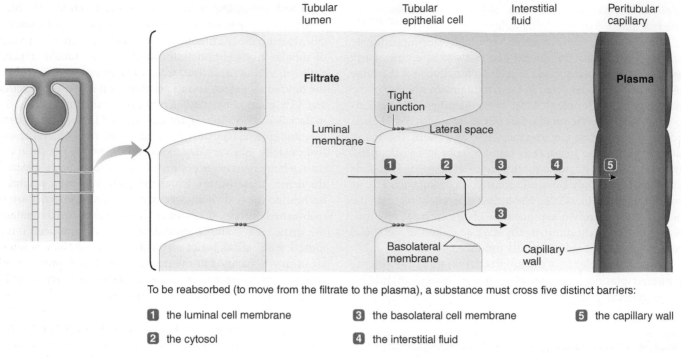

To be reabsorbed (to move from the filtrate to the plasma), a substance must cross five distinct barriers:

1 the luminal cell membrane   **3** the basolateral cell membrane   **5** the capillary wall

2 the cytosol   **4** the interstitial fluid

❘Figure 14-14 **Steps of transepithelial transport.**

---

osmotic gradients. In contrast, **active reabsorption** takes place if any step in the transepithelial transport of a substance requires energy, even if the four other steps are passive. With active reabsorption, net movement of the substance from the tubular lumen to the plasma occurs *against* an electrochemical gradient. Substances that are actively reabsorbed are of particular importance to the body, such as glucose, amino acids, and other organic nutrients, as well as Na^+ and other electrolytes, such as $PO_4{}^{3-}$. Rather than specifically describing the reabsorptive process for each of the many filtered substances returned to the plasma, we provide illustrative examples of the general mechanisms involved, after first highlighting the unique and important case of Na^+ reabsorption.

Na^+ reabsorption depends on the Na^+–K^+ ATPase pump in the basolateral membrane.

Sodium reabsorption is unique and complex. Of the total energy spent by the kidneys, 80% is used for Na^+ transport, indicating the importance of this process. Unlike most filtered solutes, Na^+ is reabsorbed throughout most of the tubule, but this occurs to varying extents in different regions. Of the Na^+ filtered, 99.5% is normally reabsorbed. Of the Na^+ reabsorbed, on average 67% is reabsorbed in the proximal tubule, 25% in the loop of Henle, and 8% in the distal and collecting tubules. Sodium reabsorption plays different important roles in each of these segments, as will become apparent as our discussion continues. Here is a preview of these roles:

■ Sodium reabsorption in the *proximal tubule* plays a pivotal role in reabsorbing glucose, amino acids, H_2O, Cl^-, and urea.

■ Sodium reabsorption in the ascending limb of the *loop of Henle,* along with Cl^- reabsorption, plays a critical role in the

kidneys' ability to produce urine of varying concentrations and volumes, depending on the body's need to conserve or eliminate H_2O.

■ Sodium reabsorption in the *distal and collecting tubules* is variable and subject to hormonal control. It plays a key role in regulating ECF volume, which is important in long-term control of arterial blood pressure, and is linked in part to K^+ secretion.

Sodium is reabsorbed throughout the tubule with the exception of the descending limb of Henle's loop. You will learn about the significance of this exception later. Throughout all Na^+-reabsorbing tubular segments, the active step in Na^+ reabsorption involves the energy-dependent Na^+–K^+ ATPase carrier located in the tubular cell's basolateral membrane (❘Figure 14-15). This carrier is the same Na^+–K^+ pump present in all cells that actively extrudes Na^+ from the cell (see p. 73). As this basolateral pump transports Na^+ out of the tubular cell into the lateral space, it keeps the intracellular Na^+ concentration low while simultaneously building up the Na^+ concentration in the lateral space—that is, it moves Na^+ against a concentration gradient. Because the intracellular Na^+ concentration is kept low by basolateral pump activity, a concentration gradient is established that favors passive movement of Na^+ from its higher concentration in the tubular lumen across the luminal border into the tubular cell. The nature of the luminal Na^+ channels and transport carriers that permit Na^+ movement from the lumen into the cell varies for different parts of the tubule, but in each case, Na^+ movement across the luminal membrane is always a passive step. For example, in the proximal tubule, Na^+ crosses the luminal border by a symport carrier that simultaneously moves Na^+ and an organic nutrient such as glucose from the lumen into the cell. You will learn more about

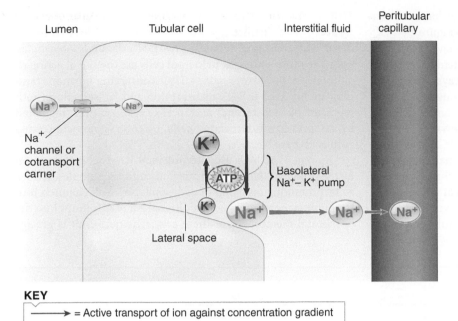

Lumen | Tubular cell | Interstitial fluid | Peritubular capillary

KEY

———➤ = Active transport of ion against concentration gradient

———➤ = Passive movement of ion down concentration gradient

Figure 14-15 Sodium reabsorption. The basolateral Na^+–K^+ pump actively transports Na^+ from the tubular cell into the interstitial fluid within the lateral space. This process establishes a concentration gradient for passive movement of Na^+ from the lumen into the tubular cell and from the lateral space into the peritubular capillary, accomplishing net transport of Na^+ from the tubular lumen into the blood at the expense of energy.

90% of the ECF's osmotic activity. Whenever we speak of Na^+ load, we tacitly mean salt load, too, because Cl^- goes along with Na^+. (NaCl is common table salt.) The Na^+ load is subject to regulation; Cl^- passively follows along. Recall that osmotic pressure can be thought of loosely as a "pulling" force that attracts and holds H_2O (see p. 67). When the Na^+ load is above normal and the ECF's osmotic activity is therefore increased, the extra Na^+ "holds" extra H_2O, expanding ECF volume. Conversely, when the Na^+ load is below normal, thereby decreasing ECF osmotic activity, less H_2O than normal can be held in the ECF, so ECF volume is reduced. Because plasma is part of the ECF, the most important result of a change in ECF volume is the matching change in blood pressure with expansion (increased blood pressure) or reduction (decreased blood pressure) of the plasma volume. Thus, long-term control of arterial blood pressure ultimately depends on Na^+-regulating mechanisms. We now turn attention to these mechanisms.

Activation of the Renin–Angiotensin–Aldosterone System The most important and best-known hormonal system involved in regulating Na^+ is the **renin–angiotensin–aldosterone system (RAAS)**. The granular cells of the *juxtaglomerular apparatus* (see **Figure 14-11**) secrete an enzymatic hormone, **renin**, into the blood in response to a fall in NaCl, ECF volume, and arterial blood pressure. This function is in addition to the role the macula densa cells of the juxtaglomerular apparatus play in autoregulation. Specifically, the following three inputs to the granular cells increase renin secretion:

1. The granular cells themselves function as *intrarenal baroreceptors*. They are sensitive to pressure changes within the afferent arteriole. When the granular cells detect a fall in blood pressure, they secrete more renin.

2. The macula densa cells in the tubular portion of the juxtaglomerular apparatus are sensitive to the NaCl moving past them through the tubular lumen. In response to a fall in NaCl, the macula densa cells trigger increased renin secretion.

3. The granular cells are innervated by the sympathetic nervous system. When blood pressure falls below normal, the baroreceptor reflex increases sympathetic activity. As part of this reflex response, increased sympathetic activity stimulates the granular cells to secrete more renin.

These interrelated signals for increased renin secretion all indicate the need to expand plasma volume to increase arterial pressure to normal in the long term. Through a complex series of events involving RAAS, increased renin secretion brings about increased Na^+ reabsorption by the distal and collecting tubules (with Cl^- passively following Na^+'s active movement). The ultimate benefit of this salt retention is osmotically induced H_2O retention, which helps restore plasma volume.

this cotransport process shortly. By contrast, in the collecting duct, Na^+ crosses the luminal border through a Na^+ leak channel (see p. 58). Once Na^+ enters the cell across the luminal border by whatever means, it is actively extruded to the lateral space by the basolateral Na^+–K^+ pump. This step is the same throughout the tubule. Na^+ continues to diffuse down a concentration gradient from its high concentration in the lateral space into the surrounding interstitial fluid and finally into the peritubular capillary blood. Thus, net transport of Na^+ from the tubular lumen into the blood occurs at the expense of energy.

First, we consider the importance and mechanism of regulating Na^+ reabsorption in the distal portion of the nephron.

Aldosterone stimulates Na^+ reabsorption in the distal and collecting tubules.

In the proximal tubule and loop of Henle, a constant percentage of the filtered Na^+ is reabsorbed regardless of the **Na^+ load** (*the total amount* of Na^+ in the body fluids, *not the concentration* of Na^+ in the body fluids). In the distal and collecting tubules, the reabsorption of a small percentage of the filtered Na^+ is subject to hormonal control. The extent of this controlled, discretionary reabsorption is inversely related to the magnitude of the Na^+ load in the body. If there is too much Na^+, little of this controlled Na^+ is reabsorbed; instead, it is lost in the urine, thereby removing excess Na^+ from the body. If Na^+ is depleted, most or all of this controlled Na^+ is reabsorbed, conserving for the body Na^+ that otherwise would be lost in the urine.

The Na^+ load in the body is reflected by ECF volume. Sodium and its accompanying anion Cl^- account for more than

Let us examine in further detail the RAAS mechanism that ultimately leads to increased Na^+ reabsorption (■Figure 14-16). Once secreted into the blood, renin acts as an enzyme to activate **angiotensinogen** into **angiotensin I**. Angiotensinogen is a plasma protein synthesized by the liver and always present in the plasma in high concentration. On passing through the lungs via the pulmonary circulation, angiotensin I is converted into **angiotensin II** by **angiotensin-converting enzyme (ACE)**, which is abundant in the pulmonary capillaries. ACE is located in small pits in the luminal surface of the pulmonary capillary endothelial cells. Angiotensin II is the main stimulus for secretion of the hormone *aldosterone* from the adrenal cortex. The *adrenal cortex* is an endocrine gland that produces several hormones, each secreted in response to different stimuli.

Functions of the Renin–Angiotensin–Aldosterone System

Two distinct types of tubular cells are located in the distal and collecting tubules: *principal cells* and *intercalated cells*. The more abundant **principal cells** are the site of action of aldosterone and vasopressin, a H_2O-conserving hormone, and thus are involved in Na^+ reabsorption and K^+ secretion (both regulated by aldosterone) and in H_2O reabsorption (regulated by vasopressin). **Intercalated cells**, by contrast, are concerned with acid–base balance.

Among its actions, **aldosterone** increases Na^+ reabsorption by the principal cells of the distal and collecting tubules. It does so by promoting insertion of additional Na^+ leak channels into the luminal membranes and additional Na^+–K^+ pumps into the basolateral membranes of these cells. The net result is greater

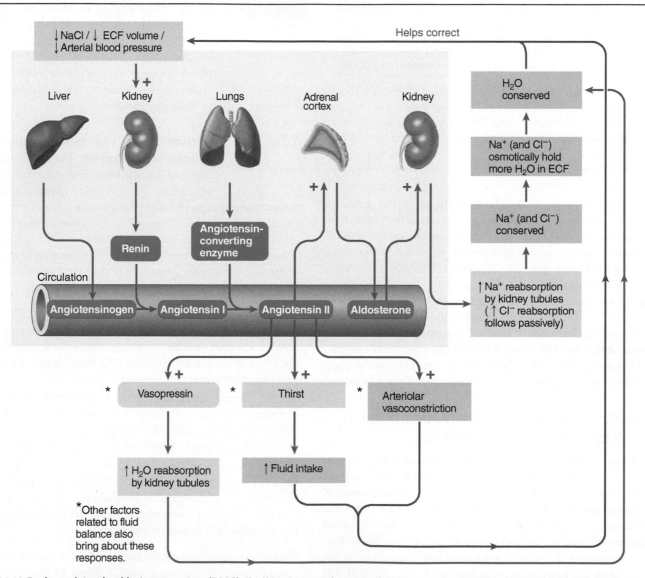

■Figure 14-16 **Renin–angiotensin–aldosterone system (RAAS).** The kidneys secrete the enzymatic hormone renin in response to reduced NaCl, ECF volume, and arterial blood pressure. Renin activates angiotensinogen, a plasma protein produced by the liver, into angiotensin I. Angiotensin I is converted into angiotensin II by angiotensin-converting enzyme (ACE) produced in the lungs. Angiotensin II stimulates the adrenal cortex to secrete the hormone aldosterone, which stimulates Na^+ reabsorption by the kidneys. The resulting retention of Na^+ exerts an osmotic effect that holds more H_2O in the ECF. Together, the conserved Na^+ and H_2O help correct the original stimuli that activated this hormonal system. Angiotensin II also exerts other effects that help rectify the original stimuli, such as by promoting arteriolar vasoconstriction.

FIGURE FOCUS: *If a person's blood pressure falls because of loss of fluid and salt through heavy sweating, summarize the short-term and long-term compensatory measures shown in this figure and Figure 14-12 to help restore blood pressure to normal.*

passive movement of Na^+ into these distal and collecting tubular cells from the lumen and increased active pumping of Na^+ out of the cells into the plasma—that is, an increase in Na^+ reabsorption, with Cl^- following passively. RAAS thus promotes salt retention and a resulting H_2O retention and rise in arterial blood pressure. Acting in a negative-feedback fashion, this system alleviates the factors that triggered the initial release of renin—namely, salt depletion, plasma volume reduction, and decreased arterial blood pressure (Figure 14-16).

In addition to stimulating aldosterone secretion, angiotensin II is a potent constrictor of the systemic arterioles, directly increasing blood pressure by increasing total peripheral resistance (see p. 349). Furthermore, it stimulates thirst (increasing fluid intake) and stimulates vasopressin (increasing H_2O retention by the kidneys), both of which contribute to plasma volume expansion and elevation of arterial pressure. (As you will learn later, other mechanisms related to long-term regulation of blood pressure and ECF osmolarity are also important in controlling thirst and vasopressin secretion.)

The opposite situation exists when the Na^+ load, ECF and plasma volume, and arterial blood pressure are above normal. Under these circumstances, renin secretion is inhibited. Therefore, because angiotensinogen is not activated to angiotensin I and II, aldosterone secretion is not stimulated. Without aldosterone, the small aldosterone-dependent part of Na^+ reabsorption in the distal segments of the tubule does not occur. Instead, this nonreabsorbed Na^+ is lost in the urine. In the absence of aldosterone, the ongoing loss of this small percentage of filtered Na^+ can rapidly remove excess Na^+ from the body. Even though only about 8% of the filtered Na^+ depends on aldosterone for reabsorption, this small loss, multiplied many times as the entire plasma volume is filtered through the kidneys many times per day, can lead to a sizable loss of Na^+.

The amount of aldosterone secreted, and consequently the relative amount of salt conserved versus salt excreted, varies depending on the body's needs. For example, an average salt consumer typically excretes nearly 10 g of salt per day in the urine, a heavy salt consumer excretes more, and someone who has lost considerable salt during heavy sweating excretes less urinary salt. With maximum aldosterone secretion, all the filtered Na^+ (and, accordingly, all the filtered Cl^-) is reabsorbed, so salt excretion in the urine is zero. By varying the amount of renin and aldosterone secreted in accordance with the salt-determined fluid load in the body, the kidneys can finely adjust the amount of salt conserved or eliminated. In doing so, they maintain the salt load, ECF volume, and arterial blood pressure at a relatively constant level despite wide variations in salt consumption and abnormal losses of salt-laden fluid.

_{Clinical Note} **Role of the Renin–Angiotensin–Aldosterone System in Various Diseases** Some cases of hypertension (high blood pressure) are the result of abnormal increases in RAAS activity. This system is also responsible in part for the fluid retention and edema accompanying congestive heart failure. Because of the failing heart, cardiac output is reduced and blood pressure is low despite a normal or even expanded plasma volume. When a fall in blood pressure is the result of a failing heart rather than a reduced salt and fluid load in the body, the salt- and fluid-retaining reflexes triggered by the low blood pressure are inappropriate. Sodium excretion may fall to zero despite continued salt ingestion and accumulation in the body. The resulting ECF expansion produces edema and intensifies the congestive heart failure because the weakened heart cannot pump the additional plasma volume.

_{Clinical Note} **Drugs that Affect Na^+ Reabsorption** Because their salt-retaining mechanisms are being inappropriately triggered, patients with congestive heart failure are placed on low-salt diets. Often they are treated with **diuretics**, therapeutic agents that cause **diuresis** (increased urinary output) and thus promote fluid loss from the body. Many of these drugs function by inhibiting tubular reabsorption of Na^+. For example, *thiazide diuretics* such as hydrochlorothiazide inhibit Na^+ reabsorption in the distal tubule. As more Na^+ is excreted, more H_2O is also lost from the body, helping remove excess ECF. **ACE inhibitor drugs**, which block the action of ACE, and **aldosterone receptor blockers (ARBs)**, which block the binding of aldosterone with its renal receptors, are both also beneficial in treating hypertension and congestive heart failure. These two classes of drugs halt the ultimate salt- and fluid-conserving actions and arteriolar constrictor effects of RAAS.

The natriuretic peptides inhibit Na^+ reabsorption.

Whereas RAAS exerts the most powerful influence on renal handling of Na^+, this Na^+-retaining, blood pressure–raising system is opposed by a Na^+-losing, blood pressure–lowering system that involves the hormones **atrial natriuretic peptide (ANP)** and **brain natriuretic peptide (BNP)**. These peptides produce **natriuresis**, or excretion of large amounts of sodium in the urine. The heart, in addition to its pump action, produces ANP and BNP. As its name implies, ANP is produced in the atrial cardiac muscle cells. BNP was first discovered in the brain (hence its name) but is produced primarily in the ventricular cardiac muscle cells. ANP and BNP are stored in granules and released when the heart muscle cells are mechanically stretched by an expansion of the circulating plasma volume when ECF volume is increased. This expansion, which occurs as a result of Na^+ and H_2O retention, increases blood pressure. In turn, the NPs promote natriuresis and accompanying diuresis, decreasing the plasma volume, and also directly influence the cardiovascular system to lower blood pressure (Figure 14-17).

The main action of ANP and BNP is to directly inhibit Na^+ reabsorption in the distal parts of the nephron, thus increasing Na^+ excretion and accompanying osmotic H_2O excretion in the urine. They further increase Na^+ excretion in the urine by inhibiting two steps of the Na^+-conserving RAAS. The NPs inhibit renin secretion by the kidneys and act on the adrenal cortex to inhibit aldosterone secretion. In addition, they inhibit the secretion and actions of vasopressin, the H_2O-conserving hormone. ANP and BNP also promote natriuresis and accompanying diuresis by increasing the GFR.

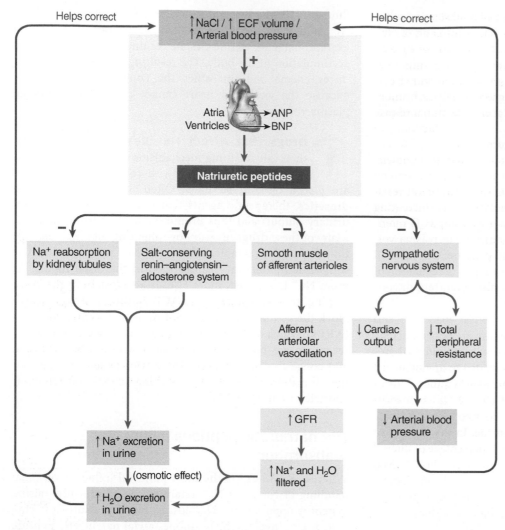

Atrial and brain natriuretic peptide. The cardiac atria secrete the hormone atrial natriuretic peptide (ANP) and the cardiac ventricles secrete brain natriuretic peptide (BNP) in response to being stretched by Na⁺ retention, expansion of the ECF volume, and increase in arterial blood pressure. ANP and BNP, in turn, promote natriuretic, diuretic, and hypotensive effects to help correct the original stimuli that resulted in their release.

They dilate the afferent arterioles and constrict the efferent arterioles, thus raising glomerular capillary blood pressure and increasing the GFR. They further increase the GFR by relaxing the glomerular mesangial cells, leading to an increase in K_f. As more salt and water are filtered, more salt and water are excreted in the urine. Besides indirectly lowering blood pressure by reducing the Na⁺ load and hence the fluid load in the body, ANP and BNP directly lower blood pressure by decreasing cardiac output and reducing total peripheral resistance by inhibiting sympathetic nervous activity to the heart and blood vessels, respectively.

The relative contributions of ANP and BNP in maintaining salt and H_2O balance and blood pressure regulation are presently being intensively investigated. A deficiency of the counterbalancing natriuretic system may underlie some cases of long-term hypertension by leaving the powerful Na⁺-conserving system unopposed. The resulting salt retention, especially in association with high salt intake, could expand ECF volume and elevate blood pressure.

We now shift attention to the reabsorption of other filtered solutes. Nevertheless, we continue to discuss Na⁺ reabsorption because the reabsorption of many other solutes is linked in some way to Na⁺ reabsorption.

Glucose and amino acids are reabsorbed by Na⁺-dependent secondary active transport.

Large quantities of nutritionally important organic molecules such as glucose and amino acids are filtered each day. Because these molecules normally are completely reabsorbed into the blood by energy- and Na⁺-dependent mechanisms located in the proximal tubule, none of these nutrients are usually excreted in the urine, thus protecting against their loss.

Glucose and amino acids are reabsorbed by **secondary active transport**. With this process, specialized *symport carriers*, such as the *sodium and glucose cotransporter (SGLT)*, simultaneously transfer both Na⁺ and the specific organic molecule from the lumen into the cell (see IFigure 3-18, p. 76). Within the kidney, SGLT is located only in the proximal tubule. This luminal cotransport carrier is the means by which Na⁺ passively crosses the luminal membrane in the proximal tubule. Once transported into the tubular cell, glucose and amino acids passively diffuse down their concentration gradients across the basolateral membrane into the plasma, facilitated by a carrier, such as the *glucose transporter (GLUT)*, which does not depend on energy (see p. 72).

In general, actively reabsorbed substances exhibit a tubular maximum.

All actively reabsorbed substances bind with plasma membrane carriers that transfer them across the membrane against a concentration gradient. Each carrier is specific for the types of substances it can transport; for example, SGLT can transport glucose but not amino acids. Because a limited number of each carrier type is present in the tubular cells, an upper limit exists on how much of a particular substance can be actively transported from the tubular fluid in a given period. The maximum reabsorption rate is reached when all of the carriers specific for a particular substance are fully occupied or saturated so that they cannot handle additional passengers at that time (see p. 71). This maximum reabsorption rate is designated as the

tubular maximum, or T_m.[1] Any quantity of a substance filtered beyond its T_m is not reabsorbed and escapes instead into the urine. With the exception of Na^+, all actively reabsorbed substances have a T_m. (Even though individual Na^+ transport carriers can become saturated, the tubules as a whole do not display a T_m for Na^+ because aldosterone promotes the insertion of more active Na^+–K^+ carriers in the distal and collecting tubular cells as needed.)

The plasma concentrations of some but not all substances that display carrier-limited reabsorption are regulated by the kidneys. How can the kidneys regulate some actively reabsorbed substances but not others, when the renal tubules limit the quantity of each of these substances that can be reabsorbed and returned to the plasma? We compare glucose, a substance that has a T_m but *is not regulated* by the kidneys, with phosphate, a T_m-limited substance that *is regulated* by the kidneys.

Glucose is an actively reabsorbed substance not regulated by the kidneys.

The normal plasma concentration of glucose is 100 mg of glucose for every 100 mL of plasma. Because glucose is freely filterable at the glomerulus, it passes into Bowman's capsule at the same concentration it has in the plasma. Accordingly, 100 mg of glucose are present in every 100 mL of plasma filtered. With 125 mL of plasma normally being filtered each minute (average GFR = 125 mL/min), 125 mg of glucose pass into Bowman's capsule with this filtrate every minute. The quantity of any substance filtered per minute, known as its **filtered load**, can be calculated as follows:

Filtered load of a substance = plasma concentration × GFR
of the substance

Filtered load of glucose = 100 mg/100 mL × 125 mL/min

= 125 mg/min

At a constant GFR, the filtered load of glucose is directly proportional to the plasma glucose concentration. Doubling the plasma glucose concentration to 200 mg/100 mL doubles the filtered load of glucose to 250 mg/min, and so on (I Figure 14-18).

Tubular Maximum for Glucose The T_m for glucose averages 375 mg/min—that is, the glucose carrier mechanism is capable of actively reabsorbing up to 375 mg of glucose per minute before it reaches its maximum transport capacity. At a normal plasma glucose concentration of 100 mg/100 mL, the 125 mg of glucose filtered per minute can readily be reabsorbed by the glucose carrier mechanism because the filtered load is well below the T_m for glucose. Ordinarily, therefore, no glucose appears in the urine. Not until the filtered load of glucose exceeds 375 mg/min is the T_m reached. When more

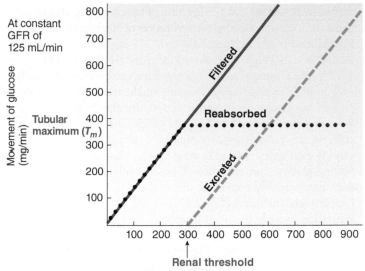

I Figure 14-18 **Renal handling of glucose as a function of plasma glucose concentration.** At a constant GFR, the quantity of glucose filtered per minute is directly proportional to the plasma concentration of glucose. All the filtered glucose can be reabsorbed up to the tubular maximum (T_m). If the amount of glucose filtered per minute exceeds the T_m, the maximum amount of glucose is reabsorbed (a T_m worth) and the rest stays in the filtrate to be excreted in urine. The renal threshold is the plasma concentration at which the T_m is reached and glucose first starts appearing in the urine.

FIGURE FOCUS: *Use the graph to determine how much glucose is filtered, reabsorbed, and excreted at plasma glucose concentrations of (1) 200 mg/100 mL, (2) 300 mg/100mL, and (3) 400 mg/100mL.*

glucose is filtered per minute than can be reabsorbed because the T_m has been exceeded, the maximum amount is reabsorbed, and the rest stays in the filtrate to be excreted. Accordingly, the plasma glucose concentration must be greater than 300 mg/100 mL—more than three times normal—before the amount filtered exceeds 375 mg/min and glucose starts spilling into the urine.

Renal Threshold for Glucose The plasma concentration at which the T_m of a particular substance is reached and the substance first starts appearing in the urine is called the **renal threshold**. At the average T_m of 375 mg/min and GFR of 125 mL/min, the renal threshold for glucose is 300 mg/100 mL.[2] Beyond the T_m, reabsorption stays constant at its maximum rate, and any further increase in the filtered load leads to a directly proportional increase in the amount of the substance excreted. For example, at a plasma glucose concentration of 400 mg/100 mL, the filtered load of glucose is 500 mg/min, 375 mg/min of which can be reab-

---

[1]For clarification, although both are designated as T_m, *transport maximum* refers to the upper limit on transport of a particular substance across a cell's plasma membrane that occurs when all of the carriers specific for the substance are saturated (see p. 71), whereas *tubular maximum* refers to the upper limit on transepithelial transport across the kidney tubules when all of the carriers specific for the substance are saturated.

[2]This is an idealized situation. In reality, glucose often starts spilling into the urine at glucose concentrations of 180 mg/100 mL and above. Glucose is often excreted before the average renal threshold of 300 mg/100 mL is reached for two reasons. First, not all nephrons have the same T_m, so some nephrons may have exceeded their T_m and be excreting glucose while others have not yet reached their T_m. Second, the efficiency of the glucose cotransport carrier may not be working at its maximum capacity at elevated values less than the true T_m, so some of the filtered glucose may fail to be reabsorbed and spill into the urine even though the average renal threshold has not been reached.

sorbed (a T_m worth) and 125 mg/min of which are excreted in the urine. At a plasma glucose concentration of 500 mg/100 mL, the filtered load is 625 mg/min, still only 375 mg/min can be reabsorbed, and 250 mg/min spill into the urine (Figure 14-18).

Clinical Note The plasma glucose concentration can become extremely high in *diabetes mellitus,* an endocrine disorder involving inadequate insulin action. Insulin is a pancreatic hormone that facilitates transport of glucose into many body cells. When cellular glucose uptake is impaired, glucose that cannot gain entry into cells stays in the plasma, elevating the plasma glucose concentration. Consequently, although glucose does not normally appear in urine, it is found in the urine of people with untreated diabetes when the plasma glucose concentration exceeds the renal threshold, even though renal function has not changed.

What happens when plasma glucose concentration falls below normal? The renal tubules reabsorb all filtered glucose because the glucose reabsorptive capacity is far from being exceeded. The kidneys cannot do anything to raise a low plasma glucose level to normal. They simply return all filtered glucose to the plasma.

Reason Why the Kidneys Do Not Regulate Glucose The kidneys do not influence plasma glucose concentration over a range of values, from abnormally low levels up to three times the normal level. Because the T_m for glucose is well above the normal filtered load, the kidneys usually conserve all the glucose, thereby protecting against loss of this important nutrient in the urine. The kidneys do not regulate glucose because they do not maintain glucose at some specific plasma concentration. Instead, this concentration is normally regulated by endocrine and liver mechanisms, with the kidneys merely maintaining whatever plasma glucose concentration is set by these other mechanisms (except when excessively high levels overwhelm the kidneys' reabsorptive capacity). The same principle holds true for other organic plasma nutrients, such as amino acids and water-soluble vitamins.

Phosphate is an actively reabsorbed substance regulated by the kidneys.

The kidneys do directly contribute to the regulation of many electrolytes, such as phosphate and calcium, because the renal thresholds of these inorganic ions equal their normal plasma concentrations. The transport carriers for these electrolytes are located in the proximal tubule. We use PO_4^{3-} as an example. Our diets are generally rich in PO_4^{3-}, but because the tubules can reabsorb up to the normal plasma concentration's worth of PO_4^{3-} and no more, the excess ingested PO_4^{3-} is quickly spilled into the urine, restoring the plasma concentration to normal. The greater the amount of PO_4^{3-} ingested beyond the body's needs, the greater the amount excreted. In this way, the kidneys maintain the desired plasma PO_4^{3-} concentration while eliminating any excess PO_4^{3-} ingested.

Unlike the reabsorption of organic nutrients, the reabsorption of PO_4^{3-} and Ca^{2+} is also subject to hormonal control. Parathyroid hormone can alter the renal thresholds for PO_4^{3-} and Ca^{2+}, thus adjusting the quantity of these electrolytes conserved, depending on the body's momentary needs (see Chapter 19).

Active Na^+ reabsorption is responsible for passive reabsorption of Cl^-, H_2O, and urea.

Not only is secondary active reabsorption of glucose and amino acids linked to the basolateral Na^+–K^+ pump, but passive reabsorption of Cl^-, H_2O, and urea also depends on this active Na^+ reabsorption mechanism.

Chloride Reabsorption Negatively charged Cl^- is passively reabsorbed down the electrical gradient created by active reabsorption of positively charged Na^+. For the most part, Cl^- passes between, not through, the tubular cells (through "leaky" tight junctions). The amount of Cl^- reabsorbed is determined by the rate of active Na^+ reabsorption instead of being directly controlled by the kidneys.

Water Reabsorption Water is passively reabsorbed throughout the length of the tubule as H_2O osmotically follows actively reabsorbed Na^+. Of the H_2O filtered, 65%—117 liters per day—is passively reabsorbed in the proximal tubule. Another 15% of the filtered H_2O is passively reabsorbed in the loop of Henle. This 80% of the filtered H_2O is obligatorily reabsorbed in the early parts of the nephron regardless of the H_2O load in the body and is not subject to regulation. Variable amounts of the remaining 20% are reabsorbed in the distal portions of the tubule; the extent of reabsorption in the distal and collecting tubules is under direct hormonal control, depending on the body's state of hydration. No part of the tubule directly requires energy for this tremendous reabsorption of H_2O.

During reabsorption, H_2O passes primarily through **aquaporins (AQPs)**, or **water channels**, formed by specific plasma membrane proteins in the tubular cells. Different types of water channels are present in various parts of the nephron. The water channels in the proximal tubule, AQP-1, are always open, accounting for the high H_2O permeability of this region. The AQP-2 channels in the principal cells in the distal parts of the nephron, in contrast, are regulated by the hormone *vasopressin,* accounting for the variable H_2O reabsorption in this region.

The main driving force for H_2O reabsorption in the proximal tubule is a compartment of hypertonicity in the lateral spaces between the tubular cells established by the basolateral pump's active extrusion of Na^+ (Figure 14-19). As a result of this pump activity, the concentration of Na^+ rapidly diminishes in the tubular fluid and tubular cells while it simultaneously increases in the localized region within the lateral spaces. This osmotic gradient induces the passive net flow of H_2O from the lumen into the lateral spaces, either through the cells or intercellularly through "leaky" tight junctions. The accumulation of fluid in the lateral spaces results in a buildup of hydrostatic (fluid) pressure, which flushes H_2O out of the lateral spaces into the interstitial fluid and finally into the peritubular capillaries. Water also osmotically follows other preferentially reabsorbed solutes such as glucose (which is also Na^+ dependent), but the direct influence of Na^+ reabsorption on passive H_2O reabsorption is quantitatively more important.

This return of filtered H_2O to the plasma is enhanced by the fact that the plasma-colloid osmotic pressure is greater in the peritubular capillaries than elsewhere. The concentration of plasma proteins, which is responsible for π_P, is elevated in the blood entering the peritubular capillaries because of the extensive filtration of H_2O through the glomerular capillaries upstream. The plasma proteins left behind in the glomerulus are concentrated into a smaller volume of plasma H_2O, increasing π_P of the unfiltered blood that leaves the glomerulus and enters the peritubular capillaries. This force tends to "pull" H_2O into the peritubular capillaries simultaneously with the "push" of the hydrostatic pressure in the lateral spaces that drives H_2O toward the capillaries. By these means, 65% of the filtered H_2O—117 liters per day—is passively reabsorbed by the end of the proximal tubule.

The mechanisms responsible for H_2O reabsorption beyond the proximal tubule will be described later.

Urea Reabsorption Passive reabsorption of urea, in addition to Cl^- and H_2O, is indirectly linked to active Na^+ reabsorption. **Urea** is a waste product from the breakdown of protein. The osmotically induced reabsorption of H_2O in the proximal tubule secondary to active Na^+ reabsorption produces a concentration gradient for urea that favors passive reabsorption of this waste (❚Figure 14-20). Extensive reabsorption of H_2O in the proximal tubule gradually reduces the original 125 mL/min of filtrate until only 44 mL/min of fluid remain in the lumen by the end of the proximal tubule (65% of the H_2O in the original filtrate, or 81 mL/min, has been reabsorbed). Substances that have been filtered but not reabsorbed become progressively more concentrated in the tubular fluid as H_2O is reabsorbed while they are left behind. Urea is one such substance. Urea's concentration as it is filtered at the glomerulus is identical to its concentration in the plasma entering the peritubular capillaries. The quantity of urea present in the 125 mL of filtered fluid at the beginning of the proximal tubule, however, is concentrated almost threefold in the 44 mL left at the end of the proximal tubule. As a result, the urea concentration within the tubular fluid becomes greater than the urea concentration in the adjacent capillaries. Therefore, a concentration gradient is created for urea to passively diffuse from the tubular lumen into the peritubular capillary plasma. Because the walls of the proximal tubules are only somewhat permeable to urea, only about 50% of the filtered urea is passively reabsorbed by this means.

 Even though only half of the filtered urea is eliminated from the plasma with each pass through the nephrons, this removal rate is adequate. The urea concentration

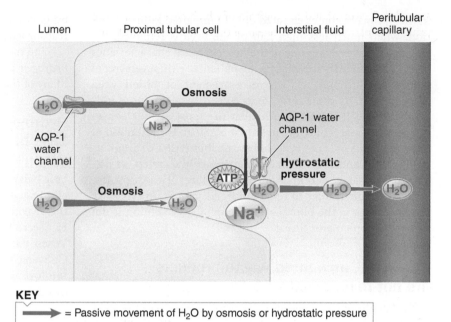

KEY

⟶ = Passive movement of H_2O by osmosis or hydrostatic pressure

→ = Active transport of ion

❚Figure 14-19 **Water reabsorption in the proximal tubule.** The force for osmotic H_2O reabsorption is the compartment of hypertonicity in the lateral spaces established by active extrusion of Na^+ by the basolateral pump. The resultant accumulation of H_2O in the lateral spaces creates a hydrostatic pressure that drives the H_2O into the peritubular capillaries.

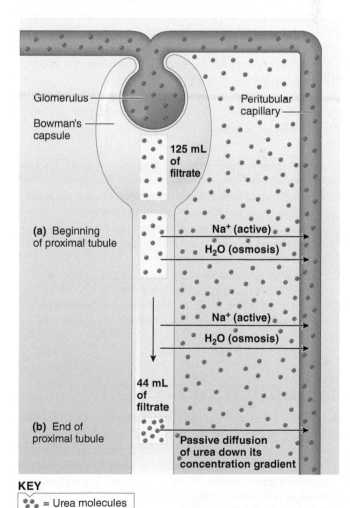

❚Figure 14-20 **Passive reabsorption of urea at the end of the proximal tubule.** (a) In Bowman's capsule and at the beginning of the proximal tubule, urea is at the same concentration as in the plasma and surrounding interstitial fluid. (b) By the end of the proximal tubule, 65% of the original filtrate has been reabsorbed, concentrating the filtered urea in the remaining filtrate. This establishes a concentration gradient favoring passive reabsorption of urea.

FIGURE FOCUS: *Explain why urea is passively reabsorbed in the late part but not the early part of the proximal tubule.*

KEY

⦙⦙ = Urea molecules

in the plasma becomes elevated only in impaired kidney function, when much less than half of the urea is removed. An elevated urea level was one of the first chemical characteristics to be identified in the plasma of patients with severe renal failure. Accordingly, clinical measurement of **blood urea nitrogen (BUN)** came into use as a crude assessment of kidney function. It is now known that the most serious consequences of renal failure are not attributable to the retention of urea, which itself is not especially toxic, but rather to accumulation of H^+ and K^+, which are inadequately secreted (as discussed in a later section about renal failure). Health professionals still often refer to renal failure as *uremia,* in reference to excess urea in the blood, even though urea retention is not this condition's major threat.

In general, unwanted waste products are not reabsorbed.

The other filtered waste products besides urea, such as *uric acid, creatinine,* and *phenols* (derived from many foods) are likewise concentrated in the tubular fluid as H_2O leaves the filtrate to enter the plasma. But urea molecules, being the smallest of the waste products, are the only wastes passively reabsorbed by this concentrating effect. The other wastes cannot leave the lumen down their concentration gradients to be passively reabsorbed because they cannot permeate the tubular wall. Therefore, these waste products generally remain in the tubules and are excreted in the urine in highly concentrated form. This excretion of metabolic wastes is not subject to physiologic control, but when renal function is normal, the excretory processes proceed at a satisfactory rate.

Having completed discussion of tubular reabsorption, we now shift to the other basic renal process carried out by the tubules—tubular secretion.

Check Your Understanding 14.3

1. Show the steps of transepithelial transport on a sketch you make of a kidney tubule and associated peritubular capillary.

2. Describe the sequence of events that take place in the renin–angiotensin–aldosterone system in response to a fall in NaCl, ECF volume, and arterial blood pressure.

3. If the plasma concentration of a substance is 200 mg/100 mL, the substance's T_m is 200 mg/min, and the GFR is 125 mL/min, (1) what is the filtered load of this substance, (2) how much of the substance is reabsorbed, and (3) how much of it is excreted?

14.4 Tubular Secretion

Like tubular reabsorption, tubular secretion involves transepithelial transport, but now the steps are reversed. By providing a second route of entry into the tubules for selected substances, *tubular secretion,* the discrete transfer of substances from the peritubular capillaries into the tubular lumen, is a supplemental mechanism that hastens elimination of these compounds from the body. Anything that gains entry to the tubular fluid, whether by glomerular filtration or tubular secretion, and fails to be reabsorbed is eliminated in the urine.

The most important substances secreted by the tubules are *hydrogen ion, potassium ion,* and *organic anions and cations.* Many of the latter are compounds foreign to the body.

Hydrogen ion secretion is important in acid–base balance.

Renal H^+ secretion is extremely important in regulating acid–base balance in the body. H^+ secreted into the tubular fluid is eliminated from the body in the urine. H^+ can be secreted by the proximal, distal, and collecting tubules, with the extent of H^+ secretion depending on the acidity of the body fluids. When the body fluids are too acidic, H^+ secretion increases. Conversely, when the H^+ concentration in the body fluids is too low, H^+ secretion decreases. (See Chapter 15 for further details.)

Potassium ion secretion is controlled by aldosterone.

Potassium is one of the most abundant cations in the body, but about 98% of the K^+ is in the intracellular fluid because the Na^+–K^+ pump actively transports K^+ into the cells. Because only a relatively small amount of K^+ is in the ECF, even slight changes in the ECF K^+ load can have a pronounced effect on the plasma K^+ concentration. Changes in the plasma K^+ concentration have a marked influence on membrane excitability. Therefore, plasma K^+ concentrations are tightly controlled, primarily by the kidneys.

Renal handling of K^+ is complex. K^+ is selectively moved in opposite directions in different parts of the tubule; it is actively reabsorbed in the proximal tubule and actively secreted by principal cells in the distal and collecting tubules. Furthermore, one type of intercalated cell actively secretes K^+ and another type actively reabsorbs K^+ in the distal and collecting tubules in conjunction with H^+ transport (see Chapter 15). Early in the tubule, K^+ is constantly reabsorbed without regulation, whereas K^+ secretion later in the tubule by the principal cells is variable and subject to regulation. Because the filtered K^+ is almost completely reabsorbed in the proximal tubule, most K^+ in the urine is derived from controlled K^+ secretion in the distal parts of the nephron rather than from filtration.

During K^+ depletion, K^+ secretion in the distal parts of the nephron is reduced to a minimum, so only the small percentage of filtered K^+ that escapes reabsorption in the proximal tubule is excreted in the urine. In this way, K^+ that normally would have been lost in urine is conserved for the body. Conversely, when plasma K^+ levels are elevated, K^+ secretion is adjusted so that just enough K^+ is added to the filtrate for elimination to reduce the plasma K^+ concentration to normal. Thus, K^+ secretion, not the filtration or reabsorption of K^+, is varied in a controlled fashion to regulate the rate of K^+ excretion and maintain the desired plasma K^+ concentration.

Mechanism Of K⁺ Secretion K^+ secretion in the principal cells of the distal and collecting tubules is coupled to Na^+ reab-

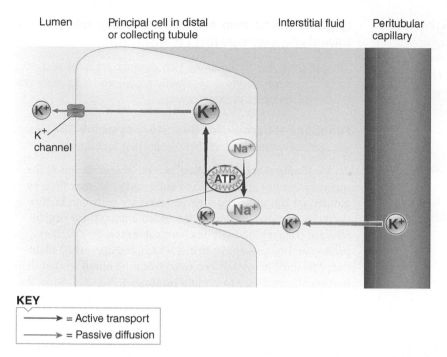

Lumen Principal cell in distal Interstitial fluid Peritubular
           or collecting tubule   capillary

K⁺ channel

KEY

→ = Active transport

→ = Passive diffusion

▮Figure 14-21 **Potassium ion secretion.** The basolateral pump simultaneously transports Na⁺ into the lateral space and K⁺ into the tubular cell. In the parts of the tubule that secrete K⁺, this ion leaves the cell through channels located in the luminal border, thus being secreted. (In the parts of the tubule that do not secrete K⁺, the K⁺ pumped into the cell during Na⁺ reabsorption leaves the cell through channels located in the basolateral border, thus being retained in the body.)

FIGURE FOCUS: *(1) What happens to the Na⁺ transported by the basolateral Na⁺–K⁺ pump during K⁺ secretion? (2) What happens to the K⁺ transported by the basolateral Na⁺–K⁺ pump during Na⁺ reabsorption in the segments of the tubule that do not secrete K⁺?*

sorption by the energy-dependent basolateral Na⁺–K⁺ pump (▮Figure 14-21). This pump not only moves Na⁺ out of the cell into the lateral space, but also transports K⁺ from the lateral space into the tubular cells. The resulting high intracellular K⁺ concentration favors net movement of K⁺ from the cells into the tubular lumen. Movement across the luminal membrane occurs passively through the large number of K⁺ leak channels in this barrier in the distal and collecting tubules. By keeping the interstitial fluid concentration of K⁺ low as it transports K⁺ into the tubular cells from the surrounding interstitial fluid, the basolateral pump encourages passive movement of K⁺ out of the peritubular capillary plasma into the interstitial fluid. A potassium ion leaving the plasma in this manner is later pumped into the cells, from which it passively moves into the lumen. In this way, the basolateral pump actively induces the net secretion of K⁺ from the peritubular capillary plasma into the tubular lumen in the distal parts of the nephron.

Because K⁺ secretion is linked with Na⁺ reabsorption by the Na⁺–K⁺ pump, why isn't K⁺ secreted throughout the Na⁺-reabsorbing segments of the tubule instead of taking place only in the distal parts of the nephron? The answer lies in the location of the passive K⁺ leak channels. In the principal cells of the distal and collecting tubules, the K⁺ channels are concentrated in the luminal membrane, providing a route for K⁺ pumped into the cell to exit into the tubular lumen, thus being secreted. In the proximal tubule, the K⁺ leak channels are located primarily in the basolateral membrane. As a result, K⁺ pumped

into the cell from the lateral space by the Na⁺–K⁺ pump simply moves back out into the lateral space through these channels. This K⁺ recycling permits the ongoing operation of the Na⁺–K⁺ pump to accomplish Na⁺ reabsorption with no local net effect on K⁺.

Control of K⁺ Secretion Several factors can alter the rate of K⁺ secretion, the most important being aldosterone. This hormone stimulates K⁺ secretion by the principal tubular cells late in the nephron while simultaneously enhancing these cells' reabsorption of Na⁺. A rise in plasma K⁺ concentration directly stimulates the adrenal cortex to increase its output of aldosterone, which in turn promotes the secretion and ultimate urinary excretion and elimination of excess K⁺. Conversely, a decline in plasma K⁺ concentration causes a reduction in aldosterone secretion and a corresponding decrease in aldosterone-stimulated renal K⁺ secretion. The amount of K⁺ excreted in the urine varies from 80% to 1% of the filtered quantity, depending on the body's momentary needs.

Note that a rise in plasma K⁺ concentration *directly* stimulates aldosterone secretion by the adrenal cortex, whereas a fall in plasma Na⁺ concentration stimulates aldosterone secretion by means of the complex RAAS pathway. Thus, aldosterone secretion can be stimulated by two separate pathways (▮Figure 14-22). No matter what the stimulus, however, increased aldosterone secretion always promotes simultaneous Na⁺ reabsorption and K⁺ secretion. For this reason, K⁺ secretion can be inadvertently stimulated as a result of increased aldosterone activity brought about by Na⁺ depletion, ECF volume reduction, or a fall in arterial blood pressure totally unrelated to K⁺ balance. The resulting inappropriate loss of K⁺ can lead to K⁺ deficiency.

Effect of H⁺ Secretion on K⁺ Secretion Another factor that can inadvertently alter the magnitude of K⁺ secretion is the acid–base status of the body. The intercalated cells in the distal portions of the nephron secrete either K⁺ or H⁺. An increased rate of secretion of either K⁺ or H⁺ is accompanied by a decreased rate of secretion of the other ion. Normally the kidneys secrete a preponderance of K⁺, but when the body fluids are too acidic and H⁺ secretion is increased as a compensatory measure, K⁺ secretion is correspondingly reduced. This reduced secretion leads to inappropriate K⁺ retention in the body fluids.

Importance of Regulating Plasma K⁺ Concentration
Except in the overriding circumstances of K⁺ imbalances inadvertently induced during renal compensations for Na⁺ or ECF volume deficits or acid–base imbalances, the kidneys usually exert a fine degree of control over plasma K⁺ concentration. This is extremely important because even minor fluctuations in plasma K⁺ concentration can have detrimental consequences.

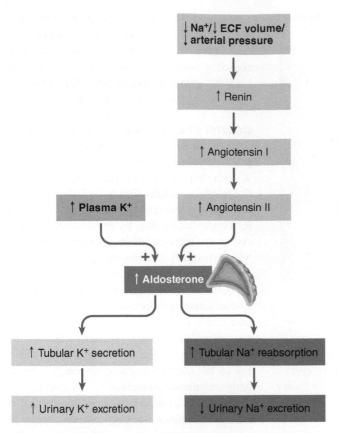

Figure 14-22 **Dual control of aldosterone secretion by K⁺ and Na⁺.**

Clinical Note K⁺ plays a key role in the membrane electrical activity of excitable tissues. Both increases and decreases in the plasma (ECF) K⁺ concentration can alter the intracellular-to-extracellular K⁺ concentration gradient, which in turn can change the resting membrane potential. The most serious consequences of both K⁺ excess and K⁺ deficiency are related to their effect on the heart. Both conditions result in decreased cardiac excitability, for different reasons. A *rise* in ECF K⁺ concentration reduces resting potential (makes it less negative), which decreases the excitability of neurons; skeletal muscle cells; and, most importantly, cardiac muscle cells, by keeping the voltage-gated Na⁺ channels responsible for the rising phase of the cardiac action potential in their inactive (closed and not capable of opening) state (see p. 92). The cell membrane is unable to repolarize completely after depolarization to return the channel to its closed and capable of opening conformation. Some Na⁺ channels are more sensitive than others to the depolarizing effect. As more and more Na⁺ channels are inactivated by rising K⁺ levels, cardiac excitability progressively decreases. A *fall* in ECF K⁺ concentration results in hyperpolarization of nerve and muscle cell membranes, which also reduces their excitability. A greater depolarization than normal is needed to bring the membrane to threshold potential. Thus, both low and high ECF K⁺ concentrations can lead to abnormalities in cardiac rhythm and even death.

Organic anion and cation secretion hastens elimination of foreign compounds.

The proximal tubule contains two distinct types of secretory carriers, one for secretion of organic anions and a separate system for secretion of organic cations.

Functions of Organic Ion Secretory Systems The organic ion secretory systems serve three important functions:

1. By adding more of a particular type of organic ion to the quantity that has already gained entry to the tubular fluid by glomerular filtration, the organic secretory pathways facilitate excretion of these substances. Included among these organic ions are certain blood-borne chemical messengers such as prostaglandins and epinephrine, which, having served their purpose, must be rapidly removed from the blood so that their biological activity is not unduly prolonged.

2. In some important instances, organic ions are poorly soluble in water. To be transported in blood, they are extensively but not irreversibly bound to plasma proteins. Because they are attached to plasma proteins, these substances cannot be filtered through the glomeruli. Tubular secretion facilitates elimination of these nonfilterable organic ions in urine. Even though a given organic ion is largely bound to plasma proteins, a small percentage of the ion always exists in free or unbound form in the plasma. Removal of this free organic ion by secretion permits "unloading" of some of the bound ion, which is then free to be secreted. This, in turn, encourages the unloading of even more organic ion, and so on.

3. Most important, the proximal tubule organic ion secretory systems play a key role in eliminating many foreign compounds from the body. These systems can secrete a large number of different organic ions, both those produced within the body and those foreign organic ions that have gained access to the body fluids. This nonselectivity permits the organic ion secretory systems to hasten removal of many foreign organic chemicals, including food additives, environmental pollutants (for example, pesticides), drugs, and other nonnutritive organic substances that have entered the body. Even though this mechanism helps rid the body of potentially harmful foreign compounds, it is not subject to physiologic adjustments. The carriers cannot pick up their secretory pace when confronting an elevated load of these organic ions.

The liver plays an important role in helping rid the body of foreign compounds. Many foreign organic chemicals are not ionic in their original form, so they cannot be secreted by the organic ion systems. The liver converts these foreign substances into an anionic form that facilitates their secretion by the organic anion system and thus accelerates their elimination.

Clinical Note Many drugs, such as penicillin and nonsteroidal antiinflammatory drugs (NSAIDs), are eliminated from the body by the organic ion secretory systems. To keep the plasma concentration of these drugs at effective levels, the dosage must be repeated frequently to keep pace with the rapid removal of these compounds in the urine.

Because the organic ion secretory carriers are not very selective, different drugs can compete for binding sites on the same carrier. For example, cimetidine (a drug used to treat stomach ulcers; see p. 589) and procainamide (a drug used to treat cardiac arrhythmias) both are secreted by the organic cation secretory carriers. If these drugs were given to the same patient, the urinary excretion rate of both substances would be decreased because they would compete for elimination by the secretory carriers. Thus coadministration of these drugs would lead to much higher blood concentrations of both substances than when each is given alone. Thus, to avoid potential drug toxicity, drugs eliminated by the same secretory pathway should not be taken together.

Summary of Reabsorptive and Secretory Processes

This completes our discussion of the reabsorptive and secretory processes that occur across the proximal and distal portions of the nephron. These processes are summarized in Table 14-2. To generalize, the proximal tubule does most of the reabsorbing. This mass reabsorber transfers much of the filtered water

and needed solutes back into the blood in unregulated fashion. Similarly, the proximal tubule is the major site of secretion, with the exception of K^+ secretion. The distal and collecting tubules then determine the final amounts of H_2O, Na^+, K^+, and H^+ excreted in the urine and thus eliminated from the body. They do so by fine-tuning the amount of Na^+ and H_2O reabsorbed and the amount of K^+ and H^+ secreted. These processes in the distal part of the nephron are all subject to control, depending on the body's momentary needs. The unwanted filtered waste products are left behind to be eliminated in the urine, along with excess amounts of filtered or secreted nonwaste products that fail to be reabsorbed.

We next focus on the end result of the basic renal processes—what's left in the tubules to be excreted in urine, and, as a consequence, what has been cleared from plasma.

TABLE 14-2 Summary of Transport across Proximal and Distal Portions of the Nephron

PROXIMAL TUBULE

Reabsorption	Secretion
67% of filtered Na^+ actively reabsorbed, not subject to control; Cl^- follows passively	Variable H^+ secretion, depending on acid–base status of body
All filtered glucose and amino acids reabsorbed by secondary active transport; not subject to control	Organic ion secretion; not subject to control
Variable amounts of filtered PO_4^{3-} and other electrolytes reabsorbed; subject to control	
65% of filtered H_2O osmotically reabsorbed; not subject to control	
50% of filtered urea passively reabsorbed; not subject to control	
Almost all filtered K^+ reabsorbed; not subject to control	

DISTAL TUBULE AND COLLECTING DUCT

Reabsorption	Secretion
Variable Na^+ reabsorption, controlled by aldosterone; Cl^- follows passively	Variable H^+ secretion, depending on acid–base status of body
Variable H_2O reabsorption, controlled by vasopressin	Variable K^+ secretion, controlled by aldosterone

14.5 Urine Excretion and Plasma Clearance

Of the 125 mL of plasma filtered per minute, typically 124 mL/min are reabsorbed, so the final quantity of urine formed averages 1 mL/min. Thus, of the 180 liters filtered per day, 1.5 liters of urine are excreted.

Urine contains high concentrations of various waste products plus variable amounts of the substances regulated by the kidneys, with any excess quantities having spilled into the urine. Useful substances are conserved by reabsorption, so they do not appear in the urine.

A relatively small change in the quantity of filtrate reabsorbed can bring about a large change in the volume of urine formed. For example, a reduction of less than 1% in the total reabsorption rate, from 124 to 123 mL/min, increases the urinary excretion rate by 100%, from 1 to 2 mL/min.

Plasma clearance is the volume of plasma cleared of a particular substance per minute.

Compared to plasma entering the kidneys through the renal arteries, plasma leaving the kidneys through the renal veins lacks the materials that were left behind to be eliminated in the urine. By excreting substances in the urine, the kidneys clean or "clear" the plasma flowing through them of these substances. The **plasma clearance** of any substance is defined as the volume of plasma completely cleared of that substance by the kidneys per minute.[3] It refers not to the *amount of the substance* removed but to the *volume of plasma* from which that amount was removed. Plasma clearance is actually a more useful measure than urine excretion; it is more important to know what effect urine excretion has on removing materials from body fluids than to know the volume and composition of discarded urine. Plasma clearance expresses the kidneys' effectiveness in removing various substances from the internal fluid environment.

Plasma clearance can be calculated for any plasma constituent as follows:

$$\text{Clearance rate of a substance (mL/min)} = \frac{\begin{array}{c}\text{urine concentration}\\\text{of the substance}\\\text{(quantity/mL urine)}\end{array} \times \begin{array}{c}\text{urine flow rate}\\\text{(mL/min)}\end{array}}{\begin{array}{c}\text{plasma concentration of the substance}\\\text{(quantity/mL plasma)}\end{array}}$$

The plasma clearance rate varies for different substances, depending on how the kidneys handle each substance. Let us consider how three common patterns of renal handling influence clearance rates for the involved substances.

Plasma Clearance Rate for a Substance Filtered But Not Reabsorbed or Secreted

Assume that a plasma constituent, substance X, is freely filterable at the glomerulus but is not reabsorbed or secreted. As 125 mL/min of plasma are filtered and subsequently reabsorbed, the quantity of substance X originally contained within the 125 mL is left behind in the tubules to be excreted. Thus, 125 mL of plasma are cleared of substance X each minute (Figure 14-23a). (Of the 125 mL/min of plasma filtered, 124 mL/min of the filtered fluid are returned, through reabsorption, to the plasma minus substance X, thus clearing this 124 mL/min of substance X. In addition, the 1 mL/min of fluid lost in urine is eventually replaced by an equivalent volume of ingested H_2O that is already clear of substance X. Therefore, 125 mL of plasma cleared of substance X are, in effect, returned to the plasma for every 125 mL of plasma filtered per minute.) Thus, *the plasma clearance rate of a substance filtered but not reabsorbed or secreted always equals the GFR.*

Clinical Note No normally occurring chemical in the body has the characteristics of substance X. All substances naturally present in the plasma, even wastes, are reabsorbed or secreted to some extent. However, **inulin** (do not confuse with insulin), a harmless foreign carbohydrate produced abundantly by Jerusalem artichokes and to a lesser extent by other root vegetables such as onions and garlic, is freely filtered and not reabsorbed or secreted—an ideal substance X. Inulin can be injected and its plasma clearance determined as a clinical means of finding out the GFR. Because all glomerular filtrate formed is cleared of inulin, the volume of plasma cleared of

inulin per minute equals the volume of plasma filtered per minute—that is, the GFR, as the following example illustrates:

$$\text{Clearance rate for inulin} = \frac{30 \text{ mg/mL urine} \times 1.25 \text{ mL urine/min}}{0.30 \text{ mg/mL plasma}}$$

$$= 125 \text{ mL plasma/min}$$

Although determination of inulin plasma clearance is accurate and straightforward, it is not very convenient because inulin must be infused continuously throughout the determination to maintain a constant plasma concentration. Therefore, the plasma clearance of an endogenous substance, **creatinine**, is often used instead to find a rough estimate of the GFR. Creatinine, an end product of muscle metabolism, is produced at a relatively constant rate. It is freely filtered and not reabsorbed but is slightly secreted. Accordingly, creatinine clearance is not a completely accurate reflection of the GFR, but it does provide a close approximation and can be more readily determined than inulin clearance.

Plasma Clearance Rate for a Substance Filtered and Reabsorbed

Some or all of a reabsorbable substance that has been filtered is returned to the plasma. Because less than the filtered volume of plasma is cleared of the substance, *the plasma clearance rate of a reabsorbable substance is always less than the GFR.* For example, the plasma clearance for glucose is normally zero. All the filtered glucose is reabsorbed with the rest of the returning filtrate, so none of the plasma is cleared of glucose (Figure 14-23b).

For a substance that is partially reabsorbed, such as urea, only part of the filtered plasma is cleared of that substance. With about 50% of the filtered urea being passively reabsorbed, only half of the filtered plasma, or 62.5 mL, is cleared of urea each minute (Figure 14-23c).

Clearance Rate for a Substance Filtered and Secreted

Tubular secretion allows the kidneys to clear certain materials from the plasma more efficiently. Only 20% of the plasma entering the kidneys is filtered. The remaining 80% passes unfiltered into the peritubular capillaries. The only means by which this unfiltered plasma can be cleared of any substance during the trip through the kidneys before being returned to the general circulation is by secretion. An example is H^+. Not only is filtered plasma cleared of nonreabsorbable H^+, but the plasma from which H^+ is secreted is also cleared of H^+. For example, if the quantity of H^+ secreted is equivalent to the quantity of H^+ present in 25 mL of plasma, the clearance rate for H^+ will be 150 mL/min at the normal GFR of 125 mL/min. Every minute 125 mL of plasma loses its H^+ through filtration and failure of reabsorption, and an additional 25 mL of plasma loses its H^+ through secretion. Thus, *the plasma clearance rate for a secreted substance is always greater than the GFR* (Figure 14-23d).

Clinical Note Just as inulin can be used to determine the GFR, plasma clearance of another foreign compound, the organic anion **para-aminohippuric acid (PAH)**, can be used to measure renal plasma flow. Like inulin, PAH is freely filterable and nonreabsorbable. It differs, however, in that all the PAH in

[3]Actually, plasma clearance is an artificial concept because when a particular substance is excreted in the urine, that substance's concentration in the plasma as a whole is uniformly decreased as a result of thorough mixing in the circulatory system. However, it is useful for comparative purposes to consider clearance in effect as the volume of plasma that would have contained the total quantity of the substance (at the substance's concentration prior to excretion) that the kidneys excreted in one minute—that is, the hypothetical volume of plasma completely cleared of that substance per minute.

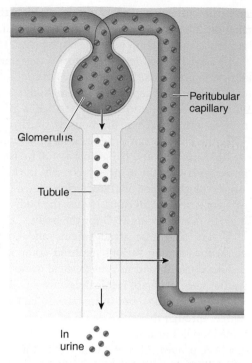

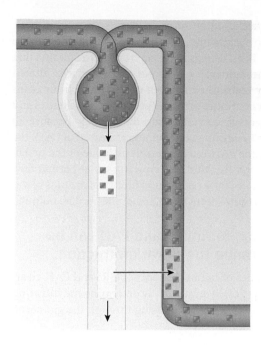

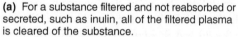

Glomerulus

Peritubular capillary

Tubule

In urine

(a) For a substance filtered and not reabsorbed or secreted, such as inulin, all of the filtered plasma is cleared of the substance.

(b) For a substance filtered, not secreted, and completely reabsorbed, such as glucose, none of the filtered plasma is cleared of the substance.

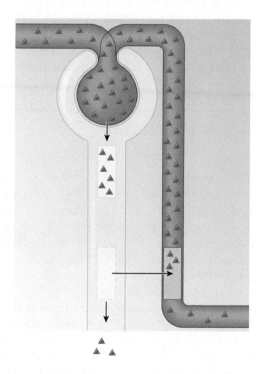

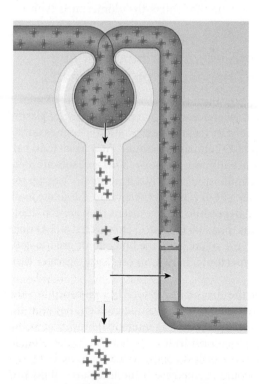

(c) For a substance filtered, not secreted, and partially reabsorbed, such as urea, only a portion of the filtered plasma is cleared of the substance.

(d) For a substance filtered and secreted but not reabsorbed, such as hydrogen ion, all of the filtered plasma is cleared of the substance, and the peritubular plasma from which the substance is secreted is also cleared.

❙Figure 14-23 **Plasma clearance for substances handled in different ways by the kidneys.**

the plasma that escapes filtration is secreted from the peritubular capillaries by the organic anion secretory pathway in the proximal tubule. Thus, PAH is removed from all the plasma that flows through the kidneys—both from plasma that is filtered and subsequently reabsorbed without its PAH and from unfiltered plasma that continues on in the peritubular capillaries and loses its PAH by active secretion into the tubules. Because all the plasma that flows through the kidneys is cleared of PAH, the plasma clearance for PAH is a reasonable estimate of the rate of plasma flow through the kidneys. Typically, renal plasma flow averages 625 mL/min, for a renal blood flow (plasma plus blood cells) of 1140 mL/min—more than 20% of the cardiac output.

Clearance rates for inulin and PAH can be used to determine the filtration fraction.

If you know the rates of inulin clearance (GFR) and PAH clearance (renal plasma flow) you can easily determine the **filtration fraction**, the fraction of plasma flowing through the glomeruli that is filtered into the tubules:

$$\text{Filtration fraction} = \frac{\text{GFR (plasma inulin clearance)}}{\text{renal plasma flow (plasma PAH clearance)}}$$

$$= \frac{125\ \text{mL/min}}{625\ \text{mL/min}} = 20\%$$

Thus, 20% of the plasma that enters the glomeruli is typically filtered.

The kidneys can excrete urine of varying concentrations depending on body needs.

Having considered how the kidneys deal with a variety of solutes in the plasma, we now concentrate on renal handling of plasma H_2O. The ECF osmolarity (solute concentration) depends on the relative amount of H_2O compared to solute. At normal fluid balance and solute concentration, the body fluids are **isotonic** at an osmolarity of 300 milliosmols per liter (mOsm/L) (see pp. 69 and A-7). If too much H_2O is present relative to the solute load, the body fluids are **hypotonic**, which means they are too dilute at an osmolarity less than 300 mOsm/L. However, if a H_2O deficit exists relative to the solute load, the body fluids are too concentrated or are **hypertonic**, having an osmolarity greater than 300 mOsm/L.

Knowing that the driving force for H_2O reabsorption the entire length of the tubules is an osmotic gradient between the tubular lumen and the surrounding interstitial fluid, you would expect, given osmotic considerations, that the kidneys could not excrete urine more or less concentrated than the body fluids. Indeed, this would be the case if the interstitial fluid surrounding the tubules in the kidneys were identical in osmolarity to the rest of the body fluids. Water reabsorption would proceed only until the tubular fluid equilibrated osmotically with the interstitial fluid, and the body would have no way to eliminate excess H_2O when the body fluids were hypotonic or to conserve H_2O in the presence of hypertonicity.

Fortunately, a **vertical osmotic gradient** is uniquely maintained in the medullary interstitial fluid of each kidney. The concentration of the interstitial fluid progressively increases

from the cortical boundary down through the depth of the renal medulla until it reaches a maximum of 1200 mOsm/L in humans at the junction with the renal pelvis (▮Figure 14-24).

By a mechanism described shortly, this gradient enables the kidneys to produce urine that ranges in concentration from 100 to 1200 mOsm/L, depending on the body's state of hydration. When the body is in ideal fluid balance, 1 mL/min of isotonic urine is formed. When the body is overhydrated (too much H_2O), the kidneys can produce a large volume of dilute urine (up to 25 mL/min and hypotonic at 100 mOsm/L), eliminating the excess H_2O in the urine. Conversely, the kidneys can put out a small volume of concentrated urine (down to 0.3 mL/min and hypertonic at 1200 mOsm/L) when the body is dehydrated (too little H_2O), conserving H_2O for the body.

Unique anatomic arrangements and complex functional interactions among various nephron components in the renal medulla establish and use the vertical osmotic gradient. Recall that the hairpin loop of Henle dips only slightly into the medulla in cortical nephrons, but in juxtamedullary nephrons the loop plunges through the entire depth of the medulla so that the tip of the loop lies near the renal pelvis (see ▮Figures 14-1c, p. 493, and 14-5, p. 496). Also, the vasa recta of juxtamedullary nephrons follow the same deep hairpin loop as the long loop of Henle. Flow in both the long loops of Henle and the vasa recta is considered *countercurrent* because the flow in the two closely adjacent limbs of the loop moves in opposite directions. Also running through the medulla in the descending direction only, on their way to the renal pelvis, are the collecting ducts that serve both types of nephrons. This arrangement, coupled with the permeability and transport characteristics of these tubular

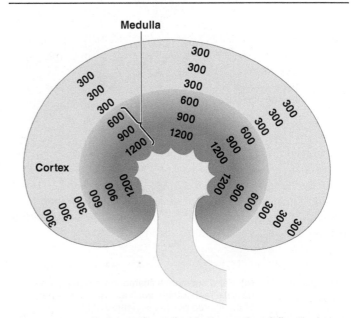

▮Figure 14-24 **Vertical osmotic gradient in the renal medulla.** All values are in mOsm/L. The osmolarity of the interstitial fluid throughout the renal cortex is isotonic at 300 mOsm/L, but the osmolarity of the interstitial fluid in the renal medulla increases progressively from 300 mOsm/L at the boundary with the cortex to a maximum of 1200 mOsm/L at the junction with the renal pelvis. (The kidney is rotated 90° from its normal position in an upright person for better visualization of the vertical osmotic gradient in the renal medulla.)

segments, plays a key role in the kidneys' ability to produce urine of varying concentrations, depending on the body's needs for water conservation or elimination. Briefly, the juxtamedullary nephrons' long loops of Henle *establish* the vertical osmotic gradient by means of countercurrent multiplication, their vasa recta *preserve* this gradient while providing blood to the renal medulla by means of countercurrent exchange, and the collecting ducts of all nephrons *use* the gradient, in conjunction with the hormone vasopressin, to produce urine of varying concentrations. We next examine each of these processes in greater detail.

Long Henle's loops establish the vertical osmotic gradient by countercurrent multiplication.

We now follow the filtrate through a long-looped nephron to see how this structure establishes a vertical osmotic gradient in the renal medulla via **countercurrent multiplication** during which an active concentrating mechanism's effect is multiplied as a result of countercurrent flow. Immediately after the filtrate is formed, uncontrolled osmotic reabsorption of filtered H_2O occurs in the proximal tubule secondary to active Na^+ reabsorption. As a result, by the end of the proximal tubule, about 65% of the filtrate has been reabsorbed, but the 35% remaining in the tubular lumen still has the same osmolarity as the body fluids. Therefore, the fluid entering the loop of Henle is still isotonic. An additional 15% of the filtered H_2O is obligatorily reabsorbed from the loop of Henle during the establishment and maintenance of the vertical osmotic gradient, with the osmolarity of the tubular fluid being altered in the process.

Properties of the Descending and Ascending Limbs of a Long Henle's Loop The following functional distinctions between the descending limb of a long Henle's loop (which carries fluid from the proximal tubule down into the depths of the medulla) and the ascending limb (which carries fluid up and out of the medulla into the distal tubule) are crucial for establishing the incremental osmotic gradient in the medullary interstitial fluid.

The *descending limb* (1) is highly permeable to H_2O (via abundant, always-open AQP-1 water channels) and (2) does not actively extrude Na^+—that is, it does not reabsorb Na^+. (It is the only segment of the entire tubule that does not do so.)

The *ascending limb* (1) actively transports NaCl out of the tubular lumen into the surrounding interstitial fluid and (2) is always impermeable to H_2O, so salt leaves the tubular fluid without H_2O osmotically following along.

Mechanism of Countercurrent Multiplication The close proximity and countercurrent flow of the two limbs allow important interactions between them. Even though the flow of fluids is continuous through the loop of Henle, we can visualize what happens step by step, much like an animated film run so slowly that each frame can be viewed.

Initially, before the vertical osmotic gradient is established, the medullary interstitial fluid concentration is uniformly 300 mOsm/L, as are the rest of the body fluids (I Figure 14-25, Initial scene).

The active salt pump in the ascending limb can transport NaCl out of the lumen until the surrounding interstitial fluid is 200 mOsm/L more concentrated than the tubular fluid in this limb. When the ascending limb pump starts actively extruding NaCl, the medullary interstitial fluid becomes hypertonic. Water cannot follow osmotically from the ascending limb because this limb is impermeable to H_2O. However, net diffusion of H_2O does occur from the descending limb into the interstitial fluid. The tubular fluid entering the descending limb from the proximal tubule is isotonic. Because the descending limb is highly permeable to H_2O, net diffusion of H_2O occurs by osmosis out of the descending limb into the more concentrated interstitial fluid. The passive movement of H_2O out of the descending limb continues until the osmolarities of the fluid in the descending limb and the interstitial fluid become equilibrated. Thus, the tubular fluid entering the loop of Henle immediately starts to become more concentrated as it loses H_2O. At equilibrium, the osmolarity of the ascending limb fluid is 200 mOsm/L and the osmolarities of the interstitial fluid and descending limb fluid are equal at 400 mOsm/L (I Figure 14-25, step **1**).

If we now advance the entire column of fluid in the loop several frames (step **2**), a mass of 200 mOsm/L fluid exits from the top of the ascending limb into the distal tubule, and a new mass of isotonic fluid at 300 mOsm/L enters the top of the descending limb from the proximal tubule. At the bottom of the loop, a comparable mass of 400 mOsm/L fluid from the descending limb moves forward around the tip into the ascending limb, placing it opposite a 400 mOsm/L region in the descending limb, but the 200 mOsm/L concentration difference has been lost at both the top and the bottom of the loop.

The ascending limb pump again transports NaCl out while H_2O passively leaves the descending limb until a 200 mOsm/L difference is reestablished between the ascending limb and both the interstitial fluid and the descending limb at each horizontal level (step **3**). Note, however, that the concentration of tubular fluid is progressively increasing in the descending limb and progressively decreasing in the ascending limb.

As the tubular fluid is advanced still farther (step **4**), the 200 mOsm/L concentration gradient is disrupted again at all horizontal levels. Again, active extrusion of NaCl from the ascending limb, coupled with the net diffusion of H_2O out of the descending limb, reestablishes the 200 mOsm/L gradient at each horizontal level (step **5**).

As the fluid flows slightly forward again and this stepwise process continues (step **6**), the fluid in the descending limb becomes progressively more hypertonic until it reaches a maximum concentration of 1200 mOsm/L at the bottom of the loop, four times the normal concentration of body fluids. Because the interstitial fluid always achieves equilibrium with the descending limb, an incremental vertical concentration gradient ranging from 300 to 1200 mOsm/L is likewise established in the medullary interstitial fluid. In contrast, the concentration of the tubular fluid progressively decreases in the ascending limb as NaCl is pumped out but H_2O is unable to follow. In fact, the tubular fluid even becomes hypotonic before leaving the ascending limb to enter the distal tubule at a concentration of 100 mOsm/L, one third the normal concentration of body fluids.

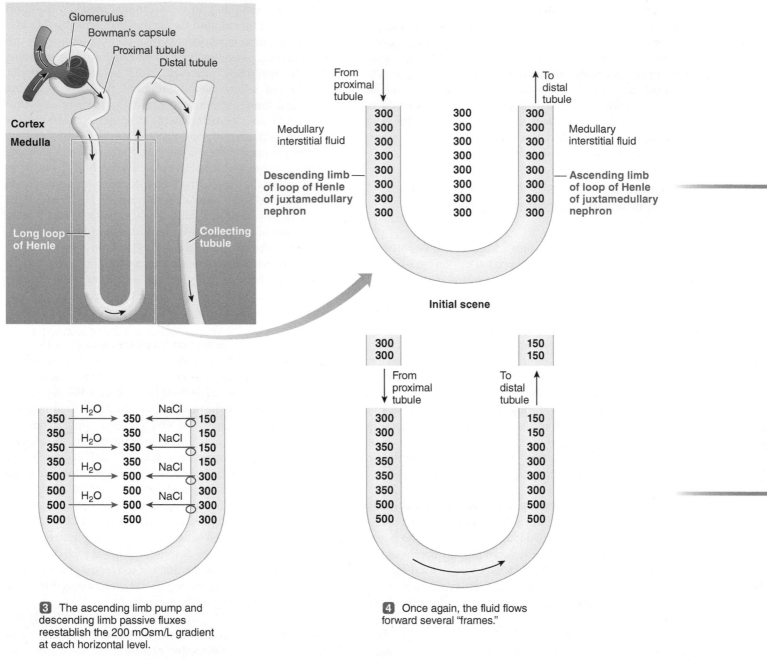

Figure 14-25 Countercurrent multiplication in the renal medulla. All values are in mOsm/L.

Note that although a gradient of only 200 mOsm/L exists between the ascending limb and the surrounding fluids at each medullary horizontal level, a larger vertical gradient exists from the top to the bottom of the medulla. Even though the ascending limb pump can generate a gradient of only 200 mOsm/L, this effect is multiplied into a large vertical gradient because of the countercurrent flow within the loop. Thus, this concentrating mechanism accomplished by the loop of Henle is known as **countercurrent multiplication**.

We have artificially described countercurrent multiplication in a stop-and-flow, stepwise fashion to facilitate understanding. However realize that once the incremental medul-

lary gradient is established, it stays constant because of the continuous flow of fluid, coupled with the ongoing ascending limb active transport and the accompanying descending limb passive fluxes.

Benefits of Countercurrent Multiplication If you consider only what happens to the tubular fluid as it flows through the loop of Henle, the whole process seems an exercise in futility. The isotonic fluid that enters the loop becomes progressively more concentrated as it flows down the descending limb, achieving a maximum concentration of 1200 mOsm/L, only to become progressively more diluted as it

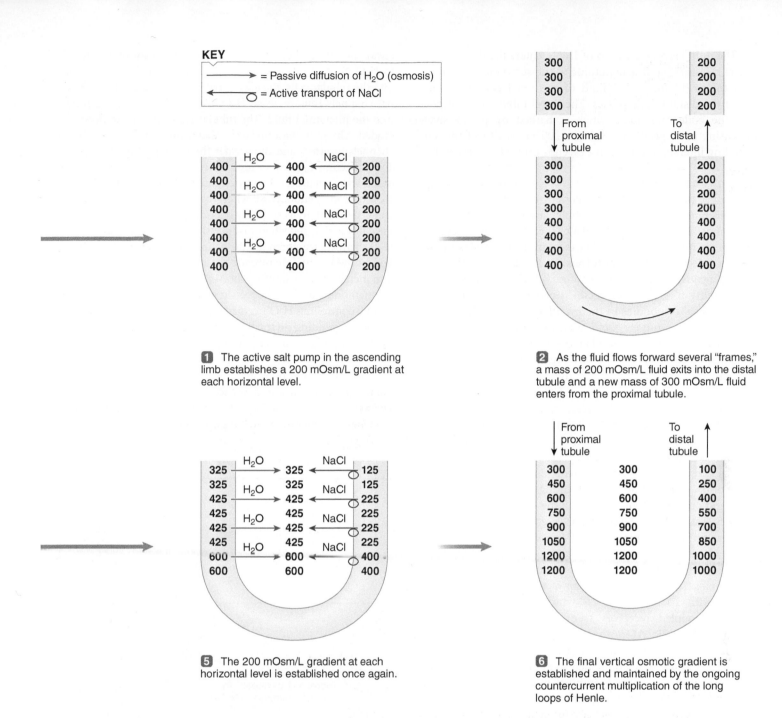

KEY

⟶ = Passive diffusion of H₂O (osmosis)

⟵ ○ = Active transport of NaCl

1 The active salt pump in the ascending limb establishes a 200 mOsm/L gradient at each horizontal level.

2 As the fluid flows forward several "frames," a mass of 200 mOsm/L fluid exits into the distal tubule and a new mass of 300 mOsm/L fluid enters from the proximal tubule.

5 The 200 mOsm/L gradient at each horizontal level is established once again.

6 The final vertical osmotic gradient is established and maintained by the ongoing countercurrent multiplication of the long loops of Henle.

flows up the ascending limb, finally leaving the loop at a minimum concentration of 100 mOsm/L. What is the point of concentrating the fluid fourfold and then turning around and diluting it until it leaves at one third the concentration at which it entered? Such a mechanism offers two benefits. First, it establishes a vertical osmotic gradient in the medullary interstitial fluid. This gradient, in turn, is used by the collecting ducts to concentrate the tubular fluid so that a urine *more concentrated* than normal body fluids can be excreted. Second, because the fluid is hypotonic as it enters the distal parts of the tubule, the kidneys can excrete a urine *more dilute* than normal body fluids. Let us see how.

Vasopressin controls variable H₂O reabsorption in the final tubular segments.

After obligatory H₂O reabsorption from the proximal tubule (65% of the filtered H₂O) and loop of Henle (15% of the filtered H₂O), 20% of the filtered H₂O remains in the lumen to enter the distal and collecting tubules for variable reabsorption under hormonal control. This is still a large volume of filtered H₂O subject to regulated reabsorption; 20% × GFR (180 L/day) = 36 L/day to be reabsorbed to varying extents, depending on the body's state of hydration. This is more than 13 times the amount of plasma H₂O in the entire circulatory system.

The fluid leaving the loop of Henle enters the distal tubule at 100 mOsm/L, so it is hypotonic to the surrounding isotonic (300 mOsm/L) interstitial fluid of the renal cortex through which the distal tubule passes. The distal tubule then empties into the collecting duct, which is bathed by progressively increasing concentrations (300 to 1200 mOsm/L) of the surrounding interstitial fluid as it descends through the medulla.

Role of Vasopressin For H_2O absorption to occur across a segment of the tubule, two criteria must be met: (1) an osmotic gradient must exist across the tubule, and (2) the tubular segment must be permeable to H_2O. The distal and collecting tubules are *impermeable* to H_2O except in the presence of **vasopressin**, also known as **antidiuretic hormone** (*antidiuretic* means "against increased urine output"),[4] which increases their permeability to H_2O. Vasopressin is produced by several specific neuronal cell bodies in the *hypothalamus* and then stored in the *posterior pituitary gland,* which is attached to the hypothalamus by a thin stalk (see Figure 18-4, p. 647). The hypothalamus controls release of vasopressin from the posterior pituitary into the blood. In negative-feedback fashion, vasopressin secretion is stimulated by a H_2O deficit when the ECF is too concentrated (that is, hypertonic) and H_2O must be conserved for the body, and it is inhibited by a H_2O excess when the ECF is too dilute (that is, hypotonic) and surplus H_2O must be eliminated in urine.

Vasopressin reaches the basolateral membrane of the principal tubular cells lining the distal and collecting tubules through the circulatory system. Here, it binds with V_2 receptors specific for it (Figure 14-26). (Vasopressin binds with different V_1 *receptors* on vascular smooth muscle to exert its vasoconstrictor effects; see p. 350.) Binding of vasopressin with its V_2 receptors, which are G-protein-coupled receptors (see p. 117), activates the cyclic AMP (cAMP) second-messenger system within these tubular cells (see p. 123). This binding ultimately increases permeability of the opposite luminal membrane to H_2O by promoting insertion of aquaporins (specifically, *AQP-2*) in this membrane by means of exocytosis. Without these aquaporins, the luminal membrane is impermeable to H_2O. Once H_2O enters the tubular cells from the filtrate through these vasopressin-regulated luminal water channels, it passively leaves the cells down the osmotic gradient across the cells' basolateral membrane to enter the interstitial fluid. The aquaporins in the basolateral membrane of the distal and collecting tubule (*AQP-3* and *AQP-4*) are always present and open, so this membrane is always permeable to H_2O. By permitting more H_2O to permeate from the lumen into the tubular cells, the additional vasopressin-regulated luminal channels thus increase H_2O reabsorption from the filtrate into the interstitial fluid. The tubular response to vasopressin is graded: The more vasopressin present, the more luminal water channels inserted, and the greater the permeability of the distal and collecting tubules to H_2O. The increase in luminal membrane water channels is not permanent, however. The channels are retrieved by endocytosis when vasopressin secretion decreases and cAMP activity is similarly decreased. Accordingly, H_2O permeability is reduced when vasopressin secretion decreases. These H_2O channels are stored in internalized vesicles ready for reinsertion in the luminal membrane the next time vasopressin secretion increases. This shuttling of AQP-2 into and out of the luminal membrane under vasopressin command provides a means of rapidly controlling H_2O permeability of the distal and collecting tubules, depending on the body's momentary needs.

Vasopressin influences H_2O permeability only in the distal and collecting tubules. It has no influence over the 80% of the filtered H_2O that is obligatorily reabsorbed without control in the proximal tubule and descending limb of the loop of Henle. The ascending limb of Henle's loop is always impermeable to H_2O, even in the presence of vasopressin.

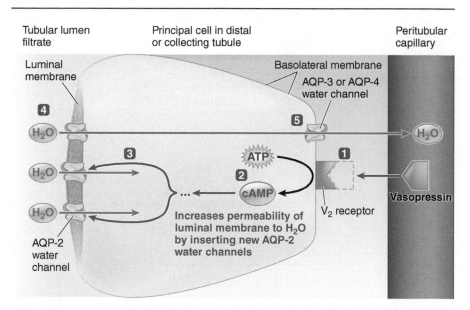

1 Blood-borne vasopressin binds with its receptor sites on the basolateral membrane of a principal cell in the distal or collecting tubule.

2 This binding activates the cyclic AMP (cAMP) second-messenger pathway within the cell.

3 Cyclic AMP increases the opposite luminal membrane's permeability to H_2O by promoting the insertion of vasopressin-regulated AQP-2 water channels into the membrane. This membrane is impermeable to water in the absence of vasopressin.

4 Water enters the tubular cell from the tubular lumen through the inserted water channels.

5 Water exits the cell through different, always open water channels (either AQP-3 or AQP-4) permanently positioned at the basolateral border, and then enters the blood, in this way being reabsorbed.

Figure 14-26 **Mechanism of action of vasopressin.**

[4]Even though textbooks traditionally have tended to use the name *antidiuretic hormone* for this hormone, especially when discussing its actions on the kidney, investigators in the field now prefer *vasopressin*.

Regulation of H₂O Reabsorption in Response to a H₂O Deficit When vasopressin secretion increases in response to a H₂O deficit and the permeability of the distal and collecting tubules to H₂O accordingly increases, the hypotonic tubular fluid entering the distal part of the nephron can lose progressively more H₂O by osmosis into the interstitial fluid as the tubular fluid first flows through the isotonic cortex and then is exposed to the ever-increasing osmolarity of the medullary interstitial fluid as it plunges toward the renal pelvis (■Figure 14-27a). As the 100 mOsm/L tubular fluid enters the distal tubule and is exposed to a surrounding interstitial fluid of 300 mOsm/L, H₂O leaves the tubular fluid by osmosis across the now-permeable tubular cells until the tubular fluid reaches a maximum concentration of 300 mOsm/L by the end of the distal tubule. As this 300 mOsm/L tubular fluid progresses farther into the collecting duct, it is exposed to even higher osmolarity in the surrounding medullary interstitial fluid. Consequently, the tubular fluid loses more H₂O by osmosis and becomes further concentrated; only to move farther forward, be exposed to an even higher interstitial fluid osmolarity, and lose even more H₂O; and so on.

Under the influence of maximum levels of vasopressin, the tubular fluid can be concentrated up to 1200 mOsm/L by the end of the collecting ducts. The fluid is not modified any further beyond the collecting duct, so what remains in the tubules at this point is urine. As a result of this extensive vasopressin-promoted reabsorption of H₂O in the late segments of the tubule, a small volume of urine concentrated up to 1200 mOsm/L can be excreted. As little as 0.3 mL of urine may be formed each minute, less than one third the normal urine flow rate of 1 mL/min. The reabsorbed H₂O entering the medullary interstitial fluid is picked up by the peritubular capillaries and returned to the general circulation, thus being conserved for the body.

Although vasopressin promotes H₂O conservation by the body, it cannot halt urine production, even when a person is not taking in any H₂O, because a minimum volume of H₂O must be excreted with the solute wastes. Collectively, the waste products and other constituents eliminated in the urine average 600 milliosmols each day. Because the maximum urine concentration is 1200 mOsm/L, the minimum volume of urine required to excrete these wastes is 500 mL per day (600 milliosmols of wastes per day ÷ 1200 milliosmols per liter of urine = 0.5 liter, [500 mL] per day, or 0.3 mL/min). Thus, under maximal vasopressin influence, 99.7% of the 180 liters of plasma H₂O filtered per day is returned to the blood, with an obligatory H₂O loss of 0.5 liter.

The kidneys' ability to tremendously concentrate urine to minimize H₂O loss when necessary is possible only because of the presence of the vertical osmotic gradient in the medulla. If this gradient did not exist, the kidneys could not produce a urine more concentrated than the body fluids no matter how much vasopressin was secreted because the only driving force for H₂O reabsorp-

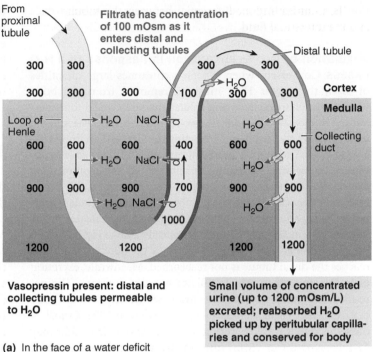

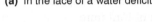

Vasopressin present: distal and collecting tubules permeable to H₂O

Small volume of concentrated urine (up to 1200 mOsm/L) excreted; reabsorbed H₂O picked up by peritubular capillaries and conserved for body

(a) In the face of a water deficit

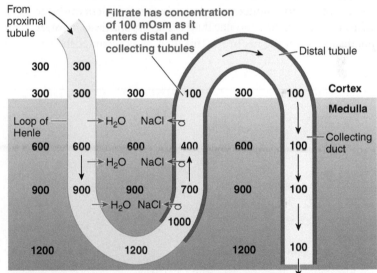

No vasopressin present: distal and collecting tubules impermeable to H₂O

Large volume of dilute urine; (as low as 100 mOsm/L) excreted; no H₂O reabsorbed in distal portion of nephron; excess H₂O eliminated from body in urine

(b) In the face of a water excess

KEY

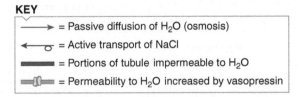

⟶	= Passive diffusion of H₂O (osmosis)
⟵ₒ	= Active transport of NaCl
▬▬	= Portions of tubule impermeable to H₂O
▬◖▬	= Permeability to H₂O increased by vasopressin

■Figure 14-27 **Excretion of urine of varying concentration depending on the body's needs.** All values are in mOsm/L.
FIGURE FOCUS: *(1) Furosemide is a loop diuretic that acts on the loop of Henle to block NaCl transport by the ascending limb. Explain how this action promotes diuresis (increased urine output). (2) Diabetes insipidus is a disease characterized by a deficiency of vasopressin. Explain how this condition promotes diuresis.*

tion is a concentration differential between the tubular fluid and the interstitial fluid.

Regulation of H_2O Reabsorption in Response to a H_2O Excess

Conversely, when a person consumes large quantities of H_2O, the excess H_2O must be removed from the body without simultaneously losing solutes that are critical for maintaining homeostasis. Under these circumstances, no vasopressin is secreted, so the distal and collecting tubules remain impermeable to H_2O. The tubular fluid entering the distal tubule is hypotonic (100 mOsm/L), having lost salt without an accompanying loss of H_2O in the ascending limb of Henle's loop. As this hypotonic fluid passes through the distal and collecting tubules (Figure 14-27b), the medullary osmotic gradient cannot exert any influence because the late tubule is impermeable to H_2O. Thus, in the absence of vasopressin, the 20% of the filtered fluid that reaches the distal tubule is not reabsorbed. Meanwhile, excretion of wastes and other urinary solutes remains constant. The net result is a large volume of dilute urine, which helps rid the body of excess H_2O. Urine osmolarity may be as low as 100 mOsm/L—the same as in the fluid entering the distal tubule. Urine flow may be increased up to 25 mL/min in the absence of vasopressin, compared to the normal urine production of 1 mL/min.

The ability to produce urine less concentrated than the body fluids depends on the tubular fluid being hypotonic as it enters the distal part of the nephron. This dilution is accomplished in the ascending limb, as NaCl is actively extruded but H_2O cannot follow. Therefore, the loop of Henle, by simultaneously establishing the medullary osmotic gradient and diluting the tubular fluid before it enters the distal segments, plays a key role in allowing the kidneys to excrete urine that ranges in concentration from 100 to 1200 mOsm/L.

The vasa recta preserve the vertical osmotic gradient by countercurrent exchange.

The renal medulla must be supplied with blood to nourish the tissues in this area and to transport water that is reabsorbed by the loops of Henle and collecting ducts back to the general circulation. In doing so, however, it is critical that circulation of blood through the medulla does not disturb the vertical gradient of hypertonicity established by the loops of Henle. Consider the situation if blood were to flow straight through from the cortex to the inner medulla and then directly into the renal vein (Figure 14-28a). Because capillaries are freely permeable to NaCl and H_2O, the blood would progressively pick up salt and lose H_2O through passive fluxes down concentration and osmotic gradients as it flowed through the depths of the medulla. Isotonic blood entering the medulla, on equilibrating with each medullary level, would leave the medulla very hypertonic at 1200 mOsm/L. It would be impossible to establish and maintain the medullary hypertonic gradient because the NaCl

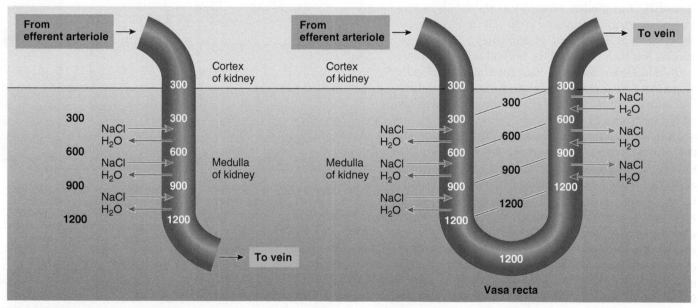

(a) Hypothetical pattern of blood flow   **(b)** Actual pattern of blood flow

KEY

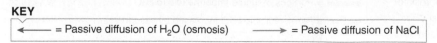

= Passive diffusion of H_2O (osmosis)   = Passive diffusion of NaCl

Figure 14-28 **Countercurrent exchange in the renal medulla.** All values are in mOsm/L. (a) If the blood supply to the renal medulla flowed straight through from the cortex to the inner medulla, the blood would be isotonic on entering but very hypertonic on leaving, having picked up salt and lost H_2O as it equilibrated with the surrounding interstitial fluid at each incremental horizontal level. It would be impossible to maintain the vertical osmotic gradient because the salt pumped out by the ascending limb of Henle's loop would be continuously flushed away by blood flowing through the medulla. (b) Blood equilibrates with the interstitial fluid at each incremental horizontal level in both the descending limb and the ascending limb of the vasa recta, so blood is isotonic as it enters and leaves the medulla. This countercurrent exchange prevents dissolution of the medullary osmotic gradient while providing blood to the renal medulla.

pumped into the medullary interstitial fluid would continuously be carried away by the circulation.

This dilemma is avoided by the hairpin construction of the vasa recta, which, by looping back through the concentration gradient in reverse, allows the blood to leave the medulla and enter the renal vein essentially isotonic to incoming arterial blood (Figure 14-28b). As blood passes down the descending limb of the vasa recta, equilibrating with the progressively increasing concentration of the surrounding interstitial fluid, it picks up salt and loses H_2O until it is very hypertonic by the bottom of the loop. Then, as blood flows up the ascending limb, salt diffuses back out into the interstitial fluid, and H_2O reenters the vasa recta as progressively decreasing concentrations are encountered in the surrounding interstitial fluid. This passive exchange of solutes and H_2O between the two limbs of the vasa recta and the interstitial fluid is known as **countercurrent exchange**. Unlike countercurrent multiplication, it does not *establish* the concentration gradient. Rather, it *preserves* (prevents the dissolution of) the gradient. Because blood enters and leaves the medulla at the same osmolarity as a result of countercurrent exchange, the medullary tissue is nourished with blood, yet the incremental gradient of hypertonicity in the medulla is preserved.

Water reabsorption is only partially linked to solute reabsorption.

It is important to distinguish between H_2O reabsorption that mandatorily follows solute reabsorption and reabsorption of "free" H_2O not linked to solute reabsorption.

- *In the tubular segments permeable to H_2O, solute reabsorption is* always *accompanied by comparable H_2O reabsorption because of osmotic considerations.* Therefore, the total volume of H_2O reabsorbed is determined in large part by the total mass of solute reabsorbed; this is especially true of NaCl because it is the most abundant solute in the ECF.

- *Solute excretion is* always *accompanied by comparable H_2O excretion because of osmotic considerations.* This fact is responsible for the obligatory excretion of at least a minimal volume of H_2O to accompany waste excretion, even when a person is severely dehydrated. For the same reason, when excess unreabsorbed solute is present in the tubular fluid, its presence exerts an osmotic effect to hold excessive H_2O in the lumen, leading to osmotic diuresis. There are two types of diuresis: osmotic diuresis and water diuresis. **Osmotic diuresis** involves increased excretion of both H_2O and solute caused by excess unreabsorbed solute in the tubular fluid, such as occurs in untreated diabetes mellitus. The large quantity of unreabsorbed glucose that remains in the tubular fluid in people with diabetes osmotically drags H_2O with it into the urine. **Water diuresis**, in contrast, is increased urinary output of H_2O with little or no increase in excretion of solutes.

- *A loss or gain of pure H_2O that is not accompanied by comparable solute deficit or excess in the body* (that is, "free" H_2O) *leads to changes in ECF osmolarity.* Such an imbalance between H_2O and solute is corrected by partially dissociating H_2O reabsorption from solute reabsorption in the distal portions of

the nephron through the combined effects of vasopressin secretion and the medullary osmotic gradient. Through this mechanism, free H_2O can be reabsorbed without comparable solute reabsorption to correct for hypertonicity of the body fluids. Conversely, a large quantity of free H_2O can be excreted unaccompanied by comparable solute excretion (that is, water diuresis) to rid the body of excess pure H_2O, thus correcting for hypotonicity of the body fluids. Water diuresis is normally a compensation for ingesting too much H_2O.

Excessive water diuresis follows alcohol ingestion. Because alcohol inhibits vasopressin secretion, the kidneys inappropriately lose too much H_2O. Typically, more fluid is lost in the urine than is consumed in the alcoholic beverage, so the body becomes dehydrated despite substantial fluid ingestion.

Table 14-3 summarizes how various tubular segments of the nephron handle Na^+ and H_2O and the significance of these processes.

Renal failure has wide-ranging consequences.

Urine excretion and the resulting clearance of wastes and excess electrolytes from the plasma are crucial for maintaining homeostasis. When the functions of both kidneys are so disrupted that they cannot perform their regulatory and excretory functions sufficiently to maintain homeostasis, **renal failure** has set in. Renal failure has a variety of causes, some of which begin elsewhere in the body and affect renal function secondarily. Among the causes are the following:

1. *Infectious organisms,* either blood-borne or gaining entrance to the urinary tract through the urethra

2. *Toxic agents,* such as lead, arsenic, pesticides, or even long-term exposure to high doses of aspirin

3. *Inappropriate immune responses,* such as **glomerulonephritis**, which occasionally follows streptococcal throat infections as antigen–antibody complexes leading to localized inflammatory damage are deposited in the glomeruli (see p. 419)

4. *Obstruction of urine flow* by kidney stones, tumors, or an enlarged prostate gland, with back pressure reducing glomerular filtration and damaging renal tissue

5. An *insufficient renal blood supply* that leads to inadequate filtration pressure, which can occur secondary to circulatory disorders such as heart failure, hemorrhage, shock, or narrowing and hardening of the renal arteries by atherosclerosis

The glomeruli or tubules may be independently affected, or both may be dysfunctional. Regardless of cause, renal failure can manifest itself either as *acute renal failure,* characterized by a sudden onset with rapidly reduced urine formation until less than the essential minimum of around 500 mL of urine is being produced per day, or as *chronic renal failure,* characterized by slow, progressive, insidious loss of renal function. A person may die from acute renal failure, or the condition may be reversible and lead to full recovery. Chronic renal failure, in contrast, is not reversible. Gradual, permanent destruction of renal tissue eventually proves fatal. Chronic renal failure is insidious because up to 75% of the kidney tissue can be destroyed before the loss of kidney function is even noticeable. Because of the

Tubular Segment	Na$^+$ REABSORPTION		H$_2$O REABSORPTION	
	Percentage of Reabsorption in This Segment	Distinguishing Features	Percentage of Reabsorption in This Segment	Distinguishing Features
Proximal tubule	67	Active; uncontrolled; plays a pivotal role in the reabsorption of glucose, amino acids, Cl$^-$, H$_2$O, and urea	65	Passive; obligatory osmotic reabsorption following active Na$^+$ reabsorption
Loop of Henle	25	Active, uncontrolled; NaCl reabsorption from the ascending limb helps establish the medullary interstitial vertical osmotic gradient, which is important in the kidneys' ability to produce urine of varying concentrations and volumes, depending on the body's needs	15	Passive; obligatory osmotic reabsorption from the descending limb as the ascending limb extrudes NaCl into the interstitial fluid (that is, reabsorbs NaCl)
Distal and collecting tubules	8	Active; variable and subject to aldosterone control; important in the regulation of ECF volume and long-term control of blood pressure; linked to K$^+$ secretion and H$^+$ secretion	20	Passive; not linked to solute reabsorption; variable quantities of "free" H$_2$O reabsorption subject to vasopressin control; driving force is the vertical osmotic gradient in the medullary interstitial fluid established by the long loops of Henle; important in regulating ECF osmolarity

abundant reserve of kidney function, only 25% of kidney tissue is needed to adequately maintain all the essential renal excretory and regulatory functions. With less than 25% of functional kidney tissue remaining, however, renal insufficiency becomes apparent. *End-stage renal failure* results when 90% of kidney function has been lost. More than 26 million people in the United States have some extent of kidney disease, which leads to more than 80,000 deaths per year.

We will not sort out the stages and symptoms associated with various renal disorders, but Table 14-4, which summarizes the potential consequences of renal failure, gives you an idea of the broad effects that kidney impairment can have. When the kidneys cannot maintain a normal internal environment, widespread disruption of cell activities can bring about abnormal function in other organ systems as well. By the time end-stage renal failure occurs, literally every body system has become impaired to some extent. The most life-threatening consequences of renal failure are retention of H$^+$ (causing metabolic acidosis) and K$^+$ (leading to cardiac malfunction) because these ions are not adequately secreted and eliminated in the urine.

Because chronic renal failure is irreversible and eventually fatal, treatment is aimed at maintaining renal function by alternative methods, such as dialysis and kidney transplantation. (For further explanation of these procedures, see the boxed feature on p. 530, Concepts, Challenges, and Controversies.)

This finishes our discussion of kidney function. For the remainder of the chapter, we focus on the plumbing that stores and carries the urine formed by the kidneys to the outside.

Urine is temporarily stored in the bladder, from which it is emptied by micturition.

Once urine has been formed by the kidneys, it is transmitted through the smooth–muscle walled ureters to the urinary bladder. Urine does not flow through the ureters by gravitational pull alone. Peristaltic (forward-pushing) contractions of the smooth muscle within the ureteral wall propel the urine forward from the kidneys to the bladder. The ureters penetrate the wall of the bladder obliquely, coursing through the wall several centimeters before they open into the bladder cavity. This anatomic arrangement prevents backflow of urine from the bladder to the kidneys when pressure builds up in the bladder. As the bladder fills, the ureteral ends within its wall are compressed closed. Urine can still enter, however, because ureteral contractions generate enough pressure to overcome the resistance and push urine through the occluded ends.

Role of the Bladder The bladder can accommodate large fluctuations in urine volume. The bladder wall consists of smooth muscle lined by a special type of epithelium. It was once assumed that the bladder was an inert sac. However, both the epithelium and the smooth muscle actively participate in the bladder's ability to accommodate large changes in urine volume. The epithelial lining can increase and decrease in surface area by the orderly process of membrane recycling as the bladder alternately fills and empties. Membrane-enclosed cytoplasmic vesicles are inserted by exocytosis into the surface area

TABLE 14-4 Potential Ramifications of Renal Failure

Uremic toxicity caused by retention of waste products

  Nausea, vomiting, diarrhea, and ulcers caused by a toxic effect on the digestive system

  Bleeding tendency arising from a toxic effect on platelet function

  Mental changes—such as reduced alertness, insomnia, and shortened attention span, progressing to convulsions and coma—caused by toxic effects on the central nervous system

  Abnormal sensory and motor activity caused by a toxic effect on the peripheral nerves

Metabolic acidosis caused by the inability of the kidneys to adequately secrete H^+ that is continually being added to the body fluids as a result of metabolic activity (*among most life-threatening consequences of renal failure*)

  Altered enzyme activity caused by the action of too much acid on enzymes

  Depression of the central nervous system caused by the action of too much acid interfering with neuronal excitability

Potassium retention resulting from inadequate tubular secretion of K^+ (*among most life-threatening consequences of renal failure*)

  Altered cardiac and neural excitability as a result of changing the resting membrane potential of excitable cells

Sodium imbalances caused by inability of the kidneys to adjust Na^+ excretion to balance changes in Na^+ consumption

  Elevated blood pressure, generalized edema, and congestive heart failure if too much Na^+ is consumed

  Hypotension and, if severe enough, circulatory shock if too little Na^+ is consumed

Phosphate and calcium imbalances arising from impaired reabsorption of these electrolytes

  Disturbances in skeletal structures caused by abnormalities in deposition of calcium phosphate crystals, which harden bone

Loss of plasma proteins as a result of increased "leakiness" of the glomerular membrane

  Edema caused by a reduction in plasma-colloid osmotic pressure

Inability to vary urine concentration as a result of impairment of the countercurrent system

  Hypotonicity of body fluids if too much H_2O is ingested

  Hypertonicity of body fluids if too little H_2O is ingested

Hypertension arising from the combined effects of salt and fluid retention and vasoconstrictor action of excess angiotensin II

Anemia caused by inadequate erythropoietin production

Depression of the immune system caused by toxic levels of wastes and acids

  Increased susceptibility to infections

---

during bladder filling; then the vesicles are withdrawn by endocytosis to shrink the surface area following emptying (see pp. 29, 31 and 77). As is characteristic of smooth muscle, bladder muscle can stretch tremendously without building up bladder wall tension (see p. 293). In addition, the highly folded bladder wall flattens out during filling to increase bladder storage capacity. Because the kidneys continuously form urine, the bladder must have enough storage capacity to preclude the need to continuously get rid of the urine.

Bladder smooth muscle is richly supplied by parasympathetic fibers, stimulation of which causes bladder contraction. If the passageway through the urethra to the outside is open, contraction empties urine from the bladder. The exit from the bladder, however, is guarded by two sphincters, the *internal urethral sphincter* and the *external urethral sphincter*.

Role of the Urethral Sphincters A *sphincter* is a ring of muscle that can variably close off or permit passage through an opening (see p. 264). The **internal urethral sphincter** is smooth muscle and, accordingly, under involuntary control. It is not really a separate muscle but instead consists of the last part of the bladder. When the bladder is relaxed, the anatomic arrangement of the internal urethral sphincter region closes the bladder outlet.

Farther down the passageway, the urethra is encircled by a layer of skeletal muscle, the **external urethral sphincter**. This sphincter is reinforced by the entire **pelvic diaphragm**, a skeletal muscle sheet that forms the floor of the pelvis and helps support the pelvic organs (see Figure 14-2, p. 494). The motor neurons that supply the external sphincter and pelvic diaphragm fire continuously at a moderate rate unless they are inhibited, keeping these muscles tonically contracted so that they prevent urine from escaping through the urethra. Normally, when the bladder is relaxed and filling, both the internal and the external urethral sphincters are closed to keep urine from dribbling out. Furthermore, because the external sphincter and pelvic diaphragm are skeletal muscle and thus under voluntary control, the person can deliberately tighten them to prevent urination from occurring even when the bladder is contracting and the internal sphincter is open.

Dialysis: Cellophane Tubing or Abdominal Lining as an Artificial Kidney

ECAUSE CHRONIC RENAL FAILURE IS irreversible and eventually fatal, treatment is aimed at maintaining renal function by alternative methods, such as dialysis and kidney transplantation. More than 300,000 people in the United States are currently undergoing dialysis, and this number is expected to climb as the population ages and the incidence of diabetes mellitus, one of the leading causes of kidney failure, continues to rise. End-stage renal failure (less than 10% kidney function) caused by diabetes mellitus is increasing at a rate of more than 11% annually.

The process of dialysis bypasses the kidneys to maintain normal fluid and electrolyte balance and remove wastes artificially. In the original method of dialysis, **hemodialysis**, a patient's blood is pumped through cellophane tubing that is surrounded by a large volume of fluid similar in composition to normal plasma. After dialysis, the blood is returned to the patient's circulatory system. During hemodialysis, about 250 mL of blood is outside of the body at any given time.

Like capillaries, cellophane is highly permeable to most plasma constituents but is impermeable to plasma proteins. As blood flows through the tubing, solutes move across the cellophane down their individual concentration gradients; plasma proteins, however, stay in the blood. Urea and other wastes, which are absent in the dialysis fluid, diffuse out of the plasma into the surrounding fluid, cleaning the blood of these wastes. Plasma constituents that are not regulated by the kidneys and are at normal concentration, such as glucose, do not move across the cellophane into the dialysis fluid because there is no driving force to produce their movement. (The dialysis fluid's glucose concentration is the same as normal plasma glucose concentration.) Electrolytes, such as K^+ and PO_4^{3-}, which are higher than their normal plasma concentrations because the diseased kidneys cannot eliminate excess quantities of these substances, move out of the plasma until equilibrium is achieved between the plasma and the dialysis fluid. Because the dialysis fluid's solute concentrations are maintained at normal plasma values, the solute concentration of the blood returned to the patient after dialysis is essentially normal.

Hemodialysis is repeated as often as necessary to maintain the plasma composition within an acceptable level. Conventionally, it is done three times per week for up to five hours at each session at a treatment center, but newer, more user-friendly, at-home methods dialyze the blood up to six times per week during the day or at night while the person is sleeping. The more frequent methods maintain better stability in plasma constituents than the less frequent methods do.

Another method of dialysis, **continuous ambulatory peritoneal dialysis (CAPD)**, uses the peritoneal membrane (the lining of the abdominal cavity) as the dialysis membrane. With this method, 2 liters of dialysis fluid are inserted into the patient's abdominal cavity through a permanently implanted catheter. Urea, K^+, and other wastes and excess electrolytes diffuse from the plasma across the peritoneal membrane into the dialysis fluid, which is drained off and replaced several times a day. The CAPD method offers several advantages: The patient can self-administer it, the patient's blood is continuously purified and adjusted, and the patient can engage in normal activities while dialysis is being accomplished. One drawback is increased risk of peritoneal infections.

Although dialysis can remove metabolic wastes and foreign compounds and help maintain fluid and electrolyte balance within acceptable limits, this plasma-cleansing technique cannot make up for the failing kidneys' reduced ability to produce hormones (erythropoietin and renin) and to activate vitamin D. One promising new technique under investigation incorporates living kidney cells derived from pigs within a dialysislike machine. Standard ultrafiltration technology like that used in hemodialysis purifies and adjusts the plasma as usual. Importantly, the living cells not only help maintain even better control of plasma constituents, especially K^+, but also add the deficient renal hormones to the plasma passing through the machine and activate vitamin D.

For now, transplanting a healthy kidney from a donor is another option for treating chronic renal failure. A kidney is one of the few transplants that can be provided by a living donor. Because 25% of the total kidney tissue can maintain the body, both the donor and the recipient have ample renal function with only one kidney each. The biggest problem with transplants is the possibility that the patient's immune system rejects the organ. Risk of rejection can be minimized by matching the tissue types of the donor and the recipient as closely as possible (the best donor choice is usually a close relative), coupled with immunosuppressive drugs. More than 15,000 kidney transplants are performed in the United States each year, with 60,000 more people on waiting lists for a donor kidney.

Another new technique on the horizon for treating end-stage renal failure is a continuously functioning artificial kidney that mimics natural renal function. Using nanotechnology (very small-scale devices), researchers are working on a device that contains two membranes, the first for filtering blood like the glomerulus does and the second for mimicking the renal tubules by selectively altering the filtrate. The device, which will directly process the blood on an ongoing basis without using dialysis fluid, will return important substances to the body while discharging unneeded substances to a disposable bag that will serve as an external bladder. Scientists have developed computer models for such a device and thus far have created the filtering membrane.

Micturition Reflex Micturition, or **urination**, the process of bladder emptying, is governed by two mechanisms: the micturition reflex and voluntary control. The **micturition reflex** is initiated when stretch receptors within the bladder wall are stimulated (▮ Figure 14-29). The bladder in an adult can accommodate 250 to 400 mL of urine before the tension within its walls begins to rise sufficiently to activate the stretch receptors (▮ Figure 14-30). The greater the distension beyond this, the greater the extent of receptor activation. Afferent fibers from the stretch receptors carry impulses into the spinal cord and eventually, via interneurons, stimulate the parasympathetic supply to the bladder and inhibit the motor-neuron supply to the external sphincter. Parasympathetic stimulation of the bladder causes it to contract. No special mechanism is required to open the internal sphincter; changes in bladder shape during contraction mechanically pull the internal sphincter open. Simultaneously, the external sphincter relaxes as its motor neuron supply is inhibited. Now both sphincters are open, and urine is expelled through the urethra by the force of bladder contraction. This micturition reflex, which is entirely a spinal reflex, governs bladder emptying in infants. As soon as the bladder fills enough to trigger the reflex, the baby automatically wets.

Voluntary Control of Micturition In addition to triggering the micturition reflex, bladder filling gives rise to the conscious urge to urinate. The perception of bladder fullness appears before the external sphincter reflexly relaxes, warning that micturition is imminent. As a result, voluntary control of micturition, learned during toilet training in early childhood, can override the micturition reflex so that bladder emptying can take place at your convenience rather than when bladder filling first activates the stretch receptors. If the time when the micturition reflex is initiated is inopportune for urination, you can voluntarily prevent bladder emptying by deliberately tightening your external sphincter and pelvic diaphragm. Voluntary excitatory impulses from the cerebral cortex override the reflex inhibitory input from the stretch receptors to the involved motor neurons (the relative balance of excitatory and inhibitory postsynaptic potentials [EPSPs and IPSPs]; see p. 106), keeping these muscles contracted so that no urine is expelled.

Urination cannot be delayed indefinitely. As the bladder continues to fill, reflex input from the stretch receptors increases with time. Finally, reflex inhibitory input to the external sphincter motor neuron becomes so powerful that it can no longer be overridden by voluntary excitatory input, so the sphincter relaxes and the bladder uncontrollably empties.

Micturition can also be deliberately initiated, even though the bladder is not distended, by voluntarily relaxing the external sphincter and pelvic diaphragm. Lowering of the pelvic floor allows the bladder to drop downward, which simultaneously pulls open the internal urethral sphincter and stretches the bladder wall. The subsequent activation of the stretch receptors brings about bladder contraction by the micturition reflex. Voluntary bladder emptying may be further assisted by contracting the abdominal wall and respiratory diaphragm. The resulting

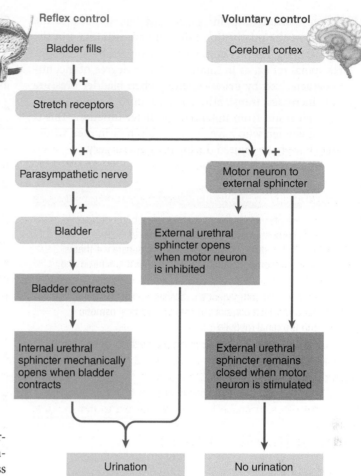

I Figure 14-29 **Reflex and voluntary control of micturition.**

increase in intra-abdominal pressure squeezes down on the bladder to facilitate its emptying.

Clinical Note **Urinary Incontinence** **Urinary incontinence**, or inability to prevent discharge of urine, occurs when descending pathways in the spinal cord that mediate voluntary control of the external sphincter and pelvic dia-

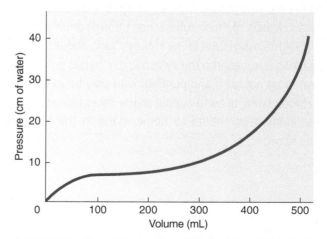

I Figure 14-30 **Pressure changes within the urinary bladder as the bladder fills with urine.**

phragm are disrupted, as in spinal-cord injury. Because the components of the micturition reflex arc are still intact in the lower spinal cord, bladder emptying is governed by an uncontrollable spinal reflex, as in infants. A lesser degree of incontinence characterized by urine escaping when bladder pressure suddenly increases transiently, such as during coughing or sneezing, can result from impaired sphincter function. This is common in women who have borne children or in men whose sphincters have been injured during prostate surgery.

Check Your Understanding 14.5

1. State how the plasma clearance rate for each of the following substances compares with the GFR: (1) a substance that is filtered but not reabsorbed or secreted, (2) a substance that is filtered and reabsorbed, and (3) a substance that is filtered and secreted.
2. Tell which nephron component establishes, which component preserves, and which component uses the vertical osmotic gradient in the renal medulla.
3. Explain how vasopressin increases the permeability of the distal and collecting tubules to H_2O.
4. Describe the micturition reflex.

Homeostasis: Chapter in Perspective

The kidneys contribute to homeostasis more extensively than any other single organ. They regulate the electrolyte composition, volume, osmolarity, and pH of the internal environment and eliminate all the waste products of bodily metabolism except for respiration-removed CO_2. They accomplish these regulatory functions by eliminating in the urine substances the body doesn't need, such as metabolic wastes and excess quantities of ingested salt or water, while conserving useful substances. The kidneys can maintain the plasma constituents they regulate within the narrow range compatible with life, despite wide variations in intake and losses of these substances through other avenues. Illustrating the magnitude of the kidneys' task, about a quarter of the blood pumped into the systemic circulation goes to the kidneys to be adjusted and purified, with only three quarters of the blood being used to supply all the other tissues.

The kidneys contribute to homeostasis in the following specific ways:

Regulatory Functions

■ The kidneys regulate the quantity and concentration of most ECF electrolytes, including those important in maintaining proper neuromuscular excitability.

■ They help maintain proper pH by eliminating excess H^+ (acid) or HCO_3^- (base) in the urine.

■ They help maintain proper plasma volume, which is important in long-term regulation of blood pressure, by controlling the body's salt balance. The ECF volume, including plasma volume, reflects total salt load in the ECF because Na^+ and its attendant anion, Cl^-, are responsible for more than 90% of the ECF's osmotic (water-holding) activity.

■ The kidneys maintain water balance in the body, which is important in maintaining proper ECF osmolarity (concentration of solutes). This role is essential in maintaining stable cell volume by keeping water from osmotically moving into or out of the cells, thus preventing them from swelling or shrinking, respectively.

Excretory Functions

■ The kidneys excrete metabolic end products in the urine. If allowed to accumulate, these wastes are toxic to cells.

■ The kidneys excrete many foreign compounds that enter the body.

Hormonal Functions

■ The kidneys produce erythropoietin, the hormone that stimulates bone marrow to produce red blood cells. This action contributes to homeostasis by helping maintain the optimal O_2 content of blood. More than 98% of O_2 in the blood is bound to hemoglobin within red blood cells.

■ They produce renin, the hormone that initiates the renin–angiotensin–aldosterone pathway for controlling renal tubular Na^+ reabsorption, which is important in long-term maintenance of plasma volume and blood pressure.

Metabolic Functions

■ The kidneys help activate vitamin D, which is essential for Ca^{2+} absorption from the digestive tract. Calcium, in turn, exerts a variety of homeostatic functions.

Review Exercises
Answers begin on p. A-42

Reviewing Terms and Facts

1. Part of the kidneys' energy supply is used to accomplish glomerular filtration. (*True or false?*)

2. Sodium reabsorption is under hormonal control throughout the length of the tubule. (*True or false?*)

3. Glucose and amino acids are reabsorbed by secondary active transport. (*True or false?*)

4. Solute excretion is always accompanied by comparable H_2O excretion. (*True or false?*)

5. Water excretion can occur without comparable solute excretion. (*True or false?*)

6. The functional unit of the kidneys is the _____.

7. _____ is the only ion actively reabsorbed in the proximal tubule and actively secreted in the distal and collecting tubules.

8. The daily minimum volume of obligatory H_2O loss that must accompany excretion of wastes is _____ mL.

9. Indicate whether each of the following factors would (a) increase or (b) decrease the GFR if everything else remained constant.

   1. a rise in Bowman's capsule pressure resulting from ureteral obstruction by a kidney stone
   2. a fall in plasma protein concentration resulting from loss of these proteins from a large burned surface of skin
   3. a dramatic fall in arterial blood pressure following severe hemorrhage (<80 mm Hg)
   4. afferent arteriolar vasoconstriction
   5. tubuloglomerular feedback response to decreased salt delivery to the distal tubule
   6. myogenic response of an afferent arteriole stretched as a result of an increased driving blood pressure
   7. ↑ sympathetic activity to the afferent arterioles
   8. contraction of mesangial cells
   9. contraction of podocytes

10. Which of the following filtered substances is normally *not* present in the urine at all?

   a. Na^+
   b. PO_4^{3-}
   c. urea
   d. H^+
   e. glucose

11. Reabsorption of which of the following substances is *not* linked in some way to active Na^+ reabsorption?

   a. glucose
   b. PO_4^{3-}
   c. H_2O
   d. urea
   e. Cl^-

In questions 12–14, indicate, by writing the identifying letters in the proper order in the blanks, the proper sequence through which fluid flows as it traverses the structures in question.

12. a. ureter   ___ ___ ___ ___ ___
    b. kidney
    c. urethra
    d. bladder
    e. renal pelvis

13. a. efferent arteriole   ___ ___ ___ ___ ___ ___
    b. peritubular capillaries
    c. renal artery
    d. glomerulus
    e. afferent arteriole
    f. renal vein

14. a. loop of Henle   ___ ___ ___ ___ ___ ___ ___
    b. collecting duct
    c. Bowman's capsule
    d. proximal tubule
    e. renal pelvis
    f. distal tubule
    g. glomerulus

15. Using the answer code on the right, indicate what the osmolarity of the tubular fluid is at each of the designated points in a nephron:

   1. Bowman's capsule
   2. end of proximal tubule
   3. tip of Henle's loop of juxtamedullary nephron (at the bottom of the U-turn)
   4. end of Henle's loop of juxtamedullary nephron (before entry into distal tubule)
   5. end of collecting duct

   (a) isotonic (300 mOsm/L)
   (b) hypotonic (100 mOsm/L)
   (c) hypertonic (1200 mOsm/L)
   (d) ranging from hypotonic to hypertonic (100 mOsm/L to 1200 mOsm/L)

Understanding Concepts
(Answers at www.cengagebrain.com)

1. List the functions of the kidneys.

2. Describe the anatomy of the urinary system. Describe the components of a nephron.

3. Describe the three basic renal processes; indicate how they relate to urine excretion.

4. Distinguish between *secretion* and *excretion*.

5. Discuss the forces involved in glomerular filtration. What is the average GFR?

6. How is GFR regulated as part of the baroreceptor reflex?

7. Why do the kidneys receive a seemingly disproportionate share of the cardiac output? What percentage of renal blood flow is normally filtered?

8. List the steps in transepithelial transport.

9. Distinguish between active and passive reabsorption.

10. Describe all the tubular transport processes that are linked to the basolateral Na^+–K^+ ATPase carrier.

11. Describe the renin–angiotensin–aldosterone system. What are the functions of aldosterone and angiotensin II?

12. Discuss the source and functions of ANP and BNP.

13. Compare two substances that display a T_m, one that *is* and one that *is not* regulated by the kidneys.

14. What is the importance of tubular secretion? What are the most important secretory processes?

15. What is the average rate of urine formation?

16. Define *plasma clearance*.

17. What establishes a vertical osmotic gradient in the medullary interstitial fluid? Of what importance is this gradient?

18. Discuss vasopressin's function and mechanism of action.

19. Compare countercurrent multiplication and countercurrent exchange.

20. Describe the transfer of urine to, the storage of urine in, and the emptying of urine from the bladder.

Solving Quantitative Exercises

1. Two patients are voiding protein in their urine. To determine whether this proteinuria indicates a serious problem, a physician injects small amounts of inulin and PAH into each patient. Recall that inulin is freely filtered and neither secreted nor reabsorbed in the nephron and that PAH at this concentration is completely removed from the blood by tubular secretion. The data collected are given in the following table, where $[I]_u$ and $[PAH]_u$ are the concentrations of inulin or PAH in the urine (in mM), respectively; $[I]_p$ and $[PAH]_p$ are the concentrations of these substances in the plasma; and v_u is the flow rate of urine (in mL/min).

Patient	$[I]_u$	$[I]_p$	$[PAH]_u$	$[PAH]_p$	v_u
1	25	2	186	3	10
2	31	1.5	300	4.5	6

a. Calculate each patient's GFR and renal plasma flow.

b. Calculate the renal blood flow for each patient, assuming both have a hematocrit of 0.45.

c. Calculate the filtration fraction for each patient.

d. Which of the values calculated for each patient are within the normal range?

e. Which values are abnormal? What could be causing these deviations from normal?

2. What is the filtered load of sodium if inulin clearance is 125 mL/min and the sodium concentration in plasma is 145 mM?

3. Calculate a patient's rate of urine production, given that his inulin clearance is 125 mL/min and his urine and plasma concentrations of inulin are 300 mg/liter and 3 mg/liter, respectively.

4. If the urine concentration of a substance is 7.5 mg/mL of urine, its plasma concentration is 0.2 mg/mL of plasma, and the urine flow rate is 2 mL/min, what is the clearance rate of the substance? Is the substance being reabsorbed or secreted by the kidneys?

Applying Clinical Reasoning

Marcus T. has noted a gradual decrease in his urine flow rate and is now experiencing difficulty in initiating micturition. He needs to urinate frequently, and often he feels as if his bladder is not empty even though he has just urinated. Analysis of Marcus's urine reveals no abnormalities. Are his urinary tract symptoms most likely caused by kidney disease, a bladder infection, or prostate enlargement?

Thinking at a Higher Level

1. What would the clinical implications be on finding each of the following substances in a 72-year-old, sedentary person's urine: (1) glucose, (2) protein, (3) sodium?

2. The juxtamedullary nephrons of animals adapted to survive with minimal water consumption, such as desert rats, have relatively much longer loops of Henle than humans have. Of what benefit would these longer loops be?

3. *Conn's syndrome* is an endocrine disorder brought about by a tumor of the adrenal cortex that secretes excessive aldosterone in uncontrolled fashion. Given what you know about the functions of aldosterone, describe what the most prominent features of this condition would be.

4. Because of a mutation, a child was born with an ascending limb of Henle that was water permeable. What would be the minimum/maximum urine osmolarities (in units of mOsm/L) the child could produce?

    a. 100/300

    b. 300/1200

    c. 100/100

    d. 1200/1200

    e. 300/300

5. An accident victim suffers permanent damage of the lower spinal cord and is paralyzed from the waist down. Describe what governs bladder emptying in this individual.

To access the course materials and companion resources for this text, please visit
www.cengagebrain.com

Fluid and Acid–Base Balance

Yellow Dog Productions/Photographer's Choice/
Getty Images

Maintaining fluid balance. This woman running a road race grabs a cup of water to replace fluid lost in sweat. A thirst center in the hypothalamus drives fluid ingestion. Input must equal output to keep a body constituent such as water in balance.

Homeostasis Highlights

Homeostasis depends on maintaining a balance between the input and the output of all constituents in the internal fluid environment. Regulation of **fluid balance** involves two separate components: control of extracellular fluid (ECF) volume, of which circulating plasma volume is a part, and control of ECF osmolarity (solute concentration). The kidneys control ECF volume by maintaining **salt balance** and control ECF osmolarity by maintaining **water balance**. The kidneys maintain this balance by adjusting the output of salt and water in the urine as needed to compensate for variable input and abnormal losses of these constituents.

Similarly, the kidneys help maintain **acid–base balance** by adjusting the urinary output of hydrogen ion (acid) and bicarbonate ion (base) as needed. Also contributing to acid–base balance are the buffer systems in the body fluids, which chemically compensate for changes in hydrogen ion concentration, and the lungs, which can adjust the rate at which they excrete hydrogen ion–generating CO_2.

15.1 Balance Concept

The cells of complex multicellular organisms are able to survive and function only within a narrow range of composition of the ECF, the internal fluid environment that bathes them.

The internal pool of a substance is the amount of that substance in the ECF.

The quantity of any particular substance in the ECF is a readily available internal **pool**. The amount of the substance in the pool may be increased either by transferring more in from the external environment (usually by ingestion) or by metabolically producing it within the body (▮Figure 15-1). Substances may be removed from the body by being excreted to the outside or by being used up in a metabolic reaction. If the quantity of a substance is to remain stable within the body, its **input** through ingestion or metabolic production must be balanced by an equal **output** through excretion or metabolic consumption. This relationship, known as the **balance concept**, is extremely important in maintaining homeostasis. Not all input and output pathways apply to every body-fluid constituent. For example, salt is not synthesized or used up by the body, so maintaining a stable salt concentration in the body fluids depends entirely on a balance between salt ingestion and salt excretion.

For some ECF constituents, the ECF pool is further altered by transferring this specific constituent into or out of storage within the body. If the body as a whole has a surplus or deficit of a particular stored substance, the storage site can be expanded or partially depleted to maintain the ECF concentration of the substance within homeostatically prescribed limits. For example, after absorption of a meal, when more glucose is entering the plasma than is being consumed by the cells, the extra glu-

cose can be temporarily stored, in the form of glycogen, in muscle and liver cells. This storage depot can then be tapped between meals as needed to maintain the plasma glucose level when no new nutrients are being added to the blood by eating.

Another possible internal exchange between the pool and the rest of the body is reversible incorporation of a plasma constituent into a more complex molecular structure to serve a specific purpose. For example, iron is incorporated into hemoglobin within the red blood cells but is released intact back into the body fluids when the red blood cells degenerate.

To maintain stable balance of an ECF constituent, its input must equal its output.

When total body input of a particular substance equals its total body output, a **stable balance** exists. When the gains via input for a substance exceed its losses via output, a **positive balance** exists. The result is an increase in the total amount of the substance in the body. In contrast, when losses for a substance exceed its gains, a **negative balance** exists and the total amount of the substance in the body decreases.

Changing the magnitude of any input or output pathways for a given substance can alter its plasma concentration. To maintain homeostasis, any change in input must be balanced by a corresponding change in output (for example, increased salt intake must be matched by a corresponding increase in salt output in the urine), and conversely, increased losses must be compensated for by increased intake. Thus, maintaining a stable balance requires control. However, not all input and output pathways are regulated to maintain balance. Generally, input of various plasma constituents is poorly controlled or not controlled at all. We frequently ingest salt and H_2O, for example, not because we need them but because we want them, so the

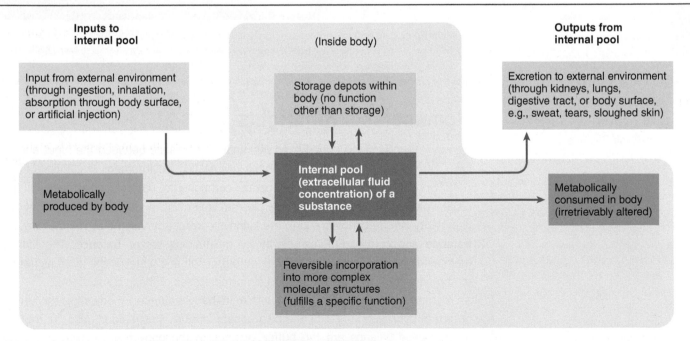

▮Figure 15-1 **Inputs to and outputs from the internal pool of a body constituent.**
FIGURE FOCUS: *Indicate which of the pathways into and out of the internal pool shown in the figure are applicable for phosphate (PO_4^{3-}). Review on pp. 512 and 529 what you have already learned about phosphate handling in the body.*

intake of salt and H_2O is highly variable. Likewise, hydrogen ion (H^+) is uncontrollably generated internally and added to the body fluids. Salt, H_2O, and H^+ can also be lost to the external environment to varying degrees through the digestive tract (vomiting), skin (sweating), and elsewhere without regard for the salt, H_2O, or H^+ balance in the body. Compensatory adjustments in the urinary excretion of these substances maintain the body fluids' volume and salt and acid composition within the extremely narrow homeostatic range compatible with life despite the wide variations in input and unregulated losses of these plasma constituents.

The rest of this chapter is devoted to discussing the regulation of fluid balance (maintaining salt and H_2O balance) and acid–base balance (maintaining H^+ balance).

| **Check Your Understanding 15.1** |

1. List the possible inputs to and outputs from the internal pool of a given body constituent.
2. Define *stable balance, positive balance,* and *negative balance.*

15.2 Fluid Balance

Water is by far the most abundant component of the body, averaging 60% of body weight but ranging from 40% to 80%. The H_2O content of an individual remains fairly constant because the kidneys efficiently regulate H_2O balance, but the percentage of body H_2O varies from person to person. The main reason for the wide range in body H_2O among individuals is their variable amount of adipose tissue (fat). Adipose tissue has a low H_2O percentage compared to other tissues. Plasma, as you might suspect, is more than 90% H_2O. Even the soft tissues such as skin, muscles, and internal organs consist of 70% to 80% H_2O. The relatively drier skeleton is only 22% H_2O. Fat, however, is the driest tissue of all, having only 10% H_2O content. Accordingly, a high body H_2O percentage is associated with leanness and a low body H_2O percentage with obesity because a larger proportion of the overweight body consists of relatively dry fat.

Body water is distributed between the ICF and the ECF compartments.

Body H_2O is distributed between two major fluid compartments: fluid within the cells, or **intracellular fluid (ICF)**, and fluid surrounding the cells, or **extracellular fluid (ECF)** (Table 15-1). (The terms *water [H_2O]* and *fluid* are commonly used interchangeably. Although this usage is not entirely accurate, because it ignores the solutes in body fluids, it is acceptable when discussing total volume of fluids because the major proportion of these fluids consists of H_2O.)

Proportion of H_2O in the Major Fluid Compartments
The ICF compartment composes about two thirds of the total body H_2O. Even though each cell contains a unique mixture of constituents, these trillions of minute fluid compartments are

TABLE 15-1 Major Body Fluid Compartments

Compartment	Volume of Fluid (in Liters)	Percentage of Body Fluid	Percentage of Body Weight
Total body fluid	42	100	60
Intracellular fluid (ICF)	28	67	40
Extracellular fluid (ECF)	14	33	20
Plasma	2.8	6.6 (20% of ECF)	4
Interstitial fluid	11.2	26.4 (80% of ECF)	16

similar enough to be considered collectively as one large fluid compartment.

The remaining third of the body H_2O found in the ECF compartment is further subdivided into plasma and interstitial fluid. **Plasma**, the fluid portion of blood, makes up about a fifth of the ECF volume. **Interstitial fluid**, the fluid that lies in the spaces between cells and makes exchanges with the cells, represents the other four fifths of the ECF compartment.

Minor ECF Compartments Two other minor categories are included in the ECF: lymph and transcellular fluid. **Lymph** is the fluid being returned from the interstitial fluid to the plasma by means of the lymphatic system, where it is filtered through lymph nodes for immune defense purposes (see pp. 358 and 405). **Transcellular fluid** consists of a number of small, specialized fluid volumes, all of which are secreted by specific cells into a particular body cavity to perform some specialized function. Transcellular fluid includes *cerebrospinal fluid* (surrounding, cushioning, and nourishing the brain and spinal cord); *intraocular fluid* (maintaining the shape of and nourishing the eye); *synovial fluid* (lubricating and serving as a shock absorber for the joints); *pericardial, intrapleural,* and *peritoneal fluids* (lubricating movements of the heart, lungs, and intestines, respectively); and the *digestive juices* (digesting ingested foods).

Although these fluids are extremely important functionally, they represent an insignificant fraction of total body H_2O. Furthermore, the transcellular compartment as a whole usually does not reflect changes in the body's fluid balance. For example, cerebrospinal fluid does not decrease in volume when the body as a whole is experiencing a negative H_2O balance. This is not to say that these fluid volumes never change. Localized changes in a particular transcellular fluid compartment can occur pathologically (such as too much intraocular fluid accumulating in the eyes of people with glaucoma; see p. 193), but such a localized fluid disturbance does not affect the fluid balance of the body. Therefore, the transcellular compartment can usually be ignored when dealing with problems of fluid balance. The main exception to this generalization occurs when digestive juices are abnormally lost from the body during heavy vomiting or diarrhea, which can bring about a fluid imbalance.

Plasma and interstitial fluid are similar in composition, but ECF and ICF differ markedly.

Several barriers separate the body-fluid compartments, limiting the movement of H_2O and solutes among the various compartments to differing degrees.

The Barrier Between Plasma and Interstitial Fluid: Blood Vessel Walls
The two components of the ECF—plasma and interstitial fluid—are separated by the walls of the blood vessels. However, H_2O and all plasma constituents except for plasma proteins are continuously and freely exchanged between plasma and interstitial fluid by passive means across the thin, pore-lined capillary walls. Accordingly, plasma and interstitial fluid are nearly identical in composition, except that interstitial fluid lacks plasma proteins. Any change in one of these ECF compartments is quickly reflected in the other compartment because they are constantly mixing.

The Barrier Between ECF And ICF: Cellular Plasma Membranes
In contrast to the similar composition of plasma and interstitial fluid, the composition of the ECF differs considerably from that of the ICF (Figure 15-2). Each cell is surrounded by a highly selective plasma membrane that permits passage of certain materials while excluding others. Movement through the membrane barrier occurs by both passive and active means and may be highly discriminating (see Table 3-2, p. 78). Among the major differences between ECF and ICF are (1) the presence of cell proteins in the ICF that cannot permeate the enveloping membranes to leave the cells and (2) the unequal distribution of Na^+ and K^+ and their attendant anions (negatively charged ions) as a result of the action of the membrane-bound Na^+–K^+ pump present in all cells. Because this pump actively transports Na^+ out of and K^+ into cells, Na^+ is the primary ECF cation (positively charged ion) and K^+ is the primary ICF cation (see p. 73; also see Table 3-3, p. 80).

Except for the extremely small, electrically unbalanced portion of the total intracellular and extracellular ions involved in membrane potential, most ECF and ICF ions are electrically balanced. In the ECF, Na^+ is accompanied primarily by the anion chloride (Cl^-) and to a lesser extent by bicarbonate (HCO_3^-). The major intracellular anions are phosphate (PO_4^{3-}) and the negatively charged proteins trapped within the cell.

The movement of H_2O between plasma and interstitial fluid across capillary walls is governed by relative imbalances between capillary blood pressure (a fluid, or hydrostatic, pressure) and colloid osmotic pressure (see pp. 356–357). The net transfer of

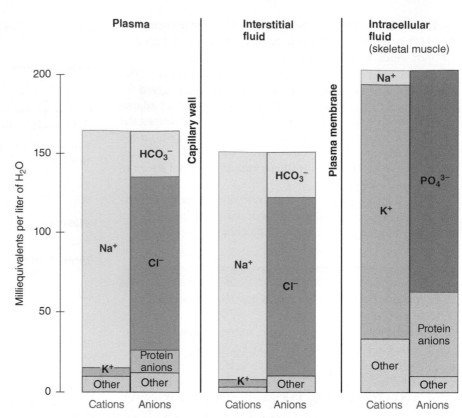

Figure 15-2 Ionic composition of the major body-fluid compartments.

H_2O between the interstitial fluid and the ICF across the cellular plasma membranes occurs as a result of osmotic effects alone (see p. 68). The hydrostatic pressures of the interstitial fluid and ICF are both extremely low and fairly constant. All cells are freely permeable to H_2O.

Fluid balance is maintained by regulating ECF volume and osmolarity.

All exchanges of H_2O and other constituents between the ICF and the external world must occur through the ECF, so the ECF serves as an intermediary between the cells and the external environment. Water added to the body fluids always enters the ECF first, and fluid always leaves the body via the ECF.

Plasma is the only fluid that can be acted on directly to control its volume and composition. This fluid circulates through all the reconditioning organs that perform homeostatic adjustments (see p. 336). However, because of the free exchange across the capillary walls, if the volume and composition of the plasma are regulated, the volume and composition of the interstitial fluid bathing the cells are likewise regulated. Thus, any control mechanism that operates on plasma in effect regulates the entire ECF. The ICF in turn is influenced by changes in the ECF to the extent permitted by the permeability of membrane barriers surrounding the cells.

Two factors are regulated to maintain **fluid balance** in the body: ECF volume and ECF osmolarity. Although regulation of these two factors is interrelated, both depending on the relative NaCl and H_2O loads in the body, the reasons why and the mechanisms by which they are controlled are notably different:

1. *ECF volume* must be closely regulated to help *maintain blood pressure*. Maintaining *salt balance* is of primary importance in the long-term regulation of ECF volume.

2. *ECF osmolarity* must be closely regulated to *prevent swelling or shrinking of cells*. Maintaining *water balance* is of primary importance in regulating ECF osmolarity.

Control of ECF volume is important in the long-term regulation of blood pressure.

A reduction in ECF volume causes a fall in arterial blood pressure by decreasing plasma volume. Conversely, expanding ECF volume raises arterial blood pressure by increasing plasma volume. Two compensatory measures come into play to transiently adjust blood pressure until the ECF volume can be restored to normal. Let us review them.

Short-Term Control Measures to Maintain Blood Pressure

1. *The baroreceptor reflex alters both cardiac output and total peripheral resistance* to adjust blood pressure in the proper direction through autonomic nervous system effects on the heart and blood vessels (see p. 367). Cardiac output and total peripheral resistance are both increased to raise blood pressure when it falls too low, and conversely, both are decreased to reduce blood pressure when it rises too high.

2. *Fluid shifts occur temporarily and automatically between plasma and interstitial fluid* as a result of changes in the balance of hydrostatic and osmotic forces acting across the capillary walls that arise when plasma volume deviates from normal (see p. 357). A reduction in plasma volume is partially compensated for by a shift of fluid out of the interstitial compartment into the blood vessels, expanding the circulating plasma volume at the expense of the interstitial compartment. Conversely, when plasma volume is too large, much of the excess fluid shifts into the interstitial compartment.

These two measures provide temporary relief to help keep blood pressure fairly constant, but they are not long-term solutions. Furthermore, these short-term compensatory measures have a limited ability to minimize a change in blood pressure. For example, if plasma volume is too inadequate, blood pressure remains too low no matter how vigorous the pump action of the heart, how constricted the resistance vessels, or what proportion of interstitial fluid shifts into the blood vessels.

Long-Term Control Measures to Maintain Blood Pressure
In the long run, other compensatory measures come into play to restore ECF volume to normal. Long-term regulation of blood pressure rests with the kidneys and the thirst mechanism, which control urinary output and fluid intake, respectively. In so doing, they make needed fluid exchanges between the ECF and the external environment to regulate the body's total fluid volume. Accordingly, they have an important long-term influence on arterial blood pressure. Of these measures, control of urinary output by the kidneys is the most crucial for maintaining blood pressure. You will see why as we discuss these long-term mechanisms in more detail.

Control of salt balance is primarily important in regulating ECF volume.

To review, sodium (Na^+) and its accompanying anion chloride (Cl^-) account for more than 90% of the ECF osmotic activity. As the kidneys conserve salt (NaCl) by actively reabsorbing Na^+, with Cl^- passively following, they automatically conserve H_2O because H_2O comes along osmotically. This retained salt solution is isotonic (see p. 69). The more salt in the ECF, the more H_2O in the ECF. The *concentration* of salt is not changed by changing the *total amount* of salt in the body (that is, by changing the Na^+ load) because H_2O always follows salt to maintain osmotic equilibrium—in other words, to maintain the normal concentration of salt. A reduced salt load leads to decreased H_2O retention, so the ECF remains isotonic but reduced in volume. The *Na^+ load* therefore determines the ECF volume, and appropriately, regulation of ECF volume depends primarily on controlling salt balance.

To maintain salt balance at a set level, salt input must equal salt output, thus preventing salt accumulation or deficit in the body. We now look at control of salt input and output.

Poor Control of Salt Intake The only avenue for salt input is ingestion, which typically is well in excess of the body's need for replacing obligatory salt losses. In our example of a typical daily salt balance (Table 15-2), salt intake is 10 g per day; yet 0.5 g of salt per day is adequate to replace the small amounts of salt usually lost in sweat and feces. (The average American salt intake is about 8.5 to 10 g per day, although many people are consciously reducing their salt intake.)

Because humans typically consume salt in excess of our needs, obviously our salt intake is not well controlled. Carnivores (meat eaters) and omnivores (eaters of meat and plants, like humans), which naturally get enough salt in fresh meat (meat contains an abundance of salt-rich ECF), normally do not display a physiological appetite to seek additional salt. In contrast, herbivores (plant eaters), which lack salt naturally in their diets, develop salt hunger and will travel miles to a salt lick. Humans have a hedonistic (pleasure-seeking) rather than a regulatory appetite for salt; we consume salt because we like it rather than because we have a physiological need.

TABLE 15-2 Daily Salt Balance

SALT INPUT		SALT OUTPUT	
Avenue	Amount (g/day)	Avenue	Amount (g/day)
Ingestion	10.0	Obligatory loss in sweat and feces	0.5
		Controlled excretion in urine	9.5
Total input	10.0	**Total output**	10.0

Precise Control of Salt Output in the Urine To maintain salt balance, excess ingested salt must be excreted in the urine. The three avenues for salt output are obligatory loss of salt in *sweat* and *feces* and controlled excretion of salt in *urine* (Table 15-2). The total amount of sweat produced is unrelated to salt balance, being determined instead by factors that control body temperature. The small salt loss in feces is not subject to control. Except when sweating heavily or during diarrhea, the body uncontrollably loses only about 0.5 g of salt per day. This amount is the only salt that normally needs to be replaced by salt intake.

Because salt consumption is typically far more than the meager amount needed to compensate for uncontrolled losses, the kidneys precisely excrete the excess salt in the urine to maintain salt balance. In our example, 9.5 g of salt are eliminated in the urine per day so that total salt output exactly equals salt input. By regulating the rate of urinary salt excretion (that is, by regulating the rate of Na^+ excretion, with Cl^- following along), the kidneys normally keep the total Na^+ load (tacitly including the total Cl^- load) in the ECF constant despite any notable changes in dietary intake of salt or unusual losses through sweating or diarrhea. As a reflection of keeping the total Na^+ load in the ECF constant, the ECF volume, in turn, is maintained within the narrowly prescribed limits essential for normal circulatory function.

Deviations in ECF volume accompanying changes in the salt load trigger renal compensatory responses that quickly bring the Na^+ load and ECF volume back into line. Na^+ is freely filtered at the glomerulus and actively reabsorbed, but it is not secreted by the tubules, so the amount of Na^+ excreted in the urine represents the amount of Na^+ filtered but not subsequently reabsorbed:

$$Na^+ \text{ excreted} = Na^+ \text{ filtered} - Na^+ \text{ reabsorbed}$$

The kidneys accordingly adjust the amount of salt excreted by controlling two processes: (1) the glomerular filtration rate (GFR) and (2) more important, the tubular reabsorption of Na^+. You have already learned about these regulatory mechanisms, but we are pulling them together here as they relate to long-term control of ECF volume and blood pressure.

■ *The amount of Na^+ filtered is controlled by regulating the GFR.* The amount of Na^+ filtered is equal to the plasma Na^+ concentration times the GFR. At any given plasma Na^+ concentration, any change in the GFR correspondingly changes the amount of Na^+ and accompanying fluid that are filtered. Thus, control of the GFR can adjust the amount of Na^+ filtered each minute. Recall that the GFR is deliberately changed to alter the amount of salt and fluid filtered, as part of the general baroreceptor reflex response to a change in blood pressure (see Figure 14-12, p. 503). Changes in Na^+ load in the body are not sensed as such; instead, they are monitored indirectly through the effect that the Na^+ load ultimately has on blood pressure. It is fitting that baroreceptors that monitor fluctuations in blood pressure bring about adjustments in the amounts of Na^+ filtered and eventually excreted.

■ *The amount of Na^+ reabsorbed is controlled through the renin–angiotensin–aldosterone system.* The amount of Na^+ re-absorbed also depends on regulatory systems that play an important role in controlling blood pressure. Although Na^+ is reabsorbed throughout most of the tubule's length, only its reabsorption in the late parts of the tubule is subject to control. The main factor controlling the extent of Na^+ reabsorption in the distal and collecting tubules is the powerful renin–angiotensin–aldosterone system (RAAS), which promotes Na^+ reabsorption and thereby Na^+ retention. Sodium retention, in turn, promotes osmotic retention of H_2O and subsequent expansion of plasma volume and elevation of arterial blood pressure. Appropriately, this Na^+-conserving system is activated by a reduction in NaCl, ECF volume, and arterial blood pressure (see Figure 14-16, p. 508).

Thus, control of GFR and Na^+ reabsorption are interrelated, and both are intimately tied in with long-term regulation of ECF volume as reflected by blood pressure. For example, a fall in arterial blood pressure brings about (1) a reflex reduction in the GFR to decrease the amount of Na^+ filtered and (2) a hormonally adjusted increase in the amount of Na^+ reabsorbed (Figure 15-3). Together, these effects reduce the amount of Na^+ excreted, thereby conserving for the body the Na^+ and accompanying H_2O needed to compensate for the fall in arterial pressure. (To look at how exercising muscles and cooling mechanisms compete for an inadequate plasma volume, see the boxed feature on p. 542, A Closer Look at Exercise Physiology.)

Controlling ECF osmolarity prevents changes in ICF volume.

Maintaining fluid balance depends on regulating both ECF volume and ECF osmolarity. Regulating ECF osmolarity is important in preventing changes in cell volume. The **osmolarity** of a fluid is a measure of the concentration of the individual solute particles dissolved in it. The higher the osmolarity, the higher the concentration of solutes or, to look at it differently, the lower the concentration of H_2O. Recall that water tends to move by osmosis down its own concentration gradient from an area of lower solute (higher H_2O) concentration to an area of higher solute (lower H_2O) concentration (see p. 66).

Ions Responsible for ECF and ICF Osmolarity Osmosis occurs across the cellular plasma membranes only when a difference in concentration of nonpenetrating solutes exists between the ECF and the ICF. Solutes that can penetrate a barrier separating two fluid compartments quickly become equally distributed between the two compartments and thus do not contribute to osmotic differences.

Na^+ and accompanying Cl^-, being by far the most abundant solutes in the ECF in terms of numbers of particles, account for most ECF osmotic activity. In contrast, K^+ and its accompanying intracellular anions are responsible for ICF osmotic activity. Even though small amounts of Na^+ and K^+ passively diffuse across the plasma membrane all the time, these ions behave as if they were nonpenetrating because of Na^+–K^+ pump activity. Any Na^+ that passively diffuses down its electrochemical gradient into the cell is promptly pumped back outside, so the result is the same as if Na^+ were barred from the cells. In reverse, K^+ in effect remains trapped within the cells.

Normally, the osmolarities of the ECF and ICF are the same because the total concentration of K^+ and other effectively nonpenetrating solutes inside the cells is equal to the total concentration of Na^+ and other effectively nonpenetrating solutes in the fluid surrounding the cells. Even though nonpenetrating solutes in the ECF and ICF differ, their concentrations are normally identical, and the number (not the nature) of the unequally distributed particles per volume determines the fluid's osmolarity. Because the osmolarities of the ECF and ICF are normally equal, no net movement of H_2O usually occurs into or out of the cells. Therefore, cell volume normally remains constant.

Importance of Regulating ECF Osmolarity Any circumstance that results in a loss or gain of *free H_2O* (that is, loss or gain of H_2O that is not accompanied by comparable solute deficit or excess) leads to changes in ECF osmolarity. If there is a deficit of free H_2O in the ECF, the solutes become too concentrated and ECF osmolarity becomes abnormally high—that is, it becomes *hypertonic* (see p. 69). If there is excess free H_2O in the ECF, the solutes become too dilute and ECF osmolarity becomes abnormally low—that is, it becomes *hypotonic*. When ECF osmolarity changes with respect to ICF osmolarity, osmosis takes place, with H_2O either leaving or entering the cells,

depending on whether the ECF is more concentrated or less concentrated, respectively, than the ICF.

The osmolarity of the ECF must therefore be regulated to prevent these undesirable shifts of H_2O out of or into the cells. As far as the ECF itself is concerned, the concentration of its solutes does not really matter. However, it is crucial that ECF osmolarity be maintained within narrow limits to prevent the cells from shrinking (by osmotically losing water to the ECF) or swelling (by osmotically gaining fluid from the ECF).

Let us examine the fluid shifts that occur between the ECF and the ICF when ECF osmolarity becomes hypertonic or hypotonic relative to the ICF.

During ECF hypertonicity, cells shrink as H_2O leaves them.

Hypertonicity of the ECF, the excessive concentration of ECF solutes, is usually associated with **dehydration**, or a negative free H_2O balance.

Causes of Hypertonicity (Dehydration) Dehydration with accompanying hypertonicity can be brought about in three ways:

1. *Insufficient H_2O intake*, such as might occur during desert travel or might accompany difficulty in swallowing

2. *Excessive H_2O loss*, such as might occur during heavy sweating, vomiting, or diarrhea (even though both H_2O and solutes are lost during these conditions, relatively more H_2O is lost, so the remaining solutes become more concentrated)

3. *Diabetes insipidus, a disease characterized by a deficiency of vasopressin*

Clinical Note *Vasopressin (antidiuretic hormone)* increases the permeability of the distal and collecting tubules to H_2O and thus enhances water conservation by reducing urinary output of water (see p. 524). Without adequate vasopressin in **diabetes insipidus**, the kidneys cannot conserve H_2O because they cannot reabsorb H_2O from the late parts of the nephron. Such patients typically produce up to 20 liters of very dilute urine daily, compared to the normal average of 1.5 liters per day. Unless H_2O intake keeps pace with this tremendous loss of H_2O in the urine, the person quickly dehydrates. Such patients complain that they spend an extraordinary amount of time day and night going the bathroom and getting drinks. Fortunately, they can be treated with replacement vasopressin administered by nasal spray.

See Figure 14-12 for details of mechanism.

See Figure 14-16 for details of mechanism.

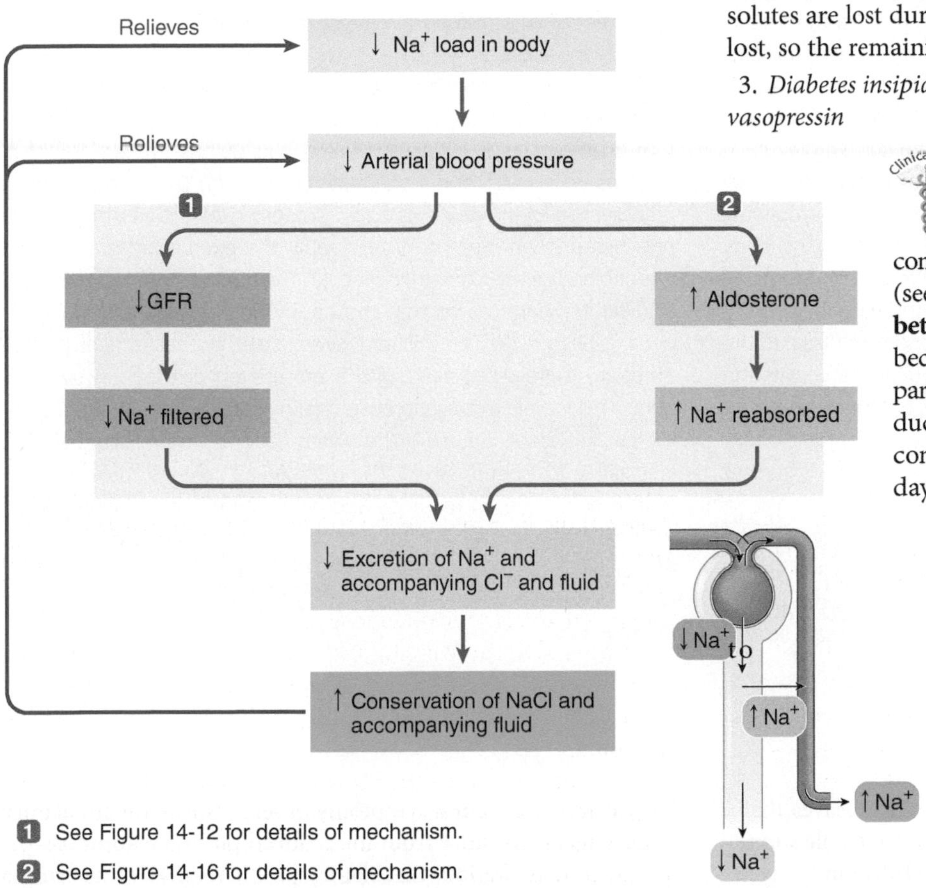

Figure 15-3 **Dual effect of a fall in arterial blood pressure on renal handling of Na^+.**
FIGURE FOCUS: *Describe the two pathways by which an excess Na^+ load is eliminated from the body.*

Clinical Note **Direction and Resulting Symptoms of Water Movement During Hypertonicity** Whenever the ECF compartment becomes hypertonic, H_2O moves out of the cells by osmosis into the more concentrated ECF until the

A Potentially Fatal Clash: When Exercising Muscles and Cooling Mechanisms Compete for an Inadequate Plasma Volume

AN INCREASING NUMBER OF PEOPLE of all ages are participating in walking or jogging programs to improve their level of physical fitness and decrease their risk of cardiovascular disease. For people living in environments that undergo seasonal temperature changes, fluid loss can make exercising outdoors dangerous during the transition from the cool days of spring to the hot, humid days of summer. If exercise intensity is not modified until the participant gradually adjusts to the hotter environmental conditions, dehydration and salt loss can indirectly lead to heat cramps, heat exhaustion, or ultimately heat stroke and death.

The term **acclimatization** refers to the gradual adaptations the body makes to maintain long-term homeostasis in response to a prolonged physical change in the surrounding environment, such as a change in temperature. When a person exercises in the heat without gradually adapting to the hotter environment, the body faces a terrible dilemma. During exercise, large amounts of blood must be delivered to the muscles to supply O_2 and nutrients and to remove the wastes that accumulate from their high rate of activity. Exercising muscles also produce heat. To maintain body temperature in the face of this extra heat, blood flow to the skin is increased so that heat from the warmed blood can be lost through the skin to the surrounding environment. If the environmental temperature is hotter than the body temperature, heat cannot be lost from the blood to the surrounding environment despite maximal skin vasodilation. Instead, the body gains heat from its warmer surroundings, further adding to the dilemma. Because extra blood is diverted to both the muscles and the skin when a person exercises in the heat, less blood is returned to the heart, and the heart pumps less blood per beat in accordance with the Frank–Starling mechanism (see p. 322). Therefore, the heart must beat faster than it would in a cool environment to deliver the same amount of blood per minute. The increased rate of cardiac pumping further contributes to heat production.

The sweat rate also increases so that evaporative cooling can take place to help maintain the body temperature during periods of excessive heat gain. In an unacclimatized person, maximal sweat rate is about 1.5 liters per hour. During sweating, water-retaining salt, as well as water, is lost. The resulting loss of plasma volume through sweating further depletes the blood supply available for muscular exercise and for cooling through skin vasodilation.

The heart has a maximum rate at which it can pump. If exercise continues at a high intensity and this maximal rate is reached, the exercising muscles win the contest for blood supply. Cooling is sacrificed as skin blood flow decreases. If exercise continues, body heat continues to rise, and heat exhaustion (rapid, weak pulse; hypotension; profuse sweating; and disorientation) or heat stroke (failure of the temperature control center in the hypothalamus; hot, dry skin; extreme confusion or unconsciousness; and possibly death) can occur (see p. 634). Every year people die of heat stroke running in marathons during hot, humid weather. (Some people make matters worse by adding a caffeine-containing energy drink to their workout or competition. Caffeine may provide a jolt of energy, but it also acts as a diuretic and can lead to performance-reducing dehydration, the opposite effect of what the people might think they are accomplishing by drinking these beverages.)

By contrast, if a person exercises in the heat for 2 weeks at reduced, safe intensities, the body makes the following adaptations so that after acclimatization the person can do the same amount of work as was possible in a cool environment: (1) The plasma volume is increased by as much as 12%. Expansion of the plasma volume provides enough blood to both supply the exercising muscles and direct blood to the skin for cooling. (2) The person begins sweating at a lower temperature so that the body does not get so hot before cooling begins. (3) The maximal sweat rate increases nearly three times, to 4 liters per hour, with a more even distribution over the body. This increase in evaporative cooling reduces the need for cooling by skin vasodilation. (4) The sweat becomes more dilute so that less salt is lost in the sweat. The retained salt exerts an osmotic effect to hold water in the body and help maintain circulating plasma volume. These adaptations take 14 days and occur only if the person exercises in the heat. Being patient until these changes take place can enable the person to exercise safely throughout the summer months.

ICF osmolarity equilibrates with the ECF. As H_2O leaves them, the cells shrink. Of particular concern is that considerable shrinking of brain neurons disturbs brain function, which can be manifested as mental confusion and irrationality in moderate cases and delirium, convulsions, or coma in more severe hypertonic conditions.

Rivaling the neural symptoms in seriousness are circulatory disturbances that arise from the reduced plasma volume associated with dehydration. Circulatory problems may range from a slight lowering of blood pressure to circulatory shock and death.

Other more common symptoms become apparent even in mild cases of dehydration. For example, dry skin and sunken eyeballs indicate loss of H_2O from the underlying soft tissues, and the tongue becomes dry and parched because salivary secretion is suppressed.

During ECF hypotonicity, the cells swell as H_2O enters them.

Hypotonicity of the ECF is associated with **overhydration**—that is, excess free H_2O. When a positive free H_2O balance exists, the ECF is less concentrated (more dilute) than normal.

Causes of Hypotonicity (Overhydration) Usually, any surplus free H_2O is promptly excreted in the urine, so hypotonicity generally does not occur. However, hypotonicity arises in three ways:

1. Patients with *renal failure* who cannot excrete dilute urine become hypotonic when they consume relatively more H_2O than solutes.

2. Hypotonicity occurs transiently in healthy people if *H_2O is rapidly ingested* to such an excess that the kidneys cannot respond quickly enough to eliminate the extra H_2O.

3. Hypotonicity occurs when excess H_2O without solute is retained in the body as a result of the *syndrome of inappropriate vasopressin secretion.*

 Vasopressin is normally secreted in response to a H_2O deficit, which is relieved by increasing H_2O reabsorption in the distal part of the nephrons. However, vasopressin secretion, and therefore hormonally controlled tubular H_2O reabsorption, can be increased in response to pain, trauma, and other stressful situations, even when the body has no H_2O deficit. The increased vasopressin secretion and resulting H_2O retention elicited by stress are appropriate in anticipation of potential blood loss in the stressful situation. The extra retained H_2O could minimize the effect a loss of blood volume would have on blood pressure. However, because modern-day stressful situations generally do not involve blood loss, the increased vasopressin secretion is inappropriate as far as the body's fluid balance is concerned, leading to the **syndrome of inappropriate vasopressin secretion**. The resultant retention of too much H_2O dilutes the body's solutes. In addition to stress inappropriately promoting vasopressin secretion from its normal source, some types of lung cancer secrete vasopressin, inappropriately diluting the body fluids.

 Direction and Resulting Symptoms of Water Movement During Hypotonicity Excess free H_2O retention first dilutes the ECF, making it hypotonic. The resulting difference in osmotic activity between the ECF and the ICF causes H_2O to move by osmosis from the more dilute ECF into the cells, with the cells swelling as H_2O moves into them. Like the shrinking of cerebral neurons, pronounced swelling of brain cells also leads to brain dysfunction. Symptoms include confusion, irritability, lethargy, headache, dizziness, vomiting, drowsiness, and in severe cases, convulsions, coma, and death.

Nonneural symptoms of overhydration include weakness caused by swelling of muscle cells and circulatory disturbances, including hypertension and edema, caused by expansion of plasma volume.

The condition of overhydration, hypotonicity, and cellular swelling resulting from excess free H_2O retention is known as **water intoxication**. It should not be confused with the fluid retention that occurs with excess salt retention. In the latter case, the ECF is still isotonic because the increase in salt is matched by a corresponding increase in H_2O. Because the interstitial fluid is still isotonic, no osmotic gradient exists to drive the extra H_2O into the cells. The excess salt and H_2O burden is therefore confined to the ECF compartment, with circulatory consequences being the most important concern. In water intoxication, in addition to any circulatory disturbances, symptoms caused by cell swelling become a problem.

We now contrast the situations of hypertonicity and hypotonicity with what happens when isotonic fluid is gained or lost.

No water moves into or out of cells during an ECF isotonic fluid gain or loss.

 An example of an isotonic fluid gain is therapeutic intravenous administration of an isotonic solution, such as isotonic saline. When isotonic fluid is injected into the ECF compartment, ECF volume increases, but the concentration of ECF solutes remains unchanged; in other words, the ECF is still isotonic. Because the ECF osmolarity has not changed, the ECF and ICF are still in osmotic equilibrium, so no net fluid shift occurs between the two compartments. The ECF compartment has increased in volume without shifting H_2O into the cells.

Similarly, in an isotonic fluid loss such as hemorrhage, the loss is confined to the ECF, with no corresponding loss of fluid from the ICF. Fluid does not shift out of the cells because the ECF remaining within the body is still isotonic, so no osmotic gradient draws H_2O out of the cells. Many other mechanisms counteract loss of blood, but the ICF compartment is not directly affected by the loss.

We now consider how free H_2O balance and consequently ECF osmolarity are normally maintained to minimize harmful changes in cell volume.

Vasopressin control of free H_2O balance is important in regulating ECF osmolarity.

Control of free H_2O balance is crucial for regulating ECF osmolarity. Because increases in free H_2O cause the ECF to become too dilute and deficits of free H_2O cause the ECF to become too concentrated, the osmolarity of the ECF must be immediately corrected by restoring stable free H_2O balance to avoid harmful osmotic fluid shifts into or out of the cells.

To maintain a stable H_2O balance, H_2O input must equal H_2O output.

Sources of H_2O Input
- In a person's typical daily H_2O balance (Table 15-3), a little more than a liter of H_2O is added to the body by *drinking liquids*.

- Surprisingly, an amount almost equal to that is obtained from *eating solid food*. Muscles consist of about 75% H_2O;

ITABLE 15-3 Daily Water Balance

WATER INPUT		WATER OUTPUT	
Avenue	Quantity (mL/day)	Avenue	Quantity (mL/day)
Fluid intake	1250	Insensible loss (from lungs and nonsweating skin)	900
H_2O in food intake	1000	Sweat	100
Metabolically produced H_2O	350	Feces	100
		Urine	1500
Total input	2600	**Total output**	2600

meat (animal muscle) is therefore 75% H_2O. Fruits and vegetables consist of 60% to 96% H_2O. Therefore, people normally get almost as much H_2O from solid foods as from liquids.

■ The third source of H_2O input is *metabolically produced H_2O*. Chemical reactions within cells convert food and O_2 into energy, producing CO_2 and H_2O in the process. This **metabolic H_2O** produced during cell metabolism and released into the ECF averages about 350 mL per day.

The average H_2O intake from these three sources totals 2600 mL per day. Another source of H_2O often employed therapeutically is intravenous infusion of fluid.

Sources of H₂O Output

■ On the output side of the H_2O balance tally, you lose nearly a liter of H_2O daily without being aware of it. This **insensible loss** (loss of which the person has no sensory awareness) occurs from the *lungs* and *nonsweating skin*. During respiration, inspired air becomes saturated with H_2O within the airways. This H_2O is lost when the moistened air is subsequently expired (see p. 447). Normally, you are not aware of this H_2O loss, but you can recognize it on cold days, when H_2O vapor condenses so that you can "see your breath." The other insensible loss is continual loss of H_2O from the skin even in the absence of sweating. Water molecules can diffuse through skin cells and evaporate without being noticed. Fortunately, the skin is fairly waterproofed by its keratinized exterior layer, which protects against a greater loss of H_2O by this avenue (see p. 440). When this protective surface layer is lost, such as when a person has extensive burns, increased fluid loss from the burned surface can cause serious problems with fluid balance.

■ Sensible loss (loss of which the person is aware) of H_2O from the skin occurs through *sweating*, which represents another avenue of H_2O output. At an air temperature of 68°F, an average of 100 mL of H_2O is lost daily through sweating. Loss of water from sweating can vary substantially, depending on the environmental temperature and humidity and the degree

of physical activity; it may range from zero up to as much as several liters per hour in very hot weather.

■ Another passageway for H_2O loss from the body is through the *feces*. Normally, only about 100 mL of H_2O are lost this way each day. During fecal formation in the large intestine, most H_2O is absorbed out of the digestive tract lumen into the blood, thereby conserving fluid and solidifying the digestive tract's contents for elimination. Additional H_2O can be lost from the digestive tract through vomiting or diarrhea.

■ By far the most important output mechanism is *urine excretion*, with 1500 mL (1.5 liters) of urine being produced daily on average.

The total H_2O output is 2600 mL/day, the same as the volume of H_2O input in our example. This balance is not by chance. Normally, H_2O input matches H_2O output so that the H_2O in the body remains in balance.

Factors Regulated to Maintain Water Balance Of the many sources of H_2O input and output, only two can be regulated to maintain H_2O balance. On the intake side, thirst influences the amount of fluid ingested; on the output side, the kidneys can adjust how much urine is formed. Controlling H_2O output in the urine is the most important mechanism in controlling H_2O balance.

Some of the other factors are regulated, but not for maintaining H_2O balance. Food intake is subject to regulation to maintain energy balance, and control of sweating is important in maintaining body temperature. Metabolic H_2O production and insensible losses are unregulated.

Control of Water Output in the Urine by Vasopressin Fluctuations in ECF osmolarity caused by imbalances between H_2O input and H_2O output are quickly compensated for by adjusting urinary excretion of H_2O without changing the usual excretion of salt—that is, H_2O reabsorption and excretion are partially dissociated from solute reabsorption and excretion, so the amount of free H_2O retained or eliminated can be varied to quickly restore ECF osmolarity to normal. Free H_2O reabsorption and excretion are adjusted through changes in vasopressin secretion (see IFigure 14-27, p. 525). Throughout most of the nephron, H_2O reabsorption is important in regulating ECF volume because salt reabsorption is accompanied by comparable H_2O reabsorption. In the distal and collecting tubules, however, variable free H_2O reabsorption can take place without comparable salt reabsorption because of the vertical osmotic gradient in the renal medulla to which this part of the tubule is exposed. Vasopressin increases the permeability of this late part of the tubule to H_2O. Depending on the amount of vasopressin present, the amount of free H_2O reabsorbed can be adjusted as necessary to restore ECF osmolarity to normal.

Vasopressin is produced by the hypothalamus, stored in the posterior pituitary gland, and released from the posterior pituitary on command from the hypothalamus.

Control of Water Input by Thirst Thirst is the subjective sensation that drives you to ingest H_2O. The **thirst center** is located in the hypothalamus close to the vasopressin-secreting cells.

We now elaborate on the mechanisms that regulate vasopressin secretion and thirst.

Vasopressin secretion and thirst are largely triggered simultaneously.

The hypothalamic control centers that regulate vasopressin secretion (and thus urinary output) and thirst (and thus drinking) act in concert. Vasopressin secretion and thirst are both stimulated by a free H_2O deficit and suppressed by a free H_2O excess. Thus, appropriately, the same circumstances that call for reducing urinary output to conserve body H_2O give rise to the sensation of thirst to replenish body H_2O.

Role of Hypothalamic Osmoreceptors The predominant excitatory input for both vasopressin secretion and thirst comes from **hypothalamic osmoreceptors** located near the vasopressin-secreting cells and thirst center. These osmo-receptors monitor the osmolarity of fluid surrounding them, which in turn reflects the concentration of the entire internal fluid environment. As ECF osmolarity increases (too little H_2O) and the need for H_2O conservation increases, vasopressin secretion and thirst are both stimulated (❙Figure 15-4). As a result, reabsorption of H_2O in the distal and collecting tubules is increased so that urinary output is reduced and H_2O is conserved, while H_2O intake is simultaneously encouraged. These actions restore depleted H_2O stores, thus relieving the hypertonic condition by diluting the solutes to normal concentration. In contrast, H_2O excess, manifested by reduced ECF osmolarity, prompts increased urinary output (through decreased vasopressin release) and suppresses thirst, which together reduce the water load in the body.

Role of Left Atrial Volume Receptors Although the major stimulus for vasopressin secretion and thirst is an increase in ECF osmolarity, the vasopressin-secreting cells and thirst center are both influenced to a moderate extent by changes in ECF volume (and therefore plasma volume) mediated by input from the **left atrial volume receptors**. Located in the left atrium, these volume receptors respond to pressure-induced stretch caused by blood flowing through, which reflects the ECF volume—that is, they monitor the "fullness" of the vascular system. In contrast, the aortic arch and carotid sinus baroreceptors monitor the mean driving pressure in the vascular system (see p. 367). In response to a major reduction in ECF volume (>7% loss of volume), and accordingly in arte-

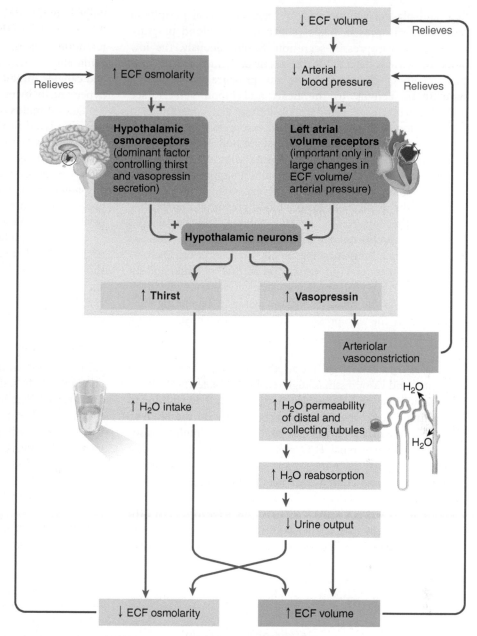

❙Figure 15-4 **Control of increased vasopressin secretion and thirst during a H_2O deficit.**

rial pressure, as during hemorrhage, the left atrial volume receptors reflexly stimulate both vasopressin secretion and thirst. (By comparison, a change as small as a 1% increase in ECF osmolarity triggers increased vasopressin secretion, and an increase in osmolarity of 2% or more produces a strong desire to drink, indicative of the greater influence of the hypothalamic osmoreceptors than the left atrial volume receptors in controlling vasopressin secretion and thirst.) In the face of a marked reduction in ECF volume, the outpouring of vasopressin and the increased thirst lead to decreased urine output and increased fluid intake, respectively. Furthermore, vasopressin, at the circulating levels elicited by a large decline in ECF volume and arterial pressure, exerts a potent vasoconstrictor (that is, a "vaso" "pressor") effect on arterioles (thus giving rise to its name; see p. 350). Both by helping expand the

ECF and plasma volume and by increasing total peripheral resistance, vasopressin helps relieve the low blood pressure that elicited vasopressin secretion. Simultaneously, the low blood pressure is detected by the aortic arch and carotid sinus baroreceptors, which help raise the pressure by increasing sympathetic activity to the heart and blood vessels (see p. 368). Furthermore, sympathetic activity also contributes to the sensation of thirst and increased vasopressin activity.

Conversely, vasopressin and thirst are both inhibited when ECF volume (and, accordingly, plasma volume) and arterial blood pressure are elevated. The resultant suppression of H_2O intake, coupled with elimination of excess ECF and plasma volume in the urine, helps restore blood pressure to normal.

Recall that low NaCl, low ECF volume, and low arterial blood pressure also reflexly increase aldosterone secretion via RAAS. The resulting increase in Na^+ reabsorption leads to osmotic retention of H_2O, expansion of ECF volume, and an increase in arterial blood pressure. Aldosterone-controlled Na^+ reabsorption is the most important factor in regulating ECF volume, with the vasopressin and thirst mechanism playing only a supportive role.

Role of Angiotensin II Yet another stimulus for increasing both thirst and vasopressin is angiotensin II (Table 15-4). When RAAS is activated to conserve Na^+, angiotensin II, in addition to stimulating aldosterone secretion, acts directly on the brain to give rise to the urge to drink and concurrently stimulates vasopressin to enhance renal H_2O reabsorption (see p. 508). The resultant increased H_2O intake and decreased urinary output help correct the reduction in ECF volume that triggered RAAS.

Regulatory Factors that Do Not Link Vasopressin and Thirst Several factors affect vasopressin secretion but not thirst. As described earlier, vasopressin is stimulated by stress-related inputs such as pain and trauma that have nothing directly to do with maintaining H_2O balance. In fact, H_2O retention from the inappropriate secretion of vasopressin can bring about a hypotonic H_2O imbalance. In contrast, alcohol and caffeine inhibit vasopressin secretion and can lead to ECF hypertonicity by promoting excessive free H_2O excretion.

One stimulus that promotes thirst but not vasopressin secretion is a direct effect of dryness of the mouth. Nerve endings in the mouth are directly stimulated by dryness, which causes an intense sensation of thirst that can often be relieved merely by moistening the mouth even though no H_2O is actually ingested. A dry mouth can exist when salivation is suppressed by factors unrelated to the body's H_2O content, such as nervousness, excessive smoking, or certain drugs.

Factors that affect vasopressin secretion or thirst but have nothing directly to do with the body's need for H_2O are usually short-lived. The dominant, long-standing control of vasopressin and thirst is directly correlated with the body's state of H_2O—namely, by the status of ECF osmolarity and, to a lesser extent, by ECF volume.

Oral Metering Some kind of "oral H_2O metering" appears to exist, at least in animals. A thirsty animal will rapidly drink only enough H_2O to satisfy its H_2O deficit. It stops drinking before the ingested H_2O has had time to be absorbed from the digestive tract and return the ECF compartment to normal. Receptors in the mouth, pharynx (throat), and upper digestive tract signal that enough H_2O has been consumed. This mechanism seems to be less effective in humans because we frequently drink more than is necessary to meet the needs of our bodies or, conversely, may not drink enough to make up a deficit.

Nonphysiological Influences on Fluid Intake Although the thirst mechanism exists to control H_2O intake, fluid con-

| TABLE 15-4 Factors Controlling Vasopressin Secretion and Thirst |

Factor	Effect on Vasopressin Secretion	Effect on Thirst	Comment
↑ ECF osmolarity	↑	↑	Major stimulus for vasopressin secretion and thirst
↓ ECF volume	↑	↑	Important only for large changes in ECF volume/arterial blood pressure
Angiotensin II	↑	↑	Part of dominant pathway for promoting compensatory salt and H_2O retention when ECF volume and arterial blood pressure are reduced
Pain, trauma, and other stress-related inputs	Inappropriate ↑ unrelated to body's H_2O balance	No effect	Promotes excess H_2O retention and ECF hypotonicity (resultant H_2O retention of potential value in maintaining arterial blood pressure in case of blood loss in the stressful situation)
Alcohol and caffeine	Inappropriate ↓ unrelated to body's H_2O balance	No effect	Promote excess H_2O loss and ECF hypertonicity
Dry mouth	No effect	↑	Nerve endings in the mouth that ultimately give rise to the sensation of thirst are directly stimulated by dryness

TABLE 15-5 Summary of the Regulation of ECF Volume and Osmolarity

Regulated Variable	Need to Regulate the Variable	Outcomes If the Variable Is Not Normal	Mechanism for Regulating the Variable
ECF volume	Important in the long-term control of arterial blood pressure	↓ ECF volume → ↓ arterial blood pressure ↑ ECF volume → ↑ arterial blood pressure	Maintenance of salt balance; salt osmotically "holds" H_2O, so the Na^+ load determines the ECF volume. Accomplished primarily by aldosterone-controlled adjustments in urinary Na^+ excretion
ECF osmolarity	Important to prevent detrimental osmotic movement of H_2O between the ECF and ICF	↑ ECF osmolarity (hypertonicity) → H_2O leaves the cells → cells shrink ↓ ECF osmolarity (hypotonicity) → H_2O enters the cells → cells swell	Maintenance of free H_2O balance. Accomplished primarily by vasopressin-controlled adjustments in excretion of H_2O in the urine

sumption by humans is often influenced more by habit and sociological factors than by the need to regulate H_2O balance. Thus, even though H_2O intake is critical in maintaining fluid balance, it is not precisely controlled in humans, who err especially on the side of excess H_2O consumption. We usually drink when we are thirsty, but we often drink even when we are not thirsty because, for example, we are on a coffee break.

With H_2O intake being inadequately controlled and indeed even contributing to H_2O imbalances in the body, the primary factor involved in maintaining H_2O balance is urinary output regulated by the kidneys. Accordingly, *vasopressin-controlled H_2O reabsorption is of primary importance in regulating ECF osmolarity.*

Before we shift to acid–base balance, examine ▌Table 15-5, which summarizes the regulation of ECF volume and osmolarity, the two factors important in maintaining fluid balance.

Check Your Understanding 15.2

1. Make a chart showing the percentage of body H_2O distributed among the major body fluid compartments.
2. Compare how ECF volume and ECF osmolarity are regulated, and discuss the importance of each being regulated.
3. What factors are regulated to maintain salt balance and to maintain water balance?
4. Compare the effect of ECF hypertonicity and hypotonicity on cell volume.

15.3 Acid–Base Balance

The term **acid–base balance** refers to the precise regulation of **free** (that is, unbound) **hydrogen ion (H^+) concentration** in the body fluids. To indicate the concentration of a chemical, its symbol is enclosed in square brackets. Thus, $[H^+]$ designates H^+ concentration.

Acids liberate free hydrogen ions, whereas bases accept them.

Acids are a special group of hydrogen-containing substances that *dissociate,* or separate, when in solution to liberate free H^+ and anions. Many other substances (for example, carbohydrates) also contain hydrogen, but they are not classified as acids because the hydrogen is tightly bound within their molecular structure and is never liberated as free H^+.

A strong acid has a greater tendency to dissociate in solution than a weak acid does—that is, a greater percentage of a strong acid's molecules separate into free H^+ and anions. Hydrochloric acid (HCl) is an example of a strong acid; every HCl molecule dissociates into free H^+ and chloride (Cl^-) when dissolved in H_2O. With a weaker acid such as carbonic acid (H_2CO_3), only a portion of the molecules dissociates in solution into H^+ and bicarbonate anions (HCO_3^-). The remaining H_2CO_3 molecules remain intact. Because only free H^+ contributes to the acidity of a solution, H_2CO_3 is a weaker acid than HCl because H_2CO_3 does not yield as many free H^+ per number of acid molecules present in solution (▌Figure 15-5).

The extent of dissociation for a given acid is always constant—that is, when in solution, the same proportion of a particular acid's molecules always separate to liberate free H^+, with the other portion always remaining intact. The constant degree of dissociation for a particular acid (in this example, H_2CO_3) is expressed by its **dissociation constant (K)** as follows:

$$[H^+][HCO_3^-]/[H_2CO_3] = K$$

where

$[H^+][HCO_3^-]$ represents the concentration of ions resulting from H_2CO_3 dissociation and $[H_2CO_3]$ represents the concentration of intact (undissociated) H_2CO_3.

The dissociation constant varies for different acids.

A **base** is a substance that can combine with a free H^+ and thus remove it from solution. A strong base can bind H^+ more readily than a weak base can.

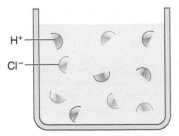

(a) Strong acid (HCl)

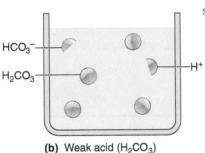

(b) Weak acid (H_2CO_3)

KEY

    = Undissociated acid

    = Free H^+ = Free anion

❙Figure 15-5 **Comparison of a strong and a weak acid.** (a) Five molecules of a strong acid. A strong acid such as HCl (hydrochloric acid) completely dissociates into free H^+ and anions in solution. (b) Five molecules of a weak acid. A weak acid such as H_2CO_3 (carbonic acid) only partially dissociates into free H^+ and anions in solution.

The greater the $[H^+]$, the larger the number by which 1 must be divided and the lower the pH.

   2. *Every unit change in pH actually represents a 10-fold change in $[H^+]$* because of the logarithmic relationship. A log to the base 10 indicates how many times 10 must be multiplied by itself to produce a given number. For example, the log of 10 = 1, whereas the log of 100 = 2. The number 10 must be multiplied by itself twice to yield 100 ($10 \times 10 = 100$). Numbers less than 10 have logs less than 1. Numbers between 10 and 100 have logs between 1 and 2, and so on. Accordingly, each unit of change in pH indicates a 10-fold change in $[H^+]$. For example, a solution with a pH of 7 has a $[H^+]$ 10 times less than that of a solution with a pH of 6 (a 1 pH-unit difference) and 100 times less than that of a solution with a pH of 5 (a 2 pH-unit difference).

Acidic and Basic Solutions in Chemistry The pH of pure H_2O is 7.0, which is considered chemically neutral. An extremely small proportion of H_2O molecules dissociate into H^+ and hydroxyl (OH^-) ions. Because an equal number of acidic H^+ and basic OH^- are formed, H_2O is neutral, being neither acidic nor basic. Solutions having a pH less than 7.0 contain a higher $[H^+]$ than pure H_2O and are considered **acidic**. Solutions having a pH value greater than 7.0 have a lower $[H^+]$ than pure

The pH designation is used to express $[H^+]$.

The $[H^+]$ in the ECF is normally 4×10^{-8} or 0.00000004 equivalents per liter, about 3 million times less than the $[Na^+]$ in the ECF. The concept of pH was developed to express the low value of $[H^+]$ more conveniently. Specifically, **pH** equals the logarithm (log) to the base 10 of the reciprocal of $[H^+]$.

$$pH = \log 1/[H^+]$$

Two important points should be noted about this formula:

   1. Because $[H^+]$ is in the denominator, *a high $[H^+]$ corresponds to a low pH and a low $[H^+]$ corresponds to a high pH.*

❙Figure 15-6 **pH considerations in chemistry and physiology.** (a) Relationship of pH to the relative concentrations of H^+ and base (OH^-) under chemically neutral, acidic, and alkaline conditions. (b) Blood pH range under normal, acidotic, and alkalotic conditions.

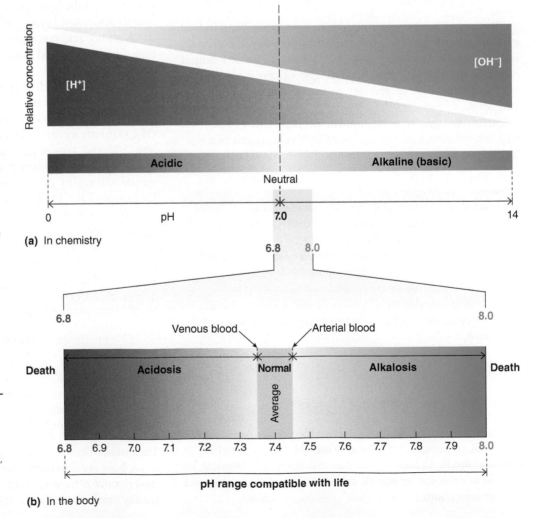

(a) In chemistry

(b) In the body

H$_2$O and are considered **basic**, or **alkaline** (❙Figure 15-6a). ❙Figure 15-7 compares the pH values of common solutions.

Acidosis and Alkalosis in the Body The pH of arterial blood is normally 7.45 and the pH of venous blood is 7.35, for an average blood pH of 7.4. The pH of venous blood is slightly lower (more acidic) than that of arterial blood because H$^+$ is generated by the formation of H$_2$CO$_3$ from CO$_2$ picked up at the tissue capillaries. **Acidosis** exists whenever blood pH falls below 7.35, whereas **alkalosis** occurs when blood pH is above 7.45 (see ❙Figure 15-6b). Note that the reference point for determining the body's acid–base status is not the chemically neutral pH of 7.0 but the normal blood pH of 7.4. Thus, a blood pH of 7.2 is considered acidotic even though in chemistry a pH of 7.2 is considered basic.

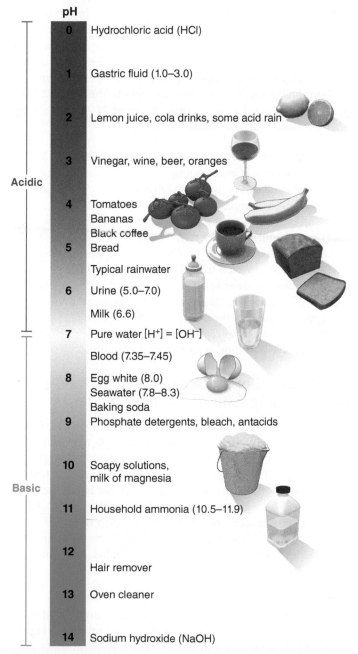

pH	
0	Hydrochloric acid (HCl)
1	Gastric fluid (1.0–3.0)
2	Lemon juice, cola drinks, some acid rain
3	Vinegar, wine, beer, oranges
4	Tomatoes Bananas Black coffee
5	Bread
	Typical rainwater
6	Urine (5.0–7.0)
	Milk (6.6)
7	Pure water [H$^+$] = [OH$^-$]
	Blood (7.35–7.45)
8	Egg white (8.0) Seawater (7.8–8.3) Baking soda
9	Phosphate detergents, bleach, antacids
10	Soapy solutions, milk of magnesia
11	Household ammonia (10.5–11.9)
12	
	Hair remover
13	Oven cleaner
14	Sodium hydroxide (NaOH)

Acidic (pH 0–7), Basic (pH 7–14)

❙Figure 15-7 **Comparison of pH values of common solutions.**

An arterial pH of less than 6.8 or greater than 8.0 is not compatible with life. Because death occurs if arterial pH falls outside the range of 6.8 to 8.0 for more than a few seconds, [H$^+$] in the body fluids must be carefully regulated.

Fluctuations in [H$^+$] alter nerve, enzyme, and K$^+$ activity.

Only a narrow pH range is compatible with life because even small changes in [H$^+$] have dramatic effects on normal cell function, as the following consequences indicate:

1. *Changes in excitability of nerve and muscle cells are among the major clinical manifestations of pH abnormalities.*

   ■ The major clinical effect of increased [H$^+$] (acidosis) is depression of the central nervous system (CNS). Acidotic patients become disoriented and, in more severe cases, eventually die in a state of coma.

   ■ In contrast, the major clinical effect of decreased [H$^+$] (alkalosis) is overexcitability of the nervous system, first the peripheral nervous system and later the CNS. Peripheral nerves become so excitable that they fire even in the absence of normal stimuli. Such overexcitability of the afferent (sensory) nerves gives rise to abnormal "pins-and-needles" tingling sensations. Overexcitability of efferent (motor) nerves brings about muscle twitches and, in more pronounced cases, severe muscle spasms. Death may occur in extreme alkalosis because spasm of the respiratory muscles seriously impairs breathing. Alternatively, patients with severe alkalosis may die of convulsions resulting from overexcitability of the CNS. In less serious situations, CNS overexcitability is manifested as extreme nervousness.

2. *Hydrogen ion concentration exerts a marked influence on enzyme activity.* Even slight deviations in [H$^+$] alter the shape and activity of protein molecules. Because enzymes are proteins, a shift in the body's acid–base balance disturbs the normal pattern of metabolic activity catalyzed by these enzymes.

3. *Changes in [H$^+$] influence K$^+$ levels in the body.* When reabsorbing Na$^+$ from the filtrate, the renal tubular cells secrete either K$^+$ or H$^+$ in exchange. Normally, they secrete a preponderance of K$^+$ compared to H$^+$. Because of the intimate relationship between secretion of H$^+$ and that of K$^+$ by the kidneys, when H$^+$ secretion increases to compensate for acidosis, less K$^+$ than usual can be secreted; conversely, when H$^+$ secretion is reduced during alkalosis, more K$^+$ is secreted than normal. The resulting changes in ECF [K$^+$] can lead to cardiac abnormalities, among other detrimental consequences (see p. 516).

Hydrogen ions are continually added to the body fluids as a result of metabolic activities.

As with any other constituent, input of hydrogen ions must be balanced by an equal output to maintain a constant [H$^+$] in the body fluids. On the input side, only a small amount of acid capable of dissociating to release H$^+$ is taken in with food, such as the weak citric acid found in oranges. Most H$^+$ in the body fluids is generated internally from metabolic activities.

Sources of H⁺ in the Body Normally, H^+ is continuously added to the body fluids from the three following sources:

1. *Carbonic acid formation.* The major source of H^+ is from metabolically produced CO_2. Cellular oxidation of nutrients yields energy, with CO_2 and H_2O as end products. Without catalyst influence, CO_2 and H_2O slowly form H_2CO_3, which then rapidly partially dissociates to liberate free H^+ and HCO_3^-:

$$CO_2 + H_2O \xrightleftharpoons{slow} H_2CO_3 \xrightleftharpoons{fast} H^+ + HCO_3^-$$

The slow first reaction is the rate-limiting step in the plasma, but the hydration (combination with H_2O) of CO_2 is greatly accelerated by the enzyme *carbonic anhydrase,* which is abundant in red blood cells (see p. 477), some special secretory cells of the stomach and pancreas (see pp. 584 and 591), and kidney tubular cells. Under the influence of carbonic anhydrase (represented by *ca* in the next equation), these cells directly convert CO_2 and H_2O into H^+ and HCO_3^- (with no intervening production of H_2CO_3) as follows:

Step 1. Carbonic anhydrase catalyzes the formation of HCO_3^- from metabolically produced CO_2 in the reaction:

$$CO_2 + OH^- \xrightleftharpoons{ca} HCO_3^-$$

Step 2. Water dissociates, forming more OH^- that can be used in Step 1, yielding H^+ in the process:

$$H_2O \rightleftharpoons OH^- + H^+$$

Collectively, these steps can be summarized as:

$$CO_2 + H_2O \xrightleftharpoons{ca} H^+ + HCO_3^-$$

The OH^- used up in step 1 is generated by step 2; as a result, there's no net loss or gain of OH^-, so we can ignore it in this summary equation.

These reactions are reversible because they can proceed in either direction, depending on the concentrations of the substances involved as dictated by the *law of mass action* (see p. 472). Within the systemic capillaries, the CO_2 level in the blood increases as metabolically produced CO_2 enters from the tissues. This drives the reaction (with or without carbonic anhydrase) to the H^+ side. In the lungs, the reaction is reversed: CO_2 diffuses from the blood flowing through the pulmonary capillaries into the alveoli (air sacs), from which it is expired to the atmosphere. The resultant reduction in blood CO_2 drives the reaction toward the CO_2 side. H^+ and HCO_3^- form CO_2 and H_2O again. The CO_2 is exhaled while the hydrogen ions generated at the tissue level are incorporated into H_2O molecules.

When the respiratory system can keep pace with the rate of metabolism, there is no net gain or loss of H^+ in the body fluids from metabolically produced CO_2. When the rate of CO_2 removal by the lungs does not match the rate of CO_2 production at the tissue level, however, the resulting accumulation or deficit of CO_2 leads to an excess or shortage, respectively, of free H^+ in the body fluids.

2. *Inorganic acids produced during breakdown of nutrients.* Dietary proteins found abundantly in meat contain a large quantity of sulfur and phosphorus. When these nutrient molecules are broken down, sulfuric acid and phosphoric acid are produced as by-products. Being moderately strong acids, these two inorganic acids largely dissociate, liberating free H^+ into the body fluids. Acids are likewise generated during breakdown of the proteins in grains and dairy products. In contrast, breakdown of fruits and vegetables produces bases that, to some extent, neutralize acids derived from meat, grain, and dairy protein metabolism. Generally, however, more acids than bases are produced during breakdown of ingested food, leading to an excess of these acids.

3. *Organic acids resulting from intermediary metabolism.* Numerous organic acids are produced during normal intermediary metabolism. For example, fatty acids are produced during fat metabolism, and muscles produce lactic acid (lactate) during heavy exercise (see p. 272). These acids partially dissociate to yield free H^+.

Hydrogen ion generation therefore normally goes on continuously, as a result of ongoing metabolic activities. In certain disease states, additional acids may be produced that further contribute to the total body pool of H^+. For example, in diabetes mellitus, large quantities of keto acids may be produced by abnormal fat metabolism (see pp. 689 and 694). Some types of acid-producing medications may also add to the total H^+ load that the body must handle. Thus, input of H^+ is unceasing, highly variable, and essentially unregulated.

Three Lines of Defense Against Changes in [H⁺] The key to H^+ balance is maintaining normal alkalinity of the ECF (pH 7.4) despite this constant onslaught of acid. The generated free H^+ must be largely removed from solution while in the body and ultimately must be eliminated so that the pH of body fluids can remain within the narrow range compatible with life. Mechanisms must also exist to compensate rapidly for the occasional situation in which the ECF becomes too alkaline.

Three lines of defense against changes in [H^+] operate to maintain [H^+] of body fluids at a nearly constant level despite unregulated input: (1) the *chemical buffer systems,* (2) the *respiratory mechanism of pH control,* and (3) the *renal mechanism of pH control.* We now look at each of these methods.

Chemical buffer systems minimize changes in pH by binding with or yielding free H⁺.

A **chemical buffer system** is a mixture in a solution of two chemical compounds that minimize pH changes when either an acid or a base is added to or removed from the solution. A buffer system consists of a pair of substances involved in a reversible reaction—one substance that can yield free H^+ as the [H^+] starts to fall and another that can bind with free H^+ (thus removing it from solution) when [H^+] starts to rise.

An important example of such a buffer system is the carbonic acid–bicarbonate (H_2CO_3:HCO_3^-) buffer pair, which is involved in the following reversible reaction:

$$H_2CO_3 \rightleftharpoons H^+ + HCO_3^-$$

When a strong acid such as HCl is added to an unbuffered solution, all the dissociated H^+ remains free in the solution (I Figure 15-8a). In contrast, when HCl is added to a solution containing the H_2CO_3:HCO_3^- buffer pair, the HCO_3^- immediately binds with the free H^+ to form H_2CO_3 (Figure 15-8b). This weak H_2CO_3 dissociates only slightly compared to the marked reduction in pH that occurred when the buffer system was not present and the additional H^+ remained unbound. In the opposite case, when the pH of the solution starts to rise from the addition of base or loss of acid, the H^+-yielding member of the buffer pair, H_2CO_3, releases H^+ to minimize the rise in pH.

The body has four buffer systems: (1) the H_2CO_3:HCO_3^- buffer system, (2) the protein buffer system, (3) the hemoglobin buffer system, and (4) the phosphate buffer system. Each buffer system serves an important role (Table 15-6).

The H_2CO_3:HCO_3^- buffer pair is the primary ECF buffer for noncarbonic acids.

The H_2CO_3:HCO_3^- buffer pair is the most important buffer system in the ECF for buffering pH changes brought about by causes other than fluctuations in CO_2-generated H_2CO_3.

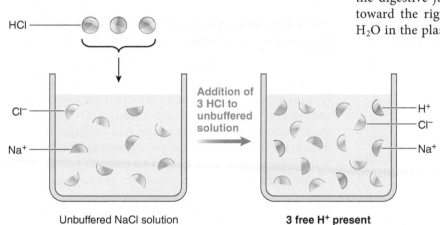

(a) Addition of HCl to an unbuffered solution

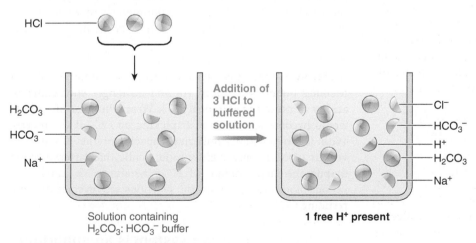

(b) Addition of HCl to a buffered solution

Figure 15-8 Action of chemical buffers. (a) Addition of HCl to an unbuffered solution. All the added hydrogen ions (H^+) remain free and contribute to the acidity of the solution. (b) Addition of HCl to a buffered solution. Bicarbonate ions (HCO_3^-), the basic member of the buffer pair, bind with some of the added H^+ and remove them from solution so that they do not contribute to its acidity.

It is an effective ECF buffer system for two reasons. First, H_2CO_3 and HCO_3^- are abundant in the ECF, so this system is readily available to resist changes in pH. Second, and more importantly, each component of this buffer pair is closely regulated. The kidneys regulate HCO_3^-, and the respiratory system regulates CO_2, which generates H_2CO_3. Thus, in the body the H_2CO_3:HCO_3^- buffer system includes involvement of CO_2 via the following reaction, with which you are already familiar:

$$CO_2 + H_2O \rightleftharpoons H_2CO_3 \rightleftharpoons H^+ + HCO_3^-$$

When new H^+ is added to the plasma from any source other than CO_2 (for example, through lactic acid released into the ECF from exercising muscles), the preceding reaction is driven toward the left side of the equation. As the extra H^+ binds with HCO_3^-, it no longer contributes to the acidity of body fluids, so the rise in $[H^+]$ abates. In the converse situation, when the plasma $[H^+]$ occasionally falls below normal for some reason other than a change in CO_2 (such as the loss during vomiting of plasma-derived HCl in the digestive juices in the stomach), the reaction is driven toward the right side of the equation. Dissolved CO_2 and H_2O in the plasma form H_2CO_3, which generates additional H^+ to make up for the H^+ deficit. In so doing, the H_2CO_3:HCO_3^- buffer system resists the fall in $[H^+]$.

This system cannot buffer changes in pH induced by fluctuations in H_2CO_3. A buffer system cannot buffer itself. Consider, for example, the situation in which the plasma $[H^+]$ is elevated by CO_2 retention from a respiratory problem. The rise in CO_2 drives the reaction to the right according to the law of mass action, elevating $[H^+]$. The increase in $[H^+]$ occurs as a result of the reaction being driven to the *right* by an increase in CO_2, so the elevated $[H^+]$ cannot drive the reaction to the *left* to buffer the increase in $[H^+]$. Only if the increase in $[H^+]$ is brought about by some mechanism other than CO_2 accumulation can this buffer system be shifted to the CO_2 side of the equation and effectively reduce $[H^+]$. Likewise, in the opposite situation, the H_2CO_3:HCO_3^- buffer system cannot compensate for a reduction in $[H^+]$ from a deficit of CO_2 by generating more H^+-yielding H_2CO_3 when the problem in the first place is a shortage of $H_2CO_3^-$-forming CO_2. Other mechanisms, to be described shortly, are available for resisting fluctuations in pH caused by changes in CO_2 levels.

Henderson–Hasselbalch Equation

The relationship between $[H^+]$ and the members of a buffer pair can be expressed according to the **Henderson–**

TABLE 15-6 Chemical Buffers and Their Primary Roles

Buffer System	Major Functions
H_2CO_3:HCO_3^- buffer system	Primary ECF buffer against noncarbonic acid changes
Protein buffer system	Primary ICF buffer; also buffers ECF
Hemoglobin buffer system	Primary buffer against carbonic acid changes
Phosphate buffer system	Important urinary buffer; also buffers ICF

Hasselbalch equation, which, for the H_2CO_3:HCO_3^- buffer system is as follows:

$$pH = pK + log[HCO_3^-]/[H_2CO_3]$$

Although you do not need to know the mathematical manipulations involved, it is helpful to understand how this formula is derived. Recall that the dissociation constant K for H_2CO_3 acid is

$$[H^+][HCO_3^-]/[H_2CO_3] = K$$

and that the relationship between pH and $[H^+]$ is

$$pH = log\ 1/[H^+]$$

Then, by solving the dissociation constant formula for $[H^+]$ (that is, $[H^+] = K \times [H_2CO_3]/[HCO_3^-]$) and replacing this value for $[H^+]$ in the pH formula, one comes up with the Henderson–Hasselbalch equation.

Practically speaking, $[H_2CO_3]$ directly reflects the concentration of dissolved CO_2, from now on referred to as $[CO_2]$, because most of the CO_2 in the plasma is converted into H_2CO_3. (The dissolved CO_2 concentration is equivalent to P_{CO_2}, as described in Chapter 13.) Therefore, the equation becomes

$$pH = pK + log[HCO_3^-]/[CO_2]$$

The pK is the logarithm of l/K, and like K, pK always remains a constant for any given acid. For H_2CO_3, pK is 6.1. Because pK is always a constant, changes in pH are associated with changes in the ratio between $[HCO_3^-]$ and $[CO_2]$.

- Normally, the ratio between $[HCO_3^-]$ and $[CO_2]$ in the ECF is 20 to 1—that is, there is 20 times more HCO_3^- than CO_2. We plug this ratio into our formula:

$$pH = pK + log[HCO_3^-]/[CO_2]$$
$$= 6.1 + log\ 20/1$$

The log of 20 is 1.3. Therefore, pH = 6.1 + 1.3 = 7.4, which is the normal pH of plasma.

- When the ratio of $[HCO_3^-]$ to $[CO_2]$ increases above 20/1, pH increases. Accordingly, either a rise in $[HCO_3^-]$ or a fall in $[CO_2]$, both of which increase the $[HCO_3^-]/[CO_2]$ ratio if the other component remains constant, shifts the acid–base balance toward the alkaline side.

- In contrast, when the $[HCO_3^-]/[CO_2]$ ratio decreases below 20/1, pH decreases toward the acid side. This can occur either if the $[HCO_3^-]$ decreases or if the $[CO_2]$ increases while the other component remains constant.

Because $[HCO_3^-]$ is regulated by the kidneys and $[CO_2]$ by the lungs, the pH of the plasma can be shifted up and down by kidney and lung influences. The kidneys and lungs regulate pH (and thus free $[H^+]$) largely by controlling plasma $[HCO_3^-]$ and $[CO_2]$, respectively, to restore their ratio to normal. Accordingly,

$$pH \propto \frac{[HCO_3^-]\ \text{controlled by kidney function}}{[CO_2]\ \text{controlled by respiratory function}}$$

Because of this relationship, not only do both the kidneys and lungs normally participate in pH control, but renal or respiratory dysfunction can also induce acid–base imbalances by altering the $[HCO_3^-]/[CO_2]$ ratio. We will build on this principle when we examine respiratory and renal control of pH and acid–base abnormalities later in the chapter. For now, we are going to continue our discussion of the roles of the different buffer systems.

The protein buffer system is primarily important intracellularly.

The most plentiful buffers of the body fluids are the proteins, including the intracellular proteins and the plasma proteins. Proteins are excellent buffers because they contain both acidic and basic groups that can give up or take up H^+. Quantitatively, the protein system is most important in buffering changes in $[H^+]$ in the ICF because of the sheer abundance of the intracellular proteins. The more limited number of plasma proteins reinforces the H_2CO_3:HCO_3^- system in extracellular buffering.

The hemoglobin buffer system buffers H^+ generated from CO_2.

Hemoglobin (Hb) buffers the H^+ generated from metabolically produced CO_2 in transit between the tissues and the lungs. At the systemic capillary level, CO_2 continuously diffuses into the blood from the tissue cells where it is being produced. The greatest percentage of this CO_2, along with H_2O, forms H^+ and HCO_3^- under the influence of carbonic anhydrase within the red blood cells. Most H^+ generated from CO_2 at the tissue level becomes bound to reduced Hb and no longer contributes to acidity of body fluids (see p. 477). Were it not for Hb, blood would become too acidic after picking up CO_2 at the tissues. With the tremendous buffering capacity of the Hb system, venous blood is only slightly more acidic than arterial blood despite the large volume of H^+-generating CO_2 carried in venous blood. At the lungs, the reactions are reversed and the resulting CO_2 is exhaled.

The phosphate buffer system is an important urinary buffer.

The phosphate buffer system consists of an acid phosphate salt (NaH_2PO_4) that can donate a free H^+ when the $[H^+]$ falls and a basic phosphate salt (Na_2HPO_4) that can accept a free H^+ when the [H$^+$] rises. Basically, this buffer pair can alternately switch a H$^+$ for a Na$^+$ as demanded by the [H$^+$]:

$$Na_2HPO_4 + H^+ \rightleftharpoons NaH_2PO_4 + Na^+$$

Even though the phosphate pair is a good buffer, its concentration in the ECF is rather low, so it is not very important as an ECF buffer. Because phosphates are most abundant within the cells, this system contributes significantly to intracellular buffering, being rivaled only by the more plentiful intracellular proteins.

Even more important, the phosphate system serves as an excellent urinary buffer. Humans normally consume more phosphate than needed. The excess phosphate filtered through the kidneys is not reabsorbed but remains in the tubular fluid to be excreted (because the renal threshold for phosphate is exceeded; see p. 512). This excreted phosphate buffers urine as it forms by removing from solution the H$^+$ secreted into the tubular fluid. None of the other body-fluid buffer systems are present in the tubular fluid to play a role in buffering urine during its formation. Most or all of the filtered HCO$_3$$^-$ and CO$_2$ (alias H$_2$CO$_3$) are reabsorbed, whereas Hb and plasma proteins are not even filtered.

Chemical buffer systems act as the first line of defense against changes in [H$^+$].

All chemical buffer systems act immediately, within fractions of a second, to minimize changes in pH. When [H$^+$] is altered, the reversible chemical reactions of the involved buffer systems shift at once (by the law of mass action) to compensate for the change in [H$^+$]. Accordingly, the buffer systems are the *first line of defense* against changes in [H$^+$] because they are the first mechanism to respond.

Through the mechanism of buffering, most hydrogen ions seem to disappear from the body fluids from the time they are generated until the time they are eliminated. Recognize, however, that none of the chemical buffer systems actually eliminate H$^+$ from the body. The buffer systems merely remove hydrogen ions from solution by incorporating them within one member of the buffer pair, thus preventing them from contributing to body-fluid acidity. Because each buffer system has a limited capacity to soak up H$^+$, the H$^+$ that is unceasingly produced must ultimately be removed from the body by another means. If H$^+$ were not eventually eliminated, soon all the body-fluid buffers would already be bound with H$^+$ and there would be no further buffering ability.

The respiratory and renal mechanisms of pH control actually eliminate acid from the body instead of merely suppressing it, but they respond more slowly than chemical buffer systems. We now turn to these other defenses against changes in acid–base balance.

The respiratory system regulates [H$^+$] by controlling the rate of CO$_2$ removal.

The respiratory system plays an important role in acid–base balance through its ability to alter pulmonary ventilation and consequently to alter excretion of H$^+$-generating CO$_2$. The level of respiratory activity is governed in part by arterial [H$^+$], as follows (Table 15-7):

- An increase in arterial [H$^+$] as the result of a *nonrespiratory* (or *metabolic*) cause brings about reflex stimulation of the respiratory center in the brain stem (see p. 479) via the peripheral chemoreceptors to increase pulmonary ventilation (the rate at which gas is exchanged between the lungs and the atmosphere; see p. 484). As the rate and depth of breathing increase, more CO$_2$ than usual is blown off. Because hydration of CO$_2$ generates H$^+$, removal of CO$_2$ in essence removes acid from this source from the body, offsetting extra H$^+$ present from a nonrespiratory source.

- Conversely, when arterial [H$^+$] falls because of a nonrespiratory cause, pulmonary ventilation is reflexly reduced. As a result of slower, shallower breathing, metabolically produced CO$_2$ diffuses from the cells into the blood faster than it is removed from the blood by the lungs, so higher-than-usual amounts of acid-forming CO$_2$ accumulate in the blood, thus restoring [H$^+$] toward normal.

The lungs are extremely important in maintaining [H$^+$]. Every day they remove from body fluids what amounts to 100 times more H$^+$ derived from CO$_2$ than the kidneys remove from sources other than CO$_2$–H$^+$. Furthermore, the respiratory system, through its ability to regulate arterial [CO$_2$], can adjust the amount of H$^+$ added to body fluids from this source as needed to restore pH toward normal when fluctuations occur in [H$^+$] from sources other than CO$_2$–H$^+$.

The respiratory system serves as the second line of defense against changes in [H$^+$].

Respiratory regulation acts at a moderate speed, coming into play only when chemical buffer systems alone cannot minimize [H$^+$] changes. When deviations in [H$^+$] occur, the buffer sys-

TABLE 15-7 Respiratory Adjustments to Nonrespiratory Acidosis and Alkalosis

Respiratory Compensations	Normal pH	Nonrespiratory (Metabolic) Acidosis	Nonrespiratory (Metabolic) Alkalosis
Ventilation	Normal	↑	↓
Rate of CO$_2$ removal	Normal	↑	↓
Rate of H$^+$ generation from CO$_2$	Normal	↓	↑

tems respond immediately, whereas adjustments in ventilation require a few minutes to be initiated. If a deviation in $[H^+]$ is not swiftly and completely corrected by the buffer systems, the respiratory system comes into action a few minutes later, thus serving as the *second line of defense* against changes in $[H^+]$.

The respiratory system alone can return the pH only 50% to 75% of the way toward normal. Two reasons contribute to the respiratory system's inability to fully compensate for a nonrespiratory-induced acid–base imbalance. First, during respiratory compensation for a deviation in pH, the peripheral chemoreceptors, which increase ventilation in response to an elevated arterial $[H^+]$, and the central chemoreceptors, which increase ventilation in response to a rise in $[CO_2]$ (by monitoring CO_2-generated H^+ in the brain ECF; see p. 483), work at odds. Consider what happens in response to acidosis arising from a non-respiratory cause. When the peripheral chemoreceptors detect an increase in arterial $[H^+]$, they reflexly *stimulate* the respiratory center to step up ventilation, causing more acid-forming CO_2 to be blown off. In response to the resultant fall in CO_2, however, the central chemoreceptors start to *inhibit* the respiratory center. By opposing the action of the peripheral chemoreceptors, the central chemoreceptors stop the compensatory increase in ventilation short of restoring pH all the way to normal.

Second, the driving force for the compensatory increase in ventilation is diminished as the pH moves toward normal. Ventilation is increased by the peripheral chemoreceptors in response to a rise in arterial $[H^+]$, but as the $[H^+]$ is gradually reduced by stepped-up removal of H^+-forming CO_2, the enhanced ventilatory response is also gradually reduced.

When changes in $[H^+]$ stem from $[CO_2]$ fluctuations that arise from respiratory abnormalities, the respiratory mechanism cannot contribute to pH control. For example, if acidosis exists because of CO_2 accumulation caused by lung disease, the impaired lungs cannot possibly compensate for acidosis by increasing the rate of CO_2 removal. The buffer systems (other than the H_2CO_3:HCO_3^- pair) plus renal regulation are the only mechanisms available for defending against respiratory-induced acid–base abnormalities.

We next see how the kidneys help maintain acid–base balance.

The kidneys adjust their rate of H^+ excretion by varying the extent of H^+ secretion.

The kidneys control the pH of body fluids by adjusting three interrelated factors: (1) H^+ excretion, (2) HCO_3^- excretion, and (3) ammonia (NH_3) secretion. We examine each of these mechanisms in further detail, starting with H^+ excretion.

Acids are continuously added to body fluids as a result of metabolic activities, yet the generated H^+ must not be allowed to accumulate. Although the body's buffer systems can resist changes in pH by removing H^+ from solution, the persistent production of acidic metabolic products would eventually overwhelm the limits of this buffering capacity. Therefore, the constantly generated H^+ must ultimately be eliminated from the body. The lungs can remove only CO_2-generated H^+ by eliminating CO_2. The task of eliminating H^+ derived from sulfuric,

phosphoric, lactic, and other acids rests with the kidneys. Furthermore, the kidneys can eliminate extra H^+ derived from CO_2.

All of the filtered H^+ is excreted because the kidney tubules are not able to reabsorb H^+, but most excreted H^+ enters the urine via secretion. Recall that the filtration rate of H^+ equals plasma $[H^+]$ times GFR. Because plasma $[H^+]$ is extremely low (less than in pure H_2O except during extreme acidosis, when pH falls below 7.0), the filtration rate of H^+ is likewise extremely low. This minute amount of filtered H^+ is excreted in the urine. However, most excreted H^+ gains entry into the tubular fluid by being actively secreted by the tubular cells from the peritubular capillary plasma into the tubular lumen. The proximal, distal, and collecting tubules all secrete H^+. Because the kidneys normally excrete H^+, urine is usually acidic, having an average pH of 6.0.

The H^+ secretory process begins in the tubular cells with CO_2 from three sources: CO_2 diffused into the tubular cells from (1) plasma or (2) tubular fluid or (3) CO_2 metabolically produced within the tubular cells. Catalyzed by carbonic anhydrase within the tubular cells, CO_2 and H_2O form H^+ and HCO_3^-. To secrete H^+, an energy-dependent carrier in the luminal membrane then transports H^+ out of the cell into the tubular lumen. The luminal-membrane carrier differs in different parts of the nephron.

Mechanism of Renal H^+ Secretion in the Proximal Tubule In the proximal tubule, H^+ is secreted by both primary active transport via H^+ ATPase pumps (see p. 73) and by secondary active transport via Na^+–H^+ antiporters (see p. 74). The antiporters transport Na^+ derived from glomerular filtrate in the opposite direction of H^+ secretion, so H^+ secretion and Na^+ reabsorption are partially linked in the proximal tubule.

Mechanism of Renal H^+ Secretion in the Distal and Collecting Tubules Recall that two types of cells are located in the distal and collecting tubules, *principal cells* and *intercalated cells* (see p. 508). Principal cells are the ones with which you are already familiar. These are the cells that play an important role in Na^+ (and subsequently Cl^-—that is, salt) balance and in K^+ balance under the influence of aldosterone. They are also the cells involved in maintaining H_2O balance under the influence of vasopressin.

Intercalated cells, which are interspersed among the principal cells, are involved in fine regulation of acid–base balance. There are two types of intercalated cells, Type A (most abundant) and Type B:

■ **Type A intercalated cells** are H^+-secreting, HCO_3^--reabsorbing, K^+-reabsorbing cells. They actively secrete H^+ into the tubular lumen via two types of primary active transport mechanisms: H^+ ATPase pumps and H^+–K^+ ATPase pumps. The latter secrete H^+ in exchange for uptake (reabsorption) of K^+. Both of these types of carriers are located at the luminal membrane in Type A cells (❚ Figure 15-9). The HCO_3^- generated in the process of forming H^+ from CO_2 under the influence of carbonic anhydrase enters the blood (is reabsorbed) in exchange for Cl^- at the basolateral membrane via Cl^-–HCO_3^- antiporters.

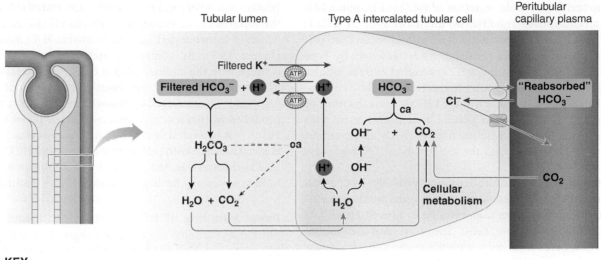

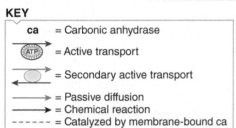

ca = Carbonic anhydrase

= Active transport

= Secondary active transport

= Passive diffusion

= Chemical reaction

----- = Catalyzed by membrane-bound ca

Figure 15-9 Hydrogen ion secretion coupled with bicarbonate reabsorption in a Type A intercalated cell. The H^+-secreting pumps are located at the luminal membrane, and the HCO_3^--reabsorbing antiporters are located at the basolateral membrane. Because the disappearance of a filtered HCO_3^- from the tubular fluid is coupled with the appearance of another HCO_3^- in the plasma, HCO_3^- is considered "reabsorbed."

■ **Type B intercalated cells** are HCO_3^--secreting, H^+-reabsorbing, K^+-secreting cells, just the opposite actions of the Type A intercalated cells. In contrast to Type A cells, the active H^+ ATPase pumps and H^+-K^+ ATPase pumps are located at the basolateral membrane and the Cl^-–HCO_3^- antiporters are located at the luminal membrane. In this case, when H^+ and HCO_3^- are generated from the hydration of CO_2 under the influence of carbonic anhydrase, HCO_3^- moves into the tubular lumen (is secreted) in exchange for Cl^-, and H^+ is reabsorbed into the plasma in exchange for K^+ across the basolateral membrane (■ Figure 15-10). Even though the Type B intercalated cells actively secrete K^+, the principal cells under the control of aldosterone actively secrete quantitatively much more K^+.

Type A intercalated cells are more active than Type B intercalated cells under normal circumstances, and their activity increases even more during acidosis. Type B intercalated cells become more active during alkalosis.

The kidneys conserve or excrete HCO_3^- depending on the plasma [H^+].

Before being eliminated by the kidneys, H^+ generated from noncarbonic acids is buffered to a large extent by plasma

HCO_3^-. Appropriately, therefore, renal handling of acid–base balance also involves adjustment of HCO_3^- excretion, depending on the H^+ load in the plasma.

The kidneys regulate plasma [HCO_3^-] by three interrelated mechanisms: (1) variable reabsorption of filtered HCO_3^- back into the plasma in conjunction with H^+ secretion, (2) variable addition of new HCO_3^- to the plasma in conjunction with H^+

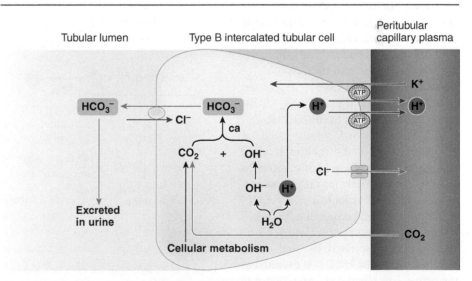

Figure 15-10 Bicarbonate secretion coupled with hydrogen ion reabsorption in a Type B intercalated cell. The HCO_3^--secreting antiporters are located at the luminal membrane, and the H^+-reabsorbing pumps are located at the basolateral membrane.

FIGURE FOCUS: Comparing this figure with Figure 15-9, explain why H^+ is secreted by Type A intercalated cells, whereas H^+ is reabsorbed by Type B intercalated cells when identical chemical reactions that produce H^+ and HCO_3^- take place in both cells.

secretion, and (3) variable secretion of HCO_3^- in conjunction with H^+ reabsorption. The first two mechanisms of renal handling of HCO_3^- are inextricably linked with H^+ secretion, primarily by proximal tubular cells and to a lesser extent by Type A intercalated cells. Every time a H^+ is secreted into the tubular fluid, a HCO_3^- is simultaneously transferred into the peritubular capillary plasma. Whether a filtered HCO_3^- is reabsorbed or a new HCO_3^- is added to the plasma in accompaniment with H^+ secretion depends on whether filtered HCO_3^- is present in the tubular fluid to react with the secreted H^+.

Coupling of HCO_3^- Reabsorption with H^+ Secretion

Bicarbonate is freely filtered, but because the luminal membranes of tubular cells are impermeable to filtered HCO_3^-, it cannot diffuse into these cells. Therefore, reabsorption of HCO_3^- must occur indirectly. We will use the Type A intercalated cell as an example (see ❘Figure 15-9). H^+ secreted into the tubular fluid combines with filtered HCO_3^- to form H_2CO_3. Under the influence of a form of carbonic anhydrase that is located on the surface of the luminal membrane, H_2CO_3 decomposes into CO_2 and H_2O within the filtrate. Unlike HCO_3^-, CO_2 and H_2O can easily penetrate tubular cell membranes. Within the cells, CO_2 and H_2O, under the influence of intracellular carbonic anhydrase, form H^+ and HCO_3^-. Because HCO_3^- can permeate these tubular cells' basolateral membrane by means of the Cl^-–HCO_3^- antiporter, it diffuses out of the cells and into the peritubular capillary plasma. Meanwhile, the generated H^+ is actively secreted. Because the disappearance of a HCO_3^- from the tubular fluid is coupled with the appearance of another HCO_3^- in the plasma, a HCO_3^- has, in effect, been "reabsorbed." Even though the HCO_3^- entering the plasma is not the same HCO_3^- that was filtered, the net result is the same as if HCO_3^- were directly reabsorbed.

The same steps are involved in HCO_3^- reabsorption in the proximal tubular cells, except in addition to having basolateral Cl^-–HCO_3^- antiporters, these cells also have more abundant basolateral Na^+–HCO_3^- symporters that simultaneously reabsorb Na^+ and HCO_3^-.

Normally, slightly more H^+ is secreted into the tubular fluid than HCO_3^- is filtered. Accordingly, all the filtered HCO_3^- is usually reabsorbed because secreted H^+ is available in the tubular fluid to combine with it to form highly reabsorbable CO_2 and H_2O. By far the largest part of the secreted H^+ combines with HCO_3^- and is not excreted because it is "used up" in HCO_3^- reabsorption. However, the slight excess of secreted H^+ that is not matched by filtered HCO_3^- is excreted in the urine. This normal H^+ excretion rate keeps pace with the normal rate of noncarbonic acid H^+ production.

To emphasize what you just learned, secreted H^+ that is coupled with HCO_3^- reabsorption is not excreted. Instead of being excreted, the secreted H^+ combines with filtered HCO_3^- and ultimately becomes incorporated into reabsorbable H_2O molecules (see ❘Figure 15-9). By contrast, secreted H^+ that is excreted is coupled with the addition of new HCO_3^- to the plasma. When all the filtered HCO_3^- has been reabsorbed and additional secreted H^+ is generated by dissociation of H_2CO_3, the HCO_3^- produced by this reaction diffuses into the plasma as a "new" HCO_3^-. It is termed "new" because its appearance in plasma is not associated with reabsorption of filtered HCO_3^- (❘Figure 15-11). Meanwhile, the secreted H^+ combines with urinary buffers, especially basic phosphate (HPO_4^{2-}) and is excreted.

Renal Handling of H^+ During Acidosis and Alkalosis

The kidneys are able to exert a fine degree of control over body pH. Renal handling of H^+ and HCO_3^- depends primarily on a direct effect of the plasma's acid–base status on the kidney's tubular cells. Under normal circumstances, the proximal tubular cells and Type A intercalated cells are predominantly active, promoting net H^+ secretion and HCO_3^- reabsorption. This pattern of activity is adjusted when pH deviates from the set point.

Let us look first at the influence of acidosis and alkalosis on H^+ secretion (❘Figure 15-12):

■ When the $[H^+]$ of the plasma passing through the peritubular capillaries is elevated above normal, the proximal tubular cells and Type A intercalated cells respond by secreting greater-than-usual amounts of H^+ from the plasma into the tubular fluid to be excreted in the urine.

■ Conversely, when plasma $[H^+]$ is lower than normal, the kidneys conserve H^+ by reducing its secretion by proximal cells and Type A intercalated cells. Also, Type B intercalated

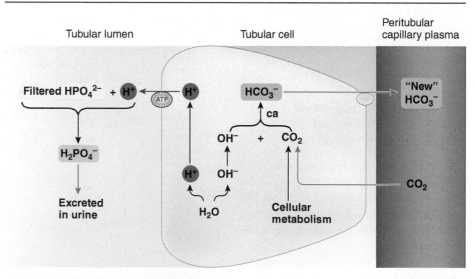

❘**Figure 15-11 Hydrogen ion secretion and excretion coupled with the addition of new HCO_3^- to the plasma.** Secreted H^+ does not combine with filtered HPO_4^{2-} and is not subsequently excreted until all the filtered HCO_3^- has been "reabsorbed," as depicted in Figure 15-9. Once all the filtered HCO_3^- has combined with secreted H^+, further secreted H^+ is excreted in the urine, primarily in association with urinary buffers such as basic phosphate. Excretion of H^+ is coupled with the appearance of new HCO_3^- in the plasma. The "new" HCO_3^- represents a net gain rather than merely a replacement for filtered HCO_3^-.

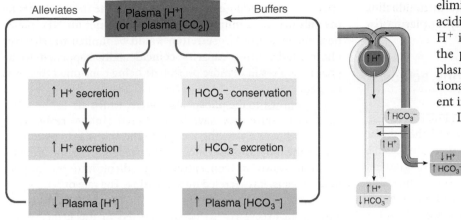

Figure 15-12 Control of the rate of tubular H⁺ secretion and HCO₃⁻ reabsorption.

cells become more active to compensate for alkalosis by increasing H⁺ reabsorption. Together these actions decrease H⁺ excretion in the urine.

Because chemical reactions for H⁺ secretion begin with CO_2, the rate at which they proceed is also influenced by $[CO_2]$.

■ When plasma $[CO_2]$ increases, the rate of H⁺ secretion speeds up (Figure 15-12).

■ Conversely, the rate of H⁺ secretion slows when plasma $[CO_2]$ falls below normal.

These responses are especially important in renal compensations for acid–base abnormalities involving a change in H_2CO_3 caused by respiratory dysfunction. Thus, the kidneys can adjust H⁺ excretion to compensate for changes in both carbonic and noncarbonic acids.

Renal Handling of HCO₃⁻ During Acidosis and Alkalosis When plasma [H⁺] is elevated during acidosis, more H⁺ is secreted than normal. At the same time, less HCO_3^- is filtered than normal because more of the plasma HCO_3^- is used up in buffering the excess H⁺ in the ECF. This greater-than-usual inequity between filtered HCO_3^- and secreted H⁺ has two consequences. First, more of the secreted H⁺ is excreted in the urine because more hydrogen ions are entering the tubular fluid at a time when fewer are needed to reabsorb the reduced quantities of filtered HCO_3^-. In this way, extra H⁺ is

eliminated from the body, making the urine more acidic than normal. Second, because excretion of H⁺ is linked with the addition of new HCO_3^- to the plasma, more HCO_3^- than usual enters the plasma passing through the kidneys. This additional HCO_3^- is available to buffer excess H⁺ present in the body.

In the opposite situation of alkalosis, the rate of H⁺ secretion diminishes, whereas the rate of HCO_3^- filtration increases compared to normal. When plasma [H⁺] is below normal, a smaller proportion of the HCO_3^- pool than usual is tied up buffering H⁺, so plasma $[HCO_3^-]$ is elevated above normal. As a result, the rate of HCO_3^- filtration correspondingly increases. Not all the filtered HCO_3^- is reabsorbed because bicarbonate ions are in excess of secreted hydrogen ions in the tubular fluid and HCO_3^- cannot be reabsorbed without first reacting with H⁺. Excess HCO_3^- is left in the tubular fluid to be excreted in the urine, thus reducing plasma $[HCO_3^-]$ while making the urine alkaline. Furthermore, Type B intercalated cells come into play during alkalosis, further decreasing the excess HCO_3^- load in the body by secreting HCO_3^- into the urine.

In short, when plasma [H⁺] increases above normal during acidosis, renal compensation includes the following (Table 15-8):

1. Increased secretion and subsequent increased excretion of H⁺ in the urine, thereby eliminating the excess H⁺ and decreasing plasma [H⁺]

2. Reabsorption of all filtered HCO_3^-, plus addition of new HCO_3^- to the plasma, resulting in increased plasma $[HCO_3^-]$

When plasma [H⁺] falls below normal during alkalosis, renal responses include the following:

1. Decreased secretion and subsequent reduced excretion of H⁺ in the urine, conserving H⁺ and increasing plasma [H⁺]

2. Incomplete reabsorption of filtered HCO_3^- coupled with secretion of HCO_3^-, leading to increased excretion of HCO_3^- and reduced plasma $[HCO_3^-]$

Note that to compensate for acidosis, the kidneys acidify urine (by getting rid of extra H⁺) and alkalinize plasma (by conserving HCO_3^-) to bring plasma pH to normal. In the

TABLE 15-8 Renal Responses to Acidosis and Alkalosis

Acid–Base Abnormality	H⁺ Secretion	H⁺ Excretion	HCO₃⁻ Reabsorption and Addition of New HCO₃⁻ to Plasma	HCO₃⁻ Excretion	pH of Urine	Compensatory Change in Plasma pH
Acidosis	↑	↑	↑	Normal (zero; all filtered is reabsorbed)	Acidic	Alkalinization toward normal
Alkalosis	↓	↓	↓	↑	Alkaline	Acidification toward normal

opposite case—alkalosis—the kidneys make urine alkaline (by eliminating excess HCO_3^-) while acidifying plasma (by conserving H^+).

The kidneys secrete ammonia during acidosis to buffer secreted H^+.

The energy-dependent H^+ carriers in the tubular cells can secrete H^+ against a concentration gradient until the tubular fluid (urine) becomes 800 times more acidic than the plasma. At this point, further H^+ secretion stops because the gradient becomes too great for the secretory process to continue. The kidneys cannot acidify urine beyond a gradient-limited urinary pH of 4.5. If left unbuffered as free H^+, only about 1% of the excess H^+ typically excreted daily would produce a urinary pH of this magnitude at normal urine flow rates, and elimination of the other 99% of the usually secreted H^+ load would be prevented—a situation that would be intolerable. For H^+ secretion to proceed, most secreted H^+ must be buffered in the tubular fluid so that it does not exist as free H^+ and, accordingly, does not contribute to tubular acidity.

Bicarbonate cannot buffer urinary H^+ as it does H^+ in the ECF because HCO_3^- is not excreted in the urine simultaneously with H^+. (Whichever of these substances is in excess in the plasma is excreted in the urine.) There are, however, two important urinary buffers: (1) filtered phosphate buffers and (2) secreted ammonia.

Filtered Phosphate as a Urinary Buffer Normally, secreted H^+ is first buffered by the phosphate buffer system, which is in the tubular fluid because excess ingested phosphate has been filtered but not reabsorbed. The basic member of the phosphate buffer pair binds with secreted H^+. When H^+ secretion is high, the buffering capacity of urinary phosphates is exceeded. The kidneys can only control the quantity of phosphate reabsorbed (under the influence of parathyroid hormone; see pp. 512 and 707). They can do nothing about the quantity of phosphate filtered and available for reabsorption; that depends on how much phosphate has been consumed. As soon as all basic phosphate ions that are coincidentally excreted (because of dietary excess) have soaked up H^+, the acidity of the tubular fluid quickly rises as more H^+ ions are secreted. Without additional buffering capacity from another source, H^+ secretion would soon halt abruptly as the free $[H^+]$ in the tubular fluid quickly rose to the critical limiting level.

Secreted NH_3 as a Urinary Buffer When acidosis exists, the tubular cells secrete **ammonia (NH_3)** into the tubular fluid once the normal urinary phosphate buffers are saturated. This NH_3 enables the kidneys to continue secreting additional H^+ ions because NH_3 combines with free H^+ in the tubular fluid to form **ammonium ion (NH_4^+)** as follows:

$$NH_3 + H^+ \rightleftharpoons NH_4^+$$

The tubular membranes are not very permeable to NH_4^+, so the ammonium ions remain in the tubular fluid and are lost in the urine, each one taking a H^+ with it. Thus, NH_3 secreted during acidosis buffers excess H^+ in the tubular fluid so that

large amounts of H^+ can be secreted into the urine before the pH falls to the limiting value of 4.5. Were it not for NH_3 secretion, the extent of H^+ secretion would be limited to whatever phosphate-buffering capacity coincidentally happened to be present as a result of more phosphate being consumed than was needed.

In contrast to the phosphate buffers, which are in the tubular fluid because they have been filtered but not reabsorbed, NH_3 is deliberately synthesized from the amino acid glutamine within the tubular cells. Once synthesized, NH_3 readily diffuses passively down its concentration gradient into the tubular fluid—that is, it is secreted into the urine. The rate of NH_3 secretion is controlled by a direct effect on the tubular cells of the amount of excess H^+ to be transported in the urine. When someone has been acidotic for more than two or three days, the rate of NH_3 production increases substantially. This extra NH_3 provides additional buffering capacity to allow H^+ secretion to continue after the normal phosphate-buffering capacity is overwhelmed during renal compensation for acidosis.

The kidneys are a powerful third line of defense against changes in $[H^+]$.

The kidneys require hours to days to compensate for changes in body-fluid pH, compared to the immediate responses of the buffer systems and the few minutes of delay before the respiratory system responds. Therefore, they are the *third line of defense* against $[H^+]$ changes in body fluids. Although not responding as quickly as the other means of pH control, the kidneys are the most potent acid–base regulatory mechanism; they not only can variably remove H^+ from any source, but they also can variably conserve or eliminate HCO_3^- depending on the acid–base status of the body. By simultaneously removing acid (H^+) from and adding base (HCO_3^-) to body fluids, the kidneys are able to restore the pH toward normal more effectively than the lungs, which can adjust only the amount of H^+-forming CO_2 in the body.

Also contributing to the kidneys' acid–base regulatory potency is their ability to return pH almost exactly to normal. By comparison to the respiratory system's inability to fully compensate for a pH abnormality, the kidneys can continue to respond to a change in pH until compensation is essentially complete.

Acid–base imbalances can arise from either respiratory or metabolic disturbances.

Clinical Note Deviations from normal acid–base status are divided into four categories, depending on the source and direction of the abnormal change in $[H^+]$. These categories are *respiratory acidosis, respiratory alkalosis, metabolic acidosis,* and *metabolic alkalosis.*

Because of the relationship between $[H^+]$ and concentrations of the members of a buffer pair, changes in $[H^+]$ are reflected by changes in the ratio of $[HCO_3^-]$ to $[CO_2]$. Recall that the normal ratio is 20/1. Using the Henderson–Hasselbalch equation and with pK being 6.1 and the log of 20 being 1.3, normal pH = 6.1 + 1.3 = 7.4. Determinations of $[HCO_3^-]$ and $[CO_2]$ provide more meaningful information about the underlying factors responsible for a particular acid–base status than do direct measurements of $[H^+]$ alone. The following rules of thumb apply when examining acid–base imbalances before any compensations take place:

1. A change in pH that has a *respiratory* cause is associated with an *abnormal [CO2]*, giving rise to a change in $H_2CO_3^-$-generated H^+. In contrast, a pH deviation of *metabolic* origin is associated with an *abnormal [HCO3⁻]* resulting from an inequality between the amount of HCO_3^- available and the amount of H^+ generated from noncarbonic acids that the HCO_3^- must buffer.

2. Anytime the $[HCO_3^-]/[CO_2]$ ratio falls *below 20/1*, an *acidosis* exists. The log of any number lower than 20 is less than 1.3 and, when added to the pK of 6.1, yields an acidotic pH below 7.4. Anytime the ratio *exceeds 20/1*, an *alkalosis* exists. The log of any number greater than 20 is more than 1.3 and, when added to the pK of 6.1, yields an alkalotic pH above 7.4.

Let us put these two points together:

- Respiratory acidosis has a ratio of less than 20/1 arising from an increase in $[CO_2]$.
- Respiratory alkalosis has a ratio greater than 20/1 because of a decrease in $[CO_2]$.
- Metabolic acidosis has a ratio of less than 20/1 associated with a fall in $[HCO_3^-]$.
- Metabolic alkalosis has a ratio greater than 20/1 arising from an elevation in $[HCO_3^-]$.

We will examine each of these categories separately in more detail. For comparison, Figure 15-13 presents each of these categories by two means—the Henderson–Hasselbalch equation and the "balance beam" or "seesaw" concept—to help you better visualize the contributions of the lungs and kidneys to the causes of and compensations for various acid–base disorders. The normal situation is represented in Figure 15-13a.

Respiratory acidosis arises from an increase in [CO2].

Respiratory acidosis is the result of abnormal CO_2 retention arising from *hypoventilation* (see p. 478). As less-than-normal amounts of CO_2 are lost through the lungs, the resulting increase in CO_2 generates more H^+ from this source.

Causes of Respiratory Acidosis Possible causes include lung disease, depression of the respiratory center by drugs or disease, nerve or muscle disorders that reduce respiratory muscle ability, or (transiently) even the simple act of holding one's breath.

In uncompensated respiratory acidosis (Figure 15-13b, left), $[CO_2]$ is elevated (in our example, it is doubled), whereas $[HCO_3^-]$ is normal, so the ratio is 20/2 (10/1) and pH is reduced. Let us clarify a potentially confusing point. You might wonder why when $[CO_2]$ is elevated and drives the reaction $CO_2 + H_2O \rightleftharpoons H^+ + HCO_3^-$ to the right, we say that $[H^+]$ becomes elevated but $[HCO_3^-]$ remains normal, although the same quantities of H^+ and HCO_3^- are produced

by this reaction. The answer lies in the fact that normally $[HCO_3^-]$ is 600,000 times $[H^+]$. For every one hydrogen ion and 600,000 bicarbonate ions present in the ECF, the generation of one additional H^+ and one HCO_3^- doubles $[H^+]$ (a 100% increase) but only increases $[HCO_3^-]$ 0.00017% (from 600,000 to 600,001 ions). Therefore, an elevation in $[CO_2]$ brings about a pronounced increase in $[H^+]$, but $[HCO_3^-]$ remains essentially normal.

Compensations for Respiratory Acidosis Compensatory measures act to restore pH to normal.

1. The chemical buffers immediately take up additional H^+.

2. The respiratory mechanism usually cannot respond with compensatory increased ventilation because impaired respiration is the problem in the first place.

3. Thus, the kidneys are most important in compensating for respiratory acidosis. They conserve all the filtered HCO_3^- and add new HCO_3^- to the plasma while simultaneously secreting and, accordingly, excreting more H^+.

As a result, HCO_3^- stores in the body become elevated. In our example (Figure 15-13b, right), the plasma $[HCO_3^-]$ is doubled, so the $[HCO_3^-]/[CO_2]$ ratio is 40/2 rather than 20/2 as it was in the uncompensated state. A ratio of 40/2 is equivalent to a normal 20/1 ratio, so pH is once again the normal 7.4. Enhanced renal conservation of HCO_3^- has fully compensated for CO_2 accumulation, thus restoring pH to normal, although both $[CO_2]$ and $[HCO_3^-]$ are now distorted. Note that maintaining a normal pH depends on preserving a normal ratio between $[HCO_3^-]$ and $[CO_2]$, no matter what the absolute values of each of these buffer components are. (Bear in mind that the values used are only representative. Deviations in pH actually occur over a range, and the degree to which compensation can be accomplished varies.)

Respiratory alkalosis arises from a decrease in [CO2].

Respiratory alkalosis occurs when excessive CO_2 is lost from the body as a result of *hyperventilation* (see p. 478). When pulmonary ventilation increases out of proportion to the rate of CO_2 production, too much CO_2 is blown off. Consequently, less $[H^+]$ is formed from this source.

Causes of Respiratory Alkalosis Possible causes of respiratory alkalosis include fever, anxiety, and aspirin poisoning, all of which excessively stimulate ventilation without regard to the status of O_2, CO_2, or H^+ in the body fluids. Respiratory alkalosis also occurs as a result of physiological mechanisms at high altitude. When the low concentration of O_2 in arterial blood reflexly stimulates ventilation to obtain more O_2, too much CO_2 is blown off, inadvertently leading to an alkalotic state (see p. 480).

If we look at the biochemical abnormalities in uncompensated respiratory alkalosis (Figure 15-13c, left), the increase in pH reflects a reduction in $[CO_2]$ (half the normal value in our example), whereas $[HCO_3^-]$ remains normal. This yields an alkalotic ratio of 20/0.5, which is comparable to 40/1.

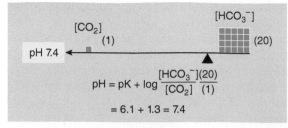

$$pH = pK + \log \frac{[HCO_3^-]}{[CO_2]} \frac{(20)}{(1)}$$

$$= 6.1 + 1.3 = 7.4$$

(a) Normal acid–base balance

‖Figure 15-13 Relationship of [HCO$_3^-$] and [CO$_2$] to pH in various acid–base statuses. Each of the different acid–base statuses are shown mathematically as a solution to the Henderson–Hasselbalch equation and visually as a balance beam or seesaw, where the beam tips down when the relative concentration of one member of the seesaw pair increases (or gets "heavier") and tips up when the relative concentration of a member of the pair decreases (gets "lighter"). (Note that the lengths of the arms of the balance beam analogy are not drawn to scale. The [CO$_2$] arm should be 20 times longer than the [HCO$_3^-$] arm, because, when the beam is in balance, force 1 × distance 1 from the pivot point = force 2 × distance 2 from the pivot point.) **FIGURE FOCUS**: *Alkalosis is always present when [HCO$_3^-$] is elevated.* (True or false?)

Uncompensated acid–base disorders   **Compensated acid–base disorders**

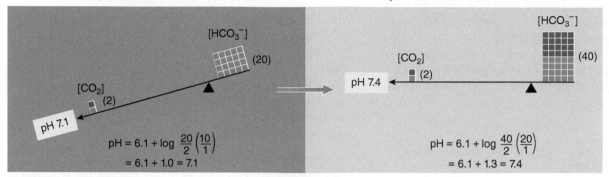

$$pH = 6.1 + \log \frac{20}{2} \left(\frac{10}{1}\right)$$
$$= 6.1 + 1.0 = 7.1$$

$$pH = 6.1 + \log \frac{40}{2} \left(\frac{20}{1}\right)$$
$$= 6.1 + 1.3 = 7.4$$

(b) Respiratory acidosis

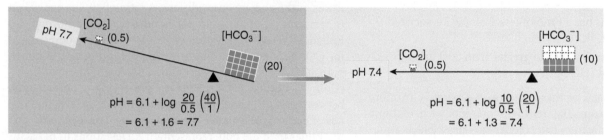

$$pH = 6.1 + \log \frac{20}{0.5} \left(\frac{40}{1}\right)$$
$$= 6.1 + 1.6 = 7.7$$

$$pH = 6.1 + \log \frac{10}{0.5} \left(\frac{20}{1}\right)$$
$$= 6.1 + 1.3 = 7.4$$

(c) Respiratory alkalosis

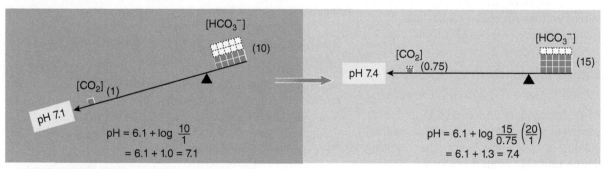

$$pH = 6.1 + \log \frac{10}{1}$$
$$= 6.1 + 1.0 = 7.1$$

$$pH = 6.1 + \log \frac{15}{0.75} \left(\frac{20}{1}\right)$$
$$= 6.1 + 1.3 = 7.4$$

(d) Metabolic acidosis

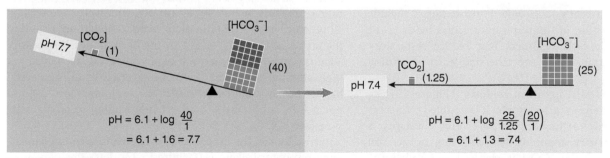

$$pH = 6.1 + \log \frac{40}{1}$$
$$= 6.1 + 1.6 = 7.7$$

$$pH = 6.1 + \log \frac{25}{1.25} \left(\frac{20}{1}\right)$$
$$= 6.1 + 1.3 = 7.4$$

(e) Metabolic alkalosis

Compensations for Respiratory Alkalosis Compensatory measures act to shift pH back toward normal.

- The chemical buffer systems liberate H^+ to diminish the severity of the alkalosis.
- As plasma $[CO_2]$ and $[H^+]$ fall below normal because of excessive ventilation, two of the normally potent stimuli for driving ventilation are removed. This effect tends to "put brakes" on the extent to which some nonrespiratory factors such as fever or anxiety can overdrive ventilation. Therefore, hyperventilation does not continue completely unabated.
- If the situation continues for a few days, the kidneys compensate by conserving H^+ and excreting more HCO_3^-. If, as in our example (Figure 15-13c, right), HCO_3^- stores are reduced by half by loss of HCO_3^- in the urine, the $[HCO_3^-]/[CO_2]$ ratio becomes 10/0.5, equivalent to the normal 20/1. Therefore, the pH is restored to normal by reducing the HCO_3^- load to compensate for the CO_2 loss.

Metabolic acidosis is associated with a fall in $[HCO_3^-]$.

Metabolic acidosis (also known as **nonrespiratory acidosis**) encompasses all types of acidosis besides that caused by excess CO_2 in body fluids. In the uncompensated state (Figure 15-13d, left), metabolic acidosis is always characterized by a reduction in plasma $[HCO_3^-]$ (in our example it is halved), whereas $[CO_2]$ remains normal, producing an acidotic ratio of 10/1. The problem may arise from excessive loss of HCO_3^--rich fluids from the body or from an accumulation of noncarbonic acids. In the latter case, plasma HCO_3^- is used up in buffering the additional H^+.

Causes of Metabolic Acidosis Metabolic acidosis is the type of acid–base disorder most frequently encountered. Here are its most common causes:

1. *Severe diarrhea.* During digestion, a HCO_3^--rich digestive juice is normally secreted into the digestive tract by the pancreas and is later reabsorbed back into the plasma when digestion is completed. During diarrhea, this HCO_3^- is lost from the body rather than reabsorbed. Because of the loss of HCO_3^-, less HCO_3^- is available to buffer H^+, leading to more free H^+ in the body fluids. Looking at the situation differently, loss of HCO_3^- shifts the $CO_2 + H_2O \rightleftharpoons H^+ + HCO_3^-$ reaction to the right to compensate for the HCO_3^- deficit, increasing $[H^+]$ above normal.

2. *Diabetes mellitus.* Abnormal fat metabolism resulting from the inability of cells to preferentially use glucose because of inadequate insulin action leads to formation of excess keto acids whose dissociation increases plasma $[H^+]$.

3. *Strenuous exercise.* When muscles resort to anaerobic glycolysis during strenuous exercise, excess lactic acid (lactate) is produced, raising plasma $[H^+]$.

4. *Uremic acidosis.* In severe renal failure (uremia), the kidneys cannot rid the body of even the normal amounts of H^+ generated from noncarbonic acids formed by ongoing metabolic processes, so H^+ starts to accumulate in the body fluids.

Also, the kidneys cannot conserve an adequate amount of HCO_3^- for buffering the normal acid load.

Compensations for Metabolic Acidosis Except in uremic acidosis, metabolic acidosis is compensated for by both respiratory and renal mechanisms as well as by chemical buffers.

- The buffers take up extra H^+.
- The lungs blow off additional H^+-generating CO_2.
- The kidneys excrete more H^+ and conserve more HCO_3^-.

In our example (Figure 15-13d, right), these compensatory measures restore the ratio to normal by reducing $[CO_2]$ to 75% of normal and by raising $[HCO_3^-]$ halfway back toward normal (up from 50% to 75% of the normal value). This brings the ratio to 15/0.75 (equivalent to 20/1).

Note that in compensating for metabolic acidosis, the lungs deliberately displace $[CO_2]$ from normal in an attempt to restore $[H^+]$ toward normal. Whereas in respiratory-induced acid–base disorders an abnormal $[CO_2]$ is the *cause* of the $[H^+]$ imbalance, in metabolic acid–base disorders $[CO_2]$ is intentionally shifted from normal as an important *compensation* for the $[H^+]$ imbalance.

When kidney disease causes metabolic acidosis, complete compensation is not possible because the renal mechanism is not available for pH regulation. Recall that the respiratory system can compensate only up to 75% of the way toward normal. Uremic acidosis is very serious because the kidneys cannot help restore pH all the way to normal.

Metabolic alkalosis is associated with an elevation in $[HCO_3^-]$.

Metabolic (or **nonrespiratory**) **alkalosis** is a reduction in plasma $[H^+]$ caused by a relative deficiency of noncarbonic acids. This acid–base disturbance is associated with an increase in $[HCO_3^-]$, which, in the uncompensated state, is not accompanied by a change in $[CO_2]$. In our example (Figure 15-13e, left), $[HCO_3^-]$ is doubled, producing an alkalotic ratio of 40/1.

Causes of Metabolic Alkalosis This condition arises most commonly from the following:

1. *Vomiting* causes abnormal loss of H^+ from the body as a result of lost acidic gastric (stomach) juices. HCl is secreted into the stomach lumen during digestion. In the course of gastric HCl secretion, HCO_3^- is added to the plasma. This HCO_3^- is neutralized by H^+ as the gastric secretions are eventually reabsorbed back into the plasma, so normally there is no net addition of HCO_3^- to the plasma from this source. However, when the secreted acid is lost from the body during vomiting instead of being reabsorbed, not only is plasma $[H^+]$ decreased, but also reabsorbed H^+ is no longer available to neutralize the extra HCO_3^- added to the plasma during gastric HCl secretion. Thus, loss of HCl in effect increases plasma $[HCO_3^-]$. (In contrast, with "deeper" vomiting, HCO_3^- in the digestive juices secreted into the upper intestine may be lost in the vomit, resulting in acidosis instead of alkalosis.)

2. *Ingestion of alkaline drugs* can produce alkalosis, such as when baking soda ($NaHCO_3$, which dissociates in solution into Na^+ and HCO_3^-) is used as a self-administered remedy for treating gastric hyperacidity. By neutralizing excess acid in the stomach, HCO_3^- relieves the symptoms of stomach irritation and heartburn; but when more HCO_3^- than needed is ingested, the extra HCO_3^- is absorbed from the digestive tract and increases plasma $[HCO_3^-]$. The extra HCO_3^- binds with some of the free H^+ normally present in plasma from noncarbonic acid sources, reducing free $[H^+]$. (In contrast, commercial alkaline products for treating gastric hyperacidity are not absorbed from the digestive tract to any extent and therefore do not alter the body's acid–base status.)

Compensations for Metabolic Alkalosis

■ In metabolic alkalosis, chemical buffer systems immediately liberate H^+.

■ Ventilation is reduced so that extra H^+-generating CO_2 is retained in the body fluids.

■ If the condition persists for several days, the kidneys conserve H^+ and excrete the excess HCO_3^- in the urine.

The resultant compensatory increase in $[CO_2]$ (up 25% in our example— ❚ Figure 15-13e, right) and the partial reduction in $[HCO_3^-]$ (75% of the way back down toward normal in our example) together restore the $[HCO_3^-]/[CO_2]$ ratio back to the equivalent of 20/1 at 25/1.25.

Overview of Compensated Acid–Base Disorders An individual's acid–base status cannot be assessed on the basis of pH alone. Uncompensated acid–base abnormalities can readily be distinguished on the basis of deviations of either $[CO_2]$ or

$[HCO_3^-]$ from normal (❚ Table 15-9). However, when compensation has been accomplished and pH is essentially normal, determinations of $[CO_2]$ and $[HCO_3^-]$ can reveal an acid–base disorder, but the type of disorder cannot be distinguished. For example, in both compensated respiratory acidosis and compensated metabolic alkalosis, $[CO_2]$ and $[HCO_3^-]$ are both above normal. With respiratory acidosis, the original problem is an abnormal increase in $[CO_2]$, and a compensatory increase in $[HCO_3^-]$ restores the $[HCO_3^-]/[CO_2]$ ratio to 20/1. Metabolic alkalosis, by contrast, is characterized by an abnormal increase in $[HCO_3^-]$ in the first place; then a compensatory rise in $[CO_2]$ restores the ratio to normal. Similarly, compensated respiratory alkalosis and compensated metabolic acidosis share similar patterns of $[CO_2]$ and $[HCO_3^-]$. Respiratory alkalosis starts out with reduced $[CO_2]$, which is compensated by a reduction in $[HCO_3^-]$. With metabolic acidosis, $[HCO_3^-]$ falls below normal, followed by a compensatory decrease in $[CO_2]$. Thus, in compensated acid–base disorders, the original problem must be determined by clinical signs and symptoms other than deviations in $[CO_2]$ and $[HCO_3^-]$ from normal.

Check Your Understanding 15.3

1. Explain why only a narrow pH range is compatible with life.
2. Discuss in what ways H^+ is continuously added to the body fluids.
3. State the Henderson–Hasselbalch equation and use it to explain why the pH of the plasma can be shifted up or down by both kidney and respiratory influences.
4. If a person has severe diarrhea, tell what type of acid–base abnormality will likely result and describe the compensatory responses of the three lines of defense against this change in $[H^+]$.

❚**TABLE 15-9** Summary of $[CO_2]$, $[HCO_3^-]$, and pH in Uncompensated and Compensated Acid–Base Abnormalities

Acid–Base Status	pH	$[CO_2]$ (Compared to Normal)	$[HCO_3^-]$ (Compared to Normal)	$[HCO_3^-]/[CO_2]$
Normal	Normal	Normal	Normal	20/1
Uncompensated respiratory acidosis	Decreased	Increased	Normal	20/2 (10/1)
Compensated respiratory acidosis	Normal	Increased	Increased	40/2 (20/1)
Uncompensated respiratory alkalosis	Increased	Decreased	Normal	20/0.5 (40/1)
Compensated respiratory alkalosis	Normal	Decreased	Decreased	10/0.5 (20/1)
Uncompensated metabolic acidosis	Decreased	Normal	Decreased	10/1
Compensated metabolic acidosis	Normal	Decreased	Decreased	15/0.75 (20/1)
Uncompensated metabolic alkalosis	Increased	Normal	Increased	40/1
Compensated metabolic alkalosis	Normal	Increased	Increased	25/1.25 (20/1)

Homeostasis: Chapter in Perspective

Homeostasis depends on maintaining a balance between the input and the output of all constituents in the internal fluid environment. Regulation of fluid balance involves two separate components: control of salt balance and control of H_2O balance. Control of salt balance is important in the long-term regulation of arterial blood pressure because the body's salt load osmotically holds H_2O, thereby determining the ECF volume, of which plasma volume is a part. An increased salt load in the ECF leads to an expansion in ECF volume, including plasma volume, which in turn causes a rise in blood pressure. Conversely, a reduction in the ECF salt load brings about a fall in blood pressure. Salt balance is maintained by constantly adjusting salt output in the urine to match unregulated, variable salt intake.

Control of H_2O balance is important in preventing changes in ECF osmolarity, which would induce detrimental osmotic shifts of H_2O between the cells and the ECF. Such shifts of H_2O into or out of the cells would cause the cells to swell or shrink, respectively. Cells, especially brain neurons, do not function normally when swollen or shrunken. Water balance is largely maintained by controlling the volume of free H_2O (H_2O not accompanied by solute) lost in the urine to compensate for uncontrolled losses of variable volumes of H_2O from other avenues, such as through sweating or diarrhea, and for poorly regulated H_2O intake. Even though a thirst mechanism exists to control H_2O intake based on need, the amount a person drinks is often influenced by social custom and habit instead of thirst alone.

A balance between input and output of H^+ is critical to maintaining the body's acid–base balance within the narrow limits compatible with life. Deviations in the internal fluid environment's pH lead to altered neuromuscular excitability, to changes in enzymatically controlled metabolic activity, and to K^+ imbalances, which can cause cardiac arrhythmias. These effects are fatal if the pH falls outside the range of 6.8 to 8.0.

Hydrogen ions are uncontrollably and continuously added to the body fluids as a result of ongoing metabolic activities, yet the ECF pH must be kept constant at a slightly alkaline level of 7.4 for optimal body function. Like salt and H_2O balance, control of H^+ output by the kidneys is the main regulatory factor in achieving H^+ balance. The lungs, which can adjust their rate of excretion of H^+-generating CO_2, also help eliminate H^+ from the body. Furthermore, chemical buffer systems can take up or liberate H^+, transiently keeping its concentration constant within the body until its output can be brought into line with its input. Such a mechanism is not available for salt or H_2O balance.

Review Exercises Answers begin on p. A-43

Reviewing Terms and Facts

1. The only avenue by which materials can be exchanged between the cells and the external environment is the ECF. (*True or false?*)

2. Water is driven into the cells when the ECF volume is expanded by an isotonic fluid gain. (*True or false?*)

3. Salt balance in humans is poorly regulated because of our hedonistic salt appetite. (*True or false?*)

4. An unintentional increase in CO_2 is a cause of respiratory acidosis, but a deliberate increase in CO_2 compensates for metabolic alkalosis. (*True or false?*)

5. Secreted H^+ that is coupled with HCO_3^- reabsorption is not excreted, whereas secreted H^+ that is excreted is linked with the addition of new HCO_3^- to plasma. (*True or false?*)

6. The largest body-fluid compartment is the _____.

7. Of the two members of the H_2CO_3:HCO_3^- buffer system, _____ is regulated by the lungs and _____ is regulated by the kidneys.

8. Which of the following factors does *not* increase vasopressin secretion?
   a. ECF hypertonicity
   b. alcohol
   c. stressful situations
   d. an ECF volume deficit
   e. angiotensin II

9. Indicate all correct answers: pH
   a. equals log $1/[H^+]$
   b. equals pK + log $[CO_2]/[HCO_3^-]$
   c. is high in acidosis
   d. falls lower as $[H^+]$ increases
   e. is normal when the $[HCO_3^-]/[CO_2]$ ratio is 20/1

10. Indicate all correct answers: Acidosis
   a. causes overexcitability of the nervous system
   b. exists when the plasma pH falls below 7.35
   c. occurs when the $[HCO_3^-]/[CO_2]$ ratio exceeds 20/1
   d. occurs when CO_2 is blown off more rapidly than it is being produced by metabolic activities
   e. occurs when excessive HCO_3^- is lost from the body, as in diarrhea

11. Indicate all correct answers: The kidney tubular cells secrete NH_3

    a. when the urinary pH becomes too high

    b. when the body is in a state of alkalosis

    c. to enable further renal secretion of H^+ to occur

    d. to buffer excess filtered HCO_3^-

    e. when there is excess NH_3 in the body fluids

12. Complete the following chart:

$\dfrac{[HCO_3^-]}{[CO_2]}$	Uncompensated Abnormality	Possible Cause	pH
10/1	1. _____	2. _____	3. _____
20/0.5	4. _____	5. _____	6. _____
20/2	7. _____	8. _____	9. _____
40/1	10. _____	11. _____	12. _____

Understanding Concepts

(Answers at www.cengagebrain.com)

1. Explain the balance concept.

2. Outline the distribution of body H_2O.

3. Define *transcellular fluid*, and identify its components. Does the transcellular compartment as a whole reflect changes in the body's fluid balance?

4. Compare the ionic composition of plasma, interstitial fluid, and intracellular fluid.

5. What factors are regulated to maintain the body's fluid balance?

6. Why is regulation of ECF volume important? How is it regulated?

7. Why is regulation of ECF osmolarity important? How is it regulated? What are the causes and consequences of ECF hypertonicity and ECF hypotonicity?

8. Outline the sources of input and output in a daily salt balance and a daily H_2O balance. Which are subject to control to maintain the body's fluid balance?

9. Distinguish between an acid and a base.

10. What is the relationship between $[H^+]$ and pH?

11. What is the normal pH of body fluids? How does this compare to the pH of H_2O? Define *acidosis* and *alkalosis*.

12. What are the consequences of fluctuations in $[H^+]$?

13. What are the body's sources of H^+?

14. Describe the three lines of defense against changes in $[H^+]$ in terms of their mechanisms and speed of action.

15. List and indicate the functions of each of the body's chemical buffer systems.

16. Compare the means by which H^+ and HCO_3^- are handled in the proximal tubules and in the Type A and Type B intercalated cells of the distal and collecting tubules.

17. What are the causes of the four categories of acid–base imbalances?

18. Why is uremic acidosis so serious?

Solving Quantitative Exercises

1. Given that plasma pH = 7.4, arterial P_{CO_2} = 40 mm Hg, and each mm Hg partial pressure of CO_2 is equivalent to a plasma $[CO_2]$ of 0.03 mM, what is the value of plasma $[HCO_3^-]$?

2. Death occurs if the plasma pH falls outside the range of 6.8 to 8.0 for an extended time. What is the concentration range of H^+ represented by this pH range?

3. A person drinks 1 liter of distilled water. Use the data in Table 15-1, p. 537, to calculate the resulting percent increase in total body water (TBW), ICF, ECF, plasma, and interstitial fluid. Repeat the calculations for ingestion of 1 liter of isotonic NaCl. Which solution would be better at expanding plasma volume in a patient who has just hemorrhaged?

Applying Clinical Reasoning

Marilyn Y. has had pronounced diarrhea for more than a week as a result of having acquired salmonellosis, a bacterial intestinal infection, from improperly handled food. What effect has this prolonged diarrhea had on her fluid balance and acid–base balance? In what ways has Marilyn's body been trying to compensate for these imbalances?

Thinking at a Higher Level

1. Alcoholic beverages inhibit vasopressin secretion. Given this fact, predict the effect of alcohol on the rate of urine formation. Predict the actions of alcohol on ECF osmolarity. Explain why a person still feels thirsty after excessive consumption of alcoholic beverages.

2. If a person loses 1500 mL of salt-rich sweat and drinks 1000 mL of water during the same time period, what will happen to vasopressin secretion? Why is it important to replace both the water and the salt?

3. If a solute that can penetrate the plasma membrane, such as dextrose (a type of sugar), is dissolved in sterile water at a concentration equal to that of normal body fluids and then is injected intravenously to provide nourishment, what is the effect on the body's fluid balance?

4. Explain why it is safer to treat gastric hyperacidity with antacids that are poorly absorbed from the digestive tract than with baking soda, which is a good buffer for acid but is readily absorbed.

5. Which of the following reactions would buffer the acidosis accompanying severe pneumonia?

    a. $H^+ + HCO_3^- \rightarrow H_2CO_3 \rightarrow CO_2 + H_2O$

    b. $CO_2 + H_2O \rightarrow H_2CO_3 \rightarrow H^+ + HCO_3^-$

    c. $H^+ + Hb \rightarrow HHb$

    d. $HHb \rightarrow H^+ + Hb$

    e. $NaH_2PO_4 + Na^+ \rightarrow Na_2HPO_4 + H^+$

To access the course materials and companion resources for this text, please visit www.cengagebrain.com

The Digestive System

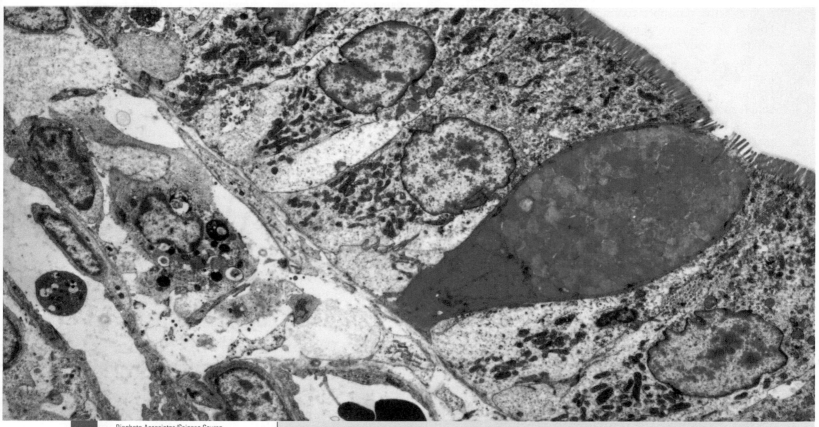

Biophoto Associates/Science Source

A transmission electron micrograph of small-intestine villus cells. Shown here is a section of a villus, one of myriad fingerlike projections from the small-intestine lining. Even smaller hairlike projections, the microvilli (*blue*) arise from the surface of the elongated epithelial cells (*yellow with purple nuclei*) that form the villus surface. The villi and microvilli greatly increase the surface area available for processing and absorbing nutrients. Also seen is a mucous cell (*red*), which secretes protective mucus.

Homeostasis Highlights

To maintain homeostasis, nutrient molecules used for energy production must continually be replaced by new, energy-rich nutrients. Also, nutrient molecules, especially proteins, are needed for ongoing synthesis of new cells and cell parts in the course of growth and tissue turnover. Similarly, water and electrolytes constantly lost in urine and sweat and through other avenues must be replenished regularly. The **digestive system** contributes to homeostasis by transferring nutrients, water, and electrolytes from the external environment to the internal environment. The digestive system does not directly regulate the concentration of any of these constituents in the internal environment. It does not vary nutrient, water, or electrolyte uptake based on body needs (with few exceptions); rather, it optimizes conditions for digesting and absorbing what is ingested.

16.1 General Aspects of Digestion

The primary function of the **digestive** (**gastrointestinal** or **GI**) **system** (*gastro* means "stomach") is to transfer nutrients, water, and electrolytes from the food we eat into the body's internal environment. Ingested food is essential as an energy source, or fuel, from which the cells can generate adenosine triphosphate (ATP) to carry out their particular energy-dependent activities, such as active transport, contraction, synthesis, and secretion. Food is also a source of building supplies for the renewal and addition of body tissues.

The act of eating does not automatically make the pre-formed organic molecules in food available to body cells. Food first must be digested, or chemically broken down, into small, simple molecules that can be absorbed from the digestive tract into the circulatory system for distribution to the cells. Normally, about 95% of the ingested food is made available for the body's use.

We provide an overview of the digestive system, examining the common features of the various components of the system, before we begin a detailed tour of the tract from beginning to end.

The digestive system performs four basic digestive processes.

There are four basic digestive processes: *motility, secretion, digestion,* and *absorption.*

Motility The term **motility** refers to the muscular contractions that mix and move forward the contents within the tract. Although the smooth muscle in the walls of the digestive tract is phasic smooth muscle that displays action potential–induced bursts of contraction (see p. 289), it also maintains a constant low level of contraction known as **tone**. Tone is important in maintaining a steady pressure on the contents of the digestive tract and in preventing its walls from remaining permanently stretched following distension.

Two basic types of phasic digestive motility are superimposed on this ongoing tonic activity: propulsive movements and mixing movements. *Propulsive movements* propel or push the contents forward through the digestive tract. *Mixing movements* have a twofold function. First, by mixing food with the digestive juices, these movements promote digestion of the food. Second, they facilitate absorption by exposing all parts of the intestinal contents to the absorbing surfaces of the digestive tract.

Smooth muscle contraction within the walls of the digestive organs accomplishes movement of material through most of the digestive tract. The exceptions are at the ends of the tract: the mouth through the early part of the esophagus at the beginning and the external anal sphincter at the end. In these regions, motility involves skeletal muscle rather than smooth muscle activity. Accordingly, the acts of chewing, swallowing, and defecation have voluntary components because skeletal muscle is under voluntary control. By contrast, motility accomplished by smooth muscle throughout the rest of the tract is controlled by complex involuntary mechanisms.

Secretion The digestive system produces both exocrine and endocrine secretions. Digestive exocrine gland cells are specialized epithelial cells found in the lining of the digestive tract and in accessory digestive organs like the exocrine pancreas that secrete digestive juices into the digestive tract lumen on appropriate neural or hormonal stimulation. Each **digestive secretion** consists of water, electrolytes, and specific organic constituents important in the digestive process, such as enzymes, bile salts, or mucus. The secretory cells extract from the plasma large volumes of water and the raw materials necessary to produce their particular secretion. Secretion of all digestive juices requires energy, both for active transport of some of the raw materials into the cell (others diffuse in passively) and for synthesis of secretory products. Normally, the digestive secretions are reabsorbed in one form or another back into the blood after their participation in digestion. Failure to do so (because of vomiting or diarrhea, for example) results in loss of this fluid that has been "borrowed" from the plasma.

The digestive system is considered the largest endocrine organ in the body. Whereas peripheral endocrine tissues typically are organized into distinct glands, the endocrine tissue of the gastrointestinal tract is organized as single, individual cells scattered throughout the length of the tract. These specialized epithelial cells produce a range of signal proteins, which are classified as either **GI hormones** or **GI peptides**, that enter the blood and are carried to targets within the tract and outside of the tract. Regardless of their classification, these endocrine secretions regulate digestive function.

Digestion Humans consume three primary categories of energy-rich foodstuffs: *carbohydrates, proteins,* and *fats* (❙Figure 16-1). These large molecules cannot cross plasma membranes intact to be absorbed from the lumen of the digestive tract into the blood or lymph. Therefore, the purpose of **digestion** is to chemically break down the structurally complex foodstuffs of the diet into smaller, absorbable units as follows:

1. The simplest **carbohydrates** are the simple sugars or **monosaccharides** ("one-sugar" molecules), such as **glucose, fructose**, and **galactose**, very few of which are normally found in the diet (❙Figure 16-1a; also see pp. A-9–A-10). Most ingested carbohydrate is in the form of **polysaccharides** ("many-sugar" molecules), which consist of chains of interconnected glucose molecules. The most common polysaccharide consumed is **starch**, consisting of the polysaccharides **amylose** (unbranched chain of glucose) and **amylopectin** (branched chain of glucose) derived from plant sources. In addition, meat contains **glycogen**, the more highly branched polysaccharide storage form of glucose in muscle. Indigestible dietary polysaccharides found in plant walls include **insoluble fiber** such as *cellulose* and **soluble fiber** such as *pectin*, which cannot be digested into their constituent monosaccharides by digestive juices humans secrete; thus, indigestible fiber represents the "bulk" of our diets.

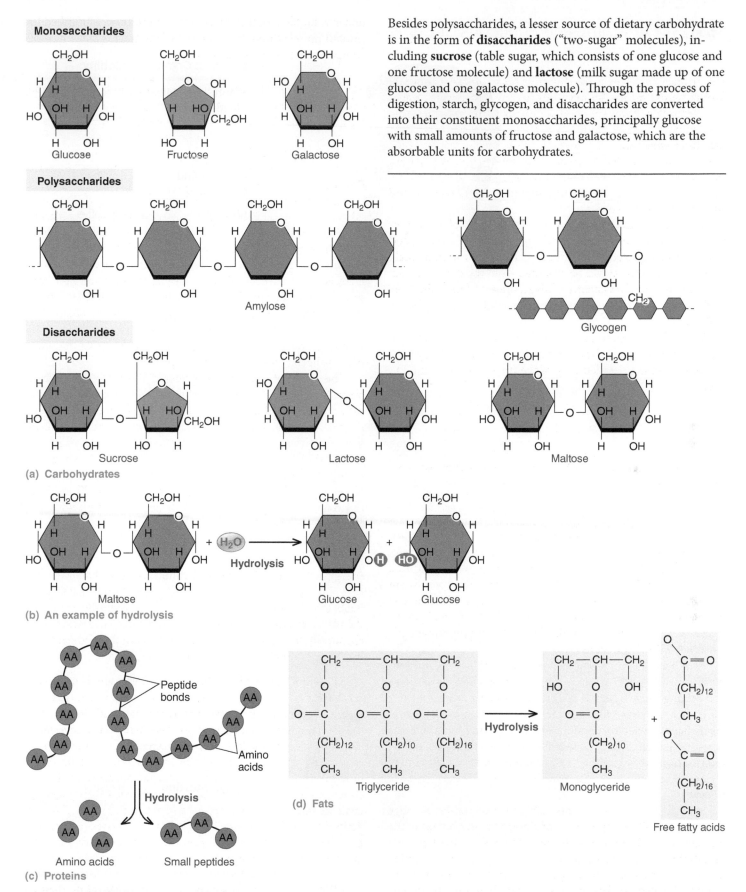

Besides polysaccharides, a lesser source of dietary carbohydrate is in the form of **disaccharides** ("two-sugar" molecules), including **sucrose** (table sugar, which consists of one glucose and one fructose molecule) and **lactose** (milk sugar made up of one glucose and one galactose molecule). Through the process of digestion, starch, glycogen, and disaccharides are converted into their constituent monosaccharides, principally glucose with small amounts of fructose and galactose, which are the absorbable units for carbohydrates.

Monosaccharides

Glucose

Fructose

Galactose

Polysaccharides

Amylose

Glycogen

Disaccharides

Sucrose

Lactose

Maltose

(a) **Carbohydrates**

Maltose + H₂O → Glucose + Glucose

Hydrolysis

(b) **An example of hydrolysis**

AA — Peptide bonds

AA — Amino acids

Hydrolysis

Amino acids   Small peptides

(c) **Proteins**

Triglyceride → Monoglyceride + Free fatty acids

Hydrolysis

(d) **Fats**

Figure 16-1 Energy-rich nutrients and hydrolysis. In the example of hydrolysis in part (b), the disaccharide maltose (the intermediate breakdown product of polysaccharides) is broken down into two glucose molecules by the addition of H₂O at the bond site. The structural details of amino acids and peptide bonds in proteins in part (b) can be found in Figures A-13 and A-14 on p. A-12.

TABLE 16-1 Anatomy and Functions of Components of the Digestive System

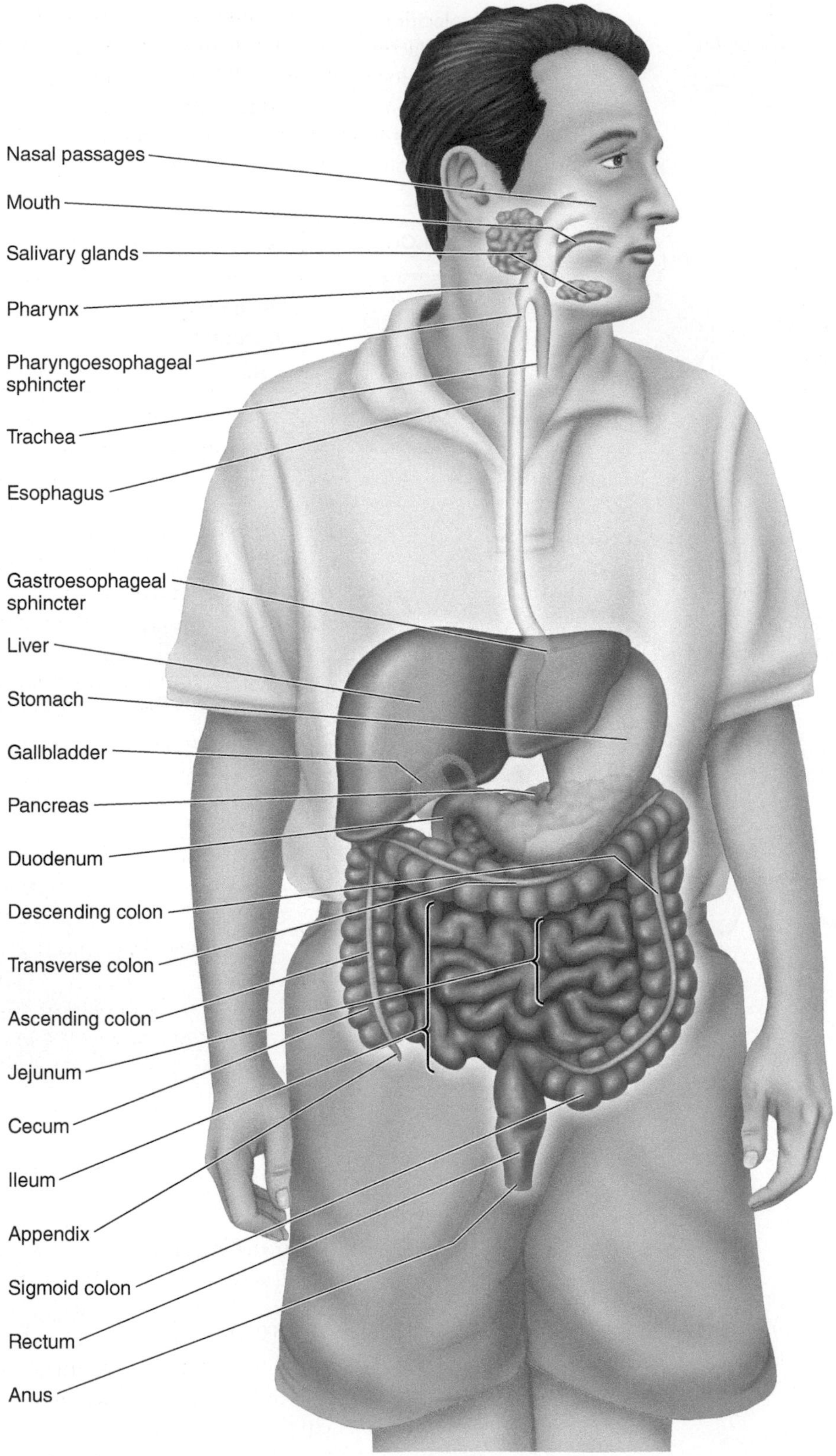

Nasal passages
Mouth
Salivary glands
Pharynx
Pharyngoesophageal sphincter
Trachea
Esophagus
Gastroesophageal sphincter
Liver
Stomach
Gallbladder
Pancreas
Duodenum
Descending colon
Transverse colon
Ascending colon
Jejunum
Cecum
Ileum
Appendix
Sigmoid colon
Rectum
Anus

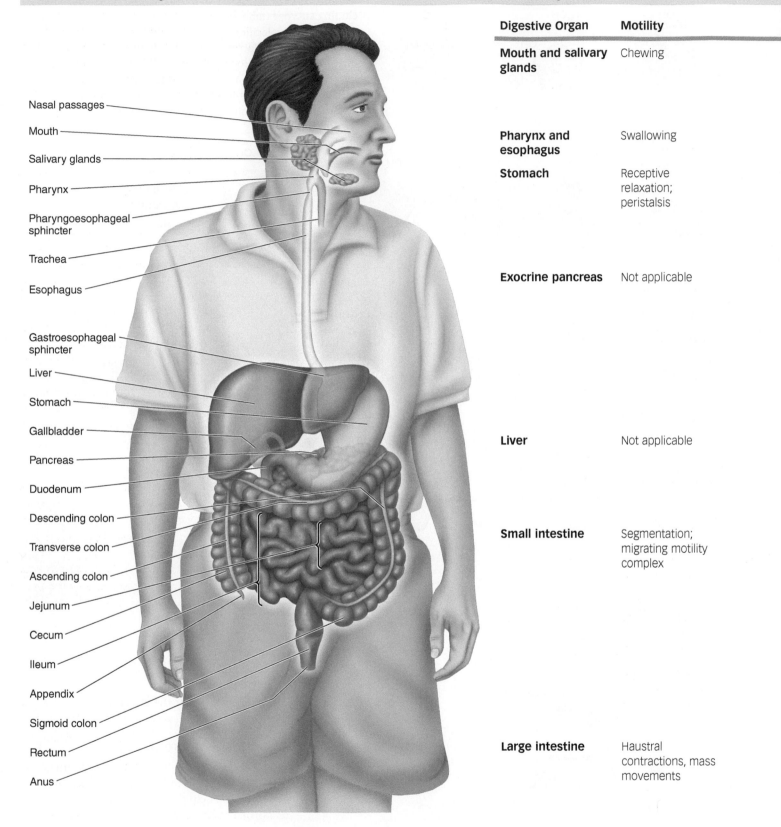

Digestive Organ	Motility
Mouth and salivary glands	Chewing
Pharynx and esophagus	Swallowing
Stomach	Receptive relaxation; peristalsis
Exocrine pancreas	Not applicable
Liver	Not applicable
Small intestine	Segmentation; migrating motility complex
Large intestine	Haustral contractions, mass movements

Secretion	Digestion	Absorption
Saliva ■ Amylase ■ Mucus ■ Lysozyme	Carbohydrate digestion begins	No foodstuffs; a few medications—for example, nitroglycerin
Mucus	None	None
Gastric juice ■ HCl ■ Pepsin ■ Mucus ■ Intrinsic factor	Carbohydrate digestion continues in body of stomach; protein digestion begins in antrum of stomach	No foodstuffs; a few lipid-soluble substances, such as alcohol and aspirin
Pancreatic digestive enzymes ■ Trypsin, chymotrypsin, carboxypeptidase ■ Amylase ■ Lipase Pancreatic aqueous NaHCO₃ secretion	These pancreatic enzymes accomplish digestion in duodenal lumen	Not applicable
Bile ■ Bile salts ■ Alkaline secretion ■ Bilirubin	Bile does not digest anything, but bile salts facilitate fat digestion and absorption in duodenal lumen	Not applicable
Succus entericus ■ Mucus ■ Salt (Small intestine enzymes—disaccharidases and aminopeptidases—are not secreted but function within the brush-border membrane)	In lumen, under influence of pancreatic enzymes and bile, carbohydrate and protein digestion continues and fat digestion is completely accomplished; in brush border, carbohydrate and protein digestion completed	All nutrients, most electrolytes, and water
Mucus	None by human enzymes; bacterial enzymes digest some fiber	Salt and water, converting contents to feces; small amount of nutrients made available by microbial activity

2. Dietary **proteins** consist of various combinations of **amino acids** held together by peptide bonds (❙Figure 16-1c; also see pp. A-11–A-12). Through the process of digestion, proteins are degraded primarily into their constituent amino acids and a few **small polypeptides** (several amino acids linked by peptide bonds), both of which are the absorbable units for protein.

3. Most dietary **fats** are in the form of **triglycerides**, which are neutral fats, each consisting of a glycerol (an alcohol) with three **fatty acid** molecules attached (*tri* means "three") (❙Figure 16-1d; also see pp. A-10–A-11). Enzymatic digestion of neutral fats splits two of the fatty acid molecules from the triglyceride, leaving a **monoglyceride**, a glycerol molecule with one fatty acid molecule attached (*mono* means "one"). Thus, the end products of fat digestion are monoglycerides and free fatty acids, which are the absorbable units of fat.

Digestion of all dietary foodstuffs is accomplished by enzymatic **hydrolysis** ("breakdown by water"; see p. A-14). By adding H_2O at the bond site, enzymes in the digestive secretions break down the bonds that hold the small molecular subunits within the nutrient molecules together, thus setting the small molecules free (❙Figure 16-1b). The removal of H_2O at the bond sites originally joined these small subunits to form nutrient molecules. Hydrolysis replaces the H_2O and frees the small absorbable units. Digestive enzymes are specific in the bonds they can hydrolyze. As food moves through the digestive tract, it is subjected to various enzymes, each of which breaks down the food molecules even further. In this way, large food molecules are converted to simple absorbable units in a progressive, stepwise fashion, like an assembly line in reverse, as the digestive tract contents are propelled forward.

Absorption In the small intestine, digestion is completed and most absorption occurs. Through the process of **absorption**, the small absorbable units that result from digestion, along with water, vitamins, and electrolytes, are transferred from the digestive tract lumen into the blood or lymph.

As we examine the digestive tract from beginning to end, we discuss the four processes of motility, secretion, digestion, and absorption as they take place within each digestive organ (❙Table 16-1).

The digestive tract and accessory digestive organs make up the digestive system.

The digestive system consists of the digestive tract plus the accessory digestive organs. The **accessory digestive organs** include the *salivary glands*, the *exocrine pancreas*, and the *biliary system*, which is composed of the *liver* and *gallbladder*. These exocrine organs lie outside the digestive tract and empty their secretions through ducts into the digestive tract lumen.

The **digestive tract** is essentially a tube about 4.5 m (15 feet) in length in its normal contractile state.[1] Running through the

---

[1]Because the uncontracted digestive tract in a cadaver is about twice as long as the contracted tract in a living person, anatomy texts indicate that the digestive tract is 30 feet long compared to the length of 15 feet indicated in physiology texts.

middle of the body, the digestive tract includes the following organs (Table 16-1): *mouth, pharynx* (throat), *esophagus, stomach, small intestine* (consisting of the *duodenum, jejunum,* and *ileum*), *large intestine* (the *cecum, appendix, colon,* and *rectum*), and *anus.* Although these organs are continuous with one another, they are considered as separate entities because of their regional modifications, which allow them to specialize in particular digestive activities.

Because the digestive tract is continuous from the mouth to the anus, the lumen of this tube, like the lumen of a straw, is continuous with the external environment. As a result, the contents within the lumen of the digestive tract are technically outside the body, just as the soda you suck through a straw is not a part of the straw. A substance is considered within the body only after it moves from the lumen into the absorptive epithelial cells lining the intestine. This is important because conditions essential to the digestive process can be tolerated in the digestive tract lumen that could not be tolerated in the body proper. Consider the following examples:

- The pH of the stomach contents falls as low as 2 as a result of gastric secretion of hydrochloric acid (HCl), yet in the body fluids the range of pH compatible with life is 6.8 to 8.0.

- The digestive enzymes that hydrolyze the protein in food could also destroy the body tissues that produce them. (Protein is the main structural component of cells.) Therefore, once these enzymes are synthesized in inactive form, they are not activated until they reach the lumen, where they actually attack the food outside the body (that is, within the lumen), thereby protecting the body tissues against self-digestion.

- In the lower part of the intestine exist quadrillions of living microorganisms that are normally harmless and even beneficial, yet if these same microorganisms enter the body proper (as may happen with a ruptured appendix), they may be extremely harmful or even lethal.

- Foodstuffs are complex foreign particles that would be attacked by the immune system if they were in contact with the body proper. However, the foodstuffs are digested within the lumen into absorbable units such as glucose, amino acids, and fatty acids that are indistinguishable from these simple energy-rich molecules already present in the body.

The digestive tract wall has four layers.

The digestive tract wall has the same general structure throughout most of its length from the esophagus to the anus, with some local variations characteristic for each region. A cross section of the digestive tube reveals four major tissue layers (Figure 16-2). From the innermost layer outward, they are the *mucosa,* the *submucosa,* the *muscularis externa,* and the *serosa.*

Mucosa The **mucosa** lines the luminal surface of the digestive tract. It is divided into three layers:

- The primary component of the mucosa is a **mucous membrane**, an inner epithelial layer that serves as a protective surface. It is also modified in particular areas for secretion and absorption. The mucous membrane contains *exocrine gland cells* for secretion of digestive juices, *endocrine gland cells* for secretion of blood-borne GI hormones, and *epithelial cells* specialized for absorbing digested nutrients.

- The **lamina propria** is a thin middle layer of connective tissue on which the epithelium rests. It houses the **gut-associated lymphoid tissue (GALT)**, which is important in defense against disease-causing intestinal bacteria (see p. 405).

- The **muscularis mucosa** is a sparse outermost mucosal layer of smooth muscle.

In some parts of the tract, such as the small intestine (the main site of digestion and absorption), the mucosal surface is highly folded, with many ridges and valleys that greatly increase the surface area available to maximize nutrient, water, and electrolyte absorption. In contrast, the esophagus exhibits little mucosal folding because it functions primarily as a transit tube. The pattern of surface folding can be modified by contraction of the muscularis mucosa. This is important in exposing different areas of the absorptive surface to the luminal contents.

Submucosa The **submucosa** ("under the mucosa") is a thick layer of connective tissue that provides the digestive tract with its distensibility and elasticity. It contains the larger blood and lymph vessels, both of which send branches inward to the mucosal layer and outward to the surrounding thick muscle layer. Also, a nerve network known as the **submucosal plexus** lies within the submucosa (*plexus* means "network").

Muscularis Externa The **muscularis externa**, the major smooth muscle coat of the digestive tube, surrounds the submucosa. In most parts of the tract, the muscularis externa consists of two layers: an *inner circular layer* and an *outer longitudinal layer.* The fibers of the inner smooth muscle layer (adjacent to the submucosa) run circularly around the tube. Contraction of these circular fibers decreases the diameter of the lumen, constricting the tube at the point of contraction. Contraction of the fibers in the outer layer, which run longitudinally along the length of the tube, shortens the tube. Together, contractile activity of these smooth muscle layers produces the propulsive and mixing movements. Another nerve network, the **myenteric plexus**, lies between the two muscle layers (*myo* means "muscle"; *enteric* means "intestine"). Together the submucosal and myenteric plexuses, along with GI hormones and local chemical mediators, help regulate local gut activity.

Serosa The outer connective tissue covering of the digestive tract is the **serosa**, which secretes a watery, slippery fluid **(serous fluid)** that lubricates and prevents friction between the digestive organs and the surrounding viscera. Throughout much of the tract, the serosa is continuous with the **mesentery**, which suspends the digestive organs from the inner wall of the abdominal cavity like a sling (Figure 16-2). This attachment provides relative fixation, supporting the digestive organs in proper position, while still allowing them freedom for mixing and propulsive movements.

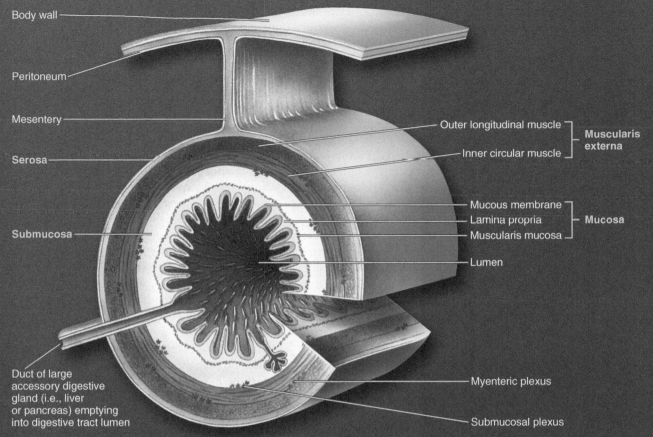

Body wall

Peritoneum

Mesentery

Serosa

Submucosa

Duct of large
accessory digestive
gland (i.e., liver
or pancreas) emptying
into digestive tract lumen

Outer longitudinal muscle ⎤ **Muscularis**
Inner circular muscle ⎦ **externa**

Mucous membrane ⎤
Lamina propria ⎬ **Mucosa**
Muscularis mucosa ⎦

Lumen

Myenteric plexus

Submucosal plexus

❙Figure 16-2 Layers of the digestive tract wall. The digestive tract wall consists of four major layers: from the innermost out, they are the mucosa, submucosa, muscularis externa, and serosa.

Regulation of digestive function is complex and synergistic.

Digestive motility and secretion are carefully regulated to maximize digestion and absorption of ingested food. Four factors are involved in regulating digestive system function: (1) autonomous smooth muscle function, (2) intrinsic nerve plexuses, (3) extrinsic nerves, and (4) GI hormones.

Autonomous Smooth Muscle Function Smooth muscle of the digestive tract undergoes spontaneous, rhythmic cycles of depolarization and repolarization. The prominent type of self-induced electrical activity in digestive smooth muscle is **slow-wave potentials** (see p. 291), alternatively referred to as the digestive tract's **basic electrical rhythm (BER)**. Located throughout the layers of the muscularis externa are pacemaker cells known as the **interstitial cells of Cajal**. These pacemakers generate the slow-wave potentials that propagate via gap junctions (see p. 62) to adjacent smooth muscle cells. Slow waves are not action potentials and do not directly induce muscle contraction; they are rhythmic, wavelike fluctuations in membrane potential that cyclically bring the membrane closer to or farther from threshold potential. If these waves reach threshold at the peaks of depolarization, a volley of action potentials is triggered

at each peak, resulting in rhythmic cycles of muscle contraction (see ❙Figure 8-32b, p. 291).

Slow waves propagate quickly from cell-to-cell throughout a sheet of digestive smooth muscle by gap junctions through which charge-carrying ions can flow, similar to pacemaker potentials propagating through cardiac muscle. Thus, the whole muscle sheet behaves like a functional syncytium, becoming excited and contracting as a unit when threshold is reached (see p. 291). If threshold is not achieved, the oscillating slow-wave electrical activity continues to sweep across the muscle sheet without being accompanied by contractile activity.

Whether threshold is reached depends on the effect of various mechanical, neural, and hormonal factors that influence the starting point around which the slow-wave rhythm oscillates. If the starting point is nearer the threshold level, as it is when food is present in the digestive tract, the depolarizing slow-wave peak reaches threshold, so action potential frequency and its accompanying contractile activity increase. Conversely, if the starting point is farther from threshold, as when no food is present, threshold will less likely be reached, so action potential frequency and contractile activity are reduced.

The *rate* (frequency) of self-induced rhythmic digestive contractile activities depends on the inherent rate established by the involved pacemaker cells. (Specific details about these rhythmic contractions will be discussed when we examine the organs involved.) The *intensity* (strength) of these contractions depends on the number of action potentials that occur when the slow-wave potential reaches threshold, which in turn depends on how long threshold is sustained. At threshold, voltage-gated Ca^{2+} channels are activated (see p. 89), resulting in Ca^{2+} influx into the smooth muscle cell. The resultant Ca^{2+} entry has two effects: (1) It is responsible for the rising phase of an action potential, with the falling phase being brought about as usual by K^+ efflux; and (2) it triggers a contractile response (see p. 290). The greater the number of action potentials, the higher the cytosolic Ca^{2+} concentration, the greater the cross-bridge activity, and the stronger the contraction. Other factors that influence contractile activity also do so by altering the cytosolic Ca^{2+} concentration. Thus, the level of contractility can range from low-level tone to vigorous mixing and propulsive movements by varying the cytosolic Ca^{2+} concentration.

Intrinsic Nerve Plexuses The **intrinsic nerve plexuses** are the two major nerve fiber networks—the *submucosal plexus* and the *myenteric plexus*—that lie entirely within the digestive tract wall and run its entire length. Thus, unlike any other body system, the digestive tract has its own intramural ("within-wall") nervous system, which contains as many neurons as the spinal cord (about 100 million neurons) and endows the tract with a considerable degree of self-regulation. Together, these two plexuses are termed the **enteric nervous system** (see p. 135).

The intrinsic plexuses influence all facets of digestive tract activity. Various types of neurons are present in the intrinsic plexuses. Sensory neurons called **intrinsic primary afferent neurons** respond to specific local stimuli in the digestive tract. **Intrinsic efferent neurons** innervate and control smooth muscle and exocrine and endocrine cells of the digestive tract. Similar to the connections within the central nervous system, interneurons receive synaptic input from intrinsic primary afferent neurons and modulate output of the intrinsic efferent neurons.

The intrinsic efferent neurons can directly affect digestive tract motility, secretion of digestive juices, and secretion of GI hormones through excitatory or inhibitory interactions. For example, neurons that release *acetylcholine (ACh)* as a neurotransmitter promote contraction of digestive tract smooth muscle, whereas the neurotransmitters *nitric oxide* and *vasoactive intestinal peptide* act in concert to cause its relaxation. These intrinsic nerve networks primarily coordinate local activity within the digestive tract. To illustrate, if a large piece of food gets stuck in the esophagus, the intrinsic plexuses coordinate local responses to push the food forward. Adding to the complexity of control and accomplishing extensive coordination throughout the digestive tract, intrinsic nerve activity can be influenced by a vast array of endocrine, paracrine, and extrinsic nerve signals.

Extrinsic Nerves The **extrinsic nerves** are the nerve fibers from both branches of the autonomic nervous system that originate outside the digestive tract and regulate digestive tract function. The autonomic nerves influence digestive tract motility and secretion either by modifying ongoing activity in the intrinsic plexuses, altering the level of GI hormone secretion, or acting directly on the smooth muscle and glands.

Recall that, in general, the sympathetic and parasympathetic nerves supplying any given tissue exert opposing actions on that tissue. The sympathetic system, which dominates in "fight-or-flight" situations, tends to inhibit or slow down digestive tract contraction and secretion. This action is appropriate, considering that digestive processes are not of highest priority when the body faces an emergency. The parasympathetic nervous system, by contrast, dominates in quiet, "rest-and-digest" situations, when general maintenance types of activities such as digestion can proceed optimally. Accordingly, the parasympathetic nerve fibers supplying the digestive tract, which arrive primarily by way of the vagus nerve, tend to increase smooth muscle motility and promote secretion of digestive enzymes and hormones. Unique to the parasympathetic nerve supply to the digestive tract, the postganglionic parasympathetic nerve fibers are actually a part of the intrinsic nerve plexuses. They are the ACh-secreting output neurons within the plexuses. Thus, ACh is released in response to local reflexes coordinated entirely by the intrinsic plexuses as well as to vagal stimulation, which acts through the intrinsic plexuses.

In addition to being called into play during generalized sympathetic or parasympathetic discharge, the autonomic nerves, especially the vagus nerve, can be discretely activated to modify only digestive activity. One of the major purposes of specific activation of extrinsic innervation is to coordinate activity among different regions of the digestive system. For example, the act of chewing food reflexly increases not only salivary secretion but also stomach, pancreatic, and liver secretion via vagal reflexes in anticipation of the arrival of food.

GI Hormones The GI hormones produced by specialized endocrine cells tucked within the mucosa of certain regions of the digestive tract exert either excitatory or inhibitory influences on digestive smooth muscle and exocrine gland cells.

Of note, many of these same hormones are released from neurons in the brain, where they act as neurotransmitters and neuromodulators. During embryonic development, certain cells of the developing neural tissue migrate to the digestive system, where they become endocrine cells.

Receptor activation alters digestive activity through neural and hormonal pathways.

The digestive tract wall contains three types of sensory receptors that respond to local changes in the digestive tract: (1) *chemoreceptors* sensitive to chemical components within the lumen, (2) *mechanoreceptors* (pressure receptors) sensitive to stretch or tension within the wall, and (3) *osmoreceptors* sensitive to the osmolarity of the luminal contents.

Stimulation of these receptors elicits neural reflexes or secretion of hormones, both of which alter the activity level in the digestive system's effector cells. These effector cells include smooth muscle cells (for modifying motility), exocrine gland cells (for controlling secretion of digestive juices), and endocrine gland cells (for varying secretion of GI hormones; Figure 16-3). Receptor activation may bring about two types of neural reflexes—short reflexes and long reflexes. A **short reflex** takes place when all elements of the reflex are located within the wall of the digestive tract itself—that is, when the intrinsic nerve networks influence local motility or secretion in response to specific local stimulation. Extrinsic autonomic nervous activity can be superimposed on the local controls to modify smooth muscle and glandular responses, either to correlate activity between different regions of the digestive system or to modify digestive system activity in response to external influences. Because the autonomic reflexes involve long pathways between the central nervous system and digestive system, they are known as **long reflexes**.

In addition to the sensory receptors within the digestive tract wall that monitor luminal content and wall tension, the plasma membranes of the digestive system's effector cells have receptor proteins that bind with and respond to GI hormones, neurotransmitters, and local chemical mediators.

From this overview, you can see that regulation of GI function is complex, being influenced by many synergistic, interrelated pathways designed to ensure that the appropriate responses occur to digest and absorb the ingested food. Nowhere else in the body experiences so much overlapping control.

We are now going to take a "tour" of the digestive tract, beginning with the mouth and ending with the anus. We examine the four basic digestive processes of motility, secretion, digestion, and absorption at each digestive organ along the way. Table 16-1 summarizes these activities and serves as a useful reference throughout the rest of the chapter.

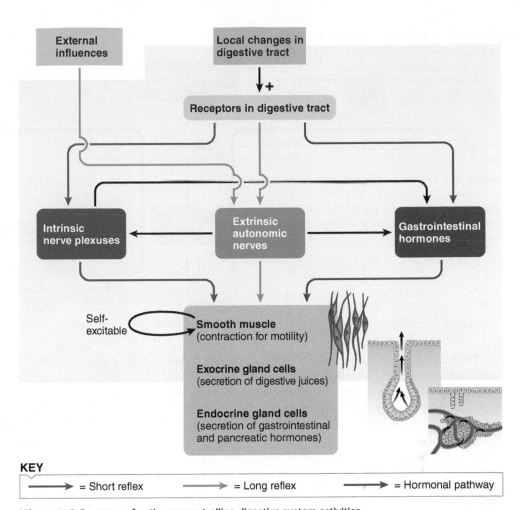

KEY

⟶ = Short reflex   ⟶ = Long reflex   ⟶ = Hormonal pathway

Figure 16-3 **Summary of pathways controlling digestive system activities.**
FIGURE FOCUS: *Describe the means by which the extrinsic autonomic nerves that supply the digestive system are activated and the means by which these nerves control digestive activity*

Check Your Understanding 16.1

1. List the three categories of energy-rich foodstuffs and the absorbable units of each.

2. Draw a cross section of the digestive tract and label the following: mucosa, submucosa, muscularis externa, serosa, submucosal plexus, and myenteric plexus.

3. Describe how pacemaker activity affects smooth muscle function.

16.2 Mouth

The oral cavity is the entrance to the digestive tract.

Entry to the digestive tract is through the **mouth**, or **oral cavity**. The opening is formed by the muscular **lips**, which help procure, guide, and contain the food in the mouth. The lips also have nondigestive functions; they are important in speech (articulation of many sounds depends on a particular lip formation) and as a sensory receptor in interpersonal relationships (for example, as in kissing). The lips are endowed with especially well-developed tactile (touch) sensation.

The **palate**, which forms the arched roof of the oral cavity, separates the mouth from the nasal passages. Its presence allows breathing and chewing or sucking to take place simultaneously. Hanging from the palate in the rear of the throat is a dangling projection, the **uvula**, which plays an important role in sealing off the nasal passages during swallowing. (The uvula is the structure you elevate when you say "ahhh" so that your healthcare provider can better see your throat.)

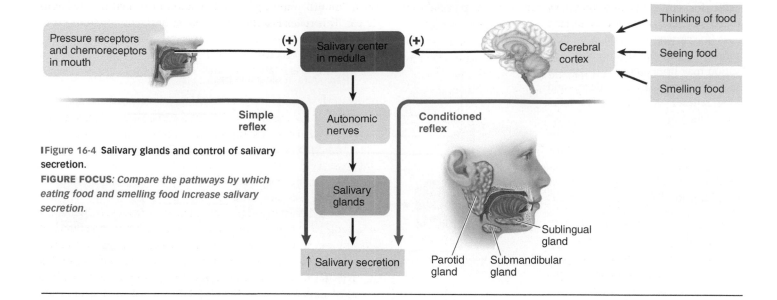

I Figure 16-4 **Salivary glands and control of salivary secretion.**
FIGURE FOCUS: *Compare the pathways by which eating food and smelling food increase salivary secretion.*

The **tongue**, which forms the floor of the oral cavity, is composed of voluntarily controlled skeletal muscle. The tongue guides food within the mouth during chewing and swallowing and also plays an important role in speech. Furthermore, the major **taste buds** are located on the tongue (see p. 224).

The teeth mechanically break down food.

The first step in the digestive process is **mastication**, or **chewing**, the mouth motility that involves the slicing, tearing, grinding, and mixing of ingested food by the **teeth**. The teeth are firmly embedded in and protrude from the jawbones. The teeth can exert forces much greater than those necessary to eat ordinary food. For example, an adult man can exert a crushing force of up to 200 pounds with his molars, which is sufficient to crack a hard nut, but ordinarily these powerful forces are not used. The degree of **occlusion** (how well the upper and lower teeth fit together when the jaw is closed) is more important than the force of the bite in determining the efficiency of chewing. *Clinical Note* When the upper and lower teeth do not make proper contact with one another, they cannot accomplish their normal cutting and grinding action adequately. Such **malocclusion** is usually caused either by overcrowding of teeth too large for the available jaw space or by one jaw being displaced in relation to the other. Malocclusions can often be corrected by applying braces, which exert prolonged gentle pressure against the teeth to move them gradually to the desired position.

The exposed part of a tooth is covered by **enamel**, the hardest structure of the body. Enamel forms before the tooth's eruption by special cells that are lost as the tooth erupts. *Clinical Note* Because enamel cannot be regenerated after the tooth has erupted, any defects (**dental caries**, or "cavities") that develop in the enamel must be patched by artificial fillings, or else the surface continues to erode into the underlying living pulp.

The act of chewing can be voluntary, but most chewing during a meal is a rhythmic reflexlike activity accomplished without conscious effort (see p. 277). The functions of chewing are (1) to mechanically break food into smaller pieces to facilitate swallowing and to increase the food surface area on which salivary enzymes can act, (2) to mix food with saliva, and (3) to expose food to the taste buds. Taste bud stimulation not only gives rise to the pleasurable sensation of taste, but also, in feedforward fashion, reflexly increases salivary, gastric, pancreatic, and bile secretion to prepare for the arrival of food.

Saliva begins carbohydrate digestion and helps swallowing, speech, taste, and oral health.

Saliva, the secretion associated with the mouth, is produced largely by three major pairs of salivary glands that lie outside the oral cavity and discharge saliva through short ducts into the mouth (I Figure 16-4).

Saliva is about 99.5% H_2O and 0.5% electrolytes and protein. The salivary salt (NaCl) concentration is only one seventh of that in the plasma, which is important in perceiving salty tastes. Similarly, discrimination of sweet tastes is enhanced by the absence of glucose in the saliva. The most important salivary proteins are *amylase, mucus,* and *lysozyme*. They contribute to the functions of saliva, which are as follows:

1. Saliva begins digestion of dietary starches through action of the enzyme **salivary amylase**. The products of digestion include **maltose**, a disaccharide consisting of two glucose molecules (see I Figure 16-1b), and **α-limit dextrins**, a branched polysaccharide resulting from amylopectin digestion.

2. Saliva facilitates swallowing by moistening food particles, thereby holding them together, and by providing lubrication through the presence of **mucus**, which is thick and slippery.

3. Saliva exerts some antibacterial action by a fourfold effect—first, by **lysozyme**, a salivary enzyme that lyses, or destroys, certain bacteria by breaking down their cell walls; second, by salivary *IgA antibodies* (see p. 417); third, by salivary *lactoferrin*, which tightly binds to iron that bacteria need to

multiply (see p. 411); and fourth, by rinsing away material that may be a food source for bacteria.

4. Saliva serves as a solvent for molecules that stimulate the taste buds. Only molecules in solution can react with taste bud receptors. The flow of saliva also flushes away the food particles on taste buds so that you can taste the next bite of food.

5. Saliva aids speech by facilitating movements of the lips and tongue. It is difficult to talk when your mouth feels dry.

6. Saliva plays an important role in oral hygiene by helping keep the mouth and teeth clean. The constant flow of saliva helps flush away food residues, foreign particles, and old epithelial cells that have shed from the oral mucosa. Saliva's contribution in this regard is apparent when you have a foul taste in your mouth when salivation is suppressed for a while, such as during a fever or when you are experiencing prolonged anxiety.

7. Saliva is rich in bicarbonate buffers, which neutralize acids in food and acids produced by bacteria in the mouth, thereby helping prevent dental caries.

Despite these many functions, saliva is not essential for digesting and absorbing foods because enzymes produced by the pancreas and small intestine can complete food digestion even in the absence of salivary and gastric secretion. The main problems associated with diminished salivary secretion, or **xerostomia**, are difficulty in chewing and swallowing, inarticulate speech unless frequent sips of water are taken when talking, and a rampant increase in dental caries unless special precautions are taken.

Salivary secretion is continuous and can be reflexly increased.

On average, about 1 to 2 liters of saliva are secreted per day, ranging from a continuous spontaneous basal rate of 0.5 mL/min to a maximum flow rate of about 5 mL/min in response to a potent stimulus such as sucking on a lemon. In the absence of food-related stimuli, low-level parasympathetic stimulation induces production of basal salivary secretion. This basal secretion is important in keeping the mouth and throat moist at all times. In addition to this continuous, low-level secretion, salivary secretion may be increased by two types of salivary reflexes, simple and conditioned (❚ Figure 16-4).

Simple and Conditioned Salivary Reflexes The **simple salivary reflex** occurs when chemoreceptors and pressure receptors within the oral cavity respond to the presence of food. On activation, these receptors initiate impulses in afferent nerve fibers that carry the information to the **salivary center**, which is located in the medulla of the brain stem, as are all the brain centers that control digestive activities. The salivary center, in turn, sends impulses via the extrinsic autonomic nerves to the salivary glands to promote increased salivation. Dental procedures, by activating pressure receptors in the mouth, promote salivary secretion in the absence of food.

With the **conditioned**, or **acquired**, **salivary reflex**, salivation occurs without oral stimulation. Just thinking about, seeing, smelling, or hearing the preparation of pleasant food initiates salivation through this reflex. All of us have experienced such "mouth watering" in anticipation of something delicious to eat. This reflex is a learned response based on previous experience. The cerebral cortex stimulates the medullary salivary center when it receives inputs that arise outside the mouth and are mentally associated with the pleasure of eating.

Autonomic Influence on Salivary Secretion The salivary center controls the degree of salivary output by means of the autonomic nerves that supply the salivary glands. Unlike the autonomic nervous system elsewhere in the body, sympathetic and parasympathetic responses in the salivary glands are not antagonistic. Both sympathetic and parasympathetic stimulation increase salivary secretion, but the quantity and characteristics differ. Parasympathetic stimulation, which exerts the dominant role in salivary secretion, produces a prompt and abundant flow of watery saliva that is rich in enzymes. Sympathetic stimulation, by contrast, produces a smaller volume of thick saliva that is rich in mucus. Because sympathetic stimulation elicits a smaller volume of saliva, the mouth feels drier than usual when the sympathetic system is dominant, such as in stressful situations. For example, people often experience a dry feeling in the mouth when they are nervous about giving a speech.

Salivary secretion is the only digestive secretion entirely under neural control. All other digestive secretions are regulated by both nervous system reflexes and hormones.

Digestion in the mouth is minimal; no absorption of nutrients occurs.

Digestion in the mouth involves the hydrolysis of polysaccharides by amylase. However, most digestion by this enzyme is accomplished in the body of the stomach after the food mass and saliva have been swallowed. Acid inactivates amylase, but in the center of the food mass, where stomach acid has not yet reached, this salivary enzyme continues to function for several more hours.

No absorption of foodstuff occurs from the mouth. Importantly, some drugs can be absorbed by the oral mucosa, a prime example being *nitroglycerin*, a vasodilator drug sometimes used by cardiac patients to relieve angina attacks (see p. 330) associated with myocardial ischemia (see p. 314).

Check Your Understanding 16.2

1. State the functions of salivary mucus, amylase, and lysozyme.

2. Distinguish between simple and conditioned salivary reflexes.

3. Compare the effects of parasympathetic versus sympathetic stimulation of the salivary glands.

16.3 Pharynx and Esophagus

The **pharynx** is the cavity at the rear of the throat. It acts as a common passageway for both the digestive system (by serving as the link between the mouth and esophagus, for food) and the respiratory system (by providing access between the nasal passages and trachea, for air). This arrangement necessitates mechanisms (to be described shortly) to guide food and air into the proper passageways beyond the pharynx. Housed within the side walls of the pharynx are the **tonsils**, lymphoid tissues that are part of the body's defense team.

The motility associated with the pharynx and esophagus is swallowing. Most of us think of swallowing as the limited act of moving food out of the mouth into the esophagus. However, **swallowing** is the entire process of moving food from the mouth through the esophagus into the stomach.

Swallowing is a sequentially programmed all-or-none reflex.

Swallowing is initiated when a **bolus**, or ball of chewed or liquid food, is voluntarily forced by the tongue to the rear of the mouth and into the pharynx (❙ Figure 16-5, step **1**). The pressure of the bolus stimulates pharyngeal pressure receptors, which send afferent impulses to the **swallowing center** located in the medulla of the brain stem. The swallowing center then reflexly activates in the appropriate sequence the muscles involved in swallowing. Swallowing is the most complex reflex in the body, with multiple highly coordinated responses being triggered in a specific all-or-none pattern over a period of time. Swallowing is initiated voluntarily, but once begun it cannot be stopped. Perhaps you have experienced this when a large piece of hard candy inadvertently slipped to the rear of your throat, triggering an unintentional swallow.

Next we describe the two stages of swallowing: the *oropharyngeal stage* and the *esophageal stage*.

During swallowing, food is prevented from entering the wrong passageways.

The **oropharyngeal stage** consists of moving the bolus from the mouth through the pharynx and into the esophagus. When the tongue propels the bolus into the pharynx (❙ Figure 16-5, step **2**), the following coordinated activities prevent the bolus from entering the respiratory passageways and direct it into the esophagus:

- The swallowing center temporarily inhibits the respiratory center (step **3**); thus the person does not attempt futile respiratory efforts while the airways are briefly sealed off.

- The uvula is elevated and lodges against the back of the throat, sealing off the nasal passage from the pharynx so that food does not enter the nose (step **4**).

- The tongue's position against the hard palate keeps food from reentering the mouth during swallowing (step **5**).

- Food is prevented from entering the trachea primarily by elevation of the larynx and tight closure of the vocal folds across the laryngeal opening, or **glottis** (step **6**). The first part of the trachea is the *larynx,* or *voice box,* across which the *vocal folds* are stretched. During swallowing, the vocal folds serve a purpose unrelated to speech. Contraction of laryngeal muscles aligns the vocal folds in tight apposition to each other, thus sealing the glottis entrance (see ❙ Figure 13-3, p. 448). Last, the **epiglottis** (*epi* means "upon"), a flap of cartilaginous tissue an-

terior to the glottis, folds backward down over the closed glottis as further protection from food entering the respiratory airways (step **7**).

- With the glottis closed, pharyngeal muscles contract to force the bolus into the esophagus (step **8**).

The pharyngoesophageal sphincter prevents air from entering the digestive tract.

The **esophagus** is a fairly straight muscular tube that extends between the pharynx and the stomach (see ❙ Table 16-1, p. 568). Lying mostly in the thoracic cavity, it penetrates the diaphragm and joins the stomach in the abdominal cavity a few centimeters below the diaphragm.

The esophagus is guarded at both ends by sphincters. A sphincter is a ringlike muscular structure that, when closed, prevents passage through the tube it guards. The upper esophageal sphincter is the *pharyngoesophageal sphincter,* and the lower esophageal sphincter is the *gastroesophageal sphincter.* We first discuss the role of the pharyngoesophageal sphincter, then the process of esophageal transit of food, and finally the importance of the gastroesophageal sphincter.

Because the esophagus is exposed to subatmospheric intrapleural pressure as a result of respiratory activity (see p. 450), a pressure gradient exists between the atmosphere and the esophagus. Except during a swallow, the **pharyngoesophageal sphincter** remains closed as a result of neurally induced contraction of the sphincter's circular skeletal muscle. Tonic contraction of this upper esophageal sphincter prevents large volumes of air from entering the esophagus and stomach during breathing, thus averting excessive *eructation* (burping). Instead, air is directed only into the respiratory airways. During swallowing, this sphincter opens and allows the bolus to pass into the esophagus (❙ Figure 16-5, step **8**). Once the bolus has entered the esophagus, the pharyngoesophageal sphincter closes, the respiratory airways are opened, and breathing resumes (step **9**). The oropharyngeal stage is complete, and about 1 second has passed since the swallow was initiated.

Peristaltic waves push food through the esophagus.

The **esophageal stage** of the swallow now begins. The swallowing center triggers a **primary peristaltic wave** that sweeps from the beginning to the end of the esophagus, forcing the bolus ahead of it toward the stomach. The term **peristalsis** refers to ringlike contractions of the circular smooth muscle that move progressively forward, pushing the bolus into a relaxed area ahead of the contraction (❙ Figure 16-5, step **10**). The peristaltic wave takes about 5 to 9 seconds to reach the lower end of the esophagus. Progression of the wave is controlled by the swallowing center, with innervation by means of the vagus.

If a large or sticky swallowed bolus, such as a bite of peanut butter sandwich, fails to be carried along to the stomach by the primary peristaltic wave, the lodged bolus distends the esophagus, stimulating stretch receptors within its walls. In response to the stimulus, the intrinsic nerve plexus at the point of distension

(a) Position of the oropharyngeal structures at rest

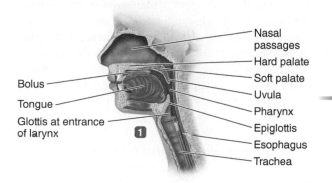

- Nasal passages
- Hard palate
- Soft palate
- Uvula
- Pharynx
- Epiglottis
- Esophagus
- Trachea
- Bolus
- Tongue
- Glottis at entrance of larynx
- **1**

1 Swallowing is initiated voluntarily. At start of swallow, tongue presses bolus against hard palate.

(b) Oral part of oropharyngeal stage of swallowing

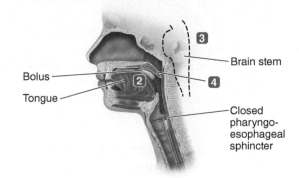

- **3**
- Brain stem
- Bolus
- Tongue
- **2**
- **4**
- Closed pharyngo-esophageal sphincter

2 Tongue propels bolus to pharynx.

3 Swallowing center inhibits respiratory center in brain stem.

4 Elevation of uvula prevents food from entering nasal passageways.

(d) Beginning of esophageal stage of swallowing

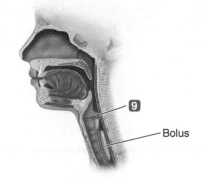

- **9**
- Bolus

9 Pharyngoesophageal sphincter closes, oropharyngeal structures return to resting position, and breathing resumes.

(c) Pharyngeal part of oropharyngeal stage

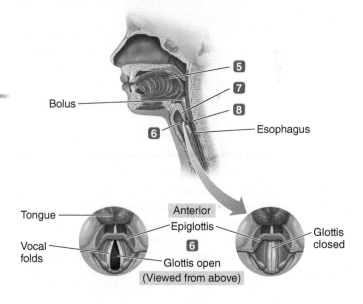

- **5**
- **7**
- **8**
- Bolus
- **6**
- Esophagus

- Tongue
- Vocal folds
- Anterior
- Epiglottis
- **6**
- Glottis open
- Glottis closed
- (Viewed from above)

5 Position of tongue prevents food from reentering mouth.

6 Tight alignment of vocal cords prevents food from entering trachea.

7 Epiglottis folds over closed glottis.

8 Contraction of pharyngeal muscles pushes bolus through opened pharyngoesophageal sphincter into esophagus.

(e) Completion of esophageal stage

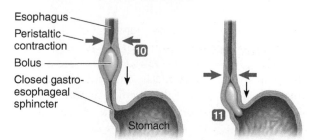

- Esophagus
- Peristaltic contraction
- Bolus
- Closed gastro-esophageal sphincter
- **10**
- Stomach
- **11**

10 Peristalsis propels bolus down length of esophagus.

11 Gastroesophageal sphincter relaxes as peristalsis pushes bolus into stomach. Swallow is complete. Sphincter again contracts.

I Figure 16-5 Oropharyngeal and esophageal stages of swallowing.
FIGURE FOCUS: *On occasion, vomit may accidentally be inhaled, or aspi-rated. Follow the route by which vomit leaves the stomach and enters the trachea.*

initiates additional peristaltic waves to clear the lodged bolus. These **secondary peristaltic waves** do not involve the swallowing center, nor is the person aware of their occurrence. Distension of the esophagus also reflexly increases salivary secretion. The trapped bolus is eventually dislodged and moved forward through the combined effects of lubrication by the extra swallowed saliva and the forceful secondary peristaltic waves. Esophageal peristalsis is so effective you could eat an entire meal while you were upside down and it would all promptly be pushed to the stomach.

The gastroesophageal sphincter prevents reflux of gastric contents.

Except during swallowing, the **gastroesophageal sphincter**, which is smooth muscle in contrast to the upper esophageal sphincter, stays tonically contracted by means of myogenic activity (see p. 291). Contraction also increases during inspiration, reducing the chance of reflux of acidic gastric contents into the esophagus during the time when the subatmospheric intrapleural pressure would favor backward movement of gastric contents. If gastric contents do flow backward despite the sphincter, the acidity of these contents irritates the esophagus, causing the esophageal discomfort known as **heartburn**. (The heart itself is not involved.)

As the peristaltic wave sweeps down the esophagus, the gastroesophageal sphincter relaxes so that the bolus can pass into the stomach (Figure 16-5, step **11**). After the bolus has entered the stomach, the swallow is complete and this lower esophageal sphincter again contracts.

Esophageal secretion is entirely protective.

Esophageal secretion is entirely mucus, which lubricates passage of food, thereby lessening the likelihood of esophageal damage by any sharp edges on food. Also, mucus helps protect the esophagus from damage by acid and enzymes in gastric juice should gastric reflux occur. (In fact, protective mucus is secreted throughout the length of the digestive tract.)

The entire transit time in the pharynx and esophagus averages a mere 6 to 10 seconds, too short a time for any digestion or absorption in this region. We now move on to our next stop, the stomach.

Check Your Understanding 16.3

1. Describe how the pharynx prevents food entry into the trachea during a swallow.

2. State the functions of the pharyngoesophageal and the gastroesophageal sphincters.

16.4 Stomach

The **stomach** is a J-shaped saclike chamber lying between the esophagus and the small intestine. It is divided into three sections based on structural and functional distinctions (Figure 16-6). The **fundus** is the part of the stomach that lies above the esophageal opening. The middle or main part of the stomach is

the **body**. The smooth muscle layers in the fundus and body are relatively thin, but the lower part of the stomach, the **antrum**, has heavier musculature. This difference in muscle thickness plays an important role in gastric motility in these two regions, as you will see shortly. There are also glandular differences in the mucosa of these regions, as described later. The terminal portion of the stomach is the **pyloric sphincter**, which acts as a barrier between the stomach and the upper part of the small intestine, the duodenum.

The stomach stores food and begins protein digestion.

The stomach performs three main functions:

1. The stomach's most important function is to store ingested food until it can be emptied into the small intestine at a rate appropriate for optimal digestion and absorption. It takes hours to digest and absorb a meal that was consumed in only a matter of minutes. Because the small intestine is the primary site for this digestion and absorption, it is important that the stomach store the food and meter it into the duodenum at a rate that does not exceed the small intestine's capacities.

2. The stomach secretes hydrochloric acid (HCl) and enzymes that begin protein digestion.

3. Through the stomach's mixing movements, the ingested food is pulverized and mixed with gastric secretions to produce a thick liquid mixture known as **chyme**. The stomach contents must be converted to chyme before they can be emptied into the duodenum.

Next we discuss how the stomach accomplishes these functions as we examine the four basic digestive processes as they relate

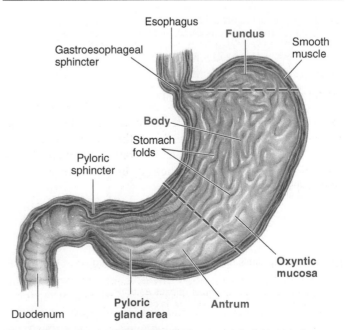

|Figure 16-6 **Anatomy of the stomach.** The stomach is divided into three sections based on structural and functional distinctions—the fundus, body, and antrum. The mucosal lining of the stomach is divided into the oxyntic mucosa and the pyloric gland area based on differences in glandular secretion.

to the stomach, starting with motility. Gastric motility is complex and subject to multiple regulatory inputs. The four aspects of gastric motility are (1) filling, (2) storage, (3) mixing, and (4) emptying.

Gastric filling involves receptive relaxation.

When empty, the stomach has a volume of about 50 mL, but it can expand up to 20-fold to a capacity of about 1 liter (1000 mL) during a meal. Here's how. The interior of the stomach is thrown into deep folds. During a meal, the folds get smaller and nearly flatten out as the stomach relaxes slightly with each mouthful, much like the gradual expansion of a collapsed ice bag as it is being filled. This vagally mediated response, called **receptive relaxation**, allows the stomach to accommodate the meal with little change in intragastric pressure. If more than a liter of food is consumed, however, the stomach becomes overdistended, intragastric pressure rises, and the person experiences discomfort.

Gastric storage takes place in the body of the stomach.

A group of pacemaker cells (interstitial cells of Cajal) located in the upper fundus region of the stomach generate slow-wave potentials that sweep down the length of the stomach toward the pyloric sphincter at a rate of three per minute. This rhythmic pattern of spontaneous depolarizations—the basic electrical rhythm, or BER, of the stomach—occurs continuously and may or may not be accompanied by contraction of the stomach's circular smooth muscle layer. Depending on the level of excitability in the smooth muscle, it may be brought to threshold by this flow of current and undergo action potentials, which in turn initiate peristaltic waves that sweep over the stomach in pace with the BER at a rate of three per minute.

Once initiated, a peristaltic wave spreads over the fundus and body to the antrum and pyloric sphincter. Because the muscle layers are thin in the fundus and body, the peristaltic contractions in this region are weak. When the waves reach the antrum, they become stronger and more vigorous because the muscle there is thicker.

Because only feeble mixing movements occur in the body and fundus, food delivered to the stomach from the esophagus is stored in the relatively quiet body without being mixed. The fundus usually does not store food but contains only a pocket of gas. Food is gradually fed from the body into the antrum, where mixing does take place.

Gastric mixing takes place in the antrum of the stomach.

The strong antral peristaltic contractions mix the food with gastric secretions to produce chyme. Each antral peristaltic wave propels chyme distally toward the pyloric sphincter. Tonic contraction of the pyloric sphincter normally keeps it almost, but not completely, closed. The opening is large enough for water and other fluids to pass through with ease, although particles larger than 2 mm in diameter typically do not leave. As the peristaltic wave reaches the pyloric sphincter and closes it

tightly, the large particles are forced backward toward the body of the stomach (Figure 16-7). The bulk of the antral chyme that is forced backward is again propelled forward and then tumbled back as the next peristaltic wave advances. This churning action is called **retropulsion**, which thoroughly shears and grinds the chyme until the particles are small enough for emptying, mixing the contents in the process.

Gastric emptying is largely controlled by factors in the duodenum.

In addition to mixing gastric contents, the antral peristaltic contractions are the driving force for gastric emptying (Figure 16-7). The amount of chyme that escapes into the duodenum with each peristaltic wave before the pyloric sphincter tightly closes depends largely on the strength of antral peristalsis. The intensity of antral peristalsis and thus the rate of gastric emptying can vary markedly under the influence of various signals from both the stomach and duodenum (Table 16-2). These factors influence the stomach's excitability by slightly depolarizing or hyperpolarizing the gastric smooth muscle. The greater the excitability is, the more frequently the BER generates action potentials, the greater the strength of antral peristalsis, and the faster the rate of gastric emptying.

Factors in the Stomach that Influence the Rate of Gastric Emptying The main gastric factor that influences the strength of contraction is the amount of chyme in the stomach. Other things being equal, the stomach empties at a rate proportional to the volume of chyme in it at any given time. Stomach distension triggers increased gastric motility through a direct effect of stretch on the smooth muscle and through involvement of the intrinsic plexuses, the vagus nerve, and the stomach hormone *gastrin*. (The source, control, and other functions of this hormone will be described later.)

Furthermore, the degree of fluidity of the chyme influences gastric emptying. The stomach contents must be converted into a finely divided, thick liquid form before emptying. The sooner the appropriate degree of fluidity can be achieved, the more rapidly the contents are ready to be evacuated.

Factors in the Duodenum that Influence the Rate of Gastric Emptying Despite these gastric influences, factors in the duodenum are of primary importance in controlling the rate of gastric emptying. The duodenum must be ready to receive the chyme and can delay gastric emptying by reducing the strength of antral peristalsis until the duodenum is ready to accommodate more chyme. The four most important duodenal factors that influence gastric emptying are *fat, acid, hypertonicity,* and *distension*. The presence of one or more of these stimuli in the duodenum activates appropriate duodenal receptors, triggering neural and hormonal responses that put brakes on antral peristaltic activity, thereby slowing the rate of gastric emptying:

■ The *neural response* is mediated through both the intrinsic plexuses (short reflex) and the autonomic nerves (long reflex). Together these constitute the **enterogastric reflex.**

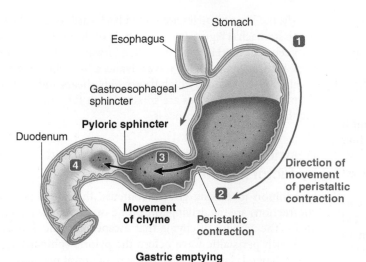

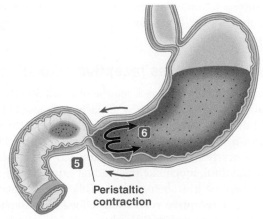

Gastric emptying

1 A peristaltic contraction originates in the upper fundus and sweeps down toward the pyloric sphincter.

2 The contraction becomes more vigorous as it reaches the thick-muscled antrum.

3 The strong antral peristaltic contraction propels the chyme forward.

4 A small portion of chyme is pushed through the partially open sphincter into the duodenum. The stronger the antral contraction, the more chyme is emptied with each contractile wave.

Gastric mixing

5 When the peristaltic contraction reaches the pyloric sphincter, the sphincter is tightly closed and no further emptying takes place.

6 When chyme that was being propelled forward hits the closed sphincter, it is tossed back into the antrum. Mixing of chyme is accomplished as chyme is propelled forward and tossed back into the antrum with each peristaltic contraction, a process called retropulsion.

❙Figure 16-7 **Gastric emptying and mixing as a result of antral peristaltic contractions.**

■ The *hormonal response* involves the release from the small-intestine mucosa into the blood of several hormones collectively known as **enterogastrones**. The blood carries these hormones to the stomach, where they inhibit antral contractions to reduce gastric emptying. The two most important enterogastrones are **secretin** and **cholecystokinin (CCK)**. Secretin was the first hormone discovered (in 1902). Because it was a secretory product that entered the blood, it was termed *secretin*. The name *cholecystokinin* derives from this same hormone also causing contraction of the bile-containing gallbladder (*chole* means "bile";

❙TABLE 16-2 Factors Regulating Gastric Motility and Emptying

Factors	Mode of Regulation	Effects on Gastric Motility and Emptying
Within the Stomach		
Volume of chyme	Distension has a direct effect on gastric smooth muscle excitability, as well as acting through the intrinsic plexuses, the vagus nerve, and gastrin	Increased volume stimulates motility and emptying
Degree of fluidity	Direct effect; contents must be in a fluid form to be evacuated	Increased fluidity allows more rapid emptying
Within the Duodenum		
Presence of fat, acid, hypertonicity, or distension	Initiates the enterogastric reflex or triggers the release of enterogastrones (secretin, cholecystokinin)	These factors in the duodenum inhibit further gastric motility and emptying until the duodenum has coped with factors already present
Outside the Digestive System		
Emotion	Alters autonomic balance	Stimulates or inhibits motility and emptying
Intense pain	Increases sympathetic activity	Inhibits motility and emptying

Pregame Meal: What's In and What's Out?

MANY COACHES AND ATHLETES BELIEVE intensely in special food rituals before a competitive event. For example, a football team may always eat steak before a game. Another may always include bananas in their pregame meal. Do these rituals work?

Many studies have been done to determine the effect of the pregame meal on athletic performance. Although laboratory studies have shown that substances such as caffeine improve endurance, no food substance that greatly enhances performance has been identified. The athlete's prior training is the most important determinant of performance. Even though no particular food confers a special benefit before an athletic contest, some food choices can actually hinder the competitors. For example, a meal of steak is high in fat and could take so long to digest that it might impair the football team's performance and thus should be avoided. However, food rituals that do not impair performance, such as eating bananas, but give the athletes a morale boost or extra confidence, are harmless and should be respected. People may attach special meanings to eating certain foods, and their faith in these practices can make the difference between winning and losing a game.

The greatest benefit of the pregame meal is to prevent hunger during competition. Because the stomach can take from 1 to 4 hours to empty, an athlete should eat at least 3 to 4 hours before competition begins. Excessive quantities of food should not be consumed before competition. Food that remains in the stomach during competition may cause nausea and possibly vomiting. This condition can be aggravated by nervousness, which slows digestion and delays gastric emptying by means of the sympathetic nervous system.

The best choices are foods that are high in carbohydrate and low in fat and protein. The goal is to maintain blood glucose levels and carbohydrate stores in the body and to not have much undigested food in the stomach during the event. High-carbohydrate foods are recommended because they are emptied from the stomach more quickly than fat or protein is. Carbohydrates do not inhibit gastric emptying by means of enterogastrone release, whereas fats and proteins do. Fats in particular delay gastric emptying and are slowly digested. Metabolic processing of proteins yields nitrogenous wastes such as urea whose osmotic activity draws water from the body and increases urine volume, both of which are undesirable during an athletic event. Good choices for a pregame meal include breads, pasta, rice, potatoes, gelatins, and fruit juices. Not only are these complex carbohydrates emptied from the stomach if consumed 1 to 4 hours before a competitive event, but they also help maintain the blood glucose level during the event.

Although it might seem logical to consume something sugary immediately before a competitive event to provide an "energy boost," beverages and foods high in sugar should be avoided because they trigger insulin release. Insulin is the hormone that enhances glucose entry into most body cells. Once the person begins exercising, insulin sensitivity increases (see p. 72), which lowers blood glucose level. A lowered blood glucose level induces feelings of fatigue and an increased use of muscle glycogen stores, which can limit performance in endurance events such as a marathon. Therefore, sugar consumption just before a competition can actually impair performance instead of giving the sought-after energy boost.

Within an hour of competition, it is best for athletes to drink only plain water, to ensure adequate hydration.

---

cysto means "bladder"; and *kinin* means "contraction"). Secretin and CCK are major GI hormones that perform other important functions in addition to serving as enterogastrones.

Let us examine why it is important that each of these stimuli in the duodenum (fat, acid, hypertonicity, and distension) delays gastric emptying:

■ *Fat.* Among the different nutrients that we consume, fat is most effective in delaying gastric emptying. This effect is important because fat digestion and absorption take more time than for the other nutrients and take place only in the small-intestine lumen. Triglycerides strongly stimulate duodenal release of CCK. This hormone inhibits antral contractions and also induces contraction of the pyloric sphincter, which both slow gastric emptying. This delay in emptying ensures that the small intestine has enough time to digest and absorb the fat already there before more fat enters from the stomach. That fat is the most potent inhibitor of gastric emptying is evident when you compare the rate of emptying of a high-fat meal (after 6 hours, some of a bacon-and-eggs meal may still be in the stomach) with that of a protein and carbohydrate meal (a meal of lean meat and potatoes may empty in 3 hours). (For a discussion of the pregame meal before participation in an athletic event, see the accompanying boxed feature, ▮A Closer Look at Exercise Physiology.)

■ *Acid.* Because the stomach secretes HCl, highly acidic chyme empties into the duodenum, where it is neutralized by sodium bicarbonate ($NaHCO_3$) secreted into the duodenum primarily from the pancreas. Unneutralized acid may damage the duodenal mucosa and inactivate the pancreatic digestive enzymes secreted into the duodenum. Appropriately, unneutralized acid in the duodenum induces the release of secretin, a hormone that slows emptying of acidic gastric contents until complete neutralization can be accomplished.

■ *Hypertonicity.* As molecules of protein and starch are digested in the duodenum, large numbers of amino acid and glucose molecules are released. If absorption of these amino acid and glucose molecules does not keep pace with the rate at which protein and carbohydrate digestion proceeds, these large numbers of molecules remain in the chyme and increase the osmolarity of the duodenal contents. Osmolarity depends on the number of molecules present, not on their size, and one protein molecule may be split into several hundred amino acid molecules, each of which has the same osmotic activity as the original protein molecule. The same holds true for one large starch molecule, which yields many smaller but equally osmotically active glucose molecules. Because water is freely diffusible across the duodenal wall, it enters the duodenal lumen from the plasma as the duodenal osmolarity rises. Large volumes of water entering the intestine from the plasma lead to intestinal distension, and, more important, circulatory disturbances result because of the reduction in plasma volume. To prevent these effects, gastric emptying is reflexly inhibited when the osmolarity of the duodenal contents starts to rise. Thus, the amount of food entering the duodenum for further digestion into a multitude of additional osmotically active particles is reduced until absorption processes have had an opportunity to catch up.

■ *Distension.* Too much chyme in the duodenum inhibits the emptying of even more gastric contents, giving the distended duodenum time to cope with the excess volume of chyme it already contains before it gets any more.

Emotions can influence gastric motility.

Other factors unrelated to digestion, such as emotions, can alter gastric motility by acting through the autonomic nerves to influence the degree of gastric smooth muscle excitability. Even though the effect of emotions on gastric motility varies among people and is not always predictable, sadness and fear generally tend to decrease motility, whereas anger and aggression tend to increase it. In addition to emotional influences, intense pain from any part of the body tends to inhibit motility throughout the digestive tract. This response is brought about by increased sympathetic activity.

The stomach does not actively participate in vomiting.

Clinical Note **Vomiting**, or **emesis**, the forceful expulsion of gastric contents out through the mouth, is not accomplished by reverse peristalsis in the stomach, as might be predicted. Actually, the stomach itself does not actively participate in vomiting. The stomach, the esophagus, and associated sphincters are all relaxed during vomiting. The major force for expulsion comes, surprisingly, from contraction of the respiratory muscles—namely, the diaphragm (the major inspiratory muscle) and the abdominal muscles (the muscles of active expiration) (see ▮Figure 13-11, p. 454).

The complex act of vomiting is coordinated by a **vomiting center** in the medulla of the brain stem. Vomiting is usually preceded by profuse salivation, sweating, rapid heart rate, and sensation of nausea. Vomiting begins with a deep inspiration and closure of the glottis. The contracting diaphragm descends downward on the stomach while simultaneous contraction of the abdominal muscles compresses the abdominal cavity, increasing

the intra-abdominal pressure and forcing the abdominal viscera upward. As the flaccid stomach is squeezed between the diaphragm from above and the compressed abdominal cavity from below, the gastric contents are forced upward through the relaxed sphincters and esophagus and out through the mouth. The glottis is closed, so vomited material does not enter the trachea. Also, the uvula is raised to close off the nasal cavity. The vomiting cycle may be repeated several times until the stomach is emptied.

Causes of Vomiting Vomiting can be initiated by afferent input to the vomiting center from various receptors throughout the body, including the following:

■ Tactile (touch) stimulation of the back of the throat, which is one of the most potent stimuli. For example, sticking a finger in the back of the throat or even the presence of a tongue depressor or dental instrument in the back of the mouth can trigger gagging and even vomiting in some people.

■ Irritation or distension of the stomach and duodenum.

■ Elevated intracranial pressure, such as that caused by cerebral hemorrhage. Thus, vomiting after a head injury is considered a bad sign; it suggests swelling or bleeding within the cranial cavity.

■ Rotation or acceleration of the head producing dizziness, such as in motion sickness.

■ Chemical agents, including drugs or noxious substances that initiate vomiting (that is, **emetics**) either by acting in the upper parts of the GI tract or by stimulating chemoreceptors in a specialized **chemoreceptor trigger zone** next to the vomiting center in the brain. For example, chemotherapeutic agents used in treating cancer often cause vomiting by acting on the chemoreceptor trigger zone.

■ Psychogenic vomiting induced by emotional factors, including those accompanying nauseating sights and odors and anxiety, as before taking an examination.

Effects of Vomiting With excessive vomiting, the body experiences large losses of secreted fluids and acids that normally would be reabsorbed. The resulting reduction in plasma volume can lead to dehydration and circulatory problems, and the loss of acid from the stomach can lead to metabolic alkalosis (see p. 561).

We have completed our discussion of gastric motility and now shift to gastric secretion.

Gastric digestive juice is secreted by glands located at the base of gastric pits.

Each day, the stomach secretes about 2 liters of gastric juice. The cells that secrete gastric juice are located in the gastric mucosa, which is divided into two distinct areas: (1) the **oxyntic mucosa**, which lines the body and fundus, and (2) the **pyloric gland area (PGA)**, which lines the antrum. The luminal surface of the stomach is pitted with deep pockets formed by infoldings of the gastric mucosa. The first parts of these invaginations are called **gastric pits**, at the base of which lie the **gastric glands**. A variety of secretory cells line these invaginations, some exocrine and some endocrine or paracrine (▮Table 16-3). Let us look at the gastric exocrine secretory cells first.

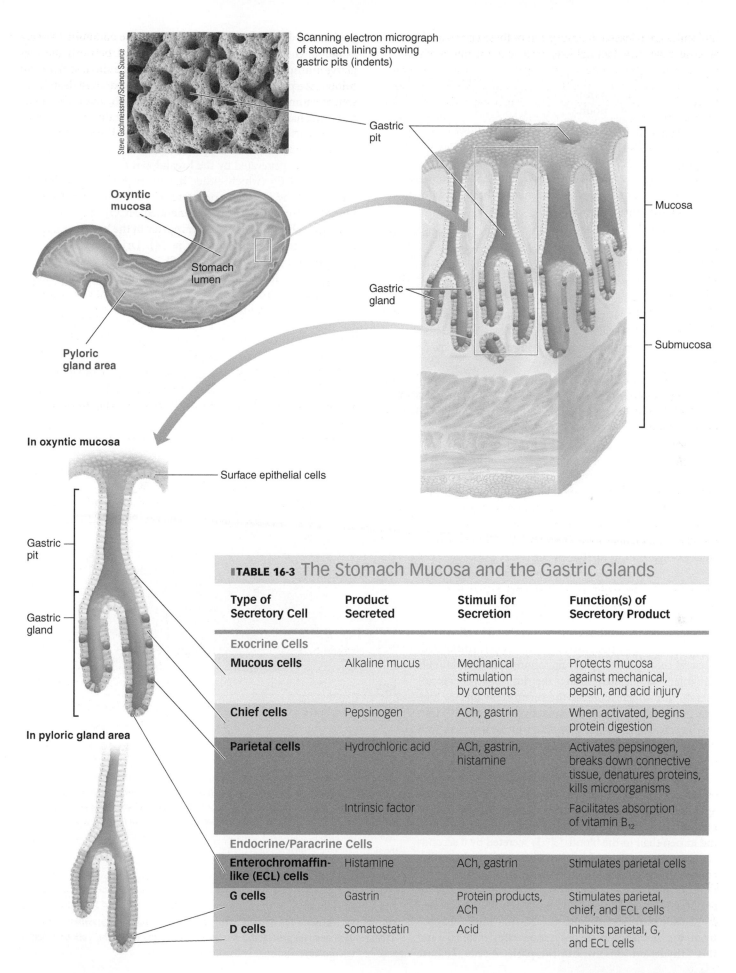

Scanning electron micrograph of stomach lining showing gastric pits (indents)

Steve Gschmeissner/Science Source

Gastric pit

Mucosa

Submucosa

Oxyntic mucosa

Stomach lumen

Gastric gland

Pyloric gland area

In oxyntic mucosa

Surface epithelial cells

Gastric pit

Gastric gland

In pyloric gland area

■TABLE 16-3 The Stomach Mucosa and the Gastric Glands

Type of Secretory Cell	Product Secreted	Stimuli for Secretion	Function(s) of Secretory Product
Exocrine Cells			
Mucous cells	Alkaline mucus	Mechanical stimulation by contents	Protects mucosa against mechanical, pepsin, and acid injury
Chief cells	Pepsinogen	ACh, gastrin	When activated, begins protein digestion
Parietal cells	Hydrochloric acid	ACh, gastrin, histamine	Activates pepsinogen, breaks down connective tissue, denatures proteins, kills microorganisms
	Intrinsic factor		Facilitates absorption of vitamin B_{12}
Endocrine/Paracrine Cells			
Enterochromaffin-like (ECL) cells	Histamine	ACh, gastrin	Stimulates parietal cells
G cells	Gastrin	Protein products, ACh	Stimulates parietal, chief, and ECL cells
D cells	Somatostatin	Acid	Inhibits parietal, G, and ECL cells

Three types of gastric exocrine secretory cells are found in the walls of the pits and glands in the oxyntic mucosa:

- **Mucous cells** line the gastric pits and the entrance of the glands. They secrete a thin, watery *mucus*. (*Mucous* is the adjective; *mucus* is the noun.)
- The deeper parts of the gastric glands are lined by chief and parietal cells. The more numerous **chief cells** secrete the enzyme precursor *pepsinogen*.
- The **parietal** (or **oxyntic**) **cells** secrete *HCl* and *intrinsic factor* (*oxyntic* means "sharp," a reference to these cells' potent HCl secretory product).

These exocrine secretions are all released into the gastric lumen. Collectively, they make up the gastric digestive juice.

In contrast to the oxyntic mucosa, the gastric glands of the PGA primarily secrete mucus and a small amount of pepsinogen; no acid is secreted in this area.

Between the gastric pits, the gastric mucosa is covered by **surface epithelial cells**, which secrete a sticky, alkaline mucus that forms a visible layer several millimeters thick over the surface of the mucosa.

A few **stem cells** are also found in the gastric pits. These cells rapidly divide and are the parent cells of all new cells of the gastric mucosa. The daughter cells that result from cell division either migrate out of the pit to become surface epithelial cells or migrate deeper to the gastric glands where they differentiate into chief or parietal cells. Through this activity, the entire stomach mucosa is replaced about every 3 days. This frequent turnover is important because the harsh acidic stomach contents expose the mucosal cells to lots of wear and tear.

Let us consider these exocrine products and their roles in digestion in further detail.

Hydrochloric acid is secreted by parietal cells and activates pepsinogen.

Parietal cells are scattered among the chief cells in the epithelial lining of the gastric glands. When stimulated, parietal cells form deep invaginations called **canaliculi** (singular *canaliculus*) along the luminal (or apical) membrane, which increase the membrane surface area bearing transport proteins that actively secrete HCl into the lumen of the gastric pits. Each pit drains acid into the lumen of the stomach, which can cause the luminal pH to fall as low as 2.

Hydrogen ion (H^+) and chloride ion (Cl^-) are actively secreted by separate pumps. H^+ is actively transported against a tremendous concentration gradient, with the H^+ concentration being as much as 3 million times greater in the lumen than in the blood. Cl^- is secreted by a secondary active-transport mechanism against a much smaller concentration gradient of only 1.5 times.

Mechanism of H^+ and Cl^- Secretion The secreted H^+ is not transported from the plasma but is derived from the breakdown of H_2O molecules into H^+ and OH^- (hydroxyl ions) within the parietal cells (❙Figure

16-8). This H^+ is secreted into the lumen by a H^+–K^+ ATPase pump in the parietal cell's luminal membrane. This primary active-transport carrier also pumps K^+ into the cell from the lumen (see p. 73). The transported K^+ then passively leaks back into the lumen through luminal K^+ channels, thus leaving K^+ levels unchanged by the process of H^+ secretion.

The parietal cells contain an abundance of the enzyme carbonic anhydrase (ca) (see pp. 477 and 550). In the presence of ca, the OH^- generated by the breakdown of H_2O readily combines with CO_2 (which either has been produced within the parietal cell by metabolic processes or has diffused in from the blood) to form HCO_3^-. The generated HCO_3^- is moved into the plasma by a Cl^-–HCO_3^- antiporter in the basolateral membrane of the parietal cells (see p. 74). Driven by the HCO_3^- gradient, this carrier moves HCO_3^- out of the cell into the plasma down its electrochemical gradient and simultaneously transports Cl^- from the plasma into the parietal cell against its electrochemical gradient. By building up the concentration of Cl^- inside the parietal cell, the Cl^-–HCO_3^- antiporter establishes a Cl^- concentration gradient between the parietal cell and gastric lumen. Because of this concentration gradient and because the cell interior is negative compared to the luminal contents, the negatively charged Cl^- pumped into the cell by the basolateral antiporter diffuses out of the cell down its elec-

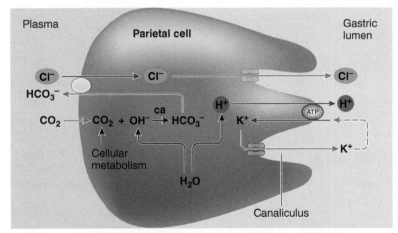

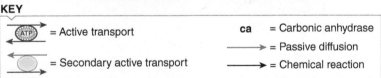

KEY

❙(ATP)❙ = Active transport	**ca** = Carbonic anhydrase
	⟶ = Passive diffusion
⬭ = Secondary active transport	⟶ = Chemical reaction

❙Figure 16-8 **Mechanism of HCl secretion.** The stomach's parietal cells actively secrete H^+ and Cl^- by the actions of two separate pumps. H^+ is secreted into the lumen by a primary H^+–K^+ ATPase active-transport pump at the parietal cell's luminal border. The H^+ that is secreted, as well as HCO_3^-, is formed within the parietal cell from H_2O and CO_2 in a reaction catalyzed by carbonic anhydrase. Cl^- is secreted by secondary active transport. Driven by the HCO_3^- concentration gradient, a Cl^-–HCO_3^- antiporter in the basolateral membrane transports HCO_3^- down its concentration gradient into the plasma and simultaneously transports Cl^- into the parietal cell against its concentration gradient. Cl^- secretion is completed as the Cl^- that entered from the plasma diffuses out of the cell down its electrochemical gradient through a luminal Cl^- channel into the lumen.

FIGURE FOCUS: *What effect does a drug that blocks the parietal cells' H^+–K^+ ATPase pump have on gastric HCl secretion and on the pH of the venous blood leaving the stomach?*

trochemical gradient through Cl^- channels in the luminal membrane into the gastric lumen, completing the Cl^- secretory process. In the meantime, the blood leaving the stomach is alkaline because HCO_3^- has been added to it.

Functions of HCl Although HCl does not actually digest anything (that is, it does not break apart nutrient chemical bonds), it performs these specific functions that aid digestion:

1. HCl activates the enzyme precursor pepsinogen to an active enzyme, *pepsin,* and provides an acid environment optimal for pepsin action.

2. It aids in the breakdown of connective tissue and muscle fibers, reducing large food particles into smaller particles.

3. It denatures protein—that is, it uncoils proteins from their highly folded final form, thus exposing more of the peptide bonds for enzymatic attack (see p. A-14).

4. Along with salivary lysozyme, HCl kills most of the microorganisms ingested with food, although some escape and then grow and multiply in the large intestine.

Pepsinogen is activated to pepsin, which begins protein digestion.

The major digestive constituent of gastric secretion is **pepsinogen**, an inactive enzymatic molecule produced by the chief cells. Pepsinogen, once activated to the enzyme **pepsin**, begins protein digestion. Pepsinogen is stored in the chief cells' cytoplasm within secretory vesicles known as **zymogen granules**, from which it is released by exocytosis on appropriate stimulation (see p. 29). When pepsinogen is secreted into the gastric lumen, HCl cleaves off a small fragment of the molecule, converting it to the active form of pepsin. Once formed, pepsin acts on other pepsinogen molecules to produce more pepsin, a mechanism called an **autocatalytic process** (*autocatalytic* means "self-activating").

Pepsin initiates protein digestion by splitting certain amino acid linkages in proteins to yield peptide fragments (small amino acid chains); it works most effectively in the acid environment provided by HCl. Because pepsin can digest protein, it must be stored and secreted in an inactive form so that it does not digest the proteins of the cells in which it is formed. Therefore, pepsin is maintained in the inactive form of pepsinogen until it reaches the gastric lumen, where it is activated by HCl secreted into the lumen by a different cell type.

Mucus is protective.

The surface of the gastric mucosa is covered by a layer of mucus derived from the surface epithelial cells and mucous cells. This mucus is a protective barrier against several forms of potential injury to the gastric mucosa:

■ Through its lubricating properties, mucus protects the gastric mucosa against mechanical injury.

■ It helps protect the stomach wall from self-digestion because pepsin is inhibited when it comes in contact with the layer of mucus coating the stomach lining. (However, mucus does not affect pepsin activity in the lumen, where digestion of dietary protein proceeds without interference.)

■ Being alkaline, mucus helps protect against acid injury by neutralizing HCl in the vicinity of the gastric lining, but it does not interfere with the function of HCl in the lumen. Whereas the pH in the lumen may be as low as 2, the pH in the layer of mucus adjacent to the mucosal cell surface is about 7.

Intrinsic factor is essential for absorption of vitamin B_{12}.

Intrinsic factor, another secretory product of the parietal cells in addition to HCl, is necessary for absorption of vitamin B_{12}. Binding of intrinsic factor with vitamin B_{12} triggers receptor-mediated endocytosis of this complex in the terminal ileum, the last part of the small intestine (see p. 31). Vitamin B_{12} is essential for normal formation of red blood cells.

 Clinical Note In the absence of intrinsic factor, vitamin B_{12} is not absorbed, so erythrocyte production is defective and *pernicious anemia* results (see p. 386). Pernicious anemia is typically caused by an autoimmune attack against the parietal cells (see p. 432).

Multiple regulatory pathways influence the parietal and chief cells.

In addition to the gastric exocrine secretory cells, other secretory cells in the gastric glands release endocrine and paracrine regulatory factors instead of products involved in the digestion of nutrients in the gastric lumen (see p. 114). These other secretory cells are also shown in ❙Table 16-3:

■ Endocrine cells known as **G cells,** found in the gastric pits only in the PGA, secrete the hormone *gastrin* into the blood.

■ **Enterochromaffin-like (ECL) cells,** dispersed among the parietal and chief cells in the gastric glands of the oxyntic mucosa, secrete the paracrine *histamine.*

■ **D cells,** which are scattered in gastric glands near the pylorus but are more numerous in the duodenum, secrete the paracrine *somatostatin.*

These three regulatory factors from the gastric pits, along with the neurotransmitter *ACh,* primarily control secretion of gastric digestive juices. Parietal cells have separate receptors for each of these chemical messengers. Three of them—ACh, gastrin, and histamine—stimulate HCl secretion. The fourth regulatory agent—somatostatin—inhibits HCl secretion. ACh and gastrin also increase pepsinogen secretion through their stimulatory effect on the chief cells. We now consider each of these chemical messengers in further detail (❙Table 16-3).

■ **ACh** is a neurotransmitter released from the intrinsic nerve plexuses in response to both short local reflexes and vagal stimulation. ACh stimulates not only the parietal and chief cells but also the G cells and ECL cells.

■ The G cells secrete the hormone **gastrin** into the blood in response to protein products in the stomach lumen and in response to ACh. Like secretin and CCK, gastrin is a major GI hormone. After being carried by the blood back to the stomach mucosa, gastrin stimulates the parietal and chief cells, promoting secretion of a highly acidic gastric juice. In addition to directly stimulating the parietal cells, gastrin indirectly promotes HCl secretion by stimulating the ECL cells to release histamine, which also stimulates the parietal cells. Gastrin is the main factor that brings about increased HCl secretion during meal digestion. Gastrin is also *trophic* (growth promoting) to the mucosa of the stomach and small intestine, thereby maintaining their secretory capabilities.

■ **Histamine**, a paracrine, is released from the ECL cells in response to ACh and gastrin. Histamine acts locally on nearby parietal cells to speed up HCl secretion and potentiates (makes stronger) the actions of ACh and gastrin.

ACh and gastrin both operate through IP_3/Ca^{2+} second-messenger pathways; histamine activates a cAMP second-messenger pathway to bring about its effects (see p. 123). These messengers all bring about increased secretion of HCl by promoting the insertion of additional $H^+–K^+$ ATPases into the parietal cells' plasma membrane. A pool of these pumps is stored within the parietal cell in intracellular vesicles, which fuse with the luminal membrane via exocytosis, forming the deep inward folding canaliculi as the surface membrane expands in the process of adding more of these active carriers to the membrane as needed to increase HCl secretion.

■ **Somatostatin** is released from the D cells in response to acid. It acts locally as a paracrine in negative-feedback fashion to inhibit secretion by the parietal cells, G cells, and ECL cells, thus turning off the HCl-secreting cells and their most potent stimulatory pathway.

From this list, it is obvious not only that multiple chemical messengers influence the parietal and chief cells but also that these chemicals influence one another. As we examine the phases of gastric secretion, you will see under what circumstances each of these regulatory agents is released.

Control of gastric secretion involves three phases.

The rate of gastric secretion can be influenced by (1) factors arising before food reaches the stomach, (2) factors resulting from food in the stomach, and (3) factors in the duodenum after food has left the stomach. Accordingly, gastric secretion is divided into three phases: cephalic, gastric, and intestinal.

Cephalic Phase The cephalic phase of gastric secretion refers to increased secretion of HCl and pepsinogen that occurs in feedforward fashion in response to stimuli acting in the head even before food reaches the stomach (*cephalic* means "head"). Thinking about, seeing, smelling, tasting, chewing, and swallowing food increases gastric secretion via vagal stimulation in two ways. First, vagal stimulation of the intrinsic plexuses promotes increased secretion of ACh, which leads to increased secretion of HCl and pepsinogen by the secretory cells. Second, direct vagal stimulation of the G cells induces gastrin release, which in turn further enhances secretion of HCl and pepsinogen, with the effect on HCl being potentiated by gastrin promoting the release of histamine (▮Table 16-4).

Gastric Phase The gastric phase of gastric secretion begins when food reaches the stomach. Stimuli acting on the stomach— namely *protein, distension, caffeine,* and *alcohol*—increase gastric secretion by overlapping efferent pathways. For example, the most potent stimulus, protein (especially short peptide fragments) in the stomach lumen, stimulates chemoreceptors that activate intrinsic-plexus pathways that induce gastric secretion. Furthermore, protein brings about activation of the extrinsic vagal fibers to the stomach. Vagal activity further enhances intrinsic nerve stimulation of the secretory cells and triggers the release of gastrin. Last, protein also directly stimulates the release of gastrin. Gastrin, in turn, is a powerful stimulus for further HCl and pepsinogen secretion. Through these synergistic and overlapping pathways, protein induces secretion of a highly acidic, pepsin-rich gastric juice, which continues digestion of the protein that first initiated the process (▮Table 16-4).

When the stomach is distended with protein-rich food that needs to be digested, these secretory responses are appropriate. Caffeine and, to a lesser extent, alcohol also stimulate secretion of a highly acidic gastric juice, even when no food is present. This unnecessary acid can irritate the linings of the stomach and duodenum. For this reason, people with ulcers or gastric hyperacidity should avoid caffeinated and alcoholic beverages.

▮**TABLE 16-4** Stimulation of Gastric Secretion

Phase	Stimuli	Excitatory Mechanism for Enhancing Gastric Secretion
Cephalic phase of gastric secretion	Stimuli in the head— seeing, smelling, tasting, chewing, swallowing food	+ Vagus → + Intrinsic nerves → ↑ACh → + Chief and parietal cells → ↑Gastric secretion; + G cells → ↑Gastrin; + ECL cells → ↑Histamine
Gastric phase of gastric secretion	Stimuli in the stomach— protein, (peptide fragments), distension, caffeine, alcohol	+ Vagus → + Intrinsic nerves → ↑ACh → + Chief and parietal cells → ↑Gastric secretion; + G cells → ↑Gastrin; + ECL cells → ↑Histamine

Intestinal Phase The intestinal phase of gastric secretion encompasses the factors originating in the small intestine that influence gastric secretion. Whereas the other phases are excitatory, this phase is inhibitory. The intestinal phase is important in helping shut off the flow of gastric juices as chyme begins to be emptied into the small intestine, a topic to which we now turn.

Gastric secretion gradually decreases as food empties from the stomach into the intestine.

You now know what factors turn on gastric secretion before and during a meal, but how is the flow of gastric juices shut off when they are no longer needed? Gastric secretion is gradually reduced in three ways as the stomach empties (Table 16-5):

1. As the meal gradually empties into the duodenum, the major stimulus for enhanced gastric secretion—protein in the stomach—is withdrawn.

2. After foods leave the stomach, gastric juices accumulate to such an extent that gastric pH falls very low. This fall in pH within the stomach lumen comes about largely because food proteins that had been buffering HCl are no longer present in the lumen as the stomach empties. (Recall that proteins serve as excellent buffers; see p. 552.) Somatostatin is released in response to this high gastric acidity (pH less than 3). As a result of somatostatin's inhibitory effects, gastric secretion declines.

3. The same stimuli that inhibit gastric motility (fat, acid, hypertonicity, or distension in the duodenum brought about by emptying of stomach contents into the duodenum) inhibit gastric secretion also. The enterogastric reflex and the enterogastrones suppress the gastric secretory cells while they simultaneously reduce the strength of antral peristalsis. This inhibitory response is the intestinal phase of gastric secretion.

The gastric mucosal barrier protects the stomach lining from gastric secretions.

How can the stomach contain strong acid contents and proteolytic enzymes without destroying itself? In addition to mucus providing a protective physical coating, the surface mucous cells secrete HCO_3^- that is trapped in the mucus and neutralizes acid in the vicinity.

Other barriers to mucosal acid damage are provided by the mucosal lining itself. First, the luminal membranes of the gastric mucosal cells are essentially impermeable to H^+, so acid cannot penetrate into the cells and damage them. Second, the lateral edges of these cells are joined near their luminal borders by tight junctions, so acid cannot diffuse between the cells from the lumen into the underlying submucosa (see p. 61). The properties of the gastric mucosa that enable the stomach to contain acid without injuring itself constitute the **gastric mucosal barrier** (Figure 16-9). These protective mechanisms are further enhanced by replacement of the entire stomach lining every 3 days. Because of rapid mucosal turnover, cells are usually replaced before they are exposed to the wear and tear of harsh gastric conditions long enough to suffer damage.

Clinical Note Despite the protection provided by mucus, by the gastric mucosal barrier, and by frequent turnover of cells, the barrier occasionally is broken and the gastric wall is injured by its acidic and enzymatic contents, producing an erosion, or **peptic ulcer**, of the stomach wall. Excessive gastric reflux into the esophagus and dumping of excessive acidic gastric contents into the duodenum can lead to peptic ulcers in these locations also. (For a further discussion of ulcers, see the boxed feature on p. 589, Concepts, Challenges, and Controversies.)

We now turn to the remaining two digestive processes in the stomach: gastric digestion and absorption.

TABLE 16-5 Inhibition of Gastric Secretion

Region	Stimuli	Inhibitory Mechanism for Gastric Secretion
Body and antrum	Removal of protein and distension as the stomach empties	→ Intrinsic nerves (−) → Vagus (−) → G cells → ↓Gastrin → ↓Histamine ⟹ ↓Gastric secretion
Antrum and duodenum	Accumulation of acid	(+) D cells → ↑Somatostatin → Parietal cells (−), G cells (−), ECL cells → ↓Gastric secretion
Duodenum (intestinal phase of gastric secretion)	Fat Acid Hypertonicity Distension	(+) Enterogastric reflex ↑Enterogastrones (cholecystokinin and secretin) → Parietal cells (−), Chief cells (−), Smooth muscle cells → ↓Gastric secretion and motility

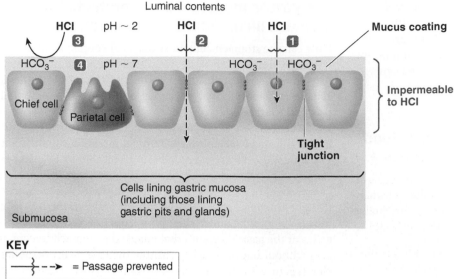

Luminal contents

HCl pH ~ 2 **HCl** **HCl** Mucus coating

HCO₃⁻ **4** pH ~ 7 HCO₃⁻ HCO₃⁻

Chief cell

Parietal cell

Impermeable to HCl

Tight junction

Cells lining gastric mucosa (including those lining gastric pits and glands)

Submucosa

The components of the gastric mucosal barrier enable the stomach to contain acid without injuring itself:

1 The luminal membranes of the gastric mucosal cells are impermeable to H⁺ so that HCl cannot penetrate into the cells.

2 The cells are joined by tight junctions that prevent HCl from penetrating between them.

3 A mucus coating over the gastric mucosa serves as a physical barrier to acid penetration.

4 The HCO₃⁻-rich mucus also serves as a chemical barrier that neutralizes acid in the vicinity of the mucosa. Even when luminal pH is 2, the mucus pH is 7.

KEY

—|‑ - ‑► = Passage prevented

I Figure 16-9 **Gastric mucosal barrier.**

Carbohydrate digestion continues in the body of the stomach; protein digestion begins in the antrum.

Two separate digestive processes take place within the stomach. In the body of the stomach, food remains in a semisolid mass because peristaltic contractions in this region are too weak for mixing to occur. Because food is not mixed with gastric secretions, little protein digestion occurs here. In the interior of the mass, however, carbohydrate digestion continues under the influence of salivary amylase. Even though acid inactivates salivary amylase, the unmixed interior of the food mass is free of acid.

Chemical digestion by the gastric juice itself occurs in the antrum of the stomach, where the food is thoroughly mixed with pepsin and HCl via retropulsion.

The stomach absorbs alcohol and aspirin but no food.

No food or water is absorbed into the blood through the stomach mucosa. However, two noteworthy nonnutrient substances are absorbed directly by the stomach—*ethyl alcohol* and *aspirin*. Alcohol is somewhat lipid soluble, so it can diffuse through the lipid membranes of the epithelial cells that line the stomach and can enter the blood through the submucosal capillaries. However, it can be absorbed even more rapidly by the small-intestine mucosa because of the greater surface area of the small-intestine mucosa. Thus, alcohol absorption occurs more slowly if gastric emptying is delayed so that the alcohol remains in the more slowly absorbing stomach longer. Because fat is the most potent duodenal stimulus for inhibiting gastric motility, consuming fat-rich foods (for example, whole milk, pizza, or nuts) before or during alcohol ingestion delays gastric emptying and prevents the alcohol from producing its effects as rapidly.

Clinical Note Another category of substances absorbed by the gastric mucosa includes weak acids, most notably *acetylsalicylic acid* (aspirin). In the highly acidic environment of the stomach lumen, weak acids are almost totally undissociated—that is, the H⁺ and associated anion of the acid are bound together. In an undissociated, or intact, form, these weak acids are lipid soluble, so they can be absorbed quickly by crossing the plasma membranes of the epithelial cells that line the stomach. Most other drugs are not absorbed until they reach the small intestine, so they do not begin to take effect as quickly.

Having completed our coverage of the stomach, we move to the next part of the digestive tract, the small intestine and the accessory digestive organs that release their secretions into the small-intestine lumen.

Check Your Understanding 16.4

1. Describe the process of retropulsion and explain what it accomplishes.

2. Name the secretions of these gastric cells: mucous cells, chief cells, parietal cells, ECL cells, G cells, and D cells.

3. Discuss how food-related stimuli induce gastric secretions during the cephalic phase of gastric secretion.

4. Describe the mechanisms that protect the gastric mucosa from acid damage.

16.5 Pancreatic and Biliary Secretions

When gastric contents empty into the duodenum, they are mixed not only with juice secreted by the small-intestine mucosa, but also with the secretions of the exocrine pancreas

Ulcers: When Bugs Break the Barrier

PEPTIC ULCERS ARE EROSIONS THAT typically begin in the mucosal lining of the stomach and may penetrate into the deeper layers of the stomach wall. They occur when the gastric mucosal barrier is disrupted, and thus pepsin and HCl act on the stomach wall as well as on food in the lumen. Frequent backflow of acidic gastric juices into the esophagus or excess unneutralized acid from the stomach in the duodenum can lead to peptic ulcers in these sites also.

In a surprising discovery in the early 1990s, the bacterium *Helicobacter pylori* was pinpointed as the cause of more than 80% of all peptic ulcers. Thirty percent of the population in the United States harbor *H. pylori*. Those who have this slow bacterium have a 3 to 12 times greater risk of developing an ulcer within 10 to 20 years of acquiring the infection than those without the bacterium. They are also at increased risk of developing stomach cancer.

For years, scientists had overlooked the possibility that ulcers could be triggered by an infectious agent because bacteria typically cannot survive in a strongly acidic environment such as the stomach lumen. An exception, *H. pylori* exploits several strategies to survive in this hostile environment. First, these organisms are motile, being equipped with four to six flagella (whiplike appendages; see the accompanying figure), which enable them to tunnel through and take up residence under the stomach's thick layer of alkaline mucus. Here they are protected from the highly acidic gastric contents. Furthermore, *H. pylori* preferentially settles in the antrum, which has no acid-producing parietal cells, although HCl from the upper parts of the stomach does reach the antrum. Also, these bacteria produce *urease,* an enzyme that breaks down urea, an end product of protein metabolism, into ammonia (NH_3) and CO_2. Ammonia serves as a buffer (see p. 558) that neutralizes stomach acid locally in the vicinity of the *H. pylori.*

H. pylori contributes to ulcer formation in part by secreting toxins that cause a persistent inflammation, or *chronic superficial gastritis,* at the site it colonizes. *H. pylori* further weakens the gastric mucosal barrier by disrupting the tight junctions between the gastric epithelial cells, thereby making the gastric mucosa leakier than normal.

Alone or in conjunction with this infectious culprit, other factors are known to contribute to ulcer formation. Frequent exposure to some chemicals can break the gastric mucosal barrier; the most important

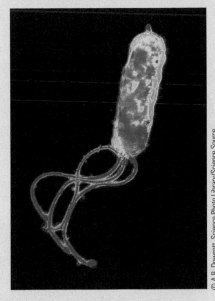

Helicobacter pylori. *Helicobacter pylori,* the bacterium responsible for most cases of peptic ulcers, has flagella that enable it to tunnel beneath the protective layer of mucus that coats the stomach lining.

of these are ethyl alcohol and nonsteroidal anti-inflammatory drugs (NSAIDs), such as aspirin, ibuprofen, or more potent medications for the treatment of arthritis or other chronic inflammatory processes. The barrier frequently breaks in patients with preexisting debilitating conditions, such as severe injuries or infections. Persistent stressful situations are frequently associated with ulcer formation, presumably because emotional response to stress can stimulate excessive gastric secretion.

When the gastric mucosal barrier is broken, acid and pepsin diffuse into the mucosa and underlying submucosa, with serious pathophysiologic consequences. The surface erosion, or ulcer, progressively enlarges as increasing levels of acid and pepsin continue to damage the stomach wall. Two of the most serious consequences of ulcers are (1) hemorrhage resulting from damage to submucosal capillaries and (2) perforation, or complete erosion through the stomach wall, resulting in the escape of potent gastric contents into the abdominal cavity.

Treatment of ulcers includes antibiotics, H-2 histamine receptor blockers, and proton pump inhibitors. With the discovery of the infectious component of most ulcers, antibiotics are now a treatment of choice. The other drugs are also used alone or in combination with antibiotics.

Two decades before the discovery of *H. pylori,* researchers discovered an antihistamine (*cimetidine*) that specifically blocks H-2 receptors, the type of receptors that bind histamine released from the stomach. These receptors differ from H-1 receptors, which bind the histamine involved in allergic respiratory disorders. Accordingly, traditional antihistamines used for respiratory allergies (such as hay fever and asthma) are not effective against ulcers, nor is cimetidine useful for respiratory problems. Because histamine potentiates the acid-promoting actions of ACh and gastrin, treatment with H-2 histamine blockers significantly suppresses acid secretion despite the fact that they do not directly interfere with the actions of these other two stimulatory messengers.

Another newer class of drugs used in treating ulcers inhibits acid secretion by directly blocking the pump that transports H^+ into the stomach lumen. These so-called *proton-pump inhibitors* (H^+ is a naked proton without its electron), for example *Prilosec,* help reduce the corrosive effect of HCl on the exposed tissue.

and liver that are released into the duodenum. We discuss the roles of each of these accessory digestive organs before we examine the contributions of the small intestine itself.

The pancreas is a mixture of exocrine and endocrine tissue.

The **pancreas** is an elongated gland that lies behind and below the stomach, above the first loop of the duodenum (Figure 16-10). This mixed gland contains both exocrine and endocrine tissue. The predominant exocrine part consists of grape-like clusters of secretory cells that form sacs known as **acini**, which connect to ducts that eventually empty into the duodenum. The smaller endocrine part consists of isolated islands of endocrine tissue, the **islets of Langerhans**, which are dispersed throughout the pancreas. The most important hormones secreted by the islet cells are insulin and glucagon (Chapter 19). The exocrine and endocrine tissues of the pancreas are derived from different tissues during embryonic development and share only their location in common. Although both are involved with the metabolism of nutrient molecules, they have different functions under the control of different regulatory mechanisms.

The exocrine pancreas secretes digestive enzymes and an alkaline fluid.

The **exocrine pancreas** secretes a pancreatic juice consisting of two components: (1) *pancreatic enzymes* actively secreted by the *acinar cells* that form the acini and (2) an *aqueous alkaline solution* actively secreted by the *duct cells* that line the pancreatic ducts. The aqueous (watery) alkaline component is rich in sodium bicarbonate ($NaHCO_3$).

Like pepsinogen, pancreatic enzymes are stored within zymogen granules (secretory vesicles) after being produced and then are released by exocytosis as needed. The acinar cells secrete three types of pancreatic enzymes capable of digesting all three categories of foodstuffs: (1) *proteolytic enzymes* for protein digestion, (2) *pancreatic amylase* for carbohydrate digestion, and (3) *pancreatic lipase* for fat digestion. Pancreatic enzymes can almost completely digest food in the absence of all other digestive secretions.

Pancreatic Proteolytic Enzymes The three major pancreatic **proteolytic enzymes** are *trypsinogen*, *chymotrypsinogen*, and *procarboxypeptidase*, each of which is secreted in an inactive form. When **trypsinogen** is secreted into the duodenal lumen, it is activated to its active enzyme form, **trypsin**, by **enteropeptidase** (formerly known as **enterokinase**), an enzyme embedded in the luminal membrane of the cells that line the duodenal mucosa. Like pepsinogen, trypsinogen must remain inactive within the pancreas to prevent this proteolytic enzyme from digesting the proteins of the cells in which it is formed. Trypsinogen remains inactive, therefore, until it reaches the duodenal lumen, where enteropeptidase triggers the activation process. Trypsin then autocatalytically activates more trypsinogen. As further protection, the pancreas also produces a chemical known as **trypsin inhibitor**, which blocks trypsin's actions if spontaneous activation of trypsinogen inadvertently occurs within the pancreas.

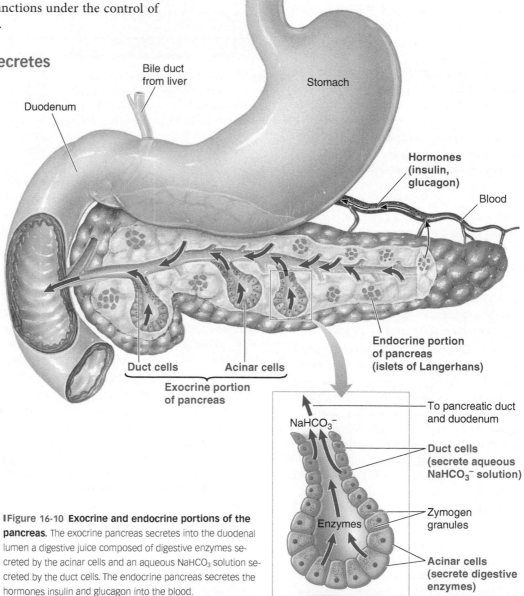

Figure 16-10 Exocrine and endocrine portions of the pancreas. The exocrine pancreas secretes into the duodenal lumen a digestive juice composed of digestive enzymes secreted by the acinar cells and an aqueous $NaHCO_3$ solution secreted by the duct cells. The endocrine pancreas secretes the hormones insulin and glucagon into the blood.

Chymotrypsinogen and **procarboxypeptidase** are converted by trypsin to their active forms, **chymotrypsin** and **carboxypeptidase**, respectively, within the duodenal lumen. Thus, once enteropeptidase has activated some of the trypsin, trypsin carries out the rest of the activation process.

Each of these proteolytic enzymes attacks different peptide linkages. The end products that result from this action are a mixture of small peptide chains and amino acids. Mucus secreted by the intestinal cells protects against digestion of the small-intestine wall by the activated proteolytic enzymes.

Pancreatic Amylase Like salivary amylase, **pancreatic amylase** contributes to carbohydrate digestion by converting dietary starches (amylose and amylopectin) into the disaccharide maltose and the branched polysaccharide α-limit dextrins. Amylase is secreted in the pancreatic juice in an active form because active amylase does not endanger the secretory cells. These cells do not contain any polysaccharides.

Pancreatic Lipase **Pancreatic lipase** is extremely important because it is the only enzyme secreted throughout the entire digestive system that can digest fat. (In humans, insignificant amounts of lipase are secreted in the saliva and gastric juice—*lingual lipase* and *gastric lipase*.) Pancreatic lipase hydrolyzes dietary triglycerides into monoglycerides and free fatty acids, which are the absorbable units of fat (see Figure 16-1d, p. 567). Like amylase, lipase is secreted in its active form because there is no risk of pancreatic self-digestion by lipase. Triglycerides are not a structural component of pancreatic cells.

Clinical Note **Pancreatic Insufficiency** When pancreatic enzymes are deficient, digestion of food is incomplete. Because the pancreas is the only significant source of lipase, pancreatic enzyme deficiency results in serious maldigestion and malabsorption of dietary fat. The main clinical manifestation of pancreatic exocrine insufficiency is **steatorrhea**, or excessive undigested fat in the feces. Up to 60% to 70% of the ingested fat may be excreted in the feces. Digestion of protein and carbohydrates is impaired to a lesser degree because salivary, gastric, and small-intestinal enzymes contribute to the digestion of these two foodstuffs.

Pancreatic Aqueous Alkaline Secretion Pancreatic enzymes function best in a neutral or slightly alkaline environment, yet the highly acidic gastric contents empty into the duodenum in the vicinity of pancreatic enzyme entry into the duodenum. This acidic chyme must be neu-

tralized quickly in the duodenal lumen, not only to allow optimal functioning of the pancreatic enzymes but also to prevent acid damage to the duodenal mucosa. The alkaline (NaHCO$_3$-rich) fluid secreted by the pancreatic duct cells into the duodenum serves the important function of neutralizing the acidic chyme that empties into the duodenum from the stomach. This aqueous NaHCO$_3$ secretion is by far the largest component of pancreatic secretion. The volume of pancreatic secretion ranges between 1 and 2 liters per day, depending on the type and degree of stimulation.

Following is the current model of pancreatic NaHCO$_3$ secretion based on recent evidence (Figure 16-11). The HCO$_3^-$ that is secreted into the duct lumen comes from two sources: (1) Some of the secreted HCO$_3^-$ is derived within the pancreatic duct cell from CO$_2$ that has either diffused in from the plasma or been produced by cellular metabolism. Under the influence of carbonic anhydrase, CO$_2$ in the duct cell combines with OH$^-$ generated from H$_2$O to produce HCO$_3^-$, just as in the gastric parietal cell. The H$^+$ simultaneously generated from H$_2$O enters the plasma across the basolateral border by a Na$^+$–H$^+$ antiporter, thus acidifying the venous blood that leaves the pancreas. (2) However, most of

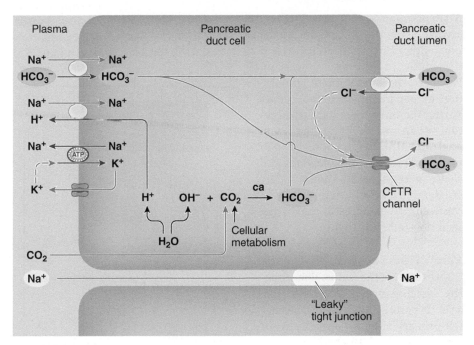

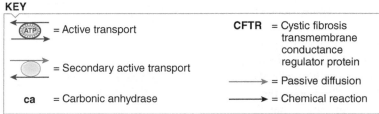

Figure 16-11 Mechanism of NaHCO$_3$ secretion. Most of the to-be-secreted HCO$_3^-$ enters the pancreatic duct cell by means of a Na$^+$–HCO$_3^-$ symporter in the basolateral membrane, but some is generated within the duct cell through carbonic-anhydrase catalyzed formation of HCO$_3^-$ and H$^+$ from H$_2$O and CO$_2$. HCO$_3^-$ is secreted from the pancreatic duct cell into the pancreatic duct lumen by two avenues: via a HCO$_3^-$–Cl$^-$ antiporter and through a CFTR channel, both in the luminal membrane. Na$^+$ diffuses down its electrochemical gradient through "leaky" tight junctions between the pancreatic duct cells to complete NaHCO$_3$ secretion.

the secreted HCO_3^- is not generated within the pancreatic duct cell. Instead the bulk of the HCO_3^- that exits across the luminal membrane enters the cell from the plasma via a Na^+–HCO_3^- symporter at the basolateral membrane. This is the active step in $NaHCO_3$ secretion. The basolateral Na^+–K^+ pump provides the driving energy for the secondary active transport mechanisms.

HCO_3^- is secreted (that is, exits the pancreatic duct cell across the luminal membrane to enter the duct lumen) by two means: (1) Cl^-–HCO_3^- antiporters in the luminal membrane move HCO_3^- into the lumen in exchange for Cl^-, and (2) HCO_3^- also enters the lumen by diffusing through a cystic fibrosis transmembrane conductance regulator (CFTR) channel. This is the same channel that serves as a Cl^- channel and is absent in the lungs and pancreas in cystic fibrosis (see p. 59). Na^+ does not exit the pancreatic duct cells to be secreted. Instead, Na^+ diffuses down an electrochemical gradient via paracellular transport through the "leaky" tight junctions between the duct cells into the lumen (see p. 62). Together these actions accomplish $NaHCO_3$ secretion.

Pancreatic exocrine secretion is regulated by secretin and CCK.

A small amount of parasympathetically induced pancreatic exocrine secretion occurs during the cephalic phase of digestion, with a further token increase occurring during the gastric phase in response to gastrin. However, the predominant stimulation of pancreatic secretion occurs during the intestinal phase of digestion when chyme is in the small intestine. The release of the two major enterogastrones, secretin and CCK, in response to chyme in the duodenum plays the central role in controlling pancreatic exocrine secretion (Figure 16-12).

Role of Secretin in Pancreatic Secretion
Of the factors that stimulate enterogastrone secretion (fat, acid, hypertonicity, and distension), the primary stimulus specifically for secretin release into the blood from the duodenal mucosa is acid in the duodenal lumen. The blood carries secretin to the pancreas, where it stimulates the duct cells to markedly increase their secretion of a $NaHCO_3$-rich aqueous fluid into the duodenum. It is appropriate that the most potent stimulus for secretin release is acid in the small intestine lumen because the resulting alkaline pancreatic secretion neutralizes the acid. The amount of secretin released is proportional to the amount of acid that enters the duodenum, so the amount of $NaHCO_3$ secreted parallels duodenal acidity.

Role of CCK in Pancreatic Secretion
CCK is important in regulating pancreatic digestive enzyme secretion. The main stimulus for CCK release into the blood from the duodenal mucosa is the presence of fat and, to a lesser extent, products of protein digestion in the lumen. The blood transports CCK to the pancreas where it stimulates the pancreatic acinar cells to increase digestive enzyme secretion. Among these enzymes are pancreatic lipase and the proteolytic enzymes, which appropriately further digest the fat and protein that initiated the response. In contrast to fat and protein, carbohydrate does not directly influence pancreatic digestive enzyme secretion.

All three types of pancreatic digestive enzymes are packaged together in the zymogen granules, so all the pancreatic enzymes are released together during exocytosis. Therefore, even though the *total amount* of enzymes released varies depending on the type of meal consumed (the most being secreted in response to fat), the *proportion* of enzymes released does not vary on a meal-to-meal basis. That is, a high-protein meal does not cause the release of a greater proportion of proteolytic enzymes.

Just as gastrin is trophic to the stomach and small intestine, CCK and secretin exert trophic effects on the exocrine pancreas to maintain its integrity.

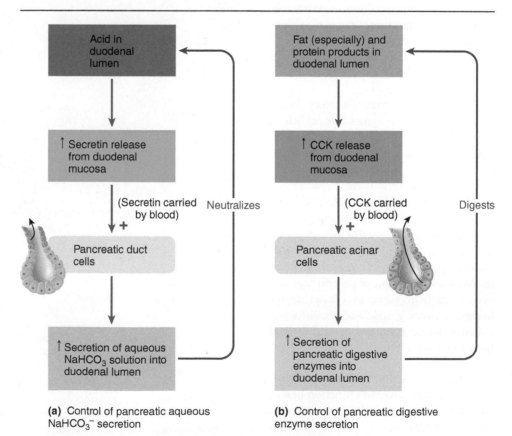

(a) Control of pancreatic aqueous $NaHCO_3^-$ secretion

(b) Control of pancreatic digestive enzyme secretion

Figure 16-12 **Hormonal control of pancreatic exocrine secretion.**

FIGURE FOCUS: *If a person has bacon, eggs, toast, and orange juice for breakfast, discuss which GI hormones are released from the duodenal mucosa as a result of each component of the meal.*

We now look at the contributions of the remaining accessory digestive unit, the liver and gallbladder.

The liver performs various important functions, including bile production.

Besides pancreatic juice, the other secretory product emptied into the duodenal lumen is **bile**. The **biliary system** includes the *liver*, the *gallbladder*, and associated ducts.

Liver Functions The **liver** is the largest and most important metabolic organ in the body; it can be viewed as the body's major biochemical factory. Its functions include the following:

1. Secretion of *bile salts*, which aid fat digestion and absorption. This is the only liver function directly related to digestion.

2. Metabolic processing of the major categories of nutrients (carbohydrates, proteins, and lipids) after their absorption from the digestive tract (see p. 609)

3. Detoxifying or degrading body wastes and hormones, as well as drugs and other foreign compounds (see p. 27)

4. Synthesizing plasma proteins, including those needed for blood clotting (see p. 397), those that transport steroid and thyroid hormones and cholesterol in the blood (see pp. 122 and 328), and angiotensinogen important in the salt-conserving renin–angiotensin–aldosterone system (see p. 508)

5. Storing glycogen, fats, iron, copper, and many vitamins (see p. 44)

6. Activating vitamin D, which the liver does in conjunction with the kidneys (see p. 709)

7. Secreting the hormones thrombopoietin (stimulates platelet production; see p. 395), hepcidin (inhibits iron uptake from the intestine; see p. 603), and insulin-like growth factor-I (stimulates growth; see p. 654)

8. Producing acute phase proteins important in inflammation (see p. 411)

9. Excreting cholesterol (see p. 329) and bilirubin, the latter being a breakdown product derived from the destruction of worn-out red blood cells (see p. 597)

10. Removing bacteria and worn-out red blood cells, thanks to its resident macrophages.

Given this range of complex functions, there is amazingly little specialization among cells within the liver. Each liver cell, or **hepatocyte**, performs the same wide variety of metabolic and secretory tasks (*hepato* means "liver"; *cyte* means "cell"). The specialization comes from the highly developed organelles within each hepatocyte. The only liver function not accomplished by hepatocytes is the phagocytic activity carried out by the resident macrophages, which are known as **Kupffer cells**.

Liver Blood Flow To carry out these wide-ranging tasks, the anatomic organization of the liver permits each hepatocyte to be in direct contact with blood from two sources: arterial blood coming from the heart and venous blood coming directly from the digestive tract. Like other cells, the hepatocytes receive fresh arterial blood via the hepatic artery, which supplies their oxygen and delivers blood-borne metabolites for hepatic process-

ing. Venous blood also enters the liver by the **hepatic portal system**, a unique and complex vascular connection between the digestive tract and the liver (Figure 16-13). The veins draining the digestive tract do not directly join the inferior vena cava, the large vein that returns blood to the heart. Instead, the veins

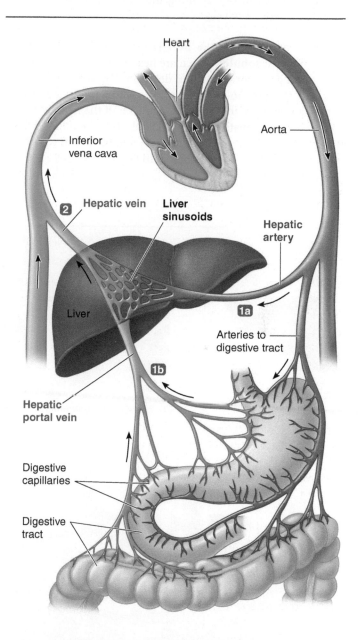

The liver receives blood from two sources:

1a Arterial blood, which provides the liver's O_2 supply and carries blood-borne metabolites for hepatic processing, is delivered by the **hepatic artery**.

1b Venous blood draining the digestive tract is carried by the **hepatic portal vein** to the liver for processing and storage of newly absorbed nutrients.

2 Blood leaves the liver via the **hepatic vein**.

Figure 16-13 Schematic representation of liver blood flow.
FIGURE FOCUS: *State the vessel through which (1) O_2 enters the liver, (2) just absorbed glucose enters the liver, and (3) ingested alcohol that has been detoxified by the liver enters the systemic circulation.*

from the stomach and intestine enter the *hepatic portal vein,* which carries the products absorbed from the digestive tract directly to the liver for processing, storage, or detoxification before they gain access to the general circulation. Within the liver, the portal vein again breaks up into a capillary network (the liver *sinusoids)* to permit exchange between the blood and the hepatocytes before draining into the hepatic vein, which joins the inferior vena cava.

Liver Organization The liver is organized into functional units known as **lobules**, which are hexagonal columns of tissue surrounding a central vein and delineated by vascular and bile channels (❙Figure 16-14a and b). At each of the six outer corners of a lobule are three vessels: a branch of the hepatic artery, a branch of the hepatic portal vein, and a bile duct. Blood from the branches of both the hepatic artery and the portal vein flows from the periphery of the lobule into large, expanded capillary

spaces called **sinusoids** (see p. 354), which run between rows of liver cells to the central vein like spokes on a bicycle wheel (❙Figure 16-14b). The Kupffer cells line the sinusoids and engulf and destroy old red blood cells and bacteria that pass through in the blood. The hepatocytes are arranged between the sinusoids in plates that are two cell layers thick so that each lateral edge faces a sinusoidal pool of blood. The central veins of all the liver lobules converge to form the hepatic vein, which carries blood away from the liver. The thin bile-carrying channel, a **bile canaliculus**, runs between the cells within each hepatic plate (❙Figure 16-14c). Hepatocytes continuously secrete bile into these thin channels, which carry the bile to a bile duct at the periphery of the lobule. The bile ducts from the various lobules converge to eventually form the **common bile duct**, which transports the bile from the liver to the duodenum. Each hepatocyte is in contact with a sinusoid on one side and a bile canaliculus on the other side.

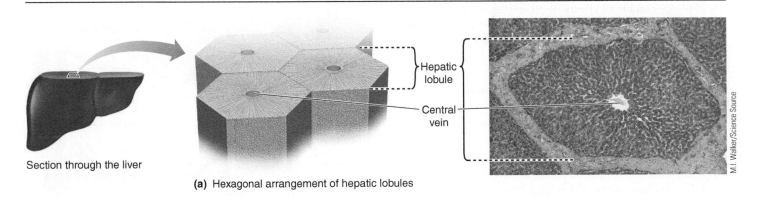

Section through the liver

(a) Hexagonal arrangement of hepatic lobules

M.I. Walker/Science Source

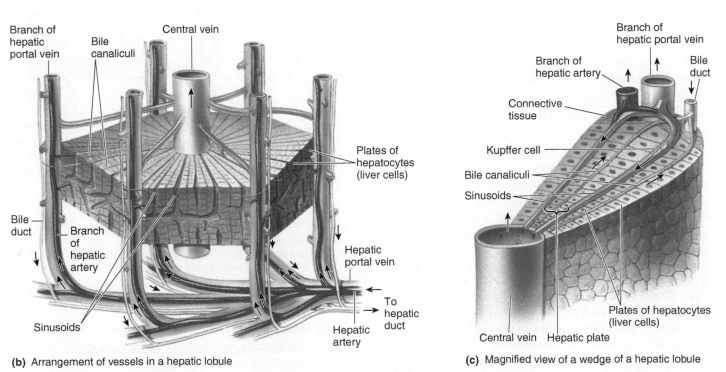

(b) Arrangement of vessels in a hepatic lobule

(c) Magnified view of a wedge of a hepatic lobule

❙Figure 16-14 **Anatomy of the liver.** The photomicrograph is of a transverse section of a lobule in a pig liver, often used in teaching because the lobules are more clearly demarcated by connective tissue in pigs than in humans so that the hexagonal arrangement of the lobule is readily apparent.

Bile is continuously secreted by the liver and is diverted to the gallbladder between meals.

The liver continuously secretes bile, even between meals. The opening of the bile duct into the duodenum is guarded by the **sphincter of Oddi**, which prevents bile from entering the duodenum except during digestion of meals (❙ Figure 16-15). When this sphincter is closed, bile secreted by the liver hits the closed sphincter and is diverted back up into the **gallbladder**, a small, saclike structure tucked beneath but not directly connected to the liver. Thus, bile is not transported directly from the liver to the gallbladder. Bile is subsequently stored and concentrated in the gallbladder between meals. Active transport of salt out of the gallbladder, with water following osmotically, results in a 5 to 10 times greater concentration of the organic constituents.

After a meal, bile enters the duodenum as a result of the combined effects of relaxation of the sphincter of Oddi, gallbladder contraction, and increased bile secretion by the liver. The amount of bile secreted per day ranges from 250 mL to 1 liter, depending on the degree of stimulation.

Clinical Note Because the gallbladder stores concentrated bile, it is the primary site for precipitation of concentrated bile constituents into **gallstones**. Fortunately, the gallbladder does not play an essential digestive role, so its removal as a treatment for gallstones presents no particular problem. After gallbladder removal, the bile secreted between meals is stored instead in the common bile duct, which becomes dilated.

Bile salts are recycled through the enterohepatic circulation.

Bile contains several organic constituents, namely, *bile salts, cholesterol, lecithin* (a phospholipid), and *bilirubin* (all derived from hepatocyte activity) in an *aqueous alkaline fluid* (added by the duct cells). Even though bile does not contain any digestive enzymes, it is important for the digestion and absorption of fats, primarily through the activity of bile salts.

Bile salts are derivatives of cholesterol. They are actively secreted into the bile and eventually enter the duodenum, along with the other biliary constituents. Following their participation in fat digestion and absorption, most bile salts are reabsorbed into the blood by special active-transport mechanisms located only in the terminal ileum. From here, bile salts are returned by the hepatic portal system to the liver, which resecretes them into the bile. This recycling of bile salts (and some of the other biliary constituents) between the small intestine and the liver is called the **enterohepatic circulation** (*entero* means "intestine"; *hepatic* means "liver") (❙ Figure 16-15).

The total amount of bile salts in the body averages about 3 to 4 g, yet 3 to 15 g of bile salts may be emptied into the duode-

❶ Secreted bile salts consist of 95% old, recycled bile salts and 5% newly synthesized bile salts.

❸ Reabsorbed bile salts are recycled by enterohepatic circulation.

Liver

Bile salts ← Cholesterol

Common bile duct

Stomach

Gallbladder

Sphincter of Oddi

Duodenum

Duct from pancreas

Hepatic portal vein

❹ 5% of bile salts are lost in feces.

Colon

KEY

⟵ = Enterohepatic circulation of bile salts

Terminal ileum

❷ 95% of bile salts are reabsorbed by torminal ileum.

❙ Figure 16-15 **Enterohepatic circulation of bile salts.** Most bile salts are recycled between the liver and small intestine through the enterohepatic circulation (*blue arrows*). After participating in fat digestion and absorption, most bile salts are reabsorbed by active transport in the terminal ileum and returned through the hepatic portal vein to the liver, which resecretes them in the bile.

num in a single meal. On average, bile salts cycle between the liver and the small intestine twice during the digestion of a typical meal. Usually, only about 5% of the secreted bile escapes into the feces daily. These lost bile salts are replaced by new bile salts synthesized by the liver; thus, the size of the pool of bile salts is kept constant.

Bile salts aid fat digestion and absorption.

Bile salts aid fat digestion through their detergent action (emulsification) and facilitate fat absorption by participating in the formation of micelles. Both functions are related to the structure of bile salts. Let us see how.

Detergent Action of Bile Salts The term **detergent action** refers to bile salts' ability to convert large fat globules into a **lipid emulsion** consisting of many small fat droplets suspended in the aqueous chyme. Breaking up the large fat globule into small, stabilized droplets increases the surface area available for attack by pancreatic lipase. To digest fat, lipase must come into direct contact with the triglyceride molecule. Because triglycerides are not soluble in water, they tend to aggregate into large droplets in the watery environment of the small-intestine lumen. If bile salts did not emulsify these large droplets, lipase could act on the triglyceride molecules only at the surface of the large droplets, and fat digestion would be greatly prolonged.

Bile salts emulsify fats similar to the detergent you use to break up grease when you wash dishes. A bile salt molecule contains a lipid-soluble part (a steroid derived from cholesterol) plus a negatively charged, water-soluble part. Bile salts *adsorb* on the surface of a fat droplet—that is, the lipid-soluble part of the bile salt dissolves in the fat droplet, leaving the charged water-soluble part projecting from the surface of the droplet (❚ Figure 16-16a). Intestinal mixing movements break up large fat droplets into smaller ones. These small droplets would quickly recoalesce were it not for bile salts adsorbing on their surface and creating a shell of water-soluble negative charges on the surface of each little droplet. Because like charges repel, these negatively charged groups on the droplet surfaces cause the small fat droplets to repel one another (❚ Figure 16-16b) and prevent them from recoalescing into large droplets. The small emulsified fat droplets range in diameter from 200 to 5000 nm, with an average of about 1000 nm (1 μm).

Although bile salts increase the surface area available for attack by pancreatic lipase, lipase alone cannot penetrate the layer of bile salts adsorbed on the surface of the small emulsified fat droplets. To solve this dilemma, the pancreas secretes the polypeptide **colipase** along with lipase. Like bile salts, colipase has both a lipid-soluble part and a water-soluble part. Colipase displaces some bile salts and lodges at the surface of the fat droplets, where it binds to lipase, thus anchoring this enzyme to its site of action amid the bile-salt coating.

Formation of Micelles Bile salts— along with cholesterol and lecithin, which are also constituents of bile— play an important role in facilitating fat absorption through formation of micelles. Like bile salts, lecithin (a phospholipid similar to the ones in the lipid bilayer of the plasma mem-

brane) has both a lipid-soluble and a water-soluble part, whereas cholesterol is almost totally insoluble in water. In a **micelle**, the bile salts and lecithin aggregate in small clusters with their fat-soluble parts huddled together in the middle to form a hydrophobic ("water-fearing") core, while their water-soluble parts form an outer hydrophilic ("water-loving") shell (❚ Figure 16-17). A micelle is 3 to 10 nm in diameter, compared to an average diameter of 1000 nm for an emulsified lipid droplet. Micelles are water soluble because of their hydrophilic shells, but they can dissolve water-insoluble (and hence lipid-soluble)

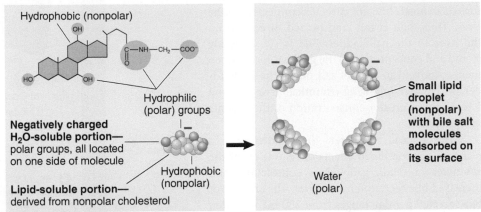

Negatively charged H₂O-soluble portion— polar groups, all located on one side of molecule

Lipid-soluble portion— derived from nonpolar cholesterol

(a) Structure of bile salts and their adsorption on the surface of a small lipid droplet

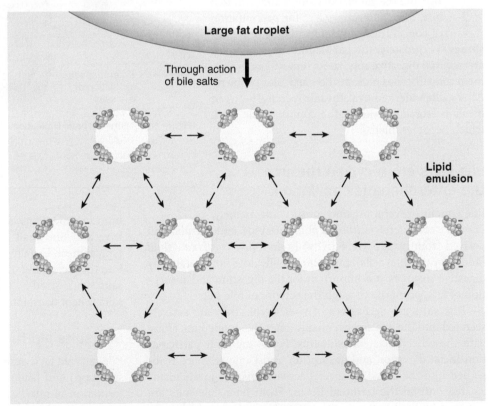

(b) Formation of a lipid emulsion through the action of bile salts

❚ **Figure 16-16 Schematic structure and function of bile salts.** (a) A bile salt consists of a lipid-soluble part that dissolves in the fat droplet and a negatively charged, water-soluble part that projects from the surface of the droplet. (b) When a large fat droplet is broken up into smaller fat droplets by intestinal contractions, bile salts adsorb on the surface of the small droplets, creating shells of negatively charged, water-soluble bile salt components that cause the fat droplets to repel one another. This emulsifying action holds the fat droplets apart and prevents them from recoalescing, increasing the surface area of exposed fat available for digestion by pancreatic lipase.

substances in their lipid-soluble cores. Micelles thus provide a handy vehicle for carrying water-insoluble substances through the watery luminal contents. The most important lipid-soluble substances carried within micelles are the products of fat digestion (monoglycerides and free fatty acids), and fat-soluble vitamins, which are all transported to their sites of absorption by this means. If they did not hitch a ride in the water-soluble micelles, these nutrients would float on the surface of the aqueous chyme (just as oil floats on top of water), never reaching the absorptive surfaces of the small intestine. In addition, cholesterol, a highly water-insoluble substance, dissolves in the micelle's hydrophobic core.

Bile salts stimulate bile secretion; CCK promotes gallbladder emptying.

Any substance that increases bile secretion is called a **choleretic**. The most potent choleretic is bile salts themselves. Between meals bile is stored in the gallbladder, but during a meal bile is emptied into the duodenum as the gallbladder contracts. After bile salts participate in fat digestion and absorption, they are reabsorbed and returned by the enterohepatic circulation to the liver, where they act as potent choleretics to stimulate further bile secretion. Therefore, during a meal, when bile salts are needed and being used, bile secretion by the liver is enhanced.

Vagal stimulation of the liver plays a minor role in bile secretion during the cephalic phase of digestion, promoting an increase in liver bile flow before food ever reaches the intestine.

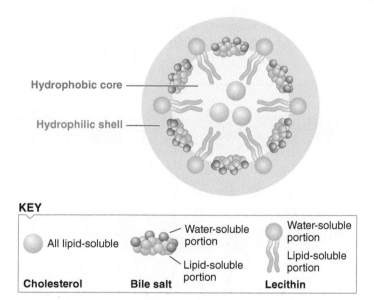

KEY

All lipid-soluble	Water-soluble portion / Lipid-soluble portion	Water-soluble portion / Lipid-soluble portion
Cholesterol	**Bile salt**	**Lecithin**

IFigure 16-17 **A micelle.** Bile constituents (bile salts, lecithin, and cholesterol) aggregate to form micelles that consist of a hydrophilic (water-soluble) shell and a hydrophobic (lipid-soluble) core. Because the outer shell of a micelle is water soluble, the products of fat digestion, which are not water soluble, can be carried through the watery luminal contents to the absorptive surface of the small intestine by dissolving in the micelle's lipid-soluble core. This figure is not drawn to scale compared to the lipid emulsion droplets in IFigure 16-16b. An emulsified fat droplet ranges in diameter from 200 to 5000 nm (average 1000 nm) compared to a micelle, which is 3 to 10 nm in diameter.

When chyme reaches the small intestine, the presence especially of fat products in the duodenal lumen triggers release of CCK. This hormone stimulates contraction of the gallbladder and relaxation of the sphincter of Oddi, so bile is discharged into the duodenum, where it appropriately aids in the digestion and absorption of the fat that initiated CCK release.

Bilirubin is a waste product excreted in the bile.

Bilirubin, the other major constituent of bile, does not play a role in digestion but instead is a waste product excreted in the bile. Bilirubin is the primary bile pigment derived from the breakdown of worn-out red blood cells, which are removed from the blood by the macrophages that line the liver sinusoids and reside in other areas in the body. Bilirubin is the end product from degradation of the heme (iron-containing) part of the hemoglobin contained within these old red blood cells (see p. 383). Hepatocytes take up bilirubin from the plasma, slightly modify the pigment to increase its solubility, then actively excrete it into the bile.

Bilirubin is a yellow pigment that gives bile its color. Within the intestinal tract, this pigment is modified by bacterial enzymes, giving rise to the characteristic brown color of feces. When bile secretion does not occur, as when the bile duct is completely obstructed by a gallstone, the feces are grayish white. A small amount of bilirubin is normally reabsorbed by the intestine back into the blood, and when it is eventually excreted in the urine, it is largely responsible for the urine's yellow color. The kidneys cannot excrete bilirubin until after it has been modified during its passage through the liver and intestine.

Clinical Note If bilirubin is formed more rapidly than it can be excreted, it accumulates in the body and causes **jaundice**. Patients with this condition appear yellowish, with this color being seen most easily in the whites of their eyes. Jaundice can be brought about in three ways:

1. *Prehepatic* (the problem occurs "before the liver"), or *hemolytic, jaundice* arises from excessive breakdown (hemolysis) of red blood cells, which results in the liver being presented with more bilirubin than it is capable of excreting.

2. *Hepatic* (the problem is the "liver") *jaundice* occurs when the liver is diseased and cannot deal with even the normal load of bilirubin.

3. *Posthepatic* (the problem occurs "after the liver"), or *obstructive, jaundice* occurs when the bile duct is obstructed, such as by a gallstone, so that bilirubin cannot be eliminated in the feces.

Hepatitis and cirrhosis are the most common liver disorders.

Clinical Note **Hepatitis** is an inflammatory disease of the liver that results from a variety of causes, including viral infection, obesity (*fatty liver disease*), or most commonly exposure to toxic agents, including alcohol, carbon tetrachloride, and certain tranquilizers. Hepatitis ranges in severity from mild, reversible symptoms to acute massive liver damage, with possible imminent death resulting from acute hepatic failure.

Repeated or prolonged hepatic inflammation, usually in association with chronic alcoholism, can lead to **cirrhosis**, a condition in which damaged hepatocytes are permanently replaced by connective tissue. Liver tissue has the ability to regenerate, normally undergoing a gradual turnover of cells. If part of the hepatic tissue is destroyed, the lost tissue can be replaced by an increase in the rate of cell division. There is a limit, however, to how rapidly hepatocytes can be replaced. In addition to hepatocytes, a small number of fibroblasts (connective tissue cells) are dispersed between the hepatic plates and form a supporting framework for the liver. Chronic or high exposure to alcohol impairs hepatocyte replacement such that the sturdier fibroblasts take advantage of the situation and overproduce. This extra connective tissue leaves little space for the hepatocytes' regrowth. Thus, as cirrhosis develops slowly over time, active liver tissue is gradually reduced, leading eventually to chronic liver failure.

Having looked at the accessory digestive organs that empty their exocrine products into the small-intestine lumen, we now examine the contributions of the small intestine itself.

Check Your Understanding 16.5

1. State the functions of the pancreatic enzymes.
2. Explain the significance of some pancreatic enzymes being stored as precursors in zymogen granules.
3. Describe how bile salts contribute to dietary fat digestion.

16.6 Small Intestine

The **small intestine** is the site where most digestion and absorption takes place. The small intestine lies coiled within the abdominal cavity, extending between the stomach and the large intestine. It is arbitrarily divided into three segments—the **duodenum**, the **jejunum**, and the **ileum**.

As usual, we examine motility, secretion, digestion, and absorption in the small intestine in that order. Small-intestine motility includes *segmentation* and the *migrating motility complex*.

Segmentation contractions mix and slowly propel the chyme.

Segmentation, the small intestine's primary motility during digestion of a meal, both mixes and slowly propels the chyme. Segmentation consists of oscillating, ringlike contractions of the circular smooth muscle along the small intestine's length; between the contracted segments are relaxed areas containing a small bolus of chyme. The contractile rings occur every few centimeters, dividing the small intestine into segments like a chain of sausages. These contractile rings do not sweep along the length of the intestine as peristaltic waves do. Rather, after a brief period, the contracted segments relax, and ringlike contractions appear in the previously relaxed areas (❙ Figure 16-18).

The new contraction forces the chyme in a previously relaxed segment to move in both directions into the now relaxed adjacent segments. A newly relaxed segment therefore receives chyme from both the contracting segment immediately ahead of it and the one immediately behind it. Shortly thereafter, the areas of contraction and relaxation alternate again. In this way, the chyme is chopped, churned, and thoroughly mixed. These contractions can be compared to squeezing a pastry tube with your hands to mix the contents.

Initiation and Control of Segmentation Segmentation contractions are initiated by a BER produced by small intestine's pacemaker cells. If the small-intestine BER brings the circular smooth muscle layer to threshold, segmentation contractions are induced, with the frequency of segmentation following the frequency of the BER.

The circular smooth muscle's degree of responsiveness and thus the intensity of segmentation contractions can be influenced by distension of the intestine, by the hormone gastrin, and by extrinsic nerve activity. All these factors influence the excitability of the small-intestine smooth muscle cells by moving the starting potential around which the BER oscillates closer to or farther from threshold. Segmentation is slight or absent between meals but becomes vigorous immediately after a meal. Both the duodenum and the ileum start to segment simultaneously when the meal first enters the small intestine. The duodenum starts to segment primarily in response to local distension caused by the presence of chyme. Segmentation of the empty ileum, in contrast, is brought about by gastrin secreted in response to the presence of chyme in the stomach, a mechanism known as the **gastroileal reflex**. Extrinsic nerves can modify the strength of these contractions. Parasympathetic stimulation enhances segmentation, whereas sympathetic stimulation depresses segmental activity.

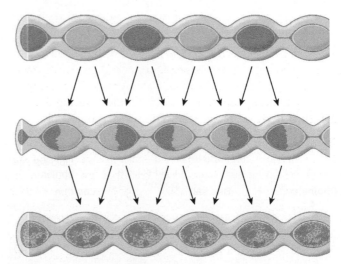

❙Figure 16-18 **Segmentation.** Segmentation consists of ringlike contractions along the length of the small intestine. Within a matter of seconds, the contracted segments relax and the previously relaxed areas contract. These oscillating contractions thoroughly mix the chyme within the small-intestine lumen.

Functions of Segmentation The mixing accomplished by segmentation serves the dual functions of (1) mixing the chyme with the digestive juices secreted into the small-intestine lumen and (2) exposing all the chyme to the absorptive surfaces of the small-intestine mucosa.

Segmentation not only accomplishes mixing but also slowly moves chyme through the small intestine. How can this be, when each segmental contraction propels chyme both forward and backward? The chyme slowly progresses forward because the frequency of segmentation declines along the length of the small intestine. The pacemaker cells in the duodenum spontaneously depolarize faster than those farther down the tract, with segmentation contractions occurring in the duodenum at a rate of 12 per minute, compared to only 9 per minute in the terminal ileum. Because segmentation occurs with greater frequency in the upper part of the small intestine than in the lower part, more chyme, on average, is pushed forward than is pushed backward. As a result, chyme is moved slowly from the upper to the lower part of the small intestine, being shuffled back and forth to accomplish thorough mixing and absorption in the process. This slow propulsive mechanism is advantageous because it allows ample time for the digestive and absorptive processes to take place. The contents usually take 3 to 5 hours to move through the small intestine.

The migrating motility complex sweeps the intestine clean between meals.

During periods of short fasting, when most of the meal has been absorbed, the stomach and small intestine exhibit a unique motor activity. Intestinal segmentation contractions cease and are replaced by the **migrating motility complex (MMC)**, or **"intestinal housekeeper"** activity. The MMC cycles through the following phases in a repetitive pattern about every 1.5 hours as long as a person is fasting:

1. Phase I: A long period lasting about 40 to 60 minutes of relative quiet with very few contractions

2. Phase II: A 20- to 30-minute period with some peristaltic contractions, with the time varying between contractions

3. Phase III: The shortest phase, where intense peristaltic contractions begin in the upper stomach and propagate (migrate) through to the end of the small intestine. The contractions rhythmically repeat for 5 to 10 minutes. During this period, the pyloric sphincter relaxes and opens completely.

The motor activity of the MMC is thought to sweep any remnants of the preceding meal plus mucosal debris and bacteria forward toward the colon, just like a good "intestinal housekeeper." If a person continues to fast, the MMC motor activity repeats itself, beginning again at Phase I. While fasting, some individuals become acutely aware of the MMC because Phase III contractions cause gurgling noises that are often thought of as the stomach "growling." The MMC is regulated between meals by the hormone *motilin,* which is secreted during the unfed state by endocrine cells of the small-intestine mucosa. When the next meal arrives, the MMC ceases and the motor activity associated with a meal takes over. Motilin release is inhibited by feeding.

The ileocecal juncture prevents contamination of the small intestine by colonic bacteria.

At the juncture between the small and the large intestines, the last part of the ileum empties into the cecum. Two factors contribute to this region's ability to act as a barrier between the small and the large intestines. First, the anatomic arrangement is such that valvelike folds of tissue protrude from the ileum into the lumen of the cecum. When the ileal contents are pushed forward, this **ileocecal valve** is easily pushed open, but the folds of tissue are forcibly closed when the cecal contents attempt to move backward. Second, the smooth muscle within the last several centimeters of the ileal wall is thickened, forming a sphincter that is under neural and hormonal control. Most of the time, this **ileocecal sphincter** remains at least mildly constricted. Pressure on the cecal side of the sphincter causes it to contract more forcibly; distension of the ileal side causes the sphincter to relax, a reaction mediated by the intrinsic plexuses in the area. In this way, the ileocecal juncture prevents the bacteria-laden contents of the large intestine from contaminating the small intestine yet lets the ileal contents pass into the colon. If the colonic bacteria gained access to the nutrient-rich small intestine, they would multiply rapidly. Relaxation of the sphincter is enhanced by release of gastrin at the onset of a meal, when increased gastric activity is taking place. This relaxation allows the undigested fibers and unabsorbed solutes from the preceding meal to be moved forward as the new meal enters the tract.

Small-intestine secretions do not contain any digestive enzymes.

Each day, the exocrine gland cells in the small-intestine mucosa secrete into the lumen about 1.5 liters of an aqueous salt and mucus solution called **succus entericus** ("juice of intestine"). Secretion increases after a meal in response to local stimulation of the small-intestine mucosa by the presence of chyme.

The mucus in the secretion provides protection and lubrication. Furthermore, this aqueous secretion provides plenty of H_2O to participate in the enzymatic digestion of food. Recall that digestion involves hydrolysis—bond breakage by reaction with H_2O—which proceeds most efficiently when all the reactants are in solution.

No digestive enzymes are secreted into this intestinal juice. The small intestine does synthesize digestive enzymes, but they act intracellularly within the brush-border membrane of the epithelial cells that line the lumen instead of being secreted directly into the lumen.

The small-intestine enzymes complete digestion within the brush-border membrane.

Pancreatic enzymes are responsible for most of the digestion within the small-intestine lumen, with fat digestion being enhanced by bile secretion. As a result of pancreatic enzymatic activity, fats are completely reduced to their absorbable units of monoglycerides and free fatty acids, proteins are broken down into small peptide fragments and some amino acids, and carbohydrates are reduced to disaccharides, α-limit dextrins, and some monosaccharides. Thus, fat digestion is completed within the small-intestine lumen, but carbohydrate and protein digestion have not been brought to completion.

Special hairlike projections on the luminal surface of the small-intestine epithelial cells, the **microvilli**, form the **brush border** (see ❙Figure 2-26, p. 51 and chapter opener photo, p. 565). The brush-border plasma membrane contains three categories of membrane-spanning proteins that function as membrane-bound enzymes:

1. **Enteropeptidase**, which activates the pancreatic proteolytic enzyme trypsinogen

2. The **disaccharidases** (**maltase**, **sucrase-isomaltase**, and **lactase**), which target maltose, α-limit dextrins, and dietary disaccharides. Maltose (which is a product of salivary and pancreatic amylase) is broken down to glucose by maltase or sucrase-isomaltase activity. However, the other product of starch digestion, the α-limit dextrins is only broken down by sucrase-isomaltase. The end digestion of the dietary disaccharides sucrose and lactose is completed by sucrase-isomaltase and lactase, respectively.

3. The **aminopeptidases**, which hydrolyze most of the small peptide fragments into their amino acid components, thereby completing protein digestion

Thus, carbohydrate and protein digestion are completed within the confines of the brush border. (❙Table 16-6 provides a summary of the digestive processes for the three major categories of nutrients.)

Clinical Note A fairly common disorder, **lactose intolerance**, involves a deficiency of lactase, the disaccharidase specific for the digestion of lactose, or milk sugar. Most children younger than 4 years of age have adequate lactase, but this may be gradually lost so that, in many adults, lactase activity is diminished or absent. When lactose-rich milk or dairy products are consumed by a person with lactase deficiency, the undigested lactose remains in the lumen and has several related consequences. First, accumulation of undigested lactose creates an osmotic gradient that draws H_2O into the intestinal lumen. Second, bacteria living in the large intestine have lactose-splitting ability, so they eagerly attack the lactose as an energy source, producing large quantities of CO_2 and methane gas in the process. Distension of the intestine by both fluid and gas produces pain (cramping) and diarrhea. Infants with lactose intolerance may also suffer from malnutrition.

Finally, we are ready to discuss absorption of nutrients. Up to this point, no food, water, or electrolytes have been absorbed.

The small intestine is remarkably well adapted for its primary role in absorption.

All products of carbohydrate, protein, and fat digestion, and most of the ingested electrolytes, vitamins, and water, are normally absorbed by the small intestine indiscriminately. Usually, only the absorption of calcium and iron is adjusted to the body's

❙**TABLE 16-6** Digestive Processes for the Three Major Categories of Nutrients

Nutrients	Enzymes for Digesting Nutrient	Source of Enzymes	Site of Action of Enzymes	Action of Enzymes	Absorbable Units of Nutrients
Carbohydrate	Amylase	Salivary glands	Mouth and (mostly) body of stomach	Hydrolyzes polysaccharides to disaccharides and α-limit dextrins	
		Exocrine pancreas	Small-intestine lumen		
	Disaccharidases (maltase, sucrase-isomaltase, lactase)	Small-intestine epithelial cells	Small-intestine brush border	Hydrolyze disaccharides to monosaccharides	Monosaccharides, especially glucose
Protein	Pepsin	Stomach chief cells	Stomach antrum	Hydrolyzes protein to peptide fragments	
	Trypsin, chymotrypsin carboxypeptidase	Exocrine pancreas	Small-intestine lumen	Attack different peptide fragments	
	Aminopeptidases	Small-intestine epithelial cells	Small-intestine brush border	Hydrolyze peptide fragments to amino acids	Amino acids and a few small peptides
Fat	Lipase	Exocrine pancreas	Small-intestine lumen	Hydrolyzes triglycerides to fatty acids and monoglyerides	Fatty acids and monoglycerides
	Bile salts (not an enzyme)	Liver	Small-intestine lumen	Emulsify large fat globules for attack by pancreatic lipase	

needs. Thus, the more food consumed, the more that is digested and absorbed, as people who are trying to control their weight are all too painfully aware.

Most absorption occurs in the duodenum and jejunum; very little occurs in the ileum, not because the ileum does not have absorptive capacity but because most absorption has already been accomplished before the intestinal contents reach the ileum. The small intestine has an abundant reserve absorptive capacity. About 50% of the small intestine can be removed with little interference to absorption—with one exception. If the terminal ileum is removed, vitamin B_{12} and bile salts are not properly absorbed because the specialized transport mechanisms for these two substances are located only in this region. All other substances can be absorbed throughout the small intestine's length.

The mucous lining of the small intestine is remarkably well adapted for its special absorptive function for two reasons: (1) it has a large surface area, and (2) the epithelial cells in this lining have a variety of specialized transport mechanisms.

Adaptations that Increase the Small Intestine's Surface Area The following special modifications of the small-intestine mucosa greatly increase the surface area available for absorption (Figure 16-19):

■ The inner surface of the small intestine is thrown into permanent **circular folds** that are visible to the naked eye and increase the surface area threefold.

■ Extending from this folded surface are microscopic, fingerlike projections known as **villi**, which give the lining a velvety appearance and increase the surface area another 10 times (Figure 16-20). The surface of each villus is covered by epithelial cells interspersed occasionally with mucous cells.

■ Even smaller hairlike projections, the *microvilli* or *brush border*, arise from the luminal surface of these epithelial cells, increasing the surface area another 20-fold. Each epithelial cell has as many as 3000 to 6000 of these microvilli, which are visible only with an electron microscope (see chapter opener photo). The small-intestine enzymes perform their functions within the membrane of this brush border.

Altogether, the folds, villi, and microvilli provide the small intestine with a luminal surface area 600 times greater than if it were a tube of the same length and diameter lined by a flat surface. If the surface area of the small intestine were spread out flat, it would cover an entire tennis court.

Clinical Note **Malabsorption** (impairment of absorption) may be caused by damage to or reduction of the surface area of the small intestine. One of the most common causes is **gluten enteropathy**, also known as **celiac disease**. In this condition, the person's small intestine is abnormally sensitive to *gluten,* a protein constituent of wheat, barley, and rye. These grain products are widely prevalent in processed foods. This condition is a complex immunological disorder in which exposure to gluten erroneously activates a T-cell response (see p. 423) that damages the intestinal villi: The normally luxuriant array of villi is reduced, the mucosa becomes flattened, and the brush border becomes short and stubby (Figure 16-21). Because this loss of villi and microvilli decreases the surface

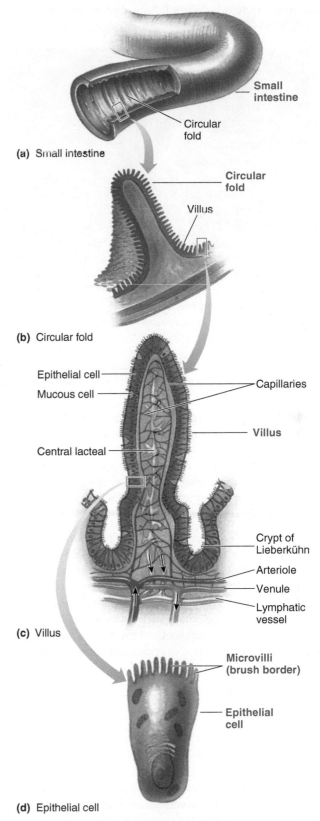

(a) Small intestine

(b) Circular fold

(c) Villus

(d) Epithelial cell

Figure 16-19 Small-intestine absorptive surface. (a) Gross structure of the small intestine. (b) The circular folds of the small-intestine mucosa collectively increase the absorptive surface area threefold. (c) Microscopic fingerlike projections known as villi collectively increase the surface area another 10-fold. (d) Each epithelial cell on a villus has microvilli on its luminal border; the microvilli increase the surface area another 20-fold. Together, these surface modifications increase the small intestine's absorptive surface area 600-fold.

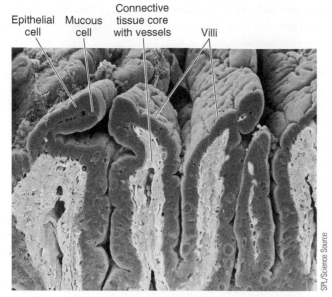

Epithelial cell   Mucous cell   Connective tissue core with vessels   Villi

SPL/Science Source

❙Figure 16-20 Villi projecting from the small-intestine mucosa. This scanning electron micrograph of a freeze-fractured surface of the small-intestine mucosa shows the fingerlike surface projections known as villi that have been sectioned longitudinally.

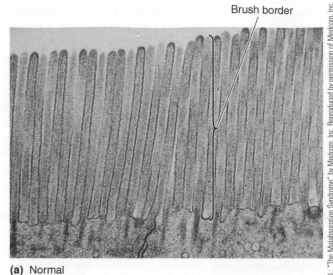

Brush border

(a) Normal

Brush border

(b) Gluten enteropathy

Thomas W. Sheehy, M.D.; Robert L. Slaughter, M.D.: "The Malabsorption Syndrome" by Medcom, Inc. Reproduced by permission of Medcom, Inc.

❙Figure 16-21 Reduction in the brush border with gluten enteropathy. (a) Electron micrograph of the brush border of a small-intestine epithelial cell in a normal individual. (b) Electron micrograph of the short, stubby brush border of a small-intestine epithelial cell in a patient with gluten enteropathy.

area available for absorption, absorption of all nutrients is impaired. The condition is treated by not eating gluten.

Structure of a Villus Absorption across the digestive tract wall involves transepithelial transport similar to movement of material across the kidney tubules (see p. 505). Each villus has the following major components (see ❙Figures 16-19c and 16-20):

■ *Epithelial cells that cover the surface of the villus.* The epithelial cells are joined at their lateral borders by tight junctions, which limit passage of luminal contents between the cells, although the tight junctions in the small intestine are leakier than those in the stomach. Within their luminal brush borders, these epithelial cells have carriers for absorption of specific nutrients and electrolytes from the lumen, and the membrane-bound digestive enzymes that complete carbohydrate and protein digestion.

■ *A connective tissue core.* This core is formed by the lamina propria.

■ *A capillary network.* Each villus is supplied by an arteriole that breaks up into a capillary network within the villus core. The capillaries rejoin to form a venule that drains away from the villus.

■ *A terminal lymphatic vessel.* Each villus is supplied by a single blind-ended lymphatic vessel known as the **central lacteal**, which occupies the center of the villus core.

During the process of absorption, digested substances enter the capillary network or the central lacteal. To be absorbed, a substance must pass completely through the epithelial cell, diffuse through the interstitial fluid within the connective tissue core of the villus, and then cross the wall of a capillary or lymph vessel. Like renal transport, intestinal absorption may be active or passive, with active absorption involving energy expenditure during at least one of the transepithelial transport steps.

The mucosal lining experiences rapid turnover.

Dipping down into the mucosal surface between the villi are shallow invaginations known as the **crypts of Lieberkühn** (see ❙Figure 16-19c). Unlike the gastric pits, these intestinal crypts do not secrete digestive enzymes, but they do secrete water and salt, which, along with the mucus secreted by the cells on the villus surface, constitute the succus entericus.

Furthermore, the crypts function as nurseries. The epithelial cells lining the small intestine slough off and are replaced at a rapid rate as a result of high mitotic activity of *stem cells* in the crypts. New cells that are continually being produced in the crypts migrate up the villi and, in the process, push off the older cells at the tips of the villi into the lumen. In this manner, more than 100 million intestinal cells are shed per minute. The entire trip from crypt to tip averages about 3 days, so the mucosal epithelium is replaced approximately every 3 days. Because of this high rate of cell division, the crypt stem cells are very sensitive to damage by radiation and anticancer drugs, both of which may inhibit cell division.

The new cells undergo several changes as they migrate up the villus. The concentration of brush-border enzymes increases and the capacity for absorption improves, so the cells at the tip of the villus have the greatest digestive and absorptive capability. Just at their peak, these cells are pushed off by the newly migrating cells. Thus, the luminal contents are constantly exposed to cells that are optimally equipped to complete the digestive and absorptive functions efficiently. Furthermore, just as in the stomach, the rapid turnover of cells in the small intestine is essential because of the harsh luminal conditions. Cells exposed to the abrasive and corrosive luminal contents are easily damaged and cannot live for long, so they must be continually replaced by a fresh supply of newborn cells.

The old cells sloughed off into the lumen are not entirely lost to the body. These cells are digested, with the cell constituents being absorbed into the blood and reclaimed for synthesis of new cells, among other things.

In addition to stem cells, defensive **Paneth cells** are found in the crypts. Paneth cells produce two chemicals that thwart bacteria: (1) *lysozyme,* the bacteria-lysing enzyme also found in saliva; and (2) *defensins,* small proteins with antimicrobial powers (see p. 438).

We now turn attention to the ways in which the epithelial lining of the small intestine is specialized to accomplish absorption of luminal contents and the mechanisms through which the specific dietary constituents are normally absorbed.

Energy-dependent Na⁺ absorption drives passive H₂O absorption.

Na^+ may be absorbed both passively and actively. When the electrochemical gradient favors movement of Na^+ from the lumen to the blood, passive diffusion of Na^+ can occur by paracellular transport *between* the intestinal epithelial cells through the "leaky" tight junctions into the interstitial fluid within the villus. Movement of Na^+ *through* the cells is energy dependent and involves different carriers or channels at the luminal and basolateral membranes, similar to the process of Na^+ reabsorption across the kidney tubules (see pp. 506 and 510). Na^+ enters the epithelial cells across the luminal border either by itself passively through Na^+ channels or in the company of another ion or a nutrient molecule by secondary active transport via three different carriers: Na^+–Cl^- symporter, Na^+–H^+ antiporter, or Na^+–glucose (or amino acid) symporter. Na^+ is actively pumped out of the cell by the Na^+–K^+ pump at the basolateral membrane into the interstitial fluid in the lateral spaces between the cells where they are not joined by tight junctions. From the interstitial fluid, Na^+ diffuses into the capillaries.

As with the renal tubules in the early part of the nephron, the absorption of Cl^-, H_2O, glucose, and amino acids from the small intestine is linked to this energy-dependent Na^+ absorption. Cl^- passively follows down the electrical gradient created by Na^+ absorption and also can be absorbed by secondary active transport if needed. Most H_2O absorption in the digestive tract depends on the active carrier that pumps Na^+ into the lateral spaces, resulting in a concentrated area of high osmotic pressure in that localized region between the cells, similar to the situation in the kidneys (see p. 512). This localized high osmotic pressure induces H_2O to move from the lumen through the cell (and possibly from the lumen through the leaky tight junction) into the lateral space. Water entering the space reduces the osmotic pressure but raises the hydrostatic (fluid) pressure. The elevated hydrostatic pressure flushes H_2O out of the lateral space into the interior of the villus, where it is picked up by the capillary network. Meanwhile, more Na^+ is pumped into the lateral space to encourage more H_2O absorption.

Digested carbohydrates and proteins are both absorbed by secondary active transport and enter the blood.

Absorption of the digestion end products of both carbohydrates and proteins is accomplished by Na^+-dependent symport, and both categories of end products are absorbed into the blood.

Carbohydrate Absorption Dietary carbohydrates are presented to the small intestine for absorption mainly in the forms of the disaccharides maltose, sucrose, and lactose (and to a lesser extent in the form of the short polysaccharide a-limit dextrins) (Figure 16-22a). The disaccharidases located in the brush-border membrane of the small intestine cells further reduce these disaccharides and polysaccharides into the absorbable monosaccharide units of glucose (mostly), galactose, and fructose.

Glucose and galactose are both absorbed by secondary active transport, in which symport carriers, such as the *sodium and glucose cotransporter* (SGLT; see Figure 3-18, p. 76) on the luminal membrane transport both the monosaccharide and Na^+ from the lumen into the interior of the intestinal cell (Figure 16-22b). The operation of these symporters, which do not directly use energy themselves, depends on the Na^+ concentration gradient established by the energy-consuming basolateral Na^+–K^+ pump (see p. 73). Glucose (or galactose), having been concentrated in the cell by these symporters, leaves the cell down its concentration gradient by facilitated diffusion (passive carrier-mediated transport; see p. 72) via the *glucose transporter GLUT-2* in the basal border to enter the blood within the villus. In addition to glucose being absorbed through the cells by means of the symporter, recent evidence suggests that a significant amount of glucose crosses the epithelial barrier through the leaky tight junctions between the epithelial cells.

Fructose enters the epithelial cells from the lumen via *GLUT-5* using facilitated diffusion. This process involves the higher concentration of luminal fructose driving the monosaccharide into the cell. Like the other monosaccharides, fructose exits via GLUT-2 and enters the blood (Figure 16-22b).

Protein Absorption Both ingested proteins and endogenous (within the body) proteins that have entered the digestive tract lumen from the following sources are digested and absorbed:

1. Digestive enzymes, all of which are proteins, that have been secreted into the lumen

2. Proteins within the cells that are pushed off from the villi into the lumen during the process of mucosal turnover

3. Small amounts of plasma proteins that normally leak from the capillaries into the digestive tract lumen

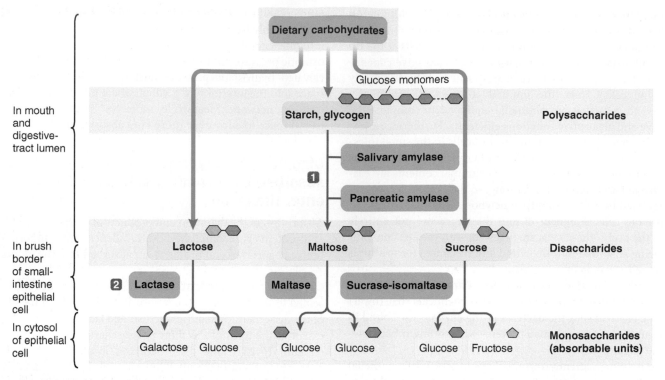

(a) Carbohydrate digestion

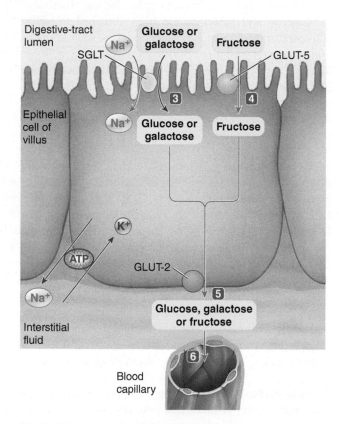

(b) Carbohydrate absorption

■ The dietary polysaccharides starch and glycogen are converted into the disaccharide maltose through the action of salivary and pancreatic amylase.

■ Maltose and the dietary disaccharides lactose and sucrose are converted to their respective monosaccharides by the disaccharidases (maltase, lactase, and sucrase-isomaltase) located in the brush borders of the small-intestine epithelial cells.

■ The monosaccharides glucose and galactose are absorbed into the epithelial cells by Na^+- and energy-dependent secondary active transport (via the symporter SGLT) located at the luminal membrane.

■ The monosaccharide fructose enters the cell by passive facilitated diffusion via GLUT-5.

■ Glucose, galactose, and fructose exit the cell at the basal membrane by passive facilitated diffusion via GLUT-2.

■ These monosaccharides enter the blood by simple diffusion.

KEY

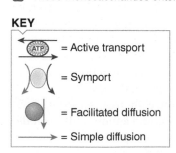

ATP = Active transport
= Symport
= Facilitated diffusion
→ = Simple diffusion

IFigure 16-22 Carbohydrate digestion and absorption.

About 20 to 40 g of endogenous proteins enter the lumen each day from these three sources. This quantity can amount to more than the quantity of proteins in ingested food. All endogenous proteins must be digested and absorbed, along with the dietary proteins, to prevent depletion of the body's protein stores. The amino acids absorbed from both food and endogenous proteins are used primarily to synthesize new proteins in the body.

The proteins presented to the small intestine for absorption are in the form of amino acids and a few small peptide fragments (Figure 16-23a). Amino acids are absorbed into the intestinal cells by symporters, similar to glucose and galactose absorption (Figure 16-23b). The sugar symporters are distinct from the amino-acid symporters, and the amino-acid symporters are selective for different amino acids. Small peptides gain entry by means of yet another Na^+-dependent carrier in a process known as **tertiary active transport** (*tertiary* meaning "third," in reference to a third linked step ultimately being driven by energy used in the first step). In this case, the symporter simultaneously transports both H^+ and the peptide from the lumen into the cell, driven by H^+ moving down its concentration gradient and the peptide moving against its concentration gradient (Figure 16-23b). The H^+ gradient is established by an antiporter in the luminal membrane that is driven by Na^+ moving into the cell down its concentration gradient and H^+ moving out of the cell against its concentration gradient. The Na^+ concentration gradient that drives the antiporter in turn is established by the energy-dependent Na^+–K^+ pump at the basolateral membrane. Thus, glucose, galactose, amino acids, and small peptides all get a "free ride" in on the energy expended for Na^+ transport. The small peptides are broken down into their constituent amino acids by the aminopeptidases in the brush-border membrane or by intracellular peptidases (Figure 16-23a). Like monosaccharides, amino acids leave the intestinal cells by facilitated diffusion and enter the capillary network within the villus.

Digested fat is absorbed passively and enters the lymph.

Fat absorption is different from carbohydrate and protein absorption because the insolubility of fat in water presents a special problem. Fat must be transferred from the watery chyme through the watery body fluids, even though fat is not water soluble. Therefore, fat must undergo a series of physical and chemical transformations to circumvent this problem during its digestion and absorption (Figure 16-24).

A Review of Fat Emulsification and Digestion When the stomach contents are emptied into the duodenum, the ingested fat is aggregated into large, oily triglyceride droplets that float in the chyme. Recall that through the bile salts' detergent action in the small-intestine lumen, the large droplets are dispersed into a lipid emulsification of small droplets, exposing a greater surface area of fat for digestion by pancreatic lipase (Figure 16-24, step **1**). The products of lipase digestion (monoglycerides and free fatty acids; step **2**) are also not very water soluble, so little of these end products of fat digestion can diffuse through the aqueous chyme to reach the absorptive lining.

However, biliary components facilitate absorption of these fatty end products by forming micelles.

Fat Absorption Remember that micelles are water-soluble particles that can carry the end products of fat digestion within their lipid-soluble interiors (Figure 16-24, step **3**). Once these micelles reach the luminal membranes of the epithelial cells, the monoglycerides and free fatty acids passively diffuse from the micelles through the lipid component of the epithelial cell membranes to enter the interior of these cells (step **4**).

Bile salts continuously repeat their fat-solubilizing function down the length of the small intestine until all fat is absorbed. Then the bile salts themselves are reabsorbed in the terminal ileum by special active transport. This is an efficient process because relatively small amounts of bile salts can facilitate digestion and absorption of large amounts of fat, with each bile salt performing its ferrying function repeatedly before it is reabsorbed.

Once within the interior of the epithelial cells, the monoglycerides and free fatty acids are resynthesized into triglycerides (step **5**). These triglycerides conglomerate into droplets and are coated with a layer of lipoprotein (synthesized by the endoplasmic reticulum of the epithelial cell), which makes the fat droplets water soluble (step **6**). The large, coated fat droplets, known as **chylomicrons**, are extruded by exocytosis from the epithelial cells into the interstitial fluid within the villus (step **7**). Chylomicrons are 75 to 500 nm in diameter, compared to micelles, which are 3 to 10 nm in diameter. The chylomicrons subsequently enter the central lacteals rather than the capillaries because of the structural differences between these two vessels (step **8**). Capillaries have a basement membrane (an outer layer of polysaccharides) (see p. 352) that prevents the chylomicrons from entering, but the lymph vessels do not have this barrier. Thus, fat can be absorbed into the lymphatics but not directly into the blood.

The actual absorption of monoglycerides and free fatty acids from the chyme across the luminal membrane of the small-intestine epithelial cells is traditionally considered a passive process because the lipid-soluble fatty end products merely dissolve in and pass through the lipid part of the membrane. However, the overall sequence of events needed for fat absorption requires energy. For example, bile salts are actively secreted by the liver, the resynthesis of triglycerides and formation of chylomicrons within the epithelial cells are active processes, and the exocytosis of chylomicrons requires energy.

Vitamin absorption is largely passive.

Water-soluble vitamins are primarily absorbed passively with water, whereas fat-soluble vitamins are carried in the micelles and absorbed passively with the end products of fat digestion. Some of the vitamins can also be absorbed by carriers, if necessary. Vitamin B_{12} is unique in that it must be in combination with gastric intrinsic factor for absorption by receptor-mediated endocytosis in the terminal ileum.

Iron and calcium absorption is regulated.

In contrast to the almost complete, unregulated absorption of other ingested electrolytes, dietary iron and calcium may not be absorbed completely because their absorption is subject to

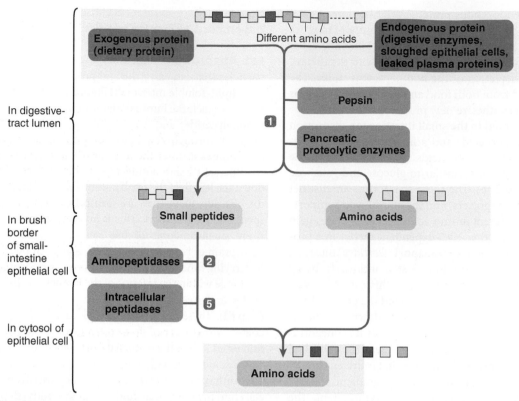

(a) Protein digestion

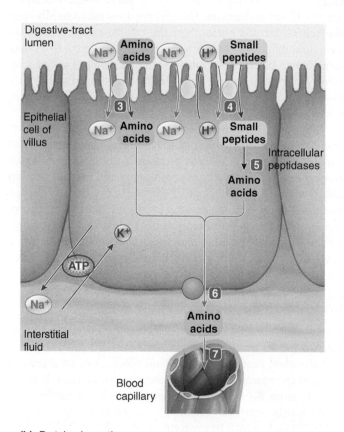

Digestive-tract lumen

Epithelial cell of villus

Interstitial fluid

Blood capillary

(b) Protein absorption

1 Dietary and endogenous proteins are hydrolyzed into their constituent amino acids and a few small peptide fragments by gastric pepsin and the pancreatic proteolytic enzymes.

2 Many small peptides are converted into their respective amino acids by the aminopeptidases located in the brush borders of the small-intestine epithelial cells.

3 Amino acids are absorbed into the epithelial cells by means of Na$^+$- and energy-dependent secondary active transport via a symporter. Various amino acids are transported by carriers specific for them.

4 Some small peptides are absorbed by a different type of symporter driven by H$^+$, Na$^+$-, and energy-dependent tertiary active transport.

5 Most absorbed small peptides are broken down into their amino acids by intracellular peptidases.

6 Amino acids exit the cell at the basal membrane via various passive carriers.

7 Amino acids enter the blood by simple diffusion. (A small percentage of di- and tripeptides also enter the blood intact.)

KEY

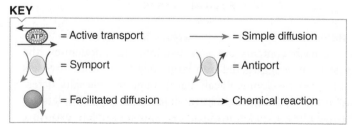

= Active transport   = Simple diffusion

= Symport   = Antiport

= Facilitated diffusion   = Chemical reaction

IFigure 16-23 Protein digestion and absorption.

FIGURE FOCUS: *Which monosaccharide is absorbed passively and which require energy to be absorbed?*

Figure 16-24 Fat digestion and absorption. Because fat is not soluble in water, it must undergo a series of transformations to be digested and absorbed.

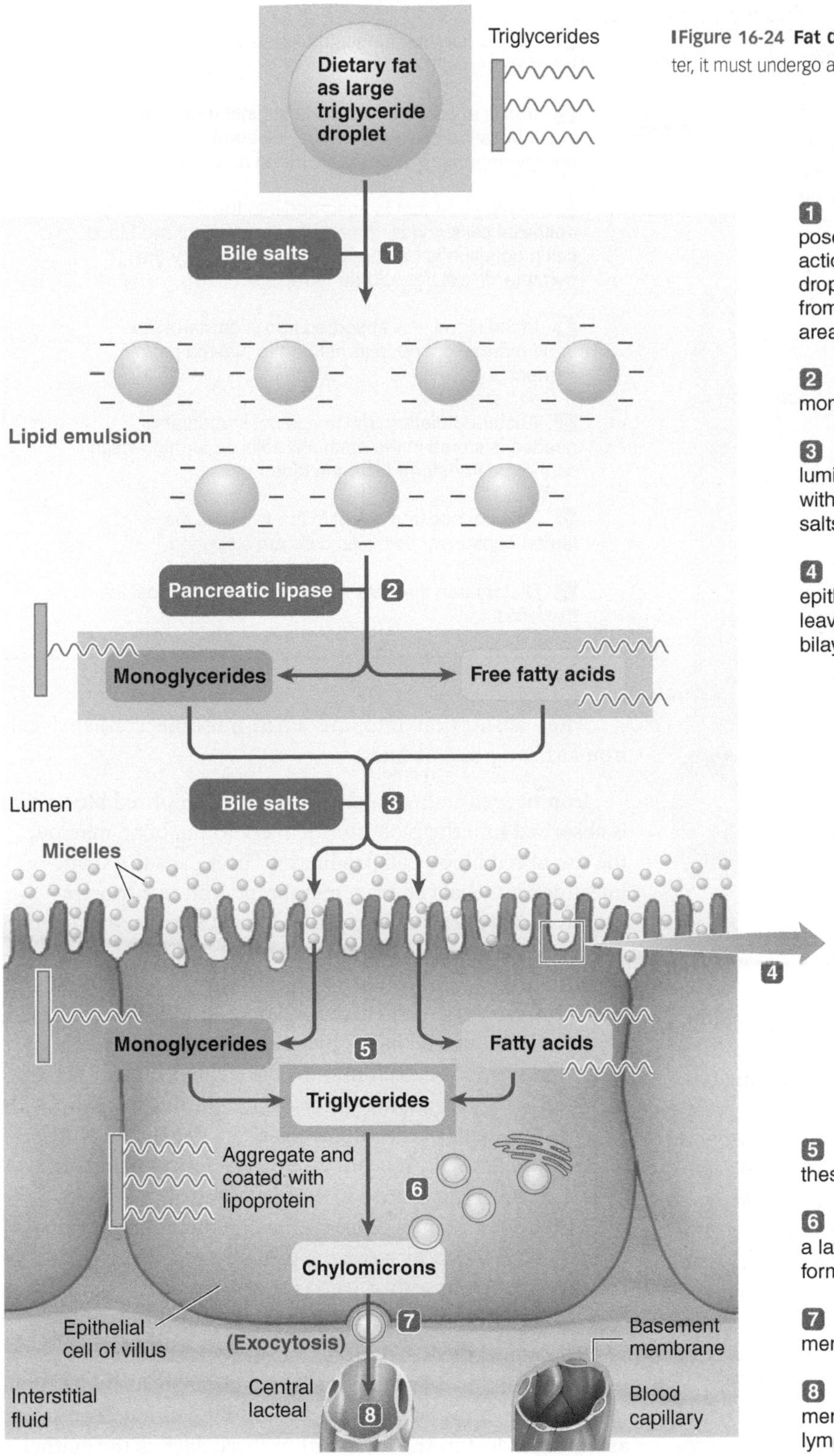

Dietary fat as large triglyceride droplet
Triglycerides
Bile salts
1
Lipid emulsion
Pancreatic lipase
2
Monoglycerides
Free fatty acids
Lumen
Bile salts
3
Micelles
4
Monoglycerides
5
Fatty acids
Triglycerides
Aggregate and coated with lipoprotein
6
Chylomicrons
7
Epithelial cell of villus
(Exocytosis)
Basement membrane
Interstitial fluid
Central lacteal
8
Blood capillary

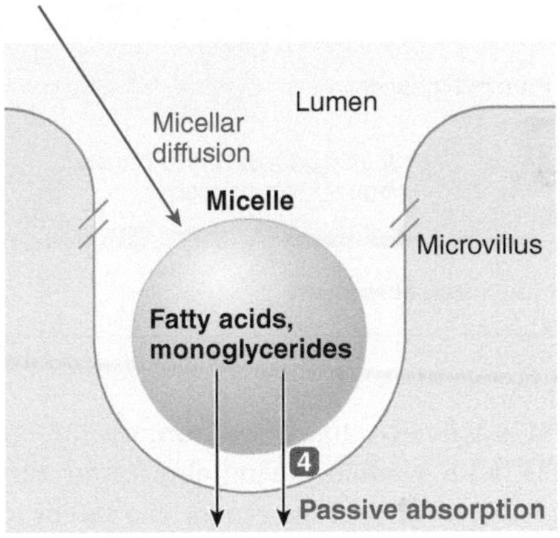

Micellar diffusion
Lumen
Micelle
Microvillus
Fatty acids, monoglycerides
4
Passive absorption

1 Dietary fat in the form of large fat globules composed of triglycerides is emulsified by the detergent action of bile salts into a suspension of smaller fat droplets. This lipid emulsion prevents the fat droplets from coalescing and thereby increases the surface area available for attack by pancreatic lipase.
2 Lipase hydrolyzes the triglycerides into monoglycerides and free fatty acids.
3 These water-insoluble products are carried to the luminal surface of the small-intestine epithelial cells within water-soluble micelles, which are formed by bile salts and other bile constituents.
4 When a micelle approaches the absorptive epithelial surface, the monoglycerides and fatty acids leave the micelle and passively diffuse through the lipid bilayer of the luminal membranes.
5 The monoglycerides and free fatty acids are resynthesized into triglycerides inside the epithelial cells.
6 These triglycerides aggregate and are coated with a layer of lipoprotein from the endoplasmic reticulum to form water-soluble chylomicrons.
7 Chylomicrons are extruded through the basal membrane of the cells by exocytosis.
8 Chylomicrons are unable to cross the basement membrane of capillaries, so instead they enter the lymphatic vessels, the central lacteals.

regulation, depending on the body's needs for these electrolytes. Normally, only enough iron and calcium are actively absorbed into the blood to maintain the homeostasis of these electrolytes, with excess ingested quantities being lost in the feces.

Iron Absorption Iron is essential for hemoglobin production. The normal iron intake is typically 15 to 20 mg/day, yet a man usually absorbs about 0.5 to 1 mg/day into the blood, and a woman takes up slightly more, at 1.0 to 1.5 mg/day. (Women need more iron because they periodically lose iron in menstrual blood flow.)

Two main steps are involved in absorption of iron into the blood: (1) absorption of iron from the lumen into the small-intestine epithelial cells and (2) absorption of iron from the epithelial cells into the blood (Figure 16-25).

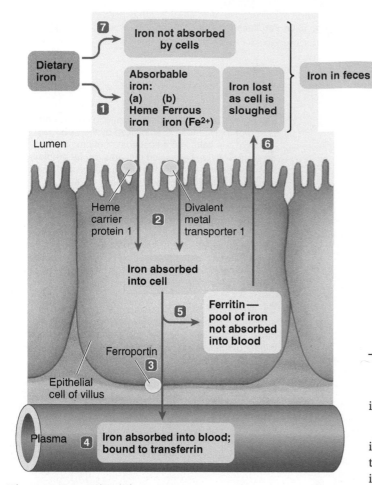

7 Iron not absorbed by cells

Dietary iron

Absorbable iron:
(a) **(b)**
Heme **Ferrous**
iron **iron (Fe²⁺)**
1

Iron lost as cell is sloughed

Iron in feces

Lumen

Heme carrier protein 1

Divalent metal transporter 1
2

Iron absorbed into cell

Ferritin— pool of iron not absorbed into blood
5

Ferroportin
3

Epithelial cell of villus

Plasma **4** **Iron absorbed into blood; bound to transferrin**

ⅠFigure 16-25 **Iron absorption.**

1 Only a portion of ingested iron is in a form that can be absorbed, either heme iron or ferrous iron (Fe^{2+}).

2 Iron is absorbed across the luminal membrane of small-intestine epithelial cells by different energy-dependent carriers for heme and Fe^{2+}.

3 Dietary iron that is absorbed into the small-intestine epithelial cells and is immediately needed for red blood cell production is transferred into the blood by the membrane iron transporter ferroportin.

4 In the blood, the absorbed iron is carried to the bone marrow bound to transferrin, a plasma protein carrier.

5 Absorbed dietary iron that is not immediately needed is stored in the epithelial cells as ferritin, which cannot be transferred into the blood.

6 This unused iron is lost in the feces as the ferritin-containing epithelial cells are sloughed.

7 Dietary iron that was not absorbed is also lost in the feces.

Iron is actively transported from the lumen into the epithelial cells, with women having about four times more active-transport sites for iron than men. The extent to which ingested iron is taken up by the epithelial cells depends on the type of iron consumed. Dietary iron exists in two forms: *heme iron,* in which iron is bound as part of a heme group found in hemoglobin (see p. 383) and is present in meat, and *inorganic iron,* which is present in plants. Dietary heme is absorbed more efficiently than inorganic iron is. Dietary inorganic iron exists primarily in the oxidized ferric iron (Fe^{3+}) form, but the reduced ferrous iron (Fe^{2+}) form is absorbed more easily. Dietary Fe^{3+} is reduced to Fe^{2+} by a membrane-bound enzyme at the luminal membrane before absorption. The presence of other substances in the lumen can either promote or reduce iron absorption. For example, vitamin C increases iron absorption, primarily by reducing Fe^{3+} to Fe^{2+}. Phosphate and oxalate, in contrast, combine with ingested iron to form insoluble iron salts that cannot be absorbed.

Heme iron and Fe^{2+} are transported across the luminal membrane by separate energy-dependent carriers in the brush border: Heme iron enters the intestinal cell by *heme carrier protein 1* and Fe^{2+} is carried in via *divalent metal transporter 1,* which also transports other metals that have a valence of +2. An enzyme within the cell frees iron from the heme complex.

After absorption into the small-intestine epithelial cells, iron has two possible fates:

1. Iron needed immediately for production of red blood cells is absorbed into the blood for delivery to the bone marrow, the site of red blood cell production. Iron exits the small-intestine epithelial cell via a membrane iron transporter known as **ferroportin.** Iron absorption is largely controlled by a recently discovered hormone, **hepcidin,** which is released from the liver when iron levels in the body become too high. Hepcidin prevents further iron export from the small-intestine epithelial cell into the blood by binding with ferroportin and promoting its internalization into the cell by endocytosis and its subsequent degradation by lysosomes. Thus, hepcidin is the primary regulator of iron homeostasis. A deficiency of hepcidin leads to tissue iron overload because ferroportin continues to transfer iron into the body without control.

Iron that exits the small-intestine epithelial cell is transported in the blood by a plasma protein carrier known as **transferrin.** The absorbed iron is then used in the synthesis of hemoglobin for the newly produced red blood cells.

2. Iron not immediately needed is irreversibly stored within the small-intestine epithelial cells in a granular form called **ferritin,** which cannot be absorbed into the blood. Iron stored as ferritin is lost in the feces within three days as the epithelial cells containing these granules are sloughed off during mucosal regeneration. Large amounts of iron in the feces give them a dark, almost black color.

Calcium Absorption The amount of calcium (Ca^{2+}) absorbed is also regulated. Calcium enters the luminal membrane of the small-intestine epithelial cells down its electrochemical gradient through a specialized Ca^{2+} channel; is ferried within the cell by a Ca^{2+}-binding protein, **calbindin;** and exits the basolateral membrane by two energy-dependent mechanisms: a primary active transport Ca^{2+} ATPase pump and a secondary active transport $Na^+–Ca^{2+}$ antiporter. Vitamin D greatly enhances all of these steps in Ca^{2+} absorption. Vitamin D can exert this effect only after it has been activated in the liver and kidneys, a process that is enhanced by parathyroid hormone. Appropriately, secretion of parathyroid hormone increases in response to a fall in Ca^{2+} concentration in the blood. Normally, of the average 1000 mg of Ca^{2+} taken in daily, only about two thirds is absorbed in the small intestine, with the rest passing out in the feces.

Most absorbed nutrients immediately pass through the liver for processing.

The venules that leave the small-intestine villi, along with those from the rest of the digestive tract, empty into the hepatic portal vein, which carries the blood to the liver. Consequently, anything absorbed into the digestive capillaries first must pass through the hepatic biochemical factory before entering the general circulation. Thus, the products of carbohydrate and protein digestion are channeled into the liver, where many of these energy-rich products are subjected to immediate metabolic processing. Furthermore, harmful substances that may have been absorbed are detoxified by the liver before gaining access to the general circulation. After passing through the portal circulation, the venous blood from the digestive system empties into the vena cava and returns to the heart to be distributed throughout the body, carrying glucose and amino acids for use by the tissues.

Fat, which cannot penetrate the intestinal capillaries, is picked up by the central lacteal and enters the lymphatic system instead, bypassing the hepatic portal system. Contractions of the villi, accomplished by the muscularis mucosa, periodically compress the central lacteal and "milk" the lymph out of this vessel. The smaller lymph vessels converge and eventually form the *thoracic duct,* a large lymph vessel that empties into the venous system within the chest. In this way, fat ultimately gains access to the blood. The absorbed fat is carried by the systemic circulation to the liver and to other tissues of the body. Therefore, the liver does have a chance to act on the digested fat, but not until the fat has been diluted by the blood in the general circulatory system. This dilution of fat protects the liver from being inundated with more fat than it can handle at one time.

Extensive absorption by the small intestine keeps pace with secretion.

The small intestine normally absorbs about 9 liters of fluid per day in the form of H_2O and solutes, including the absorbable units of nutrients, vitamins, and electrolytes. How can that be, when humans normally ingest only about 1250 mL of fluid and consume 1250 g of solid food (80% of which is H_2O) per day (see p. 544)? ▮Table 16-7 illustrates the tremendous daily absorption performed by the small intestine. Each day, about 9500 mL of H_2O and solutes enter the small intestine. Note that of this 9500 mL, only 2500 mL are ingested from the external environment. The remaining 7000 mL (7 liters) of fluid are digestive juices derived from the plasma. Recall that plasma is

▮**TABLE 16-7** Volumes Absorbed by the Small and Large Intestine per Day

Volume entering the small intestine per day

Sources	Ingested	Food eaten	1250 g*
		Fluid drunk	1250 mL
	Secreted from the plasma	Saliva	1500 mL
		Gastric juice	2000 mL
		Pancreatic juice	1500 mL
		Bile	500 mL
		Intestinal juice	1500 mL

	9500 mL
Volume absorbed by the small intestine per day	9000 mL
Volume entering the colon from the small intestine per day	500 mL
Volume absorbed by the colon per day	350 mL
Volume of feces eliminated from the colon per day	150 g*

*Because 1 mL of H_2O weighs 1 g, and a high percentage of food and feces is H_2O, we can roughly equate grams with milliliters of fluid.

the ultimate source of digestive secretions because the secretory cells extract from the plasma the necessary raw materials for their secretory product. Considering that the entire plasma volume is only about 2.75 liters, absorption must closely parallel secretion to keep the plasma volume from falling sharply.

Of the 9500 mL of fluid entering the small-intestine lumen per day, about 95%, or 9000 mL of fluid, is normally absorbed by the small intestine back into the plasma, with only 500 mL of the small-intestine contents passing on into the colon. Thus, the body normally does not lose the digestive juices. After the constituents of the juices are secreted into the digestive tract lumen and perform their function, they are returned to the plasma. The only secretory product that escapes from the body is bilirubin, a waste product that must be eliminated.

Biochemical balance among the stomach, pancreas, and small intestine is normally maintained.

Production of gastric and pancreatic digestive secretions typically does not alter the acid–base status of the body because the amount of H^+ secreted by the gastric parietal cells is usually matched by the amount of HCO_3^- secreted by the pancreatic duct cells. Also, the by-products generated during these secretory processes—HCO_3^- by the parietal cells and H^+ by the pancreatic duct cells—are normally transported back into the plasma in equal amounts.

Furthermore, the acid–base balance is not altered as the secreted juices are absorbed back into the plasma. Within the small-intestine lumen, the HCl secreted by the parietal cells of the stomach is neutralized by the $NaHCO_3$ secreted by the pancreatic duct cells:

$$HCl + NaHCO_3 \rightarrow NaCl + H_2CO_3$$

The resultant H_2CO_3 decomposes into $CO_2 + H_2O$:

$$H_2CO_3 \rightarrow CO_2 + H_2O$$

The end products of these reactions—NaCl (ionized as Na^+ and Cl^-), CO_2, and H_2O—are all absorbed by the small intestine into the blood. Thus, through these interactions, the body normally does not experience a net gain or loss of acid or base during digestion.

Diarrhea results in loss of fluid and electrolytes.

Clinical Note When secretion and absorption do not parallel each other, however, acid–base abnormalities can result because these normal neutralization processes cannot take place. We have already described vomiting and the subsequent loss of acidic gastric contents leading to metabolic alkalosis. The other common digestive tract disturbance that can lead to a loss of fluid and an acid–base imbalance is **diarrhea**. This condition is characterized by passage of a highly fluid fecal matter, often with increased frequency of defecation. Not only are some of the ingested materials lost but some of the secreted materials that normally would have been reabsorbed are lost also. Excessive loss of intestinal contents causes dehydration, loss of nutrient material, and metabolic acidosis resulting from loss of HCO_3^- (see p. 561). The abnormal fluidity of the feces usually occurs because the small intestine is unable to absorb fluid as extensively as normal. This extra unabsorbed fluid passes out in the feces.

The causes of diarrhea are as follows:

1. The most common cause of diarrhea is excessive small-intestinal motility, which arises either from local irritation of the gut wall by bacterial or viral infection of the small intestine or from emotional stress. Rapid transit of the small-intestine contents does not allow enough time for adequate absorption of fluid to occur.

2. Diarrhea also occurs when excess osmotically active particles, such as those found in lactase deficiency, are present in the digestive tract lumen. These particles cause excessive fluid to enter and be retained in the lumen, thus increasing the fluidity of the feces.

3. Toxins of the bacterium *Vibrio cholera* (the causative agent of cholera) and certain other microorganisms promote the secretion of excessive amounts of fluid by the small-intestine mucosa, resulting in profuse diarrhea. Diarrhea produced in response to toxins from infectious agents is the leading cause of death of small children in developing nations. Fortunately, a low-cost, effective *oral rehydration therapy* that takes advantage of the intestine's glucose symport carrier is saving the lives of millions of children. (For details about oral rehydration therapy, see the accompanying boxed feature, ❚Concepts, Challenges, and Controversies.)

Check Your Understanding 16.6

1. Explain how segmentation accomplishes both mixing and propulsion.

2. Describe the structural features that increase the surface area of the small intestine and explain the significance of increasing the surface area.

3. Discuss how the $Na^+–K^+$ pump of mucosal epithelial cells facilitates nutrient absorption.

16.7 Large Intestine

The **large intestine** consists of the colon, cecum, appendix, and rectum (❚Figure 16-26). The **cecum** forms a blind-ended pouch below the junction of the small and large intestines at the ileocecal valve. The small, fingerlike projection at the bottom of the cecum is the **appendix**, a lymphoid tissue that houses lymphocytes (see p. 405). The **colon**, which makes up most of the large intestine, is not coiled like the small intestine but consists of three relatively straight parts—the *ascending colon,* the *transverse colon,* and the *descending colon.* The end part of the descending colon becomes S shaped, forming the *sigmoid colon* (*sigmoid* means "S shaped"), and then straightens out to form the rectum (meaning "straight").

The large intestine is primarily a drying and storage organ.

The colon normally receives about 500 mL of chyme from the small intestine each day. Because most digestion and absorption have been accomplished in the small intestine, the contents delivered to the colon consist of indigestible food residues (such as cellulose), unabsorbed biliary components, and the remaining fluid. The colon extracts more H_2O and salt, drying and compacting the contents to form a firm mass known as **feces** for elimination from the body. The primary function of the large intestine is to store feces before defecation. Cellulose and other

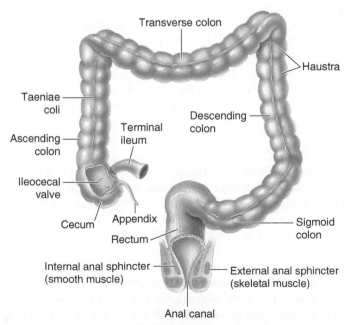

❚Figure 16-26 **Anatomy of the large intestine.**

Oral Rehydration Therapy: Sipping a Simple Solution Saves Lives

DIARRHEA-INDUCING MICROORGANISMS SUCH AS *Vibrio cholera*, which causes cholera, are the leading cause of death in children younger than age 5 worldwide. The problem is especially pronounced in developing countries, refugee camps, and elsewhere where poor sanitary conditions encourage the spread of the microorganisms, and medical supplies and health-care personnel are scarce. Fortunately, a low-cost, easily obtainable, uncomplicated remedy—**oral rehydration therapy (ORT)**—has been developed to combat potentially fatal diarrhea. This treatment exploits the symporters located at the luminal border of the villus epithelial cells.

Let us examine the pathophysiology of life-threatening diarrhea and then see how simple ORT can save lives. During digestion of a meal, the crypt cells of the small intestine normally secrete succus entericus, a salt and mucus solution, into the lumen. These cells actively transport Cl^- into the lumen, promoting the parallel passive transport of Na^+ and H_2O from the blood into the lumen. The fluid provides the watery environment needed for enzymatic breakdown of ingested nutrients into absorbable units. Glucose and amino acids, the absorbable units of dietary carbohydrates and proteins, respectively, are absorbed by secondary active transport. This absorption mechanism uses the Na^+–glucose (or amino acid) cotransport carriers (SGLT) located at the luminal membrane of the villus epithelial cells (see p. 75). In addition, separate active Na^+ carriers not linked with nutrient absorption transfer Na^+, passively accompanied by Cl^- and H_2O, from the lumen into the blood.

The net result of these various carrier activities is absorption of the secreted salt and H_2O along with the digested nutrients. Normally, absorption of salt and H_2O exceeds their secretion, so not only are the secreted fluids salvaged, but also additional ingested salt and H_2O are absorbed.

Cholera and most diarrhea-inducing microbes cause diarrhea by stimulating the secretion of Cl^- or impairing the absorption of Na^+. As a result, more fluid is secreted from the blood into the lumen than is subsequently transferred back into the blood. The excess fluid is lost in the feces, producing the watery stool characteristic of diarrhea. More important, the loss of fluids and electrolytes that came from the blood leads to dehydration. The subsequent reduction in effective circulating plasma volume can cause death in a matter of days or even hours, depending on the severity of the fluid loss.

In the middle of the past century, physicians learned that replacing the lost fluids and electrolytes intravenously saves the lives of most patients with diarrhea. In many parts of the world, however, adequate facilities, equipment, and personnel are not available to administer intravenous rehydration therapy. Consequently, millions of children still succumbed to diarrhea annually.

In 1966 researchers learned that SGLT is not affected by diarrhea-causing microbes. This discovery led to the development of ORT. When both Na^+ and glucose are present in the lumen, this symporter transports them both from the lumen into the villus epithelial cells, from which they enter the blood. Because H_2O osmotically follows the absorbed Na^+, ingestion of a glucose and salt solution promotes the uptake of fluid into the blood from the intestinal tract without the need for intravenous replacement of fluids.

The first proof of ORT's life-saving ability in the field came in 1971 when several million refugees poured into India from war-ravaged Bangladesh. Of the thousands of refugees who fell victim to cholera and other diarrheal diseases, more than 30% died because of the scarcity of sterile fluids and needles for intravenous therapy. In one refugee camp, however, under the supervision of a group of scientists who had been experimenting with ORT, families were taught to administer ORT to people with diarrhea, most of whom were small children. The scarce intravenous solutions were reserved for those unable to drink. Death from diarrhea was reduced to 3% in this camp, compared with a 10-fold higher mortality among refugees elsewhere.

Based on this evidence, the World Health Organization (WHO) started aggressively promoting ORT. Packets of dry ingredients for ORT are now manufactured locally in more than 60 countries. The WHO estimates that about 30% of the world's children who contract diarrhea are treated with the prepackaged mixture or home-prepared versions. In the United States, commercially prepared oral solutions are widely available at pharmacies and supermarkets. An estimated 1 million children worldwide are saved annually as a result of ORT.

---

indigestible substances in the diet provide bulk and help maintain regular bowel movements by contributing to the volume of the colonic contents.

Haustral contractions slowly shuffle the colonic contents back and forth.

The outer longitudinal smooth muscle layer does not completely surround the large intestine. Instead, it consists only of three separate, conspicuous, longitudinal bands of muscle, the **taeniae coli**, which run the length of the large intestine. These taeniae coli are shorter than the underlying circular smooth muscle and mucosal layers would be if these layers were stretched out flat. Because of this, the underlying layers are gathered into pouches or sacs called **haustra**, much as the material of a full skirt is gathered at the narrower waistband. The haustra are not merely passive permanent gathers, however; they actively change location as a result of contraction of the circular smooth muscle layer.

Most of the time, movements of the large intestine are slow and nonpropulsive, as is appropriate for its absorptive and storage functions. The colon's main motility is **haustral contrac-**

tions initiated by the BER of colonic smooth muscle cells. These contractions, which throw the large intestine into haustra, are oscillating ringlike contractions similar to small-intestine segmentations but occur less frequently. Thirty minutes may elapse between haustral contractions, whereas segmentation contractions in the small intestine occur at rates of between 9 and 12 per minute. The location of the haustral sacs gradually changes as a relaxed segment that has formed a sac slowly contracts while a previously contracted area simultaneously relaxes to form a new sac. These movements are nonpropulsive; they slowly shuffle the contents in a back-and-forth mixing movement that exposes the colonic contents to the absorptive mucosa. Haustral contractions are largely controlled by locally mediated reflexes involving the intrinsic plexuses.

Mass movements propel feces long distances.

Three to four times a day, generally after meals, a marked increase in motility takes place during which large segments of the ascending and transverse colon contract simultaneously, driving the feces one third to three fourths of the length of the colon in a few seconds. These massive contractions, appropriately called **mass movements**, drive the colonic contents into the distal part of the large intestine, where material is stored until defecation occurs.

When food enters the stomach, mass movements are triggered in the colon primarily by the **gastrocolic reflex**, which is mediated from the stomach to the colon by gastrin and by parasympathetic innervation. In many people, this reflex is most evident after the first meal of the day and is often followed by the urge to defecate. Thus, when a new meal enters the digestive tract, reflexes are initiated to move the existing contents farther along the tract to make way for the incoming food. The gastroileal reflex moves the remaining small-intestine contents into the large intestine, and the gastrocolic reflex pushes the colonic contents into the rectum, triggering the defecation reflex.

Feces are eliminated by the defecation reflex.

When mass movements move feces into the rectum, the resultant distension of the rectum stimulates stretch receptors in the rectal wall, initiating the **defecation reflex**. This reflex causes the **internal anal sphincter** (which is smooth muscle) to relax and the rectum and sigmoid colon to contract more vigorously. If the **external anal sphincter** (which is skeletal muscle) is also relaxed, defecation occurs. Being skeletal muscle, the external anal sphincter is under voluntary control. The initial rectal distension is accompanied by the conscious urge to defecate. If circumstances are unfavorable for defecation, voluntary tightening of the external anal sphincter can prevent defecation despite the defecation reflex. If defecation is delayed, the distended rectal wall gradually relaxes, and the urge to defecate subsides until the next mass movement propels more feces into the rectum, again distending the rectum and triggering the defecation reflex. During periods of inactivity, both anal sphincters remain contracted to ensure fecal continence.

When defecation does occur, it is usually assisted by voluntary straining movements that involve simultaneous contraction of the abdominal muscles and a forcible expiration against a closed glottis. This maneuver greatly increases intra-abdominal pressure, which helps expel the feces.

Constipation occurs when the feces become too dry.

If defecation is delayed too long, **constipation** may result. When colonic contents are retained for longer periods than normal, more than the usual amount of H_2O is absorbed from the feces, so they become hard and dry. Normal variations in frequency of defecation among individuals range from after every meal to up to once a week. When the frequency is delayed beyond what is normal for a particular person, constipation and its attendant symptoms may occur. These symptoms include abdominal discomfort, dull headache, loss of appetite sometimes accompanied by nausea, and mental depression. Contrary to popular belief, these symptoms are not caused by toxins absorbed from the retained fecal material. Although bacterial metabolism produces some potentially toxic substances in the colon, these substances normally pass through the portal system and are removed by the liver before they can reach the systemic circulation. Instead, the symptoms associated with constipation are caused by prolonged distension of the large intestine, particularly the rectum; the symptoms promptly disappear after relief from distension.

Possible causes for delayed defecation that might lead to constipation include (1) ignoring the urge to defecate; (2) decreased colon motility accompanying aging, emotion, or a low-bulk diet; (3) obstruction of fecal movement in the large bowel caused by a local tumor or colonic spasm; and (4) impairment of the defecation reflex, such as through injury of the nerve pathways involved.

Clinical Note If hardened fecal material becomes lodged in the appendix, it may obstruct normal circulation and mucus secretion in this narrow, blind-ended appendage. This blockage leads to **appendicitis**. The inflamed appendix often becomes swollen and filled with pus, and the tissue may die as a result of local circulatory interference. If not surgically removed, the diseased appendix may rupture, spewing its infectious contents into the abdominal cavity.

Intestinal gases are absorbed or expelled.

Occasionally, instead of feces passing from the anus, intestinal gas, or **flatus**, passes out. This gas is derived from two sources: (1) swallowed air (as much as 500 mL of air may be swallowed during a meal) and (2) gas produced by bacterial fermentation in the colon. The presence of gas percolating through the luminal contents gives rise to gurgling sounds. Burping removes most of the swallowed air from the stomach, but some passes on into the intestine. Usually, very little gas is present in the small intestine because the gas is either quickly absorbed or passes on into the colon. Most gas in the colon is the result of bacterial activity, with the quantity and nature of the gas depending on the type of food eaten and the characteristics of the colonic bacteria. Much of the gas is absorbed through the intestinal mucosa. The rest is expelled through the anus.

To selectively expel gas when feces are also present in the rectum, the person voluntarily contracts the abdominal muscles and external anal sphincter at the same time. When abdominal contraction raises the pressure against the contracted anal sphincter sufficiently, the pressure gradient forces air out at a high velocity through a slitlike anal opening that is too narrow for solid feces to escape through. This passage of air at high velocity causes the edges of the anal opening to vibrate, giving rise to the characteristic low-pitched sound accompanying passage of gas.

Large-intestine secretion is entirely protective.

Colonic secretion consists of an alkaline ($NaHCO_3$) mucus solution, whose function is to protect the large-intestine mucosa from mechanical and chemical injury. The mucus provides lubrication to facilitate passage of feces, whereas the $NaHCO_3$ neutralizes irritating acids produced by local bacterial fermentation. Secretion increases in response to mechanical and chemical stimulation of the colonic mucosa mediated by short reflexes and parasympathetic innervation.

The large intestine does not secrete any digestive enzymes. Digestion of food constituents that humans have the ability to digest is completed within the small intestine. Of the ingested energy-rich nutrients, only indigestible fiber reaches the colon. However, the colon contains an abundance of bacteria that break down undigested fiber for their own and our use.

The colon contains myriad beneficial bacteria.

Because of slow colonic movement, bacteria have time to grow and accumulate in the large intestine. Not all ingested bacteria are destroyed by antimicrobial agents earlier in the digestive tract; the surviving bacteria continue to thrive in the large intestine. About 10 times more bacteria live in the human colon than the human body has cells. About 2000 species of bacteria have been identified in the human large intestine. Although bacteria are by far the most numerous microscopic colonic residents, fungi and viruses of various sorts also live in the large intestine. The gut microbes weigh up to 4 pounds in total.

Microbes also inhabit other parts of our bodies in contact with the external environment, such as the skin, nose, mouth, pharynx, and vagina. Collectively, the community of microbes that coexist peacefully and usefully with their human host is called the **microbiota**, and the aggregate collection of genomes of our microbiota is known as the **microbiome**. Researchers have identified 8 million nonredundant microbial genes in the microbiome, a number 400 times greater than the 20,000 genes in our human genome (see p. 24). Each person has a unique microbiota. We normally harbor no microbes in the sterile uterus but are seeded with bacteria from our mother's birth canal during birth and acquire further resident microorganisms through variable exposure to people, animals, and objects in our environment after birth. The composition of our gut microbiota is further shaped by our behaviors, such as what we eat (different dietary habits support different populations of microorganisms) and the type and extent of antibiotics we use (antibiotics inadvertently destroy intestinal microbes).

We do not just provide a home for the microbiota. In a mutually beneficial relationship, the diverse mix of microbes

that reside with us plays important roles in our well-being, as will be described shortly. Because we and our microbiota codevelop and are interdependent, some scientists boldly propose that we should broaden the concept of "self" to start thinking about our resident microbial community as part of us—that is, to view a person as a superorganism consisting of an assembly of human cells and many symbiotic microbial species with a vast combined inventory of genes.

The indigenous colonic microbes:

1. *Promote colonic motility.*

2. *Help maintain colonic mucosal integrity.*

3. *Aid immune function.* In interplay fashion, the immune system helps shape the composition of the colon microbiota, and these microorganisms in turn promote the normal development and activities of the immune system. Some bacteria also help rein in localized intestinal inflammation and help calm immune system overreactivity.

4. *Compete with potentially pathogenic microbes for nutrients and space* (see p. 441). By crowding out infectious microbes, the normal resident microbiota make it hard for these disease-causing microorganisms to establish themselves in the intestine. Taking an oral antibiotic for an infection elsewhere in the body can sometimes lead to an intestinal infection by disrupting the normal protective gut microbial community.

5. *Help digest food and make nutritional contributions.* Gut microbes increase digestive efficiency by producing enzymes that break down dietary fiber that human digestive enzymes cannot hydrolyze. Colonic bacteria ferment fiber primarily to short-chain fatty acids, with gases being produced as a by-product. This metabolic processing not only nourishes the microbes but also provides their host with an additional source of nutrition that is otherwise lost in the feces. Being fat soluble, some of the fatty acids are absorbed by simple diffusion by the colonic mucosa, accounting on average for 5% to 10% of daily caloric intake. Gut microbes make other nutritional contributions in addition to digesting fiber. For example, bacteria synthesize vitamin K, vitamin B_{12}, and folate, which can be absorbed by the colon and serve as endogenous sources of these vitamins. Furthermore, the microbiota releases products that raise colonic acidity, thereby promoting the absorption of calcium, magnesium, and zinc.

6. *Influence the brain and behavior.* A new "bowel-to-brain" concept is emerging from recent studies. Products encoded by the gut microbiome and released into the digestive tract lumen can enter the blood and have a far-reaching effect on the brain and elsewhere. Furthermore, bacterial products in the digestive tract lumen can stimulate the vagus nerve, which has an important role in signaling the brain from the digestive tract. For instance, gut bacteria have been shown to mediate behavior and mood in animal studies, and other investigations have hinted that an altered distribution of specific bacterial species in the colon might be linked with autism in humans.

Although a good balance of colonic microbes can confer health benefits, sometimes the landscape of the microbiota can contribute to health disorders, an example being obesity. When food is scarce, the gut microbes' supplemental caloric contribution by gleaning energy from fiber is valuable. On the other hand, when food is plentiful, extra energy obtained from bacterial processing of indigestible food may contribute to the development of obesity. Making matters worse, obese individuals compared to their leaner counterparts host a greater proportion of bacteria that are highly efficient at extracting energy from food.

Less than ideal balance in the gut microbiota has been implicated in a variety of other diseases, including diabetes, asthma, irritable bowel disease, colon cancer, atherosclerosis, and rheumatoid arthritis.

Scientists are still only at the beginning stages of studying the microbiome and its implications for human health and disease susceptibility. In 2007 the National Institutes of Health launched an ambitious **Human Microbiome Project (HMP)** to catalog all of the bacterial genes that make up the microbiome. By sampling bacteria that live in multiple body sites and comparing the differences among healthy and unhealthy people, the investigators hope to use their findings to ultimately manipulate the microbial inhabitants to improve health and to fight a variety of diseases. Unlike the human genome, which is essentially permanent in a given individual, the person's microbiome can readily be changed, a feature that holds promise for favorable intervention. For example, the composition of the gut microbiota can be altered through consumption of **probiotics** (foods or dietary supplements that contain live bacteria that confer a health benefit on the host) or **prebiotics** (nondigestible dietary supplements that stimulate growth and activity of beneficial resident colonic bacteria) or by instilling a desired mix of bacteria directly into the colon (such as by fecal transplant). Furthermore, a change in the types of foods a person predominantly eats (for example, shifting from an abundance of meat to an abundance of plant foods) can quickly shift the composition and activity of microbes in the gut. Through further studies in this hot area of science, dietary recommendations will continue to be forthcoming to improve the microbiome for the host's benefit.

The large intestine absorbs salt and water, converting the luminal contents into feces.

Some absorption takes place within the colon but not to the same extent as in the small intestine. Because the luminal surface of the colon is fairly smooth, it has considerably less absorptive surface area than the small intestine. Furthermore, the colon is not equipped with extensive specialized transport mechanisms like the small intestine. When excessive small-intestine motility delivers the contents to the colon before absorption of nutrients has been completed, the colon cannot absorb most of these materials and they are lost in diarrhea.

The colon normally absorbs salt and H_2O. Na^+ is actively absorbed, Cl^- follows passively down the electrical gradient, and H_2O follows osmotically. The colon absorbs token amounts of other electrolytes, as well as the short-chain fatty acids and vitamins produced by colonic bacteria.

Through absorption of salt and H_2O, a firm fecal mass is formed. Of the 500 mL of material entering the colon per day from the small intestine, the colon normally absorbs about 350 mL, leaving 150 g of feces to be eliminated from the body each day (see ▌Table 16-7, p. 609). This fecal material normally consists of 100 g of H_2O and 50 g of solid, including undigested cellulose, bilirubin, bacteria, and small amounts of salt. Thus, contrary to popular thinking, the digestive tract is not a major excretory passageway for eliminating wastes from the body. The main waste product excreted in the feces is bilirubin. The other fecal constituents are unabsorbed food residues and bacteria, which were never actually a part of the body. Bacteria account for nearly one third of the dry weight of feces.

> **Check Your Understanding 16.7**
>
> 1. Compare haustral contractions of the large intestine to segmentation contractions of the small intestine.
> 2. State the role of $NaHCO_3$ secretions by the large intestine mucosa. Compare the function of this secretion with that of pancreatic $NaHCO_3$ secretion.
> 3. List the contributions of the gut microbiota.

16.8 Overview of the GI Hormones

Throughout our discussion of digestion, we have repeatedly mentioned different functions of the three major GI hormones: gastrin, secretin, and CCK. We now fit all of these functions together so that you can appreciate the overall adaptive importance of these interactions. Furthermore, we introduce a more recently identified GI hormone, *glucose-dependent insulinotropic peptide* (*GIP*). All of these hormones are small peptides that perform their functions by binding to G-protein-coupled receptors on the plasma membrane of their target cells, thereby activating second-messenger pathways that bring about the desired responses (see p. 117).

Gastrin Protein in the stomach stimulates the release of gastrin, which performs the following functions:

1. It acts in multiple ways to increase secretion of HCl and pepsinogen, two substances of primary importance in initiating digestion of the protein that promoted their secretion.

2. It enhances gastric motility, stimulates ileal motility, relaxes the ileocecal sphincter, and induces mass movements in the colon—all functions aimed at keeping the contents moving through the tract on arrival of a new meal.

3. It also is trophic to both the stomach mucosa and the small-intestine mucosa, helping maintain a well-developed, functionally viable digestive tract lining.

Predictably, gastrin secretion is inhibited by an accumulation of acid in the stomach and by the presence in the duodenal lumen of acid and other constituents that necessitate a delay in gastric secretion.

Secretin As the stomach empties into the duodenum, the presence of acid in the duodenum stimulates the release of secretin, which performs the following interrelated functions:

1. It inhibits gastric emptying to prevent further acid from entering the duodenum until the acid already present is neutralized.

2. It inhibits gastric secretion to reduce the amount of acid being produced.

3. It stimulates the pancreatic duct cells to produce a large volume of aqueous $NaHCO_3$ secretion, which is emptied into the duodenum to neutralize the acid. Neutralization of the acidic chyme in the duodenum helps prevent damage to the duodenal walls and provides a suitable environment for optimal functioning of the pancreatic digestive enzymes, which are inhibited by acid.

4. Secretin and CCK are both trophic to the exocrine pancreas.

CCK As chyme empties from the stomach, fat and other nutrients enter the duodenum. These nutrients—especially fat and, to a lesser extent, protein products—cause the release of CCK, which performs the following interrelated functions:

1. It inhibits gastric motility and secretion, thereby allowing adequate time for the nutrients already in the duodenum to be digested and absorbed.

2. It stimulates the pancreatic acinar cells to increase secretion of pancreatic enzymes, which continue the digestion of these nutrients in the duodenal lumen (this action is especially important for fat digestion because pancreatic lipase is the only enzyme that digests fat).

3. It causes contraction of the gallbladder and relaxation of the sphincter of Oddi so that bile is emptied into the duodenum to aid fat digestion and absorption. Bile salts' detergent action is particularly important in enabling pancreatic lipase to perform its digestive task. Again, the multiple effects of CCK are remarkably well adapted to dealing with the fat whose presence in the duodenum triggered this hormone's release.

4. Besides facilitating digestion of ingested nutrients, CCK is an important regulator of food intake. It plays a key role in satiety, the sensation of having had enough to eat (see p. 623).

GIP A more recently recognized hormone released by the duodenum, GIP, helps promote metabolic processing of the nutrients once they are absorbed. This hormone was originally named *gastric inhibitory peptide (GIP)* for its presumed role as an enterogastrone. It was believed to inhibit gastric motility and secretion, similar to secretin and CCK. Its contribution in this regard is now considered minimal. Instead, this hormone stimulates insulin release by the pancreas, so it is now called **glucose-dependent insulinotropic peptide** (again, **GIP**). This action is remarkably adaptive. As soon as the meal is absorbed, the body has to shift its metabolic gears to use and store the newly arriving nutrients. The metabolic activities of this absorptive phase are largely under the control of insulin (see pp. 688 and 690–692). Stimulated by the presence of a meal, especially glucose, in the digestive tract, GIP initiates the release of insulin in anticipation of absorption of the meal, in a feedforward fashion. Insulin is especially important in promoting the uptake and storage of glucose.

This overview of the multiple, integrated, adaptive functions of the GI hormones provides an excellent example of the remarkable efficiency of the human body.

Check Your Understanding 16.8

1. Explain the significance of some GI hormones being trophic.
2. Name the targets of secretin and of CCK.
3. State the function of GIP.

Homeostasis: Chapter in Perspective

 To maintain constancy in the internal environment, materials that are used up in the body (such as energy-rich nutrients and O_2) or uncontrollably lost from the body (such as evaporative H_2O loss from the airways or salt loss in sweat) must constantly be replaced by new supplies of these materials from the external environment. All these replacement supplies except O_2 are acquired through the digestive system. Fresh supplies of O_2 are transferred to the internal environment by the respiratory system, but all the nutrients, H_2O, and various electrolytes needed to maintain homeostasis are acquired through the digestive system. The large, complex food that is ingested is broken down by the digestive system into small absorbable units. These small, energy-rich nutrient molecules are transferred across the small-intestine epithelium into the blood for delivery to the cells to replace the nutrients constantly used for ATP production and for repair and growth of body tissues. Likewise, ingested H_2O, salt, and other electrolytes are absorbed by the intestine into the blood.

Unlike regulation in most body systems, regulation of digestive system activities is not aimed at maintaining homeostasis. The quantity of nutrients and H_2O ingested is subject to control, but the quantity of ingested materials absorbed by the digestive tract is not subject to control, with few exceptions. The hunger mechanism governs food intake to help maintain energy balance (see Chapter 17), and the thirst mechanism controls H_2O intake to help maintain H_2O balance (see Chapter 15). However, we often do not heed these control mechanisms, eating and drinking even when we are not hungry or thirsty. Once these materials are in the digestive tract, the digestive system does not vary its rate of nutrient, H_2O, or electrolyte uptake according to body needs (with the exception of iron and calcium); rather, it optimizes conditions for digesting and absorbing what is ingested. Truly, what you eat is what you get. The digestive system is subject to many regulatory processes, but these are not influenced by the nutritional or hydration state of the body. Instead, these control mechanisms are governed by the composition and volume of digestive tract contents so that the rate of motility and secretion of digestive juices are optimal for digestion and absorption of the ingested food.

If excess energy-rich nutrients are ingested and absorbed, the extra nutrients are placed in storage, such as in adipose tissue (fat), so that the blood level of nutrient molecules is kept at a constant level. Excess ingested H_2O and electrolytes are eliminated in the urine to homeostatically maintain the blood levels of these constituents.

Review Exercises
Answers begin on p. A-46

Reviewing Terms and Facts

1. The extent of nutrient uptake from the digestive tract depends on the body's needs. (*True or false?*)

2. The stomach is relaxed during vomiting. (*True or false?*)

3. Acid cannot normally penetrate into or between the cells lining the stomach, which enables the stomach to contain acid without injuring itself. (*True or false?*)

4. Protein is continually lost from the body through digestive secretions and sloughed epithelial cells, which pass out in the feces. (*True or false?*)

5. Digested foodstuffs not absorbed by the small intestine are absorbed by the large intestine. (*True or false?*)

6. The endocrine pancreas secretes secretin and CCK. (*True or false?*)

7. A digestive reflex involving the autonomic nerves is known as a _____ reflex, whereas a reflex in which all elements of the reflex arc are located within the gut wall is known as a _____ reflex.

8. The salivary center, swallowing center, and vomiting center are all located in the _____.

9. When food is mechanically broken down and mixed with gastric secretions, the resultant thick, liquid mixture is known as _____.

10. The entire lining of the small intestine is replaced approximately every _____ days.

11. The two substances absorbed by specialized transport mechanisms located only in the terminal ileum are _____ and _____.

12. The most potent choleretic is _____.

13. Match the following:

1. prevents reentry of food into the mouth during swallowing	(a) closure of the pharyngo-esophageal sphincter
2. triggers the swallowing reflex	(b) elevation of the uvula
3. seals off the nasal passages during swallowing	(c) position of the tongue against the hard palate
4. prevents air from entering the esophagus during breathing	(d) closure of the gastro-esophageal sphincter
5. closes off the respiratory airways during swallowing	(e) bolus pushed to the rear of the mouth by the tongue
6. prevents gastric contents from backing up into the esophagus	(f) tight apposition of the vocal folds

14. Which of the following is *not* a function of saliva?
   a. begins digestion of carbohydrate
   b. facilitates absorption of glucose across the oral mucosa
   c. facilitates speech
   d. exerts an antibacterial effect
   e. plays an important role in oral hygiene

15. Use the answer code on the right to identify the characteristics of the listed substances:

1. activates pepsinogen	(a) pepsin
2. inhibits amylase	(b) mucus
3. is essential for vitamin B_{12} absorption	(c) HCl
4. can act autocatalytically	(d) intrinsic factor
5. is a potent stimulant for acid secretion	(e) histamine
6. denatures protein	
7. begins protein digestion	
8. serves as a lubricant	
9. kills ingested bacteria	
10. is alkaline	
11. is deficient in pernicious anemia	
12. coats the gastric mucosa	

Understanding Concepts
(Answers at www.cengagebrain.com)

1. Describe the four basic digestive processes.

2. List the components of the digestive system. Describe the cross-sectional anatomy of the digestive tract.

3. What four general factors are involved in regulating digestive system function? What is the role of each?

4. Describe the types of motility in each component of the digestive tract. What factors control each type of motility?

5. State the composition of the digestive juice secreted by each component of the digestive system. Describe the factors that control each digestive secretion.

6. List the enzymes involved in digesting each category of foodstuff. Indicate the source and control of secretion of each of the enzymes.

7. Why are some digestive enzymes secreted in inactive form? How are they activated?

8. What absorption processes take place within each component of the digestive tract? What special adaptations of the small intestine enhance its absorptive capacity?

9. Describe the absorptive mechanisms for salt, water, carbohydrate, protein, and fat.

10. What are the contributions of the accessory digestive organs? What are the nondigestive functions of the liver?

11. Summarize the functions of each of the three major GI hormones.

12. What waste product is excreted in the feces?

13. How is vomiting accomplished? What are the causes and consequences of vomiting, diarrhea, and constipation?

14. Describe the process of mucosal turnover in the stomach and small intestine.

Solving Quantitative Exercises

1. Suppose a lipid droplet in the gut is essentially a sphere with a diameter of 1 cm.

   a. What is the surface area–to–volume ratio of the droplet? (*Hint:* The area of a sphere is $4\pi r^2$, and the volume is $4/3\pi r^3$.)

   b. Now, suppose that this sphere were emulsified into 100 essentially equal-sized droplets. What is the average surface area–to–volume ratio of each droplet?

   c. How much greater is the total surface area of these 100 droplets compared to the original single droplet?

   d. How much did the total volume change as a result of emulsification?

Applying Clinical Reasoning

Thomas W. experiences a sharp pain in his upper right abdomen after eating a high-fat meal. Also, he has noted that his feces are grayish-white instead of brown. What is the most likely cause of his symptoms? Explain why each of these symptoms occurs with this condition.

Thinking at a Higher Level

1. Why do patients who have had a large part of their stomachs removed for treatment of stomach cancer or severe peptic ulcer disease have to eat small quantities of food frequently instead of consuming three meals a day?

2. The number of immune cells in the *gut-associated lymphoid tissue (GALT)* housed in the mucosa is estimated to be equal to the total number of these defense cells in the rest of the body. Speculate on the adaptive significance of this extensive defense capability of the digestive system.

3. How would defecation be accomplished in a patient paralyzed from the waist down by a lower spinal-cord injury?

4. After bilirubin is extracted from the blood by the liver, it is conjugated (combined) with glucuronic acid by the enzyme *glucuronyl transferase* within the liver. Only when conjugated can bilirubin be actively excreted into the bile. For the first few days of life, the liver does not make adequate quantities of glucuronyl transferase. Explain how this transient enzyme deficiency leads to the common condition of jaundice in newborns.

5. Explain why removal of either the stomach or the terminal ileum leads to pernicious anemia.

17 Energy Balance and Temperature Regulation

Examples of energy in and energy out. To maintain energy balance and body weight, energy input (food intake) must equal energy output, which encompasses external work (such as cycling) and internal work (such as pumping blood). The hypothalamus governs food intake to maintain energy balance. The young adults shown here are enjoying El Capitan State Beach in California.

Venture Media Group/Aurora Open/Jupiter Images

Homeostasis Highlights

Food intake is essential to power cell activities. For body weight to remain constant, the caloric value of food (energy input) must equal total energy needs (energy output or expenditure). **Energy balance** and thus body weight are maintained primarily by controlling food intake.

Energy expenditure generates heat, which is important in **temperature regulation.** Humans, usually in environments cooler than their bodies, must constantly generate heat to maintain their body temperatures. Also, they must have mechanisms to cool the body if it gains too much heat from heat-generating skeletal muscle activity or from a hot external environment. Body temperature must be regulated because the rate of cellular chemical reactions depends on temperature and because overheating damages cell proteins.

The hypothalamus is the major integrating center for maintaining both energy balance and body temperature.

17.1 Energy Balance

Each cell in the body needs energy to perform the functions essential for the cell's survival (such as active transport and cellular repair) and to carry out its specialized contributions toward maintaining homeostasis (such as gland secretion or muscle contraction). All energy used by cells is ultimately provided by food intake.

Most food energy is ultimately converted into heat in the body.

According to the **first law of thermodynamics**, energy can be neither created nor destroyed. Therefore, energy is subject to the same kind of input–output balance as are the chemical components of the body, such as water and salt (see p. 536).

Energy Input and Output The energy in ingested food constitutes *energy input* to the body. Chemical energy locked in the bonds that hold the atoms together in nutrient molecules is released when these molecules are broken down in the body. Cells capture a portion of this nutrient energy in the high-energy phosphate bonds of adenosine triphosphate (ATP; see pp. 34 and A-15). Energy harvested from biochemical processing of ingested nutrients is either used immediately to perform biological work or stored in the body for later use as needed during periods when food is not being digested and absorbed.

Energy output or *expenditure* by the body falls into two categories: **external work** and **internal work** (❚ Figure 17-1). **External work** is the energy expended when skeletal muscles contract to move external objects or to move the body in relation to the environment. **Internal work** constitutes all other forms of biological energy expenditure that do not accomplish mechanical work outside the body. Internal work encompasses two types of energy-dependent activities: (1) skeletal muscle activity used for purposes other than external work, such as the contractions associated with postural maintenance and shivering, and (2) all energy-expending activities that must go on all the time just to sustain life. The latter include the work of pumping blood and breathing, the energy required for active transport of critical materials across plasma membranes, and the energy used during synthetic reactions essential for the maintenance, repair, and growth of cellular structures—in short, the "metabolic cost of living."

Conversion of Nutrient Energy to Heat Not all energy in nutrient molecules can be harnessed to perform biological work. Energy cannot be created or destroyed, but it can be converted from one form to another. The energy in nutrient molecules not used to energize work is transformed into **thermal energy**, or **heat**. During biochemical processing, only about 50% of the energy in nutrient molecules is transferred to ATP; the other 50% of nutrient energy is immediately lost as heat. When the cells expend ATP, another 25% of the energy derived from ingested food becomes heat. Because the body is not a heat engine, it cannot convert heat into work. Therefore, not more than 25% of nutrient energy is available for work, either external or internal. The remaining 75% is lost as heat.

Furthermore, of the energy actually captured for use by the body, almost all expended energy eventually becomes heat. To exemplify, energy expended by the heart to pump blood is gradually changed into heat by friction as blood flows through the vessels. Likewise, energy used in synthesizing structural protein eventually appears as heat when that protein is degraded during the normal course of turnover of bodily constituents. Even in performing external work, skeletal muscles convert chemical energy into mechanical energy inefficiently; as much as 75% of the expended energy is lost as heat. Thus, all energy liberated from ingested food that is not directly used for moving external objects or stored in fat (adipose tissue) deposits (or, in the case of growth, as protein) eventually becomes body heat. This heat is not entirely wasted energy, however, because much of it is used to maintain body temperature.

The metabolic rate is the rate of energy use.

The rate at which energy is expended by the body during both external and internal work is known as the **metabolic rate**:

$$\text{Metabolic rate} = \text{energy expenditure/unit of time}$$

Because most of the body's energy expenditure eventually appears as heat, the metabolic rate is normally expressed in terms of the rate of heat production in kilocalories per hour. The basic unit of heat energy is the **calorie**, which is the amount of heat required to raise the temperature of 1 g of H_2O by 1°C. This unit is too small to be convenient when discussing the human body because of the magnitude of heat involved, so the **kilocalorie** or **Calorie**, which is equivalent to 1000 calories, is used. When nutritionists speak of "calories" in quantifying the energy content of various foods, they are actually referring to kilocalories or Calories. Four kilocalories of heat energy are released when 1 g of glucose is oxidized or "burned," whether the oxidation takes place inside or outside the body.

Energy input

Food energy → Metabolic pool in body → Internal work / External work → Thermal energy (heat)

Energy storage

❚Figure 17-1 **Energy input and output.**

(Photos: *left*, © Brian Chase/Shutterstock.com; *center*, Ed Reschke/Photolibrary/Getty Images; *right*, © Val Thoermer/Shutterstock.com.)

Conditions for Measuring the Basal Metabolic Rate

The metabolic rate and, consequently, the rate of heat production vary depending on several factors, such as exercise, anxiety, shivering, and food intake. Increased skeletal muscle activity is the factor that can increase metabolic rate to the greatest extent. Even slight increases in muscle tone notably elevate the metabolic rate, and various levels of physical activity markedly alter energy expenditure and heat production (Table 17-1). For this reason, a person's metabolic rate is determined under standardized basal conditions established to control as many variables that can alter metabolic rate as possible. In this way, the metabolic activity necessary to maintain the basic body functions at rest can be determined. The **basal metabolic rate (BMR)** is a reflection of the body's "idling speed," or the minimal waking rate of internal energy expenditure. The BMR is measured under the following specified conditions:

1. The person should be at physical rest, having refrained from exercise for at least 30 minutes to eliminate any contribution of muscular exertion to heat production.

2. The person should be at mental rest to minimize skeletal muscle tone (people "tense up" when they are nervous) and to prevent a rise in epinephrine, a hormone secreted in response to stress that increases metabolic rate.

3. The measurement should be performed at a comfortable room temperature so that the person does not shiver. Shivering can markedly increase heat production.

4. The subject should not have eaten any food within 12 hours before the BMR determination to avoid **diet-induced thermogenesis** (*thermo* means "heat"; *genesis* means "production"), or the obligatory, short-lived (less than 12-hour) rise in metabolic rate that occurs as a result of the increased metabolic activity associated with processing and storing ingested nutrients.

Methods of Measuring the Basal Metabolic Rate The rate of heat production in BMR determinations can be measured directly or indirectly. To directly measure heat production, the person sits in an insulated chamber with water circulating through the walls. The difference in the temperature of the water entering and leaving the chamber reflects the amount of heat liberated by the person and picked up by the water as it passes through the chamber. A more convenient method of indirectly determining the rate of heat production requires only measuring the person's O_2 uptake per unit of time, which is a simple task using minimal equipment. Recall that

$$Food + O_2 \rightarrow CO_2 + H_2O + energy \text{ (mostly transformed into heat)}$$

Accordingly, a direct relationship exists between the volume of O_2 used and the quantity of heat produced. This relationship also depends on the type of food being oxidized. Although carbohydrates, proteins, and fats require different amounts of O_2 for their oxidation and yield different amounts of kilocalories when oxidized, an average estimate can be made of the quantity of heat produced per liter of O_2 consumed on a typical mixed American diet. This approximate value, known as the **energy equivalent of O_2**, is 4.8 kcal of energy liberated per liter of O_2 consumed. Using this method, the metabolic rate of a person consuming 15 liters/hour of O_2 can be estimated as follows:

15 liters/hr	= O_2 consumption
$\times$ 4.8 kilocalories/liter	= energy equivalent of O_2
72 kilocalories/hr	= estimated metabolic rate

Once the rate of heat production is determined under the prescribed basal conditions, it must be compared with normal values for people of the same sex, age, height, and weight because these factors all affect the basal rate of energy expenditure. For example, a large man has a higher rate of heat production than a smaller man, but expressed in terms of total surface area (which is a reflection of height and weight), the output in kilocalories per hour per square meter of surface area is normally about the same.

Factors Influencing the Basal Metabolic Rate Thyroid hormone is the primary but not sole determinant of the rate of basal metabolism. As thyroid hormone increases, the BMR increases correspondingly. As mentioned, epinephrine also increases the BMR.

Contrary to what might be expected, the BMR is not the body's lowest metabolic rate. The rate of energy expenditure during sleep is 10% to 15% lower than the BMR, presumably because of the more complete muscle relaxation that occurs during the paradoxical stage of sleep (see p. 169).

TABLE 17-1 Rate of Energy Expenditure for a 70-kg (154-Pound) Person During Different Activities

Energy Activity	Expenditure (kcal/hr)
Sleeping	65
Awake, lying still	77
Sitting at rest	100
Standing relaxed	105
Getting dressed	118
Typing	140
Walking slowly on level ground (2.6 mi/hr)	200
Carpentry or painting a house	240
Sexual intercourse	280
Bicycling on level ground (5.5 mi/hr)	304
Shoveling snow or sawing wood	480
Swimming	500
Jogging (5.3 mi/hr)	570
Rowing (20 strokes/min)	828
Walking up stairs	1100

Energy input must equal energy output to maintain a neutral energy balance.

Because energy cannot be created or destroyed, energy input must equal energy output, as follows:

$$\text{Energy input} = \text{energy output}$$

$$\begin{array}{ccccc} \text{Energy in food} & = & \text{external} & + & \text{internal heat} & \pm & \text{stored} \\ \text{consumed} & & \text{work} & & \text{production} & & \text{energy} \end{array}$$

There are three possible states of energy balance:

- *Neutral energy balance.* If the amount of energy in food intake exactly equals the amount of energy expended in performing external work plus the basal internal energy expenditure that eventually appears as body heat, then energy input and output are exactly in balance, and body weight remains constant.

- *Positive energy balance.* If the amount of energy in food intake is greater than the amount of energy expended, the extra energy taken in but not used is stored in the body, primarily as adipose tissue, so body weight increases.

- *Negative energy balance.* If the energy derived from food intake is less than the body's immediate energy requirements, the body must use stored energy to supply energy needs, and body weight decreases accordingly.

For a person to maintain a constant body weight (with the exception of minor fluctuations caused by changes in H_2O content), energy acquired through food intake must equal energy expenditure by the body. Because the average adult maintains a fairly constant weight over long periods, this implies that precise homeostatic mechanisms exist to maintain a long-term balance between energy intake and energy expenditure. Theoretically, total body energy content could be maintained at a constant level by regulating the magnitude of food intake, physical activity, or internal work and heat production. Control of food intake to match changing metabolic expenditures is the major means of maintaining a neutral energy balance. The level of physical activity is principally under voluntary control, and mechanisms that alter the degree of internal work and heat production are aimed primarily at regulating body temperature rather than total energy balance.

However, after several weeks of eating less or more than desired, small counteracting changes in metabolism may occur. For example, a compensatory increase in the body's efficiency of energy use in response to underfeeding partially explains why some dieters become stuck at a plateau after having lost the first 10 or so pounds of weight fairly easily. Similarly, a compensatory reduction in the efficiency of energy use in response to overfeeding accounts in part for the difficulty experienced by very thin people who are deliberately trying to gain weight. Despite these modest compensatory changes in metabolism, regulation of food intake is the most important factor in long-term maintenance of energy balance and body weight.

Food intake is controlled primarily by the hypothalamus.

Even though food intake is adjusted to balance changing energy expenditures over time, there are no calorie receptors per se to monitor energy input, energy output, or total body energy content. Instead, various blood-borne chemical factors that signal the body's nutritional state, such as how much fat is stored or the feeding status, are important in regulating food intake. Control of food intake does not depend on changes in a single signal but is determined by the integration of many inputs that provide information about the body's energy status, both the levels of stored and circulating nutrients. Multiple molecular signals together ensure that feeding behavior is synchronized with the body's immediate and long-term energy needs. Some information is used for short-term control of meal size and frequency. Even so, over a 24-hour period the energy in ingested food rarely matches energy expenditure for that day. The correlation between total caloric intake and total energy output is excellent, however, over long periods. As a result, the total energy content of the body—and, consequently, body weight—remains relatively constant long term.

Role of the Arcuate Nucleus: NPY and Melanocortins

The **arcuate nucleus** of the hypothalamus plays a central role in both long-term control of energy balance and body weight and short-term control of food intake from meal to meal. The arcuate nucleus is an arc-shaped collection of neurons located adjacent to the floor of the third ventricle. Multiple, highly integrated, redundant pathways crisscross into and out of the arcuate nucleus, indicative of the complex systems involved in feeding and satiety. **Feeding,** or **appetite, signals** give rise to the sensation of **hunger,** driving us to eat. By contrast, **satiety** is the feeling of being full. **Satiety signals** tell us when we have had enough and suppress the desire to eat.

The arcuate nucleus has two subsets of neurons that function in an opposing manner. One subset releases *neuropeptide Y,* and the other releases *melanocortins* derived from pro-opiomelanocortin (POMC), a precursor molecule that can be cleaved in different ways to produce several hormone products (see p. 648).[1] **Neuropeptide Y (NPY),** one of the most potent appetite stimulators ever found, leads to increased food intake, thus promoting weight gain. **Melanocortins,** a group of hormones traditionally known to be important in varying the skin color for the purpose of camouflage in some species, have been shown to exert an unexpected role in energy homeostasis in humans. Melanocortins, most notably α-*melanocyte-stimulating hormone (α-MSH)* from the hypothalamus, suppress appetite, thus leading to reduced food intake and weight loss. Melanocortins do not play a role in determining inherited skin coloration in humans, but α-MSH produced in the skin in response to ultraviolet light from the sun acts locally on melanin (pigment)-producing cells to cause tanning (see p. 441). Melanocortins' major importance in our species, however, is the role of hypothalamic α-MSH in toning down appetite.

NPY and melanocortins are not the final effectors in appetite control. These arcuate-nucleus chemical messengers influence the release of neuropeptides in other parts of the brain that exert more direct control over food intake. Scientists are trying

---

[1] The two subsets of neurons in the arcuate nucleus are the **NPY/AgRP** population and the **POMC/CART** population. *Agouti-related protein (AgRP),* like NPY, stimulates appetite, and *cocaine- and amphetamine-related transcript (CART),* like melanocortins, suppresses appetite. For simplicity's sake, we discuss only the roles of NPY and melanocortins as representative examples.

to unravel the other factors that act upstream and downstream from NPY and melanocortins to regulate appetite. Based on current evidence, the following regulatory inputs to the arcuate nucleus and beyond are important in the long-term maintenance of energy balance and the short-term control of food intake at meals (∎Figure 17-2).

Regulatory Inputs to the Arcuate Nucleus in Long-Term Maintenance of Energy Balance: Leptin and Insulin

Scientists' notion of fat cells (**adipocytes**) in adipose tissue as merely storage space for triglyceride fat underwent a dramatic change late in the past century with the discovery of their active role in energy homeostasis. Adipose tissue secretes several hormones, collectively termed **adipokines,** that play important roles in energy balance and metabolism. Thus, adipose tissue is now considered an endocrine gland and is the largest hormone-secreting organ in the body. Some adipokines are released only from adipocytes, an example being *leptin,* which plays an important role in energy balance. Some, like *tumor necrosis factor (TNF)* and *interleukin 6 (IL-6),* are released

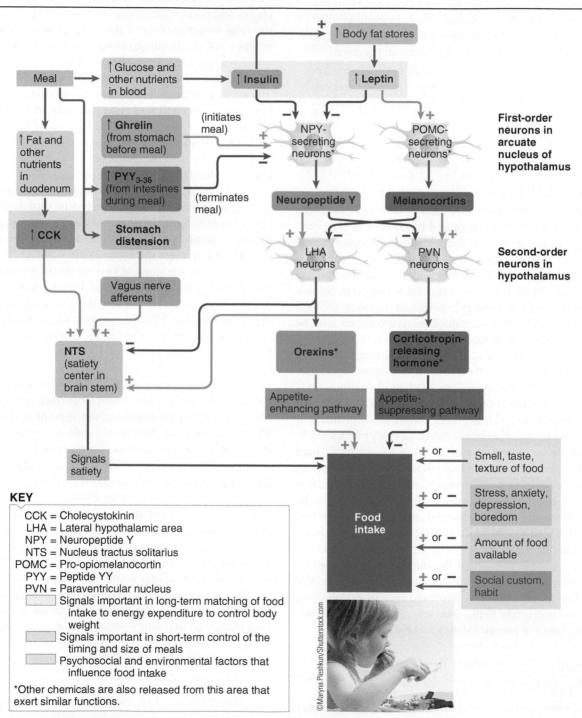

KEY

CCK = Cholecystokinin
LHA = Lateral hypothalamic area
NPY = Neuropeptide Y
NTS = Nucleus tractus solitarius
POMC = Pro-opiomelanocortin
PYY = Peptide YY
PVN = Paraventricular nucleus

☐ Signals important in long-term matching of food intake to energy expenditure to control body weight
☐ Signals important in short-term control of the timing and size of meals
☐ Psychosocial and environmental factors that influence food intake

*Other chemicals are also released from this area that exert similar functions.

∎Figure 17-2 **Factors that influence food intake.**
FIGURE FOCUS: *Follow the pathways by which increased body fat stores lead to decreased food intake.*

from adipocytes and from immune cells such as macrophages (large stationary phagocytes; see p. 411) that reside in abundance in excessive adipose tissue. These inflammation-causing adipokines contribute to inflammation in obese fat stores and to chronic low-grade systemic inflammation that alters glucose metabolism, among other detrimental metabolic effects. This newly identified link between obesity-induced inflammation and its metabolic consequences is termed **metaflammation**. Some adipokines, like *visfatin*, are released only from **visceral fat,** the deep, "bad" fat that surrounds the abdominal organs. Visceral fat is more likely to be chronically inflamed and is associated with increased heart disease and other disorders, in contrast to the more superficial and less harmful **subcutaneous fat** that is deposited under the skin. (Subcutaneous fat is the fat you can pinch.) Some, like *adiponectin*, are "good" adipokines. Adiponectin increases sensitivity to insulin (which helps protect against Type 2, or adult-onset, diabetes mellitus; see p. 696); decreases body weight; exerts anti-inflammatory actions; and promotes cell survival via its anti-apoptosis effects. Unfortunately, obesity suppresses adiponectin secretion. By contrast, some, like *resistin*, are "bad" adipokines. Resistin, which is released primarily in obesity, leads to insulin resistance (thus increasing the risk of developing Type 2 diabetes).

One of the most important adipokines is **leptin,** a hormone essential for normal body-weight regulation (*leptin* means "thin"). The amount of leptin in the blood is an excellent indicator of the total amount of triglyceride fat stored in adipose tissue: The larger the fat stores, the more leptin released into the blood. This blood-borne signal, discovered in the mid-1990s, was the first molecular satiety signal identified. This finding touched off a flurry of research responsible for greatly expanding our knowledge in recent years of the complex interplay of chemical signals that regulate food intake and body size.

The arcuate nucleus is the major site for leptin action. Acting in negative-feedback fashion, increased leptin from burgeoning fat stores serves as a "trim-down" signal. Leptin suppresses appetite, thus decreasing food consumption and promoting weight loss, by inhibiting hypothalamic output of appetite-stimulating NPY and stimulating output of appetite-suppressing melanocortins. Conversely, a decrease in fat stores and the resultant decline in leptin secretion bring about an increase in appetite, leading to weight gain. The leptin signal is generally considered the dominant factor responsible for the long-term matching of food intake to energy expenditure so that total body energy content remains balanced and body weight remains constant.

Another blood-borne signal besides leptin that plays an important role in long-term control of body weight is **insulin**. Insulin, a hormone secreted by the pancreas in response to a rise in the concentration of glucose and other nutrients in the blood following a meal, stimulates cellular uptake, use, and storage of these nutrients (see p. 690). Thus, the increase in insulin secretion that accompanies nutrient abundance, use, and storage appropriately inhibits the NPY-secreting cells of the arcuate nucleus, thus suppressing further food intake.

Beyond the Arcuate Nucleus: Orexins and Others Two hypothalamic areas are richly supplied by axons from the NPY-and melanocortin-secreting neurons of the arcuate nucleus. These second-order neuronal areas involved in energy balance and food intake are the **lateral hypothalamic area (LHA)** and **paraventricular nucleus (PVN)**. The LHA and PVN release chemical messengers in response to input from the arcuate nucleus neurons. These messengers act downstream from the NPY and melanocortin signals to regulate appetite. The LHA produces **orexins** (ore-EKS-ins), which are potent stimulators of food intake (*orexis* means "appetite"). NPY stimulates and melanocortins inhibit the release of appetite-enhancing orexins. By contrast, the PVN releases chemical messengers, for example, **corticotropin-releasing hormone**, that decrease appetite and food intake. (As its name implies, corticotropin-releasing hormone is better known for its role as a hormone; see p. 675.) Melanocortins stimulate and NPY inhibits the release of these appetite-suppressing chemicals.

In addition to the importance of leptin, insulin, and perhaps other so-called **adiposity signals** (signals related to the size of fat stores in adipose tissue) and their downstream mediators in the long-term control of body weight, other factors play a role in controlling the timing and size of meals.

Short-Term Eating Behavior: Ghrelin and PYY$_{3-36}$ Secretion Two blood-borne peptides secreted by the digestive tract that are important in regulating how often and how much we eat in a given day are *ghrelin* and *peptide YY$_{3-36}$* (*PYY$_{3-36}$*), which signify hunger and fullness, respectively. **Ghrelin** (GRELL-in), the so-called hunger hormone, is a potent appetite stimulator produced by the stomach and regulated by the feeding status (*ghrelin* is the Hindu word for "growth"). Secretion of this mealtime stimulator peaks before meals and makes people feel like eating, and then it falls once food is eaten. Ghrelin stimulates appetite by activating the hypothalamic NPY-secreting neurons.

PYY$_{3-36}$ is a counterpart of ghrelin. The secretion of PYY$_{3-36}$, which is produced by the small and large intestines, is at its lowest level before a meal but rises during meals and signals satiety. This peptide acts by inhibiting the appetite-stimulating NPY-secreting neurons in the arcuate nucleus. By thwarting appetite, PYY$_{3-36}$ is an important mealtime terminator.

The following other factors are also involved in signaling where the body is on the hunger–satiety scale.

Satiety Center In addition to the key role the hypothalamus plays in maintaining energy balance, a **satiety center** in the brain stem known as the **nucleus tractus solitarius (NTS)** processes signals important in the feeling of being full and thus contributes to short-term control of meals. Not only does the NTS receive input from the higher hypothalamic neurons involved in energy homeostasis, but it also receives afferent inputs from the digestive tract (for example, afferent input indicating the extent of stomach distension) and elsewhere that signal satiety. We now turn to cholecystokinin, one of the most important of these satiety signals.

Cholecystokinin as a Satiety Signal Cholecystokinin **(CCK)**, one of the gastrointestinal hormones released from the duodenal mucosa during digestion of a meal, is an important satiety signal for regulating meal size. CCK is secreted in response to the presence of nutrients in the small intestine. Through multiple effects on the digestive system, CCK facilitates digestion and absorption of these nutrients (see p. 615). It is appropriate that this blood-borne signal, whose rate of secretion is correlated with the amount of nutrients ingested, contributes to the sense of being filled after a meal has been consumed but before it has been digested and absorbed. We feel satisfied when adequate food to replenish the stores is in the digestive tract even though the body's energy stores are still low. This explains why we stop eating before the ingested food is made available to meet the body's energy needs. Other related, more recently discovered gut peptides released in response to a meal that serve as satiety signals include **glucagon-like peptide 1 (GLP-1)** and **oxyntomodulin.**

Table 17-2 summarizes the effect of involuntary regulatory signals on appetite.

Psychosocial and Environmental Influences Thus far, we have described involuntary signals that automatically occur to control food intake. However, as with water intake, people's eating habits are also shaped by psychological, social, and environmental factors. Often our decision to eat or stop eating is not determined merely by whether we are hungry or full, respectively. Frequently, we eat out of habit (eating three meals a day on schedule no matter what our status on the hunger–satiety continuum) or because of social custom (food often plays a prime role in entertainment, leisure, and business activities).

TABLE 17-2 Effects of Involuntary Regulatory Signals on Appetite

Regulatory Signal	Source of Signal	Effect of Signal on Appetite
Neuropeptide Y	Arcuate nucleus of hypothalamus	↑
Melanocortins	Arcuate nucleus of hypothalamus	↓
Leptin	Adipose tissue	↓
Insulin	Endocrine pancreas	↓
Orexins	Lateral hypothalamus	↑
Corticotropin-releasing hormone	Paraventricular nucleus of hypothalamus	↓
Ghrelin	Stomach	↑
Peptide YY_{3-36}	Small and large intestines	↓
Stomach distension	Stomach	↓
Cholecystokinin	Small intestine	↓

Even well-intentioned family pressure—"Clean your plate before you leave the table"—can affect the amount consumed.

Furthermore, the amount of pleasure derived from eating can reinforce feeding behavior. Eating foods with an enjoyable taste, smell, and texture can increase appetite and food intake. Stress, anxiety, depression, and boredom have also been shown to alter feeding behavior in ways unrelated to energy needs. People often eat to satisfy psychological needs rather than to satisfy hunger. Furthermore, environmental influences, such as the amount of food available, play an important role in determining the extent of food intake. Thus, any comprehensive explanation of how food intake is controlled must take into account these voluntary eating acts that can reinforce or override the internal signals governing feeding behavior.

Obesity occurs when more kilocalories are consumed than are burned.

Clinical Note Experts categorize body weight-for-height into four groups: underweight, healthy, overweight, and obese. The arbitrary boundary for being **overweight** is having a **body mass index (BMI)** (a mathematical means of assessing the proportion of body fat) of between 25 and 29.9, and for being **obese** as having a BMI of 30 or greater. (See the accompanying boxed feature, A Closer Look at Exercise Physiology, to see how to calculate the BMI.) Having excessive fat content in adipose tissue stores may have an adverse effect on health. More than two thirds of the adults in the United States weigh too much, with one third being clinically obese, and nearly one third of children are overweight or obese. To make matters worse, obesity is on the rise. The weight gain started in the 1970s and is continuing a sharp upward trend. The number of obese adults in the United States is nearly 2.5 times higher now than it was in the early 1970s, up from 14% to 34% of adults, and the incidence of childhood obesity is increasing even more rapidly. Researchers project that by 2030, if the rising rate is not curtailed, 50% of adults in the United States will be obese. Much of the world is following the same trend, leading the World Health Organization (WHO) to coin the word *globesity* to describe the worldwide situation. Obesity is now recognized as a disease state with multiple pathophysiological consequences by the WHO, the National Institutes of Health, the Food and Drug Administration, and the American Medical Association.

Obesity occurs when, over time, more kilocalories are ingested in food than are used to support the body's energy needs, with the excessive energy being stored as triglycerides in adipose tissue. Early in the development of obesity, existing fat cells get larger. An average adult has between 40 billion and 50 billion adipocytes. Each fat cell can store the maximum of about 1.2 mg of triglycerides. Once existing fat cells are full, if people continue to consume more calories than they expend, they make more adipocytes.

The causes of obesity are many, and some remain obscure. Both genetic and environmental factors, including but not limited to the following, may contribute to the development of obesity:

- *Hereditary tendencies.* Often, differences in the regulatory pathways for energy balance—either those governing food in-

BODY COMPOSITION IS THE PERCENTAGE of body weight that is composed of lean tissue and adipose tissue. Assessing body composition is an important step in evaluating a person's health status. One crude means of assessing body composition is by calculating the **body mass index (BMI)** using the following formula:

$$BMI = \frac{(\text{weight in pounds}) \times 700}{(\text{height in inches})^2}$$

A BMI of less than 25 is considered healthy, whereas a BMI of 30 or higher (being obese) places the person at increased risk for various diseases and premature death. BMIs between this range (being overweight but not obese) are considered borderline.

BMI determinations and the age–height–weight tables used by insurance companies can be misleading for determining healthy body weight. By these charts, many athletes, for example, would be considered overweight. A football player may be 6 feet 5 inches tall and weigh 300 pounds but have only 12% body fat. This player's extra weight is muscle, not fat, and therefore is not a detriment to his health. A sedentary person, in contrast, may be normal on the height–weight charts but have 30% body fat. This person should maintain body weight while increasing muscle mass and decreasing fat. Ideally, men should have 15% fat or less and women should have 20% fat or less.

The most accurate method for assessing body composition is underwater weighing. This technique is based on the fact that lean tissue is denser than water and fat tissue is less dense than water. (You can readily demonstrate this for yourself by dropping a piece of lean meat and a piece of fat into a glass of water; the lean meat sinks and the fat floats.) The results are used to determine body density using equations that take into consideration the difference between the person's weight in air and underwater. Because of the difference in density between lean and fat tissue, people who have more fat have a lower density and weigh relatively less underwater than in air compared to their lean counterparts. Body composition is then determined by means of an equation that correlates percentage fat with body density.

A more convenient and more commonly used but less accurate way to assess body composition is skinfold thickness. Because approximately half of the body's total fat content is located just beneath the skin, total body fat can be estimated from measurements of skinfold thickness taken at various sites on the body. Skinfold thickness is determined by pinching up a fold of skin at one of the designated sites and measuring its thickness by means of a caliper, a hinged instrument that fits over the fold and is calibrated to measure thickness. Mathematical equations specific for the person's age and sex can be used to predict the percentage of fat from the skinfold thickness scores. A major criticism of skinfold assessment is that accuracy depends on the investigator's skill.

There are different ways to be fat, and one way is more dangerous than the other. Obese patients can be classified into two categories—*android*, a male-type of adipose tissue distribution, and *gynoid*, a female-type distribution—based on the anatomic distribution of adipose tissue measured as the ratio of waist circumference to hip circumference. **Android obesity** is characterized by abdominal fat distribution (people shaped like "apples"), whereas **gynoid obesity** is characterized by fat distribution in the hips and thighs (people shaped like "pears"). Both sexes can display either android or gynoid obesity.

Android obesity is associated with a number of disorders, including insulin resistance, Type 2 (adult-onset) diabetes mellitus, excess blood lipid levels, high blood pressure, coronary heart disease, stroke, cancer, and dementia. "Apple" people have a greater proportion of visceral fat, which is more worrisome than accumulation of subcutaneous fat because visceral fat releases more of the bad adipokines that promote insulin resistance and boost the low-level inflammation that underlies the development of atherosclerosis (see p. 327). Gynoid obesity is not associated with the high risk of these diseases.

Research on the success of weight-reduction programs indicates that it is very difficult for people to lose weight, but when weight loss occurs, it is from the areas of increased stores. Because very-low-calorie diets are difficult to maintain, an alternative to severely cutting caloric intake to lose weight is to increase energy expenditure through physical exercise. Exercise physiologists often assess body composition as an aid in prescribing and evaluating exercise programs. Exercise generally reduces the percentage of body fat and, by increasing muscle mass, increases the percentage of lean tissue. An aerobic exercise program further helps reduce the risk of the disorders associated with android obesity.

take or those influencing energy expenditure—arise from genetic variations. Scientists have identified variations in more than 20 genes that predispose people to gaining excess weight easily. Take for example the *FTO* gene (the polite acronym now used instead of the less flattering original name "fatso gene"). There are two versions of the FTO gene: the "normal" T variant and the "faulty" A variant. Individuals who have one A version and one T version are 30% more likely to be obese than those who have two T copies of the gene. For people who have two A copies of FTO, the risk of becoming obese jumps to 70%. Recent evidence suggests ghrelin, the hunger hormone, is not properly suppressed by eating in those with A versions of the gene.

■ *Disturbances in the leptin-signaling pathway.* Some cases of obesity have been linked to leptin resistance. Some investigators suggest that the hypothalamic centers involved in maintaining energy homeostasis are "set at a higher level" in obese people. For example, the problem may lie with faulty leptin receptors in the brain that do not respond appropriately to the high levels of circulating leptin from abundant adipose stores.

Thus, the brain does not detect leptin as a signal to turn down appetite until a higher set point (and accordingly greater fat storage) is achieved. This could explain why overweight people tend to maintain their weight but at a heavier-than-normal level. Instead of faulty leptin receptors, other disturbances in the leptin pathway may be at fault, such as defective transport of leptin across the blood–brain barrier or a deficiency of one of the chemical messengers in the leptin pathway.

■ *An abundance of convenient, highly palatable, energy-dense, relatively inexpensive foods.* The current availability and convenience of food compared to most of human history make overeating easier than ever. Making matters worse, bad-for-you fast food and junk food are heavily marketed via powerful, widespread food advertising using modern mass media as a vehicle.

■ *Differences in extracting energy from food.* Another reason lean people and obese people may have dramatically different body weights despite consuming the same number of kilocalories may lie in the efficiency with which energy is extracted from food. Studies suggest that leaner individuals tend to derive less energy from the food they consume because they convert more of the food's energy into heat than into energy for immediate use or for storage. For example, slimmer individuals have more *uncoupling proteins,* which allow their cells to convert more of the nutrient calories into heat instead of fat. These are the people who can eat a lot without gaining weight. By contrast, obese people may have more efficient metabolic systems for extracting energy from food—a useful trait in times of food shortage but a hardship when trying to maintain a desirable weight when food is plentiful.

■ *Composition of colonic bacterial communities.* Studies demonstrate that obese people have a greater proportion of a type of bacteria in their colon that breaks down indigestible fiber more efficiently for absorption from the digestive tract compared to the bacterial communities in lean people (see p. 614). By making more absorbable units available for uptake from the digestive tract, the fat-promoting bacteria help their human host obtain more energy from the same number of kilocalories consumed than leaner people who have a preponderance of less energy-efficient colonic bacteria.

■ *Lack of exercise.* Numerous studies have shown that, on average, fat people do not eat more than thin people. One possible explanation is that overweight people do not overeat but underexercise—the "couch potato" syndrome. Low levels of physical activity typically are not accompanied by comparable reductions in food intake. For this reason, modern technology is partly to blame for the current obesity epidemic. Our ancestors had to exert physical effort to eke out subsistence. By comparison, we now have machines to replace much manual labor, automobiles that we use to get to locations within walking distance, remote controls to operate our machines with minimal effort, and computers that encourage long hours of sitting. We have to make a conscious effort to exercise.

■ *Differences in the "fidget factor."* **Nonexercise activity thermogenesis (NEAT),** or the "fidget factor," might explain some variation in fat storage among people. NEAT refers to energy expended by physical activities other than planned exercise.

Those who engage in toe tapping or other types of repetitive, spontaneous physical activity expend a substantial number of kilocalories throughout the day without conscious effort.

■ *Certain endocrine disorders such as hypothyroidism* (see p. 669). Hypothyroidism involves a deficiency of thyroid hormone, the main factor that bumps up the BMR so that the body burns more calories in its idling state.

■ *Emotional disturbances in which overeating replaces other gratifications.*

■ *Stress.* Evidence suggests that chronic stress leads to increased release of NPY from sympathetic nerves, in turn causing increased deposition of visceral fat.

■ *Eating out of sync with normal biological rhythms set by the "master clock"* (see p. 660). The body's daily rhythms such as the cyclic rise and fall in secretion of particular hormones, the regular fluctuations in body temperature, the sleep/wake cycle, and metabolic patterns are guided by a "master biologic clock" in the brain that operates in tune with environmental patterns of light and darkness. Consuming food at night when the body naturally shifts to a metabolic mode in anticipation of a sleeping and fasting phase may contribute to weight gain.

■ *Too little sleep.* Some studies suggest that decreased time sleeping may be a contributing factor in the recent rise of obesity. On average, people in the United States are sleeping 1 to 2 hours less per night now than they did 40 years ago. Researchers found that those who typically sleep 6 hours a night are 23% more likely to be obese, those who average 5 hours of sleep are 50% more likely to be obese, and those who sleep 4 hours nightly are 75% more likely to be obese than "traditional" sleepers who sleep for 7 to 8 hours. Studies have shown that levels of leptin (a signal to stop eating) are lower and levels of ghrelin (a signal to start eating) are higher in people who sleep less compared with those who sleep the traditional 8 hours.

■ *A possible virus link.* One intriguing proposal links a relatively common cold virus to a propensity to become overweight and may account for a portion of the current obesity epidemic. One study showed that the cold virus *adenovirus-36* might lead to obesity by transforming adult tissue-specific stem cells into fat-storing adipocytes.

■ *Heating and air conditioning.* Climate control has reduced the need for calorie-consuming activities such as shivering and sweating that the body uses to maintain body temperature.

■ *Development of an excessive number of fat cells as a result of overfeeding.* One of the problems in fighting obesity is that once fat cells are created they do not disappear with dieting and weight loss. Even if a dieter loses a large portion of the triglyceride fat stored in these cells, the depleted cells remain, ready to refill. Therefore, rebound weight gain after losing weight is difficult to avoid and discouraging for the dieter.

Despite this rather lengthy list, our knowledge about the causes and control of obesity is still rather limited, as evidenced by the number of people who are constantly trying to stabilize their weight at a more desirable level. Losing excess pounds is important from more than an aesthetic viewpoint. It is known that obesity, especially of the android type, can predispose an individual to illness and premature death from a multitude of diseases. (To learn about the differences between android and gynoid obesity, see the boxed feature on p. 625, A Closer Look at Exercise Physiology.) The burden of obesity-related problems on the U.S health care system is $147 billion annually. Scientists are working on multiple fronts to find ways to help curb the obesity epidemic.

People suffering from anorexia nervosa have a pathological fear of gaining weight.

Clinical Note The converse of obesity is generalized nutritional deficiency. The obvious causes for reduction of food intake below energy needs are lack of availability of food, interference with swallowing or digestion, and impairment of appetite.

One poorly understood disorder in which lack of appetite is a prominent feature is **anorexia nervosa,** a condition that affects 0.6% of the adult population in the United States. Patients with this disorder, most commonly adolescent girls and young women, have a morbid fear of becoming fat. They have a distorted body image, tending to visualize themselves as being heavier than they actually are. Because they have an aversion to food, they eat little and consequently lose considerable weight, perhaps even starving themselves to death. Among females 15 to 24 years old, the death rate associated with anorexia is 12 times higher than the death rate from all other causes. Other characteristics of the condition include altered secretion of many hormones, absence of menstrual periods, and low body temperature. It is unclear whether these symptoms occur secondarily as a result of general malnutrition or arise independently of the eating disturbance as a part of a primary brain malfunction. The underlying mechanisms responsible for anorexia nervosa are presently unknown, although researchers are scrambling to find answers in hopes of developing better therapies. Researchers cannot even agree on whether the causal problem is biological or psychological.

Check Your Understanding 17.1

1. Define *external work, internal work, metabolic rate, appetite signals, satiety signals, adiposity signals, adipokines, visceral fat,* and *subcutaneous fat.*

2. Explain how the basal metabolic rate can be determined indirectly.

3. Make a chart listing the involuntary regulatory signals on appetite and indicate the source and effect of each (that is, whether it increases or decreases appetite).

17.2 Temperature Regulation

Humans are usually in environments cooler than their bodies, but they constantly generate heat internally, which helps maintain body temperature. Heat production ultimately depends on oxidation of metabolic fuel derived from food.

Changes in body temperature in either direction alter cell activity—an increase in temperature speeds up cellular chemical reactions, whereas a decrease in temperature slows down these reactions. Because cell function is sensitive to fluctuations in internal temperature, humans homeostatically maintain body temperature at a level optimal for stable cellular metabolism. Overheating is more serious than cooling. Even moderate elevations of body temperature begin to cause nerve malfunction and irreversible protein denaturation. Most people suffer convulsions when the internal body temperature reaches about 106°F (41°C); 110°F (43.3°C) is considered the upper limit compatible with life.

Clinical Note By contrast, most of the body's tissues can transiently withstand substantial cooling. This characteristic is useful during cardiac surgery when the heart must be stopped. For such surgery, the patient's body temperature is deliberately lowered; the cooled tissues need less nourishment than they do at normal body temperature because of their reduced metabolic activity. However, a pronounced, prolonged fall in body temperature slows metabolism to a fatal level.

Internal core temperature is homeostatically maintained at 100°F (37.8°C).

Normal body temperature taken orally (by mouth) has traditionally been considered 98.6°F (37°C). However, more recent studies indicate that normal body temperature varies among individuals and varies throughout the day, ranging from 96.0°F (35.5°C) in the morning to 99.9°F (37.7°C) in the evening, with an overall average of 98.2°F (36.7°C).

Furthermore, there is no one body temperature because the temperature varies from organ to organ. From a thermoregulatory viewpoint, the body may conveniently be viewed as a *central core* surrounded by an *outer shell*. The temperature within the central core, which consists of the abdominal and thoracic organs, the central nervous system, and the skeletal muscles, generally remains fairly constant. This internal **core temperature** is subject to precise regulation to maintain its homeostatic constancy. The core tissues function best at a relatively constant temperature of around 100°F (37.8°C).

The skin and subcutaneous fat constitute the outer shell. In contrast to the constant high temperature in the core, the temperature within the shell is generally cooler and may vary substantially. For example, skin temperature may fluctuate between 68°F and 104°F (20°C and 40°C) without damage. As you will see, the temperature of the skin is deliberately varied as a control measure to help maintain the core's thermal constancy.

Sites for Monitoring Body Temperature Several easily accessible sites are used for monitoring body temperature. The oral and axillary (under the armpit) temperatures are comparable, whereas rectal temperature averages about 1°F (0.56°C) higher. Also available is a temperature-monitoring instrument that scans the heat generated by the eardrum and converts this temperature into an oral equivalent. A more recent device is the temporal scanner, a computerized instrument that is gently stroked across the forehead to measure the temperature of the blood in the temporal artery, which lies less than 2 mm below the skin surface in this region. Temporal temperature is the best determinant of core temperature because it is nearly identical to the temperature of the blood exiting the heart. However, none of these measurements is an absolute indication of the internal core temperature, which is a bit higher, at 100°F, than the monitored sites.

Normal Variations in Core Temperature Even though the core temperature is held relatively constant, several factors cause it to vary slightly:

1. Most people's core temperature normally varies about 1.8°F (1°C) during the day, with the lowest level early in the morning before rising (6 to 7 A.M.) and the highest point in late afternoon (5 to 7 P.M.). This variation is the result of an innate rhythm driven by the master biologic clock.

2. Women also experience a monthly rhythm in core temperature in connection with their menstrual cycle. For an undetermined reason, the "set point" is elevated (the body thermostat is turned up) so that core temperature averages 0.9°F (0.5°C) higher during the last half of the cycle from the time of ovulation to menstruation.

3. The core temperature increases during exercise because of the tremendous increase in heat production by the contracting muscles. During hard exercise, the core temperature may increase to as much as 104°F (40°C). In a resting person, this temperature would be considered a fever, but it is normal during strenuous exercise.

4. Older is colder. The elderly naturally have lower temperatures, with a midday average of 97.7°F (36.4°C).

5. Because the temperature-regulating mechanisms are not 100% effective, the core temperature may vary slightly with exposure to extremes of temperature. For example, the core temperature may fall several degrees in cold weather or rise a degree or so in hot weather.

Thus, the core temperature can vary at the extremes between about 96° and 104°F but usually deviates less than a few degrees. This relative constancy is made possible by multiple thermoregulatory mechanisms coordinated by the hypothalamus.

Heat input must balance heat output to maintain a stable core temperature.

The core temperature is a reflection of the body's total heat content. Heat input to the body must balance heat output to maintain a constant total heat content and thus a stable core temperature (❚ Figure 17-3). *Heat input* occurs by way of heat gain from the external environment and internal heat production, the latter being the most important source of heat for the body. Usually, more heat is generated than required to maintain normal body temperature, so the excess heat must be eliminated. *Heat output* occurs by way of heat loss from exposed body surfaces to the external environment.

Balance between heat input and heat output is frequently disturbed by (1) changes in internal heat production for purposes unrelated to regulation of body temperature—most notably by exercise, which markedly increases heat production—and

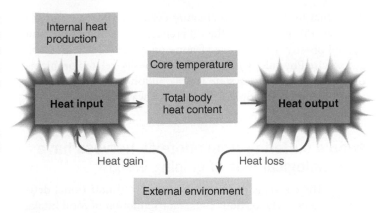

❚Figure 17-3 **Heat input and output.**

(2) changes in the external environmental temperature that influence the degree of heat gain or heat loss that occurs between the body and its surroundings. Compensatory adjustments must take place in heat-loss and heat-gain mechanisms to maintain body temperature within narrow limits, despite changes in metabolic heat production and changes in environmental temperature. We now elaborate on how these adjustments are made.

Heat exchange takes place by radiation, conduction, convection, and evaporation.

All heat loss or heat gain between the body and the external environment must take place between the body surface and its surroundings. The same physical laws of nature that govern heat transfer between inanimate objects control the transfer of heat between the body surface and the environment. The temperature of an object is a measure of the concentration of heat within the object. Heat always moves down its concentration gradient—that is, down a **thermal gradient** from a warmer to a cooler region. The body uses four mechanisms of heat transfer: *radiation, conduction, convection,* and *evaporation*.

Radiation **Radiation** is the emission of heat energy from the surface of a warm body in the form of **electromagnetic waves,** or **heat waves,** which travel through space (❚ Figure 17-4a). When radiant energy strikes an object and is absorbed, the energy of the wave motion is transformed into heat within the object. The human body both emits (source of heat loss) and absorbs (source of heat gain) radiant energy. Whether the body loses or gains heat by radiation depends on the difference in temperature between the skin surface and the surfaces of other objects in the body's environment. Because net transfer of heat by radiation is always from warmer objects to cooler ones, the body gains heat by radiation from objects warmer than the skin surface, such as the sun, burning logs, or a radiant heat system. By contrast, the body loses heat by radiation to objects in its environment whose surfaces are cooler than the surface of the skin, such as building walls, furniture, or trees. On average, humans lose close to half of their heat energy through radiation.

Conduction **Conduction** is the transfer of heat between objects of differing temperatures that are in direct contact with

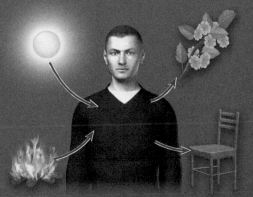

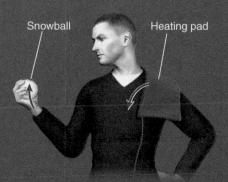

(a) **Radiation**—the transfer of heat energy from a warmer object to a cooler object in the form of electromagnetic waves ("heat waves"), which travel through space.

(b) **Conduction**—the transfer of heat from a warmer to a cooler object that is in direct contact with the warmer one. The heat is transferred through the movement of thermal energy from molecule to adjacent molecule.

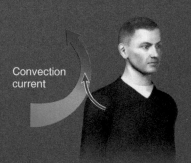

Convection current

Liquid converted to gaseous vapor

(c) **Convection**—the transfer of heat energy by air currents. Cool air warmed by the body through conduction rises and is replaced by more cool air. This process is enhanced by the forced movement of air across the body surface.

(d) **Evaporation**—conversion of a liquid such as sweat into a gaseous vapor, a process that requires heat (the heat of vaporization), which is absorbed from the skin.

❙Figure 17-4 **Mechanisms of heat transfer.** The direction of the arrows depicts the direction of heat transfer.
FIGURE FOCUS: *Describe the heat transfer mechanisms taking place when a person is jogging on a warm, sunny afternoon.*

each other, with heat moving down its thermal gradient from the warmer to the cooler object.

The rate of heat transfer by conduction depends on the *temperature difference* between the touching objects and the *thermal conductivity* of the substances involved (that is, how easily heat is conducted by the substances). Heat can be lost or gained by conduction when the skin is in contact with a good conductor (❙Figure 17-4b). When you hold a snowball, for example, your hand becomes cold because heat moves by conduction from your hand to the snowball. Conversely, when you apply a heating pad to a body part, the part is warmed up as heat is transferred directly from the pad to the body.

Similarly, you either lose or gain heat by conduction to the layer of air in direct contact with your body. The direction of heat transfer depends on whether the air is cooler or warmer, respectively, than your skin. Only a small percentage of total heat exchange between the skin and the environment takes place by conduction alone, however, because air is not a good conductor of heat. (For this reason, swimming pool water at 80°F [26.7°C] feels cooler than air at the same temperature; heat

is conducted more rapidly from the body surface into the water, which is a good conductor, than into the air, which is a poor conductor.)

Convection The term **convection** refers to the transfer of heat energy by *air* (or *water*) *currents*. As the body loses heat by conduction to the surrounding cooler air, the air in immediate contact with the skin is warmed. Because warm air is lighter (less dense) than cool air, the warmed air rises while cooler air moves in next to the skin to replace the vacating warm air. The process is then repeated (❙Figure 17-4c). These air movements, known as *convection currents,* help carry heat away from the body. Without convection currents, no further heat could be dissipated from the skin by conduction once the temperature of the layer of air immediately around the body equilibrated with skin temperature.

The combined conduction–convection process of dissipating heat from the body is enhanced by forced movement of air across the body surface, either by external air movements, such as those caused by the wind or a fan, or by movement of the body through the air, as during bicycle riding. Because forced air movement sweeps away the air warmed by conduction and replaces it with cooler air more rapidly, a greater total amount of heat can be carried away from the body over a given period. Thus, wind makes us feel cooler on hot days, and windy days in the winter are more chilling than calm days at the same cold temperature. For this reason, weather forecasters have developed the concept of *wind chill factor* (how cold it feels).

Evaporation During **evaporation** from the skin surface, the heat required to transform water from a liquid to a gaseous state is absorbed from the skin, thereby cooling the body (❚Figure 17-4d). Evaporative heat loss makes you feel cooler when your bathing suit is wet than when it is dry. Evaporative heat loss occurs continually from the linings of the respiratory airways and from the surface of the skin. Heat is continuously lost in expired air as a result of the air being humidified (gaining water vapor) during its passage through the respiratory system (see p. 447). Similarly, because the skin is not completely waterproof, water molecules constantly diffuse through the skin and evaporate. This ongoing evaporation from the skin is unrelated to the sweat glands. These passive evaporative heat-loss processes are not subject to physiological control and go on even in very cold weather, when the problem is one of conserving body heat.

Sweating is a regulated evaporative heat-loss process.

Sweating is an active evaporative heat-loss process under sympathetic nervous control. The rate of evaporative heat loss can be deliberately adjusted by varying the extent of sweating, which is an important homeostatic mechanism to eliminate excess heat as needed. In fact, when the environmental temperature exceeds the skin temperature, sweating is the only avenue for heat loss because the body is gaining heat by radiation and conduction under these circumstances. At normal temperature, an average of 100 mL of sweat is produced per day; this value increases to 1.5 liters during hot weather and climbs to 4 liters during heavy exercise.

Most **sweat** is an odorless, dilute salt solution actively extruded to the surface of the skin by **eccrine sweat glands** dispersed all over the body. This clear, salty sweat is the kind important in cooling the body. Eccrine sweat glands also produce **dermcidin**, a newly discovered antimicrobial peptide, a natural antibiotic in sweat that helps defend against potential skin infections. **Apocrine sweat glands,** which are located primarily in the armpits and genital area, produce a thick, milky sweat that is rich in organic constituents, like proteins and lipids. Apocrine sweat is initially odorless, but body odor is generated when bacteria in the vicinity break down these organic compounds into substances that have an unpleasant scent. Apocrine sweat glands are remnants of sexual scent glands found in other species. Apocrine sweat is most abundant during emotional stress and sexual excitement. Both eccrine and apocrine sweat glands are stimulated by sympathetic innervation, but the postganglionic sympathetic fibers supplying eccrine glands are unusual in that they release acetylcholine instead of norepinephrine. ACh binds with muscarinic receptors on the eccrine glands, whereas apocrine glands have the usual adrenergic receptors that bind with catecholamines (see p. 239).

Eccrine sweat must be evaporated from the skin for heat loss to occur. If sweat merely drips from the surface of skin or is wiped away, no heat loss is accomplished. The most important factor determining the extent of evaporation of sweat is the *relative humidity* of the surrounding air (the percentage of H_2O vapor actually present in the air compared to the greatest amount that the air can possibly hold at that temperature; for example, a relative humidity of 70% means that the air contains 70% of the H_2O vapor it is capable of holding). When the relative humidity is high, the air is already almost fully saturated with H_2O, so it has limited ability to take up additional moisture from the skin. Thus, little evaporative heat loss can occur on hot, humid days. The sweat glands continue to secrete, but the sweat simply remains on the skin or drips off instead of evaporating and producing a cooling effect. As a measure of the discomfort associated with combined heat and high humidity, meteorologists have devised the *temperature–humidity index,* or *heat index* (how hot it feels).

The hypothalamus integrates a multitude of thermosensory inputs.

The hypothalamus is the body's thermostat. The home thermostat keeps track of the temperature in a room and triggers a heating mechanism (the furnace) or a cooling mechanism (the air conditioner) as necessary to maintain room temperature at the indicated setting. Similarly, the hypothalamus, as the body's thermoregulatory integrating center, receives afferent information about the temperature in various regions of the body and initiates extremely complex, coordinated adjustments in heat-gain and heat-loss mechanisms as necessary to correct any deviations in core temperature from normal. The hypothalamus is far more sensitive than your home thermostat. It can respond to changes in blood temperature as small as 0.01°C.

To appropriately adjust the delicate balance between the heat-loss mechanisms and the opposing heat-producing and heat-conserving mechanisms, the hypothalamus must be apprised continuously of both the core and the skin temperature by specialized temperature-sensitive receptors called **thermoreceptors**. The core temperature is monitored by *central thermoreceptors,* which are located in the hypothalamus itself, and in the abdominal organs and elsewhere. *Peripheral thermoreceptors* monitor skin temperature throughout the body.

Two centers for temperature regulation are in the hypothalamus. The *posterior region,* activated by cold, triggers reflexes that mediate heat production and heat conservation. The *anterior region,* activated by warmth, initiates reflexes that mediate heat loss. Let us examine the means by which the hypothalamus fulfills its thermoregulatory functions (❚Figure 17-5).

Shivering is the primary involuntary means of increasing heat production.

The body can gain heat as a result of internal heat production generated by metabolic activity or from the external environment if the latter is warmer than body temperature. Because

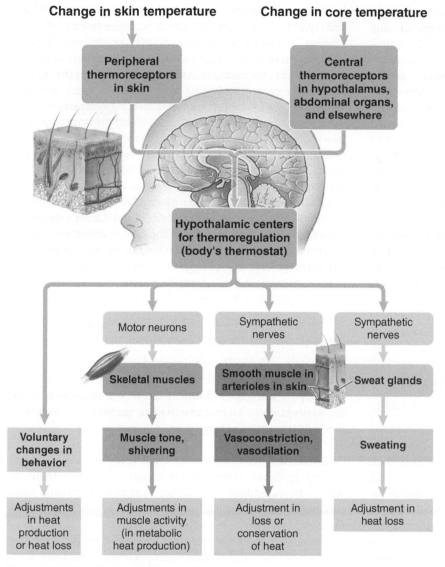

IFigure 17-5 Major thermoregulatory pathways.

FIGURE FOCUS: *Drugs that block the action of acetylcholine at muscarinic receptors (for example, oxybutynin used to treat overactive bladder) can increase a person's risk of heat exhaustion or heat stroke. Explain how.*

skeletal muscle tone. (Muscle tone is the constant level of tension within the muscles.) Soon shivering begins. **Shivering** consists of rhythmic, oscillating skeletal muscle contractions that occur at a rapid rate of 10 to 20 per second. This mechanism is efficient and effective in increasing heat production; all energy liberated during these muscle tremors is converted to heat because no external work is accomplished. Within seconds to minutes, internal heat production may increase two- to fivefold as a result of shivering.

Frequently, these reflex changes in skeletal muscle activity are augmented by increased voluntary, heat-producing actions such as bouncing up and down or hand clapping. The hypothalamus influences these behavioral responses as well as the involuntary physiological responses. As part of the limbic system, the hypothalamus is extensively involved with controlling motivated behavior (see p. 155).

In the opposite situation—a rise in core temperature caused by heat exposure—two mechanisms reduce heat-producing skeletal muscle activity: Muscle tone is reflexly decreased, and voluntary movement is curtailed. When the air becomes very warm, people often complain it is "too hot even to move."

Nonshivering Thermogenesis by Brown Fat Although reflex and voluntary changes in muscle activity are the major means of increasing the rate of heat production, **nonshivering (chemical) thermogenesis** also plays a role in thermoregulation. In most mammals, chronic cold exposure brings about an increase in metabolic heat production independent of muscle contraction by changes in heat-generating chemical activity. Nonshivering thermogenesis is mediated on cold exposure by the sympathetic nervous system, which increases heat production by stimulating **brown adipose tissue**, or **brown fat**, a special type of adipose tissue that is especially capable of converting chemical energy from food into heat. In humans nonshivering thermogenesis is most important in newborns, who have prominent deposits of brown fat. Unlike the ordinary **white adipose tissue** that stores energy in the form of triglyceride deposits, brown adipose tissue acts like a furnace that burns energy to generate heat. Newborns use brown fat to keep warm because they cannot shiver. Brown fat is brown in color because it has an abundance of mitochondria that contain iron, which causes the tissue to appear reddish brown. The mitochondria of brown fat contain a unique **uncoupling protein** called **thermogenin** ("heat producer") that uncouples the electron transport system from the process of generating ATP (see p. 37) during oxidation of glucose and fatty acids. Instead of some of the energy released by the electron transport system being harnessed in ATP by chemiosmosis, all of the energy is dissipated as heat.

body temperature usually is higher than environmental temperature, metabolic heat production is the primary source of body heat. In a resting person, most body heat is produced by the thoracic and abdominal organs as a result of ongoing, cost-of-living metabolic activities. Beyond this basal level, the rate of metabolic heat production can be variably increased primarily by changes in skeletal muscle activity or, to a lesser extent, by certain chemical activity. Thus, changes in skeletal muscle activity constitute the major way heat gain is controlled for temperature regulation.

Adjustments in Heat Production by Skeletal Muscles

In response to a fall in core temperature caused by exposure to cold, the hypothalamus takes advantage of increased skeletal muscle activity generating more heat. Acting through descending pathways that terminate on the motor neurons controlling the skeletal muscles, the hypothalamus first gradually increases

Brown fat deposits regress beyond infancy and were thought to disappear by 2 years of age. However, new radiologic imaging techniques such as PET scans (see p. 145) have revealed small, persistent brown fat reserves in adults. Brown fat is more abundant in lean people and decreases with advancing age. As expected, adult brown fat becomes more active on cold exposure. Brown fat not only is a heat-producing, but also a calorie-consuming tissue because all of the calories in nutrients burned by brown fat are turned into heat. Thus, researchers are searching for ways to fight obesity by boosting the amount or activity of adult brown fat stores or by coaxing white fat cells to behave more like brown fat cells. Fifty grams of brown fat burns 500 calories per day without the person expending any energy, as much as burned by 1 hour of aerobic exercise.

Recent studies suggest that exercise may promote "browning" of white adipose tissue. Exercising muscles release **irisin** into the blood. This hormone appears to promote synthesis of uncoupling proteins in mitochondria of white fat cells, making these cells act more like brown fat cells.

Having examined the mechanisms for adjusting heat production, we now turn to the other side of the equation: adjustments in heat loss.

The magnitude of heat loss can be adjusted by varying the flow of blood through the skin.

Heat-loss mechanisms are subject to control, again largely by the hypothalamus. When we are hot, we need to increase heat loss to the environment; when we are cold, we need to decrease heat loss. The amount of heat lost to the environment by radiation and the conduction–convection process is largely determined by the temperature gradient between the skin and the external environment. To maintain a constant core temperature, the insulative capacity and temperature of the skin can be adjusted to vary the temperature gradient between the skin and the external environment, thereby influencing the extent of heat loss.

The insulative capacity of the skin can be varied by controlling the amount of blood flowing through. Skin blood flow serves two functions. First, it provides a nutritive blood supply to the skin. Second, most skin blood flow is for the function of temperature regulation; at normal room temperature, 20 to 30 times more blood flows through the skin than is needed for skin nutrition. In the process of thermoregulation, skin blood flow can vary tremendously, from 400 to 2500 mL/min. The more blood that reaches the skin from the warm core, the closer the skin's temperature is to the core temperature. The skin's blood vessels diminish the effectiveness of the skin as an insulator by carrying heat to the surface, where it can be lost from the body by radiation and the conduction–convection process. Accordingly, vasodilation of skin arterioles increases heat loss by permitting increased flow of heated blood through the skin. Conversely, skin arteriolar vasoconstriction, which reduces skin blood flow, decreases heat loss by keeping the warm blood in the central core, where it is insulated from the external environment. This response conserves heat that otherwise would have been lost.

These skin vasomotor responses are coordinated by the hypothalamus by means of sympathetic nervous system output.

Increased sympathetic activity to the skin arterioles produces heat-conserving vasoconstriction in response to cold exposure, whereas decreased sympathetic activity produces heat-losing vasodilation of these vessels in response to heat exposure.

Recall that the cardiovascular control center in the medulla also exerts control over the skin arterioles (and arterioles throughout the body) by means of adjusting sympathetic activity to these vessels for the purpose of blood pressure regulation (see pp. 349 nd 350). Hypothalamic control over the skin arterioles for the purpose of temperature regulation takes precedence over the cardiovascular control center's control of these same vessels (see p. 369). Thus, changes in blood pressure can result from pronounced thermoregulatory skin vasomotor responses. For example, blood pressure can fall on exposure to a very hot environment because the skin vasodilator response set in motion by the hypothalamic thermoregulatory center overrides the skin vasoconstrictor response called forth by the medullary cardiovascular control center.

The hypothalamus simultaneously coordinates heat-production and heat-loss mechanisms.

Let us now pull together the coordinated adjustments in heat production and heat loss in response to exposure to either a cold or a hot environment (▌Table 17-3). (For a discussion of the effects of extreme cold or heat exposure, see the boxed feature on p. 634, ▌Concepts, Challenges, and Controversies.)

Coordinated Responses to Cold Exposure In response to cold exposure, the posterior region of the hypothalamus directs increased heat production, such as by shivering, while simultaneously decreasing heat loss (that is, conserving heat) by skin vasoconstriction and other measures.

Because there is a limit to the body's ability to reduce skin temperature through vasoconstriction, even maximum vasoconstriction is not sufficient to prevent excessive heat loss when the external temperature falls too low. Accordingly, other measures must be instituted to further reduce heat loss. In animals with dense fur or feathers, the hypothalamus, acting through the sympathetic nervous system, brings about contraction of the tiny muscles at the base of the hair or feather shafts to lift the hair or feathers off the skin surface. This puffing up traps a layer of poorly conductive air between the skin surface and the environment, thus increasing the insulating barrier between the core and the cold air and reducing heat loss. Even though the hair-shaft muscles contract in humans in response to cold exposure, this heat-retention mechanism is ineffective because of the low density and fine texture of most human body hair. The result instead is useless *goose bumps.*

After maximum skin vasoconstriction has been achieved, further heat dissipation in humans can be prevented only by behavioral adaptations, such as postural changes that reduce as much as possible the exposed surface area from which heat can escape. These heat-conserving maneuvers include hunching over or clasping the arms in front of the chest. Putting on warmer clothing further insulates the body from too much heat loss. Clothing entraps layers of poorly conductive air between the skin surface and the environment, thereby diminishing loss

IN RESPONSE TO COLD EXPOSURE (COORDINATED BY THE POSTERIOR HYPOTHALAMUS)		IN RESPONSE TO HEAT EXPOSURE (COORDINATED BY THE ANTERIOR HYPOTHALAMUS)	
Increased Heat Production	**Decreased Heat Loss (Heat Conservation)**	**Decreased Heat Production**	**Increased Heat Loss**
Increased muscle tone	Skin vasoconstriction	Decreased muscle tone	Skin vasodilation
Shivering	Postural changes to reduce exposed surface area (hunching shoulders, etc.)*	Decreased voluntary exercise*	Sweating
Increased voluntary exercise*			Cool clothing*
Nonshivering thermogenesis	Warm clothing*		

*Behavioral adaptations

of heat by conduction from the skin to the cold external air and curtailing the flow of convection currents.

Coordinated Responses to Heat Exposure During heat exposure, the anterior part of the hypothalamus reduces heat production by decreasing skeletal muscle activity and promotes increased heat loss by inducing skin vasodilation. When even maximal skin vasodilation is inadequate to rid the body of excess heat, sweating is brought into play to accomplish further heat loss through evaporation. If the air temperature rises above the temperature of maximally vasodilated skin, the temperature gradient reverses itself so that heat is gained from the environment. Sweating is the only means of heat loss under these conditions.

Humans also employ voluntary measures, such as using fans, wetting the body, drinking cold beverages, and wearing cool clothing, to further enhance heat loss. Contrary to popular belief, wearing light-colored, loose clothing is cooler than being nude. Naked skin absorbs almost all radiant energy that strikes it, whereas light-colored clothing reflects almost all radiant energy that falls on it. Thus, if light-colored clothing is loose and thin enough to permit convection currents and evaporative heat loss to occur, wearing it is actually cooler than going without any clothes.

Thermoneutral Zone Skin vasomotor activity is highly effective in controlling heat loss in environmental temperatures between the upper 60s and mid-80s. This range, within which core temperature can be kept constant by vasomotor responses without calling supplementary heat-production mechanisms (shivering) or heat-loss mechanisms (sweating) into play, is called the **thermoneutral zone.**

During a fever, the hypothalamic thermostat is "reset" at an elevated temperature.

Clinical Note The term **fever** refers to an elevation in body temperature as a result of infection or inflammation. In response to microbial invasion, macrophages release **endogenous pyrogen,** which acts on the hypothalamic thermoregula-

tory center to raise the thermostat setting (Figure 17-6; also see p. 411). The hypothalamus now maintains the temperature at the new set level instead of maintaining normal body temperature. If, for example, endogenous pyrogen raises the set point to 102°F (38.9°C), the hypothalamus senses that the normal pre-fever temperature is too cold, so it initiates cold-response mechanisms to raise the temperature to 102°F. It promotes skin vasoconstriction to rapidly reduce heat loss and initiates shivering to rapidly increase heat production, both of which drive the temperature upward. These events account for the sudden cold chills often experienced at the onset of a fever. Feeling cold, the person may voluntarily pile on more blankets to help raise body temperature by conserving body heat. Once the new temperature is achieved, body temperature is regulated as normal in response to cold and heat—but at a higher setting. Thus, fever production in response to an infection is a deliberate outcome and is not caused by a breakdown of thermoregulation. Although the physiological significance of a fever is still unclear, many medical experts believe that a rise in body temperature has a beneficial role in fighting infection. A fever augments the inflammatory response and may interfere with bacterial multiplication.

During fever production, endogenous pyrogen raises the set point of the hypothalamic thermostat by triggering the local release of *prostaglandins,* which are local chemical mediators that act directly on the hypothalamus (see p. 119). Aspirin reduces a fever by inhibiting synthesis of prostaglandins. Aspirin does not lower the temperature in a nonfebrile person because in the absence of endogenous pyrogen, prostaglandins are not present in the hypothalamus in appreciable quantities.

The exact molecular cause of a fever "breaking" naturally is unknown, although it presumably results from reduced pyrogen release or decreased prostaglandin synthesis. When the hypothalamic set point is restored to normal, the temperature at 102°F (in this example) is too high. Heat-response mechanisms are instituted to cool the body. Skin vasodilation occurs, and sweating commences. The person feels hot and throws off extra covers. The gearing up of these heat-loss mechanisms by the hypothalamus reduces the temperature to normal.

The Extremes of Heat and Cold Can Be Fatal

PROLONGED EXPOSURE TO TEMPERATURE EXTREMES in either direction can overtax the body's thermoregulatory mechanisms, leading to disorders and even death.

Cold-Related Disorders

The body can be harmed by cold exposure in two ways: frostbite and generalized hypothermia. **Frostbite** involves excessive cooling of a particular part of the body to the point where tissue in that area is damaged. If exposed tissues actually freeze, tissue damage results from disruption of the cells by formation of ice crystals or by lack of liquid water.

Hypothermia, a fall in body temperature, occurs when generalized cooling of the body exceeds the ability of the normal heat-producing and heat-conserving regulatory mechanisms to match the excessive heat loss. As hypothermia sets in, the rate of all metabolic processes slows because of the declining temperature. Higher cerebral functions are the first affected by body cooling, leading to loss of judgment, apathy, disorientation, and tiredness, all of which diminish the cold victim's ability to initiate voluntary mechanisms to reverse the falling body temperature. As body temperature continues to plummet, depression of the respiratory center occurs, reducing the ventilatory drive so that breathing becomes slow and weak. Activity of the cardiovascular system also is gradually reduced. The heart is slowed and cardiac output decreased. Cardiac rhythm is disturbed, eventually leading to ventricular fibrillation and death.

Heat-Related Disorders

At the other extreme, two disorders related to excessive heat exposure are heat exhaustion and heat stroke. **Heat exhaustion** is a state of collapse, usually manifested by fainting, that is caused by reduced blood pressure brought about as a result of overtaxing the heat-loss mechanisms. Extensive sweating reduces cardiac output by depleting the plasma volume, and pronounced skin vasodilation causes a drop in total peripheral resistance. Because blood pressure is determined by cardiac output times total peripheral resistance, blood pressure falls, an insufficient amount of blood is delivered to the brain, and fainting takes place. Thus, heat exhaustion is a consequence of overactivity of the heat-loss mechanisms rather than a breakdown of these mechanisms. Because the heat-loss mechanisms have been very active, body temperature is only mildly elevated in heat exhaustion. By forcing cessation of activity when the heat-loss mechanisms are no longer able to cope with heat gain through exercise or a hot environment, heat exhaustion serves as a safety valve to help prevent the more serious consequences of heat stroke.

Heat stroke is an extremely dangerous situation that arises from the complete breakdown of the hypothalamic thermoregulatory systems. Heat exhaustion may progress into heat stroke if the heat-loss mechanisms continue to be overtaxed. Heat stroke is more likely to occur on overexertion during a prolonged exposure to a hot, humid environment. The elderly, in whom thermoregulatory responses are generally slower and less efficient, are particularly vulnerable to heat stroke during prolonged, stifling heat waves. So too are individuals who are taking certain drugs that can hinder the body's ability to cool off, for example by hampering sweating (anticholinergics), by preventing increased blood flow to the skin (sympathomimetics), or by interfering with the hypothalamic thermoregulatory centers' neurotransmitter activity (some tranquilizers).

The most striking feature of heat stroke is a lack of compensatory heat-loss measures, such as sweating, in the face of a rapidly rising body temperature **(hyperthermia)**. No sweating occurs, despite a markedly elevated body temperature, because the hypothalamic thermoregulatory control centers are not functioning properly and cannot initiate heat-loss mechanisms. During the development of heat stroke, body temperature starts to climb as the heat-loss mechanisms are eventually overwhelmed by prolonged, excessive heat gain. Once the core temperature reaches the point at which the hypothalamic temperature-control centers are damaged by the heat, the body temperature rapidly rises even higher because of the complete shutdown of heat-loss mechanisms. Furthermore, as the body temperature increases, the rate of metabolism increases correspondingly because higher temperatures speed up the rate of all chemical reactions; the result is even greater heat production. This positive-feedback state sends the temperature spiraling upward. Heat stroke is a very dangerous situation that is rapidly fatal if untreated. Even with treatment to halt and reverse the rampant rise in body temperature, the mortality rate is still high. The rate of permanent disability in survivors is also high because of irreversible protein denaturation caused by the high internal heat.

Hyperthermia can occur unrelated to infection.

Hyperthermia denotes any elevation in body temperature above the normally accepted range. The term *fever* is usually reserved for an elevation in temperature caused by endogenous pyrogen resetting the hypothalamic set point during infection or inflammation; *hyperthermia* refers to all other imbalances between heat gain and heat loss that increase body temperature. Hyperthermia has a variety of causes, some of which are normal and harmless, others pathologic and fatal.

Exercise-Induced Hyperthermia The most common cause of hyperthermia is sustained exercise. As a physical consequence of the tremendous heat load generated by exercising muscles, body temperature rises during the initial stage of exer-

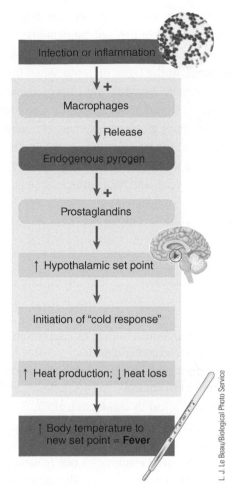

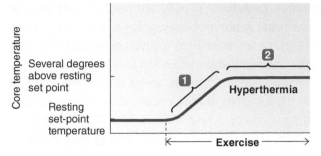

1 At the onset of exercise, the rate of heat production initially exceeds the rate of heat loss so the core temperature rises.

2 When heat loss mechanisms are reflexly increased sufficiently to equalize the elevated heat production, the core temperature stabilizes slightly above the resting point for the duration of the exercise.

I Figure 17-7 **Hyperthermia in sustained exercise.**

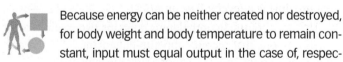

I Figure 17-6 **Fever production.**

cise because heat gain exceeds heat loss (I Figure 17-7). The elevation in core temperature reflexly triggers heat-loss mechanisms (skin vasodilation and sweating), which eliminate the discrepancy between heat production and heat loss. As soon as the heat-loss mechanisms are stepped up sufficiently to counterbalance the increased heat production, the core temperature stabilizes at a level slightly above the set point despite continued heat-producing exercise. Thus, during sustained exercise, body temperature initially rises, then is maintained at the higher level as long as the exercise continues.

Pathological Hyperthermia Hyperthermia can also be brought about in a completely different way: excessive heat production in connection with abnormally high circulating levels of thyroid hormone or epinephrine that result from dysfunctions of the thyroid gland or adrenal medulla, respectively. Both these hormones elevate the core temperature by increasing the overall rate of metabolic activity and heat production.

Hyperthermia can also result from malfunction of the hypothalamic control centers. Certain brain lesions, for example, destroy the normal regulatory capacity of the hypothalamic thermostat. When the thermoregulatory mechanisms are not functional, lethal hyperthermia may occur very rapidly. Normal metabolism produces enough heat to kill a per-

son in less than 5 hours if the heat-loss mechanisms are completely shut down. Exposure to severe, prolonged heat stress can also break down the function of hypothalamic thermoregulation (see the accompanying boxed feature, I Concepts, Challenges, and Controversies.)

Check Your Understanding 17.2

1. Describe the mechanisms of heat transfer.
2. Make a chart comparing the responses initiated by the posterior hypothalamus and the anterior hypothalamus to maintain core body temperature when the environmental temperature becomes cold or hot.
3. Define *thermoneutral zone*.

Homeostasis: Chapter in Perspective

Because energy can be neither created nor destroyed, for body weight and body temperature to remain constant, input must equal output in the case of, respectively, the body's total energy balance and its heat energy balance. If total energy input exceeds total energy output, the extra energy is stored in the body and body weight increases. Similarly, if the input of heat energy exceeds its output, body temperature increases. Conversely, if output exceeds input, body weight decreases or body temperature falls. The hypothalamus is the major integrating center for maintaining both a constant total energy balance (and thus a constant body weight) and a constant heat energy balance (and thus a constant body temperature).

Body temperature, which is one of the homeostatically regulated factors of the internal environment, must be maintained within narrow limits because the structure and reactivity of the chemicals that compose the body are temperature sensitive. Deviations in body temperature outside a limited range result in protein denaturation and death of the individual if the temperature rises too high or metabolic slowing and death if the temperature falls too low.

Body weight, in contrast, varies widely among individuals. Only the extremes of imbalances between total energy input and total energy output become incompatible with life. For example, in the face of insufficient energy input in the form of ingested food during prolonged starvation, the body resorts to breaking down muscle protein to meet its needs for energy expenditure once the adipose stores are depleted. Body weight dwindles because of this self-cannibalistic mechanism until death finally occurs as a result of loss of heart muscle, among other things. At the other extreme, when the food energy consumed greatly exceeds the energy expended, the extra energy input is stored as adipose tissue and body weight increases. The resultant gross obesity can also lead to heart failure. Not only must the heart work harder to pump blood to the excess adipose tissue, but obesity also predisposes the person to atherosclerosis and heart attacks (see p. 327).

Review Exercises Answers begin on p. A-47

Reviewing Terms and Facts

1. If more food energy is consumed than is expended, the excess energy is lost as heat. *(True or false?)*

2. All energy within nutrient molecules can be harnessed to perform biological work. *(True or false?)*

3. Each liter of O_2 contains 4.8 kcal of heat energy. *(True or false?)*

4. A body temperature greater than 98.2°F is always indicative of a fever. *(True or false?)*

5. Core temperature is relatively constant, but skin temperature can vary markedly. *(True or false?)*

6. Sweat that drips off the body has no cooling effect. *(True or false?)*

7. Production of "goose bumps" has no value in regulating body temperature. *(True or false?)*

8. The posterior region of the hypothalamus triggers shivering and skin vasoconstriction. *(True or false?)*

9. The hypothalamus is not effective in regulating body temperature during a fever. *(True or false?)*

10. The _____ of the hypothalamus contains two populations of neurons, one that secretes appetite-enhancing NPY and another that secretes appetite-suppressing melanocortins.

11. The primary means of involuntarily increasing heat production is _____.

12. Increased heat production independent of muscle contraction is known as _____.

13. The only means of heat loss when the environmental temperature exceeds the core temperature is _____.

14. Which of the following statements concerning heat exchange between the body and the external environment is *incorrect?*

   a. Heat gain is primarily by means of internal heat production.

   b. Radiation serves as a means of heat gain but not of heat loss.

   c. Heat energy always moves down its concentration gradient from warmer to cooler objects.

   d. The temperature gradient between the skin and the external air is subject to control.

   e. Little body heat is lost by conduction alone.

15. Using the answer code on the right, indicate which mechanism of heat transfer is being described:

1. sitting on a cold metal chair	(a) radiation
2. sunbathing on the beach	(b) conduction
3. being in a gentle breeze	(c) convection
4. sitting in front of a fireplace	(d) evaporation
5. sweating	
6. riding in a car with the windows open	
7. lying under an electric blanket	
8. sitting in a wet bathing suit	
9. fanning yourself	
10. immersing yourself in cold water	

Understanding Concepts
(Answers at www.cengagebrain.com)

1. Differentiate between external work and internal work.

2. Define *metabolic rate* and *basal metabolic rate*.

3. Describe the three states of energy balance.

4. By what means is energy balance primarily maintained?

5. Describe the source and role of the following in long-term regulation of energy balance and short-term control of the timing and size of meals: neuropeptide Y, melanocortins, leptin, insulin, orexins, corticotropin-releasing hormone, ghrelin, peptide YY_{3-36}, cholecystokinin, and stomach distension.

6. List the sources of heat input and heat output for the body.

7. Discuss the compensatory measures that occur in response to a fall in core temperature as a result of cold exposure and in response to a rise in core temperature as a result of heat exposure.

8. Describe the sequence of events in fever production.

Solving Quantitative Exercises

1. The basal metabolic rate (BMR) is a measure of how much energy the body consumes to maintain its "idling speed." Normal BMR = about 72 kcal/hr. Most of this energy is converted to heat. Our thermoregulatory systems function to eliminate this heat to keep body temperature constant. If our bodies were not able to lose this heat, our temperature would rise until we boiled (of course, a person would die before reaching that temperature). It is relatively easy to calculate how long it would take to reach the hypothetical boiling point. If an amount of energy ΔU is put into a liquid of mass m, the temperature change ΔT (in °C) is given by the following formula:

$$\Delta T = \Delta U/m \times C$$

In this equation, C is the specific heat of the liquid. For water, $C = 1.0$ kcal/kg-°C. Calculate how long it would take for the heat from the BMR to boil your body fluids (assume 42 liters of water in your body and a starting point of normal body temperature at 37°C). When exercising maximally, a person consumes about 1000 kcal/hr. How long would it take to boil in this case?

Applying Clinical Reasoning

Michael F., a near-drowning victim, was pulled from the icy water by rescuers 15 minutes after he fell through thin ice when skating. Michael is now alert and recuperating in the hospital. How can you explain his "miraculous" survival even though he was submerged without breathing air for 15 minutes, yet irreversible brain damage, soon followed by death, normally occurs if the brain is deprived of O_2 for more than 4 or 5 minutes?

Thinking at a Higher Level

1. Explain how drugs that selectively inhibit cholecystokinin increase feeding behavior in experimental animals.

2. What advice would you give an overweight friend who asks for your help in designing a safe, sensible, inexpensive program for losing weight?

3. Why is it dangerous to engage in heavy exercise on a hot, humid day?

4. Describe the avenues for heat loss in a person soaking in a hot bath.

5. Consider the difference between you and a fish in a local pond with regard to control of body temperature. Humans are *thermoregulators*; they can maintain a remarkably constant, rather high internal body temperature despite the body's exposure to a range of environmental temperatures. To maintain thermal homeostasis, humans physiologically manipulate mechanisms within their bodies to adjust heat production, conservation, and loss. In contrast, fish are *thermoconformers*; their body temperatures conform to the temperature of their surroundings. Thus, their body temperatures vary capriciously with changes in the environmental temperature. Even though fish produce heat, they cannot physiologically regulate internal heat production, nor can they control heat exchange with their environment to maintain a constant body temperature when the temperature in their surroundings rises or falls. Knowing this, do you think fish run a fever when they have a systemic infection? Why or why not?

To access the course materials and companion resources for this text, please visit www.cengagebrain.com

18 Principles of Endocrinology; The Central Endocrine Glands

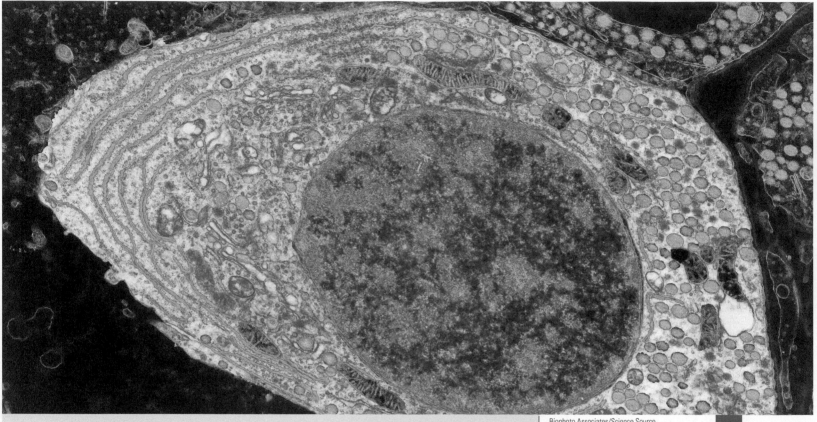

A transmission electron micrograph of a growth hormone–producing cell in the anterior pituitary. Growth hormone, synthesized by the extensive endoplasmic reticulum (*thin, light blue, curved sacs*), is stored in numerous secretory vesicles (*blue-green circles*) until it is released by exocytosis into the blood on appropriate stimulation.

Homeostasis Highlights

The endocrine system regulates activities that require duration rather than speed. Endocrine glands release hormones, blood-borne chemical messengers that act on target cells located a long distance from the endocrine gland. Most target-cell activities under hormonal control are directed toward maintaining homeostasis. The **central endocrine glands**, which are in or closely associated with the brain, include the hypothalamus, the pituitary gland, and the pineal gland. The **hypothalamus** and **posterior pituitary gland** act as a unit to release hormones essential for maintaining water balance, for giving birth, and for breast-feeding. The hypothalamus also secretes regulatory hormones that control the hormonal output of the **anterior pituitary gland**, which secretes six hormones that, in turn, largely control the hormonal output of several peripheral endocrine glands. One anterior pituitary hormone, growth hormone, promotes growth and influences nutrient homeostasis. The **pineal gland** secretes a hormone important in establishing the body's biological rhythms.

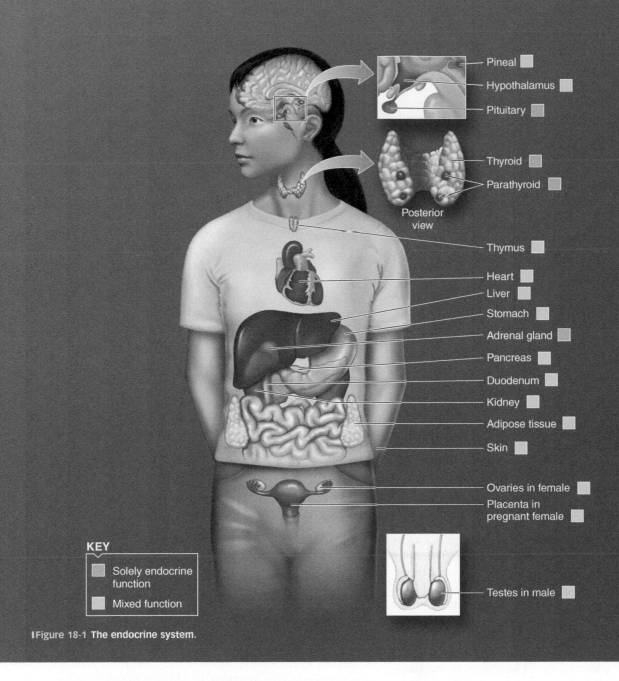

Pineal
Hypothalamus
Pituitary

Thyroid
Parathyroid

Posterior view

Thymus

Heart
Liver
Stomach
Adrenal gland
Pancreas
Duodenum
Kidney
Adipose tissue
Skin

Ovaries in female
Placenta in pregnant female

Testes in male

KEY

■ Solely endocrine function

■ Mixed function

■Figure 18-1 **The endocrine system.**

18.1 General Principles of Endocrinology

The endocrine system consists of the ductless endocrine glands (see p. 6) scattered throughout the body (■ Figure 18-1). Even though the endocrine glands for the most part are not connected anatomically, they constitute a system in a functional sense. They all accomplish their functions by secreting hormones into the blood, and many functional interactions take place among various endocrine glands. Once secreted, a hormone travels in the blood to its distant target cells, where it regulates a particular function (see p. 113). **Endocrinology** is the study of the homeostatic chemical adjustments and other activities that hormones accomplish.

Even though the blood distributes hormones throughout the body, only specific target cells can respond to each hormone because only the target cells have receptors for binding with the particular hormone. Binding of a hormone with its specific target-cell receptors initiates a chain of events within the target cells to bring about the hormone's final effect.

Recall that the means by which a hormone brings about its ultimate physiologic effect depends on whether the hormone is hydrophilic (peptide hormones, catecholamines, and indoleamines) or lipophilic (steroid hormones and thyroid hormone). *Peptide hormones,* the most abundant chemical category of hormone, are chains of amino acids of varying length. *Catecholamines,* produced by the adrenal medulla, are derived from the amino acid tyrosine. *Indoleamines* are produced by the pineal gland and are derived from the amino acid tryptophan. *Steroid hormones,* produced by the adrenal cortex and reproductive endocrine glands, are neutral lipids derived from cholesterol. *Thyroid hormone,* produced by the thyroid gland, is an iodinated tyrosine derivative. To review, hydrophilic hormones on binding with surface membrane receptors primarily act through second-messenger systems to alter the activity of preexisting proteins, such as enzymes, within the target cell to produce their physiologic response. Lipophilic steroid hormones and thyroid hormone, by contrast, activate genes on binding with receptors inside the cell, thus bringing about formation of new proteins in the target cell that carry out the desired response. Hydrophilic hormones circulate in the blood largely dissolved in the plasma, whereas lipophilic hormones are largely bound to plasma proteins. (See pp. 120–127 for further detail.)

Hormones exert a variety of regulatory effects throughout the body.

The endocrine system is one of the body's two major regulatory systems, the other being the nervous system, with which you are already familiar (Chapters 4 through 7). In general, the nervous system coordinates rapid, precise responses and is especially important in mediating the body's interactions with the external environment. The endocrine system primarily controls processes that require duration rather than speed, most of which are aimed at maintaining homeostasis, such as regulating nutrient metabolism and water and electrolyte balance; promoting growth; and facilitating reproductive capacity. Furthermore, the endocrine system works along with the autonomic nervous system to control and integrate activities of both the circulatory and the digestive systems.

Tropic Hormones Some hormones regulate the production and secretion of another hormone. A hormone that has as its primary function the regulation of hormone secretion by another endocrine gland is classified functionally as a **tropic hormone** (*tropic* means "nourishing" and is pronounced "trō-pik"). Tropic hormones not only stimulate but also maintain the structure of their endocrine target tissues. A tropic hormone's actions aimed at maintaining the structural integrity of its target gland are specifically known as *trophic* (growth promoting) actions, similar to the effects of the GI hormones in maintaining functionally viable digestive organs (see pp. 586 and 592), but for convenience we link these terms together under the common umbrella term *tropic*. For example, thyroid-stimulating hormone (TSH), a tropic hormone from the anterior pituitary, stimulates thyroid hormone secretion by the thyroid gland and maintains the structural integrity of this gland. In the absence of TSH, the thyroid gland atrophies (shrinks) and produces very low levels of its hormones.

Complexity of Endocrine Function The following factors add to the complexity of the endocrine system:

■ A single endocrine gland may produce multiple hormones. The anterior pituitary, for example, secretes six different hormones, each under a different control mechanism and having distinct functions.

■ A single hormone may be secreted by more than one endocrine gland. For example, both the hypothalamus and pancreas secrete the hormone somatostatin, and somatostatin acts as a paracrine in the stomach.

■ Frequently, a single hormone has more than one type of target cell and therefore can induce more than one type of effect, typically by binding with different subtypes of receptors. For example, vasopressin promotes H_2O reabsorption by the kidney tubules by binding with V_2 (vasopressin 2) receptors on the distal and collecting tubular cells and causes vasoconstriction of arterioles throughout the body by binding with V_1 receptors on arteriolar smooth muscle. Sometimes hormones that have multiple target-cell types can coordinate and integrate the activities of various tissues toward a common end. For example, the effects of insulin on muscle, liver, and fat all act in concert to store nutrients after absorption of a meal.

■ The rate of secretion of some hormones varies considerably over the course of time in a cyclic pattern. Therefore, endocrine systems also provide temporal (time) coordination of function. This is particularly apparent in endocrine control of reproductive cycles, such as the menstrual cycle, in which normal function requires highly specific patterns of change in the secretion of various hormones.

■ A single target cell may be influenced by more than one hormone. Some cells contain an array of receptors for responding in different ways to different hormones. To illustrate, insulin promotes the conversion of glucose into glycogen within liver cells by stimulating one particular hepatic enzyme, whereas another hormone, glucagon, by activating yet another hepatic enzyme, enhances the degradation of glycogen into glucose within liver cells.

■ The same chemical messenger may be either a hormone or a neurotransmitter, depending on its source and mode of delivery to the target cell. Norepinephrine, which is secreted as a hormone by the adrenal medulla and released as a neurotransmitter from sympathetic postganglionic nerve fibers, is a prime example.

■ Some organs are solely endocrine in function (they specialize in hormone secretion alone, the anterior pituitary being an example), whereas other organs of the endocrine system perform nonendocrine functions, in addition to secreting hormones. For example, the testes produce sperm and secrete the male sex hormone testosterone.

The effective plasma concentration of a hormone is influenced by the hormone's secretion, peripheral conversion, transport, inactivation, and excretion.

The primary function of most hormones is regulation of various homeostatic activities. Because hormones' effects are proportional to their concentrations in the plasma, these concentrations are subject to control according to homeostatic need. Furthermore, the magnitude of the hormonal response depends on the availability and sensitivity of the target cells' receptors for the hormone. We examine the factors that influence the plasma concentration of hormones before turning to the target cells' responsiveness to hormones.

The **effective plasma concentration** of the free, biologically active form of a hormone—and thus the hormone's availability to its receptors—depends on several factors:

- *The hormone's rate of secretion into the blood by the endocrine gland.* The rate of secretion, a factor that increases the plasma concentration of the hormone, is subject to control to maintain the hormone concentration at a desired set point.

- *For a few hormones, its rate of metabolic activation or conversion.* After being secreted into the blood by the endocrine gland, lipophilic hormones in particular are often modified in other organs. Sometimes this peripheral (away from the endocrine gland) modification results in a more active form of the hormone. For example, the most abundant form of thyroid hormone secreted by the thyroid gland is thyroxine (which contains four iodines), but the most powerful form of thyroid hormone in the blood is tri-iodothyronine (which contains three iodines). Once secreted, thyroxine is converted to the more active form as a result of one of its iodines being stripped peripherally, primarily by the liver and kidneys. Usually the rate of such hormone activation is itself under hormonal control. Sometimes peripheral action actually converts one hormone into a functionally different hormone. For example, a small proportion of testosterone, a potent male sex hormone, is converted peripherally by action of the enzyme *aromatase* in adipose tissue and elsewhere into estrogen, a potent female sex hormone.

- *For lipophilic hormones, its extent of binding to plasma proteins.* Because lipophilic hormones are poorly soluble in water, they circulate in the plasma bound to specific plasma proteins. Only the small, unbound portion of the hormone is free to interact with its target cells. The magnitude of this free pool rather than the total pool of hormone is monitored and adjusted to maintain normal endocrine function.

- *Its rate of removal from the blood by metabolic inactivation and excretion in the urine.* All hormones are eventually inactivated by enzymes in the liver, kidneys, blood, or target cells. The amount of time after a hormone is secreted before it is inactivated, and the means by which this takes place, differ for different classes of hormones. Hydrophilic peptides most commonly are inactivated by hydrolysis of peptide bonds (see p. 569). In the case of some peptide hormones, such as insulin, the target cell actually engulfs the bound hormone by receptor-mediated endocytosis and degrades it intracellularly (see p. 31). Catecholamines are enzymatically converted to related biologically inactive molecules. Lipophilic steroid hormones and thyroid hormone are inactivated by alteration of the active portion of the molecule by various biochemical means. After lipophilic hormones are inactivated, the liver typically adds charged groups to make them more water soluble so that they are freed from their plasma protein carriers and eliminated in the urine.

In general, the hydrophilic peptides and catecholamines are easy targets for blood and tissue enzymes, so they remain in the blood only briefly (a few minutes to a few hours) before being enzymatically inactivated. In contrast, binding of lipophilic hormones to plasma proteins makes them less vulnerable to metabolic inactivation and keeps them from escaping into the urine. Therefore, lipophilic hormones are removed from plasma more slowly: They may persist in the blood for hours (steroids) or up to a week (thyroid hormone).

Hormones and their metabolites are typically eliminated from the plasma by urinary excretion. In contrast to the tight controls on hormone secretion, hormone inactivation and excretion are not regulated.

The effective plasma concentration of a hormone is normally regulated by changes in the rate of its secretion.

Normally, the effective plasma concentration of a hormone is regulated by appropriate adjustments in the rate of its secretion. Endocrine glands do not secrete their hormones at a constant rate; the secretion rates of all hormones vary, subject to control often by a combination of several complex mechanisms. The regulatory system for each hormone is considered in detail in later sections. For now, we address these general mechanisms of controlling secretion that are common to many different hormones: negative-feedback control, neuroendocrine reflexes, and diurnal (circadian) rhythms.

Negative-Feedback Control Negative feedback is a prominent feature of hormonal control systems. Stated simply, *negative feedback exists when the output of a system counteracts a change in input,* maintaining a controlled variable within a narrow range around a desired level, or set point (see p. 16). Negative feedback maintains the plasma concentration of a hormone at a given level, like a home heating system maintains the room temperature at a given set point. For example, when the plasma concentration of free circulating thyroid hormone falls below a given set point, the anterior pituitary secretes thyroid-stimulating hormone (TSH), which stimulates the thyroid to increase its secretion of thyroid hormone. Thyroid hormone, in turn, inhibits further secretion of TSH by the anterior pituitary. Negative feedback ensures that once thyroid gland secretion has been "turned on" by TSH, it will not continue unabated but will be "turned off" when the appropriate level of free circulating thyroid hormone has been achieved. Thus, the effect of a particular hormone's actions can inhibit its own secretion. The feedback loops often become quite complex.

Neuroendocrine Reflexes Many endocrine control systems involve **neuroendocrine reflexes**, which include neural as well as hormonal components. The purpose of such reflexes is to produce a sudden increase in hormone secretion (that is, "turn up the thermostat setting") in response to a specific stimulus, frequently a stimulus external to the body. In some instances, neural input to the endocrine gland is the only factor regulating secretion of the hormone. For example, secretion of epinephrine by the adrenal medulla is solely controlled by the sympathetic nervous system. Some endocrine control systems, in contrast, include both feedback control (which maintains a constant basal level of the hormone) and neuroendocrine reflexes (which cause sudden bursts in secretion in response to a sudden increased need for the hormone). An example is the increased secretion of cortisol, the "stress hormone," by the adrenal cortex during a stress response (see ❙Figure 19-9, p. 675).

Diurnal (Circadian) Rhythms The secretion rates of many hormones rhythmically fluctuate up and down as a function of time. The most common endocrine rhythm is the **diurnal** ("day–night"), or **circadian** ("around a day") **rhythm**, which is characterized by repetitive oscillations in hormone levels that are very regular and cycle once every 24 hours. This rhythmicity is caused by endogenous oscillators whose activity level autonomously rises and falls, similar to the self-paced respiratory neurons in the brain stem that control the rhythmic motions of breathing, except the timekeeping oscillators cycle on a much longer time scale. Furthermore, unlike the rhythmicity of breathing, endocrine rhythms are locked on, or **entrained**, to external cues such as the light–dark cycle. That is, the inherent 24-hour cycles of peak and ebb of hormone secretion are set to "march in step" with cycles of light and dark. For example, cortisol secretion rises during the night, reaches its peak secretion in the morning before a person gets up, then falls throughout the day to its lowest level at bedtime (❙Figure 18-2). Inherent hormonal rhythmicity and entrainment are not accomplished by the endocrine glands themselves but result from the central nervous system changing the set point of these glands. We discuss the master biological clock further in a later section. Negative-feedback control mechanisms operate to maintain whatever set point is established for that time of day. Some endocrine cycles operate on time scales other than a circadian rhythm, a well-known example being the monthly menstrual cycle.

Endocrine disorders result from hormone excess or deficiency or decreased target-cell responsiveness.

Abnormalities in a hormone's effective plasma concentration can arise from a variety of factors (❙Table 18-1). Endocrine disorders most commonly result from abnormal plasma concentrations of a hormone caused by inappropriate rates of secretion—that is, too little hormone secreted (**hyposecretion**) or too much hormone secreted (**hypersecretion**). Occasionally, endocrine dysfunction arises because target-cell respon-

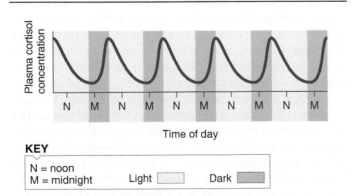

❙Figure 18-2 **Diurnal rhythm of cortisol secretion.**
(Source: Adapted from George A. Hedge, Howard D. Colby, and Robert L. Goodman, *Clinical Endocrine Physiology*, Figure 1-13, p. 28. © 1987, with permission from Elsevier.)

❙TABLE 18-1 **Means by Which Endocrine Disorders Can Arise**

Too Little Hormone Activity	Too Much Hormone Activity
Too little hormone secreted by the endocrine gland (hyposecretion)*	Too much hormone secreted by the endocrine gland (hypersecretion)*
Increased removal of the hormone from the blood	Reduced plasma protein binding of the hormone (too much free, biologically active hormone)
Abnormal tissue responsiveness to the hormone	Decreased removal of the hormone from the blood
Lack of target-cell receptors	Decreased inactivation
Lack of an enzyme essential to the target-cell response	Decreased excretion

*Most common causes of endocrine dysfunction.

siveness to the hormone is abnormally low, even though plasma concentration of the hormone is normal.

Hyposecretion *Primary hyposecretion* occurs when an endocrine gland is secreting too little of its hormone because of an abnormality within that gland. *Secondary hyposecretion* takes place when an endocrine gland is normal but is secreting too little hormone because of a deficiency of its tropic hormone.

The following are among the many different factors (each listed with an example) that may cause primary hyposecretion: (1) genetic (inborn absence of an enzyme that catalyzes synthesis of the hormone, such as the inability to synthesize cortisol because of the lack of a specific enzyme in the adrenal cortex); (2) dietary (lack of iodine, which is needed for synthesis of thyroid hormone); (3) chemical or toxic (certain insecticide residues may destroy the adrenal cortex); (4) immunologic (autoimmune antibodies sometimes destroy thyroid tissue); (5) other disease processes (cancer or tuberculosis may coincidentally destroy endocrine glands); (6) *iatrogenic* (physician induced, such as surgical removal of a cancerous thyroid gland); and (7) *idiopathic* (meaning the cause is not known).

Hypersecretion Like hyposecretion, hypersecretion by a particular endocrine gland is designated as primary or secondary depending on whether the defect lies in that gland or results from excessive stimulation from the outside, respectively. Hypersecretion may be caused by (1) tumors that ignore the normal regulatory input and continuously secrete excess hormone and (2) immunologic factors, such as excessive stimulation of the thyroid gland by an abnormal antibody that mimics the action of TSH. Excessive levels of a particular hormone may also arise from substance abuse, such as the outlawed practice among athletes of using certain steroids that increase muscle mass by promoting protein synthesis in muscle cells (see p. 276).

Abnormal Target-Cell Responsiveness Endocrine dysfunction can also occur because target cells do not respond adequately to the hormone, even though the effective plasma concentration of a hormone is normal. This unresponsiveness may be caused, for example, by an inborn lack of receptors for the hormone, as in **testicular feminization syndrome**. In this condition, receptors for testosterone, a masculinizing hormone produced by the male testes, are not produced because of a specific genetic defect. Although adequate testosterone is available, masculinization does not take place, just as if no testosterone were present.

The responsiveness of a target cell can be varied by regulating the number of hormone-specific receptors.

In contrast to endocrine dysfunction caused by *unintentional* receptor abnormalities, the target-cell receptors for a particular hormone can be *deliberately altered* as a result of physiologic control mechanisms. A target cell's response to a hormone is correlated with the number of the cell's receptors occupied by molecules of that hormone, which in turn depends not only on the plasma concentration of the hormone but also on the number of receptors in the target cell for that hormone. Thus, the response of a target cell to a given plasma concentration can be fine-tuned up or down by varying the number of receptors available for hormone binding.

Down Regulation As an illustration of this fine-tuning, when the plasma concentration of insulin is chronically elevated, the total number of target-cell receptors for insulin is gradually reduced as a direct result of the effect a sustained elevation of insulin has on the insulin receptors. This phenomenon, known as **down regulation**, constitutes an important locally acting negative-feedback mechanism that prevents the target cells from overreacting to a prolonged high concentration of insulin—that is, the target cells are *desensitized* to insulin, helping blunt the effect of insulin hypersecretion.

Down regulation of insulin is accomplished by the following mechanism. The binding of insulin to its surface receptors first triggers the dictated cellular response, then induces receptor-mediated endocytosis of the hormone receptor complex, which is subsequently attacked by intracellular lysosomal enzymes (see p. 30). This internalization serves a twofold purpose: It provides a pathway for degrading the hormone after it has exerted its effect and helps regulate the number of receptors available for binding on the target cell's surface. At high plasma insulin concentrations, the number of surface receptors for insulin is gradually reduced by the accelerated rate of receptor internalization and degradation brought about by increased hormonal binding. The rate of synthesis of new receptors within the endoplasmic reticulum and their insertion in the plasma membrane do not keep pace with their rate of destruction. Over time, this self-induced loss of target-cell receptors for insulin reduces the target cell's sensitivity to the elevated hormone concentration.

Permissiveness, Synergism, and Antagonism A given hormone's effects are influenced not only by the concentration of the hormone itself but also by the concentra-

tions of other hormones that interact with it. Because hormones are widely distributed through the blood, target cells may be exposed simultaneously to many different hormones, giving rise to numerous hormonal interactions on target cells. Hormones frequently alter the receptors for other kinds of hormones as part of their normal activity. A hormone can influence the activity of another hormone at a given target cell in one of three ways: permissiveness, synergism, and antagonism.

With **permissiveness**, one hormone must be present in adequate amounts to "permit" another hormone to exert its full effect. The first hormone enhances a target cell's responsiveness to the second hormone by increasing the number of receptors for the second hormone. For example, thyroid hormone increases the number of receptors for epinephrine in epinephrine's target cells, increasing the effectiveness of epinephrine. In the absence of thyroid hormone, epinephrine is only marginally effective.

Synergism occurs when the actions of several hormones are complementary and their combined effect is greater than the sum of their separate effects. An example is the synergistic action of follicle-stimulating hormone and testosterone, both of which are needed to maintain the normal rate of sperm production. Synergism results from each hormone's influence on the number or affinity (attraction) of receptors for the other hormone.

Antagonism occurs when one hormone causes the loss of another hormone's receptors, reducing the effectiveness of the second hormone. To illustrate, progesterone (a hormone secreted during pregnancy that decreases contractions of the uterus) inhibits uterine responsiveness to estrogen (another hormone secreted during pregnancy that increases uterine contractions). By causing loss of estrogen receptors on uterine smooth muscle, progesterone prevents estrogen from exerting its excitatory effects during pregnancy and thus keeps the uterus as a quiet (noncontracting) environment suitable for the developing fetus.

This has been a brief overview of the general features of the endocrine system. ▌Table 18-2 on pp. 644–645 summarizes the most important specific functions of the major hormones. As extensive as the table appears, it leaves out a variety of "candidate" hormones whose hormonal status has not yet been conclusively documented. Furthermore, new hormones are likely to be discovered, and additional functions may be found for known hormones. As an example, vasopressin's role in conserving H_2O during urine formation was determined first, followed later by the discovery of its constrictor effect on arterioles. More recently, vasopressin has also been found to play roles in fever, learning, memory, and behavior.

Some of the hormones listed in the table have been introduced elsewhere and are not discussed further; these are the renal hormones (erythropoietin in Chapter 11 and renin in Chapter 14), the hepatic hormones (thrombopoietin in Chapter 11 and hepcidin in Chapter 16), thymosin from the thymus (Chapter 12), atrial and brain natriuretic peptides from the heart (Chapter 14), the gastrointestinal (GI) hormones (Chapter 16), leptin and other adipokines from adipose tissue (Chapter 17), and hunger and satiety signals from the diges-

Endocrine Gland	Hormones	Target Cells	Major Functions of Hormones
Hypothalamus	Releasing and inhibiting hormones (TRH, CRH, GnRH, GHRH, somatostatin, PrRP, dopamine)	Anterior pituitary	Control release of anterior pituitary hormones
Posterior pituitary (hormones stored in)	Vasopressin (antidiuretic hormone, ADH)	Kidney tubules	Increases H_2O reabsorption
		Arterioles	Produces vasoconstriction
	Oxytocin	Uterus	Increases contractility
		Mammary glands (breasts)	Causes milk ejection
Anterior pituitary	Thyroid-stimulating hormone (TSH)	Thyroid follicular cells	Stimulates T_3 and T_4 secretion
	Adrenocorticotropic hormone (ACTH)	Zona fasciculata and zona reticularis of the adrenal cortex	Stimulates cortisol secretion
	Growth hormone (GH)	Bone and soft tissues	Is essential but not solely responsible for growth and exerts metabolic effects. By means of IGF-I, indirectly stimulates growth of bones and soft tissues; directly stimulates protein synthesis, mobilizes fat, and conserves glucose
		Liver	Stimulates IGF-I secretion
	Follicle-stimulating hormone (FSH)	Females: Ovarian follicles	Promotes follicular growth and development; stimulates estrogen secretion
		Males: Seminiferous tubules in testes	Stimulates sperm production
	Luteinizing hormone (LH)	Females: Ovarian follicle and corpus luteum	Stimulates ovulation, corpus luteum development, and estrogen and progesterone secretion
		Males: Interstitial cells of Leydig in testes	Stimulates testosterone secretion
	Prolactin (PRL)	Females: Mammary glands	Promotes breast development; stimulates milk secretion
		Males	Uncertain
Pineal gland	Melatonin	Brain, anterior pituitary, reproductive organs, and possibly others	Entrains body's biological rhythm with external cues; inhibits gonadotropins; its reduction likely initiates puberty; acts as an antioxidant
Thyroid gland follicular cells	Tetraiodothyronine (T_4, thyroxine); tri-iodothyronine (T_3)	Most cells	Increases metabolic rate; is essential for normal growth and nerve development
Thyroid gland C cells	Calcitonin	Bone	Decreases plasma Ca^{2+} concentration
Adrenal cortex			
Zona glomerulosa	Aldosterone (mineralocorticoid)	Kidney tubules	Increases Na^+ reabsorption and K^+ secretion
Zona fasciculata and zona reticularis	Cortisol (glucocorticoid)	Most cells	Increases blood glucose at the expense of protein and fat stores; contributes to stress adaptation
	Androgens (dehydroepiandrosterone)	Females: Hair follicles and brain	Promotes axillary and pubic hair growth and sex drive
Adrenal medulla	Epinephrine and norepinephrine	Sympathetic receptor sites throughout the body	Reinforce sympathetic nervous system; contribute to stress adaptation and blood pressure regulation

Endocrine Gland	Hormones	Target Cells	Major Functions of Hormones
Endocrine pancreas (islets of Langerhans)	Insulin (β cells)	Most cells	Promotes cellular uptake, use, and storage of absorbed nutrients
	Glucagon (α cells)	Most cells	Is important for maintaining nutrient levels in blood during the postabsorptive state
	Somatostatin (D cells)	Digestive system	Inhibits digestion and absorption of nutrients
Parathyroid gland	Parathyroid hormone (PTH)	Bone, kidneys, and intestine	Increases plasma Ca^{2+} and decreases plasma PO_4^{3-} concentrations; stimulates vitamin D activation
Female gonads: Ovaries	Estrogen (estradiol)	Female sex organs and body as a whole	Promotes follicular development; governs development of female secondary sexual characteristics; stimulates uterine and breast growth
		Bone	Enhances pubertal growth spurt; promotes closure of the epiphyseal plate
	Progesterone	Uterus	Prepares for pregnancy
Male gonads: Testes	Testosterone	Male sex organs and body as a whole	Stimulates sperm production; governs development of male secondary sexual characteristics; promotes sex drive
		Bone	Enhances pubertal growth spurt; promotes closure of the epiphyseal plate
Testes and ovaries	Inhibin	Anterior pituitary	Inhibits secretion of FSH
Placenta	Estrogen (estriol) and progesterone	Female sex organs	Help maintain pregnancy; prepare breasts for lactation
	Human chorionic gonadotropin (hCG)	Ovarian corpus luteum	Maintains corpus luteum of pregnancy
Kidneys	Renin (by activating angiotensin)	Zona glomerulosa of adrenal cortex (acted on by angiotensin, which is activated by renin)	Stimulates aldosterone secretion; angiotensin II is also a potent vasoconstrictor and stimulates thirst
	Erythropoietin	Bone marrow	Stimulates erythrocyte production
Stomach	Ghrelin	Hypothalamus	Signals hunger; stimulates appetite
	Gastrin	Digestive tract exocrine glands and smooth muscles; pancreas; liver; gallbladder	Control motility and secretion to facilitate digestive and absorptive processes
Small intestine	Secretin and cholecystokinin (CCK)		
	Glucose-dependent insulinotropic peptide (GIP)	Endocrine pancreas	Stimulates insulin secretion
	Peptide YY$_{3-36}$	Hypothalamus	Signals satiety; suppresses appetite
Liver	Insulin-like growth factor I (IGF-I)	Bone and soft tissues	Promotes growth
	Thrombopoietin	Bone marrow	Stimulates platelet production
	Hepcidin	Intestine	Inhibits iron absorption into blood
Skin	Vitamin D	Intestine	Increases absorption of ingested Ca^{2+} and PO_4^{3-}
Thymus	Thymosin	T lymphocytes	Enhances T lymphocyte proliferation and function
Heart	Atrial and brain natriuretic peptides (ANP; BNP)	Kidney tubules	Inhibit Na^+ reabsorption
Adipose tissue	Leptin	Hypothalamus	Suppresses appetite; is important in long-term control of body weight
	Other adipokines	Multiple sites	Play a role in metabolism and inflammation

tive tract (Chapter 17). The remainder of the hormones are described in greater detail in this and the next two chapters. We start in this chapter with the central endocrine glands—those in the brain or in close association with the brain—namely, the hypothalamus, the pituitary gland, and the pineal gland. The peripheral endocrine glands are discussed in the following chapters.

Check Your Understanding 18.1

1. Define *tropic hormone*.
2. Describe the three general mechanisms of controlling hormone secretion that are common to many different hormones.
3. Explain down regulation, permissiveness, synergism, and antagonism.

18.2 Hypothalamus and Pituitary

The **pituitary gland**, or **hypophysis**, is a small endocrine gland located in a bony cavity at the base of the brain just below the hypothalamus (Figure 18-3). The pituitary is connected to the hypothalamus by a thin connecting stalk. If you point one finger between your eyes and another finger toward one of your ears, the imaginary point where these lines would intersect is about where your pituitary is located.

The pituitary gland consists of anterior and posterior lobes.

The pituitary has two anatomically and functionally distinct lobes, the **posterior pituitary** and the **anterior pituitary** (Figure 18-3). The posterior pituitary is composed of nervous tissue and thus is also termed the **neurohypophysis**. The anterior pituitary consists of glandular epithelial tissue and accordingly is also called the **adenohypophysis** (*adeno* means "glandular"). The posterior and anterior pituitary lobes have only their location in common. They arise from different tissues embryonically, serve different functions, and are subject to different control mechanisms.

The release of hormones from both the posterior and the anterior pituitary is directly controlled by the hypothalamus, but the natures of these relationships are entirely different. The posterior pituitary connects to the hypothalamus by a neural pathway, whereas the anterior pituitary connects to the hypothalamus by a unique vascular link. We look first at the posterior pituitary.

The hypothalamus and posterior pituitary act as a unit to secrete vasopressin and oxytocin.

The hypothalamus and posterior pituitary form a neuroendocrine system that consists of a population of neurosecretory neurons whose cell bodies lie in two well-defined clusters in the hypothalamus, the **supraoptic nucleus** and the **paraventricular nucleus**. The axons of these neurons pass down through the connecting stalk to terminate on capillaries in the posterior pituitary (Figure 18-4). The posterior pituitary consists of these neuronal terminals plus glial-like supporting cells known as **pituicytes**. Functionally and anatomically, the posterior pituitary is simply an extension of the hypothalamus.

The posterior pituitary does not actually produce any hormones. It simply stores and, on appropriate stimulation, releases into the blood two small peptide hormones, *vasopressin* and *oxytocin*, which are synthesized by the neuronal cell bodies in the hypothalamus. These hydrophilic peptides are made in both the supraoptic and paraventricular nuclei, but a single neuron can produce only one of these hormones. The synthesized hormones are packaged in secretory granules (secretory vesicles) that are transported by motor proteins down the cytoplasm of the axon (see p. 48) and stored in the neuronal terminals within the posterior pituitary. Each terminal stores either vasopressin or oxytocin. Thus, these hormones can be released independently as needed. On stimulatory input to the hypothalamus, either vasopressin or oxytocin is released into the systemic blood from the posterior pituitary by exocytosis of the appropriate secretory granules. This hormonal release is triggered in

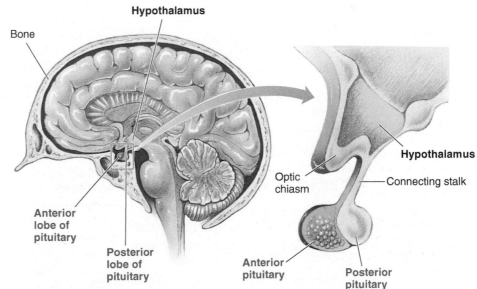

(a) Relation of pituitary gland to hypothalamus and rest of brain

(b) Enlargement of pituitary gland and its connection to hypothalamus

Figure 18-3 **Anatomy of the pituitary gland.** (a) Connected by a stalk to the base of the brain and commanded by the hypothalamus, (b) the pea-sized pituitary gland consists of the posterior pituitary (*right*), which is composed of nervous tissue, and the anterior pituitary (*left*), which consists of glandular tissue.

FIGURE FOCUS: *By examining this figure, explain why an anterior pituitary tumor sometimes causes visual disturbances.*

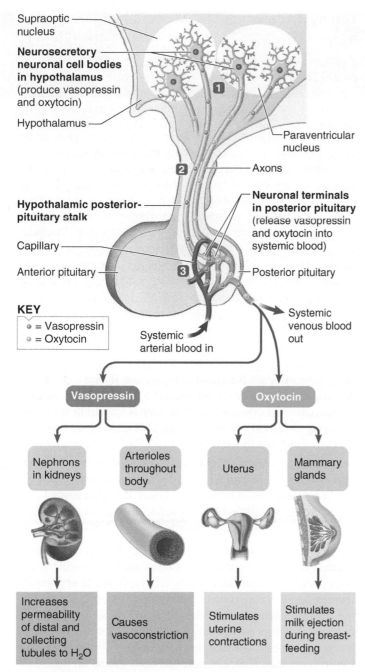

Supraoptic nucleus

Neurosecretory neuronal cell bodies in hypothalamus (produce vasopressin and oxytocin)

Hypothalamus

Paraventricular nucleus

Axons

Hypothalamic posterior-pituitary stalk

Neuronal terminals in posterior pituitary (release vasopressin and oxytocin into systemic blood)

Capillary

Anterior pituitary

Posterior pituitary

KEY
- = Vasopressin
- = Oxytocin

Systemic arterial blood in

Systemic venous blood out

Vasopressin

Oxytocin

| Nephrons in kidneys | Arterioles throughout body | Uterus | Mammary glands |

| Increases permeability of distal and collecting tubules to H_2O | Causes vasoconstriction | Stimulates uterine contractions | Stimulates milk ejection during breast-feeding |

1 The paraventricular and supraoptic nuclei both contain neurons that produce vasopressin and oxytocin. The hormone, either vasopressin or oxytocin depending on the neuron, is synthesized in the neuronal cell body in the hypothalamus.

2 The hormone travels down the axon to be stored in the neuronal terminals within the posterior pituitary.

3 When the neuron is excited, the stored hormone is released from the terminals into the systemic blood for distribution throughout the body.

Figure 18-4 **Relationship of the hypothalamus and posterior pituitary.**

response to action potentials that originate in the hypothalamic cell body and sweep down the axon to the neuronal terminal in the posterior pituitary. As in any other neuron, action potentials are generated in these neurosecretory neurons in response to synaptic input to their cell bodies.

The actions of vasopressin and oxytocin are briefly summarized here to make our endocrine story complete. They are described more thoroughly elsewhere—vasopressin in Chapters 14 and 15 and oxytocin in Chapter 20.

Vasopressin Vasopressin (antidiuretic hormone, ADH) has two major effects that correspond to its two names: (1) it conserves H_2O during urine formation by the kidney nephrons (an antidiuretic effect) and (2) it causes contraction of arteriolar smooth muscle (a vessel pressor effect). The first effect has more physiologic importance. Under normal conditions, vasopressin is the primary endocrine factor that regulates urinary H_2O loss and overall H_2O balance. In contrast, typical levels of vasopressin play only a minor role in regulating blood pressure by means of promoting arteriolar vasoconstriction.

The major control for hypothalamic-induced release of vasopressin from the posterior pituitary is input from hypothalamic osmoreceptors, which increase vasopressin secretion in response to a rise in plasma osmolarity. A less powerful input from the left atrial volume receptors increases vasopressin secretion in response to a fall in ECF volume and arterial blood pressure (see p. 545). (For further information on the importance of vasopressin secretion when exercising in the heat, see the boxed feature on p. 648, A Closer Look at Exercise Physiology.)

Oxytocin Oxytocin stimulates contraction of uterine smooth muscle to help expel the infant during childbirth, and it promotes ejection of milk from the mammary glands (breasts) during breast-feeding. Appropriately, oxytocin secretion is increased by reflexes that originate within the birth canal during childbirth and by reflexes that are triggered when the infant suckles the breast.

In addition to these two major physiologic effects, oxytocin influences a variety of behaviors, especially maternal behaviors. For example, this hormone fittingly facilitates bonding, or attachment, between a mother and her infant. For this reason, oxytocin is sometimes nicknamed the "love hormone" or "cuddle chemical." Recent studies suggest that oxytocin may play a role in other types of close human attachment, such as helping bond couples to one another.

Most anterior pituitary hormones are tropic.

Unlike the posterior pituitary, which releases hormones synthesized by the hypothalamus, the anterior pituitary synthesizes the hormones it releases into the blood. Five different cell populations within the anterior pituitary secrete six major peptide hormones. The actions of each of these hormones are described in detail in later sections. For now, this brief statement of their source and primary effects provides a rationale for their names (Figure 18-5):

1. The anterior pituitary cells known as **somatotropes** secrete **growth hormone (GH, somatotropin)**, the primary hormone responsible for regulating overall body growth (*somato* means "body"). GH also exerts important metabolic actions.

2. **Thyrotropes** secrete **thyroid-stimulating hormone (TSH, thyrotropin)**, which stimulates secretion of thyroid hormone and growth of the thyroid gland.

The Endocrine Response to the Challenge of Combined Heat and Marching Feet

WHEN ONE EXERCISES IN A hot environment, maintaining plasma volume becomes a critical homeostatic concern. Exercise in the heat results in losing large amounts of fluid through sweating. Simultaneously, blood is needed for shunting to the skin for cooling and for increased blood flow to nourish the working muscles. To maintain cardiac output, venous return must also be adequate. The hypothalamus–posterior pituitary neurosecretory system responds to these multiple, conflicting needs for fluid by releasing water-conserving vasopressin, reducing urinary fluid loss to preserve plasma volume.

Studies have generally shown that exercise in heat stimulates vasopressin release, which results in decreased urinary fluid loss. In one study conducted during an 18-mile road march in heat, the partici-

pants' average urine output dropped to 134 mL (normal urine output during the same time period would be about 2 ½ times that much), whereas sweat loss averaged 4 liters. Overhydration before exercise appears to decrease the intensity of this response, suggesting that increased vasopressin release is related to plasma osmolarity. If lost fluid is not adequately replaced, plasma osmolarity increases. When the hypothalamic osmoreceptors detect this hypertonic condition, they promote increased secretion of vasopressin from the posterior pituitary. Some investigators believe, however, that increased vasopressin release results from other factors, such as changes in blood pressure or renal blood flow. Regardless of the mechanism, vasopressin release is an important physiologic response to exercise in heat.

3. **Corticotropes** produce and release **adrenocorticotropic hormone (ACTH, adrenocorticotropin)**, the hormone that stimulates cortisol secretion by and promotes growth of the adrenal cortex.

ACTH is synthesized as part of a large precursor molecule known as **pro-opiomelanocortin (POMC)**. POMC can be cleaved into three active products: *ACTH, melanocyte-stimulating hormone (MSH)*, and *endorphin*. Several diverse cell types produce POMC and slice it in unique ways, depending on the processing enzymes they possess, to yield different active products, along with peptide "scraps" that have no known function. For example, as their major active product from this same precursor molecule, corticotropes produce ACTH; in response to UV light from the sun, keratinocytes in the skin produce α-MSH, which promotes dispersal from nearby melanocytes of the pigment melanin to cause tanning (see p. 441); appetite-suppressing neurons in the hypothalamus secrete α-MSH to control food intake (see p. 621); and other neurons in the CNS produce endorphin, an endogenous opioid that suppresses pain (see p. 192).

4. **Gonadotropes** secrete two hormones that act on the gonads (reproductive organs, namely, the ovaries and testes)—follicle-stimulating hormone and luteinizing hormone. **Follicle-stimulating hormone (FSH)** helps regulate gamete (reproductive cells, namely, ova and sperm) production in both sexes. In females, it stimulates growth and development of ovarian follicles, within which the ova, or eggs, develop. It also promotes secretion of the hormone estrogen by the ovaries. In males, FSH is required for sperm production.

5. **Luteinizing hormone (LH)**, also secreted by gonadotropes, helps control sex hormone secretion in both sexes, among other important actions in females. LH regulates ovarian secretion of the female sex hormones, estrogen and progesterone. In males, the same hormone stimulates the testes to

secrete the male sex hormone, testosterone. In females, LH is also responsible for ovulation (egg release) and luteinization (formation of a hormone-secreting corpus luteum in the ovary following ovulation). Note that both FSH and LH are named for their functions in females.

6. **Lactotropes** secrete **prolactin (PRL)**, which enhances breast development and lactation (milk production) in females. Its reproductive function in males is uncertain. Evidence suggests other functions of prolactin unrelated to the reproductive system, such as enhancing the immune system in both sexes.

GH, TSH, ACTH, FSH, and LH are all tropic hormones because they each regulate secretion of another specific endocrine gland. FSH and LH are collectively referred to as **gonadotropins** because they control secretion of the sex hormones by the gonads. Among the anterior pituitary hormones, PRL is the only one that does not stimulate secretion of another hormone. It acts directly on nonendocrine tissue to exert its effects. Of the tropic hormones, FSH, LH, and GH exert effects on nonendocrine target cells in addition to stimulating secretion of other hormones.

TSH, ACTH, FSH, and LH all act at their target organs by binding with G-protein-coupled receptors that activate the cyclic adenosine monophosphate (cAMP) second-messenger system (see p. 123). GH and PRL both exert their effects via the JAK/STAT pathway (see p. 116).

Hypothalamic releasing and inhibiting hormones help regulate anterior pituitary hormone secretion.

None of the anterior pituitary hormones are secreted at a constant rate. Even though each of these hormones has a unique control system, they have some common regulatory patterns. The two most important factors that regulate anterior pituitary hormone secretion are hypothalamic hormones and feedback by target-gland hormones.

Release of each anterior pituitary hormone is largely controlled by still other hormones produced by the hypothalamus. The secretion of these regulatory neurohormones, in turn, is controlled by a variety of neural and hormonal inputs to the hypothalamic neurosecretory cells.

Role of the Hypothalamic Releasing and Inhibiting Hormones The secretion of each anterior pituitary hormone

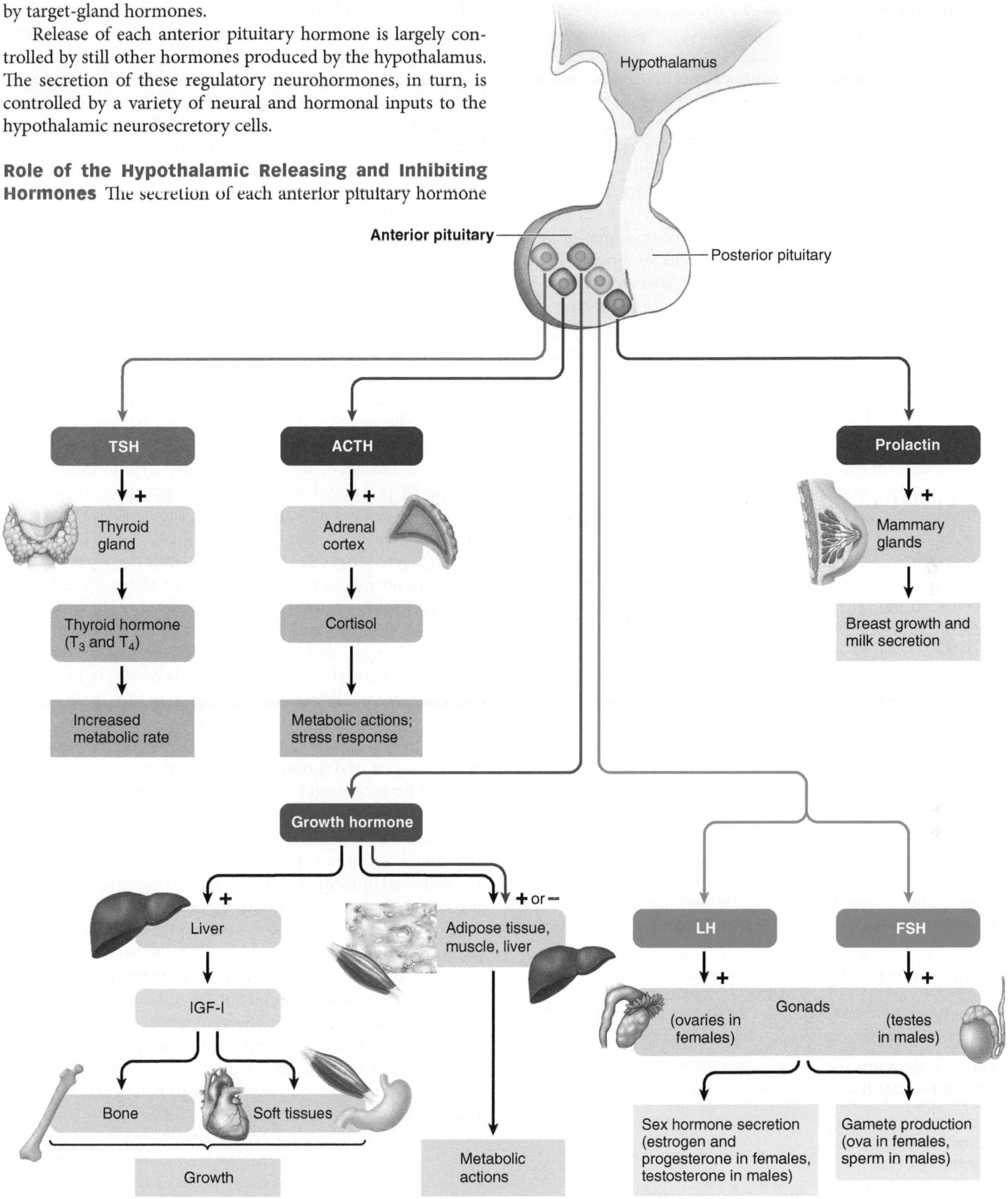

Figure 18-5 Functions of the anterior pituitary hormones. Five different endocrine cell types produce the six anterior pituitary hormones—TSH, ACTH, growth hormone, LH and FSH (produced by the same cell type), and prolactin—which exert a wide range of effects throughout the body.

FIGURE FOCUS: Use this figure to identify which anterior pituitary hormones are tropic.

is stimulated or inhibited by one or more of seven hypothalamic **hypophysiotropic hormones** (*hypophysis* means "pituitary"; *tropic* means "nourishing"). These small peptide hormones are listed in Table 18-3. Depending on their actions, these hormones are called **releasing hormones** or **inhibiting hormones**. In each case, the primary action of the hormone is apparent from its name. For example, **thyrotropin-releasing hormone (TRH)** stimulates release of TSH (alias thyrotropin) from the anterior pituitary, whereas **prolactin-inhibiting hormone (PIH)**, which is **dopamine** (the same as the neurotransmitter in the basal nuclei and in the "pleasure" pathways in the brain; see pp. 154 and 156), inhibits release of PRL from the anterior pituitary. Note that hypophysiotropic hormones in most cases are involved in a three-hormone hierarchic chain of command (Figure 18-6): The hypothalamic hypophysiotropic hormone (*hormone 1*) controls the output of an anterior-pituitary tropic hormone (*hormone 2*). This tropic hormone, in turn, regulates secretion of the target endocrine gland's hormone (*hormone 3*), which exerts the final physiologic effect. This three-hormone sequence is called an **endocrine axis**, as in the hypothalamus–pituitary–thyroid axis.

Although endocrinologists originally speculated that there was one hypophysiotropic hormone for each anterior pituitary hormone, some hypothalamic hormones have more than one effect, so their names indicate only the function first identified. Moreover, a single anterior pituitary hormone may be regulated by two or more hypophysiotropic hormones, which may even exert opposing effects. For example, **growth hormone–releasing hormone (GHRH)** stimulates growth hormone secretion, whereas **growth hormone–inhibiting hormone (GHIH)**, also known as **somatostatin**, inhibits it. The output of the anterior-pituitary somatotropes (that is, the rate of growth hormone secretion) in response to two such opposing inputs depends on the relative concentrations of these hypothalamic hormones and on the intensity of other regulatory inputs.

Chemical messengers identical in structure to the hypothalamic releasing and inhibiting hormones and to vasopressin are produced in many areas of the brain outside the hypothalamus. Instead of being released into the blood, these messengers act locally as neurotransmitters or neuromodulators in these other sites, an example being that PIH is identical to the neurotransmitter dopamine. Others modulate a variety of functions that range from motor activity (TRH) to libido (GnRH) to learning (vasopressin). These examples further illustrate the multiplicity of ways chemical messengers function.

Role of the Hypothalamic–Hypophyseal Portal System

The hypothalamic regulatory hormones reach the anterior pituitary by means of a unique vascular link. In contrast to the direct neural connection between the hypothalamus and the posterior pituitary, the anatomic and functional link between the hypothalamus and the anterior pituitary is an unusual capillary-to-capillary connection, the **hypothalamic–hypophyseal portal system**. (A portal system is a vascular arrangement in which venous blood flows directly from one capillary bed through a connecting vessel to another capillary bed, as in the hepatic portal system; see p. 593.) The hypothalamic–hypophyseal portal system provides a critical link between the brain and much of the endocrine system. It begins in the base of the hypothalamus with a group of capillaries that recombine into small portal vessels, which pass down through the connecting stalk into the anterior pituitary. Here, the portal vessels branch to form most of the anterior pituitary capillaries, which in turn drain into the systemic venous system (Figure 18-7).

As a result, almost all blood supplied to the anterior pituitary must first pass through the hypothalamus. Because materials can be exchanged between blood and surrounding tissue only at the capillary level, the hypothalamic–hypophyseal portal system provides a "private" route through which releasing and inhibiting hormones can be picked up at the hypothalamus and delivered immediately and directly to the anterior pituitary at relatively high concentrations, bypassing the general circulation. If the portal system did not exist, once the hypophysiotropic hormones were picked up in the hypothalamus, they would be returned to the heart through the systemic veins, pumped through the pulmonary circulation, then returned to the heart and finally be pumped into the systemic arteries for delivery throughout the body, including the anterior pituitary. Not only would this process take much longer, but the hypophysiotropic hormones would be considerably diluted before arriving at the anterior pituitary.

The axons of the neurosecretory neurons that produce the hypothalamic regulatory hormones terminate on the capillaries at the origin of the portal system. These hypothalamic neurons secrete their hormones in the same way as the hypothalamic neurons that produce vasopressin and oxytocin. The hormone is synthesized in the cell body and then transported in vesicles by motor proteins to the axon terminal. It is stored there until its release by exocytosis into an adjacent capillary on appropri-

┃TABLE 18-3 Major Hypophysiotropic Hormones	
Hormone	**Effect on the Anterior Pituitary**
Thyrotropin-releasing hormone (TRH)	Stimulates release of TSH (thyrotropin) and prolactin
Corticotropin-releasing hormone (CRH)	Stimulates release of ACTH (corticotropin)
Gonadotropin-releasing hormone (GnRH)	Stimulates release of FSH and LH (gonadotropins)
Growth hormone–releasing hormone (GHRH)	Stimulates release of GH
Somatostatin (growth hormone–inhibiting hormone; GHIH)	Inhibits release of GH and TSH
Prolactin-releasing peptide (PrRP)	Stimulates release of PRL
Dopamine (prolactin-inhibiting hormone; PIH)	Inhibits release of PRL

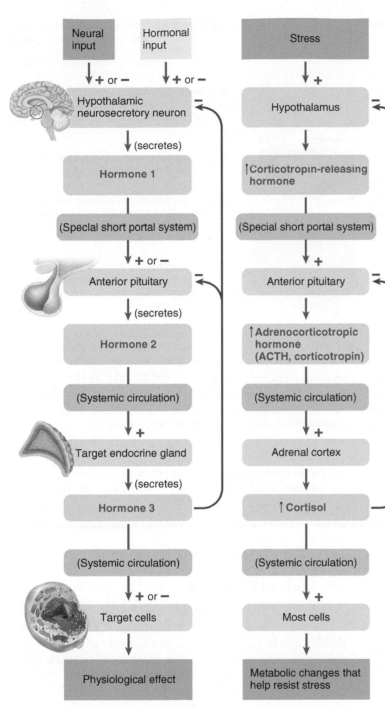

Control of Hypothalamic Releasing and Inhibiting
Hormones What regulates secretion of these hypophysiotropic hormones? Like other neurons, the neurons secreting these regulatory hormones receive abundant input of information (both neural and hormonal and both excitatory and inhibitory) that they must integrate. Studies are still in progress to unravel the complex neural input from many diverse areas of the brain to the hypophysiotropic secretory neurons. Some of these inputs carry information about a variety of environmental conditions. One example is the marked increase in secretion of corticotropin-releasing hormone (CRH) in response to stress (see Figure 18-6). Numerous neural connections also exist between the hypothalamus and the portions of the brain concerned with emotions (the limbic system; see p. 155). Thus, emotions greatly influence secretion of hypophysiotropic hormones. The menstrual irregularities sometimes experienced by women who are emotionally upset are a common manifestation of this relationship.

In addition to being regulated by different regions of the brain, the hypophysiotropic neurons are controlled by various chemical inputs that reach the hypothalamus through the blood. Unlike other regions of the brain, portions of the hypothalamus are not guarded by the blood–brain barrier, so the hypothalamus can easily monitor chemical changes in the blood. The most common blood-borne factors that influence hypothalamic neurosecretion are the negative-feedback effects of target-gland hormones, to which we now turn.

Figure 18-6 **Hierarchic chain of command and negative feedback in endocrine control.** The general pathway involved in the hierarchic chain of command in the hypothalamus–anterior pituitary–peripheral target endocrine-gland axis is depicted on the left. The pathway on the right leading to cortisol secretion provides a specific example of this endocrine chain of command. The hormone ultimately secreted by the target endocrine gland, such as cortisol, acts in negative-feedback fashion to reduce secretion of the regulatory hormones higher in the chain of command.

Target-gland hormones inhibit hypothalamic and anterior pituitary hormone secretion via negative feedback.

In most cases, hypophysiotropic hormones initiate a three-hormone sequence: hypophysiotropic hormone, anterior-pituitary tropic hormone, and hormone from the peripheral target endocrine gland. Typically, in addition to producing its physiologic effects, the target-gland hormone suppresses secretion of the tropic hormone that is driving it. This negative feedback is accomplished by the target-gland hormone acting directly on the pituitary and on the release of hypothalamic hormones, which in turn regulate anterior pituitary function (see Figure 18-6). As an example, consider the CRH–ACTH–cortisol system. Hypothalamic CRH (corticotropin-releasing hormone) stimulates the anterior pituitary to secrete ACTH (alias corticotropin), which in turn stimulates the adrenal cortex to secrete cortisol. The final hormone in the system, cortisol, inhibits the hypothalamus to reduce CRH secretion and acts directly on the corticotropes in the anterior pituitary to reduce ACTH secretion. Through this double-barreled approach, cortisol exerts negative-feedback control to stabilize its plasma concentration. If plasma cortisol levels start to rise above a prescribed level, cortisol suppresses its

ate stimulation. The major difference is that the hypophysiotropic hormones are released into the portal vessels, which deliver them to the anterior pituitary, where they control release of anterior pituitary hormones into the general circulation. In contrast, the hypothalamic hormones stored in the posterior pituitary are themselves released into the general circulation.

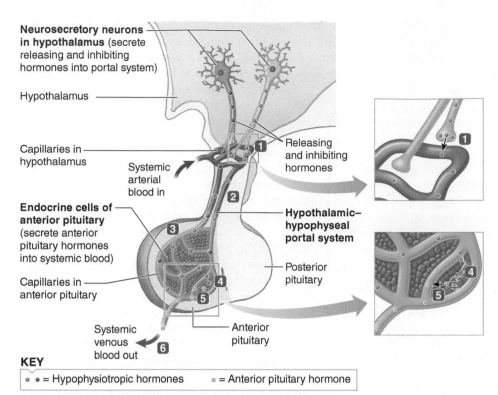

Neurosecretory neurons in hypothalamus (secrete releasing and inhibiting hormones into portal system)

Hypothalamus

Capillaries in hypothalamus

Systemic arterial blood in

Endocrine cells of anterior pituitary (secrete anterior pituitary hormones into systemic blood)

Capillaries in anterior pituitary

Systemic venous blood out

Releasing and inhibiting hormones

Hypothalamic–hypophyseal portal system

Posterior pituitary

Anterior pituitary

KEY

• • = Hypophysiotropic hormones   • = Anterior pituitary hormone

1 Hypophysiotropic hormones (releasing hormones and inhibiting hormones) produced by neurosecretory neurons in the hypothalamus enter the hypothalamic capillaries.

2 These hypothalamic capillaries rejoin to form the hypothalamic–hypophyseal portal system, a vascular link to the anterior pituitary.

3 The portal system branches into the capillaries of the anterior pituitary.

4 The hypophysiotropic hormones, which leave the blood across the anterior pituitary capillaries, control the release of anterior pituitary hormones.

5 When stimulated by the appropriate hypothalamic releasing hormone, the anterior pituitary secretes a given hormone into these capillaries.

6 The anterior pituitary capillaries rejoin to form a vein, through which the anterior pituitary hormones leave for ultimate distribution throughout the body by the systemic circulation.

▮Figure 18-7 **Vascular link between the hypothalamus and anterior pituitary.**

FIGURE FOCUS: *What hormone exchanges take place across the hypothalamic capillaries and across the anterior pituitary capillaries?*

own secretion by its inhibitory actions at the hypothalamus and anterior pituitary. This mechanism ensures that once a hormonal system is activated its secretion does not continue unabated. If plasma cortisol levels fall below the desired set point, cortisol's inhibitory actions at the hypothalamus and anterior pituitary are reduced, so the driving forces for cortisol secretion (CRH–ACTH) increase accordingly. The other target-gland hormones act by similar negative-feedback loops to maintain their plasma levels relatively constant at the set point.

Diurnal rhythms are superimposed on this type of stabilizing negative-feedback regulation—that is, the set point changes as a function of the time of day. Furthermore, other controlling inputs may break through the negative-feedback control to alter hormone secretion (that is, change the set level) at times of special need, such as stress raising the set point for cortisol secretion.

The detailed functions and control of all the anterior pituitary hormones except growth hormone are discussed elsewhere in conjunction with the target tissues that they influence; for example, thyroid-stimulating hormone is covered in the next chapter with the discussion of the thyroid gland. Accordingly, growth hormone is the only anterior pituitary hormone we elaborate on at this time.

Check Your Understanding 18.2

1. Draw a flow diagram showing the hierarchic chain of command and negative feedback in the hypothalamic–anterior pituitary–peripheral target endocrine-gland axis.

2. List the posterior pituitary hormones, the anterior pituitary hormones, and the hypophysiotropic hormones.

3. Compare the means by which the hypothalamus controls hormonal output from the posterior pituitary and from the anterior pituitary.

18.3 Endocrine Control of Growth

In growing children, continuous net protein synthesis occurs under the influence of growth hormone (GH) as the body steadily gets larger. Weight gain alone is not synonymous with growth because weight gain may occur from retaining excess water or storing fat without true structural growth of tissues. Growth requires net synthesis of proteins and includes lengthening of the long bones (the bones of the extremities) and increases in the size and number of cells in the soft tissues.

Growth depends on GH but is influenced by other factors.

Although, as the name implies, growth hormone is essential for growth, it is not wholly responsible for determining the rate and final magnitude of growth in a given individual. The following factors affect growth:

- *Genetic determination* of an individual's maximum growth capacity. Attaining this full growth potential further depends on the other factors listed here.

- *Adequate diet,* including enough total protein and ample essential amino acids to accomplish the protein synthesis necessary for growth. Malnourished children never achieve their full growth potential. By contrast, a person cannot exceed his or her genetically determined maximum by eating a more-than-adequate diet. The excess food intake produces obesity instead of growth.

■ *Freedom from chronic disease and stressful conditions.* Stunted growth under adverse circumstances is largely a result of the prolonged stress-induced secretion of cortisol from the adrenal cortex. Cortisol exerts several potent antigrowth effects, such as promoting protein breakdown, inhibiting growth in the long bones, and blocking secretion of GH.

■ *Normal levels of growth-influencing hormones.* In addition to the essential GH, other hormones including thyroid hormone, insulin, and the sex hormones play secondary roles in promoting growth.

The rate of growth is not continuous, nor are the factors responsible for promoting growth the same throughout the growth period. *Fetal growth* is promoted largely by certain hormones from the placenta (the hormone-secreting organ of exchange between the fetal and maternal circulatory systems; see p. 757), with the size at birth being determined principally by genetic and environmental factors. GH plays no role in fetal development. After birth, GH and other nonplacental hormonal factors begin to play an important role in regulating growth. Genetic and nutritional factors also strongly affect growth during this period.

Children display two periods of rapid growth—a *postnatal* ("after birth") *growth spurt* during their first 2 years of life and a *pubertal growth spurt* during adolescence (Figure 18-8). From age 2 until puberty, the *rate* of linear growth progressively declines, even though the child is still growing. Before puberty, there is little sexual difference in height or weight by age. During puberty, a marked acceleration in linear growth takes place because the long bones lengthen. Puberty begins at about age 11 in girls and 13 in boys and lasts for several years in both sexes. Both genetic and hormonal factors are involved in the pubertal growth spurt. GH secretion is elevated during puberty and contributes to growth acceleration during this time. Furthermore, the sex hormones, whose secretion increases dramatically at puberty, also directly contribute to the pubertal growth spurt and stimulate further secretion of GH. Ultimately, however, estrogen in females and testosterone-turned-estrogen (by action of aromatase) in males act on bone to halt its further growth so that full adult height is attained by the end of adolescence.

GH is essential for growth, but it also directly exerts metabolic effects not related to growth.

GH is the most abundant hormone produced by the anterior pituitary, even in adults in whom growth has already ceased, although GH secretion typically starts to decline after middle age. The continued high secretion of GH beyond the growing period implies that this hormone has important influences beyond its influence on growth, such as metabolic effects. We briefly describe GH's metabolic effects before turning to its growth-promoting actions.

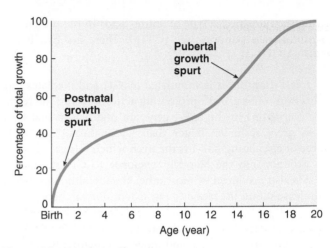

Figure 18-8 **Normal growth curve.**

To exert its metabolic effects, GH binds directly with its target organs, namely, adipose tissue, skeletal muscles, and liver. GH increases fatty acid levels in the blood by enhancing the breakdown of triglyceride fat stored in adipose tissue, and it increases blood glucose levels by decreasing glucose uptake by muscles and increasing glucose output by the liver. Muscles use the mobilized fatty acids instead of glucose as a metabolic fuel. Thus, the overall metabolic effect of GH is to mobilize fat stores as a major energy source while sparing glucose for glucose-dependent tissues such as the brain. The brain can use only glucose as its metabolic fuel, yet nervous tissue cannot store glycogen (stored glucose) to any extent. This metabolic pattern induced by GH is suitable for maintaining the body during prolonged fasting or other situations when the body's energy needs exceed available glucose stores.

GH also stimulates amino acid uptake and protein synthesis and inhibits protein degradation throughout the body, especially in protein-rich muscle, decreasing blood amino acids in the process. GH directly contributes to growth via these metabolic effects. However, GH acts indirectly by means of insulin-like growth factors to accomplish its other growth-related actions.

GH mostly exerts its growth-promoting effects indirectly by stimulating insulin-like growth factors.

GH's growth-producing actions include protein synthesis, increased cell division, and bone growth. Except for directly increasing protein synthesis, GH indirectly brings about its other growth-promoting actions by stimulating production of **insulin-like growth factors (IGFs)**, which directly act on the target cells to cause growth of both soft tissues and bones. IGFs are produced in many tissues and have endocrine, paracrine, and autocrine actions (see p. 114). Originally called **somatomedins**, these peptide mediators are now preferentially called insulin-like growth factors because they are structurally and functionally similar to insulin. Like insulin, IGFs exert their effects largely by binding with receptor-enzymes that activate designated effector proteins within the target cell by phosphorylating the tyrosines (a type of amino acid) in the protein (the tyrosine kinase pathway; see p. 116). There are two IGFs— **IGF-I** and **IGF-II**.

IGF-I IGF-I synthesis is stimulated by GH and mediates most of this hormone's growth-promoting actions. (As an interesting aside, variations in the IGF-I gene are one reason that Great Danes grow so much larger than Chihuahuas.) The major source of circulating IGF-I is the liver, which releases this peptide product into the blood in response to GH stimulation. IGF-I is also produced by most other tissues, although they do not release it into the blood to any extent. Researchers propose that IGF-I produced locally in target tissues may act through paracrine means. Such a mechanism could explain why blood levels of GH are no higher, and indeed circulating IGF-I levels are lower, during the first several years of life compared to adult values, even though growth is quite rapid during the postnatal period. Local production of IGF-I in target tissues may be more important than delivery of blood-borne IGF-I or GH during this time.

Production of IGF-I is controlled by a number of factors other than GH, including nutritional status, age, and tissue-specific factors as follows:

- IGF-I production depends on adequate nutrition. Inadequate food intake reduces IGF-I production. As a result, changes in circulating IGF-I levels do not always coincide with changes in GH secretion. For example, fasting decreases IGF-I levels even though it increases GH secretion.

- Age-related factors influence IGF-I production. A dramatic increase in circulating IGF-I levels accompanies the moderate increase in GH at puberty, which is important for the pubertal growth spurt.

- Finally, various tissue-specific stimulatory factors can increase IGF-I production in particular tissues. To illustrate, the gonadotropins and sex hormones stimulate IGF-I production within reproductive organs such as the testes in males and the ovaries and uterus in females.

Thus, control of IGF-I production is complex and subject to a variety of systemic and local factors.

IGF-II In contrast to IGF-I, GH does not influence IGF-II production. IGF-II is primarily important during fetal development. Although IGF-II continues to be produced during adulthood, its role in adults remains unclear.

We now describe GH's growth-promoting effects, mostly mediated by IGF-I.

GH, through IGF-I, promotes growth of soft tissues by stimulating hypertrophy and hyperplasia.

When tissues are responsive to its growth-promoting effects, GH stimulates growth of both the soft tissues and the skeleton. GH promotes growth of soft tissues by (1) increasing the size of cells (**hypertrophy**) and (2) increasing the number of cells (**hyperplasia**). GH increases the size of cells by favoring synthesis of proteins, the main structural component of cells. GH and

IGF-I both directly stimulate almost all aspects of protein synthesis while simultaneously inhibiting protein degradation. GH (via IGF-I) increases the number of cells by stimulating cell division and by preventing apoptosis (programmed cell death; see p. 40).

Growth of the long bones resulting in increased height is the most dramatic effect of GH. Before you can understand the means by which GH stimulates bone growth, you must become familiar with bone structure and how growth of bone is accomplished.

Bone grows in thickness and in length by different mechanisms, both stimulated by GH.

Bone is a living tissue. Being a form of connective tissue, it consists of cells and an extracellular organic matrix known as **osteoid** produced by the cells. The bone cells that produce the organic matrix are known as **osteoblasts** (*osteo* means "bone"; *blasts* means "formers"). Osteoid is composed of collagen fibers (see p. 60) in a semisolid gel. This organic matrix has a rubbery consistency and is responsible for bone's tensile strength (the resilience of bone to breakage when tension is applied). Bone is made hard by precipitation of *calcium phosphate crystals* within the osteoid. These inorganic crystals provide the bone with compressional strength (the ability of bone to hold its shape when squeezed or compressed). If bones consisted entirely of inorganic crystals, they would be brittle, like pieces of chalk. Bones have structural strength approaching that of reinforced concrete, yet they are not brittle and weigh much less because they have the structural blending of an organic scaffolding hardened by inorganic crystals. **Cartilage** is similar to bone, except that living cartilage is not calcified.

A long bone basically consists of a fairly uniform cylindrical shaft, the **diaphysis**, with a flared articulating knob at either end, an **epiphysis**. In a growing bone, the diaphysis is separated at each end from the epiphysis by a layer of cartilage known as the **epiphyseal plate** (❙ Figure 18-9a). The central cavity of the bone is filled with bone marrow, the site of blood cell production (see p. 385).

Mechanisms of Bone Growth Growth in *thickness* of bone is achieved by adding new bone on top of the outer surface of existing bone. This growth is produced by osteoblasts within the **periosteum**, a connective tissue sheath that covers the outer bone. As osteoblast activity deposits new bone on the external surface, other cells within the bone, the **osteoclasts** ("bone breakers"), dissolve the bony tissue on the inner surface next to the marrow cavity. In this way, the marrow cavity enlarges to keep pace with the increased circumference of the bone shaft.

Growth in *length* of long bones is accomplished by a different mechanism. Bones lengthen as a result of activity of the cartilage cells, or **chondrocytes**, in the epiphyseal plates (❙ Figure 18-9b). During growth, cartilage cells on the outer edge of the plate next to the epiphysis divide and multiply. As new chondrocytes are formed on the epiphyseal border, the older cartilage cells toward the diaphyseal border are enlarging. This combination of proliferation of new cartilage cells and hypertrophy of maturing chondrocytes temporarily widens the epiphyseal plate. This thickening of the intervening cartilaginous plate pushes the bony epiphysis farther away from the diaphysis. Soon the matrix surrounding the oldest hypertrophied cartilage becomes calcified. Because cartilage lacks its own capillary network, the survival of cartilage cells depends on diffusion of nutrients and O_2 through the matrix, a process prevented by the deposition of calcium salts. As a result, the old nutrient-deprived cartilage cells on the diaphyseal border die. As osteoclasts clear away dead chondrocytes and the calcified matrix that imprisoned them, the area is invaded by osteoblasts, which swarm upward from the diaphysis, trailing their capillary supply with them. These new tenants lay down bone around the persisting remnants of disin-

tegrating cartilage until bone entirely replaces the inner region of cartilage on the diaphyseal side of the plate. When this **ossification** ("bone formation") is complete, the bone on the diaphyseal side has lengthened and the epiphyseal plate has returned to its original thickness. The cartilage that bone has replaced on the diaphyseal end of the plate is as thick as the new cartilage on the epiphyseal end of the plate. Thus, bone growth is made possible by the growth and death of cartilage, which acts like a "spacer" to push the epiphysis farther out while it provides a framework for future bone formation on the end of the diaphysis.

Mature, Nongrowing Bone As the extracellular organic matrix produced by an osteoblast calcifies, the osteoblast becomes entombed by the matrix it has deposited around itself. Unlike chondrocytes, however, osteoblasts trapped within a calcified matrix do not die because they are supplied by nutrients transported to them through small canals that the osteoblasts form by sending out cytoplasmic extensions around which the bony matrix is deposited. Thus, within the final bony product, a network of permeating tunnels radiates from each

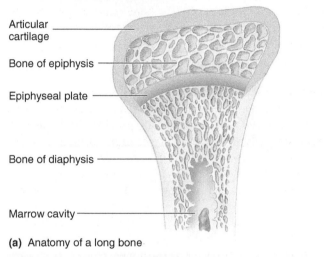

Articular cartilage

Bone of epiphysis

Epiphyseal plate

Bone of diaphysis

Marrow cavity

(a) Anatomy of a long bone

KEY

	Cartilage
	Calcified cartilage
	Bone

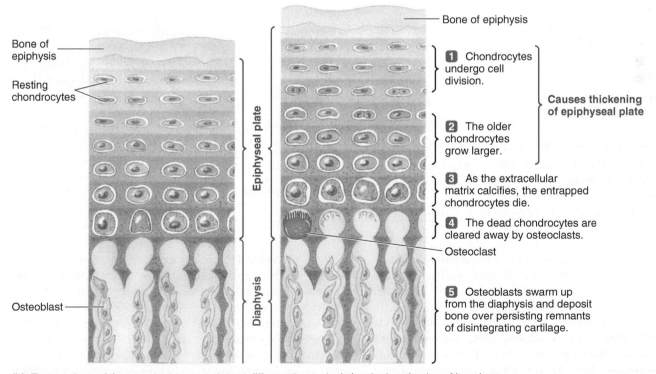

Bone of epiphysis

Bone of epiphysis

Resting chondrocytes

1 Chondrocytes undergo cell division.

Causes thickening of epiphyseal plate

2 The older chondrocytes grow larger.

3 As the extracellular matrix calcifies, the entrapped chondrocytes die.

4 The dead chondrocytes are cleared away by osteoclasts.

Osteoclast

Epiphyseal plate

Diaphysis

Osteoblast

5 Osteoblasts swarm up from the diaphysis and deposit bone over persisting remnants of disintegrating cartilage.

(b) Two sections of the same epiphyseal plate at different times, depicting the lengthening of long bones

IFigure 18-9 **Anatomy and growth of long bones.**

entrapped osteoblast, serving as a lifeline system for nutrient delivery and waste removal. The entrapped osteoblasts, now called **osteocytes**, retire from active bone-forming duty because their imprisonment prevents them from laying down new bone. However, they are involved in the hormonally regulated exchange of calcium between bone and blood. This exchange is under the control of parathyroid hormone (discussed in the next chapter), not GH.

GH Control of Bone Growth GH causes bones to grow both in length and in thickness via IGF-I, which has profound effects on cartilage and bone. IGF-I stimulates proliferation of epiphyseal cartilage, thereby making space for more bone formation, and stimulates osteoblast activity. GH/IGF-I can promote lengthening of long bones as long as the epiphyseal plate remains cartilaginous, or is "open." At the end of adolescence, sex steroids completely ossify, or "close" the epiphyseal plates so that the bones cannot lengthen any further despite the presence of GH and IGF-I. Thus, after the plates are closed, the individual does not grow any taller.

GH secretion is regulated by two hypophysiotropic hormones.

The control of GH secretion is complex, with two hypothalamic hypophysiotropic hormones playing a key role.

Growth Hormone–Releasing Hormone and Growth Hormone–Inhibiting Hormone Two antagonistic regulatory hormones from the hypothalamus are involved in controlling GH secretion: growth hormone–releasing hormone (GHRH), which is stimulatory and the dominant influence, and growth hormone–inhibiting hormone (GHIH, or somatostatin), which is inhibitory (Figure 18-10). (Note the distinctions among *somatotropin,* alias growth hormone; *somatomedin,* a liver hormone (alias IGF-I) that directly mediates the effects of GH; and *somatostatin,* which inhibits GH secretion.) Both GHRH and somatostatin act on the anterior pituitary somatotropes by binding with G-protein-coupled receptors linked to the cAMP second-messenger pathway, with GHRH increasing cAMP and somatostatin decreasing cAMP.

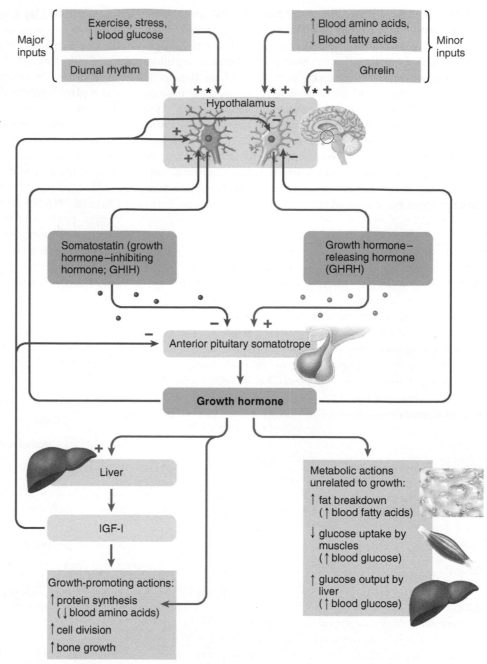

*These factors all increase growth hormone secretion, but it is unclear whether they do so by stimulating GHRH or inhibiting somatostatin (GHIH), or both.

Figure 18-10 **Control of growth hormone secretion.**
FIGURE FOCUS: *Describe the negative-feedback loops involved in control of growth hormone secretion.*

As with control of other anterior pituitary hormones, negative-feedback loops participate in controlling GH secretion. Complicating the feedback loops for the hypothalamus–pituitary–liver axis is direct regulation of GH secretion by both stimulatory and inhibitory factors. Therefore, negative-feedback loops involve both inhibition of stimulatory factors and stimulation of inhibitory factors. GH stimulates IGF-I secretion by the liver, and IGF-I in turn is the primary inhibitor of GH secretion by the anterior pituitary. IGF-I inhibits the somatotropes in the pituitary directly and further decreases GH secretion by inhibiting GHRH-secreting cells and stimulating the somatostatin-secreting cells in the hypothalamus, thus decreasing hypothalamic stimulation of the somatotropes. Furthermore, GH itself inhibits hypothalamic GHRH secretion and stimulates somatostatin release.

Factors that Influence GH Secretion A number of factors influence GH secretion by acting on the hypothalamus. GH secretion displays a well-characterized diurnal rhythm. Through most of the day, GH levels tend to be low and fairly constant. About an hour after onset of deep sleep, however, GH secretion markedly increases up to five times the low daytime value.

Superimposed on this diurnal fluctuation in GH secretion are further bursts in secretion that occur in response to exercise, stress, and low blood glucose, the major stimuli for increased secretion. The benefits of increased GH secretion during these situations when energy demands outstrip the body's glucose reserves are presumably that glucose is conserved for the brain and fatty acids are provided as an alternative energy source for muscle.

Because GH uses up fat stores and promotes synthesis of body proteins, it encourages a change in body composition away from adipose deposition toward an increase in muscle protein. Accordingly, the increase in GH secretion that accompanies exercise may at least in part mediate the effects of exercise in reducing the percentage of body fat while increasing lean body mass.

Several minor inputs also influence GH secretion. A rise in blood amino acids after a high-protein meal enhances GH secretion. In turn, GH promotes the use of these amino acids for protein synthesis. GH is also stimulated by a decline in blood fatty acids. Because GH mobilizes fat, such regulation helps maintain fairly constant blood fatty acid levels.

Finally, ghrelin, a potent appetite stimulator released from the stomach, also stimulates GH secretion (see p. 623). This "hunger hormone" may play a role in coordinating growth with nutrient acquisition.

Note that the known regulatory inputs for GH secretion are aimed at adjusting the levels of glucose, amino acids, and fatty acids in the blood. No known growth-related signals influence GH secretion. The whole issue of what really controls growth is complicated by GH levels during early childhood, a period of quite rapid growth in height, being similar to those in normal adults. As mentioned earlier, the poorly understood control of local IGF-I activity may be important in this regard. Another related question is, Why aren't adult tissues still responsive to GH's growth-promoting effects? We know we do not grow any taller after adolescence because the epiphyseal plates have closed, but why do soft tissues not continue to grow through hypertrophy and hyperplasia under the influence of GH? One speculation is that levels of GH may only be high enough to produce its growth-promoting effects during the secretion bursts that occur in deep sleep. Time spent in deep sleep is greatest in infancy and gradually declines with age. Still, even as we age, we spend some time in deep sleep, yet we do not gradually get larger. Further research is needed to unravel these mysteries.

Abnormal GH secretion results in aberrant growth patterns.

 Diseases related to both deficiencies and excesses of GH can occur. The effects on the pattern of growth are more pronounced than the metabolic consequences.

GH Deficiency GH deficiency may be caused by a pituitary defect (lack of GH) or may occur secondary to hypothalamic dysfunctions (lack of GHRH). Hyposecretion of GH in a child is one cause of **dwarfism**. The predominant feature is short stature caused by retarded skeletal growth (▮Figure 18-11). Less obvious characteristics include poorly developed muscles (reduced muscle protein synthesis) and excess subcutaneous fat (less fat mobilization).

In addition, growth may be thwarted because the tissues fail to respond normally to GH. An example is *Laron dwarfism,* which is characterized by abnormal GH receptors that are unresponsive to the hormone. The symptoms of this condition resemble those of severe GH deficiency even though blood levels of GH are actually high. In some instances, GH levels are adequate and target-cell responsiveness is normal, but IGF-I is lacking, as is the case with *African pygmies.*

▮Figure 18-11 **Examples of the effect of abnormalities in growth hormone secretion on growth.** The man at the left displays pituitary dwarfism resulting from underproduction of growth hormone in childhood. The man at the center has gigantism caused by excessive growth hormone secretion in childhood. The woman at the right is of average height.

Growth and Youth in a Bottle?

AFTER HUMAN GH BECAME AVAILABLE through genetic engineering, a new problem for the medical community was to determine under which circumstances synthetic GH treatment is appropriate. Until recently, the FDA had approved treatment only for the following uses: (1) for children with GH deficiency, (2) for adults with a pituitary tumor or other disease that causes severe GH deficiency, and (3) for patients with AIDS who have severe muscle wasting. Although not approved by the FDA for this use, GH therapy is also widely used to promote faster healing of skin in patients who have been severely burned. In 2003, amid emotionally charged debates, the FDA approved GH shots for another group, the shortest 1.2% of children who are unusually short for no apparent reason. This therapy involves multiple GH injections per week for a number of years under careful supervision of pediatric endocrinologists for an average gain in height of 1 to 3 inches. Children with GH deficiency experience more dramatic gains of about 6 to 8 inches on GH therapy.

Another group who may benefit from replacement GH therapy is the elderly. GH secretion typically peaks during a person's 20s, then in many people may start to dwindle after age 40. This decline may contribute to some of the characteristic signs of aging:

■ Decreased muscle mass (GH promotes synthesis of proteins, including muscle protein)

■ Increased fat deposition (GH promotes leanness by mobilizing fat stores for use as an energy source)

■ Reduced bone density (GH stimulates bone-forming cells)

■ Thinner, sagging skin (GH promotes proliferation of skin cells)

(However, inactivity is also believed to play a major role in age-associated reductions in muscle mass, bone density, and strength.)

Several studies in the early 1990s suggested that some of these consequences of aging may be counteracted through the use of synthetic GH in people older than age 60. Elderly men treated with supplemental GH showed increased muscle mass, reduced fatty tissue, and thickened skin. In similar studies in elderly women, supplemental GH therapy did not increase muscle mass significantly but did reduce fat mass and protect against bone loss.

Even though these early results were exciting, further studies were more discouraging. Despite increased lean body mass, many treated people surprisingly do not have increased muscle strength or exercise capabilities. Also, when GH is supplemented for an extended time or in large doses, harmful side effects include an increased likelihood of diabetes, kidney stones, high blood pressure, headaches, joint pain, and carpal tunnel syndrome (thickening and narrowing of the tunnel in the wrist through which the nerve supply to the hand muscles passes; *carpal* means "wrist"). Furthermore, synthetic GH is

---

The onset of GH deficiency in adulthood after growth is already complete produces relatively few symptoms. Adults who are GH deficient tend to have reduced skeletal muscle mass and strength (less muscle protein), and decreased bone density (less osteoblast activity during ongoing bone remodeling). Furthermore, because GH is essential for maintaining cardiac muscle mass and performance in adulthood, GH-deficient adults may be at increased risk of developing heart failure. (For a discussion of GH therapy, see the accompanying boxed feature, ▐ Concepts, Challenges, and Controversies.)

GH Excess Hypersecretion of GH is most often caused by a tumor of the GH-producing cells of the anterior pituitary. The symptoms depend on the age of the individual when the abnormal secretion begins. If overproduction of GH begins in childhood before the epiphyseal plates close, the principal manifestation is a rapid growth in height without distortion of body proportions. Appropriately, this condition is known as **gigantism** (▐ Figure 18-11). If not treated by removal of the tumor or by drugs that block the effect of GH, the person may reach a height of 8 feet or more. All the soft tissues grow correspondingly, so the body is still well proportioned.

If GH hypersecretion occurs after adolescence when the epiphyseal plates have already closed, further growth in height

is prevented. Under the influence of excess GH, however, the bones become thicker and the soft tissues, especially connective tissue and skin, proliferate. This disproportionate growth pattern produces a disfiguring condition known as **acromegaly** (*acro* means "extremity"; *megaly* means "large"). Bone thickening is most obvious in the extremities and face. A marked coarsening of the features to an almost apelike appearance gradually develops as the jaws and cheekbones become more prominent because of thickening of the facial bones and the skin (▐ Figure 18-12). The hands and feet enlarge, and the fingers and toes become greatly thickened.

Other hormones besides growth hormone are essential for normal growth.

Several other hormones in addition to GH contribute in special ways to overall growth:

■ *Thyroid hormone* is essential for growth but is not itself directly responsible for promoting growth. It plays a permissive role in skeletal growth; the actions of GH fully manifest only when enough thyroid hormone is present. As a result, growth is severely stunted in hypothyroid children, but hypersecretion of thyroid hormone does not cause excessive growth.

costly ($15,000 to $20,000 annually) and must be injected regularly. Also, some scientists worry that sustained administration of synthetic GH may raise the risk of developing cancer by promoting uncontrolled cell proliferation. For these reasons, most investigators no longer view synthetic GH as a potential "fountain of youth." Instead, they hope it can be used in a more limited way to strengthen muscle and bone sufficiently in the many elderly who have GH deficits, to help reduce the incidence of bone-breaking falls that often lead to disability. The National Institute of Aging is currently sponsoring a nationwide series of studies involving GH therapy in the elderly to help sort out potentially legitimate roles of this supplemental hormone.

An ethical dilemma is whether the drug should be used by others who have normal GH levels but want the product's growth-promoting actions for cosmetic or athletic reasons, such as normally growing teenagers who wish to attain even greater height. The drug is already being used illegally by some athletes and bodybuilders. Furthermore, a recent study found that only 4 out of 10 children who are legitimately receiving GH therapy under medical supervision are actually GH deficient or in the bottom 1.2% of height. The others are receiving the treatment because parents, physicians, and the children are being swayed by perceived cultural pressures that favor height rather than by medical factors.

Using the drug in children with normal GH levels may be problematic because synthetic GH is a double-edged product. Although it promotes growth and muscle mass, it also has negative effects, such as potentially troubling side effects. Furthermore, one study revealed that supplemental GH therapy in children who do not lack the hormone redistributes the body's fat and protein. The investigators compared two groups of otherwise-healthy 6- to 8-year-old children who were among the shortest for their age. One group consisted of children who were receiving GH and the other consisted of children who were not. At the end of 6 months, the children taking the synthetic hormone had outpaced the untreated group in growth by more than 1.5 inches per year. However, the untreated children added both muscle and fat as they grew, whereas the treated children became unusually muscular and lost up to 76% of their body fat. The loss of fat became especially obvious in their faces and limbs, giving them a rawboned, gangly appearance. It is unclear what long-term effects—either deleterious or desirable—these dramatic changes in body composition might have. Scientists also express concern that these readily observable physical changes may be accompanied by more subtle abnormalities in organs and cells. Thus, the debate about whether to use GH in normal but short children is likely to continue.

■ *Insulin* is an important growth promoter. Insulin deficiency often blocks growth, and hyperinsulinism frequently spurs excessive growth. Because insulin promotes protein synthesis, its growth-promoting effects should not be surprising. However, these effects may also arise from a mechanism other than insulin's direct effect on protein synthesis. Insulin structurally resembles the IGFs and may interact with the IGF-I receptor, which is very similar to the insulin receptor.

■ *Sex steroids* are potent growth stimulators and contribute to the pubertal growth spurt, but they ultimately stop further growth by closing the epiphyseal plates. Sex steroids include *androgens* (masculinizing hormones, such as testosterone in

Age 13　　　　　　　　　Age 21　　　　　　　　　Age 35

Dean Warchak@www.acromegalysupport.com.

I**Figure 18-12 Progressive development of acromegaly.** Note how the patient's brow bones, cheekbones, and jawbones are becoming progressively more prominent as a result of ongoing thickening of the bones and skin caused by excessive GH secretion during adulthood.

males) and *estrogen* (feminizing hormone in females). Androgens but not estrogen also powerfully stimulates protein synthesis, thereby increasing muscle mass. This action is responsible for men developing heavier musculature than women do.

Several factors contribute to the average height differences between men and women. First, because puberty occurs about 2 years earlier in girls than in boys, on the average boys have 2 more years of prepubertal growth than girls do. As a result, boys are usually several inches taller than girls at the start of their respective growth spurts. Second, boys experience a greater steroid-hormone induced growth spurt than girls before their respective gonadal steroids seal their long bones from further growth; this results in greater heights in men than in women on average. Third, evidence suggests that androgens "imprint" the brains of males during development, giving rise to a "masculine" secretory pattern of GH characterized by higher cyclic peaks, which are speculated to contribute to the greater height of males.

In addition to these hormones that exert overall effects on body growth, a number of poorly understood peptide *growth factors* have been identified that stimulate mitotic activity of specific tissues (for example, epidermal growth factor).

We now shift attention to the other endocrine gland in the brain—the pineal gland.

Check Your Understanding 18.3

1. Describe GH's metabolic effects.

2. Explain the relationship between GH and IGF-I in promoting growth.

3. Compare the means by which bone grows in thickness and in length.

18.4 Pineal Gland and Circadian Rhythms

The **pineal gland**, (PIN-ē-ul) a tiny, pinecone-shaped structure located in the center of the brain (see Figure 5-7b, p. 143, Figure 5-15, p. 154, and Figure 18-1, p. 639), secretes the hormone **melatonin**, an indoleamine hormone. (Do not confuse melatonin with the skin-darkening pigment, *melanin*.) Although melatonin was discovered in 1959, investigators have only recently begun to unravel its many functions. One of melatonin's most widely accepted roles is helping keep the body's inherent circadian rhythms in synchrony with the light–dark cycle. We examine circadian rhythms in general before looking at the role of melatonin in this regard and considering other functions of this hormone.

The suprachiasmatic nucleus is the master biological clock.

Hormone secretion rates are not the only factor in the body that fluctuates cyclically over a 24-hour period. Humans have similar biological clocks for many other bodily functions, ranging from gene expression, to physiological processes such as temperature regulation (see p. 628), to feeding and metabolic patterns (see p. 626), to behavior. The master biological clock that serves as the pacemaker for the body's circadian rhythms is the **suprachiasmatic nucleus (SCN)**. It consists of two clusters of nerve cell bodies (one on each side of the brain) in the hypothalamus above the optic chiasm, the point at which part of the nerve fibers from each eye cross to the opposite half of the brain (*supra* means "above"; *chiasm* means "cross") (see p. 207; Figure 5-7b, p. 143 and Figure 18-3, p. 646). Only 20,000 neurons populate the SCN, yet the self-induced rhythmic firing of this meager number of SCN neurons plays a major role in synchronizing all the body's inherent daily rhythms. This is analogous to the residents of a small town setting the schedule for all the residents of the world. Many tissues have independent clocks, but the SCN serves as the central timekeeper to keep these peripheral clocks in sync.

Role of Clock Proteins The underlying molecular mechanism responsible for the SCN's circadian oscillations involves autoregulatory feedback loops leading to alternating cycles of activation and repression of genes within the nuclei of SCN neurons that code for oscillator **clock proteins**. Self-starting transcription factors known as **CLOCK** and **BMAL-1**[1] in the nucleus of an SCN neuron form a complex that activates transcription of DNA's genetic code for the clock proteins **PER** and **CRY**[1], setting in motion a series of events leading to synthesis of these clock proteins in the cytosol surrounding the nucleus (Figure 18-13). As the day wears on, these clock proteins continue to accumulate, finally reaching a critical mass, at which time they are transported into the nucleus. Here, PER and CRY inhibit CLOCK and BMAL-1, thereby repressing the genetic process responsible for these clock proteins' production. After the clock proteins block synthesis of more clock proteins, the level of clock proteins gradually dwindles as they degrade within the nucleus, thus removing their inhibitory influence from the clock-protein genetic machinery. No longer being blocked, these genes again rev up the production of more clock proteins as the cycle repeats itself. Each cycle takes about a day. The fluctuating levels of clock proteins bring about cyclic changes in neural output from the SCN that, in turn, lead to cyclic changes in effector organs throughout the day. An example is the diurnal variation in cortisol secretion (see Figure 18-2, p. 642). In this way, internal timekeeping (circadian rhythms) is a self-sustaining mechanism built into the genetic makeup of the SCN neurons. The SCN provides "standard time" by which all peripheral tissue clocks are set.

Synchronization of the Biological Clock with Environmental Cues On its own, this biological clock generally cycles a bit slower than the 24-hour environmental cycle. Without any external cues, the SCN sets up cycles that average about 25 hours. The cycles are consistent for a given individual

---

[1]*CLOCK* is the acronym for "circadian rhythms of locomotor activity kaput." *BMAL-1* stands for "brain and muscle Arnt-like-1." *PER* refers to proteins known as "period." *CRY* is short for "cryptochrome" proteins.

Tinkering with Our Biological Clocks

RESEARCH SHOWS THAT THE HECTIC pace of modern life, stress, noise, and the irregular schedules many workers follow can upset internal rhythms, illustrating how a healthy external environment affects our own internal environment—and our health.

The greatest disrupter of our natural circadian rhythms is the variable work schedule common in industrialized countries. Today one out of every four working men and one out of every six working women has a variable work schedule—shifting frequently between day and night work. To spread the burden, many companies that maintain shifts round the clock alter their workers' schedules. One week, employees work the day shift. The next week, they move to the "graveyard shift" from midnight to 8 a.m. The next week, they work the night shift from 4 P.M. to midnight. Many shift workers feel tired most of the time and have trouble staying awake at the job. Work performance suffers because of the workers' fatigue. When workers arrive home for bed, they're exhausted but cannot sleep because they are trying to doze off when the body is trying to wake them up. Unfortunately, the weekly changes in schedule never permit workers' internal alarm clocks to fully adjust. Most people require 4 to 14 days to adjust to a new schedule.

People whose lifestyle operates against their circadian clock for any reason suffer more ulcers, insomnia, irritability, depression, tension, cardiovascular disease, cancer, and metabolic disorders like Type 2 diabetes mellitus and obesity than those whose daily schedule is in sync with their biological rhythms. To make matters worse, tired, irritable workers whose judgment is impaired by fatigue pose a threat to society. Consider these infamous industrial accidents that resulted from judgment errors and occurred during the night shift: in 1979, the worst nuclear accident in United States history at the Three Mile Island nuclear reactor in Pennsylvania; in 1986, the world's largest and most costly nuclear accident in Chernobyl, Ukraine; and in 1989, accidental grounding of the oil tanker *Exxon Valdez* and subsequent oil spill in Alaska. These disasters may have been the result of workers operating at a time unsuitable for clear thinking. One has to wonder how many plane crashes, auto accidents, and acts of medical malpractice can be traced to compromised vigilance and judgment resulting from working against inherent body rhythms.

Thanks to studies of biological rhythms, researchers are finding ways to reset biological clocks when warranted. For instance, one simple measure is to put shift workers on 3-week cycles to give their clocks time to adjust. Bright lights can also be used to reset the biological clock. Furthermore, use of supplemental melatonin, the hormone that sets the internal clock to march in step with environmental cycles, may prove useful in resetting the body's clock when that clock is out of sync with external cues.

---

but vary somewhat among different people. If this master clock were not continually adjusted to keep pace with the world outside, the body's circadian rhythms would become progressively out of sync with the cycles of light (periods of activity) and dark (periods of rest). Thus, the SCN must be reset daily by external cues so that biological rhythms are synchronized with the activity levels driven by the surrounding environment. For example, during the day people are typically awake and active and their metabolic activity is high, while at night they sleep and their metabolism is low. The effect of not keeping the internal clock synchronized with the environment is well known by people who experience **jet lag** when their inherent rhythm is out of step with external cues. (For a discussion of problems associated with being out of sync with environmental cues, see the accompanying boxed feature, ▌Concepts, Challenges, and Controversies.)

The SCN works with the pineal gland and its hormonal product melatonin to synchronize the various circadian rhythms with the 24-hour day–night cycle.

Melatonin helps keep the body's circadian rhythms in time with the light–dark cycle.

Daily changes in light intensity are the major environmental cue used to adjust the SCN master clock. Special photoreceptors in the retina pick up light signals and transmit them directly to the SCN. These photoreceptors are distinct from the rods and cones used to see light (see p. 200). **Melanopsin**, a protein found in a special type of retinal ganglion cell (see p. 199), is the receptor molecule for light that keeps the body in tune with external time (▌Figure 18-13). Most retinal ganglion cells receive input from the rod and cone photoreceptors. The axons of these ganglion cells form the *optic nerve*, which carries information for final processing by the visual cortex in the occipital lobe (see pp. 148 and 207). Intermingled among the visually oriented retinal ganglion cells, about 1% to 2% of the retinal ganglion cells instead form an entirely independent light-detection system that responds to levels of illumination, like a light meter on a camera, rather than the contrasts, colors, and contours detected by the image-forming visual system. The melanopsin-containing, light-detecting retinal ganglion cells cue the pineal gland about the presence or absence of light by sending their signals along the **retino-hypothalamic tract** to the SCN. This pathway is distinct from the neural systems that result in vision perception. The SCN relays the message regarding light status to the pineal gland. This is the major way the internal clock is coordinated to a 24-hour day. Melatonin is the hormone of darkness. Melatonin secretion increases up to 10-fold during the darkness of night and then falls to low levels during the light of day. Fluctuations

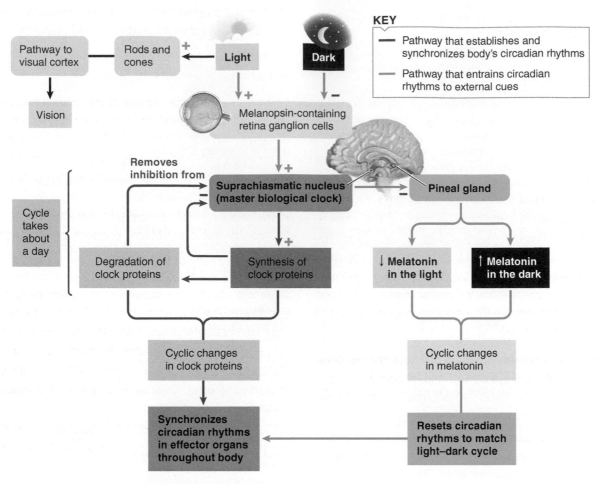

Figure 18-13 Synchronization and entrainment of circadian rhythms.

in melatonin secretion, in turn, help entrain the body's biological rhythms with the external light–dark cues.

Studies suggest the following additional roles of melatonin, besides its clock-related function:

■ Taken exogenously (in a pill), melatonin induces a natural sleep without the side effects that accompany hypnotic sedatives, so it may play a normal role in promoting sleep.

■ Melatonin inhibits the hormones that stimulate reproductive activity. Puberty may be initiated by a reduction in melatonin secretion.

■ In a related role, in some species, seasonal fluctuations in melatonin secretion associated with changes in the number of daylight hours are important triggers for seasonal breeding, migration, and hibernation.

■ Melatonin is an effective *antioxidant*, a defense tool against biologically damaging free radicals. *Free radicals* are very unstable electron-deficient particles that are highly reactive and destructive. Free radicals have been implicated in several chronic diseases, such as coronary artery disease (see p. 327) and cancer, and are believed to contribute to the aging process.

■ Melatonin may slow the aging process, perhaps by removing free radicals or by other means.

■ Melatonin enhances immunity, reduces inflammation, and has been shown to reverse some of the age-related shrinkage of the thymus, the source of T lymphocytes (see p. 416).

■ Melatonin has positive effects on the brain, such as protecting against neurodegeneration, participating in neurogenesis (production of new neurons), and combating depression.

■ Melatonin also facilitates learning, memory, and cognition.

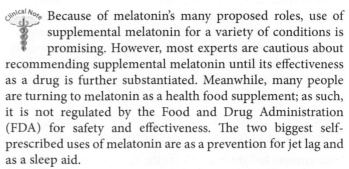 Because of melatonin's many proposed roles, use of supplemental melatonin for a variety of conditions is promising. However, most experts are cautious about recommending supplemental melatonin until its effectiveness as a drug is further substantiated. Meanwhile, many people are turning to melatonin as a health food supplement; as such, it is not regulated by the Food and Drug Administration (FDA) for safety and effectiveness. The two biggest self-prescribed uses of melatonin are as a prevention for jet lag and as a sleep aid.

1. Discuss the means by which the suprachiasmatic nucleus serves as the master biological clock.
2. Distinguish between *melanopsin* and *melatonin.*
3. Compare the type of information carried by the optic nerve pathway and by the retino-hypothalamic tract.

Homeostasis: Chapter in Perspective

The endocrine system is one of the body's two major regulatory systems; the other is the nervous system. Through its relatively slowly acting hormone messengers, the endocrine system generally regulates activities that require duration rather than speed. Most of these activities are directed toward maintaining homeostasis. The specific contributions of the central endocrine organs to homeostasis are as follows:

■ The hypothalamus–posterior pituitary unit secretes vasopressin, which acts on the kidneys during urine formation to help maintain H_2O balance. Control of H_2O balance, in turn, is essential for maintaining ECF osmolarity and proper cell volume.

■ For the most part, the hormones secreted by the anterior pituitary do not directly contribute to homeostasis. Instead, most are tropic—that is, they stimulate the secretion of other hormones.

■ However, growth hormone from the anterior pituitary, in addition to its growth-promoting actions, also exerts metabolic effects that help maintain the plasma concentration of glucose, fatty acids, and amino acids.

■ The pineal gland secretes melatonin, which helps entrain the body's circadian rhythm to the environmental cycle of light (period of activity) and dark (period of inactivity).

The peripheral endocrine glands further help maintain homeostasis in the following ways:

■ Hormones help maintain the proper concentration of nutrients in the internal environment by directing chemical reactions involved in the cellular uptake, storage, and release of these molecules. Furthermore, the rate at which these nutrients are metabolized is controlled in large part by the endocrine system.

■ Salt balance, which is important in maintaining the proper ECF volume and arterial blood pressure, is achieved by hormonally controlled adjustments in salt reabsorption by the kidneys during urine formation.

■ Likewise, hormones act on various target cells to maintain the plasma concentration of calcium and other electrolytes. These electrolytes, in turn, play key roles in homeostatic activities. For example, maintenance of calcium levels within narrow limits is critical for neuromuscular excitability and blood clotting, among other life-supporting actions.

■ The endocrine system orchestrates a wide range of adjustments that help the body maintain homeostasis in response to stressful situations.

■ The endocrine and nervous systems work in concert to control the circulatory and digestive systems, which in turn carry out important homeostatic activities.

Unrelated to homeostasis, hormones direct the growing process and control most aspects of the reproductive system.

Review Exercises Answers begin on p. A-48

Reviewing Terms and Facts

1. One endocrine gland may secrete more than one hormone. *(True or false?)*

2. One hormone may influence more than one type of target cell. *(True or false?)*

3. All endocrine glands are exclusively endocrine in function. *(True or false?)*

4. A single target cell may be influenced by more than one hormone. *(True or false?)*

5. Hyposecretion or hypersecretion of a specific hormone can occur even though its endocrine gland is perfectly normal. *(True or false?)*

6. Growth hormone levels in the blood are no higher during the early childhood growing years than during adulthood. *(True or false?)*

7. A hormone that has as its primary function the regulation of another endocrine gland is classified functionally as a _____ hormone.

8. Self-induced reduction in the number of receptors for a specific hormone is known as _____.

9. Activity within the cartilaginous layer of bone known as the _____ is responsible for lengthening of long bones.

10. The _____ in the hypothalamus is the body's master biological clock.

11. Indicate the relationships among the hormones in the hypothalamus–anterior pituitary–adrenal cortex axis by using the following answer code to identify which hormone belongs in each blank:

(a) cortisol   (b) ACTH   (c) CRH

(1) _____ from the hypothalamus stimulates the secretion of (2) _____ from the anterior pituitary. (3) _____, in turn, stimulates the secretion of (4) _____ from the adrenal cortex. In negative-feedback fashion, (5) _____ inhibits secretion of the releasing hormone (6) _____ and furthermore inhibits secretion of the tropic hormone (7) _____.

Understanding Concepts

(Answers at www.cengagebrain.com)

1. Discuss the general functions of the endocrine system.

2. How is the plasma concentration of a hormone normally regulated?

3. List and briefly state the source and functions of the posterior pituitary hormones.

4. List and briefly state the source and functions of the anterior pituitary hormones.

5. Compare the relationship between the hypothalamus and posterior pituitary with the relationship between the hypothalamus and anterior pituitary. Describe the role of the hypothalamic–hypophyseal portal system and the hypothalamic releasing and inhibiting hormones.

6. Describe the actions of growth hormone that are unrelated to growth. What are growth hormone's growth-promoting actions? What is the role of IGFs?

7. Discuss the control of growth hormone secretion.

8. Describe the role of clock proteins.

9. What are the source, functions, and stimulus for secretion of melatonin?

Applying Clinical Reasoning

At 18 years of age and 8 feet tall, Anthony O. was diagnosed with gigantism caused by a pituitary tumor. The condition was treated by surgically removing his pituitary gland. What hormonal replacement therapy would Anthony need?

Thinking at a Higher Level

1. Would you expect the concentration of hypothalamic releasing and inhibiting hormones in a systemic venous blood sample to be higher, lower, or the same as the concentration of these hormones in a sample of hypothalamic–hypophyseal portal blood?

2. Thinking about the feedback control loop among TRH, TSH, and thyroid hormone, would you expect the concentration of TSH to be normal, above normal, or below normal in a person whose diet is deficient in iodine?

3. A patient displays symptoms of excess cortisol secretion. What factors could be measured in a blood sample to determine whether the condition is caused by a defect at the hypothalamus–anterior pituitary level or the adrenal cortex level?

4. Why would males with testicular feminization syndrome be unusually tall?

5. A black market for growth hormone abuse exists among weight lifters and other athletes. What actions of growth hormone would induce a full-grown athlete to take supplemental doses of this hormone? What are the potential detrimental side effects?

To access the course materials and companion resources for this text, please visit www.cengagebrain.com

The Peripheral Endocrine Glands

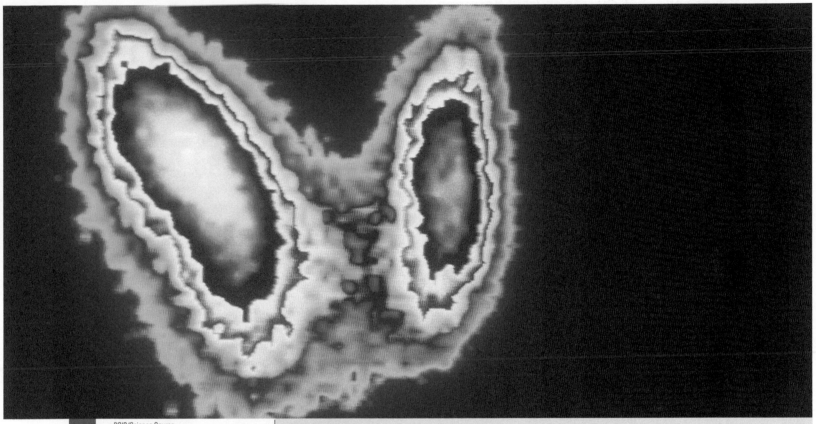

BSIP/Science Source

A scintiscan of a normal thyroid gland. In this diagnostic technique, the activity of the thyroid gland is detected by a gamma camera that shows how much of an injected radioactive chemical tracer is taken up, with *red* being the most active and *blue* being the least active areas. Note that the two lobes of this endocrine gland give it a bow tie or butterfly shape. The thyroid gland secretes hormones that control the body's basal metabolic rate (idling speed).

CHAPTER AT A GLANCE

Homeostasis Highlights

The endocrine system, by means of the blood-borne hormones it secretes, generally regulates activities that require duration rather than speed. Most target-cell activities under hormonal control are directed toward maintaining homeostasis. The peripheral endocrine glands include the **thyroid gland**, which controls the body's basal metabolic rate; the **adrenal glands**, which secrete hormones important in maintaining salt balance, in metabolizing nutrient molecules, and in adapting to stress; the **endocrine pancreas**, which secretes hormones important in metabolizing nutrient molecules; and the **parathyroid glands**, which secrete a hormone important in Ca^{2+} metabolism.

19.1 Thyroid Gland

The **thyroid gland** consists of two lobes of endocrine tissue joined in the middle by a narrow portion of the gland, the *isthmus,* giving it a bow-tie shape (alternatively described as a butterfly shape) (❙Figure 19-1a and chapter opener photo). The gland is even located in the appropriate place for a bow tie, lying in the neck over the trachea just below the larynx.

The major cells that secrete thyroid hormone are organized into colloid-filled follicles.

The major thyroid secretory cells, known as **follicular cells,** are arranged into hollow spheres, each of which forms a functional unit called a **follicle.** On microscopic section (❙Figure 19-1b), the follicles appear as rings consisting of a single layer of follicular cells enclosing an inner lumen filled with **colloid,** a substance that serves as an extracellular storage site for thyroid hormone. Note that the colloid within the follicular lumen is extracellular (that is, outside the thyroid cells), even though it is located within the interior of the follicle. Colloid is not in direct contact with the extracellular fluid (ECF) that surrounds the follicle, similar to an inland lake that is not in direct contact with the oceans that surround a continent.

The chief constituent of colloid is a large glycoprotein molecule known as **thyroglobulin (Tg),** within which are incorporated the thyroid hormones in their various stages of synthesis. The follicular cells produce two iodine-containing hormones derived from the amino acid tyrosine: **tetraiodothyronine (T_4, or thyroxine)** and **tri-iodothyronine (T_3).** The prefixes *tetra* and *tri* and the subscripts *4* and *3* denote the number of iodine atoms incorporated into each of these hormones. These two hormones, collectively referred to as **thyroid hormone,** are important regulators of overall basal metabolic rate.

Interspersed in the interstitial spaces between the follicles is another secretory cell type, the **C cells,** which secrete the peptide hormone **calcitonin.** Calcitonin plays a role in calcium (Ca^{2+}) metabolism and is not related to T_4 and T_3. We discuss T_4 and T_3 here and talk about calcitonin later in a section dealing with endocrine control of Ca^{2+} balance.

Thyroid hormone is synthesized and stored on the thyroglobulin molecule.

The basic ingredients for thyroid hormone synthesis are tyrosine and iodine, both of which must be taken up from the blood by the follicular cells. Tyrosine, an amino acid, is synthesized in sufficient amounts by the body, so it is not essential to the diet. By contrast, the iodine needed for thyroid hormone synthesis must be obtained from dietary intake. Dietary *iodine (I)* is reduced to *iodide (I^-)* before absorption by the small intestine.

Most steps of thyroid hormone synthesis take place on the thyroglobulin molecules within the colloid. Thyroglobulin itself is produced by the endoplasmic reticulum–Golgi complex of the thyroid follicular cells. The amino acid tyrosine becomes incorporated in the much larger thyroglobulin molecules as the latter are being produced. Once produced, tyrosine-containing

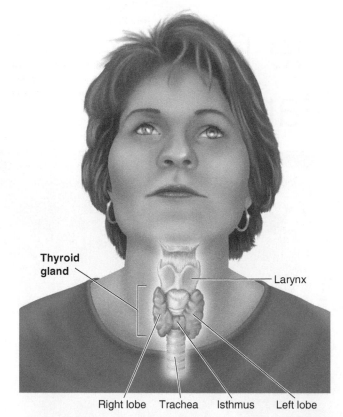

Thyroid gland — **Larynx**

Right lobe   Trachea   Isthmus   Left lobe

(a) Gross anatomy of thyroid gland

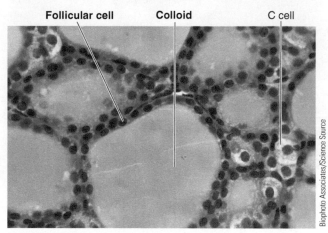

Follicular cell   **Colloid**   **C cell**

(b) Light-microscopic appearance of thyroid gland

❙Figure 19-1 **Anatomy of the thyroid gland.** (a) Gross anatomy of the thyroid gland, anterior view. The thyroid gland lies over the trachea just below the larynx and consists of two lobes connected by a thin strip called the isthmus. (b) Light-microscopic appearance of the thyroid gland. The thyroid gland is composed primarily of colloid-filled spheres enclosed by a single layer of follicular cells.

thyroglobulin is exported in vesicles from the follicular cells into the colloid by exocytosis (❙Figure 19-2, step **1**). The thyroid captures I^- from the blood and transfers it into the follicular cell by an iodide pump referred to as the **iodide trap**—the powerful, energy-requiring carrier protein in the outer membrane of a follicular cell (step **2**). The iodide trap is a symporter driven by the Na^+ concentration gradient established by the

Biophoto Associates/Science Source

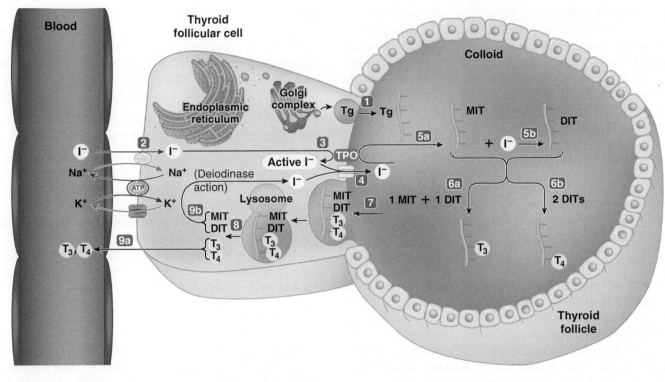

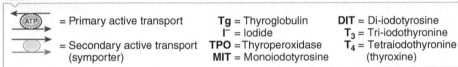

KEY

(ATP) = Primary active transport	**Tg** = Thyroglobulin	**DIT** = Di-iodotyrosine
= Secondary active transport (symporter)	**I⁻** = Iodide	T_3 = Tri-iodothyronine
	TPO = Thyroperoxidase	T_4 = Tetraiodothyronine (thyroxine)
	MIT = Monoiodotyrosine	

1 Tyrosine-containing Tg produced within the thyroid follicular cells by the endoplasmic reticulum–Golgi complex is transported by exocytosis into the colloid.

2 Iodide is carried by secondary active transport from the blood into the colloid by symporters in the basolateral membrane of the follicular cells.

3 In the follicular cell, the iodide is oxidized to active form by TPO at the luminal membrane.

4 The active iodide exits the cell through a luminal channel to enter the colloid.

5a Catalyzed by TPO, attachment of one iodide to tyrosine within the Tg molecule yields MIT.

5b Attachment of two iodides to tyrosine yields DIT.

6a Coupling of one MIT and one DIT yields T_3.

6b Coupling of two DITs yields T_4.

7 On appropriate stimulation, the thyroid follicular cells engulf a portion of Tg-containing colloid by phagocytosis.

8 Lysosomes attack the engulfed vesicle and split the iodinated products from Tg.

9a T_3 and T_4 diffuse into the blood (secretion).

9b MIT and DIT are deiodinated, and the freed iodide is recycled for synthesizing more hormone.

❙Figure 19-2 Synthesis, storage, and secretion of thyroid hormone. Note that the organelles are not drawn to scale. The endoplasmic reticulum–Golgi complex are proportionally too small.

FIGURE FOCUS: *If you were trying to develop a drug to reduce synthesis of T_3 and T_4 as a treatment for hyperthyroidism (too much thyroid hormone), which steps in this biosynthetic pathway are unique to the thyroid gland and thus would be good targets to block?*

Na⁺–K⁺ pump at the basolateral membrane (the membrane in contact with the interstitial fluid). The iodide trap transports Na⁺ into the follicular cell down its concentration gradient and I⁻ into the cell against its concentration gradient. Almost all I⁻ in the body is moved against its concentration gradient to become trapped in the thyroid for thyroid hormone synthesis. Iodide is usually about 30 times more concentrated in the thyroid follicular cells than in the blood. Iodide serves no other function in the body.

Inside the follicular cell, iodide is oxidized to "active" iodide by a membrane-bound enzyme, **thyroperoxidase (TPO)**, located at the luminal membrane (the membrane in contact with the colloid) (step **3**). This active I⁻ exits through a channel in the luminal membrane to enter the colloid (step **4**).

Within the colloid, TPO, still membrane-bound, quickly attaches I⁻ to a tyrosine within the thyroglobulin molecule. Attachment of one iodide to tyrosine yields **monoiodotyrosine (MIT)** (step **5a**). Attachment of two iodides to tyrosine yields di-iodotyrosine (DIT) (step **5b**). After MIT and DIT are formed, a coupling process occurs within the thyroglobulin molecule between the iodinated tyrosine molecules to form the thyroid hormones. Coupling of one MIT (with one iodide) and one DIT (with two iodides) yields **tri-iodothyronine**, or **T_3** (with three iodides) (step **6a**). Coupling of two DITs (each bearing two iodides) yields **tetraiodothyronine (T_4, or thyroxine),** the four-iodide form of thyroid hormone (step **6b**). Coupling does not occur between two MIT molecules. All these products remain attached to thyroglobulin by peptide bonds. Thyroid hormones remain stored in this form in the colloid until they are split off and secreted. Sufficient thyroid hormone is normally stored to supply the body's needs for several months.

To secrete thyroid hormone, the follicular cells phagocytize thyroglobulin-laden colloid.

Release of thyroid hormone into the systemic circulation is complex for two reasons. First, before their release, T_3 and T_4 are still bound within the thyroglobulin molecule. Second, these hormones are stored at an inland extracellular site, in the colloid in the follicular lumen, so they must be transported completely across the follicular cells to reach the capillaries that course through the interstitial spaces between the follicles.

The process of thyroid hormone secretion essentially involves the follicular cells "biting off" a piece of colloid, breaking the thyroglobulin molecule down into its component parts, and "spitting out" the freed T_3 and T_4 into the blood. On appropriate stimulation for thyroid hormone secretion, the follicular cells internalize a portion of the thyroglobulin-hormone complex by phagocytizing a piece of colloid (∎Figure 19-2, step **7**). Within the cells, the membrane-enclosed droplets of colloid coalesce with lysosomes, whose enzymes split off the biologically active thyroid hormones, T_3 and T_4, and the inactive MIT and DIT (step **8**). The thyroid hormones, being very lipophilic, pass freely through the outer membranes of the follicular cells and into the blood (step **9a**).

MIT and DIT are of no endocrine value. The follicular cells contain an iodide-removing enzyme, *deiodinase*, which swiftly removes the I^- from MIT and DIT, allowing the freed I^- to be recycled for synthesis of more hormone (step **9b**). This highly specific enzyme removes I^- only from the worthless MIT and DIT, not the valuable T_3 or T_4.

Once released into the blood, the highly lipophilic (and therefore water-insoluble) thyroid hormone molecules quickly bind with several plasma proteins. Most circulating T_4 and T_3 is transported by **thyroxine-binding globulin**, a plasma protein that selectively binds only thyroid hormone. Less than 0.1% of the T_4 and less than 1% of the T_3 remain in the unbound (free) form. This is remarkable, considering that only the free portion of the total thyroid hormone pool has access to the target-cell receptors and thus can exert an effect.

About 90% of the secretory product released from the thyroid gland is in the form of T_4, yet T_3 is about 10 times more biologically potent. However, most secreted T_4 is converted into T_3, or *activated,* by being stripped of one of its iodides outside the thyroid gland, primarily in the liver and kidneys. These organs contain a different deiodinase enzyme than the type found in the thyroid gland, one that only removes an I^- from T_4. About 80% of the circulating T_3 is derived from secreted T_4 that has been peripherally stripped. Therefore, T_3 is the major biologically active form of thyroid hormone at the cellular level, even though the thyroid gland secretes mostly T_4.

Thyroid hormone increases the basal metabolic rate and exerts other effects.

Thyroid hormone does not have any discrete target organs. It affects virtually every tissue in the body. Like all lipophilic hormones, thyroid hormone crosses the plasma membrane and binds with an intracellular receptor, in this case a nuclear receptor bound to the **thyroid-response element** of DNA. This binding alters the transcription of specific mRNAs and thus synthesis of specific new proteins, typically enzymes, which carry out the cellular response. The nuclear thyroid hormone receptor has a 10 times greater affinity for T_3 than for T_4. Because a hormone's potency depends on how strongly the hormone binds to its target-cell receptors, T_3 is more potent than T_4.

Compared to other hormones, thyroid hormone action is "sluggish." The response to an increase in thyroid hormone is detectable only after a delay of several hours, and the maximal response is not evident for several days. The duration of the response is also quite long, partially because thyroid hormone is not rapidly degraded but also because the response to an increase in secretion continues for days or even weeks after the plasma thyroid hormone concentrations have returned to normal.

All body cells are affected either directly or indirectly by thyroid hormone. The effects of T_3 and T_4 can be grouped into several overlapping categories.

Effect on Metabolic Rate and Heat Production Thyroid hormone increases the body's overall basal metabolic rate (BMR), or "idling speed" (see p. 620). It is the most important regulator of the body's rate of O_2 consumption and energy expenditure under resting conditions.

Closely related to thyroid hormone's metabolic effect is its **calorigenic effect** (*calorigenic* means "heat-producing"). Increased metabolic activity results in increased heat production.

Sympathomimetic Effect Any action similar to one produced by the sympathetic nervous system is known as a **sympathomimetic effect** (*sympathomimetic* means "sympathetic-mimicking"). Thyroid hormone increases target-cell responsiveness to catecholamines (epinephrine and norepinephrine), the chemical messengers used by the sympathetic nervous system and its hormonal reinforcements from the adrenal medulla. Thyroid hormone accomplishes this permissive action by causing a proliferation of catecholamine target-cell receptors (see p. 643). Because of this action, many of the effects observed when thyroid hormone secretion is elevated are similar to those that accompany activation of the sympathetic nervous system.

Effect on the Cardiovascular System Thyroid hormone increases heart rate and force of contraction, thus increasing cardiac output (see p. 319), via both its direct effect on the heart and through its effect of increasing the heart's responsiveness to catecholamines.

Effect on Growth and the Nervous System Thyroid hormone is essential for normal growth because of its effects on growth hormone (GH) and IGF-I (see p. 654). Thyroid hormone not only stimulates GH secretion and increases production of IGF-I by the liver but also promotes the effects of GH and IGF-I on the synthesis of new structural proteins and on bone growth. Thyroid-deficient children have stunted growth that can be reversed by thyroid replacement therapy, but excess thyroid hormone does not produce excessive growth.

Thyroid hormone plays a crucial role in development of the nervous system, especially the CNS, an effect impeded in children who have thyroid deficiency from birth. Thyroid hormone is also essential for normal CNS activity in adults.

Thyroid hormone is regulated by the hypothalamus–pituitary–thyroid axis.

Thyroid-stimulating hormone (TSH), the thyroid tropic hormone from the anterior pituitary, is the most important regulator of thyroid hormone secretion (⎮Figure 19-3) (see p. 647). TSH acts by increasing cAMP (see p. 123) in the thyrotropes. TSH stimulates almost every step of thyroid hormone synthesis and release. In addition to enhancing thyroid hormone secretion, TSH maintains the structural integrity of the thyroid gland. In the absence of TSH, the thyroid atrophies (decreases in size) and secretes its hormones at a very low rate. Conversely, it undergoes hypertrophy (increases the size of each follicular cell) and hyperplasia (increases the number of follicular cells) in response to excess TSH stimulation.

The hypothalamic **thyrotropin-releasing hormone (TRH)**, in tropic fashion, "turns on" TSH secretion by the anterior pituitary (see p. 650), whereas thyroid hormone, in negative-feedback fashion, "turns off" TSH secretion by inhibiting the anterior pituitary and hypothalamus. TRH functions via the IP$_3$ and DAG second-messenger pathways (see p. 124). Like other negative-feedback loops, the one between thyroid hormone and TSH tends to maintain a stable thyroid hormone output.

Negative feedback between the thyroid and anterior pituitary accomplishes day-to-day regulation of free thyroid hormone levels, whereas the hypothalamus mediates long-range adjustments. Unlike most other hormonal systems, the hormones in the hypothalamus–pituitary–thyroid axis in an adult normally do not undergo sudden, wide swings in secretion, although they do display a modest diurnal rhythm (day–night

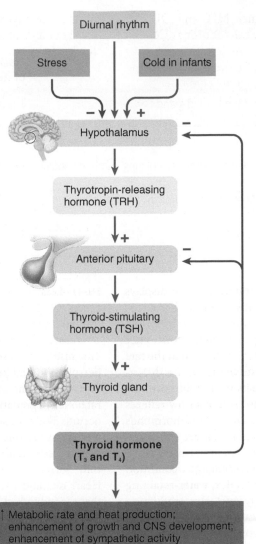

⎮Figure 19-3 **Regulation of thyroid hormone secretion.**
FIGURE FOCUS: *How would the blood concentrations of TRH, TSH, and T$_3$ and T$_4$ compare to normal in a person consuming insufficient iodine?*

fluctuations in secretion; see p. 642), with peak levels taking place as morning approaches and lowest concentrations occurring in early evening.

The only known factor that increases TRH secretion (and, accordingly, TSH and thyroid hormone secretion) is exposure to cold in newborn infants, a highly adaptive mechanism. The dramatic increase in heat-producing thyroid hormone secretion helps maintain body temperature during the abrupt drop in surrounding temperature at birth as the infant passes from the mother's warm body to the cooler environmental air. A similar immediate rise in secretion of hormones in this axis in response to cold exposure does not occur in adults, although it would make sense physiologically and does occur in many animals. Evidence suggests that on a longer-term basis during acclimatization to a cold environment, the concentration of hormones in this axis does increase as a means to increase the BMR and heat production.

Various types of stress, including physical stress, starvation, and infection, inhibit TSH and thyroid hormone secretion, presumably through neural influences on the hypothalamus, although the adaptive importance of this inhibition is unclear.

Abnormalities of thyroid function include both hypothyroidism and hyperthyroidism.

Clinical Note Abnormalities of thyroid function are among the most common endocrine disorders. They fall into two major categories—**hypothyroidism** and **hyperthyroidism**—reflecting deficient and excess thyroid hormone secretion, respectively. A number of causes can give rise to each of these conditions (⎮Table 19-1). Whatever the cause, the consequences of too little or too much thyroid hormone are largely predictable, based on knowing this hormone's functions.

Hypothyroidism Hypothyroidism can result (1) from primary failure of the thyroid gland; (2) secondary to a deficiency of TRH, TSH, or both; or (3) from inadequate dietary iodine.

Most symptoms of hypothyroidism are caused by a reduction in overall metabolic activity: A patient with hypothyroidism

Thyroid Dysfunction	Cause	Plasma Concentrations of Relevant Hormones	Goiter Present?
Hypothyroidism	Primary failure of the thyroid gland	↓ T_3 and T_4, ↑ TSH	Yes
	Secondary to hypothalamic or anterior pituitary failure	↓ T_3 and T_4, ↓ TRH and/or ↓ TSH	No
	Lack of dietary iodine	↓ T_3 and T_4, ↑ TSH	Yes
Hyperthyroidism	Abnormal presence of thyroid-stimulating immunoglobulin (TSI) (Graves' disease)	↑ T_3 and T_4, ↓ TSH	Yes
	Secondary to excess hypothalamic or anterior pituitary secretion	↑ T_3 and T_4, ↑ TRH and/or ↑ TSH	Yes
	Hypersecreting thyroid tumor	↑ T_3 and T_4, ↓ TSH	No

has a reduced BMR (less energy expenditure at rest); displays poor tolerance of cold (lack of the calorigenic effect); has a tendency to gain excessive weight (not burning fuels at a normal rate); and is easily fatigued (lower energy production). Another symptom is a slow, weak pulse (caused by a reduction in the rate and strength of cardiac contraction resulting from the diminished direct and sympathomimetic effects of thyroid hormone on the heart). A hypothyroid patient also exhibits slow reflexes and slow mental responsiveness (because of this hormone's effect on the nervous system). The mental effects are characterized by diminished alertness, slow speech, and poor memory.

Another notable characteristic is an edematous condition caused by infiltration of the skin with complex, water-retaining carbohydrate molecules (*glycosaminoglycans*), the production of which by connective tissue cells is normally suppressed by thyroid hormone. The resultant puffy appearance, primarily of the face, hands, and feet, is known as **myxedema.** If a person has hypothyroidism from birth, a condition known as **cretinism** develops. Because adequate levels of thyroid hormone are essential for normal growth and CNS development, cretinism is characterized by dwarfism and mental retardation, in addition to general symptoms of thyroid deficiency. The mental retardation is preventable if replacement therapy is started promptly, but it is not reversible once it has developed for a few months after birth, even with later treatment with thyroid hormone.

Hypothyroidism is treated by taking replacement thyroid hormone pills, except in hypothyroidism caused by iodine deficiency, in which case the remedy is adequate dietary iodine.

Hyperthyroidism The most common cause of hyperthyroidism is **Graves' disease.** This is an autoimmune disease in which the body erroneously produces **thyroid-stimulating immunoglobulin (TSI)**, also known as **long-acting thyroid stimulator (LATS)**, an antibody whose target is the TSH receptors on the thyroid cells. (An *autoimmune disease* is one in which the immune system produces antibodies for one of the body's tissues.) TSI stimulates both secretion and growth of the thyroid in a manner similar to TSH. Unlike TSH, however, TSI is not subject to negative-feedback inhibition by thyroid hormone, so thyroid secretion and growth continue unchecked (IFigure

19-4). Less frequently, hyperthyroidism occurs secondary to excess TRH or TSH or in association with a hypersecreting thyroid tumor.

As expected, the hyperthyroid patient has an elevated BMR. The resultant increase in heat production leads to excessive perspiration and poor tolerance of heat. Body weight typically falls because the body is burning fuel at an abnormally rapid rate. Net degradation of carbohydrate, fat, and protein stores occurs. The resultant loss of skeletal muscle protein results in weakness. Various cardiovascular abnormalities are associated with hyperthyroidism, caused both by the direct effects of thyroid hormone and by its interactions with catecholamines. Heart rate and strength of contraction may increase so much that the individual has palpitations (an unpleasant awareness of the heart's activity). The effects on the CNS are characterized by an excessive degree of mental alertness to the point where the patient is irritable, tense, anxious, and excessively emotional.

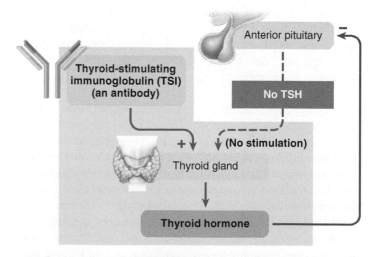

IFigure 19-4 **Role of thyroid-stimulating immunoglobulin in Graves' disease.** Thyroid-stimulating immunoglobulin (TSI), an antibody erroneously produced in the autoimmune condition of Graves' disease, binds with the TSH receptors on the thyroid gland and continuously stimulates thyroid hormone secretion outside the normal negative-feedback control system.

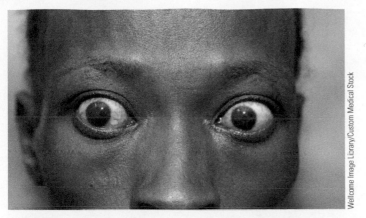

Figure 19-5 **Woman with Graves' disease displaying exophthalmos.**

Figure 19-6 **Woman with a goiter.**

A prominent feature of Graves' disease but not of the other types of hyperthyroidism is **exophthalmos** (bulging eyes) (Figure 19-5). Inflammation and swelling of the eye muscles and fat behind the eyes within the orbits (eye sockets in the skull) push the eyeballs forward so that they bulge from these bony cavities.

Three general methods of treatment can suppress excess thyroid hormone secretion: use of anti-thyroid drugs that specifically interfere with thyroid hormone synthesis (for example, drugs that block symporter uptake of I^- or drugs that inhibit thyroperoxidase); surgical removal of a portion of the oversecreting thyroid gland; or administration of radioactive iodine, which, after being concentrated in the thyroid gland by the iodide trap, selectively destroys thyroid glandular tissue.

A goiter develops when the thyroid gland is overstimulated.

Clinical Note A **goiter** is an enlarged thyroid gland. Because the thyroid lies over the trachea, a goiter is readily palpable and usually highly visible (Figure 19-6). A goiter occurs whenever either TSH or TSI excessively stimulates the thyroid gland. Note from Table 19-1 that a goiter may accompany hypothyroidism or hyperthyroidism, but it need not be present in either condition. Knowing the hypothalamus–pituitary–thyroid axis and feedback control, we can predict which types of thyroid dysfunction will produce a goiter. Let us consider hypothyroidism first:

■ Hypothyroidism secondary to hypothalamic or anterior pituitary failure is not accompanied by a goiter because the thyroid gland is not being adequately stimulated, let alone excessively stimulated.

■ With hypothyroidism caused by thyroid gland failure or lack of iodine, a goiter develops because the circulating level of thyroid hormone is so low that little negative-feedback inhibition takes place on the anterior pituitary and hypothalamus; TSH secretion is therefore elevated. TSH acts on the thyroid to increase the size and number of follicular cells and to increase their rate of secretion. If the thyroid cells cannot secrete hormone because of a lack of a critical enzyme or lack of iodine, no amount of TSH will be able to induce these cells to secrete T_3 and T_4. However, TSH can still promote hypertrophy and hyperplasia of the thyroid, with a consequent paradoxical enlargement of the gland (that is, a goiter), even though the gland is still underproducing.

Similarly, a goiter may or may not accompany hyperthyroidism:

■ Excess TSH secretion resulting from a hypothalamic or anterior pituitary defect would be accompanied by a goiter and excess T_3 and T_4 secretion because of overstimulation of thyroid growth.

■ In Graves' disease, a hypersecreting goiter occurs because TSI promotes growth of the thyroid, as well as enhancing secretion of thyroid hormone. Because the high levels of circulating T_3 and T_4 inhibit the anterior pituitary, TSH secretion itself is low.

■ Hyperthyroidism resulting from overactivity of the thyroid in the absence of overstimulation, such as caused by an uncontrolled thyroid tumor, is not accompanied by a goiter. The spontaneous secretion of excessive amounts of T_3 and T_4 inhibits TSH, so there is no stimulatory input to promote growth of the thyroid. (Even though a goiter does not develop, a tumor may cause enlargement of the thyroid, depending on the nature or size of the tumor.)

Check Your Understanding 19.1

1. Define *thyroid follicle, colloid, thyroglobulin, MIT, DIT, T_3,* and *T_4.*

2. Describe the action of the iodide trap.

3. Draw a flow diagram showing the effects and regulation of thyroid hormone secretion.

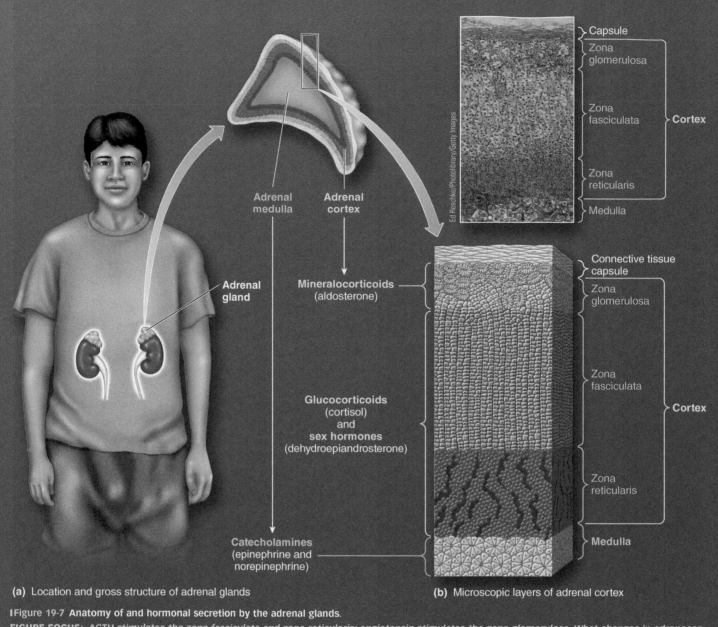

(a) Location and gross structure of adrenal glands

(b) Microscopic layers of adrenal cortex

ǀFigure 19-7 **Anatomy of and hormonal secretion by the adrenal glands.**
FIGURE FOCUS: *ACTH stimulates the zona fasciculata and zona reticularis; angiotensin stimulates the zona glomerulosa. What changes in adrenocortical output occur in response to an ACTH-secreting tumor of the lung?*

19.2 Adrenal Glands

There are two **adrenal glands,** one embedded above each kidney in a capsule of fat (*ad* means "next to"; *renal* means "kidney") (ǀFigure 19-7a).

Each adrenal gland consists of a steroid-secreting cortex and a catecholamine-secreting medulla.

Each adrenal is composed of two endocrine glands, one surrounding the other. The outer layers composing the **adrenal cortex** secrete a variety of steroid hormones; the inner portion, the **adrenal medulla,** secretes catecholamines. Thus, the adrenal cortex and medulla secrete hormones belonging to different

chemical categories, whose functions, mechanisms of action, and regulation are entirely different. We examine the adrenal cortex before turning to the adrenal medulla.

The adrenal cortex secretes mineralocorticoids, glucocorticoids, and sex hormones.

The adrenal cortex consists of three layers, or zones: the **zona glomerulosa,** the outermost layer; the **zona fasciculata,** the middle and largest portion; and the **zona reticularis,** the innermost zone (ǀFigure 19-7b). The adrenal cortex produces a number of different **adrenocortical hormones,** all of which are steroids derived from the common precursor molecule, cholesterol. All steroidogenic ("steroid-producing") tissues first convert cholesterol to *pregnenolone,* then modify this common core molecule by stepwise enzymatic reactions to produce active steroid hormones. Each steroidogenic tissue has a complement of enzymes to produce one or several but not all steroid hormones (❚ Figure 19-8). The adrenal cortex produces a greater variety of hormones than any other steroidogenic tissue. Slight variations in structure confer different functional capabilities on the various adrenocortical hormones, which can be divided into three categories based on their primary actions:

1. **Mineralocorticoids,** mainly *aldosterone,* influence mineral (electrolyte) balance, specifically Na^+ and K^+ balance.

2. **Glucocorticoids,** primarily *cortisol,* play a major role in glucose metabolism, as well as in protein and lipid metabolism and in adaptation to stress.

3. **Sex hormones** are identical or similar to those produced by the gonads (testes in males, ovaries in females). The most abundant and physiologically important of the adrenocortical sex hormones is *dehydroepiandrosterone,* an androgen, or "male" sex hormone.

The three categories of adrenal steroids are produced in anatomically distinct portions of the adrenal cortex as a result of differential distribution of the enzymes required to catalyze the different biosynthetic pathways leading to the formation of each of these steroids. Of the two major adrenocortical hormones, aldosterone is produced exclusively in the zona glo-

merulosa, whereas cortisol synthesis is limited to the two inner layers of the cortex, with the zona fasciculata being the major source of this glucocorticoid (see ❚ Figure 19-7b). No other steroidogenic tissues have the capability of producing either mineralocorticoids or glucocorticoids. In contrast, the adrenal sex hormones, also produced by the two inner cortical zones, are produced in far greater abundance in the gonads.

Because the adrenocortical hormones are all lipophilic and immediately diffuse through the plasma membrane of the steroidogenic cell into the blood after being synthesized, controlling the rate of synthesis regulates the rate of secretion.

Being lipophilic, the adrenocortical hormones are all carried in the blood extensively bound to plasma proteins. Cortisol is bound mostly to a plasma protein specific for it called **corticosteroid-binding globulin (transcortin)**, whereas aldosterone and dehydroepiandrosterone are largely bound to albumin, which nonspecifically binds a variety of lipophilic hormones.

Each of the adrenocortical steroid hormones binds with a receptor specific for it within the cytoplasm of the hormone's target cells: Mineralocorticoids bind to the **mineralocorticoid receptor (MR)**, glucocorticoids to the **glucocorticoid receptor (GR)**, and dehydroepiandrosterone to the **androgen receptor (AR)**. As is true of all steroid hormones, each hormone-receptor complex moves to the nucleus and binds with a complementary hormone-response element in DNA, namely the **mineralocorticoid response element**, **glucocorticoid response**

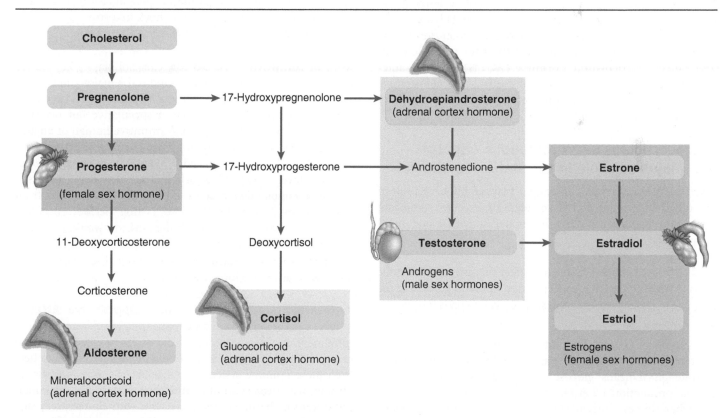

❚Figure 19-8 **Steroidogenic pathways for the major steroid hormones.** All steroid hormones are produced through a series of enzymatic reactions that modify cholesterol molecules, such as by varying the side groups attached to them. Each steroidogenic organ can produce only those steroid hormones for which it has a complete set of the enzymes needed to appropriately modify cholesterol, after first converting it to pregnenolone. The active hormones produced in the steroidogenic pathways are highlighted by screens. The intermediates that are not biologically active in humans are not screened.

FIGURE FOCUS: *"Male" sex hormones are produced in both males and females by the adrenal cortex. (True or false?)*

element, and **androgen response element**. This binding initiates specific gene transcription leading to synthesis of new proteins that carry out the effects of the hormone.

The major effects of mineralocorticoids are on Na⁺ and K⁺ balance and blood pressure homeostasis.

The actions and regulation of the primary adrenocortical mineralocorticoid, **aldosterone,** are described thoroughly elsewhere (Chapters 14 and 15). The principal site of aldosterone action is on the distal and collecting tubules of the kidney, where it promotes Na^+ retention and enhances K^+ elimination during the formation of urine. The promotion of Na^+ retention by aldosterone secondarily induces osmotic retention of H_2O, expanding the ECF volume (including the plasma volume), which is important in the long-term regulation of blood pressure.

Mineralocorticoids are *essential for life*. Without aldosterone, a person rapidly dies from circulatory shock because of the marked fall in plasma volume caused by excessive losses of H_2O-holding Na^+. With most other hormonal deficiencies, death is not imminent, even though a chronic hormonal deficiency may eventually lead to a premature death.

Aldosterone secretion is increased (1) via the complex renin–angiotensin–aldosterone system (RAAS) in response to a reduction in Na^+ and a fall in blood pressure (see Figure 14-16, p. 508), and (2) via direct stimulation of the adrenal cortex by a rise in plasma K^+ concentration (see Figure 14-22, p. 516). In tropic fashion, in addition to its effect on aldosterone secretion, angiotensin promotes growth of the zona glomerulosa. **Adrenocorticotropic hormone (ACTH)** from the anterior pituitary promotes secretion of cortisol, not aldosterone. Thus, regulation of aldosterone secretion is independent of anterior pituitary control.

Glucocorticoids exert metabolic effects and play a key role in adaptation to stress.

Cortisol, the primary glucocorticoid, plays an important role in carbohydrate, protein, and fat metabolism; executes significant permissive actions for other hormonal activities; and helps people resist stress.

Metabolic Effects The overall effect of cortisol's metabolic actions is to increase the concentration of blood glucose at the expense of protein and fat stores. Specifically, cortisol performs the following functions:

- It stimulates hepatic (liver) **gluconeogenesis,** the conversion of noncarbohydrate sources (namely, amino acids) into carbohydrate (*gluco* means "glucose"; *neo* means "new"; *genesis* means "production"). Between meals or during periods of fasting, when no new nutrients are being absorbed into the blood for use and storage, the glycogen (stored glucose) in the liver tends to become depleted as it is broken down to release glucose into the blood. Gluconeogenesis is an important factor in replenishing hepatic glycogen stores and thus in maintaining normal blood glucose levels between meals. This is essential because the brain can use only glucose as its metabolic fuel, yet nervous tissue cannot store glycogen to any extent. The concentration of glucose in the blood must therefore be maintained at an appropriate level to adequately supply the glucose-dependent brain with nutrients.

- Cortisol inhibits glucose uptake and use by many tissues, but not the brain, thus sparing glucose for use by the brain, which requires it as a metabolic fuel. This action, like gluconeogenesis, increases blood glucose.

- It stimulates protein degradation in many tissues, especially muscle. By breaking down a portion of muscle proteins into their constituent amino acids, cortisol increases the blood amino acid concentration. These mobilized amino acids are available for use in gluconeogenesis or wherever else they are needed, such as for repair of damaged tissue or synthesis of new cellular structures.

- Cortisol facilitates **lipolysis,** the breakdown of lipid (fat) stores in adipose tissue, thus releasing free fatty acids into the blood (*lysis* means "breakdown"). The mobilized fatty acids are available as an alternative metabolic fuel for tissues that can use this energy source in lieu of glucose, thereby conserving glucose for the brain.

Permissive Actions Cortisol is extremely important for its permissiveness. For example, cortisol must be present in adequate amounts to permit the catecholamines to induce vasoconstriction (blood vessel narrowing). A person lacking cortisol, if untreated, may go into circulatory shock in a stressful situation that demands immediate widespread vasoconstriction.

Role in Adaptation to Stress Cortisol plays a key role in adaptation to stress. Stress of any kind is the major stimulus for increased cortisol secretion. Although cortisol's precise role in adapting to stress is not known, a speculative but plausible explanation might be as follows: A primitive human or an animal wounded or faced with a life-threatening situation must forgo eating. A cortisol-induced shift away from protein and fat stores in favor of expanded carbohydrate stores and increased availability of blood glucose would help protect the brain from malnutrition during the imposed fasting period. Also, the amino acids liberated by protein degradation would provide a supply of building blocks for tissue repair if physical injury occurred. Thus, cortisol increases the pool of glucose, amino acids, and fatty acids for use as needed.

Anti-Inflammatory and Immunosuppressive Effects When stress is accompanied by tissue injury, inflammatory and immune responses accompany the stress response. Cortisol exerts *anti-inflammatory* and *immunosuppressive* effects to help hold these immune system responses in a check-and-balance fashion. An exaggerated inflammatory response has the potential of causing harm. Cortisol interferes with almost every step of inflammation, such as by suppressing migration of neutrophils to the injured site and interfering with their phagocytic activity (see p. 410) and by suppressing production of inflammatory cytokines (see p. 411). Cortisol inhibits immune responses by interfering with antibody production by lymphocytes.

Blurring the line between endocrine and immune control, lymphocytes have been shown to secrete ACTH, and some of the cytokines released from immune cells can stimulate the hypothalamus–pituitary–adrenal axis. In feedback fashion, cortisol in turn has a profound dampening (turning-down) impact on the immune system. These interactions between the immune system and cortisol secretion help maintain immune homeostasis, an area only beginning to be explored.

Synthetic glucocorticoids (drugs) have been developed that maximize the anti-inflammatory and immunosuppressive effects of these steroids while minimizing the metabolic effects (see p. 412). When these drugs are administered therapeutically at pharmacologic levels (that is, at higher-than-physiologic concentrations), they are effective in treating conditions in which the inflammatory response itself has become destructive, such as *rheumatoid arthritis*. Glucocorticoids used in this manner do not affect the underlying disease process; they merely suppress the body's response to the disease. Because glucocorticoids also exert multiple inhibitory effects on the overall immune process, these agents have proved useful in managing various allergic disorders (inappropriate immune attacks) and in preventing organ transplant rejections (immune attack against foreign cells).

However, these steroids should be used only when warranted, and then only sparingly, for several important reasons. First, because these drugs suppress the normal inflammatory and immune responses that form the backbone of the body's defense system, a glucocorticoid-treated person has limited ability to resist infections. Second, troublesome side effects may occur with prolonged exposure to higher-than-normal concentrations of glucocorticoids. These effects include development of gastric ulcers, high blood pressure, atherosclerosis, menstrual irregularities, and bone thinning. Third, high levels of exogenous glucocorticoids act in negative-feedback fashion to suppress the hypothalamus–pituitary–adrenal axis that drives normal glucocorticoid secretion and maintains the integrity of the adrenal cortex. Prolonged suppression of this axis can lead to irreversible atrophy (shrinkage) of the cortisol-secreting cells of the adrenal gland and thus to permanent inability of the body to produce its own cortisol. That is why *nonsteroidal anti-inflammatory drugs (NSAIDs),* such as aspirin and ibuprofen, are used as alternative anti-inflammatory therapy.

Cortisol secretion is regulated by the hypothalamus–pituitary–adrenal cortex axis.

Cortisol secretion by the adrenal cortex is regulated by a negative-feedback system involving the hypothalamus and anterior pituitary (Figure 19-9). ACTH from the anterior pituitary corticotropes, acting through the cAMP pathway, stimulates the adrenal cortex to secrete cortisol. Being tropic to the zona fasciculata and zona reticularis, ACTH stimulates both the growth and the secretory output of these two inner layers of the cortex. In the absence of adequate amounts of ACTH, these layers shrink considerably and cortisol secretion is drastically reduced. Recall that angiotensin, not ACTH, maintains the size of the zona glomerulosa. ACTH enhances many steps in the synthesis of cortisol. ACTH-producing cells, in turn, secrete only at the

command of **corticotropin-releasing hormone (CRH)** from the hypothalamus. CRH stimulates the corticotropes via the cAMP pathway. The feedback control loop is completed by cortisol's inhibitory actions on CRH and ACTH secretion by the hypothalamus and anterior pituitary, respectively.

The negative-feedback system for cortisol maintains the level of cortisol secretion relatively constant around the set point. Superimposed on the basic negative-feedback control system are two additional factors that influence plasma cortisol concentrations by changing the set point: *diurnal rhythm* and *stress,* both of which act on the hypothalamus to vary the secretion rate of CRH.

Recall that plasma cortisol concentration displays a characteristic diurnal rhythm, with the highest level occurring in the morning and the lowest level at night (see Figure 18-2, p. 642).

The other major factor that is independent of, and can override, the stabilizing negative-feedback control of cortisol is stress. Dramatic increases in cortisol secretion, mediated by the CNS through enhanced activity of the CRH–ACTH–cortisol

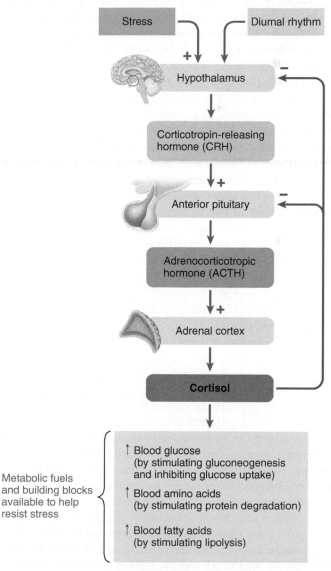

Figure 19-9 **Control of cortisol secretion.**

system, occur in response to all kinds of stressful situations. The magnitude of the increase in plasma cortisol concentration is generally proportional to the intensity of the stressful stimulation: A greater increase in cortisol levels takes place in response to severe stress than to mild stress.

The adrenal cortex secretes both male and female sex hormones in both sexes.

In both sexes, the adrenal cortex produces both *androgens,* or "male" sex hormones, and *estrogens,* or "female" sex hormones. The main site of production for the sex hormones is the gonads: the testes for androgens and the ovaries for estrogens. Accordingly, males have a preponderance of circulating androgens, whereas in females estrogens predominate. However, no hormones are unique to either males or females (except those from the placenta during pregnancy) because the adrenal cortex in both sexes produces small amounts of the sex hormone of the opposite sex. Additional small amounts of sex hormone of the opposite sex come from nonadrenal sources. Some testosterone in males is converted into estrogen by the enzyme *aromatase,* found especially in adipose tissue (see p. 726). In females, the ovaries produce androgen as an intermediate step in estrogen production (see Figure 19-8). A little of this androgen is released into the blood instead of being converted into estrogen.

Under normal circumstances, the adrenal androgens and estrogens are not sufficiently abundant or powerful to induce masculinizing or feminizing effects, respectively. The only adrenal sex hormone that has any biological importance is the androgen **dehydroepiandrosterone (DHEA)**. The testes' primary androgen product is the potent testosterone, but the most abundant adrenal androgen is the weaker DHEA. (Testosterone exerts about 100 times greater "androgenicity" than DHEA.) Adrenal DHEA is overpowered by testicular testosterone in males but is of physiologic significance in females, who otherwise have little androgens. DHEA governs androgen-dependent processes in the female such as growth of pubic and axillary (armpit) hair, enhancement of the pubertal growth spurt, and development and maintenance of the female sex drive.

In addition to controlling cortisol secretion, ACTH (not the pituitary gonadotropic hormones) controls adrenal androgen secretion. In general, cortisol and DHEA output by the adrenal cortex parallel each other. However, adrenal androgens feed back outside the hypothalamus–pituitary–adrenal axis. Instead of inhibiting CRH, DHEA inhibits gonadotropin-releasing hormone, just as testicular androgens do. Furthermore, sometimes adrenal androgen and cortisol output diverge from each other—for example, at the time of puberty adrenal androgen secretion undergoes a marked surge, but cortisol secretion does not change. This enhanced secretion initiates the development of androgen-dependent processes in females. In males the same thing is accomplished primarily by testicular androgen secretion, which is also aroused at puberty. The nature of the pubertal inputs to the adrenals and gonads is still unresolved.

A surge in DHEA secretion begins at puberty and peaks between the ages of 25 and 30. DHEA is the most abundant circulating steroid in young adults. After 30, DHEA secretion slowly tapers off until, by the age of 60, the plasma DHEA concentration is less than 15% of its peak level.

Some scientists suspect that the age-related decline of DHEA and other hormones such as GH (see p. 658) plays a role in some problems of aging. (See the accompanying boxed feature, Concepts, Challenges, and Controversies, for a discussion of theories of aging and anti-aging strategies under investigation.) DHEA supplementation as an anti-aging therapy is not recommended because of lack of evidence for its effectiveness and because of possible harmful side effects, such as raising the odds of acquiring ovarian or breast cancer in women and prostate cancer in men. Ironically, although the Food and Drug Administration (FDA) banned sales of DHEA as an over-the-counter drug in 1985 because of concerns about very real risks coupled with little proof of benefits, the product is widely available today as an unregulated food supplement. DHEA can be marketed as a dietary supplement without approval by the FDA as long as the product label makes no specific medical claims.

The adrenal cortex may secrete too much or too little of any of its hormones.

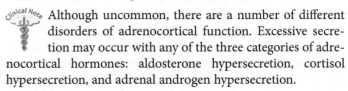 Although uncommon, there are a number of different disorders of adrenocortical function. Excessive secretion may occur with any of the three categories of adrenocortical hormones: aldosterone hypersecretion, cortisol hypersecretion, and adrenal androgen hypersecretion.

Aldosterone Hypersecretion Excess aldosterone secretion may be caused by (1) a hypersecreting adrenal tumor made up of aldosterone-secreting cells (**primary hyperaldosteronism**, or **Conn's syndrome**) or (2) inappropriately high activity of RAAS (**secondary hyperaldosteronism**). The latter may be produced by any number of conditions that cause a chronic reduction in arterial blood flow to the kidneys, thereby excessively activating RAAS. An example is atherosclerotic narrowing of the renal arteries.

Symptoms are related to the exaggerated effects of aldosterone—namely, excessive Na^+ retention (*hypernatremia*) and K^+ depletion (*hypokalemia*). Also, high blood pressure (hypertension) is generally present, at least partially because of excessive Na^+ and fluid retention.

Cortisol Hypersecretion Excessive cortisol secretion (**Cushing's syndrome**) can be caused by (1) overstimulation of the adrenal cortex by excessive amounts of CRH, ACTH, or both, or (2) adrenal tumors that uncontrollably secrete cortisol independent of ACTH, or (3) ACTH-secreting tumors located in places other than the pituitary, most commonly in the lung.

The prominent symptoms of this syndrome are related to the exaggerated effects of glucocorticoid, specifically those caused by excessive gluconeogenesis. When too many amino acids are converted into glucose, the body suffers from combined glucose excess (high blood glucose) and protein shortage. Because the resultant hyperglycemia and glucosuria (glucose in the urine) mimic diabetes mellitus, the condition is sometimes referred to as *adrenal diabetes.* For unclear reasons, some of the extra glucose is deposited as body fat in locations characteristic for this disease, typically in the abdomen, above the shoulder blades, and in the face. The abnormal fat distributions in the latter two locations are descriptively called a "buffalo hump" and a "moon face," respectively (Figure 19-10).

Besides the effects attributable to excessive glucose production, other effects arise from the widespread mobilization of amino acids from body proteins for use as glucose precursors. The appendages are thin because of muscle breakdown, in contrast to the abnormal fat deposits elsewhere. Loss of muscle protein leads to muscle weakness and fatigue. The protein-poor, thin skin of the abdomen becomes overstretched by the excessive underlying fat deposits, forming irregular, reddish purple linear streaks. Loss of structural protein within the walls of the small blood vessels leads to easy bruisability. Wounds heal poorly because formation of collagen, a major structural protein found in scar tissue, is depressed. Furthermore, loss of the collagen framework of bone weakens the skeleton, so fractures may result from little or no apparent injury.

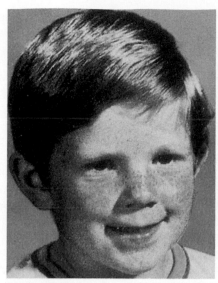 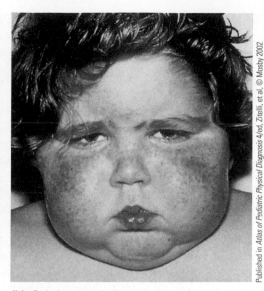

(a) Young boy prior to onset of the condition

(b) Only four months later, the same boy displaying a "moon face" characteristic of Cushing's syndrome

Figure 19-10 **Patient with Cushing's syndrome.**

Published in *Atlas of Pediatric Physical Diagnosis* 4/ed, Zitelli, et al, © Mosby 2002

Adrenal Androgen Hypersecretion Excess adrenal androgen secretion, a masculinizing condition, is more common than the extremely rare feminizing condition of excess adrenal estrogen secretion. Either condition is referred to as **adrenogenital syndrome,** emphasizing the pronounced effects that excessive adrenal sex hormones have on the genitalia and associated sexual characteristics.

The symptoms that result from excess androgen secretion depend on the sex of the individual and the age at which the hyperactivity first begins.

■ *In adult females.* Because androgens exert masculinizing effects, a woman with this disease develops a male pattern of body hair, a condition referred to as **hirsutism.** She also acquires other male secondary sexual characteristics, such as deepening of the voice and more muscular arms and legs. The breasts become smaller, and menstruation may cease as a result of androgen suppression of the woman's hypothalamus and pituitary so that her ovaries are not stimulated to secrete the female sex hormones responsible for the menstrual cycle.

■ *In newborn females.* Female infants born with adrenogenital syndrome manifest male-type external genitalia because excessive androgen secretion occurs early enough during fetal life to induce development of their genitalia along male lines, similar to the development of males under the influence of testicular androgen (see p. 721). The clitoris, which is the female homologue of the male penis, enlarges under androgen influence and takes on a penile appearance, so in some cases it is difficult at first to determine the child's sex. This hormonal abnormality is one of the major causes of female **pseudoher-** **maphroditism,** a condition in which female gonads (ovaries) are present but the external genitalia resemble those of a male. (A true hermaphrodite has the gonads of both sexes.)

■ *In prepubertal males.* Excessive adrenal androgen secretion in prepubertal boys causes them to prematurely develop male secondary sexual characteristics—for example, deep voice, beard, enlarged penis, and sex drive. This condition is referred to as **precocious pseudopuberty** to differentiate it from true puberty, which occurs as a result of increased testicular activity. In precocious pseudopuberty, the androgen secretion from the adrenal cortex is not accompanied by sperm production or any other gonadal activity because the testes are still in their nonfunctional prepubertal state.

■ *In adult males.* Overactivity of adrenal androgens in adult males has no apparent effect because any masculinizing effect induced by the weak DHEA, even when in excess, is unnoticeable in the face of the powerful masculinizing effects of the much more abundant and potent testosterone from the testes.

The adrenogenital syndrome is most commonly caused by an inherited enzymatic defect in the cortisol steroidogenic pathway. The pathway for synthesis of androgens branches from the normal biosynthetic pathway for cortisol (see Figure 19-8). When an enzyme specifically essential for synthesis of cortisol is deficient, the result is decreased secretion of cortisol. The decline in cortisol secretion removes the negative-feedback effect on the hypothalamus and anterior pituitary so that levels of CRH and ACTH increase considerably (Figure 19-11). The defective adrenal cortex is incapable of responding to this increased ACTH secretion with cortisol output and instead shunts more of its cholesterol precursor into the androgen pathway. The result is excess DHEA production. This excess androgen does not inhibit ACTH but rather inhibits the gonadotropins. Because gamete production is not stimulated in the absence of gonadotropins,

Still a Big Question: Why Do We Age?

W E ALL RECOGNIZE OUTWARD SIGNS of aging when we see it (wrinkled faces, gray hair, slower movement). We are also aware of the increased incidence of cardiovascular disease, Alzheimer's disease, arthritis, and cataracts, to name a few, among the elderly. **Aging** is the progressive impairment of cellular, organ, and system functions characterized by loss of skin elasticity, hair color, and muscle strength; decreased immunity; metabolic dysfunction; decline in memory and other cognitive abilities; increased incidence of age-related diseases; and ultimately death. However, despite years of intensive research, no one knows for sure why and how aging happens. Many theories, some interrelated, and each with some supportive evidence, have been set forth. However, none of the theories alone can account for the inevitable progress toward senescence. Many of these theories are not mutually exclusive. Undoubtedly multiple factors contribute to the aging process.

Theories of aging can be grouped into three broad categories: (1) *programmed aging theories*, which propose that aging and lifespan are predetermined by a built-in timing mechanism; (2) *damage or error accumulation theories*, which suggest that aging occurs as a result of accumulated structural damage and functional errors arising from chance environmental insults that lead to progressive failure of cells and organs; and (3) *evolutionary theories*, which maintain that aging is an inadvertent by-product of natural selection for reproductive fitness. Following are the major theories in each category.

Programmed Aging Theories

■ The *age program theory*: We are genetically programmed to age and die. That is, aging follows an intrinsic biological timeline that is predestined, in the same way that an infant inevitably grows and matures into an adult.

■ The *limit theory*: Body cells can only divide an estimated 50 times on average after birth. Then they stop dividing, after which organs progressively fail, as cells that die are no longer replaced. Related is the *telomere shortening theory*: Telomeres, the end caps of chromosomes, get shorter with each successive cell division; when they get too short, further cell division is blocked.

■ The *endocrine theory*: Hormones control the pace of aging, much as they control growth, puberty, and reproductive capacity. For example, the natural decline of growth hormone and dehydroepiandrosterone contribute to the aging process.

■ The *immunological theory*: A programmed decline in immune function with advancing age leaves the elderly more susceptible to aging-associated diseases, including infectious diseases (for example, pneumonia) and cancer. Dysregulated immune responses are also implicated in cardiovascular disease and Alzheimer's disease, among other detrimental outcomes that lead to progressive loss of health and eventual death.

Damage or Error Accumulation Theories of Aging

■ The *free radical and oxidative stress theory*: Toxic by-products of living lead to aging and death. To sustain life, the mitochondrial electron transport chain uses O_2 to generate ATP from energy derived from nutrient molecules (see p. 37). Resultant side reactions constantly produce toxic by-products, namely *reactive oxygen species (ROS)* that can form destructive *free radicals*, highly reactive particles that damage DNA, proteins, and other biomolecules; see p. 142). Natural antioxidants in the body help curb the dangerous buildup of free radicals. *Oxidative stress* occurs when the body's ability to detoxify oxidizing agents is overwhelmed by a high ROS load. Environmental sources of oxidizing activity such as exposure to ionizing radiation or toxins in tobacco smoke can add to the body's total oxidative stress burden. The cumulative damage resulting from the inability of our antioxidant defenses to cope with generated reactive oxidative products over time gradually increases the risk of late-onset diseases. In a vicious cycle, oxidative stress promotes systemic inflammation, and inflammation intensifies oxidative stress.

■ The *accumulative waste theory*: Buildup of cellular garbage interferes with cellular functions. Molecules damaged by free radicals and other means can accumulate within a cell with advancing age to the point of interfering with cell signaling, with metabolic pathways, and with intracellular transport, causing cells to malfunction and deteriorate.

■ The *somatic DNA damage theory*: The genes that code for body proteins get messed up with time. Damage to DNA occurs continuously as a result of oxidative stress, copying errors during DNA replication, and external factors. Most DNA damage can be repaired, but some persist. The resulting defects in our body cells' genetic integrity accumulate with advancing age, leading to the progressive cellular and organ malfunctions associated with senescence.

■ The *wear and tear theory*: We are not built to last. The body gradually wears out as forces that damage the body (such as diseases, accidents, oxidative stress, and so on) outpace the body's ability to prevent, repair, overcome, or cope with these insults. Cumulative damage to irreplaceable molecules and cells eventually leads to progressive organ failure with increasing age.

■ The *cross-linkage theory*: Protein molecules literally stick together to the detriment of body function. Excess sugar in the blood can haphazardly attach to proteins, which causes cross links to form between and within proteins. Progressive formation of cross links among proteins such as collagen in the skin, tendons, and blood vessels causes these structures to lose their elasticity. Similar cross linking of other proteins, including enzymes, likewise gradually impairs cellular and organ function elsewhere.

Evolutionary Theories of Aging

■ The *mutation accumulation theory*: A deleterious gene mutation that kills the young will not be passed on to the next generation (that is, the mutation is strongly selected against), whereas a harmful gene mutation present within eggs and sperm as well as other body cells that does not manifest itself until people are past their reproductive years and have already passed the gene on to their offspring is not selected against. Over successive generations, late-acting lethal mutations accumulate, leading to an increase in mortality rates in older individuals.

- The *antagonistic pleitropy theory*: This theory is based on a benefit-while-you're-young, pay-the-price-later principle. Some genes have multiple effects (this is what *pleitropy* means). An example is the gene that codes for the *mTOR* protein, a two-faced enzyme that is beneficial early in life for cell growth and development but is harmful after maturity as a key driver of aging. That is, the gene has antagonistic actions, depending on age. This gene would have been naturally selected because it favors reproductive fitness, despite its negative effect beyond the reproductive years. In this roundabout way, therefore, senescence has inadvertently been selected during evolution.

mTOR stands for *m*ammalian *t*arget *o*f *r*apamycin. Rapamycin is a drug that extends life in yeast and experimental animals by inhibiting their versions of TOR. Rapamycin itself has too many side effects for widespread use in humans, but scientists are searching for an anti-aging alternative. Severe calorie restriction (nutritionally adequate, near-starvation diet) suppresses mTOR activity and reduces oxidative stress, among other beneficial effects. As a result, the time of youthful vigor is increased and lifespan is extended up to 50% in experimental animals. Investigators are seeking milder ways to mimic these longevity effects. The hope is to buy us quality time by delaying or preventing the diseases of aging, not just to extend our years of senility. (Even now, healthy lifestyle choices are recommended to add some quality years to your life.)

Scientists are searching for other clues about what causes aging and possible ways to intercede by studying what goes wrong when children are on an incorrect aging timetable—either aging ahead of or behind schedule.

Progeria: Aging Ahead of Schedule

Progeria is a rare disorder (about 100 confirmed cases worldwide) where children manifest signs of aging and aging-related diseases (wrinkled skin, atherosclerosis, and stiff joints) along with other characteristics of the condition, including limited growth and a distinctive appearance (relatively large head, narrow face, no hair) (see the accompanying figure) (*pro* means "in favor of"; *geras* means "aging"). Mental and motor development is usually normal. Children with progeria age at a rate 8 to 10 times faster than normal and typically die in their early teens from heart attacks or strokes.

Progeria is a genetic condition that arises as a result of a new (not inherited) mutation. The most common mutation leads to production of a recently discovered, shortened form of a normal protein. This mutant form, dubbed *progerin*, causes significant disfigurement of the nuclear membrane and leads to cell dysfunction. Progerin has been associated not only with abnormal aging in progeria but also with normal aging. Progerin is produced in skin on exposure to ultraviolet light from the sun and tanning beds and is speculated to accelerate photoaging of skin, and a new finding suggests that during normal aging cells produce progerin when their telomeres become short.

Perpetual Childhood: Aging Behind Schedule

Even rarer, a handful of cases of "perpetual childhood" have been identified around the world. These individuals are moving along the

Seven-year-old girl with progeria. Seen here with her mother.

aging timeline at an incredibly slow pace, each apparently from a different underlying cause. One, a girl from Montana, at 8 years of age weighs only 11 pounds and still has the skin, dependency, and appearance of a baby, except that her hair, nails, and teeth are growing normally (see the accompanying figure). An aging specialist studying her case has identified some specific DNA flaws in her fully sequenced genes, but many questions remain. Might further analysis lead to a better understanding not only of this child's condition but of the aging process in general?

Eight-year-old "baby" girl. Seen here with her mother.

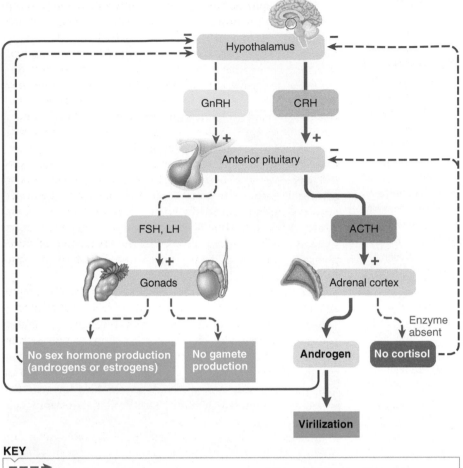

- - - -> = Normal pathways that do not occur
ACTH = Adrenocorticotropic hormone
GnRH = Gonadotropin-releasing hormone

FSH = Follicle-stimulating hormone
LH = Luteinizing hormone
CRH = Corticotropin-releasing hormone

Figure 19-11 Hormonal interrelationships in adrenogenital syndrome. The adrenocortical cells that are supposed to produce cortisol produce androgens instead because of a deficiency of a specific enzyme essential for cortisol synthesis. Because no cortisol is secreted to act in negative-feedback fashion, CRH and ACTH levels are elevated. The adrenal cortex responds to increased ACTH by further increasing androgen secretion. The excess androgen produces virilization and inhibits the gonadotropin pathway, with the result that the gonads stop producing sex hormones and gametes.

people with adrenogenital syndrome are sterile. Of course, they also exhibit symptoms of cortisol deficiency.

The symptoms of adrenal virilization (having physical characteristics befitting a man in a person other than an adult male), sterility, and cortisol deficiency are all reversed by glucocorticoid therapy. Administration of exogenous glucocorticoid replaces the cortisol deficit and, more dramatically, inhibits the hypothalamus and pituitary so that ACTH secretion is suppressed. Once ACTH secretion is reduced, the profound stimulation of the adrenal cortex ceases and androgen secretion declines markedly. Removing the large quantities of adrenal androgens from circulation allows masculinizing characteristics to gradually recede and normal gonadotropin secretion to resume. Without understanding how these hormonal systems are related, it would be very difficult to comprehend how glucocorticoid administration could dramatically reverse symptoms of masculinization and sterility.

Adrenocortical Insufficiency If one adrenal gland is nonfunctional or removed, the other healthy organ can take over the function of both through hypertrophy and hyperplasia. Therefore, both glands must be affected before adrenocortical insufficiency occurs.

In **primary adrenocortical insufficiency (Addison's disease)**, all layers of the adrenal cortex are undersecreting. This condition is most commonly caused by autoimmune destruction of the cortex by erroneous production of adrenal cortex–attacking antibodies, in which case both aldosterone and cortisol are deficient. **Secondary adrenocortical insufficiency** may occur because of a pituitary or hypothalamic abnormality, resulting in insufficient ACTH secretion. In this case, only cortisol is deficient because aldosterone secretion does not depend on ACTH stimulation.

The symptoms associated with aldosterone deficiency in Addison's disease are the most threatening. If severe enough, the condition is fatal because aldosterone is essential for life. However, the loss of adrenal function may develop slowly and insidiously so that aldosterone secretion may be subnormal but not totally lacking. Patients with aldosterone deficiency display K^+ retention (*hyperkalemia*), caused by reduced K^+ loss in the urine, and Na^+ depletion (*hyponatremia*), caused by excessive urinary loss of Na^+. The former disturbs cardiac rhythm. The latter reduces ECF volume, including circulating blood volume, which in turn lowers blood pressure (hypotension).

Symptoms of cortisol deficiency are as would be expected: poor response to stress, hypoglycemia (low blood glucose) caused by reduced gluconeogenic activity, and lack of permissive action for many metabolic activities. The primary form of the disease also produces hyperpigmentation (darkening of the skin) resulting from excessive secretion of ACTH. Because the pituitary is normal, the decline in cortisol secretion brings about an uninhibited elevation in ACTH output (resulting from reduced negative feedback). Recall that ACTH and α melanocyte-stimulating hormone (α-MSH, a skin-darkening hormone that promotes dispersion of the pigment melanin) can both be cleaved from the same pro-opiomelanocortin precursor molecule (but not at the same time nor in the same organ; see p. 648). However, being closely related, at high levels ACTH can also bind with α-MSH's receptors in the skin and cause darkening of the skin.

Having completed discussion of the adrenal cortex, we now shift attention to the adrenal medulla.

The adrenal medulla consists of modified sympathetic postganglionic neurons.

The adrenal medulla is actually a modified part of the sympathetic nervous system. A sympathetic pathway consists of two neurons in sequence. A preganglionic neuron originating in the CNS has an axonal fiber that terminates on a second peripherally located postganglionic neuron, which in turn terminates on the effector organ (see p. 234). The neurotransmitter released by sympathetic postganglionic fibers is norepinephrine, which interacts locally with the innervated organ by binding with specific target receptors known as adrenergic receptors.

The adrenal medulla consists of modified postganglionic sympathetic neurons called **chromaffin cells** because of their staining preference for chromium ions. Unlike ordinary postganglionic sympathetic neurons, chromaffin cells do not have axonal fibers that terminate on effector organs. Instead, on stimulation by the preganglionic fiber the chromaffin cells release their chemical transmitter directly into the blood (see Figure 7-2, p. 235). In this case, the transmitter qualifies as a hormone instead of a neurotransmitter. Like sympathetic fibers, the adrenal medulla does release norepinephrine, but its most abundant secretory output is a similar chemical messenger known as **epinephrine**. Both epinephrine and norepinephrine are catecholamines that possess the same structure except for epinephrine having a methyl group.

Epinephrine and norepinephrine are synthesized almost entirely within the cytosol of the adrenomedullary secretory cells. Once produced, these catecholamines are stored in **chromaffin granules**, which are similar to the transmitter storage vesicles found in sympathetic nerve endings. Segregation of catecholamines in chromaffin granules protects them from being destroyed by cytosolic enzymes during storage.

Secretion of Catecholamines from the Adrenal Medulla

Catecholamines are secreted into the blood by exocytosis of chromaffin granules. Their release is analogous to the release of stored peptide hormones from secretory vesicles or the release of norepinephrine at sympathetic postganglionic terminals.

Of the total adrenomedullary catecholamine output, epinephrine accounts for 80% and norepinephrine for 20%. Epinephrine is produced exclusively by the adrenal medulla, but the bulk of norepinephrine is produced by sympathetic postganglionic fibers. Adrenomedullary norepinephrine is secreted in quantities too small to exert significant effects on target cells. Therefore, for practical purposes we can assume that norepinephrine effects are predominantly mediated directly by the sympathetic nervous system and that epinephrine effects are brought about exclusively by the adrenal medulla.

Epinephrine and norepinephrine vary in their affinities for different receptor types.

Epinephrine and norepinephrine have differing affinities for four distinctive receptor types: α_1, α_2, β_1, and β_2 adrenergic receptors (see p. 240). (See Table 7-1, p. 238 and Table 7-2, p. 240 to review the distribution of these receptor types among target organs.)

Norepinephrine binds predominantly with α and β_1 receptors located near postganglionic sympathetic-fiber terminals. Hormonal epinephrine, which can reach all α and β_1 receptors via its circulatory distribution, interacts with these same receptors. Norepinephrine has a little greater affinity than epinephrine for the α receptors, and the two hormones have approximately the same potency at the β_1 receptors. Thus, epinephrine and norepinephrine exert similar effects in many tissues, with epinephrine generally reinforcing sympathetic nervous activity. In addition, epinephrine activates β_2 receptors, over which the sympathetic nervous system exerts no influence. Many of the epinephrine-exclusive β_2 receptors are located at tissues not even supplied by the sympathetic nervous system but reached by epinephrine through the blood. Examples include skeletal muscle, where epinephrine exerts metabolic effects such as promoting the breakdown of stored glycogen, and bronchiolar smooth muscle, where it causes bronchodilation.

Sometimes epinephrine, through its exclusive β_2-receptor activation, brings about a different action from that elicited by norepinephrine and epinephrine action through their mutual activation of other adrenergic receptors. As an example, norepinephrine and epinephrine bring about a generalized vasoconstrictor effect mediated by α_1-receptor stimulation. By contrast, epinephrine promotes vasodilation of the blood vessels that supply skeletal muscles and the heart through β_2-receptor activation (see p. 350).

Realize, however, that epinephrine functions only at the bidding of the sympathetic nervous system, which is solely responsible for stimulating its secretion from the adrenal medulla. Epinephrine secretion always accompanies a generalized sympathetic nervous system discharge, so sympathetic activity indirectly controls actions of epinephrine. By having the more versatile circulating epinephrine at its call, the sympathetic nervous system has a means of reinforcing its own neurotransmitter effects plus a way of executing additional actions on tissues that it does not directly innervate.

Catecholamines exert their effects via second-messenger pathways. Effects arising from binding to β receptor types are mediated via increased cAMP, to the α_2 receptors via decreased cAMP, and to the α_1 receptors via increased IP$_3$ and DAG.

Epinephrine reinforces the sympathetic nervous system and exerts metabolic effects.

Adrenomedullary hormones are not essential for life, but virtually all organs in the body are affected by these catecholamines. They play important roles in mounting stress responses, regulating arterial blood pressure, and controlling fuel metabolism. The following sections discuss epinephrine's major effects, which it achieves either in collaboration with the sympathetic transmitter norepinephrine or alone to complement direct sympathetic response.

Effects on Organ Systems Together, the sympathetic nervous system and adrenomedullary epinephrine mobilize the body's resources to support peak physical exertion in emergency or stressful situations. The sympathetic and epinephrine actions constitute a fight-or-flight response that prepares the person to combat an enemy or flee from danger (see p. 238). Specifically, the sympathetic system and epinephrine increase cardiac output by increasing the rate and strength of cardiac contraction, and their generalized vasoconstrictor effects increase total peripheral resistance. Together, these effects raise arterial blood pressure, thus ensuring an appropriate driving pressure to force blood to the organs most vital for meeting the emergency. Meanwhile, vasodilation of coronary and skeletal muscle blood vessels induced by epinephrine and local metabolic factors shifts blood to the heart and skeletal muscles from other vasoconstricted regions of the body, thus shunting blood to the areas of most immediate need. Because of their profound influence on the heart and blood vessels, the sympathetic system and epinephrine also play an important role in ongoing maintenance of arterial blood pressure, even in the absence of an emergency.

Epinephrine (but not norepinephrine) dilates the respiratory airways to reduce the resistance encountered in moving air in and out of the lungs. Epinephrine and norepinephrine also reduce digestive activity and inhibit bladder emptying, both activities that can be "put on hold" during a fight-or-flight situation.

Metabolic Effects Epinephrine exerts some important metabolic effects. In general, it prompts the mobilization of stored carbohydrate and fat so that extra energy is available for use as needed to fuel muscular work. Specifically, epinephrine increases blood glucose by several different mechanisms. First, it stimulates both hepatic gluconeogenesis and **glycogenolysis**, the latter being the breakdown of stored glycogen into glucose, which is released into the blood. Epinephrine also stimulates glycogenolysis in skeletal muscles. Because of the difference in enzyme content between liver and muscle, however, muscle glycogen cannot be converted directly to glucose. Instead, the breakdown of muscle glycogen releases lactate into the blood. The liver removes lactate from the blood and converts it into glucose, so epinephrine's actions on skeletal muscle indirectly help raise blood glucose levels. Epinephrine and the sympathetic system further add to this hyperglycemic effect by inhibiting secretion of insulin, the pancreatic hormone primarily responsible for removing glucose from the blood, and by stimulating glucagon, another pancreatic hormone that promotes hepatic glycogenolysis and gluconeogenesis. In addition to increasing blood glucose levels, epinephrine also increases the level of blood fatty acids by promoting lipolysis.

Epinephrine's metabolic effects are appropriate for fight-or-flight situations. The elevated levels of glucose and fatty acids provide additional fuel to power the muscular movement required by the situation and also assure adequate nourishment for the brain during the crisis when no new nutrients are being consumed. Muscles can use fatty acids for energy production, but the brain cannot.

Because of its other widespread actions, epinephrine also increases the overall metabolic rate. Under the influence of epinephrine, many tissues metabolize faster. For example, the work of the heart and respiratory muscles increases, and the pace of liver metabolism steps up. Thus, epinephrine and thyroid hormone both increase the metabolic rate.

Other Effects Epinephrine acts on the CNS to promote a state of arousal and increased alertness. This permits "quick thinking" to help cope with the impending emergency. Many drugs used as stimulants or sedatives exert their effects by altering catecholamine levels in the CNS.

Both epinephrine and norepinephrine cause sweating, which helps the body rid itself of extra heat generated by increased muscular activity. Also, epinephrine acts on smooth muscles within the eyes to dilate the pupil and flatten the lens. These actions adjust the eyes for more encompassing vision so that the whole threatening scene can be quickly viewed.

Epinephrine is released only on sympathetic stimulation of the adrenal medulla.

Catecholamine secretion by the adrenal medulla is controlled entirely by sympathetic input to the gland. When the sympathetic system is activated under conditions of fear or stress, it simultaneously triggers a surge of adrenomedullary catecholamine release. The concentration of epinephrine in the blood may increase up to 300 times normal, with the amount of epinephrine released depending on the type and intensity of the stressful stimulus.

Because both parts of the adrenal gland are important in responding to stress, this is an appropriate place to pull together the major factors involved in the stress response, the topic of the next section.

Check Your Understanding 19.2

1. List the three categories of adrenocortical hormones, name the primary hormone in each category, and state the functions of each of these hormones.
2. Discuss the effect of ACTH on the adrenal cortex.
3. Name the two catecholamines secreted by the adrenal medulla and describe how they are stored and released.

19.3 Integrated Stress Response

Stress is the generalized, nonspecific response of the body to any factor that overwhelms, or threatens to overwhelm, the body's compensatory abilities to maintain homeostasis. Contrary to popular usage, the agent inducing the response is correctly called a *stressor*, whereas *stress* refers to the state induced by the stressor. The following types of noxious stimuli illustrate the range of factors that can induce a stress response: *physical* (trauma, surgery, intense heat or cold); *chemical* (reduced O_2 supply, acid–base imbalance); *physiologic* (heavy exercise, hemorrhagic shock, pain); *infectious* (bacterial invasion); *psychological* or *emotional* (anxiety, fear, sorrow); and *social* (personal conflicts, change in lifestyle).

The stress response is a generalized pattern of reactions to any situation that threatens homeostasis.

Different stressors may produce some specific responses characteristic of that stressor; for example, the body's specific response to cold exposure is shivering and skin vasoconstriction, whereas the specific response to bacterial invasion includes increased phagocytic activity and antibody production. In addition to their specific response, however, all stressors produce a similar nonspecific, generalized response (Figure 19-12). This set of responses common to all noxious stimuli is called the **general adaptation syndrome.** When a stressor is recognized, both nervous and hormonal responses bring about defensive measures to cope with the emergency. The result is a state of intense readiness and mobilization of biochemical resources.

To appreciate the value of the multifaceted stress response, imagine a primitive cave dweller who has just seen a large wild beast lurking in the shadows. We consider both the neural and the hormonal responses that would take place in this scenario. The body responds in the same way to modern-day stressors. We cover all these responses in further detail elsewhere. At this time, we just examine how they work together.

Roles of the Sympathetic Nervous System and Epinephrine in Stress The major neural response to such a stressful stimulus is generalized activation of the sympathetic nervous system, which prepares the body for a fight-or-flight response. Simultaneously, the sympathetic system calls forth a massive outpouring of epinephrine from the adrenal medulla. Epinephrine strengthens sympathetic responses and mobilizes carbohydrate and fat stores.

Roles of the CRH–ACTH–Cortisol System In Stress Besides epinephrine, a number of other hormones are involved in the overall stress response. The predominant hormonal response is activation of the CRH–ACTH–cortisol system. Recall that cortisol's role in helping the body cope with stress is presumed to be related to its metabolic effects, namely increasing the pool of glucose, amino acids, and fatty acids in the blood for use as needed, such as to sustain nourishment to the brain and provide building blocks for repair of damaged tissues.

In addition to the effects of cortisol in the hypothalamus–pituitary–adrenal cortex axis, ACTH may also play a role in resisting stress. ACTH is one of several peptides that facilitate learning and behavior. Thus, an increase in ACTH during psychosocial stress may help the body cope more readily with similar stressors in the future by facilitating the learning of appropriate behavioral responses.

Role of Other Hormonal Responses in Stress Besides the CRH–ACTH–cortisol system, other hormonal systems play key roles in the stress response, as follows:

■ *Elevation of blood glucose and fatty acids through decreased insulin and increased glucagon.* The sympathetic nervous system and the epinephrine secreted at its bidding both inhibit insulin and stimulate glucagon. These hormonal changes act in concert to elevate blood glucose and blood fatty acids. Epinephrine and glucagon, both of which are increased during stress, promote hepatic glycogenolysis and (along with cortisol) hepatic gluconeogenesis. However, insulin, whose secretion is suppressed during stress, opposes the breakdown of liver glycogen stores. All these effects help increase blood glucose. The primary stimulus for insulin secretion is a rise in blood glucose; in turn, a primary effect of insulin is to lower blood glucose. If it were not for the deliberate inhibition of insulin during the stress response, the hyperglycemia caused by stress would stimulate secretion of glucose-lowering insulin. As a result, the elevation in blood glucose could not be sustained. Stress-related hormonal responses also promote release of fatty acids from fat stores because lipolysis is favored by epinephrine, glucagon, and cortisol but opposed by insulin.

■ *Maintenance of blood volume and blood pressure through increased renin–angiotensin–aldosterone and vasopressin activity.* In addition to the hormonal changes that mobilize energy stores during stress, other hormones are simultaneously called into play to sustain plasma volume and blood pressure during the emergency. The sympathetic system and epinephrine play major roles in acting directly on the heart and blood vessels to improve circulatory function. In addition, RAAS is activated as a consequence of increased sympathetic activity to the kidneys (see p. 507). Vasopressin secretion is also increased during stressful situations (see p. 543). Collectively, these hormones expand the plasma volume by promoting retention of salt and H_2O. Presumably, the enlarged plasma volume is a protective measure to help sustain blood pressure should acute loss of plasma fluid occur through hemorrhage or heavy sweating during the impending period of danger. Vasopressin and angiotensin also have direct vasopressor effects, which would be of benefit in maintaining an adequate blood pressure in the event of acute blood loss (see p. 350). Vasopressin is further believed to facilitate learning, which has implications for future adaptation to stress.

The multifaceted stress response is coordinated by the hypothalamus.

All the individual responses to stress just described are either directly or indirectly influenced by the hypothalamus (Figure 19-13). The hypothalamus receives input concerning physical

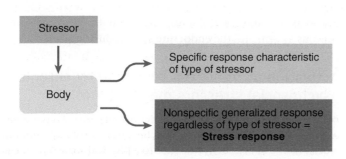

Figure 19-12 **Action of a stressor on the body.**

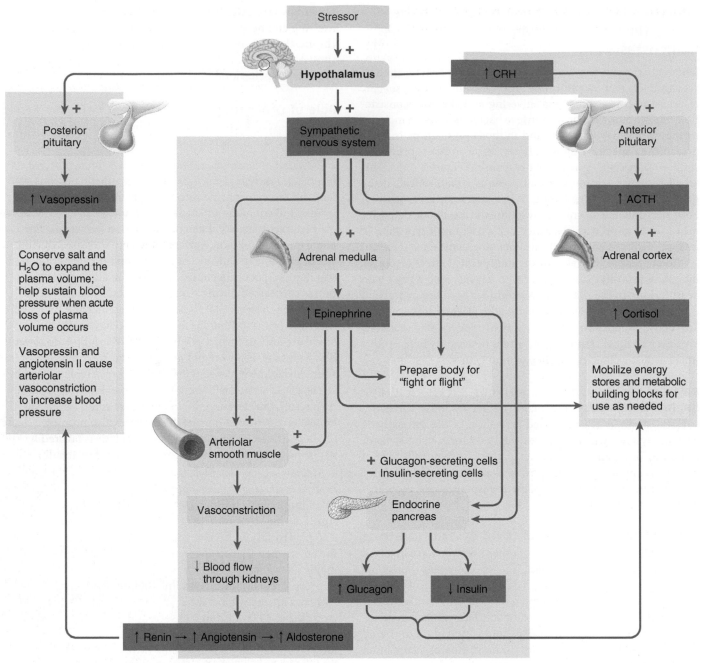

I Figure 19-13 **Integration of the stress response by the hypothalamus.**
FIGURE FOCUS: *(1) Trace all of the pathways that lead to mobilization of energy stores and metabolic building blocks during the stress response. (2) Likewise, trace all of the pathways that lead to salt and water conservation.*

and emotional stressors from virtually all areas of the brain and from many receptors throughout the body. In response, the hypothalamus directly activates the sympathetic nervous system, secretes CRH to stimulate ACTH and cortisol release, and triggers vasopressin release. Sympathetic stimulation, in turn, brings about secretion of epinephrine, with which it has a conjoined effect on the pancreatic secretion of insulin and glucagon. Furthermore, vasoconstriction of the renal afferent arterioles by the catecholamines indirectly triggers secretion of renin by reducing blood flow through the kidneys (a stimulus for renin secretion). Renin, in turn, sets in motion RAAS. Thus, the hypothalamus integrates the responses of both the sympathetic nervous system and the endocrine system during stress.

Activation of the stress response by chronic psychosocial stressors may be harmful.

Acceleration of cardiovascular and respiratory activity, retention of salt and H_2O, and mobilization of metabolic fuels and building blocks can be of benefit in response to a physical stressor, such as an athletic competition. Most of the stressors in our everyday lives are psychosocial in nature; however, they induce these same magnified responses. Stressors such as anxiety about an exam, conflicts with loved ones, or impatience while sitting in a traffic jam can elicit a stress response. Although the rapid mobilization of body resources is appropriate in the face of real or threatened physical injury, it is generally inappropriate in response to nonphysical stress. If no extra energy is demanded, no tissue is damaged, and no blood lost, body stores are being broken down and fluid retained needlessly, probably to the detriment of the emotionally stressed individual. In fact, strong circumstantial evidence suggests a link between chronic exposure to psychosocial stressors and development of pathological conditions such as high blood pressure, although no definitive cause-and-effect relationship has been ascertained. As a result of "unused" stress responses, could hypertension result from too much sympathetic vasoconstriction? From too much salt and H_2O retention? From too much vasopressin and angiotensin pressor activity? A combination of these? Other factors? Recall that hypertension can develop with prolonged exposure to pharmacological levels of glucocorticoids. Could long-standing lesser elevations of cortisol, such as might occur in the face of continual psychosocial stressors, do the same thing, only more slowly? Considerable work remains to be done to evaluate the contributions that the stressors in our everyday lives make toward disease production.

Check Your Understanding 19.3

1. Discuss the major hormonal changes and the purposes served by each change during the stress response.
2. Explain how the hypothalamus brings about each facet of the stress response.

19.4 Endocrine Pancreas and Control of Fuel Metabolism

We have just discussed the metabolic changes elicited during the stress response. Now we concentrate on the metabolic patterns that occur in the absence of stress, including the hormonal factors that govern this normal metabolism.

Fuel metabolism includes anabolism, catabolism, and interconversions among energy-rich organic molecules.

The term **metabolism** refers to all the chemical reactions that occur within body cells. Those reactions involving the degradation, synthesis, and transformation of the three classes of energy-rich organic molecules—protein, carbohydrate, and fat—are collectively known as **intermediary metabolism,** or **fuel metabolism** (Table 19-2).

During the process of digestion, large nutrient molecules **(macromolecules)** are broken down into their smaller absorbable subunits as follows: Proteins are converted into amino acids, complex carbohydrates into monosaccharides (mainly glucose), and triglycerides (dietary fats) into monoglycerides and free fatty acids. These absorbable units are transferred from the digestive tract lumen into the blood, either directly or by way of the lymph (Chapter 16).

Anabolism and Catabolism Once absorbed, the organic subunits are constantly exchanged between the blood and the body cells. The chemical reactions in which these small organic molecules participate within the cells are categorized into two metabolic processes: anabolism and catabolism (Figure 19-14). **Anabolism** is the buildup or synthesis of larger organic macromolecules from small organic molecular subunits. Anabolic reactions generally require energy input in the form of adenosine triphosphate (ATP). These reactions result in either (1) the manufacture of materials needed by the cell, such as cellular structural proteins or secretory products, or (2) the storage of excess ingested nutrients not immediately needed for energy production or as cellular building blocks. Storage is in the form of glycogen (the storage form of glucose) or fat reservoirs. **Catabolism** is the breakdown, or degradation, of large, energy-rich organic molecules within cells. Catabolism encompasses two levels of breakdown: (1) hydrolysis (see pp. 31 and A-14) of large cellular organic macromolecules into their smaller subunits, similar to the process of digestion except that the reactions take place within the body cells instead of within the digestive tract lumen (for example, release of glucose by the catabolism of stored glycogen), and (2) oxidation of the smaller subunits, such as glucose, to yield energy for ATP production (see p. 37).

TABLE 19-2 Summary of Reactions in Fuel Metabolism

Metabolic Process	Reaction	Consequence
Glycogenesis	Glucose → glycogen	↓ Blood glucose
Glycogenolysis	Glycogen → glucose	↑ Blood glucose
Gluconeogenesis	Amino acids → glucose	↑ Blood glucose
Protein synthesis	Amino acids → protein	↓ Blood amino acids
Protein degradation	Protein → amino acids	↑ Blood amino acids
Fat synthesis (lipogenesis or triglyceride synthesis)	Fatty acids and glycerol → triglycerides	↓ Blood fatty acids
Fat breakdown (lipolysis or triglyceride degradation)	Triglycerides → fatty acids and glycerol	↑ Blood fatty acids

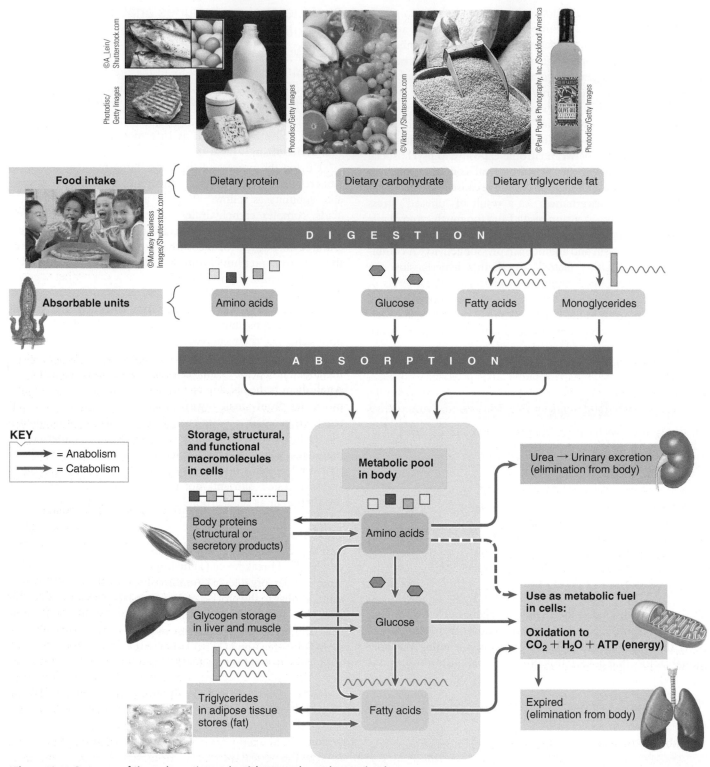

Food intake
Dietary protein · Dietary carbohydrate · Dietary triglyceride fat

D I G E S T I O N

Absorbable units
Amino acids · Glucose · Fatty acids · Monoglycerides

A B S O R P T I O N

KEY
→ = Anabolism
→ = Catabolism

Storage, structural, and functional macromolecules in cells

Body proteins (structural or secretory products)

Glycogen storage in liver and muscle

Triglycerides in adipose tissue stores (fat)

Metabolic pool in body

Amino acids

Glucose

Fatty acids

Urea → Urinary excretion (elimination from body)

Use as metabolic fuel in cells:

Oxidation to $CO_2 + H_2O$ + ATP (energy)

Expired (elimination from body)

I Figure 19-14 Summary of the major pathways involving organic nutrient molecules.

As an alternative to energy production, the smaller, multipotential organic subunits derived from intracellular hydrolysis may be released into the blood. These mobilized glucose, fatty acid, and amino acid molecules can then be used as needed for energy production or cellular synthesis elsewhere in the body.

In an adult, the rates of anabolism and catabolism are generally in balance, so the adult body remains in a dynamic steady state and appears unchanged even though the organic molecules that determine its structure and function are continuously being turned over. During growth, anabolism exceeds catabolism.

Interconversions Among Organic Molecules In addition to being able to resynthesize catabolized organic molecules back into the same type of molecules, many cells of the body, especially liver cells, can convert most types of small organic molecules into other types—as in, for example, transforming

| TABLE 19-3 Stored Metabolic Fuel in the Body

Metabolic Fuel	Circulating Form	Storage Form	Major Storage Site	Percentage of Total Body Energy Content (and Calories*)	Reservoir Capacity	Role
Carbohydrate	Glucose	Glycogen	Liver, muscle	1% (1500 calories)	Less than a day's worth of energy	First energy source; essential for the brain
Fat	Free fatty acids	Triglycerides	Adipose tissue	77% (143,000 calories)	About two months' worth of energy	Primary energy reservoir; energy source during a fast
Protein	Amino acids	Body proteins	Muscle	22% (41,000 calories)	Death results long before capacity is fully used because of structural and functional impairment	Source of glucose for the brain during a fast; last resort to meet other energy needs

*Actually refers to kilocalories; see p. 619.

amino acids into glucose or fatty acids. Because of these interconversions, adequate nourishment can be provided by a range of molecules present in different types of foods. There are limits, however. **Essential nutrients,** such as the essential amino acids and vitamins, cannot be formed in the body by conversion from another type of organic molecule.

The major fate of ingested carbohydrates and fats is catabolism to yield energy. Amino acids are predominantly used for protein synthesis but can be used to supply energy after being converted to carbohydrate or fat. Thus, all three categories of foodstuff can be used as fuel, and excesses of any foodstuff can be deposited as stored fuel, as you will see shortly.

At a superficial level, fuel metabolism appears relatively simple: The amount of nutrients in the diet must be sufficient to meet the body's needs for energy production and cellular synthesis. This apparently simple relationship is complicated, however, by two important considerations: (1) nutrients taken in at meals must be stored and then released between meals, and (2) the brain must be continuously supplied with glucose. Let us examine the implications of each.

Because food intake is intermittent, nutrients must be stored for use between meals.

Dietary fuel intake is intermittent, not continuous. As a result, excess energy must be absorbed during meals and stored for use during fasting periods between meals, when dietary sources of metabolic fuel are not available. Despite intermittent energy intake, the body cells' demand for energy is ever-present and fluctuating. That is, energy must constantly be available for cells to use as needed no matter what the status of food intake is. Stored energy fills in the gaps between meals. Energy storage takes three forms (▮Table 19-3):

■ *Excess circulating glucose* is stored in the liver and muscle as *glycogen,* a large molecule consisting of interconnected glucose molecules (see ▮Figures 2-18, p. 44, and 16-1, p. 567). About

twice as much glycogen is stored in the skeletal muscles collectively as in the liver. Because glycogen is a relatively small energy reservoir, less than a day's energy needs can be stored in this form. Once the liver and muscle glycogen stores are "filled up," additional glucose is transformed into fatty acids and glycerol, which are used to synthesize *triglycerides* (glycerol with three fatty acids attached), primarily in adipose tissue (fat).

■ *Excess circulating fatty acids* derived from dietary intake also become incorporated into triglycerides.

■ *Excess circulating amino acids* not needed for protein synthesis are not stored as extra protein but are converted to glucose and fatty acids, which ultimately end up being stored as triglycerides.

Thus, the major site of energy storage for excess nutrients of all three classes is adipose tissue. Normally, enough triglyceride is stored to provide energy for about 2 months, with more stored in an overweight person. Consequently, during any prolonged period of fasting, the fatty acids released from triglyceride catabolism are the primary source of energy for most tissues. Catabolism of stored triglycerides yields 90% fatty acids and 10% glycerol by weight. Glycerol (but not fatty acids) can be converted to glucose by the liver and contributes in a small way to maintaining blood glucose during a fast.

As a third energy reservoir, a substantial amount of energy is stored as *structural protein,* primarily in muscle, the most abundant protein mass in the body. Protein is not the first choice to tap as an energy source, however, because it serves other essential functions; in contrast, the glycogen and triglyceride reservoirs are solely energy depots.

The brain must be continuously supplied with glucose.

The second factor complicating fuel metabolism (besides intermittent nutrient intake and the resultant necessity of storing nutrients) is that the brain normally depends on delivery of adequate blood glucose as its sole source of energy. Consequently, the blood glucose concentration must be maintained above a critical level. The blood glucose concentration is typically 100 mg of glucose for every 100 mL of plasma and is normally kept within the narrow limits of 70 to 110 mg per 100 mL.[1] Liver glycogen is an important reservoir for maintaining blood glucose levels during a short fast. However, liver glycogen is depleted relatively rapidly, so during a longer fast other mechanisms must meet the energy requirements of the glucose-dependent brain. First, when no new dietary glucose is entering the blood, tissues not obligated to use glucose shift their metabolic gears to burn fatty acids instead, sparing glucose for the brain. Fatty acids are made available by catabolism of triglyceride stores as an alternative energy source for tissues that are not glucose dependent. Second, amino acids can be converted to glucose by gluconeogenesis, whereas fatty acids cannot. Thus, once glycogen stores are depleted despite glucose sparing, new glucose supplies for the brain are provided by the catabolism of body proteins and conversion of the freed amino acids into glucose.

Metabolic fuels are stored during the absorptive state and mobilized during the postabsorptive state.

The preceding discussion should make clear that how the body deals with organic molecules depends on the body's metabolic state. The two functional metabolic states—the *absorptive state* and the *postabsorptive state*—are related to eating and fasting cycles, respectively (❙Table 19-4).

Absorptive State After a meal, ingested nutrients are being absorbed and entering the blood during the **absorptive**, or **fed**, **state**. During this time, glucose is plentiful and is the major energy source. Very little absorbed fat and amino acids are used for energy during the absorptive state because most cells preferentially use glucose when it is available. Extra nutrients not immediately used for energy or structural repairs are channeled into storage as glycogen or triglycerides.

Postabsorptive State The average meal is completely absorbed in about 4 hours. Therefore, on a typical three-meals-a-day diet, no nutrients are being absorbed from the digestive tract during late morning and late afternoon and throughout the night. These times constitute the **postabsorptive**, or **fasting**, **state**. During this state, endogenous energy stores are mobilized to provide energy, whereas gluconeogenesis and glucose sparing maintain the blood glucose at an adequate level to nourish the brain. Synthesis of protein and fat is curtailed. Instead, stores of these organic molecules are catabolized for glucose formation and energy production, respectively. Carbohydrate synthesis does occur through gluconeogenesis, but the use of glucose for energy is greatly reduced.

Note that the blood concentration of nutrients does not fluctuate markedly between the absorptive and postabsorptive states. During the absorptive state, the glut of absorbed nutrients is swiftly removed from the blood and placed into storage; during the postabsorptive state, these stores are catabolized to maintain the blood concentrations at levels necessary to fill tissue energy demands.

Roles of Key Tissues in Metabolic States During these alternating metabolic states, various tissues play different roles, as summarized here:

■ The *liver* plays the primary role in maintaining normal blood glucose levels. It stores glycogen when excess glucose is available, releases glucose into the blood when needed, and is the principal site for metabolic interconversions such as gluconeogenesis.

■ *Adipose tissue* serves as the primary energy storage site and is important in regulating fatty acid levels in the blood.

■ *Muscle* is the primary site of amino acid storage and is the major energy user.

---

[1]Blood glucose concentration is sometimes given in terms of molarity, with normal blood glucose concentration hovering around 5 mM (see p. A-7).

❙TABLE 19-4 Comparison of Absorptive and Postabsorptive States

Metabolic Fuel	Absorptive State	Postabsorptive State
Carbohydrate	Glucose providing major energy source	Glycogen degradation and depletion
	Glycogen synthesis and storage	Glucose sparing to conserve glucose for the brain
	Excess converted and stored as triglyceride fat	Production of new glucose through gluconeogenesis
Fat	Triglyceride synthesis and storage	Triglyceride catabolism
		Fatty acids providing major energy source for non-glucose–dependent tissues
Protein	Protein synthesis	Protein catabolism
	Excess converted and stored as triglyceride fat	Amino acids used for gluconeogenesis

- The *brain* normally can use only glucose as an energy source, yet it does not store glycogen, making it mandatory that adequate blood glucose levels be maintained.

Lesser energy sources are tapped as needed.

Several other organic intermediates play a lesser role as energy sources—namely, glycerol, lactate, and ketone bodies.

- As mentioned earlier, glycerol derived from triglyceride hydrolysis can be converted to glucose by the liver.
- Similarly, lactate, which is produced by the incomplete catabolism of glucose via glycolysis in muscle (see p. 272), can also be converted to glucose in the liver.
- **Ketone bodies** (namely *acetone, acetoacidic acid,* and *β-hydroxybutyric acid*) are a group of compounds produced by the liver during glucose sparing. Unlike other tissues, when the liver uses fatty acids as an energy source, it oxidizes them only to acetyl coenzyme A (acetyl CoA), which it is unable to process through the citric acid cycle for further energy extraction (see p. 35). Thus, the liver does not degrade fatty acids all the way to CO_2 and H_2O for maximum energy release. Instead, it partially extracts the available energy and converts the remaining energy-bearing acetyl CoA molecules into ketone bodies, which it releases into the blood. Ketone bodies serve as an alternative energy source for tissues capable of oxidizing them further by means of the citric acid cycle.

During long-term starvation, the brain starts using ketones instead of glucose as a major energy source. Because death resulting from starvation is usually the result of protein wasting rather than hypoglycemia (low blood glucose), prolonged survival without any caloric intake requires that gluconeogenesis be kept to a minimum as long as the energy needs of the brain are not compromised. A sizable portion of cell protein can be catabolized without serious cellular malfunction, but a point is finally reached at which a cannibalized cell can no longer function adequately. To ward off the fatal point of failure as long as possible during prolonged starvation, the brain starts using ketones as a major energy source, correspondingly decreasing its use of glucose. Use by the brain of this fatty acid "table scrap" left over from the liver's "meal" limits the necessity of mobilizing body proteins for glucose production to nourish the brain. Both the major metabolic adaptations to prolonged starvation—a decrease in protein catabolism and use of ketones by the brain—are attributable to the high levels of ketones in the blood at the time. The brain uses ketones only when blood ketone level is high. The high blood levels of ketones also directly inhibit protein degradation in muscle. Thus, ketones spare body proteins while satisfying the brain's energy needs.

The pancreatic hormones, insulin and glucagon, are most important in regulating fuel metabolism.

How does the body "know" when to shift its metabolic gears from a system of net anabolism and nutrient storage to one of net catabolism and glucose sparing? The flow of organic nutrients along metabolic pathways is influenced by a variety of hormones, including insulin, glucagon, epinephrine, cortisol, and GH. Under most circumstances, the pancreatic hormones, insulin and glucagon, are the dominant hormonal regulators that shift the metabolic pathways from net anabolism to net catabolism and glucose sparing, or vice versa, depending on whether the body is feasting or fasting, respectively.

Islets of Langerhans The **pancreas** is an organ composed of both exocrine and endocrine tissues. The exocrine portion secretes a watery, alkaline solution and digestive enzymes through the pancreatic duct into the digestive tract lumen (see p. 590). Scattered throughout the pancreas between the exocrine cells are about a million clusters, or "islands," of endocrine cells known as the **islets of Langerhans** (Figure 19-15a). The islets make up about 1% to 2% of the total pancreatic mass. The pancreatic endocrine cell types, in decreasing order of abundance, are:

- **β (beta) cells** (constituting about 60% of islet cells), which simultaneously secrete *insulin*, the most important pancreatic hormone, and *amylin*, the newest discovered pancreatic hormone. β cells secrete 100 times more insulin than amylin.
- **α (alpha) cells** (about 25% of islet cells), which produce the hormone *glucagon*
- **delta, or D, cells** (about 10% of islet cells), the pancreatic site of *somatostatin* synthesis
- **gamma, or F, cells** (about 4% of islet cells), which secrete *pancreatic polypeptide,* a hormone that plays a possible role in reducing appetite and food intake, is poorly understood, and will not be discussed any further.
- **epsilon cells** (<1% of islet cells), which are newly found cells that release *ghrelin*, the "hunger hormone." Most ghrelin is secreted by the stomach before meals (see p. 623).

The β cells are concentrated centrally in the islets, with the other cells clustered around the periphery (Figure 19-15b). We briefly highlight somatostatin now, and then we focus on insulin (and accompanying amylin) and glucagon, the most important hormones in the regulation of fuel metabolism.

Somatostatin Acting as a hormone, pancreatic **somatostatin** inhibits the digestive system in a variety of ways, the overall effects of which are to inhibit digestion of nutrients and to decrease nutrient absorption. Somatostatin is released from the pancreatic D cells in direct response to an increase in blood glucose and blood amino acids during absorption of a meal. By exerting its inhibitory effects, pancreatic somatostatin acts in negative-feedback fashion to put the brakes on the rate at which the meal is being digested and absorbed, thereby preventing excessive plasma levels of nutrients. Pancreatic somatostatin also acts as a paracrine in regulating pancreatic hormone secretion. The local presence of somatostatin decreases the secretion of insulin, glucagon, and somatostatin itself, but the physiologic importance of such paracrine function has not been determined.

Cells lining the digestive tract also produce somatostatin, where it acts locally as a paracrine to inhibit most digestive processes (see p. 585). Furthermore, somatostatin (alias GHIH)

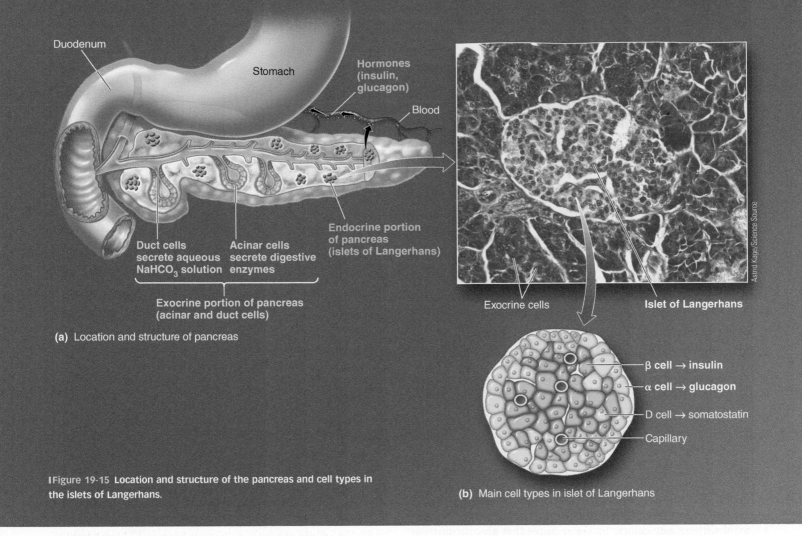

(a) Location and structure of pancreas

Duodenum

Stomach

Hormones (insulin, glucagon)

Blood

Duct cells secrete aqueous NaHCO₃ solution

Acinar cells secrete digestive enzymes

Endocrine portion of pancreas (islets of Langerhans)

Exocrine portion of pancreas (acinar and duct cells)

Exocrine cells

Islet of Langerhans

β cell → insulin

α cell → glucagon

D cell → somatostatin

Capillary

Astrid Kage/Science Source

❙Figure 19-15 **Location and structure of the pancreas and cell types in the islets of Langerhans.**

(b) Main cell types in islet of Langerhans

is produced by the hypothalamus, where it inhibits secretion of GH and TSH (see p. 650).

We next consider insulin and then glucagon, followed by a discussion of how insulin and glucagon function as an endocrine unit to shift metabolic gears between the absorptive and the postabsorptive states.

Insulin lowers blood glucose, fatty acid, and amino acid levels and promotes their storage.

Insulin has important effects on carbohydrate, fat, and protein metabolism. It lowers the blood levels of glucose, fatty acids, and amino acids and promotes their storage. As these nutrient molecules enter the blood during the absorptive state, insulin promotes their cellular uptake and conversion into glycogen, triglycerides, and protein, respectively. Insulin exerts its many effects by altering either transport of specific blood-borne nutrients into cells or activity of the enzymes involved in specific metabolic pathways. To accomplish its effects, in some instances insulin increases the activity of an enzyme, for example *glycogen synthase,* the enzyme that synthesizes glycogen from glucose molecules. In other cases, however, insulin decreases the activity of an enzyme, for example by inhibiting *hormone-sensitive lipase,* the enzyme that catalyzes the breakdown of stored triglycerides back to free fatty acids and glycerol.

Actions on Carbohydrates Maintaining blood glucose homeostasis is a particularly important function of the pancreas. The balance among the following processes determines circulating glucose concentrations (❙Figure 19-16): glucose absorption from the digestive tract, transport of glucose into cells, hepatic glucose production, and (abnormally) urinary excretion of glucose. Among these factors, only glucose transport into cells and hepatic glucose production are subject to control.

Insulin exerts four effects that lower blood glucose levels and promote carbohydrate storage:

1. Insulin facilitates glucose transport into most cells. (The mechanism of this increased glucose uptake is explained after insulin's other blood-glucose lowering effects are listed.)

2. Insulin stimulates **glycogenesis,** the production of glycogen from glucose, in both skeletal muscle and the liver.

3. Insulin inhibits glycogenolysis, the breakdown of glycogen into glucose.

4. Insulin inhibits gluconeogenesis, the conversion of amino acids into glucose in the liver. Insulin does so by decreasing the amount of amino acids in the blood available to the liver for gluconeogenesis and by inhibiting the hepatic enzymes required for converting amino acids into glucose.

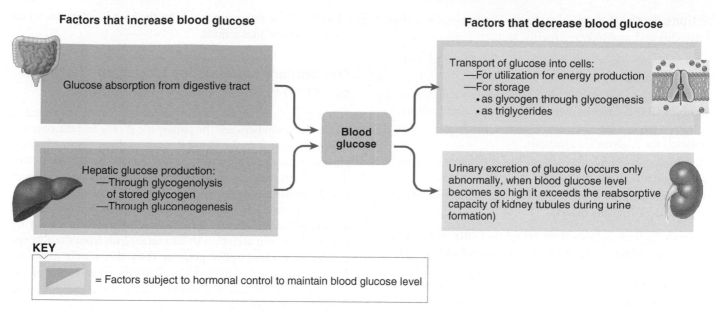

Factors that increase blood glucose

Glucose absorption from digestive tract

Hepatic glucose production:
—Through glycogenolysis of stored glycogen
—Through gluconeogenesis

Blood glucose

Factors that decrease blood glucose

Transport of glucose into cells:
—For utilization for energy production
—For storage
  • as glycogen through glycogenesis
  • as triglycerides

Urinary excretion of glucose (occurs only abnormally, when blood glucose level becomes so high it exceeds the reabsorptive capacity of kidney tubules during urine formation)

KEY

= Factors subject to hormonal control to maintain blood glucose level

IFigure 19-16 **Factors affecting blood glucose concentration.**
FIGURE FOCUS: *Name the two factors that are subject to hormonal control to maintain blood glucose.*

Thus, insulin decreases the concentration of blood glucose by promoting the cells' uptake of glucose from the blood for use and storage while simultaneously blocking the two mechanisms by which the liver releases glucose into the blood (glycogenolysis and gluconeogenesis). Insulin is the only hormone capable of lowering blood glucose.

Insulin promotes uptake of glucose from the blood by most cells through recruitment of glucose transporters. A **glucose transporter (GLUT)** is a plasma membrane carrier that accomplishes passive facilitated diffusion of glucose across the plasma membrane (see p. 72). Once GLUT facilitates entry of glucose into a cell down this nutrient's concentration gradient, an enzyme within the cell immediately phosphorylates glucose to **glucose-6-phosphate**. Glucose-6-phosphate has no means out of the cell, unlike "plain" glucose, which could exit through the bidirectional glucose transporter. Phosphorylation of glucose as it enters the cell not only "traps" the glucose inside the cell but also keeps the intracellular concentration of plain glucose low so that a gradient favoring the facilitated diffusion of glucose into the cell is maintained.

Fourteen forms of glucose transporters have been identified, named in the order they were discovered—GLUT-1, GLUT-2, and so on. Each member of the GLUT family performs slightly different functions. For example, *GLUT-1* transports glucose across the blood–brain barrier, *GLUT-2* transfers into the adjacent bloodstream the glucose that has entered the kidney and intestinal cells by means of the sodium and glucose cotransporter (SGLT; see p. 75), and *GLUT-3* is the main transporter of glucose into neurons. The transporter responsible for glucose uptake by most body cells is *GLUT-4,* which operates only at the bidding of insulin. Glucose molecules cannot readily penetrate most cell membranes in the absence of insulin, making most tissues highly dependent on insulin for uptake of glucose from the blood and for its subsequent use. GLUT-4 is especially abundant in the tissues that account for the bulk of glucose uptake from the blood during the absorptive state, namely, resting skeletal muscle and adipose tissue cells.

GLUT-4 is the only type of transporter that responds to insulin. Unlike the other types of GLUT molecules, which are always present in the plasma membranes at the sites where they perform their functions, GLUT-4 is not present in the plasma membrane in the absence of insulin. Insulin promotes glucose uptake by **transporter recruitment.** Insulin-dependent cells maintain a pool of intracellular vesicles containing GLUT-4. When insulin binds with its receptor (a receptor that acts as a tyrosine kinase enzyme; see p. 116) on the surface membrane of the target cell, the subsequent signaling pathway induces these vesicles to move to the plasma membrane and fuse with it, thus inserting GLUT-4 molecules into the plasma membrane. In this way, increased insulin secretion promotes a rapid 10- to 30-fold increase in glucose uptake by insulin-dependent cells. When insulin secretion decreases, these GLUTs are retrieved from the membrane by endocytosis and returned to the intracellular pool.

Several tissues do not depend on insulin for their glucose uptake—namely, the brain, working muscles, and liver. The brain, which requires a constant supply of glucose for its minute-to-minute energy needs, is freely permeable to glucose at all times by means of GLUT-1 and GLUT-3 molecules. Skeletal muscle cells do not depend on insulin for their glucose uptake during exercise, even though they are dependent at rest. Muscle contraction triggers insertion of GLUT-4 into the plasma membranes of exercising muscle cells in the absence of insulin. This fact is important in managing diabetes mellitus (insulin deficiency), as described later. The liver also does not depend on insulin for glucose uptake because it does not use GLUT-4. However, insulin does enhance the metabolism of glucose by the liver by stimulating the first step in glucose metabolism, the phosphorylation of glucose to form glucose-6-phosphate.

Actions on Fat Insulin exerts multiple effects to lower blood fatty acids and promote triglyceride storage:

1. It enhances entry of fatty acids from the blood into adipose tissue cells.

2. It increases transport of glucose into adipose tissue cells by means of GLUT-4 recruitment. Glucose serves as a precursor for formation of fatty acids and glycerol, which are the raw materials for triglyceride synthesis.

3. It promotes chemical reactions that ultimately use fatty acids and glucose derivatives for triglyceride synthesis.

4. It inhibits lipolysis, reducing the release of fatty acids from adipose tissue into the blood.

Collectively, these actions favor removal of fatty acids and glucose from the blood and promote their storage as triglycerides.

Actions on Protein Insulin lowers blood amino acid levels and enhances protein synthesis through several effects:

1. It promotes active transport of amino acids from the blood into muscles and other tissues. This effect decreases the blood amino acid level and provides the building blocks for protein synthesis within the cells.

2. It increases the rate of amino acid incorporation into protein by stimulating the cells' protein-synthesizing machinery.

3. It inhibits protein degradation.

The collective result of these actions is a protein anabolic effect. For this reason, insulin is essential for normal growth.

Summary of Insulin's Actions In short, insulin primarily exerts its effects by acting on the liver, adipose tissue, and non-working skeletal muscle. It stimulates biosynthetic pathways that lead to increased glucose use, increased carbohydrate and fat storage, and increased protein synthesis. In so doing, this hormone lowers the blood glucose, fatty acid, and amino acid levels. This metabolic pattern is characteristic of the absorptive state. Indeed, insulin secretion rises during this state and shifts metabolic pathways to net anabolism.

When insulin secretion is low, the opposite effects occur. The rate of glucose entry into cells is reduced, and net catabolism occurs rather than net synthesis of glycogen, triglycerides, and protein. This pattern is reminiscent of the postabsorptive state; indeed, insulin secretion is reduced during the postabsorptive state. However, the other major pancreatic hormone, glucagon, also plays an important role in shifting from absorptive to postabsorptive metabolic patterns, as described later.

Role of Amylin Amylin is cosecreted with insulin from pancreatic β cells at a ratio of 1:100. Unlike insulin, amylin does not act on peripheral tissues but instead acts in the central nervous system to enhance satiety (feeling of being full), delay gastric emptying (thereby slowing digestion of food), and suppress secretion of glucagon (a hormone that raises blood glucose). Thus amylin acts as a partner hormone to insulin to lower blood glucose. In complementary fashion, amylin reduces glucose influx into the blood from the digestive tract after a meal while insulin removes glucose from the blood and promotes its use and storage after a meal.

The primary stimulus for increased insulin secretion is an increase in blood glucose.

The primary control of insulin secretion is a direct negative-feedback system between the pancreatic β cells and the concentration of glucose in the blood flowing to them. Elevated blood glucose, such as during absorption of a meal, directly stimulates the β cells to secrete insulin. Increased insulin, in turn, reduces blood glucose to normal and promotes use and storage of this nutrient. Conversely, a fall in blood glucose below normal, such as during fasting, directly inhibits insulin secretion. Lowering the rate of insulin secretion shifts metabolism from the absorptive to the postabsorptive pattern. Thus, this simple negative-feedback system can maintain a relatively constant supply of glucose to the tissues without requiring the participation of nerves or other hormones.

Glucose stimulates insulin secretion by means of an **excitation–secretion coupling** process. That is, glucose initiates a chain of events that changes the β cell's membrane potential, leading to secretion of insulin. This is one of the few known examples where cells other than nerve or muscle cells undergo functionally related changes in membrane potential. Specifically, glucose enters the β cell by means of GLUT-2 (❙ Figure 19-17, step ❶). Once inside, glucose is immediately phosphorylated to glucose-6-phosphate (step ❷), which is oxidized by the β cell to yield ATP (step ❸). A β cell has two types of channels: an **ATP-sensitive K⁺ channel,** which is a leak channel that remains open unless ATP binds to it, and a **voltage-gated Ca²⁺ channel,** which is closed at resting potential. The ATP-sensitive K⁺ channel closes when ATP generated from glucose-6-phosphate binds to it (step ❹). The resultant decrease in K⁺ permeability leads to depolarization of the β cell (because of less outward movement of positively charged K⁺) (step ❺). This depolarization causes the voltage-gated Ca²⁺ channels to open (step ❻). The subsequent Ca²⁺ entry (step ❼) triggers exocytosis of secretory vesicles containing insulin (step ❽), resulting in insulin secretion (step ❾).

In addition to blood glucose concentration, which is the major controlling factor, other inputs are involved in regulating insulin secretion, as follows (❙ Figure 19-18):

■ An elevated blood amino acid level, such as after a high-protein meal, directly stimulates the β cells to increase insulin secretion. In negative-feedback fashion, increased insulin enhances the entry of these amino acids into the cells, lowering the blood amino acid level while promoting protein synthesis. Amino acids increase insulin secretion in the same way as glucose does, by generating ATP that leads to excitation–secretion coupling.

■ Gastrointestinal hormones secreted by the digestive tract in response to the presence of food, especially *glucose-dependent insulinotropic peptide (GIP)* (see p. 615) and *glucagon-like peptide 1 (GLP-1)* (see p. 624), stimulate pancreatic insulin secretion, in addition to having direct regulatory effects on the digestive system. Through this control, insulin secretion is

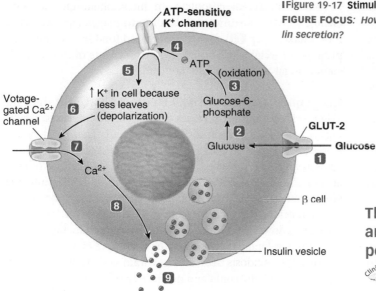

1. Glucose enters β cell by facilitated diffusion via GLUT-2.

2. Glucose is phosphorylated to glucose-6-phosphate.

3. Oxidation of glucose-6-phosphate generates ATP.

4. ATP acts on ATP-sensitive K+ channel, closing it.

5. Reduced exit of K+ depolarizes membrane.

6. Depolarization opens voltage-gated Ca²⁺ channels.

7. Ca²⁺ enters β cell.

8. Ca²⁺ triggers exocytosis of insulin vesicles.

9. Insulin is secreted.

cAMP. The fall in insulin level allows the blood glucose level to rise, an appropriate response to the circumstances under which generalized sympathetic activation occurs—namely, stress (fight-or-flight) and exercise. In both situations, extra fuel is needed for increased muscle activity.

The symptoms of diabetes mellitus are characteristic of an exaggerated postabsorptive state.

Clinical Note **Diabetes mellitus** is by far the most common of all endocrine disorders. The acute symptoms of diabetes mellitus are attributable to inadequate insulin action. Because insulin is the only hormone capable of lowering blood glucose levels, one of the most prominent features of diabetes mellitus is elevated blood glucose levels, or *hyperglycemia*. *Diabetes* literally means "siphon" or "running through," a reference to the large urine volume accompanying this condition. A large urine volume occurs in both diabetes mellitus (a result of insulin insufficiency) and diabetes insipidus (a result of vasopressin deficiency; see p. 541). *Mellitus* means "sweet"; *insipidus* means "tasteless." The urine of patients with diabetes mellitus acquires

increased in "feedforward," or anticipatory, fashion even before nutrient absorption increases the blood concentration of glucose and amino acids. Hormones released from the digestive tract that "notify" the pancreatic β cell of the impending rise in blood nutrients (primarily blood glucose) are termed **incretins**. Incretins increase insulin secretion by increasing cAMP, which enhances Ca²⁺-induced release of insulin.

■ The autonomic nervous system also directly influences insulin secretion. The islets are richly innervated by both parasympathetic (vagal) and sympathetic nerve fibers. The increase in parasympathetic activity that occurs in response to food in the digestive tract stimulates insulin release, with the parasympathetic neurotransmitter acetylcholine acting through the IP₃–Ca²⁺ pathway. This, too, is a feedforward response in anticipation of nutrient absorption. In contrast, sympathetic stimulation and the concurrent increase in epinephrine both inhibit insulin secretion by decreasing

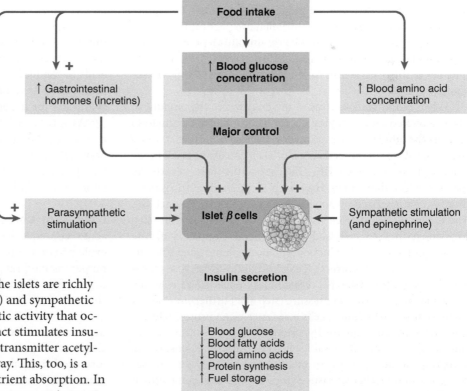

Figure 19-18 **Factors controlling insulin secretion.**

its sweetness from excess blood glucose that spills into the urine, whereas the urine of patients with diabetes insipidus contains no sugar, so it is tasteless. (Aren't you glad you were not a health professional at the time when these two conditions were distinguished on the basis of the taste of the urine?)

Diabetes mellitus has two major variants, differing in the capacity for pancreatic insulin secretion: *Type 1 diabetes*, characterized by a lack of insulin secretion, and *Type 2 diabetes,* characterized by normal or even increased insulin secretion but reduced sensitivity of insulin's target cells to this hormone. (For a further discussion of the distinguishing features of these two types of diabetes mellitus, see the boxed feature on pp. 696–697, Concepts, Challenges, and Controversies.)

The acute consequences of diabetes mellitus can be grouped according to the effects of inadequate insulin action on carbohydrate, fat, and protein metabolism (Figure 19-19). The figure may look overwhelming, but the numbers, which correspond to the numbers in the following discussion, help you work your way through this complex disease step by step.

Consequences Related to Effects on Carbohydrate Metabolism Because the postabsorptive metabolic pattern is induced by low insulin activity, the changes that occur in diabetes mellitus are an exaggeration of this state, with the exception of hyperglycemia. In the usual fasting state, the blood glucose level is slightly below normal. Hyperglycemia, the hallmark of diabetes mellitus, arises from reduced glucose uptake by cells, coupled with increased output of glucose from the liver (Figure 19-19, step **1**). As the glucose-yielding processes of glycogenolysis and gluconeogenesis proceed unchecked in the absence of insulin, hepatic output of glucose increases. Because many of the body's cells cannot use glucose without the help of insulin, an ironic extracellular glucose excess occurs coincident with an intracellular glucose deficiency—"starvation in the midst of plenty." Even though the non–insulin-dependent brain is adequately nourished during diabetes mellitus, further consequences of the disease lead to brain dysfunction, as you will see shortly.

When blood glucose rises to the level where the amount of glucose filtered by the kidney nephrons during urine formation exceeds the tubular cells' capacity for reabsorption (that is, when the T_m for glucose is exceeded; see p. 511), glucose appears in the urine *(glucosuria)* (step **2**). Glucose in the urine exerts an osmotic effect that draws H_2O with it, producing an osmotic diuresis characterized by *polyuria* (frequent urination) (step **3**). The excess fluid lost from the body leads to dehydration (step **4**), which in turn can lead to peripheral circulatory failure because of the marked reduction in blood volume (step **5**). Circulatory failure, if uncorrected, can lead to death because of low cerebral blood flow (step **6**) or secondary renal failure resulting from inadequate filtration pressure (step **7**). Furthermore, as the body becomes dehydrated, cells lose water by an osmotic shift of water from the cells into the hypertonic (too concentrated) extracellular fluid (step **8**). Brain cells are especially sensitive to shrinking, so nervous system malfunction ensues (step **9**) (see p. 542). Another characteristic symptom of diabetes mellitus is *polydipsia* (excessive thirst) (step **10**), which is actually a compensatory mechanism to counteract the dehydration.

The story is not complete. In intracellular glucose deficiency, appetite is stimulated, leading to *polyphagia* (excessive food intake) (step **11**). Despite increased food intake, however, progressive weight loss occurs from the effects of insulin deficiency on fat and protein metabolism.

Consequences Related to Effects on Fat Metabolism Triglyceride synthesis decreases while lipolysis increases, resulting in large-scale mobilization of fatty acids from triglyceride stores (step **12**). The cells largely use the increased blood fatty acids as an alternative energy source. Increased liver use of fatty acids results in the release of excessive ketone bodies into the blood, causing *ketosis* (step **13**). Ketone bodies include several different acids, such as acetoacetic acid, that result from incomplete breakdown of fat during hepatic energy production. Therefore, this developing ketosis leads to progressive metabolic acidosis (step **14**). Acidosis depresses the brain and, if severe enough, can lead to diabetic coma and death (step **15**).

A compensatory measure for metabolic acidosis is increased ventilation to blow off extra, acid-forming CO_2 (step **16**). Exhalation of one of the ketone bodies, acetone, causes a "fruity" breath odor that smells like a combination of Juicy Fruit gum and nail polish remover. Sometimes because of this odor, passersby unfortunately mistake a patient collapsed in a diabetic coma for a "wino" passed out in a state of drunkenness. (This situation illustrates the merits of medical alert identification tags.) People with Type 1 diabetes are more prone to develop ketosis than are Type 2 diabetics.

Consequences Related to Effects on Protein Metabolism The effects of a lack of insulin on protein metabolism result in a net shift toward protein catabolism. The net breakdown of muscle proteins leads to muscle wasting and weakness, as well as weight loss (step **17**) and, in child diabetics, a reduction in overall growth. Reduced amino acid uptake coupled with increased protein degradation results in excess amino acids in the blood (step **18**). The increased circulating amino acids can be used for additional gluconeogenesis, which further aggravates the hyperglycemia (step **19**).

As you can readily appreciate from this overview, diabetes mellitus is a complicated disease that can disturb not only carbohydrate, fat, and protein metabolism but also fluid and acid–base balance. Moreover it can have repercussions on the circulatory system, kidneys, respiratory system, and nervous system.

Long-Term Complications In addition to these potential acute consequences of untreated diabetes, which can be explained on the basis of insulin's short-term metabolic effects, numerous long-range complications of this disease frequently occur after 15 to 20 years despite treatment to prevent the short-term effects. These chronic complications, which account for the shorter life expectancy of diabetics, primarily involve degenerative disorders of the blood vessels and nervous system. Cardiovascular lesions are the most common cause of premature death in diabetics. Heart disease and strokes occur with greater incidence than in nondiabetics. Because vascular lesions often develop in the kidneys and retinas of the eyes, diabetes is the leading cause of both kidney failure and blindness in the

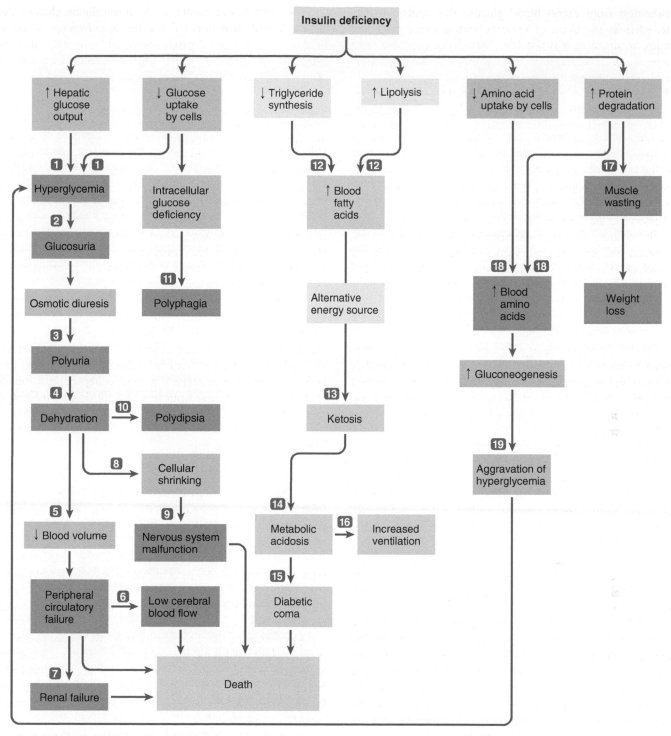

IFigure 19-19 Acute effects of diabetes mellitus. The acute consequences of diabetes mellitus can be grouped according to the effects of inadequate insulin action on carbohydrate, fat, and protein metabolism. These effects ultimately cause death through a variety of pathways. See p. 694 for an explanation of the numbers.

United States. Impaired delivery of blood to the extremities may cause these tissues to become gangrenous, and toes or even whole limbs may have to be amputated. In addition to circulatory problems, degenerative lesions in nerves lead to multiple neuropathies that result in dysfunction of the brain, spinal cord, and peripheral nerves. The latter is most often characterized by pain, numbness, and tingling, especially in the extremities.

Regular exposure of tissues to excess blood glucose over a prolonged time leads to tissue alterations responsible for the development of these long-range vascular and neural degenerative complications. Thus, the best management for diabetes mellitus is to continuously keep blood glucose levels within normal limits to diminish the incidence of these chronic abnormalities. However, the blood glucose levels of diabetic patients on traditional therapy typically fluctuate over a broader range than normal, exposing their tissues to a moderately elevated blood glucose level during a portion of each day. Fortunately, recent advances in understanding and learning how to

Diabetics and Insulin: Some Have It and Some Don't

THERE ARE TWO DISTINCT TYPES of diabetes mellitus. **Type 1 (insulin-dependent**, or **juvenile-onset) diabetes mellitus (T1DM)** is characterized by a lack of insulin. Because their pancreatic β cells do not secrete insulin, Type 1 diabetics require exogenous insulin for survival. In **Type 2 (non–insulin-dependent**, or **maturity-onset) diabetes mellitus (T2DM)**, insulin secretion may be normal or even increased, but insulin's target cells are less sensitive than normal to this hormone. That is, Type 2 diabetics are *insulin resistant*. Ninety percent of diabetics have the Type 2 form. Although either type can first be manifested at any age, T1DM occurs more in children, whereas T2DM more generally arises in adulthood, hence the age-related designations. Diabetes of both types currently affects more than 25 million people in the United States, costing this country an estimated $245 billion annually in health-care expenses and accounting for 10% of the health-care dollars spent. The U.S. diabetes-related death rate has increased by 30% since 1980, largely because the incidence of the disease has been rising.

Underlying Defect in Type 1 Diabetes

Type 1 diabetes mellitus is an autoimmune process involving the erroneous, selective destruction of pancreatic β cells by inappropriately activated T lymphocytes (see p. 432). The precise cause of this self-attack is unclear. Some have a genetic susceptibility to acquiring T1DM. Environmental triggers also appear to play a role, but investigators have not been able to definitively pin down any culprits.

Underlying Defect in Type 2 Diabetes

Various genetic and lifestyle factors appear important in the development of T2DM. Obesity is the biggest risk factor; 90% of Type 2 diabetics are obese.

Many Type 2 diabetics have *metabolic syndrome* as a forerunner of diabetes. **Metabolic syndrome** encompasses a cluster of features that predispose the person to developing T2DM and atherosclerosis (see p. 327). These features include obesity, large waist circumference ("apple" shapes; see p. 625), high triglyceride levels, low HDL (the "good" cholesterol; see p. 328), high blood glucose, and high blood pressure. An estimated 20% of the U.S. population has metabolic syndrome (up to 45% for those older than age 50).

The ultimate cause of T2DM remains elusive despite intense investigation, but researchers have identified a number of possible links between obesity and reduced insulin sensitivity. Circulating adipokines (hormones secreted by adipose cells) modulate the responsiveness of skeletal muscle and liver to insulin. For example, adipose tissue, especially the troublesome, inflamed visceral fat (see p. 623), secretes the hormone *resistin,* which promotes insulin resistance by interfering with insulin action. Resistin production increases in obesity. By contrast, the adipokine *adiponectin* increases insulin sensitiv-

ity by enhancing insulin's effects, but its production is decreased in obesity. Compounding the problem, the abundant macrophages attracted to inflamed fat secrete inflammatory signals that amplify the inflammation. Furthermore, excess free fatty acids released from overloaded adipose tissue can abnormally accumulate in muscle and interfere with insulin action in muscle. Excess free fatty acids deposited in the liver can contribute to hyperglycemia by stimulating inappropriate hepatic glucose production. Also, excess free fatty acids can indirectly trigger apoptosis of β cells. High levels of glucose and free fatty acids can also promote inflammation and progressive failure of β cells in pancreatic islets by triggering local production of IL-1β. And this is only part of the tangled web of mechanisms involved in T2DM that investigators are trying to unravel.

Early in the development of the disease, the resulting decrease in sensitivity to insulin is overcome by secretion of additional insulin. However, the sustained overtaxing of the genetically weak β cells eventually exceeds their reserve secretory capacity. Even though insulin secretion may be normal or even elevated, symptoms of insulin insufficiency develop because the amount of insulin is still inadequate to prevent hyperglycemia.

Treatment of Diabetes

The treatment for T1DM is a controlled balance of regular insulin injections timed around meals, management of the amounts and types of food consumed, and exercise. Insulin is injected because if it were swallowed, protein-digesting enzymes in the stomach and small intestine would digest this peptide hormone. Exercise is useful in managing both types of diabetes because working muscles are not insulin dependent. Exercising muscles take up some of the excess glucose in the blood, reducing the overall need for insulin.

Whereas Type 1 diabetics are permanently insulin dependent, dietary control and weight reduction may be all that is necessary to completely reverse the symptoms in Type 2 diabetics. Seven classes of blood-glucose lowering oral medications are currently available for use if needed for treating T2DM in conjunction with a dietary and exercise regime. These pills help the patient's body use its own insulin more effectively, each by a different mechanism, as follows:

1. By suppressing liver output of glucose (*metformin*; for example, Glucophage). The American Diabetes Association recommends metformin as the firstline therapy.

2. By stimulating the β cells to secrete more insulin than they do on their own (*sulfonylureas*; for example, Glucotrol)

3. By blocking enzymes that digest complex carbohydrates, thus slowing glucose absorption into the blood from the digestive tract and blunting the surge of glucose immediately after a meal (*alpha-glycosidase inhibitors*; for example, Precose)

4. By making muscle and fat cells more receptive to insulin (*thiazolidinediones*; for example, Avandia)

5. By mimicking naturally occurring incretins, which are hormones released by the gut in response to food that act in feedforward fashion on the endocrine pancreas to reduce the anticipated rise in blood glucose (*incretin mimetics*; for example, Byetta, which mimics the gut-released hormone glucagon-like peptide 1 [GLP-1; see p. 624]). Like GLP-1, Byetta, which must be injected, stimulates insulin secretion when blood glucose is high but not when glucose is in the normal range. Drugs that mimic the action of amylin, which is produced by β cells rather than intestinal cells, act in a manner similar to incretin mimetics (injectable *amylin analogs*, such as Symlin).

6. By increasing endogenous GLP-1 levels (*dipeptidyl peptidase-4, or DPP-4, inhibitors*; for example, Januvia). This class of drugs increases endogenous GLP-1 levels by blocking DPP-4, an enzyme that breaks down GLP-1, thus prolonging action of this incretin. Prolonged activity of GLP-1 boosts insulin secretion until glucose levels return to normal. Januvia also suppresses release of glucose by the liver and slows digestion.

7. By blocking a sodium and glucose cotransporter (SGLT2) found almost exclusively in the kidney that is responsible for 90% of glucose reabsorption during urine formation (*SGLT2 inhibitors,* such as Invokana). This newest class of approved diabetic drugs suppresses reabsorption of glucose by the renal tubules, thereby reducing blood glucose by increasing urinary excretion of the surplus glucose. It also promotes weight loss because the glucose lost in the urine is no longer available as an energy source for the body.

Nearly 400 new drugs for diabetes, mostly for the Type 2 form of the disease, are under development.

In the future a surgical approach might also be used to treat T2DM. Obese individuals with Type 2 diabetes who undergo *bariatric (weight loss) surgery* frequently experience a lessening or disappearance of their diabetes symptoms even before they begin to shed pounds. The mechanism responsible for the improved glucose metabolism is still being researched, but apparently gastric bypass surgery precludes the stomach from releasing food-induced blood-borne chemicals that affect blood glucose levels.

New Approaches to Managing Insulin-Dependent Diabetes

Because none of the available oral diabetic drugs deliver new insulin to the body, they cannot replace insulin therapy for people with T1DM. Furthermore, sometimes the weakened β cells in T2DM eventually burn out and can no longer produce insulin, in which case the previously non–insulin-dependent patient must be placed on insulin therapy.

Several newer approaches are currently available for insulin-dependent diabetics that preclude the need for the one or more insulin injections daily.

- *Implanted insulin pumps* can deliver a prescribed amount of insulin on a regular basis, but the recipient must time meals with care to match the automatic insulin delivery.

- *Pancreas transplants* are also being performed more widely now, with increasing success. On the downside, transplant recipients must take immunosuppressive drugs for life to prevent rejection of their donated organs. Also, donor organs are in short supply.

The following new treatments for T1DM are on the horizon:

- Some new delivery methods circumvent the need for daily insulin injections by using alternative routes of administration that bypass the destructive digestive tract enzymes. Examples include, using inhaled powdered insulin (*Afrezza*, recently approved but not yet being manufactured) or using an oral insulin spray product that can be absorbed in the mouth. A related approach is to protect swallowed insulin from destruction by the digestive tract, for example, by attaching oral insulin to vitamin B_{12}, which protects the insulin from digestive enzymes until the vitamin–insulin complex is absorbed by intrinsic factor-induced endocytosis in the terminal ileum (see p. 585).

- A new vaccine that selectively removes only the T cells responsible for the autoimmunity that destroys β cells (leaving the other useful T cells intact to do their job) is in late-stage clinical trials.

- Another hope is pancreatic islet transplants. Scientists have developed several types of devices that isolate donor islet cells from the recipient's immune system. Such immunoisolation of islet cells permits use of grafts from other animals, circumventing the shortage of human donor cells. Pig islet cells are an especially good source because pig insulin is nearly identical to human insulin.

- Some researchers have coaxed stem cells to develop into insulin-secreting cells that hopefully can be implanted.

- In a related approach, others are turning to genetic engineering to develop surrogates for pancreatic β cells. An example is the potential reprogramming of the small-intestine endocrine cells that produce the GI hormone GIP (see p. 615). The goal is to cause these non-β cells to cosecrete both insulin and GIP in response to a meal.

- Another approach under development is an implanted, glucose-detecting, insulin-releasing "artificial pancreas" that would continuously monitor the patient's blood glucose level and deliver insulin in response to need.

- One of the latest leads is the discovery of a gene in the liver and fat tissue that codes for a newly identified hormone, *betatrophin*, which can prompt production of more pancreatic β cells.

manipulate underlying molecular defects in diabetes offer hope that more effective therapies will soon be developed to better manage or even cure existing cases and perhaps to prevent new cases of this devastating disease. (See the boxed feature on diabetes on pp. 696–697 for current and potential future treatment strategies for this disorder.)

Insulin excess causes brain-starving hypoglycemia.

Clinical Note Let us now look at the opposite of diabetes mellitus, insulin excess, which is characterized by *hypoglycemia* (low blood glucose) and can arise in two ways. First, insulin excess can occur in a diabetic patient when too much insulin has been injected for the person's caloric intake and exercise level, resulting in **insulin shock.** Second, blood insulin level may rise abnormally high in a nondiabetic individual whose β cells are overresponsive to glucose, a condition called **reactive hypoglycemia.** Such β cells "overshoot" and secrete more insulin than necessary in response to elevated blood glucose after a high-carbohydrate meal. The excess insulin drives too much glucose into the cells, resulting in hypoglycemia.

The consequences of insulin excess are primarily manifestations of the effects of hypoglycemia on the brain. Recall that the brain relies on a continuous supply of blood glucose for its nourishment and that glucose uptake by the brain does not depend on insulin. With insulin excess, more glucose than necessary is driven into the other insulin-dependent cells. The result is a lowering of the blood glucose level so that not enough glucose is left in the blood to be delivered to the brain. In hypoglycemia, the brain literally starves. The symptoms, therefore, are primarily referable to depressed brain function, which, if severe enough, may rapidly progress to unconsciousness and death. People with overresponsive β cells do not become sufficiently hypoglycemic to manifest these more serious consequences, but they do show milder symptoms of depressed CNS activity.

The true incidence of reactive hypoglycemia is a subject of intense controversy because it is difficult to diagnose without confirming the presence of low blood glucose during the time of symptoms. Mild symptoms of depressed CNS function, such as tremor, fatigue, sleepiness, and inability to concentrate, are nonspecific and could also be attributable to emotional problems or other factors. Therefore, a definitive diagnosis based on symptoms alone is impossible to make.

The treatment of hypoglycemia depends on the cause. At the first indication of a hypoglycemic attack with insulin overdose, the diabetic person should eat or drink something sugary. Prompt treatment of severe hypoglycemia is imperative to prevent brain damage. Note that a diabetic can lose consciousness and die from either diabetic ketoacidotic coma caused by prolonged insulin deficiency or acute hypoglycemia caused by insulin shock. Fortunately, the other accompanying signs and symptoms differ sufficiently between the conditions to enable medical caretakers to administer appropriate therapy, either insulin or glucose. For example, ketoacidotic coma is accompanied by deep, labored breathing (in compensation for the metabolic acidosis) and fruity breath (from exhaled ketone bodies), whereas insulin shock is not.

Ironically, even though reactive hypoglycemia is characterized by low blood glucose, people with this disorder are treated by limiting their intake of sugar and other glucose yielding carbohydrates to prevent their β cells from overresponding to a high glucose intake. Giving a symptomatic individual with reactive hypoglycemia something sugary temporarily alleviates the symptoms. The blood glucose level is transiently restored to normal so that the brain's energy needs are again satisfied. However, as soon as the extra glucose triggers further insulin release, the situation is merely aggravated.

Glucagon in general opposes the actions of insulin.

Even though insulin plays a central role in controlling metabolic adjustments between the absorptive and the postabsorptive states, the secretory product of the pancreatic islet α cells, **glucagon,** is also important. Many physiologists view the insulin-secreting β cells and the glucagon-secreting α cells as a coupled endocrine system whose combined secretory output is a major factor in regulating fuel metabolism.

Glucagon affects many of the same metabolic processes that insulin influences, but in most cases glucagon's actions are opposite to those of insulin. The major site of action of glucagon is the liver, where it exerts a variety of effects on carbohydrate, fat, and protein metabolism. Glucagon acts by increasing cAMP.

Actions on Carbohydrate The overall effects of glucagon on carbohydrate metabolism result in an increase in hepatic glucose production and release and thus in an increase in blood glucose. Glucagon exerts its hyperglycemic effects by decreasing glycogenesis, promoting glycogenolysis, and stimulating gluconeogenesis.

Actions on Fat Glucagon also antagonizes the actions of insulin with regard to fat metabolism by promoting lipolysis and inhibiting triglyceride synthesis, thus increasing blood levels of fatty acids. Glucagon enhances hepatic ketone production (ketogenesis) by promoting conversion of fatty acids to ketone bodies.

Actions on Protein Glucagon inhibits hepatic protein synthesis and promotes degradation of hepatic protein. Stimulation of gluconeogenesis further contributes to glucagon's catabolic effect on hepatic protein metabolism. Glucagon promotes protein catabolism in the liver, but it does not have any significant effect on blood amino acid levels because it does not affect muscle protein, the major protein store in the body.

Glucagon secretion is increased during the postabsorptive state.

Glucagon secretion increases during the postabsorptive state and decreases during the absorptive state, just the opposite of insulin secretion. In fact, insulin is sometimes referred to as a "hormone of feasting" and glucagon as a "hormone of fasting." Insulin tends to put nutrients in storage when their blood levels are high, such as after a meal, whereas glucagon promotes catabolism of nutrient stores between meals to keep up the blood nutrient levels, especially blood glucose.

As in insulin secretion, the major factor regulating glucagon secretion is a direct effect of the blood glucose concentration on the endocrine pancreas. In this case, the pancreatic α cells increase glucagon secretion in response to a fall in blood glucose. Glucagon's hyperglycemic actions tend to raise blood glucose back to normal. Conversely, an increase in blood glucose concentration, such as after a meal, inhibits glucagon secretion, which tends to drop blood glucose back to normal.

Insulin and glucagon work as a team to maintain blood glucose and fatty acid levels.

Thus, a direct negative-feedback relationship exists between blood glucose concentration and both β cells' and α cells' rates of secretion, but in opposite directions. An elevated blood glucose stimulates insulin secretion but inhibits glucagon secretion, whereas a fall in blood glucose leads to decreased insulin secretion and increased glucagon secretion (Figure 19-20). Because insulin lowers and glucagon raises blood glucose, the changes in secretion of these pancreatic hormones in response to deviations in blood glucose work together homeostatically to restore blood glucose to normal.

Similarly, a fall in blood fatty acid concentration directly inhibits insulin output and stimulates glucagon output by the pancreas, both of which are negative-feedback control mechanisms to restore the blood fatty acid level to normal.

The opposite effects exerted by blood concentrations of glucose and fatty acids on the pancreatic α and β cells are appropriate for regulating the circulating levels of these nutrient molecules because the actions of insulin and glucagon on carbohydrate and fat metabolism oppose one another. The effect of blood amino acid concentration on the secretion of these two hormones is a different story. A rise in blood amino acid concentration stimulates both insulin and glucagon secretion. Why this seeming paradox because glucagon does not exert any effect on blood amino acid concentration? The identical effect of high blood amino acid levels on both insulin and glucagon secretion makes sense if you consider the concomitant effects these two hormones have on blood glucose levels. If, during absorption of a protein-rich meal, the rise in blood amino acids stimulated only insulin secretion, hypoglycemia might result. Because little carbohydrate is available for absorption following consumption of a high-protein meal, the amino acid–induced increase in insulin secretion would drive too much glucose into the cells, causing a sudden, inappropriate drop in blood glucose. However, the simultaneous increase in glucagon secretion elicited by elevated blood amino acid levels increases hepatic glucose production. Because the hyperglycemic effects of glucagon counteract the hypoglycemic actions of insulin, the net result is maintenance of normal blood glucose levels (and pre-

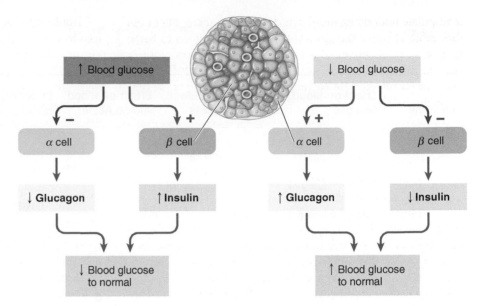

Figure 19-20 **Complementary interactions of glucagon and insulin.**

vention of hypoglycemic starvation of the brain) during absorption of a meal that is high in protein but low in carbohydrates.

Glucagon excess can aggravate the hyperglycemia of diabetes mellitus.

Clinical Note No known clinical abnormalities are caused by glucagon deficiency or excess per se. However, diabetes mellitus is frequently accompanied by excess glucagon secretion because insulin is required for glucose to gain entry into the α cells, where it can exert control over glucagon secretion. As a result, diabetics frequently have a high rate of glucagon secretion concurrent with their insulin insufficiency because the elevated blood glucose cannot inhibit glucagon secretion as it normally would. Because glucagon is a hormone that raises blood glucose, its excess intensifies the hyperglycemia of diabetes mellitus. For this reason, some insulin-dependent diabetics respond best to a combination of insulin and somatostatin therapy. By inhibiting glucagon secretion, somatostatin indirectly helps achieve better reduction of the elevated blood glucose than can be accomplished by insulin therapy alone.

Epinephrine, cortisol, and growth hormone also exert direct metabolic effects.

The pancreatic hormones are the most important regulators of normal fuel metabolism. However, several other hormones exert direct metabolic effects, even though control of their secretion is keyed to factors other than transitions in metabolism between feasting and fasting states (Table 19-5).

The stress hormones, epinephrine and cortisol, both increase blood glucose and blood fatty acids through a variety of metabolic effects. In addition, cortisol mobilizes amino acids by promoting protein catabolism. Neither hormone plays an important role in regulating fuel metabolism under resting

| TABLE 19-5 Summary of Hormonal Control of Fuel Metabolism

	MAJOR METABOLIC EFFECTS				CONTROL OF SECRETION	
Hormone	Effect on Blood Glucose	Effect on Blood Fatty Acids	Effect on Blood Amino Acids	Effect on Muscle Protein	Major Stimuli for Secretion	Primary Role in Metabolism
Insulin	↓ +Glucose uptake +Glycogenesis − Glycogenolysis − Gluconeogenesis	↓ +Triglyceride synthesis − Lipolysis	↓ +Amino acid uptake	↑ +Protein synthesis − Protein degradation	↑ Blood glucose ↑ Blood amino acids	Primary regulator of absorptive and postabsorptive cycles
Glucagon	↑ +Glycogenolysis +Gluconeogenesis − Glycogenesis	↑ +Lipolysis − Triglyceride synthesis	No effect	No effect	↓ Blood glucose ↑ Blood amino acids	Regulation of absorptive and postabsorptive cycles in concert with insulin; protection against hypoglycemia
Epinephrine	↑ +Glycogenolysis +Gluconeogenesis − Insulin secretion +Glucagon secretion	↑ +Lipolysis	No effect	No effect	Sympathetic stimulation during stress and exercise	Provision of energy for emergencies and exercise
Cortisol	↑ +Gluconeogenesis − Glucose uptake by tissues other than brain; glucose sparing	↑ +Lipolysis	↑ +Protein degradation	↓ +Protein degradation	Stress	Mobilization of metabolic fuels and building blocks during adaptation to stress
Growth Hormone	↑ − Glucose uptake by muscles; glucose sparing	↑ +Lipolysis	↓ +Amino acid uptake	↑ +Protein synthesis − Protein degradation +Synthesis of DNA and RNA	Deep sleep Stress Exercise Hypoglycemia	Promotion of growth; normally little role in metabolism; mobilization of fuels plus glucose sparing in extenuating circumstances

conditions, but both are critical for the metabolic responses to stress.

GH (itself and acting through IGF-I) has protein anabolic effects in muscle. In fact, this is one of its growth-promoting features. Although GH can elevate the blood levels of glucose and fatty acids, it is normally of little importance to the overall regulation of fuel metabolism. Deep sleep (responsible for the marked nighttime diurnal increase in GH), exercise, stress, and severe hypoglycemia stimulate GH secretion, possibly to provide fatty acids as an energy source and spare glucose for the brain under these circumstances.

Although thyroid hormone increases the overall metabolic rate and has both anabolic and catabolic actions, changes in thyroid hormone secretion are usually not important for fuel homeostasis, for two reasons. First, control of thyroid hormone secretion is not directed toward maintaining nutrient levels in the blood. Second, the onset of thyroid hormone action is too slow to have any significant effect on the rapid adjustments required to maintain normal blood levels of nutrients.

Note that, with the exception of the anabolic effects of GH on protein metabolism, all the metabolic actions of these other hormones are opposite to those of insulin. Insulin alone can reduce blood glucose and blood fatty acid levels, whereas glucagon, epinephrine, cortisol, and GH all increase blood levels of these nutrients. These other hormones are therefore considered **insulin antagonists**. The main reason diabetes mellitus has such devastating metabolic consequences is that no other control mechanism is available to pick up the slack to promote anabolism when insulin activity is insufficient, so the catabolic reactions promoted by other hormones proceed unchecked. The only exception is protein anabolism stimulated by GH.

The hypothalamus plays a role in controlling glucose homeostasis.

In addition to the hormones that play key roles in peripherally regulating blood glucose, recent evidence indicates that the CNS, in particular the hypothalamus, directly senses blood-borne nutrients (glucose and fatty acids) and hormones associated with nutrient management (insulin, leptin, and GLP-1) and uses this information to directly or indirectly influence the three main regulators of glucose homeostasis—pancreatic insulin and glucagon secretion, hepatic glucose output, and glucose uptake by skeletal muscles—to help maintain stable blood glucose levels in response to ingestion of food. A recent surge in research related to the role of the CNS in maintaining glucose homeostasis suggests that the pathways involved overlap considerably with the CNS circuits that maintain energy balance and body weight. For example, the arcuate nucleus in the hypothalamus is involved in both sensing glucose (after it has been converted to pyruvate) and in regulating hepatic glucose production, in addition to housing the appetite-stimulating NPY neurons and appetite-suppressing POMC neurons (see p. 621). Unraveling these pathways could lead to new therapies for combating obesity-related factors in the development of Type 2 diabetes.

Check Your Understanding 19.4

1. Define *glycogenesis, glycogenolysis,* and *gluconeogenesis.*
2. Describe the metabolic effects of insulin and glucagon.
3. Compare the effect of increased blood glucose on pancreatic β cells and α cells.
4. Distinguish between Type 1 and Type 2 diabetes mellitus.

19.5 Parathyroid Glands and Control of Calcium Metabolism

Besides regulating the concentration of organic nutrient molecules in the blood by manipulating anabolic and catabolic pathways, the endocrine system regulates the plasma concentration of a number of inorganic electrolytes. As you already know, aldosterone controls Na^+ and K^+ concentrations in the ECF. Three other hormones—*parathyroid hormone, calcitonin,* and *vitamin D*—control calcium (Ca^{2+}) and phosphate (PO_4^{3-}) metabolism. These hormonal agents concern themselves with regulating plasma Ca^{2+}; in the process, plasma PO_4^{3-} is maintained. Plasma Ca^{2+} concentration is one of the most tightly controlled variables in the body. The need for the precise regulation of plasma Ca^{2+} stems from its critical influence on so many body activities.

Plasma Ca^{2+} must be closely regulated to prevent changes in neuromuscular excitability.

About 99% of the Ca^{2+} in the body (about 1000 g) is in crystalline form within the skeleton and teeth. Of the remaining Ca^{2+}, about 0.9% (9 g) is found intracellularly within the soft tissues; less than 0.1% (1 g) is present in the ECF. Approximately half of the ECF Ca^{2+} is bound to plasma proteins and therefore restricted to the plasma or is complexed with PO_4^{3-} and not free to participate in chemical reactions. The other half of the ECF Ca^{2+} is freely diffusible and can readily pass from the plasma into the interstitial fluid and interact with the cells. Only this free ECF Ca^{2+} is biologically active and subject to regulation; it constitutes less than one thousandth of the total Ca^{2+} in the body.

This small, free fraction of ECF Ca^{2+} plays a vital role in a number of essential activities, including the following:

1. *Neuromuscular excitability.* Even minor variations in the concentration of free ECF Ca^{2+} can have a profound and immediate effect on the sensitivity of excitable tissues. A fall in free Ca^{2+} results in overexcitability of nerves and muscles; conversely, a rise in free Ca^{2+} depresses neuromuscular excitability. These effects result from the influence of Ca^{2+} on membrane permeability to Na^+. A decrease in free Ca^{2+} increases Na^+ permeability, with the resultant influx of Na^+ moving the resting potential closer to threshold. Consequently, in the presence of *hypocalcemia* (low blood Ca^{2+}), excitable tissues may be brought to threshold by normally ineffective physiologic stimuli so that skeletal muscles discharge and contract (go into spasm) "spontaneously" (in the absence of normal stimulation). If severe enough, spastic contraction of the respiratory muscles results in death by asphyxiation. *Hypercalcemia* (elevated blood Ca^{2+}) is also life threatening because it causes cardiac arrhythmias and generalized depression of neuromuscular excitability.

2. *Excitation–contraction coupling in cardiac and smooth muscle.* Entry of ECF Ca^{2+} into cardiac and phasic smooth muscle cells, resulting from increased Ca^{2+} permeability in response to an action potential, triggers the contractile mechanism. Calcium is also necessary for excitation–contraction coupling in skeletal muscle fibers, but in this case the Ca^{2+} is released from intracellular Ca^{2+} stores in response to an action potential. A significant part of the increase in cytosolic Ca^{2+} in cardiac muscle cells also derives from internal stores.

Note that a rise in *cytosolic Ca^{2+}* within a muscle cell causes contraction, whereas an increase in *free ECF Ca^{2+}* decreases neuromuscular excitability and reduces the likelihood of contraction. Unless one keeps this point in mind, it is difficult to understand why low plasma Ca^{2+} levels bring about muscle hyperactivity when Ca^{2+} is necessary to switch on the contractile apparatus. We are talking about two different Ca^{2+} pools, which exert different effects.

3. *Stimulus–secretion coupling.* The entry of Ca^{2+} into secretory cells, which results from increased permeability to Ca^{2+} in response to appropriate stimulation, triggers the release of the secretory product by exocytosis. This process is important

for the secretion of neurotransmitters by nerve cells and for secretion of hydrophilic hormones by endocrine cells.

4. *Excitation–secretion coupling.* In pancreatic β cells, Ca^{2+} entry from the ECF in response to membrane depolarization leads to insulin secretion.

5. *Maintenance of tight junctions between cells.* Calcium forms part of the intercellular cement that holds particular cells tightly together.

6. *Clotting of blood.* Calcium serves as a cofactor in several steps of the cascade of reactions that leads to clot formation.

In addition to these functions of free ECF Ca^{2+}, intracellular Ca^{2+} serves as a second messenger in many cells and is involved in cell motility and cilia action. Finally, the Ca^{2+} in bone and teeth is essential for the structural and functional integrity of these tissues.

Because of the profound effects of deviations in free ECF Ca^{2+}, especially on neuromuscular excitability, the plasma concentration of this electrolyte is regulated with extraordinary precision. Let us see how.

Control of Ca^{2+} metabolism includes regulation of both Ca^{2+} homeostasis and Ca^{2+} balance.

Maintaining the proper plasma concentration of free Ca^{2+} differs from the regulation of Na^+ and K^+ in two important ways: (1) Na^+ and K^+ homeostasis is maintained primarily by regulating the urinary excretion of these electrolytes so that controlled output matches uncontrolled input. Although urinary excretion of Ca^{2+} is hormonally controlled, in contrast to Na^+ and K^+, not all ingested Ca^{2+} is absorbed from the digestive tract; instead, the extent of absorption is hormonally controlled and depends on the Ca^{2+} status of the body. (2) Bone serves as a large Ca^{2+} reservoir that can be drawn on to maintain the free plasma Ca^{2+} concentration within the narrow limits compatible with life should dietary intake become too low. Exchange of Ca^{2+} between the ECF and bone is also subject to hormonal control. Similar in-house stores are not available for Na^+ and K^+.

Regulation of Ca^{2+} metabolism depends on hormonal control of exchanges between the ECF and three other compartments: *bone, kidneys,* and *intestine.* Control of Ca^{2+} metabolism encompasses two aspects:

■ Regulation of **calcium homeostasis** involves the immediate adjustments required to maintain a constant free plasma Ca^{2+} concentration on a minute-to-minute basis. This is largely accomplished by rapid exchanges between bone and ECF and to a lesser extent by modifications in urinary excretion of Ca^{2+}.

■ Regulation of **calcium balance** involves the more slowly responding adjustments required to maintain a constant total amount of Ca^{2+} in the body. Control of Ca^{2+} balance ensures that Ca^{2+} intake is equivalent to Ca^{2+} excretion over the long term (weeks to months). Calcium balance is maintained by adjusting the extent of intestinal Ca^{2+} absorption and urinary Ca^{2+} excretion.

Parathyroid hormone, the principal regulator of Ca^{2+} metabolism, acts directly or indirectly on all three of these effector sites. It is the primary hormone responsible for maintaining Ca^{2+} homeostasis and is essential for maintaining Ca^{2+} balance, although vitamin D also contributes in important ways to Ca^{2+} balance. The third Ca^{2+}-influencing hormone, calcitonin, is not essential for maintaining either Ca^{2+} homeostasis or balance. It serves a backup function during the rare times of extreme hypercalcemia. We now examine the specific effects of each of these hormonal systems in more detail.

Parathyroid hormone raises free plasma Ca^{2+}, a life-saving effect.

Parathyroid hormone (PTH) is a peptide hormone secreted by the **parathyroid glands**, four rice grain–sized glands located on the back surface of the thyroid gland, one in each corner (❙Figure 19-21). Like aldosterone, PTH *is essential for life.* The overall effect of PTH is to increase the Ca^{2+} concentration of plasma (and, accordingly, of the entire ECF), thereby preventing hypocalcemia. In the complete absence of PTH, death ensues within a few days, usually because of asphyxiation caused by hypocalcemic spasm of respiratory muscles. By its actions on bone, kidneys, and intestine, PTH raises plasma Ca^{2+} concentration when it starts to fall, so hypocalcemia and its effects are normally avoided. This hormone also lowers plasma PO_4^{3-} concentration. We consider each of these mechanisms, beginning with an overview of bone remodeling and the actions of PTH on bone.

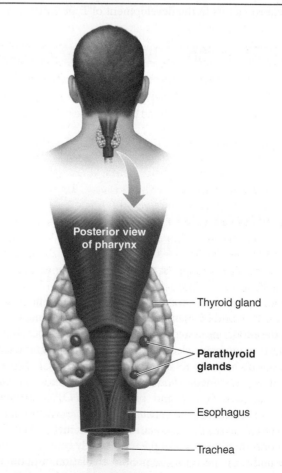

Posterior view of pharynx

Thyroid gland

Parathyroid glands

Esophagus

Trachea

❙Figure 19-21 **Anatomy of parathyroid glands.**

Bone continuously undergoes remodeling.

Because 99% of the body's Ca^{2+} is in bone, the skeleton serves as a storage depot for Ca^{2+}. Bone is a living tissue composed of an organic extracellular matrix or *osteoid* (see p. 654) made hard by **hydroxyapatite crystals** consisting primarily of precipitated calcium phosphate salts $(Ca_3(PO_4)_2)$. Normally, $Ca_3(PO_4)_2$ salts are in solution in the ECF, but conditions within bone are suitable for these salts to precipitate (crystallize) around the collagen fibers in the matrix. By mobilizing some of these Ca^{2+} stores in bone, PTH raises plasma Ca^{2+} concentration when it starts to fall.

Bone Remodeling Despite the apparent inanimate nature of bone, its constituents are continually being turned over. **Bone deposition** (formation) and **bone resorption** (removal) normally go on concurrently so that bone is constantly being remodeled, much as people remodel buildings by tearing down walls and replacing them. Through remodeling, the adult human skeleton is completely regenerated an estimated every 10 years. Bone remodeling serves two purposes: (1) it keeps the skeleton appropriately "engineered" for maximum effectiveness in its mechanical uses, and (2) it helps maintain the plasma Ca^{2+} level. Let us examine in more detail the underlying mechanisms and controlling factors for each of these purposes.

Recall that three types of bone cells are present in bone (see pp. 654 and 656). The *osteoblasts* secrete the extracellular organic matrix within which the $Ca_3(PO_4)_2$ crystals precipitate. The *osteocytes* are the retired osteoblasts imprisoned within the bony wall they have deposited around themselves. The *osteoclasts* resorb bone in their vicinity. The large, multinucleated osteoclasts attach to the organic matrix and form a "ruffled membrane" that increases its surface area in contact with the bone. Thus attached, the osteoclast actively secretes hydrochloric acid that dissolves the $Ca_3(PO_4)_2$ crystals and enzymes that break down the organic matrix. After it has created a cavity, an osteoclast moves on to an adjacent site to burrow another hole. Osteoblasts move into the vacated cavity and secrete osteoid to fill in the hole. Subsequent mineralization of this organic matrix results in new bone to replace the bone dissolved by the osteoclast. Thus, a constant cellular tug-of-war goes on in bone, with bone-forming osteoblasts countering the efforts of the bone-destroying osteoclasts. These construction and demolition crews, working side by side, continuously remodel bone. At any given time, about a million microscopic sites throughout the skeleton are undergoing resorption or deposition. Throughout most of adult life, the rates of bone formation and bone resorption are about equal, so total bone mass remains fairly constant during this period.

Osteocytes, despite their boney confinement, influence ongoing bone formation by secreting **sclerostin**, a paracrine that inhibits osteoblast activity in a check-and-balance way to prevent excessive bone growth that might end up clamping off nearby nerves or fusing the spinal column. Parathyroid hormone and mechanical stress, both of which favor bone formation, inhibit sclerostin, whereas calcitonin, which favors bone resorption, stimulates sclerostin.

Osteoblasts and osteoclasts both trace their origins to the bone marrow. Osteoblasts are derived from a type of connective tissue cell in the bone marrow, whereas osteoclasts differentiate from macrophages, which are tissue-bound derivatives of monocytes, a type of white blood cell that originates in the bone marrow (see p. 393). In a unique communication system, osteoblasts and their immature precursors produce two chemical signals that govern osteoclast development and activity in opposite ways—*RANK ligand* and *osteoprotegerin*—as follows (❙ Figure 19-22):

■ **RANK ligand (RANKL)** revs up osteoclast action. (A ligand is a small molecule that binds with a larger protein molecule; an example is an extracellular chemical messenger binding with a plasma membrane receptor.) As its name implies, RANK ligand binds to **RANK** (for *receptor activator of NF-κB*), a protein receptor on the membrane surface of nearby macrophages. This binding induces the macrophages to differentiate into osteoclasts and helps them live longer by suppressing apoptosis. As a result, bone resorption is stepped up and bone mass decreases.

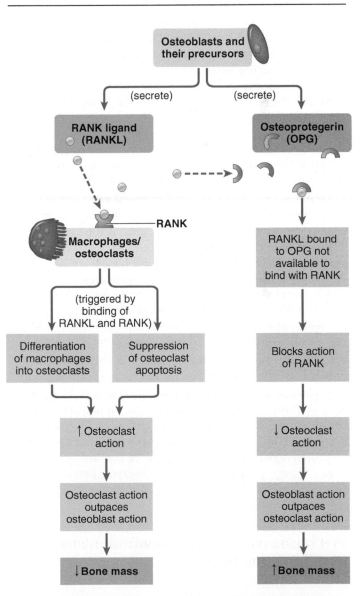

❙Figure 19-22 **Role of osteoblasts in governing osteoclast development and activity.**

- Alternatively, neighboring osteoblasts can secrete **osteoprotegerin (OPG)**, which by contrast suppresses osteoclast activity. OPG secreted into the matrix serves as a freestanding decoy receptor that binds with RANKL. By taking RANKL out of action so that it cannot bind with its intended RANK receptors, OPG prevents RANKL from revving up osteoclasts' bone-resorbing activity. As a result, the matrix-making osteoblasts are able to outpace the matrix-removing osteoclasts, so bone mass increases. The balance between RANKL and OPG thus is an important determinant of bone density. If osteoblasts produce more RANKL, the more osteoclast action, the lower the bone mass. If osteoblasts produce more OPG, the less osteoclast action, the greater the bone mass. Scientists are currently unraveling the influence of various factors on this balance. For example, the female sex hormone estrogen stimulates activity of the OPG-producing gene in osteoblasts and also promotes apoptosis of osteoclasts, both mechanisms by which this hormone preserves bone mass.

Mechanical stress favors bone deposition.

As a child grows, the bone builders keep ahead of the bone destroyers under the influence of GH and IGF-I (see pp. 654–656).

Mechanical stress also tips the balance in favor of bone deposition, causing bone mass to increase and the bones to strengthen. Mechanical factors adjust the strength of bone in response to the demands placed on it. The greater the physical stress and compression to which a bone is subjected, the greater the rate of bone deposition. For example, the bones of athletes are stronger and more massive than those of sedentary people.

By contrast, bone mass diminishes and the bones weaken when bone resorption gains a competitive edge over bone deposition in response to removal of mechanical stress. For example, bone mass decreases in people who undergo prolonged bed confinement or those in spaceflight. Early astronauts lost up to 20% of their bone mass during their time in orbit. Therapeutic exercise can limit or prevent such loss of bone.

Clinical Note Bone mass also decreases as a person ages. Bone density peaks when a person is in the 30s and then starts to decline after age 40. By 50 to 60 years of age, bone resorption often exceeds bone formation. The result is a reduction in bone mass known as **osteoporosis** (meaning "porous bones"). This bone-thinning condition is characterized by a diminished laying down of organic matrix as a result of reduced osteoblast activity, increased osteoclast activity, or both, rather than abnormal bone calcification. The underlying cause of osteoporosis is uncertain. Plasma Ca^{2+} and PO_4^{3-} levels are normal, as is PTH. Osteoporosis occurs with greatest frequency in postmenopausal women because of the associated withdrawal of bone-preserving estrogen. (For more details on osteoporosis, see the boxed feature on pp. 706–707, A Closer Look at Exercise Physiology.)

PTH raises plasma Ca^{2+} by withdrawing Ca^{2+} from the bone bank.

In addition to the factors geared toward controlling the mechanical effectiveness of bone, throughout life PTH uses bone as a "bank" from which it withdraws Ca^{2+} as needed to maintain plasma Ca^{2+} level. PTH has two major effects on bone that raise plasma Ca^{2+} concentration. First, it induces a fast Ca^{2+} efflux into the plasma from the small *labile pool* of Ca^{2+} in the bone fluid. Second, by stimulating bone dissolution, it promotes a slow transfer into the plasma of both Ca^{2+} and PO_4^{3-} from the *stable pool* of bone minerals in bone itself. Let us examine more thoroughly PTH's actions in mobilizing Ca^{2+} from its labile and stable pools in bone.

PTH's immediate effect is to promote transfer of Ca^{2+} from bone fluid into plasma.

Compact bone forms the dense outer portion of a bone. Interconnecting spicules of **trabecular bone** make up the more lacy-appearing inner core of a bone (Figure 19-23a). The inner honeycomb network allows bones to be sturdy without being too heavy. Compact bone is organized into **osteon** units, each of which consists of a **central canal** surrounded by concentrically arranged **lamellae** (Figure 19-23b). Lamellae are layers of osteocytes entombed within the bone they have deposited around themselves (Figure 19-23b and c). The osteons typically run parallel to the long axis of the bone. Blood vessels penetrate the bone from either the outer surface or the marrow cavity and run through the central canals. Osteoblasts are present along the outer surface of the bone and along the inner surfaces lining the central canals. Osteoclasts are also located on bone surfaces undergoing resorption. The surface osteoblasts and entombed osteocytes are connected by an extensive network of small, fluid-containing canals, the **canaliculi**, which permit exchange of substances between the trapped osteocytes and the circulation. These small canals also contain long, filmy cytoplasmic extensions, or "arms," of osteocytes and osteoblasts that are connected to one another, much as if these cells were "holding hands." The "hands" of adjacent cells are connected by gap junctions, which permit communication and exchange of materials among these bone cells. The interconnecting cell network, which is called the **osteocytic–osteoblastic bone membrane**, separates the mineralized bone itself from the blood vessels within the central canals (Figure 19-24a, p. 708). The small, labile pool of Ca^{2+} is in the **bone fluid** that lies between this bone membrane and the adjacent bone, both within the canaliculi and along the surface of the central canal.

PTH exerts its effects via cAMP. The earliest effect of PTH is to activate membrane-bound Ca^{2+} pumps located in the plasma membranes of the cytoplasmic extensions of osteocytes and osteoblasts. These pumps promote movement of Ca^{2+}, without the accompaniment of PO_4^{3-}, from the bone fluid into these cells, which in turn transfer the Ca^{2+} into the plasma within the central canal. Thus, PTH stimulates the transfer of Ca^{2+} from the bone fluid across the osteocytic–osteoblastic bone membrane into the plasma. Movement of Ca^{2+} out of the labile pool across the bone membrane accounts for the fast exchange between bone and plasma (Figure 19-24b). Because of the large surface area of the bone membrane, small movements of Ca^{2+} across individual cells are amplified into large Ca^{2+} fluxes between the bone fluid and plasma.

After Ca^{2+} is pumped out, the bone fluid is replenished with Ca^{2+} from partially mineralized bone along the adjacent bone

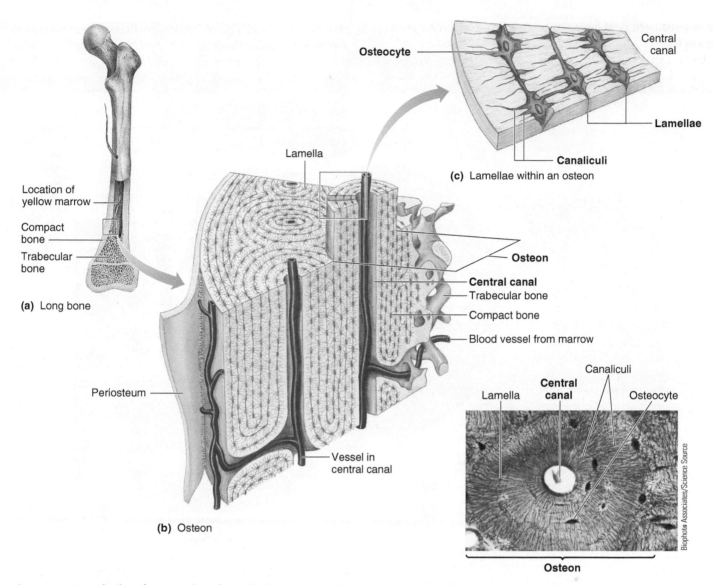

Figure 19-23 Organization of compact bone into osteons. (a) Structure of a long bone showing location of compact bone and trabecular bone. (b) An osteon, the structural unit of compact bone, consists of concentric lamellae (layers of osteocytes entombed by the bone they have deposited around themselves) surrounding a central canal containing a small blood vessel branch. The light micrograph is of compact bone in a human femur (thigh bone). (c) A magnification of lamellae.

(Source: Modified and redrawn with permission from *Human Anatomy and Physiology*, 3rd Edition, by A. Spence and E. Mason. Copyright © 1987 by The Benjamin/Cummings Publishing Company. Reprinted by permission of Pearson Education, Inc.)

surface. Thus, the fast exchange of Ca^{2+} does not involve resorption of completely mineralized bone, and bone mass is not decreased. Through this means, PTH draws Ca^{2+} out of the "quick-cash branch" of the bone bank and rapidly increases plasma Ca^{2+} without actually entering the bank (that is, without breaking down mineralized bone itself). Under normal conditions, this exchange is sufficient for maintaining plasma Ca^{2+} concentration.

PTH's chronic effect is to promote localized dissolution of bone to release Ca^{2+} into plasma.

Under conditions of chronic hypocalcemia, such as may occur with dietary Ca^{2+} deficiency, PTH stimulates localized dissolution of bone, promoting a slower transfer into the plasma of both Ca^{2+} and PO_4^{3-} from the minerals within the bone itself. Osteoblasts have PTH receptors, but osteoclasts do not. PTH promotes localized bone dissolution by acting on osteoblasts, causing them to secrete RANKL, thereby indirectly stimulating osteoclasts to gobble up bone. PTH also transiently inhibits the bone-forming activity of the osteoblasts. Bone contains so much Ca^{2+} compared to the plasma (more than 1000 times as much) that even when PTH tips the balance in favor of increased bone resorption, no immediate effects on the skeleton are discernible because such a tiny amount of bone is affected. Yet the negligible amount of Ca^{2+} "borrowed" from the bone bank can be lifesaving in terms of restoring the free plasma Ca^{2+} level to normal. The borrowed Ca^{2+} is then redeposited in the bone at another time when Ca^{2+} supplies are more abundant. Meanwhile, the plasma Ca^{2+} level has been maintained without sacrificing bone integrity. However, prolonged excess

OSTEOPOROSIS, A DECREASE IN BONE density resulting from reduced deposition of the bone's organic matrix (see the accompanying figure), is a major health problem affecting 200 million people worldwide. In the United States, 8 million women and 2 million men already have osteoporosis, with another 34 million having a low bone mass that makes them at high risk for developing this disorder. The condition is especially prevalent among women following menopause (permanent cessation of menstruation), owing to the marked drop in bone-preserving estrogen. Thirty percent of postmenopausal women have osteoporosis. Following menopause, women start losing 1% or more bone density each year. Skeletons of elderly women are typically only 50% to 80% as dense as at their peak at about age 35, whereas elderly men's skeletons retain 80% to 90% of their youthful density. Similar to estrogen, testosterone also helps preserve bone density, but unlike women's programmed withdrawal of estrogen at menopause, men do not experience a similar built-in loss of testosterone secretion.

Osteoporosis is responsible for the greater incidence of bone fractures among women older than age 50 years than among the population at large. One in three women with osteoporosis ends up with a fractured bone, most commonly of the hip or spine, which may lead to permanent disability or even death. Because bone mass is reduced, osteoporotic bones are more brittle and more susceptible to fracture in response to a fall, blow, or lifting action that normally would not strain stronger bones. For every 10% loss of bone mass, the risk of fracture doubles. In the United States, osteoporosis is the underlying cause of approximately 1.5 million fractures each year, and the attendant medical and rehabilitation cost is $17 billion per year. The cost in pain, suffering, and loss of independence is not measurable. Half of all American women have spinal pain and deformity by age 75.

Drug Therapy for Osteoporosis

Ca^{2+} and vitamin D supplementation and a regular weight-bearing exercise program are long-standing therapeutic approaches used to prevent or treat osteoporosis. Additionally, the following drugs can minimize or reverse bone loss:

■ *Estrogen replacement therapy* was used to treat osteoporosis in the past. Estrogen slows bone loss by promoting apoptosis (cell suicide) of osteoclasts and by enhancing activity of osteoblasts. However, the Food and Drug Administration (FDA) no longer approves es-

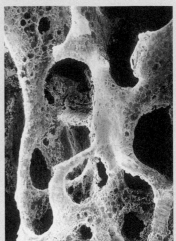

Dr. P. Motta, Department of Anatomy, University "La Sapienza" Rome/ Science Photo Library/Science Source

Normal bone   Osteoporotic bone

Comparison of normal and osteoporotic bone. Note the reduced density of osteoporotic trabecular bone compared to normal trabecular bone.

trogen therapy for treating osteoporosis because this drug has been linked to an increased risk of breast cancer and cardiovascular disease. However, the following newer classes of drugs are already approved by the FDA or are under investigation.

■ *Alendronate* (Fosamax), a bisphosphonate, was the first nonhormonal osteoporosis drug. It works by blocking osteoclasts' bone-destroying actions. Alendronate pills have to be taken daily, or a newer version can be taken weekly. Even newer bisphosphonates can be taken at longer intervals, such as *ibondronate* (Boniva) (once-a-month pill) and *zoledronic acid* (Reclast) (once-a-year intravenous infusion).

■ *Calcitonin* (Miacalcin), the thyroid C-cell hormone that slows osteoclast activity, is used to treat advanced osteoporosis, but traditionally it had to be injected daily, a deterrent to patient compliance. Now calcitonin is available in a more patient-friendly nasal spray (Fortical).

■ *Raloxifene* (Evista) belongs to a new class of drugs known as *selective estrogen receptor modulators (SERMs)*. Raloxifene does not

PTH secretion over months or years eventually leads to the formation of holes throughout the skeleton, which are filled with very large, overstuffed osteoclasts.

When PTH promotes dissolution of $Ca_3(PO_4)_2$ crystals in bone to harvest their Ca^{2+} content, both Ca^{2+} and PO_4^{3-} are released into the plasma. An elevated plasma PO_4^{3-} is undesirable, but PTH deals with this dilemma by its actions on the kidneys.

PTH acts on the kidneys to conserve Ca^{2+} and eliminate PO_4^{3-}.

PTH promotes Ca^{2+} conservation and PO_4^{3-} elimination by the kidneys during urine formation. Under the influence of PTH, the kidneys can reabsorb more of the filtered Ca^{2+}, so less Ca^{2+} escapes into the urine. This effect increases plasma Ca^{2+} and decreases urinary Ca^{2+} losses. (It would be counterproduc-

bind with estrogen receptors in reproductive organs, but it does bind with estrogen receptors outside the reproductive system, such as in bone. Through this selective receptor binding, raloxifene mimics estrogen's beneficial effects on bone to provide protection against osteoporosis by keeping osteoclasts in check while avoiding estrogen's potentially harmful effects on reproductive organs, such as increased risk of breast cancer.

■ *Teriparatide* (Forteo) was the first approved treatment that stimulates bone formation instead of acting to prevent bone loss, as the other drugs do. Teriparatide, which must be injected, is an active fragment of parathyroid hormone (PTH). Even though continuous exposure to PTH, as with hyperparathyroidism, increases osteoclast activity and thereby promotes the breakdown of bone, evidence suggests that, by contrast, intermittent administration of PTH (or its active teriparatide fragment) increases osteoblast formation and prolongs survival of these bone builders by blocking osteoblast apoptosis.

• *Denosumab* (Prolia) is the newest drug for treatment of osteoporosis. This injectable drug reduces destruction of bone by binding to and inhibiting RANKL, the protein that promotes maturation, function, and survival of bone-resorbing osteoclasts. Denosumab mimics the natural action of osteoprotegerin (see p. 704).

■ The *statins* (for example, Lipitor) are another group of drugs with some promise for treating osteoporosis. The statins are already commonly used as cholesterol-lowering agents. They also stimulate osteoblast activity, promoting bone formation and reducing the fracture rate, which are side benefits to their favorable cholesterol actions. They still have not been approved specifically for use in preventing bone loss.

■ *ANGELS (activators of nongenomic estrogen-like signaling)* is a new class of osteoporosis drug under development. Most of estrogen's effects are brought about by estrogen binding with its receptors in the target cell's nucleus, thereby turning on specific genes, just as all steroids do (see p. 126). However, scientists recently discovered that estrogen blocks apoptosis among osteoblasts by using a different pathway. In this alternative cytoplasmic-signaling pathway, estrogen binds with a cytoplasmic receptor instead of binding with its nuclear receptor to bring about its effect. ANGELS drugs trigger estrogen's cytoplasmic signaling pathway to block osteoblast apoptosis. The term ANGELS refers to activation of this nongene pathway, in contrast to SERMs, which trigger estrogen's traditional nuclear gene pathway in bone.

■ *Romosozumab* is the latest class of investigational osteoporosis drug. It is an antibody developed to bind to and block the action of sclerostin, the osteocyte-derived protein that naturally inhibits bone-forming osteoblasts (see p. 703). By this means, romosozumab favors bone rebuilding by allowing osteoblasts to proceed uninterrupted.

Benefits of Exercise on Bone

Despite advances in osteoporosis therapy, treatment is still often less than satisfactory, and all the current therapeutic agents are associated with some undesirable side effects. Therefore, prevention is by far the best approach to managing this disease. Development of strong bones to begin with before menopause through a good, Ca^{2+}-rich diet and adequate exercise appears to be the best preventive measure. A large reservoir of bone at midlife may delay the clinical manifestations of osteoporosis in later life. Continued physical activity throughout life appears to retard or prevent bone loss, even in the elderly.

It is well documented that osteoporosis can result from disuse—that is, from reduced mechanical loading of the skeleton. Space travel has clearly shown that lack of gravity results in a decrease in bone density. Studies of athletes, by contrast, demonstrate that weight-bearing physical activity increases bone density. Within groups of athletes, bone density correlates directly with the load the bone must bear. If one looks at athletes' femurs (thigh bones), the greatest bone density is found in weight lifters, followed in order by throwers, runners, soccer players, and finally swimmers. In fact, the bone density of swimmers does not differ from that of nonathletic controls. Swimming does not place any strain on bones. The bone density in the playing arm of male tennis players has been found to be as much as 35% greater than in their other arm; female tennis players have been found to have 28% greater density in their playing arm than in their other arm. One study found that very mild activity in nursing-home patients, whose average age was 82 years, not only slowed bone loss but even resulted in bone buildup over a 36-month period. Thus, exercise is a good defense against osteoporosis.

---

tive to dissolve bone to obtain more Ca^{2+} only to lose it in urine.) By contrast, PTH decreases PO_4^{3-} reabsorption, thus increasing urinary PO_4^{3-} excretion. As a result, PTH reduces plasma PO_4^{3-} at the same time it increases plasma Ca^{2+}.

This PTH-induced urinary removal of extra PO_4^{3-} from the body fluids is essential for preventing reprecipitation of the Ca^{2+} freed from bone. Both Ca^{2+} and PO_4^{3-} are released from bone when PTH promotes bone dissolution. Because PTH is secreted only when plasma Ca^{2+} falls below normal, the released Ca^{2+} is needed to restore plasma Ca^{2+} to normal, yet the released PO_4^{3-} tends to raise plasma PO_4^{3-} above normal. Because of the solubility characteristics of $Ca_3(PO_4)_2$ salt, the product of the plasma concentration of Ca^{2+} times the plasma concentration of plasma PO_4^{3-} must remain roughly constant. Therefore, an inverse relationship exists between the plasma concentrations of Ca^{2+} and PO_4^{3-}. If plasma PO_4^{3-} levels were

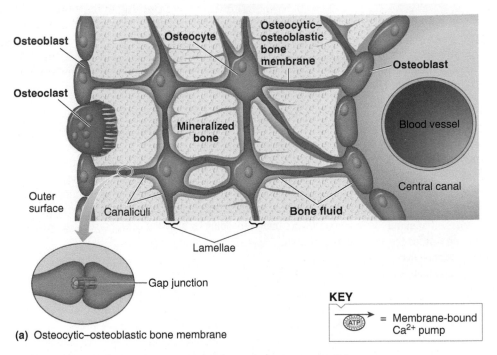

Osteoblast • **Osteocyte** • **Osteocytic–osteoblastic bone membrane** • **Osteoblast**

Osteoclast

Mineralized bone

Blood vessel

Central canal

Outer surface • Canaliculi • **Bone fluid**

Lamellae

Gap junction

(a) Osteocytic–osteoblastic bone membrane

KEY

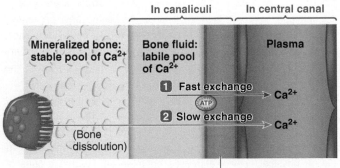

In canaliculi | In central canal

Mineralized bone: stable pool of Ca²⁺

Bone fluid: labile pool of Ca²⁺

Plasma

1 Fast exchange → Ca²⁺

ATP

2 Slow exchange → Ca²⁺

(Bone dissolution)

Osteocytic–osteoblastic bone membrane (formed by filmy cytoplasmic extensions of interconnected osteocytes and osteoblasts)

1 In a fast exchange, Ca^{2+} is moved from the labile pool in the bone fluid into the plasma by PTH-activated Ca^{2+} pumps located in the osteocytic–osteoblastic bone membrane.

2 In a slow exchange, Ca^{2+} is moved from the stable pool in the mineralized bone into the plasma through PTH-induced dissolution of the bone by osteoclasts.

(b) Fast and slow exchange of Ca²⁺ between bone and plasma

┃Figure 19-24 Fast and slow exchanges of Ca²⁺ between bone and plasma. (a) Entombed osteocytes and surface osteoblasts are interconnected by long cytoplasmic processes that extend from these cells and connect to one another within the canaliculi. This interconnecting cell network, the osteocytic–osteoblastic bone membrane, separates the mineralized bone from the plasma in the central canal. Bone fluid lies between the membrane and the mineralized bone. (b) Fast exchange of Ca²⁺ between the bone and plasma is accomplished by Ca²⁺ pumps in the osteocytic–osteoblastic bone membrane that transport Ca²⁺ from the bone fluid into these bone cells, which transfer the Ca²⁺ into the plasma. Slow exchange of Ca²⁺ between the bone and plasma is accomplished by osteoclast dissolution of bone.

FIGURE FOCUS: *Use one of these terms to fill in each blank in the following statements regarding PTH action: (a) labile, (b) stable, (c) osteoclasts, (d) bone-membrane Ca²⁺ pumps. In a fast exchange, PTH moves Ca²⁺ from the ___ Ca²⁺ pool in bone by stimulating ___. In a slow exchange, PTH moves Ca²⁺ from the ___ Ca²⁺ pool in bone by increasing activity of ___.*

though extra PO_4^{3-} is being released from bone into the plasma.

The third important action of PTH on the kidneys (besides increasing Ca^{2+} reabsorption and decreasing PO_4^{3-} reabsorption) is to enhance activation of vitamin D by the kidneys.

PTH indirectly promotes absorption of Ca²⁺ and PO₄³⁻ by the intestine.

Although PTH has no direct effect on the intestine, it indirectly increases both Ca^{2+} and PO_4^{3-} absorption from the small intestine by helping activate vitamin D. This vitamin, in turn, directly increases intestinal absorption of Ca^{2+} and PO_4^{3-}, a topic we discuss more thoroughly shortly.

The primary regulator of PTH secretion is plasma concentration of free Ca²⁺.

All the effects of PTH raise plasma Ca^{2+} levels. Appropriately, PTH secretion increases when plasma Ca^{2+} falls and decreases when plasma Ca^{2+} rises. The secretory cells of the parathyroid glands are directly and exquisitely sensitive to changes in free plasma Ca^{2+}. Because PTH regulates plasma Ca^{2+} concentration, this relationship forms a simple negative-feedback loop for controlling PTH secretion without involving any nervous or other hormonal intervention (┃Figure 19-25).

Calcitonin lowers plasma Ca²⁺ concentration but is not important in the normal control of Ca²⁺ metabolism.

Calcitonin, the hormone produced by the C cells of the thyroid gland (see ┃Figure 19-1b, p. 666), also exerts an influence on plasma Ca^{2+} levels. Like PTH, calcitonin has two effects on bone, but in this case both effects

decrease plasma Ca^{2+} levels: Short term, calcitonin decreases Ca^{2+} movement from the bone fluid into the plasma. Long term, calcitonin decreases bone resorption by inhibiting the activity of osteoclasts via the cAMP pathway. The suppression of bone resorption reduces plasma Ca^{2+} concentration and lowers plasma PO_4^{3-} levels. Calcitonin also inhibits Ca^{2+} and PO_4^{3-} reabsorption from the kidney tubules during urine formation, further reinforcing its hypocalcemic and hypophos-

allowed to rise above normal, some of the released Ca^{2+} would be forced along with the PO_4^{3-} back into bone through hydroxyapatite crystal formation to keep the calcium phosphate product constant. This self-defeating redeposition of Ca^{2+} would lower plasma Ca^{2+}, just the opposite of the needed effect. Therefore, PTH acts on the kidneys to decrease the reabsorption of PO_4^{3-} by the renal tubules. This action increases urinary excretion of PO_4^{3-} and lowers its plasma concentration, even

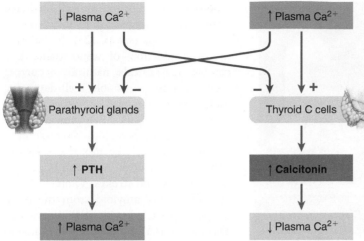

Figure 19-25 Negative-feedback loops controlling parathyroid hormone (PTH) and calcitonin secretion.

phatemic effects. Calcitonin has no effect on the intestine or on vitamin D.

As with PTH, the primary regulator of calcitonin release is free plasma Ca^{2+} concentration, but unlike with PTH, an increase in plasma Ca^{2+} stimulates calcitonin secretion and a fall in plasma Ca^{2+} inhibits calcitonin secretion (Figure 19-25). Because calcitonin reduces plasma Ca^{2+} levels, this system constitutes a second simple negative-feedback control over plasma Ca^{2+} concentration, one opposed to the PTH system.

However, calcitonin plays little or no role in the normal control of Ca^{2+} or PO_4^{3-} metabolism. Although calcitonin protects against hypercalcemia, this condition rarely occurs under normal circumstances. Moreover, neither thyroid removal nor calcitonin-secreting tumors alter circulating levels of Ca^{2+} or PO_4^{3-}, implying that this hormone is not normally essential for maintaining Ca^{2+} or PO_4^{3-} homeostasis. Calcitonin may, however, play a role in protecting skeletal integrity when there is a large Ca^{2+} demand, such as during pregnancy or breast-feeding. Furthermore, some experts speculate that calcitonin may hasten the storage of newly absorbed Ca^{2+} following a meal.

Vitamin D is actually a hormone that increases Ca^{2+} absorption in the intestine.

The final factor involved in regulating Ca^{2+} metabolism is **cholecalciferol,** or **vitamin D,** a steroidlike compound essential for Ca^{2+} absorption in the intestine. Strictly speaking, vitamin D should be considered a hormone because the skin (specifically, the keratinocytes; see p. 441) can produce it from a precursor related to cholesterol (7-dehydrocholesterol) on exposure to sunlight. Skin-produced vitamin D is subsequently released into the blood to act at a distant target site, the intestine. The skin, therefore, is actually an endocrine gland and vitamin D is a hormone. Traditionally, however, this chemical messenger has been considered a vitamin for two reasons. First, it was originally discovered and isolated from a dietary source and tagged as a vitamin. Second, even though the skin would be an adequate source of vitamin D if it were exposed to sufficient sun-

light, indoor dwelling and clothing in response to cold weather and social customs preclude significant exposure of the skin to sunlight in the United States and many other parts of the world most of the time. At least part of the essential vitamin D must therefore be derived from dietary sources.

Activation of Vitamin D Regardless of its source, vitamin D is biologically inactive when it first enters the blood from either the skin or the digestive tract. It must be activated by two sequential biochemical alterations that involve the addition of two hydroxyl (—OH) groups (Figure 19-26). The first of these reactions occurs in the liver, and the second takes place in the kidneys. The end result is production of the active form of vitamin D, *1,25-(OH)$_2$-vitamin D$_3$*, also known as *calcitriol*. PTH stimulates the kidney enzymes involved in the second step of vitamin D activation in response to a fall in plasma Ca^{2+}. To a lesser extent, a fall in plasma PO_4^{3-} also enhances the activation process. Vitamin D in its various forms circulates in the blood primarily bound to **vitamin D–binding protein**.

Function of Vitamin D The most dramatic effect of activated vitamin D is to increase Ca^{2+} absorption by the intestine. Unlike most dietary constituents, dietary Ca^{2+} is not indiscriminately absorbed by the digestive system. In fact, the majority of ingested Ca^{2+} is typically not absorbed but is lost in the feces. When needed, more dietary Ca^{2+} is absorbed into the plasma under the influence of vitamin D. Independently of its effects on Ca^{2+} absorption, the active form of vitamin D increases intestinal PO_4^{3-} absorption. Furthermore, vitamin D increases responsiveness of bone to PTH. Thus, vitamin D and PTH are closely interdependent. Like steroid hormones, vitamin D exerts its effects by binding with a nuclear **vitamin D receptor (VDR)**, with this complex regulating gene transcription in the target cells by binding with the **vitamin D–response element** in DNA. (Because vitamin D is the last major hormone to be introduced, for your convenience we summarize in Table 19-6 the signal transduction pathways used by the major hormones. Review these pathways on pp. 116–117, 123–124, and 126–127.)

PTH is principally responsible for controlling Ca^{2+} homeostasis because the actions of vitamin D are too sluggish for it to contribute substantially to the minute-to-minute regulation of plasma Ca^{2+} concentration. However, both PTH and vitamin D are essential to Ca^{2+} balance, the process ensuring that, over the long term, dietary Ca^{2+} input into the body is equivalent to Ca^{2+} output in the urine. When dietary Ca^{2+} intake is reduced, the resultant transient fall in plasma Ca^{2+} level stimulates PTH secretion. The increased PTH has two effects important for maintaining Ca^{2+} balance: (1) It stimulates Ca^{2+} reabsorption by the kidneys, thereby decreasing Ca^{2+} output; and (2) it activates vitamin D, which increases the efficiency of uptake of ingested Ca^{2+}. Because PTH also promotes bone resorption, a substantial loss of bone minerals occurs if Ca^{2+} intake is reduced for a prolonged period, even though bone is not directly involved in maintaining Ca^{2+} input and output in balance.

A torrent of studies in recent decades indicates that vitamin D's functions are more far reaching than its effects on uptake of

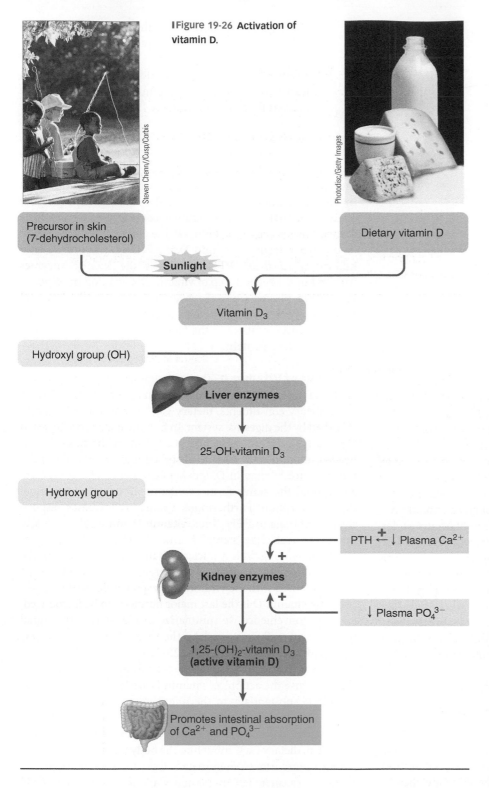

| Figure 19-26 **Activation of vitamin D.**

Precursor in skin (7-dehydrocholesterol)

Dietary vitamin D

Sunlight

Vitamin D_3

Hydroxyl group (OH)

Liver enzymes

25-OH-vitamin D_3

Hydroxyl group

PTH $\xleftarrow{+}$ ↓ Plasma Ca^{2+}

Kidney enzymes +

↓ Plasma PO_4^{3-} +

1,25-$(OH)_2$-vitamin D_3 (active vitamin D)

Promotes intestinal absorption of Ca^{2+} and PO_4^{3-}

blood cells responsible for cell-mediated immunity that targets virally invaded cells and cancer cells (see p. 423). It also promotes production of antioxidants that combat free radicals, naturally occurring highly reactive, unstable, cell-damaging molecules (see p. 142). Studies further suggest that vitamin D helps thwart development of diabetes mellitus, counter autoimmune diseases like multiple sclerosis, and lower the risk of high blood pressure, heart attacks, and strokes. Vitamin D may help clear beta amyloid from the brain plaques associated with Alzheimer's disease (see p. 164). Researchers continue to seek the underlying mechanisms by which vitamin D exerts these protective effects.

Phosphate metabolism is controlled by the same mechanisms that regulate Ca^{2+} metabolism.

Intracellular PO_4^{3-} is important in the high-energy phosphate bonds of ATP, plays a key regulatory role in phosphorylating designated proteins in second-messenger pathways, and helps form the backbone of DNA molecules. Excreted PO_4^{3-} is an important urinary buffer. In the ECF, plasma PO_4^{3-} concentration is not as tightly controlled as plasma Ca^{2+} concentration. Phosphate is regulated directly by vitamin D and indirectly by the plasma Ca^{2+}–PTH feedback loop. To illustrate, a fall in plasma PO_4^{3-} concentration exerts a twofold effect to help raise the circulating PO_4^{3-} level back to normal (| Figure 19-27). First, because of the inverse relationship between the plasma concentrations of PO_4^{3-} and Ca^{2+}, a fall in plasma PO_4^{3-} increases plasma Ca^{2+}, which directly suppresses PTH secretion. In the presence of reduced PTH, PO_4^{3-} reabsorption by the kidneys increases, returning plasma PO_4^{3-} concentration toward normal. Second, a fall in plasma PO_4^{3-} also increases activation of vitamin D, which then promotes PO_4^{3-} absorption in the intestine. This further helps alleviate the initial hypophosphatemia. Note that these changes do not compromise Ca^{2+} balance. Although the increase in activated vitamin D stimulates Ca^{2+} absorption from the intestine, the concurrent fall in PTH produces a compensatory increase in urinary Ca^{2+} excretion because less of the filtered Ca^{2+} is reabsorbed. Therefore, plasma Ca^{2+} remains unchanged while plasma PO_4^{3-} increases to normal.

ingested Ca^{2+} and PO_4^{3-}. Vitamin D's broad effect results from its activation of VDRs that have been found in many organs throughout the body. Vitamin D, at higher blood concentrations than those sufficient to protect bone, appears to bolster muscle strength and thereby help prevent falls by improving leg strength. It also seems to be an important force in energy metabolism and immune health. Vitamin D reduces inflammation and must be present for activation of T cells, the white

TABLE 19-6 Signal Transduction Pathways Used by Major Hormones

Signal Transduction Pathway	Hormones That Use This Pathway
↑ Cyclic AMP/PKA	TSH, ACTH, FSH, LH, CRH, GHRH, somatostatin, vasopressin, epinephrine (for β adrenergic receptor induced actions), secretin, glucagon, PTH, calcitonin
↓ Cyclic AMP/PKA	Dopamine, melatonin, epinephrine (for α_2 adrenergic receptor induced actions), PPY_{3-36}
IP_3/Ca^{2+} and DAG/PKC	TRH, GnRH, oxytocin, ghrelin, gastrin, cholecystokinin, epinephrine (for α_1 adrenergic receptor induced actions)
Tyrosine kinase	Insulin, IGF-I, and IGF-II
JAK/STAT	Growth hormone, prolactin, erythropoietin, leptin
Hormone-response elements on DNA	All lipophilic hormones: thyroid hormone, cortisol, aldosterone, testosterone, estrogen, progesterone, vitamin D

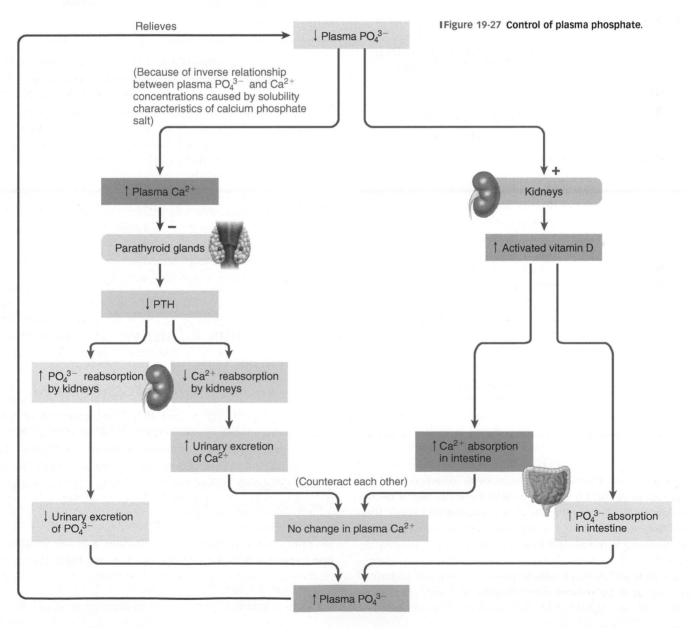

Figure 19-27 Control of plasma phosphate.

Disorders in Ca²⁺ metabolism may arise from abnormal levels of PTH or vitamin D.

 Clinical Note The primary disorders that affect Ca^{2+} metabolism are too much or too little PTH or a deficiency of vitamin D.

PTH Hypersecretion Excess PTH secretion, or **hyperparathyroidism,** which is usually caused by a hypersecreting tumor in one of the parathyroid glands, is characterized by *hypercalcemia* and *hypophosphatemia.* The affected individual can be asymptomatic or symptoms can be severe, depending on the magnitude of the problem. The following are among the possible consequences:

- Hypercalcemia reduces the excitability of muscle and nervous tissue, leading to muscle weakness and neurologic disorders, including decreased alertness, poor memory, and depression. Cardiac disturbances may also occur.
- Excessive mobilization of Ca^{2+} and PO_4^{3-} from skeletal stores leads to thinning of bone, which may result in skeletal deformities and increased incidence of fractures.
- An increased incidence of Ca^{2+}-containing kidney stones occurs because the excess quantity of Ca^{2+} being filtered through the kidneys may precipitate and form stones. These stones may impair renal function. Passage of the stones through the ureters causes extreme pain. Because of these potential multiple consequences, hyperparathyroidism has been called a disease of "bones, stones, and abdominal groans."
- To further account for the "abdominal groans," hypercalcemia can cause peptic ulcers, nausea, and constipation.

PTH Hyposecretion Because of the parathyroid glands' close anatomic relation to the thyroid, the most common cause of deficient PTH secretion, or **hypoparathyroidism,** used to be inadvertent removal of the parathyroid glands (before doctors knew about their existence) during surgical removal of the thyroid gland (to treat thyroid disease). If all the parathyroid tissue was removed, these patients died, of course, because PTH is essential for life. Physicians were puzzled why some patients died soon after thyroid removal but others did not. Now that the location and importance of the parathyroid glands are known, surgeons are careful to leave parathyroid tissue during thyroid removal. Rarely, PTH hyposecretion results from an autoimmune attack against the parathyroid glands.

Hypoparathyroidism leads to *hypocalcemia* and *hyperphosphatemia.* The symptoms are mainly caused by increased neuromuscular excitability from the reduced level of free plasma Ca^{2+}. In the complete absence of PTH, death is imminent. With a relative deficiency of PTH, milder symptoms of increased neuromuscular excitability become evident. Muscle cramps and twitches occur from spontaneous activity in the motor nerves, whereas tingling and pins-and-needles sensations result from spontaneous activity in the sensory nerves. Mental changes include irritability and paranoia.

Vitamin D Deficiency The major consequence of vitamin D deficiency is impaired intestinal absorption of Ca^{2+}. In the face of reduced Ca^{2+} uptake, PTH maintains the plasma Ca^{2+} level

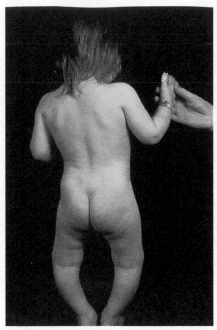

IFigure 19-28 **Rickets.**

Biophoto Associates/Science Source

at the expense of the bones. As a result, the bone matrix is not properly mineralized because Ca^{2+} salts are not available for deposition. The demineralized bones become soft and deformed, bowing under the pressures of weight bearing, especially in children. This condition is known as **rickets** in children and **osteomalacia** in adults (**I**Figure 19-28).

Check Your Understanding 19.5

1. Describe the distribution of Ca^{2+} in the body and the functions of free ECF Ca^{2+}.

2. Distinguish between calcium homeostasis and calcium balance.

3. Discuss the effects of PTH on bone, kidneys, and the intestine.

Homeostasis: Chapter in Perspective

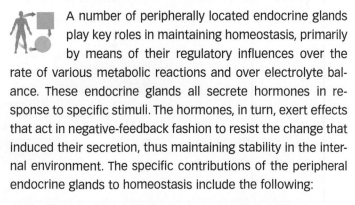 A number of peripherally located endocrine glands play key roles in maintaining homeostasis, primarily by means of their regulatory influences over the rate of various metabolic reactions and over electrolyte balance. These endocrine glands all secrete hormones in response to specific stimuli. The hormones, in turn, exert effects that act in negative-feedback fashion to resist the change that induced their secretion, thus maintaining stability in the internal environment. The specific contributions of the peripheral endocrine glands to homeostasis include the following:

- Two closely related hormones secreted by the thyroid gland, tetraiodothyronine (T_4) and tri-iodothyronine (T_3), increase the overall metabolic rate. Not only does this action influence the rate at which cells use nutrient molecules and O_2 within the internal environment, but it also produces heat, which helps maintain body temperature.

■ The adrenal cortex secretes three classes of hormones. Aldosterone, the primary mineralocorticoid, is essential for Na^+ and K^+ balance. Because of Na^+'s osmotic effect, Na^+ balance is critical to maintaining the proper ECF volume and arterial blood pressure. This action is essential for life. Without aldosterone's Na^+- and H_2O-conserving effect, so much plasma volume would be lost in the urine that death would quickly ensue. Maintaining K^+ balance is essential for homeostasis because changes in extracellular K^+ profoundly affect neuromuscular excitability, jeopardizing normal heart function, among other detrimental effects.

■ Cortisol, the primary glucocorticoid secreted by the adrenal cortex, increases the plasma concentrations of glucose, fatty acids, and amino acids above normal. Although these actions destabilize the concentrations of these molecules in the internal environment, they indirectly contribute to homeostasis by making the molecules readily available as energy sources or building blocks for tissue repair to help the body adapt to stressful situations.

■ The sex hormones secreted by the adrenal cortex do not contribute to homeostasis.

■ The major hormone secreted by the adrenal medulla, epinephrine, generally reinforces activities of the sympathetic nervous system. It contributes to homeostasis directly by its role in blood pressure regulation. Epinephrine also contributes to homeostasis indirectly by helping prepare the body for peak physical responsiveness in fight-or-flight situations. This includes increasing the plasma concentrations of glucose and fatty acids above normal, which provides additional energy sources for increased physical activity.

■ The two major hormones secreted by the endocrine pancreas, insulin and glucagon, are important in shifting metabolic pathways between the absorptive and postabsorptive states, which maintains the appropriate plasma levels of nutrient molecules.

■ Parathyroid hormone from the parathyroid glands is critical to maintaining plasma concentration of Ca^{2+}. PTH is essential for life because of Ca^{2+}'s effect on neuromuscular excitability. In the absence of PTH, death rapidly occurs from asphyxiation caused by pronounced spasms of the respiratory muscles.

Review Exercises Answers begin on p. A-50

Reviewing Terms and Facts

1. The response to thyroid hormone is detectable within a few minutes after its secretion. (*True or false?*)

2. Adrenal androgen hypersecretion is caused by a deficit of an enzyme crucial to cortisol synthesis. (*True or false?*)

3. Excess glucose and amino acids as well as fatty acids can be stored as triglycerides. (*True or false?*)

4. Insulin is the only hormone that can lower blood glucose levels. (*True or false?*)

5. The most life-threatening consequence of hypocalcemia is reduced blood clotting. (*True or false?*)

6. All ingested Ca^{2+} is indiscriminately absorbed in the intestine. (*True or false?*)

7. The $Ca_3(PO_4)_2$ bone crystals form a labile pool from which Ca^{2+} can rapidly be extracted under the influence of PTH. (*True or false?*)

8. The lumen of the thyroid follicle is filled with _____, the chief constituent of which is a large protein molecule known as _____.

9. _____ is the conversion of glucose into glycogen. _____ is the conversion of glycogen into glucose. _____ is the conversion of amino acids into glucose.

10. The three major tissues that do not depend on insulin for their glucose uptake are _____, _____, and _____.

11. The three compartments with which ECF Ca^{2+} is exchanged are _____, _____, and _____.

12. Among the bone cells, _____ are bone builders, _____ are bone dissolvers, and _____ are entombed.

13. Which of the following hormones does *not* exert a direct metabolic effect?
    a. epinephrine
    b. growth hormone
    c. aldosterone
    d. cortisol
    e. thyroid hormone

14. Which of the following are characteristic of the postabsorptive state? (Indicate all that apply.)
    a. glycogenolysis
    b. gluconeogenesis
    c. lipolysis
    d. glycogenesis
    e. protein synthesis
    f. triglyceride synthesis
    g. protein degradation
    h. increased insulin secretion
    i. increased glucagon secretion
    j. glucose sparing

Understanding Concepts

(Answers at www.cengagebrain.com)

1. Describe the steps of thyroid hormone synthesis.

2. What are the effects of T_3 and T_4? Which is the more potent? What is the source of most circulating T_3?

3. Describe the regulation of thyroid hormone secretion.

4. Discuss the causes and symptoms of both hypothyroidism and hyperthyroidism. For each cause, indicate whether a goiter occurs, and explain why.

5. What hormones are secreted by the adrenal cortex? What are the functions and control of each of these hormones?

6. Discuss the causes and symptoms of each type of adrenocortical dysfunction.

7. What is the relationship of the adrenal medulla to the sympathetic nervous system? What are the functions of epinephrine? How is epinephrine release controlled?

8. Define *stress*. Describe the neural and hormonal responses to a stressor.

9. Define *fuel metabolism, anabolism,* and *catabolism.*

10. Indicate the primary circulating form and storage form of each of the three classes of organic nutrients.

11. Distinguish between the absorptive and postabsorptive states with regard to the handling of nutrient molecules.

12. Name the two major cell types of the islets of Langerhans, and indicate the primary hormonal product of each.

13. Compare the functions and control of insulin secretion with those of glucagon secretion.

14. What are the consequences of diabetes mellitus?

15. Why must plasma Ca^{2+} be closely regulated?

16. Explain how osteoblasts influence osteoclast function.

17. Discuss the contributions of parathyroid hormone, calcitonin, and vitamin D to Ca^{2+} metabolism. Describe the source and control of each of these hormones.

18. Discuss the major disorders in Ca^{2+} metabolism.

Applying Clinical Reasoning

Najma G. sought medical attention after her menstrual periods ceased and she started growing excessive facial hair. Also, she had been thirstier than usual and urinated more frequently. A clinical evaluation revealed that Najma was hyperglycemic. Her physician told her that she had an endocrine disorder dubbed "diabetes of bearded ladies." What underlying defect do you think is responsible for Najma's condition?

Thinking at a Higher Level

1. Iodine is naturally present in salt water and is abundant in soil along coastal regions. Fish and shellfish living in the ocean and plants grown in coastal soil take up iodine from their environment. Fresh water does not contain iodine, and the soil becomes more iron poor the farther inland it is. Knowing this, explain why the midwestern United States was once known as an endemic goiter belt. Why is this region no longer an endemic goiter belt even though the soil is still iodine poor?

2. Why do doctors recommend that people who are allergic to bee stings and thus at risk for anaphylactic shock (see p. 438) carry a vial of epinephrine for immediate injection in case of a sting?

3. Why would an infection tend to raise the blood glucose level of a diabetic individual?

4. Tapping the facial nerve at the angle of the jaw in a patient with moderate hyposecretion of a particular hormone elicits a characteristic grimace on that side of the face. What endocrine abnormality could give rise to this so-called *Chvostek's sign*?

5. Soon after a technique to measure plasma Ca^{2+} levels was developed in the 1920s, physicians observed that hypercalcemia accompanied a broad range of cancers. Early researchers proposed that malignancy-associated hypercalcemia arose from metastatic (see p. 432) tumor cells that invaded and destroyed bone, releasing Ca^{2+} into the blood. This conceptual framework was overturned when physicians noted that hypercalcemia often appeared in the absence of bone lesions. Furthermore, cancer patients often manifested hypophosphatemia in addition to hypercalcemia. This finding led investigators to suspect that the tumors might be producing a PTHlike substance. Explain how they reached this conclusion. In 1987, this substance was identified and named *parathyroid hormone–related peptide (PTHrP)*, which binds to and activates PTH receptors.

To access the course materials and companion resources for this text, please visit
www.cengagebrain.com

The Reproductive System

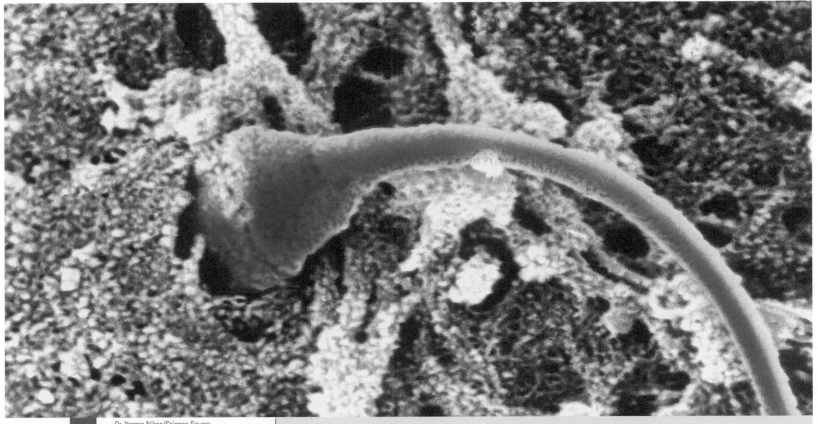

Dr. Yorgos Nikas/Science Source

A scanning electron micrograph of a human sperm penetrating an egg. This image, taken in a fertility clinic, shows one spermatozoon penetrating an ovum (egg) by means of enzymes at its head end. The enzymes from many sperm are needed to break down the outer barrier before one victorious sperm penetrates into the egg cytoplasm to accomplish fertilization.

Homeostasis Highlights

Normal functioning of the reproductive system is not aimed at homeostasis and is not necessary for survival of an individual, but it is essential for survival of the species. Only through reproduction can the complex genetic blueprint of each species survive beyond the lives of individual members of the species.

20.1 Uniqueness of the Reproductive System

The central theme of this book has been the physiologic processes aimed at maintaining homeostasis to ensure survival of the individual. We are now going to leave this theme to discuss the **reproductive system**, which serves primarily the purpose of perpetuating the species.

Unique among body systems, the reproductive system does not contribute to homeostasis but plays other roles.

Even though the reproductive system does not contribute to homeostasis and is not essential for survival of an individual, it still plays an important role in a person's life. The manner in which people relate as sexual beings contributes in significant ways to psychosocial behavior and has important influences on how people view themselves and how they interact with others. Reproductive function also has a profound effect on society. The universal organization of societies into family units provides a stable environment that is conducive for perpetuating our species.

Reproductive capability depends on intricate relationships among the hypothalamus, anterior pituitary, reproductive organs, and target cells of the sex hormones. These relationships employ many of the regulatory mechanisms used by other body systems for maintaining homeostasis, such as negative-feedback control. In addition to these basic biological processes, sexual behavior and attitudes are deeply influenced by emotional factors and the sociocultural mores of the society in which the individual lives. We concentrate on the basic sexual and reproductive functions under nervous and hormonal control and do not examine the psychological and social ramifications of sexual behavior.

The reproductive system includes the gonads, reproductive tract, and accessory sex glands, all of which differ in males and females.

Reproduction depends on the union of male and female **gametes** (**reproductive**, or **germ**, **cells**), each with a half set of chromosomes, to form a new individual with a full, unique set of chromosomes. Unlike other body systems, which are essentially identical in the two sexes, the reproductive systems of males and females are markedly different, befitting their different roles in the reproductive process. The **male** and **female reproductive systems** are designed to enable union of genetic material from the two sexual partners, and the female system is equipped to house and nourish the offspring to the developmental point at which it can survive independently in the external environment.

The **primary reproductive organs**, or **gonads**, consist of a pair of **testes** in the male and a pair of **ovaries** in the female. In both sexes, the mature gonads perform the dual function of (1) producing gametes (**gametogenesis**), that is, **spermatozoa** (**sperm**) in the male and **ova** (**eggs**) in the female; and (2) secreting sex hormones, specifically, **testosterone** in males and **estrogen** and **progesterone** in females. (The term *estrogen* refers to a group of closely related compounds, namely *estradiol, estrone,* and *estriol,* of which estradiol is the principal estrogen secreted by the ovaries.)

In addition to the gonads, the reproductive system in each sex includes a **reproductive tract** encompassing a system of ducts specialized to transport or house the gametes after they are produced, plus **accessory sex glands** that empty their supportive secretions into these passageways. In females, the breasts are also considered accessory reproductive organs. The externally visible portions of the reproductive system are known as **external genitalia**.

Secondary Sexual Characteristics The **secondary sexual characteristics** are the many external characteristics not directly involved in reproduction that distinguish males and females, such as body configuration and hair distribution. In humans, for example, males have broader shoulders, whereas females have curvier hips, and males have beards, whereas females do not. Testosterone in the male and estrogen in the female govern the development and maintenance of these characteristics. Progesterone has no influence on secondary sexual characteristics.

In some species, the secondary sexual characteristics are of great importance in courting and mating behavior; for example, the rooster's comb attracts the female's attention, and the stag's antlers are useful to ward off other males. In humans, the differentiating marks between males and females do attract the opposite sex, but attraction is also strongly influenced by the complexities of human society and cultural behavior.

Overview of Male Reproductive Functions and Organs The essential male reproductive functions are as follows:

1. Production of sperm (*spermatogenesis*)
2. Delivery of sperm to the female

The sperm-producing organs, the testes, are suspended outside the abdominal cavity in a skin-covered sac, the **scrotum**, which lies within the angle between the legs. The male reproductive system is designed to deliver sperm to the female reproductive tract in a liquid vehicle, *semen,* which is conducive to sperm viability. The major **male accessory sex glands**, whose secretions provide the bulk of the semen, are the *seminal vesicles, prostate gland,* and *bulbourethral glands* (❚ Figure 20-1). The **penis** is the organ used to deposit semen in the female. The **glans penis** (the cap at the distal end of the penis) is exquisitely sensitive erotic tissue important in sexual arousal. Sperm exit each testis through the **male reproductive tract**, consisting on each side of an *epididymis, ductus (vas) deferens,* and *ejaculatory duct.* These pairs of reproductive tubes, along with the secretions from the accessory sex glands, empty into a single *urethra,* the canal that runs the length of the penis and empties to the exterior. All of these parts are described more thoroughly when their functions are discussed.

Overview of Female Reproductive Functions and Organs The female's role in reproduction is more complicated

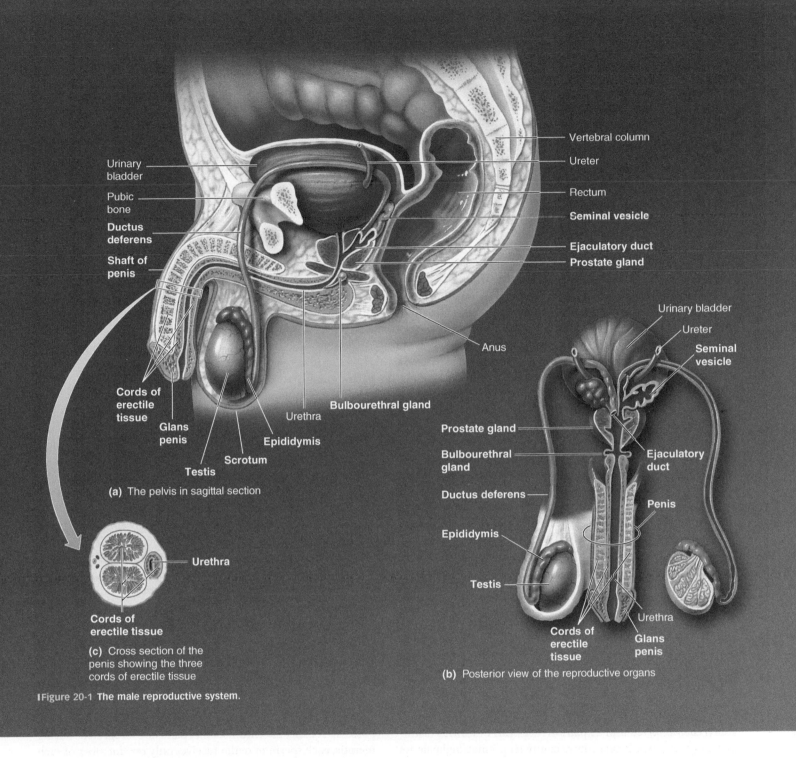

Urinary bladder

Pubic bone

Ductus deferens

Shaft of penis

Vertebral column

Ureter

Rectum

Seminal vesicle

Ejaculatory duct

Prostate gland

Anus

Bulbourethral gland

Urethra

Cords of erectile tissue

Glans penis

Epididymis

Scrotum

Testis

(a) The pelvis in sagittal section

Urethra

Cords of erectile tissue

(c) Cross section of the penis showing the three cords of erectile tissue

Urinary bladder

Ureter

Seminal vesicle

Prostate gland

Bulbourethral gland

Ductus deferens

Ejaculatory duct

Penis

Epididymis

Testis

Cords of erectile tissue

Glans penis

Urethra

(b) Posterior view of the reproductive organs

▮Figure 20-1 The male reproductive system.

than the male's. The essential female reproductive functions include the following:

1. Production of ova (*oogenesis*)

2. Reception of sperm

3. Transport of the sperm and ovum to a common site for union (*fertilization,* or *conception*)

4. Maintenance of the developing fetus until it can survive in the outside world (*gestation,* or *pregnancy*), including formation of the *placenta,* the organ of exchange between mother and fetus

5. Giving birth to the baby (*parturition*)

6. Nourishing the infant after birth by milk production (*lactation*)

The product of fertilization is known as an **embryo** during the first 2 months of intrauterine development, when tissue differentiation is taking place. Beyond this time, the developing living being is recognizable as human and is known as a **fetus** during the remainder of gestation. Although no further tissue differentiation takes place during fetal life, it is a time of tremendous tissue growth and maturation.

The ovaries and female reproductive tract lie within the pelvic cavity. The female reproductive tract consists of the following components (❙Figure 20-2a and b): Two **oviducts (uterine,** or **fallopian tubes)**, which are in close association with the two ovaries, pick up ova on ovulation (ovum release from an ovary) and serve as the site for fertilization. The thick-walled, hollow **uterus** is primarily responsible for maintaining the fetus during its development and expelling it at the end of pregnancy. The **vagina** is a muscular, expandable tube that connects the uterus to the external environment. The lowest portion of the uterus, the **cervix**, projects into the vagina and contains a single, small opening, the **cervical canal**. Sperm are deposited in the vagina by the penis during sexual intercourse. The cervical canal is a pathway for sperm through the uterus to the site of fertilization in the oviduct and, when greatly dilated during parturition, is the passageway for delivery of the baby from the uterus.

The vaginal opening is located in the **perineum** (the external region that bounds the pelvic outlet) between the urethral opening anteriorly and the anal opening posteriorly (❙Figure 20-2c). It is partially covered by a thin mucous membrane, the **hymen**, which typically is physically disrupted by the first sexual intercourse. The vaginal and urethral openings are surrounded laterally by two pairs of skin folds, the **labia minora** and **labia majora**. The smaller labia minora are located medially to the more prominent labia majora. The **clitoris** in females is derived from the same embryonic tissue as is the penis in males. Most of the clitoris is located internally, except for the externally visible **glans clitoris**, an erotic structure composed of tissue similar to that of the glans penis (❙Figure 20-2d). The glans clitoris is roughly the size and shape of a pea and lies at the anterior end of the folds of the labia minora (❙Figure 20-2c). The female external genitalia are collectively referred to as the **vulva**.

Reproductive cells each contain a half set of chromosomes.

The deoxyribonucleic acid (DNA) molecules that carry the cell's genetic code are not randomly crammed into the nucleus but, along with associated nuclear proteins, are precisely organized into **chromosomes**. Each chromosome consists of a different DNA molecule that contains a unique set of genes. Somatic (body) cells contain 46 chromosomes (the **diploid number**), which can be sorted into 23 pairs on the basis of various distinguishing features. Chromosomes composing a matched pair are termed **homologous chromosomes**, one member of each pair having been derived from the individual's mother and the other member from the father. Gametes (that is, sperm and eggs) contain only one member of each homologous pair for a total of 23 chromosomes (the **haploid number**).

Gametogenesis is accomplished by meiosis, resulting in genetically unique sperm and ova.

Most cells in the human body have the ability to reproduce themselves, a process important in growth, replacement, and repair of tissues. Cell division involves two components: division of the nucleus and division of the cytoplasm. Nuclear division in somatic cells is accomplished by **mitosis**. In mitosis, the chromosomes replicate (make duplicate copies of themselves); then, the identical chromosomes are separated so that a complete set of genetic information (that is, a diploid number of chromosomes) is distributed to each of the two new daughter cells. Nuclear division in the specialized case of gametes is accomplished by **meiosis**, in which only a half set of genetic information (that is, a haploid number of chromosomes) is distributed to each of four new daughter cells.

During meiosis, a specialized diploid germ cell undergoes one chromosome replication followed by two nuclear divisions. In the first meiotic division, the replicated chromosomes do not separate into two individual, identical chromosomes but remain joined. The doubled chromosomes sort themselves into homologous pairs, and the pairs separate so that each of two daughter cells receives a half set of doubled chromosomes. During the second meiotic division, the doubled chromosomes within each of the two daughter cells separate and are distributed into two cells, yielding four daughter cells, each containing a half set of chromosomes, a single member of each pair. During this process, the maternally and paternally derived chromosomes of each homologous pair are distributed to the daughter cells in random assortments containing one member of each chromosome pair without regard for its original derivation. That is, not all of the mother-derived chromosomes go to one daughter cell and the father-derived chromosomes to the other cell. More than 8 million (2^{23}) mixtures of the 23 paternal and maternal chromosomes are possible. This genetic mixing provides novel combinations of chromosomes. Crossing-over contributes even further to genetic diversity. *Crossing-over* refers to the physical exchange of chromosome material between the homologous pairs before their separation during the first meiotic division.

Thus, sperm and ova each have a unique haploid number of chromosomes. During fertilization, a sperm and ovum fuse to form the start of a new individual with 46 chromosomes, one member of each chromosomal pair having been inherited from the mother and the other member from the father.

The sex of an individual is determined by the combination of sex chromosomes.

Whether individuals are destined to be males or females is a genetic phenomenon determined by the sex chromosomes they possess. As the 23 chromosome pairs are separated during meiosis, each sperm or ovum receives only one member of each chromosome pair. Of the chromosome pairs, 22 are **autosomal chromosomes** that code for general human characteristics and for specific traits such as eye color. The remaining pair of chromosomes consists of the **sex chromosomes**, of which there are two genetically different types—a larger **X chromosome** and a smaller **Y chromosome** (❙Figure 20-3).

Sex determination depends on the combination of sex chromosomes: **Genetic males** have both an X and a Y sex chromosome; **genetic females** have two X sex chromosomes. Thus, the genetic difference responsible for all the anatomic and functional distinctions between males and females is the single Y chromosome. Males have it; females do not.

As a result of meiosis during gametogenesis, all chromosome pairs are separated so that each daughter cell contains

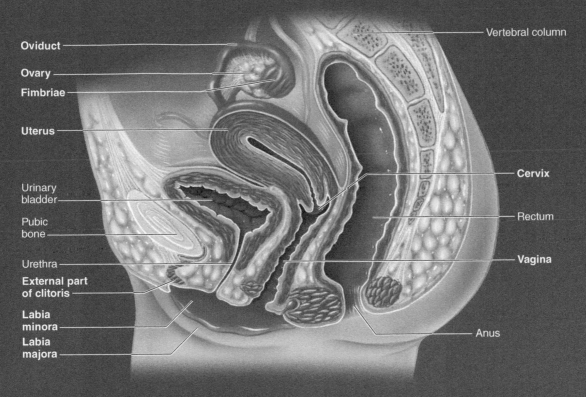

(a) The pelvis in sagittal section

Oviduct

Ovary

Fimbriae

Uterus

Urinary bladder

Pubic bone

Urethra

External part of clitoris

Labia minora

Labia majora

Vertebral column

Cervix

Rectum

Vagina

Anus

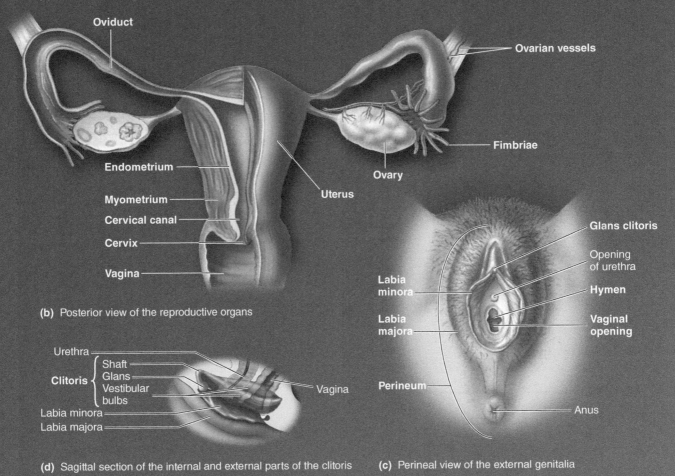

Oviduct

Ovarian vessels

Endometrium

Myometrium

Cervical canal

Cervix

Vagina

Fimbriae

Ovary

Uterus

(b) Posterior view of the reproductive organs

Urethra

Clitoris { Shaft / Glans / Vestibular bulbs

Labia minora

Labia majora

Vagina

(d) Sagittal section of the internal and external parts of the clitoris

Glans clitoris

Opening of urethra

Hymen

Vaginal opening

Labia minora

Labia majora

Perineum

Anus

(c) Perineal view of the external genitalia

| Figure 20-2 **The female reproductive system.**

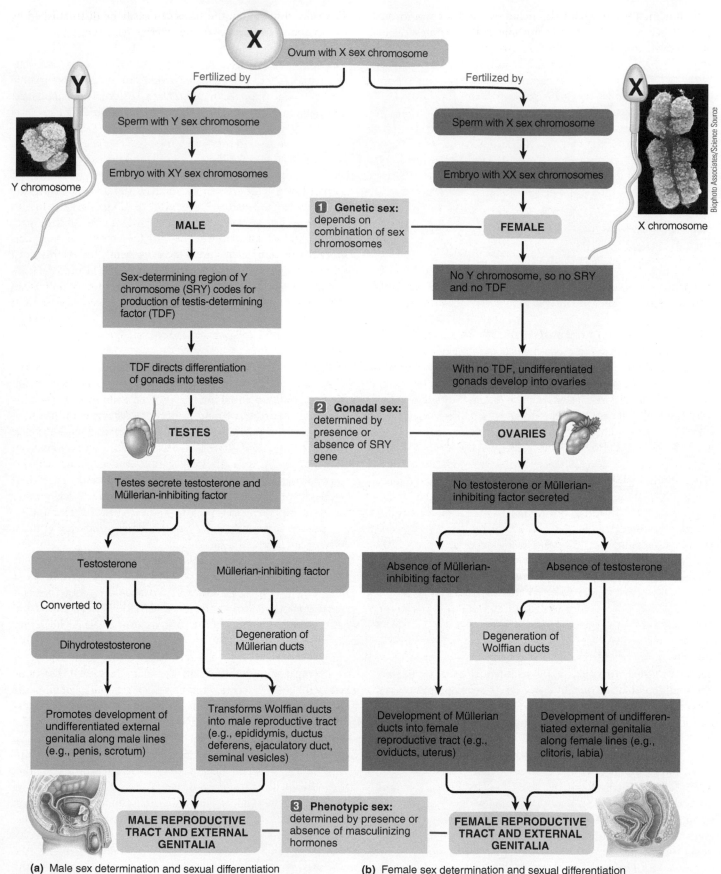

(a) Male sex determination and sexual differentiation

(b) Female sex determination and sexual differentiation

❙Figure 20-3 **Sex determination and sexual differentiation.**

FIGURE FOCUS: *Describe the gonads, reproductive tract, and external genitalia that would be present in a newborn with an XY sex chromosome who lacks 5α-reductase, the enzyme that converts testosterone to dihydrotestosterone.*

only one member of each pair, including the sex chromosome pair. When the XY sex chromosome pair separates during sperm formation, half the sperm receive an X chromosome and the other half a Y chromosome. In contrast, during oogenesis, every ovum receives an X chromosome because separation of the XX sex chromosome pair yields only X chromosomes. During fertilization, combination of an X-bearing sperm with an X-bearing ovum produces a genetic female, XX, whereas union of a Y-bearing sperm with an X-bearing ovum results in a genetic male, XY. Thus, genetic sex is determined at the time of conception and depends on which type of sex chromosome is contained within the fertilizing sperm.

Sexual differentiation along male or female lines depends on the presence or absence of masculinizing determinants.

Differences between males and females exist at three sex levels: genetic, gonadal, and phenotypic (anatomic) (❚ Figure 20-3).

Genetic and Gonadal Sex **Genetic sex**, which depends on the combination of sex chromosomes at the time of conception, in turn determines **gonadal sex**, that is, whether testes or ovaries develop. The presence or absence of a Y chromosome determines gonadal differentiation. For the first month and a half of gestation, all embryos have the potential to differentiate along either male or female lines because the developing reproductive tissues of both sexes are identical and indifferent. Gonadal specificity appears during the seventh week of intrauterine life when the **gonadal ridge** (undifferentiated gonadal tissue present in both males and females) of a genetic male begins to differentiate into testes under the influence of the **sex-determining region** of the Y chromosome (**SRY**), the gene solely responsible for sex determination. This gene triggers a chain of reactions that leads to physical development of a male. SRY "masculinizes" the gonads by coding for production of **testis-determining factor (TDF)** (also known as **SRY protein**) within primitive gonadal cells. TDF directs a series of events that leads to differentiation of the gonads into testes.

Because genetic females lack the SRY gene and consequently do not produce TDF, their gonadal cells never receive a signal for testes formation, so by default during the ninth week the undifferentiated gonadal tissue starts developing into ovaries instead, guided by female-specific gene products whose action is blocked by TDF in male embryos.

Phenotypic Sex **Phenotypic sex**, the apparent anatomic sex of an individual, is hormonally mediated and depends on the genetically determined gonadal sex. The term **sexual differentiation** refers to the embryonic development of the external genitalia and reproductive tract along either male or female lines. As with the undifferentiated gonads, embryos of both sexes have the potential to develop male or female external genitalia and reproductive tracts. Differentiation into a male-type reproductive system is induced by **androgens**, which are masculinizing hormones secreted by the developing testes. The absence of these testicular hormones in female fetuses results in the development of a female-type reproductive system. By 10 to 12 weeks of gestation, the sexes can easily be distinguished by the anatomic appearance of the external genitalia.

Sexual Differentiation of the External Genitalia Male and female external genitalia develop from the same embryonic tissue. In both sexes, the undifferentiated external genitals consist of a *genital tubercle*; paired *urethral folds* surrounding a urethral groove; and, more laterally, *genital (labioscrotal) swellings* (❚ Figure 20-4). The **genital tubercle** gives rise in males to the penis and in females to the clitoris (see ❚ Figures 20-1a and 20-2d). In males, the **urethral folds** fuse around the urethral groove to form the cord of erectile tissue that encircles the urethra within the penis (see ❚ Figure 20-1c). The **genital swellings** similarly fuse to form the scrotum; the skin that covers the penis; and the **prepuce**, a fold of skin that extends over the end of the penis and more or less completely covers the glans penis (unless this extra skin is cut off during **circumcision**). In females, the urethral folds and genital swellings do not fuse at midline but develop instead into the labia minora and labia majora, respectively. The urethral groove remains open, providing access to the interior through both the urethral opening and the vaginal opening.

Sexual Differentiation of the Reproductive Tract Although the male and female external genitalia develop from the same multipotential undifferentiated embryonic tissue, this is not the case with the reproductive tracts. Two primitive duct systems—the Wolffian ducts and the Müllerian ducts—develop in all embryos. In males, the reproductive tract develops from the **Wolffian ducts** and the Müllerian ducts degenerate, whereas in females the **Müllerian ducts** differentiate into the reproductive tract and the Wolffian ducts regress (❚ Figure 20-5). Because both duct systems are present before sexual differentiation occurs, the early embryo has the potential to develop either a male or a female reproductive tract.

Development of the reproductive tract along male or female lines is determined by the presence or absence of two hormones secreted by two different cell types in the fetal testes—*testosterone* produced by the newly developed Leydig cells and **Müllerian-inhibiting factor** (also known as **anti-Müllerian hormone**), produced by the early Sertoli cells (see ❚ Figure 20-3). (You will learn shortly about the location and functions of Leydig and Sertoli cells in the adult testes.) A hormone released by the placenta, *human chorionic gonadotropin,* is the stimulus for this early testicular secretion. A portion of testosterone secreted by the testes is converted peripherally by the enzyme **5α-reductase** into **dihydrotestosterone (DHT)**, a closely related, potent androgen. Testosterone induces development of the Wolffian ducts into the male reproductive tract (epididymis, ductus deferens, ejaculatory duct, and seminal vesicles), whereas its derivative DHT promotes differentiation of the external genitalia into the penis and scrotum. Meanwhile, Müllerian-inhibiting factor causes regression of the Müllerian ducts. The other two male accessory reproductive organs, the prostate gland and the bulbourethral glands, arise from the **urogenital sinus**, an embryonic structure from which the urinary bladder and urethra are also derived.

In the absence of testosterone and Müllerian-inhibiting factor in females, the Wolffian ducts regress, the Müllerian ducts

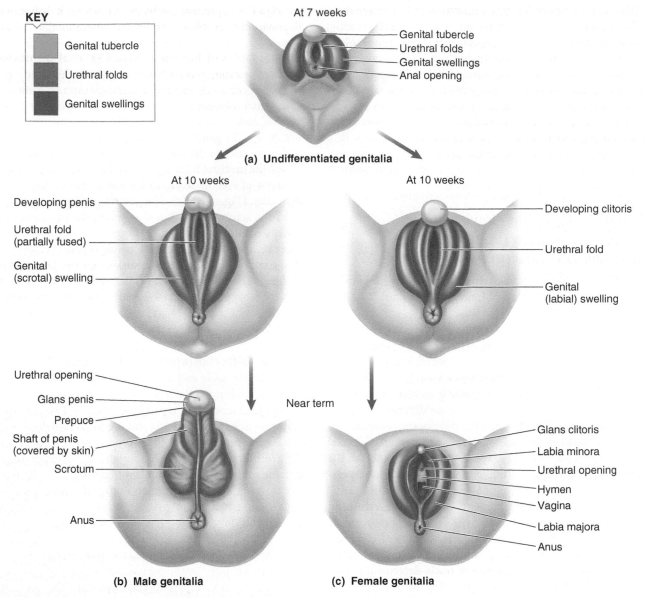

Genital tubercle

Urethral folds

Genital swellings

At 7 weeks

Genital tubercle
Urethral folds
Genital swellings
Anal opening

(a) Undifferentiated genitalia

At 10 weeks

Developing penis

Urethral fold
(partially fused)

Genital
(scrotal) swelling

At 10 weeks

Developing clitoris

Urethral fold

Genital
(labial) swelling

Near term

Urethral opening

Glans penis

Prepuce

Shaft of penis
(covered by skin)

Scrotum

Anus

Glans clitoris
Labia minora
Urethral opening
Hymen
Vagina
Labia majora
Anus

(b) Male genitalia

(c) Female genitalia

Figure 20-4 Sexual differentiation of the external genitalia.

develop into the female reproductive tract (oviducts, uterus, cervix, and upper vagina), and the external genitalia differentiate into the clitoris and labia. The lower vagina is derived from the urogenital sinus, as is the urinary bladder and urethra.

Note that the undifferentiated embryonic reproductive tissue develops into a female structure unless actively acted on by masculinizing factors that inhibit female development. In the absence of male testicular hormones, a female reproductive tract and external genitalia develop regardless of the genetic sex of the individual. For feminization of the fetal genital tissue, ovaries do not even need to be present. Such a control pattern for determining sexual differentiation is appropriate, considering that fetuses of both sexes are exposed to high concentrations of female sex hormones throughout gestation. If female sex hormones influenced the development of the reproductive tract and external genitalia, all fetuses would be feminized.

Clinical Note **Errors in Sexual Differentiation** Genetic sex and phenotypic sex are usually compatible—that is, a genetic male anatomically appears to be a male and functions as a male, and the same compatibility holds true for females. Occasionally, however, discrepancies occur between genetic and anatomic sexes because of errors in sexual differentiation, as the following examples illustrate:

- If testes in a genetic male fail to properly differentiate and secrete hormones, the result is the development of an apparent anatomic female, who will be sterile, in a genetic male. Similarly, genetic males whose target cells lack receptors for testosterone are feminized, even though their testes secrete adequate testosterone (see p. 643, *testicular feminization syndrome*, also known as *androgen insensitivity syndrome*).

- Because testosterone acts on the Wolffian ducts to convert them into a male reproductive tract but the testosterone derivative DHT is responsible for masculinization of the external genitalia, a genetic deficiency of the enzyme that converts testosterone into DHT results in a genetic male with testes and a male reproductive tract but with female external genitalia.

■ The adrenal gland normally secretes a weak androgen, *dehydroepiandrosterone (DHEA)*, in insufficient quantities to masculinize females. However, pathologically excessive secretion of this hormone in a genetically female fetus during critical developmental stages imposes differentiation of the reproductive tract and genitalia along male lines (see p. 677, *adrenogenital syndrome*).

Sometimes these discrepancies between genetic sex and apparent anatomic sex are not recognized until puberty, when the discovery produces a psychologically traumatic gender identity crisis. For example, a masculinized genetic female with ovaries but with male-type external genitalia may be reared as a boy until puberty, when breast enlargement (caused by estrogen secretion by the awakening ovaries) and lack of beard growth (caused by lack of testosterone secretion in the absence of tes-

tes) signal an apparent problem. Therefore, it is important to diagnose any problems in sexual differentiation in infancy. Once a sex has been assigned, it can be reinforced, if necessary, with surgical and hormonal treatment so that psychosexual development can proceed as normally as possible. Less dramatic cases of inappropriate sexual differentiation often appear as sterility problems.

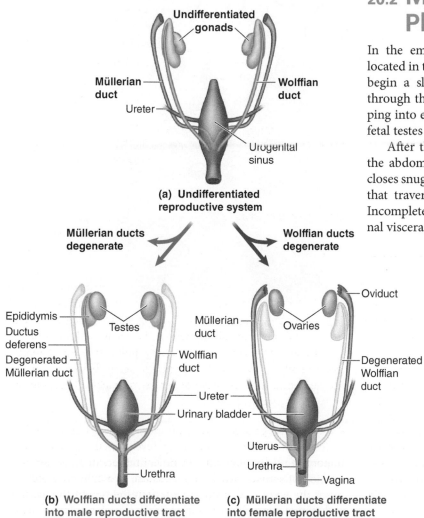

(a) Undifferentiated reproductive system

(b) Wolffian ducts differentiate into male reproductive tract (shown before descent of testes into scrotum)

(c) Müllerian ducts differentiate into female reproductive tract

⏐Figure 20-5 Sexual differentiation of the reproductive tract.

20.2 Male Reproductive Physiology

In the embryo, the testes develop from the gonadal ridge located in the rear of the abdominal cavity. Late in fetal life, they begin a slow descent, passing out of the abdominal cavity through the **inguinal canal** into the scrotum, one testis dropping into each pocket of the scrotal sac. Testosterone from the fetal testes induces descent of the testes into the scrotum.

After the testes descend into the scrotum, the opening in the abdominal wall through which the inguinal canal passes closes snugly around the sperm-carrying duct and blood vessels that traverse between each testis and the abdominal cavity. Incomplete closure or rupture of this opening permits abdominal viscera to slip through, resulting in an **inguinal hernia**.

Although the time varies somewhat, descent is usually complete by the seventh month of gestation. As a result, descent is complete in 98% of full-term baby boys.

Clinical Note However, in a substantial percentage of premature male infants the testes are still within the inguinal canal at birth. In most instances of retained testes, descent occurs naturally before puberty or can be encouraged with administration of testosterone. Rarely, a testis remains undescended into adulthood, a condition known as **cryptorchidism** (*crypt* means "hidden"; *orchid* means "testis").

The scrotal location of the testes provides a cooler environment for spermatogenesis.

The temperature within the scrotum averages several degrees Celsius less than normal body (core) temperature. Descent of the testes into

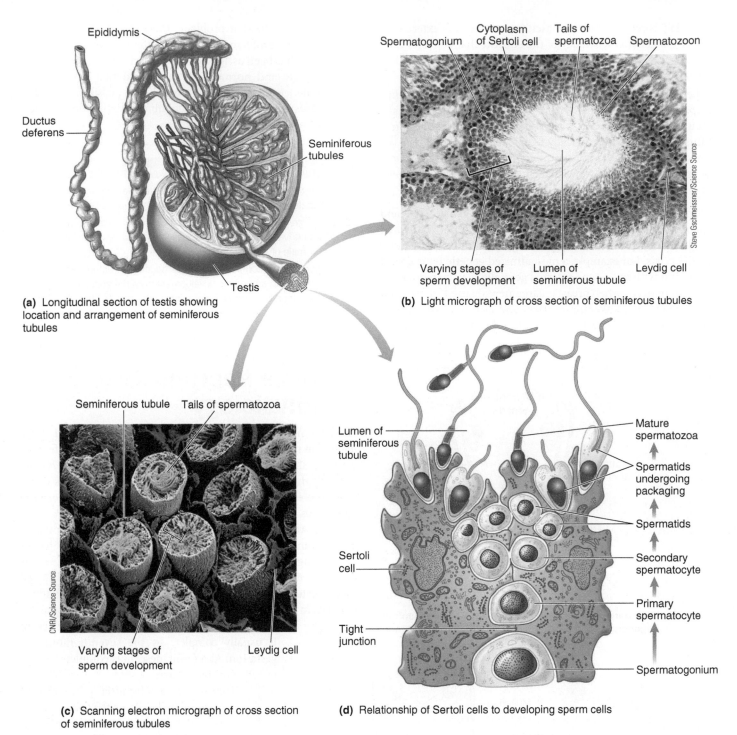

(a) Longitudinal section of testis showing location and arrangement of seminiferous tubules

(b) Light micrograph of cross section of seminiferous tubules

(c) Scanning electron micrograph of cross section of seminiferous tubules

(d) Relationship of Sertoli cells to developing sperm cells

❙Figure 20-6 **Anatomy of testis depicting the site of spermatogenesis.** (a) The seminiferous tubules are the sperm-producing portion of the testis. (b) The undifferentiated germ cells (the spermatogonia) lie in the periphery of the tubule, and the differentiated spermatozoa are in the lumen, with the various stages of sperm development in between. (c) Note the presence of the highly differentiated spermatozoa (recognizable by their tails) in the lumen of the seminiferous tubules. (d) Relationship of the Sertoli cells to the developing sperm cells.

this cooler environment is essential because spermatogenesis is temperature sensitive and cannot occur at normal body temperature. Therefore, a cryptorchid is unable to produce viable sperm.

The position of the scrotum in relation to the abdominal cavity can be varied by a spinal reflex mechanism that plays an important role in regulating testicular temperature. Reflex contraction of scrotal muscles on exposure to a cold environment raises the scrotal sac to bring the testes closer to the warmer abdomen. Conversely, relaxation of the muscles on exposure to heat permits the scrotal sac to become more pendulous, moving the testes farther from the warm core of the body.

The testicular Leydig cells secrete masculinizing testosterone.

The testes perform the dual function of producing sperm and secreting testosterone. About 80% of the testicular mass consists of highly coiled **seminiferous tubules**, within which spermatogenesis takes place (Figure 20-6a). The endocrine cells that produce testosterone—the **Leydig**, or **interstitial**, **cells**—lie in the connective tissue (interstitial tissue) between the seminiferous tubules (Figure 20 6b). Thus, the portions of the testes that produce sperm and secrete testosterone are structurally and functionally distinct.

Testosterone is a steroid hormone derived from a cholesterol precursor molecule, as are the female sex hormones, estrogen and progesterone (see Figure 19-8, p. 673). Once produced, some of the testosterone is secreted into the blood, where it is transported to its target sites of action. A substantial portion of the newly synthesized testosterone goes into the lumen of the seminiferous tubules, where it plays an important role in sperm production.

To exert its effects, testosterone binds with androgen receptors in the cytoplasm of target cells. The androgen-receptor complex moves to the nucleus, where it binds with the androgen response element on DNA, leading to transcription of genes that direct synthesis of new proteins that carry out the desired cellular response.

Most of testosterone's actions ultimately function to ensure delivery of sperm to the female. The effects of testosterone can be grouped into five categories: (1) effects on the reproductive system before birth, (2) effects on sex-specific tissues after birth, (3) other reproduction-related effects, (4) effects on secondary sexual characteristics, and (5) nonreproductive actions (Table 20-1).

Effects on the Reproductive System Before Birth

Before birth, testosterone secretion by the Leydig cells of the fetal testes masculinizes the reproductive tract and external genitalia and promotes descent of the testes into the scrotum, as already described. After birth, testosterone secretion ceases, and the testes and remainder of the reproductive system remain small and nonfunctional until puberty.

Effects on Sex-Specific Tissues After Birth

Puberty is the period of arousal and maturation of the previously nonfunctional reproductive system, culminating in sexual maturity and the ability to reproduce. It usually begins some time between the ages of 10 and 14 in males (on average, earlier, between the ages of 9 and 13, in females). Usually lasting three to five years, puberty encompasses a complex sequence of endocrine, physical, and behavioral events. **Adolescence** is a broader concept that refers to the entire transition period between childhood and adulthood, not just to sexual maturation. In both sexes, the reproductive changes that take place during puberty are (1) enlargement and maturation of gonads, (2) development of secondary sexual characteristics, (3) achievement of fertility (gamete production), (4) growth and maturation of the reproductive tract, and (5) attainment of libido (sex drive). The pubertal growth spurt also occurs.

TABLE 20-1 Effects of Testosterone
Effects before Birth
Masculinizes the reproductive tract and external genitalia
Promotes descent of the testes into the scrotum
Effects on Sex-Specific Tissues after Birth
Promotes growth and maturation of the reproductive system at puberty
Is essential for spermatogenesis
Maintains the reproductive tract throughout adulthood
Other Reproduction-Related Effects
Develops the sex drive at puberty
Controls gonadotropin hormone secretion
Effects on Secondary Sexual Characteristics
Induces the male pattern of hair growth (e.g., beard)
Causes the voice to deepen because vocal folds thicken
Promotes muscle growth responsible for the male body configuration
Nonreproductive Actions
Exerts a protein anabolic effect
Promotes bone growth at puberty
Closes the epiphyseal plates after being converted to estrogen by aromatase
May induce aggressive behavior

At puberty in males, the Leydig cells start secreting testosterone once again. Testosterone is responsible for growth and maturation of the entire male reproductive system. Under the influence of the pubertal surge in testosterone secretion, the testes enlarge and start producing sperm for the first time, the accessory sex glands enlarge and become secretory, and the penis and scrotum enlarge.

Ongoing testosterone secretion is essential for spermatogenesis and for maintaining a mature male reproductive tract throughout adulthood. Once initiated at puberty, testosterone secretion and spermatogenesis occur continuously throughout the male's life, although testicular efficiency gradually declines after 45 to 50 years of age. However, men in their 70s and beyond may continue to enjoy an active sex life, and some even father a child at this late age. The gradual reduction in circulating testosterone levels and in sperm production is not caused by a decrease in stimulation of the testes but arises instead from degenerative changes associated with aging that occur in small testicular blood vessels. This age-related gradual decline is sometimes mistakenly termed "male menopause" or "andropause," but it is not comparable to female menopause, which is preprogrammed and results in complete, abrupt cessation of reproductive capability. The androgen decline in males is more aptly termed **androgen deficiency in aging males (ADAM)**.

Following **castration** (surgical removal of the testes) or testicular failure caused by disease, the other sex organs regress in size and function.

Other Reproduction-Related Effects Testosterone governs development of sexual libido at puberty and helps maintain the sex drive in the adult male. Stimulation of this behavior by testosterone is important for facilitating delivery of sperm to females. In humans, libido is also influenced by many interacting social and emotional factors.

In another reproduction-related function, testosterone participates in the normal negative-feedback control of gonadotropin hormone secretion by the anterior pituitary, a topic covered more thoroughly later.

Effects on Secondary Sexual Characteristics All male secondary sexual characteristics depend on testosterone for their development and maintenance. These nonreproductive male characteristics induced by testosterone include (1) the male pattern of hair growth (for example, beard and chest hair and, in genetically predisposed men, baldness); (2) a deep voice caused by enlargement of the larynx and thickening of the vocal folds; (3) thick skin; and (4) the male body configuration (for example, broad shoulders and heavy arm and leg musculature) as a result of protein deposition. A male castrated before puberty (a **eunuch**) does not mature sexually, nor does he develop secondary sexual characteristics.

Nonreproductive Actions Testosterone exerts several important effects not related to reproduction. It has a general protein anabolic (synthesis) effect, thus contributing to the more muscular physique of males. It also plays a role in the pubertal growth spurt. Ironically, testosterone not only stimulates bone growth, but also eventually prevents further growth by sealing the growing ends of the long bones (that is, ossifying, or "closing," the epiphyseal plates; see p. 656).

In animals, testosterone induces aggressive behavior, but its influence on human behavior other than in the area of sexual behavior is unresolved. Even though some athletes and bodybuilders who (illegally) take testosteronelike anabolic androgenic steroids to increase muscle mass display more aggressive behavior (see p. 277), to what extent general behavioral differences between the sexes are hormonally induced or result from social conditioning is unclear.

Conversion of Testosterone to Estrogen in Males Although testosterone is classically considered the male sex hormone and estrogen a female sex hormone, the distinctions are not as clear-cut as once thought. In addition to the minuscule amount of estrogen produced by the adrenal cortex (see p. 676), a small portion of testosterone secreted by the testes is converted to estrogen outside the testes by the enzyme **aromatase**, which is widely distributed but most abundant in adipose tissue. (Remember that part of the secreted testosterone is also converted to dihydrotestosterone. Unlike testosterone, DHT cannot be converted to estrogen.) It is sometimes difficult to distinguish effects of testosterone and those of testosterone-turned-estrogen. For example, closure of the epiphyseal plates in males is induced not by testosterone per se but by testosterone turned into estrogen by aromatization. Estrogen receptors have been identified in the testes, prostate, bone, and elsewhere in males. Estrogen even plays an essential role in male reproductive health; for example, it is important in spermatogenesis and, surprisingly, contributes to male heterosexuality. The depth, breadth, and mechanisms of action of estrogen in males are only beginning to be explored. (Likewise, in addition to the weak androgenic hormone DHEA produced by the adrenal cortex in both sexes, the ovaries in females secrete a small amount of testosterone, the functions of which remain unclear.)

We now shift attention from testosterone secretion to the other function of the testes—sperm production.

Spermatogenesis yields an abundance of highly specialized, mobile sperm.

About 250 m (800 ft) of sperm-producing seminiferous tubules are packed within the testes (▮Figure 20-6a, b, and c). Two functionally important cell types are present in these tubules: *germ cells,* most of which are in various stages of sperm development, and *Sertoli cells,* which provide crucial support for spermatogenesis (▮Figure 20-6b and d). **Spermatogenesis** is a complex process by which relatively undifferentiated primordial (primitive or initial) germ cells, the **spermatogonia** (each of which contains a diploid complement of 46 chromosomes), proliferate and are converted into extremely specialized, motile spermatozoa (sperm), each bearing a randomly distributed haploid set of 23 chromosomes.

Microscopic examination of a seminiferous tubule reveals layers of germ cells in a progression of sperm development, starting with the least differentiated in the outer layer and moving inward through various stages of division to the lumen, where the highly differentiated sperm are ready for exit from the testis (▮Figure 20-6b, c, and d). Spermatogenesis takes 64 days for development from a spermatogonium to a mature sperm. At any given time, different seminiferous tubules are in different stages of spermatogenesis. Up to several hundred million sperm may reach maturity daily. Spermatogenesis encompasses three major stages: *mitotic proliferation, meiosis,* and *packaging* (▮Figure 20-7).

Mitotic Proliferation Spermatogonia located in the outermost layer of the tubule continuously divide mitotically, with all new cells bearing the full complement of 46 chromosomes identical to those of the parent cell. Such proliferation provides a continual supply of new germ cells. Following mitotic division of a spermatogonium, one of the daughter cells remains at the outer edge of the tubule as an undifferentiated spermatogonium, thus maintaining the germ-cell line. The other daughter cell starts moving toward the lumen while undergoing the various steps required to form sperm, which will be released into the lumen. In humans, the sperm-forming daughter cell divides mitotically twice more to form four identical **primary spermatocytes**. After the last mitotic division, the primary spermatocytes enter a resting phase during which the chromosomes are duplicated and the doubled strands remain together in preparation for the first meiotic division.

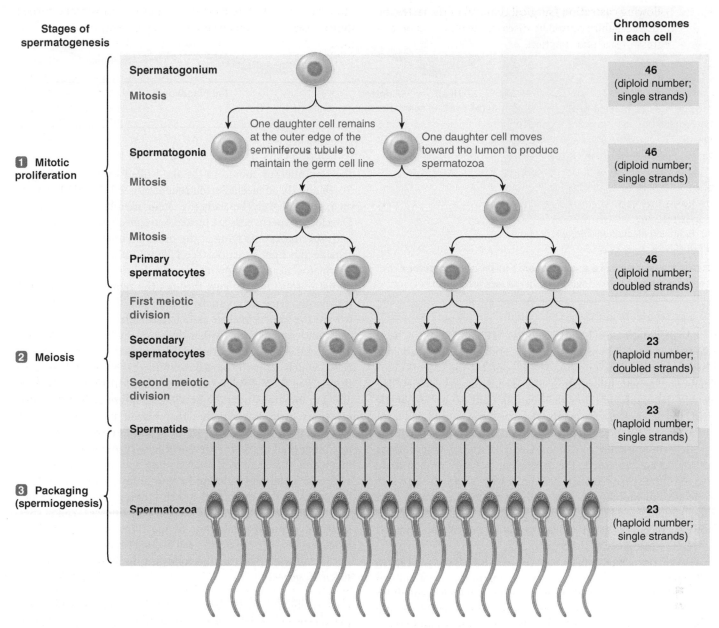

Stages of spermatogenesis

Chromosomes in each cell

Spermatogonium
Mitosis

46 (diploid number; single strands)

1 Mitotic proliferation

Spermatogonia

One daughter cell remains at the outer edge of the seminiferous tubule to maintain the germ cell line

One daughter cell moves toward the lumen to produce spermatozoa

46 (diploid number; single strands)

Mitosis

Mitosis

Primary spermatocytes

46 (diploid number; doubled strands)

First meiotic division

Secondary spermatocytes

2 Meiosis

23 (haploid number; doubled strands)

Second meiotic division

Spermatids

23 (haploid number; single strands)

3 Packaging (spermiogenesis)

Spermatozoa

23 (haploid number; single strands)

I Figure 20-7 **Spermatogenesis.**
FIGURE FOCUS: *(1) How many spermatozoa are produced from a single spermatogonium? (2) Compare the chromosome composition in these two stages of gamete development.*

Meiosis During meiosis, each primary spermatocyte (with a diploid number of 46 doubled chromosomes) forms two **secondary spermatocytes** (each with a haploid number of 23 doubled chromosomes) during the first meiotic division, finally yielding four **spermatids** (each with 23 single chromosomes) as a result of the second meiotic division.

No further division takes place beyond this stage of spermatogenesis. Each spermatid is remodeled into a single spermatozoon. Because each sperm-producing spermatogonium mitotically produces four primary spermatocytes and each primary spermatocyte meiotically yields four spermatids (spermatozoa-to-be), the spermatogenic sequence in humans can theoretically produce 16 spermatozoa each time a spermatogonium initiates this process. Usually, however, some cells

are lost at various stages, so the efficiency of productivity is rarely this high.

Packaging Even after meiosis, spermatids still resemble undifferentiated spermatogonia structurally, except for their half complement of chromosomes. Production of extremely specialized, mobile spermatozoa from spermatids requires extensive remodeling, or **packaging**, of cell elements, a process alternatively known as **spermiogenesis**. Sperm are essentially "stripped-down" cells in which most of the cytosol and any organelles not needed for delivering the sperm's genetic information to an ovum have been extruded. Thus, sperm travel lightly, taking with them only the bare essentials to accomplish fertilization.

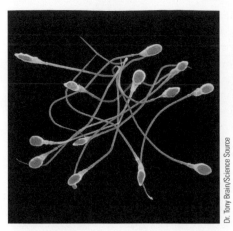

(a) Scanning electron micrograph of human spermatozoa

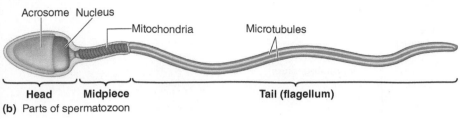

(b) Parts of spermatozoon

❙Figure 20-8 **Anatomy of a spermatozoon.** A spermatozoon has three functional parts: a head with its acrosome "cap," a midpiece, and a tail.

A **spermatozoon** has three parts (❙Figure 20-8): a head capped with an acrosome, a midpiece, and a tail. The **head** consists primarily of the nucleus, which contains the sperm's complement of genetic information. The **acrosome**, an enzyme-filled vesicle that caps the tip of the head, is used as an "enzymatic drill" for penetrating the ovum. The acrosome, a modified lysosome (see p. 30), is formed by aggregation of vesicles produced by the endoplasmic reticulum–Golgi complex before these organelles are discarded. The acrosomal enzymes remain inactive until the sperm contacts an egg, at which time the enzymes are released. Mobility for the spermatozoon is provided by a long, whiplike **tail** (a flagellum; see p. 48), movement of which is powered by energy generated by the mitochondria concentrated within the **midpiece** of the sperm.

Until sperm maturation is complete, the developing germ cells arising from a single primary spermatocyte remain joined by cytoplasmic bridges. These connections, which result from incomplete cytoplasmic division, permit the four developing sperm to exchange cytoplasm. This linkage is important because the X chromosome, but not the Y chromosome, contains genes that code for cell products essential for sperm development. (Whereas the large X chromosome contains several thousand genes, the small Y chromosome has only a few dozen, the most important of which are the SRY gene and others that play critical roles in male fertility.) During meiosis, half the sperm receive an X and the other half a Y chromosome. Were it not for the sharing of cytoplasm so that all the haploid cells are provided with the products coded for by X chromosomes until sperm development is complete, the Y-bearing, male-producing sperm could not develop and survive.

Throughout their development, sperm remain intimately associated with Sertoli cells.

The seminiferous tubules house **Sertoli cells** in addition to the developing sperm cells. Sertoli cells, which are epithelial cells, lie side by side and form a ring around the tubule lumen, with each Sertoli cell spanning the entire distance from the outer surface of the seminiferous tubule to the fluid-filled lumen (see ❙Figure 20-6b and d). Adjacent Sertoli cells are joined by tight junctions at a point slightly beneath the outer membrane (see p. 61). Developing sperm cells are tucked between adjacent Sertoli cells, with spermatogonia lying at the outer perimeter of the tubule, outside the tight junction (see ❙Figure 20-6b and d). During spermatogenesis, developing sperm cells arising from spermatogonial mitotic activity pass through the tight junctions, which transiently separate to make a path for them, then migrate toward the lumen in close association with the adjacent Sertoli cells, undergoing their further divisions during this migration. The lateral edges of the Sertoli cells envelop the migrating sperm cells, which remain buried within these cavelike recesses throughout their development. Sertoli cells form tight junctions and gap junctions with the developing sperm cells. Unlike gap junctions in excitable tissues that permit passage of charge-carrying ions, gap junctions in the seminiferous tubules serve a role other than transfer of electrical activity. At all stages of spermatogenic maturation, the developing sperm and Sertoli cells exchange small molecules and communicate with one another by means of this direct cell-to-cell binding. Final release of a mature spermatozoa from the Sertoli cell, a process called **spermiation**, requires breakdown of the tight junctions and gap junctions between the Sertoli cell and spermatozoa.

Sertoli cells perform the following functions essential for spermatogenesis:

1. The tight junctions between adjacent Sertoli cells form a **blood–testes barrier** that prevents blood-borne substances from passing between the cells to gain entry to the lumen of the seminiferous tubule. Because of this barrier, only selected molecules that can pass through Sertoli cells reach the intratubular fluid. As a result, the composition of the intratubular fluid varies considerably from that of the blood. The unique composition of this fluid that bathes the germ cells is critical for later stages of sperm development. The blood–testes barrier also prevents the antibody-producing cells in the ECF from reaching the tubular sperm factory, thus preventing the formation of antibodies against the highly differentiated spermatozoa.

2. Because the secluded developing sperm cells do not have direct access to blood-borne nutrients, the "nurse" Sertoli cells provide nourishment for them. Developing sperm cells cannot efficiently use glucose. Sertoli cells take up glucose, metabolize the glucose to lactate, then transfer the lactate to the sperm cells, which can use lactate as an energy source.

3. Sertoli cells have an important phagocytic function: They engulf the cytoplasm extruded from spermatids during their

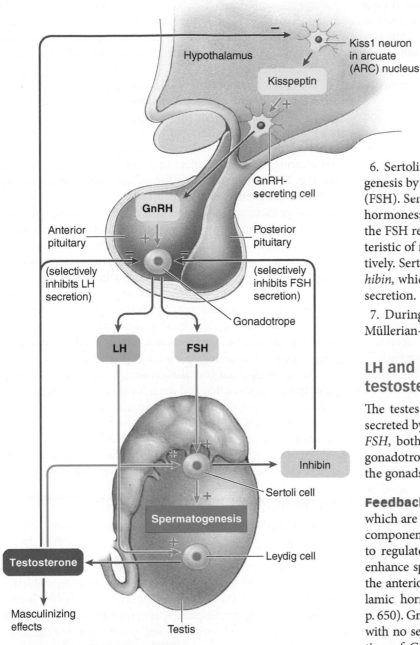

IFigure 20-9 **Control of testicular function.**

FIGURE FOCUS: *(1) What is LH's effect on the testes? (2) By what means is LH secretion inhibited? (3) What is FSH's effect on the testes? (4) By what means is FSH secretion inhibited?*

the seminiferous tubule fluid than in the blood. This high local concentration of testosterone is essential for sustaining sperm production. Androgen-binding protein is necessary to retain testosterone within the lumen because this steroid hormone is lipid soluble and could easily diffuse across the plasma membranes and leave the lumen. Testosterone itself stimulates production of androgen-binding protein.

6. Sertoli cells are the site of action for control of spermatogenesis by both testosterone and follicle-stimulating hormone (FSH). Sertoli cells have distinct receptors for each of these hormones: The receptors for testosterone are intracellular and the FSH receptors are on the membrane surface, as is characteristic of receptors for steroid and peptide hormones, respectively. Sertoli cells themselves release another hormone, *inhibin*, which acts in negative-feedback fashion to regulate FSH secretion.

7. During fetal development, Sertoli cells also secrete Müllerian-inhibiting factor.

LH and FSH from the anterior pituitary control testosterone secretion and spermatogenesis.

The testes are controlled by the two gonadotropic hormones secreted by the anterior pituitary, *luteinizing hormone (LH)* and *FSH*, both of which are produced by the same cell type, the gonadotrope (see p. 648). Both hormones in both sexes act on the gonads by activating cAMP.

Feedback Control of Testicular Function LH and FSH, which are named for their functions in females, act on separate components of the testes (IFigure 20-9). LH acts on Leydig cells to regulate testosterone secretion. FSH acts on Sertoli cells to enhance spermatogenesis. Secretion of both LH and FSH from the anterior pituitary is stimulated in turn by a single hypothalamic hormone, *gonadotropin-releasing hormone (GnRH)* (see p. 650). GnRH is released in bursts once every two to three hours, with no secretion occurring in between. The blood concentration of GnRH depends on the frequency of these bursts in secretion. This pulsatile secretion of GnRH stimulates ongoing LH and FSH secretion. However, LH and FSH are segregated to a large extent into separate secretory vesicles in the gonadotrope and are not secreted in equal amounts because other regulatory factors influence how much of each gonadotropin is secreted.

Two factors—*testosterone* and *inhibin*—differentially influence the secretory rate of LH and FSH. Testosterone, the product of LH stimulation of Leydig cells, acts in negative-feedback fashion to inhibit LH secretion in two ways. The predominant negative-feedback effect of testosterone is to decrease GnRH release by acting on the hypothalamus, thus indirectly decreasing both LH and FSH release by the anterior pituitary. In addition, testosterone acts directly on the anterior pituitary to selectively reduce LH secretion. The latter action explains why testosterone exerts a greater inhibitory effect on LH secretion than on FSH secretion.

remodeling, and they destroy defective germ cells that fail to successfully complete all stages of spermatogenesis.

4. Sertoli cells secrete into the lumen **seminiferous tubule fluid**, which "flushes" the released sperm from the tubule into the epididymis for storage and further processing.

5. An important component of this Sertoli secretion is **androgen-binding protein**. As the name implies, this protein binds androgens (predominantly testosterone), thus maintaining a very high level of this hormone within the seminiferous tubule lumen. Testosterone is 100 times more concentrated in

The testicular inhibitory signal specifically directed at controlling FSH secretion is the peptide hormone **inhibin**, which is secreted by the Sertoli cells in response to FSH. Inhibin acts directly on the anterior pituitary to selectively inhibit FSH secretion. This feedback inhibition of FSH by a Sertoli cell product is appropriate because FSH stimulates spermatogenesis by acting on Sertoli cells.

Recent studies indicate that control of testicular function begins farther upstream than GnRH. **Kiss1 neurons** in the **arcuate (ARC) nucleus** of the hypothalamus (the same region involved in control of food intake and body weight; see p. 621) release **kisspeptins**, which are peptide neurotransmitters that stimulate GnRH secretion. (The researchers used the name "kiss" after the location where they made their discovery— Hershey, Pennsylvania, famous for chocolate Hershey Kisses.) GnRH-secreting neurons are also located in the ARC nucleus. Testosterone exerts its negative-feedback effects at the hypothalamus on the kiss1 neurons, not directly on the GnRH-secreting neurons. GnRH-secreting neurons do not have androgen receptors (or estrogen or progesterone receptors in females), but kiss1 neurons do have these receptors. By directly inhibiting kiss1 neurons, testosterone indirectly inhibits GnRH-secreting neurons by blocking kiss1 neurons' excitatory action on GnRH-secreting neurons. Kisspeptin signaling appears to be critical for integrating central and peripheral input to regulate GnRH (and thus FSH and LH and sex steroid hormone) output, for initiating puberty, and for maintaining normal reproductive function.

Roles of Testosterone and FSH in Spermatogenesis

Both testosterone and FSH play critical roles in controlling spermatogenesis, each exerting its effect by acting on Sertoli cells. Testosterone is essential for both mitosis and meiosis of the germ cells, whereas FSH is needed for spermatid remodeling. Only the high concentration of testicular testosterone maintained in the intratubular fluid (as a result of a substantial portion of this hormone being complexed with androgen-binding protein) is adequate to sustain sperm production.

GnRH activity increases at puberty.

Even though the fetal testes secrete testosterone, which directs masculine development of the reproductive system, after birth the testes become dormant until puberty. During the prepubertal period, LH and FSH are not secreted at adequate levels to stimulate any significant testicular activity. The prepubertal delay in the onset of reproductive capability allows time for the individual to mature physically (although not necessarily psychologically) enough to handle child rearing. (This physical maturation is especially important in the female, whose body must support the developing fetus.)

During the prepubertal period, GnRH activity is inhibited. The pubertal process is initiated by an increase in GnRH activity. Early in puberty, GnRH secretion occurs only at night, causing a brief nocturnal increase in LH secretion and, accordingly, testosterone secretion. The extent of GnRH secretion gradually increases as puberty progresses until the adult pattern of GnRH, FSH, LH, and testosterone secretion is established. Under the influence of the rising levels of testosterone during puberty, the physical changes that encompass the secondary sexual characteristics and reproductive maturation become evident.

The factors responsible for initiating puberty in humans remain unclear. The hormone *melatonin,* which is secreted by the *pineal gland* within the brain, appears to play an important role (see p. 662). Melatonin, whose secretion decreases during exposure to the light and increases during exposure to the dark, has an antigonadotropic effect in many species. Light striking the eyes inhibits the nerve pathways that stimulate melatonin secretion. In many seasonally breeding species, the overall decrease in melatonin secretion in connection with longer days and shorter nights initiates the mating season. A reduction in the overall rate of melatonin secretion coincides with the onset of puberty in humans—particularly during the night, when the peaks in GnRH secretion first occur. Furthermore, *leptin,* the hormone released from adipose (fat) stores (see p. 623), plays an important role in the onset of puberty, especially in females. In evolutionary history, this mechanism could have been a way to ensure that females had sufficient energy stored to sustain a pregnancy when food supplies were unpredictable.

Recent studies suggest that known signals for triggering puberty, such as circadian and nutritional cues, converge on the ARC nucleus kiss1 neurons, which activate the neuroendocrine reproductive axis by initiating pulsatile secretion of GnRH. Therefore, puberty might begin with a "kiss."

Having completed discussing testicular function, we now shift to the other components of the male reproductive system.

The reproductive tract stores and concentrates sperm and increases their fertility.

The remainder of the male reproductive system (besides the testes) is designed to deliver sperm to the female reproductive tract. It consists of (1) a tortuous pathway of tubes (the reproductive tract), which transports sperm from the testes to outside the body; (2) several accessory sex glands, which contribute secretions important to the viability and motility of the sperm; and (3) the penis, which is designed to penetrate and deposit sperm within the female vagina. We examine each of these parts in greater detail, beginning with the reproductive tract.

Components of the Male Reproductive Tract

A comma-shaped **epididymis** is loosely attached to the rear surface of each testis (see ❙Figures 20-1, p. 717, and 20-6a, p. 724). After sperm are produced in the seminiferous tubules, they are swept into the epididymis as a result of the pressure created by the continual secretion of tubular fluid by the Sertoli cells. The epididymal ducts from each testis converge to form a large, thick-walled, muscular duct called the **ductus (vas) deferens**. The ductus deferens from each testis passes up out of the scrotal sac and runs back through the inguinal canal into the abdominal cavity, where it eventually empties into the urethra at the neck of the bladder (see ❙Figure 20-1). The urethra carries sperm out of the penis during ejaculation, the forceful expulsion of semen from the body.

Functions of the Epididymis and Ductus Deferens

These ducts perform several important functions. The epididymis and ductus deferens serve as the exit route for sperm from the testis. As they leave the testis, sperm are not capable of either moving or fertilizing. They gain both capabilities during their passage through the epididymis. This maturational process is stimulated by the testosterone retained within the tubular fluid bound to androgen-binding protein. The capacity of sperm to fertilize is further enhanced by exposure to secretions of the female reproductive tract. This enhancement of sperm's capacity in the male and female reproductive tracts is known as **capacitation**. The epididymis also concentrates sperm 100-fold by absorbing most of the fluid that enters from the seminiferous tubules. The maturing sperm are slowly moved through the epididymis into the ductus deferens by rhythmic contractions of the smooth muscle in the walls of these tubes.

The ductus deferens is an important site for sperm storage. Because the tightly packed sperm are relatively inactive and their metabolic needs are accordingly low, they can be stored in the ductus deferens for up to 2 months, even though they have no nutrient blood supply and are nourished only by simple sugars present in the tubular secretions.

Clinical Note **Vasectomy** In a **vasectomy**, a common sterilization procedure in males, a small segment of each ductus deferens (alias vas deferens, hence the term *vasectomy*) is surgically removed after it passes from the testis but before it enters the inguinal canal, thus blocking exit of sperm from the testes. Sperm that build up behind the tied-off testicular end of the severed ductus are removed by phagocytosis. Although this procedure blocks sperm exit, it does not interfere with testosterone activity because the Leydig cells secrete testosterone into the blood, not through the ductus deferens. Thus, vasectomy does not diminish testosterone-dependent masculinity or libido.

The accessory sex glands contribute the bulk of the semen.

Several accessory sex glands—the seminal vesicles and prostate—empty their secretions into the duct system before it joins the urethra (see ▌Figure 20-1). A pair of saclike *seminal vesicles* empties into the last portion of the two ductus deferens, one on each side. The short segment of duct that passes beyond the entry point of the seminal vesicle to join the urethra is called the *ejaculatory duct*. The *prostate* is a large, single gland that completely surrounds the ejaculatory ducts and urethra. Another pair of accessory sex glands, the *bulbourethral glands*, drains into the urethra after this canal has passed through the prostate and just before it enters the penis. Numerous mucus-secreting glands also lie along the length of the urethra.

Clinical Note In a significant number of men, the prostate enlarges in middle to older age (a condition called **benign prostatic hypertrophy**, or **BPH**). Difficulty in urinating often occurs

as the enlarging prostate impinges on the portion of the urethra that passes through the prostate.

Semen During ejaculation, the accessory sex glands contribute secretions that provide support for the continuing viability of sperm inside the female reproductive tract. These secretions constitute the bulk of the **semen**, which is a mixture of accessory sex gland secretions, sperm, and mucus. Sperm make up only a small percentage of the total ejaculated fluid.

Functions of the Male Accessory Sex Glands Although the accessory sex gland secretions are not absolutely essential for fertilization, they do greatly facilitate the process.

■ The **seminal vesicles** (1) supply fructose, which is the primary energy source for ejaculated sperm; (2) secrete prostaglandins (see p. 119), which stimulate contractions of the smooth muscle in both the male and the female reproductive tracts, thereby helping transport sperm from their storage site in the male to the site of fertilization in the female oviduct; (3) provide about 60% of the semen volume, which helps wash the sperm into the urethra and dilutes the thick mass of sperm, enabling them to become mobile; and (4) secrete fibrinogen, a precursor of fibrin, which forms the meshwork of a clot (see p. 397).

■ The **prostate gland** (1) secretes an alkaline fluid that neutralizes the acidic vaginal secretions, an important function because sperm are more viable in a slightly alkaline environment; (2) provides clotting enzymes; and (3) releases prostate-specific antigen. The prostatic clotting enzymes act on fibrinogen from the seminal vesicles to produce fibrin, which "clots" the semen, thus helping keep the ejaculated sperm in the female reproductive tract during withdrawal of the penis. Shortly thereafter, the seminal clot is broken down by **prostate-specific antigen (PSA)**, a fibrin-degrading enzyme from the prostate, thus releasing mobile sperm within the female tract.

Clinical Note Because PSA is produced only in the prostate gland, measurement of PSA levels in the blood is used as one type of screening test for possible prostate cancer. Elevated levels of PSA in the blood are associated with prostate cancer, benign prostatic hypertrophy, or prostate infections.

A new study suggests another role for the prostate: releasing **prostasomes**, which are vesicles that fuse with and transfer to sperm a molecular "tool kit" that contains components needed for Ca^{2+} signal transduction. For example, prostasomes pass ryanodine receptors (Ca^{2+}-release channels; see p. 258) and an enzyme that opens these channels. These materials passed to sperm from the prostate gland let the stripped-down sperm that cannot make these components themselves use enhanced Ca^{2+} signaling needed for optimal motility and efficient fertilization.

■ During sexual arousal, the **bulbourethral glands** secrete mucus that provides lubrication for sexual intercourse. However, females provide most lubrication for sex.

▌Table 20-2 summarizes the locations and functions of the components of the male reproductive system.

Next, before considering the female in greater detail, we examine the means by which males and females come together via sexual intercourse to accomplish reproduction.

Component	Number and Location	Functions
Testis	Pair; located in the scrotum, a skin-covered sac suspended within the angle between the legs	Produce sperm Secrete testosterone
Epididymis and ductus deferens	Pair; one epididymis attached to the rear of each testis; one ductus deferens travels from each epididymis up out of the scrotal sac through the inguinal canal and empties into the urethra at the neck of the bladder	Serve as the sperm's exit route from the testis Serve as the site for maturation of the sperm for motility and fertility Concentrate and store the sperm
Seminal vesicle	Pair; both empty into the last portion of the ductus deferens, one on each side	Supply fructose to nourish the ejaculated sperm Secrete prostaglandins that stimulate motility to help transport the sperm within the male and female Provide the bulk of the semen Provide precursors for the clotting of semen
Prostate gland	Single; completely surrounds the urethra at the neck of the bladder	Secretes an alkaline fluid that neutralizes the acidic vaginal secretions Triggers clotting of the semen to keep the sperm in the vagina during penis withdrawal
Bulbourethral gland	Pair; both empty into the urethra, one on each side, just before the urethra enters the penis	Secrete mucus for lubrication

---

Check Your Understanding 20.2

1. List the functions of testosterone.
2. Define *seminiferous tubules, Leydig cells, Sertoli cells, spermatogenesis, spermiogenesis, spermiation, spermatogonia, primary spermatocytes, secondary spermatocytes, spermatids,* and *spermatozoa.*
3. Draw a flow diagram showing the control of testicular function.
4. State the functions of the male accessory sex glands.

20.3 Sexual Intercourse between Males and Females

Ultimately, union of male and female gametes to accomplish reproduction in humans requires delivery of sperm-laden semen into the female vagina through the **sex act**, also known as **sexual intercourse**, **coitus**, or **copulation**.

The male sex act is characterized by erection and ejaculation.

The male sex act involves two components: (1) **erection**, or hardening of the normally flaccid penis to permit its entry into the vagina, and (2) **ejaculation**, or forceful expulsion of semen into the urethra and out of the penis (Table 20-3). In addition to these strictly reproduction-related components, the **sexual**

response cycle encompasses broader physiologic responses that can be divided into four phases:

1. The *excitement phase* includes erection and heightened sexual awareness.

2. The *plateau phase* is characterized by intensification of these responses, plus more generalized body responses, such as steadily increasing heart rate, blood pressure, respiratory rate, and muscle tension.

3. The *orgasmic phase* includes ejaculation and other responses that culminate the mounting sexual excitement and are collectively experienced as an intense physical pleasure.

4. The *resolution phase* returns the genitalia and body systems to their prearousal state.

The human sexual response is a multicomponent experience that, in addition to these physiologic phenomena, encompasses emotional, psychological, and sociological factors. We examine only the physiologic aspects of sex.

Erection is accomplished by penis vasocongestion.

Erection is accomplished by engorgement of the penis with blood. The penis consists almost entirely of **erectile tissue** made up of three columns or cords of spongelike vascular spaces extending the length of the organ (see Figure 20-1a, b, and c). In the absence of sexual excitation, the erectile tissues contain little blood because the arterioles that supply these vascular chambers are constricted. As a result, the penis remains small

| TABLE 20-3 Components of the Male Sex Act

Components of the Male Sex Act	Definition	Accomplished by
Erection	Hardening of the normally flaccid penis to permit its entry into the vagina	Engorgement of the penis erectile tissue with blood as a result of marked parasympathetically induced vasodilation of the penile arterioles and mechanical compression of the veins
Ejaculation		
Emission phase	Emptying of sperm and accessory sex-gland secretions (semen) into the urethra	Sympathetically induced contraction of the smooth muscle in the walls of the reproductive ducts and accessory sex glands
Expulsion phase	Forceful expulsion of semen from the penis	Motor neuron–induced contraction of the skeletal muscles at the base of the penis

and flaccid. During sexual arousal, these arterioles reflexly dilate and the erectile tissue fills with blood, causing the penis to enlarge both in length and in width and to become more rigid. The veins that drain the erectile tissue are mechanically compressed by this engorgement and expansion of the vascular spaces, reducing venous outflow and thereby contributing even further to the buildup of blood, or *vasocongestion*. These local vascular responses transform the penis into a hardened, elongated organ capable of penetrating the vagina.

Erection Reflex The erection reflex is a spinal reflex triggered by stimulation of highly sensitive mechanoreceptors located in the glans penis, which caps the tip of the penis. An **erection-generating center** lies in the lower spinal cord. Tactile stimulation of the glans reflexly triggers, by means of this center, increased parasympathetic vasodilator activity and decreased sympathetic vasoconstrictor activity to the penile arterioles. The result is rapid, pronounced vasodilation of these arterioles and an ensuing erection (Figure 20-10). As long as this spinal reflex arc remains intact, erection is possible even in men paralyzed by a higher spinal-cord injury.

This parasympathetically induced vasodilation is the major instance of direct parasympathetic control over blood vessel diameter in the body. Parasympathetic stimulation brings about relaxation of penile arteriolar smooth muscle by *nitric oxide (NO)*, which causes arteriolar vasodilation in response to local tissue changes elsewhere in the body (see p. 346). Only sympathetic nerves typically supply arterioles, with increased sympathetic activity producing vasoconstriction and decreased sympathetic activity resulting in vasodilation (see p. 349). Concurrent parasympathetic stimulation and sympathetic inhibition of penile arterioles accomplish vasodilation more rapidly and

in greater magnitude than is possible in other arterioles supplied only by sympathetic nerves. Through this efficient means of rapidly increasing blood flow into the penis, the penis can become completely erect in as little as 5 seconds. At the same time, parasympathetic impulses promote secretion of lubricat-

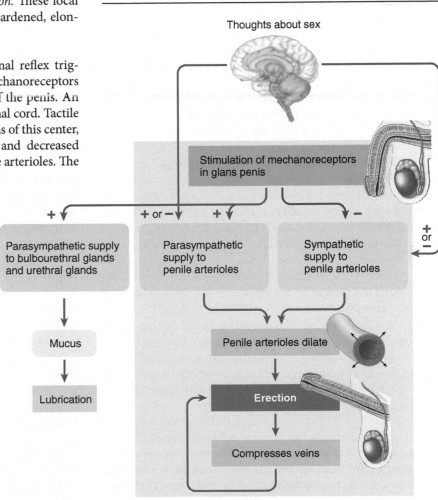

Thoughts about sex

Stimulation of mechanoreceptors in glans penis

Parasympathetic supply to bulbourethral glands and urethral glands

Parasympathetic supply to penile arterioles

Sympathetic supply to penile arterioles

Mucus

Lubrication

Penile arterioles dilate

Erection

Compresses veins

Figure 20-10 **Erection reflex.**

ing mucus from the bulbourethral glands and the urethral glands in preparation for coitus.

Numerous regions throughout the brain can influence the male sexual response. The erection-influencing brain sites appear extensively interconnected and function as a unified network to either facilitate or inhibit the basic spinal erection reflex, depending on the momentary circumstances. As an example of facilitation, mental stimuli, such as viewing something sexually exciting, can induce an erection in the absence of tactile stimulation of the penis. In contrast, failure to achieve an erection despite appropriate stimulation may result from inhibition of the erection reflex by higher brain centers. Let us examine erectile dysfunction in more detail.

Erectile Dysfunction A pattern of failing to achieve or maintain an erection suitable for sexual intercourse—**erectile dysfunction (ED)** or **impotence**—may be attributable to psychological or physical factors. An occasional episode of a failed erection does not constitute impotence, but a man who becomes overly anxious about his ability to perform the sex act may well be on his way to chronic failure. Anxiety can lead to ED, which fuels the man's anxiety level and thus perpetuates the problem. Impotence may also arise from physical limitations, including nerve damage, certain medications that interfere with autonomic function, and problems with blood flow through the penis.

ED is widespread. More than 50% of men between ages 40 and 70 experience some impotence, climbing to nearly 70% by age 70. No wonder, then, that more prescriptions were written for the much-publicized drug *sildenafil* (*Viagra*) during its first year on the market after its approval in 1998 for treating ED than for any other new drug in history. Sildenafil does not produce an erection, but it amplifies and prolongs an erectile response triggered by usual means of stimulation. Here is how the drug works: Nitric oxide released in response to parasympathetic stimulation activates a membrane-bound enzyme, **soluble guanylate cyclase (sGC)**, within nearby arteriolar smooth muscle cells. This enzyme activates **cyclic guanosine monophosphate (cyclic GMP or cGMP)**, an intracellular second messenger similar to cAMP. Cyclic GMP, in turn, leads to relaxation of the penile arteriolar smooth muscle, bringing about pronounced local vasodilation. Under normal circumstances, once cGMP is activated and brings about an erection, this second messenger is broken down by the intracellular enzyme **phosphodiesterase 5 (PDE5)**. Sildenafil inhibits PDE5. As a result, cGMP remains active longer so that penile arteriolar vasodilation continues and the erection is sustained long enough for a formerly impotent man to accomplish the sex act. Just as pushing a pedal on a piano will not cause a note to be played but will prolong a played note, sildenafil cannot cause the release of NO and subsequent activation of erection-producing cGMP, but it can prolong the triggered response. The drug has no benefit for those who do not have ED, but its success rate has been high among sufferers of the condition. Side effects have been limited because the drug concentrates in the penis, thus having more effect on this organ than elsewhere in the body.

When NO formation is severely impaired and PDE5 inhibitors are not effective, two new classes of drugs are under investigation for treatment of ED: *sGC stimulators* and *sGC activators* that act independently of NO in slightly different ways to increase cGMP formation and promote erection.

Ejaculation includes emission and expulsion.

The second component of the male sex act is ejaculation. Like erection, ejaculation is a spinal reflex. The same types of tactile and psychic stimuli that induce erection cause ejaculation when the level of excitation intensifies to a critical peak. The overall ejaculatory response occurs in two phases: *emission* and *expulsion* (see Table 20-3).

Emission First, sympathetic impulses cause sequential contraction of smooth muscles in the prostate, reproductive ducts, and seminal vesicles. This contractile activity delivers prostatic fluid, then sperm, and finally seminal vesicle fluid (collectively, semen) into the urethra. This phase of the ejaculatory reflex is called **emission**. During this time, the sphincter at the neck of the bladder is tightly closed to prevent semen from entering the bladder and urine from being expelled along with the ejaculate through the urethra.

Expulsion Second, filling of the urethra with semen triggers nerve impulses that activate skeletal muscles at the base of the penis. Rhythmic contractions of these muscles occur at 0.8-second intervals and increase the pressure within the penis, forcibly expelling the semen through the urethra to the exterior. This is the **expulsion** phase of ejaculation.

Orgasm and resolution complete the sexual response cycle.

The third phase of the sexual response cycle, orgasm, accompanies the expulsion part of the ejaculatory response and is followed by the resolution phase of the cycle.

Orgasm The rhythmic contractions that occur during semen expulsion are accompanied by involuntary rhythmic throbbing of pelvic muscles and peak intensity of the overall body responses that were climbing during the earlier phases. Heavy breathing, a heart rate of up to 180 beats per minute, marked generalized skeletal muscle contraction, and heightened emotions are characteristic. These pelvic and overall systemic responses that culminate the sex act are associated with an intense pleasure characterized by a feeling of release and complete gratification, an experience known as **orgasm**.

Resolution During the resolution phase following orgasm, sympathetic vasoconstrictor impulses slow the inflow of blood into the penis, causing the erection to subside. A deep relaxation ensues, often accompanied by a feeling of fatigue. Muscle tone returns to normal, while the cardiovascular and respiratory systems return to their prearousal level of activity. Once ejaculation has occurred, a temporary refractory period of variable duration ensues before sexual stimulation can trigger another erection. Males therefore cannot experience multiple orgasms within a matter of minutes, as females sometimes do.

Volume and sperm content of the ejaculate vary.

The volume and sperm content of the ejaculate depend on the length of time between ejaculations. The average volume of semen is 2.75 mL, ranging from 2 to 6 mL, the higher volumes following periods of abstinence. An average human ejaculate contains about 165 million sperm (60 million/mL), but some ejaculates contain as many as 400 million sperm.

Clinical Note Both quantity and quality of sperm are important determinants of fertility. A man is considered clinically **infertile** if his sperm concentration falls below 20 million/mL of semen. Even though only one spermatozoon actually fertilizes the ovum, large numbers of accompanying sperm are needed to provide sufficient acrosomal enzymes to break down the barriers surrounding the ovum until the victorious sperm penetrates into the ovum's cytoplasm (▌Figure 20-11). The quality of sperm also must be taken into account when assessing the fertility potential of a semen sample. The presence of substantial numbers of sperm with abnormal motility or structure, such as sperm with distorted tails, reduces the chances of fertilization. (For a discussion of how environmental estrogens may be decreasing sperm counts and negatively affecting the male and female reproductive systems in other ways, see the boxed feature on pp. 736–737, ▌Concepts, Challenges, and Controversies.)

The female sexual cycle is similar to the male cycle.

Both sexes experience the same four phases of the sexual cycle—excitement, plateau, orgasm, and resolution. Furthermore, the physiologic mechanisms responsible for orgasm are fundamentally the same in males and females.

The excitement phase in females can be initiated by either physical or psychological stimuli. Tactile stimulation of the glans clitoris and surrounding perineal area is an especially powerful sexual stimulus. These stimuli trigger spinal reflexes that bring about parasympathetically induced vasodilation of arterioles throughout the vagina and external genitalia, especially the clitoris. The resultant inflow of blood becomes evident as swelling of the labia and erection of the clitoris. The latter—like its male homolog, the penis—is composed largely of erectile tissue. Contrary to a common misconception, the clitoris is much larger than its externally visible portion. Most of the clitoris is located internally and consists mainly of large, highly vascular **vestibular bulbs** that surround the urethra and vagina (see ▌Figure 20-2d, p. 719). These bulbs engorge with blood during erection. The functional significance of this vasocongestion is unclear. Scientists speculate that it may (1) squeeze the urethra closed to prevent contamination of the urinary tract during sexual activity, (2) support the vaginal wall and squeeze the penis during sexual intercourse, and (3) increase pleasure signaling.

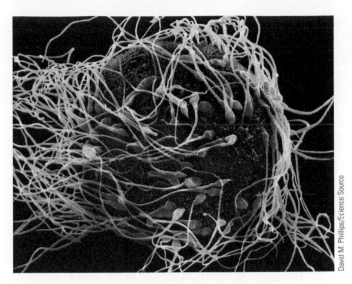

David M. Phillips/Science Source

▌Figure 20-11 **Scanning electron micrograph of sperm amassed at the surface of an ovum**.

Vasocongestion of vaginal capillaries forces fluid out of these vessels into the vaginal lumen. This fluid, which is the first positive indication of sexual arousal, serves as the primary lubricant for intercourse. Mucus secretions from the male and mucus released during sexual arousal from glands located at the outer opening of the vagina provide additional lubrication. Also during the excitement phase in the female, the nipples become erect and the breasts enlarge as a result of vasocongestion.

During the plateau phase, the changes initiated during the excitement phase intensify, while systemic responses similar to those in the male (such as increased heart rate, blood pressure, respiratory rate, and muscle tension) occur. Further vasocongestion of the lower third of the vagina during this time (accompanied by vasocongestion of the surrounding vestibular bulbs) reduces the inner capacity of the vagina so that it tightens around the thrusting penis, heightening tactile sensation for both the female and the male. Simultaneously, the uterus raises upward, lifting the cervix and enlarging the upper two thirds of the vagina. This ballooning, or **tenting effect**, creates a space for ejaculate deposition.

If erotic stimulation continues, the sexual response culminates in orgasm as sympathetic impulses trigger rhythmic contractions of pelvic musculature at 0.8-second intervals, the same rate as in males. The contractions occur most intensely in the engorged lower third of the vaginal canal. Systemic responses identical to those of the male orgasm also occur. In fact, the orgasmic experience in females parallels that of males with two exceptions. First, there is no female counterpart to ejaculation. Second, females do not become refractory following an orgasm, so they can respond immediately to continued erotic stimulation and achieve multiple orgasms.

During resolution, pelvic vasocongestion and the systemic manifestations gradually subside. As with males, this is a time of great physical relaxation for females.

We now examine how females fulfill their part of the reproductive process.

Environmental "Estrogens": Bad News for the Reproductive System

UNKNOWINGLY, DURING THE PAST 70 years humans have been polluting our environment with synthetic endocrine-disrupting chemicals as an unintended side effect of industrialization. Known as **endocrine disrupters,** these hormonelike pollutants bind with the receptor sites normally reserved for the naturally occurring hormones. Depending on how they interact with the receptors, these disrupters can either mimic or block normal hormonal activity. Many of these environmental contaminants mimic or alter the action of estrogen, the feminizing steroid hormone produced by the female ovaries.

Estrogenic pollutants are everywhere. They contaminate our food, drinking water, and air. Proved feminizing synthetic compounds are found in (1) certain weed killers and insecticides, (2) some detergent breakdown products, (3) petroleum by-products found in car exhaust, (4) a common food preservative used to retard rancidity, (5) dental sealants, (6) cash register receipts, and (7) softeners that make plastics flexible. These plastic softeners are commonly found in food packaging and can readily leach into food with which they come in contact, especially during heating. The softeners were also found to leach from some babies' plastic teething toys into the saliva. They are in numerous medical products, such as the bags in which blood is stored. Plastic softeners are among the most plentiful industrial contaminants in our environment.

Investigators are only beginning to identify and understand the implications for reproductive health of myriad synthetic chemicals that are an integral part of modern societies. An estimated 87,000 synthetic chemicals are already in our environment. Scientists suspect that the estrogen-mimicking chemicals among these may underlie a spectrum of reproductive disorders that have been rising in the past 70 years—the same period during which large amounts of these pollutants have been introduced into our environment. Here are examples of male reproductive dysfunctions that may be circumstantially linked to exposure to environmental **estrogen disrupters:**

■ *Falling sperm counts.* The average sperm count has fallen from 113 million sperm per milliliter of semen in the 1940s to 60 million/ mL now. Making matters worse, the volume of a single ejaculate has declined from 3.40 to 2.75 mL. This means that men, on average, are now ejaculating less than half the number of sperm as men did about 70 years ago—a drop from more than 380 million sperm to about 165 million sperm per ejaculate. Furthermore, the number of motile sperm has dipped. Of note, the sperm count has not declined in the less polluted areas of the world during the same period.

■ *Increased incidence of testicular and prostate cancer.* Cases of testicular cancer have tripled since the 1940s, and the rate continues to climb. Prostate cancer has also been on the rise over the same period.

■ *Rising number of male reproductive tract abnormalities at birth.* The incidence of cryptorchidism (undescended testis) nearly doubled from the 1950s to the 1970s. The number of cases of *hypospadia,* a malformation of the penis, more than doubled between the mid-1960s and the mid-1990s. Hypospadia results when the urethral fold fails to fuse closed during development of a male fetus.

■ *Evidence of reduced masculinization in animals.* Some fish and wild animal populations severely exposed to environmental estrogens—such as those living in or near water heavily polluted with hormone-mimicking chemical wastes—display a high rate of grossly impaired reproductive systems. Examples include male fish that are hermaphrodites (having both male and female reproductive parts) and male alligators with abnormally small penises. Similar reproductive abnormalities have been identified in land mammals. Presumably, excessive estrogen exposure is emasculating these populations.

■ *Decline in male births.* Many countries are reporting a slight decline in the ratio of baby boys to baby girls being born. In the United States, 17 fewer males were born per 10,000 births in 2007 compared to 1970, and Japan has seen an overall drop of 37 males per 10,000 births during the same period. Although several other plausible explanations have been put forth (such as late-age childbearing,

Check Your Understanding 20.3

1. List and describe the four phases of the sexual response cycle.
2. Define the components of the male sex act and explain how each is accomplished.
3. Compare orgasm in males and females.

20.4 Female Reproductive Physiology

Female reproductive physiology is more complex than male reproductive physiology.

Complex cycling characterizes female reproductive physiology.

Unlike the continuous sperm production and essentially constant testosterone secretion characteristic of the male, release of ova is intermittent, and secretion of female sex hormones displays wide cyclic swings. The tissues influenced by these sex hormones also undergo cyclic changes, the most obvious of which is the monthly menstrual cycle (*menstruus* means "monthly"). During each cycle, the female reproductive tract is prepared for the fertilization and implantation of an ovum released from the ovary at ovulation. If fertilization does not occur, the cycle repeats. If fertilization does occur, the cycles are interrupted while the female system adapts to nurture and pro-

increasing obesity, and greater use of reproductive technologies), many researchers attribute this troubling trend primarily to disruption of normal male fetal development by environmental estrogens. In one compelling piece of circumstantial evidence, people inadvertently exposed to the highest level of an endocrine-disrupting agent during an industrial accident subsequently had all daughters and no sons, whereas those least exposed had the normal ratio of girls and boys. Similarly, a 2004 study in the Russian Arctic found a remarkable ratio of 2.5 to 1 female-to-male births among women who had blood concentrations of 4 mg/L or greater of a known estrogen-mimicking pollutant.

Environmental estrogens are also implicated in the rising incidence of breast cancer in females. Breast cancer is 25% to 30% more prevalent now than in the 1940s. Many of the established risk factors for breast cancer, such as starting to menstruate earlier than usual and undergoing menopause later than usual, are associated with an elevation in the total lifetime exposure to estrogen. Because increased exposure to natural estrogen bumps up the risk for breast cancer, prolonged exposure to environmental estrogens may be contributing to the rising prevalence of this malignancy among women (and men).

In addition to the estrogen disrupters, scientists more recently identified a new class of chemical offenders—**androgen disrupters** that either mimic or suppress the action of male hormones. For example, studies suggest that bacteria in wastewater from pulp mills can convert the sterols in pine pulp into androgens. By contrast, anti-androgen compounds have been found in the fungicides commonly sprayed on vegetable and fruit crops. Yet another cause for concern comes from the androgens used by the livestock industry to enhance the production of muscle (that is, meat) in feedlot cattle. (Androgens have a protein anabolic effect.) These drugs do not end up in the meat, but they can get into drinking water and other food as hormone-laden feces contaminate rivers and streams.

Under the *Toxic Substances Control Act (TSCA)*, which became law in the United States in 1976, chemicals are presumed safe unless proven otherwise. The Environmental Protection Agency (EPA) must show that a chemical is dangerous after it is already in use. In response to the growing evidence that has emerged circumstantially linking numerous environmental pollutants to disturbing reproductive abnormalities, the U.S. Congress legally mandated the EPA in 1996 to determine which synthetic chemicals might be endocrine disrupters. In response, the EPA formed an advisory committee, which in 1998 proposed an ambitious plan to begin comprehensive testing of manufactured compounds for their potential to disrupt hormones in humans and wildlife. Although eventually all the 87,000 existing synthetic compounds will be tested, the initial screening was narrowed to evaluate the endocrine-disrupting potential of widely used chemicals. Declaring this a national health priority, the government has allocated millions of dollars for this research. Yet in this time-consuming process, only a few thousand chemicals were tested in the first 10 years of investigation as the EPA's *Toxic Release Inventory* slowly grew and other chemicals were deemed safe.

To more quickly, more efficiently, and more effectively screen chemicals for their potential toxicity, the *Toxicology in the 21st Century (Tox21) program* was established in 2008. Tox21 is a federal collaboration involving the EPA, National Institutes of Health (NIH), and Food and Drug Administration (FDA). In 2011, the Tox21 consortium introduced a high-speed, cost-effective robot screening system that is currently testing 10,000 chemicals to predict their potential for disrupting physiologic pathways that may negatively affect human health.

In the meantime, increasingly, environmental watchdogs are calling for additional measures to limit exposure to synthetic chemicals (such as switching from plastic to glass baby bottles) and better labeling from manufacturers so that consumers can make more informed decisions about products they use.

tect the newly conceived human until it has developed into an individual capable of living outside the maternal environment. Furthermore, the female continues her reproductive functions after birth by producing milk (lactation) for the baby's nourishment. Thus, the female reproductive system is characterized by complex cycles that are interrupted by even more complex changes should pregnancy occur.

The ovaries perform the dual function of producing ova (oogenesis) and secreting the female sex hormones, estrogen and progesterone. These hormones act together to promote fertilization of the ovum and to prepare the female reproductive system for pregnancy. Estrogen in the female governs many functions similar to those carried out by testosterone in the male, such as maturation and maintenance of the entire female reproductive system and establishment of female secondary sexual characteristics. In general, the actions of estrogen are important to preconception events. Estrogen is essential for ova maturation and release, development of physical characteristics that are sexually attractive to males, and transport of sperm from the vagina to the site of fertilization in the oviduct. Furthermore, estrogen contributes to breast development in anticipation of lactation. The other ovarian steroid, progesterone, sometimes called the "hormone of pregnancy," is important in preparing a suitable environment for nourishing a developing embryo and then fetus and for contributing to the breasts' ability to produce milk.

Being steroids, estrogen and progesterone exert their multiple effects by binding with their respective receptors in the cytoplasm of their target cells, with the hormone-receptor complex moving to the nucleus where it binds with a hormone-specific DNA hormone-response element. This binding leads to gene transcription and synthesis of designated proteins that exert the hormone's dictated response in the target cells. Estrogen has two different cytoplasmic receptors, which have a differential distribution in various tissues and permit selective actions in specific target tissues. **Selective estrogen receptor modulators (SERMs)**, like *raloxifene,* are drugs that selectively bind with a specific estrogen receptor. Raloxifene is approved to treat osteoporosis because it selectively binds with estrogen receptors in bone, where it mimics estrogen's beneficial effect on maintaining bone density, while not exerting any effect on reproductive organs, where extra estrogenlike influence could increase the risk of cancer (see pp. 704 and 706). Estrogen also binds with surface membrane receptors, where it acts via cAMP to elicit rapid, nongenomic effects (see p. 127).

As in males, reproductive capability begins at puberty in females, but unlike males, who have reproductive potential throughout life, female reproductive potential ceases during middle age at menopause.

The steps of gametogenesis are the same in both sexes, but the timing and outcome differ sharply.

Even though the identical steps of chromosome replication and division take place during gamete production in both sexes, **oogenesis** contrasts sharply with spermatogenesis in several important aspects. The undifferentiated primordial germ cells in the fetal ovaries, the **oogonia** (comparable to spermatogonia), divide mitotically to produce about 7 million oogonia by the fifth month of gestation, when mitotic proliferation ceases.

Formation of Primary Oocytes and Primordial Follicles During the last part of fetal life, the oogonia begin the early steps of the first meiotic division but do not complete it. Known now as **primary oocytes**, they contain the diploid number of 46 replicated chromosomes, which are gathered into homologous pairs but do not separate. The primary oocytes remain in this state of **meiotic arrest** for years until they are prepared for ovulation.

Before birth, each primary oocyte is surrounded by a single layer of connective tissue–derived **granulosa cells**. Together, an oocyte and surrounding granulosa cells make up a **primordial follicle**. Primary oocytes that are not incorporated into follicles self-destruct by apoptosis (cell suicide). At birth, only about 2 million primordial follicles remain, each containing a single primary oocyte capable of producing a single ovum. No new oocytes or follicles appear after birth, with the follicles already present in the ovaries at birth serving as a reservoir from which all ova throughout the reproductive life of a female arise. The follicular pool gradually dwindles away as a result of processes that "use up" the oocyte-containing follicles. Even before puberty, the pool of primordial follicles gives rise to an ongoing trickle of developing follicles, stimulated by poorly understood

paracrine factors produced by both oocytes and granulosa cells. Once it starts to develop, a follicle is destined for one of two fates: It will reach maturity and ovulate, or it will degenerate to form scar tissue, a process known as **atresia**. Until puberty, all the follicles that start to develop undergo atresia in the early stages without ever ovulating. For the first few years after puberty, many of the cycles are anovulatory (that is, no ovum is released). Of the original total pool of follicles at birth, about 300,000 remain at puberty, and only about 400 will mature and release ova; 99.97% never ovulate but instead undergo atresia at some stage of development. By menopause, which occurs on average in a woman's early 50s, few primordial follicles remain, having either already ovulated or become atretic. From this point on, the woman's reproductive capacity ceases.

This limited gamete potential in females is in sharp contrast to the continual process of spermatogenesis in males, who have the potential to produce several hundred million sperm in a single day. Furthermore, considerable chromosome wastage occurs in oogenesis compared with spermatogenesis. Let us see how.

Formation of Secondary Oocytes The primary oocyte within a primordial follicle is still a diploid cell that contains 46 doubled chromosomes. From puberty until menopause, a portion of the developing follicles in the pool progress cyclically into more advanced follicles. The mechanisms determining which follicles in the pool develop during a given cycle are unknown. Further development of a follicle is characterized by growth of the primary oocyte and by expansion and differentiation of the surrounding cell layers. We focus now on oogenesis before we discuss the accompanying changes in the follicle. The oocyte enlarges about a thousandfold. This oocyte enlargement is caused by a buildup of cytoplasmic materials that are needed by the early embryo.

Just before ovulation, the primary oocyte, whose nucleus has been in meiotic arrest for years, completes its first meiotic division. This division yields two daughter cells, each receiving a haploid set of 23 doubled chromosomes, analogous to the formation of secondary spermatocytes (❙ Figure 20-12). However, almost all the cytoplasm remains with one of the daughter cells, now called the **secondary oocyte**, which is destined to become the ovum. The chromosomes of the other daughter cell, together with a small share of cytoplasm, form the **first polar body**. In this way, the ovum-to-be loses half of its chromosomes to form a haploid gamete but retains all of its nutrient-rich cytoplasm. The nutrient-poor polar body soon degenerates.

Formation of a Mature Ovum Actually, the secondary oocyte, and not the mature ovum, is ovulated and fertilized, but common usage refers to the developing female gamete as an *ovum* even in its primary and secondary oocyte stages. Sperm entry into the secondary oocyte is needed to trigger the second meiotic division. Unfertilized secondary oocytes never complete this final division. During this division, a half set of chromosomes, along with a thin layer of cytoplasm, is extruded as the **second polar body**. The other half set of 23 unpaired chromosomes remains behind in what is now the **mature ovum** (sometimes called the *ootid,* comparable to the spermatid, until

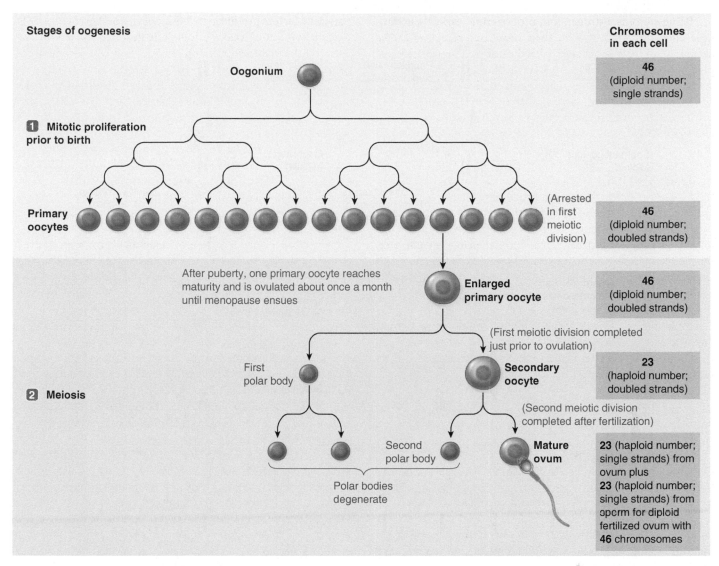

Stages of oogenesis

Chromosomes in each cell

Oogonium
46 (diploid number; single strands)

1 Mitotic proliferation prior to birth

Primary oocytes — (Arrested in first meiotic division)
46 (diploid number; doubled strands)

After puberty, one primary oocyte reaches maturity and is ovulated about once a month until menopause ensues

Enlarged primary oocyte
46 (diploid number; doubled strands)

(First meiotic division completed just prior to ovulation)

2 Meiosis

First polar body

Secondary oocyte
23 (haploid number; doubled strands)

(Second meiotic division completed after fertilization)

Second polar body

Mature ovum
23 (haploid number; single strands) from ovum plus 23 (haploid number; single strands) from sperm for diploid fertilized ovum with 46 chromosomes

Polar bodies degenerate

❘Figure 20-12 **Oogenesis.** Compare with Figure 20-7, p. 727, spermatogenesis.

the polar bodies disintegrate and the mature ovum alone remains). These 23 maternal chromosomes unite with the 23 paternal chromosomes of the penetrating sperm to complete fertilization. If the first polar body has not already degenerated, it too undergoes the second meiotic division at the same time the fertilized secondary oocyte is dividing its chromosomes.

Comparison of Steps in Oogenesis and Spermatogenesis The steps involved in chromosome distribution during oogenesis parallel those of spermatogenesis, except that the cytoplasmic distribution and time span for completion differ sharply (❘Figure 20-13). Just as four haploid spermatids are produced by each primary spermatocyte, four haploid daughter cells are produced by each primary oocyte (if the first polar body does not degenerate before it completes the second meiotic division). In spermatogenesis, each daughter cell develops into a highly specialized, motile spermatozoon unencumbered by unessential cytoplasm and organelles, its only destiny being to supply half of the genes for a new individual. In oogenesis,

however, of the four daughter cells, only the one destined to become the mature ovum receives cytoplasm. This uneven distribution of cytoplasm is important because the ovum, in addition to providing half the genes, provides all of the cytoplasmic components needed to support early development of the fertilized ovum. The large, relatively undifferentiated ovum contains numerous nutrients, organelles, and structural and enzymatic proteins. The three other cytoplasm-scarce daughter cells, the polar bodies, rapidly degenerate, their chromosomes being deliberately wasted.

Note also the considerable difference in time to complete spermatogenesis and oogenesis. A spermatogonium develops into fully remodeled spermatozoa in about 2 months. In contrast, development of an oogonium (present before birth) to a mature ovum takes anywhere from 11 years (beginning of ovulation at onset of puberty) to 50 years (end of ovulation at onset of menopause). The length of the active steps in meiosis is the same in both males and females, but in females the developing eggs remain in meiotic arrest for a variable number of years.

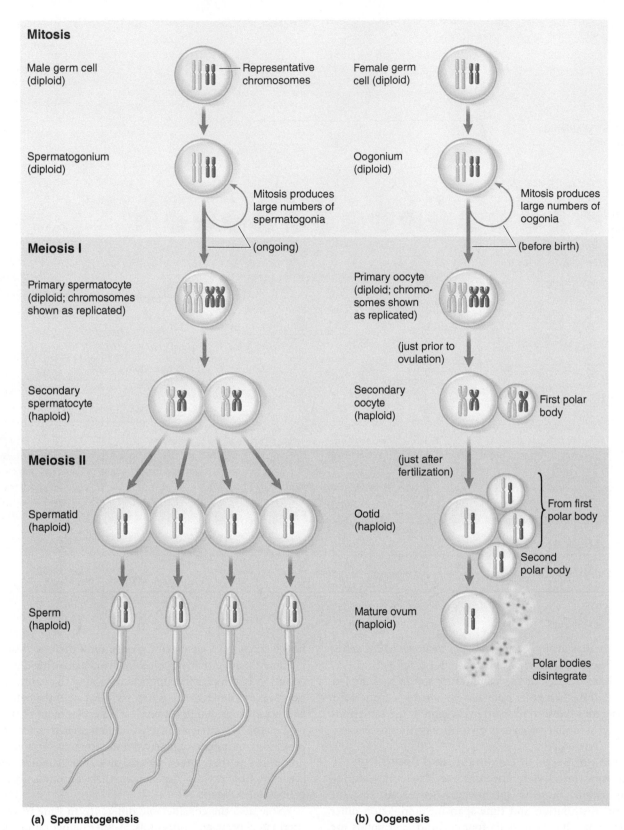

Mitosis

Male germ cell (diploid)

Representative chromosomes

Female germ cell (diploid)

Spermatogonium (diploid)

Mitosis produces large numbers of spermatogonia

Oogonium (diploid)

Mitosis produces large numbers of oogonia

Meiosis I

(ongoing)

(before birth)

Primary spermatocyte (diploid; chromosomes shown as replicated)

Primary oocyte (diploid; chromosomes shown as replicated)

(just prior to ovulation)

Secondary spermatocyte (haploid)

Secondary oocyte (haploid)

First polar body

Meiosis II

(just after fertilization)

Spermatid (haploid)

Ootid (haploid)

From first polar body

Second polar body

Sperm (haploid)

Mature ovum (haploid)

Polar bodies disintegrate

(a) **Spermatogenesis**

(b) **Oogenesis**

‖Figure 20-13 **Comparison of mitotic and meiotic divisions producing spermatozoa and eggs from germ cells.**

The older age of ova released by women in their late 30s and 40s is believed to account for the higher incidence of genetic abnormalities, such as Down syndrome, in children born to women in this age range.

The ovarian cycle consists of alternating follicular and luteal phases.

After the onset of puberty, the ovary constantly alternates between two phases: the **follicular phase**, which is dominated by the presence of *maturing follicles*; and the **luteal phase**, which is characterized by the presence of the *corpus luteum* (to be described shortly). Normally, this cycle is interrupted only if pregnancy occurs and is finally terminated by menopause. The average ovarian cycle lasts 28 days, but this varies among women and among cycles in any particular woman. The follicle operates in the first half of the cycle to produce a mature egg ready for ovulation at midcycle. The corpus luteum takes over during the last half of the cycle to prepare the female reproductive tract for pregnancy in case fertilization of the released egg occurs.

The follicular phase is characterized by development of maturing follicles.

At any given time throughout the cycle, a portion of the primordial follicles (❙ Figure 20-14b, step **1**) is starting to develop under paracrine influence. However, only those that reach a certain stage of development during the follicular phase, when the gonadotropin hormone environment is right to promote their final maturation, continue beyond the earlier stages of development. The others, lacking hormonal support, undergo atresia. During follicular development, as the primary oocyte is synthesizing and storing materials for future use if fertilized, important changes take place in the cells surrounding the reactivated oocyte in preparation for the egg's release from the ovary.

Preantral Follicular Development The first stage of follicular development is conversion of selected primordial follicles into **preantral follicles** (❙ Figure 20-14, step **2**, and ❙ Table 20-4). A preantral follicle is a follicle that has begun to grow but has not yet formed an *antrum*, a fluid-filled cavity within the follicle's interior. When a primordial follicle begins to develop into a preantral follicle, the single layer of granulosa cells thickens and then proliferates to create several layers that surround the oocyte. The oocyte and granulosa cell secrete glycoproteins that form a thick, gel-like "rind" that covers the oocyte and separates it from the surrounding granulosa cells. This intervening membrane is known as the **zona pellucida**. Gap junctions penetrate the zona pellucida and extend between the oocyte and surrounding granulosa cells. Glucose, amino acids, and other important molecules are delivered to the oocyte from the granulosa cells through these connecting tunnels, enabling the egg to stockpile these critical nutrients. Also, signaling molecules pass through the gap junctions in both directions, helping coordinate the changes that take place in the oocyte and surrounding cells as both mature and prepare for ovulation. The nurturing relationship between granulosa cells and a developing egg is similar in many ways to the relationship between Sertoli cells and developing sperm.

At the same time as the oocyte is enlarging and granulosa cells are proliferating, specialized ovarian connective tissue cells in contact with the expanding granulosa cells proliferate and differentiate to form an outer layer of **thecal cells** in response to paracrines secreted by the granulosa cells. Thecal and granulosa cells, collectively known as **follicular cells**, have the ability to function as a unit to secrete estrogen, although they do not do so at this early stage of follicular development. Preantral follicular development takes several months to complete and occurs without gonadotropin influence.

Formation of an Antral Follicle; Estrogen Secretion The next stage of follicular development is gonadotropin-dependent and involves formation of the antrum and conversion of the preantral follicle into an **antral follicle** that secretes estrogen (❙ Figure 20-14, step **3**). During this stage of follicular development, a fluid-filled cavity, or **antrum**, forms in the middle of the granulosa cells (❙ Figure 20-15). The follicular fluid originates partially from transudation (passage through capillary pores) of plasma and partially from follicular cell secretions. As the follicular cells start producing estrogen, some of this hormone is secreted into the blood for distribution throughout the body. However, a portion of the estrogen collects in the hormone-rich antral fluid. Of the three physiologically important estrogens in order of potency—*estradiol, estrone,* and *estriol*—**estradiol** is the principal ovarian estrogen.

The oocyte reaches full size during early development of the antrum. The shift from a preantral follicle to an antral follicle initiates a period of rapid follicular growth. Part of the follicular growth is the result of continued proliferation of the granulosa and thecal cells, but most results from a dramatic expansion of the antrum. As the follicle grows, estrogen is produced in increasing quantities.

Early antral development depends on the presence of gonadotropins, but the fluctuating levels of these hormones that occur during the monthly reproductive cycle do not influence early antral follicles. This early antral development takes another 45 days and, like preantral development, is not part of the follicular phase of the ovarian cycle. Only antral follicles that have developed to the point of becoming extremely sensitive to FSH are "recruited" for further rapid development at the beginning of the follicular phase when FSH levels rise. Typically during each cycle, about 15 to 20 follicles are recruited (see ❙ Figure 20-14, step **4**). Whereas the diameter of a preantral follicle is still less than 1 mm, that of a recruitable antral follicle is 2 to 5 mm, and that of a recruited mature follicle reaches 15 to 20 mm shortly before ovulation.

Formation of a Mature Follicle Of the follicles recruited, one, the "dominant" follicle, usually grows more rapidly than the others, developing into a **mature (preovulatory, tertiary,** or **Graafian) follicle** within about 14 days after being recruited (step **5**). Rapid growth of recruited follicles and development of a mature follicle is the only stage of follicular development that takes place during the follicular phase of the ovarian cycle under the influence of FSH. The dominant follicle that develops into a mature follicle generally has the most FSH receptors and

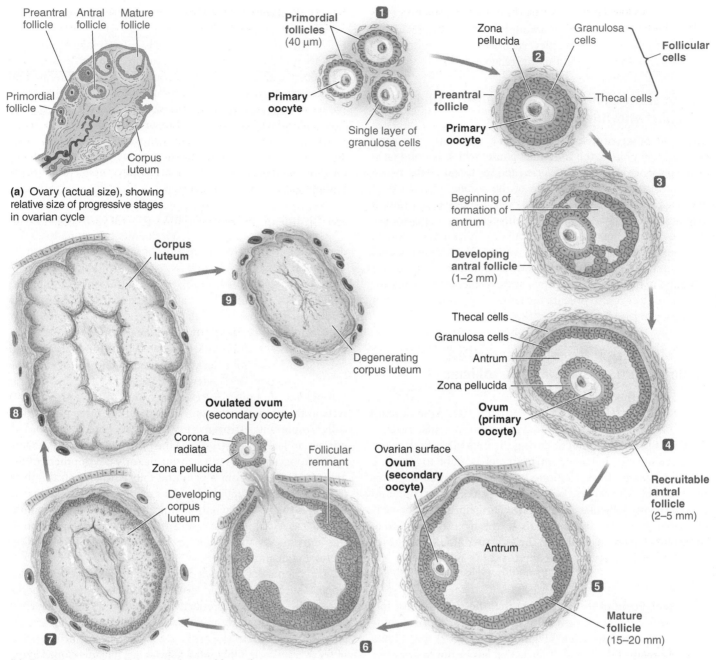

(a) Ovary (actual size), showing relative size of progressive stages in ovarian cycle

Preantral follicle
Antral follicle
Mature follicle
Primordial follicle
Corpus luteum

1 Primordial follicles (40 μm)
Primary oocyte
Single layer of granulosa cells

2 Zona pellucida
Granulosa cells
Follicular cells
Preantral follicle
Thecal cells
Primary oocyte

3 Beginning of formation of antrum
Developing antral follicle (1–2 mm)

Thecal cells
Granulosa cells
Antrum
Zona pellucida
Ovum (primary oocyte)

4 **Recruitable antral follicle** (2–5 mm)

Ovarian surface
Ovum (secondary oocyte)
Antrum

5 **Mature follicle** (15–20 mm)

6 Follicular remnant

Corpus luteum

9 Degenerating corpus luteum

Ovulated ovum (secondary oocyte)
Corona radiata
Zona pellucida
Developing corpus luteum

8

7

(b) Development of a follicle, ovulation, and formation and degeneration of a corpus luteum

1 In a primordial follicle, a primary oocyte is surrounded by a single layer of granulosa cells.

2 During development of a preantral follicle, under the influence of local paracrines, granulosa cells proliferate, the zona pellucida forms around the oocyte, and surrounding ovarian connective tissue cells differentiate into thecal cells.

3 During early development of an antral follicle, an estrogen-rich antrum starts to form and the follicle continues to enlarge.

4 Antral follicles that have reached a given size at the beginning of the follicular phase of

the ovarian cycle are recruited for further rapid development and antrum expansion under the influence of FSH.

5 After about 2 weeks of rapid growth under the influence of FSH, the follicle has developed into a mature follicle, which has a greatly expanded antrum; the oocyte, which by now has developed into a secondary oocyte, is displaced to one side.

6 At midcycle, in response to a burst in LH secretion, the mature follicle, bulging on the ovarian surface, ruptures and releases the oocyte, resulting in ovulation and ending the follicular phase.

7 Ushering in the luteal phase, the ruptured follicle develops into a corpus luteum under the influence of LH.

8 The corpus luteum continues to grow and secrete progesterone and estrogen that prepare the uterus for implantation of a fertilized ovum.

9 After 14 days, if a fertilized ovum does not implant in the uterus, the corpus luteum degenerates, the luteal phase ends, and a new follicular phase begins under the influence of a changing hormonal milieu.

|Figure 20-14 **Ovarian cycle.** (a) Ovary showing progressive stages in one ovarian cycle. All of these stages occur sequentially at one site, but the stages are represented in a loop in the periphery of the ovary so that all of the stages can be seen in progression simultaneously. (b) Enlarged view of the stages in one ovarian cycle.
◄ **FIGURE FOCUS:** *(1) Which of these stages of follicular development take(s) place without any influence by the ovarian cycle? (2) Which stages take(s) place during the follicular phase? (3) Which take(s) place during the luteal phase?*

therefore is most responsive to hormonal stimulation. The antrum occupies most of the space in a mature follicle. The oocyte, surrounded by the zona pellucida and a single layer of granulosa cells, is displaced asymmetrically at one side of the growing follicle, in a little mound that protrudes into the antrum.

Ovulation The greatly expanded mature follicle bulges on the ovarian surface, creating a thin area that ruptures to release the oocyte at ovulation (|Figure 20-16). Rupture of the follicle is facilitated by release from the follicular cells of enzymes (triggered by a burst in LH secretion, which is described later) that digest the connective tissue in the follicular wall. The bulging wall, thus weakened, balloons out even farther to the point that it can no longer contain the rapidly expanding follicular contents.

Just before ovulation, the oocyte completes its first meiotic division. The ovum (secondary oocyte), still surrounded by its

|TABLE 20-4 Stages of Follicular Development

Phase	Size of Follicle (in mm)	Status of Follicle	Influence of Gonadotropins on Follicle	Hormone Production by	Status of Gamete
Resting primordial follicle	0.04	Gamete is surrounded by single layer of granulosa cells.	None	None	The primary oocyte has entered the first meiotic division and is in meiotic arrest.
Preantral follicle	0.04–1	Granulosa cells thicken and proliferate; zona pellucida forms; thecal cells develop.	None (Early growth of selected follicles occurs under paracrine influence.)	None (Follicular cells [granulosa cells and thecal cells collectively] develop capacity to secrete estrogen but do not do so.)	The primary oocyte begins to grow and stockpile nutrients.
Early developing antral follicle	1–2	The antrum starts to form, contributing to follicular growth.	The follicle is dependent on both FSH and LH for estrogen secretion but is not influenced by cyclical fluctuations in these hormones.	Estrogen production begins; part is secreted into the blood; part remains in the follicle, expanding the antrum.	The primary oocyte grows rapidly and reaches near full size.
Recruitable antral follicle	2–5	After a follicle is recruited, the antrum continues to expand and the follicle grows rapidly.	The follicle is extremely sensitive to FSH and is recruited by the increase in FSH at the onset of the follicular phase of the ovarian cycle for further rapid growth and development.	The follicle continues to produce increasing amounts of estrogen.	The primary oocyte grows slowly and remains in meiotic arrest.
Dominant, mature follicle	15–20	The antrum occupies most of the greatly expanded follicle, which bulges on the ovarian surface. The oocyte and its surrounding layers are displaced to one side.	Rapid growth of the dominant follicle is stimulated by FSH during the follicular phase of the ovarian cycle.	A marked increase in estrogen secretion takes place, which triggers the LH surge responsible for ovulation.	The oocyte completes its first meiotic division and becomes a secondary oocyte just before ovulation.

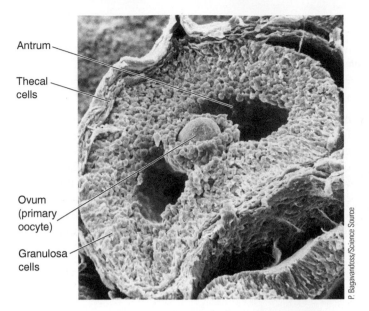

Antrum

Thecal cells

Ovum (primary oocyte)

Granulosa cells

P. Bagavandoss/Science Source

❙Figure 20-15 **Scanning electron micrograph of an early developing antral follicle.**

tightly adhering zona pellucida and granulosa cells (now called the **corona radiata**, meaning "radiating crown"), is swept out of the ruptured follicle into the abdominal cavity by the leaking antral fluid (see ❙Figure 20-14, step **6**). The released ovum is quickly drawn into the oviduct, where fertilization may or may not take place.

The other developing follicles that failed to reach maturation and ovulate undergo degeneration, never to be reactivated. Occasionally, two (or perhaps more) follicles reach maturation and ovulate around the same time. If both are fertilized, **fraternal twins** result. Because fraternal twins arise from separate ova fertilized by separate sperm, they share no more in common than any other two siblings except for the same birth date. **Identical twins**, in contrast, develop from a single fertilized ovum that completely divides into two separate, genetically identical embryos at an early stage of development.

Rupture of the follicle at ovulation signals the end of the follicular phase and ushers in the luteal phase.

The luteal phase is characterized by the presence of a corpus luteum.

The ruptured follicle left behind in the ovary after release of the ovum changes rapidly as the granulosa and thecal cells remaining in the remnant follicle undergo a dramatic structural and functional transformation.

Formation of the Corpus Luteum; Estrogen and Progesterone Secretion These old follicular cells form the **corpus luteum (CL)**, a process called *luteinization* (step **7**). The follicular-turned-luteal cells enlarge and are converted into very active steroid hormone–producing tissue. Abundant storage of cholesterol, the steroid precursor molecule, in lipid droplets within the corpus luteum gives this tissue a yellowish appearance, hence its name (*corpus* means "body"; *luteum* means "yellow").

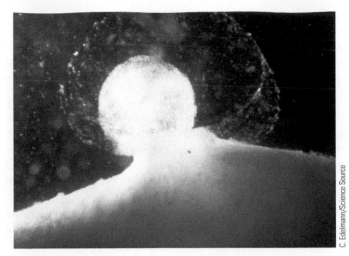

C. Edelmann/Science Source

❙Figure 20-16 **Ovulation.** A photomicrograph of an ovum emerging from the ovary surrounded by the cloudy halo of the corona radiata.

The CL secretes into the blood abundant quantities of progesterone, along with smaller amounts of estrogen. Estrogen secretion in the follicular phase followed by progesterone secretion in the luteal phase is essential for preparing the uterus for implantation of a fertilized ovum. The CL becomes fully functional within 4 days after ovulation, but it continues to increase in size for another 4 or 5 days (step **8**).

Degeneration of the Corpus Luteum If the released ovum is not fertilized and does not implant, the CL degenerates within about 14 days after its formation (step **9**). The luteal cells degenerate and are phagocytized, and connective tissue rapidly fills in to form a fibrous tissue mass. The luteal phase is now over, and one ovarian cycle is complete. A new wave of development of recruited follicles, which begins when degeneration of the old CL is completed, signals the onset of a new follicular phase.

Corpus Luteum of Pregnancy If fertilization and implantation do occur, the corpus luteum continues to grow and produce increasing quantities of progesterone and estrogen instead of degenerating. Now called the *corpus luteum of pregnancy,* this ovarian structure persists until pregnancy ends. It provides the hormones essential for maintaining pregnancy until the developing placenta can take over this crucial function. You will learn more about the role of these structures later.

The ovarian cycle is regulated by complex hormonal interactions.

The ovary has two related endocrine units: (1) the estrogen-secreting follicle during the first half of the cycle and (2) the corpus luteum, which secretes both progesterone and estrogen, during the last half of the cycle. These units are sequentially triggered by complex cyclic hormonal relationships among the hypothalamus, anterior pituitary, and these two ovarian endocrine units.

As in the male, gonadal function in the female is directly controlled by the anterior pituitary gonadotropic hormones—namely, follicle-stimulating hormone and luteinizing hormone.

▌Figure 20-17 Correlation between hormonal levels and cyclic ovarian and uterine changes. Anterior pituitary gonadotropin hormone secretion governs the ovarian cycle and ovarian hormone secretion, which in turn drives cyclical uterine changes. Follow each horizontal strip across to see the cyclical, sequential changes that take place in the plasma concentrations of FSH and LH from the anterior pituitary, the resultant follicular and luteal changes in the ovary, the subsequent cyclical changes in the plasma concentrations of estrogen and progesterone from the ovaries, the consequential endometrial changes that occur during the uterine cycle, and the corresponding phases of the ovarian cycle that occur concurrently with and cause the phases of the uterine cycle. Follow vertically downward to see the concurrent interrelationships among these factors at any given time during the monthly female reproductive cycle. See pp. 746–750 for a detailed explanation of the numbered points.

FIGURE FOCUS: *Compare the relationship among the concentration of the gonadotropic hormones, the status of ovarian follicle/corpus luteum development, the concentration of ovarian hormones, the status of the endometrial lining, and the phases of the uterine cycle and the phases of the ovarian cycle at day 8 and day 21 of the female reproductive cycle.*

These hormones are regulated by hypothalamic gonadotropin-releasing hormone. The GnRH-secreting neurons, in turn, are stimulated by kisspeptin released by higher-level hypothalamic kiss1 neurons. Feedback actions of gonadal hormones at the anterior pituitary and hypothalamus complete the control loop. Unlike in the male, however, control of the female gonads is complicated by the cyclic nature of ovarian function. For example, the effects of FSH and LH on the ovaries depend on the stage of the ovarian cycle. Furthermore, as you will see, estrogen exerts negative-feedback effects during part of the cycle and positive-feedback effects during another part of the cycle, depending on the concentration of estrogen. Also in contrast to the male, FSH is not strictly responsible for gametogenesis, nor is LH solely responsible for gonadal hormone secretion. We consider control of follicular function, ovulation, and the corpus luteum separately, using ▌Figure 20-17 as a means of integrating the various concurrent

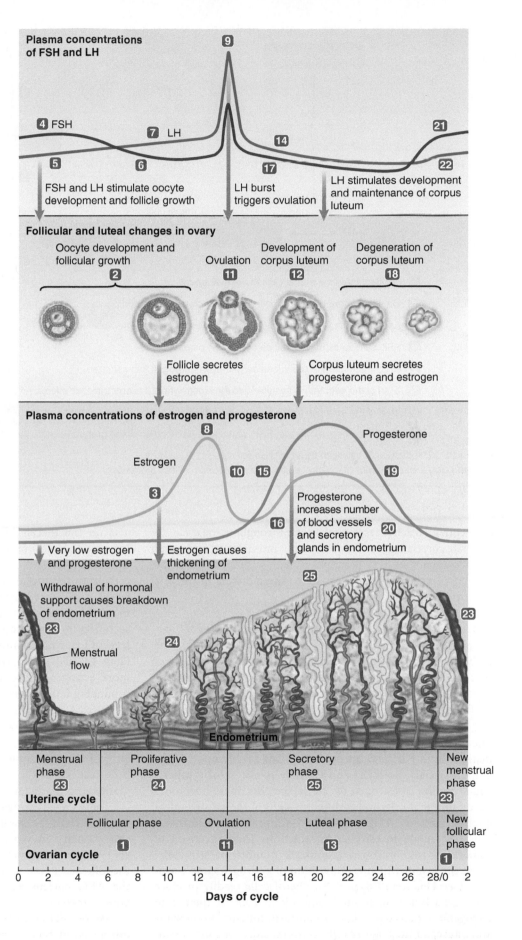

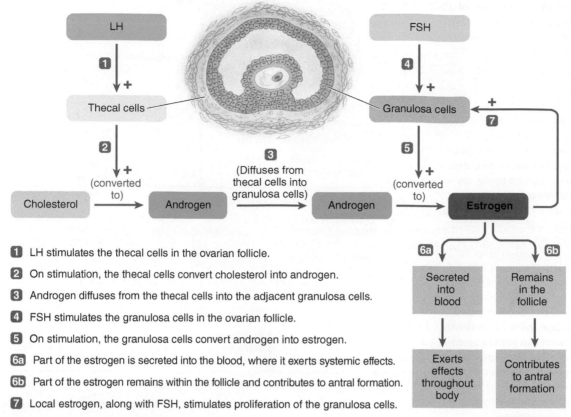

1 LH stimulates the thecal cells in the ovarian follicle.

2 On stimulation, the thecal cells convert cholesterol into androgen.

3 Androgen diffuses from the thecal cells into the adjacent granulosa cells.

4 FSH stimulates the granulosa cells in the ovarian follicle.

5 On stimulation, the granulosa cells convert androgen into estrogen.

6a Part of the estrogen is secreted into the blood, where it exerts systemic effects.

6b Part of the estrogen remains within the follicle and contributes to antral formation.

7 Local estrogen, along with FSH, stimulates proliferation of the granulosa cells.

❙Figure 20-18 **Production of estrogen by an ovarian follicle.**
FIGURE FOCUS: *Deduce from this figure which type of follicular cell has aromatase.*

and sequential activities that take place throughout the cycle. To facilitate correlation between this rather intimidating figure and the accompanying text description of this complex cycle, the numbers in the figure correspond to the numbers in the text description.

Control of Follicular Function We begin with the follicular phase of the ovarian cycle (❙Figure 20-17, step 1). The early stages of preantral follicular growth and oocyte maturation precede the follicular phase and do not require gonadotropic stimulation. Hormonal support is required, however, for further follicular development and antrum formation (step 2), and for estrogen secretion (step 3). Estrogen, FSH (step 4), and LH (step 5) are all needed. FSH induces antrum formation. Both FSH and estrogen stimulate proliferation of the granulosa cells. Both LH and FSH are required for synthesis and secretion of estrogen by the follicle, but these hormones act on different cells and at different steps in the estrogen production pathway (❙Figure 20-18). Both granulosa and thecal cells participate in estrogen production. The conversion of cholesterol into estrogen requires a number of sequential steps, the last of which is conversion of androgens into estrogens (see ❙Figure 19-8, p. 673). Thecal cells readily produce androgens but have limited capacity to convert them into estrogens. Granulosa cells, in contrast, contain the enzyme aromatase, so they can readily convert androgens into estrogens, but they cannot produce androgens in the first place. LH

acts on the thecal cells to stimulate androgen production, whereas FSH acts on the granulosa cells to promote conversion of thecal androgens (which diffuse into the granulosa cells from the thecal cells) into estrogens. Because low basal levels of FSH (see ❙Figure 20-17, step 6) are sufficient to promote this final conversion to estrogen, the rate of estrogen secretion by the follicle primarily depends on the circulating level of LH, which continues to rise during the follicular phase (step 7). Furthermore, as the follicle continues to grow, more estrogen is produced simply because more estrogen-producing follicular cells are present.

Part of the estrogen produced by the growing follicle is secreted into the blood and is responsible for the steadily increasing plasma estrogen levels during the follicular phase (step 8). The remainder of the estrogen remains within the follicle, contributing to the antral fluid and stimulating further granulosa cell proliferation (see ❙Figure 20-18).

The secreted estrogen, in addition to acting on sex-specific tissues such as the uterus, inhibits the hypothalamus and anterior pituitary in typical negative-feedback fashion (❙Figure 20-19). The rising, moderate levels of estrogen characterizing the follicular phase act directly on the hypothalamus to inhibit the ARC nucleus kiss1 neurons, thus indirectly inhibiting GnRH secretion and thereby suppressing GnRH-prompted release of FSH and LH from the anterior pituitary. However, estrogen's primary effect is directly on the pituitary. Estrogen selectively inhibits FSH secretion by the gonadotropes.

This differential secretion of FSH and LH by the gonadotropes induced by estrogen is in part responsible for the declining plasma FSH level, unlike the rising plasma LH concentration, during the follicular phase as the estrogen level rises (see ❙Figure 20-17, step **6**). Another contributing factor to the fall in FSH during the follicular phase is secretion of *inhibin* by the follicular cells. Inhibin preferentially inhibits FSH secretion by acting at the anterior pituitary, just as it does in the male (see ❙Figure 20-19). The decline in FSH secretion brings about atresia of all but the single dominant, most mature of the developing follicles.

In contrast to FSH, LH secretion continues to rise slowly during the follicular phase (see ❙Figure 20-17, step **7**) despite inhibition of GnRH (and thus, indirectly, LH) secretion. This seeming paradox occurs because estrogen alone cannot completely suppress **tonic** (low-level, ongoing) **LH secretion**; to completely inhibit tonic LH secretion, both estrogen and progesterone are required. Because progesterone does not appear until the luteal phase of the cycle, the basal level of circulating LH slowly increases during the follicular phase under incomplete inhibition by estrogen alone.

Control of Ovulation Ovulation and subsequent luteinization of the ruptured follicle are triggered by an abrupt, massive increase in LH secretion (step **9**). This **LH surge** brings about four major changes in the follicle:

1. It halts estrogen synthesis by the follicular cells (step **10**).

2. It reinitiates meiosis in the mature follicle's oocyte by blocking release of **oocyte maturation inhibitor** produced by the granulosa cells. This substance is responsible for arresting meiosis in the primary oocytes once they are wrapped within granulosa cells in the fetal ovary.

3. It triggers production of local prostaglandins, which induce ovulation by promoting vascular changes that cause rapid follicular swelling while inducing enzymatic digestion of the follicular wall. Together, these actions lead to rupture of the weakened wall that covers the bulging follicle (step **11**).

4. It causes differentiation of follicular cells into luteal cells. Because the LH surge triggers both ovulation and luteinization, formation of the corpus luteum automatically follows ovulation (step **12**). Thus, the midcycle burst in LH secretion is a dramatic point in the cycle; it terminates the follicular phase and initiates the luteal phase (step **13**).

The two modes of LH secretion—tonic LH secretion (step **7**) that promotes ovarian hormone secretion and the LH surge (step **9**) that causes ovulation—not only occur at different

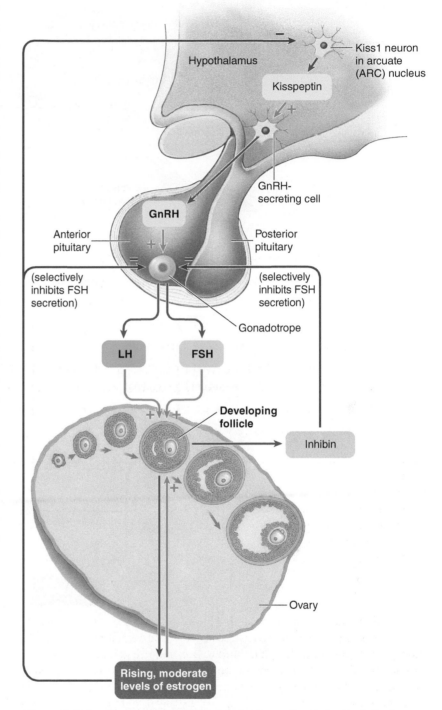

❙Figure 20-19 **Feedback control of FSH and tonic LH secretion during the follicular phase.**

times and produce different effects but also are controlled by different mechanisms. Tonic LH secretion is partially suppressed (step **7**) by the inhibitory action of the rising, moderate levels of estrogen (step **3**) during the follicular phase and is completely suppressed (step **14**) by the increasing levels of progesterone during the luteal phase (step **15**). Because tonic LH secretion stimulates both estrogen and progesterone secretion, this is a typical negative-feedback effect. Estrogen and progesterone both suppress LH secretion by inhibiting the ARC nucleus kiss1 neurons.

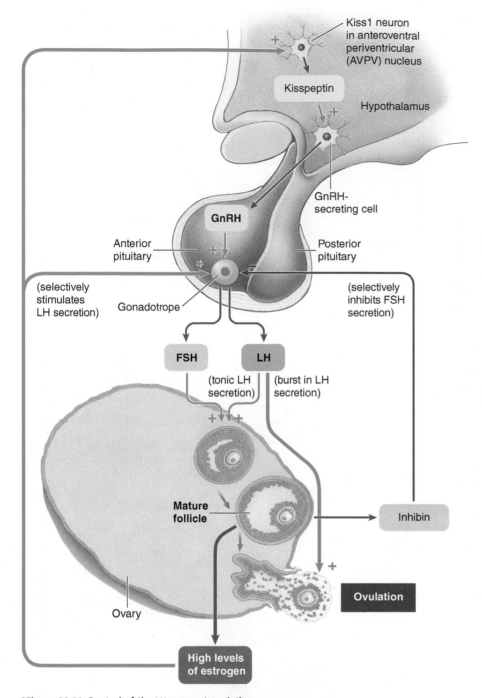

third ventricular cavity. Thus, females have two sets of kiss1 neurons, one set in the ARC nucleus (the same as in males) that is inhibited by estrogen (and progesterone in females or testosterone in males) for negative feedback and one set in the AVPV nucleus that is stimulated by high levels of estrogen for positive feedback. The high plasma concentration of estrogen acts directly on the AVPV nucleus kiss1 neurons to increase kisspeptin and thereby GnRH release, which increases both LH and FSH secretion. Thus, LH enhances estrogen production by the follicle, and the resultant peak estrogen concentration brings about increased LH secretion. A high estrogen level also acts directly on the anterior pituitary to specifically increase LH secretion by the gonadotropes. The latter effect largely accounts for the much greater surge in LH secretion compared to FSH secretion at midcycle (see ❙Figure 20-17, step ⑨). Also, continued inhibin secretion by the follicular cells preferentially inhibits FSH secretion, keeping the FSH levels from rising as high as the LH levels. There is no known role for the modest midcycle surge in FSH that accompanies the pronounced and pivotal LH surge. Because only a mature, preovulatory follicle, not follicles in earlier stages of development, can secrete high-enough levels of estrogen to trigger the LH surge, ovulation is not induced until a follicle has reached the proper size and degree of maturation. In a way, then, the follicle lets the hypothalamus know when it is ready to be stimulated to ovulate. The LH surge lasts for about a day at midcycle, just before ovulation.

Control of the Corpus Luteum LH "maintains" the corpus luteum—that is, after triggering development of the CL, LH stimulates ongoing steroid hormone secretion by this ovarian structure. Under the influence of LH, the CL secretes both progesterone (see ❙Figure 20-17, step ⑮) and

❙Figure 20-20 **Control of the LH surge at ovulation.**
FIGURE FOCUS: *Compare the feedback effect of rising, moderate levels of estrogen in Figure 20-19 with the feedback effect of high levels of estrogen in Figure 20-20.*

In contrast, the LH surge is triggered by a *positive-feedback effect*. Whereas the rising, moderate level of estrogen early in the follicular phase *inhibits* LH secretion, the high level of estrogen that occurs during peak estrogen secretion late in the follicular phase (step ⑧) *stimulates* LH secretion and initiates the LH surge (❙Figure 20-20). A high estrogen level generates the LH surge by stimulating another set of kisspeptin-releasing neurons unique to females located in the **anteroventral periventricular (AVPV) nucleus**. The AVPV nucleus is positioned in the hypothalamus along the anterior part of the wall of the

estrogen (step ⑯), with progesterone being its most abundant hormonal product. The plasma progesterone level increases for the first time during the luteal phase. No progesterone is secreted during the follicular phase (except for a small amount a few hours before ovulation). Therefore, the follicular phase is dominated by estrogen and the luteal phase by progesterone.

A transitory drop in the level of circulating estrogen occurs at midcycle (step ⑩) as the estrogen-secreting follicle meets its demise at ovulation. The estrogen level climbs again during the luteal phase because of the CL's activity, although it does not reach the same peak as during the follicular phase. What keeps the modestly high estrogen level during the luteal phase from triggering another LH surge? Progesterone. Even though a high level of estrogen stimulates LH secretion, progesterone, which dominates the luteal phase, powerfully inhibits LH secretion (step **14**) as well as FSH secretion (step **17**) by acting at both the hypothalamic ARC nucleus and the anterior pituitary (**Figure 20-21**). Furthermore, the luteal cells secrete inhibin, which selectively inhibits FSH secretion. Inhibition of FSH and LH prevents new follicular maturation and ovulation during the luteal phase. Under progesterone's influence, the reproductive system is gearing up to support the just-released ovum, should it be fertilized, instead of preparing other ova for release.

The corpus luteum functions for an average of 2 weeks and then degenerates if fertilization does not occur (see **Figure 20-17**, step **18**). The mechanisms that govern degeneration of the CL are not fully understood. The declining level of circulating LH (step **14**), driven down by inhibitory actions of progesterone, undoubtedly contributes to the CL's downfall. Prostaglandins and estrogen released by the luteal cells themselves may play a role. Demise of the CL terminates the luteal phase and sets the stage for a new follicular phase. As the CL degenerates, plasma progesterone (step **19**) and estrogen (step **20**) levels fall rapidly because these hormones are no longer being produced. Withdrawal of the inhibitory effects of these hormones on the hypothalamus allows FSH (step **21**) and tonic LH (step **22**) secretion to modestly increase again. Under the influence of these gonadotropic hormones, another batch of early developing follicles (step **2**) is recruited and induced to mature as a new follicular phase begins (step **1**).

Cyclic uterine changes are caused by hormonal changes during the ovarian cycle.

The fluctuations in circulating levels of estrogen and progesterone during the ovarian cycle induce profound changes in the uterus, giving rise to the **menstrual**, or **uterine**, **cycle**. Because it reflects hormonal changes during the ovarian cycle, the menstrual cycle averages 28 days, as does the ovarian cycle, although even normal adults vary considerably from this mean. The outward manifestation of the cyclic changes in the uterus is the menstrual bleeding once during each menstrual cycle (that is, once a month). Less obvious changes take place throughout the cycle, however, as the uterus is prepared for implantation, should a released ovum be fertilized, and then is stripped clean of its prepared lining (menstruation) if implantation does not occur, only to repair itself and start preparing for the ovum that will be released during the next cycle.

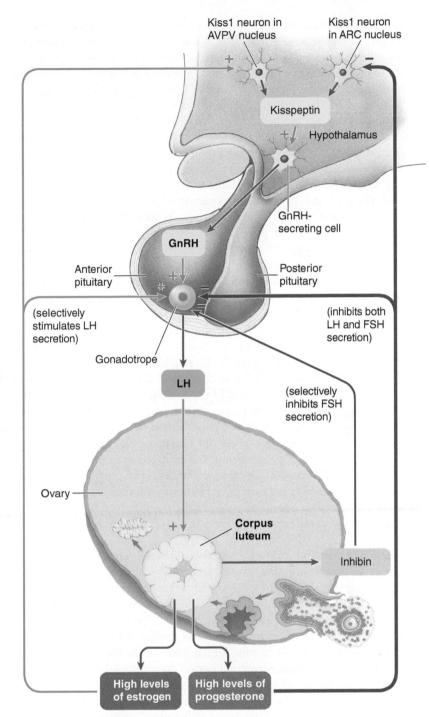

Figure 20-21 **Feedback control during the luteal phase.**

We briefly examine the influences of estrogen and progesterone on the uterus and then consider the effects of cyclic fluctuations of these hormones on uterine structure and function.

Influences of Estrogen and Progesterone on the Uterus The uterus consists of two main layers: the **myometrium**, the outer smooth muscle layer, and the **endometrium**, the inner lining that contains numerous blood vessels and glands. Estrogen stimulates growth of both the myometrium and the endometrium. It also induces synthesis of progesterone receptors in the endometrium. Thus, progesterone can exert an effect on the endometrium only after it has been "primed" by estrogen. Progesterone acts on the estrogen-primed endometrium to convert it into a hospitable and nutritious lining suitable for implantation of a fertilized ovum. Under the influence of progesterone, the endometrial connective tissue becomes loose and edematous as a result of an accumulation of electrolytes and water, which facilitates implantation of the fertilized ovum. Progesterone further prepares the endometrium to sustain an early-developing embryo by stimulating the endometrial glands to secrete and store large quantities of glycogen (stored glucose) and by causing tremendous growth of the endometrial blood vessels. Progesterone also reduces the contractility of the uterus to provide a quiet environment for implantation and embryonic growth.

The menstrual cycle consists of three phases: the *menstrual phase*; the *proliferative phase*; and the *secretory, or progestational, phase.*

Menstrual Phase The **menstrual phase** is the most overt phase, characterized by discharge of blood and endometrial debris from the vagina (see ❚ Figure 20-17, step **23**). By convention, the first day of menstruation is considered the start of a new cycle. It coincides with the end of the ovarian luteal phase and onset of a new follicular phase. As the corpus luteum degenerates because fertilization and implantation of the ovum released during the preceding cycle did not take place (step **18**), circulating levels of progesterone and estrogen drop precipitously (steps **19** and **20**). Because the net effect of progesterone and estrogen is to prepare the endometrium for implantation of a fertilized ovum, withdrawal of these steroids deprives the highly vascular, nutrient-rich uterine lining of its hormonal support.

The fall in ovarian hormone levels also stimulates release of a uterine prostaglandin that causes vasoconstriction of the endometrial vessels, disrupting the blood supply to the endometrium. The subsequent reduction in O_2 delivery causes death of the endometrium, including its blood vessels. The resulting bleeding through the disintegrating vessels flushes the dying endometrial tissue into the uterine lumen. Most of the uterine lining sloughs during each menstrual period except for a deep, thin layer of epithelial cells and glands, from which the endometrium regenerates. The same local uterine prostaglandin also stimulates mild rhythmic contractions of the uterine myometrium. These contractions help expel the blood and endometrial debris from the uterine cavity out through the vagina as **menstrual flow**. Excessive uterine contractions caused by prostaglandin overproduction produce the **dysmenorrhea** (*menstrual cramps*) some women experience.

The average blood loss during a single menstrual period is 50 to 150 mL. Blood that seeps slowly through the degenerating endometrium clots within the uterine cavity, then is acted on by fibrinolysin, a fibrin dissolver that breaks down the fibrin forming the meshwork of the clot. Therefore, blood in the menstrual flow usually does not clot because it has already clotted and the clot has been dissolved before it passes out of the vagina. When blood flows rapidly through the leaking vessels, however, it may not be exposed to sufficient fibrinolysin, so when the menstrual flow is most profuse, blood clots may appear. In addition to the blood and endometrial debris, large numbers of leukocytes are found in the menstrual flow. These white blood cells play an important defense role in helping the raw endometrium resist infection.

Menstruation typically lasts for about 5 to 7 days after degeneration of the CL, coinciding in time with the early portion of the ovarian follicular phase (steps **23** and **1**). Withdrawal of progesterone and estrogen (steps **19** and **20**) on degeneration of the CL leads simultaneously to sloughing of the endometrium (menstruation) (step **23**) and rapid development of newly recruited antral follicles in the ovary (step **2**) under the influence of rising gonadotropic hormone levels (steps **21** and **22**). The drop in gonadal hormone secretion removes inhibitory influences from the hypothalamus and anterior pituitary, so FSH and LH secretion increases and a new follicular phase begins (step **1** again). After 5 to 7 days under the influence of FSH and LH, the newly recruited, rapidly growing antral follicles are secreting enough estrogen (step **3**) to promote repair and growth of the endometrium.

Proliferative Phase Thus, menstrual flow ceases, and the **proliferative phase** of the uterine cycle begins concurrent with the last portion of the ovarian follicular phase as the endometrium starts to repair itself and proliferate (step **24**) under the influence of estrogen from the newly recruited antral follicles. When the menstrual flow ceases, a thin endometrial layer less than 1 mm thick remains. Estrogen stimulates proliferation of epithelial cells, glands, and blood vessels in the endometrium, increasing this lining to a thickness of 3 to 5 mm. The estrogen-dominant proliferative phase lasts from the end of menstruation to ovulation. Peak estrogen levels (step **8**) trigger the LH surge (step **9**) responsible for ovulation (step **11**).

Secretory, or Progestational, Phase After ovulation, when a new CL is formed (step **12**), the uterus enters the **secretory**, or **progestational**, **phase** (step **25**), which coincides in time with the ovarian luteal phase (step **13**). The CL secretes large amounts of progesterone (step **15**) and estrogen (step **16**). Progesterone converts the thickened, estrogen-primed endometrium to a richly vascularized, glycogen-filled tissue. This period is called either the *secretory phase* because the endometrial glands are actively secreting glycogen into the uterine lumen for early nourishment of a developing embryo before it implants, or the *progestational* ("before pregnancy") *phase,* referring to the development of a lush endometrial lining capable of supporting an early embryo after implantation. If fertilization and implantation do not occur, the CL degenerates and a new follicular phase and menstrual phase begin again.

Various factors can interfere with the delicate balance of the hypothalamic–pituitary–ovarian–peripheral target organ axis, leading to menstrual irregularities and fertility problems. Among these factors are starvation (a problem for example with anorexia nervosa; see p. 627), stress, and heavy exercise. (For the effects of exercise on this cycle, see the accompanying boxed feature, ❚ A Closer Look at Exercise Physiology).

Menstrual Irregularities: When Cyclists and Other Female Athletes Do Not Cycle

SINCE THE 1970S, AS WOMEN in growing numbers have participated in a variety of sports requiring vigorous training regimens, researchers have become increasingly aware that many women experience changes in their menstrual cycles as a result of athletic participation. These changes are referred to as **athletic menstrual cycle irregularity (AMI)**. The menstrual cycle dysfunction can vary in severity from amenorrhea (cessation of menstrual periods) to oligomenorrhea (cycles at irregular or infrequent intervals) to cycles that are normal in length but are anovulatory (no ovulation) or that have a short or inadequate luteal phase.

In early research studies using surveys and questionnaires to determine the prevalence of the problem, the frequency of these sport-related disorders varied from 2% to 51%. In contrast, the rate of occurrence of menstrual cycle dysfunction in females of reproductive age in the general population is 2% to 5%. A major problem of using surveys to determine the frequency of menstrual cycle irregularity is the questionable accuracy of recall of menstrual periods. Furthermore, without blood tests to determine hormone levels throughout the cycle, a woman would not know whether she was anovulatory or had had a shortened luteal phase. Studies in which hormone levels have been determined throughout the menstrual cycle have demonstrated that seemingly normal cycles in athletes frequently have a short luteal phase (less than two days long with low progesterone levels).

In a study conducted to determine whether strenuous exercise spanning two menstrual cycles would induce menstrual disorders, 28 initially untrained college women with documented ovulation and luteal adequacy served as subjects. The women participated in an eight-week exercise program in which they initially ran 4 miles per day and progressed to 10 miles per day by the fifth week. They were expected to participate daily in 3.5 hours of moderate-intensity sports. Only four women had normal menstrual cycles during the training. Abnormalities resulting from training included abnormal bleeding, delayed menstrual periods, abnormal luteal function, and loss of LH surge. All the women returned to normal cycles within six months after training. The results of this study suggest that the frequency of AMI with strenuous exercise may be much greater than indicated by questionnaire alone. In other studies using low-intensity exercise regimens, AMI was much less frequent.

The mechanisms of AMI are unknown at present, although studies have implicated rapid weight loss, decreased percentage of body fat, dietary insufficiencies, prior menstrual dysfunction, stress, age at onset of training, and the intensity of training as factors that play a role. Epidemiologists have shown that if girls participate in vigorous sports before **menarche** (the first menstrual period), menarche is delayed. On average, athletes have their first period when they are about three years older than their nonathletic counterparts. Furthermore, females who participate in sports before menarche seem to have a higher frequency of AMI throughout their athletic careers than those who begin to train after menarche. Hormonal changes found in female athletes include (1) severely depressed FSH levels, (2) elevated LH levels, (3) low progesterone during the luteal phase, (4) low estrogen levels in the follicular phase, and (5) an FSH–LH environment totally unbalanced as compared to those of age-matched nonathletic women. The preponderance of evidence indicates that cycles return to normal once vigorous training is stopped.

The major problem associated with athletic amenorrhea is a reduction in bone mineral density. Studies have shown that the mineral density in the vertebrae of the lower spine of those with athletic amenorrhea is lower than in athletes with normal menstrual cycles and lower than in age-matched nonathletes. However, amenorrheic runners have higher bone mineral density than amenorrheic nonathletes, presumably because the mechanical stimulus of exercise helps retard bone loss. Studies have shown that amenorrheic athletes are at higher risk for stress-related fractures than athletes with normal menstrual cycles. One study, for example, found stress fractures in 6 of 11 amenorrheic runners but in only 1 of 6 runners with normal menstrual cycles. The mechanism for bone loss is probably the same as is found in postmenopausal osteoporosis—lack of estrogen (see p. 706). The problem is serious enough that an amenorrheic athlete should discuss the possibility of estrogen replacement therapy with her physician.

There may be some positive benefits of athletes' menstrual dysfunction. A recent epidemiological study to determine if the long-term reproductive and general health of women who had been college athletes differed from that of college nonathletes showed that former athletes had less than half the lifetime occurrence rate of cancers of the reproductive system and half the breast cancer occurrence compared to nonathletes. Because these are hormone-sensitive cancers, the delayed menarche and lower estrogen levels found in women athletes may play a key role in decreasing the risk of cancer of the reproductive system and breast.

Fluctuating estrogen and progesterone levels produce cyclical changes in cervical mucus.

Hormonally induced changes also take place in the cervix during the ovarian cycle. Under the influence of estrogen during the follicular phase, the mucus secreted by the cervix becomes abundant, clear, and thin. This change, which is most pronounced when estrogen is at its peak and ovulation is approaching, facilitates passage of sperm through the cervical canal. After ovulation, under the influence of progesterone from the CL, the mucus becomes thick and sticky, essentially plugging up the cervical opening. This plug is an important defense mechanism, preventing bacteria (that might threaten a possible pregnancy) from entering the uterus from the vagina. Sperm also cannot penetrate this thick mucus barrier.

Pubertal changes in females are similar to those in males.

Regular menstrual cycles are absent in both young and aging females, but for different reasons. The female reproductive system does not become active until puberty. Unlike the fetal testes, the fetal ovaries need not be functional because in the absence of fetal testosterone secretion in a female, the reproductive system is automatically feminized, without requiring the presence of female sex hormones. Puberty occurs in females when hypothalamic GnRH activity increases for the first time. As in the male, the mechanisms that govern the onset of puberty are not clearly understood but are believed to involve melatonin's and leptin's influence on the ARC nucleus kiss1 neurons. Leptin's action might account at least in part for the fact that, on average, heavier girls tend to enter puberty earlier than their leaner counterparts do.

GnRH begins stimulating release of anterior pituitary gonadotropic hormones, which in turn stimulate ovarian activity. The resulting secretion of estrogen by the activated ovaries induces growth and maturation of the female reproductive tract, and development of the female secondary sexual characteristics. Estrogen's prominent action in the latter regard is to promote fat deposition in strategic locations, such as the breasts, buttocks, and thighs, giving rise to the typical curvaceous female figure. Enlargement of the breasts at puberty is the result primarily of fat deposition in the breast tissue, not functional development of the mammary glands. The pubertal rise in estrogen also closes the epiphyseal plates, halting further growth in height, similar to the effect of testosterone-turned-estrogen in males. Three other pubertal changes in females—growth of axillary and pubic hair, the pubertal growth spurt, and development of libido—are attributable to a spurt in adrenal androgen secretion at puberty, not to estrogen.

Menopause is unique to females.

The cessation of a woman's menstrual cycles at **menopause** sometime between the ages of 45 and 55 has traditionally been attributed to the limited supply of ovarian follicles present at birth. According to this proposal, once this reservoir is depleted, ovarian cycles, and hence menstrual cycles, cease. Thus, the termination of reproductive potential in a middle-aged woman is "preprogrammed" at her own birth. Recent evidence suggests, however, that a midlife hypothalamic change instead of aging ovaries may trigger the onset of menopause. Evolutionarily, menopause may have developed as a mechanism that prevented pregnancy in women beyond the time that they could likely rear a child before their own death.

Males do not experience complete gonadal failure as females do, for two reasons. First, a male's germ cell supply is unlimited because mitotic activity of the spermatogonia continues. Second, gonadal hormone secretion in males is not inextricably dependent on gametogenesis, as it is in females. If female sex hormones were produced by separate tissues unrelated to those governing gametogenesis, as are male sex hormones, estrogen and progesterone secretion would not automatically stop when oogenesis stopped.

Menopause is preceded by a period of progressive ovarian failure characterized by increasingly irregular cycles and dwindling estrogen levels. During the period of transition from sexual maturity to cessation of reproductive capability, ovarian estrogen production declines from as much as 300 mg per day to essentially nothing. Postmenopausal women are not completely devoid of estrogen, however, because adipose tissue, the liver, and the adrenal cortex continue to produce up to 20 mg of estrogen per day. In addition to the ending of ovarian and menstrual cycles, the loss of ovarian estrogen following menopause brings about many physical and emotional changes. These changes include vaginal dryness, which can cause discomfort during sex, and gradual atrophy of the genital organs. However, postmenopausal women still have a sex drive because of their adrenal androgens.

Clinical Note Because estrogen has widespread physiological actions beyond the reproductive system, the dramatic loss of ovarian estrogen in menopause affects other body systems, most notably the skeleton and the cardiovascular system. Estrogen helps build strong bones, shielding premenopausal women from the bone-thinning condition of osteoporosis (see p. 706). The postmenopausal reduction in estrogen increases activity of the bone-dissolving osteoclasts and diminishes activity of the bone-building osteoblasts. The result is decreased bone density and a greater incidence of bone fractures.

Estrogen also helps modulate the actions of epinephrine and norepinephrine on the arteriolar walls by promoting local release of the vasodilator nitric oxide. The menopausal diminution of estrogen leads to unstable control of blood flow, especially in the skin vessels. Transient increases in the flow of warm blood through these superficial vessels are responsible for the "hot flashes" that frequently accompany menopause. Vasomotor stability is gradually restored in postmenopausal women so that hot flashes eventually cease.

You have now learned about the events that take place if fertilization does not occur. Because the primary function of the reproductive system is, of course, reproduction, we next turn attention to the sequence of events that take place when fertilization does occur.

The oviduct is the site of fertilization.

Fertilization, the union of male and female gametes, normally occurs in the **ampulla**, the upper third of the oviduct (❙Figure 20-22). Thus, both the ovum and the sperm must be transported from their gonadal site of production to the ampulla.

Ovum Transport to the Oviduct Unlike the male reproductive tract, which has a continuous lumen from the site of sperm production in the seminiferous tubules to exit of the sperm from the urethra at ejaculation, the ovaries are not in direct contact with the reproductive tract. The ovum is released into the abdominal cavity at ovulation. Normally, however, the oviduct quickly picks up the egg. The dilated end of the oviduct

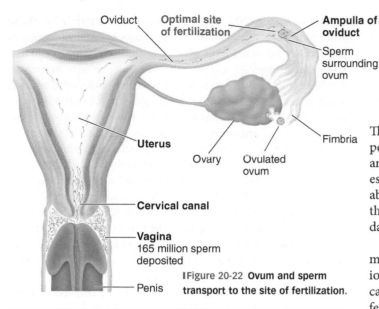

Oviduct — Optimal site of fertilization — **Ampulla of oviduct** — Sperm surrounding ovum

Uterus

Ovary — Ovulated ovum — Fimbria

Cervical canal

Vagina
165 million sperm deposited

Penis

❙Figure 20-22 **Ovum and sperm transport to the site of fertilization.**

cups around the ovary and contains **fimbriae,** fingerlike projections that contract in a sweeping motion to guide the released ovum into the oviduct (see ❙Figures 20-2b, p. 719, and 20-22). Furthermore, the fimbriae are lined by cilia—fine, hairlike projections that beat in waves toward the interior of the oviduct—further assuring the ovum's passage into the oviduct (see p. 48). Within the oviduct, the ovum is rapidly propelled by peristaltic contractions and ciliary action to the ampulla.

Conception can take place during a limited time span in each cycle (the **fertile period**). If not fertilized, the ovum begins to disintegrate within 12 to 24 hours and is subsequently phagocytized by cells that line the reproductive tract. Fertilization must therefore occur within 24 hours after ovulation, when the ovum is still viable. Sperm typically survive about 48 hours but can survive up to 5 days in the female reproductive tract, so sperm deposited from 5 days before ovulation to 24 hours after ovulation may be able to fertilize the released ovum, although these times vary considerably.

Clinical Note Occasionally, an ovum fails to be transported into the oviduct and remains instead in the abdominal cavity. Rarely, such an ovum gets fertilized, resulting in an **ectopic abdominal pregnancy,** in which the fertilized egg implants in the rich vascular supply to the digestive organs rather than in its usual site in the uterus (*ectopic* means "out of place"). An abdominal pregnancy often leads to life-threatening hemorrhage because the digestive organ blood supply is not primed to respond appropriately to implantation as the endometrium is. If this unusual pregnancy proceeds to term, the baby must be delivered surgically because the normal vaginal exit is not available. The probability of maternal complications at birth is greatly increased because the digestive vasculature is not designed to "seal itself off" after birth as the endometrium does.

Sperm Transport to the Oviduct After sperm are deposited in the vagina at ejaculation, they must travel through the cervical canal, through the uterus, and then up to the egg in the upper third of the oviduct (❙Figure 20-22). The first sperm

arrive in the oviduct within half an hour after ejaculation. Even though sperm are mobile by means of whiplike contractions of their tails, 30 minutes is too soon for a sperm's mobility to transport it to the site of fertilization. To make this formidable journey, sperm need the help of the female reproductive tract.

The first hurdle is passage through the cervical canal. Throughout most of the cycle, the cervical mucus is too thick to permit sperm penetration. The cervical mucus becomes thin and watery enough to permit sperm to penetrate only when estrogen levels are high, as in the presence of a mature follicle about to ovulate. Sperm migrate up the cervical canal under their own power. The canal remains penetrable for only 2 or 3 days during each cycle, around the time of ovulation.

Once sperm have entered the uterus, contractions of the myometrium churn them around in "washing-machine" fashion. This action quickly disperses sperm throughout the uterine cavity. When sperm reach the oviduct, they are propelled to the fertilization site in the upper end of the oviduct by upward contractions of the oviduct smooth muscle. These myometrial and oviduct contractions that facilitate sperm transport are induced by the high estrogen level just before ovulation, aided by seminal prostaglandins.

New research indicates that when sperm reach the ampulla, ova are not passive partners in conception. Sperm have a specific olfactory receptor (OR) called **hOR17-4** that is identical to one found in the nose for smell perception (see p. 227). This receptor binds to the odorant **bourgeonal,** a molecule that gives rise to the floral odor of lilies of the valley. In the ampulla, bourgeonal serves as a chemoattractant or chemotaxin (see p. 410), attracting sperm and causing them to propel themselves toward the waiting female gamete. Thus sperm "smell" their way to the egg. The source of bourgeonal in the human female reproductive tract appears to be the layer of follicular cells (corona radiata) surrounding the egg at ovulation. Activation of hOR17-4 on binding with bourgeonal triggers a cAMP second-messenger pathway in sperm that brings about intracellular Ca^{2+} release. This Ca^{2+} turns on the microtubule sliding that brings about tail movement and sperm swimming in the direction of a higher concentration of bourgeonal, toward the waiting "perfumed" egg (see p. 49).

Progesterone released into the oviduct from the follicular cells that surround the egg at ovulation is another major chemoattractant. This progesterone binds with fast-responding nongenomic surface membrane receptors on the sperm, unlike this steroid's usual binding to slow-responding intracellular receptors in other target cells. On binding, progesterone opens Ca^{2+}-permeable cation channels called **CatSper** channels found exclusively in the plasma membrane of a sperm tail. The resultant, swift Ca^{2+} entry is crucial for the following fertilization-related events in sperm: (1) capacitation, (2) hyperactivated motility, and (3) the acrosome reaction. Thus CatSper activation is essential for male fertility. You already know about capacitation and will soon learn about the acrosome reaction. We focus now on the motility changes that occur. When Ca^{2+} floods into the cell on progesterone-induced opening of CatSper channels, sperm switch from their usual smooth swimming motion to a highly asymmetric, frantic beating of the tail called

hyperactivated motility. This more powerful type of motility generates the extra thrust needed for sperm to penetrate the corona radiata and zona pellucida to enter the egg.

Fertilization Even around ovulation time, when sperm can penetrate the cervical canal, of the 165 million sperm typically deposited in a single ejaculate, only a few thousand make it to the site of fertilization. That only a very small percentage of the deposited sperm ever reach their destination is one reason sperm concentration must be so high (20 million/mL of semen) for a man to be fertile. The other reason is that the acrosomal enzymes of many sperm are needed to break down the barriers surrounding the ovum.

The tail of the sperm is used to maneuver for final penetration of the ovum. To fertilize an ovum, a sperm must first pass through the corona radiata and zona pellucida surrounding it. The sperm penetrates the corona radiata by means of membrane-bound enzymes in the surface membrane that surrounds the head (▮ Figure 20-23a, step ❶ and chapter opener photo, p. 715). Sperm can penetrate the zona pellucida only after binding with specific binding sites on the surface of this layer. As for the binding partners, *fertilin,* a plasma membrane protein on the sperm head, binds with *ZP3,* a glycoprotein in the outer layer of the zona pellucida. Only sperm of the same species can bind to these zona pellucida sites and pass through. Binding of the sperm head to ZP3 triggers the Ca^{2+}-dependent **acrosome reaction**, in which the acrosomal membrane disrupts and the acrosomal enzymes are released (▮ Figure 20-23 part a, step ❷, and part b). Calcium that enters the sperm tail through the opened CatSper channels rapidly moves within a few seconds to the head, where it participates in the acrosome reaction. The acrosomal enzymes digest the zona pellucida, enabling the sperm, with its tail still beating, to tunnel a path through this protective barrier (step ❸). The first sperm to reach the ovum itself fuses with the plasma membrane of the ovum (actually a secondary oocyte), and its head (bearing its DNA) enters the ovum's cytoplasm (step ❹). The sperm's tail is frequently lost in this process, but the head carries the crucial genetic information. Sperm–egg fusion triggers a chemical change in the ovum's surrounding membrane that makes this outer layer impenetrable to the entry of any more sperm. Following is how this phenomenon, known as **block to polyspermy** ("many sperm") is accomplished. The outermost, or cortical, region of the ovum contains enzyme-filled *cortical granules.* Fertilization-

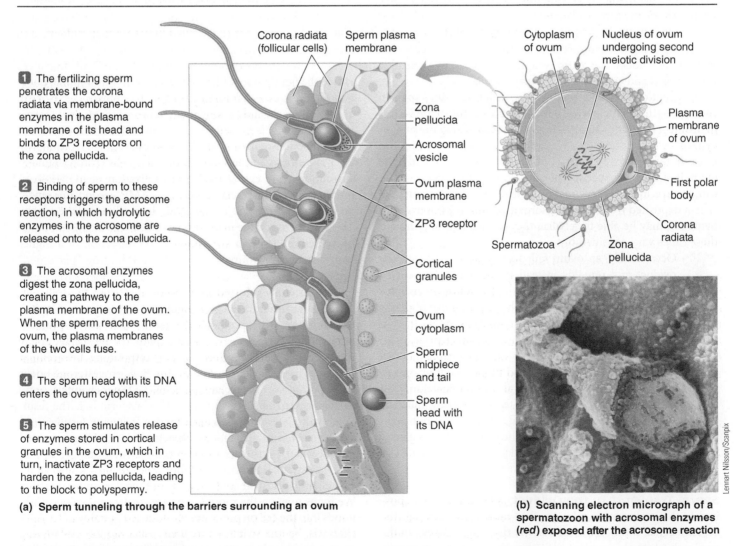

❶ The fertilizing sperm penetrates the corona radiata via membrane-bound enzymes in the plasma membrane of its head and binds to ZP3 receptors on the zona pellucida.

❷ Binding of sperm to these receptors triggers the acrosome reaction, in which hydrolytic enzymes in the acrosome are released onto the zona pellucida.

❸ The acrosomal enzymes digest the zona pellucida, creating a pathway to the plasma membrane of the ovum. When the sperm reaches the ovum, the plasma membranes of the two cells fuse.

❹ The sperm head with its DNA enters the ovum cytoplasm.

❺ The sperm stimulates release of enzymes stored in cortical granules in the ovum, which in turn, inactivate ZP3 receptors and harden the zona pellucida, leading to the block to polyspermy.

Corona radiata (follicular cells)
Sperm plasma membrane
Zona pellucida
Acrosomal vesicle
Ovum plasma membrane
ZP3 receptor
Cortical granules
Ovum cytoplasm
Sperm midpiece and tail
Sperm head with its DNA

Cytoplasm of ovum
Nucleus of ovum undergoing second meiotic division
Plasma membrane of ovum
First polar body
Corona radiata
Zona pellucida
Spermatozoa

(a) Sperm tunneling through the barriers surrounding an ovum

(b) Scanning electron micrograph of a spermatozoon with acrosomal enzymes (*red*) exposed after the acrosome reaction

Lennart Nilsson/Scanpix

▮Figure 20-23 **Process of fertilization.**

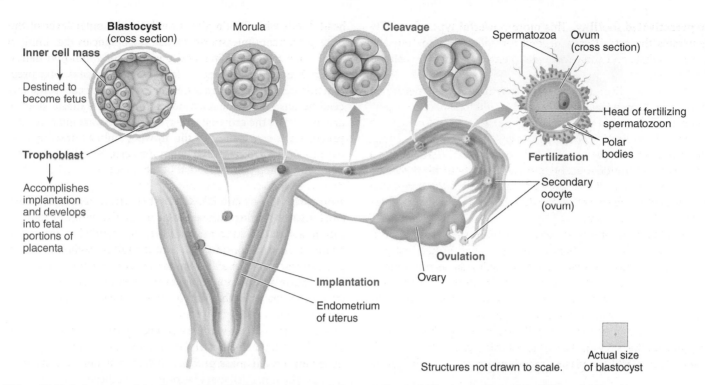

Blastocyst (cross section)
Inner cell mass — Destined to become fetus
Trophoblast — Accomplishes implantation and develops into fetal portions of placenta

Morula

Cleavage

Spermatozoa   Ovum (cross section)
Head of fertilizing spermatozoon
Polar bodies
Fertilization
Secondary oocyte (ovum)
Ovulation
Ovary
Implantation
Endometrium of uterus

Structures not drawn to scale.   Actual size of blastocyst

▮Figure 20-24 **Early stages of development from fertilization to implantation.** Note that the fertilized ovum progressively divides and differentiates into a blastocyst as it moves from the site of fertilization in the upper oviduct to the site of implantation in the uterus.

induced release of intracellular Ca^{2+} into the ovum cytosol triggers the exocytosis of these cortical granules into the space between the egg membrane and the zona pellucida (step ⑤). These enzymes diffuse into the zona pellucida, where they inactivate the ZP3 receptors so that other sperm reaching the zona pellucida cannot bind with it. The enzymes also harden the zona pellucida and seal off tunnels in progress to keep other penetrating sperm from advancing.

Furthermore, the released Ca^{2+} in the ovum cytosol triggers the second meiotic division of the egg, which is now ready to unite with the sperm to complete the fertilization process.

Within an hour, the sperm and egg nuclei fuse, thanks to a centrosome (microtubule organizing center; see p. 46) provided by the sperm that forms microtubules to bring the male and female chromosome sets together for uniting. In addition to contributing its half of the chromosomes to the fertilized ovum, now called a **zygote**, the victorious sperm also activates ovum enzymes essential for the early embryonic developmental program. Thus, fertilization accomplishes the dual events of combining genes from the two parents to form a genetically unique organism and setting in motion the development of that organism.

The blastocyst implants in the endometrium by means of its trophoblastic enzymes.

During the first 3 to 4 days following fertilization, the zygote remains within the ampulla because a constriction between the ampulla and the remainder of the oviduct canal prevents further movement of the zygote toward the uterus.

The Beginning Steps in the Ampulla The zygote is not idle during this time. It rapidly undergoes a number of mitotic cell divisions to form a solid ball of cells called the *morula* (▮Figure 20-24). Meanwhile, the rising levels of progesterone from the newly developed CL that formed after ovulation stimulate release of glycogen from the endometrium into the reproductive tract lumen for use as energy by the early embryo. The nutrients stored in the cytoplasm of the ovum can sustain the embryo for less than a day. The concentration of secreted nutrients increases more rapidly in the small confines of the ampulla than in the uterine lumen.

Descent of the Morula to the Uterus About 3 to 4 days after ovulation, progesterone is being produced in sufficient quantities to relax the oviduct constriction, thus permitting the morula to be rapidly propelled into the uterus by oviductal peristaltic contractions and ciliary activity. The temporary delay before the developing embryo passes into the uterus lets enough nutrients accumulate in the uterine lumen to support the embryo until implantation can take place. If the morula arrives prematurely, it dies.

When the morula descends to the uterus, it floats freely within the uterine cavity for another 3 to 4 days, living on endometrial secretions and continuing to divide. During the first 6 to 7 days after ovulation, while the developing embryo is in transit in the oviduct and floating in the uterine lumen, the uterine lining is simultaneously being prepared for implantation under the influence of luteal-phase progesterone. During this time, the uterus is in its secretory, or progestational phase, storing up glycogen and becoming richly vascularized.

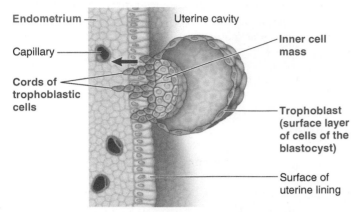

Endometrium — Uterine cavity

Capillary —

Cords of trophoblastic cells —

Inner cell mass

Trophoblast (surface layer of cells of the blastocyst)

Surface of uterine lining

1 When the free-floating blastocyst adheres to the endometrial lining, cords of trophoblastic cells begin to penetrate the endometrium.

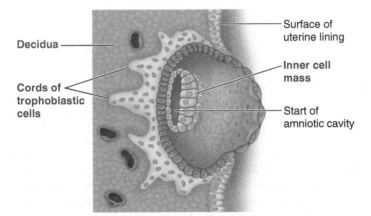

Surface of uterine lining

Decidua —

Cords of trophoblastic cells —

Inner cell mass

Start of amniotic cavity

2 Advancing cords of trophoblastic cells tunnel deeper into the endometrium, carving out a hole for the blastocyst. The boundaries between the cells in the advancing trophoblastic tissue disintegrate.

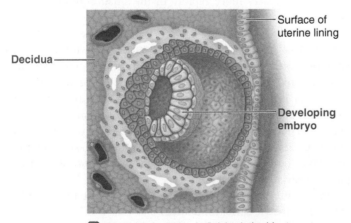

Surface of uterine lining

Decidua —

Developing embryo

3 When implantation is finished, the blastocyst is completely buried in the endometrium.

Figure 20-25 Implantation of the blastocyst.

Implantation of the Blastocyst in the Prepared Endometrium By the time the endometrium is suitable for implantation (about a week after ovulation), the morula has descended to the uterus and continued to proliferate and differentiate into a *blastocyst* capable of implantation. The week's delay after fertilization and before implantation allows time for both the endometrium and the developing embryo to prepare for implantation.

A **blastocyst** is a single-layer hollow ball of about 50 cells encircling a fluid-filled cavity, with a dense mass of cells known as the **inner cell mass** grouped together at one side (Figure 20-24). The inner cell mass becomes the embryo and then fetus. The rest of the blastocyst is never incorporated into the fetus, instead serving a supportive role during intrauterine life. The thin outermost layer, the **trophoblast**, accomplishes implantation, after which it develops into the fetal portion of the placenta.

When the blastocyst is ready to implant, its surface becomes sticky. By this time, the endometrium is ready to accept the early embryo and it too has become more adhesive through increased formation of cell adhesion molecules (CAMs) that help "Velcro" the blastocyst when it first contacts the uterine lining. The blastocyst adheres to the uterine lining on the side of its inner cell mass (Figure 20-25, step **1**). **Implantation** begins when, on contact with the endometrium, the trophoblastic cells overlying the inner cell mass release protein-digesting enzymes. These enzymes digest pathways between the endometrial cells, permitting fingerlike cords of trophoblastic cells to penetrate into the depths of the endometrium, where they continue to digest uterine cells (step **2**). Through its cannibalistic actions, the trophoblast performs the dual functions of accomplishing implantation as it carves out a hole in the endometrium for the blastocyst and making metabolic fuel and raw materials available for the developing embryo as the advancing trophoblastic projections break down the nutrient-rich endometrial tissue. The plasma membranes of the advancing trophoblastic cells degenerate, forming a multinucleated syncytium that eventually becomes the fetal portion of the placenta.

Stimulated by the invading trophoblast, the endometrial tissue at the contact site undergoes dramatic changes that enhance its ability to support the implanting embryo. In response to a chemical messenger released by the blastocyst, the underlying endometrial cells secrete prostaglandins, which locally increase vascularization, produce edema, and enhance nutrient storage. The endometrial tissue so modified at the implantation site is called the **decidua**. It is into this super-rich decidual tissue that the blastocyst becomes embedded. After the blastocyst burrows into the decidua by means of trophoblastic activity, a layer of endometrial cells covers over the surface of the hole, completely burying the blastocyst within the uterine lining (step ❸). The trophoblastic layer continues to digest the surrounding decidual cells, providing energy for the embryo until the placenta develops.

Contraception Couples wishing to engage in sexual intercourse but avoid pregnancy have available a number of methods of **contraception** ("against conception"). These methods act by blocking one of three major steps in the reproductive process: sperm transport to the ovum, ovulation, or implantation. (See the boxed feature on pp. 758 and 759, Concepts, Challenges, and Controversies, for further details on the ways and means of contraception.)

Preventing Rejection of the Embryo–Fetus What prevents the mother from immunologically rejecting the embryo–fetus, which is actually a "foreigner" to the mother's immune system, being half derived from genetically different paternal chromosomes? Numerous, considerably redundant, and incompletely understood pathways are involved in the fetomaternal immune cross-talk that confers maternal tolerance of the embryo–fetus. Following are examples of pathways currently thought to play a role:

- *Natural killer (NK) cells,* the cells of the innate immune system that kill foreign cells such as bacteria (see p. 413), accumulate in the maternal decidua. However, *decidual NK (dNK) cells* have an altered function: They are unable to kill fetal cells, unlike NK cell's usual cell-killing role. Instead, they promote fetal tolerance through various means. For example, dNKs lock *decidual dendritic cells* in a tolerance producing state. Dendritic cells are special immune cells that normally play a central role in triggering adaptive immunity (see p. 427). Synergistically, dNKs and decidual dendritic cells induce apoptosis of *maternal cytotoxic T cells* capable of destroying the embryo–fetus, sparing it from immune rejection. Cytotoxic T cells are the killer cells of the adaptive immune system that are programmed to destroy specific foreign cells (see p. 423). Also, dNKs and decidual dendritic cells both promote production of *regulatory T cells* (see p. 427).
- Production of regulatory T cells doubles or triples in pregnancy. Regulatory T cells suppress maternal cytotoxic T cells that might target the fetus.
- Trophoblastic cells express a special type of surface identity marker (a unique class I MHC glycoprotein; see p. 428) that allows them to evade interactions with killer cells.
- Trophoblastic cells produce *Fas ligand,* which binds with *Fas,* a specialized receptor on the surface of approaching activated maternal cytotoxic T cells. This binding triggers apoptosis of these immune cells that are targeted to destroy the developing foreigner.
- Furthermore, the fetal portion of the placenta, which is derived from trophoblasts, produces an enzyme, *indoleamine 2,3-dioxygenase (IDO),* which destroys tryptophan. Tryptophan, an amino acid, is a critical factor in activation of maternal cytotoxic T cells. Thus, the embryo–fetus, through its trophoblast connection, defends itself against rejection in part by shutting down the activity of the mother's cytotoxic T cells within the placenta that would otherwise attack the developing foreign tissues.

Next, we examine the placenta in further detail.

The placenta is the organ of exchange between maternal and fetal blood.

The glycogen stores in the endometrium are sufficient to nourish the embryo only during its first few weeks. To sustain the growing embryo and then fetus for the duration of intrauterine life, the **placenta**, a specialized organ of exchange between maternal and fetal blood, rapidly develops (Figure 20-26, p. 760). The placenta is derived from both trophoblastic and decidual tissue. It is an unusual organ because it is composed of tissues of two organisms: the embryo–fetus and the mother.

Formation of the Placenta and Amniotic Sac By day 12, the embryo is completely embedded in the decidua. By this time, the trophoblastic layer is two cell layers thick and is called the **chorion**. As the chorion continues to release enzymes and expand, it forms an extensive network of cavities within the decidua. As the expanding chorion erodes decidual capillary walls, maternal blood leaks from the capillaries and fills these cavities. The blood is kept from clotting by an anticoagulant produced by the chorion. Fingerlike projections of chorionic tissue extend into the pools of maternal blood. Soon the developing embryo sends out capillaries into these chorionic projections to form **placental villi**. Some villi extend completely across the blood-filled spaces to anchor the fetal portion of the placenta to the endometrial tissue, but most simply project into the pool of maternal blood.

Each placental villus contains embryonic (later fetal) capillaries surrounded by a thin layer of chorionic tissue, which separates the embryonic–fetal blood from the pools of maternal blood in the intervillous ("between villi") spaces. Maternal and fetal blood do not actually mingle, but the barrier between them is extremely thin. To visualize this relationship, think of your hands (the fetal capillary blood vessels) in rubber gloves (the chorionic tissue) immersed in water (the pool of maternal blood). Only the rubber gloves separate your hands from the water. In the same way, only the thin chorionic tissue (plus the capillary wall of the fetal vessels) separates the fetal and maternal blood. All exchanges between these two bloodstreams take place across this extremely thin barrier. This entire system of interlocking maternal (decidual) and fetal (chorionic) structures makes up the placenta. When fully developed, the placental interface for exchange between mother and fetus would be more than 12 m^2 if stretched out flat.

Even though not fully developed, the placenta is well established and operational by 5 weeks after implantation. By this time, the heart of the developing embryo is pumping blood into the placental villi and to the embryonic tissues. Throughout gestation, fetal blood continuously traverses between the placental villi and the circulatory system of the fetus by means of two *umbilical arteries* and one *umbilical vein,* which are wrapped

The Ways and Means of Contraception

THE TERM *CONTRACEPTION* REFERS TO the process of avoiding pregnancy while engaging in sexual intercourse. A number of methods of contraception are available that range in ease of use and effectiveness (see the accompanying table). These methods can be grouped into three categories based on the means by which they prevent pregnancy: blocking sperm transport to the ovum, preventing ovulation, or blocking implantation. After examining the most common ways in which contraception can be accomplished by each of these means, we will take a glimpse at future contraceptive possibilities on the horizon before concluding with a discussion of terminating unwanted pregnancies.

Blocking Sperm Transport to the Ovum

■ *Natural contraception* or the *rhythm method* of birth control relies on abstinence from intercourse during the woman's fertile period. The woman can predict when ovulation is to occur based on keeping careful records of her menstrual cycles. Because of variability in cycles, this technique is only partially effective. The time of ovulation can be determined more precisely by recording body temperature each morning before getting up. Body temperature rises slightly about a day after ovulation has taken place. The temperature rhythm method is not useful in determining when it is safe to engage in intercourse before ovulation, but it can be helpful in determining when it is safe to resume sex after ovulation.

■ *Coitus interruptus* involves withdrawal of the penis from the vagina before ejaculation occurs. This method is only moderately effective, however, because timing is difficult, and some sperm may pass out of the urethra before ejaculation.

■ *Chemical contraceptives,* such as spermicidal ("sperm-killing") jellies, foams, creams, and suppositories, when inserted into the vagina are toxic to sperm for about an hour after application.

Average Failure Rate of Various Contraceptive Techniques

Contraceptive Method	Average Failure Rate (Annual Pregnancies/100 Women)
None	90
Natural (rhythm) methods	20–30
Coitus interruptus	23
Chemical contraceptives	20
Barrier methods	10–20
Intrauterine device	4
Oral contraceptives	2–2.5
Implanted contraceptives	1

■ *Barrier methods* mechanically prevent sperm transport to the oviduct. For males, the *condom* is a thin, strong rubber or latex sheath placed over the erect penis before ejaculation to prevent sperm from entering the vagina. For females, the *diaphragm* or smaller *cervical cap,* both of which must be fitted by a trained professional, are flexible rubber domes that are inserted through the vagina and positioned over the cervix to block sperm entry into the cervical canal. The *female condom* (or *vaginal pouch*), the latest barrier method, is a long, polyurethane, cylindrical pouch that is closed on one end and open on the other end, with a flexible ring at both ends. The ring at the closed end of the device is inserted into the vagina and fits over the cervix while the ring at the open end of the pouch is positioned outside the vagina over the external genitalia. Barrier methods are often used in conjunction with spermicidal agents for increased effectiveness.

■ *Sterilization,* which involves surgical disruption of either the ductus deferens (*vasectomy*) in men or the oviduct (*tubal ligation*) in women, is considered a permanent method of preventing sperm and ovum from uniting.

Preventing Ovulation

■ *Oral contraceptives,* or *birth control pills,* available only by prescription, prevent ovulation primarily by suppressing gonadotropin secretion. These pills, which contain synthetic estrogenlike and progesteronelike steroids, are taken for three weeks, either in combination or in sequence, and then are withdrawn for one week. These steroids, like the natural steroids produced during the ovarian cycle, inhibit kisspeptin and GnRH and thus FSH and LH secretion. As a result, follicle maturation and ovulation do not take place, so conception is impossible. The endometrium responds to the exogenous steroids by thickening and developing secretory capacity, just as it would respond to the natural hormones. When these synthetic steroids are withdrawn after three weeks, the endometrial lining sloughs and menstruation occurs, as it normally would on degeneration of the corpus luteum (CL). In addition to blocking ovulation, oral contraceptives prevent pregnancy by increasing the viscosity of cervical mucus, which makes sperm penetration more difficult, and by decreasing muscular contractions in the female reproductive tract, which reduces sperm transport to the oviduct. Oral contraceptives have been shown to increase the risk of intravascular clotting, especially in women who also smoke tobacco. Birth control pills have been available for more than 50 years, with only incremental improvements during that time.

■ Several other contraceptive methods contain synthetic female sex hormones and act similarly to birth control pills to prevent ovulation. These include long-acting *subcutaneous* ("under the skin") *implantation* of hormone-containing capsules that gradually release hormones at a nearly steady rate for five years and *birth control patches* impregnated with hormones that are absorbed through the skin.

Blocking Implantation

Medically, pregnancy is not considered to begin until implantation. According to this view, any mechanism that interferes with implantation is said to prevent pregnancy. Not all hold this view, however. Some consider pregnancy to begin at time of fertilization. To them, any interference with implantation is a form of abortion. Therefore, methods of contraception that rely on blocking implantation are more controversial than methods that prevent fertilization from taking place.

Blocking implantation is most commonly accomplished by a physician inserting a small *intrauterine device (IUD)* into the uterus. The presence of this foreign object in the uterus induces a local inflammatory response that prevents implantation of a fertilized ovum.

Emergency Contraception

Emergency contraception is aimed at preventing pregnancy if used within the immediate days following unplanned unprotected sexual intercourse. Emergency contraception is for emergency use only—for instance, if a condom breaks or in the case of rape—and should not be used as a substitute for ongoing contraceptive methods.

The two means of emergency contraception are taking *morning-after pills* and insertion of a *copper IUD*. To greatly reduce the likelihood of an unwanted pregnancy following unprotected intercourse or known contraceptive failure, morning-after pills must be taken within 3 days (not just the morning after) or an IUD must be inserted within 5 days. Following are the mechanisms of action of these emergency contraceptives:

- Hormonal morning-after pills (such as *Plan B One-Step*) consist of high doses of the same hormones found in birth control pills, either progesteronelike steroids alone or in combination with estrogenlike steroids. These pills, now available over the counter, work by suppressing ovulation and stopping fertilization by affecting sperm motility.

- An alternative nonhormonal morning-after pill (*Ella*), available by prescription only, is a progesterone receptor modulator that delays or inhibits ovulation.

- Timely insertion by a health-care professional of a copper IUD prevents fertilization primarily via copper's inhibitory effect on sperm function and also interferes with implantation.

Future Possibilities

- A future birth control technique is *immunocontraception*—the use of vaccines that prod the immune system to produce antibodies targeted against a particular protein critical to the reproductive process. The contraceptive effects of the vaccines are expected to last about a year. For example, in the testing stage is a vaccine that induces formation of antibodies against human chorionic gonadotropin

(see p. 762) so that this essential CL-supporting hormone is not effective if pregnancy occurs. Another promising immunocontraception approach is aimed at blocking the acrosomal enzymes so that sperm could not enter the ovum.

- Some scientists are searching for a better *male contraceptive* beyond condoms and irreversible vasectomy. For example, they are seeking ways to manipulate hormones via a "male birth control pill" to block sperm production without depriving the man of testosterone. Still others are trying to interfere with the bond between Sertoli cells and developing sperm so that spermatogenesis cannot proceed to completion. Another approach is blocking testicular synthesis of retinoic acid (a derivative of vitamin A), which is essential for spermatogenesis. Yet another strategy under investigation is injection into the vas deferens of an exit-blocking polymer gel that can later be dissolved if desired—in essence a reversible "vasectomy."

- One interesting avenue being explored holds hope for a unisex contraceptive that would stop sperm in their tracks and could be used by either males or females. The idea is to use Ca^{2+}-blocking drugs to prevent entry of motility-inducing Ca^{2+} into sperm tails. With no Ca^{2+}, sperm would not be able to maneuver to accomplish fertilization. With the recent discovery of CatSper channels comes an opportunity to interfere with these sperm-specific Ca^{2+} channels, disrupting sperms' fertilizing capacity without having any effect on the female, who lacks these channels.

Terminating Unwanted Pregnancies

- When contraceptive practices fail or are not used and an unwanted pregnancy results, women often turn to *abortion* to terminate the pregnancy. More than half of the approximately 6.4 million pregnancies in the United States each year are unintended, and about 1.6 million of them end with an abortion. Although surgical removal of an embryo–fetus is legal in the United States, the practice of abortion is fraught with emotional, ethical, and political controversy.

- In late 2000 the "abortion pill," *RU 486,* or *mifepristone,* was approved for use in the United States amid considerable controversy, even though it had been available in other countries since 1988. This drug terminates an early pregnancy by chemical interference rather than by surgery. RU 486, a progesterone antagonist, binds tightly with the progesterone receptors on the target cells but does not evoke progesterone's usual effects and prevents progesterone from binding and acting. Deprived of progesterone activity, the highly developed endometrial tissue sloughs off, carrying the implanted embryo with it. RU 486 administration is followed in 48 hours by a prostaglandin that induces uterine contractions to help expel the endometrium and embryo.

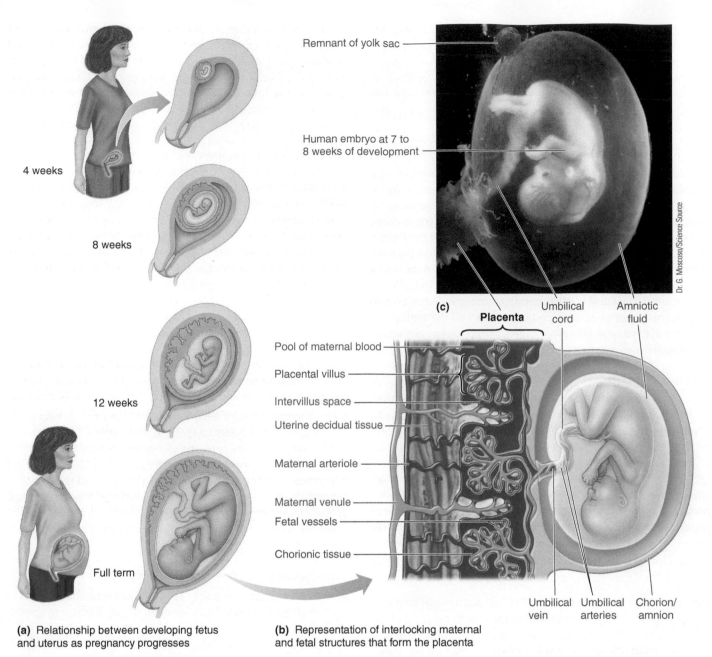

(a) Relationship between developing fetus and uterus as pregnancy progresses

4 weeks

8 weeks

12 weeks

Full term

Remnant of yolk sac

Human embryo at 7 to 8 weeks of development

Dr. G. Moscoso/Science Source

(c)

Placenta

Umbilical cord

Amniotic fluid

Pool of maternal blood

Placental villus

Intervillus space

Uterine decidual tissue

Maternal arteriole

Maternal venule

Fetal vessels

Chorionic tissue

Umbilical vein

Umbilical arteries

Chorion/ amnion

(b) Representation of interlocking maternal and fetal structures that form the placenta

❙ Figure 20-26 **Developing embryo–fetus, placenta, and amniotic fluid.** (a) The uterus progressively enlarges to accommodate the growing embryo–fetus during pregnancy. (b) During placentation, fingerlike projections of chorionic (fetal) tissue form the placental villi, which protrude into a pool of maternal blood. Decidual (maternal) capillary walls are broken down by the expanding chorion so that maternal blood oozes through the spaces between the placental villi. Fetal placental capillaries branch off the umbilical arteries and project into the placental villi. Fetal blood flowing through these vessels is separated from the maternal blood by only the capillary wall and thin chorionic layer that forms the placental villi. Maternal blood enters through the maternal arterioles, then percolates through the pool of blood in the intervillus spaces. Here, exchanges are made between the fetal and maternal blood before the fetal blood leaves through the umbilical vein and maternal blood exits through the maternal venules. (c) The embryo–fetus floats in a sac that forms during development and is filled with cushioning amniotic fluid.

within the **umbilical cord**, a lifeline between the fetus and the placenta (❙ Figure 20-26). The maternal blood within the placenta is continuously replaced as fresh blood enters through uterine arterioles; percolates through the intervillous spaces, where it exchanges substances with fetal blood in the surrounding villi; and then exits through uterine venules.

Meanwhile, during the time of implantation and early placental development, the inner cell mass forms a fluid-filled **amniotic cavity** between the trophoblast–chorion and the portion of the inner cell mass destined to become the fetus (see ❙ Figure 20-25, step ❷). The epithelial layer that encloses the amniotic cavity is called the **amniotic sac, or amnion.** As it continues to develop, the amniotic sac eventually fuses with the chorion, forming a single combined membrane that surrounds the embryo–fetus. The fluid in the amniotic cavity, the **amniotic fluid**, which is similar in composition to normal ECF, surrounds and cushions the fetus throughout gestation (❙ Figure 20-26).

Functions of the Placenta During intrauterine life, the placenta performs the functions of the digestive system, the respiratory system, and the kidneys for the "parasitic" fetus. The fetus has these organ systems, but within the uterine environment they cannot (and do not need to) function. Nutrients and O_2 move from the maternal blood across the thin placental barrier into the fetal blood, whereas CO_2 and other metabolic wastes simultaneously move from the fetal blood into the maternal blood. The nutrients and O_2 brought to the fetus in the maternal blood are acquired by the mother's digestive and respiratory systems, and the CO_2 and wastes transferred into the maternal blood are eliminated by the mother's lungs and kidneys, respectively. Thus, the mother's digestive tract, respiratory system, and kidneys serve the fetus's needs and her own.

The means by which materials move across the placenta depends on the substance. Some substances that can permeate the placental membrane, such as O_2, CO_2, water, and electrolytes, cross by simple diffusion. Some traverse the placental barrier by special mediated-transport systems in the placental membranes, such as glucose by facilitated diffusion and amino acids by secondary active transport. Other substances such as cholesterol in the form of LDL (see p. 328) move across by receptor-mediated endocytosis.

Clinical Note Unfortunately, many drugs, environmental pollutants, other chemical agents, and microorganisms in the mother's bloodstream also can cross the placental barrier, and some of them may harm the developing fetus. Individuals born limbless as a result of exposure to *thalidomide,* a tranquilizer prescribed for pregnant women before this drug's devastating effects on the growing fetus were known, serve as a grim reminder of this fact. Similarly, newborns who become "addicted" during gestation by their mother's abuse of a drug such as heroin suffer withdrawal symptoms after birth. Even more common chemical agents such as aspirin, alcohol, and agents in cigarette smoke can reach the fetus and have adverse effects. Furthermore, some infectious agents can cross the placenta, an example being *Listeria monocytogenes,* a bacterium that can contaminate food and cause food poisoning with potentially devastating results for the pregnancy and the fetus. Pregnant women should therefore be cautious about potentially harmful exposure from any source.

The placenta has yet another important responsibility—it becomes a temporary endocrine organ during pregnancy, a topic to which we now turn. During pregnancy, three endocrine systems interact to support and enhance the growth and development of the fetus, to coordinate the timing of parturition, and to prepare the mammary glands for nourishing the baby after birth: placental hormones, maternal hormones, and fetal hormones.

Hormones secreted by the placenta play a critical role in maintaining pregnancy.

The fetally derived portion of the placenta has the remarkable capacity to secrete a number of peptide and steroid hormones essential for maintaining pregnancy. The most important are *human chorionic gonadotropin, estrogen,* and *progesterone* (Table 20-5). Serving as the major endocrine organ of pregnancy, the placenta is unique among endocrine tissues in two regards. First, it is a transient tissue. Second, secretion of its hormones is not subject to extrinsic control, in contrast to the stringent, often complex mechanisms that regulate secretion of other hormones. Instead, the type and rate of placental hormone secretion depend primarily on the stage of pregnancy.

ITABLE 20-5 Placental Hormones

Hormone	Function
Human chorionic gonadotropin (hCG)	Maintains the corpus luteum of pregnancy
	Stimulates secretion of testosterone by the developing testes in XY embryos
Estrogen (also secreted by the corpus luteum of pregnancy)	Stimulates growth of the myometrium, increasing uterine strength for parturition
	Helps prepare the mammary glands for lactation
Progesterone (also secreted by the corpus luteum of pregnancy)	Suppresses uterine contractions to provide a quiet environment for the fetus
	Promotes formation of a cervical mucus plug to prevent uterine contamination
	Helps prepare the mammary glands for lactation
Human chorionic somatomammotropin (has a structure similar to that of both growth hormone and prolactin)	Reduces maternal use of glucose and promotes the breakdown of stored fat (similar to growth hormone) so that greater quantities of glucose and free fatty acids may be shunted to the fetus
	Helps prepare the mammary glands for lactation (similar to prolactin)
Relaxin (also secreted by the corpus luteum of pregnancy)	Softens the cervix in preparation for cervical dilation at parturition
	Loosens the connective tissue between the pelvic bones in preparation for parturition
Placental PTHrp (parathyroid hormone–related peptide)	Increases maternal plasma Ca^{2+} level for use in calcifying fetal bones; if necessary, promotes localized dissolution of maternal bones, mobilizing their Ca^{2+} stores for use by the developing fetus

Secretion of Human Chorionic Gonadotropin One of the first endocrine events is secretion by the developing chorion of **human chorionic gonadotropin (hCG)**, a peptide hormone that prolongs the life span of the corpus luteum. Recall that, during the ovarian cycle, the CL degenerates and the highly prepared, luteal-dependent uterine lining sloughs off if fertilization and implantation do not occur. When fertilization does occur, the implanted blastocyst saves itself from being flushed out in menstrual flow by producing hCG. This hormone, which is similar to LH and binds to the same receptor as LH, stimulates and maintains the CL so that it does not degenerate. Now called the **corpus luteum of pregnancy**, this ovarian endocrine unit grows even larger and produces increasingly greater amounts of estrogen and progesterone for an additional 10 weeks until the placenta takes over secretion of these steroid hormones. Because of the persistence of estrogen and progesterone, the thick, pulpy endometrial tissue is maintained instead of sloughing. Accordingly, menstruation ceases during pregnancy. Stimulation by hCG is necessary to maintain the CL of pregnancy because LH, which maintains the CL during the normal luteal phase of the uterine cycle, is suppressed through feedback inhibition by the high levels of progesterone.

Maintenance of a normal pregnancy depends on high concentrations of progesterone and estrogen. Thus, hCG production is critical during the first trimester to maintain ovarian output of these hormones. In a male fetus, hCG also stimulates the precursor Leydig cells in the fetal testes to secrete testosterone, which masculinizes the developing reproductive tract.

The secretion rate of hCG increases rapidly during early pregnancy to save the CL from demise. Peak secretion of hCG occurs about 60 days after the end of the last menstrual period (❙ Figure 20-27). By the 10th week of pregnancy, hCG output declines to a low rate of secretion that is maintained for the duration of gestation. The fall in hCG occurs because the placenta has begun to secrete substantial quantities of estrogen and progesterone, which inhibit hCG secretion. By this time, the CL of pregnancy is no longer needed for its steroid hormone output because the placenta is secreting sufficient estrogen and progesterone. Therefore the CL of pregnancy is the source of estrogen and progesterone during the first trimester of gestation, and the placenta takes over this role during the last two trimesters. The CL of pregnancy partially regresses as hCG secretion dwindles, but it is not converted into scar tissue until after delivery of the baby.

Clinical Note Human chorionic gonadotropin is eliminated from the body in the urine. Pregnancy diagnosis tests can detect hCG in urine as early as the first month of pregnancy, about 2 weeks after the first missed menstrual period. Because this is before the growing embryo can be detected by physical examination, the test permits early confirmation of pregnancy.

A frequent early clinical sign of pregnancy is **morning sickness**, a daily bout of nausea and vomiting that often occurs in the morning but can take place at any time of day. Because this condition usually appears shortly after implantation and coincides with the time of peak hCG production, scientists speculate that this early placental hormone may trigger the symptoms, perhaps by acting on the chemoreceptor trigger zone next to the vomiting center (see p. 582).

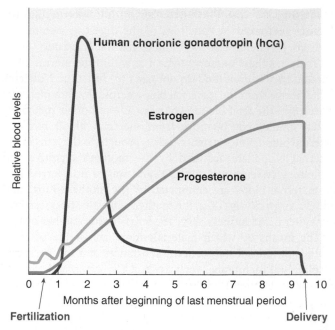

❙Figure 20-27 **Secretion rates of placental hormones.**
FIGURE FOCUS: *(1) Why do estrogen and progesterone levels continue to climb after the first trimester of pregnancy even though human chorionic gonadotropin output has declined substantially? (2) Why do estrogen and progesterone levels drop precipitously at delivery?*

Secretion of Estrogen and Progesterone Why does the developing placenta not start producing estrogen and progesterone in the first place instead of secreting hCG, which in turn stimulates the CL to secrete these two critical hormones? The answer is that, for different reasons, the placenta cannot produce enough estrogen or progesterone in the first trimester of pregnancy. In the case of estrogen, the placenta does not have all the enzymes needed for complete synthesis of this hormone. Estrogen synthesis requires a complex interaction between the placenta and the fetus (❙ Figure 20-28). The placenta converts the androgen hormone produced by the fetal adrenal cortex, dehydroepiandrosterone (DHEA), into estrogen. The placenta cannot produce estrogen until the fetus has developed to the point that its adrenal cortex is secreting DHEA into the blood. The placenta extracts DHEA from the fetal blood and converts it into estrogen, which it then secretes into the maternal blood.

Estrogen comes in several variants. The primary estrogen synthesized by the placenta is **estriol**, in contrast to the main estrogen product of the ovaries, estradiol. Because estriol can only be synthesized from fetal DHEA, measurement of estriol levels in the maternal urine can be used clinically to assess the viability of the fetus.

In the case of progesterone, the placenta can synthesize this hormone soon after implantation. Even though the early placenta has the enzymes necessary to convert cholesterol extracted from the maternal blood into progesterone, it does not produce much of this hormone because the amount of progesterone produced is proportional to placental weight. The placenta is simply too small in the first 10 weeks of pregnancy to produce enough progesterone to maintain the endometrial tissue. The notable increase in circulating progesterone in the last 7 months of gestation reflects placental growth during this period.

Roles of Estrogen and Progesterone During Pregnancy

As noted earlier, high concentrations of estrogen and progesterone are essential to maintain a normal pregnancy. Estrogen stimulates growth of the myometrium, which increases in size throughout pregnancy. The stronger uterine musculature is needed to expel the fetus during labor. Estriol also promotes development of the ducts within the mammary glands, through which milk is ejected during lactation.

Progesterone performs various roles throughout pregnancy. Its main function is to prevent miscarriage by suppressing contractions of the uterine myometrium. Progesterone also promotes formation of a thick mucus plug in the cervical canal to prevent vaginal contaminants from reaching the uterus. Finally, placental progesterone stimulates development of milk glands in the breasts in preparation for lactation.

Maternal body systems respond to the increased demands of gestation.

The period of **gestation (pregnancy)** is about 38 weeks from conception (40 weeks from the end of the last menstrual period). During gestation, the embryo–fetus develops and grows to the point of being able to leave its maternal life-support system. Meanwhile, a number of physical changes within the mother accommodate the demands of pregnancy. The most obvious change is uterine enlargement. The uterus expands and increases in weight more than 20 times, exclusive of its contents. The breasts enlarge and develop the ability to produce milk. Body systems other than the reproductive system also make needed adjustments. The volume of blood increases by 30%, and the cardiovascular system responds to the increasing demands of the growing placental mass. Weight gain during pregnancy is the result only in part of the weight of the fetus. The remainder is mostly from increased weight of the uterus, including the placenta, and increased blood volume. Respiratory activity increases by about 20% to handle the additional fetal requirements for O_2 use and CO_2 removal. Urinary output increases, and the kidneys excrete the additional wastes from the fetus.

The increased metabolic demands of the growing fetus increase nutritional requirements for the mother. In general, the fetus takes what it needs from the mother, even if this leaves the mother with a nutritional deficit. For example, the placental hormone **human chorionic somatomammotropin (hCS)** decreases use of glucose by the mother and mobilizes free fatty acids from maternal adipose stores, similar to the actions of growth hormone (see p. 653). (In fact, hCS has a structure similar to that of both growth hormone and prolactin and exerts similar actions.) The hCS-induced metabolic changes in the mother make available greater quantities of glucose and fatty acids for shunting to the fetus. Also, if the mother does not consume enough Ca^{2+}, yet another placental hormone similar to parathyroid hormone, **parathyroid hormone-related peptide (PTHrp)**, mobilizes Ca^{2+} from the maternal bones to ensure adequate calcification of the fetal bones (❚ Table 20-5).

Changes during late gestation prepare for parturition.

Parturition (labor, delivery, or birth) requires (1) dilation of the cervical canal to accommodate passage of the fetus from the uterus through the vagina and to the outside and (2) contractions of the uterine myometrium that are sufficiently strong to expel the fetus.

Several changes take place during late gestation in preparation for the onset of parturition. During the first two trimesters of gestation, the uterus remains relatively quiet because of the inhibitory effect of the high levels of

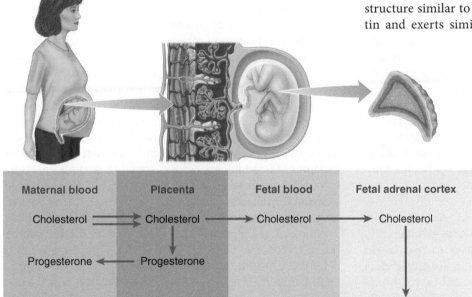

Maternal blood	Placenta	Fetal blood	Fetal adrenal cortex
Cholesterol ⇌	Cholesterol	→ Cholesterol	→ Cholesterol
Progesterone ←	Progesterone		
	DHEA ←	DHEA ←	Dehydroepiandrosterone (DHEA)
Estrogen ←	Estrogen		

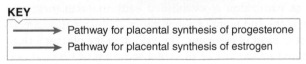

KEY

→ Pathway for placental synthesis of progesterone
→ Pathway for placental synthesis of estrogen

❚Figure 20-28 **Secretion of estrogen and progesterone by the placenta.** The placenta secretes increasing quantities of progesterone and estrogen into the maternal blood after the first trimester. The placenta can convert cholesterol into progesterone (*orange pathway*) but lacks some of the enzymes necessary to convert cholesterol into estrogen. However, the placenta can convert DHEA derived from cholesterol in the fetal adrenal cortex into estrogen when DHEA reaches the placenta by means of the fetal blood (*blue pathway*).

progesterone on the uterine muscle. During the last trimester, however, the uterus becomes progressively more excitable, so mild contractions (**Braxton–Hicks contractions**) are experienced with increasing strength and frequency. Sometimes these contractions become regular enough to be mistaken for the onset of labor, a phenomenon called "false labor."

Throughout gestation, the exit of the uterus remains sealed by the rigid, tightly closed cervix. As parturition approaches, the cervix begins to soften (or "ripen") as a result of the dissociation of its tough connective tissue (collagen) fibers. Because of this softening, the cervix becomes malleable so that it can gradually yield, dilating the exit, as the fetus is forcefully pushed against it during labor. This cervical softening is caused largely by **relaxin**, a peptide hormone produced by the CL of pregnancy and by the placenta. Other factors to be described shortly contribute to cervical softening. Relaxin also "relaxes" the birth canal by loosening the connective tissue between pelvic bones.

Meanwhile, the fetus shifts downward (the baby "drops") and is normally oriented so that the head is in contact with the cervix in preparation for exiting through the birth canal. In a **breech birth**, any part of the body other than the head approaches the birth canal first.

Scientists are closing in on the factors that trigger the onset of parturition.

Rhythmic, coordinated contractions, usually painless at first, begin at the onset of labor. As labor progresses, the contractions increase in frequency, intensity, and discomfort. These strong, rhythmic contractions force the fetus against the cervix, dilating the cervix. Then, after having dilated the cervix enough for the fetus to pass through, these contractions force the fetus out through the birth canal.

The exact factors triggering the increase in uterine contractility and thus initiating parturition are not fully established, although much progress has been made in recent years in unraveling the sequence of events. Let us look at what is known about this process.

Role of High Estrogen Levels During early gestation, maternal estrogen levels are relatively low, but as gestation proceeds, placental estrogen secretion continues to rise. In the immediate days before the onset of parturition, soaring levels of estrogen bring about changes in the uterus and cervix to prepare them for labor and delivery (Figures 20-27 and 20-29). First, high levels of estrogen promote synthesis of connexons within the uterine smooth muscle cells. These myometrial cells are not functionally linked to any extent throughout most of gestation. The newly manufactured connexons are inserted in the myometrial plasma membranes to form gap junctions that electrically link together the uterine smooth muscle cells so that they become able to contract as a coordinated unit (see p. 62).

Simultaneously, high levels of estrogen dramatically and progressively increase the concentration of myometrial receptors for oxytocin. Together, these myometrial changes collectively bring about the increased uterine responsiveness to oxytocin that ultimately initiates labor.

In addition to preparing the uterus for labor, the increasing estrogen levels promote production of local prostaglandins that contribute to cervical ripening by stimulating cervical enzymes that degrade local collagen fibers. These prostaglandins also increase uterine responsiveness to oxytocin.

Role of Oxytocin **Oxytocin** is a peptide hormone produced by the hypothalamus, stored in the posterior pituitary, and released into the blood from the posterior pituitary on nervous stimulation by the hypothalamus (see p. 647). Oxytocin exerts its effects via the $IP_3/Ca^{2+}/DAG$ pathway. A powerful uterine muscle stimulant, oxytocin plays the key role in the progression of labor. However, this hormone was once discounted as the trigger for parturition because circulating levels of oxytocin remain constant before the onset of labor. The discovery that uterine responsiveness to oxytocin is 100 times greater at term than in nonpregnant women (because of the connexons and increased concentration of myometrial oxytocin receptors) led to the now widely accepted conclusion that labor begins when myometrial responsiveness to oxytocin reaches a critical threshold that permits onset of strong, coordinated contractions in response to ordinary levels of circulating oxytocin.

Role of Corticotropin-Releasing Hormone For years scientists were baffled by the factors that raise levels of placental estrogen secretion. Recent studies suggest that *corticotropin-releasing hormone* (CRH) secreted by the fetal portion of the placenta into both the maternal and the fetal circulations not only drives the manufacture of placental estrogen, thus ultimately dictating the timing of the onset of labor, but also promotes changes in the fetal lungs needed for breathing air (Figure 20-29). Recall that CRH is normally secreted by the hypothalamus and regulates the output of ACTH by the anterior pituitary (see pp. 650 and 675). In turn, ACTH stimulates production of both cortisol and DHEA by the adrenal cortex. In the fetus, much of the CRH comes from the placenta rather than solely from the fetal hypothalamus. The additional cortisol secretion summoned by the extra CRH promotes fetal lung maturation. Specifically, cortisol stimulates synthesis of pulmonary surfactant, which facilitates lung expansion and reduces the work of breathing (see p. 459).

The bumped-up rate of DHEA secretion by the adrenal cortex in response to placental CRH leads to the rising levels of placental estrogen secretion because the placenta converts DHEA from the fetal adrenal gland into estrogen, which enters the maternal bloodstream (see Figure 20-28). When sufficiently high, this estrogen sets in motion the events that initiate labor. Thus, pregnancy duration and delivery timing are determined largely by the placenta's rate of CRH production. That is, a **"placental clock"** ticks out the length of time until parturition. The timing of parturition is established early in pregnancy, with delivery at the end point of a maturational process that extends throughout most of gestation. The ticking of the placental clock is measured by the rate of placental secretion of CRH. As the pregnancy progresses, CRH levels in maternal plasma rise. Researchers can accurately predict the timing of parturition by measuring the maternal plasma levels of CRH as early as the end of the first trimester. Higher-than-normal levels are associated

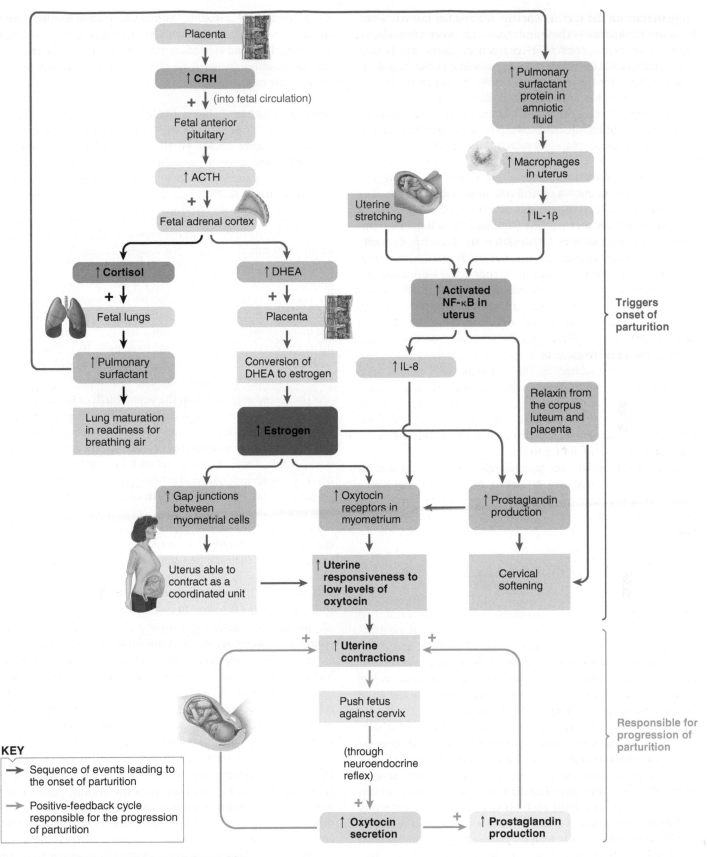

KEY

→ Sequence of events leading to the onset of parturition

→ Positive-feedback cycle responsible for the progression of parturition

I Figure 20-29 Initiation and progression of parturition.

with premature deliveries, whereas lower-than-normal levels indicate late deliveries. These and other data suggest that when a critical level of placental CRH is reached, parturition is triggered. This critical CRH level ensures that when labor begins the infant is ready for life outside the womb. It does so by concurrently increasing the fetal cortisol needed for lung maturation and the estrogen needed for the uterine changes that bring on labor. The remaining unanswered puzzle regarding the placental clock: What controls placental secretion of CRH?

Role of Inflammation Surprisingly, new evidence suggests that inflammation plays a central role in the labor process. Key to this inflammatory response is activation of **nuclear factor κB (NF-κB)** in the uterus. NF-κB boosts production of inflammatory cytokines such as *interleukin 8 (IL-8)* (see p. 411) and prostaglandins that increase the sensitivity of the uterus to contraction-inducing chemical messengers and help soften the cervix. What activates NF-κB, setting off an inflammatory cascade that helps prompt labor? Various factors associated with the onset of full-term labor and premature labor can cause an upsurge in NF-κB. These include stretching of the uterine muscle and the presence of a specific pulmonary surfactant protein *SP-A* (stimulated by the action of CRH on the fetal lungs) in the amniotic fluid from the fetus. SP-A promotes migration of fetal macrophages (see p. 393) to the uterus. These macrophages, in turn, produce the inflammatory cytokine *interleukin 1β (IL-1β)* that activates NF-κB. In this way, fetal lung maturation contributes to the onset of labor.

Clinical Note Bacterial infections and allergic reactions can lead to premature labor by activating NF-κB. Also, multiple-fetus pregnancies are at risk for premature labor, likely because the increased uterine stretching triggers earlier activation of NF-κB.

Parturition is accomplished by a positive-feedback cycle.

Once uterine responsiveness to oxytocin reaches a critical level and regular uterine contractions begin, myometrial contractions progressively increase in frequency, strength, and duration throughout labor until they expel the uterine contents. At the beginning of labor, contractions lasting 30 seconds or less occur about every 25 to 30 minutes; by the end, they last 60 to 90 seconds and occur every 2 to 3 minutes.

As labor progresses, a positive-feedback cycle involving oxytocin and prostaglandin ensues, incessantly increasing myometrial contractions (■Figure 20-29). Each uterine contraction begins at the top of the uterus and sweeps downward, forcing the fetus toward the cervix. Pressure of the fetus against the cervix does two things. First, the fetal head pushing against the softened cervix wedges open the cervical canal. Second, stimulation of receptors in the cervix in response to fetal pressure sends a neural signal up the spinal cord to the hypothalamus, which in turn triggers oxytocin release from the posterior pituitary. This additional oxytocin promotes more powerful uterine contractions. As a result, the fetus is pushed more forcefully against the cervix, stimulating the release of even more oxytocin, and so on. This cycle is reinforced as oxytocin stimulates prostaglandin production by the decidua.

As a powerful myometrial stimulant, prostaglandin further enhances uterine contractions. Oxytocin secretion, prostaglandin production, and uterine contractions continue to increase in positive-feedback fashion throughout labor until delivery relieves the pressure on the cervix.

Stages Of Labor Labor is divided into three stages: (1) cervical dilation, (2) delivery of the baby, and (3) delivery of the placenta (■Figure 20-30). At the onset of labor or sometime during the first stage, the amniotic sac, or "bag of waters," ruptures. As amniotic fluid escapes out of the vagina, it helps lubricate the birth canal.

1. During the *first stage,* the cervix is forced to dilate to accommodate the diameter of the baby's head, usually to a maximum of 10 cm. This stage is the longest, lasting from several hours to as long as 24 hours in a first pregnancy. If another part of the fetus's body other than the head is oriented against the cervix, it is generally less effective than the head as a wedge. The head has the largest diameter of the baby's body. If the baby approaches the birth canal feet first, the feet may not dilate the cervix enough to let the head pass. In such a case, without medical intervention the baby's head would remain stuck behind the too-narrow cervical opening.

2. The *second stage* of labor, the actual birth of the baby, begins once cervical dilation is complete. When the infant begins to move through the cervix and vagina, stretch receptors in the vagina activate a neural reflex that triggers contractions of the abdominal wall in synchrony with the uterine contractions. These abdominal contractions greatly increase the force pushing the baby through the birth canal. The mother can help deliver the infant by voluntarily contracting the abdominal muscles at this time in unison with each uterine contraction (that is, by "pushing" with each "labor pain"). Stage 2 is usually shorter than the first stage, lasting 30 to 90 minutes. The infant is still attached to the placenta by the umbilical cord at birth. The cord is tied and severed, with the stump shriveling up in a few days to form the **umbilicus (navel)**.

3. Shortly after delivery of the baby, a second series of uterine contractions separates the placenta from the myometrium and expels it through the vagina. Delivery of the placenta, or **afterbirth**, constitutes the *third stage* of labor, typically the shortest stage, being completed within 15 to 30 minutes after the baby is born. After the placenta is expelled, continued contractions of the myometrium constrict the uterine blood vessels supplying the site of placental attachment to prevent hemorrhage.

Uterine Involution After delivery, the uterus shrinks to its pregestational size, a process known as **involution**, which takes 4 to 6 weeks to complete. During involution, the remaining endometrial tissue not expelled with the placenta gradually disintegrates and sloughs off, producing a vaginal discharge called **lochia** that continues for 3 to 6 weeks following parturition. After this period, the endometrium is restored to its nonpregnant state.

Involution occurs largely because of the precipitous fall in circulating estrogen and progesterone when the placental source of these steroids is lost at delivery. The process is facilitated in mothers who breast-feed their infants because oxytocin

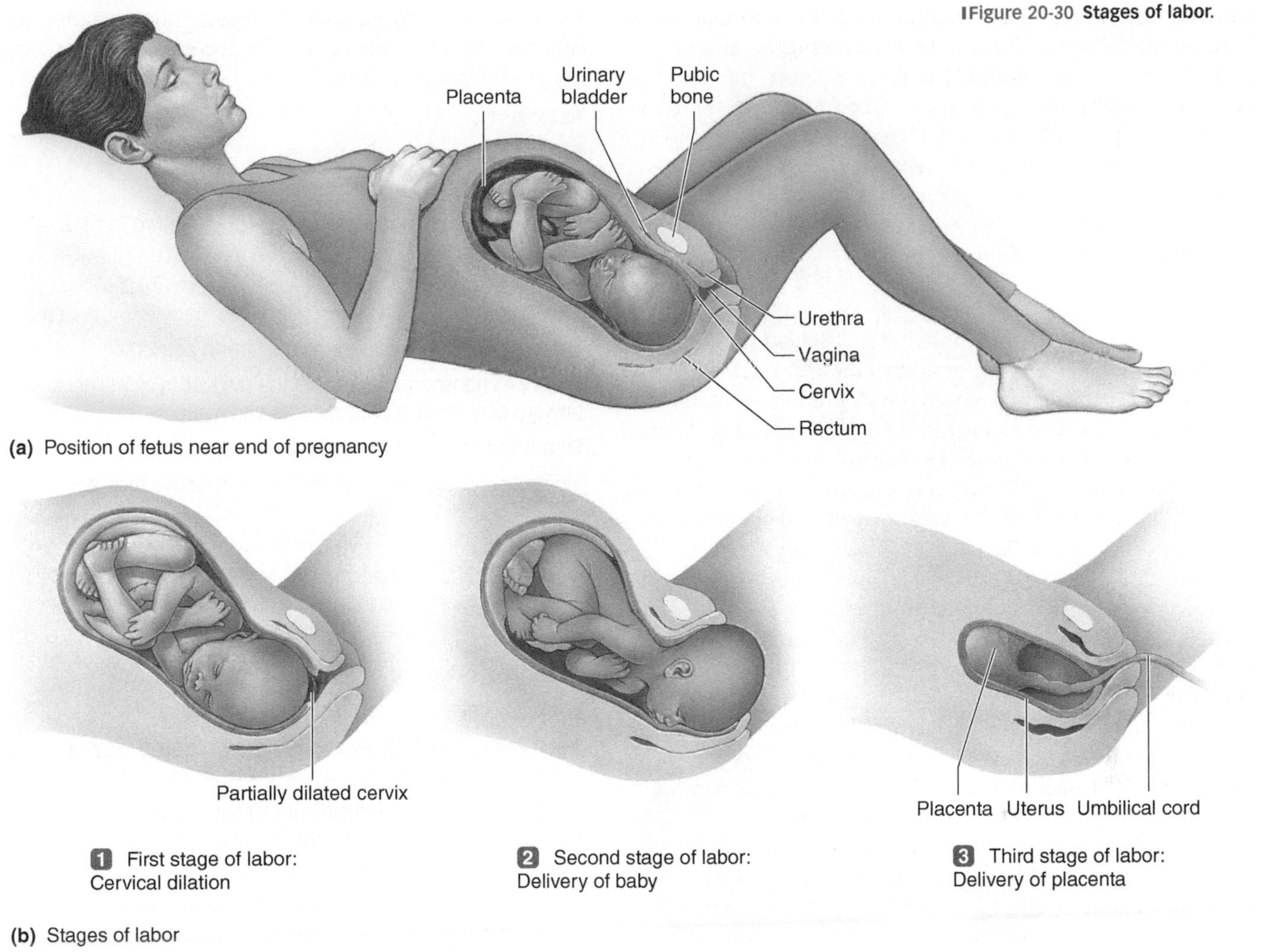

|Figure 20-30 **Stages of labor.**

Placenta — Urinary bladder — Pubic bone

— Urethra
— Vagina
— Cervix
— Rectum

(a) Position of fetus near end of pregnancy

Partially dilated cervix

Placenta Uterus Umbilical cord

1 First stage of labor: Cervical dilation

2 Second stage of labor: Delivery of baby

3 Third stage of labor: Delivery of placenta

(b) Stages of labor

---

is released in response to suckling. In addition to playing an important role in lactation, this periodic nursing-induced release of oxytocin promotes myometrial contractions that help maintain uterine muscle tone, enhancing involution. Involution is usually complete in about 4 weeks in nursing mothers but takes about 6 weeks in those who do not breast-feed.

Lactation requires multiple hormonal inputs.

The female reproductive system supports the new being from the moment of conception through gestation and continues to nourish it during its early life outside the supportive uterine environment. Milk (or its equivalent) is essential for survival of the newborn. Accordingly, during gestation the **mammary glands**, or **breasts**, are prepared for **lactation** (milk production).

The breasts in nonpregnant females consist mostly of adipose tissue and a rudimentary duct system. Breast size is determined by the amount of adipose tissue, which has nothing to do with the ability to produce milk.

Preparation of the Breasts for Lactation Under the hormonal environment present during pregnancy, the mammary glands develop the internal glandular structure and func-

tion necessary for milk production. A breast capable of lactating has a network of progressively smaller ducts that branch out from the nipple and terminate in lobules (|Figure 20-31a). Each lobule is made up of a cluster of saclike epithelial-lined, milk-producing glands known as **alveoli**. (Both milk-producing sacs in the breast and air sacs in the lungs are called *alveoli,* which means "little cavities.") Milk is synthesized by the alveolar epithelial cells and then secreted into the alveolar lumen, which is drained by a milk-collecting duct that transports the milk to the surface of the nipple (|Figure 20-31b).

During pregnancy, the high concentration of *estrogen* promotes extensive duct development, whereas the high level of *progesterone* stimulates abundant alveolar–lobular formation. Elevated concentrations of *prolactin* (an anterior pituitary hormone stimulated by the rising levels of estrogen) and *human chorionic somatomammotropin (hCS,* a placental hormone that has a structure similar to that of both growth hormone and prolactin) also contribute to mammary gland development by inducing the synthesis of enzymes needed for milk production. So great is the commitment to preparing the breasts for infant nutrition that the pituitary gland doubles or triples in size during pregnancy as a result of the estrogen-induced increase in the number of prolactin-secreting cells.

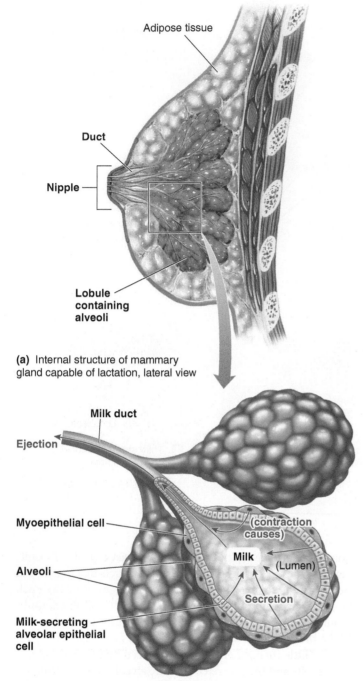

Adipose tissue

Duct

Nipple

Lobule
containing
alveoli

(a) Internal structure of mammary
gland capable of lactation, lateral view

Milk duct

Ejection

Myoepithelial cell

Alveoli

Milk-secreting
alveolar epithelial
cell

(contraction
causes)

Milk

(Lumen)

Secretion

(b) Alveoli within mammary gland

IFigure 20-31 Mammary gland anatomy. The alveolar epithelial cells secrete milk into the lumen. Contraction of the surrounding myoepithelial cells ejects the secreted milk out through the duct.

TABLE 20-6 Actions of Estrogen and Progesterone

Estrogen

Effects on sex-specific tissues

Is essential for egg maturation and release

Stimulates growth and maintenance of the entire female reproductive tract

Stimulates granulosa cell proliferation, which leads to follicle maturation

Thins the cervical mucus to permit sperm penetration

Enhances transport of sperm to the oviduct by stimulating upward contractions of the uterus and oviduct

Stimulates growth of the endometrium and myometrium

Induces synthesis of endometrial progesterone receptors

Triggers onset of parturition by increasing uterine responsiveness to oxytocin during late gestation through a twofold effect: by inducing synthesis of myometrial oxytocin receptors and by increasing myometrial gap junctions so that the uterus can contract as a coordinated unit in response to oxytocin

Other reproductive effects

Promotes development of secondary sexual characteristics

If at a low or moderate level, inhibits secretion of kisspeptin, GnRH, and gonadotropins

If at a high level, is responsible for triggering the LH surge by stimulating secretion of kisspeptin and GnRH

Stimulates duct development in the breasts during gestation

Inhibits milk-secreting actions of prolactin during gestation

Nonreproductive effects

Promotes fat deposition

Increases bone density

Closes the epiphyseal plates

Improves blood cholesterol profile by increasing HDL and decreasing LDL

Promotes vasodilation by increasing nitric oxide production in arterioles (cardioprotective)

Progesterone

Prepares a suitable environment for nourishment of a developing embryo and then fetus

Causes a thick mucus plug to form in the cervical canal

Inhibits secretion of kisspeptin, GnRH, and gonadotropins

Inhibits uterine contractions during gestation

Stimulates alveolar development in the breasts during gestation

Inhibits milk-secreting actions of prolactin during gestation

In addition to preparing the mammary glands for lactation, prolactin and hCS also promote fetal growth by stimulating production of the insulin-like growth factors, IGF-I and IGF-II (see p. 654). Surprisingly, growth hormone secreted by the fetal anterior pituitary does not control growth of the fetus.

Prevention of Lactation During Gestation Most of these changes in the breasts occur during the first half of gestation, so the mammary glands are fully capable of producing milk by the middle of pregnancy. However, milk secretion does not occur until parturition. The high estrogen and progesterone concentrations during the last half of pregnancy prevent lactation by blocking prolactin's stimulatory action on milk secretion. Prolactin is the primary stimulant of milk secretion. Thus, even though the high levels of placental steroids promote development of the milk-producing machinery in the breasts, they prevent these glands from becoming operational until the baby is born and milk is needed.

The abrupt decline in estrogen and progesterone that occurs with loss of the placenta at parturition initiates lactation.

We have now completed our discussion of the functions of estrogen and progesterone during gestation and lactation, as well as throughout the reproductive life of females. These functions are summarized in ❙ Table 20-6.

Stimulation of Lactation via Suckling Once milk production begins after delivery, two hormones are critical for maintaining lactation: (1) *oxytocin,* which causes milk ejection, and (2) *prolactin,* which promotes milk secretion. **Milk ejection,** or **milk letdown,** refers to the forced expulsion of milk from the lumen of the alveoli out through the ducts. A neuroendocrine reflex triggered by suckling stimulates release of both of these hormones (❙ Figure 20-32).

■ *Oxytocin release and milk ejection.* The infant cannot directly suck milk out of the alveolar lumen. Instead, milk must be actively squeezed out of the alveoli into the ducts, and hence toward the nipple, by contraction of specialized **myoepithelial cells** (smooth-musclelike epithelial cells) that surround each alveolus (see ❙ Figure 20-31b). The infant's suckling of the breast stimulates sensory nerve endings in the nipple, initiating action potentials that travel up the spinal cord to the hypothalamus. Thus activated, the hypothalamus triggers a burst of oxytocin release from the posterior pituitary. Oxytocin in turn stimulates contraction of the myoepithelial cells in the breasts to bring about milk ejection. Milk letdown continues only as long as the infant continues to nurse. In this way, the milk ejection reflex ensures that the breasts release milk only when required and in the amount needed by the baby. Even though the alveoli may be full of milk, the milk cannot be released without oxytocin. The reflex can become conditioned to stimuli other than suckling, however. For example, the infant's cry can trigger milk letdown, causing a spurt of milk to leak from the nipples. In contrast, psychological stress, acting through the hypothalamus, can easily inhibit milk ejection. For this reason, a positive attitude toward breast-feeding and a relaxed environment are essential for successful breast-feeding.

■ *Prolactin release and milk secretion.* Suckling not only triggers oxytocin release but also stimulates prolactin secretion. Prolactin output by the anterior pituitary is controlled by two hypothalamic secretions: **prolactin-inhibiting hormone (PIH)** and **prolactin-releasing peptide (PrRP)**. PIH is now known to be *dopamine,* which also serves as a neurotransmitter in the brain.

Throughout most of the female's life, PIH is the dominant influence, so prolactin concentrations normally remain low. During lactation, a burst in prolactin secretion occurs each time the infant suckles. Afferent impulses initiated in the nipple on suckling are carried by the spinal cord to the hypothalamus. This

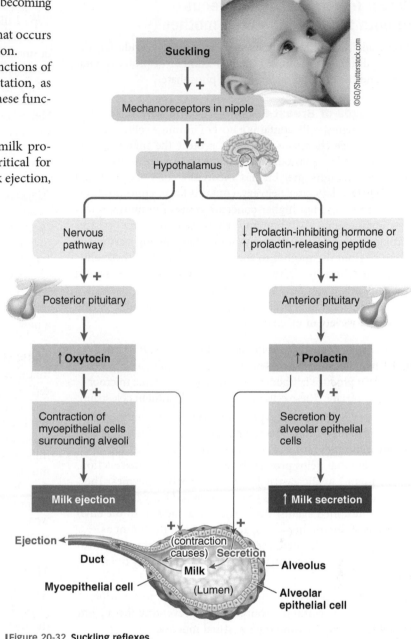

❙Figure 20-32 **Suckling reflexes.**

reflex ultimately leads to prolactin release by the anterior pituitary, although it is unclear whether this is from inhibition of PIH secretion, stimulation of PrRP secretion, or both. Prolactin then acts on the alveolar epithelium to promote secretion of milk to replace the ejected milk (❙ Figure 20-32). Prolactin exerts its effect by means of the JAK/STAT signaling pathway (see p. 116).

Concurrent stimulation by suckling of both milk ejection and milk production ensures that the rate of milk synthesis keeps pace with the baby's needs for milk. The more the infant nurses, the more milk is removed by letdown and the more milk is produced for the next feeding.

In addition to prolactin, which is the most important factor controlling synthesis of milk, at least four other hormones are essential for their permissive role in ongoing milk production: cortisol, insulin, parathyroid hormone, and growth hormone.

Breast-feeding is advantageous to both the infant and the mother.

Nutritionally, **milk** is composed of water, triglyceride fat, the carbohydrate lactose (milk sugar), a number of proteins, vitamins, and the minerals calcium and phosphate.

Advantages of Breast-Feeding for the Infant In addition to nutrients, milk contains a host of immune cells, antibodies, and other chemicals that help protect the infant against infection until it can mount an effective immune response on its own a few months after birth. **Colostrum**, the milk produced for the first 5 days after delivery, contains lower concentrations of fat and lactose but higher concentrations of immunoprotective components. All human babies acquire some passive immunity during gestation by antibodies passing across the placenta from the mother to the fetus (see p. 421). These antibodies are short lived, however, and they often do not persist until the infant can fend for itself immunologically. Breast-fed babies gain additional protection during this vulnerable period through a variety of mechanisms:

■ Breast milk contains an abundance of *immune cells*—both B and T lymphocytes, macrophages, and neutrophils (see pp. 392–393)—that produce antibodies and destroy pathogenic microorganisms outright. These cells are especially plentiful in colostrum.

■ *Secretory IgA,* a special type of antibody, is present in great amounts in breast milk. Secretory IgA consists of two IgA antibody molecules (see p. 417) joined with a so-called secretory component that helps protect the antibodies from destruction by the infant's acidic gastric juice and digestive enzymes. The collection of IgA antibodies that a breast-fed baby receives is specifically aimed against the particular pathogens in the environment of the mother—and, accordingly, of the infant as well. Appropriately, therefore, these antibodies protect against the infectious microbes that the infant is most likely to encounter.

■ Some components in mother's milk, such as *mucus,* adhere to potentially harmful microorganisms, preventing them from attaching to and crossing the intestinal mucosa.

■ *Lactoferrin* is a breast-milk constituent that thwarts growth of harmful bacteria by decreasing the availability of iron, a mineral needed for multiplication of these pathogens (see p. 411).

■ *Bifidus factor* in breast milk promotes multiplication of the nonpathogenic microorganism *Lactobacillus bifidus* in the infant's digestive tract. Growth of this harmless bacterium helps crowd out potentially harmful bacteria.

■ Other components in breast milk promote maturation of the baby's digestive system so that it is less vulnerable to diarrhea-causing bacteria and viruses.

■ Still other factors in breast milk hasten development of the infant's immune capabilities.

Thus, breast milk helps protect infants from disease in a variety of ways.

Some studies hint that in addition to the benefits of breast milk during infancy, breast-feeding may reduce the risk of developing certain serious diseases later in life. Examples include allergies such as asthma, autoimmune diseases such as Type 1 diabetes mellitus, and cancers such as lymphoma.

Infants who are bottle-fed on a formula made from cow's milk or another substitute do not have the protective advantage provided by human milk and, accordingly, have a higher incidence of infections of the digestive tract, respiratory tract, and ears than breast-fed babies do. Also, the digestive system of a newborn is better equipped to handle human milk than cow milk–derived formula, so bottle-fed babies tend to have more digestive upsets.

Advantages of Breast-Feeding for the Mother Breast-feeding is also advantageous for the mother. Oxytocin release triggered by nursing hastens uterine involution. In addition, suckling suppresses the menstrual cycle because prolactin (sometimes termed "nature's contraceptive") inhibits GnRH, thereby suppressing FSH and LH secretion. Lactation, therefore, tends to prevent ovulation, decreasing the likelihood of another pregnancy (although it is not a reliable means of contraception). This mechanism permits all the mother's resources to be directed toward the newborn instead of being shared with a new embryo.

Cessation of Milk Production at Weaning When the infant is weaned, two mechanisms contribute to cessation of milk production. First, without suckling, prolactin secretion is not stimulated, removing the main stimulus for continued milk synthesis and secretion. Also, in the absence of suckling, oxytocin is not released and milk letdown does not occur. Because milk production does not immediately shut down, milk accumulates in the alveoli, engorging the breasts. The resulting pressure buildup acts directly on the alveolar epithelial cells to suppress further milk production. Cessation of lactation at weaning therefore results from a lack of suckling-induced stimulation of both prolactin and oxytocin secretion.

The end is a new beginning.

Reproduction is an appropriate way to end our discussion of physiology from cells to systems. The single cell resulting from the union of male and female gametes divides mitotically and differentiates into a multicellular individual made up of a number of body systems that interact cooperatively to maintain homeostasis (that is, stability in the internal environment). All the life-supporting homeostatic processes introduced throughout this book begin again at the start of a new life.

Check Your Understanding 20.4

1. List and describe the stages of follicular development and indicate the status of the gamete in each of these stages.

2. Tell what ovarian hormones the follicle and the corpus luteum secrete, state the effects of these hormones on the uterus, and indicate during which phase of the ovarian cycle each of the phases of the uterine cycle takes place.

3. Define *zygote, blastocyst, inner cell mass, trophoblast, decidua, chorion, placenta, embryo,* and *fetus.*

4. Discuss the role of oxytocin during parturition and during breast-feeding.

Homeostasis: Chapter in Perspective

 The reproductive system is unique in that it is not essential for homeostasis or for survival of the individual but is essential for sustaining the thread of life from generation to generation. Reproduction depends on the union of male and female gametes (reproductive cells), each with a half set of chromosomes, to form a new individual with a full, one-of-a-kind set of chromosomes. Unlike the other body systems, which are essentially identical in the two sexes, the reproductive systems of males and females are remarkably different, befitting their different roles in the reproductive process.

The male system is designed to continuously produce huge numbers of mobile spermatozoa that are delivered to the female during the sex act. Male gametes must be produced in abundance for two reasons: (1) Only a small percentage of them survive the hazardous journey through the female reproductive tract to the site of fertilization, and (2) the cooperative effort of many spermatozoa is required to break down the barriers surrounding the female gamete (ovum or egg) to enable one spermatozoon to penetrate and unite with the ovum.

The female reproductive system undergoes complex changes on a cyclic monthly basis. During the first half of the cycle, a single nonmotile ovum is prepared for release. During the second half, the reproductive system is geared toward preparing a suitable environment for supporting the ovum if fertilization (union with a spermatozoon) occurs. If fertilization does not occur, the prepared supportive environment within the uterus sloughs off, and the cycle starts again as a new ovum is prepared for release. If fertilization occurs, the female reproductive system adjusts to support growth and development of the new individual until it can survive on its own on the outside.

There are three important parallels in the male and female reproductive systems, even though they differ considerably in structure and function. First, the same set of undifferentiated reproductive tissues in the embryo can develop into either a male or a female system, depending on the presence or absence, respectively, of male-determining factors. Second, the same hormones—namely, hypothalamic kisspeptins and GnRH and anterior pituitary FSH and LH—control reproductive function in both sexes. In both cases, gonadal steroids and inhibin act in negative-feedback fashion to control hypothalamic and anterior pituitary output, with the exception that a high estrogen level in females induces an ovulation-triggering surge in LH secretion in positive-feedback fashion. Third, the same events take place in the developing gamete's nucleus during sperm formation and egg formation, although males produce millions of sperm in a day, whereas females produce only about 400 ova in a lifetime.

Review Exercises Answers begin on p. A-53

Reviewing Terms and Facts

1. It is possible for a genetic male to have the anatomic appearance of a female. *(True or false?)*

2. Testosterone secretion essentially ceases from birth until puberty. *(True or false?)*

3. Females do not experience erection. *(True or false?)*

4. Most of the lubrication for sexual intercourse is provided by the female. *(True or false?)*

5. If a follicle does not reach maturity during one ovarian cycle, it can finish maturing during the next cycle. *(True or false?)*

6. Rising, moderate levels of estrogen inhibit tonic LH secretion, whereas high levels of estrogen stimulate the LH surge. *(True or false?)*

7. Spermatogenesis takes place within the _____ of the testes, stimulated by the hormones _____ and _____.

8. During estrogen production by the follicle, the _____ cells under the influence of the hormone _____ produce androgens, and the _____ cells under the influence of the hormone _____ convert these androgens into estrogens.

9. The source of estrogen and progesterone during the first 10 weeks of gestation is the _____. The source of these hormones during the last two trimesters of gestation is the _____.

10. Detection of _____ in the urine is the basis of pregnancy diagnosis tests.

11. Which of the following statements concerning chromosomal distribution is *incorrect*?

    a. All human somatic cells contain 23 chromosomal pairs for a total diploid number of 46 chromosomes.

    b. Each gamete contains 23 chromosomes, one member of each chromosomal pair.

    c. During meiotic division, the members of the chromosome pairs regroup themselves into the original combinations derived from the individual's mother and father for separation into haploid gametes.

    d. Sex determination depends on the combination of sex chromosomes, an XY combination being a genetic male and XX a genetic female.

    e. The sex chromosome content of the fertilizing sperm determines the sex of the offspring.

12. When the corpus luteum degenerates,
   a. circulating levels of estrogen and progesterone rapidly decline.
   b. FSH and LH secretion start to rise as the inhibitory effects of the gonadal steroids are withdrawn.
   c. the endometrium sloughs off.
   d. both (a) and (b) occur.
   e. all of the above

13. Match the following:
   1. secrete(s) prostaglandins
   2. increase(s) motility and fertility of sperm
   3. secrete(s) an alkaline fluid
   4. provide(s) fructose
   5. act(s) as the storage site for sperm
   6. concentrate(s) the sperm a 100-fold
   7. secrete(s) fibrinogen
   8. provide(s) clotting enzymes
   9. contain(s) erectile tissue

   (a) epididymis and ductus deferens
   (b) prostate gland
   (c) seminal vesicles
   (d) bulbourethral glands
   (e) penis

14. Using the following answer code, indicate when each event takes place during the ovarian cycle:
   1. rapid development of recruited antral follicles
   2. secretion of estrogen
   3. secretion of progesterone
   4. menstruation
   5. repair and proliferation of the endometrium
   6. increased vascularization and glycogen storage in the endometrium

   (a) occurs during the follicular phase
   (b) occurs during the luteal phase
   (c) occurs during both the follicular and the luteal phases

Understanding Concepts
(Answers at www.cengagebrain.com)

1. What constitutes the primary reproductive organs, gametes, sex hormones, reproductive tract, accessory sex glands, external genitalia, and secondary sexual characteristics in males and in females?

2. List the essential reproductive functions of the male and of the female.

3. Discuss what determines genetic sex, gonadal sex, and phenotypic sex in males and in females.

4. Of what functional significance is the scrotal location of the testes?

5. Discuss the source and functions of testosterone.

6. Describe the three major stages of spermatogenesis. Discuss the functions of each part of a spermatozoon. What are the roles of Sertoli cells?

7. Discuss the control of testicular function.

8. Compare the sex act in males and females.

9. Compare oogenesis with spermatogenesis.

10. Discuss the structure of, gamete status in, stimulation of, and hormonal secretion (if any) by each of the following: primordial follicle, preantral follicle, early antral follicle, recruitable antral follicle, mature follicle, and corpus luteum.

11. How are the ovum and spermatozoa transported to the site of fertilization? Describe the process of fertilization.

12. Describe the process of implantation and placenta formation.

13. What are the functions of the placenta? What hormones does the placenta secrete?

14. What is the role of human chorionic gonadotropin?

15. What factors contribute to the initiation of parturition? What are the stages of labor? What is the role of oxytocin?

16. Describe the hormonal factors that play a role in lactation.

17. Summarize the actions of estrogen and progesterone.

Applying Clinical Reasoning

Maria A., who is in her second month of gestation, has been experiencing severe abdominal cramping. Her physician has diagnosed her condition as a tubal pregnancy: The developing embryo is implanted in the oviduct instead of in the uterine endometrium. Why must this pregnancy be surgically terminated?

Thinking at a Higher Level

1. The hypothalamus releases GnRH in pulsatile bursts once every 2 to 3 hours, with no secretion occurring between bursts. The blood concentration of GnRH depends on the frequency of these bursts of secretion. A promising line of research for a new method of contraception involves administration of GnRHlike drugs. In what way could such drugs act as contraceptives when GnRH is the hypothalamic hormone that triggers the chain of events leading to ovulation? (Hint: The anterior pituitary is "programmed" to respond only to the normal pulsatile pattern of GnRH.)

2. Occasionally, testicular tumors composed of interstitial cells of Leydig may secrete up to 100 times the normal amount of testosterone. When such a tumor develops in young children, they grow up much shorter than their genetic potential. Explain why. What other symptoms would be present?

3. What type of sexual dysfunction might arise in men taking drugs that inhibit sympathetic nervous system activity as part of the treatment for high blood pressure?

4. Explain the physiologic basis for administering a posterior pituitary extract to induce or facilitate labor.

5. The symptoms of menopause are sometimes treated with supplemental estrogen and progesterone. Why wouldn't treatment with GnRH or FSH and LH also be effective?

A Review of Chemical Principles

By Spencer Seager, Weber State University, and Lauralee Sherwood

A.1 Chemical Level of Organization in the Body

Matter is anything that occupies space and has mass, including all living and nonliving things in the universe. **Mass** is the amount of matter in an object. **Weight,** in contrast, is the effect of gravity on that mass. The more gravity exerted on a mass, the greater the weight of the mass. An astronaut has the same mass whether on Earth or in space but is weightless in the zero gravity of space.

Atoms

All matter is made up of tiny particles called **atoms.** These particles are too small to be seen individually, even with the most powerful electron microscopes available today.

Even though extremely small, atoms consist of three types of even smaller subatomic particles. The types of atoms vary in the numbers of these subatomic particles they contain. **Protons** and **neutrons** are particles of nearly identical mass, with protons carrying a positive charge and neutrons having no charge. **Electrons** have a much smaller mass than protons and neutrons and are negatively charged. An atom consists of two regions—a dense, central *nucleus* made of protons and neutrons surrounded by a three-dimensional *electron cloud,* where electrons move rapidly around the nucleus in orbitals (I Figure A-1). The magnitude of the charge of a proton exactly matches that of an electron, but it is opposite in sign, being positive. In all atoms, the number of protons in the nucleus is equal to the number of electrons moving around the nucleus, so their charges balance and the atoms are neutral.

Elements and atomic symbols

A pure substance composed of only one type of atom is called an **element.** A pure sample of the element carbon contains only carbon atoms, even though the atoms might be arranged in the form of diamond or in the form of graphite (pencil "lead"). Each element is designated by an **atomic symbol,** a one- or two-letter chemical shorthand form of the element's name. Usually these symbols are easy to follow, because they are derived from the English name for the element. Thus, H stands for *hydrogen,* C for *carbon,* and O for *oxygen.* In a few cases, the atomic symbol is based on the element's Latin name—for example, Na for *sodium* (*natrium* in Latin) and K for *potassium*

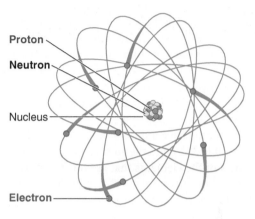

I Figure A-1 **The atom.** The atom consists of two regions. The central nucleus contains protons and neutrons and makes up 99.9% of the mass. Surrounding the nucleus is the electron cloud, where the electrons move rapidly around the nucleus. (Figure not drawn to scale.)

(kalium). Of the 109 known elements, 26 are normally found in the body. Four elements—hydrogen, carbon, oxygen, and nitrogen—comprise 96% of the body's mass.

Compounds and molecules

Pure substances composed of more than one type of atom are known as **compounds.** Pure water, for example, is a compound that contains atoms of hydrogen and atoms of oxygen in a 2:1 ratio, regardless of whether the water is in the form of liquid, solid (ice), or vapor (steam). A **molecule** is the smallest unit of a pure substance that has the properties of that substance and is capable of a stable, independent existence. For example, a molecule of water consists of two atoms of hydrogen and one atom of oxygen, held together by chemical bonds.

Atomic number

Exactly what are we talking about when we refer to a "type" of atom? That is, what makes hydrogen, carbon, and oxygen atoms different? The answer is the number of protons in the nucleus. Regardless of where they are found, all hydrogen atoms have 1 proton in the nucleus, all carbon atoms have 6, and all oxygen atoms have 8. These numbers also represent the number of electrons moving around each nucleus, because the number of electrons and number of protons in an atom are equal. The number of protons in the nucleus of an atom of an element is called the **atomic number** of the element.

Atomic weight

As expected, tiny atoms have tiny masses. For example, the actual mass of a hydrogen atom is 1.67×10^{-24} g, that of a carbon atom is 1.99×10^{-23} g, and that of an oxygen atom is 2.66×10^{-23} g. These very small numbers are inconvenient to work with in calculations, so a system of relative masses has been developed. These relative masses simply compare the actual masses of the atoms with one another. Suppose the actual masses of two people were determined to be 45.50 and 113.75 kg. Their relative masses are determined by dividing each mass by the smaller mass of the two: 45.50/45.50 = 1.00, and 113.75/45.50 = 2.50. Thus, the relative masses of the two people are 1.00 and 2.50; these numbers simply express that the mass of the heavier person is 2.50 times that of the other person. The relative masses of atoms are called **atomic masses,** or **atomic weights,** and they are given in *atomic mass units (amu).* In this system, hydrogen atoms, the least massive of all atoms, have an atomic weight of 1.01 amu. The atomic weight of carbon atoms is 12.01 amu, and that of oxygen atoms is 16.00 amu. Thus, oxygen atoms have a mass about 16 times that of hydrogen atoms. ▮Table A-1 gives the atomic weights and some other characteristics of the elements that are most important physiologically.

A.2 Chemical Bonds

Because all matter is made up of atoms, atoms must somehow be held together to form matter. The forces holding atoms together are called **chemical bonds.** Not all chemical bonds are formed in the same way, but all involve the electrons of atoms. Whether one atom bonds with another depends on the number and arrangement of its electrons. An atom's electrons are arranged in electron shells, to which we now turn our attention.

▮TABLE A-1 Characteristics of Selected Elements

Name and Symbol	Number of Protons	Atomic Number	Atomic Weight (amu)
Hydrogen (H)	1	1	1.01
Carbon (C)	6	6	12.01
Nitrogen (N)	7	7	14.01
Oxygen (O)	8	8	16.00
Sodium (Na)	11	11	22.99
Magnesium (Mg)	12	12	24.31
Phosphorus (P)	15	15	30.97
Sulfur (S)	16	16	32.06
Chlorine (Cl)	17	17	35.45
Potassium (K)	19	19	39.10
Calcium (Ca)	20	20	40.08

Electron shells

Electrons tend to move around the nucleus in a specific pattern. The orbitals, or pathways traveled by electrons around the nucleus, are arranged in an orderly series of concentric layers known as **electron shells,** which consecutively surround the nucleus. Each electron shell can hold a specific number of electrons. The first shell, closest to the nucleus (innermost), can contain a maximum of only 2 electrons, no matter what the element is. The second shell can hold a total of 8 more electrons. The third shell also can hold a maximum of 8 electrons. As the number of electrons increases with increasing atomic number, still more electrons occupy successive shells, each at a greater distance from the nucleus. Each successive shell from the nucleus has a higher **energy level.** Because the negatively charged electrons are attracted to the positively charged nucleus, it takes more energy for an electron to overcome the nuclear attraction and orbit farther from the nucleus. Thus, the first electron shell has the lowest energy level and the outermost shell of an atom has the highest energy level.

In general, electrons belong to the lowest energy shell possible, up to the maximum capacity of each shell. For example, hydrogen atoms have only 1 electron, so it is in the first shell. Helium atoms have 2 electrons, which are both in the first shell and fill it. Carbon atoms have 6 electrons, 2 in the first shell and 4 in the second shell, whereas the 8 electrons of oxygen are arranged with 2 in the first shell and 6 in the second shell.

Bonding characteristics of an atom and valence

Atoms tend to undergo processes that result in a filled outermost electron shell. Thus, the electrons of the outer or higher-energy shell determine the bonding characteristics of an atom and its ability to interact with other atoms. Atoms that have a vacancy in their outermost shell tend to give up, accept, or share electrons with other atoms (whichever is most favorable energetically) so that all participating atoms have filled outer shells. For example, an atom that has only 1 electron in its outermost shell may empty this shell so that its remaining shells are full. By contrast, another atom that lacks only 1 electron in its outer shell may acquire the deficient electron from the first atom to fill all its shells to the maximum. The number of electrons an atom loses, gains, or shares to achieve a filled outer shell is known as the atom's **valence.** A **chemical bond** is the force of attraction that holds participating atoms together as a result of an interaction between their outermost electrons.

Consider sodium atoms (Na) and chlorine atoms (Cl) (▮Figure A-2). Sodium atoms have 11 electrons: 2 in the first shell, 8 in the second shell, and 1 in the third shell. Chlorine atoms have 17 electrons: 2 in the first shell, 8 in the second shell, and 7 in the third shell. Because 8 electrons are required to fill the second and third shells, sodium atoms have 1 electron more than is needed to provide a filled second shell, whereas chlorine atoms have 1 fewer electron than is needed to fill the third shell. Each sodium atom can lose an electron to a chlorine atom, leaving each sodium atom with 10 electrons; 8 of these are in the second shell, which is full and is now the outer shell occupied by electrons. By accepting 1 electron, each chlorine atom now

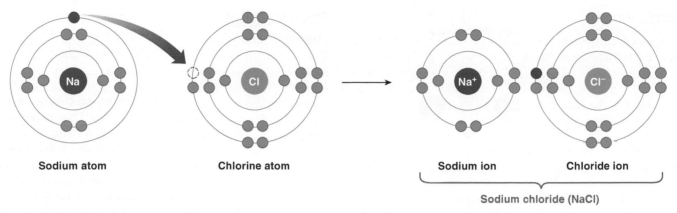

Sodium atom   Chlorine atom   Sodium ion   Chloride ion

Sodium chloride (NaCl)

❙Figure A-2 Ions and ionic bonds. Sodium (Na) and chlorine (Cl) atoms both have partially filled outermost shells. Therefore, sodium tends to give up its lone electron in the outer shell to chlorine, filling chlorine's outer shell. As a result, sodium becomes a positively charged ion and chlorine becomes a negatively charged ion known as chloride. The oppositely charged ions attract each other, forming an ionic bond.

has a total of 18 electrons, with 8 of them in the third, or outer, shell, which is now full.

Ions; ionic bonds

Recall that atoms are electrically neutral because they have an identical number of positively charged protons and negatively charged electrons. By giving up and accepting electrons, the sodium atoms and chlorine atoms have achieved filled outer shells, but now each atom is unbalanced electrically. Although each sodium atom now has 10 electrons, it still has 11 protons in the nucleus and a net electrical charge, or valence, of +1. Similarly, each chlorine atom now has 18 electrons but only 17 protons. Thus, each chlorine atom has a −1 charge. Such charged atoms are called **ions.** Positively charged ions are called **cations;** negatively charged ions are called **anions.** As a helpful hint to keep these terms straight, imagine the "t" in *cation* as standing for a "+" sign and the first "n" in *anion* as standing for "negative."

Note that both a cation and an anion are formed whenever an electron is transferred from one atom to another. Because opposite charges attract, sodium ions (Na^+) and charged chlorine atoms, now called *chloride* ions (Cl^-), are attracted toward each other. This electrical attraction that holds cations and anions together is known as an **ionic bond.** Ionic bonds hold Na^+ and Cl^- together in the compound **sodium chloride, NaCl,** which is common table salt. A sample of sodium chloride actually contains sodium and chloride ions in a three-dimensional geometrical arrangement called a *crystal lattice.* The ions of opposite charge occupy alternate sites within the lattice (❙Figure A-3).

Covalent bonds

It is not favorable, energywise, for an atom to give up or accept more than three electrons. Nevertheless, carbon atoms, which have four electrons in their outer shell, form compounds. They do so by another bonding mechanism, *covalent bonding.* Atoms that would have to lose or gain four or more electrons to achieve outer-

shell stability usually bond by *sharing* electrons. Shared electrons actually orbit around *both* atoms. Thus, a carbon atom can share its 4 outer electrons with the 4 electrons of 4 hydrogen atoms, as shown in Equation A-1, where the outer-shell electrons are shown as dots around the symbol of each atom. (The resulting compound is methane, CH_4, a gas made up of individual CH_4 molecules.)

$$\cdot \overset{\cdot}{\underset{\cdot}{C}} \cdot + 4 \cdot H \rightarrow H : \overset{\cdot\cdot}{\underset{\cdot\cdot}{C}} : H \qquad \text{Eq. A-1}$$

Shared electron pairs

Shared electron pairs

Crystals of sodium chloride (NaCl)

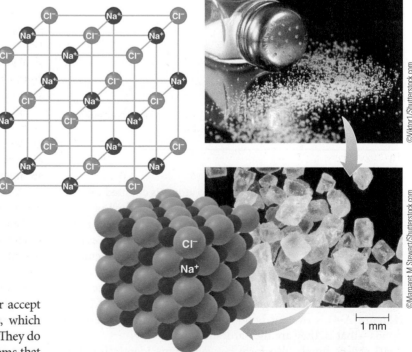

❙Figure A-3 Crystal lattice for sodium chloride (NaCl; table salt).

Molecular formula	Atomic structure	Structural formula with covalent bond

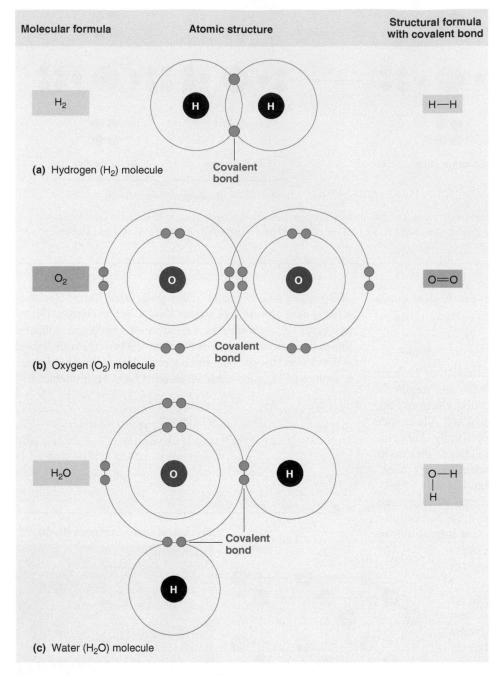

(a) Hydrogen (H_2) molecule

(b) Oxygen (O_2) molecule

(c) Water (H_2O) molecule

▌Figure A-4 **A covalent bond.** A covalent bond is formed when atoms that share a pair of electrons are both attracted toward the shared pair.

Each electron that is shared by two atoms is counted toward the number of electrons needed to fill the outer shell of each atom. Thus, each carbon atom shares four pairs, or 8 electrons, and thus has 8 in its outer shell. Each hydrogen atom shares one pair, or 2 electrons, and thus has a filled outer shell. (Remember, hydrogen atoms need only two electrons to complete their outer shell, which is the first shell.) The sharing of a pair of electrons by atoms binds them together by means of a **covalent bond** (▌Figure A-4). Covalent bonds are the strongest of chemical bonds—that is, they are the hardest to break.

Covalent bonds also form between some identical atoms. For example, two hydrogen atoms can complete their outer

shells by sharing one electron pair made from the single electrons of each atom, as shown in Equation A-2:

$$H\cdot + \cdot H \rightarrow H{:}H \qquad \text{Eq. A-2}$$

Thus, hydrogen gas consists of individual H_2 molecules (▌Figure A-4a). (A subscript following a chemical symbol indicates the number of that type of atom present in the molecule.) Several other nonmetallic elements also exist as molecules because covalent bonds form between identical atoms; oxygen (O_2) is an example (▌Figure A-4b).

Often, an atom can form covalent bonds with more than one atom. One of the most familiar examples is water (H_2O), consisting of two hydrogen atoms each forming a single covalent bond with one oxygen atom (▌Figure A-4c). Equation A-3 represents the formation of water's covalent bonds:

$$\begin{array}{c} H\cdot \\ + \cdot \ddot{O}{:} \rightarrow H{:}\ddot{O}{:} \qquad \text{Eq. A-3} \\ H\cdot \qquad\qquad H \end{array}$$

The water molecule is sometimes represented as

$$\begin{array}{c} H-O \\ | \\ H \end{array}$$

where the nonshared electron pairs are not shown and the covalent bonds, or shared pairs, are represented by dashes.

Nonpolar and polar molecules

The electrons between two atoms in a covalent bond are not always shared equally. When the atoms sharing an electron pair are identical, such as two oxygen atoms, the electrons are attracted equally by both atoms and so are shared equally. The result is a **nonpolar molecule.** The term *nonpolar* implies no difference at the two ends (two "poles") of the bond. Because both atoms within the molecule exert the same pull on the shared electrons, each shared electron spends the same amount of time orbiting each atom. Thus, both atoms remain electrically neutral in a nonpolar molecule such as O_2.

When the sharing atoms are not identical, unequal sharing of electrons occurs because atoms of different elements do not exert the same pull on shared electrons. For example, an oxygen atom strongly attracts electrons when it is bonded to other atoms. A **polar molecule** results from the unequal sharing of electrons between different types of atoms covalently bonded together. The water molecule is a good example of a polar molecule. The oxygen atom pulls the shared electrons more strongly than do the

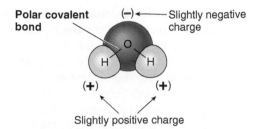

I Figure A-5 **A polar molecule.** A water molecule is an example of a polar molecule, in which the distribution of shared electrons is not uniform. Because the oxygen atom pulls the shared electrons more strongly than the hydrogen atoms do, the oxygen side of the molecule is slightly negatively charged and the hydrogen sides are slightly positively charged.

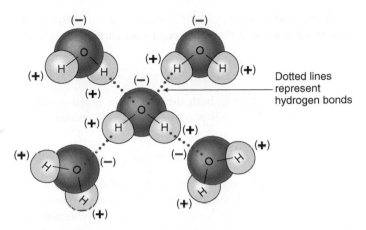

I Figure A-6 **A hydrogen bond.** A hydrogen bond is formed by the attraction of a positively charged hydrogen end of a polar molecule to the negatively charged end of another polar molecule.

hydrogen atoms within each of the two covalent bonds. Consequently, the electron of each hydrogen atom tends to spend more time away orbiting around the oxygen atom than at home around the hydrogen atom. Because of this nonuniform distribution of electrons, the oxygen side of the water molecule where the shared electrons spend more time is slightly negative, and the two hydrogens that are visited less frequently by the electrons are slightly more positive (I Figure A-5). Note that the entire water molecule has the same number of electrons as it has protons, so as a whole it has no net charge. This is unlike ions, which have an electron excess or deficit. Polar molecules have a balanced number of protons and electrons but an unequal distribution of the shared electrons among the atoms making up the molecule.

Hydrogen bonds

Polar molecules are attracted to other polar molecules. In water, for example, an attraction exists between the positive hydrogen ends of some molecules and the negative oxygen ends of others. Hydrogen is not a part of all polar molecules, but when it is covalently bonded to an atom that strongly attracts electrons to form a covalent molecule, the attraction of the positive (hydrogen) end of the polar molecule to the negative end of another polar molecule is called a **hydrogen bond** (I Figure A-6). Thus, the polar attractions of water molecules to each other are an example of hydrogen bonding.

A.3 Chemical Reactions

Processes in which chemical bonds are broken, formed, or both are called **chemical reactions.** Reactions are represented by equations in which the reacting substances (**reactants**) are typically written on the left, the newly produced substances (**products**) are written on the right, and an arrow meaning "yields" points from the reactants to the products. These conventions are illustrated in Equation A-4:

$$A + B \rightarrow C + D \qquad \text{Eq. A-4}$$
$$\underset{\text{Reactants}}{} \quad \underset{\text{Products}}{}$$

Balanced equations

A chemical equation is a "chemical bookkeeping" ledger that describes what happens in a reaction. By the **law of conservation of mass,** the total mass of all materials entering a reaction

equals the total mass of all the products. Thus, the total number of atoms of each element must always be the same on the left and right sides of the equation, because no atoms are lost. Such equations in which the same number of atoms of each type appears on both sides are called **balanced equations.** When writing a balanced equation, the number *preceding* a chemical symbol designates the number of independent (unjoined) atoms, ions, or molecules of that type, whereas a number written as a subscript *following* a chemical symbol denotes the number of a particular atom within a molecule. The absence of a number indicates "one" of that particular chemical. Let us look at a specific example, the oxidation of glucose (the sugar that cells use as fuel), as shown in Equation A-5:

$$\underset{\text{glucose}}{C_6H_{12}O_6} + \underset{\text{oxygen}}{6\ O_2} \rightarrow \underset{\substack{\text{carbon}\\\text{dioxide}}}{6\ CO_2} + \underset{\text{water}}{6\ H_2O} \qquad \text{Eq. A-5}$$

According to this equation, 1 molecule of glucose reacts with 6 molecules of oxygen to produce 6 molecules of carbon dioxide and 6 molecules of water. Note the following balance in this reaction:

- 6 carbon atoms on the left (in 1 glucose molecule) and 6 carbon atoms on the right (in 6 carbon dioxide molecules)

- 12 hydrogen atoms on the left (in 1 glucose molecule) and 12 on the right (in 6 water molecules, each containing 2 hydrogen atoms)

- 18 oxygen atoms on the left (6 in 1 glucose molecule plus 12 more in the 6 oxygen molecules) and 18 on the right (12 in 6 carbon dioxide molecules, each containing 2 oxygen atoms, and 6 more in the 6 water molecules, each containing 1 oxygen atom)

Reversible and irreversible reactions

Under appropriate conditions, the products of a reaction can be changed back to the reactants. For example, carbon dioxide gas dissolves in and reacts with water to form carbonic acid, H_2CO_3:

$$CO_2 + H_2O \rightarrow H_2CO_3 \qquad \text{Eq. A-6}$$

Carbonic acid is not very stable, however, and as soon as some is formed, part of it decomposes to give carbon dioxide and water:

$$H_2CO_3 \rightarrow CO_2 + H_2O \qquad \text{Eq. A-7}$$

Reactions that go in both directions are called **reversible reactions.** Double arrows pointing in both directions usually represent them:

$$CO_2 + H_2O \rightleftharpoons H_2CO_3 \qquad \text{Eq. A-8}$$

Theoretically, every reaction is reversible. Often, however, conditions are such that a reaction, for all practical purposes, goes in only one direction; such a reaction is called **irreversible.** For example, an irreversible reaction takes place when an explosion occurs, because the products do not remain in the vicinity of the reaction site to get together to react.

Catalysts; enzymes

The rates (speeds) of chemical reactions are influenced by a number of factors, of which catalysts are one of the most important. A **catalyst** is a "helper" molecule that speeds up a reaction without being used up in the reaction. Living organisms use catalysts known as **enzymes.** These enzymes exert amazing influence on the rates of chemical reactions that take place in the organisms. Reactions that take weeks or even months to occur under normal laboratory conditions take place in seconds under the influence of enzymes in the body. One of the fastest-acting enzymes is **carbonic anhydrase (ca),** which catalyzes the reaction between carbon dioxide and water to form carbonic acid. This reaction is important in the transport of carbon dioxide from tissue cells, where it is produced metabolically, to the lungs, where it is excreted. The equation for the reaction was shown in Equation A-6. (Carbonic anhydrase indirectly catalyzes this reaction by converting $CO_2 + H_2O$ directly to $H^+ + HCO_3^-$, which can form H_2CO_3. In the absence of ca, $CO_2 + H_2O$ slowly, directly form H_2CO_3. The reactions both with and without catalyst are commonly shown as in Equation A-6; see p. 550 for specific details.) Each molecule of carbonic anhydrase catalyzes the conversion of 36 million CO_2 molecules per minute! Enzymes are important in essentially every chemical reaction that takes place in living organisms.

A.4 Molecular and Formula Weight and the Mole

Because molecules are made up of atoms, the relative mass of a molecule is simply the sum of the relative masses (atomic weights) of the atoms found in the molecule. The relative masses of molecules are called **molecular masses** or **molecular weights.** The molecular weight of water, H_2O, is thus the sum of the atomic weights of two hydrogen atoms and one oxygen atom, or 1.01 amu + 1.01 amu + 16.00 amu = 18.02 amu.

Not all compounds exist in the form of molecules. Ionically bonded substances such as sodium chloride consist of three-dimensional arrangements of sodium ions (Na^+) and chloride ions (Cl^-) in a 1:1 ratio. The formulas for ionic compounds reflect only the ratio of the ions in the compound and should not be interpreted in terms of molecules. Thus, the formula for sodium chloride, NaCl, indicates that the ions combine in a 1:1 ratio. It is convenient to apply the concept of relative masses to ionic compounds even though they do not exist as molecules. The **formula weight** for such compounds is defined as the sum of the atomic weights of the atoms found in the formula. Thus, the formula weight of NaCl is equal to the sum of the atomic weights of one sodium atom and one chlorine atom, or 22.99 amu + 35.45 amu = 58.44 amu.

As you have seen, chemical reactions can be represented by equations and discussed in terms of numbers of molecules, atoms, and ions reacting with one another. To carry out reactions in the laboratory, however, a scientist cannot count out numbers of reactant particles but instead must be able to weigh the correct amount of each reactant. Using the mole concept makes this task possible. A **mole** (abbreviated *mol*) of a pure element or compound is the amount of material contained in a sample of the pure substance that has a mass in grams equal to the substance's atomic weight (for elements) or the molecular weight or formula weight (for compounds). Thus, 1 mol of potassium, K, would be a sample of the element with a mass of 39.10 g. Similarly, 1 mol of H_2O would have a mass of 18.02 g, and 1 mol of NaCl would be a sample with a mass of 58.44 g.

Atomic weights, molecular weights, and formula weights are relative masses, which leads to a fundamental characteristic of moles. For example, 1 mol of oxygen atoms has a mass of 16.00 g, and 1 mol of hydrogen atoms has a mass of 1.01 g. Thus, the ratio of the masses of 1 mol of each element is 16.00 to 1.01, the same as the ratio of the atomic weights for the two elements. Recall that these atomic weights compare the relative masses of oxygen and hydrogen. Accordingly, the number of oxygen atoms present in 16 g of oxygen (1 mol of oxygen) is the same as the number of hydrogen atoms present in 1.01 g of hydrogen. Therefore, 1 mol of oxygen contains exactly the same number of oxygen atoms as the number of hydrogen atoms in 1 mol of hydrogen. Thus, it is possible and sometimes useful to think of a mole as a specific number of particles. This number, called **Avogadro's number,** is equal to 6.02×10^{23}.

A.5 Solutions, Colloids, and Suspensions

In contrast to a compound, a **mixture** consists of two or more types of elements or molecules physically blended together (intermixed) instead of being linked by chemical bonds. A compound has very different properties from the individual elements of which it is composed. For example, the solid, white NaCl (table salt) crystals you use to flavor your food are different from either sodium (a silvery white metal) or chlorine (a poisonous yellow–green gas found in bleach). By comparison, each component of a mixture retains its chemical properties. If you mix salt and sugar together, each retains a distinct taste and other individual properties. The constituents of a compound can only be separated by chemical means—bond breakage. By contrast, the components of a mixture can be separated by physical means, such as filtration or evaporation. The most common mixtures in the body are mixtures of water and various other substances. These mixtures are categorized as *solutions, colloids,* or *suspensions,* depending on the size and nature of the substance mixed with water.

Solutions

Most chemical reactions in the body take place between reactants that have dissolved to form solutions. **Solutions** are homogenous mixtures containing a relatively large amount of one substance called the **solvent** (the dissolving medium) and smaller amounts of one or more substances called **solutes** (the dissolved particles). Saltwater, for example, contains mostly water, which is thus the solvent, and a smaller amount of salt, which is the solute. Water is the solvent in most solutions found in the human body.

Electrolytes versus nonelectrolytes

When ionic solutes are dissolved in water to form solutions, the resulting solution will conduct electricity. This is not true for most covalently bonded solutes. For example, a salt–water solution conducts electricity, but a sugar–water solution does not. When salt dissolves in water, the solid lattice of Na^+ and Cl^- is broken down, and the individual ions are separated and distributed uniformly throughout the solution. These mobile, charged ions conduct electricity through the solution. Solutes that form ions in solution and conduct electricity are called **electrolytes.** When sugar dissolves, however, individual covalently bonded sugar molecules leave the solid and become uniformly distributed throughout the solution. These uncharged molecules cannot conduct a current. Solutes that do not form conductive solutions are called **nonelectrolytes.**

Measures of concentration

The amount of solute dissolved in a specific amount of solution can vary. For example, a salt–water solution might contain 1 g of salt in 100 mL of solution, or it could contain 10 g of salt in 100 mL of solution. Both solutions are salt–water solutions, but they have different concentrations of solute. The **concentration** of a solution indicates the relationship between the amount of solute and the amount of solution. Concentrations can be given in various units.

Molarity Concentrations given in terms of **molarity (M)** give the number of moles of solute in exactly 1 liter of solution. Thus, a half molar (0.5 M) solution of NaCl would contain one-half mole, or 29.22 g, of NaCl in each liter of solution.

Normality When the solute is an electrolyte, it is sometimes useful to express the concentration of the solution in a unit that gives information about the amount of ionic charge in the solution. This is done by expressing concentration in terms of **normality (N).** The normality of a solution gives the number of equivalents of solute in exactly 1 liter of solution. An **equivalent** of an electrolyte is the amount that produces 1 mole of positive (or negative) charges when it dissolves. The number of equiva-

lents of an electrolyte can be calculated by multiplying the number of moles of electrolyte by the total number of positive charges produced when one formula unit of the electrolyte dissolves. Consider NaCl and calcium chloride ($CaCl_2$) as examples. The ionization reactions for one formula unit of each solute are

$$NaCl \rightarrow Na^+ + Cl^- \qquad \text{Eq. A-9}$$

$$CaCl_2 \rightarrow Ca^{2+} + 2\,Cl^- \qquad \text{Eq. A-10}$$

Thus, 1 mol of NaCl produces 1 mole of positive charges (Na^+) and so contains 1 equivalent:

$$(1 \text{ mol NaCl}) \times 1 = 1 \text{ equivalent}$$

where the number 1 used to multiply the 1 mol of NaCl came from the +1 charge on Na^+.

One mole of $CaCl_2$ produces 1 mol of Ca^{2+}, which is 2 moles of positive charge. Thus, 1 mol of $CaCl_2$ contains 2 equivalents:

$$(1 \text{ mol CaCl}_2) \times 2 = 2 \text{ equivalents}$$

where the number 2 used in the multiplication came from the +2 charge on Ca^{2+}.

If two solutions were made such that one contained 1 mol of NaCl per liter and the other contained 1 mol of $CaCl_2$ per liter, the NaCl solution would contain 1 equivalent of solute per liter and would be 1 normal (1 N). The $CaCl_2$ solution would contain 2 equivalents of solute per liter and would be 2 normal (2 N).

Osmolarity Another expression of concentration frequently used in physiology is **osmolarity (Osm/L)**, which indicates the total *number* of solute particles in a liter of solution instead of the relative weights of the specific solutes. The osmolarity of a solution is the product of molarity (M) and *n*, where *n* is the number of moles of solute particles obtained when 1 mole of solute dissolves. Because nonelectrolytes such as glucose do not dissociate in solution, $n = 1$ and the osmolarity ($n \times M$) is equal to the molarity of the solution. For electrolyte solutions, the osmolarity exceeds the molarity by a factor equal to the number of ions produced on dissociation of each molecule in solution. For example, because a NaCl molecule dissociates into two ions, Na^+ and Cl^-, the osmolarity of a 1 M solution of NaCl is $2 \times 1 \text{ M} = 2 \text{ Osm/L}$.

Colloids and suspensions

In solutions, solute particles are ions or small molecules. By contrast, the particles in colloids and suspensions are much larger than ions or small molecules. In colloids and suspensions, these particles are known as **dispersed-phase particles** instead of *solutes.* When the dispersed-phase particles are no more than about 100 times the size of the largest solute particles found in a solution, the mixture is called a **colloid.** The dispersed-phase particles of colloids generally do not settle out. All dispersed-phase particles of colloids carry electrical charges of the same sign. Thus, they repel each other. The constant buffeting from these collisions keeps the particles from settling. The most abundant colloids in the body are small functional proteins that are dispersed in the body fluids. An example is the colloidal dispersion of the plasma proteins in the blood (see p. 357).

When dispersed-phase particles are larger than those in colloids, if the mixture is left undisturbed the particles settle out because of the force of gravity. Such mixtures are usually called **suspensions.** The major example of a suspension in the body is the mixture of blood cells suspended in the plasma (see p. 381). The constant movement of blood as it circulates through the blood vessels keeps the blood cells rather evenly dispersed within the plasma. However, if a blood sample is placed in a test tube and treated to prevent clotting, the heavier blood cells gradually settle to the bottom of the tube.

A.6 Inorganic and Organic Chemicals

Chemicals are commonly classified into two categories: inorganic and organic.

Distinction between inorganic and organic chemicals

The original criterion used for this classification was the origin of the chemicals. Those that came from living or once-living sources were *organic,* and those that came from other sources were *inorganic.* Today, the basis for classification is the element carbon. **Organic** chemicals are generally those that contain carbon. All others are classified as **inorganic.** A few carbon-containing chemicals are also classified as inorganic; the most common are pure carbon in the form of diamond and graphite, carbon dioxide (CO_2), carbon monoxide (CO), carbonates such as limestone ($CaCO_3$), and bicarbonates such as baking soda ($NaHCO_3$).

The unique ability of carbon atoms to bond to one another and form networks of carbon atoms results in an interesting fact. Even though organic chemicals all contain carbon, millions of these compounds have been identified. Some were isolated from natural plant or animal sources, and many have been synthesized in laboratories. Inorganic chemicals include all the other 108 elements and their compounds. The number of known inorganic chemicals made up of all these other elements is estimated to be about 250,000, compared to millions of organic compounds made up predominantly of carbon.

Monomers and polymers

Another result of carbon's ability to bond to itself is the large size of some organic molecules. Organic molecules range in size from methane (CH_4), a small, simple molecule with one carbon atom, to molecules such as DNA that contain as many as a million carbon atoms. Organic molecules that are essential for life are called **biological molecules,** or **biomolecules** for short. Some biomolecules are rather small organic compounds, including *simple sugars, fatty acids, amino acids,* and *nucleotides.* These small, single units, known as **monomers** (meaning "single unit"), are building blocks for the synthesis of larger biomolecules, including *complex carbohydrates, lipids, proteins,* and *nucleic acids,* respectively. These larger organic molecules are called **polymers** (meaning "many units"), reflecting that they are made by the bonding together of a number of smaller monomers. For example, starch is formed by linking many glucose molecules together. Very large organic polymers are often referred to as **macromolecules,** reflecting their large size (*macro* means "large"). Macromolecules include many naturally occurring molecules, such as DNA and structural proteins, as well as many molecules that are synthetically produced, such as synthetic textiles (for example, nylon) and plastics.

A.7 Acids, Bases, and Salts

Acids, bases, and salts may be inorganic or organic compounds.

Acids and bases

Acids and bases are chemical opposites, and salts are produced when acids and bases react with each other. In 1887, Swedish chemist Svante Arrhenius proposed a theory defining acids and bases. He said that an *acid* is any substance that dissociates, or breaks apart, when dissolved in water and in the process releases a hydrogen ion (H^+). Similarly, *bases* are substances that dissociate when dissolved in water and in the process release a hydroxyl ion (OH^-). Hydrogen chloride (HCl) and sodium hydroxide ($NaOH$) are examples of Arrhenius acids and bases; their dissociations in water are represented in Equations A-11 and A-12, respectively:

$$HCl \rightarrow H^+ + Cl^- \qquad \text{Eq. A-11}$$

$$NaOH \rightarrow Na^+ + OH^- \qquad \text{Eq. A-12}$$

Note that the hydrogen ion is a bare proton, the nucleus of a hydrogen atom. Also note that both HCl and $NaOH$ would behave as electrolytes.

Arrhenius did not know that free hydrogen ions cannot exist in water. They covalently bond to water molecules to form hydronium ions, as shown in Equation A-13:

$$H^+ + :\overset{..}{O}-H \rightarrow \left[H-\overset{..}{O}-H \right]^+ \qquad \text{Eq. A-13}$$
$$\qquad\quad | \qquad\qquad\qquad |$$
$$\qquad\quad H \qquad\qquad\qquad H$$

In 1923 Johannes Brønsted in Denmark and Thomas Lowry in England proposed an acid–base theory that took this behavior into account. They defined an **acid** as any hydrogen-containing substance that donates a proton (hydrogen ion) to another substance (an acid is a *proton donor*) and a **base** as any substance that accepts a proton (a base is a *proton acceptor*). According to these definitions, the acidic behavior of HCl given in Equation A-11 is rewritten as shown in Equation A-14:

$$HCl + H_2O \rightleftharpoons H_3O^+ + Cl^- \qquad \text{Eq. A-14}$$

Note that this reaction is reversible and the hydronium ion is represented as H_3O^+. In Equation A-14, HCl acts as an acid (hydrochloric acid) in the forward (left-to-right) reaction, whereas H_2O acts as a base. In the reverse reaction (right-to-left), H_3O^+ gives up a proton and thus is an acid, whereas Cl^- accepts the proton and so is a base. It is still a common practice to use equations such as Equation A-11 to simplify the representation of the dissociation of an acid, even though scientists recognize that equations like Equation A-14 are more correct.

When acids or bases are used as solutes in solutions, the concentrations can be expressed as normalities the same as for salts. An equivalent of acid is the amount that gives up 1 mol of H^+ in solution. Thus, 1 mol of HCl is also 1 equivalent, but 1 mol of H_2SO_4 (sulfuric acid) is 2 equivalents. Bases are described in a similar way, but an equivalent is the amount of base that gives 1 mol of OH^-.

Salts; neutralization reactions

At room temperature, **inorganic salts** are crystalline solids that contain the positive ion (cation) of an Arrhenius base such as NaOH and the negative ion (anion) of an acid such as HCl. Salts can be produced by mixing solutions of appropriate acids and bases, allowing a **neutralization reaction** to occur. In neutralization reactions, the acid and base react to form a salt and water. Most salts that form are water soluble and can be recovered by evaporating the water. Equation A-15 is a neutralization reaction:

$$HCl + NaOH \rightarrow NaCl + H_2O \qquad \text{Eq. A-15}$$

See Chapter 15 for a discussion of acid–base balance in the body.

A.8 Functional Groups of Organic Molecules

Organic molecules consist of carbon and one or more additional elements covalently bonded to one another in "Tinker Toy" fashion. The simplest organic molecules, **hydrocarbons** (such as methane and petroleum products), have only hydrogen atoms attached to carbon backbones of varying lengths. All biomolecules always have elements besides hydrogen added to the carbon backbone. The carbon backbone forms the stable portion of most biomolecules. Other atoms covalently bonded to the carbon backbone, either alone or in clusters, form functional groups. **Functional groups** are specific combinations of atoms that generally react in the same way, regardless of the number of carbon atoms in the molecule to which they are attached. For example, all *aldehydes* contain a functional group that includes one carbon atom, one oxygen atom, and one hydrogen atom covalently bonded in a specific way:

$$\begin{array}{c} O \\ \parallel \\ (-C-H) \end{array}$$

The carbon atom in an aldehyde group forms a single covalent bond with the hydrogen atom and a **double bond** (a bond in which two covalent bonds are formed between the same atoms, designated by a double line between the atoms) with the oxygen atom. The aldehyde group is attached to the rest of the molecule by a single covalent bond extending to the left of the carbon atom. Most aldehyde reactions are the same regardless of the size and nature of the rest of the molecule to which the aldehyde group is attached. Reactions of physiological importance often occur between two functional groups or between one functional group and a small molecule such as water.

A.9 Carbohydrates

Carbohydrates are organic compounds of tremendous biological and commercial importance. They are widely distributed in nature and include such familiar substances as starch, table sugar, and cellulose. Carbohydrates have five important functions in living organisms: They provide energy, serve as a stored form of chemical energy, provide dietary fiber, supply carbon atoms for the synthesis of cell components, and form part of the structural elements of cells.

Chemical composition of carbohydrates

Carbohydrates contain carbon, hydrogen, and oxygen. They acquired their name because most of them contain these three elements in an atomic ratio of one carbon to two hydrogens to one oxygen. This ratio suggests that the general formula is CH_2O and that the compounds are simply carbon hydrates ("watered" carbons), or carbohydrates. It is now known that they are not hydrates of carbon, but the name persists. All carbohydrates have a large number of functional groups per molecule. The most common functional groups in carbohydrates are *alcohol, ketone,* and *aldehyde*

$$\begin{array}{ccc} & O & O \\ & \parallel & \parallel \\ (-OH), & (-C-), & (-C-H) \\ \text{Alcohol} & \text{Ketone} & \text{Aldehyde} \end{array}$$

or functional groups formed by reactions between pairs of these three.

Types of carbohydrates

The simplest carbohydrates are simple sugars, also called **monosaccharides.** As their name indicates, they consist of single, simple-sugar units called saccharides (*mono* means "one"). The molecular structure of glucose, an important monosaccharide, is shown in Figure A-7a. In solution, most glucose molecules assume the ring form shown in Figure A-7b. Other common monosaccharides are *fructose, galactose,* and *ribose* (see p. 567).

Disaccharides are sugars formed by linking two monosaccharide molecules together through a covalent bond (*di* means "two"). Some common examples of disaccharides are *sucrose*

(a) Chain form of glucose   (b) Ring form of glucose

Figure A-7 **Forms of glucose.**

(common table sugar) and *lactose* (milk sugar). Sucrose molecules are formed from one glucose and one fructose molecule. Lactose molecules each contain one glucose and one galactose unit.

Because of the many functional groups on carbohydrate molecules, large numbers of simple carbohydrate molecules are able to bond together and form long chains and branched networks. The resultant substances, **polysaccharides,** contain many saccharide units (*poly* means "many"). Three common polysaccharides made up entirely of glucose units are glycogen, starch, and cellulose:

- *Glycogen* is a storage carbohydrate found in animals. It is a highly branched polysaccharide that averages a branch every 8 to 12 glucose units. The structure of glycogen is represented in ❙Figure A-8, where each circle represents one glucose unit.

- *Starch,* a storage carbohydrate of plants, consists of two fractions, amylose and amylopectin. Amylose consists of long, essentially unbranched chains of glucose units. Amylopectin is a highly branched network of glucose units averaging 24 to 30 glucose units per branch. Thus, it is less highly branched than glycogen.

- *Cellulose,* a structural carbohydrate of plants, exists in the form of long, unbranched chains of glucose units. The bonding between the glucose units of cellulose is slightly different from the bonding between the glucose units of glycogen and starch. Humans have digestive enzymes that catalyze the breaking (hydrolysis) of the glucose-to-glucose bonds in starch but lack the necessary enzymes to hydrolyze cellulose glucose-to-glucose bonds. Thus, starch is a food for humans, but cellulose is not. Cellulose is the indigestible fiber in our diets.

A.10 Lipids

Lipids are a diverse group of organic molecules made up of substances with widely different compositions and molecular structures. Unlike carbohydrates, which are classified on the basis of their *molecular structure,* substances are classified as lipids on the basis of their *solubility.* **Lipids** are insoluble in water but soluble in nonpolar solvents such as alcohol.

Lipids are the waxy, greasy, or oily compounds found in plants and animals. These compounds repel water, a useful characteristic of the protective wax coatings found on some plants. Fats and oils are energy rich and have relatively low densities. These properties account for the use of fats and oils as stored energy in plants and animals. Still other lipids occur as structural components, especially in cellular membranes. The oily plasma membrane that surrounds each cell serves as a barrier that separates the intracellular contents from the surrounding extracellular fluid (see pp. 4, 22, and 56).

Simple lipids

Simple lipids contain just two types of components: fatty acids and alcohols. **Fatty acid molecules** consist of a hydrocarbon chain with a *carboxyl* functional group (—COOH) on the end. The hydrocarbon chain can be of variable length, but natural fatty acids always contain an even number of carbon atoms. The

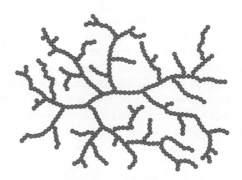

❙Figure A-8 **A simplified representation of glycogen.** Each circle represents a glucose molecule.

hydrocarbon chain can also contain one or more double bonds between carbon atoms. Fatty acids with no double bonds are called **saturated fatty acids,** whereas those with double bonds are called **unsaturated fatty acids.** The more double bonds present, the higher the degree of unsaturation. Saturated fatty acids predominate in dietary animal products (for example, meat, eggs, and dairy products), whereas unsaturated fatty acids are more prevalent in plant products (for example, grains, vegetables, and fruits). Consumption of a greater proportion of saturated than unsaturated fatty acids is linked with a higher incidence of cardiovascular disease (see p. 329).

The most common alcohol found in simple lipids is **glycerol** (glycerin), a three-carbon alcohol that has three alcohol functional groups (—OH).

Simple lipids called fats and oils are formed by a reaction between the carboxyl group of three fatty acids and the three alcohol groups of glycerol. The resulting lipid is an E-shaped molecule called a **triglyceride.** Such lipids are classified as fats or oils on the basis of their melting points: *fats* are solids at room temperature, whereas *oils* are liquids. Their melting points depend on the degree of unsaturation of the fatty acids of the triglyceride. The melting point goes down with increasing degree of unsaturation. Thus, oils contain more unsaturated fatty acids than fats do. Examples of the components of fats and oils and a typical triglyceride molecule are shown in ❙Figure A-9.

When triglycerides form, a molecule of water is released as each fatty acid reacts with glycerol. Adipose tissue in the body contains triglycerides. When the body uses adipose tissue as an energy source, the triglycerides react with water to release free fatty acids into the blood. The fatty acids can be used as an immediate energy source by many organs. In the liver, free fatty acids are converted into compounds called **ketone bodies** (see p. 689). Two of the ketone bodies are acids, and one is *acetone* (found in nail polish remover). Excess ketone bodies are produced during diabetes mellitus, a condition in which most cells resort to using fatty acids as an energy source because the cells are unable to take up adequate amounts of glucose in the face of inadequate insulin action (see p. 694).

Complex lipids

Complex lipids have more than two types of components. The different complex lipids usually contain three or more of the following components: glycerol, fatty acids, a phosphate group,

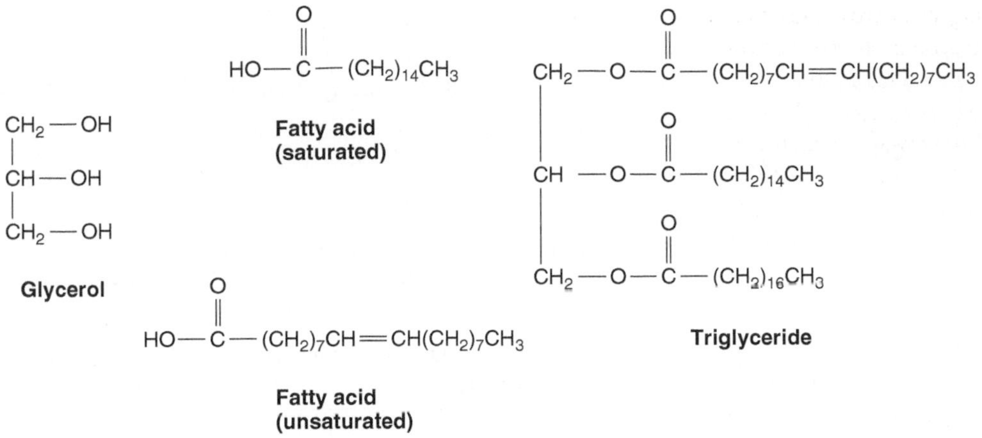

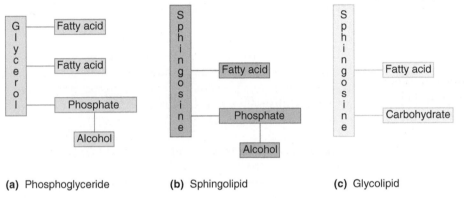

HO—C(=O)—(CH$_2$)$_{14}$CH$_3$

Fatty acid (saturated)

CH$_2$—OH
|
CH—OH
|
CH$_2$—OH

Glycerol

HO—C(=O)—(CH$_2$)$_7$CH=CH(CH$_2$)$_7$CH$_3$

Fatty acid (unsaturated)

CH$_2$—O—C(=O)—(CH$_2$)$_7$CH=CH(CH$_2$)$_7$CH$_3$
|
CH—O—C(=O)—(CH$_2$)$_{14}$CH$_3$
|
CH$_2$—O—C(=O)—(CH$_2$)$_{16}$CH$_3$

Triglyceride

❙Figure A-9 **Triglyceride components and structure.**

Glycerol — Fatty acid
Glycerol — Fatty acid
Glycerol — Phosphate — Alcohol

(a) Phosphoglyceride

Sphingosine — Fatty acid
Sphingosine — Phosphate — Alcohol

(b) Sphingolipid

Sphingosine — Fatty acid
Sphingosine — Carbohydrate

(c) Glycolipid

❙Figure A-10 **Examples of complex lipids.** In parts (b) and (c), sphingosine is an alcohol similar to glycerol.

an alcohol other than glycerol, and a carbohydrate. Those that contain phosphate are called **phospholipids.** ❙Figure A-10 contains representations of a few complex lipids; it emphasizes the components but does not give details of the molecular structures.

Steroids are lipids that have a unique structural feature consisting of a fused carbon ring system that contains three six-membered rings and a single five-membered ring (❙Figure A-11). Different steroids possess this characteristic ring structure but have different functional groups and carbon chains attached.

Cholesterol, a steroidal alcohol, is the most abundant steroid in the human body. It is a component of cell membranes and is used by the body to produce other important steroids that include bile salts, male and female sex hormones, and adrenocortical hormones. The structures of cholesterol and cortisol, an important adrenocortical hormone, are given in ❙Figure A-12.

A.11 Proteins

The name *protein* is derived from the Greek word *proteios,* which means "of first importance." It is certainly an appropriate term for these important biological compounds. Proteins

are indispensable components of all living things, where they play crucial roles in all biological processes. Proteins are the main structural component of cells, and enzymes, all of which are proteins, catalyze all chemical reactions in the body.

Chemical composition of proteins

Proteins are macromolecules made up of monomers called **amino acids.** Hundreds of different amino acids, both natural and synthetic, are known, but only 20 are commonly found in natural proteins. From this limited pool of 20 amino acids, cells build thousands of types of proteins, each with a distinct function, in much the same way that composers create diverse music from a relatively small number of notes. Different proteins are constructed by varying the types and numbers of amino acids used and by varying the order in which they are linked together. However, proteins are not built haphazardly, by randomly linking together amino acids. Every protein in the body is deliberately and precisely synthesized under the direction of the blueprint laid down in the person's genes. Thus, amino acids are assembled in a specific pattern to produce a given protein that can accomplish a particular structural or functional task in the body. (See pp. 23–24. More information about protein synthesis can be found at the book's Web site at **www.cengagebrain.com.**)

Peptide bonds

Each amino acid molecule has three important parts: an amino functional group (—NH$_2$), a carboxyl functional group (—COOH), and a characteristic side chain or R group. These components are shown in expanded form in ❙Figure A-13. Amino acids form long chains as a result of reactions between the amino group of one amino acid and the carboxyl group of another amino acid. This reaction is illustrated in Equation A-16:

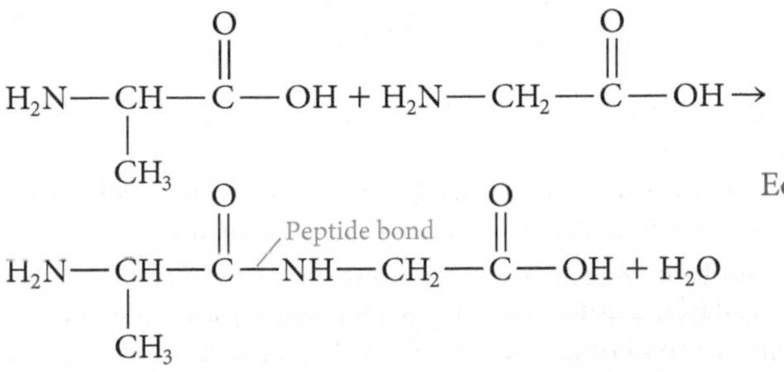

$$H_2N—CH(CH_3)—C(=O)—OH + H_2N—CH_2—C(=O)—OH \rightarrow$$

$$H_2N—CH(CH_3)—C(=O)—NH—CH_2—C(=O)—OH + H_2O$$

Peptide bond

Eq. A-16

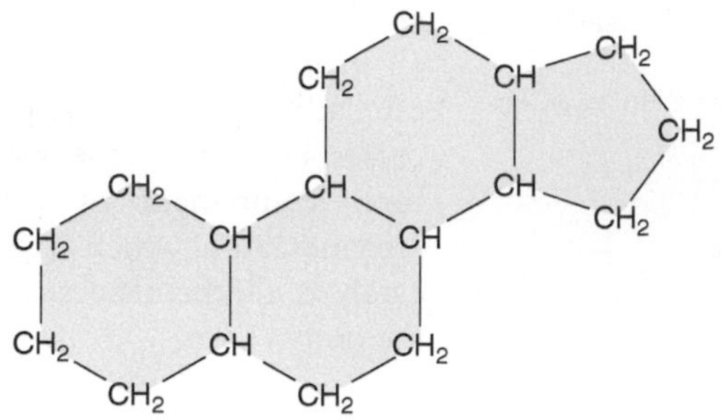

(a) Detailed steroid ring system

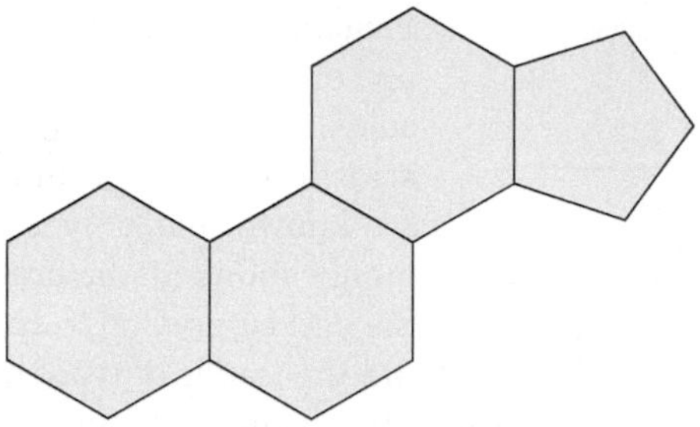

(b) Simplified steroid ring system

▮Figure A-11 **The steroid ring system.**

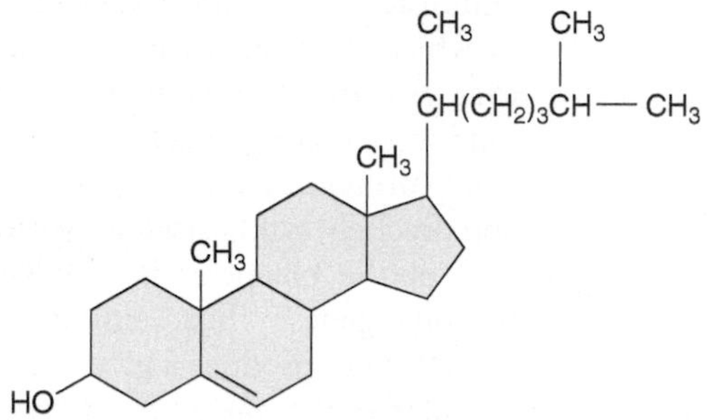

Cholesterol

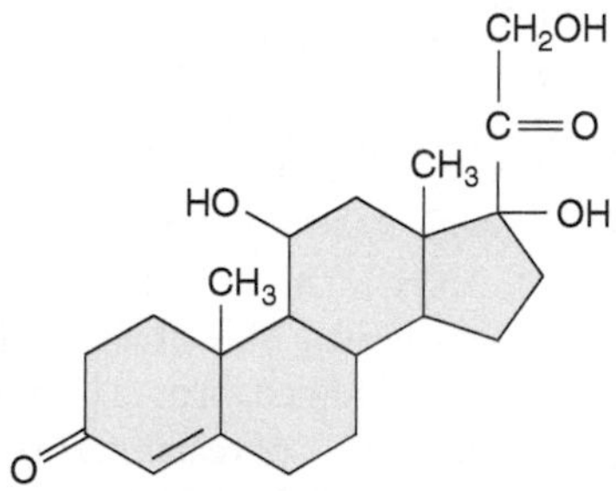

Cortisol

▮Figure A-12 **Examples of steroidal compounds.**

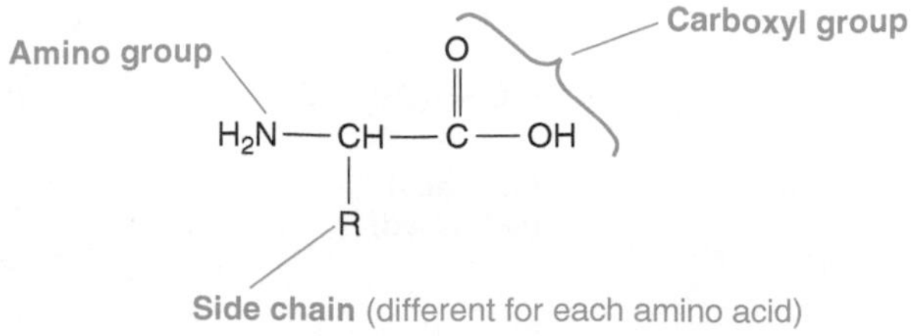

The covalent bond formed in the reaction is called a **peptide bond** (▮Figure A-14). Notice that after the two molecules react, the ends of the product still have an amino group and a carboxyl group that can react to extend the chain length.

On a molecular scale, proteins are immense molecules. Their size can be illustrated by comparing a glucose molecule to a mol-

▮Figure A-13 **The general structure of amino acids.**

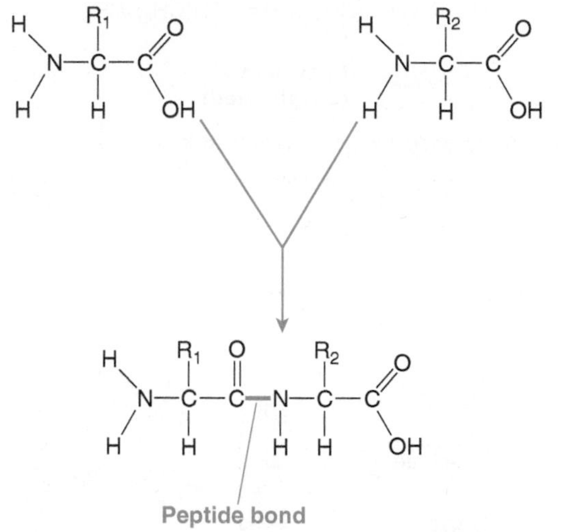

▮Figure A-14 **A peptide bond.** In forming a peptide bond, the carboxyl group of one amino acid reacts with the amino group of another amino acid.

Thr—Lys—Pro—Thr—Tyr—Phe—Phe—Gly—Arg— · · · · ·

▮Figure A-15 **A portion of the primary protein structure of human insulin.**

ecule of hemoglobin, a protein. Glucose has a molecular weight of 180 amu and a molecular formula of $C_6H_{12}O_6$. Hemoglobin, a relatively small protein, has a molecular weight of 65,000 amu and a molecular formula of $C_{2952}H_{4664}O_{832}N_{812}S_8Fe_4$.

Levels of protein structure

The many atoms in a protein are not arranged randomly. In fact, proteins have a high degree of structural organization that plays an important role in their behavior in the body.

Primary Structure The first level of protein structure is called the **primary structure.** It is simply the order in which amino acids are bonded together to form the protein chain. Amino acids are frequently represented by three-letter abbreviations, such as Gly for glycine and Arg for arginine. When this practice is followed, the primary structure of a protein can be represented as in ▮Figure A-15, which shows part of the primary structure of human insulin, or as in ▮Figure A-16a, which depicts a portion of the primary structure of hemoglobin.

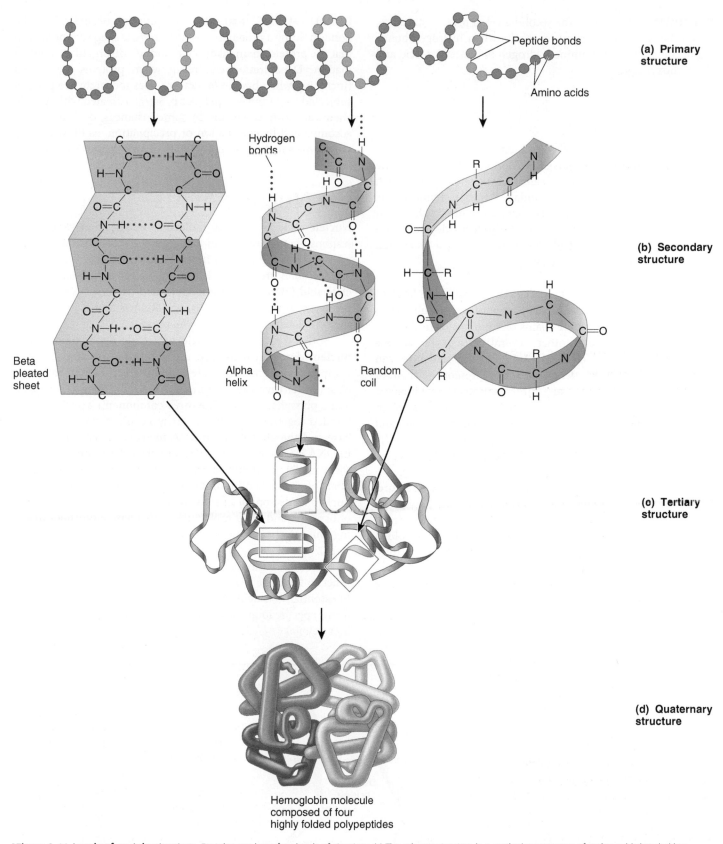

(a) Primary structure

Peptide bonds

Amino acids

(b) Secondary structure

Hydrogen bonds

Beta pleated sheet

Alpha helix

Random coil

(c) Tertiary structure

(d) Quaternary structure

Hemoglobin molecule composed of four highly folded polypeptides

❚Figure A-16 **Levels of protein structure.** Proteins can have four levels of structure. (a) The primary structure is a particular sequence of amino acids bonded in a chain. (b) At the secondary level, hydrogen bonding occurs between various amino acids within the chain, causing the chain to assume a particular shape. The most common secondary protein structure in the body is the alpha helix. (c) The tertiary structure is formed by the folding of the secondary structure into a functional three-dimensional configuration. (d) Many proteins have a fourth level of structure composed of several polypeptides, as exemplified by hemoglobin.

Secondary Structure The second level of protein structure, called the **secondary structure,** results when hydrogen bonding occurs between the amino hydrogen of one amino acid and the carboxyl oxygen

$$(-\overset{\overset{\displaystyle O}{\|}}{C}-)$$

of another amino acid in the same chain. As a result of this hydrogen bonding, the involved portion of the chain typically assumes a coiled, helical shape called an *alpha* (α) *helix,* which is by far the most common secondary structure found in the body (❙Figure A-16b). Other secondary structures such as *beta* (β) *pleated sheets* and *random coils* can also form, depending on the pattern of hydrogen bonding between amino acids located in different parts of the same chain.

Tertiary and Quaternary Structure The third level of structure in proteins is the **tertiary structure.** It results when functional groups of the side chains of amino acids in the protein chain react with each other. Several types of interactions are possible, as shown in ❙Figure A-17. Tertiary structures can be visualized by letting a length of wire represent the chain of amino acids in the primary structure of a protein. Next imagine that the wire is wound around a pencil to form a helix, which represents the secondary structure. The pencil is removed, and the helical structure is now folded back on itself or carefully wadded into a ball. Such folded or spherical structures represent the tertiary structure of a protein (see ❙Figure A-16c).

All functional proteins exist in at least a tertiary structure. Sometimes, several polypeptides interact with one another to form a fourth level of protein structure, the **quaternary structure.** For example, hemoglobin contains four highly folded polypeptide chains (the **globin** portion) (see ❙Figure A-16d). Four iron-containing *heme* groups, one tucked within the interior of each of the folded polypeptide subunits, complete the quaternary structure of hemoglobin (see ❙Figure 11-2, p. 383).

Hydrolysis and denaturation

In addition to serving as enzymes that catalyze the many essential chemical reactions of the body, proteins can undergo reactions themselves. Two of the most important are hydrolysis and denaturation.

Hydrolysis Notice that according to Equation A-16, the formation of peptide bonds releases water molecules. Under appropriate conditions, it is possible to reverse such reactions by adding water to the peptide bonds and breaking them. **Hydrolysis** ("breakdown by H_2O") reactions of this type convert large proteins into smaller fragments or even into individual amino acids. Hydrolysis is the means by which digestive enzymes break down ingested food into small units that can be absorbed from the digestive tract lumen into the blood.

Denaturation **Denaturation** of proteins occurs when the bonds holding a protein in its characteristic shape are broken so that the protein chain takes on a random, disorganized conformation. Denaturation can result when proteins are heated (including when body temperature rises too high; see p. 627), subjected to extremes of pH (see p. 549), or treated with specific chemicals such as alcohol. In some instances, denaturation is accompanied by coagulation or precipitation, as illustrated by the changes that occur in the white of an egg as it is fried.

A.12 Nucleic Acids

Nucleic acids are high-molecular-weight macromolecules responsible for storing and using genetic information in living cells and passing it on to future generations. These important biomolecules are classified into two categories: **deoxyribonucleic acid (DNA)** and **ribonucleic acid (RNA).** DNA is found primarily in the cell's nucleus, and RNA is found primarily in the cytoplasm that surrounds the nucleus.

Both types of nucleic acid are made up of units called **nucleotides,** which in turn are composed of three simpler components: Each nucleotide contains an organic nitrogenous base (either thymine, adenine, cytosine, or guanine), a sugar, and a phosphate group. The three components are chemically bonded together, with the sugar molecule lying between the base and the phosphate. In RNA the sugar is ribose, whereas in DNA it is deoxyribose. When nucleotides bond together to form nucleic acid chains, the bonding is between the phosphate of one nucleotide and the sugar of another. The resulting nucleic acids consist of polynucleotide strands of alternating phosphates and sugar molecules, with a base molecule extending out of the strand from each sugar molecule (❙Figure A-18).

DNA takes the form of two strands that coil around each other to form the well-known double helix. Some RNA occurs in essentially straight chains, whereas in other types the chain forms specific loops or helices (see the book's Web site for further details).

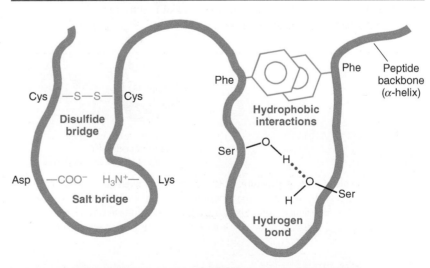

❙Figure A-17 Side chain interactions leading to the tertiary protein structure.

A.13 High-Energy Biomolecules

Not all nucleotides are used to construct nucleic acids. One important nucleotide—**adenosine triphosphate (ATP)**—is used as the body's primary energy carrier. Certain bonds in ATP temporarily store energy that is harnessed during the metabolism of foods and make it available to the parts of the cells where it is needed to do specific cellular work (see pp. 34–40). Let us see how ATP functions in this role. Structurally, ATP is a modified RNA (ribose-containing) nucleotide that has adenine as its base and two additional phosphates bonded in sequence to the original nucleotide phosphate. Thus, as its name implies, adenosine triphosphate has three phosphates attached in a string to *adenosine,* the composite of ribose and adenine (Figure A-19). Attaching these additional phosphates requires considerable energy input. The high-energy input used to create these **high-energy phosphate bonds** is "stored" in the bonds for later use. Most energy transfers in the body involve ATP's terminal phosphate bond. When energy is needed, the third phosphate is cleaved off by hydrolysis, yielding *adenosine diphosphate (ADP)* and an inorganic phosphate (P_i) and releasing energy in the process (Equation A-17):

$$ATP \rightarrow ADP + P_i + \text{energy for use by cell} \quad \text{Eq. A-17}$$

Why use ATP as an energy currency that cells can cash in by splitting the high-energy phosphate bonds as needed? Why not just directly use the energy released during the oxidation of nutrient molecules such as glucose? If all the chemical energy stored in glucose were to be released at once, most of the energy would be squandered because the cell could not capture much of the energy for immediate use. Instead, the energy trapped within the glucose bonds is gradually released and harnessed as cellular "bite-size pieces" in the form of the high-energy phosphate bonds of ATP.

Under the influence of an enzyme, ATP can be converted to a cyclic form of adenosine monophosphate, which contains only one phosphate group, the other two having been cleaved off. The resultant molecule, called **cyclic AMP** or **cAMP,** serves as an intracellular messenger, affecting the activities of a number of enzymes involved in important reactions in the body (see p. 122).

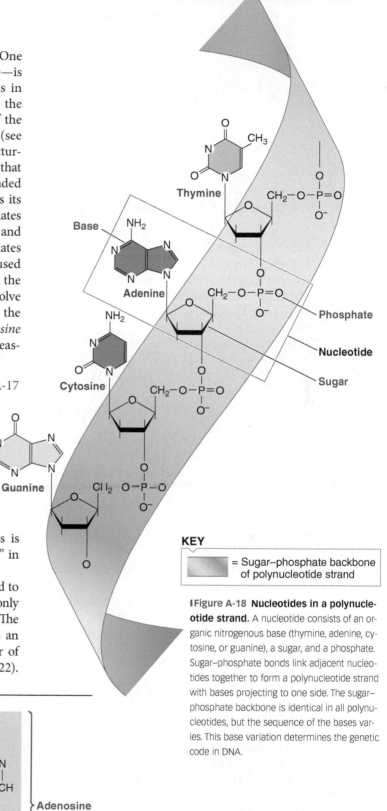

KEY

= Sugar–phosphate backbone of polynucleotide strand

Figure A-18 **Nucleotides in a polynucleotide strand.** A nucleotide consists of an organic nitrogenous base (thymine, adenine, cytosine, or guanine), a sugar, and a phosphate. Sugar–phosphate bonds link adjacent nucleotides together to form a polynucleotide strand with bases projecting to one side. The sugar–phosphate backbone is identical in all polynucleotides, but the sequence of the bases varies. This base variation determines the genetic code in DNA.

Figure A-19 **The structure of ATP.**

Text References to Exercise Physiology

A Closer Look at Exercise Physiology Boxed Features by Chapter

Exercise References by Topic

Answers

Chapter 1 Introduction to Physiology and Homeostasis

Check Your Understanding

1.1 (Questions on p. 2.)

1. *Physiology* is the study of body functions.

2. to increase the surface area across which nutrients can be absorbed from the digestive tract into the blood

1.2 (Questions on p. 7.)

1. *Chemical level:* Various atoms and molecules make up all body structures. *Cellular level:* Specific chemicals are organized into living cells, which are the basic units of both structure and function. *Tissue level:* Groups of cells of similar specialization are organized into tissues. *Organ level:* An organ is made up of several tissue types that act together as a unit to perform a particular function or functions. *Body system level:* A body system is a collection of related organs that interact to accomplish a common activity essential for survival of the whole body. *Organism level:* The body systems are structurally and functionally packaged together into the whole body, which is a single, multicellular organism capable of living independently in the surrounding external environment.

2. Every cell performs *basic cell functions* essential for its survival, including (1) obtaining food and O_2 from the environment surrounding the cell; (2) performing chemical reactions using food and O_2 to provide energy for the cell; (3) eliminating to the surrounding environment wastes produced by these reactions; (4) synthesizing proteins and other components needed by the cell; (5) largely controlling exchanges between the cell and surrounding environment; (6) moving materials within the cell or, in the case of some cells, moving the cell; (7) being sensitive and responsive to changes in the surrounding environment; and (8) reproducing (except for nerve cells and muscle cells).

3. *muscle tissue* (skeletal muscle attached to bones, cardiac muscle in the heart, and smooth muscle in the walls of hollow organs); *nervous tissue* (in the brain, spinal cord, nerves, and special sense organs); *epithelial tissue* (skin, linings of hollow organs, exocrine glands, and endocrine glands); and *connective tissue* (loose connective tissue, tendons, bone, and blood)

1.3 (Questions on p. 16.)

1. The *external environment* is the surrounding environment in which an organism lives. The *internal environment* is the fluid inside the body and outside the cells in which the cells live. The *intracellular fluid* is the fluid collectively within all body cells. The *extracellular fluid* is the fluid inside the body and outside the cells and constitutes the internal environment. The extracellular fluid is made up of the *plasma*, the fluid portion of the blood, and *interstitial fluid*, the fluid that surrounds and bathes the cells.

2. *Homeostasis* is the maintenance of a relatively stable internal environment.

3. See Figure 1-7, p 12.

1.4 (Questions on p. 18.)

1. *Intrinsic controls* are inherent compensatory responses that act locally in an organ; *extrinsic controls* are systemic controls initiated outside an organ by the regulatory systems (nervous or endocrine systems) to alter the organ's activity.

2. In *negative feedback,* the output of a control system drives a controlled variable in the opposite direction of an initial change, thus counteracting the change. In *positive feedback,* the output of a control system drives a controlled variable in the same direction as the initial change, thus enhancing the change.

3. See Figure 1-9a, p. 17.

Figure Focus

FIGURE 1-4 (P. 6): Milk-secreting glands are exocrine; they secrete through ducts to the outside. An endocrine gland secretes the hormone oxytocin into the blood.

FIGURE 1-8 (P. 14): The urinary, digestive, endocrine, and skeletal systems all contribute to maintaining the proper concentration of calcium in the blood. The circulatory system merely transports this electrolyte.

FIGURE 1-9 (P. 17): go up

Reviewing Terms and Facts

(Questions on p. 19.)

1. e **2.** b **3.** c **4.** T **5.** F **6.** T **7.** muscle tissue, nervous tissue, epithelial tissue, connective tissue **8.** secretion **9.** exocrine, endocrine, hormones **10.** intrinsic, extrinsic **11.** 1.d, 2.g, 3.a, 4.e, 5.b, 6.j, 7.h, 8.i, 9.c, 10.f

Applying Clinical Reasoning

(Questions on p. 19.)

Loss of fluids threatens the maintenance of proper plasma volume and blood pressure. Loss of acidic digestive juices threatens the maintenance of the proper pH in the internal fluid environment. The urinary system helps restore the proper plasma volume and pH by reducing the amount of water and acid eliminated in the urine. The respiratory system helps restore the pH by adjusting the rate of removal of acid-forming CO_2. Adjustments in the circulatory system help maintain blood pressure despite fluid loss. Increased thirst encourages increased fluid intake to help restore plasma volume. These compensatory changes in the urinary, respiratory, and circulatory systems, and the sensation of thirst, are all regulated by the two regulatory systems, the nervous and endocrine systems. Furthermore, the endocrine system makes internal adjustments to help maintain the concentration of nutrients in the internal environment even though no new nutrients are being absorbed from the digestive system.

Thinking at a Higher Level

(Questions on p. 20.)

1. The respiratory system eliminates internally produced CO_2 to the external environment. A decrease in CO_2 in the internal environment brings about a reduction in respiratory activity (that is, slower, shallower breathing) so that CO_2 produced within the body is allowed to accumulate instead of being blown off as rapidly as normal to the external environment. The extra CO_2 retained in the body increases the CO_2 levels in the internal environment to normal.

2. (b) (c) (b)

3. b

4. immune defense system

5. When a person is engaged in strenuous exercise, the temperature-regulating center in the brain brings about widening of the blood vessels of the skin. The resultant increased blood flow through the skin carries the extra heat generated by the contracting muscles to the body surface, where it can be lost to the surrounding environment.

Chapter 2 Cell Physiology

Check Your Understanding

2.1 (Questions on p. 22.)

1. See Table 2-1, p. 22.

2. These cells are all about the same size.

2.2 (Questions on p. 25.)

1. *DNA* provides the genetic code for protein synthesis and serves as a genetic blueprint during cell replication. The DNA code is transcribed into *messenger RNA (mRNA)*, which is translated into the specified protein by ribosomes that contain *ribonucleic RNA (rRNA)*. *Transfer RNA (tRNA)* delivers the appropriate amino acids to their designated site in the protein under construction.

2. The *genome* is all of the genetic information coded in a complete single set of DNA in a typical body cell. The *proteome* is the complete set of proteins coded for by the genome. *Epigenetics* refers to the environmentally induced modifications that influence gene activity without altering the gene's DNA code.

3. *Cytoplasm* is the portion of the cell interior not occupied by the nucleus. It consists of organelles, cytosol, and cytoskeleton. *Organelles* are distinct, highly organized structures that perform specialized functions within the cell. The *cytosol* is the gel-like portion of cytoplasm that surrounds the organelles. The *cytoskeleton* is a scaffolding of proteins within the cytoplasm that serves as the cell's "bone" and "muscle" by providing support and enabling movement.

2.3 (Questions on p. 28.)

1. The endoplasmic reticulum (ER) is one continuous, extensive organelle. The *rough ER* consists of stacks of relatively flattened interconnected sacs studded with ribosomes that synthesize proteins. The *smooth ER* is a meshwork of tiny interconnected tubules that in most cells serves as a central packaging and discharge site for products synthesized by the ER.

2. A *ribosome* is a nonmembranous organelle consisting of a large and a small subunit that are brought together to serve as the "workbench" for protein synthesis. When messenger RNA moves through a groove formed between the two subunits, the ribosome translates the mRNA into chains of amino acids in the ordered sequence dictated by the DNA code.

3. secreted out of the cell or used for construction of new membrane

4. Misfolded, damaged, or unneeded intracellular proteins are tagged with *ubiquitin*. *Proteasomes* break down ubiquinated proteins into recyclable building blocks.

2.4 (Questions on p. 30.)

1. The *Golgi complex* consists of a stack of flattened, slightly curved, membrane-enclosed sacs. It (1) processes raw materials synthesized by the ER into their final form and (2) sorts and directs the finished products to their final destinations.

2. *Secretion* is release to the cell's exterior by exocytosis of a specific product synthesized by the cell for a particular function.

3. The v-SNARE docking marker of a secretory vesicle can bind lock-and-key fashion only with the t-SNARE docking-marker acceptor on the targeted plasma membrane, thus ensuring that secretory vesicles can dock only with the plasma membrane to release their contents to the cell's exterior.

2.5 (Questions on p. 33.)

1. *Hydrolytic enzymes* catalyze hydrolysis, the breakdown of organic molecules by the addition of water at a bond site.

2. See Figure 2-9, p. 32.

3. *Autophagy* is selective self-digestion of dysfunctional organelles by lysosomes.

2.6 (Questions on p. 33.)

1. Peroxisomes detoxify various wastes and foreign compounds within the cell by means of *oxidative enzymes* that use oxygen to strip hydrogen from these organic molecules.

2. hydrogen peroxide (H_2O_2)

2.7 (Questions on p. 41.)

1. See Figure 2-10a, p. 34.

2. The three stages of *cellular respiration* are (1) glycolysis in the cytosol, (2) the citric acid cycle in the mitochondrial matrix, and (3) oxidative phosphorylation (consisting of the electron transport system and chemiosmosis) at the mitochondrial inner membrane.

3. In *anaerobic conditions* 2 molecules of ATP are produced (by glycolysis) and in *aerobic conditions* 32 molecules of ATP are produced (2 by glycolysis, 2 by the citric acid cycle, and 28 by oxidative phosphorylation) from one glucose molecule.

4. *Apoptosis* is intentional suicide of a cell that is no longer useful. *Necrosis* is uncontrolled, accidental death of useful, injured cells.

2.8 (Questions on p. 42.)

1. When closed, *vaults* are hollow, octagonal-shaped barrels. When open, each half looks like unfolded flowers with eight "petals" attached to a central ring.

2. Vaults are thought to serve as cellular "trucks" that carry cargo (either mRNA or the ribosomal subunits) from the nucleus to cytoplasmic sites of protein synthesis.

2.9 (Questions on p. 44.)

1. (1) intermediary metabolism, (2) protein synthesis by free ribosomes, and (3) storage of nutrients (as glycogen and fat) and secretory vesicles

2. *Intermediary metabolism* encompasses the intracellular chemical reactions involving the degradation, synthesis, and transformation of simple sugars, amino acids, and fatty acids.

2.10 (Questions on p. 51.)

1. (1) *microtubules* (maintain asymmetric cell shapes, serve as highways for transport of secretory vesicles, provide movement of cilia and flagella, form mitotic spindle); (2) *microfilaments* (play a key role in cellular contractile systems, serve as mechanical stiffeners); and (3) *intermediate filaments* (help resist mechanical stress)

2. A *motor protein* attaches to a particle to be transported, then walks along a microtubular "highway" by alternately attaching and releasing its "feet" as it cyclically swings the rear foot ahead of the front foot, stepping on one tubulin molecule after another.

3. A duplicated *centriole* moves from the centrosome to a position just under the plasma membrane, where microtubules grow outward from the centriole to form a cilium or flagellum. The centriole remains at the base of this motile appendage as the basal body of the structure.

4. In a process called *treadmilling* in amoeboid movement, actin filaments extend forward at the leading edge through the addition of actin molecules that have been removed from the rear of the filament and transferred to the front of the filament, an action that pushes a pseudopod (a fingerlike protrusion) forward. In treadmilling, the filament does not get any longer, it just advances forward.

Figure Focus

FIGURE 2-3 (P. 27): A *transport vesicle* contains a mixture of proteins that have been newly synthesized by the rough ER. A *secretory vesicle* contains a specific finished protein product that has been modified and sorted by the Golgi complex.

FIGURE 2-6 (P. 29): The surface membrane increases during exocytosis and decreases during endocytosis.

FIGURE 2-13 (P. 36): 3 NADH and 1 $FADH_2$ are generated for each "turn" of the citric acid cycle. Because glycolysis splits glucose into two pyruvate molecules, each of which enters one "turn" of the citric acid cycle, 6 NADH and 2 $FADH_2$ are produced by this cycle for every glucose molecule processed.

FIGURE 2-14 (P. 39): from air breathed in

FIGURE 2-15 (P. 39): 32 ATPs with O_2, 2 ATPs without O_2

FIGURE 2-21 (P. 47): Without its microtubule "highway" intact, the damaged neuron cannot transport secretory products from the cell body to the axon terminal for release to the ECF or debris from the axon terminal to the cell body for degradation by lysosomes. Failure to transport these materials to their sites of elimination from the cell leads to their accumulation within the cell, which can lead to disruption of cellular activities and ultimately death of the neuron.

Reviewing Terms and Facts

(Questions on p. 52.)

1. plasma membrane **2.** deoxyribonucleic acid (DNA), nucleus **3.** nucleus, organelles, cytoplasm **4.** organelles, cytosol, cytoskeleton **5.** endoplasmic reticulum, Golgi complex **6.** oxidative **7.** adenosine triphosphate (ATP) **8.** F **9.** F **10.** 1.b, 2.c, 3.c, 4.a, 5.b, 6.c, 7.a, 8.c, 9.c **11.** 1.b, 2.a, 3.b, 4.b

Solving Quantitative Exercises

(Questions on p. 53.)

1. b

2.

$$24 \text{ moles } O_2/\text{day} \times 6 \text{ moles ATP/mole } O_2 = 144 \text{ moles ATP/day}$$
$$144 \text{ moles ATP/day} \times 507 \text{ g ATP/mole} = 73{,}000 \text{ g ATP/day}$$
$$1000 \text{ g}/2.2 \text{ lb} = 73{,}000 \text{ g}/x \text{ lb}$$
$$1000 \, x = 160{,}600$$
$$x = \text{approximately } 160 \text{ lb}$$

3. 144 mol/day (7300 cal/mol) = 1,051,200 cal/day (1051 kilocal/day)

4. About 2/3 of the water in the body is intracellular. Because a person's mass is about 60% water, for a 150-pound (68-kg) person,

$$68 \text{ kg}(0.6)(2/3) = 27.2 \text{ kg}$$

is the mass of water. Assume that 1 mL of body water weighs 1 g. Then the total volume in the person's cells is about 27.2 liters. The volume of an average cell is

$$\frac{4}{3}\pi(1 \times 10^{-3} \text{ cm})^3 \approx 4.2 \times 10^{-9} \text{ cm}^3 = 4.2 \times 10^{-9} \text{ mL}$$

So, the number of cells in a 68-kg person is about

$$27.2 \text{ liters} \left(\frac{1000 \text{ mL}}{1 \text{ L}}\right)\left(\frac{1 \text{ cell}}{4.2 \times 10^{-9} \text{ mL}}\right) = 6.476 \times 10^{12} \text{ cells}$$

5. $150 \text{ mg}\left(\dfrac{1 \text{ mL}}{0.015 \text{ mg}}\right) = 10{,}000 \text{ mL} \, (10 \text{ L})$

Applying Clinical Reasoning

(Question on p. 53.)

Some hereditary forms of male sterility involving nonmotile sperm have been traced to defects in the cytoskeletal components of the sperm's flagella. These same individuals usually also have long histories of recurrent respiratory tract disease because the same types of defects are present in their respiratory cilia, which are unable to clear mucus and inhaled particles from the respiratory system.

Thinking at a Higher Level

(Questions on p. 53.)

1. The *chief cells* have an extensive rough endoplasmic reticulum, with this organelle being responsible for synthesizing these cells' protein secretory product, namely, pepsinogen. Because the parietal cells do not secrete a protein product to the cells' exterior, they do not need an extensive rough endoplasmic reticulum.

2. With *cyanide poisoning*, the cellular activities that depend on ATP expenditure, such as synthesis of new chemical compounds, membrane transport, and mechanical work, could not continue. The resultant inability of the heart to pump blood and failure of the respiratory muscles to accomplish breathing would lead to imminent death.

3. catalase

4. ATP is required for muscle contraction. Muscles are able to store limited supplies of nutrient fuel for use in the generation of ATP. During *anaerobic exercise*, muscles generate ATP from these nutrient stores by means of glycolysis, which yields 2 molecules of ATP per glucose molecule processed. During *aerobic exercise*, muscles can generate ATP by means of oxidative phosphorylation, which yields 32 molecules of ATP per glucose molecule processed. Because glycolysis inefficiently generates ATP from nutrient fuels, it rapidly depletes the muscle's limited stores of fuel, and ATP can no longer be produced to sustain the muscle's contractile activity. Aerobic exercise, in contrast, can be sustained for prolonged periods. Not only does oxidative phosphorylation use far less nutrient fuel to generate ATP, but it can be supported by nutrients delivered to the muscle by means of the blood instead of relying on stored fuel in the muscle. Intense anaerobic exercise outpaces the ability to deliver supplies to the muscle by the blood, so the muscle must rely on stored fuel and inefficient glycolysis, thus limiting anaerobic exercise to brief periods of time before energy sources are depleted.

5. skin. The mutant keratin weakens the skin cells of patients with *epidermolysis bullosa* so that the skin blisters in response to even a light touch.

Chapter 3 The Plasma Membrane and Membrane Potential

Check Your Understanding

3.1 (Questions on p. 60.)

1. See Figure 3-2, p. 57.

2. Under an electron microscope, the plasma membrane has a "sandwich" appearance with two dark layers separated by a light middle layer. The two dark layers are the hydrophilic polar regions of the lipid and protein molecules that take up a stain, whereas the light middle layer is the poorly stained hydrophobic core made up of the nonpolar regions of these molecules.

3. (1) form channels, (2) serve as carriers, (3) serve as docking-marker acceptors, (4) function as membrane-bound enzymes, (5) serve as receptors, (6) serve as cell adhesion molecules (CAMs), and (7) are important in "self" recognition

3.2 (Questions on p. 63.)

1. the biological "glue" that holds neighboring cells together; consists of an intricate meshwork of proteins in a watery, gel-like substance (interstitial fluid)

2. (1) *desmosome* (adhering junction that spot-rivets two adjacent but nontouching cells, anchoring them together in tissues subject to considerable stretching); (2) *tight junction* (impermeable junction that joins the lateral edges of epithelial cells near their luminal borders, thus preventing movement of materials between the cells); and (3) *gap junction* (communicating junction made up of small connecting tunnels that permit movement of charge-carrying ions and small molecules between two adjacent cells)

3. See Figure 3-4, p. 61.

3.3 (Questions on p. 63.)

1. Lipid-soluble substances of any size can permeate the plasma membrane without assistance by dissolving in the lipid bilayer. Small water-soluble substances (ions) can pass through the membrane without assistance through open channels specific for them.

2. *Passive forces* do not require energy and *active forces* require energy to produce movement across the plasma membrane.

3.4 (Questions on p. 70.)

1. movement down a concentration gradient (including osmosis) and movement along an electrical gradient

2. *Osmotic pressure* is a "pulling" pressure; it is a measure of the tendency for osmotic flow of water into a solution resulting from its relative concentration of nonpenetrating solutes and water. *Hydrostatic (fluid) pressure* is a "pushing" pressure; it is the pressure exerted by a stationary fluid on an object.

3. See Figure 3-13, p. 69.

3.5 (Questions on p. 77.)

1. See Figure 3-15, p. 71.

2. In *facilitated diffusion,* the carrier undergoes spontaneous changes in shape as a result of thermal energy. In *primary active transport,* phosphorylation (binding of the phosphate group derived from the carrier splitting ATP) increases affinity of the carrier for its passenger ion; this binding causes the carrier to change its shape. In *secondary active transport,* the change in shape of a cotransport carrier that binds both Na^+ and the transported solute is driven by a Na^+ concentration gradient established by a primary active transport mechanism.

3. Symport and antiport are both secondary active transport mechanisms. In *symport,* the cotransported solute moves uphill in the same direction the driving ion moves downhill. In *antiport,* the cotransported solute moves uphill in the opposite direction the driving ion moves downhill.

3.6 (Questions on p. 84.)

1. a separation of opposite charges across the membrane, or a difference in the relative number of cations and anions in the ECF and ICF

2. Because the resting membrane is 25 to 30 times more permeable to K^+ than to Na^+, K^+ passes through more readily than Na^+. The substantially larger movement of K^+ out of the cell influences the resting membrane potential to a much greater extent than the smaller movement of Na^+ into the cell does. As a result, the resting potential (-70 mV) is closer to the equilibrium potential for K^+ (-90 mV) than to the equilibrium potential for Na^+ ($+60$ mV). The resting potential is less than the K^+ equilibrium potential because the limited entry of Na^+ neutralizes some of the potential that would be created by K^+ alone.

3. In a *steady state,* opposing passive and active forces exactly counterbalance each other. In a *dynamic equilibrium,* opposing passive forces exactly counterbalance each other. In both cases, no net change takes place, but energy is used to maintain this constancy in a steady state, but no energy is needed in a dynamic equilibrium.

Figure Focus

FIGURE 3-5 (P. 62): Tight junctions prevent specialized membrane proteins from migrating between the luminal and basolateral parts of the plasma membrane of epithelial cells.

FIGURE 3-13 (P. 69): As the penetrating solutes enter the cell, ICF osmolarity increases and the osmolarity of the solution surrounding the cell decreases As water enters the cell down the resulting osmotic gradient, cell volume increases. Thus the solution is hypotonic; even though it has the same beginning osmolarity as the red blood cells, it has a lower concentration of nonpenetrating solutes than the cells do.

FIGURE 3-16 (P. 74): If insufficient ATP were available to run the Na^+–K^+ pump, the Na^+ concentration would fall in the ECF and rise in the ICF, whereas the K^+ concentration would rise in the ECF and fall in the ICF. Passive movement of these ions down their concentration gradients across the plasma membrane through leak channels would not be adequately counterbalanced by active pumping of these ions across the membrane against their concentration gradients.

FIGURE 3-18 (P. 76): When both Na^+ and glucose are present in the digestive tract lumen, they are cotransported via SGLT across the digestive tract wall into the blood. Water follows osmotically, helping rehydrate a child who has dehydrating diarrhea.

FIGURE 3-20 (P. 81): more negative. Because the ECF K^+ concentration is lower but its ICF concentration is the same, the concentration gradient for K^+ to exit the cell is greater than normal. Therefore the opposing electrical gradient at E_{K^+} must be greater than normal to exactly counterbalance the larger concentration gradient. That is, E_{K^+} must be more negative. The same conclusion can be reached by using the Nernst equation. Plugging a value less than the normal 5mM extracellular concentration of K^+ into the equation yields a value more negative than -90mV.

Reviewing Terms and Facts

(Questions on p. 85.)

1. T **2.** T **3.** T **4.** negative, positive **5.** 1.b, 2.a, 3.b, 4.a, 5.c, 6.b, 7.a, 8.b **6.** 1.a, 2.a, 3.b, 4.a, 5.b, 6.a, 7.b **7.** 1.c, 2.b, 3.a, 4.a, 5.c, 6.b, 7.c, 8.a, 9.b

Solving Quantitative Exercises

(Questions on p. 86.)

1. $E = \dfrac{61 \text{ mV}}{z} \log \dfrac{C_o}{C_i}$

a. $\dfrac{61 \text{ mV}}{2} \log \dfrac{1 \times 10^{-3}}{100 \times 10^{-9}} = +122 \text{ mV}$

b. $\dfrac{61 \text{ mV}}{-1} \log \dfrac{110 \times 10^{-3}}{10 \times 10^{-3}} = -63.5 \text{ mV}$

2. $I_x = G_x(V_m - E_x)$

$E_{Na^+} = 61 \text{ mV} \log \dfrac{145 \text{ mM}}{15 \text{ mM}} = 60.1 \text{ mV}$

a. $= 1 \text{ ns } (-70 \text{ mV} - 60.1 \text{ mV})$

$= 1 \text{ ns } (-130 \text{ mV})$

$= -130 \text{ pA (A = amperes)}$

b. entering

c. with concentration gradient; with electrical gradient

3.
$$V_m = 61 \log \frac{P_{K^+}[K^+]_o + P_{Na^+}[Na^+]_o}{P_{K^+}[K^+]_i + P_{Na^+}[Na^+]_i}$$

$$= 61 \log \frac{(1)(10) + (0.04)(150)}{(1)(150) + (0.04)(15)}$$

$$= 61 \log \frac{10 + 6}{150 + 0.6}$$

$$= 61 \log 0.1062$$

Because the log of 0.106 is -0.974,

$$V_m = 61\,(-0.974) = -59 \text{ mV}$$

Therefore, the resting membrane potential is less than normal (that is, slightly depolarized compared to normal).

Applying Clinical Reasoning

(Questions on p. 86.)

As Cl⁻ is secreted by the intestinal cells into the intestinal tract lumen, Na⁺ follows passively along the established electrical gradient. Water passively accompanies this salt (Na⁺ and Cl⁻) secretion by osmosis. The toxin produced by the cholera pathogen prevents the normal inactivation of the mechanism (cAMP pathway; see p. 123) that opens the Cl⁻ channels in the luminal membranes of the intestinal cells. Increased, ongoing secretion of Cl⁻ and the subsequent passively induced secretion of Na⁺ and water are responsible for the severe diarrhea that characterizes cholera.

Thinking at a Higher Level

(Questions on p. 86.)

1. d. active transport. Leveling off of the curve designates saturation of a carrier molecule, so carrier-mediated transport is involved. The graph indicates that active transport is being used instead of facilitated diffusion because the concentration of the substance in the intracellular fluid is greater than the concentration in the extracellular fluid at all points until after the transport maximum is reached. Thus, the substance is being moved *against* a concentration gradient, so active transport must be the method of transport being used.

2. c. As Na⁺ moves from side 1 to side 2 down its concentration gradient, Cl⁻ remains on side 1, unable to permeate the membrane. The resultant separation of charges produces a membrane potential, negative on side 1 because of unbalanced chloride ions and positive on side 2 because of unbalanced sodium ions. Sodium does not continue to move to side 2 until its concentration gradient is dissipated because of the development of an opposing electrical gradient.

3. Osmolarity refers to the concentration of all particles in a solution, both penetrating and nonpenetrating, yet only nonpenetrating solutes contribute to the tonicity of a solution. Therefore, a solution with a mixture of penetrating and nonpenetrating solutes may have an osmolarity of 300 mOsm/L, the same as in the ICF, but be hypotonic to the cells because the solution's concentration of nonpenetrating solutes is less than the concentration of nonpenetrating solutes inside the cells. The entire osmolarity of the ICF at 300 mOsm/L is attributable to nonpenetrating solutes.

4. more positive. Because the electrochemical gradient for Na⁺ is inward, the membrane potential would become more positive as a result of an increased influx of Na⁺ into the cell if the membrane were more permeable to Na⁺ than to K⁺. (Indeed, this is what happens during the rising phase of an action potential once threshold potential is reached; see Chapter 4.)

5. vesicular transport. The maternal antibodies in the infant's digestive tract lumen are taken up by the intestinal cells by endocytosis and are extruded on the opposite side of the cell into the interstitial fluid by exocytosis. The antibodies are picked up from the intestinal interstitial fluid by the blood supply to the region.

Chapter 4 Principles of Neural and Hormonal Communication

Check Your Understanding

4.1 (Questions on p. 89.)

1. nerve and muscle

2. See Figure 4-1, p. 88.

3. *Voltage-gated channels* open or close in response to changes in membrane potential. *Chemically gated channels* change conformation in response to binding of a specific extracellular chemical messenger to a surface membrane receptor. *Mechanically gated channels* respond to mechanical deformation such as stretching. *Thermally gated channels* respond to heat or cold.

4.2 (Questions on p. 90.)

1. The stronger a triggering event, the larger the graded potential. The longer the duration of a triggering event, the longer the duration of the graded potential.

2. An increase in the difference in potential increases current flow. An increase in resistance decreases current flow.

3. Spread of graded potentials is decremental because leakage of charge-carrying ions through open channels in the plasma membrane results in progressive loss of current with increasing distance from the initial site of the change in potential.

4.3 (Questions on p. 102.)

1. See Figure 4-6, p. 94.

2. (1) *closed but capable of opening* (at resting potential to threshold); (2) *open* (*activated*) (from threshold to the peak of an action potential—that is, throughout the rising phase); and (3) *closed and not capable of opening* (*inactivated*) (from the peak of an action potential until return to resting potential—that is, throughout the falling phase)

3. See Figure 4-7a, p. 95.

4. Saltatory conduction is faster than contiguous conduction because the action potential jumps from one node of Ranvier to the next in saltatory conduction, skipping over the myelinated sections of the axon, whereas the action potential must be regenerated within every section of an unmyelinated axon from beginning to end during contiguous conduction.

4.4 (Questions on p. 113.)

1. Because the presynaptic terminal releases the neurotransmitter and the subsynaptic membrane of the postsynaptic neuron has receptor-channels for the neurotransmitter, the synapse can operate only in the direction from presynaptic to postsynaptic neuron.

2. See Figure 4-15, p. 106.

3. *Temporal summation* is the summing of several EPSPs occurring very close together in time as a result of rapid, repetitive firing of a single presynaptic neuron. *Spatial summation* is the summing of EPSPs originating simultaneously from several different presynaptic neurons.

4. A *neurotransmitter* is a chemical messenger released from a presynaptic neuron that binds to and alters the permeability and thereby the potential of a postsynaptic neuron at a synapse. Neurotransmitters produce EPSPs or IPSPs. A *neuromodulator* is a chemical messenger that binds to nonsynaptic sites on a neuron and does not produce EPSPs or IPSPs but instead acts slowly to bring about long-term changes that subtly modulate (depress or enhance) synaptic activity. For example, neuromodulators may alter the synthesis and release of neurotransmitter from a presynaptic neuron or vary formation of receptors for the neurotransmitter in the postsynaptic neuron.

4.5 (Questions on p. 118.)

1. a cell influenced by a particular extracellular chemical messenger

2. *Paracrines* are secreted by local cells and exert short-range effects on neighboring target cells. *Neurotransmitters* are secreted by neurons and exert short-range effects on the target cells they innervate. *Hormones* are secreted into the blood by endocrine-gland cells and exert long-range effects on distant target cells. *Neurohormones* are released into the blood by neurons and exert long-range effects on distant target cells.

3. (1) by opening or closing chemically gated *receptor-channels,* (2) by activating *receptor-enzyme complexes,* or (3) by activating second-messenger pathways via *G-protein-coupled receptors.*

4. *Protein kinases* transfer a phosphate group from ATP to a particular intracellular protein (phosphorylate the protein), thereby activating it. *Protein phosphatases* remove phosphate groups from a designated protein (dephosphorylate the protein), thereby inactivating it.

4.6 (Questions on p. 120.)

1. *Cytokines* are *protein* signal molecules secreted by immune and other cells that largely act locally to regulate immune responses. *Eicosanoids* are *lipid* signal molecules derived from arachidonic acid in the plasma membrane of most cells that act locally to regulate diverse cell activities throughout the body.

2. The enzyme *phospholipase A₂* splits arachidonic acid from the plasma membrane. The enzyme *cyclooxygenase* leads to formation of prostaglandins and thromboxanes from arachidonic acid. The enzyme *lipooxygenase* leads to formation of leukotrienes from arachidonic acid.

3. NSAIDs provide pain relief by inhibiting cyclooxygenase, thus blocking conversion of arachidonic acid into pain-intensifying prostaglandins.

4.7 (Questions on p. 127.)

1. See Table 4-5, p. 120.

2. Cyclic AMP always brings about the cellular response by modifying a designated, preexisting protein within the target cell. The type of protein altered by cAMP depends on the unique specialization of a particular cell type. In this way, a common second messenger such as cAMP can induce widely differing responses in different cells because it modifies different proteins that lead to different cell events.

3. When the complex consisting of a lipophilic hormone bound with its intracellular receptor binds with DNA at a specific attachment site known as the *hormone response element,* this binding activates a specific gene that leads to synthesis of a new protein that carries out the cellular response.

4.8 (Questions on p. 129)

1. Neural specificity depends on the anatomic proximity of the neurotransmitter-secreting neuronal terminal to the target organ to which it is "wired." Specificity of the "wireless" endocrine system depends on specialization of target cell receptors for a specific circulating hormone.

2. The nervous system enables you to turn the pages of this book, and the endocrine system maintains your blood glucose levels.

Figure Focus

FIGURE 4-3 (P. 91): by K⁺ leaking out down its electrochemical gradient through K⁺ leak channels

FIGURE 4-4 (P. 91): False. The magnitude at the peak of an action potential is 30 mV (positive inside), whereas the magnitude at resting potential is 70 mV (negative inside).

FIGURE 4-6 (P. 94): During the *rising phase,* the Na⁺ channel is open and activated (both its activation and inactivation gates are open) and the K⁺ channel is closed (its activation gate is closed). During the falling phase, the Na⁺ channel is closed and inactivated (its activation gate is open but its inactivation gate is closed) and the K⁺ channel is open (its activation gate is open).

FIGURE 4-8 (P. 97): A graded potential produced in the dendrites and cell body in response to a triggering event spreads to the axon hillock. If the graded potential is of sufficient magnitude at the axon hillock to bring this region to threshold, an action potential is initiated here; that is, the axon hillock becomes an "active area." Local current flow between an active area and adjacent inactive area reduces the potential in the inactive area to threshold, triggering an action potential in the previously inactive area. Simultaneously, the old active area returns to resting potential. Local current flow between the new active area and the adjacent inactive area next to it triggers an action potential in this next area, and so on as the action potential propagates to the end of the axon.

FIGURE 4-17 (P. 111): As a result of presynaptic facilitation, the potential in postsynaptic cell C would depolarize to a greater extent (have a larger EPSP) in response to stimulation by excitatory terminal D than it would have if terminal D itself was not simultaneously stimulated by excitatory terminal E. Stimulation of terminal D by terminal E causes greater release of excitatory neurotransmitter from D.

FIGURE 4-25 (P. 123): No. Protein kinase activated by the cAMP pathway phosphorylates and thereby activates different designated proteins in different cell types.

FIGURE 4-28 (P. 127): A lipophilic hormone diffuses into the cell and binds to the steroid hormone receptor in the cytoplasm (or in the case of some hormones in the nucleus). Subsequent events that take place in the nucleus include binding of the hormone receptor complex with DNA's hormone response element, activation of the gene by this binding, and subsequent transcription of mRNA by the activated gene. The new mRNA exits the nucleus to enter the cytoplasm, where ribosomes "read" mRNA to synthesize the designated protein, which, once released from the ribosome and processed into its final folded form, brings about the desired response.

Reviewing Terms and Facts

(Questions on p. 130.)

1. T 2. F 3. F 4. F 5. T 6. F 7. nerve and muscle 8. refractory period 9. axon hillock 10. synapse 11. temporal summation 12. spatial summation 13. convergence, divergence 14. G protein 15. receptor-channel, receptor-enzyme, G-protein-coupled receptor 16. prostaglandins, thromboxanes, leukotrienes 17. 1.b, 2.a, 3.a, 4.b, 5.b, 6.a 18. 1.a, 2.b, 3.a, 4.b, 5.d, 6.b, 7.b, 8.b, 9.a, 10.b, 11.a, 12.c

Solving Quantitative Exercises

(Questions on p. 131.)

1. **a.** 0.6 m (1 sec/0.7 m) = 0.8571 sec
   b. 0.6 m (1 sec/120 m) = 0.005 sec
   c. unmyelinated: 0.8591 sec; myelinated: 0.007 sec
   d. unmyelinated: 0.8621 sec; myelinated: 0.01 sec

2. Total conduction time for the single axon is 1/60 sec. Let v m/sec be the unknown conduction velocity for the three neurons. Our equation for the total conduction time then is

$$\frac{1}{60} \text{ sec} = \left(\frac{1}{v} \times 1 \text{ m} \right) + 0.002 \text{ sec}$$

Solving for v, we obtain

$$v \text{ m/sec} = 1 \text{ m}/(1/60 \text{ sec} - 0.002 \text{ sec}) = 68.18 \text{ m/sec}$$

3. $25 \times 10^{-3} \text{V} \left[\dfrac{3.3\ \mu\text{S/cm}^2 (240\ \mu\text{S/cm}^2)}{(3.3 + 240)\ \mu\text{S/cm}^2} \right] \log \dfrac{240(145)}{3.3(4)}$

$= 25 \times 10^{-3}(11.1361)\text{V} \times \mu\text{S/cm}^2 = 0.2784\ \mu\text{A/cm}^2$

Applying Clinical Reasoning

(Question on p. 131.)

Initiation and propagation of action potentials does not occur in nerve fibers acted on by local anesthetic because blockage of Na$^+$ channels by the local anesthetic prevents the massive opening of voltage-gated Na$^+$ channels at threshold potential. As a result, pain impulses (action potentials in nerve fibers that carry pain signals) are not initiated and propagated to the brain and therefore do not reach the level of conscious awareness.

Thinking at a Higher Level

(Questions on p. 132.)

1. accelerate. During an action potential, Na$^+$ enters and K$^+$ leaves the cell. Repeated action potentials would eventually "run down" the Na$^+$ and K$^+$ concentration gradients were it not for the Na$^+$–K$^+$ pump returning the Na$^+$ that entered back to the outside and the K$^+$ that left back to the inside. Indeed, the rate of pump activity is accelerated by the increase in both ICF Na$^+$ and ECF K$^+$ concentrations that occurs as a result of action potential activity, thus hastening the restoration of the concentration gradients.

2. c. The action potentials would stop as they met in the middle. As the two action potentials moving toward each other both reached the middle of the axon, the two adjacent patches of membrane in the middle would be in a refractory period, so further propagation of either action potential would be impossible.

3. The hand could be pulled away from the hot stove by flexion of the elbow accomplished by summation of EPSPs at the cell bodies of the neurons controlling the biceps muscle, thus bringing these neurons to threshold. The subsequent action potentials generated in these neurons would stimulate contraction of the biceps. Simultaneous contraction of the triceps muscle, which would oppose the desired flexion of the elbow, could be prevented by generation of IPSPs at the cell bodies of the neurons controlling this muscle. These IPSPs would keep the triceps neurons from reaching threshold and firing so that the triceps would not be stimulated to contract.

The arm could deliberately be extended despite a painful finger prick by voluntarily generating EPSPs to override the reflex IPSPs at the neuronal cell bodies controlling the triceps while simultaneously generating IPSPs to override the reflex EPSPs at the neuronal cell bodies controlling the biceps.

4. An EPSP, being a graded potential, spreads decrementally from its site of initiation in the postsynaptic neuron. If presynaptic neuron A (near the axon hillock of the postsynaptic cell) and presynaptic neuron B (on the opposite side of the postsynaptic cell body) both initiate EPSPs of the same magnitude and frequency, the EPSPs from A are of greater strength when they reach the axon hillock than the EPSPs from B. An EPSP from B decreases more in magnitude as it travels farther before reaching the axon hillock, the region of lowest threshold and thus the site of action potential initiation. Temporal summation of the larger EPSPs from A may bring the axon hillock to threshold and initiate an action potential in the postsynaptic neuron, whereas temporal summation of the weaker EPSPs from B at the axon hillock may not be sufficient to bring this region to threshold. Thus, the proximity of a presynaptic neuron to the axon hillock can bias its influence on the postsynaptic cell.

5. Estrogen acting on estrogen-dependent breast cancer cells promotes survival of these cells. By interfering with the ability of estrogen

to bind with its receptors in the breast cancer cells, selective estrogen receptor modulators (SERMs) prevent estrogen from promoting survival of these cells. SERMs are taken for a number of years following removal of cancerous breast tissue with the goal of thwarting any cancer cells that may remain in the body.

Because SERMs interfere with estrogen (a lipophilic steroid hormone) binding with its receptors, which are located inside the target cell, one can infer that SERMs must also enter the target cell. Therefore, SERMs must be lipophilic, so they could be taken orally without risk of being destroyed by protein-digesting enzymes in the digestive tract.

Chapter 5 The Central Nervous System

Check Your Understanding

5.1 (Questions on p. 138.)

1. See Figure 5-1, p. 134.

2. An *afferent neuron* has a sensory receptor at its peripheral ending, a long peripheral axon (afferent fiber), a cell body devoid of presynaptic inputs located adjacent to the spinal cord, and a short central axon that terminates in the spinal cord. Afferent neurons relay signals from the periphery to the CNS. The cell body of an *efferent neuron* lies in the CNS and has many presynaptic inputs converging on it. Its long peripheral axon (efferent fiber) branches into axon terminals at the effector organ. Efferent neurons carry instructions from the CNS to effector organs. In contrast to afferent and efferent neurons, which lie primarily in the PNS, *interneurons* lie entirely in the CNS. The cell body of an interneuron receives converging input from afferent neurons and other interneurons and its diverging output terminates on efferent neurons or other interneurons. Interneurons are important in integrating afferent information and formulating an efferent response and for accomplishing all higher mental functions associated with the "mind."

3. astrocytes, oligodendrocytes, microglia, and ependymal cells

5.2 (Questions on p. 141.)

1. dura mater, arachnoid mater, and pia mater

2. CSF is a shock-absorbing fluid that surrounds and cushions the brain and spinal cord.

3. Tight junctions anatomically prevent transport between the cells that form the walls of brain capillaries, and highly selective membrane-bound carriers physiologically restrict transport through these cells. Together, these mechanisms constitute the blood–brain barrier.

5.3 (Questions on p. 144.)

1. act or process of "knowing," including both awareness and judgment

2. 1. brain stem
   2. cerebellum
   3. forebrain
     a. diencephalon
      (1) hypothalamus
      (2) thalamus
     b. cerebrum
      (1) basal nuclei
      (2) cerebral cortex

3. Without its convolutions, the human cortex would take up to three times the area it does and would not fit like a cover over the underlying structures.

5.4 (Questions on p. 153.)

1. See Figure 5-11, p. 148.

2. primary motor cortex, supplementary motor area, premotor cortex, and posterior parietal cortex

3. the ability of the brain to change or be functionally remodeled in response to the demands placed on it

4. *Broca's area* governs speaking ability. It commands facial and tongue muscles to speak words. *Wernicke's area* is concerned with language comprehension. It is important in understanding spoken and written messages and plans coherent content of spoken words.

5.5 (Questions on p. 155.)

1. (1) inhibit muscle tone, (2) maintain purposeful motor activity while suppressing useless patterns of movement, and (3) coordinate slow, sustained movements related to posture

2. All sensory input on its way to the higher cortex synapses in the thalamus, which screens out insignificant signals and routes the important signals to appropriate areas of the cortex. In this way the thalamus serves as a relay station for preliminary processing of sensory input.

3. hypothalamus

5.6 (Questions on p. 157.)

1. emotions, basic survival and sociosexual behavioral patterns, motivation, and learning

2. amygdala

3. norepinephrine, dopamine, and serotonin

5.7 (Questions on p. 163.)

1. the process of transferring and fixing short-term memory traces into long-term memory stores

2. *Short-term memory* involves transient modifications in the function of preexisting synapses, such as increased neurotransmitter release from a presynaptic neuron or increased responsiveness of a postsynaptic neuron to neurotransmitter. *Long-term memory* involves gene activation and protein synthesis that leads to relatively permanent structural or functional changes, such as formation of new synapses.

3. With *long-term potentiation*, in response to increased use at a given preexisting synapse, modifications take place in the postsynaptic neuron and/or presynaptic neuron that enhance the future ability of the presynaptic neuron to excite the postsynaptic neuron.

4. The hippocampus is important for *declarative memories*, the "what" memories of specific people, places, objects, facts, and events that often result after only one experience. The cerebellum plays an essential role in the "how to" *procedural memories* involving motor skills gained through repetitive training. The prefrontal association cortex is the major orchestrator of *working memory*, which temporarily holds currently relevant data—both new information and knowledge retrieved from memory stores—and manipulates and relates them to accomplish complex reasoning functions.

5.8 (Questions on p. 166.)

1. The *vestibulocerebellum* maintains balance and controls eye movements. The *spinocerebellum* enhances muscle tone and coordinates skilled, voluntary movements. The *cerebrocerebellum* plays a role in planning and initiating voluntary activity and stores procedural memories.

2. A *resting tremor* is an involuntary, useless, or unwanted movement, such as hands rhythmically shaking, and is often associated with Parkinson's disease. An *intention tremor* is characterized by oscillating to-and-fro movements of a limb as it approaches an intended destination, and is often associated with cerebellar disease.

5.9 (Questions on p. 172.)

1. (1) origin of most cranial nerves; (2) contains centers for control of circulation, respiration, and digestion; (3) helps regulate muscle reflexes involved in equilibrium and posture; (4) receives all incoming sensory synaptic input and is the origin of the reticular activating system; and (5) contains some of the centers that govern sleep.

2. refers to subjective awareness of the external world and self

3. (1) an *arousal system,* which is regulated by a group of neurons in the hypothalamus and involves the reticular activating system in the brain stem; (2) a *slow-wave sleep center* in the hypothalamus that contains sleep-on neurons that induce slow-wave sleep; and (3) a *paradoxical sleep center* in the brain stem, which houses REM sleep-on neurons that switch from slow-wave to REM sleep

5.10 (Questions on p. 178.)

1. See Figure 5-24, p. 174.

2. A *tract* is a bundle of nerve fibers (axons of long interneurons) with similar function that travel up (ascending tract) or down (descending tract) in the white matter of the spinal cord. A *ganglion* is a collection of neuronal cell bodies located outside the CNS. A *center,* or *nucleus,* is a functional collection of neuronal cell bodies located within the CNS. A *nerve* is a bundle of peripheral axons (both afferent and efferent fibers), enclosed by a connective-tissue covering and following the same pathway.

3. receptor, afferent pathway, integrating center, efferent pathway, effector

4. (1) as *spinal* or *cranial*, (2) as *innate* or *conditioned*, (3) as *somatic* or *autonomic*, and (4) as *monosynaptic* or *polysynaptic*

Figure Focus

FIGURE 5-1 (P. 134): When you are *taking a walk,* the afferent division of the PNS receives sensory stimuli, and the CNS issues commands to skeletal muscles acting via the somatic nervous system component of the efferent division of the PNS. When you are *digesting a meal,* the afferent division of the PNS receives sensory stimuli (such as the smell and taste of the food) and visceral stimuli (for example, the protein and fat content of the food and distention of the digestive tract). The CNS issues commands to the smooth muscle of the digestive tract and the exocrine glands that secrete digestive juices via the autonomic nervous system component of the efferent division of PNS.

FIGURE 5-9 (P. 147): (1) pyramidal cells, (2) stellate cells

FIGURE 5-11 (P. 148): (1) occipital lobe, (2) frontal lobe, (3) parietal lobe, (4) temporal lobe

FIGURE 5-13 (P. 151): (1) no, (2) yes

FIGURE 5-17 (P. 161): No. Even though binding of glutamate to an NMDA receptor-channel opens this channel's gate, Mg^{2+} still blocks the channel. Mg^{2+} must be driven out of the channel by depolarization before Ca^{2+} can enter through the NMDA receptor-channel to bring about long-term potentiation. The single EPSP produced by Na^+ entry when glutamate binds to and opens the AMPA receptor-channel does not depolarize the postsynaptic neuron sufficiently to drive Mg^{2+} out. Other EPSPs must occur concurrently via input from this or other presynaptic inputs to depolarize the postsynaptic membrane enough to drive Mg^{2+} out, thereby allowing long-term potentiation to occur.

FIGURE 5-19 (P. 167): Auditory impulses initiated by your alarm going off are carried by afferent fibers that synapse within the reticular formation in the brain stem. Ascending fibers that originate in the reticular formation and compose the reticular activating system subsequently carry signals upward to arouse and activate the cerebral cortex.

FIGURE 5-21 (P. 170): Most stage 4 deep, slow-wave sleep occurs during the first three hours of the sleep period. Slow-wave sleep occupies progressively less time and only passes through its lighter stages, whereas REM sleep occupies progressively more time, throughout the remainder of this cyclical sleep pattern.

Reviewing Terms and Facts

(Questions on p. 179.)

1. F **2.** F **3.** T **4.** F **5.** T **6.** F **7.** F **8.** F **9.** habituation **10.** consolidation **11.** dorsal, ventral **12.** 1.a, 2.c, 3.a and b, 4.b, 5.a, 6.c, 7.c **13.** 1.d, 2.c, 3.f, 4.e, 5.a, 6.b

Applying Clinical Reasoning

(Question on p. 180.)

The deficits following the stroke—numbness and partial paralysis on the upper right side of the body and inability to speak—are indicative of damage to the left somatosensory cortex and left primary motor cortex in the regions devoted to the upper part of the body plus Broca's area.

Thinking at a Higher Level

(Questions on p. 180.)

1. Only the left hemisphere has language ability. When sharing of information between the two hemispheres is prevented as a result of severance of the corpus callosum, the left hemisphere cannot verbally identify visual information presented only to the right hemisphere because the left hemisphere is unaware of the information. However, the information can be recognized by nonverbal means, of which the right hemisphere is capable.

2. c. A severe blow to the back of the head is most likely to traumatize the visual cortex in the occipital lobe.

3. Insulin excess drives too much glucose into insulin-dependent cells so that the blood glucose falls below normal and insufficient glucose is delivered to the non-insulin-dependent brain. Therefore, the brain, which depends on glucose as its energy source, does not receive adequate nourishment.

4. Salivation when seeing or smelling food, striking the appropriate letter on the keyboard when typing, and many of the actions involved in driving a car are conditioned reflexes. You undoubtedly will have many other examples.

5. Strokes occur when a portion of the brain is deprived of its vital O_2 and glucose supply because the cerebral blood vessel supplying the area either is blocked by a clot or has ruptured. Although a clot-dissolving drug could be helpful in restoring blood flow through a cerebral vessel blocked by a clot, such a drug would be detrimental in the case of a ruptured cerebral vessel sealed by a clot. Dissolution of a clot sealing a ruptured vessel would lead to renewed hemorrhage through the vessel and make the problem worse.

Chapter 6 The Peripheral Nervous System: Afferent Division; Special Senses

Check Your Understanding

6.1 (Questions on p. 189.)

1. A *stimulus* is a change detectable by the body. A *receptor potential* is a graded potential change in a receptor in response to a stimulus. A *labeled line* is a committed, incoming neural pathway carrying information regarding a particular sensory modality detected by a specialized receptor type at a specific site in the periphery and delivered to a defined area in the somatosensory cortex. *Perception* is the conscious interpretation of the external world as created by the brain from the sensory input it receives.

2. See Figure 6-4, p. 185.

3. The receptive field size for a sensory neuron on your tongue would be smaller than for a sensory neuron on your back because your tongue has greater discriminative ability than your back does.

6.2 (Questions on p. 192.)

1. Pain is a multidimensional experience because the sensation of pain is accompanied by motivated behavioral responses and emotional reactions.

2. *A-delta fibers* constitute a fast pain pathway that carries signals arising from mechanical and thermal nociceptors. *C fibers* constitute a slow pain pathway that carries impulses from polymodal nociceptors.

3. *Endogenous opioids* serve as endogenous analgesics by binding with opiate receptors at the synaptic knob of afferent pain fibers where they inhibit release of the pain neurotransmitter, substance P, thereby blocking further transmission of the pain signal.

6.3 (Questions on p. 211.)

1. See Figure 6-19a, p. 198.

2. When a photopigment absorbs light, retinal changes to the all-*trans* form, activating the photopigment. The activated photopigment activates the G protein transducin, which then activates the intracellular enzyme phosphodiesterase. In the dark the second messenger cGMP had been keeping chemically gated Na^+ channels open, resulting in a passive, inward, depolarizing Na^+ leak (dark current). Activated phosphodiesterase degrades cGMP, permitting these chemically gated Na^+ channels to close, stopping the depolarizing Na^+ leak and causing hyperpolarization of the photoreceptor (the receptor potential).

3. See Table 6-3, p. 204.

4. All of the axons of the ganglion cells in the retina of one eye are bundled together into the *optic nerve* that exits the eye. At the *optic chiasm*, located underneath the hypothalamus, the fibers in the two optic nerves carrying information from the medial halves of each retina cross to the opposite side, but those from the lateral halves remain on the original side. This partial crossover brings together information from the same half of the visual field as viewed by both eyes. These reorganized bundles of fibers that leave the optic chiasm are known as *optic tracts*, which synapse in the thalamus. The thalamus separates the visual information it receives and relays it via fiber bundles known as *optic radiations* to different zones in the visual cortex.

6.4 (Questions on p. 224.)

1. The *middle ear* amplifies the tympanic membrane vibrations and converts them into wavelike movements in the inner ear fluid at the same frequency as the original sound waves.

2. *Pitch discrimination* depends on which region of the basilar membrane naturally vibrates maximally with a given sound frequency. *Loudness discrimination* depends on the amplitude of the vibrations. *Timbre discrimination* depends on overtones of varying frequencies causing many points along the basilar membrane to vibrate simultaneously but less intensely than the fundamental tone.

3. The stereocilia of an auditory hair cell are organized into rows of graded heights, from short to tall. *Tip links* that connect the mechanically gated channel of a stereocilium to its next tallest member are stretched when the stereocilia all bend toward their tallest member, thereby pulling the channels open and permitting K^+ entry into the hair cell. When the tip links slacken as the stereocilia bend away from their tallest member, the channels close and K^+ entry into the hair cell ceases.

4. See Figure 6-40, p. 222.

6.5 (Questions on p. 230.)

1. *salty* (stimulated by chemical salts); *sour* (caused by free H^+ in acids); *sweet* (evoked by the particular configuration of glucose or by artificial sweeteners, which are organic molecules similar in structure to glucose but have no calories); *bitter* (elicited by alkaloids and poisonous substances, thus discouraging ingestion of potentially dangerous compounds); *umami* (triggered by amino acids, especially glutamate, as in meat)

2. *Gustducin* is the G protein activated in the second-messenger pathways involved in perception of bitter, sweet, and umami tastes.

3. An odorant is dissected into various components, and each olfactory receptor responds to a particular odor component that may be shared in common by multiple scents. Each glomerulus in the olfactory bulb receives signals only from receptors that detect a particular odor component. Odor discrimination is based on different patterns of glomeruli (the "smell files") activated by various scents.

Figure Focus

FIGURE 6-3 (P. 184): An acid-monitoring neuron's receptor potential, frequency of action potentials in its afferent fiber, and rate of neurotransmitter release at its axon terminals would all increase if body fluids became too acidic. These neuronal responses would all decrease if body fluids became too alkaline.

FIGURE 6-9 (P. 191): Oxycontin binds to the opiate receptors at the afferent pain-fiber terminal and suppresses the terminal's release of substance P, a pain neurotransmitter.

FIGURE 6-20 (P. 199): yes for rods, no for cones

FIGURE 6-25 (P. 202): A high concentration of cGMP in the photoreceptor in the dark keeps Na^+ channels open, leading to a depolarizing dark current. On absorption of light, the photoreceptor's adequate stimulus, a chain of events is initiated that leads to decreased concentration of cGMP, thereby closing the Na^+ channels and subsequently bringing about hyperpolarization. Hyperpolarization of the photoreceptor causes decreased release of neurotransmitter. The on-center ganglion cells' response to decreased neurotransmitter release is depolarization and subsequent initiation and propagation of action potentials.

FIGURE 6-35 (P. 216): The spiral shape of the cochlea steers the low-frequency sound waves toward the region of the basilar membrane that responds maximally to these bass sounds.

FIGURE 6-37 (P. 218): Uniquely, the endolymph that surrounds the stereocilia has a higher K^+ concentration than inside the hair cell. Therefore, when the stereocilia's K^+ channels open, K^+ enters the cell, leading to depolarization. In sharp contrast, a postsynaptic neuron at an inhibitory synapse is surrounded by ECF that has a lower K^+ concentration than inside the cell. As a result, when K^+ channels open, K^+ leaves the neuron, leading to hyperpolarization (an IPSP).

FIGURE 6-40 (P. 222): As you somersault forward, the endolymph moves backward, in the opposite direction of the forward roll.

FIGURE 6-44 (P. 229): Each separate odor component of a cologne's fragrance (such as floral, fruity, or woody) is detected by one of a thousand different olfactory receptor types, each of which sends this information to a specific glomerulus (or "smell file") in the olfactory bulb for further processing.

Reviewing Terms and Facts

(Questions on p. 231.)

1. transduction **2.** adequate stimulus **3.** F **4.** T **5.** T **6.** T **7.** F **8.** F **9.** T **10.** T **11.** F **12.** 1.f, 2.h, 3.l, 4.d, 5.i, 6.e, 7.b, 8.j, 9.a, 10.g, 11.c, 12.k **13.** 1.a, 2.b, 3.c, 4.c, 5.c, 6.a, 7.b, 8.b

Solving Quantitative Exercises

(Questions on p. 232.)

1. The slow pain pathway takes about (1.3 m) (1 sec/12 m) = 0.1083 sec. The fast pathway takes (1.3 m) (1 sec/30 m) = 0.0433 sec. The difference is 0.1083 sec − 0.0433 sec = 0.065 sec = 65 msec.

2. a. The amount of light entering the eye is proportional, approximately, to the area of the open pupil. Recall that the area of a circle is πr^2. Let r be the pupil radius and A_1 be the original pupil area. Halving the diameter also halves the radius, so the new pupil area is

$$\pi\left(\frac{1}{2}r\right)^2 = \frac{1}{4}\pi r^2 = \frac{1}{4}A_1$$

Therefore, the amount of light allowed into the eye is a quarter of what it was originally.

b. The area of a rectangle is hw, where h is the height and w the width. Halving either dimension halves the area and hence the amount of light allowed into the eye.

c. The cat's pupil can be considered more precise. Think about the coarse and fine adjustments on a microscope. Fine adjustment translates rotations of the knob into much smaller movement of the stage than coarse adjustment does.

3. a. Solve the following for I:

$$\beta = (10\ dB)\log_{10}(I/I_0)$$
$$I = I_0 10^{B/10}\ W/m^2$$

Therefore,

$$I_1 = 10^{-12}(10^{20/10}) = 10^{-12}(10^2) = 10^{-10}\ W/m^2$$
$$I_2 = 10^{-12}(10^{70/10}) = 10^{-12}(10^7) = 10^{-5}\ W/m^2$$
$$I_3 = 10^{-12}(10^{120/10}) = 10^{-12}(10^{12}) = 1\ W/m^2$$
$$I_4 = 10^{-12}(10^{170/10}) = 10^{-12}(10^{17}) = 10^5\ W/m^2$$

b. Because of the logarithm in the definition of decibel, the sound intensity increases exponentially with respect to sound level. This fact should be clear from the definition of dB solved for I. This result implies that the human ear performs well throughout an enormous range of sound intensities.

Applying Clinical Reasoning

(Questions on p. 232.)

Syncope most frequently occurs as a result of inadequate delivery of blood carrying sufficient oxygen and glucose supplies to the brain. Possible causes include circulatory disorders such as impaired pumping of the heart or low blood pressure; respiratory disorders resulting in poorly oxygenated blood; anemia, in which the oxygen-carrying capacity of the blood is reduced; or low blood glucose resulting from improper endocrine management of blood glucose levels. *Vertigo*, in contrast, typically results from a dysfunction of the vestibular apparatus, arising, for example, from viral infection or trauma, or abnormal neural processing of vestibular information, as, for example, with a brain tumor.

Thinking at a Higher Level

(Questions on p. 232.)

1. Pain is a conscious warning that tissue damage is occurring or about to occur. A patient unable to feel pain because of a nerve disorder does not consciously take measures to withdraw from painful stimuli and thus prevent more serious tissue damage.

2. Pupillary dilation (mydriasis) can be deliberately induced by ophthalmic instillation of either an adrenergic drug (such as epinephrine or related compound) or a cholinergic blocking drug (such as atropine or related compound). Adrenergic drugs produce mydriasis by causing contraction of the sympathetically supplied radial (dilator) muscle of the iris. Cholinergic blocking drugs cause pupillary dilation by blocking parasympathetic activity to the circular (constrictor) muscle of the iris so that action of the adrenergically controlled radial muscle of the iris is unopposed.

3. The defect would be in the left optic tract or optic radiation.

4. Rod deterioration associated with *retinitis pigmentosa* leads to tunnel vision, in which the person's field of vision is a constricted circular view as from within a tunnel looking out, because of loss of peripheral vision (where most rods are located) and retention of central vision (where most cones are located).

5. Fluid accumulation in the middle ear in accompaniment with middle ear infections impedes the normal movement of the tympanic membrane, ossicles, and oval window in response to sound. All these structures vibrate less vigorously in the presence of fluid, causing temporary hearing impairment. Chronic fluid accumulation in the middle ear is sometimes relieved by surgical implantation of drainage tubes in the eardrum. Hearing is restored to normal as the fluid drains to the exterior. Usually, the tube "falls out" as the eardrum heals and pushes out the foreign object.

Chapter 7 The Peripheral Nervous System: Efferent Division

Check Your Understanding

7.1 (Questions on p. 241.)

1. See Figure 7-2, p. 235.

2. The *sympathetic nervous system* dominates in emergency or stressful (fight-or-flight) situations and promotes responses that prepare the body for strenuous physical activity. The *parasympathetic nervous system* dominates in quiet, relaxed (rest-and-digest) situations and promotes "general housekeeping" activities such as digestion.

3. The *adrenal medulla* is a modified sympathetic ganglion that does not give rise to postganglionic fibers but instead, on stimulation by the preganglionic fiber, secretes the hormones epinephrine and norepinephrine into the blood.

7.2 (Questions on p. 244.)

1. The *autonomic nervous system* innervates cardiac muscle, smooth muscle, most exocrine glands, some endocrine glands, and adipose tissue. The *somatic nervous system* innervates skeletal muscles.

2. Motor neurons are considered the *final common pathway* because the only way any other parts of the nervous system can influence skeletal muscle activity is by acting, in common, on these motor neurons.

7.3 (Questions on p. 248.)

1. *Acetylcholine (ACh)* is the neuromuscular junction neurotransmitter. When it is released from the motor-neuron terminal button in response to an action potential, it binds with and opens nonspecific cation receptor-channels in the motor end plate of the muscle fiber. The resultant ion movement leads to an end-plate potential, which initiates a contraction-inducing action potential that is propagated throughout the muscle fiber. *Acetylcholinesterase (AChE)* is an enzyme in the motor end-plate membrane that inactivates ACh. By removing ACh, AChE permits the choice of allowing relaxation to take place (no more ACh released) or keeping the contraction going (more ACh released), depending on the body's momentary needs.

2. An EPP is larger than an EPSP. Because of its magnitude, an EPP is normally large enough to promote sufficient local current flow to bring the muscle membrane adjacent to the motor end plate to threshold, thus initiating an action potential. Thus, one-to-one transmission of action potentials occurs between a motor neuron and a muscle fiber at a neuromuscular junction. By contrast, one EPSP is not of sufficient magnitude to bring the postsynaptic neuron to threshold. Summation of EPSPs arising from multiple presynaptic action potentials is needed to initiate an action potential in the postsynaptic neuron.

3. *Myasthenia gravis* is an autoimmune condition caused by erroneous production of antibodies against the motor end-plate ACh receptor-channels. As a result, insufficient ACh receptor-channels are available for binding with contraction-inducing ACh released at a neuromuscular junction, leading to the extreme muscular weakness characteristic of the disease. The condition is treated with a drug such as neostigmine that temporarily inhibits AChE, thus prolonging activity of ACh at the neu-

romuscular junction so that an EPP of sufficient magnitude to initiate an action potential and subsequent contraction can develop through binding of ACh at the limited number of available ACh receptor-channels.

Figure Focus

FIGURE 7-2 (P. 235): (1) all autonomic preganglionic fibers and parasympathetic postganglionic fibers, (2) both parasympathetic and sympathetic, (3) only parasympathetic

FIGURE 7-4 (P. 244): An *axon terminal* is the small end branch of a motor neuron that terminates on a single muscle cell (muscle fiber). A *neuromuscular junction* consists of both the neuronal (axon terminal) and muscle cell (motor end plate) components of this junction. Within a neuromuscular junction the axon terminal divides into fine branches, each of which ends in an enlarged *terminal button*.

FIGURE 7-5 (P. 245): by (1) interfering with release of ACh from the synaptic vesicles, (2) blocking binding of ACh to the receptor-channels, or (3) inhibiting AChE action

Reviewing Terms and Facts

(Questions on p. 248.)

1. T **2.** F **3.** c **4.** c **5.** sympathetic, parasympathetic **6.** adrenal medulla **7.** motor end plate **8.** 1.a, 2.b, 3.a, 4.b, 5.a, 6.a, 7.b **9.** 1.b, 2.b, 3.a, 4.a, 5.b, 6.b, 7.a **10.** 1.c,f, 2.a, 3.d,f, 4.e, 5.e, 6.b,f

Solving Quantitative Exercises

(Question on p. 249.)

$$1.\ t = \frac{x^2}{2D} = \frac{(200\ \text{nm})^2}{2 \times 10^{-5}\ \text{cm}^2/\text{sec}}$$

$$= \frac{4 \times 10^{-14}\ \text{m}^2 \cdot \text{sec}}{2 \times 10^{-5}\ \text{cm}^2}\left(\frac{10^4\ \text{cm}^2}{\text{m}^2}\right) = 20\ \mu\text{sec}$$

Applying Clinical Reasoning

(Question on p. 249.)

Drugs that block β_1 receptors are useful for prolonged treatment of angina pectoris because they interfere with sympathetic stimulation of the heart during exercise or emotionally stressful situations. By preventing increased cardiac metabolism and thus an increased need for oxygen delivery to the cardiac muscle during these situations, beta blockers can reduce the frequency and severity of angina attacks.

Thinking at a Higher Level

(Questions on p. 250.)

1. By promoting arteriolar constriction, epinephrine administered in conjunction with local anesthetics reduces blood flow to the region and thus helps the anesthetic stay in the region instead of being carried away by the blood.

2. No. Atropine blocks the effect of acetylcholine at muscarinic receptors but does not affect nicotinic receptors. Nicotinic receptors are present on the motor end plates of skeletal muscle fibers.

3. The voluntarily controlled external urethral sphincter is composed of skeletal muscle and supplied by the somatic nervous system.

4. By interfering with normal acetylcholine activity at the neuromuscular junction, α-bungarotoxin leads to skeletal muscle paralysis, with death ultimately occurring as a result of an inability to contract the diaphragm and breathe.

5. If the motor neurons that control the respiratory muscles, especially the diaphragm, are destroyed by poliovirus or amyotrophic lateral sclerosis, the person is unable to breathe and dies (unless breathing is assisted by artificial means).

Chapter 8 Muscle Physiology

Check Your Understanding

8.1 (Questions on p. 256.)

1. A *muscle fiber* is composed of *myofibrils* that extend the entire length of the muscle fiber; in general, larger diameter muscle fibers have a greater number of myofibrils. A *whole muscle* is composed of muscle fibers that extend the entire length of the muscle; in general, larger diameter muscles have more muscle fibers.

2. See Figure 8-2c, d, and e, p. 253.

3. The regulatory protein, troponin, binds to both actin and tropomyosin. In the relaxed state, troponin assumes a conformation that causes tropomyosin to cover the myosin cross-bridge binding sites on the actin molecules.

8.2 (Questions on p. 262.)

1. See Figure 8-7, p. 257.

2. The dihydropyridine receptors serve as voltage-gated sensors that are activated by an action potential as it propagates along the T tubule. The activated dihydropyridine receptors trigger opening of Ca^{2+}-release channels (ryanodine receptors) in the adjacent lateral sacs of the sarcoplasmic reticulum, thereby permitting Ca^{2+} release from the lateral sacs. This released Ca^{2+} repositions the troponin–tropomyosin complex so that actin and the myosin cross bridges can interact to accomplish contraction.

3. See the cross-bridge cycle in Figure 8-12, p. 261. ATP binds to the myosin head and causes the head to detach from the actin molecule. During the cocking of the myosin head, ATP is hydrolyzed to ADP and P_i. When the myosin head binds to actin, P_i is released from the head during the power stroke. ADP is released from the myosin head after the power stroke.

4. During *rigor mortis*, the cytosolic Ca^{2+} concentration rises owing to leaky membranes, bringing about binding of previously energized myosin cross bridges to actin. However, no fresh ATP is available to bind with the myosin cross bridge and permit its detachment from actin. Therefore myosin and actin remain locked together, producing a "stiffness of death."

8.3 (Questions on p. 269.)

1. Muscle tension generated by sarcomere shortening stretches the series-elastic component (tendon). The stretched tendon transmits the generated tension to the bone to which it is attached.

2. Refer to Figure 8-17, p. 265. Changing the power arm from 5 cm to 10 cm would reduce the force needed to support the 5 kg load; a 17.5 kg muscle force would be required (at the new lever ratio of 1:3.5). The trade-off for this improved mechanical advantage would be slower movement of the hand (at 3.5 cm/unit of time) as the biceps shortens at 1 cm/unit of time.

3. Greater strength of contraction can be achieved through motor unit recruitment, twitch summation–tetanus, positioning the muscle at its optimal length, the absence of fatigue, and hypertrophy of a muscle (strength training).

4. In twitch summation, the level of cytosolic Ca^{2+} is increased by repeated release of Ca^{2+} from the lateral sacs. In addition, with repeated excitation of a skeletal muscle cell, insufficient time is available between action potentials for the sarcoplasmic reticulum to pump all of the released Ca^{2+} back into the lateral sacs. The sustained, elevated cytosolic Ca^{2+} leads to prolonged exposure of myosin cross-bridge binding sites for interaction with actin and thus greater power stroke opportunities.

8.4 (Questions on p. 275.)

1. ATP is expended (1) to cock the head of the myosin molecule to permit a power stroke, (2) to pump released Ca^{2+} back into the lateral sacs against a concentration gradient, and (3) to run the Na^+–K^+ pump to maintain the concentrations of these ions in the ECF and ICF following an action potential in the skeletal muscle cell. Also, (4) binding (but not splitting) of a fresh molecule of ATP lets the cross bridge detach from actin at the end of a power stroke.

2. The leg (drumstick) muscles of a turkey consist primarily of red muscle fibers, which have a large number of mitochondria, high levels of myoglobin, low glycogen content, and relatively few glycolytic enzymes. The turkey's leg muscles are built for endurance, not for speed or power. In contrast, the turkey's breast muscles, composed primarily of white muscle fibers, have relatively few mitochondria, low levels of myoglobin, high glycogen content, and an abundance of glycolytic enzymes. The breast muscles are built for speed and power, but lack endurance (for example, turkeys can fly only a very short distance).

3. ATP comes from all of these sources during the course of the race, but oxidative phosphorylation is responsible for generating the largest amount of ATP expended in this event.

8.5 (Questions on p. 286.)

1. (1) *Somatic reflexes* are automatic, purposeful responses brought about by skeletal muscles without conscious effort. They are integrated by the spinal cord or brain stem. (2) *Voluntary movements* are goal-directed movements initiated and terminated consciously and integrated by the cerebral cortex. (3) *Rhythmic movements* are stereotypical movements repeated in a general pattern, like walking. They are initiated and terminated consciously by the cerebral cortex, but their reflexlike execution is accomplished without conscious effort by lower CNS levels.

2. The following have direct input to motor neurons: (1) afferent neurons, (2) the primary motor cortex, and (3) the brain stem as part of the multineuronal motor system.

3. See Figure 8-25, p. 284, which depicts this monosynaptic reflex; components include muscle spindle receptors, afferent neuron, efferent neuron (alpha motor neuron), and quadriceps femoris muscle.

8.6 (Questions on p. 294.)

1. In both skeletal and smooth muscle, ATP-powered cross-bridge stroking causes the thin filaments to slide in relation to the stationary thick filaments. In skeletal muscle, the myosin molecules are arranged so that no heads are located in the bare H zone in the middle of the sarcomere, and all cross-bridge stroking pulls the surrounding thin filaments toward the center of the sarcomere. In smooth muscle, the heads of the myosin molecules are located throughout the entire length of the thick filaments (no bare zone), and the heads are arranged to pull half of the surrounding thin filaments in one direction and the other half in the opposite direction within the same region of the thick filament. In skeletal muscle, the thin filaments are made up of actin, troponin, and tropomyosin. In smooth muscle, the thin filaments are made up of only actin and tropomyosin; the filaments lack the regulatory protein, troponin. Smooth muscle has 10 to 15 thin filaments per thick filament; in skeletal muscle the ratio is 2 to 1. The thick filaments are considerably longer in smooth muscle than in skeletal muscle, thereby allowing smooth muscle to develop tension when stretched to 2½ times its resting muscle length.

2. *Multiunit smooth muscle* is neurogenic and consists of multiple discrete muscle fiber units that must be separately stimulated by nerves and function independently of one another. *Single-unit smooth muscle* is myogenic and organized into a sheet of muscle fibers interconnected via gap junctions to form a functional syncytium that becomes excited and contracts as a single unit.

3. In *skeletal muscle,* the sarcoplasmic reticulum is the sole source of contraction-inducing Ca^{2+}. Contraction of *cardiac muscle* is a result of Ca^{2+} influx from the ECF and intracellular Ca^{2+} release from the sarcoplasmic reticulum. The source of Ca^{2+} in *smooth muscle* is predominantly from the ECF; the sarcoplasmic reticulum is poorly developed in smooth muscle.

Figure Focus

FIGURE 8-2 (P. 253): A cross section through the H zone would have only thick filaments; a cross section through the I band would have only thin filaments; the cross section through the A band in part (c) has both thick and thin filaments.

FIGURE 8-7 (P. 257): The H zone disappears.

FIGURE 8-12 (P. 261): Each cross-bridge cycle uses one ATP molecule. The fresh ATP molecule that binds to the cross bridge to permit its detachment from actin is the same ATP that is split and energizes the cross bridge for the subsequent power stroke.

FIGURE 8-17 (P. 265): 42 kg. The power arm of the lever is 4 cm, and the load arm is 28 cm for a lever ratio of 1:7 (4 cm:28 cm). Thus, to lift a 6 kg backpack with one hand, the child must generate an upward applied force in the biceps muscle of 42 kg. (With a lever ratio of 1:7, the muscle must exert seven times the force of the load; 7×6 kg $= 42$ kg.)

FIGURE 8-19 (P. 267): A muscle fiber develops considerably more tension during tetanus than during a single twitch. Because of repetitive action potentials during tetanus, the cytosolic Ca^{2+} concentration remains elevated, thereby keeping the troponin–tropomyosin complex pulled aside so that more of the cross bridges can participate in the cycling process for a longer time.

FIGURE 8-20 (P. 269): 70%

FIGURE 8-32 (P. 291): (1) true, (2) false

Reviewing Terms and Facts

(Questions on p. 295.)

1. F **2.** F **3.** F **4.** T **5.** concentric, eccentric **6.** alpha, gamma **7.** denervation atrophy, disuse atrophy, age-related atrophy (sarcopenia) **8.** tone **9.** functional syncytium **10.** Latch phenomenon **11.** b **12.** 1.f, 2.d, 3.c, 4.e, 5.b, 6.g, 7.a **13.** d, e, f

Solving Quantitative Exercises

(Questions on p. 296.)

1. a. For the weekend athlete, the lever ratio is 70 cm/9 cm. So the velocity at the end of the arm is 2.6 m/sec (70/9) = 20.2 m/sec (about 45 mph).

b. For the professional ballplayer, the lever ratio is 90 cm/9 cm. So

$$10x = 85 \text{ mph}$$
$$x = 8.5 \text{ mi/hr}(1609 \text{ m/mi})(1 \text{ hr}/3600 \text{ sec})$$
$$= 3.8 \text{ m/sec}$$

2. The force–velocity curve is as follows:

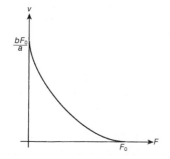

a. The shape of the curve indicates that it takes time to develop force and that the greater the force developed, the more time is needed.

b. The maximum velocity does not change when F_0 is increased, but the muscle is able to lift heavier loads or to generate more force. The maximum load does not increase when the cross-bridge cycling rate increases, but the muscle is able to lift lighter loads faster. If the muscle increases in size, b increases, and the entire curve shifts up with respect to the v axis.

Applying Clinical Reasoning

(Questions on p. 296.)

The muscles in the immobilized leg have undergone disuse atrophy. The physician or physical therapist can prescribe regular resistance-type exercises that specifically use the atrophied muscles to help restore them to their normal size.

Thinking at a Higher Level

(Questions on p. 296.)

1. By placing increased demands on the heart to sustain increased delivery of O_2 and nutrients to working skeletal muscles, regular aerobic exercise induces changes in cardiac muscle that enable it to use O_2 more efficiently, such as increasing the number of capillaries supplying blood to the heart muscle. Intense exercise of short duration, such as weight training, in contrast, does not induce cardiac efficiency. Because this type of exercise relies on anaerobic glycolysis for ATP formation, no demands are placed on the heart for increased delivery of blood to the working muscles.

2. The length of the thin filaments is represented by the distance between a Z line and the edge of the adjacent H zone. This distance remains the same in a relaxed and contracted myofibril, leading to the conclusion that the thin filaments do not change in length during muscle contraction.

3. Regular bouts of anaerobic, short-duration, high-intensity resistance training would be recommended for competitive downhill skiing. By promoting hypertrophy of the fast-glycolytic fibers, such exercise better adapts the muscles to activities that require intense strength for brief periods, such as a swift, powerful descent downhill. In contrast, regular aerobic exercise would be more beneficial for competitive cross-country skiers. Aerobic exercise induces metabolic changes within the oxidative fibers that enable the muscles to use O_2 more efficiently. These changes, which include an increase in mitochondria and capillaries within the oxidative fibers, adapt the muscles to better endure the prolonged activity of cross-country skiing without fatiguing.

4. (a) If the gamma motor neurons are activated and the alpha motor neurons are not activated, the intrafusal fibers contract and stretch the noncontractile portion of the muscle spindle. This causes the primary and secondary receptors to increase their rate of firing. **(b)** If the gamma motor neurons are not activated and the alpha motor neurons are activated, the receptors decrease their rate of firing or stop firing because the muscle spindle slackens as the whole muscle shortens.

5. Because the site of voluntary control to overcome the micturition reflex is at the external urethral sphincter and not the bladder, the external urethral sphincter must be skeletal muscle, which is innervated by the voluntarily controlled somatic nervous system, and the bladder must be smooth muscle, which is innervated by the involuntarily controlled autonomic nervous system. The only other type of involuntarily controlled muscle besides smooth muscle is cardiac muscle, which is found only in the heart. Therefore, the bladder must be smooth, not cardiac, muscle.

Chapter 9 Cardiac Physiology

Check Your Understanding

9.1 (Questions on p. 303.)

1. See Figure 9-1, p. 298.

2. The *right* and *left atrioventricular (AV valves)* let blood flow from the atria to the ventricles during ventricular filling but prevent backflow of blood from the ventricles into the atria during ventricular emptying. The *aortic* and *pulmonary semilunar valves* let blood flow from the ventricles into the aorta and pulmonary arteries during ventricular emptying but prevent backflow of blood from these major arteries into the ventricles during ventricular filling.

3. Adjacent cardiac muscle cells are joined end to end at specialized structures called *intercalated discs,* within which are two types of membrane junctions: desmosomes that hold the cells together mechanically and gap junctions that permit spread of electrical current between the cells.

9.2 (Questions on p. 314.)

1. See Figure 9-7, p. 304 and Figure 9-10, p. 308.

2. *SA node* (70–80 action potentials/min), *AV node* (40–60 action potentials/min), and *Bundle of His and Purkinje fibers* (20–40 action potentials/min)

3. refers to excitation-induced entry of a small amount of Ca^{2+} into the cytosol from the ECF through voltage-gated surface membrane Ca^{2+} channels triggering the opening of Ca^{2+}-release channels in the sarcoplasmic reticulum, thereby inducing a much larger release of Ca^{2+} into the cytosol from these intracellular stores.

4. See Figure 9-14, p. 312. The electrical activity associated with atrial repolarization occurs simultaneously with ventricular depolarization and is masked by the QRS complex on a normal ECG.

9.3 (Questions on p. 318.)

1. *Systole* is the period of contraction and emptying and *diastole* is the period of relaxation and filling during the cardiac cycle.

2. (1) aortic pressure > atrial pressure > ventricular pressure, (2) aortic pressure > ventricular pressure > atrial pressure, (3) ventricular pressure > aortic pressure > atrial pressure, (4) aortic pressure > ventricular pressure > atrial pressure

3. *End-diastolic volume (EDV)* is the volume of blood in the ventricle at the end of diastole when filling is complete. *End-systolic volume (ESV)* is the volume of blood in the ventricle at the end of systole when emptying is complete. *Stroke volume (SV)* is the volume of blood ejected by each ventricle with each contraction. EDV − ESV = SV.

9.4 (Questions on p. 325.)

1. *Parasympathetic stimulation* decreases heart rate and has no effect on stroke volume. *Sympathetic stimulation* increases heart rate and increases stroke volume by increasing the contractile strength of the heart.

2. See Figure 9-21 on p. 322.

3. The extent of ventricular filling is referred to as the *preload* because it is the workload imposed on the heart before contraction begins. The arterial blood pressure is called the *afterload* because it is the workload imposed on the heart after the contraction has begun.

9.5 (Questions on p. 331.)

1. The heart receives most of its blood supply during diastole because blood flow through the coronary vessels is substantially reduced during systole as a result of (1) the contracting myocardium compressing the coronary arteries and (2) the open aortic valve partially blocking the entrance to the coronary arteries.

2. When the heart is more active metabolically (that is, when it is pumping harder) and needs more O_2, the cardiac cells form and release more adenosine from ATP during this stepped-up metabolic activity.

Adenosine promotes vasodilation of the coronary vessels, allowing more O_2-rich blood to flow to the more active cardiac cells to match their increased O_2 demand.

3. An *atherosclerotic plaque* consists of a lipid-rich core covered by an abnormal overgrowth of smooth muscle cells, topped off by a collagen-rich connective tissue cap.

Figure Focus

FIGURE 9-1 (P. 298): right, systemic, pulmonary, left, pulmonary, systemic

FIGURE 9-7 (P. 304): The rate of slow depolarization would be decreased. A lower ECF K^+ concentration would increase the K^+ concentration gradient between a cardiac pacemaker cell and the ECF. The resultant increased movement of K^+ outward down this increased gradient would oppose the diminished outflow of K^+ caused by the slow closure of K^+ channels that contributes to development of the pacemaker potential.

FIGURE 9-10 (P. 308): Entry of Ca^{2+} through L-type Ca^{2+} channels in cardiac contractile cells is responsible for the plateau phase of the action potential, whereas its entry through these channels in autorhythmic cells is responsible for the rising phase of the action potential.

FIGURE 9-14 (P. 312): during the *PR segment*: atria completely depolarized, ventricles still at resting potential; during the *ST segment*: atria completely repolarized, ventricles completely depolarized; during the *TP segment*: atria and ventricles both completely depolarized

FIGURE 9-16 (P. 316): No. When diastolic time is reduced in half as a result of increased heart rate, filling is nearly the same as at normal heart rate because much of ventricular filling occurs early in diastole during the rapid filling phase.

FIGURE 9-17 (P. 317): The pressure difference between points 4 and 6 would be less in the right-sided loop than in the left-sided loop because the ventricular pressure that must be developed to open the pulmonary valve is less than the pressure needed to open the aortic valve. The volume difference between points 6 and 8 would be the same in both loops because both ventricles eject the same stroke volume.

FIGURE 9-24 (P. 324): A transplanted heart that does not have any innervation adjusts the cardiac output to meet the body's changing needs by means of both intrinsic control (the Frank–Starling mechanism) and extrinsic hormonal influences, such as the effect of epinephrine on the rate and strength of cardiac contraction.

Reviewing Terms and Facts

(Questions on p. 332.)

1. F **2.** F **3.** F **4.** F **5.** T **6.** F **7.** heart, blood vessels, blood **8.** endothelium, myocardium, epicardium **9.** intercalated discs, desmosomes, gap junctions **10.** 80% **11.** adenosine **12.** d **13.** e **14.** 1.e, 2.a, 3.d, 4.b, 5.f, 6.c **15.** AV, systole, semilunar, diastole **16.** 1.b, 2.c, 3.a, 4.b, 5.a, 6.b, 7.b, 8.c, 9.c, 10.a, 11.b, 12.c

Solving Quantitative Exercises

(Questions on p. 333.)

1. CO = HR × SV

40 liters/min = HR × 0.07 liter
HR = (40 liters/min)/(0.07 liter) = 571 beats/min

This rate is not physiologically possible.

2. ESV = EDV − SV
= 125 mL − 85 mL
= 40 mL

3. Ejection fraction = SV/EDV
   a. Ejection fraction = 70/135 = 52%
   b. Ejection fraction = 100/135 = 74%
   c. Ejection fraction = 140/175 = 80%

Applying Clinical Reasoning

(Questions on p. 333.)

The most likely diagnosis is atrial fibrillation. This condition is characterized by rapid, irregular, uncoordinated depolarizations of the atria. Many of these depolarizations reach the AV node at a time when it is not in its refractory period, thus bringing about frequent ventricular depolarizations and a rapid heartbeat. However, because impulses reach the AV node erratically, the ventricular rhythm and thus the heartbeat are also very irregular as well as being rapid.

Ventricular filling is only slightly reduced despite the fact that the fibrillating atria are unable to pump blood because most ventricular filling occurs during diastole before atrial contraction. Because of the erratic heartbeat, variable lengths of time are available between ventricular beats for ventricular filling. However, most ventricular filling occurs early in ventricular diastole after the AV valves first open, so even though the filling period may be shortened, the extent of filling may be near normal. Only when the ventricular filling period is very short is ventricular filling substantially reduced.

Cardiac output, which depends on stroke volume and heart rate, usually is not seriously impaired with atrial fibrillation. Because ventricular filling is only slightly reduced during most cardiac cycles, stroke volume, as determined by the Frank–Starling mechanism, is likewise only slightly reduced. Only when the ventricular filling period is very short and the cardiac muscle fibers are operating on the lower end of their length–tension curve is the resultant ventricular contraction weak. When the ventricular contraction becomes too weak, the ventricles eject a small or no stroke volume. During most cardiac cycles, however, the slight reduction in stroke volume is often offset by the increased heart rate so that cardiac output is usually near normal. Furthermore, if the mean arterial blood pressure falls because the cardiac output does decrease, increased sympathetic stimulation of the heart brought about by the baroreceptor reflex helps restore cardiac output to normal by shifting the Frank–Starling curve to the left.

On those cycles when ventricular contractions are too weak to eject enough blood to produce a palpable wrist pulse, if the heart rate is determined directly, either by the apex beat or via the ECG, and the pulse rate is taken concurrently at the wrist, the heart rate exceeds the pulse rate, producing a pulse deficit.

Thinking at a Higher Level

(Questions on p. 333.)

1. Because, at a given heart rate, the interval between a premature beat and the next normal beat is longer than the interval between two normal beats, the heart fills for a longer period of time following a premature beat before the next period of contraction and emptying begins. Because of the longer filling time, the end-diastolic volume is larger, and, according to the Frank–Starling law of the heart, the subsequent stroke volume is also correspondingly larger.

2. Trained athletes' hearts are stronger and can pump blood more efficiently so that the resting stroke volume is larger than in an untrained person. For example, if the resting stroke volume of a strong-hearted athlete is 100 mL, a resting heart rate of only 50 beats/minute produces a normal resting cardiac output of 5000 mL/minute. An untrained individual with a resting stroke volume of 70 mL, in contrast, must have a heart rate of about 70 beats/minute to produce a comparable resting cardiac output.

3. The direction of flow through a patent ductus arteriosus is the reverse of the flow that occurs through this vascular connection during fetal life. With a patent ductus arteriosus, some of the blood present in the aorta is shunted into the pulmonary artery because, after birth, the aortic pressure is greater than the pulmonary artery pressure. This abnormal blood flow produces a "machinery murmur," which lasts throughout the cardiac cycle but is more intense during systole and less intense during diastole. Thus, the murmur waxes and wanes with each beat of the heart, sounding somewhat like a washing machine as the agitator rotates back and forth. The murmur is present throughout the cardiac cycle because a pressure differential between the aorta and pulmonary artery is present during both systole and diastole. The murmur is more intense during systole because more blood is diverted through the patent ductus arteriosus as a result of the greater pressure differential between the aorta and pulmonary artery during ventricular systole than during ventricular diastole. Typically, the systolic aortic pressure is 120 mm Hg, and the systolic pulmonary arterial pressure is 24 mm Hg, for a pressure differential of 96 mm Hg. By contrast, the diastolic aortic pressure is normally 80 mm Hg, and the diastolic pulmonary arterial pressure is 8 mm Hg, for a pressure differential of 72 mm Hg.

4. In left bundle-branch block, the right ventricle becomes completely depolarized more rapidly than the left ventricle. As a result, the right ventricle contracts before the left ventricle, and the right AV valve is forced closed before closure of the left AV valve. Because the two AV valves do not close in unison, the first heart sound is "split"—that is, two distinct sounds in close succession can be detected as closure of the left valve lags behind closure of the right valve.

5. Lub-whistle-dup-swish-lub-whistle-dup-swish, and so on

Chapter 10 The Blood Vessels and Blood Pressure

Check Your Understanding

10.1 (Questions on p. 339.)

   1. digestive tract, kidneys, and skin
   2. $F = \Delta P/R$
   $R \propto 1/r^4$
   3. *arteries* (passageway from heart to organs, serve as pressure reservoir); *arterioles* (primary resistance vessels, determine distribution of cardiac output); *capillaries* (site of exchange, determine distribution of ECF between plasma and interstitial fluid); and *veins* (passageway to heart from organs, serve as blood reservoir); microcirculation = arterioles, capillaries, and venules

10.2 (Questions on p. 343.)

   1. an abundance of elastin fibers that allow the arteries to stretch to accommodate the extra volume of blood pumped into them during systole, then to recoil and drive the extra blood forward into the remaining vasculature during diastole
   2. See Figure 10-7a, p. 342.
   3. 105 mm Hg

10.3 (Questions on p. 350.)

   1. See Figure 10-9, p. 344.
   2. See Figure 10-11a, p. 347. *Active hyperemia* refers to the local increase in blood flow to an organ to meet increased local needs when the organ is more active metabolically. More active cells need more blood to bring in extra O_2 and nutrients and to remove the additional wastes they generate. Active hyperemia occurs when the endothelial cells that line the arterioles in the vicinity release nitric oxide in response to local chemical changes that occur in the more active organ (such as decreased O_2). Nitric oxide causes local arteriolar vasodilation

by relaxing the arteriolar smooth muscle, allowing more blood to flow into the organ.

    3. local arteriolar myogenic and chemical mechanisms that keep tissue blood flow fairly constant despite rather wide deviations in mean arterial driving pressure

    4. (determinants of mean arterial pressure):

Mean arterial pressure = cardiac output × total peripheral resistance

    (calculation of mean arterial pressure):

Mean arterial pressure = diastolic pressure + 1/3 pulse pressure

10.4 (Questions on p. 360.)

    1. *Diffusion* down concentration gradients is the primary means by which individual solutes cross the capillary walls. *Bulk flow* is responsible for the distribution of ECF between the plasma and the interstitial fluid.

    2. The forces that tend to move fluid out of the capillary are capillary blood pressure (main force) and interstitial fluid–colloid osmotic pressure (if plasma proteins are abnormally present in the interstitial fluid). The forces that tend to move fluid into the capillary are plasma-colloid osmotic pressure (main force) and interstitial fluid hydrostatic pressure (minor contributor). At the arteriolar end of the capillary, the outward forces exceed the inward forces, causing ultrafiltration, but at the venular end of the capillary, the inward forces exceed the outward forces, causing reabsorption. The reason for this shift in balance is a steady decline of the outward capillary blood pressure along the capillary's length while the inward forces remain constant.

    3. When interstitial fluid enters a lymphatic vessel, it is called *lymph*.

10.5 (Questions on p. 365.)

    1. Because veins have little elastin and little inherent myogenic tone, they are highly distensible and have little elastic recoil (that is, they stretch easily and do not tend to snap back when stretched). As a result, they easily distend to accommodate an extra volume of blood with only a small increase in venous pressure.

    2. volume of blood being returned to and pumped out by the heart

    3. Sympathetically induced venous vasoconstriction, skeletal muscle pump, respiratory pump, and cardiac-suction effect all enhance venous return. Venous valves prevent blood from moving backward in the veins.

10.6 (Questions on p. 377.)

    1. See Figure 10-33b, p. 368.

    2. *Secondary hypertension* is high blood pressure that occurs secondary to another known primary problem. *Primary hypertension* is high blood pressure for which the underlying cause is unknown. *Orthostatic hypotension* results from insufficient compensatory responses to gravitational shifts in blood when a person moves from lying down to standing up. *Circulatory shock* occurs when blood pressure falls so low that adequate blood flow to the tissues can no longer be maintained.

    3. The baroreceptors do not respond to bring blood pressure back to normal during hypertension because they adapt (are "reset") to maintain blood pressure at the new, higher level.

Figure Focus

FIGURE 10-3 (P. 338): Because $F = \Delta P/R$ and $R \propto 1/r^A$, F in vessel 3 is 2 times that in vessel 2 and 32 times that in vessel 1, and F in vessel 4 is 16 times that in vessel 2 and 256 times that in vessel 1.

FIGURE 10-7 (P. 342): (1) 125 mm Hg, (2) 77 mm Hg, (3) 48 mm Hg, (4) 93 mm Hg, (5) no, (6) yes, (7) no

FIGURE 10-11 (P. 347): *Triggering cause:* (a) ↑ local metabolic activity, (b) occlusion of local blood supply, (c) ↓ mean arterial (driving) pressure. *Local mediators:* (a) and (b) vasodilating paracrines, (c) myogenic relaxation. *Compensatory result:* (a) ↑ O_2 delivery and CO_2 re-

moval, (b) ↑ O_2 delivery and CO_2 removal following removal of occlusion, (c) blood flow to tissue normal despite ↓ driving pressure.

FIGURE 10-20 (P. 356): ultrafiltration = 13 mm Hg, reabsorption = 7 mm Hg. The volume of the interstitial fluid would increase compared to normal, resulting in edema.

FIGURE 10-26 (P. 363): (c), (a), (b), (d)

FIGURE 10-29 (P. 366): ↓ RAAS activity → ↓ salt and water conservation → ↓ blood volume → ↓ venous return → ↓ stroke volume → ↓ cardiac output → ↓ mean arterial pressure. Also, ↓ angiotensin II → ↓ extrinsic vasoconstrictor effect → ↑ arteriolar radius → ↓ total peripheral resistance → ↓ mean arterial pressure.

FIGURE 10-35 (P. 376): A compensatory ↑ cardiac output (CO) in response to hemorrhage is brought about by ↑ heart rate (HR), ↑ stroke volume (SV), and ↑ plasma volume. (1) HR increases as a result of (a) ↑ sympathetic and (b) ↓ parasympathetic activity to the heart. (2) SV increases in two ways: (a) ↑ sympathetic activity to the heart → ↑ contractility of the heart → ↑ SV, and (b) ↑ sympathetic activity to the veins → venous vasoconstriction → ↑ venous return → ↑ SV. (3) Plasma volume increases by three means: (a) ↑ sympathetic activity to the arterioles → arteriolar vasoconstriction → ↓ renal blood flow → ↓ urine output → plasma volume conserved, (b) fluid shift from interstitial fluid into plasma across capillaries → ↑ plasma volume, and (c) ↑ thirst → ↑ plasma volume.

Reviewing Terms and Facts

(Questions on p. 377.)

    1. T **2.** F **3.** T **4.** T **5.** F **6.** T **7.** resistance **8.** autoregulation **9.** flow rate, velocity of flow **10.** vasoconstriction, vasodilation **11.** a, c, d, e, f **12.** 1.a, 2.a, 3.b, 4.a, 5.b, 6.a

Solving Quantitative Exercises

(Questions on p. 378.)

    1. (120 mm Hg)/(30 L/min) = 4 PRU

    2. a. 90 mm Hg + (180 mm Hg − 90 mm Hg)/3 = 120 mm Hg

    b. Because the other forces acting across the capillary wall, such as plasma-colloid osmotic pressure, typically do not change with age, one would suspect fluid loss from the capillaries into the tissues as a result of the increase in capillary blood pressure.

    3. systemic: (95 mm Hg)/(19 PRU) = 95 mm Hg/(19 mm Hg/liters/min) = 5 L/min

    pulmonary: (20 mm Hg)/(4 PRU) = 5 L/min

    4. e

Applying Clinical Reasoning

(Question on p. 379.)

    The abnormally elevated levels of epinephrine found with a *pheochromocytoma* bring about secondary hypertension by (1) increasing the heart rate; (2) increasing cardiac contractility, which increases stroke volume; (3) causing venous vasoconstriction, which increases venous return and subsequently stroke volume by means of the Frank–Starling mechanism; and (4) causing arteriolar vasoconstriction, which increases total peripheral resistance. Increased heart rate and stroke volume both lead to increased cardiac output. Increased cardiac output and increased total peripheral resistance both lead to increased arterial blood pressure.

Thinking at a Higher Level

(Questions on p. 379.)

    1. An elastic support stocking increases external pressure on the remaining veins in the limb to produce a favorable pressure gradient that promotes venous return to the heart and minimizes swelling that would result from fluid retention in the extremity.

2. The classmate has apparently fainted because of insufficient blood flow to the brain as a result of pooling of blood in the lower extremities brought about by standing still for a prolonged time. When the person faints and assumes a horizontal position, the pooled blood quickly returns to the heart, improving cardiac output and blood flow to his brain. Trying to get the person up would be counterproductive, so the classmate trying to get him up should be advised to let him remain lying down until he recovers on his own.

3. The drug is apparently causing the arteriolar smooth muscle to relax by causing the release of a local vasoactive chemical mediator from the endothelial cells that induces relaxation of the underlying smooth muscle.

4. Decreased synthesis of plasma proteins as a result of insufficient dietary protein intake reduces the plasma-colloid osmotic pressure, leading to increased ultrafiltration and decreased reabsorption across the capillaries. Accumulation in the abdominal cavity of the resultant excess fluid that exits the capillaries is responsible for the markedly distended abdomen in *kwashiorkor*.

5. When sitting relatively immobile for a prolonged period in cramped seating during an airplane flight, blood tends to pool in the deep veins of the legs, leading to an increased risk of *deep venous thrombosis (DVT)* because blood clots tend to form in pooled blood that moves through the veins slowly. To reduce your risk of developing a DVT during a long flight, exercise your leg muscles while seated (for example, lifting your bended legs up and down, pushing the floor with the balls of your feet, stretching your legs out beneath the seat in front of you), getting up and walking in the aisle when allowed to do so, and staying hydrated (to maintain adequate circulating blood volume).

Chapter 11 The Blood

Check Your Understanding

11.1 (Questions on p. 383.)
1. See Figure 11-1, p. 381.
2. The *hematocrit* is the packed cell volume in a centrifuged blood sample; it essentially represents the percentage of erythrocytes in the total blood volume.
3. Plasma proteins (1) exert an osmotic effect important in the distribution of ECF between the vascular and interstitial compartments, (2) buffer pH changes, (3) transport many substances that are poorly soluble in plasma, (4) include clotting factors, (5) include inactive precursor molecules, and (6) include antibodies.

11.2 (Questions on p. 392.)
1. Four anatomic features of erythrocytes contribute to the efficiency with which they transport O_2. (1) Most importantly, they are stuffed full of O_2-carrying hemoglobin, with 98.5% of the O_2 in the blood being bound to hemoglobin. (2) Their unique biconcave shape provides a larger surface area for diffusion of O_2 across the membrane than a spherical cell of the same volume would. (3) Their thinness enables O_2 to diffuse rapidly from the exterior to the innermost regions of the cell. (4) Their membrane flexibility enables them to deform so that they can squeeze through capillaries less than half their diameter on their O_2 delivery route.
2. O_2, CO_2, H^+, CO, and NO
3. *Erythropoietin,* which the kidneys secrete into the blood in response to reduced O_2 delivery, stimulates increased erythrocyte production by the red bone marrow.

11.3 (Questions on p. 394.)
1. *Immunity* is the body's ability to resist or eliminate potentially harmful foreign material or abnormal cells. The *immune system* defends against disease-producing microorganisms, paves the way for tissue repair by removing worn-out cells and tissue debris, and defends against cancer cells that arise.
2. See Figure 11-8, p. 393.
3. *Neutrophils* are phagocytic specialists, act like suicide bombers by releasing neutrophil extracellular traps (NETs) during a unique type of programmed cell death, are the first defenders on the scene of bacterial invasion, and are important in inflammatory responses. *Eosinophils* are important in allergic reactions and attack parasitic worms. *Basophils* release histamine that is important in allergic reactions and heparin that speeds up removal of fat particles from the blood. *Monocytes* leave the blood and set up residence in tissues throughout the body, where they mature into large tissue phagocytes known as macrophages. *B lymphocytes* produce antibodies. *T lymphocytes* accomplish cell-mediated immunity by releasing chemicals that punch holes in virally invaded body cells and in cancer cells.

11.4 (Questions on p. 400.)
1. The plasma protein von Willebrand factor (vWF) adheres to exposed collagen at a vessel defect. Circulating platelets attach to binding sites on vWF. Collagen activates the bound platelets, which release ADP that causes passing-by platelets to become sticky and pile on top of the growing *platelet plug* at the defect site in a positive-feedback fashion.
2. See Figure 11-13, p. 398.
3. The same factor (factor XII, Hageman factor) that activates the fast cascade of reactions that leads to clot formation also activates a fast cascade that leads to plasmin activation. Activated plasmin, a fibrinolytic enzyme, is trapped in the clot, which it later dissolves by slowly breaking down the clot's fibrin meshwork.

Figure Focus

FIGURE 11-1 (P. 381): 63%, below normal

FIGURE 11-4 (P. 385): The rate of erythropoiesis would be decreased because of the failing kidneys' reduced ability to secrete erythropoietin.

FIGURE 11-7 (P. 389): (1) A transfusion reaction would occur. The type B donor red blood cells would be attacked by the anti-B antibodies in the recipient's plasma, leading to agglutination or hemolysis of the donor cells. (2) Nothing would happen to the type O donor red blood cells because type O cells bear no antigens and therefore would not be attacked by the anti-A antibodies in the type B recipient's plasma.

FIGURE 11-9 (P. 394): platelets, basophils, eosinophils, neutrophils, erythrocytes, and monocytes

FIGURE 11-13 (P. 398): (1) by a cascade of reactions initiated by activation of factor XII (Hageman factor) on exposure to collagen in a damaged vessel or to a foreign surface (intrinsic pathway) or (2) by tissue thromboplastin released from damaged tissue (extrinsic pathway)

Reviewing Terms and Facts

(Questions on p. 401.)
1. T **2.** F **3.** T **4.** T **5.** F **6.** lymphocytes **7.** liver **8.** d **9.** a **10.** 1.c, 2.f, 3.b, 4.a, 5.g, 6.d, 7.h, 8.e, 9.f **11.** 1.e, 2.c, 3.b, 4.d, 5.g, 6.f, 7.a, 8.h

Solving Quantitative Exercises

(Questions on p. 402.)
1. a. $(15 \text{ g})/(100 \text{ mL}) = (150 \text{ g/L})$

$$(150 \text{ g/L}) \times (1 \text{ mole}/66 \times 10^3 \text{ g}) = 2.27 \text{ mM}$$

b. $(2.27 \text{ mM}) \times (4 \text{ } O_2/\text{Hb}) = 9.09 \text{ mM}$

c. $(9.09 \times 10^{-3} \text{ moles } O_2/\text{L blood}) \times$
$(22.4 \text{ liters } O_2/1 \text{ mole } O_2) = 204 \text{ mL } O_2/\text{L blood}$

2. Normal blood contains 5×10^9 RBCs/mL.

Normal blood volume is 5 liters.

Thus, a normal person has $(5 \times 10^9$ RBCs/mL$) \times (5000$ mL$) = 25 \times 10^{12}$ RBCs.

The normal hematocrit (Hct) is 45%, whereas the anemic has a Hct of 30%. This represents a loss of 1/3 of the RBCs, that is, 8.3×10^{12} RBCs. If RBCs are produced at a rate of 3×10^6 RBCs/sec, then the time to reestablish the Hct is 8.3×10^{12} RBCs/$(3 \times 10^6$ RBCs/sec$) = 2.77 \times 10^6$ sec = 32 days.

Thus, it takes about a month to replace a hemorrhagic loss of RBCs of this magnitude.

3. $v = 1.5 \times \exp(2h)$; calculate v for $h = 0.4$ and $h = 0.7$.

When $h = 0.4$, $v = 1.5 \times \exp (0.8) = 3.3$.

When $h = 0.7$, $v = 1.5 \times \exp (1.4) = 6.1$.

$6.1/3.3 = 1.85$, that is, an 85% increase in viscosity. Because resistance is directly proportional to viscosity, the resistance also increases by 85%.

Applying Clinical Reasoning

(Questions on p. 402.)

1. Heather's firstborn Rh-positive child did not have hemolytic disease of the newborn because the fetal and maternal blood did not mix during gestation. Consequently, Heather did not produce any maternal antibodies against the fetus's Rh factor during gestation.

2. Because a small amount of the infant's blood likely entered the maternal circulation during the birthing process, Heather would produce antibodies against the Rh factor as she was first exposed to it at that time. During any subsequent pregnancies with Rh-positive fetuses, Heather's maternal antibodies against the Rh factor could cross the placental barrier and bring about destruction of fetal erythrocytes.

3. If any Rh factor that accidentally mixed with the maternal blood during the birthing process were immediately tied up by Rh immunoglobulin administered to the mother, the Rh factor would not be available to induce maternal antibody production. Thus, no anti-Rh antibodies would be present in the maternal blood to threaten the RBCs of an Rh-positive fetus in a subsequent pregnancy. (The exogenously administered Rh immunoglobulin, being a passive form of immunity, is short-lived. In contrast, the active immunity that would result if Heather were exposed to Rh factor would be long-lived.)

4. Rh immunoglobulin must be administered following the birth of every Rh-positive child Heather bears to sop up any Rh factor before it can induce antibody production. Once an immune attack against Rh factor is launched, subsequent treatment with Rh immunoglobulin does not reverse the situation. Thus, if Heather were not treated with Rh immunoglobulin following the birth of a first Rh-positive child, and a second Rh-positive child developed hemolytic disease of the newborn, administration of Rh immunoglobulin following the second birth would not prevent the condition in a third Rh-positive child. Nothing could be done to eliminate the long-lived maternal antibodies already present.

Thinking at a Higher Level

(Questions on p. 402.)

1. No. You cannot conclude that a person with a hematocrit of 62 definitely has polycythemia. With 62% of the whole-blood sample consisting of erythrocytes (normal being 45%), the number of erythrocytes compared to the plasma volume is definitely elevated. However, the person *may* have polycythemia, in which the number of erythrocytes is abnormally high, or may be dehydrated, in which case a normal number of erythrocytes is concentrated in a smaller-than-normal plasma volume.

2. If the genes that direct fetal hemoglobin F synthesis could be reactivated in a patient with sickle cell anemia, a portion of the abnormal hemoglobin S that causes the erythrocytes to warp into defective sickle-shaped cells would be replaced by "healthy" hemoglobin F, thus sparing a portion of the RBCs from premature rupture. Hemoglobin F would not completely replace hemoglobin S because the gene for synthesis of hemoglobin S would still be active.

3. A person with type A positive blood could safely receive a transfusion of type A positive, type A negative, type O positive, or type O negative blood (considering the ABO and Rh blood systems).

4. Most heart-attack deaths are attributable to the formation of abnormal clots that prevent normal blood flow. The sought-after chemicals in the "saliva" of bloodsucking creatures are agents that break up or prevent the formation of these abnormal clots.

Although genetically engineered tissue–plasminogen activator (tPA) is already being used as a clot-busting drug, this agent brings about degradation of fibrinogen as well as fibrin. Thus, even though the life-threatening clot in the coronary circulation is dissolved, the fibrinogen supplies in the blood are depleted for up to 24 hours until the liver synthesizes new fibrinogen. If the patient sustains a ruptured vessel in the interim, insufficient fibrinogen might be available to form a blood-staunching clot. For example, many patients treated with tPA suffer hemorrhagic strokes within 24 hours of treatment due to incomplete sealing of a ruptured cerebral vessel. Therefore, scientists are searching for better alternatives to combat abnormal clot formation by examining the naturally occurring chemicals produced by bloodsucking creatures that permit them to suck a victim's blood without the blood clotting.

5. When considering the symptoms of porphyria, one could imagine how tales of vampires—blood-craving, hairy, fanged, monstrous-looking creatures who roamed in the dark and were warded off by garlic—might easily have evolved from people's encounters with victims of this condition. This possibility is especially likely when considering how stories are embellished and distorted as word of mouth passes them along.

Chapter 12 Body Defenses

Check Your Understanding

12.1 (Questions on p. 407.)

1. *Immunity* is the body's ability to resist or eliminate potentially harmful foreign materials or abnormal cells.

2. *Innate immune responses* are inherent defense mechanisms that nonselectively defend against foreign or abnormal material of any type, even on initial exposure to it. *Adaptive immune responses* are selectively targeted against a particular foreign material to which the body has already been exposed and has had an opportunity to prepare for an attack aimed discriminatingly at the enemy.

3. *TLRs*, *RLRs*, and *NLRs* are a trio of pattern recognition receptors that recognize unique, telltale patterns characteristic of particular pathogens or other targets of the innate immune system. TLRs (toll-like receptors) function at the surface of phagocytes to recognize *PAMPs* (pathogen-associated molecular patterns) and *DAMPs* (damage-associated molecular patterns), in response to which the phagocyte engulfs the pathogen or damaged cell and secretes chemicals that enhance inflammation. Similarly, RLRs and NLRs are pattern recognition receptors located inside cells that recognize viral nucleic acid and intracellular PAMPs, respectively, leading to responses appropriate for the invader recognized.

12.2 (Questions on p. 415.)

1. (1) *Inflammation* is a nonspecific response to foreign invasion or tissue damage mediated largely by phagocytes and macrophages that destroy or incapacitate the invaders, remove debris, and prepare for subsequent healing and repair. (2) *Interferon* is released by virally invaded cells and nonspecifically, transiently interferes with replication of the same or unrelated viruses in other host cells. (3) *Natural killer cells* nonspecifically destroy virally invaded cells and cancer cells by releasing chemicals that directly lyse these cells on first exposure to them. (4) The *complement system* is a group of inactive plasma proteins that, when sequentially activated, destroy foreign cells by forming holes in their plasma membranes.

2. *Chemotaxins* are chemical mediators that attract phagocytic cells to a site of tissue damage or invasion by microorganisms. *Opsonins* are chemical mediators that enhance phagocytosis by linking a bacterium to a phagocyte.

3. When the complement cascade is activated by exposure to microorganism-specific carbohydrates or to antibodies produced against the microorganism, the final five complement components assemble into a large, doughnut-shaped protein complex, the *membrane attack complex,* which embeds itself in the surface membrane of the nearby microorganisms. This hole-punching lets water leak into the victim cell, causing it to swell and burst.

12.3 (Questions on p. 416.)

1. *antibody mediated immunity* accomplished by B lymphocytes and *cell-mediated* immunity accomplished by T lymphocytes

2. An *antigen* is a large, foreign, unique molecule that triggers a specific immune response against itself.

12.4 (Questions on p. 422.)

1. An *antibody* is Y-shaped. The *Fab* regions on the tip of each arm are variable among antibodies and determine with which specific antigen the antibody can bind lock-and-key fashion. The *Fc* tail region, which is constant for a given antibody class, binds with a particular antibody mediator, thereby determining what the antibody does once it binds with antigen.

2. neutralization of bacterial toxins, agglutination, activation of the complement system, enhancement of phagocytosis by acting as opsonins, and stimulation of natural killer cells

3. See Figure 12-12, p. 420.

12.5 (Questions on p. 435.)

1. *Cytotoxic,* or *killer, T cells* destroy virus-invaded host cells and cancer cells by secreting (1) perforin, which forms hole-punching complexes in the victim cell; or (2) granzymes, which trigger the victim cell to self-destruct through apoptosis. *Helper T cells* secrete cytokines that amplify the activities of other immune cells. *Regulatory T cells* suppress immune responses; they inhibit both innate and adaptive immune responses in check-and-balance fashion to minimize harmful immune pathology.

2. *Antigen-presenting cells* process and present antigen, complexed with MHC molecules (self-antigens), on their surface to T cells. T cells cannot interact with antigen without this "formal introduction."

3. clonal deletion, clonal anergy, active suppression by regulatory T cells, receptor editing, and immunological tolerance

4. In the process of *immune surveillance,* natural killer cells, cytotoxic T cells, macrophages, and the interferon they collectively secrete normally eradicate newly arisen cancer cells before they have a chance to multiply and spread.

12.6 (Questions on p. 438.)

1. autoimmune responses, immune complex diseases, and allergies

2. See Table 12-5, p. 439.

3. In contrast to *IgG antibodies,* which freely circulate and immediately amplify innate defense mechanisms on binding with their specific antigen, *IgE antibodies* specific for different antigens attach by their tail portions to mast cells and basophils in the absence of antigen. Binding of an appropriate antigen triggers the rupture of the cell's granules, which contain histamine and other inflammatory chemical mediators that spew forth into the surrounding tissue, where they cause the allergic response.

12.7 (Questions on p. 442.)

1. The skin consists of an outer epidermis and inner dermis. The *epidermis* has an inner layer of living cube-shaped cells and an outer keratinized layer of dead, flattened cells. The epidermis has no direct blood supply, instead getting its nourishment from the underlying *dermis,* a connective tissue layer that has an abundance of blood vessels. The skin is anchored to muscle or bone by the *hypodermis,* a layer of loose connective tissue that often contains an abundance of fat cells, in which case it is called *adipose tissue.*

2. (1) *Melanocytes* produce the pigment melanin, which is responsible for different skin colors. (2) *Keratinocytes* produce keratin, which gives rise to the outer protective keratinized layer of the epidermis. (3) *Langerhans cells* present antigen to helper T cells. (4) *Granstein cells* are immune-suppressive cells.

3. The *mucus escalator* refers to the cilia lining the respiratory airways constantly beating to move up and out the bacteria- and dust-laden mucus coating the airway interior.

Figure Focus

FIGURE 12-2 (P. 408): (1) activated resident macrophages; (2) by squeezing out amoeboid fashion between adjacent endothelial cells, a process called *diapedesis*

FIGURE 12-3 (P. 409): *Histamine* promotes (1) local arteriolar vasodilation, which brings about increased blood delivery to the injured tissue, and (2) increased local capillary permeability, which leads to local accumulation of fluid. Collectively, the extra blood and associated inflammatory mediators in the vicinity give rise to the undesirable gross manifestations of *inflammation*—swelling, redness, heat, and pain—as well as to unpleasant systemic responses such as fever. The desirable outcome is designated by the box "Defense against foreign invader; tissue repair."

FIGURE 12-5 (P. 413): *Interferon* released by virally invaded cells binds with all uninvaded cells and causes them to produce inactive enzymes capable of breaking down viral messenger RNA and of inhibiting virally directed protein synthesis. These viral-blocking enzymes are activated if these forewarned cells are invaded by any virus.

FIGURE 12-11 (P. 419): by (1) activating the complement system, which can destroy the bacteria via membrane attack complexes; (2) serving as opsonins to enhance phagocytosis; and (3) stimulating natural killer cells, which release chemicals that destroy the bacteria

FIGURE 12-15 (P. 425): The *membrane attack complex* is a porelike channel formed in victim cells by activated complement proteins C5 through C9. The resulting osmotic flux of water into the cell causes it to rupture. Perforin molecules released by a killer cell on binding with a target cell form similar lethal pores in the victim cell.

FIGURE 12-20 (PP. 430-431): *Helper T cells* bind (1) with an *APC,* which processes and presents the antigen to helper T cells with matching TCRs and secretes cytokines that activate these T cells, and (2) with *B cells* that have internalized and display the same antigen, whereupon the helper T cells secrete cytokines that stimulate B cell proliferation to produce a clone of these selected effector cells.

FIGURE 12-24 (P. 439): Only the *dermis* is supplied by blood. The epidermis, which has no direct blood supply, is nourished by diffusion of nutrients from the underlying dermis. As newly forming cells in the inner layer of the epidermis push the older cells in its outer layer farther from their nutrient supply, these older cells die and become the outermost keratinized layer of the skin.

Reviewing Terms and Facts

(Questions on p. 442.)

1. F 2. F 3. F 4. F 5. T 6. toll-like receptors 7. membrane attack complex 8. pus 9. inflammation 10. opsonin 11. cytokines 12. perforin, granzymes 13. b 14. 1.c, 2.d, 3.a, 4.b 15. 1.a, 2.a, 3.b, 4.b, 5.c, 6.c, 7.b, 8.a, 9.b, 10.b, 11.a, 12.b 16. 1.b, 2.a, 3.a, 4.b, 5.a, 6.a, 7.a, 8.b

Solving Quantitative Exercises

(Question on p. 444.)

1. *NEP* = net outward pressure − net inward pressure

$$NEP = (P_C + \pi_{IF}) - (P_{IF} + \pi_P)$$

Note for this problem, $(P_{IF} + \pi_P) = (25 \text{ mm Hg} + 1 \text{ mm Hg}) = 26 \text{ mm Hg}$, is constant for all cases.

Normal ($\pi_{IF} = 0$ mm Hg)

Arteriolar end NEP	$(37 + 0) - 26 = +11$ mm Hg
Venular end NEP	$(17 + 0) - 26 = -9$ mm Hg
Average NEP	$(+11 - 9)/2 = +1$ mm Hg (outward)

a. ($\pi_{IF} = 5$ mm Hg)

Arteriolar end NEP	$(37 + 5) - 26 = +16$ mm Hg
Venular end NEP	$(17 + 5) - 26 = -4$ mm Hg
Average NEP	$(+16 - 4)/2 = +6$ mm Hg (outward)
Condition	mild edema

b. ($\pi_{IF} = 10$ mm Hg)

Arteriolar end NEP	$(37 + 10) - 26 = +21$ mm Hg
Venular end NEP	$(17 + 10) - 26 = +1$ mm Hg
Average NEP	$(+21 + 1)/2 = +11$ mm Hg (outward)
Condition	Extreme edema

Applying Clinical Reasoning

(Questions on p. 444.)

If offending allergens bind with IgG antibodies instead of IgE antibodies, no allergy symptoms result because IgG antibodies do not attach to mast cells and basophils like IgE antibodies do.

Thinking at a Higher Level

(Questions on p. 444.)

1. See Table 12-3, p. 430 for a summary of immune responses to bacterial invasion and Table 12-2, p. 426 for a summary of defenses against viral invasion.

2. A vaccine against a particular microbe can be effective only if it induces formation of antibodies or activated T cells against a stable antigen that is present on all microbes of this type. It has not been possible to produce a vaccine against HIV because it frequently mutates. Specific immune responses induced by vaccination against one form of HIV may prove to be ineffective against a slightly modified version of the virus.

3. Failure of the thymus to develop would lead to an absence of T lymphocytes and no cell-mediated immunity after birth. This outcome would seriously compromise the individual's ability to defend against viral invasion and cancer.

4. Researchers are working on ways to "teach" the immune system to view foreign tissue as "self" as a means of preventing the immune systems of organ-transplant patients from rejecting the foreign tissue while leaving the patients' immune defense capabilities fully intact. The immunosuppressive drugs now used to prevent transplant rejection cripple the recipients' immune defense systems and leave the patients more vulnerable to microbial invasion.

5. The skin cells visible on the body's surface are all dead.

Chapter 13 The Respiratory System

Check Your Understanding

13.1 (Questions on p. 450.)

1. (1) ventilation or gas exchange between the atmosphere and alveoli in the lungs, (2) exchange of O_2 and CO_2 between air in the alveoli and blood in the pulmonary capillaries, (3) transport of O_2 and CO_2 in the blood between the lungs and the tissues, and (4) exchange of O_2 and CO_2 between the blood in the systemic capillaries and the tissue cells. The respiratory system is involved in steps 1 and 2; the circulatory system is involved in steps 2, 3, and 4.

2. Only 0.5 μm separates the air in the alveoli from the blood in the pulmonary capillaries, and the alveolar air–blood interface presents a tremendous surface area (75 m^2) for exchange. The thinness and extensive surface area of the alveolar membrane facilitate gas exchange because the rate of diffusion is inversely proportional to the thickness and directly proportional to the surface area of this interface.

3. *Type I alveolar cells* form the walls of the alveoli, *Type II alveolar cells* secrete pulmonary surfactant, and wandering *alveolar macrophages* are phagocytic specialists that scavenge within the lumen of the alveoli.

13.2 (Questions on p. 465.)

1. During *normal quiet breathing*, the diaphragm and external intercostal muscles contract during inspiration, expanding the chest wall and thoracic cavity. The lungs passively follow along, with the intra-alveolar pressure dropping from 760 to 759 mm Hg as the lungs enlarge. Air moves into the lungs down its pressure gradient from the atmosphere until the intra-alveolar pressure equilibrates with the atmospheric pressure of 760 mm Hg. When these inspiratory muscles relax, the chest and lungs passively recoil to their preinspiratory size, increasing the intra-alveolar pressure to 761 mm Hg. Air moves out of the lungs down its pressure gradient to the atmosphere as a passive expiration occurs. During *strenuous exercise*, the diaphragm and external intercostal muscles contract more vigorously and the accessory inspiratory muscles (muscles in the neck) come into play to expand the chest and lungs even more than during quiet breathing. During this forceful inspiration, the intra-alveolar pressure drops even farther, for example to 758 mm Hg, so more air flows into the lungs before equilibration with atmospheric pressure is achieved. During active, or forced, expiration, the inspiratory muscles relax and the expiratory muscles (the abdominal muscles and the internal intercostal muscles) contract to reduce the volume of the chest even more than during passive expiration, allowing the lungs to recoil to a greater extent. As a result, the intra-alveolar pressure increases even more than during quiet breathing, for example to 762 mm Hg, so more air leaves the lungs before equilibrating with atmospheric pressure.

2. *Compliance* is a measure of how much effort is required to stretch the lungs; specifically, it is a measure of the change in lung volume resulting from a given change in the transmural pressure gradient. *Elastic recoil* refers to how readily the lungs rebound after having been stretched.

3. The forces that keep the alveoli open are the transmural pressure gradient and pulmonary surfactant (which opposes alveolar surface tension), and the forces that promote alveolar collapse are elastic recoil and alveolar surface tension.

4. See Figure 13-15b, p. 461.

13.3 (Questions on p. 471.)

1. *Partial pressure* is the individual pressure exerted independently by a particular gas within a mixture of gases.

2. See Figure 13-22, p. 469.

3. See Table 13-5, p. 470.

13.4 (Questions on p. 478.)

1. See Table 13-6, p. 472.

2. See Figure 13-24, p. 473.

3. *Hypoventilation* results in accumulation of CO_2 because less CO_2 is blown off to the atmosphere than is produced. Because CO_2 generates acid, the excess H^+ leads to respiratory acidosis. *Hyperventilation* leads to a reduction in CO_2 because CO_2 is blown off more rapidly than it is produced. As a result, less H^+ is generated from CO_2 than normal, leading to respiratory alkalosis.

13.5 (Questions on p. 487.)

1. The *medullary respiratory center* is the primary respiratory control center. It contains the dorsal respiratory group (DRG) and ventral respiratory group (VRG). The *DRG* consists mostly of inspiratory neurons that alternately fire to cause inspiration and cease firing to bring about expiration during quiet breathing. The *VRG* is composed of inspiratory neurons and expiratory neurons that remain inactive during quiet breathing but are called into play by the DRG as an overdrive mechanism to cause forceful inspiration and active expiration during periods when demands for ventilation are increased. The *pneumotaxic center* in the pons helps switch off the DRG inspiratory neurons, and the *apneustic center* in the pons prevents these inspiratory neurons from being switched off, in a check-and-balance system. The *pre-Bötzinger complex* generates the basic respiratory rhythm and drives the rhythmic firing of the DRG inspiratory neurons.

2. The *peripheral chemoreceptors* are stimulated when the arterial P_{O_2} falls to the point of being life threatening (<60 mm Hg). In turn, they stimulate the medullary inspiratory neurons as an emergency mechanism to maintain ventilation during a time when such a low arterial P_{O_2} directly depresses the respiratory center. The peripheral chemoreceptors are also responsive to changes in arterial H^+ concentration and adjust ventilation accordingly to help maintain acid–base balance by altering the rate at which H^+-generating CO_2 is eliminated from the body. The peripheral chemoreceptors are weakly stimulated by increased arterial P_{CO_2}.

3. The major mechanism controlling ventilation under resting conditions is specifically aimed at regulating the brain-ECF H^+ concentration, which in turn directly reflects changes in arterial P_{CO_2} because CO_2 that enters the brain ECF generates H^+ to which the central chemoreceptors are highly responsive. The central chemoreceptors adjust ventilation accordingly to keep the brain-ECF H^+ and thus arterial P_{CO_2} normal.

Figure Focus

FIGURE 13-4 (P. 449): from alveolar air across the alveolar fluid lining, through a Type I alveolar cell forming the wall of an alveolus, through the interstitial fluid, across an endothelial cell forming the wall of a pulmonary capillary and into the blood

FIGURE 13-12 (P. 455): Airflow would double if intra-alveolar pressure falls to 758 mm Hg instead of 759 mm Hg at the same resistance because $F = \Delta P/R$.

FIGURE 13-17 (P. 463): VC is reduced in *obstructive lung disease* because the lungs cannot empty as completely as normal. Because residual volume is increased, less additional air can be inspired beyond the tidal volume on maximal inspiratory effort before the normal total lung capacity is reached, and less air than usual can be expired on maximal expiratory effort. VC is reduced in *restrictive lung disease* because the lungs cannot expand and fill as completely as normal. Less air than usual can be inspired beyond the tidal volume on maximal inspiratory effort before the reduced total lung capacity is reached. The lungs can empty as normal on maximal expiratory effort, but not as much air is available to be expired during a VC measurement because of the reduced inspiratory capacity.

FIGURE 13-20 (P. 467): (1) bottom, (2) bottom, (3) top

FIGURE 13-21 (P. 467): $P_{O_2} = 630$ mm Hg $\times$ 0.21 = 132 mm Hg

FIGURE 13-24 (P. 473): (1) If P_{CO_2} declines from 100 mm Hg to 75 mm Hg, % Hb saturation decreases from 97.5% to 95%. (2) If P_{CO_2} declines from 40 mm Hg to 15 mm Hg, % Hb saturation decreases from 75% to 19%.

FIGURE 13-27 (P. 476): O_2, CO_2, H^+

Reviewing Terms and Facts

(Questions on p. 488.)

1. F **2.** F **3.** T **4.** F **5.** F **6.** F **7.** F **8.** transmural pressure gradient, pulmonary surfactant action **9.** pulmonary elasticity, alveolar surface tension **10.** compliance **11.** elastic recoil **12.** carbonic anhydrase **13.** a **14.** a.<, b. >, c. =, d. =, e. =, f. =, g. >, h. <, i. approximately =, j. approximately =, k. =, l. = **15.** 1.d, 2.a, 3.b, 4.a, 5.b, 6.a

Solving Quantitative Exercises

(Questions on p. 489.)

For general reference for questions 1 and 2:

$$P_{AO_2} = P_{IO_2} - (V_{O_2}/V_A) \times 863 \text{ mm Hg}$$
$$P_{ACO_2} = (V_{CO_2}/V_A) \times 863 \text{ mm Hg}$$

1. $V_A = 3$ L/min
$V_{O_2} = 0.3$ L/min, RQ = 1, therefore $V_{CO_2} = 0.3$ L/min
$P_{ACO_2} = (0.3 \text{ L/min}/3 \text{ L/min}) \times 863 \text{ mm Hg}$

$\qquad = 86.3$ mm Hg

2. a. 380 mm Hg $\times$ 0.21 = 79.8 mm Hg
b. $P_{AO_2} = 79.8$ mm Hg $-$ (0.06) $\times$ 431.5 mm Hg
$\qquad = 79.8$ mm Hg $-$ 25.8 mm Hg
$\qquad = 54$ mm Hg

c. $P_{ACO_2} = (0.2 \text{ L/min}/4.2 \text{ L/min}) \times 431.5 \text{ mm Hg}$
$\qquad = 20.5$ mm Hg

3. $TV = 350$ mL, BR = 12/min, $V_A = 0.8 \times V_E$, DS = ?
$V_A = BR \times (TV - DS)$
$V_E = BR \times TV$
$0.8 = V_A/V_E = [BR \times (TV - DS)]/(BR \times TV)$
$\qquad = 1 - (DS/TV)$
$0.8 = 1 - (DS/350 \text{ mL})$
$DS/350 \text{ mL} = 0.2$
$DS = 0.2(350 \text{ mL}) = 70$ mL

Applying Clinical Reasoning

(Questions on p. 489.)

Emphysema is characterized by a collapse of smaller respiratory airways and a breakdown of alveolar walls. Because of the collapse of smaller airways, airway resistance is increased with emphysema. As with other chronic obstructive pulmonary diseases, expiration is impaired to a greater extent than inspiration because airways are naturally dilated slightly more during inspiration than expiration as a result of the greater transmural pressure gradient during inspiration. Because airway resistance is increased, a patient with emphysema must produce larger-than-normal intra-alveolar pressure changes to accomplish a normal tidal volume. Unlike quiet breathing in a normal person, the accessory inspiratory muscles (neck muscles) and the muscles of active expiration (abdominal muscles and internal intercostal muscles) must be brought into play to inspire and expire a normal tidal volume of air.

The spirogram would be characteristic of chronic obstructive pulmonary disease. Because the patient experiences more difficulty emptying the lungs than filling them, the total lung capacity would be essentially normal, but the functional residual capacity and the residual volume would be elevated as a result of the additional air trapped in the lungs following expiration. Because the residual volume is increased, the inspiratory capacity and vital capacity are reduced. Also, the FEV_1 is markedly reduced because the airflow rate is decreased by the airway obstruction. The FEV_1-to–vital capacity ratio is much lower than the normal 80%.

Because of the reduced surface area for exchange as a result of a breakdown of alveolar walls, gas exchange would be impaired. Therefore, arterial P_{CO_2} would be elevated and arterial P_{O_2} reduced compared to normal.

Ironically, administering O_2 to this patient to relieve his hypoxic condition would markedly depress his drive to breathe by elevating the arterial P_{O_2} and removing the primary driving stimulus for respiration. Because of this danger, O_2 therapy should either not be administered or administered extremely cautiously.

Thinking at a Higher Level

(Questions on p. 490.)

1. Total atmospheric pressure decreases with increasing altitude, yet the percentage of O_2 in the air remains the same. At an altitude of 30,000 feet, the atmospheric pressure is only 226 mm Hg. Because 21% of atmospheric air consists of O_2, the P_{O_2} of inspired air at 30,000 feet is only 47.5 mm Hg, and alveolar P_{O_2} is even lower at about 20 mm Hg. At this low P_{O_2}, hemoglobin is only about 30% saturated with O_2—much too low to sustain tissue needs for O_2.

The P_{O_2} of inspired air can be increased by two means when flying at high altitude. First, by pressurizing the plane's interior to a pressure comparable to that of atmospheric pressure at sea level, the P_{O_2} of inspired air within the plane is 21% of 760 mm Hg, or the normal 160 mm Hg. Accordingly, alveolar and arterial P_{O_2} and percent hemoglobin saturation are likewise normal. In the emergency situation of failure to maintain internal cabin pressure, breathing pure O_2 can raise the P_{O_2} considerably above that accomplished by breathing normal air. When a person is breathing pure O_2, the entire pressure of inspired air is attributable to O_2. For example, with a total atmospheric pressure of 226 mm Hg at an altitude of 30,000 feet, the P_{O_2} of inspired pure O_2 is 226 mm Hg, which is more than adequate to maintain normal arterial hemoglobin saturation.

2. (a) normal, (b) normal, (c) normal, (d) below normal, (e) below normal, (f) below normal

3. a. Hypercapnia would not accompany the hypoxia associated with cyanide poisoning. In fact, CO_2 levels decline because oxidative metabolism is blocked by the tissue poisons so that CO_2 is not being produced.

b. Hypercapnia could but may not accompany the hypoxia associated with pulmonary edema. Pulmonary diffusing capacity is reduced in pulmonary edema, but O_2 transfer suffers more than CO_2 transfer because the diffusion constant for CO_2 is 20 times that for O_2. As a result, hypoxia occurs much more readily than hypercapnia in these circumstances. Hypercapnia does occur, however, when pulmonary diffusing capacity is severely impaired.

c. Hypercapnia would accompany the hypoxia associated with restrictive lung disease because ventilation is inadequate to meet the metabolic needs for both O_2 delivery and CO_2 removal. Both O_2 and CO_2 exchange between the lungs and atmosphere are equally affected.

d. Hypercapnia would not accompany the hypoxia associated with high altitude. In fact, arterial P_{CO_2} levels actually decrease. One of the compensatory responses in acclimatization to high altitudes is reflex stimulation of ventilation as a result of the reduction in arterial P_{O_2}. This compensatory hyperventilation to obtain more O_2 blows off too much CO_2 in the process, so arterial P_{CO_2} levels decline below normal.

e. Hypercapnia would not accompany the hypoxia associated with severe anemia. Reduced O_2-carrying capacity of the blood has no influence on blood CO_2 content, so arterial P_{CO_2} levels are normal.

f. Hypercapnia would accompany the circulatory hypoxia associated with congestive heart failure. Just as the diminished blood flow fails to deliver adequate O_2 to the tissues, it also fails to remove sufficient CO_2.

g. Hypercapnia would accompany the hypoxic hypoxia associated with obstructive lung disease because ventilation would be inadequate to meet the metabolic needs for both O_2 delivery and CO_2 removal. Both O_2 and CO_2 exchange between the lungs and atmosphere would be equally affected.

4. Voluntarily hyperventilating before going underwater lowers the arterial P_{CO_2} but does not increase the O_2 content in the blood. Because the P_{CO_2} is below normal, the person can hold his or her breath longer than usual before the arterial P_{CO_2} increases to the point that he or she is driven to surface for a breath. Therefore, the person can stay underwater longer. The risk, however, is that the O_2 content of the blood, which was normal, not increased, before going underwater, continues to fall. Therefore, the O_2 level in the blood can fall dangerously low before the CO_2 level builds to the point of driving the person to take a breath. Low arterial P_{O_2} does not stimulate respiratory activity until it has plummeted to 60 mm Hg. Meanwhile, the person may lose consciousness and drown due to inadequate O_2 delivery to the brain. If the person does not hyperventilate so that both the arterial P_{CO_2} and O_2 content are normal before going underwater, the buildup of CO_2 drives the person to the surface for a breath before the O_2 levels fall to a dangerous point.

5. c. The arterial P_{O_2} is less than the alveolar P_{O_2}, and the arterial P_{CO_2} is greater than the alveolar P_{CO_2}. Because pulmonary diffusing capacity is reduced, arterial P_{O_2} and P_{CO_2} do not equilibrate with alveolar P_{O_2} and P_{CO_2}. O_2 exchange is affected to a greater extent than CO_2 exchange because of CO_2's greater diffusion constant.

If the person is administered 100% O_2, the alveolar P_{O_2} increases and the arterial P_{O_2} increases accordingly. Even though arterial P_{O_2} does not equilibrate with alveolar P_{O_2}, it is higher than when the person is breathing atmospheric air.

The arterial P_{CO_2} remains the same whether the person is administered 100% O_2 or is breathing atmospheric air. The alveolar P_{CO_2} and thus the blood-to-alveolar P_{CO_2} gradient are not changed by breathing 100% O_2 because the P_{CO_2} in atmospheric air and 100% O_2 are both essentially zero (P_{CO_2} in atmospheric air = 0.23 mm Hg).

Chapter 14 The Urinary System

Check Your Understanding

14.1 (Questions on p. 498.)

1. The functional unit of the kidney is the *nephron,* which is made up of a vascular component and a tubular component that are intimately related structurally and functionally. The vascular component consists of an afferent arteriole that carries blood to the glomerulus; the glomerulus, which is a tuft of capillaries that filters a protein-free plasma into the tubular component; an efferent arteriole that carries blood from the glomerulus; and peritubular capillaries that wrap around and exchange materials with the tubular component. The tubular component begins with Bowman's capsule that cups around the glomerulus and collects the filtrate; the proximal tubule across which uncontrolled exchanges take place between the tubular fluid and peritubular capillary blood; the loop of Henle, which plays a key role in producing urine of varying concentration depending on the body's needs; and the distal tubule and collecting duct across which controlled exchanges take place between the tubular fluid and blood. The fluid exiting the collecting duct is urine.

2. See Figure 14-6, p. 497.

3. *Cortical nephrons* are the most abundant type of nephron. The *juxtamedullary nephrons* are important in establishing the medullary vertical osmotic gradient. The glomeruli of cortical nephrons lie in the outer cortex, whereas the glomeruli of juxtamedullary nephrons lie in the inner part of the cortex next to the medulla. The loops of Henle of cortical nephrons dip only slightly into the medulla, but the juxtamedullary nephrons have long loops of Henle that plunge deep into the medulla. The juxtamedullary nephrons' peritubular capillaries form hairpin loops known as vasa recta. See Figure 14-5, p. 496.

14.2 (Questions on p. 505.)

1. See Table 14-1, p. 500.

2. Mechanisms of autoregulation of the GFR: (1) a *myogenic* mechanism, which responds to changes in pressure within the nephron's vascular component; and (2) a *tubuloglomerular feedback* mechanism, which senses changes in the salt level in the fluid flowing through the nephron's tubular component. Autoregulation prevents inappropriate fluctuations in the GFR in response to changes in the mean arterial driving pressure within the range of 80 to 180 mm Hg.

3. Extrinsic control of the GFR is accomplished by sympathetic nervous system input to the afferent arterioles aimed at regulating arterial blood pressure. When arterial blood pressure falls, sympathetically induced afferent arteriolar vasoconstriction decreases glomerular capillary blood pressure, causing a decrease in GFR. The resultant reduction in urine volume conserves fluid and salt that would have been lost from the body in urine. This saved fluid helps to restore plasma volume and blood pressure to normal in the long term. The opposite compensatory effect takes place when arterial blood pressure is too high.

14.3 (Questions on p. 514.)

1. See Figure 14-14, p. 506.

2. The kidneys secrete the enzymatic hormone renin in response to reduced NaCl, ECF volume, and arterial blood pressure. Renin activates angiotensinogen, a plasma protein produced by the liver, into angiotensin I, which is converted into angiotensin II by angiotensin-converting enzyme produced in the lungs. Angiotensin II stimulates the adrenal cortex to secrete the hormone aldosterone, which stimulates Na^+ reabsorption by the kidneys. The resulting retention of Na^+ exerts an osmotic effect that holds more H_2O in the ECF. Together, the conserved Na^+ and H_2O help correct the original stimuli that activated this hormonal system.

3. (1) 250 mg/min filtered, (2) 200 mg/min reabsorbed (a T_m worth of the substance), (3) 50 mg/min excreted

14.4 (Questions on p. 517.)

1. secretion of H^+, K^+, and organic anions and cations

2. The *proximal tubule cells* actively reabsorb K^+ in constant, unregulated fashion. The *principal cells* of the distal and collecting tubules actively secrete K^+. Potassium secretion by the late tubule is varied in a controlled fashion by aldosterone to regulate the rate of K^+ excretion and maintain the desired plasma K^+ concentration.

3. The removal of foreign organic compounds from the body is hastened via secretion by the organic ion secretory systems in the proximal tubule. The liver helps rid the body of foreign compounds that are not ionic in their original form by converting these compounds into an anionic form.

14.5 (Questions on p. 532.)

1. (1) clearance rate = GFR, (2) clearance rate < GFR, (3) clearance rate > GFR

2. The juxtamedullary nephrons' long loop of Henle establishes the medullary vertical osmotic gradient, their vasa recta preserves the gradient, and the collecting ducts of all nephrons use the gradient, in conjunction with vasopressin, to produce urine of varying concentrations.

3. *Vasopressin,* which is secreted into the blood by the posterior pituitary in response to a H_2O deficit, increases the permeability of the distal and collecting tubules to H_2O. Vasopressin activates the cAMP pathway within the principal cells lining these tubules when it binds with V_2 receptors specific for it at the basolateral membrane of these cells. cAMP increases the opposite luminal membrane's permeability to H_2O by promoting the insertion of AQP-2 water channels into the membrane. Water enters the cell from the tubular lumen through the inserted water channels and exits the cell through different, permanently positioned water channels at the basolateral border to enter the blood, in this way being reabsorbed to correct the H_2O deficit.

4. The *micturition reflex* is initiated when bladder filling activates stretch receptors in the bladder wall. Afferent input from the stretch receptors, via interneurons, stimulates the parasympathetic nerve to the bladder, leading to bladder contraction and mechanical opening of the internal urethral sphincter, and simultaneously inhibits the motor neuron to the external urethral sphincter, leading to opening of this sphincter. Bladder contraction expels urine through the open sphincters to the outside.

Figure Focus

FIGURE 14-4 (P. 495): The afferent arteriole has a larger diameter than the efferent arteriole has.

FIGURE 14-6 (P. 497): Substances can *enter* the tubular fluid via glomerular filtration or tubular secretion and can leave the tubular fluid by means of tubular reabsorption or urine excretion.

FIGURE 14-16 (P. 508): When blood pressure falls as a result of excessive fluid loss via sweating, in the short term the baroreceptor reflex increases sympathetic activity to the heart and arterioles, leading respectively to increased cardiac output and increased total peripheral resistance, which together produce a compensatory increase in blood pressure. An increase in ECF volume, particularly blood volume, is important in the long term for restoring blood pressure to normal without having to sustain the adjustments in cardiac output and total peripheral resistance. Multiple pathways increase blood volume. Afferent arteriolar vasoconstriction occurring as part of the baroreceptor reflex brings about a reduction in the GFR, which leads to conservation of fluid and salt that would have been lost in the urine. More importantly, the original decrease in NaCl, ECF volume, and arterial blood pressure resulting from heavy sweating collectively trigger increased activity of the renin–angiotensin–aldosterone system, which causes increased Na^+ reabsorption, with H_2O osmotically following along. This extra fluid reabsorption expands the blood volume. Furthermore, angiotensin II activated as part of the RAAS stimulates vasopressin secretion, which leads to increased water reabsorption; promotes thirst, which leads to increased fluid intake; and also causes arteriolar vasoconstriction, which contributes to short-term compensation for the fall in blood pressure.

FIGURE 14-18 (P. 511): (1) at *200* mg/100 mL: 250 mg/min filtered, 250 mg/min reabsorbed, 0 mg/min excreted; (2) at *300* mg/100 mL: 375 mg/min filtered, 375 mg/min reabsorbed, 0 mg/min excreted; (3) at *400* mg/100 mL: 500 mg/min filtered, 375 mg/min reabsorbed, 125 mg/min excreted

FIGURE 14-20 (P. 513): In the early part of the proximal tubule, the urea concentration in the tubular fluid is the same as its concentration in the surrounding interstitial fluid and peritubular capillary blood, so no concentration gradient exists for passive diffusion of urea out of the tubule. By the late part of the proximal tubule, however, the urea concentration in the tubular fluid has increased above its concentration in the surrounding fluids as a result of extensive reabsorption of water by the proximal tubule. Therefore, urea is passively reabsorbed down this concentration gradient in the late part of the proximal tubule.

FIGURE 14-21 (P. 515): (1) This Na^+ is reabsorbed into the peritubular capillary blood. (2) This K^+ is recycled as it diffuses from the tubular cell (into which it had been pumped) back into the lateral space through K^+ leak channels located in the basolateral membrane.

FIGURE 14-27 (P. 525): (1) By blocking NaCl transport (reabsorption) by the ascending limb, *loop diuretics* prevent generation of the vertical osmotic gradient in the medullary interstitial fluid, abolishing the driving force for osmotic free water reabsorption in the collecting duct. Because the tubular fluid cannot be concentrated and water conserved, more urine than usual is formed. (2) Because of the vasopressin deficiency associated with *diabetes insipidus*, the kidneys are unable to increase the water permeability of the distal and collecting tubules sufficiently to permit free water reabsorption down the medullary vertical osmotic gradient to conserve water and concentrate urine as needed. As a result, more urine than usual is formed.

Reviewing Terms and Facts

(Questions on p. 533.)

1. F **2.** F **3.** T **4.** T **5.** T **6.** nephron **7.** potassium **8.** 500 **9.** 1.b, 2.a, 3.b, 4.b, 5.a, 6.b, 7.b, 8.b, 9.b **10.** e **11.** b **12.** b, e, a, d, c **13.** c, e, d, a, b, f **14.** g, c, d, a, f, b, e **15.** 1.a, 2.a, 3.c, 4.b, 5.d

Solving Quantitative Exercises

(Questions on p. 534.)

1.	Patient 1	Patient 2
GFR	125 mL/min	124 mL/min
RPF	620 mL/min	400 mL/min
RBF	1127 mL/min	727 mL/min
FF	0.20	0.31

All of Patient 1's values are within the normal range. Patient 2's *GFR* is normal, but he has a low renal plasma flow and a high filtration fraction. Therefore, his *GFR* is too high for that *RPF*. This might imply enlarged filtration slits or a "leaky" glomerulus in general. The low *RPF* may imply low renal blood pressure, perhaps from a partially blocked renal artery.

2. filtered load = *GFR* × plasma concentration
$$= (0.125\ L/min) \times (145\ mmol/L)$$
$$= 18.125\ mmol/min$$

3. $GFR = (U \times [I]_U)/[I]_B$
$$U = (GFR \times [I]_B)/[I]_U$$
$$= (125\ mL/min)(3\ mg/L)/(300\ mg/L)$$
$$= 1.25\ mL/min$$

4. $$\text{Clearance rate of a substance} = \frac{\text{urine concentration of a substance} \times \text{urine flow rate}}{\text{plasma concentration of the substance}}$$
$$= \frac{7.5\ mg/mL \times 2\ mL/min}{0.2\ mg/mL}$$
$$= 75\ mL/min$$

Because a clearance rate of 75 mL/min is less than the average GFR of 125 mL/min, the substance is being reabsorbed.

Applying Clinical Reasoning

(Question on p. 534.)

prostate enlargement

Thinking at a Higher Level

(Questions on p. 534.)

1. (1) Glucose in the urine is indicative of diabetes mellitus. (2) Protein in the urine is indicative of kidney disease. (3) Na^+ in the urine is normal.

2. The longer loops of Henle in desert rats (known as *kangaroo rats*) permit a greater magnitude of countercurrent multiplication and thus a larger medullary vertical osmotic gradient. As a result, these rodents can produce urine that is concentrated up to an osmolarity of almost 6000 mOsm/L, which is five times more concentrated than maximally concentrated human urine at 1200 mOsm/L. Because of this tremendous concentrating ability, kangaroo rats never have to drink; the H_2O produced metabolically within their cells during oxidation of foodstuff (food + $O_2 \rightarrow CO_2 + H_2O$ + energy) is sufficient for their needs.

3. Aldosterone stimulates Na^+ reabsorption and K^+ secretion by the renal tubules. Therefore, the most prominent features of Conn's syndrome (hypersecretion of aldosterone) are hypernatremia (elevated Na^+ levels in the blood) caused by excessive Na^+ reabsorption, hypophosphatemia (below-normal K^+ levels in the blood) caused by excessive K^+ secretion, and hypertension (elevated blood pressure) caused by excessive salt and water retention.

4. e. 300/300. If the ascending limb were permeable to water, it would not be possible to establish a vertical osmotic gradient in the interstitial fluid of the renal medulla, nor would the ascending limb fluid become hypotonic before entering the distal tubule. As the ascending limb pumped NaCl into the interstitial fluid, water would osmotically follow, so both the interstitial fluid and the ascending limb would remain isotonic at 300 mOsm/L. With the tubular fluid entering the distal tubule being 300 mOsm/L instead of the normal 100 mOsm/L, it would not be possible to produce urine with an osmolarity less than 300 mOsm/L. Likewise, in the absence of the medullary vertical osmotic gradient, it would not be possible to produce urine more concentrated than 300 mOsm/L, no matter how much vasopressin was present.

5. Because the descending pathways between the brain and the motor neurons supplying the external urethral sphincter and pelvic diaphragm are no longer intact, the accident victim can no longer voluntarily control micturition. Therefore, bladder emptying in this individual is governed entirely by the micturition reflex.

Chapter 15 Fluid and Acid–Base Balance

Check Your Understanding

15.1 (Questions on p. 537.)

1. *inputs* = ingestion or metabolic production; *outputs* = excretion or metabolic consumption

2. *stable balance:* when total body input of a substance equals its total body output; *positive balance:* when the gains via input for a substance exceed its losses via output; *negative balance:* when losses for a substance exceed its gains

15.2 (Questions on p. 547.)

1. See Table 15-1, p. 537.

2. *ECF volume* is regulated by maintaining salt balance and is important to help maintain blood pressure. *ECF osmolarity* is regulated by maintaining free water balance and is important to prevent swelling or shrinking of cells.

3. Maintenance of salt balance is accomplished by controlling salt output in the urine. Salt excretion is controlled primarily by the renin–angiotensin–aldosterone system regulating the amount of Na^+ reabsorbed and to a lesser extent by the baroreceptor reflex controlling the GFR and thus of the amount of Na^+ filtered. Maintenance of water balance is accomplished primarily by controlling water output in the urine via vasopressin regulating the amount of free H_2O reabsorbed, and to a lesser extent by controlling water input via thirst.

4. When the ECF is hypertonic, water osmotically leaves the cells, causing them to shrink. When the ECF is hypotonic, water osmotically enters the cells, causing them to swell.

15.3 (Questions on p. 562.)

1. Only a narrow pH range is compatible with life because even small changes in $[H^+]$ have dramatic effects on normal cell function. Changes in $[H^+]$ (1) cause changes in excitability of nerve and muscle cells, which can lead to death by depressing the CNS in the case of severe acidosis or causing spasm of the respiratory muscles in the case of severe alkalosis; (2) alter enzyme activity, thus disturbing life-supporting metabolic activity catalyzed by these enzymes; and (3) influence $[K^+]$, which can lead to fatal cardiac abnormalities.

2. The following sources continually add H^+ to the body fluids: (1) carbonic acid formed from continuously generated CO_2; (2) inorganic acids (phosphoric and sulfuric acid) produced during the breakdown of dietary proteins that contain phosphorus and sulfur; and (3) organic acids (such as fatty acids and lactic acid) resulting from intermediary metabolism.

3. *Henderson–Hasselbalch equation:* $pH = pK + \log[HCO_3^-]/[H_2CO_3]$

Because $[HCO_3^-]$ is controlled by kidney function, and $[H_2CO_3]$, which is a reflection of $[CO_2]$, is controlled by respiratory function, pH can be shifted up or down by kidney and respiratory influences.

4. metabolic acidosis. To compensate, the ECF buffers take up extra H^+, the lungs blow off additional H^+-generating CO_2, and the kidneys excrete more H^+ and conserve more HCO_3^-.

Figure Focus

FIGURE 15-1 (P. 536): Input from the external environment via ingestion, excretion to the external environment in the urine, and storage within the body as calcium phosphate crystals, which harden bone.

FIGURE 15-3 (P. 541): Excess Na^+ is eliminated from the body by increasing urinary excretion of Na^+ by (1) increasing the GFR via the baroreceptor reflex, which increases the amount of Na^+ filtered, and (2) decreasing the amount of Na^+ reabsorbed via the renin–angiotensin–aldosterone system.

FIGURE 15-10 (P. 555): In *Type A intercalated cells*, the H^+ pumps are located at the luminal membrane where they accomplish H^+ secretion; in *Type B intercalated cells*, the H^+ pumps are located at the basolateral membrane where they accomplish H^+ reabsorption.

FIGURE 15-13 (P. 560): false. $[HCO_3^-]$ increases not only during metabolic alkalosis but also during compensation for respiratory acidosis.

Reviewing Terms and Facts

(Questions on p. 563.)

1. T **2.** F **3.** F **4.** T **5.** T **6.** intracellular fluid **7.** $[H_2CO_3]$, $[HCO_3^-]$ **8.** b **9.** a, d, e **10.** b, e **11.** c **12.** 1. metabolic acidosis, 2. diabetes mellitus, 3. pH = 7.1, 4. respiratory alkalosis, 5. anxiety, 6. pH = 7.7, 7. respiratory acidosis, 8. pneumonia, 9. pH = 7.1, 10. metabolic alkalosis, 11. vomiting, 12. pH = 7.7

Solving Quantitative Exercises

(Questions on p. 564.)

1. $pH = 6.1 + \log [HCO_3^-]/(0.03 \text{ mM/mm Hg} \times 40 \text{ mm Hg})$
$7.4 = 6.1 + \log [HCO_3^-]/1.2 \text{ mM}$
$\log [HCO_3^-]/1.2 \text{ mM} = 7.4 - 6.1 = 1.3$
$[HCO_3^-] = 1.2 \text{ mM} \times (10^{1.3}) = 24 \text{ mM}$

2. $pH = -\log [H^+]$, $[H^+] = 10^{-pH}$
$[H^+] = 10^{-6.8} = 158 \text{ nM for pH} = 6.8$
$[H^+] = 10^{-8.0} = 10 \text{ nM for pH} = 8.0$

3. Note that distilled water can permeate all barriers, so it distributes equally among all compartments. Because saline does not enter cells, it stays in the ECF. The resultant distributions are summarized in the chart at the bottom of the page. Clearly, saline is better at expanding the plasma volume.

Ingested Fluid	Compartment	Size of Compartment before Ingestion (liters)	Size of Compartment after Ingestion (liters)	% Increase in Size of Compartment after Ingestion
Distilled water	TBW	42	43	2%
	ICF (2/3 TBW)	28	28.667	2%
	ECF (1/3 TBW)	14	14.333	2%
	plasma (20% ECF)	2.8	2.866	2%
	ISF (80% ECF)	11.2	11.466	2%
Saline	TBW	42	43	2%
	ICF	28	28	0%
	ECF	14	15	7%
	plasma	2.8	3	7%
	ISF	11.2	12	7%

Applying Clinical Reasoning

(Questions on p. 564.)

The resultant prolonged diarrhea leads to dehydration and metabolic acidosis as a result of, respectively, excessive loss in the feces of fluid and $NaHCO_3$ that normally would have been absorbed into the blood.

Compensatory measures for dehydration have included increased vasopressin secretion, resulting in increased water reabsorption by the distal and collecting tubules and a subsequent reduction in urine output. Simultaneously, fluid intake has been encouraged by increased thirst. The metabolic acidosis has been combated by removal of excess H^+ from the ECF by the HCO_3^- member of the $H_2CO_3:HCO_3^-$ buffer system, by increased ventilation to reduce the amount of acid-forming CO_2 in the body fluids, and by the kidneys excreting extra H^+ and conserving HCO_3^-.

Thinking at a Higher Level

(Questions on p. 564.)

1. The rate of urine formation increases when alcohol inhibits vasopressin secretion and the kidneys are unable to reabsorb water from the distal and collecting tubules. Because extra free water that normally would have been reabsorbed from the distal parts of the tubule is lost from the body in the urine, the body becomes dehydrated and the ECF osmolarity increases following alcohol consumption. That is, more fluid is lost in the urine than is consumed in the alcoholic beverage as a result of alcohol's action on vasopressin. Thus, the imbibing person experiences a water deficit and still feels thirsty, despite the recent fluid consumption.

2. If a person loses 1500 mL of salt-rich sweat and drinks 1000 mL of water without replacing the salt during the same time period, there is still a volume deficit of 500 mL, and the body fluids are hypotonic (the remaining salt in the body is diluted by the ingestion of 1000 mL of free H_2O). As a result, the hypothalamic osmoreceptors (the dominant input) signal the vasopressin-secreting cells to *decrease* vasopressin secretion and thus increase urinary excretion of the extra free water that is making the body fluids too dilute. Simultaneously, the left atrial volume receptors signal the vasopressin-secreting cells to *increase* vasopressin secretion to conserve water during urine formation and thus help relieve the volume deficit. These two conflicting inputs to the vasopressin-secreting cells are counterproductive. For this reason, it is important to replace both water and salt following heavy sweating or abnormal loss of other salt-rich fluids. If salt is replaced along with water intake, the ECF osmolarity remains close to normal and the vasopressin-secreting cells receive signals only to increase vasopressin secretion to help restore the ECF volume to normal.

3. When a dextrose solution equal in concentration to that of normal body fluids is injected intravenously, the ECF volume is expanded but the ECF and ICF are still osmotically equal. Therefore, no net movement of water occurs between the ECF and ICF. When the dextrose enters the cell and is metabolized, however, the ECF becomes hypotonic as this solute leaves the plasma. If the excess free water is not excreted in the urine rapidly enough, water moves into the cells by osmosis.

4. Because baking soda ($NaHCO_3$) is readily absorbed from the digestive tract, treatment of gastric hyperacidity with baking soda can lead to metabolic alkalosis as too much HCO_3^- is absorbed. Treatment with antacids that are poorly absorbed is safer because these products remain in the digestive tract and do not produce an acid–base imbalance.

5. c. The hemoglobin buffer system buffers CO_2-generated H^+. In the case of respiratory acidosis accompanying severe pneumonia, the $H^+ + Hb \rightarrow HHb$ reaction shifts toward the HHb side, thus removing some of the extra free H^+ from the blood.

Chapter 16 The Digestive System

Check Your Understanding

16.1 (Questions on p. 573.)

1. (1) *carbohydrates*, absorbable units = monosaccharides, principally glucose, (2) *proteins*, absorbable units = primarily amino acids and a few small polypeptides, (3) *fats*, absorbable units = monoglycerides and free fatty acids

2. See Figure 16-2, p. 571.

3. Pacemaker cells called interstitial cells of Cajal (ICC) generate slow-wave potentials that cyclically bring the membrane closer to or farther from threshold potential. Slow-wave potentials spread via gap junctions from ICCs to the surrounding smooth muscle cells, which themselves are interconnected by gap junctions into functional syncytia. If the slow waves reach threshold at the peaks of depolarization, a volley of action potentials is triggered at each peak, resulting in rhythmic cycles of contraction in the sheet of smooth muscle cells driven by the pacemaker. The slow-wave frequency varies regionally in the digestive tract. Contractions within the tract occur at the rate of slow-wave frequency within a given section of tract.

16.2 (Questions on p. 575.)

1. *mucus:* facilitates swallowing by moistening food particles to hold them together and by providing lubrication; *amylase:* begins digestion of dietary starches; *lysozyme:* exerts antibacterial actions by lysing bacteria, binding IgA antibodies, secreting lactoferrin that binds iron needed for bacterial multiplication, and rinsing away material that may serve as a food source for bacteria

2. A *simple salivary reflex* occurs when receptors in the oral cavity are activated by the presence of food in the mouth and send impulses to the salivary center to trigger salivation. A *conditioned salivary reflex* occurs without oral stimulation via inputs that arise outside the mouth (such as sight and smell of food) that act through the cerebral cortex to stimulate the salivary center.

3. *Parasympathetic stimulation* induces production of a large volume of watery saliva that is rich in enzymes. *Sympathetic stimulation* produces a small volume of thick saliva that is rich in mucus.

16.3 (Questions on p. 578.)

1. During swallowing, respiration is temporarily inhibited while the entrance to the trachea is closed off to prevent food from entering the respiratory airways. Contraction of the laryngeal muscles tightly aligns the vocal folds across the glottis. For further protection, the epiglottis folds downward over the closed glottis.

2. The *pharyngoesophageal sphincter* at the upper end of the esophagus separates the pharynx from the esophagus and prevents air from entering the esophagus and stomach during breathing. This sphincter remains tonically contracted except during a swallow, when it opens. The *gastroesophageal sphincter* at the lower end of the esophagus separates the esophagus from the stomach and prevents reflux of gastric contents into the esophagus. It relaxes during a swallow.

16.4 (Questions on p. 588.)

1. A strong peristaltic contraction in the stomach antrum pushes the luminal contents downward toward a slightly open pyloric sphincter. A small portion of the chyme is forced through the sphincter before the peristaltic wave reaches the sphincter and closes it tightly. When food that is moving forward hits the closed sphincter, it is tossed backward, only to be propelled forward again by the next peristaltic wave. This *retropulsion* (cycles of food being moved forward and backward in the antrum) shears and grinds the food into smaller pieces and thoroughly mixes it with gastric secretions, converting it into chyme before emptying.

2. *Mucous cells*: mucus; *chief cells*: pepsinogen; *parietal cells*: HCl, intrinsic factor; *ECL cells*: histamine; *G cells*: gastrin; *D cells*: somatostatin

3. The cephalic phase of gastric secretion occurs in response to food-related stimuli "in the head." Such stimuli include thinking about, tasting, smelling, seeing, chewing, and swallowing food. The response is mediated by vagal activation of the chief and parietal secretory cells (which secrete HCl and pepsinogen, respectively) and the endocrine G cells (which secrete the hormone gastrin that further stimulates the chief and parietal cells).

4. The following components of the gastric mucosal barrier enable the stomach to contain potent HCl without injuring itself: (1) The luminal membranes of the gastric mucosal cells are impermeable to H^+ so that HCl cannot penetrate into the cells, (2) tight junctions that join the cells prevent HCl from penetrating between them, (3) the surface mucous cells secrete mucus that serves as a physical barrier to acid penetration, and (4) HCO_3^- secreted with the mucus serves as a chemical barrier that neutralizes acid in the vicinity of the mucosa. Finally, the entire stomach lining undergoes significant turnover, with the cells being replaced every 3 days.

16.5 (Questions on p. 598.)

1. The *pancreatic proteolytic enzymes* (trypsinogen, chymotrypsinogen, and procarboxypeptidase), when activated, digest dietary proteins and peptide fragments to small peptide chains and amino acids. *Pancreatic amylase* converts dietary starches into the disaccharide maltose (mostly) and the polysaccharide α-limit dextrins. *Pancreatic lipase* hydrolyzes dietary triglycerides into monoglycerides and free fatty acids, the absorbable units of fat.

2. The pancreas produces proteolytic enzymes for digesting food proteins, but these enzymes could also digest the proteins of the cells that produce them if they were not stored in inactive form until they are secreted. These inactive enzymes are stored in secretory zymogen granules in the pancreatic acinar cells until appropriate stimuli induce their secretion. They are activated to protein-digesting enzymes (trypsin, chymotrypsin, and carboxypeptidase) only when they reach the duodenal lumen, where they act on food, not the pancreatic cells.

3. Bile salts contribute to dietary fat digestion by their detergent action (fat emulsification) in the small intestine. Intestinal mixing movements break up large fat globules into smaller droplets. The droplets do not recoalesce into the large globule because bile salts adsorb on the surface of the smaller droplets, creating a *lipid emulsion* consisting of many small fat droplets suspended in the watery chyme. This action increases the surface area of fat available for attack by pancreatic lipase.

16.6 (Questions on p. 610.)

1. The alternating *segmentation* contractions chop and thoroughly mix the chyme as they squeeze the small intestine contents both forward and backward to adjacent segments with each round of contraction. Segmentation also slowly propels the contents forward; because the frequency of segmentation declines along the length of the small intestine, more chyme on average is pushed forward than is pushed backward.

2. Three major structural features increase the surface area of the small intestine available for absorption: (1) Large *circular folds* on the small-intestine luminal surface increase the surface area threefold, (2) microscopic fingerlike projections known as *villi* that sit atop the circular folds increase the surface area another 10-fold, and (3) a multitude of hairlike protrusions referred to as *microvilli* or the *brush border* on each villus surface collectively increase the surface area another 20-fold. Together these three adaptations increase the absorptive surface area of the small intestine approximately 600 times greater compared to a completely smooth tube of the same length and diameter.

3. Absorption of most dietary carbohydrate and protein is accomplished by secondary active transport that involves the cotransport of Na^+ and the nutrient molecule into the absorptive cell. Although the cotransporter does not use ATP directly, the Na^+–K^+ pump (which directly uses energy) establishes a Na^+ concentration gradient that drives the cotransporter to move Na^+ downhill and move the nutrient molecule uphill into the cell across the luminal membrane. The nutrient molecules move out of the cell into the blood across the basal membrane by facilitated diffusion, completing the absorptive process.

16.7 (Questions on p. 614.)

1. Large-intestine *haustral contractions* are very similar to small-intestine *segmentation contractions* in that both are rhythmic, autonomous, oscillating ringlike contractions initiated by pacemaker cells, but haustral contractions occur much less frequently (once every 30 minutes) than segmentation contractions (9 to 12 per minute). Haustral contractions, which are nonpropulsive, shuffle the colonic content back and forth, facilitating salt and water absorption and thereby compacting the content and forming a firm fecal mass. In contrast, segmentation in the small intestine is both a slowly propulsive and mixing movement.

2. The $NaHCO_3$ produced by the *large intestine* mainly serves to protect the large intestine from acid produced during colonic fermentation by the intestinal bacteria. The primary role of *pancreatic* $NaHCO_3$ secreted into the duodenal lumen is to neutralize the acidic chyme that empties from the stomach, thereby protecting the small intestine from acid damage in addition to creating an alkaline environment that optimizes pancreatic digestive enzyme activity.

3. The gut *microbiota* (1) promote colon motility, (2) enhance mucosal integrity, (3) aid immune function, (4) compete with potential gut pathogens, (5) help digest food, and (6) influence the brain and behavior.

16.8 (Questions on p. 615.)

1. *Trophic hormones* induce growth of their target tissues. For example, gastrin is trophic to the gastric mucosa in addition to stimulating secretion by the parietal and chief cells. The significance is that the hormone helps maintain the gastric mucosa, despite harsh conditions, thereby maintaining the secretory capabilities of this stomach lining.

2. *secretin*: stomach smooth muscle and parietal cells (acts as an enterogastrone to inhibit gastric motility and acid secretion), pancreatic duct cells (stimulates $NaHCO_3$ secretion); *CCK*: stomach smooth muscle and parietal cells (acts as an enterogastrone), pancreatic acinar cells (stimulates secretion of pancreatic digestive enzymes), gallbladder (causes contraction), sphincter of Oddi (causes relaxation), and brain (signals satiety)

3. *GIP* stimulates insulin secretion in feedforward fashion.

Figure Focus

FIGURE 16-3 (P. 573): Extrinsic autonomic nerves can be activated via receptors in the digestive tract that respond to local changes in the digestive tract and by external influences arising from outside the digestive tract. Extrinsic autonomic nerves directly control smooth muscle motility, digestive exocrine gland secretion, and gastrointestinal hormone secretion and further indirectly control these effectors by acting through the intrinsic nerve plexuses and GI hormones.

FIGURE 16-4 (P. 574): Eating food increases salivary secretion via the simple salivary reflex, whereas smelling food increases salivary secretion by means of a conditioned salivary reflex.

FIGURE 16-5 (P. 577): After being pushed out of the stomach and up the esophagus into the pharynx, vomit can accidentally pass through an open glottis into the trachea (that is, be aspirated) instead of passing into the mouth and out. This can happen if a person tries to inspire while vomit is in the pharynx.

FIGURE 16-8 (P. 584): Gastric HCl secretion is reduced and the venous blood leaving the stomach is not as alkaline as normal.

FIGURE 16-12 (P. 592): The fat (especially) and protein in the bacon and eggs stimulate release of CCK. The acid in the orange juice stimulates release of secretin. Even more importantly, HCl secreted by the stomach in response to protein-triggered gastrin release is emptied into the duodenum, where it potently stimulates secretin secretion. The carbohydrate in the toast has no direct effect on GI hormones.

FIGURE 16-13 (P. 593): (1) hepatic artery, (2) hepatic portal vein, (3) hepatic vein

FIGURE 16-23 (P. 606): Fructose is absorbed passively by facilitated diffusion. Glucose and galactose are absorbed by secondary active transport.

Reviewing Terms and Facts

(Questions on p. 616.)

1. F **2.** T **3.** T **4.** F **5.** F **6.** F **7.** long, short **8.** medulla **9.** chyme **10.** three **11.** vitamin B_{12}, bile salts **12.** bile salts **13.** 1.c, 2.e, 3.b, 4.a, 5.f, 6.d **14.** b **15.** 1.c, 2.c, 3.d, 4.a, 5.e, 6.c, 7.a, 8.b, 9.c, 10.b, 11.d, 12.b

Solving Quantitative Exercises

(Questions on p. 617.)

1. a. $r = 0.5$ cm, area $= 4\pi (0.25) = \pi$ cm^2

volume $= \left(\frac{4}{3}\pi\right)(0.5)^3 = 0.5236$ cm^3

area/volume $= 6$

b. Each new droplet's volume is 5.236×10^{-3} cm^3, so the average radius is therefore 0.1077 cm. The area of a sphere with that radius is

$4\pi (0.1077$ cm$)^2 = 0.1458$ cm^2

area/volume $= 27.8$

c. Area emulsified/area droplet $= (100)(0.1458$ cm$^2)/\pi$ cm^2

$= 4.64$

Thus, the total surface area of all 100 emulsified droplets is 4.64 times the area of the original larger lipid droplet.

d. Volume emulsified/volume droplet $= (100)(5.236 \times 10^{-3}$ cm$^3)/$ 0.5236 cm$^3 = 1.0$.

Thus, the total volume did not change as a result of emulsification, as would be expected because the total volume of the lipid is conserved during emulsification. The volume originally present in the large droplet is divided up among the 100 emulsified droplets.

Applying Clinical Reasoning

(Question on p. 617.)

A person whose bile duct is blocked by a gallstone experiences a painful "gallbladder attack" after eating a high-fat meal because the ingested fat triggers the release of cholecystokinin, which stimulates gallbladder contraction. As the gallbladder contracts and bile is squeezed into the blocked bile duct, the duct becomes distended before the blockage. This distension is painful.

The feces are grayish white because no bilirubin-containing bile enters the digestive tract when the bile duct is blocked. Bilirubin, when acted on by bacterial enzymes, is responsible for the brown color of feces, which are grayish white in its absence.

Thinking at a Higher Level

(Questions on p. 617.)

1. Patients who have had their stomachs removed must eat small quantities of food frequently instead of consuming the typical three meals a day because they have lost the ability to store food in the stomach and meter it into the small intestine at an optimal rate. If a person without a stomach consumed a large meal that entered the small intestine all at once, the luminal contents would quickly become too hypertonic as digestion of the large nutrient molecules into a multitude of small, osmotically active, absorbable units outpaced the more slowly acting process of absorption of these units. As a consequence of this increased luminal osmolarity, water would enter the small-intestine lumen from the plasma by osmosis, resulting in circulatory disturbances and intestinal distension. To prevent this "dumping syndrome" from occurring, the patient must "feed" the small intestine only small amounts of food at a time so that absorption of the digestive end products can keep pace with their rate of production. The person has to consciously take over metering the delivery of food into the small intestine because the stomach is no longer present to assume this responsibility.

2. The gut-associated lymphoid tissue launches an immune attack against any pathogenic (disease-causing) microorganisms that enter the readily accessible digestive tract and escape destruction by salivary lysozyme or gastric HCl. This action defends against entry of these potential pathogens into the body proper. The large number of immune cells in the gut-associated lymphoid tissue is adaptive as a first line of defense against foreign invasion when considering that the surface area of the digestive tract lining represents the largest interface between the body proper and the external environment.

3. Defecation would be accomplished entirely by the defecation reflex in a patient paralyzed from the waist down because of lower spinal-cord injury. Voluntary control of the external anal sphincter would be impossible because of interruption in the descending pathway between the primary motor cortex and the motor neuron supplying this sphincter.

4. When insufficient glucuronyl transferase is available in the neonate to conjugate all of the bilirubin produced during erythrocyte degradation with glucuronic acid, the extra unconjugated bilirubin cannot be excreted into the bile. Therefore, this extra bilirubin remains in the body, giving rise to mild jaundice in the newborn.

5. Removal of the stomach leads to pernicious anemia because of the resultant lack of intrinsic factor, which is necessary for absorption of vitamin B_{12}. Removal of the terminal ileum leads to pernicious anemia because this is the only site where vitamin B_{12} can be absorbed.

Chapter 17 Energy Balance and Temperature Regulation

Check Your Understanding

17.1 (Questions on p. 627.)

1. *External work* is the energy expended by skeletal muscles to move external objects or to move the body in relation to the environment. *Internal work* constitutes all biological energy expenditure that does not accomplish mechanical work outside the body; it includes skeletal muscle activity associated with postural maintenance and shivering and all energy-expending activities essential for sustaining life, such as pumping blood or breathing. *Metabolic rate* is the rate at which energy is expended during both external and internal work: Metabolic rate = energy expenditure/unit of time. *Appetite signals* give the sensation of hunger, driving us to eat. *Satiety signals* give the sensation of being full, suppressing the desire to eat. *Adiposity signals* are indicative of the size of fat stores in adipose tissue and are important in the long-term control of body weight. *Adipokines* refer to hormones secreted by adipocytes. *Visceral fat* is the deep, "bad" fat that surrounds the abdominal organs and is likely to be chronically inflamed and secrete harmful adipokines. *Subcutaneous fat* is deposited under the skin (it is the fat you can pinch) and is less harmful than visceral fat.

2. The basal metabolic rate can be determined indirectly by measuring a person's O_2 uptake per unit of time and multiplying this value by the energy equivalent of O_2, which is 4.8 kilocalories of energy liberated per liter of O_2 consumed. This technique is based on the fact that a direct relationship exists between the volume of O_2 used and the quantity of heat produced as follows: Food $+ O_2 \rightarrow CO_2 + H_2O +$ energy (mostly transformed into heat).

3. See Table 17-2, p. 624.

17.2 (Questions on p. 635.)

1. *radiation:* transfer of heat energy from a warmer object to a cooler object in the form of heat waves that travel through space; *conduction:* transfer of heat between objects of differing temperatures that are in direct contact with each other; *convection:* transfer of heat energy by air currents; *evaporation:* conversion of a liquid such as sweat to a gaseous vapor, a process that requires heat, which is absorbed from the skin

2. See Table 17-3, p. 633.

3. the environmental temperature range within which core temperature can be kept constant by varying skin blood flow without calling into play supplementary heat-production mechanisms such as shivering or heat-loss measures such as sweating.

Figure Focus

FIGURE 17-2 (P. 622): Increased leptin released by expanded body fat stores inhibits the appetite-enhancing pathway and stimulates the appetite-suppressing pathway. It does so by inhibiting secretion of NPY, which is a potent appetite stimulator, and by stimulating secretion of melanocortins, which are appetite suppressors. Farther downstream in the appetite-enhancing pathway, orexins, which are potent stimulators of food intake, are inhibited by the higher level of melanocortins, and their excitation by NPY is suppressed by the decline in NPY. Simultaneously, downstream in the appetite-suppressing pathway, corticotropin-releasing hormone and similar appetite suppressing mediators are stimulated by the stepped-up secretion rate of melanocortins, and their inhibition by NPY is reduced as NPY's rate of secretion drops. Furthermore, the decrease in NPY and increase in melanocortins work together at the satiety center to decrease food intake.

FIGURE 17-4 (P. 629): Assume environmental temperature is lower than body temperature, which it would be on a warm (not hot) afternoon. The person gains heat by radiation from the sun and by heat generated internally by jogging. The person loses heat to the surrounding air by radiation and conduction. Further loss of heat takes place via convection as a result of the person's movement during jogging, coupled with the effect of any wind. Also, if the person sweats, evaporative heat loss occurs.

FIGURE 17-5 (P. 631): Because eccrine sweat glands are stimulated to secrete by binding at their muscarinic receptors of acetylcholine released from sympathetic nerve fibers, drugs that block this action prevent secretion of evaporative heat-losing sweat, leading to increased risk of heat-related disorders.

Reviewing Terms and Facts

(Questions on p. 636.)

1. F **2.** F **3.** F **4.** F **5.** T **6.** T **7.** T **8.** T **9.** F **10.** arcuate nucleus **11.** shivering **12.** nonshivering thermogenesis **13.** sweating **14.** b **15.** 1.b, 2.a, 3.c, 4.a, 5.d, 6.c, 7.b, 8.d, 9.c, 10.b

Solving Quantitative Exercises

(Questions on p. 637.)

1. From physics we know that $\Delta T(°C) = \Delta U/(C \times m)$. Also note that $\Delta U/t = BMR$ (i.e., the rate of using energy is the basal metabolic

rate). m represents the mass of body fluid; for a typical person, this is 42 L.

$$(42 \text{ L}) \times (1 \text{ kg/L}) = 42 \text{ kg}$$
$$C = 1.0 \text{ kcal/(kg-°C)}$$

Given that water boils at 100°C and normal body temperature is 37°C, we need to change the temperature by 63°C. Thus,

$$t = (\Delta T \times C \times m)/BMR = (63°C)[1.0 \text{ kcal/(kg-°C)}] (42 \text{ kg})/ (75 \text{ kcal/hr}) - 35 \text{ hr}$$

At the higher metabolic rate during exercise,

$$t = (63°C)[1.0 \text{ kcal/(kg-°C)}](42 \text{ kg})/(1000 \text{ kcal/hr}) = 2.6 \text{ hr}$$

Applying Clinical Reasoning

(Question on p. 637.)

Cooled tissues need less nourishment than they do at normal body temperature because of their pronounced reduction in metabolic activity. The lower O_2 need of cooled tissues accounts for the occasional survival of drowning victims who have been submerged in icy water considerably longer than one could normally survive without O_2.

Thinking at a Higher Level

(Questions on p. 637.)

1. CCK serves as a satiety signal: It serves as a signal to stop eating when enough food has been consumed to meet the body's energy needs, even though the food is still in the digestive tract. Therefore, when drugs that inhibit CCK release are administered to experimental animals, the animals overeat because this satiety signal is not released.

2. Do not go on a "crash diet." Be sure to eat a nutritionally balanced diet that provides all essential nutrients, but reduce total caloric intake, especially by cutting down on high-fat foods. Spread out consumption of the food throughout the day instead of just eating several large meals. Avoid bedtime snacks. Burn more calories through a regular exercise program.

3. Engaging in heavy exercise on a hot day is dangerous because of problems arising from trying to eliminate the extra heat generated by the exercising muscles. First, there are conflicting demands for distribution of the cardiac output—temperature-regulating mechanisms trigger skin vasodilation to promote heat loss from the skin surface, whereas metabolic changes within the exercising muscles induce local vasodilation in the muscles to match the increased metabolic needs with increased blood flow. Further aggravating the problem of conflicting demands for blood flow is the loss of effective circulating plasma volume resulting from the loss of a large volume of fluid through another important cooling mechanism, sweating. Therefore, it is difficult to maintain an effective plasma volume and blood pressure and simultaneously keep the body from overheating when engaging in heavy exercise in the heat, so heat exhaustion is likely to ensue.

4. When a person is soaking in a hot bath, loss of heat by radiation, conduction, convection, and evaporation is limited to the small surface area of the body exposed to the cooler air. Heat is being gained by conduction at the larger skin surface area exposed to the hotter water.

5. The thermoconforming fish would not run a fever when it has a systemic infection because it has no mechanisms for regulating internal heat production or for controlling heat exchange with its environment. The fish's body temperature varies capriciously with the external environment no matter whether it has a systemic infection or not. It is not able to maintain body temperature at a "normal" set point or an elevated set point (i.e., a fever).

Chapter 18 Principles of Endocrinology; The Central Endocrine Glands

Check Your Understanding

18.1 (Questions on p. 646.)

1. The primary function of a *tropic hormone* is to regulate hormone secretion by another endocrine gland.

2. (1) *Negative feedback* maintains the plasma concentration of a hormone at a given set level because the output of the hormonal control system counteracts a change in the input. For example when the concentration of a hormone falls, the control system triggers increased secretion of the hormone, which feeds back to shut off the stimulatory response when the set level is achieved. (2) *Neuroendocrine reflexes* produce a sudden increase in hormone secretion (that is, "turn up the thermostat setting") in response to a specific stimulus, which is often external to the body. (3) *Diurnal (circadian) rhythms* are rhythmic fluctuations up and down in the secretion rate of many hormones as a function of the time of day.

3. *Down regulation* is a reduction in the number of target-cell receptors in the face of a prolonged increase in a hormone. *Permissiveness* refers to one hormone having to be present in adequate amounts to permit another hormone to fully exert its effects. *Synergism* results when the combined effect of two hormones is greater than the sum of their separate effects. *Antagonism* takes place when one hormone decreases the effectiveness of another hormone.

18.2 (Questions on p. 652.)

1. See Figure 18-6, p. 651.

2. *posterior pituitary hormones:* vasopressin (antidiuretic hormone, ADH) and oxytocin; *anterior pituitary hormones:* growth hormone (GH, somatotropin); thyroid-stimulating hormone (TSH, thyrotropin); adrenocorticotropic hormone (ACTH, corticotropin); follicle-stimulating hormone (FSH); luteinizing hormone (LH); and prolactin (PRL); *hypophysiotropic hormones:* thyrotropin-releasing hormone (TRH); corticotropin-releasing hormone (CRH); gonadotropin-releasing hormone (GnRH); growth hormone–releasing hormone (GHRH); somatostatin (growth hormone–inhibiting hormone, GHIH); prolactin-releasing peptide (PrRP); and dopamine (prolactin-inhibiting hormone, PIH)

3. The hypothalamus controls hormonal output from the posterior pituitary by a neural connection and controls hormonal output from the anterior pituitary by a vascular connection. The posterior pituitary is a neural extension of the hypothalamus. Neuronal cell bodies in the hypothalamus produce vasopressin and oxytocin, which are transported down the axons that pass through the connecting stalk to the posterior pituitary, where these hormones are stored in the neuronal terminals. When stimulated by the hypothalamus, these hormones are independently released into the systemic blood. The hypothalamus secretes hypophysiotropic hormones into the hypothalamic–hypophyseal portal system, a capillary-to-capillary vascular link that transports them through the connecting stalk to the anterior pituitary, where they control the secretion of hormones produced by the anterior pituitary into the systemic blood.

18.3 (Questions on p. 660.)

1. GH directly ↑ fatty acid levels in the blood by enhancing the breakdown of triglyceride stores in adipose tissue and ↑ blood glucose levels by decreasing glucose uptake by muscles and increasing glucose output by the liver, thus mobilizing fat stores as a major energy source for muscle while conserving glucose for the glucose-dependent brain. GH also directly and indirectly (via insulin-like growth factor-I [IGF-I]) brings about protein synthesis, ↓ blood amino acids in the process.

2. GH does not act directly to bring about most of its growth-producing actions (increased cell division, enhanced protein synthesis, and bone growth). Instead, GH stimulates the liver to release IGF-I, which directly mediates these growth-promoting actions. GH's only direct growth-promoting action is to stimulate protein synthesis, which it does in conjunction with IGF-I.

3. Bone grows in *thickness* by osteoblasts adding new bone on top of the outer surface of existing bone simultaneously with osteoclasts expanding the marrow cavity. Bone grows in *length* as chondrocytes (cartilage cells) in the epiphyseal plate form a new-growth template that is replaced by bone through osteoblast activity after the old chondrocytes die.

18.4 (Questions on p. 663.)

1. The self-induced rhythmic firing of the neurons in the *suprachiasmatic nucleus (SCN)* (brought about by cyclic synthesis and degradation of clock proteins) establishes many of the body's inherent daily rhythms. The fluctuating levels of clock proteins bring about cyclic changes in neural output from the SCN that lead to cyclic changes in effector organs throughout the day. Accordingly, the SCN is the *master biological clock* or pacemaker for the body's diurnal (circadian) rhythms.

2. *Melanopsin* is the receptor for light found in special retinal ganglion cells that informs the SCN about the absence or presence of light. The SCN relays the message regarding the light status to the pineal gland, which increases secretion of the hormone *melatonin* in the dark. Fluctuations in melatonin secretion help entrain the body's biological rhythm established by the SCN (which on its own cycles a bit slower than the 24-hour environmental cycle) to stay in sync with the external light–dark cycle.

3. *The optic nerve pathway* carries visual information from the rods and cones in the eyes to the visual cortex in the occipital lobes. The *retino-hypothalamic tract* carries circadian rhythm-related information from the light-detecting melanopsin-secreting retinal ganglion cells to the SCN.

Figure Focus

FIGURE 18-3 (P. 646): Because the pituitary gland lies beneath the optic chiasm (the point at which half of the nerve fibers passing from the eyes cross over on their way to the visual cortex), enlargement of this gland as a result of an anterior pituitary tumor can cause visual disturbances by pressing against this part of the visual pathway.

FIGURE 18-5 (P. 649): TSH, ACTH, GH, LH, and FSH

FIGURE 18-7 (P. 652): *across hypothalamic capillaries:* hypophysiotropic hormones enter blood; *across anterior pituitary capillaries;* hypophysiotropic hormones leave blood, anterior pituitary hormones enter blood

FIGURE 18-10 (P. 656): GHRH stimulates and somatostatin (GHIH) inhibits growth hormone (GH) secretion by the anterior pituitary. GH stimulates IGF-I secretion by the liver. In negative-feedback fashion, GH and IGF-I both inhibit GHRH secretion and stimulate somatostatin secretion by the hypothalamus. IGF-I also inhibits GH secretion by the anterior pituitary.

Reviewing Terms and Facts

(Questions on p. 663.)

1. T 2. T 3. F 4. T 5. T 6. T 7. tropic 8. down regulation 9. epiphyseal plate 10. suprachiasmatic nucleus 11. 1.c, 2.b, 3.b, 4.a, 5.a, 6.c, 7.b

Applying Clinical Reasoning

(Question on p. 664.)

Hormonal replacement therapy following pituitary gland removal should include thyroid hormone (the thyroid gland does not produce sufficient thyroid hormone in the absence of TSH) and glucocorticoid (because of the absence of ACTH), especially in stress situations. If indicated, male or female sex hormones can be replaced, even though these hormones are not essential for survival. For example, testosterone in males plays an important role in libido. Growth hormone and prolactin need not be replaced because their absence produces no serious consequences in this individual. Vasopressin may have to be replaced if the blood at the hypothalamus picks up insufficient quantities of this hormone in the absence of the posterior pituitary.

Thinking at a Higher Level

(Questions on p. 664.)

1. The concentration of hypothalamic releasing and inhibiting hormones would be considerably lower (in fact, almost nonexistent) in a systemic venous blood sample compared to the concentration of these hormones in a sample of hypothalamic–hypophyseal portal blood. These hormones are secreted into the portal blood for local delivery between the hypothalamus and anterior pituitary. Any portion of these hormones picked up by the systemic blood at the anterior pituitary capillary level is greatly diluted by the much larger total volume of systemic blood compared to the extremely small volume of blood within the portal vessel.

2. Above normal. Without sufficient iodine, the thyroid gland is unable to synthesize enough thyroid hormone. The resultant reduction in negative-feedback activity by the reduced level of thyroid hormone would lead to increased TSH secretion. Despite the elevated TSH, however, the thyroid gland still could not secrete adequate thyroid hormone because of the iodine deficiency.

3. If CRH and/or ACTH are elevated in accompaniment with the excess cortisol secretion, the condition is secondary to a defect at the hypothalamic/anterior pituitary level. If CRH and ACTH levels are below normal in accompaniment with the excess cortisol secretion, the condition is caused by a primary defect at the adrenal cortex level, with the excess cortisol inhibiting the hypothalamus and anterior pituitary in negative-feedback fashion.

4. Males with *testicular feminization syndrome* would be unusually tall because of the inability of testosterone (aromatized to estradiol) to promote closure of the epiphyseal plates of the long bones in the absence of testosterone receptors.

5. Full-grown athletes sometimes illegally take supplemental doses of growth hormone because it promotes increased skeletal muscle mass through its and IGF-I's-protein anabolic effect. However, excessive growth hormone can have detrimental side effects, such as possibly causing diabetes or high blood pressure.

Chapter 19 The Peripheral Endocrine Glands

Check Your Understanding

19.1 (Questions on p. 671.)

1. A *thyroid follicle,* the functional unit of the thyroid gland, is a hollow sphere consisting of a single layer of follicular cells enclosing an inner lumen filled with *colloid,* which serves as an extracellular storage site for thyroid hormone. Colloid is filled with *thyroglobulin,* a large glycoprotein within which are incorporated the thyroid hormones in their various stages of synthesis. During synthesis, attachment of one iodide to tyrosine yields *MIT* (monoiodotyrosine), and attachment of two iodides to tyrosine yields *DIT* (di-iodotyrosine). Coupling of one MIT and one DIT yields T_3 (tri-iodothyronine), and coupling of two DITs yields T_4 (tetraiodothyronine or thyroxine). T_3 and T_4 are collectively referred to as thyroid hormone.

2. The *iodide trap* is the powerful symporter that transports iodide against its concentration gradient from the blood into the follicular cell.

3. See Figure 19-3, p. 669.

19.2 (Questions on p. 682.)

1. (1) *mineralocorticoids:* aldosterone (promotes Na^+ retention and K^+ elimination during urine formation); (2) *glucocorticoids:* cortisol (increases blood glucose at the expense of protein and fat stores and helps the body adapt to stress); and (3) *sex hormones:* dehydroepiandrosterone (governs androgen-dependent processes in the female, such as growth of pubic and axillary hair, enhancing pubertal growth spurt, and maintaining the sex drive)

2. ACTH stimulates the adrenal cortex to secrete cortisol but has no effect on aldosterone.

3. *epinephrine* and *norepinephrine,* which are stored in chromaffin granules, from which they are secreted into the blood by exocytosis on stimulation by preganglionic sympathetic fibers

19.3 (Questions on p. 685.)

1. (1) ↑ epinephrine: reinforces sympathetic nervous system in "fight-or-flight responses"; mobilizes energy stores (↑ blood glucose and blood fatty acids); (2) ↑ CRH–ACTH–cortisol: mobilizes energy stores and metabolic building blocks (↑ blood glucose, blood fatty acids, and blood amino acids); (3) ↑ glucagon and ↓ insulin: act in concert to ↑ blood glucose and blood fatty acids; (4) ↑ renin–angiotensin–aldosterone and ↑ vasopressin: conserve salt and H_2O to expand plasma volume; angiotensin II and vasopressin cause arteriolar vasoconstriction. These actions help sustain blood pressure if acute loss of plasma volume occurs.

2. The hypothalamus (1) increases sympathetic nervous system activity, which (a) prepares the body for "fight or flight"; (b) promotes arteriolar vasoconstriction that reduces blood flow to the kidneys, thereby triggering ↑ renin–angiotensin–aldosterone activity; and (c) stimulates the adrenal medulla to ↑ epinephrine secretion. Epinephrine reinforces the sympathetic nervous system and additionally acts on the pancreas to ↑ glucagon and ↓ insulin secretion. The hypothalamus further (2) secretes CRH, thereby ↑ CRH–ACTH–cortisol activity; and (3) stimulates the posterior pituitary to ↑ vasopressin secretion.

19.4 (Questions on p. 701.)

1. *Glycogenesis* is the conversion of glucose to glycogen, *glycogenolysis* is the conversion of glycogen to glucose, and *gluconeogenesis* is the conversion of amino acids to glucose.

2. *Insulin* ↓ blood glucose and promotes carbohydrate storage by facilitating glucose transport into most cells via transporter recruitment and by stimulating glycogenesis and inhibiting glycogenolysis and gluconeogenesis. It ↓ blood fatty acids and promotes triglyceride storage by increasing the entry of fatty acids and glucose into adipose cells, promoting triglyceride synthesis, and inhibiting lipolysis. It ↓ blood amino acids and enhances protein synthesis by promoting active transport of amino acids into muscles, stimulating protein synthesis, and inhibiting protein degradation. *Glucagon* ↑ blood glucose and reduces carbohydrate stores by increasing hepatic glucose production via decreasing glycogenesis and by promoting glycogenolysis and gluconeogenesis. It ↑ blood fatty acids and reduces fat stores by promoting lipolysis and inhibiting triglyceride synthesis. It ↑ blood ketone levels by enhancing conversion of fatty acids to ketone bodies by the liver. Glucagon inhibits protein synthesis and promotes protein degradation in the liver but not in muscle, the body's major protein store, so it does not have any significant effect on blood amino acid levels.

3. Increased blood glucose stimulates the pancreatic β cells, thus increasing insulin secretion, and inhibits the pancreatic α cells, thus decreasing glucagon secretion.

4. *Type 1 diabetes mellitus* is caused by autoimmune destruction of β cells, so it is characterized by a lack of insulin secretion. Its onset most commonly begins in childhood. *Type 2 diabetes mellitus,* which is characterized by reduced sensitivity of insulin's target cells, most commonly arises in adulthood.

19.5 (Questions on p. 712.)

1. Following is the distribution of body Ca^{2+}: 99% in the skeleton and teeth, 0.9% in the cells, and 0.1% in the ECF. Of the Ca^{2+} in the ECF, half is bound in complexes and not available to participate in chemical reactions, and the other half is free Ca^{2+} that is biologically active. This free ECF Ca^{2+} plays a vital role in neuromuscular excitability, excitation–contraction coupling in cardiac and smooth muscle, stimulus–secretion coupling, excitation–secretion coupling, maintenance of tight junctions, and blood clotting.

2. *Calcium homeostasis* refers to maintenance of a constant free plasma Ca^{2+} concentration. *Calcium balance* refers to maintenance of a constant total amount of Ca^{2+} in the body.

3. PTH acts on bone, kidneys, and the intestine to increase plasma Ca^{2+} as follows: PTH stimulates a fast exchange of Ca^{2+} from the small labile pool of Ca^{2+} in the bone fluid into the plasma by activating Ca^{2+} pumps in the osteocytic–osteoblastic bone membrane. It induces a slow exchange of Ca^{2+} from the stable pool of Ca^{2+} in the bone minerals of the bone itself into the plasma by stimulating osteoclasts to dissolve bone. PTH acts on the kidneys to conserve Ca^{2+} and eliminate PO_4^{3-} during urine formation and to activate vitamin D. PTH indirectly increases Ca^{2+} and PO_4^{3-} absorption from the small intestine via its role in activating vitamin D, which directly promotes the intestinal absorption of these ingested electrolytes.

Figure Focus

FIGURE 19-2 (P. 667): Potential targets for blocking synthesis of T_3 and T_4 are inhibiting the iodide trap (symporter) or blocking TPO action. Indeed, drugs are available for both targets. However, drugs that inhibit the iodide trap are rarely used because of toxic side effects. Drugs that block TPO action are in common use. These so-called *thionamides* prevent conversion of iodide to active I^-, inhibit attachment of active I^- to tyrosine in thyroglobulin, and halt coupling of iodotyrosines into thyroid hormone.

FIGURE 19-3 (P. 669): T_3 and T_4 concentrations would be decreased because insufficient iodide would be available for their synthesis. The resultant reduction in negative feedback by thyroid hormone at the hypothalamus and anterior pituitary would lead to increased TRH and TSH concentrations.

FIGURE 19-7 (P. 672): Cortisol and dehydroepiandrosterone secretion would increase; aldosterone secretion would not be affected.

FIGURE 19-8 (P. 673): True

FIGURE 19-13 (P. 684): (1) pathways that lead to mobilization of energy stores and metabolic building blocks: hypothalamus stimulates ↑ CRH–ACTH–cortisol secretion → ↑ blood glucose, blood fatty acids, and blood amino acids; hypothalamus stimulates sympathetic nervous system, which stimulates the adrenal medulla to increase epinephrine secretion, which acts on the endocrine pancreas to ↑ glucagon and ↓ insulin → ↑ blood glucose and blood fatty acids. (2) pathways that lead to salt and H_2O conservation: hypothalamus stimulates posterior pituitary to release vasopressin → ↑ H_2O reabsorption during urine formation; hypothalamus stimulates the sympathetic nervous system, which causes arteriolar vasoconstriction, including the afferent arterioles, leading to decreased blood flow through the kidneys, which triggers ↑ renin–angiotensin–aldosterone → ↑ salt and osmotic H_2O reabsorption during urine formation.

FIGURE 19-16 (P. 691): hepatic glucose production and transport of glucose into cells

FIGURE 19-17 (P. 693): Glucose that enters the β cell by facilitated diffusion is phosphorylated to glucose-6-phosphate, whose oxidation generates ATP, which closes the ATP-sensitive K^+ channels. The resultant reduced exit of K^+ depolarizes the β cell, an action that opens voltage-gated Ca^{2+} channels. Ca^{2+} entry triggers exocytosis of insulin vesicles, in this way causing insulin secretion.

FIGURE 19-24 (P. 708): (a), (d), (b), (c)

Reviewing Terms and Facts

(Questions on p. 713.)

1. F **2.** T **3.** T **4.** T **5.** F **6.** F **7.** F **8.** colloid, thyroglobulin **9.** glycogenesis, glycogenolysis, gluconeogenesis **10.** brain, working muscles, liver **11.** bone, kidneys, digestive tract **12.** osteoblasts, osteoclasts, osteocytes **13.** c **14.** a, b, c, g, i, j

Applying Clinical Reasoning

(Question on p. 714.)

"Diabetes of bearded ladies" is descriptive of both excess cortisol and excess adrenal androgen secretion. Excess cortisol secretion causes hyperglycemia and glucosuria. Glucosuria promotes osmotic diuresis, which leads to dehydration and a compensatory increased sensation of thirst. All these symptoms—hyperglycemia, glucosuria, polyuria, and polydipsia—mimic diabetes mellitus. Excess adrenal androgen secretion in females promotes masculinizing characteristics, such as beard growth. Simultaneous hypersecretion of both cortisol and adrenal androgen most likely occurs secondary to excess CRH–ACTH secretion because ACTH stimulates both cortisol and androgen production by the adrenal cortex.

Thinking at a Higher Level

(Questions on p. 714.)

1. The midwestern United States is no longer an endemic goiter belt, even though the soil is still iodine poor because individuals living in this region obtain iodine from iodine-supplemented nutrients, such as iodinated salt, and from seafood and other naturally iodine-rich foods shipped from coastal regions.

2. Anaphylactic shock is an extremely serious allergic reaction brought about by massive release of chemical mediators in response to exposure to a specific allergen—such as one associated with a bee sting—to which the individual has been highly sensitized. These chemical mediators bring about circulatory shock (severe hypotension) through a twofold effect: (1) by relaxing arteriolar smooth muscle, thus causing widespread arteriolar vasodilation and a resultant fall in total peripheral resistance and arterial blood pressure; and (2) by causing a generalized increase in capillary permeability, resulting in a shift of fluid from the plasma into the interstitial fluid. This shift decreases the effective circulating volume, further reducing arterial blood pressure. Additionally, these chemical mediators bring about pronounced bronchoconstriction, making it impossible for the victim to move sufficient air through the narrowed airways. Because these responses take place rapidly and can be fatal, people allergic to bee stings are advised to keep injectable epinephrine in their possession. By promoting arteriolar vasoconstriction through its action on α_1 receptors in arteriolar smooth muscle and promoting bronchodilation through its action on β_2 receptors in bronchiolar smooth muscle (see Table 7-2, p. 240), epinephrine counteracts the life-threatening effects of the anaphylactic reaction to the bee sting.

3. An infection elicits the stress response, which brings about increased secretion of cortisol and epinephrine, both of which increase the blood glucose level. This becomes a problem for diabetic patients

who have to bring down the elevated blood glucose by injecting additional insulin or, preferably, by reducing carbohydrate intake and/or exercising (difficult to do when sick) to use up some of the extra blood glucose. In a normal individual, the check-and-balance system between insulin and the other hormones that oppose insulin's actions helps maintain the blood glucose within reasonable limits during the stress response.

4. The presence of Chvostek's sign results from increased neuromuscular excitability caused by moderate hyposecretion of parathyroid hormone.

5. If malignancy-associated hypercalcemia arose from metastatic tumor cells that invaded and destroyed bone, both hypercalcemia and hyperphosphatemia would result as $Ca_3(PO_4)_2$ salts were released from the destructed bone. The fact that hypophosphatemia, not hyperphosphatemia, often accompanies malignancy-associated hypercalcemia led investigators to rule out bone destruction as the cause of the hypercalcemia. Instead, they suspected that the tumors produced a substance that mimics the actions of PTH in promoting concurrent hypercalcemia and hypophosphatemia.

Chapter 20 The Reproductive System

Check Your Understanding

20.1 (Questions on p. 723.)

1. *testes* in males and *ovaries* in females. In both sexes, these primary reproductive organs perform the dual function of producing gametes (sperm in males and ova in females) and secreting sex hormones (testosterone in males and estrogen and progesterone in females).

2. *Genetic sex* depends on the combination of sex chromosomes at fertilization, with males having an XY and females having an XX combination. *Gonadal sex* depends on whether testes or ovaries develop. The indifferent gonadal ridge develops into testes during the seventh week of intrauterine life under the influence of the testis-determining factor (TDF) secreted under direction of the sex-determining region of the Y chromosome (SRY). Because female embryos do not have a Y chromosome, they lack the masculinizing SRY and TDF, so by default the gonadal ridge develops into ovaries during the ninth week of gestation. *Phenotypic sex,* the apparent anatomic sex of an individual, is hormonally mediated and depends on the genetically determined gonadal sex. Embryos of both sexes have the same undifferentiated exter-

nal genitals and two duct systems, the Wolffian ducts and the Müllerian ducts. The fetal testes secrete testosterone, which transforms the Wolffian ducts into a male reproductive tract and, when converted into dihydrotestosterone, promotes development of the external genitals along male lines. Müllerian-inhibiting factor causes degeneration of the Müllerian ducts. In the absence of these masculinizing secretions, the Wolffian ducts degenerate, the *Müllerian* ducts develop into a female reproductive tract, and the undifferentiated external genitalia develop along female lines. The sex of the developing fetus can easily be distinguished by 10 to 12 weeks of gestation.

3. See chart below.

20.2 (Questions on p. 732.)

1. See Table 20-1, p. 725

2. The *seminiferous tubules* are the highly coiled tubular component of the testes where spermatogenesis takes place. The *Leydig cells,* which lie in the interstitial spaces between the seminiferous tubules, secrete testosterone. *Sertoli cells* are epithelial cells that lie in close association with and protect, nurse, and enhance the developing sperm cells throughout their development in the seminiferous tubules. Sertoli cells also secrete inhibin and are the site of action for control of spermatogenesis by both testosterone and FSH. *Spermatogenesis* is the sequence of steps by which relatively undifferentiated primordial germ cells proliferate and are converted into extremely specialized, motile spermatozoa. *Spermiogenesis* refers to the remodeling or packaging of the haploid spermatids into spermatozoa. *Spermiation* is the final release of a mature spermatozoon from the Sertoli cell to which it has been attached throughout development. *Spermatogonia* are the undifferentiated primordial germ cells that have a diploid number of single strand chromosomes. Two mitotic divisions of a spermatogonium yield four *primary spermatocytes,* each with a diploid number of double strand chromosomes. The first meiotic division of these primary spermatocytes yields eight *secondary spermatocytes,* each with a haploid number of double strand chromosomes. The second meiotic division of these secondary spermatocytes yields 16 *spermatids,* each with a haploid number of single strand chromosomes. The spermatids are packaged into highly specialized *spermatozoa,* each with a haploid number of single strand chromosomes.

3. See Figure 20-9, p. 729

4. See Table 20-2, p. 732

20.3 (Questions on p. 736.)

1. (1) *excitement phase* (erection and heightened sexual awareness); (2) *plateau phase* (intensification of sexual responses and systemic responses, such as increased heart rate, blood pressure, respiratory rate, and muscle tension); (3) *orgasmic phase* (ejaculation in

Embryonic Structure	Reproductive Structure Derived from This Embryonic Structure in Males	Reproductive Structure Derived from This Embryonic Structure in Females
Gonadal ridge	Testes	Ovaries
Genital tubercle	Penis	Clitoris
Urethral folds	Cord of erectile tissue that surrounds the urethra in the penis	Labia minora
Genital swellings	Scrotum and prepuce	Labia majora
Wolffian ducts	Epididymis, ductus deferens, ejaculatory duct, and seminal vesicles	Nothing; ducts degenerate
Müllerian ducts	Nothing; ducts degenerate	Oviducts, uterus, cervix, and upper vagina
Urogenital sinus	Prostate gland, bulbourethral gland	Lower vagina

males, orgasm in both sexes); and (4) *resolution phase* (return to pre-arousal state)

2. The *male sex act* includes erection and ejaculation. *Erection,* hardening of the normally flaccid penis to permit its entry into the vagina, is accomplished by engorgement of the penis erectile tissue with blood as a result of marked parasympathetically induced vasodilation of the penile arterioles and mechanical compression of the veins. *Ejaculation* occurs in two phases: The *emission phase,* emptying of sperm and accessory sex gland secretions (semen) into the urethra, is accomplished by sympathetically induced contraction of smooth muscle in the walls of the reproductive ducts and accessory sex glands. The *expulsion phase,* the forceful expulsion of semen from the penis, is accomplished by motor-neuron-induced contraction of the skeletal muscles at the base of the penis.

3. *Orgasm*, which culminates the sex act, is characterized in both sexes by involuntary rhythmic contractions of pelvic muscles at 0.8-second intervals and identical peak systemic responses, such as heavy breathing, markedly elevated heart rate, and generalized skeletal muscle contraction, all of which are associated with a feeling of intense pleasure, release, and complete gratification. The only differences between male and female orgasms are no accompanying ejaculation in females, and males become refractory to further sexual stimulation following an orgasm but females do not.

20.4 (Questions on p. 770.)

1. See Table 20-4, p. 743

2. The ovarian *follicle* secretes estrogen; the *corpus luteum* secretes progesterone (most) and estrogen. *Estrogen* stimulates growth of the endometrium and myometrium and induces synthesis of progesterone receptors in the endometrium. *Progesterone* acts on the estrogen-primed endometrium to convert it into a hospitable and nutritious lining (with loose, edematous connective tissue, glycogen stores, and increased blood supply) suitable for implantation. The *menstrual phase* of the uterine cycle takes place during the first half of the ovarian follicular phase, after the end of the last cycle when the estrogen and progesterone supply was withdrawn on degeneration of the corpus luteum. The highly developed uterine lining sloughs (menstruation) simultaneous with the beginning of a new follicular phase under the influence of rising FSH and LH on withdrawal of the inhibitory actions of estrogen and progesterone. The *proliferative phase* of the uterine cycle occurs during the last half of the ovarian follicular phase as the endometrium stops sloughing and starts to repair itself and proliferate under the influence of rising levels of estrogen from the newly recruited, rapidly growing antral follicles. Peak estrogen secretion causes the LH surge, which triggers ovulation, ending both the ovarian follicular phase and the uterine proliferative phase. The *secretory,* or *progestational, phase* of the uterine cycle begins concurrent with the ovarian luteal phase as progesterone from the corpus luteum converts the thickened, estrogen-primed endometrium to a lush environment capable of supporting an early embryo should the released egg be fertilized and implant.

3. *zygote:* fertilized ovum, following union of the male and female chromosomes; *blastocyst:* a single-layer hollow ball of about 50 cells resulting from mitotic cell divisions of the zygote; the blastocyst is the developmental stage that implants in the endometrium; *inner cell mass:* a dense mass on one side of the blastocyst that is destined to become the fetus; *trophoblast:* the thin outermost layer of the blastocyst that accomplishes implantation, after which it develops into the fetal part of the placenta; *decidua:* the endometrial tissue modified at the implantation site to enhance its ability to support the implanting embryo; *chorion:* the thickened trophoblastic layer after implantation is complete; *placenta:* the specialized organ of exchange between the maternal and fetal blood that is derived from both trophoblastic embryonic tissue and decidual maternal tissue and that secretes peptide and steroid hormones essential for maintaining pregnancy; *embryo:* the product of fertilization during the first 2 months of intrauterine development when tissue differentiation is taking place; *fetus:* the product of fertilization during the last 7 months of gestation after differentiation is complete and tremendous tissue growth and maturation occur

4. During *parturition,* a positive-feedback loop involving *oxytocin,* a powerful uterine muscle stimulant, is responsible for the progression of labor. Oxytocin promotes uterine muscle contractions that force the fetus against the cervix, dilating it and triggering a neuroendocrine reflex that results in secretion of even more oxytocin, which stimulates even stronger contractions, and so on as labor progresses until the cervix is dilated sufficiently for the baby to be pushed out. During *breast-feeding,* oxytocin causes milk ejection (milk letdown) by stimulating contraction of the myoepithelial cells surrounding the milk-secreting alveoli.

Figure Focus

FIGURE 20-3 (P. 720): The individual would have testes; a male reproductive tract (epididymis, ductus deferens, ejaculatory duct); seminal vesicles; and female external genitalia (clitoris, labia).

FIGURE 20-7 (P. 727): (1) 16 (2) *Spermatogonium* has 46 chromosomes; spermatozoon has 23 chromosomes.

FIGURE 20-9 (P. 729): (1) *LH* stimulates Leydig cells to secrete testosterone. (2) LH secretion is inhibited by testosterone inhibiting kisspeptin and subsequently GnRH secretion at the hypothalamus and selectively inhibiting LH secretion at the anterior pituitary. (3) *FSH* acts on Sertoli cells to stimulate spermatogenesis and secretion of inhibin. (4) FSH secretion is inhibited via testosterone's inhibition of kisspeptin and subsequently GnRH at the hypothalamus. Furthermore, inhibin selectively inhibits FSH secretion at the anterior pituitary.

FIGURE 20-14 (P. 743): (1) stages *without influence by ovarian cycle*: resting primordial follicle, preantral follicle, early antral follicle (steps 1, 2, and 3 in figure); (2) stages during follicular phase: recruitable antral follicle, dominant mature follicle (steps 4 and 5); (3) stages during luteal phase: developing and degenerating corpus luteum (steps 7, 8, and 9)

FIGURE 20-17 (P. 745): *Day 8:* FSH is high but declining and LH secretion is on the rise. FSH and LH stimulate oocyte development and follicular growth; the ovarian cycle is in the follicular phase. Increased estrogen secretion by the follicle is just beginning. No progesterone is secreted. The endometrium is repairing itself and starting to thicken under the influence of rising estrogen levels. The uterine cycle is in the proliferative phase, following the preceding menstrual phase. *Day 21:* FSH and LH secretion are declining under the influence of peak progesterone levels from the mature corpus luteum. The ovarian cycle is in the luteal phase. The estrogen level is elevated but not as high as at its peak at the end of the follicular phase when it triggered the LH surge responsible for ovulation and formation of the CL. Under the influence of high progesterone levels, the endometrium is undergoing vascular and secretory changes to prepare for implantation should the released ovum be fertilized. The uterine cycle is in the secretory phase.

FIGURE 20-18 (P. 746): granulosa cells

FIGURE 20-20 (P. 748): In negative-feedback fashion, rising, moderate levels of estrogen inhibit kisspeptin and subsequently GnRH at the hypothalamus and selectively inhibit FSH secretion at the anterior pituitary, thus suppressing secretion of the factors promoting estrogen secretion (although LH continues to slowly rise [tonic LH secretion] under incomplete suppression by estrogen). By contrast, in positive-feedback fashion, LH-induced high levels of estrogen stimulate a different population of kisspeptin-secreting neurons at the hypothalamus and selectively stimulate LH secretion at the anterior pituitary, giving rise to the LH surge that triggers ovulation.

FIGURE 20-27 (P. 762): (1) The placenta secretes estrogen and progesterone after the first trimester. (2) The concentrations of these hormones drop precipitously at delivery when the placenta is delivered and no longer secreting hormones.

Reviewing Terms and Facts

(Questions on p. 771.)

1. T 2. T 3. F 4. T 5. F 6. T 7. seminiferous tubules, FSH, testosterone 8. thecal, LH, granulosa, FSH 9. corpus luteum of pregnancy, placenta 10. human chorionic gonadotropin 11. c 12. e 13. 1.c, 2.a, 3.b, 4.c, 5.a, 6.a, 7.c, 8.b, 9.e 14. 1.a, 2.c, 3.b, 4.a, 5.a, 6.b

Applying Clinical Reasoning

(Question on p. 772.)

The first warning of a tubal pregnancy is pain caused by stretching of the oviduct by the growing embryo. A tubal pregnancy must be surgically terminated because the oviduct cannot expand as the uterus does to accommodate the growing embryo. If not removed, the enlarging embryo ruptures the oviduct, causing possibly lethal hemorrhage.

Thinking at a Higher Level

(Questions on p. 772.)

1. The anterior pituitary responds only to the normal pulsatile pattern of GnRH and does not secrete gonadotropins in response to continuous exposure to GnRH. In the absence of FSH and LH secretion, ovulation and other events of the ovarian cycle do not ensue, so continuous GnRH administration may find use as a contraceptive technique.

2. Testosterone hypersecretion in a young boy causes premature closure of the epiphyseal plates so that he stops growing before he reaches his genetic potential for height. The child would also display signs of precocious pseudopuberty, characterized by premature development of secondary sexual characteristics, such as deep voice, beard, enlarged penis, and sex drive.

3. A potentially troublesome side effect of drugs that inhibit sympathetic nervous system activity as part of the treatment for high blood pressure is males' inability to carry out the sex act. Both divisions of the autonomic nervous system are required for the male sex act. Parasympathetic activity is essential for accomplishing erection, and sympathetic activity is important for ejaculation.

4. Posterior pituitary extract contains an abundance of stored oxytocin, which can be administered to induce or facilitate labor by increasing uterine contractility. Exogenous oxytocin is most successful in inducing labor if the woman is near term because of the increasing concentration of myometrial oxytocin receptors and formation of gap junctions between the uterine smooth muscle cells at that time.

5. GnRH or FSH and LH are not effective in treating the symptoms of menopause because the ovaries are no longer responsive to the gonadotropins. Thus, treatment with these hormones would not cause estrogen and progesterone secretion. In fact, GnRH, FSH, and LH levels are already elevated in postmenopausal women because of lack of negative feedback by the ovarian hormones.

Glossary

A band One of the dark bands that alternate with light (I) bands to create a striated appearance in a skeletal or cardiac muscle fiber when these fibers are viewed with a light microscope

absorption The transfer of digested nutrients and ingested liquids from the digestive tract lumen into the blood or lymph

absorptive state The metabolic state following a meal when nutrients are being absorbed and stored; fed state

accessory digestive organs Exocrine organs outside the wall of the digestive tract that empty their secretions through ducts into the digestive tract lumen

accessory sex glands Glands that empty their secretions into the reproductive tract

accommodation The ability to adjust the strength of the lens in the eye so that both near and far sources can be focused on the retina

acetylcholine (ACh) (ass′-uh-teal-KŌ–lēn) The neurotransmitter released from all autonomic preganglionic fibers, parasympathetic postganglionic fibers, and motor neurons

acetylcholinesterase (AChE) (ass′-uh-teal-kō-luh-NES-tuh-rās) An enzyme present in the motor end-plate membrane of a skeletal muscle fiber that inactivates acetylcholine

ACh See *acetylcholine*

AChE See *acetylcholinesterase*

acid A hydrogen-containing substance that yields a free hydrogen ion and anion on dissociation

acidosis (ass-i-DŌ-sus) Blood pH of less than 7.35

acini (ĀS-i-ni) The secretory component of saclike exocrine glands, such as digestive enzyme–producing pancreatic glands or milk-producing mammary glands

acquired immune responses Responses that are selectively targeted against particular foreign material to which the body has previously been exposed; see also *antibody-mediated immunity* and *cell-mediated immunity*

ACTH See *adrenocorticotropic hormone*

actin The contractile protein forming the backbone of the thin filaments in muscle fibers

action potential A brief, rapid, large change in membrane potential that serves as a long-distance electrical signal in an excitable cell

active expiration Emptying of the lungs more completely than when at rest by contracting the expiratory muscles; also called *forced expiration*

active force A force that requires expenditure of cellular energy (ATP) in the transport of a substance across the plasma membrane

active reabsorption When any one of the five steps in the transepithelial transport of a substance reabsorbed across the kidney tubules requires energy expenditure

active transport Active carrier-mediated transport involving transport of a substance against its concentration gradient across the plasma membrane

acuity Discriminative ability; the ability to discern between two different points of stimulation

adaptation A reduction in receptor potential despite sustained stimulation of the same magnitude

adenosine diphosphate (ADP) (uh-DEN-uh-sēn) The two-phosphate product formed from the splitting of ATP to yield energy for the cell's use

adenosine triphosphate (ATP) The body's common energy "currency," which consists of an adenosine with three phosphate groups attached; splitting of the high-energy, terminal phosphate bond provides energy to power cellular activities

adenylyl cyclase (ah-DEN-il-il si-klās) The membrane-bound enzyme activated by a G-protein intermediary in response to binding of an extracellular messenger with a surface membrane receptor that, in turn, activates cyclic AMP, an intracellular second messenger

adequate stimulus The type of stimulus to which a specific receptor type responds, such as a photoreceptor responding to light

ADH See *vasopressin*

adipocytes Fat cells in adipose tissue; store triglyceride fat and secrete hormones termed *adipokines*

adipokines Hormones secreted by adipose tissue that play important roles in energy balance and metabolism; some trigger inflammation in fat

adipose tissue The tissue specialized for storage of triglyceride fat; found under the skin in the hypodermis or surrounding the abdominal viscera

ADP See *adenosine diphosphate*

adrenal cortex (uh-DRĒ-nul) The outer portion of the adrenal gland; secretes three classes of steroid hormones: glucocorticoids, mineralocorticoids, and sex hormones

adrenal medulla (muh-DŪL-uh) The inner portion of the adrenal gland; secretes the hormones epinephrine and norepinephrine into the blood in response to sympathetic stimulation

adrenergic fibers (ad′-ruh-NUR-jik) Nerve fibers that release norepinephrine as their neurotransmitter

adrenocorticotropic hormone (ACTH) (ad-rē′-nō-kor′-tuh-kō-TRŌP-ik) An anterior pituitary hormone that stimulates cortisol secretion by the adrenal cortex and promotes growth of the adrenal cortex

aerobic Referring to a condition in which oxygen is available

aerobic exercise Exercise that can be supported by ATP formation accomplished by oxidative phosphorylation because adequate O_2 is available to support the muscle's modest energy demands; also called *endurance-type exercise*

afferent arteriole (AF-er-ent ar-TIR-ē-ōl) The vessel that carries blood into the glomerulus of the kidney's nephron

afferent division The portion of the peripheral nervous system that carries information from the periphery to the central nervous system

afferent neuron Neuron that possesses a sensory receptor at its peripheral ending and carries information to the central nervous system

after hyperpolarization (hī′-pur-pō-luh-ruh-ZĀ-shun) A slight, transient hyperpolarization that sometimes occurs at the end of an action potential

agonist A substance that binds to a neurotransmitter's receptors and mimics the neurotransmitter's response

agranulocytes (ā-GRAN-yuh-lō-sits′) Leukocytes that do not contain granules, including lymphocytes and monocytes

albumin (al-BEW-min) The smallest and most abundant plasma proteins; binds and transports many water-insoluble substances in the blood; contributes extensively to plasma-colloid osmotic pressure

aldosterone (al-dō-steer-OWN) or (al-DOS-tuh-rōn) The adrenocortical hormone that stimulates Na^+ reabsorption and K^+ secretion by the distal and collecting tubules of the kidney's nephron during urine formation

alkalosis (al′-kuh-LŌ-sus) Blood pH of greater than 7.45

allergy Acquisition of an inappropriate specific immune reactivity to a normally harmless environmental substance

all-or-none law An excitable membrane either responds to a stimulus with a maximal action potential that spreads nondecrementally throughout the membrane or does not respond with an action potential at all

alpha (α) cells The endocrine pancreatic cells that secrete the hormone glucagon

alpha motor neuron A motor neuron that innervates ordinary skeletal muscle fibers

alveolar surface tension (al-VĒ-ō-lur) The surface tension of the fluid lining the alveoli in the lungs; see *surface tension*

alveolar ventilation The volume of air exchanged between the atmosphere and alveoli per minute; alveolar ventilation = (tidal volume − dead space volume) × respiratory rate

alveoli (of lungs) (al-VĒ-ō-li) The air sacs across which O_2 and CO_2 are exchanged between the blood and air in the lungs

alveoli (of mammary glands) Clusters of saclike, epithelial-lined milk-producing glands at the terminal end of the milk ducts in the breasts

amines (AH-mēnz) Hormones derived from the amino acid tyrosine, including thyroid hormone and catecholamines

amoeboid movement (uh-MĒ-boid) "Crawling" movement of white blood cells, similar to the means by which amoebas move

AMPA receptor One of two types of receptor-channels on a postsynaptic membrane to which the neurotransmitter glutamate binds, this one leading to the formation of EPSPs

anabolism (ah-NAB-ō-li-zum) The buildup, or synthesis, of larger organic molecules from small organic molecular subunits

anaerobic (an′-uh-RŌ-bik) Referring to a condition in which oxygen is not present

anaerobic exercise High-intensity exercise that can be supported by ATP formation accomplished by anaerobic glycolysis for brief periods of time when O_2 delivery to a muscle is inadequate to support oxidative phosphorylation

analgesic (an-al-JĒ-zic) Pain relieving

anatomy The study of body structure

androgen A masculinizing "male" sex hormone; includes testosterone from the testes and dehydro-epiandrosterone from the adrenal cortex

anemia A reduction below normal in O_2-carrying capacity of the blood

anion (AN-ī-on) Negatively charged ion that has gained one or more electrons in its outer shell

ANP See *atrial natriuretic peptide*

antagonism Actions opposing each other; in the case of hormones, when one hormone causes the loss of another hormone's receptors, reducing the effectiveness of the second hormone

antagonist A substance that blocks a neurotransmitter's receptor, thus preventing the neurotransmitter from binding and producing a response

anterior pituitary The glandular portion of the pituitary that synthesizes, stores, and secretes growth hormone, TSH, ACTH, FSH, LH, and prolactin

antibody An immunoglobulin produced by a specific activated B lymphocyte (plasma cell) against a particular antigen; binds with the specific antigen against which it is produced and promotes the antigenic invader's destruction by augmenting nonspecific immune responses already initiated against the antigen

antibody-mediated immunity A specific immune response accomplished by antibody production by B cells

antidiuretic hormone (ADH) (an′-ti-dī′-yū-RET-ik) See *vasopressin*

antigen A large, complex molecule that triggers a specific immune response against itself when it gains entry into the body

antioxidant A substance that helps inactivate biologically damaging free radicals

antiport The form of secondary active transport in which the driving ion and transported solute move in opposite directions across the plasma membrane; also called *countertransport* or *exchange*

antral follicle A developing ovarian follicle that is secreting estrogen and forming an antrum

antrum (of ovary) The fluid-filled cavity formed within a developing ovarian follicle

antrum (of stomach) The lower portion of the stomach

aorta (a-OR-tah) The large vessel that carries blood from the left ventricle

aortic valve A one-way valve that permits the flow of blood from the left ventricle into the aorta during ventricular emptying but prevents the backflow of blood from the aorta into the left ventricle during ventricular relaxation

apoptosis (ā-pop-TŌ-sis) Programmed cell death; deliberate self-destruction of a cell

aquaporin Water channel

aqueous humor (Ā-kwē-us) or (AK-we-us) The clear, watery fluid in the anterior chamber of the eye; provides nourishment for the cornea and lens

arcuate nucleus (ARE-kyou-it′) The subcortical brain region housing neurons that secrete appetite-enhancing neuropeptide Y and those that secrete appetite-suppressing melanocortins

aromatase An enzyme that converts testosterone to estrogen outside of the testes

arterioles (ar-TIR-ē-ōlz) The highly muscular, high-resistance vessels that branch from arteries, the caliber of which can be changed to regulate the amount of blood distributed to the various tissues

artery A vessel that carries blood away from the heart

ascending tract A bundle of nerve fibers of similar function that travels up the spinal cord to transmit signals derived from afferent input to the brain

astrocyte A type of glial cell in the brain; major functions include holding the neurons together in proper spatial relationship, inducing the brain capillaries to form tight junctions important in the blood–brain barrier, and enhancing synaptic activity

atmospheric pressure The pressure exerted by the weight of the air in the atmosphere on objects on Earth's surface; equals 760 mm Hg at sea level

ATP See *adenosine triphosphate*

ATPase An enzyme that can split ATP

ATP synthase The mitochondrial enzyme that catalyzes the synthesis of ATP from ADP and inorganic phosphate

atrial natriuretic peptide (ANP) (Ā-trē-al NĀ-tree-ur-eh′tik) A peptide hormone released from the cardiac atria that promotes urinary loss of Na^+

atrioventricular (AV) node (ā′-trē-ō-ven-TRIK-yuh-lur) A small bundle of specialized cardiac cells at the junction of the atria and ventricles that is the only site of electrical contact between the atria and ventricles

atrioventricular (AV) valve A one-way valve that permits the flow of blood from the atrium to the ventricle during filling of the heart but prevents the backflow of blood from the ventricle to the atrium during emptying of the heart

atrium (**atria,** plural) (Ā-tree-um) An upper chamber of the heart that receives blood from the veins and transfers it to the ventricle

atrophy (AH-truh-fē) Decrease in mass of an organ

autoimmune disease Disease characterized by erroneous production of antibodies against one of the body's own tissues

autonomic nervous system The portion of the efferent division of the peripheral nervous system that innervates smooth and cardiac muscle and exocrine glands; subdivided into the sympathetic and the parasympathetic nervous systems

autophagy Selective self-digestion of cell parts such as worn-out organelles by lysosomes

autoregulation The ability of an organ to adjust its own rate of blood flow despite changes in the driving mean arterial blood pressure

autorhythmicity The ability of an excitable cell to rhythmically initiate its own action potentials

AV nodal delay The delay in impulse transmission between the atria and ventricles at the AV node, which allows enough time for the atria to become completely depolarized and contract, emptying their contents into the ventricles, before ventricular depolarization and contraction occur

AV valve See *atrioventricular (AV) valve*

axon A single, elongated tubular extension of a neuron that conducts action potentials away from the cell body; also known as a *nerve fiber*

axon hillock The first portion of a neuronal axon plus the region of the cell body from which the axon leaves; the site of action potential initiation in most neurons

axon terminals The branched endings of a neuronal axon, which release a neurotransmitter that influences target cells in close association with the axon terminals

balance concept The balance between input of a substance through ingestion or metabolic production and its output through excretion or metabolic consumption

baroreceptor reflex An autonomically mediated reflex response that influences the heart and blood vessels to oppose a change in mean arterial blood pressure

baroreceptors Receptors located within the circulatory system that monitor blood pressure

basal ganglia See *basal nuclei*

basal metabolic rate (BMR) (BĀ-sul) The minimal waking rate of internal energy expenditure; the body's "idling speed"

basal nuclei Several masses of gray matter located deep within the white matter of the cerebrum of the brain; play an important inhibitory role in motor control

base A substance that can combine with a free hydrogen ion and remove it from solution

basic electrical rhythm (BER) Self-induced electrical activity of the digestive tract smooth muscle

basilar membrane (BAS-ih-lar) The membrane that forms the floor of the middle compartment of the cochlea and bears the organ of Corti, the sense organ for hearing

basophils (BĀY-sō-fills) White blood cells that synthesize, store, and release histamine, which is important in allergic responses, and heparin, which hastens the removal of fat particles from the blood

BER See *basic electrical rhythm*

beta (β) cells The endocrine pancreatic cells that secrete the hormone insulin

bicarbonate (HCO_3^-) The anion resulting from dissociation of carbonic acid, H_2CO_3

bile An alkaline solution containing bile salts and bilirubin secreted by the liver, stored in the gallbladder, and emptied into the small-intestine lumen

bile salts Cholesterol derivatives secreted in the bile that facilitate fat digestion through their detergent action and facilitate fat absorption through their micellar formation

biliary system (BIL-ē-air′-ē) The bile-producing system, consisting of the liver, gallbladder, and associated ducts

bilirubin (bill-eh-RŪ-bin) A bile pigment that is a waste product derived from the degradation of hemoglobin during the breakdown of old red blood cells

bipolar cells Neurons in the retina on which photoreceptors terminate and which in turn terminate on retinal ganglion cells

blastocyst The developmental stage of the fertilized ovum by the time it is ready to implant; consists of a single-layered sphere of cells encircling a fluid-filled cavity

blood–brain barrier (BBB) Special structural and functional features of the brain capillaries that limit access of materials from the blood into the brain tissue

B lymphocytes (B cells) White blood cells that produce antibodies against specific targets to which they have been exposed

body of the stomach The main, or middle, part of the stomach

body system A collection of organs that perform related functions and interact to accomplish a common activity that is essential for survival of the whole body; for example, the digestive system

bone marrow The soft, highly cellular tissue that fills the internal cavities of bones and is the source of most blood cells

Bowman's capsule The beginning of the tubular component of the kidney's nephron that cups around the glomerulus and collects the glomerular filtrate as it is formed

Boyle's law (boils) At any constant temperature, the pressure exerted by a gas varies inversely with the volume of the gas

brain stem The portion of the brain that is continuous with the spinal cord, serves as an integrating link between the spinal cord and higher brain levels, and controls many life-sustaining processes, such as breathing, circulation, and digestion

bronchioles (BRONG-kē-ōlz) The small, branching airways within the lungs

bronchoconstriction Narrowing of the respiratory airways

bronchodilation Widening of the respiratory airways

brown fat A special type of adipose tissue especially capable of converting the chemical energy from food into heat

brush border The collection of microvilli projecting from the luminal border of epithelial cells lining the digestive tract and kidney tubules

buffer See *chemical buffer system*

bulbourethral glands (bul-bō-you-RĒTH-ral) Male accessory sex glands that secrete mucus for lubrication

bulk flow Movement in bulk of a protein-free plasma across the capillary walls between the blood and surrounding interstitial fluid; encompasses ultrafiltration and reabsorption

bundle of His (hiss) A tract of specialized cardiac cells that rapidly transmits an action potential down the interventricular septum of the heart

C cells The thyroid cells that secrete calcitonin

calcitonin (kal'-suh-TŌ-nun) A hormone secreted by the thyroid C cells that lowers plasma Ca^{2+} levels

calcium balance Maintenance of a constant total amount of Ca^{2+} in the body; accomplished by slowly responding adjustments in intestinal Ca^{2+} absorption and in urinary Ca^{2+} excretion

calcium homeostasis Maintenance of a constant free plasma Ca^{2+} concentration, accomplished by rapid exchanges of Ca^{2+} between the bone and ECF and, to a lesser extent, by modifications in urinary Ca^{2+} excretion

calcium-induced calcium release When in cardiac muscle cells the excitation-induced entry of a small amount of Ca^{2+} from the ECF through voltage-gated surface membrane receptors triggers the opening of Ca^{2+}-release channels in the sarcoplasmic reticulum, causing a much larger release of Ca^{2+} into the cytosol from this intracellular store

calmodulin (kal'-MA-jew-lin) An intracellular Ca^{2+} binding protein that, on activation by Ca^{2+}, induces a change in structure and function of another intracellular protein; especially important in smooth muscle excitation–contraction coupling

cAMP See *cyclic adenosine monophosphate*

CAMs See *cell adhesion molecules*

capillaries The thin-walled, pore-lined smallest of blood vessels, across which exchange between the blood and surrounding tissues takes place

carbonic anhydrase (an-HĪ-drās) The enzyme found in erythrocytes, kidney tubular cells, and a few other specialized cells that catalyzes the direct conversion of CO_2 and H_2O into H^+ and HCO_3^-

cardiac cycle One period of systole and diastole

cardiac muscle The specialized muscle found only in the heart

cardiac output (CO) The volume of blood pumped by each ventricle each minute; cardiac output = stroke volume × heart rate

cardiovascular control center The integrating center located in the medulla of the brain stem that controls mean arterial blood pressure

carrier-mediated transport Transport of a substance across the plasma membrane facilitated by a carrier molecule

carrier molecules Membrane proteins that, by undergoing reversible changes in shape so that specific binding sites are alternately exposed at either side of the membrane, can bind with and transfer particular substances unable to cross the plasma membrane on their own

cascade A series of sequential reactions that culminates in a final product, such as a clot

catabolism (kuh-TAB-ō-li-zum) The breakdown, or degradation, of large, energy-rich molecules within cells

catalase (KAT-ah-lās) An antioxidant enzyme found in peroxisomes that decomposes potent hydrogen peroxide (H_2O_2) into harmless H_2O and O_2

catecholamines (kat'-uh-KŌ-luh-means) Amine hormones derived from tyrosine and secreted largely by the adrenal medulla

cations (KAT-i-onz) Positively charged ions that have lost one or more electrons from their outer shell

CatSper channels Ca^{2+}-permeable channels found exclusively in the plasma membrane of sperm tails, which when opened in response to egg-released progesterone permit Ca^{2+} entry that triggers hyperactivated motility of sperm

CCK See *cholecystokinin*

cell The smallest unit capable of carrying out the processes associated with life; the basic unit of both structure and function in living organisms

cell adhesion molecules (CAMs) Proteins that protrude from the surface of the plasma membrane and form loops or other appendages that the cells use to grip one another and the surrounding connective tissue fibers

cell body The portion of a neuron that houses the nucleus and organelles

cell-mediated immunity A specific immune response accomplished by activated T lymphocytes, which directly attack unwanted cells

cellular respiration The entire series of chemical reactions involving the intracellular breakdown of nutrient-rich molecules to yield energy, using O_2 and producing CO_2 in the process

center A functional collection of cell bodies within the central nervous system

central chemoreceptors (kē-mō-rē-SEP-turz) Receptors located in the medulla near the respiratory center that respond to changes in ECF H^+ concentration resulting from changes in arterial P_{CO_2} and adjust respiration accordingly

central lacteal (LAK-tē-ul) The initial lymphatic vessel that supplies each of the small-intestinal villi

central nervous system (CNS) The brain and spinal cord

central sulcus (SUL-kus) A deep infolding of the brain surface that runs roughly down the middle of the lateral surface of each cerebral hemisphere and separates the parietal and frontal lobes

centrioles (SEN-trē-ōlz) A pair of short, cylindrical structures within a cell that form the mitotic spindle during cell division

centrosome The cell's microtubule organizing center located near the nucleus and consisting of the pair of centrioles surrounded by an amorphous mass of proteins; also called the *cell center*

cerebellum (ser'-uh-BEL-um) The part of the brain attached at the rear of the brain stem and concerned with maintaining proper position of the body in space and subconscious coordination of motor activity

cerebral cortex The outer shell of gray matter in the cerebrum; site of initiation of all voluntary motor output and final perceptual processing of all sensory input as well as integration of most higher neural activity

cerebral hemispheres The cerebrum's two halves, which are connected by a thick band of neuronal axons

cerebrospinal fluid (CSF) (ser'-uh-brō-SPĪ-nul) or (sah-REE-brō-SPĪ-nul) A special cushioning fluid that is produced by, surrounds, and flows through the central nervous system

cerebrum (SER-uh-brum) or (sah-REE-brum) The division of the brain that consists of the basal nuclei and cerebral cortex

channels Small, water-filled passageways through the plasma membrane; formed by membrane proteins that span the membrane and provide highly selective passage for small water-soluble substances such as ions

chemical bonds The forces holding atoms together

chemical buffer system A mixture in a solution of two or more chemical compounds that minimize pH changes when either an acid or a base is added to or removed from the solution

chemically gated channels Channels in the plasma membrane that open or close in response to the binding of a specific chemical messenger with a membrane receptor site that is in close association with the channel

chemical mediator A chemical secreted by a cell that influences an activity outside the cell

chemical synapse The most abundant type of junction between two neurons in which an action potential in a presynaptic neuron alters the postsynaptic neuron's potential by means of a chemical messenger, a neurotransmitter

chemiosmosis (kē-ma-OS-mō-sis) ATP production in mitochondria catalyzed by ATP synthase, which is activated by flow of H⁺ down a concentration gradient established by the electron transport system

chemoreceptor (kē-mō-rē-sep′-tur) A sensory receptor sensitive to specific chemicals

chemotaxin (kē-mō-TAK-sin) A chemical released at an inflammatory site that attracts phagocytes to the area

chief cells The cells in the gastric pits that secrete pepsinogen

cholecystokinin (CCK) (kō′-luh-sis-tuh-kī-nun) A hormone released from the duodenal mucosa primarily in response to the presence of fat; inhibits gastric motility and secretion, stimulates pancreatic enzyme secretion, stimulates gallbladder contraction, and acts as a satiety signal

cholesterol A type of fat molecule that serves as a precursor for steroid hormones and bile salts and is a stabilizing component of the plasma membrane

cholinergic fibers (kō′-lin-ER-jik) Nerve fibers that release acetylcholine as their neurotransmitter

chromaffin granules The granules that store catecholamines in adrenomedullary cells

chyme (kim) A thick liquid mixture of food and digestive juices

cilia (SILL-ē-ah) Motile, hairlike protrusions from the surface of cells lining the respiratory airways and the oviducts

ciliary body The portion of the eye that produces aqueous humor and contains the ciliary muscle

ciliary muscle A circular ring of smooth muscle within the eye whose contraction increases the strength of the lens to accommodate for near vision

circadian rhythm (sir-KĀ-dē-un) Repetitive oscillations in the set point of various body activities, such as hormone levels and body temperature, that are very regular and have a frequency of one cycle every 24 hours, usually linked to light–dark cycles; diurnal rhythm; biological rhythm

circulatory shock When mean arterial blood pressure falls so low that adequate blood flow to the tissues can no longer be maintained

citric acid cycle A cyclic series of biochemical reactions that processes the intermediate breakdown products of nutrient molecules, generating carbon dioxide and preparing hydrogen carrier molecules for entry into the high-energy-yielding electron transport system

clot a mass of activated fibrin threadlike molecules that form a netlike meshwork that traps blood cells, thereby transforming blood in the vicinity from a flowing liquid into a nonflowing gel

CNS See *central nervous system*

cochlea (KOK-lē-uh) The snail-shaped portion of the inner ear that houses the receptors for sound

cognition The act of "knowing," including both awareness and judgment

collecting tubule The last portion of tubule in the kidney's nephron that empties into the renal pelvis

colloid (KOL-oid) The thyroglobulin-containing substance enclosed within the thyroid follicles

colloid osmotic pressure The osmotic force across the capillary wall resulting from the uneven colloidal dispersion of plasma proteins between the blood and interstitial fluid

competition When several closely related substances compete for the same carrier binding sites

complement system A collection of plasma proteins that are activated in cascade fashion on exposure to invading microorganisms, ultimately producing a membrane attack complex that destroys the invaders

compliance The distensibility of a hollow, elastic structure, such as a blood vessel or the lungs; a measure of how easily the structure can be stretched

concave surface Curved in, as a surface of a lens that diverges light rays

concentration gradient A difference in concentration of a particular substance between two adjacent areas

conditioned reflex A reflex response that occurs as a result of learning; also called *acquired reflex*

conduction The transfer of heat between objects of differing temperature that are in direct contact

cones The eye's photoreceptors used for color vision in the light

connective tissue Tissue that serves to connect, support, and anchor various body parts; distinguished by relatively few cells dispersed within an abundance of extracellular material

contiguous conduction The means by which an action potential is propagated throughout a nonmyelinated nerve fiber; local current flow between an active and adjacent inactive area brings the inactive area to threshold, triggering an action potential in a previously inactive area

contractility (of heart) The strength of contraction of the heart at any given end-diastolic volume

control center See *integrator*

controlled variable Some factor that can vary but is held within a narrow range by a control system

convection Transfer of heat energy by air or water currents

convergence The converging of many presynaptic terminals from thousands of other neurons on a single neuronal cell body and its dendrites so that activity in the single neuron is influenced by the activity of many other neurons

convex surface Curved out, as a surface in a lens that converges light rays

core temperature The temperature within the inner core of the body (abdominal and thoracic organs, central nervous system, and skeletal muscles) that is homeostatically maintained at about 100°F

cornea (KOR-nē-ah) The clear, anteriormost outer layer of the eye through which light rays pass to the interior of the eye

coronary circulation The blood vessels that supply the heart muscle

corpus callosum (ka-LŌ-sum) The thick band of nerve fibers that connects the two cerebral hemispheres structurally and functionally

corpus luteum (LOO-tē-um) The ovarian structure that develops from a ruptured follicle after ovulation

cortical nephrons The most abundant type of nephrons, whose glomeruli lie in the outer layer of the renal cortex and whose short loops of Henle dip only slightly into the renal medulla

corticotropes Anterior pituitary cells that secrete adrenocorticotropic hormone

cortisol (KORT-uh-sol) The adrenocortical hormone that plays an important role in carbohydrate, protein, and fat metabolism and helps the body resist stress

cotransport See *symport*

countercurrent multiplication The means by which long loops of Henle establish the vertical osmotic gradient in the renal medulla, making it possible to put out urine of variable concentration depending on the body's needs

countertransport See *antiport*

cranial nerves The 12 pairs of peripheral nerves, most of which arise from the brain stem

cross bridges The myosin molecules' globular heads that protrude from a thick filament within a muscle fiber and interact with the actin molecules in the thin filaments to bring about shortening of the muscle fiber during contraction

CSF See *cerebrospinal fluid*

current The flow of electrical charge, such as by movement of positive charges toward a more negatively charged area

cyclic adenosine monophosphate (cyclic AMP or cAMP) An intracellular second messenger derived from ATP

cyclic guanosine monophosphate (cyclic GMP or cGMP) An intracellular second messenger similar to cAMP

cytokines Protein signal molecules secreted by immune and other cells that largely act locally to regulate immune responses

cytoplasm (SĪ-tō-plaz′-um) The portion of the cell interior not occupied by the nucleus

cytoskeleton The complex intracellular protein network that acts as the "bone and muscle" of the cell

cytosol (SĪ-tuh-sol′) The semiliquid portion of the cytoplasm not occupied by organelles

cytotoxic T cells (sī-tō-TOK-sik) The population of T cells that destroys host cells bearing foreign antigen, such as body cells invaded by viruses or cancer cells

DAG See *diacylglycerol*

dead-space volume The volume of air that occupies the respiratory airways as air is moved in and out and that is not available to participate in exchange of O₂ and CO₂ between the alveoli and atmosphere

dehydration A water deficit in the body

dehydroepiandrosterone (DHEA) (dē-HĪ-drō-epi-and-ro-steer-own) The androgen secreted by the adrenal cortex in both sexes

dendrites Projections from the surface of a neuron's cell body that carry signals toward the cell body

deoxyribonucleic acid (DNA) (dē-OK-sē-rī-bō-new-klā-ik) The cell's genetic material, which is found within the nucleus and provides codes for protein synthesis and serves as a blueprint for cell replication

depolarization (de′-pō-luh-ruh-ZĀ-shun) A reduction in membrane potential from resting potential; movement of the potential from resting toward 0 mV

dermis The connective tissue layer that lies under the epidermis in the skin; contains the skin's blood vessels and nerves

descending tract A bundle of nerve fibers of similar function that travels down the spinal cord to relay messages from the brain to efferent neurons

desmosome (dez′-muh-sōm) An adhering junction between two adjacent but nontouching cells formed by the extension of filaments between the cells' plasma membranes; most abundant in tissues that are subject to considerable stretching

DHEA See *dehydroepiandrosterone*

diacylglycerol (DAG) (die-ACE-sul-gli-sir-all) A component cleaved from phosphatidylinositol bisphosphate (PIP_2) in the plasma membrane that serves as a second messenger in response to binding of an extracellular (first) messenger with a G-protein-coupled receptor

diaphragm (DIE-uh-fram) A dome-shaped sheet of skeletal muscle that forms the floor of the thoracic cavity; the major inspiratory muscle

diastole (dī-AS-tō-lē) The period of cardiac relaxation and filling

diencephalon (dī'-un-SEF-uh-lan) The division of the brain that consists of the thalamus and hypothalamus

differentiation The process of each type of cell becoming specialized during development of a multicellular organism to carry out a particular function

diffusion Random collisions and intermingling of molecules as a result of their continuous thermally induced random motion

digestion The process by which the structurally complex foodstuffs of the diet are broken down into smaller absorbable units by the enzymes produced within the digestive system

dihydropyridine receptors (die-HIGH-dro-PEER-ih-deen) Voltage-gated receptors on the T tubules that trigger the opening of adjoining ryanodine receptors on the sarcoplasmic reticulum during excitation–contraction coupling

diploid number (DIP-loid) A complete set of 46 chromosomes (23 pairs), as found in all human somatic cells

discriminative ability See *acuity*

distal tubule A highly convoluted tubule that extends between the loop of Henle and the collecting duct in the kidney's nephron

diurnal rhythm (dī-URN'-ul) Repetitive oscillations in hormone levels that are very regular and have a frequency of one cycle every 24 hours, usually linked to the light–dark cycle; circadian rhythm; biological rhythm

divergence The diverging, or branching, of a neuron's axon terminals so that activity in this single neuron influences the many other cells with which its terminals synapse

DNA See *deoxyribonucleic acid*

dorsal root ganglion A cluster of afferent neuronal cell bodies located adjacent to the spinal cord

down regulation A reduction in the number of receptors for (and thereby the target cell's sensitivity to) a particular hormone as a direct result of the effect that an elevated level of the hormone has on its receptors

dynamic equilibrium When two opposing passive movements exactly counterbalance each other so that no net movement takes place, with no energy needed to maintain this constancy

dynein (DIE-neen) The motor protein that "walks" along microtubular "highways" toward the cell center, such as in transporting debris from the axon terminal to the cell body for destruction by lysosomes

ECG See *electrocardiogram*

ECM See *extracellular matrix*

edema (i-DĒ-muh) Swelling of tissues as a result of excess interstitial fluid

EDV See *end-diastolic volume*

EEG See *electroencephalogram*

effector The component of a control system that accomplishes the output commanded by the integrator

effector organs The muscles or glands innervated by the nervous system that carry out the nervous system's orders to bring about a desired effect, such as a particular movement or secretion

efferent division (EF-er-ent) The portion of the peripheral nervous system that carries instructions from the central nervous system to effector organs

efferent neuron Neuron that carries information from the central nervous system to an effector organ

efflux (Ē-flux) Movement out of the cell

eicosanoids (ī-KŌ-sa-noydz) A group of lipid signal molecules derived from arachidonic acid in the phospholipid tails of the plasma membrane, including the prostaglandins, prostacyclins, thromboxanes, and leukotrienes, that act locally to regulate diverse cellular processes throughout the body

elastic recoil Rebound of the lungs after having been stretched

electrical gradient A difference in charge between two adjacent areas

electrical synapse The least common type of junction between two neurons in which an action potential in a presynaptic neuron spreads directly to the postsynaptic neuron via gap junctions

electrocardiogram (ECG) The graphic record of the electrical activity that reaches the surface of the body as a result of cardiac depolarization and repolarization

electrochemical gradient The simultaneous existence of an electrical gradient and concentration (chemical) gradient for a particular ion

electroencephalogram (EEG) (i-lek'-trō-in-SEF-uh-luh-gram') A graphic record of the collective postsynaptic potential activity in the cell bodies and dendrites located in the cortical layers of the brain under a recording electrode

electrolytes Solutes that form ions in solution and conduct electricity

electron transport system The series of electron carriers in the mitochondrial inner membrane that transfer electrons from higher to lower energy levels, with the released energy being used to establish the H^+ concentration gradient in the mitochondria that powers ATP synthesis

embolus (EM-bō-lus) A freely floating clot

embryo The product of fertilization during the first two months of intrauterine life when tissue differentiation is taking place

embryonic stem cells (ESCs) Undifferentiated cells resulting from the early divisions of a fertilized egg that ultimately give rise to all the mature, specialized cells of the body while at the same time renewing themselves

end-diastolic volume (EDV) The volume of blood in the ventricle at the end of diastole, when filling is complete

endocrine axis A three-hormone sequence consisting of (1) a hypothalamic hypophysiotropic hormone that controls the output of (2) an anterior pituitary tropic hormone that regulates secretion of (3) a target endocrine gland hormone, which exerts the final physiological effect

endocrine glands Ductless glands that secrete hormones into the blood

endocytic vesicle A small, intracellular, membrane-enclosed vesicle in which extracellular material is trapped

endocytosis (en'-dō-sī-TŌ-sis) Internalization of extracellular material within a cell as a result of the plasma membrane forming a pouch that contains the extracellular material, then sealing at the surface of the pouch to form an endocytic vesicle

endogenous opiates (en-DAJ'-eh-nus Ō'-pē-ātz) Endorphins and enkephalins, which bind with opiate receptors and are important in the body's natural analgesic system

endogenous pyrogen (pī'-ruh-jun) A chemical released from macrophages during inflammation that acts by means of local prostaglandins to raise the set point of the hypothalamic thermostat to produce a fever

endolymph The fluid within the cochlear duct and vestibular organs in the inner ear

endometrium (en'-dō-MĒ-trē-um) The lining of the uterus

endoplasmic reticulum (ER) (en'-dō-PLAZ-mik ri-TIK-yuh-lum) An organelle consisting of a continuous membranous network of fluid-filled tubules (smooth ER) and flattened sacs, partially studded with ribosomes (rough ER); synthesizes proteins and lipids for formation of new cell membrane and manufactures products for secretion

endothelium (en'-dō-THĒ-lē-um) The thin, single-celled layer of epithelial cells lining the entire circulatory system

end-plate potential (EPP) The graded receptor potential that occurs at the motor end plate of a skeletal muscle fiber in response to binding with acetylcholine

end-systolic volume (ESV) The volume of blood in the ventricle at the end of systole, when emptying is complete

endurance-type exercise See *aerobic exercise*

energy balance The balance between energy input by means of food intake and energy output by means of external work and internal work

enteric nervous system The extensive network of nerve fibers consisting of the myenteric plexus and submucous plexus within the digestive tract wall that endows the tract with considerable self-regulation

enterogastrones (ent'-uh-rō-GAS-trōnz) Hormones secreted by the duodenal mucosa that inhibit gastric motility and secretion; include secretin and cholecystokinin

enterohepatic circulation (en'-tur-ō-hi-PAT-ik) The recycling of bile salts and other bile constituents between the small intestine and liver by means of the hepatic portal vein

enzyme A special protein molecule that speeds up a particular chemical reaction in the body

eosinophils (ē'-uh-SIN-uh-fils) White blood cells that are important in allergic responses and in combating internal parasite infestations

ependymal cells (eh-PEN-dim-ul) The glial cells lining the ventricles of the brain, which serve as neural stem cells

epidermis (ep'-uh-DER-mus) The outer layer of the skin, consisting of numerous layers of epithelial cells, with the outermost layers being dead and flattened

epididymis The male accessory reproductive organ that stores sperm and increases their motility and fertility before ejaculation

epigenetics Environmentally induced modifications of gene activity that do not alter the gene's DNA code

epinephrine (ep′-uh-NEF-rin) The primary hormone secreted by the adrenal medulla; important in preparing the body for "fight-or-flight" responses and in regulating arterial blood pressure; adrenaline

epiphyseal plate (eh-pif-i-SĒ-al) A layer of cartilage that separates the diaphysis (shaft) of a long bone from the epiphysis (flared end); the site at which bones grow longer before the cartilage ossifies (turns into bone)

epithelial tissue (ep′-uh-THĒ-lē-ul) A functional grouping of cells specialized in the exchange of materials between the cell and its environment; lines and covers various body surfaces and cavities and forms secretory glands

EPSP See *excitatory postsynaptic potential*

equilibrium The sense of body orientation and motion

equilibrium potential (E_x) The potential that exists when the concentration gradient and opposing electrical gradient for a given ion exactly counterbalance each other so that there is no net movement of the ion

erythrocytes (i-RITH-ruh-sīts) Red blood cells, which are plasma membrane–enclosed bags of hemoglobin that transport O_2 and, to a lesser extent, CO_2 and H^+ in the blood; RBCs

erythropoiesis (i-rith′-rō-poi-Ē-sus) Erythrocyte production by the bone marrow

erythropoietin The hormone released from the kidneys in response to a reduction in O_2 delivery to the kidneys; stimulates the bone marrow to increase erythrocyte production

ESCs See *embryonic stem cells*

esophagus (i-SOF-uh-gus) A straight muscular tube that extends between the pharynx and stomach

estrogen Feminizing "female" sex hormone

ESV See *end-systolic volume*

evaporation The transfer of heat from the body surface by the transformation of water from a liquid to a gaseous state

exchange See *antiport*

excitable tissue Tissue capable of producing electrical signals when excited; includes nervous and muscle tissue

excitation–contraction coupling The series of events linking muscle excitation (the presence of an action potential) to muscle contraction (filament sliding and sarcomere shortening)

excitatory postsynaptic potential (EPSP) (pōst′-si-NAP-tik) A small depolarization of the postsynaptic membrane in response to neurotransmitter binding, bringing the membrane closer to threshold

excitatory synapse (SIN-aps′) Synapse in which the postsynaptic neuron's response to neurotransmitter release is a small depolarization of the postsynaptic membrane, bringing the membrane closer to threshold

exercise physiology The study of both the functional changes that occur in response to a single session of exercise and the adaptations that result from regular, repeated exercise sessions

exocrine glands Glands that secrete through ducts to the outside of the body or into a cavity that communicates with the outside

exocytosis (eks′-Ō-sī-TŌ-sis) Fusion of a membrane-enclosed intracellular vesicle with the plasma membrane, followed by the opening of the vesicle and the emptying of its contents to the outside

expiration A breath out

expiratory muscles The skeletal muscles whose contraction reduces the size of the thoracic cavity and lets the lungs recoil to a smaller size, bringing about movement of air from the lungs to the atmosphere

external environment The environment surrounding the body

external genitalia The externally visible portions of the reproductive system

external intercostal muscles Inspiratory muscles whose contraction elevates the ribs, thereby enlarging the thoracic cavity

external work Energy expended by contracting skeletal muscles to move external objects or to move the body in relation to the environment

extracellular fluid (ECF) All the fluid outside the cells of the body; consists of interstitial fluid and plasma

extracellular matrix (ECM) An intricate meshwork of fibrous proteins embedded in the interstitial fluid secreted by local cells

extrinsic controls Regulatory mechanisms initiated outside an organ that alter the activity of the organ; accomplished by the nervous and endocrine systems

extrinsic nerves The nerves originating outside the digestive tract that innervate the various digestive organs

facilitated diffusion Passive carrier-mediated transport involving transport of a substance down its concentration gradient across the plasma membrane

fatigue Inability to maintain muscle tension at a given level despite sustained stimulation

feedback A response that occurs after a change has been detected; may be *negative feedback* or *positive feedback*

feedforward A response designed to prevent an anticipated change in a controlled variable

feeding signals Appetite signals that give rise to the sensation of hunger and promote the desire to eat

fetus The product of fertilization during the last 7 months of gestation when differentiation is complete and tremendous growth and maturation occur

fibrinogen (fī-BRIN-uh-jun) A large, soluble plasma protein that when converted into an insoluble, threadlike molecule forms the meshwork of a clot during blood coagulation

Fick's law of diffusion The rate of net diffusion of a substance across a membrane is directly proportional to the substance's concentration gradient, the lipid solubility of the substance, and the surface area of the membrane and inversely proportional to the substance's molecular weight and the diffusion distance

fight-or-flight response The changes in activity of the various organs innervated by the autonomic nervous system in response to sympathetic stimulation, which collectively prepare the body for strenuous physical activity in the face of an emergency or stressful situation, such as a physical threat from the outside environment

firing When an excitable cell undergoes an action potential

first messenger An extracellular messenger, such as a hormone, that binds with a surface membrane receptor and activates an intracellular second messenger to carry out the desired cellular response

flagellum (fluh-JEL-um) The single, long, whip-like appendage that serves as the tail of a spermatozoon

flow rate (of blood or air) The volume of blood or air passing through a blood vessel or airway, respectively, per unit of time

fluid balance Maintenance of ECF volume (for long-term control of blood pressure) and ECF osmolarity (for maintaining normal cell volume)

follicle (of ovary) A developing ovum and the surrounding specialized cells

follicle-stimulating hormone (FSH) An anterior pituitary hormone that stimulates ovarian follicular development and estrogen secretion in females and stimulates sperm production in males

follicular cells (of ovary) (fah-LIK-you-lar) Collectively, the granulosa and thecal cells

follicular cells (of thyroid gland) The cells that form the walls of the colloid-filled follicles in the thyroid gland and secrete thyroid hormone

follicular phase The phase of the ovarian cycle dominated by the presence of maturing antral follicles before ovulation

Frank–Starling law of the heart Intrinsic control of the heart such that increased venous return resulting in increased end-diastolic volume leads to an increased strength of contraction and increased stroke volume; that is, the heart normally pumps out all the blood returned to it

free radicals Very unstable electron-deficient particles that are highly reactive and destructive

frontal lobes The lobes of the cerebral cortex at the top of the brain in front of the central sulcus, which are responsible for voluntary motor output, speaking ability, and elaboration of thought

FSH See *follicle-stimulating hormone*

fuel metabolism See *intermediary metabolism*

functional syncytium (sin-sish′-ē-um) A group of smooth or cardiac muscle cells that are interconnected by gap junctions and function electrically and mechanically as a single unit

functional unit The smallest component of an organ that can perform all the functions of the organ

gametes (GAM-ētz) Reproductive, or germ, cells, each containing a haploid set of chromosomes; sperm and ova

gamma motor neuron A motor neuron that innervates the fibers of a muscle–spindle receptor

ganglion (GAN-glē-un) A collection of neuronal cell bodies located outside the central nervous system

ganglion cells The nerve cells in the outermost layer of the retina whose axons form the optic nerve

gap junction A communicating junction formed between adjacent cells by small connecting tunnels that permit passage of charge-carrying ions between the cells so that electrical activity in one cell is spread to the adjacent cell

gastrin A hormone secreted by the pyloric gland area of the stomach that stimulates the parietal and chief cells to secrete a highly acidic gastric juice

gastrointestinal hormone Hormones secreted into the blood by the endocrine cells in the digestive tract mucosa that control motility and secretion in other parts of the digestive system

gene expression The multistep process by which information encoded in a gene is used to direct synthesis of a protein

genome All of the genetic information coded in a complete single set of DNA in a typical body cell

gestation Pregnancy

ghrelin (GRELL-in) The "hunger" hormone, a potent appetite stimulator secreted by the empty stomach

glands Epithelial tissue derivatives that are specialized for secretion

glial cells (glē-ul) Connective tissue cells of the CNS, which support the neurons both physically and metabolically, including astrocytes, oligodendrocytes, ependymal cells, and microglia

gliotransmitters Chemical mediators released from glial cells that influence neurons and other glial cells

glomerular filtration (glow-MAIR-yū-lur) Filtration of a protein-free plasma from the glomerular capillaries into the tubular component of the kidney's nephron as the first step in urine formation

glomerular filtration rate (GFR) The rate at which glomerular filtrate is formed

glomerulus (in kidney) (glow-MAIR-yū-lus) A ball-like tuft of capillaries in the kidney's nephron that filters water and solute from the blood as the first step in urine formation

glomerulus (in olfactory bulb) A ball-like neural junction within the olfactory bulb that serves as a "smell file" sorting different scent components

glucagon (GLOO-kuh-gon) The pancreatic hormone that raises blood glucose and blood fatty-acid levels

glucocorticoids (gloo'-kō-KOR-ti-koidz) Adrenocortical hormones that are important in intermediary metabolism and in helping the body resist stress; primarily cortisol

gluconeogenesis (gloo'-kō-nē-ō-JEN-uh-sus) The conversion of amino acids into glucose

glycogen (GLĪ-kō-jen) The storage form of glucose in the liver and muscle

glycogenesis (glī'-kō-JEN-i-sus) The conversion of glucose into glycogen

glycogenolysis (glī'-kō-juh-NOL-i-sus) The conversion of glycogen to glucose

glycolysis (glī-KOL-uh-sus) A biochemical process taking place in the cell's cytosol that breaks down glucose into pyruvate molecules

GnRH See *gonadotropin-releasing hormone*

Golgi complex (GOL-jē) An organelle consisting of sets of stacked, flattened membranous sacs; processes raw materials transported to it from the endoplasmic reticulum into finished products and sorts and directs the finished products to their final destination

gonadotropes Anterior pituitary cells that secrete gonadotropin-releasing hormone

gonadotropin-releasing hormone (GnRH) (gō-nad'-uh-TRŌ-pin) The hypothalamic hormone that stimulates the release of FSH and LH from the anterior pituitary

gonadotropins FSH and LH; hormones that control secretion of the sex hormones by the gonads

gonads (GŌ-nadz) The primary reproductive organs, which produce the gametes and secrete the sex hormones; testes and ovaries

G protein A membrane-bound intermediary, which, when activated on binding of an extracellular first messenger to a surface receptor, activates effector proteins on the intracellular side of the membrane in the cAMP second-messenger system

G-protein-coupled receptor (GPCR) A type of receptor that activates the associated G protein on binding with an extracellular chemical messenger

gradation of contraction Variable magnitudes of tension produced in a single whole muscle

graded potential A local change in membrane potential that occurs in varying grades of magnitude; serves as a short-distance signal in excitable tissues

grand postsynaptic potential (GPSP) The total composite potential in a postsynaptic neuron resulting from the sum of all EPSPs and IPSPs occurring at the same time

granulocytes (gran'-yuh-lō-sīts) Leukocytes that contain granules, including neutrophils, eosinophils, and basophils

granulosa cells (gran'-yuh-LŌ-suh) The layer of cells immediately surrounding a developing oocyte within an ovarian follicle

gray matter The portion of the central nervous system composed primarily of densely packaged neuronal cell bodies and dendrites

growth factors Cytokines important in development

growth hormone (GH) An anterior pituitary hormone that is primarily responsible for regulating overall body growth and is also important in intermediary metabolism; somatotropin

H⁺ See *hydrogen ion*

haploid number (HAP-loid) The number of chromosomes found in gametes; a half set of chromosomes, one member of each pair, for a total of 23 chromosomes in humans

Hb See *hemoglobin*

hCG See *human chorionic gonadotropin*

helper T cells The population of T cells that enhances the activity of other immune-response effector cells

hematocrit (hi-MAT'-uh-krit) The percentage of blood volume occupied by erythrocytes as they are packed down in a centrifuged blood sample

hemoglobin (Hb) (HĒ-muh-glō'-bun) A large iron-bearing protein molecule in erythrocytes that binds with and transports most O_2 in the blood; also carries some of the CO_2 and H^+ in the blood

hemolysis (hē-MOL-uh-sus) Rupture of red blood cells

hemostasis (hē'-mō-STĀ-sus) The stopping of bleeding from an injured vessel

hepatic portal system (hi-PAT-ik) A complex vascular connection between the digestive tract and liver such that venous blood from the digestive system drains into the liver for processing of absorbed nutrients before being returned to the heart

hippocampus (hip-ō-CAM-pus) The elongated, medial portion of the temporal lobe that is a part of the limbic system and is especially crucial for forming long-term memories

histamine A chemical released from mast cells or basophils that brings about vasodilation and increased capillary permeability; important in allergic responses and inflammation

homeostasis (hō'-mē-ō-STĀ-sus) Maintenance by the highly coordinated, regulated actions of the body systems of relatively stable chemical and physical conditions in the internal fluid environment that bathes the body cells

homeostatic control system A regulatory system that includes a sensor, integrator, and effectors that work together to bring about a corrective adjustment that opposes an original deviation from a normal set point

hormone A long-distance chemical mediator secreted by an endocrine gland into the blood, which transports it to its target cells

hormone response element (HRE) The specific attachment site on DNA for a given steroid hormone and its nuclear receptor

host cell A body cell infected by a virus

HRE See *hormone response element*

human chorionic gonadotropin (hCG) (kō-rē-ON-ik gō-nad'-uh-TRŌ-pin) A hormone secreted by the developing placenta that stimulates and maintains the corpus luteum of pregnancy

hydrogen ion (H⁺) The cationic portion of a dissociated acid

hydrolysis (hī-DROL-uh-sis) The digestion of a nutrient molecule by the addition of water at a bond site

hydrostatic (fluid) pressure (hī-drō-STAT-ik) The pressure exerted by fluid on the walls that contain it

hyperglycemia (hī'-pur-glī-SĒ-mē-uh) Elevated blood glucose concentration

hyperplasia (hī-pur-PLĀ-zē-uh) An increase in the number of cells

hyperpolarization An increase in membrane potential from resting potential; potential becomes even more negative than at resting potential

hypersecretion Too much of a particular hormone secreted

hypertension (hī'-pur-TEN-chun) Sustained, above-normal mean arterial blood pressure

hypertonic solution (hī'-pur-TON-ik) A solution with osmolarity greater than that of normal body fluids; more concentrated than normal

hypertrophy (hī-PUR-truh-fē) Increase in the size of an organ as a result of an increase in the size of its cells

hyperventilation Overbreathing; when the rate of ventilation is in excess of the body's metabolic needs for CO_2 removal

hypophysiotropic hormones (hi-PŌ-fiz-ē-ō-TRŌ-pik) Hormones secreted by the hypothalamus that regulate the secretion of anterior pituitary hormones; see also *releasing hormone* and *inhibiting hormone*

hyposecretion Too little of a particular hormone secreted

hypotension (hī-pō-TEN-chun) Sustained, below-normal mean arterial blood pressure

hypothalamic–hypophyseal portal system (hī-pō-thuh-LAM-ik hī-pō-FIZ-ē-ul) The vascular connection between the hypothalamus and anterior pituitary gland used for the pickup and delivery of hypophysiotropic hormones

hypothalamus (hī'-pō-THAL-uh-mus) The brain region beneath the thalamus that regulates many aspects of the internal fluid environment, such as water and salt balance and food intake; serves as an important link between the autonomic nervous system and endocrine system

hypotonic solution (hī'-pō-TON-ik) A solution with osmolarity less than that of normal body fluids; more dilute than normal

hypoventilation Underbreathing; ventilation inadequate to meet the metabolic needs for O_2 delivery and CO_2 removal

hypoxia (hī-POK-sē-uh) Insufficient O_2 at the cellular level

I band One of the light bands that alternate with dark (A) bands to create a striated appearance in a skeletal or cardiac muscle fiber when these fibers are viewed with a light microscope

ICF See *intracellular fluid*

IGF See *insulin-like growth factor*

immune surveillance Recognition and destruction of newly arisen cancer cells by the immune system

immunity The body's ability to resist or eliminate potentially harmful foreign materials or abnormal cells

immunoglobulins (im´-ū-nō-GLOB-yū-lunz) Antibodies; gamma globulins

impermeable Prohibiting passage of a particular substance through the plasma membrane

implantation The burrowing of a blastocyst into the endometrial lining

inclusion A nonpermanent mass of stored material, such as glycogen or triglycerides (fat), in a cell

incretin A hormone released by the digestive tract that stimulates insulin secretion by the pancreas

indoleamine A type of amine hormone derived from the amino acid tryptophan and secreted by the pineal gland

inflammation An innate, nonspecific series of highly interrelated events, especially involving neutrophils, macrophages, and local vascular changes, that are set into motion in response to foreign invasion or tissue damage

influx Movement into the cell

inhibin (in-HIB-un) A hormone secreted by the Sertoli cells of the testes or by the ovarian follicles that inhibits FSH secretion

inhibiting hormone A hypothalamic hormone that inhibits the secretion of a particular anterior pituitary hormone

inhibitory postsynaptic potential (IPSP) (pōst´-si-NAP-tik) A small hyperpolarization of the postsynaptic membrane in response to neurotransmitter binding, thereby moving the membrane farther from threshold

inhibitory synapse (SIN-aps´) Synapse in which the postsynaptic neuron's response to neurotransmitter release is a small hyperpolarization of the postsynaptic membrane, moving the membrane farther from threshold

innate immune responses Inherent defense responses that nonselectively defend against foreign or abnormal material, even on initial exposure to it; see also *inflammation, interferon, natural killer cells,* and *complement system*

innate reflex A built-in, unlearned reflex response

inorganic Referring to substances that do not contain carbon; from nonliving sources

inositol trisphosphate (IP₃) A component cleaved from phosphatidylinositol bisphosphate (PIP₂) in the plasma membrane that mobilizes the Ca^{2+} second-messenger system in response to binding of an extracellular (first) messenger to a G-protein-coupled receptor

insensible loss Loss of water of which the person is not aware from the lungs or nonsweating skin

inspiration A breath in

inspiratory muscles The skeletal muscles whose contraction enlarges the thoracic cavity, bringing about lung expansion and movement of air into the lungs from the atmosphere

insulin (IN-suh-lin) The pancreatic hormone that lowers blood levels of glucose, fatty acids, and amino acids and promotes their storage

insulin-like growth factor (IGF) Synonymous with *somatomedin;* hormone secreted by the liver into the blood on stimulation by growth hormone that acts directly on target cells to promote growth, with other tissues producing IGF that acts locally as a paracrine to promote growth

integral protein A plasma-membrane protein that extends through the thickness of the membrane

integrator A region that determines efferent output based on processing of afferent input; also called a *control center*

integument (in-TEG-yuh-munt) The skin and underlying connective tissue

intercalated cells Cells in the distal and collecting tubules of the kidney important in renal control of acid–base balance

intercostal muscles (int-ur-KOS-tul) The muscles that lie between the ribs; see also *external intercostal muscles* and *internal intercostal muscles*

interferon (in´-tur-FĒR-on) A chemical released from virus-invaded cells that provides nonspecific resistance to viral infections by transiently interfering with replication of the same or unrelated viruses in other host cells

intermediary metabolism The collective set of intracellular chemical reactions that involve the degradation, synthesis, and transformation of small nutrient molecules; also known as *fuel metabolism*

intermediate filaments Threadlike cytoskeletal elements that play a structural role in parts of the cells subject to mechanical stress

internal environment The body's aqueous extracellular environment, which consists of the plasma and interstitial fluid and which must be homeostatically maintained for the cells to make life-sustaining exchanges with it

internal intercostal muscles Expiratory muscles whose contraction pulls the ribs downward and inward, thereby reducing the size of the thoracic cavity

internal respiration The intracellular metabolic processes carried out within the mitochondria that use O_2 and produce CO_2 during the derivation of energy from nutrient molecules

internal work All forms of biological energy expenditure that do not accomplish mechanical work outside the body

interneuron Neuron that lies entirely within the central nervous system and is important for integrating peripheral responses to peripheral information as well as for the abstract phenomena associated with the "mind"

interstitial fluid (in´-tur-STISH-ul) The portion of the extracellular fluid that surrounds and bathes all the body cells

intra-alveolar pressure (in´-truh-al-VĒ-uh-lur) The pressure within the alveoli

intracellular fluid (ICF) The fluid collectively contained within all the body cells

intrapleural pressure (in´-truh-PLOOR-ul) The pressure within the pleural sac

intrinsic controls Local control mechanisms inherent to an organ

intrinsic factor A special substance secreted by the parietal cells of the stomach that must be combined with vitamin B_{12} for this vitamin to be absorbed by the intestine; deficiency produces pernicious anemia

intrinsic nerve plexuses Interconnecting networks of nerve fibers within the digestive tract wall

involuntary muscle Muscle innervated by the autonomic nervous system and not subject to voluntary control; cardiac and smooth muscle

ion An atom that has gained or lost one or more of its electrons, so it is not electrically balanced

IP₃ See *inositol trisphosphate*

IPSP See *inhibitory postsynaptic potential*

iris A pigmented smooth muscle that forms the colored portion of the eye and controls pupillary size

islets of Langerhans (LAHNG-er-honz) The endocrine portion of the pancreas that secretes the hormones insulin and glucagon into the blood

isometric contraction (ī´-sō-MET-rik) A muscle contraction in which muscle tension develops at constant muscle length

isotonic contraction A muscle contraction in which muscle tension remains constant as the muscle fiber changes length

isotonic solution (ī´-sō-TON-ik) A solution with osmolarity equal to that of normal body fluids

janus kinase (JAK) An enzyme attached to the cytosolic side of a surface-membrane receptor for several specific hormones; binding of the hormone to the receptor activates the JAK enzymes, which phosphorylate STAT that turns on gene transcription resulting in the synthesis of new proteins that carry out the cellular response dictated by the hormone

juxtaglomerular apparatus (juks´-tuh-glō-MAIR-ū-lur) A cluster of specialized vascular and tubular cells at a point where the ascending limb of the loop of Henle passes through the fork formed by the afferent and efferent arterioles of the same nephron in the kidney

juxtamedullary nephrons (juks´-tuh-MED-you-lair-ee) Nephrons whose glomeruli lie in the renal cortex next to the medulla and whose long loops of Henle dip deeply into the medulla; establish the medullary vertical osmotic gradient

keratin (CARE-uh-tin) The protein found in the intermediate filaments in skin cells that give the skin strength and help form a waterproof outer layer

killer (K) cells Cells that destroy a target cell that has been coated with antibodies by lysing its membrane

kinesin (kī-NÊ´-sin) The motor protein that transports secretory vesicles along the microtubular highway within neuronal axons by "walking" along the microtubule to the end of the axon

kisspeptin A neurotransmitter released from kiss1 neurons in the hypothalamus that controls gonadotropin-releasing hormone (GnRH) secretion, thus regulating the reproductive hormonal axis; may play a role in the onset of puberty

lactate (lactic acid) An end product formed from pyruvate (pyruvic acid) during the anaerobic process of glycolysis

lactation Milk production by the mammary glands

larynx (LARE-inks) The "voice box" at the entrance of the trachea; contains the vocal cords

lateral inhibition The phenomenon in which the most strongly activated signal pathway originating from the center of a stimulus area inhibits the less excited pathways from the fringe areas by means of lateral inhibitory connections within sensory pathways

lateral sacs The expanded saclike regions of a muscle fiber's sarcoplasmic reticulum; store and release calcium, which plays a key role in triggering muscle contraction

law of mass action If the concentration of one of the substances involved in a reversible reaction is increased, the reaction is driven toward the opposite side, and if the concentration of one of the substances is decreased, the reaction is driven toward that side

leak channels Unregulated, ungated channels that are open all the time

left ventricle The heart chamber that pumps blood into the systemic circulation

length–tension relationship The relationship between the length of a muscle fiber at the onset of contraction and the tension the fiber can achieve on a subsequent contraction

lens A transparent, biconvex structure of the eye that refracts (bends) light rays and whose strength can be adjusted to accommodate for vision at different distances

leptin A hormone released from adipose tissue that plays a key role in long-term regulation of body weight by acting on the hypothalamus to suppress appetite

leukocytes (LOO-kuh-sīts) White blood cells, which are the immune system's mobile defense units

leukotrienes (loo-ko-TRĪ-eenz) Locally acting eicosanoids derived from the plasma membrane that are especially important in development of asthma

Leydig cells (LĪ-dig) The interstitial cells of the testes that secrete testosterone

LH See *luteinizing hormone*

LH surge The burst in LH secretion that occurs at midcycle of the ovarian cycle and triggers ovulation

limbic system (LIM-bik) A functionally interconnected ring of forebrain structures that surrounds the brain stem and is concerned with emotions, basic survival and sociosexual behavioral patterns, motivation, and learning

lipase (LĪ-payz) An enzyme secreted primarily by pancreatic acinar cells that digests dietary fat

lipid emulsion A suspension of small fat droplets held apart as a result of adsorption of bile salts on their surface

loop of Henle (HEN-lē) A hairpin loop that extends between the proximal and distal tubule of the kidney's nephron

lumen (LOO-men) The interior space of a hollow organ or tube

luteal phase (LOO-tē-ul) The phase of the ovarian cycle dominated by the presence of a corpus luteum

luteinization (loot′-ē-un-uh-ZĀ-shun) Formation of a postovulatory corpus luteum in the ovary

luteinizing hormone (LH) An anterior pituitary hormone that stimulates ovulation, luteinization, and secretion of estrogen and progesterone in females and testosterone secretion in males

lymph Interstitial fluid that is picked up by the lymphatic vessels and returned to the venous system, meanwhile passing through the lymph nodes for defense purposes

lymphocytes White blood cells that provide immune defense against targets for which they are specifically programmed

lymphoid tissues Tissues that produce and store lymphocytes, such as lymph nodes and tonsils

lysosomes (LĪ-sō-sōmz) Organelles consisting of membrane-enclosed sacs containing powerful hydrolytic enzymes that destroy unwanted material within the cell, such as internalized foreign material or cellular debris

macrophages (MAK-ruh-fājs) Large, tissue-bound phagocytes

mast cells Cells located within connective tissue that synthesize, store, and release histamine, as during allergic responses

mature follicle The dominant recruited follicle that grows the largest and is ovulated

mean arterial blood pressure The average pressure responsible for driving blood forward through the arteries into the tissues throughout the cardiac cycle; mean arterial blood pressure = cardiac output × total peripheral resistance

mechanically gated channels Channels that open or close in response to stretching or other mechanical deformation

mechanoreceptor (meh-CAN-ō-rē-SEP-tur) or (mek′-uh-nō-rē-SEP-tur) A sensory receptor sensitive to mechanical energy, such as stretching or bending

medullary respiratory center (MED-you-LAIR-ē) Several aggregations of neuronal cell bodies within the medulla that provide output to the respiratory muscles and receive input important for regulating the magnitude of ventilation

megakaryocyte A large bone-marrow bound cell that sheds off blood-borne platelets from its outer edges

meiosis (mī-ō-sis) Cell division in which the chromosomes replicate followed by two nuclear divisions so that only a half set of chromosomes is distributed to each of four new daughter cells

melanocyte-stimulating hormone (MSH) (mel-AH-nō-sīt) A hormone produced by the pituitary in lower vertebrates that regulates skin coloration for camouflage in these species; in humans, is secreted as a paracrine by the hypothalamus for control of food intake and by keratinocytes in the skin to control dispersion of melanin granules from melanocytes during tanning

melatonin (mel-uh-TŌ-nin) A hormone secreted by the pineal gland during darkness that helps entrain the body's biological rhythms with the external light and dark cues

membrane attack complex A collection of the five final activated components of the complement system that aggregate to form a porelike channel in the plasma membrane of an invading microorganism, with the resultant leakage leading to destruction of the invader

membrane potential A separation of charges across the membrane; a slight excess of negative charges lined up along the inside of the plasma membrane and separated from a slight excess of positive charges on the outside

memory cells B or T cells that are newly produced in response to a microbial invader but that do not participate in the current immune response against the invader; instead, they remain dormant, ready to launch a swift, powerful attack should the same microorganism invade again in the future

meninges (men-IN-geez) Three membranes that wrap the brain and spinal cord

menstrual cycle (men′-stroo-ul) The cyclic changes in the uterus that accompany the hormonal changes in the ovarian cycle

menstrual phase The phase of the menstrual cycle characterized by sloughing of endometrial debris and blood out through the vagina

messenger RNA (mRNA) Carries the transcribed genetic blueprint for synthesis of a particular protein from nuclear DNA to the cytoplasmic ribosomes where the protein is synthesized

metabolic acidosis (met-uh-bol′-ik) Acidosis resulting from any cause other than excess accumulation of carbonic acid in the body

metabolic alkalosis (al′-kuh-LŌ-sus) Alkalosis caused by a relative deficiency of noncarbonic acid

metabolic rate Energy expenditure per unit of time

metabolism All chemical reactions that occur within the body cells

micelle (mī-SEL) A water-soluble aggregation of bile salts, lecithin, and cholesterol that has a hydrophilic shell and a hydrophobic core; carries the water-insoluble products of fat digestion to their site of absorption

microbiome The collective genomes of the microbiota that inhabit the human body

microbiota The community of microbes that live peacefully and usefully with their human host

microfilaments Cytoskeletal elements made of actin molecules (and myosin molecules in muscle cells); play a major role in various cellular contractile systems and serve as a mechanical stiffener for microvilli

microglia Glial cells that serve as the immune defense cells of the CNS

microtubules Cytoskeletal elements made of tubulin molecules arranged into long, slender, unbranched tubes that help maintain asymmetric cell shapes and coordinate complex cell movements

microvilli (mī′-krō-VIL-ī) Actin-stiffened, nonmotile, hairlike projections from the luminal surface of epithelial cells lining the digestive tract and kidney tubules; tremendously increase the surface area of the cell exposed to the lumen

micturition (mik-too-RISH-un) or (mik-chuh-RISH-un) The process of bladder emptying; urination

milk ejection The squeezing out of milk produced and stored in the alveoli of the breasts by means of contraction of the myoepithelial cells that surround each alveolus

mineralocorticoids (min-uh-rul-ō-KOR-ti-koidz) The adrenocortical hormones that are important in Na$^+$ and K$^+$ balance; primarily aldosterone

mitochondria (mī-tō-KON-drē-uh) The energy organelles, which contain the enzymes for oxidative phosphorylation

mitosis (mī-TŌ-sis) Cell division in which the chromosomes replicate before nuclear division so that each of the two daughter cells receives a full set of chromosomes

mitotic spindle The system of microtubules assembled during mitosis along which the replicated chromosomes are moved away from each other toward opposite sides of the cell before cell division

modality The energy form to which sensory receptors respond, such as heat, light, pressure, and chemical changes

molecule A chemical substance formed by the linking of atoms; the smallest unit of a given chemical substance

monocytes (MAH-nō-sīts) White blood cells that emigrate from the blood, enlarge, and become tissue macrophages

monosaccharides (mah′-nō-SAK-uh-rīdz) Simple sugars, such as glucose; the absorbable unit of digested carbohydrates

motility Muscular contractions of the digestive tract wall that mix and propel forward the luminal contents

motor activity Movement of the body accomplished by contraction of skeletal muscles

motor end plate The specialized portion of a skeletal muscle fiber that lies immediately underneath the terminal button of the motor neuron and possesses receptor sites for binding acetylcholine released from the terminal button

motor neurons The neurons that innervate skeletal muscle and whose axons constitute the somatic nervous system

motor protein Specialized protein molecule with "feet" that can be alternately swung forward enabling the molecule to "walk" along a microtubular highway, carrying cargo from one part of the cell to another

motor unit One motor neuron plus all the muscle fibers it innervates

motor unit recruitment The progressive activation of a muscle fiber's motor units to accomplish increasing gradations of contractile strength

mucosa (mew-KŌ-sah) The innermost layer of the digestive tract that lines the lumen

multiunit smooth muscle A smooth muscle mass consisting of multiple discrete units that function independently of one another and that must be separately stimulated by autonomic nerves to contract

muscarinic receptor (MUS-ka-rin′-ik) Type of cholinergic receptor found at the effector organs of all parasympathetic postganglionic fibers

muscle fiber A single muscle cell, which is relatively long and cylindrical in shape

muscle tension See *tension*

muscle tissue A functional grouping of cells specialized for contraction and force generation

myelin (MĪ-uh-lun) An insulative lipid covering that surrounds myelinated nerve fibers at regular intervals along the axon's length; each patch of myelin is formed by a separate myelin-forming cell that wraps itself jelly-roll fashion around the neuronal axon

myelinated fibers Neuronal axons covered at regular intervals with insulative myelin

myocardial ischemia (mī-ō-KAR-dē-ul is-KĒ-mē-uh) Inadequate blood supply to the heart tissue

myocardium (mī′-ō-KAR-dē-um) The cardiac muscle within the heart wall

myofibril (mī′-ō-FĪB-rul) A specialized intracellular structure of muscle cells that contains the contractile apparatus

myogenic activity Muscle-produced, nerve-independent contractile activity

myometrium (mī′-ō-mē-TRĒ-um) The smooth muscle layer of the uterus

myosin (MĪ-uh-sun) The protein forming the thick filaments in muscle fibers

Na⁺–K⁺ pump A carrier that actively transports Na⁺ out of the cell and K⁺ into the cell

Na⁺ load See *sodium (Na⁺) load*

natural killer (NK) cells Naturally occurring, lymphocytelike cells that nonspecifically destroy virus-infected cells and cancer cells by directly lysing their membranes on first exposure to them

negative balance Situation in which the losses for a substance exceed its gains so that the total amount of the substance in the body decreases

negative feedback A regulatory mechanism in which a change in a controlled variable triggers a response that opposes the change, thus maintaining a relatively steady set point for the regulated factor

nephron (NEF-ron′) The functional unit of the kidney; consisting of an interrelated vascular and tubular component, it is the smallest unit that can form urine

nerve A bundle of peripheral neuronal axons, some afferent and some efferent, enclosed by a connective tissue covering and following the same pathway

nerve fiber See *axon*

nervous system One of the two major regulatory systems of the body; in general, coordinates rapid activities of the body, especially those involving interactions with the external environment

nervous tissue A functional grouping of cells specialized for initiation and transmission of electrical signals

net diffusion The difference between the opposing movements of two types of molecules in a solution

net filtration pressure The net difference in the hydrostatic and osmotic forces acting across the glomerular membrane that favors the filtration of a protein-free plasma into Bowman's capsule

NETs See *neutrophil extracellular traps*

neurogenic activity Contractile activity in muscle cells initiated by nerves

neuroglia See *glial cells*

neurohormones Hormones released into the blood by neurosecretory neurons

neuromodulators (ner′ō-MA-jew-lā′-torz) Chemical messengers that bind to neuronal receptors at nonsynaptic sites and bring about long-term changes that subtly depress or enhance synaptic effectiveness

neuromuscular junction The juncture between a motor neuron and a skeletal muscle fiber

neuron (NER-on) A nerve cell specialized to initiate, propagate, and transmit electrical signals, typically consisting of a cell body, dendrites, and an axon

neuropeptide Y (NPY) A potent appetite stimulator secreted by the hypothalamic arcuate nucleus

neuropeptides Large, slow-acting peptide molecules released from axon terminals along with classical neurotransmitters; most neuropeptides function as neuromodulators

neurotransmitter The chemical messenger released from the axon terminal of a neuron in response to an action potential that influences another neuron or an effector with which the neuron is anatomically linked

neutrophil extracellular traps (NETs) A web of prepared fibers released into the ECF by neutrophils when they undergo an unusual type of programmed cell death; bind with bacteria and contain bacteria-killing chemicals

neutrophils (new′-truh-filz) White blood cells that are phagocytic specialists and important in inflammatory responses and defense against bacterial invasion

nicotinic receptor (nick′-ō-TIN-ik) Type of cholinergic receptor found at all autonomic ganglia and the motor end plates of skeletal muscle fibers

nitric oxide A local chemical mediator released from endothelial cells and other tissues; its effects range from causing local arteriolar vasodilation to acting as a toxic agent against foreign invaders to serving as a unique type of neurotransmitter

NMDA receptor One of two types of receptor-channels on a postsynaptic membrane to which the neurotransmitter glutamate binds, this one being both chemically mediated and voltage dependent and permitting Ca^{2+} entry when open

nociceptor (nō-sē-SEP-tur) A pain receptor, sensitive to tissue damage

nodes of Ranvier (RAN-vē-ā) The portions of a myelinated neuronal axon between the segments of insulative myelin; the axonal regions where the axonal membrane is exposed to the ECF and membrane potential exists

norepinephrine (nor′-ep-uh-NEF-run) The neurotransmitter released from sympathetic postganglionic fibers; noradrenaline

NPY See *neuropeptide Y*

nucleus (of brain) (NŪ-klē-us) A functional aggregation of neuronal cell bodies within the brain

nucleus (of cells) A distinct spherical structure, usually located near the center of a cell, that contains the cell's genetic material, deoxyribonucleic acid (DNA)

O₂–Hb dissociation curve A graphic depiction of the relationship between arterial P_{O_2} and percent hemoglobin saturation

occipital lobes (ok-SIP′-ut-ul) The posterior lobes of the cerebral cortex, which initially process visual input

oligodendrocytes (ol-i-gō′-DEN-drō-sitz) The myelin-forming cells of the central nervous system

oogenesis (ō′-ō-JEN-uh-sus) Egg production

opsonin (OP′-suh-nun) Body-produced chemical that links bacteria to macrophages, thereby making the bacteria more susceptible to phagocytosis

optic nerve The bundle of nerve fibers leaving the retina that relay information about visual input

optimal length (l_o) The length before the onset of contraction of a muscle fiber at which maximal force can be developed on a subsequent contraction

organ A distinct structural unit composed of two or more types of primary tissue organized to perform one or more particular functions; for example, the stomach

organ of Corti (KOR-tē) The sense organ of hearing within the inner ear that contains hair cells whose hairs are bent in response to sound waves, setting up action potentials in the auditory nerve

organelles (or′-gan-ELZ) Distinct, highly organized, membrane-bound intracellular compartments, each containing a specific set of chemicals for carrying out a particular cellular function

organic Referring to substances that contain carbon; originally from living or once-living sources

organism A living entity, whether unicellular or multicellular

osmolarity (oz′-mo-LAIR-ut-ē) A measure of the concentration of a solution given in terms of milliosmoles/liter (mOsm/L), the number of millimoles of solute particles in a liter of solution

osmosis (os-MŌ-sis) Movement of water across a membrane down its own concentration gradient toward the area of higher solute concentration

osmotic pressure (os-MAH-tic) A measure of the tendency for osmotic flow of water into a solution resulting from its relative concentration of nonpenetrating solutes and water

osteoblasts (OS-tē-ō-blasts') Bone cells that produce the organic matrix of bone

osteoclasts Bone cells that dissolve bone in their vicinity

osteocytes Retired osteoblasts entombed within the bone that they have laid down around themselves that continue to participate in calcium and phosphate exchange between the bone fluid and plasma

otolith organs (ŌT'-ul-ith) Sense organs in the inner ear that provide information about rotational changes in head movement; include the utricle and saccule

oval window The membrane-covered opening that separates the air-filled middle ear from the upper compartment of the fluid-filled cochlea in the inner ear

overhydration Water excess in the body

ovulation (ov'-yuh-LĀ-shun) Release of an ovum from a mature ovarian follicle

oxidative phosphorylation (fos'-for-i-LĀ-shun) The entire sequence of mitochondrial biochemical reactions that uses oxygen to extract energy from the nutrients in food and transforms it into ATP, producing CO_2 and H_2O in the process; includes the electron transport system and chemiosmosis

oxyhemoglobin (ok-si-HĒ-muh-glō-bun) Hemoglobin combined with O_2

oxyntic mucosa (ok-SIN-tic) The mucosa lining the body and fundus of the stomach, which contains gastric pits that lead to the gastric glands lined by mucous neck cells, parietal cells, and chief cells

oxytocin (ok'-sē-TŌ-sun) A hypothalamic hormone stored in the posterior pituitary that stimulates uterine contraction and milk ejection

pacemaker activity Self-excitable activity of an excitable cell in which its membrane potential gradually depolarizes to threshold on its own

pacemaker potential A self-induced slow depolarization to threshold occurring in a pacemaker cell as a result of shifts in passive ionic fluxes across the membrane accompanying automatic changes in channel permeability

pancreas (PAN-krē-us) A mixed gland composed of an exocrine portion that secretes digestive enzymes and an aqueous alkaline secretion into the duodenal lumen and an endocrine portion that secretes the hormones insulin and glucagon into the blood

paracellular transport Between-cell movement across an epithelial sheet

paracrine (PEAR-uh-krin) A local chemical messenger whose effect is exerted only on neighboring cells in the immediate vicinity of its site of secretion

parasympathetic nervous system (pear'-uh-sim-puh-THET-ik) The subdivision of the autonomic nervous system that dominates in quiet, relaxed situations and promotes body maintenance activities such as digestion and emptying of the urinary bladder

parathyroid glands (pear'-uh-THĪ-roid) Four small glands located on the posterior surface of the thyroid gland that secrete parathyroid hormone

parathyroid hormone (PTH) A hormone that raises plasma Ca^{2+} levels

parietal cells (puh-RĪ-ut-ul) The stomach cells that secrete hydrochloric acid and intrinsic factor

parietal lobes The lobes of the cerebral cortex that lie at the top of the brain behind the central sulcus, which contain the somatosensory cortex

partial pressure (P_x) The individual pressure exerted independently by a particular gas within a mixture of gases

partial pressure gradient A difference in the partial pressure of a gas between two regions that promotes the movement of the gas from the region of higher partial pressure to the region of lower partial pressure

parturition (par'-too-RISH-un) Delivery of a baby

passive expiration Expiration accomplished during quiet breathing as a result of elastic recoil of the lungs on relaxation of the inspiratory muscles, with no energy expenditure required

passive force A force that does not require expenditure of cellular energy to accomplish transport of a substance across the plasma membrane

passive reabsorption Reabsorption when none of the steps in the transepithelial transport of a substance across the kidney tubules requires energy expenditure

pathogens (PATH-uh-junz) Disease-causing microorganisms, such as bacteria or viruses

pathophysiology (path'-ō-fiz-ē-OL-ō-gē) Abnormal functioning of the body associated with disease

pepsin; pepsinogen (pep-SIN-uh-jun) An enzyme secreted in inactive form by the stomach that, once activated, begins protein digestion

peptide hormones Hormones that consist of a chain of specific amino acids of varying length

peptide YY$_{3-36}$ A satiety signal secreted by the small and large intestines that inhibits appetite and serves as a mealtime terminator

percent hemoglobin saturation A measure of the extent to which the hemoglobin present is combined with O_2

perception The conscious interpretation of the external world as created by the brain from a pattern of nerve impulses delivered to it from sensory receptors

peripheral chemoreceptors (kē'-mō-rē-SEP-turz) The carotid and aortic bodies, which respond to changes in arterial P_{O_2}, P_{CO_2}, and H^+ and adjust respiration accordingly

peripheral nervous system (PNS) Nerve fibers that carry information between the central nervous system and other parts of the body

peripheral protein A plasma-membrane protein that studs the surface instead of penetrating the membrane

peristalsis (per'-uh-STOL-sus) Ringlike contractions of the circular smooth muscle of a tubular organ that move progressively forward with a stripping motion, pushing the contents of the organ ahead of the contraction

peritubular capillaries (per'-i-TŪ-bū-lur) Capillaries that intertwine around the tubules of the kidney's nephron; they supply the renal tissue and participate in exchanges between the tubular fluid and blood during the formation of urine

permeable Permitting passage of a particular substance

permissiveness When one hormone must be present in adequate amounts for the full exertion of another hormone's effect

peroxisomes (puh-ROK'-suh-sōmz) Saclike organelles containing powerful oxidative enzymes that detoxify various wastes produced within the cell or foreign compounds that have entered the cell

pH The logarithm to the base 10 of the reciprocal of the hydrogen ion concentration; $pH = \log 1/[H^+]$

phagocytosis (fag'-ō-sī-TŌ-sus) A type of endocytosis in which large, multimolecular, solid particles are engulfed by a cell

pharynx (FARE-inks) The back of the throat, which serves as a common passageway for the digestive and respiratory systems

phasic smooth muscle Smooth muscle that contracts in bursts, triggered by the generation of action potentials

phosphorylation (fos'-for-i-LĀ-shun) Addition of a phosphate group to a molecule

photoreceptor A sensory receptor responsive to light

phototransduction The mechanism of converting light stimuli into electrical activity by the rods and cones of the eye

physiology (fiz-ē-OL-ō-gē) The study of body functions

pineal gland (PIN-ē-ul) A small endocrine gland located in the center of the brain that secretes the hormone melatonin

pinocytosis (pin-ō-cī-TŌ-sus) A type of endocytosis in which the cell internalizes fluid

pitch The tone of a sound, determined by the frequency of vibrations (that is, whether a sound is a C or G note)

pituitary gland (pih-TWO-ih-tair-ee) A small endocrine gland connected by a stalk to the hypothalamus; consists of the anterior pituitary and posterior pituitary

placenta (plah-SEN-tah) The organ of exchange between the maternal and fetal blood; also secretes hormones that support the pregnancy

plaque A deposit of cholesterol and other lipids, perhaps calcified, and thickened, abnormal smooth-muscle cells within blood vessel walls as a result of atherosclerosis

plasma The liquid portion of the blood

plasma cell An antibody-producing derivative of an activated B lymphocyte

plasma clearance The volume of plasma that is completely cleared of a given substance by the kidneys per minute

plasma-colloid osmotic pressure (KOL-oid os-MOT-ik) The force caused by the unequal distribution of plasma proteins between the blood and surrounding fluid that encourages fluid movement into the capillaries

plasma membrane A protein-studded lipid bilayer that encloses each cell, separating it from the extracellular fluid

plasma proteins The proteins in the plasma, which perform a number of important functions; include albumins, globulins, and fibrinogen

plasticity (plas-TIS-uh-tē) The ability of portions of the brain to assume new responsibilities in response to the demands placed on it

platelets (PLĀT-lets) Specialized cell fragments in the blood that participate in hemostasis by forming a plug at a vessel defect

pleural sac (PLOOR-ul) A double-walled, closed sac that separates each lung from the thoracic wall

pluripotent stem cells Precursor cells; for example, those that reside in the bone marrow and continuously divide and differentiate to give rise to each of the types of blood cells

polarization The state of having membrane potential

polycythemia (pol-ē-sī-THĒ-mē-uh) Excess circulating erythrocytes, accompanied by an elevated hematocrit

polysaccharides (pol′-ē-SAK-uh-rīdz) Complex carbohydrates, consisting of chains of interconnected glucose molecules

pool (of a substance) Total quantity of any particular substance in the ECF

positive balance Situation in which the gains via input for a substance exceed its losses via output so that the total amount of the substance in the body increases

positive feedback A regulatory mechanism in which the input and the output in a control system continue to enhance each other so that the controlled variable is progressively moved further from a steady state

postabsorptive state The metabolic state after a meal is absorbed during which endogenous energy stores must be mobilized and glucose must be spared for the glucose-dependent brain; fasting state

posterior pituitary The neural portion of the pituitary that stores and releases into the blood on hypothalamic stimulation two hormones produced by the hypothalamus, vasopressin and oxytocin

postganglionic fiber (pōst′-gan-glē-ON-ik) The second neuron in the two-neuron autonomic nerve pathway; originates in an autonomic ganglion and terminates on an effector organ

postsynaptic neuron (pōst′-si-NAP-tik) The neuron that conducts its action potentials away from a synapse

power stroke The ATP-powered cross-bridge binding and bending that pulls the thin filaments in closer together between the thick filaments during contraction of a muscle fiber

preganglionic fiber The first neuron in the two-neuron autonomic nerve pathway; originates in the central nervous system and terminates on an autonomic ganglion

pressure gradient A difference in pressure between two regions that drives the movement of blood or air from the region of higher pressure to the region of lower pressure

presynaptic facilitation Enhanced release of neurotransmitter from a presynaptic axon terminal as a result of excitation of another neuron that terminates on the axon terminal

presynaptic inhibition A reduction in the release of a neurotransmitter from a presynaptic axon terminal as a result of excitation of another neuron that terminates on the axon terminal

presynaptic neuron (prē-si-NAP-tik) The neuron that conducts its action potentials toward a synapse

primary active transport A carrier-mediated transport system in which energy is directly required to operate the carrier and move the transported substance against its concentration gradient

primary motor cortex The portion of the cerebral cortex that lies anterior to the central sulcus and is responsible for voluntary motor output

primodial follicle A primary oocyte surrounded by a single layer of granulosa cells in the ovary

principal cells Cells in the distal and collecting tubules of the kidney that are the site of action of aldosterone and vasopressin

progestational phase See *secretory phase*

prolactin (PRL) (prō-LAK-tun) An anterior pituitary hormone that stimulates breast development and milk production in females

proliferative phase The phase of the menstrual cycle during which the endometrium repairs itself and thickens following menstruation; lasts from the end of the menstrual phase until ovulation

pro-opiomelanocortin (prō-ōp′-Ē-ō-ma-LAN-ō-kor′-tin) A large precursor molecule that can be variably cleaved into adrenocorticotropic hormone, melanocyte-stimulating hormone, and endorphin

proprioception (prō′-prē-ō-SEP-shun) Awareness of position of body parts in relation to one another and to surroundings

prostaglandins (pros′-tuh-GLAN-dins) Local chemical mediators that are derived from a component of the plasma membrane, arachidonic acid

prostate gland A male accessory sex gland that secretes an alkaline fluid, which neutralizes acidic vaginal secretions

proteasome A nonmembranous organelle that degrades misfolded, damaged, or unneeded intracellular proteins that have been tagged with ubiquitin

protein kinase (KĪ-nase) An enzyme that phosphorylates and thereby induces a change in the shape and function of a particular intracellular protein

proteolytic enzymes (prōt′-ē-uh-LIT-ik) Enzymes that digest protein

proteome The complete set of proteins that can be synthesized using the genetic codes in the genome

proximal tubule (PROKS-uh-mul) A highly convoluted tubule that extends between Bowman's capsule and the loop of Henle in the kidney's nephron

PTH See *parathyroid hormone*

pulmonary artery (PULL-mah-nair-ē) The large vessel that carries blood from the right ventricle to the lungs

pulmonary circulation The closed loop of blood vessels carrying blood between the heart and lungs

pulmonary surfactant (sur-FAK-tunt) A phospholipoprotein complex secreted by the Type II alveolar cells that intersperses between the water molecules that line the alveoli, thereby lowering the surface tension within the lungs

pulmonary valve A one-way valve that permits the flow of blood from the right ventricle into the pulmonary artery during ventricular emptying but prevents the backflow of blood from the pulmonary artery into the right ventricle during ventricular relaxation

pulmonary veins The large vessels that carry blood from the lungs to the heart

pulmonary ventilation The volume of air breathed in and out in one minute; pulmonary ventilation = tidal volume × respiratory rate

pupil An adjustable round opening in the center of the iris through which light passes to the interior portions of the eye

Purkinje fibers (pur-KIN-jē) Small terminal fibers that extend from the bundle of His and rapidly transmit an action potential throughout the ventricular myocardium

pyloric gland area (PGA) (pī-LŌR-ik) The specialized region of the mucosa in the antrum of the stomach that secretes gastrin

pyloric sphincter (pī-LŌR-ik SFINGK-tur) The juncture between the stomach and duodenum

PYY$_{3-36}$ See *peptide YY$_{3-36}$*

RAAS See *renin–angiotensin–aldosterone system*

radiation Emission of heat energy from the surface of a warm body in the form of electromagnetic waves

reabsorption The net movement of interstitial fluid into the capillary

receptive field The circumscribed region surrounding a sensory neuron within which the neuron responds to stimulus information

receptor See *sensory receptor* or *receptor (in membrane)*

receptor (in membrane) Membrane protein that binds with a specific extracellular chemical messenger, bringing about membrane and intracellular events that alter the activity of the particular cell

receptor potential The graded potential change that occurs in a sensory receptor in response to a stimulus; generates action potentials in the afferent neuron fiber

receptor-channel A type of receptor that is an integral part of a channel that opens (or closes) on binding with an extracellular messenger

receptor-enzyme A type of receptor that functions as an enzyme on binding with an extracellular chemical messenger

receptor-mediated endocytosis Import of a particular large molecule from the ECF into a cell by formation and pinching off of an endocytic pouch in response to binding of the molecule to a surface membrane receptor specific for it

reduced hemoglobin Hemoglobin that is not combined with O_2

reflex Any response that occurs automatically without conscious effort; the components of a reflex arc include a receptor, afferent pathway, integrating center, efferent pathway, and effector

refraction Bending of a light ray

refractory period (rē-FRAK-tuh-rē) The time period when a recently activated patch of membrane is refractory (unresponsive) to further stimulation, which prevents the action potential from spreading backward into the area through which it has just passed and ensures the unidirectional propagation of the action potential away from the initial site of activation

regulatory proteins Troponin and tropomyosin, which play a role in regulating muscle contraction by either covering or exposing the sites of interaction between actin and the myosin cross bridges

regulatory T cells A class of T lymphocytes that suppresses the activity of other lymphocytes

releasing hormone A hypothalamic hormone that stimulates the secretion of a particular anterior pituitary hormone

renal cortex An outer granular-appearing region of the kidney

renal medulla (RĒ-nul muh-DUL-uh) An inner striated-appearing region of the kidney

renal threshold The plasma concentration at which the T_m of a particular substance is reached and the substance first starts appearing in the urine

renin (RĒ-nin) An enzymatic hormone released from the kidneys in response to a decrease in NaCl or ECF volume or arterial blood pressure; activates angiotensinogen

renin–angiotensin–aldosterone system (RAAS) (an′jē-ō-TEN-sun al-dō-steer-OWN) The salt-conserving system triggered by release of renin from the kidneys, which activates angiotensin, stimulating aldosterone secretion and Na$^+$ reabsorption by the kidney tubules during the formation of urine

repolarization (rē'-pō-luh-ruh-ZĀ-shun) Return of membrane potential to resting potential following a depolarization

reproductive tract The system of ducts specialized to transport or house the gametes after they are produced

residual volume The minimum volume of air remaining in the lungs even after a maximal expiration

resistance Hindrance of blood or air flow through a blood vessel or respiratory airway, respectively

respiration The sum of processes that accomplish ongoing passive movement of O_2 from the atmosphere to the tissues, as well as the continual passive movement of metabolically produced CO_2 from the tissues to the atmosphere

respiratory acidosis (as-i-DŌ-sus) Acidosis resulting from abnormal retention of CO_2 arising from hypoventilation

respiratory airways The system of tubes that conducts air between the atmosphere and the alveoli of the lungs

respiratory alkalosis (al'-kuh-LŌ-sus) Alkalosis caused by excessive loss of CO_2 from the body as a result of hyperventilation

respiratory rate Breaths per minute

resting membrane potential The membrane potential that exists when an excitable cell is not displaying an electrical signal

reticular activating system (RAS) (ri-TIK-ū-lur) Ascending fibers that originate in the reticular formation and carry signals upward to arouse and activate the cerebral cortex

reticular formation A network of interconnected neurons that runs throughout the brain stem and initially receives and integrates all synaptic input to the brain

retina The innermost layer in the posterior region of the eye that contains the eye's photoreceptors (rods and cones)

rhythmic activities Stereotypical movements that repeat in a general manner, like walking; started and stopped consciously by the cerebral cortex but executed subconsciously by lower CNS levels

ribonucleic acid (RNA) (rī-bō-new-KLĀ-ik) A nucleic acid that exists in three forms (messenger RNA, ribosomal RNA, and transfer RNA), which participate in gene transcription and protein synthesis

ribosomes (RĪ-bō-sōmz) Special ribosomal RNA–protein complexes that synthesize proteins under the direction of nuclear DNA

right atrium (Ā-trē'-um) The heart chamber that receives venous blood from the systemic circulation

right ventricle The heart chamber that pumps blood into the pulmonary circulation

RNA See *ribonucleic acid*

rods The eye's photoreceptors used for night vision

rough ER The flattened, ribosome-studded sacs of the endoplasmic reticulum that synthesize proteins for export or for use in membrane construction

round window The membrane-covered opening that separates the lower chamber of the cochlea in the inner ear from the middle ear

ryanodine receptors (rye-ah-NO-deen) Receptors on the sarcoplasmic reticulum that bind with dihydropyridine receptors on the adjoining T tubule and serve as Ca^{2+}-release channels during excitation–contraction coupling

SA node See *sinoatrial node*

salivary amylase (AM-uh-lās') An enzyme produced by the salivary glands that begins carbohydrate digestion in the mouth and continues it in the stomach after the food and saliva have been swallowed

salt balance Balance between salt intake and salt output; important in controlling ECF volume

saltatory conduction (SAL-tuh-tōr'-ē) The means by which an action potential is propagated throughout a myelinated fiber, with the impulse jumping over the myelinated regions from one node of Ranvier to the next

sarcomere (SAR-kō-mir) The functional unit of skeletal muscle; the area between two Z lines within a myofibril

sarcoplasmic reticulum (ri-TIK-yuh-lum) A fine meshwork of interconnected tubules that surrounds a muscle fiber's myofibrils; contains expanded lateral sacs, which store calcium that is released into the cytosol in response to a local action potential

satiety signals (suh-TĪ-ut-ē) Signals that lead to the sensation of fullness and suppress the desire to eat

saturation The condition in which all binding sites on a carrier molecule are occupied

Schwann cells (shwahn) The myelin-forming cells of the peripheral nervous system

sclera (SKLAIR-a) The visible, white, outer layer of the eye

second messenger An intracellular chemical that is activated by binding of an extracellular first messenger to a surface receptor site, triggering a preprogrammed series of biochemical events that alter activity of intracellular proteins controlling a particular cellular activity

secondary active transport A transport mechanism in which a carrier molecule for glucose or an amino acid is driven by a Na^+ concentration gradient established by the energy-dependent Na^+–K^+ pump to transfer the glucose or amino acid uphill without directly expending energy to operate the carrier

secondary sexual characteristics The many external characteristics that are not directly involved in reproduction but that distinguish males and females

secretin (si-KRĒT-n) A hormone released from the duodenal mucosa primarily in response to the presence of acid; inhibits gastric motility and secretion and stimulates secretion of $NaHCO_3$ solution from the pancreas and liver

secretion Release to a cell's exterior, on appropriate stimulation, of substances produced by the cell

secretory phase The phase of the menstrual cycle characterized by the development of a lush endometrial lining capable of supporting a fertilized ovum; also known as the *progestational phase*

secretory vesicles (VES-i-kuls) Membrane-enclosed sacs containing proteins that have been synthesized and processed by the endoplasmic reticulum and Golgi complex of the cell and which are released to the cell's exterior by exocytosis on appropriate stimulation

segmentation The small intestine's primary method of motility; consists of oscillating, ringlike contractions of the circular smooth muscle along the small intestine's length

selectively permeable membrane A membrane that permits some particles to pass through while excluding others

self-antigens Antigens that are characteristic of a person's cells

semen (SĒ-men) A mixture of accessory sex gland secretions and sperm

semicircular canal Sense organ in the inner ear that detects rotational or angular acceleration or deceleration of the head

semilunar valves (sem'-ī-LEW-nur) The aortic and pulmonary valves

seminal vesicles (VES-i-kuls) Male accessory sex glands that supply fructose to ejaculated sperm and secrete prostaglandins

seminiferous tubules (sem'-uh-NIF-uh-rus) The highly coiled tubules within the testes that produce spermatozoa

sensor The component of a control system that monitors the magnitude of the controlled variable

sensory afferent Pathway into the central nervous system carrying information that reaches the level of consciousness

sensory input Input from somatic sensation and special senses

sensory receptor An afferent neuron's peripheral ending, which is specialized to respond to a particular stimulus in its environment

sensory transduction The conversion of stimulus energy into a receptor potential

Sertoli cells (sur-TŌL-lē) Cells located in the seminiferous tubules that support spermatozoa during their development

serum Plasma minus fibrinogen and other clotting precursors

set point The desired level at which homeostatic control mechanisms maintain a controlled variable

sex hormones The steroid hormones secreted by the gonads that govern reproductive function and are responsible for the development of masculine and feminine characteristics; testosterone in males and estrogens and progesterone in females

signal molecule An extracellular chemical messenger that initiates signal transduction in a cell

signal transducers and activators of transcription (STAT) A cytosolic protein that, when phosphorylated by JAK enzymes, moves to the nucleus and turns on gene transcription resulting in synthesis of designated proteins that carry out the effect dictated by the hormone that sets the JAK/STAT signal transduction pathway in motion

signal transduction The sequence of events in which incoming signals from extracellular chemical messengers are conveyed into a target cell where they are transformed into the dictated cellular response

single-unit smooth muscle The most abundant type of smooth muscle; made up of muscle fibers interconnected by gap junctions so that they become excited and contract as a unit; also known as *visceral smooth muscle*

sinoatrial (SA) node (sī-nō-Ā-trē-ul) A small specialized autorhythmic region in the right atrial wall of the heart that has the fastest rate of spontaneous depolarizations and serves as the normal pacemaker of the heart

skeletal muscle Striated muscle, which is attached to the skeleton and is responsible for movement of the bones in purposeful relation to one another; innervated by the somatic nervous system and under voluntary control

slow-wave potentials Self-excitable activity of an excitable cell in which its membrane potential undergoes gradually alternating depolarizing and hyperpolarizing swings

smooth ER The tubules of the endoplasmic reticulum that package newly synthesized proteins in transport vesicles

smooth muscle Involuntary muscle in the walls of hollow organs and tubes innervated by the autonomic nervous system

sodium (Na⁺) load The total amount of Na^+ in the body, which determines the ECF volume through its osmotic effect

somatic cells (sō-MAT-ik) Body cells, as contrasted with reproductive cells

somatic nervous system The portion of the efferent division of the peripheral nervous system that innervates skeletal muscles; consists of the axonal fibers of the alpha motor neurons

somatic sensation Sensory information arising from the body surface, including somesthetic sensation and proprioception

somatosensory cortex The region of the parietal lobe immediately behind the central sulcus; the site of initial processing of somesthetic and proprioceptive input

somatotropes Anterior pituitary cells that secrete growth hormone

somesthetic sensation (SŌ-mes-THEH-tik) Awareness of sensory input such as touch, pressure, temperature, and pain from the body's surface

sound waves Traveling vibrations of air in which regions of high pressure caused by compression of air molecules alternate with regions of low pressure caused by rarefaction of the molecules

spatial summation The summing of several postsynaptic potentials arising from the simultaneous activation of several excitatory (or several inhibitory) synapses

special senses Vision, hearing, equilibrium, taste, and smell

specificity Ability of carrier molecules to transport only specific substances across the plasma membrane

spermatogenesis (spur′-mat-uh-JEN-uh-sus) Sperm production

sphincter (sfink-tur) A voluntarily controlled ring of skeletal muscle that controls passage of contents through an opening into or out of a hollow organ or tube

spinal reflex A reflex that is integrated by the spinal cord

spleen A lymphoid tissue in the upper left part of the abdomen that stores lymphocytes and platelets and destroys old red blood cells

STAT See *signal transducers and activators of transcription*

state of equilibrium State of a system in which no net change is occurring

steady state State of a system in which no net movement occurs because passive forces and active forces exactly counterbalance each other, with energy being used to maintain the constancy

stem cells Relatively undifferentiated cells that can give rise to highly differentiated, specialized cells while at the same time making new stem cells

stereocilia The auditory and vestibular hair cells that transduce mechanical movements into electrical signals

steroids (STEER-oidz) Hormones derived from cholesterol

stimulus A detectable physical or chemical change in the environment of a sensory receptor

stress The generalized, nonspecific response of the body to any factor that overwhelms, or threatens to overwhelm, the body's compensatory abilities to maintain homeostasis

stretch reflex A monosynaptic reflex in which an afferent neuron originating at a stretch-detecting receptor in a skeletal muscle terminates directly on the efferent neuron supplying the same muscle to cause it to contract and counteract the stretch

stroke volume (SV) The volume of blood pumped out of each ventricle with each contraction, or beat, of the heart

subcortical regions The brain regions that lie under the cerebral cortex, including the basal nuclei, thalamus, and hypothalamus

submucosa The connective tissue layer of the digestive tract that lies under the mucosa and contains the larger blood and lymph vessels and a nerve network

substance P The neurotransmitter released from pain fibers

subsynaptic membrane (sub-sih-NAP-tik) The portion of the postsynaptic cell membrane that lies immediately underneath a synapse and contains receptor sites for the synapse's neurotransmitter

suprachiasmatic nucleus (soup′-ra-kī-as-MAT-ik) A cluster of nerve cell bodies in the hypothalamus that serves as the master biological clock, acting as the pacemaker establishing many of the body's circadian rhythms

surface tension The force at the liquid surface of an air–water interface resulting from the greater attraction of water molecules to the surrounding water molecules than to the air above the surface; a force that tends to decrease the area of a liquid surface and resists stretching of the surface

sympathetic nervous system The subdivision of the autonomic nervous system that dominates in emergency ("fight-or-flight") or stressful situations and prepares the body for strenuous physical activity

symport The form of secondary active transport in which the driving ion and transported solute move in the same direction across the plasma membrane; also called *cotransport*

synapse (SIN-aps) The specialized junction between two neurons where an action potential in the presynaptic neuron influences the membrane potential of the postsynaptic neuron, typically by releasing a chemical messenger that diffuses across the small cleft between the neurons

synergism (SIN-er-jiz′-um) The result of several complementary actions in which the combined effect is greater than the sum of the separate effects

systemic circulation (sis-TEM-ik) The closed loop of blood vessels carrying blood between the heart and body systems

systole (SIS-tō-lē) The period of cardiac contraction and emptying

T₃ See *tri-iodothyronine*

T₄ See *thyroxine*

T lymphocytes (T cells) White blood cells that accomplish cell-mediated immune responses against targets to which they have been previously exposed; see also *cytotoxic T cells, helper T cells,* and *regulatory T cells*

T tubule See *transverse tubule*

tactile (TACK-til) Referring to touch

target cells The cells that a particular extracellular chemical messenger, such as a hormone or a neurotransmitter, influences

target-cell receptors Receptors located on a target cell that are specific for a particular chemical mediator

temporal lobes The lateral lobes of the cerebral cortex, which initially process auditory input

temporal summation The summing of several postsynaptic potentials that occur very close together in time because of successive firing of a single presynaptic neuron

tension The force produced during muscle contraction by shortening of the sarcomeres, resulting in stretching and tightening of the muscle's elastic connective tissue and tendon, which transmit the tension to the bone to which the muscle is attached

terminal button A motor neuron's enlarged knoblike ending that terminates near a skeletal muscle fiber and releases acetylcholine in response to an action potential in the neuron

testosterone (tes-TOS-tuh-rōn) The male sex hormone, secreted by the Leydig cells of the testes

tetanus (TET′-n-us) A smooth, maximal muscle contraction that occurs when the fiber is stimulated so rapidly that it does not have a chance to relax at all between stimuli

tetraiodothyronine See *thyroxine*

thalamus (THAL-uh-mus) The brain region that serves as a synaptic integrating center for preliminary processing of all sensory input on its way to the cerebral cortex

thecal cells (THĀY-kel) The outer layer of specialized ovarian connective tissue cells in a maturing follicle

thermoreceptor (thur′-mō-rē-SEP-tur) A sensory receptor sensitive to heat and cold

thick filaments Specialized cytoskeletal structures within skeletal muscle that are made up of myosin molecules and interact with the thin filaments to shorten the fiber during muscle contraction

thin filaments Specialized cytoskeletal structures within skeletal muscle that are made up of actin, tropomyosin, and troponin molecules and interact with the thick filaments to shorten the fiber during muscle contraction

thoracic cavity (thō-RAS-ik) Chest cavity

threshold potential The critical potential that must be reached before an action potential is initiated in an excitable cell

thrombus An abnormal clot attached to the inner lining of a blood vessel

transcellular transport Transport across an epithelial barrier through, not between, the cells

thymus (THĪ-mus) A lymphoid gland located midline in the chest cavity that processes T lymphocytes and produces the hormone thymosin, which maintains the T-cell lineage

thyroglobulin (thī′-rō-GLOB-yuh-lun) A large, complex molecule on which all steps of thyroid hormone synthesis and storage take place

thyroid gland A bilobed endocrine gland that lies over the trachea and secretes the hormones thyroxine and tri-iodothyronine, which regulate overall basal metabolic rate, and calcitonin, which contributes to control of calcium balance

thyroid hormone Collectively, the hormones secreted by the thyroid follicular cells, namely, thyroxine and tri-iodothyronine

thyroid-stimulating hormone (TSH) An anterior pituitary hormone that stimulates secretion of thyroid hormone and promotes growth of the thyroid gland; thyrotropin

thyrotropes Anterior pituitary cells that secrete thyroid-stimulating hormone

thyroxine (thī-ROCKS-in) The most abundant hormone secreted by the thyroid gland; important in the regulation of overall metabolic rate; also known as *tetraiodothyronine* or T_4

tidal volume The volume of air entering or leaving the lungs during a single breath

tight junction An impermeable junction between two adjacent epithelial cells formed by the sealing together of the cells' lateral edges near their luminal borders; prevents passage of substances between the cells

tip links Cell adhesion molecules that link the tips of stereocilia of hair cells in adjacent rows in the cochlea and vestibular organs and that pull open mechanically gated channels in the stereocilia during signal transduction

tissue (1) A functional aggregation of cells of a single specialized type, such as nerve cells forming nervous tissue; (2) the aggregate of various cellular and extracellular components that make up a particular organ, such as lung tissue

tissue-specific stem cells Partially differentiated cells that can generate the highly differentiated, specialized cell types composing a particular tissue

titin A giant, highly elastic protein that extends in both directions from the M line along the length of the thick filament to the Z lines and is responsible for the parallel elastic component of the muscle fiber

T_m See *transport maximum* and *tubular maximum*

tone The ongoing baseline of activity in a given system or structure, as in muscle tone, sympathetic tone, or vascular tone

tonic smooth muscle Smooth muscle that is partially contracted at all times in the absence of action potentials

tonicity A measure of the effect a solution has on cell volume when the solution surrounds the cell

total peripheral resistance The resistance offered by all the peripheral blood vessels, with arteriolar resistance contributing most extensively

trachea (TRĀ-kē-uh) The "windpipe"; the conducting airway that extends from the pharynx and branches into two bronchi, each entering a lung

tract A bundle of nerve fibers (axons of long interneurons) with a similar function within the spinal cord

transduction Conversion of stimuli into action potentials by sensory receptors

transepithelial transport (tranz-ep-i-THĒ-lē-al) The entire sequence of steps involved in the transfer of a substance across the epithelium between either the renal tubular lumen or digestive tract lumen and the blood

transmural pressure gradient The pressure difference across the lung wall (intra-alveolar pressure is greater than intrapleural pressure) that stretches the lungs to fill the thoracic cavity, which is larger than the unstretched lungs

transport maximum (T_m) The maximum rate of a substance's carrier-mediated transport across the membrane when the carrier is saturated; known as *tubular maximum* in transepithelial transport across the kidney tubules

transport vesicle Membranous sac enclosing newly synthesized proteins that buds off the smooth endoplasmic reticulum and moves the proteins to the Golgi complex for further processing and packaging for their final destination

transporter recruitment The phenomenon of inserting additional transporters (carriers) for a particular substance into the plasma membrane, thereby increasing membrane permeability to the substance, in response to an appropriate stimulus

transverse (T) tubule A perpendicular infolding of the surface membrane of a muscle fiber; rapidly spreads surface electric activity into the central portions of the muscle fiber

triglycerides (trī-GLIS-uh-rīdz) Neutral fats composed of one glycerol molecule with three fatty acid molecules attached

tri-iodothyronine (T_3) (trī-ī-ō-dō-THĪ-ro-nēn) The most potent hormone secreted by the thyroid follicular cells; important in the regulation of overall metabolic rate

trophic hormone A hormone that helps maintain the structural integrity of its target gland

trophoblast (TRŌF-uh-blast′) The outer layer of cells in a blastocyst that is responsible for accomplishing implantation and developing the fetal portion of the placenta

tropic hormone (TRŌ-pik) A hormone whose primary function is to regulate the secretion of another hormone

tropomyosin (trōp′-uh-MĪ-uh-sun) One of the regulatory proteins in the thin filaments of muscle fibers

troponin (trō-PŌ-nun) One of the regulatory proteins in the thin filaments of muscle fibers

TSH See *thyroid-stimulating hormone*

tubular maximum (T_m) The maximum amount of a substance that the renal tubular cells can actively transport within a given time period; the kidney cells' equivalent of transport maximum

tubular reabsorption The selective transfer of substances from tubular fluid into peritubular capillaries during urine formation

tubular secretion The selective transfer of substances from peritubular capillaries into the tubular lumen during urine formation

tunneling nanotubes Long, thin, hollow filaments between cells that can transfer larger cargo considerably longer distances from cell to cell than gap junctions can

twitch A brief, weak contraction that occurs in response to a single action potential in a muscle fiber

twitch summation The addition of two or more muscle twitches as a result of rapidly repetitive stimulation, resulting in greater tension in the fiber than that produced by a single action potential

tympanic membrane (tim-PAN-ik) The eardrum, which is stretched across the entrance to the middle ear and which vibrates when struck by sound waves funneled down the external ear canal

Type I alveolar cells (al-VĒ-ō-lur) The single layer of flattened epithelial cells that forms the wall of the alveoli within the lungs

Type II alveolar cells The cells within the alveolar walls that secrete pulmonary surfactant

tyrosine kinase A type of receptor-enzyme that brings about the dictated cellular response on binding with an extracellular chemical messenger by phosphorylating the amino acid tyrosine in designated intracellular proteins

ultrafiltration The net movement of a protein-free plasma out of the capillary into the surrounding interstitial fluid

umami A meaty or savory taste

ureter (yū-RĒ-tur) A duct that transmits urine from the kidney to the bladder

urethra (yū-RĒ-thruh) A tube that carries urine from the bladder out of the body

urine excretion The elimination of substances from the body in the urine; anything filtered or secreted and not reabsorbed is excreted

vagus nerve (VĀ-gus) The tenth cranial nerve, which serves as the major parasympathetic nerve

varicosities Swellings in autonomic postganglionic fibers that simultaneously release neurotransmitter over a large area of an innervated organ

vascular tone The state of partial constriction of arteriolar smooth muscle that establishes a baseline of arteriolar resistance

vasoconstriction (vā′-zō-kun-STRIK-shun) The narrowing of a blood vessel lumen as a result of contraction of the vascular circular smooth muscle

vasodilation The enlargement of a blood vessel lumen as a result of relaxation of the vascular circular smooth muscle

vasopressin (vā-zō-PRES-sin) A hormone secreted by the hypothalamus, then stored and released from the posterior pituitary; increases the permeability of the distal and collecting tubules of the kidneys to water and promotes arteriolar vasoconstriction; also known as *antidiuretic hormone (ADH)*

vaults Nonmembranous organelles shaped like octagonal barrels; believed to serve as transporters for messenger RNA or the ribosomal subunits from the nucleus to sites of protein synthesis

vein A vessel that carries blood toward the heart

vena cava (venae cavae, plural) (VĒ-nah CĀV-ah; VĒ-nē cāv-ē) A large vein that empties blood into the right atrium

venous return (VĒ-nus) The volume of blood returned to each atrium per minute from the veins

ventilation The mechanical act of moving air in and out of the lungs; breathing

ventricle (of brain) (VEN-tri-kul) One of four interconnected chambers within the brain through which cerebrospinal fluid flows

ventricle (of heart) A lower chamber of the heart that pumps blood into the arteries

vertical osmotic gradient A progressive increase in the concentration of the interstitial fluid in the renal medulla from the cortical boundary down to the renal pelvis; important in the ability of the kidneys to put out urine of variable concentration, depending on the body's needs

vesicle (VES-i-kul) A small, intracellular, fluid-filled, membrane-enclosed sac

vesicular transport Movement of large molecules or multimolecular materials into or out of the cell within a vesicle, as in endocytosis or exocytosis

vestibular apparatus (veh-STIB-yuh-lur) The component of the inner ear that provides information essential for the sense of equilibrium and for coordinating head movements with eye and postural movements; consists of the semicircular canals, utricle, and saccule

villus (villi, plural) (VIL-us) Microscopic finger-like projections from the inner surface of the small intestine

virulence (VIR-you-lentz) The disease-producing power of a pathogen

visceral afferent A pathway into the central nervous system that carries subconscious information derived from the internal viscera

visceral smooth muscle (VIS-uh-rul) See *single-unit smooth muscle*

viscosity (vis-KOS-i-tē) The friction developed between molecules of a fluid as they slide over each other during flow of the fluid; the greater the viscosity, the greater the resistance to flow

visible light The portion of the electromagnetic spectrum to which the eyes' photoreceptors are responsive (wavelengths between 400 and 700 nanometers)

vital capacity The maximum volume of air that can be moved out during a single breath following a maximal inspiration

vitreous humor The jellylike substance in the posterior cavity of the eye between the lens and retina

voltage-gated channels Channels in the plasma membrane that open or close in response to changes in membrane potential

voluntary movement Goal-directed movement initiated and terminated at will and integrated by the cerebral cortex

voluntary muscle Muscle innervated by the somatic nervous system and subject to voluntary control; skeletal muscle

water balance The balance between water intake and water output; important in controlling ECF osmolarity

white matter The portion of the central nervous system composed of myelinated nerve fibers

Z line A flattened disclike cytoskeletal protein that connects the thin filaments of two adjoining sarcomeres

zona fasciculata (zō-nah fa-SIK-ū-lah-ta) The middle and largest layer of the adrenal cortex; major source of cortisol

zona glomerulosa (glō-MAIR-yū-lō-sah) The outermost layer of the adrenal cortex; sole source of aldosterone

zona reticularis (ri-TIK-yuh-lair-us) The innermost layer of the adrenal cortex; produces cortisol, along with the zona fasciculata

zygote A fertilized ovum

Index

Corticospinal motor system, 279, 281
Corticosteroid-binding globulin, 673
Corticosteroids. *See* Glucocorticoids
Corticotropes, 648
Corticotropin-releasing hormone (CRH), 650, 651–652
    appetite and, 623
    cortisol secretion and, 675
    parturition and, 764–766
    stress response and, 683
Cortisol
    in adaptation to stress, 674, 675–676, 683
    antigrowth effects, 653
    anti-inflammatory effects, 674
    bound to plasma proteins, 673
    deficiency of, 680
    diurnal rhythm of secretion, 642, 675
    effects of, summarized, 674–675
    hypersecretion of, 676–677
    immunosuppressive effects, 674–675
    metabolic effects, 674, 699–700
    milk production and, 769
    permissive actions of, 674
    regulation of secretion, 651–652, 675–676
    structure, A-12
    synthesis of, 673
Cotransmitters, autonomic, 236
Cotransport, 74, 75. *See also* Sodium and glucose cotransporter
Cough, 442
Countercurrent exchange, 526–527
Countercurrent multiplication, 521–523
Countertransport, 74–75
Coupled-clock system, 304, 305
Covalent bonds, A-3–A-4
COX (cyclooxygenase), 119
CPR (cardiopulmonary resuscitation), 299
Cranial nerves, 166, 167
Cranial reflexes, 177, 277
C-reactive protein (CRP), 329, 411
Creatine kinase, 271
Creatine phosphate, 269–271, 272–273
Creatinine, 518
CREB (cAMP responsive element binding protein), 162
Crenated red blood cells, 69
Cretinism, 670
CRH. *See* Corticotropin-releasing hormone
Cristae, 33–34
Cross bridges
    ATP power and, 259, 261
    calcium and, 255–256, 259, 260, 261, 267–268
    cardiac muscle, 309, 322–323
    contraction time and, 261–262
    optimal muscle length and, 268
    power stroke and, 256–258, 259, 260, 261
    smooth muscle, 288–289, 292, 293
    structural features of, 253, 254, 255–256
    in twitch summation, 267–268
Crossed extensor reflex, 285–286
Crossing-over, 718
CRP (C-reactive protein), 329, 411
Cryptorchidism, 723, 724, 736
Crypts of Lieberkühn, 602
Crystal lattice of sodium chloride, A-3
CSF (cerebrospinal fluid), 139, 140
    spinal tap and, 173
Cupula, 221, 222
Curare, 246–247
Current, electric
    defined, 89
    direction of, 89
    during graded potential, 89–90
Cushing's syndrome, 676–677
CVA (cerebrovascular accident), 142. *See also* Stroke
Cyanide poisoning, 478
Cyclic adenosine monophosphate (cAMP)
    calcitonin and, 708
    degradation by phosphodiesterase, 124
    formation of, A-15

glucagon and, 698
growth hormone secretion and, 656
heart activity and, 319, 320
hormones using pathway of, 711
insulin secretion and, 693
in long-term potentiation, 161
in memory consolidation, 162
parathyroid hormone and, 704
as second messenger, 122–125
in sensitization, 160
in sperm, 753
Cyclic guanosine monophosphate (cGMP)
    erection and, 734
    in photoreceptors, 201, 203
    as second messenger, 124
Cyclooxygenase (COX), 119
Cyclosporin, 429, 431
Cysteinuria, 71
Cystic fibrosis, 59
Cystic fibrosis transmembrane conductance regulator (CFTR), 59
Cytochrome *c*
    apoptosis activated by, 40–41, 42–43
    in electron transport, 37, 38
Cytokines
    defined, 411
    helper T-cell secretion of, 425–426, 427
    inflammation and, 410, 411–412
    nervous system and, 434, 435
    overview, 118
Cytokinesis, 49, 50
Cytoplasm, 24–25. *See also* Cytoskeleton; Cytosol; Organelles
    summary of components, 45
Cytoskeleton, 25, 44–51
    functions of, 44
    as integrated whole, 51
    intermediate filaments, 44, 45, 51
    microfilaments, 44, 45, 49–51
    microtubules, 44, 45, 46–49
    of muscle fiber, 254–256
    summary of, 45
Cytosol, 25, 42–44
    glycolysis in, 35, 39
    intermediary metabolism in, 43
    ribosomal protein synthesis in, 26, 43
    storage of fat, glycogen, and secretory vesicles in, 43–44
    summary of, 45
Cytotoxic T cells, 423–424, 425
    cancer cells and, 433–434
    class I MHC glycoproteins and, 428
    embryo–fetus defense against, 757

D cells. *See* Delta cells
DAG (diacylglycerol), 124
Damage-associated molecular patterns (DAMPs), 407
Dark adaptation, 206
Dark current, 201
Dead space
    alveolar, 465
    anatomic, 462
Deafness, 220–221
Death receptors, 43
Decibels (dB), 212, 214
Decidua, 756–757
Declarative memories, 162
Decompression sickness, 481
Deep-sea diving, 481
Default mode network (DMN), 152–153
Defecation reflex, 612
Defenses, body. *See also* Immune system
    of digestive system, 441
    of respiratory system, 441–442
    of urogenital system, 441
Defensins, 59, 438, 603
Degranulation, 392
Dehydration
    causes of, 541
    diarrhea causing, 610, 611
    hematocrit associated with, 387, 388

hypertonic ECF in, 541–543
    symptoms of, 541–543
Dehydroepiandrosterone (DHEA), 673, 676, 723, 762, 764
Deiodinase, 668
Delayed hypersensitivity, 436, 438, 439
Delta (D) cells, 583, 585, 586, 689, 690
Delta waves, 169
Denaturation of proteins, A-14
    by gastric acid, 585
Dendrites, 95–96
Dendritic cells, 427–428, 429
    decidual, 757
Dendritic spines, 95–96
Denervation atrophy, 275
Denosumab, 707
Dense bodies, 288
Dense-core vesicles, 110–111
Dental caries, 574, 575
Deoxyhemoglobin, 472
Deoxyribonucleic acid. *See* DNA
Depolarization
    in action potential, 91–92, 93, 96, 99
    defined, 88
    in excitatory postsynaptic potential (EPSP), 106
    in graded potential, 89
    in pacemaker potential, 304
Depolarization block, 246
Depression, 156–157
Depth perception, 209
Dermatomes, 176
Dermcidin, 630
Dermis, 440
Descending tracts, spinal cord, 173–174, 175
Desmosomes, 61
    of intercalated discs, 303
Detergent action of bile salts, 595–596
Detoxification
    by liver, 27
    by peroxisomes, 33
Development
    apoptosis in, 42
    of embryo, 755–757
Dextrins, α-limit, 574, 600
DHEA (dehydroepiandrosterone), 673, 676, 723, 762, 764
DHT (dihydrotestosterone), 721, 723
Diabetes insipidus, 541
Diabetes mellitus, 693–698
    acute consequences of, 694
    as autoimmune disease, 696, 697
    carbohydrate metabolism in, 694
    fat metabolism in, 694
    glucagon excess in, 699
    glucose in urine due to, 512
    glucose transport and, 72
    long-term complications of, 694–695, 698
    metabolic acidosis caused by, 561
    protein metabolism in, 694
    treatment of, 696–697
    Type 1, 72, 432, 694, 696, 697
    Type 2, 72, 694, 696–697
    Underlying defects in, 696
Diacylglycerol (DAG), 124
Dialysis, 530
Diapedesis, 410
Diaphragm (contraceptive), 758
Diaphragm (muscle), 449, 453, 454
    vomiting and, 582
Diaphysis, 654
Diarrhea, 610
    metabolic acidosis caused by, 561
    oral rehydration therapy for, 610, 611
Diastole, 314–315, 316, 317
Diastolic heart failure, 324, 325
Diastolic murmur, 318
Diastolic pressure, 341
Dicrotic notch, 315, 316
Diencephalon, 143, 154
Diet-induced thermogenesis, 620
Differentiation, cell, 4

Hypocretin, 170, 172
Hypodermis, 440
Hypoglycemia
  insulin excess leading to, 698
  reactive, 698
Hypokalemia, 676
Hyponatremia, 680
Hypoparathyroidism, 712
Hypophosphatemia, 710, 712
Hypophysiotropic hormones, 650–652
Hypophysis. *See* Pituitary gland
Hyposecretion, hormonal, 642
Hypospadias, 736
Hypotension
  circulatory shock associated with, 374–375, 377
  defined, 369
  orthostatic (postural), 374
Hypothalamic osmoreceptors, 369, 545, 647, 648
Hypothalamic–hypophyseal portal system, 650–651
Hypothalamus. *See also* Arcuate nucleus
  anterior pituitary and, 647–652
  autonomic responses regulated by, 241
  AVPV nucleus in, 748
  basic behavioral patterns and, 155
  blood–brain barrier and, 141
  fever and, 411
  food intake control by, 621–623
  glucose homeostasis and, 701
  homeostasis and, 154–155
  hyperthermia due to lesions of, 635
  lactation and, 769
  limbic system and, 155
  ovarian cycle and, 746, 747, 748, 749
  overview, 143, 144–145
  posterior pituitary and, 646–647
  releasing and inhibiting hormones of, 648–652
  sleep and, 170, 172
  smell and, 228
  stress response coordinated by, 683–684
  suprachiasmatic nucleus in, 660–661
  temperature regulation and, 369, 630, 631, 632–635
  thirst center in, 544, 545
Hypothalamus–pituitary–adrenal axis, 675
Hypothalamus–pituitary–thyroid axis, 669
Hypothermia, 634
Hypothyroidism, 669–670
  goiter in, 671
  obesity and, 626
Hypotonic body fluids, 520, 524
Hypotonic solution, 69
Hypotonicity of ECF, 543
Hypoventilation, 478, 559
Hypovolemic shock, 374, 375
Hypoxia, 477–478
Hypoxic hypoxia, 477, 480
Hz (Hertz), 212

I band, 253, 254, 256, 258
IC (inspiratory capacity), 461, 462
ICF. *See* Intracellular fluid
Identical twins, 744
IDO (indoleamine 2,3-dioxygenase), 757
IEGs (immediate early genes), 162
I_f (funny) channels, 304
IgA, 417, 421, 442, 574
  secretory, 770
IgD, 417
IgE, 417, 418, 426, 436–438
IgG, 417, 418, 420, 421
IgM, 417, 420, 421
IGF-I, 654, 656, 657, 659, 669, 768
IGF-II, 654, 768
ILCs (innate lymphoid cells), 415
Ileocecal sphincter, 599
Ileocecal valve, 599
Ileum, 598
  absorption of vitamin B_{12} and bile salts in, 601
Immediate early genes (IEGs), 162
Immediate hypersensitivity, 436–438, 439
Immune complex disease, 419–420, 436

Immune diseases. *See also* Autoimmune diseases; HIV (human immunodeficiency virus); Immune complex disease
  allergies (hypersensitivities), 436–438
  immunodeficiency, 435–436
Immune surveillance, for cancer, 405, 432, 433–434
Immune system. *See also* Adaptive (acquired) immune system; Innate immune system
  activities attributable to, 405
  apoptosis in, 42
  colon microbiota and, 613
  components of, 9, 392–393
  cortisol and, 675
  cytokines and, 118
  effector cells of, 405–406
  embryo–fetus tolerated by, 757
  endocrine system and, 434, 435
  exercise effect on, 435
  functions of, 392
  homeostasis and, 13, 15, 404, 442
  inappropriate attacks by, 436
  intercellular communication by, 113
  nervous system and, 434–435
  skin cells involved in, 441
  targets of, 405
Immunity
  active, 421
  defined, 405
  passive, 421
Immunization, 421, 422
Immunocontraception, 759
Immunodeficiency diseases, 435–436
  AIDS as, 137–138, 426, 436
Immunoglobulin, 417. *See also* Antibodies
  IgA, 417, 421, 442, 574, 770
  IgD, 417
  IgE, 417, 418, 426, 436–438
  IgG, 417, 418, 420, 421
  IgM, 417, 420, 421
Immunological ignorance, 431
Immunosuppressive effects, of cortisol, 674–675
Impermeable membrane, 63
Implantation, 756–757
  blocking, to prevent pregnancy, 759
  preparation of endometrium for, 750, 755
Impotence, 734
Inactivation gate, 92
Inclusions, cellular, 43–44
Incontinence, urinary, 531–532
Incretin mimetics, 697
Incretins, 693
Incus, 214
Indoleamine 2,3-dioxygenase (IDO), 757
Indoleamines, 120, 121, 639. *See also* Melatonin
Induced pluripotent stem cells (iPSCs), 11
Infertility, male, 735
Inflammasomes, 407
Inflammation. *See also* Nonsteroidal anti-inflammatory drugs
  in adipose tissue, 623
  atherosclerosis and, 327, 328, 329
  capillary permeability increased in, 409
  chronic illnesses associated with, 412
  complement system augmentation of, 414, 418
  cytokines in mediation of, 410, 411–412
  defined, 408
  drugs for suppression of, 412
  edema in, 409
  glucocorticoids and, 674, 675
  goal of, 408–409
  gross manifestations of, 409
  histamine in, 348, 354, 409, 411, 414
  leukocyte emigration in, 410
  leukocyte proliferation in, 410
  leukocytic destruction of bacteria in, 410–411
  molecular pattern recognition and, 407
  opsonins in, 410, 411
  parturition and, 766
  sequence of events in, 408–412
  stress and, 434–435
  tissue macrophages in, 409

  tissue repair following, 412
  vasodilation in, 409
  walling off of inflamed area, 410
Inguinal canal, 723
Inhibin
  in female, 747, 748, 749
  in male, 729, 730
Inhibiting hormones, 649–652
Inhibitory postsynaptic potential (IPSP), 107–108, 109, 111
Inhibitory synapse, 106–107
Initial lymphatics, 358
Initial segment, of axon, 96
Innate immune system, 408–415
  amplified by antibodies, 418–419
  complement in, 408, 413–415, 418
  components of, 408
  inflammation and, 407, 408–412
  interferon in, 408, 412–413, 433
  linked to adaptive immune system, 407, 415, 424
  natural killer (NK) cells in, 408, 413, 418, 424, 425, 433, 757
  overview, 406
  pathogen receptors of, 407
  response to bacterial invasion, 430
Innate lymphoid cells (ILCs), 415
Innate reflexes, 177
  overridden in sports, 178
Innate response activator (IRA) B cells, 415
Inner cell mass, 756, 760
Inner ear, 211, 212
  cochlea, 214–219
  vestibular apparatus, 221–224
Inner hair cells, 217–219
Innervation, defined, 102
Inorganic chemicals, A-8
Inorganic salts, A-9
Inositol triphosphate (IP_3), 124
  hormones using pathway of, 711
Input to extracellular fluid, 536–537
Insensible water loss, 544
Insertion, muscle, 263
Inspiration, 452–453, 454
  defined, 452
  intra-alveolar pressure and, 452
Inspiratory capacity (IC), 461, 462
Inspiratory muscles
  accessory, 453
  major, 452–453, 454
Inspiratory neurons, 479, 480
Inspiratory reserve volume (IRV), 461, 462
Insufficient (incompetent) valve, 318
Insulin. *See also* Diabetes mellitus
  actions of, 690–692
  administered for diabetes type 1, 696, 697
  amylin as partner of, 689, 692
  blood glucose as stimulus for secretion of, 692–693
  carbohydrates and, 690–691
  in control of food intake, 623
  diabetes mellitus and, 693–695, 696–697
  down regulation of, 643
  excess of, 698
  excitation-secretion coupling and, 692
  factors controlling secretion of, 692–693
  fats and, 692
  feedforward control of, 18
  glucagon in relation to, 698–699
  glucose uptake and, 72
  growth promotion by, 659
  milk production and, 769
  pancreatic secretion of, 689, 690, 702
  protein and, 692
  stress response and, 683
  summary of actions, 692
  tyrosine kinase pathway and, 116
Insulin antagonists, 700–701
Insulin pumps, 697
Insulin resistance, 696
Insulin shock, 698

Macrophages
  alveolar, 442, 448, 449
  angry, 426
  antigen presentation by, 427, 429
  atherosclerosis and, 328
  cancer cells and, 433
  cytokine secretion by, 411
  fever and, 633
  inflammation and, 409, 410–411, 412
  Kupffer cells of liver, 593, 594
  life span, 393
  monocytes maturing into, 393, 405
  osteoclasts derived from, 703
  as phagocytes, 393, 405
Macula densa, 502, 507
Macula lutea, 200
Macular degeneration, 200
Magnetic resonance imaging, functional (fMRI), 145
Major histocompatibility complex (MHC), 428.
    See also MHC molecules
Malabsorption, 601–602
Malaria, 387–388
Male menopause, 725
Male reproductive system. See also Testes;
    Testosterone
  accessory sex glands, 716, 717, 731, 732
  anatomy, 716, 717, 724, 730–731, 732
  components of, location and function, 732
  ejaculation, 732, 733, 734
  erection, 732–734
  overview, 716
  puberty and, 725–726, 730
  semen, 716, 731, 734, 735
  sexual response cycle, 732
  spermatogenesis, 723–724, 725, 726–730
Males, genetic, 718, 720, 721
Malignant tumor, 432–433
Malleus, 214
Malocclusion, 574
Maltase, 600, 604
Maltose
  hydrolysis of, 567, 600
  from starch digestion, 574
Mammary glands. See also Lactation
  anatomy, 767, 768
  preparation for lactation, 767–768
MAOIs (monoamine oxidase inhibitors), 157
MAP. See Mean arterial pressure (MAP)
Margination, 410
Mass, A-1
  molecular, A-6
Mass action, law of, 472
Mass movements, 612
Mast cells, 393
  histamine released by, 409, 414, 436
  immediate hypersensitivity and, 436, 437, 438
Mastication, 574
Matrix, mitochondrial, 33–34
  beta oxidation of fatty acids in, 39–40
  citric acid cycle in, 35–37
Matter, A-1
Mature follicle, 741, 742, 743
Mature ovum, 738–739
Maximal O_2 consumption (max VO_2), 487
Mean arterial pressure (MAP). See also Blood
    pressure
  autoregulation of tissue blood flow and, 347, 348
  defined, 341, 343
  determinants of, 365–366
  regulation of, 349–350, 365–369
Mechanical nociceptors, 189
Mechanical work, ATP required for, 40
Mechanically gated channels, 89
Mechanoreceptors
  adequate stimuli for, 182
  digestive tract, 572–573
  hair cells as, 217, 221
  tactile, 185, 187
Medial geniculate nucleus, 220
Medulla, 143, 144–145. See also Brain stem
  autonomic output controlled by, 241
  cardiovascular control center in, 321, 350, 367, 632

central chemoreceptors in, 483–484
respiratory center in, 479–481, 484, 486
salivary center in, 575
swallowing center in, 576
vomiting center in, 582
Medulla, renal, 494
  vertical osmotic gradient in, 520–523, 525,
    526–527, 544
Medullary respiratory center, 479–481, 484,
    486
Megakaryocytes, 395
Meiosis
  in oogenesis, 738, 739, 743, 747, 755
  overview, 718
  in spermatogenesis, 727
Meiotic arrest, 738
Meissner's corpuscle, 185
Melanin, 440, 441
Melanocortins, 621–622, 623
α-Melanocyte stimulating hormone (α-MSH),
    441, 621, 648, 680
Melanocytes, 440–441
Melanopsin, 210, 661
Melatonin, 660, 661–662
  puberty and, 730, 752
Membrane. See Plasma membrane
Membrane attack complex, 414, 418
Membrane carbohydrate, 57
Membrane clock mechanism, 304, 305
Membrane permeability changes
  action potentials and, 92–94
  triggering event for, 88, 89
Membrane potential
  action potentials, 91–102
  changes in, due to ion movement, 88–89
  chloride and, 84
  defined, 77, 79
  equilibrium potential for K^+ (EK^+), 80–81
  equilibrium potential for Na^+ (ENa^+), 81–82
  Goldman-Hodgkin-Katz equation for, 82–83
  graded potentials, 89–91
  Na^+-K^+ ATPase pump and, 83–84
  resting, 79–84
  specialized use in nerve and muscle cells, 88
  threshold potential, 91–92
  types of changes in, 88
  units of measurement, 77
  voltage-gated channels and, 89
Membrane proteins. See also Carrier-mediated
    transport; Channels
  carbohydrates bound to, 57, 60
  fluid mosaic model and, 56
  functions, 58–60
  in lipid rafts, 57
  types of, 56, 58–60
Membrane transport. See also Active transport;
    Diffusion
  active forces in, 63
  active transport, overview of, 73–75
  assisted, 70–77
  ATP required for, 40
  carrier-mediated, 70–75, 78
  diffusion, overview of, 63–66
  along electrical gradient, 66
  down electrochemical gradient, 66
  facilitated diffusion, 72–73
  of lipid soluble substances, 63, 65, 70
  osmosis, 66–69
  overview, 63
  passive forces in, 63
  summary of, 77, 78
  unassisted, 63–70
  vesicular, 75–77, 78
Membranous organelles, 24–25, 45
Memory, 157–163
  amnesia and, 159
  brain regions involved in, 162–163
  consolidation, 158, 162
  declarative, 162
  defined, 157
  episodic, 162
  habituation as, 159, 160

long-term, 157–159, 161–162
long-term potentiation and, 160–161
procedural, 162
reconsolidation, 158
semantic, 162
sensitization as, 159–160
short-term, 157–161
sleep and, 171
working, 152, 158
Memory cells, 417, 420–421, 423
Memory trace, defined, 157
Menarche, 751
Ménière's disease, 224
Meninges, 139, 140
Meningiomas, 138
Menopause, 752
Menstrual cycle, 749–752
  age of onset, 751
  average length of, 749
  cessation at menopause, 752
  core temperature variations in, 628
  cramps associated with, 750
  estrogen and progesterone during, 745,
    749–750, 751
  irregularities in, 750, 751
  overview, 736–737
  phases of, 750
Menstrual flow, 750
Menstrual phase, of uterine cycle, 750
Merkel's disc, 185
Mesangial cells, 504, 510
Mesentery, 570
Messenger RNA (mRNA)
  in protein synthesis, 23, 24, 25
  vaults and, 41
Metabolic acidosis, 559, 560, 561, 562
  causes, 561
  in diabetes mellitus, 694, 698
  diarrhea causing, 610
  in renal failure, 528, 561
  respiratory adjustments to, 553–554, 561
Metabolic alkalosis, 559, 560, 561–562
  causes, 561–562
  respiratory adjustments to, 553–554, 562
Metabolic rate, 619–620
  epinephrine and, 682
Metabolic states, 688–689
Metabolic syndrome, 696
Metabolic water, 544
Metabolism. See also Fuel metabolism
  cortisol effects on, 674
  defined, 685
  epinephrine effects on, 682
  glucagon effects on, 698
  growth hormone effects on, 653
  insulin effects on, 690–692
  skeletal muscle, 269–273
  summary of hormonal effects on, 700
Metaflammation, 623
Metarteriole, 354, 355
Metastasis, 432
Metformin, 696
Metoprolol, 241
MHC molecules, 423
  antigen-presenting cells and, 427–428
  class I MHC glycoproteins, 428
  class II MHC glycoproteins, 428–429
  tissue rejection and, 429
Micelles, 596–597
Microbiome, 613
Microbiota, 613
Microcirculation, 339
Microfilaments, 44, 45, 49–51
  in amoeboid movement, 50
  in cell contractile systems, 50
  in cytokinesis, 49, 50
  as mechanical stiffeners, 50–51
  in microvilli, 50–51
  of muscle fiber, 254–256
  structure, 49
Microglia, 137–138
MicroRNA (miRNA), 24

Potassium (K$^+$). *See also* Sodium-potassium (Na$^+$-K$^+$) pump
   arteriolar radius and, 345
   astrocytes taking up cerebral excess of, 137
   cardiac malfunction and, 516, 528
   equilibrium potential for, 80–81
   importance of regulating plasma concentration, 515–516
   tubular secretion of, 514–516, 549
Potassium channels, 58, 80
   ATP-sensitive, 692
   voltage-gated, 92–93, 99, 100, 304–305, 308–309
Potassium ion secretion
   aldosterone and, 515
   hydrogen ion secretion affecting, 515, 549
   importance of controlling, 515–516
   inadvertent effects on, 515
   mechanism of, 514–515
Power arm, 265
Power stroke, 256–258, 259, 260, 261
PPAR-δ (peroxisome proliferator-activated receptor delta), 277
PR segment, 311, 312
Preantral follicle, 741, 742, 743
Prebiotics, 614
Pre-Bötzinger complex, 480
Precapillary sphincters, 354–355
Precocious pseudopuberty, 677
Prefrontal association cortex, 152, 156
   autonomic activity influenced by, 241
   executive functions and, 163
   working memory and, 163
Pregame meals, 581
Preganglionic fiber, 234
Pregnancy
   corpus luteum of, 744, 762, 764
   ectopic, 753, 756
   edema of lower extremities in, 360
   estrogen in, 761, 762–763
   human chorionic gonadotropin in, 761, 762
   implantation for, 756–757
   lactation prevention during, 768–769
   length of, 763
   maternal body systems in, 763
   morning sickness in, 762
   placenta in, 757, 760–763
   preparation for parturition in, 763–764
   preventing immune rejection of embryo-fetus in, 757
   prevention of, 757, 758–759
   progesterone in, 761, 762–763
   termination of, 759
Pregnenolone, 673
Prehypertension, 374
Preload, 322
Premature ventricular contraction (PVC), 307, 313
Premotor cortex, 150, 280, 281
Preovulatory follicle, 741
Preprohormones, 121
Prepuce, 721
Presbycusis, neural, 220–221
Presbyopia, 197
Pressure. *See also* Blood pressure; Hydrostatic pressure
   atmospheric (barometric), 450
   Bowman's capsule hydrostatic pressure, 500, 501
   Boyle's law, 452
   hydrostatic, 66–67
   important in ventilation, 450
   interstitial fluid hydrostatic pressure, 357
   interstitial fluid–colloid osmotic pressure, 357
   intra-alveolar (intrapulmonary), 450, 452, 453, 455
   intrapleural, 450, 451, 453, 455
   net filtration, 500, 501
   osmotic, 67, 357
   plasma–colloid osmotic pressure, 357, 382, 500, 501
   transmural pressure gradient, 450–451

Pressure gradient, 337, 338, 450, 456
Pressure reservoir, 340
Pressure–volume loop, 317
Presynaptic inhibition/facilitation, 111–112
Presynaptic neuron, 103–104, 112
PRH (prolactin-releasing hormone), 650, 769
Primary active transport, 73
Primary auditory cortex, 220
Primary cilium, 48
Primary (annulospiral) endings, 282
Primary gustatory (taste) cortex, 226
Primary hypertension, 369, 372–373
Primary immune response, 421, 423
Primary motor cortex, 148, 149, 278–279, 280, 281
Primary olfactory cortex, 228
Primary oocytes, 738, 739
Primary peristaltic wave, 576
Primary polycythemia, 388
Primary reproductive organs, 716
Primary spermatocytes, 726–727, 728
Primary structure, of protein, A-12, A-13
Primary visual cortex, 206, 208–209
   multiple senses and, 210–211
Primordial follicle, 738, 742, 743
Principal cells, 508
   potassium secretion and, 514, 515
   V$_2$ receptors on, 524
Prions, 165
PRL. *See* Prolactin
Probiotics, 614
Procarboxypeptidase, 590, 591
Procedural memories, 162, 277
Products of reaction, A-5
Progeria, 679
Progestational phase, of uterine cycle, 750
Progesterone
   cervical mucus and, 751–752
   from corpus luteum, 744, 747, 748–749, 755
   mammary gland development and, 767
   ovarian cycle and, 745
   ovarian secretion of, 716, 737
   placental secretion of, 761, 762–763
   in pregnancy, 761, 762–763
   sperm receptors binding to, 753
   summary of actions, 768
   uterine cycle and, 745, 749–750, 751
Programmed cell death. *See* Apoptosis; Neutrophil extracellular traps (NETs)
Projection on neural pathways, 147
   of sensory input, 186
Prolactin (PRL)
   fetal growth and, 768
   JAK/STAT pathway and, 116–117, 769
   mammary glands and, 767
   milk secretion and, 769
   overview, 648, 649
Prolactin-inhibiting hormone (PIH), 650, 769
Prolactin-releasing peptide (PrRP), 650, 769
Proliferative phase, of uterine cycle, 750
Pro-opiomelanocortin (POMC), 621, 648, 680
Proprioception, 186
   in sports, 187
Proprioceptors, muscle, 281–284
Propulsive movements, 566
Prostacyclin, 396
Prostaglandins
   actions of, 119
   in fever production, 411, 633
   at implantation site, 756
   menstrual cycle and, 750
   nociceptor sensitization by, 189
   nomenclature, 119
   ovulation and, 747
   in parturition, 764, 765, 766
   seminal vesicles' secretion of, 731, 753
   structure, 119
Prostasomes, 731
Prostate gland
   anatomy, 716, 717
   benign hypertrophy of, 731
   cancer of, 736

   functions of, 731, 732
   urethra in relation to, 493, 732
Prostate-specific antigen (PSA), 731
Proteasomes, 27–28
Protective reflexes, 276
Protein. *See also* Membrane proteins; Plasma proteins
   absorption in small intestine, 603, 605, 606
   chemical composition of, A-11
   cortisol actions on, 674, 700
   degraded in the cell, 27–28, 31
   denaturation, A-14
   denatured by gastric acid, 585
   diabetes mellitus and, 694
   digestion of, 567, 569, 585, 588, 590–591, 600, 603, 605, 606
   endogenous, to be digested and absorbed, 603, 605
   as energy source, 687
   of extracellular matrix, 60–61
   functions in body, A-11
   gastric secretion stimulated by, 586, 587
   glucagon actions on, 698, 699, 700
   growth hormone actions on, 653, 700
   hydrolysis of, 567, 570, A-14
   insulin actions on, 692
   membrane potential and, 80
   overview, A-11–A414
   peptide bonds of, A-11–A-12
   in plasma membrane, 56, 58–60
   processing in Golgi complex, 29
   storage of, 687
   structural levels of, A-12–A-14
Protein buffer system, 552
Protein kinase A (PKA), 123, 125
Protein kinase C (PKC), 124
Protein kinases, 116, 117, 118
Protein phosphatases, 118
Protein synthesis
   ATP used for, 40
   in cytosol, 26, 43
   destruction of defective products of, 27–28
   DNA and RNA in, 23–24, 25
   growth hormone and, 653, 654, 657
   insulin acting for enhancement of, 692
   processing of newly synthesized proteins, 25, 29
   ribosomes in, 25–26
   rough endoplasmic reticulum and, 25–26
Proteinuria, 499
Proteolytic enzymes, pancreatic, 590–591
Proteome, 24, 26
Prothrombin, 397, 398
Proton acceptor, A-8
Proton donor, A-8
Proton pump inhibitors, 589
Protons, A-1
Proximal tubule
   bicarbonate reabsorption in, 556
   glucose and amino acid reabsorption in, 510
   hydrogen ion secretion in, 514, 554, 556–557
   organic ion secretion in, 516–517
   phosphate reabsorption in, 512
   sodium reabsorption in, 506, 507, 510, 512, 528
   structure, 495
   summary of processes carried out by, 517
   urea reabsorption in, 513
   water reabsorption in, 512–513, 528
PrRP (prolactin-releasing peptide), 650, 769
PSA (prostate-specific antigen), 731
Pseudohermaphroditism, female, 677
Pseudomonas aeruginosa, 59
Pseudonephritis, athletic, 499
Pseudopods, 31, 32, 50
Psoriasis, 432
Psychiatric disorders, 156
Psychoactive drugs, 156, 157
PTH. *See* Parathyroid hormone
PTHrp (parathyroid hormone–related peptide), 761
Pubertal growth spurt, 653, 654, 659–660

Respiratory quotient, 446
Respiratory rate, 446, 462, 463, 464
Respiratory rhythm, generation of, 479–480
Respiratory system. *See also* Lungs; Ventilation
  airflow matched to blood flow, 465
  airway resistance, 456–458, 460, 465, 466
  anatomy, 447–450
  blood-gas abnormalities, 477–478
  central chemoreceptors and, 483–484
  chronic obstructive pulmonary disease
    (COPD), 457–458, 460
  cigarette smoking effect on, 442
  cilia in, 48, 441–442
  clinical terminology related to, 477
  components of, 8
  conditions affecting function of, 462
  control of, 479–487
  defenses of, 441–442
  gas exchange and, 446, 466–471, 473–475
  gas transport and, 471–478
  heights and depths, effect of, 480–481
  homeostasis and, 12, 14, 445, 488
  mechanics, 450–466
  mucus escalator of, 442
  nonrespiratory functions, 447
  peripheral chemoreceptors, 482–483, 484, 487
  pH regulation by, 553–554
  steps carried out by, 446–447
Resting membrane potential, 79–84
  balance of passive leaks and active pumping
    at, 83–84
  chloride movement at, 84
  defined, 79
  equilibrium potentials of ions and, 80–83
  ions responsible for, 80
  return to, 88, 93
Resting tremor, 154
Restrictive lung disease, 462, 463
Reticular activating system (RAS), 167–168
  sensory processing by, 182
  sleep and, 170
Reticular formation, 167
Reticulocytes, 386
Retina. *See also* Photoreceptors
  anatomical relations of, 193
  cortical maps of, 209
  focus of image on, 194, 195, 198–199
  layers of, 193, 199–200
  on-center and off-center cells in, 203
  photoreceptors in, 193, 199, 200–203,
    204–206
  processing of light input, 203, 206–207
Retinal, 200–201, 203, 204, 206
Retino-hypothalamic tract, 661
Retinoic acid inducible gene I (RIG-I)-like
    receptors (RLRs), 407
Retrograde amnesia, 159
Retropulsion, 579, 588
Reverse axonal transport, 48
Reversible reactions, A-5–A-6
Reward center, 156
Rh blood-group system, 389
Rheumatic fever, 318, 432
Rheumatoid arthritis, 420, 432, 675
Rhodopsin, 200–201, 204, 206
Rhythm, abnormalities of heart, 313–314
Rhythm method, 758
Rhythmic activities, 277
Ribonucleic acid. *See* RNA
Ribosomal RNA (rRNA), 23, 25
Ribosomes
  free in cytosol, 25, 26
  protein synthesis and, 23, 25–26
  rough endoplasmic reticulum and, 25–26
  structure, 25
  vaults and, 41
Ribs, 449
Rickets, 712
Right cerebral hemisphere, 152
Rigor mortis, 259, 261
RLRs (retinoic acid inducible gene I (RIG-I)-like
    receptors), 407

RNA (ribonucleic acid)
  in protein synthesis, 23, 24, 25
  types of, 23–24
  vaults and, 41
RNA bandage, for muscular dystrophy, 279
RNA interference (RNAi), 24
Rods. *See also* Photoreceptors
  acuity, 204
  night vision and, 204
  parts of, 200
  photopigment of, 200–201, 203, 204
  properties of, 204
  retinal layer containing, 193, 199
  sensitivity, 204
  shades of gray provided by, 204
Romosozumab, 707
Rotational acceleration of head, 221, 222, 223
Rough endoplasmic reticulum, 25–26
Round window, 215, 217
rRNA (ribosomal RNA), 23, 25
RU 486, 759
Ruffini endings, 185
Runner's high, 192
RV (residual volume), 461, 462
Ryanodine receptors, 258
  for sperm, 731

SA node. *See* Sinoatrial (SA) node
Saccule, 222, 223, 224
Sacral nerves, 173
Salbutamol, 241
Saliva, 574–575
Salivary amylase, 574, 575, 588
Salivary center, 575
Salivary glands, 568–569, 574
  innervation of, 239, 575
Salivary reflexes, 575
Salt. *See also* Sodium chloride
  absorption in large intestine, 614
  absorption in small intestine, 603
  balance, 539–540
  in extracellular fluid, 12
  hypertension and, 372, 373
  ingestion of, 539
  tubuloglomerular feedback and, 502
SALT (skin-associated lymphoid tissue), 441
Salt balance
  extracellular fluid volume and, 539–540
  intake in, 539
  urinary output in, 540
Saltatory conduction, 100
Salts, inorganic, A-9
Salty taste, 226
Sarcolemma, 258
Sarcomere, 253, 254
  banding pattern changes, 256, 257
  as contractile component, 262
  sliding filament mechanism, 256, 258
  widths of, 269
Sarcopenia, 275
Sarcoplasmic reticulum (SR)
  calcium release from, 258–259, 260
  cardiac muscle, 294, 304, 309
  as smooth endoplasmic reticulum, 27
  smooth muscle, 290
  structure, 258
Sarcoplasmic/endoplasmic reticulum Ca^{2+}–
    ATPase (SERCA) pump, 261, 268, 272, 304,
    309
Satellite cells, 275, 279
Satiety, 621
Satiety center, 623
Satiety signals, 621, 623, 624, 692
Saturated fatty acids, A-10, A-11
Saturation of carriers, 71
Scala media (cochlear duct), 214, 215, 217
Scala tympani, 214, 215
Scala vestibuli, 214, 215
Scar tissue, 412
Schwann cells, 100, 104
Sclera, 193
Sclerostin, 703

Scrotum, 716, 721, 723
Scurvy, 60
Sebaceous glands, 440
Sebum, 440
Second messengers. *See also* Cyclic adenosine
    monophosphate
  amplification by, 124–125
  calcium as, 122–123, 124
  of G-protein-coupled receptors, 116, 117
  of hydrophilic hormone action, 122–126
  reversal by protein phosphatases, 118
  at "slow" synapses, 118
Second polar body, 738
Secondary active transport, 74–75
  in carbohydrate/protein absorption, 603–605
  defined, 73
  in glucose and amino acid reabsorption, 510
  hydrogen ion secretion in proximal tubule,
    554
  sodium and glucose cotransporter (SGLT), 75,
    510, 603, 604, 611, 697
  sodium-dependent, 512
Secondary (flower-spray) endings, 282
Secondary hypertension, 369, 372
Secondary immune response, 421, 423
Secondary oocyte, 738, 739, 742, 743, 754
Secondary peristaltic wave, 578
Secondary polycythemia, 388
Secondary sexual characteristics, 716
  androgens and, 726, 752
  estrogen and, 752
Secondary spermatocytes, 727
Secondary structure, of protein, A-13, A-14
Second-order sensory neuron, 186
Secretin, 580–581
  overview, 614–615
  pancreatic secretion and, 592
Secretion
  ATP used for, 40
  defined, 6, 29
  by exocytosis, 29, 59, 701–702
Secretion, in digestive system, 566, 569
  bile, 593, 594, 595
  esophageal, 578
  gastric, 582–587
  large intestine, 613
  pancreatic, 568–569, 590–592
  salivary, 574–575
  small intestine, 599
  volumes of, 609
Secretion, in kidney. *See* Tubular secretion
Secretory glands, as epithelial tissue, 6
Secretory IgA, 770
Secretory phase, of uterine cycle, 750
Secretory vesicles, 29–30, 44
  axonal transport of, 47–48
  membrane proteins docking with, 58–59
Segmentation, small intestine, 598–599
Selectins, 410
Selective estrogen receptor modulators (SERMs),
    706–707, 738
Selective serotonin reuptake inhibitors (SSRIs),
    108, 157
Selectively permeable membrane, 63
  osmosis and, 66–69
Self-tolerance, 431
Self-antigens
  autoimmune diseases and, 432
  MHC molecules and, 423, 428
  tolerance of, 431
Self-excitable smooth muscle, 291–292
Self-identity markers, 60
Semantic memories, 162
Semen, 716, 731, 734, 735
Semicircular canals, 221–223
  damage to, 224
Semilunar valves, 301
  malfunctioning, 318
  second heart sound and, 317
Seminal vesicles, 716, 731, 732
Seminiferous tubule fluid, 729
Seminiferous tubules, 724, 725, 726

Volume receptors, left atrial, 369, 545–546
Voluntary movements, 277
   cerebellum and, 277
Voluntary muscle, 252
Vomeronasal organ, 229–230
Vomiting, 582
   metabolic alkalosis caused by, 561
Vomiting center, 582
Von Willebrand factor (vWF), 395–396
v-SNAREs, 29–30
Vulva, 718

Waking state, 168
Walking, 277
Warfarin, 400
Waste products
   excreted in urine, 514
   homeostasis and, 12
Water. See also Fluid balance
   absorption in large intestine, 614
   absorption in small intestine, 603
   aquaporins, 66, 512, 524
   body percentage of, 537
   distribution among fluid compartments,
      537–538
   forces governing movement between
      compartments, 538
   homeostatic regulation of, 12
   insensible losses of, 544
   metabolic, 544
   in osmosis, 66–69
   from oxidative phosphorylation, 37, 38, 39
   tubular reabsorption of, 512–513, 523–526,
      527, 528, 544, 545, 546, 547
   in various tissues, 537

Water balance
   angiotensin II and, 546
   daily quantities, 544
   dehydration, 541–543
   hypothalamic osmoreceptors and, 545, 648
   left atrial volume receptors and, 545
   oral metering of, 546
   overhydration, 543
   regulated by thirst, 544, 545, 546–547
   regulated by urine output, 544, 545, 546, 547
   sources of input, 543–544
   sources of output, 544
   thirst and, 544, 545, 546–547
   vasopressin in control of, 544–547
Water channels (aquaporins), 66, 512, 524
Water diuresis, 527
Water intoxication, 543
Wavelength, 194
WBCs. See Leukocytes (white blood cells)
Weaning, 770
Weight, A-1
   molecular, A-6
Weight control, 621, 623
Wernicke's area, 151, 152
White blood cell count, 393–394
White blood cells. See Leukocytes (white blood
   cells)
White fibers, 273–274
White matter
   brain, 144–145
   in learning and memory, 162
   spinal cord, 173–174, 175
   Whooping cough toxin, 126
Wigger's diagram, 314–317
Wind chill factor, 630

Windpipe, 447
Withdrawal reflex, 284–285
Withdrawal symptoms, in addiction, 112
Wolffian ducts, 721, 722
Work
   of breathing, 460
   defined, 264
   external, 619
   internal, 619
Work capacity, 487
Working memory, 158
   prefrontal cortex and, 163
   World Anti-Doping Agency (WADA), 277

X chromosome, 718, 720, 721
   in spermatogenesis, 728
Xerostomia, 575

Y chromosome, 718, 720, 721
Yellow bone marrow, 386

Z line, 253, 254, 256, 268
Zona fasciculata, 672, 673, 675
Zona glomerulosa, 672, 673, 674
Zona pellucida, 741, 742, 743, 744, 754, 755
Zona reticularis, 672, 675
Zygote, 755
Zymogen granules
   gastric, 585
   pancreatic, 590

1.1 Introduction to Physiology (p. 2)

■ *Physiology* is the study of body functions.

■ Physiologists explain body function in terms of the mechanisms of action involving cause-and-effect sequences of physical and chemical processes.

■ Physiology and anatomy are closely interrelated because body functions highly depend on the structure of the body parts that carry them out.

1.2 Levels of Organization in the Body (pp. 2–7)

■ The human body is a complex combination of specific *atoms* and *molecules*.

■ These nonliving chemicals are organized in a precise way to form *cells,* the smallest entities capable of carrying out life processes. Cells are the body's structural and functional living building blocks. *(Review Figure 1-2.)*

■ The *basic functions* performed by each cell for its own survival include (1) obtaining O_2 and nutrients, (2) performing energy-generating chemical reactions, (3) eliminating wastes, (4) synthesizing proteins and other cell components, (5) controlling movement of materials between the cell and its environment, (6) moving materials throughout the cell, (7) responding to the environment, and (8) reproducing.

■ In addition to its basic functions, each cell in a multicellular organism performs a *specialized function* essential for survival of the organism.

■ Cells of similar structure and specialized function combine to form the four *primary tissues* of the body: muscle, nervous, epithelial, and connective. *(Review Figure 1-3.)*

■ *Glands* are derived from epithelial tissue and specialized for secretion. *Exocrine glands* secrete through ducts to the body surface or into a cavity that communicates with the outside; *endocrine glands* secrete hormones into the blood. *(Review Figure 1-4.)*

■ *Organs* are combinations of two or more types of tissues that act together to perform one or more functions. An example is the stomach. *(Review Figure 1-3.)*

■ *Body systems* are collections of organs that perform related functions and interact to accomplish a common activity essential for survival of the whole body. An example is the digestive system. *(Review Figure 1-5.)*

■ Body systems combine to form the *organism,* or whole body.

1.3 Concept of Homeostasis (pp. 7–16)

■ The fluid inside the cells of the body is *intracellular fluid (ICF);* the fluid outside the cells is *extracellular fluid (ECF).*

■ Because most body cells are not in direct contact with the external environment, cell survival depends on maintaining a relatively stable *internal fluid environment* with which the cells directly make life-sustaining exchanges.

■ The ECF serves as the body's internal environment. It consists of the *plasma* and the *interstitial fluid*. *(Review Figure 1-6.)*

■ *Homeostasis* is the maintenance of a dynamic steady state in the internal environment.

■ The factors of the internal environment that must be homeostatically maintained are its (1) concentration of nutrients; (2) concentration of O_2 and CO_2; (3) concentration of waste products; (4) pH; (5) concentrations of water, salt, and other electrolytes; (6) volume and pressure; and (7) temperature. *(Review Figure 1-8.)*

■ The functions performed by the 11 body systems are directed toward maintaining homeostasis. These functions ultimately depend on the specialized activities of the cells that make up the system. Thus, *homeostasis is essential for each cell's survival, and each cell contributes to homeostasis. (Review Figures 1-7 and 1-8.)*

1.4 Homeostatic Control Systems (pp. 16–18)

■ A *homeostatic control system* is a network of body components working together to maintain a controlled variable in the internal environment relatively constant near an optimal set point despite changes in the variable.

■ Homeostatic control systems can be classified as (1) *intrinsic (local) controls,* which are inherent compensatory responses of an organ to a change, and (2) *extrinsic (systemic) controls,* which are responses of an organ triggered by factors external to the organ—namely, by the nervous and endocrine systems.

■ Both intrinsic and extrinsic control systems generally operate on the principle of *negative feedback:* A change in a controlled variable triggers a response that drives the variable in the opposite direction of the initial change, thus opposing the change. *(Review Figure 1-9.)*

■ In *positive feedback,* a change in a controlled variable triggers a response that drives the variable in the same direction as the initial change, thus amplifying the change. Positive feedback is uncommon in the body but is important in several instances, such as during childbirth.

■ *Feedforward mechanisms* respond in anticipation of a change in a regulated variable.

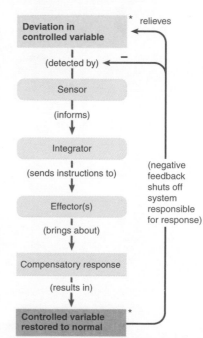

(a) Components of a negative-feedback control system

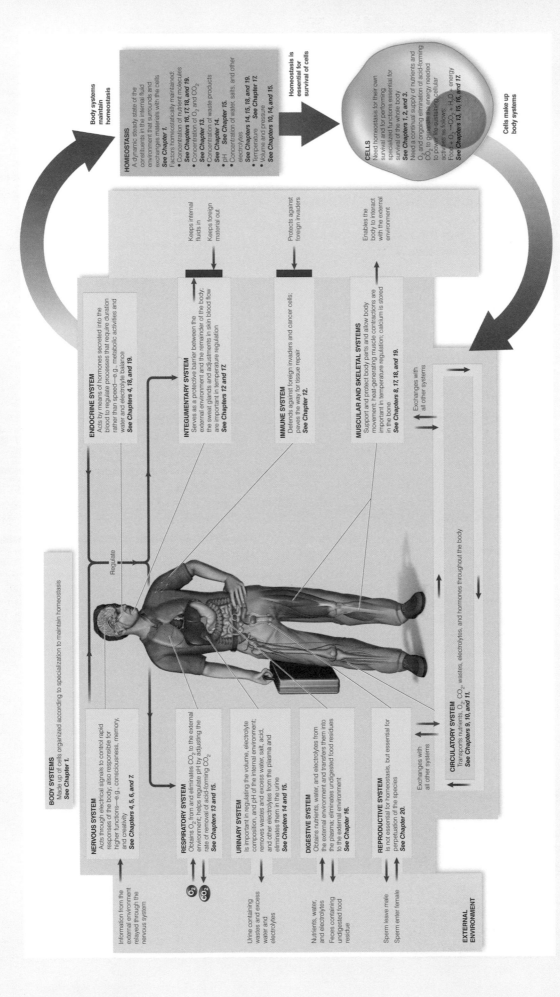

BODY SYSTEMS
Made up of cells organized according to specialization to maintain homeostasis
See Chapter 1.

NERVOUS SYSTEM
Acts through electrical signals to control rapid responses of the body; also responsible for higher functions—e.g., consciousness, memory, and creativity
See Chapters 4, 5, 6, and 7.

RESPIRATORY SYSTEM
Obtains O_2 from and eliminates CO_2 to the external environment; helps regulate pH by adjusting the rate of removal of acid-forming CO_2
See Chapters 13 and 15.

URINARY SYSTEM
Is important in regulating the volume, electrolyte composition, and pH of the internal environment; removes wastes and excess water, salt, acid, and other electrolytes from the plasma and eliminates them in the urine
See Chapters 14 and 15.

DIGESTIVE SYSTEM
Obtains nutrients, water, and electrolytes from the external environment and transfers them into the plasma; eliminates undigested food residues to the external environment
See Chapter 16.

REPRODUCTIVE SYSTEM
Is not essential for homeostasis, but essential for perpetuation of the species
See Chapter 20.

Information from the external environment relayed through the nervous system

O_2
CO_2

Urine containing wastes and excess water and electrolytes

Nutrients, water, and electrolytes

Feces containing undigested food residue

Sperm leave male
Sperm enter female

EXTERNAL ENVIRONMENT

Regulate

ENDOCRINE SYSTEM
Acts by means of hormones secreted into the blood to regulate processes that require duration rather than speed—e.g., metabolic activities and water and electrolyte balance
See Chapters 4, 18, and 19.

INTEGUMENTARY SYSTEM
Serves as a protective barrier between the external environment and the remainder of the body; the sweat glands and adjustments in skin blood flow are important in temperature regulation
See Chapters 12 and 17.

IMMUNE SYSTEM
Defends against foreign invaders and cancer cells; paves the way for tissue repair
See Chapter 12.

MUSCULAR AND SKELETAL SYSTEMS
Support and protect body parts and allow body movement; heat-generating muscle contractions are important in temperature regulation; calcium is stored in the bone
See Chapters 8, 17, 18, and 19.

Keeps internal fluids in
Keeps foreign material out

Protects against foreign invaders

Enables the body to interact with the external environment

Exchanges with all other systems

CIRCULATORY SYSTEM
Transports nutrients, O_2, CO_2, wastes, electrolytes, and hormones throughout the body
See Chapters 9, 10, and 11.

Exchanges with all other systems

HOMEOSTASIS
A dynamic steady state of the constituents in the internal fluid environment that surrounds and exchanges materials with the cells
See Chapter 1.
Factors homeostatically maintained:
• Concentration of nutrient molecules
  See Chapters 16, 17, 18, and 19.
• Concentration of O_2 and CO_2
  See Chapter 13.
• Concentration of waste products
  See Chapter 14.
• pH *See Chapter 15.*
• Concentration of water, salts, and other electrolytes
  See Chapters 14, 15, 18, and 19.
• Temperature *See Chapter 17.*
• Volume and pressure
  See Chapters 10, 14, and 15.

Body systems maintain homeostasis

Homeostasis is essential for survival of cells

CELLS
Need homeostasis for their own survival and for performing specialized functions essential for survival of the whole body
See Chapters 1, 2, and 3.
Need a continual supply of nutrients and O_2 and ongoing elimination of acid-forming CO_2 to generate the energy needed to power life-sustaining cellular activities as follows:
Food + $O_2 \rightarrow CO_2 + H_2O$ + energy
See Chapters 13, 15, 16, and 17.

Cells make up body systems

Figure 1-8 Role of the body systems in maintaining homeostasis.

2.1 Cell Theory and Discovery (p. 22)

■ The complex organization and interaction of the chemicals within a cell confer the unique characteristics of life. The cell is the smallest unit capable of carrying out life processes.

■ Cells are the living building blocks of the body. The structure and function of a *multicellular organism* ultimately depend on the structural and functional capabilities of its cells. *(Review Table 2-1.)*

■ Cells are too small for the unaided eye to see.

■ Using early microscopes, investigators learned that all plant and animal tissues consist of individual cells.

■ Scientists now know that a cell is a complex, highly organized, compartmentalized structure.

2.2 An Overview of Cell Structure (pp. 22–25)

■ Cells have three major subdivisions: the plasma membrane, the nucleus, and the cytoplasm. *(Review Figure 2-1.)*

■ The *plasma membrane* encloses the cell and separates the intracellular fluid (ICF) and extracellular fluid (ECF). It also controls movement into and out of the cell.

■ The *nucleus* contains *deoxyribonucleic acid (DNA),* the cell's genetic material. Three types of *ribonucleic acid (RNA)* play a role in the protein synthesis coded by DNA: *messenger RNA (mRNA),* which transcribes DNA's genetic code for delivery to a ribosome; *ribosomal RNA (rRNA),* which is part of a ribosome that reads the mRNA code and translates it into a designated protein; and *transfer RNA (tRNA),* which delivers the appropriate amino acids to the protein under construction. The *genome* is all of the genetic information coded in a complete single set of DNA. The *proteome* is the complete set of proteins that can be expressed by protein-coding genes in the genome.

■ The *cytoplasm* consists of cytosol, a complex gel-like mass laced with organelles and a cytoskeleton.

■ *Organelles* are highly organized structures that serve a specific function. *Membranous organelles* are bound by a membrane that separates the organelle's contents from the surrounding cytosol.

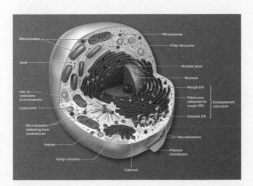

They include the endoplasmic reticulum, Golgi complex, lysosomes, peroxisomes, and mitochondria. *Nonmembranous organelles* are not surrounded by membrane and include ribosomes, proteasomes, vaults, and centrioles. *(Review Figure 2-1 and Table 2-2, p. 45.)*

2.3 Endoplasmic Reticulum and Segregated Synthesis (pp. 25–28)

■ The *endoplasmic reticulum (ER)* is a single, complex, membranous network that encloses a fluid-filled lumen.

■ The primary function of the ER is to synthesize proteins and lipids that are either (1) secreted to the exterior of the cell, such as enzymes and hormones, or (2) used to produce new cell components, particularly plasma membrane.

■ The two types of ER are *rough ER* (flattened interconnected sacs studded with ribosomes) and *smooth ER* (interconnected tubules with no ribosomes). *(Review Figure 2-2a and b.)*

■ During protein synthesis, a large and a small ribosomal subunit unite to form a *ribosome* that translates mRNA. *(Review Figure 2-2c.)* The rough-ER ribosomes synthesize proteins that are released into the ER lumen, where they are separated from the cytosol. Also entering the lumen are lipids produced within the membranous walls of the ER.

■ Synthesized products move from the rough ER to the smooth ER, where they are packaged and discharged as *transport vesicles.* Transport vesicles are formed as a portion of the smooth ER "buds off." *(Review Figure 2-3.)*

■ Misfolded proteins in the ER lumen are tagged with *ubiquitin* and moved out to a *proteasome,* which degrades them. Other unwanted intracellular proteins are also marked with ubiquitin for destruction by proteasomes. *(Review Figure 2-4.)*

2.4 Golgi Complex and Exocytosis (pp. 28–30)

■ Transport vesicles move to and fuse with the *Golgi complex,* which consists of a stack of separate, flattened, membrane-enclosed sacs. *(Review Figures 2-3 and 2-5 and chapter opener.)*

■ The Golgi complex serves a twofold function: (1) to modify into finished products the newly synthesized molecules delivered to it in crude form from the ER, and (2) to sort, package, and direct molecular traffic to appropriate intracellular and extracellular destinations.

■ *Secretion* refers to the release of a specific product synthesized by the cell for a particular purpose. The Golgi complex of secretory cells packages proteins for export from the cell in *secretory vesicles* released by *exocytosis* on appropriate stimulation. *(Review Figures 2-3, 2-6a, and 2-7.)*

2.5 Lysosomes and Endocytosis (pp. 30–33)

■ *Lysosomes* are membrane-enclosed sacs that contain powerful *hydrolytic (digestive) enzymes. (Review Figure 2-8.)*

■ Serving as the intracellular "digestive system," lysosomes destroy foreign materials such as bacteria that have been internalized by the cell and demolish worn-out cell parts to make way for replacement parts.

■ Extracellular material is brought into the cell by *endocytosis* for attack by lysosomal enzymes. The three forms of endocytosis are *pinocytosis* (nonselective uptake of ECF; "cell drinking"), receptor-mediated endocytosis (selective import of a specific large molecule), and *phagocytosis* (engulfment of a large multi-molecular particle; "cell eating"). (*Review Figures 2-6b and 2-9.*)

■ *Autophagy* is selective self-digestion of cell parts such as worn-out organelles by lysosomes.

2.6 Peroxisomes and Detoxification (p. 33)

■ *Peroxisomes* are small membrane-enclosed sacs containing powerful *oxidative enzymes*. (*Review Figure 2-8.*)

■ They carry out particular oxidative reactions that detoxify various wastes and toxic foreign compounds that have entered the cell. During these detoxification reactions, peroxisomes generate potent *hydrogen peroxide* (H_2O_2), which they decompose into harmless water and oxygen by means of *catalase*.

2.7 Mitochondria and ATP Production (pp. 33–41)

■ The rod-shaped *mitochondria* are enclosed by two membranes, a smooth outer membrane and an inner membrane consisting of a series of shelves, the *cristae*, which project into an interior gel-filled cavity, the *matrix*. (*Review Figure 2-10a and chapter opener, p. 21.*)

■ Mitochondria are the energy organelles of the cell. They efficiently convert the energy in food molecules to the usable energy stored in ATP molecules. Cells use ATP as an energy source for synthesis of new chemical compounds, for membrane transport, and for mechanical work.

■ *Cellular respiration* refers collectively to the intracellular reactions in which energy-rich molecules are broken down to form ATP, using O_2 and producing CO_2 in the process. Cellular respiration includes the sequential dismantling of nutrient molecules and subsequent ATP production in three stages: (1) *glycolysis* in the cytosol, (2) the *citric acid cycle* in the mitochondrial matrix, and (3) *oxidative phosphorylation* at the mitochondrial inner membrane. (*Review Figure 2-11.*)

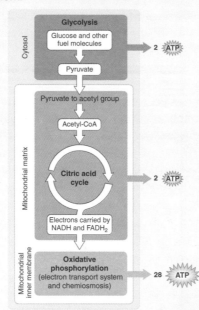

IFigure 2-11 **Stages of cellular respiration.**

■ *Oxidative phosphorylation* includes the electron transport system and chemiosmosis by ATP synthase. The *electron transport system* extracts high-energy electrons from hydrogens released during nutrient breakdown during glycolysis and the citric acid cycle and transfers them to successively lower energy levels. The free energy released during this process is used to create a hydrogen ion (H^+) gradient across the mitochondrial inner membrane. The flow of H^+ down this gradient activates *ATP synthase*,

an enzyme that synthesizes ATP in a process called *chemiosmosis*. (*Review Figures 2-12 through 2-15.*)

■ A cell is more efficient at converting food energy into ATP when O_2 is available. Without O_2 (an *anaerobic* condition), a cell can produce only 2 molecules of ATP for every glucose molecule processed by glycolysis. With O_2 (an *aerobic* condition), the mitochondria can yield another 30 molecules of ATP for every glucose molecule processed (2 from the citric acid cycle and 28 from oxidative phosphorylation). (*Review Figure 2-15.*)

2.8 Vaults as Cellular Trucks (pp. 41–42)

■ *Vaults* are hollow, octagonal structures that are the same shape and size as the nuclear pores. (*Review Figure 2-17.*) They are believed to be cellular trucks that dock at the nuclear pores and pick up cargo for transport from the nucleus.

■ The leading proposals are that vaults may transport mRNA or the ribosomal subunits from the nucleus to the cytoplasmic sites of protein synthesis.

2.9 Cytosol: Cell Gel (pp. 42–44)

■ The *cytosol* contains the enzymes involved in intermediary metabolism and the ribosomal machinery essential for synthesis of these enzymes and other cytosolic proteins.

■ Many cells store unused nutrients within the cytosol as glycogen granules or fat droplets. These nonpermanent masses of stored material are called *inclusions*. (*Review Figure 2-18.*)

■ Also present in the cytosol are various secretory, transport, and endocytic vesicles.

2.10 Cytoskeleton: Cell "Bone and Muscle" (pp. 44–51)

■ The *cytoskeleton*, which extends throughout the cytosol, serves as the "bone and muscle" of the cell. (*Review Table 2-2.*)

■ The three types of cytoskeletal elements—microtubules, microfilaments, and intermediate filaments—each consist of different proteins and perform various roles. (*Review Figure 2-19.*)

■ The *centrosome* (cell center) consists of a pair of *centrioles* surrounded by an amorphous mass. (*Review Figure 2-20.*) The centrosome and its centrioles form and organize microtubules as needed to serve various functions.

■ *Microtubules,* made of tubulin, maintain asymmetric cell shapes, serve as highways for intracellular transport by motor proteins, are the main component of cilia and flagella, and make up the mitotic spindle. (*Review Figures 2-19a and 2-21 through 2-23.*)

■ *Microfilaments,* made of actin in most cells, are important in various cellular contractile systems, including amoeboid movement and muscle contraction. They also serve as mechanical stiffeners for microvilli. (*Review Figures 2-19b and 2-24 through 2-26.*)

■ *Intermediate filaments* are irregular threadlike proteins that help cells resist mechanical stress. Different proteins make up intermediate filaments in various cell types. Intermediate filaments are especially abundant in skin cells, where they are composed of keratin. (*Review Figure 2-19c.*)

Chapter 3
Study Card

3.1 Membrane Structure and Functions (pp. 56–60)

■ All cells are bounded by a *plasma membrane,* a thin lipid bilayer that is interspersed with proteins and has carbohydrates attached on the outer surface.

■ The appearance of the plasma membrane in an electron microscope as a trilaminar structure (two dark lines separated by a light interspace) is produced by the arrangement of its molecules. The phospholipids orient themselves to form a bilayer with a hydrophobic interior (light interspace) sandwiched between hydrophilic outer and inner surfaces (dark lines). *(Review Figures 3-1, 3-2, and 3-3 and chapter opener.)*

■ The *lipid bilayer* forms the structural boundary of the cell, serving as a barrier for water-soluble substances and being responsible for the fluid nature of the membrane. Cholesterol molecules tucked between the phospholipids contribute to the fluidity and stability of the membrane.

■ According to the *fluid mosaic model* of membrane structure, the lipid bilayer is embedded with proteins. *(Review Figure 3-3.)* *Membrane proteins,* which vary in type and distribution among cells, serve as (1) channels for passage of small ions across the membrane; (2) carriers for transport of specific substances into or out of the cell; (3) docking-marker acceptors where secretory vesicles dock and release their contents; (4) membrane-bound enzymes that govern specific chemical reactions; (5) receptors for detecting and responding to chemical messengers that alter cell function; and (6) cell adhesion molecules that help hold cells together and are a structural link between the plasma membrane and the intracellular cytoskeleton.

■ *Membrane carbohydrates* on the outer surface of the cell serve as self-identity markers. *(Review Figure 3-3.)* They are important in recognition of "self" in cell-to-cell interactions such as tissue formation and tissue growth.

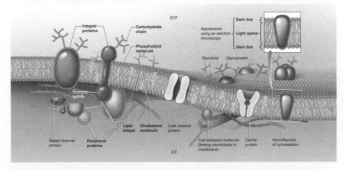

I Figure 3-3 **Fluid mosaic model of plasma membrane structure.**

3.2 Cell-to-Cell Adhesions (pp. 60–63)

■ The *extracellular matrix (ECM)* serves as a biological "glue" between the cells of a tissue. The ECM consists of a watery, gel-like substance interspersed with three major types of protein fibers: *collagen* (provides tensile strength), *elastin* (permits stretch and recoil), and *fibronectin* (promotes cell adhesion).

■ Many cells are further joined by specialized cell junctions, of which there are three types: desmosomes, tight junctions, and gap junctions.

■ *Desmosomes* serve as adhering junctions to hold cells together mechanically and are especially important in tissues subject to a great deal of stretching. *(Review Figure 3-4.)*

■ *Tight junctions* actually fuse cells to each other, preventing the passage of materials between cells and thereby permitting only regulated passage of materials through the cells. These impermeable junctions are found in the epithelial sheets that separate compartments with very different chemical compositions. *(Review Figure 3-5.)*

■ *Gap junctions* are communicating junctions between two adjacent, but not touching, cells. They form small tunnels that permit exchange of ions and small molecules between the cells. Such movement of ions plays a key role in the spread of electrical activity to synchronize contraction in heart and smooth muscle. *(Review Figure 3-6.)*

3.3 Overview of Membrane Transport (p. 63)

■ Materials can pass between the ECF and ICF by *unassisted* and *assisted* means.

■ Transport mechanisms may also be *passive* (the particle moves across the membrane without the cell expending energy) or *active* (the cell expends energy to move the particle across the membrane). *(Review Table 3-2, p. 78.)*

3.4 Unassisted Membrane Transport (pp. 63–70)

■ Nonpolar (lipid-soluble) molecules of any size cross the membrane unassisted by dissolving in and passively moving through the lipid bilayer down concentration gradients. *(Review Figures 3-7 and 3-8.)* Small ions cross the membrane unassisted by passively moving along electrochemical gradients through open protein channels specific for the ion. *(Review Figure 3-3.)*

■ In *osmosis,* water moves passively down its own concentration gradient across a selectively permeable membrane to an area of higher concentration of nonpenetrating solutes. Penetrating solutes do not have an osmotic effect. *(Review Figures 3-9 through 3-12.)*

■ The *osmolarity* of a solution is a measure of its total number of solute particles, both penetrating and nonpenetrating, both molecules and ions, per liter. The *osmotic pressure* of a solution is the pressure that must be applied to the solution to completely stop osmosis. The *tonicity* of a solution refers to the effect the solution has on cell volume and depends on the solution's relative concentration of nonpenetrating solutes compared to the concentration of nonpenetrating solutes in the cell it surrounds. *(Review Figure 3-13.)*

3.5 Assisted Membrane Transport (pp. 70–77)

■ In *carrier-mediated transport,* small polar molecules and selected ions are transported across the membrane by specific membrane carrier proteins. *Carriers* open to one side of the membrane where a passenger binds to a binding site specific for it and then change shape so that the binding site is exposed to the opposite side of the membrane where the passenger is released. Carrier-mediated transport may be passive and move the particle down its concentration gradient (*facilitated diffusion*) (*review Figure 3-14*) or active and move the particle against its concentration gradient (*active transport*). Carriers exhibit a *transport maximum* (T_m) when saturated. (*Review Figure 3-15.*)

■ The two forms of active transport include primary active transport and secondary active transport. *Primary active transport* requires the direct use of ATP to drive the pump. One of the most important examples of primary active transport is the Na^+–K^+ pump, which concentrates Na^+ in the ECF and K^+ in the ICF. (*Review Figure 3-16.*) *Secondary active transport* is driven by an ion (usually Na^+) concentration gradient established by a primary active-transport system. The two types of secondary active transport are symport (or cotransport) and antiport (or countertrans-port or exchange). In *symport* both the cotransported solute and driving ion (Na^+) are moved in the same direction (both moving into the cell), with the cotransported solute moving uphill and the driving ion moving downhill. In *antiport* the coupled solute and driving ion are moved in opposite directions (solute moving out of and Na^+ moving into the cell), with the solute moving uphill and the driving ion moving downhill. (*Review Figures 3-17 and 3-18.*)

■ Large polar molecules and multimolecular particles can leave or enter the cell by being wrapped in a piece of membrane to form vesicles that can be internalized (*endocytosis*) or externalized (*exocytosis*). (*Review Figures 2-6, 2-7, and 2-9.*)

■ Cells are differentially selective in what enters or leaves because they possess varying numbers and kinds of channels, carriers, and mechanisms for vesicular transport.

■ Large polar molecules (too large for channels and not lipid soluble) for which there are no special transport mechanisms are unable to cross the membrane.

3.6 Membrane Potential (pp. 77–84)

■ All cells have a *membrane potential,* which is a separation of opposite charges across the plasma membrane. (*Review Figure 3-19.*)

■ The *Na^+–K^+ pump* makes a small direct contribution to membrane potential because it transports more Na^+ out than K^+ in. (*Review Figure 3-16.*) However, the primary role of the Na^+–K^+ pump is to actively maintain a greater concentration of Na^+ outside and a greater concentration of K^+ inside the cell. These concentration gradients tend to passively move K^+ out of and Na^+ into the cell. (*Review Table 3-3 and Figures 3-20 and 3-21.*)

■ Because the resting membrane is 25 to 30 times more permeable to K^+ than to Na^+, substantially more K^+ leaves the cell than Na^+ enters, resulting in an excess of positive charges outside the cell. This leaves an unbalanced excess of negative charges, in the form of large protein anions (A^-), trapped inside the cell. (*Review Table 3-3 and Figure 3-22.*)

■ When the resting membrane potential of –70 mV is achieved, no further net movement of K^+ and Na^+ takes place because any further leaking of these ions down their concentration gradients is quickly reversed by the Na^+–K^+ pump.

■ The distribution of Cl^- across the membrane is passively driven by the established membrane potential so that Cl^- is concentrated in the ECF.

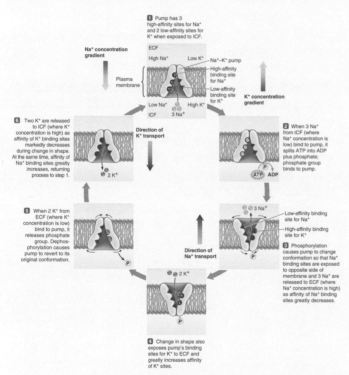

IFigure 3-16 Na^+–K^+ pump, an example of primary active transport.

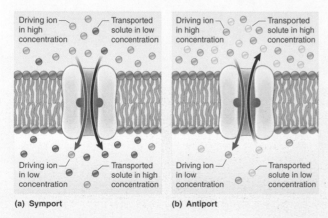

(a) Symport (b) Antiport

IFigure 3-17 Secondary active transport.

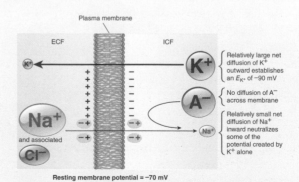

IFigure 3-22 Effect of concurrent K^+ and Na^+ movement on establishing the resting membrane potential.

4.1 Introduction to Neural Communication (pp. 88–89)

■ Nerve and muscle cells are *excitable tissues* because they can rapidly alter their membrane permeabilities and undergo transient membrane potential changes when excited. These rapid changes in potential serve as electrical signals.

■ Compared to resting potential, a membrane becomes *depolarized* when the magnitude of its negative potential is reduced (becomes less negative) and *hyperpolarized* when the magnitude of its negative potential is increased (becomes more negative). (Review Figure 4-1.)

■ Changes in potential are brought about by *triggering events* that alter membrane permeability, in turn leading to changes in ion movement across the membrane.

■ The two kinds of potential change are: (1) graded potentials, the short-distance signals, and (2) action potentials, the long-distance signals. (Review Table 4-1, p. 98.)

4.2 Graded Potentials (pp. 89–91)

■ *Graded potentials* are local changes in membrane potential that occur in varying degrees of magnitude. A graded potential, usually a depolarization, occurs in a small, specialized region of an excitable cell membrane. The site undergoing a potential change is called an active area. Graded potentials spread decrementally by local current flow between the active area and adjacent inactive areas and die out over a short distance. (Review Figures 4-2 and 4-3.)

■ The magnitude of a graded potential varies directly with the magnitude of the triggering event.

4.3 Action Potentials (pp. 91–102)

■ *Action potentials* are brief, rapid, large changes in membrane potential during which the potential reverses.

■ During an action potential, depolarization of the membrane from resting potential (-70 mV) to *threshold potential* (-50 mV) triggers sequential changes in permeability caused by conformational changes in voltage-gated Na^+ and K^+ channels. These permeability changes bring about a brief reversal of membrane potential, with Na^+ influx causing the *rising phase* (from threshold to $+30$ mV), followed by K^+ efflux causing the *falling phase* (from peak back to resting). (Review Figures 4-4 through 4-6.)

■ Before an action potential returns to resting, it regenerates an identical new action potential in the area next to it by means of current flow that brings the previously inactive area to threshold. This self-perpetuating cycle continues until the action potential spreads undiminished throughout the cell membrane.

■ There are two types of action potential propagation: (1) *contiguous conduction* in unmyelinated fibers, in which the action potential spreads along every portion of the membrane; and

(2) the more rapid, *saltatory conduction* in myelinated fibers, in which the impulse jumps from one node of Ranvier to the next over sections of the fiber covered with insulating myelin. (Review Figures 4-8, 4-11, and 4-12.)

■ The Na^+–K^+ pump gradually restores the ions that moved during propagation of the action potential to their original location, to maintain the concentration gradients.

■ It is impossible to restimulate the portion of the membrane where the impulse has just passed until it has recovered from its *refractory period,* ensuring the one-way propagation of action potentials. (Review Figures 4-9 and 4-10.)

■ Action potentials occur either maximally in response to stimulation or not at all (*all-or-none law*).

■ Variable strengths of stimuli are coded by varying the frequency of action potentials, not their magnitude, in an activated nerve fiber.

4.4 Synapses and Neuronal Integration (pp. 102–113)

■ One neuron directly interacts with another neuron primarily through a *chemical synapse. (Review Figures 4-13 and 4-14 and chapter opener, p. 87.)*

■ Most neurons have four different functional parts. (Review Figure 4-7.)

1. The *dendrite/cell body* region (the input zone) serves as the *postsynaptic* component that binds with and responds to neurotransmitters released from other neurons.

2. The *axon hillock* (the trigger zone) is where action potentials are initiated because it has the lowest threshold and thus reaches threshold first in response to an excitatory graded potential change.

3. The *axon,* or *nerve fiber* (the conducting zone), conducts action potentials in undiminished fashion from the axon hillock to the axon terminals.

4. The *axon terminals* (the output zone) serve as the *presynaptic* component, releasing a *neurotransmitter* that influences other postsynaptic cells in response to action potential propagation down the axon.

■ The released neurotransmitter combines with receptor-channels on the postsynaptic neuron. (Review Figure 4-14.) (1) If nonspecific cation channels that permit passage of both Na^+ and K^+ are opened, the resultant ionic fluxes cause an *EPSP*, a small depolarization that brings the postsynaptic cell closer to threshold. (2) If either K^+ or Cl^- channels are opened, the likelihood that the postsynaptic neuron reaches threshold is diminished when an *IPSP*, a small hyperpolarization, is produced. (Review Figure 4-15.)

■ If the dominant activity is in its excitatory inputs, the postsynaptic cell is likely to be brought to threshold and have an action potential. This can be accomplished by either (1) *temporal summation* (EPSPs from a single, repetitively firing, presynaptic input occurring so close together in time that they add together) or (2) *spatial summation* (adding of EPSPs occurring simultaneously from several different presynaptic inputs). (Review Figure 4-16.) If inhibitory inputs dominate, the postsynaptic potential is brought further than usual from threshold. If excitatory and inhibitory activity to the postsynaptic neuron is balanced, the membrane remains close to resting.

■ Even though there are a number of different neurotransmitters, each synapse always releases the same neurotransmitter to produce a given response when combined with a particular receptor. *(Review Table 4-2.)*

■ Synaptic pathways between neurons are incredibly complex as a result of *convergence* of neuronal input and *divergence* of its output. Usually, many presynaptic inputs converge on a single neuron and jointly control its level of excitability. This same neuron, in turn, diverges to synapse with and influence the excitability of many other cells. *(Review Figure 4-18.)*

4.5 Intercellular Communication and Signal Transduction (pp. 113–118)

■ Intercellular communication is accomplished *directly* via (1) gap junctions or (2) transient direct linkup of cells' complementary surface markers. *(Review Figure 4-19.)*

■ More commonly cells communicate *indirectly* with one another to carry out various coordinated activities by dispatching *extracellular chemical messengers,* which act on particular *target cells* to bring about the desired response. There are four types of extracellular chemical messengers, which differ in their source and in the distance and means by which they get to their site of action: (1) *paracrines* (local chemical messengers); (2) *neurotransmitters* (short-range chemical messengers released by neurons); (3) *hormones* (long-range chemical messengers secreted into the blood by endocrine glands); and (4) *neurohormones* (long-range chemical messengers secreted into the blood by neurosecretory neurons). *(Review Figure 4-19.)*

■ Transfer of the signal carried by the extracellular messenger into the cell for execution is known as *signal transduction*.

■ An extracellular chemical messenger that cannot gain entry to the cell, such as a peptide hormone (the first messenger), triggers the desired cellular response by binding to the target cell membrane and (1) opening *receptor-channels;* (2) activating *receptor-enzymes,* such as tyrosine kinase; or (3) activating an intracellular *second messenger* via *G-protein-coupled receptors*. *(Review Table 4-3 and Figures 4-21 and 4-22.)*

4.6 Introduction to Paracrine Communication (pp. 118–120)

■ *Paracrines* act on neighboring cells; *autocrines* act on the cell that secretes them. Most of these local signal molecules are either cytokines or eicosanoids.

■ *Cytokines* are protein signal molecules secreted mostly by immune cells but also by some nonimmune cells for the primary purpose of regulating immune responses. Some cytokines help control cell growth and differentiation, in which case they are specifically termed *growth factors*.

■ *Eicosanoids* are lipid signal molecules derived from arachidonic acid, a fatty acid constituent in the plasma membrane. The three main classes of eicosanoids are *prostaglandins, thrombocytes,* and *leukotrienes*, which collectively exert a multitude of local effects throughout the body. *(Review Figure 2-23 and Table 4-4.)*

4.7 Introduction to Hormonal Communication (pp. 120–127)

■ *Hormones* are long-distance chemical messengers secreted by the endocrine glands into the blood, which transports them to specific target sites where they control a particular function by altering protein activity within the target cells.

■ Hormones are grouped into two categories based on their solubility differences: (1) *hydrophilic (water-soluble) hormones,* which include peptides (most hormones) and catecholamines (secreted by the adrenal medulla); and (2) *lipophilic (lipid-soluble) hormones,* which include steroid hormones (the sex hormones and those secreted by the adrenal cortex) and thyroid hormone. *(Review Table 4-5.)*

■ Hydrophilic *peptide hormones* are synthesized and packaged for export by the endoplasmic reticulum–Golgi complex, stored in secretory vesicles, and released by exocytosis on appropriate stimulation. They dissolve freely in the blood for transport to their target cells.

■ At their target cells, hydrophilic hormones bind with surface membrane receptors, triggering a chain of intracellular events by means of a second-messenger pathway (typically either the cAMP pathway or the IP_3–Ca^{2+}/DAG pathway) that ultimately alters preexisting cell proteins, usually enzymes, leading to the target cell's response to the hormone. *(Review Figures 4-25 and 4-26.)* Through this cascade of reactions, the initial signal is greatly amplified. *(Review Figure 4-27.)*

■ *Steroids* are synthesized by modifying stored cholesterol via enzymes specific for each steroidogenic tissue. Steroids are not stored in endocrine cells. Being lipophilic, they diffuse out through the lipid membrane barrier as soon as they are synthesized. Control of steroids is directed at their synthesis.

■ Lipophilic steroids and thyroid hormone are both transported in the blood largely bound to carrier plasma proteins, with only free, unbound hormone being biologically active.

■ Lipophilic hormones readily cross the lipid membrane barriers of their target cells and bind with receptors inside the cell. Once the hormone binds with the receptor, the hormone receptor complex binds with DNA and activates a gene, which leads to the synthesis of new enzymatic or structural intracellular proteins that carry out the hormone's effect on the target cell. *(Review Figure 4-28.)*

4.8 Comparison of the Nervous and Endocrine Systems (pp. 127–129)

■ The nervous and endocrine systems are the two main regulatory systems of the body. *(Review Table 4-6.)* The nervous system is anatomically "wired" to its target organs, whereas the "wireless" endocrine system secretes blood-borne hormones that reach distant target organs.

■ Specificity of neural action depends on the anatomic proximity of the neurotransmitter-releasing neuronal terminal to its target organ. Specificity of endocrine action depends on specialization of target cell receptors for a specific circulating hormone.

■ In general, the nervous system coordinates rapid responses, whereas the endocrine system regulates activities that require duration rather than speed.

5.1 Organization and Cells of the Nervous System (pp. 134–138)

■ The nervous system consists of the *central nervous system (CNS),* which includes the brain and spinal cord, and the *peripheral nervous system (PNS),* which includes the nerve fibers carrying information to (afferent division) and from (efferent division) the CNS. *(Review Figure 5-1.)*

■ Three functional classes of neurons—afferent neurons, efferent neurons, and interneurons—compose the excitable cells of the nervous system. *(Review Figure 5-2.) Afferent neurons* inform the CNS about conditions in both the external and internal environment. *Efferent neurons* carry instructions from the CNS to effector organs (muscles and glands). *Interneurons* are responsible for integrating afferent information and formulating an efferent response, and for all higher mental ("mind") functions.

■ *Glial cells* form the connective tissue within the CNS and physically, metabolically, and functionally support the neurons. The glial cells are *astrocytes, oligodendrocytes, microglia,* and *ependymal cells. (Review Figures 5-3 and 5-4.)*

5.2 Protection and Nourishment of the Brain (pp. 139–142)

■ Neurons cannot divide to replace damaged cells, but several protective devices shelter the brain: (1) The brain is wrapped in three layers of protective membranes—the *meninges*—and is further surrounded by a hard, bony covering. (2) *Cerebrospinal fluid (CSF)* flows within and around the brain to cushion it against physical jarring. *(Review Figure 5-6.)* (3) Protection against chemical injury is conferred by a *blood–brain barrier (BBB)* that limits access of blood-borne substances to the brain.

■ The brain depends on a constant blood supply for delivery of O_2 and glucose because it cannot generate ATP in the absence of either of these substances.

5.3 Overview of the Central Nervous System (pp. 142–144)

■ The parts of the brain from the lowest, most primitive level to the highest, most sophisticated level are the brain stem, cerebellum, hypothalamus, thalamus, basal nuclei, and cerebral cortex. *(Review Figure 5-7 and Table 5-1.)*

5.4 Cerebral Cortex (pp. 144–153)

■ The *cerebral cortex* is the outer shell of gray matter that caps an underlying core of white matter. *Gray matter* consists primarily of neuronal cell bodies, dendrites, and glial cells. The cerebral cortex is organized into six layers, each containing varying numbers of *stellate cells* (neurons that process sensory input) and *pyramidal cells* (neurons that send output to motor neurons).

The *white matter* consists of bundles of nerve fibers that interconnect various areas. *(Review Figures 5-9 and 5-14.)*

■ Ultimate responsibility for many discrete functions is localized in particular regions of the cortex as follows: (1) the *occipital lobes* house the visual cortex; (2) the auditory cortex is in the *temporal lobes;* (3) the *parietal lobes* process somatosensory (somesthetic and proprioceptive) input; and (4) voluntary motor movement is set into motion by the motor areas in the *frontal lobes. (Review Figures 5-8 and 5-10 through 5-12.)*

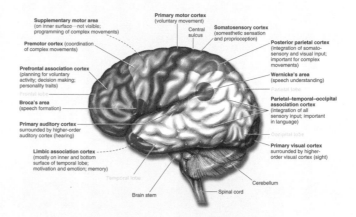

IFigure 5-11 Functional areas of the cerebral cortex.

■ *Language* ability depends on the integrated activity of two primary language areas—*Broca's area* and *Wernicke's area*— typically located only in the left cerebral hemisphere. *(Review Figures 5-11 and 5-13.)*

■ The *association areas* are regions of the cortex not specifically assigned to processing sensory input or commanding motor output or language ability. These areas provide an integrative link between diverse sensory information and purposeful action; they also play a key role in higher brain functions such as memory and decision making. They include the *prefrontal association cortex, the parietal–temporal–occipital association cortex,* and the *limbic association cortex. (Review Figure 5-11.)*

5.5 Basal Nuclei, Thalamus, and Hypothalamus (pp. 153–155)

■ The subcortical brain structures include the basal nuclei, thalamus, and hypothalamus. *(Review Figures 5-14 and 5-15.)*

■ The *basal nuclei* inhibit muscle tone; coordinate slow, sustained postural contractions; and suppress useless patterns of movement.

■ The *thalamus* serves as a relay station for preliminary processing of sensory input. It also accomplishes a crude awareness of sensation and some degree of consciousness.

■ The *hypothalamus* regulates body temperature, thirst, urine output, and food intake; extensively controls the autonomic nervous system and endocrine system; and is part of the limbic system.

5.6 Emotion, Behavior, and Motivation (pp. 155–157)

■ The *limbic system,* which includes portions of the hypothalamus and other structures that encircle the brain stem, plays an important role in emotion, basic behavioral patterns, motivation, and learning. (Review Figure 5-16.)

■ *Emotion* refers to subjective feelings and moods and the physical responses associated with these feelings.

■ *Basic behavioral patterns* triggered by the limbic system are aimed at survival (such as attack) and perpetuation of the species (such as mating behavior). Higher cortical centers can reinforce, modify, or suppress these basic behaviors.

■ *Motivation* is the ability to direct behavior toward specific goals.

■ *Norepinephrine, dopamine,* and *serotonin* are the key neurotransmitters in pathways for emotions and behavior.

5.7 Learning and Memory (pp. 157–163)
■ *Learning* refers to acquiring knowledge or skills as a result of experience, instruction, or both. *Memory* is storage of acquired knowledge for later recall and use.

■ There are two types of memory: (1) *short-term memory* with limited capacity and brief retention, coded by modifying activity at preexisting synapses; and (2) *long-term memory* with large storage capacity and enduring retention, involving relatively permanent structural or functional changes, such as forming new synapses between existing neurons. Enhanced protein synthesis underlies these long-term changes. (Review Table 5-2.)

■ *The hippocampus* plays a role in *consolidation,* the transfer of short-term memory to long-term memory. *Long-term potentiation (LTP),* a prolonged increase in the strength of existing synaptic connections in activated pathways, might be the link between short-term memory and consolidation of long-term memory. (Review Figure 5-17.)

■ The *hippocampus* and associated structures are especially important in *declarative,* or "what," *memories* of specific objects, facts, and events. The *cerebellum* and associated structures are especially important in *procedural,* or "how to," *memories* of motor skills gained through repetitive training.

■ The *prefrontal association cortex* is the site of *working memory,* which temporarily holds currently relevant data—both new information and knowledge retrieved from memory stores—and manipulates and relates them to accomplish the higher-reasoning processes of the brain.

5.8 Cerebellum (pp. 163–166)
■ The *cerebellum,* attached at the back of the brain stem beneath the cortex, consists of three functionally distinct parts. (Review Figure 5-18.)

■ The *vestibulocerebellum* helps maintain balance and controls eye movements. The *spinocerebellum* enhances muscle tone and helps coordinate voluntary movement, especially fast, phasic motor activities. The *cerebrocerebellum* plays a role in initiating voluntary movement and in storing procedural memories.

5.9 Brain Stem (pp. 166–172)
■ The *brain stem* is an important link between the spinal cord and higher brain levels.

■ The brain stem is the origin of the *cranial nerves.* (Review Table 5-3.) It also contains *centers* that control cardiovascular, respiratory, and digestive function. It further regulates postural muscle reflexes. Finally, it controls the overall degree of cortical alertness via the *reticular activating system,* or *RAS,* and plays a key role in the sleep–wake cycle. (Review Figure 5-19.)

■ *Consciousness* is the subjective awareness of the external world and self. The normal states of consciousness are wakefulness and sleep.

■ *Sleep* is an active process, not just the absence of wakefulness. While sleeping, a person cyclically alternates between slow-wave sleep and paradoxical (REM) sleep. *Slow-wave sleep* is characterized by slow waves on the EEG and little change in behavior pattern from the waking state except for not being consciously aware of the external world. *Paradoxical,* or *REM, sleep* is characterized by an EEG pattern similar to that of an alert, awake individual; rapid eye movements, dreaming, and abrupt changes in behavior pattern occur. (Review Table 5-4 and Figures 5-20 and 5-21.)

■ The prevailing *state of consciousness* depends on the cyclical interplay among an *arousal system* originating in the brain stem (RAS) and commanded by hypocretin-secreting neurons in the hypothalamus, (2) a *slow-wave sleep center* consisting of sleep-on neurons in the hypothalamus, and (3) an *REM sleep center* consisting of REM sleep-on neurons in the brain stem.

■ The leading theories of why we need sleep fall into the categories of (1) restoration and recovery, (2) memory consolidation, and (3) synaptic homeostasis.

5.10 Spinal Cord (pp. 172–178)
■ Extending from the brain stem, the *spinal cord* descends through a canal formed by surrounding protective vertebrae. (Review Figures 5-22 and 5-23.)

■ The spinal cord has two functions. (1) It serves as the link between the brain and the peripheral nervous system. All communication up and down the spinal cord is located in *ascending* and *descending tracts* in the cord's outer white matter. (Review Figures 5-25 and 5-26.) (2) It is the integrating center for spinal reflexes. The basic *reflex arc* includes a receptor, an afferent pathway, an integrating center, an efferent pathway, and an effector.

■ The centrally located gray matter of the spinal cord contains the cell bodies of efferent neurons and the interneurons interposed between afferent input and efferent output. (Review Figures 5-24 and 5-27.)

■ A *nerve* is a bundle of peripheral neuronal axons, both afferent and efferent, wrapped in connective tissue and following the same pathway. *Spinal nerves* supply specific body regions and are attached to the spinal cord in paired fashion throughout its length. (Review Figures 5-22 through 5-24 and 5-28.)

■ The 31 pairs of spinal nerves along with the 12 pairs of cranial nerves that arise from the brain stem constitute the peripheral nervous system. (Review Figure 5-23.)

Chapter 6
Study Card

6.1 Receptor Physiology (pp. 182–189)

■ The *afferent division* of the PNS carries information about the internal and the external environment to the CNS.

■ *Sensory receptors* are specialized peripheral endings of afferent neurons. *(Review Figure 6-1.)* Each receptor type (photoreceptor, mechanoreceptor, thermoreceptor, osmoreceptor, chemoreceptor, or nociceptor) responds to its *adequate stimulus* (a change in the energy form, or *modality,* to which it is responsive), translating the stimulus energy form into electrical signals.

■ A stimulus typically brings about a graded, depolarizing *receptor potential* by opening channels that allow net Na^+ entry. Receptor potentials, if of sufficient magnitude, ultimately generate action potentials in the afferent fiber next to the receptor. These action potentials self-propagate along the afferent fiber to the CNS. *(Review Figures 6-1 and 6-2.)* The strength of the stimulus determines the magnitude of the receptor potential, which in turn determines the frequency of action potentials generated. *(Review Figure 6-3 and Table 6-1.)*

■ The magnitude of the receptor potential is also influenced by the extent of receptor *adaptation,* which is a reduction in receptor potential despite sustained stimulation. (1) *Tonic receptors* adapt slowly or not at all and thus provide continuous information about the stimuli they monitor. (2) *Phasic receptors* adapt rapidly and frequently exhibit off responses, thereby providing information about changes in the energy form they monitor. *(Review Figure 6-4.)*

■ *Visceral afferent information* remains mostly subconscious. *Sensory afferent information* reaches the level of conscious awareness, including (1) somatic sensation (somesthetic sensation and proprioception) and (2) special senses.

■ Discrete *labeled-line* pathways lead from the receptors to the CNS so that the CNS can decipher information about the type and location of stimuli. *(Review Table 6-1.)*

■ The term *receptive field* refers to the area surrounding a receptor within which the receptor can detect stimuli. The *acuity,* or *discriminative ability,* of a body region varies inversely with the size of its receptive fields and also depends on the extent of *lateral inhibition* in the afferent pathways arising from receptors in the region. *(Review Figures 6-6 and 6-7.)*

■ *Perception* is the conscious interpretation of the external world that the brain creates from sensory input. What the brain perceives from its input is an abstraction and not reality. *(Review Figure 6-8.)* The only stimuli that can be detected are those for which receptors are present. Furthermore, as sensory signals ascend through progressively more complex processing, parts of the information may be suppressed or enhanced.

6.2 Pain (pp. 189–192)

■ Painful experiences are elicited by *nociceptors* responding to noxious stimuli and consist of two components: the perception of pain coupled with emotional and behavioral responses to it. The three categories of pain receptors are *mechanical, thermal,* and *polymodal* nociceptors. The latter respond to all kinds of damaging stimuli, including chemicals released from injured tissues.

■ Pain signals are transmitted over two afferent pathways: a *fast pathway* via A-delta fibers that carry sharp, prickling pain signals and a *slow pathway* via C fibers that carry dull, aching, persistent pain signals. *(Review Table 6-2.)*

■ Afferent pain fibers terminate in the spinal cord on ascending pathways that transmit the signal to the brain for processing. Descending pathways from the brain use *endogenous opioids* to suppress the release of *substance P,* a pain-signaling neurotransmitter from the afferent pain-fiber terminal. Thus, these descending pathways block further transmission of the pain signal and serve as a built-in *analgesic system. (Review Figure 6-9.)*

6.3 Eye: Vision (pp. 192–211)

■ Light is a form of electromagnetic radiation, with *visible light* being only a small band in the total electromagnetic spectrum. *(Review Figures 6-12 and 6-13.)*

■ The *eye* houses the light-sensitive *photoreceptors* essential for vision—the rods and cones found in its *retina* layer. *(Review Figures 6-10; 6-20, p. 199; and 6-24, p. 201; and Tables 6-3, p. 204, and 6-4, p. 208.)*

■ The *iris* controls the size of the *pupil* to adjust the amount of light permitted to enter the eye. *(Review Figure 6-11.)*

■ The *cornea* and *lens* are the primary *refractive* structures that bend incoming light rays to focus the image on the retina. The cornea contributes most to the total refractive ability of the eye. The strength of the lens can be adjusted through action of the *ciliary muscle* to *accommodate* for differences in near and far vision. *(Review Figures 6-14 through 6-19.)*

■ *Rods* and *cones* have three parts: a photopigment-containing outer segment, a metabolically specialized inner segment, and a neurotransmitter-secreting synaptic terminal. *(Review Figures 6-20, 6-24, and 6-25.)*

■ Rods and cones secrete neurotransmitter in the dark. They are activated when their *photopigments* differentially absorb various wavelengths of light. Photopigments consist of *opsin,* a membrane protein, and *retinal,* a vitamin A derivative. During *phototransduction,* light absorption by retinal causes a biochemical change in the photopigment that, through a series of steps, hyperpolarizes the photoreceptor, leading to decreased neurotransmitter release. Further retinal processing by *on-center* and *off-center bipolar and ganglion cells* eventually converts this light-induced signal into a change in the rate of action potential propagation in the visual pathway leaving the retina. *(Review Figures 6-24, 6-25, and 6-26.)*

■ Cones display high acuity but can be used only for *day vision* because of their low sensitivity to light. According to the *trichromatic theory*, different ratios of stimulation of three cone types (red, blue, and green) by varying wavelengths of light lead to *color vision*. According to the *opponent-process theory*, color vision depends on further processing of three opponent pairs of colors that are mutually exclusive (red/green, blue/yellow, and black/white). *(Review Table 6-3 and Figures 6-27 through 6-29.)*

■ Rods provide only indistinct vision in shades of gray, but because they are very sensitive to light, they can be used for *night vision*. *(Review Table 6-3.)*

■ The visual message is transmitted via a complex crossed and uncrossed pathway to the *visual cortex* in the occipital lobe of the brain for perceptual processing. *(Review Figure 6-30.)*

6.4 Ear: Hearing and Equilibrium (pp. 211–224)

■ The *ear* performs two unrelated functions: (1) *hearing*, which involves the *external ear, middle ear,* and *cochlea of the inner ear*; and (2) *sense of equilibrium*, which involves the *vestibular apparatus of the inner ear*. The ear receptor cells located in the inner ear—the *hair cells* in the cochlea and vestibular apparatus—are *mechanoreceptors*. *(Review Figure 6-31 and Table 6-6, p. 225.)*

■ *Hearing* depends on the ear's ability to convert airborne sound waves into mechanical deformations of auditory hair cells, thereby initiating neural signals. *Sound waves* consist of high-pressure regions of compression alternating with low-pressure regions of rarefaction of air molecules. The *pitch (tone)* of a sound is determined by the frequency of its waves, the *loudness (intensity)* by the amplitude of the waves, and the *timbre (quality)* by its characteristic overtones. *(Review Figures 6-32 and 6-33 and Table 6-5.)*

■ Sound waves are funneled through the *ear canal* to the *tympanic membrane*, which vibrates in sync with the waves. *Middle ear bones* bridging the gap between the tympanic membrane and inner ear amplify the tympanic movements and transmit them to the *oval window*, whose movement sets up traveling waves in the cochlear fluid. *(Review Figures 6-34 and 6-35.)*

■ These waves, which are at the same frequency as the original sound waves, set the *basilar membrane* in motion. Various regions of this membrane selectively vibrate more vigorously in response to different frequencies of sound. Its narrow, stiff end near the oval window vibrates best with high-frequency pitches, and its wide, flexible end near the helicotrema vibrates best with low-frequency pitches. *(Review Figure 6-35.)*

■ On top of the basilar membrane are the receptive inner hair cells of the *organ of Corti*, whose *stereocilia* ("hairs") are bent as the basilar membrane is deflected up and down in relation to the overhanging stationary *tectorial membrane*, which the hairs contact. *(Review Figures 6-34, 6-36, and 6-37.)*

■ *Pitch discrimination* depends on which region of the basilar membrane naturally vibrates maximally with a given frequency.

Loudness discrimination depends on the amplitude of the vibrations. Hair bending in the region of maximal basilar membrane vibration is transduced into neural signals that are transmitted to the *auditory cortex* in the temporal lobe of the brain for sound perception. *(Review Figure 6-38.)*

■ The vestibular apparatus in the inner ear consists of (1) the *semicircular canals*, which detect rotational acceleration or deceleration in any direction; and (2) the *utricle* and the *saccule*, which collectively detect changes in the rate of linear movement in any direction and provide information important for determining head position in relation to gravity. Neural signals are generated in response to the mechanical deformation of vestibular hair cells by specific movement of fluid and related structures within these sense organs. *(Review Figures 6-40 and 6-41.)*

■ Vestibular input goes to the *vestibular nuclei* in the brain stem and to the *cerebellum* for use in maintaining balance and posture, controlling eye movement, and perceiving motion and orientation.

6.5 Chemical Senses: Taste and Smell (pp. 224–230)

■ *Taste* and *smell* are chemical senses. In both cases, attachment of specific dissolved molecules to binding sites on a *chemoreceptor* causes receptor potentials that, in turn, set up neural impulses signaling the presence of the chemical.

■ *Taste receptors* are housed in *taste buds* on the tongue; *olfactory receptors* are located in the *olfactory mucosa* in the upper part of the nasal cavity. *(Review Figures 6-42 and 6-43.)*

■ Both sensory pathways include two routes: one to the *limbic system* for emotional and behavioral processing and one to the *cortex* for conscious perception and fine discrimination.

■ Taste and olfactory receptors are continuously renewed, unlike visual and hearing receptors, which are irreplaceable.

■ The five *primary tastes* are salty, sour, sweet, bitter, and umami (a meaty, "amino-acid" taste). *Taste discrimination* beyond the primary tastes depends on the patterns of stimulation of the taste buds by *tastants*. Each taste receptor responds to one of the primary tastes. Salty and sour tastants bring about receptor potentials in taste buds by directly affecting membrane channels, whereas the other three categories of tastants act through second-messenger pathways to bring about receptor potentials.

■ The 1000 different types of olfactory receptors each respond to only one discrete component of an odor, an *odorant*. Odorants act through second-messenger pathways to trigger receptor potentials. The afferent signals arising from the olfactory receptors are sorted according to scent component by the *glomeruli* within the *olfactory bulb*. *Odor discrimination* depends on the patterns of activation of the glomeruli. *(Review Figure 6-44.)*

7.1 Autonomic Nervous System (pp. 234–241)

■ The CNS controls muscles and glands by transmitting signals to these effector organs through the *efferent division* of the PNS.

■ There are two types of efferent output: the *autonomic nervous system,* which is under involuntary control and supplies cardiac and smooth muscle and most exocrine and some endocrine glands, and the *somatic nervous system,* which is subject to voluntary control and supplies skeletal muscle. *(Review Table 7-4, p. 242, and Table 7-5, p. 243.)*

■ The autonomic nervous system consists of two subdivisions—the *sympathetic* and *parasympathetic nervous systems. (Review Figures 7-2 and 7-3 and Tables 7-1 and 7-3.)*

■ An autonomic nerve pathway consists of a two-neuron chain. The *preganglionic fiber* originates in the CNS and synapses with the cell body of the *postganglionic fiber* in a ganglion outside the CNS. The postganglionic fiber ends on the effector organ. *(Review Figures 7-1, 7-2, and 7-3 and Table 7-3.)*

■ All preganglionic fibers and parasympathetic postganglionic fibers are *cholinergic,* that is, release *acetylcholine (ACh).* Sympathetic postganglionic fibers are *adrenergic,* that is, release *norepinephrine. (Review Figure 7-2 and Table 7-2.)*

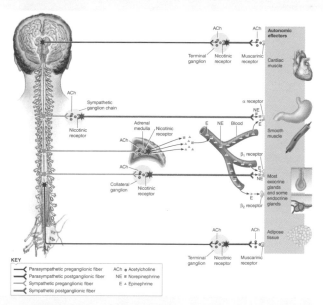

IFigure 7-2 Autonomic nervous system.

■ Postganglionic fibers have numerous swellings, or *varicosities,* that simultaneously release neurotransmitter over a large area of the innervated organ. *(Review Figures 7-1 and 8-33, p. 292.)*

■ The *adrenal medulla,* an endocrine gland, is a modified sympathetic ganglion that secretes the hormones epinephrine and to a lesser extent norepinephrine into the blood in response to stimulation by the sympathetic preganglionic fiber that innervates it. *(Review Figure 7-2.)*

■ The same neurotransmitter elicits different responses in different tissues. Thus, the response depends on specialization of the tissue cells, not on the properties of the messenger. *(Review Tables 7-1 and 7-2.)*

■ Tissues innervated by the autonomic nervous system possess one or more of several different receptor types for the postganglionic chemical messengers (and for the related adrenomedullary hormone epinephrine). Cholinergic receptors include *nicotinic* and *muscarinic receptors;* adrenergic receptors include α_1, α_2, β_1, and β_2 receptors. *(Review Figure 7-2 and Tables 7-1, 7-2, and 7-3.)*

ITABLE 7-2 Properties of Autonomic Receptor Types

Receptor Type	Neurotransmitter Affinity	Effector(s) with Receptor Type	Mechanism of Action at Effector	Effect on Effector
Nicotinic	ACh from autonomic preganglionic fibers	All autonomic postganglionic cell bodies; adrenal medulla	Opens nonspecific cation receptor-channels	Excitatory
	ACh from motor neurons	Motor end plates of skeletal muscle fibers	Opens nonspecific cation receptor-channels	Excitatory
Muscarinic	ACh from parasympathetic postganglionic fibers	Cardiac muscle, smooth muscle, glands	Activates various G-protein-coupled receptor pathways, depending on effector	Excitatory or inhibitory, depending on effector
α_1	Greater affinity for NE (from sympathetic postganglionic fibers) than for E (from the adrenal medulla)	Most sympathetic target tissues	Activates IP_3–Ca^{2+} second-messenger pathway	Excitatory
α_2	Greater affinity for NE than for E	Digestive organs	Inhibits cAMP	Inhibitory
β_1	Equal affinity for NE and for E	Heart	Activates cAMP	Excitatory
β_2	Affinity for E only	Smooth muscles of arterioles and bronchioles	Activates cAMP	Inhibitory

■ A given autonomic fiber either excites or inhibits activity in the organ it innervates. *(Review Tables 7-1 and 7-2.)*

■ Most visceral organs are innervated by both sympathetic and parasympathetic fibers, which in general produce opposite effects in a particular organ. *Dual innervation* of organs by both branches of the autonomic nervous system permits precise control over an organ's activity. *(Review Figure 7-2 and Table 7-1.)*

■ The sympathetic system dominates in emergency or stressful *("fight-or-flight")* situations and promotes responses that prepare the body for strenuous physical activity. The parasympathetic system dominates in quiet, relaxed *("rest-and-digest")* situations and promotes body-maintenance activities such as digestion. *(Review Tables 7-1 and 7-3.)*

ITABLE 7-3 Comparison of the Sympathetic and the Parasympathetic Nervous System

Feature	Sympathetic System	Parasympathetic System
Origin of preganglionic fiber	Thoracic and lumbar regions of the spinal cord	Brain and sacral region of the spinal cord
Origin of postganglionic fiber	Sympathetic ganglion chain (near the spinal cord) or collateral ganglia (about halfway between spinal cord and effector organs)	Terminal ganglia (in or near effector organs)
Fiber length	Short preganglionic fibers, long postganglionic fibers	Long preganglionic fibers, short postganglionic fibers
Neurotransmitter released	Preganglionic: ACh	Preganglionic: ACh
	Postganglionic: NE	Postganglionic: ACh
Types of receptors for neurotransmitters	For preganglionic neurotransmitter: nicotinic	For preganglionic neurotransmitter: nicotinic
	For postganglionic neurotransmitter: α_1, α_2, β_1, β_2	For postganglionic neurotransmitter: muscarinic
Dominance	Dominates in "fight-or-flight" situations	Dominates in "rest-and-digest" situations

■ Visceral afferent input is used by the CNS to direct appropriate autonomic output to maintain homeostasis. Autonomic activities are controlled by multiple areas of the CNS, including the spinal cord, medulla, hypothalamus, and prefrontal association cortex.

7.2 Somatic Nervous System (pp. 242–244)

■ The somatic nervous system consists of the axons of *motor neurons*, which originate in the spinal cord or brain stem and end on skeletal muscle. *(Review Figure 7-4 and Table 7-4.)*

■ *ACh*, the neurotransmitter released from a motor neuron, stimulates muscle contraction.

■ Motor neurons are the *final common pathway* by which various regions of the CNS exert control over skeletal muscle activity. The areas of the CNS that influence skeletal muscle activity by acting through the motor neurons are the spinal cord, motor regions of the cortex, basal nuclei, cerebellum, and brain stem.

TABLE 7-4 Comparison of the Autonomic and the Somatic Nervous System		
Feature	**Autonomic Nervous System**	**Somatic Nervous System**
Site of origin	Sympathetic: lateral horn of thoracic and lumbar spinal cord Parasympathetic: brain and sacral spinal cord	Ventral horn of spinal cord for most; those supplying muscles in head originate in brain
Number of neurons from CNS to effector organ	Two-neuron chain (preganglionic and postganglionic)	Single neuron (motor neuron)
Organs innervated	Cardiac muscle, smooth muscle, most exocrine and some endocrine glands	Skeletal muscle
Type of innervation	Most effector organs dually innervated by the two antagonistic branches of this system (sympathetic and parasympathetic)	Effector organs innervated only by motor neurons
Neurotransmitter at effector organs	May be ACh (parasympathetic terminals) or NE (sympathetic terminals)	Only ACh
Effects on effector organs	Either stimulation or inhibition (antagonistic actions of two branches)	Stimulation only (inhibition possible only centrally through IPSPs on dendrites and cell body of motor neuron)
Type of control	Under involuntary control	Subject to voluntary control; much activity subconsciously coordinated
Higher centers involved in control	Spinal cord, medulla, hypothalamus, prefrontal association cortex	Spinal cord, motor cortex, basal nuclei, cerebellum, brain stem

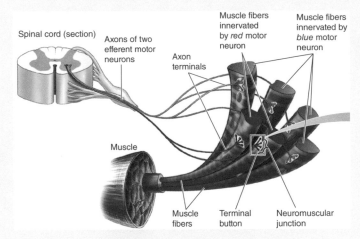

IFigure 7-4 **Motor neuron innervating skeletal muscle cells.**

7.3 Neuromuscular Junction (pp. 244–248)

■ When a motor neuron reaches a muscle, it branches into axon terminals. Each axon terminal forms a *neuromuscular junction* with a single muscle cell (fiber). The axon terminal splits into multiple fine branches, each of which ends in an enlarged *terminal button*. *(Review Figure 7-4 and chapter opener, p. 233.)*

■ The specialized region of the muscle cell membrane underlying the axon-terminal complex is called the *motor end plate*. Because these structures do not make direct contact, signals are passed between a terminal button and muscle fiber by chemical means. *(Review Figure 7-5.)*

■ An action potential in the axon terminal causes the release of ACh from its storage vesicles in the terminal button. The released ACh diffuses across the space separating the nerve and muscle cells and binds to special receptor-channels on the underlying motor end plate. This binding triggers the opening of these nonspecific cation channels. The subsequent ion movements depolarize the motor end plate, producing the *end-plate potential (EPP)*. *(Review Figure 7-5.)*

■ Local current flow between the depolarized end plate and adjacent muscle cell membrane brings these adjacent areas to threshold, initiating an action potential that is propagated throughout the muscle fiber. This muscle action potential triggers muscle contraction. *(Review Figure 7-5.)*

■ Membrane-bound *acetylcholinesterase*, an enzyme in the motor end plate, inactivates ACh, ending the EPP and, subsequently, the action potential and resultant contraction. *(Review Figure 7-5.)*

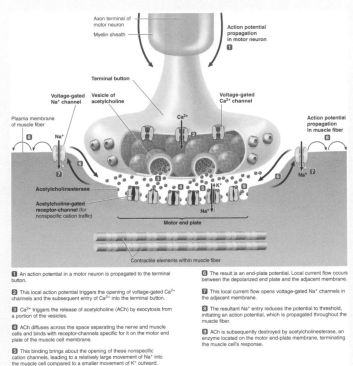

1 An action potential in a motor neuron is propagated to the terminal button.

2 This local action potential triggers the opening of voltage-gated Ca²⁺ channels and the subsequent entry of Ca²⁺ into the terminal button.

3 Ca²⁺ triggers the release of acetylcholine (ACh) by exocytosis from a portion of the vesicles.

4 ACh diffuses across the space separating the nerve and muscle cells and binds with receptor-channels specific for it on the motor end plate of the muscle cell membrane.

5 This binding brings about the opening of these nonspecific cation channels, leading to a relatively large movement of Na⁺ into the muscle cell compared to a smaller movement of K⁺ outward.

6 The result is an end-plate potential. Local current flow occurs between the depolarized end plate and the adjacent membrane.

7 This local current flow opens voltage-gated Na⁺ channels in the adjacent membrane.

8 The resultant Na⁺ entry reduces the potential to threshold, initiating an action potential, which is propagated throughout the muscle fiber.

9 ACh is subsequently destroyed by acetylcholinesterase, an enzyme located on the motor end-plate membrane, terminating the muscle cell's response.

IFigure 7-5 **Events at a neuromuscular junction.**

Chapter 8
Study Card

8.1 Structure of Skeletal Muscle (pp. 252–256)

■ *Muscles*, contraction specialists, can develop tension, shorten, produce movement, and accomplish work.

■ The three types of muscle are *skeletal, cardiac* and *smooth*. Skeletal muscle and cardiac muscle are *striated*, whereas smooth muscle is *unstriated*. Skeletal muscle is *voluntary*, whereas cardiac muscle and smooth muscle are *involuntary*. (*Review Figure 8-1.*)

■ *Skeletal muscles* are made up of bundles of long, thick, cylindrical muscle cells, called *muscle fibers*, wrapped in connective tissue. Muscle fibers are packed with *myofibrils*, each myofibril consisting of alternating, slightly overlapping stacked sets of thick and thin filaments. This arrangement leads to a skeletal muscle fiber's striated microscopic appearance, which consists of alternating dark *A bands* and light *I bands*. A *sarcomere*, the area between two Z lines, is the *functional unit* of skeletal muscle. (*Review Figures 8-2 and 8-3 and chapter opener.*)

■ *Thick filaments* consist of the protein *myosin*. *Cross bridges* made up of the myosin molecules' globular heads project from each thick filament toward the surrounding thin filaments. A cross bridge has an actin binding site and an ATPase site. (*Review Figures 8-2 and 8-4.*)

■ *Thin filaments* consist primarily of the protein *actin*, which can bind and interact with the myosin cross bridges to bring about contraction. In the resting state two other proteins, *tropomyosin* and *troponin*, lie across the thin filament surface and cover actin's binding site for attachment with a cross bridge, thus preventing this cross-bridge interaction. (*Review Figures 8-2 and 8-5.*)

8.2 Molecular Basis of Skeletal Muscle Contraction (pp. 256–262)

■ During *excitation-contraction coupling*, excitation of a skeletal muscle fiber by its motor neuron brings about contraction through a series of events resulting in the thin filaments sliding closer together between the thick filaments. (*Review Figure 8-7.*)

■ This *sliding filament mechanism* of muscle contraction is switched on by Ca^{2+} release from the *lateral sacs of the sarcoplasmic reticulum* in response to the spread of a muscle fiber action potential into the central portions of the fiber via the *T tubules*. (*Review Figures 8-9, 8-10, and 8-11.*)

■ *Released Ca^{2+}* binds to troponin, slightly repositioning tropomyosin to uncover actin's cross-bridge binding sites. (*Review Figures 8-6 and 8-11.*)

■ Binding of actin to a myosin cross bridge triggers cross-bridge stroking, powered by energy stored in the myosin head from prior splitting of ATP by myosin ATPase. During a *power stroke* the cross bridge bends toward the thick filament's center, "rowing" in the thin filament to which it is attached. (*Review Figures 8-8, 8-11, and 8-12.*)

■ When a fresh ATP attaches to the cross bridge, myosin and actin detach, the cross bridge returns to its original shape, and the cycle is repeated. Repeated cycles of cross-bridge activity slide the thin filaments inward step by step. (*Review Figures 8-8 and 8-12.*)

■ When the action potential ends, the lateral sacs actively take up the Ca^{2+}, troponin and tropomyosin slip back into their blocking position, and *relaxation* occurs. (*Review Figure 8-11.*)

■ The entire contractile response lasts up to 100 times longer than the action potential. (*Review Figure 8-13.*)

8.3 Skeletal Muscle Mechanics (pp. 262–269)

■ *Tension* is generated within a muscle by the *contractile component* (sarcomere shortening brought about by cross-bridge cycling). To move the bone to which the muscle's insertion is attached, this internal tension is transmitted to the bone as the contractile component stretches and tightens the muscle's *series-elastic component* (tendon). The *load* is the opposing force exerted on the muscle by the weight of an object. (*Review Figure 8-14.*)

■ The three primary types of muscle contraction are (1) *isotonic* (literally constant tension) in which the load remains constant as the muscle changes length (shortens in a *concentric* contraction and lengthens in an *eccentric* contraction), (2) *isokinetic* (constant velocity) in which the velocity of shortening remains constant as the muscle changes length, and (3) *isometric* (constant length) in which the muscle length does not change as tension increases.

■ The *velocity*, or speed, of shortening is inversely proportional to the load. (*Review Figure 8-16.*)

■ The amount of *work* accomplished by a contracting muscle equals the magnitude of the load times the distance the load is moved. Many muscles are part of a *lever system* that amplifies the distance and velocity of shortening of a muscle at the expense of the muscle having to exert increased force to move a load. (*Review Figure 8-17.*) The amount of energy consumed by a contracting muscle that is realized as external work varies from 0% to 25%; the remaining energy is converted to heat.

■ *Gradation* of whole-muscle contraction can be accomplished by (1) varying the number of muscle fibers contracting within the muscle and (2) varying the tension developed by each contracting fiber. (*Review Table 8-2, p. 275.*)

■ The number of fibers contracting depends on (1) size of the muscle (number of muscle fibers present), (2) extent of *motor unit recruitment* (how many motor neurons supplying the muscle are active), and (3) size of each motor unit (how many muscle fibers are activated simultaneously by a single motor neuron). A *motor unit* is a motor neuron plus all of the muscle fibers it innervates. (*Review Figure 8-18 and Table 8-2.*)

■ Two variable factors that affect fiber tension are (1) frequency of stimulation, which determines the extent of twitch summation; and (2) length of the fiber before the onset of contraction (*length–tension relationship*). (*Review Table 8-2.*)

■ *Twitch summation* is an increase in tension accompanying repetitive stimulation of a muscle fiber. After undergoing an action potential, a muscle cell membrane recovers from its refractory period and can be restimulated while some contractile activity remains so that the twitches induced by the two rapidly successive action potentials sum. If the muscle fiber is stimulated so rapidly that it does not have a chance to start relaxing between stimuli, a smooth, sustained maximal contraction known as *tetanus* occurs. (*Review Figure 8-19.*)

■ Tension also depends on fiber length at the onset of contraction. At *optimal length* (l_o), maximal opportunity for cross-bridge interaction occurs because of optimal overlap of thick and thin filaments, so the greatest tension can develop. Less tension can develop at shorter or longer lengths. (*Review Figure 8-20.*)

8.4 Skeletal Muscle Metabolism and Fiber Types (pp. 269–275)

■ Three pathways furnish the ATP needed for muscle contraction and relaxation: (1) the transfer of high-energy phosphates from stored *creatine phosphate* to ADP, providing the first source of ATP at the onset of exercise; (2) *oxidative phosphorylation,* which efficiently extracts large amounts of ATP from nutrients if enough O_2 is available to support this system; and (3) *glycolysis,* which can synthesize ATP in the absence of O_2 but uses large amounts of stored glycogen and produces lactate in the process. (*Review Figure 8-21.*)

■ The three types of skeletal muscle fibers are classified by the pathways they use for ATP synthesis (oxidative or glycolytic) and the rapidity with which they split ATP and subsequently contract (slow twitch or fast twitch): (1) *slow-oxidative fibers,* (2) *fast-oxidative fibers,* and (3) *fast-glycolytic fibers.* (*Review Table 8-1 and Figure 8-22.*)

8.5 Control of Motor Movement (pp. 276–286)

■ Motor activity is classified as (1) involuntary *somatic reflex responses,* (2) *voluntary movements,* and (3) *rhythmic activities* voluntarily initiated but subconsciously coordinated.

■ Control of motor movement depends on activity in three types of presynaptic inputs that converge on the motor neurons supplying various muscles: (1) *afferent neurons* as part of spinal reflex pathways; (2) the *corticospinal (pyramidal) motor system,* which originates in the primary motor cortex and is concerned with discrete, intricate movements of the hands; and (3) the *multineuronal (extrapyramidal) motor system,* which originates in the brain stem and is involved with postural adjustments and involuntary movements of the trunk and limbs. The final motor output from the brain stem is influenced by the cerebellum, basal nuclei, and cerebral cortex. (*Review Table 8-3.*)

■ Initiation and adjustment of motor commands depend on continuous afferent input, especially feedback about changes in muscle length (monitored by *muscle spindles*) and tension (monitored by *Golgi tendon organs*). (*Review Figure 8-23.*)

■ When a whole muscle is stretched, the stretch of its muscle spindles triggers the *stretch reflex,* which results in reflex contraction of that muscle. This reflex resists any passive changes in muscle length. (*Review Figures 8-24 and 8-25.*)

8.6 Smooth and Cardiac Muscle (pp. 286–294)

■ *Smooth muscle cells* are spindle shaped and much smaller than skeletal muscle fibers. The thick and thin filaments of smooth muscle are oriented diagonally in a diamond-shaped lattice instead of running longitudinally, so the fibers are not striated. (*Review Figures 8-28 and 8-29.*)

■ In smooth muscle, cytosolic Ca^{2+}, which enters from the ECF and is also released from sparse intracellular stores, activates cross-bridge cycling by initiating a series of biochemical reactions that result in phosphorylation of the *light chains* of the myosin cross bridges to enable them to bind with actin. (*Review Figures 8-30 and 8-31.*)

■ Smooth muscle in different organs is highly diversified and can be classified in various ways: phasic or tonic, multiunit or single-unit, and neurogenic or myogenic.

■ *Phasic smooth muscle* displays bursts of pronounced contraction in response to action potentials. *Tonic smooth muscle* is partially contracted at all times in the absence of action potentials because of ongoing Ca^{2+} entry through open surface-membrane Ca^{2+} channels.

■ *Multiunit smooth muscle* is *neurogenic,* requiring stimulation of individual muscle fibers by its autonomic nerve supply to trigger contraction. *Single-unit smooth muscle* is *myogenic;* it can initiate its own contraction. A few specialized, self-excitable, noncontractile cells in phasic, single-unit smooth muscle spontaneously depolarize to threshold as a result of *pacemaker potentials* (spontaneous drift to threshold) or *slow-wave potentials* (spontaneous alternating depolarizing and hyperpolarizing swings in potential). When threshold is reached and an action potential is initiated, this electrical activity spreads by means of gap junctions to the surrounding contractile cells within the *functional syncytium,* so the entire sheet of smooth muscle becomes excited and contracts as a single unit. (*Review Figure 8-32.*)

■ The level of tension in smooth muscle depends on the level of cytosolic Ca^{2+}. The autonomic nervous system (*review Figure 8-33*), along with hormones and local metabolites, can modify the rate and strength of contractions by altering cytosolic Ca^{2+} concentration.

■ Smooth muscle contractions are slow and energy efficient, enabling this type of muscle to economically sustain long-term contractions without fatigue. This economy, coupled with single-unit smooth muscle's ability to exist at various lengths with little change in tension, makes single-unit smooth muscle ideally suited for its task of forming the walls of hollow organs that can distend.

■ *Cardiac muscle* is found only in the heart. It has highly organized striated fibers, like skeletal muscle. Like single-unit smooth muscle, some specialized, self-excitable cardiac muscle fibers can generate action potentials, which spread throughout the heart with the aid of gap junctions. (*Review Table 8-4.*)

9.1 Anatomy of the Heart (pp. 298–303)

■ The circulatory system is the transport system of the body.

■ The three basic components of the circulatory system are the *heart* (the pump), the *blood vessels* (the passageways), and the *blood* (the transport medium).

■ The heart is positioned midline in the thoracic cavity at an angle with its wide base lying toward the right and its pointed apex toward the left.

■ The heart is a dual pump that provides the driving pressure for blood flow through the *pulmonary circulation* (between the heart and lungs) and *systemic circulation* (between the heart and other body systems). (*Review Figures 9-1 and 9-2.*)

■ The heart has four chambers: Each half of the heart consists of an *atrium,* or venous input chamber, and a *ventricle,* or arterial output chamber. The right atrium receives O_2-poor blood from the systemic circulation and the right ventricle pumps it into the pulmonary circulation. The left atrium receives O_2-rich blood from the pulmonary circulation and the left ventricle pumps it into the systemic circulation. (*Review Figures 9-1 and 9-2.*)

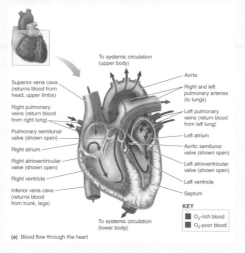

To systemic circulation (upper body)

Superior vena cava (returns blood from head, upper limbs)

Right pulmonary veins (return blood from right lung)

Pulmonary semilunar valve (shown open)

Right atrium

Right atrioventricular valve (shown open)

Right ventricle

Inferior vena cava (returns blood from trunk, legs)

Aorta

Right and left pulmonary arteries (to lungs)

Left pulmonary veins (return blood from left lung)

Left atrium

Aortic semilunar valve (shown open)

Left atrioventricular valve (shown open)

Left ventricle

Septum

To systemic circulation (lower body)

KEY
- O_2-rich blood
- O_2-poor blood

(a) Blood flow through the heart

■ Four heart valves direct blood in the right direction and keep it from flowing in the other direction. The *right* and *left atrioventricular (AV) valves* direct blood from the atria to the ventricles during diastole and prevent backflow of blood from the ventricles to the atria during systole. The *aortic* and *pulmonary semilunar valves* direct blood from the ventricles to the aorta and pulmonary artery, respectively, during systole and prevent backflow of blood from these major vessels to the ventricles during diastole. (*Review Figures 9-3, 9-4, and 9-5.*)

■ Contraction of the spirally arranged cardiac muscle fibers produces a wringing effect important for efficient pumping. Also important for efficient pumping is that the muscle fibers in each chamber contract as a coordinated unit. The branching cardiac muscle fibers are interconnected by *intercalated discs,* which contain (1) desmosomes that hold the cells together mechanically and (2) gap junctions that permit spread of electrical current between cells joined together into a functional syncytium. (*Review Figure 9-6.*)

9.2 Electrical Activity of the Heart (pp. 303–314)

■ The self-excitable heart initiates its rhythmic contractions. *Autorhythmic cells* are 1% of the cardiac muscle cells; they do not contract but are specialized to initiate and conduct action potentials. The other 99% of cardiac cells are *contractile cells* that contract in response to the spread of an action potential initiated by autorhythmic cells.

■ Autorhythmic cells display a *pacemaker potential,* a slow drift to threshold potential, as a result of a complex interplay of inherent changes in ion movement across the membrane. The first half of the pacemaker potential results from opening of unique *funny channels* that permit entry of Na^+ at the same time K^+ channels slowly close so that exit of K^+ slowly declines. Both of these actions gradually depolarize the membrane toward threshold. The final boost to threshold results from Ca^{2+} entry on opening of *T-type Ca^{2+} channels.* The rising phase of the action potential is the result of further Ca^{2+} entry on opening of *L-type Ca^{2+} channels* at threshold. The falling phase results from K^+ efflux on opening of K^+ channels at the peak of the action potential. Slow closure of these K^+ channels at the end of repolarization contributes to the next pacemaker potential. (*Review Figure 9-7.*)

■ The cardiac impulse originates at the *SA node,* the *pacemaker* of the heart, which has the fastest rate of spontaneous depolarization to threshold. (*Review Figures 9-8 and 9-9.*)

■ Once initiated, the action potential spreads throughout the right and left atria, partially facilitated by specialized conduction pathways but mostly by cell-to-cell spread of the impulse through gap junctions. (*Review Figure 9-8.*)

■ The impulse passes from the atria into the ventricles through the *AV node,* the only point of electrical contact between these chambers. The action potential is delayed briefly at the AV node, ensuring that atrial contraction precedes ventricular contraction to allow complete ventricular filling. (*Review Figure 9-8.*)

■ The impulse then travels rapidly down the interventricular septum via the *bundle of His* and rapidly disperses throughout the myocardium by means of the *Purkinje fibers.* The rest of the ventricular cells are activated by cell-to-cell spread of the impulse through gap junctions. (*Review Figure 9-8.*)

■ Thus, the atria contract as a single unit, followed after a brief delay by a synchronized ventricular contraction.

■ The action potentials of cardiac contractile cells exhibit a prolonged positive phase, or *plateau,* accompanied by a prolonged period of contraction, which ensures adequate ejection time. This plateau is primarily the result of activation of slow L-type Ca^{2+} channels. (*Review Figure 9-10.*)

■ Ca^{2+} entry though the L-type channels in the T tubules triggers a much larger release of Ca^{2+} from the sarcoplasmic reticulum. This *Ca^{2+}-induced Ca^{2+} release* leads to cross-bridge cycling and contraction. (*Review Figure 9-11.*)

■ Because a long refractory period occurs in conjunction with this prolonged plateau phase, summation and tetanus of cardiac muscle are impossible, ensuring the alternate periods of contraction and relaxation essential for pumping of blood. (Review Figure 9-12.)

■ The spread of electrical activity throughout the heart can be recorded from the body surface. In this *electrocardiogram (ECG)*, the *P wave* represents atrial depolarization; the *QRS complex*, ventricular depolarization; and the *T wave*, ventricular repolarization. Atrial repolarization is masked by the QRS complex. (Review Figures 9-13, 9-14, and 9-15.)

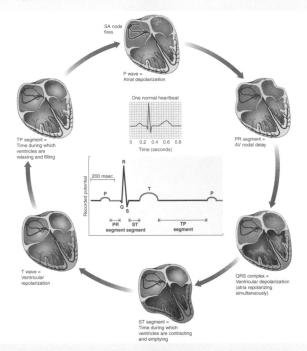

|Figure 9-14 Electrocardiogram waveforms in lead II and electrical status of heart associated with each waveform.

9.3 Mechanical Events of the Cardiac Cycle (pp. 314–318)

■ The *cardiac cycle* consists of three important events (Review Figure 9-16):

1. The generation of electrical activity as the heart autorhythmically depolarizes and repolarizes. (Review Figure 9-14).

2. Mechanical activity consisting of alternate periods of *systole* (contraction and emptying) and *diastole* (relaxation and filling), which are initiated by the rhythmic electrical cycle.

3. Directional flow of blood through the heart chambers, guided by valve opening and closing induced by pressure changes that are generated by mechanical activity.

■ The atrial pressure curve remains low throughout the entire cardiac cycle, with only minor fluctuations (normally varying between 0 and 8 mm Hg). The aortic pressure curve remains high the entire time, with moderate fluctuations (normally varying between a systolic pressure of 120 mm Hg and a diastolic pressure of 80 mm Hg). The left ventricular pressure curve fluctuates dramatically because ventricular pressure must be below the low atrial pressure to allow the AV valve to open during the *ventricular filling phase*, and, to force the aortic valve open to allow emptying, it must be above the high aortic pressure during the *ejection phase*. Therefore, ventricular pressure normally varies from 0 mm Hg during diastole to slightly more than 120 mm Hg during systole. During the

periods of *isovolumetric ventricular contraction* and *relaxation*, ventricular pressure is above the low atrial pressure and below the high aortic pressure, so all valves are closed and no blood enters or leaves the ventricles. (Review Figures 9-16 and 9-17.)

■ The *end-diastolic volume (EDV)* is the volume of blood in the ventricle when filling is complete at the end of diastole. The *end-systolic volume (ESV)* is the volume of blood remaining in the ventricle when ejection is complete at the end of systole. The *stroke volume (SV)* is the volume of blood pumped out by each ventricle each beat. (Review Figures 9-16 and 9-17.)

■ Valve closing gives rise to two normal heart sounds. The *first heart sound* is caused by closing of the AV valves and signals the onset of ventricular systole. The *second heart sound* is the result of closing of the aortic and pulmonary valves at the onset of diastole. (Review Figure 9-16.)

■ Defective valve function produces turbulent blood flow, which is audible as a *heart murmur*. Abnormal valves may be either *stenotic* and not open completely or *insufficient (leaky)* and not close completely. (Review Figure 9-18.)

9.4 Cardiac Output and Its Control (pp. 319–325)

■ *Cardiac output (CO)*, the volume of blood ejected by each ventricle each minute, is determined by heart rate times stroke volume. (Review Figure 9-24, p. 324.)

■ *Heart rate (HR)*, the number of beats per minute, varies by altering the balance of parasympathetic and sympathetic influence on the SA node. Parasympathetic stimulation slows HR, and sympathetic stimulation speeds it up. (Review Table 9-1 and Figure 9-19.)

■ *Stroke volume* depends on (1) the extent of ventricular filling, with an increased EDV resulting in a larger SV by means of the length–tension relationship (*Frank–Starling law of the heart,* a form of intrinsic control); and (2) the extent of sympathetic stimulation, with increased sympathetic stimulation resulting in increased *contractility* of the heart, that is, increased strength of contraction and increased SV at a given EDV (extrinsic control). (Review Table 9-1 and Figures 9-20 through 9-23.)

■ The *preload* of the heart (the workload imposed on the heart before contraction begins) is the extent of filling. The *afterload* of the heart (the workload imposed on the heart after contraction has begun) is the arterial blood pressure.

9.5 Nourishing the Heart Muscle (pp. 326–331)

■ Cardiac muscle is supplied with O_2 and nutrients by blood delivered to it by the *coronary circulation,* not by blood within the heart chambers. (Review Figure 9-29 and chapter opener, p. 297.)

■ Most coronary blood flow occurs during diastole because during systole the contracting heart muscle compresses the coronary vessels.

■ Coronary blood flow is normally varied to keep pace with cardiac oxygen needs. (Review Figure 9-26.)

■ Coronary blood flow may be compromised by development of *atherosclerotic plaques,* which can lead to *ischemic heart disease* ranging in severity from mild chest pain on exertion to fatal *heart attacks*. (Review Figures 9-27 through 9-29.)

10.1 Patterns and Physics of Blood Flow (pp. 336–339)

- Materials can be exchanged between various parts of the body and with the external environment by means of the blood vessel network that transports blood to and from all organs. (Review Figure 10-1 and chapter opener.)

- *Reconditioning organs* that replenish nutrient supplies and remove metabolic wastes from the blood receive a greater percentage of the cardiac output (CO) than is warranted by their metabolic needs. These reconditioning organs can better tolerate reductions in blood supply than can organs that receive blood solely for meeting their metabolic needs. The reconditioning organs are the digestive organs, kidneys, and skin.

- The brain is especially vulnerable to reductions in its blood supply. Therefore, maintaining adequate flow to this vulnerable organ is a high priority in circulatory function.

- The *flow rate* of blood through a vessel (in volume per unit of time) is directly proportional to the pressure gradient and inversely proportional to the resistance ($F = \Delta P/R$). The *pressure gradient* (ΔP) is the difference in pressure between the beginning and end of a vessel. The pressure imparted to the blood by cardiac contraction establishes the higher pressure at the beginning of a vessel. The lower pressure at the end is the result of frictional losses as flowing blood rubs against the vessel wall. (Review Figure 10-2.)

- *Resistance (R),* the hindrance to blood flow through a vessel, is influenced most by the vessel's radius. Resistance is inversely proportional to the fourth power of the radius ($R \propto 1/r^4$), so small changes in radius profoundly influence flow. As the radius increases, resistance decreases and flow increases, and vice versa. (Review Figure 10-3.)

- Blood flows in a closed loop between the heart and the organs. The *arteries* transport blood from the heart throughout the body. The *arterioles* regulate the amount of blood that flows through each organ. The *capillaries* are the actual site where materials are exchanged between blood and surrounding tissue cells. The *veins* return blood from the tissue level back to the heart. (Review Figure 10-4 and Table 10-1.)

10.2 Arteries (pp. 339–343)

- Arteries are large-radius, low-resistance *passageways from the heart to the organs*. They also serve as a *pressure reservoir*. Because of their elasticity, owing to their abundant elastin fibers, arteries expand to accommodate the extra volume of blood pumped into them by cardiac contraction and then recoil to continue driving the blood forward when the heart is relaxing. (Review Table 10-1 and Figures 10-5 and 10-6.)

- *Systolic pressure* (average 120 mm Hg) is the peak pressure exerted by the ejected blood against the vessel walls during car-

diac systole. *Diastolic pressure* (average 80 mm Hg) is the minimum pressure in the arteries when blood is draining off into the vessels downstream during cardiac diastole. When blood pressure is 120/80, *pulse pressure* (the difference between systolic and diastolic pressures) is 40 mm Hg. The average driving pressure throughout the cardiac cycle is the *mean arterial pressure (MAP)*, which can be estimated using the following equation: MAP = diastolic pressure + 1/3 pulse pressure. (Review Figure 10-7.)

10.3 Arterioles (pp. 343–350)

- Arterioles are the *major resistance vessels*. Their high resistance produces a large drop in mean pressure between the arteries and capillaries. This decline enhances blood flow by contributing to the pressure differential between the heart and organs. (Review Figure 10-8.)

- Arterioles have a thick layer of circular smooth muscle, variable contraction of which alters arteriolar caliber and resistance. *Tone,* a baseline of contractile activity, is maintained in arterioles at all times. Arteriolar *vasodilation* (expansion of arteriolar caliber above tone level) decreases resistance and increases blood flow through the vessel, whereas *vasoconstriction* (narrowing of the vessel) increases resistance and decreases flow. (Review Table 10-1 and Figure 10-9.) Arteriolar caliber is subject to both local (intrinsic) controls and extrinsic controls.

- *Local controls* primarily involve local chemical changes associated with changes in the level of metabolic activity in an organ, such as local changes in O_2, which cause the release of *vasoactive paracrines* from the *endothelial cells* in the vicinity. Examples include vasodilating *nitric oxide* and vasoconstricting *endothelin*. These vasoactive mediators act on the underlying arteriolar smooth muscle to bring about an appropriate change in the caliber of the arterioles supplying the organ. By adjusting the resistance to blood flow, the local control mechanism *matches an organ's blood flow to its momentary metabolic needs*. (Review Figures 10-9, 10-10, 10-11, and 10-13 and Tables 10-2 and 10-3.)

- Other local influences include (1) *histamine release* (important in inflammatory and allergic reactions); (2) *myogenic response to stretch* (important in *autoregulation,* which keeps tissue blood flow fairly constant despite changes in mean arterial driving pressure) (review Figures 10-11 and 10-12); (3) *chemical response to shear stress* (which resists changes in the force exerted parallel to the vessel surface by flowing blood); and (4) *local application of heat or cold* (important therapeutically). (Review Figure 10-13).

- Local control factors can adjust arteriolar caliber independently in different organs. Such adjustments are important in *variably distributing cardiac output*. (Review box figure on p. 371.)

- *Extrinsic control* is accomplished primarily by sympathetic and to a lesser extent by hormonal influence over arteriolar smooth muscle. Extrinsic controls are important in determining *total peripheral resistance (TPR)*, the total resistance offered by all systemic vessels, most of which is due to arteriolar resistance. TPR in turn plays a key role in *maintaining MAP*: MAP = CO × TPR. Arterioles are richly supplied with *sympathetic nerve fibers,* whose increased activity produces generalized vasoconstriction and a subsequent increase in TPR, thus increasing MAP. Decreased sympathetic activity produces generalized arteriolar vasodilation, which lowers MAP. These extrinsically controlled adjustments of arteriolar caliber help maintain the appropriate pressure head for driving blood forward to the tissues. Most arterioles are not supplied by parasympathetic nerves. Hormones that extrinsically influence arteriolar radius are *norepinephrine, epinephrine, vasopressin,* and *angiotensin II,* all of which cause generalized arteriolar vasoconstriction. *(Review Figure 10-13.)*

10.4 Capillaries (pp. 350–360)

■ The thin-walled, small-radius, extensively branched capillaries are ideally suited to serve as *sites of exchange* between the blood and surrounding tissue cells. Anatomically, the surface area for exchange is maximized and diffusion distance is minimized in the capillaries. Furthermore because of their large total cross-sectional area, the *velocity of blood flow* through capillaries (in distance per unit of time) is relatively slow, providing adequate time for exchanges to take place. *(Review Figures 10-14 through 10-16 and Table 10-1, p. 339.)*

■ Two types of passive exchanges—diffusion and bulk flow—take place across capillary walls.

■ Individual solutes are exchanged primarily by *diffusion down concentration gradients*. Lipid-soluble substances pass directly through the single layer of endothelial cells lining a capillary, whereas water-soluble substances pass through water-filled pores between the endothelial cells. Plasma proteins generally do not escape. *(Review Figures 10-17 and 10-19.)*

■ Imbalances in physical pressures acting across capillary walls are responsible for *bulk flow* of fluid through the pores. (1) Fluid is forced out of the first portion of the capillary *(ultrafiltration),* where outward pressures (mainly capillary blood pressure; [P_C]) exceed inward pressures (mainly plasma-colloid osmotic pressure [π_P]). (2) Fluid is returned to the capillary *(reabsorption)* along its last half, when outward pressures fall below inward pressures. The reason for the shift in balance down the capillary's length is the continuous decline in P_C while π_P remains constant. Bulk flow is responsible for the distribution of the ECF between plasma and interstitial fluid. *(Review Figures 10-8, p. 343, and 10-20.)*

■ Normally, slightly more fluid is filtered than is reabsorbed. The extra fluid, any leaked proteins, and bacteria in the tissue are picked up as *lymph* by the *lymphatic system.* Bacteria are destroyed as lymph passes through lymph nodes on the way to being returned to the venous system. *(Review Figures 10-20 through 10-22.)*

10.5 Veins (pp. 360–365)

■ Veins are large-radius, low-resistance *passageways from the organs to the heart*. In addition, the thin-walled, highly distensible veins, as *capacitance vessels,* can passively stretch to store a larger volume of blood and therefore act as a *blood reservoir.* The capacity of veins to hold blood can change markedly with little change in venous pressure. At rest, the veins contain more than 60% of the total blood volume. *(Review Table 10-1, p. 339, and Figure 10-24.)*

■ The primary force that produces *venous return* is the *pressure gradient* between the veins and atrium (that is, what remains of the driving pressure imparted to the blood by cardiac contraction). *(Review Figures 10-8, p. 343, and 10-25.)*

■ Venous return is enhanced by sympathetically induced *venous vasoconstriction* and by external compression of the veins from contraction of surrounding skeletal muscles *(skeletal muscle pump),* both of which drive blood out of the veins. These actions help counter the effects of gravity on the venous system. *(Review Figures 10-25 and 10-26.)*

■ One-way *venous valves* ensure that blood is driven toward the heart and kept from flowing back toward the tissues. *(Review Figure 10-27.)*

■ Venous return is also enhanced by the *respiratory pump* and the *cardiac suction* effect. Respiratory activity produces a less-than-atmospheric pressure in the chest cavity, thus establishing an external pressure gradient that encourages flow from the lower veins that are exposed to atmospheric pressure to the chest veins that empty into the heart. In addition, slightly negative pressures created within the atria during ventricular systole and within the ventricles during ventricular diastole exert a suctioning effect that further enhances venous return and facilitates cardiac filling. *(Review Figures 10-25 and 10-28.)*

10.6 Blood Pressure (pp. 365–377)

■ Regulation of MAP depends on control of its two main determinants, CO and TPR. Control of CO, in turn, depends on regulation of HR and SV, whereas TPR is determined primarily by the degree of arteriolar vasoconstriction. *(Review Figure 10-29.)*

■ Short-term regulation of blood pressure is accomplished mainly by the *baroreceptor reflex. Carotid sinus and aortic arch baroreceptors* continuously monitor MAP. When they detect a deviation from normal, they signal the medullary *cardiovascular center,* which responds by adjusting autonomic output to the heart and blood vessels to restore blood pressure to normal. *(Review Figures 10-30 through 10-33.)*

■ Long-term control of blood pressure involves maintaining proper plasma volume through the kidneys' control of salt and water balance. *(Review Figure 10-29.)*

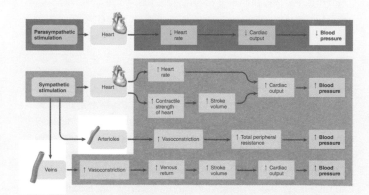

I Figure 10-32 **Summary of the effects of the parasympathetic and sympathetic nervous systems on factors that influence mean arterial blood pressure.**

11.1 Plasma (pp. 381–383)

■ Blood consists of three types of cellular elements—*erythrocytes (red blood cells), leukocytes (white blood cells),* and *platelets (thrombocytes)*—suspended in the liquid plasma. *(Review chapter opener, Figure 11-1, and Table 11-1.)*

■ The 5- to 5.5-L volume of blood in an adult consists of 42% to 45% erythrocytes, less than 1% leukocytes and platelets, and 55% to 58% plasma. The percentage of whole-blood volume occupied by erythrocytes is the *hematocrit. (Review Figure 11-1.)*

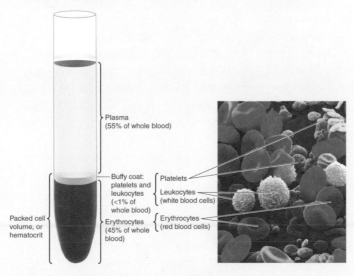

IFigure 11-1 **Hematocrit and types of blood cells.**

■ *Plasma* is a complex liquid consisting of 90% water that serves as a transport medium for substances being carried in the blood. The most abundant inorganic constituents in plasma are Na^+ and Cl^-, and the most plentiful organic constituents are *plasma proteins*. All plasma constituents are freely diffusible across the capillary walls except the plasma proteins, which remain in the plasma where they perform a variety of important functions. Plasma proteins include the *albumins, globulins (α, β, and γ),* and *fibrinogen. (Review Table 11-1.)*

11.2 Erythrocytes (pp. 383–392)

■ Erythrocytes are specialized for their primary function of O_2 *transport* in the blood. Their biconcave shape maximizes the surface area available for diffusion of O_2 into cells of this volume. *(Review Figure 11-1 and chapter opener.)* Erythrocytes do not contain a nucleus or organelles (these are extruded during development) but instead are packed full of *hemoglobin,* an iron-containing molecule that can loosely and reversibly bind with O_2. Because O_2 is poorly soluble in blood, hemoglobin is indispensable for O_2 transport. Each hemoglobin molecule can carry four O_2 molecules. *(Review Figures 11-2 and 11-3.)*

■ Hemoglobin also contributes to CO_2 transport and buffering of blood by reversibly binding with CO_2 and H^+.

■ Unable to replace cell components, erythrocytes are destined to a short life span of about 120 days.

■ Undifferentiated *pluripotent stem cells* in the *red bone marrow* give rise to all cellular elements of the blood. *(Review Figures 11-3 and 11-9, p. 394.)* Erythrocyte production *(erythropoiesis)* by the red marrow normally keeps pace with the rate of erythrocyte loss, keeping the red cell count constant. Erythropoiesis is stimulated by *erythropoietin,* a hormone secreted by the kidneys in response to reduced O_2 delivery. *(Review Figure 11-4.)*

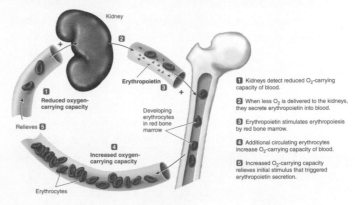

IFigure 11-4 **Control of erythropoiesis.**

■ The major *ABO blood types* depend on the presence of specific antigens on the surface of erythrocytes. The red blood cells of *type A* blood have A antigen; those of *type B* blood have B antigen, those of *type AB* blood have both A and B antigen, and those of *type O* blood have no A or B antigen. Type A blood has anti-B antibodies, type B blood has anti-A antibodies, type AB blood has no anti-A or anti-B antibodies, and type O blood has both anti-A and anti-B antibodies. These antibodies cause the RBCs with the corresponding antigens to agglutinate (clump) or rupture, causing a *transfusion reaction* if incoming donor cells are exposed to corresponding antibodies in recipient blood. *(Review Table 11-2 and Figure 11-7.)*

11.3 Leukocytes (pp. 392–394)

■ Leukocytes are the *defense corps* of the body. They attack foreign invaders (the most common of which are bacteria and viruses), clean up cellular debris, and destroy cancer cells that arise in the body. Leukocytes and certain plasma proteins make up the *immune system.*

■ The five types of leukocytes are categorized microscopically by differences in nuclear shape, presence or absence of granules, and staining properties. The *polymorphonuclear granulocytes* include neutrophils, eosinophils, and basophils. The *mononuclear agranulocytes* include monocytes and lymphocytes. *(Review Figure 11-8.)*

■ Each of the leukocyte types has a different task: (1) *Neutrophils,* the phagocytic specialists, are important in engulfing bacteria and debris. (2) *Eosinophils* specialize in attacking parasitic worms and play a role in allergic responses. (3) *Basophils* release two chemicals: *histamine,* which is also important in allergic responses; and *heparin,* which helps clear fat particles from the blood. (4) *Monocytes,* on leaving the blood, set up residence in the tissues and greatly enlarge to become the large tissue phagocytes known as *macrophages*. (5) *Lymphocytes* provide immune defense against bacteria, viruses, and other targets for which they are specifically programmed. Their defense tools include production of antibodies that mark the victim for destruction by phagocytosis or other means (for *B lymphocytes*) and release of chemicals that punch holes in the victim (for *T lymphocytes*). *(Review Figure 11-8 and Table 11-1, p. 382.)*

■ Leukocytes are present in the blood only while in transit from their site of production and storage in the bone marrow (and also in the lymphoid tissues in the case of the lymphocytes) to their site of action in the tissues. *(Review Figure 11-9.)* At any given time, most leukocytes are out in the tissues on surveillance or performing actual combat missions.

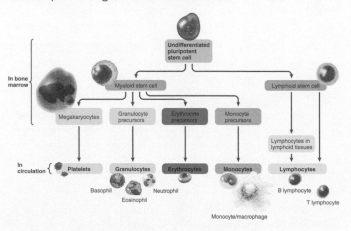

IFigure 11-9 **Blood cell production (hemopoiesis).**

■ All leukocytes have a limited life span and must be replenished by ongoing differentiation and proliferation of precursor cells. The total number and percentage of each of the different types of leukocytes are produced at variable rates depending on the momentary defense needs of the body. Factors that regulate production of the different types of leukocyte are released from invaded or injured tissues and from activated leukocytes.

11.4 Platelets and Hemostasis (pp. 395–400)

■ Platelets are cell fragments derived from large *megakaryocytes* in the bone marrow. *(Review Figures 11-9 and 11-10.)*

■ Platelets play a role in *hemostasis,* the arrest of bleeding from an injured vessel. The three main steps in hemostasis are (1) *vascular spasm,* which reduces blood flow through an injured vessel; (2) *platelet plugging;* and (3) *clot formation.*

■ Platelet aggregation at the site of vessel injury quickly plugs the defect. Platelets start to aggregate by adhering to *von Willebrand factor,* which binds to *exposed collagen* in the damaged vessel wall. These aggregated platelets secrete *ADP* and *thromboxane A₂,* which together cause other passing-by platelets to pile on, setting up a positive-feedback cycle as the platelet plug grows to fill in the defect. Normal adjacent endothelium secretes inhibitory chemicals that prevent platelets from adhering to the surrounding undamaged parts of the vessel. *(Review Figure 11-11.)*

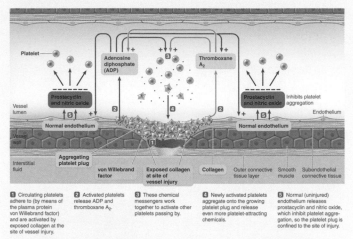

IFigure 11-11 **Platelet activation, formation of a platelet plug, and prevention of platelet aggregation at adjacent normal vessel lining.**

■ *Clot formation* reinforces the platelet plug and converts blood in the vicinity of a vessel injury into a nonflowing gel.

■ Most factors necessary for clotting are always present in the plasma in inactive precursor form. When a vessel is damaged, exposed collagen initiates a *clotting cascade* of reactions involving successive activation of these clotting factors, ultimately converting fibrinogen into fibrin via the *intrinsic clotting pathway*. *(Review Figure 11-13.)*

■ *Fibrin,* an insoluble threadlike molecule, is laid down as the meshwork of the clot; the meshwork in turn entangles blood cellular elements to complete clot formation. *(Review Figure 11-12.)*

■ Blood that has escaped into the tissues clots on exposure to *tissue thromboplastin,* which sets the *extrinsic clotting pathway* into motion. *(Review Figure 11-13.)*

■ Clots form quickly. When no longer needed, they are slowly dissolved by *plasmin,* a fibrinolytic factor also activated by exposed collagen. *(Review Figure 11-14.)*

■ Thus, exposed collagen simultaneously initiates platelet aggregation and clot formation and sets the stage for later clot dissolution.

12.1 Immune System: Targets, Effectors, Components (pp. 405–407)

■ Foreign invaders and newly arisen mutant cells are immediately confronted with multiple interrelated defense mechanisms aimed at destroying and eliminating anything that is not part of the normal self. These mechanisms, collectively referred to as *immunity,* include both innate and adaptive immune responses. *Innate immune responses* are nonspecific responses that nonselectively defend against foreign material even on initial exposure to it. *Adaptive immune responses* are specific responses that selectively target particular invaders for which the body has been specially prepared after a prior exposure.

■ The most common invaders are bacteria and viruses. *Bacteria* are self-sustaining, single-celled organisms that produce disease by virtue of the destructive chemicals they release. *Viruses* are protein-coated nucleic acid particles that invade host cells and take over the cellular metabolic machinery for their own survival to the detriment of the host cell.

■ *Leukocytes* and their derivatives are the major effector cells of the immune system and are reinforced by a number of different *plasma proteins*. Leukocytes include neutrophils, eosinophils, basophils, monocytes, and lymphocytes.

■ Immune cells also clean up cellular debris, preparing the way for tissue repair, and help defend against cancer.

12.2 Innate Immunity (pp. 408–415)

■ Innate immune responses include inflammation, interferon, natural killer (NK) cells, and the complement system.

■ *Inflammation* is a nonspecific response to foreign invasion or tissue damage mediated largely by the professional phagocytes (neutrophils and monocytes-turned-macrophages). Phagocytes have receptors, such as *toll-like receptors (TLRs),* which recognize and bind with a foreign cell's telltale markers, for example, carbohydrates found in bacterial cell walls but not in human cells. The phagocytic cells destroy foreign and damaged cells both by phagocytosis and by release of lethal chemicals. *(Review Figures 12-2 and 12-3 and chapter opener.)* Phagocytic secretions also augment inflammation, induce systemic manifestations such as fever, and enhance adaptive immune responses.

■ Histamine-induced vasodilation and increased permeability of local capillaries at the site of invasion or injury permit enhanced delivery of more phagocytic leukocytes and inactive plasma protein precursors crucial to defense, such as complement components. These vascular changes also largely produce the observable local manifestations of inflammation—swelling, redness, heat, and pain. *(Review Figure 12-3.)*

■ *Interferon* is nonspecifically released by virus-infected cells and transiently inhibits viral multiplication in other cells to which it binds. *(Review Figure 12-5.)*

■ *NK cells* nonspecifically lyse and destroy virus-infected cells and cancer cells on first exposure to them.

■ On being activated by microbes at the site of invasion or by antibodies produced against the microbes, the *complement system* directly destroys the foreign invaders by lysing their membranes and also augments other aspects of the inflammatory process, such as by acting as *opsonins* that enhance phagocytosis. The complement system lyses the targeted cells by forming a hole-punching *membrane attack complex* that inserts into the victim cell's membrane, leading to osmotic rupture of the cell. *(Review Figures 12-4 and 12-6.)*

12.3 Adaptive Immunity: General Concepts (pp. 415–416)

■ Not only is the adaptive immune system able to recognize foreign molecules as different from self-molecules—so that destructive immune reactions are not unleashed against the body itself—but it can also distinguish between millions of different foreign molecules. *Lymphocytes*, the effector cells of adaptive immunity, are each uniquely equipped with surface membrane receptors that can bind with only one specific complex foreign molecule, known as an *antigen*.

■ The two classes of adaptive immune responses are *antibody-mediated immunity* accomplished by plasma cells derived from *B lymphocytes (B cells),* and *cell-mediated immunity* accomplished by *T lymphocytes (T cells). (Review Table 12-4, p. 434.)*

■ B cells develop from a lineage of lymphocytes that originally matured within the bone marrow. The T-cell lineage comes from lymphocytes that migrated from the bone marrow to the thymus to complete their maturation. New B and T cells arise from lymphocyte colonies in *lymphoid tissues. (Review Figures 12-1, p. 406, and 12-7 and Table 12-1.)*

12.4 B Lymphocytes: Antibody-Mediated Immunity (pp. 416–422)

■ Each B cell recognizes specific free extracellular antigen, such as that found on the surface of bacteria.

■ After being activated by binding of its receptor (a *B-cell receptor,* or *BCR*) with its specific antigen, a B cell rapidly proliferates, producing a *clone* that can specifically wage battle against the invader. Most lymphocytes in the expanded B-cell clone become antibody-secreting *plasma cells* that participate in the *primary response* against the invader. Some of the new lymphocytes do not participate in the attack but become *memory cells* that lie in wait, ready to launch a swifter and more forceful *secondary response* should the same foreigner ever invade the body again. *(Review Figures 12-8, 12-9, 12-12, and 12-13.)*

■ *Antibodies* are **Y**-shaped molecules. The antigen-binding sites on the tips of each arm determine with what specific antigen the antibody can bind. Properties of the antibody's tail portion determine what the antibody does once it binds with antigen. There are five subclasses of antibodies, depending on differences in the biological activity of their tail portion: IgM, IgG, IgE, IgA, and IgD immunoglobulins. *(Review Figure 12-10.)*

■ Antibodies do not directly destroy antigenic material. Instead, they exert their protective effect by physically hindering antigens through neutralization or agglutination or by intensifying lethal innate immune responses already called into play by the foreign invasion. Antibodies activate the complement system, enhance phagocytosis, and stimulate NK cells. (*Review Figure 12-11 and Table 12-3, p. 430.*)

12.5 T Lymphocytes: Cell-Mediated Immunity (pp. 422–435)

■ T cells accomplish cell-mediated immunity by being in direct contact with their targets and by releasing *cytokines,* which are protein signal molecules other than antibodies released by leukocytes that regulate immune function.

■ There are three types of T cells: cytotoxic, helper, and regulatory T cells.

■ The targets of *cytotoxic (CD8⁺) T cells* are virally invaded cells and cancer cells, which they destroy by releasing *perforin* molecules that form a lethal hole-punching complex that inserts into the membrane of the victim cell or by releasing *granzymes* that trigger the victim cell to undergo apoptosis. (*Review Figures 12-14 and 12-15 and Table 12-2.*)

■ *Helper (CD4⁺) T cells* bind with other immune cells and release cytokines that augment the activity of these other cells. B cells cannot convert to plasma cells and produce antibodies in response to T-dependent antigen without the help of helper cells. (*Review Figure 12-20.*)

■ *Regulatory (CD4⁺CD25⁺) T cells (T_regs)* secrete cytokines that suppress other immune cells, putting the brake on immune responses in check-and-balance fashion.

■ Like B cells, T cells bear receptors (*T-cell receptors* or *TCRs*) that are antigen specific (*review Figure 12-8, p. 416*), undergo clonal selection, exert primary and secondary responses, and form memory pools for long-lasting immunity against targets to which they have already been exposed.

■ Helper T cells can recognize and bind with antigen only when it has been processed and presented to them by *antigen-presenting cells (APCs),* such as macrophages and dendritic cells. (*Review Figures 12-17 and 12-18.*)

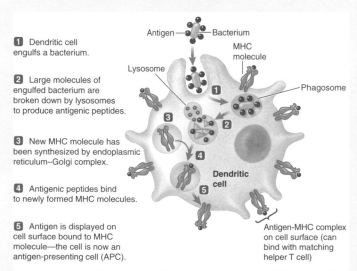

1 Dendritic cell engulfs a bacterium.

2 Large molecules of engulfed bacterium are broken down by lysosomes to produce antigenic peptides.

3 New MHC molecule has been synthesized by endoplasmic reticulum–Golgi complex.

4 Antigenic peptides bind to newly formed MHC molecules.

5 Antigen is displayed on cell surface bound to MHC molecule—the cell is now an antigen-presenting cell (APC).

Antigen-MHC complex on cell surface (can bind with matching helper T cell)

❙ Figure 12-18 **Generation of an antigen-presenting cell when a dendritic cell engulfs a bacterium.**

■ Lymphocytes produced by chance that can attack the body's cells are eliminated or suppressed so that they are prevented from functioning. In this way, the body is able to "tolerate" (not attack) its own antigens, a process known as *self-tolerance.*

■ B and T cells have different targets because their requirements for antigen recognition differ. B cells recognize freely circulating antigen, such as bacteria, that can lead to antigen destruction at long distances via antibodies. T cells, in contrast, have a dual binding requirement of foreign antigen in association with self-antigens on the surface of one of the body's cells. (*Review Figures 12-19 and 12-20.*)

■ The *self-antigens* on cell surfaces are *class I* or *class II MHC molecules,* which are unique for each individual. Cytotoxic T cells can bind only with virus-infected host cells or cancer cells, which always bear class I MHC self-antigen in association with foreign or abnormal antigen. Helper T cells can bind only with APCs and B cells that bear the class II MHC self-marker in association with foreign antigen. The APCs activate helper T cells, and helper T cells activate B cells. Thus, such differential binding ensures that the appropriate specific immune response ensues. (*Review Figures 12-19 and 12-20.*)

■ In the process of *immune surveillance,* natural killer cells, cytotoxic T cells, macrophages, and the interferon they collectively secrete normally eradicate newly arisen cancer cells before they have a chance to spread. (*Review Figure 12-22.*)

12.6 Immune Diseases (pp. 435–438)

■ *Immune diseases* are of two types: *immunodeficiency diseases* (insufficient immune responses) or *inappropriate immune attacks* (excessive or mistargeted immune responses).

■ Inappropriate attacks include *autoimmune diseases, immune complex diseases,* and *allergies (hypersensitivities),* of which there are two types: (1) *immediate hypersensitivities* involving the production of IgE antibodies by B cells that trigger release of histamine from mast cells and basophils to bring about a swift response to the allergen, and (2) *delayed hypersensitivities* involving a more slowly responding cell-mediated, symptom-producing response by T cells against the allergen. (*Review Figure 12-23 and Table 12-5.*)

12.7 External Defenses (pp. 438–442)

■ The body surfaces exposed to the outside environment—both the outer covering of skin and the linings of internal cavities that communicate with the external environment—serve not only as mechanical barriers to deter would-be pathogenic invaders but also play an active role in thwarting entry of bacteria and other unwanted materials.

■ The *skin* consists of two layers: an outer keratinized *epidermis* and an inner vascularized, connective tissue *dermis.* The epidermis contains pigment-producing *melanocytes,* keratin-producing *keratinocytes,* immune-enhancing *Langerhans cells,* and immune-suppressive *Granstein cells.* (*Review Figure 12-24.*)

■ The other main routes by which potential pathogens enter the body are the digestive system, the urogenital system, and the respiratory system, which are all defended by various antimicrobial strategies.

13.1 Respiratory Anatomy (pp. 446–450)

■ *Cellular respiration* refers to the intracellular metabolic reactions that use O_2 and produce CO_2 during energy-yielding oxidation of nutrient molecules. *External respiration* refers to the transfer of O_2 and CO_2 between the external environment and tissue cells. The respiratory and circulatory systems function together to accomplish external respiration. (*Review Figure 13-1.*)

■ The *respiratory system* exchanges air between the atmosphere and lungs. The airways conduct air from the atmosphere to the *alveoli*, across which O_2 and CO_2 are exchanged between air in these air sacs and blood in the surrounding *pulmonary capillaries*. The extremely thin alveolar walls are formed by *Type I alveolar cells. Type II alveolar cells* secrete pulmonary surfactant. (*Review Figures 13-2 and 13-4 and chapter opener.*)

■ The *lungs* consist of the smaller airways, alveoli, pulmonary arteries, and highly elastic connective tissue. They are housed within the closed compartment of the *thorax,* the volume of which can be changed by contractile activity of surrounding *respiratory muscles.* Each lung is surrounded by a double-walled, closed *pleural sac*. (*Review Figure 13-5.*)

13.2 Respiratory Mechanics (pp. 450–466)

■ *Ventilation,* or *breathing,* is the process of cyclically moving air in and out of the lungs so that old alveolar air that has given up O_2 and picked up CO_2 can be exchanged for fresh air.

■ Ventilation is mechanically accomplished by alternately shifting the direction of the pressure gradient for airflow between the atmosphere and alveoli through the cyclic expansion and recoil of the lungs. When *intra-alveolar pressure* decreases as a result of lung expansion during *inspiration,* air flows into the lungs from the higher atmospheric pressure. When intra-alveolar pressure increases as a result of lung recoil during *expiration,* air flows out of the lungs toward the lower atmospheric pressure. (*Review Figures 13-6, 13-9, 13-12, and 13-13.*)

■ Alternate contraction and relaxation of the inspiratory muscles (primarily the *diaphragm*) indirectly produce periodic inflation and deflation of the lungs by cyclically expanding and compressing the thoracic cavity, with the lungs passively following its movements. (*Review Figures 13-10 and 13-11.*)

■ The lungs follow the movements of the thoracic cavity by virtue of the *transmural pressure gradient* across the lung wall resulting from the *intrapleural pressure* being subatmospheric and thus less than the intra-alveolar pressure. (*Review Figures 13-6, 13-7, and 13-13.*)

■ Because energy is required for contracting the inspiratory muscles, inspiration is an active process, but expiration is passive during quiet breathing because it is accomplished by elastic recoil of the lungs when the inspiratory muscles relax, at no energy expense. (*Review Figure 13-11a, b, and c.*)

■ For more forceful active expiration, contraction of the expiratory muscles (namely, the *abdominal muscles*) further decreases the size of the thoracic cavity and lungs, which further increases the intra-alveolar–to–atmospheric-pressure gradient. (*Review Figures 13-10 and 13-11d.*)

■ The larger the *pressure gradient* between the alveoli and atmosphere in either direction, the larger the *airflow rate* because air flows until intra-alveolar pressure equilibrates with atmospheric pressure. (*Review Figures 13-12 and 13-13.*)

■ Besides being directly proportional to the pressure gradient, airflow rate is also inversely proportional to *airway resistance*. (*Review Table 13-1.*) Because airway resistance, which depends on the caliber of the conducting airways, is normally very low, airflow rate usually depends primarily on the pressure gradient between the alveoli and atmosphere.

■ The lungs can be stretched to varying degrees during inspiration and then recoil to their preinspiratory size during expiration because of their elastic behavior. *Pulmonary compliance* refers to the distensibility of the lungs—how much they stretch in response to a given change in the transmural pressure gradient. *Elastic recoil* refers to the snapping back of the lungs to their resting position during expiration.

■ Pulmonary elastic behavior depends on the *elastin fibers* in the lung and on *alveolar surface tension*. Alveolar surface tension, which is the result of attractive forces between the surface water molecules lining each alveolus, tends to resist the alveolus being stretched on inflation (decreases compliance) and tends to return it back to a smaller surface area during deflation (increases lung rebound). (*Review Table 13-2.*)

■ If the alveoli were lined by water alone, the surface tension would be so great that the lungs would be poorly compliant and would tend to collapse. *Pulmonary surfactant* intersperses between the water molecules and lowers alveolar surface tension, thereby increasing compliance and counteracting the tendency for alveoli to collapse. (*Review Figure 13-14 and Table 13-2.*)

■ The lungs can be filled to about 5.5 liters on maximal inspiration or emptied to about 1 liter on maximal expiration. Normally the lungs operate at "half full." Lung volume typically varies from about 2 to 2.5 liters as an average *tidal volume* of 500 mL of air is moved in and out with each breath. (*Review Figures 13-15 and 13-16.*)

■ The amount of air moved in and out of the lungs in one minute, the *pulmonary ventilation,* is equal to tidal volume times respiratory rate. Not all the air moved in and out is available for gas exchange with the blood because part occupies the conducting airways (*anatomic dead space*). *Alveolar ventilation,* the volume of air exchanged between the atmosphere and alveoli in one minute, is a measure of the air actually available for gas exchange with the blood. Alveolar ventilation equals (tidal volume minus dead space volume) times respiratory rate. (*Review Figure 13-18 and Table 13-3.*)

■ Even though ventilation and perfusion rates are both higher at the bottom than at the top of the lungs (with the ventilation–

perfusion ratio being higher at the top), local controls act on bronchiolar and pulmonary arteriolar smooth muscle to match airflow and blood flow to each area of the lung as much as possible. (See Figures 13-19 and 13-20.)

13.3 Gas Exchange (pp. 466–471)

■ O_2 and CO_2 move across body membranes by passive diffusion down partial pressure gradients. The *partial pressure* of a gas in air is that portion of the total atmospheric pressure contributed by this individual gas, which in turn is directly proportional to the percentage of this gas in the air. The partial pressure of a gas in blood depends on the amount of this gas dissolved in the blood. (*Review Figure 13-21.*)

■ Net diffusion of O_2 occurs first between the alveoli and blood and then between the blood and tissues as a result of the O_2 partial pressure gradients created by continuous use of O_2 in the cells and continuous replenishment of fresh alveolar O_2 provided by ventilation. Net diffusion of CO_2 occurs first between the tissues and blood and then between the blood and alveoli, as a result of the CO_2 partial pressure gradients created by continuous production of CO_2 in the cells and continuous removal of alveolar CO_2 through ventilation. (*Review Figure 13-22.*)

■ Other factors that influence the rate of gas exchange are surface area and thickness of the membrane across which the gas is diffusing and the diffusion constant of the gas in the membrane (Fick's law of diffusion). (*Review Table 13-5.*)

13.4 Gas Transport (pp. 471–478)

■ Because O_2 and CO_2 are not very soluble in blood, they must be transported primarily by mechanisms other than simply being physically dissolved. Only 1.5% of the O_2 is *physically dissolved* in the blood, with 98.5% chemically *bound to hemoglobin (Hb)*. (*Review Table 13-6.*)

■ The primary factor that determines the extent to which Hb and O_2 are combined (the % *Hb saturation*) is the P_{O_2} of the blood, depicted by an S-shaped curve known as the O_2–Hb dissociation curve. In the P_{O_2} range of the pulmonary capillaries (the plateau portion of the curve), Hb is still almost fully saturated even if the blood P_{O_2} falls as much as 40%. This provides a margin of safety by ensuring near-normal O_2 delivery to the tissues despite a substantial reduction in arterial P_{O_2}. In the P_{O_2} range in the systemic capillaries (the steep portion of the curve), Hb unloading increases greatly in response to a small local decline in blood P_{O_2} associated with increased cellular metabolism. In this way, more O_2 is provided to match the increased tissue needs. (*Review Figure 13-24.*)

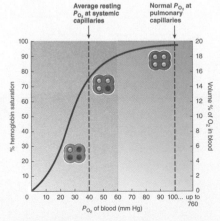

I Figure 13-24 Oxygen–hemoglobin (O_2–Hb) dissociation (saturation) curve.

■ Increased P_{CO_2}, increased acid, and increased temperature at the tissue level shift the O_2–Hb curve to the right, facilitating the unloading of O_2 from Hb for tissue use. (*Review Figure 13-26.*)

■ *Hemoglobin* facilitates a large net transfer of O_2 between alveoli and blood and between blood and tissue cells by acting as a storage depot to keep P_{O_2} (that is, dissolved O_2 concentration) low, despite a considerable increase in the total O_2 content of the blood. (*Review Figure 13-25.*)

■ CO_2 picked up at the systemic capillaries is transported in blood by three means: (1) 10% is *physically dissolved,* (2) 30% is *bound to Hb,* and (3) 60% takes the form of *bicarbonate (HCO_3^-).* The erythrocyte enzyme *carbonic anhydrase* catalyzes conversion of CO_2 to HCO_3^- according to the reaction $CO_2 + H_2O \rightarrow H^+ + HCO_3^-$. These reactions are all reversed in the lungs as CO_2 is eliminated to the alveoli. (*Review Table 13-6 and Figure 13-27.*)

13.5 Control of Respiration (pp. 479–488)

■ Ventilation involves two aspects, both subject to neural control: (1) rhythmic cycling between inspiration and expiration and (2) regulation of ventilation magnitude, which depends on control of respiratory rate and depth of tidal volume.

■ Respiratory rhythm is established by the *pre-Bötzinger complex,* which displays pacemaker activity and drives the inspiratory neurons located in the *dorsal respiratory group (DRG)* of the *medullary respiratory control center*. When these neurons fire, impulses ultimately reach the inspiratory muscles to bring about inspiration. (*Review Figure 13-29.*)

■ When the inspiratory neurons stop firing, the inspiratory muscles relax and passive expiration takes place. For active expiration, the expiratory muscles are activated at this time by expiratory neurons in the *ventral respiratory group (VRG)* of the medullary respiratory control center. (*Review Figure 13-29.*)

■ This basic rhythm is smoothed out by the apneustic and pneumotaxic centers located in the pons. The *apneustic center* prolongs inspiration; the more powerful *pneumotaxic center* limits inspiration. (*Review Figure 13-29.*)

■ Three chemical factors play a role in determining the magnitude of ventilation: P_{CO_2}, P_{O_2}, H$^+$ concentration of the arterial blood. (*Review Table 13-8.*)

■ The dominant factor in the ongoing regulation of ventilation is arterial P_{CO_2}, an increase of which is the most potent chemical stimulus for increasing ventilation. Changes in arterial P_{CO_2} alter ventilation by bringing about corresponding changes in the brain-ECF H$^+$ concentration, to which the *central chemoreceptors* are very sensitive. (*Review Figure 13-31.*)

■ The *peripheral chemoreceptors* are responsive to an increase in arterial H$^+$ concentration, which likewise reflexly brings about increased ventilation. The resulting adjustment in arterial H$^+$-generating CO_2 is important in maintaining the acid–base balance of the body. (*Review Figure 13-30.*)

■ The peripheral chemoreceptors also reflexly increase ventilation in response to a marked reduction in arterial P_{O_2} (<60 mm Hg), serving as an emergency mechanism to increase respiration when arterial P_{O_2} levels fall below the safety range provided by the plateau portion of the O_2–Hb curve.

■ Respiration activity can also be voluntarily modified.

14.1 Kidneys: Anatomy, Functions, and Basic Processes (pp. 492–498)

■ Each of the pair of *kidneys* consists of an outer *renal cortex* and inner *renal medulla*. The kidneys form *urine*. They eliminate unwanted plasma constituents in the urine while conserving materials of value to the body. Urine from each kidney is collected in the *renal pelvis,* then transmitted from both kidneys through the pair of *ureters* to the single *urinary bladder,* where urine is stored until emptied through the *urethra* to the outside. *(Review Figures 14-1 and 14-2.)*

■ The urine-forming functional unit of the kidneys, the *nephron,* is composed of interrelated vascular and tubular components. The *vascular component* consists of two capillary networks in series, the first being the *glomerulus,* a tuft of capillaries that filters large volumes of protein-free plasma into the tubular component. The second capillary network consists of the *peritubular capillaries,* which nourish the renal tissue and participate in exchanges between the tubular fluid and plasma. *(Review Figures 14-3 and 14-4 and chapter opener.)*

■ The tubular component begins with *Bowman's capsule,* which cups around the glomerulus to catch the filtrate, then continues a specific tortuous course to ultimately empty into the renal pelvis. *(Review Figure 14-3.)* As the filtrate passes through various regions of the tubule, cells lining the tubules modify it, returning to the plasma only those materials necessary for maintaining proper ECF composition and volume. What is left behind in the tubules is excreted as urine.

■ The kidneys perform three basic processes: (1) *glomerular filtration,* the nondiscriminating movement of protein-free plasma from the blood into the tubules; (2) *tubular reabsorption,* the selective transfer of specific constituents in the filtrate back into peritubular capillary blood; and (3) *tubular secretion,* the highly specific movement of selected substances from peritubular capillary blood into the tubular fluid. Everything filtered or secreted but not reabsorbed is excreted as urine. *(Review Figure 14-6.)*

14.2 Glomerular Filtration (pp. 498–505)

■ Glomerular filtrate is produced when part of the plasma flowing through each glomerulus is passively forced under pressure through the *glomerular membrane* into the underlying Bowman's capsule. The *net filtration pressure* causing filtration results from a high glomerular capillary blood pressure that favors filtration outweighing the combined opposing forces of plasma-colloid osmotic pressure and Bowman's capsule hydrostatic pressure. *(Review Figures 14-7 and 14-8 and Table 14-1.)*

■ Of the cardiac output, 20% to 25% is delivered to the kidneys to be acted on by renal regulatory and excretory processes. Of the plasma flowing through the kidneys, normally 20% is filtered through the glomeruli, for an average *glomerular filtration rate (GFR)* of 125 mL/min.

■ The *juxtaglomerular apparatus* consists of specialized vascular and tubular cells next to the glomerulus where the tubular and vascular components come into close proximity. *Myogenic mechanisms* and *tubuloglomerular feedback,* triggered by the juxtaglomerular apparatus, *autoregulate* glomerular blood flow and the GFR despite transient changes in the driving mean arterial blood pressure in the range of 80 to 180 mm Hg. *(Review Figures 14-9 through 14-11.)*

■ The GFR can be deliberately altered by changing the glomerular capillary blood pressure via sympathetic influence on the *afferent arterioles* as part of the baroreceptor reflex response that compensates for changed arterial blood pressure. When blood pressure falls too low, sympathetically induced afferent arteriolar vasoconstriction lowers glomerular blood pressure and GFR. When blood pressure rises too high, reduced sympathetic activity causes afferent arteriolar vasodilation, leading to a rise in GFR. As the GFR is altered, the amount of fluid lost in urine changes correspondingly, adjusting plasma volume as needed to help restore blood pressure to normal on a long-term basis. *(Review Figures 14-4, p. 495, 14-10, and 14-12.)*

14.3 Tubular Reabsorption (pp. 505–514)

■ After the filtrate is formed, the tubules handle each filtered substance discretely, so that even though the initial glomerular filtrate is identical to plasma (with the exception of plasma proteins), the concentrations of different constituents are variously altered as the filtered fluid flows through the tubular system. *(Review Table 14-2, p. 517.)*

■ The reabsorptive capacity of the tubular system is tremendous. More than 99% of the filtered plasma is returned to the blood through reabsorption. On average, 124 mL out of the 125 mL filtered per minute are reabsorbed.

■ Tubular reabsorption involves *transepithelial transport* from the tubular lumen into the peritubular capillary plasma. This process may be active (requiring energy) or passive (using no energy). *(Review Figure 14-14.)*

■ The pivotal event to which most reabsorptive processes are linked is the *active reabsorption of* Na^+, driven by the energy-dependent Na^+–K^+ pump in the *basolateral membrane* of the tubular cells. The transport of Na^+ out of the cells into the *lateral spaces* between adjacent cells by this carrier induces the net reabsorption of Na^+ from the tubular lumen to the peritubular capillary plasma. *(Review Figure 14-15.)*

■ Most Na^+ reabsorption takes place early in the nephron in constant unregulated fashion, but in the distal and collecting tubules, the reabsorption of a small percentage of the filtered Na^+ is variable and controlled, mostly by the *renin–angiotensin–aldosterone system (RAAS)*. *(Review Figure 14-16 and Table 14-3, p. 528.)*

■ Because Na^+ and its attendant anion, Cl^-, are the major osmotically active ions in the ECF, the ECF volume is determined by the *Na^+ load (total amount of Na^+)* in the body. In turn, the plasma volume, which reflects the total ECF volume, is important in the long-term determination of arterial blood pressure. Whenever the Na^+ load, ECF volume, plasma volume, and arterial blood pressure are below normal, the juxtaglomerular apparatus secretes *renin,* an enzymatic hormone that triggers RAAS, ultimately leading to increased aldosterone secretion by the adrenal cortex. *Aldosterone* increases Na^+ reabsorption from the distal portions of the tubule, thus correcting for the original reduction in Na^+, ECF volume, and blood pressure. *(Review Figures 14-11 and 14-16.)*

■ By contrast, Na^+ reabsorption is inhibited by the natriuretic peptides, *ANP* and *BNP,* hormones released from the cardiac atria and ventricles, respectively, in response to expansion of the ECF volume and a subsequent increase in blood pressure. *(Review Figure 14-17.)*

■ In addition to driving the reabsorption of Na^+, the energy used by the Na^+–K^+ pump is ultimately responsible for the reabsorption of *organic nutrients* (glucose or amino acids) from the proximal tubule by secondary active transport. *(Review Figure 3-18, p. 76.)*

■ Other electrolytes actively reabsorbed by the tubules, such as PO_4^{3-} and Ca^{2+}, have independently functioning carrier systems within the proximal tubule.

■ Because the electrolyte and nutrient carriers can become saturated, each exhibits a *maximal carrier-limited transport capacity (T_m)*. Once the filtered load of an actively reabsorbed substance exceeds the T_m, reabsorption proceeds at a constant maximal rate, and any additional filtered quantity of the substance is excreted in the urine. *(Review Figure 14-18.)*

■ Active Na^+ reabsorption also drives the passive reabsorption of Cl^- (via an electrical gradient), H_2O (by osmosis), and *urea* (down a urea concentration gradient created as a result of extensive osmotic-driven H_2O reabsorption). Sixty-five percent of the filtered H_2O is reabsorbed from the proximal tubule in unregulated fashion, driven by active Na^+ reabsorption. *(Review Figure 14-19 and Table 14-3, p. 528.)* Reabsorption of H_2O increases the concentration of other substances remaining in the tubular fluid, most of which are filtered waste products. The small urea molecules are the only waste products that can passively permeate the tubular membranes, so urea is the only waste product partially (50%) reabsorbed as a result of being concentrated. *(Review Figure 14-20.)*

■ The other waste products, which are not reabsorbed, remain in the urine in highly concentrated form.

14.4 Tubular Secretion (pp. 514–517)

■ Tubular secretion involves transepithelial transport from the peritubular capillary plasma into the tubular lumen. By tubular secretion, the kidney tubules can selectively add some substances to the quantity already filtered. Secretion of substances hastens their excretion in the urine.

■ The most important secretory systems are for (1) H^+ (helps regulate acid–base balance); (2) K^+ (keeps the plasma K^+ concentration at the level needed to maintain normal membrane ex-

citability in the heart, other muscles, and nerves); and (3) *organic ions* (accomplishes more efficient elimination of foreign organic compounds from the body). H^+ is secreted in the proximal, distal, and collecting tubules. K^+ is secreted only in the distal and collecting tubules under control of aldosterone. Organic ions are secreted only in the proximal tubule. *(Review Figures 14-21 and 14-22 and Table 14-2.)*

14.5 Urine Excretion and Plasma Clearance (pp. 517–532)

■ Of the 125 mL/min of glomerular filtrate formed, normally only 1 mL/min remains in the tubules to be excreted as urine. Only wastes and excess electrolytes not wanted by the body are left behind, dissolved in a given volume of H_2O to be eliminated in the urine.

■ Because the excreted material is removed or "cleared" from the plasma, the term *plasma clearance* refers to the volume of plasma cleared of a particular substance each minute by renal activity. *(Review Figure 14-23.)*

■ The kidneys can excrete urine of varying volumes and concentrations to either conserve or eliminate H_2O, depending on whether the body has a H_2O deficit or excess, respectively. The kidneys can produce urine ranging from 0.3 mL/min at 1200 mOsm/L to 25 mL/min at 100 mOsm/L by reabsorbing variable amounts of H_2O from the distal portions of the nephron.

■ This variable reabsorption is made possible by a *vertical osmotic gradient* in the medullary interstitial fluid, established by the *long loops of Henle* of the *juxtamedullary nephrons* via *countercurrent multiplication* and preserved by the *vasa recta* of these nephrons via *countercurrent exchange*. *(Review Figures 14-5, p. 496; 14-24; 14-25; and 14-28.)* This vertical osmotic gradient, to which the hypotonic (100 mOsm/L) tubular fluid is exposed as it passes through the distal portions of the nephron, establishes a passive driving force for progressive reabsorption of H_2O from the tubular fluid, but the actual extent of H_2O reabsorption depends on the amount of *vasopressin (antidiuretic hormone)* secreted. *(Review Figure 14-27.)*

■ Vasopressin increases the permeability of the distal and collecting tubules to H_2O; they are impermeable to H_2O in its absence. *(Review Figure 14-26.)* Vasopressin secretion increases in response to a H_2O deficit, increasing H_2O reabsorption. Its secretion is inhibited in response to a H_2O excess, reducing H_2O reabsorption. Thus vasopressin-controlled H_2O reabsorption helps correct any fluid imbalances.

■ Once formed, *urine* is propelled by peristaltic contractions through the ureters from the kidneys to the urinary bladder for temporary storage.

■ The *bladder* can accommodate up to 250 to 400 mL of urine before stretch receptors within its wall initiate the *micturition reflex. (Review Figure 14-30.)* This reflex causes involuntary emptying of the bladder by simultaneous bladder contraction and opening of both the *internal* and the *external urethral sphincters*. Micturition can transiently be voluntarily prevented by deliberately tightening the external sphincter and pelvic diaphragm. *(Review Figures 14-2, p. 494, and 14-29.)*

15.1 Balance Concept (pp. 536–537)

■ The *internal pool* of a substance is the quantity of that substance in the ECF. The *inputs* to the pool are by way of ingestion or metabolic production of the substance. The *outputs* from the pool are by way of excretion or metabolic consumption of the substance. *(Review Figure 15-1.)*

■ Input must equal output to maintain a *stable balance* of the substance.

15.2 Fluid Balance (pp. 537–547)

■ On average, the body fluids compose 60% of total body weight. This figure varies, depending on how much fat (a tissue with a low H_2O content) a person has. Two thirds of the body H_2O is in the ICF. The remaining third, in the ECF, is distributed between plasma (20% of ECF) and interstitial fluid (80% of ECF). *(Review Table 15-1.)*

■ Because all plasma constituents are freely exchanged across the capillary walls, the plasma and interstitial fluid are nearly identical in composition, except for the lack of plasma proteins in the interstitial fluid. In contrast, the ECF and ICF have markedly different compositions because the plasma membrane barriers are highly selective as to what materials are transported into or out of the cells. *(Review Figure 15-2.)*

■ The components of fluid balance are *control of ECF volume* by maintaining salt balance and *control of ECF osmolarity* by maintaining water balance. *(Review Tables 15-2, 15-3, and 15-5, p. 547.)*

■ Because of the osmotic holding power of Na^+, the major ECF cation, a change in the body's total Na^+ content, or load, causes a corresponding change in ECF volume, including plasma volume, which alters arterial blood pressure in the same direction. Appropriately, in the long run, *Na^+-regulating mechanisms* compensate for changes in ECF volume and arterial blood pressure. *(Review Table 15-5.)*

■ Salt intake is not controlled in humans, but control of salt output in the urine is closely regulated to maintain *salt balance*. Blood pressure–regulating mechanisms can vary the GFR, and thus the amount of Na^+ filtered, by adjusting the radius of the afferent arterioles supplying the glomeruli. Blood pressure–regulating mechanisms can also vary aldosterone secretion to adjust Na^+ reabsorption by the renal tubules. Varying Na^+ filtration and Na^+ reabsorption can adjust how much Na^+ is excreted in the urine to regulate plasma volume and thus arterial blood pressure in the long term. *(Review Figure 15-3.)*

■ ECF osmolarity must be closely regulated to prevent osmotic shifts of H_2O between the ECF and ICF because cell swelling or shrinking is harmful, especially to brain neurons. Excess free H_2O in the ECF dilutes ECF solutes; the resulting ECF hypotonicity

drives H_2O into the cells. An ECF free H_2O deficit, by contrast, concentrates ECF solutes, so H_2O leaves the cells to enter the hypertonic ECF. *(Review Table 15-5.)*

■ To prevent these harmful fluxes, changes in ECF osmolarity are primarily detected and corrected by the systems that maintain *free H_2O balance* (H_2O without accompanying solute).

■ Free H_2O balance is regulated largely by vasopressin and, to a lesser degree, by thirst. Both of these factors are governed primarily by *hypothalamic osmoreceptors*, which monitor ECF osmolarity, and to a lesser extent by *left atrial volume receptors*, which monitor vascular "fullness." The amount of *vasopressin* secreted determines the extent of free H_2O reabsorption by distal portions of the nephrons, thereby determining the volume of urinary output. *(Review Figure 15-4 and Table 15-4.)*

■ Simultaneously, intensity of *thirst* controls the volume of fluid intake. However, because the volume of fluid drunk is often not directly correlated with the intensity of thirst, control of urinary output by vasopressin is the most important regulatory mechanism for maintaining H_2O balance.

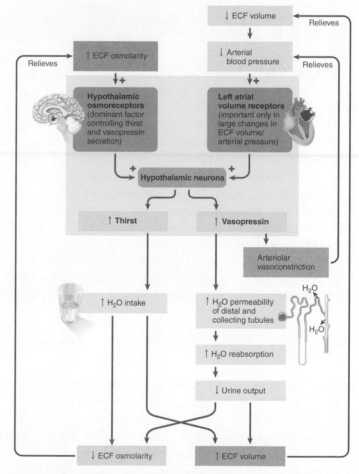

ⅠFigure 15-4 **Control of increased vasopressin secretion and thirst during a H_2O deficit.**

15.3 Acid–Base Balance (pp. 547–562)

■ *Acids* liberate *free hydrogen ions (H^+)* into solution; *bases* bind with free hydrogen ions and remove them from solution. *(Review Figure 15-5.)*

- *Acid–base balance* refers to regulation of [H$^+$] in the body fluids. To precisely maintain [H$^+$], input of H$^+$ by metabolic production of acids within the body must continually be matched with H$^+$ output by urinary excretion of H$^+$ and respiratory removal of H$^+$-generating CO_2. Furthermore, between the time of this generation and its elimination, H$^+$ must be buffered within the body to prevent marked fluctuations in [H$^+$].

- Hydrogen ion concentration is often expressed in terms of *pH*, which is the logarithm of 1/[H$^+$].

- The *normal pH* of the plasma is *7.4*, slightly alkaline compared to neutral H$_2$O, which has a pH of 7.0. A pH lower than normal (higher [H$^+$] than normal) indicates a state of *acidosis*. A pH higher than normal (lower [H$^+$] than normal) characterizes a state of *alkalosis. (Review Figure 15-6.)*

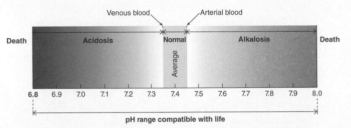

- Fluctuations in [H$^+$] have profound effects, most notably (1) changes in neuromuscular excitability, with acidosis depressing excitability, especially in the CNS, and alkalosis producing overexcitability of both the PNS and CNS; (2) disruption of normal metabolic reactions by altering the structure and function of all enzymes; and (3) alterations in plasma [K$^+$] (which affect cardiac function) brought about by H$^+$-induced changes in the rate of K$^+$ elimination by the kidneys.

- The primary challenge in controlling acid–base balance is maintaining normal plasma alkalinity despite continual addition of H$^+$ to the plasma from ongoing metabolic activity. The major source of H$^+$ is from *CO$_2$-generated H$^+$*.

- The *three lines of defense* for resisting changes in [H$^+$] are *first* the chemical buffer systems, *second* respiratory control of pH, and *third* renal control of pH.

- *Chemical buffer systems* each consist of a pair of chemicals involved in a reversible reaction, one that can liberate H$^+$ and the other that can bind H$^+$. By acting according to the law of mass action, a buffer pair acts immediately to minimize any changes in pH. The four chemical buffers are (1) H_2CO_3: HCO_3^-, (2) *proteins*, (3) *hemoglobin*, and (4) *phosphate. (Review Figure 15-8 and Table 15-6.)*

- The relationship between pH and the members of the H_2CO_3:HCO_3^- buffer pair is represented in the *Henderson–Hasselbalch equation*: pH = pK + log [HCO$_3^-$]/[CO$_2$], with [CO$_2$] reflecting [H$_2$CO$_3$]. [HCO$_3^-$] is controlled by the kidneys; [CO$_2$] is controlled by the lungs. pK is a constant at 6.1, and the normal ratio of [HCO$_3^-$]/[CO$_2$] is 20/1 (the log of which is 1.3), for a normal pH of 7.4.

- The *respiratory system* normally eliminates metabolically produced CO$_2$ so that CO$_2$-generated H$^+$ does not accumulate in the body fluids.

- When chemical buffers alone have been unable to immediately minimize a pH change, the respiratory system responds within a few minutes by altering its rate of CO$_2$ removal. An increase in [H$^+$] from sources other than CO$_2$ stimulates respiration so that more H$^+$-forming CO$_2$ is blown off, compensating for acidosis by reducing generation of CO$_2$-associated H$^+$. Conversely, a fall in [H$^+$] depresses respiratory activity so that CO$_2$ and thus H$^+$ generated from this source can accumulate in the body fluids to compensate for alkalosis. *(Review Table 15-7.)*

- The *kidneys* are the most powerful line of defense. They require hours to days to compensate for a deviation in body-fluid pH. However, not only can they eliminate the normal amount of H$^+$ produced from non-CO$_2$ sources, but they can also alter their rate of H$^+$ removal in response to changes in both non-CO$_2$ and CO$_2$-generated acids. In contrast, the lungs can adjust only H$^+$ generated from CO$_2$. Furthermore, the kidneys can regulate [HCO$_3^-$] in body fluids.

- The kidneys compensate for acidosis by secreting the excess H$^+$ in the urine while adding new HCO$_3^-$ to the plasma to expand the HCO$_3^-$ buffer pool. During alkalosis, the kidneys conserve H$^+$ by reducing its secretion in urine. They also eliminate HCO$_3^-$, which is in excess because less HCO$_3^-$ than usual is tied up buffering H$^+$ when H$^+$ is in short supply. *(Review Figures 15-9 through 15-12 and Table 15-8.)*

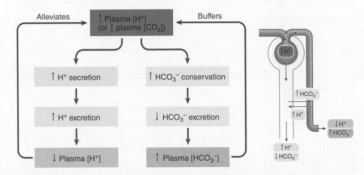

IFigure 15-12 **Control of the rate of tubular H$^+$ secretion and HCO$_3^-$ reabsorption.**

- Secreted H$^+$ must be buffered in the tubular fluid to prevent the H$^+$ concentration gradient from becoming so great that it blocks further H$^+$ secretion. Normally, H$^+$ is buffered by the urinary phosphate buffer pair, which is abundant in the tubular fluid because excess dietary phosphate spills into the urine to be excreted from the body. In acidosis, when all the phosphate buffer is already used up in buffering the extra secreted H$^+$, the kidneys secrete *NH$_3$* into the tubular fluid to serve as a buffer so that H$^+$ secretion can continue.

- The four types of acid–base imbalances are *respiratory acidosis, respiratory alkalosis, metabolic acidosis,* and *metabolic alkalosis*. Respiratory acid–base disorders stem from deviations from normal [CO$_2$], whereas metabolic acid–base imbalances include all deviations in pH other than those caused by abnormal [CO$_2$] and are always accompanied by deviations from normal [HCO$_3^-$]. *(Review Figure 15-13 and Table 15-9.)*

Chapter 16
Study Card

16.1 General Aspects of Digestion (pp. 566–573)

■ The four basic digestive processes are *motility, secretion, digestion,* and *absorption.*

■ Three classes of energy-rich nutrients are digested into absorbable units: (1) Dietary *carbohydrates* in the form of the polysaccharides starch (amylose and amylopectin) and glycogen are digested into monosaccharides, mostly glucose. (2) Dietary *proteins* are digested into amino acids and a few small polypeptides. (3) Dietary *fats* (triglycerides) are digested into monoglycerides and free fatty acids. *(Review Figure 16-1.)*

■ The digestive system consists of the *digestive tract* and *accessory digestive organs* (salivary glands, exocrine pancreas, and biliary system). *(Review Table 16-1.)*

■ The *lumen* of the digestive tract (a tube that runs from the mouth to the anus) is continuous with the external environment, so its contents are technically outside the body; this arrangement permits digestion of food without self-digestion occurring in the process.

■ The *digestive tract wall* has four layers. From innermost outward, they are the *mucosa, submucosa, muscularis externa,* and *serosa.* *(Review Figure 16-2.)*

■ Digestive activities are carefully regulated by synergistic autonomous, neural (both intrinsic and extrinsic), and hormonal mechanisms to ensure that ingested food is maximally made available to the body. *(Review Figure 16-3.)*

16.2 Mouth (pp. 573–575)

■ *Motility:* Food enters the digestive system through the mouth, where it is chewed and mixed with saliva.

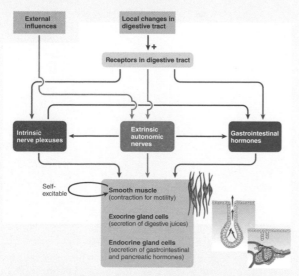

▌Figure 16-3 **Summary of pathways controlling digestive system activities.**

■ *Secretion and digestion:* The salivary enzyme, *amylase,* begins to digest polysaccharides into the disaccharide *maltose,* a process that continues in the stomach after swallowing. Salivary secretion is controlled by a *salivary center* in the medulla, mediated by autonomic nerves to the salivary glands. *(Review Figures 16-1 and 16-4.)*

■ *Absorption:* No food is absorbed from the mouth.

16.3 Pharynx and Esophagus (pp. 575–578)

■ *Motility:* The tongue propels the bolus of food to the rear of the throat, which initiates the *swallowing reflex.* The *swallowing center* in the medulla coordinates a complex group of activities that result in closure of the respiratory passages and propulsion of food through the pharynx and esophagus into the stomach. *(Review Figure 16-5.)*

■ *Secretion, digestion, and absorption:* The esophageal secretion, *mucus,* is protective. No nutrient digestion or absorption occurs here.

16.4 Stomach (pp. 578–588)

■ *Motility:* Gastric motility includes filling, storage, mixing, and emptying. *Gastric filling* is facilitated by vagally mediated *receptive relaxation* of the stomach. *Gastric storage* takes place in the body of the stomach, where peristaltic contractions of the thin muscle walls are too weak to mix the contents. *Gastric mixing* in the thick-muscled antrum results from vigorous peristaltic contractions and retropulsion. *(Review Figures 16-6 and 16-7.)*

■ *Gastric emptying* is influenced by factors in both the stomach and duodenum. (1) Increased volume and fluidity of chyme in the stomach promote emptying. (2) Fat, acid, hypertonicity, and distension in the duodenum (the dominant factors controlling gastric emptying) delay gastric emptying until the duodenum is ready to process more chyme. They do so by inhibiting stomach peristaltic activity via the *enterogastric reflex* and the enterogastrones, *secretin* and *cholecystokinin (CCK),* which are secreted by the duodenal mucosa. *(Review Figure 16-7 and Table 16-2.)*

■ *Secretion:* Gastric secretions into the stomach lumen include (1) *HCl* (from the *parietal cells*), which activates pepsinogen; (2) *pepsinogen* (from the *chief cells*), which, once activated, initiates protein digestion; (3) *mucus* (from the *mucous cells*), which provides a protective coating; and (4) *intrinsic factor* (from the parietal cells), which is needed for vitamin B_{12} absorption. *(Review Table 16-3 and Figures 16-8 and 16-9.)*

■ The stomach also secretes the hormone *gastrin,* which plays a dominant role in stimulating gastric secretion, and the paracrines *histamine* and *somatostatin,* which stimulate and inhibit gastric secretion, respectively. *(Review Table 16-3.)*

■ Gastric secretion is increased before and during a meal via excitatory vagal and intrinsic nerve responses along with the stimulatory actions of gastrin and histamine. After the meal empties, gastric secretion is reduced by withdrawal of stimulatory factors, release of inhibitory somatostatin, and inhibitory actions of the enterogastric reflex and enterogastrones. *(Review Tables 16-4 and 16-5.)*

■ *Digestion and absorption:* Carbohydrate digestion continues by swallowed salivary amylase in the body of the stomach. Protein digestion is initiated by pepsin in the antrum of the stomach, where vigorous peristaltic contractions mix the food with gastric secretions, converting it to a thick liquid mixture known as *chyme. (Review Table 16-6, p. 600.)* No nutrients are absorbed from the stomach.

16.5 Pancreatic and Biliary Secretions (pp. 588–598)

■ Pancreatic exocrine secretions and bile from the liver both enter the duodenal lumen.

■ *Pancreatic exocrine secretions* include (1) potent *digestive enzymes* from the *acinar cells,* which digest all three categories of foodstuff; and (2) an *aqueous NaHCO₃ solution* from the *duct cells,* which neutralizes the acidic contents emptied into the duodenum from the stomach. *Secretin* stimulates the pancreatic duct cells, and *CCK* stimulates the acinar cells. *(Review Figures 16-10 through 16-12.)*

■ The pancreatic digestive enzymes include (1) the proteolytic enzymes *trypsinogen, chymotrypsinogen,* and *procarboxypeptidase,* which are secreted in inactive form and are activated in the duodenal lumen on exposure to *enteropeptidase* and activated trypsin; (2) *pancreatic amylase,* which continues carbohydrate digestion; and (3) *lipase,* which accomplishes fat digestion. *(Review Table 16-6.)*

■ The *liver* performs many metabolic functions. Its contribution to digestion is secretion of bile-salt containing *bile. Bile salts* aid fat digestion through their detergent action (forming a *lipid emulsion*) and facilitate fat absorption by forming water-soluble *micelles* that carry the water-insoluble products of fat digestion to their absorption site. *(Review Figures 16-14 through 16-17 and 16-24, p. 607.)*

■ Between meals, bile is stored and concentrated in the *gallbladder,* which is stimulated by CCK to contract and empty into the duodenum during meal digestion. After participating in fat digestion and absorption, bile salts are reabsorbed and returned via the *hepatic portal system* to the liver, where they are resecreted and also act as a potent *choleretic* to stimulate secretion of more bile. *(Review Figures 16-13 and 16-15.)*

■ Bile also contains *bilirubin,* a derivative of degraded hemoglobin, which is the major excretory product in the feces.

16.6 Small Intestine (pp. 598–610)

■ *Motility: Segmentation,* the small intestine's primary motility during digestion of a meal, thoroughly mixes the chyme with digestive juices to facilitate digestion; it also exposes the products of digestion to the absorptive surfaces. *(Review Figure 16-18.)* Between meals, the *migrating motility complex (MMC)* sweeps the lumen clean.

■ *Secretion:* The juice secreted by the small intestine does not contain any digestive enzymes. The enzymes synthesized by the small intestine act within the *brush-border membrane* of the epithelial cells. *(Review Figures 16-22a and 16-23a.)*

■ *Digestion:* The small intestine is the *main site for digestion and absorption.* Carbohydrate and protein digestion continues in the small-intestine lumen by the pancreatic enzymes and is completed by the small-intestine brush-border enzymes (*disaccharidases* and *aminopeptidases,* respectively). Fat is digested entirely in the small-intestine lumen, by pancreatic lipase. *(Review Table 16-6.)*

■ *Absorption:* The small-intestine lining is remarkably adapted to its digestive and absorptive function. Its folds bear a rich array of fingerlike projections, the *villi,* which have a multitude of even smaller hairlike protrusions, the *microvilli* (brush border). Together, these surface modifications tremendously increase the area available to house the membrane-bound enzymes and to accomplish absorption. *(Review Figures 16-19 through 16-21 and chapter opener, p. 565.)* This lining is replaced about every three days to ensure it is optimally healthy despite harsh lumen conditions.

■ The energy-dependent process of Na^+ absorption provides the driving force for Cl^-, water, glucose, and amino acid absorption. All these absorbed products enter the blood. *(Review Figures 16-22b and 16-23b.)*

■ Because they are not soluble in water, the products of fat digestion must undergo a series of transformations that enable them to be passively absorbed, eventually entering the lymph. *(Review Figure 16-24.)*

■ The small intestine absorbs almost everything presented to it, from ingested food to digestive secretions to sloughed epithelial cells. In contrast to the almost complete, unregulated absorption of ingested nutrients, water, and most electrolytes, the amount of iron and calcium absorbed is variable and subject to control. Only a small amount of fluid and indigestible food residue passes on to the large intestine. *(Review Figure 16-25 and Table 16-7.)*

16.7 Large Intestine (pp. 610–614)

■ *Motility:* The colon *(review Figure 16-26)* concentrates and stores undigested food residues (fiber—that is, plant cellulose) and bilirubin until they can be eliminated in the feces. *Haustral contractions* slowly shuffle the colonic contents back and forth to mix and facilitate absorption of most of the remaining fluid and electrolytes. *Mass movements* several times a day, usually after meals, propel the feces long distances. Movement of feces into the rectum triggers the *defecation reflex.*

■ *Secretion, digestion, and absorption:* The alkaline mucus secretion is protective. The colon secretes no digestive enzymes. However, colonic bacteria (the *microbiota*) ferment for their own use some of the fiber that cannot be digested by human digestive enzymes. The colon absorbs some of these bonus nutrients. Absorption of some of the remaining salt and water converts the colonic contents into *feces.*

16.8 Overview of the GI Hormones (pp. 614–615)

■ The three major GI hormones are gastrin from the stomach mucosa and secretin and cholecystokinin (CCK) from the duodenal mucosa. *Gastrin* is released primarily in response to protein in the stomach, and its effects promote digestion of protein. *Secretin* is released primarily in response to acid in the duodenum, and its effects neutralize the acid. *CCK* is released primarily in response to fat in the duodenum, and its effects optimize conditions for digesting fat.

Chapter 17
Study Card

17.1 Energy Balance (pp. 619–627)

■ *Energy input* to the body in the form of *food energy* must equal energy output because energy cannot be created or destroyed. *Energy output* or expenditure includes (1) *external work,* performed by skeletal muscles to move an external object or move the body through the external environment; and (2) *internal work,* which consists of all other energy-dependent activities that do not accomplish external work, including active transport, smooth and cardiac muscle contraction, glandular secretion, and protein synthesis. *(Review Figure 17-1.)*

■ Only about 25% of the chemical energy in food is harnessed to do biological work. The rest is immediately converted to heat. Furthermore, all the energy expended to accomplish internal work is eventually converted into heat, and 75% of the energy expended by working skeletal muscles is lost as heat. Therefore, most of the energy in food ultimately appears as *body heat.*

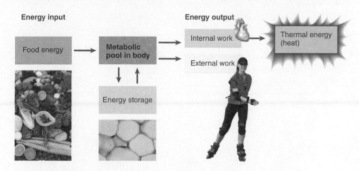

■ The *metabolic rate* (energy expenditure per unit of time) is measured in kilocalories of heat produced per hour.

■ The *basal metabolic rate (BMR)* is a measure of the body's minimal waking rate of internal energy expenditure.

■ For a neutral *energy balance,* the energy in ingested food must equal energy expended in performing external work and transformed into heat. If more energy is consumed than is expended, the extra energy is stored in the body, primarily as adipose tissue, so body weight increases. If more energy is expended than is available in the food, body energy stores are used to support energy expenditure, so body weight decreases.

■ Usually, *body weight* remains fairly constant over a prolonged period of time (except during growth) because food intake is adjusted to match energy expenditure on a long-term basis. *Food intake* is controlled primarily by the hypothalamus by means of complex regulatory mechanisms in which hunger and satiety are important components. *Feeding* or *appetite signals* give rise to the sensation of hunger and promote eating, whereas *satiety signals* lead to the sensation of fullness and suppress eating. *(Review Table 17-2, p. 624.)*

■ The *arcuate nucleus* of the hypothalamus plays a key role in energy homeostasis by virtue of the two clusters of appetite-regulating neurons it contains: neurons that secrete *neuropeptide Y (NPY),* which increases appetite and food intake; and neurons that secrete *melanocortins,* which suppress appetite and food intake. *(Review Figure 17-2.)*

■ *Adipocytes* in fat stores secrete the hormone *leptin,* which reduces appetite and decreases food consumption by inhibiting the NPY-secreting neurons and stimulating the melanocortin-secreting neurons. This mechanism is important in the long-term matching of energy intake with energy output, thus maintaining body weight. *(Review Figure 17-2.)* Adipocytes and macrophages in adipose tissue also secrete other adipokines that largely cause inflammation in obese fat stores and set off a low-grade systemic inflammation that has detrimental metabolic consequences.

■ *Insulin* released by the endocrine pancreas in response to increased glucose and other nutrients in the blood also inhibits the NPY-secreting neurons and contributes to long-term control of energy balance and body weight.

■ NPY and melanocortins bring about their effects by acting on the *lateral hypothalamus area (LHA)* and *paraventricular nucleus (PVN)* to alter the release of chemical messengers from these areas. The LHA secretes *orexins,* which are potent stimulators of food intake, whereas the PVN releases neuropeptides such as *corticotropin-releasing hormone,* which decrease food intake. *(Review Figure 17-2.)*

■ Short-term control of the timing and size of meals is mediated primarily by the actions of two peptides secreted by the digestive tract. (1) *Ghrelin,* a mealtime initiator, is secreted by the stomach before a meal and signals hunger. Its secretion drops

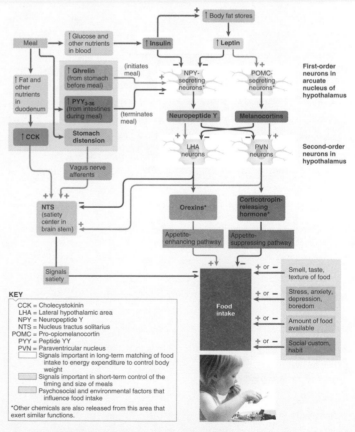

I Figure 17-2 Factors that influence food intake.

when food is consumed. Ghrelin stimulates appetite and promotes feeding behavior by stimulating the NPY-secreting neurons. (2) *PYY$_{3-36}$*, a mealtime terminator, is secreted by the small and large intestines during a meal and signals satiety. Its secretion is lowest before a meal. PYY$_{3-36}$ inhibits the NPY-secreting neurons. *(Review Figure 17-2.)*

■ The *nucleus tractus solitarius (NTS)* in the brain stem serves as the *satiety center* and in this capacity also plays a key role in short-term control of meals. The NTS receives input from the higher hypothalamic areas concerned with control of energy balance and food intake and input from the digestive tract. Satiety signals acting through the NTS to inhibit further food intake include *stomach distension* and increased *cholecystokinin (CCK),* a hormone released from the duodenum in response to nutrients in the digestive tract lumen. *(Review Figure 17-2.)*

■ Psychosocial and environmental factors can also influence food intake above and beyond the internal signals that govern feeding behavior. *(Review Figure 17-2.)*

17.2 Temperature Regulation (pp. 627–635)

■ The body can be thought of as a heat-generating *core* (internal organs, CNS, and skeletal muscles) surrounded by a *shell* of variable insulating capacity (the skin).

■ The *skin* exchanges heat energy with the external environment, with the direction and amount of heat transfer depending on the environmental temperature and the momentary insulating capacity of the shell. The four physical means by which heat is exchanged are (1) *radiation* (net movement of heat energy via electromagnetic waves); (2) *conduction* (exchange of heat energy by direct contact); (3) *convection* (transfer of heat energy by means of air currents); and (4) *evaporation* (extraction of heat energy from the body by the heat-requiring conversion of liquid H_2O to H_2O vapor). Because heat energy moves from warmer to cooler objects, radiation, conduction, and convection can be channels for either heat loss or heat gain, depending on whether surrounding objects are cooler or warmer, respectively, than the body surface. Normally, they are avenues for heat loss, along with evaporation resulting from sweating. *(Review Figure 17-4.)*

■ To prevent serious cell malfunction, the core temperature must be held constant at about *100°F* (equivalent to an average oral temperature of *98.2°F*) by continuously balancing heat gain and heat loss despite changes in environmental temperature and variation in internal heat production. *(Review Figure 17-3.)*

■ This thermoregulatory balance is controlled by the *hypothalamus*. The hypothalamus is apprised of the skin temperature by *peripheral thermoreceptors* and of the core temperature by *central thermoreceptors,* the most important of which are located in the hypothalamus itself. *(Review Figure 17-5.)*

■ The primary means of *heat gain* is heat production by metabolic activity, the biggest contributor being skeletal muscle contraction. *(Review Figure 17-5.)*

■ *Heat loss* is adjusted by sweating and by controlling to the greatest extent possible the temperature gradient between the skin and surrounding environment. The latter is accomplished by regulating the caliber of the skin's arterioles. (1) *Skin vasoconstriction* reduces the flow of warmed blood through the skin so that skin temperature falls. The layer of cool skin between the core and environment increases the insulating barrier between the warm core and the external air. (2) *Skin vasodilation* brings more warmed blood through the skin so that skin temperature approaches the core temperature, thus reducing the insulative capacity of the skin. *(Review Figure 17-5.)*

■ On exposure to cool surroundings, the core temperature starts to fall as heat loss increases because of the larger-than-normal skin-to-air temperature gradient. The *posterior hypothalamus* responds to reduce the heat loss by inducing skin vasoconstriction while simultaneously increasing heat production through heat-generating *shivering.* *(Review Table 17-3.)*

■ Conversely, in response to a rise in core temperature (resulting either from excessive internal heat production accompanying exercise or from excessive heat gain on exposure to a hot environment), the *anterior hypothalamus* triggers heat-loss mechanisms, such as skin vasodilation and *sweating*, while simultaneously decreasing heat production, such as by reducing muscle tone. *(Review Table 17-3.)*

■ In both cold and heat responses, voluntary behavioral actions also help maintain thermal homeostasis.

■ A *fever* occurs when *endogenous pyrogen* released from macrophages in response to infection raises the hypothalamic set point. An elevated core temperature develops as the hypothalamus initiates cold-response mechanisms to raise the core temperature to the new set point. *(Review Figure 17-6.)*

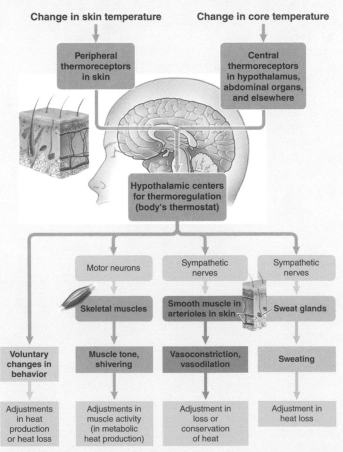

Figure 17-5 Major thermoregulatory pathways.

Chapter 18
Study Card

18.1 General Principles of Endocrinology (pp. 639–646)

- *Hormones* are long-distance chemical messengers secreted by the ductless *endocrine glands* into the blood, which transports the hormones to specific target cells where they control a particular function by altering protein activity.

- Hormones are grouped into two categories based on differences in their solubility and are further grouped according to their chemical structure—*hydrophilic hormones* (peptide hormones, catecholamines, and indoleamines) and *lipophilic hormones* (steroid hormones and thyroid hormone).

- The *endocrine system* is especially important in regulating organic metabolism, H_2O and electrolyte balance, growth, and reproduction and in helping the body cope with stress. *(Review Figure 18-1 and Table 18-2, pp. 644–645.)*

- Some hormones are *tropic*, meaning their function is to stimulate and maintain other endocrine glands.

- The *effective plasma concentration* of each hormone is normally controlled by regulated changes in the rate of hormone secretion. Secretory output of endocrine cells is primarily influenced by two types of direct regulatory inputs: (1) neural input, which increases hormone secretion in response to a specific need and governs diurnal variations in secretion; and (2) input from another hormone, which involves either stimulatory input from a tropic hormone or inhibitory input from a target-cell hormone in negative-feedback fashion. *(Review Figures 18-2 and 18-6, p. 651.)*

- The effective plasma concentration of a hormone can also be influenced by its rate of removal from the blood by metabolic inactivation and excretion and, for some hormones, by its rate of peripheral activation or its extent of binding to plasma proteins.

- *Endocrine dysfunction* arises when too much or too little of any particular hormone is secreted or when there is decreased target-cell responsiveness to a hormone. *(Review Table 18-1.)*

- Target-cell sensitivity to a given plasma concentration of a hormone to which the target cell is responsive can be modified by (1) *down regulation* (number of receptors decreases in the face of a prolonged increase in the hormone), (2) *permissiveness* (one hormone increases the effectiveness of another hormone), (3) *synergism* (combined effect of two hormones is greater than the sum of their separate effects), and (4) *antagonism* (one hormone decreases the effectiveness of another hormone).

18.2 Hypothalamus and Pituitary (pp. 646–652)

- The *pituitary gland* consists of two distinct lobes, the posterior pituitary and the anterior pituitary. *(Review Figure 18-3.)*

- The *hypothalamus*, a portion of the brain, secretes nine peptide hormones. Two are stored in the posterior pituitary, and seven are carried through a special vascular link—the *hypothalamic–hypophyseal portal system*—to the anterior pituitary, where they regulate the release of particular anterior pituitary hormones. *(Review Figures 18-4 and 18-7.)*

- The *posterior pituitary* is a neural extension of the hypothalamus. Cell bodies of neurosecretory neurons in the hypothalamus synthesize two small peptide hormones, *vasopressin* and *oxytocin*, which pass down the axon to be stored in nerve terminals within the posterior pituitary. These hormones are independently released from the posterior pituitary into the blood in response to action potentials originating in the hypothalamus. *(Review Figure 18-4.)*

- The *anterior pituitary* secretes six different peptide hormones produced by five distinct cell types in the gland. Five anterior pituitary hormones are tropic. (1) *Thyroid-stimulating hormone (TSH)* stimulates secretion of thyroid hormone. (2) *Adrenocorticotropic hormone (ACTH)* stimulates secretion of cortisol by the adrenal cortex. (3 and 4) The *gonadotropic hormones—follicle-stimulating hormone (FSH)* and *luteinizing hormone (LH)*—stimulate production of gametes (eggs and sperm) and secretion of sex hormones. (5) *Growth hormone (GH)* largely stimulates growth indirectly by stimulating liver secretion of IGF-I, which in turn promotes growth. GH exerts metabolic effects also. (6) *Prolactin (PRL)* stimulates milk secretion and is not tropic. *(Review Figure 18-5 and chapter opener, p. 638.)*

- The anterior pituitary releases its hormones into the blood at the bidding of *releasing hormones* and *inhibiting hormones* from the hypothalamus. The hypothalamus, in turn, is influenced by a variety of neural and hormonal controlling inputs. *(Review Table 18-3 and Figures 18-6 and 18-7.)*

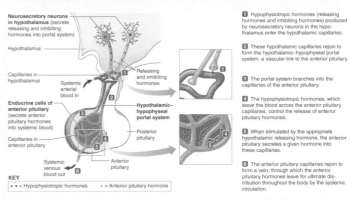

Figure 18-7 Vascular link between the hypothalamus and anterior pituitary.

- Both the hypothalamus and anterior pituitary are inhibited in negative-feedback fashion by the product of the target endocrine gland in the *hypothalamus–anterior pituitary–target-gland axis*. *(Review Figure 18-6.)*

18.3 Endocrine Control of Growth (pp. 652–660)

- *Growth* depends not only on growth hormone and other growth-influencing hormones such as thyroid hormone, insulin, and the sex hormones but also on genetic determination, an adequate diet, and freedom from chronic disease or stress. Major growth spurts occur the first few years after birth and during puberty. *(Review Figure 18-8.)*

- GH directly promotes growth by stimulating protein synthesis. GH indirectly promotes growth by triggering secretion of *IGF-I*, which causes growth by stimulating protein synthesis and cell division in soft tissues and lengthening and thickening of bones. *(Review Figures 18-9 and 18-10.)*

- GH also directly exerts metabolic effects unrelated to growth, such as conserving carbohydrates and mobilizing fat stores. *(Review Figure 18-10.)*

- GH secretion by the anterior pituitary is regulated by two hypothalamic hormones, *growth hormone–releasing hormone (GHRH)* and *growth hormone–inhibiting hormone (somatostatin)*. In negative-feedback fashion, IGF-I and GH both inhibit GHRH and stimulate somatostatin. *(Review Figure 18-10.)*

- GH levels are not highly correlated with periods of rapid growth. The primary signals for increased GH secretion are related to metabolic needs rather than growth, namely, deep sleep (during diurnal rhythm), exercise, stress, and low blood glucose.

18.4 Pineal Gland and Circadian Rhythms (pp. 660–663)

- The *suprachiasmatic nucleus (SCN)* is the body's *master biological clock*. Self-induced cyclic variations in the concentration of SCN clock proteins bring about cyclic changes in neural discharge from this area. Each cycle takes about a day and drives the body's *circadian (daily) rhythms*. *(Review Figure 18-13.)*

- The inherent rhythm of this endogenous oscillator is a bit longer than 24 hours. Therefore, each day the body's circadian rhythms must be *entrained* or adjusted to keep pace with environmental cues so that the internal rhythms are synchronized with the external light–dark cycle.

- In the eyes, special *melanopsin*-containing photoreceptors that respond to light but are not involved in vision send input to the SCN. Acting through the SCN, the *pineal gland* secretes the hormone melatonin in rhythmic cycles that fluctuate with the light–dark cycle, decreasing in the light and increasing in the dark. *Melatonin*, in turn, resets the body's natural circadian rhythms, such as diurnal (day–night) variations in hormone secretion and body temperature, to march in step with external cues such as the light–dark cycle. *(Review Figure 18-13.)*

- Other proposed roles for melatonin include (1) promoting sleep; (2) influencing reproductive activity, including the onset of puberty; (3) acting as an antioxidant to remove damaging free radicals; and (4) enhancing immunity.

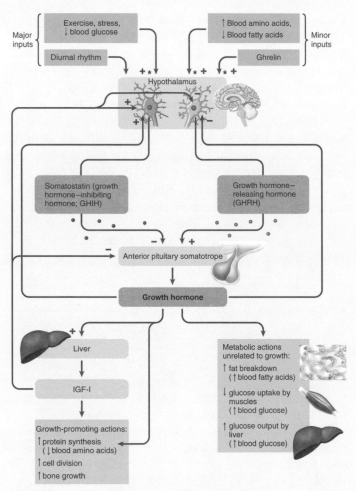

*These factors all increase growth hormone secretion, but it is unclear whether they do so by stimulating GHRH or inhibiting somatostatin (GHIH), or both.

I Figure 18-10 **Control of growth hormone secretion.**

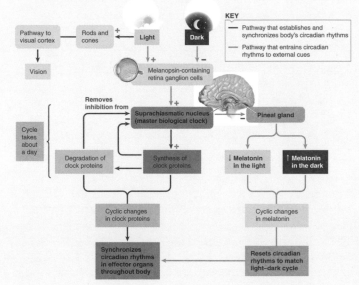

I Figure 18-13 **Synchronization and entrainment of circadian rhythms.**

19.1 Thyroid Gland (pp. 666–671)

■ The bow-tie-shaped *thyroid gland* contains two types of endocrine secretory cells: (1) *follicular cells*, which produce the iodide-containing hormones, T_4 (*thyroxine* or *tetraiodothyronine*) and T_3 (*tri-iodothyronine*), collectively known as *thyroid hormone*; and (2) *C cells*, which synthesize a Ca^{2+}-regulating hormone, *calcitonin*. (*Review Figure 19-1 and chapter opener.*)

■ Most steps of thyroid hormone synthesis take place on large *thyroglobulin* molecules within the *colloid*, an "inland" extracellular site within the interior of the spherical thyroid follicles. Dietary iodine is transported as *iodide* (I^-) from the blood into the follicular cells by the iodide pump, an energy-dependent symporter. From the follicular cells, I^- enters the colloid where it iodinates the amino acid *tyrosine* within thyroglobulin, yielding *monoiodothyronine (MIT)* and *di-iodothyronine (DIT)*. Coupling of MIT and DIT produces T_3; coupling of two DITs produces T_4. Thyroid hormone is secreted by follicular cells phagocytizing a piece of colloid and freeing T_4 and T_3, which enter the blood. (*Review Figure 19-2.*)

■ Thyroid hormone is the primary determinant of the basal metabolic rate of the body. By accelerating metabolic rate, it increases heat production. It also enhances the actions of the sympathetic catecholamines and is essential for normal growth and for development and function of the nervous system.

■ Thyroid hormone secretion is regulated by a negative-feedback system between hypothalamic TRH, anterior pituitary TSH, and thyroid gland T_3 and T_4. The feedback loop maintains thyroid hormone levels relatively constant. Cold exposure in newborn infants is the only input for increasing TRH and thus thyroid hormone secretion. (*Review Figure 19-3.*)

19.2 Adrenal Glands (pp. 672–682)

■ Each *adrenal gland* (of the pair) consists of two separate endocrine organs—an outer, steroid-secreting adrenal cortex and an inner, catecholamine-secreting adrenal medulla. (*Review Figure 19-7.*)

■ Each steroid hormone is produced by stepwise modifications of cholesterol via specific enzymes present in a given steroidogenic endocrine gland. The *adrenal cortex* has enzymes to produce three categories of steroid hormones: *mineralocorticoids* (primarily aldosterone), *glucocorticoids* (primarily cortisol), and *adrenal sex hormones* (primarily the weak androgen, dehydroepiandrosterone). (*Review Figure 19-8.*)

■ *Aldosterone* regulates Na^+ and K^+ balance and is important for blood pressure homeostasis, which is achieved secondarily by the osmotic effect of Na^+ in maintaining the plasma volume, a lifesaving effect. Control of aldosterone secretion is related to Na^+ and K^+ balance and to blood pressure regulation and is not influenced by ACTH. Aldosterone is controlled by the renin–angiotensin–aldosterone system (RAAS) and by a direct effect of K^+ on the adrenal cortex. (*Review Figure 14-22, p. 516*).

■ *Cortisol* helps regulate fuel metabolism and is important in stress adaptation. It increases blood levels of glucose, amino acids, and fatty acids and spares glucose for use by the glucose-dependent brain. The mobilized organic molecules are available for use as needed for energy or repair. Cortisol secretion is regulated by a negative-feedback loop involving hypothalamic CRH and pituitary ACTH. Stress is the most potent stimulus for increasing activity of the CRH–ACTH–cortisol axis. Cortisol also displays a characteristic diurnal rhythm. (*Review Figures 18-2, p. 642; 18-6, p. 651; 19-9, p. 675; and 19-13, p. 684.*)

■ *Dehydroepiandrosterone (DHEA)* governs the sex drive and growth of axillary and pubertal hair in females. It has no observable effect in males, in whom it is overpowered by testosterone. DHEA is under control of CRH–ACTH but negatively feeds back to the gonadotropin loop.

■ The *adrenal medulla* consists of modified sympathetic postganglionic neurons known as *chromaffin cells*, which secrete the catecholamine epinephrine into the blood in response to sympathetic stimulation. (*Review Figure 7-2, p. 235.*) Epinephrine reinforces the sympathetic system in mounting fight-or-flight responses and in maintaining arterial blood pressure. It also increases blood glucose and blood fatty acids. The primary stimulus for increased epinephrine secretion is activation of the sympathetic system by stress. (*Review Figure 19-13.*)

19.3 Integrated Stress Response (pp. 682–685)

■ The term *stress* refers to the generalized nonspecific response of the body to any factor that overwhelms, or threatens to overwhelm, the body's compensatory ability to maintain homeostasis. The term *stressor* refers to any noxious stimulus that elicits the stress response. (*Review Figure 19-12.*)

■ In addition to specific responses to various stressors, all stressors produce a similar *generalized stress response*: (1) increased sympathetic and epinephrine activity, which prepares the body for fight-or-flight; (2) activation of the CRH–ACTH–cortisol axis, which helps the body cope with stress by mobilizing metabolic resources; (3) elevation of blood glucose and fatty acids through decreased insulin and increased glucagon secretion; and (4) maintenance of blood volume and blood pressure through increased activity of RAAS and vasopressin. All these actions are coordinated by the hypothalamus. (*Review Figure 19-13.*)

19.4 Endocrine Pancreas and Control of Fuel Metabolism (pp. 685–701)

■ *Intermediary*, or *fuel*, *metabolism* is, collectively, the synthesis (anabolism), breakdown (catabolism), and transformations of the three classes of energy-rich organic nutrients—carbohydrate, fat, and protein—within the body. Glucose and fatty acids derived from carbohydrates and fats, respectively, are primarily used as metabolic fuels, whereas amino acids derived from proteins are primarily used for synthesis of structural and enzymatic proteins. (*Review Figure 19-14 and Tables 19-2 and 19-3.*)

- During the *absorptive state* following a meal, excess absorbed nutrients not immediately needed for energy production or protein synthesis are stored to a limited extent as glycogen in the liver and muscle but mostly as triglycerides in adipose tissue. During the *postabsorptive state* between meals when no new nutrients are entering the blood, the glycogen and triglyceride stores are catabolized to release nutrient molecules into the blood. If necessary, body proteins are degraded to release amino acids for conversion into glucose (*gluconeogenesis*). The blood glucose concentration must be maintained above a critical level even during the postabsorptive state because the brain depends on blood-delivered glucose as its energy source. Tissues not dependent on glucose switch to fatty acids as their metabolic fuel, sparing glucose for the brain. *(Review Table 19-4.)*

- Blood glucose concentration is controlled by factors that regulate glucose uptake by cells and glucose output by the liver. *(Review Figure 19-16.)*

- The shifts in metabolic pathways between the absorptive and postabsorptive state are controlled by hormones, the most important being insulin. *Insulin* is secreted by the β cells of the islets of Langerhans, the endocrine portion of the pancreas. *(Review Figure 19-15 and Table 19-5, p. 700.)*

- Insulin is an anabolic hormone; it promotes the cellular uptake of glucose, fatty acids, and amino acids and enhances their conversion into glycogen, triglycerides, and proteins, respectively. In so doing, it lowers the blood concentrations of these small organic molecules. Insulin secretion is increased during the absorptive state, primarily by a direct effect of an elevated blood glucose on the β cells via excitation–secretion coupling. Insulin directs nutrients into cells during this state. *(Review Figures 19-17 through 19-20.)*

- *Glucagon*, secreted by the pancreatic α cells, mobilizes the energy-rich molecules from their stores during the postabsorptive state. Glucagon, which is secreted in response to a direct effect of a fall in blood glucose on the α cells, in general opposes the actions of insulin. *(Review Figures 19-15 and 19-20.)*

19.5 Parathyroid Glands and Control of Calcium Metabolism (pp. 701–702)

- Changes in the concentration of *free, diffusible plasma Ca^{2+}*, the biologically active form of this ion, produce profound and life-threatening effects, most notably on neuromuscular excitability. Hypercalcemia reduces excitability, whereas hypocalcemia brings about overexcitability of nerves and muscles. If the overexcitability is severe enough, fatal spastic contractions of respiratory muscles can occur.

- Control of Ca^{2+} metabolism involves two aspects—regulation of Ca^{2+} homeostasis and regulation of Ca^{2+} balance—and depends on hormonal control of exchanges between the ECF and three compartments: *bone*, *kidneys*, and *intestine*. Regulation of Ca^{2+} *homeostasis* (maintenance of a constant free plasma Ca^{2+} concentration) involves rapid exchanges between bone and the ECF and to a lesser extent to adjustments in urinary Ca^{2+} excretion. Regulation of *Ca^{2+} balance* (maintenance of a constant total amount of Ca^{2+} in the body) is accomplished by adjustments in Ca^{2+} absorption from the intestine and urinary Ca^{2+} excretion.

- *Bone* consists of an *organic extracellular matrix*, the *osteoid*, which is hardened by precipitation of *$Ca_3(PO_4)_2$ (calcium phosphate)* crystals. Bone constantly undergoes remodeling by means of bone-dissolving *osteoclasts* and bone-building *osteoblasts*. Entombed *osteocytes* are "retired" osteoblasts that have deposited bone around themselves. Osteoblasts and osteocytes are interconnected by long cytoplasmic arms that extend through the tiny canals that permeate the hardened bone, forming a continuous *osteocytic–osteoblastic bone membrane*. *(Review Figures 19-22 through 19-24.)*

- Three hormones regulate the plasma concentration of Ca^{2+} (and concurrently regulate PO_4^{3-})—*parathyroid hormone (PTH)* secreted by the parathyroid glands, *calcitonin*, and *vitamin D*. *(Review Figure 19-21.)*

- *PTH*, whose secretion is directly increased by a fall in plasma Ca^{2+} concentration, acts directly on bone and kidneys and indirectly on the intestine to raise plasma Ca^{2+}. In so doing, it is essential for life by preventing the fatal consequences of hypocalcemia. PTH promotes Ca^{2+} movement across the osteocytic–osteoblastic bone membrane from the bone fluid into the plasma in the short term and promotes localized dissolution of bone in the long term by enhancing osteoclasts and suppressing osteoblasts. *(Review Figures 19-24 and 19-25.)*

- Dissolution of the $Ca_3(PO_4)_2$ bone crystals releases Ca^{2+} and PO_4^{3-} into the plasma. PTH acts on the kidneys to enhance the reabsorption of filtered Ca^{2+} thereby reducing the urinary excretion of Ca^{2+} and increasing its plasma concentration. Simultaneously, PTH reduces renal PO_4^{3-} reabsorption, in this way increasing PO_4^{3-} excretion and lowering plasma PO_4^{3-} levels. This is important because a rise in plasma PO_4^{3-} would force the redeposition of some of the plasma Ca^{2+} back into the bone. *(Review Figure 19-27.)*

- PTH facilitates activation of *vitamin D*, which in turn stimulates Ca^{2+} and PO_4^{3-} absorption from the intestine. The skin can synthesize vitamin D from cholesterol when exposed to sunlight, but frequently this endogenous source is inadequate, so vitamin D must be supplemented by dietary intake. From either source, vitamin D must be activated first by the liver and then by the kidneys (the site of PTH regulation of vitamin D activation) before it can exert its effect. *(Review Figure 19-26.)*

- *Calcitonin*, a hormone produced by the C cells of the thyroid gland, is secreted in response to an increase in plasma Ca^{2+} and lowers plasma Ca^{2+} by inhibiting activity of bone osteoclasts. Calcitonin is unimportant except during the rare condition of hypercalcemia. *(Review Figures 19-1 and 19-25.)*

20.1 Uniqueness of the Reproductive System (pp. 716–723)

■ Both sexes produce *gametes* (reproductive cells), *sperm* in males and *ova (eggs)* in females, each of which bears one member of each of the 23 pairs of chromosomes present in human cells. Union of a sperm and an ovum at fertilization results in the beginning of a new individual with 23 complete pairs of chromosomes, half from each parent.

■ The *reproductive system* is anatomically and functionally distinct in males and females. Males produce sperm and deliver them into the female. Females produce ova, accept sperm delivery, and provide a suitable environment for supporting development of a fertilized ovum until the new individual can survive on its own in the external world.

■ In both sexes, the reproductive system consists of (1) a pair of *gonads*, testes in males and ovaries in females, which are the primary reproductive organs that produce the gametes and secrete sex hormones; (2) a *reproductive tract* composed of a system of ducts that transport or house the gametes after they are produced; and (3) *accessory sex glands* that provide supportive secretions for the gametes. The *external genitalia* are the externally visible parts of the reproductive system. (*Review Figures 20-1 and 20-2.*) *Secondary sexual characteristics* are the distinguishing features between males and females not directly related to reproduction.

■ *Sex determination* is a genetic phenomenon dependent on the combination of sex chromosomes at the time of fertilization: An XY combination is a genetic male, and an XX combination, a genetic female. *Sexual differentiation* refers to the embryonic development of the gonads, reproductive tract, and external genitalia along male or female lines, which gives rise to the apparent anatomic sex of the individual. In the presence of masculinizing factors, a male reproductive system develops; in their absence, a female system develops. (*Review Figures 20-3 through 20-5.*)

20.2 Male Reproductive Physiology (pp. 723–732)

■ The *testes* are located in the *scrotum*. The cooler temperature in the scrotum than in the abdominal cavity is essential for spermatogenesis (sperm production), which occurs in the testes' highly coiled *seminiferous tubules*. *Leydig cells* in the interstitial spaces between these tubules secrete the male sex hormone testosterone into the blood. (*Review Figures 20-6 and 20-7.*)

■ *Testosterone* is secreted before birth to masculinize the developing reproductive system; then its secretion ceases until puberty, at which time it begins once again and continues throughout life. Testosterone is responsible for maturation and maintenance of the entire male reproductive tract, for development of secondary sexual characteristics, and for stimulating libido. (*Review Table 20-1.*)

■ The testes are regulated by the anterior pituitary gonadotropic hormones, luteinizing hormone (LH) and follicle-stimulating hormone (FSH), which are under control of hypothalamic gonadotropin-releasing hormone (GnRH), which itself is under control of arcuate nucleus kisspeptins. (*Review Figure 20-9.*)

■ Testosterone secretion is regulated by LH stimulation of the Leydig cells, and in negative-feedback fashion, testosterone inhibits LH secretion. (*Review Figure 20-9.*)

■ *Spermatogenesis* requires both testosterone and FSH. Testosterone stimulates the mitotic and meiotic divisions required to transform the undifferentiated diploid germ cells, the spermatogonia, into undifferentiated haploid spermatids. FSH stimulates the remodeling of spermatids into highly specialized motile spermatozoa. (*Review Figures 20-6, 20-7, 20-9, and 20-13, p. 740.*)

■ A *spermatozoon* consists only of a DNA-packed head bearing an enzyme-filled acrosome at its tip for penetrating the ovum, a midpiece containing the mitochondria for energy production, and a whiplike motile tail. (*Review Figure 20-8 and chapter opener, p. 715.*)

■ Also present in the seminiferous tubules are *Sertoli cells*, which protect, nurse, and enhance the germ cells throughout their development. FSH stimulates Sertoli cells to secrete *inhibin*, a hormone that inhibits FSH secretion, completing the negative-feedback loop. (*Review Figures 20-6b and d and 20-9.*)

■ The still-immature sperm are flushed out of the seminiferous tubules into the epididymis by fluid secreted by the Sertoli cells. The *epididymis* and *ductus deferens* store and concentrate the sperm and increase their motility and fertility before ejaculation. During ejaculation, the sperm are mixed with secretions released by the accessory glands. (*Review Figure 20-6a and Table 20-2.*)

■ The *seminal vesicles* supply fructose for energy and prostaglandins, which promote smooth muscle motility in both the male and female reproductive tracts to enhance sperm transport. The seminal vesicles also contribute the bulk of the semen. The *prostate gland* contributes an alkaline fluid to neutralize the acidic vaginal secretions. The *bulbourethral glands* release lubricating mucus. (*Review Table 20-2.*)

20.3 Sexual Intercourse between Males and Females (pp. 732–736)

■ The *male sex act* consists of erection and ejaculation, which are part of a much broader systemic *sexual response cycle*. (*Review Table 20-3.*)

■ *Erection* is a hardening of the normally flaccid *penis* that enables it to penetrate the female vagina. Erection is accomplished by marked vasocongestion of the penis brought about by reflexly induced vasodilation of the arterioles supplying the penile erectile tissue. (*Review Figures 20-1, p. 717, and 20-10.*)

■ When sexual excitation reaches a critical peak, *ejaculation* occurs. It consists of two stages: (1) *emission*, the emptying of semen (sperm and accessory sex gland secretions) into the urethra; and (2) *expulsion* of semen from the penis. The latter is accompanied by a set of characteristic systemic responses and intense pleasure referred to as *orgasm*. (*Review Table 20-3.*)

■ Females experience a sexual response cycle similar to that of males, with both having excitation, plateau, orgasmic, and resolution phases. Like the penis, the highly vascular *clitoris* undergoes erection (but not ejaculation). *(Review Figure 20-2d, p. 719.)* During sexual response, the outer portion of the vagina constricts to grip the penis, and the inner part expands to create space for sperm deposition.

20.4 Female Reproductive Physiology (pp. 736–770)

■ In the nonpregnant state, female reproductive function is controlled by a complex, cyclic, both negative- and positive-feedback system among the hypothalamus (kisspeptins and GnRH), anterior pituitary (FSH and LH), and ovaries (estrogen, progesterone, and inhibin).

■ The *ovaries* perform the dual and interrelated functions of oogenesis (producing ova) and secretion of estrogen and progesterone. *(Review Table 20-6, p. 768.)* Two related ovarian endocrine units sequentially accomplish these functions: the follicle and the corpus luteum.

■ The same steps in chromosome replication and division take place in *oogenesis* as in spermatogenesis, but the timing and end result are markedly different. Spermatogenesis is accomplished within two months, but the similar steps in oogenesis take anywhere from 12 to 50 years to complete on a cyclic basis from the onset of puberty until menopause. A female is born with a limited, nonrenewable supply of germ cells, whereas postpubertal males can produce several hundred million sperm each day. Each primary oocyte yields only one cytoplasm-rich ovum along with three doomed cytoplasm-poor polar bodies that disintegrate, whereas each primary spermatocyte yields four equally viable spermatozoa. *(Review Figures 20-12, 20-13, and 20-7, p. 727.)*

■ Oogenesis and estrogen secretion take place within an *ovarian follicle* during the first half of each reproductive cycle (the *follicular phase*) under the influence of FSH, LH, and estrogen. *(Review Figures 20-14 through 20-19.)*

■ At approximately midcycle, the maturing follicle releases a single ovum *(ovulation).* Ovulation is triggered by an *LH surge* brought about by the high level of estrogen produced by the mature follicle. *(Review Figures 20-14, 20-16, 20-17, and 20-20.)*

■ LH converts the empty follicle into a *corpus luteum (CL),* which produces progesterone and estrogen during the last half of the cycle (the *luteal phase*). This endocrine unit prepares the uterus for implantation if the released ovum is fertilized. *(Review Figures 20-14, 20-17, and 20-21.)*

■ If fertilization and implantation do not occur, the CL degenerates, withdrawing hormonal support for the highly developed uterine lining, causing it to disintegrate and slough, producing *menstrual flow.* Simultaneously, a new follicular phase is initiated. *(Review Figures 20-14 and 20-17.)*

■ Menstruation ceases and the uterine lining *(endometrium)* repairs itself under the influence of rising estrogen levels from the newly maturing follicle. *(Review Figure 20-17.)*

■ If *fertilization* does take place, it occurs in the *oviduct* as the released egg and sperm deposited in the vagina are both transported to this site. *(Review Figures 20-22 and 20-23.)*

■ The fertilized ovum begins to divide mitotically. Within a week it grows and differentiates into a *blastocyst* capable of implantation. *(Review Figure 20-24.)*

■ Meanwhile, the endometrium has become richly vascularized and stocked with stored glycogen under the influence of luteal-phase progesterone. *(Review Figure 20-17, p. 745.)* Into this especially prepared lining the blastocyst *implants* by means of enzymes released by the *trophoblasts*, which form the blastocyst's outer layer. These enzymes digest the nutrient-rich endometrial tissue, accomplishing the dual function of carving a hole in the endometrium for implantation of the blastocyst while simultaneously releasing nutrients from the endometrial cells for use by the developing embryo. *(Review Figure 20-25.)*

■ After implantation, an interlocking combination of fetal and maternal tissues, the placenta, develops. The *placenta* is the organ of exchange between maternal and fetal blood and also acts as a transient, complex endocrine organ that secretes a number of hormones essential for pregnancy. *Human chorionic gonadotropin (hCG), estrogen,* and *progesterone* are the most important of these hormones. hCG maintains the *CL of pregnancy*, which secretes estrogen and progesterone during the first trimester of gestation until the placenta takes over this function in the last two trimesters. High levels of estrogen and progesterone are essential for maintaining a normal pregnancy. *(Review Figures 20-26 through 20-28 and Table 20-5.)*

■ At *parturition*, rhythmic contractions of increasing strength, duration, and frequency accomplish the *three stages of labor*: dilation of the cervix, birth of the baby, and delivery of the placenta (afterbirth). *(Review Figure 20-30.)*

■ Parturition is initiated by a complex interplay of multiple maternal and fetal factors. Once the contractions are initiated at the onset of labor, a positive-feedback cycle is established that progressively increases their force. As contractions push the fetus against the cervix, secretion of *oxytocin*, a powerful uterine muscle stimulant, is reflexly increased. The extra oxytocin causes stronger contractions, giving rise to even more oxytocin release, and so on. This positive-feedback cycle progressively intensifies until cervical dilation and delivery are complete. *(Review Figure 20-29.)*

■ During gestation, the *breasts (mammary glands)* are specially prepared for *lactation*. The elevated levels of placental estrogen and progesterone, respectively, promote development of the ducts and alveoli in the mammary glands. *(Review Figure 20-31.)*

■ *Prolactin* stimulates the synthesis of enzymes essential for milk production by the alveolar epithelial cells. However, the high gestational level of estrogen and progesterone prevents prolactin from promoting milk production. Withdrawal of the placental steroids at parturition initiates lactation.

■ Lactation is sustained by suckling, which triggers the release of oxytocin and prolactin. Oxytocin causes *milk ejection (letdown)* by stimulating the *myoepithelial cells* surrounding the alveoli to squeeze the secreted milk out through the ducts. Prolactin stimulates secretion of more milk to replace the milk the baby nurses. *(Review Figures 20-31 and 20-32.)*

ANATOMICAL TERMS USED TO INDICATE DIRECTION AND ORIENTATION

Anterior	situated in front of or in the front part of
Posterior	situated behind or toward the rear
Ventral	toward the belly or front surface of the body; synonymous with anterior
Dorsal	toward the back surface of the body; synonymous with posterior
Medial	denoting a position nearer the midline of the body or a body structure
Lateral	denoting a position toward the side or farther from the midline of the body or a body structure
Superior	toward the head
Inferior	away from the head
Proximal	closer to a reference point
Distal	farther from a reference point
Sagittal section	a vertical plane that divides the body or a body structure into right and left sides
Longitudinal section	a plane that lies parallel to the length of the body or a body structure
Cross section	a plane that runs perpendicular to the length of the body or a body structure
Frontal or coronal section	a plane parallel to and facing the front part of the body

WORD DERIVATIVES COMMONLY USED IN PHYSIOLOGY

a; an-	absence or lack	epi-	above; over	osteo-	bone
ad-; af-	toward	erythro-	red	oto-	ear
adeno-	glandular	gastr-	stomach	para-	near
angi-	vessel	-gen; -genic	produce	pariet-	wall
anti-	against	gluc-; glyc-	sweet	peri-	around
archi-	old	hemi-	half	phago-	eat
-ase	splitter	hemo-	blood	-pod	footlike
auto-	self	hepat-	liver	-poiesis	formation
bi-	two; double	homeo-	sameness	poly-	many
-blast	former	hyper-	above; excess	post-	behind; after
brady-	slow	hypo-	below; deficient	pre-	ahead of; before
cardi-	heart	inter-	between	pro-	before
cephal-	head	intra-	within	pseudo-	false
cerebr-	brain	kal-	potassium	pulmon-	lung
chondr-	cartilage	leuko-	white	rect-	straight
-cide	kill; destroy	lip-	fat	ren-	kidney
contra-	against	macro-	large	reticul-	network
cost-	rib	mamm-	breast	retro-	backward
crani-	skull	mening-	membrane	sacchar-	sugar
-crine	secretion	micro-	small	sarc-	muscle
crypt-	hidden	mono-	single	semi-	half
cutan-	skin	multi-	many	-some	body
-cyte	cell	myo-	muscle	sub-	under
de-	lack of	natr-	sodium	supra-	upon; above
di-	two; double	neo-	new	tachy-	rapid
dys-	difficult; faulty	nephr-	kidney	therm-	temperature
ecto-; exo-; extra-	outside; away from	neuro-	nerve	-tion	act or process of
ef-	away from	oculo-	eye	trans-	across
-elle	tiny; miniature	-oid	resembling	tri-	three
-emia	blood	ophthalmo-	eye	vaso-	vessel
encephalo-	brain	oral-	mouth	-uria	urine
endo-	within; inside				